Stammtafel

(English version of this Genealogical Table: see endpapers of this book)

ulie Thieben
? – 1884

anuel Goldscheider
? – ?

Heinrich Goldscheider
1853 – 1853

Klara Goldscheider
1856 – ?

Henriette Goldscheider
? – ?

WIEN (Friedrich und Alois; ab 1897 Regine und Alois)

ab 1880 Goldscheider'sche Porzellan-Manufaktur und Porzellanmalerei
ab 1885 – offiziell – Goldscheider'sche Porzellan-Manufaktur und Majolica-Fabrik
ab 1894 Friedrich Goldscheider Erste Wiener Terrakottafabrik
ab 1907 Erste Wiener Terrakottafabrik und Atelier für Künstlerische Fajencen Friedrich Goldscheider

Walter Goldscheider ⚭ Lilly Kuschnitzky
1879 – 1962 ? – 1951

WIEN (Walter und Marcell bis 1928)

ab 1921 Wiener Manufaktur Friedrich Goldscheider
ab 1928 Walter Alleininhaber von Wiener Manufaktur Friedrich Goldscheider

WIEN (unter Josef Schuster nach Arisierung)

ab 1939 Wiener Manufaktur Friedrich Goldscheider Nachf., Inhaber Josef Schuster
ab 1941 Wiener Manufaktur Josef Schuster, vorm. Friedrich Goldscheider

Erwin Goldscheider ⚭ Charlotte Stein
1912 – 1984 1906 – 1999

TRENTON, USA (Erwin und Walter)

ab 1940 American Goldscheider Inc.
ab 1942 Goldscheider-Everlast Corporation
ab 1950 Goldscheider Inc.
ab 1952 Goldscheider of Vienna Inc.

WIEN

ab 1950 erneut in Wien unter Walter: Wiener Manufaktur Friedrich Goldscheider

FREDELSLOH unter Majolikafabrik Carstens (Walters Fortsetzung)

ab 1953 Kunstkeramische Manufaktur Goldscheider GmbH
1960 Produktion eingestellt

Marcel(l) Goldscheider ⚭ Rosemarie ?
1887 – 1964 1890 – 1964

WIEN (Marcell)

ab 1928 Vereinigte Ateliers für Kunst und Keramik
ab 1931 Marcell Goldscheider Keramik

STOKE-ON-TRENT, GB (Marcel)

ab 1940 Zusammenarbeit mit Myott, Son & Co., Staffordshire
ab 1950 Goldscheider Art Pottery
ab 1953 Goldscheider (Staffordshire) Pottery, Ltd.
1960 Produktion eingestellt

GOLDSCHEIDER

Firmengeschichte und Werkverzeichnis
History of the Company and Catalogue of Works

ARNOLDSCHE

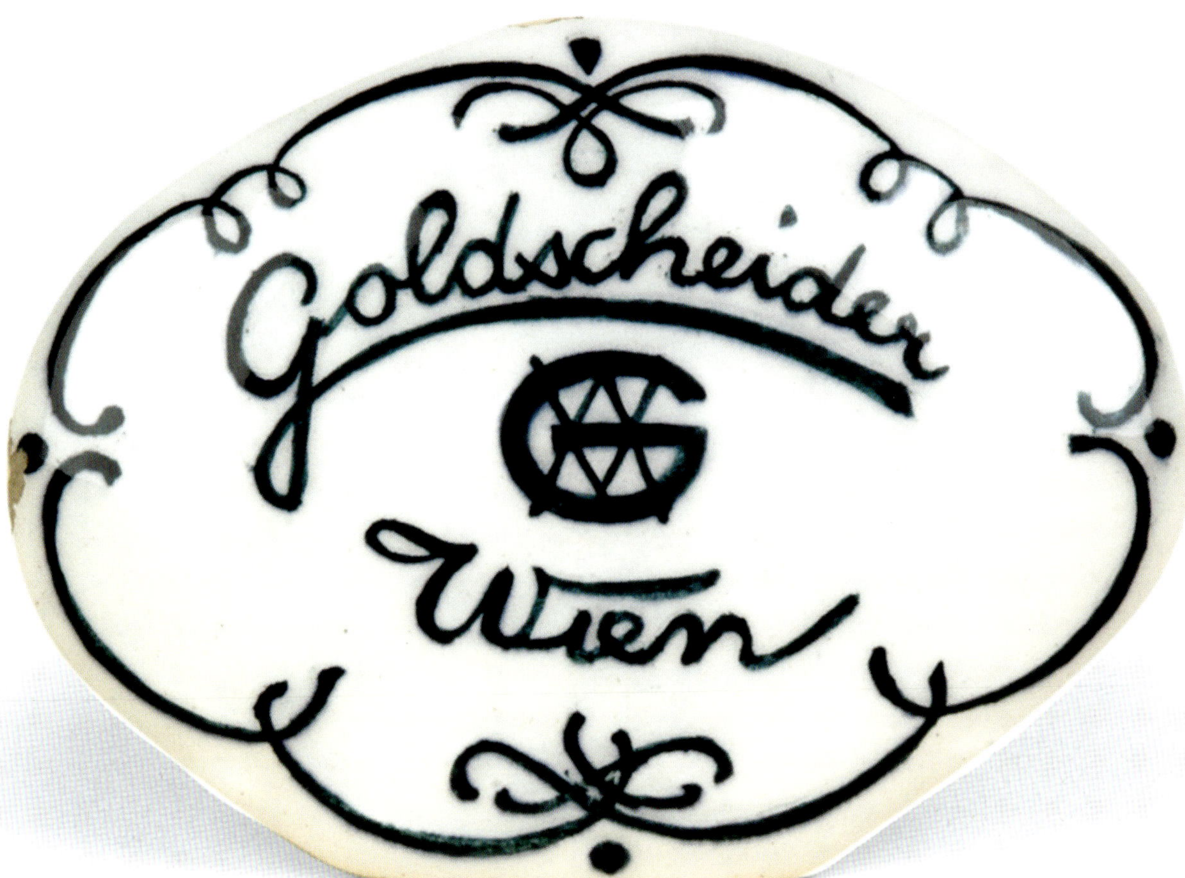
Goldscheider
Wien

Robert E. Dechant | Filipp Goldscheider

GOLDSCHEIDER

Firmengeschichte und Werkverzeichnis
Historismus, Jugendstil, Art Déco, 1950er Jahre

History of the Company and Catalogue of Works
Historicism, Art Nouveau, Art Déco, The Fifties

ARNOLDSCHE
Art Publishers

Inhalt | Contents

Vorwort | Preface Robert E. Dechant 6

Vorwort | Preface Filipp Goldscheider 8

Einführung | Introduction Peter Goldscheider 10

Firmengeschichte | History of the Company
Filipp Goldscheider

Die Anfänge in Pilsen (1867–1892) | The Beginnings in Pilsen (1867–1892) 14

„Moritz Goldscheider's Erben" (1867–1878) | "Moritz Goldscheider's Erben" (1867–1878) 14

Eduard Goldscheider und die „Goldscheidrovka" (1878–1892) | Eduard Goldscheider and the "Goldscheidrovka" (1878–1892) 18

Friedrich Goldscheider und die Entstehung einer internationalen Manufaktur in der Zeit des Historismus in Wien (1878–1897) | Friedrich Goldscheider and the Creation of an International Manufactory in the Period of Historicism in Vienna (1878–1897) 22

Die Blüte des Jugendstils unter Regine Goldscheider (1897–1918) | The Flowering of Art Nouveau under Regine Goldscheider (1897–1918) 46

Dekorative Plastik von Camillo Goldscheider in Wien (1909–1910) | Ornamental Sculpture by Camillo Goldscheider in Vienna (1909–1910) 78

Art Déco Meisterstücke der Brüder Walter und Marcell Goldscheider (1918–1928) | Art Déco Masterpieces by the Brothers Walter and Marcell Goldscheider (1918–1928) 84

Populäre und moderne Keramik der dreißiger Jahre unter Walter Goldscheider (1928–1938) | Popular and Modern 1930s Ceramics under Walter Goldscheider (1928–1938) 106

Die Arisierung der Wiener Manufaktur Friedrich Goldscheider (1938) | The Aryanization of the Viennese Manufactory Friedrich Goldscheider (1938) 154

Die Manufaktur Josef Schuster in den Kriegsjahren (1939–1944) | The Josef Schuster Manufactory during the War Years (1939–1944) 170

Emigration und Aufbau von „American Goldscheider" (1939–1952) | Emigration and Building up "American Goldscheider" (1939–1952) 182

Die Nachkriegsjahre in Wien (1945–1953) | The Postwar Years in Vienna (1945–1953) 194

Die Weltmarke der Keramik – West Germany (1953–1960) | World Mark in Ceramics – West Germany (1953–1960) 220

Dekorative Skulpturen von Arthur Goldscheider in Paris (1900 – ca. 1935) | Ornamental Sculpture from Arthur Goldscheider in Paris (1900 – ca. 1935) 228

Marcell Goldscheiders Werkstätten in Wien und Stoke-on-Trent | Marcell Goldscheider's Workshops in Vienna and Stoke-on-Trent 248

Die Vereinigten Ateliers für Kunst und Keramik (1928–1939) | United Studios of Art and Ceramic (1928–1939) 248

Emigration und Art Pottery in England (1939–1960) | Emigration and Art Pottery in England (1939–1960) 262

Die Fortsetzung der Tradition durch Peter Goldscheider in Stoob, Österreich (1987–1994) | The Tradition Continued by Peter Goldscheider in Stoob, Austria (1987–1994) 274

Werkverzeichnis | Catalogue of Works

Robert E. Dechant

Anmerkungen zum Werkverzeichnis | Notes on the Catalogue of Works 284

Wiener Manufaktur Friedrich Goldscheider | Viennese Manufactory Friedrich Goldscheider 288

Goldscheider „Moderne Keramik" und Gebrauchskeramik | Goldscheider "Modern Ceramics" and Utilitarian Ceramics 517

Goldscheider Modelle „ohne Nummerierung" | "Unnumbered" Goldscheider Models 531

Goldscheider USA 536

Goldscheider West-Germany 541

Arthur Goldscheider, Paris – La Stèle 545

Marcell Goldscheider, Wien | Marcell Goldscheider, Vienna 546

Marcel Goldscheider, Stoke-on-Trent 550

Anhang | Appendix

Künstlerverzeichnis | List of Artists 554

Markenverzeichnis | Marks Table 604

Tipps für Sammler | Pointers for Collectors 610

Bibliografie | Bibliography 615

Abkürzungen | Abbreviations 626

Register | Index 628

Vorwort | Preface

Oft werde ich gefragt, wie ich zur Wiener Keramik und, im Speziellen, zum Thema Goldscheider kam. Bereits in meiner Kindheit fiel mir in der Auslage eines Trödelladens eine Uhr durch die aufgesetzten Figuren auf dem Gehäuse auf, und seitdem hegte ich den Wunsch, später einmal auch solch eine Uhr zu besitzen.

Meine regelmäßigen Flohmarktbesuche ab Mitte der 80er Jahre ließen mich im Jahr 1993 auf eine ebensolche Uhr stoßen, deren aufgesetzte Figuren eine „Allegorie des Lebens" darstellten. Der Händler pries die Uhr als eine wunderschöne Arbeit des Wiener Jugendstils an, und ich kratzte all mein Geld zusammen, um diese Uhr mein Eigen nennen zu können. Zuhause untersuchte ich die Keramik näher und entdeckte das Goldscheider-Firmensignet, die Künstlersignatur und weitere Zeichen und Zahlen. Von da an ließ mich mein kunsthistorischer Forscherdrang – ich hatte mich schon einige Zeit mit dem Münchner Hofmaler Rudolf Wimmer und der grandiosen Maltechnik von Berta Hummel beschäftigt – nicht mehr los.

Wie ich feststellte, bestand die Literatur zum Thema Goldscheider hauptsächlich aus zwei Veröffentlichungen von Waltraud Neuwirth sowie dem Ausstellungskatalog des Historischen Museums der Stadt Wien (heute Wien Museum) und war ohne ein Werk- oder Produktionsverzeichnis als reichlich dünn zu bezeichnen. Ich begann, alle Informationen über Goldscheider zu sammeln und mir u.a. durch Auktionskataloge einen Marktüberblick zu verschaffen. Die spannende und wechselvolle Firmengeschichte, der Formenreichtum in der Produktion und das schier unerschöpfliche Modellreservoir stachelten mich immer mehr an, so dass ich Sammler, Händler und Fachleute zum Thema kennen lernen wollte und selbst zu sammeln begann.

Die erste Kontaktaufnahme war ein wunderbarer Glücksfall. Ich durfte den bekannten Wiener Sammler und „Kunstguru" Dipl.-Kfm. Rudi Schmutz besuchen, woraus sich eine bis heute bestehende Freundschaft entwickelte. Rückblickend haben wir es ihm zu verdanken, dass aus meinen ursprünglich rein privat gesammelten Informationen dieses Buch entstanden ist. Ich darf ihn nicht nur als Freund, sondern auch als Mentor bezeichnen, war er es doch, der zu einem frühen Zeitpunkt einen ersten Kontakt zu einem Verlag herstellte.

Im Laufe der Zeit lernte ich weitere interessante Wiener Persönlichkeiten kennen. Genannt seien hier folgende Sammler: Dr. Peter Goldscheider, der Schauspieler und Regisseur Otto Schenk sowie Prof. Arthur Weilinger. Besonders erwähnt sei auch Frau Mag. Uta M. Matschiner. Sie unterstützte mein Vorhaben nicht nur durch ihr Wissen über Stefan Dakon und Adolf Prischl sowie durch ihre Sammelleidenschaft, sondern zeigte sich seit unserer ersten Begegnung an meiner Arbeit sehr interessiert und war durch

I am often asked how I came to be associated with Viennese ceramics and specifically Goldscheider. Back in childhood days, a clock in a junk shop window caught my eye because of the applied figures on the case and from then on I wanted to own a clock like that myself someday.

Regular visits to flea markets from the mid-1980s led to my finally coming across a clock like that in 1993. Its applied figures represented an "Allegory of Life." The dealer touted the clock as a fine Viennese Jugendstil work and I scraped together all the money I could get hold of to be able to call that clock my own. Once at home with my purchase, I examined it more closely from the ceramics standpoint and discovered the Goldscheider firm mark, the artist signature and other marks and numerals. Since then, the art historical itch to investigate – I had already been preoccupied for some time with the Munich court painter Rudolf Wimmer and the superb painting technique of Berta Hummel – has never left me.

As I soon realized, publications on Goldscheider comprised two articles by Waltraud Neuwirth and an exhibition catalogue issued by the Historisches Museum der Stadt Wien (now the Wien Museum) and, since all three were without catalogues of works or products, the bibliography could at best be called extremely scanty. I began to collect all the information I could on Goldscheider and to gain some insight into the market through auction catalogues among other things. The exciting and turbulent history of the firm, the wealth of forms comprising the product range and the inexhaustible reserves of models spurred me on so that I wanted to meet collectors, dealers and specialists in the field and began to collect pieces myself.

The first contact I made was a felicitous coincidence. I was fortunate in being able to visit the well known Viennese collector and "art guru" Dipl.-Kfm. Rudi Schmutz and from that visit a friendship developed that has lasted until this day. In retrospect, we owe it to him that this book has grown out of what was originally a store of information collected for purely private purposes. I am privileged not only to call him friend but also my mentor since it was he who early on made the first contact to a publishing house.

Over time I met other interesting and distinguished people in Vienna. The following collectors will be mentioned here: Dr. Peter Goldscheider, the actor and director Otto Schenk and Professor Arthur Weilinger. Special mention is also due to Magistra Uta M. Matschiner. She supported my project not only with her knowledge of Stefan Dakon and Adolf Prischl and her passion for collecting; she has also been very interested in my work from our first encounter and has continued to be involved with it by contributing to a stimulating exchange of ideas.

During frequent stays in Vienna as well as several trips to Paris, New York, Rome, London, etc, I sought out both the appropriate famous antiques fairs as well as specialist dealers and was thus gradually able to round out the knowledge I gleaned from contemporary journals and newspapers, thousands of auction catalogues and related publications of all kinds. When I realized that Goldscheider had had more than ten thousand models on the books for decades, it soon became clear to me why no one had ventured on drawing up a comprehensive, let alone complete, catalogue of works.

The many trips I took and the contact with the many people I met while on them not only made work on this book steadily progress but also rewarded my efforts and uniquely enriched my life. It would, however, be impossible to name here all those who have had a share in ensuring the successful completion of this book. Nevertheless, I should like to single out for mention Wilhelm Amann, Dr. Thomas Arlt, Dr. Jolana Ascherl, Othmar Barnert (of the library of the Österreichisches Theatermuseum, Vienna), Kurt Dechant, Olga Dechant, Dr. Edina Deme (Szépmuvészeti Múzeum, Budapest), Dr. Hans Flach, Dr. Alexander Goldscheider, Mag. Christian Kircher, Mag. Michael Kostka, Dr. Sylvia Mattl-Wurm, Mag. Peter Matzelt (Naples), Dr. Frank-Manuel Peter (Deutsches Tanzarchiv, Cologne), Karl Täuber and Professor Andreas Zendler (PH Ludwigsburg). My particular thanks go to Christa Roßmann, whose friendship and collaboration have contributed decisively to my being able to finish this comprehensive book with Filipp Goldscheider. With him I am now linked by close ties of friendship that have ripened over years of collaboration and experiences shared at museums, at our desks or in the dusty, damp cellar in Staudgasse.

Robert E. Dechant

regen Gedankenaustausch ständig involviert.

Während häufigen Aufenthalten in Wien sowie auf mehreren Reisen nach Paris, New York, Rom, London usw. suchte ich neben den entsprechenden berühmten Antikmärkten auch die einschlägigen Händler auf und konnte mein Wissen, das ich aus zeitgenössischen Zeitschriften und Zeitungen, aus tausenden von Auktionskatalogen und vielseitiger Literatur bezog, nach und nach vervollständigen. Die Erkenntnis, dass Goldscheider über die Jahrzehnte mehr als 10.000 Modelle im Programm hatte, machte mir schnell klar, weshalb sich bislang niemand an ein ausführliches oder gar komplettes Werkverzeichnis gewagt hatte.

Die zahlreichen Reisen und der Kontakt mit den vielen Menschen, die ich unterwegs kennen lernte, ließen nicht nur die Arbeit an diesem Buch beständig voranschreiten, sondern belohnten meine Mühen und bereicherten mein Leben auf eine einzigartige Weise. All jene, die am Gelingen dieses Buches beteiligt waren, an dieser Stelle zu nennen, wäre ein Ding der Unmöglichkeit. Dankend erwähnt seien jedoch Wilhelm Amann, Dr. Thomas Arlt, Dr. Jolana Ascherl, Othmar Barnert (Bibliothek des Österreichischen Theatermuseums, Wien), Kurt Dechant, Olga Dechant, Dr. Edina Deme (Szépmuvészeti Múzeum, Budapest), Dr. Hans Flach, Dr. Alexander Goldscheider, Mag. Christian Kircher, Mag. Michael Kostka, Dr. Sylvia Mattl-Wurm, Mag. Peter Matzelt (Neapel), Dr. Frank-Manuel Peter (Deutsches Tanzarchiv, Köln), Karl Täuber sowie Prof. Andreas Zendler (PH Ludwigsburg). Besonders herzlich danken möchte ich Christa Roßmann, deren Freundschaft und Mitarbeit erheblich dazu beitrug, dass ich dieses umfangreiche Buch nun zusammen mit Filipp Goldscheider abschließen kann – mit dem mich durch die mehrjährige Zusammenarbeit und die gemeinsamen Erlebnisse im Museum, am Schreibtisch oder im verstaubten, feuchten Keller der Staudgasse nun eine herzliche Freundschaft verbindet.

Robert E. Dechant

Vorwort | Preface

Als im Jahr 2001 der erste Internet-Boom vorbei war und meine Karriere als New Media Manager ins Stocken zu geraten schien, wartete auf mich ein Projekt, dessen Thematik latent auch den persönlichen Hauch einer midlife crisis und Selbstfindung innehatte. Auf einer Party in München lernte ich den Kunsthändler Ron Krausz kennen, der bei meinem Namen sofort fragte, ob ich mit der berühmten Keramik-Manufaktur familiär verbunden sei. Leider musste ich entgegnen, dass ich mich weder mit der Geschichte und den Erzeugnissen der Manufaktur noch mit meiner eigenen Familiengeschichte besonders gut auskennen würde. Aus dieser Begegnung mit Ron wurde aber eine Freundschaft, und er gab mir den Anstoß, mich mit beiden Themen doch ein wenig intensiver zu beschäftigen. Ich studierte die begrenzt vorhandene Literatur, recherchierte zunächst im Internet, befragte meine Verwandten und gewann nach ein paar Monaten doch an Wissen zum Thema „Goldscheider". Von Peter Goldscheider aus Wien erfuhr ich, dass es einen Experten gäbe, der sich seit Jahren mit der Manufaktur Goldscheider beschäftigen würde: Robert Dechant.

Ich traf Robert und war beeindruckt von seinem Wissen und den Unterlagen, die er mit einer unvergleichlichen Gründlichkeit zusammengetragen hatte: umfangreiche Modellverweise, Literaturquellen, Dokumentationsmaterial. Unsere Gespräche nahmen mit der Zeit die Form vollkommener „Fachtrottelei" an, denn wir sprachen nur noch in Modellnummern miteinander. Wir beschlossen, unsere Stärken – Roberts Spürnase für Details sowie mein pragmatisches Organisationsvermögen – zusammenzulegen und das anhand der Modellvielfalt und fehlenden Inventars schier als unmöglich erscheinende Projekt eines Goldscheider-Standardwerks gemeinsam anzugehen. Robert konzentrierte sich auf das Werk- und Künstlerverzeichnis, während ich aufgrund meines persönlichen Interesses an der Familiengeschichte die Entwicklung der Manufaktur recherchierte.

Meine Recherche-Reisen führten mich nach Pilsen, Prag, Karlsbad, selbstverständlich nach Wien, Paris, Leipzig, Florenz, London, Stoke-on-Trent und andere Orte der Goldscheiderischen Odyssee, deren geographische Lage mir vorher nicht mal bekannt gewesen war. Ich traf liebenswerte Archivarinnen, stocksteife Beamte, lebenslustige Friedhofsverwalter, leidenschaftliche Sammler, skurrile Händler: Menschen, die mir in den über fünf Jahren der Arbeit an diesem Buch geholfen haben. All diesen Leuten gilt mein persönlicher Dank.

Selbstverständlich wäre ohne meine Familie und Freunde dieses mich auch finanziell belastende Unterfangen nicht möglich gewesen. Meiner Mutter Alexandra Herrmann-Goldscheider danke ich als gut erzogener Sohn an erster Stelle: für den Glauben an mich, ihre immerwährende Unterstützung

When the first internet boom was over in 2001 and my career as a new media manager seemed to be in the doldrums, a project awaited me on a subject that also, latently at least, had something to do with a personal whiff of mid-life crisis and self-actualization. At a party in Munich, I met Ron Krausz, an antiques dealer who, on hearing my name, immediately asked whether I had familial ties with the celebrated ceramics factory. Unfortunately, I was constrained to reply that I did not know much about the history and products of the factory nor did I know all that much about my own family history. A friendship grew out of that chance encounter with Ron, however, and he urged me to delve a bit deeper into both those subjects. I studied the scant literature available, did some research online, interviewed my relatives and, even after a few months, had gained some knowledge of "Goldscheider." From Peter Goldscheider in Vienna I heard that there was an expert who had been studying the Goldscheider manufactory for years: Robert Dechant.

I met Robert and was impressed with his knowledge and the material he had amassed with inimitable thoroughness: extensive references to models, sources in writings, documentary material. With time, our talks assumed the form of absolute "nerd speak" as we had long since descended to communicating solely in terms of model numbers. We decided to join forces, each contributing his best side – Robert's fine nose for detail and my pragmatic talent for organization – and embark jointly on the project of writing what would be the standard work on Goldscheider, which, considering the vast number and diversity of the models and the lack of inventory lists, seemed an utterly impossible undertaking. Robert concentrated on the catalogue of works and the index of artists while I, due to my personal interest in my family history, researched the development of the manufactory.

My research took me to Pilsen, Prague, Carlsbad, to Vienna, of course, to Paris, Leipzig, Florence, London, Stoke-on-Trent and other stations along the way in the Goldscheider odyssey, whose geographic parameters I had previously not even known. I met kindly lady archivists, officials whose shirts were solidly stuffed, jolly cemetery administrators, avid collectors, Dickensian antique dealers: all people who helped me in the more than five years during which this book was a work in progress. My personal thanks to all those helpful people.

It goes without saying that without my family and friends, an undertaking incurring such a heavy financial burden would not have been possible. I thank first and foremost, as a well-brought up son should, my mother, Alexandra Herrmann-Goldscheider: for her faith in me, her unswerving support and her healthily realistic and down-to-earth skepticism. To my wife, Ilana Goldscheider, I can only express my admiration for not only putting up with me during that difficult time but even consenting to marry me. To Dr. Peter Goldscheider I am indebted for supporting this project in every possible way. I thank Peter and his wife, Gitti Goldscheider, Dr. Arlette Leupold-Löwenthal and all my other "new-found relatives" for so warmly welcoming me to the family in Vienna.

Filipp Goldscheider

***

The wider family for us as co-authors has come to embrace the many collectors whose knowledge, years of committed support and invaluable help in providing photographs of the models have made this book possible. Our particular thanks go to:

Bernard Amigon, Karl Aumann, Gary Bachman, Franz Bogner, Hans Bundy, Luc van Cauwenbergh, Joan Ferguson, Bärbel Fischer-Daber, Françoise & Jean Pierre Girard, Dr. Alexander Goldscheider, H. A. Gregg, Rudolf Hansl, Heinz Hoever, Dr. Peter Hieke, Paul Howarth, Ursula Huck, Peter Irsigler, John Jesse, Doris Knörlein, Dr. Barbara Kramer-Egghart, Rosa Krenn, Uta Matschiner, Gisela Möller, Dieter Perl, Anne Myott & Philip Pollitt, John Nash, Wade T. Sanders, Rudi Schmutz, the Sellitsch family, Robert Senft, Professor Arthur Weilinger and all the many other Goldscheider collectors.

We owe a great debt of thanks to Karl Täuber for his eye-witness information and help in discovering the Goldscheider Archives.

In addition, we should like to thank Victor Arwas, Karen Benegmos, Dr. Julia Blaha, Joanne Cooper, Christopher Dagleish, Dr. Irene Etzersdorfer, Peter F. Goldscheider (USA), Hans Hagen Hottenroth, Franziska Mathis, Katja Miksovsky, Eva-Maria Orosz, Sarah O'Brien, Tim O'Connell, Magda Pfabigan, Cherica Schreyer-Hartmann, and Professor Jiří Štaif.

In ARNOLDSCHE Art Publishers we have found the perfect partner for this momentous undertaking: their profound knowledge of, and passionate enthusiasm for, ceramics, the exceptional quality of their books and the extremely pleasant personal experience of collaborating with Elke Thode and Gabi Koch have made the book possible in the form it has been realized. We are deeply grateful to Dirk Allgaier, who has not just been the ideal partner for the project but has also become a friend.

Finally, we should like to express our thanks to all those enthusiastic lovers of Goldscheider ceramics without whom this comprehensive and unprecedented work could not have seen the light of day.

R. E. D. & F. G.

und ihr gesundes, realitätsnahes Maß an Skepsis. Meiner Frau Ilana Goldscheider kann ich nur Bewunderung aussprechen, dass sie mich in dieser schwierigen Zeit nicht nur ertragen, sondern auch noch geheiratet hat. Dr. Peter Goldscheider danke ich in jeglicher Hinsicht für die Unterstützung dieses Projektes. Peter und seiner Frau Gitti Goldscheider, Dr. Arlette Leupold-Löwenthal und allen anderen „neuen Verwandten" danke ich für die herzliche Aufnahme in die Wiener Familie.

Filipp Goldscheider

***

Zur erweiterten Familie wurden für uns Autoren die zahlreichen Sammler, deren Kenntnisse, langjährige Unterstützung und das Abbilden der Modelle dieses Buch ermöglichten. Insbesondere danken wir:

Bernard Amigon, Karl Aumann, Gary Bachman, Franz Bogner, Hans Bundy, Luc van Cauwenbergh, Joan Ferguson, Bärbel Fischer-Daber, Françoise & Jean Pierre Girard, Dr. Alexander Goldscheider, H.A. Gregg, Rudolf Hansl, Heinz Hoever, Dr. Peter Hieke, Paul Howarth, Ursula Huck, Peter Irsigler, John Jesse, Doris Knörlein, Dr. Barbara Kramer-Egghart, Rosa Krenn, Uta Matschiner, Gisela Möller, Dieter Perl, Anne Myott & Philip Pollitt, John Nash, Wade T. Sanders, Rudi Schmutz, Familie Sellitsch, Robert Senft, Prof. Arthur Weilinger und den zahlreichen anderen Goldscheider Sammlern.

Besonderer Dank gilt Karl Täuber für seine Zeitzeugen-Informationen und die Hilfe bei der Aufdeckung des Goldscheider Archivs.

Außerdem danken wir Victor Arwas, Karen Benegmos, Dr. Julia Blaha, Joanne Cooper, Christopher Dagleish, Dr. Irene Etzersdorfer, Peter F. Goldscheider (USA), Hans Hagen Hottenroth, Franziska Mathis, Katja Miksovsky, Eva-Maria Orosz, Sarah O'Brien, Tim O'Connell, Magda Pfabigan, Cherica Schreyer-Hartmann, und Prof. Jiří Štaif.

In der ARNOLDSCHEN Verlagsanstalt haben wir den perfekten Partner für dieses große Unterfangen gefunden: Die tiefe Kenntnis und leidenschaftliche Begeisterung für das Thema Keramik, die hervorragende Qualität der Bücher und die äußerst angenehme Zusammenarbeit mit Elke Thode und Gabi Koch haben dieses Werk in seiner Form erst möglich gemacht. Herzlichst danken wir Dirk Allgaier, der nicht nur der ideale Betreuer war, sondern auch ein Freund geworden ist.

Zuletzt gilt unser Dank allen enthusiastischen Liebhabern der Goldscheider Keramik, ohne die dieses umfangreiche und einmalige Werk nie möglich gewesen wäre.

R. E. D. & F. G.

Einführung I
Introduction

Mein Dank gilt den beiden Autoren. Speziell Robert Dechant hat einen Durchbruch in der Aufarbeitung des umfangreichen Materials erzielt. Darüber hinaus ist ein erheblicher Teil seines Lebens der Aufgabe gewidmet, eine möglichst vollständige Inventarisierung der Goldscheider-Keramik in ihren vielfältigen Ausprägungen und Etappen der Familien- und Firmengeschichte zu schaffen. Filipp Goldscheider ist ihm zu einem unentbehrlichen Wegbegleiter geworden. Es gibt inzwischen eine Reihe von Publikationen über Goldscheider. Vollständigkeit, Genauigkeit in der Recherche sowie Liebe und Hingabe zur gemeinsamen Aufgabe sorgen für einen Alleinstellungsanspruch. Dieses Buch wird auch eine Orientierungshilfe für den Leser sein, sich ein eigenes Urteil über den Stellenwert von Goldscheider-Keramik zu bilden.

Meine Frau und ich haben Jahrzehnte unseres Lebens gesammelt. Ich durfte Sammlerkollegen in der ganzen Welt kennen lernen, Händler und Trödler. Ich verdanke diesen Jahren viele köstliche, interessante und bewegende Momente. Normalerweise fuhr ich Freitag nach Büroschluss nach Paris einkaufen: 1.400 km, nicht alles Autobahn; samstags verhandeln, kaufen, einladen und sonntags zurück. Nie vergessen werde ich, wie ich ein besonders großes, schweres Stück in einem Bürgerhaus in Paris in den dort angebrachten Lift bringen wollte, mich zwischen den Flügeltüren des Lifts eingesperrt fand und mich weder vor- noch zurücktraute, aus Angst, die Keramik zu beschädigen. Einmal war ich am Walserberg (Zollgrenze zwischen Deutschland und Österreich) und hatte den Wagen wie immer randvoll; neben mir saß ein lebensgroßer (Keramik-)Negerbub. Nach Erledigung aller Formalitäten erklärte mir der Zöllner, ich könne alles einführen, nur den Negerbuben nicht, weil er sicher wäre, seine Frau wolle den haben. Bei dieser Gelegenheit möchte ich ein Loblied auf die österreichischen Beamten anstimmen, die mir oft unbürokratisch geholfen haben, mich in dem Dickicht der Vorschriften nicht zu verirren. Einmal war ich mit dem Flugzeug in Paris und hatte am Samstagvormittag überraschend reiche Beute am Flohmarkt Porte de Clignancourt gemacht. Ich wollte diese einem Bekannten bringen, um sie dort zwischenzulagern. Aber wir waren schon viel zu spät dran für meinen Flug. Wie heute sehe ich uns im Stau in der Rivoli, als ich den Beschluss fasste, in ein mir völlig unbekanntes Restaurant hineinzugehen und dort zu

My thanks to the two authors. Robert Dechant in particular has achieved a breakthrough in working up the extensive source material. Moreover, a considerable part of his life has been dedicated to the task of ensuring that Goldscheider ceramics in all their diverse forms and through all stages in the history of both the family and the firm have been inventoried as completely as possible. Filipp Goldscheider has been an indispensable companion to him on this journey. By now there is quite a number of publications on Goldscheider. Completeness, accurate scholarship as well as love for, and devotion to the task, the two authors set themselves have ensured that they stand unique. This book will also be an invaluable aid to readers in getting their bearings and forming their own opinions on the status of Goldscheider ceramics.

My wife and I have spent decades of our lives collecting. I probably know fellow collectors everywhere in the world. To those years I owe many precious, interesting and stirring moments. I would usually leave for Paris on Fridays when the office closed: 1400 km, not all of it modern motorway; Saturdays were spent negotiating, buying and inviting and Sundays I would drive back. I shall never forget how, in an attempt to squeeze a particularly large and heavy piece into the lift in a Paris town house, I found myself wedged between the two halves of the lift doors and dared not move backwards or forwards lest I damage the ceramic piece. One day I was at the Walserberg (the customs border between Germany and Austria) and had the car, as always, packed to the roof; next to me sat a life-size (ceramic) black boy. After getting through all the formalities, the customs official explained to me I could bring everything into Austria except for the black boy because he was sure his wife would want to have him herself. At this juncture I should like to raise my voice in a paean of praise to the Austrian officials who so often in an unbureaucratic approach prevented me from losing my way in the jungle of regulations. Once I was due for the plane from Paris, where I had made a surprisingly rich haul at the Porte de Clignancourt flea market. I wanted to take it to a friend for temporary storage. But we were already much too late for my flight. As if it were today, I see us in a traffic jam in the rue de Rivoli when I decided to go into a restaurant I had never set eyes on before and beg the owners to store the things for me. No possibility of putting all safely down in writing, jokes about what they would do with it. Of course I found everything unharmed when I stopped by weeks later to pick it up.

Another advantage of collecting has been the opportunity it has provided for meeting many interesting people and seeing many beautiful flats and houses. The penthouse of a Greek shipping magnate in New York, celebrated singers, both men and women and, in London especially beautiful people around the King's Road and Portobello Road in London. Great excitement when a particularly important piece came up for auction; cautious calls placed to fellow collectors about whether they might refrain from bidding this time; when everything had been so carefully prepared, some big unknown quantity was sure to turn up and keep on bidding up and up.

Collectors often fight with no holds barred; once a highly reputable Salzburg dealer, a woman, rang me up and asked me, utterly bemused, why I wasn't interested in a huge stele (which we still have at home). She had after all requested Frau ... to bring photos to me in Vienna and to ask me whether I wanted to buy it. After a volley of questions on my part, Frau ... actively repented and let me bid for the stele alone without trying to bid it up herself. I had no desire to go to the Dorotheum and bid in person. My father-in-law often went there for me and often had a very attractive lady there as his rival, who had her own reasons for specializing primarily in figures representing dancers. This irritating bidder and her husband have in the meantime become very close friends of ours and have placed a great deal of picture material for this book at our disposal. Another collector, known throughout Vienna, specialized in "captured birds" (Gefangener Vogel/Captured Bird: Niddy Impekoven). A celebrated couple, both of them actors, once gave me a beautiful piece as a present.

On the occasion of the 1985 exhibition at the Wien Museum, I wrote in the accompanying catalogue about collectors and essayed a rather theoretical excursus on motifs, collector types and addictive behavior. This time I wanted to give some practical examples. The collector's life has also changed through ebay, MMS, laptops and PDAs. It has become simpler but, on the other hand, it is no longer as easy as it was to take advantage of a head start in information. There are many important collections of Goldscheider ceramics. We are delighted that we have been able to establish important endowments in two museums (MAK – Österreichisches Museum für angewandte Kunst / Gegenwartskunst, Vienna, and the Wien Museum).

This material that is so interesting to work with because of its tactile, visual and aesthetic qualities, the diversity of products

bitten, die Sachen für mich zu lagern. Keine Möglichkeit, das irgendwie schriftlich festzuhalten, Witze, was sie alles damit machen werden. Natürlich habe ich Wochen später alles unbeschadet in Empfang nehmen können.

Ein weiterer Vorteil der Sammlertätigkeit war, viele interessante Menschen und viele schöne Wohnungen und Häuser kennen zu lernen. Ein Penthouse eines griechischen Reeders in New York, berühmte Sänger, Sängerinnen, vor allem in London, beautiful people rund um King's Road und Portobello in London. Große Aufregung, wenn ein besonders wichtiges Stück in Auktionen kam; vorsichtige Anrufe bei Sammlerkollegen, ob sie dieses eine Mal nicht bieten würden; wenn dann alles vorbereitet war, kam der große Unbekannte und bot weiter und weiter.

Sammler kämpfen oft mit allen Mitteln; einmal hat mich eine sehr bekannte Salzburger Händlerin angerufen und mich völlig erstaunt gefragt, warum mich eine riesige Stele (die wir immer noch zuhause haben) nicht interessiere. Sie hätte doch Frau ... gebeten, mir Fotos nach Wien mitzubringen und mich zu fragen, ob ich kaufen wolle. Auf heftiges Befragen meinerseits übte Frau ... tätige Reue und ließ mich allein für diese Stele verhandeln, ohne mitzubieten. Ich wollte ungern ins Dorotheum gehen und selbst bieten. Mein Schwiegervater ging oft für mich hin und hatte oft eine sehr attraktive Dame als Widersacherin, die aus gegebenem Anlass sich hauptsächlich auf Figuren, die Tänzerinnen darstellten, spezialisierte. Diese lästige Bieterin und ihr Mann gehören inzwischen zu unseren engsten Freunden und haben auch viel Bildmaterial für dieses Buch zur Verfügung gestellt. Ein anderer stadtbekannter Wiener Sammler hat sich auf „Gefangene Vögel" (Niddy Impekoven) spezialisiert. Ein berühmtes Schauspieler-Ehepaar hat mir einmal eine schöne Figur geschenkt.

Ich habe aus Anlass der Ausstellung 1985 im Wien-Museum im Begleitkatalog über die Sammler geschrieben und mich eher an einem theoretischen Exkurs über Motive, Sammler-Typen und Suchtverhalten versucht. Dieses Mal wollte ich ein paar Beispiele aus der Praxis geben. Mit Ebay, MMS, Laptops und PDAs hat sich auch das Sammlerleben verändert. Es ist einfacher geworden, aber es ist auch nicht mehr so leicht, einen Informationsvorsprung zu nutzen. Es gibt viele bedeutende Sammlungen von Goldscheider-Keramik. Wir freuen uns, dass wir bedeutende Stiftungen in zwei Museen (MAK – Österreichisches Museum für angewandte Kunst / Gegenwartskunst, Wien und Wien Museum) etablieren durften.

Der haptisch, visuell und ästhetisch so interessante Werkstoff, die Vielfältigkeit der Produkte (neben Massenware in bestechender Qualität immer auch herausragende

Beispiele künstlerischer Schaffenskraft), die Kontinuität über verschiedene Stilepochen wie Historismus, Jugendstil und Art Déco haben unser Interesse aufrecht erhalten. Nicht zu vergessen die enorme kaufmännische Leistung, dieses Unternehmen über Jahrzehnte als wirtschaftlich sinnvolle Tätigkeit aufrecht zu erhalten: 1. Weltkrieg, Depression, ständige Anpassung an Stile, Moden, Verbraucherverhalten; all das wurde erfolgreich bewältigt. Die Nazi-Zeit hat dieses Stück österreichischer Kultur- und Wirtschaftsgeschichte beendet.

Ausgehend von meiner Sammeltätigkeit habe ich Ende der siebziger Jahre noch einen Versuch gestartet, diesem Unternehmen ein zweites Leben zu geben. Weder die Versuche, alte Modelle in limitierten Auflagen herzustellen, noch die Zusammenarbeit mit Professor Wander Bertoni und Maître Leherb haben zu kommerziell vertretbaren Ergebnissen geführt. Nach 20 Jahren in Großunternehmen war es für mich eine kostbare Erfahrung, einen größeren Handwerksbetrieb mit aufbauen zu dürfen. Ich werde der burgenländischen Unternehmerfamilie um Michael und Elfi Sommer immer dankbar sein, diesen Traum mit mir geträumt zu haben.

Wie so oft im Leben, hat dieser Sprung ins ungewisse Unternehmerleben auf einem ganz anderen Gebiet zu erfreulichen Resultaten geführt. Ohne den Wunsch, Goldscheider wiederzubeleben, wäre ich nie Unternehmer geworden. Die Welt kann sicherlich ohne dekorative Keramik und ohne Goldscheider existieren. Kanonisches Bilderverbot hin, Adolph Loos, Ornament und Verbrechen her, unser Leben wäre ein wenig ärmer ohne Dekoration. Loos hielt übrigens jeden Menschen, der ein Tattoo trägt, für latent verbrecherisch. Ist es nicht interessant, dass jede Zeitströmung, die auf totale Reduktion drängt (wir erleben das gerade in Architektur und Interior Design), ihre Gegenbewegung produziert? Die Menschheit bewegt sich auch auf diesem Gebiet in dialektischen Wellen. Ich wünsche diesem Buch viel Erfolg und allen Sammlern und Interessierten viel Freude und interessante Erkenntnisse bei der Lektüre.

Peter Goldscheider

(apart from mass-produced wares there are still outstanding examples of artistic creative powers of stunning quality) and continuity through various period styles such as Historicism, Art Nouveau and Art Déco have sustained our interest. Nor should we forget the enormous commercial achievement in keeping this business alive for decades as an economically sustainable proposition: the First World War, the Great Depression, the need to accommodate continually to styles, fashions, consumer behavior; all that was successfully accomplished. It was the Nazi era that put paid to this piece of Austrian cultural and economic history.

Taking my collecting activities as my point of departure, I launched an attempt in the late 1970s at bringing this business back to life. Neither experimenting with producing old models in limited editions nor collaborating with Professor Wander Bertoni and Maître Leherb led to commercially viable results. After twenty years in large businesses, being able to help build up quite a large crafts business was, however, an invaluable experience for me. I shall always be grateful to the entrepreneurial family around Michael and Elfi Sommer in Burgenland for having dreamed this dream with me.

As so often happens in life, that leap into an uncertain entrepreneurial life nevertheless led to gratifying outcomes. Without the desire to revive Goldscheider I would never have become an entrepreneur. Granted, the world can exist without decorative ceramics and without Goldscheider. Despite canonical iconoclasm, Adolph Loos, ornament and crime, our lives would, however, be just that much more impoverished without decoration. Incidentally, Loos views any person sporting a tattoo as latently criminal. Isn't it interesting that every trend urging total minimalism (right now we are in the throes of that in architecture and interior design) produces a backlash? Humanity acts in that area, too, in dialectical waves. I wish this book great success and all collectors and interested parties much joy and interesting insights from reading it.

Peter Goldscheider

Firmengeschichte I
History of the Company

Filipp Goldscheider

Die Anfänge in Pilsen (1867–1892)

The Beginnings in Pilsen (1867–1892)

„Moritz Goldscheider's Erben" (1867–1878)

Anfang des 19. Jahrhunderts muss sich Samuel Goldscheider von seiner Familie trennen und seinen Geburtsort Brennporitschen (Spálené Poríčí), etwa 25 km südöstlich von der böhmischen Stadt Pilsen gelegen, verlassen. Die Goldscheiders, deren Vorfahre Isaac im Jahr 1560 von Rudolf II. im Rahmen der Goldschmiedearbeiten und der Alchemie am Hof zum „obersten Goldscheider von Prag"[1] berufen worden war, sind seit dem 17. Jahrhundert in Brennporitschen ansässig.[2] Doch die von Karl VI. erlassenen Familiantengesetze aus dem Jahr 1726 gewähren nur dem ältesten Sohn einer jüdischen Familie das Recht auf die Heirat und Gründung einer eigenen Familie in seiner Geburtsstadt. Den jüngeren Söhnen, wie Samuel, bleibt nichts anderes übrig, als wegzuziehen. Sie lassen sich teilweise in umliegenden Gemeinden wie Blowitz (Blovice) oder Miröschau (Mirošov) nieder, manche versuchen in größeren Städten oder im Ausland Fuß zu fassen. Den ersten Halt finden sie bei bereits dort lebenden Verwandten oder in den jüdischen Gemeinden. Die repressiven Familiantengesetze fördern somit indirekt die Mobilität, Flexibilität und den Aufbau von Netzwerken, die sich geschäftlich als effizient erweisen.

Der 1786 geborene Samuel, der bis 1811 in Brennporitschen als Lehrer tätig ist, zieht wie die meisten Dorflehrer von einem Ort zum anderen. Er lebt mit seiner Frau Rosina Zirel, die aus dem Dorf Tymákov (ebenfalls in der Nähe von Pilsen) stammt, und seinem Sohn Moritz, der 1815 geboren wird. Um 1818 wird Samuel als Lehrer nach Brennporitschen zurückberufen. Aufgrund seines Wissens wird er zum „Religionsführer" ernannt und lehrt später unter dem Namen Samuel ben Salomon als Rabbiner. Er erwirbt den Status eines „Inwohners" und „Schutzjuden" und darf sich in Brennporitschen niederlassen. Moritz' Geschwister werden hier zwischen 1819 und 1822 geboren.[3] Samuel stirbt 1832 in Schönlind (Krásná Lípa), einem Städtchen im Norden Böhmens, wo er als Landesrabbiner tätig war. Sein Leben ist, wie das Schicksal vieler Juden dieser Zeit, durch fortwährende Ortswechsel bestimmt.

"Moritz Goldscheider's Erben" (1867–1878)

Early in the 19th century, Samuel Goldscheider had to part from his family and leave his birthplace, Brennporitschen (Spálené Poríčí), some 25 km south-east of the Bohemian city of Pilsen. The Goldscheiders, whose ancestor, Isaac, had been appointed "foremost gold-refiner [Goldscheider] in Prague"[1] by Rudolf II in 1560 in the context of goldsmithing and alchemy at the imperial court, had been domiciled at Brennporitschen since the 17th century.[2] However, the Familiantengesetze issued by Charles VI in 1726 granted to only the eldest son of a Jewish family to marry and start a family of his own in the city of his birth. Younger sons, like Samuel, could only move away. Some of them settled in nearby communities such as Blowitz (Blovice) and Miröschau (Mirošov). Many tried to gain a footing in larger cities or even abroad. They were initially supported by relatives already living there or in the Jewish congregations. Thus the repressive Familiantengesetze indirectly promoted mobility, flexibility and the building up of networks that would facilitate the conduct of business.

Born in 1786, Samuel, who worked as a teacher in Brennporitschen until 1811, moved about from place to place as most village school teachers did. He lived with his wife, Rosina Zirel, who came from the village of Tymákov (also near Pilsen) and his son Moritz, who was born in 1815. Around 1818, Samuel was summoned back to Brennporitschen to teach there. Due to his knowledge, he was appointed "religious leader" and later taught as a rabbi, using the name Samuel ben Salomon. He acquired "resident" and protected Jew" status and was permitted to settle permanently in Brennporitschen. Moritz's brothers and sisters were born there between 1819 and 1822.[3] Samuel died in 1832 at Schönlind (Krásná Lípa), a small town in northern Bohemia, where he officiated as district rabbi. His life was, like those of so many Jews of that time, defined by constant moving from place to place.

His son Moritz also moved from city to city. Unlike Samuel and his grandfather, Salomon, Moritz was a merchant rather than a rabbi. He married Julie, daughter of Jakob Thieben, merchant, from Rausinau (Rosenov) near Rakonitz (Rakovník), a place that some years later would become famous for the ceramics it produced.[4] The Jewish family of Thieben is said to have dealt in pottery from the late 18th century.[5] Moritz's eldest son, Siegfried Friedrich Goldscheider, was born in the nearby village of Slabetz (Slabce), 11 km south of Rakonitz, on 6 November 1845.[6] In the early 1850s, Moritz moved with his wife and sons Friedrich, Alois, Eduard and Emanuel to Pilsen, where he acquired in 1852 house No. 251 in Jungferngasse, thus becoming one of the first Jews to own a house in Pilsen.[7] His son Heinrich, who died in 1853 shortly after his birth, was born in that house and so was his daughter Klara, on 14 April 1856.[8] Klara was followed by a younger daughter and youngest child, Henriette.

After having been expelled from the royal city of Pilsen in the early 16th century, Jews had only been allowed to settle there since the early 19th century and then only under certain conditions.[9] The revolutionary upheavals of 1848 and the proclamation of the first Basic Law brought equality for the Jews. The ghettos were disbanded while the Familiantengesetze and other restrictions were revoked. These social strivings for liberty, equality and fraternity, however, also aroused Anti-Semitic tendencies since many merchants and peasants feared competition from the Jews. At that time, most Jews traded in leather, ironmongery, goose-down, regional products and produce, haberdashery, etc.[10] Within the Jewish community, they also worked as butchers, tailors, distillers and bookbinders, although the practice of most crafts by Jews was still forbidden by the guilds. From the mid-19th century, industrialization set in and Jewish merchants and traders began to manufacture in larger quantities the goods they were familiar with from merchandising: leather, textiles, ironmongery as well as paper, matches and spirits.[11] Many of them became important industrialists in Pilsen.

Moritz Goldscheider was the successful proprietor of a shop that sold a variety of wares and in 1863 had a house built at No. 285 Hussgasse, in what was known as the Reichsvorstadt.[12] Moritz died two years later and was buried in the Jewish cemetery in Lochotýn, a suburb of Pilsen.[13] Apart from the two houses, he bequeathed a barn at 301 Hussstraße as well as fields and meadows on the outskirts of the city to his wife Julie and his children.[14]

His heirs were registered as a firm, "M. Goldscheider's Erben" [M. Goldscheider, Heirs] and are listed in the 1872 Pilsen direc-

Auch sein Sohn Moritz zieht von einer Stadt zur anderen. Im Gegensatz zu Samuel und dem Großvater Salomon ist er nicht Rabbiner, sondern Händler. Er heiratet Julie, die Tochter des Händlers Jakob Thieben aus Rausinau (Rosenov) bei Rakonitz (Rakovník), einem Ort, der einige Jahre später für seine keramischen Produkte bekannt wird.[4] Die jüdische Familie Thieben soll seit Ende des 18. Jahrhunderts mit Tonwaren gehandelt haben.[5] Im nahe gelegenen Dorf Slabetz (Slabce), 11 km südlich von Rakonitz, kommt Moritz' ältester Sohn Siegfried Friedrich Goldscheider am 6. November 1845 zur Welt.[6] Anfang der 1850er Jahre zieht Moritz mit seiner Frau und den Kindern Friedrich, Alois, Eduard und Emanuel nach Pilsen, wo er im Jahr 1852 das Haus Nr. 251 in der Jungferngasse erwirbt und somit einer der ersten jüdischen Hausbesitzer in Pilsen ist.[7] In diesem Haus werden auch Sohn Heinrich, der 1853 kurz nach seiner Geburt stirbt, und am 14. April 1856 die Tochter Klara geboren.[8] Nach Klara wird als jüngste Tochter Henriette geboren.

In der königlichen Stadt Pilsen dürfen sich erst seit Beginn des 19. Jahrhunderts unter bestimmten Bedingungen Juden niederlassen, nachdem sie Anfang des 16. Jahrhunderts von dort vertrieben worden sind.[9] Die revolutionären Ereignisse des Jahres 1848 und die Proklamierung des ersten Grundgesetzes bringen eine Gleichberechtigung der Juden. Die Ghettos, die Familiantengesetze und andere Einschränkungen werden aufgelöst. Dieses gesellschaftliche Streben nach Freiheit, Gleichheit und Brüderlichkeit weckt jedoch gleichzeitig antisemitische Tendenzen, da sich viele Kaufleute und Bauern vor der jüdischen Konkurrenz fürchten. Die meisten Juden betreiben zu dieser Zeit Handel mit Leder, Eisenwaren, Gänsedaunen, regionalen Landesprodukten, Schnittwaren usw.[10] Innerhalb der jüdischen Gemeinden betätigen sie sich auch als Fleischer, Schneider, Schnapsbrenner oder Buchbinder, da ihnen die offizielle Ausübung der meisten Handwerksberufe von den Zünften bislang verboten war. Ab Mitte des 19. Jahrhunderts setzt sich in Pilsen die Industrialisierung durch und die jüdischen Händler beginnen die ihnen durch den Handel vertrauten Waren in größeren Mengen selbst zu fabrizieren: Leder, Textilien,

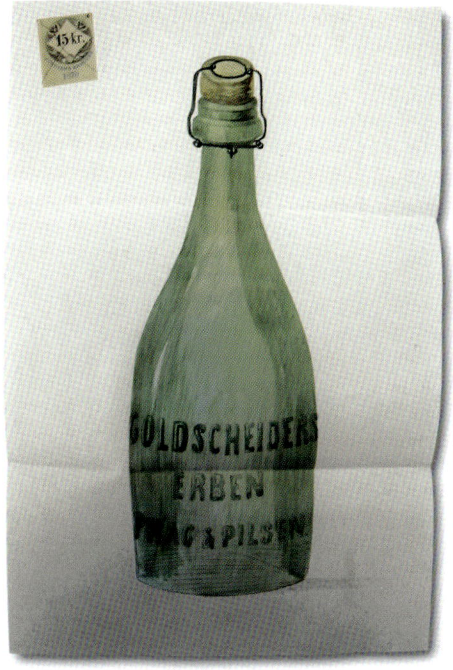

Ill. 1
Im Mai 1871 und Januar 1873 erhält die Erbengemeinschaft ein Privilegium für die exklusive Produktion von Flaschen mit einem Bügelverschluss. (Quelle: Privilegium 1873/000717)

In May 1871 and January 1873 the heirs were granted privilege for the sole rights to produce bottles with a swing cap. (Source: Privilege 1873/000717)

Eisenwaren, aber auch Papier, Streichhölzer oder Schnaps.[11] Viele von ihnen werden zu wichtigen Industriellen in Pilsen.

Moritz Goldscheider handelt erfolgreich mit gemischten Waren und lässt im Jahre 1863 in der Hussgasse Nr. 285, in der so genannten Reichsvorstadt, ein Haus erbauen.[12] Zwei Jahre später stirbt er und wird auf dem jüdischen Friedhof in der Pilsner Vorstadt Lochotýn begraben.[13] Er hinterlässt seiner Frau Julie und den Kindern neben den beiden Häusern eine Scheune in der Hussstraße 301 sowie Felder und Wiesen am Stadtrand.[14]

Die Erben treten unter der Firmierung „M. Goldscheider's Erben" auf und sind im Pilsner Adressbuch aus dem Jahr 1872 als Besitzer einer „gemischten Waarenhandlung", die allerdings seit 1869 nicht mehr betrieben wird, und der „ersten Pilsner Fabrik von kohlensauerer Wässer und moussirender Getränke" zu finden.[15] Die Hauptniederlassung ist in der Pilsner Jungferngasse 251, eine weitere Filiale befindet sich in Prag in der Ferdinandstraße 38. Bei der Fabrik handelt es sich um eine offene Handelsgesellschaft der ältesten Brüder Siegfried Friedrich und Alois Goldscheider, der ein schriftlicher Gesellschaftsvertrag vom 15. November 1867 zugrunde liegt und bei der nur Friedrich zeichnungsberechtigt ist.[16] Das Mineralwasser wird in Flaschen aus Steinzeug abgefüllt, die Verschlüsse sind teilweise aus Porzellan. Dies ist vermutlich auch einer der ersten Berührungspunkte der Familie Goldscheider mit dem Bereich der Keramik (Ill. 1 und 2).

Um 1872 wird auf dem Grundstück am Stadtrand von Pilsen eine Ziegelbrennerei

tory as the proprietors of a "shop selling a variety of wares" (which, however, had not been in operation since 1869) and as the "first Pilsen factory for sparkling waters and carbonated beverages."[15] The firm's headquarters was at 251 Jungferngasse in Pilsen. There was also a branch at 38 Ferdinandstraße in Prague. The factory was established as a partnership (OHG) owned by the two eldest brothers Siegfried Friedrich and Alois Goldscheider, and based on a written partnership contract dated 15 November 1867; only Friedrich had power of attorney.[16] The mineral water was bottled in stoneware and some bottles had porcelain caps. This is probably one of the first contacts the Goldscheider family had with ceramics (Ill. 1 and 2).

Around 1872 a brickyard was built on the property on the outskirts of Pilsen, where paving bricks and fire bricks were to be made (Ill. 3 and 4). The clay for simple pottery products such as bricks could be obtained on site. For products containing kaolin, the raw materials were later obtained in the Pilsen region, where kaolin is mined, for instance at Chlumčany or Horní Bříza. By 1873 the heirs were already represented at the Viennese World's Fair with "China clay, ochre and fire bricks."[17]

Siegfried Friedrich Goldscheider married Regine Schwarz in the synagogue at Marienbad on 15 May 1873. Regine came from a rich family[18] from the pottery city of Teplitz (Teplice) and her dowry is said to have added 40 000 guilders (Fl.) to the family coffers.[19] The young married couple lived in the house at 285 Reichsvorstadt, Pilsen, where their

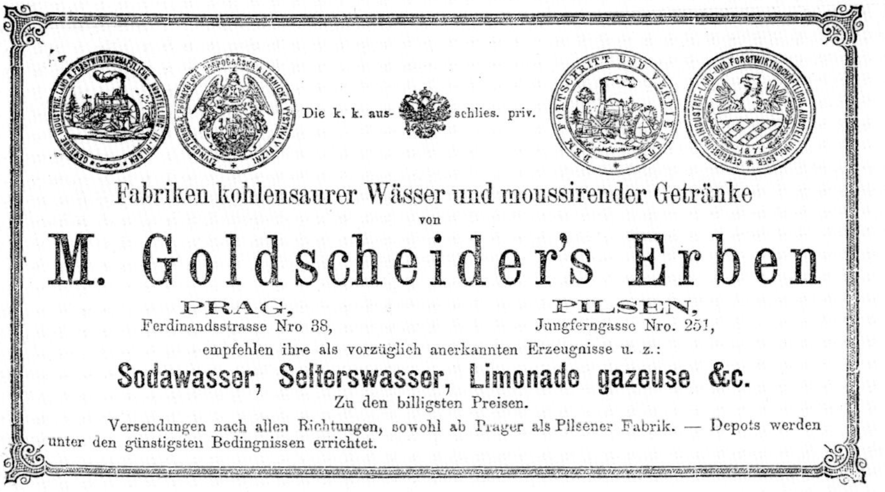

Ill. 2
Anzeige aus dem Jahr 1872 für die Sodawasserfabriken der Erbengemeinschaft in Pilsen und Prag. (Quelle: Bloch, Anzeigenteil des Pilsner Adressbuchs von 1872)

An advertisement from 1872 of the soda-water factories owned by the heirs in Pilsen and Prague (Source: Bloch, advertising section of the 1872 Pilsen directory)

sons Arthur (6 June 1874) and Camillo (28 May 1875; his godfather was Alois) were born. Their third son, Alphons (27 January 1877), died soon after he was born.[20] The year 1873 was generally marred by severe economic crisis. Although the beverage factory had been awarded a bronze medal in 1869 at the "Second Commercial, Industrial, Agricultural and Forestry Trade Fair in Pilsen,"[21] and later also received awards for quality, the firm holdings were sequestrated when "M. Goldscheider's Erben" went into receivership in April 1874. The holdings of Siegfried Friedrich and Alois, who were personally liable as partners in the Pilsen factory, as well as the personal fortunes of the Prague owners, Eduard und Emanuel, were affected.[22] Nonetheless, bankruptcy seems to have been averted because the 1876 Pilsen directory lists the soda-water factory of "M. Goldscheider's Erben" at 454 Klattauerstraße, and the houses at 251 Jungferngasse as well as 285 and 301 Hussstraße continued to belong to the heirs.[23] In addition, a "Schnitt- & Thonwaarengeschäft" ["Haberdashery & Pottery Wares Shop"] is listed as owned and run by the brothers Siegfried Friedrich, Eduard, Emanuel and Alois.[24] The production of carbonated waters was finally expunged from the commercial register on 19 June 1877 and the firm was re-registered only as "M. Goldscheider's Heirs, Pottery and Artificial Stone Manufactory, Moldau Works."[25] The designation "Moldau Works" was probably chosen for advertising purposes in order to compete on the international market with a better known name since the Moldau is far away from Pilsen.

errichtet, in der Pflastersteine und Schamottziegel hergestellt werden sollen (Ill. 3 und 4). Der Ton kann für einfache keramische Erzeugnisse wie Ziegel vor Ort geschöpft werden, für kaolinhaltige Produkte werden später die Rohstoffe aus dem Umkreis von Pilsen bezogen, wo Kaolin z.B. in Chlumčany oder Horní Bříza abgebaut wird. Im Jahr 1873 ist die Erbengemeinschaft bereits auf der Wiener Weltausstellung mit „Carlin, Oker und feuerfesten Ziegeln" vertreten.[17]

Am 15. Mai 1873 heiratet Siegfried Friedrich Goldscheider in der Synagoge in Marienbad Regine Schwarz. Die aus der Töpferstadt Teplitz (Teplice) stammende Regine kommt aus einer vermögenden Familie[18] und bringt angeblich 40.000 Gulden (Fl.) in die Ehe ein.[19] Das junge Ehepaar lebt im Haus in der Pilsner-Reichsvorstadt 285, wo auch die Söhne Arthur (6. Juni 1874) und Camillo (28. Mai 1875, sein Pate ist Alois) geboren werden. Der dritte Sohn Alphons (27. Januar 1877) stirbt kurz nach seiner Geburt.[20] Das Jahr 1873 ist allgemein durch eine schwere wirtschaftliche Krise gekennzeichnet. Obwohl die Getränkefabrik im Jahr 1869 auf der „Zweiten Gewerbe, Industrie, Land und Forstwirtschaft Ausstellung in Pilsen" die bronzene Medaille gewinnen kann[21] und später auch Qualitäts-Auszeichnungen erhält, wird über das Vermögen der Firma „M. Goldscheider's Erben" im April 1874 der Konkurs eröffnet. Auch das Vermögen der persönlich haftenden Gesellschafter der Pilsner Fabrik, Siegfried Friedrich und Alois, sowie das Vermögen der Prager Besitzer Eduard und Emanuel ist hiervon betroffen.[22] Doch der Konkurs kann anscheinend abgewendet werden, denn im Pilsner Adressbuch des Jahres 1876 wird die Sodawasserfabrik der „M. Goldscheider's Erben" mit Klattauerstraße 454 angegeben und auch die Häuser in der Jungferngasse 251 und der Hussstraße 285 und 301 gehören weiterhin zum Besitz der Erbengemeinschaft.[23] Zusätzlich wird ein „Schnitt- & Thonwaarengeschäft" der Brüder Siegfried Friedrich, Eduard, Emanuel und Alois aufgeführt.[24] Die Erzeugung kohlensaurer Wässer wird letztendlich am 19. Juni 1877 aus dem Handelsregister gelöscht und man firmiert nur noch unter „M. Goldscheider's Erben, Thon- und Kunststeinmanufactur, Werk Moldau".[25] Die Bezeichnung Werk Moldau hat wahrscheinlich werbetechnische Gründe, um international mit einem bekannteren Namen aufzutreten, denn die Moldau befindet sich fern ab von Pilsen.

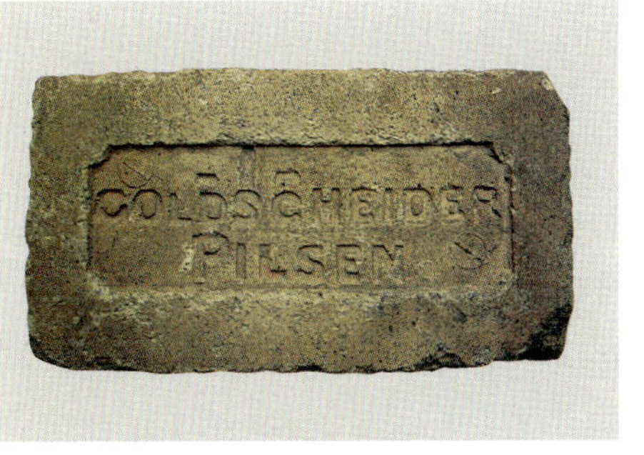

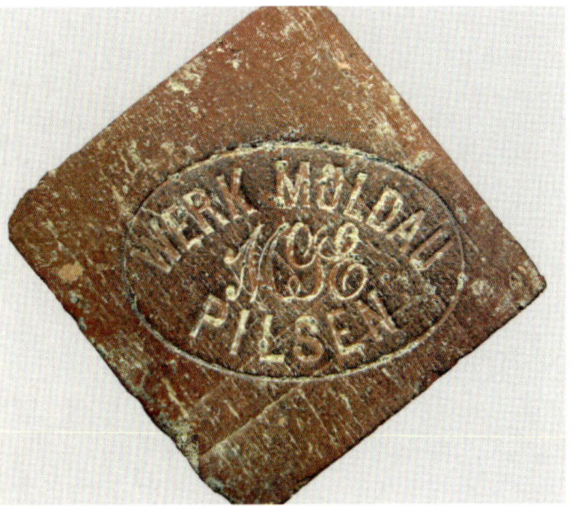

Ill. 3, 4
Die Anfänge der keramischen Produktion bilden bei Goldscheider Ziegel- und Pflastersteine. Das MGE-Zeichen auf dem Pflasterstein aus den Jahren um 1875 steht für „Moritz Goldscheider's Erben". Der Schamott-Ziegelstein (oben) ist aus den Jahren um 1879. (Quelle: Sammlung f@goldscheider.de)

The manufacture of bricks and paving bricks marked the beginnings of ceramic production at Goldscheider. The MGE mark on the paving brick from the years before and after 1875 stands for "Moritz Goldscheider's Erben." The fire brick (above) dates from around 1879. (Source: Collection f@goldscheider.de)

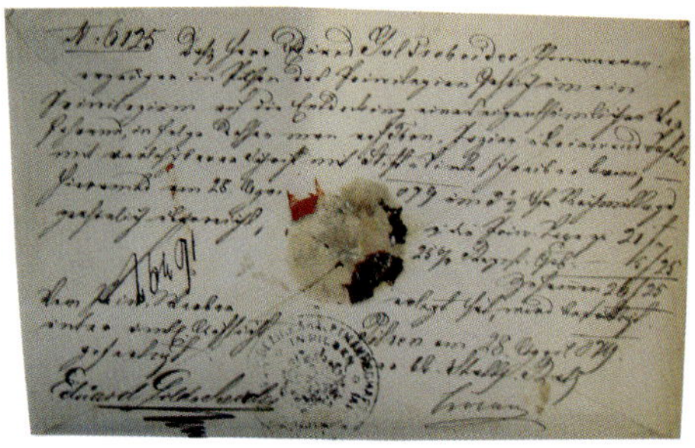

Ill. 5
Dieser Umschlag enthält das von Eduard Goldscheider beantragte Privilegium aus dem Jahr 1879 zur Erzeugung von Schreibtafeln aus Ton, Papier und Leinwand. (Quelle: Privilegium 1879/001145)

This envelope contains the privilege applied for by Eduard Goldscheider and granted in 1879 for the manufacture of slates made of clay, paper and canvas (Source: Privilege 1879/001145)

Eduard Goldscheider und die „Goldscheidrovka" (1878–1892)

Als Siegfried Friedrich Goldscheider um 1878 nach Wien zieht, ändert sich in Pilsen der Firmenname und anstatt der Erbengemeinschaft tritt man nur noch unter „Brüder Goldscheider" auf.[26] Den Betrieb leitet Eduard Goldscheider, der die Produktpalette erweitert und am 7. April 1879 das Privilegium zur Erzeugung von künstlichen Schleif- und Putzsteinen, Pflastersteinen und ähnlichen Erzeugnissen sowie am 4. Juli des gleichen Jahres ein Privilegium zur Herstellung von Schreibtafeln „aus Ton, Papier und Leinwand" erhält[27] (Ill. 5).

Diese Produkte präsentiert er bereits 1879 auf der Industrieausstellung in Teplitz, zusammen mit Schamottziegelsteinen und Geschirr aus Steingut.[28] Auch Emanuel Thieben, ein Verwandter von Julie Goldscheider (geb. Thieben), ist auf die Herstellung von Tonprodukten spezialisiert und arbeitet im „Werk Moldau" in Pilsen.[29] Das Gelände, auf dem die Ziegelbrennerei steht, wird im Volksmund und auf Stadtkarten ab 1880 allgemein als „Goldscheidrovka" bezeichnet.[30] In einer Werbeanzeige aus dem Jahr 1888 preist Eduard Goldscheider seine „Spezialitäten in hochfeuerfestem Sanitäts-Kochgeschirr (ohne jeden Bleigehalt) in verschiedenen Ausführungen", die er zu „billigsten Preisen" in seiner „Thonwaarenfabrik" in Pilsen erzeugt.[31] Im Jahr 1891 stellt er auf der Jubiläumsausstellung in Prag „gute Handels-

Eduard Goldscheider and the "Goldscheidrovka" (1878–1892)

When Siegfried Friedrich Goldscheider moved to Vienna around 1878, the firm name changed in Pilsen and, instead of being registered as Goldscheider, Heirs, now appeared solely as "Goldscheider Brothers."[26] Eduard Goldscheider managed the business, enlarging the product range. On 7 April 1879 privilege was granted for the manufacture of artificial grinding and polishing stones, paving brick and similar products and, on 4 July of the same year, another privilege was granted for the making of slates "of clay, paper and canvas"[27] (Ill. 5).

In 1879 he presented these products at the Teplitz Industrial Trade Fair along with fire brick and stoneware dishes.[28] Emanuel Thieben, a relative of Julie Goldscheider's (née Thieben), also specialized in the manufacture of ceramic products and worked at the "Moldau Works" in Pilsen.[29] The premises of the brickyard were popularly known and marked on city maps from 1880 as "Goldscheidrovka."[30] In an 1888 advertisement, Eduard Goldscheider promotes his "specialities in highly fire-proof, hygienic cookware (without any lead content) in various models," which he produced for "lowest prices" at his "pottery wares factory" in Pilsen.[31] At the 1891 Jubilee Exhibition in Prague, he showed "good commercial wares in majolica, faience tableware, sanitary ware and artificial pumice stones"[32] (Ill. 6–8).

Ill. 6–8
Neben schlichten Haushaltsartikeln, wie dem kleinen blauen Krug (10 cm Höhe, mit eingeprägtem Schriftzug GOLDSCHEIDER und PILSEN, Quelle: Národopisné Muzeum Plzeňska, Inventar-Nr. 20079), wurden auch dekorative Vasen hergestellt. Diese Vase mit 3 Maskerons trägt die Nummer 152 sowie den Stempel mit den Initialen EGP (11 cm hoch, Quelle: Národopisné Muzeum Plzeňska, Inventar-Nr. 47466).

Alongside simple household articles such as this little blue jug (H 10 cm, with GOLDSCHEIDER and PILSEN stamped on it; source: Národopisné Muzeum Plzenska, inventory no. 20079), ornamental vases were also made. This vase with three mascarons bears the number 152 and the stamp with the initials EGP (H 11 cm; source: Národopisné Muzeum Plzeňska, inventory no. 47466).

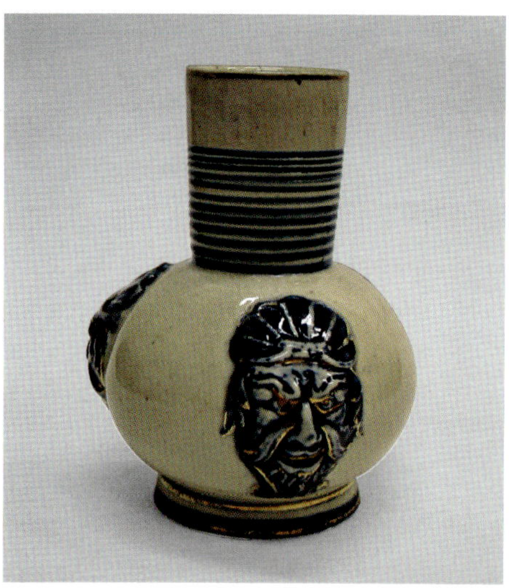

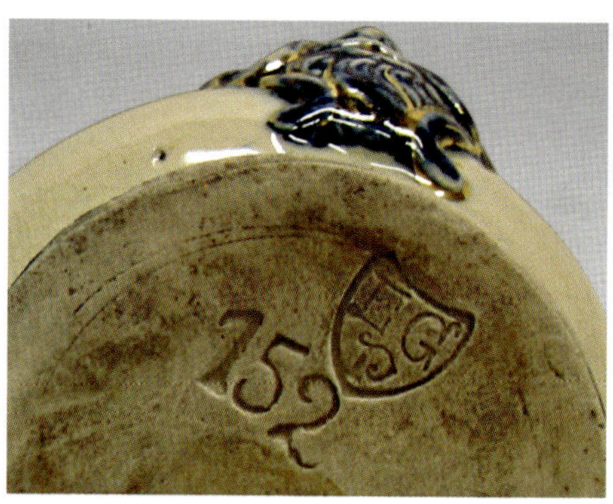

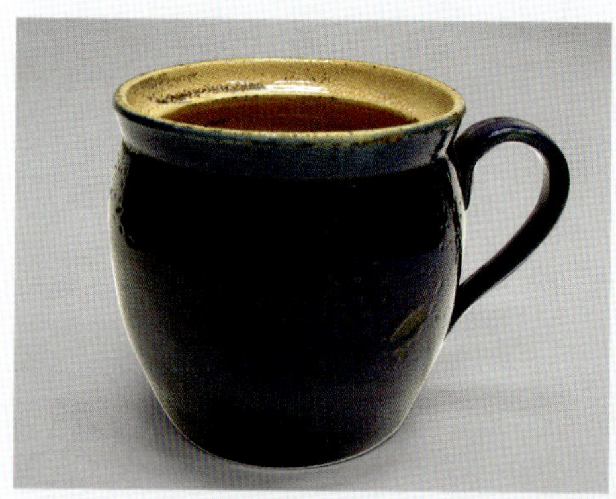

Towards the end of the year, Eduard was looking for a molder,[33] yet his factory had long since seen better days. Although he had started out with a relatively large capital base of his own, the factory was burdened with heavy debt – there was talk of 60 000 guilders. Sensing he was no longer capable of coping, Eduard committed suicide at the age of thirty-nine on 23 May 1892. He left a young wife and their one-year-old daughter, Frida.[34] The local paper, Nové Plzeňské Noviny, describes the events dramatically: Eduard was said to have disappeared several days before and was suspected to have fled. However, he returned by train to Pilsen, looked in at his home and went to the factory, which had already been officially sealed off by the authorities. Nonetheless he sat down in his office and had begun to write a letter when a workman suddenly rushed into the room and demanded the rest of his wages. Eduard gave him the last coins he had and told him to make up the difference by taking goods from the warehouse. When the man returned from the warehouse and the managing director had also arrived, Eduard handed the letter to the managing director, saying: "Give this letter to Mr S.; I have taken poison, nothing can be done for me now!" He turned pale soon afterwards and died.[35]

Eduard Goldscheider was buried next to his father, Moritz, and his mother, Julie; she had died in 1884. His estate was sequestrated.[36] For the warehouse sales, wash basins with accessories, majolica coffee and tea services (Ill. 9–11), jugs and plates in the

waaren in Majolika, Fayencegeschirr, Sanitätsgeschirr und künstliche Bimsteine [sic]" aus[32] (Ill. 6–8).

Noch Ende des Jahres sucht Eduard einen Formengießer[33], doch seine Fabrik ist längst nicht mehr mit Erfolg gesegnet. Obwohl er mit relativ viel Eigenkapital begonnen hatte, lasten auf der Fabrik hohe Schulden – von 60.000 Gulden ist die Rede. Er sieht sich der Lage nicht mehr gewachsen und nimmt sich mit 39 Jahren am 23. Mai 1892 das Leben. Er hinterlässt eine junge Frau und die einjährige Tochter Frida.[34] Die Lokalzeitung „Nové Plzeňské Noviny" beschreibt dramatisch die Geschehnisse: Eduard soll mehrere Tage nicht auffindbar gewesen sein und man vermutete bereits, dass er geflohen sei. Dann kehrte er jedoch mit dem Zug nach Pilsen zurück, schaute kurz zu Hause vorbei und ging in die Fabrik, die bereits amtlich abgeriegelt worden war. Er setzte sich dennoch in die Schreibstube und begann einen Brief zu schreiben, als ein Arbeiter plötzlich in das Zimmer stürmte und seinen restlichen Lohn verlangte. Eduard gab ihm seine letzten Münzen und bat ihn, sich die Differenz in Form von Waren aus dem Lager zu holen. Als dieser aus dem Lager zurückkehrte und auch der Betriebsleiter eintraf, übergab Eduard dem Betriebsleiter den Brief und sagte: „Geben Sie diesen Brief Herrn S., ich habe Gift genommen, mir ist nicht mehr zu helfen!" Kurz darauf wurde er bleich und starb.[35]
Eduard Goldscheider wird neben seinem Vater Moritz und seiner Mutter Julie, die bereits 1884 starb, begraben. Über seine Hinterlassenschaft wird der Konkurs verhängt.[36]

Ill. 9–11
Tee- und Kaffeeservice gehörten zu den populären Produkten der Pilsner Fabrik. Dieser Unterteller aus dem Japon-Service trägt den Eduard Goldscheider Adler-Stempel. (Quelle: Národopisné Muzeum Plzeňska, Inventar-Nr. 16754)

Tea and coffee services were among the most popular products made by the Pilsen factory. This saucer from the Japon service bears the Eduard Goldscheider eagle stamp. (Source: Národopisné Muzeum Plzeňska, inventory no. 16754)

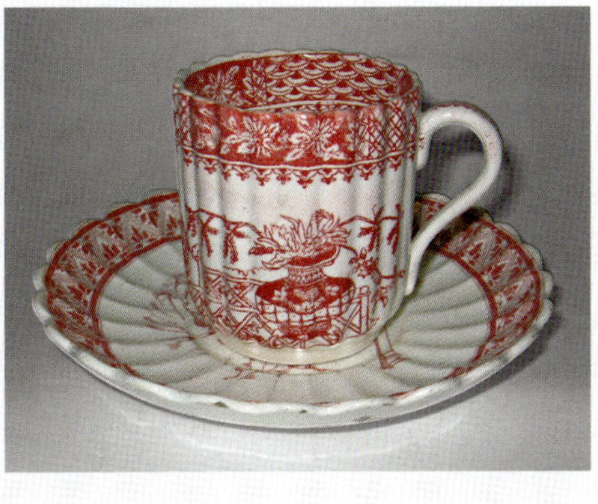

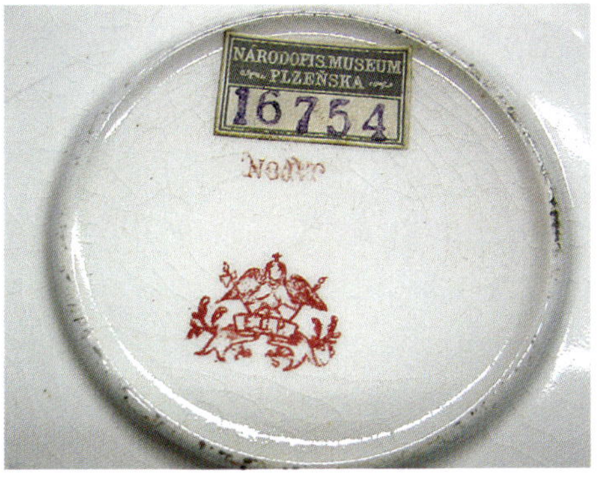

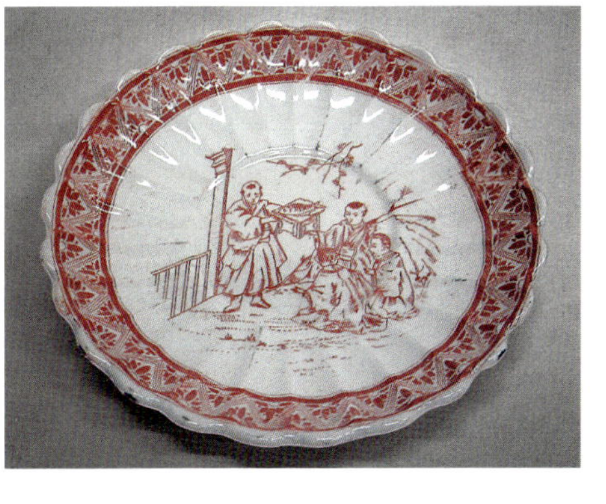

Im Ausverkauf des Lagerbestands werden Waschbecken mit Zubehör, Kaffee- und Teeservice (Ill. 9–11) aus Majolika, Krüge und Teller der „neuesten Pariser, Londoner und New Yorker Façon" angepriesen und die Leute zum Kauf dieser Waren aufgefordert, die „früher zu dem Mehrfachen der Preise" nur aus England bezogen werden konnten.[37]

Nach dem Tod von Eduard ist in den Pilsner Adressbüchern kein Mitglied der Goldscheider-Familie mehr zu finden. Doch auf der „Goldscheidrovka" wird weiterhin Keramik produziert: zunächst Tafelgeschirr in der Fabrik von Leopold Löwidt, ab Mitte der 1920er Jahre in der Fabrik von Adéla Rappová[38] (Ill. 12). Zudem dienen der anliegende Garten und die Gastwirtschaft den Pilsnern als beliebter Ausflugsort mit einem schönen Blick über die Stadt.[39] Die keramische Produktion wird dort erst um 1973 endgültig eingestellt, die Bezeichnung „Goldscheidrovka" erinnert aber heute noch an die Anfänge der berühmten Keramik-Manufaktur.

"newest Paris, London and New York fashion" were advertised and the public was encouraged to buy these wares, which previously could only be procured from England "for many times the price."[37]

After Eduard's death, no member of the Goldscheider family is listed in the Pilsen directories. Nevertheless, ceramic wares continued to be produced at the "Goldscheidrovka." At first tableware in Leopold Löwidt's factory, and from the mid-1920s in Adéla Rappová's factory[38] (Ill. 12). Moreover, the adjacent garden and inn were used by the residents of Pilsen as a popular place for day trips with a beautiful view over the city.[39] Ceramics production there did not cease forever until 1973. The name "Goldscheidrovka," however, still recalls today the beginnings of the celebrated ceramics factory.

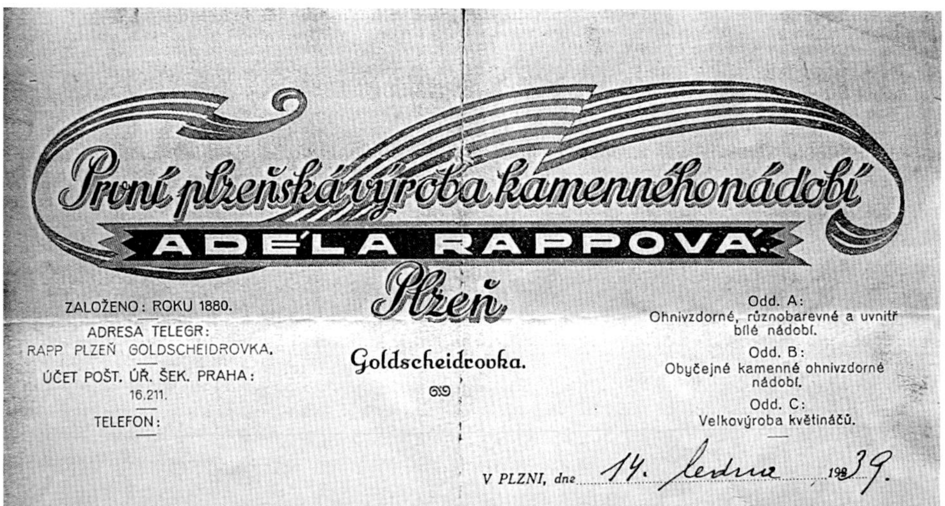

Ill. 12
Jahrzehnte nach dem Ende der Goldscheider-Produktion in Pilsen wird das Gelände am Rande der Stadt weiterhin „Goldscheidrovka" genannt. Die auf dem Grundstück sich befindende keramische Fabrik von Adéla Rappová setzt auf ihrem Briefpapier aus dem Jahr 1939 diese Bezeichnung ein. (Quelle: Archiv Goldscheider)

For decades after production ceased at Goldscheider in Pilsen, the premises on the outskirts of the city continued to be called "Goldscheidrovka." Adéla Rappová's ceramics factory located on the premises bears this name on the letter paper of 1939. (Source: Goldscheider Archives)

Notes

[1] Samuel Kurinsky: Gold and Silver Smithing; A Judaic Tradition. Part II – Europe and America. Hebrew History Federation, Factpaper 17-II, www.hebrewhistory.info
[2] Rosenzweig, p. 60. The family is said to have gone to Brennporitschen through the efforts of Mordechai Goldscheider, a Cabbalist.
[3] Geburtsmatrikel Brennporitschen, 1785–1835, Státní ústřední archív Praha, HBMa 1917
[4] Geburts-, Trauungs- und Sterbematrikel Pilsen, 1841–1869, Státní ústřední archív Praha, HBMa 1558
[5] Cf Bük (1908), p. 582
[6] Richemond (1999), p. 74
[7] Osobní Rejstřík, Historické Muzeum Plzeň (Registry of Names, Historical Museum, Pilsen)
[8] Geburts-, Trauungs- und Sterbematrikel Pilsen, 1841–1869, Státní ústřední archív Praha, HBMa 1558
[9] Špírková, p. 12f. A census conducted in 1850 does not list a single Jew; by 1854 there were 41 Jewish families comprising 249 individuals. Ibid, p. 18
[10] Rosenzweig, p. 60
[11] Špírková, p. 22
[12] Pilsner Bote, 8 Nov. 1863
[13] Geburts-, Trauungs- und Sterbematrikel Pilsen, 1841–1869, Státní ústřední archív Praha, HBMa 1558. This house was later used as a Jewish school. See Špírková, p. 36. Moritz Goldscheider died at his house, No. 285 Husstraße, on 25 March 1865.
[14] Osobní Rejstřík, Historické Muzeum Plzeň (Registry of Names, Historical Museum, Pilsen)
[15] Bloch (1872), p. 16. The haberdashery had already been "expunged," on 3 Aug. 1869.
[16] Ibid. In Prague the factory moved to 11 Langegasse (cf Allgemeines Adressen-Buch der königl. Hauptstadt Prag und vollständiges Verzeichnis handelsgerichtlich protokollierter Firmen, Bohemia Verlag, Prague 1875).
[17] Officieller General-Catalog, 1873 Vienna World's Fair, p. 513, No. 173, quoted in Neuwirth (1974), p. 25. In Vienna, the products were shown under the heading "Werk Moldau."
[18] Trauungsmatrikel Pilsen, 1868–1878, Státní ústřední archív Praha, HBMa 1549
[19] Lawsuit brought by Camillo Goldscheider, represented by Dr. Emil Frischauer, for admission as a registered partner in the firm of Friedrich Goldscheider with holdings amounting to 20 per cent, before the k.k. Handelsgericht [Royal and Imperial Trades Court] Vienna; filed on 28 October 1908. This sum was to be, according to his son, Camillo, paid out to Regine on Friedrich's death.
[20] Isrl. Geburtsmatrik vom 1. Jänner 1874 bis 31. Dezember 1900, Pilsen (Archív Hlavního Města Prahy; HBMa 1546)
[21] Hruška, p. 1088
[22] Pilsen reform, 18 April 1874
[23] Bloch (1876), p. 24 and p. 29
[24] Ibid, p. 44
[25] Konečný, p. 404
[26] Ibid. After some street and property names were changed in Pilsen in the 1870s, the changes were written down in 1879 in the "Orientirungs-Schematismus der kön. Kreisstadt Pilsen" and this shows the properties owned by the Goldscheider heirs. The house at 285 Hussstraße was divided into two houses: 2 Hussstraße and 1 Stefanspromenade. What had been the soda-water factory at 454 Klattauer Straße now went under the house number 11. The properties outside the city (513 In den Bory-Feldern) were amalgamated with those in Klattauer Straße under number 44 and the brickworks (No. 613) became 42 Klattauer Straße. Cf Orientirungs-Schematismus der kön. Kreisstadt Pilsen, Schiebl, Pilsen 1879, p. 44, p. 50, p. 52 and p. 57.
[27] Ibid
[28] Ibid
[29] Ibid. Late in 1880 Thieben was also granted privilege for manufacturing white and colored writing slates.
[30] N.N.: Goldscheidrovka – Plzeňská Hrnčírna, in: Nová Doba, 24 Dec. 1939
[31] Advertisement in: Sprechsaal, year 21, 1888, p. 134
[32] Julius Bük: Jubiläumsausstellung in Prag, in: Sprechsaal, No. 35, 26 Aug. 1891, p. 695
[33] Post advertised in: Sprechsaal, No. 43, 22 Oct. 1891, p. 867
[34] Isrl. Geburtsmatrik vom 1. Jänner 1874 bis 31. Dezember 1900, Pilsen (Archív Hlavního Města Prahy; HBMa 1546). They lived together at 4 Hussstraße.
[35] Cf Nové Plzeňské Noviny, 25 May 1892
[36] Plzeňské Listy, 31 May 1892
[37] Nové Plzeňské Noviny, 30 July 1892
[38] Konecný, pp. 404–405
[39] Goldscheidrovka – Plzeňská Hrnčírna, in: Nová Doba, 24 Dec. 1939

Anmerkungen

[1] Samuel Kurinsky: Gold and Silver Smithing; A Judaic Tradition. Part II – Europe and America. Hebrew History Federation, Factpaper 17-II, www.hebrewhistory.info
[2] Rosenzweig, S. 60. Nach Brennporitschen soll die Familie durch den Kabbalisten Mordechai Goldscheider gekommen sein.
[3] Geburtsmatrikel Brennporitschen, 1785–1835, Státní ústřední archív Praha, HBMa 1917
[4] Geburts-, Trauungs- und Sterbematrikel Pilsen, 1841–1869, Státní ústřední archív Praha, HBMa 1558
[5] Vgl. Bük (1908), S. 582
[6] Richemond (1999), S.74
[7] Osobní Rejstřík, Historické Muzeum Plzeň (Personenregister des Historischen Museums in Pilsen)
[8] Geburts-, Trauungs- und Sterbematrikel Pilsen, 1841–1869, Státní ústřední archív Praha, HBMa 1558
[9] Špírková, S. 12ff. Bei einer Volkszählung im Jahr 1850 werden in Pilsen keine Juden vermerkt, im Jahr 1854 sind es 41 jüdische Familien mit 249 Familienmitgliedern. Ebenda, S. 18
[10] Rosenzweig, S. 60
[11] Špírková, S. 22
[12] Pilsner Bote, 8.11.1863
[13] Geburts-, Trauungs- und Sterbematrikel Pilsen, 1841–1869, Státní ústřední archív Praha, HBMa 1558. Dieses Haus dient später als jüdische Schule. Siehe Špírková, S. 36. Moritz Goldscheider stirbt am 25.3.1865 in seinem Haus in der Hussstraße Nr. 285.
[14] Osobní Rejstřík, Historické Muzeum Plzeň (Personenregister des Historischen Museums in Pilsen).
[15] Bloch (1872), S. 16. Die gemischte Warenhandlung wurde bereits am 3.8.1869 „gelöscht".
[16] Ebenda. In Prag zieht die Fabrik dann in die Langegasse 11 (vgl. Allgemeines Adressen-Buch der königl. Hauptstadt Prag und vollständiges Verzeichnis handelsgerichtlich protokollierter Firmen, Bohemia Verlag, Prag 1875).
[17] Officieller General-Catalog, Weltausstellung 1873 in Wien, S. 513, Nr. 173, zitiert nach Neuwirth (1974), S. 25. In Wien stellt man unter der Bezeichnung Werk Moldau aus.
[18] Trauungsmatrikel Pilsen, 1868–1878, Státní ústřední archív Praha, HBMa 1549
[19] Klage von Camillo Goldscheider, vertreten durch Dr. Emil Frischauer um Aufnahme als öffentlicher Gesellschafter in die Firma Friedrich Goldscheider mit einer Beteiligung von 20 Prozenten, beim k.k. Handelsgericht Wien, eingelangt am 28. Oktober 1908. Dieser Betrag soll, laut dem Sohn Camillo, Regine nach dem Tod Friedrichs wieder ausbezahlt worden sein.
[20] Isrl. Geburtsmatrik vom 1. Jänner 1874 bis 31. Dezember 1900, Pilsen (Archív Hlavního Města Prahy; HBMa 1546)
[21] Hruška, S. 1088
[22] Pilsner Reform, 18.4.1874
[23] Bloch (1876), S. 24 und 29
[24] Ebenda, S. 44
[25] Konečný, S. 404
[26] Ebenda. Nachdem in Pilsen in den 1870er Jahren einige Straßen und Grundstücke umbenannt worden sind, werden 1879 im „Orientirungs-Schematismus der kön. Kreisstadt Pilsen" die Änderungen festgehalten, und es ist daraus der Immobilienbesitz der Goldscheider Erben ersichtlich. Die Hussstraße 285 wurde in zwei Häuser aufgeteilt: die Hussstraße 2 und Stefanspromenade 1. Die ehemalige Sodawasserfabrik in der Klattauer Straße 454 trägt nun die Hausnummer 11. Die Grundstücke außerhalb der Stadt (In den Bory-Feldern 513) werden in die Klattauer Straße mit der Nummer 44 eingegliedert und die Ziegelbrennerei mit Ringofen (Nr. 613) wird zur Klattauer Straße 42. Vgl. Orientirungs-Schematismus der kön. Kreisstadt Pilsen, Schiebl, Pilsen 1879, S. 44, 50, 52 und 57.
[27] Ebenda
[28] Ebenda
[29] Ebenda. Ende 1880 erhält auch Thieben das Privilegium zur Erzeugung von weißen und farbigen Schreibtafeln.
[30] N.N.: Goldscheidrovka – Plzeňská Hrnčírna, in: Nová Doba, 24.12.1939
[31] Reklame in: Sprechsaal, Jg. 21, 1888, S. 134
[32] Bük , Julius: Jubiläumsausstellung in Prag, in: Sprechsaal, Nr. 35, 26.8.1891, S. 695
[33] Stellenanzeige in: Sprechsaal, Nr. 43, 22.10.1891, S. 867
[34] Isrl. Geburtsmatrik vom 1. Jänner 1874 bis 31. Dezember 1900, Pilsen (Archív Hlavního Města Prahy; HBMa 1546). Gemeinsam wohnen sie in der Hussstraße 4.
[35] Vgl. Nové Plzeňské Noviny, 25.5.1892
[36] Plzeňské Listy, 31.5.1892
[37] Nové Plzeňské Noviny, 30.7.1892
[38] Konečný, S. 404–405
[39] Goldscheidrovka – Plzeňská Hrnčírna, in: Nová Doba, 24.12.1939

Friedrich Goldscheider und die Entstehung einer internationalen Manufaktur in der Zeit des Historismus in Wien (1878–1897)

Friedrich Goldscheider and the Creation of an International Manufactory in the Period of Historicism in Vienna (1878–1897)

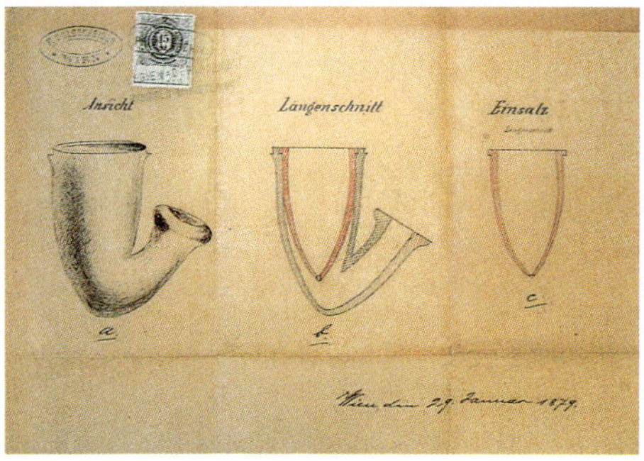

Ill. 13
Friedrich Goldscheider produziert in Wien auch Porzellanpfeifen, für die er im Jahr 1879 ein Privileg erhält. Auf dieser Skizze ist links oben einer der ersten Goldscheider-Stempel zu sehen, der je nach Herstellungsort die Aufschrift Wien oder Carlsbad trägt. (Quelle: Privileg 1879/001146)

In Vienna, Friedrich Goldscheider also produced porcelain pipes, for which he was granted a privilege in 1879. This sketch shows, top left, one of the first Goldscheider stamps, which bore, depending on where the piece was manufactured, the Vienna or Carlsbad address. (Source: Privilege 1879/001146)

Die Wiener Keramik ist in der zweiten Hälfte des 19. Jahrhunderts durch Stagnation gekennzeichnet. Das keramische Schaffen beschränkt sich auf die Porzellanmalerei, die zum größten Teil auf böhmischem Porzellan erfolgt und noch unter dem Einfluss der seit 1864 geschlossenen Wiener Porzellanmanufaktur steht.[40] Bevor der älteste der Goldscheider Brüder, Siegfried Friedrich, um 1878 nach Wien zieht und dort eine Porzellanhandlung im 9. Bezirk eröffnet, lässt er schon in den Jahren zuvor Porzellan in Karlsbad produzieren und handelt mit diesem.[41] Zusammen mit seiner Frau Regine bietet er in der Wiener Lichtensteinstraße 9 eine „reiche Auswahl in Kaffee-, Tafel-, Wasch-Service etc. – stets Nouveautés und Spezialitäten in articles de Luxues [sic]" an, die größtenteils wohl aus Karlsbad und Pilsen bezogen werden.[42] Man produziert aber auch Pfeifen aus Porzellan (Ill. 13) und erhält hierfür im Jahr 1879 ein Privilegium.[43]

Die Anfänge in Wien sind durch viele Wechsel gekennzeichnet. Um 1880 dient die Lichtensteinstraße auch als Niederlassung der „Goldscheider'schen Porzellan-Manufaktur und Porzellanmalerei", die im Jahr darauf in die Wipplingerstraße 39 umzieht. Die Porzellanhandlung und Porzellanmalerei werden dann allerdings nicht mehr aufgeführt,[44] das Porzellan wird vermutlich direkt in Karlsbad bemalt.[45] Laut dem Adressbuch der Keramischen Industrie aus dem Jahr 1887 besteht die Goldscheider Fabrik seit 1882 und die Produkte werden mit „F.G." oder „Vindabona" gekennzeichnet.[46] Gerade in der An-

Viennese ceramics stagnated noticeably in the latter half of the 19th century. Limited to porcelain painting, most of which was done on Bohemian porcelain, Viennese ceramics were still under the sway of the Vienna State Porcelain Factory, which had shut down in 1864.[40] The eldest Goldscheider brother, Siegfried Friedrich, moved to Vienna in 1878, where he opened a porcelain shop in the 9th district. In the years before he had had porcelain manufactured in Carlsbad for trade.[41] Working with his wife, Regine, at 9 Lichtensteinstraße, Vienna, he sold a "wide selection of coffee services, tableware, wash basins and ewers, etc – always nouveautés and specialities in articles de luxues [sic]," most of which probably came from Carlsbad and Pilsen.[42] Porcelain pipes were also made (Ill. 13); a privilege was granted for manufacturing them in 1879.[43]

The beginnings in Vienna were marked by change. In 1880, Lichtensteinstraße was also the address of the "Goldscheider'sche Porzellan-Manufaktur und Porzellanmalerei" [Goldscheider Manufactory and Porcelain Painting Workshop], which moved a year later to 39 Wipplingerstraße. The porcelain shop and porcelain painting workshop are no longer listed after that;[44] the porcelain was probably painted in Carlsbad before shipment to Vienna.[45] According to the 1887 Ceramics Industry Directory, the Goldscheider Manufactory existed from 1882 and its products were marked either "F.G." or "Vindabona."[46] In the early period especially, the marking of the products was still relatively

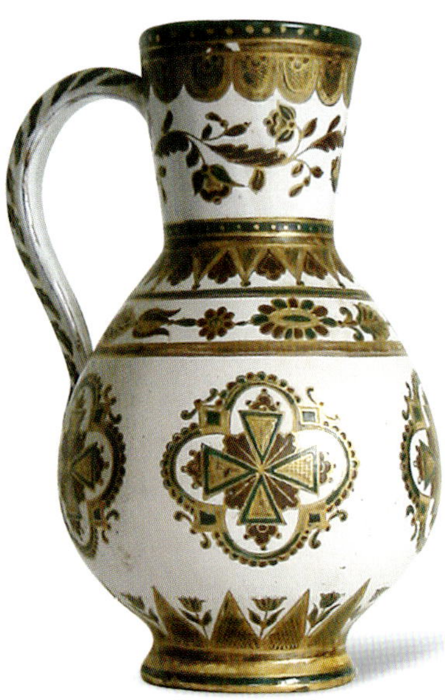

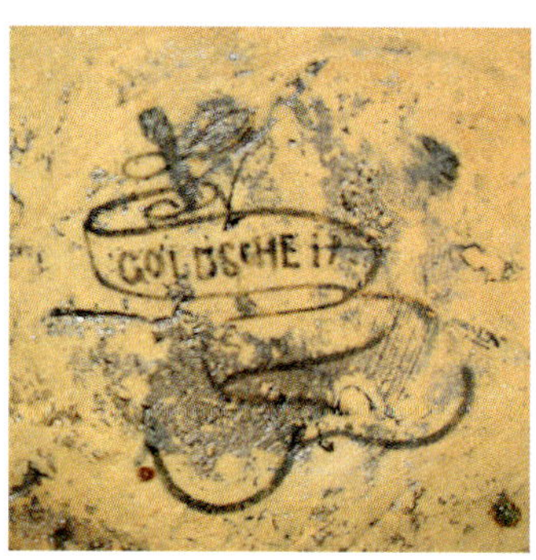

Ill. 14, 15
Zu den frühesten Wiener Erzeugnissen gehören auch Krüge. Dieser mit Blumenranken dekorierte und 28 cm hohe Krug ist wie die meisten frühen Modelle ohne eine Modellnummer markiert und trägt den so genannten Anker-Stempel. (Sammlung f@goldscheider.de)

Jugs are among the earliest Viennese products. This 28-cm- high jug decorated with flower sprays is, like most of the early models, marked without a model number and bears what is known as the anchor stamp. (Collection f@goldscheider.de)

unstandardized. Alongside the mentioned marks, the anchor stamp (Ill. 14 and 15) and the stamp "F. Goldscheider" used on the porcelain pipe were also common (Ill. 13).

Officially, however, 1885 was recorded in the annals of the firm as the year it was established. That year the "Goldscheider'sche Porzellan-Manufaktur und Majolica-Fabrik" ["Goldscheider Porcelain Manufactory and Majolica Factory"] showed porcelain services, porcelain luxury items, household tableware, majolica pieces and pottery at the Antwerpen exhibition.[47] The porcelain articles were procured in Carlsbad and marked "Friedrich Goldscheider Porzellan-Manufactur Carlsbad;" the other wares were made in Vienna (Ill. 16). The Goldscheider range also included snuff-boxes.[48] These products were stamped but not numbered.

By then Friedrich – in Vienna he left off his first name, Siegfried – already had three sons: Arthur and Camillo, born in Pilsen, and Walter, who was born in Vienna on 18 March 1879. His youngest son, Marcell, was born almost a decade later, on 25 May 1887.[49]

Since Friedrich Goldscheider's motto was "Standing still is the beginning of the end," he devoted an enormous amount of energy to building up his business.[50] He not only mastered manufacturing; he also had the necessary aesthetic sense and far-sighted commercial acumen needed. He was the first in Vienna to believe that artistically made and painted terracotta figurines would be successful.[51] By around 1887, therefore, he was calling his manufactory "Terracotta and Majolica Factory," or "First Viennese Terracotta Factory," and moved into more spacious premises at Johannesgasse, Vienna-Währing.[52] This building did not house the factory alone; the Goldscheider family also lived there.

When works are made of clay, the object must – unlike cast metal pieces or sculpture in stone – be completely modelled before firing. Firing is what gives the piece hardness, gloss and durability so that it can no longer be deformed.[53] Whereas the material is regarded as relatively inexpensive, the

fangszeit ist aber die Markierung der Produkte noch relativ bunt gemischt, neben den erwähnten Kennzeichnungen sind auch der Anker-Stempel (Ill. 14 und 15) sowie der auf der Porzellanpfeife eingesetzte Stempel „F. Goldscheider" üblich (Ill. 13).

Offiziell geht aber das Jahr 1885 als Gründungsjahr in die Firmenannalen ein und die „Goldscheider'sche Porzellan-Manufactur und Majolica-Fabrik" stellt im selben Jahr noch auf der Antwerpener Ausstellung Porzellanservice, Porzellanluxusgegenstände, Gebrauchsgeschirr, Majoliken und Tonwaren aus.[47] Die Porzellanwaren werden unter der Bezeichnung „Friedrich Goldscheider Porzellan-Manufactur Carlsbad" aus Karlsbad bezogen, die anderen Gegenstände stammen aus der Wiener Produktion (Ill. 16). Aber auch Tabakbehälter gehören zum Sortiment.[48] Die Erzeugnisse erhalten zwar einen Stempel, werden jedoch nicht nummeriert.

Zu dieser Zeit hat Friedrich – in Wien lässt er seinen ersten Vornamen Siegfried weg – bereits drei Söhne: die in Pilsen geborenen Söhne Arthur und Camillo sowie Walter, der am 18. März 1879 in Wien zur Welt kommt. Fast ein Jahrzehnt später, am 25. Mai 1887, wird sein jüngster Sohn Marcell geboren.[49]

Das Motto von Friedrich Goldscheider ist „Stillstand ist der Anfang vom Ende", und so treibt er energisch den Aufbau seines Unternehmens voran.[50] Er beherrscht nicht nur die Fabrikation, sondern hat auch das notwendige künstlerische Verständnis und den kaufmännischen Weitblick. Er glaubt als Erster in Wien an den Erfolg von künstlerisch ausgeführten und bemalten Terrakotta-Figuren.[51] Bereits um 1887 bezeichnet er deshalb seine Manufaktur als „Terracotta und Majolikafabrik" bzw. als „Erste Wiener Terracottafabrik" und bezieht größere Räumlichkeiten in der Johannesgasse 11, Wien-Währing.[52] In diesem Haus sind nicht nur die Fabrikräume, auch die Familie Goldscheider lebt hier.

Bei der Herstellung von tönernen Werken muss – im Gegensatz zum Metallgussverfahren oder der Steinbildhauerei – das Objekt vor dem Brand vollständig modelliert sein. Denn mit dem Brand erhält das Stück seine Härte, Glanz und Beständigkeit und

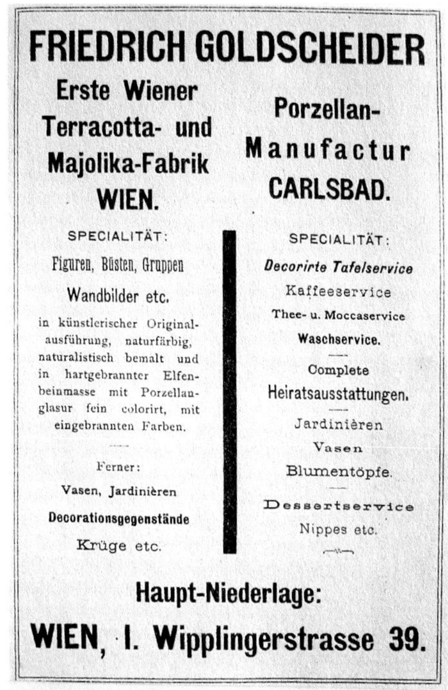

Ill. 16
Aus einer Anzeige zur Wiener Jubiläumsgewerbeausstellung im Jahr 1888 wird die Produktion der Karlsbader „Porzellan-Manufactur" ersichtlich. Seit den 1870er Jahren werden aus Karlsbad verschiedenste Produkte bezogen und bemalt sowie mit einem eigenen Stempel versehen. (Quelle: Jubiläumskatalog Gewerbeausstellung Wien 1888)

An advertisement for the Vienna 1888 Jubilee Trade Fair reveals that porcelain was being made at the Carlsbad "Porcelain Manufactory." Since the 1870s, products of many kinds had been procured from Carlsbad, painted and stamped with a proprietary stamp. (Source: Jubiläumskatalog Gewerbeausstellung Vienna 1888)

kann nicht mehr verformt werden.[53] Während das Material als relativ preiswert gilt, ist der Arbeitsprozess aufwendig und lohnintensiv. Um optimale Leistungen zu erzielen, ist deshalb auch eine intensive Zusammenarbeit zwischen Künstlern und Technikern notwendig. Einen wichtigen Beitrag leistet hierbei Friedrichs Bruder Alois, der seit 1884 mit ihm in Wien als technischer und kommerzieller Leiter zusammenarbeitet. Später wird Alois Prokurist und erhält eine Umsatzbeteiligung.[54]

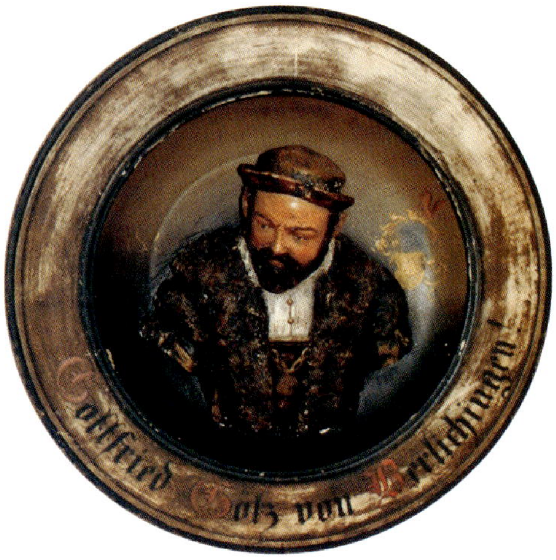

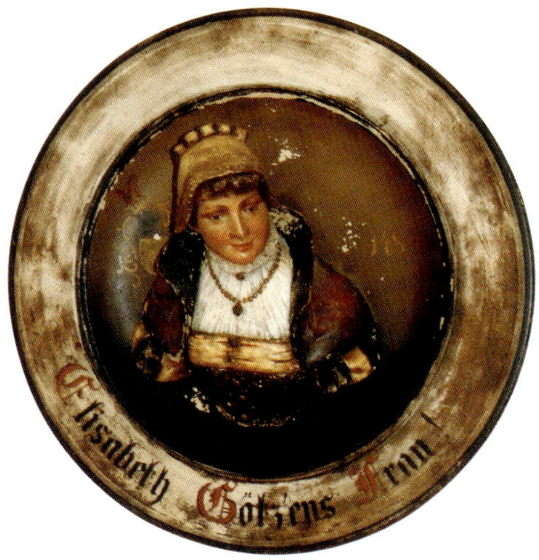

Ill. 17, 18

Die mit historischen Motiven oder Reliefs ausgestatteten Wandteller dienen auch als Wandbilder. Auf diesen großen Wandtellern mit 40 cm Durchmesser werden Götz von Berlichingen und seine Frau Elisabeth gezeigt. (Sammlung Perl, Nürnberg)

Chargers decorated with historical motifs are also made to be hung on walls as wall plates. Götz von Berlichingen and his wife Elisabeth are depicted on these large chargers, which are 40 cm in diameter. (Perl Collection, Nürnberg)

Anfangs werden bei Goldscheider Terrakotta-Figuren, Büsten, Bilder, Vasen, Schalen und „Decorations-Fantasiegegenstände" wie auch die so genannten „Wiener Typen" produziert. Der neuen Manufaktur fehlen noch geschulte Fachkräfte und die Modelle wirken noch etwas fad, die Bemalung ist zu „dick und klebrig, teils zu licht, teils zu dunkel".[55] Dennoch werden die Erzeugnisse auf der Leipziger Michaelis-Messe im Jahr 1887 ausgestellt.[56] Genauso wie ein Jahr später auf der Jubiläumsausstellung in Wien, wo sie Anklang beim Publikum finden. Friedrich beschickt konsequent weitere Messen, vor allem die Teilnahme an der Pariser Ausstellung im Jahr 1889 zeigt einen Wandel in Plastik und Malerei, in der Wahl der Modelle und in der Darstellung.[57] In dieser Zeit setzt auch die Nummerierung der Goldscheider Erzeugnisse ein, die mit einer fortlaufenden Fabriknummer (später Modellnummer genannt) und der jeweiligen Stückzahl versehen werden. Die Nummern werden zunächst eingeritzt, später mit einem Stempel in die lederharte Masse eingeprägt. Manche Modelle sind gesetzlich geschützt und dementsprechend markiert: so ist neben der Beschriftung „Gesetzlich geschützt" (s. ModNr. 214) auch „Vervielfältigung vorbehalten" üblich, für den Export werden die Modelle mit „Bréveté S.G.D.G" oder „Reproduction Réservée" gemarkt Im Gegensatz zu Büsten wird bei der Fabriknummer von vielen Vasen, Schalen und Reliefs die Ziffer 0 zur Unterscheidung vorgestellt.

*Originaltitel der Modelle stehen stets in Anführungszeichen. Neu vergebene, beschreibende Titel sind ohne Anführungszeichen eingefügt.

*Only the original titles of the models are given here in double quotation marks. New, describing titles have been used without being placed in quotation marks.

process of working it is time-consuming and causes high labor-cost. Intensive collaboration between artists and technicians is necessary for achieving optimum results. Friedrich's brother Alois, who had worked with him in Vienna since 1884 as technical and managing director, contributed substantially to this side of the business. Alois later became his deputy with powers of attorney and was entitled to receiving his share of the sales proceeds.[54]

Initially, Goldscheider produced terracotta figurines, busts, pictures, vases, bowls and "ornamental fantasy objects" as well as what were known as "Viennese Types". Since the new factory lacked trained specialists, the models still look rather insipid; the painting is too "thick and sticky, partly too light, partly too dark."[55] Nevertheless, the products were exhibited at the 1887 Michaelis-Messe [Michaelmas Trade Fair] in Leipzig.[56] They were also shown a year later at the Jubilee Exhibition in Vienna, where they made a hit with the public. Friedrich continued to pursue a consistent policy of exhibiting pieces at other trade fairs. The work he showed at the 1889 Paris Exhibition especially shows changes in both sculpture and painting, in the selection of models and in the handling of representation.[57] This was the time when numbering of the Goldscheider products set in: from now on they are marked with the consecutive factory number (later called "model number") and the individual piece number. The numbers were at first incised, but later impressed with a stamp in the body while it was in the leather-hard state. Some models are copyright and marked accordingly: "Vervielfältigung vorbehalten" ["reproduction reserved"] is usually next to "Gesetzlich geschützt" ["protected by law"]. Models for export are marked "Bréveté S.G.D.G" or "Reproduction Réservée." The factory number of many vases, bowls and reliefs is preceded by the numeral "0" to differentiate them, but this is not the case with the busts.

Professor Alexander Schmidt, a contemporary, described the objects exhibited under the auspices of the 1889 Paris Universal Exposition as follows: "Goldscheider has gathered round himself quite a number of young, fresh artisans, who provide him with the models for the Viennese and the numerous national types of the Austro-Hungarian monarchy. Even though the Viennese figures, the gentlemen of fashion, mocked as 'fops' in vernacular parlance, and the ladies from the Ringstrasse reveal a satirical treatment of the ludicrous whimsies of fashion, the artificial poses and the affected demeanor, – the representations of the national types adhere entirely to perceptive characterization and realistic truth. These itinerant tinkers of the Slovak tribe, gypsies, shepherds from the Hungarian Puszta, the Hanna from Moravia, the Bosnians and many other costume figures are all superbly modelled, captured from life, as they stand and move, without academic composition and posturing. Naturalness and truth are the aim and in this the painting, for which pure oil paint is used, also helps considerably. Unabashedly and without reservations, the coloring approaches the full appearance of reality."[58]

A desire for sculpture painted in colors was a primary concern of sculptors of naturalistic bent at that time. In addition, the eclectic character of late Historicism, which permitted diversity both in choice of materials and design, made it possible for Friedrich to experiment with all sorts of different materials, including terracotta, hard-paste and phosphatic soft-paste porcelain. With this broad range of art objects, Friedrich Goldscheider won a gold medal in Paris just as he had been distinguished the year before in Barcelona and Vienna (Ill. 19).[59]

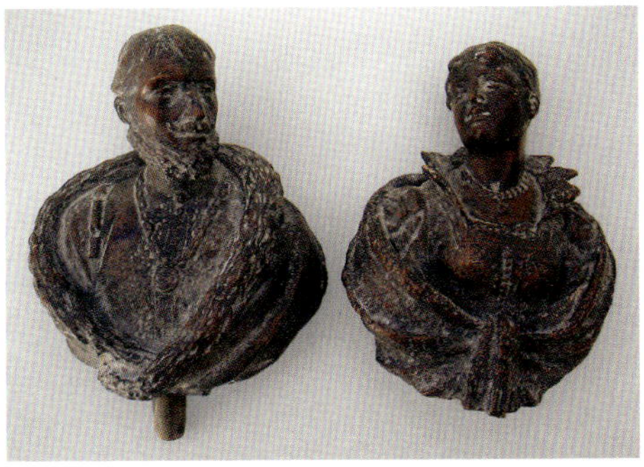

ModNr. 213, 214* Henry VIII und I and Anne Boleyn, ca. 1890, Collection f@goldscheider.de
Die beiden Büsten mit der Modellnummer 213 und 214 zeigen Henry VIII. und seine Frau Anne Boleyn. Auf der Rückseite der Büste von Anne Boleyn wird die Firmenmarke ersichtlich, die geritzte Modellnummer 214 sowie auch die Stückzahlnummer 3. Zudem ist noch „Gesetzlich Geschützt" eingeritzt. Die Signatur mit Künstlernamen ist in der Anfangsphase von Goldscheider sehr selten. Bei beiden Büsten fehlt der sonst übliche Holzsockel.

The two busts marked with model numbers 213 and 214 depict Henry VIII and Anne Boleyn, one of his wives. The firm mark, the model number 214 and the piece number 3 are discernible on the back of the Anne Boleyn bust. In addition, "Gesetzlich Geschützt" ["Copyright"] is incised. Signatures with artists' names are very rare in the early phase at Goldscheider. The two busts are without the otherwise usual wooden pedestal.

Professor Alexander Schmidt beschreibt die ausgestellten Objekte im Rahmen der 1889 in Paris stattfindenden Weltausstellung wie folgt: „Goldscheider hat eine ganze Anzahl junger, frischer Kunstkräfte um sich gesammelt, welche ihm die Modelle zu den wienerischen und den zahlreichen nationalen Typen der österreich-ungarischen Monarchie liefern. Lassen die Wiener Figuren, die Modeherren, vom spöttischen Volksmund ‚Gigerln' genannt, und die Damen der Ringstraße eine satyrische Behandlung der lächerlichen Modethorheiten, der gezierten Haltung und des affectirten Wesens erkennen, so halten sich die Darstellungen der nationalen Typen allein an scharfe Charakteristik und lebensvolle Wahrheit. Diese Rastelbinder slovakischen Stammes, die Zigeuner, die Hirten der ungarischen Puszta, die Hannaken aus Mähren, die Bosniaken und viele andere Costümfiguren sind alle vortrefflich modellirt, aus dem Leben

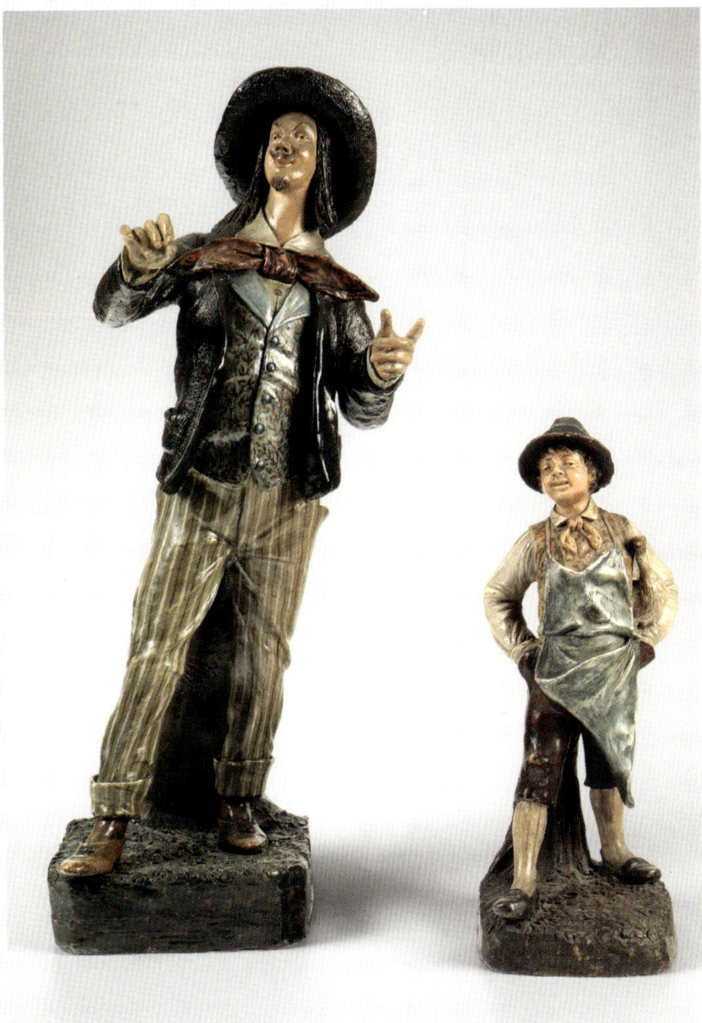

ModNr. 259, 333* „Wiener Typen" I "Viennese Types" ca. 1890, © Wien Museum

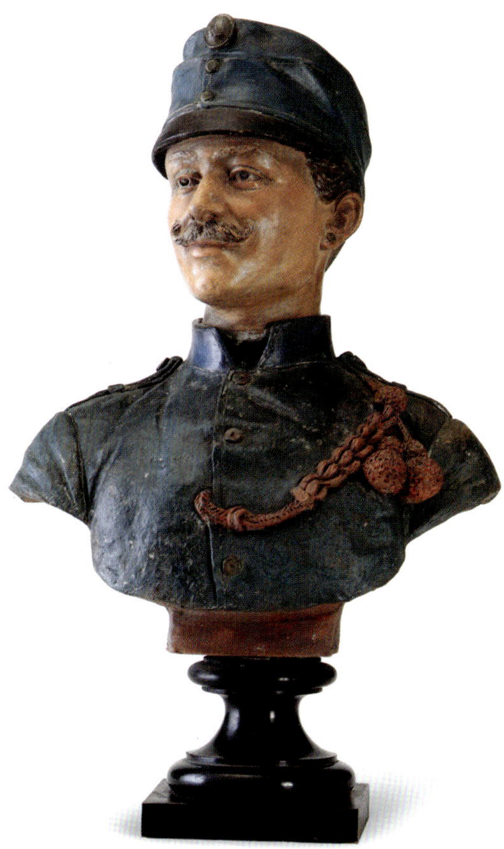

ModNr. 237 Offizier I Officer, ca. 1888, MAK – Österreichisches Museum für angewandte Kunst / Gegenwartskunst, Wien
Die Repräsentanten der k. u. k Monarchie, wie dieser Offizier auf Holzsockel, gehören genauso zum beliebten Sortiment wie die „Wiener Typen".

The representatives of the Royal and Imperial monarchy, like this officer on a wooden pedestal, are along with the "Viennese Types" part of the popular range.

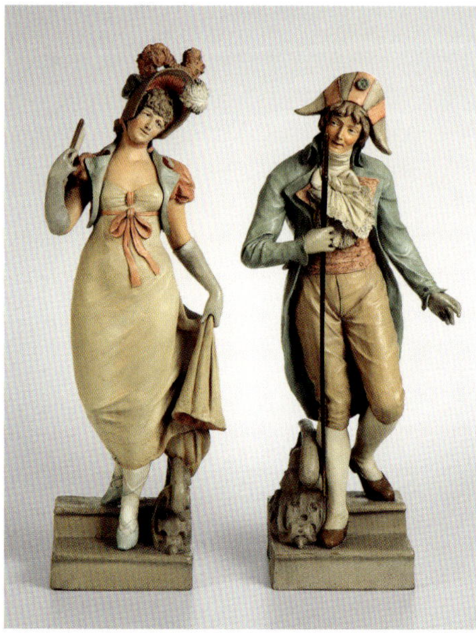

ModNr. 660, 649
Dame und Herr I
Lady and Gentleman,
ca. 1892,
© Wien Museum

herausgegriffen und hingestellt, wie sie stehen und gehen, ohne akademische Composition und Pose. Natürlichkeit, Wahrheit sind das Ziel und dazu hilft nun sehr wesentlich auch die Bemalung, zu welcher reine Oelfarbe dient. Die Farbe geht unerschrocken und rückhaltlos an die volle farbige Erscheinung der Wirklichkeit heran."[58]

Der Wunsch nach farbig bemalten Skulpturen ist eines der Hauptanliegen der damaligen Bildhauer naturalistischer Richtungen. Hinzu kommt die eklektizistische Charakteristik des späten Historismus, die eine Vielfalt an Materialien und Gestaltungsweisen erlaubt und die es Friedrich ermöglicht, mit verschiedenen Materialien wie Terrakotta, Porzellan- und Elfenbeinmasse zu experimentieren. Mit dieser breit gefassten Palette von Kunstgegenständen gewinnt Friedrich Goldscheider in Paris eine Goldene Medaille, genauso wie er im Jahr zuvor schon in Barcelona und Wien ausgezeichnet worden ist (Ill. 19).[59]

Im Jahr 1890 erhält er die Goldene Medaille in Edinburgh[60], ein Jahr später kommt die Goldene Medaille in Triest hinzu, genauso wie der Erste Preis der keramischen Ausstellung in Dresden[61]. Hier erzielen vor allem bronzierte, polychromierte Figuren Aufmerksamkeit, die nach einem besonderen Verfahren hergestellt werden.[62] Für dieses

In 1890 he was awarded a gold medal in Edinburgh.[60] It was followed a year later by a gold medal in Triest as well as first prize at the Dresden ceramics exhibition.[61] In Dresden, bronzed, polychromed figures made by a special process attracted the most attention.[62] On 19 November 1892 Friedrich Goldscheider was granted a privilege (Ill. 20) for this process, which lent objects of clay, terracotta, porcelain, faience, plaster of Paris and even concrete the "appearance of patinated bronze objects."[63] Clay was relatively inexpensive and disparaged by many as inferior. With this new, patented process he succeeded in making clay the appearance of a more "noble material." Because it was so much more inexpensive, the middle class could now afford art objects as part of their interior decoration schemes. Thus Friedrich Goldscheider opened up a broad, new clientele for himself – both in the Austro-Hungarian monarchy and abroad. After all, regular participation in international trade

Ill. 19
Die Leipziger Messe gehört zu den wichtigen Ausstellungen für Friedrich Goldscheider. In den Jahren 1888 bis 1891 gewinnt Friedrich Goldscheider international mehrere wichtige Auszeichnungen. (Quelle: Anzeige in Sprechsaal, Nr. 38, vom 17.9.1891, S. 765)

The Leipzig Trade Fair was one of the most important exhibition venues for Friedrich Goldscheider. In the years from 1888 to 1891 Friedrich Goldscheider won several important international awards. (Source: advertisement in Sprechsaal, No. 38, of 17 Sept. 1891, p. 765)

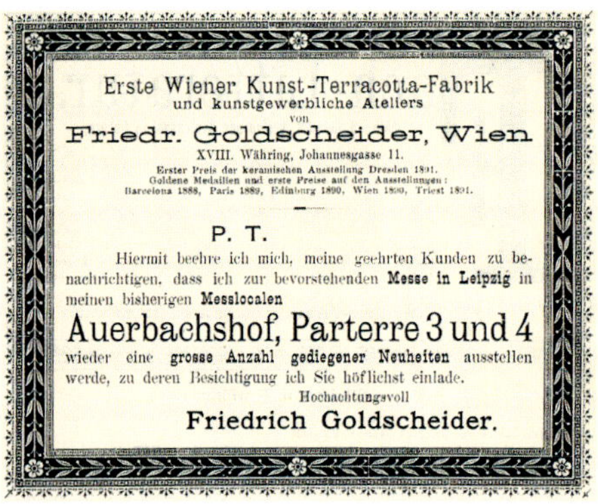

Erste Wiener Kunst-Terracotta-Fabrik
und kunstgewerbliche Ateliers
von
Friedr. Goldscheider, Wien
XVIII. Währing, Johannesgasse 11.
Erster Preis der keramischen Ausstellung Dresden 1891.
Goldene Medaillen und erste Preise auf den Ausstellungen:
Barcelona 1888, Paris 1889, Edinburg 1890, Wien 1890, Triest 1891.

P. T.

Hiermit beehre ich mich, meine geehrten Kunden zu benachrichtigen, dass ich zur bevorstehenden **Messe in Leipzig** in meinen bisherigen **Messlocalen**

Auerbachshof, Parterre 3 und 4

wieder eine **grosse Anzahl gediegener Neuheiten** ausstellen werde, zu deren Besichtigung ich Sie höflichst einlade.

Hochachtungsvoll
Friedrich Goldscheider.

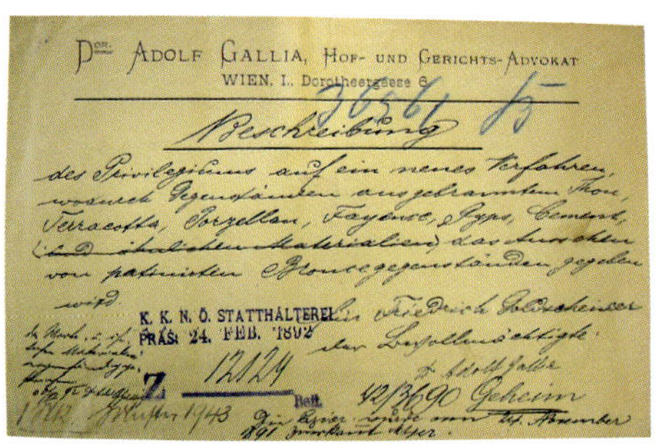

Dr. Adolf Gallia, Hof- und Gerichts-Advokat
Wien, I., Dorotheergasse 6

Beschreibung

des Privilegiums auf ein neues Verfahren, wodurch Gegenständen aus gebranntem Thon, Terracotta, Porzellan, Fayence, Gyps, Cement, das Aussehen von patinirten Bronzegegenständen gegeben wird.

K. K. N. Ö. STATTHALTEREI
PRÄS: 24. FEB. 1892
Z. 12124

Ill. 20
Ein neues, patentiertes Verfahren, das Gegenständen aus verschiedenen Materialien das Aussehen von patinierten Bronzegegenständen verleiht, erweist sich als besonderer Erfolg. (Quelle: Privileg 1892/003079)

A new, patented process for giving objects made of various materials the appearance of patinated bronzes proved particularly successful. (Source: Privilege 1892/003079)

fairs was not just necessary for building up the Goldscheider brand; it also made it possible to continuously broaden the reach of the export market.

A rising sales volume and success after success rewarded Friedrich Goldscheider's entrepreneurial drive and creativity. The progress he was making achieved recognition at the 1891 Viennese Kunstgewerbeverein exhibition: "Artistic and technical progress is discernible; new models and original products follow on each other in rapid succession, proving the extraordinary spate of activity that has necessitated enlarging the business premises. We again encounter numerous new figurines and groups designed by aspiring young sculptors who are glad to turn to industry. Among them are: the Tyrolese couple bought by the widowed Crown Princess, Sarah Bernhard as Theodora, King Lear, busts of women, and the leading figures from Blaas's paintings. The oil paintings on pottery chargers executed after

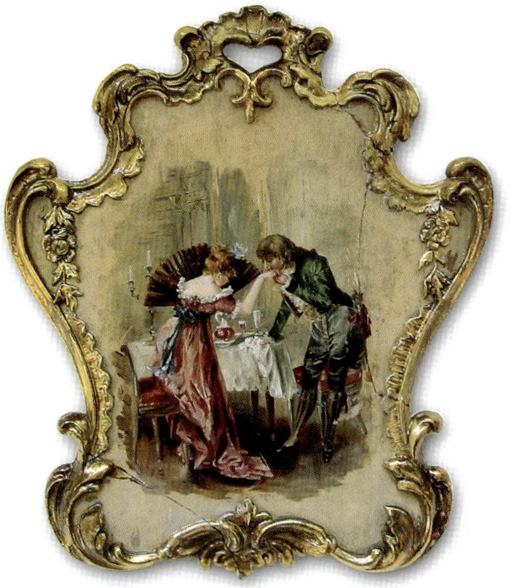

Ill. 21
Wandbilder gehören zu den populären Objekten der frühen Produktionszeit. Diese Handkuss-Szene mit goldfarbenem Rahmen ist typisch für die Darstellung traditioneller Sujets. (Sammlung f@goldscheider.de)

Murals were among the most popular objects in the early days of production. This kiss-your-hand Madame scene with a golden frame is typical of the way traditional motifs were handled. (Collection f@goldscheider.de)

Verfahren, das Gegenständen aus Ton, Terrakotta, Porzellan, Fayence, Gips und Zement das „Aussehen von patinierten Broncegegenständen" verleiht, erhält Friedrich Goldscheider am 19. November 1892 ein Privilegium (Ill. 20).[63] Es gelingt ihm somit, dem relativ preiswerten und von vielen als minderwertig bezeichneten Ton das Aussehen eines „edleren" Materials zu verleihen. Aufgrund des weitaus niedrigeren Preises kann sich jetzt auch die Mittelschicht Kunstgegenstände für die dekorative Einrichtung ihrer Wohnungen leisten. Friedrich Goldscheider erschließt sich somit eine neue, breite Käuferschicht – sowohl im Inland wie auch Ausland. Denn die regelmäßige Teilnahme an internationalen Ausstellungen ist nicht nur für den Aufbau des Markennamens Goldscheider notwendig, sie ermöglicht auch die stete Ausweitung des Exports.

Steigende Umsätze und Erfolge sind die Belohnung für Friedrich Goldscheiders unternehmerische Vitalität und Kreativität. Im Rahmen der Ausstellung des Wiener Kunstgewerbevereins im Jahr 1891 werden seine Fortschritte anerkannt: „Künstlerischer

ModNr. 607
Tabaksdose I Snuffbox, ca. 1895, Sammlung.Schmutz@chello.at
Diese realistische und zugleich makabre Darstellung eines Totenschädels als Tabaksdose ist ein besonderes Stück der frühen Goldscheider-Produktion. Als Knauf des Deckels dieser Tabakdose dient ein Knochen.

This realistic and even macabre representation of a skull as a snuffbox is a prime piece of early Goldscheider. The knop of the cover is made of bone.

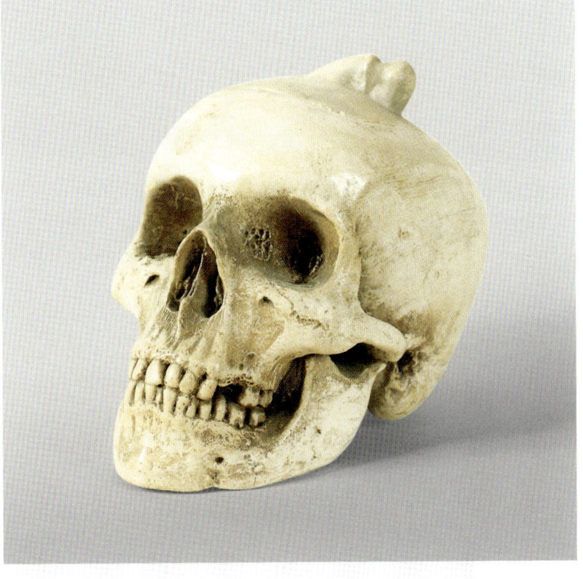

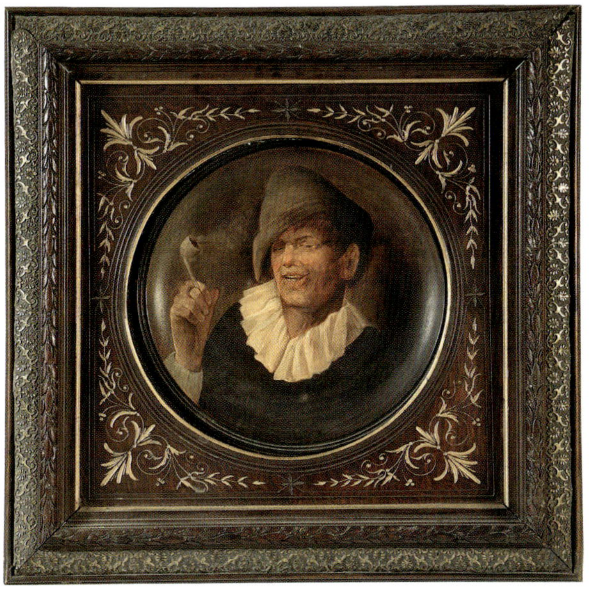

Ill. 22
Ebenso sind auch die Wandteller, hier die Modellnummer 30 im Holzrahmen, bis zur Jahrhundertwende beliebt. (© Wien Museum)

Chargers made to be mounted on walls were equally popular up to the turn of the century: here model number 30 in a wooden frame. (© Wien Museum)

rechts | right

ModNr. 850
Säule (Postament) |
Column (pedestal),
Griser, ca. 1895,
© Wien Museum

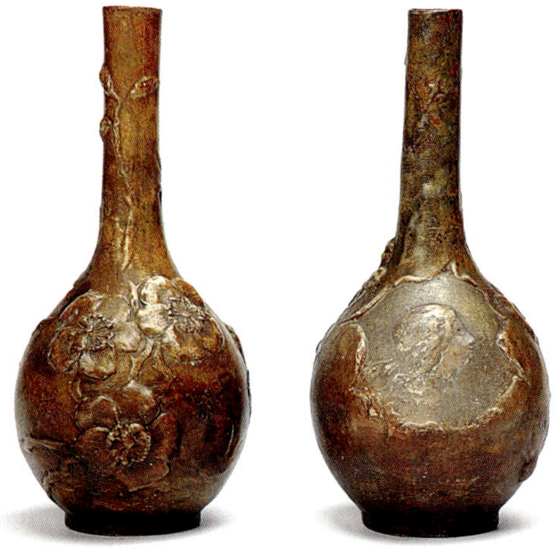

ModNr. 692, 693
Etruskische Vasen |
Etruscan Vases „Eglantine" und | and „Chrysanthème", Jacques Callot, ca. 1892, PB Wien 19

ModNr. 726
Schale | Bowle, Simon, Ausführung | execution ca. 1898, PB Wien 19

Um 1895 sind die ersten Anzeichen der neuen künstlerischen Epoche des Jugendstils erkennbar.

The first signs of the new Art Nouveau period style show up around 1895.

und technischer Fortschritt ist bemerkbar, neue Modelle und eigenartige Erzeugnisse wechseln in rascher Folge und beweisen eine außerordentliche Rührigkeit, die eine Vergrößerung der Betriebsräume erforderte. Wir finden wieder zahlreiche neue Figuren und Gruppen von aufstrebenden jungen Plastikern, die sich gerne der Industrie zuwenden. So das Tirolerpaar, das von der Kronprinzessin-Witwe angekauft wurde, Sarah Bernhard als Theodora, König Lear, weibliche Büsten, die Hauptfiguren aus den Blaas'schen Gemälden. Recht gute comercielle Erfolge werden auch die Oelgemälde auf Thonplatten haben, die nach hervorragenden Originalen moderner Meister von jungen Wiener Akademikern hergestellt werden."[64] Die Beschäftigung junger Künstler ist für Friedrich in vielerlei Hinsicht vorteilhaft: Sie sind hervorragend ausgebildet, näher am Zeitgeist, leichter zu kontrollieren und vor allem preiswerter als etablierte Künstler.

Als Maßstab für Qualität und künstlerische Darstellung gelten französische Terra-

excellent originals by modern masters among the young Viennese academicians are also bound to be commercially quite successful."[64] Employing young artists was advantageous for Friedrich in many respects: they were exceptionally well trained, closer to the *Zeitgeist*, easier to manage and, above all, less expensive than established artists.

French terracottas represented the standard against which quality and artistic skill were measured. It was, therefore, complimentary when critics wrote: "Some cupids and busts can well stand comparison with those celebrated reddish Paris terracottas à la Chéret. One might almost think that some of Goldscheider's artists studied in Paris."[65] Paris was the measure of all things so it marked a consistent development of Friedrich's plans for expansion when he opened the "Frédéric Goldscheider" bronze factory at 45 rue de Paradis, Paris, in December 1891.[66] Considering how nationally aware the French market was, it was, moreover, an advantage to be on the spot and represented in the marketplace by French artisans. From

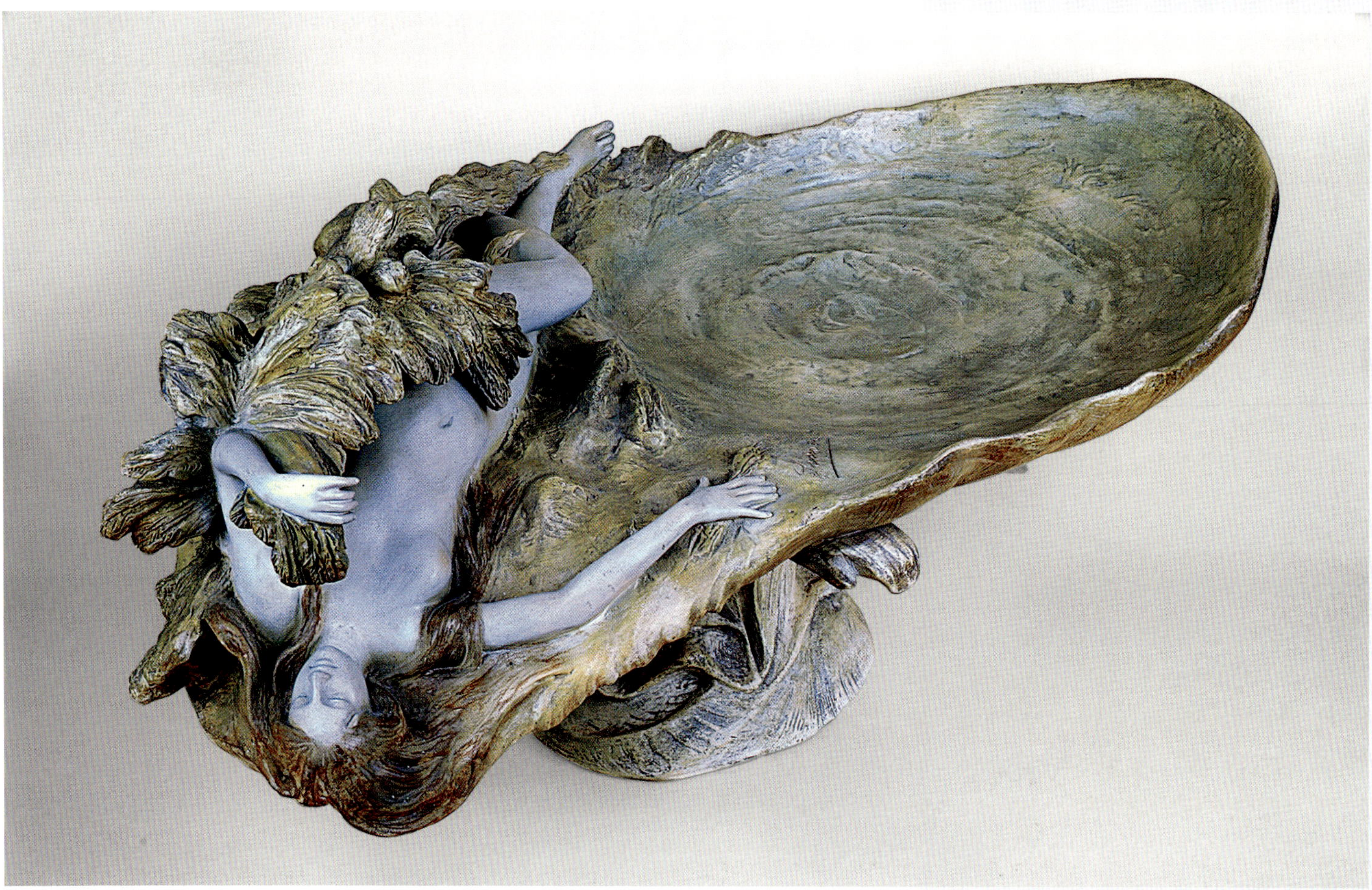

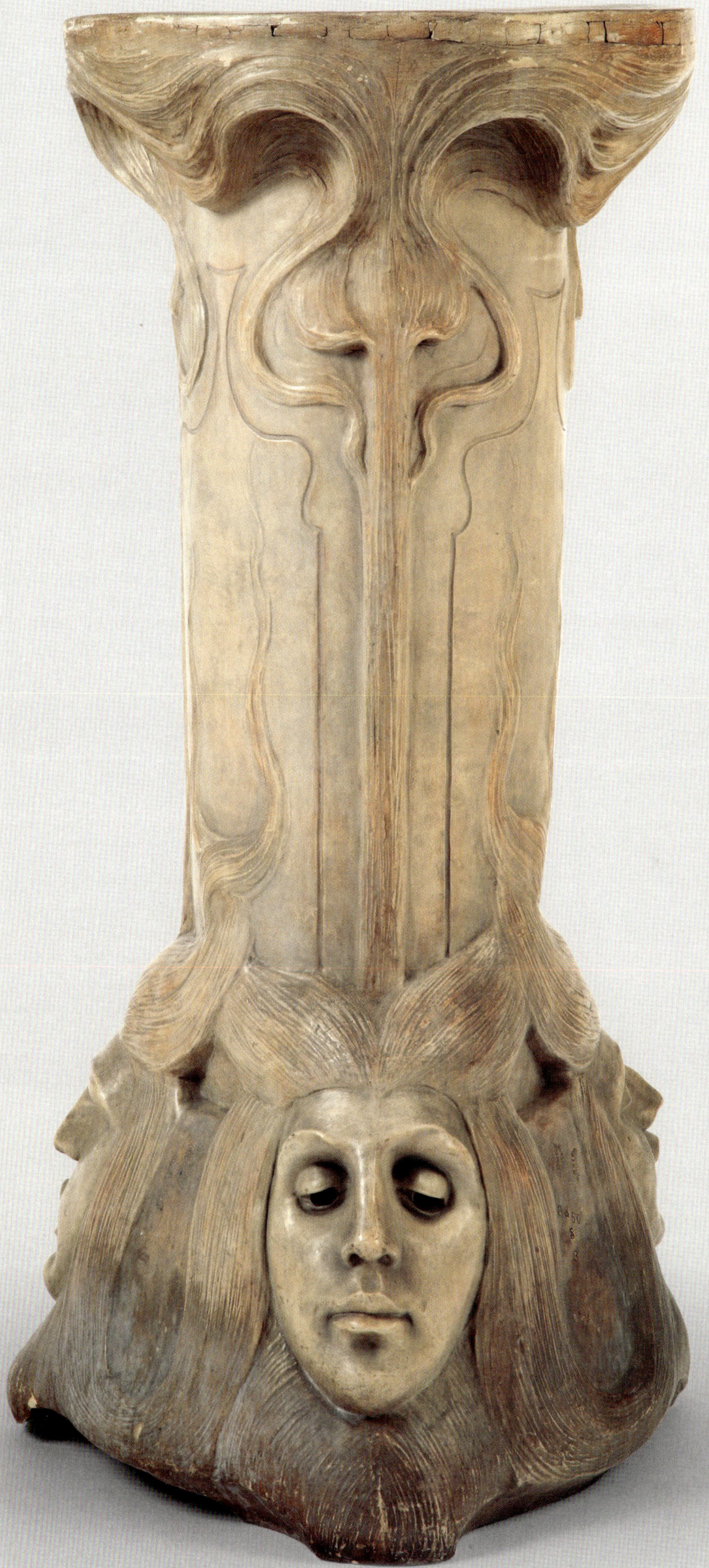

ModNr. 878
Vase „Am Brunnen" I "At the Well", Simon, Ausführung I execution ca. 1898, PB Wien 19
Diese Vase wird 1902 auch in Turin ausgestellt und von der Presse gelobt.

This vase was shown in Turin as well in 1902 and received a good press.

kotten und so gilt es als Kompliment, wenn die Kritiker schreiben: „Einige Putten und Büsten können mit jenen berühmten röthlichen Pariser Terracotten à la Chéret ganz gut rivalisieren. Man möchte fast annehmen, dass einige Künstler von Goldscheider ihre Studien in Paris gemacht haben."[65] Paris ist der Maßstab aller Dinge und so ist es eine konsequente Fortsetzung der Expansionspläne, als Friedrich im Dezember 1891 in Paris, in der Rue de Paradis 45, die Bronzefabrik „Frédéric Goldscheider" eröffnet.[66] Für den nationalbewussten französischen Markt ist es außerdem vorteilhaft, vor Ort zu sein und mit französischen Künstlern am Markt aufzutreten. Dazu gehören von nun an Allouard, Bouval, Carrièr-Belleuse, Charpentier oder Levasseur. Zusätzlich ermöglicht die Zusammenarbeit zwischen den Goldscheider-Fabriken in Paris (Bronze) und Wien (Terrakotta) einerseits die gegenseitige Reproduktion von hochwertigen und populären Modellen in den jeweiligen Materialien, andererseits auch den direkten Verkauf der

then on, they would include Allouard, Bouval, Carrièr-Belleuse, Charpentier and Levasseur. In addition, collaboration between the Paris Goldscheider factories (bronze) and Vienna (terracotta) facilitated, on the one hand, the reproduction of both the expensive and the popular models in the materials used for each; on the other, the sale of the Viennese products at the French branch. To promote acceptance and circumvent the traditional anti-German attitude prevailing in France due to the Franco-Prussian war, German artists' names were duly Gallicized for the French market: Czapek became "Capèque," Schoerk "Cherc" and Fischer was simply translated into "Pecheur."[67] The Paris models were numbered in a new sequence (also see the catalogue of works on this). Apart from reliefs, vases and bowls show the first signs of the coming spread of Art Nouveau.

At the 1893 Chicago World's Fair both bronzes and bronzed objects were shown, "superb reproductions of sculptural works of art, some of them after his own process."[68] By then the workforce amounted to 145 and 90 per cent of output was exported to "all countries and continents."[69]

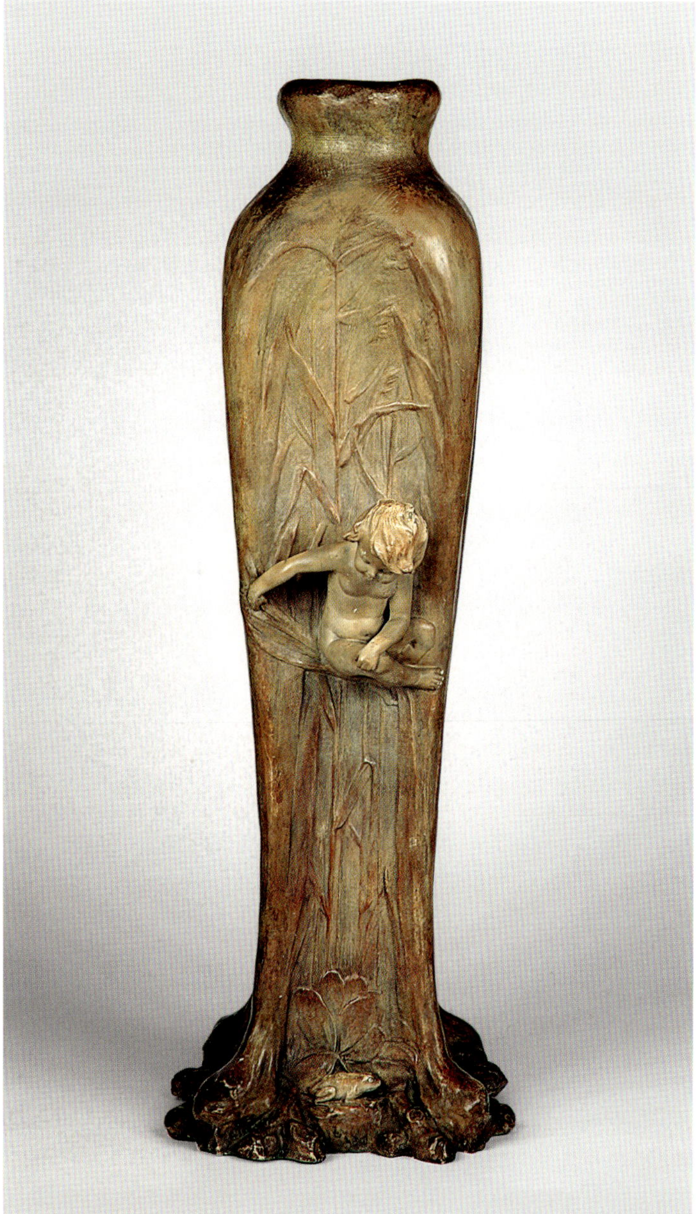

ModNr. 745
Maske (Jardiniere) I Mask (Jardinière), Léon Noël Delagrange, französische Ausführung I execution French ca. 1900, PB Wien 19

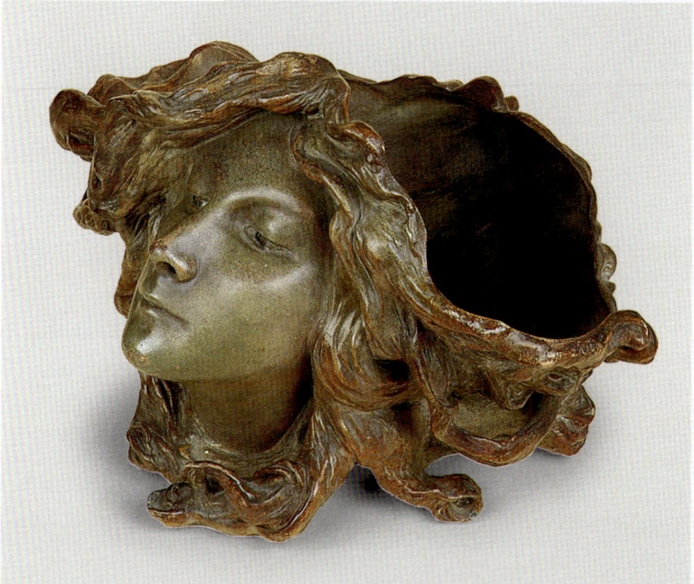

In spring 1894 operations were increased on the German market as well and in March a branch with showrooms was opened in the department store at 9/19 Neumarkt in Leipzig.[70] In the late 19th century, department stores were the big consumer attraction; they were indeed cathedrals of luxury. However, the luxury goods sold were by then available to more than just the upper-middle and upper classes. A broader clientele was being targeted: the middle and lower-middle classes. Advertising gained in importance in the context of reaching out to a wider target market; the marketing potential of a new consumer class was to be activated.

Leipzig played a significant role for Friedrich Goldscheider. On the one hand, Leipzig was important as the place for registering copyrights on German designs, which explains why he had already registered copyrights – for a term of three years – for several sculpture products by 1893. They included two wall chargers (Rococo style) and a gilded wall charger with a set-in mirror (factory numbers 6–8)[71] as well as seven frames, a bowl (factory numbers 12–19) and a figurine with "distinctive painting and color compo-

Wiener Erzeugnisse unter französischem Dach. Um die Akzeptanz zu fördern und die kriegsbedingte, traditionell antideutsche Haltung in Frankreich zu umgehen, werden sogar die deutschen Künstlernamen für den französischen Markt angepasst: so wird aus Czapek „Capèque", aus Schoerk „Cherc" und aus dem Fischer ein „Pecheur".[67] Die Nummerierung der Pariser Modelle erfolgt nach neuer Reihenfolge (vgl. hierzu auch das Werkverzeichnis). Neben Reliefs zeigen vor allem Vasen und Schalen erste Anzeichen der bevorstehenden Verbreitung des Jugendstils.

Auf der Weltausstellung in Chicago werden 1893 sowohl Bronzen wie auch bronzierte Objekte ausgestellt, „vortreffliche Reproductionen plastischer Kunstwerke, theilweise nach eigenem Verfahren"[68]. Zu dieser Zeit werden 145 Arbeiter beschäftigt und man exportiert zu 90 Prozent nach "allen Ländern und Welttheilen"[69].

Im Frühjahr 1894 werden die Aktivitäten auch auf dem deutschen Markt verstärkt und im März eine Niederlassung mit Schauräumen im Kaufhaus am Neumarkt 9/19 in Leipzig eröffnet.[70] Kaufhäuser sind Ende des 19. Jahrhunderts die Attraktion des Konsums, sie sind die Kathedralen des Luxus. Allerdings sind die dort angebotenen Luxusgüter nicht mehr nur für die gehobene Schicht zugänglich, ein breiterer Kundenkreis wird erschlossen: die Mittelschicht. Die Werbung gewinnt im Rahmen der erweiterten Kundenansprache an Bedeutung, die potenziellen Konsumenten sollen zum Kauf animiert werden.

Leipzig spielt für Friedrich Goldscheider eine bedeutende Rolle. Einerseits ist Leipzig als Meldeort für den deutschen Musterschutz wichtig, weshalb er bereits 1893 dort Musterschutz für mehrere plastische Erzeugnisse – mit einer Schutzfrist von 3 Jahren – angemeldet hat. Unter anderem sind dies zwei Wandteller (Rokoko-Stil) und ein vergoldeter Wandteller mit Spiegeleinsatz (Fabriknummern 6–8)[71] sowie sieben Rahmen, eine

ModNr. 842
Vase „Sirenenspiel" I "Play of the Sirens", Eugène Emile Moulin, französische Ausführung I execution French ca. 1898, © Wien Museum

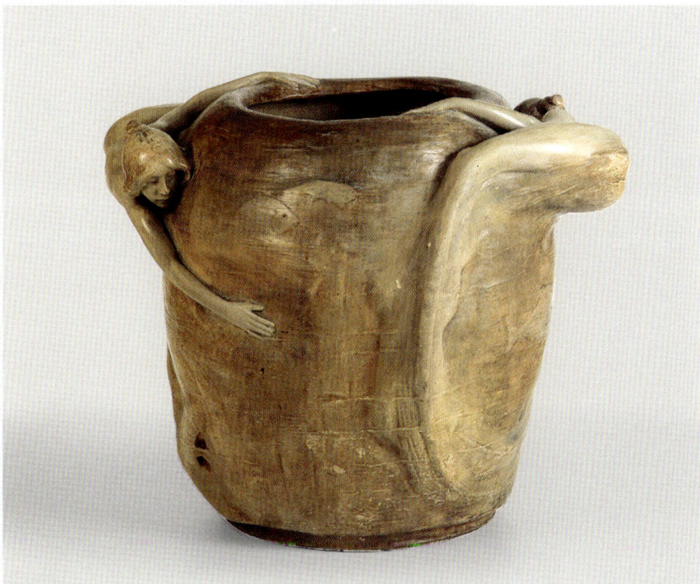

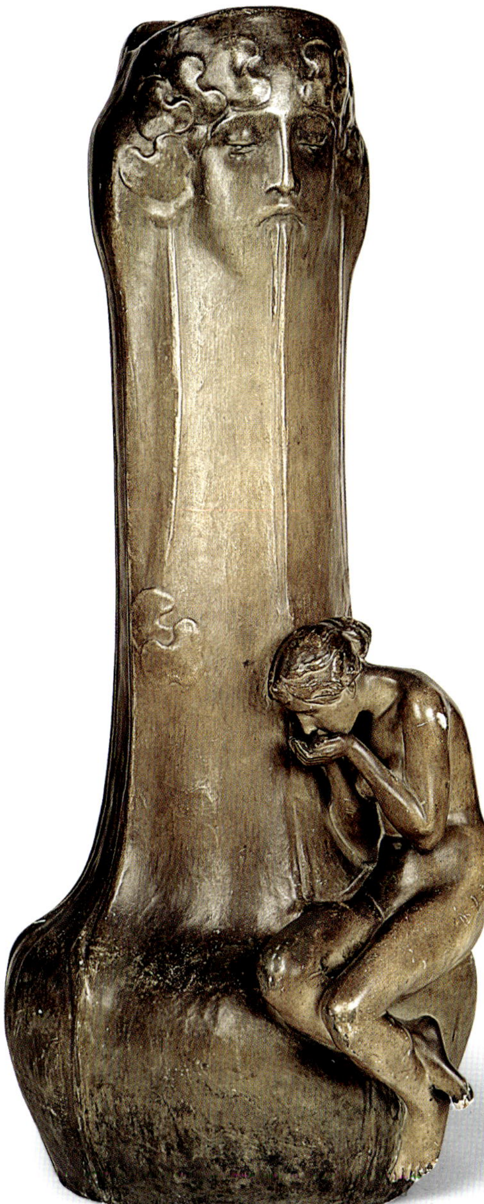

ModNr. 896
Vase, J. Louger, ca. 1897, PB Wien 19

Schale (Fabriknummern 12–19) und eine Figur mit „eigenartiger Bemalung und Farbenzusammenstellung (Decorationsmuster)" mit der Fabriknummer 512/60[72]. Die Anmeldung von Musterschutz für das Deutsche Reich findet auch in den darauf folgenden Jahren für ausgewählte Kunstgegenstände in Leipzig statt.

Vor allem ist es aber die Nähe zu den wichtigen Leipziger Messen, auf denen Friedrich seit Jahren ausstellt. Durch die neue Dependance hat er eine feste Basis im Deutschen Reich und kann die Leipziger Filiale zu den stattfindenden Frühjahrs- und Herbst-Messen als Ausstellungsraum umdekorieren. Die im Luxuskaufhaus in Szene gesetzten Kunstobjekte erregen höchste Begeisterung und Lob: „Das Sensationsstück der kürzlich abgelaufenen Ostermesse in Leipzig war die Ausstellung der Kunstterracotten von Friedrich Goldscheider in Wien in dem Neubau des alten Gewandhauses am Neumarkt. [...] [Das] große, hohe und helle Local glich

sition (decoration patterns)" bearing factory number 512/60.[72] Applications for German Empire copyrights were also made in Leipzig in subsequent years for selected art objects.

Above all, however, Leipzig meant proximity to the important Leipzig trade fairs, where Friedrich had shown products for years. Through the new Leipzig branch, he had a strong base in the German Empire so that he could redecorate the Leipzig store as a showroom for the spring and autumn trade fairs. The art objects displayed in the luxury department store met with enormous enthusiasm and praise: "The sensation of the recently finished Easter Fair in Leipzig was the display of art terracottas from Friedrich Goldscheider in Vienna in the new building of the old Gewandhaus on Neumarkt. [...] [The] large, high-ceilinged and bright setting resembled an art exhibition in content and arrangement. [...] Every figure, every bust and group was positioned on a pedestal, a column, a frame, a modelling stand at the

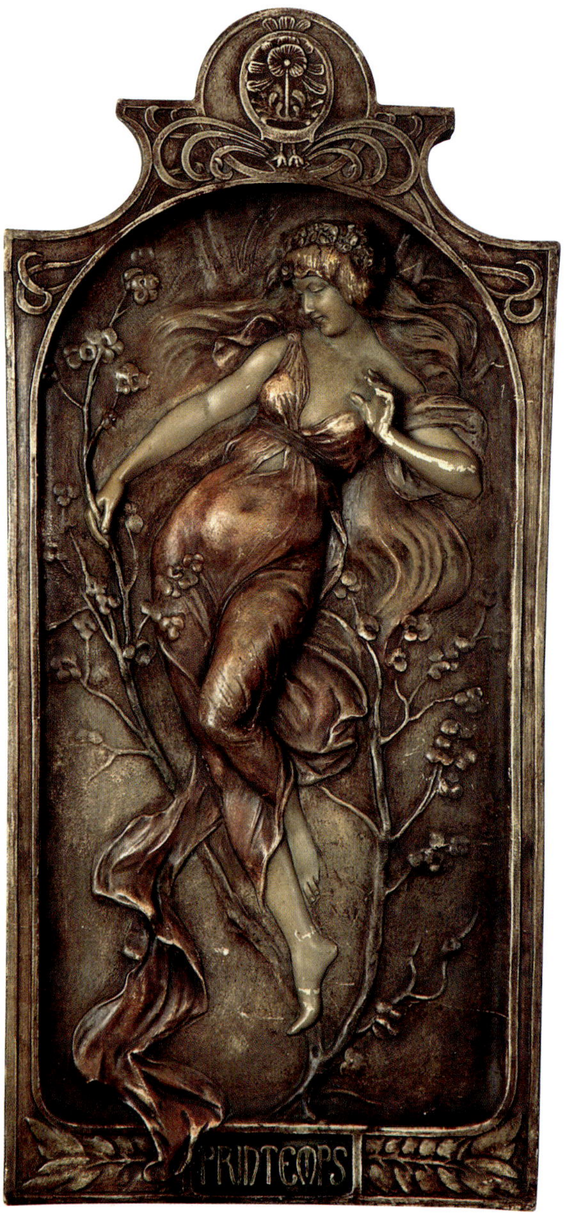

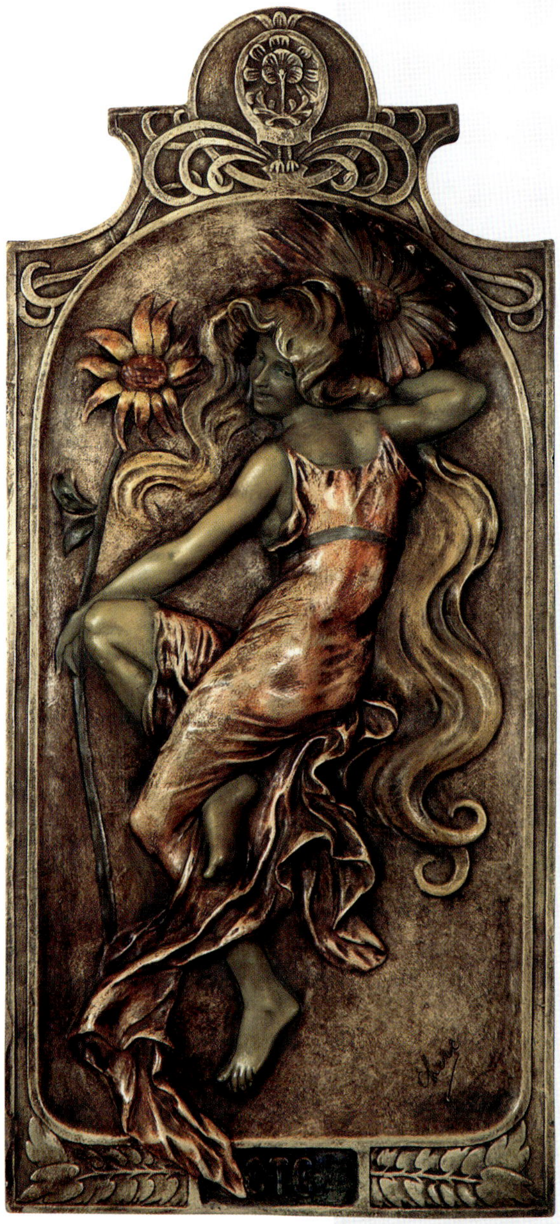

ModNr. 766
Relief „Printemps", Cherc, ca. 1895, PB Wien 19

ModNr. 768
Relief „Été", Cherc, ca. 1895, PB Wien 19

correct viewing level; blue velvet and silk fabrics corresponded to the terracotta colors, the imitated bronzes and green plants; palm fronds formed the backgrounds for the attractive sculptural works."[73] The show met such a positive response that Friedrich had the venue brightly lit up in the evenings so that passers-by could share in this "sensational art exhibition." Apart from their masterly workmanship, "immaculate" firing, tinting and polychromy, the models were praised especially for their originality as art works.[74] The handling of the "carnation" as well as the rendering of cloth textures and apparel, which was applied after the figures were fired, were a mass hit.[75] The exhibits included large figures, such as "Judith," "Gesang" ["Song"], "Die Wahrheit" ["Truth"], the group "Das Erwachen der Liebe" ["Love Awakened"], polychrome Arabs and busts representing a Mulatto couple. Friedrich Goldscheider, who had previously been represented at the Antwerp Fair with work from

einer Kunstausstellung nach Inhalt und Arrangement. [...] Jede Figur, jede Büste und Gruppe war auf einem Postament, einer Säule, einem Gestell, einem Modellirstuhl in rechter Sichthöhe aufgestellt, farbige Tücher aus Samt und Seide correspondirten mit den Terracottafarben, den Bronceimitationen, und grüne Pflanzen, Palmenwedel, bildeten Hintergründe für die reizvollen plastischen Arbeiten."[73] Die Ausstellung findet einen solch großen Anklang, dass Friedrich auch abends das Lokal hell erleuchten lässt und Schaulustige an dieser „Aufsehen erregenden Kunstausstellung" teilhaben lässt. Bei den Modellen wird neben der meisterhaften Technik, dem „tadellosen" Brand, den Tönungen und Polychromierungen vor allem die künstlerische Originalität gelobt.[74] Die Behandlung der „Fleischformen" sowie die Wiedergabe des Stofflichen, der Gewänder, die erst nachträglich der Figur aufgetragen werden, begeistern die Massen.[75] Darunter sind große Figuren wie „Judith", „Gesang",

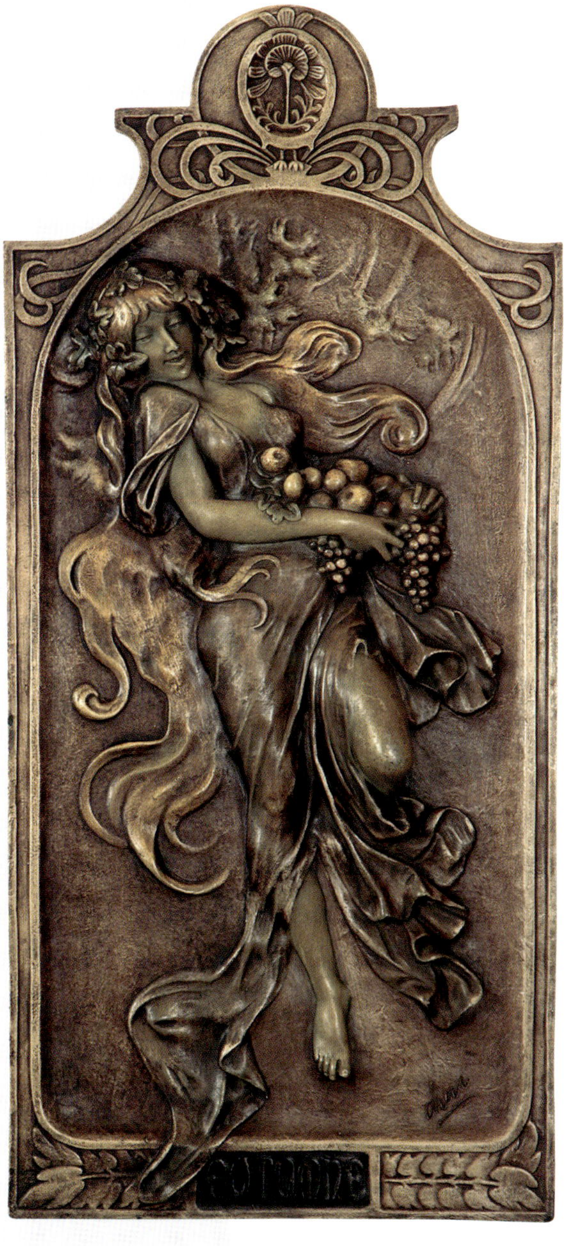

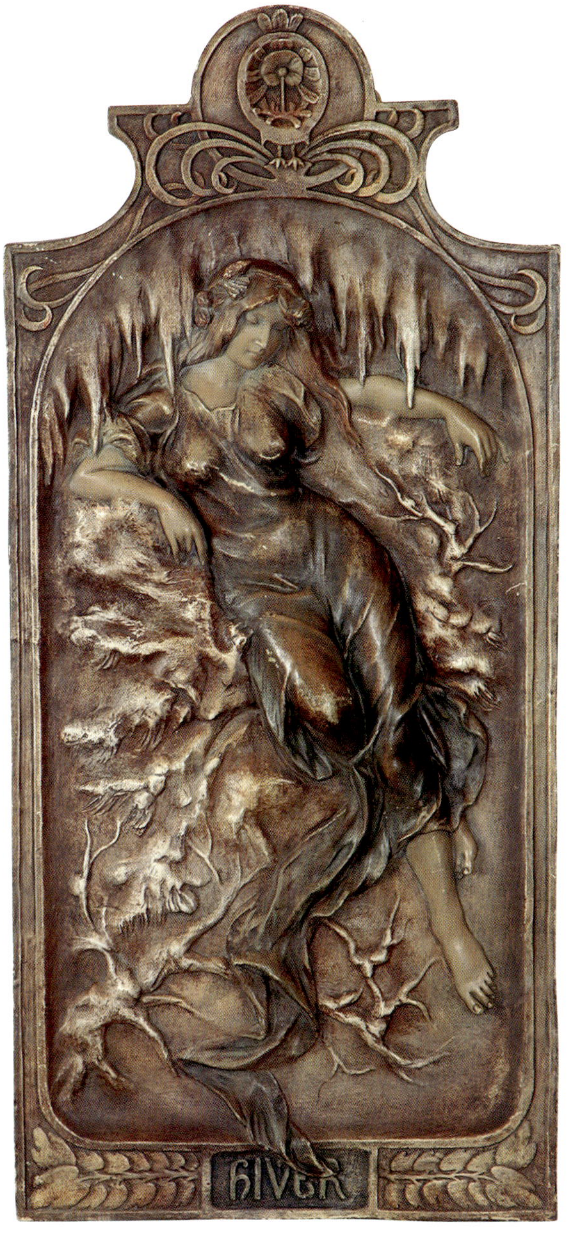

ModNr. 770
Relief „Autumne", Cherc, ca. 1895, PB Wien 19

ModNr. 772
Relief „Hiver", Cherc, ca. 1895, PB Wien 19

„Die Wahrheit", die Gruppe „Das Erwachen der Liebe", polychrome Araber wie auch Büsten eines Mulattenpaares. Friedrich Goldscheider, der zuvor auf der Antwerpener Ausstellung sowohl mit der Wiener wie auch der Pariser Fabrik vertreten war und als Jurymitglied nur außerhalb des Wettbewerbs ausgestellt hat, ist auf dem Höhepunkt seiner Karriere.[76]

Beliebte Motive dieser Zeit werden außereuropäische Völker, vor allem Araber und Mohren, die die europäischen Ängste und Sehnsüchte widerspiegeln. Gerade in Mitteleuropa, wo die Kolonialisierung ferner Länder im Vergleich zu Spanien, den Niederlanden oder Großbritannien erst später einsetzte, symbolisierte das Exotische einerseits das Wilde und Primitive im Menschen, das es zu zivilisieren galt, andererseits diente es als Projektionsfläche für das befreite, glückliche Leben ohne Zwänge. Das aufgeklärte Europa mitsamt der propagierten Sittsamkeit und Prüderie des 19. Jahrhunderts findet in der Darstellung des exotisch Unzivilisierten und Nackten zum Teil sein Ventil für erotische Phantasien. Arthur Strasser, ein renommierter Wiener Bildhauer und Professor an der Wiener Kunstgewerbeschule, arbeitet eng mit Friedrich Goldscheider zusammen und entwirft einige naturalistisch farbig gefasste Skulpturen im Stil des Orientalismus. Er gilt um die Jahrhundertwende als Erneuerer der polychromen Salonplastik. Ab 1890 nimmt auch die Bedeutung der böhmischen Manufakturen wie Bernhard Bloch, Johann Maresch, Riessner oder Stellmacher & Kessel zu, doch keine andere Manufaktur bietet so hochwertige Terrakotta-Ware wie Goldscheider.

both the Viennese and the Paris factories, was a member of the jury so he could only exhibit *hors concours*. He was at the pinnacle of his career.[76]

At that time, non-European peoples, especially Arabs and Moors, were popular motifs. They reflected European fears and yearnings. In Central Europe especially, where the colonization of distant lands set in late compared to Spain, the Netherlands and Great Britain, the exotic symbolized, on the one hand, what was wild and primitive in man, which had to be civilized; on the other, it served as a platform on which to project longings for a liberated, happy life without constraints. Enlightened Europe, with all the modesty and prudery propagated in the 19th century, found in part an outlet for erotic fantasizing in representing what was exotically uncivilized and naked. Arthur Strasser, a renowned Viennese sculptor and professor at the Vienna Kunstgewerbeschule, collaborated closely with Friedrich Goldscheider on designing some naturalistically polychromed works of sculpture in the Orientalizing style. At the turn of the century, Strasser was regarded as having recreated polychrome salon sculpture. From 1890, the importance of such Bohemian manufactories as Bernhard Bloch, Johann Maresch, Riessner, and Stellmacher & Kessel increased; yet no other factory produced terracotta wares of such high quality as Goldscheider did.

ModNr. 319, 318
Büsten auf Holzsockel I Busts on wooden pedestal, ca. 1890,
© Wien Museum
Die Büsten und Figuren von Arabern, Nordafrikanern und „Mohren" sind typisch für die Epoche des Orientalismus. Sie spiegeln die europäischen Ängste und Sehnsüchte wider.

The busts and figures of Arabs, North Africans and "Blackamoors" are typical of the Orientalism period. They reflect European fears and longings.

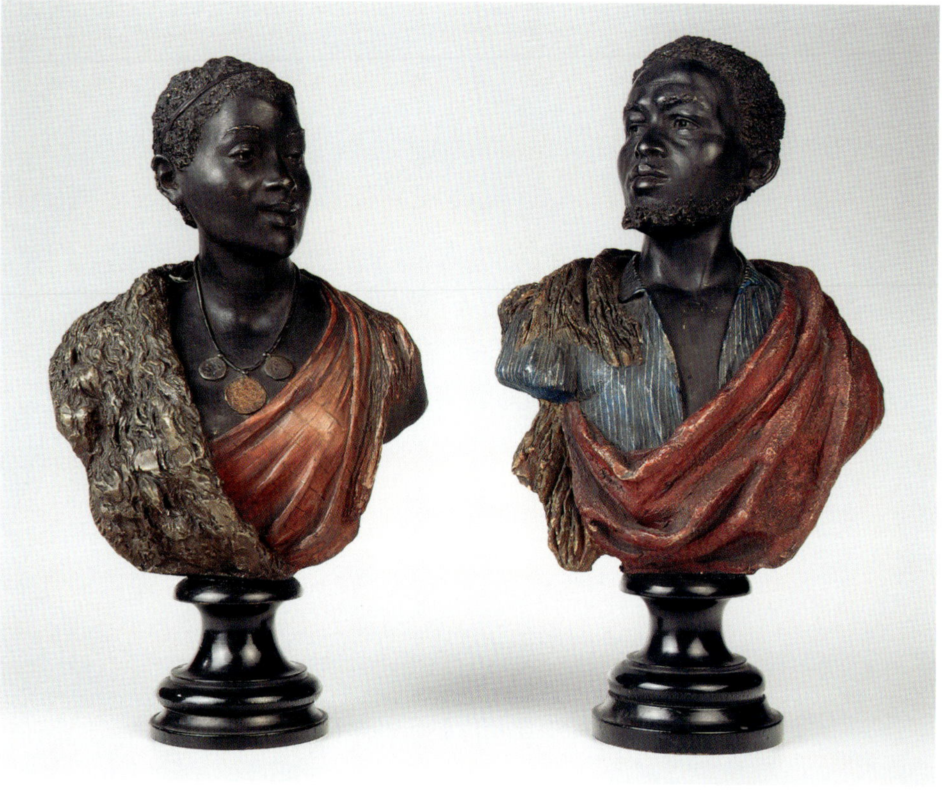

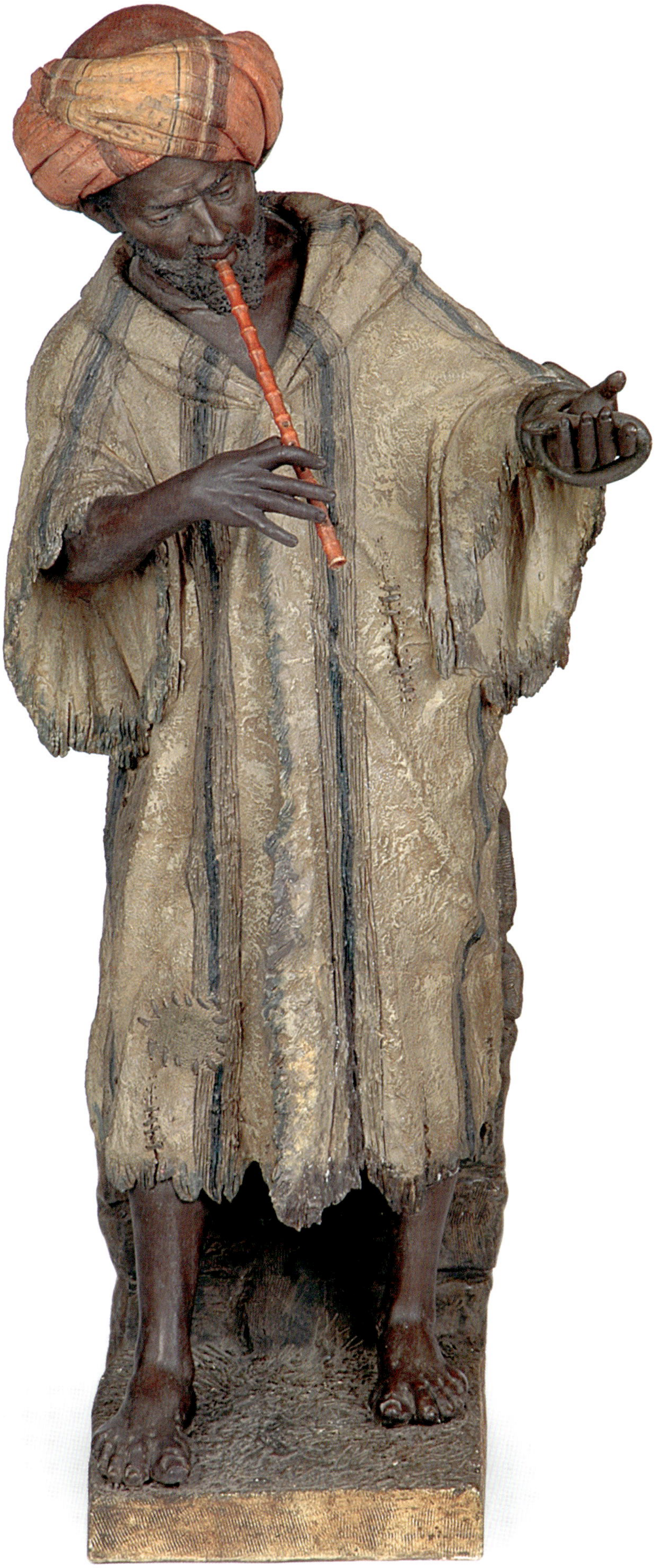

ModNr. 331
„Schlangenbändiger" |
"Snake Charmer",
Arthur Strasser zugeschrieben | attributed to,
ca. 1890, PB Wien 19

Bei der Modellnummer 1224 handelt es sich um die linke Figur aus der Gruppe der drei auf einer Mauer sitzenden „Neger"-Jungen (unten). Der Stuhl ist aus Bambus und betont die naturalistische Absicht dieser Figuren.

Model number 1224 is the left-hand figure from the group of three African boys seated on a wall (below). The chair is made of bamboo, emphasizing the naturalistic intention behind these figures.

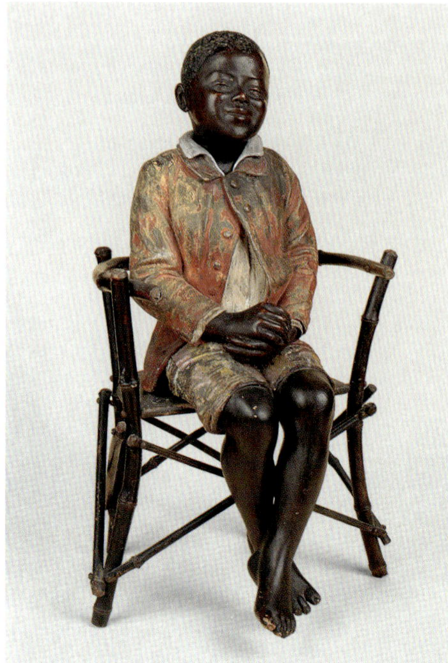

ModNr. 1224
„Beim Photographen" | "At the Photographer's", ca. 1897, PB Wien 19

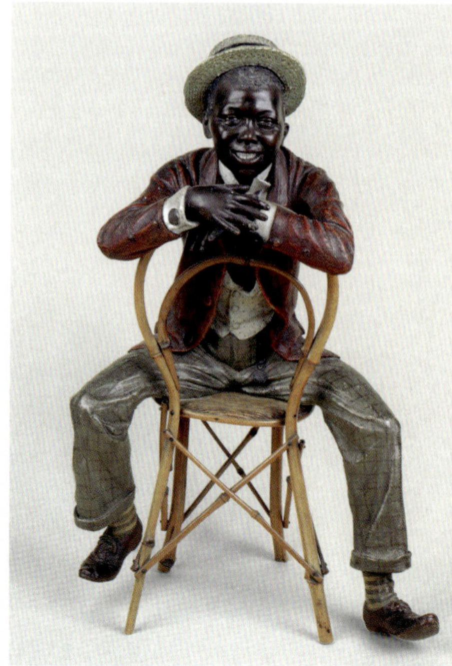

ModNr. 1338
ca. 1897, PB Wien 19

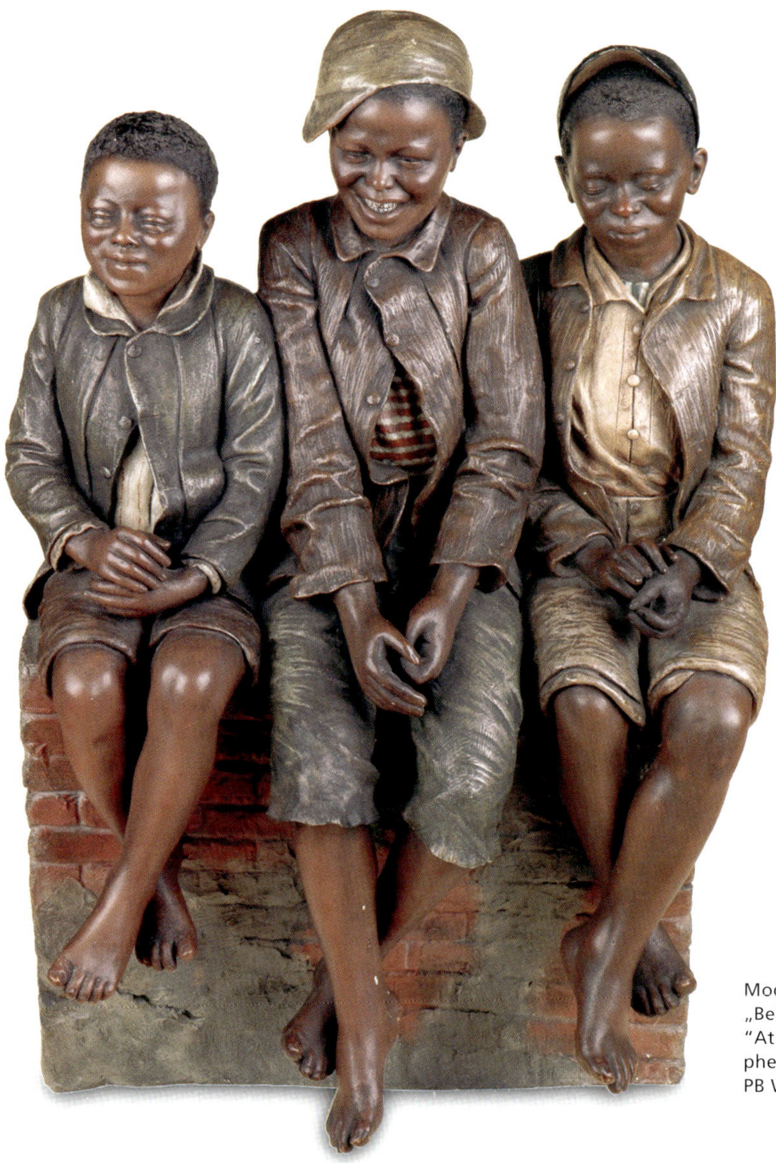

ModNr. 1190
„Beim Photographen" | "At the Photographer's", ca. 1897, PB Wien 19

ModNr. 628
Banjospieler | Banjo Player, ca. 1895, © Wien Museum
Der Banjospieler ist aus Terrakotta, das Banjo aus Holz gefertigt.

The banjo player is terracotta but his banjo is made of wood.

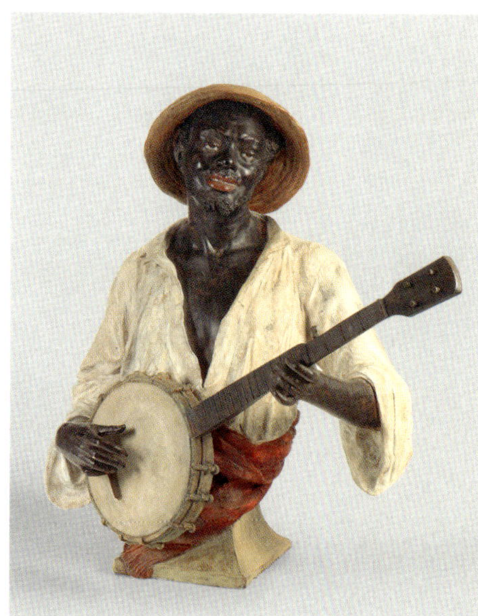

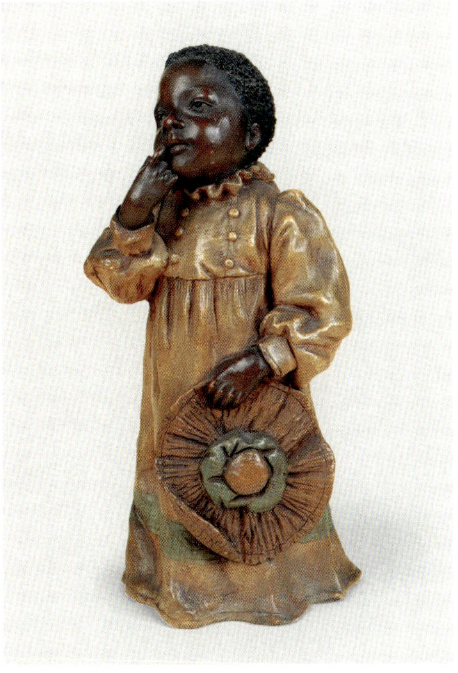

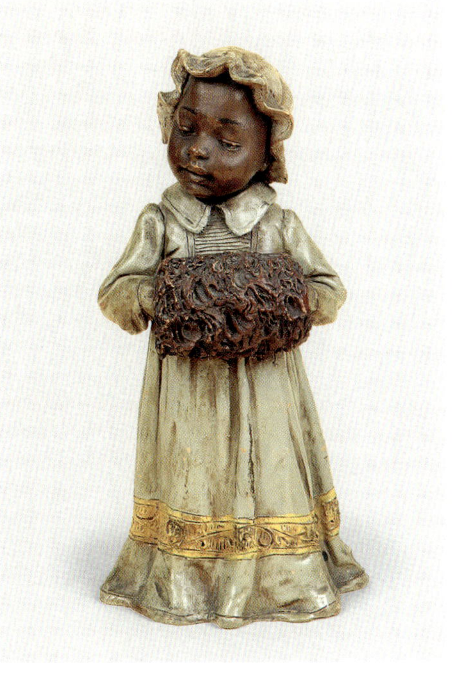

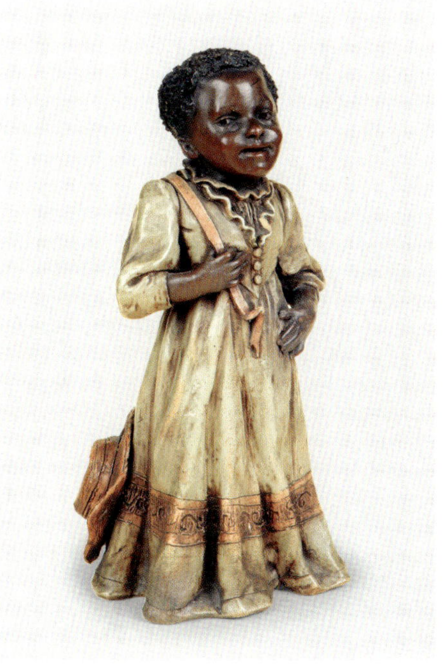

ModNr. 944, 946, 947
ca. 1895, PB Wien 19

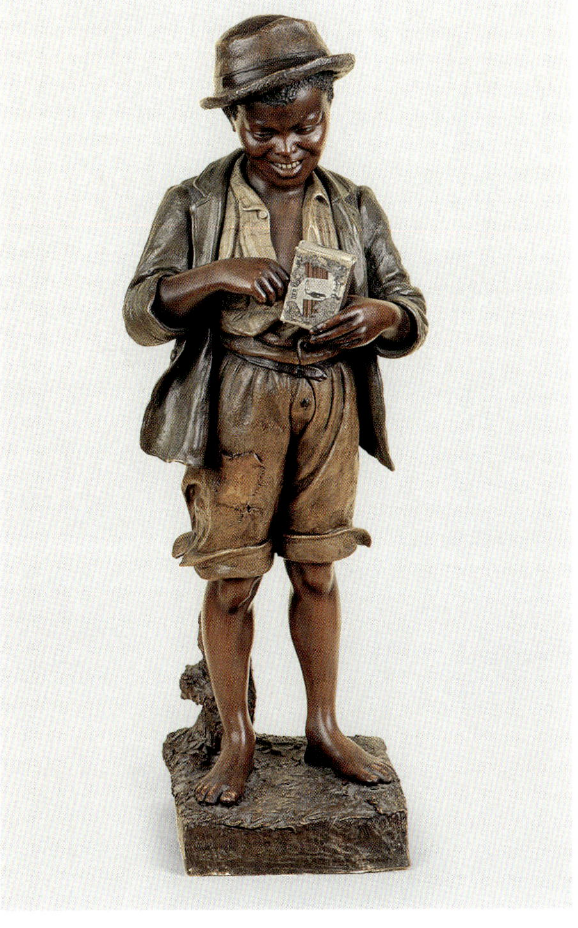

ModNr. 1471

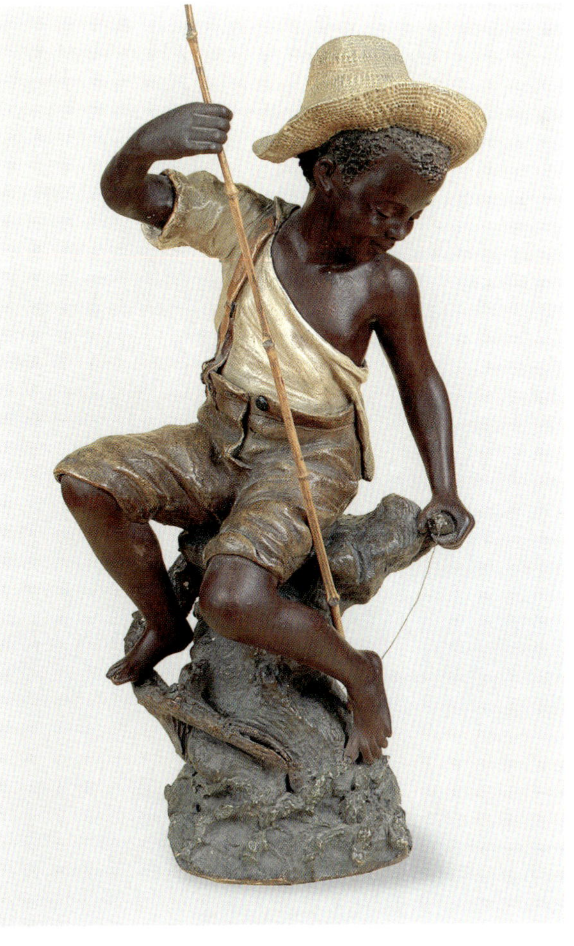

ModNr. 1559

ModNr. 1471
„Negerknabe, rauchend" | "Negro Boy, Smoking", Cherc, ca. 1899, PB Wien 19
Das Modell 1471 ist von der Figur her identisch mit Modell 1470 („Negerknabe, fischend"). Die Attribute Zündholz und Zündholzschachtel wurden lediglich als Variante zu Angel und Fisch produziert. Mit der Modellnummer 1472 war ein „Knabe mit zerbrochenem Krug" erhältlich.

The figure in model 1471 is identical with model 1470 ("Negro Boy, Fishing"). Only the attributes – a match and box of matches – were produced as variants of fishing tackle and fish. A "Knabe mit zerbrochenem Krug" ["Boy with Broken Jug"] was available as model number 1472.

ModNr. 1559
„Fischerknabe (Neger)" | "Boy Fishing (Negro)", Cherc, ca. 1899, PB Wien 19

rechts I right

ModNr. 1179
Araber mit Odaliske I Arab with Odalisque, ca. 1897, PB Wien 19
Eine seltene Gruppen-Darstellung eines alten Arabers mit Odaliske.

A rare group representation of an old Arab with an odalisque.

ModNr. 1101
„Sklavin" I "Slave Woman", Auguste Maurice Lévèque, ca. 1897, PB Wien 19

ModNr. 1341
„Mohrin" I "Blackamoor Woman", Auguste Maurice Lévèque, ca. 1898, PB Wien 19

Die Büste mit der Modellnummer 1101 konnte anhand eines Modellschilds als „Bedouine" dem Künstler Lévèque zugeschrieben werden. In einem zeitgenössischen Goldscheider Katalog wird sie allerdings als „Esclave" bezeichnet und dem Künstler Thiele zugesprochen. Das Gleiche trifft auch auf das Modell 1341 zu. Solche unterschiedlichen Zuweisungen erfolgen manchmal auch in den späteren Epochen der Manufaktur Goldscheider.

The bust with model number 1101 has been attributed to the artist Lévèque on the grounds of a model sign labelling it as "Bedouin". However, in a contemporaneous Goldscheider catalogue, it is listed as "Esclave" ["Slave"] and attributed to a different artist: Thiele. The same holds for model 1341. Attributions and classifications also occasionally vary in this way in the later years of the Goldscheider factory.

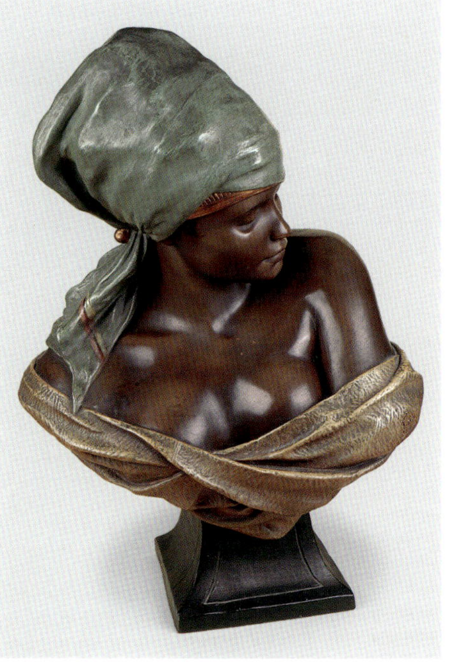

ModNr. 1101

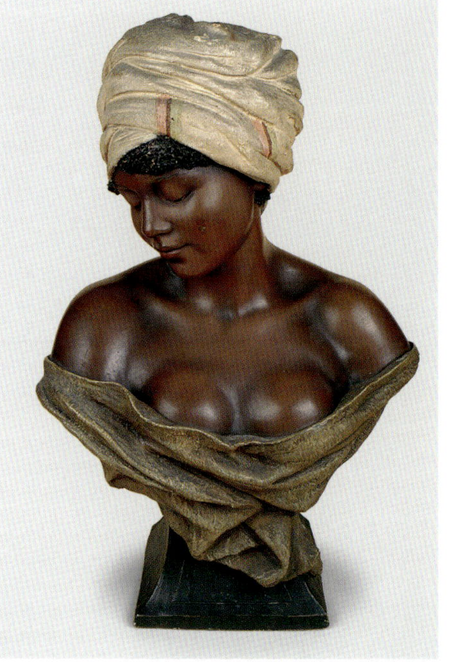

ModNr. 1341

ModNr. 1334
„Wollverkäufer" I "Wool Seller", Arthur Strasser, ca. 1898, PB Wien 19

ModNr. 1600
„Wasserträger" I "Water Carrier", Arthur Strasser, ca. 1898, PB Wien 19

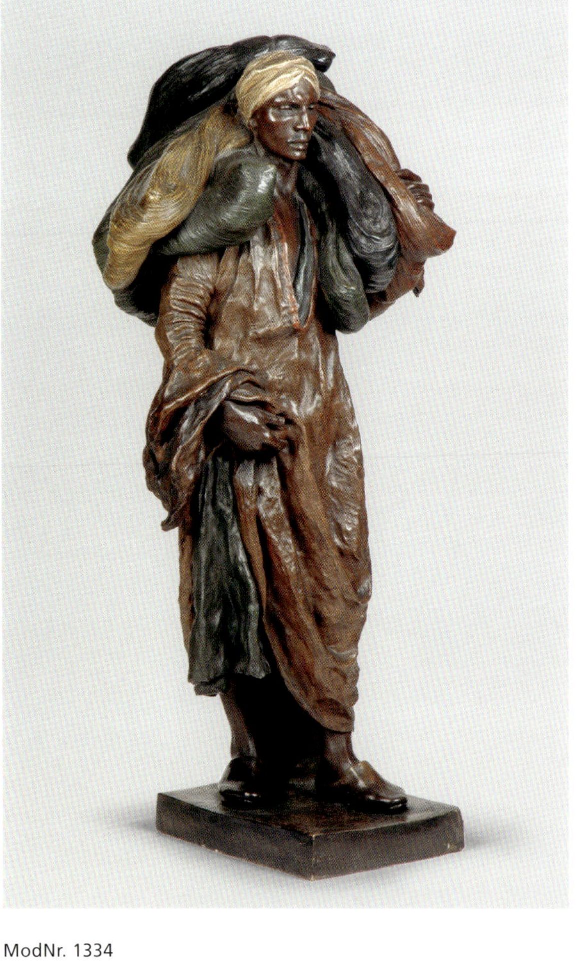

ModNr. 1334

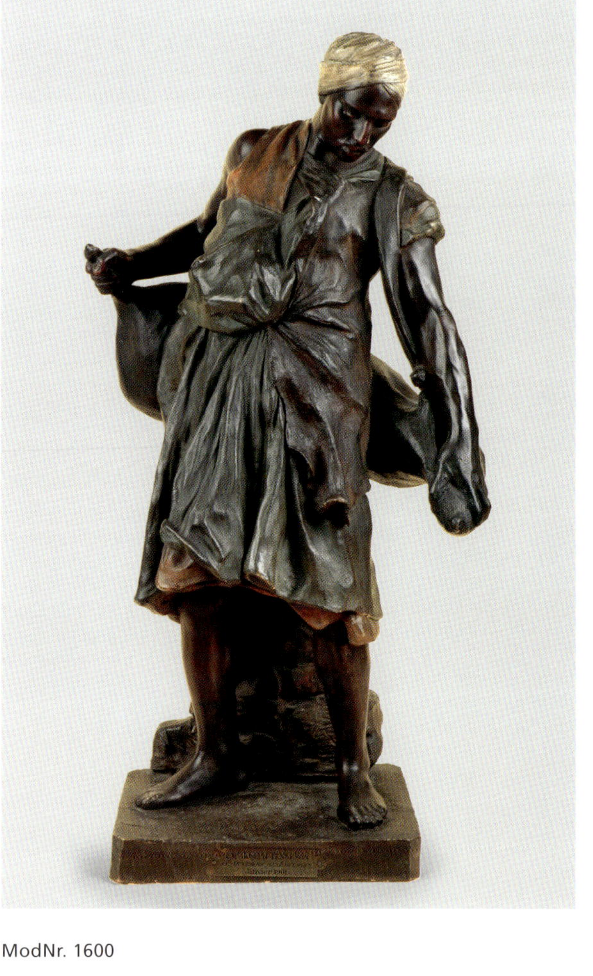

ModNr. 1600

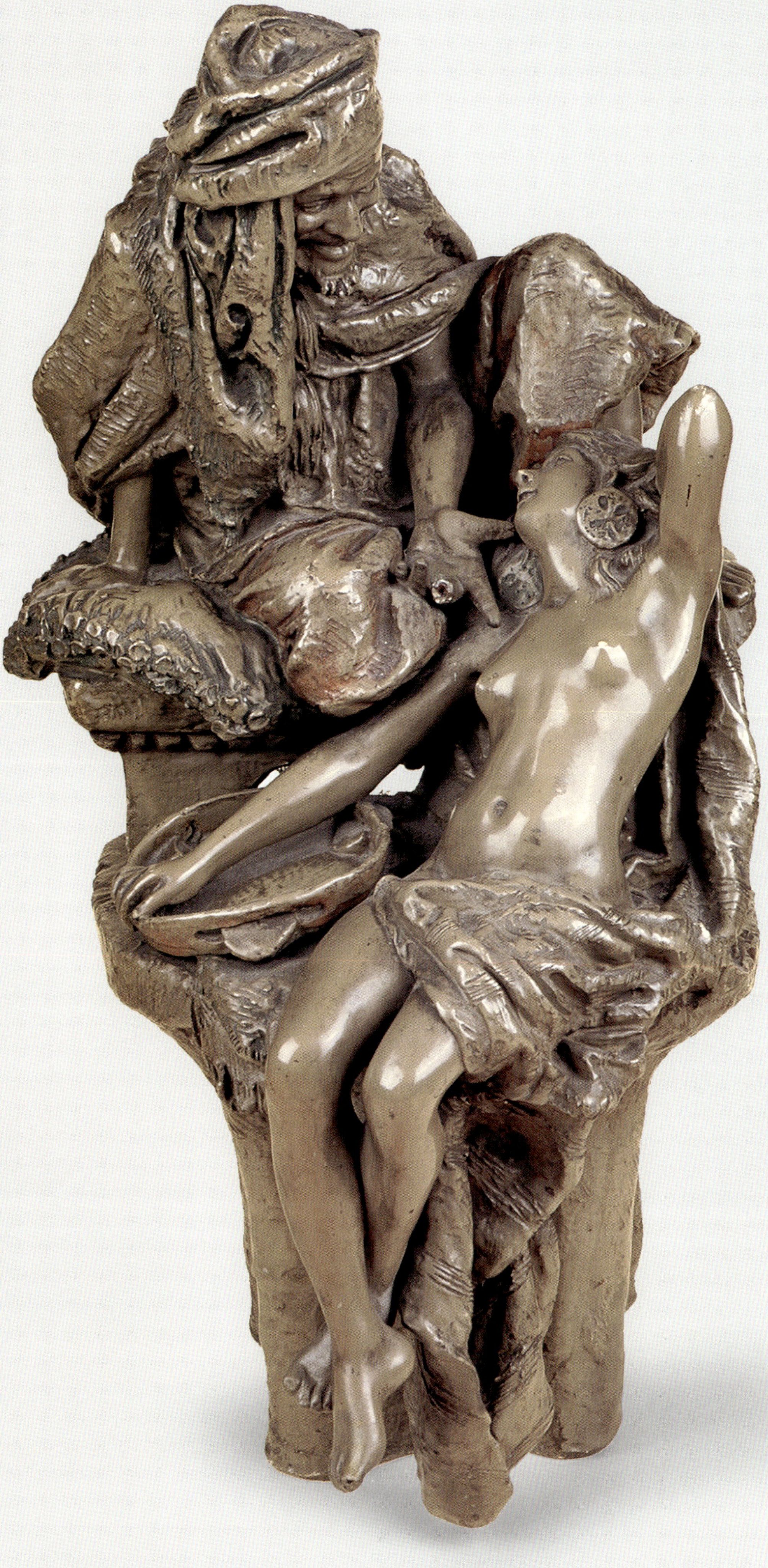

ModNr. 419
„Singhalesin" I
"Sinhalese Woman",
ca. 1890, PB Wien 19

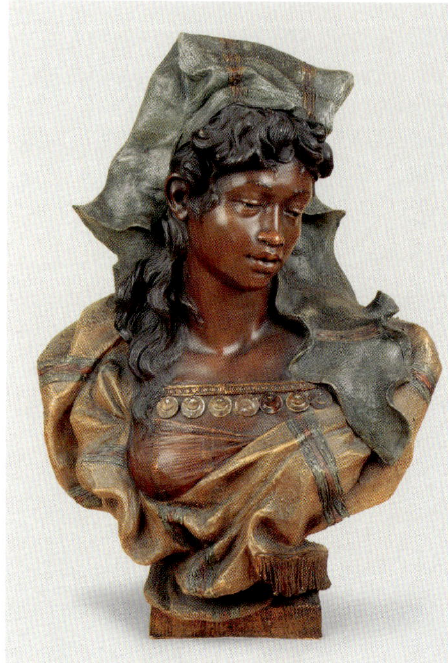

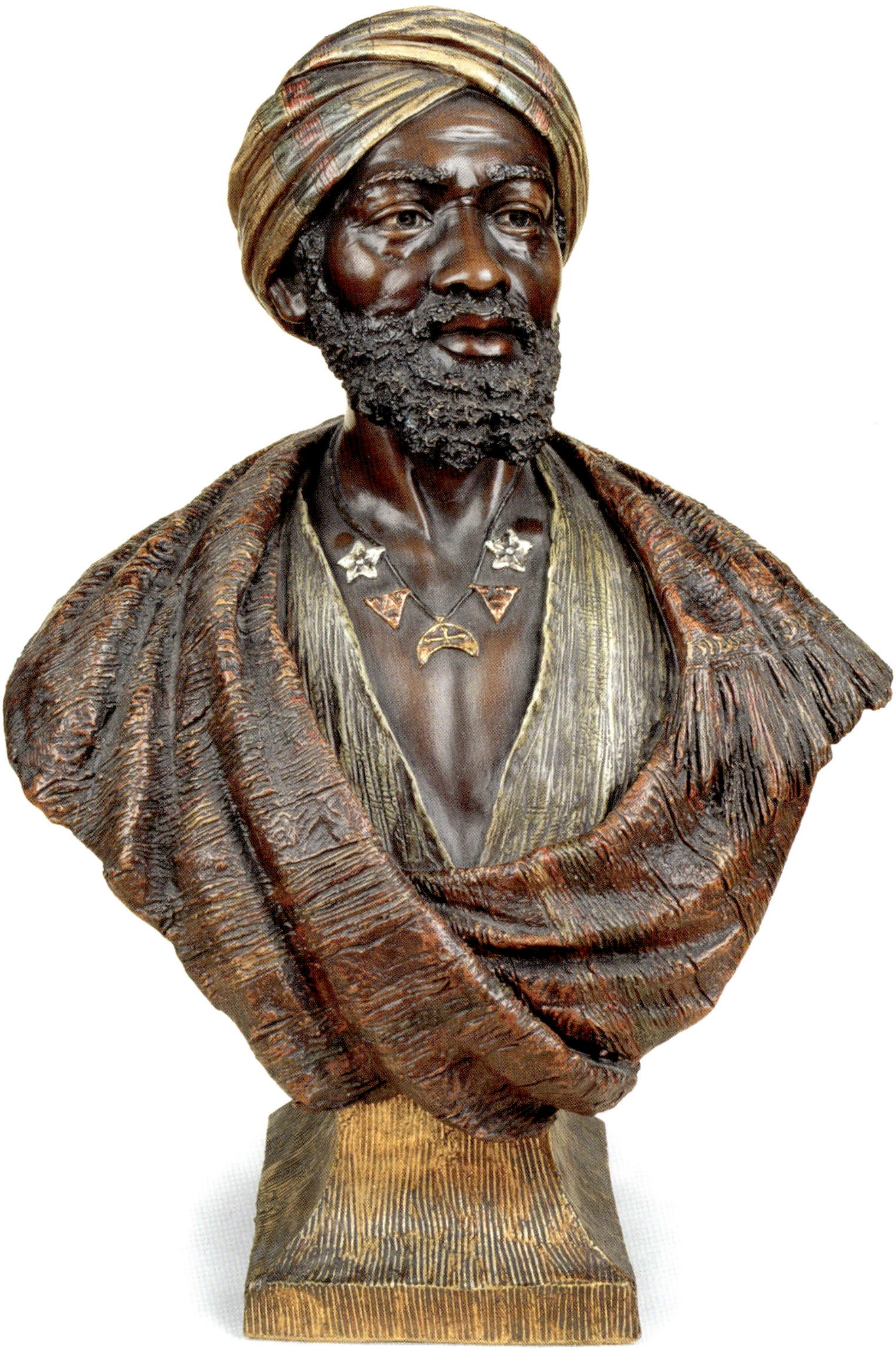

ModNr. 1060
Araber mit Turban I
Arab with Turban,
Arthur Strasser zuge-
schrieben I attributed to,
ca. 1897, PB Wien 19

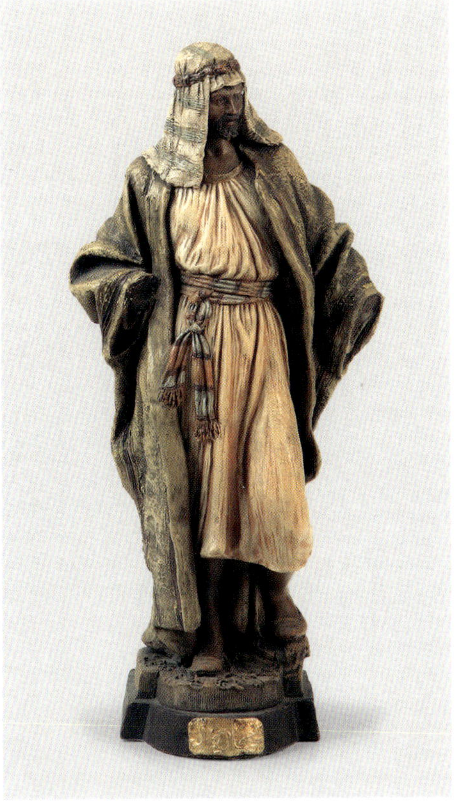

ModNr. 926

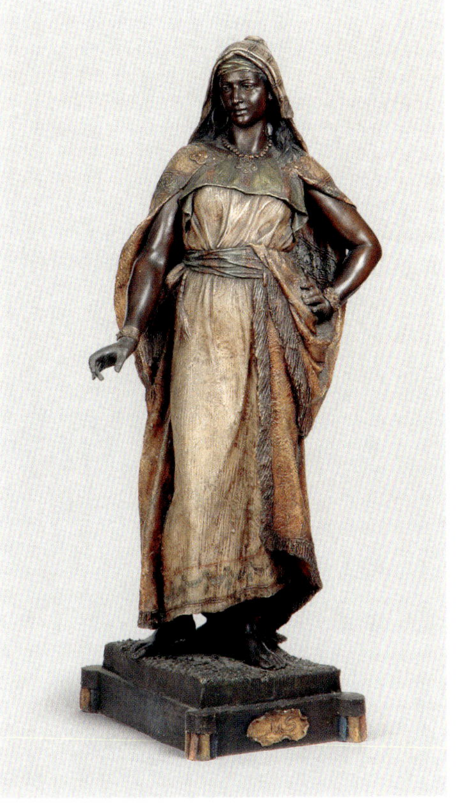

ModNr. 904

ModNr. 926
„Kabyle" | "Kabyle Berber", Arthur Strasser, ca. 1895, © Wien Museum

ModNr. 904
„Kabylin" | "Kabyle Berber Woman", Désurmont, ca. 1895, PB Wien 19

Das Kabylen-Paar "Antar" und „Abla" wurde in verschiedenen Größen hergestellt. Interessant ist hierbei, dass die Figur des Antar von Arthur Strasser entworfen wurde, das weibliche Pendant Abla hingegen von Désurmont. Die Figuren wurden unter verschiedenen Modellnummern in unterschiedlichen Größen produziert. In diesem Falle Antar mit 43 cm und Abla mit 97 cm Höhe.

"Antar" and "Abla," a Kabyle Berber couple, were made in several sizes. What is interesting in this connection is that the figure of Antar was designed by Arthur Strasser whereas the female counterpart was designed by Désurmont. These figures were produced in several sizes and marketed under different model numbers. In the present instance, Antar is 43 cm high and Abla 97 cm.

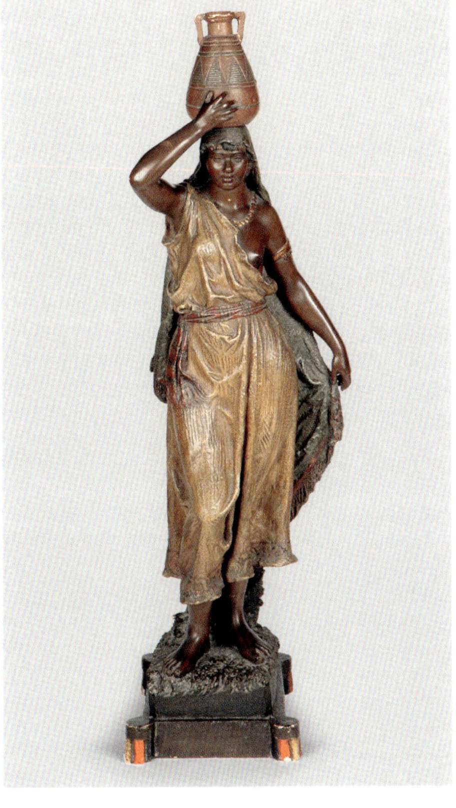

ModNr. 1183
Krugträgerin | Jug Bearer, ca. 1897, PB Wien 19

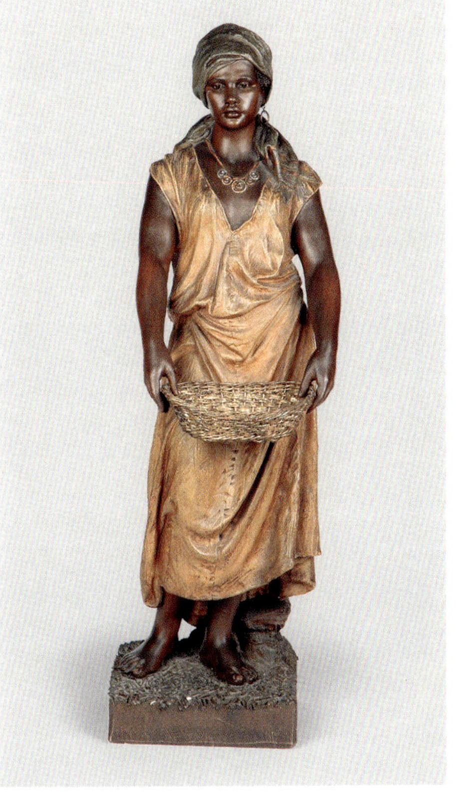

ModNr. 1356
„Mohrin" | "Blackamoor Woman", ca. 1898, PB Wien 19

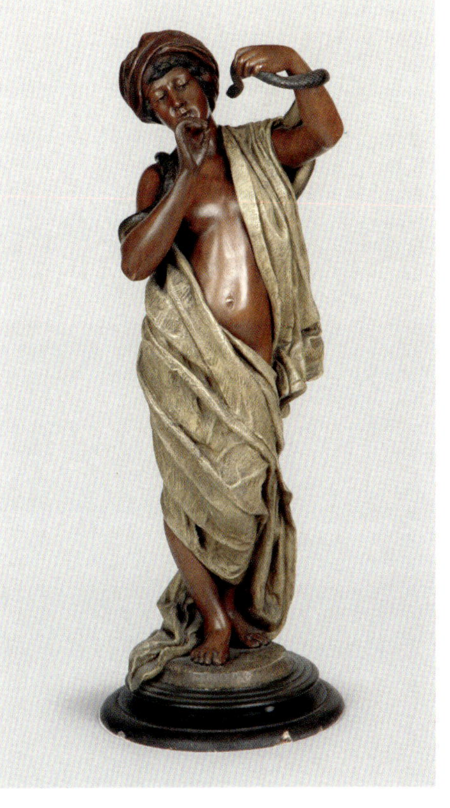

ModNr. 1399
„Schlangenbändiger" | "Snake Charmer", Josef Wind, ca. 1898, PB Wien 19

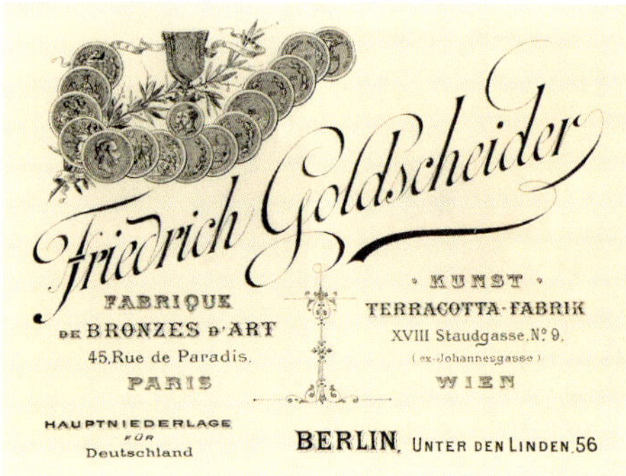

Ill. 23
Wie diese Postkarte aus dem Jahr 1895 zeigt, ist Friedrich Goldscheider in mehreren Ländern mit eigenen Dependancen ansässig. Zahlreiche Auszeichnungen bestätigen seinen internationalen Erfolg. (Sammlung a@goldscheider.net)

As this 1895 postcard shows, Friedrich Goldscheider had branches in several countries. Numerous awards confirmed his success internationally. (Collection a@goldscheider.net)

Ill. 24
Ein Abzeichen mit den Initialen von Friedrich Goldscheider und dem kaiserlichen Spruch „Viribus Unitis" bezeugt seine Kaisertreue. (PB Wien 19)

A badge with Friedrich Goldscheider's initials and the imperial motto "viribus unitis" attests to his allegiance to the monarchy. (PB Vienna 19)

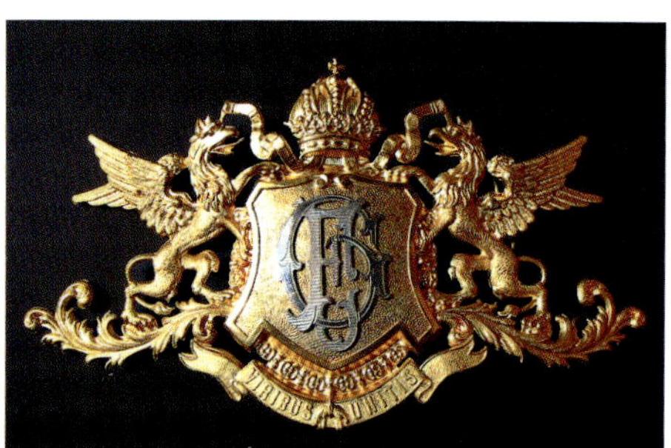

Parallel zur Eröffnung der Leipziger Filiale erfolgt im April 1894 die Einweihung der Hauptniederlassung für Deutschland in Berlin.[77] Am 24. Juli 1894 wird das Wiener Unternehmen offiziell als „Friedrich Goldscheider Erste Wiener Terrakottafabrik" ins Registerblatt der Wirtschaftskammer eingetragen. Inhaber ist Friedrich Goldscheider, die Prokuristen sind Regine und Alois Goldscheider.[78] In Adressbüchern bezeichnet Friedrich seine Fabrik unter den protokollierten Firmen als „Thon.- und Porzellanwaarenfabrik", unter seinem privaten Namenseintrag weiterhin als „Terracotta- u. Majolikafabrik".[79] Friedrich Goldscheider hat somit innerhalb eines Jahrzehnts den Markennamen Goldscheider nicht nur zu internationalem Ansehen gebracht, sondern das Unternehmen expansiv in verschiedenen Märkten positionieren können (Ill. 23). Seine Filialen befinden sich in Wien (Staudgasse 9), Paris (45 Rue de Paradis), Leipzig (Neumarkt/Gewandhaus) und Berlin (Kommandantenstraße 86).[80]

Das Unternehmen arbeitet trotz der Expansion und damit verbundener Kosten in den Jahren 1894 und 1895 mit Gewinn. Nach der Bilanz des Jahres 1894 beläuft sich das Reinvermögen der Firma Friedrich Goldscheider per 31. Dezember 1894 auf 144.775,86 Fl. (Gulden), zuzüglich des Gewinnes von 2.716 Fl. Demgegenüber stehen Forderungen in Höhe von 49.930,85 Fl. Im darauf folgenden Jahr wird ein Gewinn von 3.455,16 Fl. erwirtschaftet. Dank der Umsatzbeteiligung stehen Alois Goldscheider für die Jahre 1894 und 1895 Bezüge in Höhe von 5.737,04 Fl. zu, d.h. das Unternehmen hat in den beiden Jahren einen Umsatz von ca. 300.000 Fl. erreicht. Das Reinvermögen des Unternehmens beträgt zum 31. Dezember 1895 ca. 95.000 Fl.[81]

Die positive Geschäftsentwicklung soll durch eine Verstärkung der Präsenz auf den internationalen Märkten fortgesetzt werden. Im Oktober 1896 wird zu der bestehenden Bronzefabrik in Paris eine Detailniederlassung (also ein üppiges Verkaufsgeschäft mit Schauräumen) auf der vornehmen Einkaufsmeile Avenue de l'Opera 28 eröffnet.[82] Friedrich Goldscheider ist somit in allen Bereichen des Handels vertreten: Den Großhandel betreibt er von Wien und Paris aus, die Verkaufsräume sind in Paris, Berlin und Leipzig.

Die internationale Präsenz der Goldscheider Manufaktur wird um 1896 auch durch ein Atelier in Florenz erweitert. Im „Atelier für Marmor Sculpturen Rotonde Barbetti" werden die Modelle in Marmor ausgeführt, die direkt an die Niederlassungen geschickt und von Wien aus beglichen werden. Es handelt sich allerdings eher um eine kleine Werkstatt, die im Auftrag der Wiener Zentrale arbeitet.[83] Insgesamt sind für die vielbeschäftigte, internationale Manufaktur Friedrich Goldscheider über 200 Arbeiter tätig.[84] Im Jahr 1896 tritt auch Walter Goldscheider in das Unternehmen ein, nachdem er die

In parallel with the opening of the Leipzig branch, the German headquarters were inaugurated in Berlin in April 1894.[77] On 24 July 1894 the Viennese business was registered officially in the registry of the chamber of commerce as "Friedrich Goldscheider Erste Wiener Terrakottafabrik" ["Friedrich Goldscheider First Viennese Terracotta Factory"]. Friedrich Goldscheider was the owner and Regine and Alois Goldscheider were the head clerks with power of attorney.[78] In directories, Friedrich listed his factory under registered firms as a "Thon.- und Porzellanwaarenfabrik" ["Pottery and Porcelain Wares Manufactory"] but under his own name continued to enter it as a "Terracotta- u. Majolikafabrik" ["Terracotta and Majolica Manufactory"].[79] Thus Friedrich Goldscheider not only consolidated the reputation of the Goldscheider brand internationally but was also able to position the business to expand on several different markets within a decade (Ill. 23). His branches were in Vienna (9 Staudgasse), Paris (45 rue de Paradis), Leipzig (Neumarkt/Gewandhaus) and Berlin (86 Kommandantenstraße).[80]

Despite expansion and the concomitant costs incurred, the business was operating with profit margins in 1894 and 1895. After accounts had been settled for 1894, the net assets as of 31 December 1894 amounted to 144 775.86 Fl. (guilders), plus gains of 2716 Fl. This was offset by outstanding debts to the tune of 49 930.85 Fl. The following year profits amounted to 3455.16 Fl. Due to his share of sales proceeds, Alois Goldscheider received sums of 5737.04 Fl. for the years 1894 and 1895, respectively; ie, turnover had reached approx 300 000 Fl. each year. The net assets of the business amounted to approx 95 000 Fl. as of 31 December 1895.[81]

This positive development was to be followed up by consolidating the Goldscheider presence on the international market. In October 1896, a mounts, fittings and accessories branch (that is, a lavish retail branch with showrooms) was added to the existing bronze factory in Paris at No. 28 in the elegant shopping boulevard avenue de l'Opera.[82] Friedrich Goldscheider was now represented in all areas of the trade: he operated the wholesale trade from Vienna and Paris; his retail branches were in Paris, Berlin und Leipzig.

The international presence of the Goldscheider manufactory was enlarged around 1896 by the addition of a Florence studio.

At the "Atelier für Marmor Sculpturen Rotonde Barbetti" ["Rotonde Barbetti Studio for Marble Sculpture"], models were executed in marble and sent directly to the international branches; accounts were settled from Vienna. This newest acquisition was, however, a small workshop that handled orders from the Viennese headquarters.[83] A workforce of over 200 was employed by the internationally busy Friedrich Goldscheider manufactory[84] – Walter Goldscheider also entered the business after finishing higher secondary school, commercial college and devoting himself to "specialist arts and crafts studies."[85] The Goldscheiders continued to live at 9 Staudgasse; enlargement of the production facilities necessitated extending the factory to 7 Staudgasse.[86]

On a visit to Paris, Friedrich Goldscheider suffered a stroke. To aid convalescence, he went to Nice on the coast, where, suspecting he had not long to live, he wrote his will: the children were to inherit the business and they were to continue the family tradition, according to their talents and inclinations, as artistic or managing directors.[87] Friedrich Goldscheider died at fifty-two in Nice on 19 January 1897. An obituary written by Professor Alexander Schmidt praises not only Friedrich Goldscheider's innovative and creative activity; as an entrepreneur he was also "highly esteemed" for being so "obliging" and was also said to have been a friendly, invariably stimulating and interesting person in his dealings with others.[88] In the obituaries written about Friedrich, who was the first to launch artistically painted terracottas in Vienna, it is obvious that even then both the technical and aesthetic importance of his work for the future of art ceramics was recognized.

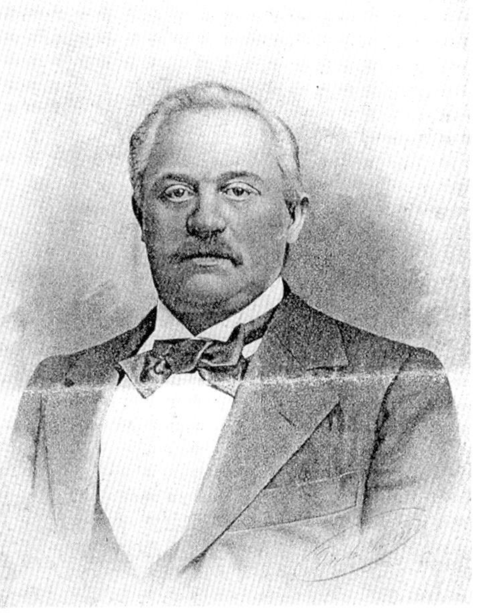

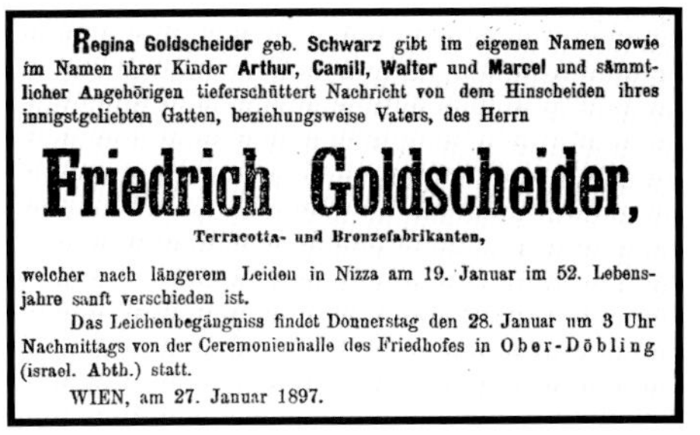

Regina Goldscheider geb. **Schwarz** gibt im eigenen Namen sowie im Namen ihrer Kinder **Arthur, Camill, Walter** und **Marcel** und sämmtlicher Angehörigen tieferschüttert Nachricht von dem Hinscheiden ihres innigstgeliebten Gatten, beziehungsweise Vaters, des Herrn

Friedrich Goldscheider,

Terracotta- und Bronzefabrikanten,

welcher nach längerem Leiden in Nizza am 19. Januar im 52. Lebensjahre sanft verschieden ist.

Das Leichenbegängniss findet Donnerstag den 28. Januar um 3 Uhr Nachmittags von der Ceremonienhalle des Friedhofes in Ober-Döbling (israel. Abth.) statt.

WIEN, am 27. Januar 1897.

Ill. 25, 26
Nach kaum zwölf Jahren des Aufbaus des Unternehmens zu einer weltweit renommierten Marke für Terrakotta und Bronze, stirbt Friedrich Goldscheider mit 52 Jahren in Nizza. (Quelle: Neue Freie Presse vom 27.1.1897, Foto aus Pudor 1903)

After barely twelve years spent building up the business into a brand renowned worldwide, Friedrich Goldscheider died at fifty-two in Nice. (Source: Neue Freie Presse of 27 Jan. 1897; photo from Pudor 1903)

Realschule und die Handelsakademie absolviert hat und sich „kunstgewerblichen Fachstudien" gewidmet hat.[85] Die Goldscheiders wohnen weiterhin in der Staudgasse 9, die Fabrik wird aufgrund der Erweiterung der Produktionsanlagen auf die Staudgasse 7 ausgeweitet.[86]

Bei einem Besuch in Paris erleidet Friedrich Goldscheider einen Schlaganfall. Zur Erholung fährt er ans Meer nach Nizza und schreibt dort unter Todesahnungen sein Testament nieder: Den Kindern solle das Geschäft zufallen und sie sollen, ihren Neigungen entsprechend, als künstlerische oder kommerzielle Leiter die Familientradition fortführen.[87] Am 19. Januar 1897 stirbt Friedrich Goldscheider mit 52 Jahren in Nizza. In einem Nachruf von Professor Alexander Schmidt wird nicht nur seine innovative und schöpferische Tätigkeit gelobt, auch als Geschäftsmann soll er wegen seiner „Coulanz hochgeschätzt" und im Umgang ein freundlicher, immer anregender und interessanter Mensch gewesen sein.[88] In den Nachrufen auf Friedrich, der in Wien als Erster künstlerisch bemalte Terrakotten herausbrachte, wird deutlich, dass man schon damals die Bedeutung seines Wirkens in technischer und künstlerischer Hinsicht für die Zukunft der Kunstkeramik erkannt hatte.

Anmerkungen

[40] Neuwirth (1973), S. 54
[41] Nach Marcel Goldscheiders Schilderung der Geschichte der Manufaktur Goldscheider („An authentic report of its origin and development"), die als Beigabe zu einem Brief an die Kunden bzgl. eines Umzugs in Hanley, vermutlich aus dem Jahr 1951 stammt. Im Stadtarchiv und dem Porzellanarchiv in Karlsbad sind keine Hinweise auf die Tätigkeit Friedrich Goldscheiders in Karlsbad nachweisbar, weder in Verbindung mit einer Porzellanhandlung, noch mit einer Malerei. Es ist eher davon auszugehen, dass er in Karlsbad bei anderen Manufakturen Porzellan aufkaufte, selbst bemalte und mit einem eigenen Stempel versah. Somit konnte er die gute Karlsbader Ware und Qualität für den Aufbau des Qualitätsnamens Goldscheider einsetzen.
[42] Lehmann's Adressbuch Wien, 1881. Das Porzellangeschäft ist bereits im Adressbuch des Jahres 1879 erwähnt, die Privatadresse der Goldscheiders ist die Hörlgasse 14 (später 16). Regine und Friedrich ziehen um 1883 in die Nähe ihrer Firma, in die Wipplingerstraße 15 (vgl. Lehmann's Adressbuch Wien, 1884).
[43] Privilegium 1879/001146, Österreichisches Patentamt, Wien.
[44] Lehmann's Adressbuch Wien, 1881 bis 1883. Im Branchenverzeichnis erscheint nur noch die Firmierung „Goldscheider'sche Porzellan-Manufaktur", diese Manufaktur wird allerdings nicht unter den protokollierten Firmen geführt. Ein Verwandter aus Brennporitschen, Karl Goldscheider, betreibt in Prag zwischen 1891 und 1897 eine eigene Porzellanmalerei (vgl. Konecný, S. 456).
[45] Schmidt, A. (1897), S. 121
[46] Ebenda, S. 40
[47] Neuwirth (1974), S. 25. Zunächst als Einzelfirma, ab 1910 als Offene Handelsgesellschaft
[48] Privilegium 1885/002296, Österreichisches Patentamt, Wien
[49] Verlassenschaftsabhandlung Friedrich Goldscheider, Bezirksgericht Währing (Stadt- und Landesarchiv Wien, 38867). Die Schreibweisen der Namen ändern sich häufig, vor allem Camillo wird mal Kamillo oder Camill geschrieben.
[50] PGGTR, February 1947, S. 140. Laut seinem Sohn Marcel heißt es übersetzt „To stand still is to decay".
[51] Vgl. Bük (1908), S. 582
[52] Lehmann's Adressbuch Wien, 1890. Vgl. auch Anzeige in: Sprechsaal, Nr. 36, 8.9.1887, S. 733. Als Adresse für die Erste Wiener Terracottafabrik ist die Johannesgasse No. 11 angegeben. Die Johannesgasse wird um 1895 in Staudgasse umbenannt. Aus einer Werbung aus dem Jahr 1888 geht noch die Wipplingerstraße 39 als Haupt-Niederlassung hervor.
[53] N.N. in: Keramische Monatshefte, 1904, S. 111. Beim Trocken- und Brennprozess tritt eine Schwindung des Tones auf, die beim Modellieren einbezogen wird und gleichmäßig auf alle Teile des Modells übertragen werden muss, damit es nicht zum Verziehen oder zu Rissen kommt. Es muss darauf Rücksicht genommen werden, dass alle Teile der Figur aus gleichem Material und von gleicher Feuchtigkeit sind. Besondere Vorsichtsmaßnahmen während des Trockenprozesses sorgen dafür, dass keine Teile, also auch nicht die ganz freistehenden Körperteile schneller trocknen als die übrigen Teile der Figur. Deshalb werden freistehende Teile durch Belegen mit beispielsweise Papier gegen zu schnelle Entweichung der Feuchtigkeit geschützt oder mit aufgelegten nassen Tüchern befeuchtet. Nach dem völligen Austrocknen werden sie gebrannt, wobei auch hier erst nach vielen Probebränden die notwendigen Erfahrungen zum jeweiligen Material gesammelt werden können. Bei zu niedrigem Feuer ist die Ware weniger haltbar, bei zu starkem Brand können die Waren durch „Verziehen oder gar durch teilweisen Schmolz minderwertig oder gänzlich unbrauchbar werden", siehe N.N., Figürliche Terrakotten, in: Keramische Monatshefte, 1904, S. 139.
[54] Stellungnahme von Dr. Herrmann Fialla, Hof- und Gerichtsadvokat, als Vertreter der Vormundschaft von Marcell Goldscheider, an das Bezirksgericht Währing, vom 31.3.1909. In einem Engagementbrief vom 18. Mai 1890 wird Alois eine Beteiligung von 2 Prozent vom Umsatz zugesprochen.
[55] Vgl. Bük (1908), S. 582
[56] Reklame in: Sprechsaal, Nr. 36, 8.9.1887, S. 733
[57] Vgl. Bük (1908), S. 582
[58] Schmidt, A. (1890), S. 77. Die meisten der jungen Künstler sind Absolventen der k.k. Fachschule für Keramik in Znaim, der Fachschule für Keramik und Kunstgewerbe in Teplitz-Schönau, der Fachschule für Tonindustrie in Bechyn oder der seit 1868 bestehenden Wiener Kunstgewerbeschule.
[59] Reklame in: Sprechsaal, Nr. 38, 18.9.1890, S. 753. In dieser Anzeige hebt Friedrich auch hervor, dass seine Firma als „Erste Kunstanstalt dieser Art in Bezug auf Schaffung wirklich künstlerischer Originalmodelle und deren künstlerischer Reproduction" anerkannt ist.
[60] Reklame in: Sprechsaal, Nr. 13, 26.3.1891, S. 247. In dieser Anzeige zur Ostermesse in Leipzig steht die Unternehmensbezeichnung „Die erste Wiener Kunst-Terracottafabrik und Atelier für Kunstgewerbe."
[61] Reklame in: Sprechsaal, Nr. 38, 17.9.1891, S. 765
[62] N.N.: Die keramische Ausstellung in Dresden 1891. In: Sprechsaal, Nr. 41, 8.10.1891, S. 816. Für Österreich-Ungarn wurde das Patent am 10.4.1891 angemeldet, später auch für Deutschland.
[63] Privilegium 1892/003079, Österreichisches Patentamt, Wien.
[64] N.N.: Ausstellung des Kunstgewerbevereins in Wien. In: Sprechsaal, Nr. 4, 22.1.1891, S. 63
[65] N.N.: Die keramische Ausstellung in Dresden 1891. In: Sprechsaal, Nr. 41, 8.10.1891, S. 816
[66] Verlassenschaftsabhandlung Friedrich Goldscheider, Bezirksgericht Währing (Stadt- und Landesarchiv Wien, 38867)
[67] Weitere Beispiele siehe Künstlerverzeichnis, S. 552ff
[68] Officieller Bericht der k.k. Öster. Central-Commission für die Weltausstellung in Chicago im Jahre 1893. Wien 1894, S. 17
[69] Amtlicher Special-Katalog der Österreichischen Abteilung auf der Weltausstellung in Chicago 1893. S. 70. Hier ist auch ein Hinweis, dass Arthur Goldscheider zu dieser Zeit als internationaler Vertreter der Manufaktur fungiert.
[70] N.N: Messräume der feinkeramischen Industrie, in: Keramos, Heft 6, Juni 1923, S. 217. Übrigens ist dies die erste Filiale, die von einer keramischen Firma in Leipzig gegründet wurde.
[71] Muster-Register, in: Keramische Rundschau, 1893, Nr. 16, 20.7.1893
[72] Muster-Register, in: Keramische Rundschau, 1893, Nr. 28, 11.10.1893
[73] Schmidt, A. (1894), S. 435
[74] N.N.: Friedrich Goldscheider, in: Sprechsaal, 27. Jahrgang, 1894, S. 1111
[75] Ebenda
[76] Ebenda. In Antwerpen ist die Wiener Fabrik mit Adresse Johannesgasse 11, die Pariser Bronzefabrik wie folgt angegeben: Frédéric Goldscheider, ateliers d'objets d'art, éditeur et fabricant de chefs-d'œuvre d'art plastique en bronze, bronze d'art etc., Fabrique et dépôt à Paris, 45, rue de Paradis. Vgl. Exposition universelle d'Anvers 1894, Brüssel, S. 76 und 301
[77] Verlassenschaftsabhandlung Friedrich Goldscheider, Bezirksgericht Währing (Stadt- und Landesarchiv Wien, 38867)
[78] Registerblatt I für Einzel- und Gesellschaftsfirmen, Wirtschaftskammer Österreich-Archiv, Gewerbearchiv
[79] Lehmann's Adressbuch Wien, 1895 bis 1897
[80] Reklame in: Sprechsaal, 1894, S. 797. Die Berliner Filiale soll zunächst in der Kommandantenstraße 86 angesiedelt gewesen sein, später dann Unter den Linden 56 sowie zum Schluss in der Leipziger Straße 119/120. In dieser Zeit wird auch die Johannesgasse in Staudgasse umbenannt, die Fabrik wird fortan mit Staudgasse 9 im XVIII. Bezirk angegeben (siehe Lehmann's Adressbuch Wien, 1896).
[81] Stellungnahme von Dr. Herrmann Fialla, Hof- und Gerichtsadvokat, als Vertreter der Vormundschaft von Marcell Goldscheider, an das Bezirksgericht Währing, vom 31.3.1909. Die Bilanzen wurden durch den Buchhalter Adolf Pfundhaller nach doppelter Buchhaltung erstellt.
[82] Verlassenschaftsabhandlung Friedrich Goldscheider, Bezirksgericht Währing (Stadt- und Landesarchiv Wien, 38867) Vermutlich erhofft man sich auch von der im Jahr 1900 in Paris statt findenden Weltausstellung zunehmende Absatzmöglichkeiten.
[83] Stellungnahme von Dr. Herrmann Fialla, Hof- und Gerichtsadvokat, als Vertreter der Vormundschaft von Marcell Goldscheider, an das Bezirksgericht Währing, vom 31.3.1909. Regine Goldscheider beschreibt hier das Florenzer Atelier wie folgt: „Dem gegenüber diene zur Aufklärung, dass zur Zeit der Errichtung der Verlassenschaftsinventur diese sogenannte Filiale ein einfaches, von dem Erblasser kurz vor seinem Ableben ins Leben gerufenes Atelier war, woselbst Figuren in Marmor ausgeführt wurden, deren jede fertige unverzüglich an eine unserer Niederlagen abging und von Wien aus Stück für Stück bezahlt werden musste. [...] Dieses Atelier hatte bis Jänner 1897 im Ganzen für rund Lire 6000,– Modelle geliefert, welche, soweit noch unverkaufte Stücke vorhanden waren, in den Lagern und Inventuren nicht unter dem ganz ungerechtfertigten Titel ‚Filiale Florenz' sondern unter der Bezeichnung ‚Marmor' figurierten."
[84] Vgl. Bük (1908), S. 582
[85] Vgl. Planer
[86] Lehmann's Adressbuch Wien, 1897. Alois Goldscheider zieht um 1897 in die Hasenauerstraße 36 (vgl. Lehmann's Adressbuch Wien, 1898).
[87] Klage von Camillo Goldscheider, vertreten durch Dr. Emil Frischauer um Aufnahme als öffentlicher Gesellschafter in die Firma Friedrich Goldscheider mit einer Beteiligung von 20 Prozenten, beim k.k. Handelsgericht Wien, eingelangt am 28. Oktober 1908.
[88] Schmidt, A. (1897), S.121

Notes

[40] Neuwirth (1973), p. 54

[41] According to Marcel Goldscheider's description of the history of the Goldscheider manufactory ("An authentic report of its origin and development"), which was enclosed in a letter to clients informing them of a move in Hanley, presumably dating from 1951. The Carlsbad Municipal Archives and Porcelain Archives seem to contain no reference to Friedrich Goldscheider's operations in Carlsbad in connection with a porcelain or a painting dealership. It is more likely that he bought porcelain from other factories in Carlsbad, had it painted in his own factory and marked with his proprietary stamp. Assuming that this was so, he would have been able to take advantage of the good Carlsbad wares and quality to build up the Goldscheider reputation for quality.

[42] Lehmann's Adressbuch Wien [Vienna], 1881: The porcelain business is listed in the 1879 directory; the Goldscheiders' private address was 14 (later 16) Hörlgasse. Around 1883, Regine and Friedrich moved to 15 Wipplingerstraße (cf Lehmann's Adressbuch Wien, 1884) to be close to the firm.

[43] Privilegium 1879/001146, Österreichisches Patentamt [Austrian Patent Office], Vienna

[44] Lehmann's Adressbuch Wien, 1881 to 1883. All that is listed in the business directory from then on is "Goldscheider'sche Porzellan-Manufaktur;" however, a manufactory of this name is not listed among the registered firms. Karl Goldscheider, a relative from Brennporitschen, ran a porcelain-painting business of his own in Prague between 1891 and 1897 (cf Konecný, p. 456).

[45] Schmidt, A. (1897), p. 121

[46] Ibid, p. 40

[47] Neuwirth (1974), p. 25. First as a business, from 1910 as an Offene Handelsgesellschaft (OHG = partnership).

[48] Privilegium 1885/002296, Österreichisches Patentamt, Vienna

[49] Verlassenschaftsabhandlung [estate settlement] Friedrich Goldscheider, Bezirksgericht Währing (Stadt- und Landesarchiv Wien, 38867). The spellings of the names often vary: Camillo, for instance, is also spelt Kamillo or Camill.

[50] PGGTR, February 1947, p. 140. "Stillstand ist der Anfang vom Ende," which means, translated literally: "Standing still is the beginning of the end."According to his son Marcel, it means in translation "To stand still is to decay."

[51] Cf Bük (1908), p. 582

[52] Lehmann's Adressbuch Wien, 1890. Cf also the advertisement in: Sprechsaal, No. 36, 8 Sept. 1887, p. 733: No. 11 Johannesgasse is given as the adress of the Erste Wiener Terracottafabrik. Around 1895 Johannesgasse was renamed Staudgasse. An advertisement dating from Jahr 1888 still gives 39 Wipplingerstraße as the main branch.

[53] N.N. in: Keramische Monatshefte, 1904, p. 111. During the process of drying and firing, the clay body of all applied parts contracts and shrinkage must be equalized for all parts of a model to avoid warping or cracking. It must be ensured that all parts of a figure are made of the same material and have the same degree of wetness. Special precautions taken during drying ensure that no parts – not even the entirely free-standing parts of the body – dry more quickly than the other parts of the figure. Free-standing parts are, therefore, protected by being covered, for instance, with paper, to prevent moisture escaping too rapidly or are moistened by having damp cloths laid on them. After fully drying, they are fired; here the necessary experience with a particular material can only be gained from many trial firings. Fired at a temperature that is too low, the ware is less durable; firing temperatures that are too high can make wares "of inferior quality or entirely useless through warping or even through being partly melted;" see N.N.: Figürliche Terrakotten, in: Keramische Monatshefte, 1904, p. 139.

[54] Deposition made by Dr. Herrmann Fialla, Hof- und Gerichtsadvokat, as representing the guardianship of Marcell Goldscheider, to the Bezirksgericht Währing, on 31 Mar. 1909. In a confirming letter dated 18 May 1890 Alois is granted a 2 per cent share of sales proceeds.

[55] Cf Bük (1908), p. 582

[56] Advertisement in: Sprechsaal, No. 36, 8 Sept. 1887, p. 733

[57] Cf Bük (1908), p. 582

[58] Schmidt, A. (1890), p. 77. Most of the young artists were graduates of the k.k. Fachschule für Keramik in Znaim, the Fachschule für Keramik und Kunstgewerbe in Teplitz-Schönau, der Fachschule für Tonindustrie in Bechyn or the Viennese Kunstgewerbeschule (est. 1868).

[59] Advertisement in: Sprechsaal, No. 38, 18 Sept. 1890, p. 753. In this advertisement, Friedrich also emphasies that his firm has achieved recognition as the "First art institution of this kind with regard to creating really artistic original models and reproducing them artistically".

[60] Advertisement in: Sprechsaal, No. 13, 26 Mar. 1891, p. 247. In this advertisement for the Leipzig Easter Fair, the firm is called "The First Vienna Art and Terracotta Manufactory and Studio for Arts and Crafts".

[61] Advertisement in: Sprechsaal, No. 38, 17 Sept. 1891, p. 765

[62] N.N.: Die keramische Ausstellung in Dresden 1891. In: Sprechsaal, No. 41, 8 Oct. 1891, p. 816. The patent was applied for on 10. April 1891 for the Austro-Hungarian monarchy, later also for Germany.

[63] Privilegium 1892/003079, Österreichisches Patentamt, Vienna

[64] N.N.: Ausstellung des Kunstgewerbeverein in Wien. In: Sprechsaal, No. 4, 22 Jan. 1891, p. 63

[65] N.N.: Die keramische Ausstellung in Dresden 1891. In: Sprechsaal, No. 41, 8 Oct. 1891, p. 816

[66] Verlassenschaftsabhandlung [Estate Settlement] Friedrich Goldscheider, Bezirksgericht Währing (Stadt- und Landesarchiv Wien, 38867)

[67] For more examples see the list of artists, p. 552f

[68] Officieller Bericht der k.k. Öster. Central-Commission für die Weltausstellung in Chicago im Jahre 1893. Vienna 1894, p. 17

[69] Amtlicher Special-Katalog der Österreichischen Abteilung auf der Weltausstellung in Chicago 1893. p. 70. Here there is also an indication that Arthur Goldscheider was by that time functioning as the Manufactory representative abroad.

[70] N.N: Messräume der feinkeramischen Industrie, in: Keramos, No. 6, Juni 1923, p. 217. Incidentally, this was the first branch established by a ceramics firm in Leipzig.

[71] Muster-Register [Pattern Index], in: Keramische Rundschau, 1893, No. 16, 20 Jul. 1893

[72] Muster-Register, in: Keramische Rundschau, 1893, No. 28, 11 Oct. 1893

[73] A. Schmidt (1894), p. 435

[74] N.N.: Friedrich Goldscheider, in: Sprechsaal, 27th year, 1894, p. 1111

[75] Ibid

[76] Ibid. In Antwerp, the factory addresses were given as follows: the Viennese factory, 11 Johannesgasse; the Paris bronze factory: Frédéric Goldscheider, ateliers d'objets d'art, éditeur et fabricant de chefs-d'œuvre d'art plastique en bronze, bronze d'art etc., Fabrique et dépôt à Paris, 45, rue de Paradis. Cf Exposition universelle d'Anvers 1894, Brussels, p. 76 and p. 301.

[77] Verlassenschaftsabhandlung Friedrich Goldscheider, Bezirksgericht Währing (Stadt- und Landesarchiv Wien, 38867)

[78] Registerblatt I [Registry Sheet I] für Einzel- und Gesellschaftsfirmen, Wirtschaftskammer Österreich-Archiv, Gewerbearchiv

[79] Lehmann's Adressbuch Wien, 1895 to 1897

[80] Advertisement in: Sprechsaal, 1894, p. 797. The Berlin branch is said to have opened at 86 Kommandantenstraße 86 but is given later as 56 Unter den Linden and finally 119/120 Leipziger Straße. During that time, Johannesgasse was renamed Staudgasse; the factory address would be listed thenceforward as 9 Staudgasse in the XVIIIth District (see Lehmann's Adressbuch Wien, 1896).

[81] Deposition made by Dr. Herrmann Fialla, Hof- und Gerichtsadvokat, representing the guardianship of Marcell Goldscheider, given at Bezirksgericht Währing, on 31 Mar. 1909. The accounts were settled by bookkeeper Adolf Pfundhaller according to the double-entry bookkeeping system.

[82] Verlassenschaftsabhandlung Friedrich Goldscheider, Bezirksgericht Währing (Stadt- und Landesarchiv Wien, 38867). Presumably, there were hopes of a growth market following the 1900 Paris Universal Exposition.

[83] Deposition made by Dr. Herrmann Fialla, Hof- und Gerichtsadvokat, as representing the wardship of Marcell Goldscheider, at Bezirksgericht Währing, on 31 Mar. 1909. Regine Goldscheider describes the Florence studio here as follows: "As concerns that, may it serve to clarify that, at the time the estate inventory was drawn up, this so-called branch was a simple studio established by the testator, where figures were executed in marble, of which each finished piece went without delay to one of our branches and had to be paid by the piece from Vienna. [...] By January 1897, that studio had delivered models to the overall sum of some 6000 Lires, which, in so far as pieces that had remained unsold were in existence, figured in the warehouses and inventories under the heading 'marble' rather than under the entirely unjustifiable entry 'Florence Branch'."

[84] Cf Bük (1908), p. 582

[85] Cf Planer

[86] Lehmann's Adressbuch Wien, 1897. Alois Goldscheider moved to 36 Hasenauerstraße around 1897 (cf Lehmann's Adressbuch Wien, 1898).

[87] Petition submitted by Camillo Goldscheider, represented by Dr. Emil Frischauer for admission as partner in the firm of Friedrich Goldscheider with a share of 20 per cent to the k.k. Handelsgericht Wien, registered on 28 Oct. 1908.

[88] Schmidt, A. (1897), p. 121

Die Blüte des Jugendstils unter Regine Goldscheider (1897–1918)

The Flowering of Art Nouveau under Regine Goldscheider (1897–1918)

Ill. 27
Das Erbe von Friedrich Goldscheider wird unter seinen Söhnen Arthur, Camillo, Walter und Marcell sowie der Witwe Regine aufgeteilt. Der endgültige Erbvertrag wird erst am 24. Dezember 1899, also fast drei Jahre nach Friedrichs Tod unterzeichnet. (Quelle: Stadt- und Landesarchiv Wien, 38867)

Friedrich Goldscheider's legacy was shared between his sons Arthur, Camillo, Walter and Marcell and his widow, Regine. The final probate contract was not signed until 24 December 1899, that is, nearly three years after Friedrich's death. (Source: Stadt- und Landesarchiv Wien, 38867)

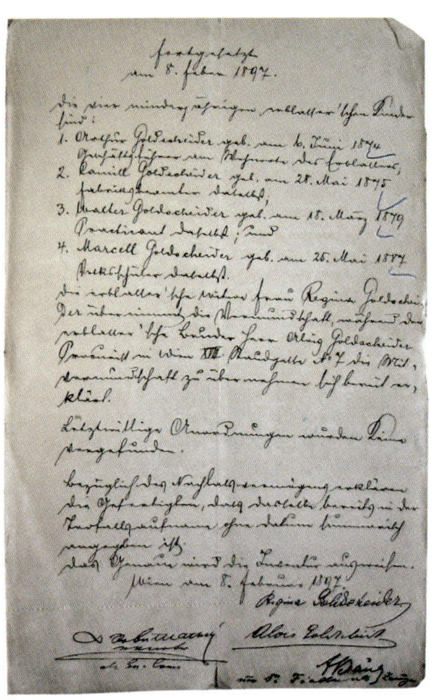

Als Alleininhaber der Manufaktur besaß Friedrich Goldscheider im Jahr 1897 Filialen oder Werkstätten in Wien, Paris, Florenz, Berlin und Leipzig. Bis zur Klärung der Hinterlassenschaft führen nach seinem Tod Regine und Alois Goldscheider die Geschäfte. Regine zieht in diesen schweren Zeiten auch zu ihrem Schwager in die Hasenauerstraße 36.[89] Doch bereits 1898 verfügt der Betrieb nicht mehr über genügend liquide Mittel. Alois Goldscheider berichtet, dass „mangels finanzieller Mittel und bei dem Missverhältnisse zwischen der Beschaffenheit der Aktiva und Passiva es schon im Jahre 1898 schier unmöglich erschien, den Betrieb weiterzuführen und den Verpflichtungen nachzukommen."[90]

Zur Begleichung der Schulden und zur Sanierung müssen Geldmittel gefunden werden. Alois und Regine beschließen, „dass die stets Geld und Waren verzehrende und nichts reproduzierende Filiale in Berlin aufgelöst, desgleichen ein grosses Warenlager bei unzulänglicher Verwertung absorbierende Niederlage in Paris abgestossen und schliesslich das in Wien übrig bleibende Geschäft nebst der Leipziger Verkaufsstelle verkauft, bezw. von einer dritten Person [...] mit allen ihm anhaftenden Geldlosigkeit und Belastung übernommen werden."[91]

So wird die Berliner Filiale am 10. September 1899, für 63.000 Mark an den dortigen Vertreter Carl Denecke verkauft.[92] Er soll weiterhin Goldscheider Objekte verkaufen und erhält hierfür 10 Prozent Vermittlungsprovision für die im Geschäftslokal Leipzigerstraße 119/120 aufgenommenen und von der Firma ausgeführten Aufträge.

Die Dependance und Fabrik in Paris sowie das Atelier in Florenz sollen an den 25-jährigen Arthur Goldscheider übergehen, der den Kauf teilweise mit seinem Erbanteil begleichen will. Der Nachlass von Friedrich Goldscheider besteht vor allem in der Firma, so dass der Unternehmenswert per Bilanz vom 30. September 1899 bestimmt und das Vermögen laut dem Erbteilungsvertrag (Ill. 27) vom 24. Dezember 1899 wie folgt aufgeteilt wird:[93] Die minderjährigen Söhne Walter und Marcell, die unter die vormundschaftliche Obhut von Regine und Alois Goldscheider fallen, erhalten jeweils 18.000 Fl. (Gulden),

As sole proprietor of the Manufactory, Friedrich Goldscheider had branches or workshops in Vienna, Paris, Florence, Berlin and Leipzig in 1897. After his death, Regine and Alois Goldscheider managed the business until the estate was settled. In that difficult time, Regine moved to live with her brother-in-law at 36 Hasenauerstraße.[89] By 1898 the business had insufficient liquid assets. Alois Goldscheider reported that "for lack of financial means considering the discrepancy between the assets and liabilities, even in the year 1898 it seemed simply impossible to continue with the business and meet [our] commitments."[90]

Funds had to be found to pay off the debts and restructure the business. Alois and Regine decided "that the branch in Berlin, that always consumed money and wares and produced nothing, should be dissolved; also a large, costly and insufficiently used warehouse in Paris should be got rid of and, finally, the business remaining in Berlin as well as the Leipzig retail outlet should be sold or taken over by a third party [...] with all the lack of funds and burden of debt."[91]

The Berlin branch was subsequently sold to Carl Denecke, the firm representative there, for 63 000 marks on 10 September 1899.[92] Denecke would continue to sell Goldscheider objects and received a 10 per cent middleman's commission on orders accepted in the outlet at 119/120 Leipzigerstraße and executed by the firm.

The Paris branch and factory as well as the Florence studio would go to twenty-five-year-old Arthur Goldscheider, who proposed financing the acquisition partly from his share of the legacy. Friedrich Goldscheider's legacy consisted mainly in the firm; consequently, the value of the business was estimated according to the balance of 30 September 1899 and the legacy was to be divided up as set out in the probate contract (Ill. 27) of 24 December 1899[93] as follows: The sons Walter and Marcell, who were still minors and under the guardianship of Regine and Alois Goldscheider, were each to receive 18 000 Fl. (guilders) that were to be kept under court sequestration until they should come of age. Camillo was to receive the

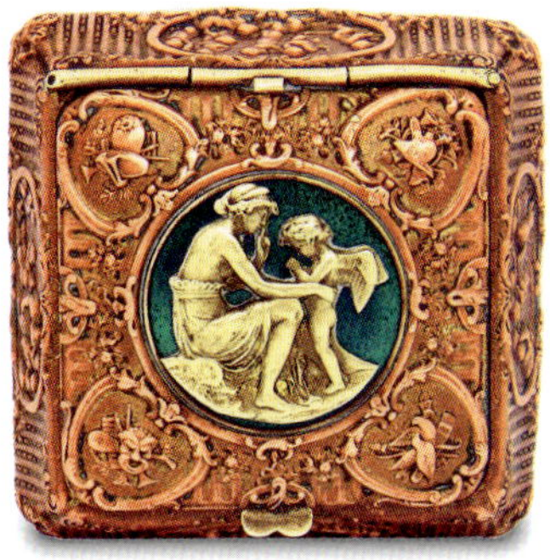
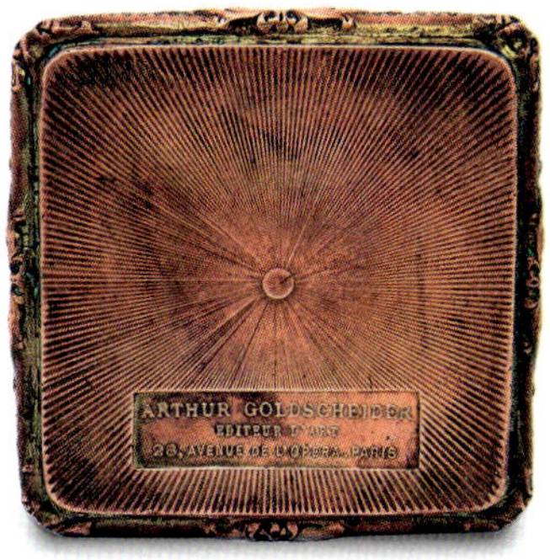

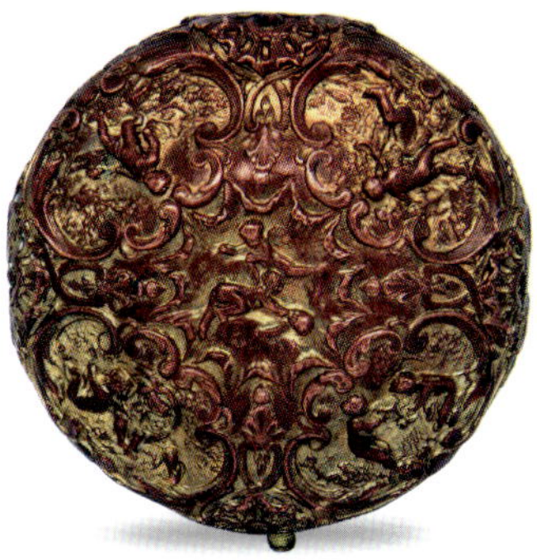
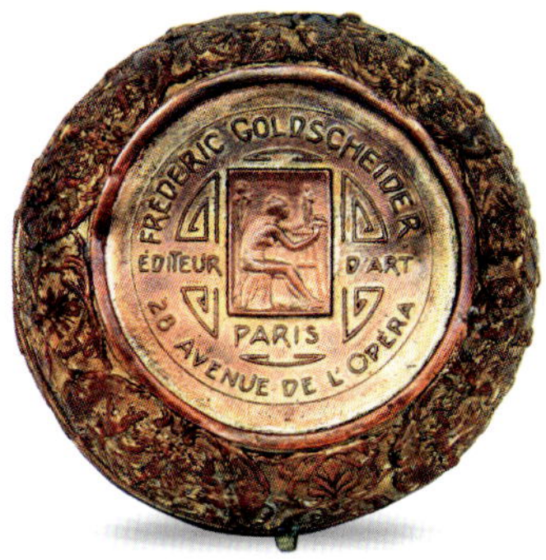

Ill. 28, 29
Die beiden Dosen zeigen die als Werbefläche eingesetzte Unterseite für die Pariser Dependance, Avenue de l'Opera 28: einmal für die Firma Frédéric Goldscheider und in einer späteren Fassung für Arthur Goldscheider, Editeur d'Art. (Sammlung a@goldscheider.net)

These two boxes show the underside used as a surface on which the Paris branch at 28 avenue de l'Opera is advertised: one advertising the firm of Frédéric Goldscheider and, in a later version, advertising Arthur Goldscheider, Editeur d'Art. (Sammlung a@goldscheider.net)

same sum as his two younger brothers, to be paid out to him from the year 1904 in annual payments of 3000 Fl. Arthur, as the eldest son and heir was to receive as his share the Paris and Florence branches (see also the chapter "Ornamental Sculpture from Arthur Goldscheider in Paris," p. 228f), including the "mounts, fittings and accessories outlet" in avenue de l'Opera, the storehouses and the flat in rue de Paradis as well as the workshops and the studio in Florence. However, he would have to advance 70 000 Fl. for this in several monthly payments. All the above would be handed over to Arthur after payment had been completed so that he could continue to operate the branches from 1 January 1900 as "sole proprietor at his own expense, cost, advantage and at his own risk." In addition, he was obliged to pay his brother Camillo a minimum annual salary of 1800 Fl. and, if possible and Camillo so desired, employ him in his business.

The future collaboration between the Vienna and the Paris factories was also settled: "Re the further production and reproduction of the models accordingly shared and the further exercise of the acquired production rights it is hereby determined that all hitherto mutually produced models, in particular all those in stock in Vienna and Paris arrived by 31 December 1899, also the Viennese models existing in Paris and Florence may continue in future to be mutually produced free of charge. Further, it is determined that, as far as granting future production rights elsewhere for the Viennese models in marble, the licensing fee is to amount to 5 % (five per cent) and in bronze

die bis zur Volljährigkeit in gerichtlicher Verwahrung bleiben. Camillo erhält die gleiche Summe wie seine beiden jüngeren Brüder, die ihm ab dem Jahr 1904 in jährlichen Raten von 3.000 Fl. ausgezahlt werden soll. Der älteste Sohn Arthur erhält als Erbe die Filialen in Paris und Florenz (siehe auch Kapitel „Dekorative Skulpturen von Arthur Goldscheider in Paris", S. 228ff), inklusive des „Detail-Verkaufslokals" in der Avenue de l'Opera, der Magazine und der Wohnung in der Rue de Paradis sowie der Werkstätten bzw. des Ateliers in Florenz. Er muss hierfür jedoch eine Abtrittszahlung von 70.000 Fl. in mehreren Monatsraten leisten. All dies soll nach vollständiger Zahlung an Arthur übertragen werden, so dass er die Filialen ab dem 1. Januar 1900 als „alleiniger Inhaber auf eigene Rechnung, Kosten, Nutzen, und Gefahr" weiterführen kann. Zusätzlich wird er verpflichtet, seinem Bruder Camillo ein Mindestjahresgehalt von 1.800 Fl. zu zahlen und ihn gegebenenfalls in seinem Geschäft zu beschäftigen.[94]

Auch die künftige Zusammenarbeit zwischen der Wiener und Pariser Manufaktur wird geregelt: „Betreffs der ferneren Reproduktion und Vervielfältigung der hiernach aufgeteilten Modelle und der ferneren Ausübung erworbener Reproduktionsrechte wird festgesetzt, dass alle bisher gegenseitig reproduzierten Modelle, speziell alle in Wien vorhandenen und bis 31. Dezember 1899 noch eintreffenden Pariser Modelle, desgleichen die in Paris und Florenz vorhandenen Wiener Modelle auch künftighin unentgeltlich gegenseitig reproduziert werden dürfen. Weiter wird fixiert, dass bezüglich der Einräumung anderweitiger künftiger Reproduktionsrechte die Lizenzgebühr für Wiener Modelle in Marmor 5 % (fünf Prozent) und in Bronce 10 % (zehn Prozent) desgleichen für Pariser Modelle zur Vervielfältigung in Terracotta 10 % (zehn Prozent) von dem für jedes Stück erzielten Netto-Verkaufserlöse beträgt."[95]

Regine Goldscheider erhält als Witwe alle „Aktiva und Passiva der Firma Friedrich Goldscheider in Wien, Berlin und Leipzig" und wird von den Erben ermächtigt, die Firma in dem damaligen Wortlaut „Friedrich Goldscheider in Wien, Berlin und Leipzig" weiterzuführen.[96] Die Erben verpflichten sich

10 % (ten per cent) and also for reproducing Paris models in terracotta 10 % (ten per cent) of the net sales proceeds from each piece."[95]

As the widow, Regine Goldscheider received all "assets and liabilities of the firm of Friedrich Goldscheider in Vienna, Berlin and Leipzig" and was empowered by the heirs to continue to manage the firm as designated at the time "Friedrich Goldscheider in Vienna, Berlin and Leipzig."[96] The heirs pledged not to establish any businesses of their own that might pursue the manufacture of terracotta nor to have any share, direct or indirect, in such a business, niether as representatives nor as agents. Moreover, Camillo was to be able to work according to his abilities in any "working environment adapted [to his needs] and for commensurate remuneration according to an agreement that was to be concluded."[97]

ModNr. 1072
„Schaukel" I "See-saw", Pecheur / Simon, ca. 1898, PB Wien 19
Dieses Modell wird zwei unterschiedlichen Künstlern zugeordnet. Einerseits konnte ein signiertes Exemplar von Simon bestimmt werden, andererseits wird in der zeitgenössischen Literatur der Künstler Pecheur angegeben (siehe Pudor 1902).

This model is attributed to two different artists. On the one hand, a signed piece by Simon has been identified; on the other, the artist named in publications of the day is Pecheur (see Pudor 1902).

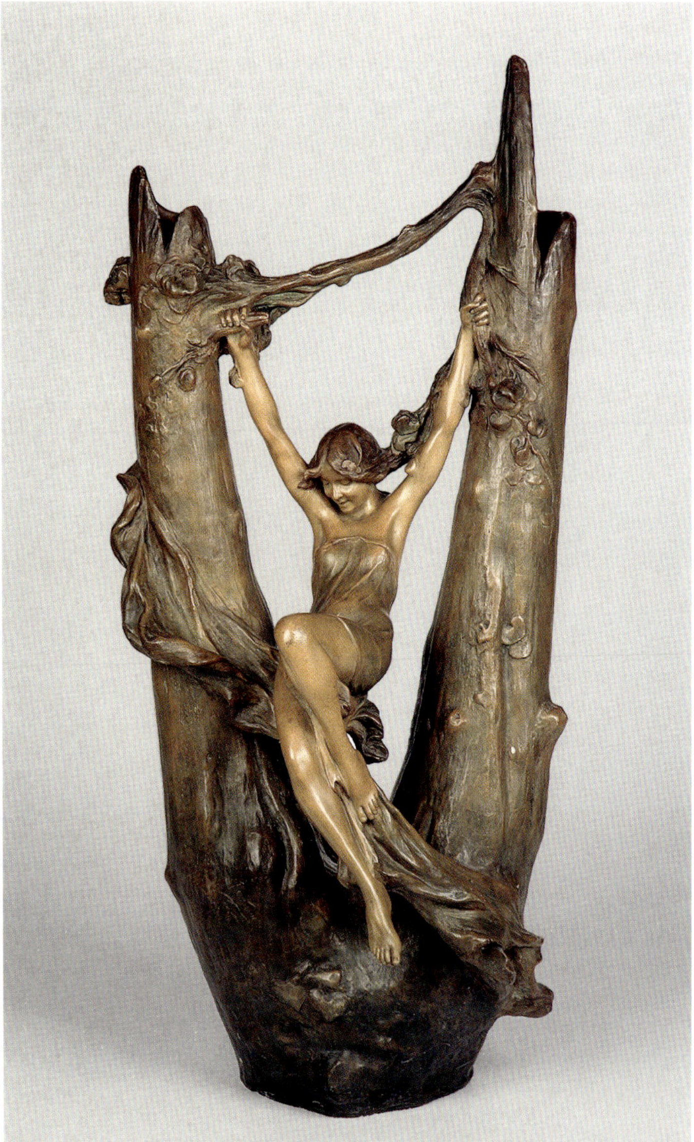

ModNr. 1067
„Pfeilschnitzer" I "Arrow Maker", Lemoine, ca. 1898, © Wien Museum

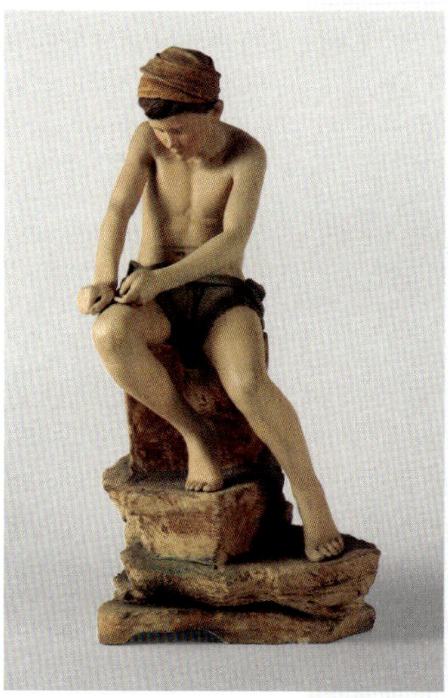

ModNr. 1973
„Vasenmaler" I "Vase Painter", Chavanne, ca. 1900, PB Wien 19

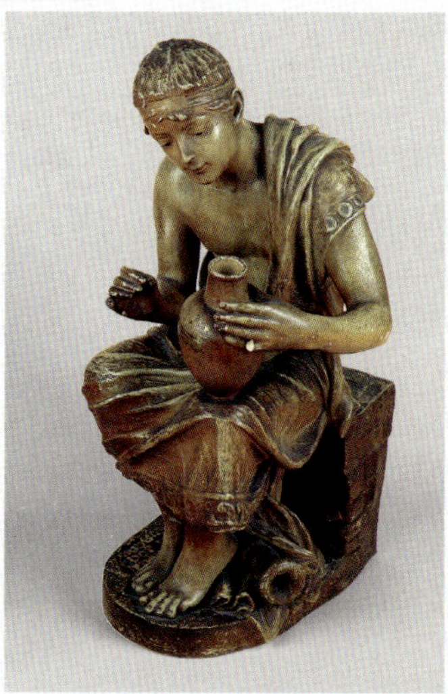

Although the liquid assets acquired by selling off the branches were insufficient to cover the debts,[98] Regine decided to continue production in Vienna and to keep the Leipzig branch: "Nevertheless, despite warnings to the contrary on the part of my brother-in-law – co-guardian – and my brother, Herr Heinrich Schwarz in Leipzig, who had been familiar with the circumstances from the outset, I summoned up the courage to take on this difficult and risky task because, more than any interest whatsoever in the money, a powerful urge grew in me to make any sacrifice to honor the memory of my spouse as far as preserving his work and his and his sons' reputation and to secure to the latter at least the available share of the legacy."[99] Another risk to be incurred by production in Vienna was the circumstance that the rental contract for the factory building in Staudgasse expired in 1899. Despite the alleged liquidity deficit, Regine bought the property at No. 9 Staudgasse on 31 July 1899 for 55 000 Fl.[100] The business year 1899 finished with losses amounting to 15 363 Fl. due to the expenses thus incurred.[101]

In those difficult years between 1897 and 1900, the accustomed drive for innovation was seen to be lacking at Goldscheider. In July 1897 a "Decor Ebène chromé" and a "Décor Chromo nouveau with distinctive painting" were introduced, followed in September 1897 by a "distinctive painting in wood brown and fresh green" but these do not represent anything special.[102] The terracottas continued to be elaborately worked after firing to ensure the customary Goldscheider quality standards and safeguard the reputation of the brand but the splendor of decoration that would have been in keeping with the times was lacking: "The figures and groups of figures made of clay body that fires light yellow are, as is well known, painstakingly retouched after being removed from the molds so that all details are artistically emphasized and each individual figure reveals the distinctive touch of the sculptor who has retouched it. After they have been dried and fired, they are painted free-hand in oil paint, whereby once again the man or woman painting it lends the wares their distinctiveness. Thus it is made possible that, although each model is cast many times, each individual finished piece creates its own particular artistic effect. Unfortunately, the oil paints are only weatherproof to a slight degree so that all these terracottas can only be used inside buildings; even the peculiar

zudem kein eigenes Unternehmen zu gründen, das der Fabrikation von Terrakotta nachgeht, oder sich an einem solchen irgendwie direkt oder indirekt zu beteiligen, auch nicht als Vertreter oder Kommissionäre. Camillo soll zudem in Wien gemäß seinen Fähigkeiten im jeweils „angepassten Wirkungskreis und den entsprechenden Entlohnungen nach einer zu treffenden Abmachung" arbeiten können.[97]

Obwohl die Beschaffung liquider Mittel durch den Verkauf der Filialen nicht zur Bewältigung der Schulden ausreicht[98], beschließt Regine, die Produktion in Wien fortzusetzen und die Filiale in Leipzig beizubehalten: „Gleichwohl gewann ich trotz Abmahnungen seitens meines Schwagers – des Mitvormundes – und meines mit den Verhältnissen seit jeher genauestens vertrauten Bruders Herrn Heinrich Schwarz in Leipzig den Wagemut, die schwierige und bedenkliche Aufgabe auf mich zu nehmen, weil mehr als jegliches Geldinteresse der Impuls mächtig in mir wirkte, dem Andenken meines Gatten betreff Erhaltung seiner Schöpfung und der Ehre seines und seiner Söhne Namens jegliches Opfer zu bringen und den letzteren für alle Fälle mindestens den vorhandenen Erbanteil zu sichern."[99] Ein weiteres Risiko für die Produktion in Wien bildet der im Jahr 1899 auslaufende Mietvertrag für das Fabrikhaus in der Staudgasse. Trotz der angeblichen Liquiditätsschwäche erwirbt am 31. Juli 1899 Regine die Liegenschaft Staudgasse Nr. 9 für 55.000 Fl.[100] Das Geschäftsjahr 1899 wird auch wegen dieser Sonderaufwendungen mit einem Verlust von 15.363 Fl. abgeschlossen.[101]

In den schwierigen Jahren zwischen 1897 und 1900 wird auch der gewohnte Innovationsschub bei Goldscheider vermisst. Zwar wird im Juli 1897 ein „Decor Ebène chromé" und „Decor Chromo nouveau mit eigenartiger Bemalung" eingeführt und im September 1897 eine „eigenartige Bemalung in holzbraun und neugrün", diese stellen aber nichts Besonderes dar.[102] Die Terrakotten werden nach dem Brennvorgang zwar weiterhin aufwändig bearbeitet, um die gewohnte Qualität und den Ruf der Marke Goldscheider zu sichern, es fehlt aber der Glanz zeitgemäßer Dekorierungen: „Die aus hellgelbbrennenden Thonen ausgeführten Figuren und Figurengruppen werden nach Herausnahme aus den Formen bekanntlich sorgfältig nachgearbeitet, so dass alle Details künstlerisch zur Erscheinung kommen und jede einzelne Figur die Eigenart des nachar-

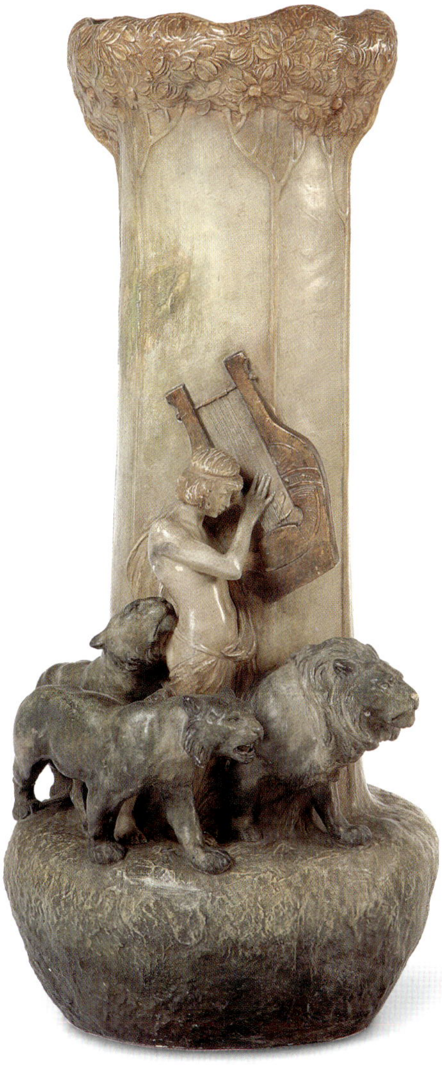

ModNr. 1251
Vase „Lautenschläger" I "Lute Player", Adolphe-Jean Lavergne, ca. 1898, © Wien Museum

ModNr. 1741
„Amazonenkönigin" |
"Queen of the Amazons", Cherc, ca. 1898,
© Wien Museum

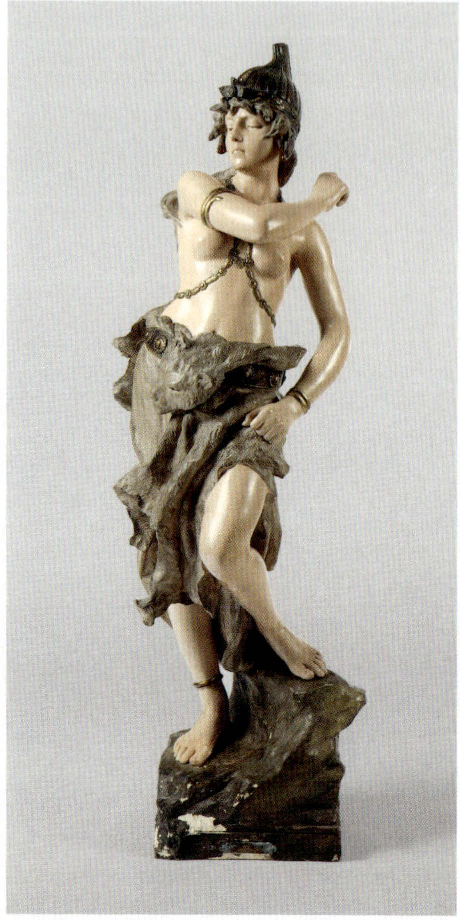

beitenden Bildhauers erkennen lässt. Nachdem sie dann getrocknet und gebrannt sind, werden sie freihändig mit Oelfarbe bemalt, wobei wiederum der ausführende Maler oder die Malerin ihre Eigenarten den Waren verleihen. Hierdurch wird es möglich, dass, obgleich jedes Modell sehr oft abgeformt wird, doch jedes einzelne fertige Stück seine besondere künstlerische Wirkung ausübt. Leider sind die Oelfarben nur in geringem Masse wetterbeständig, so dass alle diese Terrakotten nur im Innern von Gebäuden zur Verwendung gelangen können; auch der eigenartige Glanz, welchen keramische Glasuren den mit denselben versehenen Waren verleihen, fehlt diesen Stücken."[103]

Erst Anfang 1900 werden bei Goldscheider wieder neue und maßgebliche Herstellungs- und Dekorierungsverfahren entwickelt. Am 30. März 1900 wird ein Patent zur „Decorierung plastischer Gegenstände" angemeldet. Dieses neue Verfahren soll vor allem bei der Dekorierung plastischer Gegenstände aus Terrakotta oder anderen Materialien, wie Gips, künstlicher Marmor und Kunst-

brilliance, which ceramic glazes lend wares decorated with them, is lacking in these pieces."[103]

Major innovations in manufacturing and decorating were not developed again at Goldscheider until early in 1900. A patent was applied for on 30 March 1900 for "decorating sculptural objects." This new process was to be used primarily for decorating sculptural objects of terracotta or other materials such as plaster of Paris, synthetic marble and stone composite to make the surface of the "object hard, matt, durable and translucent-looking."[104]

Particularly striking, however, are some terracottas that boast decoration protected by another patent. "Patent Algen" covers, as the name suggests, subtle nuances in the representation of aquatic plants: "the

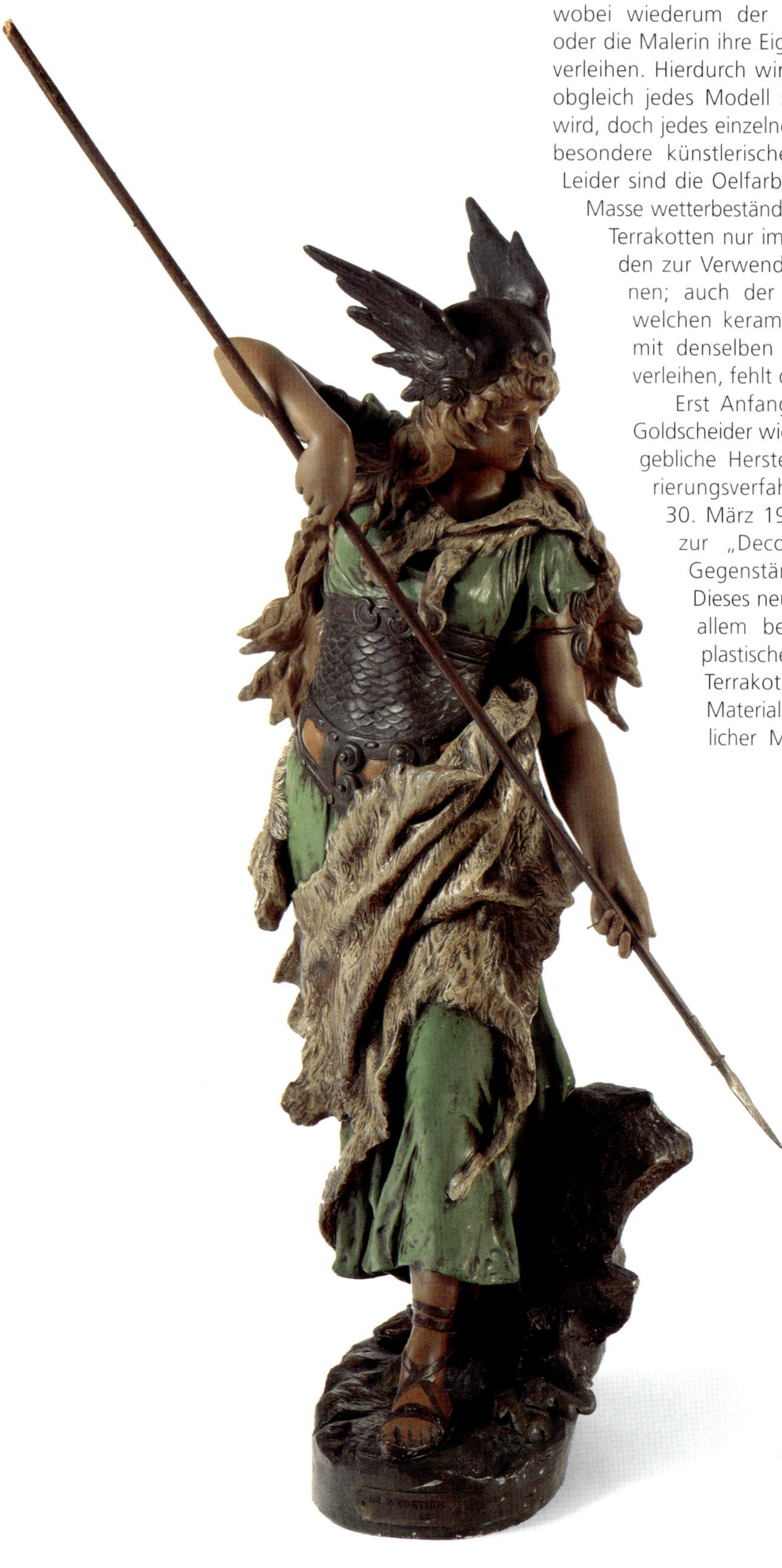

ModNr. 1861
„Walküre" | "Valkyrie",
Cherc, ca. 1899,
© Wien Museum

most matt yellowish green, which darkens into brownish green, light grey that shades through lavender into golden bronze tones. The harmonious effect achieved by the colors of these decorations is further enlivened by reddish brown highlights sporadically added in places. These basic tones are always varied individually among themselves on the various art objects for which algae decoration is used so that each piece boasts decoration of a distinctive character."[105] These models include the vases "Amour poursuivi" by Croisy, "Affen und Krokodil" ["Monkeys and Crocodile"] by Salzmann and a vase designed by Simon decorated with a scene of children dancing a round dance. The algae decoration also occurs on mirror frames by Baudry, the figure entitled "Wassernymphe" ["Water Nymph"] by Montenaves as well as other objects.

Another new decoration element introduced at that time was "Turquoise," a "vibrant, blue-green turquoise color with vivid green nuances and in two variants, ie, light and dark."[106] A perching owl conceived

steinmassen, eingesetzt werden, um die Oberfläche des „Gegenstandes hart, matt, widerstandsfähig zu machen und transparent erscheinen zu lassen."[104]

Besonders auffallend sind jedoch einige Terrakotten, die mit einem anderen Patent geschützte Dekore aufweisen. Dieses „Patent Algen" umfasst, wie der Name schon andeutet, variierende Nuancen der Wasserpflanzen: „mattestes Gelbgrün, das sich bis zu Bräunlichgrün verstärkt, lichtes Grau, das durch Lila in goldige Bronzetöne übergeht. Die harmonische Farbenwirkung dieser Dekorationen wird noch belebt durch vereinzelt an den Stellen aufgesetzte rötlich-braune Reflexe. Diese Grundtöne sind bei den verschiedenen Kunstgegenständen, bei denen Algendekor verwendet ist, stets noch individuell unter sich variiert, so dass jedes Stück einen speziellen Charakter des Dekors aufweist."[105] Zu diesen Modellen gehören die Vasen „Amour poursuivi" von Croisy, „Affen und Krokodil" von Salzmann oder die von Simon entworfene Vase mit einem Kinderringelreigen. Das Algendekor kommt aber

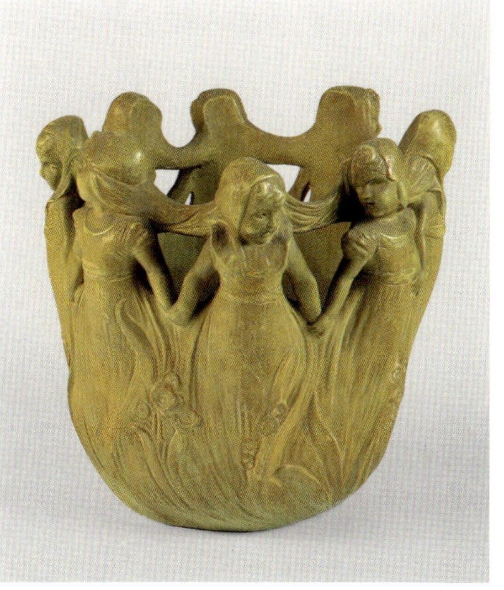

ModNr. 712
Vase, Simon, ca. 1898, © Wien Museum
Das Modell 712 wird im Algendekor produziert. Die zeitgenössische Presse schreibt: „Eine von Simon modellierte Vase stellt Ringelreigen spielende kleine Mädchen dar, deren unten zusammenfließende Kleider die Wände der Vase bilden. Die Füßchen verschwinden in dem hohen Grase der Wiese, auf welcher die Kinder spielend gedacht sind. Jedes Köpfchen der singenden Kleinen ist liebevoll und sorgsam durchgeführt, die ganze Gruppe von frappierender Charakteristik". (R., F.: Wiener Kunstkeramik, in: KM, 2.1902, S. 89, 90, m. Abb.)

Model 712 was produced with algae decoration. The press commentary at the time was: "A vase modelled by Simon represents little girls doing a round dance, whose dresses flowing together at the bottom form the wall of the vase. Their little feet disappear in the high grass of the meadow on which the children are conceived as playing. Each little head of the singing little girls is lovingly and painstakingly executed, the whole group of a striking appearance." (R., F.: Wiener Kunstkeramik, in: KM 2 1902, p. 89, p. 90, with ill.)

ModNr. 1772
„Chriemhilde" I "Kriemhild", Albert-Ernest Carrièr-Belleuse, ca. 1899, © Wien Museum
Durch den Brandschwund haben gleiche Modelle manchmal unterschiedliche Maße: Hier eine Büste des französischen Bildhauers Carrièr-Belleuse, links mit 77 cm sowie rechts mit 75 cm Höhe. Ein fast identisches Modell gibt es unter der Modell-Nummer 1647.

Shrinkage in firing often caused identical models to be turned out in different sizes: hence a bust by the French sculptor Carrièr-Belleuse might be either 77 cm high (left) or 75 cm high (right). The model number 1647 is almost identical with model 1861.

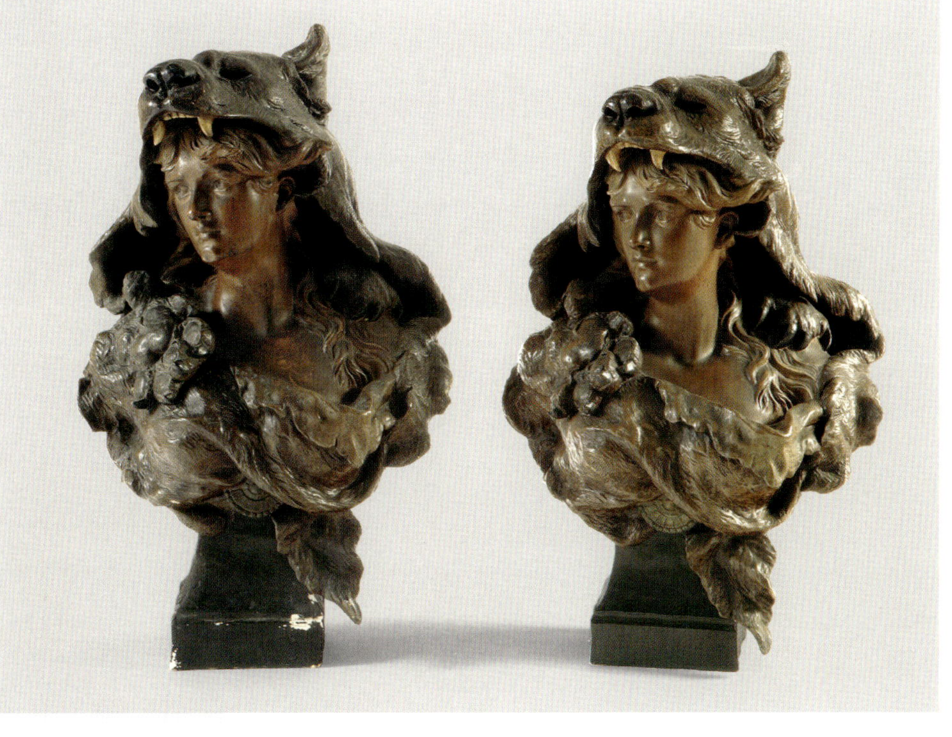

auch bei Spiegelfassungen von Baudry, der Figur „Wassernymphe" von Montenaves und anderen Gegenständen vor.

Unter den neu eingeführten Dekoren dieser Zeit ist auch „Türkis", eine „leuchtende blaugrüne Türkisfarbe mit lebhaften grünen Nuancierungen und in zwei Varianten, nämlich hell und dunkel."[106] Im dunklen Türkisdekor wird eine als Beleuchtungskörper gedachte hockende Eule verkauft, in der hellen Türkisdekoration die von Ledru entworfene Vase „Betrachtung".

Ein dritter Dekorationsstil, das „Holzdekor", eignet sich vor allem für figürliche Darstellungen. In hellem Holzdekor wird eine Muschelträgerin dargestellt, die einen „kupfrigen, leicht patinierten Bronzeton aufweist. Einige Goldreflexe, die verstreut aufgesetzt sind, machen den Eindruck, als habe sich die Patina an einigen Stellen abgelöst."[107] Haniroffs „Krugträgerin" zeigt hingegen ein dunkles Golddekor, hat somit den Anschein der schwärzlichen Farbe der Eisenbronze, und aufgesetzte Lichter, etwa wie gelblichen Messington.

Mit zarten lila Farbenvariationen spielt das Dekor „Heliotrop", zum Beispiel bei Tells „Muschelträgerin". Keramische Produkte im Dekor „Grès" tragen die natürliche Farbe des französischen Steinzeugs, dazu gehören die Schalen „Der Mann mit der Schlange" von Podolsk oder Montenaves „Wellenschlag". Das Dekor „Secession" findet man bei Reliefs, deren Motive sich in „licht gelblicher Kolorierung" von einem lederbraunen Hintergrund abheben.[108] Dieses Dekor findet man auch bei einer Sonnenblumenvase von Petri. Die Modelle werden in mehreren dieser Dekorarten und verschiedenen Mate-

as a lamp base and shaft was marketed in the dark turquoise decoration; "Betrachtung" ["Contemplation"], a lamp designed by Ledru, was produced in the light turquoise.

A third style of decoration, "Wood Décor," was particularly suited to figurative representation. A shell bearer was represented in the light wood décor, which "boasts a coppery, lightly patinated bronze tone. Some gold highlights dotted here and there create the impression that the patina has flaked off in places."[107] Haniroff's "Krugträgerin" ["Jug Bearer"], on the other hand, has dark golden decoration, which looks like the blackish color of architectural bronze (a copper alloy with iron content), and is touched with highlights approximating the yellowish tint of brass.

The decoration called "Heliotrope," plays with delicate lavender color variations, for instance, on Tell's "Muschelträgerin" ["Shell Bearer"]. Ceramics products featuring the "Grès" decoration are in the natural color of French stoneware. Examples include the bowls "Der Mann mit der Schlange" ["Man with a Serpent"] by Podolsk and Montenave's "Wellenschlag" ["Breakers"]. "Secession" is a decoration element that occurs in reliefs, with the motifs picked

Das Modell „Rebecca" wird in verschiedenen Dekoren und Größen produziert: Modell 1903 in ca. 93 cm, Modell 2171 in 63 cm und Modell 2537 in 45 cm Höhe.

The "Rebecca" model was produced featuring various decoration schemes and in several sizes: model 1903 ca 93 cm high; model 2171 63 cm and model 2537 45 cm high.

ModNr. 1903
„Rebecca", Cherc,
ca. 1900, PB Wien 19

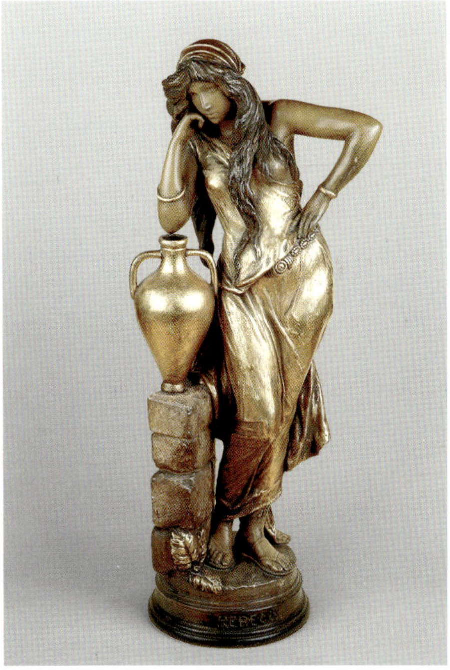

ModNr. 2171
„Rebecca", Cherc,
ca. 1901, PB Wien 19

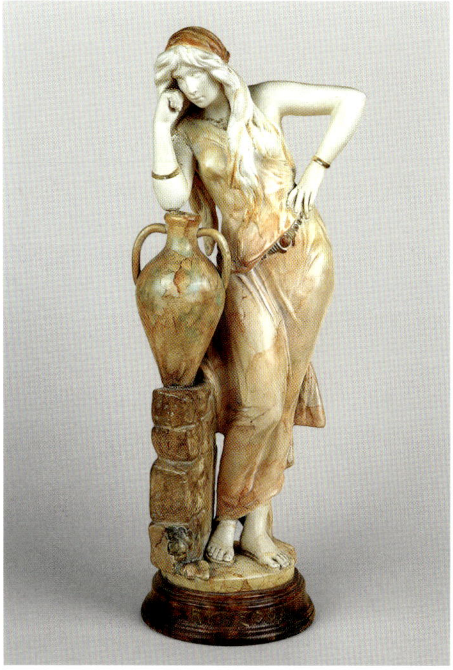

ModNr. 2537
„Rebecca", Cherc,
ca. 1902, PB Wien 19

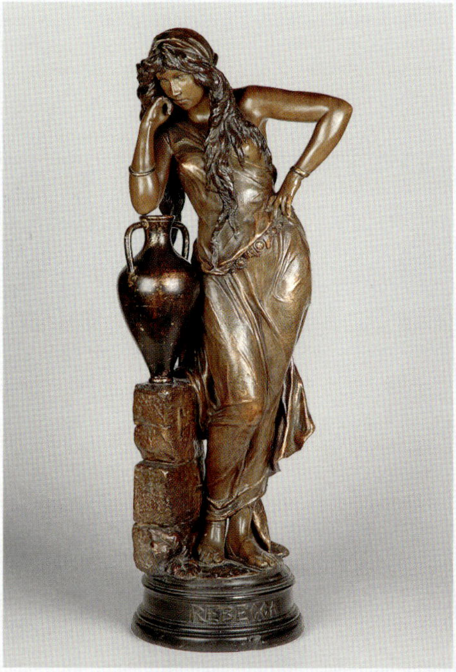

out in "light yellowish coloring" against a leather-brown background.[108] This decoration also occurs in a sunflower vase by Petri. Models were marketed in several of these decoration types as well as a wide variety of materials, including terracotta, marble, alabaster, stone composite, bronze and even pewter and zinc.

At the ground-breaking 1900 Paris Exposition Internationale Universelle, Goldscheider not only took the public by surprise with innovative exhibits. The "consummate artistry" of the sculpture shown was just as acclaimed as the "original manner of the sculptural and decorative tinting."[109] In the modern Salon, designed by the architect Neukomm, numerous statues, busts and vases were shown, including "Pariserin" ["Parisienne"] by Moreau-Vauthier, a bust of Napoleon II as "l'Aiglon," a bust of the emperor of the Austro-Hungarian monarchy by Hegenbarth and several works of Impressionist sculpture by Delagrange. The other Goldscheider Salon rooms, the Moorish-Orientalizing and the Ancient Greek exhibition space, showcased sculptures in those styles respectively.[110]

rialien aus Terrakotta, Marmor, Alabaster, Kunststein, Bronze, sogar Zinn und Zink angeboten.

Auf der richtungweisenden Pariser Weltausstellung im Jahr 1900 (Exposition Internationale Universelle) überrascht Goldscheider zwar nicht mit Neuheiten, aber die „künstlerische Vollendung" der ausgestellten Skulpturen wird genauso anerkannt wie die „originelle Art der plastischen und dekorativen Tönungen".[109] Im modernen Salon, entworfen vom Architekten Neukomm, werden zahlreiche Statuen, Büsten und Vasen gezeigt, unter anderem die „Pariserin" von Moreau-Vauthier, eine Büste von Napoleon II als „L'Aiglon", von Hegenbarth die Büste des Kaisers von Österreich-Ungarn sowie mehrere impressionistische Plastiken von Delagrange. In den anderen Salonräumen von Goldscheider, dem maurisch-orientalischen und griechisch-antiken, werden Plastiken in den jeweiligen Stilen gezeigt.[110]

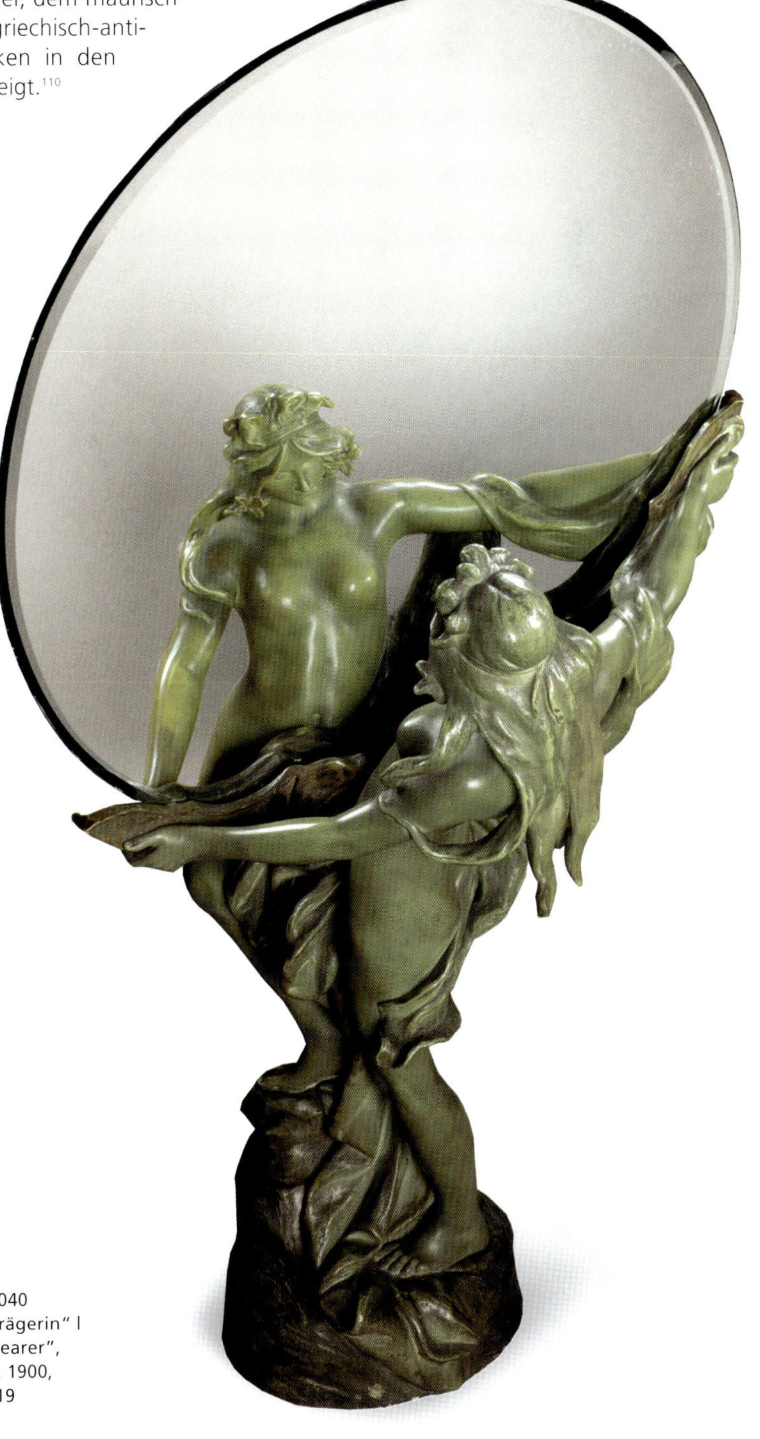

ModNr. 2040
„Spiegelträgerin" I
"Mirror Bearer",
E. Tell, ca. 1900,
PB Wien 19

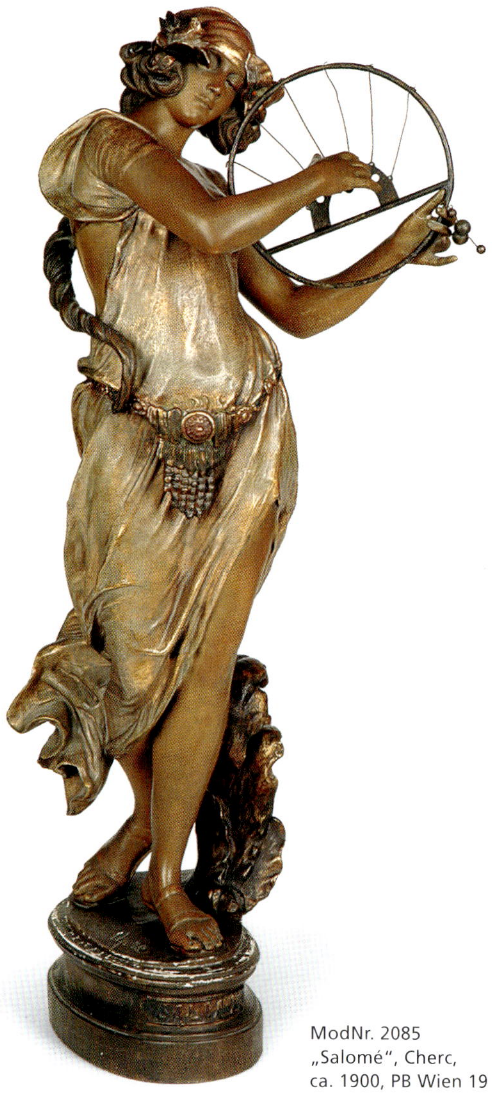

ModNr. 2085
„Salomé", Cherc,
ca. 1900, PB Wien 19

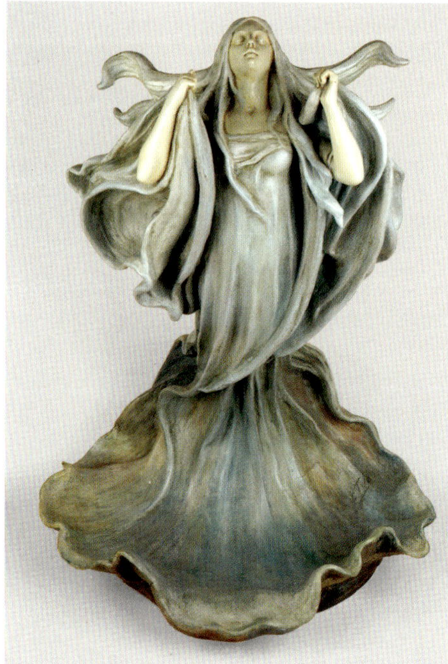

ModNr. 2149
„Die Nacht" | "Night",
E. Tell, ca. 1900,
PB Wien 19

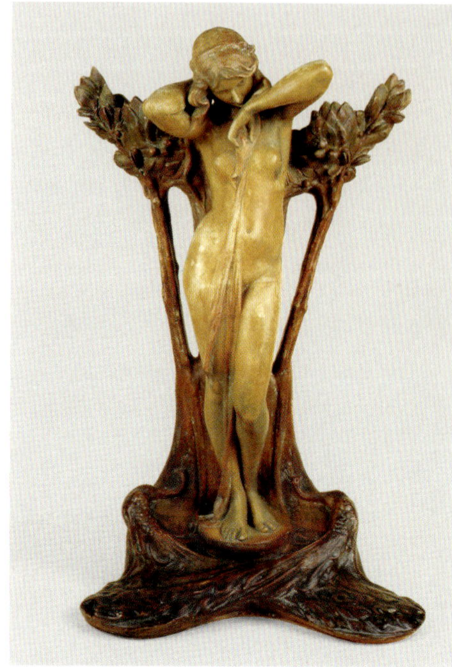

ModNr. 2543
Spiegel | Mirror „Iris",
ca. 1902, PB Wien 19

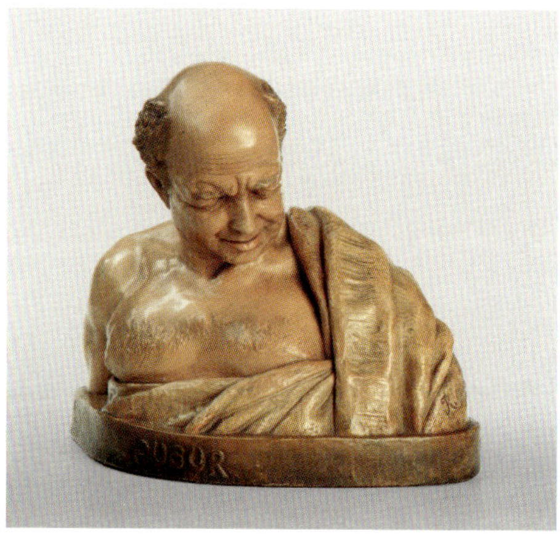

ModNr. 2147
„Augur", Rudolf
Maison, ca. 1900,
© Wien Museum

rechts | right

ModNr. 2223
„Die Jagdtrophäe" |
"Hunting Trophy",
Haniroff, ca. 1900,
© Wien Museum

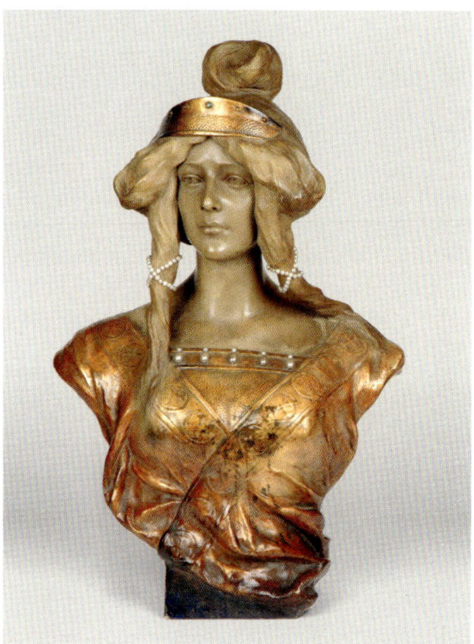

ModNr. 2440
„Byzanz" | "Byzantium",
Haniroff, ca. 1903, PB Wien 19

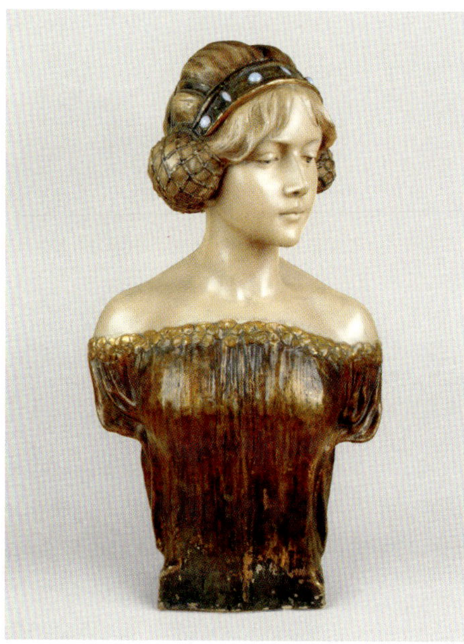

ModNr. 2664
Büste Mädchen | Bust Girl,
Montenave, ca. 1903, PB Wien 19

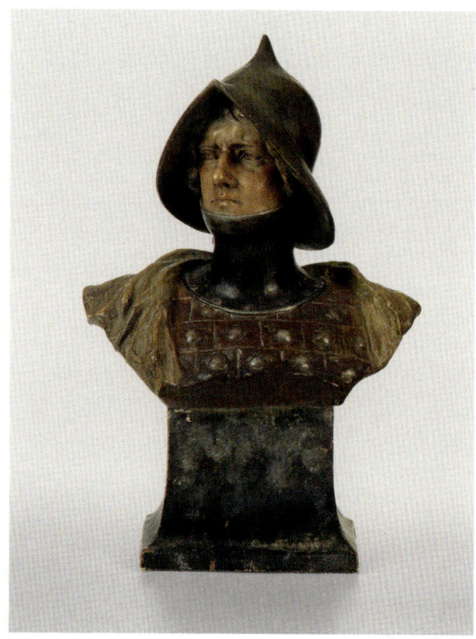

ModNr. 2854
„Jeanne d'Arc", Ciseau,
ca. 1905, © Wien Museum

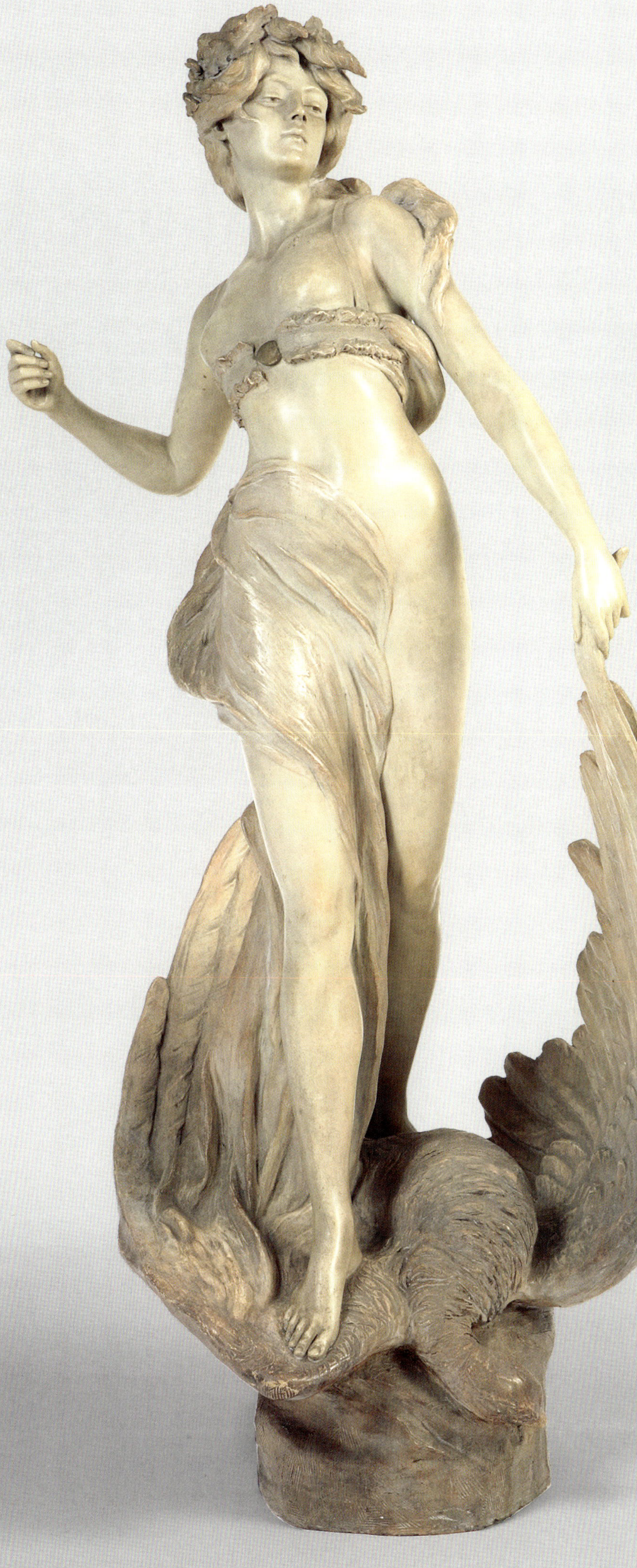

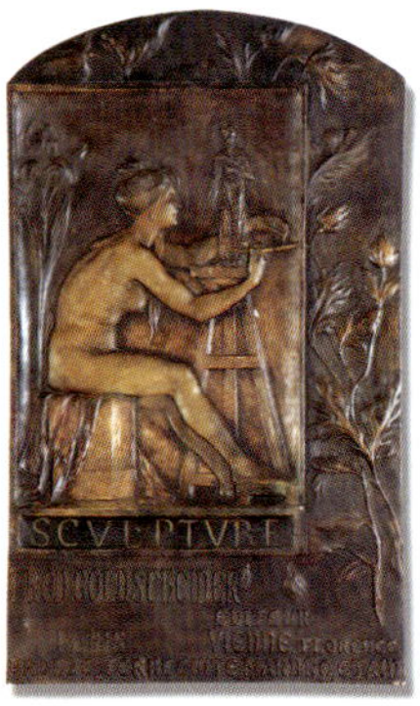

ModNr. 672
„Sculpture", Maurice Bouval, ca. 1898, MAK – Österreichisches Museum für angewandte Kunst / Gegenwartskunst, Wien
Die von Maurice Bouval als Allegorie der Bildhauerei entworfene „Sculpture" dient auch zur Kennzeichnung der Goldscheider Skulpturen.

"Sculpture," designed by Maurice Bouval as an allegorical personification, was also used as the mark for Goldscheider sculpture.

Die Firma beschäftigt um 1900 über 200 Arbeiter in Wien, darunter „kunstgewerblich und künstlerisch geschulte Kräfte ersten Ranges".[111] Unter den Modelleuren, Formgießern und Figurenretoucheuren sind: Josef Biško, Franz Fuka, Rudolf Krajc, Josef Pítra, Jaroslav Richter, Jindřich Skalák, Johann Studenovský, Anton Strakatý, František Vaňek, Hugo Vondrák, Franz Warzel.[112] In diesem prägenden Jahr meldet die Wiener Manufaktur eine neue Firmenmarke an, die auf die Tonwaren reliefartig aufgelegt und mitgebrannt wird: die von Maurice Bouval entworfene „Sculpture", auch töpfernde Frau genannt.[113] Diese Marke wurde schon seit längerem in Paris eingesetzt und zeigt eine sitzende Frau, die an einer Skulptur arbeitet, sowie darunter den Schriftzug „Friedrich Goldscheider Wien".

Die seit 1901 mit dem jüngeren Hof- und Gerichtsadvokaten Dr. S. Lewit verheiratete Regine[114] stellt im Jahr 1902 außer auf der in London stattfindenden „Exhibition of

By 1900 the firm employed a workforce of over 200 in Vienna, including "top-ranking personnel schooled in the applied and fine arts."[111] The modellers, molders and retouchers included: Josef Biško, Franz Fuka, Rudolf Krajc, Josef Pítra, Jaroslav Richter, Jindřich Skalák, Johann Studenovský, Anton Strakatý, František Vanek, Hugo Vondrák and Franz Warzel.[112] During those formative years, the Viennese factory registered a new firm trademark, which was applied to pottery wares in relief before firing: "Sculpture," also called "Woman Making Pots," designed by Maurice Bouval.[113] Already in use for some time in Paris, this trademark depicts a seated woman working on a sculpture and below her in lettering "Friedrich Goldscheider Wien."

Regine had been married since 1901 to Dr. S. Lewit, a younger man who was a Counsellor to the Court.[114] In 1902 Regine showed work at both the London Exhibition of Austrian Fine Art and Decorative Furnishings and the important Prima Esposizione d'Arte Decorativa Moderna in Turin. At the latter show, which represented Art Nouveau at its zenith, Friedrich Goldscheider's "arts and crafts workshop" was represented in the Austrian Pavilion alongside the Riessner, Stellmacher & Co ("Amphora") and Josef Böck manufactories as well as the k.k. Fach-

ohne I without ModNr.
Amor I Cupid, Francesco Flora, ca. 1900, Artimo Fine Arts, Brüssel
Dieser Amor aus Marmor wurde in der Florentiner Werkstatt hergestellt.

This marble Cupid was made in the Florence workshop.

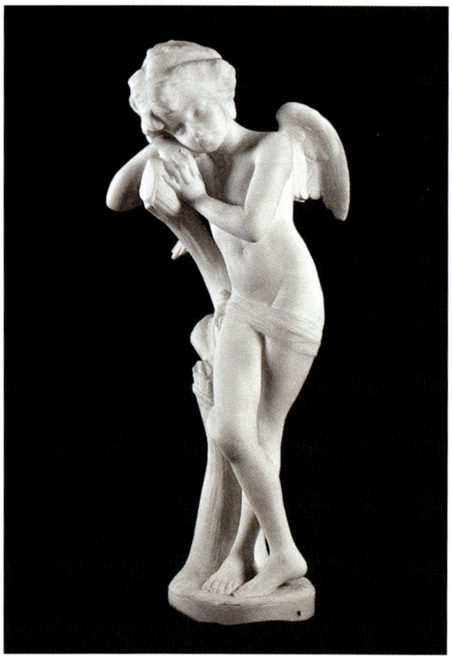

Auf der umfangreich bestückten Jugendstil-Ausstellung in Turin 1902 sind auch diese Steinzeugvasen mit sechs Maskerons nach einem Modell von Larroux vertreten. Das linke Modell 933 ist 47 cm hoch und in Terrakotta ausgearbeitet, die rechte Vase 1097 ist 38 cm hoch und aus Steinzeug produziert. (Quelle: Gmelin, L.: Die 1. internationale Ausstellung für moderne dekorative Kunst in Turin 1902. In: Kunst und Handwerk, 1901/1902, S. 334)

This stoneware vases with six mascarons after a model by Larroux were shown at the large 1902 Art Nouveau exhibition in Turin. Model 933 on the left is 47 cm high and executed in terracotta; Vase 1097 on the right is 38 cm high and is stoneware. (Source: L. Gmelin: Die 1. internationale Ausstellung für moderne dekorative Kunst in Turin 1902. In: Kunst und Handwerk, 1901/1902, p. 334)

ModNr. 933
Vase „Masques", Antonin Larroux, ca. 1895, PB Wien 19

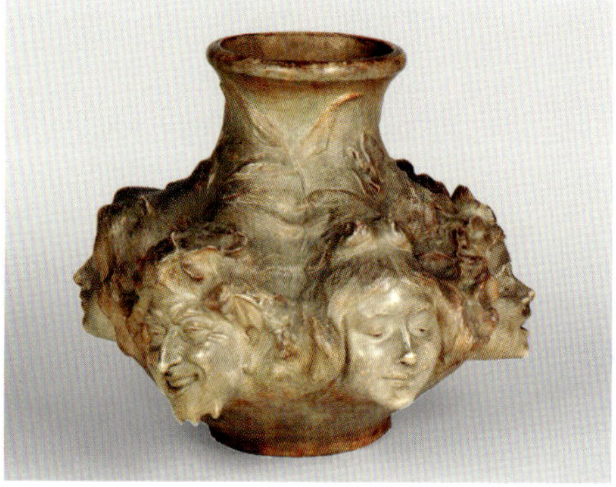

ModNr. 1097
Vase „Masques", Antonin Larroux, ca. 1895, Collection f@goldscheider.de

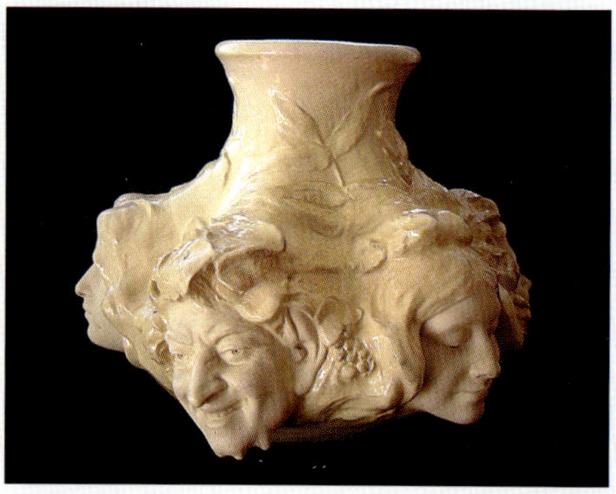

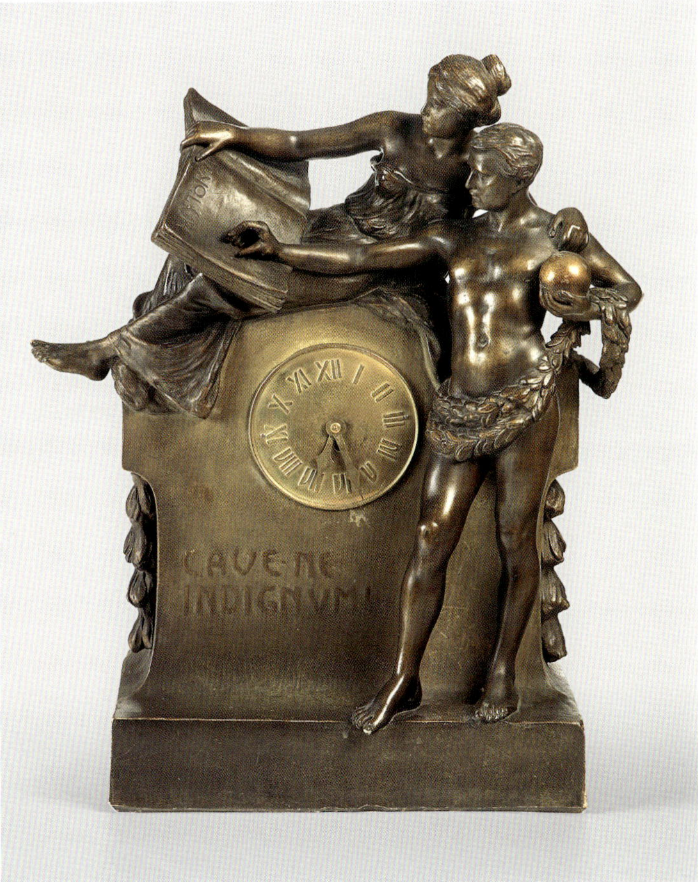

ModNr. 3396
Uhr I Clock „Historia", Adolphe-Jean Lavergne, ca. 1906, © Wien Museum

Austrian Fine Art and Decorative Furnishing" auch auf der bedeutenden „Prima Esposizione d'Arte Decorativa Moderna" in Turin aus. Auf dieser Ausstellung, die den Höhepunkt des Jugendstils bildet, ist im österreichischen Pavillon neben den Manufakturen Riessner, Stellmacher & Co („Amphora"), Josef Böck und der k.u.k. Fachschule in Teplitz auch die „kunstgewerbliche Werkstatt" Friedrich Goldscheider vertreten. Besonders hervorgehoben wird hierbei die Nachbildung einer Beethoven-Maske[115] sowie die Uhr „Amicitia vincit horas" (Die Freundschaft besiegt die Stunden) von Simon.[116]

Im Herbst kehrt Camillo Goldscheider, der seit Ende Februar 1900 bei seinem Bruder Arthur in Paris beschäftigt war, nach Wien zurück und wird zum 18. November 1902 als „Mitleiter der Fabrik, mit K 6.000 Jahresgehalt und 3 % Gewinnbeteiligung, kurzum mit Bezügen, die sich allmählich auf mehr als K 12.000,– jährlich steigerten", angestellt.[117] Camillo entwirft auch Modelle für die Manufaktur und sieht sich als tragende Kraft des Familienunternehmens: „Ich war damals schon der künstlerisch technische Leiter des

schule in Teplitz. Particularly noteworthy in this connection was a replica of a Beethoven mask[115] as well as a clock, "Amicitia vincit horas" ["Friendship Conquers Time"] by Simon.[116]

Camillo Goldscheider, who had been working for his brother Arthur in Paris since late February 1900, returned to Vienna in autumn. On 18 November 1902 he entered employment as "co-director of the factory, with an annual salary of K 6000 and a 3 % share of profits, in brief, with emoluments that gradually rose to over K 12 000 per annum."[117] Camillo also designed models for the factory and viewed himself as the pillar of the family firm: "By then I was artistic technical director of the business; the models I created in the year 1900 are what in part made the house world-renowned and I can certainly say that the vast majority of them are my own invention."[118]

In 1903 the Keller & Reiner art salon in Berlin exhibited numerous Goldscheider objects. The specialist press was full of praise for a faun mask modelled by Larroux (model

ModNr. 2678
Uhr I Clock, Capèque, ca. 1902, © Wien Museum
Diese Uhr wurde in drei Größen gefertigt und auch in der Gestaltung des Gehäusedekors kommen unterschiedliche Entwürfe zur Ausführung. Die Inschrift „Vivos voco, mortuos plango, fulgura frango" ist ein Zitat des Leitspruchs von Friedrich Schillers Dichtung „Das Lied von der Glocke" (1799).

This clock was made in three sizes and the case was decorated according to several different designs. The inscription "Vivos voco, mortuos plango, fulgura frango" is the motto of Friedrich Schiller's poem "Das Lied von der Glocke" ["Song of the Bell"] (1799).

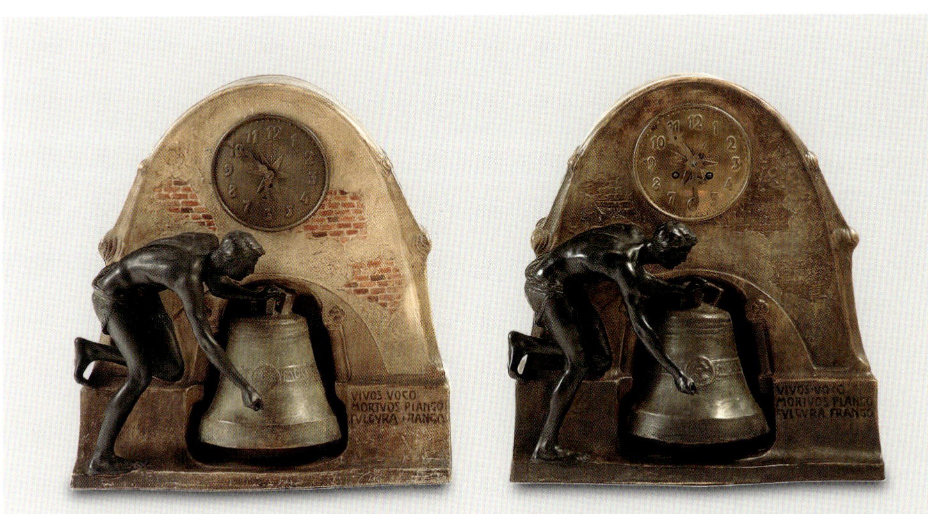

ModNr. 2452
„Der Windstoß" I
"A Gust of Wind",
Haniroff, ca. 1902,
PB Wien 19

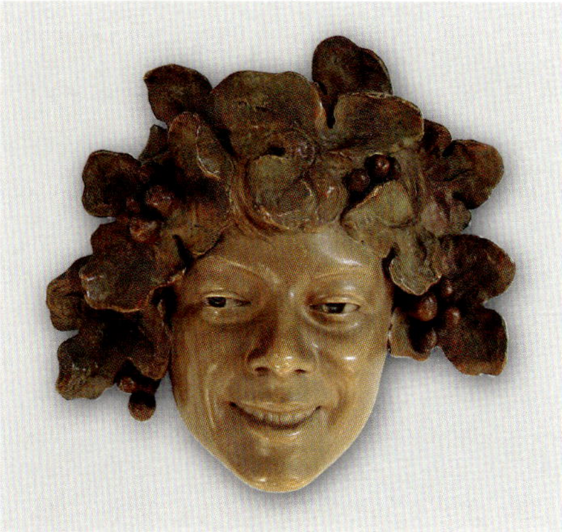

ModNr. 2288
„Bacchantin" I
"Bacchantae",
Antonin Larroux,
ca. 1901, Collection
f@goldscheider.de

Unternehmens; die Modelle die ich im Jahre 1900 geschaffen hatte, begründeten zum Teile den Weltruf des Hauses, und ich kann wohl sagen, dass sie zum weitaus grössten Teile meine eigene Erfindung sind"[118].

Im Jahr 1903 werden in Berlin im Kunstsalon von Keller & Reiner zahlreiche Objekte ausgestellt: vor allem eine von Larroux modellierte Faunmaske (Modell-Nr. 2288), die Figuren „Schmetterlingsweibchen" von Tell, „Die Zeit" von Powoley (gemeint ist Michael Powolny), die „Fischerin" von Congers und die Vase „Am Brunnen" von Simon werden von der Fachpresse in höchsten Tönen gewürdigt. „Alle diese Darstellungen stehen technisch auf der Höhe und gehören mit zu dem Besten, was die gesamte, kunstgewerbliche Terrakottenindustrie in letzter Zeit hervorgebracht hat."[119] Auf der Leipziger Ostermesse werden die „schönen Modelle" und vor allem die Büste „Pureté" wegen ihrer Einfachheit gelobt.[120] Die auf der Ersten Kärntnerischen Ausstellung in Klagenfurt ausgestellten Objekte, z.B. „Eisbär", „Hahn", „Eule" und die Jardinière von F. Gornik, werden hingegen als im „bisherigen Genre in allzu weichen süsslichen Formen und Farben" von der Kritik verurteilt.[121] Bei manchen Modellen wird das „Barockwirkende" und „an italienische Marktware" Erinnernde kritisiert.[122] Die Kunstgegenstände von Goldscheider sind beim Publikum jedoch derart populär, dass sie auch von anderen Manufakturen stilistisch kopiert werden. Der Wiener Kunstterrakotta-Fabrikant J. Pilar hebt sogar in seiner Anzeige hervor, dass er eine „eigene Kunstrichtung und keine Nachahmung von Goldscheider" betreibt.[123]

Im Frühjahr 1903 wird ein neuer Mattlack zum Patent angemeldet[124], im Januar 1904 wird ein Patent zur Herstellung von Mosaikimitation beantragt, bei dem das Verfahren wie folgt von statten geht: „Die Oberfläche des zu dekorierenden Gegenstandes ist in unregelmäßig begrenzte, unebene, sowie verschieden geneigte Flächen zerlegt, welche verschiedenartig bemalt sind. Zunächst wird ein Modell durch Gravieren aus freier Hand mit solchen Flächen versehen, worauf eine negative Abdruckform herge-

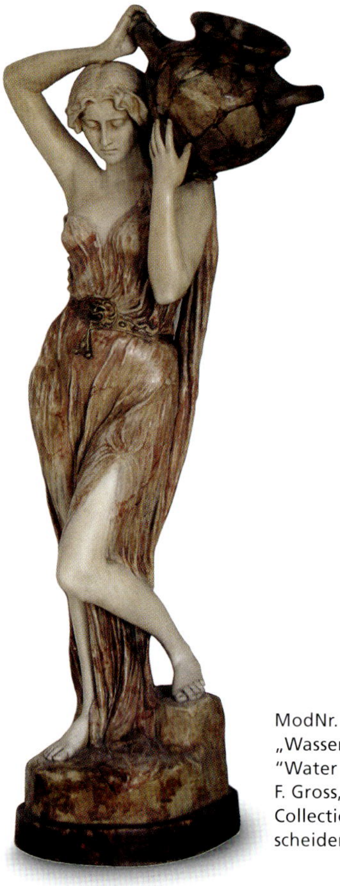

ModNr. 2770
„Wasserträgerin" I
"Water Carrier",
F. Gross, ca. 1904,
Collection a@goldscheider.net

Das Sujet der „Wasserträgerin" oder „Krugträgerin" ist äußerst beliebt und wird von verschiedenen Künstlern und in mehreren Größen produziert.

"The Water Carrier" or "Jug Bearer" motif was extremely popular and was produced by more than one artist and in several sizes.

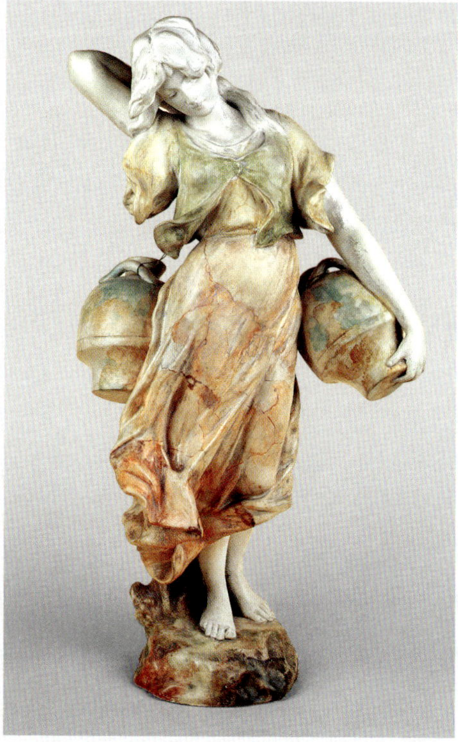

no. 2288), the figures "Schmetterlingsweibchen" ["Butterfly Woman"] by Tell, "Die Zeit" ["Time"] by Powoley (Michael Powolny), "Fischerin" ["Woman Fishing"] by Congers and the Simon vase "Am Brunnen" ["At the Well"]. "All these representations are technically first rate and are among the best things the entire arts and crafts terracotta industry has produced in recent years."[119] At the Leipzig Easter Fair, the "beautiful models" and especially the bust "Pureté" were praised for their simplicity.[120] Objects shown at the Erste Kärntnerische Ausstellung in Klagenfurt, including "Eisbär" ["Polar Bear"], "Hahn" ["Cock"], "Eule" ["Owl"] and a jardinière by F. Gornik, were, on the other hand, panned by the critics as in "the genre hitherto in forms and colors that are all too soft and cloying."[121] Some models were criticized for their "Baroque look" and being reminiscent "of Italian market wares."[122] Nonetheless, the Goldscheider art objects were so popular with the public that they were copied stylistically by other manufactories. The Viennese art terracotta manufacturer J. Pilar even felt called upon to make a point of stating flatly in his advertisement that he worked with a "style of his own in art and not a Goldscheider imitation."[123]

rechts I right

ModNr. 1156
Relief, Alois Hans
Schram, ca. 1898,
© Wien Museum
Dieses Wandrelief im Stile von Gustav Klimt ist eine Mosaikimitation.

This mural relief in the manner of Gustav Klimt is an imitation mosaic.

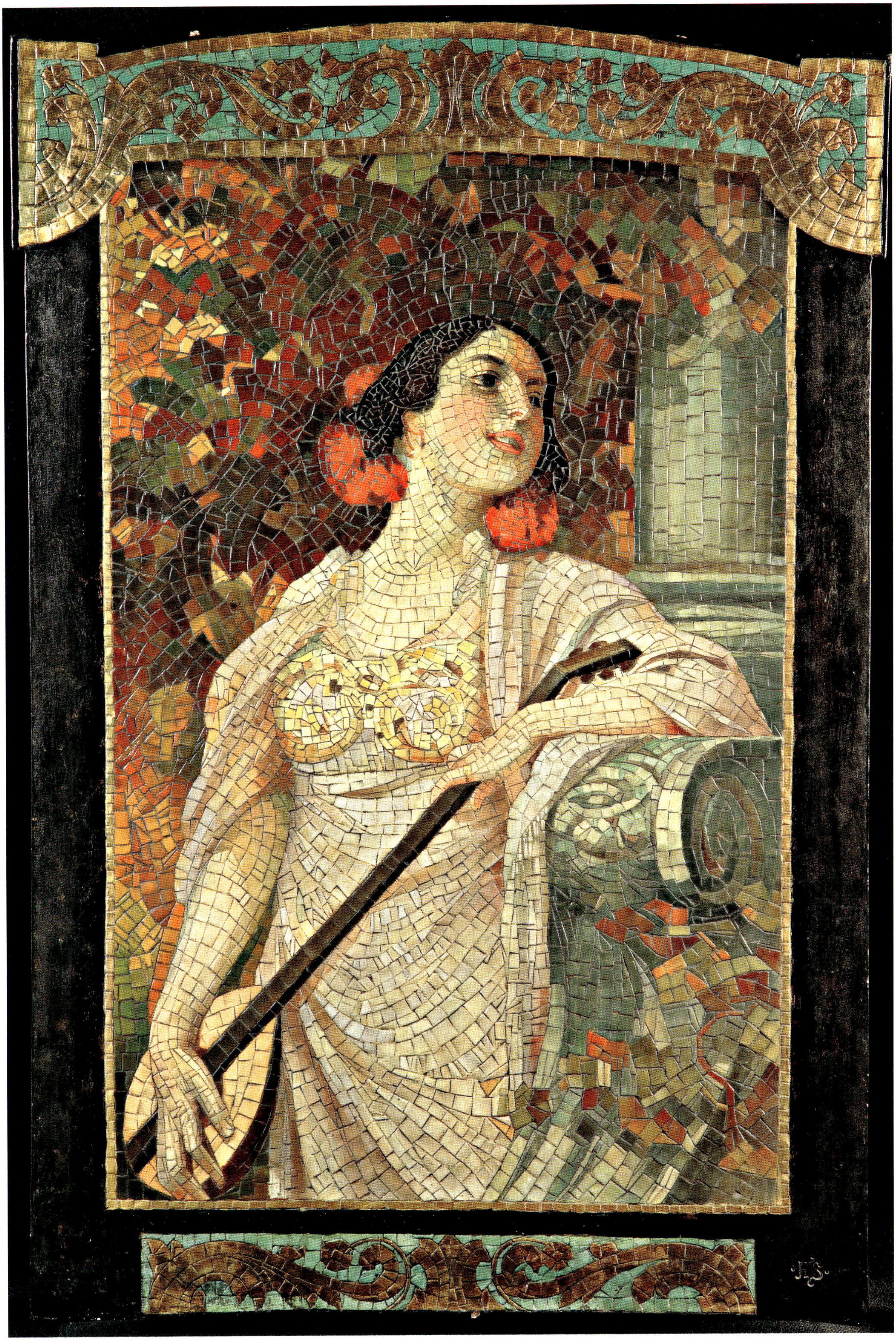

rechts I right

ModNr. 2647
Lampe mit Maiglöckchen I Lamp with Lilies of the Valley, ca. 1903, Collection Brigantine 1900, München

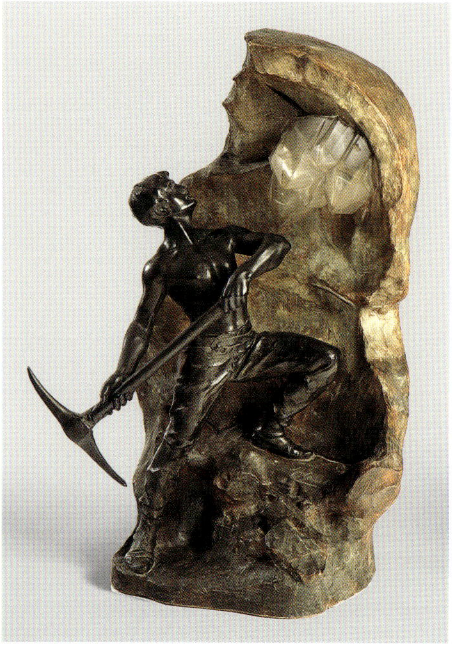

ModNr. 2859
Lampe „Bergmann mit Pickel" I Lamp "Miner with Pickaxe", Leclancher, ca. 1904, © Wien Museum

stellt wird, nach welcher die Vervielfältigungen gebildet werden. Es kann auch von einer Zeichnung, welche die Steinchenteilung darstellt, durch Lichtdruck ein Quellrelief hergestellt werden, welches nach Erhärtung in das noch plastische Matrizenmaterial eingedrückt wird"[125]. Die Mosaikimitationen werden beispielsweise bei Wandbildern, Platten, Vasen, Säulen und Sockeln angewendet.

Der Ruf der Goldscheider Manufaktur ist hervorragend und hunderte von Modellen werden auf den Markt gebracht, so dass man auch finanziell an die früheren Erfolge anknüpfen kann. Regine erwirbt am 8. Februar 1904 das zweite Fabrikgebäude in der Staudgasse 7, zum Kaufpreis von 122.000 Kronen.[126] Auch privat bezieht sie neue Räumlichkeiten. Sie erwirbt ein Haus in der Hasenauerstraße 16, im sogenannten Cottage-Viertel, und erweitert es im Folgejahr um einen Zubau.[127] In der Fachpresse wird ihre Arbeit gewürdigt und es wird gelobt, dass sie nicht nur hervorragende Bildhauer für sich gewinnen kann, „die lebenswahr ihre Bildwerke herstellen, sondern auch Former und Modelleure heranzubilden [versteht], welche die Originalwerke in Ton nachzubilden wissen, und zwar völlig unter Berücksichtigung des Umstandes, dass die geformten Bildwerke noch dem Feuer ausgesetzt werden, wobei sie ihre Dauer und Haltbarkeit erhalten."[128] Das Unternehmen beschäftigt an die 200 Arbeiter, die Niederlassungen befinden sich in Leipzig am Neumarkt und in Paris in der Rue de Paradis 45 sowie der Avenue de l'Opéra 28.[129]

Die zunehmende Versorgung der Haushalte mit Strom führt auch zu einer starken Erweiterung des Goldscheider-Sortiments im Bereich der elektrifizierten Lampen, die zur Leipziger Ostermesse 1904 neben „Hervorragenden Neuheiten" als Spezialität („figurale Objekte für elektrische Beleuchtung") geführt werden.[130] Die Lampen sind zwar originell, können sich aber langfristig nicht als bahnbrechende Geschäftsidee durchset-

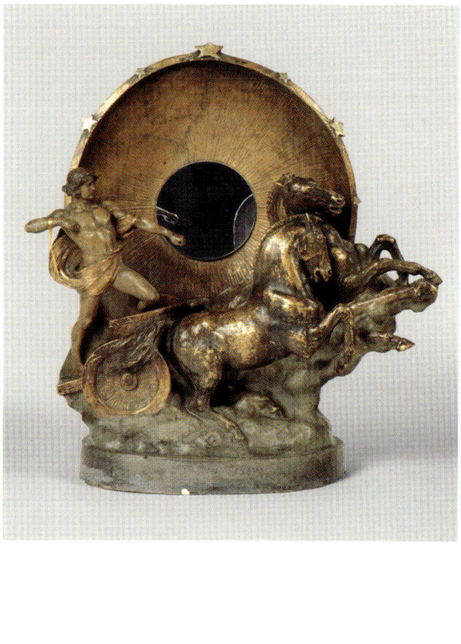

ModNr. 2998
Lampe „Sonnenwagen" I Lamp "Chariot of the Sun", Rosé, ca. 1905, © Wien Museum

In spring 1903 an application was made for a patent on a new matt lacquer.[124] In January 1904 an application was made for another process for making imitation mosaic, which is described as follows: "The surface of the object to be decorated is divided up into irregularly delimited, uneven surfaces of varying degrees of inclination, which are variously painted. First a model is given such surfaces by being engraved free hand, from which a negative mold is made, after which the replicas are formed. There can also be a drawing which represents the distribution of the tessarae; through an offset print an emulsion relief can be made which, after hardening, can be pressed into the matrix material while it is still malleable."[125] These imitation mosaics would be used for murals, chargers, vases, columns and bases.

The Goldscheider manufactory had an excellent reputation and launched hundreds of models on the market so that it was possible to link up with the earlier successes. On 8 February 1904, Regine bought the second factory building at 7 Staudgasse, paying 122 000 Kronen for it.[126] She also moved into a new home. After buying a house at 16 Hasenauerstraße, in what was then known as the Cottage Quarter, she had it enlarged the following year by building an annexe.[127] The specialist press gave her work positive reviews and she was praised for being able to attract personnel such as outstanding sculptors "who make lifelike sculpture but also [was able to] train molders and modellers who can replicate original works in clay, taking fully into consideration the circumstance that the molded sculptures will be exposed to firing to ensure that they are durable and long-lasting."[128] The business employed a workforce of 200. There were branches in Leipzig am Neumarkt and in Paris, 45 rue de Paradis as well as 28 avenue de l'Opéra.[129]

Households being provided with electricity were also a contributing factor to a considerable expansion of the Goldscheider range to included electric lamps. They were launched at the 1904 Leipzig Easter Fair alongside "outstanding novelties" as a speciality ("figurative objects for electric lighting").[130] The lamps were original in design yet they were not successful in the long term. After all, as Pazaurek, the doyen of arts and crafts critics of the day wrote, they would never dominate the lighting industry due to the "nature of the ceramic material" since it was unsuitable for "pendant fix-

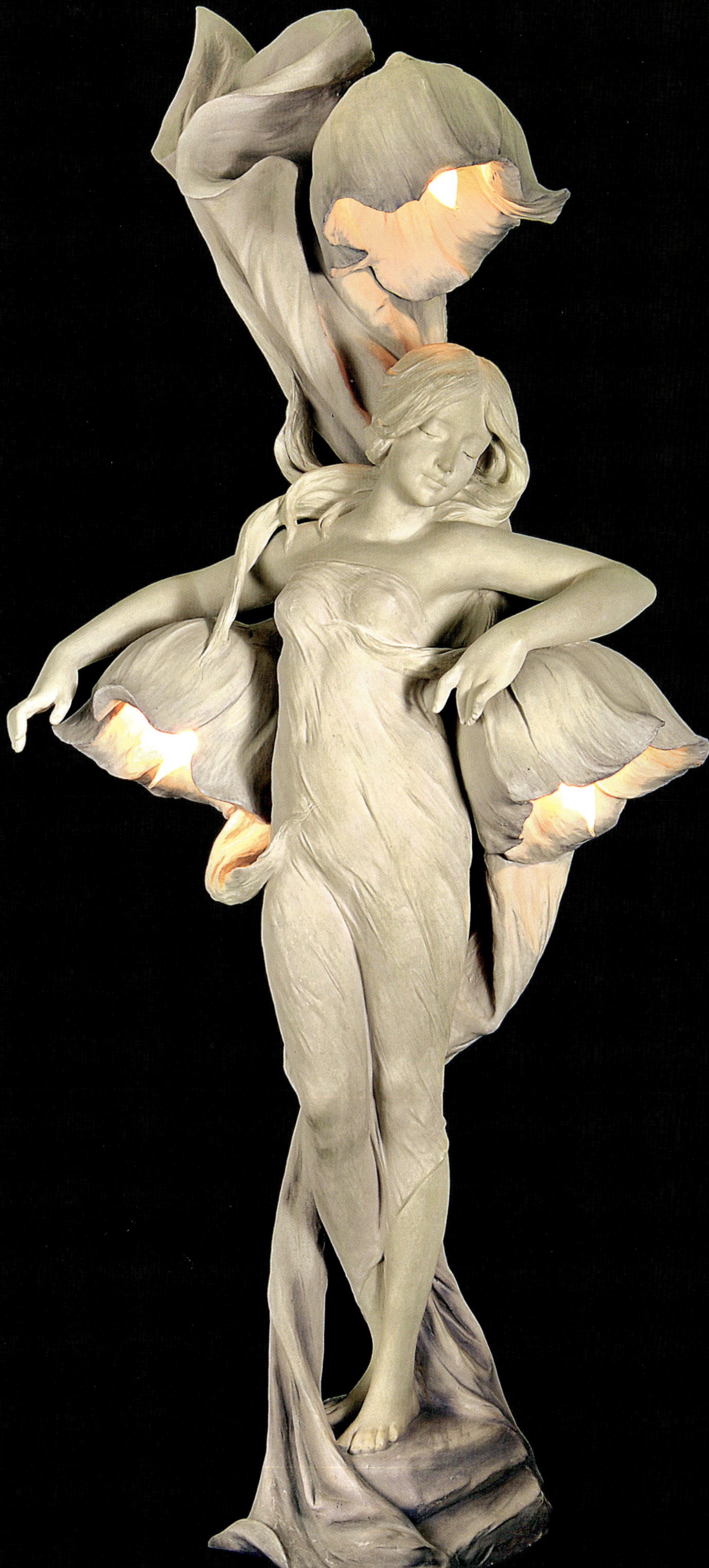

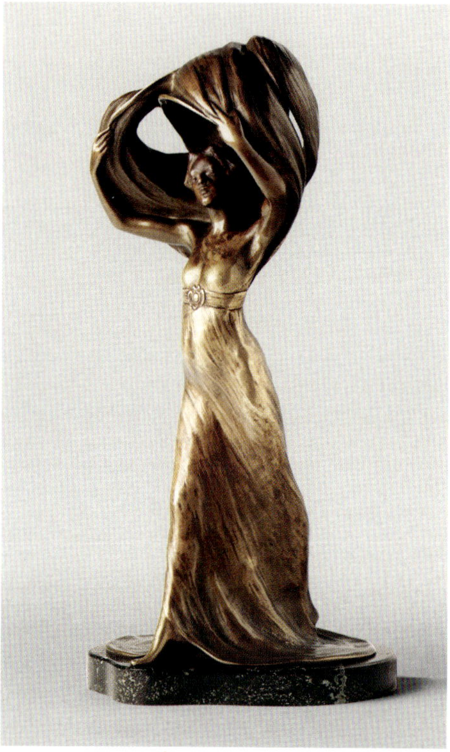

ohne I without ModNr.
Lampe I Lamp,
Gottfrid Julius Andreas Larsson, ca. 1900,
Collection John Jesse, London.
Diese bronzene Lampe mit Marmorsockel zeigt die Tänzerin Loie Fuller. Das Leuchtelement ist im Tuch versteckt.

This bronzed lamp with a marble foot depicts the dancer Loie Fuller. The lighting element is concealed in her veil.

zen, denn wie der damalige Kunstgewerbe-Experte Pazaurek schreibt, kann aufgrund der „Natur des keramischen Stoffes" dieser nie die Beleuchtungsindustrie dominieren, da er für „Hängekörper" nicht geeignet sei[131]. Für Lampen wurden eigene Sujets entworfen oder man verwendete bestehende Modelle, z.B. Figuren, als Lampenfuß.

Die Goldscheider Objekte werden allgemein als „Salon-Plastiken par excellence" bezeichnet, zu den ausgestellten Modellen auf der Leipziger Michaelis-Messe im Jahr 1904, auch Herbstmesse genannt, gehören unter anderem ein „Tanzendes Paar" und von Simon die „Topfguckerin" mit Beleuchtungseffekt.[132] Auch eine Figur aus Hulsberg-Marmor sowie eine Büste auf einem Durabit-Sockel mit majolikaähnlichem Dekor fallen auf.[133] Eine richtige Neuheit neben den „Plastiken mit und ohne elektrischem Licht, dekorativen Mosaik-Imitationen, originellen Uhren und Spiegeln" sind jedoch die zum ersten Mal beworbenen künstlerischen Fayencen.[134]

Als Fayence gilt zu dieser Zeit „jede Töpferware, bei der farbige Tonscherben von einer durchsichtigen Glasur überdeckt wird", und Majolika ist hingegen „eine solche, bei der der gefärbte Tonscherben eine undurchsichtige Schmelzglasur" trägt.[135] Die Goldscheider-Fayencen werden nach einem Verfahren, das der Porzellanmalerei ähnelt, hergestellt. D.h. die Bemalung mit keramischen Farben erfolgt auf einem bereits eingebrannten Glasurüberzug: „Das Goldscheiderische Erzeugnis ist also eine reine Fayence, die in der Art der Porzellane durch ‚Malerei auf der Glasur' dekoriert wird. Die Glasierung und Bemalung ist dabei als ein integrierender Bestandteil des Herstellungsverfahrens anzusprechen, weil die Stücke ohne Glasurüberzug überhaupt unverwendbar und durch ihre Formgebung auf die ergänzende Bemalung angewiesen sind. Da der Fayence-Glasuruntergrund viel weicher ist als die Porzellanglasur, verhalten sich die keramischen Farben beim Brande auf ihm etwas anders als auf dem glasierten Porzellanscherben"[136]. Aufgrund dieses Unterschiedes können Porzellanmaler nicht ohne eine Fortbildung bei Goldscheider anfangen, sie werden von der Firma selbst ausgebildet. Die neu eingeführ-

tures."[131] Motifs were designed for lamps or existing models, including figures, were used as lamp bases.

Goldscheider objects were universally designated as "salon sculpture par excellence." The models shown at the 1904 Leipzig Michaelmas Fair (also simply called the Autumn Fair), included a "Tanzendes Paar" ["Dancing Couple"] and the "Topfguckerin" ["Woman Potter"] by Simon with lighting effects.[132] Other notable pieces were a figure made of Hulsberg marble and a bust on a Durabit pedestal with decoration similar to majolica.[133] The real novelties, however, apart from "sculptures with and without electric light, ornamental imitation mosaics, original clocks and mirrors" were the art faiences, which were being advertised for the first time.[134]

At that time, faience was defined as "all pottery in which the colored clay body is coated with a transparent glaze." Majolika, on the other hand, was "such that the colored clay body [bears] a fired glaze that is not transparent."[135] The Goldscheider faiences were made by a process similar to that used for porcelain painting, ie, the painting was done with ceramic colors on a previously fired coating of glaze: "The Goldscheider product is, therefore, pure faience, which is decorated like porcelain with 'overglaze painting.' The glazing and painting must be viewed as an integral part of the manufacturing process because the pieces are useless without the coating of glaze and depend for their form on the supplementary painting. Since the faience that is the ground for the glaze is much softer than the porcelain glaze, the ceramic colors look different on it after firing to the way they do on glazed porcelain body."[136] Due to this difference, porcelain painters could not simply start working at

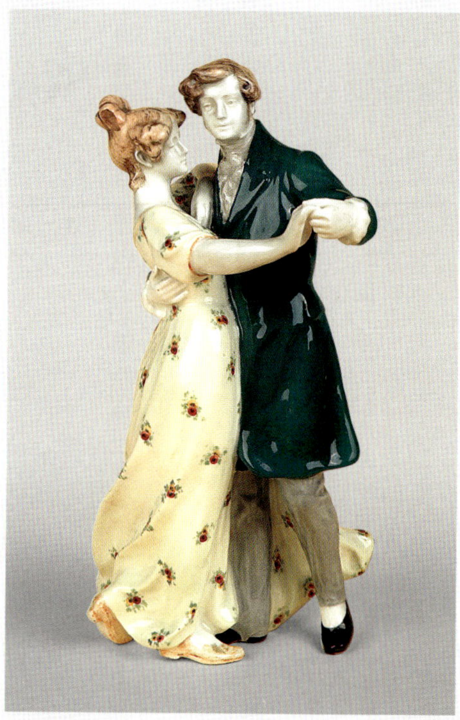

ModNr. 2839
Tanzendes Biedermeierpaar I Dancing Biedermeier Couple, ca. 1905, PB Wien 19

Goldscheider without further training; so they were trained on the job. The recently introduced faiences, which were made of a clay body with a high kaolin content that fired white and were later also decorated with underglaze painting, developed subsequently into the most important Goldscheider product line. In addition, however, figures, such as statues of nymphs and women as well as busts that were often used as mantel piece decoration continued in production.[137]

In general, woman in all her wiles was the dominant theme in the turn-of-the century decorative and applied arts. Art Nouveau and the enlightened approach taken by psychoanalysis to gender roles within middle-class society informed numerous representations of both women as lascivious types and figures of innocent girls – femme fatale and femme enfant – were the motifs of the day. Alongside those representations of women, crinoline figures and dancers, animal sculpture also became popular.[138] The production of oriental busts and statues was followed by a growth in the making of statuary from antiquity, masks and historical figures, which rubbed shoulders with Rococo and Biedermeier figures and groups. Vases, bowls, lamps and clocks were also marketed as ornaments for grand upper-middle-class salons. These products especially were informed by French Art Nouveau, with its elongated decorative elements based on plant motifs. The Art Nouveau style and its elegantly mannered handling of line held sway until 1905. After that a counter-movement that was emphatically naturalistic in approach, determined what was produced. Small statuary became more important.

At the 1906 Austrian Exhibition in London, Goldscheider had a stand measuring

ten Fayencen, die aus kaolinhaltigem, weißbrennendem Ton sind, und später auch in Unterglasur dekoriert werden, entwickeln sich sukzessive zur wichtigsten Produktreihe. Daneben gibt es aber aus Terrakotta weiterhin Figuren, Statuen, Spiegelmotive mit Nymphen und Frauengestalten sowie Büsten, die oft als Kaminaufsätze dienen.[137]

Allgemein dominiert um die Jahrhundertwende das Thema der Weiblichkeit das Kunsthandwerk. Bedingt durch den Jugendstil und die aufklärerische Auseinandersetzung der Psychoanalyse mit den Geschlechterrollen innerhalb der bürgerlichen Gesellschaft, bestimmen zahlreiche Darstellungen von lasziven Frauentypen und unschuldigen Mädchenfiguren – femme fatale und femme enfant – die Motive dieser Zeit. Neben diesen Mädchen, Krinolinen-Figuren und Tänzerinnen gewinnen auch Tierplastiken an Popularität.[138] Auf die Produktion orientalischer Büsten und Statuen folgt verstärkt die Herstellung antiker Figuren, Masken und historischer Persönlichkeiten, die neben Rokoko- und Biedermeier-Figuren und -Gruppen ihren Platz finden. Zur Dekoration großbürgerlicher Salons verkauft man auch Vasen, Schalen sowie Lampen und Uhren. Gerade diese Produkte sind durch den französischen Jugendstil (Art Nouveau) geprägt, mit länglichen, an Pflanzen orientierten Ornamenten. Dieser Stil dominiert mit seiner elegant-manieristischen Linienführung bis 1905. Eine neue Gegenbewegung, die sich betont naturalistisch gibt, bestimmt fortan die Produktion. Die Kleinplastik gewinnt an Bedeutung.

Auf der „Österreichischen Ausstellung" in London im Jahr 1906 belegt Goldscheider einen 60 Quadratmeter großen Stand.[139] Im gleichen Jahr stellt man auch auf der kunstgewerblichen Ausstellung in Bukarest aus.

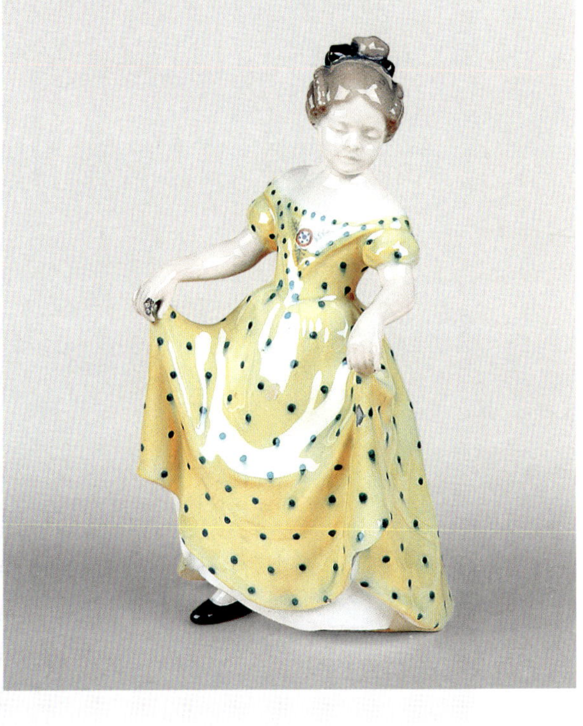

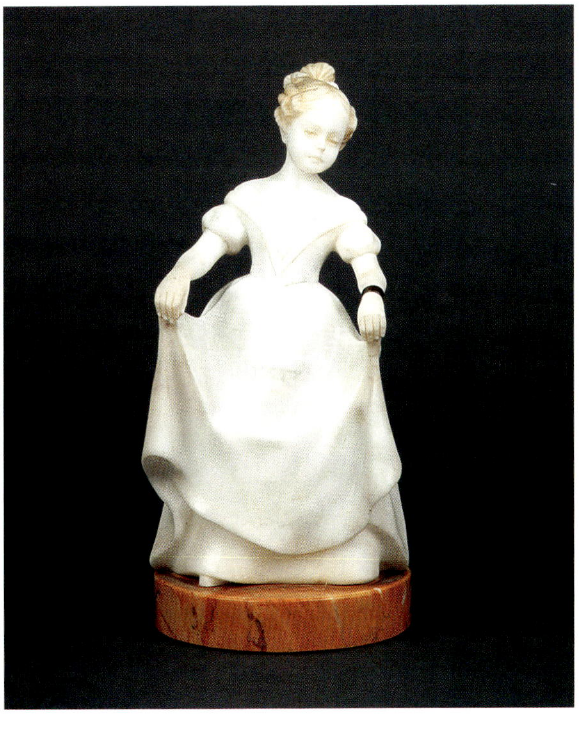

ModNr. 3009
Biedermeiermädchen I Biedermeier Girl, E. Tell, ca. 1905, beide I both PB Wien 19
Die Goldscheider-Modelle wurden nicht nur in unterschiedlichen Größen produziert, sondern auch in unterschiedlichen Materialien. Hier das Beispiel eines Biedermeiermädchens, links in Fayence und rechts in Alabaster.

The Goldscheider models were produced both in varying sizes as well as several different materials. Here for example a Biedermeier girl: left in faience and right in alabaster.

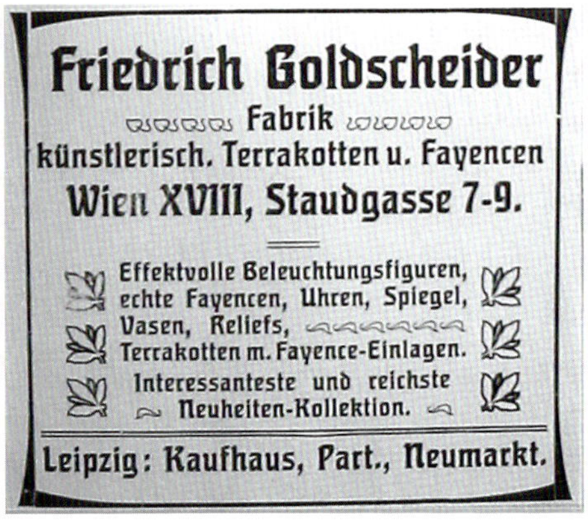

Friedrich Goldscheider
Fabrik
künstlerisch. Terrakotten u. Fayencen
Wien XVIII, Staudgasse 7-9.

Effektvolle Beleuchtungsfiguren,
echte Fayencen, Uhren, Spiegel,
Vasen, Reliefs,
Terrakotten m. Fayence-Einlagen.
Interessanteste und reichste
Neuheiten-Kollektion.

Leipzig: Kaufhaus, Part., Neumarkt.

Ill. 30
Um 1907 stellt die „Erste Wiener Terrakottafabrik und Atelier für künstlerische Fayencen" neben Figuren, die auch als Lampen dienen, vor allem Uhren, Spiegel, Vasen und Reliefs her. Als Spezialität gelten Terrakotten mit Fayence-Einlagen (Quelle: Keramische Rundschau, 1907)

Around 1907 the "First Viennese Terracotta Manufactory and Studio for Art Faiences" made, apart from figures that also served as lamps, mainly clocks, mirrors, vases and reliefs. Terracottas with faience inlay were a speciality (Source: Keramische Rundschau, 1907)

ModNr. 3285
„Mozart", Rosé, ca. 1905, © Wien Museum

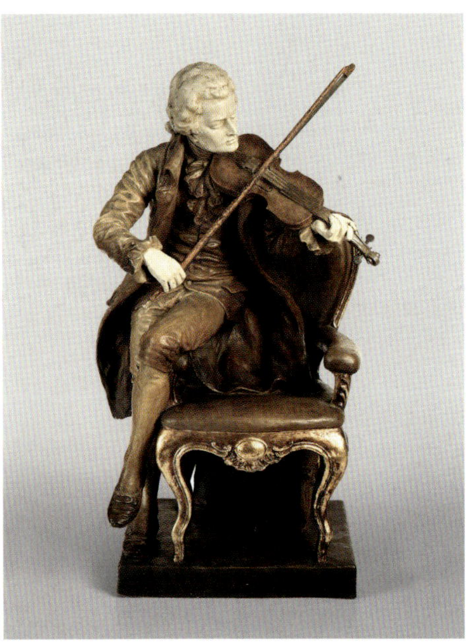

Ill. 31
Firmenaushang aus Leder um 1910. (Quelle: Archiv Goldscheider)

Firm sign made of leather, ca 1910. (Source: Goldscheider Archives)

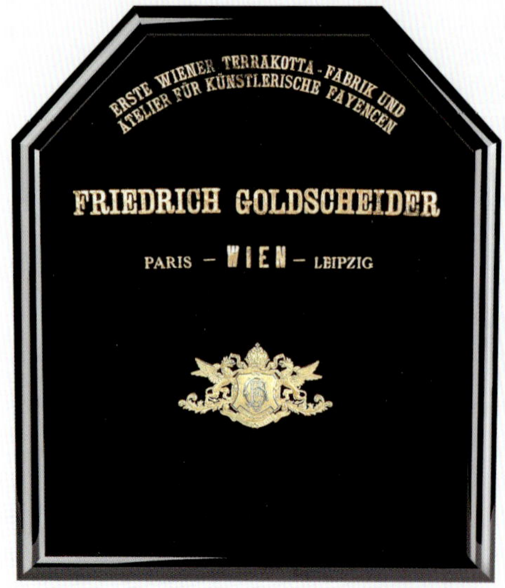

Auf die Frage des Ausstellungskomitees, wer der „wesentlichste künstlerische Mitarbeiter der Firma sei", wird Camillo Goldscheider genannt, der daraufhin die Mitarbeitermedaille erhält und mit der Jubiläumsmedaille Carols I. ausgezeichnet wird.[140] Doch zwischen Regine und Camillo kommt es schon seit Längerem zu starken Auseinandersetzungen, da Camillo überzeugt ist, dass ihm als Sohn Friedrich Goldscheiders eine Aufnahme als Mitgesellschafter der Wiener Fabrik samt Beteiligung zustehe. Außerdem seien es seine Verdienste, dass die Manufaktur erfolgreiche Modelle produziere: „Ich machte mehrere Erfindungen – neue Unterglasurfarben wurden von mir dargestellt –, erfand Neuerungen in der Fabrikation, welche mit ausserordentlichen Vorteilen verbunden waren – so wurden z.B. von den von mir erfundenen Terrakotta- und Fayencesachen bereits für 400.000 Kronen Ware verkauft."[141]

Regine vertröstet ihn, dass wenn er verheiratet und etwas ruhiger, stabiler geworden sei, sie ihn als Gesellschafter in die Firma aufnehmen werde. Diese Zusage soll sie auch Camillos Schwiegervater Jakob Hecht aus Czernowitz gegeben haben, der daraufhin die Heirat mit seiner Tochter Jolanthe erlaubt.[142] Doch Regine weigert sich weiterhin, Camillo als Gesellschafter aufzunehmen. Sie empfindet ihren Sohn als übertrieben selbstgefällig und geringschätzig gegenüber anderen Personen und Leistungen, sie sieht ihn an einer Art „Verfolgungswahn gleichendem Mistrauen [sic]" leidend und ist überzeugt, dass er wegen seiner ungezügelten Ausbrüche „absolut ungeeignet sei und seine Berufung auf eine leitende, selbständige, mit Dispositionsrechten ausgestattete Stellung eine Gefährdung der Gesamt- und sonach auch seiner eigenen Interessen bedeuten würde."[143]

Nach mehreren Jahren der vergeblichen Versuche um eine Aufnahme als Gesellschafter beschließt Camillo, seine vermeintlichen Rechte einzuklagen. Am 28. Oktober 1908 legt er beim k.u.k. Handelsgericht Wien Klage gegen seine Mutter ein und ersucht die Aufnahme als öffentlicher Gesellschafter in die Firma mit einer Beteiligung von 20 Prozent. Camillo, der mit seiner Frau und zwei Kindern fast jährlich den Wohnort wechselt[144], bezeichnet sich selbst immer als Fabrikdirektor, obwohl er als technischer Leiter angestellt ist. Mit einem, seiner Meinung nach, minimalen Gehalt von jährlich 10.000 Kronen: „Bildhauer, welche in dem Geschäfte der Firma engagiert sind, verdienen 8.400,–

60 square metres.[139] That same year Goldscheider also exhibited at the Arts and Crafts show in Bucharest. When the exhibition committee enquired who was "the most important artistic employee of the firm," Camillo Goldscheider was named and he was awarded the Carol I Celebration Medal.[140] However, for some time there had been severe disagreements between Regine and Camillo since Camillo was convinced that, as Friedrich Goldscheider's son, he was entitled to being made an official partner in the Viennese factory with the share of proceeds that this status implied. Further, he maintained, it was largely due to his efforts that the factory produced such successful models: "I came up with several inventions – new underglaze colors were made by me –, introduced innovations in manufacturing, which were linked with extraordinary advantages – the terracotta and faience things invented by me were already 400 000 Krone worth."[141]

Regine tried to put him off by promising that, when he was married and had become somewhat calmer and more stable, she would make him a partner in the firm. She supposedly made the same promise to Camillo's prospective father-in-law, Jakob Hecht of Czernowitz, who accordingly allowed his daughter, Jolanthe, to marry Camillo.[142] Regine, however, continued to refuse Camillo a partnership. She found her son grossly complacent and contemptuous of other people and their achievements. Indeed, she felt he suffered from a sort of "distrust amounting to a persecution complex" and was convinced that he was, due to his unfettered outbursts of temper "absolutely unsuited and his appointment to a managerial, independent position replete with powers of attorney would jeopardize the whole company and, therefore, his own interests."[143]

After several years of vain attempts at being admitted to a partnership, Camillo decided to bring a lawsuit to obtain what he saw as his rights. On 28 October 1908 he opened a lawsuit before the Handelsgericht [commercial court] in Vienna against his mother in an attempt to be made a partner in the firm with a 20 per cent share of proceeds. Camillo, who moved with his wife and two children on an almost yearly basis,[144] had always styled himself director of the factory although he was employed as technical director, with a minimum salary, in his opinion, of 10 000 Krone per annum: "Sculptors employed in the firm's earn 8400 Krone and more!"[145] A few days later, on 6 November 1908, he filed an additional "suit against Mme Regine Lewit, guardian, and Alois Goldscheider, co-guardian to be relieved of the guardianship of Marcell Goldscheider, minor, and appointment of the plaintiff as guardian," in which Camillo charged his mother and his uncle Alois with deliberately neglecting the interests of Marcell Goldscheider in the distribution of the legacy.[146] Camillo's partnership lawsuit, was, however, dismissed by the court on 19 January 1909. Camillo then left the family business to set up on his own[147] (see also the chapter on "Ornamental Sculpture by Camillo Goldscheider," p. 78f).

The dispute outlined above overshadowed the successful development of the firm. On 24 Dezember 1907 the firm name was officially enlarged to "Erste Wiener Terrakottafabrik und Atelier für künstlerische Fayencen Friedrich Goldscheider" ["First Viennese Terracotta Manufactory and Studio for Art Faiences Friedrich Goldscheider"].[148] It marketed "industrially manufactured products of clay and porcelain wares" and employed a workforce of over 300 artisanal and technical personnel in the two three-storeyed buildings in Staudgasse.[149] Well-known artists such as Wernekinck, Schmidt-Kestner and Gemignani, graduates of the Vienna Kunstgewerbeschule, including Ida Lehmann, Fritz Pollak and Mallener as well as their professors Arthur Strasser, Breitner, Marschall and Müllner provided Goldscheider with models, which were fired in the seven manufactory's muffle kilns.[150] The Paris branch run by Arthur at 45 rue de Paradis also served the Viennese branch as the general transshipment warehouse for goods destined for France, England, Belgium and French Switzerland.[151] In Berlin the firm was represented by Hugo Lippmann & Co, who maintained a sample showroom at 41 Ritterstraße, where the Goldscheider products were on display along with the wares of many other glass and ceramics firms.[152] At Lippmann the current Goldscheider faiences especially were showcased, German "Intellectual Heroes," such as Goethe, Schiller and Wagner as seated figures.[153]

At the 1909 Leipzig Spring Fair, Regine Goldscheider launched a new material: a composite stone called "Pierre de Vienne," which was supposed to be durable and color-fast.[154] In autumn 1909, an application was made for a patent for "Decorating

ModNr. 3839
„Nathan der Weise" I "Nathan the Wise", Léon Jules Butensky, ca. 1910, PB Deutschland

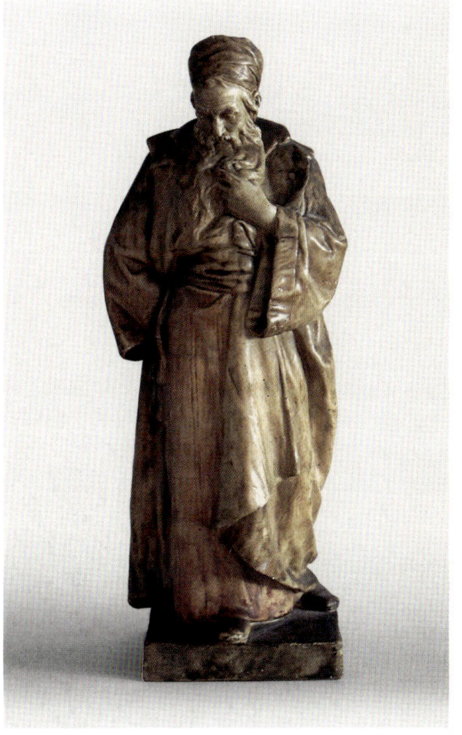

Kronen und noch mehr!"[145] Wenige Tage später, am 6. November 1908, stellt er zusätzlich einen „Antrag auf Enthebung der Vormünderin Frau Regine Lewit und des Mitvormundes Alois Goldscheider von der Vormundschaft des mj. Marcell Goldscheider und Bestellung des Einschreiters zum Vormunde", in dem er seiner Mutter und seinem Onkel Alois eine beabsichtigte Vernachlässigung der Interessen von Marcell Goldscheider bei der Erbverteilung vorwirft.[146] Camillos Gesellschafterklage wird jedoch am 19. Januar 1909 abgewiesen, woraufhin er das Familienunternehmen verlässt und sich selbständig macht[147] (siehe auch Kapitel „Dekorative Plastik von Camillo Goldscheider", S. 78ff).

Dieser Streit überschattet die erfolgreiche Entwicklung des Unternehmens. Am 24. Dezember 1907 wird der Firmenname erweitert, die Firma heißt fortan offiziell „Erste Wiener Terrakottafabrik und Atelier für künstlerische Fayencen Friedrich Goldscheider"[148], betreibt „fabriksmäßige Erzeugung von Ton- und Porzellanwaren" und beschäftigt in den beiden dreigeschossigen Häusern in der Staudgasse über 300 künstlerische und technische Kräfte[149]. Namhafte Künstler wie Wernekinck, Schmidt-Kestner, Gemignani, Schüler der Wiener Kunstgewerbeschule wie Ida Lehmann, Fritz Pollak, Mallener aber auch ihre Professoren wie Arthur Strasser, Breitner, Marschall oder Müllner liefern Modelle für Goldscheider, die in den sieben Muffelöfen der Manufaktur gebrannt werden.[150] Die Pariser Dependance von Arthur in der Rue de Paradis 45 dient auch als General-Depot der Wiener Filiale für die Länder Frankreich, England, Belgien und die französische Schweiz.[151] In Berlin wird die Firma von Hugo Lippmann & Co vertreten, der in der Ritterstraße 41, in der viele Firmen der Glas- und Keramikbranche ihre Produkte zeigen, ein Musterlager nicht nur mit Goldscheider-Objekten unterhält.[152] Bei Lippmann werden vor allem die aktuellen Fayencen von Goldscheider hervorgehoben, die deutsche „Geistesheroen" wie Goethe, Schiller und Wagner als sitzende Figuren zeigen.[153]

Auf der Leipziger Frühjahrsmesse 1909 stellt Regine Goldscheider ein neues Material vor: die Steinmasse „Pierre de Vienne", die unveränderlich haltbar in der Farbenwirkung sein soll.[154] Im Herbst 1909 wird auch ein Patent zur „Dekorierung keramischer Gegenstände durch Tränken oder Durchfärben mit Metallsalzlösungen" beantragt, bei dem weitere Variationen der Färbung durch „partielles Ätzen mit Säuren" herbeigeführt werden.[155]

ModNr. 3943
Knabe I Boy, Rosé, ca. 1911, PB Wien 19
Neben den neuen Fayence-Produkten sind weiterhin klassische Sujets aus Terrakotta im Sortiment.

Alongside the new faience products, classic motifs in terracotta continued in production.

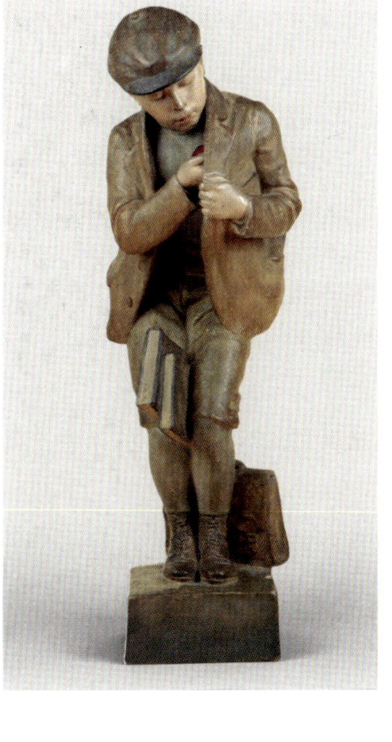

ModNr. 3697
„Helena", Ernst Seger, ca. 1909, PB Wien 19

ModNr. 3698
„Helena", Ernst Seger, ca. 1909, PB Wien 19

Das Modell „Helena" ist gleichzeitig in 56 cm sowie 40 cm Höhe produziert worden. Die zeitgenössische Presse schrieb: „Alles hervorragende Arbeiten. Bei allen diesen Plastiken werden die Körperteile, soweit sie nicht vom Gewand bedeckt sind, ganz matt-glänzend gehalten, während das Gewand sich in glänzender Frische abhebt." (Quelle: N.N.: Aus Berliner Musterlagern, in: Sprechsaal, 42.1909, S. 308)

The "Helena" model was produced in two sizes: 56 cm and 40 cm. The press of the day wrote: "All outstanding works. In all these sculptures, the body parts, in so far as they are not covered by the apparel, are rendered all in semi-gloss, while the apparel stands out brilliantly fresh." (Source: N.N.: Aus Berliner Musterlagern, in: Sprechsaal 42, 1909, p. 308)

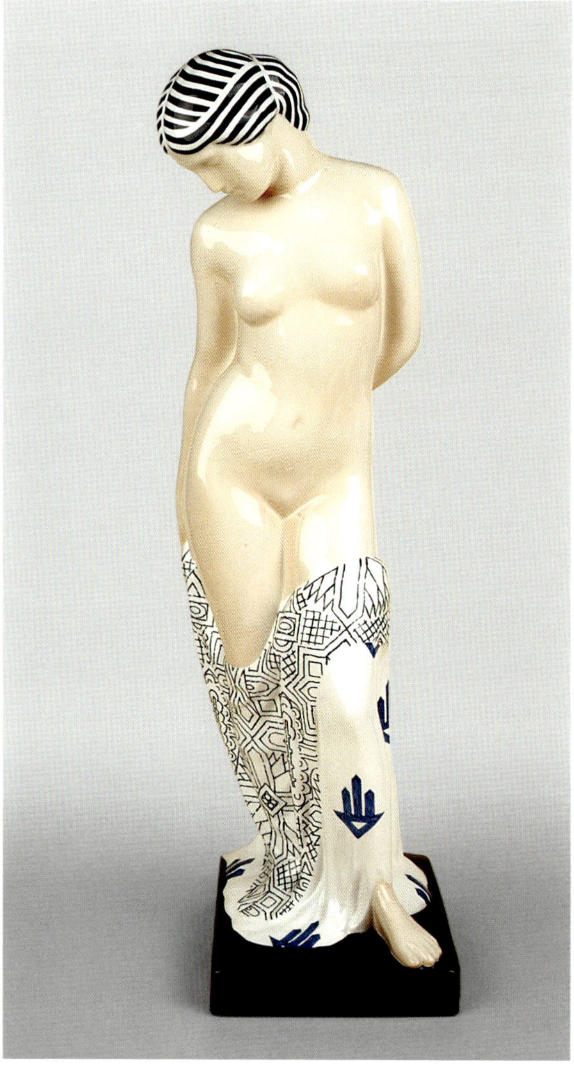

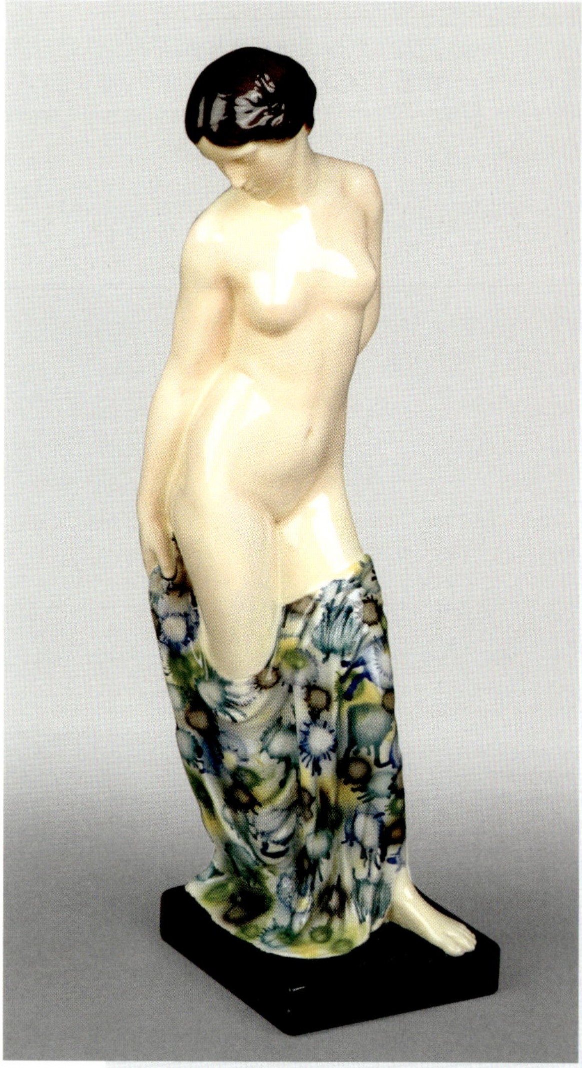

Der Vorteil dieses Verfahrens soll sein, dass im Gegensatz zu früheren Farbanstrichen, die nicht dauerhaft und unzerstörbar sind, die Farbeffekte nun stark in die Masse eindringen und „unzerstörbar" werden. Das Färben erfolgt durch die Metallsalze, so ergeben zum Beispiel „Kupfersalze blaue oder grünliche Effekte (je nach Konzentration), Eisenoxydsalze braune, gelbe, rote, graue, ja selbst tiefschwarze Effekte, einerseits je nach Konzentration, andererseits je nach Wahl des Salzes; in ähnlicher Weise kann man Stron-

ceramic objects by steeping or staining them with solutions of metallic salts," a process whereby further variations on staining could be induced by "partly biting them with acids."[155] The advantage of this process was to be that, unlike the earlier method of applying coats of paint, which were neither durable and imperishable, the new process caused the color effects thus created to soak deep into the paste and thus become "imperishable." The staining was done with metallic salts. Salts of copper, for example produced "blue or greenish effects (depending on the concentration); iron oxide salts brown, yellow, red, grey, even deep black effects, on the one hand, depending on the concentration, on the other, the salts used;

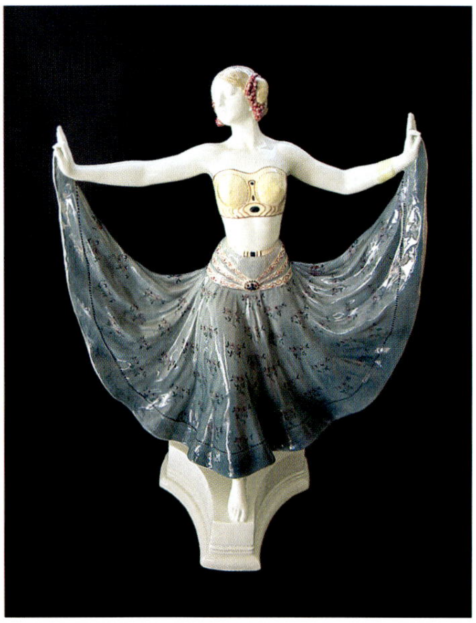

ModNr. 3992
„Ruth", Rosé, ca. 1911, Collection f@goldscheider.de

Die beliebte Figur der „Ruth", nach der Tänzerin Ruth Saint Denis benannt, wurde in fünf verschiedenen Größen und zahlreichen Dekoren produziert.

"Ruth," a popular figure named after the dancer Ruth Saint Denis, was produced in five sizes and numerous decoration schemes.

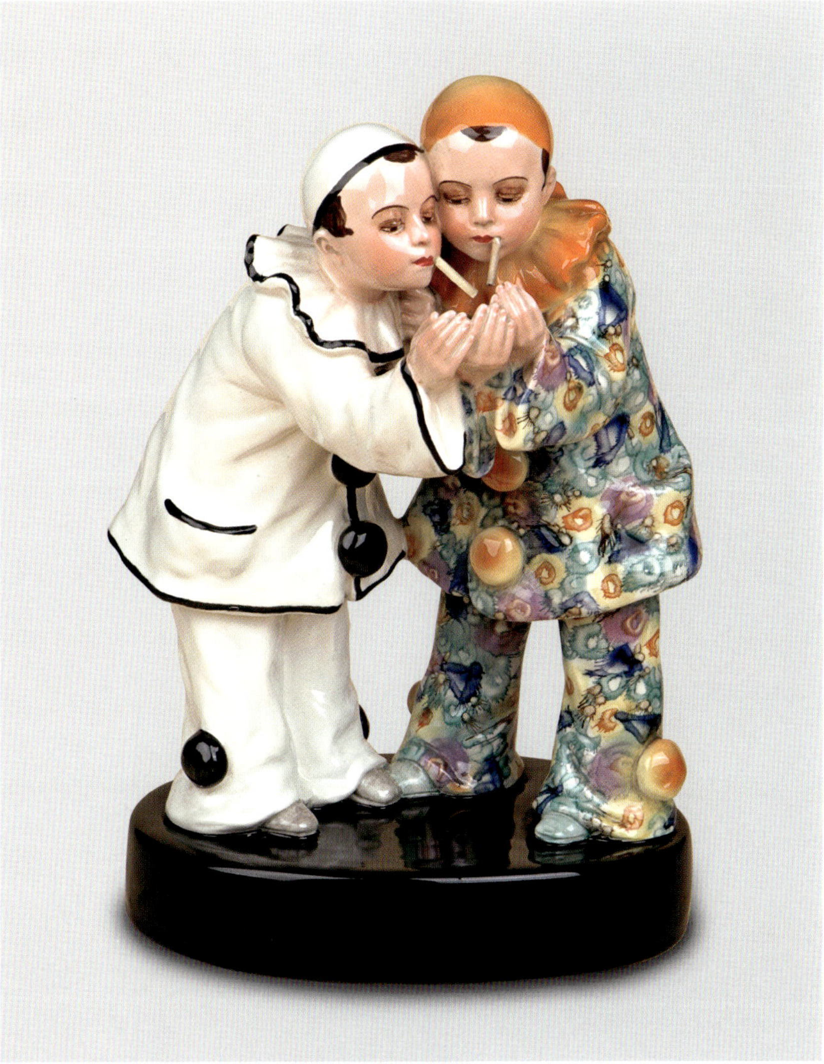

ModNr. 3721
Gruppe | Group,
Adolphe-Jean Lavergne,
ca. 1909, PB Wien 19
Diese Gruppe wurde auch als Lampe hergestellt, wobei sich die Leuchte in den Händen der Harlekine befand.

This group was also made as a lamp, with the light held in the harlequin's hands.

salts of strontium, manganese, silver, tin, zinc, platinum, chrome, etc could be similarly" used.[156]

At Lippmann in Berlin in 1909 the new "Pozzos aus Fayence" ["faience Cupids"] with flowerpots were particularly featured, as were sculptures in which the "body parts [...] were rendered all in semi-gloss, while the apparel stood out brilliantly fresh."[157]

tium-, Mangan-, Silber-, Zinn-, Zink-, Platin-, Chrom- usw. Salze" anwenden.[156]

Bei Lippmann in Berlin werden 1909 vor allem die neuen „Pozzos aus Fayence" mit Blumenkübeln angepriesen, genauso wie Plastiken, bei denen die „Körperteile [...] ganz matt-glänzend gehalten, während das Gewand sich in glänzender Frische abhebt."[157]

ModNr. 4028
„Dachshund",
Alfred Raum, ca. 1911,
Collection a@goldscheider.net

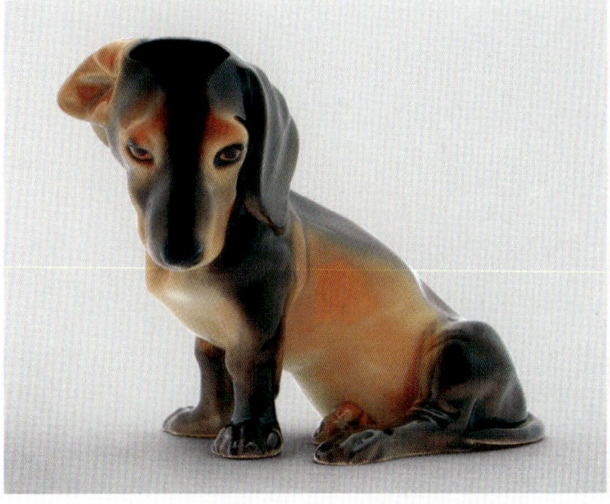

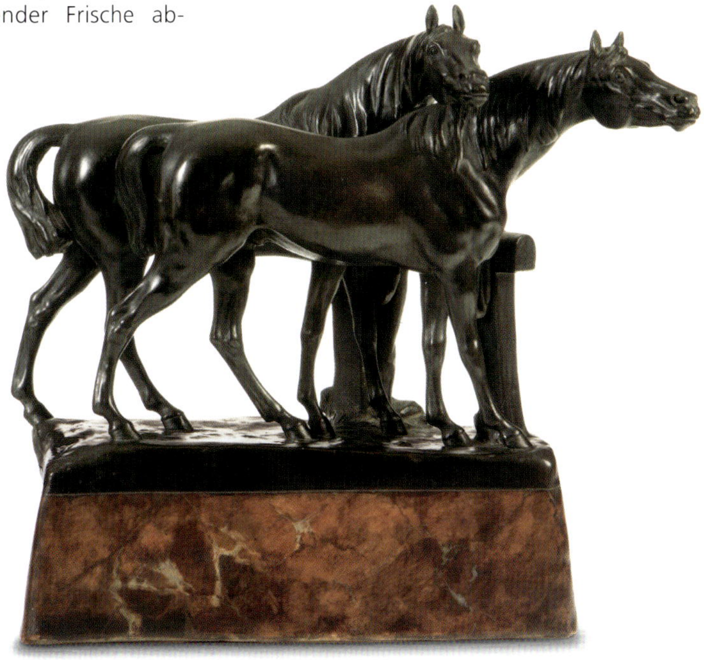

ModNr. 3625
„Pferde" | "Horses",
Capèque, ca. 1908,
© Wien Museum

Am 8. Juli 1910 wird neben Regine Lewit-Goldscheider auch Walter Goldscheider, der bislang als Mitarbeiter im Familienunternehmen tätig war, als Gesellschafter der Firma „Erste Wiener Terrakottafabrik und Atelier für künstlerische Fayencen Friedrich Goldscheider" mit der Hauptniederlassung in Wien und Zweigniederlassung in Leipzig eingetragen.[158] Alois Goldscheider wird auch unter der erweiterten Führung als Prokurist eingesetzt.[159] Die Firma beschäftigt an die 250 Arbeiter.[160] Regine Lewit wohnt zu dieser Zeit in der Hasenauerstraße 16, Walter in der Gentzgasse 57.[161] Im Jahr 1911 heiratet Walter die aus einer Gleiwitzer Industriellenfamilie stammende Lilly Kuschnitzky, ein Jahr später wird ihr einziger Sohn Erwin geboren.[162]

Auch die Kunstgegenstände der Jahre 1911 und 1912 finden großen Anklang: „Zurzeit ist das ernstere Genre besonders beliebt, wir sahen prächtige Gladiatorengestalten in Terrakotta (Bronzedekor), der ‚Triumph des Achilles', die ‚Phryne vor Gericht' aus Pierre de Vienne zeigte sehr originelle Farbentönungen. In Wiener Fayencen ist mehr das leichte Genre beliebt. Hübsche Barock- und Biedermeierfiguren gehörten dieser Kollektion an."[163]

On 8 July 1910 Walter Goldscheider, who until then had been an employee of the family business, was made a partner in the "Erste Wiener Terrakottafabrik und Atelier für künstlerische Fayencen Friedrich Goldscheider," based in Vienna with a branch in Leipzig, along with Regine Lewit-Goldscheider.[158] Alois Goldscheider continued to have power of attorney even after the management was enlarged.[159] By now the firm employed a workforce of around 250.[160] Regine Lewit was then living at 16 Hasenauerstraße and Walter at 57 Gentzgasse.[161] In 1911 Walter married Lilly Kuschnitzky, the daughter of an industrialist in Gleiwitz. A year later Erwin, their only son, was born.[162]

The art objects produced in 1911 and 1912 also met with a very positive reception: "Currently the more serious genre is especially popular; we saw splendid figures of gladiators in terracotta (bronze décor), the 'Triumph of Achilles'; 'Phryne before the Areopagus' in Pierre de Vienne revealed very original color tints. In Viennese faience, the light genre tends to be popular. Pretty Baroque and Biedermeier figures were in this collection."[163]

ModNr. 4318
Knabe in Tracht |
Boy Wearing Costume, Theodor Mallener, ca. 1912, © Wien Museum
Die Figur des stehenden Knaben in Tracht mit Hut und zwei Blumenkörben erfreut sich hoher Beliebtheit und wird auch in der internationalen Presse erwähnt (Ill. 32). (Quelle: Anzeige aus The Studio, 1914)

This figure of a standing boy wearing Austrian costume with a hat and two baskets of flowers was extremely popular and was also mentioned in the international press (Ill. 32). (Source: an advertisement from The Studio, 1914)

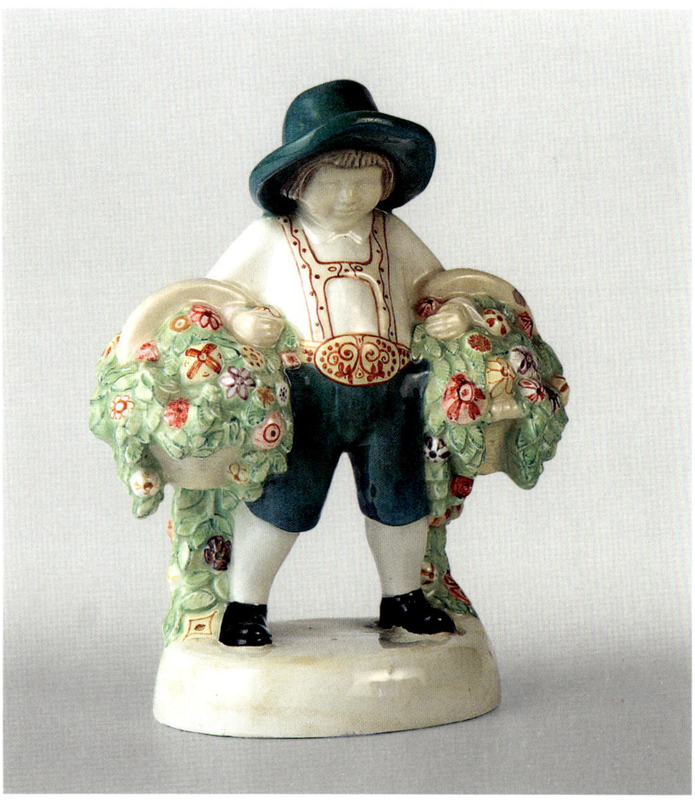

Ill. 32

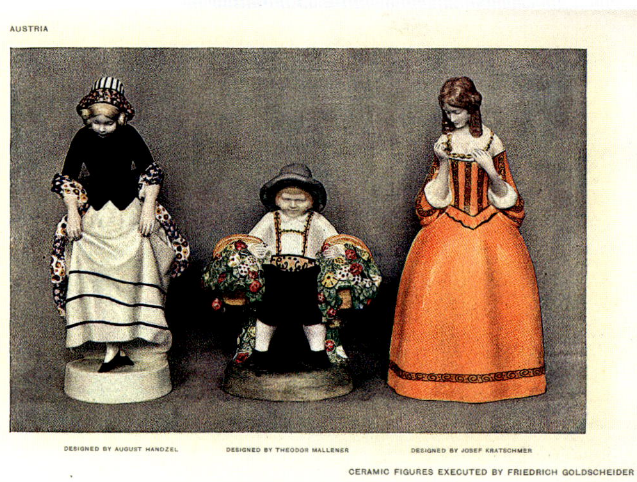

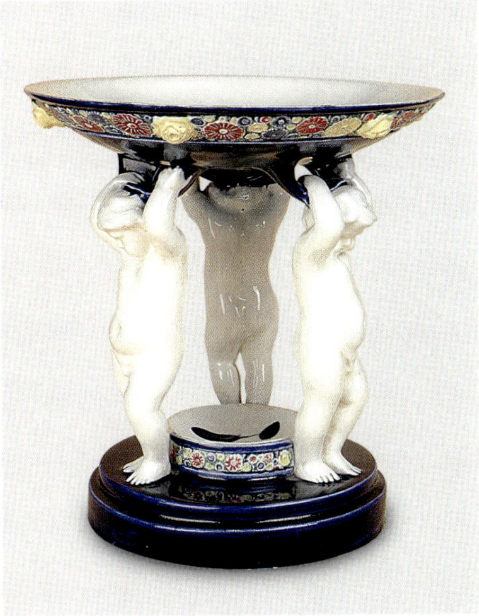

ModNr. 3831
Schale | Bowl, ca. 1910,
PB Wien 19

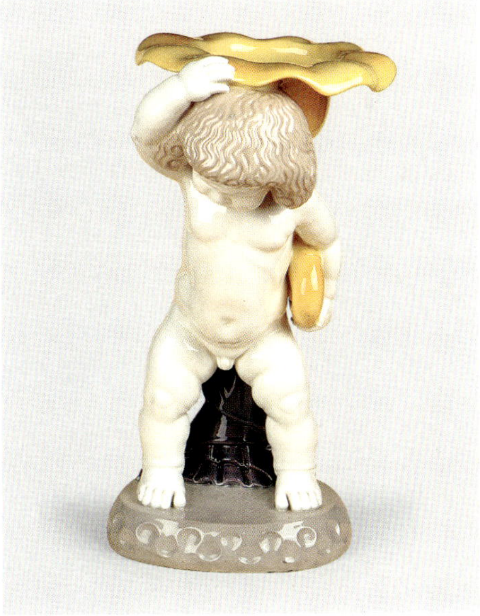

ModNr. 3856
Putto, Armin Müller,
ca. 1910, PB Wien 19

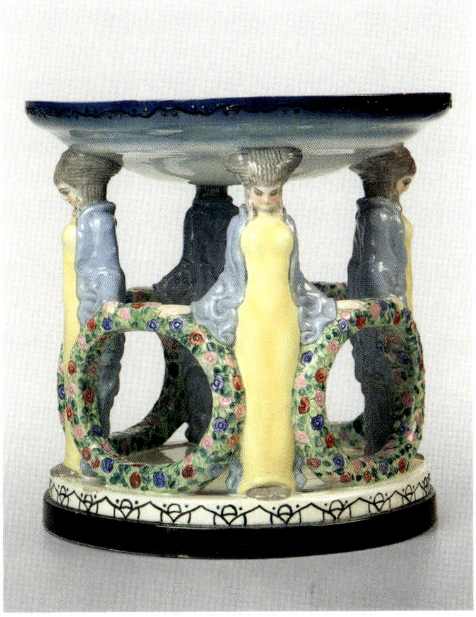

ModNr. 4387
Schale | Bowl,
Fritz Pollak, ca. 1913,
© Wien Museum

ModNr. 4103
Lampe Biedermeier-mädchen | Lamp Biedermeier Girl, Rosé,
ca. 1912, Collection
a@goldscheider.net

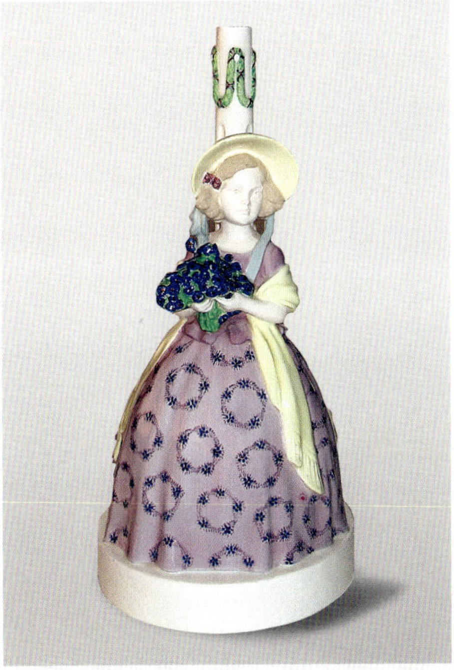

ModNr. 4221
Biedermeiermädchen |
Biedermeier Girl,
Adolphe-Jean
Lavergne, ca. 1912, PB

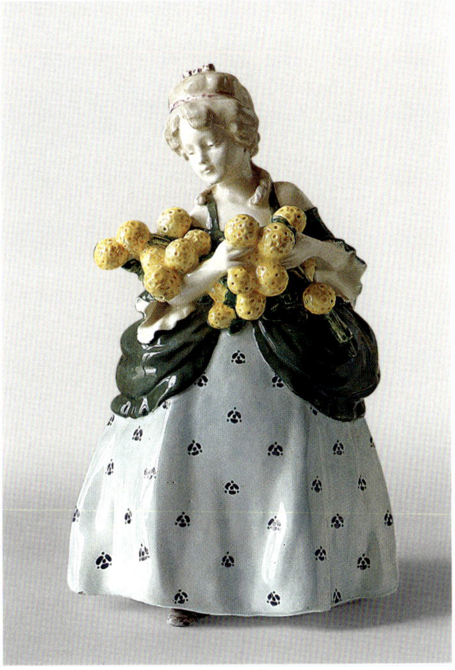

ModNr. 4390
„Im Regen" | "In the
Rain", Hentschel,
ca. 1913, PB Wien 19

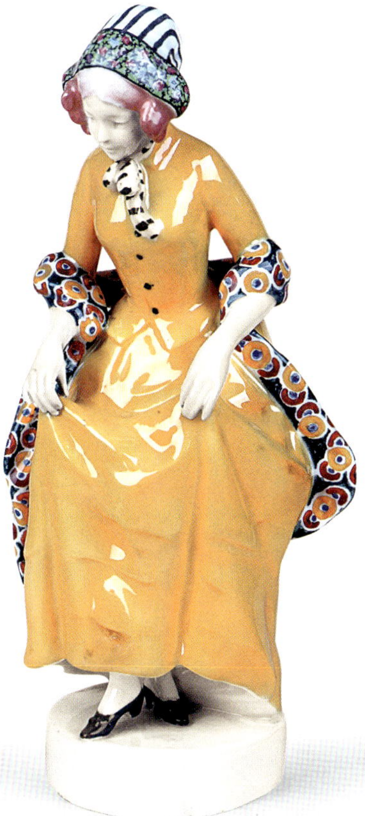

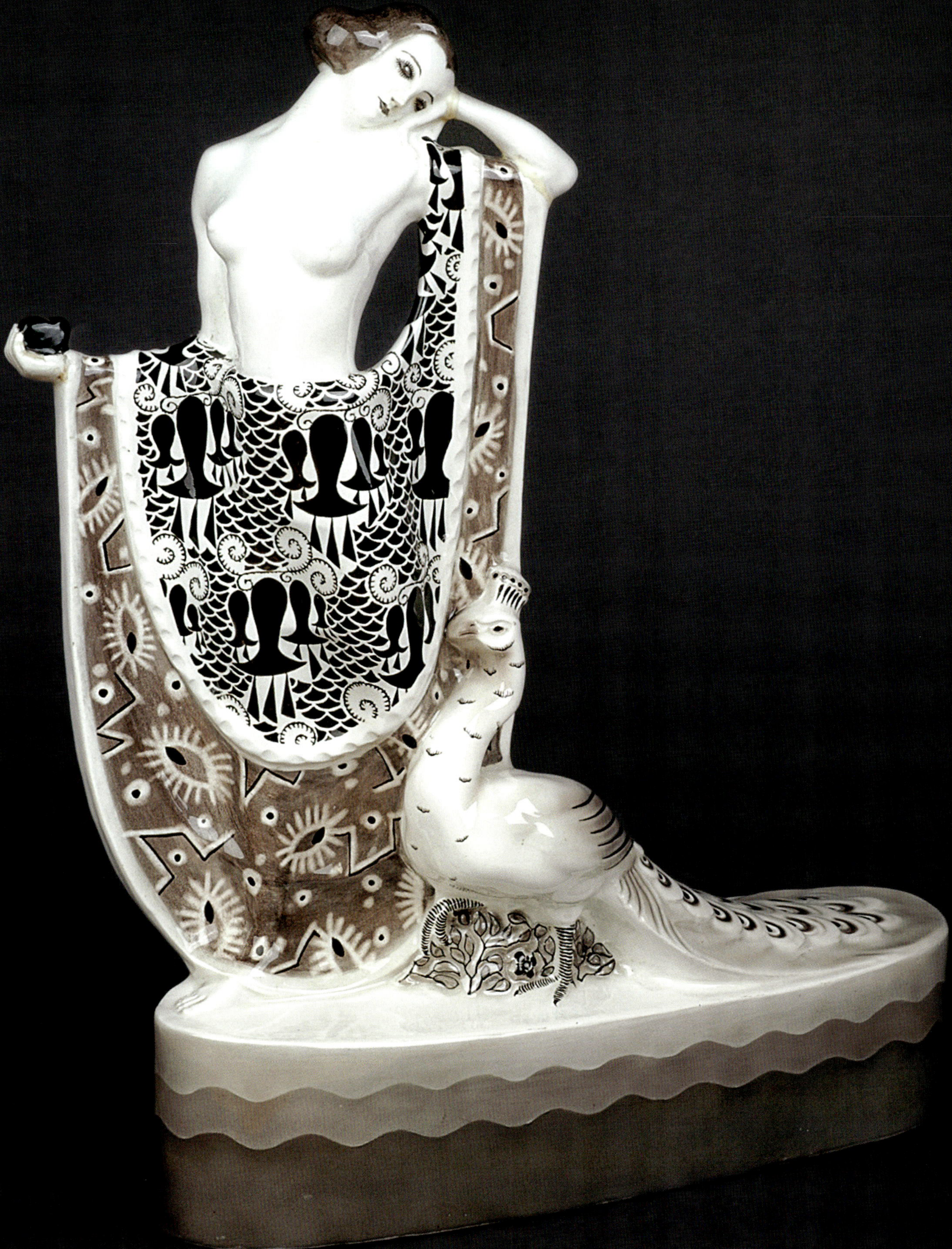

links I left

ModNr. 4483 „Juno mit Pfau" I "Juno and the Peacock", Ruß, ca. 1913, PB Wien 19
Unter dem Einfluss des Stils der Wiener Werkstätte entstehen einige Figuren im Schwarz-Weiß-Dekor. Die thematischen Vorbilder liefern oft Theaterszenen, die beispielsweise Postkarten-Motiven entnommen werden (vgl. Ill. 33); hier ein Beispiel mit Lo Hesse und Joachim von Seewitz. (Quelle: PB Deutschland)

Some figures were created in black-and-white décor under the influence of the Wiener Werkstätte style. They were often modelled thematically on theatre scenes, which in turn were taken from postcard motifs (cf Ill. 33); here an example with Lo Hesse and Joachim von Seewitz. (Source: PB Deutschland)

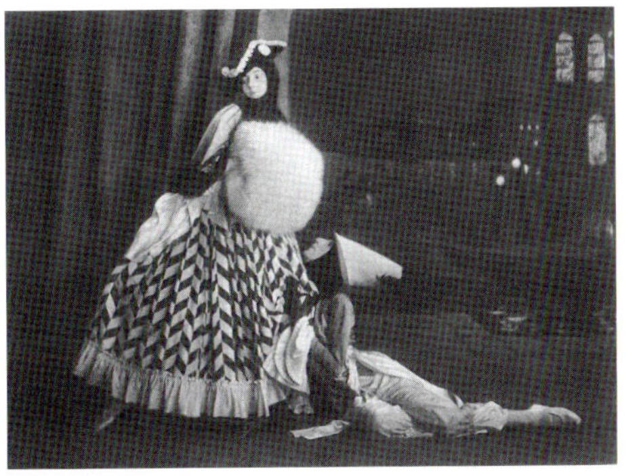

Ill. 33

The following observations were made on the objects shown at the 1912 Leipzig Pre-Easter Fair: "Friedrich Goldscheider in Vienna has thought up a new excavation décor to very naturalistic effect; the piece called 'Strength', a group of lion tamers, made the impression of having lain in the earth for many centuries. We saw many terracottas treated with a bronze impregnation that looks deceptively like genuine bronze. Further, the firm has produced figures of Pierre de Vienne, a faience paste that is chemically treated to make it look like colored marble. We focus on some particularly fine motifs from the collection, such as Tolstoy as a pilgrim in terracotta, Napoleon retreating from Moscow (1812), an oriental bead dancer with electric light effects and more of the like."[164] At the autumn fair, Goldscheider also showed animal figures and objects revealing the influence of the Wiener Werkstätte style: "We saw [...] faiences in this style, figures and vases executed in black-and-white to exquisite effect. From the new marble collection, the Madonna bust, further a Judith and a bowls (boccia) player are particularly noteworthy."[165]

The faiences were successful and gradually replaced the terracotta models. The manufacture of "crackle decoration on objects" was created in 1912 by means of a coat of lacquer, which before it was entirely dry, was given a "coat of tinted isinglass solution and finally dried, either naturally or by artificial means" so that "fissures are created in the outer coat, through which the ground of the object appears."[166] The patent was, however, revoked in 1913.

In 1912 Marcell Goldscheider, who was by then twenty-five years old, returned to Vienna to work in the family business, where he became technical and artistic director.[167] He was the only member of the family to have decided on training as a potter at the academy; his brothers, by contrast, were all businessmen. At eighteen Marcell went to the k.k. Kunstgewerbliche Fachschule Teplitz, where he attended the ceramics classes and later trained at the Kunstgewerbeschule in Vienna. After three years' training in Austria,

Über die auf der Leipziger Ostervormesse 1912 ausgestellten Objekte heißt es: „Friedrich Goldscheider in Wien hatte einen neuen Ausgrabungsdekor von sehr naturalistischer Wirkung ersonnen, so machte die Gruppe Löwenbezwinger ‚Kraft' den Eindruck, als hätte das Stück viele Jahrhunderte in der Erde gelegen. Terrakotta sahen wir vielfach mit imprägnierter Bronze behandelt, die echter Bronze täuschend ähnlich sieht. Ferner brachte die Firma Figuren aus Pierre de Vienne, einer Fayencemasse, die auf chemischem Wege den bunten Marmorcharakter erhält. Wir heben aus der Kollektion einige besonders schöne Sujets hervor, so Tolstoi als Pilger in Terrakotta, Napoleon auf der Rückkehr aus Moskau (1812), eine orientalische Perlentänzerin mit elektrischen Lichteffekten u.a.m."[164] Auf der Herbstmesse findet man bei Goldscheider auch Tierfiguren und Objekte unter dem Einfluss des Stils der Wiener Werkstätte: „Wir sahen [...] Fayencen in diesem Stil, Figuren und Vasen in Schwarz-weiss-Ausführung von feiner Wirkung. Aus der neuen Marmorkollektion sei die Büste Madonna, ferner eine Judith und eine Boggiaspielerin hervorgehoben."[165]

Die Fayencen setzen sich durch und verdrängen sukzessive die Terrakottamodelle. Die Erzeugung „kraquelierter Verzierungen auf Gegenständen" entsteht 1912 durch einen Lackanstrich, der vor dem vollständigen Trocknen mit einem „Anstrich von gefärbter Wasserglaslösung versehen und schließlich auf natürlichem oder künstlichem Weg getrocknet" wird, wobei „im äußeren Anstrich Risse entstehen, durch die der Grund des Gegenstandes zum Vorschein kommt."[166] Das Patentaufgebot wird allerdings 1913 zurückgezogen.

Im Jahr 1912 kehrt auch der fünfundzwanzigjährige Marcell Goldscheider nach Wien zurück, um im Familienunternehmen zu arbeiten, in dem er die technische und künstlerische Leitung übernimmt.[167] Er hat sich als Einziger aus der Familie für eine akademische Töpferausbildung entschieden, denn seine Brüder sind alle Kaufleute. Mit 18 Jahren ist er an die k.u.k. Kunstgewerbliche Fachschule Teplitz gegangen, besuchte dort auch die Klassen für Keramik sowie später die Wiener Kunstgewerbeschule. Nach der dreijährigen Ausbildung in Österreich setzte er sein Studium in Frankreich

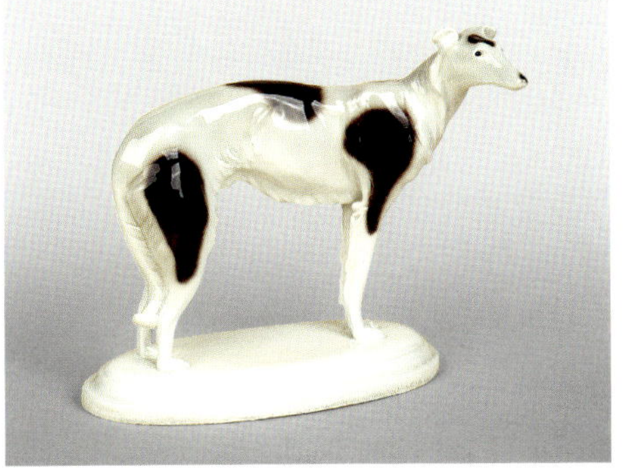

ModNr. 4687 Windhund I Whippet, Valmore Gemignani, Entwurf I design um 1911, Ausformung 20er Jahre, PB Wien 19
Gemignani hat etliche Tierfiguren für Goldscheider entworfen. Zwischen 1910 bis Ende der 30er Jahre waren vor allem Windhunde, auch Borsoi genannt, beliebt.

Gemignani designed quite a few animal figures for Goldscheider. Between 1910 and the late 1930s, Russian Wolfhounds, also called by their Russian name, Borzoi, were particular favorites.

ModNr. 4130 Sitzendes Mädchen I Sitting Girl, Polter, ca. 1912, PB Wien 19

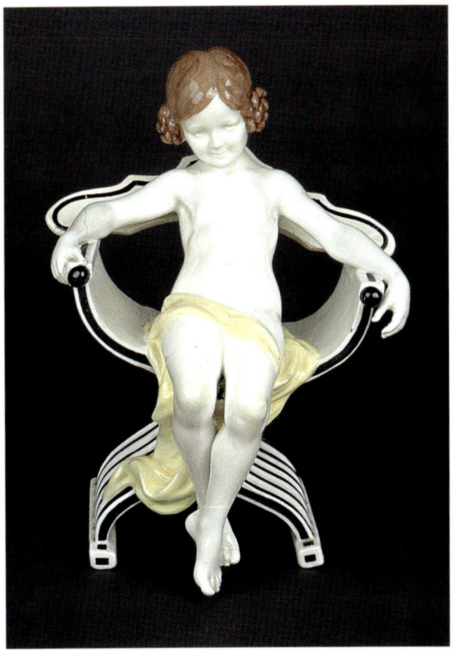

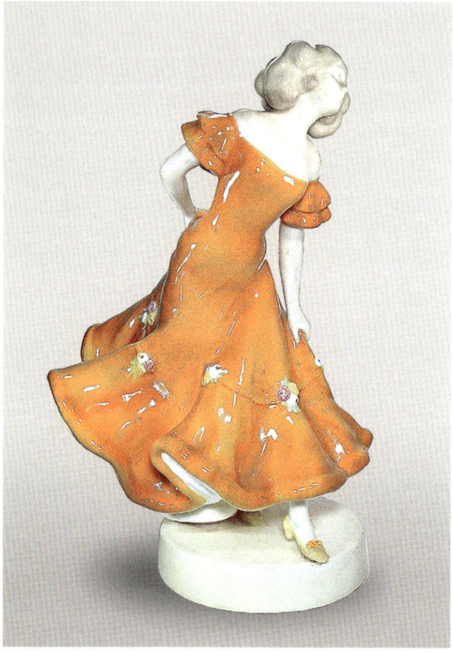

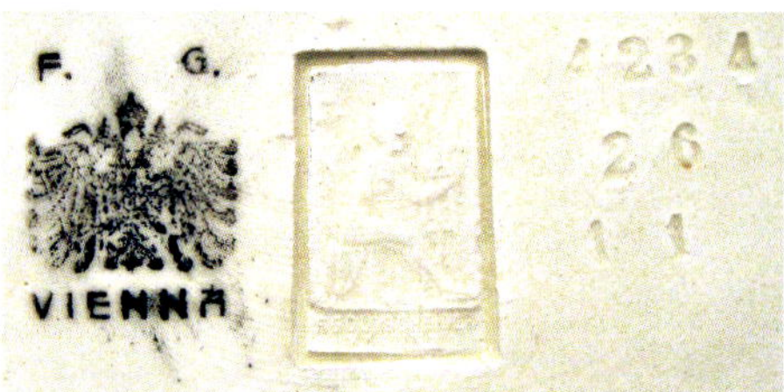

ModNr. 4234
Tänzerin I Danseuse, Latour / Rosé, ca. 1912, Collection a@gold-scheider.net
Der international bekannte Sculpture-Stempel wird sowohl auf Terrakotten wie auch Fayencen verwendet. Der Adler-Stempel erscheint fast ausschließlich auf Fayencen, häufig in Kombination mit der Sculpture-Marke.

The internationally known "Sculpture" stamp was used both on terracotta and faience. The eagle stamp occurs almost exclusively on faience, often in combination with the "Sculpture" mark.

fort, wo er über ein Jahr blieb und anschließend nach England wechselte. Zeitweise war er in London, aber auch in den so genannten „Potteries" – den traditionellen Töpfereien um Stoke-on-Trent (Staffordshire). In Wien gibt Marcell sein international erworbenes Wissen an die ausführenden Mitarbeiter weiter und verbessert die Qualität der Erzeugnisse.

In dieser Zeit werden 180 Arbeiter beschäftigt, die Pariser Filiale befindet sich nun in der Rue de Paradis 34. Die Fabrikate sind vor allem Blumentöpfe und -kübel, Bonbonnieren, Blumenkrippen, Jardinieren, Figuren, Aquarien, Säulen, Büsten, Reliefbilder, Uhrgehäuse, Vasen, Wandbilder sowie Tierstücke aus Kunstterrakotta und Fayence. Das international eingeführte Markenzeichen „Sculpture" wird sowohl für Terrakotten wie auch Fayencen eingesetzt. Zusätzlich werden viele Fayencen mit dem Adler-Stempel und der Bezeichnung Vienna oder Wien gemarkt.[168]

he went to France for further study. Following a year in France he went to England, where he spent some time in London. He also "went to the potteries" – the traditional potteries around Stoke-on-Trent in Staffordshire. In Vienna Marcell passed the knowledge and skills he had acquired abroad on to the employees engaged in ceramics-making and thus improved product quality.

By that time the business employed a workforce of 180. The Paris branch had moved to 34 rue de Paradis. The chief products were flower pots and cachepots, bonbonnières, flower crèches, jardinières, figurines, aquariums, columns, busts, reliefs, clock cases, vases, murals and animals made of artificial terracotta and faience. The trademark "Sculpture" was introduced internationally for both terracotta and faience. In addition, many faience pieces were marked with the eagle stamp and "Vienna" or "Wien."[168]

ModNr. 4286
Dame I Lady, Polter, ca. 1912, PB Wien 19

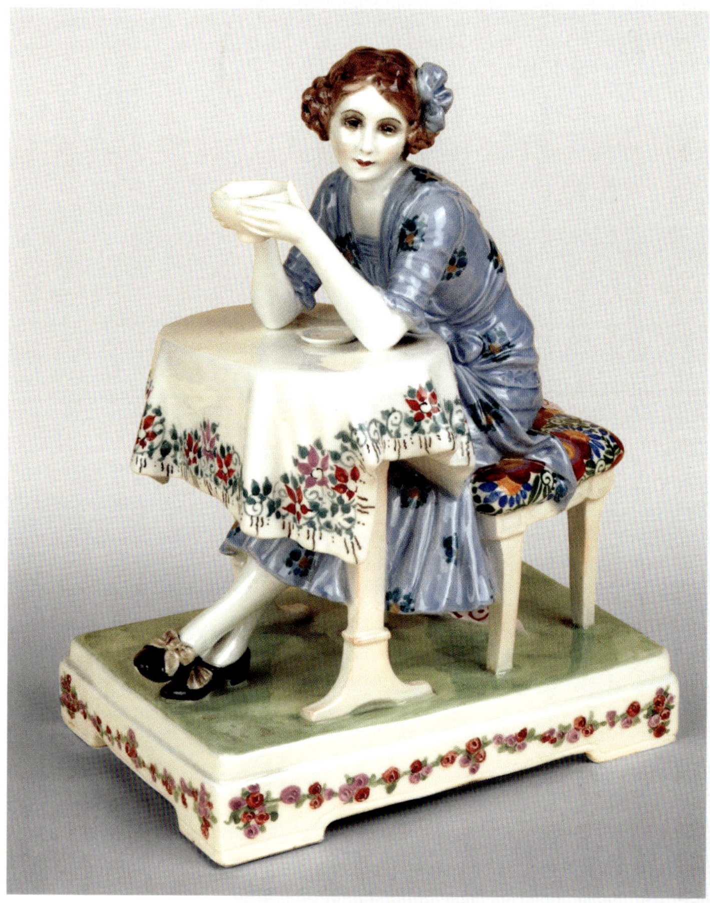

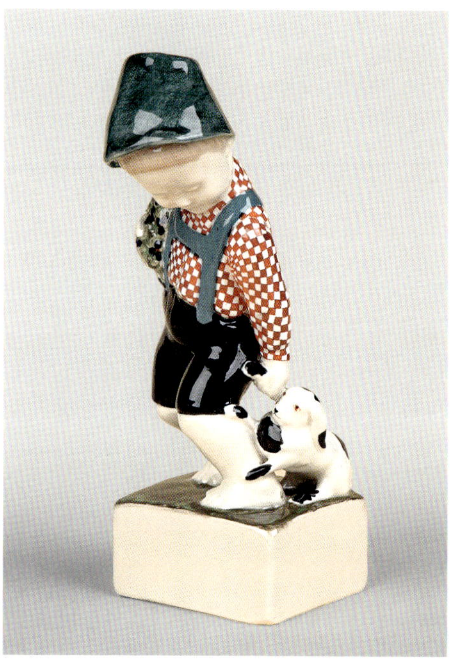

ModNr. 4319
Knabe mit Hase I Boy with Rabbit, Theodor Mallener, ca. 1912, PB Wien 19

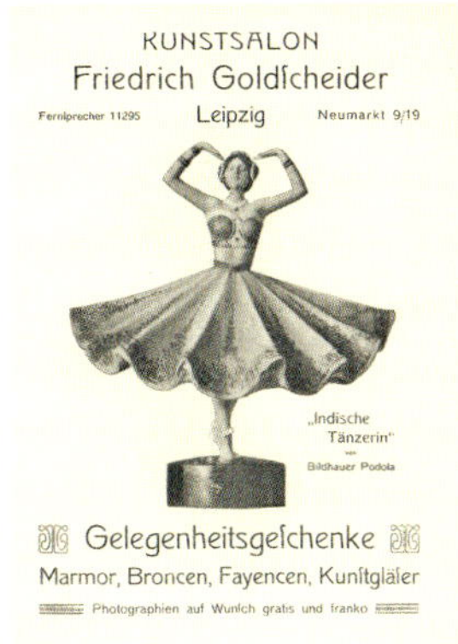

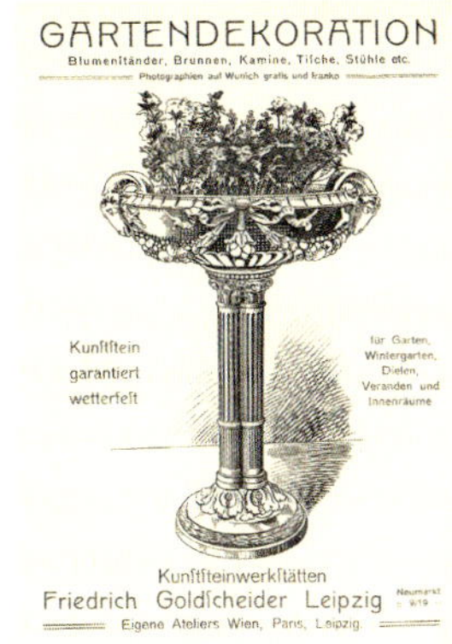

Ill. 34, 35
In Leipzig wird neben dem Kunstsalon im Kaufhof am Neumarkt auch ein Atelier betrieben, das auf Objekte für Gartendekoration spezialisiert ist. (Quelle: Heimatkalender für Thüringen und das Osterland 1914)

In Leipzig there was an art salon in the Neumarkt department store as well as a studio specialized in garden ornaments. (Source: Heimatkalender für Thüringen und das Osterland 1914)

At the "Internationale Baufach-Ausstellung" in Leipzig in 1913, Goldscheider showed a large number of small pieces of art sculpture, including typical figures in regional costume and animal figures. Of the latter, Wernekinck's "Great Crested Grebe" received particular mention. The manufactory was awarded a gold medal for the objects shown.[169] This was also the period of "Dessins" in vivid colors since faience in pale tints had gone "entirely out of fashion."[170] The Leipzig branch, however, did not just sell products from Paris and Vienna. There was also a proprietary workshop, in which flower stands, fountains, chimney-pieces, tables and chairs were made (Ill. 35).[171]

Bei der „Internationalen Baufach-Ausstellung" in Leipzig im Jahr 1913 stellt man eine Fülle von künstlerischer Kleinplastik vor, darunter charakteristische Trachtenfiguren und Tierfiguren, von denen der „Haubentaucher" von Wernekinck besondere Erwähnung findet. Für die ausgestellten Objekte erhält die Manufaktur die Goldene Medaille.[169] In dieser Zeit fallen vor allem die „Dessins" mit lebhaften Farben auf, da blasse Fayencen „ganz aus der Mode" sind.[170] In der Leipziger Dependance werden jedoch nicht nur Erzeugnisse aus Paris und Wien verkauft. Es gibt ein eigenes Atelier, in dem Blumenständer, Brunnen, Kamine, Tische und Stühle hergestellt werden (Ill. 35).[171]

ModNr. 4495
„Indische Tänzerin" | "Indian Dancer",
Podola / Rosé,
ca. 1913, Collection city-antik, Wien

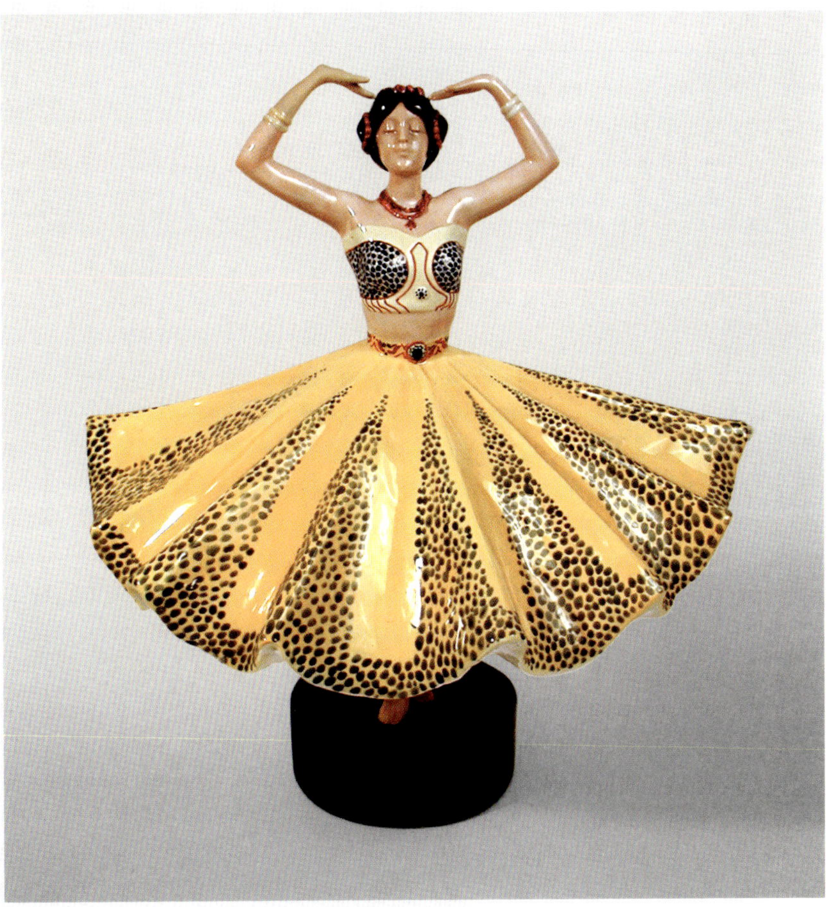

ModNr. 4609
Tänzerin | Danseuse,
Helmut Lindner,
ca. 1913, PB Wien 19

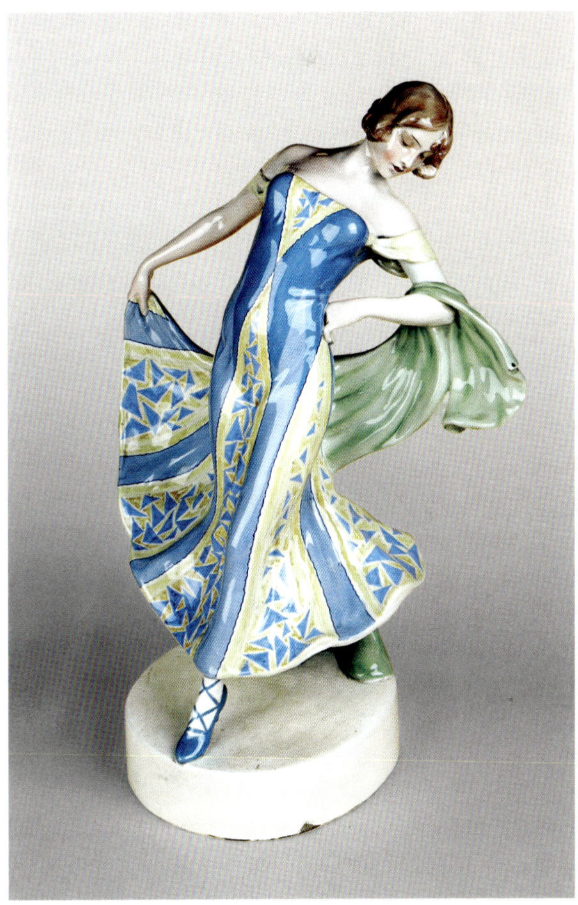

Ill. 36

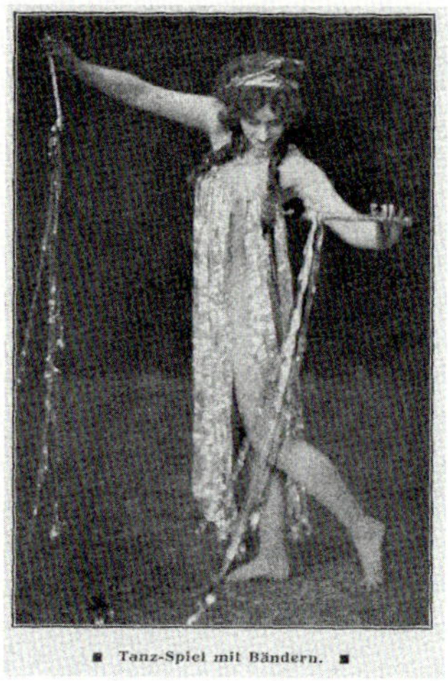

rechts | right

ModNr. 4814 „Adorée-Via Villany – Tanzspiel mit Bändern" | "– Streamer Dance", Rudolf Podany, ca. 1917, PB Wien 19
Die seinerzeit berühmt-berüchtigte (Nackt)-Tänzerin Adorée-Via Villany wird hier in ihrem „Tanzspiel mit Bändern" dargestellt (vgl. Ill. 36). Diese Plastik nimmt die Fayence-Modelle der 1920er Jahre vorweg, die oft berühmte Tänzerinnen und Schauspielerinnen abbilden. (Quelle: Postkarte aus PB Deutschland)

Notorious at the time, the (nude) dancer Adorée Via Villany is depicted here doing her "dance play with streamers" (cf Ill. 36). This sculpture anticipates the faience models of the 1920s, which often portray celebrated dancers and actresses. (Source: postcard, PB Deutschland)

Mit Ausbruch des 1. Weltkriegs wird zum 1. August 1914 der Betrieb still gelegt, doch bereits mit Beginn des Jahres 1915 wird die Arbeit wieder aufgenommen.[172] Sowohl Walter wie auch Marcell werden zum Kriegsdienst eingezogen. Walter dient von Juli 1914 bis zum November 1918 als Rittmeister der Landwehr.[173] Am 29. September 1916 scheidet er als Gesellschafter aus der Wiener Fabrik aus, und Regine Lewit führt als Alleininhaberin die Firma weiter.[174] Anfang 1917 verlässt Walter auch die Leipziger Zweigniederlassung.[175] Die Erzeugung von Ton- und Porzellanwaren wird am 24. April 1917 offi-

Production shut down on 1 August 1914 on the outbreak of the First World War but work resumed by early 1915.[172] Both Walter and Marcell were conscripted for national service. Walter served from July 1914 until November 1918 as cavalry captain in the Territorial Reserve.[173] On 29 September 1916, Walter resigned as a partner in the Viennese factory, leaving Regine Lewit to run the firm as sole proprietor.[174] Early in 1917, Walter also left the Leipzig branch.[175] On 24 April 1917, the manufacture of pottery and porcelain wares was officially reclassified as the "industrial production of sculptural objects."[176] During

ModNr. 4763 Mädchen | Girl, Johanna Meier-Michel, ca. 1916, PB Wien 19
Während des I. Weltkrieges wurde die Produktion stark reduziert, neue Modelle waren äußerst selten. Dieses Modell von Johanna Meier-Michel ist eines der wenigen, die während des Krieges aufgelegt worden sind.

During the First World War, production was considerably reduced; new models were extremely thin on the ground. This model by Johanna Meier-Michel is one of the few to have been produced during the war.

ModNr. 4834 „Bei der Toilette" | "At Her Toilette", Rudolf Podany, ca. 1918, © Wien Museum
Eines der populärsten Modelle der frühen Nachkriegszeit sind diese Balldamen im Rokokokleid bei ihrer Morgentoilette.

These ladies dressed in the Rococo style at their morning toilette after a ball were one of the most popular models of the early post-war years.

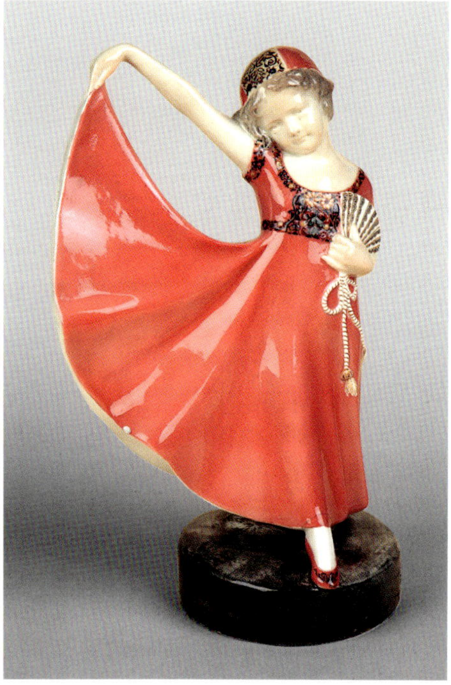

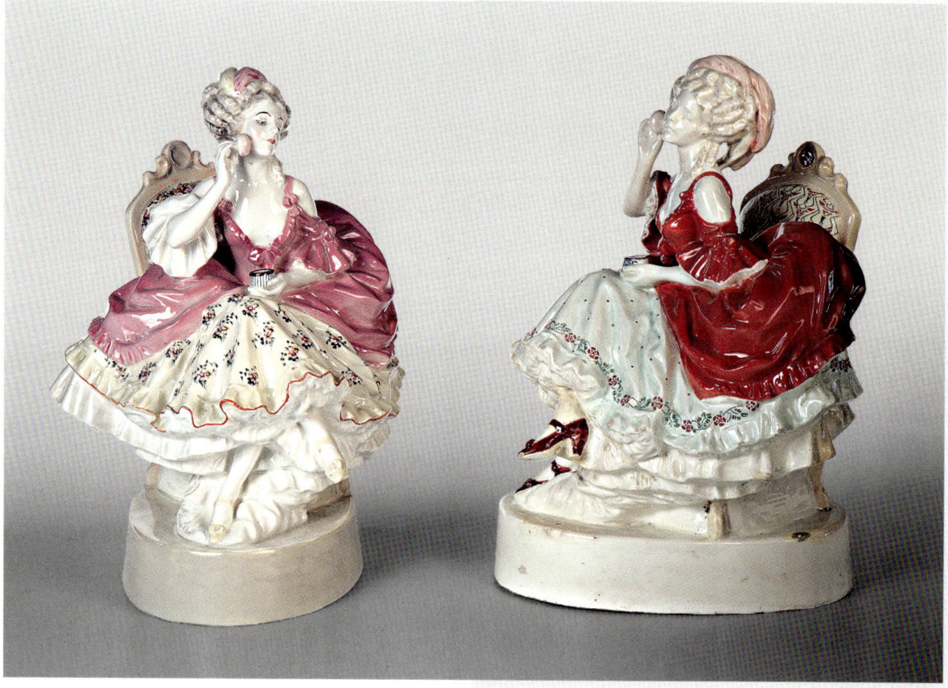

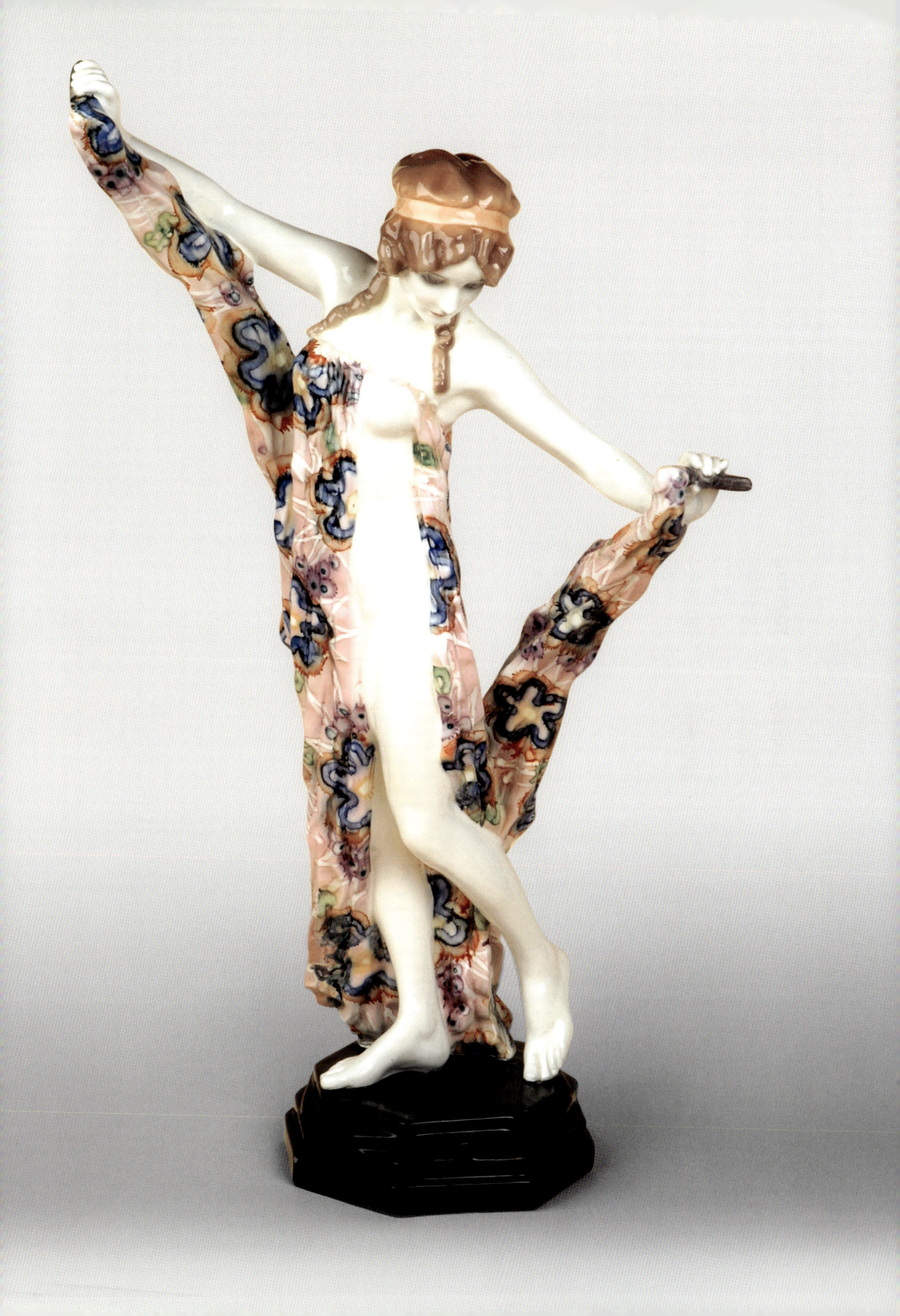

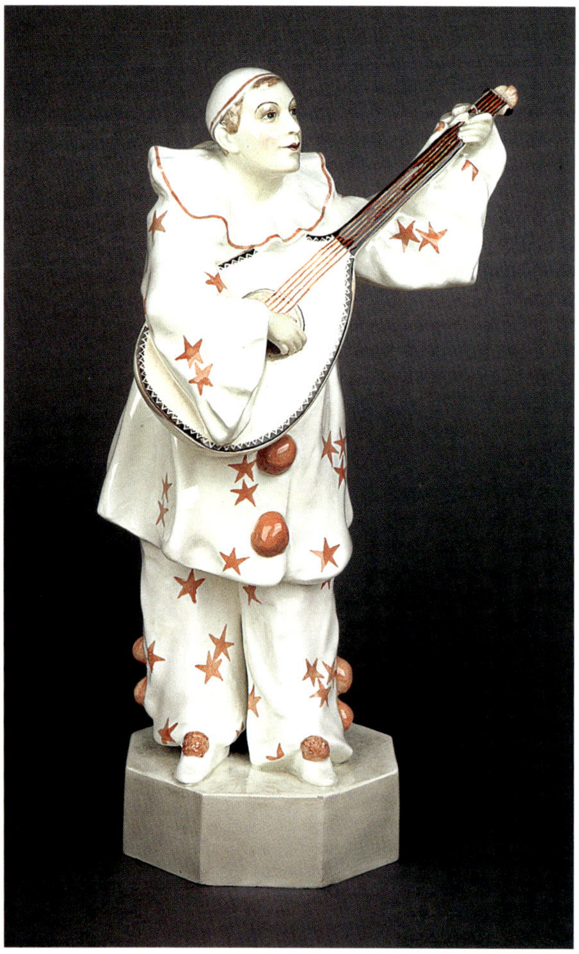

ModNr. 4935
Harlekin I Harlequin,
Podolak, ca. 1919,
PB Wien 19

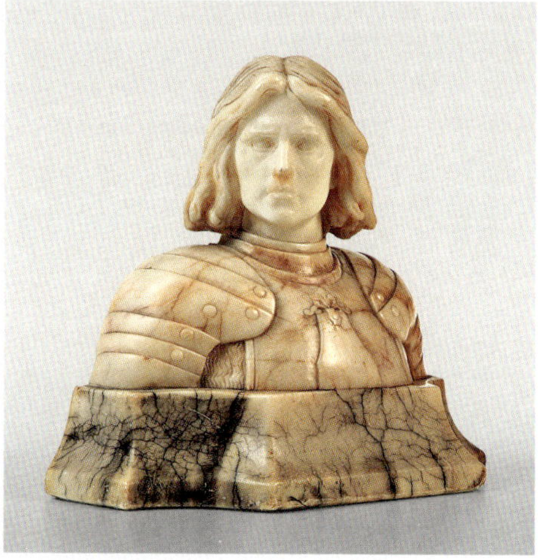

ModNr. 4807
Büste I Bust „Jeanne d'Arc", Rudolf Maison, Entwurf vor I design before 1900, Ausführung I executed ca. 1917, © Wien Museum
Während des Ersten Weltkriegs werden eher bewährte Modelle in neuen Materialien oder Größen reproduziert.

During the First World War tried and tested models tended to be reproduced in new materials or sizes.

ziell in „Fabriksmäßige Erzeugung von plastischen Gegenständen" umgeschrieben.[176] Während des Krieges werden kaum neue Modelle entworfen, die Produktion wird insgesamt reduziert.

Am 20. Oktober 1918 stirbt mit 64 Jahren Regine Lewit, verwitwete Goldscheider.[177] Sie wird neben Friedrich Goldscheider in der israelitischen Abteilung am Döblinger Friedhof begraben. Walter und Marcell erben das Haus im Cottage-Viertel, in das Marcell einzieht. Sein Bruder Walter lebt mit seiner Frau Lilly und Sohn Erwin bereits seit 1912 in der wenige Minuten entfernten Hochschulstraße 40.[178] Das Unternehmen wird für den Zeitraum der Verlassenschaftsklärung (17.12.1918 bis 30.12.1920) von Walter Goldscheider und dem Rechtsanwalt Dr. Herbert Fialla geführt, Alois Goldscheider ist weiterhin Prokurist.[179]

the war, hardly any new models were designed; production was throttled overall.

Regine Lewit, veuve Goldscheider, died at the age of sixty-four on 20 October 1918.[178] She was buried next to Friedrich Goldscheider in the Jewish section of the Döblinger Cemetery. Walter and Marcell inherited the house in the Cottage Quarter and Marcell moved into it. Since 1912, his brother Walter had been living with his wife, Lilly, and son Erwin at 40 Hochschulstraße. Until probate was declared on the will (17 Dec. 1918 bis 30 Dec. 1920), the firm was managed by Walter Goldscheider and Dr. Herbert Fialla, attorney. Alois Goldscheider continued to have power of attorney.[179]

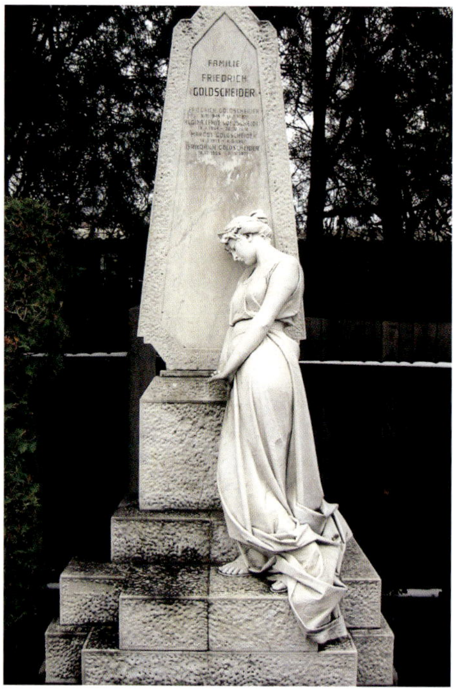

Ill. 37
Im Familiengrab von Friedrich Goldscheider sind auch seine Frau Regine sowie sein Enkel Friedrich „Fritz" Goldscheider (der Sohn von Camillo Goldscheider) und dessen Ehefrau Margot begraben.

Regine Goldscheider, Friedrich Goldscheider's widow, and his grandson, Friedrich "Fritz" Goldscheider (Camillo Goldscheider's son) and his wife, Margot, were all buried in the family vault with Friedrich Goldscheider.

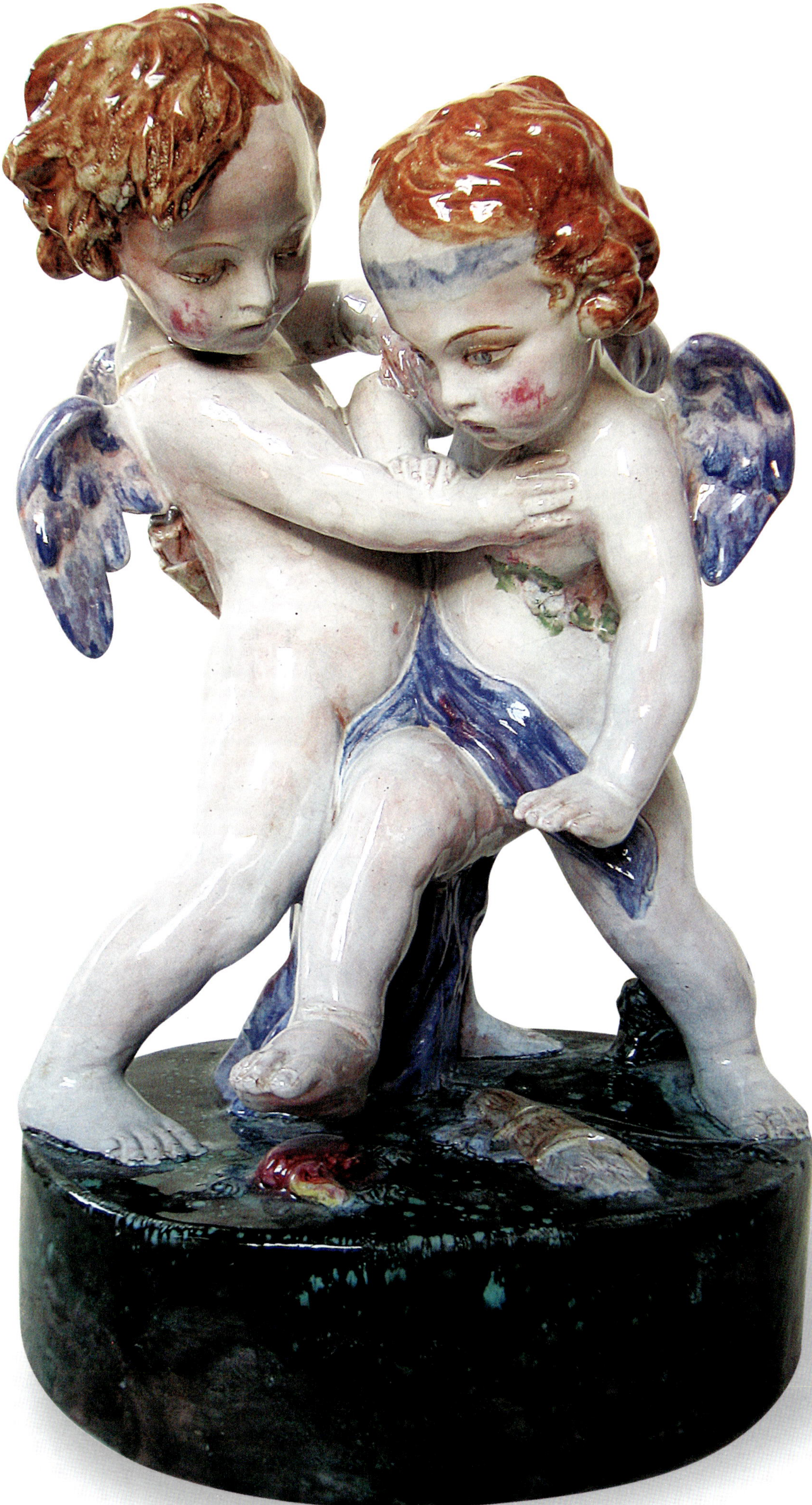

ModNr. 3762
Putti, ca. 1909, Collection Weilinger, Wien
Der Entwurf dieser Gruppe der beiden um ein Herz zankenden Putti geht auf eine Marmorskulptur von François-Joseph LeClercq, die 1780 entstand, zurück. Die Ausformung des gezeigten Exemplars ist allerdings aus den 20er Jahren.

The design for this group of two Cupids squabbling over a heart derives from a marble sculpture by François-Joseph LeClercq (1780). The cast of the model shown here, however, dates from the 1920s.

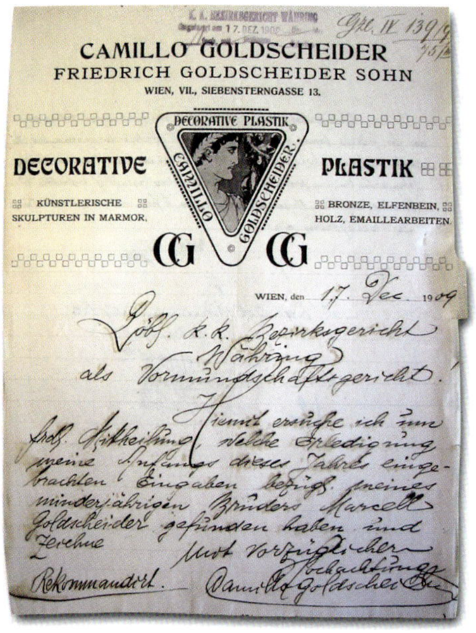

Ill. 38
Die Firma von Camillo Goldscheider währt nur ein Jahr. Der überdimensionierte Briefkopf deutet auf seinen megalomanen Charakter hin. (Quelle: Stadt und Landesarchiv Wien; Verlassenschaftsabhandlung)

Camillo Goldscheider's firm only lasted a year. The outsized letterhead is symptomatic of his megalomania. (Source: Stadt und Landesarchiv Wien; Verlassenschaftsabhandlung)

Dekorative Plastik von Camillo Goldscheider in Wien (1909–1910)

Nachdem Camillos Klage um Aufnahme als Gesellschafter ins Familienunternehmen Anfang 1909 abgewiesen wurde[180], gründet er am 27. April 1909 in der Stiftgasse 21 eine eigene Fabrik unter der Firmierung „Camillo Goldscheider, Friedrich Goldscheider Sohn" und stellt „Decorative Plastiken" in Marmor, Bronzen, Elfenbein, Holz und Emaille her[181]. In der Siebensterngasse 13 im 7. Bezirk hat er sein Büro und die Verkaufsräume.[182] Seine Firmenmarke meldet er unter der Nummer 37751 an.[183] Auch Camillo verfolgt ähnliche künstlerische Ideale wie sein Vater Friedrich und will die Kleinplastik in verschiedenen Materialien durch seine Ideen und Werke prägen.

Die Deutsche Export-Revue bezeichnet die Gründung der neuen Werkstätte als eine der „interessantesten Erscheinungen im modernen Kunstgewerbe" und rühmt die „tadellos feine [...] Ausführung der Plastiken, Uhrgehäuse, Beleuchtungskörper" in „Bildhauerei und Patinierung".[184] Auch von anderen Stimmen wird ihm „volles künstlerisches und technisches Verständnis" zugesprochen.[185] Mit Neuheiten aus Petrazit (Petrarciastein) und Beleuchtungskörpern tritt er auf der Messe am Leipziger Neumarkt im Zeissighaus auf und schaltet diesbezüglich etliche Anzeigen, unter anderem auf dem Heftumschlag der Deutschen Export-Revue Nr. 19 aus dem Jahr 1909/10 (Ill. 39 und 40).

Ornamental Sculpture by Camillo Goldscheider in Vienna (1909–1910)

After Camillo's lawsuit for a partnership in the family business was dismissed early in 1909,[180] he founded a factory of his own on 27 April 1909 at 21 Stiftgasse and had it registered as "Camillo Goldscheider, Friedrich Goldscheider Sohn." He made "ornamental sculpture" in marble, bronze, ivory, wood and enamel.[181] His office and salerooms were at 13 Siebensterngasse in the 7th District.[182] He registered his trademark under the number 37751.[183] Camillo pursued aesthetic ideals similar to those Friedrich, his father, had espoused and wanted his ideas and works to inform the small-scale statuary he produced in various materials.

The Deutsche Export-Revue described the establishment of the new workshop as one of the "most interesting débuts in modern Arts and Crafts" and praised "the exquisitely immaculate [...] execution of the sculpture, clock cases [and] lighting fixtures" in respect of "sculpture and patination."[184] Approval was also voiced by other critics, who attested to his "consummate artistry and technical skill."[185] He displayed novelties made of Petrazit (Petrarcia stone) and lighting fixtures at the Zeissighaus at Neumarkt in Leipzig and placed a good many advertisements, including on the cover of Deutsche Export-Revue No. 19 for 1909/10 (Ill. 39 and 40).

Petrazit is described as "a light, highly ductile substance that reproduces the details of line in the smallest of figures." The painting of the figures, on the other hand, was achieved by pastose "color tinting" as if injected delicately into the substance.[186] The method for making the models was innovative since this kind of tinted artifical stone had previously not been used for figurines[187] (Ill. 41–43).

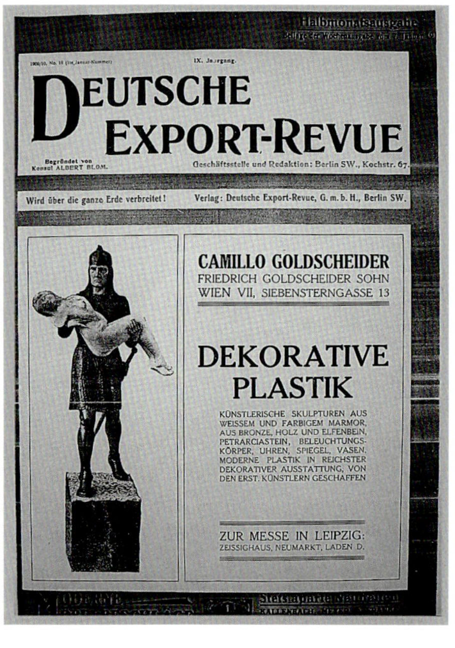

Ill. 39, 40
Auf dem Umschlag und im Innenteil der Deutschen Export-Revue bewirbt Camillo Goldscheider im Januar 1910 seine Erzeugnisse, die er in eigener Werkstatt herstellt. (Quelle: Deutsche Export-Revue No. 19, 1909/10, sowie No. 15 auf S. 628)

On the front cover and inside the January 1910 issue of Deutsche Export-Revue, Camillo Goldscheider advertised the products he made at his own workshop. (Source: Deutsche Export-Revue No. 19, 1909/10 as well as No. 15 on p. 628).

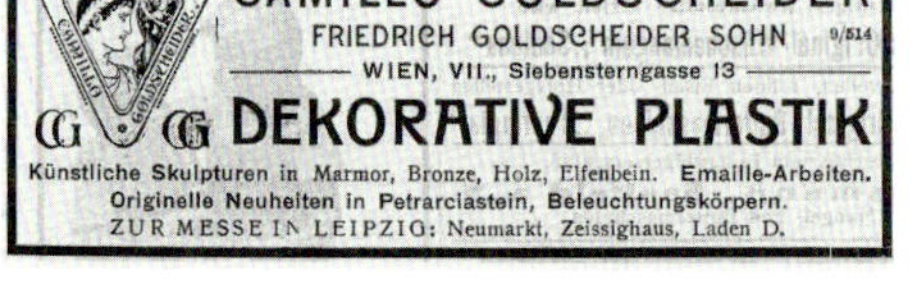

Ill. 41

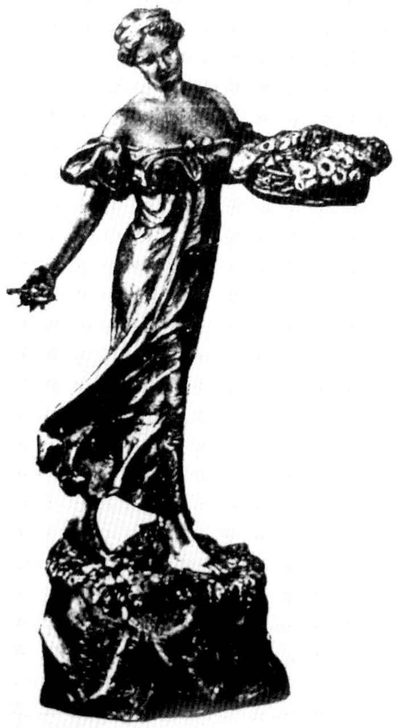

Das Petrazit wird als „eine leichte, sehr bildungsfähige Masse, die die Einzelheiten der Linienführung in den kleinsten Figürchen wiedergibt" bezeichnet, deren Bemalung eher einer pastösen, wie in die Masse eingehauchten „Tönung der Farbe" entspricht.[186] Die Herstellung der Modelle ist innovativ, da dieser kolorierte Kunststein bislang nicht für Kleinplastiken verwendet worden ist[187] (Ill. 41–43).

Camillos Weg in die Selbständigkeit ist jedoch mit dem Zusammenbruch seiner Familienbande verbunden. Zum Streit mit seiner Mutter und seinem Onkel kommt hinzu, dass sich bereits im Frühjahr 1909 seine Frau Jolanthe von ihm scheiden lässt.[188] Am 14. Januar 1910 widerruft Camillo zwar seine Vormundschaftsklage beim Bezirksgericht Währing, doch privat ist er mit allen zerstritten und psychisch in einer schlechten Verfassung. Er stellt zwar noch auf der Leipziger Ostervormesse im Jahr 1910 aus[189], doch bereits am 15. April 1910 stellt er den Betrieb ein[190]. Bis 1916 wird Camillo in den Wiener Adressbüchern geführt[191], und anschließend in ein Sanatorium eingewiesen, wo er bis Ende der dreißiger Jahre versorgt und von seinen Brüdern finanziell unterstützt wird. Vermutlich wurde Camillo Goldscheider von den Nazis im Laufe der Kriegsjahre umgebracht.

Ill. 41–43
Camillos Plastiken zeigen eine Variation an Sujets in herausragender Bearbeitung – von der jugendstilgeprägten „Dame mit Blumenkörben", der Figurengruppe „Kindersymphonie" bis hin zum galanten Tanzpaar der „Gavotte", die aus Petrazit gefertigt sind. (Quelle: Deutsche Export-Revue No. 15, 1909/10, S. 596)

Camillo's sculptures represent excellently made variations on motifs – from the Art Nouveau "Lady with Basket of Flowers," the group of figures called "Children's Symphony" to "Gavotte," an elegant dancing couple – executed in Petrazit (Source: Deutsche Export-Revue No. 15, 1909/10, p. 596)

Camillo's move towards independence was, however, linked with his breaking his family ties. Not only did he quarrel with his mother and uncle; by spring 1909, his wife Jolanthe divorced him.[188] On 14 January 1910, Camillo revoked the guardianship suit he had initiated at Währing District court but privately he was at odds with everyone and, psychologically speaking, in a bad state. He managed to show pieces at the 1910 Leipzig Pre-Easter Fair[189] but by 15 April 1910 he had shut down the factory.[190] Camillo is listed in the Viennese directories until 1916;[191] after that he was committed to a sanatorium, where he was taken care of and financially supported by his brothers until the late 1930s. Camillo Goldscheider was presumably killed by the Nazis during the war years.

Ill. 42

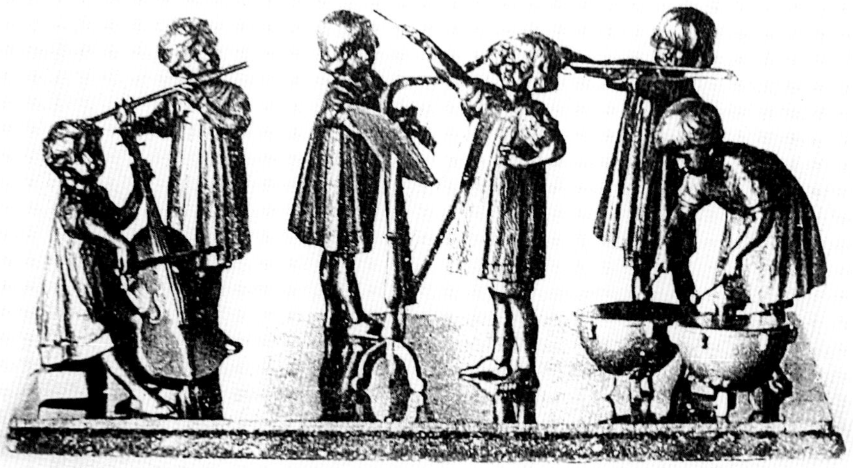

Ill. 43

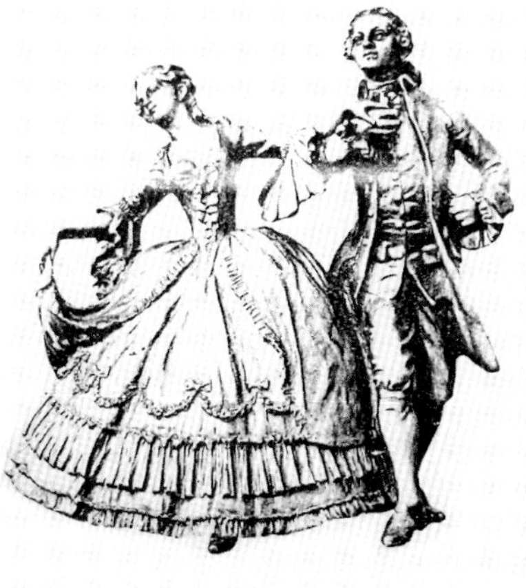

Anmerkungen

[89] Lehmann's Adressbuch Wien, 1899 bis 1901

[90] Stellungnahme von Dr. Herrmann Fialla, Hof- und Gerichtsadvokat, als Vertreter der Vormundschaft von Marcell Goldscheider, an das Bezirksgericht Währing, vom 31.3.1909. Den Kreditoren schuldete man 161.195,74 Fl. (Gulden) und diesem Posten standen nur Zahlungsmittel von 34.569,83 Fl. gegenüber, so dass sich die Schulden insgesamt auf 126.625,91 Fl. beliefen, was 253.251,82 K (Kronen) entsprach. Die Warenbestände hatten zwar einen Gesamtwert von 414.299,08 K „an denen in grösstem Ausmasse die Filiale in Paris und in nächstgrossem die Niederlage in Berlin partizipierten, aber keinerlei flottante Geldmittel gegenüberstanden," so dass trotz des Aktivüberschusses von 161.047,26 K, zu welchem „noch an anderen unveräusserlichen, nur bedingt valablen Werten, wie Modelle, Inventar und dgl. etwa K 24.000 zuzurechnen waren, somit trotz eines bücherlichen Überschusses von K 185.047,26" – also rund 92.500 Fl. – „die Gefahr imminent war, die Zahlungspflichten der Firma nicht erfüllen und einer Katastrophe nicht vorbeugen zu können." Die Währung Krone wurde 1892 eingeführt.

[91] Stellungnahme von Dr. Herrmann Fialla, Hof- und Gerichtsadvokat, als Vertreter der Vormundschaft von Marcell Goldscheider, an das Bezirksgericht Währing, vom 31.3.1909.

[92] Abschrift des Vertrags zwischen der Firma Friedrich Goldscheider, Leipzigerstraße 119/120 und dem Kaufmann Herrn Carl Denecke, Potsdamerstraße 4, vom 10.9.1899. Regine Goldscheider beschreibt die Bewertung der Berliner Filiale wie folgt: „Der Inventurwert der Berliner Filiale betrug am 19. Januar 1897 fl. 22.000,– = Mk 37.500,–, wovon Mark 24.750,– auf das Warenlager und Mark 12.750,– auf die andere Aktiva entfallen. Infolge der seither stattgehabten Verlegung der Geschäftslokalitäten in die Leipzigerstrasse und der damit verbundenen Adaptierungs- und Anschaffungskosten, sowie infolge beträchtlicher Warenanhäufung daselbst ergab sich am 10. September 1899 ein Warenbestand von Mk 59.850,– und dazu die Bewertung des aus den Adaptierungen und der Einrichtung bestehenden Inventars mit Mk 15.000,–, zusammen Mk 74.850,–, wobei von den Kreditoren der Filiale, die im ganzen fl. 56,07 ausmachten, abgesehen wird." Zu dieser Zeit entsprachen 100 Mark ungefähr 58 Fl. oder 77,5 K. Zusätzlich wird vereinbart, dass Denecke innerhalb der nächsten zehn Tage Terrakotten bis zur Fabriknummer 1891 und bis zu einem Rechnungsbetrag von 10.000 Mark mit 33,3 Prozent unter dem rabattlosen Engros-Preis, Bronzen und Marmorwaaren bis 5.000 M mit 25 Prozent Rabatt, bestellen kann.

[93] Erbteilungs-Vertrag zwischen Regine Goldscheider und den Kindern Arthur, Camillo, Walter und Marcell Goldscheider, vom 24.12.1899 (Quelle: Stadt- und Landesarchiv Wien, 38867).

[94] Ebenda

[95] Ebenda

[96] Ebenda. Der einige Monate zuvor erfolgte Verkauf der Berliner Filiale wird somit zu Gunsten von Regine Goldscheider verbucht.

[97] Ebenda

[98] Regina Goldscheider beschreibt die Beschaffung von Geldmitteln durch die Verkäufe der Filialen wie folgt: „Auf diese Weise wurden aus Berlin etwa K 73.000,– und aus Paris K 140.000, zusammen K 213.000,– mobilisiert, welche schon einigermassen gestatteten, an die Bewältigung der Schulden per K 253.000,– heranzutreten. So verblieb noch ein Schuldenbetrag von K 40.000,– offen, trotz dessen ich auch noch die drei Erbteile meiner Söhne Camillo, Walter und Marcell à K 36.000,– mit K 108.000 abzulösen mich entschloss, wodurch ich gegenüber einer finanziellen Gesamtverpflichtung von K 148.000,– ein aus toten Waren und Materialien bestehendes Aktivum von K 185.000,– abzüglich des im Jahre 1899 hauptsächlich aus der Realisierung der Filiale in Berlin hervorgegangenen Verlustes per K 30.500, K 154.500 zu übernehmen hatte. [...] Die Geldknappheit war so gross, das mein Sohn Arthur in Paris sich verpflichten musste, mir das für Marcell zum Gerichtserlage erforderliche Kapital von K 36.000,– für 1 Jahr zu beschaffen." In dieser Stellungnahme wird auch angegeben, dass am 20. Mai 1899 im Nachlassnachweis ein reines Nachlassvermögen von Fl. 92.677 = 185.355 K dargetan worden ist. Vgl. hierzu Stellungnahme von Dr. Herrmann Fialla, Hof- und Gerichtsadvokat, als Vertreter der Vormundschaft von Marcell Goldscheider, an das Bezirksgericht Währing, vom 31.3.1909.

[99] Stellungnahme von Dr. Herrmann Fialla, Hof- und Gerichtsadvokat, als Vertreter der Vormundschaft von Marcell Goldscheider, an das Bezirksgericht Währing, vom 31.3.1909.

[100] Stellungnahme von Josef Schuster an das Wiener Magistrat, Abteilung 11 Wertzuwachsabgabe, vom 21.1.1939, bezug nehmend auf die Einantwortungsurkunde des Handelsgerichts Wien vom 8.3.1921 A 188/18. Die Umrechnung erfolgt durch Josef Schuster im Jahr 1939 wie folgt: 55.000 FL = 110.000 Kronen; 1 Krone = 1,44 Schilling.

[101] Stellungnahme von Dr. Herrmann Fialla, Hof- und Gerichtsadvokat, als Vertreter der Vormundschaft von Marcell Goldscheider, an das Bezirksgericht Währing, vom 31.3.1909.

[102] Neuwirth (1974), S. 31

[103] In: Keramische Monatshefte, Halle 1901, S. 93

[104] Oesterreichische Patentschrift Nr. 3855, Classe 80: Thonwaren-, Stein- und Cementindustrie, vom 25. April 1901. Das Patent erlischt im Jahr 1906.

[105] R., F.: Wiener Kunstkeramik, in: Keramische Monatshefte, 1902, S. 88f

[106] Ebenda

[107] Ebenda

[108] Ebenda, S. 91

[109] Hecht/Barthel, S. 1584

[110] Ebenda

[111] Ebenda

[112] Neuwirth (1974), S. 31

[113] Central-Marken-Register für das Jahr 1900, Register Nr. 12167. Die Anmeldung erfolgte am 18.4.1900 für das k.k. Reich. Für das Deutsche Reich wurde die Marke am 4.12.1900 eingetragen (vgl. Waarenzeichenblatt vom Kaiserlichen Patentamt, Berlin, Heft 1, Januar 1901).

[114] Antrag auf Enthebung der Vormünderin Frau Regine Lewit und des Mitvormundes Alois Goldscheider von der Vormundschaft des mj. Marcell Goldscheider und Bestellung des Einschreiters zum Vormunde, Brief von Advokat Dr. Emil Frischauer, im Auftrag Camillo Goldscheiders, an das Bezirksgericht Währing vom 6.11.1908. Regine und Dr. Lewit wohnen seit 1901 in der Lannerstraße 20 (vgl. Lehmann's Adressbuch Wien, 1902). Alois Goldscheider zieht im Jahr 1903 in die Cottagegasse 8, wo er bis zu seinem Ableben wohnt (vgl. Lehmann's Adressbuch Wien, 1904 bis 1923).

[115] Pudor (1902a), S. 119

[116] N.N.: Von der ersten internationalen Ausstellung für dekorative Kunst in Turin 1902, in: Keramische Monatshefte, 1902, S. 162f

[117] Stellungnahme von Dr. Herrmann Fialla, Hof- und Gerichtsadvokat, als Vertreter der Vormundschaft von Marcell Goldscheider, an das Bezirksgericht Währing, vom 31.3.1909. Camillo wohnt ab ca. 1904 in der Rieglergasse 14 im 18. Bezirk und wird als Disponent in Lehmann's Adressbuch Wien, 1905 geführt. Auch Walter wird 1905 als Disponent mit Wohnort in der Nussdorferstraße 5 im 9. Bezirk gelistet, doch 1906 und 1907 sind keine Einträge verzeichnet, was auf einen Auslandsaufenthalt von Walter vermuten lässt. Walter ist erst 1908 als Disponent mit Wohnort Schulgasse 70, ein Jahr später in der Gentzgasse 57 verzeichnet (vgl. Lehmann's Adressbuch Wien, 1905 bis 1909).

[118] Klage von Camillo Goldscheider, vertreten durch Dr. Emil Frischauer um Aufnahme als öffentlicher Gesellschafter in die Firma Friedrich Goldscheider mit einer Beteiligung von 20 Prozenten, beim k.k. Handelsgericht Wien, eingelangt am 28.10.1908.

[119] R., F.: Keramische Wanderungen durch den Kunstsalon von Keller & Reiner, in: Keramische Monatshefte, 1903, S. 125f. Im Jahr 1902/1903 stellt man auch auf der Keramischen Ausstellung in Reichenberg aus. Vgl. N.N.: Unsere keramische Ausstellung, in: Mitteilungen des nordböhmischen Gewerbemuseums, Reichenberg 1902, Nr. 4, S. 88

[120] Kiessig (1904), S. 761

[121] Leisching, Eduard: Die Winterausstellung des k.k. Österr. Museums. In: Kunst und Kunsthandwerk, Nr. 7, 1904, S. 526ff. Diese museumseigene Kunstzeitschrift ist eine ergiebige Quelle für Studien der Wiener Keramik, da das Museum ihre Winter- und Frühjahrsausstellungen durch Abbildungen und Künstlerangaben dokumentierte. In Paris erfüllt der Catalogue illustré des Salons du Paris eine ähnliche Rolle. Andere in dieser Zeit entstehende Fachzeitschriften für Kunstgewerbe berichten über Goldscheider-Modelle und fördern den Bekanntheitsgrad der international renommierten Marke.

[122] Illustrierte Rundschau. In: Velhagen & Klasings Monatshefte, Bd. 17, 1902/03, S. 717

[123] Reklame in: Keramische Rundschau, 5.3.1903, S. 200

[124] Österreichisches Patentblatt, 1904, S. 499. Die Patentvergabe erfolgt zum 15.3.1904 und erhält die Patentnummer 16882. Das Patent erlischt im Jahr 1907.

[125] Österreichisches Patentblatt, Nr. 2, 1905, S. 59. Das Patent vom 21.1.1904 erhält die Nummer 20387 und erlischt 1906.

[126] Stellungnahme von Josef Schuster an das Wiener Magistrat, Abteilung 11 Wertzuwachsabgabe, vom 21.1.1939, Bezug nehmend auf die Einantwortungsurkunde des Handelsgerichts Wien vom 8.3.1921, A 188/18. Die Umrechnung erfolgt durch Josef Schuster im Jahr 1939 wie folgt: 55.000 Gulden = 110.000 Kronen; 1 Krone = 1,44 Schilling

[127] Manuskript Heidi Brunnbauer „Im Cottage von Währing/Döbling", Band II, unter anderem zur Geschichte des Hauses Hasenauerstraße 16.

[128] N.N.: Figürliche Terrakotten, in: Keramische Monatshefte, 1904, S. 138

[129] Adress-Buch der Keram-Industrie, 1904, S. 60. Die Unternehmensgründung wird hier mit 1882 angegeben, was auch den Recherchen der Autoren entspricht. Die Manufaktur Goldscheider hat allerdings in Firmenunterlagen und Werbekampagnen das Jahr 1885 als Gründungsjahr angegeben.

[130] Reklame in: Keramische Rundschau, 1904, S. 158

[131] Pazaurek (1904), S. 90

[132] N.N.: Leipziger Michaelismesse 1904, in: Keramische Rundschau, Nr. 39, 22.9.1904

[133] Leipziger Michaelis-Messe 1904, in: Centralblatt 1904, S. 1294, zitiert nach Neuwirth (1974), S. 32

[134] Reklame in: Keramische Rundschau, Wegweiser für die Leipziger Michaelismesse 1904. S. 798. Ein Jahr später erzielen als „geschützte Neuheit" Terrakotten mit Fayence-Einlagen „hervorragende Wirkung" (vgl. N.N.: Michaelismesse 1905, in: Keramische Rundschau, Nr. 37, 14.9.1905, S. 915ff)

[135] Fayence und Majolika. In: Keramische Rundschau, 1905, S. 735

Notes

[89] Lehmann's Adressbuch Wien, 1899 to 1901

[90] Deposition made by Dr. Herrmann Fialla, Hof- und Gerichtsadvokat, representing the guardianship of Marcell Goldscheider, to Bezirksgericht Währing [Währing District Court], on 31 Mar. 1909. The sum of 161 195.74 Fl. (guilders) was owed to creditors and all that was available in ready cash was 34 569.83 Fl. so that the total debt amounted to 126.625,91 Fl., which corresponded to 253 251.82 K (Kronen). The inventory in stock was worth 414 299.08 K in total "of which the bulk was shared by the branch in Paris, with the next largest stock at the Berlin branch" so that, despite a surplus of 161 047.26 K, to which "yet other inalienable, only to a limited extent saleable assets such as models, inventory and the like some K 24 000 might be added, thus, despite a bookable surplus of K 185.047.26" – that is amounting to roughly 92 500 Fl. – "there was imminent danger of being unable to meet the firm's outstanding debts and not being able to prevent disaster." The Krone was introduced as the Austrian currency in 1892.

[91] Deposition made by Dr. Herrmann Fialla, Hof- und Gerichtsadvokat, representing the guardianship of Marcell Goldscheider, to Bezirksgericht Währing on 31 Mar. 1909.

[92] Copy of the contract between the firm of Friedrich Goldscheider, 119/120 Leipzigerstraße, and the businessman Carl Denecke, 4 Potsdamerstraße, of 10 Sept. 1899. Regine Goldscheider described the financial situation of the Berlin branch as follows: "The value of the Berlin branch inventory stock was fl. 22 000 = Mk 37 500 on 19 January 1897, of which 24 750 Marks fall to the share of the warehouse and 12 750 Mark to the other liquid assets. Due to the business location having been moved in the meantime to Leipzigerstrasse and the renovation and acquisitions costs as well as the considerable accumulation of stock as a result, on 10 September 1899 the wares were worth Mk 59 850 – and in addition the valuation from the renovation and the setting up of existing inventory at Mk 15 000 – a total of Mk 74 850, apart from fl. 56.07 owed to creditors of the branch." At that time 100 Mark amounted to approx 58 Fl. or 77.5 K. In addition, it was agreed that Denecke could order terracottas up to factory number 1891 and worth up to 10 000 Marks at 33.3 per cent discount below the undiscounted wholesale price as well as bronze and marble wares worth up to 5000 Marks at 25 per cent discount.

[93] Erbteilungs-Vertrag [contract on the division of legacy at probate] between Regine Goldscheider and children Arthur, Camillo, Walter and Marcell Goldscheider, of 24 Dec. 1899 (Source: Stadt- und Landesarchiv Wien, 38867).

[94] Ibid

[95] Ibid

[96] Ibid. Consequently, the proceeds from the sale of the Berlin branch that had taken place some months previously were credited to Regine Goldscheider.

[97] Ibid

[98] Regina Goldscheider described the realization of funds through selling the branches as follows: "In this way some K 73 000 were realized from Berlin and from Paris K 140 000, together K 213 000, which already made it to some extent possible to approach paying off the debts to the tune of K 253 000. What now remained were debts amounting to K 40 000, despite which I also resolved to liquidate the three shares left to my sons Camillo, Walter and Marcell at K 36 000, amounting to K 108 000, through which I, in the face of overall financial obligations of K 148 000, had to assume an overall obligation of K 148 000 – assets consisting of unsaleable wares and materials – minus the loss of K 30 500 resulting mainly from realising the sale of the Berlin branch in 1899. [...] The lack of money was so pressing that my son Arthur in Paris had to pledge himself to procure for me the capital sum of K 36 000 needed for Marcell for the court deposit for 1 year." In this deposition, it is also mentioned that on 20 May 1899, the sum bequeathed in the estate amounted to Fl. 92 677 = 185 355 K. Cf on this the deposition made by Dr. Herrmann Fialla, Hof- und Gerichtsadvokat, representing the guardians of Marcell Goldscheider, to Währing District Court on 31 Mar. 1909.

[99] Deposition made by Dr. Herrmann Fialla, Hof- und Gerichtsadvokat, representing the guardianship of Marcell Goldscheider, to Bezirksgericht Währing, on 31 Mar. 1909

[100] Deposition made by Josef Schuster to the Viennese Magistracy, Division 11 Wertzuwachsabgabe [capital gains tax], on 21 Jan. 1939, relating to the Einantwortungsurkunde of the Handelsgericht Vienna of 8 Mar. 1921 A 188/18. The sum was converted by Josef Schuster in 1939 as follows: 55 000 FL = 110 000 Kronen; 1 Krone = 1.44 Schilling.

[101] Deposition made by Dr. Herrmann Fialla, Hof- und Gerichtsadvokat, representing the guardianship of Marcell Goldscheider, to Währing District Court on 31 Mar. 1909.

[102] Neuwirth (1974), p. 31

[103] In: Keramische Monatshefte, Halle 1901, p. 93

[104] Oesterreichische Patentschrift No. 3855, Classe 80: Thonwaren-, Stein- und Cementindustrie [pottery wares, stone and cement industry] of 25 April 1901. The patent expired in 1906.

[105] R., F.: Wiener Kunstkeramik, in: Keramische Monatshefte, 1902, p. 88f

[106] Ibid

[107] Ibid

[108] Ibid, p. 91

[109] Hecht/Barthel, p. 1584

[110] Ibid

[111] Ibid

[112] Neuwirth (1974), p. 31

[113] Central-Marken-Register [Central Trademark Register] für das Jahr 1900, Register No. 12167. The application was made on 18 Apr. 1900 for the Austro-Hungarian monarchy. The trademark was registered for the German Empire on 4 Dec. 1900 (cf Waarenzeichenblatt vom Kaiserlichen Patentamt, Berlin, [Heft] 1, January 1901).

[114] Suit for revoking the wardship of Regine Lewit and co-guardian Alois Goldscheider as guardians of Marcell Goldscheider, minor, and appointment of the plaintiff as guardian: letter from Advokat Dr. Emil Frischauer in the interest of Camillo Goldscheider to Währing District Court on 6 Nov. 1908. Regine and Dr. Lewit had lived at 20 Lannerstraße since 1901 (cf Lehmann's Adressbuch Wien, 1902). In 1903 Alois Goldscheider moved to 8 Cottagegasse, where he remained until he died (cf Lehmann's Adressbuch Wien, 1904 bis 1923).

[115] Pudor (1902a), p. 119

[116] N.N.: Von der ersten internationalen Ausstellung für dekorative Kunst in Turin 1902, in: Keramische Monatshefte, 1902, p. 162f

[117] Deposition made by Dr. Herrmann Fialla, Hof- und Gerichtsadvokat, representing the guardianship of Marcell Goldscheider, to Währing District Court on 31 Mar. 1909. From about 1904, Camillo lived at 14 Rieglergasse in the 18th District and is listed as agent in Lehmann's Adressbuch Wien, 1905. Walter is also listed as agent in 1905 and as resident at 5 Nussdorferstraße in the 9th District but there are no entries for 1906 and 1907, which suggests that Walter was abroad during those years. Walter is not listed again until 1908 as an agent and his private address is then given as 70 Schulgasse; a year later it is listed as 57 Gentzgasse (cf Lehmann's Adressbuch Wien, 1905 to 1909).

[118] Lawsuit brought by Camillo Goldscheider, represented by Dr. Emil Frischauer, for recognition as a partner [öffentlicher Gesellschafter] in the firm of Friedrich Goldscheider with a 20 per cent share before the k.k. Handelsgericht Wien, filed on 28 Oct. 1908.

[119] R., F.: Keramische Wanderungen durch den Kunstsalon von Keller & Reiner, in: Keramische Monatshefte, 1903, p. 125f. In 1902/1903 Goldscheider also showed work at the Pottery Exhibition [Keramische Ausstellung] in Reichenberg. Cf N.N.: Unsere keramische Ausstellung, in: Mitteilungen des nordböhmischen Gewerbemuseums, Reichenberg 1902, No. 4, p. 88

[120] Kiessig (1904), p. 761

[121] Eduard Leisching: Die Winterausstellung des k.k. Österr. Museums. In: Kunst und Kunsthandwerk, No. 7, 1904, p. 526f. This art journal published by the museum is a rich source of material for studies of Viennese ceramics since the museum documented its winter and spring exhibitions with illustrations and data on the artists. In Paris, the Catalogue illustré des Salons du Paris fulfilled a similar function. Other specialist journals for the decorative and applied arts of the time report on Goldscheider models, thus making the internationally renowned label even better known.

[122] Illustrierte Rundschau. In: Velhagen & Klasings Monatshefte, Vol. 17, 1902/03, p. 717

[123] Advertisement in: Keramische Rundschau, 5 Mar. 1903, p. 200

[124] Österreichisches Patentblatt, 1904, p. 499. The patent went into effect on 15 Mar. 1904 and was allocated the number 16882. The patent expired in 1907.

[125] Österreichisches Patentblatt, No. 2, 1905, p. 59. The patent went into effect on 21 Jan. 1904 under the number 20387 and expired in 1906.

[126] Deposition made by Josef Schuster before the Viennese Magistracy, Division 11 Wertzuwachsabgabe [capital gains tax] on 21 Jan. 1939, referring to the Einantwortungsurkunde of the Handelsgericht Wien of 8 Mar. 1921, A 188/18. Josef Schuster converted the currency as follows in 1939: 55 000 Guilders = 110 000 Kronen; 1 Krone = 1.44 Schilling

[127] Manuscript Heidi Brunnbauer "Im Cottage von Währing/Döbling," Vol. II, inter alia on the history of the house at 16 Hasenauerstraße.

[128] N.N.: Figürliche Terrakotten, in: Keramische Monatshefte, 1904, p. 138

[129] Adress-Buch der Keram-Industrie, 1904, p. 60. The date of establishment of the firm is given here as 1882, which also tallies with the authors' findings. The Goldscheider manufactory, however, gave the date as 1885 in firm documents and advertising campaigns.

[130] Advertisement in: Keramische Rundschau, 1904, p. 158

[131] Pazaurek (1904), p. 90

[132] N.N.: Leipziger Michaelismesse 1904, in: Keramische Rundschau, Nr. 39, 22.9.1904

[133] Leipziger Michaelis-Messe 1904, in: Centralblatt 1904, p. 1294, quoted in Neuwirth (1974), p. 32

[136] Neuwirth (1974), S. 53

[137] Nach Marcel Goldscheiders Schilderung der Geschichte der Manufaktur Goldscheider (An authentic report of its origin and development), Beigabe zu einem Brief an die Kunden bzgl. eines Umzugs in Hanley, vermutlich aus dem Jahr 1951. „After the death of Fredrick Goldscheider his wife took over until her death in 1918 and introduced a new type of glazed ware, which was called ‚Fayence', but made in a pure white porous body, decorated underglaze with liquid colours, and carrying a transparent glaze. Both these and the Terracotta figures which were made up to a life-size found a large sale everywhere."

[138] Ebenda „It is perhaps interesting to learn that Goldscheider also made Crinoline Figures (around 1906), but the public of those days preferred movement in figures to the cut-off feet and the monotony connected with that type."

[139] Heller, S. 102

[140] Klage von Camillo Goldscheider, vertreten durch Dr. Emil Frischauer um Aufnahme als öffentlicher Gesellschafter in die Firma Friedrich Goldscheider mit einer Beteiligung von 20 Prozenten, beim k.k. Handelsgericht Wien, eingelangt am 28.10.1908.

[141] Ebenda

[142] Ebenda. Laut des Registerblatts I für Einzel- und Gesellschaftsfirmen, Wirtschaftskammer Österreich-Archiv, Gewerbearchiv, war die Hochzeit mit Jolanthe am 20.3.1906.

[143] Stellungnahme von Dr. Herrmann Fialla, Hof- und Gerichtsadvokat, als Vertreter der Vormundschaft von Marcell Goldscheider, an das Bezirksgericht Währing, vom 31.3.1909.

[144] Camillo wohnt bis 1906 in der Rieglergasse 14, dann in der Gentzgasse 9. Im Jahr 1907 in der Martinsstraße 91 und um 1908 in der Gymnasiumsstraße 32 (vgl. Lehmann's Adressbuch Wien, 1905 bis 1909).

[145] Klage von Camillo Goldscheider, vertreten durch Dr. Emil Frischauer um Aufnahme als öffentlicher Gesellschafter in die Firma Friedrich Goldscheider mit einer Beteiligung von 20 Prozenten, beim k.k. Handelsgericht Wien, eingelangt am 28.10.1908.

[146] Antrag auf Enthebung der Vormünderin Frau Regine Lewit und des Mitvormundes Alois Goldscheider von der Vormundschaft des mj. Marcell Goldscheider und Bestellung des Einschreiters zum Vormunde, Brief von Advokat Dr. Emil Frischauer, im Auftrag Camillo Goldscheiders, an das Bezirksgericht Währing vom 6.11.1908.

[147] Stellungnahme von Dr. Herrmann Fialla, Hof- und Gerichtsadvokat, als Vertreter der Vormundschaft von Marcell Goldscheider, an das Bezirksgericht Währing, vom 31.3.1909.

[148] Registerblatt I für Einzel- und Gesellschaftsfirmen, Wirtschaftskammer Österreich-Archiv, Gewerbearchiv. Die Schreibweise von Fayence variiert in dieser Zeit, mal heißt es „Fajence", dann wieder „Fayenze" oder wie üblich „Fayence".

[149] Bük (1908), S. 582

[150] Neuwirth (1973), S. 57

[151] Reklame der Firma „Friedrich Goldscheider Kunstterrakotta-Fabrik Ateliers künstlerischer Fayencen" in Leuchs Adressbuch 1908

[152] Blom's Engros- und Export-Adressbuch, 1907/8

[153] N.N.: Aus Berliner Musterlagern, in: Sprechsaal, No. 40, 1907, S. 684. Auf der internationalen kunstgewerblichen Ausstellung in Petersburg im Jahr 1908 werden die historistischen Goldscheider-Modelle, vor allem ein Teilrelief mit einer germanischen Walküre mit ihrem Ross, eine Reiterstatue Napoleons I. sowie die „graziösen Kinder- und Frauengruppen" von „Kennern bewundert". Vgl. H., H.: Internationale kunstgewerbliche Ausstellung in Petersburg 1908. Keramik und Glas, in: Keramische Monatshefte, 1908, S. 168

[154] vgl. KeramRS 17.1909, S. 177

[155] Österreichisches Patentblatt, 1911, S. 91. Das Patent wird erst am 15.1.1911 erteilt und erhält die Nr. 49972. Aufgrund von unterlassener Gebührenzahlung erlischt es im Jahr 1914.

[156] Oesterreichische Patentschrift Nr. 49972, Kais. Königl. Patentamt, 25.9.1911

[157] N.N.: Aus Berliner Musterlagern 1909, in: Sprechsaal, 42.1909, S. 308

[158] Handelsregister zur Firma „Erste Wiener Terrakottafabrik und Atelier für künstlerische Fayenzen Friedrich Goldscheider" (Bestand Handelsgericht Wien, Handelsregister A 16/19). Das Unternehmen ist seit dem 1. Januar 1910 als offene Handelsgesellschaft eingetragen, Walter wird dort als Kaufmann geführt. Vgl. Keramische Rundschau, Nr. 38, 1910, S. 441

[159] Ebenda

[160] Jahrbuch der österreichischen Industrie, 1910, S. 179

[161] Zählblatt J9.6 der Handels- und Gewerbekammer in Wien, vom 7.3.1911, Wirtschaftskammer Österreich-Archiv, Gewerbearchiv

[162] Siehe Planer. Erwin ist laut Geburts-Zeugnis am 11.7.1912 in Wien geboren.

[163] N.N.: Kunstkeramik, in: DER, 11.1911/12, S. 566

[164] N.N.: Illustrierte Rundschau über die Leipziger Ostervormesse 1912, in: DER, 12.1912/1913, S. 112

[165] N.N: Die Leipziger Herbstmesse 1912, in: DER, 12.1912/13, S. 674

[166] Österreichisches Patentblatt, Nr. 5, 1913, S. 225

[167] PGGTR, June 1941, S. 486

[168] Adress-Buch der Keram-Industrie, Juni 1913, S. 479

[169] N.N.: Goldene Medaille auf der Internationalen Baufach-Ausstellung in Leipzig, in: KeramRS, 21.1913, S. 464

[170] N.N.: Die Leipziger Messe, Ostern 1913, in: DER, 13.1913/14, S. 83

[171] Anzeige in Heimatkalender für Thüringen und das Osterland 1914, Ernst Mühlbach, 1913, S. 216

[172] Undatiertes Expose „Entstehung und Werdegang der Wiener Manufaktur Friedrich Goldscheider", wahrscheinlich geschrieben von der Arbeitsgemeinschaft unter Josef Paul um 1946.

[173] Brief von Walter Goldscheider an die Vermögensverkehrsstelle, Wien, 4.8.1938

[174] Handelsregister zur Firma „Erste Wiener Terrakottafabrik und Atelier für künstlerische Fayenzen Friedrich Goldscheider" (Bestand Handelsgericht Wien, Handelsregister A 16/19)

[175] KeramRS, 25.1917, S. 29

[176] Handelsregister zur Firma „Erste Wiener Terrakottafabrik und Atelier für künstlerische Fayenzen Friedrich Goldscheider" (Bestand Handelsgericht Wien, Handelsregister A 16/19)

[177] Inschrift auf Grabstein der Familie Goldscheider, Friedhof in Döbling I-1/1, Gruft 22

[178] Vgl. Lehmann's Adressbuch 1913. Die Hochschulstraße wird ab ca. 1934 in Gregor Mendel Straße 58 umbenannt.

[179] Registerblatt I für Einzel- und Gesellschaftsfirmen,Wirtschaftskammer Österreich-Archiv, Gewerbearchiv

[180] Ebenda

[181] Ebenda. In Lehmann's Adressbuch von Wien aus dem Jahr 1910 lässt Camillo seinen Namen hervorheben und um seine Auszeichnung aus Bukarest „Bes. d. rum. Inh. Med. Carol I" ergänzen.

[182] Jahrbuch der österreichischen Industrie, 1910, S. 179. In der Gymnasiumstraße 32 im XVII. Bezirk hat er seine Wohnung (vgl. „High Life" Almanach der Österr. Gesellschaft, 5. Jg 1909, Volkswirtschaftlicher Verlag), später ist er mit Wohnsitz Stiftgasse 21 gelistet (vgl. „High Life" Almanach der Österr. Gesellschaft, 7. Jg 1911, Volkswirtschaftlicher Verlag).

[183] ZMA 1909, S. 225. Die Anmeldung erfolgte bereits am 6.3.1909.

[184] N.N.: Moderne Plastik, in: DER 9.1909/10, S. 596

[185] Dal Vecchio: Dekorative Plastik, in: IZB 24.1909, S. 3601

[186] Ebenda

[187] Ebenda

[188] Laut des Registerblatts I für Einzel- und Gesellschaftsfirmen,Wirtschaftskammer Österreich-Archiv, Gewerbearchiv, war die Hochzeit mit Jolanthe am 20.3.1906. Die Ehe wurde am 4.5.1909 geschieden.

[189] N.N: Leipziger Ostervormesse 1910, in: DER 9.1909/10, S. 1030

[190] Registerblatt I für Einzel- und Gesellschaftsfirmen, Wirtschaftskammer Österreich-Archiv, Gewerbearchiv

[191] vgl. Lehmann's Adressbuch Wien, Jahrgänge 1911 bis 1917

[134] Advertisement in: Keramische Rundschau, Wegweiser für die Leipziger Michaelismesse 1904, p. 798. A year later, terracotta with faience inlay create as a "copyright novelty" an "outstanding effect" (cf N.N.: Michaelismesse 1905, in: Keramische Rundschau, No. 37, 14 Sept. 1905, p. 915f.)
[135] Fayence und Majolika. In: Keramische Rundschau, 1905, p. 735
[136] Neuwirth (1974), p. 53
[137] According to Marcel Goldscheider's description of the history of the Goldscheider manufactory (An authentic report of its origin and development), enclosure in a letter to clients relating to a move in Hanley, presumably dating from 1951. "After the death of Fredrick Goldscheider his wife took over until her death in 1918 and introduced a new type of glazed ware, which was called 'Fayence,' but made in a pure white porous body, decorated underglaze with liquid colors, and carrying a transparent glaze. Both these and the Terracotta figures which were made up to a life-size found a large sale everywhere."
[138] Ibid. "It is perhaps interesting to learn that Goldscheider also made Crinoline Figures (around 1906), but the public of those days preferred movement in figures to the cut-off feet and the monotony connected with that type."
[139] Heller, p. 102
[140] Suit brought by Camillo Goldscheider, represented by Dr. Emil Frischauer for recognition as öffentlicher Gesellschafter [partner] in the firm of Friedrich Goldscheider with a 20 per cent share of proceeds, before the k.k. Handelsgericht Wien, filed on 28 Oct. 1908.
[141] Ibid
[142] Ibid. According to Registerblatt I für Einzel- und Gesellschaftsfirmen, Wirtschaftskammer Österreich-Archiv, Gewerbearchiv, the marriage to Jolanthe took place on 20 Mar. 1906.
[143] Deposition made by Dr. Herrmann Fialla, Hof- und Gerichtsadvokat, representing the guardians of Marcell Goldscheider, to Währing District Court, on 31 Mar. 1909.
[144] Until 1906, Camillo lived at 14 Rieglergasse, then at 9 Gentzgasse, in 1907 at 91 Martinsstraße and ca 1908 at 32 Gymnasiumsstraße (cf Lehmann's Adressbuch Wien, 1905 to 1909).
[145] Lawsuit brought by Camillo Goldscheider, represented by Dr. Emil Frischauer, for official recognition as a partner in the firm of Friedrich Goldscheider with a 20 per cent share in the proceeds, before the k.k. Handelsgericht Wien, filed on 28 Oct.1908.
[146] Suit for revoking the wardship of Regine Lewit and co-guardian Alois Goldscheider as guardians of Marcell Goldscheider, minor, and appointment of the plaintiff as guardian: letter from Advokat Dr. Emil Frischauer in the interest of Camillo Goldscheider to Währing District Court on 6 Nov. 1908.
[147] Deposition made by Dr. Herrmann Fialla, Hof- und Gerichtsadvokat, representing the guardians of Marcell Goldscheider, to Währing District Court on 31 Mar 1909.
[148] Registerblatt I für Einzel- und Gesellschaftsfirmen, Wirtschaftskammer Österreich-Archiv, Gewerbearchiv. At that time the spelling of "Fayence" varied; sometimes it was written "Fajence," then again "Fayenze" or – the usual form – "Fayence."
[149] Bük (1908), p. 582
[150] Neuwirth (1973), p. 57
[151] Advertisement of the firm: "Friedrich Goldscheider Kunstterrakotta-Fabrik Ateliers künstlerischer Fayencen" in Leuchs Adressbuch 1908
[152] Blom's Engros- und Export-Adressbuch, 1907/8
[153] N.N.: Aus Berliner Musterlagern, in: Sprechsaal, No. 40, 1907, p. 684. At the 1908 international Arts and Crafts trade fair in St Petersburg, the historical Goldscheider models, especially a part relief with a Teutonic valkyrie and her steed, an equestrian statue of Napoleon I as well as "graceful groups of children and women," were "admired by connoisseurs." Cf H., H.: Internationale kunstgewerbliche Ausstellung in Petersburg 1908. Keramik und Glas, in: Keramische Monatshefte, 1908, p. 168
[154] Cf KeramRS 17.1909, p. 177
[155] Österreichisches Patentblatt, 1911, p. 91. The patent was not granted until 15 Jan. 1911 and was given the number 49972. It expired in 1914 because the fees were not paid.
[156] Oesterreichische Patentschrift No. 49972, Kais. Königl. Patentamt, 25 Sept. 1911
[157] N.N.: Aus Berliner Musterlagern 1909, in: Sprechsaal, 42.1909, p. 308
[158] Handelsregister zur Firma "Erste Wiener Terrakottafabrik und Atelier für künstlerische Fayenzen Friedrich Goldscheider" (Bestand Handelsgericht Wien, Handelsregister A 16/19). The firm had been registered as "offene Handelsgesellschaft" [OHG: partnership] since 1 Januar 1910; Walter is listed there as a businessman. Cf Keramische Rundschau, No. 38, 1910, p. 441
[159] Ibid
[160] Jahrbuch der österreichischen Industrie, 1910, p. 179
[161] Zählblatt J9.6 der Handels- und Gewerbekammer in Wien, of 7 Mar. 1911, Wirtschaftskammer Österreich-Archiv, Gewerbearchiv
[162] See Planer. According to his birth certificate, Erwin was born in Vienna on 11 July 1912.
[163] N.N.: Kunstkeramik, in: DER, 11.1911/12, p. 566
[164] N.N.: Illustrierte Rundschau über die Leipziger Ostervormesse 1912, in: DER, 12.1912/1913, p. 112
[165] N.N: Die Leipziger Herbstmesse 1912, in: DER, 12.1912/13, p. 674
[166] Österreichisches Patentblatt, No. 5, 1913, p. 225
[167] PGGTR, June 1941, p. 486
[168] Adress-Buch der Keram-Industrie, Juni 1913, p. 479
[169] N.N.: Goldene Medaille auf der Internationalen Baufach-Ausstellung in Leipzig, in: KeramRS, 21.1913, p. 464
[170] N.N.: Die Leipziger Messe, Ostern 1913, in: DER, 13.1913/14, p. 83
[171] Advertisement in: Heimatkalender für Thüringen und das Osterland 1914, Ernst Mühlbach, 1913, p. 216
[172] Undated exposé "Entstehung und Werdegang der Wiener Manufaktur Friedrich Goldscheider," probably written by the work group under Josef Paul ca 1946.
[173] Letter from Walter Goldscheider to the Vermögensverkehrsstelle, Vienna, 4 Aug. 1938
[174] Handelsregister zur Firma "Erste Wiener Terrakottafabrik und Atelier für künstlerische Fayenzen Friedrich Goldscheider" (Bestand Handelsgericht Wien, Handelsregister A 16/19)
[175] KeramRS, 25.1917, p. 29
[176] Handelsregister zur Firma "Erste Wiener Terrakottafabrik und Atelier für künstlerische Fayenzen Friedrich Goldscheider" (Bestand Handelsgericht Wien, Handelsregister A 16/19)
[177] Inscription on the Goldscheider family gravestone, cemetery in Döbling I-1/1, vault 22
[178] Cf Lehmann's Adressbuch 1913. Around 1934, the Hochschulstraße was renamed Gregor Mendel Straße 58.
[179] Registerblatt I für Einzel- und Gesellschaftsfirmen,Wirtschaftskammer Österreich-Archiv, Gewerbearchiv
[180] Ibid
[181] Ibid. In Lehmann's Adressbuch Wien for the year 1910, Camillo has his name emphasized and his award from Bucharest added to it "Bes. d. rum. Inh. Med. Carol I."
[182] Jahrbuch der österreichischen Industrie, 1910, p. 179. He lived at 32 Gymnasiumstraße in the XVIIth District ("High Life" Almanach der Österr. Gesellschaft, 5. [Jg] 1909, Volkswirtschaftlicher Verlag); later he is listed as living at 21 Stiftgasse (cf "High Life" Almanach der Österr. Gesellschaft, 7. [Jg] 1911, Volkswirtschaftlicher Verlag).
[183] ZMA 1909, p. 225. The application had been made on 6 Mar. 1909.
[184] N.N.: Moderne Plastik, in: DER 9.1909/10, p. 596
[185] Dal Vecchio: Dekorative Plastik, in: IZB 24.1909, p. 3601
[186] Ibid
[187] Ibid
[188] According to Registerblatt I für Einzel- und Gesellschaftsfirmen, Wirtschaftskammer Österreich-Archiv, Gewerbearchiv, the marriage to Jolanthe took place on 20 Mar. 1906. The divorce was granted on 4 May 1909.
[189] N.N: Leipziger Ostervormesse 1910, in: DER 9.1909/10, p. 1030
[190] Registerblatt I für Einzel- und Gesellschaftsfirmen, Wirtschaftskammer Österreich-Archiv, Gewerbearchiv
[191] Cf Lehmann's Adressbuch Wien, 1911 to 1917

Art Déco Meisterstücke der Brüder Walter und Marcell Goldscheider (1918–1928)

Art Déco Masterpieces by the Brothers Walter and Marcell Goldscheider (1918–1928)

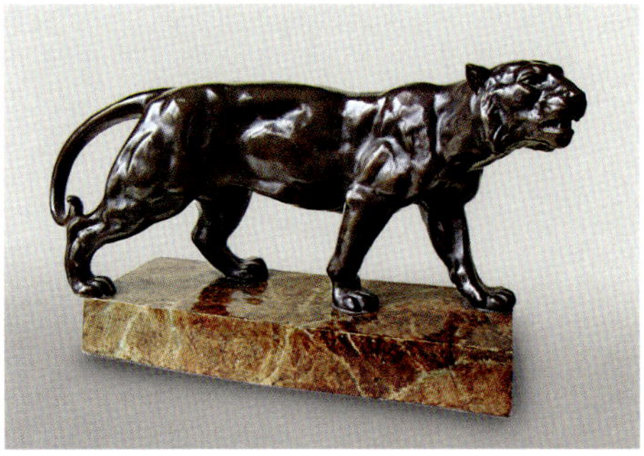

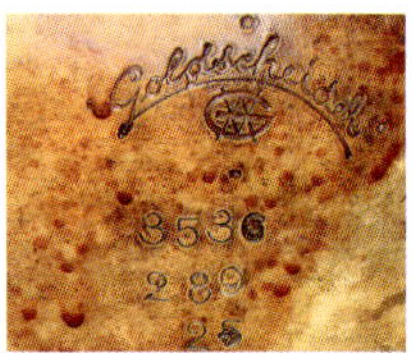

ModNr. 3536
Tiger, Adolphe-Jean Lavergne, Entwurf | design ca. 1907, Ausführung 20er Jahre | executed 1920s, Collection f@goldscheider.de

Die neue Firmenmarke nimmt den seit 1921 eingetragenen Firmennamen „Wiener Manufaktur Friedrich Goldscheider" auf und zeigt das Monogram GWM. Bereits in den Jahren seit Ende des Ersten Weltkrieges wurde ein Stempel benutzt, der die Buchstaben „W" und „M" für „Wiener Manufaktur" innerhalb des Firmenlogos zeigt.

The new firm mark subsumes the firm name registered in 1921, "Wiener Manufaktur Friedrich Goldscheider," and bears the monogram "GWM." By the post-war years, a stamp was in use that imprinted the letters "W" and "M" standing for "Wiener Manufaktur" within the firm logo.

Nach dem Tod ihrer Mutter wollen Walter und Marcell an die erfolgreichen Zeiten vor dem Krieg anknüpfen. Am 1. März 1920 werden beide Gesellschafter der Leipziger Niederlassung, die offene Handelsgesellschaft wird umgeschrieben und Alois Goldscheider als Prokurist ausgetragen.[192] Am 30. Dezember 1920 werden die Brüder im Handelsregister als Gesellschafter des Wiener Unternehmens eingetragen.[193] Walter, der außer deutsch perfekt französisch, englisch, italienisch und spanisch spricht, ist für die kaufmännische Seite verantwortlich, Marcel übernimmt die künstlerische und technische Leitung.[194] Am 5. Juli 1921 wird die Firma offiziell in „Wiener Manufaktur Friedrich Goldscheider" umbenannt, mit dem Schwerpunkt der fabrikmäßigen „Erzeugung von Ton- und plastischen Gegenständen".[195] Die Firmenbezeichnung geht nun endgültig in die neue Firmenmarke ein, die unterhalb des Schriftzugs „Goldscheider" das Monogram GWM zeigt. Die Buchstaben „W" und „M" stehen für „Wiener Manufaktur", das „G" für „Goldscheider".

After their mother's death, Walter and Marcell wanted to link up with the successful days before the First World War. On 1 March 1920 the two brothers became partners in the Leipzig branch. The OHG was transferred to them and Alois Goldscheider's powers of attorney expired.[192] The brothers were entered in the commercial register as partners in the Viennese business on 30 December 1920.[193] Walter, who, besides German, also spoke French, English, Italian and Spanish perfectly, was responsible for the business side while Marcel became artistic and technical director.[194] On 5 July 1921 the firm was officially renamed "Wiener Manufaktur Friedrich Goldscheider," specializing mainly in industrial "manufacture of pottery and sculptural objects".[195] The firm name was from then on fused with the new firm mark to bear the monogram "GWM" below "Goldscheider" in lettering. The letters "W" and "M" stand for "Wiener Manufaktur," "G" for "Goldscheider."

ModNr. 4892
„Faun mit Schnecke" | "Faun with Snail", Josef Kostial / Sigismund Wernekinck, ca. 1918, beide | both PB Wien 19

Auch bei diesem beliebten Sujet der spielenden Faune wurden versehentlich zwei unterschiedliche Künstlersignaturen eingesetzt.

Two different artists' signatures were also erroneously put on this popular motif of fauns playing.

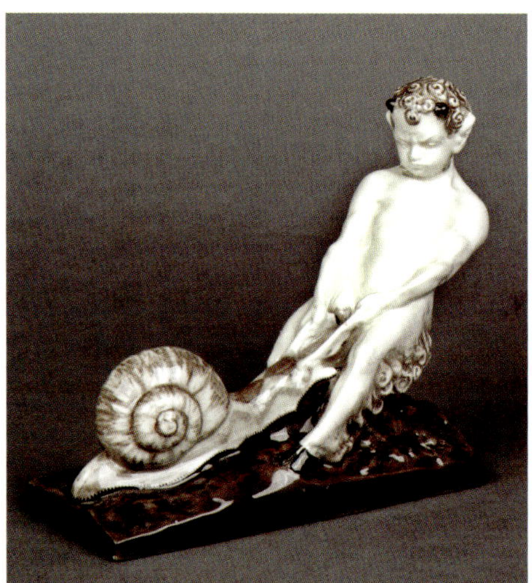

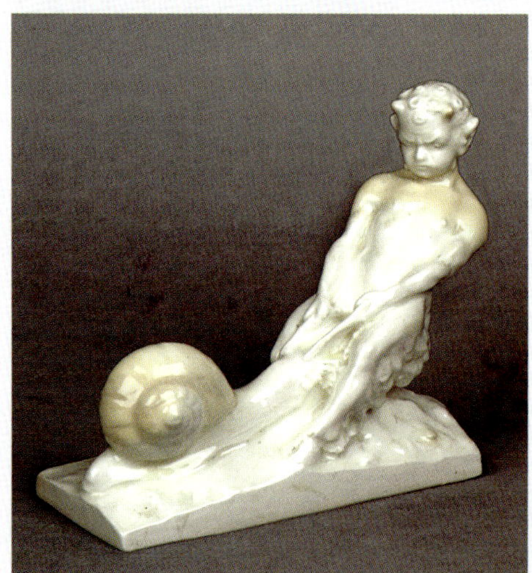

The factory remained at 7 and 9 Staudgasse in the XVIIIth District. From 1924 until ca 1927, an additional factory for the manufacture of terracottas was established at 5 Plenergasse in the same district.[196] Alois Goldscheider's powers of attorney expired on 21 March 1922[197] and thenceforward the brothers decided on the way the business would develop. Walter Goldscheider was even made a Kommerzialrat in 1922, an honorary title that indicated recognition of his status as an important Austrian entrepreneur.[198]

In the early 1920s, Goldscheider, like Rosenthal and Hertwig, launched a popular line in putti (Cupids) with animals that stood for "simple sensuous pleasure."[199] Josef Kostial, for instance, designed the models called

Die Fabrik befindet sich weiterhin in der Staudgasse 7 und 9 im XVIII. Bezirk, von 1924 bis um 1927 wird in der Plenergasse 5 im gleichen Bezirk eine zusätzliche Fabrikation von Terrakotten eingerichtet.[196] Die Prokura von Alois Goldscheider erlischt am 21. März 1922[197] und die Brüder bestimmen allein über die Entwicklung des Unternehmens. Walter Goldscheider wird 1922 sogar zum Kommerzialrat ernannt, ein Zeichen seiner Anerkennung als bedeutender österreichischer Unternehmer.[198]

Anfang der zwanziger Jahre entstehen bei Goldscheider, wie bei Rosenthal oder Hertwig, die beliebten Putten mit Tieren, die für „einfache Sinnenfreude" stehen.[199] Josef Kostial entwirft beispielsweise die Modelle „Duett" und „Faun mit Schnecke".

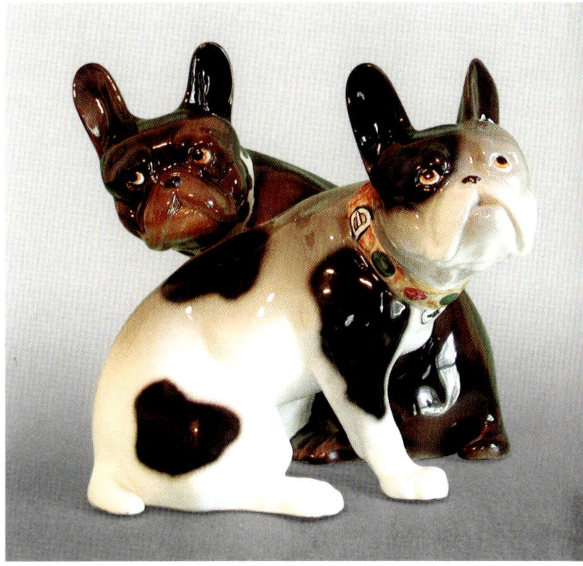

ModNr. 4997 Französische Bulldoggen I French Bulldogs, Friedrich Herkner, ca. 1921, Collection Gary Bachman, USA
Diese Hunde wurden unter derselben Modell-Nummer in zwei verschiedenen Ausführungen angeboten. Einerseits das linke Einzeltier als separate Figur, andererseits das Paar als Gruppe. Ansonsten wurde bei Goldscheider in solchen Fällen für jede Variante eher eine eigene Modellnummer vergeben.

These dogs were marketed in two different versions under the same model number. On the one hand, the individual dog as a separate figure on the left, on the other, as a pair. Otherwise Goldscheider tended in such cases to use a different model number for each variant.

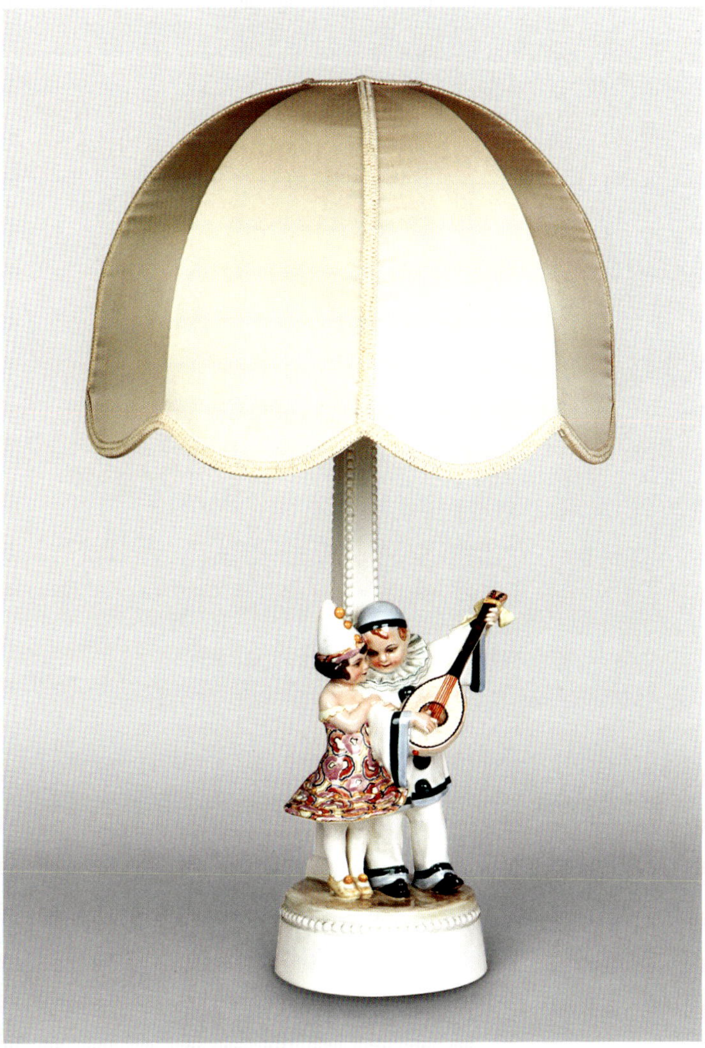

ModNr. 5057
Lampe I Lamp Pierrot und I and Pierrette, Josef Lorenzl, um 1922, PB Wien 19

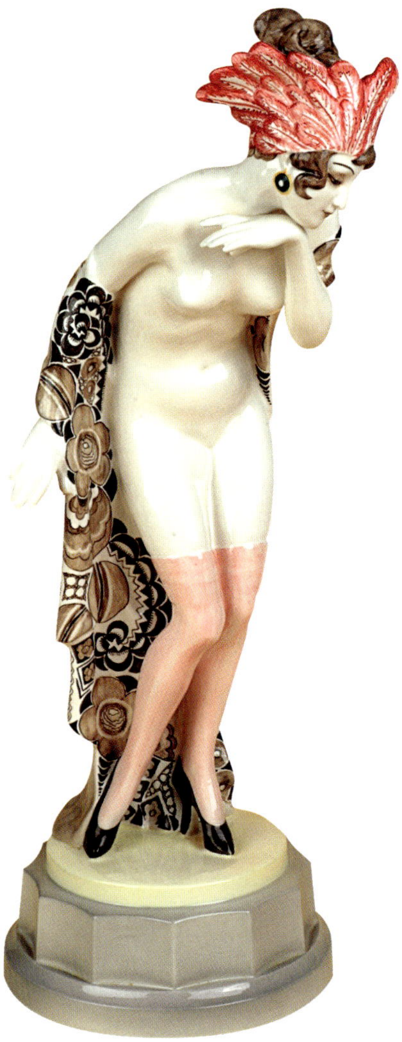

ModNr. 5060
„Faszination" I
"Fascination",
Wilhelm Thomasch,
ca. 1922, PB Wien 19

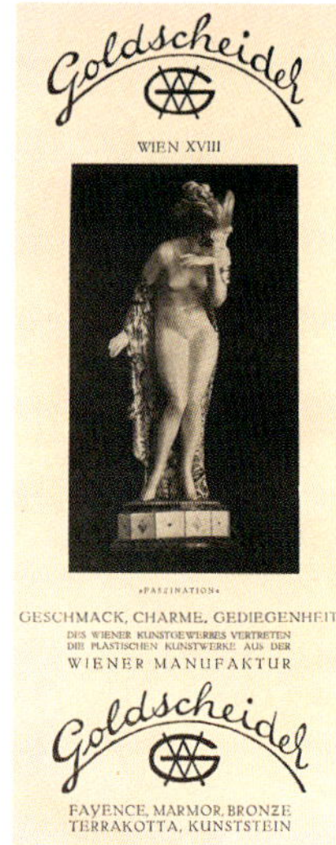

Ill. 44
Diese Anzeige aus der Zeitschrift Die Dame zeigt Ende 1922 das neue Goldscheider Logo sowie eine faszinierende Figur des Künstlers Thomasch mit der Modell-Nummer 5060. (Quelle: Die Dame, November 1922)

This advertisement from Die Dame magazine in late 1922 features the new Goldscheider logo and a figure called "Fascination" created by Thomasch that bore the model number 5060. (Source: Die Dame, November 1922)

Zur Deutschen Gewerbeschau für Keramik und Glas in München werden Modelle wie die schwarz-weiße Keramik „Maskerade" von Kostial oder „Velasquez" von Johanna Meier geschickt.[200] Man greift aber auch auf Entwürfe aus den 10er Jahren zurück. Es werden weiterhin Figuren von Mädchen und Kindern in Biedermeiertracht produziert, genauso wie zahlreiche Tiermodelle.

Und auch die von den Brüdern wieder verstärkt eingesetzten Bronzen, die eigentlich die „reinste Wiedergabe des meist in Ton oder Wachs geschaffenen Modells" gestatten, werden ins Programm genommen und von der Kritik wohlwollend beurteilt: „Auch bei dieser Art von Wiedergabe mittel- und kleinplastischer Kunstwerke vermissen wir angenehm alles Fabrikmäßige so sehr, daß ein liebevoller Beschauer aus jedem Stück die Individualität des schaffenden Künstlers wie die des ausführenden Kunsthandwerkers herauszufühlen vermeint."[201] Trotz der seriellen Produktion gilt bei Goldscheider seit jeher der hohe Anspruch, den individuellen, schöpferischen Charakter eines Kunstwerks zu erhalten. Künstlerisch sieht man sich in Konkurrenz zur Wiener Werkstätte und anderen auf Originalwerke bzw. Unikate

"Duet" and "Faun with Snail". Kostial models such as the black-and-white ceramic object "Masquerade" and "Velasquez" by Johanna Meier were sent to the Deutsche Gewerbeschau für Keramik und Glas in Munich.[200] However, there was a reversion to designs from the first decade of the century. Figures of girls and children in Biedermeier costume as well as numerous animal models continued in production.

And, the bronzes, which the brothers were making in larger numbers again because they permitted "the purest replication of the model, which was usually made in pottery or wax," figured once again in the program. They were well received by critics: "In this type of replication of medium-size and small sculptural art work, we are so pleasantly surprised to find no trace of industrialized manufacture; an observer dwelling on each of these pieces might feel he could make out the individual artist who created it as well as the craftsman who executed it."[201] Although the pieces were mass-produced, at Goldscheider the lofty aim of preserving the individual, creative character of a work of art had prevailed from the outset. On the artistic plane, Goldscheider viewed itself as competitive with the Wiener Werkstätte and other studios specializing in original works or one-offs. Whereas many of those studios were crisis-prone commercially, the Wiener Manufaktur Friedrich Goldscheider succeeded in consolidating its niche on the world market with artistically ambitious models that could be produced economically.

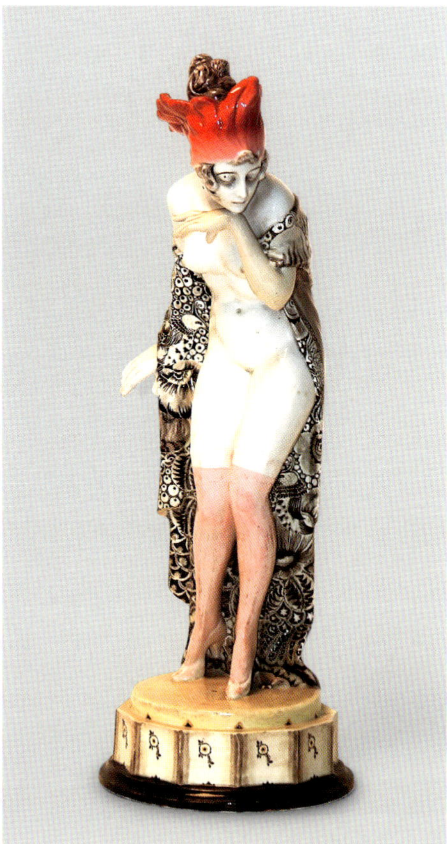

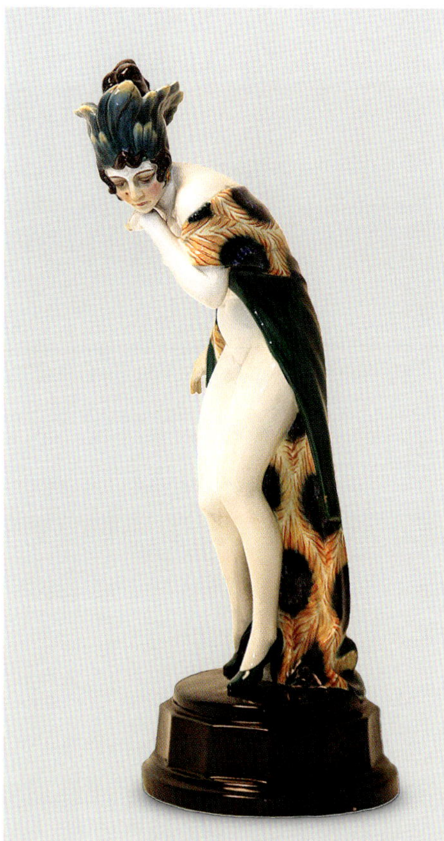

beide I both:
ModNr. 5060
„Faszination" I
"Fascination",
Wilhelm Thomasch,
ca. 1922, private collection, Wien

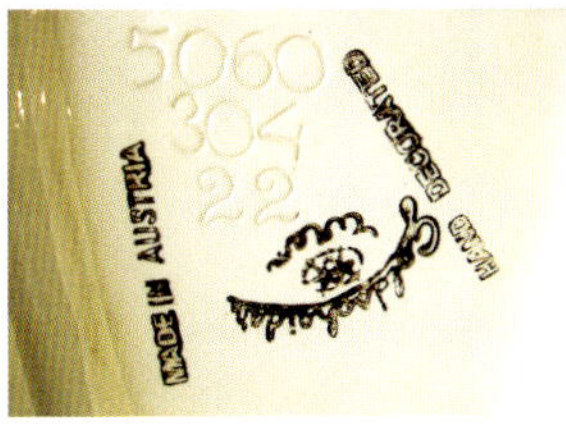

The discovery of Tutankhamen's tomb in 1922 made the Egyptianizing style popular in art. Fashion magazines, newspapers and the cinema created public awareness for this fashion and trend. Typical of it is a bold use of color, geometric forms and an attractively exotic opulence that represented a backlash against both Art Nouveau and the horrors and privation of the First World War. Extravagance and luxury informed the Zeitgeist; decorative figurative room ornaments brought the glamour of sophistication and high society into the home (Ill. 44).

spezialisierte Ateliers. Während viele dieser Ateliers in eine wirtschaftliche Krise steuern, kann sich die Wiener Manufaktur Friedrich Goldscheider mit künstlerisch anspruchsvollen und dennoch wirtschaftlich reproduzierbaren Modellen erfolgreich am Weltmarkt durchsetzen.

Mit der Entdeckung des Grabs von Tutanchamun im Jahr 1922 wird der ägyptische Kunststil populär. Modezeitschriften, Zeitungen und das Cinema kreieren ein Bewusstsein für Mode und Trend. Typisch ist der gewagte Einsatz von Farbe, geometrischen Formen und einer reizvollen exotischen Fülle, die auch als Reaktion auf Art Nouveau und die Erlebnisse des Ersten Weltkriegs zurückzuführen ist. Extravaganz und Luxus prägen den Zeitgeist, dekorativ figürlicher Raumschmuck bringt den Glanz der mondänen Welt in die eigenen vier Wände (Ill. 44).

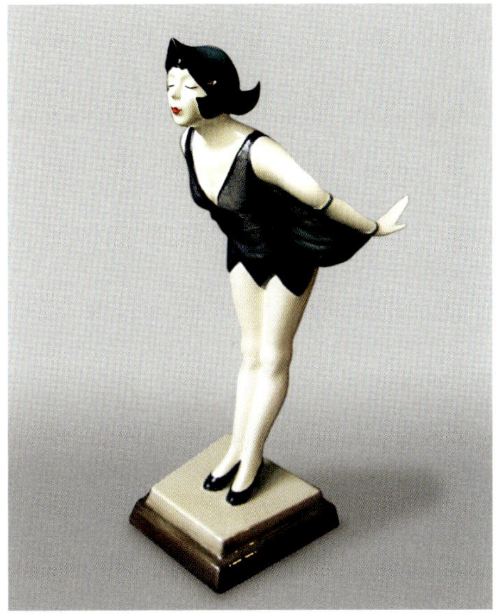

ModNr. 5060
Stephan Dakon,
ca. 1928, Collection
f@goldscheider.de
Unter der gleichen Modell-Nummer 5060 (S. 86) erscheint auch die von Stephan Dakon signierte Figur dieser Tänzerin, die eine der beiden Dolly Sisters darstellt (vgl. auch Modell 5712). Da Dakon erst ab 1924 bei Goldscheider beschäftigt ist, kann dieses Modell auf ca. 1928 datiert werden und entspricht der Modell-Nummer 5781. Es handelt sich um eine der zahlreichen Doppelbelegungen für Modell-Nummern, die aufgrund der enormen Produktionsvielfalt bei Goldscheider zu Stande kamen.

Signed by Stephan Dakon, this figure of a dancer representing one of the two Dolly Sisters went under the same model number, 5060 (see p. 86; also cf model 5712). However, since Dakon was not employed by Goldscheider until 1924, this model must be dated to around 1928 and corresponds to model number 5781. This is one of many examples of numbers being used for more than one model, a slip-up that occurred because the Goldscheider output was so diverse.

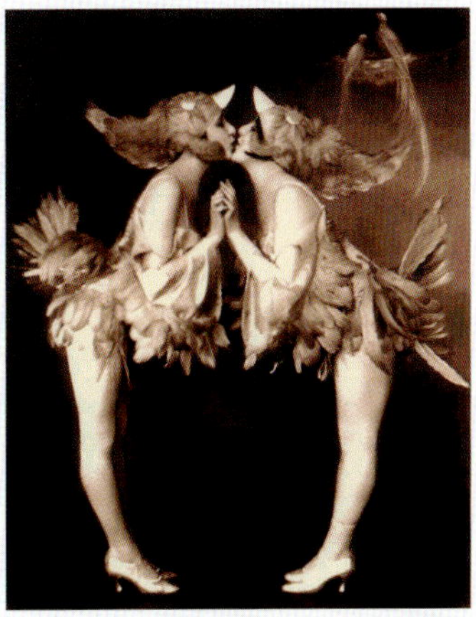

Ill. 45

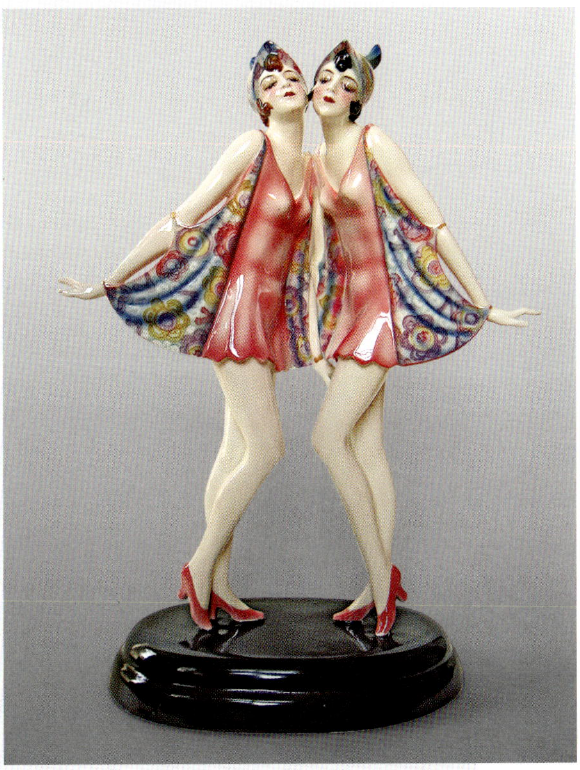

ModNr. 5712
„Turteltauben" I
"Turtle Doves",
Stephan Dakon,
ca. 1927, Collection
Weilinger, Wien
Die Dolly Sisters in einer typischen Pose aus ihren Revuetänzen. Vgl. Ill. 45 (Postkarte aus PB Deutschland)

The Dolly Sisters in a pose typical of their revue numbers. Cf Ill. 45 (postcard from PB Deutschland)

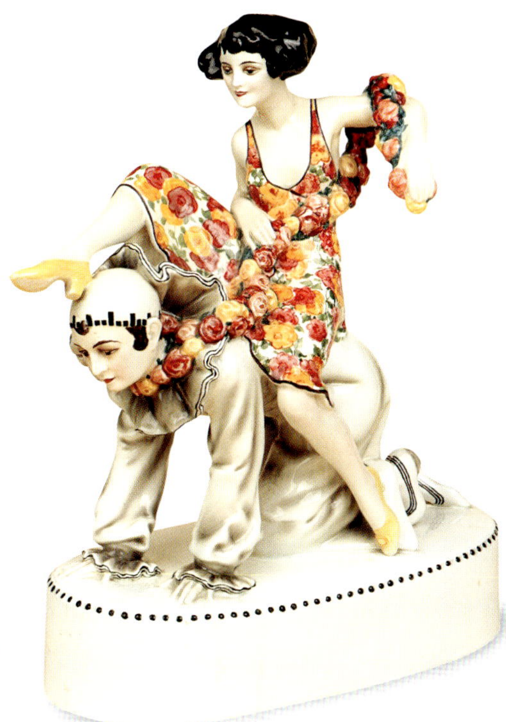

ModNr. 5214
Gruppe I Group Pierrot
und I and Pierrette,
Wilhelm Thomasch,
ca. 1923, PB Wien 19

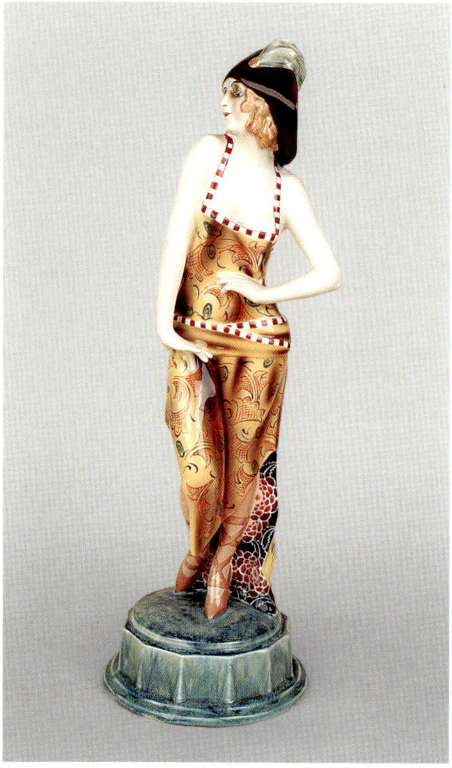

ModNr. 5178
Tänzerin I Danseuse,
Josef Lorenzl, ca. 1923,
PB Wien 19

Marcell verfeinert die Methode der Unterglasur-Dekoration und führt einen neuen niedrig brennenden Scherben ein. Dadurch kann er sukzessive die eher blassen Farben des „grand feu"-Typs mit neuen, vielseitigen und lebendigen Unterglasur-Farben ersetzen. Mit diesem neuen Dekorationsstil, der durch den Einsatz einer Spritztechnik mit Aerographen erfolgt, hebt sich die Goldscheider Manufaktur erfolgreich von der Konkurrenz ab, die diese Technik bei Weitem nicht so gut beherrschen.[202] Am 3. März 1923 wird von Goldscheider ein Patent (Ill. 46) beantragt, das ein Verfahren zur Erzeugung farbiger Muster durch Aufspritzen schützen soll.[203] Bislang wurden farbig gemusterte Waren, vor allem Fayence, zumeist „durch Aufspritzen von Farbe mittels eines Zerstäubers hergestellt", wodurch allerdings nur Muster mit scharfen Konturen ohne Farbübergänge und Abstufungen der Farbintensität möglich sind. Deshalb fehlt den farbigen Mustern auf Fayence die Weichheit und Feinheit, die man sonst bei Porzellanmalereien bewundern kann. Bei dem neuen Verfahren wird mit einem dosierten Wasserstrahl die Farbe ver-

Marcell refined the methods used for underglaze decoration and introduced a new clay body that fired at low temperatures. By using it, he could gradually replace the rather pale "grand feu" (ie, fired in a high-temperature kiln) colors with new, versatile and vivid underglaze colors. With this new style of decoration, which was made possible by spraying on the glazes with an aerograph, the Goldscheider manufactory stood head and shoulders above the competition, who lagged far behind in mastering this technique.[202] On 3 March 1923, Goldscheider applied for a patent (Ill. 46) to protect a process for creating colored design patterns by spraying them on.[203] Up to then, wares sporting colored designs, especially faience, were usually "made by spraying on the color with an atomizer," which, however, only made possible crisply contoured patterns devoid of blurring colors and varying degrees of color intensity. As a result, colored designs on faience lacked the subtlety and fineness so admired in porcelain painting. By means of the new process, a controlled jet of water blurred colors to create soft transitions. Objects made in this way are especially aesthetically satisfying and suited in form the taste of the day. The harmony thus created between artistic model, execution and decoration enhanced the world-

Ill. 46

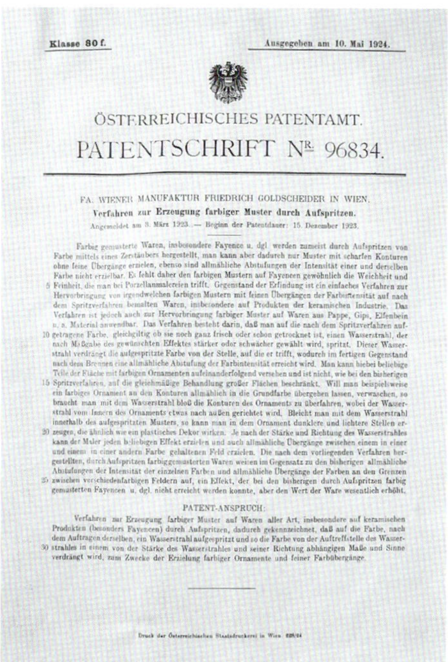

Klasse 80 f. — Ausgegeben am 10. Mai 1924.

ÖSTERREICHISCHES PATENTAMT.

PATENTSCHRIFT Nr. 96834.

FA. WIENER MANUFAKTUR FRIEDRICH GOLDSCHEIDER IN WIEN.

Verfahren zur Erzeugung farbiger Muster durch Aufspritzen.

Angemeldet am 3. März 1923. — Beginn der Patentdauer: 15. Dezember 1923.

Farbig gemusterte Waren, insbesondere Fayence u. dgl. werden zumeist durch Aufspritzen von Farbe mittels eines Zerstäubers hergestellt, man kann aber dadurch nur Muster mit scharfen Konturen ohne feine Übergänge erzielen, ebenso sind allmähliche Abstufungen der Intensität einer und derselben Farbe nicht erzielbar. Es fehlt daher den farbigen Mustern auf Fayencen gewöhnlich die Weichheit und Feinheit, die man bei Porzellanmalereien trifft. Gegenstand der Erfindung ist ein einfaches Verfahren zur Hervorbringung von irgendwelchen farbigen Mustern mit feinen Übergängen der Farbintensität auf nach dem Spritzverfahren bemalten Waren, insbesondere auf Produkten der keramischen Industrie. Das Verfahren ist jedoch auch zur Hervorbringung farbiger Muster auf Waren aus Pappe, Gips, Elfenbein u. a. Material anwendbar. Das Verfahren besteht darin, daß man auf die nach dem Spritzverfahren aufgetragene Farbe, gleichgiltig ob sie noch ganz frisch oder schon getrocknet ist, einen Wasserstrahl, der nach Maßgabe des gewünschten Effektes stärker oder schwächer gewählt wird, spritzt. Dieser Wasserstrahl verdrängt die aufgespritzte Farbe von der Stelle, auf die er trifft, wodurch im fertigen Gegenstand nach dem Brennen eine allmähliche Abstufung der Farbintensität erreicht wird. Man kann hiebei beliebige Teile der Fläche mit farbigen Ornamenten aufeinanderfolgend versehen und ist nicht, wie bei den bisherigen Spritzverfahren, auf die gleichmäßige Behandlung großer Flächen beschränkt. Will man beispielsweise ein farbiges Ornament an den Konturen allmählich in die Grundfarbe übergehen lassen, verwaschen, so braucht man mit dem Wasserstrahl bloß die Konturen des Ornaments zu überfahren, wobei der Wasserstrahl vom Innern des Ornaments etwas nach außen gerichtet wird. Bleicht man mit dem Wasserstrahl innerhalb des aufgespritzten Musters, so kann man in dem Ornament dunklere und lichtere Stellen erzeugen, die ähnlich wie ein plastisches Dekor wirken. Je nach der Stärke und Richtung des Wasserstrahles kann der Maler jeden beliebigen Effekt erzielen und auch allmähliche Übergänge zwischen einem in einer und einem in einer andern Farbe gehaltenen Feld erzielen. Die nach dem vorliegenden Verfahren hergestellten, durch Aufspritzen farbiggemusterten Waren weisen im Gegensatz zu den bisherigen allmähliche Abstufungen der Intensität der einzelnen Farben und allmähliche Übergänge der Farben an den Grenzen zwischen verschiedenfarbigen Feldern auf, ein Effekt, der bei den bisherigen durch Aufspritzen farbig gemusterten Fayencen u. dgl. nicht erreicht werden konnte, aber den Wert der Ware wesentlich erhöht.

PATENT-ANSPRUCH:

Verfahren zur Erzeugung farbiger Muster auf Waren aller Art, insbesondere auf keramischen Produkten (besonders Fayencen) durch Aufspritzen, dadurch gekennzeichnet, daß auf die Farbe, nach dem Auftragen derselben, ein Wasserstrahl aufgespritzt und so die Farbe von der Auftreffstelle des Wasserstrahles in einem von der Stärke des Wasserstrahles und seiner Richtung abhängigen Maße und Sinne verdrängt wird, zum Zwecke der Erzielung farbiger Ornamente und feiner Farbübergänge.

Druck der Österreichischen Staatsdruckerei in Wien.

Ill. 47

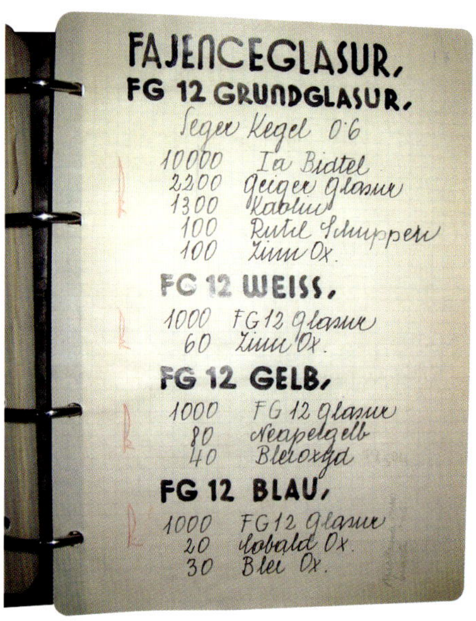

Ill. 46
Ein neues patentiertes Verfahren, mit dem weiche Farbübergänge möglich sind, bildet die Grundlage für die berühmten farbigen Muster der Art Déco Periode. (Quelle: Patentschrift 96834)

The celebrated colored designs of the Art Déco period were based on a newly patented process which made possible softly blurred colors. (Source: Patentschrift 96834)

Ill. 47
Das geheime Glasuren-Rezeptbuch wird wie ein Schatz aufbewahrt, beinhaltet es doch die den Weltruf der Goldscheider Manufaktur sichernden Zusammensetzungen der Farbmischungen. (PB Wien 19)

The glaze recipe book was a treasured, well-guarded secret since it contained the recipes for the color blends that ensured the reputation of the Goldscheider manufactory worldwide. (PB Wien 19)

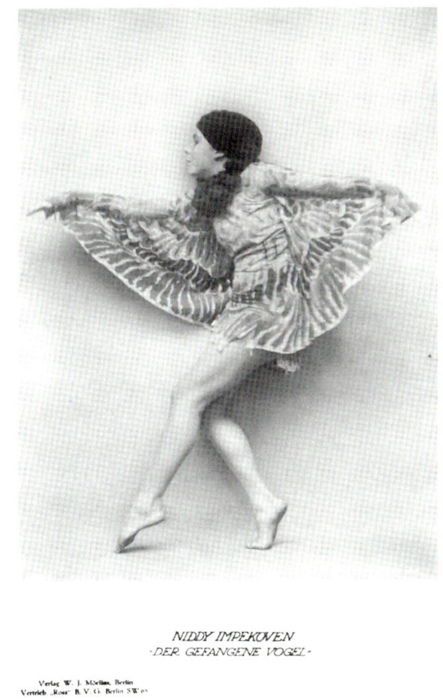

Ill. 48

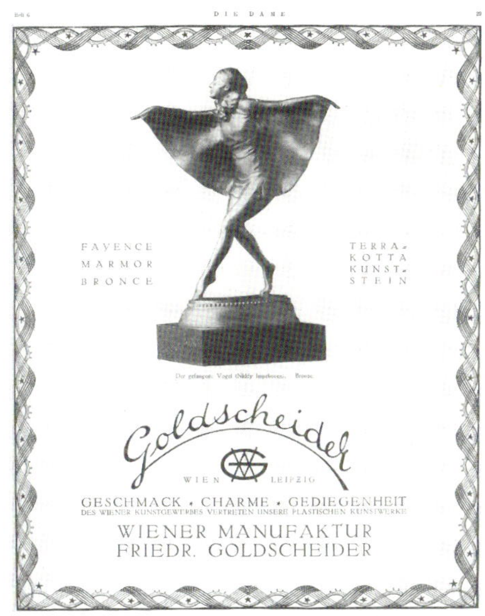

Ill. 49

wide success of the Goldscheider manufactory and led to the term "Wiener Fayence" being equated with Goldscheider products.

Thanks to this process, one of the most impressive faience models was created by the Goldscheider manufactory: "Captured Bird" after a model by Josef Lorenzl. This portrayal of the dancer Niddy Impekoven is notable for brilliantly hued, colorful decoration in a wide range of variations.

wischt, um weiche Farbübergänge zu erzeugen. Die so hergestellten Objekte erscheinen als besonders ästhetisch und in ihrer Form den Zeitgeist treffend. Diese Harmonie aus künstlerischem Modell, Ausarbeitung und Dekoration stärkt den weltweiten Erfolg der Goldscheider Manufaktur und manifestiert den Begriff „Wiener Fayence" mit Goldscheider Erzeugnissen.

Dank dieses Verfahrens entsteht auch eines der imposantesten Fayence-Modelle der Goldscheider Manufaktur: der „Gefangene Vogel" nach einem Modell von Josef Lorenzl. Diese Darstellung der Tänzerin Niddy Impekoven zeigt eine brillante und farbenfrohe Dekoration in den unterschiedlichsten Variationen.

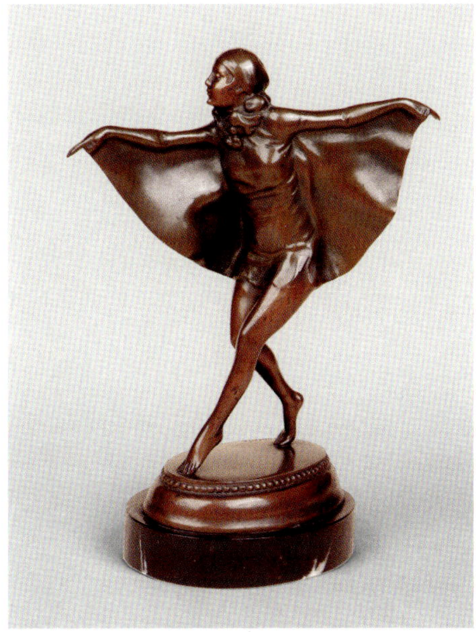

ModNr. 5054
Bronze „Gefangener Vogel" I "Captured Bird", Josef Lorenzl, ca. 1922, PB Wien 19
Die Tänzerin Niddy Impekoven gehört zu den Stars des Tanztheaters um 1920 (vgl. Ill. 48). Das Modell ihrer tänzerischen Darbietung „Der Gefangene Vogel" wird 1922 in der Zeitschrift Die Dame zuerst als Bronze beworben (Ill. 49), die die Modell-Nummer 5054 trägt. Zum berühmtesten Modell der Goldscheider Manufaktur wird allerdings die Fayence-Version mit der Modellnummer 5230, die in Kombination aus Unterglasur-Malerei und hervorragender Aerograph-Spritztechnik dekoriert wird. (Quelle: Die Dame, Dezember 1922, Nr. 6, S. 29; Postkarte aus PB Deutschland)

The dancer Niddy Impekoven was a dance-theatre star around 1920 (cf Ill. 48). The model portraying her performing "Der Gefangene Vogel" ["Captured Bird"] was first advertised in bronze in Die Dame magazine in 1922 (Ill. 49), bearing the model number 5054. The most famous model produced by the Goldscheider manufactory, however, was the faience version, model number 5230, which was decorated with a combination of underglaze painting and virtuoso aerograph spray technique. (Source: Die Dame, December 1922, No. 6, p. 29; postcard from PB Deutschland)

ModNr. 5230
„Gefangener Vogel" I "Captured Bird", Josef Lorenzl, ca. 1923, Collection f@goldscheider.de

ModNr. 5230
„Gefangener Vogel" I "Captured Bird", Josef Lorenzl, ca. 1923, beide I both Collection Weilinger, Wien

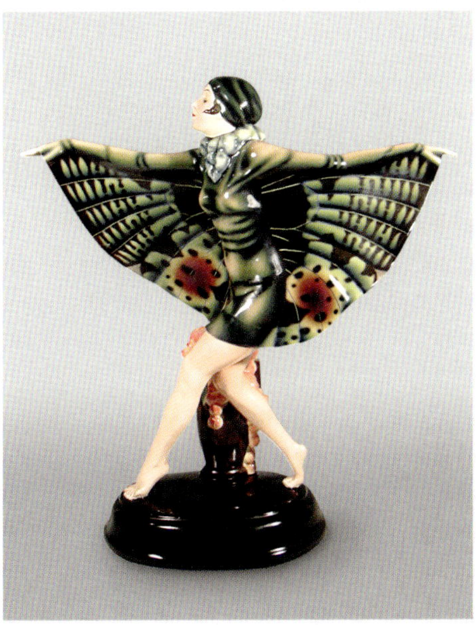

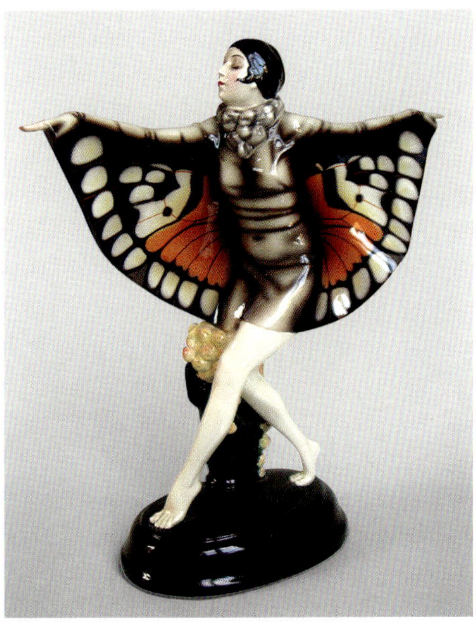

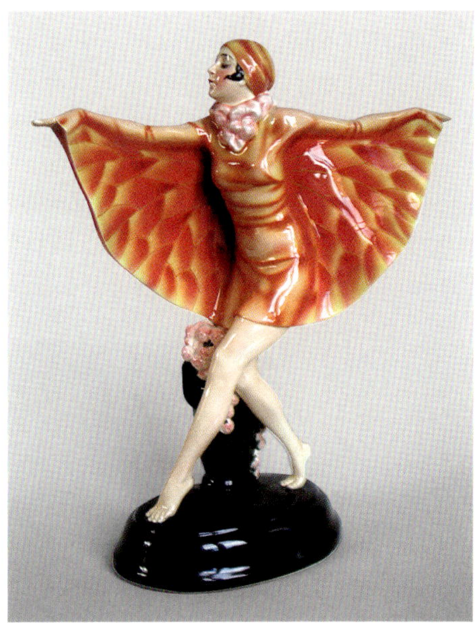

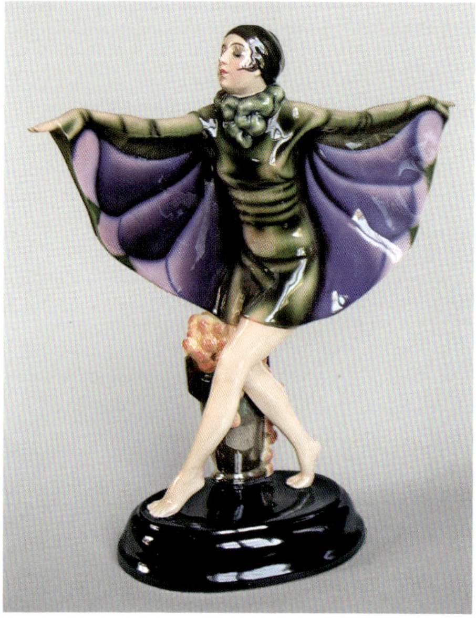
ModNr. 5230 (A)

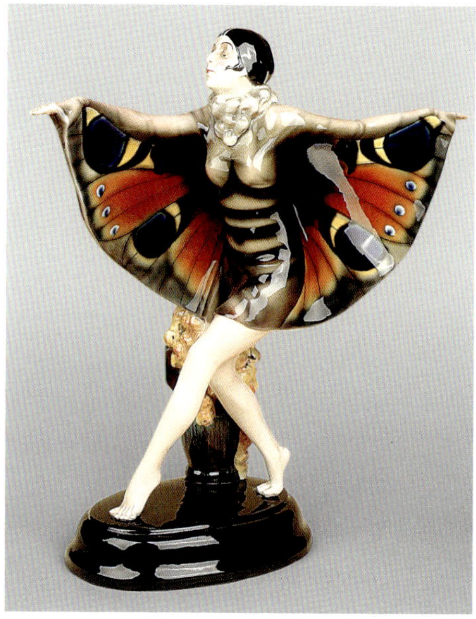
ModNr. 5230 (B)

ModNr. 5230 (A)
„Gefangener Vogel" | "Captured Bird", Josef Lorenzl, ca. 1923, Collection Weilinger, Wien

ModNr. 5230 (B – D)
„Gefangener Vogel" | "Captured Bird", Josef Lorenzl, ca. 1923, PB Wien 19

ModNr. 5230 (E + F)
„Gefangener Vogel" | "Captured Bird", Josef Lorenzl, ca. 1923, Sammlung.Schmutz@chello.at

ModNr. 5960
„Gefangener Vogel" | "Captured Bird", Josef Lorenzl, ca. 1929, Sammlung.Schmutz@chello.at

rechts | right

ModNr. 5230
„Gefangener Vogel" | "Captured Bird", Josef Lorenzl, ca. 1923, Collection f@goldscheider.de, Foto: Im Kinsky Kunstauktionen GmbH

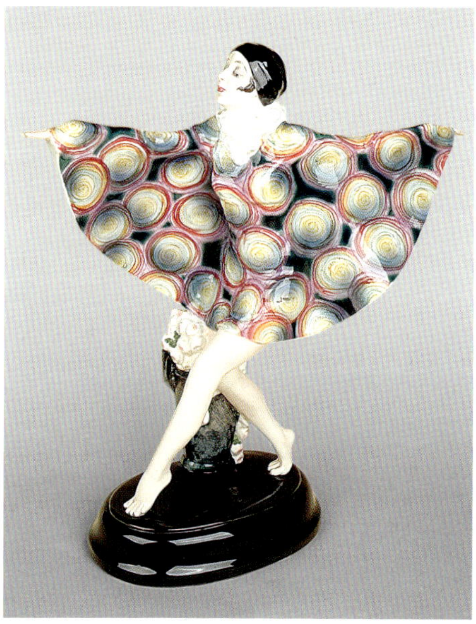
ModNr. 5230 (C)

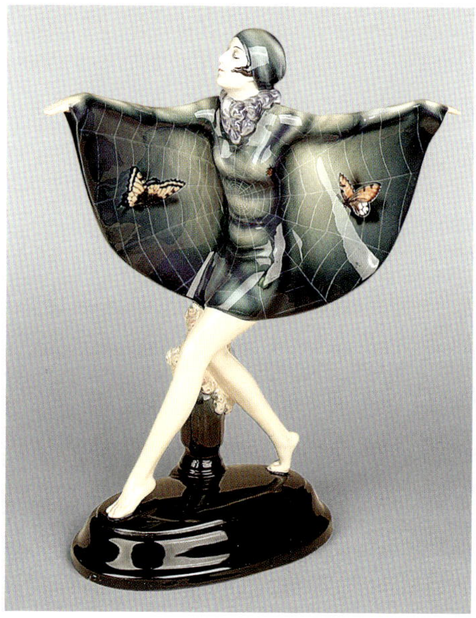
ModNr. 5230 (D)

„Der Gefangene Vogel" wird in diversen Größen produziert; die Modell-Nummer 5960 ist bspw. 38 cm hoch – im Gegensatz zur größten Figur 5230, die fast 50 cm hoch ist. Aber auch nach links schauende Modelle mit anderem Haaransatz bzw. eine Lampe mit einem Reigen aus drei Tänzerinnen werden angeboten.

"Captured Bird" was produced in several different sizes; model number 5960, for instance, is 38 cm high – the largest figure, No. 5230, by contrast, is almost 50 cm high. Models looking to the left with a different hairstyle or a lamp with three dancers in a row were also marketed.

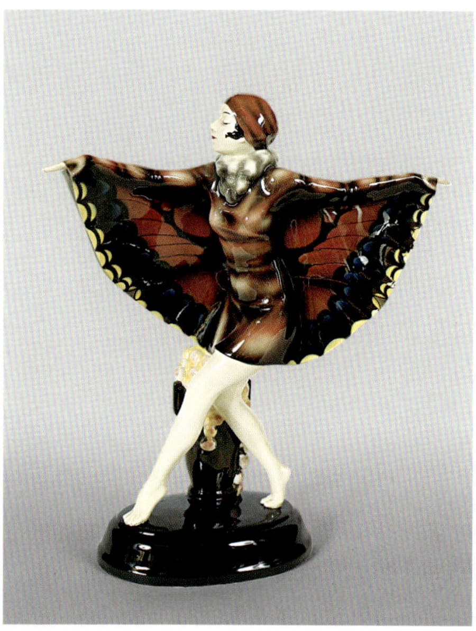
ModNr. 5230 (E)

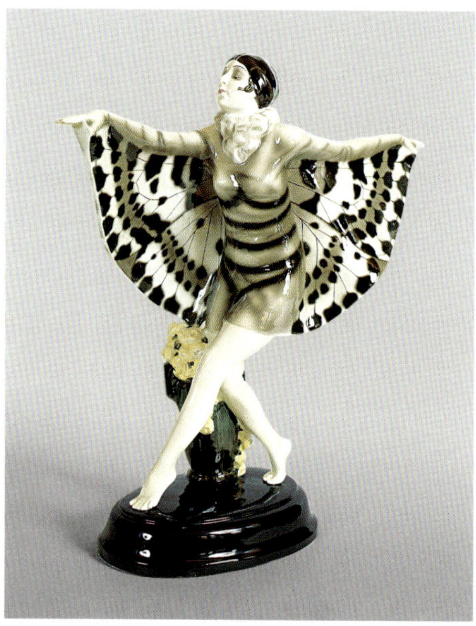
ModNr. 5230 (F)

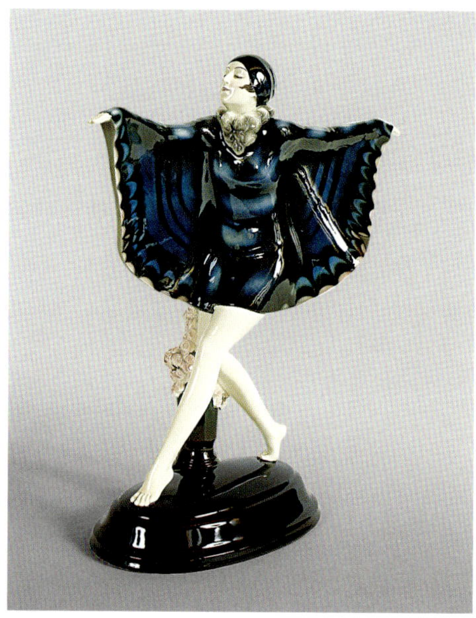
ModNr. 5960

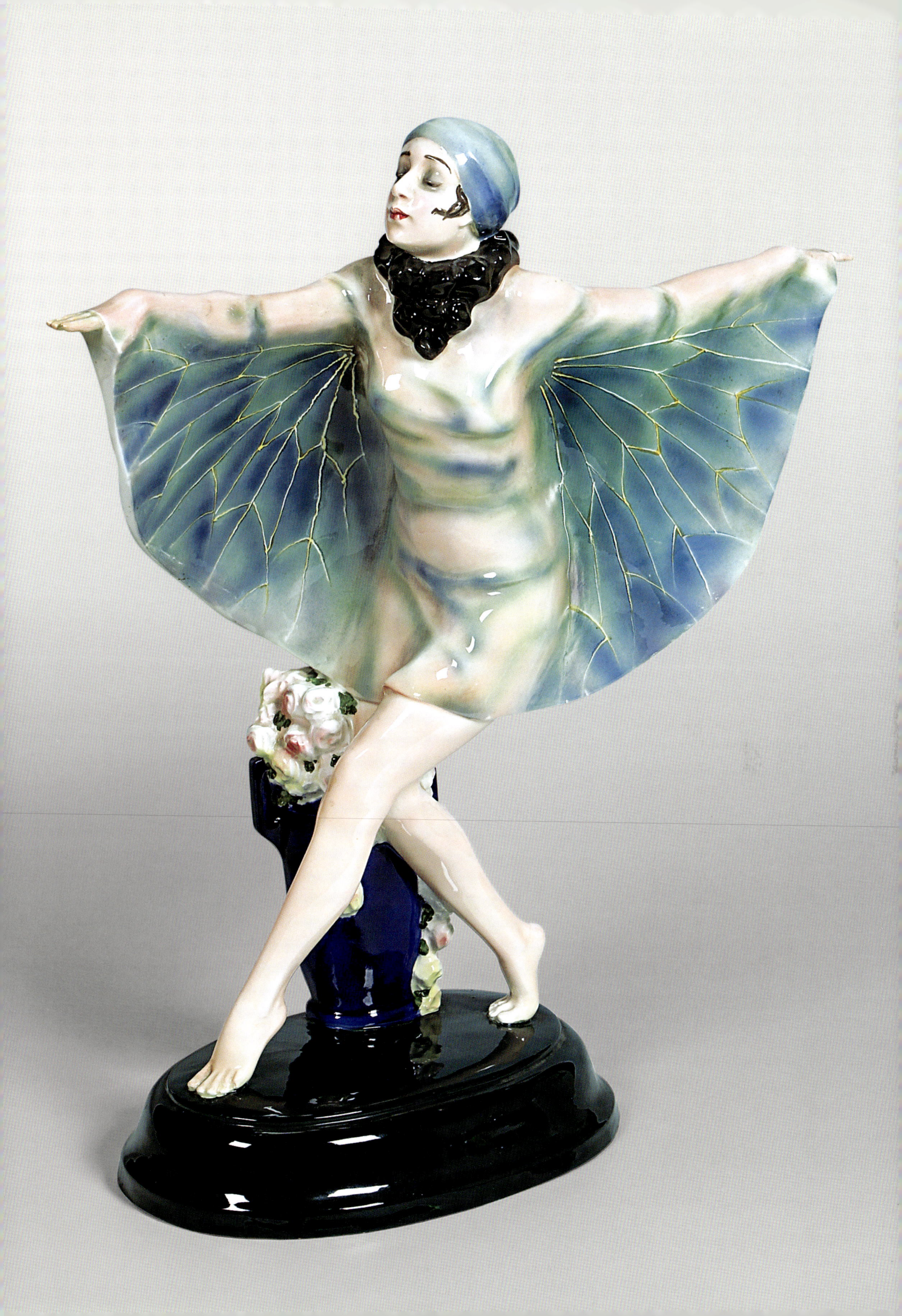

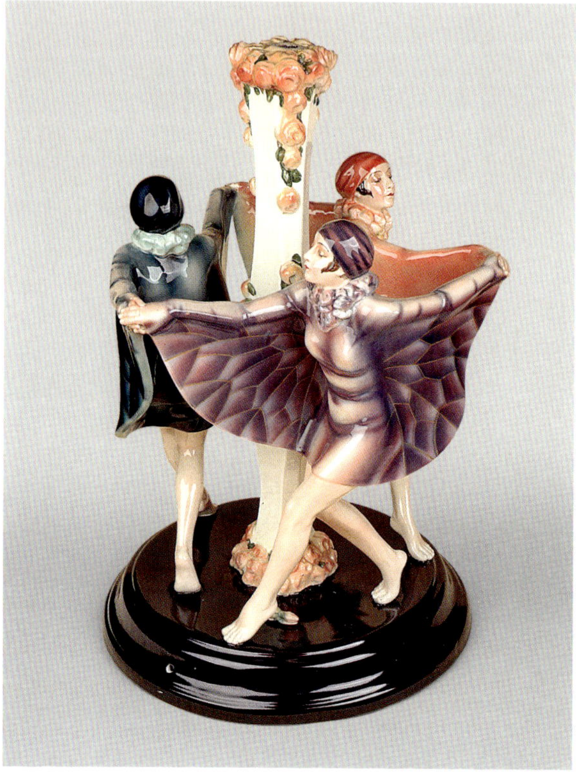

ModNr. 5530
„Gefangener Vogel" I
"Captured Bird",
Josef Lorenzl, ca. 1925,
PB Wien 19

Ende 1923 finden auch andere Figuren in Fayence und Bronze beachtlichen Anklang bei der Fachpresse: „Die Wiener Manufaktur Friedrich Goldscheider, Wien, deren Ruf und Bedeutung in aller Welt feststeht, bringt wieder einige trefflich gelungene Figuren in Fayence und Bronze, die in ihren Schrittstellungen, im ganzen Zauber des Mienenspiels, im Wurf und in der Farbe der Kostüme, den ganzen Reiz des echt Wienerischen bestrickend zum Ausdruck bringt."[204] Zu den Neuheiten gehören Lorenzls „Carmen" und „Wilder Tanz", Kostials „Tanz der Eitelkeit", „Harlekin" und „Pierrette (Modell-Nr. 5094), Perls „Seiltänzerin" sowie einige Vasen von Gult. Kostial ist zu dieser Zeit besonders dank seiner Harlekinaden, manieriert-ironischen Figuren und Gruppen, gefragt.

Die Manufaktur beschäftigt nach der Flaute des Ersten Weltkriegs wieder über 100 Mitarbeiter.[205] Betriebsleiter ist der Bildhauer Wilhelm Thomasch, der nach seiner Ausbildung an der Kunstgewerbeschule Teplitz und danach an der Akademie für Bildende Künste in Wien bei der Manufaktur Friedrich Goldscheider eingestellt wurde und eigene

Late in 1923, other figures in faience and bronze were also well received by the specialist press: "The Viennese Manufactory Friedrich Goldscheider, Vienna, whose reputation and importance are established throughout the world, has again launched some admirably caught figures in faience and bronze, which, in their poses, in all the enchanting play of feature, in design and the color of their costumes, stunningly express the charm of what is authentically Viennese."[204] The new models included Lorenzl's "Carmen" and "Wilder Tanz" ["Wild Dance"], Kostial's "Tanz der Eitelkeit" ["Dance of Vanity"], "Harlekin" ["Harlequin"] and "Pierrette" (model number 5094), Perl's "Seiltänzerin" ["Tightrope Walker"] as well as some vases by Gult. Kostial's Harlequinades, ironically mannered figures and groups, were particularly sought-after at the time.

Following the stagnation that set in after the First World War, the manufactory once again employed a workforce of over 100.[205] The sculptor Wilhelm Thomasch was managing director. After training at the Teplitz Kunstgewerbeschule and the Viennese Art Academy, he was hired by the Friedrich Goldscheider manufactory, where he contributed designs of his own.[206] Stephan Dakon, who attended a sculpture school and then served

ModNr. 5434
„Gefangener Vogel" I
"Captured Bird",
Josef Lorenzl, ca. 1924,
MAK – Österreichisches Museum für angewandte Kunst / Gegenwartskunst, Wien

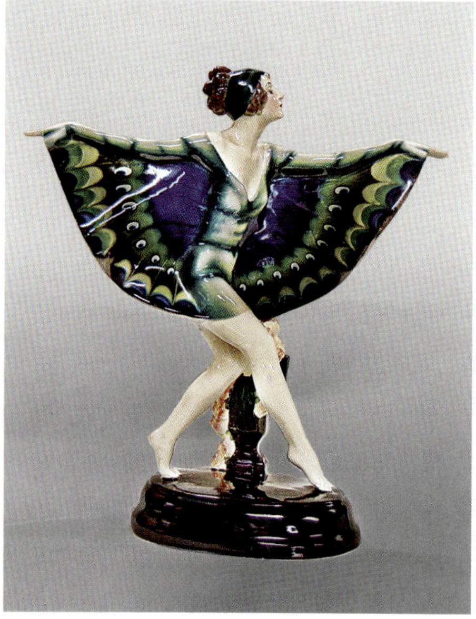

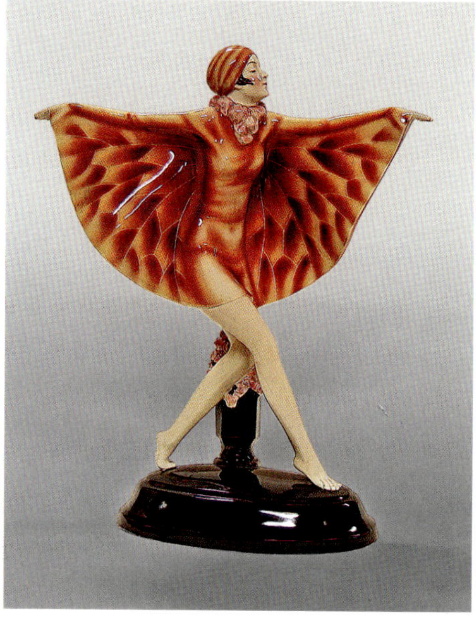

ModNr. 5964
„Gefangener Vogel" I
"Captured Bird",
Josef Lorenzl, ca. 1929,
MAK – Österreichisches Museum für angewandte Kunst / Gegenwartskunst, Wien

ModNr. 5500
„Stehende Haremstänzerin" I "Standing Harem Dancer", Josef Lorenzl, ca. 1925, private collection, Wien

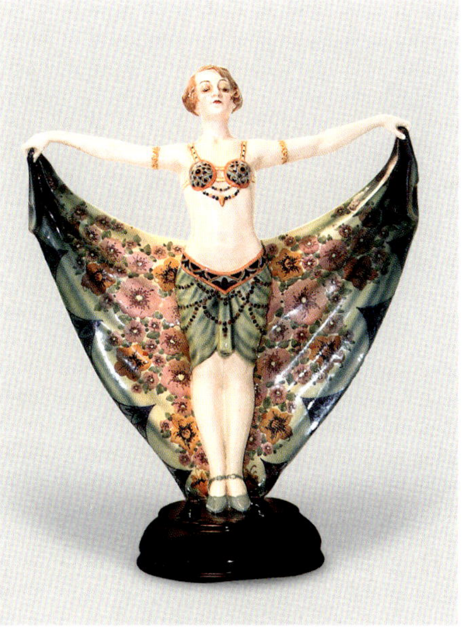

an apprenticeship at the art bronze foundry in the Vienna Arsenal, where he made the acquaintance of the sculptor Josef Lorenzl, was hired by Goldscheider around 1924 and contributed his first models.[207]

A long article by Marcell Goldscheider on the subject of "Modern oder zeitgemäß?" ["Modern or Contemporary?"] was published early in 1925. Searchingly examining the terms "Moderne" ["Modernism/Modern"] and "Kunst & Kunstgewerbe" ["fine art and the decorative and applied arts"] as used in the discourse of the day, Marcell uttered an urgent plea against the term "modern" since it resembled fashion and could dissipate within a space of a few months. He also emphatically aired his views on the "contemporary" decorative and applied arts. On the one hand, he believed it was necessary to accommodate oneself to the prevailing taste: "The exponent of the decorative and applied arts, be he ever so independent, will have to take into consideration the buyer's aesthetic feelings and taste tendency, his habits and needs, whether he

Entwürfe beisteuert.[206] Stephan Dakon, der nach der Bildhauerschule eine Lehre in der Kunsterzgießerei im Wiener Arsenal machte und dort den Bildhauer Josef Lorenzl kennen lernte, wird um 1924 bei Goldscheider angestellt und steuert die ersten Modelle bei.[207]

Anfang 1925 erscheint ein ausführlicher Artikel von Marcell Goldscheider zum Thema „Modern oder zeitgemäß?". In diesem eindringlichen Plädoyer, der sich mit dem damaligen Diskurs um die Begriffe „Moderne" sowie „Kunst & Kunstgewerbe" auseinandersetzt und in dem sich Marcell gegen die Bezeichnung „modern" wehrt, da diese einer Mode ähnelt und bereits nach Monaten verfliegen kann, werden auch seine Ansichten zum „zeitgenössischen" Kunstgewerbe deutlich. Einerseits glaubt er, dass es notwendig ist, sich dem Publikumsgeschmack anzupassen: „Der Kunstgewerbler, mag er noch so unabhängig sein, wird auf das ästhetische Gefühl und die Geschmacksrichtung des Käufers, auf seine Nationalität, seine Gebräuche und Bedürfnisse Rücksicht nehmen müssen, ob er will oder nicht."[208] Andererseits

links I left
ModNr. 5366
Kerzenhalter I Candlestick, Wilhelm Thomasch, ca. 1924, PB Wien 19

rechts I right
ModNr. 5267
Lampe I Lamp, Josef Lorenzl, ca. 1923, PB Wien 19

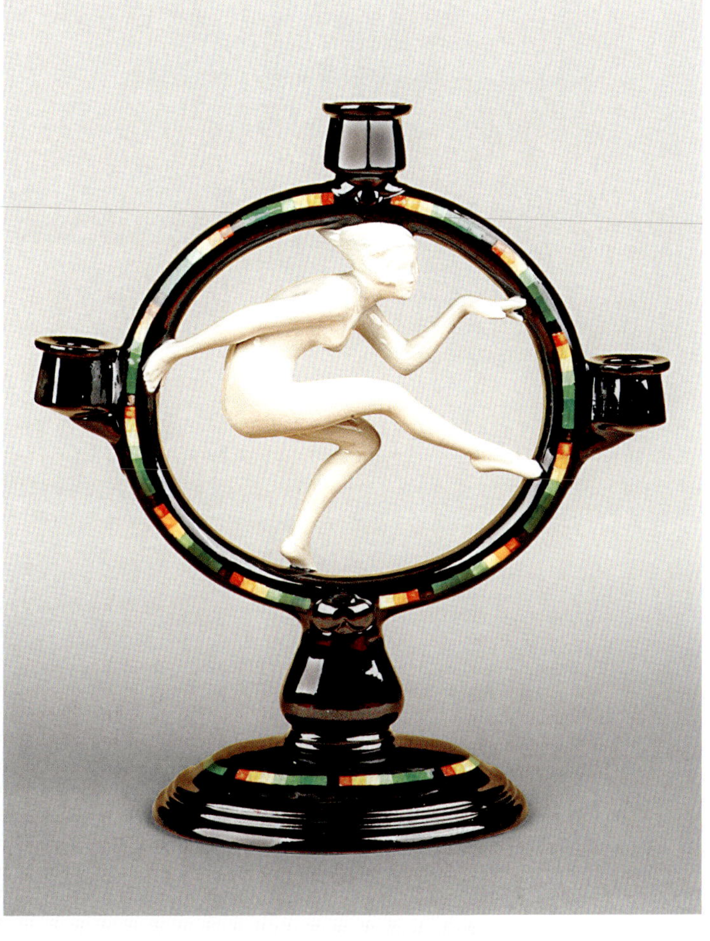

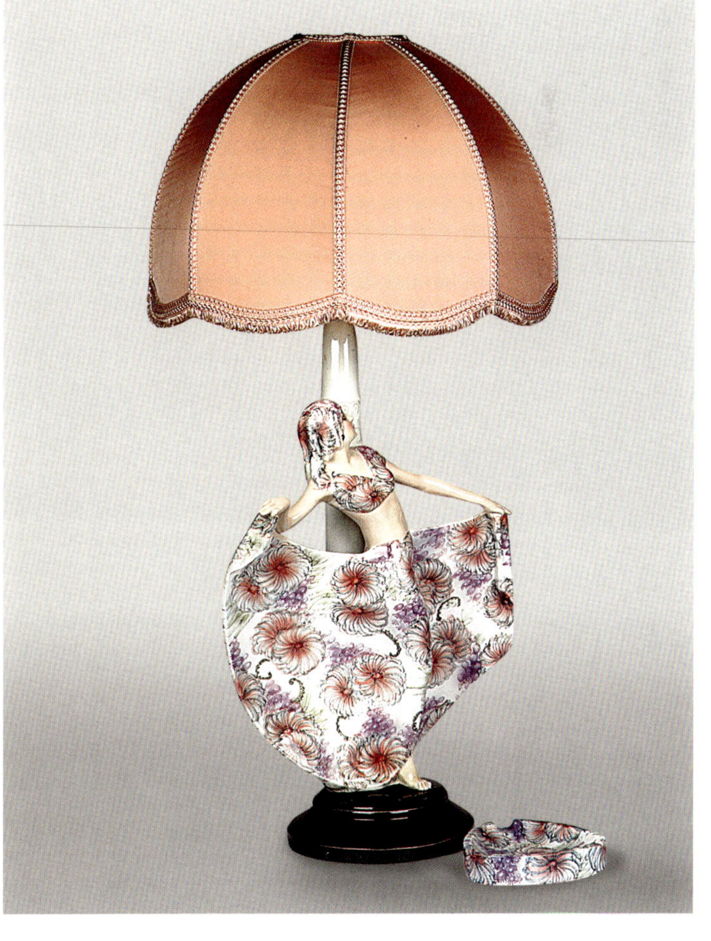

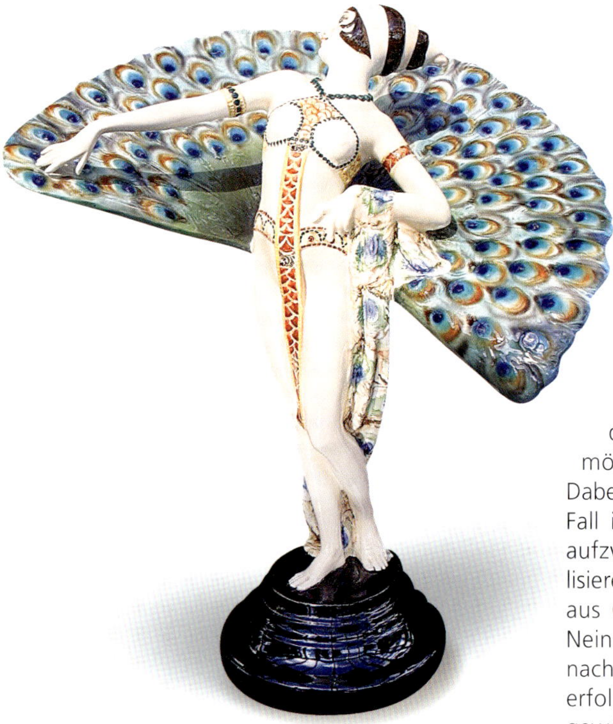

ModNr. 5405
Tänzerin | Danseuse,
Paul Philippe, ca. 1924,
Collection Robert
Senft

ist er davon überzeugt, dass der Künstler seine Überzeugung nicht dem Materialismus opfern darf: „Soll er [der Künstler] voll genommen werden, so muß er konsequent seinem Ideale zustreben und es möglichst zu verwirklichen trachten."[209] Dabei soll er nicht, wie dies bei der Mode der Fall ist, sein Kunstgewerbe dem Publikum aufzwingen und es zu einer Modesache stilisieren, bei der vor allem weibliche Käufer aus Gesellschaftsdruck die Waren kaufen. Nein, nach Marcell Goldscheider soll ein nachhaltigerer Erfolg für das Kunstgewerbe erfolgen. Es gilt, das „Publikum langsam zu gewöhnen, die herrschenden Formen und Richtungen durch Anpassung an den Zeitgeist aufzugeben."[210] Zusammenfassend sagt er, dass das zeitgemäße Kunstgewerbe im weiten Sinne Schöpfungen umfasst, die „ein Kompromiß zwischen Künstler und Publikum bedeuten und an Tradition anknüpfend auf dem Weg der Entwicklung zu einer Richtung

wants to or not."[208] On the other hand, Marcell Goldscheider was convinced that the artist need not sacrifice his convictions to materialism: "If he is to be taken seriously, he must aspire consistently to his ideal and strive as far as possible to realize it."[209] In so doing, the artist should not, as in the case of fashions in apparel, seek to foist his decorative and applied art on the public and stylize it to a fashion issue, exerting peer pressure on women customers especially to buy the wares. On the contrary, thus Marcell Goldscheider, the decorative and applied arts should be assured more long-lasting success. The aim was "to accustom the public gradually to abandon the prevailing forms and stylistic tendencies by adapting to the Zeitgeist."[210] In conclusion, he contended that the decorative and applied arts of the day encompassed in the broad sense creations that represented "a compromise between artist and public and, linking up with tradition, want to attain by developing it a style which is not tied to the times but [signifies] a lasting value."[211] He addressed the entrepreneurs dealing in the decorative and applied arts since introducing a new style naturally meant making considerable financial sacrifice and also necessitated retraining artisans yet "for all that, one need not reject some-

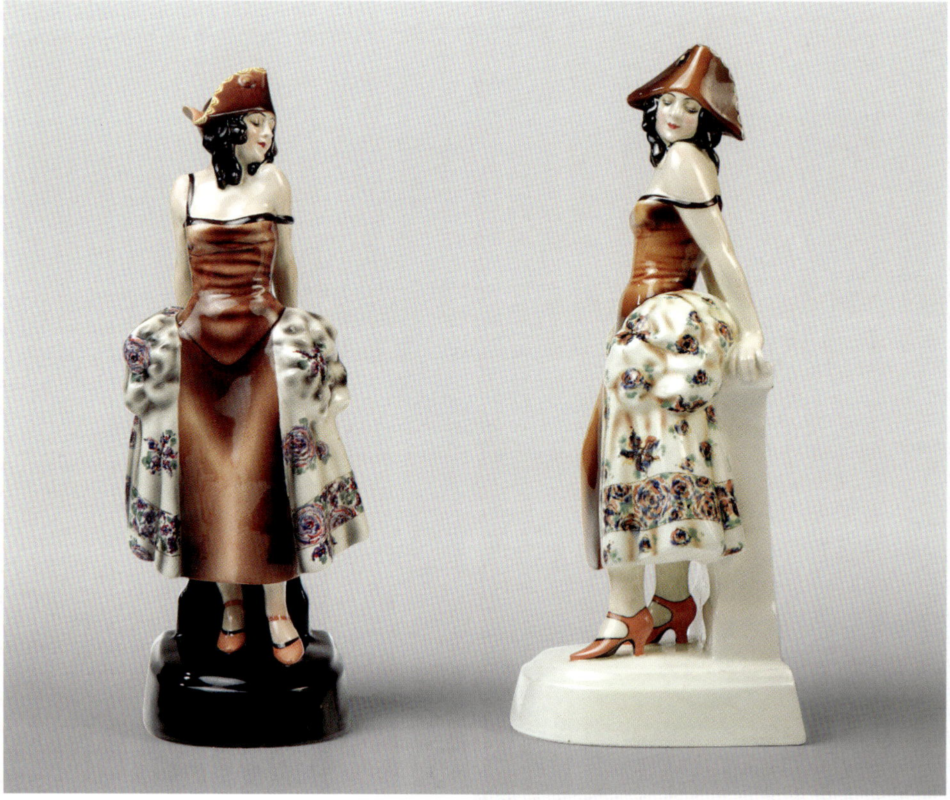

ModNr. 5486
Figur und Lampe |
Statue and Lamp,
Josef Lorenzl, ca. 1925,
© Wien Museum

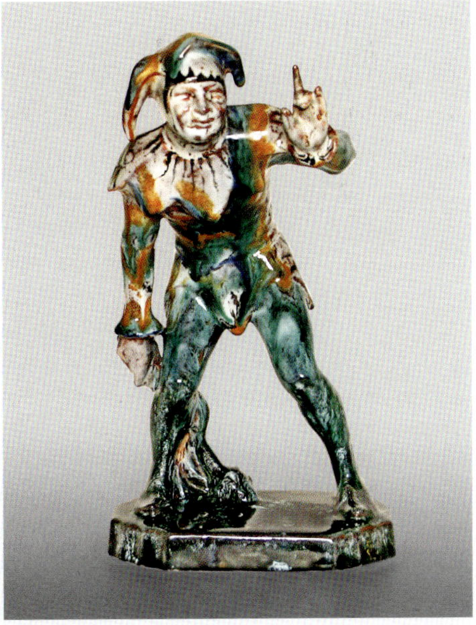

ModNr. 5558
Harlekin | Harlequin,
Josef Lorenzl, ca. 1925,
Collection Irsigler,
Wien

gelangen wollen, welche keinen an die Zeit gebundenen, sondern bleibenden Wert" bedeutet.[211] Er appelliert an die Geschäftsleute des Kunstgewerbes, da die Einführung neuer Stile naturgemäß zunächst größere finanzielle Opfer mit sich bringt und auch ein Umlernen der Handwerker erfordert, doch „dessen ungeachtet darf man etwas Neues, mag es noch so kraß und absurd erscheinen, nicht einfach deswegen verwerfen."[212] Bei all dem ist für Marcell die enge Zusammenarbeit zwischen den Entwerfern und den Handwerkern am Wichtigsten, denn nur so kann langfristig künstlerisches und qualitätsvolles Kunstgewerbe entstehen: „Wollen aber die Führer und Schöpfer neuer Richtungen, daß diese für die Entwicklung gewerblicher Edelarbeit mehr bedeuten, als eine rasch verschwindende Mode, wollen sie, daß die Besitzer von Gegenständen unserer Zeit sie mit Freude und Stolz an spätere Genera-

thing new for those reasons, may it appear ever so crass and absurd."[212] The most important point for Marcell in all this was close collaboration between designers and artisans because only through it could aesthetically satisfying decorative and applied arts objects of high quality be produced: "If, however, the leaders and creators of new styles want them to mean more for the development of

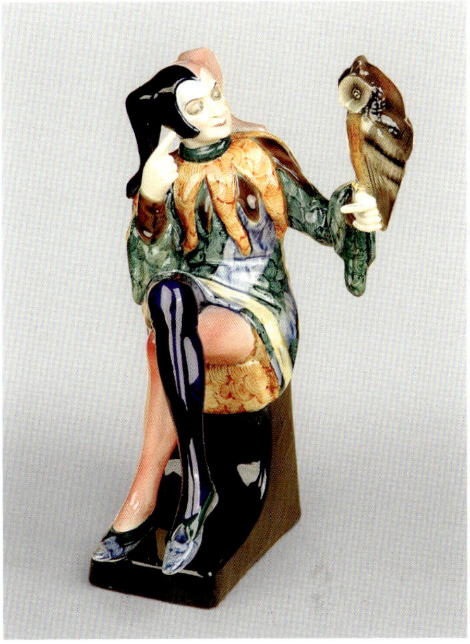

ModNr. 5509
„Till Eulenspiegel mit Eule" | "Till Eulenspiegel with Owl", Josef Lorenzl, Entwurf | design ca. 1923,
PB Wien 19

ModNr. 5616
„Carmen", Josef Lorenzl, ca. 1927,
PB Wien 19
Die Schauspielerin Dolores del Rio wurde in ihrer Rolle als Carmen in zahlreichen Modellen verewigt.

The actress Dolores del Rio was immortalized playing the role of Carmen in numerous models.

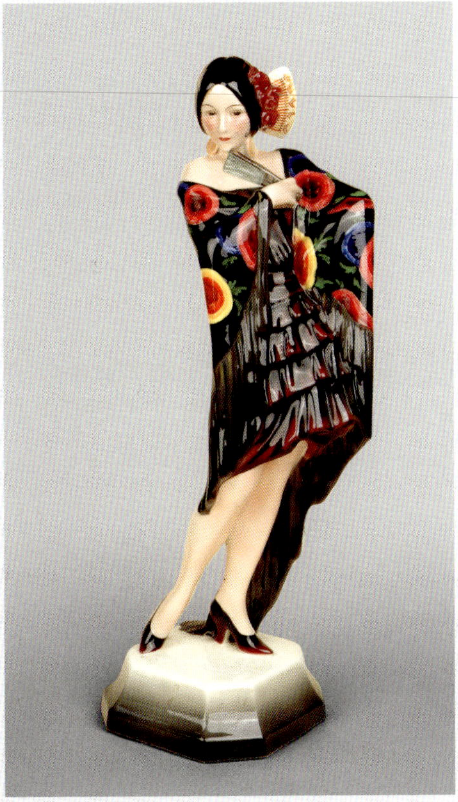

ModNr. 5516
Gruppe und Lampe |
Group and Lamp,
Karin Jarl, ca. 1925,
© Wien Museum

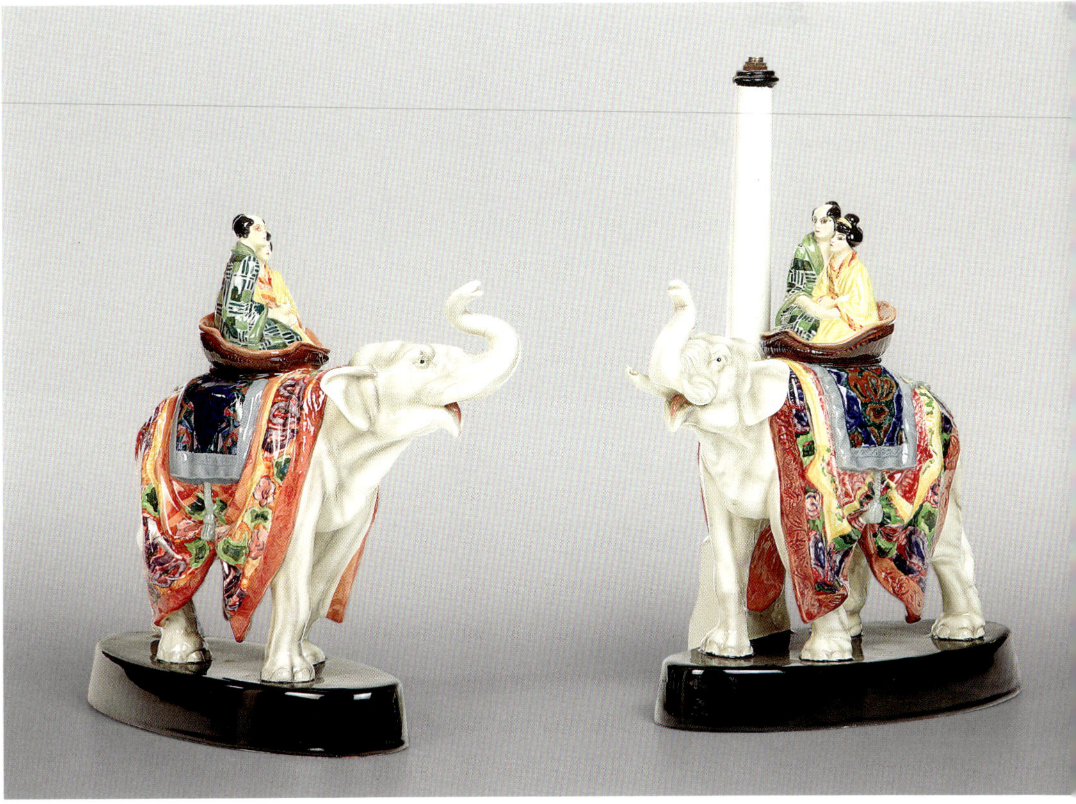

Ill. 50
Die Brüder setzen Postkarten als Werbemittel ein, hier eine Abbildung von Haniroffs Faun. (Sammlung f@goldscheider.de)

The brothers used postcards as vehicles for advertising; here a picture of Haniroff's Faun. (Collection f@goldscheider.de)

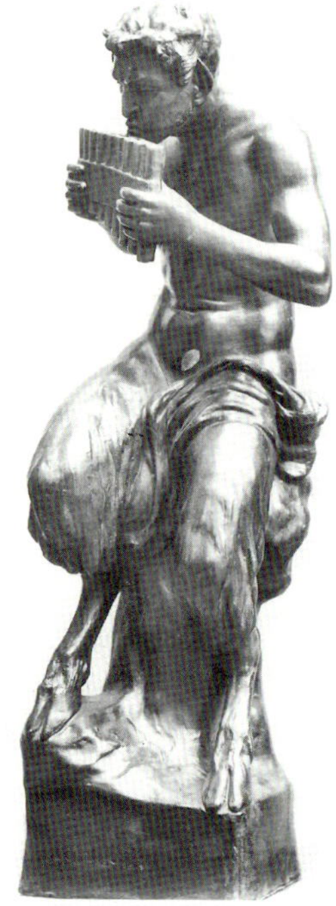

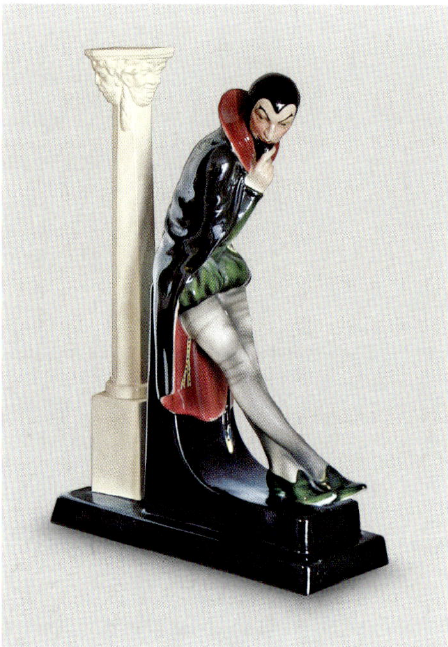

ModNr. 5537
Lampe I Lamp „Mephisto", Josef Lorenzl, ca. 1925, Collection a@goldscheider.net

tionen weitergeben, so müssen wir sie in lebendiger Zusammenarbeit mit dem Erzeuger und Handwerker, gestützt auf dessen Erfahrungen, aufbauen."[213]

Zur Vertriebsunterstützung setzen Walther und Marcell Werbemittel ein, hierbei vor allem Anzeigen und Postkarten mit Abbildungen von Modellen. Die Manufaktur stiftet auch öfters eigene Erzeugnisse für Preisausschreiben, so erhält die Fachzeitschrift „Kunst und Kunstgewerbe" beispielsweise den „Mephisto" von Lorenzl im Wert von 76 Mark.[214] Die Leipziger Filiale wird weiterhin auffällig dekoriert und selbstverständlich werden auch zahlreiche Messen und Ausstellungen beschickt. Man stellt beispielsweise 1922 auf der Deutschen Gewerbeschau in München „Moderne Fayencen (Kunstkeramik) nach Entwürfen führender Künstler Deutschlands und Österreichs"[215] sowie auf der Jubiläumsausstellung 1884–1924 des Wiener Kunstgewerbe-Vereins aus.[216]

fine arts and crafts than a rapidly vanishing fashion, if they want the owners of objects [made] in our times to hand them down to later generations with joy and pride, we must build them up in active collaboration with the maker and artisan, supported by his experience."[213]

To aid marketing and distribution, Walter and Marcell employed advertising devices, especially advertisements and postcards depicting models. The firm also donated its products to competitions as prizes: the specialist journal Kunst und Kunstgewerbe was given "Mephisto" by Lorenzl (value: 76 Marks).[214] The Leipzig branch continued to be strikingly decorated and of course objects were sent to numerous trade fairs and exhibitions. In 1922, for instance, Goldscheider sent "modern faiences (art ceramics) after designs by leading artists from Germany and Austria" to the Deutsche Gewerbeschau in Munich[215] and exhibited at the 1884–1924 Jubilee Exhibition mounted by the Viennese Kunstgewerbe-Verein.[216]

Ill. 51, 52
Im Leipziger Städtischen Kaufhaus am Neumarkt Nr. 9 werden im ständigen Verkaufsraum der Wiener Manufaktur Friedrich Goldscheider beliebte Modelle präsentiert. (Quelle: Keramos, Juni 1923, Heft 6, S. 216)

Popular models were presented at the Städtisches Kaufhaus, No. 9 Neumarkt in Leipzig in the permanent Wiener Manufaktur Friedrich Goldscheider sale-room. (Source: Keramos, June 1923 [Heft] 6, p. 216)

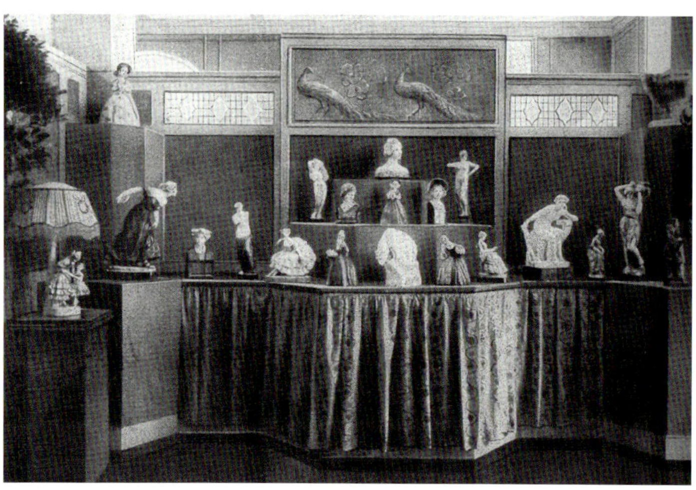

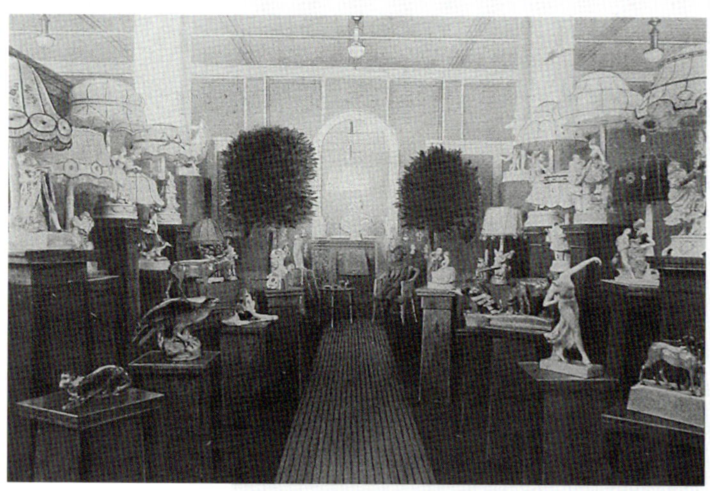

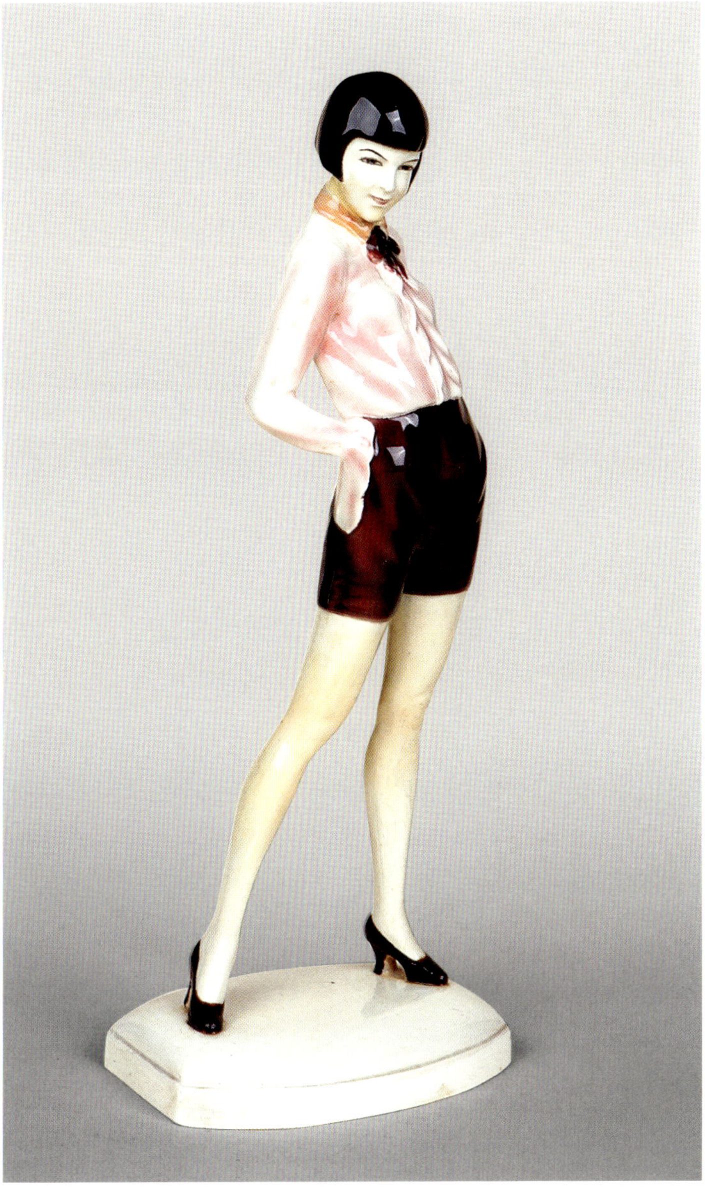

Ill. 54

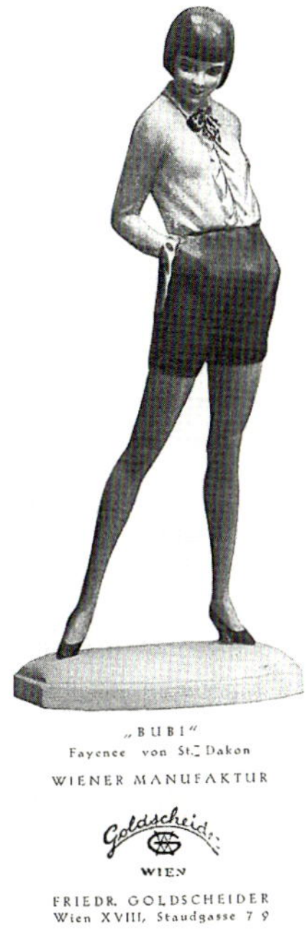

Ill. 53

ModNr. 5815 „Bubi" I "Bob", Stephan Dakon, ca. 1928, PB Wien 19
Die Tänzerin und Schauspielerin Louise Brooks wird hier im für die 20er Jahre prägenden Bubi-Stil präsentiert. Vgl. Ill. 53 (Quelle: Moderne Welt, Jahrgang 11, 1929/1930)

The dancer and Hollywood actress Louise Brooks is presented here sporting a bob, the hairstyle she made fashionable in the 1920s. Cf Ill. 53 (Source: Moderne Welt, 11, 1929/1930)

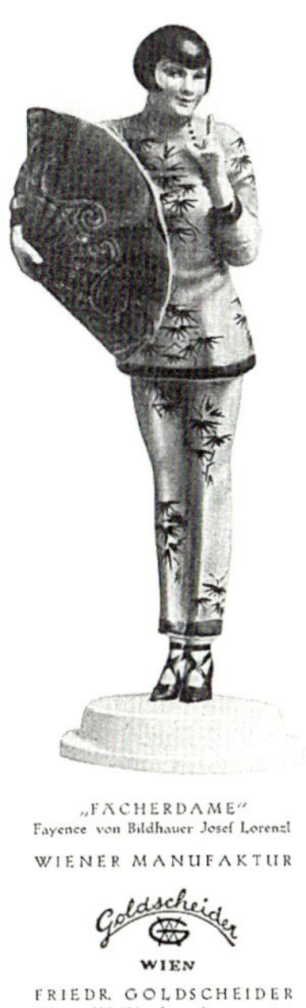

ModNr. 5870 „Fächerdame" I "Lady with Fan", Josef Lorenzl, ca. 1929, PB Wien 19
Die Zeitschrift Moderne Welt ist eines der wichtigsten Werbemedien für die Manufaktur Goldscheider, vgl. Ill 54. (Quelle: Moderne Welt, Jahrgang 11, 1929/1930)

Moderne Welt magazine was one of the most important advertising media for Goldscheider, cf Ill. 54 (Source: Moderne Welt, 11, 1929/1930)

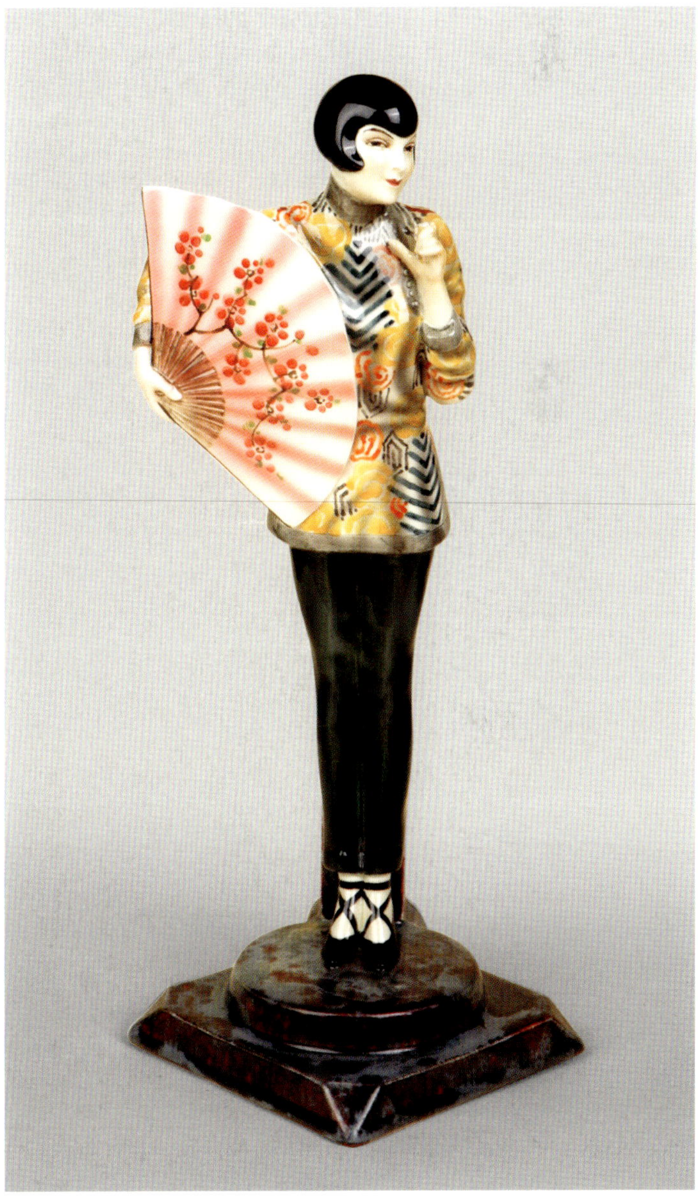

Auch auf der legendären Pariser Kunstgewerbeausstellung „Exposition Internationale des Arts Décoratifs et Industriels Modernes" im Jahr 1925 ist die Wiener Manufaktur vertreten, unter anderem mit der Fayencegruppe „Kahnfahrt" von Hertha Bucher, mit Lampenfüßen, Vasen und Leuchtern von Dina Kuhn, exotisch anmutenden Figuren in bemaltem roten Ton nach Entwürfen von Spuller, Studienköpfen in Terrakotta von Grete Neuwalder, Tierplastiken von Karin Jarl und schwarz glasierten Amorettengruppen von Herlinger-Schwarz.[217] Aber auch eine liegende weibliche Figur aus Marmor von Ena Rottenberg und ein vom Architekten Josef Berger entworfener Kamin werden von Goldscheider präsentiert.[218] In Paris stellt übrigens auch Arthur Goldscheider in einem eigenen Pavillon aus (siehe auch Kapitel „Dekorative Skulpturen von Arthur Goldscheider in Paris", S. 228ff).

The Viennese manufactory was also represented at the legendary 1925 Exposition Internationale des Arts Décoratifs et Industriels Modernes. One of the Goldscheider exhibits was the faience group "Kahnfahrt" ["Canoeing"] by Hertha Bucher, alongside lamp bases, vases and candlesticks by Dina Kuhn, exotic-looking figures in painted red pottery after designs by Spuller, head studies in terracotta by Grete Neuwalder, animal sculptures by Karin Jarl and black-glazed groups of amoretti by Herlinger-Schwarz.[217] Goldscheider also presented a reclining female figure in marble by Ena Rottenberg and a fireplace surround designed by the architect Josef Berger.[218] Incidentally, Arthur Goldscheider also exhibited in Paris in a pavilion of his own (also see the chapter "Ornamental Sculpture from Arthur Goldscheider in Paris," p. 228f).

ModNr. 5345
„Kahnfahrt" I "Canoeing", Hertha Bucher, ca. 1924, PB Wien 19
Eine der bekanntesten Wiener Keramikerinnen, Hertha Bucher, kreierte die Gruppe „Kahnfahrt" für Goldscheider.

Hertha Bucher, one of the best known Viennese ceramicists, created the "Canoeing" group for Goldscheider.

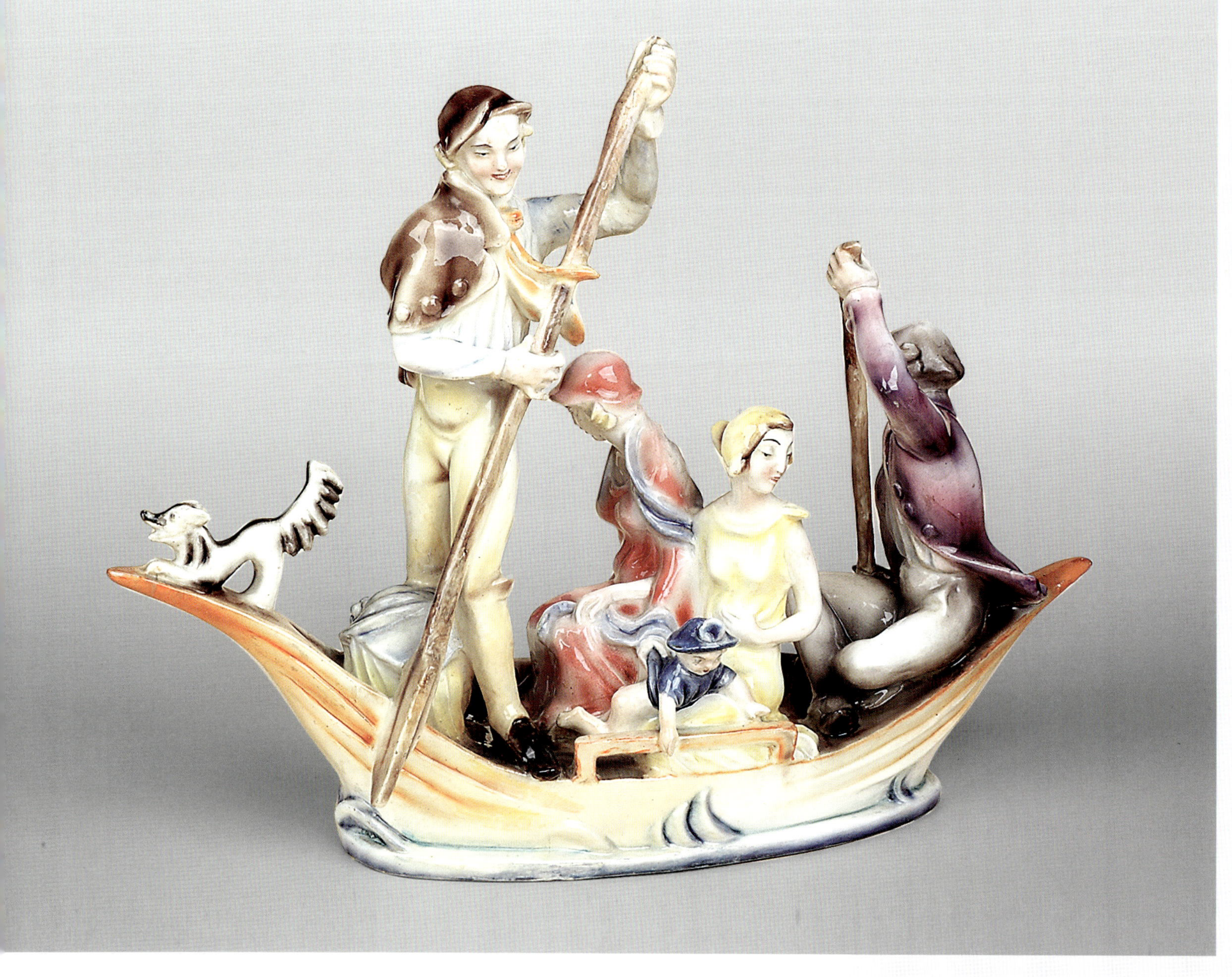

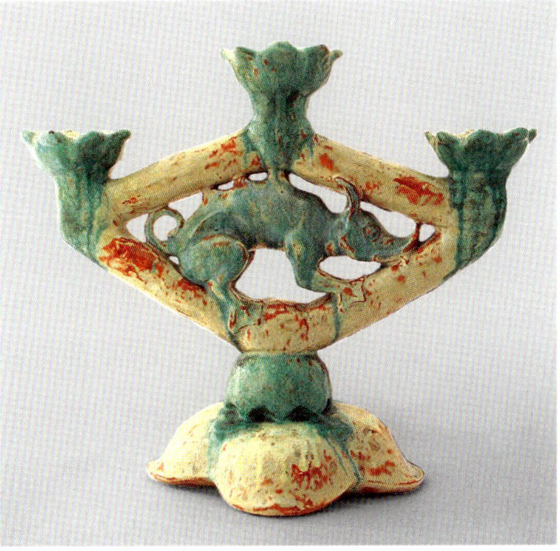

ModNr. 5693
Kerzenhalter I Candlestick, Pejic, ca. 1927, Collection a@goldscheider.net

Dina Kuhn's models in red and polychrome majolica were also exhibited at the Leipzig Fair, where Willy Bormann's "English Greyhound" attracted attention as particularly well done because it was so sinuously elegant. Other notably successful exhibits were Lorenzl's colorful "Aida" and especially a group designed by Karl Perl, "Die sieben Schwaben" ["The Seven Swabians"] because of its composition and comical air.[219] This figure was also remarked on by Professor

Dina Kuhns Modelle aus roter und bunter Majolika werden auch auf der Leipziger Messe ausgestellt, wo ansonsten Willy Bormanns „Englischer Greyhound" aufgrund seiner Geschmeidigkeit und Eleganz, Lorenzls farbenfrohe „Aida", aber vor allem die von Karl Perl entworfene Gruppe „Die sieben Schwaben" wegen ihrer Komposition und Komik als besonders gelungen auffallen.[219] Diese Figur wird von Prof. Konrad Meyer auch als entscheidender Beweis für die damalige Debatte zur Vorrangigkeit des künstlerischen Schaffens vor dem materiellen Wert eines Arbeitsstoffes genannt, mit der Schlussfolgerung, dass man Fayence und Porzellan als „gleichwertig für künstlerische Bearbeitungsmöglichkeiten ansehen" sollte.[220] Während Porzellan transparenter und dezenter wirkt, hat Fayence aufgrund der niedrigeren Brenn-

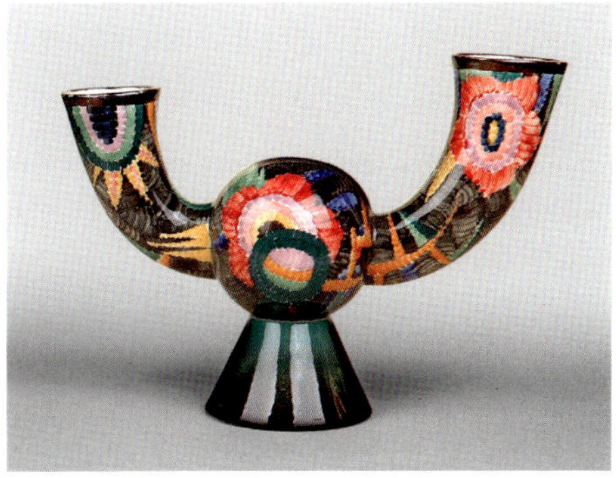

ModNr. 5343
Kerzenhalter I Candlestick, Dina Kuhn, ca. 1924, PB Wien 19

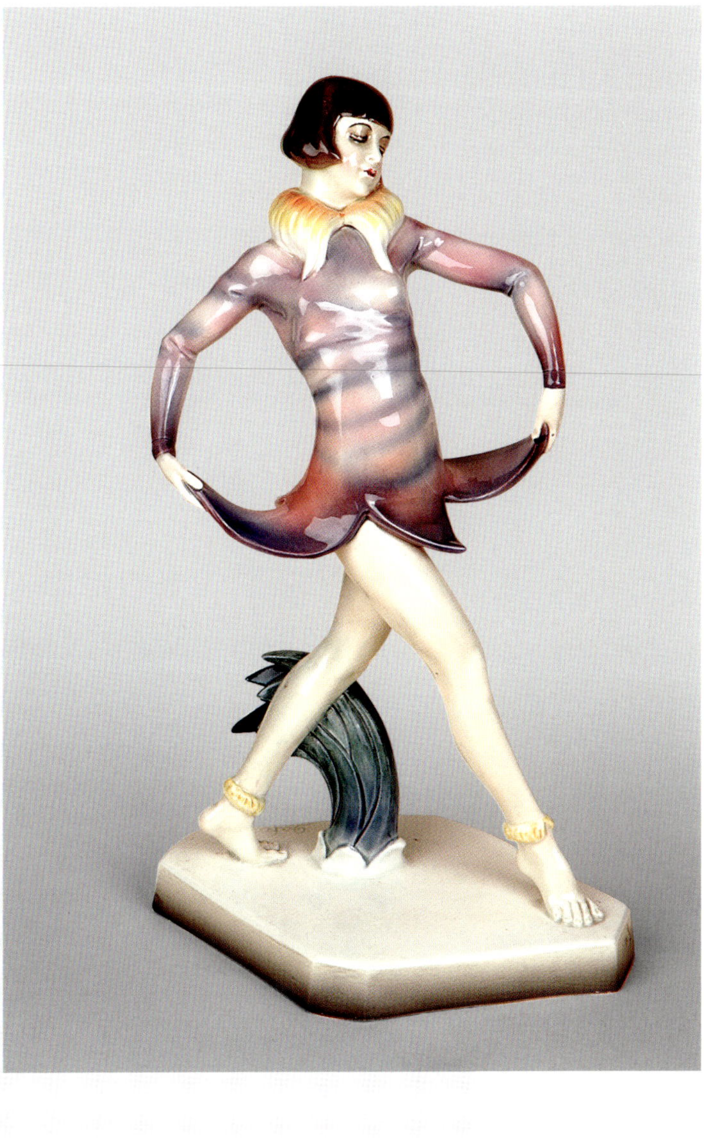

ModNr. 5649
Figur und Lampe I Statue and Lamp „Revue", Stephan Dakon, ca. 1927, beide I both PB Wien 19

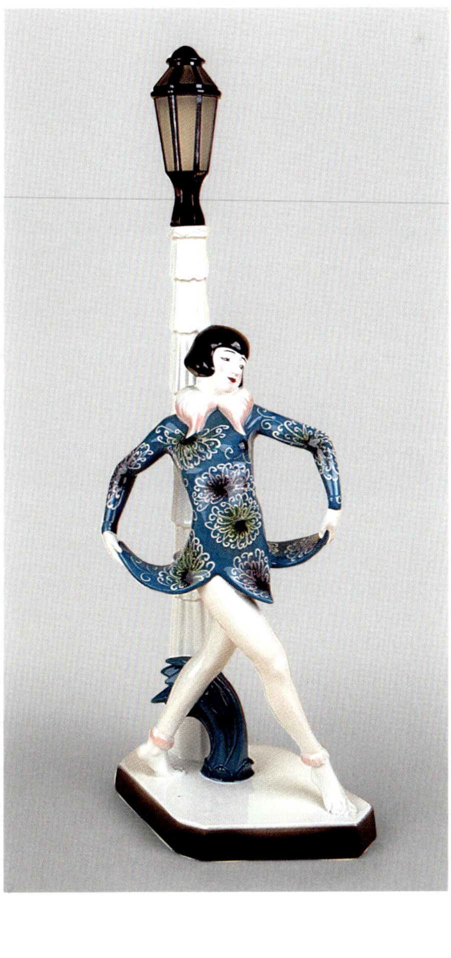

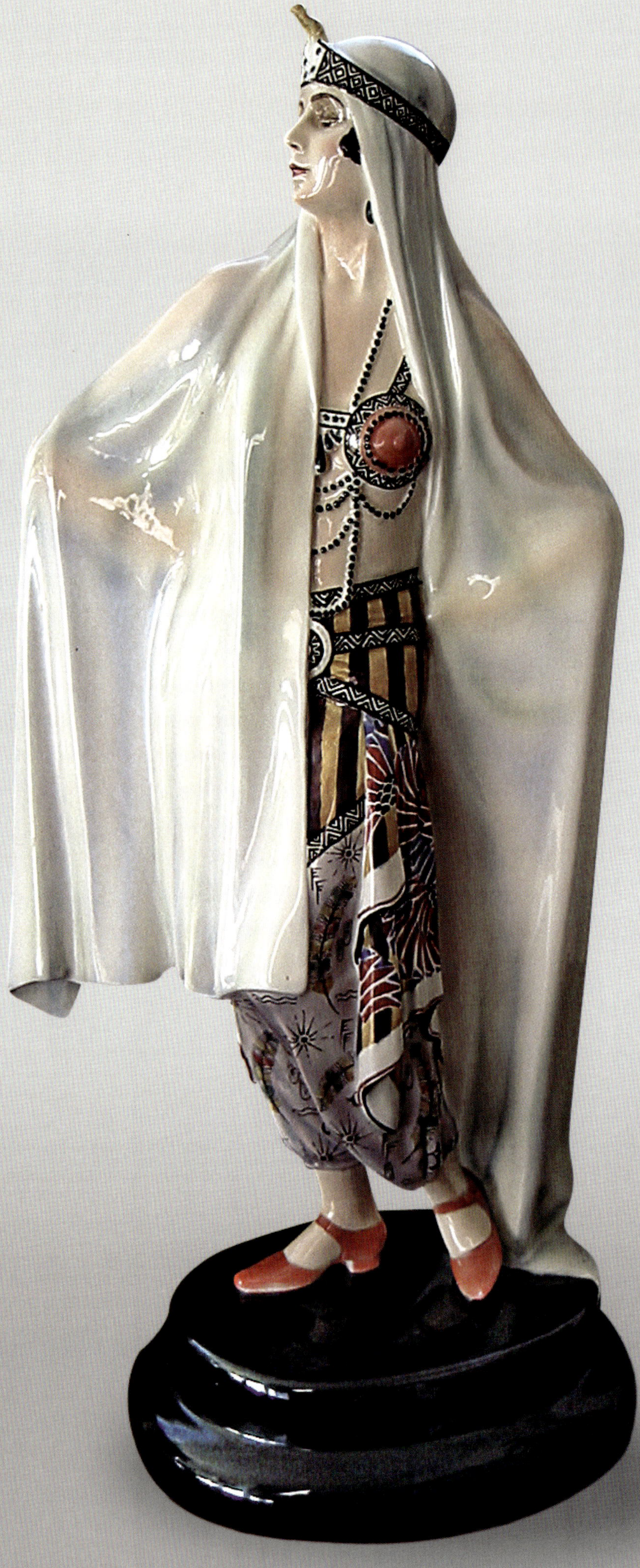

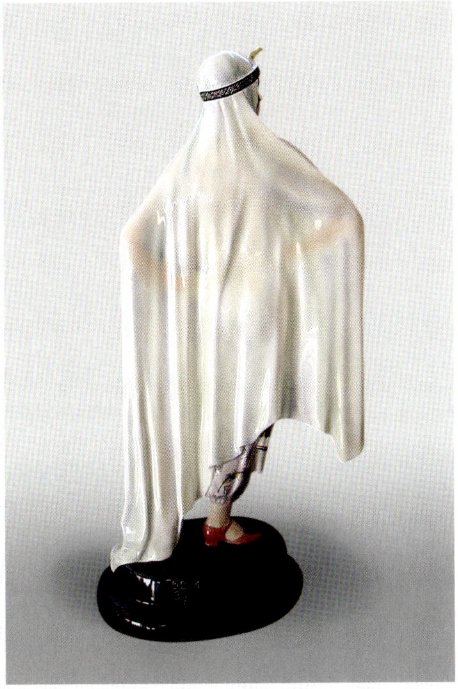

ModNr. 5281
„Aida", Josef Lorenzl, ca. 1923, Collection f@goldscheider.de
Diese Ausführung täuscht durch die Bemalung eine Durchsichtigkeit des Umhangs vor, eine fast porzellanhafte Transparenz. Der damalige Diskurs um die künstlerische Gleichwertigkeit, unabhängig von einer Wertigkeit des Arbeitsstoffes, findet hier ein Beispiel.

The painting of this version features a deceptively transparent cloak, almost like porcelain in its translucency. The discourse at the time is exemplified here, regardless of the issue of the aesthetic value of the material used.

Konrad Meyer as crucial evidence in the debate raging at the time on artistic creation taking precedence over the material value of the material used to make it, concluding that faience and porcelain should be "viewed as of equal value as possibilities for working artistically."[220] Whereas porcelain looked more translucent and elegant, faience had the advantage, due to its lower firing temperature, of being able to take on more

temperatur den Vorteil, dass sie lebendigere, kräftigere Farbtöne aufnehmen kann und deshalb beide „ihren Platz" in einem Heim nebeneinander beanspruchen können.

Im Jahr 1926 beschäftigt die Wiener Manufaktur 100 bis 120 Arbeiter.[221] Hinzu kommen weitere Arbeiter in der Leipziger und Pariser Filiale. Die kühl-elegante Linienführung der „Neuen Sachlichkeit", später der Art Déco, bestimmt die Stilrichtung; eine Vielzahl von eingesetzten Materialien kennzeichnet die Produktion. Die Wiener Manufaktur verkauft Kunstgegenstände in Fayence, Marmor, Bronze und Terrakotta,[222] aber auch Kaminverkleidungen und Objekte aus Kunststein gehören zum Sortiment[223]. Passend zu den nun extrem schlanken Proportionen werden die Figuren auch in kleineren Größen produziert. Dieser Trend setzt sich auch in den 1930er Jahren mit den so genannten „Goldscheider-Nippes" fort. Auf der Wiener Weihnachtsschau im Künstlerhaus werden Keramiken von Hertha Bucher, Herlinger-Schwarz, Kostial, Dina Kuhn, Grete Neuwalder und Karl Perl sowie Arbeiten aus Marmor und Alabaster nach Entwürfen von Ena Rottenberg-Henny und Josef Berger ausgestellt.[224]

Zwischen den Brüdern kommt es jedoch verstärkt zu Auseinandersetzungen über die Ausrichtung des Unternehmens. Während Marcell, der eine kunstkeramische Ausbildung genoss, den Künstler viel stärker in den Produktionsprozess einbeziehen will und mit expressiverer Keramik nicht einfach nur dem

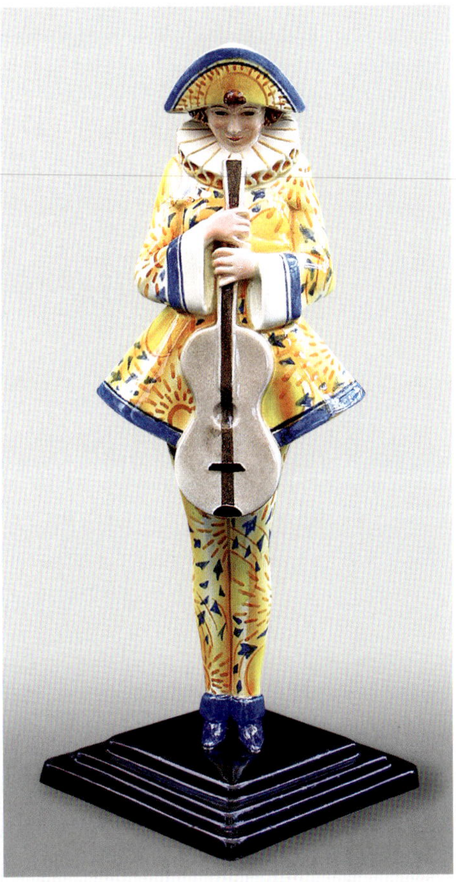

ModNr. 5903
„Aschermittwoch" I "Ash Wednesday", Stephan Dakon, ca. 1929, Collection Robert Senft

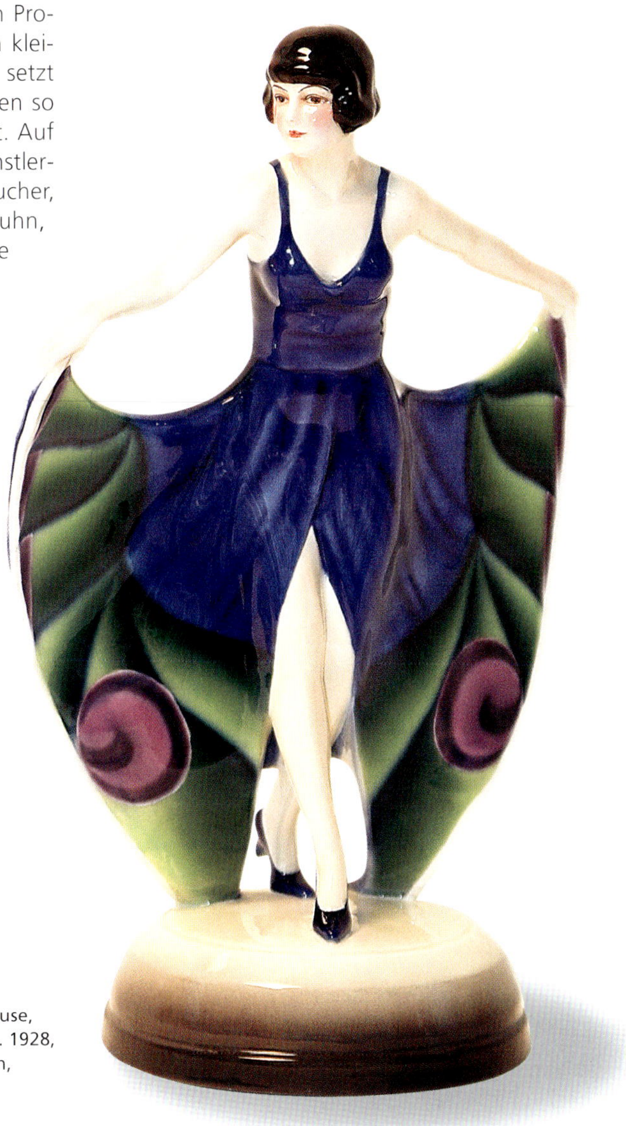

ModNr. 5715
Tänzerin I Danseuse, Josef Lorenzl, ca. 1928, private collection, Wien

rechts I right

ModNr. 5940 Dame I Lady, Josef Lorenzl, ca. 1929, Collection John Jesse, London

Marktgeschmack folgen will, steht für Walter, der eine kaufmännische Ausbildung absolvierte, der wirtschaftliche Erfolg im Vordergrund. Und dieser ist nur durch eine Befriedigung des Konsumentengeschmacks und einer seriellen Reproduzierbarkeit der Modelle möglich. Da es zwischen den Brüdern zu keinem Kompromiss kommt, beschließt man sich zu trennen. Am 18. Januar 1928 erwirbt Walter von seinem Bruder

ModNr. 5822 „Spanischer Shawl" I "Spanish Shawl", Josef Lorenzl, ca. 1928, Sammlung.schmutz@chello.at

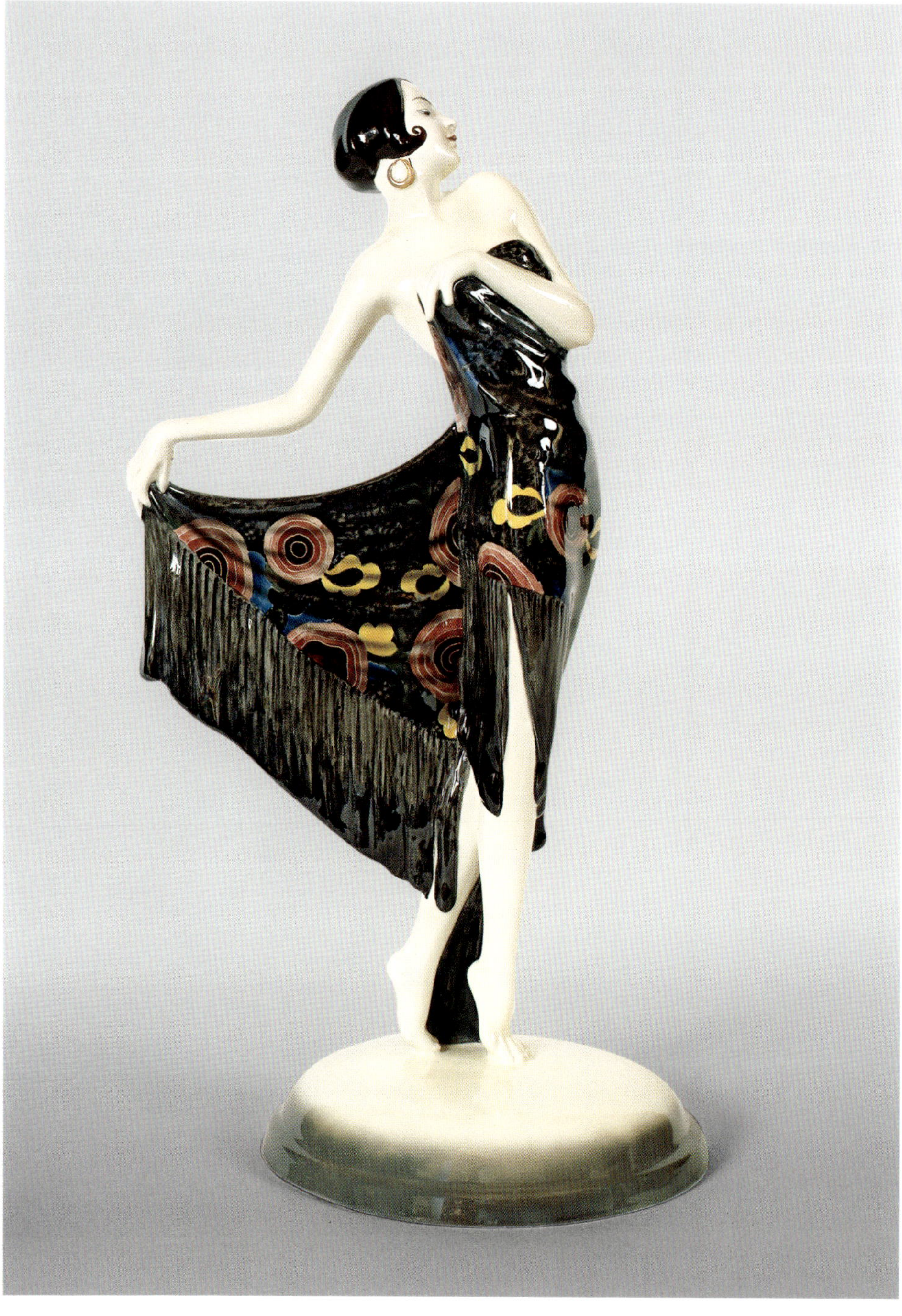

vivid, stronger color tones. Both, therefore, could lay claim on an equal footing to "a place" in the home.

In 1926 the Viennese factory employed a workforce ranging from 100 to 120.[221] In addition, there were more workmen at the Leipzig and Paris branches. The cool, elegant "New Objectivity" handling of line that would turn into Art Déco distinguished the products. The Viennese factory sold art objects in faience, marble, bronze and terracotta.[222] The product range also included fireplace surrounds as well as objects made of simulated stone.[223] Matching the extremely slender proportions now in style, figures were produced in smaller sizes. This trend continued on into the 1930s with what was called "Goldscheider Nippes" ["Goldscheider bric-à-brac"]. At the Christmas show held at the Künstlerhaus in Vienna, work was shown by Hertha Bucher, Herlinger-Schwarz, Kostial, Dina Kuhn, Grete Neuwalder and Karl Perl as well as pieces executed in marble and alabaster after designs by Ena Rottenberg-Henny and Josef Berger.[224]

Between the brothers, however, disputes were now erupting more frequently on what direction the business should be going. Marcell, who had, after all, trained as a ceramicist, wanted to involve artists more closely in the production process and make more expressive ceramics, not just pander to market taste. For Walter, on the other hand,

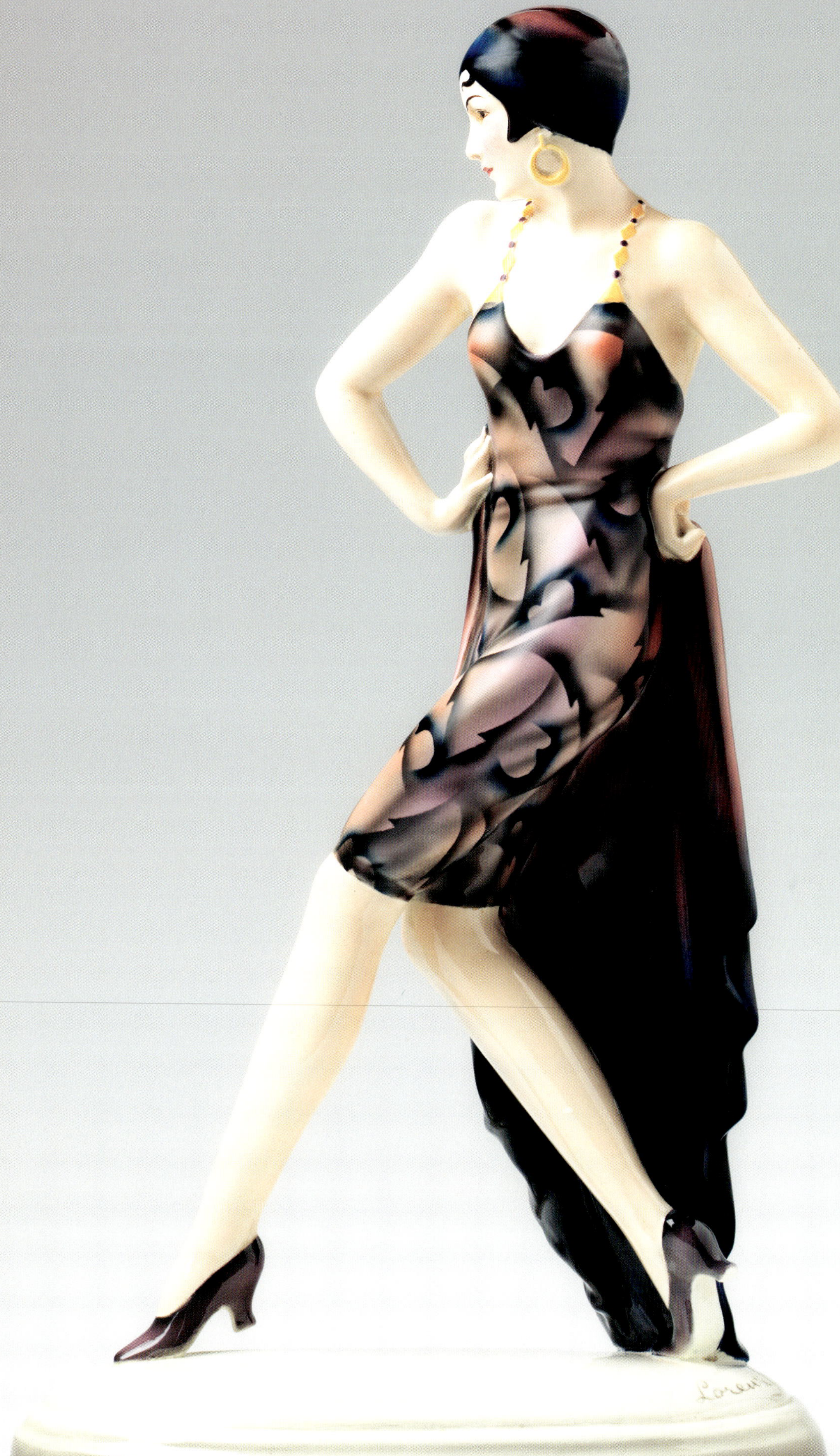

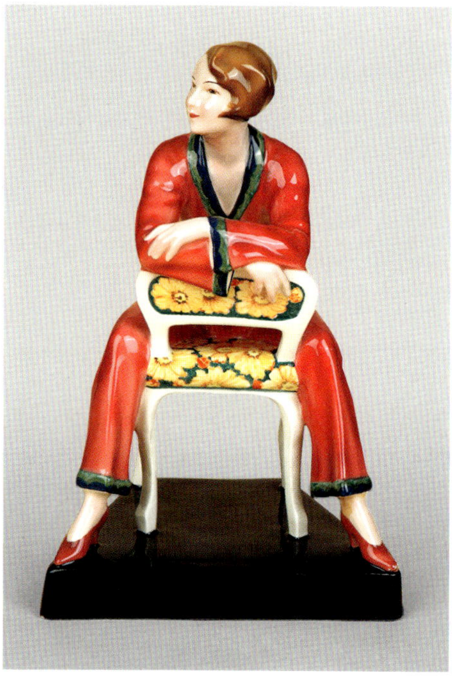

ModNr. 5780
„Pyjama" I "Pyjamas",
Josef Lorenzl, ca. 1928,
PB Wien 19

Marcell die zweite Hälfte der Fabrik in der Staudgasse 7 und 9 für den Betrag von 30.000 Schilling, sein Erbanteil an dem Wohnhaus in der Hasenauerstraße 16 geht an Marcell über.[225] Zwei Tage später wird Walter als Alleininhaber eingetragen, nachdem sein jüngerer Bruder Marcell gegen eine „sehr hohe Ablöse" seinen Geschäftsanteil an ihn verkauft hat.[226] Eine der glanzvollsten Zeiten der Wiener Manufaktur Friedrich Goldscheider, die in diesen wenigen Jahren einige der Meisterstücke der Art Déco Keramik hervorgebracht hat, geht zu Ende.

with his business training, commercial success had priority. And success of this kind was only possible if consumer tastes were taken seriously and models were mass produced. Since no compromise was arrived at between the brothers, they decided to go separate ways. On 18 January 1928, Walter bought Marcell's half of the factory at 7 and 9 Staudgasse for 30 000 Schillings. His share of the dwelling at 16 Hasenauerstraße went to Marcell.[225] Two days later, Walter was registered as sole proprietor of the firm after his younger brother Marcell had let him buy out his share of the business to him for "a very large sum."[226] Thus ended one of the most brilliant eras of the Wiener Manufaktur Friedrich Goldscheider, which in just a few years had produced some masterpieces of Art Déco ceramics.

ModNr. 5891
Buchstütze I Bookend,
Ida Meisinger,
ca. 1928, Collection
a@goldscheider.net

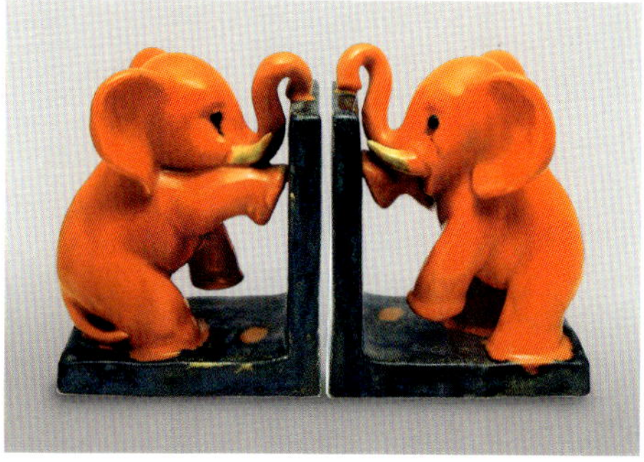

ModNr. 5811
Buchstütze I Bookend,
Ida Meisinger,
ca. 1928, PB Wien

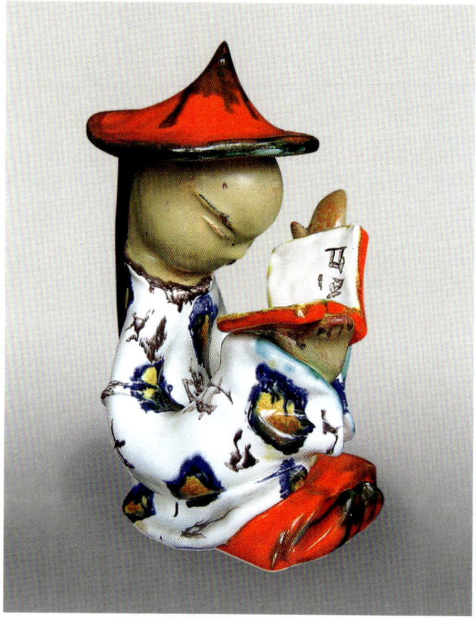

ModNr. 5773
Buchstütze I Bookend
Josef C. Motto,
ca. 1928, Collection
a@goldscheider.net

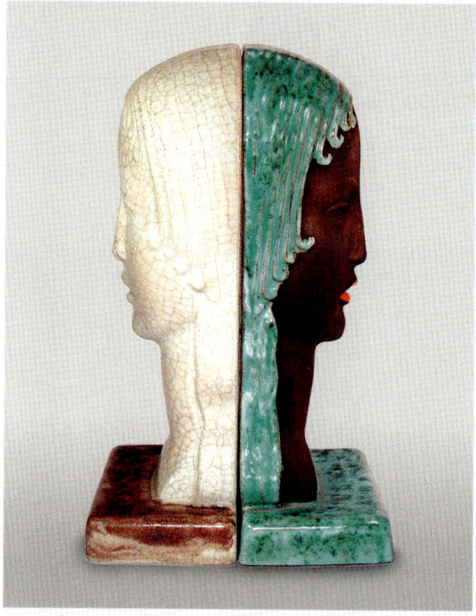

Notes

[192] See Tonwaren-, Ofen- und Töpfer-Zeitung, 46.1922, p. 54
[193] Handelsregister [commercial register] zur Firma "Erste Wiener Terrakottafabrik und Atelier für künstlerische Fayenzen Friedrich Goldscheider" (Bestand Handelsgericht Wien, Handelsregister A 16/19)
[194] Letter written by Walter Goldscheider to the Vermögensverkehrsstelle [Office for Property Tax], 4 Aug. 1938
[195] Handelsregister zur Firma "Erste Wiener Terrakottafabrik und Atelier für künstlerische Fayenzen Friedrich Goldscheider" (Bestand Handelsgericht Wien, Handelsregister A 16/19)
[196] Cf Lehmann's Adressbuch, 1925 to 1928
[197] Handelsregister zur Firma "Erste Wiener Terrakottafabrik und Atelier für künstlerische Fayenzen Friedrich Goldscheider" (Bestand Handelsgericht Wien, Handelsregister A 16/19). No entries relating to Alois Goldscheider occur in the Viennese directory from 1923, which implies that Alois died in 1922.
[198] Letter written by Walter Goldscheider to the Vermögensverkehrsstelle, 4 Aug. 1938
[199] J. A. Meisenbach: Putten mit Tieren, in: Keramos, 2.1923, p. 305
[200] W. Burger: Die Deutsche Kunstgewerbeschau in München, Keramik und Glas, in: Dekorative Kunst, 25.1921/22, p. 265
[201] N.N.: Wiener Kleinplastik einst und jetzt, in: MW, 2.1920/21, [Heft] 12, p. 29
[202] PGGTR, June 1941, p. 486. "By schooling the decorateurs of the factory, however, in improved methods of underglaze decoration, and by the introduction of a lower fired body, he was able gradually to replace the rather pale colours of the "grand feu" type, which were then in general use at the works, by the introduction of new underglaze colours of breadth and vivacity, in the utilisation of which, new styles of decoration were introduced calling for the skilful use of aerograph pencil at a time when that process had not been developed to any great extent."
[203] Patentschrift 96834, Österreichisches Patentblatt, Nr. 12, 1923, p. 382. The patent was valid as from 15 Dec. 1923.
[204] Ros, p. 16
[205] Taschenbuch für Keramiker, 1924, p. 140
[206] Makus (2000), p. 180. Thomasch left Goldscheider in 1926 and bought a pottery at Sierndorf.
[207] Ibid, p. 182
[208] Goldscheider, M.: Modern oder zeitgemäß? Stimmen zur Internationalen Kunstgewerbeausstellung in Paris, in: Das Handelsmuseum, 40. 1925, p. 5–6. Two styles dominated sculpture at that time: on the one hand, a naturalistic manner, that is, one that tended to be retro and not influenced by contemporary art, on the other, a style that was expressive and radical for the time.
[209] Ibid
[210] Ibid
[211] Ibid
[212] Ibid
[213] Ibid
[214] Als Preise gelangen zur Verteilung, in: KuKGW, 8.1927/28
[215] Deutsche Gewerbeschau München 1922, Amtlicher Führer
[216] Jubiläumsausstellung 1884–1924. Wiener Kunstgewerbe-Verein, 1924, p. 32
[217] Grell, p. 32f.
[218] N.N.: Österreich in Paris, in: ÖBWK, 1.1924/25, p. 331
[219] Meyer, p. 8f.
[220] Ibid, p. 12
[221] IC, 1926/27, p. 379
[222] Advertisement placed by the Wiener Manufaktur Friedrich Goldscheider in the catalogue to the Christmas show [Weihnachtsschau] at the Künstlerhaus in Vienna, 1926
[223] Lehmann's Adressbuch, 1926
[224] Weihnachtsschau im Künstlerhaus, Kunst und Kunsthandwerk. November-December 1926. Ausstellungsprospekt Raum VIII Keramik, p. 56 and p. 61
[225] Deposition made by Josef Schuster to the Viennese Magistracy, Division 11 Wertzuwachsabgabe on 21 Jan. 1939, with reference to the Urkundensammlung des Amtsgerichtes Döbling 392/28
[226] Exposé relating to the Aryanization of Dr. Stefan Schwartz to Vermögensverkehrsstelle, 12 Aug. 1938

Anmerkungen

[192] Siehe Tonwaren-, Ofen- und Töpfer-Zeitung, 46.1922, S. 54
[193] Handelsregister zur Firma „Erste Wiener Terrakottafabrik und Atelier für künstlerische Fayenzen Friedrich Goldscheider" (Bestand Handelsgericht Wien, Handelsregister A 16/19)
[194] Brief von Walter Goldscheider an die Vermögensverkehrsstelle, 4.8.1938
[195] Handelsregister zur Firma „Erste Wiener Terrakottafabrik und Atelier für künstlerische Fayenzen Friedrich Goldscheider" (Bestand Handelsgericht Wien, Handelsregister A 16/19)
[196] Vgl. Lehmann's Adressbuch, Jahrgänge 1925 bis 1928
[197] Handelsregister zur Firma „Erste Wiener Terrakottafabrik und Atelier für künstlerische Fayenzen Friedrich Goldscheider" (Bestand Handelsgericht Wien, Handelsregister A 16/19). Ab dem Jahr 1923 sind zu Alois Goldscheider keine Einträge im Wiener Adressbuch zu finden, so dass davon auszugehen ist, dass Alois im Jahr 1922 verstorben ist.
[198] Brief von Walter Goldscheider an die Vermögensverkehrsstelle, 4.8.1938
[199] Meisenbach, J.A.: Putten mit Tieren, in: Keramos, 2.1923, S. 305
[200] Burger, W.: Die Deutsche Kunstgewerbeschau in München, Keramik und Glas, in: Dekorative Kunst, 25.1921/22, S. 265
[201] N.N.: Wiener Kleinplastik einst und jetzt, in: MW, 2.1920/21, Heft 12, S. 29
[202] PGGTR, June 1941, S. 486. „By schooling the decorateurs of the factory, however, in improved methods of underglaze decoration, and by the introduction of a lower fired body, he was able gradually to replace the rather pale colours of the „grand feu" type, which were then in general use at the works, by the introduction of new underglaze colours of breadth and vivacity, in the utilisation of which, new styles of decoration were introduced calling for the skilful use of aerograph pencil at a time when that process had not been developed to any great extent."
[203] Patentschrift 96834, Österreichisches Patentblatt, Nr. 12, 1923, S. 382. Das Patent wird ab dem 15.12.1923 gewährt.
[204] Ros, S. 16
[205] Taschenbuch für Keramiker, 1924, S. 140
[206] Makus (2000), S. 180. Thomasch verlässt 1926 Goldscheider und erwirbt eine Töpferei in Sierndorf.
[207] Ebenda, S. 182
[208] Goldscheider, M.: Modern oder zeitgemäß? Stimmen zur Internationalen Kunstgewerbeausstellung in Paris, in: Das Handelsmuseum, 40. 1925, S. 5–6. In der damaligen Bildhauerkunst herrschten vor allem zwei Richtungen vor: einerseits die naturalistische, also eher retrospektive und unzeitgemäße, andererseits die stilisierende, expressive und für damalige Verhältnisse radikale Bildhauerei.
[209] Ebenda
[210] Ebenda
[211] Ebenda
[212] Ebenda
[213] Ebenda
[214] Als Preise gelangen zur Verteilung, in: KuKGW, 8.1927/28
[215] Deutsche Gewerbeschau München 1922, Amtlicher Führer
[216] Jubiläumsausstellung 1884–1924. Wiener Kunstgewerbe-Verein, 1924, S. 32
[217] Grell, S. 32f
[218] N.N.: Österreich in Paris, in: ÖBWK, 1.1924/25, S. 331
[219] Meyer, S. 8ff
[220] Ebenda, S. 12
[221] IC, 1926/27, S. 379
[222] Anzeige der „Wiener Manufaktur Friedrich Goldscheider" im Katalog zur Weihnachtsschau im Künstlerhaus Wien, 1926
[223] Lehmann's Adressbuch, 1926
[224] Weihnachtsschau im Künstlerhaus, Kunst und Kunsthandwerk. November-Dezember 1926. Ausstellungsprospekt Raum VIII Keramik, S. 56 sowie S. 61
[225] Stellungnahme von Josef Schuster an das Wiener Magistrat, Abteilung 11 Wertzuwachsabgabe, vom 21.1.1939, bezugnehmend auf die Urkundensammlung des Amtsgerichtes Döbling 392/28
[226] Expose betreffend die Arisierung von Dr. Stefan Schwartz an die Vermögensverkehrsstelle, 12.8.1938

Populäre und moderne Keramik der dreißiger Jahre unter Walter Goldscheider (1928–1938)

Popular and Modern 1930s Ceramics under Walter Goldscheider (1928–1938)

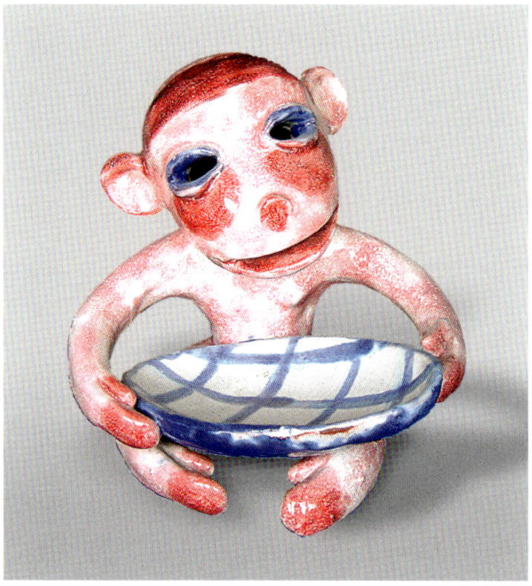

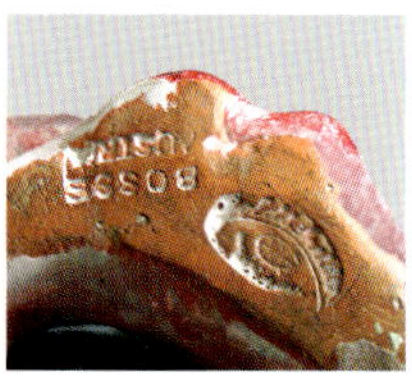

ohne | without ModNr.
Affe mit Schale |
Monkey with Bowl,
Walter Bosse, ca. 1928,
PB Wien

Nach der Trennung der beiden Brüder führt Walter 1928 einen zweiten Hauptzweig der Produktion auf dem Markt ein: die modernen Keramiken. Diese umfassen die Erzeugung von Köpfen, Masken, Tieren, Figuren, Gebrauchskeramik, Reliefs etc. und sollen den neuen Geschmack der Wohnungseinrichtungen bedienen.[227] Die Zeitschrift für Kunstgewerbe „Die Schaulade" schreibt über die Moderne Keramik von Goldscheider: „Die jüngste Zeit brachte, bedingt durch die moderne Gestaltung der Wohnräume, eine neuzeitliche Keramik, ausgezeichnet durch Klarheit und Harmonie in Form und Oberflächengestaltung. Vasen, Schalen, Leuchter, Masken und moderne Plastiken geben dieser neuen Richtung das Gepräge."[228]

Die ersten Modelle dieser neuen Kollektion werden vom 24jährigen Walter Bosse angefertigt, der an der Kunstgewerbeschule in Wien unter Michael Powolny und Franz

After the two brothers went their separate ways, Walter launched a second important product line in 1928: modern ceramics. They encompassed heads, masks, animals, figures, utilitarian household ceramics, reliefs, etc, and were intended to suit the new taste in furnishing and decorating the homes.[227] Die Schaulade, a decorative arts magazine, wrote as follows on the Goldscheider Modern Ceramics: "Recent years have brought forth, due to the modern decoration of rooms in dwelling, modern ceramics distinguished by clarity and harmony of form and surface design. Vases, bowls, candlesticks, masks and modern pieces of sculpture shape this new trend."[228]

The first models in that new collection were made by Walter Bosse, who was only twenty-four years old and had trained at the Viennese Kunstgewerbeschule with Michael Powolny and Franz Cizek. He ran a work-

Ill. 56

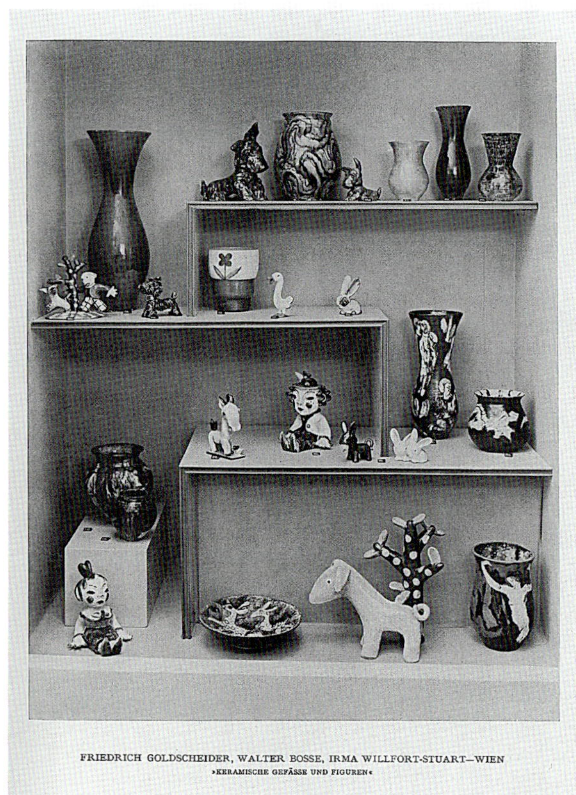

Ill. 55

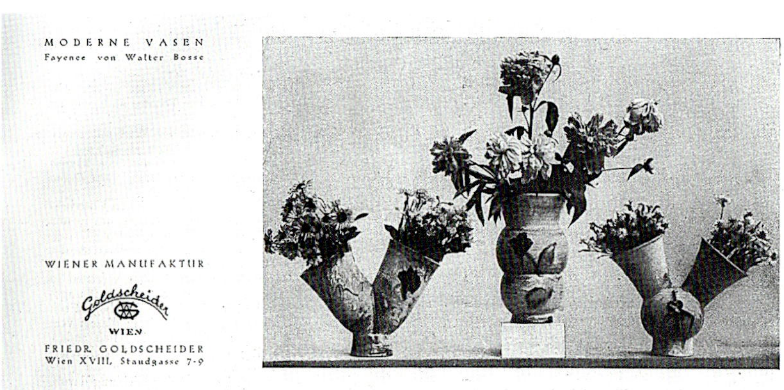

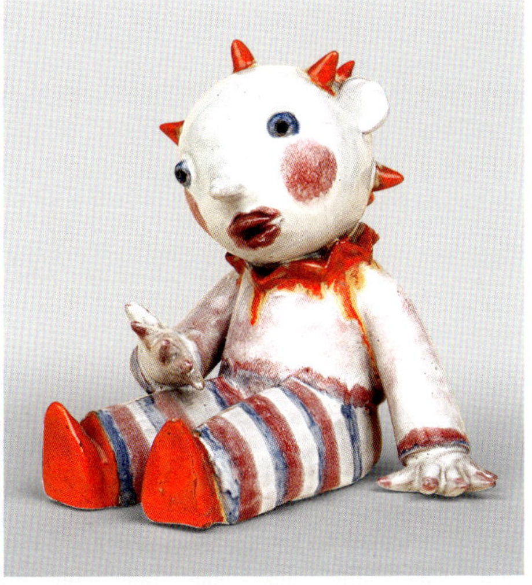

ohne I without ModNr.
Buchstütze Igelmann ohne Buch I Bookend Hedgehog Man without Book, Walter Bosse, ca. 1928, PB Wien 19

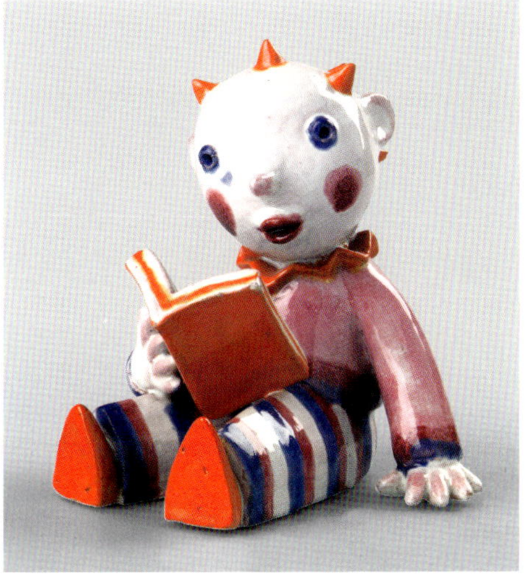

ohne I without ModNr.
Buchstütze Igelmann mit Buch I Bookend Hedgehog Man with Book, Walter Bosse, ca. 1928, Collection mad

Walter Bosse wird als externer Mitarbeiter beschäftigt und prägt um 1930 die neue Linie der „Modernen Keramik". Diese wird auch konsequent in den Zeitschriften Moderne Welt sowie Die Dame oder Deutsche Kunst und Dekoration beworben, vgl. Ill. 55 und 56. (Quelle: Moderne Welt, Jahrgang 11, 1929/30; Deutsche Kunst und Dekoration, Band 67) Die Modelle von Walter Bosse sind öfters nicht markiert und werden auch teilweise ohne den Goldscheider-Stempel in Bosses eigener Werkstatt produziert und verkauft.

Employed on a freelance basis, Walter Bosse shaped the new line in "Modern Ceramics" around 1930. It was consistently advertised in such magazines as Moderne Welt, Die Dame and Deutsche Kunst und Dekoration; cf Ill. 55 and 56. (Source: Moderne Welt, 11, 1929/30; Deutsche Kunst und Dekoration, vol. 67) Walter Bosse's models are often unmarked and some of them were produced and sold at Bosse's own workshop without the Goldscheider stamp.

shop of his own at Kufstein,[229] where he made the modern ceramics for Goldscheider and stamped them with a Goldscheider stamp placed at his disposal for the purpose.[230] Some objects were also marked with his own Bosse stamp. The collaboration between Walter Goldscheider and Walter Bosse was very productive at the time: Bosse delivered wares worth about 500 Schillings on a weekly basis.[231] Apart from modern utilitarian household ceramics, Bosse also made

Cizek studiert hat und eine eigene Werkstatt in Kufstein betreibt.[229] Dort stellt er die modernen Keramiken für Goldscheider her und prägt sie mit einem ihm zur Verfügung gestellten Goldscheider-Stempel.[230] Manche Objekte erhalten zusätzlich seinen eigenen Bosse-Stempel. Die Zusammenarbeit zwischen Walter Goldscheider und Walter Bosse ist in dieser Zeit sehr produktiv, Bosse liefert wöchentlich Ware im Wert von etwa 500 Schilling.[231] Außer der modernen Gebrauchs-

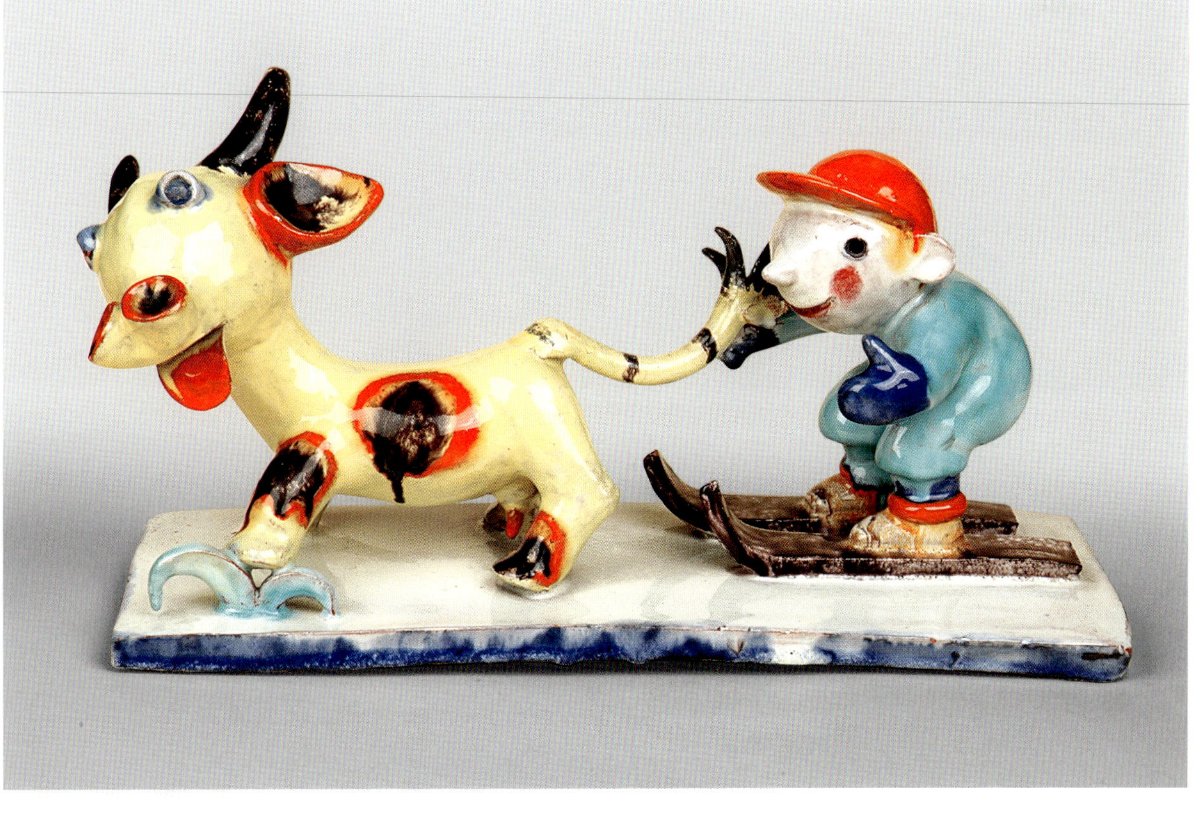

ohne I without ModNr.
Skifahrer mit Kuh I Skier with Cow, Walter Bosse, ca. 1928, PB Wien 19

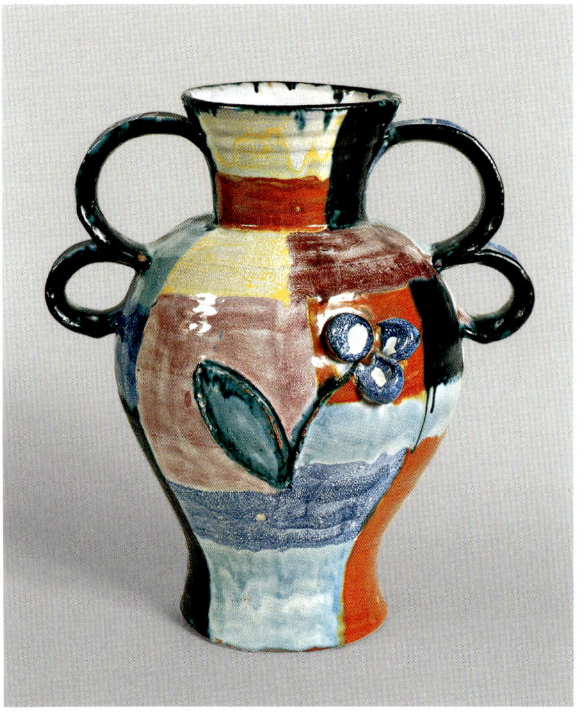

ohne I without ModNr.
Vase, Walter Bosse,
ca. 1928,
Sammlung.Schmutz@
chello.at

keramik produziert Bosse für Goldscheider auch figürliche Keramik, die von der Fachpresse als „ungequälter keramischer Humor" bezeichnet wird.[232] Im Jahr 1930 bewirbt sich Bosse bei Walter Goldscheider um die Stelle des künstlerischen Leiters, die Verhandlungen scheitern jedoch und die Zusammenarbeit lässt nach.[233]

Während der größte Teil der modernen Keramik an die bisherigen Modellnummern anschließt, wird die Gebrauchskeramik wie Vasen und Schalen zum größten Teil mit einer neuen Nummerierung versehen, die bei Null beginnt. In der Zeit vor dieser neuen Nummerierung wurde die Gebrauchskeramik nach der traditionellen Nummernvergabe geführt.

figurative ceramics for Goldscheider, which the specialist press praised as exemplifying "unstilted ceramic humor."[232] In 1930 Bosse applied to Walter Goldscheider for the position of artistic director but the negotiations were unsuccessful and the collaboration declined.[233]

Whereas most of the modern ceramics continued sequentially with the previous model numbers, household articles such as vases and dishes were usually numbered according to a new sequence beginning with zero. Before this new numbering system was implemented, utilitarian household ceramics followed the traditional numbering system.

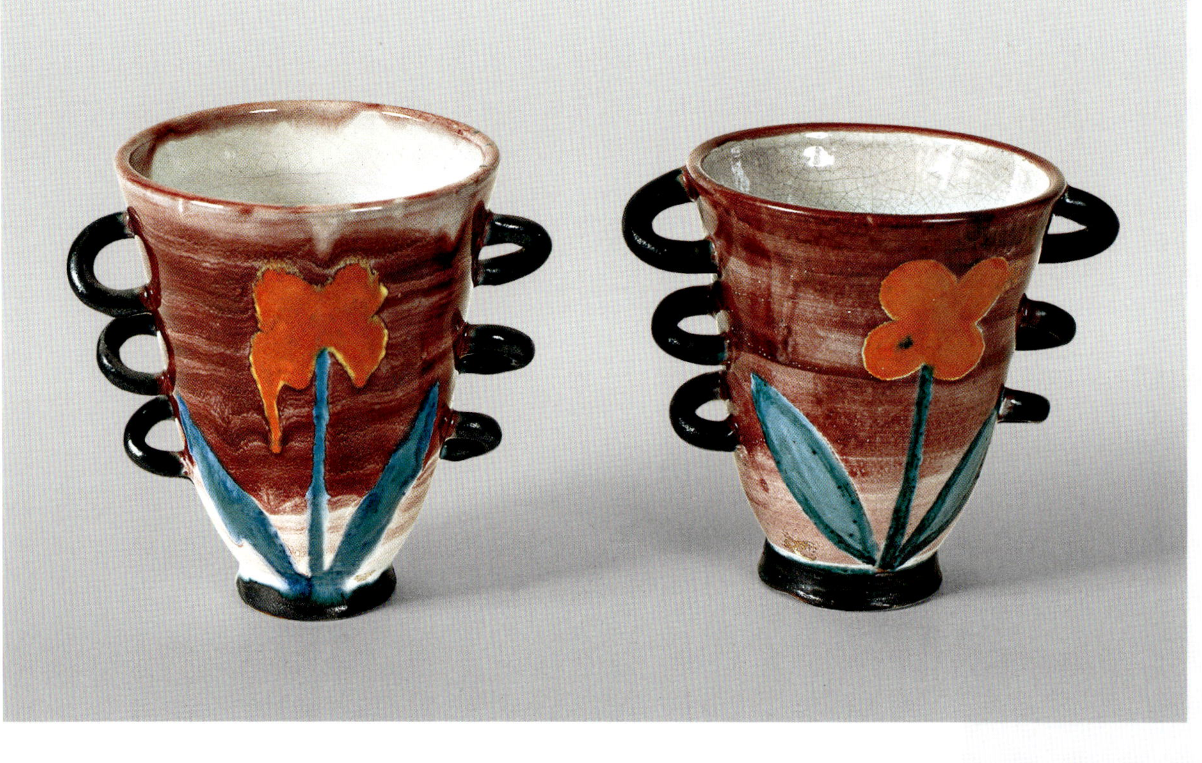

ohne I without ModNr.
Zwei Vasen I Two Vases,
Walter Bosse, ca. 1928,
Sammlung.Schmutz@
chello.at

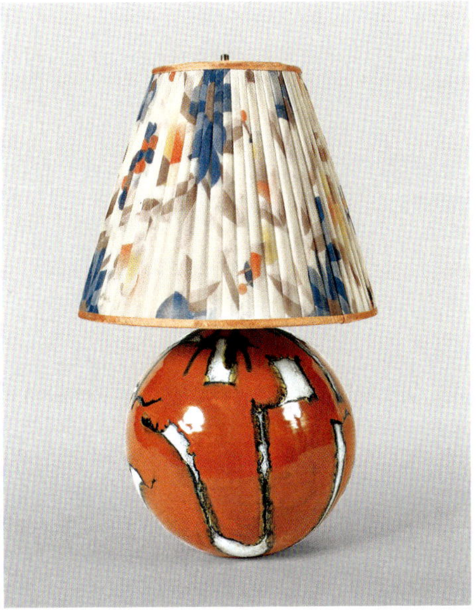

ModNr. 1

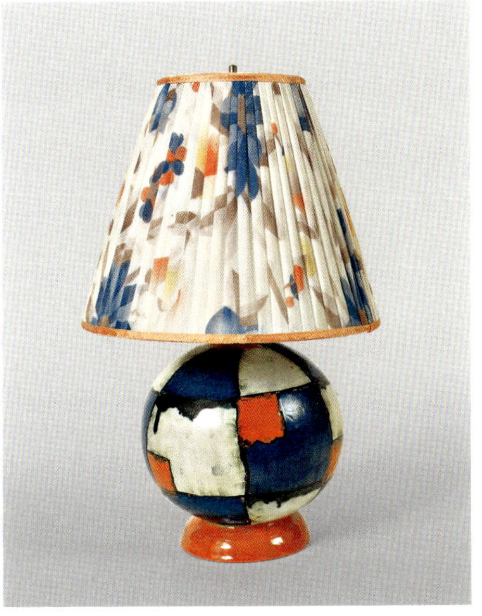

ModNr. 556

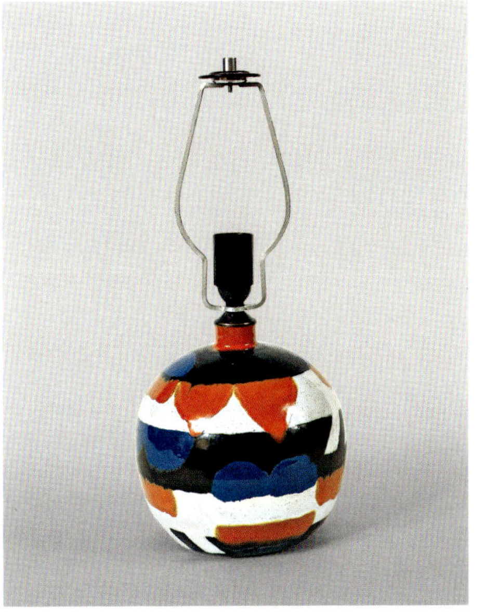

ModNr. 628

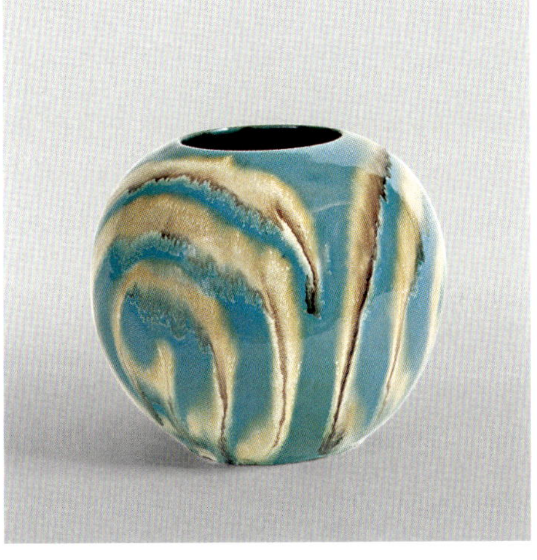

ModNr. 780

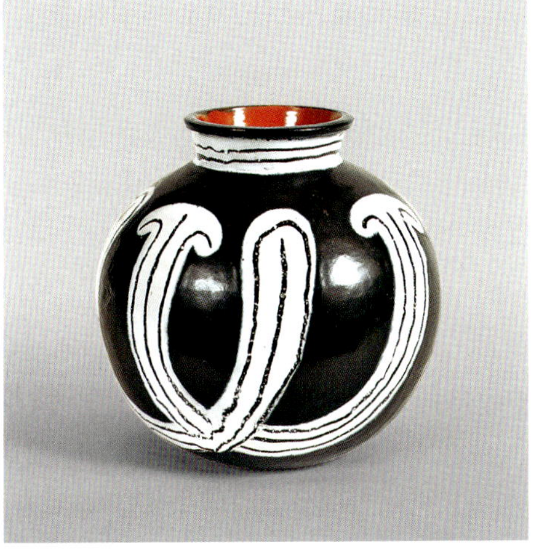

ModNr. 1039

ModNr. 1
Lampe | Lamp,
Ernst Weber,
ca. 1928–1938

ModNr. 556
Lampe | Lamp,
ca. 1928–1938

ModNr. 628
Lampe | Lamp,
ca. 1928–1938

ModNr. 780
Vase, ca. 1928–1938

ModNr. 1039
Vase, ca. 1928–1938

alle | all
Sammlung.Schmutz@chello.at

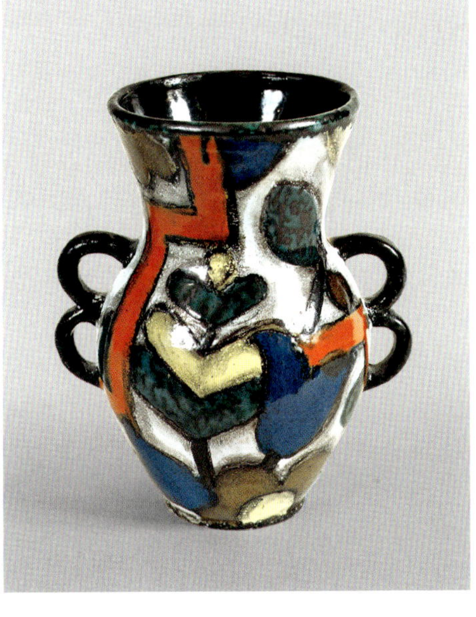

ModNr. 526

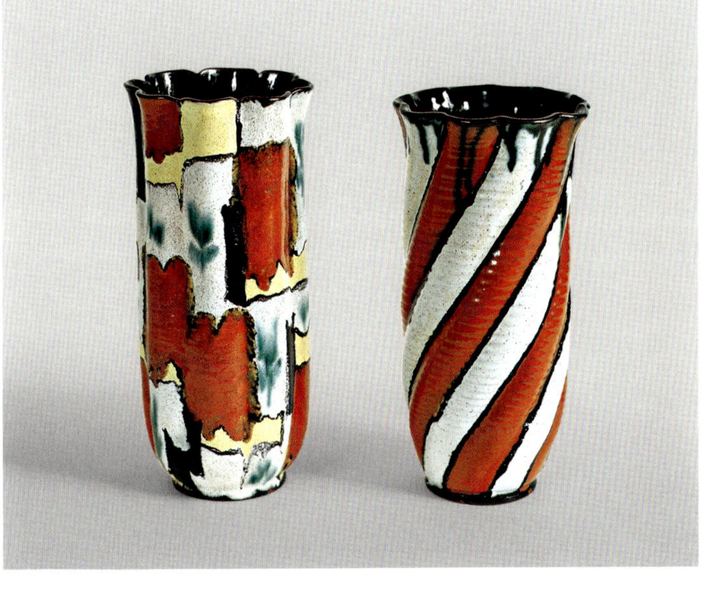

ModNr. 549 + ModNr. 552

ModNr. 526
Vase, Benno Geiger zugeschrieben | attributed to,
ca. 1928–1938

ModNr. 549
Vase, ca. 1928–1938

ModNr. 552
Vase, ca. 1928–1938

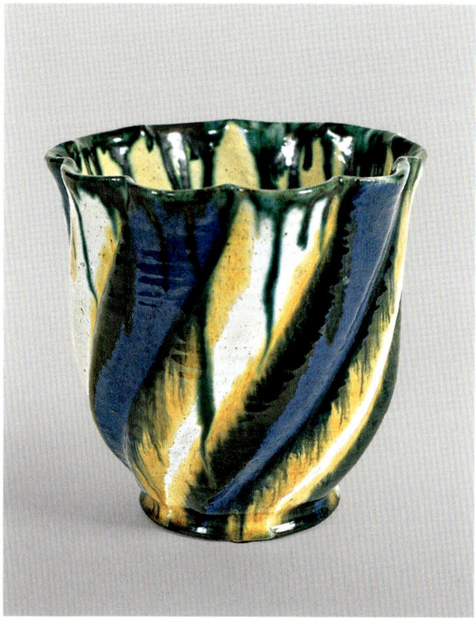

ModNr. 550

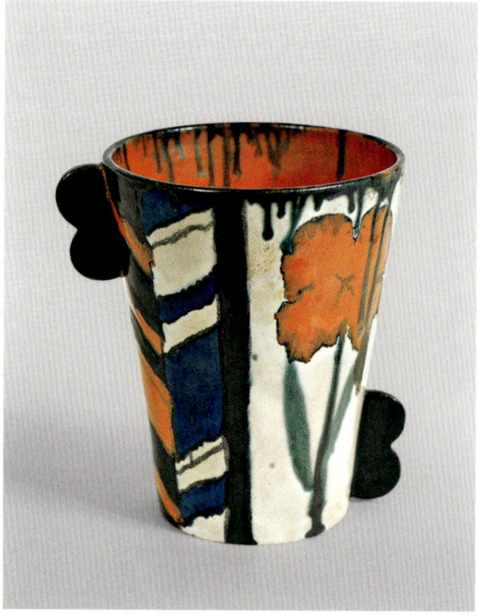

ModNr. 688

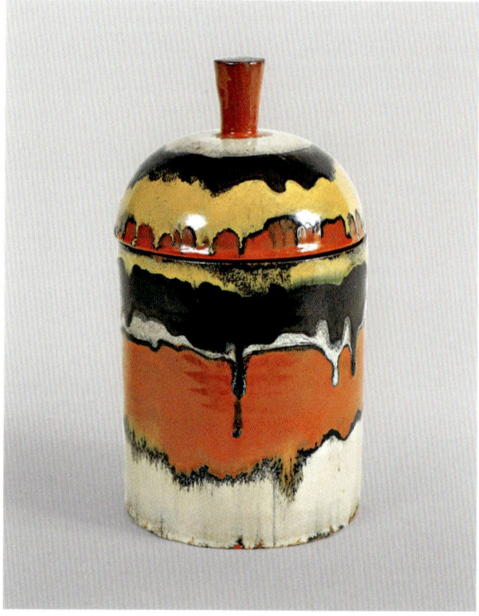

ModNr. 696

ModNr. 550
Vase, ca. 1928–1938

ModNr. 688
Vase, ca. 1928–1938

ModNr. 696
Deckeldose | Covered Box, ca. 1928–1938

ModNr. 706
Vase, ca. 1928–1938

ModNr. 1003
Vase, Benno Geiger zugeschrieben | attributed to, ca. 1928–1934

ModNr. 574
Vase, ca. 1928–1938

ModNr. 703
Vase, ca. 1928–1938

ModNr. 713
Vase, ca. 1928–1938

alle | all
Sammlung.Schmutz@chello.at

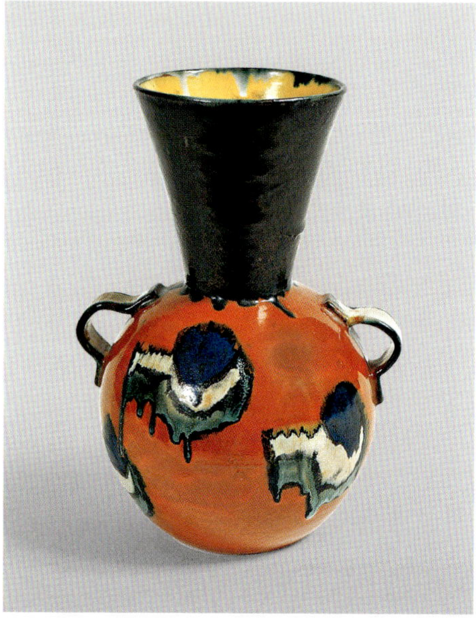

ModNr. 706

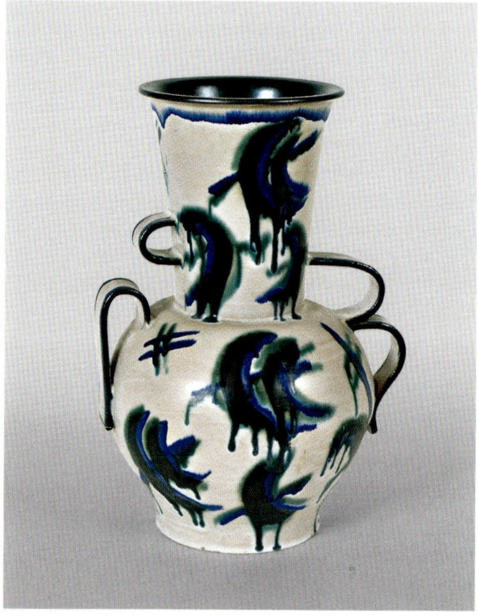

ModNr. 1003

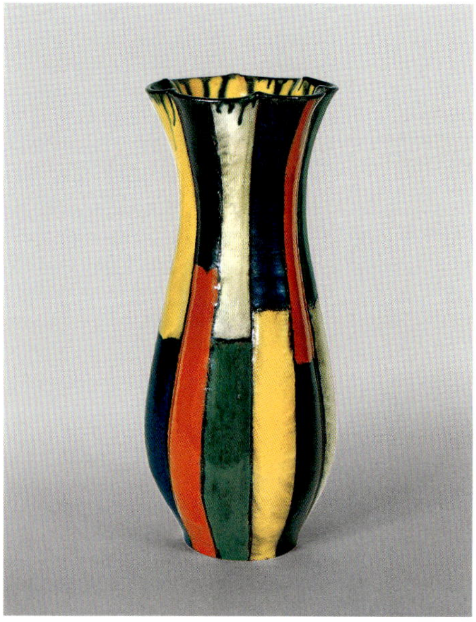

ModNr. 574

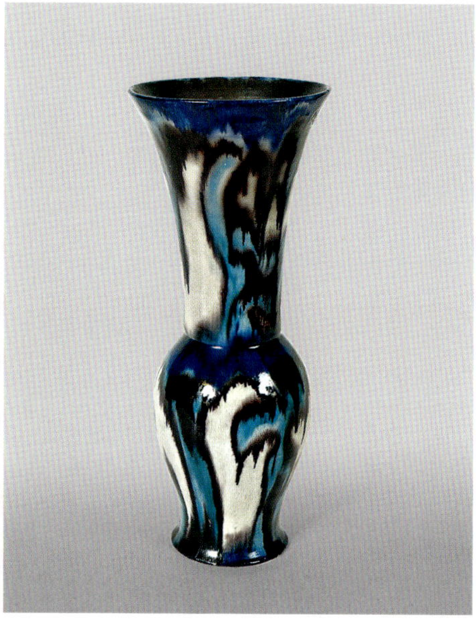

ModNr. 703

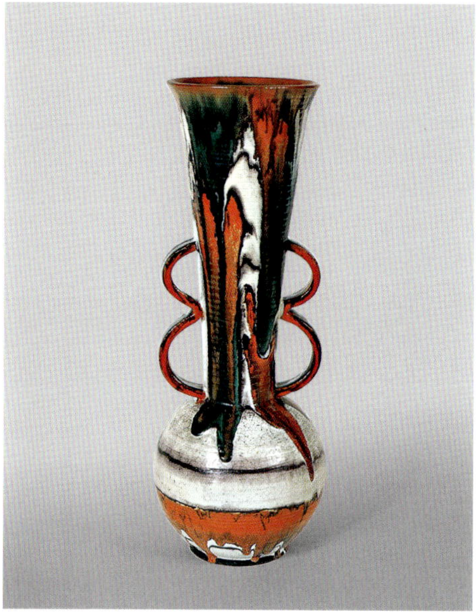

ModNr. 713

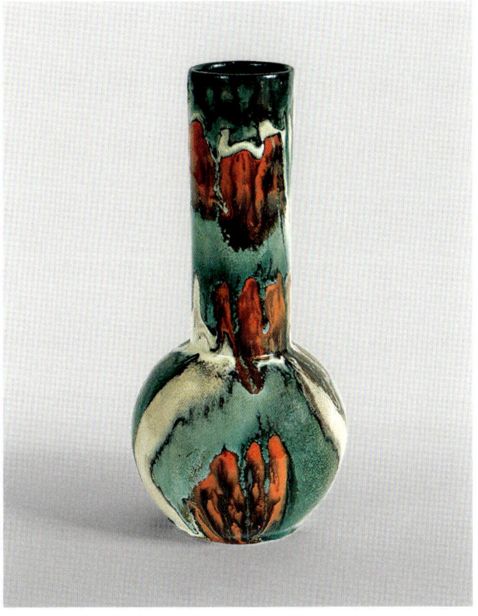

ModNr. 742

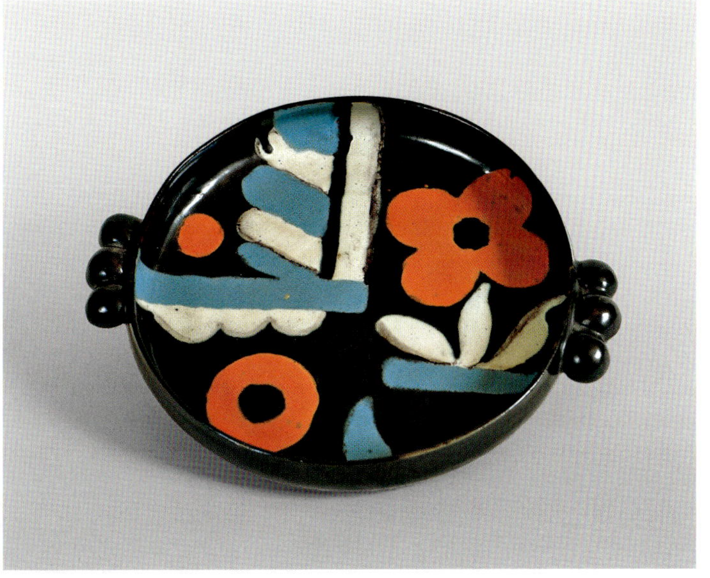

ModNr. 782

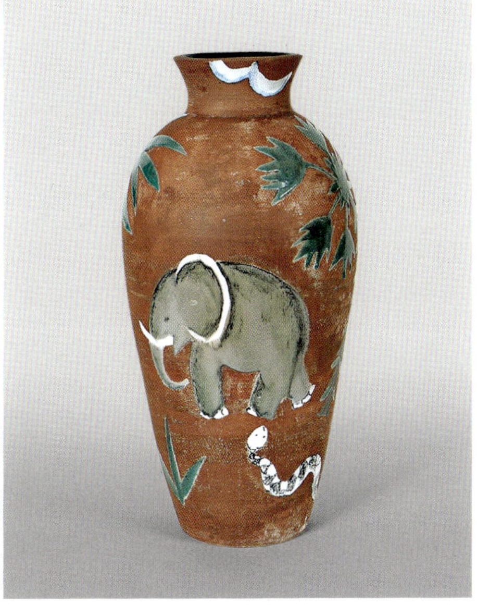

ModNr. 1166

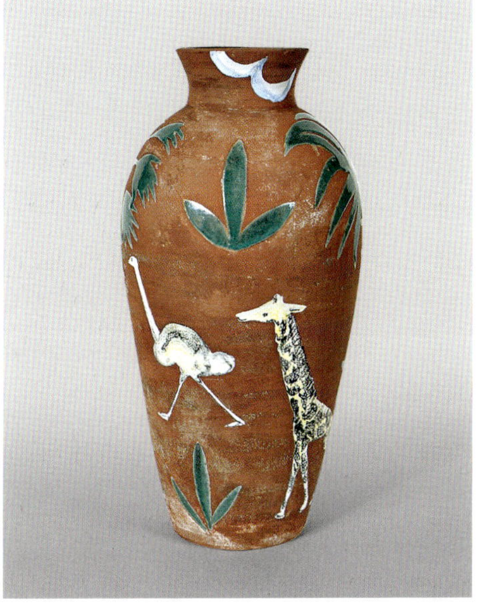

ModNr. 1166

ModNr. 742
Vase, ca. 1928–1938

ModNr. 782
Schale | Bowl,
ca. 1928–1938

ModNr. 1166
Vase, ca. 1928–1938

ModNr. 1242
Vase, ca. 1928–1938

ModNr. 1183
Zwei Vasen | Two Vases,
ca. 1928–1938

alle | all
Sammlung.Schmutz@chello.at

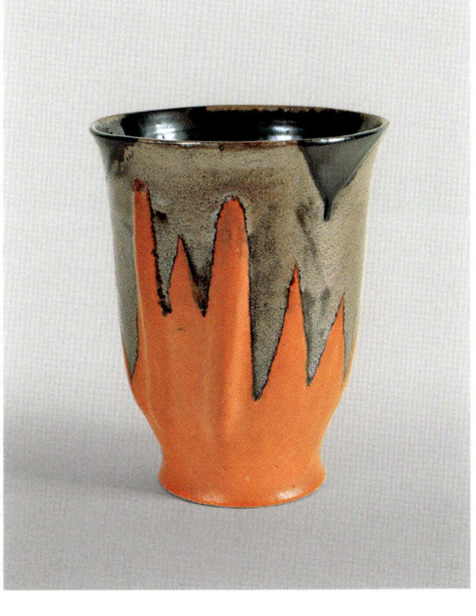

ModNr. 1242

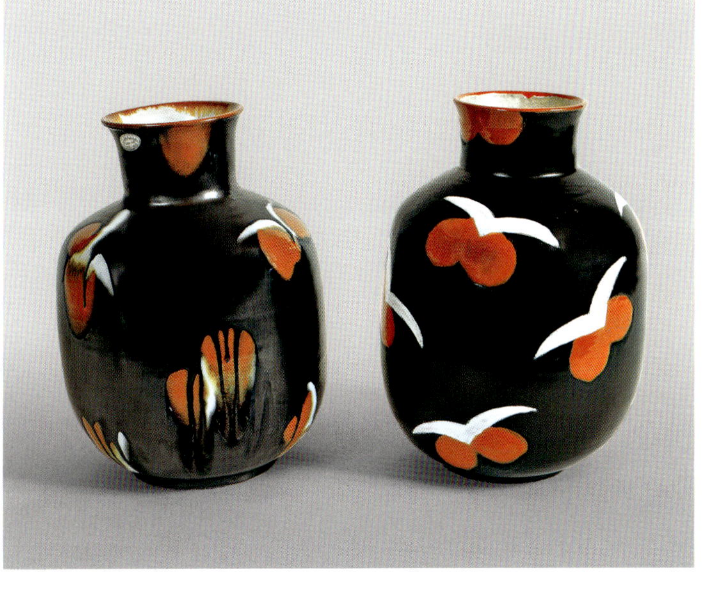

ModNr. 1183

ohne | without ModNr.
Vase, ca. 1928–1934,
Sammlung.Schmutz@chello.at

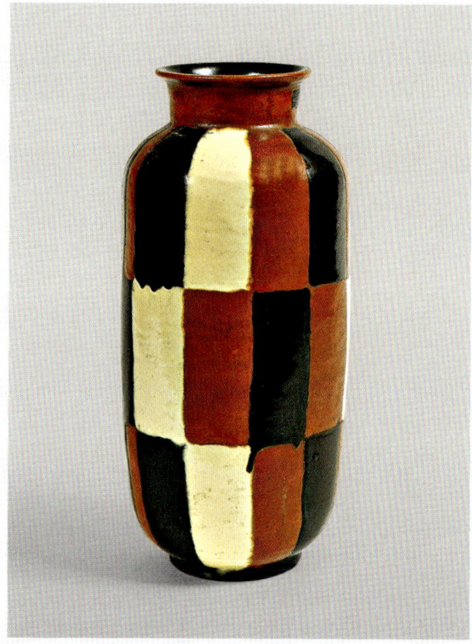

Im Jahr 1928 nimmt auch der fünfundzwanzigjährige Benno Geiger, ebenfalls ein Absolvent der Wiener Kunstgewerbeschule und Schüler Powolnys, seine Arbeit bei Walter Goldscheider auf.[234] Er gehört zu den produktivsten Mitarbeitern bei Goldscheider. Seine Werke, vor allem Vasen und Schalen, beschreibt die Fachpresse als „keramische Schilderei von leuchtender Farbigkeit".[235] In den Jahren seit 1928 entwirft er etliche Vasen und Schalen sowie figürliche Keramik für Goldscheider. Er steigt zum Abteilungsleiter auf, 1934 kehrt er in die Schweiz zurück.[236]

Walter Goldscheider ist Ende der 1920er Jahre eine angesehene Persönlichkeit des Wiener Gesellschaftslebens: Er ist nicht nur Kommerzialrat, sondern seit 1926 Vizeprä-

In 1928 Benno Geiger, also a graduate of the Vienna Kunstgewerbeschule and a pupil of Powolny's, began to work for Walter Goldscheider.[234] Twenty-five years old at the time he was hired, Geiger was one of the most prolific and versatile Goldscheider employees. His pieces, especially his vases and bowls, were described in glowing terms by the specialist press as "ceramic depiction in vibrant colors."[235] From 1928 Geiger designed numerous vases and dishes as well as figurative ceramics for Goldscheider. He was promoted to division manager but returned to his native Switzerland in 1934.[236]

By the late 1920s, Walter Goldscheider was a respected and distinguished member of Viennese society: he not only held the honorary title of Kommerzialrat but was vice-president of the Austrian Institute for the Promotion of Exports, chairman of the League of Fine Ceramics Producers of Austria, vice-

ModNr. 592
Vase, Benno Geiger,
ca. 1928–1934,
Sammlung.Schmutz@chello.at

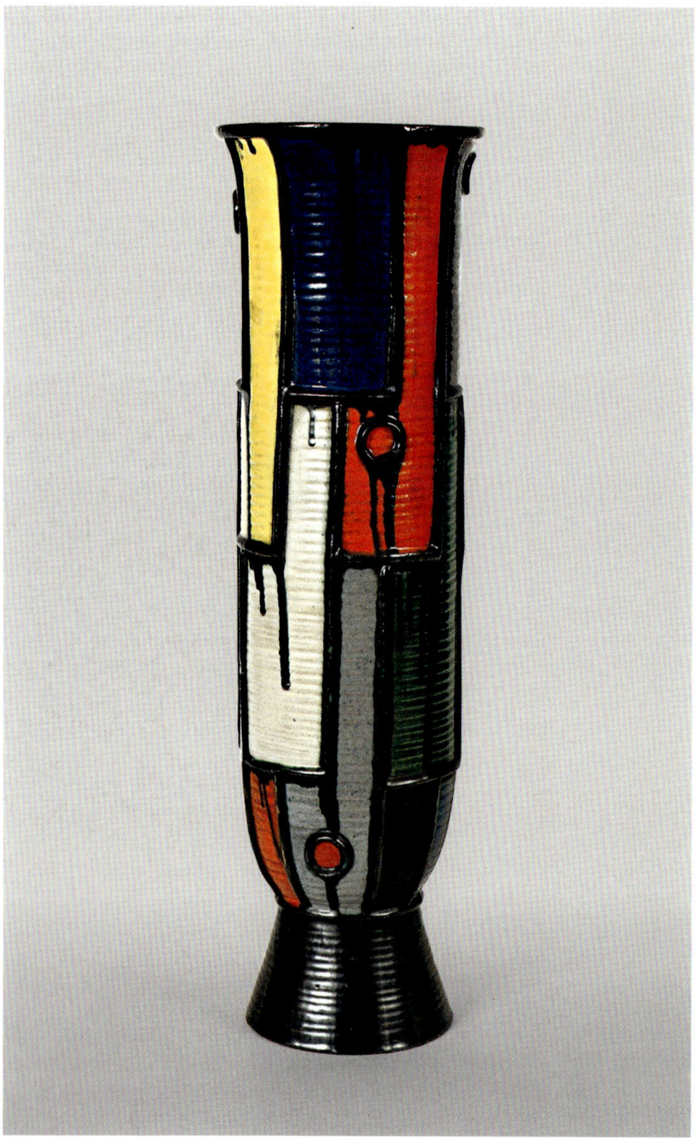

Ill. 57

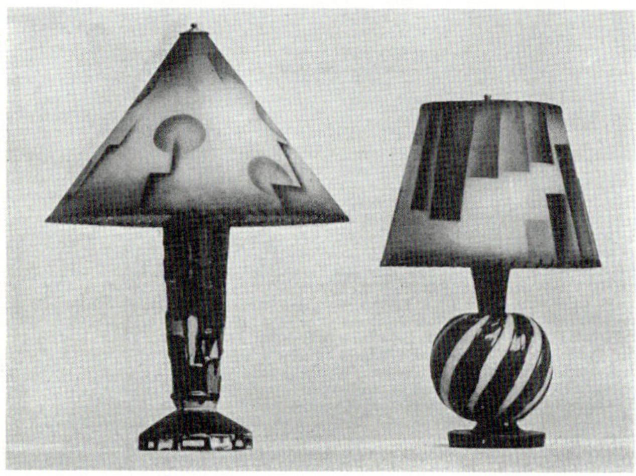

»Moderne Lampenfüße«: Links von Bildhauer Ernst Weber. Rechts von Bildhauer Benno Geiger

Goldscheider WIEN

WIENER MANUFAKTUR FRIEDRICH GOLDSCHEIDER
WIEN XVIII, STAUDGASSE 7–9. GEGRÜNDET 1885

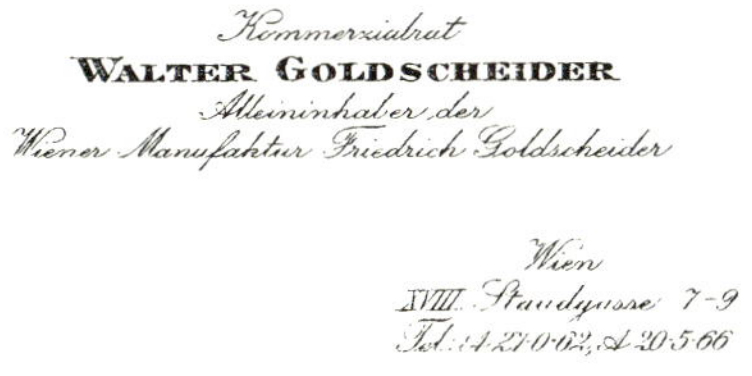
Kommerzialrat
WALTER GOLDSCHEIDER
Alleininhaber der
Wiener Manufaktur Friedrich Goldscheider
Wien
XVIII. Staudgasse 7–9
Tel.: A-27-0-62, A-20-5-66

Ill. 59

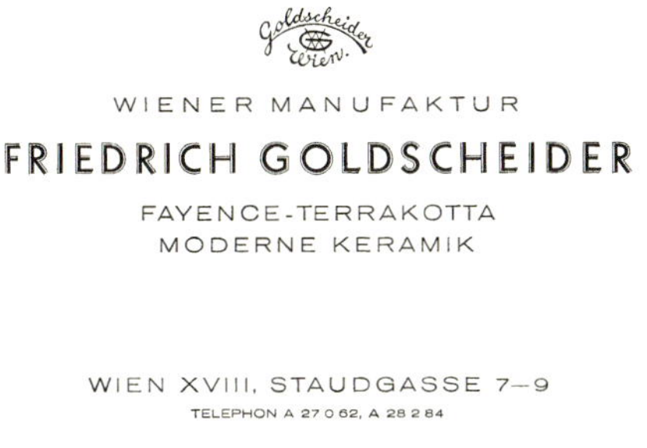
WIENER MANUFAKTUR
FRIEDRICH GOLDSCHEIDER
FAYENCE-TERRAKOTTA
MODERNE KERAMIK
WIEN XVIII, STAUDGASSE 7–9
TELEPHON A 27 0 62, A 28 2 84

Ill. 60

Ill. 59, 60
Walter Goldscheider wird in der Branche hoch geachtet und trägt seit 1922 den Titel Kommerzialrat. Unter der Visitenkarte ist eine Empfehlungskarte aus den 1930er Jahren abgebildet. (beide: Quelle: Archiv Goldscheider)

Highly respected in his field, Walter Goldscheider bore the honorary title of Kommerzialrat from 1922. Below the visiting card is a promotional card from the 1930s. (both: Source: Goldscheider Archives)

president of the Standing Committee for the Austrian Decorative and Applied Arts, vice-president of the Trade Museum and was on the administrative council of the Kunstgewerbeverein.[237] Moreover, as vice-president of the Main Association of Parties Interested in Viennese Fairs and the Trade Fair Council, he was closely associated with the Viennese Trade Fair.[238] He would later join the board of governors of the Viennese Kunstgewerbeschule, ultimately becoming "watch-dog for his trade association and indeed since its inception."[239] (See also Ill. 59 and 60)

His models were so popular and the firm marketed on such a global scale that Walter Goldscheider's business did not succumb to the prevailing worldwide economic crisis.

sident des österreichischen Exportförderungsinstituts, Vorsitzender des Verbands der feinkeramischen Betriebe Österreichs, Vizepräsident des ständigen Ausschusses des österreichischen Kunstgewerbes, Vizepräsident des Handelsmuseums und gehört dem Verwaltungsrat des Kunstgewerbevereins an.[237] Zudem ist er als Vizepräsident des Hauptverbands der Wiener Messe-Interessenten und Messebeirat eng an die Wiener Messe gebunden.[238] Später wird er Mitglied des Kuratoriums der Wiener Kunstgewerbeschule und schließlich auch „Obmann des Verbandes seiner Branche und zwar seit dessen Gründung".[239] (Siehe auch Ill. 59 und 60)

Seine Manufaktur kann dank der populären Modelle und des internationalen Vertriebs der weltweiten Wirtschaftskrise trotzen. Eine eigene Filiale in Paris wird ab dem Jahr 1929 allerdings nicht mehr betrieben, so dass davon auszugehen ist, dass spätestens ab diesem Zeitpunkt die Brüder Walter, Marcell und Arthur Goldscheider gänzlich getrennte Wege gehen.[240] Die Vertretung für Frankreich erhält Guerin Pouyat in der Rue d'Hauteville 68, ab dem Jahr 1934 die Soc. Reima in der Rue de Paradis 2.[241] Am 1. Februar 1929 wird Walters Frau Lilly als Alleinprokuristin eingetragen,[242] Anfang 1932 kommen als Kollektivprokuristen Eduard Stein, Franz Schwarzbart und Josef Paul hinzu.[243]

Ill. 58

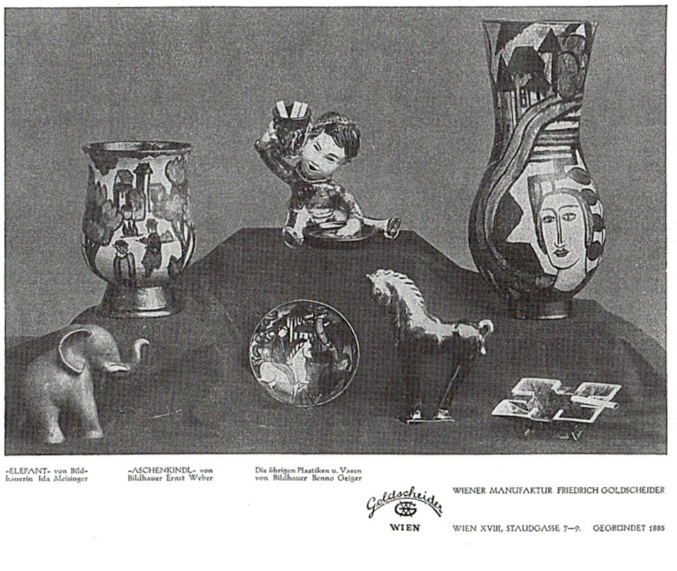
»ELEFANT« von Bildhauerin Ida Meisinger
»ASCHENKINDL« von Bildhauer Ernst Weber
Die übrigen Plastiken u. Vasen von Bildhauer Benno Geiger
Goldscheider WIEN
WIENER MANUFAKTUR FRIEDRICH GOLDSCHEIDER
WIEN XVIII, STAUDGASSE 7–9. GEGRÜNDET 1885

Ill. 57, 58
In der Zeitschrift Moderne Welt werden etliche Entwürfe von Benno Geiger, Ida Meisinger und Ernst Weber beworben. (Quelle: Moderne Welt, Jahrgang 11, 1929/1930)

Quite a few pieces designed by Benno Geiger, Ida Meisinger and Ernst Weber were advertised in Moderne Welt magazine. (Source: Moderne Welt, 11, 1929/1930)

ModNr. 5790
„Aschenkindl" I "Ash Child",
Ernst Weber, ca. 1928,
PB Wien

Diese Groteske dient als Rauchergarnitur, mit Aschenschale und Streichholztopf.

This grotesque is a smoking-set, consisting of an ashtray and a matchbox holder.

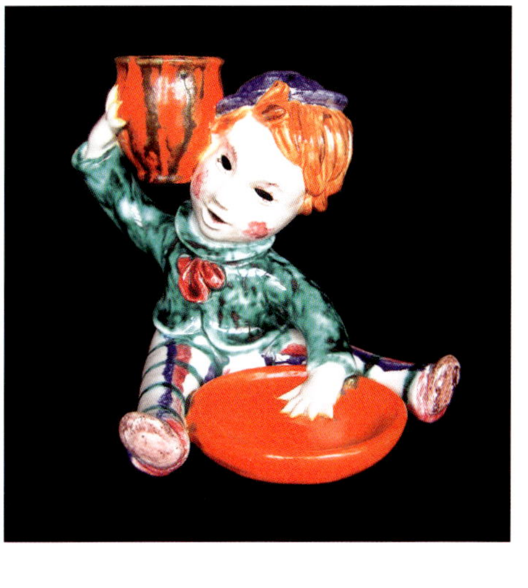

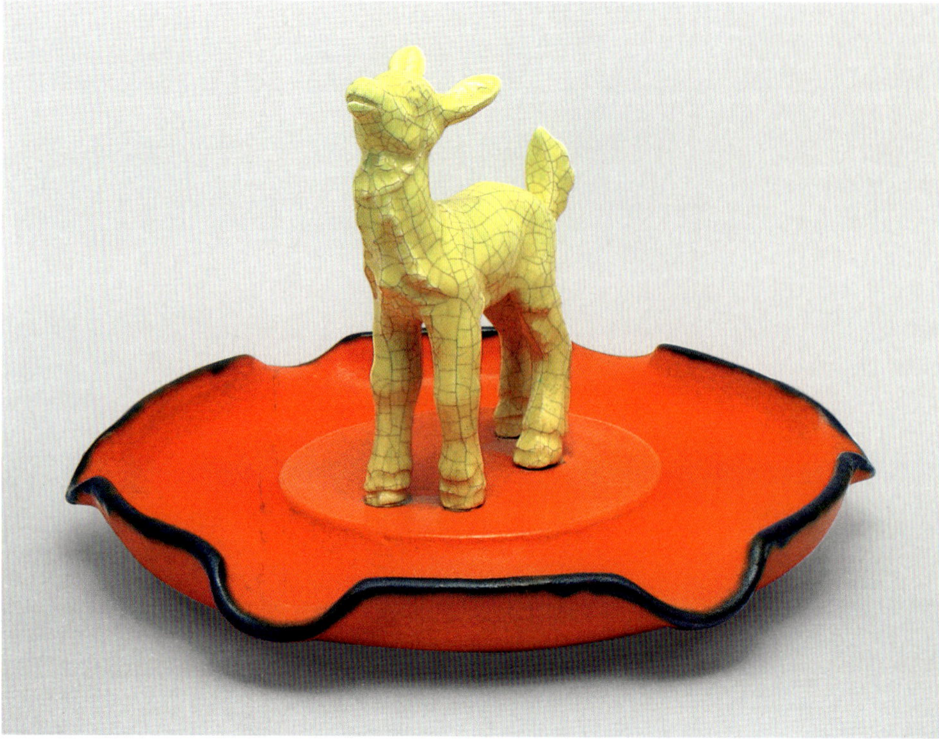

ModNr. 667
Ascher I Ashtray, Ida Meisinger, ca. 1928–1938, Collection a@goldscheider.net
Dieser Ascher von Ida Meisinger wurde mittig mit verschiedenen Tierfiguren dekoriert; hier ein Rehkitz, populär waren aber auch die Meisinger Terrier.

This ashtray by Ida Meisinger was decorated down the centre with various animal figures; here it is a fawn but Meisinger's terriers were also popular.

Auf dem deutschen Markt werden die modernen Keramiken zunächst verhalten aufgenommen, da deren „sehr" moderner Fayencestil von „typisch wienerischer Prägung" etwas abschreckt.[244] Diese Vasen, Schalen, Aschenbecher (Ascher) und Deckeldosen, die in ihren Wandungen einen „leicht ungleichmäßigen, handtöpferischen Charakter haben und Laufglasuren in großen, farbkräftigen Flächen aufweisen", vor allem im leuchtenden Weiß und saftigem Blutrot, bestehen neben den eingeführten Figurenserien.[245] Schon bald werden diese Farben jedoch als leuchtend, edel und „sehr gewählt zueinander gestimmt" empfunden.[246]

However, a branch of his in Paris was no longer in operation from 1929; the implication is that by then at the latest the brothers Walter, Marcell and Arthur Goldscheider had chosen to go their separate ways.[240] Guerin Pouyat at 68 rue d'Hauteville 68 was the French Goldscheider representative; from 1934 Goldscheider was represented there by Soc. Reima at 2 rue de Paradis.[241] Walter's wife Lilly was registered as having sole power of attorney on 1 February 1929;[242] early in 1932, power of attorney was also granted collectively to Eduard Stein, Franz Schwarzbart and Josef Paul.[243]

The German market at first had reservations about the Goldscheider modern ceramics. The "very" modern style of those faience pieces tended to act as a deterrent, exemplifying as it did "the typical Viennese manner."[244] Vases, bowls, ashtrays and covered boxes notable for having a "slightly irregular, hand-thrown character and treacle glazes on large, vibrant surfaces," especially in brilliant white and saturated Sang-de-bœuf, were made alongside the series of figures that had been introduced.[245] It would not be long, however, before those colors were perceived as brilliant, elegant and "very discerningly attuned to one another."[246]

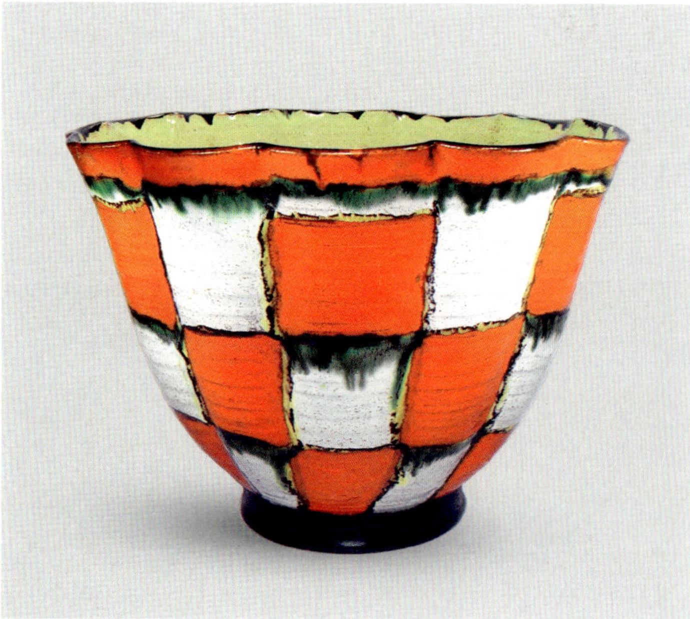

ModNr. 564
Vase, ca. 1928–1938, Collection a@goldscheider.net

Die moderne Keramik besticht durch den verlaufenden Auftrag der Glasur und die ungewöhnlich schrillen Farben; hier eine Serie von türkisen bis coelinblauen Farbtönen.

Goldscheider Modern Ceramics are notable for treacle glazes and unusually bright colors; here a series of color tones ranging from turquoise to azure.

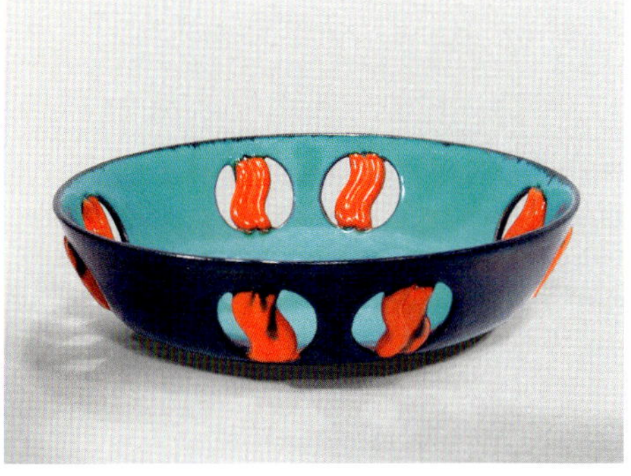

ModNr. 1026
Schale | Bowl,
ca. 1928–1938,
Collection a@goldscheider.net

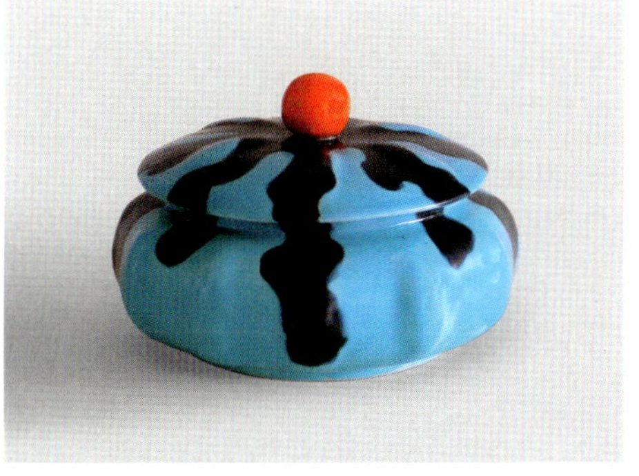

ModNr. 1061
Deckeldose | Covered Box, ca. 1928–1938,
PB Deutschland

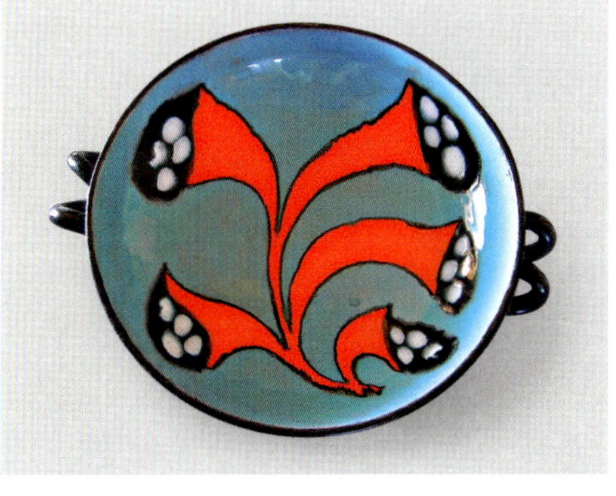

ModNr. 1078
Schale | Bowl,
ca. 1928–1938,
Sammlung.schmutz@chello.at

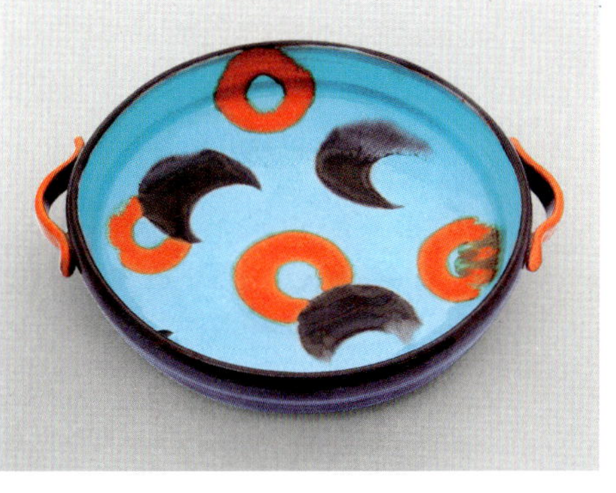

ModNr. 1209
Schale | Bowl,
ca. 1928–1938,
Collection a@goldscheider.net

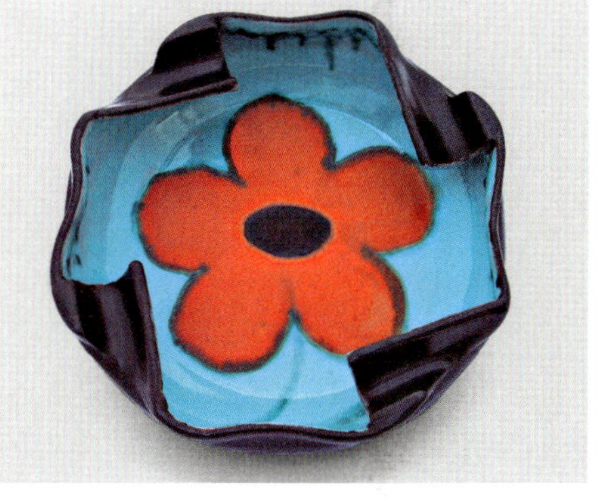

ModNr. 1230
Ascher | Ashtray, Viktor Matula, ca. 1928–1938,
Collection a@goldscheider.net

ModNr. 1074
Vase, ca. 1928–1938,
PB Wien

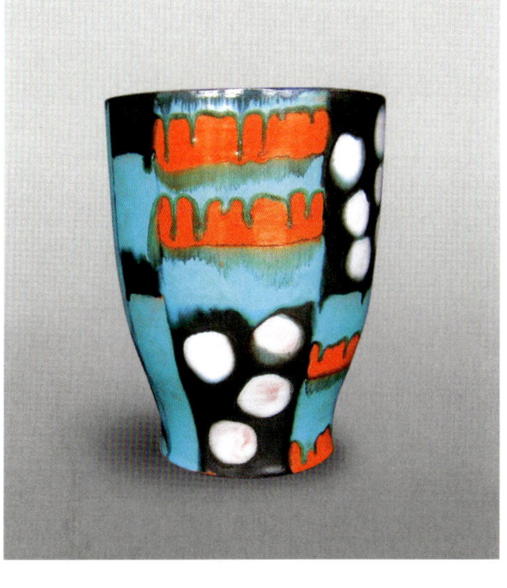

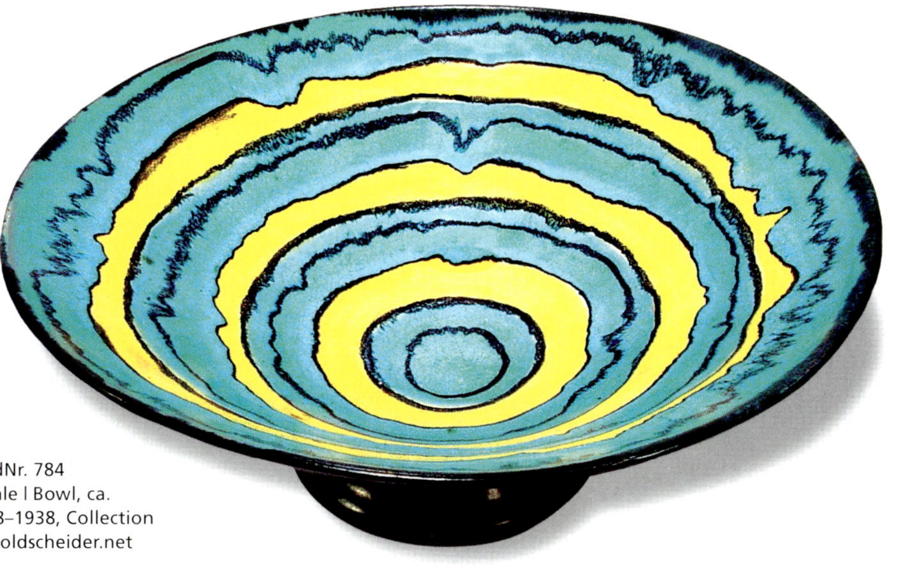

ModNr. 784
Schale | Bowl, ca. 1928–1938, Collection a@goldscheider.net

rechts | right

ModNr. 5914 „Schwestern" | "Sisters", Stephan Dakon, ca. 1929, PB Wien 19
Fayence-Gruppen waren relativ selten, dieses Modell zeigt die Vigo-Drillinge, die als Varieté-Tänzerinnen berühmt geworden sind.

Faience groups were relatively rare. This model depicts the Vigo Triplets, who were famous music-hall dancers.

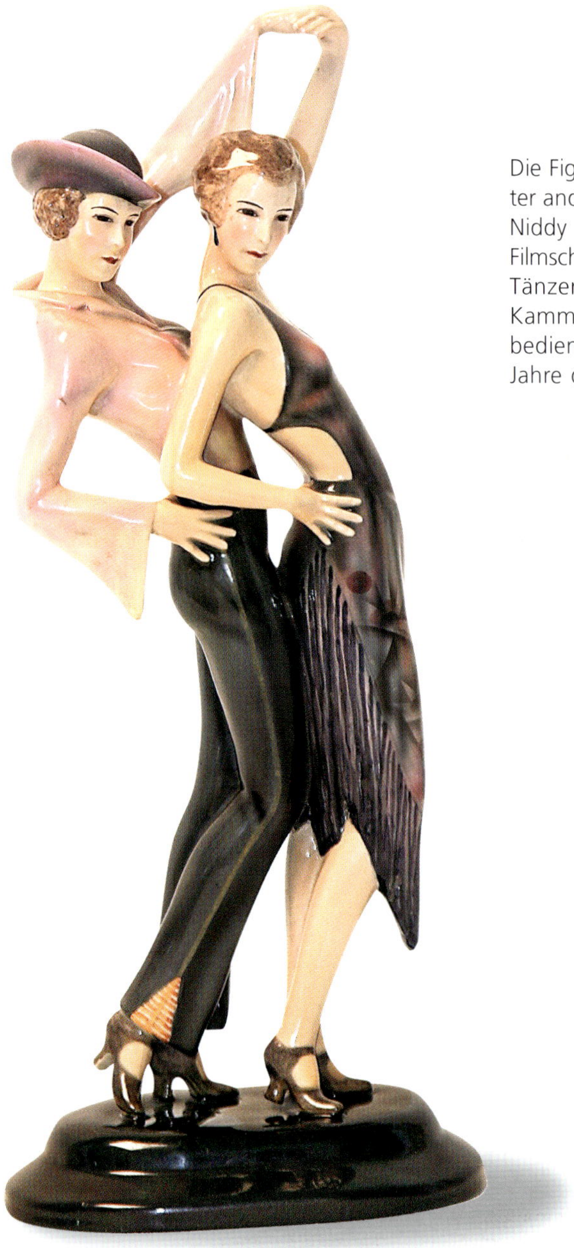

ModNr. 5975
Gruppe „Tanz" |
Group "The Dance",
Stephan Dakon,
ca. 1929, private
collection, Wien

Die Figurenserie „Berühmte Frauen", die unter anderem die moderne Ausdruckstänzerin Niddy Impekoven, die glutäugige, exotische Filmschauspielerin Dolores del Rio, die Revue-Tänzerinnen Dolly Sisters oder auch die Kammersängerin Maria Jeritza darstellen, bedienen schon seit Mitte der Zwanziger Jahre die Nachfrage der sowohl weiblichen

The "Berühmte Frauen" ["Famous Women"] series of figurines, which included the modern expressive dancer Niddy Impekoven, the exotic film actress Dolores del Rio, famous for her smoldering glance, the revue dancers known as the Dolly Sisters and the Kammersängerin Maria Jeritza among other, and had been much in demand with a male as well

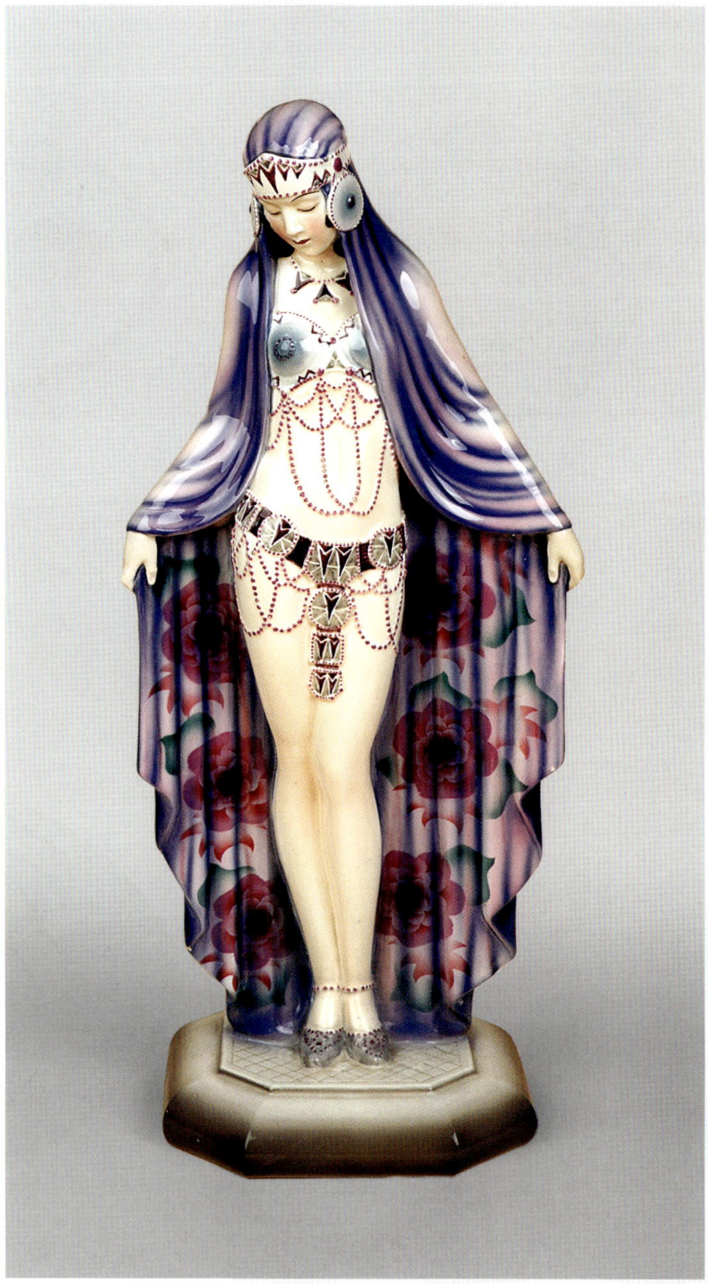

ModNr. 5927
Harems-Tänzerin |
Harem Dancer,
Josef Lorenzl,
ca. 1929, PB Wien 19

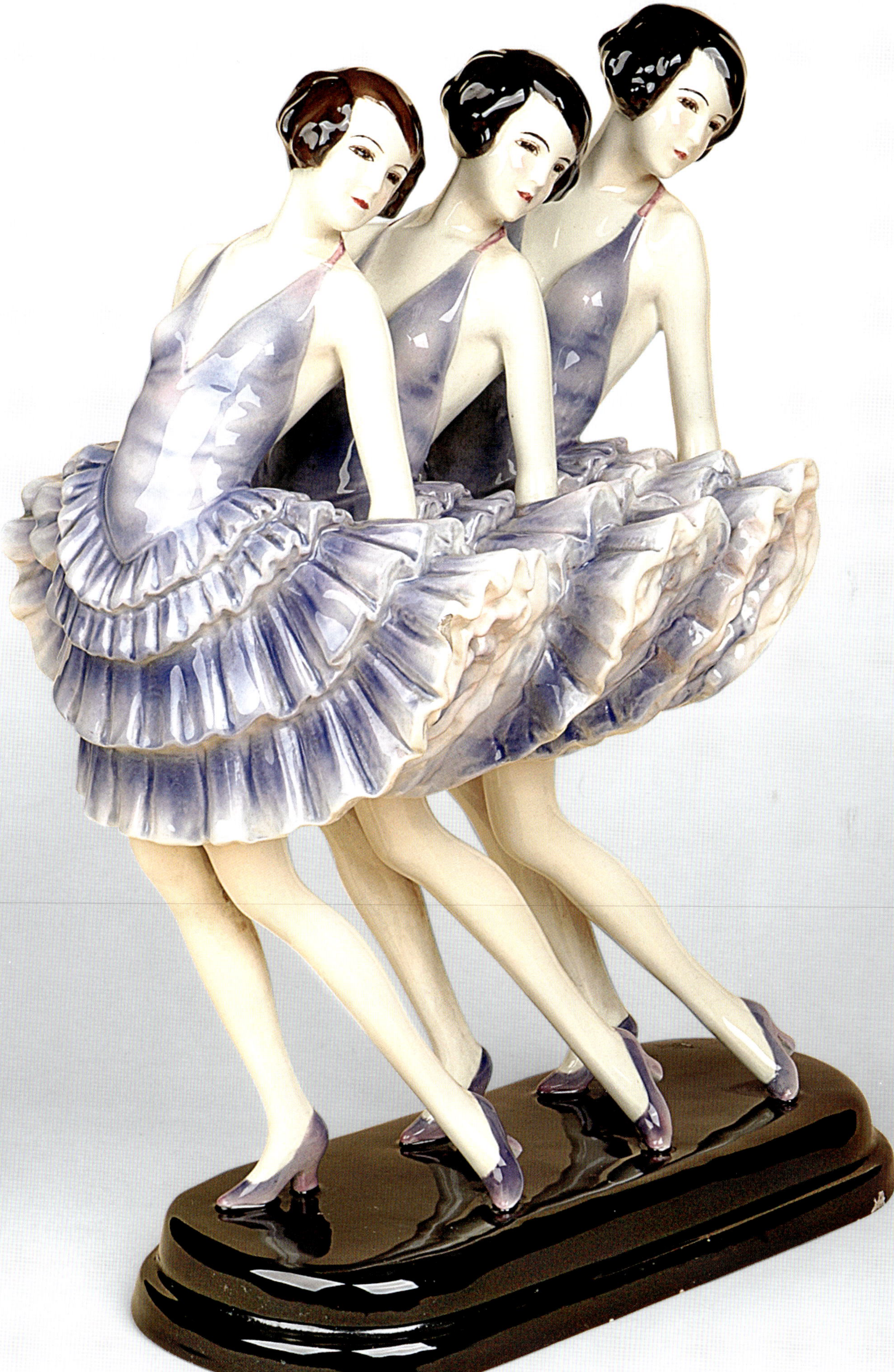

wie männlichen Käufer. Sie erreichen ihre beste Wirkung nicht in Stein oder Bronze, sondern in den farbenintensiven, handbemalten Dekoren der Fayence und bringen sozusagen den Glamour der Stars und Starlets ins heimische Wohnzimmer. Diese Figuren sind „Idole gleichsam der Sehnsucht, die in jeder Frau lebt", sie sind „ganz aus der Zeit des Jazz, der Körperkultur geboren."[247]

Ende der Zwanziger Jahre werden auch zunehmend Hundeplastiken produziert und der Hund, in allen Rassen, als innigster Freund des Menschen abgebildet. Neben einzelnen Schäferhunden, Dackeln, Pekinesen

as female clientele since the mid-1920s. They were most effective in the saturated hues of hand-painted faience rather than in stone or bronze, bringing the glamour of the stars and starlets into the home. These figures are "icons, as it were, of the longings innate in every woman;" they were "entirely children of the Jazz Age, of the physical culture era."[247]

In the late 1920s, dog figurines were produced in increasing quantities. Pedigreed dogs of all breeds were portrayed as man's best friend. Alongside solitary German Shepherds, Dachshunds, Pekinese and Newfound-

ModNr. 6003
Tänzerin I Danseuse,
Josef Lorenzl, ca. 1930,
PB Wien 19

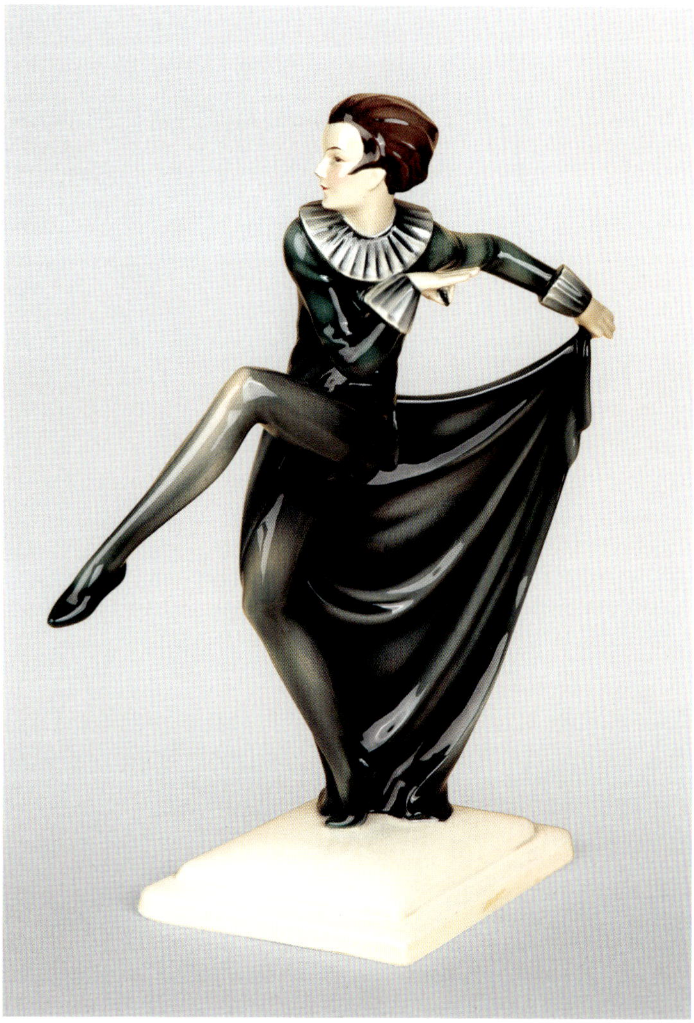

ModNr. 6606
„Blonder Traum" I
"Dreamy Blonde"
(Lilian Harvey),
Stephan Dakon,
ca. 1932, PB Wien 19

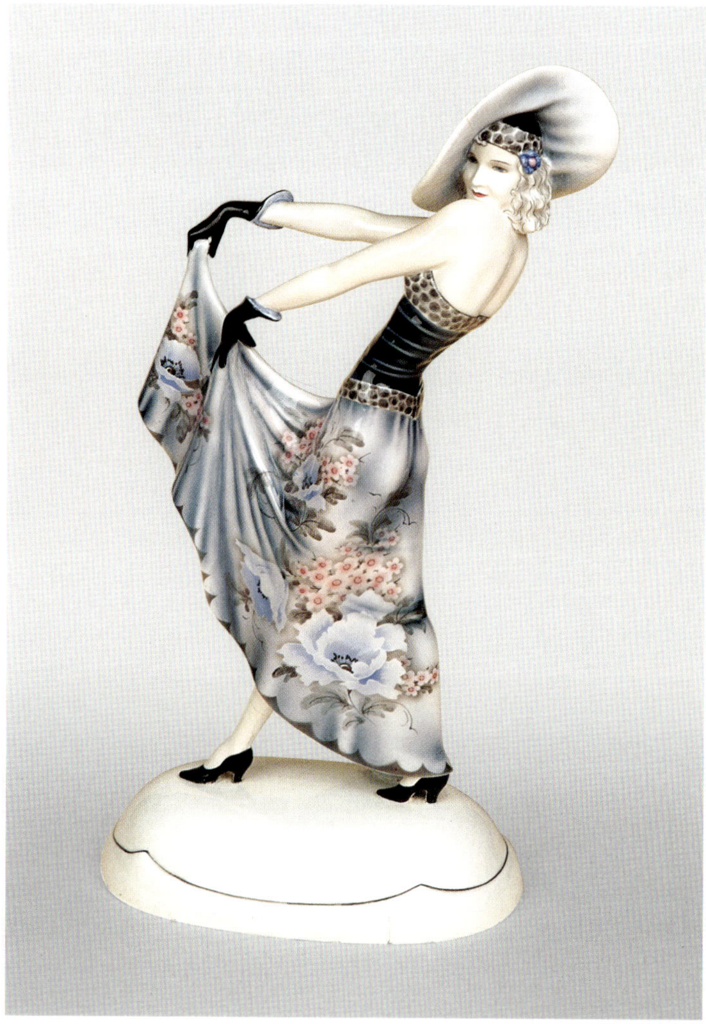

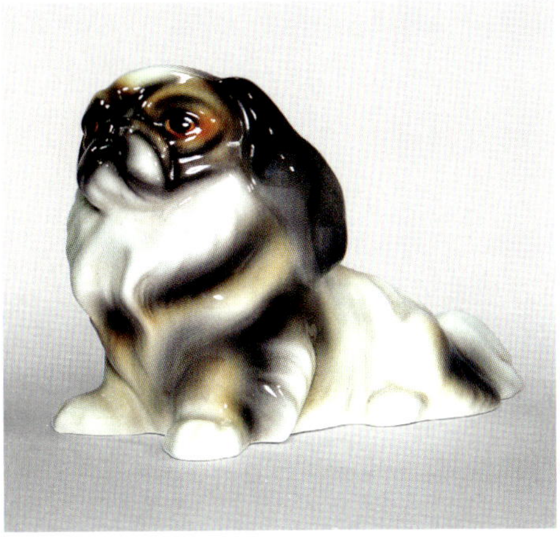

ModNr. 6136
Pekinese, Hugo Postl,
ca. 1929, Collection
a@goldscheider.net

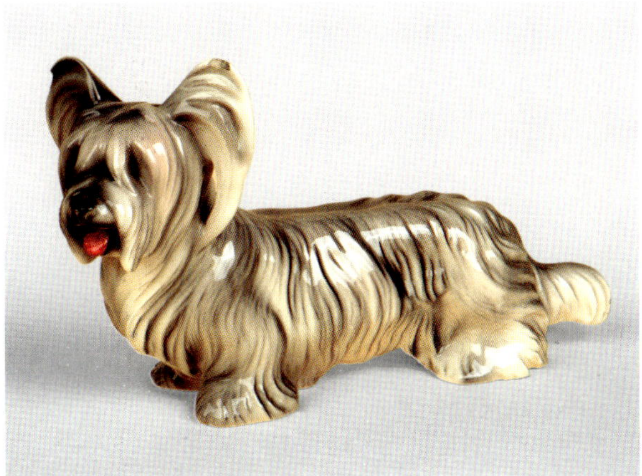

ModNr. 6503
„Skyeterrier" I "Skye Terrier", Hugo Postl,
ca. 1930, PB Wien 19

lands, combinations in groups matched the taste of the times: "The child strokes a dog's fur; the cosmopolitan society lady parades her terriers, Pekinese, Japanese Chins, just as the woman in the street is proud of her spitz or terrier. The gentleman is fond of the larger, more powerful breeds such as the Great Dane, the boxer, the sheep-dog, the doberman and, if he shoots, the pointer."[248]

One of the most popular models in the 1930s was Ida Meisinger's terrier. Produced in several versions and colors, the terrier might stand alone, be a pair designated "Pat and Patachon" (Ill. 62), assume the form of book-ends or even be executed as a miniature. It was the Goldscheider figurine that sold best of all. Schaulade magazine called the terrier the most recent "smash hit from

und Neufundländern sind es vor allem kombinierte Gruppen, die den Geschmack der Zeit treffen: „Das Kind grault dem Hund in seinem Fell, die mondäne Dame der Gesellschaft paradiert ebenso mit ihren Terriers, Pekinesen, Chins, wie die einfache Frau aus dem Volk auf ihren Spitz oder Rattler stolz ist. Der Herr liebt die größeren, mächtigeren Rassen wie Dogge, Boxer, Schäfer, Dobermann und, so er Jäger ist, Vorstehhund."[248]

Ill. 61

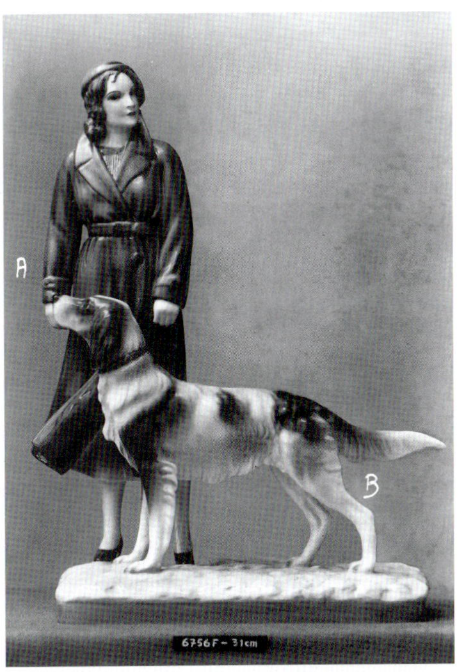

ModNr. 6756
Dame mit Setter I
Lady with Setter,
Stephan Dakon,
ca. 1933, PB Wien 19

Bei der dargestellten Dame ist eine gewisse Ähnlichkeit zu Marlene Dietrich erkennbar. Der Hund (Modell-Nr. 6804) dieser Gruppe ist vermutlich kein Entwurf Dakons, sondern wurde der stehenden Frauengestalt durch die Manufaktur beigefügt. (vgl. Ill. 61: Bezeichnung der Figuren mit A und B. Quelle: Archiv Mimi Dakon)

The lady depicted recognizably resembles Marlene Dietrich. The dog (model number 6804) in this group was probably not designed by Dakon but was added at the factory to the standing figure of the lady. (cf Ill. 61: the two figures designated with A and B. Source: Mimi Dakon Archives)

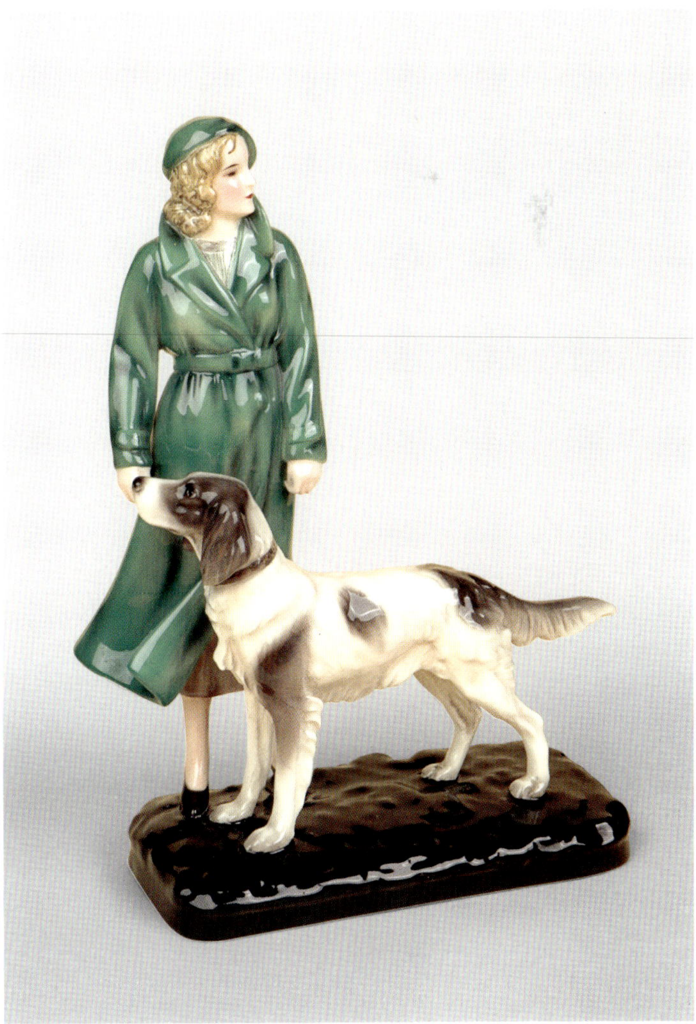

Ill. 62
Die Ausführung der sitzenden Airedale-Terrier wurde „Pat und Patachon" genannt (ModNr. 6362; Quelle: Fotoarchiv Karl Täuber)

This version of the sitting Airedale Terriers was named "Pat and Patachon" (model number 6362; source: Fotoarchiv Karl Täuber)

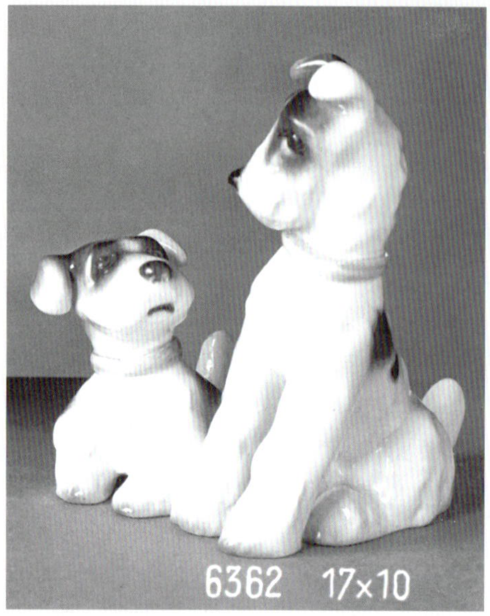

Eines der populärsten Modelle der 1930er Jahre ist der Terrier von Ida Meisinger. Der in verschiedenen Versionen und Farben produzierte Terrier, einzeln stehend oder im Paar als „Pat und Patachon" bezeichnet (Ill. 62), in Form von Buchstützen bis hin zu Miniaturen, ist die am meisten verkaufte Kleinplastik von Goldscheider. Die „Schaulade" bezeichnet den Terrier als den jüngsten „Erfolg der Firma, der sich zu einem wahren Welterfolg auswuchs. Wer hat nicht schon die mannigfaltig gestalteten, drolligen Terriers gesehen, die zu Tausenden aus ihrer Geburtsstadt Wien in aller Herren Länder gingen und vielerorts eine geradezu unzählbare Nachkommenschaft fanden."[249] Die teilweise grotesken Tierfiguren treffen den Geschmack der Leute, die sich nach der Weltwirtschaftskrise etwas Positives wünschen: „Leicht, unbeschwert, sorglos, trotz Krise und allgemeiner Missstimmung – das ist der Grundzug dieser heiteren Kleinkunst."[250] Die erfolgreiche Vermarktung der Terrier etabliert den Begriff der „Goldscheider-Nippes".

the firm, which grew into a worldwide success story. Is there anyone who has not seen the droll terrier in all its manifold guises, thousands of which have gone out from their native Vienna to the rest of the world and in many places have produced innumerable offspring."[249] These animal figures, some of them grotesque, suited people's taste, especially those who wanted something positive after the Great Depression: "Light, free-and-easy, carefree despite the crisis and the general depression – that is the basic feature of this cheerful art in miniature."[259] Successful sales of the terrier established the concept of "Goldscheider Nippes" ["Goldscheider bric-à-brac"].

ModNr. 5854
„Der Modehund" | "The Fashionable Dog", Ida Meisinger, ca. 1928, PB Wien 19
Dieses Modell gehört eindeutig zu den beliebtesten der ganzen Produktionspalette von Goldscheider. Es wurde nicht nur in hohen Stückzahlen (besonders in der mittleren Ausführung, Modell-Nr. 5879) hergestellt, sondern auch in fünf verschiedenen Größen (5,4–6,2 cm, 12 cm, 16,5 cm, 22 cm, 39 cm). Vor allem die Farben Orange und Weiß waren populär, aber auch Schwarz, Braun und sogar Türkis.

This model was beyond question one of the most popular in the entire Goldscheider product range. It was not only made in large quantities (especially the medium version, model number 5879) but also in five different sizes (5,4-6,2 cm, 12 cm, 16,5 cm, 22 cm, 39 cm). The colors orange and white were frequently used but black, brown and even turquoise were also popular.

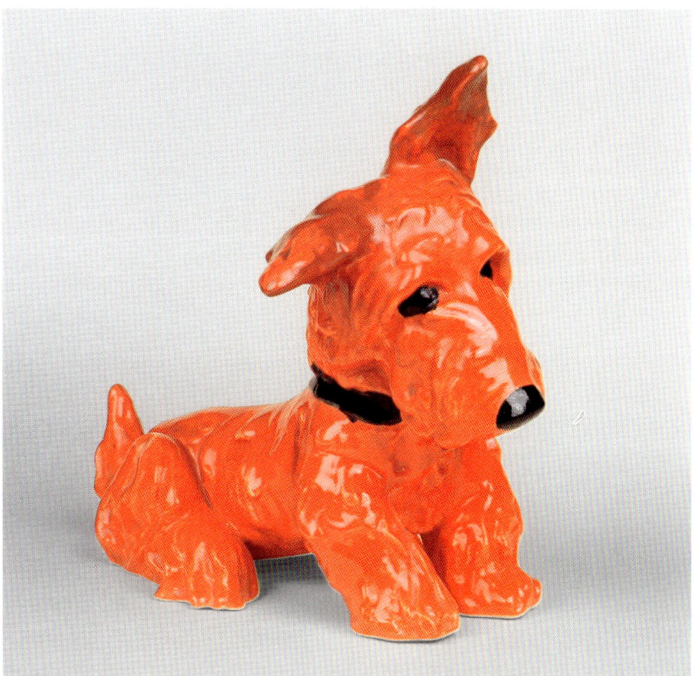

ModNr. 5879
„Der Modehund" | "The Fashionable Dog", Ida Meisinger, ca. 1928, PB Wien 19

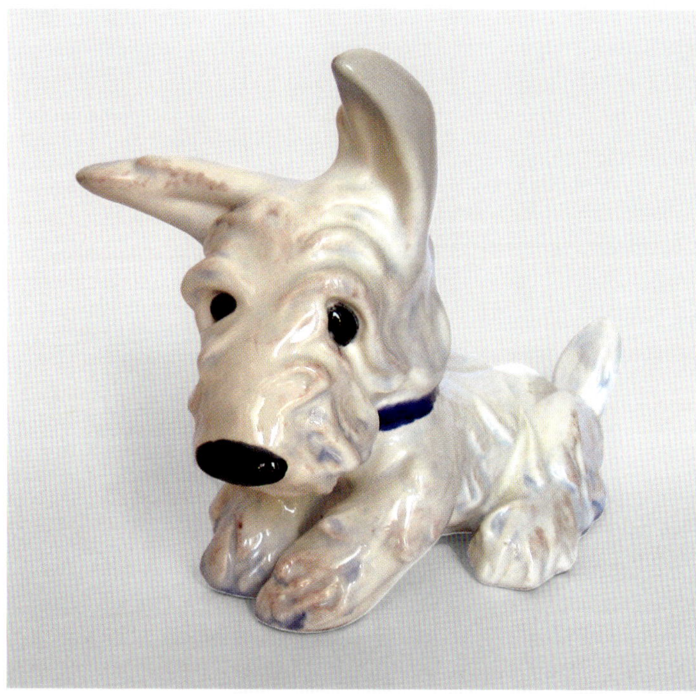

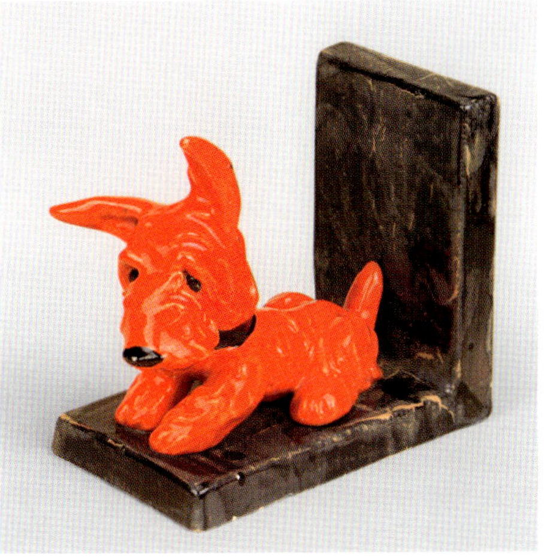

ModNr. 5879

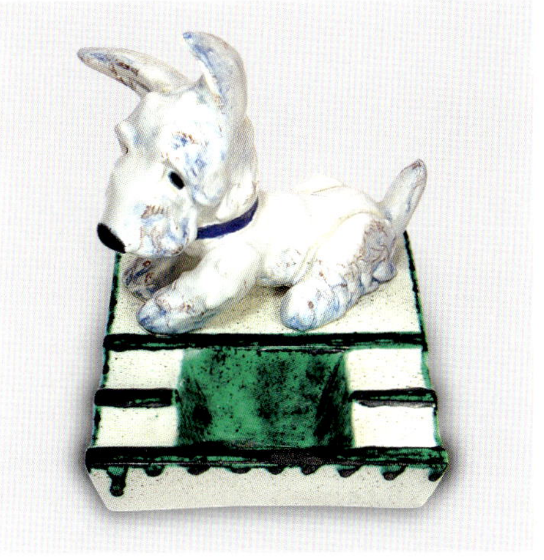

ModNr. 6043

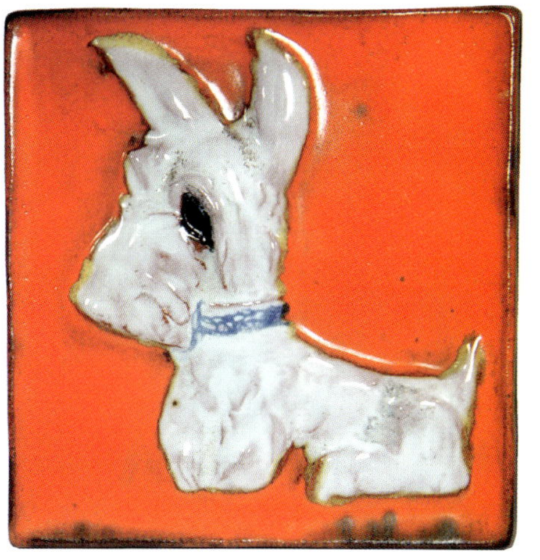

ModNr. 6285

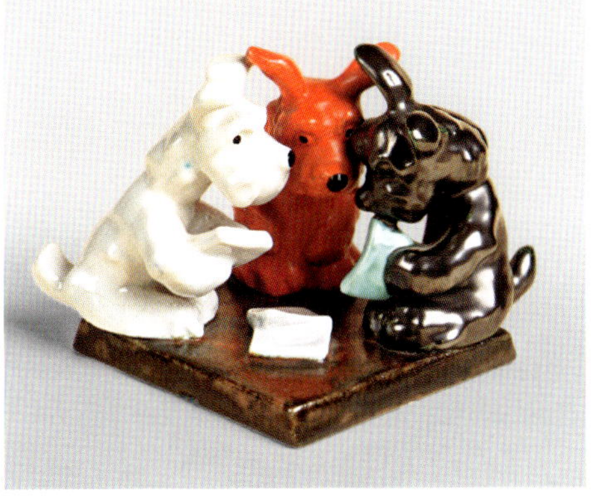

ModNr. 6596

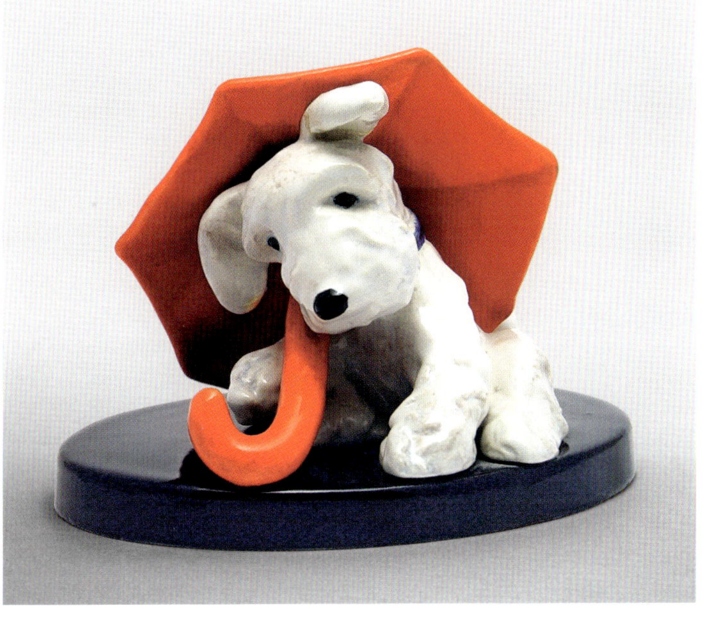

ModNr. 6449

ModNr. 6316
Deckeldose Scotch-terrier I Covered Box Scottish Terrier, Ida Meisinger, ca. 1930, beide I both Collection a@goldscheider.net

ModNr. 5879
Buchstütze „Der Mode-hund" I Bookend "The Fashionable Dog", Ida Meisinger, ca. 1928, PB Wien 19

ModNr. 6043
Ascher „Der Mode-hund" I Ashtray "The Fashionable Dog", Ida Meisinger, ca. 1929, Collection a@goldscheider.net

ModNr. 6285
Relief Scotchterrier I Relief Scottish Terrier, Ida Meisinger, ca. 1930, Collection a@goldscheider.net

ModNr. 6596
Drei Terrier beim Kartenspiel I Three Terriers Playing Cards, Ida Meisinger, ca. 1932, PB Wien 19

ModNr. 6449
„Hundewetter" I "Beastly Weather", Ida Meisinger, ca. 1931, Collection a@goldscheider.net

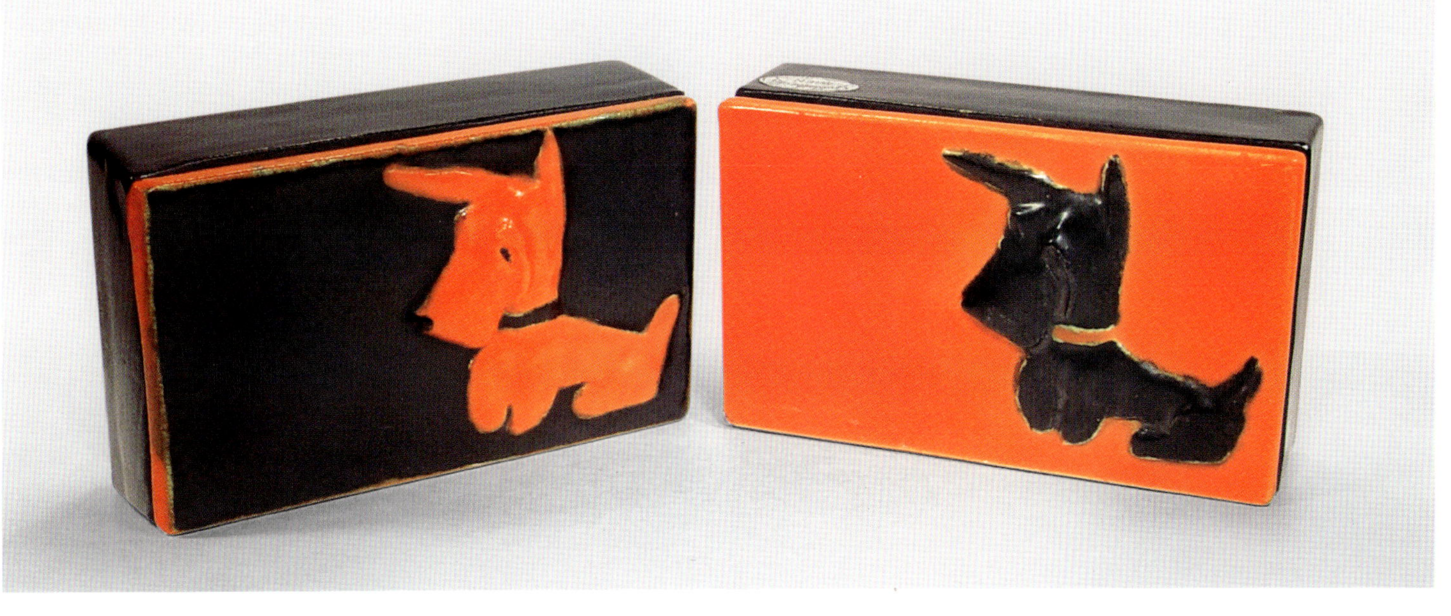

ModNr. 7070
„Zahnweh" I "Toothache", Ida Meisinger, ca. 1935, PB Wien 19

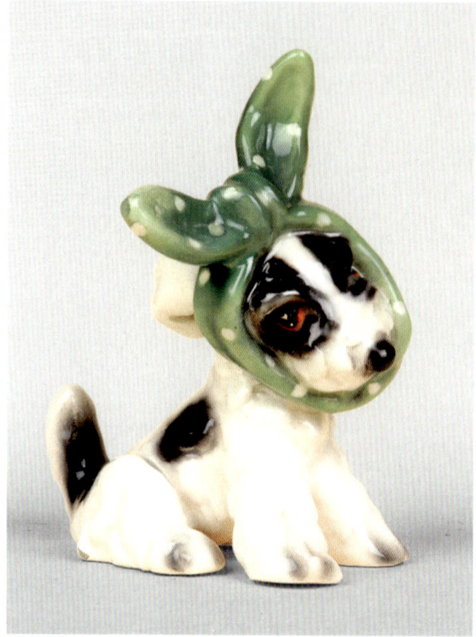

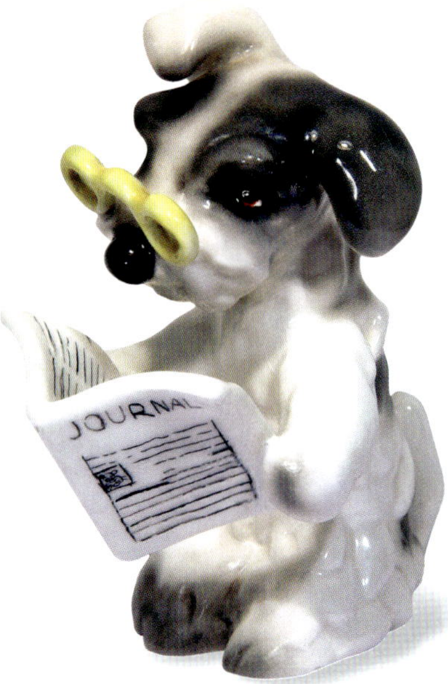

ModNr. 7186
„Zeitungsleser" I "Reading the Paper", Ida Meisinger, ca. 1935, Collection a@goldscheider.net

Fünfundzwanzig Jahre nach der Einführung der Fayence-Figuren erscheint zum Jubiläum im Jahr 1929 der Katalog „25 Jahre Wiener Fayence", der für Werbezwecke eingesetzt wird.[251] Die Erweiterung der Produktpalette um preiswerte kleinfigürliche und angewandte Keramik ist in diesen von der Weltwirtschaftskrise geprägten Zeiten sehr erfolgreich und wird auch von der Fachpresse wohlwollend aufgenommen. Die neue Ausrichtung unter Walter Goldscheider wird Anfang 1932 wie folgt beschrieben: „Mit besonderer Ambition und Energie hat die alte Wiener Manufaktur Friedrich Goldscheider den Weg zum neuen Stil gefunden. Ein Stab junger Künstler arbeitet dort und hat die leuchtend bunten Gefäße mit starken, satten Glasuren geschaffen, die besonders das Ausland schnell gewonnen haben. Köstliche Tierfiguren, hübsche blumengeschmückte Schüsseln, Dosen usw. sind aus dieser Werkstätte hervorgegangen. Die neuen Arbeiten zeigen den Ehrgeiz feiner Craquelés, gedämpfter, vornehmer Glasuren und bemühen sich, mit höheren künstlerischen Werten das heute so schwierige Publikum zu gewinnen."[252] Walter gelingt es, unterschiedliche Geschmäcker und Zielgruppen zu bedienen und mit seinen Produkten ein wesentliches Element moderner Wohngestaltung anzubieten. Die Zeitschrift „Moderne Welt" dient ab 1928 hierbei als wichtigstes Werbemedium von Goldscheider (Ill. 55, S. 106). Immer wieder werden dort Anzeigen für aktuelle Modelle geschaltet.

Im Jahr 1932 wird Rudolf Knörlein Leiter der modernen Abteilung und künstlerischer Leiter des Betriebes. Knörlein, der vorher schon für Marcell Goldscheider tätig war und

Twenty-five years after the faience figures had been introduced, the catalogue "25 Jahre Wiener Fayence" celebrating this anniversary was published in 1929 for advertising purposes.[251] The enlargement of the product range to include reasonably priced figurines and related ceramics was highly successful in an era shaped by the worldwide economic crisis and also met with a positive response from the specialist press. The new orientation under Walter Goldscheider was described as follows early in 1932: "With remarkable ambition and energy, the old Viennese Manufactory Friedrich Goldscheider has found the way to a new style. A staff of young artists works there and has created the brilliantly colorful vessels in strong, saturated glazes that have rapidly conquered the foreign market especially. Delightful animal figures, pretty bowls with floral decoration, boxes, etc, have come forth from this workshop. The new works ambitiously boast fine crackle, more subdued, elegant glazes and aspire to win over the public, which is today so hard to please with more lofty aesthetic values."[252] Walter succeeded in catering for various different tastes and target markets by offering an essential element of modern interior decoration with his products. From 1928, Moderne Welt magazine served as the most important advertising vehicle for Goldscheider in this respect (Ill. 55, p. 106). Advertisements were always being placed there for current models.

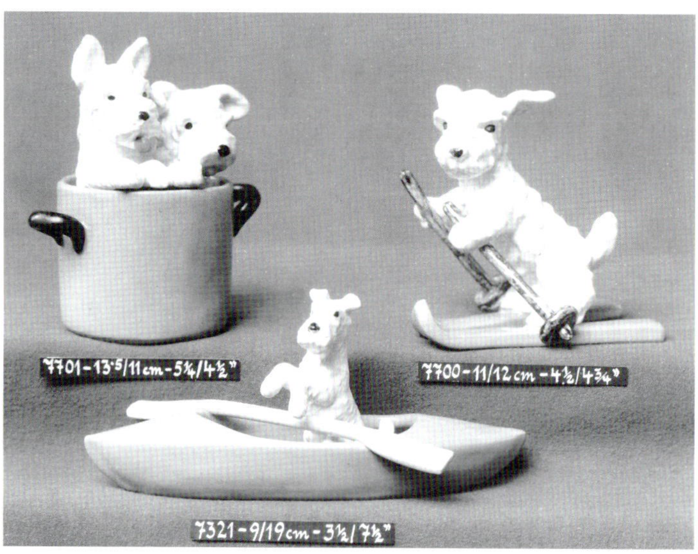

Ill. 63
Allgemein gehen die grotesken Hundefiguren als „Goldscheider-Nippes" in die Geschichte der Wiener Keramik ein. (Quelle: Fotoarchiv Karl Täuber)

On the whole, the grotesque dog figures made history in Viennese ceramics as "Goldscheider Nippes" ["Goldscheider bric-à-brac"]. (Source: Fotoarchiv Karl Täuber)

In 1932 Rudolf Knörlein became head of the Modern Department and artistic director of the business. Knörlein, who had already worked for Marcell Goldscheider and whose heads of women as wall masks and busts were successful on an international scale, moved on 6 December 1934 into the business flat at 7 Staudgasse, where he lived with his family until he left Goldscheider in February 1936.[253] Many years later, Knörlein would write a very personal letter to Walter Goldscheider about those days: "One can, in fact, really work well under & with you – even though you kept a very tight rein on one. And I must say this to you – I have never come across a proprietor of a business in our sector with whom one could discuss so much about the matter at hand – who had so much insight into the various problems – who tried to concede as much as possible to the artistic part with all the necessary feeling for the business side."[254] Although he was employed for only a relatively short time, Knörlein exerted a paramount influence on the Goldscheider output in the 1930s with his designs of slender figures and tousle-haired wall masks and heads.

dessen Frauenköpfe als Wandmasken und Büsten international erfolgreich sind, bezieht am 6. Dezember 1934 die Betriebswohnung in der Staudgasse 7, wo er bis zu seinem Weggang von Goldscheider, im Februar 1936, mit seiner Familie wohnt.[253] Viele Jahre später schreibt Knörlein an Walter Goldscheider in einem sehr persönlichen Brief über diese Zeit: „Unter Ihnen & mit Ihnen konnte man nämlich wirklich arbeiten – auch wenn Sie einen etwas sehr kurz gehalten haben. Und ich muss es Ihnen sagen – ich habe noch keinen Besitzer eines Unternehmens unserer Branche gefunden, mit dem man so über die Materie diskutieren konnte – der so viel Verständnis hatte für die verschiedenen Probleme – bei allem notwendigen Sinne für die geschäftliche Seite, dem künstlerischen Teile das möglichste zuzugestatten versuchte."[254] Knörlein hat trotz des relativ kurzen Beschäftigungsverhältnisses mit seinen feingliedrigen Figuren und den krausköpfigen Wandmasken und Stehköpfen maßgeblich die Goldscheider-Produktion der 30er Jahre beeinflusst.

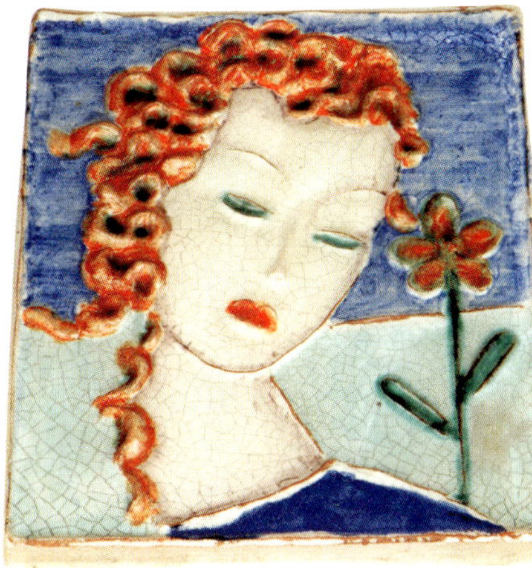

ModNr. 6661
Relief, Rudolf Knörlein, ca. 1933, PB Wien 19

Ill. 64
Das Dekor dieser Dose mit der Modell-Nummer 7430 findet sich auch auf einer Werbeschale der Manufaktur, die zum 50-jährigen Firmenjubiläum 1935 editiert wurde. Durch die eigenhändige Datierung Knörleins (1934) auf dem Trägerblatt des Originalfotos kann die Entstehung um 1934 bestimmt werden. (Quelle: Privatarchiv)

The decoration on this model number 7430 jar also occurs on a dish, produced in a limited edition for the firm's fiftieth anniversary in 1935 and used for advertising purposes. It can be dated around 1934 because Knörlein dated it in his own hand (1934) on the original photo. (Source: private archive)

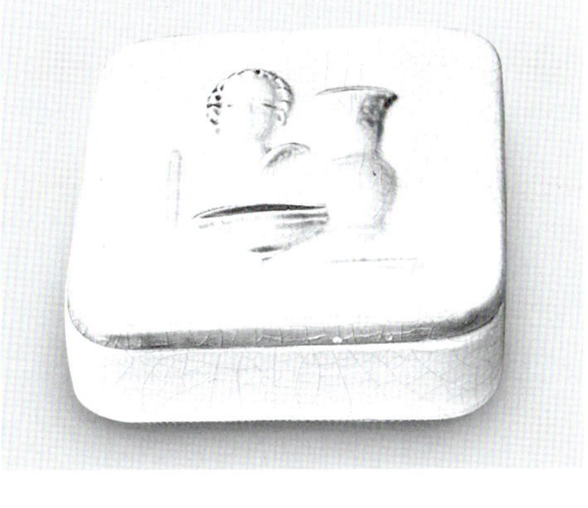

Ill. 65
Der liegende Akt (ModNr. 7717, Rudolf Knörlein zugeschrieben) entsteht ca. 1936. Die krausköpfige Frisur findet man auch häufig bei den Wandmasken. (Quelle: Fotoarchiv Karl Täuber)

The Reclining Nude (model number 7717, attributed to Rudolf Knörlein) originated ca. 1936. This tousled hairstyle also occurs frequently on the wall masks. (Source: Fotoarchiv Karl Täuber)

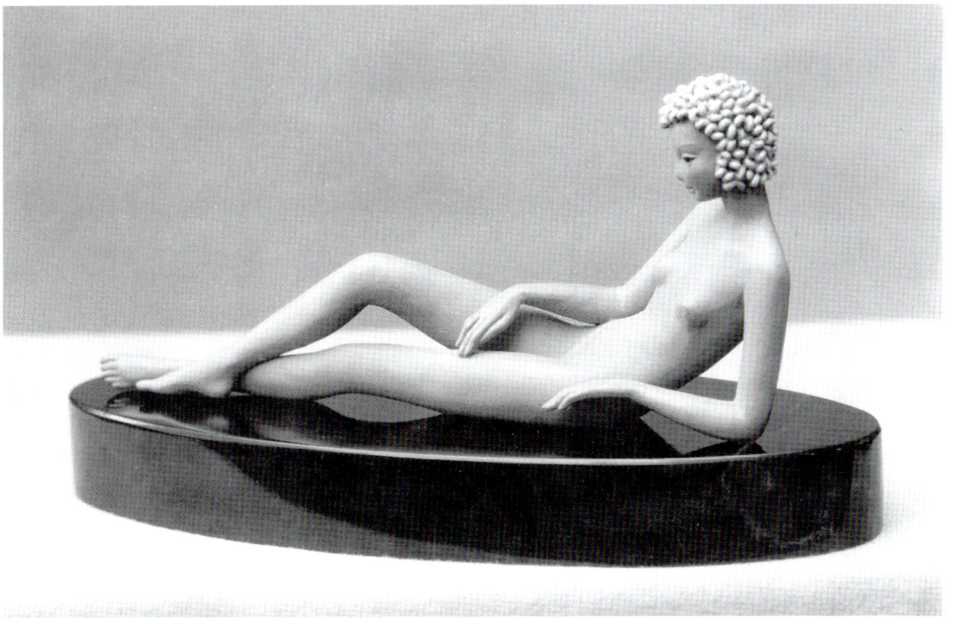

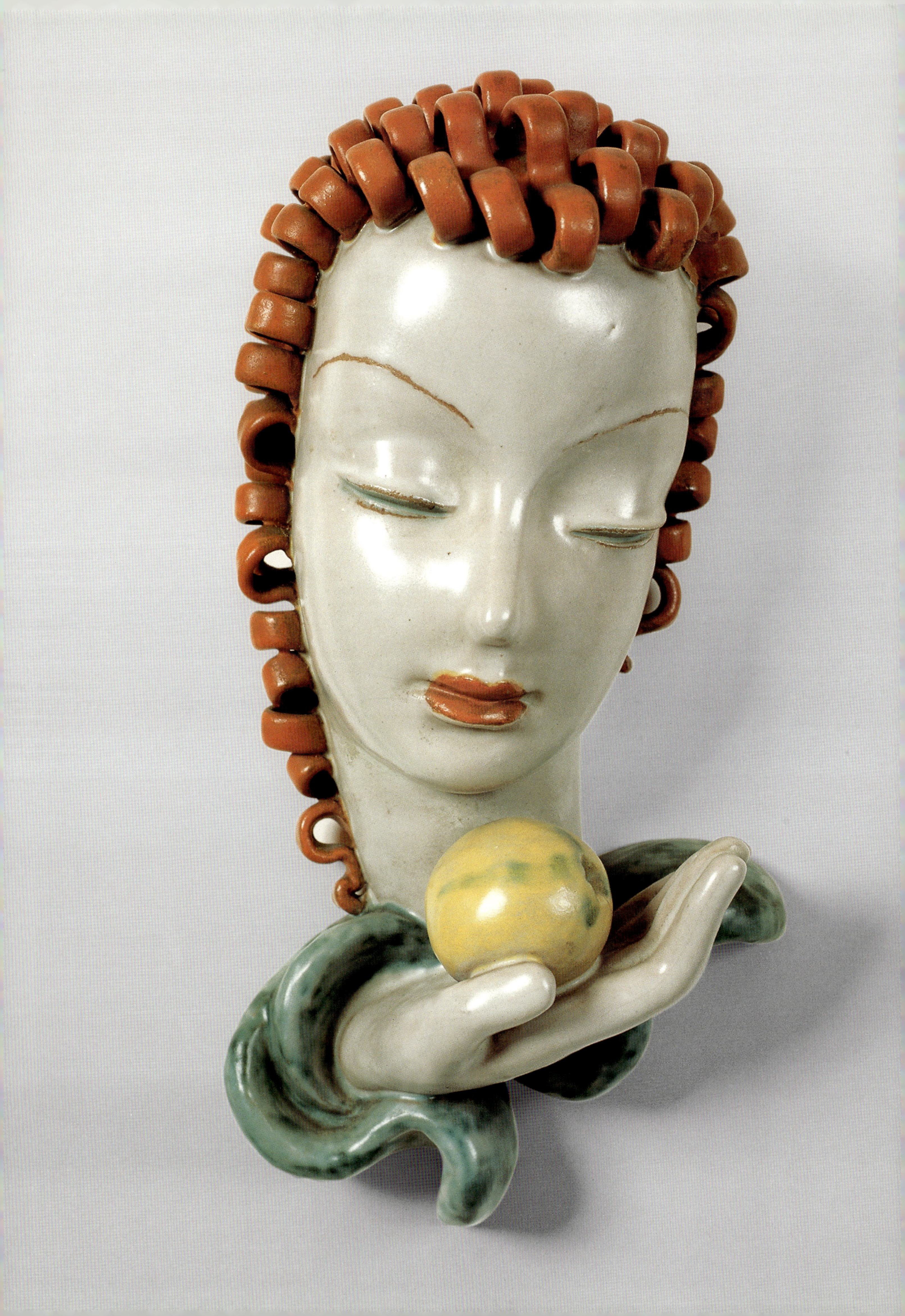

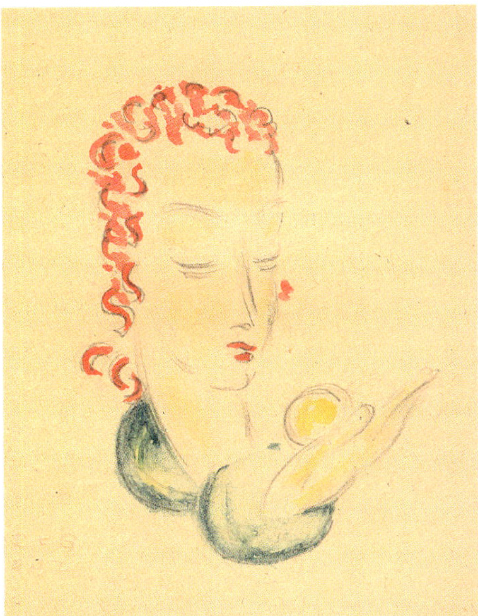

Ill. 66

ModNr. 6774
„Eva" mit rotem Haar | "Eve" with Red Hair, Rudolf Knörlein, ca. 1933, Sammlung.schmutz@chello.at

Die populärste Wandmaske von Knörlein ist „Eva", die in etlichen Variationen produziert wurde. Der Original-Entwurf von Knörlein zeigt sie mit einem Apfel in der Hand und rötlichem Haar (Ill. 66; Quelle: Privatsammlung D; Modell Sammlung.schmutz@chello.at). Es gibt sie aber auch mit grauem Haar (Sammlung f@goldscheider.de), grünem Haar (Sammlung a@goldscheider.net), schwarzem Haar (PB Wien 19), goldenem Haar, schwarzem Teint und rotem Apfel (Sammlung a@goldscheider.net) sowie mit einer Rose anstatt des Apfels (Sammlung a@goldscheider.net). Selbstverständlich darf dann auch ein Stehkopf nicht fehlen (Ill. 67; Quelle: Fotoarchiv Karl Täuber; Modell Sammlung.schmutz@chello.at)

The most popular Knörlein wall mask was "Eva," which was produced in several versions. The original Knörlein design depicts Eve with the apple in her hand and reddish hair (Ill. 66; source: private collection D; model Sammlung.schmutz@chello.at). There are also examples with grey hair (Sammlung f@goldscheider.de), green hair (Sammlung a@goldscheider.net), black hair (PB Wien 19), golden hair, black skin and a red apple (Sammlung a@goldscheider.net) as well as with a rose instead of the apple (Sammlung a@goldscheider.net).

It goes without saying that a head was also included in the range (Ill. 67; source: Fotoarchiv Karl Täuber; model Sammlung.schmutz@chello.at).

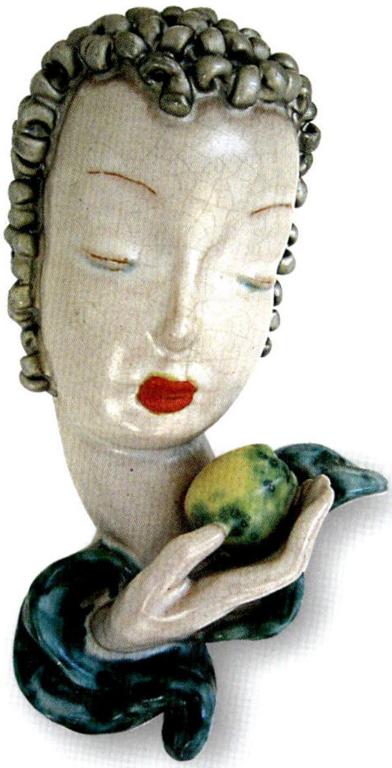

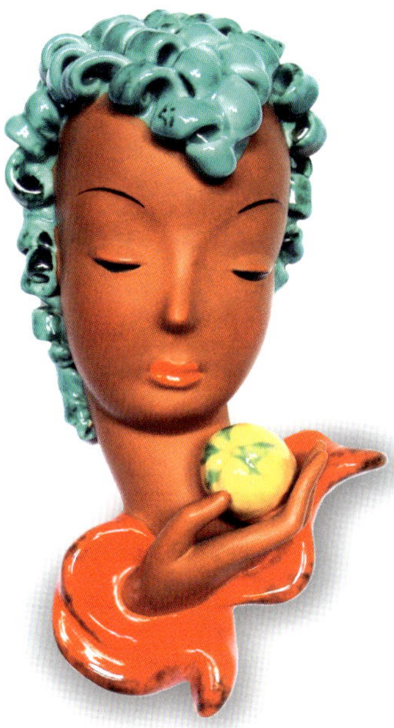

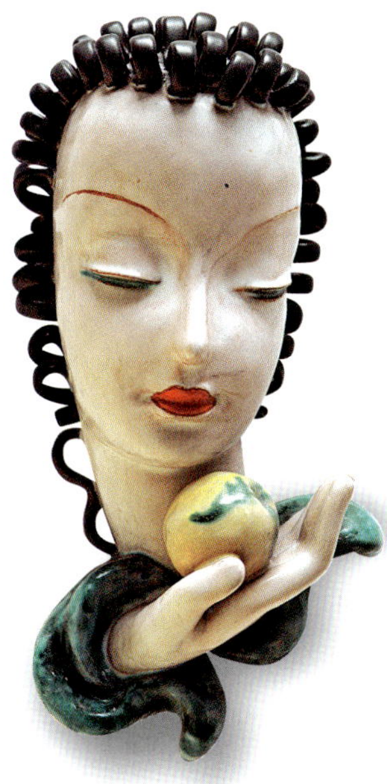

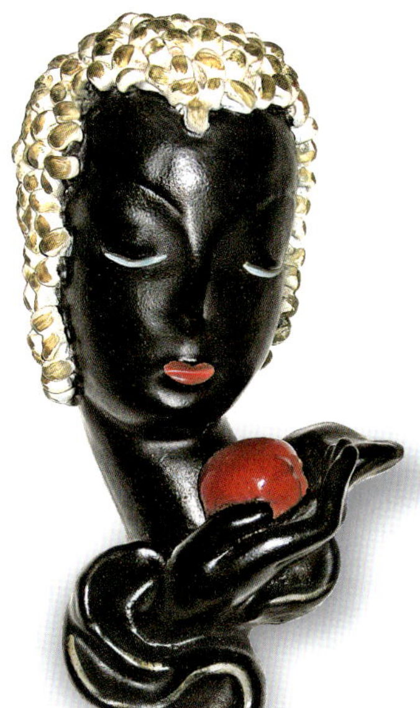

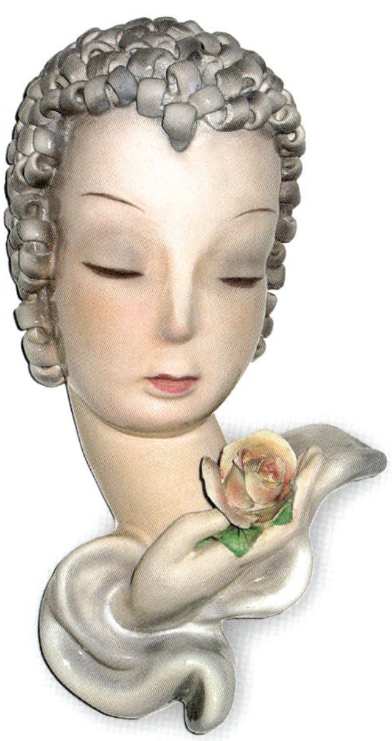

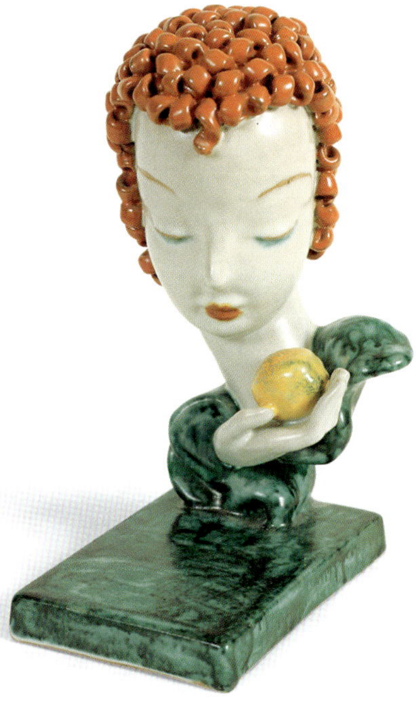

Ill. 67

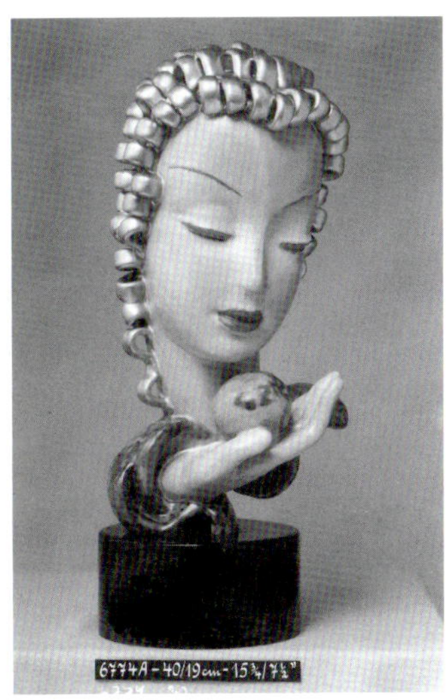

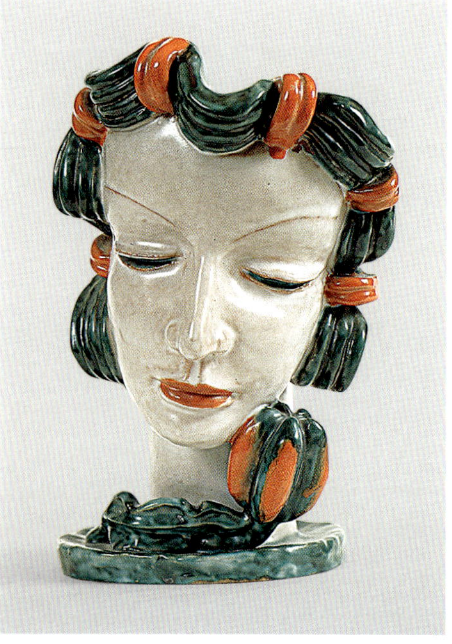

ModNr. 6305
Stehmaske | Mask Bust, Rudolf Knörlein, ca. 1931, Sammlung.schmutz@chello.at

rechts und rechte Seite | right and right side

ModNr. 6542
Wandmaske mit Fächer | Wall Mask with Fan, Rudolf Knörlein, ca. 1932, Sammlung.schmutz@chello.at

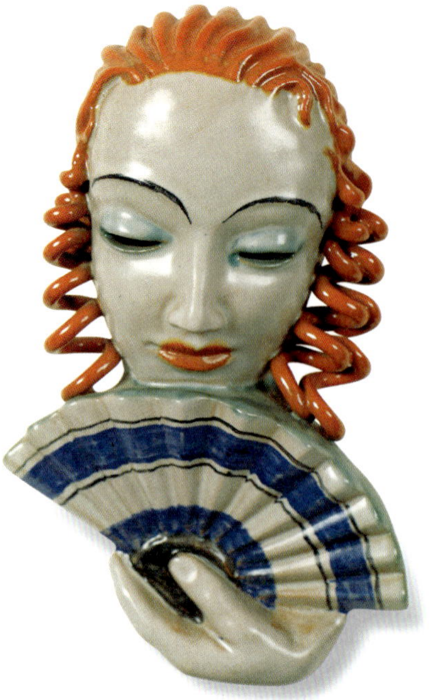

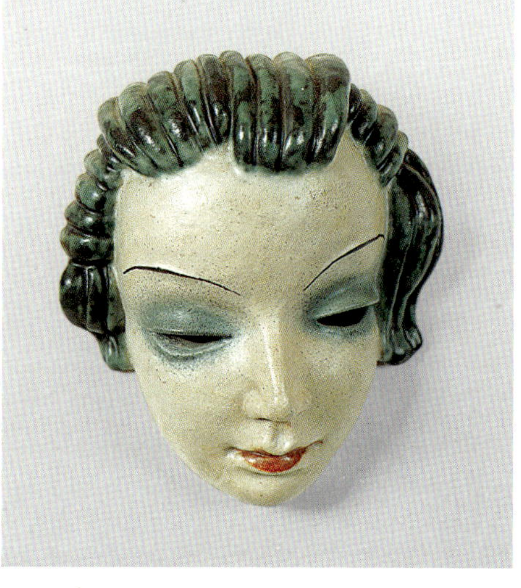

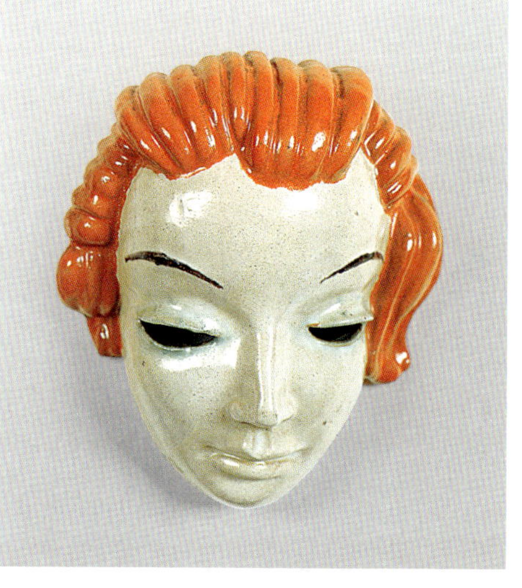

ModNr. 6306
Wandmaske | Wall Mask, Rudolf Knörlein, ca. 1931, Sammlung.schmutz@chello.at

Ill. 68

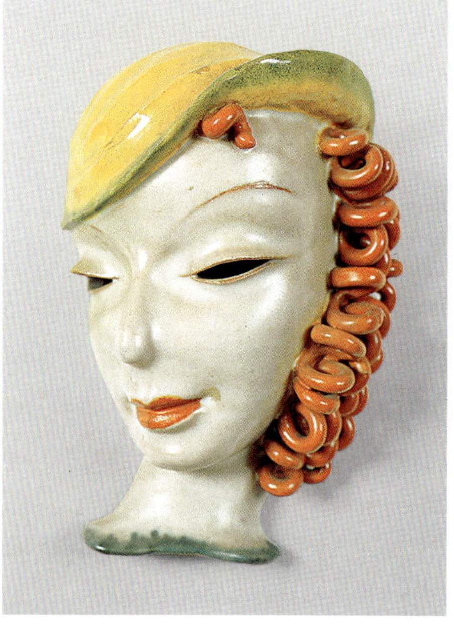

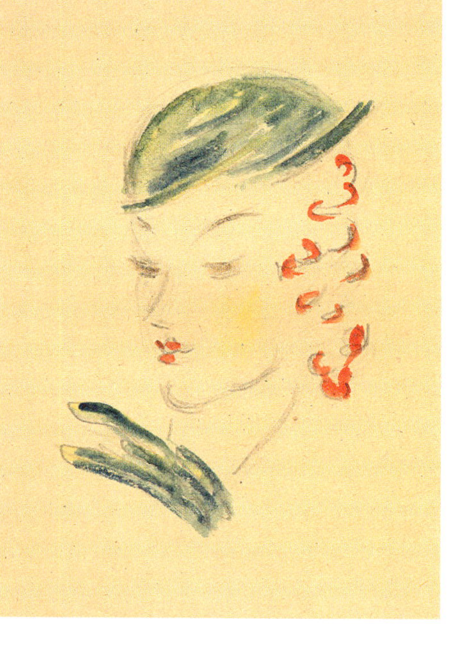

ModNr. 6428
Wandmaske | Wall Mask, Rudolf Knörlein, ca. 1931, Sammlung.schmutz@chello.at

Dank einer Originalzeichnung (Aquarell über Graphit) kann diese Maske Knörlein zugeordnet werden (vgl. Ill. 68); das Halstuch fiel vermutlich wegen der höheren Produktionskosten weg. (Quelle: Privatsammlung D)

Thanks to an original drawing (watercolor over graphite pencil), this mask can be attributed to Knörlein (cf Ill. 68); the scarf round the neck was probably eliminated to reduce production costs. (Source: private collection D)

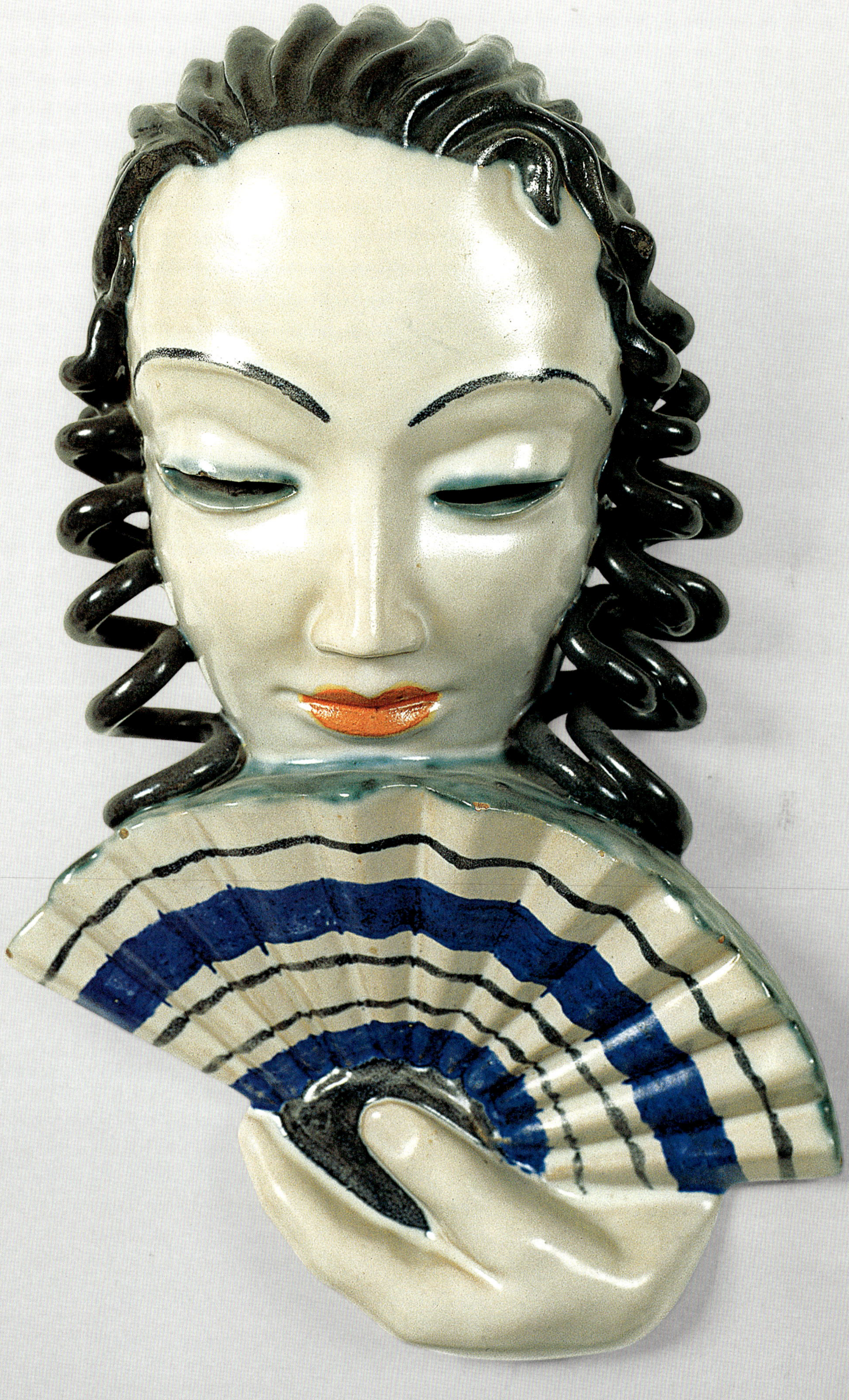

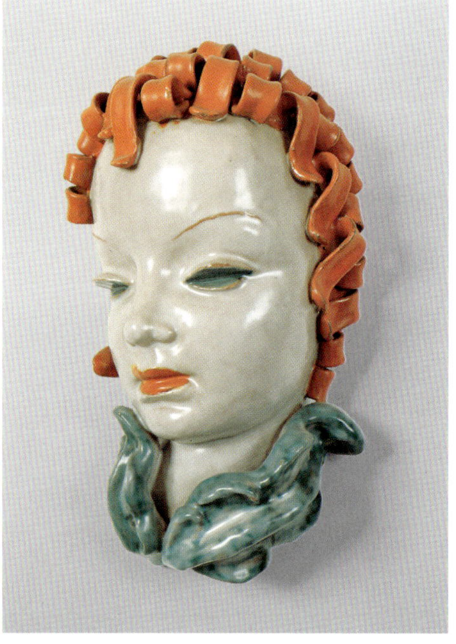

ModNr. 6735

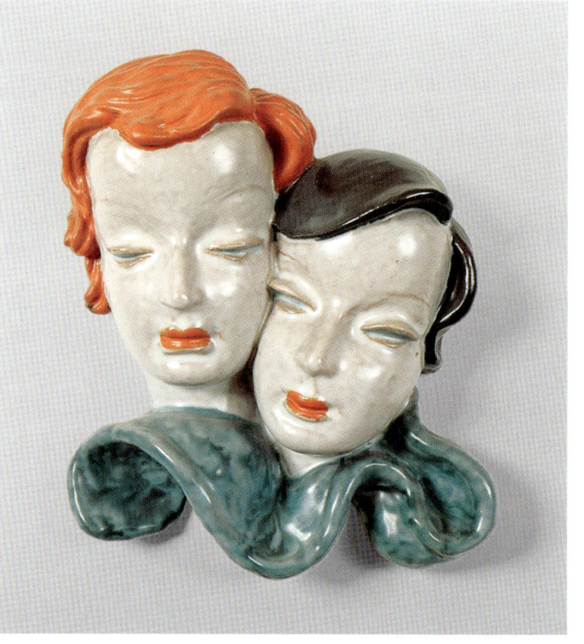

ModNr. 6883

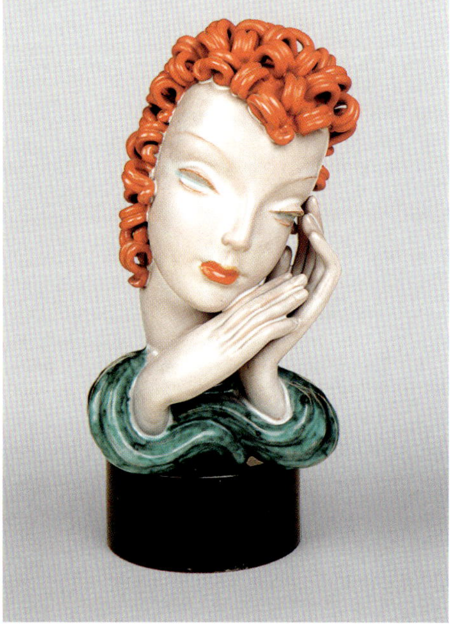

ModNr. 7495

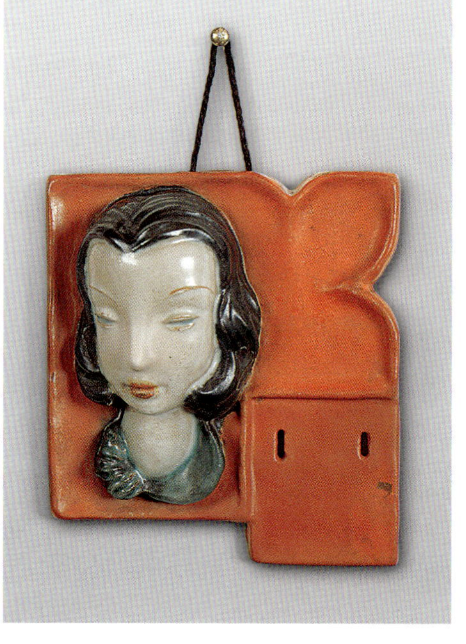

ModNr. 6880

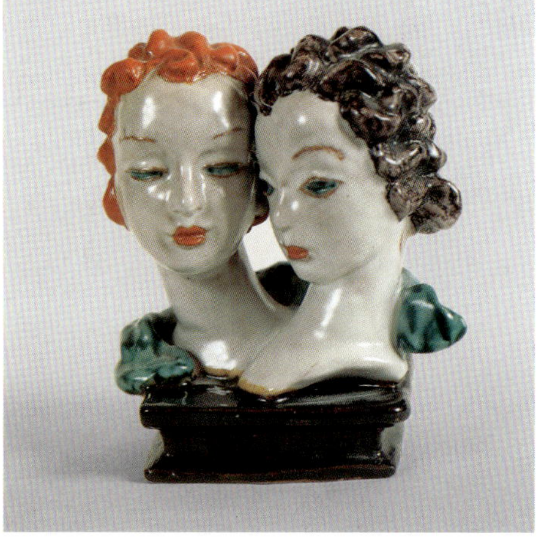

ModNr. 6739

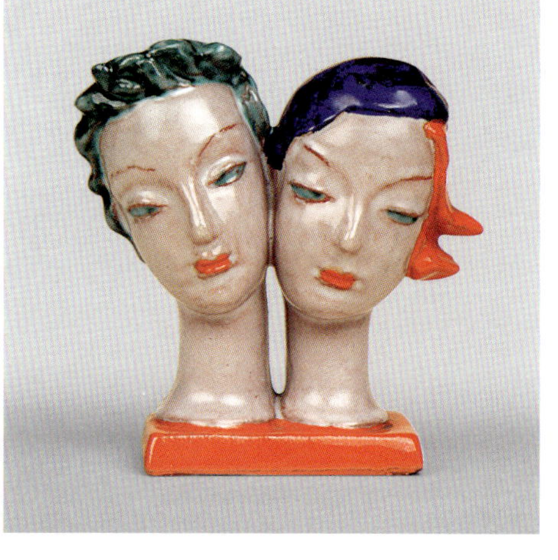

ModNr. 6560

ModNr. 6735
Wandmaske |
Wall Mask, Rudolf Knörlein, ca. 1933, Sammlung.schmutz@chello.at

ModNr. 6883
Doppelköpfe |
Two Heads, Rudolf Knörlein, ca. 1934, Sammlung.schmutz@chello.at

ModNr. 7495
Stehmaske | Mask Bust, Rudolf Knörlein, ca. 1936, PB Wien 19

ModNr. 6880
Abreißkalender |
Block Calendar, Rudolf Knörlein, ca. 1934, Sammlung.schmutz@chello.at

ModNr. 6739
Doppelstehkopf |
Two Heads, Rudolf Knörlein, ca. 1933, Sammlung.schmutz@chello.at

ModNr. 6560
Doppelstehkopf |
Two Heads, Rudolf Knörlein, ca. 1932, PB Wien 19

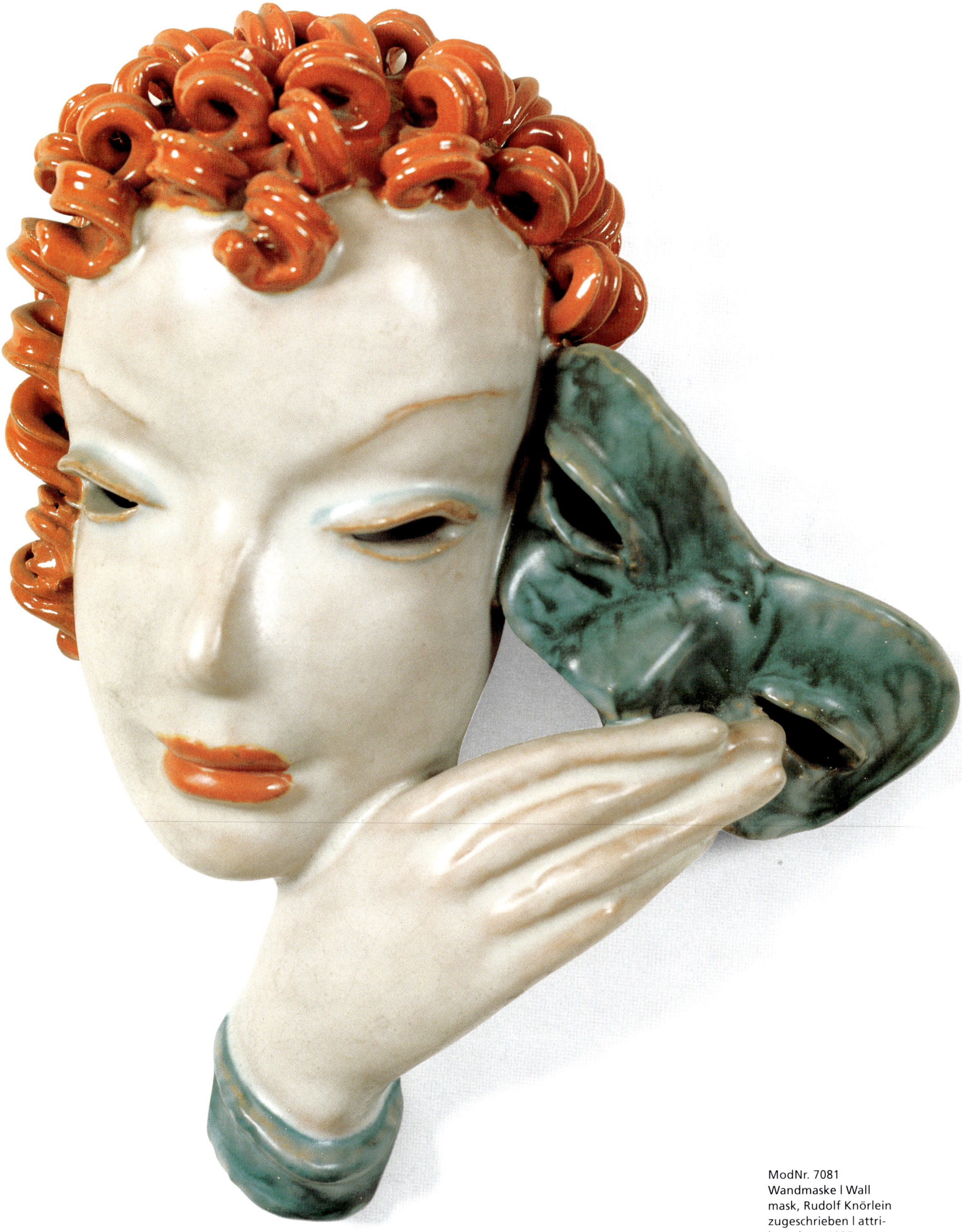

ModNr. 7081
Wandmaske | Wall mask, Rudolf Knörlein zugeschrieben | attributed, ca. 1935, Sammlung.schmutz@chello.at

rechts I right

ModNr. 7261
„Maske", ca. 1935,
Collection Rudolf
Hansl, St. Florian

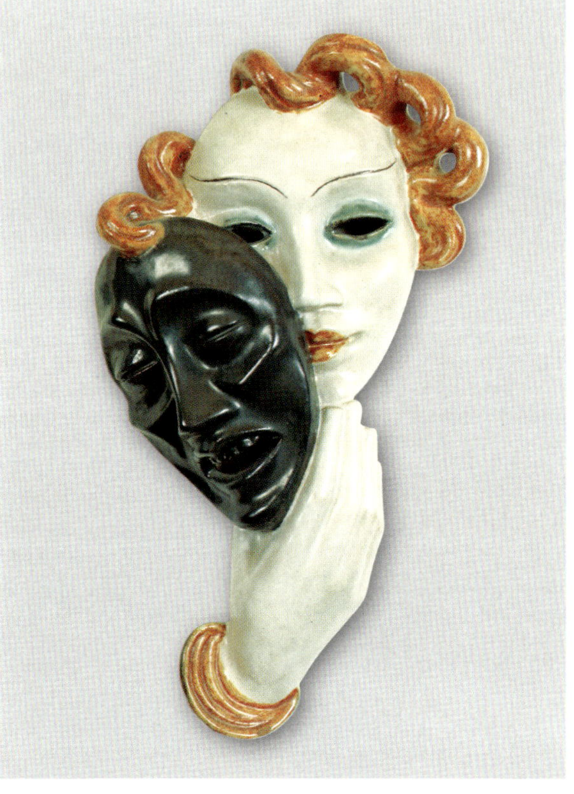

ModNr. 6288

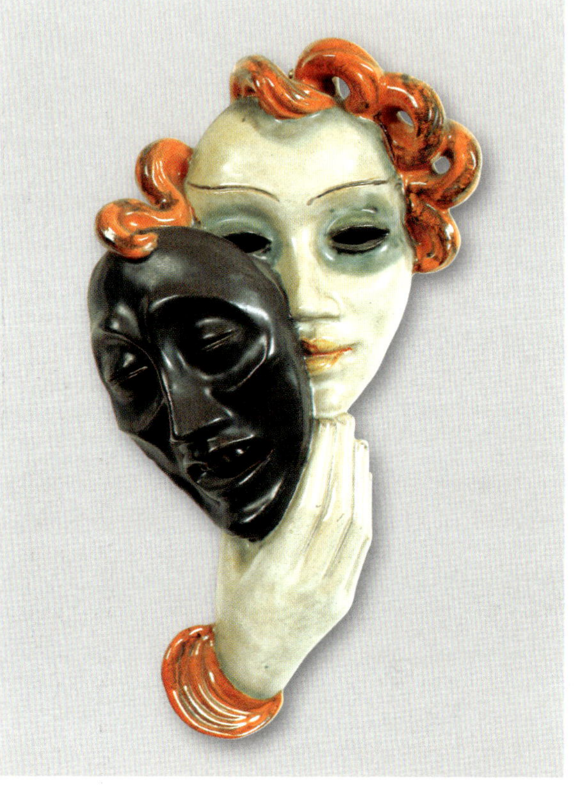

ModNr. 6288

ModNr. 6288
Frauenkopf mit Maske I
Head of a Woman
with Mask, Ernst
Weber, ca. 1930

ModNr. 7114
Frauenkopf mit Skye-
terrier I Head of a
Woman with Skye
Terrier, ca. 1935

ModNr. 7089
Frauenkopf mit
geknotetem Schal I
Head of a Woman
with Knotted Scarf,
ca. 1935

ModNr. 7102
Frauenkopf mit klei-
nem Vogel I Head of
a Woman with Little
Bird, ca. 1935

alle I all
Sammlung.Schmutz@
chello.at

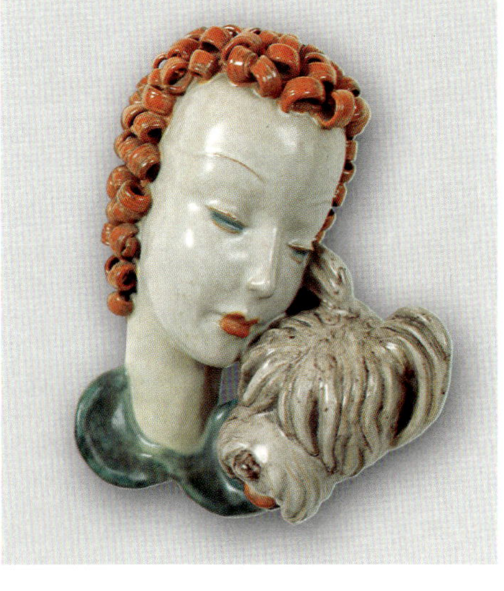

ModNr. 7114

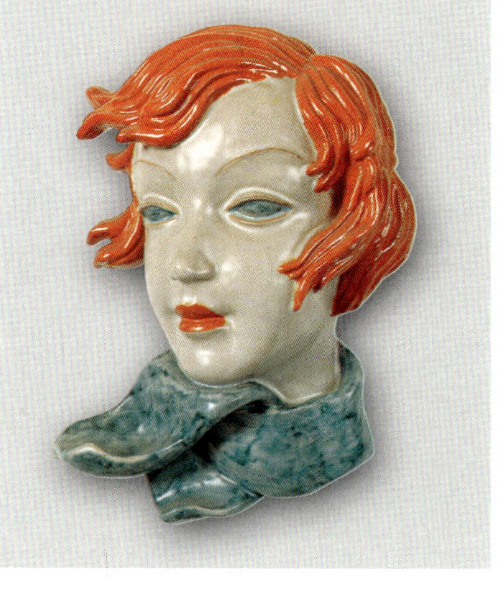

ModNr. 7089

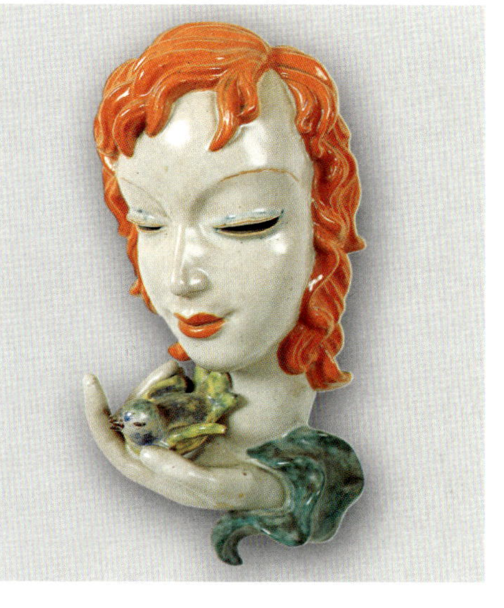

ModNr. 7102

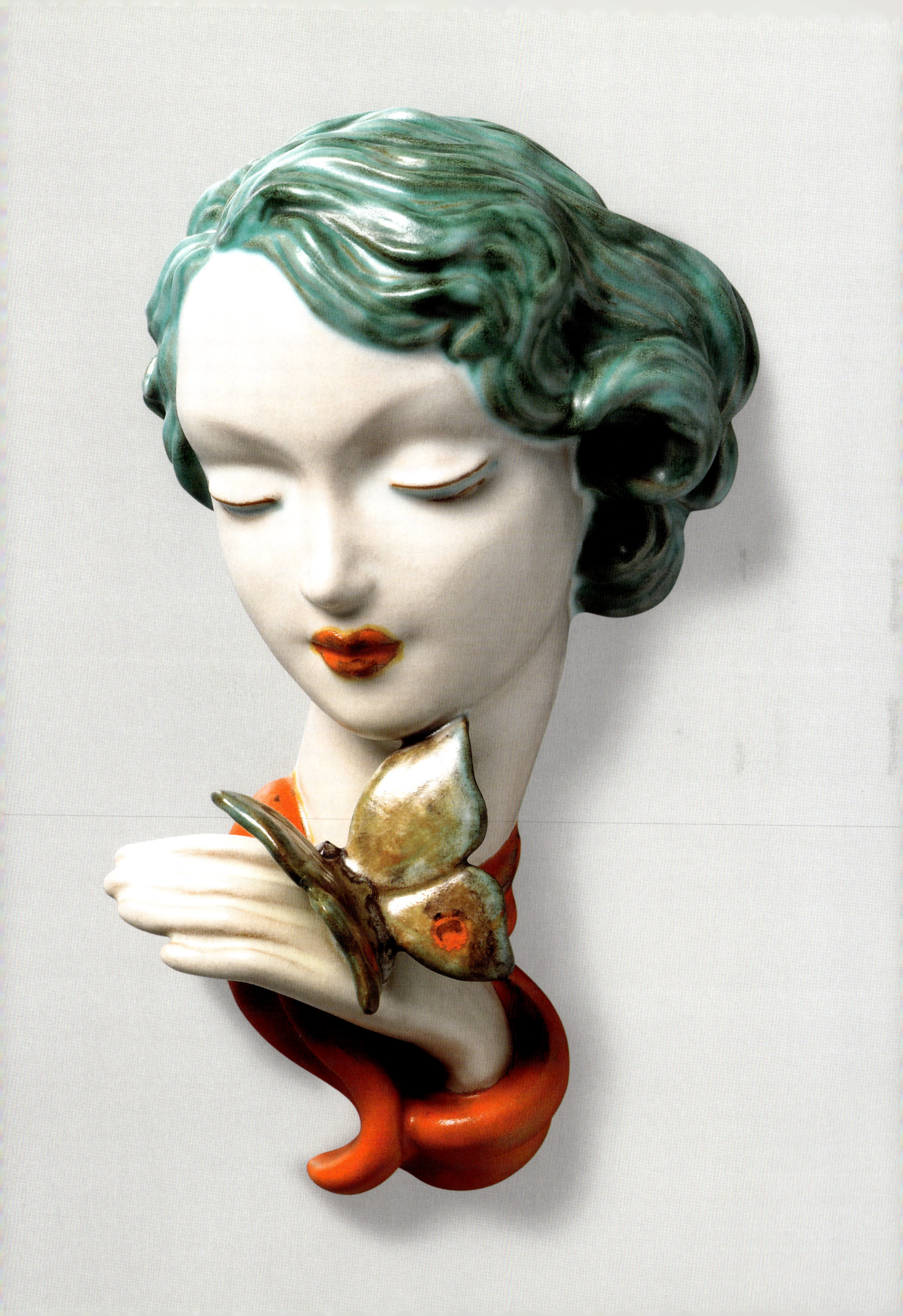

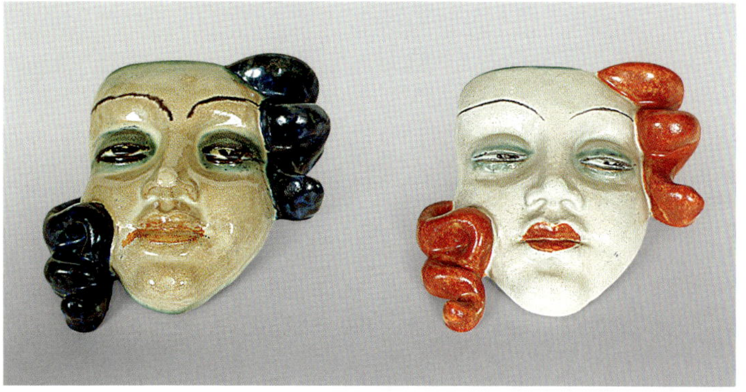

ModNr. 6086

ModNr. 6086 Wandmaske (Wandvase) | Wall Mask (Wall Vase), Ernst Weber, ca. 1930, beide | both Sammlung.schmutz@chello.at
Dieses Modell wurde in zwei Versionen produziert. Zum einen die gängige Ausführung als Wandmaske, zum anderen als Wandvase. Bei Goldscheider war dies eine durchaus übliche Praxis, durch zusätzliche Funktion Neues zu kreieren und zugleich die Modellpalette zu erweitern.

Two versions of this model were produced. On the one hand, the standard wall mask version, on the other, as a wall vase. It was definitely the usual practice at Goldscheider to enlarge the model range by widening the functional scope of available models.

rechts | right

ModNr. 7891 Doppelmaske | Double Mask, ca. 1937, Sammlung.schmutz@chello.at

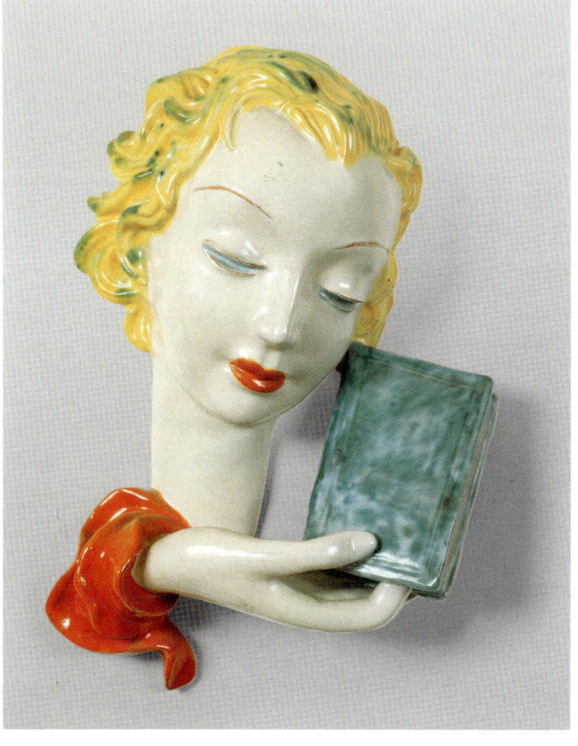

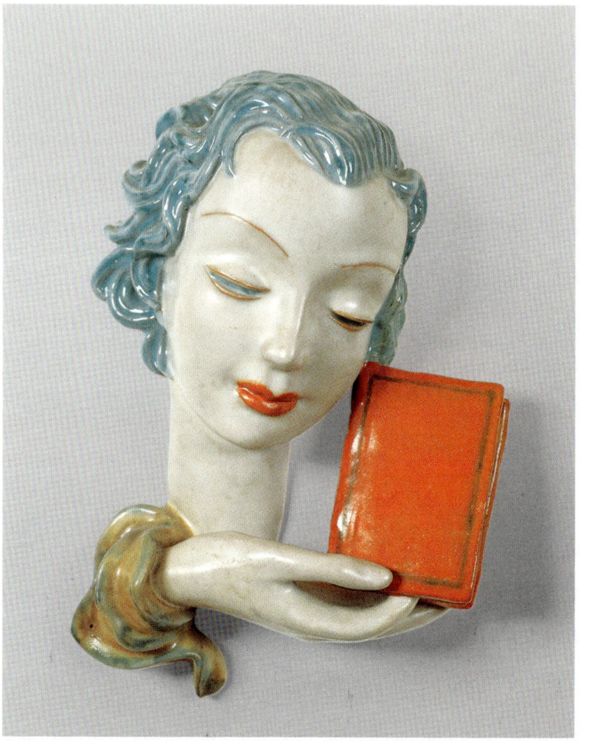

ModNr. 7419 Frauenkopf mit Buch | Head of a Woman with Book, ca. 1935, beide | both Sammlung.schmutz@chello.at

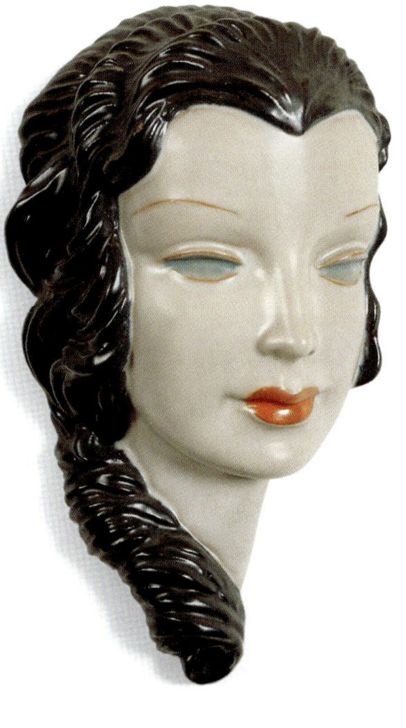

ModNr. 7695 Wandmaske | Wall Mask, ca. 1936, Sammlung.schmutz@chello.at

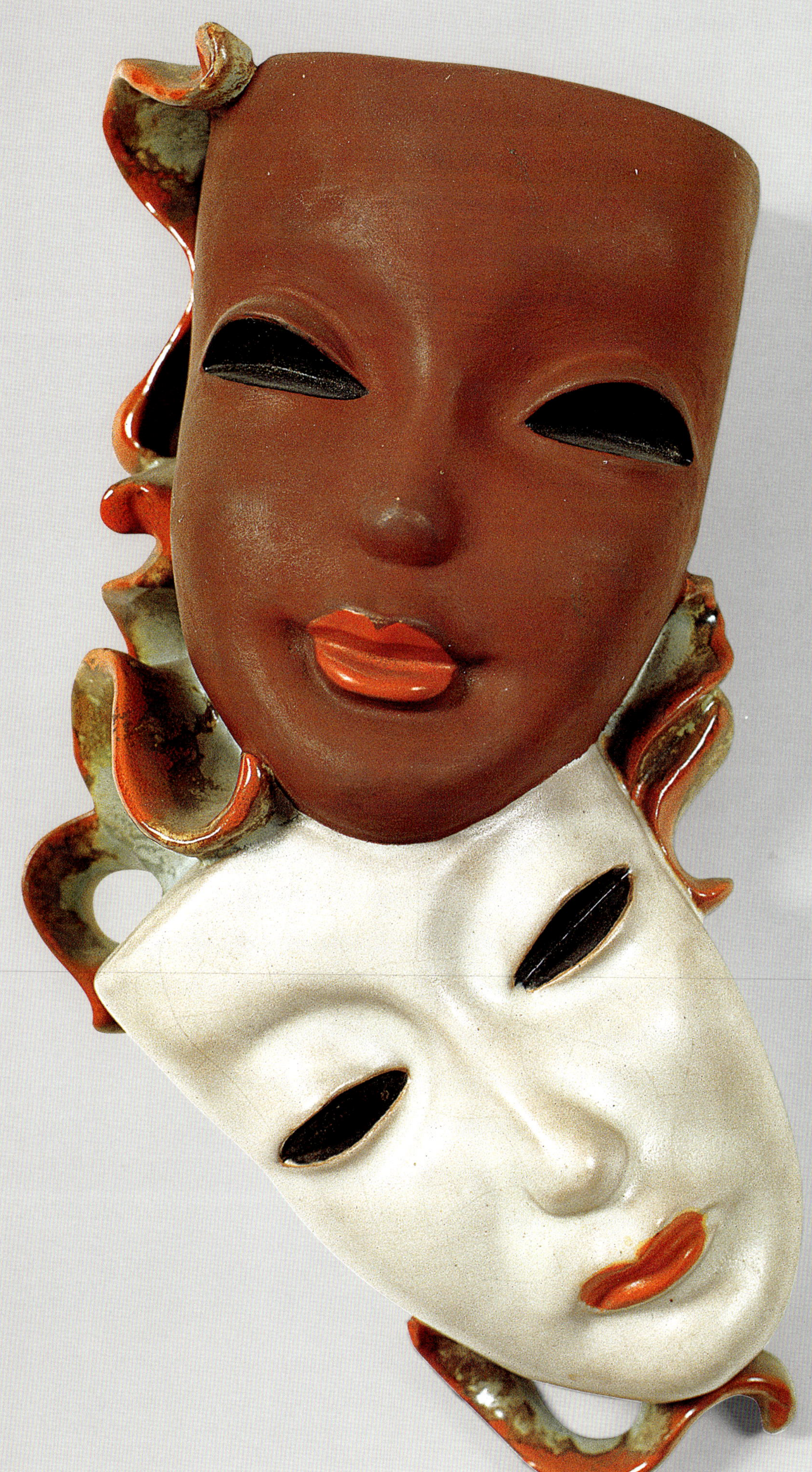

Ill. 69
Mitte der 1930er Jahre wird eine von Prof. Robert Obsieger kreierte Produktlinie eingeführt, bereits kurze Zeit später aber wegen mangelnder Nachfrage wieder eingestellt.

A product line created by Professor Robert Obsieger was launched in the mid-1930s but was soon discontinued for lack of demand.

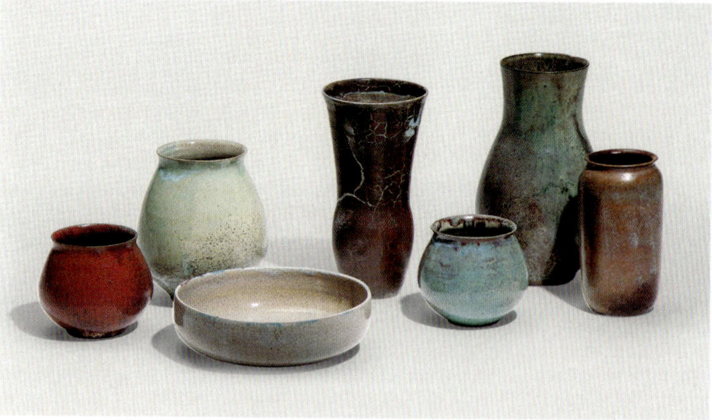

ModNr. 7195 „Beauty", Stephan Dakon, ca. 1935, Collection city antik, Wien
Eines der beliebtesten Modelle ist die von Stephan Dakon um 1935 entworfene Tänzerin mit der Bezeichnung „Beauty". Sie wird in verschiedenen Größen hergestellt und ist sogar als Lampe zu erwerben. Als Modell 7195 ist sie 40 cm hoch, unter der Modellnummer 7996 hat sie eine Höhe von 26 cm und unter 7897 nur 16 cm. Sie wird größtenteils in blauer und roter Glasur geliefert, manchmal auch in grün oder gelb, mit verschiedenen Staffierungen (Bemalungen) auf dem Kleid. (Vgl. Ill. 70, Quelle: Archiv Mimi Dakon)

One of the most popular models was the dancer called "Beauty," designed by Stephan Dakon around 1935. It was produced in several sizes and was even marketed as a lamp. As model number 7195, it was 40 cm high, as model number 7996 26 cm and as model number 7897 only 16 cm high. It was usually made in blue and red glaze but occasionally also in green or yellow, with the dress painted in several different ways. (Cf Ill. 70, source: Mimi Dakon Archives)

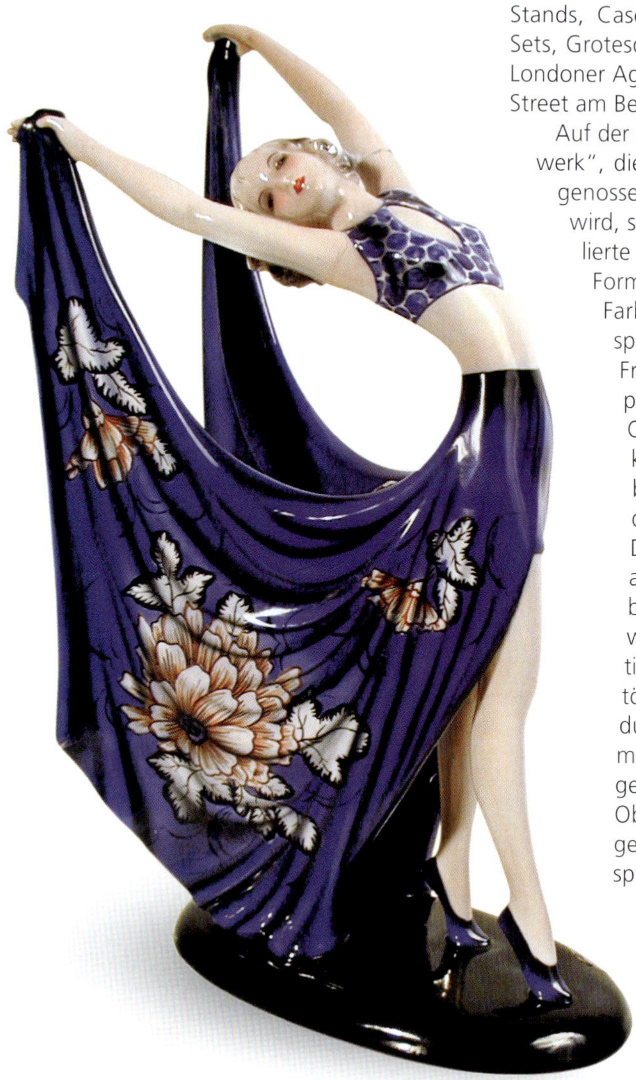

Als Betriebsleiter fungiert in den 1930er Jahren Fritz Iglauer, der Betrieb hat ca. 100 Mitarbeiter.[255] National erhält die Wiener Manufaktur Friedrich Goldscheider große Aufträge wie die keramische Dekoration der 1932 errichteten Wiener Werkbundsiedlung.[256] Die internationale Teilnahme an Ausstellungen, wie 1929 auf der Exposición Internacional in Barcelona, fördert zudem den Export.[257] Auf der Ausstellung „Austria in London" im Jahr 1934 zeigt die Firma „The Vienna Manufactory Frederick Goldscheider" Modelle des aktuellen Sortiments: „High Class Art Goods in Pottery Terra Cotta, Modern Ceramics, Statuettes, Animals, Lamp Stands, Cases, Cups, Ash Trays, Smoking Sets, Grotesque Figures of all kinds."[258] Der Londoner Agent ist J. Plichta in der 37 Store Street am Bedford Square.

Auf der Ausstellung „Das befreite Handwerk", die 1934 vom Wiener Gewerbegenossenschafts-Verband veranstaltet wird, stellt Goldscheider neue craquelierte Gefäßkeramik vor, die eine klare Formgebung sowie reduzierte Uni-Farben zeigt.[259] Wenige Monate später werden auf der Wiener Frühjahrsmesse weitere Gefäße präsentiert. Die von Prof. Robert Obsieger entworfenen Keramiken werden 1935 als nach ihm benannte, eigenständige Produktlinie eingeführt (Ill. 69). Diese künstlerischen Werke sind aus „dünn ausgedrehtem Scherben" mit „edler Oberflächenwirkung" in leuchtenden roten, tiefrotbraunen und grünen Matttönen, teilweise „heller oder dunkler gefleckt". Manche sind mit „dünn ausgedrehtem, leicht gerieftem Scherben und edler Oberflächenwirkung" in quittengelben Matttönen, dunkel gesprenkelt.[260]

In the 1930s, Fritz Iglauer was managing director; and a workforce of approx 100 was employed.[255] The Wiener Manufaktur Friedrich Goldscheider dealt with large domestic orders such as the ceramic decoration for the Wiener Werkbund settlement built in 1932.[256] Participation in trade fairs abroad, such as the 1929 Exposición Internacional in Barcelona, promoted exports.[257] At the 1934 "Austria in London" exhibition, the firm, listed as "The Vienna Manufactory Frederick Goldscheider" showed models from the current range: "High Class Art Goods in Pottery, Terra Cotta, Modern Ceramics, Statuettes, Animals, Lamp Stands, Cases, Cups, Ash Trays, Smoking Sets, Grotesque Figures of all kinds."[258] The London agent for Goldscheider was J. Plichta at 37 Store Street, Bedford Square.

At the 1934 trade fair "Das befreite Handwerk" mounted by the Viennese Gewerbegenossenschafts-Verband [Decorative and Applied Arts Co-operative Union], Goldscheider launched new crackle-glaze ceramics notable for clear vessel design and subdued monochrome.[259] More vessels were

Ill. 70

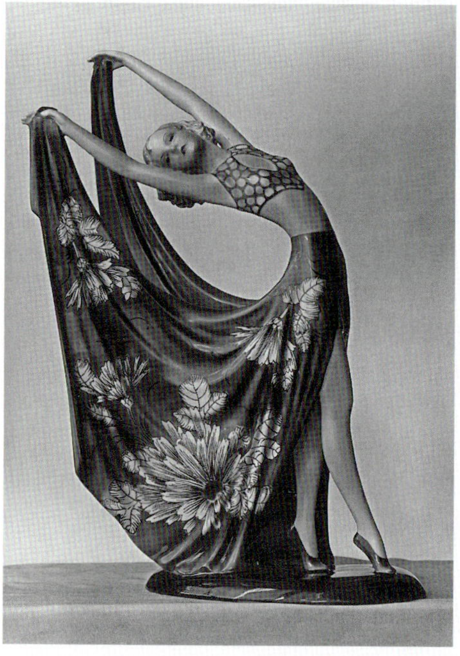

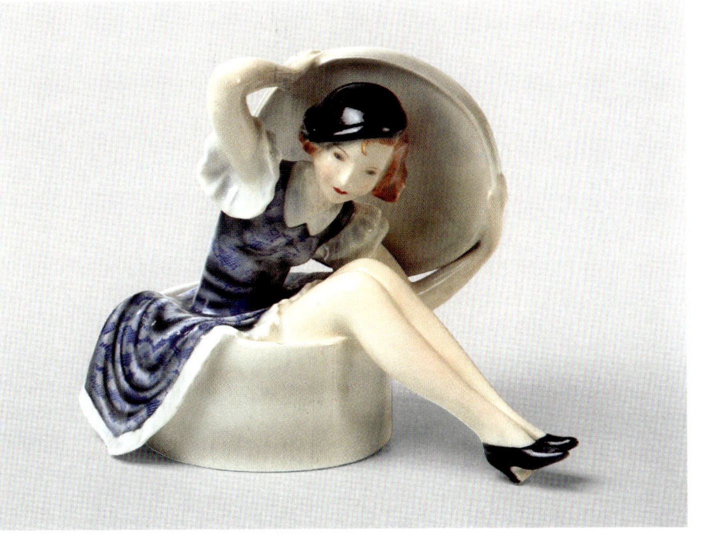

ModNr. 7200 „Surprise", Stephan Dakon, ca. 1935, © Wien Museum

shown a few months later at the Viennese Spring Fair. Designed by Professor Robert Obsieger, these ceramics were launched in 1935 as a separate product line named after him (Ill. 69). These art works are made of "thin, turned body" with an "elegant-looking surface" in vibrant red, deep reddish brown and green matt tones, some of them featuring "lighter or darker mottling." Some had "thin-walled, slightly grooved body and an elegant-looking surface" in matt tones of quince yellow with dark spots.[260]

In the 1930s the Goldscheider manufactory was again at the height of its powers and enjoyed worldwide popularity. On the occasion of the firm's fiftieth anniversary, Die Schaulade wrote in February 1935 on "Viennese faience," which had become synonymous with Goldscheider: "Models distinguished by originality, which are suitably adapted to their time, elegant glazes, choice colors, tasteful decoration, those are their distinguishing features that are also valued and respected today."[261] Typical 1930 Goldscheider models are the dancers by Josef Lorenzl, Stephan Dakon and Klára Herczeg (Claire Weiss), wall masks by Kurt Goebel, Rudolf Knörlein and Adolf Prischl, animal figurines by Karin Jarl, nude figurines by Lorenzl, who also designed some Madonnas, and figurines of children by Dakon and Weiss. Lorenzl, a sculptor with academy training, did have a studio of his own in the factory building but he also produced to a limited extent for other manufactories. Karl Sailer, who was permanently employed at Goldscheider as a retoucher, designed quite

Die Goldscheider Manufaktur erlebt in den dreißiger Jahren einen erneuten Höhepunkt ihres Schaffens und ihrer weltweiten Popularität. Zum fünfzigjährigen Bestehen der Manufaktur schreibt „Die Schaulade" im Februar 1935 über die „Wiener Fayence", die zum Synonym für Goldscheider geworden ist: „durch Originalität ausgezeichnete Modelle, die der jeweiligen Zeit richtig angepasst, edle Glasuren, erlesene Farben, geschmackvolle Dekore, das sind ihre auch heute noch in Geltung und Ansehen stehende Kennzeichen."[261] Typische Modelle der dreißiger Jahre sind die Tänzerinnen von Josef Lorenzl, Stephan Dakon und Klára Herczeg (Claire Weiss), Wandmasken von Kurt Goebel, Rudolf Knörlein, Adolf Prischl, Tierfiguren von Karin Jarl, Aktfiguren von Lorenzl, der auch einige Madonnen entwirft, Kinderfiguren von Dakon und Weiss. Der akademisch ausgebildete Bildhauer Lorenzl hat zwar ein eigenes Atelier im Fabrikgebäude, produziert in geringem Umfang aber auch für andere Manufakturen. Karl Sailer, der als Retuscheur bei Goldscheider fest angestellt ist, entwirft etliche Kinderfiguren.[262] Claire Weiss und Kurt Goebel sind als freie

ModNr. 8169 „Modedame" I "Lady of Fashion", Claire Herczeg, ca. 1937, private collection, Wien
Bei der dargestellten Person handelt es sich um die Schauspielerin Marlene Dietrich. (Vgl. Ill. 71, Quelle: Fotoarchiv Karl Täuber)

The person portrayed is the actress Marlene Dietrich. (Cf Ill. 71, source: Fotoarchiv Karl Täuber)

Ill. 71

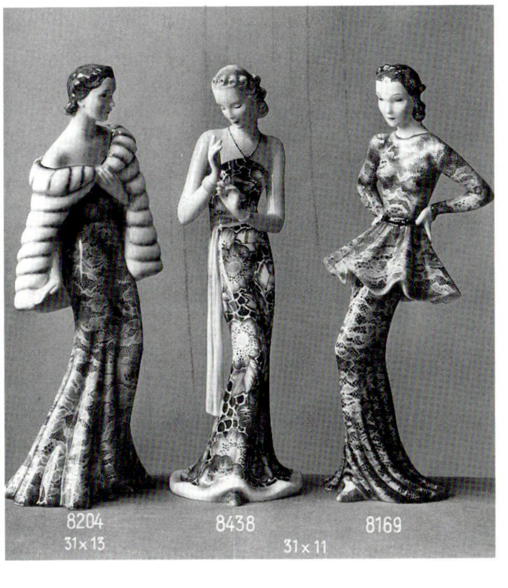

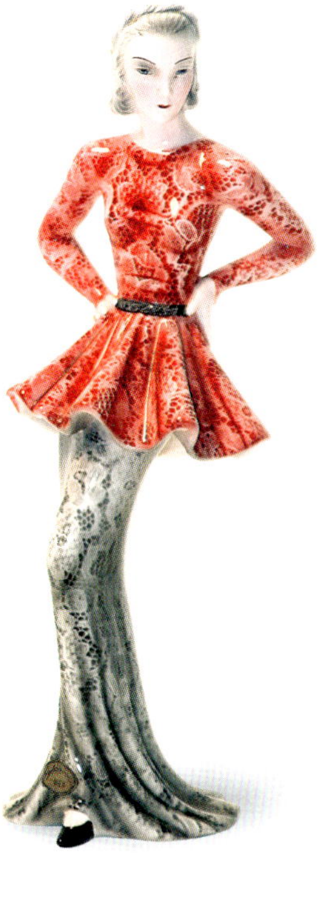

ModNr. 7655
„Skifahrer" I "Skier",
Kurt Goebel, ca. 1936,
© Wien Museum

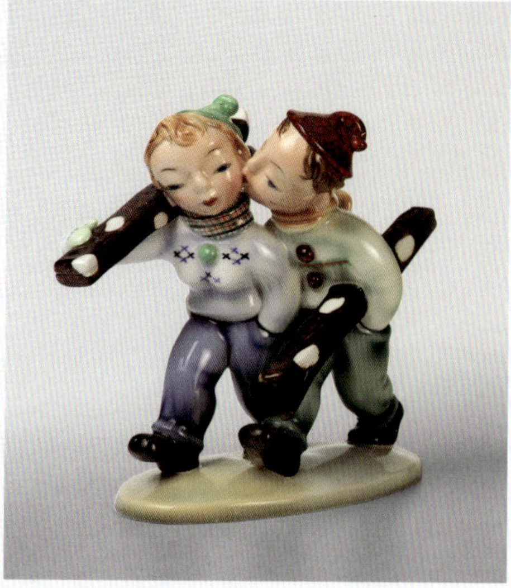

ModNr. 6922
Ascher mit Kopf I
Ashtray with Head,
Kurt Goebel, ca. 1934,
Collection a@goldscheider.net

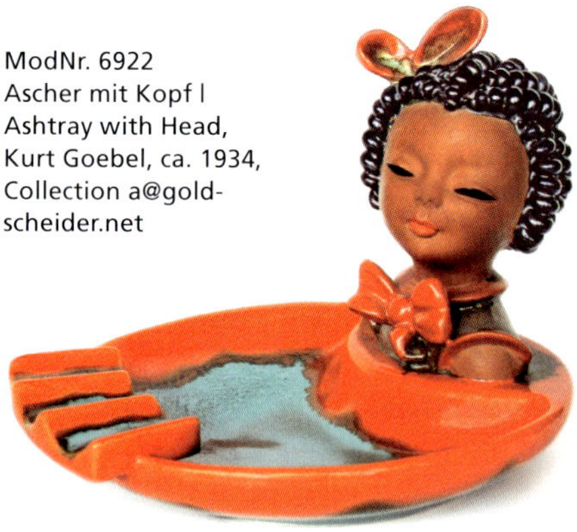

Mitarbeiter für Goldscheider und andere Manufakturen tätig. Prischl und Dakon sind hingegen fest angestellt und eng an die Goldscheider Manufaktur gebunden. Ein sehr beliebtes Modell ist beispielsweise die von Dakon entworfene Tänzerin mit der Bezeichnung „Beauty", die neben der Modellnummer 7195 in weiteren Größen erhältlich ist (Ill. 70, S. 134). Aber auch der „Gefangene Vogel" von Lorenzl, der in seiner größten Ausführung die Modellnummer 5230 trägt, ist weiterhin gefragt.

Einer der auffälligsten Keramiker ist ab dem Jahr 1936 Kurt Goebel, dessen Arbeiten, eine Stehmaske (Modell-Nr. 7616) und ein Frauenkopf (Modell-Nr. 8339) sowie die Gruppe „Skifahrer" (Modell-Nr. 7655), auf der Wiener Herbstmesse besonderen Anklang finden. Seine „Kleinplastiken, Gruppen, voll frischer Lebendigkeit im Umriß und in der Oberflächenbehandlung, voll frischer Farbigkeit in Glasur und Dekor, gewannen insbesondere durch den Humor und Charme der Darstellung auch das anspruchsvolle Publikum."[263]

a few figurines of children.[262] Claire Weiss and Kurt Goebel worked on a freelance basis for Goldscheider and other manufactories. Prischl and Dakon, on the other hand, were permanently employed and closely associated with the Goldscheider manufactory. A highly popular model was, to take one example, the dancer called "Beauty" designed by Dakon. It was available in several sizes besides model number 7195 (Ill. 70, p. 134). However, Lorenzl's "Gefangener Vogel" ["Captured Bird"], the largest version of which was model number 5230, continued to be much in demand.

One of the most remarkable ceramicists from 1936 was Kurt Goebel, whose works, a mask bust (model number 7616) and a head of a woman (model number 8339) as well as the "Skiers" group (model number 7655) met with a particularly enthusiastic reception at the Vienna Autumn Fair. Goebel's "figurines, groups, full of freshness and liveliness in silhouette and in the surface treatment, full of fresh color in glaze and decoration, enchanted even the most discerning public, so charming and humorous are they."[263]

ModNr. 7844, 7845
„Rendez-vous", Kurt
Goebel, ca. 1937,
Collection a@goldscheider.net

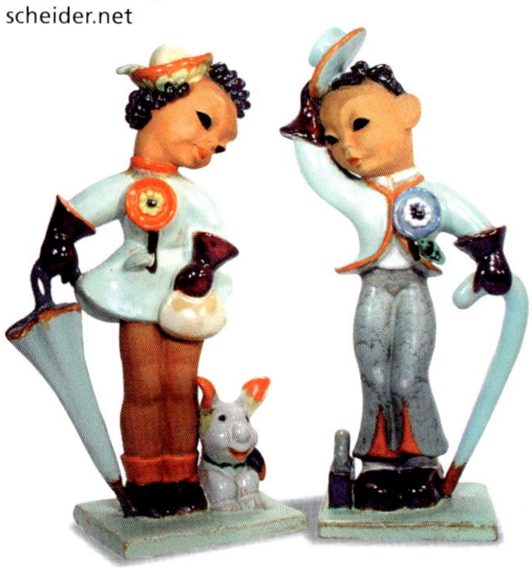

ModNr. 7464
Knabe auf Fisch I Boy
on a Fish, Kurt Goebel,
ca. 1936, Collection
a@goldscheider.net

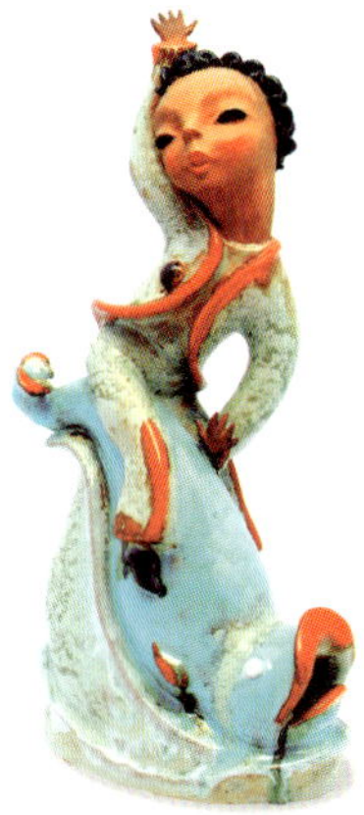

ModNr. 7466
Der Bücherwurm I
Bookworm, Kurt
Goebel, ca. 1936,
Collection John Nash

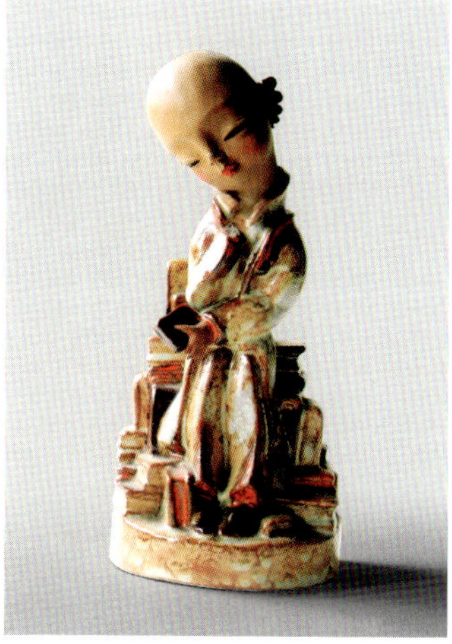

ModNr. 7686
Buchstützen I Bookends, Kurt Goebel,
ca. 1936, PB Wien 19

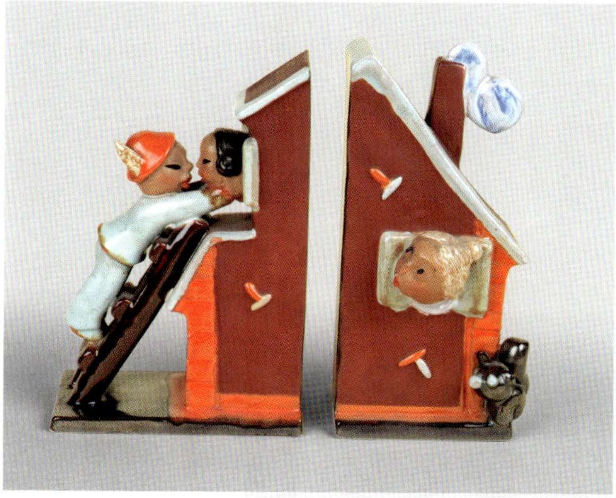

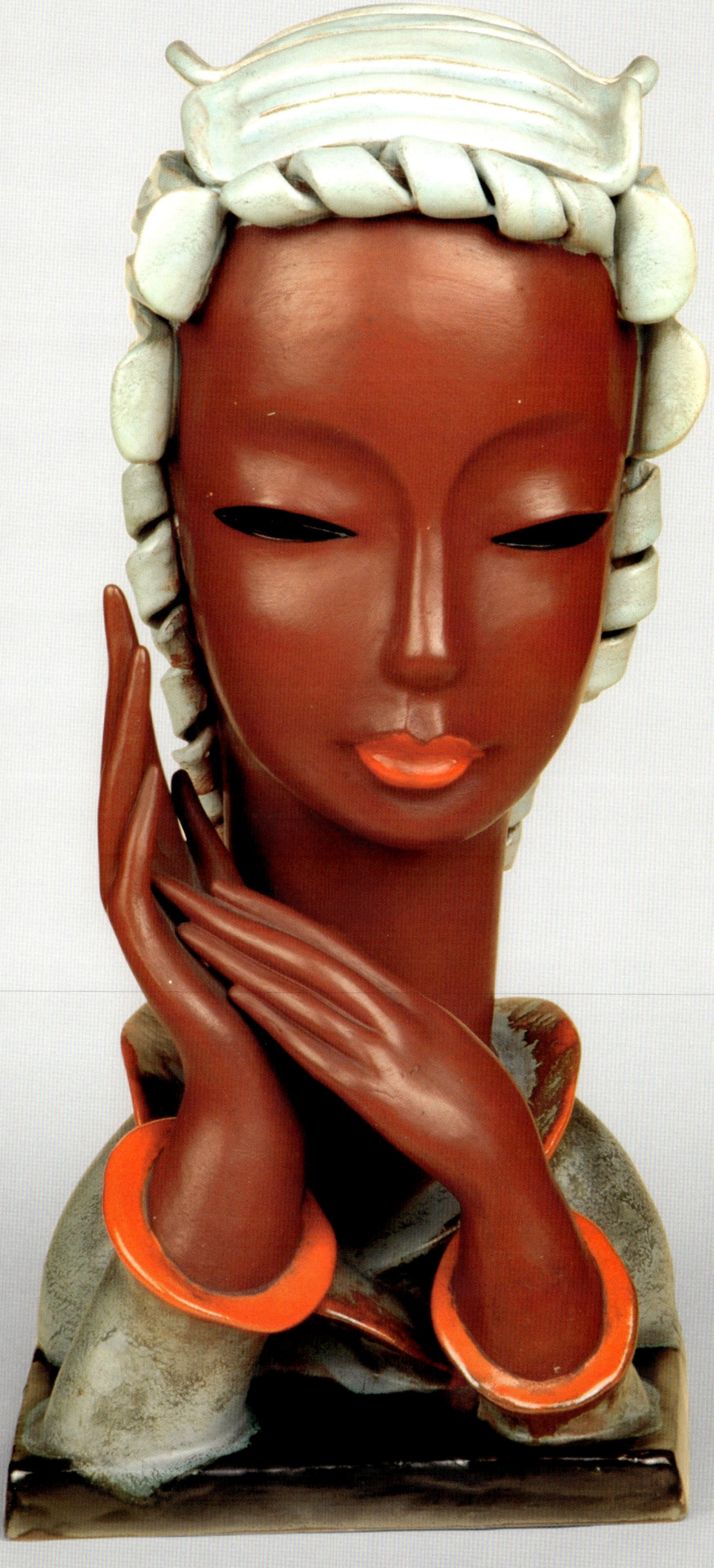

ModNr. 7653
„Kopf" | "Head",
Kurt Goebel, ca. 1936,
MAK – Österreichisches
Museum für ange-
wandte Kunst / Gegen-
wartskunst, Wien

Auch Arbeiten des bei Goldscheider beschäftigten Adolf Prischl werden wenige Monate später der Öffentlichkeit vorgestellt. Die Kritiker loben seinen Mädchenkopf (Modell-Nr. 7982) sowie ein Pferd und bezeichnen Prischl als „eine Begabung, die entwicklungsfähig ist und eigene Wege" finden wird.[264]

In der zweiten Hälfte der 1930er Jahre ist die Veränderung des populären Geschmacks ersichtlich. Es werden verstärkt liebliche Tier- und Kinderfiguren hergestellt, auch das Erscheinungsbild der Tänzerinnen hat sich von der eleganten Ausstrahlung der 20er Jahre hin zu der eher völkisch orientierten Naturalismus-Ästhetik der 40er Jahre entwickelt.

Den Prozess von der Idee bis zur Herstellung eines Modells beschreibt der damalige Mitarbeiter Karl Täuber, der im Jahr 1937 als Praktikant unter Walter Goldscheider in der Manufaktur zu arbeiten beginnt:[265] In regelmäßigen Arbeitsbesprechungen wird allgemein über erfolgreiche und weniger populäre Modelle befunden. Die Mitarbeiter tauschen bei diesen Treffen ihre Ideen aus, suchen Zeit-

Works by Adolf Prischl, who was employed at Goldscheider, were publicly launched a few months later. Critics praised his head of a girl (model number 7982) as well as a horse and called Prischl "a talent who is capable of development and [will go] his own way."[264]

By the latter half of the 1930s, the popular taste had noticeably changed. More figurines of animals and children were made. Dancers' appearance had evolved from the elegance radiated in the 1920s to the more naturalistically folkloric aesthetic that would dominate the 1940s.

The process of going from an idea to making a model has been described by Karl Täuber, a former Goldscheider employee who began as an intern at the manufactory under Walter Goldscheider in 1937:[265] At regular staff conferences, there was a general discussion on which models were successful and which were less popular. At such meetings, the staff exchanged ideas and leafed through magazines and newspapers for subject matter that was in fashion at the moment. Insight gained from trade fairs as well as models presented by the competition

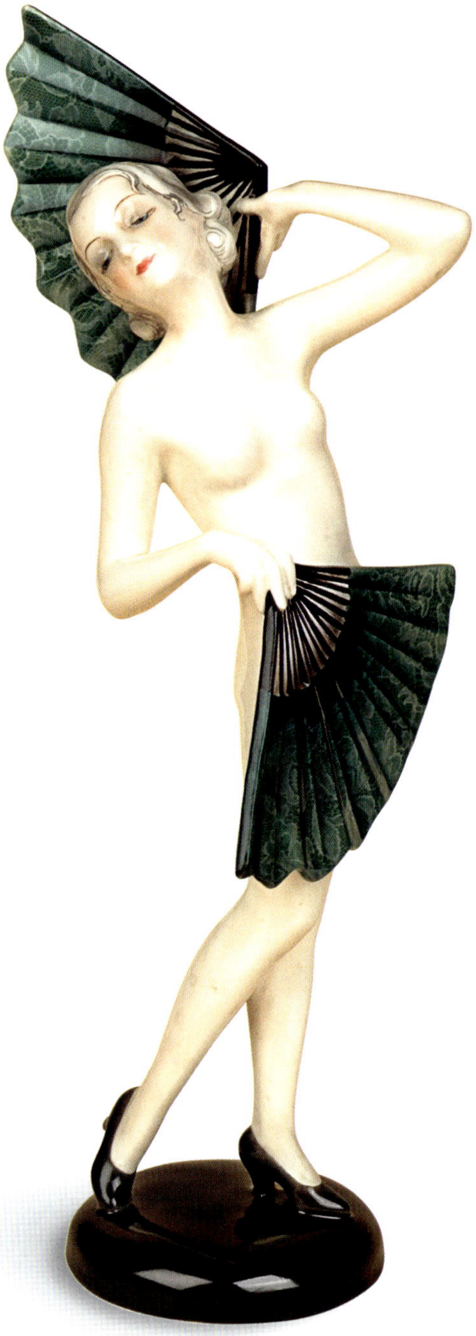

ModNr. 7688
Damenakt mit zwei Fächern I Female Nude with Two Fans, Josef Lorenzl, ca. 1937, PB Wien 19

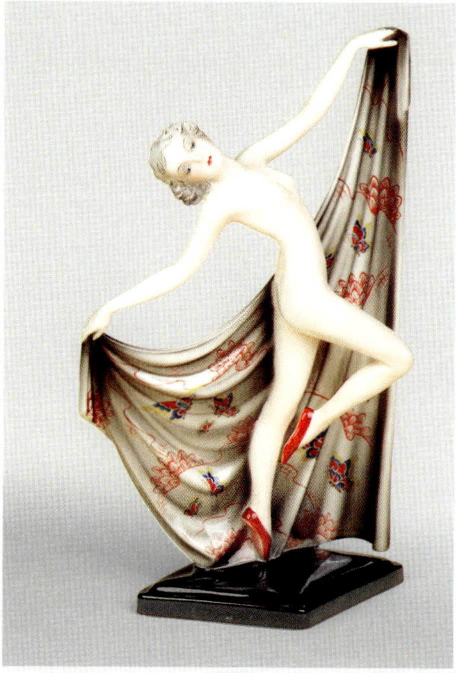

ModNr. 8174
„Nocturno", Karl Nieschlag / Josef Melcher, ca. 1937, PB Wien 19
Der Künstler kann bei diesem Modell nicht eindeutig bestimmt werden, da unterschiedlich signierte Modelle existieren.

This model cannot be attributed with assurance to one particular artist, since versions with different signatures exist.

ModNr. 7279
Dame mit Hut | Lady with Hat, Claire Weiss, ca. 1935, PB Wien 19

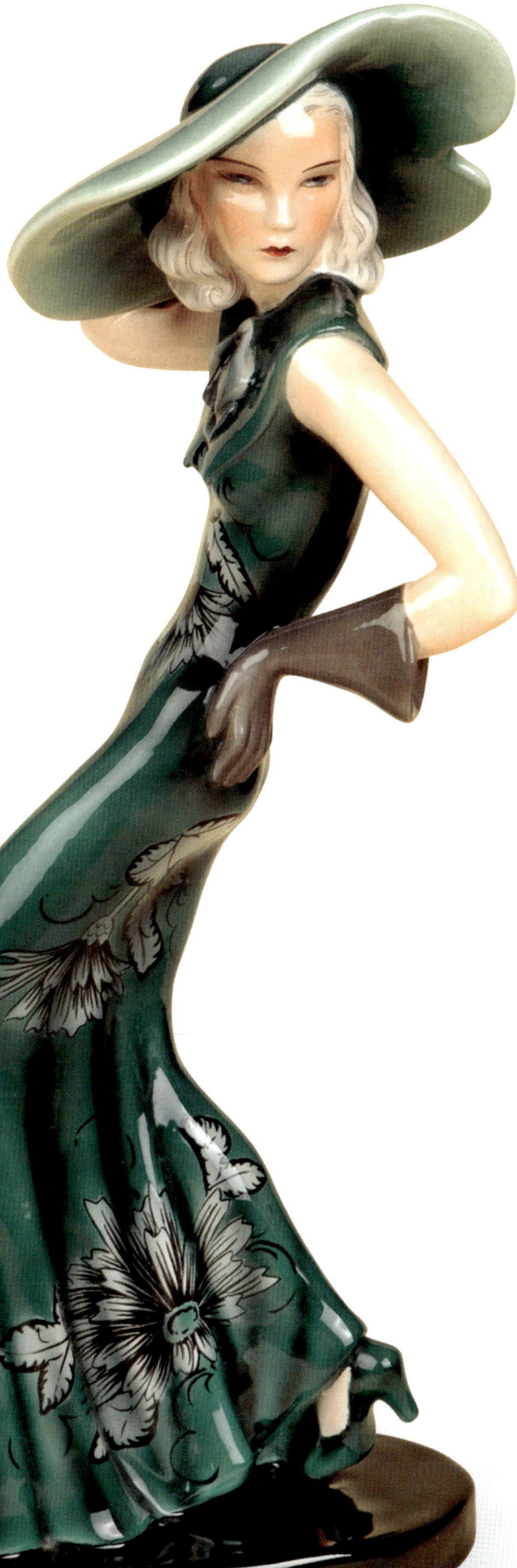

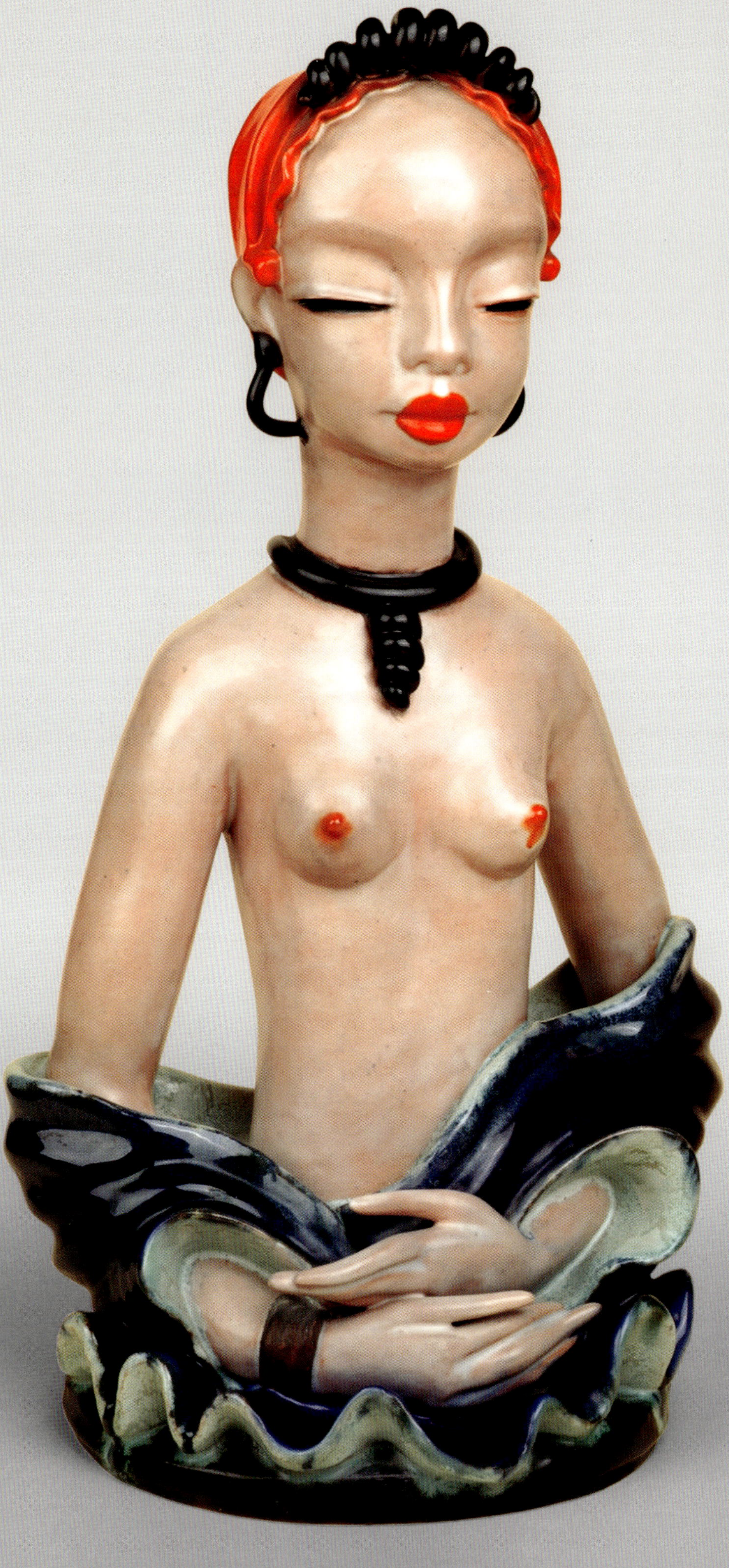

links I left

ModNr. 7617
Torso, Kurt Goebel,
ca. 1936, PB Wien 19

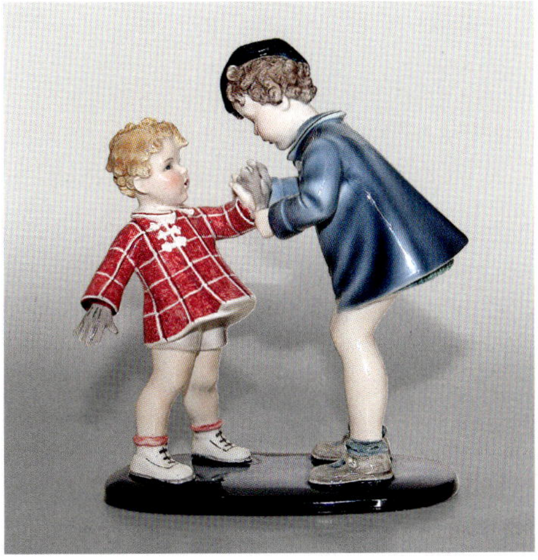

ModNr. 8365 „Kindergruppe" I "Group of Children", Germaine Bouret / Stephan Dakon, ca. 1938, Collection Bernard Amigon
Auch das Modell 8365 (Originaltitel: „Les hommes, ça donne du travail") ist aus der Kombination Bouret & Dakon entstanden. (Abbildung der Figur aus Sammlung Bernard Amigon; Skizze aus Archiv Goldscheider, Ill. 72)

Model 8365 (original title: "Les hommes, ça donne du travail") resulted from collaboration between Bouret and Dakon. (Figure reproduced from the Bernard Amigon Collection, sketch from the Goldscheider Archives, Ill. 72)

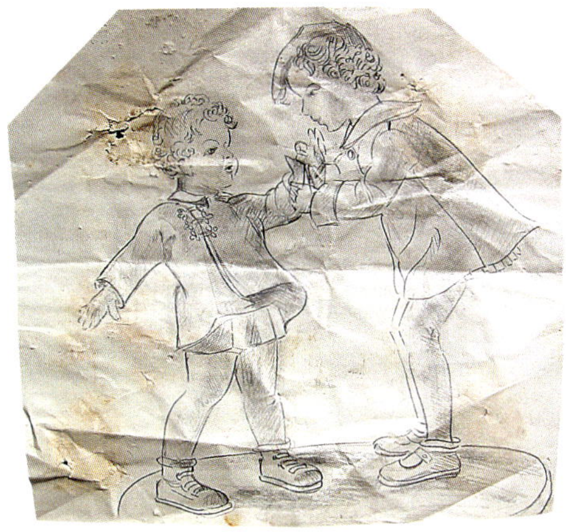

Ill. 72

were discussed and it was decided jointly with management which new models might be taken into the collections. Walter Goldscheider, who contributed many ideas of his own for producing new models, was the one to decide whether a model would be accepted and to commission the right artist to design it. Some artists, such as Karin Jarl-Sakellarios, who specialized in animal figurines, had pronounced thematic preferences.

The second step was for the sculptors themselves or the modelers, to do a trial model design from a sketch or a prototype. Here they had to ensure that technical peculiarities were also taken into consideration, and assess whether the model could be cast at all as proposed or whether it might not turn out to be too fragile. Whereas dog figurines, for instance, hardly ever produced wasters due to their form, dancers or figurines of other animals such as rearing horses required care because they were so dynamic in action or their limbs were so fine. Fragile models were said to be "shaky on their feet."[266] Designed in modeling clay, the model was taken to the casting workshop and the sculptor's commitment usually ended at that point. At the casting workshop, slip was poured into the mold of the model and the two parts of the mold were fitted together. Individual parts such as head, hands, feet, etc, were cast separately and attached afterwards with slip while retouching

schriften und Medien nach zeitgemäßen Themen ab, Erkenntnisse aus Messen sowie Modelle der Konkurrenz werden diskutiert und in Absprache mit dem Vertrieb geklärt, welche neuen Modelle in die Kollektionen passen könnten. Walter Goldscheider, der selbst viele Ideen für die Produktion neuer Modelle einbringt, beurteilt in letzter Instanz die Abnahme des Modells und beauftragt geeignete Künstler mit der Kreation des Modellentwurfs. Einige Künstler haben ihre besonderen thematischen Vorlieben, wie beispielsweise Karin Jarl-Sakellarios, die auf Tierfiguren spezialisiert ist.

Im zweiten Schritt erstellen die Bildhauer selbst oder Modelleure nach einer Skizze oder Vorlage den Modellentwurf, bei dem sie auch die technischen Eigenheiten beachten müssen, sowie allgemein ob das Modell überhaupt auf diese Art gegossen werden kann und nicht allzu bruchanfällig ist. Während zum Beispiel Hundefiguren aufgrund ihrer Form kaum zu Ausschuss führen, muss man bei Tänzerinnen oder Tierfiguren wie den aufsteigenden Pferden aufgrund der dargestellten Dynamik oder Feinheit der Gliedmaßen vorsichtig sein. Diese bruchanfälligen Modelle werden als „schwach auf den Beinen" bezeichnet.[266] Das entworfene Plastelinmodell wird in die Gießerei gegeben und das Engagement des Bildhauers hört an diesem Punkt zumeist auf. In der Gießerei wird die Modellform mit Schlicker ausgegossen und in zwei Teilen jeweils zusammengefügt. Die einzel-

ModNr. 7663 „Naschkätzchen" I "Sweet Tooth", Germaine Bouret / Stephan Dakon, ca. 1936, Collection Bernard Amigon
Nach den Vorlagen der französischen Zeichnerin Germaine Bouret entstanden mehrere Figuren, die von Stephan Dakon für die Manufaktur modelliert wurden. Hier das Modell 7663 (Gruppe „Naschkätzchen") nach einer Postkarte, der Gips-Prototyp und das fertige Exemplar. (Abbildung der Postkarte und Figur aus Sammlung Bernard Amigon; Gips-Prototyp aus Archiv Goldscheider)

Several figures modelled by Stephan Dakon for the manufactory were made after designs by the French draughtswoman Germaine Bouret. Here is model number 7663 (the "Naschkätzchen" ["Sweet Tooth"] group) after a post card, the plaster of Paris prototype and the finished piece. (Postcard and figure reproduced from the Bernard Amigon Collection; plaster of Paris prototype from the Goldscheider Archives).

Ill. 73

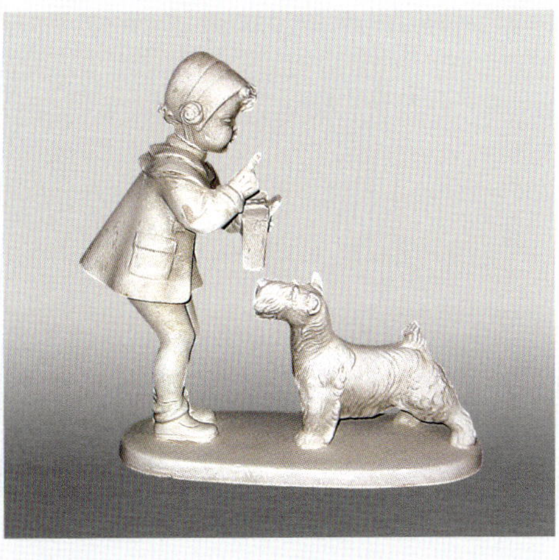

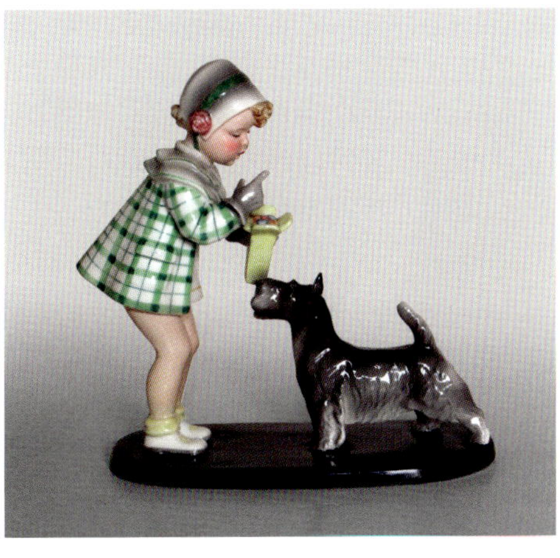

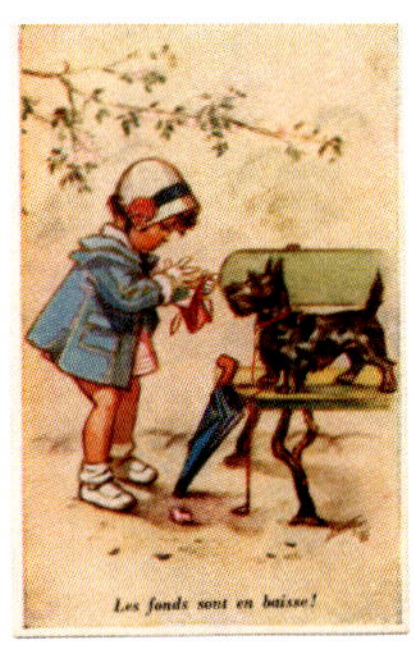

nen Bestandteile wie Kopf, Hände, Füße etc. werden dabei separat gegossen und in der Retusche anschließend mit Schlicker zusammengesetzt. Dieser Scherben kommt dann in den Rohbrand und wird nach dem Abkühlen in der Malerei gespritzt oder in der Malerei für Moderne Keramik mit Pinsel bemalt und glasiert. Die Muster auf den Kleidern der Tänzerinnen werden teilweise mit Hilfe von Textilvorlagen wie Stickereien, Spitzen oder Stofffetzen aus Gardinen aufgetragen.[267] Diese werden vor allem im Unterglasurverfahren wie Schablonen aufgesetzt und mit einem Aerographen gespritzt, was zu auffallend klaren, außergewöhnlichen Effekten führt (vgl. Ill. 74): „Feine Spitze mit ornamentaler oder figuraler Musterung (allenfalls aus Tüll) wird mit kräftiger Appretur versehen, um sie einigermaßen steif zu erhalten. Gespritzt wird nun auf Flächen, die ganz oder teilweise von der Spitze bedeckt sind und auf diese Weise ein Negativ erscheint. Durch anschließende Schattierung mit geeigneten Farben lässt sich eine phantastische Wirkung erzielen. Als Beispiel: Die Darstellung völlig naturgetreuer Musterungen auf Bekleidungen von Figuren (samt Faltenwurf) oder etwa die

was going on. The body was then glost-fired and, after it had cooled down, was sprayed in the painting workshop or painted and glazed in the painting workshop for Modern Ceramics. The designs on the dancers' apparel were sometimes applied with the aid of textiles such as embroidery, lace or scraps of curtain cloth.[267] They were applied like stencils mainly in the underglaze process and sprayed with an aerograph to create strikingly crisp, extraordinary effects (cf Ill. 74): "Fine lace with decorative or figurative designs (if need be, from tulle) is heavily starched to keep it as stiff as possible. The spray is directed at surfaces that are entirely or partly covered by the lace to make a negative appear. A fantastic effect can be created by shading afterwards with appropriate colors. To take just a few examples: the entirely faithful reproduction of patterns on figures' apparel (including folds) or reproducing a delicate chinoiserie landscape on ornamental pieces or tableware. The sprays used are similar to conventional paint and lacquer sprays but considerably smaller, often only as big as a fountain pen (retouching tip). Since the work is so delicate, it is done with

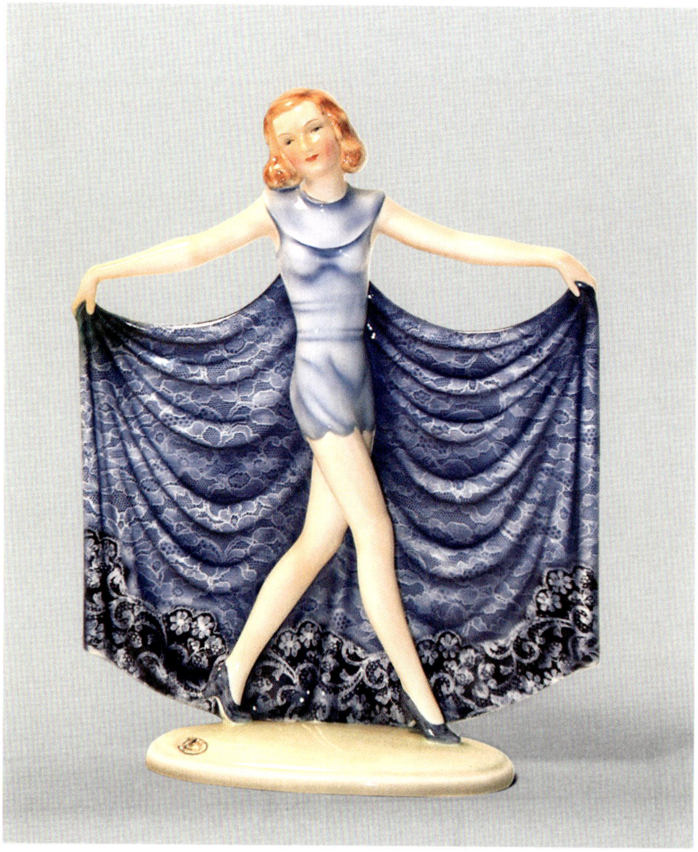

Ill. 74
Spitzen- oder Gardinenmuster werden als Schablonen eingesetzt und zeigen dank einer ausgefeilten Spritztechnik detailreiche, auffällige Effekte. (Privatsammlung, Wien)

Lace or curtain patterns were used as stencils to create strikingly detailed effects, thanks to a sophisticated spray technique. (Private collection, Vienna)

the smallest nozzle (needle), at low pressure and with highly diluted paints. The spray painting is in no respect inferior, technically or aesthetically, to drawing with pen or brush but can supplement and embellish this in many different ways."[268]

Some artists specialized in particular materials. Rudolf Knörlein, for one, worked a great deal in majolica and preferred modeling what was called Modern Ceramics. The greenish majolica clay body was often used for crackle-glazed pieces, sometimes for specific animal figures, busts or demi-nudes.[269] Many dancer figurines and animals were made in faience, which has a white body. Firing temperatures and glazing techniques depended on the material of the body.

After painting, the glaze was fired to lend the faience its distinctive gloss. Some pieces, such as the popular "Rosenkavalier" (model number 6535), were given an additional coat of glaze and were fired again. That was expensive but gave a figure an even more exquisite appearance resembling porcelain. The figure of Mozart, seated, with a violin (model numbers 3285; 3522; 5483) was also treated to a second glaze in this way.

The models made were given a model number, each piece was numbered consecutively and the retoucher's number was

Abbildung einer zarten Chinalandschaft auf Zier- oder Geschirrstücken. Die verwendeten Spritzen sind den herkömmlichen Farb- und Lackspritzen ähnlich, jedoch wesentlich kleiner, haben oft nur die Größe einer Füllfeder (Retuschierspitze). Entsprechend der feinen Arbeit wird mit kleinster Düse (Nadel), niedrigem Luftdruck und stark verdünnten Farben gearbeitet. Die Spritzmalerei steht sowohl in technischer als auch in künstlerischer Bedeutung der Feder- oder Pinselzeichnung in keiner Weise nach, sondern kann diese in vielfältiger Art ergänzen und bereichern."[268]

Manche Künstler sind auch nur auf bestimmte Materialien spezialisiert. Rudolf Knörlein arbeitet beispielsweise viel in Majolika und modelliert vorzugsweise die sogenannte Moderne Keramik. Den grünlichen Ton der Majolika findet man oft bei craquelierten Stücken, manchmal auch bei bestimmten Tierfiguren, den Büsten oder Halbakten.[269] In Fayence sind hingegen viele Tänzerinnen und Tiere, die aus einem weißen Scherben bestehen. Je nach Material sind unterschiedliche Brenntemperaturen und Glasurverfahren notwendig.

Nach der Bemalung wird die aufgetragene Glasur eingebrannt, damit sie der Fayence den besonderen Glanz verleiht. Manche Stücke, wie der beliebte „Rosenkavalier" (Modell-Nr. 6535), erhalten eine

Ill. 75

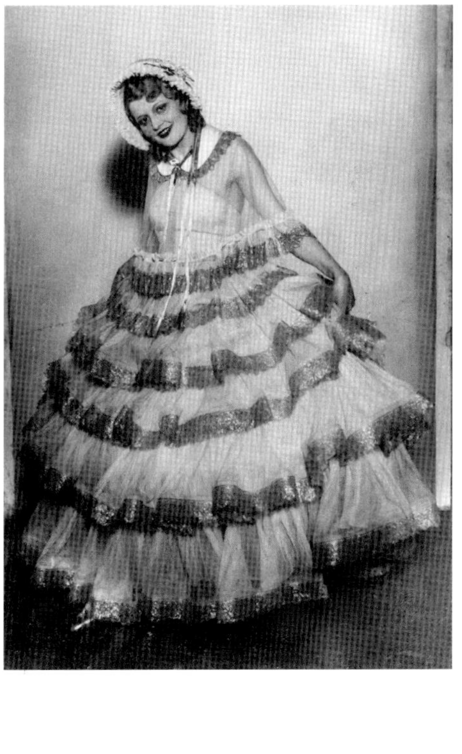

Ill. 76

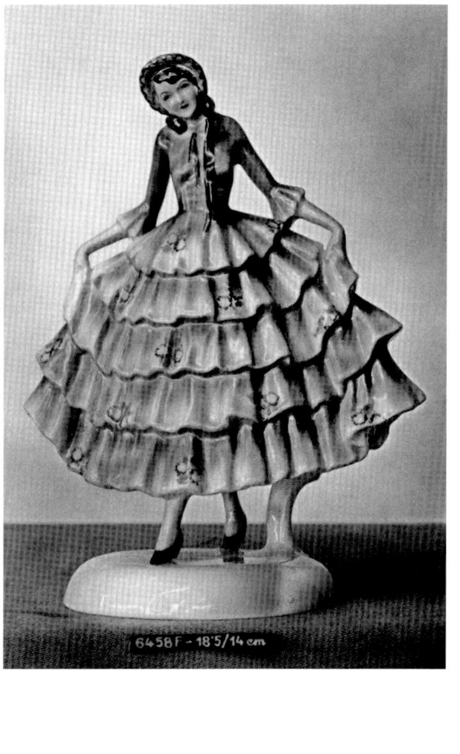

Ill. 75, 76
Das Modell 6458 von Stephan Dakon entstand nach einer Bild-Vorlage und wurde „Alt-Wien" benannt. Dieses Modell wurde in den 1940er Jahren auch in der amerikanischen Fabrik produziert. (Beide Abbildungen aus dem Archiv Mimi Dakon)

Model number 6458 by Stephan Dakon was created after a picture and was called "Alt-Wien" ["Old Vienna"]. This model was also produced in the American factory during the 1940s. (Both pictures from the Mimi Dakon Archives)

zusätzliche Glasur und werden noch einmal gebrannt. Dies ist zwar teuer, aber verschafft der Figur ein noch porzellanähnlicheres und feineres Aussehen. Auch die Figur des sitzenden Mozarts mit Geige (Modell-Nr. 3285; 3522; 5483) wird mit dieser zweiten Glasur hergestellt.

Die hergestellten Modelle erhalten eine Modellnummer, eine fortlaufende Stückzahlnummer sowie die Nummer des Retuscheurs. Meistens kommt auch ein Zeichen des Malers hinzu. Somit kann man stets den Herstellungsweg zurückverfolgen, die Fehlerquelle identifizieren und eventuelle Missstände bei der nächsten Produktion vermeiden. Zuletzt werden die Modelle auch mit dem Goldscheider-Stempel versehen, den es in mehreren Varianten gibt. Für englischsprachige Länder wird aus Wien „Vienna", „Vienne" ist für Frankreich bestimmt und auf manchen für Tschechien produzierten Stücken findet man dann beispielsweise „Víden". Oft sind aufeinander folgende Modellnummern vom selben Künstler, da dieser zumeist mehrere Modelle gleichzeitig zur einmaligen Ab-

added. There was usually also a painter's mark. Consequently, it was always possible to trace the making of a piece, identify at source any mistakes made and avoid possible problems in future batches. Finally, the models were given the Goldscheider stamp, of which there are several variants. For the English-speaking countries, "Wien" became "Vienna;" "Vienne" was destined for France and some pieces produced for Czechoslovakia are stamped "Víden." Sequential model numbers often indicate the same artist since an artist would usually sell several models at once to Walter Goldscheider for a fee in installments or on a royalty basis. Models 7653 to 7657 were, for instance, by Goebel.

When a model proved successful, it was also produced and marketed in different sizes (under different model numbers). Sometimes, however, a model was produced in various combinations under a single model number: model number 7039 occurs both as a single standing figure or as a group of two or three figures; the group of three was even made as an inkstand for use on a desk.

ModNr. 6535

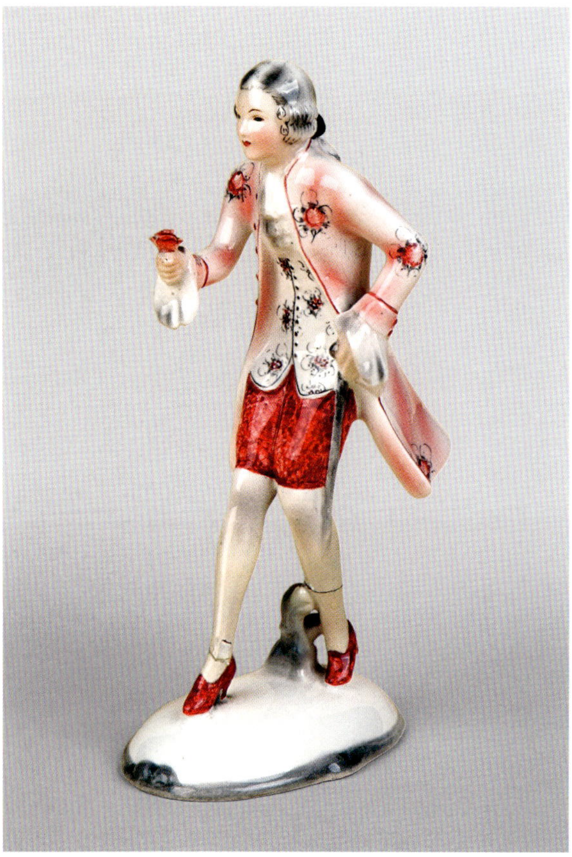

ModNr. 5483

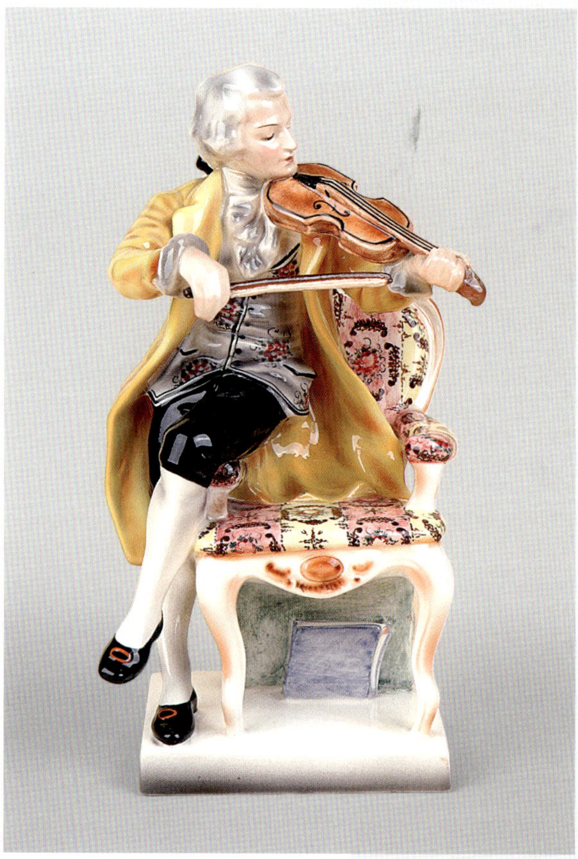

ModNr. 6535 „Rosenkavalier", Josef Lorenzl, ca. 1931, PB Wien 19

ModNr. 5483 „Mozart", Rosé, Entwurf I design ca. 1905, Ausführung 30er Jahre I execution 1930s, PB Wien 19

Die Modelle „Rosenkavalier" und „Mozart" erhalten eine zusätzliche Glasur, um sie noch filigraner und porzellanähnlicher erscheinen zu lassen.

The "Rosenkavalier" and "Mozart" models were given an additional glaze to make them look even more delicate and porcelain-like.

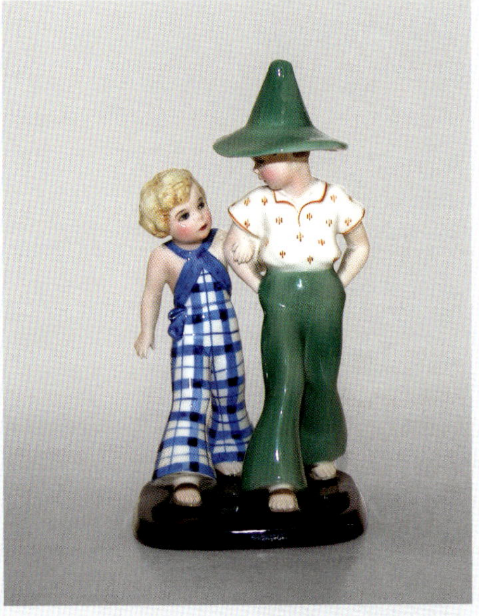

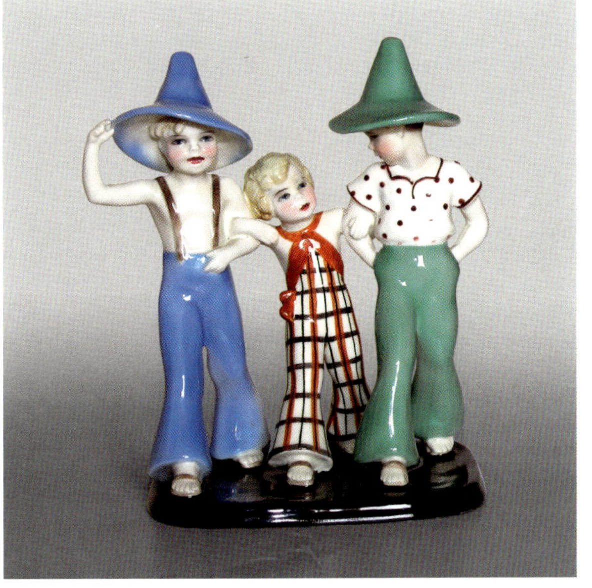

ModNr. 7039
Kinder I Children,
Germaine Bouret,
ca. 1934/35, Collection
Bernard Amigon

Manchmal erscheinen unter einer bestimmten Modellnummer verschiedene Kombinationen von Einzelfiguren und Gruppen. (Sammlung Bernard Amigon, Frankreich)

Various combinations of individual figures and groups were sometimes sold under one model number. (Bernard Amigon Collection, France)

The "Windhunde" ["Whippets"] group, model number 7054, by Robert Ullmann was sold as a group but individual dogs were also available under the same model number. The popular wall mask "Eva" ["Eve"] (model number 6774) was made in several different sizes under the same number. It was sold both as a wall mask and as a mask bust.

In the 1930s models were usually made to order since storage was too risky both economically and because the pieces were so fragile. Molds from the store-room were used to produce a model again if re-ordered by a customer. Clients could order to a certain extent the colors and patterns they wanted; with some decoration schemes, choice incurred a 5–20 per cent surcharge.[270] A model was preferably made by the same workman each time because he would be best informed on the problems associated with a particular model. Wage costs were high but the material costs were relatively low. Dealers were paid a commission when new models were involved to induce them and to test them on the market; practical market research, as it were.[271] No pieces were sold with defective body but there was a 20–25 per cent discount for what are known as seconds, in this case, with slight defects in the glaze. There is no evidence that such pieces were specially marked.[272] Clients were also entitled to send pieces they had broken directly to the manufactory for restoration.

Marketing at Goldscheider was carried out through trade fairs, travelling salesmen, representatives working on commission and through orders placed by regular customers. One of the most important trade fairs was the Leipzig Fair (in spring and autumn), where the firm had been represented since it was first established in "uninterrupted sequence" and was regarded as the oldest Austrian exhibitor (Ill. 77 and 78).[273] Leipzig had always been an important venue for the Goldscheider Manufactory since Goldschei-

schlagszahlung oder auf Tantiemen-Basis an Walter Goldscheider verkauft. Die Modelle von 7653 bis 7657 sind beispielsweise von Goebel.

Wenn sich ein Modell als erfolgreich erweist, wird es auch in anderen Größen (unter anderer Modellnummer) produziert und vertrieben. Manchmal wird allerdings unter einer Modellnummer das Modell in verschiedenen Kombinationen hergestellt: Die Modellnummer 7039 ist sowohl als allein stehende Figur wie auch als Zweier- oder Dreiergruppe erhältlich, die Dreiergruppe sogar als Schreibtischablage. Die Gruppe „Windhunde" von Robert Ullmann mit der Modellnummer 7054 wird als Gruppe verkauft, die einzelnen Hunde sind aber auch unter der gleichen Modellnummer getrennt erhältlich. Die beliebte Wandmaske „Eva" (Modell-Nr. 6774) wird unter der gleichen Nummer in verschiedenen Größen hergestellt und mal als Wand-, ein anderes mal als Stehmaske verkauft.

Die Produktion der Modelle erfolgt in den 1930er Jahren größtenteils auf Bestellung, da man wegen der Kapitalbindung und Bruchgefahr keine Lagerung betreiben will. Für die Reproduktion greift man auf im Depot liegende Formen zurück und produziert das Modell erneut für die jeweilige Bestellung des Kunden. Die Farben und das Muster können teilweise von den Kunden bestimmt werden, bei manchen Dekoren ist dies mit einem Preiszuschlag von 5–20 Prozent verbunden.[270] Die Modelle werden vorzugsweise von den gleichen Arbeitern reproduziert, da diese am besten die Problembereiche des jeweiligen Modells kennen. Die Produktion ist eher lohnintensiv, denn das Material ist relativ preiswert. Nur neue Modelle, die auf dem Markt erst eingeführt werden müssen, erhalten die Händler auf Kommission, um die Attraktivität des Modells für den Markt zu testen.[271] Es werden grundsätzlich keine vom Scherben her minderwertigen Stücke verkauft, aber die so genannte zweite Wahl mit leichten Glasurfehlern ist mit 20–25 Prozent

Rabatt erhältlich. Eine spezielle Kennzeichnung dieser Stücke ist nicht nachweisbar.[272] Die Kunden haben auch die Möglichkeit, kaputt gegangene Stücke direkt an die Manufaktur zur Restaurierung zu senden.

Der Vertrieb erfolgt bei Goldscheider durch Messen, eigene Reisende, Provisionsvertreter und durch Korrespondenzaufträge der Stammkundschaft. Zu der wichtigsten Messe zählt Leipzig (im Frühjahr und Herbst), wo die Firma seit ihrem Bestehen in „ununterbrochener Reihenfolge" vertreten ist und als ältester österreichischer Messeaussteller gilt (Ill. 77 und 78).[273] Leipzig war und ist für die Goldscheider Manufaktur seit jeher ein wichtiger Standort gewesen, denn von 1894 bis 1935 wurde in Leipzig, am Neumarkt 9, eine Niederlassung mit einem „eigenen grossen Detaillokal unterhalten". Im Jahr 1935 wird aufgrund des Boykotts jüdischer Geschäfte in Deutschland dieses Geschäftslokal von der Firma Georg Roth jun. übernommen. Der neue Besitzer stellt jedoch für Messezwecke weiterhin die Räume zur Verfügung.[274] Auch an der Wiener Messe stellt man im Frühjahr und Herbst in einem „eigenen, bestgelegenen grossen Pavillon" aus, der im Herbst 1937 noch vergrößert wird. Weiterhin präsentiert man sich auf der im April stattfindenden Mailänder Messe mit einem „bestausgestatteten Messestand im Ausmasse von 20m²".[275]

Vier Reisende sind für den Vertrieb eingestellt, die ausschließlich für die Erzeugnisse der Firma Europa und Nordamerika bereisen. Ausgestattet mit gepolsterten Musterkoffern, die zumeist die kleineren Modelle enthalten sowie Fotos der größeren Modelle, stellen sie den potenziellen Weiterverkäufern die Goldscheider-Objekte vor.[276] Die auf diese Weise gewonnenen Aufträge werden mit einer Provision von ca. 10 Prozent der erzielten Umsätze verrechnet.[277] Auch Walter Goldscheider unternimmt oft diese Geschäftsreisen, um einen persönlichen Kontakt zu seinen Kunden herzustellen und einen Eindruck des jeweiligen nationalen Marktes zu bekommen. Des Weiteren wird eine

der maintained a branch in Leipzig, at 9 Neumarkt with a "large proprietary accessories stock" from 1894 until 1935. In 1935, the Germany-wide boycott of businesses owned by Jews caused this branch to be taken over by the firm of Georg Roth jun. The new owner, however, continued to place rooms at Goldscheider's disposal when they were needed for the trade fair.[274] Goldscheider also showed pieces regularly at the Vienna Fair in spring and autumn in a "large proprietary pavilion in the best possible situation," which was even enlarged in autumn 1937. The firm also continued to exhibit at the annual April Milan at a "stall equipped to the highest standard measuring 20m2."[275]

Four traveling salesmen were hired just to market the firm's products in Europe and North America. Equipped with padded sample cases, which usually held smaller models as well as photographs of the bigger models, the salesmen presented the Goldscheider objects to potential retailers.[276] The orders received in this way were on a commission basis amounting to approx 10 per cent of net proceeds.[277] Walter Goldscheider himself often went on such business trips to make contact personally with his clientele and get an idea of the market in each country. Further, "quite a large number of representatives on commission" was employed. They were responsible mainly for the countries no one from Goldscheider traveled to personally: for instance, South and Central America, Africa and Australia. They were given an average commission of 15 per cent. Other factors contributing to turnover were clients re-ordering on a regular basis, and orders placed "by firms abroad whose inter-

Ill. 77, 78
Neben Leipzig gehört auch die Wiener Messe zu den bedeutenden und bewährten Präsentationsorten für die Objekte der Goldscheider Manufaktur. (Quelle: Archiv Goldscheider)

After Leipzig, the Vienna Fair was another of the most important and tried and tested venues for presenting Goldscheider objects. (Source: Goldscheider Archives)

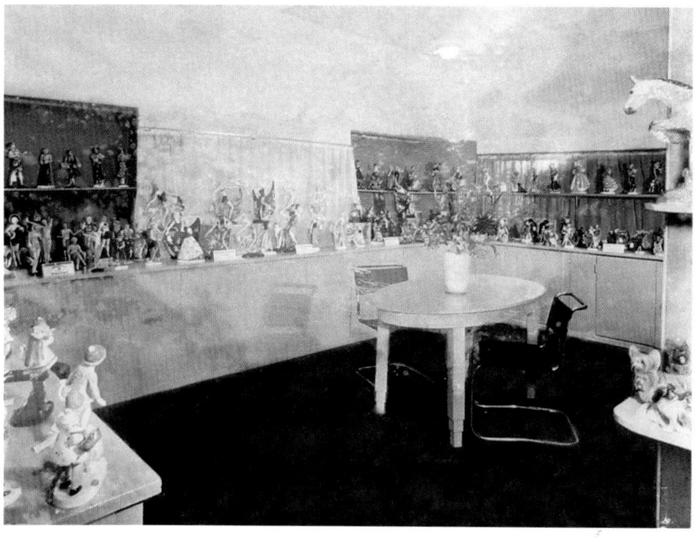

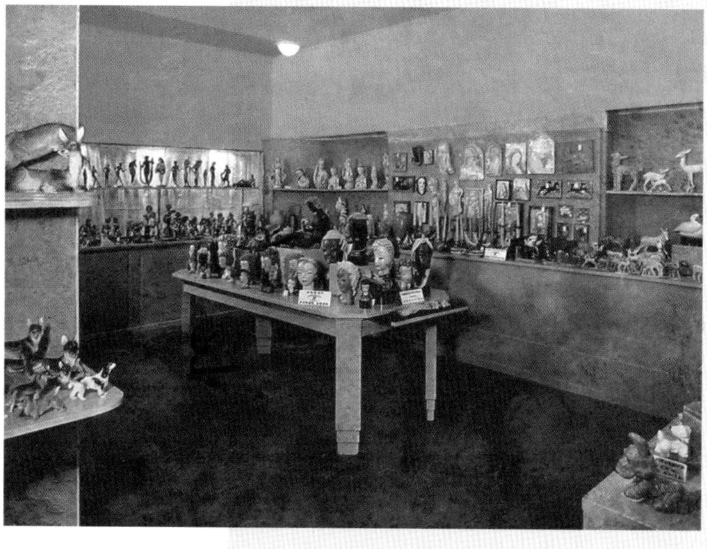

est was aroused by word of mouth and, finally, the regular business conducted in the factory sample warehouse, which, however, [was almost entirely accounted for] by the Viennese exporters with buyers going to Vienna."[278]

Goldscheider was a popular brand worldwide and demand was more than satisfactory but the challenge in marketing consisted in ensuring that wholesalers kept buying in ever larger quantities.[279] With new clients, the models were offered for sale on a commission basis to compensate for the initial outlay and forge constructive, long-term ties.[280] Aggressive advertising was unnecessary because, according to former Goldscheider employee, Karl Täuber "the product was its own best advertising." There was a possible threat of competition from two manufactories: Augarten, whose porcelain products were almost three times as expensive because of the material; and Keramos, which produced products similar to Goldscheider. Keramos models were, in fact, slightly cheaper but they rarely attained the grace and quality of Goldscheider wares and did not compare favorably with the Goldscheider simulated porcelain as far as appearance went. The Keramos manufactory was even said to copy Goldscheider.[281] This was, however, also due to so many artists' selling designs to both Goldscheider and Keramos.

Walter Goldscheider found the best marketing and distributing strategy consisted in concentrating on selling to shops specializing in porcelain, which enhanced the perceived quality and exclusiveness of the brand. Such specialist dealers were given Goldscheider signs for advertising purposes, which they were supposed to display in their windows (Ill. 79–81). In Vienna, the regular clients included such exclusive shops as Rasper, Wahliss, Haas & Czjzek and Denk am Graben.[282]

„ziemlich grosse Anzahl von Provisionsvertretern" beschäftigt. Diese sind vor allem für Länder zuständig, die man nicht selbst bereist: beispielsweise Süd- und Zentralamerika, Afrika und Australien. Der Provisionssatz beträgt hierbei durchschnittlich 15 Prozent. Zum Absatz tragen auch die regelmäßigen Nachbestellungen der Stammkundschaft bei, Bestellungen der „durch Propaganda interessierten ausländischen Firmen und schließlich der laufende Kundenverkehr im Fabriksmusterlager, der allerdings fast ausschließlich von den Wiener-Exporteuren mit den nach Wien kommenden Einkäufern" bestritten wird.[278]

Die Herausforderung im Vertrieb besteht eigentlich nur darin, den Großeinkäufern höhere Stück-Abnahmen abzuringen, denn Goldscheider ist eine populäre Weltmarke und die Nachfrage sehr zufriedenstellend.[279] Bei neuen Geschäftspartnern werden die Modelle als Kommissionsware zum Verkauf angeboten, um ihnen den Kapitalaufwand zu erleichtern und eine positive, langfristige Kundenbeziehung herzustellen.[280] Aggressive Werbung ist nicht notwendig, denn laut dem ehemaligen Mitarbeiter Karl Täuber „war das Produkt die beste Werbung" gewesen. Als Konkurrenten betrachtet wurden die beiden Manufakturen Augarten, die mit ihren Porzellanprodukten wegen des höherwertigeren Materials fast um das dreifache teurer war, sowie Keramos, die ähnliche Objekte wie Goldscheider produzierte. Keramos-Modelle waren etwas preiswerter, sie erreichten aber selten die Grazie und Qualität von Goldscheider und kamen vom Erscheinungsbild nicht an das imitierte Porzellan heran. Der Manufaktur Keramos wurde auch nachgesagt, dass sie Goldscheider kopierte.[281] Dies ist jedoch auch auf die Arbeiten vieler Künstler zurückzuführen, die Entwürfe sowohl an Goldscheider wie auch an Keramos verkauft haben.

Als Vertriebsstrategie steht für Walter Goldscheider fest, dass vor allem Fachgeschäfte für Porzellan beliefert werden sollten, um das Niveau und die Exklusivität der Marke zu stärken. Die Fachgeschäfte bekommen für Werbezwecke Goldscheider-Schilder, die sie in den Schaufenstern platzieren sollen (Ill. 79–81). In Wien gehören auch die renommierten Häuser Rasper, Wahliss, Haas & Czjzek sowie Denk am Graben zu den Stammkunden.[282]

Ill. 79

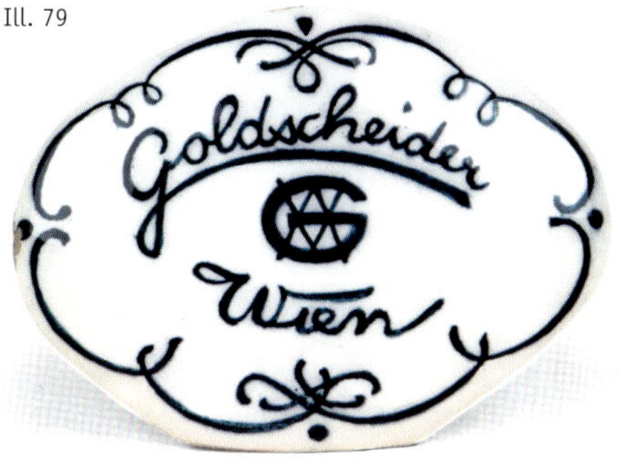

Ill. 80

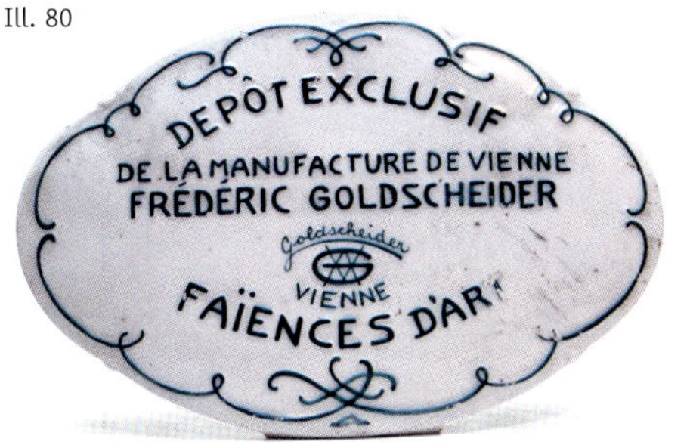

Ill. 81

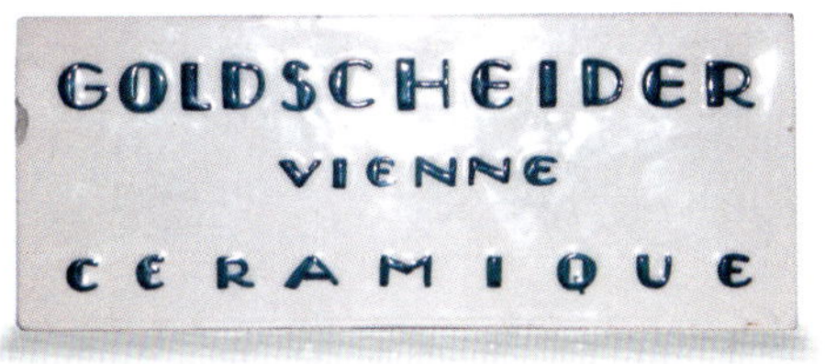

Ill. 79–81
Die Markenschilder der Goldscheider Manufaktur werden in Schaufenstern von Fachgeschäften als Werbemittel eingesetzt. Hier sind zwei Werbeschilder für den französischen sowie eines für den deutschsprachigen Markt zu sehen. (Quelle: französische Schilder aus Sammlung Bernard Amigon, Frankreich; deutsches aus PB Wien 19)

The Goldscheider manufactory signs were set up as advertising vehicles in the window displays of specialist shops. Here are two advertising signs for the French market alongside one for the German-speaking market. (Source: French signs from the Bernard Amigon Collection, France; German sign from PB Wien 19)

Das Unternehmen exportiert nicht nur in fast alle Staaten Europas, sondern auch nach Übersee. Die Hauptabsatzländer sind vor allem England, Italien, Deutschland und Nordamerika. Die Exportquote beträgt im Jahr 1937 erstaunliche 95,3 Prozent, wovon 10,8 Prozent auf das so genannte Altreich (Deutschland) fallen (vgl. Ill. 82). Im Zeitraum von 1934 bis 1937 steigt der Export nach Übersee ums Zehnfache: von 3,76 auf 32,20 Prozent.[283] Der Vertrieb tendiert eindeutig zum nordamerikanischen Markt hin, bedingt durch die wirtschaftlich und politisch angespannte Situation in Europa sowie durch die Vertriebstätigkeit von Erwin Goldscheider, dem Sohn Walters, der für USA und Kanada zuständig ist.

Die Rendite des Unternehmens liegt in den Jahren 1933 bis 1937 zwischen 5,3 und 8,2 Prozent. Die Nettoumsätze steigen von 440.653 Schilling im Jahr 1933 auf 708.891 Schilling im Jahr 1937, der Reingewinn ver-

The firm exported to most European countries as well as overseas. The main markets were England, Italy, Germany and North America. The export quota amounted in 1937 to an astonishing 95.3 per cent, with 10.8 per cent going to what was known then as the Altreich (the Old Empire, ie, Germany) (cf. Ill. 82). Between 1934 and 1937, exports overseas increased tenfold: from 3.76 to 32.2 per cent.[283] Marketing was definitely aimed at the North American market, particularly since the economic and political situation in Europe was so uncertain and Erwin Goldscheider, Walter's son, was responsible for marketing to the US and Canada.

The firm's profit margins ranged between 5.3 and 8.2 per cent in the years between 1933 and 1937. Net turnover grew from 440 653 Schillings in 1933 to 708 891 Schillings in 1937. During that period, net profits doubled from 24 300 to 51 900 Schillings.[284] At a time that was on the whole difficult, Walter Goldscheider managed with his consistent export policy to achieve remarkable growth in profits. The biggest turnover was made on the renowned faiences and the Modern Ceramics (cf Ill. 83). The bronzes and marbles, on the other hand, became less important with time as did the terracottas and Obsieger.[285] The share of Obsieger ceramics in overall turnover amounted to about 1.5 per cent in 1935 but dropped to 50 per cent the following year and by 1937 amounted to only .27 per cent of the year's turnover. Walter Goldscheider blamed the lack of growth in art bronze production

B r u t t o u m s ä t z e

in Schillingen.

Europa:	1934	%	1935	%	1936	%	1937	%
Oesterreich	37370	6.91	42630	6.7	41522	5.35	37431	4.7
Deutschland	61619	11.45	79021	12.5	100237	12.90	86248	10.8
Schweiz	22113	4.09	17752	2.9	14785	1.90	13683	1.7
C.S.R.	48577	8.99	30898	4.8	37875	4.88	27045	3.4
Italien	88649	16.40	104948	16.6	132824	17.10	95924	12.5
Frankreich	45988	8.52	36231	5.7	18398	2.36	37226	4.7
Ungarn	6953	1.29	7372	1.2	3614	-.46	6849	-.8
England	86994	16.10	130503	20.5	171163	22.05	126106	15.8
Polen	19022	3.52	18292	2.9	16219	2.08	14662	1.8
Schweden	25693	4.75	24959	3.9	36025	4.63	34802	4.4
Spanien	30713	5.69	29527	4.6	20215	2.60	--	--
Norwegen	320	-.06	--	--	1615	-.20	2968	-.5
Rumänien	9682	1.79	9902	1.6	13580	1.74	12270	1.6
Bulgarien	372	-.06	--	--	--	--	--	--
Dänemark	626	-.12	--	--	--	--	195	-.-
Portugal	3349	-.62	2163	-.3	2444	-.31	5381	-.7
Holland	10019	1.89	5306	-.8	4741	-.61	4039	-.5
Jugoslavien	10310	1.90	14759	2.3	10382	1.34	7079	-.9
Belgien	9201	1.70	14163	2.2	23297	3.07	24399	3.1
Griechenland	1968	-.35	4355	-.7	3511	-.45	1758	-.2
Litauen	179	-.04	--	--	307	-.04	478	-.-
Türkei	--	--	545	-.1	3250	-.41	1165	-.1
Finnland	--	--	--	--	--	--	348	-.-
Estland	--	--	--	--	--	--	386	-.-
	519917	96.24	573326	90.2	656004	84.48	540442	67.80
Uebersee:								
Südafrika	330	-.05	11495	1.8	11499	1.48	20834	2.6
Nordafrika	8757	1.64	2380	-.4	3657	-.48	7005	-.9
Südamerika	1372	-.25	10638	1.7	7809	1.--	13220	1.7
Nordamerika	5492	1.02	24369	3.8	96205	12.39	213203	26.8
Australien	3347	-.62	12518	2.-	1037	-.13	1101	-.1
Asien	1026	-.18	648	-.1	303	-.04	1335	-.2
	20324	3.76	62048	9.8	120510	15.52	256698	32.20
Gesamtumsatz	540241		635374		776514		797140	

Ill. 82
Bei einem Exportanteil von 95,3 Prozent sind im Jahr 1937 die Hauptabnehmer vor allem USA, England, Italien und Deutschland. (Quelle: Archiv Goldscheider)

With exports amounting to 95.3 per cent in 1937, the main markets were the US, England, Italy and Germany (Source: Goldscheider Archives)

on "insufficient investment" and also came up with "other reasons."[286] Ornamental figures were produced in bronze as they were in faience, with "the available models [made] both in ceramics and metal" whenever suitable for both materials.[287] This multiple use of models for different materials and in different sizes is in general characteristic of Goldscheider production methods. Not only could the cost of supplying models be kept down in this way but advertising expenses as well since an advertising package only had to be developed for one size. Similarly, when a model had proven its popularity in one material, the risk of not succeeding with another was lower.

Since business prospered between 1935 and 1937, the plant and facilities could be enlarged and modernized and new kilns were bought.[288] The Wiener Manufaktur Friedrich Goldscheider continued to be housed on its own premises at 7–9 Staudgasse. No. 7, at the corner of Hans Sachs Gasse, had three storeys and a cellar beneath two-thirds of the building. It had once been a domestic dwelling, "which [was] adapted to use as a factory by the removal of partitions."[289] The factory kilns stood in the courtyard, muffle kilns, a "small, single-storey annexe without a basement and a brick chimney 26 m high as well as a glazed linking room."[290] The spraying and painting workshops were on the ground floor of the building, with the office and a corner room for the boss on the first floor. The entire second floor had been turned into a sample showroom for clients. There were two

doppelt sich in dieser Zeit von 24.300 Schilling auf 51.900 Schilling.[284] In dieser allgemein schwierigen Zeit gelingt es Walter Goldscheider, mit seiner konsequenten Exportorientierung eine bemerkenswerte Gewinnsteigerung zu erzielen. Die größten Umsatzbringer sind hierbei die renommierten Fayencen und die moderne Keramik (vgl. Ill. 83). Die Bereiche Bronzen und Marmor nehmen an Bedeutung ab, genauso wie Terrakotta und Obsieger.[285] Der Anteil der Obsieger-Keramik am Gesamtumsatz beträgt im Jahr 1935 etwa 1,5 Prozent, fällt im darauf folgenden Jahr auf die Hälfte und trägt im Jahr 1937 nur zu 0,27 Prozent zum Jahresumsatz bei. Den Rückgang der Erzeugung von künstlerischen Bronzen, einem „Nebenartikel" der Goldscheider-Produktion, begründet Walter Goldscheider mit „mangelnden Investitionen" und „anderen Gründen".[286] In Bronze werden genauso wie in Fayence Dekorationsfiguren produziert; wobei „die vorhandenen Modelle sowohl in Keramik wie auch Metall" ausgeführt werden, wenn sie sich für beide Materialien eignen.[287] Diese Mehrfachverwendung von Modellen für verschiedene Materialien und Größen ist allgemein kennzeichnend für die Goldscheider-Produktion. Somit wurde nicht nur der Aufwand für die Modellbeschaffung gesenkt, sondern auch die Werbekosten, da beispielsweise Werbematerial nur für eine Größe entwickelt werden musste. Zusätzlich war auch aufgrund der bewährten Popularität das Risiko eines Flops im anderen Materialbereich niedriger.

Aufgrund der erfolgreichen Geschäftsentwicklung werden in den Jahren 1935 bis 1937 die Betriebsanlagen erweitert und modernisiert sowie neue Öfen angeschafft.[288] Die Wiener Manufaktur Friedrich Goldscheider ist weiterhin in den zwei eigenen Fabrikhäusern in der Staudgasse 7–9 untergebracht. Das Haus Nr. 7, an der Ecke zur Hans Sachs Gasse, besteht aus drei Stockwerken und ist zu zwei Dritteln unterkellert. Es ist ein

Netto Umsätze und Reingewinne
in Schillingen.

Netto Umsatz	1933	1934	1935	1936	1937
Fayence	221401	276258	348415	424364	418015
Keramik	206566	210954	200392	239742	283804
Terrakotta	2059	2426	1510	456	934
Obsiegen	--	--	8470	4060	2729
Marmor	5576	571	366	50	331
Bronze	5051	5392	4591	3951	3078
	440653	495601	563744	672623	708891
Reingewinn:	24300	33600	30000	55700	51900
	5.5 %	6.7 %	5.3 %	8.2 %	7.3 %

Ill. 83
Vor allem Fayence und die Moderne Keramik sind die Umsatzbringer der Manufaktur, der Verkauf von Marmor- und Bronzeerzeugnissen geht stark zurück. (Quelle: Archiv Goldscheider)

Faience and Modern Ceramics drove turnover at Goldscheider; sales of marble and bronze products fell sharply. (Source: Goldscheider Archives)

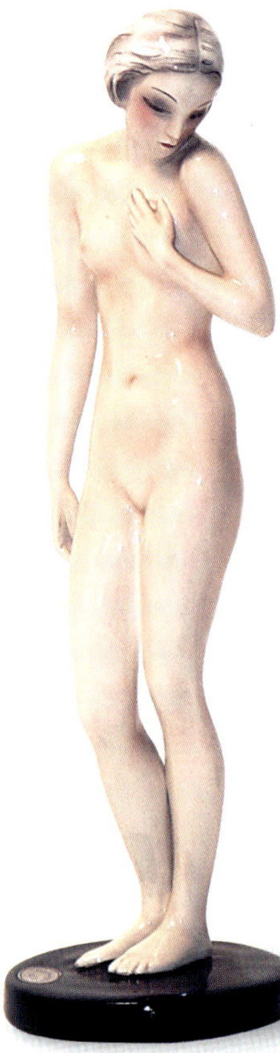

ModNr. 8070
„Akt" I "Nude",
Klára Herczeg,
ca. 1937, private
collection, Wien

ehemaliges Wohngebäude, „welches durch Entfernen von Zwischenmauern zu Fabrikszwecken adaptiert" wurde.[289] Im Hofe des Hauses steht die Brennerei mit Muffelöfen, ein „kleiner, ebenerdiger, nicht unterkellerter Anbau und ein 26 m hoher, gemauerter Schornstein, sowie ein verglaster hölzerner Verbindungsraum."[290] Im Parterre des Hauses ist die Spritzerei und Malerei, im ersten Stock das Büro mit einem Eckzimmer als Chefbüro. Der zweite Stock ist durchgehend als Musterzimmer für die Kunden eingerichtet. Im dritten Stock sind zwei Dienstwohnungen, in denen Ende der dreißiger Jahre der Betriebsleiter Fritz Iglauer und der Prokurist Franz Schwarzbart mit ihren Familien wohnen. Zusätzlich gibt es im dritten Stock vier Freizeiträume für die Belegschaft, unter anderem auch eine Bibliothek.

Die Staudgasse 9 ist ebenfalls ein dreistöckiges, ehemaliges Wohnhaus, das bis auf die Einfahrt unterkellert ist. Das Haus Nr. 7 dient als Haupteingang und die Einfahrt im Haus Nr. 9 ist für Materialtransporte vorgesehen. Beide Häuser sind im ersten Stock mit einem Durchgang verbunden, hier ist die Expedition und Kommission angesiedelt. Im zweiten Stock ist die Moderne Malerei und

service flats on the third floor, in which Fritz Iglauer, managing director, and Franz Schwarzbart, head clerk, lived with their families during the late 1930s. In addition, there were four leisure and recreation rooms, including a library, for the staff on the third floor.

The building at 9 Staudgasse was also a three-storeyed structure that had once been a private dwelling. It had a cellar up to the entry. No. 7 was the main entrance and the entry to No. 9 was reserved for transporting materials. The two buildings were linked by a passage on the first floor, where orders were filled and shipping was arranged. Modern painting was on the second floor. The workshop for making plaster casts and repairing damaged figures or retouching defects in colors were on the third. The third floor also housed the studio used by Josef Lorenzl, who by the late 1930s dropped in only occasionally and had a studio of his own outside the factory premises. The other studios used by Adolf Prischl and Viktor Matula, who were employed permanently as sculptors by Goldscheider, were on the second floor.[291] A two-storeyed building in the courtyard housed the workshops, the re-

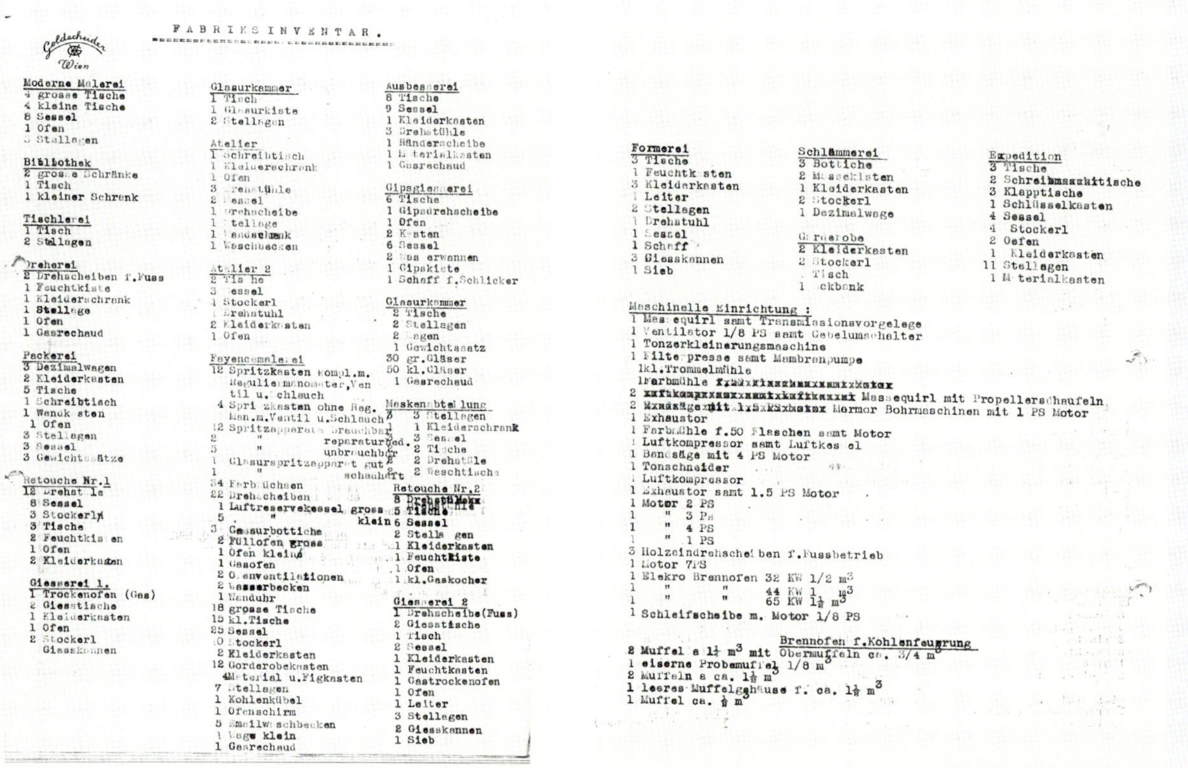

Goldscheider Wien

FABRIKSINVENTAR.

Moderne Malerei
4 grosse Tische
4 kleine Tische
8 Sessel
1 Ofen
3 Stellagen

Bibliothek
2 grosse Schränke
1 Tisch
1 kleiner Schrank

Tischlerei
1 Tisch
2 Stellagen

Dreherei
2 Drehscheiben f.Fuss
1 Feuchtkiste
1 Kleiderschrank
1 Stellage
1 Ofen
1 Gasrechaud

Packerei
3 Dezimalwagen
2 Kleiderkasten
5 Tische
1 Schreibtisch
1 Wandkasten
1 Ofen
3 Stellagen
3 Sessel
3 Gewichtssätze

Retouche Nr.1
12 Drehstühle
8 Sessel
3 Stockerl
3 Tische
2 Feuchtkisten
1 Ofen
2 Kleiderkasten

Giesserei 1.
1 Trockenofen (Gas)
2 Giesstische
1 Kleiderkasten
1 Ofen
2 Stockerl
Giesskannen

Glasurkammer
1 Tisch
1 Glasurkiste
2 Stellagen

Atelier
1 Schreibtisch
1 Kleiderschrank
1 Ofen
3 Drehstühle
2 Sessel
1 Drehscheibe
1 Stellage
1 Wandschrank
1 Waschbecken

Atelier 2
2 Tische
3 Sessel
1 Stockerl
1 Drehstuhl
2 Kleiderkasten
1 Ofen

Fayencemalerei
12 Spritzkasten kompl.m. Reguliermanometer,Ventil u.Schlauch
4 Spritzkasten ohne Reg. Man.m.Ventil u.Schlauch
12 Spritzapparate brauchbar
2 " reparaturbed.
5 " unbrauchbar
1 Glasurspritzapparat gut
1 " schadhaft
34 Farbbüchsen
22 Drehscheiben
1 Luftreservekessel gross
5 " klein
3 Gassurbottiche
2 Füllofen gross
1 Ofen klein
1 Gasofen
2 Ofenventilationen
2 Wasserbecken
1 Wanduhr
18 grosse Tische
15 kl.Tische
25 Sessel
10 Stockerl
2 Kleiderkasten
12 Gorderobekasten
4 Material u.Figkasten
7 Stellagen
1 Kohlenkübel
1 Ofenschirm
5 Emailwaschbecken
1 Wage klein
1 Gasrechaud

Ausbesserei
8 Tische
9 Sessel
1 Kleiderkasten
3 Drehstühle
1 Händerscheibe
1 Materialkasten
1 Gasrechaud

Gipsgiesserei
6 Tische
1 Gipsdrehscheibe
1 Ofen
2 Kasten
6 Sessel
2 Wasserwannen
1 Gipskiste
1 Schaff f.Schlicker

Glasurkammer
2 Tische
2 Stellagen
2 Wagen
1 Gewichtsatz
30 gr.Gläser
50 kl.Gläser
1 Gasrechaud

Maskenabteilung
3 Stellagen
1 Kleiderschrank
3 Sessel
2 Tische
2 Drehstühle
2 Waschtische

Retouche Nr.2
8 Drehstühle
5 Tische
6 Sessel
2 Stellagen
1 Kleiderkasten
1 Feuchtkiste
1 Ofen
1 kl.Gaskocher

Giesserei 2
1 Drehscheibe(Fuss)
2 Giesstische
1 Tisch
2 Sessel
1 Kleiderkasten
1 Feuchtkasten
1 Gastrockenofen
1 Ofen
1 Leiter
3 Stellagen
2 Giesskannen
1 Sieb

Formerei
3 Tische
1 Feuchtkasten
3 Kleiderkasten
1 Leiter
2 Stellagen
1 Drehstuhl
1 Sessel
1 Schaff
3 Giesskannen
1 Sieb

Schlämmerei
3 Bottiche
2 Massekisten
1 Kleiderkasten
2 Stockerl
1 Dezimalwage

Garderobe
2 Kleiderkasten
2 Stockerl
1 Tisch
1 Bank

Expedition
3 Tische
2 Schreibmaschintische
3 Klapptische
1 Schlüsselkasten
4 Sessel
4 Stockerl
2 Oefen
1 Kleiderkasten
11 Stellagen
1 Materialkasten

Maschinelle Einrichtung :
1 Massequirl samt Transmissionsvorgelege
1 Ventilator 1/8 PS samt Gebelumschalter
1 Tonzerkleinerungsmaschine
1 Filterpresse samt Membranpumpe
1 kl.Trommelmühle
1 Farbmühle
2 Massequirl mit Propellerschaufeln
2 Marmor Bohrmaschinen mit 1 PS Motor
1 Exhaustor
1 Farbmühle f.50 Flaschen samt Motor
1 Luftkompressor samt Luftkessel
1 Bandsäge mit 4 PS Motor
1 Tonschneider
1 Luftkompressor
1 Exhaustor samt 1.5 PS Motor
1 Motor 2 PS
1 " 3 PS
1 " 4 PS
1 " 1 PS
3 Holzeindrehscheiben f.Fussbetrieb
1 Motor 7PS
1 Elekro Brennofen 32 KW 1/2 m^3
1 " " 44 KW 1 m^3
1 " " 65 KW 1½ m^3
1 Schleifscheibe m. Motor 1/8 PS

Brennofen f.Kohlenfeuerung
2 Muffel a 1½ m^3 mit Obermuffeln ca. 3/4 m^3
1 eiserne Probemuffel 1/8 m^3
2 Muffeln a ca. 1½ m^3
1 leeres Muffelgehäuse f. ca. 1½ m^3
1 Muffel ca. ½ m^3

Ill. 84
Eine Übersicht des Fabrikinventars zeigt den Umfang der Wiener Manufaktur, die um 1937 insgesamt an die 125 Personen beschäftigte. (Quelle: Archiv Goldscheider)

An overview of the factory inventory reveals the scope of the Viennese manufactory, which employed a workforce of 125 in 1937. (Source: Goldscheider Archives)

ModNr. 8316
Akt mit überkreuzten Armen I Nude with Arms Crossed, Josef Melcher, ca. 1938, private collection, Wien

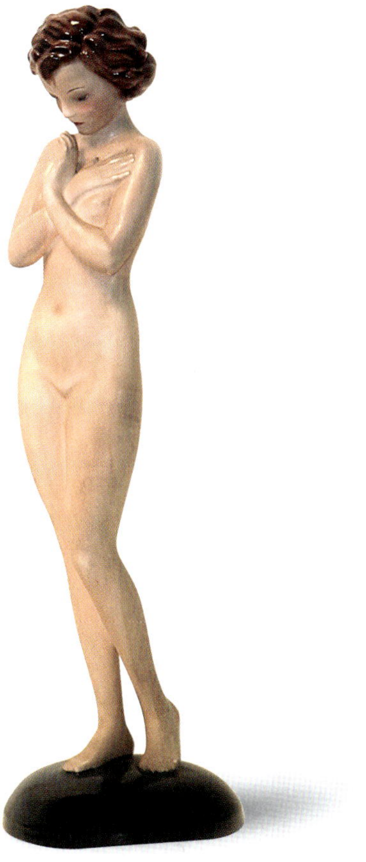

touch workshop and the "rote Retusche" ["red retouches"] for terracotta, the workshop where clay was prepared, etc. A bust of Friedrich Goldscheider, the founder of the firm, stood in the entry to the house.[292]

The Goldscheider manufactory employed a workforce of 125 in 1937 (cf Ill. 84).[293] The personnel broke down into workmen in the faience molding, casting and retouching, who were paid by the piece at a fee agreed on by contract. The retouchers earned most. Skilled workers such as sprayers were paid an hourly wage of 0.75 Schillings. Other workmen, for instance in the blank polishing workshop, received up to one Schilling an hour and those working at the kilns were paid up to 1.42 Schillings an hour. Women were paid lower wages. Trained draughtswomen were paid only 0.60 Schillings an hour; qualified workers earned only 0.65 Schillings instead of the 0.80 Schillings their male counterparts received.[294] The work week was Monday through Friday from 8 am to 12 noon and from 12.30 pm to 5.30 pm although for some tasks the workday lasted until 6 pm. On Saturday, the work day lasted until 1 pm. Former employee Karl Täuber remembers the work environment as pleasant. Walter Goldscheider is described as a strict yet just, hard-working and almost too punctilious boss, who expected a great deal of his employees.[295] Shortly before Christmas 1937, Walter's son, Erwin Friedrich Goldscheider, was registered as having collective power of attorney.[296]

The last big trade fair before the Second World War was the second 1937 Paris Exposition Arts et Techniques. The Viennese manufactory was represented there with objects by Professor Robert Obsieger, Grete Fucik-Fischmeister and other artists and was awarded a Grand Prix for the exhibits.[297]

im dritten Stock die Gipsgießerei und Ausbesserei für beschädigte Figuren oder Farbfehler. Im dritten Stock ist auch das Atelier von Josef Lorenzl, der aber Ende der dreißiger Jahre nur noch sporadisch kommt und ein eigenes Atelier außerhalb des Fabrikgebäudes hat. Die anderen Ateliers von Adolf Prischl und Viktor Matula, die als fest angestellte Bildhauer für Goldscheider tätig sind, befinden sich im zweiten Stock.[291] Im zweistöckigen Hofgebäude sind die Werkstätten, die Retusche sowie die „rote Retusche" für Terrakotta, die Schlämmerei, wo der Ton vorbereitet wird etc. In der Einfahrt des Hauses steht eine Büste des Firmengründers Friedrich Goldscheider.[292]

Die Goldscheider Manufaktur beschäftigt im Jahr 1937 insgesamt 125 Personen (vgl. Ill. 84).[293] Dazu gehören Arbeiter in der Fayenceformerei, Gießerei und Retusche, die nach Akkord-Verrechnung aufgrund der festgesetzten Stückpreise bezahlt werden, wobei die Retuscheure am besten verdienen. Ausgelernte Arbeiter wie die Spritzer erhalten einen Stundenlohn von 0,75 Schilling, andere Arbeiter, z.B. in der Rohschleiferei, erhalten bis zu einem Schilling, in der Brennerei bis zu 1,42 Schilling. Frauen erhalten einen niedrigeren Lohn; ausgelernte Dessineurinnen bekommen nur 0,60 Schilling, qualifizierte Hilfsarbeiterinnen verdienen anstatt 0,80 Schilling nur 0,65 Schilling.[294] Die Arbeitszeit ist von Montag bis Freitag von 8 bis 12 und von 12.30 bis 17.30 Uhr beziehungsweise für manche Aufgaben bis 18 Uhr. Am Samstag wird bis 13 Uhr gearbeitet. Das Betriebsklima wird von dem damaligen Mitarbeiter Karl Täuber als angenehm, Walter Goldscheider als strenger aber gerechter, fleißiger und fast überkorrekter Chef beschrieben, der viel von seinen Mitarbeitern verlangt.[295] Kurz vor Weihnachten 1937 wird auch Walters Sohn Erwin Friedrich Goldscheider als Kollektivprokurist eingetragen.[296]

Die letzte große Ausstellung vor dem zweiten Weltkrieg ist die zweite Pariser Kunstgewerbeausstellung im Jahr 1937. Die Wiener Manufaktur ist dort mit Objekten von Professor Robert Obsieger, Grete Fucik-Fischmeister und anderen Künstlern vertreten und erhält für ihre Beteiligung den Grand Prix.[297]

Anmerkungen

[227] Brief von Walter Goldscheider an die Vermögensverkehrsstelle, 4.8.1938. Nach Angaben von Walter Bosse hat er Ende 1927 für Goldscheider die erste Kollektion produziert. Weitere Informationen siehe auch Kapitel „Die Arisierung der Wiener Manufaktur Friedrich Goldscheider", S. 154ff
[228] N.N.: 50 Jahre Wiener Manufaktur Friedrich Goldscheider, in: Schaulade, 11.1935, S. 152ff
[229] Siehe Schreyer-Hartmann
[230] Brief von Walter Bosse an die Österreichische Kontrollbank in Wien, 5.11.1938
[231] Ebenda
[232] Lehmann, G.: Neuheiten, in: KuKGW, 9.1928/29, S. 126
[233] Brief von Walter Bosse an die Österreichische Kontrollbank in Wien, 5.11.1938
[234] Schnyder, S. 33
[235] Lehmann, G.: Neuheiten, in: KuKGW, 9.1928/29, S. 127
[236] Schnyder, S. 33
[237] Siehe Planer
[238] Ausstellungskatalog: Frühjahrsmesse Wien, Amtliche Ausgabe, 7. bis 13. März 1926, S. XV
[239] Brief von Walter Goldscheider an die Vermögensverkehrsstelle, 4.8.1938
[240] IC 1929/30, S. 373
[241] AdK, 1930 bis 1934
[242] Handelsregister zur Firma „Erste Wiener Terrakottafabrik und Atelier für künstlerische Fayenzen Friedrich Goldscheider" (Bestand Handelsgericht Wien, Handelsregister A 16/19)
[243] Ebenda. Die Eintragung der Prokuristen erfolgt zum 19.1.1932.
[244] Schaulade, 1930, S. 797
[245] Ebenda
[246] Heizer (1931), S. 236. Die Darstellungen sind oft Revuen der Folies Bergère oder Ziegfeld Follies entlehnt. In den Zwanziger Jahren nimmt auch die Bedeutung der Modemagazine wie Vogue und Harper's Bazaar zu.
[247] N.N.: Berühmte Frauen, Moderne Keramik, in: MW, 10.1928/29, Heft 6, S. 28
[248] Sch(illing-Schletter), A.: Hunde-Plastiken, in: MW, 11.1929/30, Heft 6, S. 12
[249] N.N.: 50 Jahre Wiener Manufaktur Friedrich Goldscheider, in: Schaulade, 11.1935, S. 154
[250] Bernd, in: Schaulade, 8.1932, S. 480
[251] GK, S. 18. Dieser Katalog ist leider verschollen.
[252] Hofmann, E.: Österreichisches Kunsthandwerk II: Keramik, in: ÖK, 3.1932, S. 20ff
[253] Auszug aus den Erinnerungen von Rudolf Knörlein, zur Verfügung gestellt von seiner Tochter Ursula Huck, 9.9.2004. Nachdem sich Walter Goldscheider und Rudolf Knörlein nicht über eine Gehaltserhöhung einigen konnten, hat Knörlein am 28.2.1936 das Unternehmen verlassen und im März 1936 als Betriebsleiter in Gmunden angefangen. Im Mai wurde er Direktor der Gmundner Keramik. Der Kontakt zwischen Knörlein und dem Besitzer der Gmundner Keramik, August Warchalowsky, kam durch Knörleins ehemaligen Lehrer, Prof. Obsieger, zustande, der bei Goldscheider eine eigene Modellreihe produzierte. Bei der Gmundner Keramik hatte Knörlein seine Lehrzeit absolviert.
[254] Brief von Rudolf Knörlein, Neuss am Rhein, an Walter Goldscheider, Wien, 20.4.1953
[255] AdK, 1934, S. 312f
[256] Neuwirth (1973), S. 62
[257] Exposición Internacional de Barcelona 1929, Catálogo Oficial, S. 156
[258] Austria in London. Dorland Hall, April 16 to May 12, 1934, Ausstellungskatalog. S. 67
[259] Labek, A.: Zur Ausstellung „Das befreite Handwerk", in: ÖK, 5.1934, Nr. 11, S. 20
[260] N.N.: 50 Jahre Wiener Manufaktur Friedrich Goldscheider, in: Schaulade, 11.1935, S. 154
[261] Ebenda
[262] Interview mit dem ehemaligen Praktikanten und Prokuristen Karl Täuber, Wien, 25.11.2003
[263] N.N.: Neue Keramik,in: ÖK, 8.1937, Nr. 7, S. 22
[264] N.N.: Neue Keramik, in: ÖK, 7.1936, Nr. 10, S. 25
[265] Interview mit Karl Täuber, Wien, 25.11.2003
[266] Ebenda
[267] Bestellung von Musterstücken bei der Firma Rudolf Hagen, Stickereien & Spitzen, Lustenau, 27.8.1952
[268] Neuwirth (1974), S. 54
[269] Interview mit Karl Täuber, Wien, 25.11.2003
[270] Siehe auch Kapitel „Die Manufaktur Josef Schuster in den Kriegsjahren", S. 170ff
[271] Interview mit Karl Täuber, Wien, 25.11.2003
[272] Ebenda
[273] Exposé zur Arisierung der Firma Goldscheider, von Dr. Stefan Schwartz an die Vermögensverkehrsstelle, 12.8.1938
[274] Ebenda
[275] Ebenda
[276] Interview mit Karl Täuber, Wien, 25.11.2003
[277] Exposé zur Arisierung der Firma Goldscheider, von Dr. Stefan Schwartz an die Vermögensverkehrsstelle, 12.8.1938
[278] Ebenda
[279] Interview mit Karl Täuber, Wien, 25.11.2003
[280] Ebenda
[281] Ebenda
[282] Ebenda
[283] Brief und Beilage von Walter Goldscheider an die Vermögensverkehrsstelle, 4.8.1938
[284] Ebenda
[285] Bericht von Dr. Schwartz zur Lage des Unternehmens an die Vermögensverkehrsstelle, 7.9.1938
[286] Exposé zur Arisierung der Firma Goldscheider, von Dr. Stefan Schwartz an die Vermögensverkehrsstelle, 12.8.1938
[287] Ebenda
[288] Brief von Siemens Austria, Wien an Walter Goldscheider, 29.5.1953. Im Jahr 1935 wurde bei Siemens der Ofen 1 mit 36 kW Anschlusswert gekauft, zwei Jahre später kam Ofen 3 mit 72 kW Anschlusswert hinzu.
[289] Exposé zur Arisierung der Firma Goldscheider, von Dr. Stefan Schwartz an die Vermögensverkehrsstelle, 12.8.1938
[290] Ebenda
[291] Ebenda
[292] Interview mit Karl Täuber, Wien, 25.11.2003
[293] Brief des kommissarischen Verwalters, Dr. Stefan Schwartz, an die Pruefsteller der keramischen Industrie, Wien, 23.9.1938
[294] Kollektivvertrag zwischen der Wiener Manufaktur Friedrich Goldscheider und der Belegschaft, 28.1.1937. Die Arbeiter können allgemein in drei Gruppen aufgegliedert werden: Spitzenarbeiter wie Retuscheure, Spritzer, Gipsgießer etc., fachliche Hilfskräfte und Hilfsarbeiter sowie so genannte Regiekräfte. Zusätzlich wird noch eine große Anzahl von Lehrlingen beschäftigt. Im Jahr 1937 und 1938 gibt es unter den Angestellten durchschnittlich 10 bis 12 kaufmännische Angestellte, die ca. 210,– Reichsmark pro Monat verdienen, 5 Fabrikangestellte mit einem Verdienst von durchschnittlich RM 190,– sowie 3 Bildhauer mit ca. RM 210,– im Monat, wobei „allerdings noch freie Bildhauer ausser Haus" beschäftigt werden. Siehe Exposé zur Arisierung der Firma Goldscheider, von Dr. Stefan Schwartz an die Vermögensverkehrsstelle, 12.8.1938
[295] Interview mit Karl Täuber, Wien, 25.11.2003. Er erinnert sich auch, dass Walter seinen Arbeitstag ab 17 Uhr mit der Erledigung der Post beendete, die ihm Täuber als Praktikant brachte. Wenn Walter Goldscheider einen Fehler oder ein Eselsohr an einem Brief gefunden hat, musste dieser Brief komplett neu geschrieben werden.
[296] Handelsregister zur Firma „Erste Wiener Terrakottafabrik und Atelier für künstlerische Fayenzen Friedrich Goldscheider" (Bestand Handelsgericht Wien, Handelsregister A 16/19). Die Eintragung erfolgte am 22.12.1938.
[297] Neuwirth (1974), S. 45

Notes

[227] Letter written by Walter Goldscheider to the Vermögensverkehrsstelle [Office for Property Tax], 4 Aug. 1938. According to Walter Bosse, he produced the first collection for Goldscheider in late 1927. For more information, see the chapter "The Aryanization of the Viennese Manufactory Friedrich Goldscheider (1938)," p. 154f
[228] N.N.: 50 Jahre Wiener Manufaktur Friedrich Goldscheider, in: Schaulade, 11.1935, p. 152f
[229] See Schreyer-Hartmann
[230] Letter written by Walter Bosse to the Österreichische Kontrollbank in Wien, 5 Nov. 1938
[231] Ibid
[232] Lehmann, G.: Neuheiten, in: KuKGW, 9.1928/29, p. 126
[233] Letter written by Walter Bosse to the Österreichische Kontrollbank in Vienna, 5 Nov. 1938
[234] Schnyder, p. 33
[235] Lehmann, G.: Neuheiten, in: KuKGW, 9.1928/29, p. 127
[236] Schnyder, p. 33
[237] See Planer
[238] Exhibition catalogue: Frühjahrsmesse Wien, Amtliche Ausgabe, 7 to 13 March 1926, p. XV
[239] Letter written by Walter Goldscheider to the Vermögensverkehrsstelle, 4 Aug. 1938
[240] IC 1929/30, p. 373
[241] AdK, 1930 to 1934
[242] Handelsregister zur Firma "Erste Wiener Terrakottafabrik und Atelier für künstlerische Fayenzen Friedrich Goldscheider" (Bestand Handelsgericht Wien, Handelsregister A 16/19)
[243] Ibid. Registered with power of attorney as from 19 Jan. 1932
[244] Schaulade, 1930, p. 797
[245] Ibid
[246] Heizer (1931), p. 236. The representations are often taken from the Folies Bergère revues or the Ziegfeld Follies. Fashion magazines such as Vogue and Harper's Bazaar became increasingly important in the 1920s.
[247] N.N.: Berühmte Frauen, Moderne Keramik, in: MW, 10.1928/29, [Heft] 6, p. 28
[248] A. Sch(illing-Schletter): Hunde-Plastiken, in: MW, 11.1929/30, [Heft] 6, p. 12
[249] N.N.: 50 Jahre Wiener Manufaktur Friedrich Goldscheider, in: Schaulade, 11.1935, p. 154
[250] Bernd, in: Schaulade, 8.1932, p. 480
[251] GK, p. 18. This catalogue has, unfortunately, been lost.
[252] E. Hofmann: Österreichisches Kunsthandwerk II: Keramik, in: ÖK, 3.1932, p. 20f
[253] Excerpt from Rudolf Knörlein's memoirs, placed at the authors' disposal by his daughter, Ursula Huck, 9 Sept. 2004. After Walter Goldscheider and Rudolf Knörlein could not agree on a salary rise, Knörlein left the business on 28 Feb. 1936 to begin as managing director in Gmunden in March 1936. In May he was made director of Gmundner Keramik. The contact between Knörlein and the owner of Gmundner Keramik, August Warchalowsky, was arranged by Knörlein's former teacher, Professor Obsieger, who produced a line of models of his own for Goldscheider. Knörlein had served his apprenticeship at Gmundner Keramik.
[254] Letter written by Rudolf Knörlein, Neuss am Rhein, to Walter Goldscheider, Vienna, 20 April 1953
[255] AdK, 1934, p. 312f
[256] Neuwirth (1973), p. 62
[257] Exposición Internacional de Barcelona 1929, Catálogo Oficial, p. 156
[258] Austria in London. Dorland Hall, April 16 to May 12, 1934, exhibition catalogue. p. 67
[259] A. Labek: Zur Ausstellung "Das befreite Handwerk," in: ÖK, 5.1934, No. 11, p. 20
[260] N.N.: 50 Jahre Wiener Manufaktur Friedrich Goldscheider, in: Schaulade, 11.1935, p. 154
[261] Ibid
[262] Interview with Karl Täuber, former intern and head clerk with power of attorney, Vienna, 25 Nov. 2003
[263] N.N.: Neue Keramik, in: ÖK, 8.1937, No. 7, p. 22
[264] N.N.: Neue Keramik, in: ÖK, 7.1936, No. 10, p. 25
[265] Interview with Karl Täuber, Vienna, 25 Nov. 2003
[266] Ibid
[267] Samples ordered from the firm of Rudolf Hagen, Stickereien & Spitzen, Lustenau, 27 Aug. 1952
[268] Neuwirth (1974), p. 54
[269] Interview with Karl Täuber, Vienna, 25 Nov. 2003
[270] See also the chapter "The Josef Schuster Manufactory during the war years (1939–1944)," p. 170f
[271] Interview with Karl Täuber, Vienna, 25 Nov. 2003
[272] Ibid
[273] Exposé on the Aryanization of the firm of Goldscheider by Dr. Stefan Schwartz to the Vermögensverkehrsstelle [Office for Property Tax], 12 Aug. 1938
[274] Ibid
[275] Ibid
[276] Interview with Karl Täuber, Vienna, 25 Nov. 2003
[277] Exposé on the Aryanization of the firm of Goldscheider by Dr. Stefan Schwartz to the Vermögensverkehrsstelle, 12 Aug. 1938
[278] Ibid
[279] Interview with Karl Täuber, Vienna, 25 Nov. 2003
[280] Ibid
[281] Ibid
[282] Ibid
[283] Letter and enclosure from Walter Goldscheider to the Vermögensverkehrsstelle, 4 Aug. 1938
[284] Ibid
[285] Report by Dr. Schwartz on the Condition of the Business Made to the Vermögensverkehrsstelle, 7 Sept. 1938
[286] Exposé on the Aryanization of the firm of Goldscheider by Dr. Stefan Schwartz to the Vermögensverkehrsstelle, 12 Aug. 1938
[287] Ibid
[288] Letter from Siemens Austria, Vienna, to Walter Goldscheider, 29 May 1953. In 1935 kiln 1, with 36 kW connected load/power requirement, was purchased from Siemens; it was followed two years later by kiln 3: 72 kW connected load/power requirement.
[289] Exposé on the Aryanization of the firm of Goldscheider by Dr. Stefan Schwartz to the Vermögensverkehrsstelle, 12 Aug. 1938
[290] Ibid
[291] Ibid
[292] Interview with Karl Täuber, Vienna, 25 Nov. 2003
[293] Letter from the provisional administrator, Dr. Stefan Schwartz to the Inspectorate of the Ceramics Industry, Vienna, 23 Sept. 1938
[294] Collective contract concluded between the Wiener Manufaktur Friedrich Goldscheider and the personnel, 28 Jan. 1937. The workmen can be divided in general into three groups: top specialist workmen such as repairers, sprayers, plaster molders, etc; specialized workers and workmen, called at the time Regiekräfte. In addition, a large number of apprentices worked at Goldscheider. In 1937 and 1938 ten to twelve commercial employees were on the staff, who earned approx 210 Reichsmarks a month as well as five factory workers earning on average RM 190 and three sculptors making approx RM 210 a month "although sculptors were also employed on a freelance basis." See exposé on the Aryanization of the firm of Goldscheider by Dr. Stefan Schwartz to the Vermögensverkehrsstelle, 12 Aug. 1938
[295] Interview with Karl Täuber, Vienna, 25 Nov. 2003. He also recalls that Walter finished his working day from 5 pm by going through the post, which Täuber brought to him as an intern. Whenever Walter Goldscheider spotted a mistake or a page looked dog-eared, a letter had to be entirely redone.
[296] Handelsregister zur Firma "Erste Wiener Terrakottafabrik und Atelier für künstlerische Fayenzen Friedrich Goldscheider" (Bestand Handelsgericht Wien, Handelsregister A 16/19). The entry was made on 22 Dec. 1938.
[297] Neuwirth (1974), p. 45

Die Arisierung der Wiener Manufaktur Friedrich Goldscheider (1938)

The Aryanization of the Viennese Manufactory Friedrich Goldscheider (1938)

Die Wiener Manufaktur Friedrich Goldscheider erfährt einen jähen Abbruch ihres Erfolgs mit dem „Anschluss" Österreichs an das Deutsche Reich. Nach dem im März 1938 erfolgten Einmarsch deutscher Truppen sieht sich die assimilierte jüdische Familie antisemitischen Übergriffen ausgesetzt. Der langjährige Buchhalter Walter Brunner kommt in SA-Uniform ins Büro, grüßt nur noch mit „Heil Hitler" und macht aus seiner nationalsozialistischen, antisemitischen Gesinnung keinen Hehl.[298] Die Arisierung des Betriebes wird angeordnet und am 5. Mai 1938 wird dem Betrieb Dr. Stefan Schwartz als kommissarischer Verwalter zugewiesen.[299]

Der Hilfsmonteur Wranovsky lässt sich aufgrund seiner Beziehungen zur Partei zum Betriebsobmann ernennen und zeigt die Leitung des Unternehmens bei der Polizei an.[300] Wenige Tage später, am 9. Mai 1938, werden Walter und Lilly Goldscheider sowie der Prokurist Franz Schwarzbart in polizeiliche Schutzhaft genommen. Obwohl ihnen keine strafbare Handlung zur Last gelegt werden kann, werden sie erst fünf Wochen später entlassen.[301] Die Schutzhaft ist ein gängiges Mittel des Arisierungsprozesses: einerseits dient sie der Einschüchterung, um die Besitzer zu einem schnellen Verkauf ihrer Unternehmen zu zwingen, andererseits entzieht sie den Besitzern die Kontrolle über die Führung der Geschäfte, so dass Termine und Auflagen nicht mehr eingehalten werden können und das Unternehmen ins Chaos verfällt und an Wert verliert. Auch Walter und Lilly müssen deshalb über ihren Rechtsanwalt, Dr. Arthur Mayer, bei der Vermögensverkehrsstelle um Terminaufschub für die von allen Juden verlangte Vermögensaufstellung bitten, denn, wie Walter schreibt, haben sich während der Zeit der Haft „grosse Rückstände in der Buchhaltung ergeben, so dass es mir, Walter Goldscheider, trotz eifrigen Bemühens nicht möglich gewesen ist, die für die Vermögensanmeldung erforderliche Zwischenbilanz über mein Unternehmen fertigzustellen."[302] Als weiteren Grund für das Aufschieben des Abgabetermins gibt er im gleichen Brief an, dass der kommissarische Verwalter „den Buchwert der Fabriksgebäude für zu niedrig findet" und deshalb eine sachverständige Schätzung der beiden Objekte stattfinden sollte. Da das Vermögen der Goldscheiders in dem Unternehmen

The "Anschluss" of Austria to the Third Reich put a quick end to the Viennese Manufactory Friedrich Goldscheider. After German troops marched into Austria in March 1938, the assimilated Jewish family was subjected to anti-Semitic persecution. Walter Brunner, who had been the firm's accountant for years, started going to the office in SA uniform, invariably greeted everyone with "Heil Hitler!" and made no bones about his sympathies with National Socialism and anti-Semitic leanings.[298] The authorities ordered the business to be Aryanized and on 5 May 1938 it was assigned to the provisional administration of Dr. Stefan Schwartz.[299]

Wranovsky, an unskilled mechanic, had himself appointed watchdog for the business on the basis of his relations with the Party and denounced the management to the police.[300] Only a few days later, on 9 May 1938, Walter and Lilly Goldscheider along with head clerk Franz Schwarzbart were taken into custody. Although no charges could be laid against them, they were not released for five weeks.[301] Taking people into custody was standard practice in the Aryanization process: on the one hand, the intention was to intimidate owners of businesses and force them to sell out quickly. On the other, this step took control of a business out of its owner's hands so that deadlines and conditions could no longer be met and the business was consequently so overwhelmed by chaos that its value depreciated. Walter and Lilly were among those who had to address a petition to the Vermögensverkehrsstelle [Office for Property Tax] through their lawyer, Dr. Arther Mayer, to have the deadline for the appraisal of assets required of all Jews delayed. As Walter wrote, during the time he had spent in prison, "the books had been greatly in arrears so that it has not been possible for me, Walter Goldscheider, despite my best efforts, to complete the interim settling of accounts necessary for declaring my business assets."[302] In the same letter, he adds as another reason for wishing to delay the deadline the fact that the provisional administrator "finds the value of the factory buildings set down as too low in the books" and suggests that an expert appraisal of the two properties should, therefore, be carried out. Since the Goldscheiders' money was tied up in the business, he wrote, it would only be possible to make a complete declaration after the business had been subjected to inspection by a chartered accountant as suggested. Dr. Schwartz, a son of the Viennese sculptor and medallion-maker Professor Stefan Schwartz,[303] and himself the owner of an art workshop at 2 Sachskrügelgasse in Vienna, confirmed under separate cover the reasons given by Walter Goldscheider and added that also due to "reduction in personnel, the interim settling of accounts mentioned" could not be completed by the deadline.[304]

The forced dismissal of Jewish or non-Aryan employees and the concomitant lack of skilled personnel can be traced in the payrolls of May and August 1938 (Ill. 85 and 86). As from 1 June 1938, all non-Aryan employees had to be dismissed on the basis of a decree issued by the NSDAP Gauleitung [district authorities]. For two of the men affected, Franz Schwarzbart and Otto Link, however, a delay of two months was applied for by the commissioner for the Inspectorate of the Ceramics Industry in Vienna, Herr von Lichtenberg. This petition reveals how important the Goldscheider manufactory was for the Austrian export market and the inflow of currency in the field of ceramics and the decorative arts: "Since changing the working arrangements within the firm is causing considerable difficulties and especially since

liegt, sei eine vollständige Anmeldung des Vermögens nur nach der genannten Prüfung des Unternehmens möglich. Dr. Schwartz, der ein Sohn des Wiener Bildhauers und Medailleurs Prof. Stefan Schwartz ist[303] und selbst Inhaber einer Kunstwerkstätte in der Sachskrügelgasse 2 in Wien, bestätigt in einem separaten Schreiben die genannten Gründe und fügt hinzu, dass auch infolge des „reduzierten Personalstandes die genannte Zwischenbilanz" noch nicht abgeschlossen werden konnte.[304]

Die erzwungene Entlassung von jüdischen bzw. nicht arischen Mitarbeitern und der damit für die Produktion verbundene Mangel an Fachpersonal lassen sich anhand der Gehälterlisten vom Mai und August 1938 nachvollziehen (Ill. 85 und 86). Zum 1. Juni 1938 sollten aufgrund einer Verordnung der Gauleitung der NSDAP alle nicht arischen Angestellten entlassen werden. Für zwei der Betroffenen, Franz Schwarzbart und Otto Link, wird jedoch vom Beauftragten der Prüfungsstelle Keramische Industrie in Wien, Herrn von Lichtenberg, ein Aufschub um zwei Monate beantragt. Anhand dieses Antrags wird ersichtlich, wie bedeutsam die Goldscheider Manufaktur für das österreichische Exportgeschäft und den Devisenzufluss im Bereich der Keramik und des Kunstgewerbes ist: „Da die Umstellung der Bearbeitung innerhalb der Firma erhebliche Schwierig-

Ill. 85, 86
Die jüdischen Angestellten mussten ab 1. Juni 1938 aufgrund der durch die NSDAP Gauleitung erlassenen Vorschriften zum nächstmöglichen Termin entlassen werden. Zwei jüdische Angestellte durften aus „exportunablässigen" Gründen für kurze Zeit weiter beschäftigt werden. (Quelle: Archiv Goldscheider)

As from 1 June 1938, Jewish employees had to be discharged at the earliest possible date in compliance with regulations issued by the NSDAP district authorities. Two Jewish employees, however, were permitted to remain on the payroll for a while longer for reasons "to keep the export business running." (Source: Goldscheider Archives)

Ill. 85

Gehalte Buero Mai 1938.

Franz Schwarbart, Prokurist Wien XVIII., Staudg.7-9	S 500.-	Nichtarier	
Erwin Goldscheider, Prokurist Wien XIX., Gregor-Mendelstr.58	" 500?-	"	
Fritz Iglauer, Betriebsleiter, Wien XVIII., Staudg.7-9	" 400.-	Arier	
Walter Brunner, Buchhalter Wien XII., Schoenbrunnerstr.226	" 300.-	"	infolge Staatsanstellung ausgetreten 7.V.38.
Viktor Matula, Bildhauer Wien XVIII., Schopenhauerstr.19	" 270.-	"	
Anton Hostasch, Maler Wien XVII., Roetzerg.31	" 357.-	"	
Adolf Prischl, Bildhauer, Wien X., Troststr;116	" 270.-	"	
Karl Koehler, Korrespondent Wien XVIII., Theresieng.70	" 300?-	"	
Egon Paul, Expedient Wien.,XVIII.,Hildebrandgasse 40	" 225.-	"	
Emilie Deutsch,Kassirin Wien,XX.,Nussdorferstr. 88	" 125.-	Nichtarierin	
Wilhelmine Schadt,Malerin Wien, XVII.,Gentzg.7	" 300.-	Arierin	
Viktoria Chachola,Buchhalterin, Wien,XVI.,Reinhartg.11	" 275.-	"	
Rosa Fischer,Fakturistin Wien, XIX.,Budinskig.4	" 175.-	"	
Hilde Schomann,Kontoristin, Wien,XVIII.,Karl-Beckg.25	" 130.-:	"	
Franz Hayek,Betriebsbeamter Wien.,XVII.,Stoeberpl.9-12	" 140.-	"	
Otto Link, Korrespodent, Wien, XIII.,Nisselg.13	" 150.-	Nichtarier	
Gottfried Popper,Expedient Wien, II.,Stuwerstr; 60	" 160.-	Mischling	
Josef Lorenzl,Bildhauer,	" 453.40	Arier	

Ill. 86

Goldscheider Wien

Gehalte Buero, August 1938.

I g l a u e r Fritz,Betr.Leiter 18.Staudg.7-[illegible]	RM	266.67	Arier
H o s t a s c h Anton,Maler 17.Roetzerg.31	RM	238.--	"
R a [illegible] d a k Franz,Inkassant 9.Galileig.4	RM	116.--	"
K o e h l e r Karl, Korrespondent 18.Theresieng.70	RM	[illegible]	"
C h o c h o l a Viktoria,Buchhalt[illegible]. 1[illegible].Reinhardtg.11	RM	183.33	"
Fischer Rosa, Fakturistin 19.Budinskyg.4	RM	116.67	"
P a u l Egon,Expedient 18.Hildebrandg.40	RM	150.--	"
P r i s c h l Adolf, Bildhauer 10.Troststr.116	RM	130.--	"
M a t u l a Viktor, Bildhauer 18.Schopenhauerstr.19	RM	130.--	"
S c h o m a n n Hilde, Kontoristin 18.Karl-Beckg.25	RM	[illegible].67	"
L e n k [illegible]ric[illegible], [illegible]tikant 17.Rokitanskyg.10	RM	[illegible].33	"
T a e u b e r Karl,Praktikant 17.Weidmanng.10	RM	15.33	"
H a y e k Franz,Betriebsbeamter 17.Stoeberpl.9	RM	93.34	"
J e r a b e k Viktor,Korrespondent 18.Waehringerguertel 67	RM	150.--	"
S c h a d t Wilhelmine,Malerin 18.Gentzg.72	RM	182.48	"
L o r e n z l Josef,Bildhauer 9.Althanpl.6	RM	302.26	"
L i n k Otto, Korrespondent 13.Nisselg.13	RM	56.--	J u d e(bis 1.X.38)
ZUSAMMEN: 17 Angestellte	RM	2.497.41	

keiten macht und insbesondere Angestellte, die in Exportfragen Bescheid wissen, ausserordentlich schwer zu bekommen sind bzw. eine Einarbeitung dieses schwierigen Gebietes längere Zeit in Anspruch nimmt, würde durch das vorzeitige Ausscheiden dieser Angestellten eine Schädigung des Exports eintreten."[305]

Sowohl Walter wie auch Lilly ist klar, dass sie das Unternehmen verlieren werden und sie versuchen zu retten, was möglich ist. In einem Brief an die Vermögensverkehrsstelle vom 15. Juli 1938 macht Lilly Goldscheider darauf aufmerksam, dass sie als „Einzelprokuristin im Unternehmen tätig gewesen" sei und noch eine Einlageforderung in Höhe von rund RM 12.000 habe.[306] Eine Tante von Lilly, Rosa Kush aus New York, die der Manufaktur ein hohes Darlehen gewährt hat, schickt ein Radiogramm, dass sie mit der Verlängerung der Stundung ihrer Kapitalforderungen über drei Jahre sowie einer Rückstehung gegenüber der Länderbank einverstanden ist.[307] Dieser Verzicht soll einen schnellen Verkauf des Unternehmens ermöglichen und die Ausreise von Walter und Lilly beschleunigen.

Einige Wochen zuvor versuchte Walter bereits mit dem „Verband deutscher Fabriken für Gebrauchs-, Zier- und Kunstporzellan und Keramik G.m.b.H" ein Arrangement zu finden. Er bot dem Verband bei einer eventuellen Umwandlung seines Unternehmens eine mögliche Kapitalbeteiligung deutscher Firmen an, denn bereits am 4. Juni 1938 schreibt ihm diesbezüglich Dr. Fischer vom besagten Verband, dass aufgrund von „Besprechung mit infrage kommenden Firmen [...] zur Beurteilung der Angelegenheit" diverse Unterlagen wie Bilanzen, Gehälterübersicht, benötigtes Betriebskapital etc. benötigt werden.[308] Bei diesem Szenario stellte sich Walter wahrscheinlich vor, dass mit seiner eigenen Minderheitsbeteiligung von unter 25 Prozent die Firma nicht mehr als jüdisch gelten würde und er weiterhin an seinem Familienunternehmen beteiligt und vom Ausland aus mit dem neuen Besitzer, dem Ariseur, in Kontakt bleiben könnte. Um den Verkauf an deutsche Unternehmen zu ermöglichen, schreibt am 23. Juli Dr. Schwartz an die Vermögensverkehrsstelle und bittet um die Genehmigung, „an reichsdeutsche Interessenten informative Anbote durch direkte Offerte und Annoncen stellen zu dürfen." Dies begründet er damit, dass österreichische „Interessenten notorisch über kein auch nur nennenswertes Kapital verfügen"

employees who are experts in export-related issues are extraordinarily difficult to come by and training new personnel in this difficult sector takes quite a long time, the export trade would be impaired by the premature withdrawal of these employees."[305]

Both Walter and Lilly realized that they would lose the business and so they attempted to save what they could. In a letter to the Vermögensverkehrsstelle dated 15 July 1938, Lilly Goldscheider pointed out that she had "worked as the sole head clerk in the business" and there was still an outstanding debt on a loan to the tune of some RM 12 000.[306] An aunt of Lilly's, Rosa Kush in New York, who had loaned the manufactory a large sum, wired that she would agree to extending the deadline for payment of the capital debt in installments over three years as well as arrears at the Länderbank.[308] This waiver was intended to expedite the sale of the business and Walter and Lilly's emigration.

A few weeks previously, Walter had attempted to come to an agreement with the Verband deutscher Fabriken für Gebrauchs-, Zier- und Kunstporzellan und Keramik G.m.b.H. [Union of German Factories for Utilitarian, Ornamental and Art Porcelain and Ceramics Ltd]. He offered the union, should it come to a change in his business, the opportunity for German firms to invest capital in it since by 4 June 1938 Dr. Fischer of the above union had written him that, on the grounds of "talks with firms that might come into question [...] in judging the matter," various documents such as balance sheets, payrolls, the capital needed to cover running expenses, were needed.[308] In opting for this scenario, Walter probably believed if his share in the business amounted to less than 25 per cent, the firm would no longer be classified as Jewish so he would be able to continue as part owner of the family business and could keep in contact from abroad with the new owner, the Aryanizer. To facilitate the sale to a German business, Dr. Schwartz wrote on 23 July to the Vermögensverkehrsstelle requesting permission "to be able to make informative offers to Reich German interested parties both directly and through advertisements." He justified this approach by noting that Austrian "interested parties as is well known lack capital worth

Ill. 87
Brief Walter Goldscheiders an die Vermögensverkehrsstelle, in dem er im Rahmen der Arisierung des Unternehmens einen Verkauf von 76 Prozent des Unternehmens anbietet, um weiterhin seinem traditionsreichen Familienunternehmen verbunden bleiben zu können. (Quelle: Archiv Goldscheider)

A letter written by Walter Goldscheider to the Vermögensverkehrsstelle, in which he offered to sell 76 per cent of the business under the scheme of Aryanization in order to be able to remain linked with his family business which was so rich in tradition. (Source: Goldscheider Archives).

mentioning" or were themselves crisis-ridden. Further, he explained that in 1938 the Goldscheider manufactory was suffering from a considerable decline in turnover due to the exchange rate from Marks that was unfavorable to the export trade and slowing demand from abroad, which was also due to the boycott instituted by some countries.[309]

On 4 August Walter Goldscheider wrote a letter of several pages to the Vermögensverkehrsstelle, in which he proposed several possibilities for Aryanizing his business (Ill. 87 and 88).[310] He also made the suggestion that he remain a minority shareholder with 24 per cent and discussed the economic problems besetting the business in 1938 that led to a downturn in sales volume of 45 per cent. Apart from the "worldwide economic depression," the "boycott by the Anglo-Saxon and other countries" was making itself felt, whereby the loss of the American market was particularly damaging. Other

beziehungsweise selbst in einer Krise stecken. Des weiteren erklärt er auch, dass im Jahr 1938 die Goldscheider Manufaktur unter starkem Umsatzrückgang leide, bedingt durch das für den Export ungünstige Mark-Umrechnungsverhältnis und den langsamen Eingang der Auslandsforderungen, der auch auf den Boykott seitens einiger Länder zurückzuführen ist.[309]

Am 4. August verfasst Walter Goldscheider einen mehrseitigen Brief an die Vermögensverkehrsstelle, in dem er diverse Möglichkeiten der Arisierung seines Unternehmens vorschlägt (Ill. 87 und 88).[310] Hierbei unterbreitet er auch den Vorschlag einer Minderheitsbeteiligung von 24 Prozent und beschreibt die wirtschaftlichen Schwierigkeiten des Jahres 1938, die zu einem Umsatzrückgang von 45 Prozent führten. Neben der „wirtschaftlichen Weltdepression" sei vor allem der „Boykott der anglikanischen und anderer Länder" fühlbar, wobei vor allem der

Ill. 88
Exposé zur Geschichte der Wiener Manufaktur, das Walter Goldscheider im Rahmen der Arisierung seines traditionsreichen Familienunternehmens an die Vermögensverkehrsstelle schickt. (Quelle: Archiv Goldscheider)

Summary of the history of the Viennese manufactory sent by Walter Goldscheider to the Vermögensverkehrsstelle in the course of the Aryanization. (Source: Goldscheider Archives)

Ill. 87

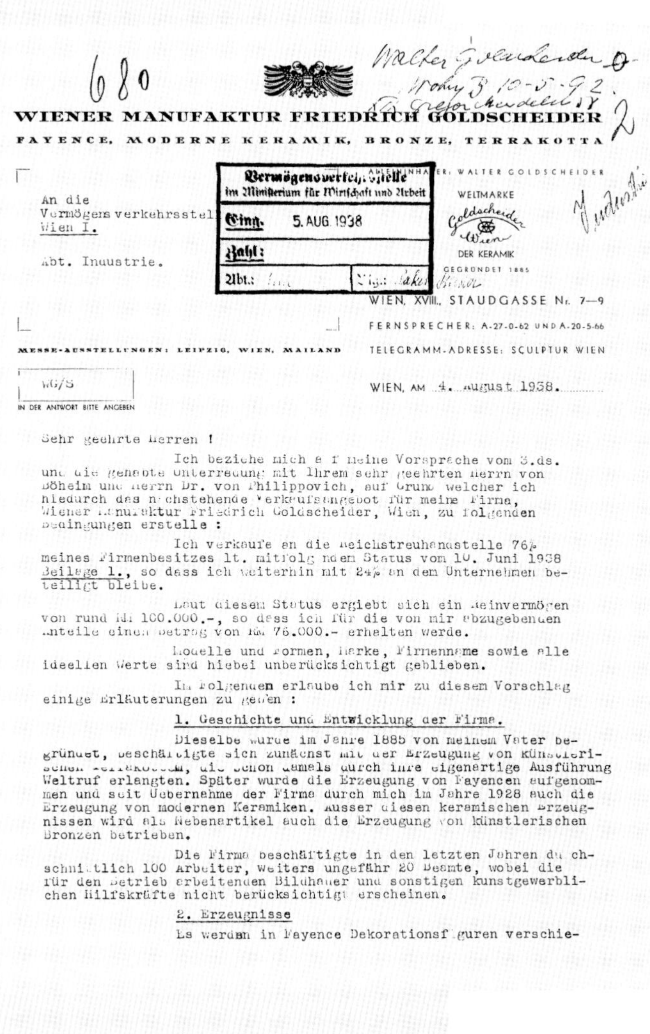

WIENER MANUFAKTUR FRIEDRICH GOLDSCHEIDER
FAYENCE, MODERNE KERAMIK, BRONZE, TERRAKOTTA

An die
Vermögensverkehrsstelle
Wien I.

Abt. Industrie.

Vermögensverkehrsstelle im Ministerium für Wirtschaft und Arbeit
Eing. 5. AUG 1938

WIEN, XVIII., STAUDGASSE Nr. 7—9
FERNSPRECHER: A-27-0-62 UND A-20-5-66
MESSE-AUSSTELLUNGEN: LEIPZIG, WIEN, MAILAND
TELEGRAMM-ADRESSE: SCULPTUR WIEN

WG/S
IN DER ANTWORT BITTE ANGEBEN

WIEN, AM 4. August 1938.

Sehr geehrte Herren !

Ich beziehe mich auf meine Vorsprache vom 3.ds. und die gehabte Unterredung mit Ihrem sehr geehrten Herrn von Böheim und Herrn Dr. von Philippovich, auf Grund welcher ich hiedurch das nachstehende Verkaufsangebot für meine Firma, Wiener Manufaktur Friedrich Goldscheider, Wien, zu folgenden Bedingungen erstelle :

Ich verkaufe an die Reichstreuhandstelle 76% meines Firmenbesitzes lt. mitfolgendem Status vom 10. Juni 1938 Beilage 1., so dass ich weiterhin mit 24% an dem Unternehmen beteiligt bleibe.

Laut diesem Status ergiebt sich ein Reinvermögen von rund RM 100.000.-, so dass ich für die von mir abzugebenden Anteile einen Betrag von RM 76.000.- erhalten werde.

Modelle und Formen, Marke, Firmenname sowie alle ideellen Werte sind hiebei unberücksichtigt geblieben.

Im Folgenden erlaube ich mir zu diesem Vorschlag einige Erläuterungen zu geben :

1. Geschichte und Entwicklung der Firma.

Dieselbe wurde im Jahre 1885 von meinem Vater begründet, beschäftigte sich zunächst mit der Erzeugung von künstlerischen Terrakotten, die schon damals durch ihre eigenartige Ausführung Weltruf erlangten. Später wurde die Erzeugung von Fayencen aufgenommen und seit Uebernahme der Firma durch mich im Jahre 1928 auch die Erzeugung von modernen Keramiken. Ausser diesen keramischen Erzeugnissen wird als Nebenartikel auch die Erzeugung von künstlerischen Bronzen betrieben.

Die Firma beschäftigte in den letzten Jahren durchschnittlich 100 Arbeiter, weiters ungefähr 20 Beamte, wobei die für den Betrieb arbeitenden Bildhauer und sonstigen kunstgewerblichen Hilfskräfte nicht berücksichtigt erscheinen.

2. Erzeugnisse

Es werden in Fayence Dekorationsfiguren verschie-

Ill. 88

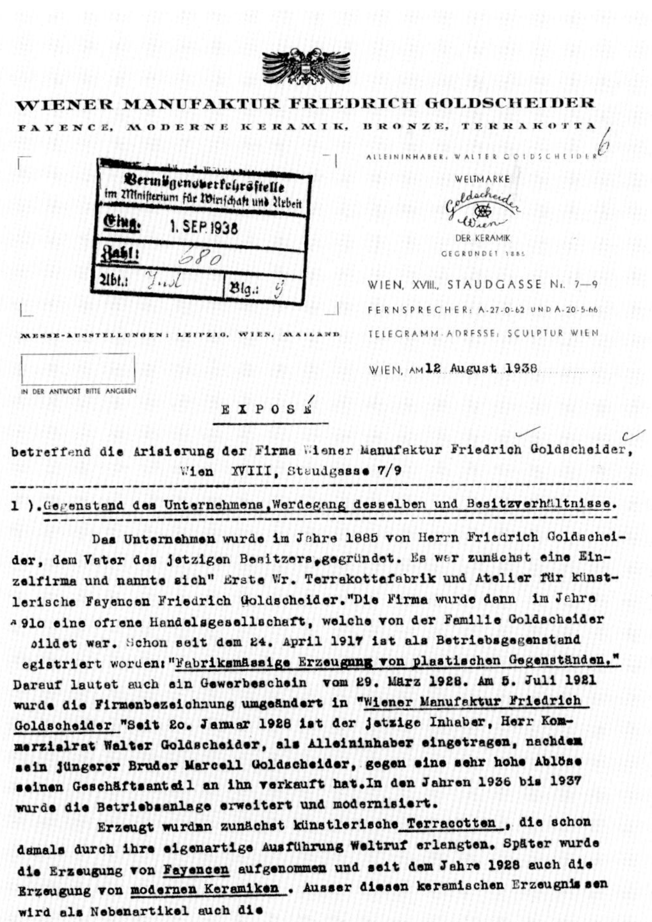

WIENER MANUFAKTUR FRIEDRICH GOLDSCHEIDER
FAYENCE, MODERNE KERAMIK, BRONZE, TERRAKOTTA

Vermögensverkehrsstelle im Ministerium für Wirtschaft und Arbeit
Eing. 1. SEP 1938
Zahl: 680

WIEN, XVIII., STAUDGASSE Nr. 7—9
FERNSPRECHER: A-27-0-62 UND A-20-5-66
MESSE-AUSSTELLUNGEN: LEIPZIG, WIEN, MAILAND
TELEGRAMM-ADRESSE: SCULPTUR WIEN

IN DER ANTWORT BITTE ANGEBEN

WIEN, AM 12 August 1938

EXPOSÉ

betreffend die Arisierung der Firma Wiener Manufaktur Friedrich Goldscheider, Wien XVIII, Staudgasse 7/9

1). Gegenstand des Unternehmens, Werdegang desselben und Besitzverhältnisse.

Das Unternehmen wurde im Jahre 1885 von Herrn Friedrich Goldscheider, dem Vater des jetzigen Besitzers, gegründet. Es war zunächst eine Einzelfirma und nannte sich" Erste Wr. Terrakottafabrik und Atelier für künstlerische Fayencen Friedrich Goldscheider."Die Firma wurde dann im Jahre 1910 eine offene Handelsgesellschaft, welche von der Familie Goldscheider gebildet war. Schon seit dem 24. April 1917 ist als Betriebsgegenstand registriert worden: "Fabriksmässige Erzeugung von plastischen Gegenständen." Darauf lautet auch ein Gewerbeschein vom 29. März 1928. Am 5. Juli 1921 wurde die Firmenbezeichnung umgeändert in "Wiener Manufaktur Friedrich Goldscheider "Seit 20. Januar 1928 ist der jetzige Inhaber, Herr Kommerzialrat Walter Goldscheider, als Alleininhaber eingetragen, nachdem sein jüngerer Bruder Marcell Goldscheider, gegen eine sehr hohe Ablöse seinen Geschäftsanteil an ihn verkauft hat. In den Jahren 1936 bis 1937 wurde die Betriebsanlage erweitert und modernisiert.

Erzeugt wurden zunächst künstlerische Terracotten , die schon damals durch ihre eigenartige Ausführung Weltruf erlangten. Später wurde die Erzeugung von Fayencen aufgenommen und seit dem Jahre 1928 auch die Erzeugung von modernen Keramiken . Ausser diesen keramischen Erzeugnissen wird als Nebenartikel auch die

Verlust des amerikanischen Marktes schmerzlich sei. Weitere Gründe sind „das schlechte Ergebnis der diesjährigen Wiener Frühjahrsmesse und die bis vor kurzem bestandene fast gänzliche Unterbindung des Geschäftes mit dem Altreich."[311] Obwohl in Deutschland die Goldscheider Keramik stets beliebt war, kommt es durch das Naziregime und die zunehmenden Restriktionen für jüdische Firmen zu einem extremen Absatzrückgang. Durch den Anschluss Österreichs an das Deutsche Reich und die anstehende Arisierung des Unternehmens wäre dieser Markt wieder offen für die Erzeugnisse des – dann nicht mehr als jüdisch geltenden – Unternehmens.

Da die Goldscheider Erzeugnisse zumeist als Geschenkartikel gekauft werden und die anstehende Herbst- und Weihnachtssaison in den Monaten September bis Dezember zu ca. 60 Prozent der Jahresumsätze beiträgt, hofft Walter auf eine schnelle Klärung der Situation. Dies sei notwendig, um „im Interesse der Erhaltung des Betriebes" sowie der rechtzeitigen „Hereinbringung der Aufträge zur Beschäftigung der Belegschaft" auch die wichtigen Messen in Leipzig und Wien organisieren und die notwendigen Vertreterreisen durchführen zu können.[312] Ein geeigneter Kaufbewerber sollte laut Walter eine Person sein, „welche neben kunstgewerblichem Verständnis auch über entsprechende kaufmännische Fähigkeiten und Erfahrungen verfügt."[313] Deshalb sieht Walter Goldscheider nur ein Szenario für die Übereignung des Unternehmens: Es soll eine vorübergehende Übernahme durch die Reichstreuhandstelle stattfinden, bei fortlaufender Geschäftsleitung durch Walter Goldscheider bis zur endgültigen Arisierung. Somit wäre auch gewährleistet, dass Walter wie bisher der „Firma durch Herbeischaffung von neuen Ideen für Modelle, neue Artikel und sonstigen Anregungen für Dekore etc." zur Verfügung steht.[314] Somit könnte er auch die jährlich zweimal zu errichtenden neuen Kollektionen unterstützen. Auch die Fortsetzung der Arbeit des kommissarischen Verwalters Dr. Schwartz sei sinnvoll, da dieser sich hervorragend eingearbeitet habe und über das notwendige Wissen verfüge.

Sein Vorschlag findet kein Gehör. Am 30. August 1938 muss sich Walter Goldscheider in einem offiziellen Dokument an die Vermögensverkehrsstelle bereit erklären, sein Unternehmen komplett samt aller Aktiva und Passiva an einen „nichtjüdischen Kaufwerber" zu übertragen. In der Beilage zu diesem Dokument listet er auch auf, dass nur

reasons were "the bad results of this year's Vienna Spring Fair and the recently enforced almost complete ban on conducting business with the Old Reich."[311] Although Goldscheider ceramics had always been popular in Germany, there was an extreme downturn in sales due to the Nazi regime and the increasing restrictions to which Jewish firms were subjected. The annexation of Austria by the Third Reich and the Aryanization of the business should – he must have reasoned – open up this market again for the products of what would no longer be classified as a Jewish commercial enterprise.

Since Goldscheider products were usually bought to be given as presents and the autumn and Christmas season during the months of September through December contributed approx. 60 per cent of the annual sales volume, Walter hoped the situation would soon be cleared up. That would be necessary "in the interests of keeping the business going" and "bringing in orders to keep the workforce busy" in time of preparing for the important fairs in Leipzig and Vienna and ensuring that the necessary salesmen's trips were carried out.[312] A suitable applicant to buy the business should be, according to Walter, a person "who, in addition to having a knowledge of the decorative and applied arts, also possessed the appropriate business skills and experience."[313] Consequently, Walter Goldscheider could envisage only one scenario for handing over the firm: a temporary takeover by the Reichstreuhandstelle [Reich Trust Authority] with Walter Goldscheider continuing to manage the business until it was wholly Aryanized. This would ensure that Walter would still have access to the "firm by contributing new ideas for models, new articles and other inspiration for decoration, etc" as hitherto.[314] Thus he could also support the making up of new biannual collections. He also felt that having Dr. Schwartz continue working as provisional administrator was a good idea because Schwartz had really learnt the ropes and possessed the necessary knowledge.

Walter's proposal fell on deaf ears. On 30 August 1938 Walter Goldscheider had to make a declaration in an official document to the Vermögensverkehrsstelle of his willingness to hand over his business in its entirety with all liquid and illiquid assets to a "non-Jewish applicant." In an enclosure with this document, he also listed 84 white-collar employees and workmen who were still working for the business as well as a Jewish employee, who had already been given notice.[315] To ensure that the business continued to run, the head of the Vermögensverkehrstelle, Count Hardegg, wrote a letter dated 1 September to the Länderbank AG, in which he requested that "no difficulties should be made [for the Goldscheider manufactory] about the payment of sums owed to it since this was a primarily export-oriented business that it seemed necessary to preserve."[316] The uncertain situation, the loss of personnel and the difficult conditions imposed led to a rapid decline in output. The business depreciated in value and the Aryanization machinery shifted into high gear.

On 7 September 1938 Dr. Schwartz sent an exhaustive report on the condition of the business to the Vermögensverkehrsstelle. He described debtors, suppliers, financial creditors, loans, supplied and the facilities, pointing out that the two buildings in Staudgasse "must be said to be undervalued at RM 20 027"[317] since the estimated value was 82 560 Reich marks according to an appraisal undertaken by the architect and urban planner Karl August Benirschke.[318] A total loss of 12 110 RM was quoted for the current year due to dwindling turnover and repayment of a credit from the Länderbank before it was actually due. In that report, the provisional administrator recommended that both Walter Goldscheider and his son Erwin, who worked in the most important market, in North America, should be kept for the good of the business. He also felt Walter Goldscheider's continuing to be a shareholder in the firm would be advantageous since "this would be the most effective way of countering the boycott movement."[319] He finished his report with the following evaluation of the business: "The firm can be described as viable. What speaks for this are the profit margins of 5–8 % achieved in the past 5 years."[320] In the report, the sum of 100 000 RM in liquid assets possessed by Walter Goldscheider, owner of the firm, which would form the basis on which the sale would be negotiated, were not mentioned despite Dr. Schwartz's calculations in his August summary.[321] On 3 October 1938 a

noch 84 Angestellte und Arbeiter für die Manufaktur tätig sind sowie ein bereits gekündigter jüdischer Angestellter.[315] Um den laufenden Betrieb des Unternehmens zu gewährleisten, schreibt am 1. September der Leiter der Vermögensverkehrsstelle, Graf Hardegg, einen Brief an die Länderbank AG, in der er darum bittet, der Manufaktur Goldscheider „in Auszahlung der für sie einlangenden Beträge keine Schwierigkeiten zu bereiten, da es sich hier um einen vorwiegend exportorientierten Betrieb handelt, dessen Erhaltung notwendig erscheint."[316] Die unsichere Situation, der Personalverlust und die erschwerten Bedingungen führen zu einem schnellen Verfall der Produktion. Das Unternehmen verliert an Wert und die Arisierungsmaschinerie kommt richtig in Fahrt.

Am 7. September 1938 schickt Dr. Schwartz einen ausführlichen Bericht zur Lage des Unternehmens an die Vermögensverkehrsstelle. Er beschreibt die Debitoren, Lieferanten, Finanzkreditoren, Darlehen, Vorräte, Anlagen und macht darauf aufmerksam, dass die beiden Häuser in der Staudgasse „mit RM 20.027,– als unterbewertet zu bezeichnen sind"[317], denn der Schätzwert liegt nach einem Gutachten des Architekten und Stadtbaumeisters Karl August Benirschke bei 82.560 Reichsmark.[318] Insgesamt wird für das laufende Jahr aufgrund der verringerten Umsätze und einer vorzeitigen Rückzahlung eines Kredits von der Länderbank ein Verlust von 12.110 Reichsmark angesetzt. In diesem Bericht empfiehlt der kommissarische Verwalter, dass sowohl Walter Goldscheider wie auch sein Sohn Erwin, der auf dem wichtigsten Markt in Nordamerika als Generalvertreter tätig ist, dem Unternehmen erhalten bleiben sollten. Auch eine weitere Beteiligung Walter Goldscheiders an der Firma wird von ihm als vorteilhaft angesehen, da „hierdurch der Boykottbewegung am wirksamsten entgegengetreten werden kann".[319] Den Bericht schließt er mit der folgenden Unternehmensbeurteilung ab: „Die Firma kann als lebensfähig bezeichnet werden. Dafür sprechen auch die bisher erzielten Reingewinne von 5–8% in den letzten 5 Jahren."[320] In diesem Bericht wird allerdings nicht mehr die Summe von 100.000 Reichsmark als Aktivvermögen des Firmeninhabers Walter Goldscheider und somit als Grundlage für die Verkaufsverhandlungen aufgeführt, wie dies von Dr. Schwartz zuvor in einem im August entworfenen Expose errechnet worden ist.[321] Am 3. Oktober 1938 wird ein erneuter Schätzungsbericht über die Liegenschaften in der Staudgasse durch den Architekten Josef Heinzle verfasst. Dieser bewertet die Gebäude als stark reparaturbedürftig und errechnet einen Verkehrswert von nur noch 68.000 Reichsmark für beide Gebäude.[322]

Trotz der negativen Berechnungen zum Wert des Unternehmens gibt es mehrere Kandidaten, die sich bei der Vermögensverkehrsstelle im Rahmen der Arisierung um die Übernahme der Goldscheider Manufaktur bewerben. Der Leiter der Wiener Porzellanfabrik Augarten, Theodor Ludwig Kraus, bewirbt sich persönlich bei Dr. Leitich von der Vermögensverkehrsstelle um den Posten eines Geschäftsführers. Für ihn setzen sich nicht nur die Barone Theodor und Gisbert von Liebig ein, auch der Kommerzialrat Stefan Rath von der Firma Lobmeyr, Ing. Siegfried Dietl als Prokurist der Generali und Rudolf Pfeiffer als Direktor der Creditanstalt unterstützen ihn – doch erfolglos.[323]

Die Genehmigung, auch an reichsdeutsche Unternehmen gezielt Informationen geben zu können, wird erteilt, führt aber zu keinen Ergebnissen. Für das Unternehmen interessiert sich aus Deutschland der Reichstagsabgeordnete Kunze, „ein persönlicher Freund des Führers, [...] dem wohl vor anderen Bewerbern der Vorrang zu geben sei", wie aus einer Notiz vom 16. September 1938 an Dr. Schumetz von der Vermögensverkehrsstelle zu entnehmen ist.[324] Doch auch hier gibt es keine weiteren Schritte.

Fachlich wäre die Bewerbung des Bildhauers Walter Bosse bevorzugt einzustufen. Aus einer Notiz von Dr. Arthur Mayer, dem Rechtsanwalt Walter Goldscheiders, geht hervor, dass Bosse bereits am 5. August 1938 Interesse an der Arisierung der Goldscheider Manufaktur geäußert hat.[325] Diese Notiz vom 16. September 1938 und ein persönlich von Walter Bosse verfasster Lebenslauf, den er am 5. November 1938 im Rahmen seines anvisierten Erwerbs der Goldscheider Manufaktur beilegt, lässt sich folgendes zu seinem Werdegang zusammenfassen: Bosse wurde am 17. November 1904 in Wien als Sohn eines Kunstmalers geboren. Er hat die Kunstgewerbeschule in Wien und München besucht, wo er die nötigen Kenntnisse für das keramische Kunstgewerbe erwarb. Im Jahr 1922 machte er sich in Kufstein mit einer kleinen keramischen Werkstätte selbständig. Er erzielte anfänglich gute Ergebnisse und beschäftigte 20–30 Arbeiter. Im Jahre 1928 war eine Erweiterung der Werkstätte notwendig und Bosse erwarb ein Fabrikobjekt. Er beschäftigte zwischen 1928 und 1937

new appraisal of the Staudgasse properties was written up by the architect Josef Heinzle. He evaluated the buildings as severely in need of repairs and calculated a sale value of only 68 000 RM for the two.[322]

Despite the negative appraisals of the value of the business, there were several candidates who applied to the Vermögensverkehrsstelle as part of the Aryanization process to take over the Goldscheider manufactory. Theodor Ludwig Kraus, director of the Augarten Viennese Porcelain Factory, applied in person to Dr. Leitich of the Vermögensverkehrsstelle for the post of managing director. Not only Barons Theodor and Gisbert von Liebig but also Kommerzialrat Stefan Rath of the firm of Lobmeyr, Ing. Siegfried Dietl, head cleark of Generali, and Rudolf Pfeiffer, director of the Creditanstalt, threw all their considerable weight behind his application – in vain.[323]

Permission was granted to give Reich German businesses as well as Austrian particulars of the sale but led to no results. Kunze, a member of the Reichstag and "a personal friend of the Führer's, [...] to whom precedence should certainly be given over other applicants," as written in a note dated 16 September 1938 to Dr. Schumetz of the Vermögensverkehrsstelle, was an applicant from Germany.[324] Nevertheless, no further steps were taken, even in this case.

As far as expertise was concerned, the application made by Walter Bosse, the sculptor, should have been given priority. From a note written by Dr. Arthur Mayer, Walter Goldscheider's lawyer, it is apparent that Bosse had already expressed interest in the Aryanization of the Goldscheider manufactory as early as 5 August 1938.[325] This note, dated 16 September 1938, and a curriculum vitae, composed by Walter Bosse himself, enclosed in his application for the right to purchase the Goldscheider manufactory, reveal the following about the applicant: Born on 17 November 1904 in Vienna, Bosse was the son of a painter. He had attended the Applied Arts Schools in Vienna and Munich, where he had acquired the knowledge and skills needed for working in ceramics. In 1922 he became self-employed, opening a small ceramics workshop in Kufstein. Successful from the outset, Bosse employed a workforce of 20–30. In 1928 he found it necessary to enlarge the workshop and acquired a factory. Between 1928 and 1937, Bosse employed between thirty and sixty men. Ninety per cent of his sales volume resulted from exports. However, he was also "continuously employed by the well-known 'Goldscheider' ceramics factory in Vienna, XVIII District."[326] Further, Dr. Mayer wrote, Bosse's assets went into receivership in 1933 and his property was sold at auction. Bosse nevertheless continued to produce but had to contend with great difficulties caused by trade restrictions and dwindling exports.[327] The business was run by Bosse's sister Gretl and his ex-wife while Bosse moved to Vienna early in 1937, where he continued production on a reduced scale with a workforce of approx 20. Bosse was described as "skilled, hard-working and in respect of his art entirely satisfactory." Dr. Mayer added that Bosse was interested in the Aryanization of the Goldscheider manufactory and, since Bosse himself had no finances to speak of, would probably be backed by "a financial source which, however, prefers to remain anonymous."

The financial backing behind Bosse was an engineer named F.R. Uffert from Berlin-Wilmersdorf, who was a specialist in industrial and communal financing. On 23 September 1938, Uffert wrote to the Austrian Kontrollbank that Walter Bosse had contacted him about taking a "Sperrmarkkredit" ["blocked-mark" or "frozen" credit], with which he "wanted to acquire the firm of Goldscheider through the Aryanization process"[328]. In this letter, Uffert mentions that the money he needed as calculated by Bosse amounted to some 160 000 to 180 000 Reich marks. He requested that informative documents and a balance-sheet of the Goldscheider factory should be sent. Although one had been drawn up by Walter Goldscheider and Dr. Schwartz and handed over to the Kontrollbank, Bosse was sent incomplete documents by the Kontrollbank. Bosse wrote a letter to the Kontrollbank dated 19 October 1938, in which he complained about the incompleteness of the documents furnished.[329]

At the same time, another interested party contacted the Kontrollbank. On 14 October 1938, Josef Schuster from Munich expressed his interest to the Austrian Kontrollbank in the Aryanization of the Viennese Manufactory Friedrich Goldscheider. He wrote that he had personally visited the business only a few days before writing. On himself, Schuster wrote confidently: "I am Aryan, politically irreproachable and have money of my own. In 1928 I founded a highly artistic business (quite a large decora-

zwischen 30 und 60 Leute, 90 Prozent des Umsatzes gingen ins Ausland. Er war aber auch gleichzeitig „ständig von der bekannten Keramik-Fabrik ‚Goldscheider' in Wien, XVIII. beschäftigt."[326] Des Weiteren führt Dr. Mayer aus, dass im Jahr 1933 über Bosses Vermögen Konkurs verhängt und sein Besitz versteigert worden ist. Bosse produzierte jedoch weiter, hatte aber mit großen Schwierigkeiten zu kämpfen, bedingt durch die Handelserschwernisse und den rückläufigen Export.[327] Das Geschäft wurde von Bosses Schwester Gretl und seiner Ex-Frau betrieben und Bosse zog Anfang 1937 nach Wien um, wo er im kleineren Umfang mit ca. 20 Beschäftigten die Produktion fortführte. Bosse wird als „fachtüchtig, fleißig und in künstlerischer Hinsicht einwandfrei" beurteilt. Ebenfalls wird von Dr. Mayer aufgeführt, dass Bosse Interesse an der Arisierung der Goldscheider Manufaktur hat und, da er selbst über keine nennenswerten Finanzmittel verfügt, wohl von einer „Finanzkraft, die aber nicht genannt sein will", unterstützt wird.

Die Finanzkraft hinter Bosse ist der auf Industrie- und Kommunalfinanzierungen spezialisierte Ingenieur F.R. Uffert aus Berlin-Wilmersdorf. Am 23. September 1938 schreibt dieser an die Österreichische Kontrollbank, dass Walter Bosse mit ihm wegen der Aufnahme eines „Sperrmarkkredites" in Verbindung steht, mit welchem er „im Arisierungsverfahren die Firma Goldscheider erwerben will"[328]. In diesem Brief gibt er an, dass der von Bosse errechnete Geldbedarf in etwa 160.000 bis 180.000 Reichsmark beträgt. Er bittet um die Zusendung von aussagekräftigen Unterlagen und einer Bilanz der Goldscheider Manufaktur. Obwohl diese von Walter Goldscheider und Dr. Schwartz erstellt worden sind und der Kontrollbank vorliegt, werden Bosse von der Kontrollbank unvollständige Unterlagen geschickt. Dies moniert Bosse am 19. Oktober 1938 in einem Brief an die Kontrollbank.[329]

Zur gleichen Zeit meldet sich ein neuer Interessent bei der Kontrollbank. Am 14. Oktober 1938 bekundet Josef Schuster aus München bei der Österreichischen Kontrollbank sein Interesse an der Arisierung der Wiener Manufaktur Friedrich Goldscheider. Er schreibt, dass er in den vergangenen Tagen den Betrieb persönlich besichtigt hätte. Über sich selbst schreibt Schuster selbstbewusst: „Ich bin arisch, politisch einwandfrei und verfüge über eigenes Vermögen. 1928 habe ich einen hochkünstlerischen Betrieb (grösseres Kunstgewerbehaus mit 5 Schaufenstern) gegründet und 10 Jahre lang mit grossem Erfolg geleitet. Ich gehöre der Reichskunstkammer (Reichskulturkammer) sowie der Reichsschriftumskammer an. Politisch bin ich verdient."[330] Im Bewerberblatt-Formular gibt er als frühere Adresse München, Vandalenstraße 15 und als jetzige Adresse bereits die Staudgasse 7/9 an, also die Adresse der Goldscheider Manufaktur. Die meisten Fragen auf dem Bewerberblatt, zum Beispiel nach finanziellen Grundlagen und seinem Angebot, lässt er unausgefüllt.

Die Österreichische Kontrollbank versucht über verschiedene Quellen weitere Informationen über Schuster zu finden, doch die Ausbeute ist mager. Man erfährt, dass Schuster etwa 1900 geboren ist und im Mai 1938 nach München-Grosshadern kam, wo er zur Miete wohnt. Er soll als Vertreter für ein größeres Unternehmen ständig unterwegs sein. Später wird noch festgestellt, dass er der Sohn des verstorbenen Gastwirtes Karl Schuster ist, der in früheren Jahren Besitzer des „Alpengasthofes zum Türken" in Obersalzberg und Pächter des Purtschellerhauses am Hohen Göll war. Die Eltern haben 1933 den Gasthof an die Partei verkauft und sind dann nach Arlaching am Chiemsee übersiedelt, wo sie ein Strandrestaurant mit Pension erwarben.[331]

Während man Erkundigungen einholt, fragt Schuster weiterhin selbstbewusst nach dem Zeitpunkt des Kaufabschlusses.[332] Walter Bosse indes bekundet am 5. November 1938 offiziell in einem Schreiben an die Kontrollbank, dass er am Erwerb der Firma Goldscheider interessiert sei (Ill. 89). Er fühlt sich aufgrund seiner bisherigen Berufstätigkeit „wie kaum ein anderer dieser Aufgabe gewachsen." Ein ungenannter Teilhaber sowie die „Bank der deutschen Arbeit" sollen die finanziellen Mittel stellen wollen, er selbst würde seine Kollektion und seinen Kufsteiner Betrieb einbringen. Bosse weist darauf hin, dass er bei der Firma Goldscheider „die moderne Kollektion eingeführt hat, welche heute den halben Umsatz der Firma abwirft."[333] Er weist aber auch darauf hin, dass er Josef Schuster auf die Arisierung von Goldscheider aufmerksam gemacht hat und dass sie ursprünglich gemeinsam das Unternehmen übernehmen wollten. Schuster hätte dann jedoch hinter seinem Rücken allein das Unternehmen erwerben wollen, um Bosse „auszuschalten".[334] Bosse weist auch spitzzüngig darauf hin, dass die von der Kontrollbank gelieferten Zahlen nicht als „stichhaltig" angesehen werden können

tive arts house with 5 display windows) and managed it with great success for ten years. I belong to the Reichskunstkammer (Reichskulturkammer) [Reich Art Chamber (Reich Culture Chamber)] and the Reichsschriftumskammer [Reich Literature Chamber]. I am politically deserving."[330] In the application form, he gives as his former address 15 Vandalenstraße, Munich and as his present address 7/9 Staudgasse, Vienna, the address of the Goldscheider manufactory. He did not even bother to answer most of the questions on the application form, for instance, about his financial situation and what he would offer for the business.

The Austrian Kontrollbank tried to glean more information on Schuster through various channels but the yield was meager. It appeared that Schuster was born in 1900 and went to Munich-Grosshadern in May 1938, where he lived in lodgings. He traveled constantly as the representative for a fairly large firm. It was later ascertained that he was the son of tavern-keeper Karl Schuster, deceased, who had for some years been the landlord of the "Alpengasthof zum Türken" in Obersalzberg and had leased the Purtschellerhaus on the Hohen Göll. Joseph Schuster's parents had sold the tavern to the Party in 1933 and then moved to Arlaching on Lake Chiemsee, where they acquired a lakeshore restaurant and guesthouse.[331]

While more information was being sought on him, Schuster continued confidently with his queries about when contracts would be concluded.[332] Walter Bosse in the meantime announced officially in a letter to the Kontrollbank dated 5 November 1938 that he was interested in buying the firm of Goldscheider (Ill. 89). He felt, on the grounds of his previous professional experience "more qualified for this than almost anyone else." An unnamed partner as well as the "Bank der deutschen Arbeit" were willing to ensure the funding; he himself would contribute his collection and the Kufstein business. Bosse pointed out that he had been instrumental at Goldscheider "in introducing the modern collection, which today accounts for half the firm's turnover."[333] He also pointed out, however, that he was the one who had made Josef Schuster aware of the Aryanization of Goldscheider and that they had originally planned to take over the business jointly. Schuster, he declared, had, however, wanted to acquire the business behind his back in order to "eliminate" Bosse.[334] Bosse also pointed out acerbically that the figures provided by the Kontrollbank could not be viewed as "sound" and that "viewed from the National Socialist standpoint, the present negotiation [looks] rather fishy" in as far as "actually qualified specialists should be employed and those unknowledgeable in the field are the ones that should be eliminated."[335] These remarks induced the civil servant processing the case at the Kontrollbank to note repeatedly in the margin: "impertinence."

At the same time, a planning session at which Dr. Hans Fischböck, the chairman of the Advisory Council at the Vermögensverkehrsstelle and economics and labor minister, was present classified the Goldscheider manufactory as a "business that should be preserved."[336] A summary dated 9 November 1938 dealing with the Aryanization of the Viennese Manufactory Friedrich Goldscheider states: the assets were appraised as of 30 June 1938 at RM 59 430, the losses per month estimated as amounting to 10 000 Reich marks and the condition of the business assessed as "deteriorated" so that "only immediate transfer to Aryan ownership can prevent complete collapse."[337] Mr and Mrs Schuster and Bosse are mentioned as potential buyers of the business.

On 10 November 1938 the council of the Vermögensverkehrsstelle decided – presumably on the one hand due to Bosse's way of expressing himself, his financial weakness and insolvency only a few years previously and, on the other, due to Schuster's mindset that implied absolute allegiance to the Party and alleged capital – that "the enterprise Viennese Manufactory

und dass „vom nationalsozialistischen Gesichtspunkt betrachtet, die gegenwärtige Verhandlung recht eigenartig" anmutet, „wo doch eigentlich berufene Fachleute eingesetzt und Branchenfremde ausgeschaltet werden sollen."[335] Diese Anmerkungen führen zu einer mehrmals wiederholten Randnotiz des bearbeitenden Beamten bei der Kontrollbank: „Frechheit".

Zur gleichen Zeit wird in einer Planungssitzung in Anwesenheit des Vorsitzenden des Beirates bei der Vermögensverkehrsstelle und Ministers für Wirtschaft und Arbeit, Dr. Hans Fischböck, die Goldscheider Manufaktur als „erhaltungswürdiges Unternehmen" eingestuft.[336] Am 9. November 1938 steht in einer Zusammenfassung zur Arisierung der Wiener Manufaktur Friedrich Goldscheider: Die Eigenmittel werden per 30. Juni 1938 mit RM 59.430,– bestimmt, der eingetretene monatliche Verlust auf 10.000 Reichsmark geschätzt und die Situation des Unternehmens als „verschlechtert" bewertet, so dass „nur eine sofortige Ueberleitung in arischen Besitz den restlosen Zusammenbruch aufhalten kann."[337] Als potenzielle Käufer des Unternehmens werden das Ehepaar Schuster und Bosse aufgeführt.

Am 10. November 1938 beschließt der Beirat der Vermögensverkehrsstelle – vermutlich einerseits durch Bosses Formulierungen, Finanzschwäche und den wenige

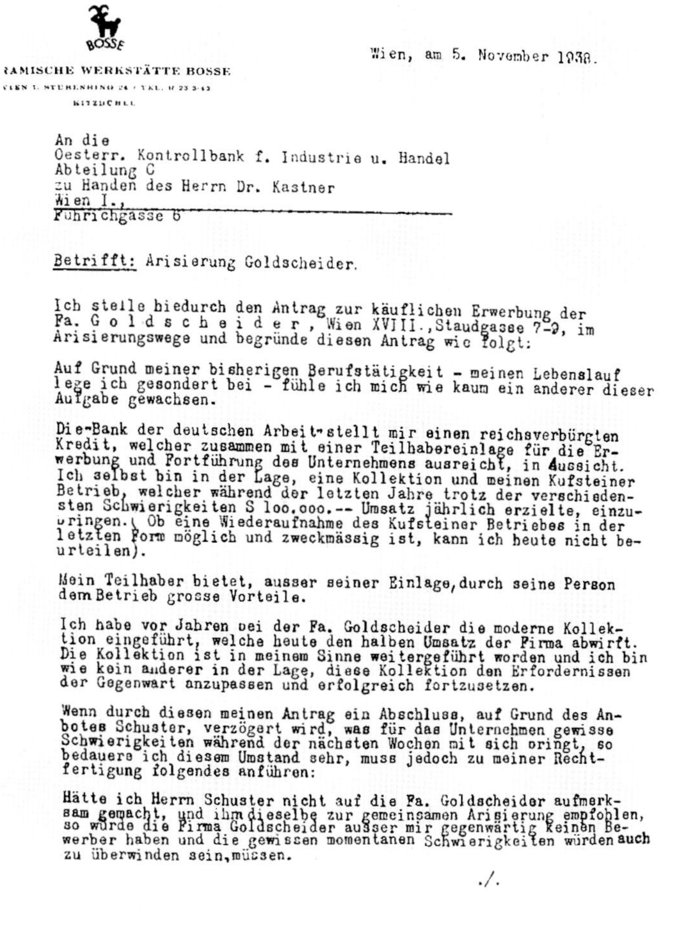

BOSSE
[KER]AMISCHE WERKSTÄTTE BOSSE
KITZBÜHEL

Wien, am 5. November 1938.

An die
Oesterr. Kontrollbank f. Industrie u. Handel
Abteilung C
zu Handen des Herrn Dr. Kastner
Wien I.,
Führichgasse 6

Betrifft: Arisierung Goldscheider.

Ich stelle hiedurch den Antrag zur käuflichen Erwerbung der Fa. G o l d s c h e i d e r , Wien XVIII., Staudgasse 7-9, im Arisierungswege und begründe diesen Antrag wie folgt:

Auf Grund meiner bisherigen Berufstätigkeit – meinen Lebenslauf lege ich gesondert bei – fühle ich mich wie kaum ein anderer dieser Aufgabe gewachsen.

Die Bank der deutschen Arbeit stellt mir einen reichsverbürgten Kredit, welcher zusammen mit einer Teilhabereinlage für die Erwerbung und Fortführung des Unternehmens ausreicht, in Aussicht. Ich selbst bin in der Lage, eine Kollektion und meinen Kufsteiner Betrieb, welcher während der letzten Jahre trotz der verschiedensten Schwierigkeiten S 100.000.-- Umsatz jährlich erzielte, einzubringen. (Ob eine Wiederaufnahme des Kufsteiner Betriebes in der letzten Form möglich und zweckmässig ist, kann ich heute nicht beurteilen).

Mein Teilhaber bietet, ausser seiner Einlage, durch seine Person dem Betrieb grosse Vorteile.

Ich habe vor Jahren bei der Fa. Goldscheider die moderne Kollektion eingeführt, welche heute den halben Umsatz der Firma abwirft. Die Kollektion ist in meinem Sinne weitergeführt worden und ich bin wie kein anderer in der Lage, diese Kollektion den Erfordernissen der Gegenwart anzupassen und erfolgreich fortzusetzen.

Wenn durch diesen meinen Antrag ein Abschluss, auf Grund des Anbotes Schuster, verzögert wird, was für das Unternehmen gewisse Schwierigkeiten während der nächsten Wochen mit sich bringt, so bedauere ich diesem Umstand sehr, muss jedoch zu meiner Rechtfertigung folgendes anführen:

Hätte ich Herrn Schuster nicht auf die Fa. Goldscheider aufmerksam gemacht, und ihm dieselbe zur gemeinsamen Arisierung empfohlen, so würde die Firma Goldscheider ausser mir gegenwärtig keinen Bewerber haben und die gewissen momentanen Schwierigkeiten würden auch zu überwinden sein, müssen.

./.

Ill. 89
Brief Walter Bosses an die Österreichische Kontrollbank, in dem er offiziell das Interesse am Erwerb der zu arisierenden Firma Goldscheider bekundet. (Quelle: Archiv Goldscheider)

Letter written by Walter Bosse to the Austrian Kontrollbank in which he officially declared his interest in acquiring the firm of Goldscheider that was to be Aryanized. (Source: Goldscheider Archives)

Jahre zurückliegenden Konkurs sowie andererseits durch Schusters parteitreue Gesinnung und angeblich vorhandenes Eigenkapital bedingt –, dass „die Unternehmung Wiener Manufaktur Friedrich Goldscheider von der Kontrollbank an die Bewerber, Eheleute Schuster, zu verkaufen ist."[338]

Die Ehepaare Josef und Lilly Schuster sowie Walter und Lilly Goldscheider treffen sich, um über die Zukunft des Unternehmens zu sprechen. Am 28. November 1938 bestätigt Josef Schuster in einem Brief ein Abkommen mit Walter und Lilly Goldscheider. Nach dem Erwerb des Unternehmens von der Österreichischen Kontrollbank durch Schuster sollen die Goldscheiders mit den Vertretungen in den westeuropäischen Ländern, wie England, Frankreich und Belgien, betraut werden sowie gegebenenfalls mit den Vertretungen für die Schweiz, Schweden, Dänemark und Italien.[339] Sie sollen für vermittelte Aufträge eine Provision von 5 Prozent erhalten. Des Weiteren sollen sie den Alleinvertrieb „sämtlicher gegenwärtiger und künftiger Erzeugnisse und Handelsartikel" für das Gebiet der Vereinigten Staaten von Nordamerika und Kanada übertragen bekommen. Sollten sich Walter und Lilly für die Errichtung einer Generalvertretung für Nordamerika entscheiden, so würde ihnen eine Provision von 22,5 Prozent für alle Lieferungen in diese Länder zustehen, in US-Dollar.[340] Zum Zwecke des Aufbaus der Vertretungen soll Walter bei seiner Ankunft in New York einen Provisionsvorschuss von 600 Dollar erhalten. Sollten Walter und Lilly das Geschäft in den Vereinigten Staaten und Kanada als so genannte „wholesaler" führen wollen, dann können sie die Waren von Schuster nach der geltenden Liste für Großverkäufer beziehen. Bei beiden Geschäftsformen sichert Schuster in seinem Abkommen zu, dass ihnen der Vertrieb exklusiv gewährt wird. Schuster verpflichtet sich, ihnen die in New York bereits vorhandenen Musterkoffer zu überlassen und bei Teilnahmen an Messen die Kosten zu übernehmen. Walter Goldscheider soll auch weiterhin Anregungen für neue Artikel beisteuern, neue Ausführungsformen (Modelle) sowie so genannte Neuausstattungen (Dekore), und für diese eine fünfprozentige Provision des damit erzielten Umsatzes erhalten. Im Gegenzug erklärt sich Walter Goldscheider einverstanden, dass er, solange Schuster seine Pflichten erfüllt, kein Konkurrenzunternehmen selbst aufbaut oder unterstützt und Josef Schuster das Wiener Unternehmen unter dem Namen „Wiener Manufaktur

Friedrich Goldscheider is to be sold by the Kontrollbank to the applicants Mr and Mrs Schuster."[338]

Josef and Lilly Schuster met Walter and Lilly Goldscheider to discuss the future of the business. On 28 November 1938 Josef Schuster confirmed in a letter an agreement with Walter and Lilly Goldscheider. After Schuster had acquired the business from the Austrian Kontrollbank, the Goldscheiders were to be entrusted with representing it in the western European countries such as England, France and Belgium as well as, if need arose, in Switzerland, Sweden, Denmark and Italy.[339] They were to receive a 5 per cent commission for each order they negotiated. Further, the sole responsibility for marketing "all present and future products and articles for trade" for the area of the United States of America and Canada was to be transferred to them. Should Walter and Lilly decide on setting up a general agency for North America, they would be entitled to a commission of 22.5 per cent, paid in US dollars, on all deliveries made to those countries.[340] For the purpose of building up the agencies, Walter was to receive a 600-dollar advance on the commission on his arrival in New York. Should Walter and Lilly want to operate the business in the United States and Canada as a wholesale [rather than a retail one], they were entitled to procure the wares from Schuster on the terms obtaining for wholesalers. In both forms of business, Schuster assured them in his agreement that they would have sole right to marketing. Schuster pledged to turn over the sample cases already in New York to them and to assume the cost of participation in trade fairs. Walter Goldscheider was to continue to contribute suggestions for new articles, new versions (of models) as well as what were termed refurbishments (decoration) and for such to receive a five per cent commission on the proceeds from the sales of such articles. For his part, Walter Goldscheider agreed that he would not build up or support another business as competition as long as Schuster fulfilled his side of the agreement and Josef could continue the Viennese business under the name "Wiener Manufaktur Friedrich Goldscheider's Nachfolger" ["Viennese Manufactory Friedrich Goldscheider, Successor"]. Walter was also to persuade his aunt, Rosa Kush, who lived in New York and had lent the business almost 20 000 dollars, to make it possible for Josef Schuster to be able to pay the debt owed to her later, with certain conditions obtaining. The agreement between Schuster and the Goldscheiders was to be valid provisionally until late 1939.[341]

An agreement with Schuster was vital for Walter and Lilly Goldscheider if they were to build up a new life as emigrants. The agreement, however, was nothing if not farcical. The following day Schuster had his lawyer, Dr. Ernst Höpler, inform the Goldscheiders that he was only willing to grant Walter Goldscheider the agency for North America and to pay him a commission for business transacted there. That was it. At the same time, Walter Goldscheider was not to be released from his responsibility for a credit that had already existed for some time.[342] In the days that followed, the situation became increasingly acute and the lawyers continued to negotiate through the mediation of the Kontrollbank. Walter Goldscheider's demands were cut down one after the other.

On 9 December 1938 Schuster's lawyer set down in minutes the terms of the agreement between the Kontrollbank and Josef Schuster: Schuster was to pay only 15 000 Reich marks for the properties at 7 and 9 Staudgasse as well as the business that operated there, the Viennese Manufactory Friedrich Goldscheider, of which 5000 were to be paid at once and the rest in monthly installments of 2000 Reich marks.[343] That same day, Schuster confirmed a new agreement with Walter Goldscheider, with himself as "sole proprietor" of the legally registered firm "Wiener Manufaktur Friedrich Goldscheider," in which he transferred to Walter and Lilly Goldscheider the sole agency for "his" products for the territory of the "United States of America and Canada." They were forbidden to represent other businesses. Walter Goldscheider was to receive a 20 per cent commission, based on the net value of the wares, from direct transactions and 10 per cent from indirect transactions. Should Walter not reach a minimum turnover of 20 000 dollars within six months of his arrival in the US, the contract could be immediately and unilaterally terminated by Schuster. Walter would have to pay for the sample collections or they could be offset from his claims to commissions. The outstanding claims by Lilly Goldscheider and Rosa Kush were no longer mentioned.[344]

Finally, the contracts were signed on 14 December 1938. Between the Austrian Kontrollbank and Walter Goldscheider it was

Friedrich Goldscheider's Nachfolger" fortführen kann. Walter soll auch seine in New York lebende Tante Rosa Kush, die dem Unternehmen fast 20.000 Dollar geliehen hat, dazu bewegen, dass Josef Schuster bei gewissen Bedingungen seine Schulden später rückzahlen kann. Das Abkommen zwischen Schuster und den Goldscheiders soll vorerst bis Ende 1939 gültig sein.[341]

Für den Aufbau einer neuen Existenz in der Emigration ist für Walter und Lilly Goldscheider ein Abkommen mit Schuster lebensnotwendig. Doch das Abkommen ist eine absolute Farce. Denn gleich am nächsten Tag lässt Schuster durch seinen Rechtsanwalt Dr. Ernst Höpler mitteilen, dass er lediglich bereit sei, Walter Goldscheider die Vertretung für Nordamerika einzuräumen und ihm eine Provision für erbrachte Geschäfte zu zahlen. Mehr aber nicht. Gleichzeitig soll Walter Goldscheider nicht aus der Haftung für einen bereits seit längerem bestehenden Kredit genommen werden.[342] In den nächsten Tagen spitzt sich die Situation zu und die Rechtsanwälte verhandeln unter der Einbeziehung der Kontrollbank weiter. Die Forderungen Walter Goldscheiders werden nach und nach gekappt.

Am 9. Dezember 1938 hält Schusters Anwalt in einem Gedächtnisprotokoll die Vereinbarungen zwischen der Kontrollbank und Josef Schuster fest: Schuster soll für die Liegenschaften in der Staudgasse 7 und 9 sowie das dort betriebene Unternehmen Wiener Manufaktur Friedrich Goldscheider nur 15.000 Reichsmark bezahlen, davon 5.000 sofort und den Rest in Monatsraten von jeweils 2.000 Reichsmark.[343] Am gleichen Tag bestätigt Schuster als „Alleineigentümer" der handelsgerichtlich eingetragenen Firma „Wiener Manufaktur Friedrich Goldscheider" ein erneutes Abkommen mit Walter Goldscheider, in dem er Walter und Lilly Goldscheider die Alleinvertretung „seiner" Erzeugnisse für das Gebiet der „Vereinigten Staaten von Nordamerika und Canada" überträgt. Es wird ihnen verboten, andere Unternehmen zu vertreten. Aus direkten Geschäften soll Walter 20 Prozent Provision erhalten, aus indirekten 10 Prozent vom Nettowarenwert. Sollte Walter innerhalb eines halben Jahres nach seiner Ankunft in den USA nicht einen Mindestumsatz von 20.000 Dollar erreichen, kann der Vertrag von Schuster sofort und einseitig gelöst werden. Die Musterkollektionen muss Walter bezahlen bzw. diese werden mit den Provisionsansprüchen verrechnet. Die ausstehenden

Forderungen von Lilly Goldscheider und Rosa Kush werden nicht mehr erwähnt.[344]

Am 14. Dezember 1938 werden letztendlich die Verträge unterschrieben. Zwischen der Österreichischen Kontrollbank für Industrie und Handel und Walter Goldscheider wird vereinbart, dass das Unternehmen mit allen Aktiva und Passiva sowie den Liegenschaften zu einem Kaufpreis von 50.000 Reichsmark an die Kontrollbank verkauft wird. Davon muss Walter Goldscheider 20.000 Reichsmark als Reichsfluchtsteuer zahlen, 8.000 Reichsmark werden für Abgaberückstände und Personalsteuer gesperrt. So bleiben ihm vom Verkauf seines traditionsreichen und erfolgreichen Familienunternehmens nur 22.000 Reichsmark übrig – das sind 31.680 Schilling, also nicht mal 5 Prozent des Umsatzes von 1937. Unter Punkt XI wird in diesem Vertrag auch festgehalten, dass der Vertrag wegen Verletzung über die Höhe des wahren Wertes nicht anzufechten sei.[345] Weiterhin wird auch festgehalten, dass die Bank berechtigt ist, alle Rechte und Ansprüche aus diesem Vertrag an Josef Schuster zu übertragen.

Ebenfalls am 14. Dezember bestätigt Schuster ein Abkommen mit Walter Goldscheider, in dem er Walter und Lilly Goldscheider die Alleinvertretung der Erzeugnisse für das Gebiet der „Vereinigten Staaten von Nordamerika und Canada" überträgt. Zum Abkommen vom 9. Dezember kommt jedoch hinzu, dass die von Walter gesandten Aufträge nach Schusters Ermessen bearbeitet werden, d.h. Schuster hat es letztendlich in der Hand, ob Walter den Mindestumsatz erreichen kann oder nicht.[346] Somit sind Walter und Lilly Goldscheider zu Vertretern ihres Wiener Unternehmens geworden, während Schuster sich durch die Arisierung vom Vertreter zum Unternehmer gemausert hat.

Am gleichen Tag erfolgt auch der Vertrag zwischen der Kontrollbank und Josef Schuster, in dem die Bank alle Rechte und Ansprüche an der Wiener Manufaktur zum Pauschalpreis von 50.000 Reichsmark an Josef Schuster verkauft. Dieser Kaufpreis soll laut Abschnitt II bereits vom Käufer, also Schuster, bezahlt worden sein. Die Liegenschaften Staudgasse 7 und 9 werden unter Abschnitt III ebenfalls mit 50.000 Reichsmark angegeben.[347] Die genannten Summen aus den verschiedenen Dokumenten sind verwirrend und deuten auf Unstimmigkeiten hin.[348] Erst aus einem Brief von Rechstanwalt Dr. Höpler an die Kontrollbank wegen Umsatzsteuerzahlung und Liegenschaftsge-

agreed that the business with all liquid and illiquid assets as well as the properties would be sold to the Kontrollbank for the sum of 50 000 Reich marks. From that sum Walter Goldscheider had to pay 20 000 Reich marks as Reichsfluchtsteuer [tax on fleeing the Reich]; 8000 Reich marks were frozen for tax debts and personal income tax. As a result, all that remained to him from the sale of what had been such a successful family business that looked back on a distinguished tradition were 22 000 Reich marks – the equivalent of 31 680 Schillings, that is, not even 5 per cent of the turnover in 1937. Under clause XI (an escape clause) of that contract, it was established that the contract could not be contested even in the case of infringements pertaining to the real value [of the business and property as assessed].[345] Further, it was also established that the Kontrollbank was entitled to transfer all rights and claims arising from the contract to Josef Schuster.

On 14 December, Schuster also confirmed an agreement with Walter Goldscheider, under the terms of which he transferred to Walter and Lilly Goldscheider the sole right to represent the wares in the territories of "the United States of America and Canada." It had, however, by then been appended to the agreement of 9 December that orders sent from Walter were to be processed as Schuster saw fit, ie, it lay with Schuster alone to determine whether Walter could attain the minimum turnover or not.[346] Thus Walter and Lilly Goldscheider became representatives for their Viennese business whereas Schuster rose from salesman to entrepreneur through Aryanization.

That same day the contract between the Kontrollbank and Josef Schuster was concluded, under the terms of which the Kontrollbank sold all rights and claims to the Viennese manufactory to Josef Schuster for the sum of 50 000 Reich marks. This price had, according to paragraph II, already been paid by the purchaser, that is, Schuster. The properties at 7 and 9 Staudgasse were also valued in paragraph II at 50 000 Reich marks.[347] The sums mentioned in the various documents are confusing and indicate that there were discrepancies.[348] Not until a letter was sent by Schuster's lawyer, Dr. Höpler, to the Kontrollbank about paying turnover tax and stamp tax did it become clear that, at the sale of the business, a sum total of 65 000 Reich marks was to be paid, of which 50 000 were to be paid for the property and

15 000 for the business.[349] After that, the letters flew back and forth between Schuster's lawyer and the Kontrollbank because Schuster did not agree to the amount of turnover tax to be paid.

By the first week in December, a letter from Walter Bosse also arrived at the Kontrollbank, in which Bosse requested to be informed on the stage the Aryanization proceedings had reached by then and demanding an explanation of why his offer had been deferred.[350] No answer to this query is recorded; however, astonishingly, as early as 15 September 1938, Dr. Mayer had recorded an interesting remark in his documentation of the Goldscheider Aryanization proceedings from an off-the-record discussion with the directors of the Austrian Kontrollbank, Dr. Georg Schumetz and Dr. Hermann Leitich. At that discussion, it was decided that Bosse could not be considered eligible to buy the firm. At best, he might be considered eligible for employment as a managing director following Aryanization or, alternatively, Dr. Schwartz, the provisional administrator, who "has been running the business now for quite some time and [has] experience in the field of bronze sculpture."[351] These notes suggest that Schuster had friends in high places and, further, with the Aryanization of the Goldscheider manufactory, other financial intentions were being pursued that would line the pockets of various parties. A further indication is that, in 1941, Ministerialrat Dr. Wacha from the finance and accounting department even initiated an investigation to clarify what had gone on in that particular Aryanization process.[352]

Early in 1939 Walter and his wife Lilly were granted a permit to emigrate to England, where they stayed for a while with friends before going on to the US. Before they emigrated, they paid 6000 Reich marks to the Psychiatrische Anstalt Am Steinhof to cover Camillo Goldscheider's board and lodging.[353] In exile, Walter Goldscheider described Aryanization as follows: "After being discharged from prison, I was forced in 1938 to 'sell' my business with all liquid assets, factory buildings, marks, names, molds, models, etc, for a lump sum of RM 50 000 set by the bank engaged in this transaction. Of this sum, after income, fleeing the Reich, being a Jew and passport taxes were deducted, a remainder of RM 6000 was actually paid out to me to cover the costs of my journey, ship ticket and baggage handling and RM 10 000 each in cash for me and my wife."[354]

bühren wird klar, dass bei dem Verkauf insgesamt 65.000 Reichsmark zu zahlen sind, davon 50.000 für die Liegenschaft und 15.000 für die Manufaktur.[349] Anschließend findet ein reger Verkehr zwischen Schusters Anwalt und der Kontrollbank statt, da Schuster nicht mit der Höhe der zu zahlenden Umsatzsteuer einverstanden ist.

Bereits in der ersten Dezemberwoche traf bei der Kontrollbank ein Brief Walter Bosses ein, in dem er um eine Mitteilung zum Stand der Arisierung bittet sowie um eine Erklärung, warum sein Angebot zurückgestellt worden sei.[350] Auf diese Anfrage ist keine Antwort dokumentiert, doch erstaunlicherweise hielt bereits am 15. September 1938 Dr. Mayer aus einer unverbindlichen Aussprache mit den Direktoren der Österreichischen Kontrollbank, Dr. Georg Schumetz und Dr. Hermann Leitich, eine interessante Anmerkung in einer Aktennotiz zur Goldscheider-Arisierung fest. Bei dieser Aussprache wurde beschlossen, dass Bosse nicht mehr als Käufer in Betracht komme, sondern höchstens zu überlegen sei, ob er nach der Arisierung als Betriebsleiter eingesetzt werden könnte oder weiterhin der kommissarische Verwalter, Dr. Schwartz, der den „Betrieb jetzt seit längerem führt und Erfahrungen im Bereich der Bronzeplastik" hat.[351] Dies sind Hinweise darauf, dass Schuster protegiert wurde und bei der Arisierung der Goldscheider Manufaktur noch andere finanzielle Absichten verfolgt wurden und zur Bereicherung verschiedener Personen führten. Im Jahr 1941 leitet der Ministerialrat Dr. Wacha von der Finanzbuchhaltung deshalb auch ein Verfahren zur Klärung des Arisierungsverfahrens ein.[352]

Walter und seine Frau Lilly erhalten Anfang 1939 die Ausreisebewilligung nach England, wo sie sich kurze Zeit bei Freunden aufhalten und dann in die USA weiterreisen. Vor der Emigration zahlten sie noch 6.000 Reichsmark an die Psychiatrische Anstalt Am Steinhof für die Verpflegungskosten von Camillo Goldscheider.[353] Walter Goldscheider beschrieb im Exil die Arisierung wie folgt: „Nach meiner Entlassung aus dem Gefängnisse wurde ich 1938 zum ‚Verkauf' meines Unternehmens mit allen Aktiven, Fabrikshäusern, Marke, Namen, Formen, Modellen etc. gegen einen seitens der bei dieser Transaktion tätigen Bank bestimmten Pauschalbetrag von RM 50.000,– gezwungen. Von dieser Summe wurde mir tatsächlich nach Abzug der Einkommens-, Reichsflucht-, Juden- und Passsteuer ein Restbetrag von RM 6.000,– zur Deckung meiner Fahrt, Schiffskarte, Gepäck ausgehändigt, und ein mitzunehmender Betrag von je RM 10.000,– für mich und meine Gattin."[354]

Anmerkungen

298 Interview mit Karl Täuber, Wien, 25.11.2003
299 Brief von Walter und Lilly Goldscheider, vertreten durch Rechtsanwalt Dr. Arthur Mayer, an die Vermögensverkehrsstelle, 15.7.1938
300 Brief der Arbeitsgemeinschaft der Manufaktur Goldscheider an das Bundesministerium für Vermögenssicherung und Wirtschaftsplanung, 25.3.1946
301 Brief von Walter und Lilly Goldscheider, vertreten durch Rechtsanwalt Dr. Arthur Mayer, an die Vermögensverkehrsstelle, 15.7.1938. Die Entlassung erfolgt am 17. Juni 1938.
302 Ebenda
303 Brief von Walter Goldscheider an die Vermögensverkehrsstelle, 4.8.1938
304 Beilage von Dr. Schwartz zum Brief von Walter und Lilly Goldscheider, vertreten durch Rechtsanwalt Dr. Arthur Mayer, an die Vermögensverkehrsstelle, 15.7.1938
305 Brief vom Beauftragten der Prüfungsstelle Keramische Industrie in Wien, Herrn von Lichtenberg, an die Gauleitung der N.S.D.A.P, 19.7.1938
306 Brief von Walter und Lilly Goldscheider, vertreten durch Rechtsanwalt Dr. Arthur Mayer, an die Vermögensverkehrsstelle, 15.7.1938
307 Radiogramm von Rosa Kush, New York an die Goldscheidersche Manufaktur, Wien, 22.8.1938
308 Brief von Dr. Fischer vom Verband Deutscher Fabriken für Gebrauchs-, Zier- und Kunstporzellan und Keramik G.m.b.H an Walter Goldscheider, 4.6.1938
309 Brief von Dr. Schwartz an die Vermögensverkehrsstelle, 23.7.1938
310 Brief von Walter Goldscheider an die Vermögensverkehrsstelle, 4.8.1938
311 Ebenda
312 Ebenda. Walter ist 1938 sowohl auf der Frühjahrsmesse wie auch der Herbstmesse in Wien vertreten, vgl. Amtlicher Ausstellerkatalog 1938
313 Ebenda
314 Ebenda
315 Einverständniserklärung von Walter Goldscheider zur Arisierung der Wiener Manufaktur Friedrich Goldscheider an die Vermögensverkehrsstelle, 30.8.1938
316 Brief von Graf Hardegg, Vermögensverkehrsstelle an Direktor Kreis von der Länderbank AG, 1.9.1938
317 Bericht von Dr. Schwartz zur Lage des Unternehmens an die Vermögensverkehrsstelle, 7.9.1938
318 Gutachten des gerichtlich beeideten Sachverständigen, Architekten und Stadtbaumeisters Karl August Benirschke 27.7.1938
319 Bericht von Dr. Schwartz zur Lage des Unternehmens an die Vermögensverkehrsstelle, 7.9.1938
320 Ebenda
321 Expose zur Arisierung von Dr. Stefan Schwartz an die Vermögensverkehrsstelle, 12.8.1938
322 Schätzungsbericht über die Liegenschaften E.Z. 814, Grundbuch Währing, von Josef Heinzle, Architekt, 3.10.1938
323 Etzersdorfer, S. 156
324 Ebenda
325 Notiz von Dr. Arthur Mayer, Rechtsanwalt von Walter Goldscheider, 16.9.1938
326 Ebenda
327 Brief von Walter Bosse an die Österreichische Kontrollbank in Wien, 5.11.1938
328 Brief von F.R. Uffert, Berlin-Wilmersdorf an die Österreichische Kontrollbank, 23.9.1938
329 Brief von Walter Bosse an die Österreichische Kontrollbank, 19.10.1938
330 Brief von Josef Schuster an die Österreichische Kontrollbank in Wien, 14.10.1938
331 Brief mit Informationen zu Josef Schuster von der Merck, Finck & Co Bank in München, an die Österreichische Kontrollbank in Wien, 7.11.1938. Aus einer bereits am 22.9.1938 erfolgten Bestätigung durch die Nationalsozialistische Deutsche Arbeiterpartei Gauleitung Franken an die Arisierungskommission beim Gau Tirol, Innsbruck, geht hervor, dass gegen „Josef Schuster und dessen Ehefrau, von April 1928 bis April 1938 wohnhaft in Dinkelsbühl, jetzt wohnhaft in München, Vandalenstrasse 15, während ihres Aufenthaltes in Dinkelsbühl in politischer Beziehung nichts nachteiliges bekannt wurde." Ob Schuster sich bereits früher für die Arisierung interessiert hat und deshalb die Nachforschungen angestellt worden sind, bleibt unbelegt. Diese Bestätigung könnte auch aus einem anderen Arisierungsverfahren stammen, das Schuster in Tirol übernehmen wollte. In einer späteren Benachrichtigung vom 7. Januar 1939 teilt die Gauleitung der NSDAP in München dem Minister für Wirtschaft und Arbeit in Wien mit, dass Erkundigungen über Josef Schuster in München zu keinem Ergebnis geführt haben und dass er in Wirtschaftskreisen nicht bekannt sei.
332 Telegramm von Josef Schuster an die Österreichische Kontrollbank in Wien, 1.11.1938
333 Brief von Walter Bosse an die Österreichische Kontrollbank in Wien, 5.11.1938. Laut anderer Quellen wurde die Moderne Keramik 1928 eingeführt.
334 Ebenda
335 Ebenda
336 Brief mit Zusammenfassung der Arisierung der Wiener Manufaktur Friedrich Goldscheider an Dr. Hans Fischböck, Minister für Wirtschaft und Arbeit, als Vorsitzender des Beirates bei der Vermögensverkehrsstelle, 9.11.1938; die Unterschrift ist unleserlich.
337 Ebenda
338 Beschlussprotokoll über die Beiratssitzung der Vermögensverkehrsstelle vom 10.11.1938 unter dem Vorsitz des Ministers Dr. Hans Fischböck.
339 Brief von Josef Schuster an Walter und Lilly Goldscheider, 28.11.1938
340 Ebenda
341 Ebenda
342 Brief von Dr. Ernst Höpler, RA von Josef Schuster an Dr. Arthur Mayer, RA von Walter Goldscheider, 29.11.1938
343 Gedächtnisprotokoll von Dr. Ernst Höpler, RA von Josef Schuster an die Österreichische Kontrollbank, zu einem Treffen zwischen Vertretern der Kontrollbank und Josef Schuster vom 9.12.1938. Die genannte und sehr niedrige Summe von 15.000 RM sorgt einige Jahre später für Nachforschungen seitens der NSDAP, da man Steuerbetrug vermutet.
344 Abkommen zwischen Josef Schuster und Walter Goldscheider vom 9.12.1938
345 Kaufvertrag zwischen der Österreichischen Kontrollbank für Industrie und Handel in Wien und Walter Goldscheider, als Alleininhaber der Manufaktur Friedrich Goldscheider in Wien, 14.12.1938
346 Abkommen zwischen Josef Schuster und Walter Goldscheider vom 14.12.1938
347 Kaufvertrag zwischen der Österreichischen Kontrollbank für Industrie und Handel, Wien und Josef Schuster, München, 14.12.1938
348 Im Gedächtnisprotokoll vom 9. Dezember wird von 15.000 RM gesprochen, im Vertrag vom 14. Dezember heißt es, der Betrag wäre bereits bezahlt. Bei einer Nachuntersuchung der Arisierungsvorgänge wird ersichtlich, dass Schuster nur 5.000 RM gezahlt hat und die restlichen 10.000 RM in Monatsraten abzahlen sollte.
349 Brief von Dr. Ernst Höpler an die Österreichische Kontrollbank, 31.1.1939
350 Brief von Walter Bosse an die Österreichische Kontrollbank, 6.12.1938
351 Aktennotiz von Dr. Arthur Mayer zur Goldscheider-Arisierung aus einer unverbindlichen Aussprache mit den Direktoren der Österreichischen Kontrollbank, Dr. Georg Schumetz und Dr. Hermann Leitich, 15.9.1938
352 Siehe Kapitel „Die Manufaktur Josef Schuster in den Kriegsjahren", S. 170ff
353 Etzersdorfer, S. 164
354 Ebenda, S. 163ff

Notes

[298] Interview with Karl Täuber, Vienna, 25 Nov. 2003
[299] Letter from Walter and Lilly Goldscheider, represented by Dr. Arthur Mayer, lawyer, to the Vermögensverkehrsstelle [Office for Property Tax], 15 July 1938
[300] Letter from the work group of Manufaktur Goldscheider to the Bundesministerium für Vermögenssicherung und Wirtschaftsplanung [Federal Ministery for Securing Wealth and Economic Planning], 25 March 1946
[301] Letter from Walter and Lilly Goldscheider, represented by Dr. Arthur Mayer, lawyer, to the Vermögensverkehrsstelle, 15 Aug. 1938. They were discharged on 17 June 1938.
[302] Ibid
[303] Letter from Walter Goldscheider to the Vermögensverkehrsstelle, 4 Aug. 1938
[304] Enclosure from Dr. Schwartz in the letter from Walter and Lilly Goldscheider, represented by Dr. Arthur Mayer, lawyer, to the Vermögensverkehrsstelle, 15 July 1938
[305] Letter from the authorized representative of the Prüfungsstelle Keramische Industrie [Inspectorate of the Ceramics Industry] in Vienna, Herr von Lichtenberg, to the Gauleitung [district authorities] of the N.S.D.A.P, 19 July 1938
[306] Letter from Walter and Lilly Goldscheider, represented by Dr. Arthur Mayer, lawyer, to the Vermögensverkehrsstelle, 15 July 1938
[307] Wire from Rosa Kush, New York, to the Goldscheider Manufaktur, Vienna, 22 Aug. 1938
[308] Letter from Dr. Fischer from the Verband Deutscher Fabriken für Gebrauchs-, Zier- und Kunstporzellan und Keramik G.m.b.H to Walter Goldscheider, 4 June 1938
[309] Letter from Dr. Schwartz to the Vermögensverkehrsstelle, 23 July 1938
[310] Letter from Walter Goldscheider to the Vermögensverkehrsstelle, 4 Aug. 1938
[311] Ibid
[312] Ibid. Walter was represented with exhibits in 1938 both at the Spring and the Autumn Fairs in Vienna: cf Amtlicher Ausstellerkatalog 1938
[313] Ibid
[314] Ibid
[315] Declaration of agreement to Aryanization made by Walter Goldscheider with respect to Aryanizing the Viennese Manufactory Friedrich Goldscheider to the Vermögensverkehrsstelle, 30 Aug. 1938
[316] Letter from Count Hardegg, Vermögensverkehrsstelle, to Director Kreis of the Länderbank AG, 1 Sept. 1938
[317] Report made by Dr. Schwartz on the condition of the business to the Vermögensverkehrsstelle, 7 Sept. 1938
[318] Expert appraisal by Karl August Benirschke, court appointed and sworn expert, architect and urban planner, on 27 July 1938
[319] Report made by Dr. Schwartz on the condition of the business to the Vermögensverkehrsstelle, 7 Sept. 1938
[320] Ibid
[321] Summary of Aryanization proceedings made by Dr. Stefan Schwartz to the Vermögensverkehrsstelle, 12 Aug. 1938
[322] Appraisal report on the properties E.Z. 814, Grundbuch Währing, by Josef Heinzle, Architect, 3 Oct. 1938
[323] Etzersdorfer, p. 156
[324] Ibid
[325] Note by Dr. Arthur Mayer, Walter Goldscheider's lawyer, 16 Sept. 1938
[326] Ibid
[327] Letter from Walter Bosse to the Österreichische Kontrollbank in Vienna, 5 Nov. 1938
[328] Letter from F.R. Uffert, Berlin-Wilmersdorf, to the Österreichische Kontrollbank, 23 Sept. 1938
[329] Letter from Walter Bosse to the Österreichische Kontrollbank, 19 Oct. 1938
[330] Letter from Josef Schuster to the Österreichische Kontrollbank in Vienna, 14 Oct. 1938
[331] Letter with information on Josef Schuster from the Merck, Finck & Co Bank in Munich to the Österreichische Kontrollbank in Vienna, 7 Nov. 1938. A statement of verification sent on 22 Sept. 1938 by the NSDAP Gauleitung Franken to the Arisierungskommission [Aryanization Commission] of Gau Tirol, Innsbruck, confirms that "nothing to their disadvantage has become known about Josef Schuster and his wife, domiciled from April 1928 to April 1938 in Dinkelsbühl, now domiciled in Munich, at 15 Vandalenstrasse 15, with respect to politics during their stay in Dinkelsbühl." There is no written evidence that Schuster had previously been interested in the Aryanization of the Goldscheider manufactory and had for that reason initiated enquiries. This confirmation might just as easily come from Aryanization proceedings conducted on another business in the Tyrol that Schuster wanted to take over. In a later message dated 7 January 1939, the NSDAP Gauleitung in Munich informs the Minister for Economics and Labor in Vienna that enquires on Josef Schuster in Munich had revealed nothing and that he was unknown in business circles.
[332] Telegram from Josef Schuster to the Österreichische Kontrollbank in Vienna, 1 Nov. 1938
[333] Letter from Walter Bosse to the Österreichische Kontrollbank in Vienna, 5 Nov. 1938. According to other sources, the Modern Ceramics were introduced at Goldscheider in 1928.
[334] Ibid
[335] Ibid
[336] Letter with the summary of the Aryanization proceedings at the Viennese Manufactory Friedrich Goldscheider to Dr. Hans Fischböck, Minister for Economics and Labor, as chairman of the Advisory council at the Vermögensverkehrsstelle, 9 Nov. 1938; the signature is illegible.
[337] Ibid
[338] Minutes on the resolutions passed at the Advisory Council meeting at the Vermögensverkehrsstelle of 10 Nov. 1938, chaired by the minister, Dr. Hans Fischböck.
[339] Letter from Josef Schuster to Walter and Lilly Goldscheider, 28 Nov. 1938
[340] Ibid
[341] Ibid
[342] Letter from Dr. Ernst Höpler, Josef Schuster's lawyer, to Dr. Arthur Mayer, Walter Goldscheider's lawyer, 29 Nov. 1938
[343] Memorandum by Dr. Ernst Höpler, Josef Schuster's lawyer, to the Österreichische Kontrollbank on a meeting between representatives of the Kontrollbank and Josef Schuster on 9 Dec. 1938. The very low sum of 15 000 RM mentioned gave rise to an investigation initiated by the NSDAP because they suspected tax fraud.
[344] Agreement between Josef Schuster and Walter Goldscheider of 9 Dec. 1938
[345] Bill of sale concluded between the Österreichische Kontrollbank für Industrie und Handel in Vienna and Walter Goldscheider, as sole proprietor of the Manufactory Friedrich Goldscheider in Vienna, 14 Dec. 1938
[346] Agreement between Josef Schuster and Walter Goldscheider of 14 Dec. 1938
[347] Bill of sale between the Österreichische Kontrollbank für Industrie und Handel, Vienna and Josef Schuster, Munich, 14 Dec. 1938
[348] In the memorandum of 9 December, RM 15 000 are mentioned; in the bill of sale of 14 December, it says the sum has already been paid. An investigation of the Aryanization proceedings revealed that Schuster had only paid RM 5000 and was supposed to pay the remainder amounting to RM 10 000 in monthly installments.
[349] Letter from Dr. Ernst Höpler to the Österreichische Kontrollbank, 31 Jan. 1939
[350] Letter from Walter Bosse to the Österreichische Kontrollbank, 6 Dec. 1938
[351] Note in Dr. Arthur Mayer's file on the Goldscheider Aryanization proceedings on an off-the-record discussion with Dr. Georg Schumetz and Dr. Hermann Leitich, directors of the Österreichische Kontrollbank, 15 Sept. 1938
[352] See chapter "The Josef Schuster Manufactory during the War Years", p. 170f
[353] Etzersdorfer, p. 164
[354] Ibid, p. 163f

Die Manufaktur Josef Schuster in den Kriegsjahren (1939–1944)

The Josef Schuster Manufactory during the War Years (1939–1944)

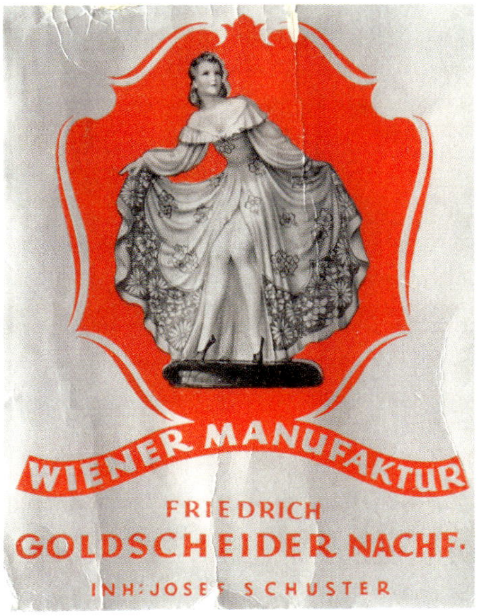

Ill. 90
Das erste unter Josef Schuster produzierte Modell ist die Figur „Jolanthe" mit der Modell-Nr. 8440, die auch auf dem Umschlag des Schuster-Katalogs abgebildet ist. (Quelle: Archiv Goldscheider)

The first model produced under Josef Schuster was the figurine "Jolanthe" as model number 8440, which was also shown on the cover of the Schuster catalogue. (Source: Goldscheider Archives)

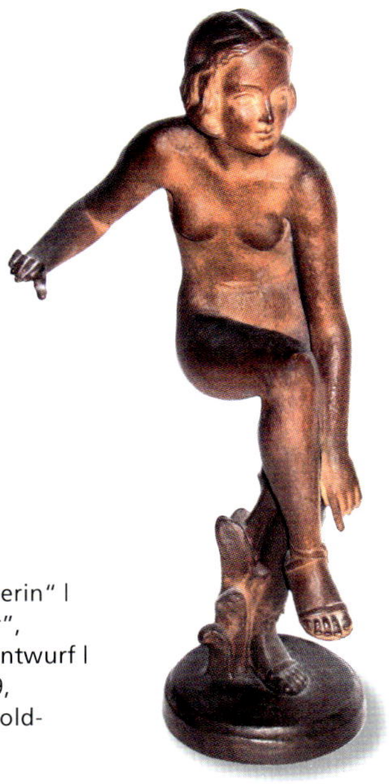

ModNr. 8702
„Sandalenbinderin" I "Sandal Binder",
Adolf Prischl, Entwurf I design ca. 1939,
Collection a@goldscheider.net

Am 27. Januar 1939 wird das Wiener Unternehmen in „Wiener Manufaktur Friedrich Goldscheider Nachf., Inhaber Josef Schuster" umbenannt und Lilly Schuster als Einzelprokuristin eingetragen.[355] Der 35-jährige Josef Schuster wird den Mitarbeitern als neuer Inhaber und kommerzieller Leiter vorgestellt, seine Frau Lilly ist für die künstlerische Leitung zuständig.[356] Einige Mitarbeiter werden entlassen sowie zum Militärdienst eingezogen und Schuster stellt neue, ihm vertraute Leute ein. Auch ein Ziehsohn der Schusters wird als Praktikant eingestellt.[357] Der seit 21. März 1922 bei Goldscheider beschäftigte Kerammaler und spätere Leiter der Unterglasurabteilung Alois Teichgrab wird unter Schuster zum Abteilungsleiter für die gesamte Spritzerei, Dessinage und den Brennhausbetrieb eingesetzt. In dieser Funktion obliegt ihm der Entwurf neuer Dekore, die Zusammenstellung der Glasuren und Farben für Aufglasur- und Unterglasurmalerei, die Einführung neuer technischer Arbeitsvorgänge sowie die „Abrichtung der Lehrlinge" und verschiedene administrative Aufgaben.[358] Im Unternehmen ist auf Anordnung von Schuster nur noch mit „Heil Hitler" zu grüßen, bei Reden von Hitler wird stets eine Betriebsversammlung einberufen und alle Mitarbeiter müssen dem Führer im Radio zuhören.[359]

Eines der ersten unter Schuster produzierten Modelle ist die Fayencefigur „Jolanthe" (Modell-Nr. 8440, Ill. 90).[360] Die neuen Modelle werden im Laufe der Kriegsjahre bevorzugt in Terrakotta hergestellt, da Fayence teuer und in der Produktion anspruchsvoller ist. Es wird auch schwieriger, den hochwertigen Quarz für den weißen Fayence-Scherben zu beziehen. Die entworfenen Modelle in der Schuster-Ära entsprechen größtenteils dem völkischen Vorbild der Nationalsozialisten: Es sind Aktfiguren von athletischen Männern, wie der „Diskuswerfer" (Modell-Nr. 8563), oder Frauen mit geflochtenen Zöpfen und zünftiger Erscheinung, wie die „Sandalenbinderin" (Modell-Nr. 8503; 8702). Einige der „zeitgemäßen" Modelle werden von Lilly Schuster persönlich entworfen, die selbst brünett ist und Zöpfe trägt und genauso wie ihr Mann nationalsozialistisch gesinnt ist.[361] Die hergestellten Objekte führen zunächst den Goldscheider-Stempel und die seit dem „Anschluss" offizielle Markung „Made in

On 27 January 1939, the Viennese business was renamed "Wiener Manufaktur Friedrich Goldscheider Nachf., Inhaber Josef Schuster" ["Viennese Manufactory Friedrich Goldscheider, Successor, Proprietor Josef Schuster"] and Lilly Schuster was registered as sole head clerk with power of attorney.[355] Josef Schuster, who was only thirty-five at the time, was introduced to the workforce as the new proprietor and managing director while his wife Lilly was artistic director.[356] Some employees were dismissed or inducted into the military and Schuster hired new people from among his familiars to replace them. An adoptive son of the Schusters' was also employed as an intern.[357] Alois Teichgrab, who had been employed at Goldscheider since 21 March 1922 as a ceramic painter and later as head of the underglaze department, was appointed division manager for the entire spray workshop, draughting and kiln operations. In this capacity, he was responsible for the designing of new decoration, the mixing of glazes and underglaze painting, implementing new technical procedures and "training apprentices" as well as various administrative tasks.[358] Schuster had issued orders that greetings had to be prefaced with "Heil Hitler!," a works council meeting was to be convened when Hitler was giving a speech and all employees had to listen to the Führer on the radio.[359]

One of the first models produced under Schuster's tenure was the faience figure "Jolanthe" (model number 8440, Ill. 90).[360] The new models were usually produced in terracotta during the war years because faience was more expensive and difficult to produce. It also became more difficult to obtain top-quality quartz to make white faience body. The models designed in the Schuster era pretty much corresponded to the purely Teutonic ideal propagated by the National Socialists: they were nude athletic male figures, such as a "Diskuswerfer" ["Discobolus"] (model number 8563) or wholesome-looking women with braided hair, such as the "Sandalenbinderin" ["Sandal Binder"] (model number 8503; 8702). Some of those "contemporary" models were designed by Lilly Schuster herself. She was a brunette but wore her hair in braids and, like her husband, had National Socialist sympathies.[361] The objects made at first bore the Goldscheider stamp and the mark "Made in Germany," which had been official since the "Anschluss." Under Schuster, however, only two numbers continued in use for the most part: the model number and the repairer number. Batch numbers were no longer used. The reason for this change is unknown but presumably this measure made it more difficult for the administration or at least not so easy to ascertain how many pieces were produced and sold. The result would have been lower taxes and lower royalty payments to artists.

In 1940 a catalogue produced by Schuster was issued featuring all models on offer (Ill. 91). It was added to in 1942 and reissued. The models were for the most part taken from the Walter Goldscheider era, especially those from the 1930s bearing model numbers from 7000 in ascending order. The few 1920s models still sold under Schuster were the faience figures "Gefangener Vogel" ["Captured Bird"] (in versions 5230, 5960, 6204, 6600 and 8176), "Kokettchen" ["Flirt"] (5802), a "Windhund" ["Whippet"] (5974) and "Wolfshund liegend" ["Alsatian, Couchant"] (5883, 5458, 5541, 5542) as well as a much older "Setter" model (3024).[362] There were also some models from the 6000 series, mainly masks and animals. The Schuster catalogue, which consisted of loose sheets, links the models to photos, informs readers on model numbers and names, gives the basic price in Schillings as well as Reich marks, English Shillings and US dollars and explains about special orders. The models were arranged by subject area. Under the heading "Dance and other figures, children, nudes in faience," for instance, ninety-one models are listed. Nude

Germany". Allerdings werden unter Schuster meistens nur noch zwei Nummern eingesetzt: die Modellnummer und die Retuscheurnummer. Die Stückzahlnummer fällt weg. Der Grund ist unbekannt, aber anzunehmen ist, dass aufgrund dieser Maßnahme die Administration erschwert worden ist bzw. nicht mehr so einfach verfolgt werden konnte, wie viel Stücke produziert und verkauft worden sind. Dies dürfte zu niedrigeren Steuerabgaben und geringeren Tantiemenzahlungen an die Künstler geführt haben.

Um das Jahr 1940 entsteht ein von Schuster produzierter Katalog (Ill. 91) mit allen angebotenen Modellen, der 1942 ergänzt und neu aufgelegt wird. Die Modelle sind größtenteils noch aus der Ära Walter Goldscheiders, vor allem aus den dreißiger Jahren mit den Modellnummern von 7000 aufwärts. Die wenigen Modelle aus den zwanziger Jahren, die auch unter Schuster verkauft werden, sind die Fayencefiguren „Gefangener Vogel" (in den Ausführungen 5230, 5960, 6204, 6600 und 8176), die Figur „Kokettchen" (5802), ein „Windhund" (5974) und „Wolfshund liegend" (5883, 5458, 5541, 5542) sowie ein viel älteres Modell eines „Setters" (3024).[362] Es finden sich auch einige Modelle aus der 6000 Reihe, vor allem Masken und Tierfiguren. Der aus einzelnen Blättern bestehende Schuster-Katalog gibt an, auf welchem Foto das Modell zu sehen ist, informiert über die Modellnummer und Bezeichnung des Modells, zeigt den Grundpreis in Schilling sowie die Preise in Reichsmark, englischen Shilling und US-Dollar und klärt über Sonderanfertigungen auf. Die Modelle sind nach mehreren Bereichen aufgeteilt. Unter „Tanz- und andere Figuren, Kinder, Akte in Fayence" sind bei-

Ill. 91
Der von Schuster eingesetzte Katalog zeigt die in den vierziger Jahren angebotene Produktpalette. Die meisten Modelle sind von Walter Goldscheider übernommen, es finden sich aber auch einige neue, der NS-Ästhetik entsprechende Objekte, die unter der künstlerischen Leitung von Lilly Schuster entstanden oder sogar selbst von ihr entworfen wurden. (Quelle: Archiv Goldscheider)

The catalogue used by Schuster shows the product range sold in the 1940s. Most models were taken over from Walter Goldscheider; there were, however also some new objects matching the National Socialist aesthetic, which were developed under Lilly Schuster as artistic director or even designed by her personally. (Source: Goldscheider Archives)

Bild Nr.	Modell Nr.	Benennung	Grundpreis	Preise in RM	engl. sh	USA.S
	6204F	Figur Gefangener Vogel	25.—	38.—	31 6	7.—
	6600F	Figur Gefangener Vogel	12.—	18.50	15 9	3.40
	8176F	Figur Gefangener Vogel	23.—	35.75	31 6	6.50

Fayenceakte können auch in „matt" hergestellt werden, wofür ein Zuschlag von 10 % berechnet wird.

Fayencefiguren können als Lampe montiert geliefert werden. Die Berechnung ist nach folgendem Schlüssel vorzunehmen:
Für den Fayence-Säulenansatz 5 % Zuschlag zum Figurenpreis.
Für die weitere Ausführung sind drei Preisgruppen vorgesehen, und zwar

	Grundpreis	RM	engl. sh	USA.S
Preisgruppe I: Preislagen bis . . .	16.—	24.25	22 -	4.50
Preisgruppe II: Höhere Preislagen bis	30.—	46.25	41 -	8.40
Preisgruppe III: Alle Preise darüber hinaus				
Zuschläge bei Lampen mit Fayencesäulenansatz und Chrommontage, jedoch **ohne** Schirm, Schnur, Stecker und Glühlampe:				
Preisgruppe I	6.—	9.—	7 6	1.70
Preisgruppe II	7.—	10.50	9 3	2.—
Preisgruppe III	14.—	21.—	18 9	3.90
Zuschläge bei Lampen mit Fayencesäulenansatz, Chrommontage und Papierfaltschirm, **ohne** Schirm, Schnur, Stecker und Glühlampe:				
Preisgruppe I	10.—	15.25	13 -	2.80
Preisgruppe II	12.—	18.50	16 6	3.40
Preisgruppe III	21.—	32.50	28 3	5.90
Zuschläge für Papierfaltschirme				
Preisgruppe I	4.—	6.—	5 -	1.10
Preisgruppe II	5.—	8.—	6 9	1.40
Preisgruppe III	7.50	10.50	9 3	2.—

Wiener-Manufaktur Friedrich Goldscheider Nachf., Wien XVIII. Staudgasse 7-9

Tanz- und andere Figuren, Kinder, Akte in Fayence

Bild Nr.	Modell Nr.	Benennung	Grundpreis	Preise in RM	engl. sh.	USA.S
1	7063F	Figur Furchtsame	11.50	17.75	15 -	3.30
	7063AF	Fig. Furchtsame ohne Hund	9.50	14.75	12 3	2.70
2	7218F	Gruppe Jugend	11.—	17.—	14 3	3.20
3	7594F	Figur Joujou	12.—	18.50	16 6	3.40
4	7663F	Gruppe Naschkätzchen	15.—	23.—	20 9	4.20
5	7664F	Gruppe Komm mit!	13.—	20.—	17 3	3.70
6	8503F	Figur Sandalenbinderin	27.50	41.50	37 -	7.75
7	6210F	Figur Bubi	20.—	31.—	25 6	5.60
8	6330F	Figur Kongreß	30.—	46.25	39 -	8.40
	6646F	Figur Kongreß	11.—	17.—	14 3	3.20
9	7290F	Figur Nora	7.—	10.50	9 -	2.—
	7285F	Figur Walzer	7.—	10.50	9 -	2.—
	7286F	Figur Alt-Wien	7.—	10.50	9 -	2.—
10	7149F	Figur Skiläuferin	19.—	28.75	24 -	5.30
	8157F	Figur Skiläuferin	14.—	21.—	18 9	3.90
11	7192F	Figur Revue	11.—	17.—	14 3	3.20
	7396F	Figur Revue	50.—	77.—	67 9	14.—
12	7195F	Figur Beauty	40.—	61.—	51 -	11.20
	7996F	Figur Beauty	28.—	43.—	37 9	7.90
	7897F	Figur Beauty	16.—	24.25	22 -	4.50
13	7280F	Figur Ballnacht	30.—	46.25	39 -	8.40
14	7327F	Gruppe Komtesse m. Hund	36.—	54.50	46 6	10.10
	8145F	Gruppe Komtesse m. Hund	20.—	31.—	26 9	5.60
15	7367F	Gruppe Maskerade	56.—	86.—	75 6	15.70
	8155F	Gruppe Maskerade	35.—	53.50	47 3	9.80
16	7390F	Figur Parisienne	34.—	51.50	45 9	9.50
	8138F	Figur Parisienne	16.—	24.25	22 -	4.50
17	7416F	Figur Golf	20.—	31.—	26 9	5.60
	8140F	Figur Golf	14.—	21.—	18 9	3.90
18	7426F	Figur Sterbender Schwan	50.—	77.—	67 9	14.—

Ill. 92
Ende der 1930er Jahre werden verstärkt Wandmasken in kleinster Ausführung von 9 cm Höhe angeboten. Diese sind preislich für jedermann erschwinglich. (Quelle: Archiv Goldscheider)

The late 1930s saw an increase in the production of wall masks in the smallest version, which was 9 cm high. Everyone could afford them. (Source: Goldscheider Archives)

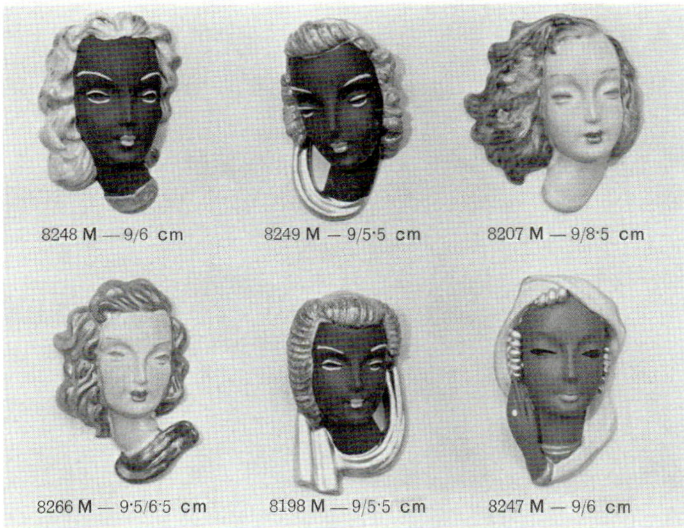

spielsweise 91 Modelle verzeichnet. Fayenceakte können bei einem Zuschlag von 10 Prozent auch „matt" hergestellt werden, Fayencefiguren können als Lampe montiert geliefert werden. Hierbei wird jedoch für den Fayence-Säulenansatz ein Zuschlag von 5 Prozent zum Figurenpreis berechnet und je nach Preisgruppe ein weiterer Zuschlag, abhängig davon, ob beispielsweise ein Papierfaltschirm mitgeliefert werden soll. Die Lampen werden stets ohne Schirm, Schnur, Stecker und Glühlampe geliefert.

Tierfiguren aus Fayence sind unter „Hunde, Pferde und andere Fayencetiere" aufgeführt und umfassen 71 Modelle, bei denen es in einer Ausführung ganz in Weiß 25 Prozent Nachlass gibt. Unter der Rubrik „Figuren und Tiere in moderner Keramik" gibt es 72 Modelle, die nicht aus Fayence, sondern Terrakotta sind. Die Aktfiguren werden größtenteils in brauner oder schwarzer Terrakotta hergestellt, wobei die Haare und

figures in faience could be ordered in "matt" but carried a 10 per cent surcharge. Faience figures were also available mounted as lamps. Here the surcharge for a faience shaft only amounted to 5 per cent of the figure price. There was another surcharge, depending on the price group and whether a folded paper shade was to be delivered. Lamps were always shipped without a shade, flex, plug and bulb.

Faience animal figures were listed under "Dogs, horses and other faience animals" and there were seventy-one different models. There was a 25 per cent discount on models sold entirely in white. Under the heading "Figures and animals in Modern Ceramics" there were seventy-two models, which were made in terracotta rather than faience. The nudes were usually made in brown or black terracotta although the hair and the base were glazed. Figures "made entirely of terracotta" were sold at a 20 per cent discount.[363] Animal figures were usually produced in "crackle glazes (brown, white, yellow, turquoise, green, etc) but also in red, green and black matt glazes. Terriers are mainly delivered in red, white or black but can also be produced in crackle glazes, chiefly white and yellow."[364]

"Masks, heads and reliefs" comprised a separate range of sixty-six models. The masks were either glazed with red, black or green hair or in "brown or black terracotta".[365] There was a 10 per cent surcharge

Einige Beispiele der unter Josef Schuster entstandenen Modelle | Some of the models produced under Schuster's tenure:

ModNr. 8477
Wandmaske | Wall Mask, ca. 1939, Collection f@goldscheider.de

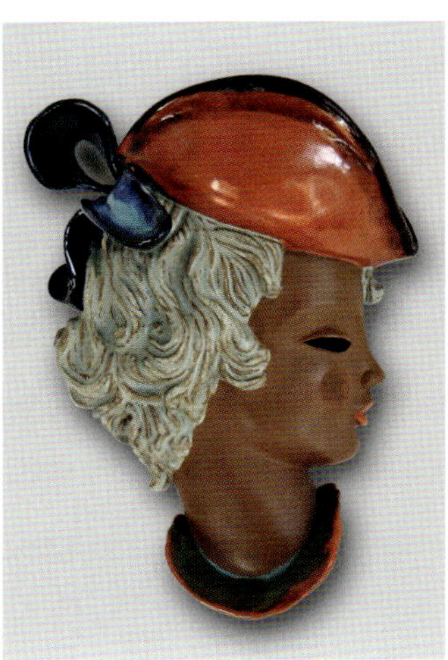

ModNr. 8613
Wandmaske „Rokoko" | Wall Mask "Rococo", ca. 1940, Collection f@goldscheider.de

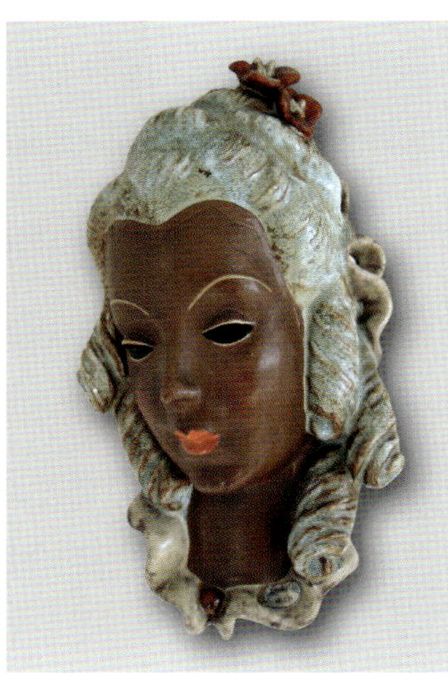

ModNr. 8720, 8721
„Max" und | and „Moritz", ca. 1941, © Wien Museum

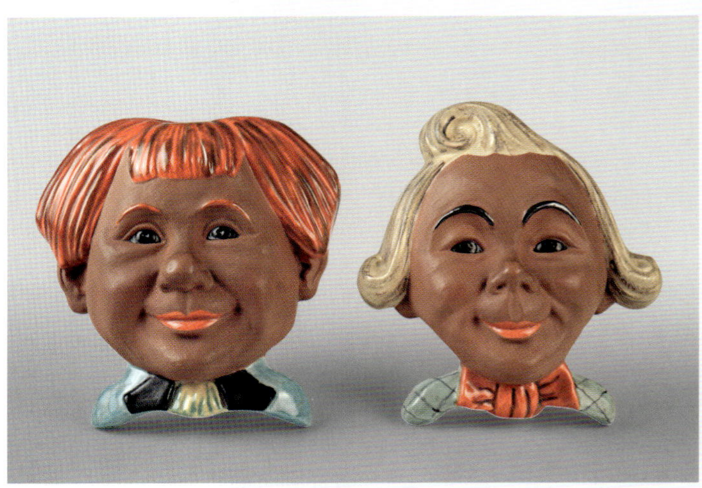

for masks, heads and busts with gold, silver or copper hair. A surcharge of 15 per cent was added for masks, heads and busts with a wooden pedestal. Reliefs, etc, with a wooden frame incurred a surcharge of 60 per cent.

There was a choice of twenty-one "Madonna" models. Special decoration schemes incurred a surcharge: 20 per cent for polychrome decoration (pink and blue, pink, blue and yellow) and for polychrome floral decoration on features such as the Madonna's cloak as well as on Madonna reliefs in cobalt or lavender. In this last case, there was a surcharge of 10 per cent; for gold, silver or copper hair it was 5 per cent.

Finally, utilitarian wares such as "ashtrays, book-ends, candlesticks, bowls and vases" were available, with forty models to choose from. Some ashtrays, such as Models 1277 to 1281, 1300 and 1322 could, if desired, be ordered in a smaller size (from approx 3.5 to 13 cm) and the basic price for such articles was 1.70 Schillings instead of 2.90. Prices for book-ends always implied a pair. On the whole, the catalogue with which Schuster first went on the market encompassed a range of three hundred and sixty-one models; later models were added that were only advertised with photographs (Ill. 93 and 94).

By spring 1939, problems were showing up with the operation of the business at Schuster. He apparently did not have enough

der Sockel glasiert werden. Bei „Herstellung in ganz Terrakotta" wird ein Preisnachlass von 20 Prozent gewährt.[363] Die Tierfiguren werden hauptsächlich in „craquelé Glasuren (braun, weiß, gelb, türkis, grün etc.), aber auch in roten, grünen und schwarzen Mattglasuren erzeugt. Die Terriers werden in der Hauptsache in Rot, Weiß oder Schwarz geliefert, können aber auch in craquelé Glasuren, vorwiegend Weiß und Gelb, erzeugt werden."[364]

Ein eigener Produktbereich sind die „Masken, Köpfe und Reliefs", die ein Sortiment aus 66 Modellen umfassen. Die Masken werden entweder mit rotem, schwarzem oder grünem Haar glasiert, oder in „braun oder schwarz Terrakotta".[365] Bei Gold-, Silber- oder Kupferhaar für Masken, Köpfe und Büsten wird ein Zuschlag von 10 Prozent berechnet. Für die Holzsockel ist bei Masken, Köpfen und Büsten ein Preiszuschlag von 15 Prozent, bei Figuren von 10 Prozent zu errichten. Die Holzrahmenpreise für Reliefs etc. bedeuten einen Preiszuschlag von 60 Prozent.

Bei „Madonnen" besteht die Auswahl aus 21 Modellen. Zuschläge werden für besondere Dekore erhoben: 20 Prozent bei bunten Dekoren (rosa-blau, rosa-blau-gelb) oder bei bunt geblumten sowie wenn der Mantel bei Madonnen und Madonnenreliefs in Kobalt oder Violett sein soll. In letzterem Fall wird ein Zuschlag von 10 Prozent erhoben, für Gold-, Silber- oder Kupferhaare sind es 5 Prozent.

Letztendlich sind auch Gebrauchsgegenstände wie „Ascher, Buchstützen, Leuchter, Schalen und Vasen" mit 40 Modellen vertreten. Einige Ascher wie die Modelle 1277 bis 1281, 1300 und 1322 können auf Wunsch auch in kleinerer Ausführung (von ca. 3,5 bis zu 13 cm) geliefert werden und kosten dann im Grundpreis anstatt 2,90 nur 1,70 Schilling.

ModNr. 8615
Gruppe Kätzchen I
Group of Kittens,
Viktor Matula,
ca. 1940, Collection
a@goldscheider.net

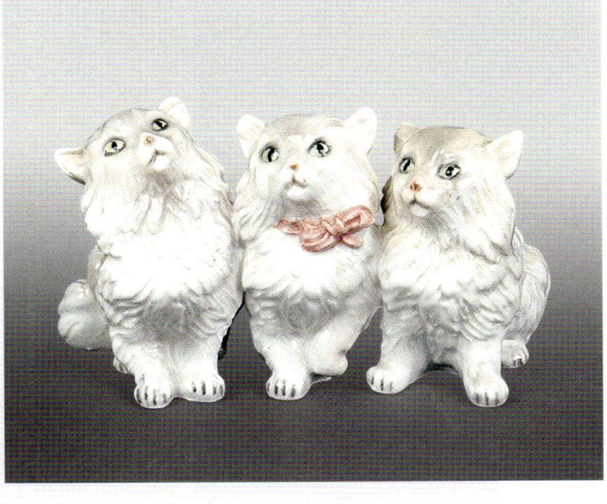

ModNr. 8525
Relief, ca. 1940, Collection f@goldscheider.de

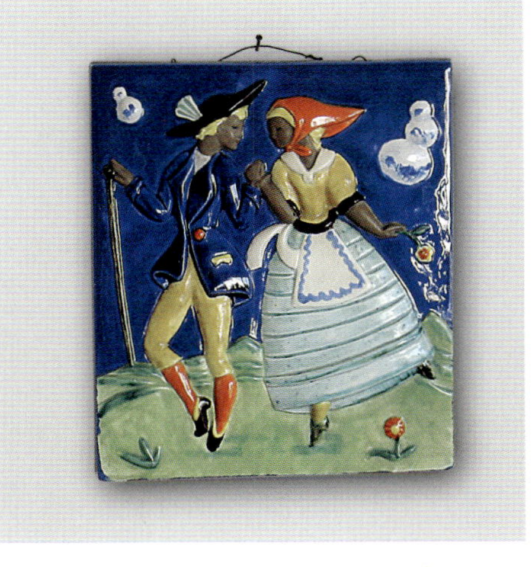

ModNr. 8784
Relief Madonna,
ca. 1942, Collection
f@goldscheider.de

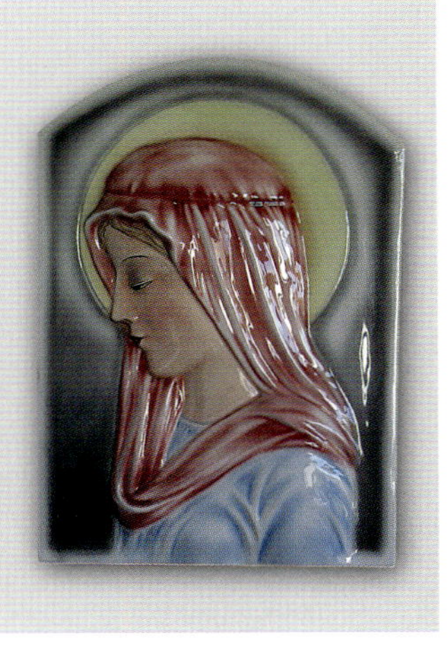

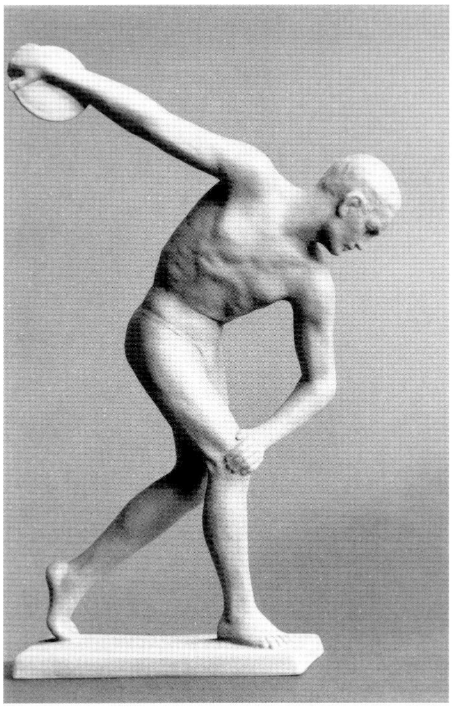

Ill. 93

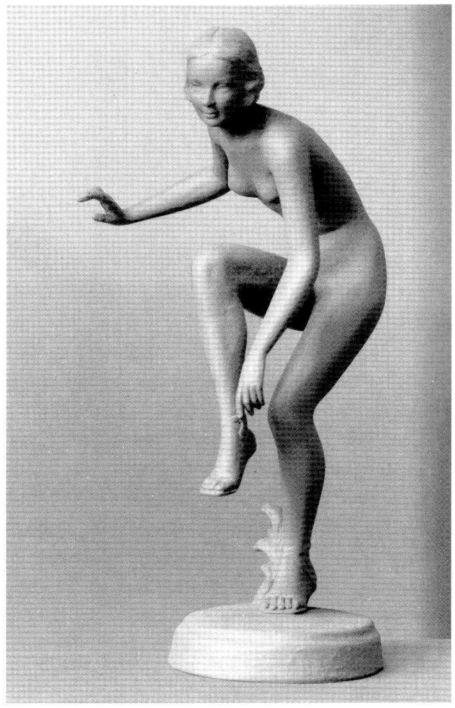

Ill. 94

Ill. 92, 93
Außer dem Katalog setzen die Vertriebsleute auch Fotos ein, vor allem von Objekten, die nicht im Katalog enthalten sind: z.B. der „Diskuswerfer" aus Terrakotta mit der Modellnummer M8563. Der Buchstabe M vor der Modellnummer steht für Moderne Keramik, im Gegensatz zu F für Fayence. Die „Sandalenbinderin" wird zunächst unter der Nummer 8503 entworfen und anschließend in verschiedenen Größen und Varianten produziert. (Quelle: Fotoarchiv Karl Täuber)

Besides the catalogue, the marketing personnel used photographs which were not in the catalogue: for instance the "Discobolus," model number M8563. The letter "M" in front of the model number stands for Modern Ceramics whereas "F" stands for faience. The "Sandal Binder" was at first designed as number 8503 and later in several different sizes and versions. (Source: Fotoarchiv Karl Täuber)

Die Preise bei den Buchstützen verstehen sich immer pro Paar. Insgesamt belegt der Katalog, dass Schuster anfänglich mit einem Sortiment von 361 Modellen am Markt aufgetreten ist, hinzu kamen nachträglich Modelle, die nur mit Fotos als Werbemittel propagiert worden sind (Ill. 93 und 94).

Bereits im Frühjahr 1939 treten bei Schuster Probleme mit der Führung des Unternehmens auf. Er hat anscheinend nicht genügend Kapital, um den laufenden Betrieb aufrechtzuerhalten und begleicht auch seine Schulden bei der Kontrollbank nicht. Über Dr. Paulick von der Prüfungsstelle der Keramischen Industrie in Wien versucht er bei der Vermögensverkehrsstelle eine Revision des Übernahmebetrages zu erzielen.[366] Dieser Revisionsanspruch wird abgelehnt, Schuster lässt aber am 23. Mai 1939 erneut über die Prüfungsstelle ein langes Schreiben an die Kontrollbank verfassen, in dem für ihn Dr. Paulick bittet, hinsichtlich des Restbetrags von 10.000 Reichsmark „das weitestgehende Entgegenkommen" zu zeigen. Als Grund gibt er an, dass die Firma Goldscheider Nachf. zu den „exportintensivsten Unternehmungen des Deutschen Reiches" gehört und daher die weit „verzweigte Auslandsorganisation dieses Unternehmens von ausserordentlicher Bedeutung" sei, „da hiedurch die Beobachtung von Märkten möglich ist, auf welchen der Absatz deutscher Waren bisher nicht möglich oder nicht mehr möglich war"[367] (Ill. 95).

In der Antwort der Kontrollbank wird Schuster vorgeworfen, dass er den ersten Fälligkeitstermin ohne Mitteilung hat verstreichen lassen und trotz des Angebots, die Ratenzahlungen von monatlich 2.000 Reichsmark auf 1.250 zu senken, er weder bezahlt

capital to keep production running nor did he pay his debts to the Kontrollbank. Through Dr. Paulick from the Inspectorate for the Ceramics Industry in Vienna, Schuster tried to persuade the Vermögensverkehrsstelle to revise the sum he was to have paid for the takeover.[366] The attempt at revision was refused but Schuster again tried on 23 May 1939 through the Inspectorate to address a long letter to the Kontrollbank, in which Dr. Paulick asked that "as large a concession as possible" be made to Schuster in the matter of the remaining sum of 10 000 RM. The reason Paulick gave for the request was that the firm of Goldscheider, Successor, was one of the most "intensive exporters in the German Reich;" hence the broadly "ramified organization abroad of this enterprise [was] of exceptional importance [...] since, through it, it was possible to keep an eye on markets on which German wares had hitherto not sold or were no longer being sold"[367] (Ill. 95).

The reply from the Kontrollbank accuses Schuster of allowing the first deadline for an installment payment to pass and, despite the offer made to him of reducing the monthly installments from 2000 RM to 1250 RM, Schuster had neither paid nor expressed any

Ill. 95
Schuster möchte die Zahlungsraten an die Kontrollbank nicht zahlen, mit Hinweis auf die notwendige „Aufbauarbeit". Im „Interesse des Reiches" verlangt er eine Stundung der Zahlungen für seinen „Exportbetrieb". (Quelle: Archiv Goldscheider)

Unwilling to pay the monthly installments to the Kontrollbank, Schuster spoke of "reconstruction work" being needed. In the "interests of the Reich", he requested that payments for his "export oriented business" be postponed. (Source: Goldscheider Archives)

opinion on this concession.[368] Schuster then wrote to the Kontrollbank on 14 June asking for the deadline for payment of the monthly installments to be delayed by one year. The Kontrollbank obliged Schuster and permitted him to postpone payment of the capital debt to 1 January 1940 in monthly installments of 1000 RM. There would, however, be 5 per cent annual interest on the outstanding 10 000 RM.[369] Since Schuster again did not respond in time to the offer from the Kontrollbank, the Kontrollbank considered taking legal measures against him.[370]

Schuster involved the Reichswirtschaftsministerium [Reich Economics Ministry], which sent a letter to both the Vermögensverkehrsstelle and the Austrian Kontrollbank on 2 August 1939. In it, a civil servant from the ministry announced that "in view of the decree issued by the civil servant for General Field Marshal Göring respecting the Four Year Plan, according to which exports were to be given absolute priority in all cases," he could under no circumstances allow "the firm of Goldscheider to neglect the export trade because of its domestic obligations"[371] (Ill. 96). After a while, the attempt at reaching an agreement between Schuster and the Kontrollbank bore fruit. Schuster wrote out a statement of liability, in which he confirmed that he owed the remainder of the purchase price amounting to 10 000 RM to the Kontrollbank and would pay said sum from 1 January 1940 in monthly installments

noch sich geäußert hat.[368] Schuster schreibt daraufhin am 14. Juni an die Kontrollbank und bittet um eine Stundung der monatlichen Abzahlungsraten um ein Jahr. Die Kontrollbank kommt Schuster entgegen und gestattet die Zahlung der Kapitalschuld erst ab dem 1. Januar 1940 in Monatsraten von 1.000 Reichsmark. Allerdings soll der Betrag von 10.000 Reichsmark mit 5 Prozent jährlich verzinst werden.[369] Da Schuster sich jedoch erneut nicht rechtzeitig zu diesem Angebot äußert, erwägt die Kontrollbank, rechtliche Schritte gegen ihn einzuleiten.[370]

Schuster schaltet das Reichswirtschaftsministerium ein, das am 2. August 1939 einen Brief an die Vermögensverkehrsstelle und die Österreichische Kontrollbank schickt. Hier teilt der Beauftragte des Ministeriums mit, dass „im Hinblick auf die vom Beauftragten für den Vierjahresplan Generalfeldmarschall Göring getroffene Anordnung, wonach in jedem Fall der Ausfuhr der unbedingte Vorrang einzuräumen ist", er es unter keinen Umständen zulassen kann, „daß die Firma Goldscheider wegen ihrer Inlandsverpflichtungen das Exportgeschäft vernachlässigt"[371] (Ill. 96). Der Versuch einer Einigung zwischen der Kontrollbank und Schuster gelingt nach einiger Zeit. Schuster fertigt einen Schuldschein aus, in dem er bestätigt, dass er der Kontrollbank den Restkaufpreis in Höhe von 10.000 Reichsmark schuldet und diesen Betrag ab dem 1. Januar 1940 in Monatsraten von 300 Reichsmark und ab dem 1. Ja-

Ill. 96
Brief des Reichsministers, in dem wegen der Bedeutung der Goldscheider Manufaktur für die Exporte und der damit verbundenen Deviseneinnahmen verlangt wird, Schuster eine Stundung seiner Schulden zu gewähren. (Quelle: Archiv Goldscheider)

A letter from the Reich minister demanding that Schuster should be permitted to postpone paying off his debt because the Goldscheider manufactory was so important for the export trade and the foreign currency revenue associated with it. (Source: Goldscheider Archives)

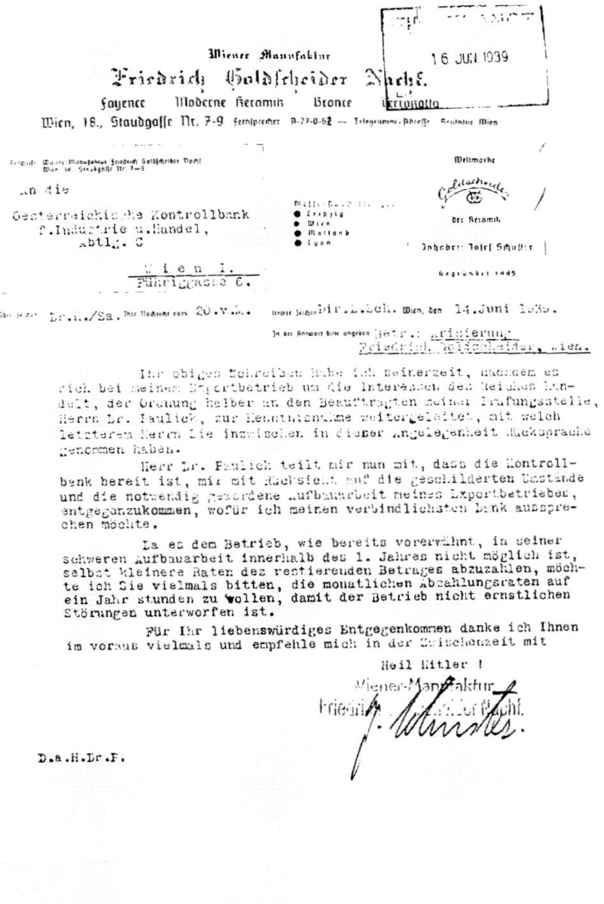
16. JUN. 1939

Wiener Manufaktur
Friedrich Goldscheider Nachf.
Fayence Moderne Keramik Bronce Terracotta
Wien, 18., Staudgasse Nr. 7-9

An die
Oesterreichische Kontrollbank
f. Industrie u. Handel,
Abtlg. C
Wien I.
Führiggasse 6.

Weltmarke Goldscheider der Keramik
Inhaber: Josef Schuster
Gegründet 1885

Ihr Zeichen Dr.K./Sa. Ihre Nachricht vom 20.V.39. Unsere Zeichen Dir.L.Sch. Wien, den 14.Juni 1939.
In der Antwort bitte angeben Betr.: Arisierung Friedrich Goldscheider, Wien.

Ihr obiges Schreiben habe ich seinerzeit, nachdem es sich bei meinem Exportbetrieb um die Interessen des Reiches handelt, der Ordnung halber an den Beauftragten meiner Prüfungsstelle, Herrn Dr. Paulick, zur Kenntnisnahme weitergeleitet, mit welch letzterem Herrn Sie inzwischen in dieser Angelegenheit Rücksprache genommen haben.

Herr Dr. Paulick teilt mir nun mit, dass die Kontrollbank bereit ist, mir mit Rücksicht auf die geschilderten Umstände und die notwendig gewordene Aufbauarbeit meines Exportbetriebes, entgegenzukommen, wofür ich meinen verbindlichsten Dank aussprechen möchte.

Da es dem Betrieb, wie bereits vorerwähnt, in seiner schweren Aufbauarbeit innerhalb des 1. Jahres nicht möglich ist, selbst kleinere Raten des restierenden Betrages abzuzahlen, möchte ich Sie vielmals bitten, die monatlichen Abzahlungsraten auf ein Jahr stunden zu wollen, damit der Betrieb nicht ernstlichen Störungen unterworfen ist.

Für Ihr liebenswürdiges Entgegenkommen danke ich Ihnen im voraus vielmals und empfehle mich in der Zwischenzeit mit

Heil Hitler !
Wiener-Manufaktur
Friedrich Goldscheider Nachf.
J. Schuster.

D.a.H.Dr.P.

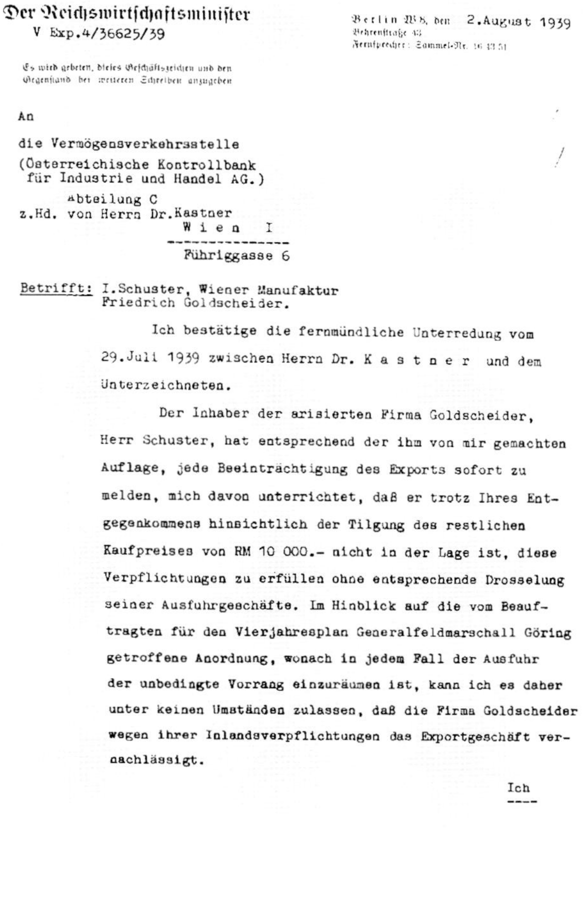
Der Reichswirtschaftsminister
V Exp.4/36625/39

Es wird gebeten, dieses Geschäftszeichen und den Gegenstand bei weiteren Schreiben anzugeben

Berlin W 8, den 2.August 1939
Behrenstraße 43

An

die Vermögensverkehrsstelle
(Österreichische Kontrollbank
für Industrie und Handel AG.)
Abteilung C
z.Hd. von Herrn Dr.Kastner
Wien I
Führiggasse 6

Betrifft: I.Schuster, Wiener Manufaktur Friedrich Goldscheider.

Ich bestätige die fernmündliche Unterredung vom 29.Juli 1939 zwischen Herrn Dr. K a s t n e r und dem Unterzeichneten.

Der Inhaber der arisierten Firma Goldscheider, Herr Schuster, hat entsprechend der ihm von mir gemachten Auflage, jede Beeinträchtigung des Exports sofort zu melden, mich davon unterrichtet, daß er trotz Ihres Entgegenkommens hinsichtlich der Tilgung des restlichen Kaufpreises von RM 10 000.- nicht in der Lage ist, diese Verpflichtungen zu erfüllen ohne entsprechende Drosselung seiner Ausfuhrgeschäfte. Im Hinblick auf die vom Beauftragten für den Vierjahresplan Generalfeldmarschall Göring getroffene Anordnung, wonach in jedem Fall der Ausfuhr der unbedingte Vorrang einzuräumen ist, kann ich es daher unter keinen Umständen zulassen, daß die Firma Goldscheider wegen ihrer Inlandsverpflichtungen das Exportgeschäft vernachlässigt.

Ich

nuar 1941 zu 500 Reichsmark bis zur Begleichung der Gesamtsumme zahlt. Zusätzlich ist der Restbetrag von 10.000 Reichsmark rückwirkend ab dem 1. Mai mit 3 Prozent zu verzinsen. In diesem Schuldschein verpflichtet sich Schuster auch, dass bei Nichtzahlung der oben genannten Beträge die Liegenschaften in der Staudgasse an die Kontrollbank verpfändet werden.[372]

Am 21. Mai 1941 wendet sich Schuster erneut an die Vermögensverkehrsstelle im Ministerium für Wirtschaft und Arbeit in Wien. Bezug nehmend auf die Erlassung eines Gesetzes vom 27. März 1941, nach dem sämtliche mit einem jüdischen Namen verbundene Firmierungen geändert werden müssen, stellt Schuster einen Antrag auf Zulassung der Beibehaltung des Namens und der Angabe eines Zusatzes, dass es sich um ein Nachfolgeunternehmen handelt (Ill. 97). Er bittet um eine Genehmigung, den jüdischen Firmennamen aufgrund seiner Bekanntheit bis zum 31. Dezember 1942 verwenden zu dürfen.[373] Aufgrund der Exportstärke des Unternehmens und im Interesse der Devisenwirtschaft soll die Firmierung lauten: „Wiener Manufaktur Josef Schuster, vormals Friedrich Goldscheider". Dies sei laut Schuster notwendig, um die kriegsbedingt nicht mehr bestehenden Kontakte zu Überseekunden wieder aufnehmen zu können. Denn unter einem vollkommen neuen Namen wäre die Wiederaufnahme dieser

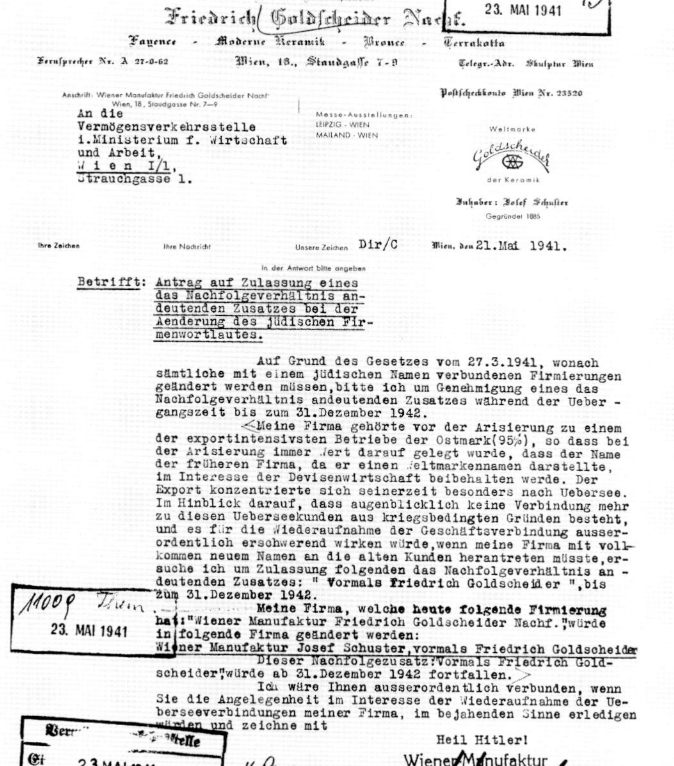

Wiener Manufaktur
Friedrich Goldscheider Nachf.
Fayence - Moderne Keramik - Bronce - Terrakotta
Fernsprecher Nr. A 27-0-62 Wien, 18., Staudgasse 7-9 Telegr.-Adr. Skulptur Wien
Postscheckkonto Wien Nr. 23520

23. MAI 1941 13

Anschrift: Wiener Manufaktur Friedrich Goldscheider Nachf.
Wien, 18., Staudgasse Nr. 7–9

Messe-Ausstellungen:
LEIPZIG - WIEN
MAILAND - WIEN

Weltmarke Goldscheider der Keramik
Inhaber: Josef Schuster
Gegründet 1885

An die
Vermögensverkehrsstelle
i.Ministerium f. Wirtschaft
und Arbeit,
Wien I/1,
Strauchgasse 1.

Ihre Zeichen Ihre Nachricht Unsere Zeichen Dir/C Wien, den 21.Mai 1941.
In der Antwort bitte angeben

Betrifft: Antrag auf Zulassung eines das Nachfolgeverhältnis andeutenden Zusatzes bei der Aenderung des jüdischen Firmenwortlautes.

Auf Grund des Gesetzes vom 27.3.1941, wonach sämtliche mit einem jüdischen Namen verbundenen Firmierungen geändert werden müssen, bitte ich um Genehmigung eines das Nachfolgeverhältnis andeutenden Zusatzes während der Uebergangszeit bis zum 31.Dezember 1942.

Meine Firma gehörte vor der Arisierung zu einem der exportintensivsten Betriebe der Ostmark(95%), so dass bei der Arisierung immer Wert darauf gelegt wurde, dass der Name der früheren Firma, da er einen Weltmarkennamen darstellte, im Interesse der Devisenwirtschaft beibehalten werde. Der Export konzentrierte sich seinerzeit besonders nach Uebersee. Im Hinblick darauf, dass augenblicklich keine Verbindung mehr zu diesen Ueberseekunden aus kriegsbedingten Gründen besteht, und es für die Wiederaufnahme der Geschäftsverbindung ausserordentlich erschwerend wirken würde, wenn meine Firma mit vollkommen neuem Namen an die alten Kunden herantreten müsste, ersuche ich um Zulassung folgenden das Nachfolgeverhältnis andeutenden Zusatzes: " Vormals Friedrich Goldscheider ", bis zum 31.Dezember 1942.

Meine Firma, welche heute folgende Firmierung hat: "Wiener Manufaktur Friedrich Goldscheider Nachf." würde in folgende Firma geändert werden:
Wiener Manufaktur Josef Schuster, vormals Friedrich Goldscheider
Dieser Nachfolgezusatz: "Vormals Friedrich Goldscheider" würde ab 31.Dezember 1942 fortfallen.

Ich wäre Ihnen ausserordentlich verbunden, wenn Sie die Angelegenheit im Interesse der Wiederaufnahme der Ueberseeverbindungen meiner Firma, im bejahenden Sinne erledigen würden und zeichne mit

Heil Hitler!
Wiener Manufaktur
Friedrich Goldscheider Nachf.

1100 9 23. MAI 1941
Et 23 MAI 1941
680
Abt.: Blg.: KB.

Ill. 97
Aufgrund der Verordnung, jüdische Namen aus Firmierungen zu entfernen, stellt Josef Schuster bei der Vermögensverkehrstelle den Antrag, den Weltmarkennamen Goldscheider „im Interesse der Devisenwirtschaft" weiterhin führen zu dürfen. (Quelle: Archiv Goldscheider)

Because of a decree according to which Jewish names were to be struck from registered firms, Josef Schuster petitioned the Vermögensverkehrsstelle for permission to continue to use the world-renowned Goldscheider brand name "in the interests of procuring foreign currency". (Source: Goldscheider Archives)

of 300 RM and from 1 January 1941 in monthly installments of 500 RM until the sum total owed should be paid. In addition, interest amounting to 3 per cent was to be paid retroactively as from 1 May on the remainder of 10 000 RM. In this statement of liability, Schuster also committed himself to forfeiting the Staudgasse properties to the Kontrollbank if the above mentioned sums were not paid.[372]

On 21 May 1941, Schuster again addressed the Vermögensverkehrsstelle in the Ministry for Economics and Labor in Vienna. Referring to a law passed on 27 March 1941, according to which all registered firms with Jewish names had to change them, Schuster petitioned for permission to retain the name with the addition that it was a successor business (Ill. 97). He requested a permit allowing him to retain the Jewish firm name until 31 December 1942 because it was so well known.[373] Since the business was so strongly engaged in the export trade, it should be registered as: "Wiener Manufaktur Josef Schuster, vormals Friedrich Goldscheider" ["Viennese Manufactory Josef Schuster, formerly Friedrich Goldscheider"]. This was, as Schuster put it, necessary for reforging the ties to the overseas clientele, which, due to the war, no longer existed. After all, it would be especially difficult to renew those contacts under an entirely different name. The addition "formerly Friedrich Goldscheider" would then be dropped as from 31 December 1942.[374]

Permission was granted on 20 June 1941 by Dr. von Philippovich at the Vermögensverkehrsstelle in Vienna; Schuster was allowed to continue to use the addition "formerly Friedrich Goldscheider" to the registered firm name until the end of the year 1942 since it referred to his renowned predecessor.[375] The firm was now registered as "Wiener Manufaktur Josef Schuster vorm. Friedrich Goldscheider" and entered under this name in the commercial register on 22 July 1941.[376] The stamp used on models also changed with the new registered firm name and the GWM stamp was replaced by a firm mark with "Wiener Manufaktur J. Schuster" in lettering (Ill. 98).

Schuster was threatened by another lawsuit in addition to his legal problems with the Kontrollbank. On 6 November 1941, a permit was issued by Ministerialrat Dr. Wacha from the finance and accounting division to initiate an investigation under his supervision of the Aryanization proceedings at the Goldscheider manufactory.

Dr. Wacha deposed that the sale price of 65 000 Reich marks (RM) had been set much too low and had not been preceded by a financial investigation. He pointed out the status as established by the provisional administrator on 10 June 1938, when liquid assets of some RM 100 000 had been estimated. The same result had also been arrived at in a "cursory report made of the clear liquid assets by an anonymous financial investigator on 30 June 1938". In both appraisals of the value of the stock, Dr. Wacha contended, "for reasons that are not clear, of warehouse stock, some RM 140 000 worth have been deducted in the individual types of wares between 60 and 95 % and, from the value resulting from this of some RM 34 000, another 40 % was deducted so that ultimately the warehouse stock was taken into the two appraisals conducted as being worth RM 20 400, respectively, RM 26 200."[377]

That meant that, at the time the Goldscheider manufactory was Aryanized in 1938 by the Kontrollbank and by Schuster, not just the difference between the purchase price (either 50 000 or 65 000 RM) and the clear liquid assets (RM 100 000) was too great and, therefore, the business had been self-evidently undervalued. The stock had also been undervalued on the basis of the appraisal at the time by more than 100 000 RM. Schuster had been able to incorporate those assets and, since they were worth far more than the sum he was to pay, he had managed to multiply his investment simply through the Aryanization proceedings. The question was whether his unwillingness to pay the remainder of the purchase price to the Kontrollbank and the allegedly dire straits in which he said he found himself represented an attempt to conceal capital gains or whether the manufactory under Schuster really had declined in output and sales volume within the briefest possible time span. In any case, Working Group 6 had on 3 December 1941 commissioned the Abwicklungsstelle [literally expedition division] at the Vermögensverkehrsstelle to investigate the "Entjudungsverfahren" ["Dejudification proceedings"] at the firm of Goldscheider and had notified the NSDAP District Finance Authority. The NSDAP, of course, was interested in taking in tax revenue it felt had been withheld rather than exposing injustice.[378]

Dr. Wacha's attention was probably drawn to the fallacious appraisal of the firm by a daughter born to Walter Goldscheider out of wedlock. Born in 1902, Elisabeth Seifert was the daughter of Anna Kowacz, who was not Jewish. She hoped to be able to claim a financial share from the Aryanization of the business. It is not known whether she did this with her father's knowledge. He had officially acknowledged her as his daughter in 1902 – that is, long before he made the acquaintance of Lilly Kuschnitzky. However, "although she handed in her claim even before the contract of purchase had been concluded with Josef Schuster, the authorities, by cleverly sending her letter back and forth between desks, misplacing and forgetting it, succeeded in not preventing the sale going through."[379]

Kontakte besonders schwer. Der Nachfolgezusatz „vormals Friedrich Goldscheider" soll dann ab 31. Dezember 1942 wegfallen.[374]

Der Antrag wird am 20. Juni 1941 von Dr. von Philippovich von der Vermögensverkehrsstelle in Wien bewilligt und Schuster darf bis Ende des Jahres 1942 in der Firmenbezeichnung den Nachfolgezusatz „vormals Friedrich Goldscheider" führen, der auf den berühmten Vorgänger hinweist.[375] Der Firmenwortlaut ist „Wiener Manufaktur Josef Schuster vorm. Friedrich Goldscheider", die Eintragung ins Handelsregister erfolgt am 22. Juli 1941.[376] Auch der Stempel auf den Modellen ändert sich mit der Umfirmierung und der GWM-Stempel wird durch eine Firmenmarke mit dem Schriftzug „Wiener Manufaktur J. Schuster" ersetzt (Ill. 98).

Außer den Auseinandersetzungen mit der Kontrollbank droht Schuster eine weitere Klage. Am 6. November 1941 wird vom Ministerialrat Dr. Wacha von der Finanzbuchhaltung eine Genehmigung bei seiner Leitung zwecks eines einzuleitenden Verfahrens zur Klärung des Arisierungsvorgangs bei der Manufaktur Goldscheider beantragt.

Dr. Wacha führt dabei auf, dass der Verkaufspreis von 65.000 Reichsmark (RM) viel zu niedrig gewesen und eine eingehende Wirtschaftsprüfung nicht vorausgegangen sei. Er macht auf den am 10. Juni 1938 erstellten Status des kommissarischen Verwalters aufmerksam, bei dem ein Reinvermögen von rund RM 100.000 ermittelt worden ist. Zum gleichen Ergebnis sei auch ein „oberflächlicher Bericht eines unbekannten Wirtschaftsprüfers mit einer per 30.6.1938 erstellten Vermögensaufstellung" gekommen. In beiden Vermögensaufstellungen wurde nämlich bei der Bewertung der Warenbestände „aus unersichtlichen Gründen vom Lagerbestand von rund RM 140.000 bei den einzelnen Warengattungen zwischen 60 u. 95 % abgeschrieben und von dem hieraus resultierenden Werte von rund RM 34.000,– weitere 40 % in Abzug gebracht und schließlich das Warenlager mit rund RM 20.400,– bzw. RM 26.200,– in die beiden Vermögensaufstellungen aufgenommen."[377]

Das bedeutet, dass bei der Arisierung der Goldscheider Manufaktur im Jahr 1938 durch die Kontrollbank und Schuster nicht nur die Differenz zwischen Kaufpreis (RM 50.000 bzw. 65.000) und Reinvermögen (RM 100.000) zu hoch und somit das Unternehmen eindeutig unterbewertet worden war. Auch die Warenbestände wurden um mehr als 100.000 Reichsmark niedriger angesetzt als ihr damaliger Wert war. Diese Wertbestände konnte sich Schuster einverleiben und da sie weitaus größer waren, als der von ihm zu zahlende Betrag, hat er allein durch den Arisierungsprozess seine Einlage vervielfachen können. Die Frage ist, ob seine Zahlungsunwilligkeit des Kaufrestbetrages an die Kontrollbank und die vermeintlich schwierige Lage den persönlichen Gewinn

Ill. 98
Ab dem Jahr 1941 ändert sich die Firmenstempelung auf den Modellen. Das bewährte Zeichen „GWM" und die Grundform werden zwar beibehalten, aber der jüdische Name Goldscheider wird durch den neuen Inhaber „J. Schuster" ersetzt. (Sammlung f@goldscheider.de)

From the year 1941, the firm stamp changes on the models. The familiar "GWM" and the basic form were retained but the Jewish name Goldscheider was replaced by that of the new owner, "J. Schuster". (Collection f@goldscheider.de)

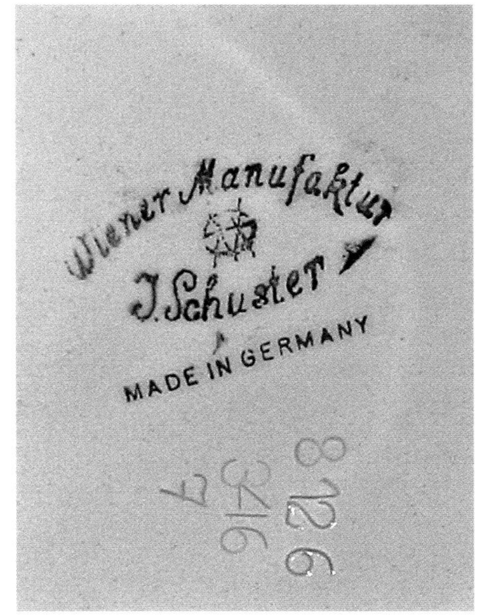

kaschieren sollten, oder ob die Manufaktur unter Schuster tatsächlich innerhalb kürzester Zeit derart an Produktions- und Vertriebsstärke verloren hat. Jedenfalls wird am 3. Dezember 1941 die Arbeitsgruppe 6 der Abwicklungsstelle der Vermögensverkehrsstelle mit der Nachprüfung des „Entjudungsverfahrens" der Firma Goldscheider beauftragt und das NSDAP Gauwirtschaftsamt benachrichtigt. Die NSDAP will natürlich nicht das Unrecht aufklären, sondern entgangene Einnahmen und Steuern kassieren.[378]

Vermutlich wurde Dr. Wacha auf die falsche Bewertung des Unternehmens durch eine uneheliche Tochter von Walter Goldscheider aufmerksam gemacht. Elisabeth Seifert, geboren 1902 und Kind der nichtjüdischen Mutter Anna Kowacz, hoffte durch die Geltendmachung ihrer Ansprüche eine finanzielle Beteiligung aus der Arisierung des Unternehmens zu erhalten. Ob sie dies mit Wissen ihres Vaters tat, der sie im Jahre 1902 – also lange bevor er Lilly Kuschnitzky kennenlernte – offiziell als Tochter anerkannte, ist nicht bekannt. Doch „obwohl sie ihre Eingabe noch vor Abschluss des Kaufvertrages mit Josef Schuster einreichte, gelang es den Behörden durch geschicktes Hin- und Herschieben, Verlegen und Vergessen ihres Ansuchens, den Kaufabschluss nicht zu behindern."[379]

Sie schreibt sogar einen Brief an Adolf Hitler und bittet ihn um Verständnis und Unterstützung: „Mein Führer! Ich bitte um Hilfe, Recht und Gerechtigkeit für meinen schwer herzleidenden sechzigjährigen Vater, der heute ein gebrochener Greis ist und für mich, die ich arbeitslos bin, schwer tuberkulös (beiliegend Bestätigung der Lungenfürsorge) und außerdem auf Grund, daß mein Vater Jude ist, bin ich vom Gatten davon gejagt, der erklärt, statt Alimente gibt es Prügel."[380] Im Mai antwortet ihr vom Reichswirtschaftsminister im Auftrag ein gewisser P. Müller: „Der dem jüdischen Vorbesitzer mit seinem Einverständnis zugebilligte Preis von M 50.000,– ist angemessen. Er stellt den tatsächlichen Wert dar, den der verkaufte Betrieb in der Hand eines Juden im Augenblick der Veräusserung bekommen hat."[381]

Der Fall Elise Seifert wird detailliert protokolliert. Am 23. Mai 1940 verklagt sie die Kontrollbank, ihre Klage wird jedoch am 18. April 1941 abgewiesen. Sie prozessiert aber weiter.[382] Am 23. Juli 1943 erstattet sie Anzeige gegen Schuster und die Vermögensverkehrsstelle mit der Begründung, es habe „Mißstände" bei der Arisierung gegeben.[383]

Elisabeth Seifert even wrote a letter to Adolf Hitler, in which she implored his understanding and support: "My Führer! I ask for help, right and justice for my sixty-year-old father, who suffers from severe heart disease and is today a broken old man, and for me, since I am unemployed and severely tubercular (enclosed is confirmation from the pulmonary examination authorities) and besides, because my father is a Jew, I have been cast off by my husband, who declares there will be blows rather than alimony."[380] In May she received a reply from the Reich economics minister signed in his stead by one P. Müller: "The price of M 50 000 granted to the previous Jewish owner with his approval is commensurate. It represents the real value which the business in the hands of a Jew at the point of sale had."[381]

The case of Elise Seifert was minutely recorded. On 23 May 1940 she sued the Kontrollbank but her suit was dismissed on 18 April 1941. She nevertheless continued to institute proceeding.[382] On 23 Juli 1943 she filed a suit against both Schuster and the Vermögensverkehrsstelle, giving as her reason there had been "abuses" in the course of Aryanization.[383] Schuster, however, refused to make any retroactive payments into her account. Nor was he willing to pay back the loan made by Rosa Kush in 1938 amounting to 20 053 dollars.[384] Finally, he refused to consider the demands made for commissions by the salesmen who had previously worked for the firm.[385]

After that, not much more is to be found on Schuster and production at the manufactory. There is an application from Schuster dated 30 June 1944 for a foreign currency permit in order to pay "the commissions due on deliveries of wares" to O. Wiederhielm, Stockholm, Sweden. Surprisingly, the firm name is given on the application as "Wiener Manufaktur Josef Schuster, vormals Friedrich Goldscheider". That means Schuster had retained the addition with the Jewish name even though he had been granted permission to do so only until the end of 1942. In November 1944 Josef Schuster and his wife fled Vienna and disappeared. There is no information on the rest of their lives.[386]

After the war, it turned out that Josef Schuster had received the sum of RM 50 000 from the Wirtschaftsgruppe Keramische Industrie, Berlin [Finance Group of the Ceramics Industry] in the course of Aryanization proceedings to be spent on new purchases although only one fifth of that sum was actually invested in the business. After the takeover, Schuster began at once to sell the warehouse stock; his profit margins between 1940 and 1943 amounted to RM 204 803.[387] Moreover, the Schusters appropriated the possessions of a deceased aunt of the Goldscheiders'. Stored in two rooms on the third floor of the house at 7 Staudgasse, this legacy consisted in "fine furnishings, valuable rugs, a great deal of silver tableware, crystal glass and valuable oil paintings."[388] In 1942, Lilly Schuster had been in a position to pay RM 92 000 for a manor with villa at Grossegg on Lake Millstätter in Carinthia to the Raaber Sparkasse with funds from the firm.[389]

The bookkeeping was not straight under the Schuster's tenure. Several sales took place without invoices and all documents pertaining to the year 1944 were alleged to have been lost through the upheavals of the war. Before they fled, the Schusters sent more than three hundred current models and molds as well as miscellaneous finished wares to Grossegg and Straubing near Munich for safekeeping in autumn 1944. The shipment included furniture and office typewriters.[390] However, most of the models and molds made under Walter Goldscheider were stolen by Schuster so that employees were confronted with a virtually empty factory and could scarcely produce anything at all.

Schuster weigert sich jedoch, irgendwelche Zahlungen nachträglich zu überweisen. Auch das im Jahr 1938 von Rosa Kush erfolgte Darlehen in Höhe von 20.053 Dollar will er nicht zurückzahlen[384], genauso wie die Provisionsforderungen bisheriger Vertreter.[385]

Über Schuster und die Produktion der Manufaktur ist dann nicht mehr viel zu finden. Am 30. Juni 1944 gibt es noch einen Antrag von Schuster auf Erteilung einer Devisengenehmigung, um die „Provisionsverrechnung mit Warenlieferungen" an O. Wiederhielm aus Stockholm, Schweden, zu begleichen. Erstaunlicherweise ist als Antragsteller die „Wiener Manufaktur Josef Schuster, vormals Friedrich Goldscheider" eingetragen. Das heißt, Schuster hat trotz der nur bis Ende 1942 gültigen Genehmigung, den Zusatz mit dem ehemaligen jüdischen Namen beibehalten. Im November 1944 flüchtet Josef Schuster mit seiner Frau aus Wien und ist nicht mehr auffindbar. Zu seinem weiteren Lebensweg gibt es keine Informationen.[386]

Nach dem Krieg stellt sich heraus, dass Josef Schuster von der Wirtschaftsgruppe Keramische Industrie, Berlin, im Rahmen der Arisierung 50.000 Reichsmark für Neuanschaffungen erhielt, wovon aber nur ein Fünftel tatsächlich ins Unternehmen geflossen ist. Schuster hat nach der Übernahme auch sofort angefangen, das Warenlager zu veräußern, und konnte von 1940 bis 1943 einen Reingewinn von RM 204.803 erzielen.[387] Außerdem haben sich die Schusters den Besitz einer verstorbenen Tante der Goldscheiders angeeignet. Dieses Erbe war im Hause Staudgasse 7 im dritten Stockwerk in zwei Räumen eingelagert, bestand aus „schönem Mobiliar, wertvollen Teppichen, viel Silbergeschirr, Kristallglas und wertvollen Oelgemälden."[388] Lilly Schuster hat im Jahr 1942 von der Raaber Sparkasse einen Gutsbesitz mit Villa in Grossegg am Millstättersee in Kärnten mit Firmengeldern für RM 92.000 erwerben können.[389]

Die Buchhaltung wurde unter den Schusters nicht ordentlich geführt, mehrere Verkäufe wurden ohne Faktura durchgeführt und für das Jahr 1944 sollen alle Unterlagen durch Kriegswirren verloren gegangen sein. Vor ihrer Flucht haben die Schusters im Herbst 1944 über dreihundert aktuelle Modelle und Formen sowie diverse Fertigwaren nach Grossegg und Straubing verschickt und in Sicherheit gebracht. Darunter auch Mobiliar und Schreibmaschinen aus dem Büro.[390] Aber auch die meisten der früher unter Walter Goldscheider geschaffenen Modelle und Formen wurden von Schuster entwendet, so dass die Mitarbeiter eine fast leere Fabrik vorfanden und kaum produzieren konnten.

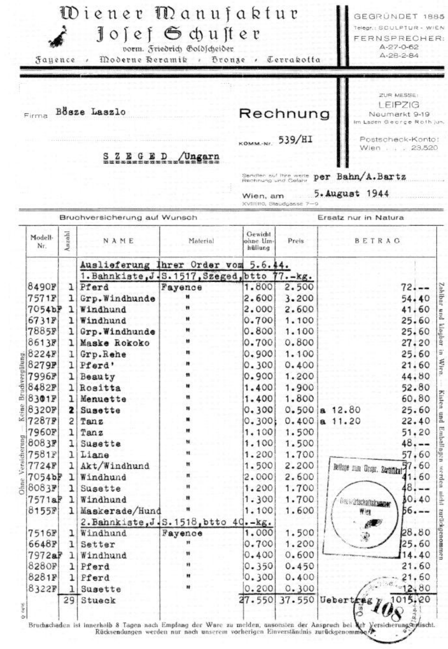

Wiener Manufaktur
Josef Schuster
vorm. Friedrich Goldscheider
Fayence · Moderne Keramik · Bronze · Terrakotta

GEGRÜNDET 1885
Telegr.: SCULPTUR - WIEN
FERNSPRECHER: A-27-0-62 A-28-2-84

ZUR MESSE: LEIPZIG Neumarkt 9-19 im Laden George Roth jun.
Postscheck-Konto: Wien ... 23.520

Firma Bösze Laszlo
S Z E G E D /Ungarn

Rechnung
KOMM.-Nr. 539/HI

Sandten auf Ihre werte Rechnung und Gefahr per Bahn/A.Bartz
Wien, am 5.August 1944
XVIII/110, Staudgasse 7–9

Bruchversicherung auf Wunsch — Ersatz nur in Natura

Modell-Nr.	Anzahl	NAME	Material	Gewicht ohne Umhüllung	Preis		BETRAG
		Auslieferung Ihrer Order vom 5.6.44.					
		1.Bahnkiste,J.S.1517,Szeged,btto 77.-kg.					
8490F	1	Pferd	Fayence	1.800	2.500		72.--
7571F	1	Grp.Windhunde	"	2.600	3.200		54.40
7054bF	1	Windhund	"	2.000	2.600		41.60
6731F	1	Windhund	"	0.700	1.100		25.60
7885F	1	Grp.Windhunde	"	0.800	1.100		25.60
8613F	1	Maske Rokoko	"	0.700	0.800		27.20
8224F	1	Grp.Rehe	"	0.900	1.100		25.60
8279F	1	Pferd'	"	0.300	0.400		21.60
7996F	1	Beauty	"	0.900	1.200		44.80
8482F	1	Rositta	"	1.400	1.900		52.80
8301F	1	Menuette	"	1.400	1.800		60.80
8320F	2	Susette	"	0.300	0.500	a 12.80	25.60
7287F	2	Tanz	"	0.300	0.400	a 11.20	22.40
7960F	1	Tanz	"	1.100	1.500		51.20
8083F	1	Susette	"	1.100	1.500		48.--
7581F	1	Liane	"	1.200	1.700		57.60
7724F	1	Akt/Windhund	"	1.500	2.200		57.60
7054bF	1	Windhund	"	2.000	2.600		41.60
8083F	1	Susette	"	1.200	1.700		48.--
7571aF	1	Windhund	"	1.300	1.700		30.40
8155F	1	Maskerade/Hund	"	1.100	1.600		56.--
		2.Bahnkiste,J.S.1518,btto 40.-kg.					
7516F	1	Windhund	Fayence	1.000	1.500		28.80
6648F	1	Setter	"	0.700	1.200		25.60
7972aF	1	Windhund	"	0.400	0.600		14.40
8280F	1	Pferd	"	0.350	0.450		21.60
8281F	1	Pferd	"	0.300	0.400		21.60
8322F	1	Susette	"	0.200	0.300		12.80
	29	Stueck		27.550	37.550	Uebertrag	1015.20

Ohne Versicherung — Keine Bruchvergütung.
Zahlbar und klagbar in Wien. — Kisten und Emballagen werden nicht zurückgenommen

Bruchschaden ist innerhalb 8 Tagen nach Empfang der Ware zu melden, ansonsten der Anspruch bei der Versicherung erlischt. Rücksendungen werden nur nach unserem vorherigen Einverständnis zurückgenommen.

Beilage zum Urspr. Zertifikat

108

Ill. 99
Eine der letzten Rechnungen, die vor Schusters Verschwinden erstellt wird. (Quelle: Archiv Goldscheider)

One of the last invoices made out before Schuster vanished. (Source: Goldscheider Archives)

Anmerkungen

[355] Handelsregister zur Firma „Erste Wiener Terrakottafabrik und Atelier für künstlerische Fayenzen Friedrich Goldscheider" (Bestand Handelsgericht Wien, Handelsregister A 16/19)
[356] Gewerbeschein, Reg.Nr. 19931/fr, Gemeindeverwaltung des Reichsgaues Wien, Wirtschaftskammer Österreich-Archiv, Gewerbearchiv. Schuster wurde am 12.4.1905 geboren.
[357] Interview mit Karl Täuber, Wien, vom 25.11.2003. Zum Militärdienst werden unter anderem Stephan Dakon, Adolf Prischl und Rudolf Knörlein eingezogen (vgl. Makus, 50er Jahre Wandmasken). Schuster hat zwischen 1939 und 1940 die meisten österreichischen kaufmännischen Beamte „durch Schikanen und Drangsalierungen förmlich gezwungen, ihre Stellung" aufzugeben, auch mehrere Bildhauer und Arbeiter wurden von ihm verklagt und diskreditiert und der Büroleiter Karl Köhler im Jahr 1940 entlassen. Vgl. Brief der Arbeitsgemeinschaft der Manufaktur Goldscheider an das Bundesministerium für Vermögenssicherung und Wirtschaftsplanung, 25.3.1946
[358] Zeugnis für Alois Teichgrab, von Walter Goldscheider vom 30. Juni 1953
[359] Interview mit Karl Täuber, Wien, 25.11.2003
[360] Ebenda
[361] Ebenda. Karl Täuber wird als Soldat eingezogen und verlässt im Jahr 1941 das Unternehmen. Erst 1945 kehrt er aus dem Krieg zurück.
[362] Katalog der Wiener Manufaktur Friedrich Goldscheider Nachf., um 1940. Anmerkung: Im Werksverzeichnis der hier vorliegenden Publikation werden die Originaltitel der Modelle in Anführungszeichen angegeben.
[363] Ebenda
[364] Ebenda
[365] Ebenda
[366] Brief von der Prüfungsstelle Keramische Industrie an die Vermögensverkehrsstelle, Wien, 15.4.1939
[367] Brief von der Prüfungsstelle Keramische Industrie an die Österreichische Kontrollbank, Wien, 23.5.1939
[368] Brief von der Österreichischen Kontrollbank an die Prüfungsstelle Keramische Industrie, Wien 24.5.1939
[369] Brief von der Österreichischen Kontrollbank an die Wiener Manufaktur Friedrich Goldscheider Nachf., 22.6.1939
[370] Brief von der Österreichischen Kontrollbank an Dr. Ernst Höpler, Rechtsanwalt von Josef Schuster, 20.7.1939
[371] Brief des Reichswirtschaftsministers, Berlin, an die Vermögensverkehrsstelle und an die Österreichische Kontrollbank, Wien, 2.8.1939
[372] Schuldschein, ausgestellt von Josef Schuster am 22.9.1939
[373] Brief von Josef Schuster an die Vermögensverkehrsstelle im Ministerium für Wirtschaft und Arbeit in Wien, 21.5.1941
[374] Schuster behält den Namenszusatz „vormals Friedrich Goldscheider" trotz der Frist bis Ende 1942 darüber hinaus bei.
[375] Brief von Dr. von Philippovich von der Vermögensverkehrsstelle an Josef Schuster, 20.6.1941
[376] Archiv des Handelsregisters Wien, HRA 7077
[377] Brief des Ministerial-Rats Dr. Wacha von der Finanzbuchhaltung, Wien an das Leitungsbüro im Hause, 6.11.1941
[378] Brief von der Abwicklungsstelle der Vermögensverkehrsstelle an das Gauwirtschaftsamt, Wien, 3.12.1941. Das Ergebnis dieser Untersuchung ist nicht bekannt.
[379] Etzersdorfer, S. 165. In einer Zusammenfassung des Arisierungsprozesses (Brief an den RA Dr. Walther Richter vom 22.5.1940) wegen der Rechtssache Elise Seifert gg. Kontrollbank werden folgende Angaben gemacht: 1. Laut Kaufvertrag vom 14.12.1938 wurden die Firma und die Liegenschaften für RM 50.000 von Walter verkauft; davon gingen 20.000 an Finanzkasse Wien Innere Stadt-Ost; 8.000 zur Deckung der Abgaberückstände; Restbetrag von 22.000 stand dem Verkäufer (also Walter Goldscheider) drei Tage nach Genehmigung des Vertrages durch die Vermögensverkehrsstelle (31.12.1938) zur Verfügung (am 5.1.1939 ausgezahlt). Am selben Tag (14.12.1938) wurde alles an Josef Schuster für RM 65.000 weiterverkauft. Der Preis wurde aufgrund der Aktiva vom 30.6.1938 berechnet, bei der die Aktiva RM 238.442 und die Passiva RM 171.039 betrug: Saldo/Sachwert daher RM 67.403. Durch den allgemeinen Boykott verbuchte die Goldscheider Manufaktur monatlich RM 4.000 Verlust, insgesamt also RM 16.000 seit dem 1. Juli, d.h. nach Abzug dieser Summe ist das Unternehmen mit nur noch rund RM 51.000 bewertet worden.
[380] Brief von Elisabeth Seifert, Gumpendorferstr. 114 III/29, Wien, an Adolf Hitler, vom 17.1.1939, in dem sie auch erwähnt, dass Walter vor seiner Ausreise noch „6.000 M für Verpflegungskosten für seinen Bruder in Steinhof gezahlt" hat. Bei dem Bruder handelt es sich um Camillo, der geistesverwirrt hospitalisiert wurde.
[381] Brief des Reichswirtschaftsministeriums an Elisabeth Seifert, 11.5.1939
[382] Elisabeth Seifert war Schneidergehilfin und wurde durch Dr. Ludwig Haydn vertreten, als bestellter Armenvertreter. Die Kontrollbank wurde durch Dr. Walther Richter vertreten. Wie aus einer Klageschrift vom 23.5.1941 hervorgeht, hat sie sich in der Verhandlung nicht anwaltlich vertreten lassen und hat die Briefe mit Verweisen auf Paragraphen und Gesetze selbst verfasst. Dies könnte ein Hinweis sein, dass sie durch Bekannte von Walter unterstützt und unter seiner Weisung das Arisierungsverfahren angefochten hat.
[383] Am 10.7.1943 bittet die Staatliche Kriminalpolizei, Kriminalpolizeistelle Wien, in einem Brief an die Abwicklungsstelle der Vermögensverkehrsstelle um die Zusendung der Akten: Arisierungsakten Industrieakt Nr. 680 mit anhängendem Exportakt Nr. 31, ferner den Liegenschaftsakt Nr. 4648 und den Sekretariatsakt 1687. Diese Akten werden gebraucht, da Elise Seifert eine Anzeige bei der Staatsanwaltschaft beim Landgericht Wien gegen die Vermögensverkehrsstelle und Josef Schuster wegen angeblicher Missstände bei der Arisierung des Betriebes ihres Vaters erstattet hat.
[384] Etzersdorfer, S. 172
[385] Brief von Alice Loeffler an die Vermögensverkehrsstelle vom 18.10.1940, in dem sie schreibt, dass ihr Ehemann Arthur Georg Loeffler, langjähriger Auslandsreisender der Firma, als Jude fristlos am 1.4.1938 entlassen wurde und Schuster die Auszahlung der Gehälter und Provisionsabrechnungen verweigert. Arthur Loeffler ist zu der Zeit in Montevideo (Uruguay), nicht geschieden und unterhaltspflichtig. Am 27.12.1940 antwortet Schuster, dass Loeffler die Manufaktur in Italien und Frankreich vertrat und von der Firma sechs große Reisekoffer und eine umfangreiche Musterkollektion zur Verfügung gestellt bekam. Da er diese nicht zurückgegeben habe, schulde Frau Loeffler ihm noch Geld.
[386] Brief der Wiener Manufaktur FG an die Bildhauerin Göhlner-Bergmann vom 11.8.1947. Vom 27.10.1944 gibt es noch eine kurze Notiz auf dem damaligen Briefpapier, die Schusters Tätigkeit nachweist.
[387] Brief der Arbeitsgemeinschaft der Manufaktur Goldscheider an das Bundesministerium für Vermögenssicherung und Wirtschaftsplanung, 25.3.1946
[388] Ebenda
[389] Ebenda
[390] Ebenda

Notes

[355] Handelsregister [commercial register] on the firm of "Erste Wiener Terrakottafabrik und Atelier für künstlerische Fayenzen Friedrich Goldscheider" (Bestand Handelsgericht Vienna, Handelsregister A 16/19)
[356] Trade licence, Reg.Nr. 19931/fr, Gemeindeverwaltung des Reichsgaues Vienna, Wirtschaftskammer Österreich-Archiv, Gewerbearchiv. Schuster was born on 12 April 1905.
[357] Interview with Karl Täuber, Vienna, on 25 Nov. 2003. Employees inducted into the military included Stephan Dakon, Adolf Prischl and Rudolf Knörlein (cf Makus, 50er Jahre Wandmasken / Wall Masks from the 1950s). Between 1939 and 1940, Schuster "literally forced" most Austrian white-collar employees "by intriguing and harassment to give up their posts;" he also brought suits against several sculptors and workmen to discredit them and Karl Köhler, the office manager, was fired in 1940. Cf letter from the works union at the Goldscheider manufactory to the Bundesministerium für Vermögenssicherung und Wirtschaftsplanung, 25 March 1946.
[358] Testimony for Alois Teichgrab given by Walter Goldscheider on 30 June 1953
[359] Interview with Karl Täuber, Vienna, 25 Nov. 2003
[360] Ibid
[361] Ibid. Karl Täuber was conscripted and left the business in 1941. He did not return from his war service until 1945.
[362] Catalogue of the Viennese Manufactory Friedrich Goldscheider, Successor, around 1940. Note: In the catalogue of works in the present publication, the original titles of the models are given in quotation marks.
[363] Ibid
[364] Ibid
[365] Ibid
[366] Letter from the Prüfungsstelle Keramische Industrie to the Vermögensverkehrsstelle, Vienna, 15 April 1939
[367] Letter from the Prüfungsstelle Keramische Industrie to the Österreichische Kontrollbank, Vienna, 23 May 1939
[368] Letter from the Österreichische Kontrollbank to the Prüfungsstelle Keramische Industrie, Vienna 24 May 1939
[369] Letter from the Österreichische Kontrollbank to the Wiener Manufaktur Friedrich Goldscheider Nachf., 22 June 1939
[370] Letter from the Österreichische Kontrollbank to Dr. Ernst Höpler, Josef Schuster's lawyer, 20 July 1939
[371] Letter from the Reichswirtschaftsminister [Reich Economics Minister], Berlin, to the Vermögensverkehrsstelle und to the Österreichische Kontrollbank, Vienna, 2 Aug. 1939
[372] Promissory note, made out by Josef Schuster on 22 Sept. 1939
[373] Letter from Josef Schuster to the Vermögensverkehrsstelle im Ministerium für Wirtschaft und Arbeit [Ministry for Economics and Labor] in Vienna, 21 May 1941
[374] Schuster retained the addition to the firm name "vormals Friedrich Goldscheider" even after the deadline had expired at the end of 1942.
[375] Letter from Dr. von Philippovich of the Vermögensverkehrsstelle to Josef Schuster, 20 June 1941
[376] Archives: Handelsregister Vienna, HRA 7077
[377] Letter from Ministerial-Rat Dr. Wacha of the Finanzbuchhaltung, Vienna, to the firm head office, 6 Nov. 1941
[378] Letter from the Abwicklungsstelle der Vermögensverkehrsstelle to the Gauwirtschaftsamt, Vienna, 3 Dec. 1941. The results of this investigation are not known.
[379] Etzersdorfer, p. 165. In a summary of the Aryanization proceedings (Letter to Dr. Walther Richter, lawyer, of 22 May 1940) in the case of Elise Seifert vs Kontrollbank, the following items of information are given: 1. According to the contract of sale of 14 Dec. 1938, the firm and the properties were sold by Walter for RM 50 000, of which RM 20 000 went to the Finanzkasse [Inland Revenue Office] Vienna Innere Stadt-Ost; RM 8000 to cover tax arrears; the remainder of RM 22 000 were due to the seller (that is, Walter Goldscheider) three days after approval of the contract by the Vermögensverkehrsstelle (31 Dec. 1938; paid out on 5 Jan. 1939). That same day (14 Dec. 1938) everything was sold to Josef Schuster for RM 65 000. The sale price was calculated on the basis of the total assets as from 30 June 1938, with the liquid assets totalling RM 238 442 and the illiquid assets RM 171 039: hence balance carried forward/real value RM 67 403. Due to the general boycott, the Goldscheider manufactory had booked RM 4000 in losses per month from 1 July, that is a total of RM 16 000; ie, after that sum was deducted, the business was evaluated at only RM 51 000, rounded off.
[380] Letter from Elisabeth Seifert, 114 III/29 Gumpendorferstraße, Vienna, to Adolf Hitler, dated 17 Jan. 1939, in which she mentions that, before he emigrated, Walter also had to pay "6000 M for board and lodgings for his brother at Steinhof." The brother was Camillo, who had been institutionalized as mentally ill.
[381] Letter from the Reichswirtschaftsministerium to Elisabeth Seifert, 11 May 1939
[382] Elisabeth Seifert was an assistant seamstress and was represented by Dr. Ludwig Haydn, appointed by the court to represent her as destitute. The Kontrollbank was represented by Dr. Walther Richter. As is indicated by a writ of 23 May 1941, she was not represented by a lawyer in court and wrote the letters with reference to laws and paragraphs herself. This might indicate that she was supported by acquaintances of Walter's and contested the Aryanization proceedings at his instruction.
[383] On 10 July 1943 the Staatliche Kriminalpolizei, Kriminalpolizeistelle [CID] Vienna, in a letter to the Abwicklungsstelle der Vermögensverkehrsstelle requesting that the following files should be sent: Arisierungsakten Industrieakt Nr. 680 with appended Exportakt Nr. 31, further Liegenschaftsakt Nr. 4648 and Sekretariatsakt 1687. These files were needed since Elise Seifert had filed a declaration with the prosecutor at Landgericht Wien [Vienna District Court] against the Vermögensverkehrsstelle and Josef Schuster alleging abuses in the course of the Aryanization proceedings against her father.
[384] Etzersdorfer, p. 172
[385] Letter from Alice Loeffler to the Vermögensverkehrsstelle dated 18 Oct. 1940, in which she writes that her husband, Arthur Georg Loeffler, who had for years been a salesman abroad for the firm, was dismissed without notice on 1 April 1938 because he was a Jew and Schuster had refused to pay the salary and commissions due to him. Arthur Loeffler was at that time in Montevideo, Uruguay, was not divorced from his wife and was required by law to support her. On 27 Dec. 1940, Schuster replied that Loeffler had represented the manufactory in Italy and France and had been given six large suitcases and an extensive collection of samples. Since, however, he had not returned them, Frau Loeffler owed him (Schuster) money.
[386] Letter of 11 Aug. 1947 from the Viennese Manufactory FG to the sculptor Göhlner-Bergmann. There is also a brief note on the letter paper of that time, dated 17 Oct. 1944, that furnishes evidence of Schuster's being active at that time.
[387] Letter from the Arbeitsgemeinschaft der Manufaktur Goldscheider to Bundesministerium für Vermögenssicherung und Wirtschaftsplanung, 25 March 1946
[388] Ibid
[389] Ibid
[390] Ibid

Emigration und Aufbau von „American Goldscheider" (1939–1952)

Emigration and Building up "American Goldscheider" (1939–1952)

Nach der Arisierung ihres Wiener Unternehmens emigrieren Walter und Lilly Goldscheider im Jahr 1939 über England nach Amerika, wo ihr Sohn Erwin seit 1938 in Miami lebt.[391] Sie setzen sich in der amerikanischen Töpferstadt Trenton in New Jersey nieder und eröffnen dort zunächst eine Porzellanhandlung.[392] Da Goldscheider-Objekte seit Ende des 19. Jahrhunderts auch in Nordamerika vertrieben werden, ist Goldscheider auf dem amerikanischen Markt eine renommierte Marke, die für eine der besten europäischen Keramikmanufakturen steht.[393]

Im Dezember 1940 wird in Trenton eine neue Fabrik eröffnet, der Walter als „president" des Unternehmens „American Goldscheider Inc." vorsteht und Ideengeber für neue Modelle ist, während sein Sohn Erwin

After their Viennese business had been Aryanized, Walter and Lilly Goldscheider emigrated in 1939 via England to America; their son, Erwin, had been living in Miami since 1938.[391] The Goldscheiders settled in the ceramics-manufacturing city of Trenton, New Jersey, where they started out by opening a porcelain shop.[392] Since Goldscheider objects had been marketed in North America as well as Europe since the late 19th century, Goldscheider was a well-known brand on the American market that stood for one of Europe's best ceramic manufactories.[393]

In December 1940, a new factory was opened in Trenton. Walter was "president" of "American Goldscheider Inc." while his son Erwin (also called Irving in America) oversaw production as "production manager"[394] (Ill. 100). The first model produced in America was a foal lying down with its head turned to look back. Another model from the earliest American output was sent to President Roosevelt, whose wife sent a letter to Walter Goldscheider thanking him.[395] Whether true or merely a public relations anecdote – Walter, who continued to be

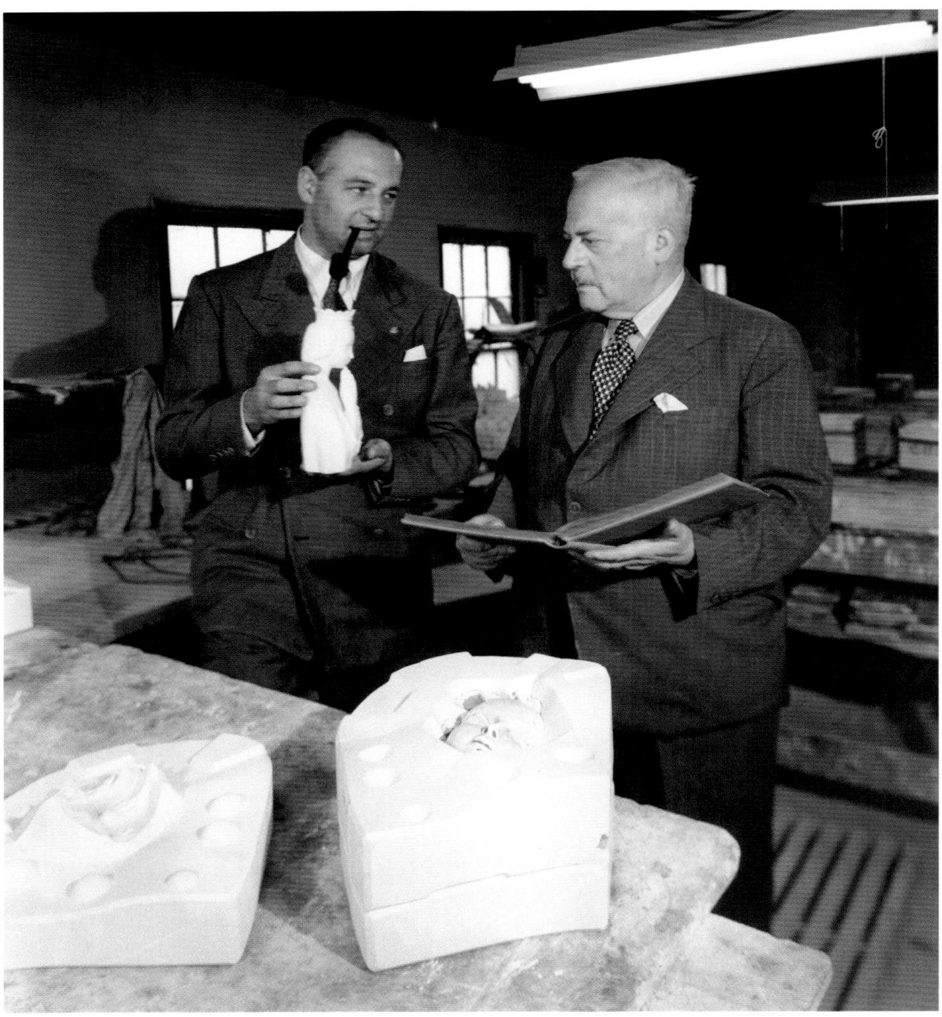

Ill. 100
Walter Goldscheider ist als Präsident der American Goldscheider Inc. für die Entwicklung neuer Modelle zuständig, während sein Sohn Erwin die Produktion leitet. (Quelle: Privatarchiv Peter F. Goldscheider, USA)

As president of American Goldscheider Inc., Walter Goldscheider was responsible for developing new models while his son, Erwin, was production manager. (Source: Peter F. Goldscheider Private Archives, USA)

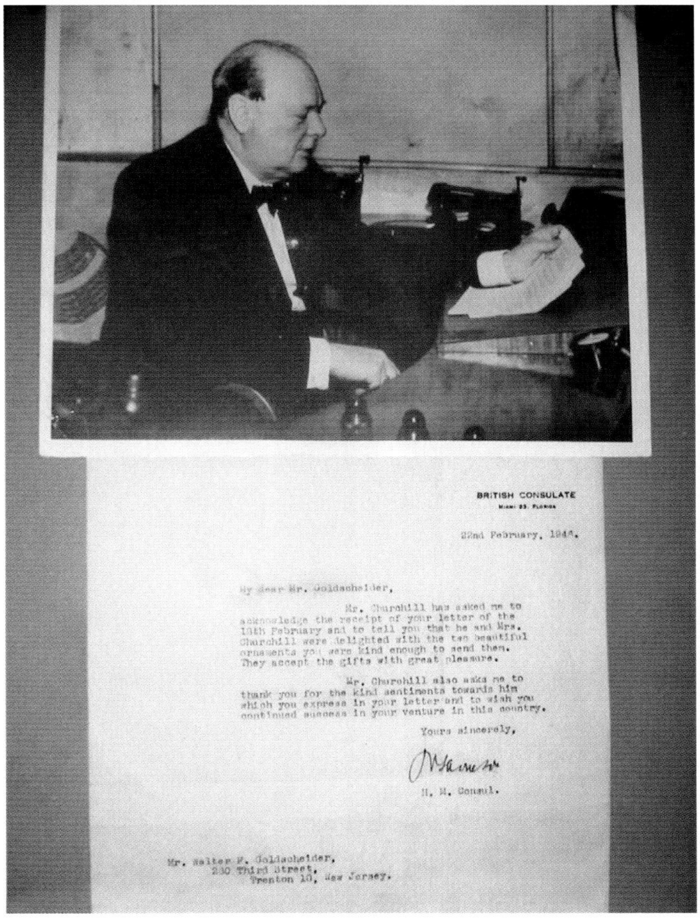

Ill. 101
Außer eines Briefs von Präsident Roosevelt liegt auch die Korrespondenz des Britischen Konsuls vor, der sich im Namen von Winston Churchill für ein Geschenk von Walter Goldscheider bedankt. (Quelle: Archiv Goldscheider)

Apart from a letter from President Roosevelt, there is correspondence with the British Consul, who thanked Walter Goldscheider on Winston Churchill's behalf for a present from Walter Goldscheider. (Source: Goldscheider Archives)

addressed as Kommerzialrat, told the press that his father, Friedrich, had been commissioned by the Emperor Franz Joseph to make a 72-piece porcelain service in cobalt and gold for the Crown Prince.[396]

Not long after the American firm was founded, Dr. Richard Kuschnitzky, a Viennese lawyer and a cousin of Lilly's, joined the business. "Kush," as he was nicknamed, worked nearly eight years for the American firm before returning to Vienna, where he represented Walter Goldscheider's interests as his lawyer.[397] Some former Viennese Goldscheider employees were hired to work at the new factory, including Franz Schwarzbart and Eduard Stein, who had been forced to emigrate because they were Jewish and who had once been head clerks at Goldscheider.[398] In the early period, models familiar from Vienna, including many figurines of dancers, Borzoi dogs and figures of children with animals were produced. Numerous new models were also created, many of which were the work of Helen Liedloff. Even the artist Vally Wieselthier, who had been living

(in Amerika auch Irving genannt) als „production manager" die Produktion beaufsichtigt[394] (Ill. 100). Das erste in Amerika produzierte Modell ist ein liegendes Fohlen, das mit dem Kopf nach hinten schaut; ein anderes Modell aus der frühesten amerikanischen Produktion wird an Präsident Roosevelt geschickt, dessen Frau sich mit einem Brief bei Walter Goldscheider bedankt.[395] Ob wahr oder nur Werbegag – Walter, der sich weiterhin mit Kommerzialrat anreden lässt, erzählt der Presse, dass früher sein Vater Friedrich vom Kaiser Franz Joseph beauftragt worden sei, für den Kronprinzen ein 72-teiliges Porzellanservice in Kobalt und Gold herzustellen.[396]

Kurz nach der Gründung tritt der Wiener Rechtsanwalt Dr. Richard Kuschnitzky, ein Cousin Lillys, dem amerikanischen Unternehmen bei. „Kush", wie er genannt wird, arbeitet fast acht Jahre für die amerikanische Manufaktur, bevor er nach Wien zurückkehrt und dort als Rechtsanwalt die Interessen von Walter Goldscheider vertritt.[397] Einige ehemalige Wiener Mitarbeiter finden in der neuen Fabrik Arbeit, beispielsweise die ehemaligen jüdischen Prokuristen Franz Schwarzbart und Eduard Stein.[398] In der Anfangszeit werden

ModNr. 777
French Bulldog,
1940er Jahre I 1940s,
Collection F. & J.P. Girard, Frankreich
In Trenton wird der Markenname Goldscheider nun um das neue Herkunftsland ergänzt: U.S.A.

In Trenton, the new country of origin was added to the Goldscheider trademark: U.S.A.

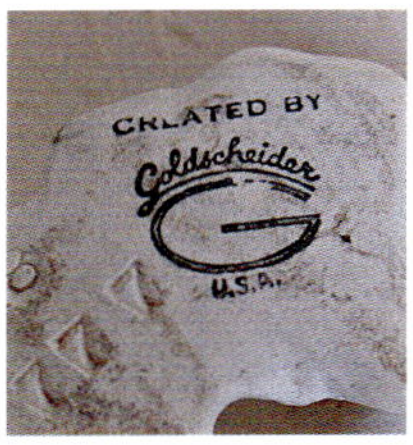

Ill. 102

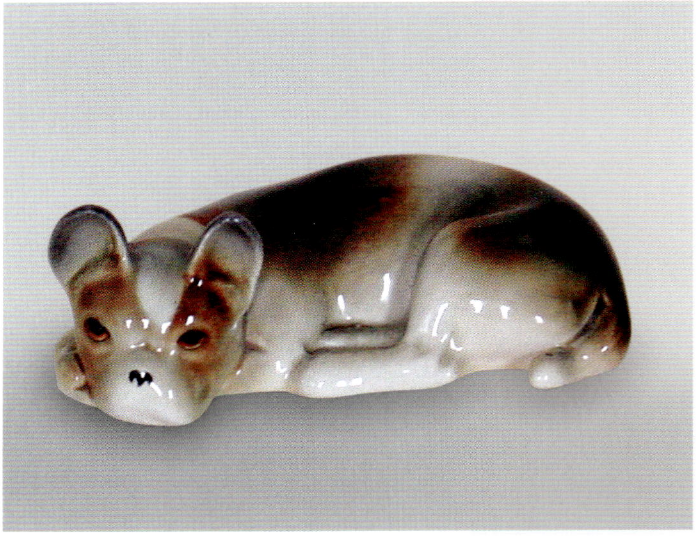

in Trenton sowohl bekannte Modelle der Wiener Produktion hergestellt, darunter viele Tänzerinnen, Borzoi-Hunde oder Kinderfiguren mit Tieren. Es entstehen aber auch zahlreiche neue Modelle, von denen viele von Helen Liedloff stammen. Sogar die seit längerem in New York lebende Künstlerin Vally Wieselthier entwirft für American Goldscheider. Wieselthier hat bereits in Wien für Goldscheider Modelle geliefert, unter anderem Wandmasken von Cherubim (Modell-Nr. 6764), die auch in Trenton produziert und für den amerikanischen Geschmack angepasst werden.

Die Fabrik wird in Trenton als „Goldscheider Pottery" bezeichnet und deren Produkte anfangs mit einer Firmenmarke gestempelt, die aus einem großen „G" und „Goldscheider U.S.A" besteht (Ill. 102). Zusätzlich erhalten einige Stücke einen Stempel mit der horizontalen Aufschrift eines Copy-

in New York for quite some time, designed for American Goldscheider. Wieselthier had made models for Goldscheider in Vienna, including Cherub wall masks (model number 6764), which were also made in Trenton but were adapted to the American taste.

The factory was known in Trenton as the "Goldscheider Pottery" and its products were initially stamped with a firm mark consisting in a large "G" and "Goldscheider U.S.A" (Ill. 102). In addition, some pieces were stamped with the horizontal replica of a copyright sign linked with "American Goldscheider Inc.". By 1942, however, the firm was registered as "Goldscheider-Everlast Corporation",[399] the copyright stamp on the models was accordingly changed to "Goldscheider-Everlast Corp.". Everlast Corporation, based in New York, was a marketing and distributing agent for decorative art works made of aluminum. It was a partner

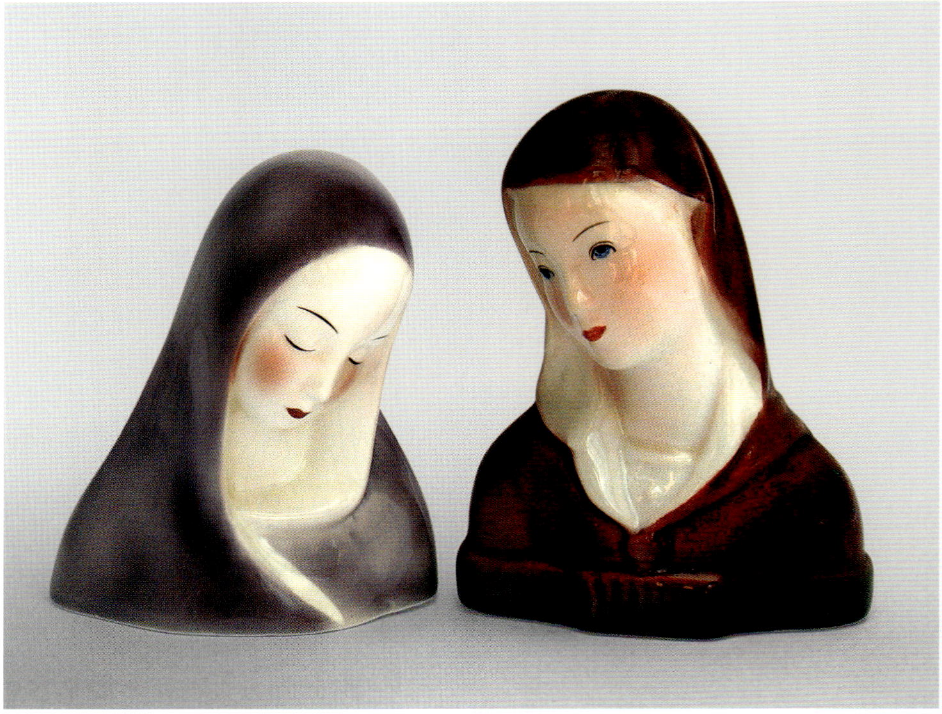

ModNr. 254, 725
Madonna, Funaro,
1940er Jahre I 1940s,
Collection a@goldscheider.net

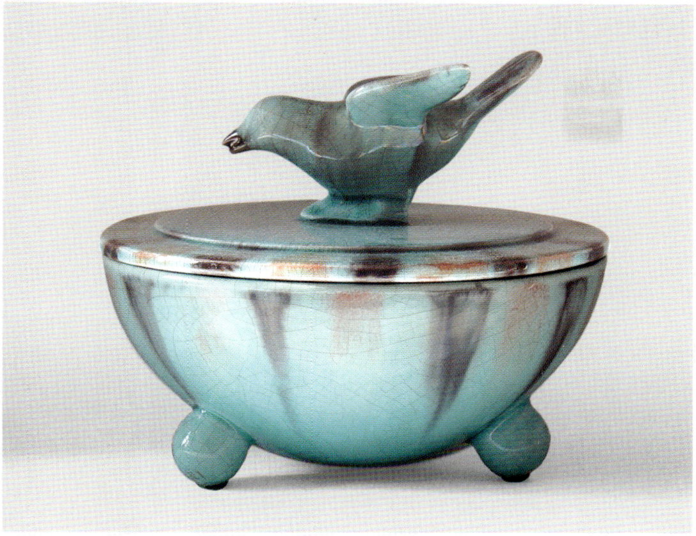

ModNr. 772
Bowl with a Bird,
1940er Jahre I 1940s,
Collection a@gold-
scheider.net

in Walter Goldscheider's factory and took over marketing and distributing the American Goldscheider products.[400] Headed by Jack Ornstein, the Everlast Corporation delivered products to more than 4000 retailers in the US and exported to twenty-two countries.[401]

In 1942 the factory employed a workforce of 20 and produced 600 objects a week costing up to 25 dollars each.[402] The American products were also fired twice. After the first firing, the decoration was sprayed on by hand and sealed with a transparent glaze.[403] The more modern American plant and machinery made it possible to

right-Zeichens verbunden mit „American Goldscheider Inc.". Doch schon um 1942 wird das Unternehmen als „Goldscheider-Everlast Corporation" geführt[399], dementsprechend wird auch die Copyright-Stempelung der Modelle in „Goldscheider-Everlast Corp." geändert. Die in New York ansässige Everlast Corporation ist eine Vertriebsagentur für dekorative Kunsterzeugnisse aus Aluminium, die sich an Walter Goldscheiders Fabrik beteiligt und den Vertrieb der amerikanischen Erzeugnisse übernimmt.[400] Die von Jack Ornstein geleitete Everlast Corporation beliefert in USA über 4.000 Geschäfte und exportiert in 22 Länder.[401]

Um 1942 werden zwanzig Mitarbeiter beschäftigt und durchschnittlich 600 Objekte pro Woche produziert, die bis zu 25 Dollar kosten können.[402] Auch die amerikanischen Produkte werden zweifach gebrannt, nach dem ersten Brand werden die Dekore von

ModNr. 506
Puppy, 1940er Jahre I
1940s, Collection
a@goldscheider.net

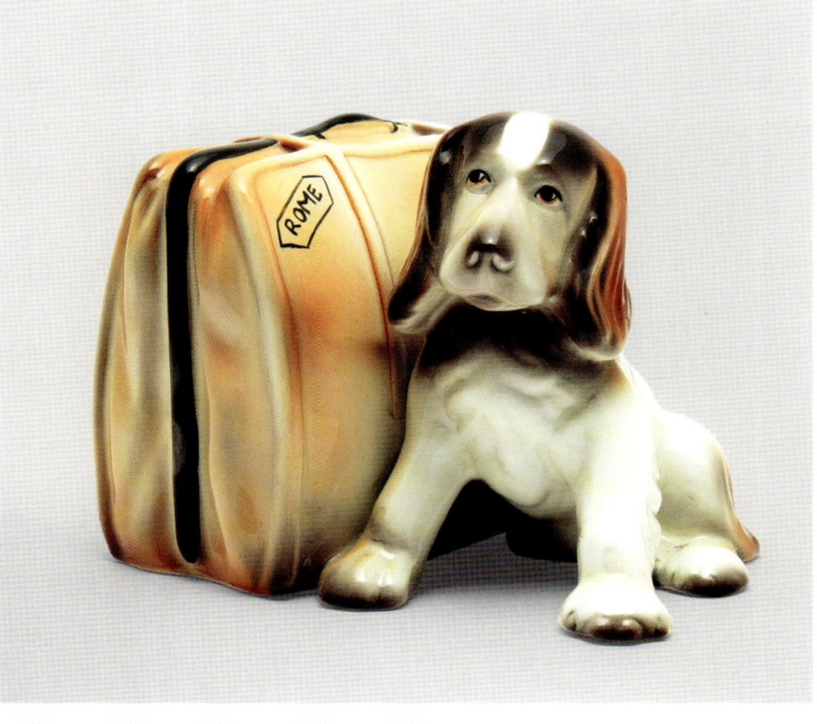

ModNr. 420
Box, 1940er Jahre I
1940s, Collection
a@goldscheider.net

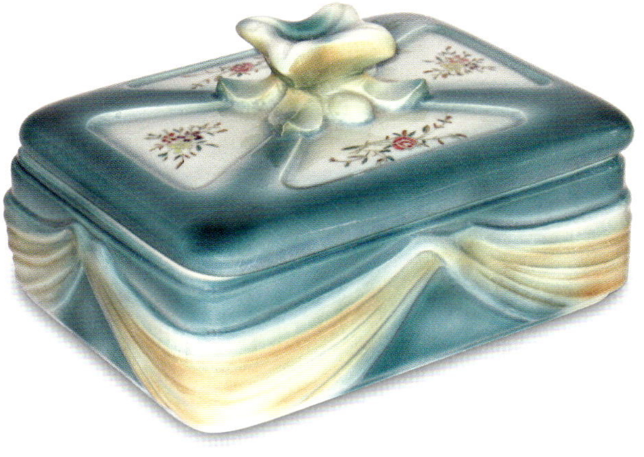

Hand aufgesprüht und mit einer durchsichtigen Glasur versiegelt.[403] Doch die moderneren amerikanischen Produktionsanlagen ermöglichen im Vergleich zur Wiener Produktion eine Reduzierung der Produktionsdauer eines Modells – der komplette Brennvorgang kann insgesamt um 10 Stunden verringert werden – sowie Arbeitszeitersparnis.[404] Die produzierten Modelle sind vielfältig: neben Gebrauchskeramik wie Dosen, Aschern und Vasen finden sich Rokoko-Statuetten im Alt-Wiener-Genre genauso wie Kostümfiguren aus der Zeit des amerikanischen Bürgerkriegs, Tierfiguren (vor allem Vögel, Hunde und exotische Tiere), balinesische Tänzerinnen sowie einige der aus den 1930er Jahren bekannte Wiener Figuren und Wandmasken. Sowohl die an den amerikanischen Stil angepassten Reproduktionen der Wiener Modelle wie auch die zeitgenössischen Trenton Modelle erhalten eine neue Nummerierung. Amerikanische Künstler wie Gloria Drew, Peggy Porcher oder A. Jacob steuern zahlreiche Modelle bei.

Im Jahr 1945 werden fast nur amerikanische Arbeiter beschäftigt, davon achtzig Prozent Frauen, die insgesamt 50.000 handgemachte Objekte produzieren.[405] Nach

reduce model production time – the entire firing process was ten hours shorter than in Vienna – and make it less labor-intensive.[404] A wide variety of models was produced: besides utilitarian household ceramics such as boxes, ashtrays and vases, there were Rococo statuettes in the Old Viennese genre as well as costume figurines from the American Civil War era, animal figurines (mainly birds, dogs and exotic animals), Balinese temple dancers and some Viennese figures and wall masks familiar from the 1930s. Both the reproductions of Viennese models adapted to the American taste and the contemporary Trenton models were numbered by a new system. American artists such as Gloria Drew, Peggy Porcher and A. Jacob contributed numerous designs for models.

By 1945 the workforce was almost exclusively American, with eighty per cent of it women, producing a total of 50 000 handmade objects.[405] After the war was over, Walter tried to target a broader export market and asked marketers and distributors and dealers who were friends of his about potential markets for his American products. In spring 1946 he received an honest answer from an English dealer in Birmingham, who

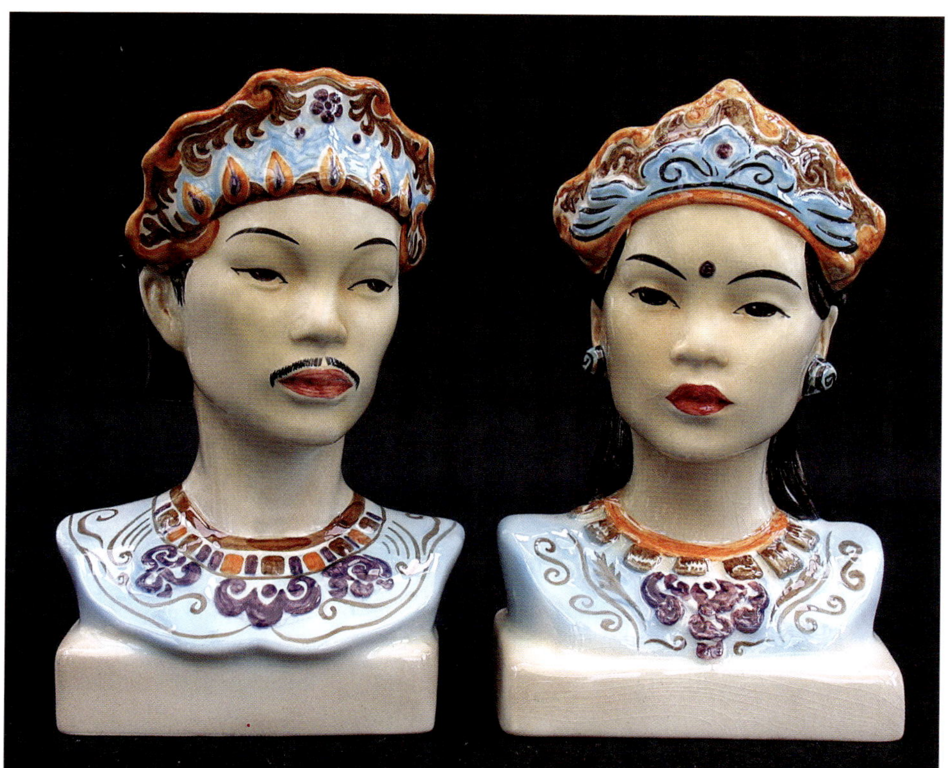

ModNr. 224, 225
„Javanese Head",
1940er Jahre I 1940s,
beide I both Collection
Robert Senft

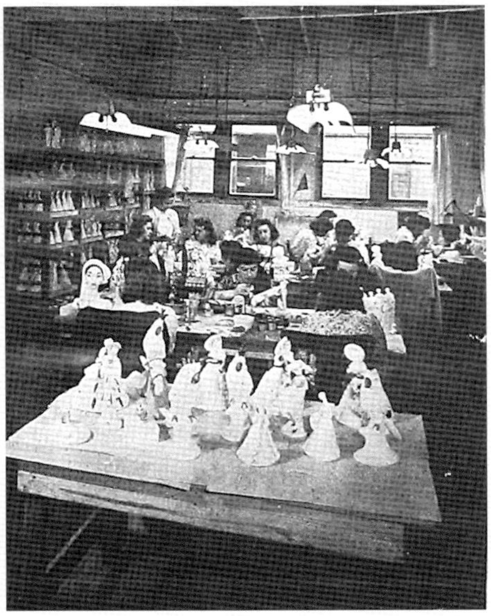

Ill. 103
Das Cover der Zeitschrift Ceramic Industry, September 1947, zeigt den „decorating shop at The Goldscheider-Everlast Corp., Trenton".

The "decorating shop at The Goldscheider-Everlast Corp., Trenton," is shown on the cover of the September 1947 issue of Ceramic Industry.

thanked him for the American-made figure sent to him but was highly skeptical about its chances of selling on the English market. His chief criticism was the coarseness of the workmanship of the figure, which too closely resembled English Royal Doulton or Beswick figurines and had none of the delicacy for which the Viennese Goldscheider ceramics were renowned.[406] Moreover, the American figures were much more expensive than their English counterparts.

Like his brother Marcel, who had emigrated to England and produced ceramics there, Walter adapted to the taste of his new country. He had an uphill struggle with the different materials and the deficient training of his American workmen, who were not capable of attaining the Viennese quality standard. In 1946 a new, more robust porcelain body was used for a separate product line, which was marked "Fine China" and was supposed to make possible underglaze decoration of a higher quality.[407] There were, however virtually no stylistic and aesthetic

Kriegsende versucht Walter, den Export auszuweiten, und erkundet sich bei befreundeten Vertriebsagenten und Händlern nach den Absatzmöglichkeiten für seine amerikanischen Erzeugnisse. Im Frühjahr 1946 erhält er eine aufrichtige Antwort eines befreundeten englischen Händlers aus Birmingham, der sich bei ihm für eine zugesandte Figur aus der amerikanischen Produktion bedankt, deren Verkaufschancen jedoch auf dem englischen Markt sehr skeptisch sieht. Er moniert vor allem die unfeine Ausarbeitung der Figur, die zu sehr den englischen Figuren von Doulton oder Beswick ähnelt und an der nichts auf den unverwechselbaren Stil und die Feinheit der Wiener Goldscheider Keramik hindeutet.[406] Zudem seien die amerikanischen Figuren viel teurer als die englischen Erzeugnisse.

Genauso wie sein Bruder Marcel, der nach England emigriert ist und dort Keramik produziert, hat sich auch Walter dem Geschmack des neuen Heimatlandes angepasst. Er hat mit den unterschiedlichen Materialien und der unzureichenden Ausbildung seiner amerikanischen Arbeiter zu kämpfen, die nicht in der Lage sind, die Wiener Qualität zu erreichen. Im Jahr 1946 wird zwar ein neues,

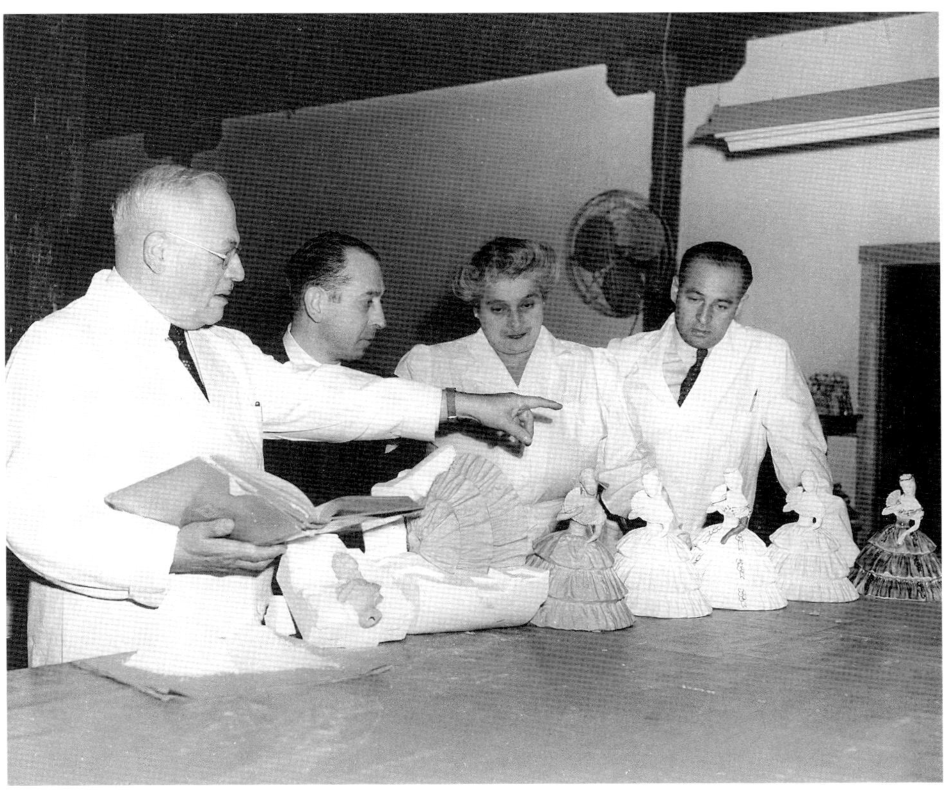

Ill. 104
Die Familie Goldscheider ist auch in Trenton in den Produktionsprozess eingebunden. Von links nach rechts: Walter Goldscheider, ein Mitarbeiter, Lilly Goldscheider und Erwin Goldscheider (Quelle: Privatarchiv Peter F. Goldscheider, USA)

As in Vienna, the Goldscheider family was also involved in the production process in Trenton. From left to right: Walter Goldscheider, an employee, Lilly Goldscheider and Erwin Goldscheider (Source: Peter F. Goldscheider Private Archives, USA)

robusteres Porzellanmaterial für eine eigene Produktionsreihe eingesetzt, das mit „Fine China" bezeichnet wird und hochwertigere Unterglasur-Dekorationen ermöglichen soll.[407] Stilistisch und künstlerisch gibt es aber kaum Entwicklungen. Die Beschriftung „Fine China" findet man ab diesem Zeitpunkt auch auf den jeweiligen Objekten, neben der Marke, die aus einem großen G und der Überschrift „ORIGINAL Goldscheider" besteht.[408] Ansonsten werden ab 1946 wieder bleihaltige Glasuren eingesetzt, die während der Kriegsjahre Mangelware waren. Es entstehen auch zahlreiche Modelle mit Überglasur, beispielsweise die auf der Gift Show in Chicago ausgestellte Tänzerin „Balinese Dancer" von Barbara Baldwin.[409] Die Überglasuren werden allgemein aufgesprüht.

Nach dem Krieg kommt es schon bald zu einer engeren Zusammenarbeit zwischen der Fabrik in Wien und Trenton, die vor allem im gegenseitigen Vertrieb der Erzeugnisse beruht. Eine neue Abteilung, die „Lamp Division", betreut die Adaptation von Plastiken wie Tänzerinnen, Büsten und exotischen Vögeln zu Lampen und Leuchtgegenständen.[410] Erwin Goldscheider entwickelt zu dieser Zeit eine matte, fleischfarbene Überglasur und experimentiert mit Dekoren aus Spitze.[411] Der im Jahr 1946 von Universal Pictures ge-

developments. From that time on, the "Fine China" marking occurs on objects alongside the mark, which consists of a large "G" and "ORIGINAL Goldscheider" in lettering above it.[408] Further, glazes with a lead content, which had been in short supply during the war years, were again used from 1946. Many overglaze models were also made, including the "Balinese Dancer" figurine by Barbara Baldwin, which featured at the Chicago Gift Show.[409] Overglazes were usually sprayed on.

Not long after the war ended, close collaboration began between the Viennese factory and Trenton, based primarily on marketing and distributing each other's products. A new division, the "Lamp Division," oversaw the adaptation of sculptures such as dances, busts and exotic birds as lamps and lighting fixtures.[410] At that time Erwin Goldscheider developed a matt, flesh-colored overglaze and was experimenting with lace decoration.[411] The short film "Pottery Poets", made by Universal Pictures in 1946, was shot in the Goldscheider factory, which Erwin used for advertising purposes and this, in turn, contributed to the growing popularity of the Goldscheider models.[412] Other media, such as the radio, magazines and newspapers, were also used by Gold-

Ill. 105
Walter ist weiterhin an der Auswahl geeigneter Modelle maßgeblich beteiligt und bereitet für die Retuscheure und Maler die historisch richtigen Kostümierungen auf. (Quelle: Privatarchiv Peter F. Goldscheider, USA)

Walter continued to play a paramount role in selecting suitable models and prepared historically accurate costumes for the retouchers and painters. (Source: Peter F. Goldscheider Private Archives, USA)

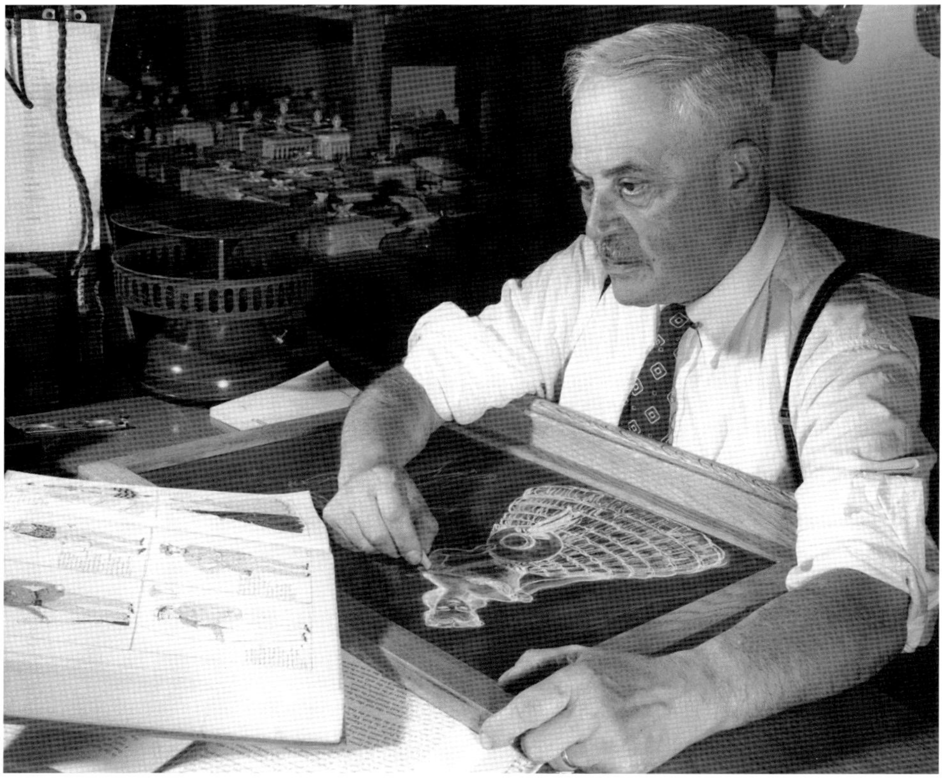

scheider to draw attention to the works produced at the American factory.[413]

Sales flourished and output rose. By 1947, a workforce of about 100 was employed at Goldscheider to make more than 500 different models.[414] In spring 1948 a new factory was opened in Trenton and seventy-five new models were announced.[415] From this time on, many American Goldscheider products were also marketed under the "Goldcrest" label – especially inexpensive figurines, animals and smoking accessories.[416] These products bear an escutcheon in which "Goldcrest" and "Fine China U.S.A." are marked. Some are marked, as they were at Goldscheider-Everlast, with an additional copyright stamp, which forms a large "C" with "Goldscheider-Everlast Corp." or now "Goldcrest Ceramics Corp." written around it. Since Walter Goldscheider and Jack Ornstein separated at that time, which is why Walter probably had to move into new factory premises, Goldcrest was probably a subsidiary of Everlast, which took some models from the collaboration with Walter and then made them itself. Kathi Urbach was one of the most versatile and prolific artists working under the Goldcrest label.

In the late 1940s, Walter Goldscheider was again caught up in a situation that made heavy demands on him. First, he had to oversee production in the new factory and see to marketing and distributing. Second, the Viennese factory was to be restored to him. After the Viennese factory was offi-

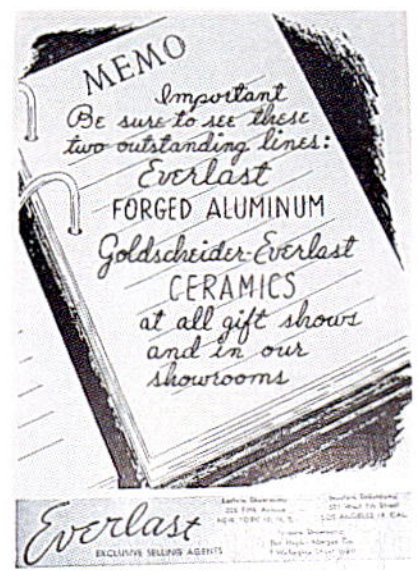

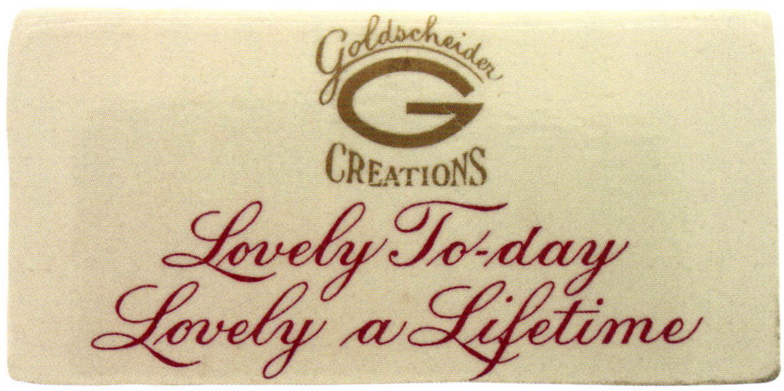

Ill. 106, 107
Auch in Amerika werden verschiedene Werbemittel eingesetzt. Hier eine Anzeige aus dem Gift Art Buyer Magazin, Mai 1946 sowie ein Werbeschild (Sammlung a@goldscheider.net)

All sorts of advertising vehicles were used in America as well. Here is an advertisement from Gift Art Buyer Magazine, May 1946, and an advertising sign. (Collection a@goldscheider.net)

drehte Kurzfilm „Pottery Poets", der in der Goldscheider Fabrik aufgenommen worden ist, wird von Erwin für Promotionszwecke eingesetzt, was zur steigenden Popularität der Goldscheider-Modelle beiträgt.[412] Auch andere Medien wie Radio, Magazine und Zeitungen werden von Goldscheider genutzt, um auf die Arbeit der amerikanischen Produktion aufmerksam zu machen.[413]

Die Produktion und der Absatz gedeihen. Im Jahr 1947 arbeiten bereits knapp 100 Mitarbeiter bei Goldscheider an über 500 verschiedenen Modellen.[414] Im Frühjahr 1948 wird eine neue Fabrik in Trenton eröffnet und 75 neue Modelle angekündigt.[415] Ab diesem Zeitpunkt werden viele der amerikanischen Goldscheider-Erzeugnisse auch unter der Marke „Goldcrest" vertrieben – vor allem preisgünstige Figuren, Tiere und Raucher-Accessoires.[416] Diese Figuren tragen ein Wappen, in dem „Goldcrest" und „Fine China U.S.A." steht. Einige sind wie schon bereits zuvor bei Goldscheider-Everlast mit einem zusätzlichen Copyright-Stempel versehen, der in einem Kreis ein großes „C" mit der Umschrift „Goldscheider-Everlast Corp." abbildet bzw. jetzt „Goldcrest Ceramics Corp.". Da sich Walter Goldscheider und Jack Ornstein zu dieser Zeit trennen und Walter wohl deshalb in die neuen Fabrikräume umziehen muss, handelt es sich bei Goldcrest wahrscheinlich um eine Tochtergesellschaft von Everlast, die aus der Zusammenarbeit mit Walter einige Modelle mitgenommen hat und diese selbst erzeugt. Zu den produktivsten Künstlerinnen unter dem Label Goldcrest gehört Kathi Urbach.

Walter Goldscheider befindet sich Ende der 1940er Jahre erneut in einer Situation, die seinen vollen Einsatz erfordert. Erstens muss er sich um die Produktion in der neuen Fabrik und den Vertrieb kümmern, zweitens soll ihm die Wiener Fabrik zurückgegeben werden. Nach der offiziellen Rückstellung der Wiener Fabrik an Walter Goldscheider und seiner Rückkehr nach Wien im Jahr 1950 findet weiterhin ein Austausch zwischen der amerikanischen und österreichischen Goldscheider Fabrik statt. Man verkauft gegenseitig die Produkte der anderen Fabrik. Die Wiener beziehen beispielsweise Madonnen aus Trenton, doch es sind vor allem die Amerikaner, die in Wien neben Tänzerinnen auch klassische Figuren wie den „Mozart" für den

cially restored to Walter Goldscheider and after he returned to Vienna in 1950, the exchange between the American and Austrian Goldscheider factories continued. They marketed each other's products. The Viennese, for instance, procured Madonnas from Trenton; however, it was mainly the Americans who ordered dancers and classic figures such as "Mozart" from Vienna for the domestic market.[417] In America, marketing and distributing were no longer done by Everlast but by a new business that was established after Goldscheider separated from Everlast Corporation: Goldscheider Inc., at 1441 Heath Avenue, Trenton, which marketed its own models under the "Goldscheider Ceramics" label.[418] As a result, the stamp on the ceramics changed; "Goldscheider Inc." once again encircled the copyright "C" sign.

Some of the old models were also produced in the new factory alongside new models and pieces to order, such as a seamstress figurine designed for the Singer Sewing Machine Company, which was conceived as a "music box" and played music

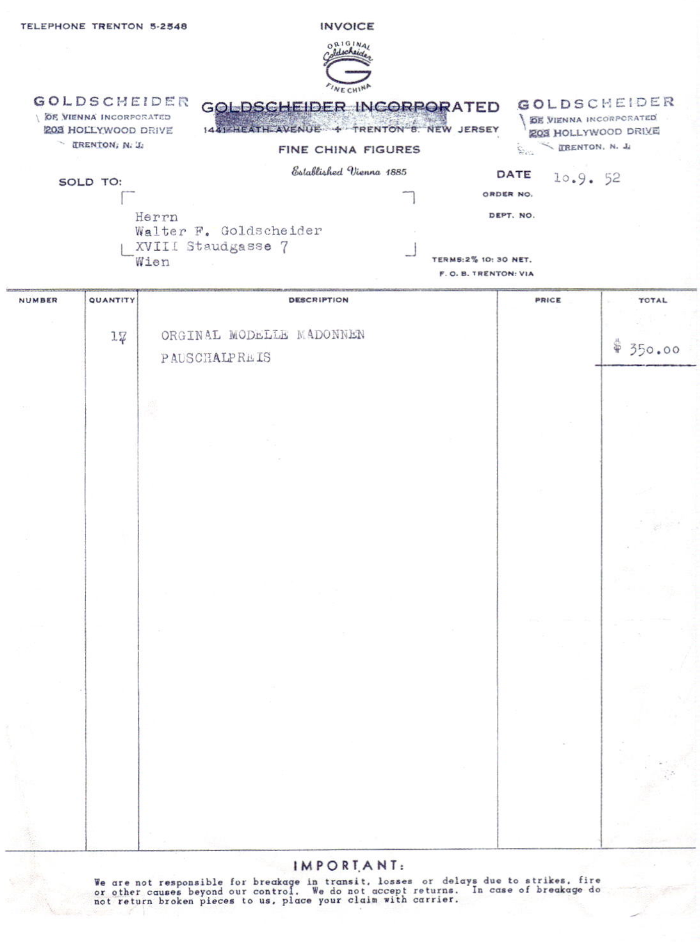
TELEPHONE TRENTON 5-2548

INVOICE

GOLDSCHEIDER OF VIENNA INCORPORATED 203 HOLLYWOOD DRIVE TRENTON, N. J.

GOLDSCHEIDER INCORPORATED
1441 HEATH AVENUE + TRENTON 8, NEW JERSEY
FINE CHINA FIGURES
Established Vienna 1885

GOLDSCHEIDER OF VIENNA INCORPORATED 203 HOLLYWOOD DRIVE TRENTON, N. J.

SOLD TO:
Herrn
Walter F. Goldscheider
XVIII Staudgasse 7
Wien

DATE 10.9. 52
ORDER NO.
DEPT. NO.
TERMS: 2% 10: 30 NET.
F. O. B. TRENTON: VIA

NUMBER	QUANTITY	DESCRIPTION	PRICE	TOTAL
	17	ORGINAL MODELLE MADONNEN PAUSCHALPREIS		$ 350.00

IMPORTANT:
We are not responsible for breakage in transit, losses or delays due to strikes, fire or other causes beyond our control. We do not accept returns. In case of breakage do not return broken pieces to us, place your claim with carrier.

Ill. 108
Im Jahr 1952 wird die amerikanische Fabrik Goldscheider Ceramics geschlossen und die Vertriebsgesellschaft Goldscheider of Vienna Incorporated gegründet, die ebenfalls in Trenton ansässig ist und den Alleinvertrieb der Wiener Erzeugnisse auf dem amerikanischen Markt durchführt. (Quelle: Archiv Goldscheider)

In 1953 the American Goldscheider Ceramics factory closed down and Goldscheider of Vienna Incorporated was founded as a marketing and distributing business, also based in Trenton, with sole rights to marketing the Viennese products on the American market. (Source: Goldscheider Archives)

when it was wound up. Singer gave this figurine to traveling salesman as a reward or to clients as advertising. This figure bears the city names "Trenton and Vienna" as the place of origin beneath "Fine China Goldscheider" and the big "G". The copyright-holder of the figure was, however, "The Singer Mfg. Co". In spring 1952, Erwin launched new models in Chicago, including vases and dishes with underglaze decoration and some figurines and animals from the "Fine China" line.[419] Sylvia Scott contributed two book-ends featuring a "Chinese Poet" and "Chinese Guitarist."

The American factory shut down half way through 1952 and a business devoted exclusively to marketing and distributing was founded: Goldscheider of Vienna Incorporated, with firm headquarters at 203 Hollywood Drive, Trenton[420] (Ill. 108). Erwin Goldscheider, his wife Charlotte and their son, Peter, lived at that address. Walter had the following to say about the fate of the American factory: "We had an art ceramics factory in America but had to give it up because of the awfully cheap Japanese as well as Italian competition. Now my son will concentrate there on selling our Viennese products."[421]

Only a year later, production ceased in Vienna as well. Erwin Goldscheider, who had been familiar with the American market for over fifteen years, was possibly planning to organize marketing and distribution in North America for Carstens, a German ceramics factory that had taken over the Viennese models and trademarks. Since realization of this plan was so uncertain, Erwin looked for other wares and firms to represent.[422] Even before the Viennese business was sold, Walter placed a classified advert under the rubric "Vertretungen" ["Representatives/Agents"] in the February issue of Der Sprechsaal: "Experienced general agent for world-renowned art ceramics factory seeks appropriate sales position for USA in luxury porcelain, crystal, etc. Meeting at the Hannover Fair possible. Detailed offers."[423]

heimischen Markt bestellen.[417] Der Vertrieb erfolgt allerdings in Amerika nicht mehr durch Everlast, sondern durch ein neues Unternehmen, das nach der Trennung von der Everlast Corporation entstanden ist: der Goldscheider Inc. in der 1441 Heath Avenue, Trenton, die unter dem Label „Goldscheider Ceramics" die selbst erzeugten Modelle vertreibt.[418] Dementsprechend ändert sich auch der Stempel auf den Keramiken, in einem Kreis rund um das Copyright-Zeichen „C" steht nun wieder „Goldscheider Inc.".

Einige der alten Modelle werden auch in der neuen Fabrik produziert, genauso wie neue Modelle und Auftragsarbeiten, beispielsweise die für die Nähmaschinenfirma Singer entworfene Figur einer Näherin, die als „music box" konzipiert ist und nach dem Aufdrehen Musik abspielt. Singer verteilt diese Figur als Belohnung für Handelsvertreter oder als Werbemittel an Kunden. Diese Figur trägt unter dem Schriftzug „Fine China Goldscheider" und dem großen „G" als Herkunftsort die Städtenamen „Trenton and Vienna". Das Copyright an dieser Figur besitzt allerdings „The Singer Mfg. Co". Im Frühjahr 1952 präsentiert Erwin in Chicago neue Modelle, wie unterglasiert dekorierte Vasen und Schalen sowie einige Figuren und Tiere aus der „Fine China"-Reihe.[419] Sylvia Scott steuert mit „Chinese Poet" und „Chinese Guitarist" zwei Buchstützen bei.

Doch Mitte des Jahres 1952 wird die amerikanische Fabrik geschlossen und es kommt zur Gründung der reinen Vertriebsgesellschaft Goldscheider of Vienna Incorporated, die ihren Sitz in 203 Hollywood Drive, Trenton, hat[420] (Ill. 108). Hier wohnt auch Erwin Goldscheider mit seiner Frau Charlotte und seinem Sohn Peter. Über das Schicksal der amerikanischen Fabrik äußert sich Walter wie folgt: „Wir hatten in Amerika eine kunstkeramische Fabrik, mussten dieselbe jedoch wegen der furchtbar billigen japanischen, aber auch italienischen Konkurrenz aufgeben. Nun wird mein Sohn sich dort auf den Verkauf unserer Wiener Erzeugnisse konzentrieren."[421]

Doch bereits ein Jahr später wird die Wiener Produktion eingestellt. Erwin Goldscheider, der den amerikanischen Markt seit über fünfzehn Jahren kennt, soll zwar möglicherweise für die deutsche Keramikfabrik Carstens, die die Wiener Modelle und Markenrechte übernommen hat, den Vertrieb in Nordamerika organisieren. Da dies aber unsicher ist, sucht er nach anderen Waren und Firmen, die er vertreten könnte.[422] Schon vor dem Verkauf des Wiener Unternehmens gab Walter daher für seinen Sohn Erwin im Februar-Heft von „Der Sprechsaal" eine Anzeige in der Rubrik „Vertretungen" auf: „Erfahrener Generalvertreter weltbekannter Kunstkeramikfabrik sucht einschlägige Vertretung für USA in Luxusporzellan, Kristall, etc. Besprechung bei Hannover-Messe möglich. Detaillierte Angebote."[423]

Ab 1953 betreibt Erwin „Goldscheider of Vienna, Inc." als Vertriebsunternehmen für europäische Kunsterzeugnisse. Im Jahr 1980 verkauft er die Firma an Leo O'Connell, der den Vertrieb fortsetzt und vor allem Devotionalien aus Spanien, Deutschland und Italien importiert. Vier Jahre später stirbt Erwin Goldscheider; seine Frau Charlotte bleibt in Trenton, sein Sohn Peter F. Goldscheider wird Rechtsanwalt in Palo Alto.[424] Leo O'Connell übergibt die Vertriebsagentur im Jahr 1991 an seinen Sohn Timothy, unter dessen Leitung sich das Unternehmen auf den Import von exklusiven christlichen Objekten aus Italien spezialisiert: Alabaster- und Porzellan-Heiligenfiguren, Madonnen, Kruzifixe und Altarbilder von Künstlern wie Ado Santini, Luciano Cazzolla oder Simona Santini.[425] Die von Goldscheider of Vienna vertriebenen Objekte werden vor allem in Devotionalien-Läden in Nordamerika verkauft. So sind der jüdische Name und die traditionelle Weltmarke Goldscheider heutzutage in den USA ein Gütezeichen für hochwertige christliche Utensilien (Ill. 109–111).

From 1953 Erwin operated "Goldscheider of Vienna, Inc." as a marketing and distributing agency for European art products. In 1980 he sold the firm to Leo O'Connell, who continued the marketing operations and imported primarily devotional objects from Spain, Germany and Italy. Erwin Goldscheider died four years later; his wife Charlotte remained in Trenton while his son, Peter F. Goldscheider, became a lawyer in Palo Alto.[424] In 1991 Leo O'Connell handed the marketing agency to his son Timothy, under whose management the business specialized in the import of exclusive Christian devotional objects from Italy: alabaster and porcelain figures of saints, Madonnas, crucifixes and portable altarpieces by artists such as Ado Santini, Luciano Cazzolla and Simona Santini.[425] The products marketed and distributed by Goldscheider of Vienna were sold mainly in shops specializing in devotional objects. As a result, the Jewish name and traditional world-famous Goldscheider label are today a quality seal for upmarket Christian devotional wares (Ill. 109–111).

Ill. 109–111
Das amerikanische Unternehmen Goldscheider of Vienna ist seit 1980 auf den Import von christlichen Gegenständen aus Europa spezialisiert. (Quelle: Goldscheider of Vienna, Inc., Katalog 2005)

Since 1980, the American company Goldscheider of Vienna has specialized in the import of Christian devotional objects from Europe. (Source: Goldscheider of Vienna, Inc., 2005 catalogue)

Ill. 109

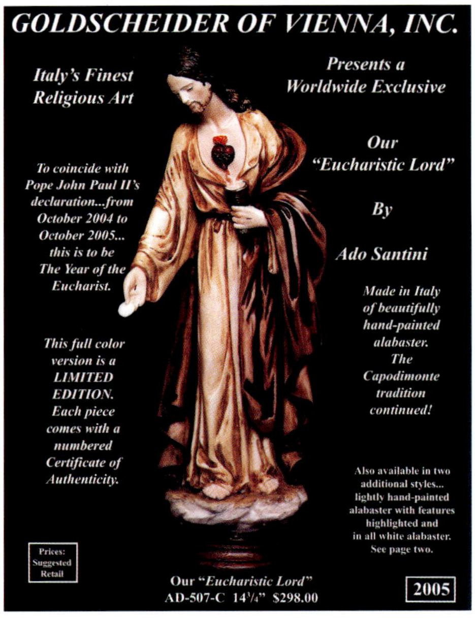

Ill. 110

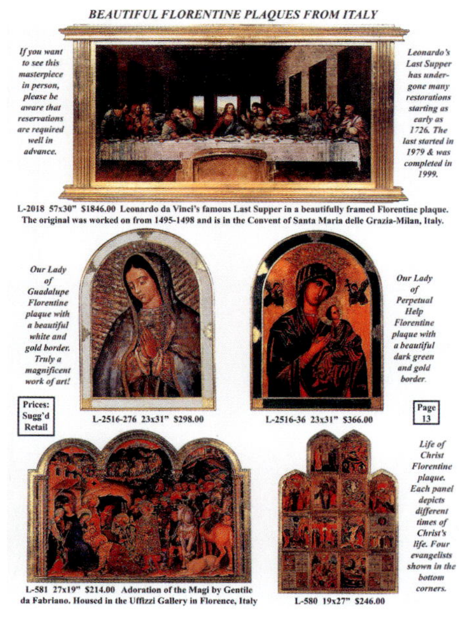

Ill. 111

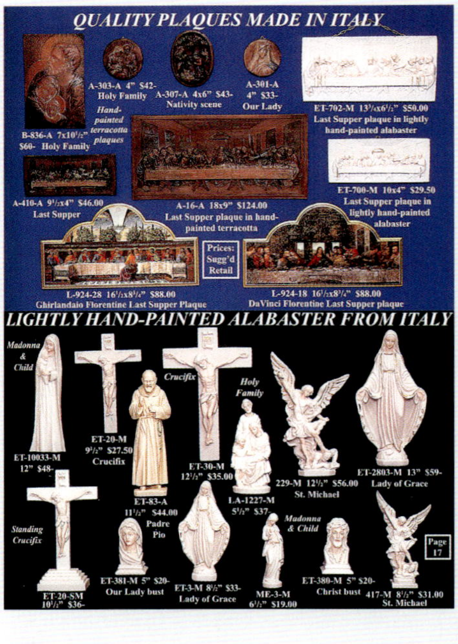

Notes

[391] Information imparted in person by George Newman, a grandson of Richard Kuschnitzky's, on 23 Sept. 2006.
[392] N.N.: Making Goldscheider Ware in America, in: Ceramic Industry, August 1942, p. 43
[393] Letter from Walter Goldscheider, Vienna, to Odette Galoppi, U.S. Department of Justice, Nuremberg, ca 1950: "Our firm is importing our line in the United States since more than 50 years, and our trade mark is well known and appreciated there as one of the best European Ceramics."
[394] N.N.: Goldscheider Vienna Plant Going Full Tilt Once More, in: Trenton Evening Times, 22 Feb. 1952
[395] N.N.: Making Goldscheider Ware in America, in: Ceramic Industry, August 1942, p. 44.
[396] Ibid
[397] N.N.: Goldscheider Vienna Plant Going Full Tilt Once More, in: Trenton Evening Times, 22 Feb. 1952. Richard Kuschnitzky lived at 199 South Broad Street, Trenton.
[398] Letter from Karl Täuber to Arturo Löffler, Buenos Aires, 7 Nov. 1951
[399] Lehner, p. 174. Excerpt from the renton City Directories 1943 to 1950
[400] See http://www.aluminet.org/manufacturers/everlast.html of 28 Sept. 2004
[401] N.N.: Goldscheiders' American Venture In 8th Year of Prosperity, in: Ceramic Industry, September 1947, p. 100
[402] N.N.: Making Goldscheider Ware in America, in: Ceramic Industry, August 1942, p. 43f
[403] Ibid
[404] Ibid
[405] N.N.: The Goldscheiders greet the Canadian market, in: Gift Buyer, December Issue 1945
[406] Letter from Hill-Ouston Co Ltd, Birmingham (UK), to the Goldscheider-Everlast Corporation, 290 Third Street, Trenton, 9 April 1946: "You see there are people over here like Fielding's, Doulton's, Beswick etc. making quite nice Figures, comparable with your Figure of 'Courtesy' at very much less money than this. To my way of thinking there is something solid and heavy about your Figure, and if this is typical of the range you are making today, I do not think they would set this country on fire! You see the Figures you made in Vienna were distinctive beyond words! They had a handwriting all their own, and one could never fail to recognise a Goldscheider Figure. They had the appearance of the very finest porcelain, as distinct from China; there was a delicacy and an 'air' about them which no-one else seemed able to get. [...] I'm obliged to say that before taking this Figure home with me, I submitted it to several customers, none of whom recognised it as a Goldscheider Figure, and I think this is what you want to know."
[407] American Ceramic Society Bulletin, Volume 25, No. 12, 1946, p. 494, quoted in S. Lehner, 174: "Erwin F. Goldscheider, vice-president of the company in charge of production, is responsible for the ceramic discovery. The direct result will be an immediate change from the present china body to an additional and stronger porcelain body. The porcelain base is a combination of American air-floated and magnetically purified kaolin feldspathic rock and silica. The Goldscheider's new fine-china body apparently assumes extra hardness when fired to vitrification, retaining the highest possible degree of pure whiteness. This permits the most delicate underglaze decoration to be applied on its surface, as well as unlimited nuances of color shading. Since it is leadless, it will retain its original sheen, a goal of ceramists for hundred of years. These new translucent porcelains will be decorated exclusively, in gold, platinum and overglaze colors for extreme delicacy of ornament, and in unlimited colors and shades for general production. Productions has begun at once for early shipment to American stores first, then global exportations to follow."
[408] Crockery and Glass Journal Directory, 1952, p. 166, quoted in S. Lehner, p. 174
[409] N.N.: Goldscheiders' American Venture In 8th Year of Prosperity, in: Ceramic Industry, September 1947, p. 100
[410] Ibid
[411] Ibid
[412] Ibid
[413] Ibid, p. 98
[414] Ibid
[415] N.N.: Goldscheiders Open New Plant, in: Ceramic Industry, June 1948, p. 55
[416] Lehner, p. 141, according to the advertisement, in the 1948 China and Glass Red Book Directory. The Goldcrest Ceramics Corporation was also based in Trenton.
[417] Invoice from Goldscheider of Vienna Incorporated, dated 9 Oct. 1952 for 17 "Original Madonna Models price for the lot" as well as a debit notice from Goldscheider Incorporated, dated 11 Dec. 1951 on Dancers (number 8126), Mozart (number 5483) and Madonnas (numbers 8743 and 8307).
[418] Lehner, p. 174
[419] Ceramic Industry, March 1952, p. 52
[420] Invoice from the Viennese factory to Goldscheider of Vienna Inc., 203 Hollywood Drive, Trenton, 7 July 1952
[421] Letter from Walter Goldscheider to Rudolf Knörlein, Lurago d'Erba, 2 July 1952
[422] Letter from Walter Goldscheider, Vienna, to Ernesto Samson, Havana, Cuba, who had been working for thirty years for Goldscheider as a representative, 29 April 1953
[423] Advertisement placed by Walter Goldscheider, Vienna, with the publishers of Der Sprechsaal, Coburg, 19 Jan. 1953
[424] Cf Death Certificate, Office of the Registrar Of Vital Statistics, Ewing Township, Trenton, New Jersey, 11 Jan. 1984. Erwin died on 10 January 1984.
[425] Letter from Timothy O'Connell, Goldscheider of Vienna, Inc., 1040 S. Clinton Avenue, Trenton, to Filipp Goldscheider, Munich, 20 Sept. 2004

Anmerkungen

[391] Persönliche Auskunft durch George Newman, einem Enkel von Richard Kuschnitzky, am 23.9.2006.
[392] N.N.: Making Goldscheider Ware in America, in: Ceramic Industry, August 1942, S. 43
[393] Brief von Walter Goldscheider, Wien, an Odette Galoppi, U.S. Department of Justice, Nürnberg, um 1950: „Our firm is importing our line in the United States since more than 50 years, and our trade mark is well known and appreciated there as one of the best European Ceramics."
[394] N.N.: Goldscheider Vienna Plant Going Full Tilt Once More, in: Trenton Evening Times, 22.2.1952
[395] N.N.: Making Goldscheider Ware in America, in: Ceramic Industry, August 1942, S. 44.
[396] Ebenda
[397] N.N.: Goldscheider Vienna Plant Going Full Tilt Once More, in: Trenton Evening Times, 22.2.1952. Richard Kuschnitzky lebte in Trenton in der 199 South Broad Street.
[398] Brief von Karl Täuber an Arturo Löffler, Buenos Aires, 7.11.1951
[399] Lehner, S. 174. Auszug aus den Trenton City Directories 1943 bis 1950
[400] Siehe http://www.aluminet.org/manufacturers/everlast.html vom 28.9.2004
[401] N.N.: Goldscheiders' American Venture In 8th Year of Prosperity, in: Ceramic Industry, September 1947, S. 100
[402] N.N.: Making Goldscheider Ware in America, in: Ceramic Industry, August 1942, S. 43ff
[403] Ebenda
[404] Ebenda
[405] N.N.: The Goldscheiders greet the Canadian market, in: Gift Buyer, December Issue 1945
[406] Brief der Hill-Ouston Co. Limited, Birmingham (UK), an Goldscheider-Everlast Corporation, 290 Third Street, Trenton, 9.4.1946: „You see there are people over here like Fielding's, Doulton's, Beswick etc. making quite nice Figures, comparable with your Figure of ‚Courtesy' at very much less money than this. To my way of thinking there is something solid and heavy about your Figure, and if this is typical of the range you are making today, I do not think they would set this country on fire! You see the Figures you made in Vienna were distinctive beyond words! They had a handwriting all their own, and one could never fail to recognise a Goldscheider Figure. They had the appearance of the very finest porcelain, as distinct from China; there was a delicacy and an ‚air' about them which no-one else seemed able to get. [...] I'm obliged to say that before taking this Figure home with me, I submitted it to several customers, none of whom recognised it as a Goldscheider Figure, and I think this is what you want to know."
[407] American Ceramic Society Bulletin, Volume 25, No. 12, 1946, S. 494, zitiert nach Lehner, S, 174: „Erwin F. Goldscheider, vice-president of the company in charge of production, is responsible for the ceramic discovery. The direct result will be an immediate change from the present china body to an additional and stronger porcelain body. The porcelain base is a combination of American air-floated and magnetically purified kaolin feldspathic rock and silica. The Goldscheider's new fine-china body apparently assumes extra hardness when fired to vitrification, retaining the highest possible degree of pure whiteness. This permits the most delicate underglaze decoration to be applied on its surface, as well as unlimited nuances of color shading. Since it is leadless, it will retain its original sheen, a goal of ceramists for hundred of years. These new translucent porcelains will be decorated exclusively, in gold, platinum and overglaze colors for extreme delicacy of ornament, and in unlimited colors and shades for general production. Productions has begun at once for early shipment to American stores first, then global exportations to follow."
[408] Crockery and Glass Journal Directory, 1952, S. 166, zitiert nach Lehner, S. 174.
[409] N.N.: Goldscheiders' American Venture In 8th Year of Prosperity, in: Ceramic Industry, September 1947, S. 100
[410] Ebenda
[411] Ebenda
[412] Ebenda
[413] Ebenda, S. 98
[414] Ebenda
[415] N.N.: Goldscheiders Open New Plant, in: Ceramic Industry, June 1948, S. 55
[416] Lehner, S. 141, laut Anzeige in China and Glass Red Book Directory aus dem Jahr 1948. Die Goldcrest Ceramics Corporation hat ihren Sitz ebenfalls in Trenton.
[417] Rechnung von Goldscheider of Vienna Incorporated, vom 9.10.1952, über 17 „Original Modelle Madonnen Pauschalpreis" sowie Debit Notice von Goldscheider Incorporated, vom 11.12.1951 über die Tänzerinnen Nr. 8126 sowie Mozart Nr. 5483 und Madonnen Nr. 8743 und 8307.
[418] Lehner, S. 174
[419] Ceramic Industry, March 1952, S. 52
[420] Rechnung der Wiener Manufaktur an Goldscheider of Vienna Inc., 203 Hollywood Drive, Trenton, 7.7.1952
[421] Brief von Walter Goldscheider an Rudolf Knörlein, Lurago d'Erba, 2.7.1952
[422] Brief von Walter Goldscheider, Wien, an den seit 30 Jahren für Goldscheider arbeitenden Vertreter Ernesto Samson, Havanna, Kuba, 29.4.1953
[423] Anzeigenauftrag von Walter Goldscheider, Wien, an den Verlag „Der Sprechsaal", Coburg, 19.1.1953
[424] Vgl. Certificate of Death, Office Registrar Of Vital Statistics Township of Ewing, Trenton, 11.1.1984. Erwin starb am 10. Januar 1984.
[425] Brief von Timothy O'Connell, Goldscheider of Vienna, Inc., 1040 S. Clinton Avenue, Trenton, an Filipp Goldscheider, München, 20.9.2004.

Die Nachkriegsjahre in Wien (1945–1953)

The Postwar Years in Vienna (1945–1953)

Die Mitarbeiter der Wiener Manufaktur stehen nach dem Verschwinden des Ehepaars Schuster Ende 1944 vor dem Ruin. Dennoch beschließen sie nach der Befreiung Österreichs durch die Alliierten und knapp sechs Monaten der Ungewissheit, die Firma bis zur Rückkehr Walter Goldscheiders als Arbeitsgemeinschaft zu übernehmen und die Produktion weiterzuführen: „Die Arbeitsgemeinschaft hat sich zum Ziel gesetzt, die Firma, die wir im Mai 1945 in sehr desolatem Zustand übernommen haben, wieder zu der Weltgeltung emporzuarbeiten, welche das Unternehmen unter Führung des Herrn Kommerzialrats Goldscheider bis zum Jahre 1938 inne hatte."[426] Der Betrieb fällt nach neu erlassenen Gesetzen der österreichischen Bundesregierung wieder an den „rechtmässigen Besitzer Herrn Kommerzialrat Walter Goldscheider" zurück. Da sich Walter jedoch in Amerika befindet, wird bis zu seiner Rückkehr vom Staatsamt für Industrie, Handel, Gewerbe und Verkehr der frühere langjährige und inzwischen siebzigjährige Prokurist und Oberbuchhalter Josef Paul zum öffentlichen Verwalter bestellt.[427] Auch Karl Täuber, der aus dem Krieg zurückkehrt, wird in dem Unternehmen als Vertreter tätig.[428] Zudem finden viele der unter Schuster entlassenen Mitarbeiter in der Manufaktur allmählich wieder Arbeit.

Der Name der Firma lautet weiterhin „Wiener Manufaktur Josef Schuster, vorm. Friedrich Goldscheider". Auch das bisherige Briefpapier findet Verwendung, genauso wie der Katalog und die Fotos. Neue Investitionen werden aufgrund der allgemeinen Notlage gemieden, die Modelle tragen weiterhin die gleichen Modellnummern und Bezeichnungen. In den Anfängen der Arbeitsgemeinschaft werden zunächst die wenigen von Schuster zurückgelassenen Lagerbestände veräußert.[429] Erst ab Mitte Oktober 1945 beginnt in ganz beschränktem Umfang die Produktion,[430] hierbei wird jedoch nur in Terrakotta produziert.[431] Die Modelle werden mit dem Vorkriegsstempel „Goldscheider WM" und „Made in Austria" versehen.

Die Arbeitsgemeinschaft hat mit verschiedenen Schwierigkeiten zu kämpfen: Es fehlen nicht nur die zur Herstellung benötigten Rohstoffe, auch sind die Formen von den wenigen, zurückgelassenen Modellen in dermaßen schlechtem Zustand, dass neue hergestellt werden müssen. Da es aber keinen Mitarbeiter für die Formenerzeugung gibt, können vorerst weder Formen noch die Modelle produziert werden. Nur die in Repa-

The employees of the Viennese factory were faced with destitution after the Schusters vanished late in 1944. Nevertheless, they decided to keep the firm operating after the liberation of Austria by the Allies. After six months of uncertainty, they resolved to take over the firm until Walter Goldscheider should return, run it as a co-operative and continue production: "The works co-operative has set itself the aim of building up the firm, which we took over in May 1945 in very decrepit condition, to the international quality standard the business represented under the management of Herr Kommerzialrat Goldscheider until the year 1938."[426] In compliance with laws recently enacted by the Austrian government, the business reverted to its "lawful owner, Herr Kommerzialrat Walter Goldscheider." Since Walter was in America, however, Josef Paul, head clerk and head accountant, who was by then seventy years old, was appointed public administrator by the State Office of Industry, Trade, Commerce and Traffic until Walter Goldscheider should return.[427] Karl Täuber, who had returned from the war, was also employed in the firm as a salesman.[428] In addition, many of the employees fired by Schuster were gradually re-hired by the factory.

The name of the firm continued to be "Wiener Manufaktur Josef Schuster, vorm. Friedrich Goldscheider." Even the old letterhead paper was still in use as were the catalogue and photographs. New investment was eschewed due to the general state of emergency. Models continued to bear the same model numbers and kept their names. In the early days of the co-operative, the few warehouse stocks left behind by Schuster were sold.[429] Not until mid-October 1945 could production begin again, albeit on a limited scale;[430] however, products were now only made in terracotta.[431] The models were marked with the pre-war stamps "Goldscheider WM" and "Made in Austria."

The works co-operative had to contend with difficulties of various kinds: not only was there a shortage of raw materials needed for production; the molds for the few models left were in such poor condition that new ones had to be made. However, since there was no employee to make molds, neither molds nor models could be produced for the time being. All that could be done was to repair figures brought in for the purpose, such as "Bäuerliche Venus" ["Peasant Venus"], sent in by Ernst Wahliss, a specialist shop.[432] By the end of 1945, the warehouses had been emptied of stock. In order to be able to begin producing faience, the co-operative placed classified advertisements seeking faience painters and other skilled employees.[433] The production of faiences resumed in January 1946. Vases and bowls were not made for the time being since there were no skilled artisans (turners) needed to make such articles.[434] The business correspondence indicates that demand for the products was high yet the co-operative was still unable to produce and deliver in large quantities.

In June 1946, after seven years of exile, Walter Goldscheider visited the Viennese factory[435] At that time, the export trade was slowly getting started again. Since there were not yet any secretaries in the business trained in foreign languages, the manufactory enquired at the Globus Language Institute what the fees were for having business correspondence translated.[436] The response from abroad was good but orders only rarely came in due to the various restrictions on imports in many places. Late in 1946, the Swedish import agent O. Wiederhielm wrote to Josef Paul, the head clerk: "I should, however, like to inform you that the clientele in this country are longing for the day when they can again receive your wares and especially of the quality they had before Herr Schuster's day. He considerably worsened quality yet I hope that the damage is not irreparable."[437] In a later letter, he complained about the collaboration with Schuster since he "proved to be extremely selfish and, in my opinion, was unreliable."[438] In the same letter, Wiederhielm also pointed out that he had business ties with Walter Goldscheider but was rarely able to sell his American products because "the prices from America are quite expensive for our collection and besides the collection Goldscheider has built up there is not so suitable for us as the Viennese collection was."[439] The manufactory successfully participated in the 1946 Autumn Viennese Fair but had to refuse some orders because the means of production were inadequate. In January and February 1947, production even had to shut down because the Energy Ministry had issued an ordinance due to the prevailing coal and energy crisis according to which all not vitally necessary Viennese industrial plants had to shut down.[440]

The postwar years were marred by hardship in all areas of life. Dwellings and

ratur eingesandten Figuren, beispielsweise die vom Fachgeschäft Ernst Wahliss eingeschickte „Bäuerliche Venus", können repariert werden.[432] Ende 1945 sind die Lagerbestände völlig aufgebraucht. Um mit der Produktion von Fayence beginnen zu können, werden Anzeigen in der Arbeiterzeitung geschaltet, in denen Fayence-Maler und andere Mitarbeiter gesucht werden.[433] Im Januar 1946 kann die Produktion von Fayencen wieder aufgenommen werden. Vasen und Schalen werden vorerst nicht hergestellt, da die geeigneten Arbeitskräfte (Dreher) für diese Artikel fehlen.[434] Die Nachfrage nach den Produkten ist laut der Geschäftskorrespondenz gut, doch die Arbeitsgemeinschaft kann noch nicht in größeren Mengen liefern.

Walter Goldscheider besichtigt im Juni 1946, nach sieben Jahren der Emigration, die Wiener Fabrik.[435] Zu dieser Zeit wird auch der Vertrieb ins Ausland langsam wieder aufgenommen. Da es im Betrieb noch keinen eigenen fremdsprachigen Korrespondenten gibt, fragt die Manufaktur beim Sprachinstitut Globus nach den Honoraren für fremdländische Korrespondenz an.[436] Die Resonanz aus dem Ausland ist gut, Bestellungen gehen aber aufgrund der diversen Importbeschränkungen nur spärlich ein. Der schwedische Import-Agent O. Wiederhielm schreibt Ende 1946 an den Prokuristen Paul: „Ich möchte Ihnen jedoch mitteilen, dass die Kunden hier im Lande sich nach der Zeit sehnen, wo sie wieder Ihre Ware erhalten können und besonders in der Qualität, die Sie vor Herrn Schusters Zeit hatten. Er verschlechterte die Qualität bedeutend, doch hoffe ich, dass der Schaden wieder repariert werden kann."[437] In einem späteren Brief beklagt er die Zusammenarbeit mit Schuster, da sich dieser „als äusserst egoistisch erwies und meiner Auffassung nach nicht zuverlässig war."[438] Wiederhielm weist in diesem Brief auch darauf hin, dass er mit Walter Goldscheider in geschäftlicher Verbindung steht, aber selten seine amerikanischen Erzeugnisse verkaufen kann, denn „die Preise von Amerika stellen sich für unsere Kollektion recht teuer, und ausserdem ist die Kollektion, die Goldscheider dort aufgebaut hat, nicht so geeignet für uns wie die Wiener Kollektion es war."[439] Die Manufaktur nimmt erfolgreich an der Herbstmesse Wien 1946 teil und muss einige Aufträge wegen ungenügender Produktionsmittel abweisen. Im Januar und Februar 1947 muss der Betrieb sogar stillgelegt werden, da das Energieministerium aufgrund der be-

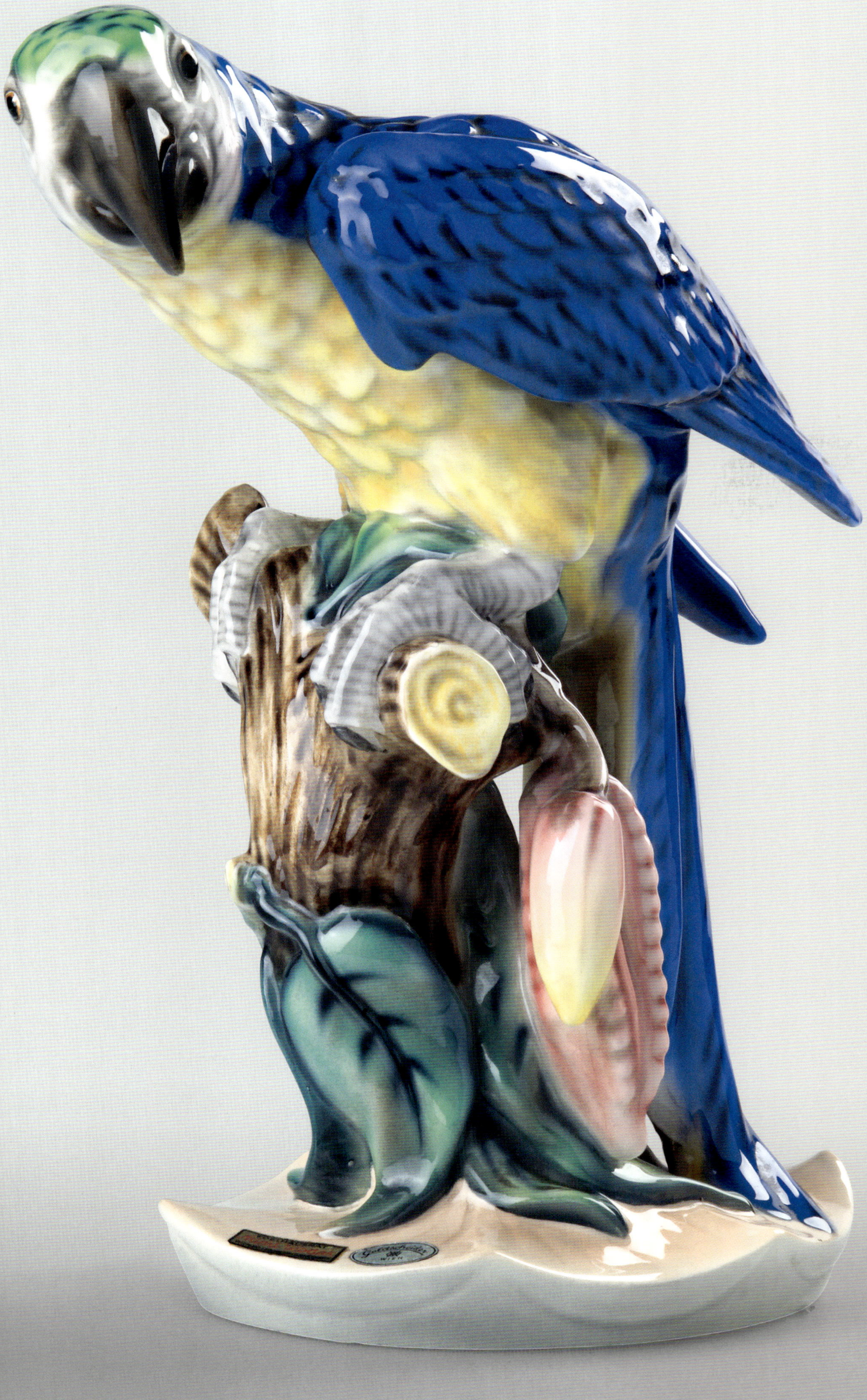

links | left

ModNr. 8734
Papagei | Parrot, Entwurf | design ca. 1941, Collection mad

businesses had been destroyed. People were jobless and hungry, needed new clothing and were dependent on aid from abroad. Not only was there a shortage of both raw materials and skilled workers on the production side. Import restrictions and bans in other countries, temporarily at least, limited sales to the Austrian domestic market. Later, despite the first trade agreements, export was still difficult since only limited export and import quotas were approved. The import and export business was often carried out on an exchange of wares basis, which was called trade by payment in kind.[441] There was no money for traveling salesmen since foreign currency was made available only in limited quantities and foreign currency was primarily reserved for paying for imported wares classified as vital necessities.

Despite the difficult prevailing conditions, demand for the products made by the Goldscheider manufactory seems to have been steady. In February 1947, Mitzi Pinget, a specialist shop in Linz, placed orders primarily for "some masks, Madonnas and other figures, mainly something in faience." The animal figurines were also popular.[442] A letter written to Wiederhielm, the Swedish agent, confirms that demand for "high-quality products in faience" was so great that "we

stehenden Kohle- und Stromkrise eine Verfügung erlässt, nach der sämtliche Industriebetriebe Wiens, so weit sie nicht lebensnotwendig sind, ihre Arbeit einstellen müssen.[440]

Die Jahre nach dem Krieg sind durch die Not in allen Bereichen gekennzeichnet. Die Wohnhäuser und Geschäfte sind zerstört, die Menschen haben keine Arbeit und hungern, brauchen neue Kleidung und sind auf Hilfe vom Ausland angewiesen. Es gibt nicht nur seitens der Produktion einen Rohstoff- und Fachkräftemangel, durch Importsperren anderer Länder ist der Verkauf vorerst auf Österreich begrenzt. Später ist trotz der ersten Handelsabkommen der Export sehr schwierig, da nur bestimmte Aus- und Einfuhrquoten für ausgewählte Produkte genehmigt werden. Oft finden die Importe und Exporte auf Basis des Warenaustauschs statt, der als Kompensationsgeschäft bezeichnet wird.[441] Für Vertreterreisen gibt es aufgrund der knappen Valutazuteilung kein Geld, da Valuta vor allem zum Bezahlen von lebenswichtigen Importwaren benötigt werden.

Trotz der schwierigen Umstände scheinen die Erzeugnisse der Goldscheider Manufaktur weiterhin gefragt zu sein. Im Februar 1947 bestellt das Linzer Fachgeschäft Mitzi Pinget bei der Wiener Manufaktur Friedrich Goldscheider vorzugsweise „einige Masken, Madonnen und sonstige Figuren, hauptsäch-

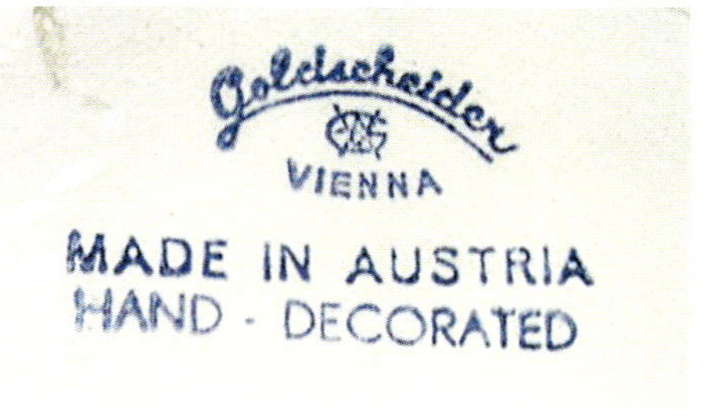

Ill. 112
Der Stempel aus den Jahren 1945–1953 ist fast identisch mit der Version aus den 30er Jahren. Allerdings kann man ab ungefähr der Modell-Nummer 8400 bei der Landesangabe „Made in Austria" von einer Nachkriegsproduktion ausgehen, während „Made in Germany" auf die Zeit des „Anschlusses", also 1938 bis 1945, zu datieren ist.

The stamp used in the years 1945–1953 is almost identical with the 1930s version of it. However, from about model number 8400, it is safe to assume when the country of origin was given as "Made in Austria" that the objects thus marked are postwar whereas "Made in Germany" indicates the "Anschluss" era, that dates an object between 1938 and 1945.

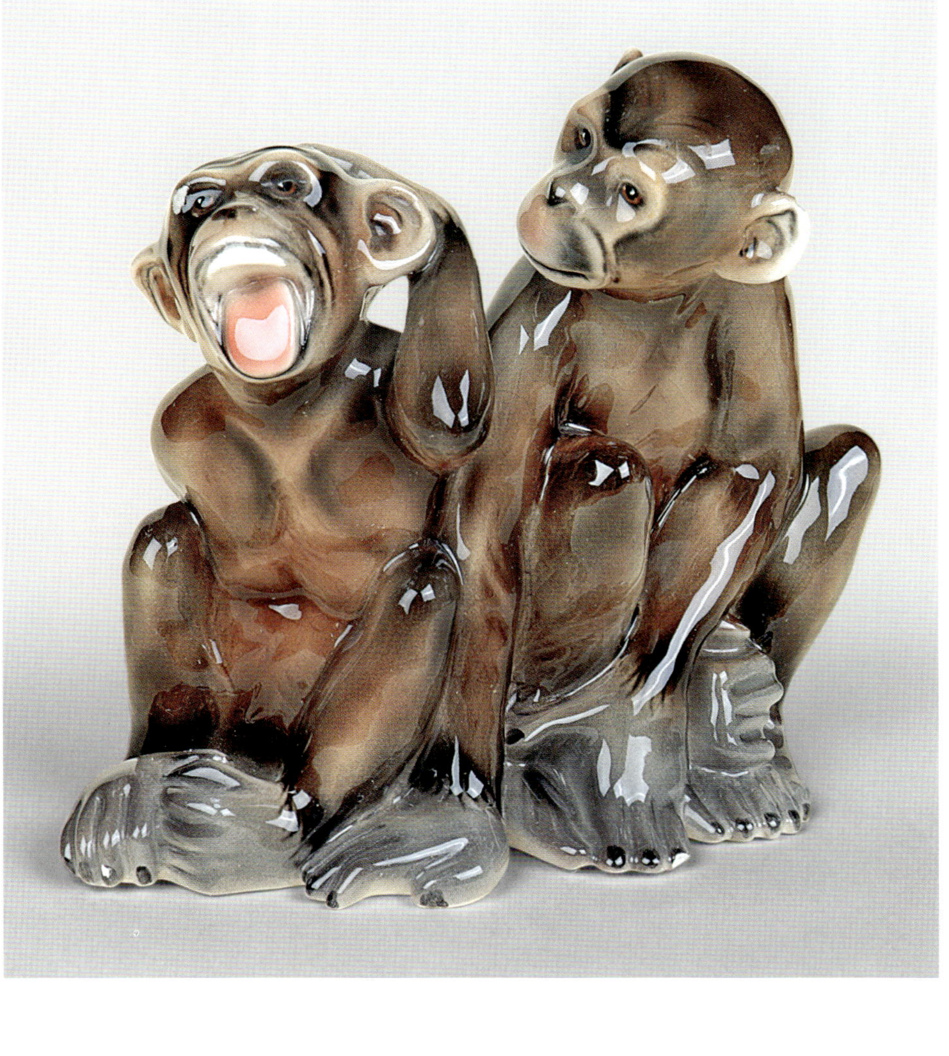

ModNr. 8793
Zwei Affen | Two Monkeys, Entwurf | design ca. 1942, PB Wien 19

lich etwas in Fajence". Beliebt sind aber auch die Tierfiguren.[442] In einem Schreiben an den schwedischen Agenten Wiederhielm wird bestätigt, dass die Nachfrage nach „hochwertigen Erzeugnissen in Fayence" hoch ist, so dass „wir mindestens 200 Arbeiter beschäftigen könnten, um alle Wünsche zu befriedigen. So haben wir unter anderem einen Auftrag von cca Dollar 100.000,– von einer amerik. Firma vorliegen. Diese Firma möchte unsere ganze Jahreserzeugung abnehmen aber auch Herr Goldscheider selbst will den ganzen Alleinverkauf für Amerika haben."[443]

Die Produktion von Fayence ist in dieser Zeit jedoch problematisch, denn Rohmaterial wie weißbrennender Ton, der für die Fayence-Herstellung notwendig ist und von Goldscheider seit Langem aus Colditz in Sachsen bezogen wurde, kann vorerst nicht importiert werden und man sucht nach Bezugsmöglichkeiten in Österreich.[444] Auch die Unterglasurfarben, die früher bei Harrison & Son aus Hanley (Stoke-on-Trent) gekauft worden sind, müssen aus anderen Quellen bezogen werden.[445] Da bisher die Materialien wie Quarz und Tone aus dem Ausland bezogen worden sind, aber dieser Vorrat nun aufgebraucht ist, wurde für die aktuelle Produktion auf inländische Materialien zurückgegriffen. Die Erfahrung mit dieser Qualität ist jedoch schlecht, wie aus einem Brief an die Firma Mitzi Pinget ersichtlich wird: „Speziell Quarz lässt sich nur sehr schwer günstig bei unserer Erzeugung einschalten und das Ergebnis ist, dass unsere jetzigen Erzeugnisse mit der neuen Masse nicht mehr ganz so in der Qualität entsprechen, wie es bis jetzt bei unserer Ware der Fall war. Bei den Fayence-

would need to employ at least two hundred workmen to meet all requests. We have currently on the books among other things an order amounting to approx 100 000 dollars from an American firm. This firm would like to take our entire year's output but Herr Goldscheider also wants to reserve to himself all sales to America."[443]

It was difficult to produce faience at the time since raw materials like clay that fires white, which is needed for making faience and had been procured by Goldscheider for a long time from Colditz, Saxony, could for the time being not be imported and sources had to be sought in Austria.[444] The underglaze colors, that had formerly been bought from Harrison & Son of Hanley (Stoke-on-Trent), also had to be bought from other suppliers.[445] Since materials such as quartz and clays had always had to be procured from abroad but the supply was now exhausted, domestic sources were drawn on for immediate production. Experience showed, however, that the quality was poor as a letter written to the firm of Mitzi Pinget indicates: "Quartz especially is hard to come by inexpensively for use in our production and the result is that the body [used] for our current products no longer meets the quality standards that have hitherto been the case with our wares. Fine hair-line cracks show up in the faience figurines that laymen cannot object to because they don't even notice them and the pieces are otherwise perfect in the color effects, etc."[446] Those pieces were, therefore, sold at a 20 per cent discount. "For us, however, since we have always been used to delivering only first-rate wares, it is extremely unpleasant even though our

Ill. 113, 114
Manche Händler werben gezielt mit den Produkten und der Marke Goldscheider. (Quelle: Archiv Goldscheider)

Some retailers targeted their markets by advertising with Goldscheider products and the Goldscheider label. (Source: Goldscheider Archives)

Ill. 113

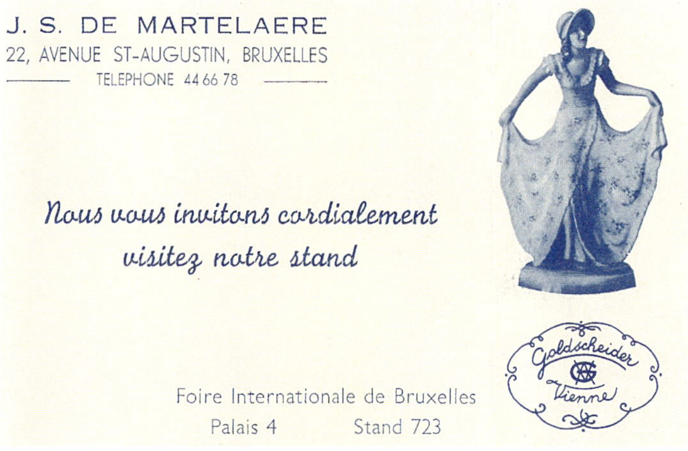

Ill. 114

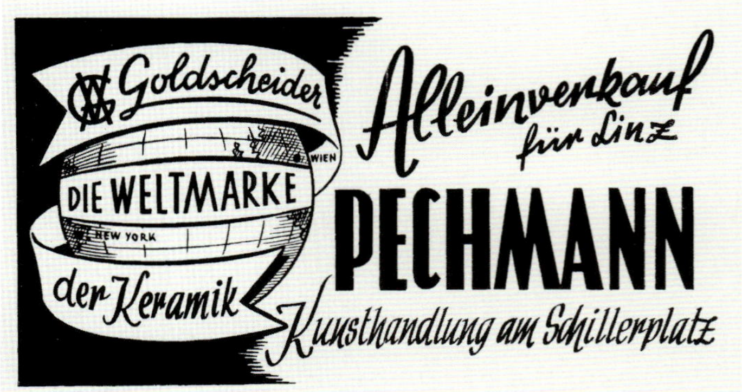

products, despite those hair-line cracks, are still better than those made by the competition."[447] They were both better and more reasonably priced than the American products.[448]

Bad as it had been, the situation deteriorated even further. As a consequence of the general wage rises and prices for various materials, electricity and gas, the factory was forced to raise the "inflation surcharge on the figurative ceramics" from 40 to 100 per cent. Since Goldscheider was even then cheaper than the competition "despite the better quality," it was hoped that, although the prices were now so much higher, the Goldscheider models would sell well.[449] However, export sales also stagnated since many countries maintained a ban on the import of luxury wares. The only positive news seems to have been that Walter Goldscheider would soon be returning to Vienna.[450] Richard Kuschnitzky, a cousin of Lilly's, returned from America in 1948 and in his capacity as a lawyer represented the interests of a clientele that also included Walter Goldscheider in the matter of the Viennese manufactory being restored to him.[451]

The factory showed work at the Autumn and Spring Vienna Fairs but in 1948 business also stagnated.[452] There were special commissions, such as decorating the new bar in the Grand Hotel at Bad Hofgastein with modern masks and figures of dancers but such orders tended to be few and far between.[453] The factory continued with the strategy of supplying only select specialist shops because it did not want to be stuck with a "reputation for making mass-produced articles." The policy was to avoid supplying too many shops in a single town. Some retailers, such as Pechmann, a specialist shop in Linz, occasionally used Goldscheider products and the Goldscheider name as advertising, showing "slides at the cinema, in reading circles and placing advertisements in newspapers"[454] (Ill. 113 and 114).

Potential clients were presented with the catalogue that had once been used by Schuster as well as numerous photographs of models with the following pointer: "Following each illustration you will find the numbers of the models shown and next to them the maximum height and width in cm. When there are several numbers for a single model, this indicates different sizes, as shown by the dimensions given. Figures from groups can also be supplied as individual pieces. For example: portfolio 2, picture 4: 7054 is the

figuren zeigen sich feine Haarrisse, die vom Laien zwar nicht beanstandet werden, weil er sie ja gar nicht bemerkt und die Stücke ja auch sonst in der Farbwirkung usw. einwandfrei sind."[446] Diese Stücke werden daher mit einem Rabatt von 20 Prozent verkauft. „Für uns, wo wir gewohnt waren stets nur 1a-Ware zu liefern ist dies aber äusserst unangenehm, obwohl unsere Erzeugnisse trotz dieser Haarrisse noch immer besser sind als die der Konkurrenz."[447] Sie sind auch besser und preiswerter als die amerikanischen Erzeugnisse.[448]

Die Situation verschlimmert sich jedoch. Denn infolge der allgemeinen Lohnerhöhungen und der Verteuerung diverser Materialien, der Strom- und Gaspreise, ist man gezwungen, den „Teuerungszuschlag bei den figuralen Keramiken" von 40 auf 100 Prozent zu erhöhen. Da man aber bislang gegenüber der Konkurrenz „trotz der besseren Qualität" immer billiger war, hofft man, dass trotz der höheren Preise die Goldscheider-Modelle gut verkauft werden können.[449] Auch der Export zieht nicht an, da in vielen Ländern Einfuhrsperren für Luxuswaren bestehen, und die einzige positive Nachricht scheint zu sein, dass Walter Goldscheider schon bald nach Wien zurückkehren soll.[450] Richard Kuschnitzky, ein Cousin Lillys, kehrt 1948 aus Amerika zurück und vertritt in Wien als Rechtsanwalt unter anderem die Interessen von Walter Goldscheider bezüglich der Rückführung der Wiener Manufaktur.[451]

Die Manufaktur stellt auf der Wiener Herbst- und Frühjahrsmesse aus, aber auch das Geschäftsjahr 1948 ist eher durch Stagnation geprägt.[452] Es gibt zwar spezielle Aufträge, wie die künstlerische Ausgestaltung der neuen Bar im Grand Hotel in Bad Hofgastein mit modernen Masken und Plastiken von Tänzerinnen, diese sind aber sehr selten.[453] Man hält weiterhin an der Strategie fest, nur ausgewählte Fachgeschäfte zu beliefern, da man nicht in den „Verruf eines Massenartikels" kommen möchte. Es werden nie zu viele Läden an einem Ort beliefert. Einige Händler, wie das Linzer Fachgeschäft Pechmann, setzen manchmal die Goldscheider Produkte und den Markennamen als Werbung ein, zeigen „Lichttransparente im Kino, Lesezirkeln und Zeitungsinserate"[454] (Ill. 113 und 114).

Sie erhalten auch den schon bei Schuster eingesetzten Katalog sowie zahlreiche Fotos der Modelle mit dem folgenden Hinweis: „Am Fussende der einzelnen Abbildungen finden Sie die Nummern der jeweiligen Modelle und daneben die maximalen Höhen-

und Breitenangaben in cm. Bei mehreren Nummern für ein Modell handelt es sich um verschiedene Grössen, was aus den Massen ersichtlich ist. Figuren von Gruppen können auch als Einzelstücke geliefert werden. Beispiel: Mappe 2, Bild 4: 7054 ist die Hundegruppe, 7054A der einzelne Windhund, stehend, und 7054B der einzelne Windhund, liegend."[455] Schon vor dem Krieg konnte man aus figuralen Gruppen auch einzelne Figuren beziehen, die dann ebenfalls alle die gleiche Modellnummer trugen.

Die Rückkehr Walter Goldscheiders lässt jedoch auf sich warten. Er und seine Frau Lilly sind bereits siebzig Jahre alt und krank, in den USA sind sie relativ etabliert und es wäre einfacher, wenn Walter nur die nordamerikanische Vertretung für die Wiener Manufaktur übernehmen würde. Andererseits fühlt er sich Wien und seinem traditionsreichen Familienbetrieb verbunden. Dem Fayencefabrikanten F. Gollhammer aus Vöcklabruck bietet er eine finanzielle Beteiligung an der Wiener Manufaktur Friedrich Goldscheider an und schlägt vor, dass dieser die Erzeugung während Walters Tätigkeit in den USA kontrollieren sollte. Gollhammer schlägt jedoch dieses Angebot aus, da er über „keine ungebundenen Vermögenswerte frei verfügen" kann und die Kontrolle des Wiener Unternehmens schwer sei, wenn Walter „nicht über einen verlässlichen Betriebsleiter" in Wien verfügt.[456] Er macht auch Walter darauf aufmerksam, dass „Sie jedenfalls in den USA unter glücklicheren Schaffensverhältnissen leben als hier, doch ich verstehe auch Ihr Gefühl, dass Sie als geborener Wiener auf Ihr Wien nicht ganz verzichten wollen."[457]

Letztendlich entscheidet sich Walter, die mit der Stadt Wien verbundene Geschichte seines Familienunternehmens in Wien fortzusetzen. Sein Sohn Erwin bleibt in den USA, Walter und Lilly Goldscheider kehren im Februar 1950 – nach über

ModNr. 8792
Revuegirl auf Schildkröte | Show Girl Riding a Turtle, Entwurf | design ca. 1942,
© Wien Museum

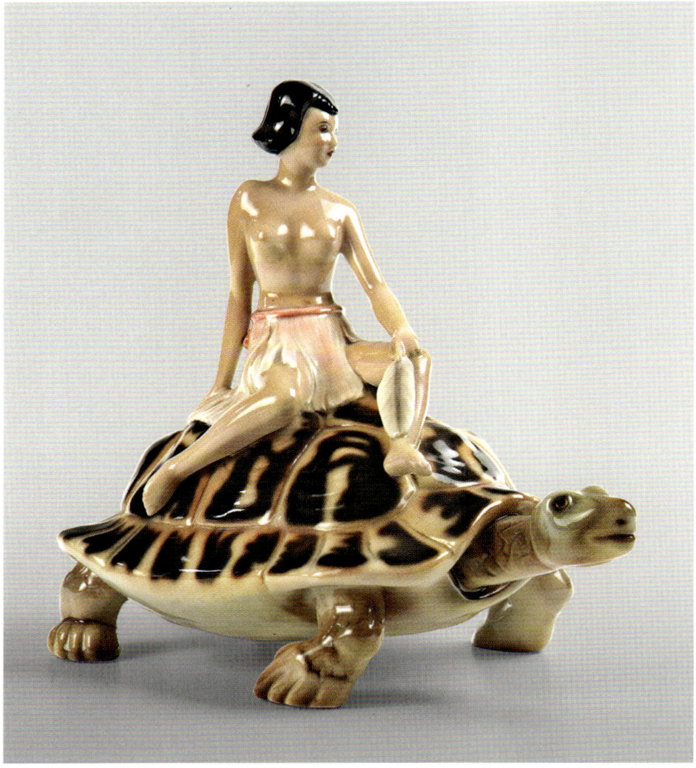

group of dogs, 7054A the individual Whippet, standing, and 7054B the individual Whippet, reclining."[455] Even before the war it had been possible to order individual figures from figurative groups, all of which then also bore the same model number.

The Walter Goldscheiders took time over their return. Walter and his wife Lilly were by now seventy years old and in poor health. They had established quite a firm footing in the US and it would have been simpler if Walter had contented himself with representing the Viennese manufactory in North America. On the other hand, he felt strong ties to Vienna and his family business with its long tradition. He offered F. Gollhammer, a faience manufacturer in Vöcklabruck a financial share in the Viennese Manufactory Friedrich Goldscheider and suggested that Gollhammer should monitor production there while Walter was operating in the US. Gollhammer, however, declined the offer since he "had no liquid assets" and it would be difficult to monitor the Viennese business if Walter "did not [have] a reliable managing director" in Vienna.[456] He also pointed out to Walter that "you in any case live under more fortunate working conditions than here but I also understand your feeling that you, as a native of Vienna, do not want to do without your Vienna entirely."[457]

Walter ultimately decided to continue the history of his family business that was linked with the city of Vienna in that city. His son Erwin remained in the US but Walter and Lilly Goldscheider returned – after over ten years in exile – to Austria in February 1950.

The public administration under Josef Paul ended on 7 April 1950 and the factory was officially restored to Walter Goldscheider.[458] The business again bore the old name "Wiener Manufaktur Friedrich Goldscheider" and Lilly Goldscheider was restored to her position as sole head clerk.[459] Dr. Richard Kuschnitzky, who had represented the manufactory for several years in his legal capacity, also had exclusive powers of attorney from August 1950.[460]

The factory was, however, in a dilapidated condition. Schuster, the "Aryanizer" had not only stolen thousands of models and molds; even the "old photographs, color sketches," the library and "two bronze busts" of Friedrich Goldscheider, founder of the firm, had vanished.[461] Walter nevertheless believed he would be able to link up with past success, especially with such outstanding personnel, who had been with Goldscheider even before the war.[462] On arriving, Walter began at once in building up the Viennese business. He contacted his former agents worldwide and looked for potential markets for the Goldscheider products, which he described as "highly artistic, figurative faiences such as: dancers, children, animals, Modern Ceramics, Madonnas, etc."[463] Marketing and distribution in Austria was done in part through Haas & Czjzek and the firm's own representatives.[464] On 4 April 1950, Walter wrote to the Swedish import-export agent, O. Wiederhielm: "I arrived here from the States 6 weeks ago to take over my factory again, which has been restored to me. Unfortunately, I have not found it in very good condition and it is now my task to re-organize it and return it to its old repute and importance. What encourages me about this task and makes me expect it to be carried out successfully are the two following advantageous circumstances: first, we still possess a root-stock of old, well-trained and experienced skilled workers and, second, the quality of our products is really excellent and possibly even better than before the war. In addition, I am going to refurbish the collection by procuring new models

zehn Jahren Emigration – nach Österreich zurück.

Die öffentliche Verwaltung unter Josef Paul wird zum 7. April 1950 beendet und die Fabrik an Walter Goldscheider offiziell zurückgegeben.[458] Das Unternehmen erhält wieder den alten Namen „Wiener Manufaktur Friedrich Goldscheider" und Lilly Goldscheider wird erneut als Einzelprokuristin eingesetzt.[459] Auch Dr. Richard Kuschnitzky, der die Manufaktur seit mehreren Jahren rechtlich vertritt, fungiert ab August 1950 als Einzelprokurist.[460]

Die Fabrik ist jedoch in einem desolaten Zustand. Der Ariseur Schuster hat vor seiner Flucht nicht nur tausende von Modellen und Formen weggebracht, auch die „alten Photographien, Farbskizzen", die Bibliothek und „zwei Bronzebüsten" des Gründers der Firma, Friedrich Goldscheider, sind verschwunden.[461] Dennoch glaubt Walter vor allem wegen der ausgezeichneten Arbeitskräfte, die schon vor dem Krieg bei Goldscheider arbeiteten, an vergangene Erfolge anknüpfen zu können.[462] Er fängt nach seiner Ankunft sofort mit dem Ausbau des Wiener Unternehmens an und kontaktiert weltweit seine früheren Agenten und sucht potenzielle Abnehmer für die Goldscheider Produkte, die er als „hochkünstlerische, figurale Fayencen wie: Tänzerinnen, Kinder, Tiere, Moderne Keramik, Madonnen etc." bezeichnet.[463] Der Vertrieb in Österreich erfolgt teilweise durch Haas & Czjzek und durch eigene Vertreter.[464] Am 4. April 1950 schreibt er an den schwedischen Import und Export Agenten O. Wiederhielm: „Ich bin vor 6 Wochen, aus den Staaten kommend, hier eingetroffen um meine Fabrik, die an mich rückgestellt wurde, wieder zu übernehmen. Leider habe ich dieselbe in keiner sehr guten Verfassung vorgefunden und ist es nun meine Auf-

Ill. 115
Im Herbst 1949 stellt die Arbeitsgemeinschaft nicht auf der Wiener Herbstmesse aus, sondern lädt potenzielle Kunden ins eigene Musterlager im Fabrikgebäude ein. Erst ab 1950 firmiert man wieder unter dem alten Namen Wiener Manufaktur Friedrich Goldscheider (Quelle: Archiv Goldscheider)

In autumn 1949, the co-operative not only showed work at the Vienna Autumn Fair but also invited potential customers to their own sample warehouse on the factory premises. Not until 1950 was the firm again registered under the old name of Wiener Manufaktur Friedrich Goldscheider. (Source: Goldscheider Archives)

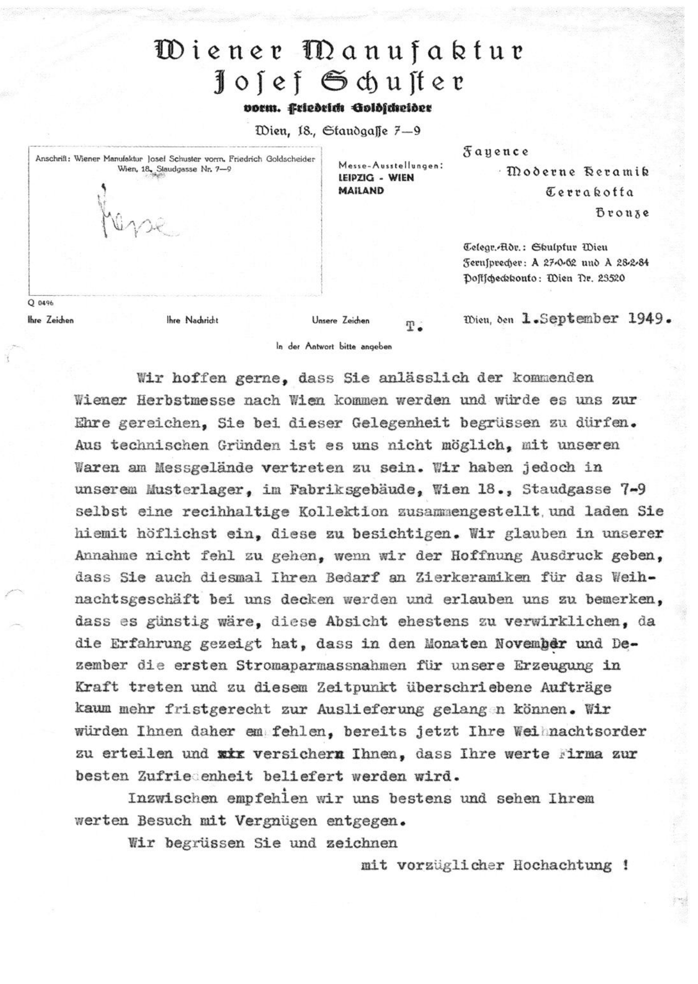
Wiener Manufaktur
Josef Schuster
vorm. Friedrich Goldscheider
Wien, 18., Staudgasse 7–9

Anschrift: Wiener Manufaktur Josef Schuster vorm. Friedrich Goldscheider
Wien, 18, Staudgasse Nr. 7–9

Messe-Ausstellungen: LEIPZIG - WIEN MAILAND

Fayence
Moderne Keramik
Terrakotta
Bronze

Telegr.-Adr.: Skulptur Wien
Fernsprecher: A 27-0-62 und A 28-2-84
Postscheckkonto: Wien Nr. 23520

Q 0496

Ihre Zeichen | Ihre Nachricht | Unsere Zeichen T. | Wien, den 1. September 1949.

In der Antwort bitte angeben

Wir hoffen gerne, dass Sie anlässlich der kommenden Wiener Herbstmesse nach Wien kommen werden und würde es uns zur Ehre gereichen, Sie bei dieser Gelegenheit begrüssen zu dürfen. Aus technischen Gründen ist es uns nicht möglich, mit unseren Waren am Messgelände vertreten zu sein. Wir haben jedoch in unserem Musterlager, im Fabriksgebäude, Wien 18., Staudgasse 7-9 selbst eine recihhaltige Kollektion zusammengestellt und laden Sie hiemit höflichst ein, diese zu besichtigen. Wir glauben in unserer Annahme nicht fehl zu gehen, wenn wir der Hoffnung Ausdruck geben, dass Sie auch diesmal Ihren Bedarf an Zierkeramiken für das Weihnachtsgeschäft bei uns decken werden und erlauben uns zu bemerken, dass es günstig wäre, diese Absicht ehestens zu verwirklichen, da die Erfahrung gezeigt hat, dass in den Monaten November und Dezember die ersten Stromsparmassnahmen für unsere Erzeugung in Kraft treten und zu diesem Zeitpunkt überschriebene Aufträge kaum mehr fristgerecht zur Auslieferung gelangen können. Wir würden Ihnen daher empfehlen, bereits jetzt Ihre Weihnachtsorder zu erteilen und versichern Ihnen, dass Ihre werte Firma zur besten Zufriedenheit beliefert werden wird.

Inzwischen empfehlen wir uns bestens und sehen Ihrem werten Besuch mit Vergnügen entgegen.

Wir begrüssen Sie und zeichnen

mit vorzüglicher Hochachtung !

gabe, dieselbe zu reorganisieren und zu ihrem alten Ansehen und Bedeutung zurückzuführen. Was mich bei dieser Aufgabe ermutigt und deren erfolgreiche Durchführung erwarten lässt, sind die beiden folgenden vorteilhaften Tatsachen: Erstens besitzen wir noch immer einen Stock von den alten, gut geschulten und erfahrenen Facharbeitern und zweitens ist die Qualität unserer Erzeugnisse eine ganz vorzügliche und womöglich noch besser als vor dem Kriege. Ausserdem werde ich die Kollektion, durch die Beschaffung von neuen Modellen, interessanter und moderner gestalten. Selbstredend bemühe ich mich gleichzeitig mit allen Exportländern die alten Verbindungen wieder aufzunehmen und deshalb wende ich mich heute an Sie mit der Anfrage ob Sie für die Vertretung meiner Firma Interesse haben, in welchem Falle ich Ihren Vorschlägen, zur raschesten Wiederaufnahme des Geschäftes ehebaldigst entgegensehe."[465]

in a more interesting and modern way. Of course I am also trying to reforge the old links with all export countries and that is why I am asking you today whether you would be interested in representing my firm, in which case I would be looking forward to your suggestions for the quickest possible resumption of business."[465]

Walter signed marketing and distributing contracts with several agents. In November 1950, the contract on representing the firm in Sweden was to be signed with O. Wiederhielm (Ill. 116). The terms of this contract regulate the duration of business, commissions and the ordering of a sample collection, which the representative had to buy from the Goldscheider manufactory at a 25 per cent discount. Photographs and catalogues would also be placed at representatives' disposal.[466] The catalogue was still the Schuster catalogue but a sticker was glued over the name the firm had had under

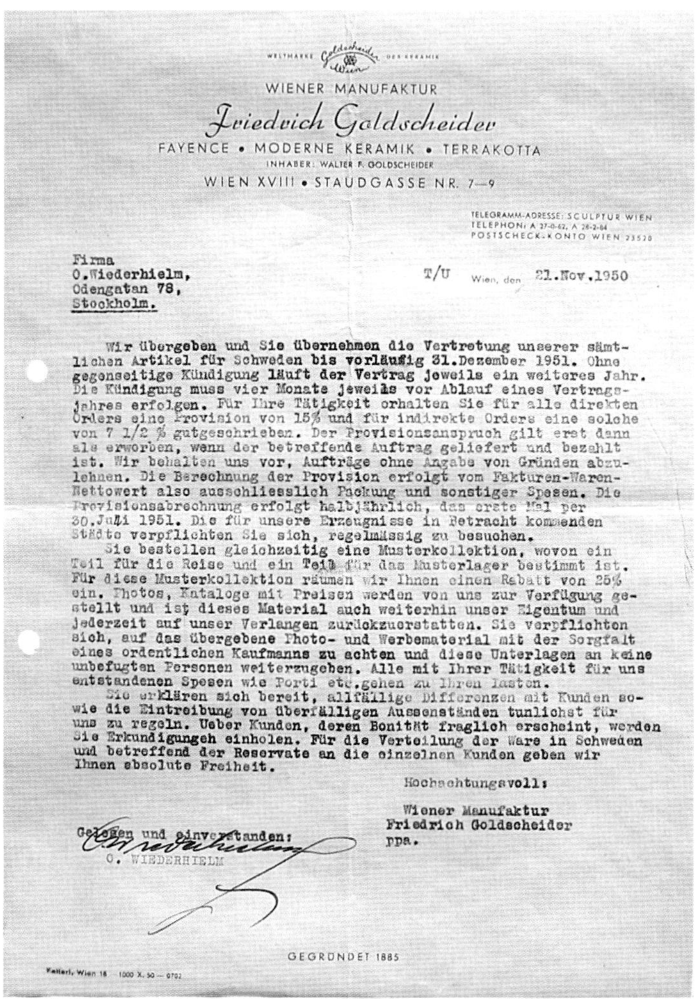

WELTMARKE Goldscheider Wien DER KERAMIK

WIENER MANUFAKTUR

Friedrich Goldscheider

FAYENCE • MODERNE KERAMIK • TERRAKOTTA

INHABER: WALTER F. GOLDSCHEIDER

WIEN XVIII • STAUDGASSE NR. 7–9

TELEGRAMM-ADRESSE: SCULPTUR WIEN
TELEPHON: A 27-0-62, A 26-2-84
POSTSCHECK-KONTO WIEN 23520

Firma
O.Wiederhielm,
Odengatan 78,
Stockholm.

T/U Wien, den 21.Nov.1950

Wir übergeben und Sie übernehmen die Vertretung unserer sämtlichen Artikel für Schweden bis vorläufig 31.Dezember 1951. Ohne gegenseitige Kündigung läuft der Vertrag jeweils ein weiteres Jahr. Die Kündigung muss vier Monate jeweils vor Ablauf eines Vertragsjahres erfolgen. Für Ihre Tätigkeit erhalten Sie für alle direkten Orders eine Provision von 15% und für indirekte Orders eine solche von 7 1/2 % gutgeschrieben. Der Provisionsanspruch gilt erst dann als erworben, wenn der betreffende Auftrag geliefert und bezahlt ist. Wir behalten uns vor, Aufträge ohne Angabe von Gründen abzulehnen. Die Berechnung der Provision erfolgt vom Fakturen-Waren-Nettowert also ausschliesslich Packung und sonstiger Spesen. Die Provisionsabrechnung erfolgt halbjährlich, das erste Mal per 30.Juli 1951. Die für unsere Erzeugnisse in Betracht kommenden Städte verpflichten Sie sich, regelmässig zu besuchen.

Sie bestellen gleichzeitig eine Musterkollektion, wovon ein Teil für die Reise und ein Teil für das Musterlager bestimmt ist. Für diese Musterkollektion räumen wir Ihnen einen Rabatt von 25% ein. Photos, Kataloge mit Preisen werden von uns zur Verfügung gestellt und ist dieses Material auch weiterhin unser Eigentum und jederzeit auf unser Verlangen zurückzuerstatten. Sie verpflichten sich, auf das übergebene Photo- und Werbematerial mit der Sorgfalt eines ordentlichen Kaufmanns zu achten und diese Unterlagen an keine unbefugten Personen weiterzugeben. Alle mit Ihrer Tätigkeit für uns entstandenen Spesen wie Porti etc.gehen zu Ihren Lasten.

Sie erklären sich bereit, allfällige Differenzen mit Kunden sowie die Eintreibung von überfälligen Aussenständen tunlichst für uns zu regeln. Ueber Kunden, deren Bonität fraglich erscheint, werden Sie Erkundigungeh einholen. Für die Verteilung der Ware in Schweden und betreffend der Reservate an die einzelnen Kunden geben wir Ihnen absolute Freiheit.

Hochachtungsvoll:

Wiener Manufaktur
Friedrich Goldscheider
ppa.

Gelesen und einverstanden:
O. WIEDERHIELM

GEGRÜNDET 1885

Kellerl, Wien 18 1000 X. 50 – 0792

Ill. 116
Der Vertrag mit dem Vertreter in Schweden sieht eine Provision von 15 Prozent für direkte sowie 7,5 Prozent für indirekte Bestellungen vor. (Quelle: Archiv Goldscheider)

The contract with the representative in Sweden provided for commissions of 15 per cent on direct orders and 7.5 per cent on indirect orders. (Source: Goldscheider Archives)

Ill. 117
Die unterschiedlichen Briefköpfe der Manufaktur dokumentieren die ungewisse und wechselhafte Zeit seit Kriegsende. Italienisch- und französischsprachige Briefe sollen Anfang der fünfziger Jahre den Export erleichtern. (Quelle: Archiv Goldscheider)

The various Goldscheider letterheads document the uncertain and turbulent years after the war ended. Letters in Italian and French were to facilitate export in the early 1950s. (Source: Goldscheider Archives)

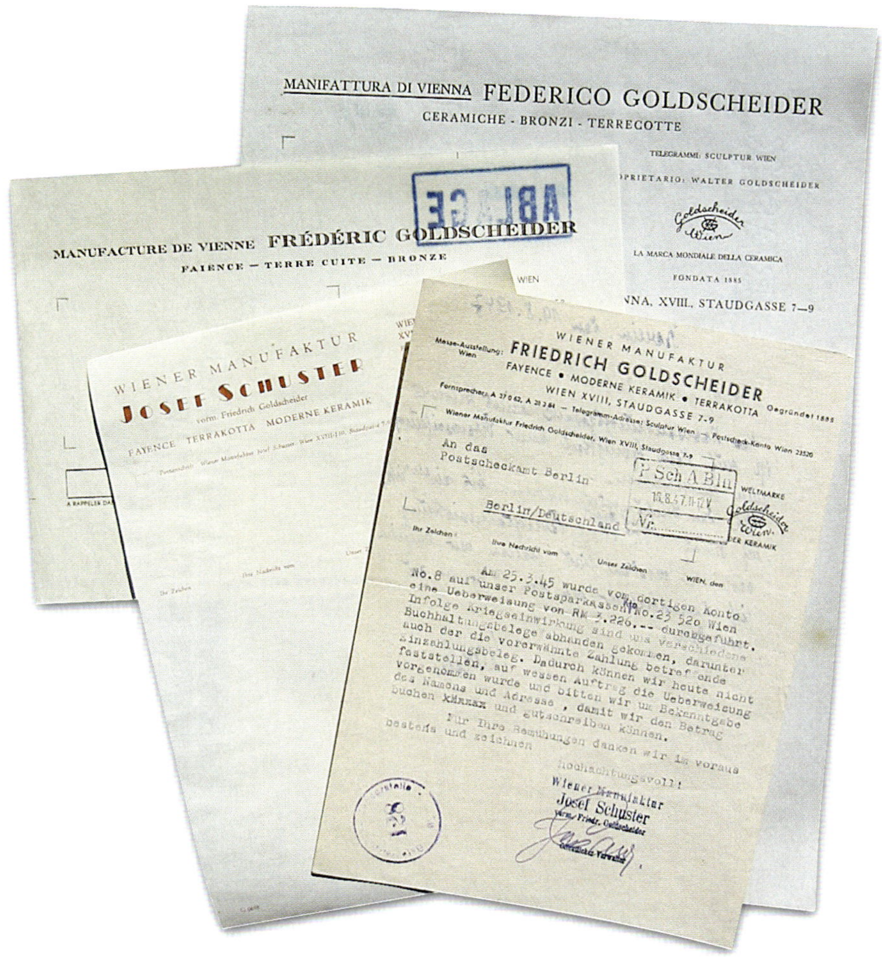

Schuster that replaced it with the new, in fact the old, name "Wiener Manufaktur Friedrich Goldscheider."[467] The Schuster letterhead paper was now only used as scrap paper by employees. Walter had new letter paper for business correspondence printed as well special versions for the Italian and French markets (Ill. 117). Retailers were given not only the catalogue with the price list but also brand signs for their shop windows; later representatives were even equipped with art prints.[468]

In a later letter, Wiederhielm drew Walter's attention to an interesting competitor: "The demand for ornamental porcelain and faiences is currently very great and the Goldscheider label from England will soon be seen in every shop. What is regrettable is that the English factory uses the same advertising signs as you have (WM) and the name Goldscheider. This circumstance will lead to much confusion"[469] (cf Ill. 118). This suggests that the brothers Walter and Marcell, who had had no contact with each other since they fell out in the late 1920s, were, moreover, still competing against each other.[470] Other members of the family supported Walter in rebuilding the Viennese factory. Dr. Fritz Goldscheider (Ill. 119), Camillo's son and a nephew of Walter's, had worked for the firm both before and during the war. He lived in Cairo and traveled from there into the Near East. After his return to Vienna, he founded the Import-Export firm IMEXA and also helped his uncle financially to rebuild the business.[471]

Walter's son Erwin represented the firm on the important American market. Lilly Goldscheider, who had been seriously ill when she returned to Vienna with Walter, underwent an operation not long after she arrived there.[472] Since her health did not improve, Walter and Lilly returned to Trenton

Walter unterzeichnet mit mehreren Agenten die Vertriebsverträge. Im November 1950 soll der Vertretungsvertrag für Schweden mit O. Wiederhielm geschlossen werden (Ill. 116). In diesem Vertrag werden die Dauer, Provisionen sowie die Bestellung einer Musterkollektion geregelt, die der Vertreter mit 25 Prozent Rabatt von der Goldscheider Manufaktur beziehen muss. Auch Fotos und Kataloge werden den Vertretern zur Verfügung gestellt.[466] Bei dem Katalog handelt es sich um den Schuster-Katalog, allerdings wird der Schuster-Firmenname überklebt und durch den neuen beziehungsweise alten Namen „Wiener Manufaktur Friedrich Goldscheider" ersetzt.[467] Das Schuster Briefpapier wird von den Mitarbeitern nur noch als Schmierzettel benutzt und Walter lässt neue Geschäftsausstattung sowie spezielle Versionen für den italienischen und französischen Markt drucken (Ill. 117). Die Händler erhalten außer dem Katalog mit Preisliste noch Markenschilder für die Schaufenster, später werden die Vertreter sogar mit Kunstdrucken ausgestattet.[468]

Wiederhielm macht in einem späteren Brief Walter auf einen interessanten Konkurrenten aufmerksam: „Die Nachfrage nach Zierporzellan und Fayancen ist augenblicklich sehr gross, und die Marke Goldscheider aus England sieht man bald in jedem Geschäft. Das Bedauerliche ist, dass die englische Fabrik die gleichen Reklameschilder verwen-

Ill. 118
Diese Korrekturfahne einer Anzeige aus dem Handelswerbeblatt Austria at Work, Ende des Jahres 1950, ähnelt stark den englischen Werbemotiven von Marcell Goldscheider. (Quelle: Archiv Goldscheider)

This galley proof of an advertisement from the commercial advertising sheet Austria at Work in late 1950 closely resembles the English advertising motifs used by Marcell Goldscheider. (Source: Goldscheider Archives)

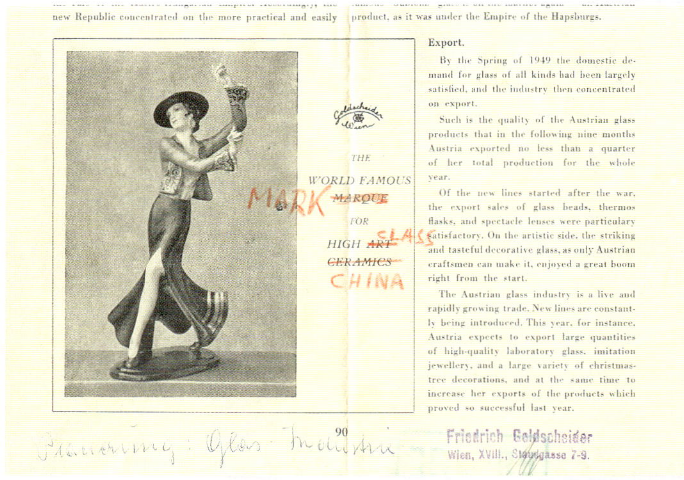

new Republic concentrated on the more practical and easily product, as it was under the Empire of the Hapsburgs.

THE WORLD FAMOUS ~~MARQUE~~ MARK FOR HIGH ~~ART~~ GLASS ~~CERAMICS~~ CHINA

Export.

By the Spring of 1949 the domestic demand for glass of all kinds had been largely satisfied, and the industry then concentrated on export.

Such is the quality of the Austrian glass products that in the following nine months Austria exported no less than a quarter of her total production for the whole year.

Of the new lines started after the war, the export sales of glass beads, thermos flasks, and spectacle lenses were particulary satisfactory. On the artistic side, the striking and tasteful decorative glass, as only Austrian craftsmen can make it, enjoyed a great boom right from the start.

The Austrian glass industry is a live and rapidly growing trade. New lines are constantly being introduced. This year, for instance, Austria expects to export large quantities of high-quality laboratory glass, imitation jewellery, and a large variety of christmas-tree decorations, and at the same time to increase her exports of the products which proved so successful last year.

90

Friedrich Goldscheider
Wien, XVIII., Stoudgasse 7-9.

det welche Sie haben (WM) und den Namen Goldscheider. Dieser Umstand wird zu vielen Verwechslungen führen"[469] (vgl. Ill. 118). Dies deutet darauf hin, dass die seit Ende der zwanziger Jahre zerstrittenen Brüder Walter und Marcell keinen Kontakt pflegten und sich nur noch als Konkurrenten gegenüberstanden.[470] Andere Familienmitglieder unterstützen Walter beim erneuten Aufbau der Wiener Manufaktur. Dr. Fritz Goldscheider (Ill. 119), der Sohn von Camillo und ein Neffe Walters, ist schon vor und während des Krieges als Vertreter für die Manufaktur tätig. Er lebt in Kairo und bereist von dort aus den Nahen Osten. Nach seiner Rückkehr nach Wien gründet er die Import und Export Handelsgesellschaft IMEXA und hilft nun finanziell seinem Onkel beim Wiederaufbau des Unternehmens.[471]

Auf dem wichtigen amerikanischen Markt fungiert Walters Sohn Erwin als Vertreter. Lilly Goldscheider, die schwer krank mit Walter nach Wien zurückkehrte, wird bald nach ihrer Ankunft operiert.[472] Da sich ihr gesundheitlicher Zustand nicht bessert, kehren Walter und Lilly Anfang 1951 nach Trenton zurück, wo Lilly im Frühjahr 1951 stirbt.[473] Walter ist vom Tod seiner Frau, mit der er über 40 Jahre zusammenlebte, schwer getroffen: „Meine Frau war die Sonne und der wahre Inhalt meines Lebens und da sie verloschen, muss mir die Welt jetzt viel dunkler erscheinen. Materielle Verluste und sonstige Bitternisse erscheinen mir dagegen unwesentlich."[474] Walter kehrt erst Anfang 1952 nach Wien zurück und lebt in der Pension Arenberg am Stubenring, wo er öfters von Karl Täuber besucht wird, der ihm als Sekretär assistiert.[475] Täuber ist neben Erwin Goldscheider auch als Einzelprokurist eingetragen.[476]

early in 1951, where Lilly died in spring that year.[473] Walter took the death of his wife of more than forty years very hard: "My wife was the Sun and real content of my life and since it has been extinguished, the world now looks much darker to me. Material losses and other bitter hardships seem inconsequential compared to this."[474] Walter did not return to Vienna until early in 1952 and lived at Pension Arenberg on the Stubenring, where Karl Täuber, who assisted him as a secretary, often visited him.[475] Täuber was also registered, like Erwin Goldscheider, as possessing power of attorney.[476]

The artists who worked most for Goldscheider around 1950 were the same ones who had done so before the war: Stephan Dakon, Kurt Goebel and Karl Sailer. What was called a "contract of cession" had been concluded with the modeler Karl Sailer, who had worked for Goldscheider before the war (Ill. 120–122). In that contract, of 25 August 1948, he ceded replication rights for fifteen models, registered under the model numbers 8805 to 8821, to the factory. In return, Sailer was to receive royalties amounting to 7 per cent of the net wholesale price on "all entirely paid deliveries." For extra costs incurred by the firm, such as "return of goods, replacements, discounts at point of sale, etc," 5 per cent was deducted from the royalty payment.[477] The royalties were to be paid every six months. On 13 February 1950, Sailer sold the rights to models 8823 to 8826 for royalties amounting to 10 per cent.[478] In July 1950, in the course of re-registration of the firm under the old name "Wiener Manufaktur Friedrich Goldscheider," the two contracts of cession were fused, with models 8833 to 8836 added, and a uniform royalty of 8 per cent on all models.[479] These model

Ill. 119
Dr. Fritz Goldscheider unterstützt finanziell seinen Onkel Walter beim Wiederaufbau der Wiener Manufaktur. (Quelle: Privatarchiv Dr. Arlette Leupold-Löwenthal)

Dr. Fritz Goldscheider financially supported his uncle Walter in rebuilding the Viennese factory. (Source: Dr. Arlette Leupold-Löwenthal Private Archives)

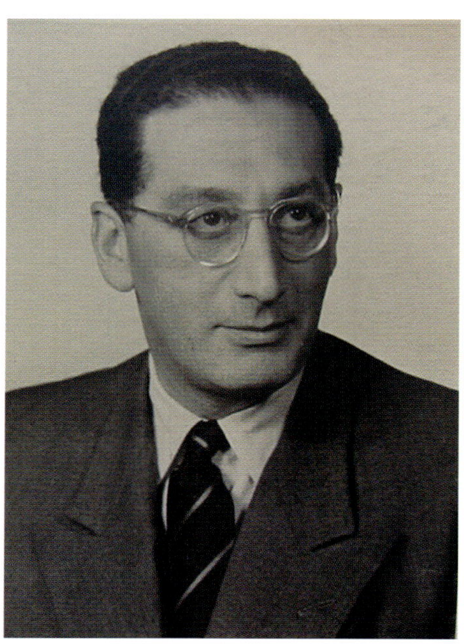

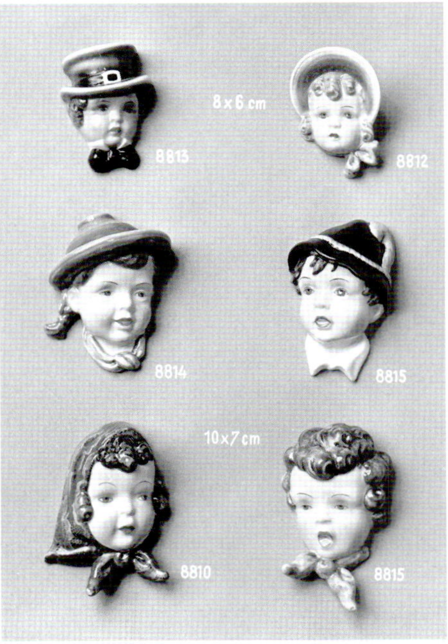

Ill. 120

Ill. 120–122
In einem Vertrag zwischen der Manufaktur und dem Modelleur Karl Sailer werden ihm für mehrere Modelle Tantiemen in Höhe von 7 Prozent zugesichert. Dazu gehören die Wandmasken mit den Modell-Nummern 8810–8815 sowie das stehende Kinderpaar 8821. (Quelle: Archiv Goldscheider)

In a contract between the factory and the modeler Karl Sailer, royalties amounting to 7 per cent were guaranteed to him. The models in question include the wall masks (models 8810-8815) and a pair of children, standing (model 8821). (Source: Goldscheider Archives)

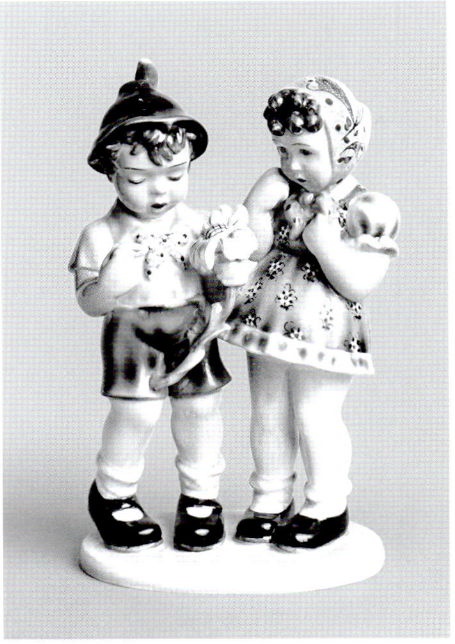

Ill. 121

Zu den meistbeschäftigten Künstlern gehören auch um 1950 Stephan Dakon, Kurt Goebel und Karl Sailer. So wird beispielsweise mit dem Modelleur Karl Sailer, der vor dem Krieg schon für Goldscheider tätig war, eine so genannte Zession geschlossen (Ill. 120–122). In diesem Vertrag vom 25. August 1948 überträgt er die Vervielfältigungsrechte für 15 Modelle an die Manufaktur, die mit den Modellnummern 8805 bis 8821 registriert werden. Hierfür erhält Sailer eine Tantieme von 7 Prozent vom Netto-Engrosverkaufspreis, von „allen voll bezahlten Lieferungen." Für Sonderkosten wie „Retouren, Vergütungen, Kassaskontis etc." werden 5 Prozent von den Tantiemenbezügen abgezogen.[477] Die Tantiemen werden halbjährlich ausgezahlt. Am 13. Februar 1950 verkauft Sailer die Rechte an den Modellen 8823 bis 8826 zu einem Tantiemensatz von 10 Prozent.[478] Im Juli 1950 werden im Rahmen der Umfirmierung der Manufaktur zu dem alten Namen „Wiener Manufaktur Friedrich Goldscheider" beide Zessionen zusammengefasst, um die Modellnummern 8833 bis 8836 ergänzt und alle Modelle einheitlich mit 8 Prozent als Tantieme verrechnet.[479] Anhand dieser Modellnummern kann

numbers make it possible to estimate that no more than three hundred and sixty models (8440–8800) were designed under Josef Schuster since, between 1945 and 1950, the years during which the factory was managed by the co-operative, only a few new models were bought in. Instead, due to the general economic stagnation and uncertain circumstances, models that had proved their worth were the ones usually reproduced.

Even in later years, the purchase of new models continued to be a matter for careful consideration. On 24 October 1952, for instance, a contract was concluded with Kurt Goebel in which eleven masks including the molds and even the glaze recipes needed were to be procured from him. Those masks bear the model numbers 8884 to 8897. This indicates that, since the contract had been concluded with Sailer in July 1950 (the last model number was 8836), not even fifty new models had been created in over two years. Most of the Goebel models are wall masks and figures of animals and children. In this contract, Goebel secured the rights from Walter Goldscheider and the factory "to repair cast masks themselves the way models are" so that it would be possible to take casts from such most precisely executed pieces."[480] Goebel committed himself to painting a couple of pieces with his glazes at the factory and to inform the workmen at Goldscheider on "producing from patterns." For the masks and these services, Goebel re-

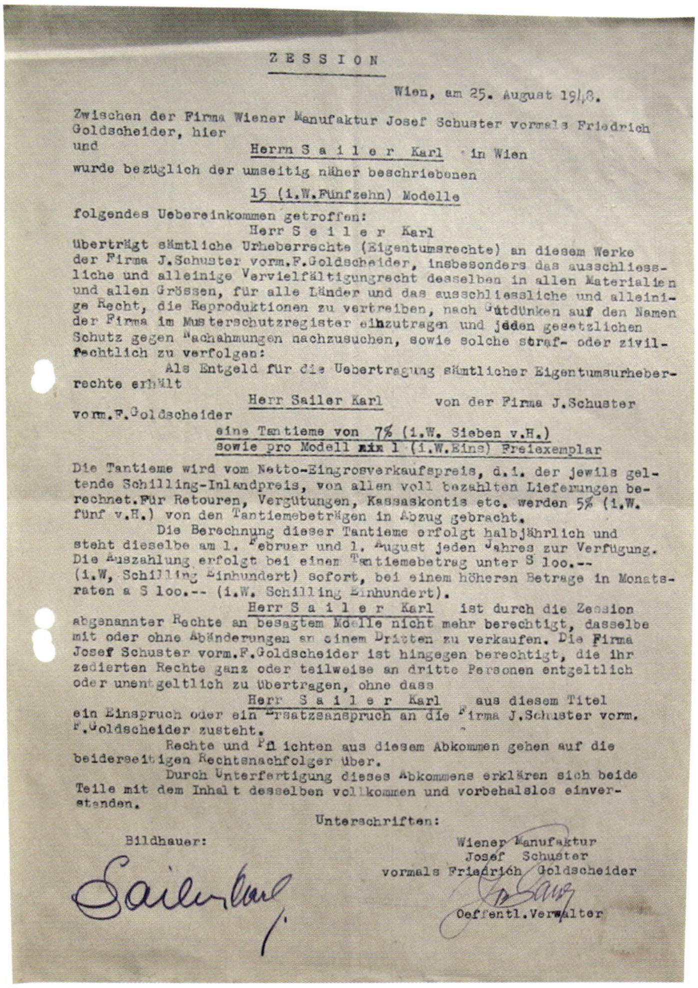

Z E S S I O N

Wien, am 25. August 1948.

Zwischen der Firma Wiener Manufaktur Josef Schuster vormals Friedrich Goldscheider, hier
und **Herrn S a i l e r Karl** in Wien
wurde bezüglich der umseitig näher beschriebenen
15 (i.W.Fünfzehn) Modelle
folgendes Uebereinkommen getroffen:
Herr S e i l e r Karl
überträgt sämtliche Urheberrechte (Eigentumsrechte) an diesem Werke der Firma J.Schuster vorm.F.Goldscheider, insbesonders das ausschliessliche und alleinige Vervielfältigungrecht desselben in allen Materialien und allen Grössen, für alle Länder und das ausschliessliche und alleinige Recht, die Reproduktionen zu vertreiben, nach Gutdünken auf den Namen der Firma im Musterschutzregister einzutragen und jeden gesetzlichen Schutz gegen Nachahmungen nachzusuchen, sowie solche straf- oder zivilrechtlich zu verfolgen:

Als Entgeld für die Uebertragung sämtlicher Eigentumsurheberrechte erhält
Herr Sailer Karl von der Firma J.Schuster vorm.F.Goldscheider
eine Tantieme von 7% (i.W. Sieben v.H.)
sowie pro Modell 1 (i.W.Eins) Freiexemplar

Die Tantieme wird vom Netto-Eingrosverkaufspreis, d.i. der jewils geltende Schilling-Inlandpreis, von allen voll bezahlten Lieferungen berechnet. Für Retouren, Vergütungen, Kassaskontis etc. werden 5% (i.W. fünf v.H.) von den Tantiemebeträgen in Abzug gebracht.

Die Berechnung dieser Tantieme erfolgt halbjährlich und steht dieselbe am 1. Februar und 1. August jeden Jahres zur Verfügung. Die Auszahlung erfolgt bei einem Tantiemebetrag unter S 100.-- (i.W, Schilling Einhundert) sofort, bei einem höheren Betrage in Monatsraten a S 100.-- (i.W. Schilling Einhundert).

Herr S a i l e r Karl ist durch die Zession abgenannter Rechte an besagtem Modelle nicht mehr berechtigt, dasselbe mit oder ohne Abänderungen an einem Dritten zu verkaufen. Die Firma Josef Schuster vorm.F.Goldscheider ist hingegen berechtigt, die ihr zedierten Rechte ganz oder teilweise an dritte Personen entgeltlich oder unentgeltlich zu übertragen, ohne dass
Herr S a i l e r Karl aus diesem Titel ein Einspruch oder ein Ersatzanspruch an die Firma J.Schuster vorm. F.Goldscheider zusteht.

Rechte und Pflichten aus diesem Abkommen gehen auf die beiderseitigen Rechtsnachfolger über.

Durch Unterfertigung dieses Abkommens erklären sich beide Teile mit dem Inhalt desselben vollkommen und vorbehalslos einverstanden.

Unterschriften:

Bildhauer: Sailer Karl

Wiener Manufaktur Josef Schuster vormals Friedrich Goldscheider
Oeffentl. Verwalter

Ill. 122

ModNr. 8850
„Geiger" | "Violinist",
Karl Sailer, ca. 1950,
PB Wien 19

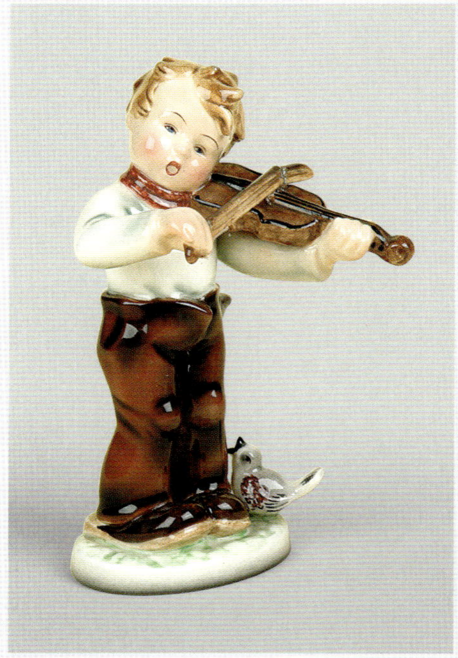

man abschätzen, dass unter Josef Schuster nicht mehr als 360 Modelle (8440–8800) entworfen worden sind. Denn zwischen 1945 und 1950, also der Verwaltung der Manufaktur unter der Arbeitsgemeinschaft, werden nur wenige neue Modelle eingekauft, sondern aufgrund der allgemeinen stagnierenden Situation und ungewisser Lage vor allem bewährte Modelle reproduziert.

Doch auch in den späteren Jahren wird der Ankauf neuer Modelle genau überlegt. Am 24. Oktober 1952 wird beispielsweise mit Kurt Goebel ein Vertrag geschlossen, in dem von ihm elf Masken samt Formen und der notwendigen Glasurrezepte bezogen werden. Diese Masken tragen die Modellnummern 8884 bis 8897, also sind seit dem im Juli 1950 geschlossenen Vertrag mit Sailer (letzte Modellnummer war 8836) in über zwei Jahren nicht mal fünfzig neue Modelle entstanden. Die meisten der Goebel-Modelle sind Wandmasken, Tier- und Kinderfiguren. Goebel sichert in diesem Vertrag gegenüber Walter Goldscheider und der Manufaktur zu, die „ausgeformten Masken selbst modellmässig zu retouchieren", damit man in der Lage ist, auf „Grund dieser genauest ausgeführten Stücke die Abgüsse herzustellen".[480] Er verpflichtet sich auch, ein oder zwei Stücke mit seinen Glasuren bei der Manufaktur zu bemalen, um die Arbeiter bei Goldscheider über die „mustergemässe Erzeugung" zu informieren. Für die Masken und all diese Leistungen erhält er eine einmalige Abschlagzahlung von 2.000 Schilling, die in Wochenraten zu 250 Schilling bezahlt werden. Mit dieser Summe sind auch alle Urheberrechte abgegolten.[481] In der Zession erteilt Goebel, genauso wie vorher Sailer, der Firma Friedrich Goldscheider das „ausschliessliche und alleinige Vervielfältigungsrecht" für die erworbenen Masken in „allen Materialien und allen Größen für alle Länder".[482]

Die Zusammenarbeit mit Kurt Goebel verläuft besonders eng. Im Herbst 1952 macht Walter Goldscheider ihn im Rahmen von Retuschearbeiten darauf aufmerksam, dass dieser vor allem die „jetzt noch ziemlich plump und etwas kurz" wirkenden Hände verfeinern solle. Denn „in allen Ländern, aber besonders in USA wird auf die Zartheit und Feinheit der Hände ganz besonders Wert gelegt."[483]

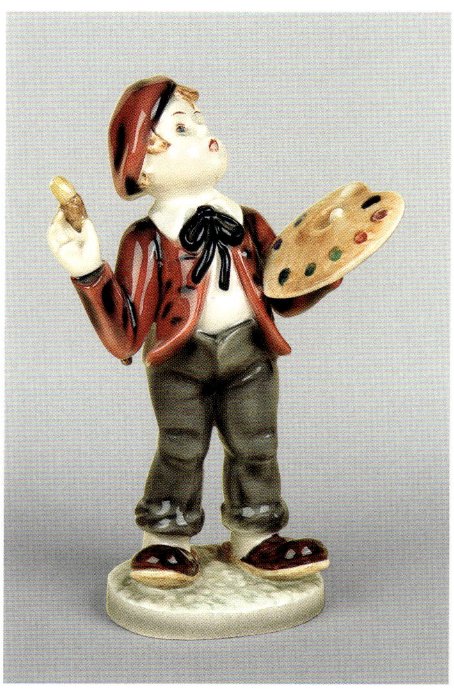

ModNr. 8851
„Maler" | "Painter",
Karl Sailer, ca. 1950,
PB Wien 19

ceived a one-off payment of 2000 Schillings. That sum also included all copyright fees.[481] In the contract of cession, Goebel, like Sailer before him, granted the firm of Friedrich Goldscheider the "exclusive and sole right of replication" for the masks acquired in "all materials and all sizes for all countries."[482]

The collaboration with Kurt Goebel was particularly close. In autumn 1952, Walter Goldscheider pointed out to Goebel in connection with repairing that he should do finer work, especially on the hands, which "at present still look rather clumsy and somewhat short." After all, "in all countries and especially the USA, particular value is attached to the delicacy and fineness of hands."[483]

Walter often exerted an influence on model design. He wrote to Dakon, for instance, in December 1952: "Today I viewed the models with engineer Guse and find the 'Mozart' and the female figure of a 'Tyrolean girl' very well caught. On the other hand, however, I think that the 'Tyrolean Boy' has legs that are too short and I request you to lengthen these somewhat. The modern head is better but I still think the hairstyle is slightly too thick so that it looks like a hood. I suggest that you should remove the back of it so that the hair looks much lighter; it might also be a good idea to remove the side curls but I don't know how that will look."[484]

Walter attached enormous importance to authenticity and correctness in details. In the case of two figures, he pointed out to Dakon that the costumes did not match and that Dakon should "make sure in the

Applied Arts Museums library [...] or the National Library" and advised him "to take tracing paper with him and trace costumes."[485] Walter consulted the Oberösterreichisches Landesmuseum in Linz in the matter of the correct costume for the figure of a woman from Linz with a golden hood.[486]

Stephan Dakon not only provided models of his own; he also repaired other sculptor's models at that time for Goldscheider. He was paid 46.09 Schillings in royalties for the first half of the year 1947.[487] For the first half of 1951, his royalties amounted to a total of 753.38 Schillings.[488] Karl Sailer, who did a lot of work for Goldscheider, was paid 3602.40 Schillings in royalties for the latter half of 1952.[489] Royalties amounting to 191.48 Schillings were paid to Njura Lorenzl, widow of Josef Lorenzl, who had died in 1950, for the first half of 1951.[490]

A disagreement arose, on the other hand, with the sculptor Adolf Prischl, who had worked for Goldscheider in the 1930s and during the war for Keramos. Since 1947 Prischl had had a workshop of his own for art ceramics.[491] Prischl made reproductions of some of the masks produced at Goldscheider in the 1930s and sold them under his own name. In the presence of lawyers, the two parties reached the following agreement: Prischl committed himself "as from immediately to cease production and further marketing in all sizes and materials of the mask that looks so similar to our 'Eve,' model 6774 that it is mistaken for it. We for our part agree not to raise any objections to the further production and further marketing of the two other mask models," which resembled the Goldscheider masks 8041 and 6427[492] (cf Ill. 123).

Even though Walter had bought in new models, there was still a lack of objects suitable for sale on the market of the time. That this was so is also clearly shown in a complaint made by the Cuban representative, Ernesto Samson: "I have been hearing all sorts of unpleasant things because I always show the same things. [...] The worst is that I cannot even say to a client that I am expecting new objects for then no one would have ordered anything but would have put me off until the new models had arrived."[493] Walter was on the look-out for models in the following areas: "Modern figures and animals, naturalistic animals, modern heads and masks, devotional models but strictly liturgical, not modern ones. Other new, original models and ideas for a new genre,

Walter übt oft Einfluss auf die Gestaltung der Modelle aus, so schreibt er an Dakon im Dezember 1952: „Ich habe heute mit Herrn Ing. Guse die Modelle besichtigt und finde den ‚Mozart' sowie die weibliche Figur ‚Tirolerin' sehr gut gelungen. Dagegen finde ich, dass der ‚Tiroler Bub' zu kurze Beine hat und ich ersuche Sie, diese etwas zu verlängern. Der moderne Kopf ist besser, aber ich glaube, dass die Frisur noch immer etwas zu stark ist, so dass sie den Eindruck einer Haube macht. Ich schlage vor, dass Sie vom Hinterkopf entsprechend wegnehmen, damit die Frisur wesentlich leichter wirkt; ev. könnte man auch die Seitenlocken wegnehmen, doch weiss ich nicht, wie dies wirkt."[484]

Er legt auch viel Wert auf Authentizität und Richtigkeit der Details. Bei zwei Figuren macht er Dakon darauf aufmerksam, dass die Kostüme nicht zueinander passen und dass Dakon „sich darüber in der Bibliothek des Museums für angewandte Kunst [...] oder in der Nationalbibliothek versichern" soll und rät ihm, „Pauspapier mitzunehmen und Kostüme abzuzeichnen."[485] Bei der Figur einer Linzerin mit Goldhaube lässt sich Walter vom Oberösterreichischen Landesmuseum in Linz wegen der passenden Tracht beraten.[486]

Stephan Dakon liefert nicht nur eigene Modelle ab, sondern retuschiert für Goldscheider in dieser Zeit auch die Modelle anderer Bildhauer. Für das erste Halbjahr des Jahres 1947 werden ihm Tantiemen in Höhe von 46,09 Schilling ausbezahlt,[487] für das 1. Halbjahr 1951 sind es insgesamt 753,38 Schilling.[488] Der viel beschäftigte Karl Sailer erhält für das 2. Halbjahr 1952 eine Abrechnung in Höhe von 3.602,40 Schilling[489] und von Josef Lorenzls Werken, der im Jahr 1950 stirbt, fällt für seine Frau Njura für das erste Halbjahr 1951 eine Tantiemenzahlung von 191,48 Schilling ab.[490]

Mit dem Bildhauer Adolf Prischl, der in den 30er Jahren bei Goldscheider und während des Krieges bei Keramos beschäftigt war und seit 1947 eine eigene Werkstätte für Kunstkeramik besitzt,[491] kommt es hingegen zu einem Streit. Prischl stellt von einigen in den 30er Jahren bei Goldscheider hergestellten Masken Reproduktionen her und verkauft sie unter seinem eigenen Namen. Im Beisein von Anwälten einigen sich beide Parteien wie folgt: Prischl verpflichtet sich, „ab sofort die Erzeugung und den weiteren Vertrieb in allen Grössen und Materialien, bei jener Maske, welche unserem Modell ‚Eva' 6774 zum Verwechseln ähnlich ist, einzustellen. Dagegen erheben wir keinen Einspruch gegen die weitere Erzeugung und den weiteren Vertrieb der anderen beiden Maskenmodelle", die den Goldscheider Masken 8041 und 6427 ähneln[492] (vgl. Ill. 123).

Trotz der von Walter neu eingekauften Modelle herrscht ein Mangel an marktgerechten Verkaufsobjekten. Dies wird auch in einer Beschwerde des kubanischen Vertreters Ernesto Samson deutlich: „Ich habe allerlei unerfreuliches zu hören bekommen, weil ich immer die gleichen Sachen zeige. [...] Das schlimmste ist, dass ich nicht einmal den Kunden sagen konnte, dass ich Neuheiten erwarte, denn dann hätte niemand bestellt, sondern man hätte mich auf das Eintreffen der Neuheiten vertröstet."[493] Walter sucht vor allem nach Modellen in den Bereichen „Moderne Figuren und Tiere, Naturalistische Tiere, Moderne Köpfe und Masken, Religiöse Modelle, jedoch nicht modern, sondern streng lithurgisch. Sonstige neue originelle Modelle und Ideen, die für einen neuen Genre besonders in kleiner Größe, cca 15 cm, und leicht reproduzierbar wären."[494] Er weiß, dass er die Modellpalette unbedingt an den

especially in a small size, approx 15 cm, and easy to reproduce."[494] He was fully aware that he definitely had to adapt his model range to more modern tastes and sounded out various artists about collaboration with him. The artists were first requested to send in pictures of their models with sizes indicated and, if they were suitable, to provide "models in Plasticine or, which we would prefer, in plaster of Paris, already repaired."[495] It cost between 300 and 500 Schillings to make such plaster of Paris models ranging in size from 15 to 22 cm, depending of course on materials and finish. Walter emphasized in this connection that he "was especially interested in models that are easy to make, that is, [have] as few places to be trimmed as possible."[496]

He chose the sculptor Ida Schwetz-Lehmann specifically for supplying Madonnas since they were popular on the North American market: "As I also mentioned to you, devotional figures are especially successful in the US [...] provided the models are strictly liturgical but still lovely and appealing."[497] They were supposed to look as if they could be prayed to. Walter commissioned several devotional figures from Schwetz-Lehmann; however, since he wanted high quality but was not willing to pay all that much, she refused most of his commissions: "Since you don't want to spend much

Ill. 123
Der Bildhauer Adolf Prischl hielt sich nicht an die Verträge und produzierte an Goldscheider verkaufte Modelle auch in seiner eigenen Werkstatt, teilweise mit leichten Änderungen. Dies führte zunächst zu einem Streit, dann jedoch zu der abgebildeten Einigung. (Quelle: Archiv Goldscheider)

The sculptor Adolf Prischl did not keep to the contracts and also produced models sold to Goldscheider in his own workshop, some of them slightly altered. This behavior led at first to a dispute but eventually the matter was settled as shown. (Source: Goldscheider Archives)

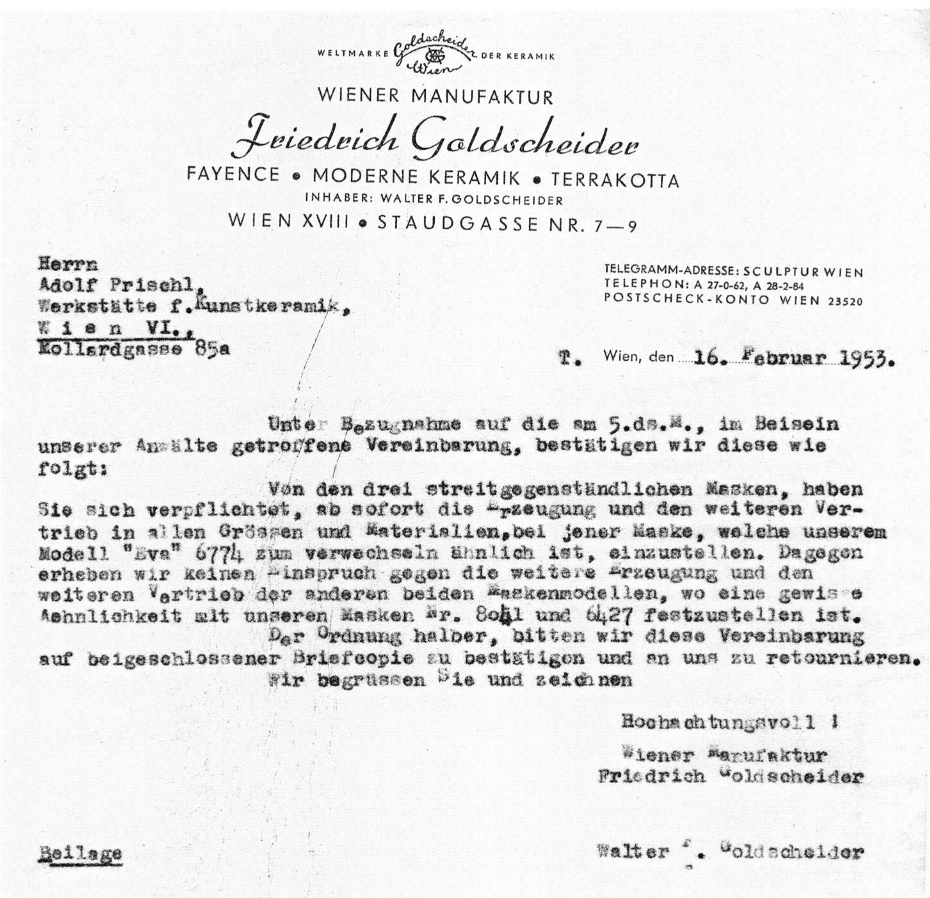

WELTMARKE Goldscheider Wien DER KERAMIK

WIENER MANUFAKTUR

Friedrich Goldscheider

FAYENCE • MODERNE KERAMIK • TERRAKOTTA

INHABER: WALTER F. GOLDSCHEIDER

WIEN XVIII • STAUDGASSE NR. 7–9

Herrn
Adolf Prischl,
Werkstätte f. Kunstkeramik,
Wien VI.,
Kollardgasse 85a

TELEGRAMM-ADRESSE: SCULPTUR WIEN
TELEPHON: A 27-0-62, A 28-2-84
POSTSCHECK-KONTO WIEN 23520

T. Wien, den 16. Februar 1953.

Unter Bezugnahme auf die am 5.ds.M., im Beisein unserer Anwälte getroffene Vereinbarung, bestätigen wir diese wie folgt:

Von den drei streitgegenständlichen Masken, haben Sie sich verpflichtet, ab sofort die Erzeugung und den weiteren Vertrieb in allen Grössen und Materialien, bei jener Maske, welche unserem Modell "Eva" 6774 zum verwechseln ähnlich ist, einzustellen. Dagegen erheben wir keinen Einspruch gegen die weitere Erzeugung und den weiteren Vertrieb der anderen beiden Maskenmodellen, wo eine gewisse Aehnlichkeit mit unseren Masken Nr. 8041 und 6427 festzustellen ist.

Der Ordnung halber, bitten wir diese Vereinbarung auf beigeschlossener Briefcopie zu bestätigen und an uns zu retournieren.

Wir begrüssen Sie und zeichnen

Hochachtungsvoll !

Wiener Manufaktur
Friedrich Goldscheider

Beilage

Walter F. Goldscheider

on smallish models (holy water vessels, little angel heads, small Madonna busts), I request you to address another sculptor as I absolutely do not want to do the work at such low prices, what is more, [paid] in installments. Fees like that, after all, take all the joy out of working. With the many corrections, the affair is also long drawn-out so I don't even have time for my own work. Yet you want to have the finest materials and finish; that should be work for a porcelain factory. It remains with the 3 models, therefore: large Madonna, smaller Madonna busts and St Francis of Assisi. I'll make those models immaculately in plaster of Paris."[498]

Walter also approached Rudolf Knörlein, who lived on Lake Como and had been managing director at Goldscheider from 1932 to 1936. Since many models and molds had been removed by Schuster, Walter asked Knörlein to recreate two popular models[499] (see Ill. 124). They were to be handed in as repaired plaster of Paris models: "We would like to have the models in the original size and whether the Stockholm head is 32 cm as we still recall it and the Tanagra head 26 cm, you are certainly bound to have a better recollection of the dimensions than we have. As you are probably aware, 10 % shrinkage of course has to be added to the finished original dimensions."500 Even with those models, as Walter put it, attention had to be paid to ensuring that they could be reproduced at low cost. They should not "have more than two or three pieces of applied trimming at most and the hair should not be applied but instead be already there on the model." Like Schwetz-Lehmann, Knörlein, too, was to design Madonnas and masks since his models (numbers 6427 and 7188; see Ill. 125 and 126) had sold particularly well.[501]

At that time Rudolf Knörlein was head of a factory that made wall chargers, and could not keep up with production of commissions for Goldscheider so that delivery lagged and no collaboration worthy of mention was again achieved. Although the Stockholm Head was actually supposed to be produced at a reasonable price and with few applied trimmings, Walter ultimately decided that the trimming on the base should not be an integral part of the plaster of Paris mold but should be applied by hand in Vienna to "preserve the original character of the model."[502] Similarly, the air was also "styled afterwards;" only the ruff round the neck was built into the mold. This factor

moderneren Geschmack anpassen muss und fragt bei diversen Künstlern die Möglichkeit einer Zusammenarbeit an. Die Künstler sollen zunächst Abbildungen mit der Größenangabe ihrer Modelle einsenden und bei Gefallen die „Modelle in Plastelin, oder was uns noch lieber wäre, bereits in Gips retouchiert" liefern.[495] Für die Herstellung solcher Gips-Modelle in der Größe zwischen 15 bis 22 cm werden zwischen 300 und 500 Schilling gezahlt, abhängig von der Ausführung. Walter macht hierbei insbesondere darauf aufmerksam, dass ihm „speziell an solchen Modellen gelegen ist, welche in der Fabrikation einfach herzustellen sind, das heißt also, möglichst wenig Garnierstellen" aufweisen.[496]

Die Bildhauerin Ida Schwetz-Lehmann bittet er insbesondere um die Lieferung von Madonnen, denn diese seien auf dem nordamerikanischen Markt beliebt: „Wie ich Ihnen auch erwähnte, sind religiöse Figuren in USA besonders erfolgreich [...] vorausgesetzt, dass die Modelle streng lithurgisch aber dennoch lieblich und ansprechend sind."[497] Sie sollen so aussehen, dass man zu ihnen beten kann. Walter gibt mehrere religiöse Figuren bei Schwetz-Lehmann in Auftrag, da er aber hohe Qualität und nicht allzu viel zahlen will, lehnt sie die meisten Aufträge ab: „Da Sie für kleinere Modelle (Weihbrunnkessel, Engelsköpferln, kleine Madonnenbüste) nicht viel ausgeben wollen, so bitte ich, sich an einen anderen Bildhauer zu wenden, da ich die Arbeiten zu so niederen Preisen, noch dazu auf Raten absolut nicht machen will. Solche Preise benehmen einem ja die ganze Lust zur Arbeit. Durch die vielen Correkturen wird die Sache auch lange hinausgezogen, da komme ich ja gar nicht zu meinen Arbeiten. Dabei wollen Sie die feinste Ausführung haben, es dürften Arbeiten für eine Porzellanfabrik sein. Es bleibt also bei den 3 Modellen: Grosse Madonna, kleinere Madonnenbüste und hl. Franz v. Assisi. Diese Modelle führe ich tadellos in Gips aus."[498]

Auch an den am Comer See lebenden Rudolf Knörlein, der von 1932 bis 1936 Betriebsleiter bei Goldscheider war, wendet sich Walter. Da durch Schuster viele Modelle bzw. Formen verschwunden sind, bittet er Knörlein zwei populäre Modelle wieder herzustellen[499] (siehe Ill. 124). Diese sollen als retuschierte Gipsmodelle eingereicht werden: „Wir möchten die Modelle in Originalgrösse haben und ist der Stockholm Kopf, wie noch in unserer Erinnerung, 32 cm und der Tanagra-Kopf 26 cm hoch, doch werden Sie

Ill. 124
Den ehemaligen Betriebsleiter Rudolf Knörlein bittet Walter Goldscheider um die Reproduktion des in den 1930er Jahren erfolgreichen Mädchenkopfs, der vom schwedischen Nationalmuseum gekauft wurde und daher auch als „Stockholm Kopf" bezeichnet wird. (Quelle: Archiv Goldscheider)

Walter Goldscheider asked former managing director Rudolf Knörlein to reproduce the head of a girl that had been so successful in the 1930s and had been bought by the Swedish National Museum, hence also known as the "Stockholm Head." (Source: Goldscheider Archives)

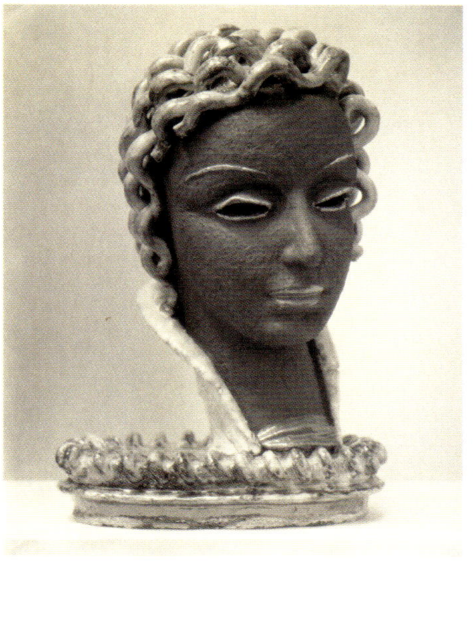

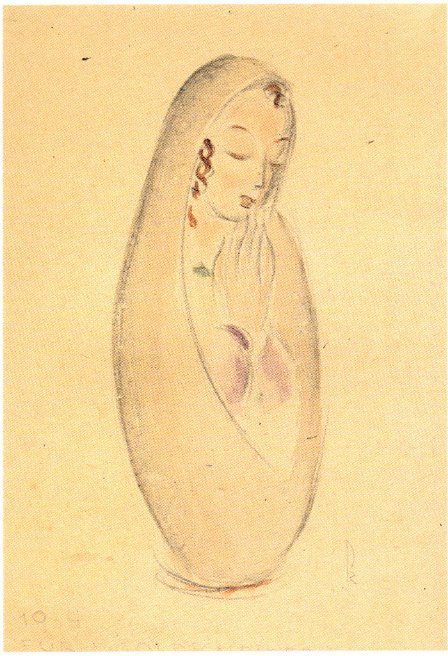

Ill. 125

Ill. 125, 126
Knörlein soll auch die Madonna mit gefalteten Händen reproduzieren; Modell 7188 „Ave Maria". (Quelle: Skizze aus Privatsammlung D; Foto aus Fotoarchiv Karl Täuber)

Knörlein was also asked to reproduce the Madonna with hands folded: model 7188, "Ave Maria." (Source: a sketch from private collection D; photograph from Fotoarchiv Karl Täuber)

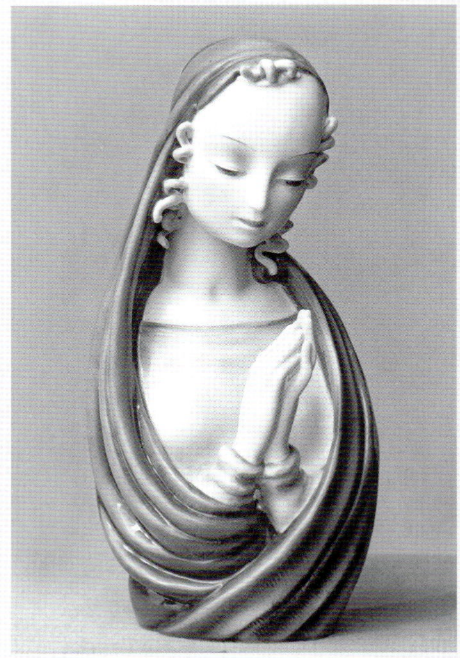

Ill. 126

die Masse sicherlich noch besser in Gedächtnis haben als wir. Zu den fertigen Original-Massen muss natürlich, wie Sie wohl wissen, die 10 % Schwindung gerechnet werden."[500] Auch bei diesen Modellen muss laut Walter darauf geachtet werden, dass sie preiswert reproduzierbar sein müssen, sie sollen „nicht mehr als höchstens zwei bis drei Garnierungsstücke haben und die Haare sollen nicht aufgesetzt werden, sondern schon im Modell vorhanden sein." Wie Schwetz-Lehmann soll auch Knörlein Madonnen und Masken entwerfen, da sich seine Modelle (Nr. 6427 und 7188, siehe Ill. 125 und 126) besonders gut verkauft haben.[501]

Rudolf Knörlein leitet zu dieser Zeit eine Wandplattenfabrik und kommt mit der Produktion für Goldscheider nicht nach, so dass sich die Abgabe hinzieht und es zu keiner nennenswerten Zusammenarbeit mehr kommt. Obwohl der Stockholm-Kopf eigentlich preiswert und mit wenigen Garnierungsstücken produziert werden soll, beschließt Walter letztendlich doch, dass die Garnierungen des Sockels nicht in der Gipsform enthalten sein sollen, sondern in Wien frei aufgelegt werden, um „den Originalcharakter des Modelles zu bewahren."[502] Genauso werden auch die Haare „nachfrisiert", nur die Halskrause wird in die Form eingebaut. Dies erhöht die Produktionskosten, entspricht aber dem traditionellen Qualitätsanspruch Walter Goldscheiders.

Neben diesen Objekten soll die Kollektion durch eine „moderne Linie von Gebrauchskeramik in guten, einfachen aber effektvollen Glasuren" erweitert werden, denn Walter erkennt „die absolute Notwendigkeit, einen neuen Genre meiner Kollektion beizufuegen, denn die Tanzfiguren sind veraltet" und verkaufen sich nur in geringem Maße.[503] Hierfür wendet sich Walter nicht nur an Prof. Robert Obsieger, sondern auch an den ehemaligen leitenden Mitarbeiter Benno Geiger, der an der Kunstgewerbeschule Bern lehrt. Doch Walters Wunsch, „dass bei dem neuen Genre die Keramik doch einfach zu reproduzieren sein soll aber dennoch originell und wirkungsvoll",[504] ist laut Geiger nicht erfüllbar: „Eine Kollektion zu machen, die modern ist, aber allen Leuten gefallen soll, gut aber billig und ohne geschulte Leute ausführbar sein soll, das ergibt einen Kompromiss, der hier in der Schweiz [...] nicht mehr üblich ist."[505]

Neben der veralteten Kollektion und den durch Handelseinschränkungen und Zölle beeinträchtigten Export gibt es noch

raised production costs but was consistent with Walter Goldscheider's traditional quality standards.

Apart from the above objects, the collection was to be enlarged by a "modern line in utilitarian ceramics with good, simple but striking glazes" because Walter was aware of "the absolute necessity of adding a new genre to my collection as the dance figures are old-fashioned" and sold poorly.[503] For this purpose, Walter not only approached Professor Robert Obsieger but also former managerial employee Benno Geiger, who was teaching at the Bern School for the Applied Arts. However, Walter's request "that the new ceramics genre should after all be easy to reproduce yet original and striking"[504] could not, as Geiger saw it, be complied with: "Making a collection that is modern but is supposed to suit everybody, good but cheap and made without trained personnel, would result in a compromise that is [...] no longer usual here in Switzerland."[505]

There were reasons for the crisis apart from the old-fashioned collection and the trade restrictions and customs duties hindering the export trade. The decorative and applied arts market had changed considerably since the war. On the whole, inexpensive, practical wares were what people were looking for. Cheap ceramic products were imported from Japan and other countries where businesses were state subsidized. Former traveling salesman Arthur Loeffler described this trend from Buenos Aires, where he was living in exile, as follows: "due to lack of understanding of art here, Italian and French objects are preferred, which cannot compete with yours, either in taste or execution (nor, naturally, as far as prices are concerned). There are many imitations of your old figures, executed in plaster of Paris, etc, and at rock bottom prices. The latter factor is, as so often, the chief consideration."[506]

By July 1952, the Goldscheider manufactory range comprised nearly 2000 different models.[507] A workforce of about 60 was employed although, according to Richard Kuschnitzky, there were allegedly so many orders that 200 workmen would have been needed. There were, on the other hand, not enough skilled workers and the political situation now made it impossible to employ skilled personnel from Czechoslovakia.[508] The Viennese Manufactory Friedrich Goldscheider is entered in the 1952 Industrie-Compass für Österreich as an exhibitor at the Vienna, Hanover, Cologne and Frankfurt am Main trade fairs. The firm of Goldscheider is entered in the list of exhibitors at the German Industrial Fair in Hanover with branches in Vienna and Frankfurt am Main, where Walter had presumably arranged short-term collaboration with Rosenthal, the glass retailer, who was a friend of his.[509] The following products are listed as exhibits: quality art faiences and terracottas such as Madonnas, animals, masks, figures of dancers and children, vases, wall pictures, ashtrays, etc – wares were supplied to all European countries as well as North and South America.[510] The United States had become the most important market for the Viennese factory, not least due to the strong commitment shown there by Walter's son Erwin. For the export trade with the English-speaking countries, 10 000 silver and black labels marked with the "GWM" sign, "Vienna" in lettering and "Made in Austria" below were ordered late in 1952.[511]

Every effort was made to increase turnover and specialist shops were requested to arrange a special "Goldscheider" display window and mark it with the Goldscheider trademark sign.[512] Special operations as advertising campaigns were mounted such as

andere Gründe für die Krise. Seit dem Krieg hat sich der Markt für Kunstgewerbe stark verändert. Allgemein suchen die Menschen vor allem billige, praktische Waren. Preiswerte keramische Erzeugnisse werden aus Japan und anderen ausländischen, staatlich subventionierten Unternehmen importiert. Der ehemalige Reisende Arthur Loeffler beschreibt dies aus seinem Exil-Ort Buenos Aires wie folgt: „infolge des mangelnden Kunstverstaendnisses hier werden italienische und franzoesische Erzeugnisse vorgezogen, die sich mit den Ihren weder an Geschmack noch nach Ausfuehrung (natuerlich auch nicht nach den Preisen) messen koennen. Es gibt viele Imitationen von Ihren alten Figuren, in Gips etc. ausgefuehrt und spottbillig. Letzteres ist wie so oft die Hauptsache."[506]

Das Angebot der Manufaktur umfasst im Juli 1952 fast 2.000 verschiedene Modelle.[507] Man beschäftigt an die 60 Arbeiter, obwohl laut Richard Kuschnitzky angeblich genug Aufträge vorliegen, um 200 Arbeiter anstellen zu können. Es gibt aber nicht genügend Fachkräfte und aufgrund der politischen Situation kann man auch nicht mehr wie früher Facharbeiter aus der Tschechoslowakei beschäftigen.[508] Im Jahr 1952 ist die Wiener Manufaktur Friedrich Goldscheider im Industrie-Compass für Österreich als Aussteller auf den Messen Wien, Hannover, Köln und Frankfurt am Main eingetragen. Auf der Deutschen Industrie-Messe in Hannover ist im Ausstellerverzeichnis die Firma Goldscheider mit Vertretungen in Wien und Frankfurt am Main angegeben, wo Walter mit dem befreundeten Glashändler Rosenthal vermutlich eine kurzfristige Zusammenarbeit arrangiert hat.[509] Als Erzeugnisse sind aufgeführt: Hochkünstlerische Fayencen und Terrakotten wie Madonnen, Tiere, Masken, Tanz- und Kinderfiguren, Vasen, Wandbilder, Ascher usw. – geliefert wird in sämtliche Staaten Europas sowie nach Nord- und Südamerika.[510] Hierbei sind vor allem die Vereinigten Staaten für die Wiener Manufaktur zum wichtigsten

Ill. 127

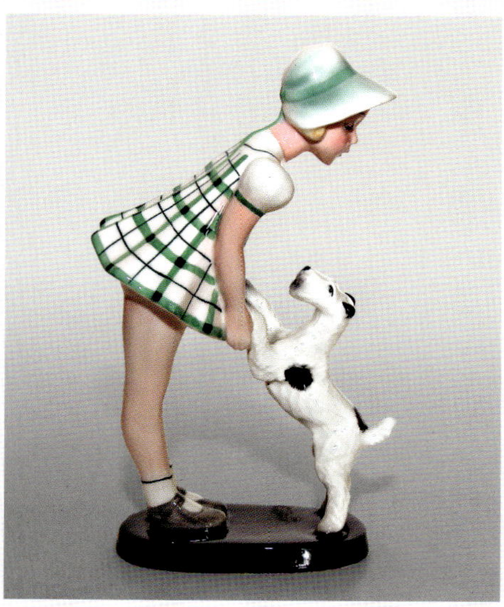

Ill. 128

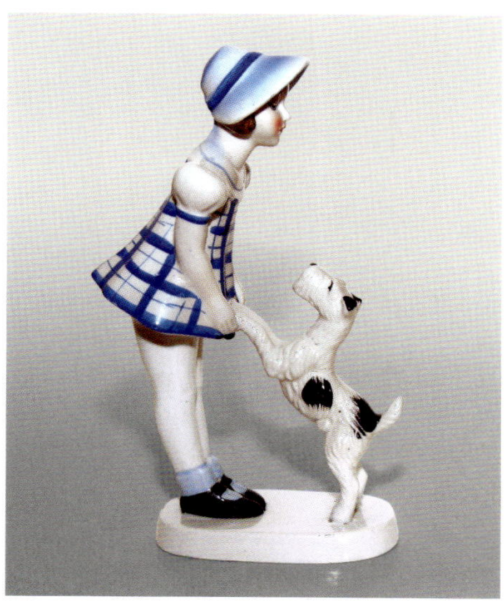

Ill. 127–129

In den 1950er Jahren bereitet den europäischen Keramikmanufakturen die Konkurrenz aus Japan Sorgen, da diese populäre Modelle kopieren und preiswerter verkaufen.

Ill. 129

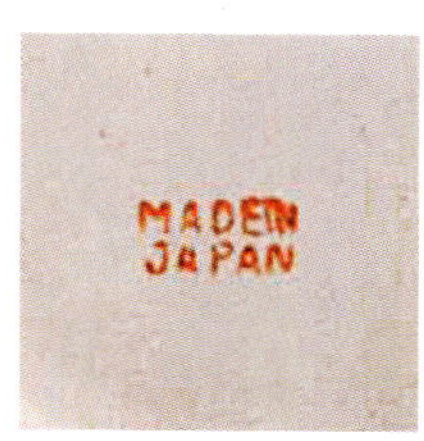

Links ein Original von Goldscheider (ModNr. 7273, Germaine Bouret, ca. 1935), rechts die japanische Kopie. (Quelle: Sammlung Bernard Amigon, Frankreich)

In the 1950s, competition from Japan caused headaches for European ceramics manufacturers since the Japanese copied popular models and sold them at lower prices. On the left, a Goldscheider original (ModNr. 7273, Germaine Bouret, ca. 1935), on the right, the Japanese copy. (Source: Bernard Amigon Collection, France)

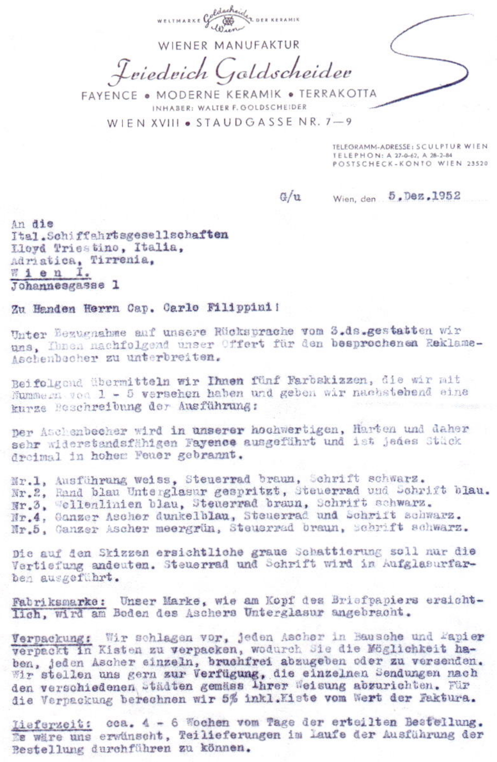

WELTMARKE Goldscheider Wien DER KERAMIK
WIENER MANUFAKTUR
Friedrich Goldscheider
FAYENCE • MODERNE KERAMIK • TERRAKOTTA
INHABER: WALTER F. GOLDSCHEIDER
WIEN XVIII • STAUDGASSE NR. 7—9

TELEGRAMM-ADRESSE: SCULPTUR WIEN
TELEPHON: A 27-0-62, A 28-2-84
POSTSCHECK-KONTO WIEN 23520

G/u Wien, den 5.Dez.1952

An die
Ital.Schiffahrtsgesellschaften
Lloyd Triestino, Italia,
Adriatica, Tirrenia,
W i e n I.
Johannesgasse 1

Zu Handen Herrn Cap. Carlo Filippini!

Unter Bezugnahme auf unsere Rücksprache vom 3.ds.gestatten wir uns, Ihnen nachfolgend unser Offert für den besprochenen Reklame-Aschenbecher zu unterbreiten.

Beifolgend übermitteln wir Ihnen fünf Farbskizzen, die wir mit Nummern von 1 - 5 versehen haben und geben wir nachstehend eine kurze Beschreibung der Ausführung:

Der Aschenbecher wird in unserer hochwertigen, Harten und daher sehr widerstandsfähigen Fayence ausgeführt und ist jedes Stück dreimal in hohem Feuer gebrannt.

Nr.1, Ausführung weiss, Steuerrad braun, Schrift schwarz.
Nr.2, Rand blau Unterglasur gespritzt, Steuerrad und Schrift blau.
Nr.3, Wellenlinien blau, Steuerrad braun, Schrift schwarz.
Nr.4, Ganzer Ascher dunkelblau, Steuerrad und Schrift schwarz.
Nr.5, Ganzer Ascher meergrün, Steuerrad braun, Schrift schwarz.

Die auf den Skizzen ersichtliche graue Schattierung soll nur die Vertiefung andeuten. Steuerrad und Schrift wird in Aufglasurfarben ausgeführt.

Fabriksmarke: Unser Marke, wie am Kopf des Briefpapiers ersichtlich, wird am Boden des Aschers Unterglasur angebracht.

Verpackung: Wir schlagen vor, jeden Ascher in Bausche und Papier verpackt in Kisten zu verpacken, wodurch Sie die Möglichkeit haben, jeden Ascher einzeln, bruchfrei abzugeben oder zu versenden. Wir stellen uns gern zur Verfügung, die einzelnen Sendungen nach den verschiedenen Städten gemäss Ihrer Weisung abzurichten. Für die Verpackung berechnen wir 5% inkl.Kiste vom Wert der Faktura.

Lieferzeit: cca. 4 - 6 Wochen vom Tage der erteilten Bestellung. Es wäre uns erwünscht, Teillieferungen im Laufe der Ausführung der Bestellung durchführen zu können.

Ill. 130
Die Aschenbecher von Goldscheider eignen sich auch als Werbeobjekte. Wie man dem Angebotsbrief aus den 1950er Jahren entnehmen kann, können Schifffahrtsgesellschaften aus verschiedenen Motiven ihren Reklame-Aschenbecher auswählen. (Quelle: Archiv Goldscheider)

Goldscheider ashtrays were also suitable advertising vehicles. As this 1950s letter illustrates, steamship lines were offered a selection of motifs for their advertising ashtrays. (Source: Goldscheider Archives)

Markt geworden, auch bedingt durch das starke Engagement vor Ort durch Walters Sohn Erwin. Für den Export in angelsächsische Länder werden Ende 1952 noch 10.000 silber-schwarze Etiketten mit dem GWM Zeichen und dem Schriftzug „Vienna" sowie darunter „Made in Austria" bestellt.[511]

Man versucht mit allen Mitteln die Umsätze zu steigern und fordert die belieferten Fachgeschäfte auf, fürs Weihnachtsgeschäft ein spezielles „Goldscheider"-Schaufenster zu arrangieren und mit dem Goldscheider-Markenschild zu versehen.[512] Auch besondere Aktionen wie die Produktion von Aschenbechern als Werbemittel für das Hotel Tyrol in Innsbruck[513] oder für italienische Schifffahrtsgesellschaften (Ill. 130) sollen größere Abnahmemengen und Umsätze bringen, doch leider scheitern diese Versuche.[514] Die schon seit dem Sommer 1951 laufenden Verhandlungen mit Walt Disney, die Walter Goldscheider eine Lizenz zur Produktion von Disney-Figuren für Österreich angeboten haben (Ill. 131), scheitern an den Vertragsbedingungen. Walter merkt diesbezüglich an, dass „die Disney Charaktere durch Ihre zu grosse Ausnützung in verschiedenen Artikeln sehr an Zugkraft einge-

producing ashtrays for the Hotel Tyrol in Innsbruck[513] or for Italian steamship lines (Ill. 130) to ensure larger orders and turnover, but these attempts, unfortunately, failed.[514] Since summer 1951, negotiations had been going on with Walt Disney, who had offered Walter Goldscheider a license to produce Disney figures for the Austrian market (Ill. 131). These talks, too, foundered on the conditions of business as set out in the proposed contract. Walter noted in this connection that "the Disney characters have very much lost their appeal because they have been too heavily exploited in so many different articles" and the Viennese manufactory would not be able "to risk the considerable cost of creating models and molds of our own at our own expense."[515] The outcome of negotiations with the Austrian tourist office was equally unsuccessful. Walter had wanted to be licensed to make group of figures based on a poster used by the tourist office that depicted two children in a sled.[516]

ModNr. 8296
Schreibzeugablage
„Übersee Post Leipzig" I Standish "The Leipzig Overseas Post", ca. 1937,
© Wien Museum

ohne I without ModNr.
Ascher Schiff I Ashtray Ship, ca. 1935,
Collection a@goldscheider.net

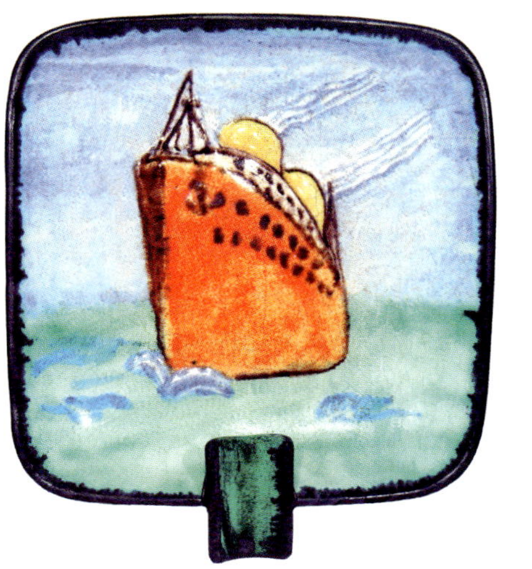

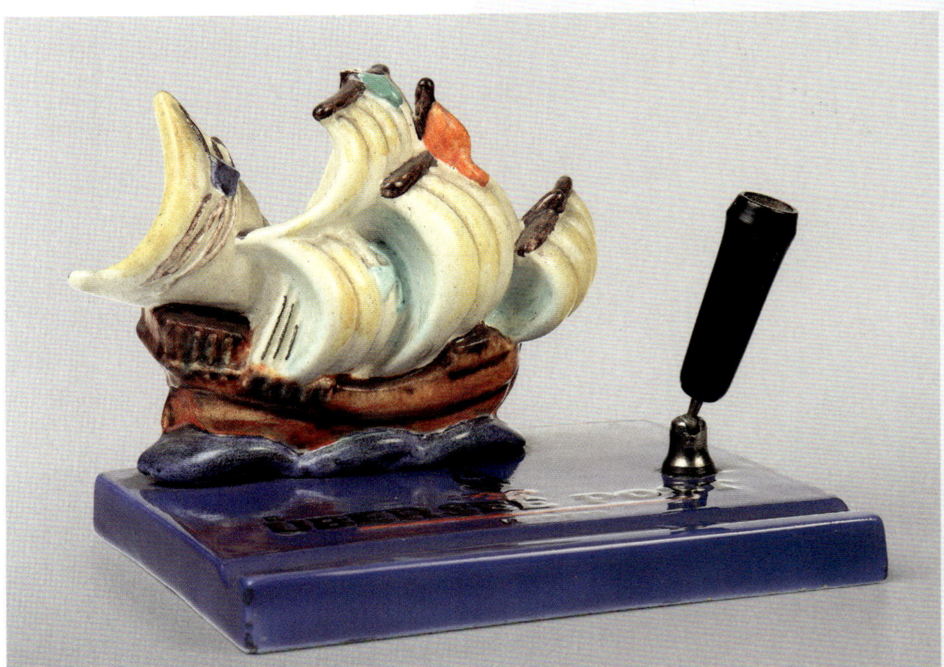

The missed opportunities for raising turnover were of essential importance to Walter because his business had for quite some time been in the throes of a severe financial crisis. By July 1952, he had for this reason placed an advertisement in Keramische Zeitschrift reading: "World-famous art ceramics factory in Austria seeks potential collaborative partner with strong capital reserves or leasing and pooling agreement, would also be interested in ties to existing porcelain factory."[517] Walter needed a credit and testimonials in order to be able to modernize the business. In November 1952, he sent a thank-you letter to Director Kramer of the firm of Ernst Wahliss to underscore his gratitude for Kramer's letter of recommendation that would shore up his efforts to be granted a credit to revamp the business: "The interest and support I have encountered in my endeavor to rebuild the business from all my clients in Austria is particularly encouraging to me."[518] The positive response he received did not, however, result in the needed funding being forthcoming. Not even a year later, he had to report in a letter to Porzellanhaus Wahliss that he had, unfortunately, come to realize that he "would not be able to carry out the rebuilding of our firm successfully, although we had tried with all our might to do so since it has been impossible for us to find the necessary backing in the sectors that would have been appropriate. We are now about to merge our production with a German art ceramics factory, which will continue both our model collection and our brand. Regarding our plan, we will soon be shutting down production at our Vienna factory."[519] He offered to supply Wahliss, as a specialist shop and Goldscheider business partner of many years' standing, in time with all the products he needed

büsst haben" und die Manufaktur nicht in der Lage ist, die „erheblichen Kosten für die Schaffung eigener Modelle und Formen aus eigenen Mitteln zu riskieren."[515] Genauso missglücken auch die Verhandlungen mit dem Österreichischen Verkehrsbüro, für das Walter gerne in Lizenz eine Figurengruppe produzieren möchte, die an ein vom Verkehrsbüro verwendetes Plakat mit zwei Kindern im Schlitten angelehnt ist.[516]

Die entgangenen Umsatzsteigerungen sind jedoch für Walter von essentieller Bedeutung, denn sein Unternehmen steckt schon seit Längerem in einer schweren finanziellen Krise. Bereits im Juli 1952 schickte er deshalb an die Keramische Zeitschrift einen Anzeigenentwurf mit dem Textlaut: „Weltbekannte kunstkeramische Fabrik in Oesterreich sucht eventuelle mittätigen, kapitalkräftigen Teilhaber oder auch Verpachtung und Interessengemeinschaft, wäre auch interessiert an Verbindung mit bestehender Porzellanfabrik."[517] Walter benötigt einen Kredit und Fürsprecher, um den Betrieb modernisieren zu können. Im November 1952 schickt er einen Dankbrief an Direktor Kramer von der Firma Ernst Wahliss und bedankt sich für ein Befürwortungsschreiben, das seine Bemühungen um den Aufbaukredit unterstreichen sollte: „Das Interesse und die Unterstützung, welche ich bei meinem Bestreben des Wiederaufbaus bei allen meinen Kunden in Österreich begegne, bedeuten für mich eine besondere Ermutigung."[518] Die positive Resonanz führt jedoch nicht zur Vergabe der notwendigen Finanzmittel, so dass er kaum ein halbes Jahr später erneut in einem Brief an das Porzellanhaus Wahliss mitteilen muss, dass er leider zu der Erkenntnis gelangt sei, „den mit aller

Ill. 131
Die Walter Goldscheider angebotene Lizenz zur Herstellung von Disney-Figuren für den österreichischen Markt scheitert an den Vertragsbedingungen. (Quelle: Archiv Goldscheider)

The license for making Disney figures for the Austrian market offered to Walter Goldscheider foundered on the terms of the contract. (Source: Goldscheider Archives)

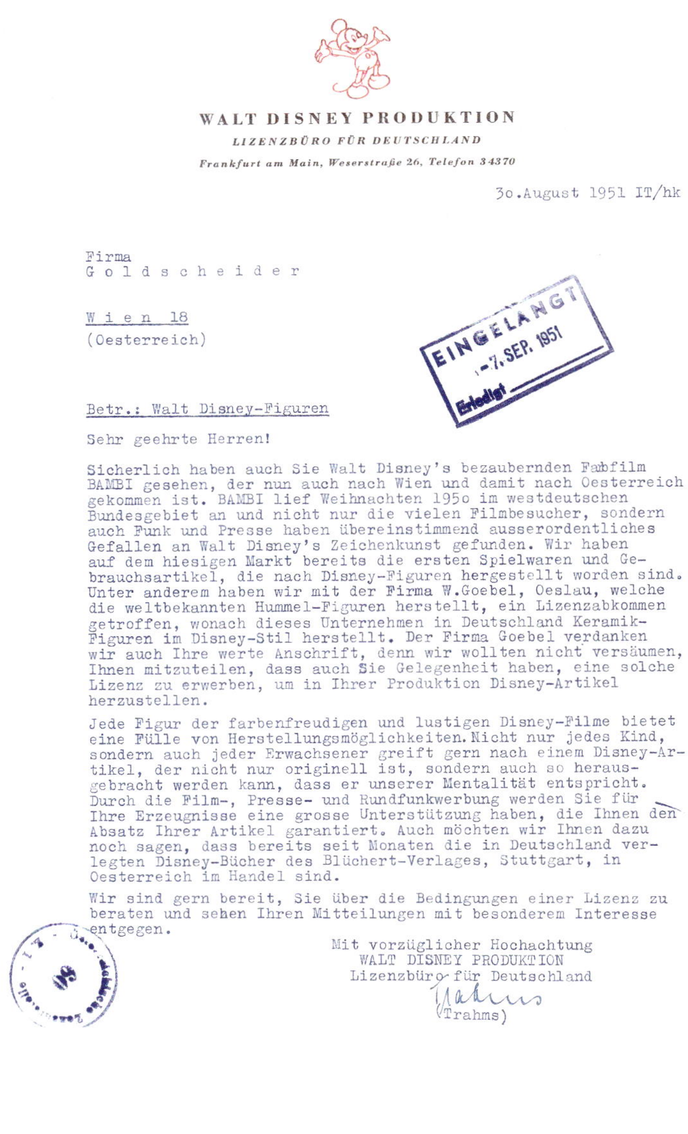

WALT DISNEY PRODUKTION
LIZENZBÜRO FÜR DEUTSCHLAND
Frankfurt am Main, Weserstraße 26, Telefon 34370

30.August 1951 IT/hk

Firma
Goldscheider

Wien 18
(Oesterreich)

EINGELANGT -7. SEP. 1951
Erledigt

Betr.: Walt Disney-Figuren

Sehr geehrte Herren!

Sicherlich haben auch Sie Walt Disney's bezaubernden Farbfilm BAMBI gesehen, der nun auch nach Wien und damit nach Oesterreich gekommen ist. BAMBI lief Weihnachten 1950 im westdeutschen Bundesgebiet an und nicht nur die vielen Filmbesucher, sondern auch Funk und Presse haben übereinstimmend ausserordentliches Gefallen an Walt Disney's Zeichenkunst gefunden. Wir haben auf dem hiesigen Markt bereits die ersten Spielwaren und Gebrauchsartikel, die nach Disney-Figuren hergestellt worden sind. Unter anderem haben wir mit der Firma W.Goebel, Oeslau, welche die weltbekannten Hummel-Figuren herstellt, ein Lizenzabkommen getroffen, wonach dieses Unternehmen in Deutschland Keramik-Figuren im Disney-Stil herstellt. Der Firma Goebel verdanken wir auch Ihre werte Anschrift, denn wir wollten nicht versäumen, Ihnen mitzuteilen, dass auch Sie Gelegenheit haben, eine solche Lizenz zu erwerben, um in Ihrer Produktion Disney-Artikel herzustellen.

Jede Figur der farbenfreudigen und lustigen Disney-Filme bietet eine Fülle von Herstellungsmöglichkeiten. Nicht nur jedes Kind, sondern auch jeder Erwachsener greift gern nach einem Disney-Artikel, der nicht nur originell ist, sondern auch so herausgebracht werden kann, dass er unserer Mentalität entspricht. Durch die Film-, Presse- und Rundfunkwerbung werden Sie für Ihre Erzeugnisse eine grosse Unterstützung haben, die Ihnen den Absatz Ihrer Artikel garantiert. Auch möchten wir Ihnen dazu noch sagen, dass bereits seit Monaten die in Deutschland verlegten Disney-Bücher des Blüchert-Verlages, Stuttgart, in Oesterreich im Handel sind.

Wir sind gern bereit, Sie über die Bedingungen einer Lizenz zu beraten und sehen Ihren Mitteilungen mit besonderem Interesse entgegen.

Mit vorzüglicher Hochachtung
WALT DISNEY PRODUKTION
Lizenzbüro für Deutschland
(Trahms)

Kraft versuchten Wiederaufbau unserer Firma nicht erfolgreich durchführen zu können, nachdem es uns unmöglich war, die hierzu notwendige Unterstützung von den in Betracht kommenden Faktoren zu finden. Wir stehen nun mehr im Begriffe, unsere Erzeugung an eine deutsche kunstkeramische Fabrik anzuschliessen, welche sowohl unsere Modellkollektion als auch unsere Marke fortsetzen wird. Mit Rücksicht auf unseren Plan, werden wir die Erzeugung in unserer Wiener Fabrik bald einstellen."[519] Als langjährigem Geschäftspartner bietet er dem Fachgeschäft Wahliss noch an, sich rechtzeitig mit den Erzeugnissen, die die alte Marke „Goldscheider Wien" tragen, einzudecken, da „das Lager geräumt wird."[520]

Bei der genannten „deutschen kunstkeramischen Fabrik" handelt es sich um die Majolikamanufaktur E. & C. Carstens aus Fredelsloh in Niedersachsen. Der Verkauf einer Produktionslizenz an Carstens erfolgt im März 1953 und soll die Verbindlichkeiten abdecken sowie die Altersrente von Walter sichern. In einem Brief an seinen Neffen, Dr. Fritz Goldscheider, schildert er die Situation einige Monate nach dem Verkauf wie folgt: „Um die notwendigen Zahlungen leisten zu können, müssen die Raten von Carstens eingehen. Weiters hänge ich von der Verwertung aller anderen Aktiva ab, Inventar, Warenlager etc. Wie dir bereits bekannt, erfolgt der Eingang der Monatsraten Carstens äusserst schleppend und ist die Firma derzeit mit 2 Monatsraten im Rückstand, trotz aller meiner dringlichsten Urgenzen. Ich glaube nicht, dass sie zahlungsunwillig sind, aber sie hat selbst mit Geldknappheit zu kämpfen. Aber auch die Verwertung des Inventars kann nur successive erfolgen, weil wir uns die wenigen keramischen Interessenten erst suchen müssen, eben um auch nicht um jeden Preis sondern im Interesse aller bestmöglichste Preise zu erzielen. Dies gilt womöglich noch mehr für das verbliebene Warenlager, das natürlich schon stark ausgesiebt ist, und wir daher mit dem Verkauf bis zum Beginn der Saison warten müssen, da jetzt im Hochsommer gar keine Kauflust in unsere Luxusware besteht."[521]

Auch die so genannte zweite Ware wird durch Walter bei der Liquidation verkauft: „Diese zweite Ware ist natürlich kein Bruch oder sonst geschädigt, sondern es sind nur solche Stücke, die von uns vom fabrikationstechnischen Standpunkt nicht als einwandfrei betrachtet werden können und daher als 2. Wahl klassifiziert werden."[522] Der Verkauf des Inventars findet ab Juli 1953 statt. Brennmuffel, Filterpresse, Kompressoren, Drehscheiben, Glasurbottiche, Spritzpistolen, Spritzkästen etc. werden verschiedenen Unternehmen angeboten.[523] Glasuren, Borsten- und Haarpinsel gehen beispielsweise an die Fayencefabrik F. Gollhammer & Co aus Vöcklabruck.[524] Den Stand für die Wiener Herbstmesse 1953 überlässt Walter der Gmundner

bearing the old "Goldscheider Wien" mark since "the warehouse is being emptied."[520]

The above mentioned "German art ceramics factory" was E. & C. Carstens of Fredelsloh, Lower Saxony, manufacturers of majolica. The license to produce under the Goldscheider name was granted to Carstens in March 1953. It was supposed to cover all obligations and secure Walter's old age pension. In a letter to his nephew, Dr. Fritz Goldscheider, Walter described the situation a few months after the sale of the license as follows: "The payments from Carstens must come in [for us] to be able to cover the necessary payments. Further, I depend on the activation of all other liquid assets from the inventory stock, warehouse, etc. As you already know, the monthly payments from Carstens are dribbling in extremely slowly and the firm is currently two months in arrears with payments despite my most urgent notification. I do not believe that they are unwilling to pay but that they also have to contend with a shortage of funds. However, liquidating the inventory, too, can only be done in installments because first we have to find out the few parties interested in ceramics, of course not at any price but to get the best possible prices in the interests of all concerned. This is probably even more true of what remains in the warehouses, which, naturally, has already been thoroughly sifted through and we therefore have to wait for the sale until the beginning of the season since now in mid-summer there is no rush to buy our luxury wares."[521]

Walter even liquidated what are known in the trade as seconds: "These seconds are, of course, not damaged by breakage or otherwise but rather they are merely pieces which cannot be regarded by us from the manufacturing standpoint as perfect and, therefore, are classified as seconds."[522] The sale of inventory took place from July 1953. Muffle kilns, filter presses, compressors, potters' wheels, glaze vats, spray-guns, spray-chests, etc, were offered to various businesses.[523] Glaze mixtures and bristle and animal hair brushes, for instance, went to F. Gollhammer & Co, a faience factory in Vöcklabruck.[524] Walter left the Goldscheider stall for the 1953 Vienna Autumn Fair to Gmundner Keramik, which had already shown an interest in a license for producing Goldscheider figures for the Austrian market.[525] On 7 August 1953, Walter sold the factory building at 9 Staudgasse and on 1 December the building at 7 Staudgasse as part of the liquidation of the firm's assets.[526] A few weeks before he had also sold the property he owned at 17 Hans Sachs Gasse.[527]

Ultimately, the sale of the business did not lead to the financial security that Walter had hoped for, as a letter he wrote to his nephew Fritz reveals: "Only one thing is certain about the whole thing, that nothing will be left for me. [...] As you know, I had really counted on it for my pension settlement. Unfortunately, from the news I have just received, that high hope has turned out to be chimerical."[528] In retrospect, he confided his entrepreneurial endeavors and personal wishes to Rudolf Knörlein: "If I had been allowed to pursue my innermost desires, I would have preferred to create only genuine crafts but here, too, as always in life and in all parts of it, one is forced to make concessions, which, unfortunately, often go so far that they suppress the actual sense of what we are striving for and want."[529]

The "Wiener Manufaktur Friedrich Goldscheider" was finally expunged from the commercial register on 26 April 1954 and thus ended the history of one of the most important Viennese ceramics factories.[530] Some employees transferred to Wilhelm Thomasch and other workshops.[531] Walter did not return to Trenton but instead retired at Rottach, where he died on 23 March 1962. As he had wished, he was buried in Trenton beside his wife Lilly.[532]

Keramik, die sich zuvor auch für die Lizenz zur Produktion von Goldscheider Figuren für Österreich interessiert hat.[525] Am 7. August 1953 verkauft Walter im Zuge der Liquidation der Firma das Fabrikshaus Staudgasse 9, zum 1. Dezember 1953 erfolgt der Verkauf der Staudgasse 7.[526] Wenige Wochen zuvor hat er auch das ihm gehörende Lokal in der Hans Sachs Gasse 17 veräußert.[527]

Der Verkauf des Unternehmens führt jedoch nicht zu der erhofften finanziellen Absicherung, wie aus dem Brief an seinen Neffen Fritz abschließend zu lesen ist: „Nur eines ist dabei sicher, dass mir dabei nichts verbleiben wird. [...] Wie Du weißt hatte ich dabei auf meine Abmachung der Lebensrente sehr gerechnet. Leider erweist sich nach den jetzt erhaltenen Nachrichten diese schöne Hoffnung als eine Chimaere."[528] Seine Bemühungen als Unternehmer und persönliche Wünsche vertraut er rückblickend Rudolf Knörlein an: „Wenn es mir gestattet gewesen wäre, meinem innersten Wunsche zu folgen, so hätte ich gerne nur echtes Kunstgewerbe geschaffen, aber auch hier wie immer im Leben und in all seinen Teilen ist man zu Konzessionen gezwungen, die leider oft so weit gehen, dass sie den eigentlichen Sinn unseres Strebens und Wollens erdrücken."[529]

Die „Wiener Manufaktur Friedrich Goldscheider" wird letztendlich am 26. April 1954 aus dem Handelsregister gelöscht, und die Geschichte einer der bedeutendsten Wiener Keramikmanufakturen geht zu Ende.[530] Einige Mitarbeiter wechseln zu Wilhelm Thomasch und in andere Werkstätten.[531] Walter reist nicht nach Trenton zurück, sondern setzt sich in Rottach zur Ruhe, wo er am 23. März 1962 stirbt. Seinem Wunsche entsprechend wird er in Trenton an der Seite seiner Frau Lilly begraben.[532]

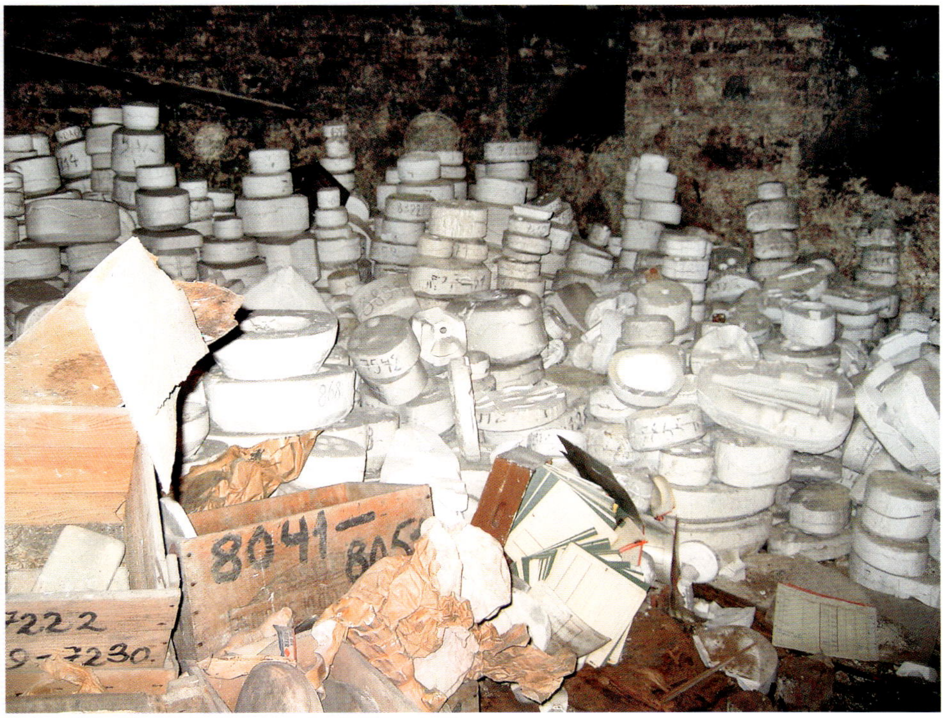

Ill. 132
Von der Manufaktur Goldscheider, der traditionsreichen Weltmarke der Keramik, bleiben in Wien lediglich einige Modellformen im Keller des Fabrikgebäudes übrig. (Quelle: Archiv Goldscheider)

All that was left in Vienna of the Goldscheider manufactory, the internationally renowned name in ceramics were some molds for models in the cellar of the factory building. (Source: Goldscheider Archives)

Anmerkungen

426 Brief der Arbeitsgemeinschaft an die Firma Porslins-Ericsson, Schweden, 14.11.1946
427 Brief der Arbeitsgemeinschaft an das Fachgeschäft Mitzi Pinget, Linz, 15.10.1945. Josef Paul wurde am 18.3.1875 geboren und wurde ab dem 25.10.1945 als öffentlicher Verwalter geführt (s. HRA 7077)
428 Interview mit Karl Täuber, Wien, 25.11.2003
429 Brief der Arbeitsgemeinschaft an das Fachgeschäft Ernst Wahliss, Wien, 14.9.1945
430 Brief der Arbeitsgemeinschaft an das Fachgeschäft Mitzi Pinget, Linz, 31.10.1945
431 Brief der Arbeitsgemeinschaft an das Fachgeschäft Mitzi Pinget, Linz, 4.12.1945
432 Brief der Arbeitsgemeinschaft an das Fachgeschäft Ernst Wahliss, Wien, 13.11.1945. Allgemein hatten die Käufer von Goldscheider Figuren seit jeher die Möglichkeit, kaputt gegangene Stücke bei der Manufaktur einzusenden, wo sie von Fachkräften wieder in Ordnung gebracht worden sind.
433 Brief der Arbeitsgemeinschaft an das Fachgeschäft Max Pilat, Attnang-Puchheim, 16.1.1946
434 Brief der Arbeitsgemeinschaft an das Fachgeschäft Max Pilat, Attnang-Puchheim, 21.5.1946. Laut eines früheren Briefes an das Fachgeschäft Pinget, vom 2.2.1946, haben sie aber „schwarze Teller" geliefert. Die Produktion wird immer wieder unterbrochen, aufgrund von Stromausfällen oder dem Ausfall von zwei Öfen im Juni 1946, die aufgrund einer Reparatur die Produktion für drei Wochen lahm legen.
435 Brief der Arbeitsgemeinschaft der Manufaktur Goldscheider an das Bundesministerium für Vermögenssicherung und Wirtschaftsplanung, 25.3.1946
436 Brief an das Sprachinstitut Globus, Wien, 22.5.1946
437 Brief von O. Wiederhielm, Stockholm, an den Prokuristen der Manufaktur Josef Paul, 27.12.1946. Wiederhielm unterstützt Josef Paul in den Jahren 1946–47 mit Nahrungsmitteln und Textilien, die er an Pauls private Adresse in der Hildebrandgasse 40/I schickt.
438 Brief von O. Wiederhielm, Stockholm an den Prokuristen der Manufaktur Josef Paul, 16.5.1947. Aus verschiedenen Briefen und Hinweisen geht hervor, dass Schuster seine Verkäufe nicht nachvollziehbar dokumentiert hat und zum Beispiel hinter dem Rücken von Verkaufsagenten direkt bei den Endkunden kassierte, um die zustehende Provision an die Agenten zu sparen.
439 Ebenda
441 Brief an das Fachgeschäft Mitzi Pinget, Linz, 17.1.1947. Auch vorher gab es schon wegen Stromausfällen Schwierigkeiten bei der Produktion. Allgemein fehlen Arbeitskräfte, und wie aus diesem Brief hervor geht, wechselten einige Arbeiter wegen höherer Löhne zur Konkurrenz. Prischl ging z.B. 1942 nach seiner Entlassung aus dem Militärdienst zu Keramos. Da Fachkräfte nicht einfach ersetzt werden können, reichen schon normale Krankheitsfälle, um die Produktion zu beeinträchtigen.
440 Brief von Walter Goldscheider an O. Wiederhielm, Stockholm, 29.8.1950. Im Gegensatz zum Clearingverfahren, dem klassischen Handelsgeschäft, werden beim Kompensationsgeschäft Waren aus diversen Branchen miteinander verrechnet: man besorgt z.B. einen österreichischen Abnehmer für chilenischen Wein im Wert von 1.000 Dollar und darf dann in Höhe dieser Summe kunstgewerbliche Gegenstände nach Chile verkaufen. Dieser Tauschhandel wird vor allem durch die Import- und Export-Agenturen geregelt, da diese die Kontakte zu verschiedenen Ländern und Branchen haben.
442 Brief von Mitzi Pinget, Linz, an die Wiener Manufaktur, 12.2.1947
443 Brief von Josef Paul an O. Wiederhielm, Stockholm, 7.6.1947
444 Brief an die Gmundner Keramik, 29.8.1946
445 Bestellung von Unterglasurfarben bei Harrison & Son (Hanley) Ltd., 10.12.1952
446 Brief an das Fachgeschäft Mitzi Pinget, Linz, 20.5.1947
447 Ebenda
448 Brief von Josef Paul an O. Wiederhielm, Stockholm, 7.6.1947. Man versucht über Wiederhielm auch die über die Kriegsjahre nicht gezahlten Schulden der schwedischen Händler einzutreiben.
449 Brief an das Fachgeschäft Mitzi Pinget, Linz, 18.8.1947
450 Ebenda
451 N.N.: Goldscheider Vienna Plant Going Full Tilt Once More, in: Trenton Evening Times, 22.2.1952
452 Brief an O. Wiederhielm, Stockholm, 5.11.1947
453 Brief des Dipl.Arch.Ing. Ferdinand Klinger, Salzburg, an die Wiener Manufaktur Friedrich Goldscheider, 25.3.1948
454 Brief von der Kunsthandlung Karl Pechmann, Linz, an die Firma Friedrich Goldscheider Wiener Manufaktur, 13.8.1949
455 Brief von Walter Goldscheider, Wien, an die Firma Hermann Melter, Pforzheim, 25.10.1950
456 Brief von F. Gollhammer von der Österreichischen Edelkeramik und Fayencefabrik F. Gollhammer & Co, Vöcklabruck, an Walter Goldscheider, Wien, 24.3.1950
457 Ebenda. Gollhammer weist auch darauf hin, dass sich einige der ehemaligen Mitarbeiter von Goldscheider selbständig gemacht haben, aber „keiner davon auch nur annähernd Ihre Qualität" erreicht hat. Dabei handelt es sich wahrscheinlich vor allem um Adolf Prischl, der sich 1947 selbständig machte und auch Modelle von Karl Sailer, Stephan Dakon und Victor Matula produzierte.
458 Brief von Walter Goldscheider an die Magistrats Abteilung 63, Wien, 3.7.1950
459 Eintrag im Handelsregister, HRA 7077, 4.5.1950
460 Eintrag im Handelsregister, HRA 7077, 30.8.1950
461 Brief von Walter Goldscheider, Wien, an Rudolf Knörlein, Lurago d'Erba, 20.8.1950
462 Brief von Walter Goldscheider, Wien, an Morris Katz von Marks & Rosenfeld Inc., New York, 11.4.1950
463 Brief von Walter Goldscheider an den ehrenamtlichen Korrespondenten Axel Wahl-Bau, Kopenhagen, 25.4.1950. Aus Amerika erhält Walter eine Anfrage von der Hudson Commercial Company, die Augarten und Keramos in USA vertreiben und gerne auch Goldscheider ins Sortiment aufnehmen möchten. Sie begründen dies damit, dass sie die Goldscheider Produkte schon immer für hochwertiger als Keramos gehalten haben. Siehe Brief von Walter Goldscheider, Wien, an Morris Katz von Marks & Rosenfeld Inc., New York, 11.4.1950
464 Brief von Haas & Czjzek, Wien, an Firma Friedrich Goldscheider Wiener Manufaktur, 15.10.1952
465 Brief von Walter Goldscheider, Wien an den Agenten O. Wiederhielm, Schweden, 4.4.1950
466 Brief von Walter Goldscheider an die Firma O. Wiederhielm, Stockholm, 21.11.1950. Die Geschäftsbeziehung zwischen Walter Goldscheider und O. Wiederhielm findet ein unfreundliches Ende. Wiederhielm sieht kaum Möglichkeiten eines Exports nach Schweden und man wird sich nicht über den Vertretungsvertrag einig. Walter beauftragt anschließend eine andere schwedische Agentur, die in der Lage ist, Goldscheider Objekte nach Schweden zu importieren. Wiederhielm weigert sich daraufhin, die noch nicht bezahlte Musterkollektion sowie das Fotomaterial zurück zu geben und verlangt, dass ihm erst die unter Schuster zustehenden Provisionen beglichen werden sollten. Die Lieferzeit für ausländische Bestellungen beträgt übrigens 2–3 Monate nach Ordereingang. Vgl. Brief von Walter Goldscheider an Dr. Fritz Goldscheider, Wien, 21.1.1953
467 Interview mit Karl Täuber, Wien, 25.11.2003
468 Brief von Walter Goldscheider an die Firma J. Fr. Pollak, Leibnitz, 28.1.1953
469 Brief von O. Wiederhielm, Stockholm, an Walter Goldscheider, 30.11.1950
470 Siehe hierzu auch das Kapitel „Marcell Goldscheiders Werkstätten in Wien und Stoke-on-Trent", S. 248ff
471 Interview mit Dr. Arlette Leupold-Löwenthal, Wien, Tochter von Dr. Friedrich „Fritz" Goldscheider, 26.11.2003
472 Brief von Walter Goldscheider an den Hospital Service Plan of New Jersey, 30.8.1950
473 Brief von Karl Täuber an Arturo Löffler, Buenos Aires, 7.11.1951
474 Brief von Walter Goldscheider an Rudolf Knörlein, Lurago d'Erba, 2.7.1952
475 Interview mit Karl Täuber, Wien, 25.11.2003
476 Eintrag im HRA 7077, vom 17.4.1951 für Erwin Friedrich Goldscheider und vom 14.6.1951 für Karl Täuber.
477 Zession zwischen der Firma Wiener Manufaktur Josef Schuster, vormals Friedrich Goldscheider und Karl Sailer, Wien, 25.8.1948
478 Zession zwischen der Firma Wiener Manufaktur Josef Schuster, vormals Friedrich Goldscheider und Karl Sailer, Wien, 13.2.1950
479 Zession zwischen der Firma Wiener Manufaktur Friedrich Goldscheider und Karl Sailer, Wien, 1.7.1950
480 Brief von Walter Goldscheider an Kurt Goebel, Wien, 24.10.1952. Auch eine Liste seiner Kunden soll Goebel zur Verfügung stellen.
481 Ebenda
482 Zession zwischen der Firma Wiener Manufaktur Friedrich Goldscheider und Kurt Goebel, Wien, 24.10.1952
483 Brief von Walter Goldscheider an Kurt Goebel, 26.11.1952
484 Brief von Walter Goldscheider an Stephan Dakon, Wien, 8.12.1952
485 Brief von Walter Goldscheider an Stephan Dakon, Wien, 17.11.1952
486 Brief des Oberösterreichischen Landesmuseum, Linz, an Firma Goldscheider, Wien, 2.7.1952
487 Tantiemenabrechung an Stephan Dakon für das 1. Halbjahr 1947, 18.8.1947
488 Tantiemenabrechnung an Stephan Dakon für das 1. Halbjahr 1951, 30.7.1951
489 Tantiemenabrechnung an Karl Sailer für das 2. Halbjahr 1952, 29.1.1953
490 Tantiemenabrechnung an Njura Lorenzl für das 1. Halbjahr 1951, 30.7.1951
491 Makus (2000), S. 176
492 Brief von Walter Goldscheider an Adolf Prischl, Wien, 16.2.1953. Eine kleine Auseinandersetzung gibt es auch mit dem Fabrikanten der berühmten Hummel-Figuren. Am 14. August 1951 sendet W. Goebel von der seit 1871 bestehenden Porzellanfabrik Goebel, in Oeslau bei Coburg, ein Einschreiben an Walter Goldscheider. Er legt zwei Fotos von Goldscheider-Figuren bei, die seiner Ansicht nach zu sehr den berühmten Hummel-Figuren aus seinem Programm ähneln und daher die Urheberrechte verletzen. Er bittet Walter diesbezüglich um eine Stellungnahme, merkt aber auch an, dass er sich „bei der mir bekannten seriösen Führung Ihres Betriebes nicht recht vorstellen" kann, dass „Sie mit

Notes

[426] Letter from the Arbeitsgemeinschaft [works co-operative] to the firm of Porslins-Ericsson, Sweden, 14 Nov. 1946
[427] Letter from the Arbeitsgemeinschaft to the specialist shop Mitzi Pinget, Linz, 15 Oct. 1945. Born on 18 March 1875, Josef Paul was registered as public administrator from 25 Oct. 1945 (see HRA 7077)
[428] Interview with Karl Täuber, Vienna, 25 Nov. 2003
[429] Letter from the Arbeitsgemeinschaft to the specialist shop Ernst Wahliss, Vienna, 14 Sept. 1945
[430] Letter from the Arbeitsgemeinschaft to the specialist shop Mitzi Pinget, Linz, 31 Oct. 1945
[431] Letter from the Arbeitsgemeinschaft to the specialist shop Mitzi Pinget, Linz, 4 Dec. 1945
[432] Letter from the Arbeitsgemeinschaft to the specialist shop Ernst Wahliss, Vienna, 13 Nov. 1945. Generally speaking, Goldscheider clients had always had the possibility of sending in pieces that had broken to the factory, where they were restored by skilled craftsmen.
[433] Letter from the Arbeitsgemeinschaft to the specialist shop Max Pilat, Attnang-Puchheim, 16 Jan. 1946
[434] Letter from the Arbeitsgemeinschaft to the specialist shop Max Pilat, Attnang-Puchheim, 21 May 1946. According to an earlier letter of 2 Feb. 1946 to the specialist shop Pinget, they delivered "black plates." Production was continually interrupted because of grid black-outs or two kilns breaking down in June 1946. Production ceased for three weeks before repairs on them could be completed.
[435] Letter from the Arbeitsgemeinschaft at Goldscheider to the Bundesministerium für Vermögenssicherung und Wirtschaftsplanung [Federal Ministry for Restitution of Moneys and Economic Planning], 25 March 1946
[436] Letter to the Globus Language Institute, Vienna, 22 May 1946
[437] Letter from O. Wiederhielm, Stockholm, to Josef Paul, head clerk at the manufactory, 27 Dec. 1946. Wiederhielm supported Josef Paul in 1946–47 with food and clothing, which he sent to Paul's home address, 40/I Hildebrandgasse.
[438] Letter from O. Wiederhielm, Stockholm, to Josef Paul, head clerk at the manufactory, 16 May 1947. Several letters and other sources indicate that Schuster did not document his sales in a manner that could be traced and, for instance, took the money directly from the retail customer, thus circumventing the middleman representing Goldscheider to save the money that otherwise would have gone to pay the commission.
[439] Ibid
[440] Letter to the specialist shop Mitzi Pinget, Linz, 17 Jan. 1947. There had been hitches in production even earlier due to black-outs. In general, workmen were in short supply and, as this letter reveals, some employees transferred to the competition for higher pay. Prischl, for instance, went to Keramos in 1942 on his discharge from the army. Since skilled craftsmen could not easily be replaced, sick-leave cases sufficed to hold up production.
[441] Letter from Walter Goldscheider to O. Wiederhielm, Stockholm, 29 Aug. 1950. Unlike the clearing house process, the classical business transaction, compensatory bartering entailed exchanging wares of approximately similar value from different sectors in lieu of cash payment: for example, one might procure Chilean wine worth 1000 dollars for an Austrian client and then sell crafts objects to Chile worth the same amount in return. This form of barter was regulated primarily by the import-export agencies since they had contacts to different countries and sectors.
[442] Letter from Mitzi Pinget, Linz, to the Viennese manufactory, 12 Feb. 1947
[443] Letter from Josef Paul an O. Wiederhielm, Stockholm, 7 June 1947
[444] Letter to Gmundner Keramik, 29 Aug. 1946
[445] Order for underglaze colors placed with Harrison & Son (Hanley) Ltd, 10 Dec. 1952
[446] Letter to the specialist shop Mitzi Pinget, Linz, 20 May 1947
[447] Ibid
[448] Letter from Josef Paul to O. Wiederhielm, Stockholm, 7 June 1947. Goldscheider tried to collect debts not paid during the war years through the Swedish agent.
[449] Letter to the specialist shop Mitzi Pinget, Linz, 18 Aug. 1947
[450] Ibid
[451] N.N.: Goldscheider Vienna Plant Going Full Tilt Once More, in: Trenton Evening Times, 22 Feb. 1952
[452] Letter to O. Wiederhielm, Stockholm, 5 Nov. 1947
[453] Letter from Dipl.Arch.Ing. Ferdinand Klinger, Salzburg, to the Wiener Manufaktur Friedrich Goldscheider, 25 March 1948
[454] Letter from the art dealer Karl Pechmann, Linz, to the Friedrich Goldscheider Wiener Manufaktur, 13 Aug. 1949
[455] Letter from Walter Goldscheider, Vienna, to the firm of Hermann Melter, Pforzheim, 25 Oct. 1950
[456] Letter from F. Gollhammer of the Österreichische Edelkeramik und Fayencefabrik F. Gollhammer & Co, Vöcklabruck, to Walter Goldscheider, Vienna, 24 March 1950
[457] Ibid. Gollhammer pointed out that some former Goldscheider employees had become self-employed but "none of them [had] even remotely [approached] your quality." The former employee concerned in this case was probably Adolf Prischl, who had become self-employed in 1947 and also produced models by Karl Sailer, Stephan Dakon and Victor Matula.
[458] Letter from Walter Goldscheider to the Magistrats Abteilung [Magistracy, Division] 63, Vienna, 3 July 1950
[459] Entry in the Handelsregister, HRA 7077, 4 May 1950
[460] Entry in the Handelsregister, HRA 7077, 30 Aug. 1950
[461] Letter from Walter Goldscheider, Vienna, to Rudolf Knörlein, Lurago d'Erba, 20 Aug. 1950
[462] Letter from Walter Goldscheider, Vienna, to Morris Katz of Marks & Rosenfeld Inc., New York, 11 April 1950
[463] Letter from Walter Goldscheider to the honorary agent Axel Wahl-Bau, Copenhagen, 25 April 1950. From America Walter received an enquiry from the Hudson Commercial Company, which marketed and distributed for Augarten and Keramos in the US and wanted to include Goldscheider in their range. They explained their interest by saying that they had always regarded Goldscheider products as superior to those of Keramos. See letter from Walter Goldscheider, Vienna, to Morris Katz of Marks & Rosenfeld Inc., New York, 11 April 1950.
[464] Letter from Haas & Czjzek, Vienna, to Friedrich Goldscheider Wiener Manufaktur, 15 Oct. 1952
[465] Letter from Walter Goldscheider, Vienna, to the agent O. Wiederhielm, Sweden, 4 April 1950
[466] Letter from Walter Goldscheider to the firm of O. Wiederhielm, Stockholm, 21 Nov. 1950. The business link between Walter Goldscheider and O. Wiederhielm was dissolved in an unfriendly manner. Wiederhielm saw virtually no possibilities for export to Sweden and no agreement was reached on a contract. Walter finally commissioned another Swedish agency that was able to import Goldscheider objects to Sweden. Wiederhielm then refused to return samples collections that had not been paid for and the photographic material and demanded that the commissions owed him by Schuster should be paid. Incidentally, the term for delivery for orders abroad was 2–3 months after placement of an order. Cf letter from Walter Goldscheider to Dr. Fritz Goldscheider, Vienna, 21 Jan. 1953.
[467] Interview with Karl Täuber, Vienna, 25 Nov. 2003
[468] Letter from Walter Goldscheider to the Firm of J. Fr. Pollak, Leibnitz, 28 Jan.1953
[469] Letter from O. Wiederhielm, Stockholm, to Walter Goldscheider, 30 Nov. 1950
[470] Also see on this the chapter "Marcell Goldscheider's workshops in Vienna and Stoke-on-Trent," p. 248f
[471] Interview with Dr. Arlette Leupold-Löwenthal, Vienna, daughter of Dr. Friedrich "Fritz" Goldscheider, 26 Nov. 2003
[472] Letter from Walter Goldscheider to the Hospital Service Plan of New Jersey, 30 Aug 1950
[473] Letter from Karl Täuber to Arturo Löffler, Buenos Aires, 7 Nov. 1951
[474] Letter from Walter Goldscheider to Rudolf Knörlein, Lurago d'Erba, 2 July. 1952
[475] Interview with Karl Täuber, Vienna, 25 Nov. 2003
[476] Entry in HRA 7077 of 17 April 1951 for Erwin Friedrich Goldscheider and of 14 June 1951 for Karl Täuber
[477] Cession between the Wiener Manufaktur Josef Schuster, vormals [formerly] Friedrich Goldscheider and Karl Sailer, Vienna, 25 Aug. 1948
[478] Cession between the Wiener Manufaktur Josef Schuster, vormals Friedrich Goldscheider and Karl Sailer, Vienna, 13 Feb. 1950
[479] Cession between the Wiener Manufaktur Friedrich Goldscheider and Karl Sailer, Vienna, 1 July 1950
[480] Letter from Walter Goldscheider to Kurt Goebel, Vienna, 24 Oct. 1952. Goebel was also asked to place a list of his clients at Goldscheider's disposal.
[481] Ibid
[482] Cession between the Wiener Manufaktur Friedrich Goldscheider and Kurt Goebel, Vienna, 24 Oct. 1952
[483] Letter from Walter Goldscheider to Kurt Goebel, 26 Nov. 1952
[484] Letter from Walter Goldscheider to Stephan Dakon, Vienna, 8 Dec. 1952
[485] Letter from Walter Goldscheider to Stephan Dakon, Vienna, 17 Nov. 1952
[486] Letter from the Oberösterreichisches Landesmuseum, Linz, to the firm of Goldscheider, Vienna, 2 July 1952
[487] Settlement of royalties account to Stephan Dakon for the 1st half year 1947, 18 Aug. 1947
[488] Settlement of royalties account to Stephan Dakon for the 1st half year 1951, 30 July 1951
[489] Settlement of royalties account to Karl Sailer for the 2nd half year 1952, 29 Jan. 1953
[490] Settlement of royalties account to Njura Lorenzl for the 1st half year 1951, 30 July 1951
[491] Makus (2000), p. 176

derartigen Figuren bewusst in fremde Urheberrechte eindringen wollten, sondern möchte annehmen, dass bei der Fertigung dieser Figuren nur versehentlich die doch allgemein bekannten Hummel'schen Urheberrechte nicht beachtet worden sind." Walter antwortet ihm am 11.9.1951, dass seit dem Bestehen der Manufaktur noch „niemals der Vorhalt einer Nachahmung" erhoben worden ist und man habe sich „immer bemüht, unsere eigenen Ideen auszuführen, um unsere Spitzenstellung durch Jahrzehnte zu bewahren." Des Weiteren macht er auch darauf aufmerksam, dass man Kinderfiguren verschiedenster Art seit mehr als dreißig Jahren produziere. Weitere Korrespondenz zu diesem Vorwurf ist nicht bekannt.

[493] Brief von Ernesto Samson, Havanna, Kuba, an Walter Goldscheider, Wien, 28.3.1953

[494] Brief von Walter Goldscheider an den akademischen Bildhauer Ottmar Obermeier, München, 3.12.1952

[495] Brief von Walter Goldscheider an Prof. Walter Kuen, Innsbruck, 27.8.1952

[496] Ebenda

[497] Brief von Walter Goldscheider an Ida Schwetz-Lehmann, Wien, 24.10.1952

[498] Brief von Ida Schwetz-Lehmann, Wien, an Walter Goldscheider, 27.1.1953

[499] Brief von Walter Goldscheider an Rudolf Knörlein, Lurago d'Erba, 6.6.1952

[500] Brief von Walter Goldscheider an Rudolf Knörlein, Lurago d'Erba, 2.7.1952

[501] Brief von Walter Goldscheider an Rudolf Knörlein, Lurago d'Erba, 4.7.1952

[502] Brief von Walter Goldscheider an Rudolf Knörlein, Lurago d'Erba, 18.8.1952

[503] Brief von Walter Goldscheider an Benno Geiger, Bern, 18.9.1952

[504] Brief von Walter Goldscheider an Benno Geiger, Bern, 4.11.1952. Der Kopf soll in roter Masse, aber nicht „in Chamottemasse" geformt werden.

[505] Brief von Walter Goldscheider an Benno Geiger, Bern, 3.12.1952

[506] Brief von Arturo G. Loeffler, Buenos Aires, an die Karl Täuber, Wien, 15.11.1951

[507] Brief von Walter Goldscheider an die Firma R. Gonnelli, Triest, 24.7.1952

[508] N.N.: Goldscheider Vienna Plant Going Full Tilt Once More, in: Trenton Evening Times, 22.2.1952

[509] Deutsche Industrie-Messe Hannover 1952, Amtlicher Messekatalog, S. 199. Goldscheider hat den Stand 616 in Halle 1. In Frankfurt befindet sich die Vertretung in der Brüder-Grimm-Str. 32.

[510] IC 1952, S. 448

[511] Auftragsbestätigung der Etikettenfabrik August Ulrich, Wien, 10.12.1952

[512] Brief von Walter Goldscheider an Rasper & Söhne, Salzburg, 3.12.1952

[513] Brief von Hotelgesellschaft Tyrol-Europa, Innsbruck, an Walter Goldscheider, Wien, 17.12.1952

[514] Brief von Walter Goldscheider an die Ital. Schiffahrtsgesellschaften Lloyd Triestino, Italia, Adriatica, Tirrenia, Wien, 5.12.1952

[515] Brief von Walter Goldscheider an das Walt Disney Lizenzbüro in Frankfurt am Main, 4.11.1952

[516] Brief von Walter Goldscheider an das Österreichische Verkehrsbüro, Wien, 14.10.1952. Bereits in den 30er Jahren hat Walter nach einem Plakat des österreichischen Verkehrsbüros eine Keramikfigur entwerfen lassen (Modell-Nr. 8345).

[517] Brief von Walter Goldscheider an die Keramische Zeitschrift, Lübeck, 16.7.1952

[518] Brief von Walter Goldscheider an Dr. Kramer von dem Fachgeschäft Ernst Wahliss, Wien, 11.11.1952

[519] Brief von Walter Goldscheider an das Fachgeschäft Ernst Wahliss, Wien, 9.4.1953

[520] Ebenda. Bereits am 12.3.1953 wurde die Einzelprokura von Lilly Goldscheider und Dr. Richard Kuschnitzky gelöscht, vgl. Eintrag im HRA 7077.

[521] Brief von Walter Goldscheider an Dr. Fritz Goldscheider, Wien, 5.8.1953

[522] Brief von Walter Goldscheider, Wien, an den seit 30 Jahren für Goldscheider arbeitenden Vertreter Ernesto Samson, Havanna, Kuba, 21.5.1953

[523] Brief von Walter Goldscheider an die Firma F. Gollhammer & Co, Vöcklabruck, 7.7.1953

[524] Postkarten der Firma F. Gollhammer & Co, Vöcklabruck, an Walter Goldscheider, 25.9.1953 und 20.10.1953

[525] Brief von Walter Goldscheider an Ing. Fritz Fink von der Gmundner Keramik, 30.6.1953. Walter erwähnt diesbezüglich auch, dass er Mitbegründer der Messe A.G. war.

[526] Brief von Walter Goldscheider an das Magistrat der Stadt Wien, 30.11.1953

[527] Brief der Liegenschaftsverwaltung Felix Brasch, Wien, an Walter Goldscheider, 2.10.1953

[528] Brief von Walter Goldscheider an Dr. Fritz Goldscheider, Wien, 5.8.1953. Für die Modellrechte, Mutterformen und Lizenzen sollen an Walter Goldscheider bis Dezember 1953 DM 69.000,– ausgezahlt worden sein. Eine monatliche Rente von DM 1.500,– wurde ihm bis zu seinem Tod zugesichert. Siehe Könecke (2002), S. 19.

[529] Brief von Walter Goldscheider an Rudolf Knörlein, Neuss am Rhein, 30.4.1953

[530] Eintrag im Handelsregister, HRA 11.511a, 26.4.1954

[531] Makus (2002), S. 35ff. Thomasch war in den 20er Jahren bei Goldscheider beschäftigt und hat zahlreiche Modelle entworfen, seit 1926 betreibt er eine eigene Manufaktur in Sierndorf. Viele der von ihm produzierten Modelle ähneln populären Goldscheider Modellen, auch der von Thomasch eingesetzte Stempel sieht dem Goldscheider GWM-Stempel zum Verwechseln ähnlich. Die bei Thomasch 1953 eingerichtete kunstkeramische Abteilung wird von Karl Sailer vorübergehend geleitet, der auch für Adolf Prischl tätig ist. Für Prischl wird übrigens auch ein anderer wichtiger Goldscheider-Künstler tätig: Stephan Dakon. Die Bildhauer Kurt Goebel, Rudolf Knörlein und Viktor Matula beliefern ebenfalls verschiedene Manufakturen mit Modellen.

[532] Brief von Erwin Goldscheider an Karl Täuber, Wien, 25.3.1962

[492] Letter from Walter Goldscheider to Adolf Prischl, Vienna, 16 Feb. 1953. There was also a minor disagreement with the maker of the famous Hummel figures. On 14 August 1951, W. Goebel of the Porzellanfabrik Goebel (est. 1871) in Oeslau near Coburg, sent a letter by registered mail to Walter Goldscheider. He enclosed two photographs of Goldscheider figures which, in his opinion, too closely resembled the famous Hummel figures from his range and, therefore, represented an infringement of copyright. He requested of Walter a statement on this but also noted that he could "not really imagine such a thing, knowing how responsibly your business is managed" and that "you deliberately set out to infringe others' copyright but would prefer to assume that in the making of these figures the generally known Hummel copyright had been unintentionally overlooked." Walter replied to him on 11 Sept. 1951 that never since the manufactory had been in existence "had an accusation of imitation" been made and the firm had "always striven to realize our own ideas in order to maintain our position as leader down through the decades." Further, he pointed out that Goldscheider had been producing figures of children of all kinds for more than thirty years. No other correspondence dealing with this charge is known.

[493] Letter from Ernesto Samson, Havana, Cuba, to Walter Goldscheider, Vienna, 28 March 1953

[494] Letter from Walter Goldscheider to the academician and sculptor Ottmar Obermeier, Munich, 3 Dec. 1952

[495] Letter from Walter Goldscheider to Professor Walter Kuen, Innsbruck, 27 Aug. 1952

[496] Ibid

[497] Letter from Walter Goldscheider to Ida Schwetz-Lehmann, Vienna, 24 Oct. 1952

[498] Letter from Ida Schwetz-Lehmann, Vienna, to Walter Goldscheider, 27 Jan. 1953

[499] Letter from Walter Goldscheider to Rudolf Knörlein, Lurago d'Erba, 6 June 1952

[500] Letter from Walter Goldscheider to Rudolf Knörlein, Lurago d'Erba, 2 July 1952

[501] Letter from Walter Goldscheider to Rudolf Knörlein, Lurago d'Erba, 4 July 1952

[502] Letter from Walter Goldscheider to Rudolf Knörlein, Lurago d'Erba, 18 Aug. 1952

[503] Letter from Walter Goldscheider to Benno Geiger, Bern, 18 Sept. 1952

[504] Letter from Walter Goldscheider to Benno Geiger, Bern, 4 Nov. 1952. The head was to be cast in red paste but not "in fire-brick paste."

[505] Letter from Walter Goldscheider to Benno Geiger, Bern, 3 Dec. 1952

[506] Letter from Arturo G. Loeffler, Buenos Aires, to Karl Täuber, Vienna, 15 Nov. 1951

[507] Letter from Walter Goldscheider to the firm of R. Gonnelli, Triest, 24 July 1952

[508] N.N.: Goldscheider Vienna Plant Going Full Tilt Once More, in: Trenton Evening Times, 22 Feb. 1952

[509] Deutsche Industrie-Messe Hannover 1952, Amtlicher Messekatalog, p. 199. Goldscheider had Stand 616 in Hall 1. In Frankfurt the representative was at 32 Brüder-Grimm-Straße.

[510] IC 1952, p. 448

[511] Confirmation of order by the August Ulrich label factory, Vienna, 10 Dec. 1952

[512] Letter from Walter Goldscheider to Rasper & Söhne, Salzburg, 3 Dec. 1952

[513] Letter from Hotelgesellschaft Tyrol-Europa, Innsbruck, to Walter Goldscheider, Vienna, 17 Dec. 1952

[514] Letter from Walter Goldscheider to the Italian steamship line Lloyd Triestino, Italy, Adriatica, Tirrenia, Vienna, 5 Dec. 1952

[515] Letter from Walter Goldscheider to the Walt Disney Licensing Office in Frankfurt am Main, 4 Nov. 1952

[516] Letter from Walter Goldscheider to the Österreichische Verkehrsbüro, Vienna, 14 Oct. 1952. Back in the 1930s Walter had had a ceramic figure designed that was based on an Austrian tourist office poster (model number 8345).

[517] Letter from Walter Goldscheider to Keramische Zeitschrift, Lübeck, 16 July 1952

[518] Letter from Walter Goldscheider to Dr. Kramer, of the specialist shop Ernst Wahliss, Vienna, 11 Nov. 1952

[519] Letter from Walter Goldscheider to the specialist shop Ernst Wahliss, Vienna, 9 April 1953

[520] Ibid. On 12 March 1953 the powers of attorney held by Lilly Goldscheider and Dr. Richard Kuschnitzky expired: cf entry in the HRA 7077.

[521] Letter from Walter Goldscheider to Dr. Fritz Goldscheider, Vienna, 5 Aug. 1953

[522] Letter from Walter Goldscheider, Vienna, to Ernesto Samson, Havana, Cuba, who had worked as a representative for Goldscheider for thirty years, 21 May 1953

[523] Letter from Walter Goldscheider to the firm of F. Gollhammer & Co, Vöcklabruck, 7 July 1953

[524] Postcards from the firm of F. Gollhammer & Co, Vöcklabruck, to Walter Goldscheider, 25 Sept. 1953 and 20 Oct. 1953

[525] Letter from Walter Goldscheider to Ing. Fritz Fink of Gmundner Keramik, 30 June 1953. Walter also mentions in this connection that he had been a cofounder of the A. G. Fair.

[526] Letter from Walter Goldscheider to the Municipal Magistracy Vienna, 30 Nov. 1953

[527] Letter from Liegenschaftsverwaltung Felix Brasch, Vienna, to Walter Goldscheider, 2 Oct. 1953

[528] Letter from Walter Goldscheider to Dr. Fritz Goldscheider, Vienna, 5 Aug. 1953. Walter Goldscheider was to receive DM 69 000 by December 1953 for the rights to the models, matrices and licenses. He was promised a monthly pension of DM 1500 until his decease. See Könecke (2002), p. 19.

[529] Letter from Walter Goldscheider to Rudolf Knörlein, Neuss am Rhein, 30 April 1953

[530] Entry in the Handelsregister, HRA 11.511a, 26 April 1954

[531] Makus (2002), p. 35f. Thomasch had been employed at Goldscheider in the 1920s and had designed numerous models. From 1926 he ran a factory of his own at Sierndorf. Many of the models he produced resemble popular Goldscheider models; similarly, the stamp used by Thomasch looks so much like the Goldscheider GWM stamp that they might easily be confused. The art ceramics division set up by Thomasch in 1953 was temporarily headed by Karl Sailer, who also worked for Adolf Prischl. Incidentally, another important Goldscheider artist also worked for Prischl: Stephan Dakon. The sculptors Kurt Goebel, Rudolf Knörlein and Viktor Matula also supplied several different factories with models.

[532] Letter from Erwin Goldscheider to Karl Täuber, Vienna, 25 March 1962

Die Weltmarke der Keramik – West Germany (1953–1960)

World Mark in Ceramics – West Germany (1953–1960)

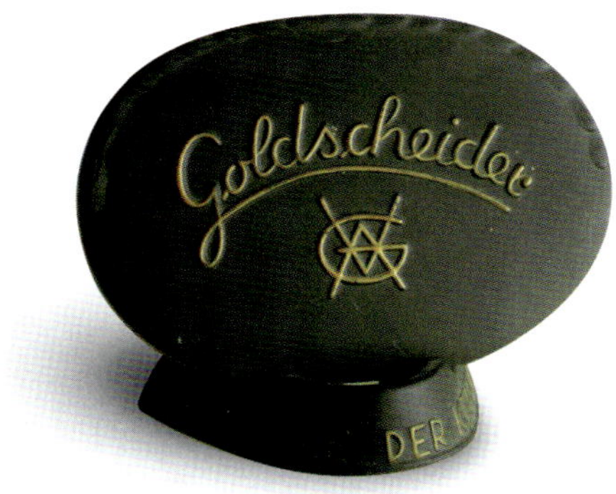

Ill. 133
Ein Werbeaufsteller aus den 1950er Jahren, der „Goldscheider – Weltmarke der Keramik" propagiert. (Quelle: Sammlung John Nash)

A 1950s sign advertising "Goldscheider – World Mark in Ceramics." (Source: John Nash Collection)

Das traditionsreiche Majolika-Unternehmen Carstens stellt nach dem Krieg im Gegensatz zu Goldscheider vor allem Artikel für den täglichen Gebrauch her, die in der Regel farblos glasiert werden. Diese werden stark nachgefragt, da durch den Krieg und die Bombenangriffe die meisten Haushalte zerstört worden sind. Carstens ist in diesem Bereich erfolgreich und beschäftigt um 1950 bereits 300 Mitarbeiter.[533] Durch den Kauf der Produktionslizenz von Goldscheider kann sich Carstens nicht nur mit der renommierten Weltmarke der Keramik schmücken, sondern seine Produktpalette um figurale Keramik und populäre Wandmasken erweitern und die Auslastung seiner Brennöfen steigern. Die Goldscheider-Produktlinie, darunter viele bewährte Modelle aus Wien, werden in einem neu errichteten Gebäude auf dem Tönnieshof, dem Fabrikgelände von Carstens, hergestellt. Da die Produktion der Goldscheider-Artikel um ein Vielfaches schwieriger ist als die bisherigen Carstens-Erzeugnisse, müssen einige der Wiener Angestellten und Arbeiter auf den Tönnieshof umziehen, um Carstens bei dem Aufbau der Produktion behilflich zu sein und die dortigen Arbeitskräfte anzulernen.[534]

Im Juni 1953 gründet die Majolikafabrik Carstens das Tochterunternehmen „Kunstkeramische Manufaktur Goldscheider GmbH" mit Sitz in Fredelsloh und startet die Produktion.[535] Da auf dem Tönnieshof die

After the war, Carstens, a venerable majolica business, made mainly utilitarian articles for household use, unlike Goldscheider. Usually sporting colorless glaze, the Carstens articles were much in demand since most households had been destroyed in the war and the air raids. Carstens was so successful in this sector that a workforce of 300 was being employed by 1950.[533] By buying the production license from Goldscheider, Carstens could not just adorn itself with the renowned world-class ceramics brand but also enlarge the scope of its own product range by adding figurative porcelain and popular wall masks, thus ensuring that its kilns were operating closer to capacity. The Goldscheider product line, including many tried and tested models from Vienna, were made in a new building erected on the Tönnieshof, the Carstens factory premises. Since producing Goldscheider articles was many times more difficult than making the products Carstens had been putting on the market up to then, some of the employees and workmen from Vienna had to transfer to the Tönnieshof to help Carstens with building up production and train the Carstens personnel.[534]

In June 1953, the Carstens majolica factory established "Kunstkeramische Manufaktur Goldscheider GmbH," a subsidiary based in Fredelsloh, and started production there.[535] Since making figurative ceramics

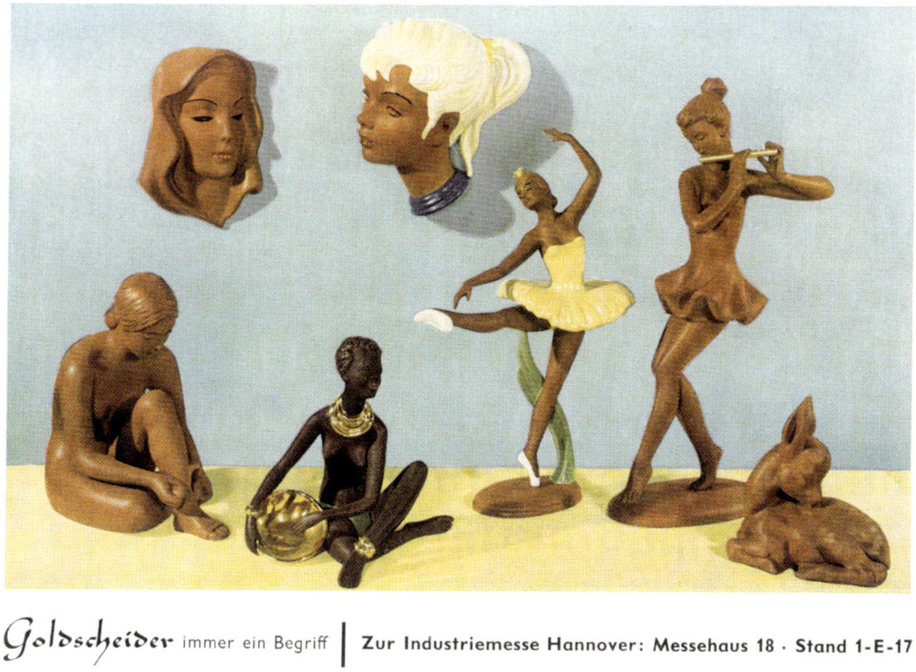

Ill. 134
Als Einladung zur Industriemesse in Hannover 1955 setzt Carstens dieses Werbeprospekt ein, das sowohl populäre Modelle aus Wiener Zeiten zeigt wie auch neue Entwürfe. (Quelle: Archiv Mimi Dakon)

As its invitation to the 1955 Hannover Industrial Fair, Carstens used this advertising prospectus showing both popular models from the Vienna days and new designs. (Source: Mimi Dakon Archives)

using the faience process had never been done at the Tönnieshof, the decision was taken to make the Goldscheider figures in "red clay body in the terracotta manner."[536] This undertaking was supported by the Goldscheider Vienna repairers Hans Sefzik, Johannes Beinhofer and Herr Förster and the sculptor Schwarz as well as other former Goldscheider Vienna employees.[537] The sculptors Prischl and Knörlein also maintained close ties to Carstens. On Walter Goldscheider's recommendation, Rudolf Knörlein became artistic director of Goldscheider production at the Tönnieshof in 1954.[538] Knörlein contributed some "Negro" masks to the Goldscheider model range, which were decorated with a new ground color he had developed. The colored glazes he had developed in 1954 and 1955 gave the 1950s Goldscheider masks their characteristic appearance.[539]

In 1955, a workforce of 20 handled the Goldscheider Production at Carstens, which comprised dance figures, nudes, animals and wall masks.[540] At the German Industrial Fair in Hanover (1955, 1956), Carstens exhibited figurative ceramics articles under the name "Kunstkeramische Manufaktur Friedrich Goldscheider GmbH."[541] The models continued to be marked with "Goldscheider" in lettering and the "G" with enclosed "W" and "M" but "Vienna" was replaced by "West Germany" as the place of manufac-

Herstellung figurativer Keramik im Fayence-Verfahren bislang unbekannt ist, beschließt man, die Figuren von Goldscheider in „rotem Ton in Terrakotta-Manier" zu fertigen.[536] Unterstützung findet man in den Wiener Retuscheuren Hans Sefzik, Johannes Beinhofer, Herrn Förster und dem Bildhauer Schwarz sowie anderen ehemaligen Mitarbeitern von Goldscheider Wien.[537] Auch die Bildhauer Prischl und Knörlein stehen in enger Beziehung zu Carstens. Rudolf Knörlein wird auf Empfehlung Walter Goldscheiders 1954 künstlerischer Direktor der Goldscheider-Herstellung auf dem Tönnieshof[538] und steuert unter anderem auch „Neger"-Masken zur Goldscheider-Modellpalette bei, die mit einer von ihm neu entwickelten Grundfarbe versehen sind. Die von ihm 1954 und 1955 entwickelten farbigen Glasuren verleihen den Goldscheider-Masken der 50er Jahre das charakteristische Erscheinungsbild.[539]

Im Jahr 1955 sind in der Goldscheider Produktion 20 Mitarbeiter beschäftigt, die Tanzfiguren, Aktfiguren, Tiere und Wandmasken produzieren.[540] Auf der Deutschen Industrie-Messe Hannover (1955, 1956) stellt man als „Kunstkeramische Manufaktur Friedrich Goldscheider GmbH" figürliche Artikel aus Keramik aus.[541] Die Modelle werden weiterhin mit dem Schriftzug Goldscheider und dem G mit eingeschlossenem W und M gemarkt, der Herstellungsort Wien wird allerdings durch West Germany ersetzt. Die

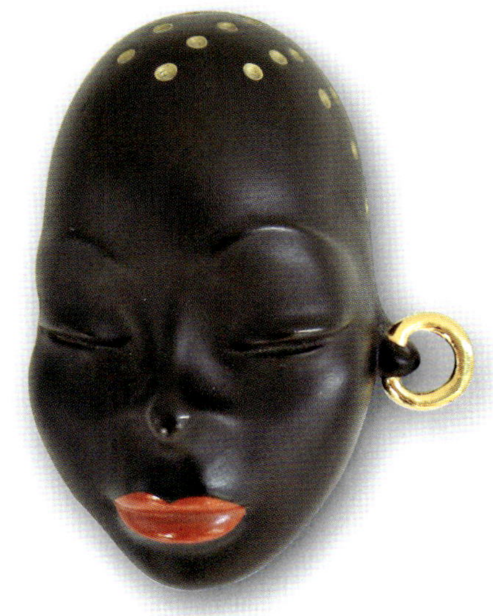

ModNr. 700 „Maske" I "Mask", Rudolf Knörlein, ca. 1955, Collection f@goldscheider.de

Rudolf Knörlein wird 1954 künstlerischer Leiter der Goldscheider-Produktion bei Carstens und steuert einige neue Modelle bei.

In 1954, Rudolf Knörlein became artistic director of Goldscheider production at Carstens and contributed some new models.

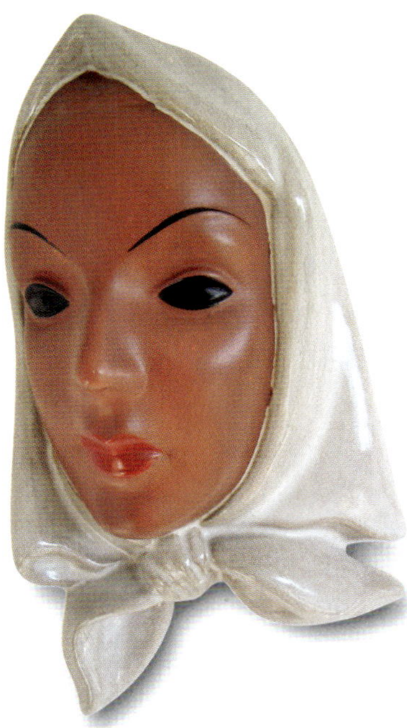

ModNr. 8328 „Maske mit Kopftuch" I "Mask with Headscarf", Entwurf I design ca. 1938, Ausführung I execution ca. 1955, Collection f@goldscheider.de

Carstens übernimmt zahlreiche Formen aus Wien und produziert beliebte alte wie auch gänzlich neue Modelle unter der Lizenzmarke „Goldscheider" bei sich auf dem Tönnieshof in „West Germany".

Carstens took over numerous molds from Vienna and produced popular old models and entirely new ones under the licensed "Goldscheider" brand at the Tönnieshof in "West Germany."

ModNr. 321
„Maske" | "Mask",
Stephan Dakon,
ca. 1955, Collection
f@goldscheider.de

Stephan Dakon steuert auf dem Tönnieshof neben Wandmasken zahlreiche „Neger"-Figuren bei, die sich in den 1950er Jahren hoher Beliebtheit erfreuen.

In addition to wall masks, Stephan Dakon contributed many "Negro" figures at the Tönnieshof, which were extremely popular in the 1950s.

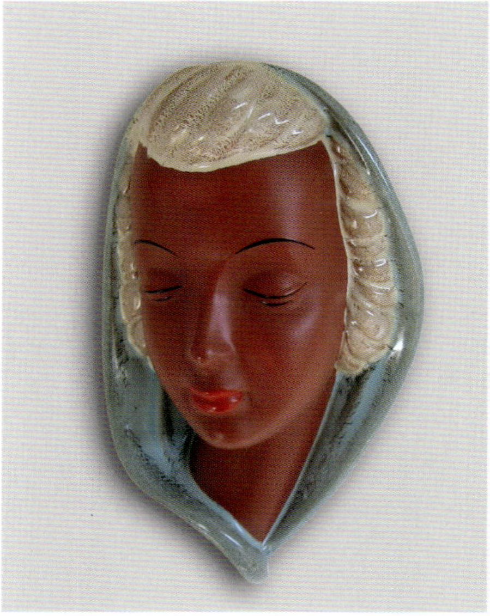

Modelle werden unterschiedlich nummeriert. Die über einhundert aus Wien übernommenen Modelle, vor allem Wandmasken, Tierfiguren, einige Madonnen, Akte und Tänzerinnen, werden mit der gleichen Nummerierung versehen wie in Wien, d.h. mit der gleichen Modellnummer, der Retuscheurnummer und manchmal mit der Stücknummer.

Bei Carstens neu entworfene Modelle erhalten eine neue Nummerierung.[542] Dabei handelt es sich vor allem um die in den 1950er Jahren beliebten Wandmasken und exotische Figuren, die Afrikaner oder Asiaten darstellen, sowie Tierfiguren. Diese ab 1955 entstandenen Modelle werden nummerisch

ture. Several different numbering systems were used for the models. The more than one hundred models taken over from Vienna, mainly wall masks, animal figures, Madonnas, nudes and dancers, were numbered just as they had been in Vienna, ie, with the same model numbers, repairer numbers and sometimes a piece number.

New models designed at Carstens were given new numbering.[542] These were mainly the wall masks so popular in the 1950s and

ModNr. 309, 303, 304, 302
Verschiedene „Neger"-Figuren | Various "Negro" figures, Stephan Dakon, ca. 1955, Collection mad

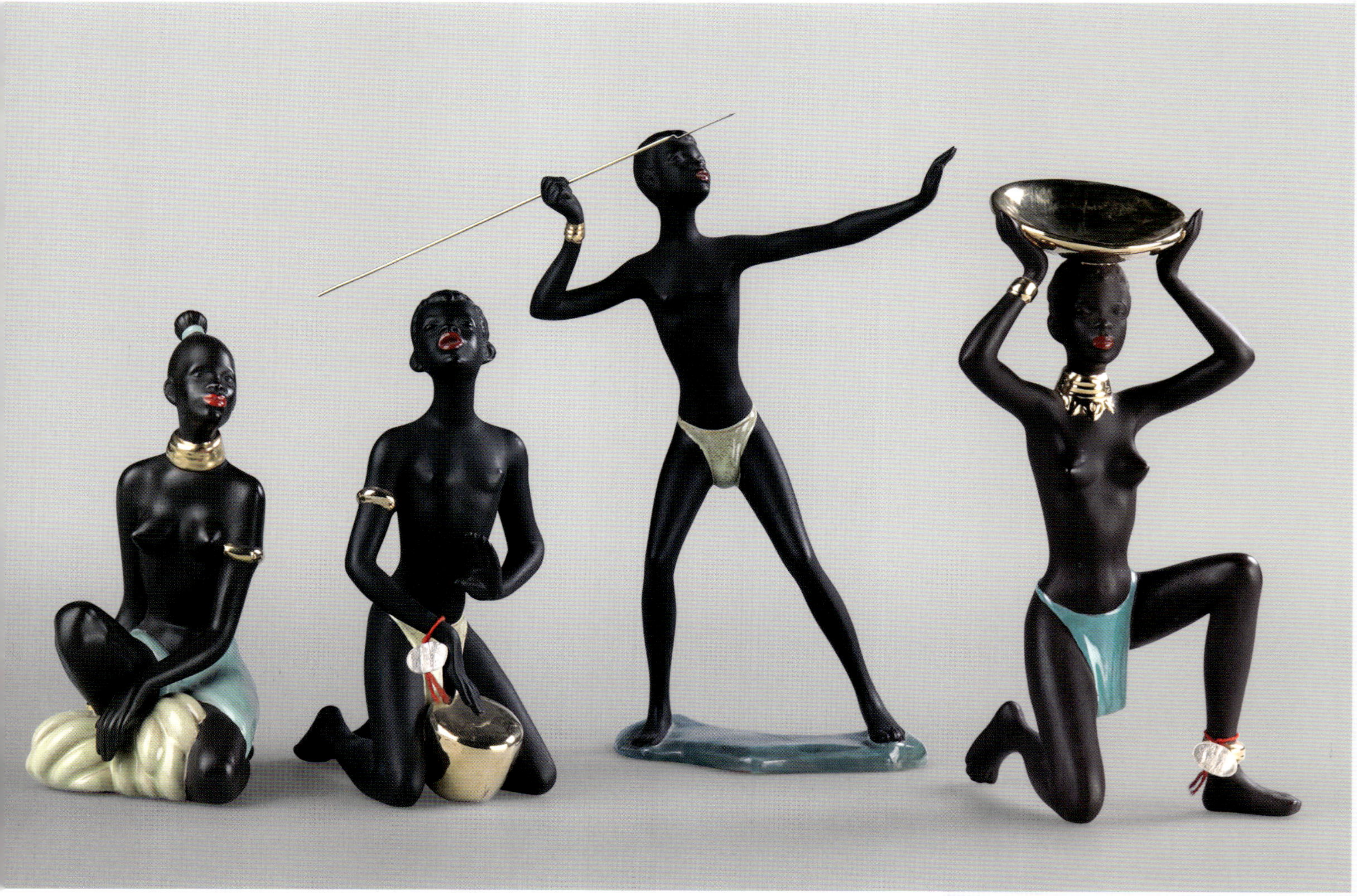

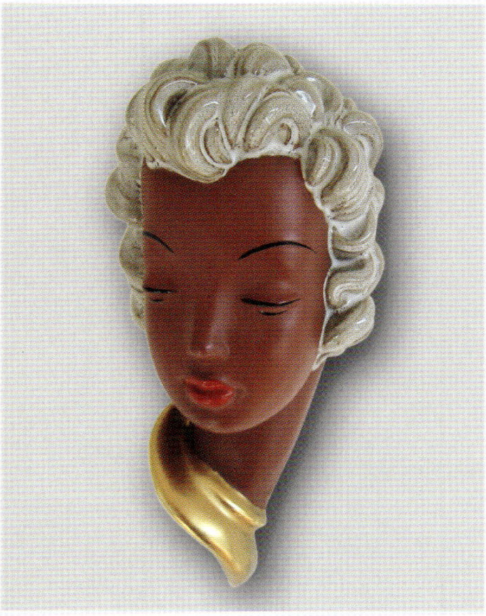

ModNr. 510
„Maske" I "Mask",
Adolf Prischl, ca. 1955,
Collection f@goldscheider.de

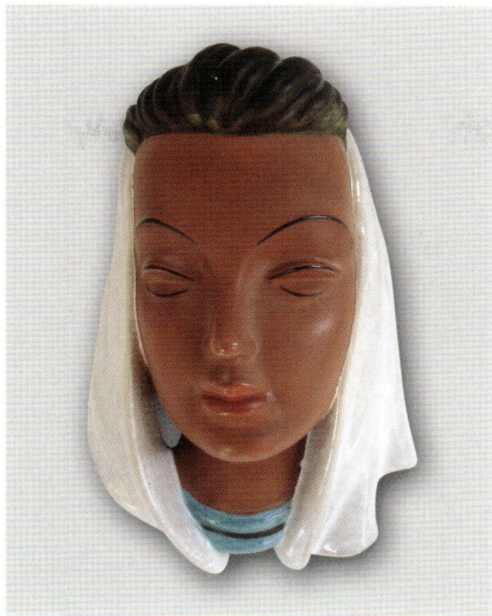

ModNr. 529
„Maske" I "Mask",
Adolf Prischl, ca. 1955,
Collection f@goldscheider.de

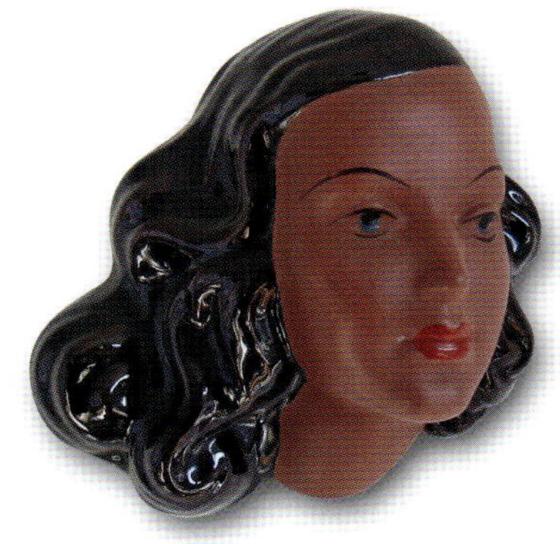

ModNr. 533
„Maske" I "Mask",
Adolf Prischl, ca. 1955,
Collection f@goldscheider.de

Viele der von Adolf Prischl an Carstens verkauften Wandmasken werden auch von ihm selbst in seiner Werkstätte „Prischl Kunstkeramik" in Wien produziert. Beispielsweise trägt das unter Carstens produzierte Modell 529 in der Prischl-Produktion die Nummer 129.

Adolf Prischl also produced many of the wall masks he sold to Carstens at his own Vienna workshop, "Prischl Kunstkeramik." To take one example, model number 529 produced at Carstens bears the number 129 in Prischl's own product line.

exotic figures, representing Africans or Asians, as well as animal figures. These numbers of the models made from 1955 indicate which artist designed them: the models in the 300 series are after designs by Stephan Dakon (exotic figures, masks, nudes); the 500 series were designed by Prischl (masks); the 700 series are by Knörlein (masks, animals, nudes) and the 1000 series were made after designs by Walter Bosse (animal figures).[543] In between are three 200 numbers by Goedtke, two 400 numbers by Müller, two 600 numbers by Chludek, two 800 numbers by Weiß and two 900 numbers by Schmidt. These new objects often marked with only an incised model number. A golden label bearing the slogan "Weltmarke der Keramik" and "Goldscheider" in lettering along with the "GWM" sign also often occurs on these models.

Producing the Goldscheider line at the Tönnieshof incurred relatively high overheads because inconvenient spatial design caused breakage of up to 25 per cent of models while they were being between the various divisions.[544] As former employees saw it, the Goldscheider products were

den Künstlern zugeordnet: die Modelle der 300er-Reihe sind nach Entwürfen von Stephan Dakon (exotische Figuren, Masken, Akte), die 500er-Reihe von Prischl (Masken), die 700er-Reihe von Knörlein (Masken, Tiere, Akte) und die 1000er-Reihe nach Entwürfen von Walter Bosse (Tierfiguren) entstanden.[543] Dazwischen finden sich drei 200er-Nummern von Goedtke, zwei 400er-Nummern von Müller, zwei 600er-Nummern von Chludek, zwei 800er-Nummern von Weiß sowie zwei 900er-Nummern von Schmidt. Bei diesen neuen Objekten wird häufig nur die Modellnummer eingeritzt. Man findet auf diesen Modellen auch öfters einen goldenen Aufkleber mit dem Slogan „Weltmarke der Keramik" und der Aufschrift „Goldscheider" mit dem „GWM"-Zeichen.

Die Goldscheider Produktion auf dem Tönnieshof ist relativ teuer, da durch die unvorteilhafte Raumgestaltung auf dem Transport zwischen verschiedenen Abteilungen bis zu 25 Prozent der Modelle zu Bruch gehen.[544] Nach der Meinung ehemaliger Mit-

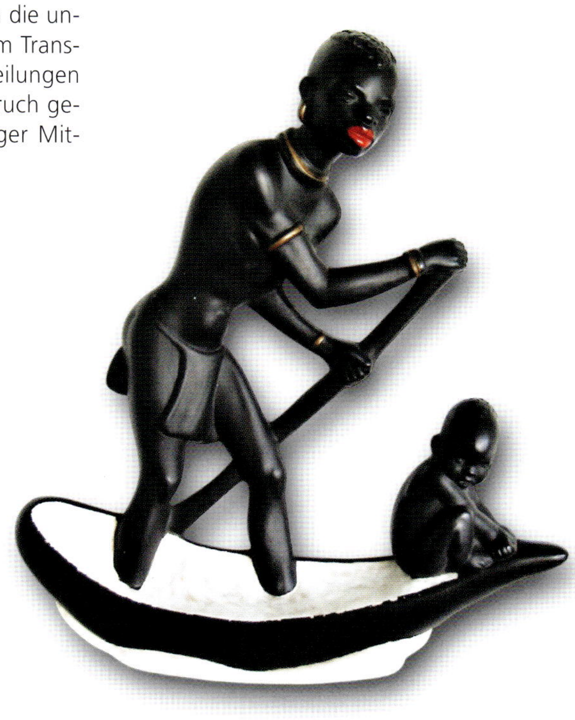

ModNr. 354
Wandplastik „Negerboot" I Wall Sculpture "Negro Boat",
Stephan Dakon,
ca. 1955, Collection f@goldscheider.de

Ill. 135–137
Ende der 1950er Jahre wird die Gefäßkeramik von Carstens ebenfalls mit einem Sticker versehen, der den renommierten Markennamen trägt. (Quelle: Sammlung a@goldscheider.net)

In the late 1950s, ceramic vessels from Carstens were labeled with a sticker that also bore the famed brand name. (Source: Collection a@goldscheider.net)

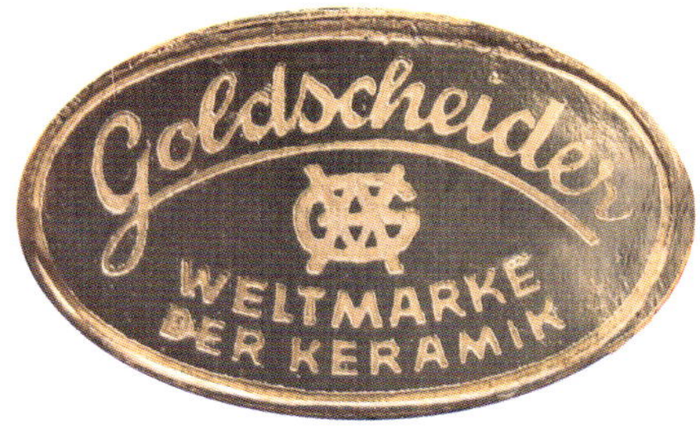

Ill. 135

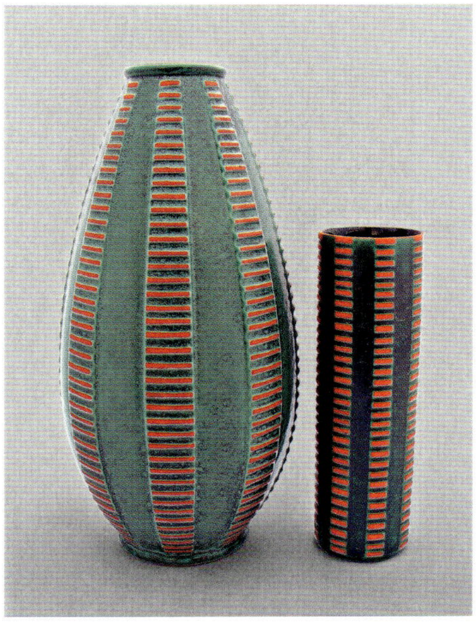

Ill. 136

rechts I right

Ill. 137

arbeiter ist die Goldscheider-Abteilung „vor allem ein Steckenpferd des Senior-Chefs, Ernst Carstens".[545] Nach seinem Tod im Jahr 1954 achtet sein Sohn Christian verstärkt auf die Wirtschaftlichkeit, so dass es gegen Ende 1954 zu den ersten Entlassungen bei der Goldscheider Produktion kommt, beziehungsweise zu Arbeitsplatzverlagerungen zur Muttergesellschaft. Außerdem wird die Produktion gestrafft: Man verzichtet ab ca. 1954/55 auf „den Schrühbrand und das Umarbeiten der sehr kleinteiligen" Wiener Modelle. Die bis zu achtteiligen Wandmaskenformen werden geändert und nur noch aus drei Teilen hergestellt.[546] Diese Reduzierung des Produktions- und Lohnaufwands führt aber zu Absatzverlusten, da die Kundschaft die zuvor ausdrucksstärkeren Modelle bevorzugt. Allgemein unterscheiden sich die Wandmasken der fünfziger Jahre von den mondänen Masken der 20er und 30er Jahre. In den 50ern wird das „Bild" der modernen Frau in den Wandmasken reflektiert. Sie stellen oft junge schöne Frauen dar, Teenager mit Zöpfen.[547]

Die Goldscheider Modelle werden weiterhin ausschließlich über den Fachhandel vertrieben und im höheren Preissegment angesetzt. Drei verschiedene Kataloge sind aus dieser Zeit zur figürlichen Produktion sowie zwei zur Gefäßproduktion bekannt.[548] Der Exportanteil ist jedoch im Gegensatz zu früher niedriger und beträgt in den Jahren 1953 bis 1960 bei fallender Tendenz zwischen 47 und 11 Prozent.[549] Die Goldscheider Produktion auf dem Tönnieshof endet im Jahr 1960, die Firma bleibt aber vermutlich aus steuerlichen Gründen weiter bestehen und vermietet die Räumlichkeiten an die Muttergesellschaft Carstens. Anfang 1963, kaum ein Jahr nach dem Tod Walter Goldscheiders, wird die „Kunstkeramische Manufaktur Friedrich Goldscheider GmbH" in die „Carstens Tönnieshof GmbH" umgewandelt.[550]

ModNr. 1004
„Pferd" I "Horse",
Walter Bosse, ca. 1955,
Collection f@goldscheider.de
Auch Walter Bosse wird in den 1950er Jahren für Carstens tätig und entwirft einige Tierfiguren.

Walter Bosse also worked for Carstens in the 1950s and designed some animal figures.

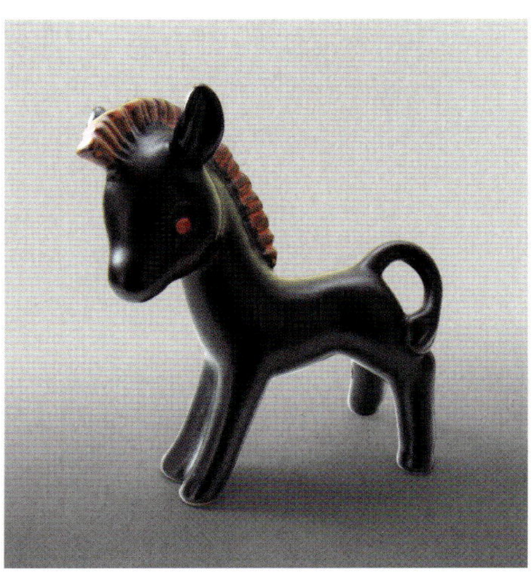

"mainly a hobby-horse for the senior boss, Ernst Carstens."[545] After he died in 1954, his son, Christian, paid more attention to economizing so that the first employees involved with the Goldscheider line were dismissed late in 1954 or were transferred to the parent company. In addition, production was streamlined: from about 1954/55, "sintering and repairing the very intricate" Viennese models were dispensed with. The wall mask molds, which had comprised up to eight pieces, were changed and made in only three pieces.[546] However, rationalizing production and, concomitantly, reducing labor costs, led to a downturn in sales because the clientele preferred the old models that had been more expressive. Generally speaking, the 1950s wall masks are unlike the stylish 1920s and 1930s masks. The "image" of the modern woman is reflected in the 1950s masks. They often depict beautiful young women or teenaged girls with their hair in pigtails.[547]

The Goldscheider models continued to be sold exclusively by specialist retailers and targeted the higher market segments. Three different catalogues listing the figures produced and two devoted to vessels are known from this period.[548] The export share of turnover was, however, lower than it had been, fluctuating between 47 and 11 per cent in the years from 1953 to 1960 on a downwards spiral.[549] The production of Goldscheider wares at the Tönnieshof ceased in 1960 but the firm continued in existence, presumably for tax purposes, and rented the rooms to Carstens, the parent company. Early in 1963, not even a year after Walter Goldscheider died, the "Kunstkeramische Manufaktur Friedrich Goldscheider GmbH" changed its name to "Carstens Tönnieshof GmbH."[550]

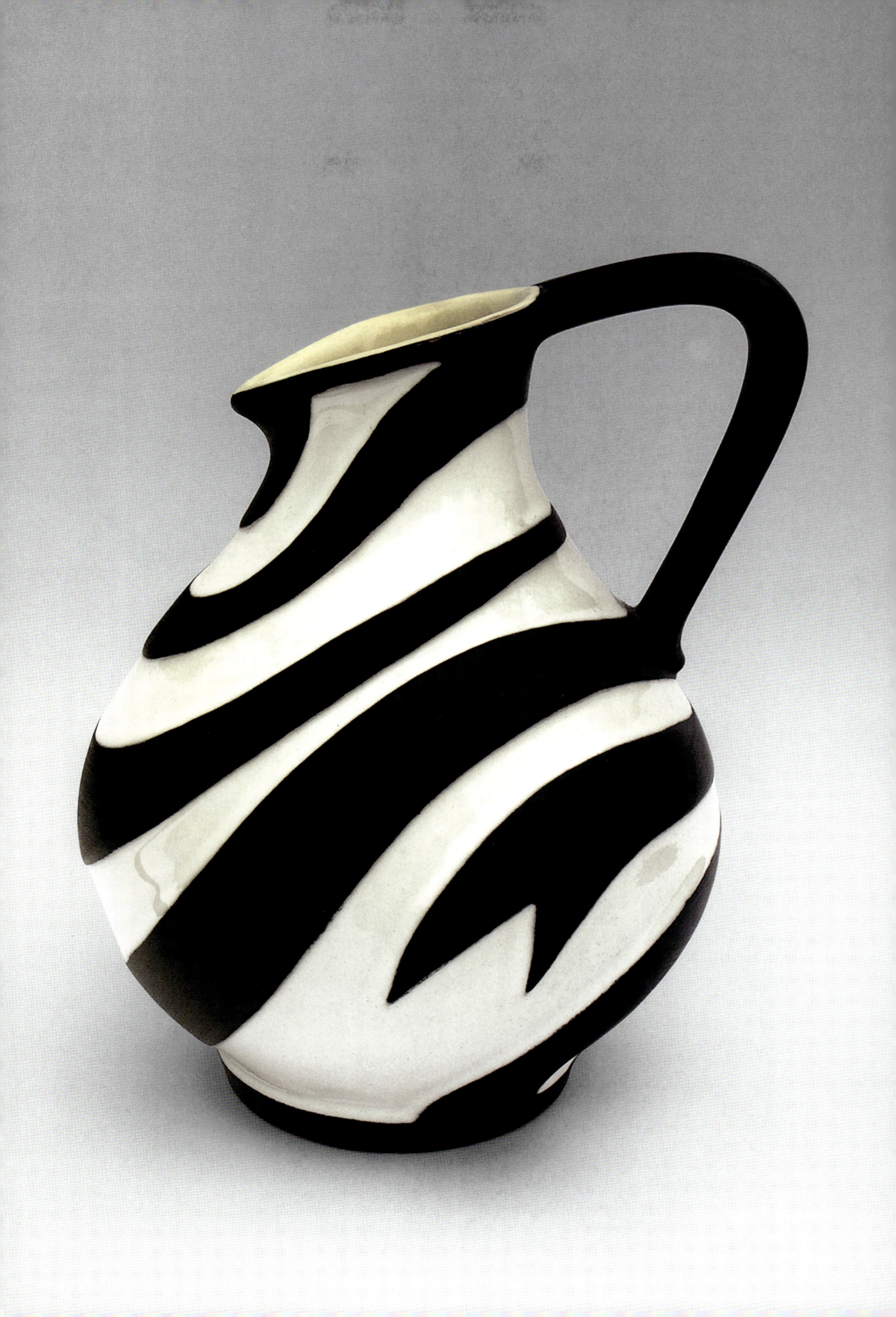

Anmerkungen

[533] Könecke (2002), S. 12. Da die Firmengeschichte der Goldscheider Manufaktur in Deutschland bereits durch Publikationen von Gerald Könecke und Horst Makus ausführlich dokumentiert ist, wird hier die Entwicklung von Goldscheider in den Jahren 1953 bis 1963 nur skizziert.
[534] Könecke (2002), S. 14. Auch der langjährige Mitarbeiter Karl Täuber fährt zu Carstens, verlässt jedoch schon bald die dortige Fabrik. Bei Carstens wird 1953 der Arbeitstag noch mit Marschmusik und einer Versammlung angefangen. Der Tönnieshof ist eine Welt für sich, einmal die Woche kommt ein Bus mit Lebensmitteln, ansonsten dreht sich alles rund um die Fabrik. S. Interview mit Karl Täuber, Wien, 25.11.2003.
[535] Könecke (2002), S. 16
[536] Porzellan und Glas, 4/1954, S. 39ff, zitiert nach Könecke (2002), S. 16. Durch den Einsatz des roten Tones kam es zu Schwierigkeiten, da die Schwindung der in Wien eingesetzten weißen Masse 6 Prozent betrug, aber bei der roten Masse es 15 Prozent waren. Die Terrakotta-Objekte wurden entweder bemalt (Haar, Gesicht) oder glasiert, manchmal mit „Glasur Bronce" oder „Mattgold". Ab 1954/55 wird der rote Engobeüberzug weitgehend durch Nitrolacke ersetzt. S. Brutto-Preisliste, in: Könecke (2002), S. 97
[537] Könecke (2002), S. 16. Nach Könecke ist Sefzik im April 1956 nach Wien zurückgekehrt, Beinhofer Ende 1955. Der Bildhauer Schwarz soll bereits Anfang 1955 zurückgegangen sein, später aber als freier Mitarbeiter für den Tönnieshof gearbeitet haben.
[538] Makus (2000), S. 158ff
[539] Ebenda
[540] AdK 1954/55, Eintragung Nr. 1379
[541] Vgl. Deutsche Industrie-Messe, Hannover 1955 bzw. 1956, Amtl. Messekatalog
[542] Hierbei findet man aber auch Modelle, z.B. Wandmasken, die von Prischl bereits in den 1930er Jahren für Goldscheider Wien entworfen wurden, aber im Lizenzvertrag zwischen Walter Goldscheider und Christian Carstens vermutlich nicht enthalten waren. Prischl hat diese wahrscheinlich im Nachhinein an Carstens verkauft.
[543] Vergleiche Auflistung der Modelle in Könecke (2002), S. 73ff
[544] Könecke (2002), S. 21
[545] Ebenda
[546] Ebenda
[547] Makus (2000), S. 16ff
[548] Ebenda, S. 20. Hinzu kommen vier Preisblätter bzw. Preislisten.
[549] Siehe Bilanzangaben Goldscheider von 1953–1960, in: Könecke (2002), S. 22
[550] Könecke (2002), S. 23

Notes

[533] Könecke (2002), p. 12. Since the history of the Goldscheider manufactory in Germany has been thoroughly documented in publications by Gerald Könecke and Horst Makus, developments at Goldscheider between 1953 and 1963 are only sketchily outlined here.
[534] Könecke (2002), p. 14. Karl Täuber, a Goldscheider employee for many years, went to Carstens but soon left the factory there. Even in 1953, the working day was ushered in at Carstens with marching music and an assembly. The Tönnieshof was a world apart; once a week a bus brought food supplies but otherwise everything revolved around the factory. See interview with Karl Täuber, Vienna, 25 Nov. 2003.
[535] Könecke (2002), p. 16
[536] Porzellan und Glas, 4/1954, p. 39f, quoted in Könecke (2002), p. 16. Using red clay caused problems because the shrinkage in firing of the white paste used in Vienna was 6 per cent but the red paste shrank by 15 per cent. Terracotta objects were either painted (hair, face) or glazed, sometimes with "Glasur Bronce" ["bronze glaze"] or "matt gold." From 1954/55 the red engobe coating was replaced with nitro-lacquers. See wholesale price list; in: Könecke (2002), p. 97.
[537] Könecke (2002), p. 16. According to Könecke, Sefzik returned to Vienna in April 1956 and Beinhofer did so in 1955. Schwarz, the sculptor, is said to have returned by early 1955 but later worked on a freelance basis for the Tönnieshof.
[538] Makus (2000), p. 158f.
[539] Ibid
[540] AdK 1954/55, entry no. 1379
[541] Cf Deutsche Industrie-Messe, Hannover 1955 and 1956, Amtl. Messekatalog
[542] There are, however, also models, including wall masks, that Prischl designed for Goldscheider Vienna in the 1930s but were presumably not covered by the licensing agreement concluded between Walter Goldscheider and Christian Carstens. Prischl probably sold them to Carstens ex post facto.
[543] Cf the list of models in Könecke (2002), p. 73f.
[544] Könecke (2002), p. 21
[545] Ibid
[546] Ibid
[547] Makus (2000), p. 16f.
[548] Ibid, p. 20. In addition, there are four price sheets or price lists.
[549] See the Goldscheider balance sheets from 1953–1960, in: Könecke (2002), p. 22.
[550] Könecke (2002), p. 23

Parallel zum Wiener Geschäftszweig der Familie Goldscheider bauen Arthur Goldscheider (in Paris) und Marcell Goldscheider (in Wien, später in Stoke-on-Trent) eigene Manufakturen auf. Einige Jahrzehnte später erfolgt die Wiederaufnahme der Produktion durch Peter Goldscheider (in Stoob).
Die Entwicklung dieser Unternehmungen wird in den folgenden Sonderkapiteln skizziert.

Paralleling the Viennese Goldscheider branch, Arthur Goldscheider (in Paris) and Marcell Goldscheider (in Vienna, later in Stoke-on-Trent) built up factories of their own. Decades later production resumed (in Stoob) under Peter Goldscheider. The three chapters that follow outline how these businesses developed.

Dekorative Skulpturen von Arthur Goldscheider in Paris (1900 – ca. 1935)

Ornamental Sculpture from Arthur Goldscheider in Paris (1900 – ca. 1935)

Arthur Goldscheider betreibt nach der Erbaufteilung ab dem Jahr 1900 das französische Unternehmen „Frédéric Goldscheider" mit dem „Dépot Général", also dem Musterlager für den Großhandel, in der 45, Rue de Paradis und dem „Maison de Vente", dem Schauraum des Einzelhandels, in der 28, Avenue de l'Opera. Die „Fabrique de Bronzes" und das „Atelier de Marbres" befinden sich ebenfalls in der Rue de Paradis, das Florenzer Atelier dient bereits ab 1900 nur noch als Zulieferbetrieb.[551] Von 1900 bis 1902 ist Arthurs Bruder Camillo in Paris beschäftigt und auch Walter und Marcell verbringen ihre Lehrjahre teilweise bei ihrem Bruder in Frankreich. Arthur ist weiterhin eng an Wien gebunden und lässt sich am 5. März 1901 als „Kunstgewerbetreibender in Paris" beim k.k. Patentamt in Wien ein Verfahren patentieren, bei dem eine neue Masse „dieselben decorativen Wirkungen hervorbringt, wie durchscheinendes und undurchsichtiges Email"[552] (Ill. 138).

Neben dem Vertrieb der Wiener Modelle und deren Reproduktionen in Bronze und Marmor ist Arthur in Paris als „Éditeur d'art" tätig, das heißt er ist Verleger bzw. Herausgeber und vertreibt die Werke diverser, zumeist französischer Künstler unter eigener Modellnummerierung. Meistens wählt er nach der Besichtigung ausgestellter Skulpturen in Ateliers und Salons von verschiedenen Künstlern Modelle aus, die er in seiner Bronzefabrik gießen lässt und anschließend international vertreibt. Um die Jahrhundert-

After probate on Friedrich Goldscheider's will in 1900, Arthur Goldscheider ran the French branch of the firm, "Frédéric Goldscheider," along with the "Dépot Général," that is, the wholesale samples warehouse, at 45 rue de Paradis and the "Maison de Vente" (the retail show-room) at 28 avenue de l'Opera. The "Fabrique de Bronzes" and the "Atelier de Marbres" were also in rue de Paradis. By 1900, the Florence studio was being drawn on only for supplies.[551] Arthur's brother, Camillo, was also employed in Paris from 1900 until 1902 and Walter and Marcell served part of their apprenticeships under their brother in France. Arthur continued to maintain close ties to Vienna. On 5 March 1901, he had a process patented in his capacity as a "seller of arts and crafts in Paris" at the k.k. Patent Office in Vienna for a new paste that "produces the same decorative effects as translucent and opaque enamel"[552] (Ill. 138).

Apart from selling the Viennese models and reproducing them in bronze and marble, Arthur worked in Paris as an "éditeur d'art," which means he was a publisher and general editor and sold works by various artists, usually French ones, using a model numbering system of his own. He usually went to look at sculpture exhibited at studios and Salons and then selected models by various artists, which he had cast at his bronze foundry and ultimately sold worldwide. Around the turn of the century, his competitors in Paris were other éditeurs d'art, including Houdebine, E. Blot and the celebrated Colin & Cie, who represented the models of artists Maurice Bouval, Léon Noël Delagrange and Louis Chalon.[553] By 1902, however, Arthur had

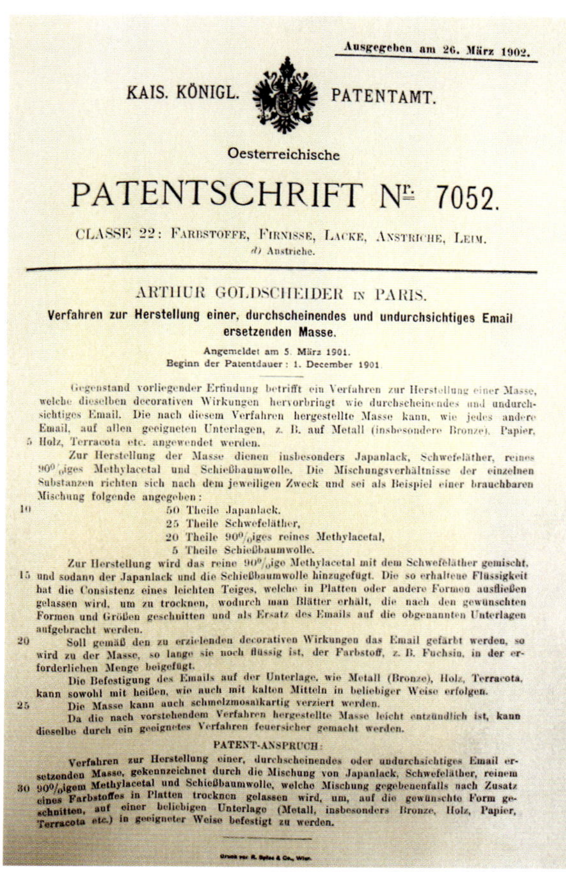

Ausgegeben am 26. März 1902.

KAIS. KÖNIGL. PATENTAMT.

Oesterreichische

PATENTSCHRIFT Nr. 7052.

CLASSE 22: Farbstoffe, Firnisse, Lacke, Anstriche, Leim.
d) Anstriche.

ARTHUR GOLDSCHEIDER in PARIS.

Verfahren zur Herstellung einer, durchscheinendes und undurchsichtiges Email ersetzenden Masse.

Angemeldet am 5. März 1901.
Beginn der Patentdauer: 1. December 1901.

Gegenstand vorliegender Erfindung betrifft ein Verfahren zur Herstellung einer Masse, welche dieselben decorativen Wirkungen hervorbringt wie durchscheinendes und undurchsichtiges Email. Die nach diesem Verfahren hergestellte Masse kann, wie jedes andere Email, auf allen geeigneten Unterlagen, z. B. auf Metall (insbesondere Bronze), Papier, Holz, Terracota etc. angewendet werden.

Zur Herstellung der Masse dienen insbesonders Japanlack, Schwefeläther, reines 90%iges Methylacetal und Schießbaumwolle. Die Mischungsverhältnisse der einzelnen Substanzen richten sich nach dem jeweiligen Zweck und sei als Beispiel einer brauchbaren Mischung folgende angegeben:

50 Theile Japanlack,
25 Theile Schwefeläther,
20 Theile 90%iges reines Methylacetal,
5 Theile Schießbaumwolle.

Zur Herstellung wird das reine 90%ige Methylacetal mit dem Schwefeläther gemischt, und sodann der Japanlack und die Schießbaumwolle hinzugefügt. Die so erhaltene Flüssigkeit hat die Consistenz eines leichten Teiges, welche in Platten oder andere Formen ausfließen gelassen wird, um zu trocknen, wodurch man Blätter erhält, die nach den gewünschten Formen und Größen geschnitten und als Ersatz des Emails auf die obgenannten Unterlagen aufgebracht werden.

Soll gemäß den zu erzielenden decorativen Wirkungen das Email gefärbt werden, so wird zu der Masse, so lange sie noch flüssig ist, der Farbstoff, z. B. Fuchsin, in der erforderlichen Menge beigefügt.

Die Befestigung des Emails auf der Unterlage, wie Metall (Bronze), Holz, Terracota, kann sowohl mit heißen, wie auch mit kalten Mitteln in beliebiger Weise erfolgen.

Die Masse kann auch schmelzmosaikartig verziert werden.

Da die nach vorstehendem Verfahren hergestellte Masse leicht entzündlich ist, kann dieselbe durch ein geeignetes Verfahren feuersicher gemacht werden.

PATENT-ANSPRUCH:

Verfahren zur Herstellung einer, durchscheinendes oder undurchsichtiges Email ersetzenden Masse, gekennzeichnet durch die Mischung von Japanlack, Schwefeläther, reinem 90%igem Methylacetal und Schießbaumwolle, welche Mischung gegebenenfalls nach Zusatz eines Farbstoffes in Platten trocknen gelassen wird, um, auf die gewünschte Form geschnitten, auf einer beliebigen Unterlage (Metall, insbesonders Bronze, Holz, Papier, Terracota etc.) in geeigneter Weise befestigt zu werden.

Druck von R. Spies & Co., Wien.

Ill. 138
Im Jahr 1901 lässt sich Arthur ein Verfahren zur Herstellung einer neuen Dekorations-Masse patentieren. (Quelle: Oesterreichische Patentschrift Nr. 7052)

In 1901, Arthur had a new process for making a paste for ornaments patented. (Source: Oesterreichische Patentschrift No. 7052)

succeeded in showing the models "Printemps de Bretagne" and "Éclosion" designed by Louis Chalon at the Salon of the Société des Artistes Français in Paris and in attracting other artists, including Marius Mars-Vallett.[554] That same year, an "Egyptian Vase" and a vase called "Les enfants" after designs by the Jugendstil artist Hans Stoltenberg Lerche were shown at the Salon of the Société Nationale des Beaux-Arts along with a statuette of a Bacchante with an inkwell designed by Gottfrid Larsson and a lighting fixture in plaster of Paris entitled "Le Vice fuyant la lumière."[555]

An advertisement Arthur placed in the specialist journal Art et Décoration (Ill. 139) indicates that in 1903 he was selling models in terracotta, bronze and marble after designs by such well-known French artists as Allouard, A. Boucher, Bouval, Croisy, Carlier, Carrièr-Belleuse, F. und A. Charpentier, Injalbert, Klein, Levasseur, Ledru, Loiseau-Rousseau, H. Lemaire, Mercié, Moreau-Vauthier, Marqueste and T. Noël as well as models by the Regensburg artist Maison and Strasser, a Viennese.[556]

In 1907, the acclaimed table decoration "Le Déjeuner fleuri" by Laporte-Blairsy was made for Arthur in Paris. It consisted in "Marquises XVIIIe siècle, portant des corbeilles."[557] Other artists were also attracting attention. The talented Paul Philippe designed several bronzes and marble figures for Arthur Goldscheider, including the marble statuette "Le Réveil" in 1907, a marble and ivory statuette of a "Femme au masque" (1908) and, in several materials, the 1909 group "Grâce et Force."[558] That same year Arthur exhibited the bronze and ivory stat-

wende konkurriert er in Paris mit anderen Verlegern wie Houdebine, E. Blot und den renommierten Colin & Cie, die Modelle der Künstler Maurice Bouval, Léon Noël Delagrange und Louis Chalon vertreten.[553] Doch bereits 1902 gelingt es Arthur, die von Louis Chalon entworfenen Modelle „Printemps de Bretagne" und „Eclosion" im Pariser Salon der Société des Artistes Français auszustellen und andere Künstler wie Marius Mars-Vallett für sich zu gewinnen.[554] Im Salon der Société Nationale des Beaux-Arts wird im gleichen Jahr eine „egyptische Vase" und die Vase „Les enfants" nach den Entwürfen des Jugendstilkünstlers Hans Stoltenberg Lerche gezeigt, genauso wie die von Gottfrid Larsson entworfene Bacchantenstatuette mit Tintenfass und der Beleuchtungskörper „Le Vice fuyant la lumière" aus Gips.[555]

Aus einer in der Fachzeitschrift „Art et Décoration" geschalteten Werbung (Ill. 139) von Arthur geht hervor, dass er 1903 Modelle aus Terrakotta, Bronze und Marmor vertreibt nach Entwürfen bekannter französischer Künstler wie Allouard, A. Boucher, Bouval, Croisy, Carlier, Carrièr-Belleuse, F. und A. Charpentier, Injalbert, Klein, Levasseur, Ledru, Loiseau-Rousseau, H. Lemaire, Mercié, Moreau-Vauthier, Marqueste, T. Noël, aber auch Modelle des Regensburger Künstlers Maison und des Wieners Strasser.[556]

Im Jahr 1907 entsteht bei Arthur in Paris die viel beachtete Tafeldekoration „Le Déjeuner fleuri" von Laporte-Blairsy, bestehend aus „Marquises XVIIIe siècle, portant des corbeilles".[557] Doch auch andere Künstler machen auf sich aufmerksam. Der talentierte Paul Philippe entwirft mehrere Bronzen und Marmorfiguren für Arthur Goldscheider,

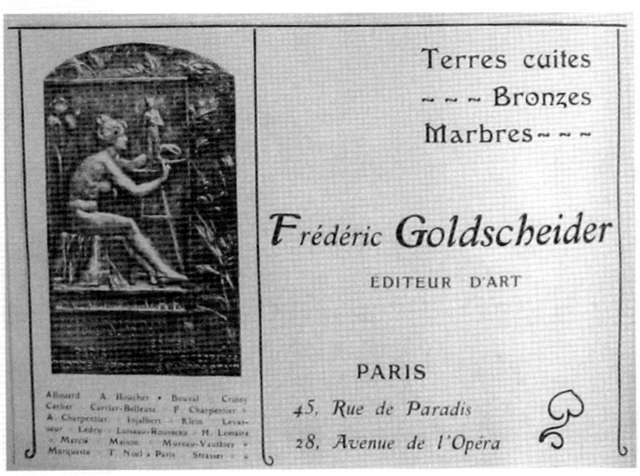

Ill. 139
Eine Anzeige in Art et Décoration aus dem Jahr 1903 zeigt die von Bouval entworfene Firmenmarke „Sculpture", die eine töpfernde Frau darstellt. (Quelle: Art et Décoration, 1903, S. 2)

A 1903 advertisement in Art et Décoration shows the firm trademark, "Sculpture" designed by Bouval and depicting a woman potter. (Source: Art et Décoration, 1903, p. 2)

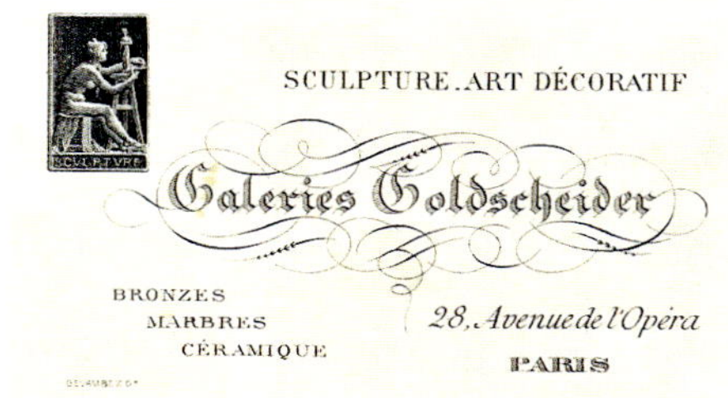

Ill. 140
Als Werbemittel legt Arthur dem Programm der Pariser Oper im Jahr 1912 seine Visitenkarte für die Galerie Goldscheider in der 28, Avenue de l'Opéra bei. (Quelle: Programm des Theatre National de l'Opéra 1912; Sammlung f@goldscheider.de)

For advertising purposes, Arthur enclosed his business card for the Goldscheider gallery at 28 avenue de l'Opéra with the Paris Opera program (Source: 1912 program from the Théâtre National de l'Opéra; Collection f@goldscheider.de)

darunter im Jahr 1907 die Marmorstatuette „Le Réveil", 1908 die Marmor-Elfenbein-Statuette „Femme au masque" sowie 1909 in verschiedenen Materialien die Gruppe „Grâce et Force".[558] Im selben Jahr stellt Arthur die Bronze-Elfenbeinstatuetten „Mimi" (statuette 1830) und „Rêverie" von Georges van der Straeten im Pariser Salon der Société des Artistes Français aus.[559] Im Jahr zuvor feierte in diesem Salon der Bildhauer Francesco La Monaca sein Debüt, der für Goldscheider die Bronzestatuetten „Danseur" und „Siffleur" lieferte.[560]

Die Pariser Dependance von Arthur in der Rue de Paradis wird in dieser Zeit auch als General-Depot der Wiener Filiale für die Länder Frankreich, England, Belgien und französische Schweiz geführt.[561] Auf der Leipziger Oster-Vormesse 1910 hat Arthur einen eigenen Stand unter der Firmierung „Frédéric Goldscheider" mit „Bronzen und Marmorwaren", daneben sind die „Erste Wiener Terrakottafabrik und Atelier für künstlerische Fayencen Friedrich Goldscheider" mit Terrakotta und Fayencen sowie „Camillo Goldscheider" mit „künstlerischen Skulpturen aus Marmor, Bronze und Elfenbein" mit jeweils einem eigenen Laden vertreten.[562] Als Werbemittel setzt Arthur neben Anzeigen auch Visitenkarten für seine Galerie in der Avenue de l'Opéra ein, die er Theaterprogrammen beilegen lässt, beispielsweise dem Programm des Theatre National de l'Opéra zur Aufführung von „Roma" sowie „Werther" in der Opéra-Comique, beide im Jahr 1912 (Ill. 140).[563]

Mit Ausbruch des Ersten Weltkriegs bricht die enge Zusammenarbeit zwischen Arthur und dem Wiener Unternehmen auseinander. In den Nachkriegsjahren verschweigt Arthur seine Wiener Wurzeln und bezeichnet seine Firma als „Maison francotchèque".[564] Offiziell vertritt er fast ausschließlich französische Künstler, zumindest dem Namen nach.[565] Nach dem Krieg stellt Arthur auf der Herbst-Messe in Leipzig 1918 aus, wo er als Inhaber der Firma „Adar" mit Sitz in Genf, 4 Boulevard James Fazy, als

uettes "Mimi" (statuette 1830) and "Rêverie" by Georges van der Straeten at the Salon of the Société des Artistes Français in Paris.[559] The previous year the sculptor Francesco La Monaca had débuted at that Salon. He provided Goldscheider with the bronze statuettes of male figures entitled "Danseur" and "Siffleur."[560]

The Paris branch run by Arthur in the rue de Paradis was at that time also the general transshipment point used by the Viennese branch for France, England, Belgium and French Switzerland.[561] At the 1910 Leipzig Pre-Easter, Arthur had a stand of his own registered as "Frédéric Goldscheider," where he displayed "bronzes and marble wares." Next to it, the "Erste Wiener Terrakottafabrik und Atelier für künstlerische Fayencen Friedrich Goldscheider" showed terracotta and faiences. "Camillo Goldscheider" also had a stand, which exhibited "art sculptures of marble, bronze and ivory." Each stand was backed up by a separate shop.[562] Apart from the usual advertisements, Arthur used other advertising tactics: for example enclosing his business cards for his gallery in the avenue de l'Opéra to theatre and opera programs, including programs for the performance of "Roma" at the Théâtre National de l'Opéra and "Werther" at the Opéra-Comique, both in 1912 (Ill. 140).[563]

The outbreak of the First World War put an end to the close collaboration between Arthur and the Viennese firm. In the years

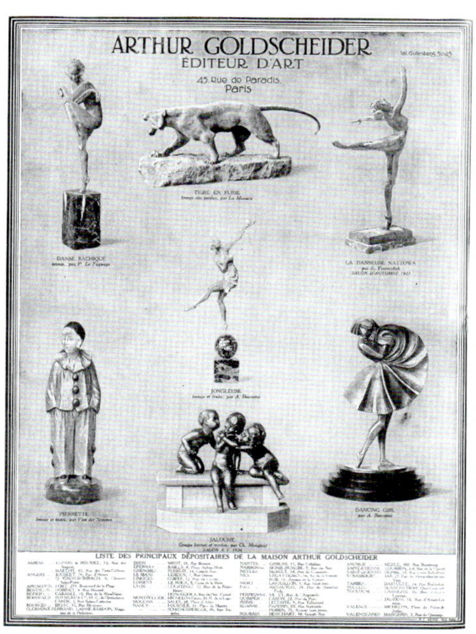

Ill. 141
Arthur Goldscheider ist mit seinen Kunstgegenständen in den wichtigsten Fachhäusern Frankreichs vertreten. (Quelle: L'Illustration vom 20.12.1924)

Arthur Goldscheider's art objects were sold at leading French specialist shops. (Source: L'Illustration of 20 Dec. 1924)

following the Great War, Arthur concealed his Viennese roots, calling his firm "Maison franco-tchèque."[564] He officially represented French artists almost exclusively, at least to judge from the names given.[565] After the war had ended, Arthur exhibited at the 1918 Leipzig Autumn Fair, where he was listed as the proprietor of "Adar," a firm based at 4 Boulevard James Fazy, Geneva, and the manufacturer of utilitarian decorative household articles (cigarette cases, boxes, canes, lamps, etc).[566] Once again he had a fair stand of his own even though the Viennese factory was also represented in Leipzig at its permanent branch in the department store at Neumarkt. In Paris, Arthur was now registered as "Arthur Goldscheider, Éditeur d'Art," based at 45, rue de Paradis.

In 1921 Céline Lepage designed the figure of a "Mendiante à Marakech" ["Beggar woman at Marrakesh"], which became a popular motif sold by Arthur in several sizes.[567] The art objects Arthur Goldscheider marketed and distributed were sold at numerous specialist shops – in cities ranging from Amiens, Bordeaux and Lyon to Toulouse (Ill. 141). The figures most in demand in late 1924 were the following: the bronzes "Tigre en furie" by La Monaca, "La danseuse Nattowa" by Serge Yourievitch (exhibited at the 1923 Salon d'Automne), "Danse Bachique" by Pierre Le Faguay, "Dancing Girl" and the bronze and ivory figures "Jongleuse" by Bouraine and "Pierrette" by van der Straeten

Fabrikant kunstgewerblicher Gebrauchsartikel (Zigarettenetuis, Schachteln, Stöcke, Lampen usw.) geführt wird.[566] Er hat erneut einen eigenen Messestand, obwohl die Wiener Fabrik dort mit der ständigen Filiale im Leipziger Kaufhaus am Neumarkt ebenfalls vertreten ist. In Paris fungiert Arthur nun unter der Firmierung „Arthur Goldscheider, Éditeur d'Art" mit Sitz in der 45, Rue de Paradis.

Im Jahr 1921 entwirft Céline Lepage die Figur „Mendiante a Marakech" (Bettlerin von Marrakesch), die zu einem beliebten Sujet wird und in verschiedenen Größen von Arthur verkauft wird.[567] Die von Arthur Goldscheider vertriebenen Kunstgegenstände werden in zahlreichen Fachhäusern angeboten – von Amiens, Bordeaux, Lyon bis Toulouse (Ill. 141). Zu den beworbenen Figuren

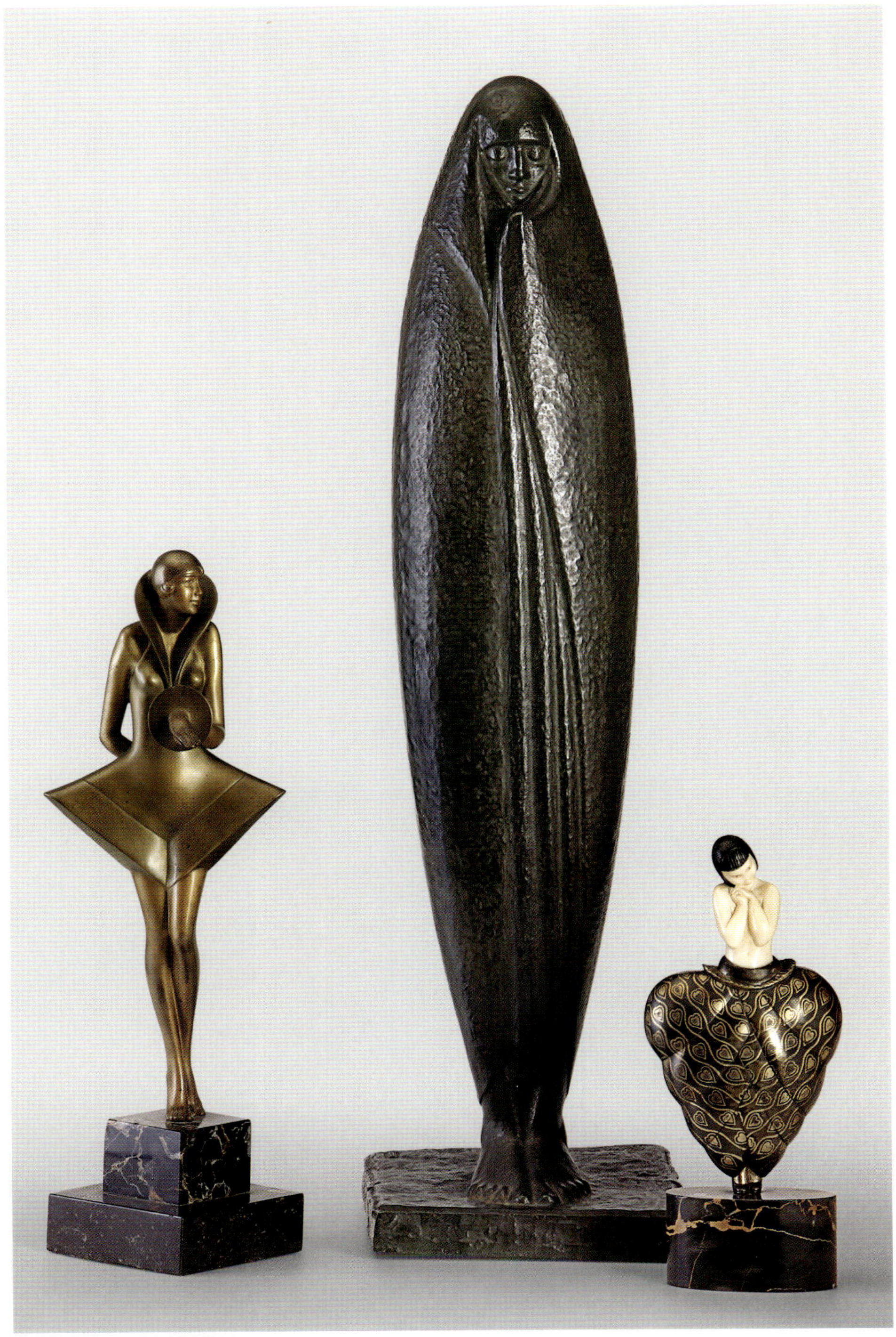

Ill. 142
Arthur Goldscheider vertritt in den 1920er Jahren etliche vielversprechende Künstler. Von links nach rechts sind folgende Modelle zu sehen: Stehendes Mädchen von Stephan Dakon, „Femme (Mendiante) de Marakech" von Céline Lepage sowie Stehendes Mädchen im herzförmigen Rock von Alexandre Kéléty.
(© Wien Museum)

In the 1920s, Arthur Goldscheider represented quite a few promising artists. From left to right, the models are as follows: Girl, standing, by Stephan Dakon, "Femme (Mendiante) de Marakech" by Céline Lepage and Girl, standing in heart-shaped skirt, by Alexandre Kéléty.
(© Wien Museum)

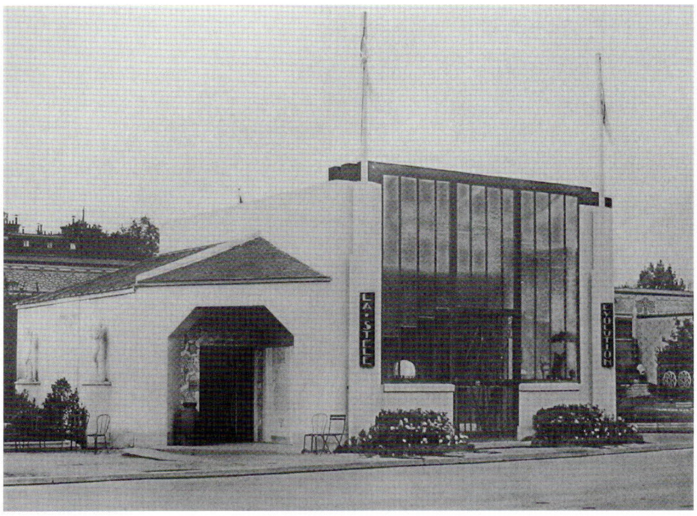

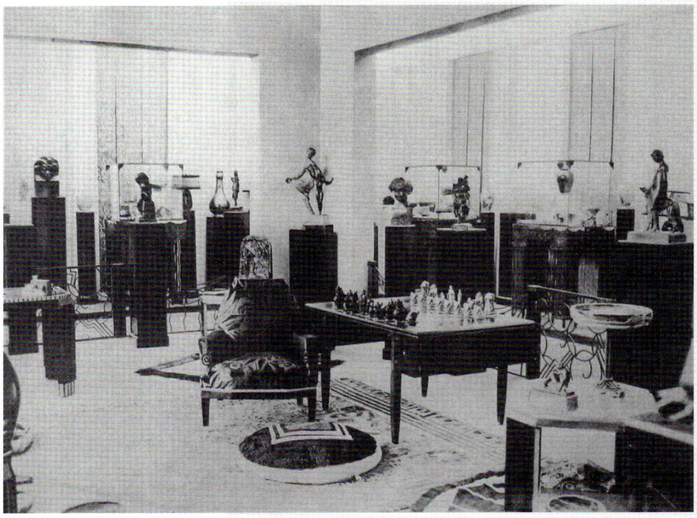

Ill. 143, 144
Auf der berühmten Ausstellung in Paris im Jahre 1925 präsentiert Arthur Goldscheider in einem eigenen Pavillon die Künstlergruppen La Stèle und Évolution. (Quelle: Artistic Evolution, Archiv Goldscheider)

At the celebrated 1925 Paris trade fair, Arthur Goldscheider presented the groups of artists known as La Stèle and Évolution. (Source: Artistic Evolution, Goldscheider Archives)

gehören Ende 1924 die Bronzen „Tigre en furie" von La Monaca, „La danseuse Nattowa" von Serge Yourievitch (ausgestellt im Salon D'Automne 1923), „Danse Bachique" von Pierre Le Faguays, „Dancing Girl" sowie die Bronze-Elfenbein-Figur „Jongleuse" von Bouraine, „Pierrette", ebenfalls aus Bronze-Elfenbein, von van der Straeten und die Bronze-Marmor-Gruppe „Jalousie" von Ch. Monginot (ausgestellt im Salon A.F. 1924).[568]

Céline Lepage gehört der von Arthur Goldscheider gegründeten Künstlergruppe „La Stèle" an, ihre „Bettlerin" stellt er auf der bedeutenden Pariser Ausstellung „Exposition des Arts Décoratifs et Industriels Modernes" im Jahr 1925 unter dem lieblicheren Namen „Femme de Marakech" (Frau aus Marrakesch; Ill. 142) aus. Arthur Goldscheiders Teilnahme an dieser – den Begriff Art Déco prägenden – Ausstellung ist von

as well as "Jalousie," a bronze and marble group by Ch. Monginot (exhibited at the 1924 Salon A. F.).[568]

Céline Lepage belonged to "La Stèle," a group of artists established by Arthur Goldscheider. He displayed her "Beggarwoman" at the paramount Paris trade fair in 1925, the "Exposition des Arts Décoratifs et Industriels Modernes" under the more appealing name "Femme de Marakech" ["Woman of Marrakesh"] (Ill. 142). Arthur Goldscheider's participation in this major trade fair – which established Art Déco as a concept – was of prime importance: in a pavilion of his own, designed by Eric Bagge (Ill. 143 and 144), Arthur showcased works by two groups of progressive artists: "La Stèle" and "Évolution." La Stèle represented modern sculptors; Évolution promoted arts and crafts decorators.

Ill. 145

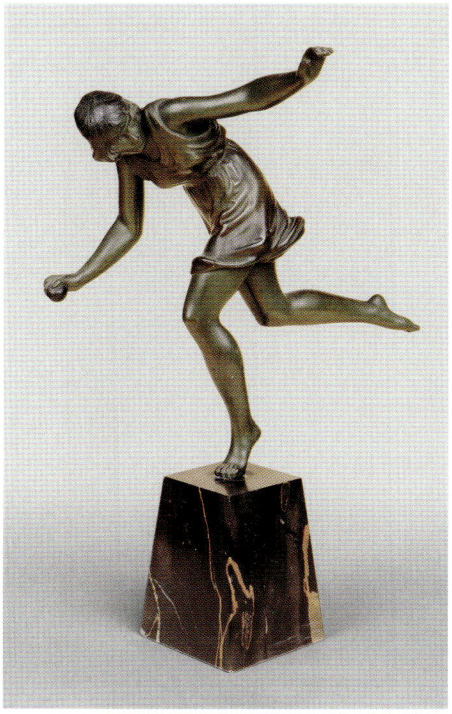

Ill. 146

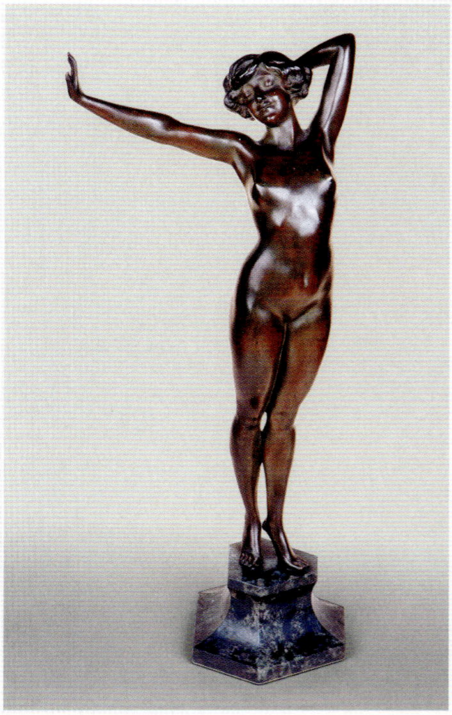

rechts | right Ill. 147

Ill. 145–147
Aus dem Nachlass von La Stèle stammen die folgenden Figuren, die auch Modell-Nummern erhielten: Laufendes Mädchen mit Ball, ModNr. 15; Damenakt von Paul Philippe, ModNr. 21;Tänzerin mit ausgebreiteten Armen, ModNr. 81. (alle aus PB Wien 19)

The following figures come from the La Stèle estate, which were also given model numbers: Girl running with ball, ModNr. 15; Female nude: Paul Philippe, ModNr. 21; Dancer with arms outspread, ModNr. 81. (all from PB Wien 19)

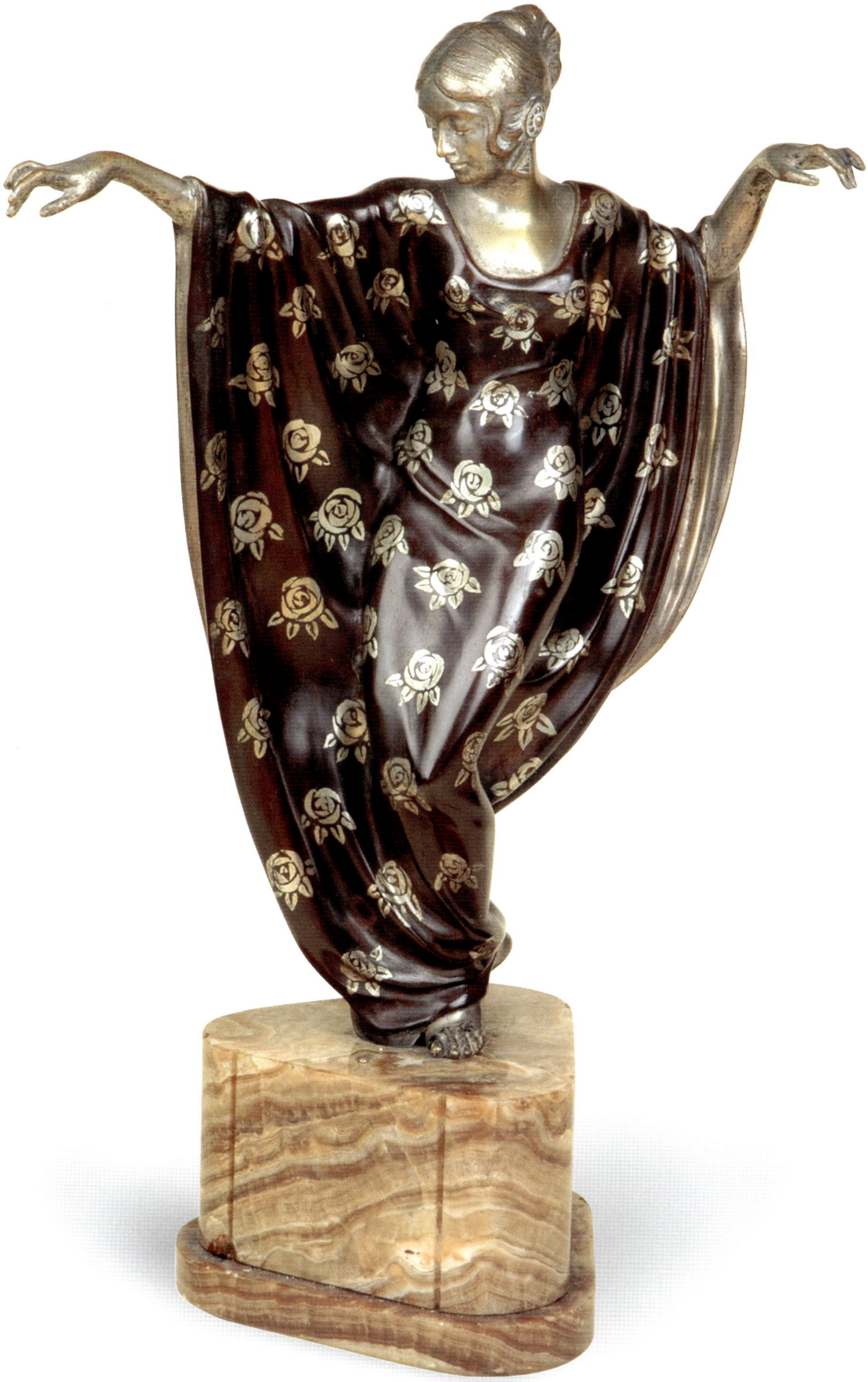

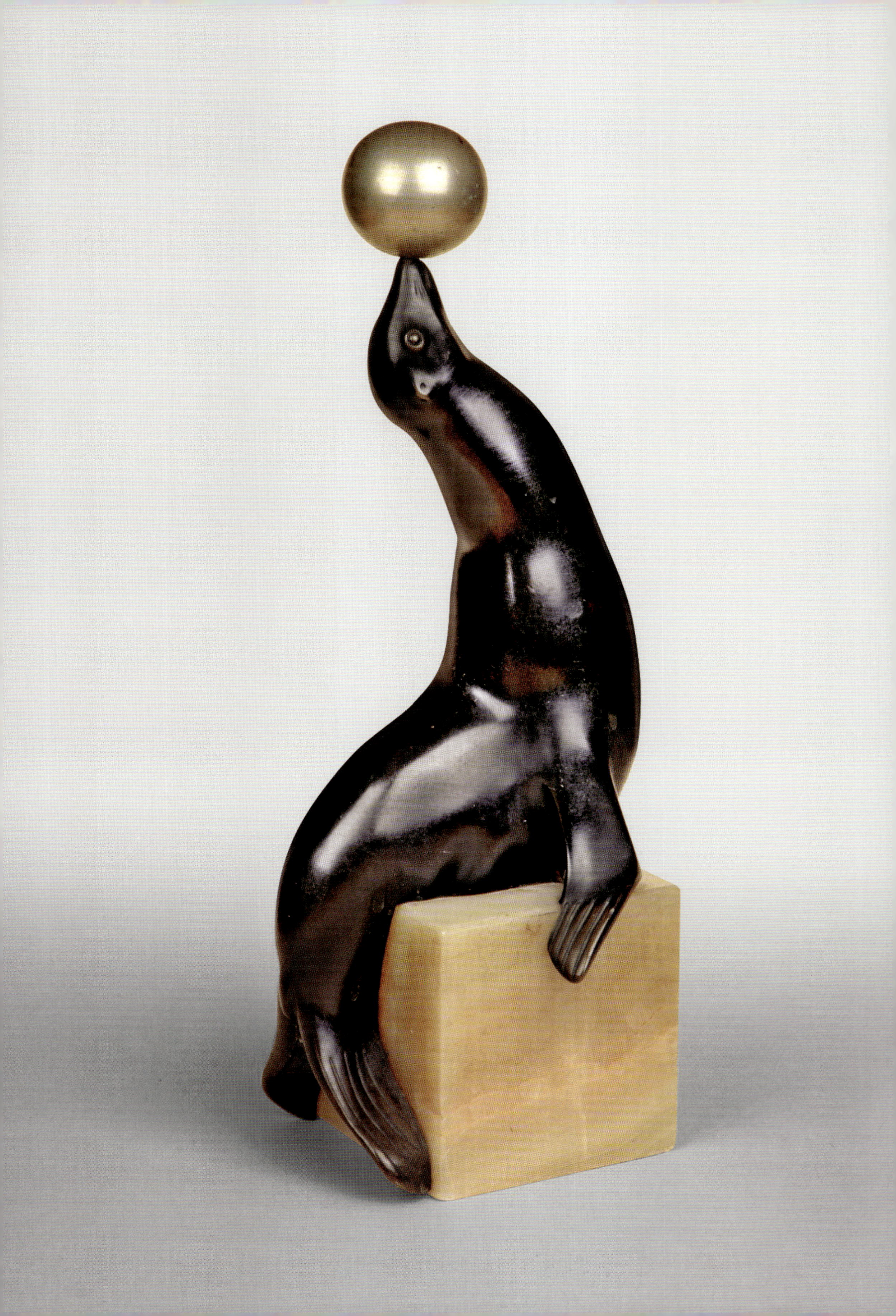

links I left:

Ill. 148
Marcel Bouraine, „Otarie" Seelöwe I Sea-lion, ModNr. 82. (PB Wien 19)

The following artists besides Céline Lepage belonged to La Stèle: E. Arnold, Daniel Bacqué, Armel Beaufils, Henri Bouchard, Max Blondat, André Marcel Bouraine, Jean-Marie Camus, Frédérique Ozeanne Cederlund, Georges Chauvel, Fernand David, Georges Halbout, Gaston Hauchecorne, Nathan Imenitoff, Raoul Lamourdedieu, Francesco La Monaca, Paul Maximilien Landowski (Ill. 150), Pierre Lenoir, Charles Malfray, Joël Martel, Henri Navarre, François-Émile Popineau, Marius Sain, Antoine Sartorio, S. Tallichet, Marcel Temporal, Pierre Traverse, Pierre Vigoureux and Jean Verschneider.[569] The members of "Évolution" specialized in making stylized figures in bronze, ivory and marble or combinations of such materials.

The following were members of the group of artists known as Évolution: the architect Eric Bagge (Ill. 157); the artisans (ceramicists, potters, goldsmiths, etc) Jean Besnard, A. Brugier, H. Brunel, Édouard Cazaux, Cochet-Ewald-Kohler, R. Drouart, H. Dubret, Dreyfus-Stern, J.-B. Gauvenet, Maurice Gensoli, L.-D. Germani, J.-Ch. Goëtz, Guénou, J. Guinard, Claudius Linossier (Ill. 151), J. Malfrey, P.-E. Mangin, Sybille May, M. Méheut, E. Membré, Alexandre Noll, S. Olesievicz, H. Pecquériaux, Ekiel Raby (who was even awarded a Grand Prix for the ivory articles he showed),[570] H. Rapin, G. Ringuet,

besonderer Bedeutung: In einem eigenen, von Eric Bagge entworfenen Pavillon (Ill. 143 und 144) präsentiert Arthur die Werke von zwei progressiven Künstlergruppen: „La Stèle" und „Évolution". Unter La Stèle sind moderne Bildhauer vertreten, unter Évolution die kunstgewerblichen Dekorateure.

Neben Céline Lepage gehören La Stèle die folgenden Bildhauer an: E. Arnold, Daniel Bacqué, Armel Beaufils, Henri Bouchard, Max Blondat, André Marcel Bouraine, Jean-Marie Camus, Frédérique Ozeanne Cederlund, Georges Chauvel, Fernand David, Georges Halbout, Gaston Hauchecorne, Nathan Imenitoff, Raoul Lamourdedieu, Francesco La Monaca, Paul Maximilien Landowski (Ill. 150), Pierre Lenoir, Charles Malfray, Joël Martel, Henri Navarre, François Emile Popineau, Marius Sain, Antoine Sartorio, S. Tallichet, Marcel Temporal, Pierre Traverse, Pierre Vigoureux und Jean Verschneider.[569] Sie stellen vor allem stilisierte Figuren in Bronze, Elfenbein, Marmor oder aus Materialkombinationen her.

Zu den Mitgliedern der Künstlergruppe Évolution gehören neben dem Architekten Eric Bagge (Ill. 157) die Kunstgewerbler (Keramiker, Töpfer, Goldschmiede etc.) und Dekorateure: Jean Besnard, A. Brugier, H. Brunel, Édouard Cazaux, Cochet-Ewald-Kohler, R. Drouart, H. Dubret, Dreyfus-Stern, J.-B. Gauvenet, Maurice Gensoli, L.-D. Ger-

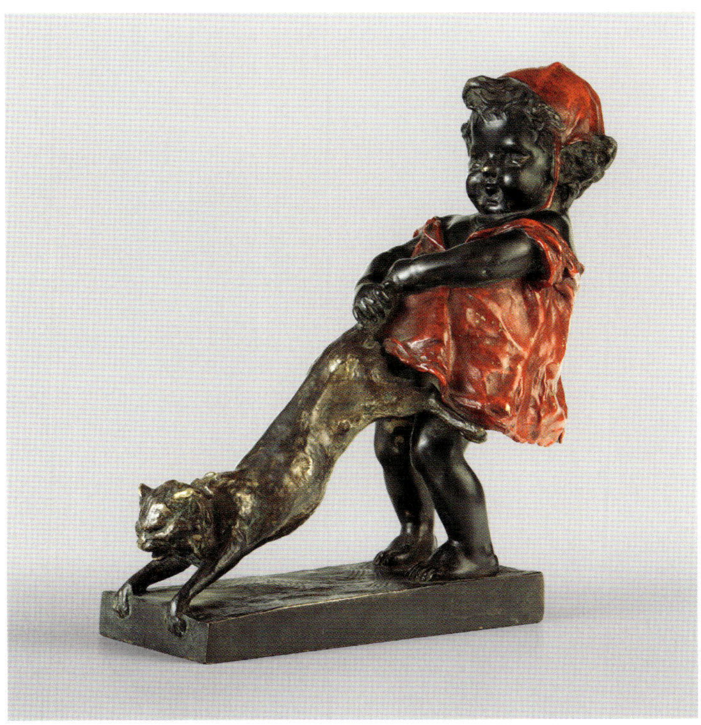

Ill. 149
Ein Exemplar des Rotkäppchens, ModNr. 61, befindet sich im Wien Museum. (© Wien Museum)

The Wien Museum collection contains an exemplar of Little Red Riding-Hood, ModNr. 61. (© Wien Museum)

mani, J.-Ch. Goëtz, Guénou, J. Guinard, Claudius Linossier (Ill. 151), J. Malfrey, P.-E. Mangin, Sybille May, M. Méheut, E. Membré, Alexandre Noll, S. Olesievicz, H. Pecquériaux, Ekiel Raby (der für seine ausgestellten Elfenbeinartikel sogar den „Großen Preis" erhält),[570] H. Rapin, G. Ringuet, A. Rivir, Jean Sala (Ill. 152), E. Schenk & fils, Jean Saint-Paul und ebenfalls Jean Verschneider sowie Victor Amalric Walter.[571]

Beide Gruppen stehen unter der künstlerischen Leitung des vielseitig begabten Marcel Temporal und beteiligen sich aktiv an der „industriellen Herstellung ihrer Kreationen."[572] Dies ist ein bedeutender Fortschritt des bisherigen Systems, bei dem der „Éditeur" bislang Modelle von Künstlern ausgewählt hat und sie ohne die „ästhetische Kontrolle" des Künstlers beliebig reproduzierte.

Im Sog des Erfolgs auf der Pariser Ausstellung ruft Arthur Goldscheider eine neue Publikation ins Leben.[573] Gemeinsam mit dem französischen Kunstkritiker Yvanhoé

A. Rivir, Jean Sala (Ill. 152), E. Schenk & fils and Jean Saint-Paul as well as Jean Verschneider and Victor Amalric Walter.[571]

Both groups were supervised by Marcel Temporal, a multi-talented artist and impresario, and participated actively in the "industrial production of their creations."[572] This represents a major advance on the previous system, whereby the "éditeur" had been the one to select models by artists and then reproduced them arbitrarily without subjecting them to "aesthetic control" exerted by the artists personally.

Following up on his success at the Paris exhibition, Arthur Goldscheider launched a new publication.[573] Jointly with the French art critic Yvanhoé Rambosson, who wrote the texts, Arthur published several issues of L'Évolution Artistique in 1926. It clearly shows the objectives of the two groups of artists.[574] The work was published in two volumes in English as Artistic Evolution for the English-language market. In those texts, Rambosson and Goldscheider made a plea for including artists more closely in the industrial manufacturing process and advocated mass-production in the interests of more reasonably priced replication. What they were calling for ran counter to the French tradition of drawing a distinction between producing one-off crafts pieces at small

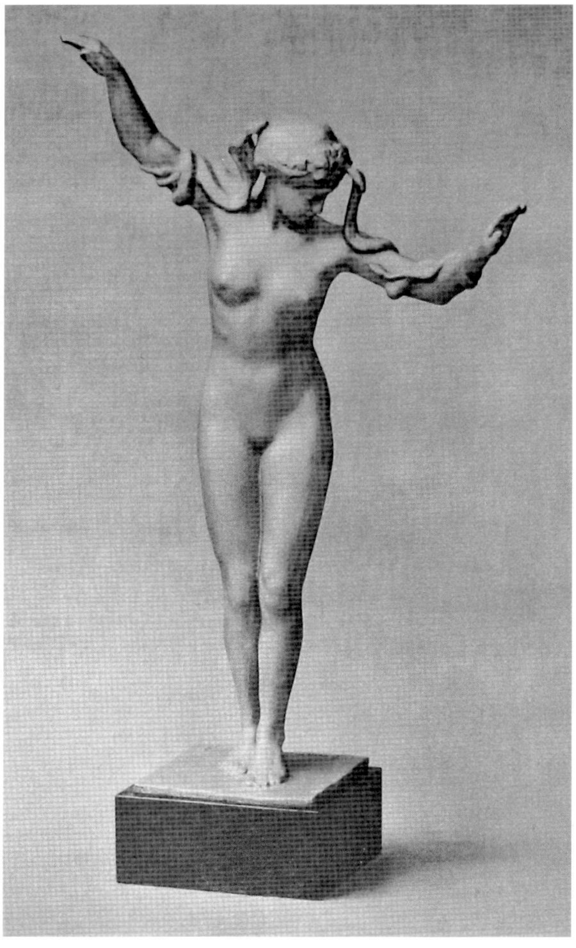

Ill. 150
Paul Landowski, der als etablierter Bildhauer mit den Entwürfen von Büsten zeitgenössischer Persönlichkeiten wie President Millerand und Marshal Pétain beauftragt wird, stellt als Mitglied von La Stèle die Bronze-Statuette „Charmeuse de serpents" (H 50 cm, B 25 cm) aus. (Quelle: Artistic Evolution, Archiv Goldscheider)

Paul Landowski, who, as an established sculptor, was commissioned to design busts of such famous contemporaries as President Millerand and Marshal Pétain, exhibited the bronze statuette "Charmeuse de serpents" (H 50 cm, W 25 cm) as a member of La Stèle. (Source: Artistic Evolution, Goldscheider Archives)

family-owned factories and industrial mass production. Up to that time, artists' wishes had not only been disregarded in industrial mass production but had often been infringed. Consequently, the aesthetic quality of the original model by the artist and that of the objects produced from it often differed greatly. Industrial manufacturers would change the colors, for instance, had the gilding laid on with a heavy hand and, ultimately, according to Rambosson, caused the artistic idea to be rendered in a shallow and superficial way, thus fostering "bad taste."[575] This was what Arthur Goldscheider was trying to prevent at his factory. He wanted to preserve the original aesthetic character of an art object. He clung so "boldly and persistently" to this idea that he joined forces with the artist Marcel Temporal to found the two groups of artists. Arthur was committed to ensuring that the artists became better acquainted with the technical process of manufacturing so that they could also optimize their models with the technical

Rambosson, der die Texte schreibt, publiziert Arthur im Jahr 1926 mehrere Hefte unter dem Titel „L'Évolution Artistique", die das Anliegen der beiden Künstlergruppen verdeutlichen.[574] Für den englischsprachigen Markt erscheint das Werk unter dem Titel „Artistic Evolution", zusammengefasst in zwei Bänden. Rambosson und Goldscheider plädieren in diesen Schriften für eine stärkere Einbeziehung der Künstler in den industriellen Produktionsprozess und sprechen sich für die Serienproduktion, also für die preiswertere Vervielfältigung aus. Ihre Forderung widerspricht der bislang üblichen französischen Tradition der Trennung der kunstgewerblichen Unikat-Erzeugnisse in kleinen Familien-Manufakturen und den industriellen Serienproduktionen der Fabriken. Die Vorgaben der Künstler wurden bislang bei einer industriellen Produktion nicht nur vernachlässigt, sondern man hat oft gegen sie verstoßen, so dass die Qualität des künstlerischen Originalmodells und der reproduzierten Objekte sich stark unterschieden. Die industriellen Produzenten änderten beispielsweise die Farben, übertrieben es mit der Vergoldung und führten letztlich, laut Rambosson, zu einer Verflachung der künstlerischen Idee und förderten den „schlechten Geschmack".[575] Arthur Goldscheider versucht genau dies in seiner Produktionsstätte zu vermeiden und den Originalcharakter eines künstlerischen Gegenstandes zu bewahren. An dieser Idee hält er „wagemutig und ausdauernd" fest und gründet deshalb gemeinsam mit dem Künstler Marcel Temporal die beiden Künstlergruppen. Arthur setzt sich dafür ein, dass die Künstler sich mit dem technischen Prozess der Herstellung

Ill. 151
Claudius Linossier, der 1925 in der Jury der Pariser Ausstellung sitzt, entwirft für Arthur Goldscheider eine Holz-Messing-Pendule (H 15 cm, L 47 cm, B 14 cm), auf der die Figur „Atlas" (H 19,5 cm, B 15 cm) von Sybille May befestigt ist. (Quelle: Artistic Evolution, Archiv Goldscheider)

Claudius Linossier, who was on the jury of the 1925 Paris exhibition, designed a pendule clock in wood and brass for Arthur Goldscheider (H 15 cm, L 47 cm, W 14 cm), with a figure of "Atlas" (H 19.5 cm, W 15 cm) by Sybille May attached to it. (Source: Artistic Evolution, Goldscheider Archives)

Ill. 152
Der Glaskünstler Jean Sala, der für seine blauen und gelben Farben bekannt ist, steuert einen Fisch aus Glas (H 31 cm, B 25 cm) bei. Sala erzeugt seine Werke in einer von ihm entwickelten Glasmasse, die er mit bloßen Händen ohne Gussform bearbeitet. (Quelle: Artistic Evolution, Archiv Goldscheider)

Jean Sala, an artist in glass famous for his blue and yellow color tones, contributed a glass fish (H 31 cm, W 25 cm). Sala executed his works in a glass paste he had developed, which he could work with bare hands without using a mold. (Source: Artistic Evolution, Goldscheider Archives)

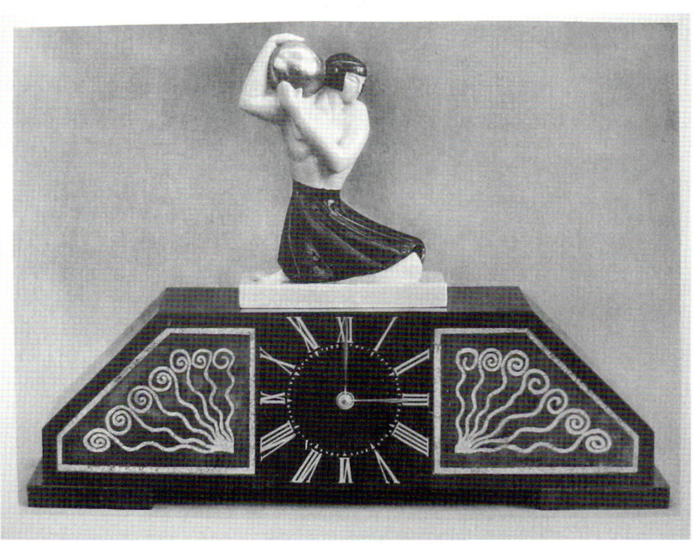

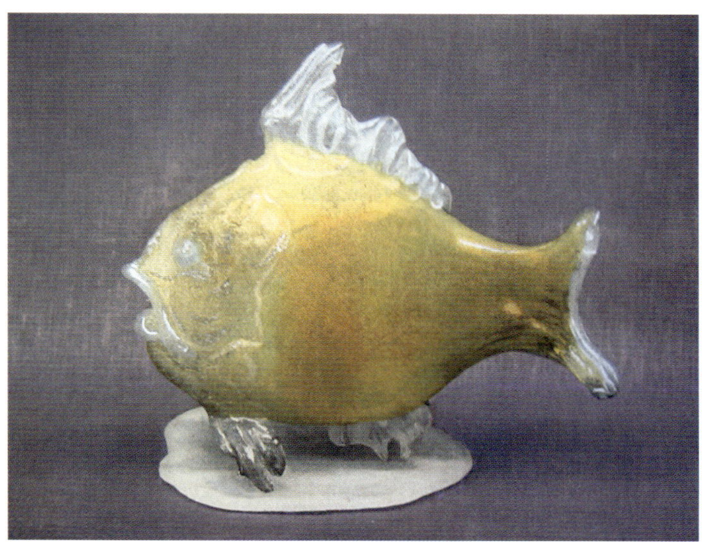

Ill. 153, 154
Marcel Bouraine, der zu den renommierten französischen Malern und Bildhauern zählt und mit Pierre Le Faguays auf der Messe 1925 für die Stadt Paris Reliefs mit einer Höhe von 3,5 Metern kreierte, zeigt bei Arthur die versilberte Bronze mit Elfenbein „Léda", ModNr. 46. (Quelle: Artistic Evolution, Archiv Goldscheider; PB Wien 19)

Marcel Bouraine, a celebrated French painter and sculptor, who created reliefs that were 3.5 m high jointly with Pierre Le Faguay for the city of Paris at the 1925 Paris Fair, showed the silvered bronze and ivory "Léda", model number 46, for Arthur Goldscheider. (Source: Artistic Evolution, Goldscheider Archives; PB Wien 19)

Ill. 153

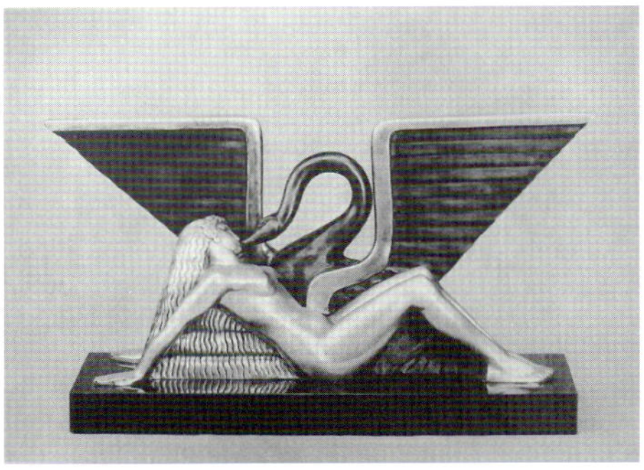

besser vertraut machen und dementsprechend auch ihre Modelle für die technischen Möglichkeiten optimieren. Durch die Serienproduktion wird einer breiteren Käuferschicht der Erwerb eines preiswerteren und dennoch hoch qualitativen Kunstgegenstandes ermöglicht, bei dem jedes Stück vom Künstler selbst für den Verkauf freigegeben worden ist. Es geht laut Rambosson um eine „Revolution" in der Methode kommerzieller Produktion.[576] Dieser Ansatz soll sowohl national wie auch international die Aktivitäten des französischen Kunstgewerbes stärken und die Zufriedenheit auf beiden Seiten, sowohl bei den Künstlern wie auch den Käufern, fördern. Rambosson und Goldscheider verfolgen hiermit auch ein aus dem Jugendstil übernommenes Gesellschaftsideal, bei dem nur durch niedrigere Preise „alle Gesellschaftsschichten" an der Schönheit des Kunstgewerbes teilhaben können.[577]

parameters in mind. Mass production made it possible for a broader-based clientele to acquire a reasonably priced object that was nonetheless of aesthetically high quality, with each piece approved for sale by the artists themselves. What was at stake, according to Rambosson, was a "revolution" in market-oriented production methods.[576] This approach would promote the operations of the French decorative and applied arts both domestically and abroad to the satisfaction of both artists and their clientele. In taking up this stance, Rambosson and Goldscheider were pursuing a social ideal promulgated by the Art Nouveau movement, according to which "all classes of society" could only share the beauty of the decorative and applied arts if prices were lower.[577]

The early publications showing only works produced in close collaboration between the artists concerned and the maker,

Ill. 154

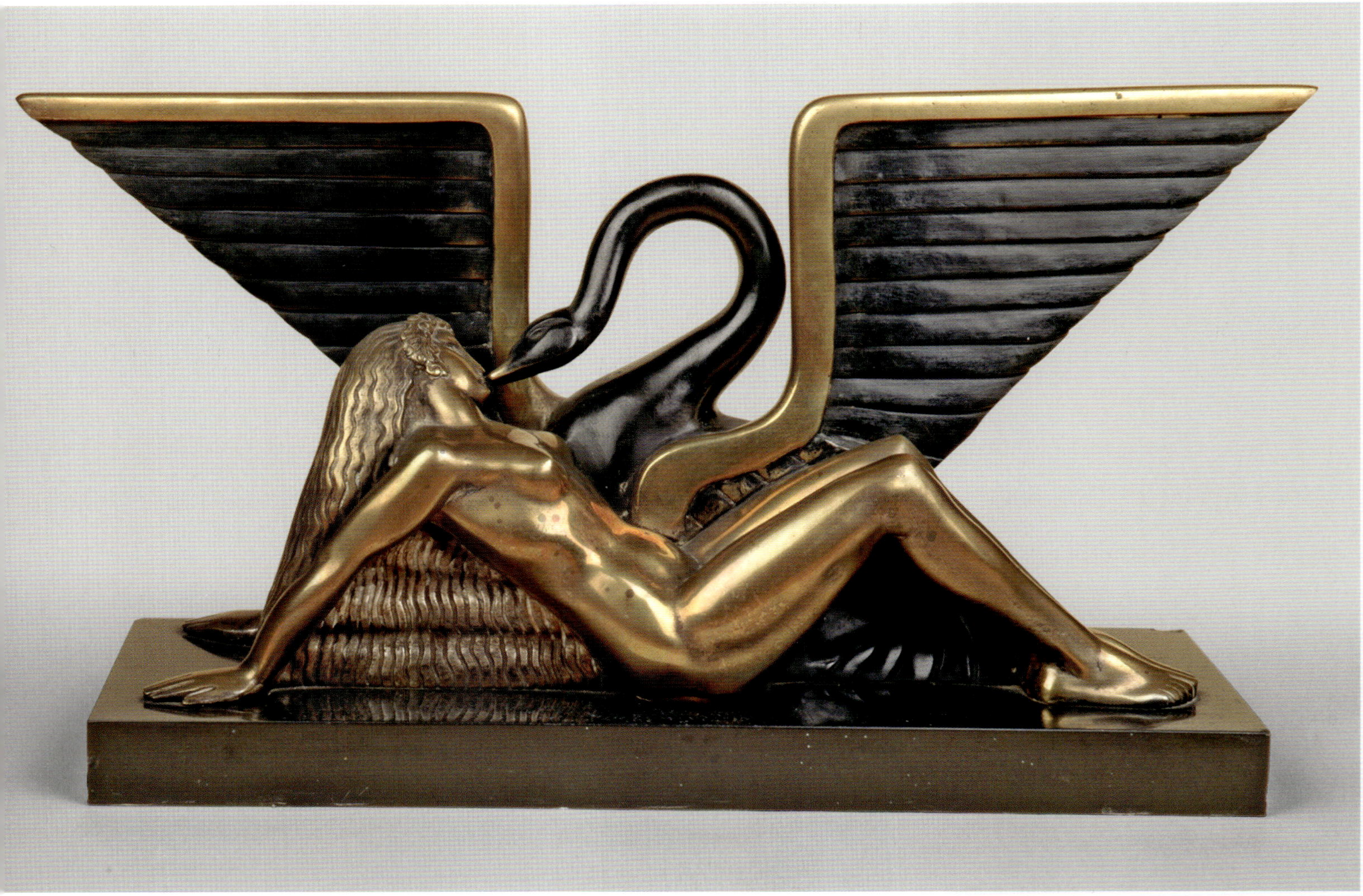

were to support this approach and enlighten the public – not just students but also the retailers who were so important if the market was to be penetrated.[578] Later issues were to be devoted to particular artists, decorative arts genres, specific target markets or subject areas.[579] But also other artists were focused on by Arthur Goldscheider in the publication, their number included Henri Lacoux, a pupil of Camille Pissaro's, who was represented by "Deer Hunting," a stoneware vase painted in underglaze pastel colors.

The second issue of Artistic Evolution, which dealt with the subject of "Art in Industry" and overtly targeted the middle class, espouses the Art Déco ideals: clear forms and the primacy of form over ornament.[580] Apart from statuettes by Lepage and Marcel Bouraine (Ill. 148, 153–156), it features figurative ceramics by May and Vigoureux and lacquered vases by the artists

Die Publikationen, die nur Werke zeigen, die in einer engen Verbindung zwischen Künstler und Produzent entstanden sind, sollen diesen Ansatz unterstützen und aufklärerisch wirken – nicht nur bei Studenten, sondern auch bei den für die Marktdurchdringung wichtigen Händlern.[578] In den ersten Heften werden die Künstler präsentiert. Später sollen Ausgaben zu besonderen Künstlern, bestimmten Kunstgewerbegattungen, einem bestimmten Markt oder Thema erscheinen.[579] Außer den auf der Pariser Messe ausgestellten Mitgliedern der beiden Künstlergruppen werden weitere Künstler von Arthur Goldscheider in der Publikation gezeigt, z.B. Henri Lacoux, ein Schüler Camille Pissaros, der mit einer in Unterglasur-Pastellfarben bemalten Steingut-Vase „Deer Hunting" vertreten ist.

Im zweiten Heft von „Artistic Evolution", das sich mit dem Thema „Art in Industry"

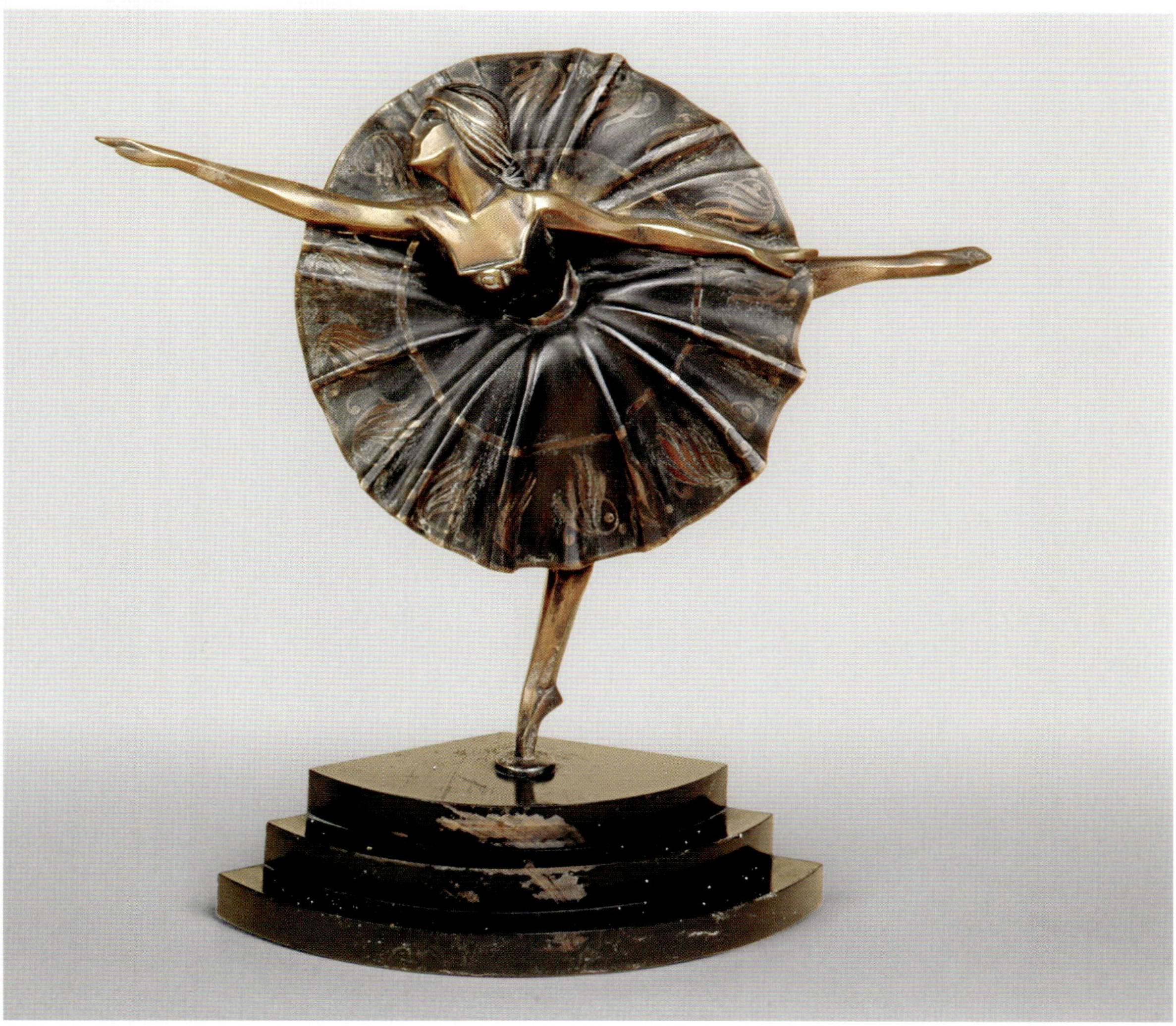

Ill. 155
Marcel Bouraine,
Ballerina, ModNr. 106.
(PB Wien 19)

beschäftigt und offiziell als Zielgruppe die Mittelschicht hat, werden die Ideale des Art Déco manifestiert: klare Formen und deren Vormachtstellung gegenüber Ornamenten.[580] Neben Statuetten von Lepage und Marcel Bouraine (Ill. 148, 153–156) sind May und Vigoureux mit figürlicher Keramik und die Künstler Savard, Cazaux, Noll, Edouard Stabi und A. Brugier mit lackierten Vasen vertreten. Sie gehören ebenfalls der Gruppe Évolution an und wollen mit klaren Formen, Dekoren und Ausführungen Kunstgewerbe in Serienproduktion für die breitere Masse herstellen, die in einer „künstlerischen Umgebung" leben will.[581]

Zu den präsentierten Künstlern des dritten Heftes, das schwerpunktmäßig Vasen

Savard, Cazaux, Noll, Édouard Stabi und A. Brugier. They, too, belonged to Évolution and wanted to industrially mass-produce arts and crafts objects with clear forms, decoration and execution for a broad market purportedly desirous of living in "aesthetic surroundings."[581]

The third issue focused on vases. The artists presented in it included Édouard Cazaux, La Monaca, Noël, Colmant, Mergier, Gilles Defraigne, Walter René Fuerst, J.-Charles Goëtz, Marek Szwarc and Jean Sala (Ill. 158).

Artistic Evolution also focused on lamps and lighting fixtures. Yvanhoé Rambosson invoked the importance of the 19th century for mankind because it saw the invention

Ill. 156
Marcel Bouraine, „Damenakt" I "Female Nude", ModNr. 147. (PB Wien 19)

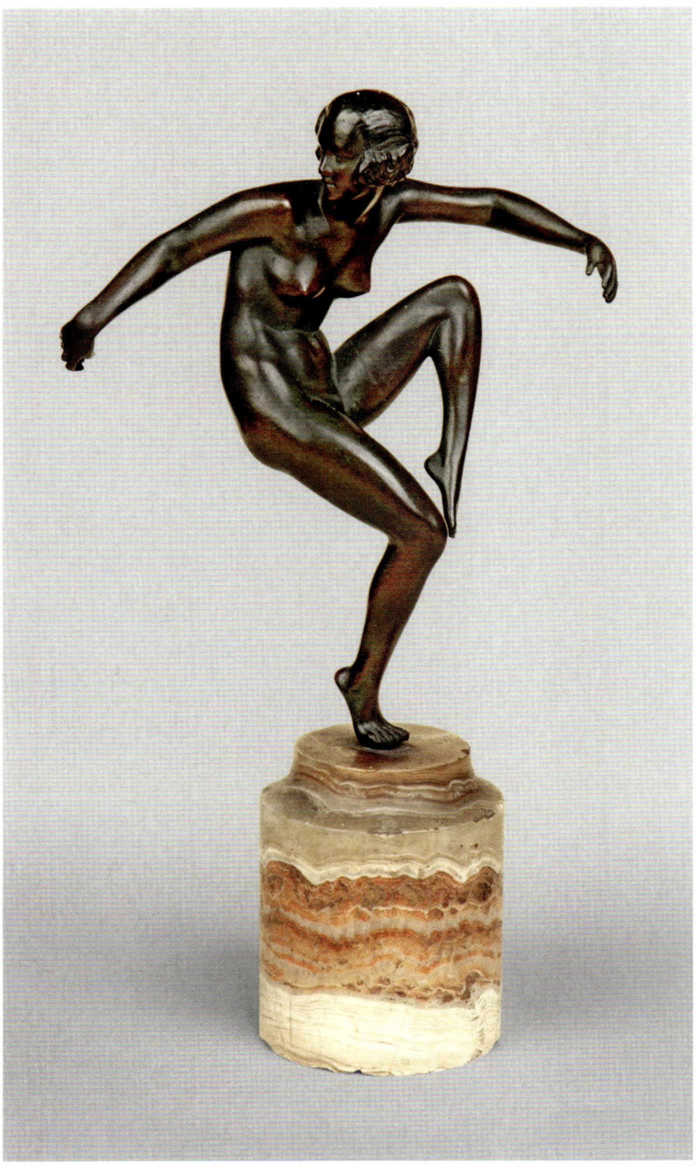

Ill. 157
Eric Bagge, der den Pavillon auf der Pariser Messe 1925 für Arthur Goldscheider entworfen hat, zeigt als Mitglied der Dekorateur-Gruppe Évolution, wie Kunstgewerbe angemessen in Vitrinen präsentiert werden sollte. (Quelle: Artistic Evolution, Archiv Goldscheider)

In his capacity as a member of the arts and crafts group called Évolution Eric Bagge, who designed the pavilion for Arthur Goldscheider at the 1925 Paris Fair, demonstrated how arts and crafts objects should be appropriately presented in glass display-cases. (Source: Artistic Evolution, Goldscheider Archives)

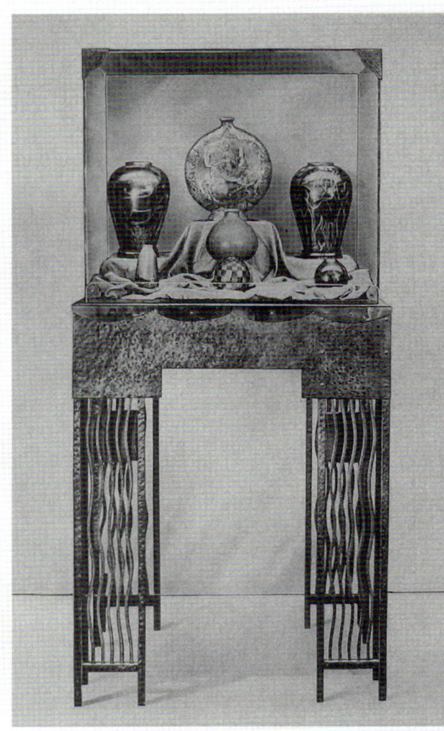

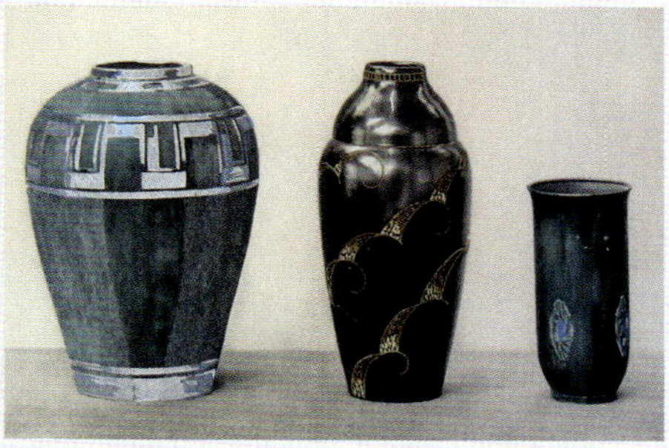

Ill. 158
Von La Monaca und Goëtz entworfene Vasen mit Dekoren von Colmant, Noll und Goëtz. (Quelle: Artistic Evolution, Archiv Goldscheider)

Vases designed by La Monaca and Goëtz with decoration by Colmant, Noll and Goëtz. (Source: Artistic Evolution, Goldscheider Archives)

of the steam engine and electricity, which changed the world. Lighting especially had a positive effect on both public and private life yet the forms of lighting fixtures, as Rambosson saw it, remained the same old, inappropriate ones.[582] Consequently, the artists in Évolution increasingly turned their attention to this new utilitarian object used in modern everyday living and tried to create a "useful and aesthetic" form of lamp in various materials. Arthur Goldscheider presented works of this kind by Magdeleine Anthouard (Ill. 159), Bouraine, Léon Fargue (Ill. 160), le Faguays, Jean Besnard and Etlin, Verschneider, et al.

Artistic Evolution also dealt with subjects such as nature and water in the form of fountains and wells. In the fifth issue, objects by Jean Verschneider, an engineer who worked as an artist in this field (Ill. 161), as

behandelt, gehören Édouard Cazaux, La Monaca, Noël, Colmant, Mergier, Gilles Defraigne, Walter René Fuerst, J.-Charles Goëtz, Marek Szwarc und Jean Sala (Ill. 158).

Einen weiteren Schwerpunkt der Artistic Evolution bilden Lampen und Leuchtkörper. Yvanhoé Rambosson beschwört die Bedeutung des 19. Jahrhunderts für die Menschheit, in dessen Verlauf die Erfindungen der Dampfmaschine und der Elektrizität die Welt veränderten. Gerade die Beleuchtung wirke sich fördernd sowohl auf das öffentliche wie private Leben aus, doch die Formen der Beleuchtungskörper blieben, laut Rambosson, die alten und unpassenden.[582] Deshalb setzen sich die Künstler von Évolution verstärkt mit diesem neuen Anwendungsgegenstand des modernen Alltags auseinander und versuchen in verschiedenen Materialien eine „nützliche und ästhetische" Form von Lampen zu kreieren. Arthur Goldscheider präsentiert diesbezüglich Werke von Magdeleine Anthouard (Ill. 159), Bouraine, Léon Fargue (Ill. 160), le Faguays, Jean Besnard und Etlin, Verschneider und anderen Künstlern.

Auch die Natur und das Element Wasser werden in der Artistic Evolution behandelt und zwar in Form von Fontänen und Brunnen. Im fünften Heft werden Objekte des auf diesem Gebiet künstlerisch tätigen Inge-

Ill. 159
Magdeleine Anthouard entwirft für Arthur Goldscheider moderne Beleuchtungskörper aus Bronze und Aragonit (H 30 cm, B 31,5 cm sowie H 35 cm, B 23,5 cm). (Quelle: Artistic Evolution, Archiv Goldscheider)

Magdeleine Anthouard designed modern lighting fixtures in bronze and Aragonite (H 30 cm, W 31.5 cm as well as H 35 cm, W 23.5 cm) for Arthur Goldscheider. (Source: Artistic Evolution, Goldscheider Archives)

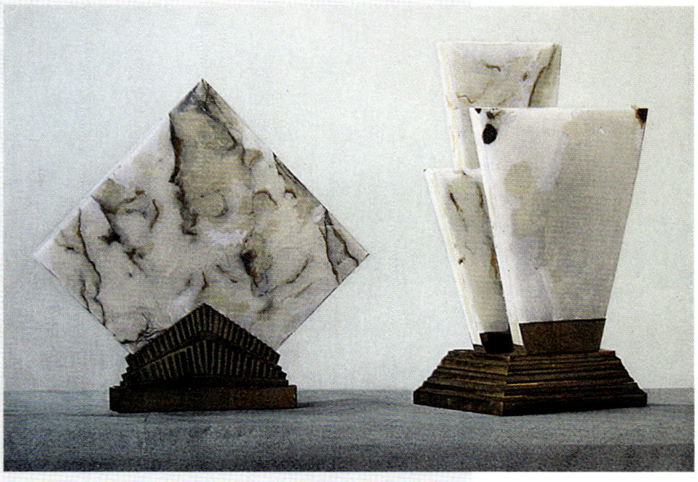

Ill. 160
Léon Fargue fällt mit einer von unten beleuchteten Schale auf (H 18 cm, B 49 cm). (Quelle: Artistic Evolution, Archiv Goldscheider)

Léon Fargue attracted attention with a bowl lit from below (H 18 cm, W 49 cm). (Source: Artistic Evolution, Goldscheider Archives)

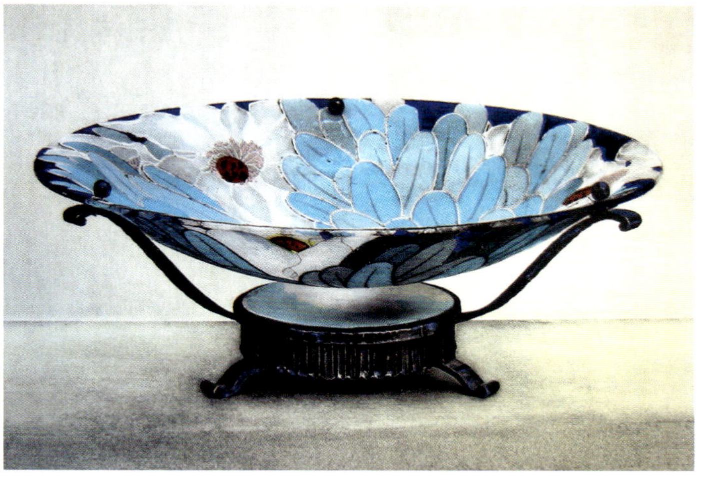

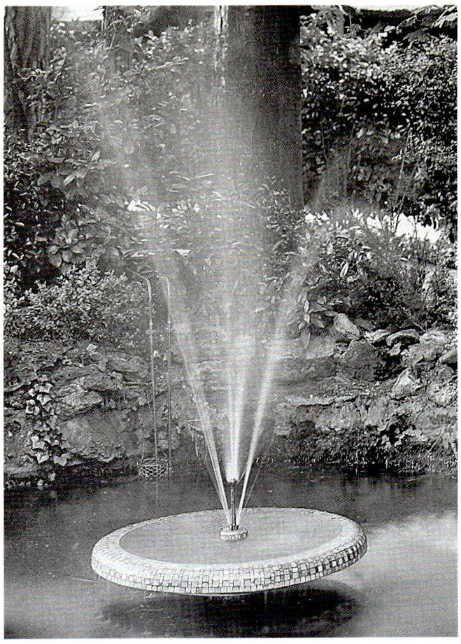

Ill. 161
Jean Verschneider, ein künstlerisch tätiger Ingenieur, kreiert für Arthur Goldscheider die Fontäne „Artesian Fountain". (Quelle: Artistic Evolution, Archiv Goldscheider)

Jean Verschneider, an engineer who worked as an artist, created this "Artesian Fountain" for Arthur Goldscheider. (Source: Artistic Evolution, Goldscheider Archives)

nieurs Jean Verschneider (Ill. 161) sowie für den Heimbereich entworfene Kleinfontänen von P. Lenoir und Goëtz gezeigt.[583]

Das sechste Heft ist dem Keramiker Édouard Cazaux gewidmet (Ill. 162). Cazaux, der alle Bereiche vom Entwurf bis Herstellung selbst durchführt, arbeitet schon seit längerem mit Arthur Goldscheider zusammen. Er stammt aus einer baskischen Töpferfamilie und erfand ein Verfahren, rote Farbtönungen in Kupfer zu fixieren – das allerdings auf Ernest Chaplets Erkenntnissen basierte.[584] Die von Cazaux hergestellten Keramiken werden unter Kennern seiner Zeit geschätzt und er selbst zu den wenigen originellen Keramikern gezählt, dessen Werke mehrere Epochen stilistisch und qualitativ überdauern sollen. Die von ihm präsentierten Objekte bestehen aus Vasen, Flaschen, Schalen und einer in Zusammenarbeit mit Marguerite Lehucher entstandenen Lampe.[585]

Im zweiten Band der englischen Ausgabe von „L'Évolution Artistique", der die Hefte vom Juli bis Dezember 1926 zusammenfasst, werden ebenfalls verschiedene Themen behandelt. Das siebte Heft widmet sich allgemein der Bildhauergruppe La Stèle. Die von Marcel Temporal und Arthur Goldscheider gegründete Gruppe gewann den „Grand Prix" auf der 1925 stattfindenden internationalen Ausstellung und beruht auf einer engen Kooperation zwischen den Künstlern und dem Éditeur.[586] Die im Heft präsentierten Objekte sind von Landowski, Blondat, Lenoir, Chauvel, Traverse, Raymond Delamarre, Bacqué sowie Jean und Joël Martel (Ill. 163).

Pierre Le Faguays Arbeiten (Ill. 164) werden im achten Heft gezeigt, seine Kunst als „leicht und klar", er selbst als „geistreich" bezeichnet.[587] Der aus Nantes stammende

well as small table fountains designed by P. Lenoir und Goëtz for use in the home are presented.[583]

The sixth issue is devoted to the ceramicist Édouard Cazaux (Ill. 162). Cazaux, who made the whole object from the drawing to the making stage himself, had worked for Arthur Goldscheider for quite some time. The scion of a Basque family of potters, Caszaux invented a process for staining copper permanently with reddish tints – based, however, on discoveries made by Ernest Chaplet.[584] The ceramics Cazaux produced were highly appreciated by connoisseurs in his day and he was one of the few original ceramicists whose works would last from the standpoint of style and quality for several design periods. The objects he presented included vases, bottles, bowls and a lamp made in collaboration with Marguerite Lehucher.[585]

The second volume of the English edition of L'Évolution Artistique, which included the July to December 1926 issues, also dealt with miscellaneous subjects. The seventh issue was devoted to the group of sculptors known as La Stèle, founded by Marcel Temporal and Arthur Goldscheider. It won the Grand Prix at the 1925 Paris exhibition and was based on close collaboration between the artists and the éditeur.[586] The objects presented in this issue are by Landowski, Blondat, Lenoir, Chauvel, Traverse, Raymond Delamarre and Bacqué as well as Jean and Joël Martel (Ill. 163).

Pierre Le Faguay's works (Ill. 164) were shown in the eighth issue. His work is described as "light and clear" and the artist himself as "witty."[587] A native of Nantes, Le Faguay was a modern artist and, even though fauns and nymphs were occasionally

Ill. 162
Édouard Cazaux, ein seit längerem mit Arthur Goldscheider verbundener Keramiker, entwarf diese Steingut-Flasche mit Teufelsfratze (H 34,5 cm, B 28,5 cm). (Quelle: Artistic Evolution, Archiv Goldscheider)

Édouard Cazaux, a ceramicist who had worked for Arthur Goldscheider for quite some time, designed this stoneware bottle with a grimacing devil mask (H 34.5 cm, W 28.5 cm). (Source: Artistic Evolution, Goldscheider Archives)

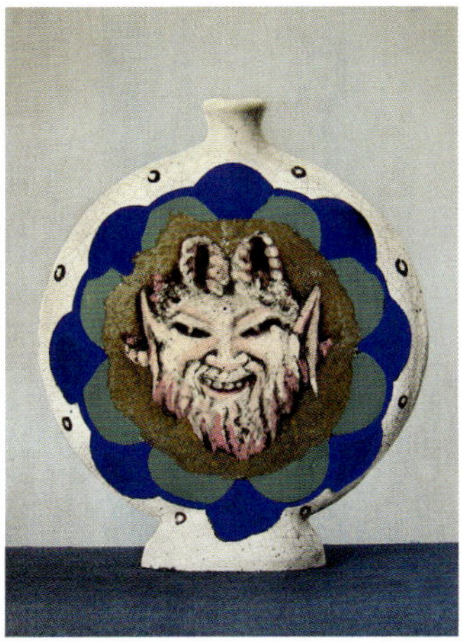

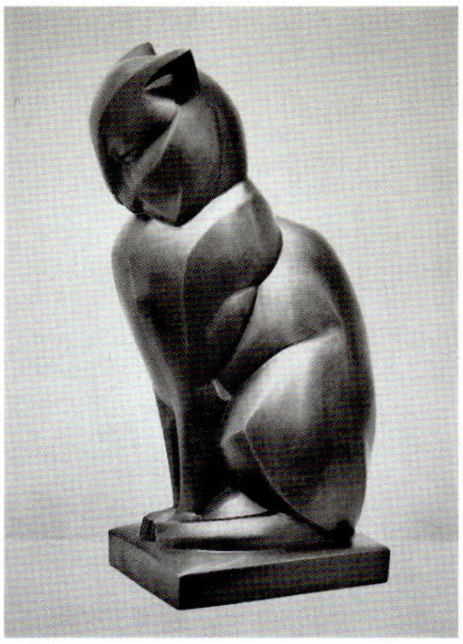

Ill. 163
Die Brüder Jean und Joël Martel entwerfen die Statuette „Le Chat" aus schwarzer Bronze (H 39 cm, B 13 cm, T 18 cm). (Quelle: Artistic Evolution, Archiv Goldscheider)

The brothers Jean and Joël Martel designed "Le Chat," a statuette in black bronze (H 39 cm, W 13 cm, D 18 cm). (Source: Artistic Evolution, Goldscheider Archives)

rechts I right

Ill. 164
Le Faguays Puppenspielerin mit zwei Handpuppen trägt die Modell-Nummer 56. (PB Wien 19)

A Le Faguay puppeteer with two hand puppets bears the model number 56. (PB Wien 19)

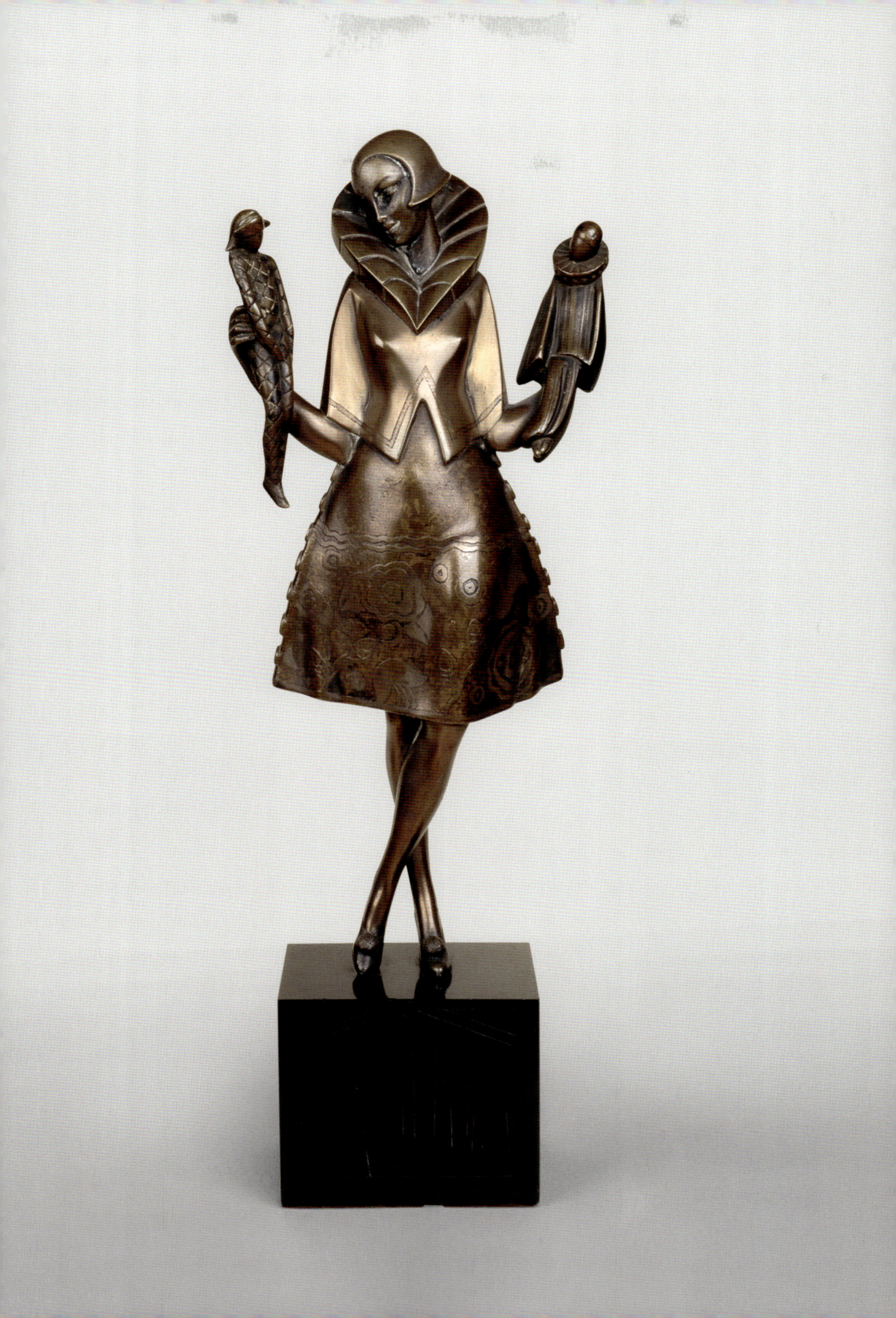

Ill. 165

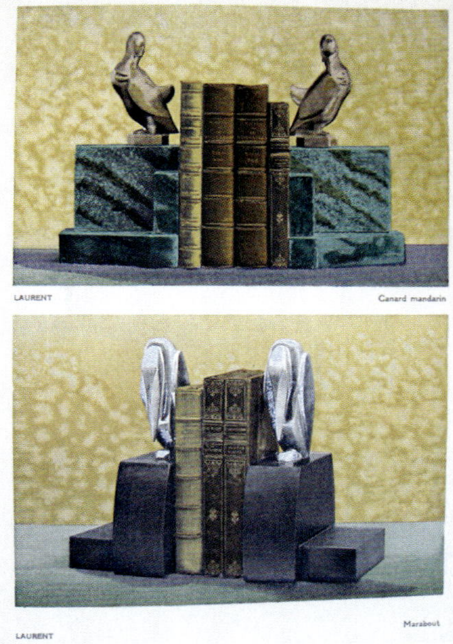

Ill. 165, 166
Arthur Goldscheider bietet die in den zwanziger Jahren beliebten Buchstützen in diversen Materialien an. Mehrere Künstler liefern ihm Modelle, darunter Georges Henri Laurent (Mandarin Ente aus versilberter Bronze; H 19 cm, B 20 cm, T 115 cm; Marabu aus vergoldeter Bronze, H 17 cm, B 16,5 cm, T 10 cm) und Marcel Bouraine (Basset Hund aus vergoldeter Bronze, H 11 cm, B 27,5 cm, T 11 cm). (Quelle: Artistic Evolution, Archiv Goldscheider)

Bookends were very popular in the 1920s. Arthur Goldscheider sold them in all sorts of materials. Several artists, including Georges Henri Laurent (Mandarin Duck in silvered bronze: H 19 cm, W 20 cm, D 11.5 cm; Marabu Stork in gilded bronze: H 17 cm, W 16.5 cm, D 10 cm) and Marcel Bouraine (Basset-hound in gilded bronze: H 11 cm, W 27.5 cm, D 11 cm). (Source: Artistic Evolution, Goldscheider Archives)

Le Faguays ist ein moderner Künstler und obwohl seine Skulpturen manchmal Faune und Nymphen zeigen, erscheinen seine Charaktere laut Rambosson nie antik.

Das neunte Heft beschäftigt sich mit dem Phänomen der Buchstützen, die seit Anfang der 1920er Jahre an Popularität gewinnen. Die vor allem auf Regalen und Schreibtischen eingesetzten Buchstützen werden benutzt, um gerade gelesene Bücher in einer ästhetischen Art „griffbereit" zu halten.[588] Aufgrund des ästhetischen Anspruchs hat man mit der Kreation der Buchstützen Bildhauer beauftragt, die in verschiedenen Stilrichtungen und Materialien tätig sind: Georges Henri Laurent (Ill. 165), J. Belville, J. Verschneider, Marcel Bouraine (Ill. 166), Alexandre Kéléty und Pierre Bouret. Seien es Tiere oder klassische Formen, die von Arthur Goldscheider veröffentlichten Buchstützen bedienen verschiedene Geschmäcker.

Der schwierig zu definierende Bereich der Kleinbronzen wird im zehnten Heft diskutiert. Marcel Temporal bezeichnet den Begriff „kleine Bronze" als „Geisteshaltung" und Yvanhoé Rambosson verdeutlicht, dass es sich zwar offensichtlich nicht um Statuen handelt, die Größe aber nicht das Maßgebende ist.[589] Vereinfacht gesagt, ist die Kleinbronze ein Ziergegenstand, der im Heim in greifbarer Nähe den Augen und Händen gefallen muss.[590] Beispielhaft werden Bronzen von Landowski, Bouchard, Fany Rozet, Guerbe, Bouraine, Rudolph Tegner, Yourievitch und Sybille May präsentiert.

Dem Bildhauer Raoul Lamourdedieu ist ein eigenes Heft gewidmet, denn laut Ram-

his subject matter, his sculpture, as Rambosson pointed out, was never in the antique manner.

The ninth issue deals with bookends, which had become phenomenally successful since the early 1920s. Placed mainly on shelves and desks, bookends were used to keep books which were being read "within immediate reach" in an aesthetically pleasing way.[588] Because bookends were supposed to be aesthetic objects, sculptors involved with various styles and working in various different materials were commissioned to design some: Georges Henri Laurent (Ill. 165), J. Belville, J. Verschneider, Marcel Bouraine (Ill. 166), Alexandre Kéléty and Pierre Bouret. Be they animals or classical forms, the bookends published by Arthur Goldscheider targeted a broad range of tastes.

The tenth issue discusses small bronzes, a field that is difficult to define. Marcel Temporal defines the term "small bronze" as "an attitude of mind" and Yvanhoé Rambosson makes it clear that, although he is obviously not dealing with statues, size is not paramount.[589] Put more simply, the small bronze is an ornamental object that, placed in the home within immediate reach, must appeal to the senses of both sight and touch.[590] Exemplary bronzes by Landowski, Bouchard, Fany Rozet, Guerbe, Bouraine, Rudolph Tegner, Yourievitch and Sybille May are presented.

An entire issue was devoted to the sculptor Raoul Lamourdedieu because Ram-

Ill. 166

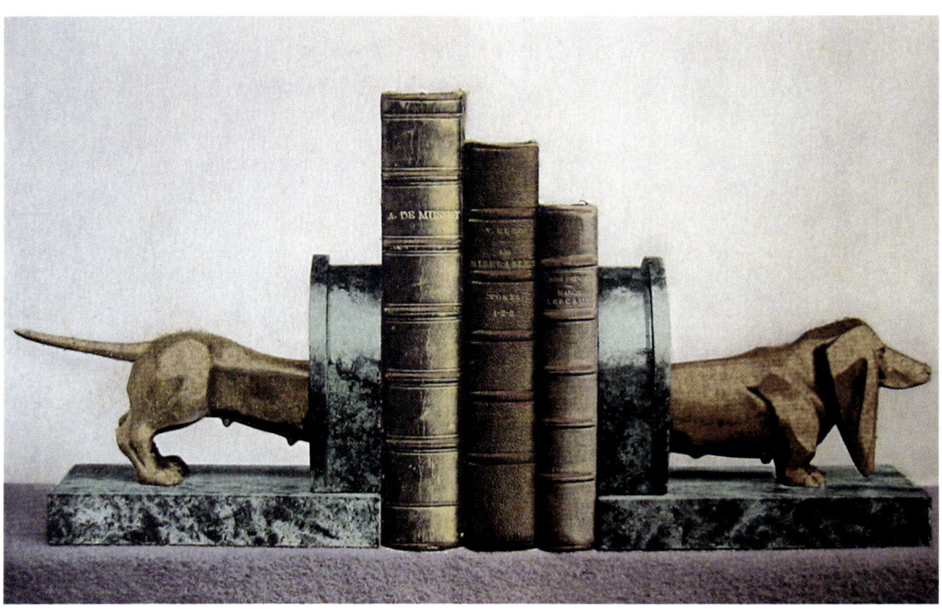

bosson felt that he exemplified the developments taking place in modern art perhaps more than any other exponent: he was versatile and had an instinct for what was modern and avant-garde.[591] He did not repeat himself in his work and was always receptive to new techniques and willing to develop them further. It was also greatly to his credit that he recognized the importance of architecture, viewing himself as its "number one collaborator."[592]

The twelfth issue deals generally with the representation of animals. Even in prehistoric times, mankind had depicted animals; domestication brought with it devotion that virtually amounted to a symbiosis. The various animal figures presented by Arthur Goldscheider in this last issue of Artistic Evolution are the work of the Martel brothers, Marie-Louise Simart (Ill. 168), Pierre Blanc (Ill. 167), Gabriel Beauvais, Alexandre Kéléty, Laurent and Lamourdedieu.[593]

Although no more issues were published in English, individual artists from the two groups, La Stèle and Évolution were successful on the market. In 1928, "Le Bouc," an ornamental table fountain by Henri Proszynski[594] and Demêtre Chiparu's bronze "La Danseuse Ta-Keo" were shown by Arthur Goldscheider at the Salon of the Société des Artistes Français. By the mid-1930s, Arthur Goldscheider was sixty years old and the firm closed down. Many of the sculptors he had represented went to a competitor, "Les Neveux de J. Lehmann."[595]

bosson verdeutlicht kaum ein anderer die Entwicklung der modernen Kunst so treffend: Er sei vielseitig und habe ein Gefühl für das Moderne und die Avantgarde.[591] Seine Werke wiederholen sich nicht und er sei stets für neue Techniken und deren Weiterentwicklung offen. Zu Gute gehalten wird ihm auch seine Anerkennung der Bedeutung von Architektur, als deren „ersten Kollaborateur" er sich sieht.[592]

Das zwölfte Heft beschäftigt sich allgemein mit der Darstellung von Tieren. Bereits in prähistorischen Zeiten haben sich Menschen mit Darstellungen von Tieren beschäftigt, durch die Domestizierung kam noch die Anhänglichkeit, fast Symbiose hinzu. Die im letzten Heft der „Artistic Evolution" von Arthur Goldscheider präsentierten unterschiedlichen Tierfiguren sind von den Brüdern Martel, Marie-Louise Simart (Ill. 168), Pierre Blanc (Ill. 167), Gabriel Beauvais, Alexandre Kéléty, Laurent sowie Lamourdedieu.[593]

Es werden keine weiteren englischen Hefte publiziert, doch einzelne Künstler der beiden Gruppen La Stèle und Évolution können sich am Markt behaupten. Im Salon der Société des Artistes Français wird von Arthur Goldscheider im Jahr 1928 Henri Proszynskis Zierbrunnen „Le Bouc" gezeigt[594] sowie Demêtre Chiparus Bronze „La Danseuse Ta-Keo". Mitte der 30er Jahre wird die Firma des sechzigjährigen Arthur Goldscheider geschlossen, viele der von ihm vertretenen Bildhauer gehen zur Konkurrenz „Les Neveux de J. Lehmann".[595]

Ill. 167

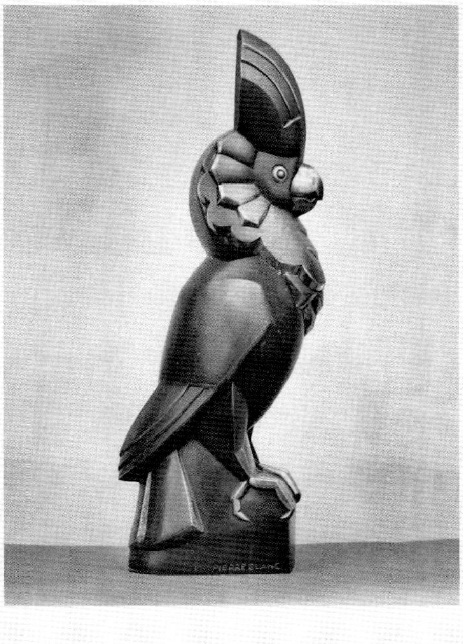

Ill. 168

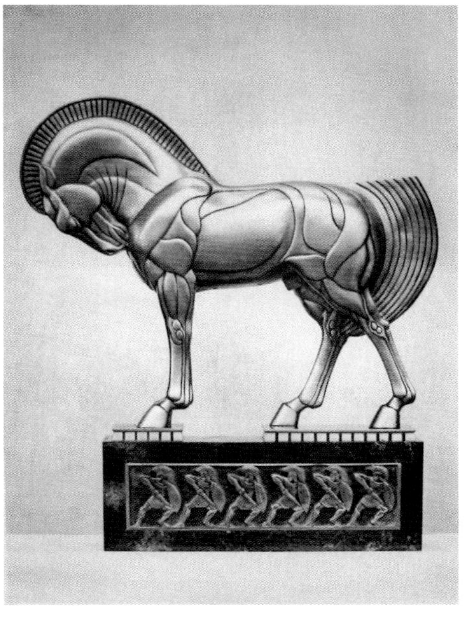

Ill. 167, 168
Tierfiguren sind ein weiteres beliebtes Sujet. Für Arthur Goldscheider entwerfen auch Pierre Blanc (Kakadu aus schwarzer Bronze mit Nickelteilen; H 40 cm, B 8 cm, T 12,5 cm) und Marie-Louise Simart (Trojanisches Pferd aus vergoldeter Bronze auf Schwarz-Marmor-Sockel, H 64 cm, B 60 cm, T 17,5 cm). (Quelle: Artistic Evolution, Archiv Goldscheider)

Animal figures were another popular subject. Pierre Blanc (Cockatoo in black bronze with nickel elements: H 40 cm, W 8 cm, D 12.5 cm) and Marie-Louise Simart (Trojan Horse in gilded bronze on a black marble plinth: H 64 cm, W 60 cm, D 17.5 cm) also designed for Arthur Goldscheider. (Source: Artistic Evolution, Goldscheider Archives)

Anmerkungen

[551] Abgebildete Anzeige aus „L'annuaire du commerce Didot Bottin" um 1900, in: Richemond (1999)
[552] ÖPB, 1901, S. 796. Das Patent erhält die Nummer 7052. Arthur Goldscheiders „Atelier für Kunstwerke in Paris" wird in Wien rechtlich durch V. Tischler vertreten, der auch die Wiener Firma vertritt. Das Patent erlischt bereits im Jahr 1903.
[553] Thomas, A.: Petits Bronzes D'Art, in: L'Art Décoratif, 3.1900/01, S. 181ff
[554] N.N.: Bronze – Elfenbein, Statuetten der Neuzeit. Louis Chalon, in: AZ, 1980, S. 379
[555] N., R.: Keramische Körperkultur, in: AZ, 1992, S. 565
[556] Art et Décoration, 1903, S. 2
[557] Neuwirth (1973), S. 60
[558] N.N.: Bronze – Elfenbein, Statuetten der Neuzeit. Paul Philippe, in: AZ, 1979, S. 581
[559] N.N.: Bronze – Elfenbein, Statuetten der Neuzeit. Georges (Joris) van der Straeten, in: AZ, 1980, S. 459
[560] N.N.: In Bronze lesen, in: AZ, 1984, S. 269
[561] Anzeige der Firma „Friedrich Goldscheider Kunstterrakotta-Fabrik Ateliers künstlerischer Fayencen" in Leuchs Adressbuch 1908
[562] Offizielles Leipziger Mess-Adressbuch, Oster-Vormesse 1910. S. 55ff sowie S. 570
[563] Archiv Goldscheider, Programme der beiden Theater
[564] Kjellberg, S. 660
[565] Arwas (1992), S. 248
[566] Herbst-Mustermesse Leipzig 1918, Amtliches Ausstellerverzeichnis
[567] Rouger, G.: Nos Artistes au Maroc, in: L'Art et les artistes, N.S. 6, 1922, S. 36
[568] Anzeige in L'Illustration, 20.12.1924, S. 14
[569] N.N.: Bronze – Elfenbein, Statuetten der Neuzeit. La Stèle, in: AZ, 1980, S. 459. Entnommen aus dem offiziellen Ausstellungskatalog zum Pavillon I-115.
[570] N.N.: Bronze – Elfenbein, Statuetten der Neuzeit. Ekiel Raby, in: AZ, 1980, S. 209
[571] Neuwirth (1974), S. 44. Entnommen aus dem offiziellen Ausstellungskatalog zum Pavillon I-115.
[572] Goldscheider, Arthur (Hg.): Artistic Evolution (L'Évolution Artistique). Texte von Yvanhoé Rambosson, Selbstverlag, Paris 1926. Originaluntertitel: Revue destinée à la diffusion des idées d'art moderne dans l'industrie et contenant des reproductions d'œuvres d'artistes des groupes „La Stèle" et „Évolution" fascicules de Janvier à Juin 1926 – Textes par Yvanhoé Rambosson. Arthur Goldscheider, Édit. 45, Rue de Paradis, Paris (Xe).
[573] Ebenda. „The incontestable and prodigious success of the Exhibition of Modern Decorative and Industriel Arts of Paris 1925 is a pledge to us that our faith in success does not rest on the desires and hopes of a limited few, but on the revolution itself of ideas and facts."
[574] Ebenda
[575] Ebenda
[576] Ebenda: „Here is a real revolution in the methods of commercial productions."
[577] Ebenda: „Nevertheless, all classes of society have a right to beauty, and the time was bound, to come when manufacturers would try to produce the conceptions of innovators in series."
[578] Ebenda: „We hope that the collection of these albums will form a most precious documentary whole for pupils, students and art amateurs, for salesman without whom modern art could not penetrate the public ..."
[579] Es erschienen in englischer Sprache nur zwei Bände, die alle Hefte des Jahres 1926 zusammenfassen. Die Publikation wurde dann eingestellt. Die beiden Künstlergruppen konnten keinen bedeutenden Einfluss ausüben, einzelne Künstler hatten sich jedoch langfristig durchsetzen können.
[580] Ebenda, 2. Heft im Band I: „It will be the honour of contemporary art to have returned to the creed of pure form and to have recognized its supremacy over ornamentation. [...] The campaigne which we shall lead with buyers and sellers as well as creators of all kinds cannot fail to bear its fruit, for it is obvious that the actual world is more and more obsessed with the desire to make for itself logical surroundings replying to the fundamental transformations introduced into customs by a scientific expansion which is without precedent in the history of civilisation. [...] In such a time as ours it is impossible that artists should live in the margin of society in an ivory castle of which the walls are sheltered from wireless. Their duty is to mingle with existence, to let themselves be inspired by that spezial sort of lyricism which emanates from the discoveries of learned men or the art of the machinist, and to dream in their trades of new technique often generating new conceptions."
[581] Ebenda, 2. Heft im Band I
[582] Ebenda, 4. Heft im Band I: „At first, the creators of models and manufacturers made the mistake of retaining for electricity shapes of fittings which had been used for other means of lightning. Do we not still see artificial electric candles of which the porcelain stem in imitation wax is surmounted by a flame-shaped bulb?"
[583] Ebenda, 5. Heft im Band I: „We may add that these home fountains will remain amongst the most original and delightful innovations of our time. They will be in demand later on like wall fixtures for washhand basins created by the manufactories of Rouen, Nevers, and other celebrated earthenware factories, and which are so eagerly sought after by collectors when offered for public sale."
[584] Ebenda, 6. Heft im Band I: „Chaplet obtained these famous red tints and all of his productions bearing them are eagerly sought after by collectors."
[585] Ebenda
[586] Ebenda, 7. Heft im Band II: „It was agreed that each artist should follow the material fabrication of his work and that the editor, even if a model were bought, could never offer for sale a copy which did not give the artist entire satisfaction."
[587] Ebenda, 8. Heft im Band II
[588] Ebenda, 9. Heft im Band II: „Under these different aspects, [...], the bookrest, an innovation of the present day, has won an important position in miniature sculpture during the last three or four years."
[589] Ebenda, 10. Heft im Band II
[590] Ebenda: „A slight and frivolous thing – at times picturesque – a ‚small bronze' is an ornament made to please the eye and hand, placed in our homes within immediate reach."
[591] Ebenda, 11. Heft im Band II: „What distinguishes M. Raoul Lamourdedieu in the crowd of contemporary producers, among whom talent is certainly not lacking, is his breadth of mind, his sense of moderness ever on alert, not the moderness of to-day which belongs already to the past, but that of to-morrow."
[592] Ebenda
[593] Ebenda, 12. Heft im Band II: „It will end by remarking what a diversity of conception we met with in eight pieces of sculpture alone. The same animal might serve as a theme to each of the artists mentioned in this number, and yet each work would be dissimilar. This shows the immense variety of the modern menagerie created by artists, a variety as great as that of temperaments and tastes."
[594] N.N.: Bronze – Elfenbein, Statuetten der Neuzeit. Henri Proszynski, in: AZ, 1980, S. 209
[595] Arwas (1987), S. 163

Notes

[551] The advertisement is from L'annuaire du commerce Didot Bottin ca 1900, in: Richemond (1999)
[552] ÖPB, 1901, p. 796. The patent was given the number 7052. Arthur Goldscheider's "Atelier für Kunstwerke in Paris" ["Studio for Art Works"] was legally represented in Vienna by V. Tischler, who also represented the Viennese firm. The patent expired in 1903.
[553] A. Thomas: Petits Bronzes d'Art, in: L'Art Décoratif, 3.1900/01, p. 181f.
[554] N.N.: Bronze – Elfenbein, Statuetten der Neuzeit. Louis Chalon, in: AZ, 1980, p. 379
[555] R. N.: Keramische Körperkultur, in: AZ, 1992, p. 555
[556] Art et Décoration, 1903, p. 2
[557] Neuwirth (1973), p. 60
[558] N.N.: Bronze – Elfenbein, Statuetten der Neuzeit. Paul Philippe, in: AZ, 1979, p. 581
[559] N.N.: Bronze – Elfenbein, Statuetten der Neuzeit. Georges (Joris) van der Straeten, in: AZ, 1980, p. 459
[560] N.N.: In Bronze lesen, in: AZ, 1984, p. 269
[561] Advertisement placed by the firm of "Friedrich Goldscheider Kunstterrakotta-Fabrik Ateliers künstlerischer Fayencen" in Leuchs Adressbuch 1908
[562] Offizielles Leipziger Mess-Adressbuch, Oster-Vormesse 1910. p. 55f. and p. 570
[563] Goldscheider Archives, programs from the two theatres
[564] Kjellberg, p. 660
[565] Arwas (1992), p. 248
[566] Herbst-Mustermesse Leipzig 1918, Amtliches Ausstellerverzeichnis
[567] G. Rouger: Nos Artistes au Maroc, in: L'Art et les artistes, N.S. 6, 1922, p. 36
[568] Advertisement in L'Illustration, 20 Dec. 1924, p. 14
[569] N.N.: Bronze – Elfenbein, Statuetten der Neuzeit. La Stèle, in: AZ, 1980, p. 459. Taken from the official exhibition catalogue for Pavilion I-115
[570] N.N.: Bronze – Elfenbein, Statuetten der Neuzeit. Ekiel Raby, in: AZ, 1980, p. 209
[571] Neuwirth (1974), p. 44. Taken from the official exhibition catalogue for Pavilion I-115
[572] Arthur Goldscheider (ed.): Artistic Evolution (L'Évolution Artistique). Texts by Yvanhoé Rambosson, published by Arthur Goldscheider, Paris 1926. Original subtitle: Revue destinée à la diffusion des idées d'art moderne dans l'industrie et contenant des reproductions d'œuvres d'artistes des groupes "La Stèle" et "Évolution" fascicules de Janvier à Juin 1926 – Textes par Yvanhoé Rambosson. Arthur Goldscheider, Édit. 45, rue de Paradis, Paris (Xe).
[573] Ibid. "The incontestable and prodigious success of the Exhibition of Modern Decorative and Industriel Arts of Paris 1925 is a pledge to us that our faith in success does not rest on the desires and hopes of a limited few, but on the revolution itself of ideas and facts."
[574] Ibid
[575] Ibid
[576] Ibid: "Here is a real revolution in the methods of commercial production."
[577] Ibid: "Nevertheless, all classes of society have a right to beauty, and the time was bound, to come when manufacturers would try to produce the conceptions of innovators in series."
[578] Ibid: "We hope that the collection of these albums will form a most precious documentary whole for pupils, students and art amateurs, for salesman without whom modern art could not penetrate the public ..."
[579] Only two volumes were published in English, containing all issues from 1926. After that publication in English ceased. The two groups of artists were not very influential although individual artists did make it over the long term.
[580] Ibid, 2nd issue in Vol. I: "It will be the honour of contemporary art to have returned to the creed of pure form and to have recognized its supremacy over ornamentation. [...] The campaign which we shall lead [sic] with buyers and sellers as well as creators of all kinds cannot fail to bear its fruit, for it is obvious that the actual world is more and more obsessed with the desire to make for itself logical surroundings replying to the fundamental transformations introduced into customs by a scientific expansion which is without precedent in the history of civilisation. [...] In such a time as ours it is impossible that artists should live in the margin of society in an ivory castle of which the walls are sheltered from wireless. Their duty is to mingle with existence, to let themselves be inspired by that spezial sort of lyricism which emanates from the discoveries of learned men or the art of the machinist, and to dream in their trades of new technique often generating new conceptions."
[581] Ibid, 2nd issue in Vol. I
[582] Ibid, 4th issue in Vol. I: "At first, the creators of models and manufacturers made the mistake of retaining for electricity shapes of fittings which had been used for other means of lightning. Do we not still see artificial electric candles of which the porcelain stem in imitation wax is surmounted by a flame-shaped bulb?"
[583] Ibid, 5th issue in Vol. I: "We may add that these home fountains will remain amongst the most original and delightful innovations of our time. They will be in demand later on like wall fixtures for washhand basins created by the manufactories of Rouen, Nevers, and other celebrated earthenware factories, and which are so eagerly sought after by collectors when offered for public sale."
[584] Ibid, 6th issue in Vol. I: "Chaplet obtained these famous red tints and all of his productions bearing them are eagerly sought after by collectors."
[585] Ibid
[586] Ibid, 7th issue in Vol. II: "It was agreed that each artist should follow the material fabrication of his work and that the editor, even if a model were bought, could never offer for sale a copy which did not give the artist entire satisfaction."
[587] Ibid, 8th issue in Vol. II
[588] Ibid, 9th issue in Vol. II: "Under these different aspects, [...], the bookrest, an innovation of the present day, has won an important position in miniature sculpture during the last three or four years."
[589] Ibid, 10th issue in Vol. II
[590] Ibid: "A slight and frivolous thing – at times picturesque – a 'small bronze' is an ornament made to please the eye and hand, placed in our homes within immediate reach."
[591] Ibid, 11th issue in Vol. II: "What distinguishes M. Raoul Lamourdedieu in the crowd of contemporary producers, among whom talent is certainly not lacking, is his breadth of mind, his sense of moderness ever on alert, not the moderness of to-day which belongs already to the past, but that of to-morrow."
[592] Ibid
[593] Ibid, 12th issue in Vol. II: "It will end by remarking what a diversity of conception we met with in eight pieces of sculpture alone. The same animal might serve as a theme to each of the artists mentioned in this number, and yet each work would be dissimilar. This shows the immense variety of the modern menagerie created by artists, a variety as great as that of temperaments and tastes."
[594] N.N.: Bronze – Elfenbein, Statuetten der Neuzeit. Henri Proszynski, in: AZ, 1980, p. 209
[595] Arwas (1987), p. 163

Marcell Goldscheiders Werkstätten in Wien und Stoke-on-Trent

Marcell Goldscheider's Workshops in Vienna and Stoke-on-Trent

Die Vereinigten Ateliers für Kunst und Keramik (1928–1939)

Nach der Trennung von seinem Bruder Walter wird im Januar 1928 die Cottage-Villa in der Hasenauerstraße 16 komplett auf Marcell übertragen. Er lebt dort bereits seit mehreren Jahren mit seiner Frau Rosemarie, die ebenfalls im Kunstgewerbe tätig ist. Rosemarie ist Glaskünstlerin, die im Jahr 1925 zwei Patente angemeldet hat: ein Verfahren zur Herstellung von Eisglas- und Glasmalerei-Imitation sowie von Seidenstickerei-Imitationen.[596] Sie betreibt das Atelier „Parivienna" in der Gymnasiumstraße 67, bevor sie es um 1927 zu Marcell in die Hasenauerstraße 16 verlegt.[597] Am gleichen Ort betreibt Marcell am Anfang seiner Selbständigkeit als Alleininhaber die „Vereinigten Ateliers für Kunst und Keramik",[598] in denen er mit 15 Mitarbeitern Fayencen, Porzellan-Figuren, Vasen und Leuchter herstellt.[599] Am 3. September 1928 meldet er den „Handel mit Porzellan- und Steingutwaren, Figuren und einschlägi-

United Studios of Art and Ceramic (1928–1939)

After the brothers Walter and Marcell separated, the villa known as the Cottage at 16 Hasenauerstraße was made over to Marcell as sole owner in January 1928. Marcell had already been living there for several years with his wife Rosemarie, who also worked in the decorative and applied arts. Rosemarie's craft was glass. In 1925 she had registered two patents: one for a process for making frosted glass and imitation stained glass and another for simulated silk embroidery.[596] She ran the "Parivienna" studio at 67 Gymnasiumstraße before it was moved around 1927 to 16 Hasenauerstraße, where Marcell worked.[597] From the time he had become self-employed, Marcell had run the "Vereinigte Ateliers für Kunst und Keramik" ["United Studios of Art and Ceramic"] there as its proprietor,[598] employing a workforce of 15 to make faience, porcelain figures, vases and candlesticks.[599] On 3 September 1928,

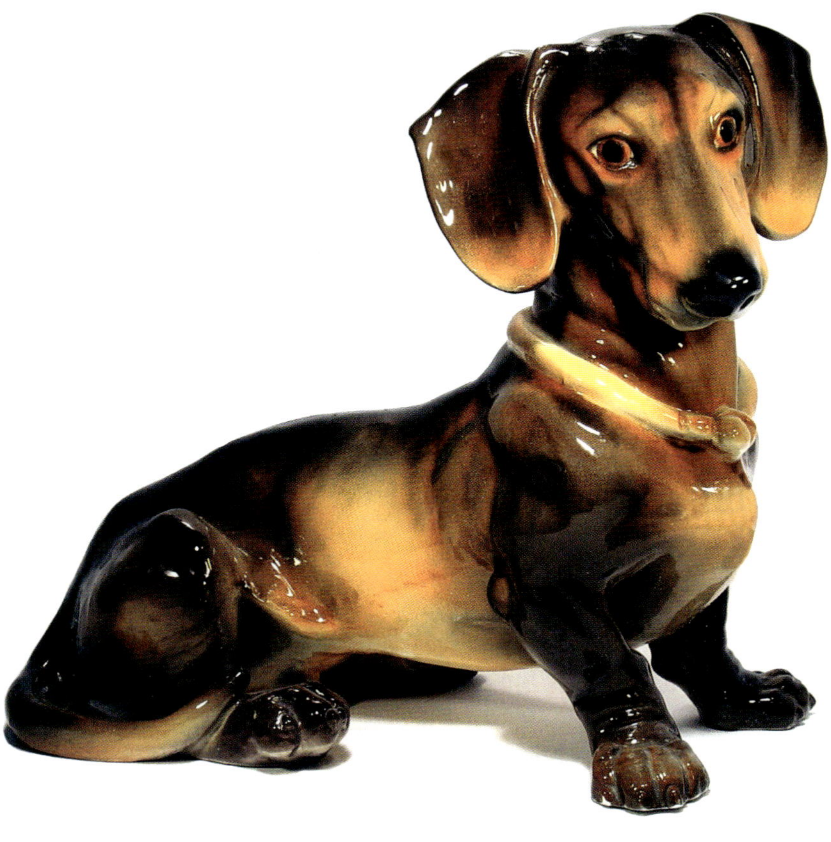

ModNr. 499
Dackel I Dachshund, ca. 1929, Collection a@goldscheider.net

Der Stempel von Marcel Goldscheider zeigt ein lachendes Gesicht in der Form des Buchstabens „M".

Marcell Goldscheider's stamp consisted of a laughing face formed by the letter "M".

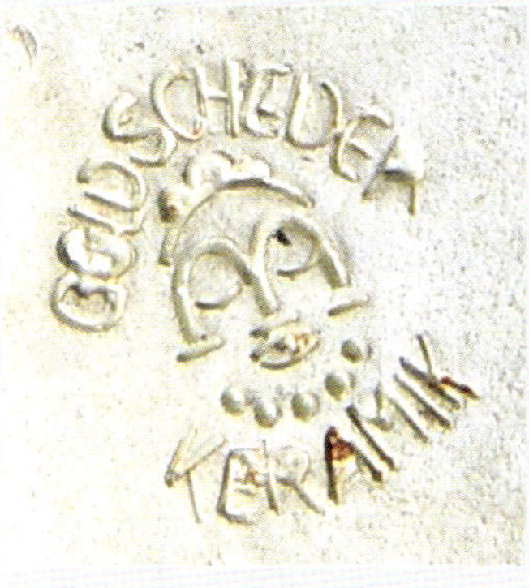

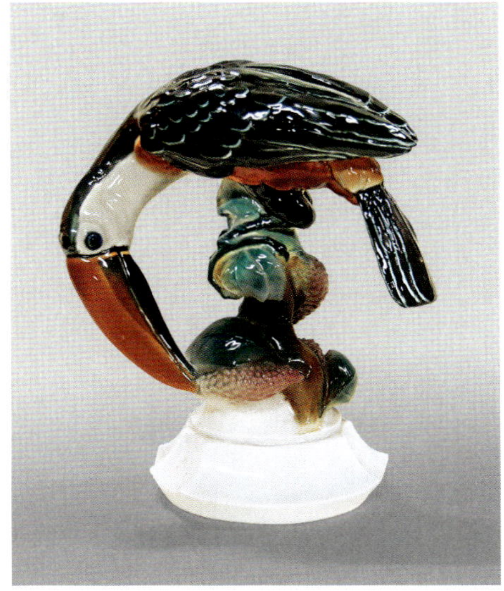

ModNr. 378
Tukan I Toucan,
Karin Jarl, um 1928/29,
MAK – Österreichisches
Museum für ange-
wandte Kunst / Gegen-
wartskunst, Wien

Marcell registered as his trade "selling porcelain and stoneware wares, figures and related articles"[600] and opened a samples warehouse and office, that he ran jointly with his wife, at 16 Getreidemarkt in the 1st District.[601] There he sold not only his own models but also art objects "after designs by leading Viennese artists."[602]

Confusion arose from the two brothers' both producing at the same time. Marcell was erroneously regarded as merely a dealer, which is why he sent a letter of explanation to the Österreichisches Museum to clarify his status. He wrote that almost all the wares he sold were his own. He felt the reason for the confusion was as follows: "The mistake may have arisen from my being accustomed to naming my artists yet my being the maker is shown in the prices."[603] Surprisingly, data on other artists or artists' signatures do not occur on later models of Marcell's. Instead, the objects are only marked with a stamp of Marcell's. This mark or rather stamp, consists of a laughing face (some may also have interpreted it as a reversed anchor) formed by the letter "M", with "Marcell" above it and

gen Artikeln" als Gewerbe an[600] und eröffnet am Getreidemarkt 16 im ersten Bezirk ein Musterlager und Büro, das er gemeinsam mit seiner Frau betreibt.[601] Dort verkauft er nicht nur seine eigenen Modelle, sondern auch Kunstgegenstände „nach Entwürfen der führenden Künstler Wiens".[602]

Die gleichzeitige Produktion der Brüder stiftet Verwirrung. Marcell wird fälschlicherweise nur für einen Händler gehalten, weshalb er einen erklärenden Brief an das Österreichische Museum schickt. Er schreibt, dass die von ihm verkauften Waren fast nahezu ausschließlich seine Erzeugnisse sind. Den Grund für die Verwirrung sieht er wie folgt: „Vielleicht ist der Irrtum dadurch entstanden, dass ich gewohnt bin, die Namen meiner Künstler zu nennen, jedoch drückt sich schon in der Preiserstellung aus, dass ich Erzeuger bin."[603] Erstaunlicherweise finden sich bei Marcel später auf den Modellen keine Angaben oder Signaturen anderer Künstler, die Objekte werden nur mit einem Stempel von Marcell gekennzeichnet. Diese Marke bzw. Stempel zeigt ein aus dem Buchstaben „M" geformtes lachendes Ge-

ModNr. 316
Akt mit Koralle, glasiert I Nude with Coral, glace, ca. 1928,
PB Wien 19

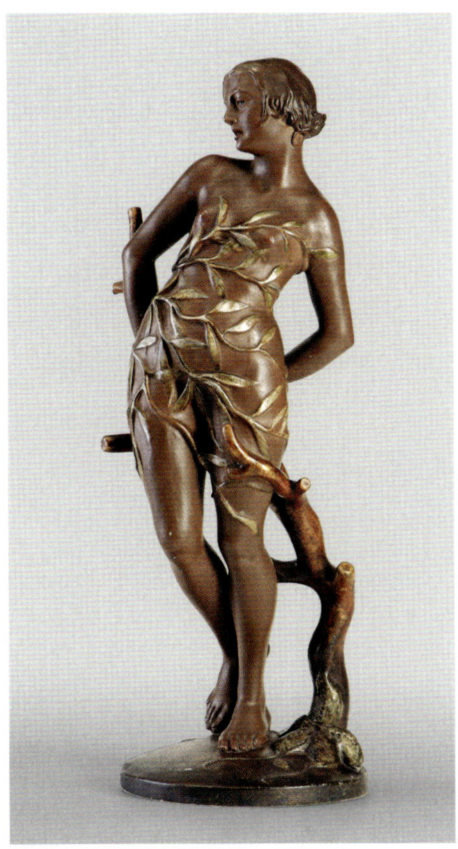

ModNr. 316
Akt mit Koralle, bronzefarben I Nude with Coral, bronze-colored, ca. 1928,
MAK – Österreichisches Museum für angewandte Kunst / Gegenwartskunst, Wien

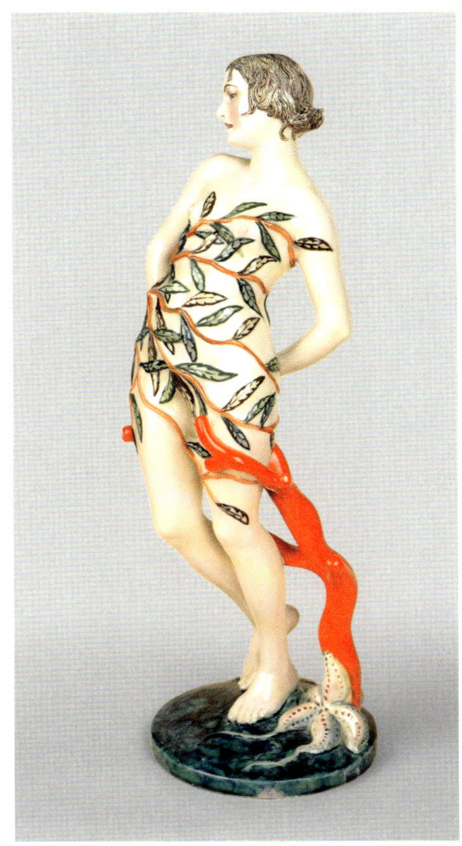

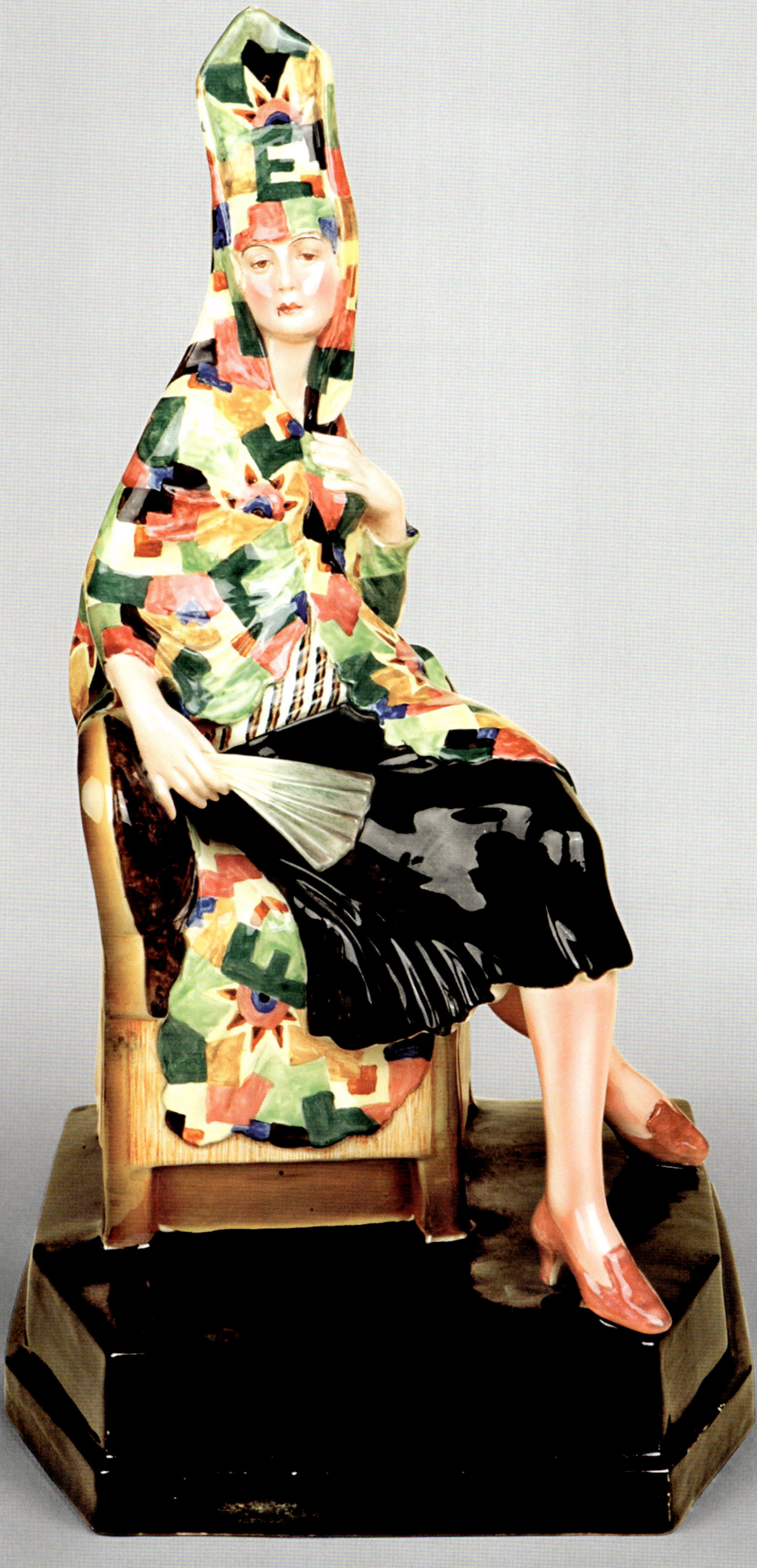

links | left

ModNr. 319
Carmen, ca. 1928,
PB Wien 19

"Goldscheider" below it. Depending on the material used, the mark is sometimes supplemented by the terms "Porzellan" or "Keramik."

The magazine Deutsche Kunst und Dekoration wrote as follows on Marcell's business: "A host of artists is at work here, among whom there are well-known, illustrious names, on a freelance basis with the entrepreneur. It might be of interest to say a few words about the way they work together. Commercially based manufactories buy models from freelance or permanently employed artists and these models are reproduced at discretion. At the 'Vereinigte Ateliers für Kunst und Keramik', the artists' activities also penetrate far into the manufacturing process. They accomplish the re-

sicht (manche mögen darin auch einen umgedrehten Anker erkennen), über dem „Marcell" und darunter „Goldscheider" steht. Je nach Material wird die Marke zusätzlich um die Bezeichnung Porzellan oder Keramik erweitert.

In der Zeitschrift „Deutsche Kunst und Dekoration" ist über Marcells Unternehmung folgendes zu lesen: „Hier arbeitet eine Künstlerschar, unter der es Namen von bekanntem und gutem Klange gibt, auf freier Basis mit dem Unternehmer zusammen. Es ist vielleicht nicht uninteressant, über diese freie Art der gemeinsamen Arbeit ein Wort zu sagen. Die Manufakturen, die auf kommerzieller Grundlage stehen, kaufen Modelle von freien oder festangestellten Künstlern an, und diese Modelle werden nach Gut-

Ill. 169

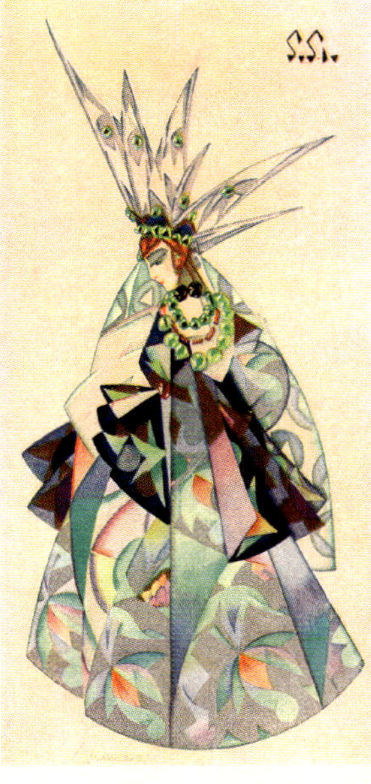

181. Poschedajeff: Figurine. Sammlung Juschni.

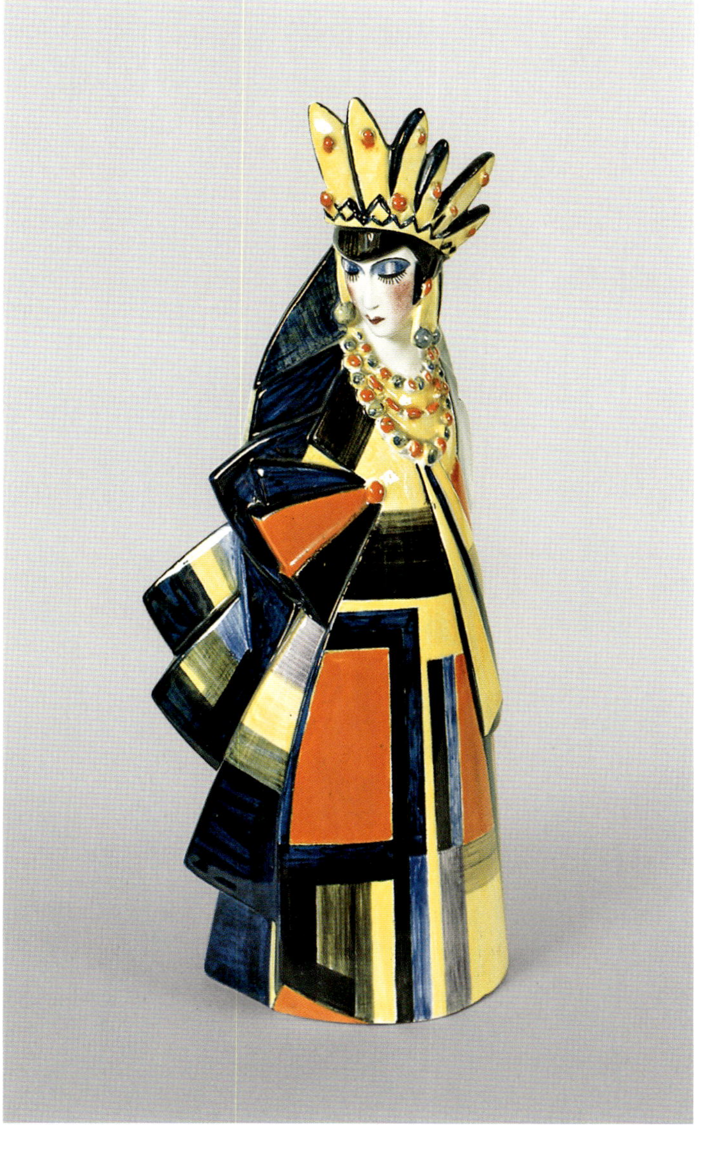

ModNr. 465
„Ballett russe",
ca. 1929–1939,
Sammlung.schmutz@chello.at
Der Kostümentwurf geht zurück auf den russischen Maler und Kostümbildner Georgi Poschedajeff (Ill. 169). (Quelle: PB Deutschland)

The design of the costume originated with the Russian painter and costume designer Georgy Poshedayev (Ill. 169). (Source: PB Deutschland)

dünken reproduziert. Bei den ‚Vereinigten Ateliers für Kunst und Keramik' reicht die Tätigkeit der Künstler auch weit in die Fabrikation hinein. Sie nehmen die Reproduktion ihrer Modelle selbst vor oder behalten doch die Leitung der Ausführung, so daß jede Reproduktion den Charakter einer Original-Ausführung erhält."[604] Marcell setzt in seinem eigenen Unternehmen die Idee eines ganzheitlich involvierten Künstlers durch. Er ist in dieser Einstellung stark von seinem Bruder Arthur geprägt, der in Paris bereits 1925 die gleichen Ideen propagierte. Selbstverständlich zeigt sich hier auch der Einfluss der Wiener Werkstätte, die Originale produzieren ließ und deren Künstler auch für Marcell Modelle entwerfen.

Als der Österreichische Werkbund im Jahr 1928 im Künstlerhaus eine Weihnachts-

production of their models themselves or at least direct the execution of them so that each reproduction is given the character of an original piece."[604] In his own enterprise, Marcell was realizing the idea of the artist being involved in the entire process. In this attitude, he was strongly influenced by his brother Arthur, who had been propagating the same ideas in Paris back in the 1920s. Undoubtedly this practice also documents the influence of the Wiener Werkstätte, where originals were produced, and whose artists also designed models for Marcell.

When the Österreichischer Werkbund organized a Christmas show at the Künstlerhaus in 1928, Marcell was represented in sale-room VI with "Ö.W.B. ceramics after designs by H. Bucher, R. Knörlein, L. Rie-Gomperz, B. Ziffer [actually Moses, Ziffer]."[605]

ModNr. 325
Deckeldose I Covered Box, ca. 1928, Collection Dr. Peter Hieke, Wien

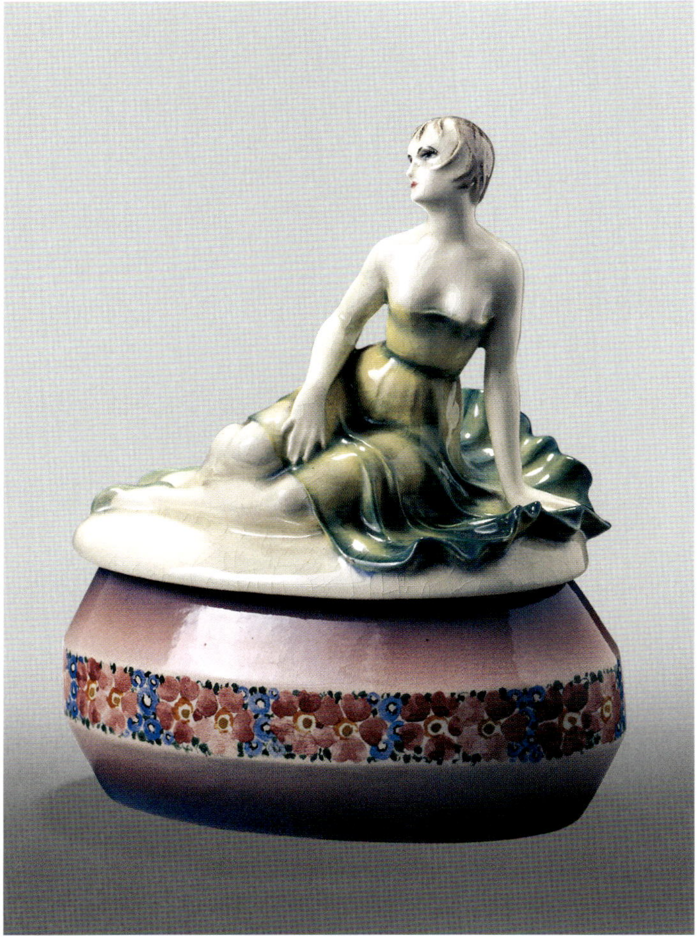

ModNr. 382
Liebespaar mit Kakadu I Lovers with Cockatoo, ca. 1928/29, © Wien Museum

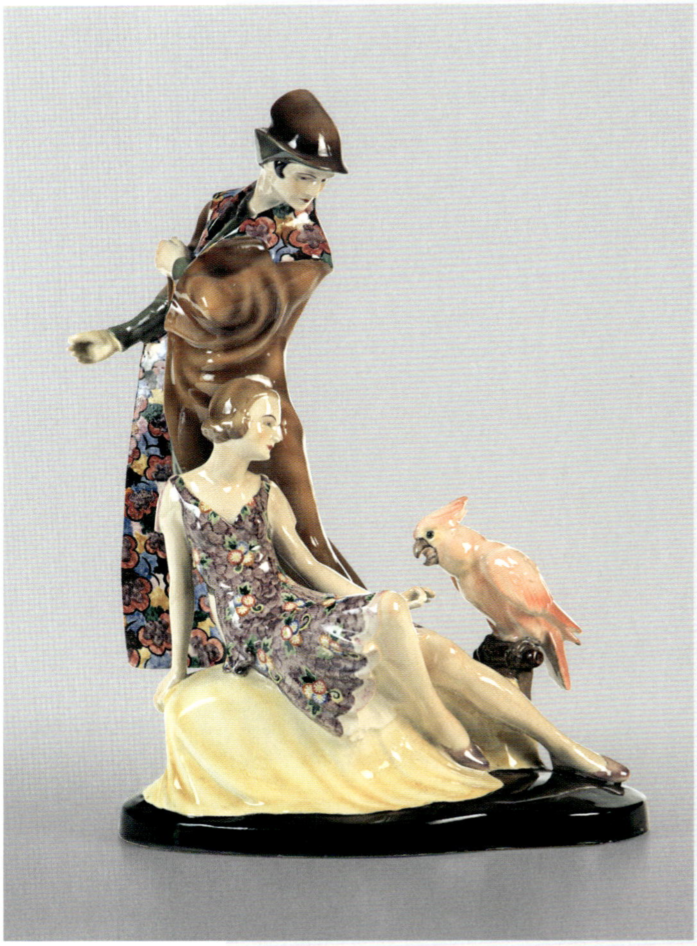

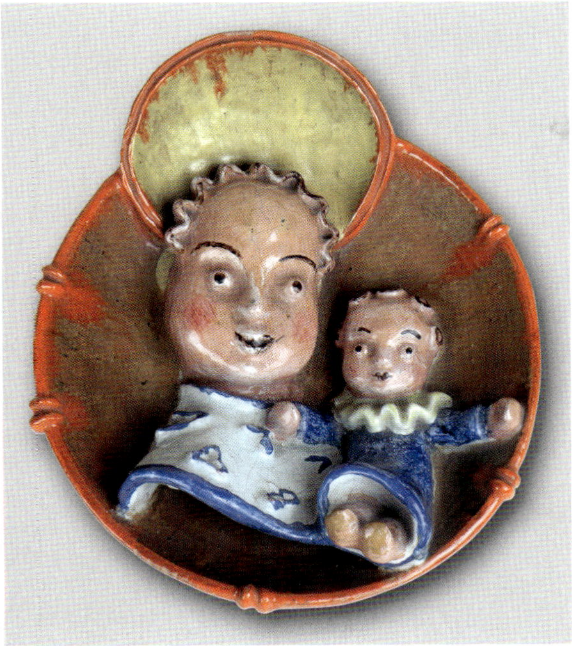

ModNr. 244
Tondo (Relief),
ca. 1928, PB Wien 19

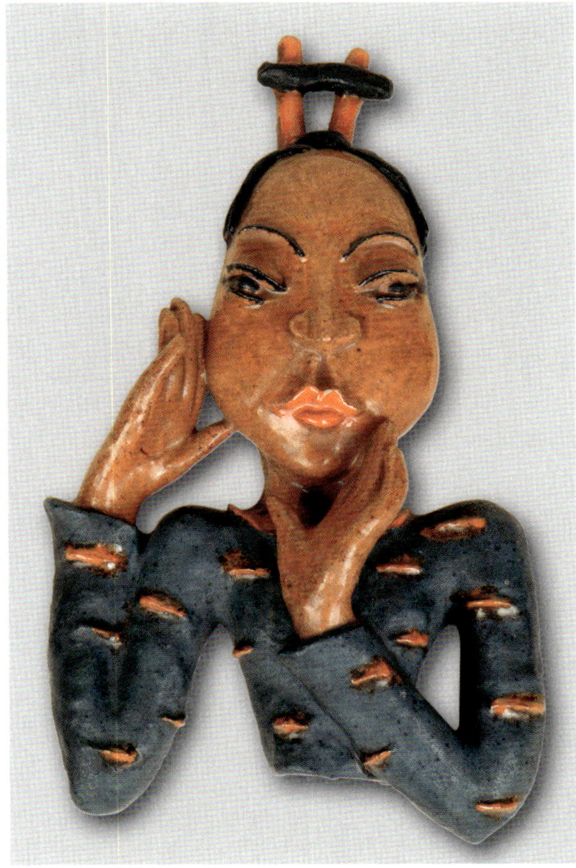

ModNr. 235
Relief, ca. 1928,
PB Wien 19

ModNr. 478
Akkordeonspieler |
Accordion Player,
ca. 1929–1939,
PB Wien 19

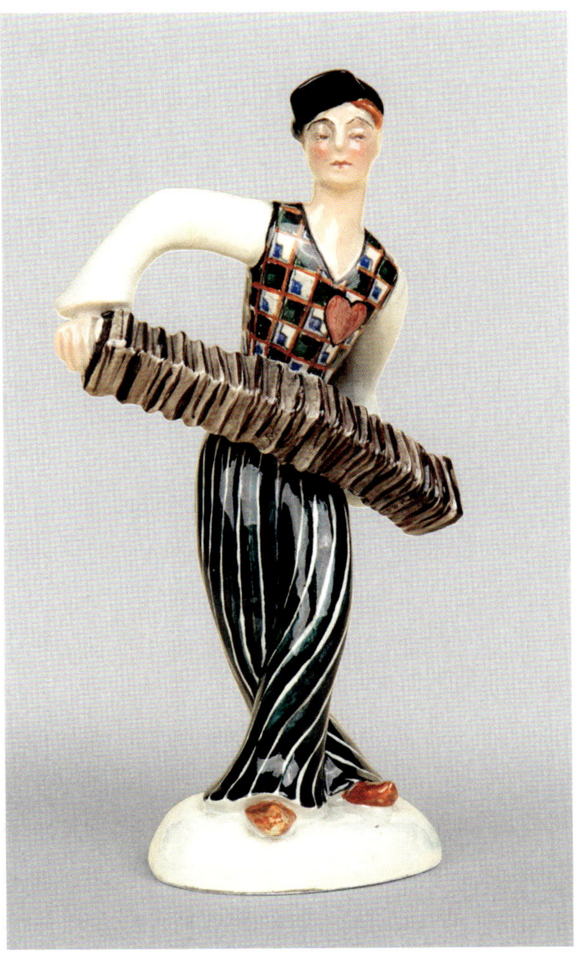

ModNr. 386,
Geisha mit Reh | with
Doe, ca. 1928/29,
Collection Weilinger,
Wien

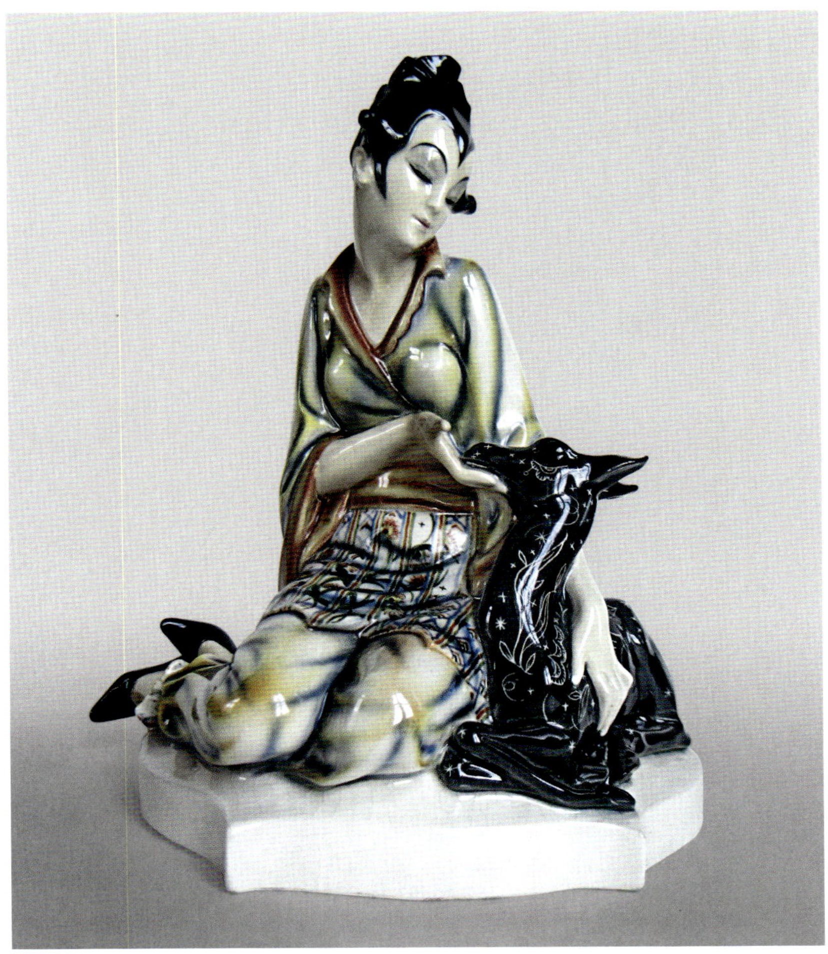

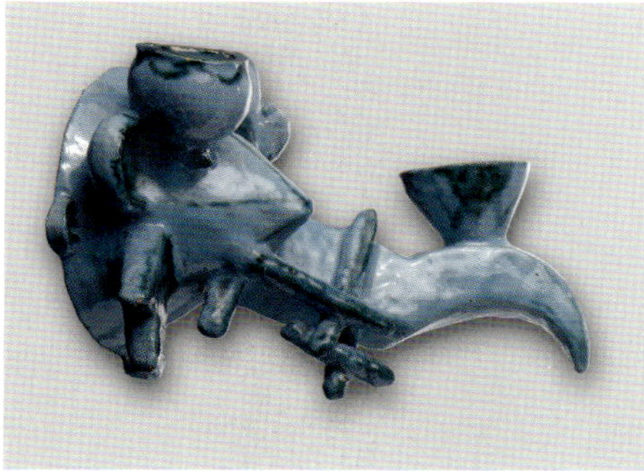

ModNr. 375
Kerzenhalter | Candlestick, ca. 1928/29, Collection Dr. Peter Hieke, Wien

rechts | right

ModNr. 484
Liliputaner auf Kugelfisch | Lilliputian with Blowfish, ca. 1929–1939, Collection Dr. Peter Hieke, Wien

schau organisiert, ist Marcell im Verkaufsraum VI mit „Ö.W.B.-Keramiken nach Entwürfen von H. Bucher, R. Knörlein, L. Rie-Gomperz, B. Ziffer [eigentlich Moses Ziffer]" vertreten.[605] Viele der bei Marcell vertretenen Künstler, wie Hertha Bucher, kennt er bereits aus dem väterlichen Betrieb, in dem er seit 1912 gearbeitet und den er seit 1920 mit seinem Bruder Walter geleitet hat. Auch Walter stellt auf dieser Ausstellung aus und präsentiert Arbeiten von W. Bosse, B. Geiger, J. Löw, E. Rottenberg und anderen.

Die von Marcell für den Österreichischen Werkbund (Ö.W.B.) erstellten Keramiken werden sehr wohlwollend aufgenommen, in der „Schaulade" steht diesbezüglich: „In diesen Arbeiten prägt sich das Wesen des Wienerischen in seiner besten Art charakteristisch aus. Die figürlichen Darstellungen sind von einer stillen, besinnlichen, fast traumhaften Art, und doch behält die Formgebung das lockere Spiel mit der Form, mit dem Formalen."[606] Und auch das Farbspiel wird gelobt: „Die Glasuren gleiten über den oft mit Absicht etwas unebenen Scherben

Marcell had met many of the artists, such as Hertha Bucher, whom he represented in his father's business, where he had worked from 1912 and which he had managed jointly with his brother Walter from 1920. Walter also showed work at that exhibition, by such artists as W. Bosse, B. Geiger, J. Löw, E. Rottenberg, et al.

The ceramics Marcell showed for the Österreichischer Werkbund (Ö.W.B.) met with a very positive reception. Schaulade noted in this connection: "The quintessence of what is Viennese at its best characteristically informs these works. The figurative representations are quiet, contemplative, almost dream-like in manner yet the formal design retains the casual interplay of form and formal."[606] And the play of color is also praised: "The glazes slide over the body, which is often intentionally rather uneven, in very free, occasionally boldly colored trickles."[607] The objects described included "Schalenträger" ["Supporting Bowls"] and figures entitled "Innehaltende" ["Reserved"] and "Niederblickende" ["Eyes Cast Down"] as well as lamp bases, vases and bowls produced by Marcell.

ohne | without ModNr.
Tafelaufsatz | Table Decoration, ca. 1928–1939, Collection Dr. Peter Hieke, Wien

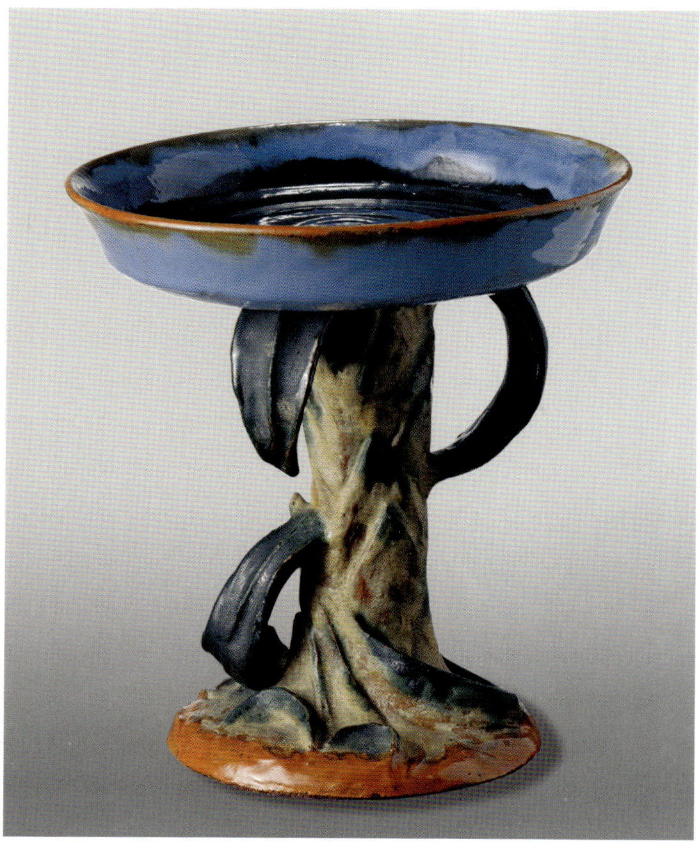

ohne | without ModNr.
Kerzenhalter Akrobaten | Candlestick Acrobats, ca. 1928–1939, PB Wien

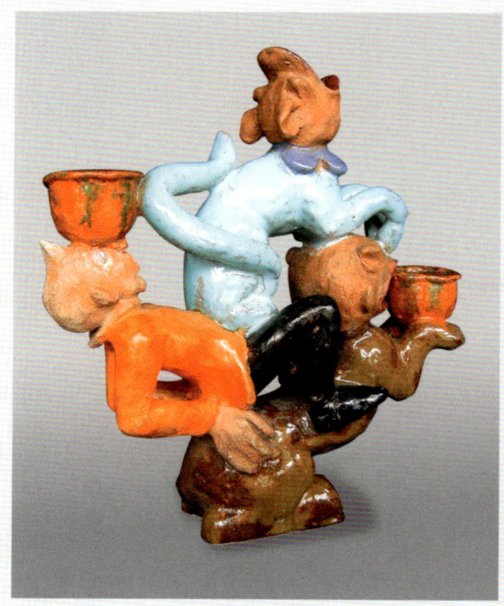

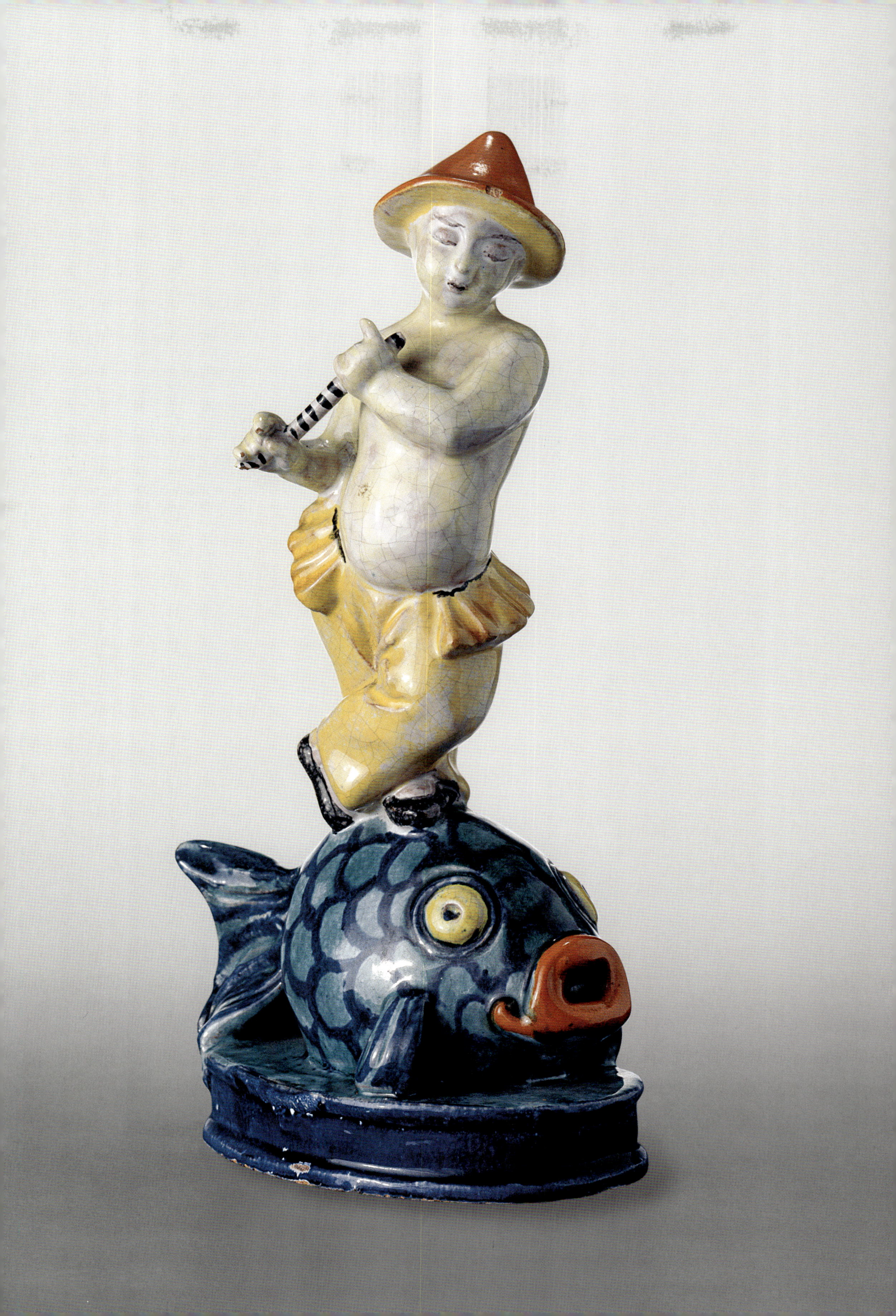

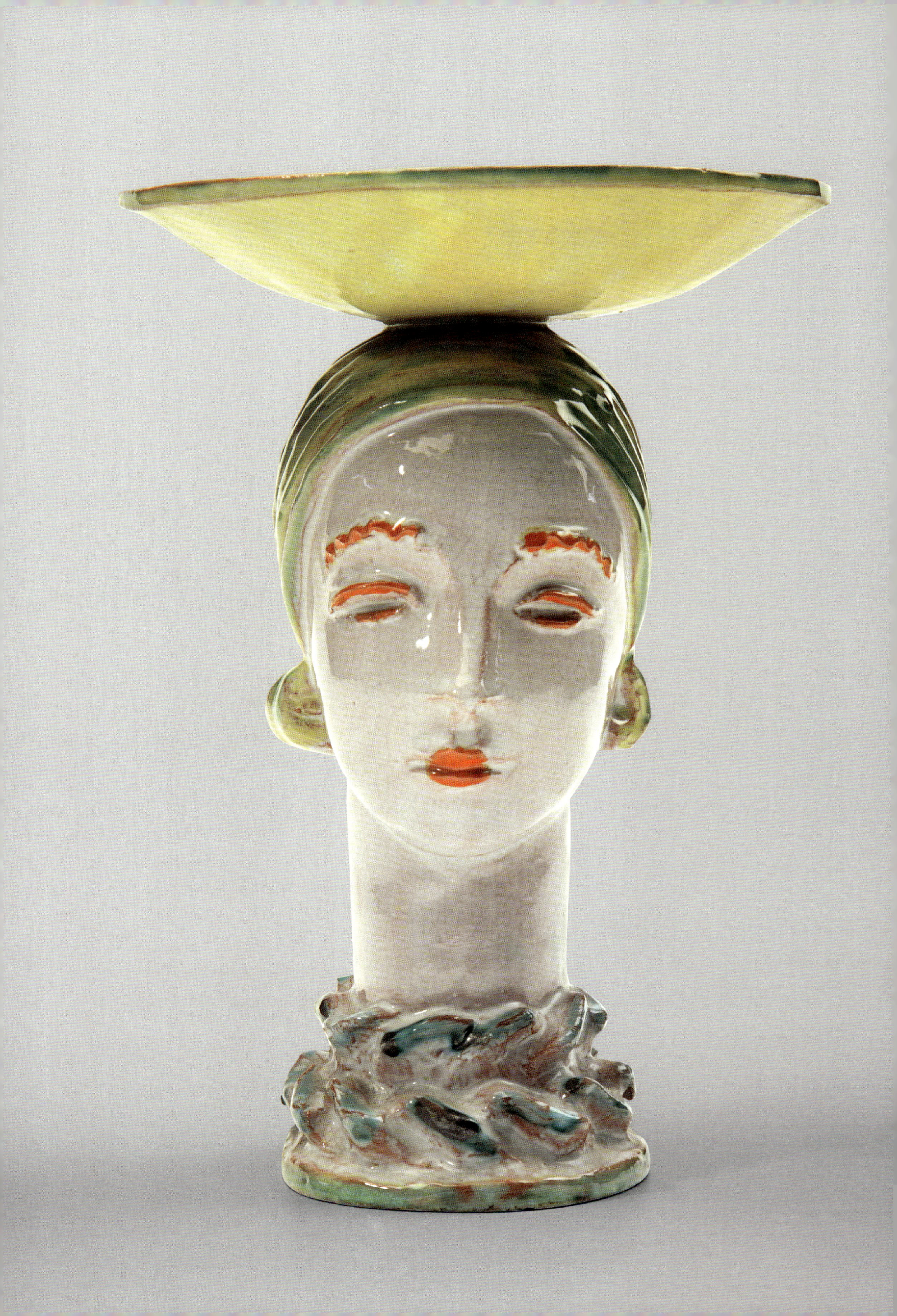

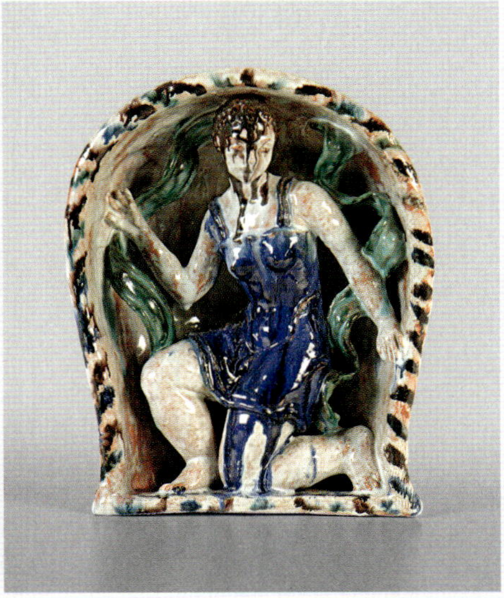

ModNr. 410 „Nische" I "Niche", Rudolf Knörlein, ca. 1929, © Wien Museum
Auf einer Entwurfszeichnung von Knörlein hat dieser das Modell auf 1931 datiert. Da dieses Exemplar aber den Stempel aus der Zusammenarbeit mit dem Österreichischen Werkbund trägt, ist es, auch wegen der Modellnummer, auf 1929 zu datieren.

Knörlein dated this model 1931 on a design sketch for it. However, since this exemplar is marked with the stamp used when Marcell collaborated with the Österreichischer Werkbund and because of the model number, it should be dated 1929.

At the 1929 exhibition "Das Bild im Raum" Marcell showed "Musikanten" ["Musicians"] ceramic figure designed by Hertha Bucher.[608] By that time, the importance of figurative ceramics for the interior decoration of the modern home was firmly established: "Figurative ceramics have almost entirely replaced sculpture in stone and bronze in our homes. But not just figurines; nowadays vases, decorative bowls and the like, even objects that are obviously utilitarian, are preferably made in ceramics. [...] Flowerpots made in ceramics are also very popular."[609] A terracotta head designed by Rudolf Knörlein (Model No. 412) is also mentioned for that year.[610]

in sehr freien, mitunter kühnen farbigen Bahnen."[607] Zu den beschriebenen Objekten zählen nicht nur der „Schalenträger" und die Figuren „Innehaltende" und „Niederblickende", es betrifft auch Lampenfüße, Vasen und Schalen, die von Marcell produziert werden.

Auf der Ausstellung „Das Bild im Raum" ist Marcell 1929 mit der von Hertha Bucher entworfenen Keramik „Musikanten" vertreten.[608] Die Bedeutung der figürlichen Keramik für die moderne Wohnungseinrichtung hat sich zu dieser Zeit bereits fest etabliert: „Die figürliche Keramik hat nahezu ganz die Plastik in Stein und Bronze aus unseren Woh-

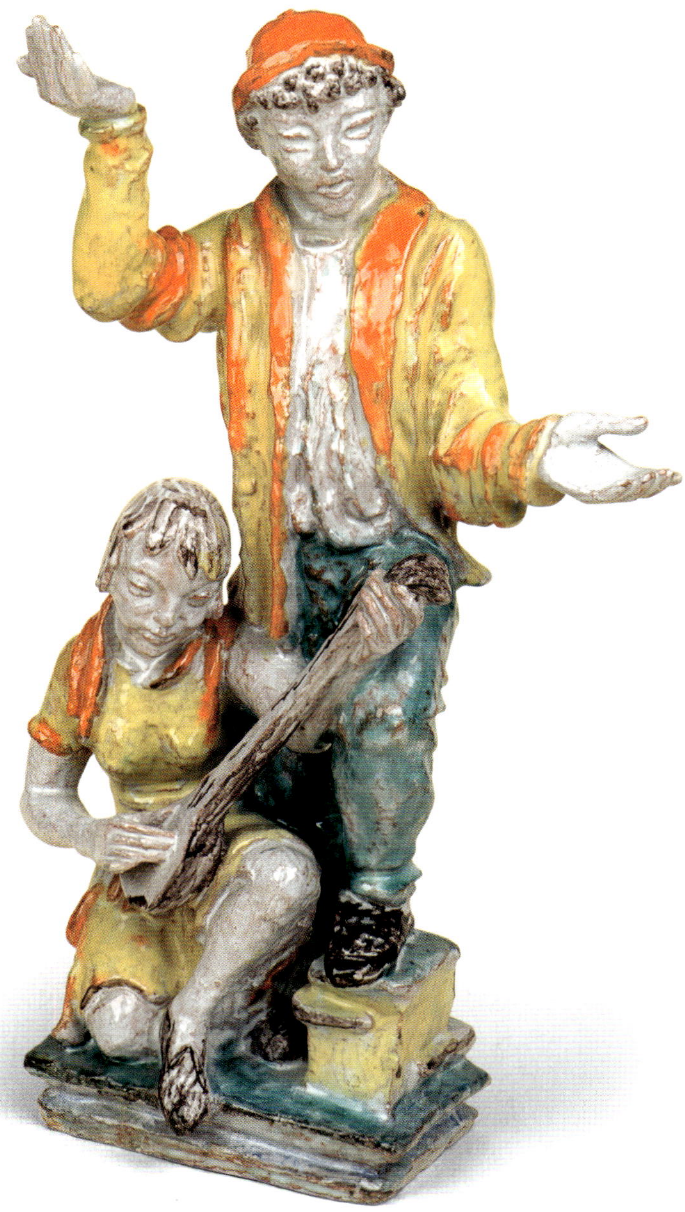

ModNr. 371 „Musikanten" I "Musicians", Hertha Bucher, ca. 1928/29, PB Wien 19

Zahlreiche Künstler, die Marcel aus der väterlichen Manufaktur kennt, entwerfen auch für sein neues, eigenes Atelier Modelle.

Numerous artists Marcell had known when his father ran the factory also designed models for Marcell's new studio.

links I left

ModNr. 400 Frauenkopf I Head of a Woman, ca. 1929, © Wien Museum

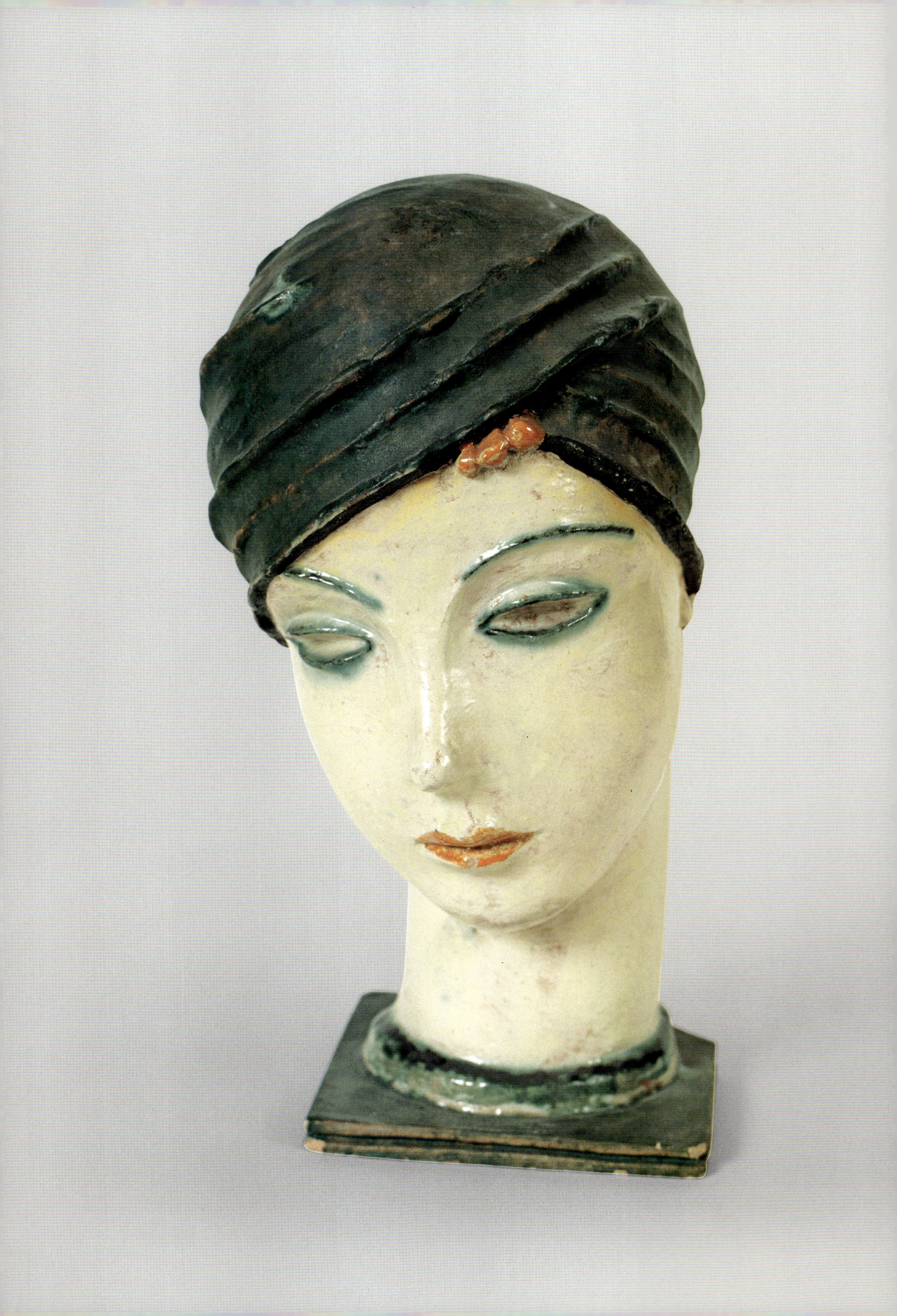

links | left

ModNr. 412
Stehkopf | Head, Rudolf Knörlein, ca. 1929, Sammlung.schmutz@chello.at
Der Bildhauer Rudolf Knörlein entwirft ab dem Jahr 1928 mehrere Modelle für Marcell Goldscheider, hier eine Terrakotta mit der Modellnummer 412. Im Jahr 1932 wechselt Knörlein zu Walter Goldscheider und wird dort künstlerischer Leiter.

From 1928, the sculptor Rudolf Knörlein designed models for Marcell Goldscheider; here a terracotta bearing the model number 412. In 1932, however, Knörlein transferred to Walter Goldscheider's business, becoming its artistic director.

nungen verdrängt. Aber nicht nur kleine figürliche Plastiken, Vasen, Zierschalen und dergleichen, auch ausgesprochene Gebrauchsgegenstände stellt man heute mit Vorliebe in Keramik her. [...] Auch Blumentöpfe, in Keramik hergestellt, erfreuen sich großer Beliebtheit."[609] Im gleichen Jahr wird auch ein von Rudolf Knörlein entworfener Terrakotta-Stehkopf (Modell-Nr. 412) erwähnt.[610]

Am 2. Mai 1930 erhält Marcell den Gewerbeschein zur „Erzeugung von kunstkeramischen Gegenständen und Erzeugung von plastischen Gegenständen aus Ton und Gips".[611] Er hat sowohl auf der Wiener Frühjahrs- wie auch auf der Herbstmesse einen Stand und stellt in der Nähe seines Bruders Walter aus, der mit zwei Ständen vertreten ist.[612] Im Künstlerhaus Wien stellt Marcell im Rahmen der Ausstellung „Buch und Raum der Gegenwart" vom Juni bis Oktober 1930 „künstlerische Keramiken" aus.[613] Auch Walter ist mit aktuellen Kunstgegenständen vertreten;[614] die beiden Brüder stehen wie schon auf den anderen Messen in Konkurrenz zueinander (Ill. 170).

Marcell erweitert seine Produktion um „Lampen aus Metall, Keramik, Holz sowie von Lampenschirmen" aus patentiertem Stoffglas, die er in seinem Musterlager am

On 2 May 1930 Marcell was granted a trade license for "producing art ceramic objects and producing sculptural objects of terracotta and plaster of Paris."[611] He had a stand at the Viennese Spring and Autumn Fairs near the two stands of his brother Walter.[612] Marcell also exhibited "art ceramics" at the "Buch und Raum der Gegenwart" exhibition that ran at the Künstlerhaus in Vienna from June to October 1930.[613] Walter was also there showing modern art objects;[614] as at so many other trade fairs, the two brothers vied with each other (Ill. 170).

Marcell broadened the scope of his product range to include "metal, ceramic, and wood lamps as well as lampshades" of patented textile-textured glass, which he presented at his samples warehouse at the Getreidemarkt.[615] In an advertisement, he touted not only "designs by leading Vien-

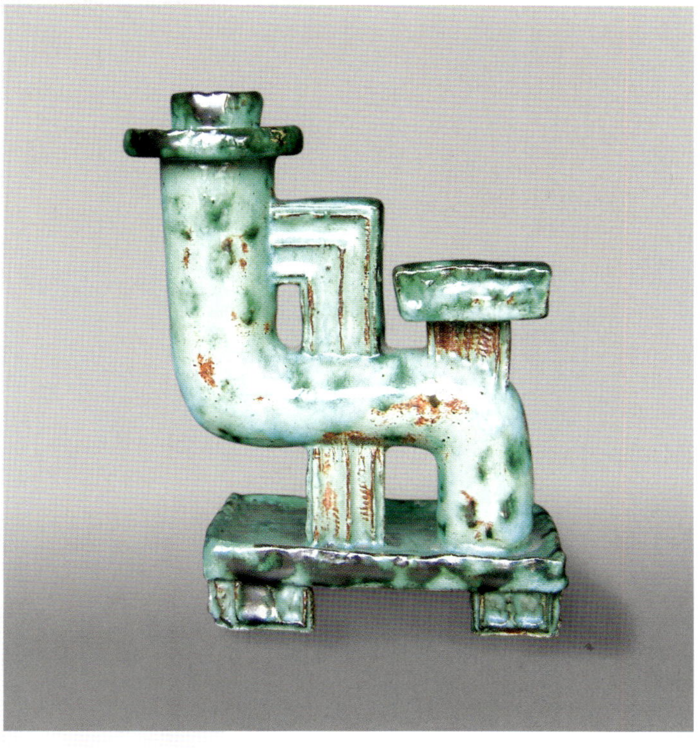

ohne | without ModNr.
Kerzenhalter | Candlestick, Rudolf Knörlein, ca. 1928–1932, PB Wien

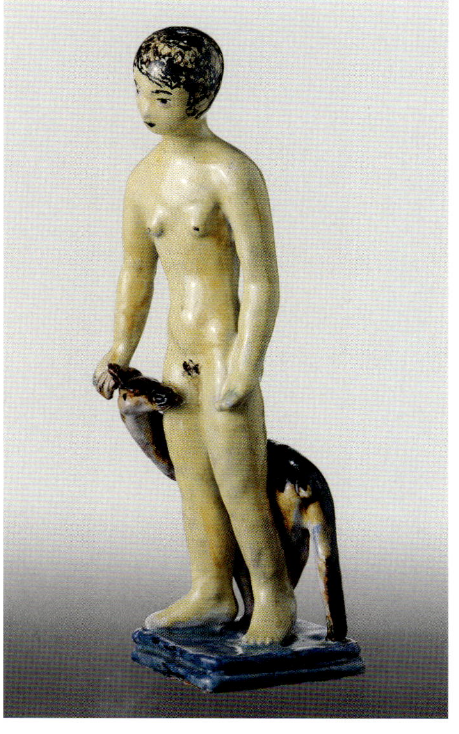

ModNr. 375
Akt mit Reh | Nude with Doe, Rudolf Knörlein, ca. 1929, Collection Dr. Peter Hieke, Wien

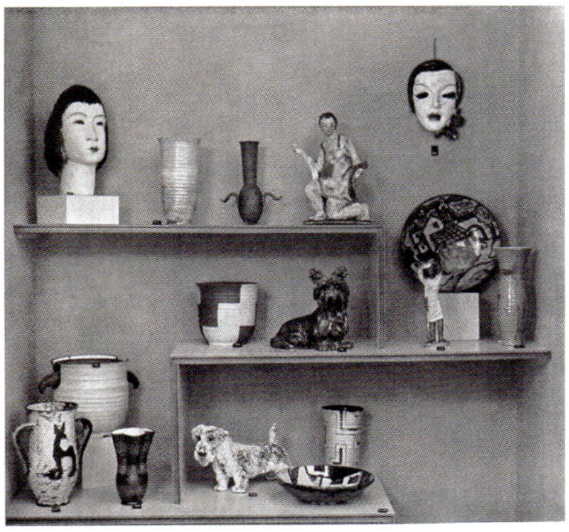

Ill. 170

Auf der Ausstellung „Buch und Raum der Gegenwart" stellen sowohl Walter wie auch Marcell Goldscheider in Konkurrenz zueinander neue Modelle vor. (Quelle: Deutsche Kunst und Dekoration, Band 67)

Both Walter and Marcell Goldscheider vied with one another in launching new models at the exhibition "Buch und Raum der Gegenwart." (Source: Deutsche Kunst und Dekoration, Vol. 67)

Ill. 171

In einer Anzeige aus dem Jahr 1930 wird anhand des englischen Firmentitels „United Studios of Art and Ceramic" auch die internationale Ausrichtung deutlich. (Quelle: Rochowanski, Führer durch das österreichische Kunstgewerbe 1930)

In a 1930 advertisement, the international scope of the firm is shown by the English translation of the firm name: "United Studios of Art and Ceramic." (Source: Rochowanski, Führer durch das österreichische Kunstgewerbe 1930)

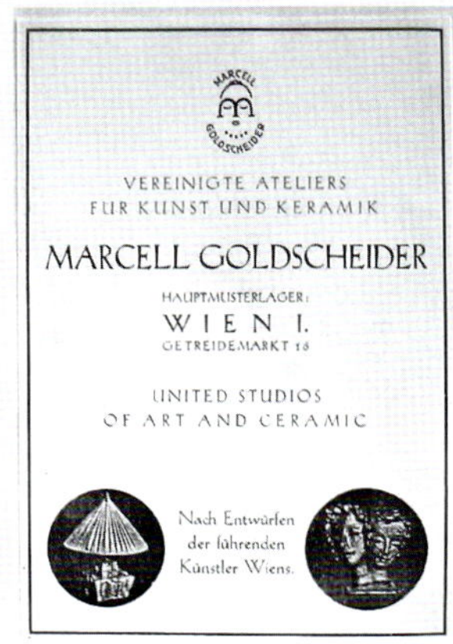

Getreidemarkt präsentiert.[615] In einer Anzeige preist er nicht nur die „Entwürfe der führenden Künstler Wiens" an, sondern übersetzt auch seinen Firmentitel ins Englische: United Studios of Art and Ceramic[616] (Ill. 171).

Für die Ausstellung „Blume und Plastik" im Jahr 1931 liefert Marcell bzw. die Vereinigten Ateliers „kostenlos" Ausstellungsarbeiten, vor allem Blumen- und Tontöpfe.[617] Doch schon im gleichen Jahr wendet Marcell die Bezeichnung „Vereinigte Ateliers für Kunst und Keramik" nicht mehr an. Er lässt sich nicht mehr als „Fabriksdirektor" betiteln, wie in den 20er Jahren, sondern schlicht als Keramiker, seine Arbeiten werden als „Marcell Goldscheider Keramik" veröffentlicht.[618]

Trotz der hohen Ablöse, die er von Walter bekam, sind Marcells finanzielle Mittel jetzt aufgebraucht. Seine unternehmerische Neugründung fällt in eine schwierige Zeit, denn die Weltwirtschaftskrise bricht aus und auch seine Geschäfte scheinen nicht gut zu laufen. Im Jahr 1932 verkauft er das Haus in der Hasenauerstraße 16[619] und zieht mit Rosemarie in die Schikanedergasse 12 im IV. Bezirk. Auch das Musterlager am Getreidemarkt wird aufgelöst, Rosemarie betreibt fortan ein Geschäft in der Friedrichstraße 3.[620]

nese artists" but also translated the name of his firm into English as "United Studios of Art and Ceramic"[616] (Ill. 171).

Marcell and the United Studios provided exhibits, mainly flowerpots and clay pots, for the 1931 exhibition "Blume und Plastik."[617] That same year, however, Marcell ceased to use the name "Vereinigte Ateliers für Kunst und Keramik." He no longer styled himself "factory director" as he had in the 1920s but rather presented his work to the public as "Marcell Goldscheider Keramik."[618]

Although Walter had paid a large sum to buy his brother out, Marcell's finances were now exhausted. He had founded his business at a difficult time because the worldwide economic crisis broke out in 1929 and his business, too, seemed to be adversely affected by it. In 1932 he sold the building at 16 Hasenauerstraße[619] and moved with Rosemarie to 12 Schikanedergasse in the IVth District. Furthermore, the samples warehouse at the Getreidemarkt was closed down and from then on Rosemarie ran a business at 3 Friedrichstraße.[620] In spring 1934 Marcell showed work at the "Austria in London" exhibition[621] and at "Das befreite Handwerk,"[622] a trade fair for the decorative and applied arts in Vienna. That same year he opened new "sales halls" at Kärntnertor.[623] At shop 23 there, Rosemarie sold "crafts" and Marcell used it as a samples warehouse for his "Ceramics Workshop."[624] By about

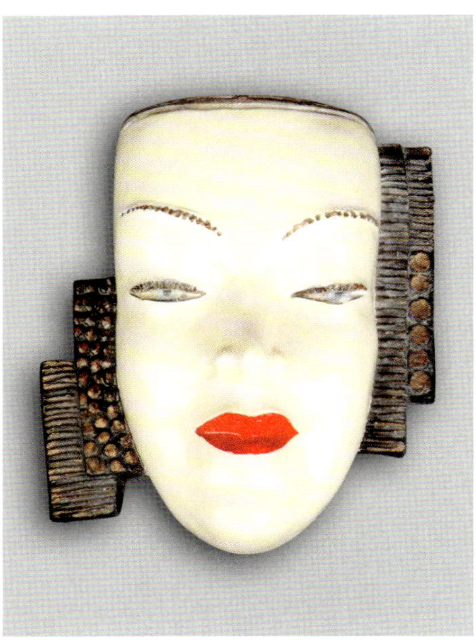

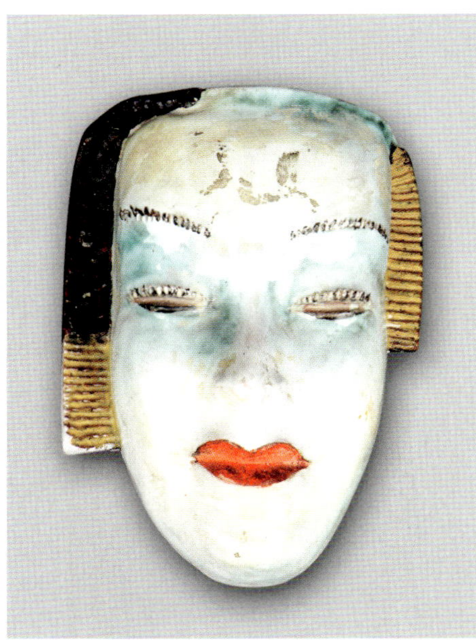

ohne | without ModNr. Wandmaske | Wall Mask, ca. 1928–1939, beide | both PB Wien 19

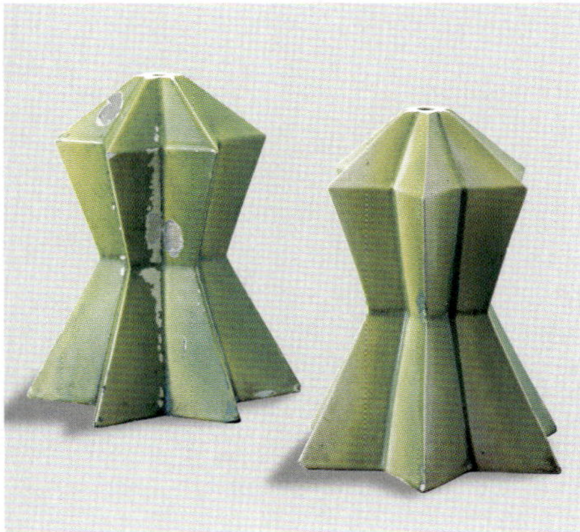

ModNr. 1013
Lampenfuß | Lamp base, ca. 1930–1939, beide | both Collection Dr. Peter Hieke, Wien

Im Frühjahr 1934 ist Marcell auf der Messe „Austria in London" vertreten,[621] auf der Kunstgewerbeausstellung „Das befreite Handwerk"[622] in Wien und er eröffnet im gleichen Jahr am Kärntnertor neue „Verkaufshallen".[623] Hier, im Laden 23, verkauft Rosemarie „Kunstgewerbe" und Marcell nutzt es als Musterlager seiner „Keramik Werkstatt".[624] Um 1937 verringert sich die Anzahl der bei Marcell tätigen Arbeiter auf nur noch zehn Personen.[625] Mit der Annektierung Österreichs wird auch Marcell Goldscheiders Arbeit immens erschwert. Am 31.8.1939 wird sein Gewerbeschein zurückgenommen,[626] bevor er einen Monat später offiziell erlischt.[627] Zu dieser Zeit befinden sich Marcell und Rosemarie jedoch nicht mehr in Wien.

1937, the workforce Marcell employed had dwindled to ten people.[625] The annexation of Austria made things enormously difficult for Marcell Goldscheider's work as well. On 31 August 1939, his trade license was revoked;[626] it had been due to expire a month from that date.[627] By that time, however, Marcell and Rosemarie were no longer in Vienna.

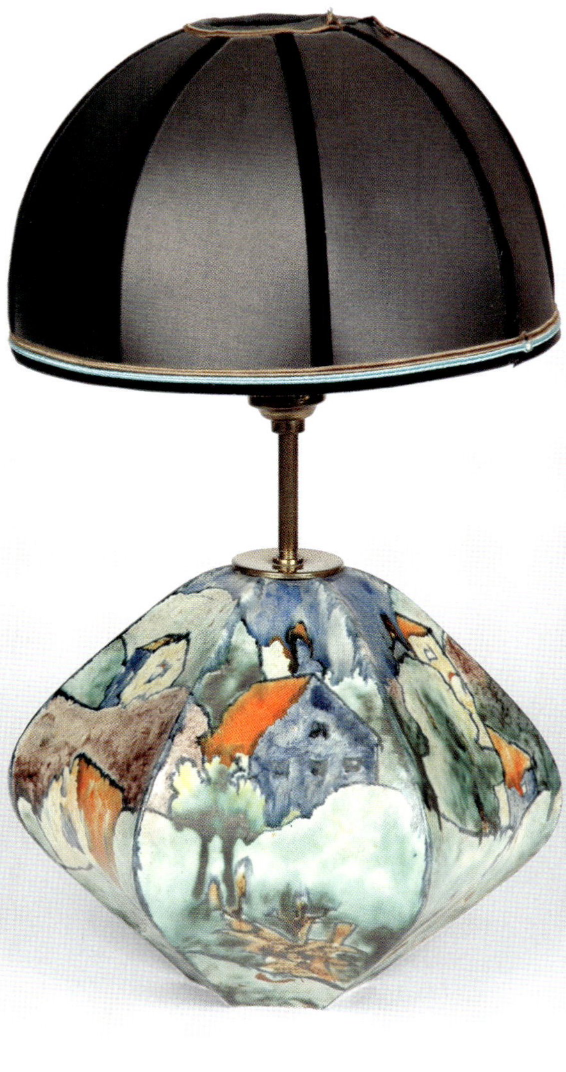

ModNr. 1024
Lampenfuß | Lamp base, ca. 1930–1939, PB Wien 19
Lampen werden zu einem wichtigen Bestandteil des Produktionssortiments.

Lamps became an important part of the product range.

ModNr. 1017
Vase, ca. 1930–1939, PB Wien 19

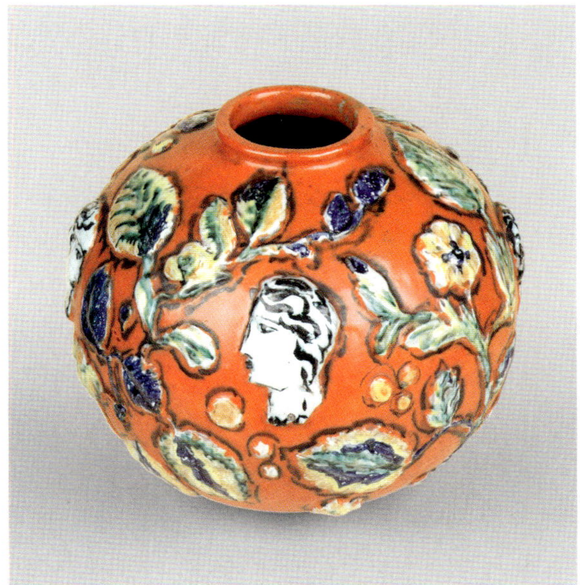

Emigration und Art Pottery in England (1939–1960)

Marcell und Rosemarie Goldscheider beschließen, nach England zu emigrieren, und verlassen Österreich Anfang 1939 gemeinsam mit ihrer Tochter Inge. In den Manufakturen in Staffordshire, die Marcel aus seiner Ausbildungszeit kennt, sehen sie die Möglichkeit, weiterhin auf hohem Niveau keramische Modelle produzieren zu können. Die technischen Bedingungen und die Facharbeiter sollen sogar besser als auf dem Kontinent sein.[628] Marcell gelingt es, vor dem Kriegsausbruch knapp 300 seiner besten Modelle nach England zu transportieren und somit den Grundstock einer Produktion zu sichern.[629]

In der Manufaktur Myott, Son & Co. findet Marcell 1940 Unterstützung. Die Brüder Myott stellen ihm für eine eigenständige Modell-Linie Produktionsräume auf ihrem Fabrikgelände, den Alexander Potteries in Hanley, einem Vorort von Stoke-on-Trent, zur Verfügung. Obwohl die Arbeitskräfte auf Marcells Stil umgelernt werden müssen, werden innerhalb von drei Monaten die ersten Modelle produziert,[630] die alle den Stempel „Made in England by Goldscheider

Emigration and Art Pottery in England (1939–1960)

Marcell and Rosemarie Goldscheider decided to emigrate to England and left Austria early in 1939 with their daughter, Inge. Marcell had trained at the Staffordshire potteries and the Goldscheiders saw a possibility there for continuing to produce high-quality ceramic models. The technical conditions and the skilled workmen were supposed to be even better there than on the Continent.[628] Before war broke out, Marcell managed to transport some three hundred of his best models to England to form the nucleus of a future product line.[629]

Marcell found backing in 1940 from Myott, Son & Co. The Myott brothers placed at his disposal rooms on their factory premises, the Alexander Potteries in Hanley, a suburb of Stoke-on-Trent, where he could

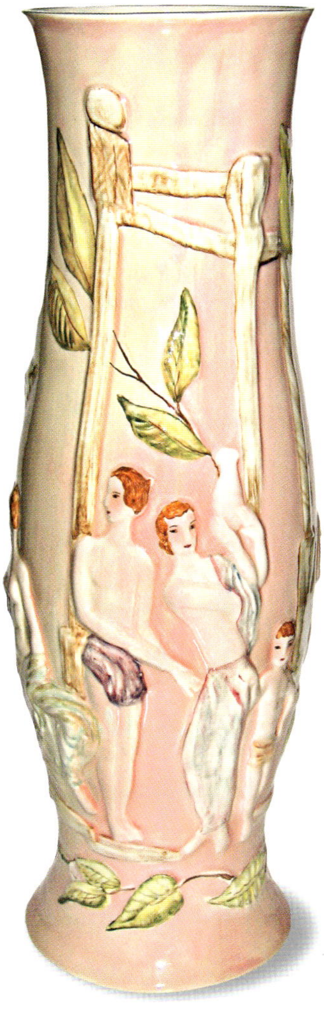

Ill. 172

Ill. 173

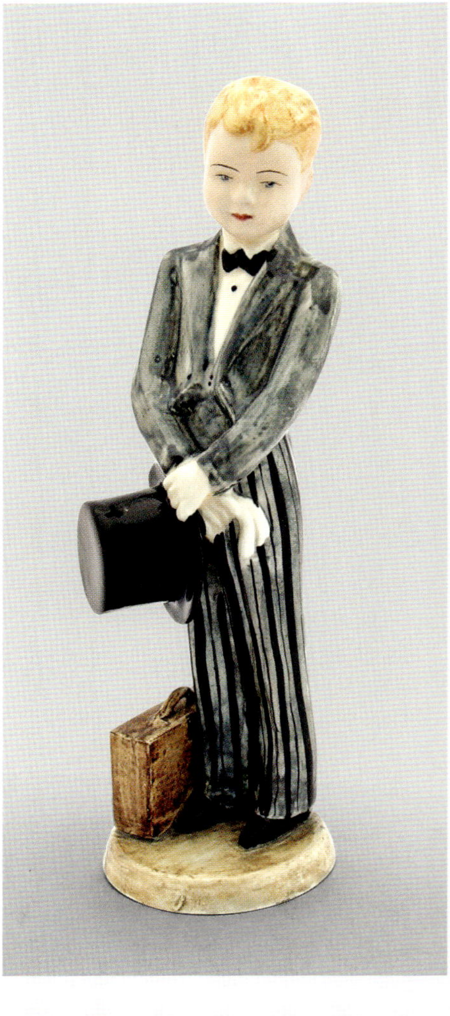

Ill. 172–175
In Zusammenarbeit mit Myott entstehen zahlreiche neue Modelle, z.B. eine Vase (ModNr. 352), die Figur eines jungen Kavaliers (ModNr. 753) oder die Figur des „George Formby in a Mexican Costume" (ModNr. 495 u. 496). In der Stempelung wird sowohl auf Myott wie auf Marcel Goldscheider verwiesen. (Sammlung a@goldscheider.net)

The collaboration with Myott saw the launch of numerous new models, including a vase (model number 352), a figure of a young Cavalier (model number 753) and a figure representing "George Formby in a Mexican Costume" (model numbers 495 u. 496). The marks stamped on referred to both Myott and Marcel Goldscheider. (Collection a@goldscheider.net)

produce an independent model line of his own. Although the employees had no idea of Marcell's style, within three months the first models were produced,[630] bearing the stamp "Made in England by Goldscheider with Myott Son & Co, Staffordshire." Marcel Goldscheider, who lopped an "l" off his name to spell it in the English way, announced that he planned to concentrate on designing new, exclusive pieces at his new place of work.[631]

Established around 1900 at Stoke-on-Trent, home to traditional potteries, Myott was run by the brothers Ashley and Sydney Myott. The most popular pieces were polychrome ones from the Art Déco period and they were exported worldwide.[632] For Myott & Son Co, which specialized in making vases, dishes and tableware, the collaboration with Marcel entailed broadening the scope of their range to embrace figurative ceramics and other ornamental models. After the outbreak of the Second World War, however, a government decree was issued that closed many potteries. Only a few factories now classified as "nucleus firms" were permitted to produce undecorated utilitarian household ceramics for the domestic market and decorated wares for the export market to

with Myott Son & Co, Staffordshire" tragen. Marcel Goldscheider, der seinen Vornamen um ein „l" auf die englische Schreibweise gekürzt hat, bekundet, vor allem neue und exklusive Arbeiten an seiner neuen Wirkungsstätte entwerfen zu wollen.[631]

Die Myott Manufaktur wird um 1900 im traditionsreichen Stoke-On-Trent gegründet und von den Brüdern Ashley und Sydney Myott geleitet. Vor allem die bunt bemalten Stücke aus der Zeit des Art Déco sind populär und werden weltweit exportiert.[632] Für die auf Vasen, Schalen und Geschirr spezialisierte Myott & Son Co. bedeutet die Zusammenarbeit mit Marcel eine Erweiterung ihres Sortiments um figurative Keramik und andere dekorative Modelle. Nach Ausbruch des Zweiten Weltkriegs müssen jedoch aufgrund eines Regierungserlasses viele Töpfereien schließen, nur wenige Manufakturen werden als „nucleus firms" klassifiziert und dürfen nichtdekorierte Gebrauchskeramik für den

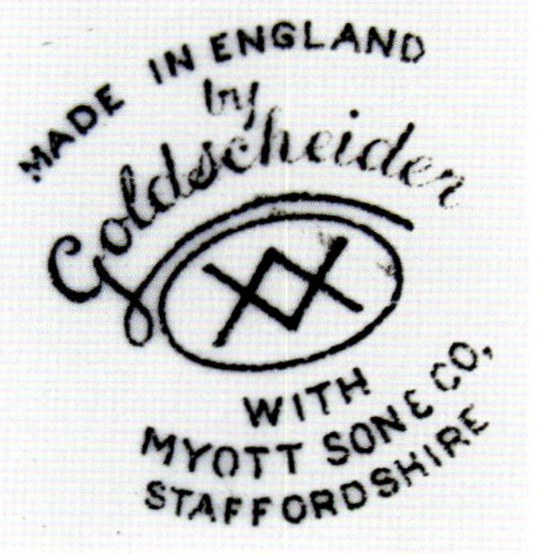

Ill. 174

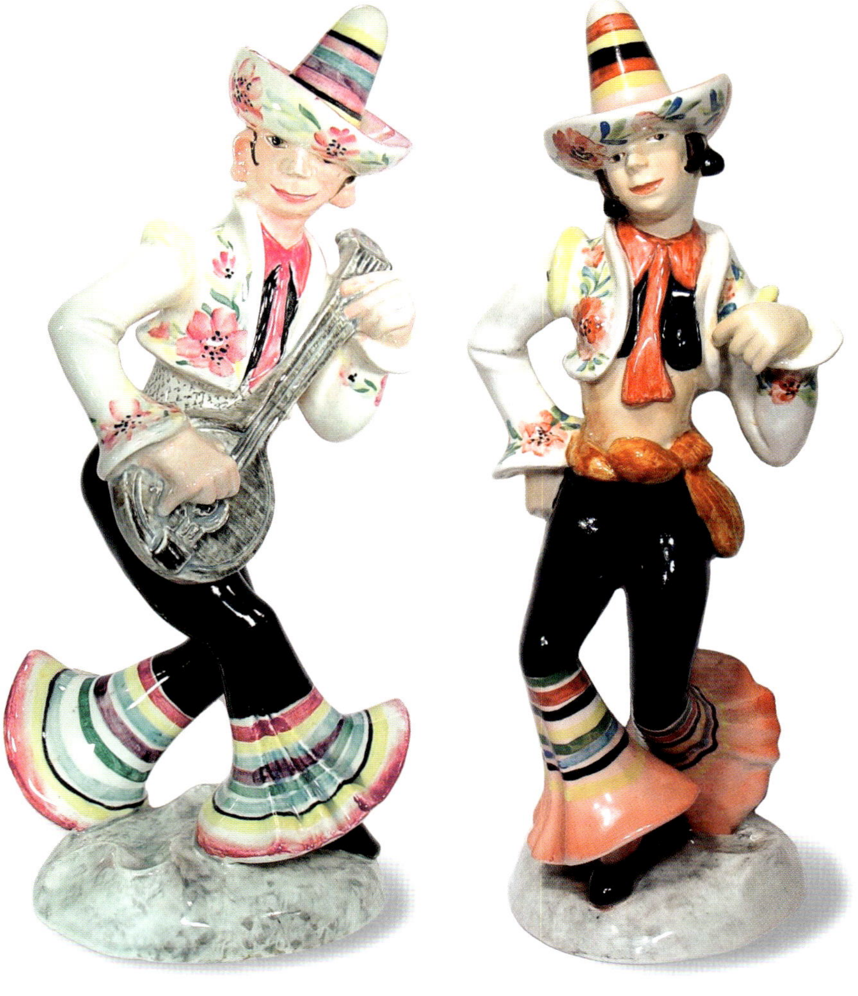

Ill. 175

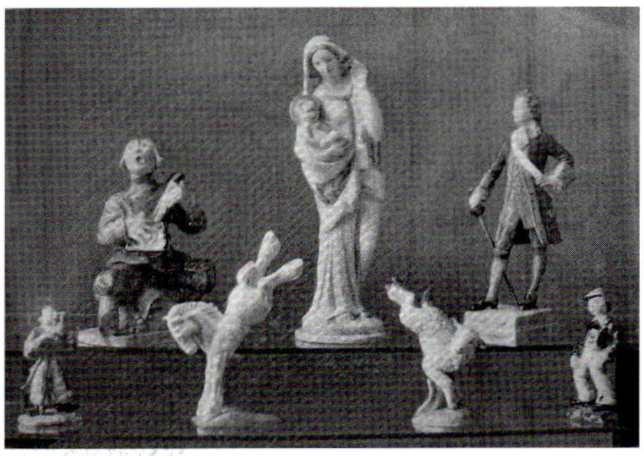

Ill. 176
Im Jahr 1942 umfasst Marcels Sortiment bereits 300 Modelle, darunter viele Tierfiguren sowie Figuren aus dem Bereich der Groteske, Geschichte und Religion. (Quelle: Pottery Gazette, Oct 1942, S. 565)

By 1942 Marcel's range encompassed three hundred models, including many animal figures as well as grotesque, historical and devotional pieces. (Source: Pottery Gazette, Oct. 1942, p. 565)

Heimmarkt sowie dekorierte Ware fürs Ausland zur Gewinnung von Devisen produzieren.[633] Auch Myott wird als „essentielles" Unternehmen eingestuft und Marcel kann deshalb seine Produktion während der Kriegsjahre fortführen.

Die Modellpalette besteht im Jahr 1941 aus Figuren, Tierskulpturen, Lampen, Wandmasken und Dekorationsobjekten für Ess- und Schreibtische. Alle hergestellten Objekte sind für den Export bestimmt, die figürliche Keramik insbesondere für den amerikanischen Markt.[634] Unter den Wandmasken befinden sich einige aus Österreich stammende Modelle im Stil Rudolf Knörleins sowie beispielsweise eine neue Wandmaske von Winston Churchill. Auch Kerzenleuchter und eine Zigarettenbox mit dem Relief eines Frauengesichts sind im Sortiment.[635] Ende 1942 umfasst das Sortiment bereits 300 Modelle und über 600 Muster. Die Figuren sind aus dem Bereich der Groteske, Geschichte und Religion und werden sowohl in farbreicher Dekoration verkauft wie auch weiß, matt oder in hellen Farben, um ein preiswerteres Segment abzudecken[636] (Ill. 176).

procure foreign currency.[633] Because Myott was classified as an "essential" business, Marcel was able to continue production even during the war years.

In 1941 the product range encompassed figurines, animal sculptures, lamps, wall masks and ornamental objects for the dining table and the desk. All objects made were destined for export, the figurative ceramics specifically for the American market.[634] The line in wall masks included some Viennese models in the manner of Rudolf Knörlein as well as new ones, for example a new wall mask of Winston Churchill. Candlesticks and a cigarette box sporting the face of a woman in relief were also made.[635] By late 1942, the range included three hundred models and more than six hundred samples. Grotesque, historical or devotional, the figures were sold featuring polychrome and white, matt and pastel-colored decoration schemes to cover the lower end of the market[636] (Ill. 176).

Within only a few years, Marcel's work enjoyed an excellent reputation in North and South America as well as on the British market. Moreover, as the Pottery Gazette wrote

Ill. 177

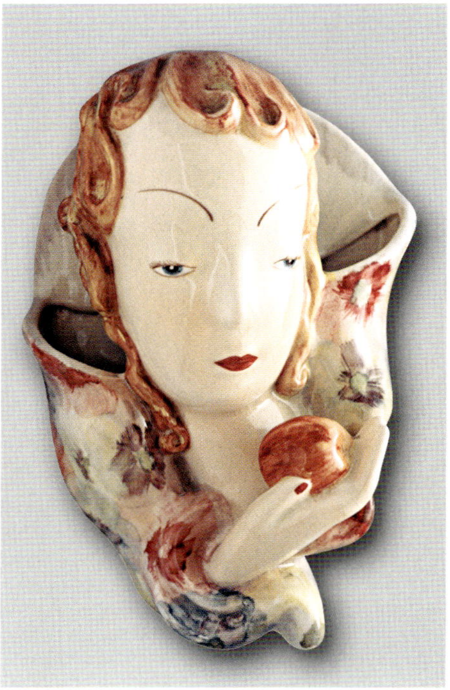

Ill. 178

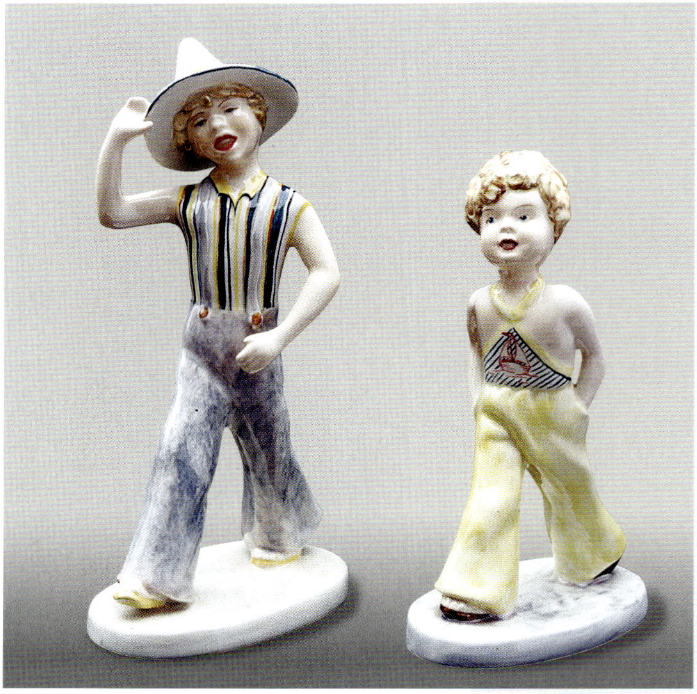

Ill. 177, 178
Etliche Modelle sind angelehnt an populäre Entwürfe der Wiener Epoche: eine Wandmaske im Stil Rudolf Knörleins (ModNr. 709) oder Figuren schreitender Kinder, die den Entwürfen Germaine Bourets bzw. Stephan Dakons nachempfunden sind (ModNr. 7039 in der Original-Produktion unter Walter Goldscheider in Wien). (Sammlung Anne Myott & Philip Pollitt)

Quite a few models borrowed from designs popular during Marcel's Viennese days: a wall mask in the manner of Rudolf Knörlein (model number 709) and striding figures of children representing an appreciation of designs by Germaine Bouret and Stephan Dakon (model number 7039 produced originally under Walter Goldscheider in Vienna). (Anne Myott & Philip Pollitt Collection)

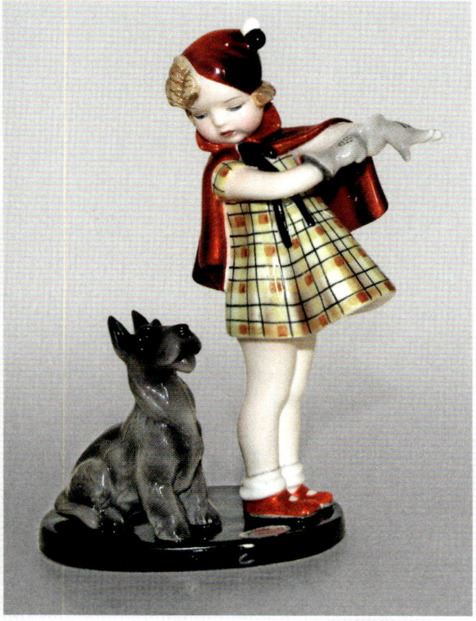

Ill. 179

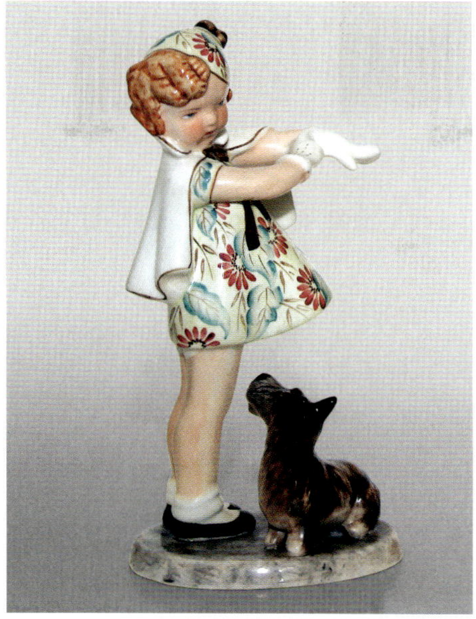

Ill. 180

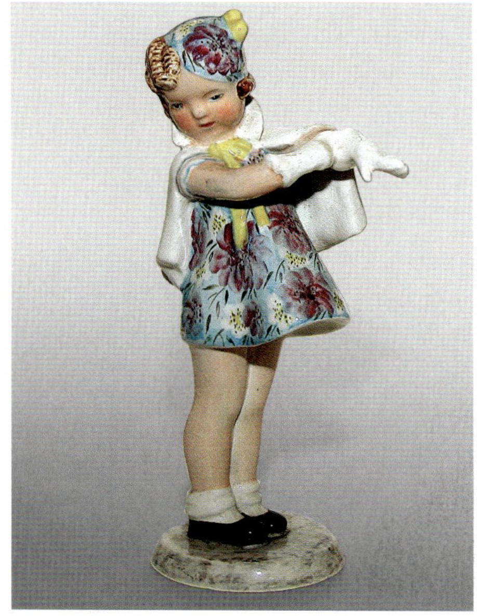

Ill. 181

in January 1945, the figures attested to "great technical daring."[637] A set entitled "Cries of Old London," for instance, comprised genre figures of women in various poses and Victorian dress as street-hawkers of various wares. They were executed in a new coloring technique (Ill. 182). Other models represented distinguished figures, usually of historic importance, vases with raised relief on the walls, bowls for fruits or salad and jugs and cups – all hand painted.[638] According to Marcel, the post-war product line was to be divided into two segments: first, "a wide variety of high-quality ceramics" and, second, lower priced objects that would cater for mass market tastes.[639]

Nach wenigen Jahren ist Marcels Arbeit sowohl in Nord- und Südamerika wie auch auf dem britischen Markt angesehen und wie die „Pottery Gazette" im Januar 1945 schreibt, sind die Figuren von „großartiger technischer Kühnheit".[637] Ein Set von Figuren unter dem Namen „Cries of Old London" zeigt beispielsweise verschiedene Damen bei diversen Tätigkeiten in der Kleidung des Viktorianismus, die in einer neuen Kolorierungstechnik entstanden sind (Ill. 182). Andere Modelle zeigen prominente, meist historische Persönlichkeiten, Vasen mit erhaben herausgearbeiteten Reliefs, Schüsseln für Früchte oder Salate sowie Krüge und Tassen – alle handbemalt.[638] Die Nachkriegsproduktion soll laut Marcel in zwei Bereiche aufgeteilt werden: erstens in „hochqualitative Keramik in großer Vielfalt" und zweitens in preiswertere Objekte, die den Massengeschmack treffen sollen.[639]

Ill. 179–181
Die aus dem Jahr 1936 stammende Kinderfigur „Komm mit!" von Germaine Bouret (links das Original mit der ModNr. 7664) wurde in der englischen Produktion kopiert und vereinfacht. (Sammlung Bernard Amigon, Frankreich)

This figure of a child dating from 1936 and entitled "Komm mit!" ["Come with me!"] by Germaine Bouret (left, the original with model number 7664) was copied and simplified for manufacture in England. (Bernard Amigon Collection, France)

Ill. 182
Im Jahr 1945 erscheint eine neue Reihe unter dem Titel „Cries of Old London". (Quelle: Pottery Gazette, January 1945, S. 37)

In 1945, a new range was launched under the title "Cries of Old London" (Source: Pottery Gazette, January 1945, p. 37)

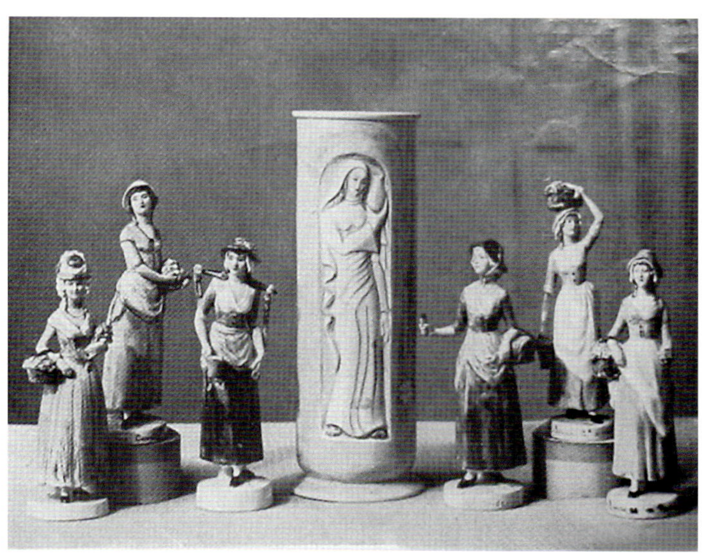

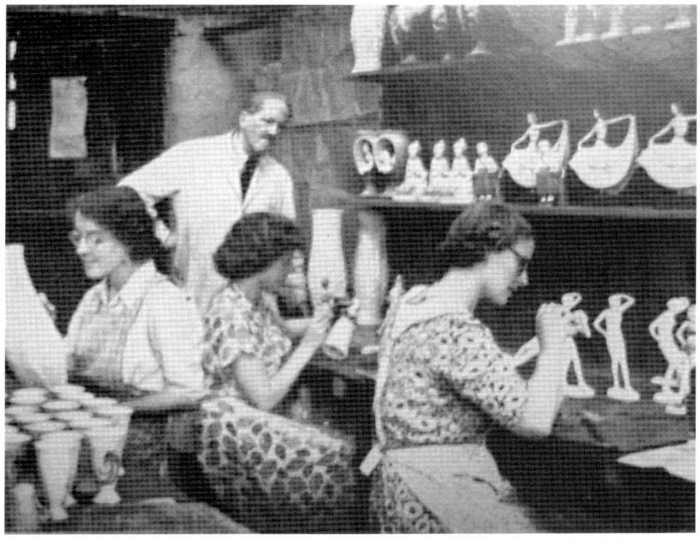

Ill. 183
Marcel Goldscheider beobachtet seine Malerinnen bei der Arbeit auf dem Myott-Gelände. (Quelle: Pottery Gazette, January 1949, S. 53)

Marcel Goldscheider watching the women painting for him on the Myott premises. (Source: Pottery Gazette, January 1949, p. 53)

Viele der bei Marcel beschäftigten Facharbeiter kommen von der Stoke-on-Trent Art School und werden von ihm im Bereich der Modellierung, der Abformtechniken und Dekorierung weitergebildet. Auch in England verfolgt Marcel das Ziel, die schöpferische Individualität des Künstlers mit einem umfassenden, fachmännischen Wissen der Herstellung zu verbinden und den Künstler in die vielfältigen Produktionsstufen einzubeziehen.[640] Die produzierten Modelle zeigen eindeutig die Einflüsse des Englischen Stils, was bei der fast ausschließlichen Beschäftigung von englischen Modelleuren und Künstlern nicht weiter verwundert. Einer der wenigen ausländischen Künstler, die bei Marcel Goldscheider in die Lehre gehen, ist Joseph Szeiler, der später eine eigene Manufaktur gründet.

Marcel glaubt auch, dass im Vergleich zu Tafelgeschirr der Markt für hochpreisige künstlerische Werke, so wie er sie anbieten will, limitiert ist. Solche Werke sollen nicht mit Preisnachlässen verkauft werden und auch nicht in Preiskonkurrenz mit Erzeugnissen aus Billiglohnländern treten. Die Qualität und der Preis sollen stimmig sein. Die Preis-

Many of the skilled workers Marcel employed came from the Stoke-on-Trent Art School and were further trained by him on the job in modeling, casting techniques and decoration. In England as in Vienna, Marcel pursued the aim of linking artists' individual creativity with a thorough specialist knowledge of manufacturing and to include artists in all the manifold stages of production.[640] The models produced unabashedly reveal the influence of the English style, which is unsurprising since only English modelers and artists were employed. One of the few non-British artists to serve an apprenticeship under Marcel Goldscheider was Joseph Szeiler, who later founded a factory of his own.

Marcel also believed that the market for high-priced artistic works of the kind he wanted to sell was limited in comparison to the market for tableware. Such upmarket works should not be sold at a discount nor should they compete with the prices charged for wares imported from countries that paid rock-bottom wages. Prices should be commensurate with quality. Items should be priced fairly to the purchaser's advantage.

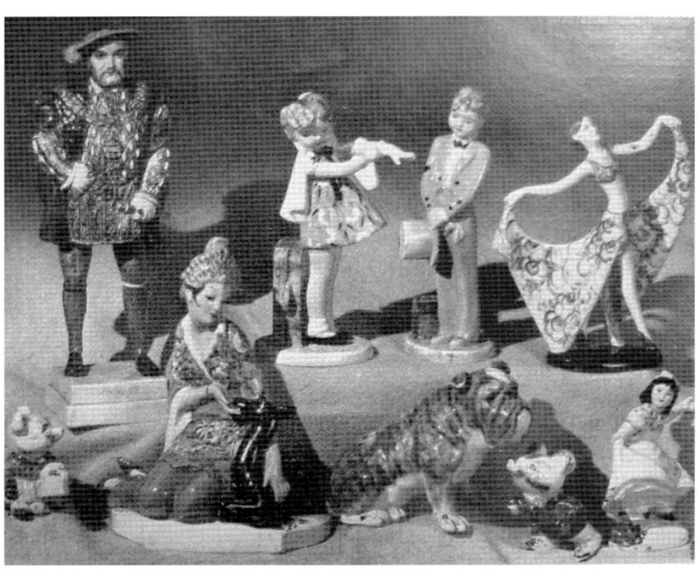

Ill. 184
Im Jahr 1949 umfasst das Sortiment bereits 750 Modelle, darunter sind etliche noch in Wien entstandene Modelle, z.B. die kniende Tänzerin mit Reh, das Mädchen mit Koffer und Handschuhen, der Junge mit Zylinder oder die Tänzerin rechts oben. (Quelle: Pottery Gazette, Jan. 1949, p. 53)

By 1949 the range comprised 750 models, including quite a few created in Vienna, such as a kneeling dancer with a doe, a little girl with a suitcase and gloves, a boy wearing a top hat and the dancer above right. (Source: Pottery Gazette, Jan. 1949, p. 53)

Marcel personally approved each piece that left his factory – quality control took place four times in the course of production. Goldscheider models were to radiate artistic individuality so that they could be distinguished at a glance from the bland objects made on a mass-production basis.[641]

The new products of the year 1947 included a colorful, comical "Lady Godiva" group, a "Nell Gwynne" figure designed by F. Halnon, animal figures such as elephants, birds and an ostrich group. Besides vases and ashtrays, the range included a reproduction in greenish black majolica of the head of a woman that had been bought by the king of Sweden at the 1931 Austrian Exhibition in Stockholm.[642] Other popular models were "Henry VIII," "Goldilocks' Bears" and figures from fairy-tales. Only two years later, in 1949, the range encompassed seven hundred and fifty models and Marcel not only showed work at the British Industry Fair (B.I.F.), which he had participated in for some time but also at the Copenhagen Fair. Besides the products made at Myott, Marcel also designed models himself that were made at his private studio[643] at 9 Moreton Parade, May Bank, Newcastle, where he lived with his wife Rosemarie, who helped him with the work.

On 24 August 1949, the Myott factory was gutted by fire.[644] On rebuilding the plant, the Myott owners decided that figurative and art ceramics would not be continued and that Myott would concentrate solely on producing their tableware, that enjoyed worldwide popularity.[645] In September 1950, the English specialist journal The Pottery and

bestimmung soll fair zum Wohle des Käufers und Verkäufers entschieden werden und Marcel begutachtet persönlich jedes Stück, das seine Manufaktur verlässt – im Laufe der Produktion erfolgt eine viermalige Qualitätskontrolle. Die Goldscheider-Modelle sollen künstlerische Individualität ausstrahlen und sich von den charakterlosen Objekten aus der Massenproduktion auf den ersten Blick unterscheiden.[641]

Zu den neuen Produkten des Jahres 1947 gehören unter anderem die farbenfrohe, humorvolle Gruppe „Lady Godiva", die von F. Halnon entworfene Figur „Nell Gwynne", Tierfiguren wie beispielsweise Elefanten, Vögel und eine Gruppe von Straußen. Neben Vasen und Aschenbechern wird auch in grünschwarzer Majolika die Reproduktion eines Frauenkopfs ins Sortiment aufgenommen, der 1931 vom schwedischen König auf der Stockholmer Österreich Ausstellung erworben wurde.[642] Weitere populäre Modelle sind: „Henry VIII", „Goldilocks Bears" sowie Figuren aus Märchen. Zwei Jahre später, 1949, umfasst das Sortiment 750 Modelle, und Marcel stellt neben der seit längerem besuchten British Industry Fair (B.I.F.) auch auf der Messe in Kopenhagen aus. Außer der Produktion bei Myott entwirft Marcel in dieser Zeit auch selbst Modelle, die in seinem privaten Studio entstehen.[643] Hier, in der Moreton Parade 9, May Bank in Newcastle wohnt er auch mit seiner Frau Rosemarie, die ihm bei der Arbeit assistiert.

Am 24. August 1949 wird die Myott Fabrik von einem Feuer zerstört.[644] Beim Wiederaufbau der Fabrik beschließen die Myott

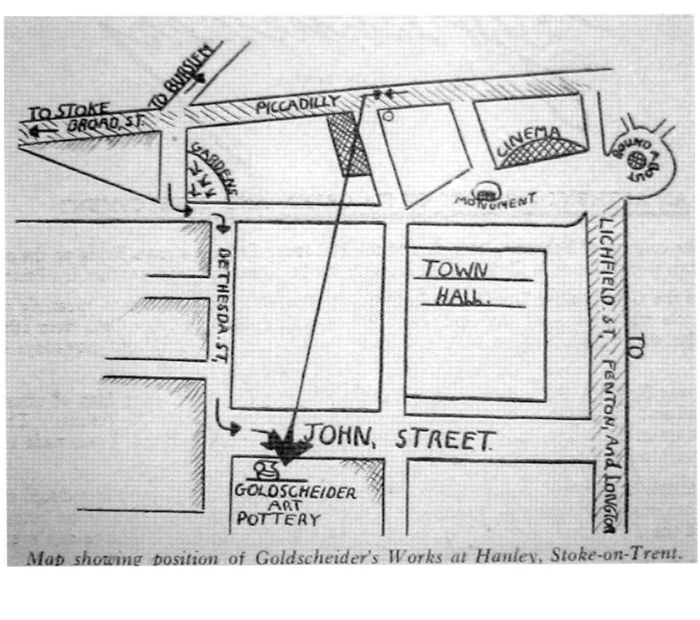

Ill. 185
Nach dem großen Feuer auf dem Myott-Gelände bezieht Marcel Ende 1950 eigene Räumlichkeiten in der John Street in Hanley. (Quelle: Firmenprospekt M. Goldscheider Staffordshire, um 1951)

After a devastating fire on the Myott premises, Marcel moved into premises of his own in John Street, Hanley. (Source: prospectus for the firm of M. Goldscheider, Staffordshire, ca 1951)

Ill. 186, 187
Nach der Gründung der eigenen Werkstatt werden mehrere Anzeigen geschaltet, die auf die selbständige Tätigkeit in den neuen Räumlichkeiten hinweisen. (Quelle: Pottery Gazette, 1950–1954)

After Marcel founded a workshop of his own, he placed several advertisements drawing the public's attention to the fact that he was now working independently on his own premises. (Source: Pottery Gazette, 1950-1954)

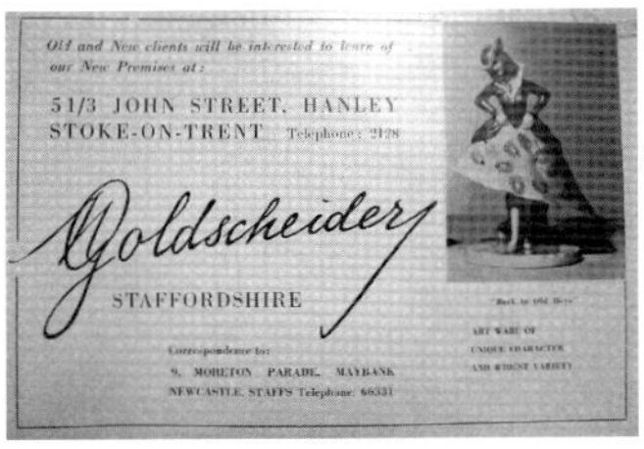

Ill. 186

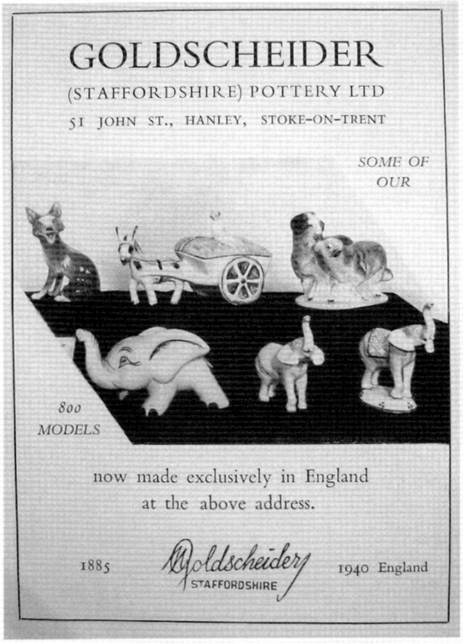

Ill. 187

Besitzer, dass der Bereich der figurativen und künstlerischen Keramik nicht fortgesetzt werden soll und man sich nur noch auf die Produktion der weltweit populären Geschirrservices von Myott konzentrieren will.[645] Im September 1950 wird daher in der englischen Fachzeitschrift „The Pottery and Glass Trade Review" bekannt gegeben, dass Marcel Goldscheider nach zehnjähriger Zusammenarbeit von nun an allein die Produktion in seinem Studio in Newcastle bestreitet[646] (Ill. 186 und 187). Im Dezember 1950 erscheint eine Anzeige, in der Marcels neue Räumlichkeiten in 51/3 John Street, Hanley, beworben werden. In dieser Anzeige ist auch der neue Goldscheider Schriftzug zu sehen, mit dem die Modelle gestempelt werden. Marcels Slogan lautet seiner geschilderten Anschauung entsprechend: „Art Ware of Unique Character and Widest Variety".[647]

In einem Brief informiert Marcel seine Kunden über die neuen Räumlichkeiten und Modelle. Dem Brief legt er ein Werbeprospekt bei, das die angebotenen Warenkategorien und eine Auswahl der inzwischen 800 Modelle enthält sowie eine Zusammenfassung der Geschichte seiner Manufaktur, die auch seine Firmen- und Schaffensphilosophie erklärt (Ill. 188). Die individuelle Behandlung der Objekte in der Produktion soll sich nach Marcel auch in einer individuellen Behandlung der Kundschaft spiegeln. So werden die Wünsche der Kunden bezüglich Ausarbeitung und Dekoration, wie schon in den Jahren zuvor, in der Herstellung voll berücksichtigt.[648] Viele der Modelle werden drei bis fünf Mal in elektrischen Öfen gebrannt. Auch kleine Porzellanfiguren (Bone China) werden

Ill. 188
Marcel Goldscheider informiert die Kundschaft Anfang der 1950er Jahre in einem Prospekt über seine neue Werkstatt in der John Street und über das Sortiment, das inzwischen knapp 800 Modelle umfasst. Er legt auch einen kurzen historischen Abriss des Familienunternehmens Goldscheider seit 1885 bei. (Quelle: Firmenprospekt M. Goldscheider Staffordshire, um 1951)

In the early 1950s, Marcel Goldscheider informed his clientele in a prospectus on his new workshop in John Street and on his product range, which by then encompassed some eight hundred models. He also included a brief historical outline of the Goldscheider family business since 1885. (Source: M. Goldscheider, Staffordshire, firm prospectus ca 1951)

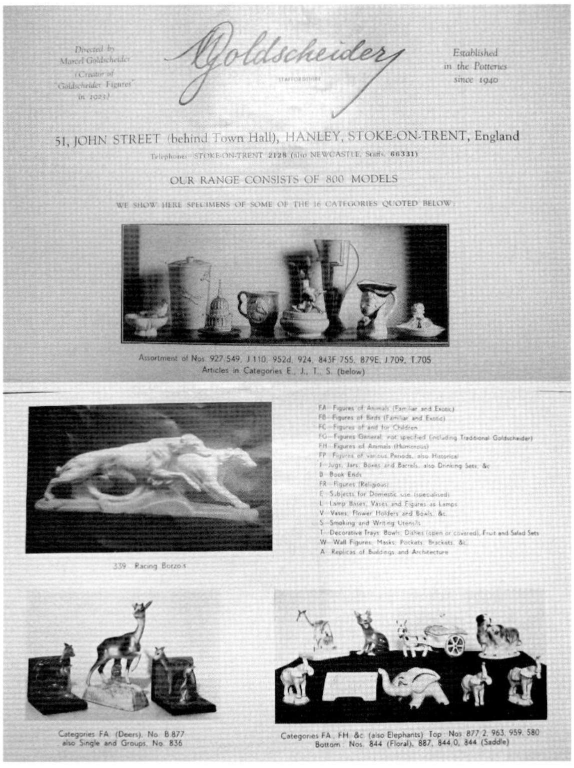

Glass Trade Review announced that Marcel Goldscheider, after collaborating with Myott for the past ten years, would thenceforward work alone from his own studio in Newcastle[646] (Ill. 186 and 187). In December 1950, an advertisement promotes Marcel's new rooms at 51/3 John Street, Hanley. The same advertisement shows the new Goldscheider mark in lettering, with which the models were stamped. Marcel's slogan reflects his thinking as outlined above: "Art Ware of Unique Character and Widest Variety."[647]

In a letter, Marcel informed his clientele of his new premises and models. He also enclosed a prospectus advertising the types of wares he was making and a selection of what by then amounted to eight hundred models as well as a brief history of his factory explaining the philosophy underlying his firm and his thinking on creativity (Ill. 188). The individual treatment accorded objects in production was, as Marcel saw it, to be reflected in individual treatment of his clientele. Client's wishes regarding the finish and decoration were faithfully complied with in manufacture, as they had been in previous years.[648] Many models were fired from three to five times in electric kilns. Small porcelain

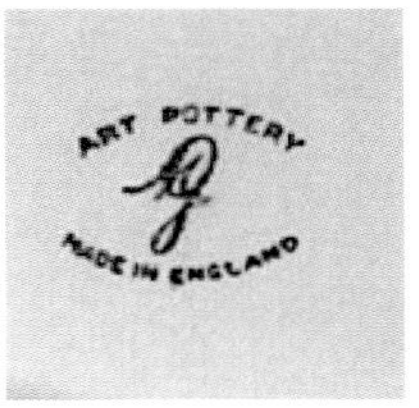

Ill. 189
Die neuen Modelle aus eigener Werkstatt erhalten einen neuen Stempel, der auf den Charakter der „Art Pottery" hinweist.

The new models from Marcel's own workshop were given a new stamp: "Art Pottery," to indicate the character of these products.

(bone china) figurines were announced, although Marcel took the opportunity of pointing out that white body was much more suited to making larger figures and art objects depicting expressive movement.[649]

Marcel, whose firm was registered as "Goldscheider Art Pottery" (Ill. 189), attracted attention with a new model series. Famous historic buildings such as "St Paul's Cathedral" (Ill. 191), "Windsor Castle," "The Tower of London" and "Tower Bridge" as well as buildings from works of literature such as "The Old Curiosity Shop" described by Charles Dickens (Ill. 192) were to be sold under the auspices of festive public ceremonies or simply as souvenirs. Marcel pointed out that such objects were not merely ornamental but were also "useful" since they could be made as snuff-boxes, fruit dishes or jewelry boxes.[650] Moreover, several figures representing characters from Dickens novels would be produced (Ill. 193–195). In addition, the figures in the new product line were to be larger, 24 inches high (approx 61 cm), and, finally, figures would be combined with lamps.

In June 1953, the name of the firm was changed and registered as "Goldscheider (Staffordshire) Pottery, Ltd." Marcel exported his "earthenware" and "bone china" models worldwide and displayed pieces at showrooms in London, Stockholm, Wellington, Montevideo, Paris, Toronto and Sydney. The Paris showrooms were at 29 rue de Paradis,[651] the very street in which his family business had had a bronze foundry for

angekündigt, obwohl Marcel gleichzeitig darauf hinweist, dass für größere Figuren und Kunstobjekte, die ausdrucksstarke Bewegungen zeigen, der weiße Scherben viel besser geeignet sei.[649]

Marcel, dessen Firma „Goldscheider Art Pottery" heißt (Ill. 189), macht auch mit einer neuen Modellreihe auf sich aufmerksam. Berühmte Gebäude wie die „St. Paul's Cathedral" (Ill. 191), „Windsor Castle", „Tower of London" und „Tower Bridge" sowie der von Charles Dickens beschriebene „Old Curiosity Shop" (Ill. 192) sollen im Rahmen von Festlichkeiten oder allgemein als Souvenirs verkauft werden. Marcel weist darauf hin, dass diese Objekte nicht nur dekorativ, sondern auch „nützlich" seien, da sie teilweise als Tabakdosen, Fruchtbehälter oder Schmuckkästchen eingesetzt werden können.[650] Nach Geschichten von Charles Dickens entstehen auch mehrere Figuren (Ill. 193–195). Außerdem soll in der neuen Produktion die Größe der Figuren auf 24 Inch (ca. 61 cm) erhöht werden sowie die Verbindung von Figuren und Lampen erfolgen.

Im Juni 1953 wird eine Umfirmierung bekannt gegeben und die Firma als „Goldscheider (Staffordshire) Pottery, Ltd." registriert. Marcel exportiert seine „earthenware" und „bone china" Modelle weltweit und stellt in Schauräumen in London, Stockholm, Wellington, Montevideo, Paris, Toronto und Sydney aus. In Paris sind die Schauräume in der Rue de Paradis 29,[651] der Straße, in der seine Familie jahrzehntelang mit einer Bronzefabrik ansässig war. Ihn vertretende

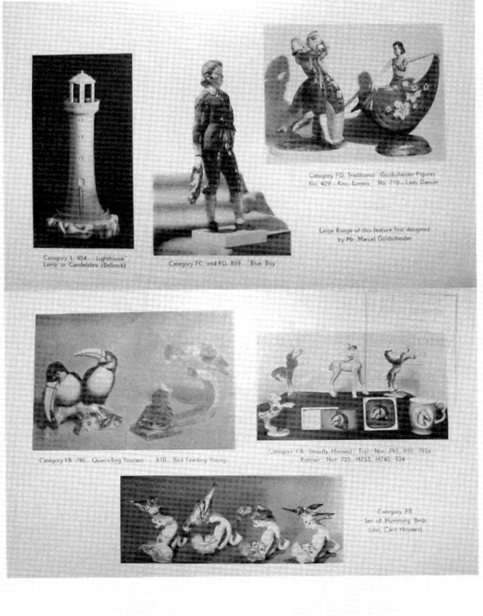

Ill. 190
Zahlreiche Tierfiguren werden um 1952 angeboten. Aber auch von Marcel Goldscheider in Wien selbst entworfene Figuren, wie die ModNr. 429 „Kiss", „Lovers" sowie die ModNr. 710 „Lady Dancer" und in England entworfene Modelle wie ModNr. 899 „Blue Boy", gehören zum Sortiment. (Quelle: Firmenprospekt M. Goldscheider Staffordshire, um 1951)

A great many animal figures were listed around 1952. Figures designed by Marcel Goldscheider in Vienna, including model number 429 "Kiss," "Lovers" and model number 710 "Lady Dancer" as well as models he had designed in England, such as model number 899 "Blue Boy," were included in the range. (Source: M. Goldscheider, Staffordshire, firm prospectus ca 1951)

Ill. 191
Das Modell der „St. Paul's Cathedral" gehört zu einer neuen Modellreihe, die berühmte Gebäude wiedergibt. (Quelle: Firmenprospekt M. Goldscheider Staffordshire, um 1951)

"St Paul's Cathedral" belonged to a new series of models representing famous historic buildings. (Source: M. Goldscheider, Staffordshire, firm prospectus ca 1951)

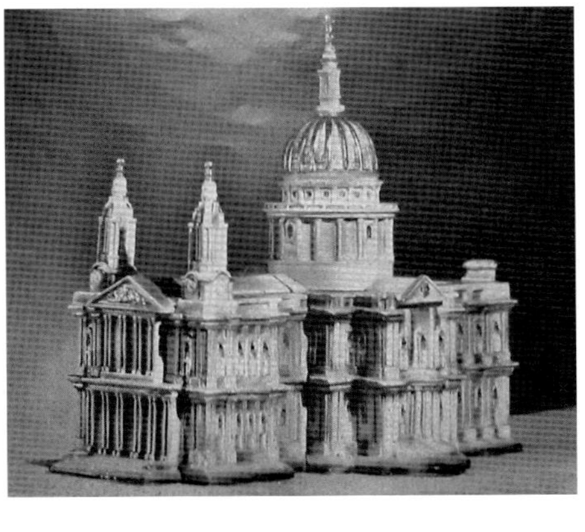

Agenten sind auch in Hongkong, Venezuela oder Kuba zu finden, wo er von Ernesto Samson vertreten wird, der schon vor dem Krieg sowohl die Modelle für Marcel wie auch Walter repräsentiert hat.[652]

Im Januar 1955 erscheint in der Fachzeitschrift „Pottery Gazette" ein interessanter Hinweis. Marcel macht hierbei vor allem Händler darauf aufmerksam, dass nur er berechtigt ist, so genannte „Goldscheider Modelle" zu produzieren, und dass die Marke Goldscheider eine geschützte Marke sei.[653] Gegen Ende des Jahres erscheint ein erneuter Hinweis, der speziell vor den aus Deutschland unter Carstens produzierten Modellen warnt.[654] Vermutlich waren die Brüder Walter und Marcel seit ihrer Trennung in den 1920er Jahren so zerstritten, dass Walter ihn nicht über den Verkauf der Namensrechte an Carstens informiert hat.

Mitte der 1950er Jahre ist die keramische Industrie in einer Krise. Billige Kopien aus Japan und anderen Ländern erobern den Markt, die Exportumsätze gehen zurück und auch die Erzeugnisse aus Plastik werden als Bedrohung betrachtet. Der Markt wird als stagnierend bezeichnet, wegen der erhöhten

decades. Agents also represented Marcel's firm Hong Kong, Venezuela and Cuba, notably Ernesto Samson, who had represented the models for Marcel and Walter before the Second World War.[652]

An interesting clue appeared in the January 1955 issue of the specialist journal Pottery Gazette. Marcel pointed out to dealers in particular that he was the sole holder of rights to reproduce what were called "Goldscheider Models" and that the Goldscheider brand was a registered trademark.[653] Towards the close of that year, a reminder of that same point was published, warning specifically about models produced in Germany under Carstens.[654] Presumably the brothers Walter and Marcel had been so completely out of touch since their ways had parted in the 1920s that Walter had neglected to inform Marcel the rights to the name had been sold to Carstens.

By the mid-1950s, the pottery industry was in the throes of a crisis. Cheap imitations from Japan and other countries were flooding the market; export sales declined and even products made of plastic were viewed as a threat. The market was de-

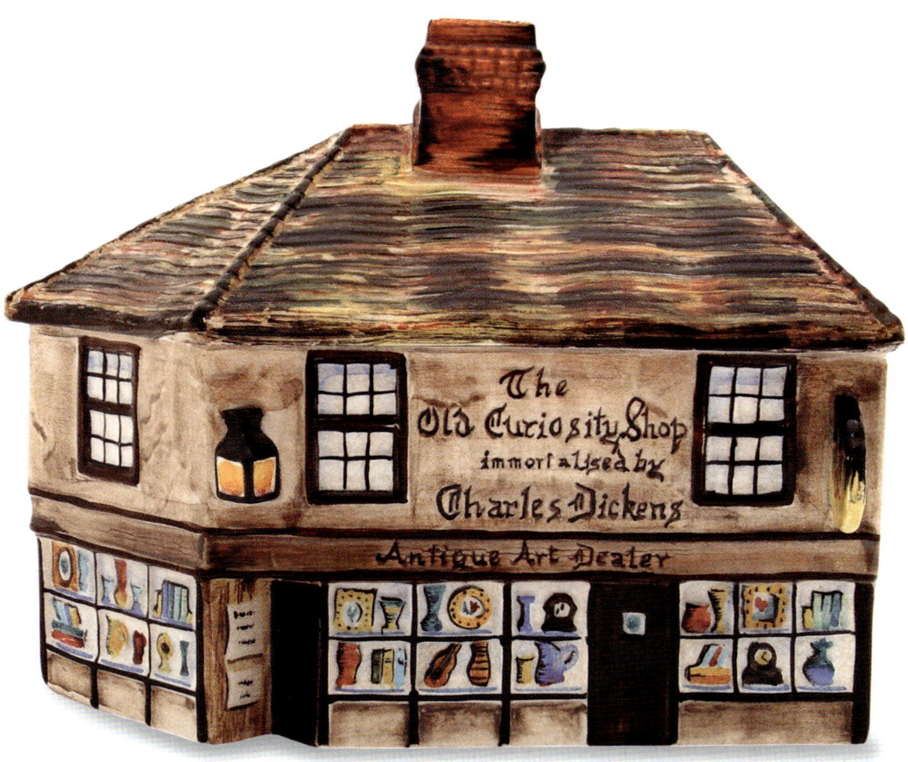

Ill. 192
Der „Old Curiosity Shop" (ModNr. 942) aus einer Geschichte von Charles Dickens. (Quelle: Sammlung a@goldscheider.net)

The "Old Curiosity Shop" (model number 942) from the Charles Dickens novel of that title. (Source: Collection a@goldscheider.net)

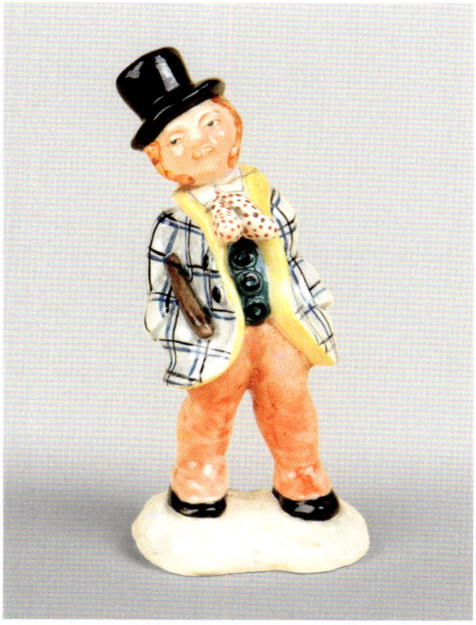

Ill. 193

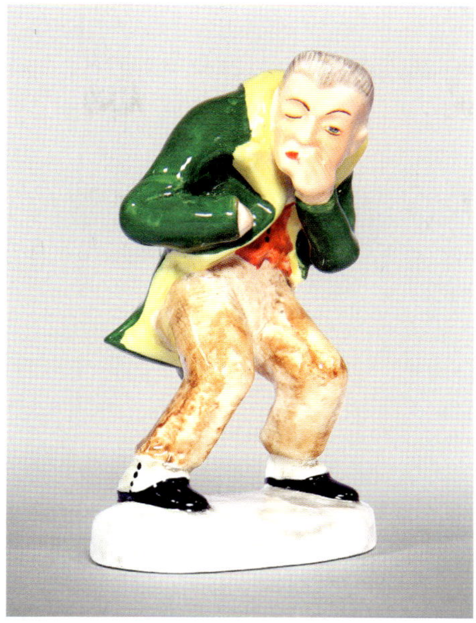

Ill. 194

Ill. 193–195
Nach Geschichten von Charles Dickens entsteht auch ein Set von Figuren, z.B. der abgebildete „Barker" (ModNr. 872) oder „Mr. Squeers" (ModNr. 914). (Quelle: Firmenprospekt M. Goldscheider Staffordshire, um 1951; Figuren aus Sammlung a@goldscheider.net)

A set of figures representing characters from Dickens novels was also produced, including "Barker" (model number 872) and "Mr. Squeers" (model number 914) shown here. (Source: M. Goldscheider, Staffordshire, firm prospectus ca 1951; figures from Collection a@goldscheider.net)

scribed as stagnating. Bone china sales especially were hit by tax increases.[655] During those parlous times, Marcel reduced his presence at trade fairs, which was probably also due to old age. At almost seventy, he was tired of traveling. In 1955 the only trade fair he participated in was the annual Harrogate fair specializing in gift articles.[656] At the trade fair in Blackpool a year later, Marcel's bookends with koala bears, figures of animals in sgraffito design and lamps with figures received a good press.[657] The new bone china figures were also shown there.[658] In January 1957 Marcel Goldscheider reduced exports worldwide and terminated contracts with some of his agents.[659] He was no longer present with articles at the 1958 "International Gifts & Fancy Goods Fair" in Blackpool. He is no longer listed as a producing potter in the 1960 specialist The Pottery Gazette and Glass Trade Review Reference Book and Directory. Marcel Goldscheider died at Worthing, West Sussex, on the south coast of England on 23 April 1964.[660]

Steuern leidet vor allem der Absatz von Bone China Figuren.[655] Die Teilnahme auf Messen reduziert Marcel in dieser schwierigen Zeit wahrscheinlich auch altersbedingt, denn mit fast 70 Jahren ist er des Reisens zu müde. Er nimmt 1955 nur noch an der spezialisierten jährlichen Messe für Geschenkartikel in Harrogate teil.[656] Auf der ein Jahr später in Blackpool stattfindenden Messe werden insbesondere die Buchstützen mit Koalabären, Tierfiguren im Sgraffito Design und Lampen mit Figuren gut aufgenommen.[657] Ausgestellt werden dort auch die neuen Bone China Figuren.[658] Im Januar 1957 reduziert Marcel Goldscheider den weltweiten Vertrieb und löst die Verträge mit einigen seiner Agenten auf.[659] Auf der 1958 stattfindenden „International Gifts & Fancy Goods Fair" in Blackpool ist er nicht mehr vertreten und im Jahr 1960 ist er auch nicht mehr in dem Fachorgan „The Pottery Gazette and Glass Trade Review Reference Book and Directory" als produzierender Töpfer eingetragen. Marcel Goldscheider stirbt am 23. April 1964 in Worthing (Sussex), an der Südküste Englands.[660]

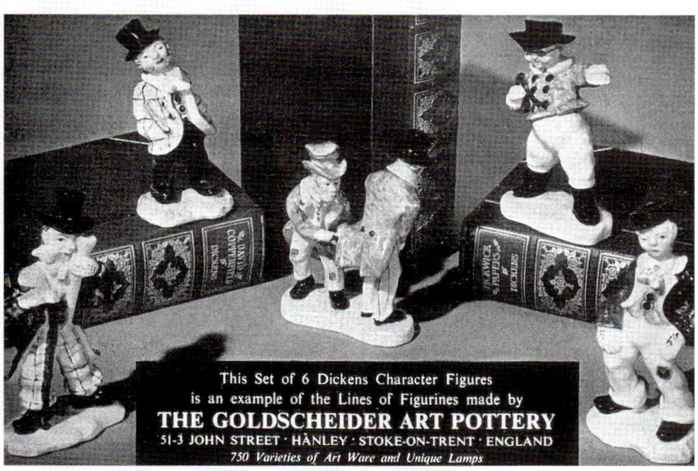

Ill. 195

Anmerkungen

[596] Patentschrift 106210 (Patent beginnt am 15.12.1926), Österreichisches Patentblatt vom 25.4.1927, sowie Patentschrift 107568 (dieses Zusatzpatent beginnt am 15.5.1927), Österreichisches Patentblatt vom 25.10.1927.
[597] Lehmann's Adressbuch, Jahrgänge 1928 bis 1932. Rosemarie Goldscheider führt das Atelier „Parivienna" bis um 1932.
[598] Lehmann's Adressbuch, 1929
[599] IC, 1929/30, S. 1589
[600] Gewerbeschein 7984/fr/XIX, Wirtschaftskammer Österreich-Archiv, Gewerbearchiv
[601] Lehmann's Adressbuch, Jahrgänge 1930 bis 1934
[602] Anzeige der Firma „Vereinigte Ateliers für Kunst und Keramik" in: Rochowanski (1930)
[603] Neuwirth (1974), S. 47
[604] Ebenda, zitiert nach Neuwirth (1974), S. 48
[605] Weihnachtsschau im Künstlerhaus, Kunst und Kunsthandwerk, November – Dezember 1928, S. 36
[606] Heizer (1929), S. 714ff
[607] Ebenda
[608] Das Bild im Raum. II. Ausstellung des Verbandes Bildender Künstlerinnen und Kunsthandwerkerinnen: Wiener Frauenkunst. Österreichisches Museum für Kunst und Industrie Wien, Februar – April 1929, S. 50. Auf Seite 79 erscheint eine Anzeige von Marcell mit einer Abbildung des „Schalenträgers".
[609] Ebenda, S. 36. In diesem Ausstellungskatalog wird auch das Verlangen nach Ästhetisierung des Wohnraums und der geistigen Stimulation dokumentiert. So ist einleitend auf Seite 5 zu lesen: „So sehr es Aufgabe unserer Zeit ist, uns selbst zu rationalisieren, unsere Verstandeskräfte immer mehr in den Dienst des tätigen Lebens zu stellen und durch zielbewusste Übung ein Gehirn zu erreichen, das mit der Präzision eines elektrischen Schaltwerks arbeitet, so sehr ist es auch Aufgabe, den inneren Menschen zu immer größerer Vollkommenheit zu bringen, das Reich der Seele in uns aufzurichten, um so in antithetischer Steigerung wahre Selbstkultur zu erringen. Ist es nun selbstverständliche Voraussetzung, daß unser Heim in einer Hingabe, wie es frühere Zeiten nie gekannt haben, dem Wohnzweck dienstbar gemacht werde, so verlangen wir, auf dieser Grundlage aufbauend, von unserem Heim noch Höheres. Denn das Heim ist ja der Ort unserer inneren Erneuerung, der Ort, wo wir ganz dem Reinmenschlichen leben. Aber nicht in einem ästhetisierenden Feinschmeckertum im Sinn des fin de siècle soll es dienen, das innere Ich gestalten zu helfen, ist seine Aufgabe. In der Art der Einrichtung eines Zimmers, in der Verteilung der Einrichtungsstücke, in den Formen der Möbel, in den Farben, die ein Zimmer beherrschen, vermögen wir einen Raum mit jenem geistigen Ausdruck zu beseelen, zu dem sich unser Ich in seinen edelsten Stunden aufzuschwingen vermag." Und auch die moderne Frau erfährt auf Seite 25 die Vorteile einer modernen Hausbibliothek: „Ein Zimmer, das wir uns als kleine Hausbücherei einrichten, wird vorwiegend Bücherstellen aufzunehmen haben. Die Bücherstellen werden wir nun nicht bis zum Himmel reichen lassen – das Bücken wäre zu unbequem – sondern stellen sie auf hohe Beine, die überdies den Vorteil haben, daß man unten gut durchfegen kann."
[610] Rochowanski (1929), S. 430
[611] Gewerbeschein 18334/fr.XVIII, Wirtschaftskammer Österreich-Archiv, Gewerbearchiv. Die Adresse der Anmeldung erfolgt auf die Währingerstraße 115.
[612] Frühjahrsmesse Wien, Amtliche Ausgabe, 9. bis 15. März 1930, S. 76 sowie Herbstmesse Wien, Amtliche Ausgabe, 7. bis 13. September 1930, S. 22ff
[613] Buch und Raum der Gegenwart. Ausstellungskat., Künstlerhaus Wien, Juni – Oktober 1930, S. 48
[614] Weihnachtsschau im Künstlerhaus, Kunst und Kunsthandwerk. Ausstellungskat., 23.11.1929–6.1.1930, S. 31 und S. 36
[615] IC, 1931/32, S. 1569
[616] Anzeige in Rochowanski (1930). In den 30er Jahren entsteht aus einer Zusammenarbeit zwischen Marcell und der Keramikschule in Teplitz eine Katzenschüssel, die mit einem U.A.T.-Stempel sowie Atelier Goldscheider gemarkt ist (gesichtet im Wiener Kunsthandel).
[617] Blume und Plastik. Ausstellungskat., Künstlerhaus Wien, 10.4.–3.5.1921, S. 7. Walter ist auf dieser Ausstellung mit einer großen Blumenvase vertreten.
[618] Lehmann's Adressbuch, Jahrgänge 1932 bis 1939
[619] Brunnbauer (2006). Das Haus wird an Max und Rifka Demuth verkauft.
[620] Lehmann's Adressbuch, Jahrgänge 1935 bis 1938. Das Geschäft befindet sich ebenfalls im IV. Bezirk.
[621] Austria in London. Dorland Hall, April 16 to May 12, 1934, Ausstellungskat., S. 70. Als Adresse dient die Friedrichstraße 3. Auch Walter nimmt an dieser Messe teil.
[622] Das befreite Handwerk. Geschmack und Wohnkultur, Ausstellungskat., 1934, S. 30
[623] Gewerbeschein 7984/fr/XIX, Wirtschaftskammer Österreich-Archiv, Gewerbearchiv. Die Eröffnung des Ladens am Kärntnertor ist am 3.5.1934. Das Geschäft in der Friedrichstraße wird aufgelöst.
[624] Lehmann's Adressbuch, 1938. Marcell und Rosemarie wohnen weiterhin in der Schikanedergasse 12. Der „Modellexport" wird mit Schottengasse 1 angegeben (vgl. Adressbuch der Keram-Industrie, Müller & Schmidt, Coburg 1937, S. 284).
[625] IC, 1937/38, S. 392
[626] Gewerbeschein 7984/fr/XIX, Wirtschaftskammer Österreich-Archiv, Gewerbearchiv. Der Laden am Kärntnertor ist bereits im April 1939 geschlossen.
[627] Gewerbeschein 7984/fr/XIX, Wirtschaftskammer Österreich-Archiv, Gewerbearchiv. Der Gewerbeschein erlischt am 21.9.1939.
[628] PGGTR, June 1941, S. 486
[629] Goldscheider, Marcel: Authentic Report of its Origin and Development, vermutlich aus dem Jahr 1951. Informative Beilage zur Firmengeschichte im Rahmen der Eröffnung neuer Räumlichkeiten in Hanley, Stoke-On-Trent.
[630] PGGTR, June 1941, S. 486ff
[631] Ebenda
[632] Myott, A. & Pollitt, P.: The Mystery of Myott, 2003, S. 7
[633] Ebenda, S. 22. In Österreich wurde die arisierte Goldscheider Manufaktur während der Kriegsjahre ebenfalls als „export unabkömmliches" Unternehmen eingestuft.
[634] PGGTR, June 1941, S. 487
[635] Ebenda
[636] PGGTR, October 1942, S. 565
[637] PGGTR, January 1945, S. 37. „... many tributes have been paid by customers in North and South America, and by people in this country, to the remarkable progress which Mr. Goldscheider has achieved in the finish and quality of his figures, both technically and artistically – a tribute we are able to endorse and enlarge by saying that figure compositions of great technical daring are now being produced."
[638] Ebenda. „Assisted by London, Staffordshire and foreign artists Marcel Goldscheider created a series of new Figures, Lamps and numerous useful lines of such popularity as London Cries, Henry VIII, Goldilocks Bears, Popular Fairy Tale subjects, etc., florally decorated humorous animals and skilfully modelled other animals and birds in natural colours and settings."
[639] Ebenda
[640] PGGTR, June 1941, S. 487
[641] PGGTR, February 1947, S. 140
[642] Ebenda, S. 141
[643] PGGTR, January 1949, S. 53. In dieser Zeit ist seine private Werkstatt bereits offiziell im Pottery Reference Book 1949 auf Seite 27 verzeichnet: Goldscheider, Marcel, The Studio, 9 Moreton Parade, May Bank, Newcastle, Staffs. Als produzierte Waren sind angegeben: Earthenware, semi-porcelain, red ware. Der Londoner Agent ist Sydney N. Folker.
[644] Myott, A. & Pollitt, P.: The Mystery of Myott, 2003, S. 24
[645] PGGTR, September 1950, S. 1360
[646] Ebenda
[647] PGGTR, December 1950, S. 1855
[648] Brief von Marcel Goldscheider, Hanley, an seine Kunden, vermutlich aus dem Jahr 1951
[649] Goldscheider, Marcel: Authentic Report of its Origin and Development, vermutlich aus dem Jahr 1951. Informative Beilage zur Firmengeschichte im Rahmen der Eröffnung neuer Räumlichkeiten in Hanley, Stoke-On-Trent.
[650] PGGTR, May 1951, S. 747
[651] The Pottery Gazette and Glass Trade Review Reference Book and Directory, 1953, S. 64. Hier werden auch die geschützten Begriffe (Trade Marks) „Goldscheider Figures" und „Popular Art Ware" festgehalten. Die angebotenen Waren sind „EARTHENWARE AND BONE CHINA ... book ends; bottles; butter coolers; cheese dishes; cigarette sundries; floral groups; flower jugs; fruit sets; ink wells; jars; musical jugs; figures and statuettes; lamp bases; plaques; salad bowls; salad sets; toby jugs; vases; wall vases."
[652] PGGTR, February 1954, S. 276
[653] PGGTR, January 1955, S. 171. „NOTICE: GOLDSCHEIDER figures and ornamental pottery and china are exclusively made by Goldscheider (Staffs) Pottery Ltd. (Incorp Goldscheider Art Pottery), 51 John Street, Hanley, Stoke-on-Trent, under the management of the creators of Goldscheider Ware, Mr. and Mrs. Marcel Goldscheider. Traders (wholesalers or retail) in Great Britain and all countries are requested to beware of any pretensions which might be made by representatives of Continental or other factories of any other models than those manufactured by above concern, being ‚former Goldscheider models'. The name Goldscheider is protected by them and the trade mark registered."
[654] PGGTR, December 1955, S. 1919. „NOTICE; It has come to our knowledge that samples have been imported of German-made figures bearing the name ‚Goldscheider'. The undersigned are the only legal successors of the former Viennese Factory and the name ‚Goldscheider' is identical with our Trade Mark Regd. Nr. 653300. We have taken steps to protect us against this infringement of our rights and would inform purchasers in their own interest of these facts. GOLDSCHEIDER (STAFFS) POTTERY LTD. Marcel GS, Rosemarie GS"
[655] PGGTR, September 1956, S. 1299 und 1311
[656] PGGTR, January 1955, S. 85
[657] Pottery and Glass Record, March 1956, S. 76. „Among the many exhibits by Goldscheider (Staffordshire) Potteries that cought the buyer's fancy were bookends especially one with two koala bears and another with Goldilock's three bears, papa supporting one end all by himself and mama and child sharing the other. They were, too, china figures and elephants decorated with a Sgraffito design and also a range of lamps with a variety of figures."
[658] PGGTR, February 1956, S. 283
[659] PGGTR, January 1957, S. 124
[660] Pinhas (2006)

Notes

[596] Patentschrift 106210 (Patent is valid as from 15 Dec. 1926), Österreichisches Patentblatt of 25 April 1927 and Patentschrift 107568 (this additional patent is valid as from 15 May 1927), Österreichisches Patentblatt of 25 Oct. 1927.
[597] Lehmann's Adressbuch, Jahrgänge 1928 bis 1932. Rosemarie Goldscheider ran the "Parivienna" studio until about 1932.
[598] Lehmann's Adressbuch, 1929
[599] IC, 1929/30, p. 1589
[600] Gewerbeschein [trade licence] 7984/fr/XIX, Wirtschaftskammer Österreich-Archiv, Gewerbearchiv
[601] Lehmann's Adressbuch, Jahrgänge 1930 bis 1934
[602] Advertisement placed by "Vereinigte Ateliers für Kunst und Keramik" in: Rochowanski (1930)
[603] Neuwirth (1974), p. 47
[604] Ibid, quoted in Neuwirth (1974), p. 48
[605] Christmas show at the Künstlerhaus, Kunst und Kunsthandwerk, November – December 1928, p. 36
[606] Heizer (1929), p. 714f
[607] Ibid
[608] Das Bild im Raum. II. Ausstellung des Verbandes Bildender Künstlerinnen und Kunsthandwerkerinnen: Wiener Frauenkunst. Österreichisches Museum für Kunst und Industrie Vienna, February – April 1929, p. 50. On p. 79 there is an advertisement placed by Marcell with a picture of the figure "Supporting Bowls."
[609] Ibid, p. 36. This exhibition catalogue also documents a desire for aestheticisation of the dwelling and intellectual stimulation. An introductory passage on p. 5 runs: "In as much as it is the task of our time to rationalise ourselves, to place our powers of ratiocination increasingly at the service of an active life and, through purposive training to attain a brain that works with the precision of an electric transformer, it is also our task to advance the inward person to ever greater perfection, to build up the realm of the soul in us in order to attain thus true self-culture in antithetical heightening. Since it is now a premise to be taken for granted that our homes are to be made useful for the purpose of dwelling in them with a dedication which earlier times did not know, we demand, building on this premise, even higher things of our homes. The home is, after all, the place of our inner renewal, the place where we live what is purely human. But it is not to serve in an aestheticising epicureanism in the fin de siècle sense; its task is to help to form the inner ego. In the manner in which a room is appointed, in the distribution of appointments, in the forms of furniture, in the colours that dominate a room, we are able to inspirit a room with that spiritual expression to which our ego can ascend in its noblest hours." And on p. 25, the modern woman is taught the advantages of a modern home library: "A room that we furnish as a small home library will in the main have to accommodate book shelves. We shall not, however, allow those book shelves to reach the sky – bending down would be too uncomfortable – but rather we shall set them on long legs, which, moreover, have the advantage that they can be easily swept out at the bottom."
[610] Rochowanski (1929), p. 430
[611] Gewerbeschein [trade licence] 18334/fr.XVIII, Wirtschaftskammer Österreich-Archiv, Gewerbearchiv. The address given on the registration is 115 Währingerstraße.
[612] Frühjahrsmesse Wien, Amtliche Ausgabe, 9.–15. March 1930, p. 76, as well as Herbstmesse Wien, Amtliche Ausgabe, 7.–13. September 1930, p. 22f
[613] Buch und Raum der Gegenwart. Exhib cat., Künstlerhaus Vienna, June – October 1930, p. 48
[614] Weihnachtsschau im Künstlerhaus, Kunst und Kunsthandwerk. Exhib. cat., 23 Nov. 1929–6 Jan. 1930, p. 31 and p. 36
[615] IC, 1931/32, p. 1569
[616] Advertisement in Rochowanski (1930). In the 1930s collaboration between Marcell and the Teplitz ceramics school resulted in a cat bowl. The bowl is marked with a U.A.T. stamp and "Atelier Goldscheider" (spotted on the Viennese art and antiques market).
[617] Blume und Plastik. Exhib. cat., Künstlerhaus Vienna, 10 April–3 May 1921, p. 7. Walter was represented by a large flower vase at that exhibition.
[618] Lehmann's Adressbuch, Jahrgänge 1932 to 1939
[619] Brunnbauer (2006). The house was sold to Max und Rifka Demuth.
[620] Lehmann's Adressbuch, Jahrgänge 1935 to 1938. The business was also in the 4th District.
[621] Austria in London. Dorland Hall, April 16 to May 12, 1934, exhib. cat., p. 70. The address given is 3 Friedrichstraße. Walter also showed pieces at that exhibition.
[622] Das befreite Handwerk. Geschmack und Wohnkultur, exhib. cat., 1934, p. 30
[623] Gewerbeschein [trade licence] 7984/fr/XIX, Wirtschaftskammer Österreich-Archiv, Gewerbearchiv. The shop at Kärntnertor opened on 3 May 1934. The business in Friedrichstraße was dissolved.
[624] Lehmann's Adressbuch, 1938. Marcell and Rosemarie continued to live at 12 Schikanedergasse. "Model export" is listed as 1 Schottengasse 1 (cf Adressbuch der Keram-Industrie, Müller & Schmidt, Coburg 1937, p. 284).
[625] IC, 1937/38, p. 392
[626] Gewerbeschein [trade licence] 7984/fr/XIX, Wirtschaftskammer Österreich-Archiv, Gewerbearchiv. By April 1939 the shop at Kärntnertor had closed.
[627] Gewerbeschein 7984/fr/XIX, Wirtschaftskammer Österreich-Archiv, Gewerbearchiv. The trade licence expired on 21 Sept. 1939.
[628] PGGTR, June 1941, p. 486
[629] Marcel Goldscheider: Authentic Report of its Origin and Development (presumably 1951). Informative enclosure on the history of the firm on the occasion of the opening of new rooms in Hanley, Stoke-On-Trent
[630] PGGTR, June 1941, p. 486f
[631] Ibid
[632] A Myott and P. Pollitt: The Mystery of Myott, 2003, p. 7
[633] Ibid, p. 22. In Austria, the Aryanized Goldscheider Manufactory was also classified during the war years as a business that was "indispensable to export."
[634] PGGTR, June 1941, p. 487
[635] Ibid
[636] PGGTR, October 1942, p. 565
[637] PGGTR, January 1945, p. 37: "[...] many tributes have been paid by customers in North and South America, and by people in this country, to the remarkable progress which Mr. Goldscheider has achieved in the finish and quality of his figures, both technically and artistically – a tribute we are able to endorse and enlarge by saying that figure compositions of great technical daring are now being produced."
[638] Ibid: "Assisted by London, Staffordshire and foreign artists Marcel Goldscheider created a series of new Figures, Lamps and numerous useful lines of such popularity as London Cries, Henry VIII, Goldilocks' Bears, Popular Fairy Tale subjects, etc., florally decorated humorous animals and skilfully modelled other animals and birds in natural colours and settings."
[639] Ibid
[640] PGGTR, June 1941, p. 487
[641] PGGTR, February 1947, p. 140
[642] Ibid, p. 141
[643] PGGTR, January 1949, p. 53. By that time, his private workshop was officially listed in the 1949 Pottery Reference Book on p. 27: Goldscheider, Marcel, The Studio, 9 Moreton Parade, May Bank, Newcastle, Staffs. The following wares are listed as produced: earthenware, semi-porcelain, red ware. The London agent was Sydney N. Folker.
[644] A Myott and P. Pollitt: The Mystery of Myott, 2003, p. 24
[645] PGGTR, September 1950, p. 1360
[646] Ibid
[647] PGGTR, December 1950, p. 1855
[648] Letter from Marcel Goldscheider, Hanley, to his clientele (presumably 1951)
[649] Marcel Goldscheider: Authentic Report of its Origin and Development (presumably 1951). Informative enclosure on the history of the firm on the occasion of the opening of new rooms in Hanley, Stoke-On-Trent
[650] PGGTR, May 1951, p. 747
[651] The Pottery Gazette and Glass Trade Review Reference Book and Directory, 1953, p. 64. Here the trademarks are also given as "Goldscheider Figures" and "Popular Art Ware." The wares sold are "EARTHENWARE AND BONE CHINA [...] book ends; bottles; butter coolers; cheese dishes; cigarette sundries; floral groups; flower jugs; fruit sets; ink wells; jars; musical jugs; figures and statuettes; lamp bases; plaques; salad bowls; salad sets; toby jugs; vases; wall vases."
[652] PGGTR, February 1954, p. 276
[653] PGGTR, January 1955, p. 171. "NOTICE: GOLDSCHEIDER figures and ornamental pottery and china are exclusively made by Goldscheider (Staffs) Pottery Ltd. (Incorp Goldscheider Art Pottery), 51 John Street, Hanley, Stoke-on-Trent, under the management of the creators of Goldscheider Ware, Mr. and Mrs. Marcel Goldscheider. Traders (wholesalers or retail) in Great Britain and all countries are requested to beware of any pretensions which might be made by representatives of Continental or other factories of any other models than those manufactured by above concern, being 'former Goldscheider models'. The name Goldscheider is protected by them and the trade mark registered."
[654] PGGTR, December 1955, p. 1919. "NOTICE; It has come to our knowledge that samples have been imported of German-made figures bearing the name 'Goldscheider'. The undersigned are the only legal successors of the former Viennese Factory and the name 'Goldscheider' is identical with our Trade Mark Regd. Nr. 653300. We have taken steps to protect us against this infringement of our rights and would inform purchasers in their own interest of these facts. GOLDSCHEIDER (STAFFS) POTTERY LTD. Marcel GS, Rosemarie GS"
[655] PGGTR, September 1956, p. 1299 and p. 1311
[656] PGGTR, January 1955, p. 85
[657] Pottery and Glass Record, March 1956, p. 76. "Among the many exhibits by Goldscheider (Staffordshire) Potteries that caught the buyer's fancy were bookends especially one with two koala bears and another with Goldilocks' three bears, papa supporting one end all by himself and mama and child sharing the other. They were, too, china figures and elephants decorated with a Sgraffito design and also a range of lamps with a variety of figures."
[658] PGGTR, February 1956, p. 283
[659] PGGTR, January 1957, p. 124
[660] Pinhas (2006)

Die Fortsetzung der Tradition durch Peter Goldscheider in Stoob, Österreich (1987–1994)

The Tradition Continued by Peter Goldscheider in Stoob, Austria (1987–1994)

Es vergehen über 30 Jahre nach der Auflösung der Wiener Manufaktur durch Walter Goldscheider, bis die erste große Retrospektive zur Goldscheider Keramik stattfindet. Im Historischen Museum der Stadt Wien wird im Dezember 1985 ein Teil des Depot-Bestandes ausgestellt, der größtenteils einer Schenkung des Urenkels von Friedrich Goldscheider zu verdanken ist. Dr. Peter Goldscheider und seine Frau Brigitte übergeben dem Museum über 300 Objekte aus ihrer über Jahrzehnte selbst erworbenen Sammlung und ermöglichen somit diese Ausstellung zum hundertjährigen Jubiläum der Goldscheider Manufaktur. Deren Exponate umfassen alle Stil-Epochen der Manufaktur: Historismus, Jugendstil und Art Déco.

Zwei Jahre später, am 8. Juli 1987, gründet Peter Goldscheider gemeinsam mit dem Industriellen Michael Sommer im burgenländischen Stoob die „Goldscheider Keramik GmbH" zur „Erzeugung und Vertrieb von keramischen Produkten".[661] In Zusammenarbeit mit der in Stoob ansässigen Keramikschule beginnt mit acht Mitarbeitern die neue Produktion von Fayencemodellen. Der in Kairo im Jahr 1945 geborene Peter Goldscheider, ein Sohn von Dr. Fritz Goldscheider, unterbricht seine Karriere als erfolgreicher Manager in der Finanzbranche und widmet sich der Familientradition, um die berühmte Weltmarke der Keramik zu neuem Leben zu erwecken. Er will keine Massenproduktion, sondern die künstlerische und technische Qualität der Manufaktur fortsetzen: „Goldscheider-Figuren werden von Kunsthandwerkern in Handarbeit gefertigt. Diese sind nicht nur vom Entwurf, sondern auch von der Ausführung her Kunstwerke."[662] Zur ersten Modellreihe gehört eine Auswahl von etwa 15 der populärsten Goldscheider Modelle aus der Zeit zwischen 1885 und 1938.

More than thirty years would elapse after the Viennese factory had been disbanded by Walter Goldscheider before the first large-scale retrospective of Goldscheider ceramics was mounted. In December 1985, the Historisches Museum der Stadt Wien showed some pieces from the Museum collections, most of which had been donated by Friedrich Goldscheider's great-grandson. Dr. Peter Goldscheider and his wife, Brigitte, gave the Museum more than three hundred objects from the collection they had amassed over decades, thus making it possible to mount the exhibition on the centenary of the Goldscheider Manufactory. The exhibits covered all period styles the manufactory worked in: Historicism, Jugendstil and Art Déco.

Two years later, on 8 July 1987, Peter Goldscheider and the industrialist Michael Sommer founded "Goldscheider Keramik GmbH" at Stoob, in Burgenland to "produce and market ceramic products."[661] Collaborating with the ceramics school at Stoob, they began to produce faience models with a workforce of 8. Born in Cairo in 1945, Peter Goldscheider, a son of Dr. Fritz Goldscheider, interrupted his career as a successful financial manager to devote himself to the family tradition in order to breathe new life into the celebrated world name in ceramics. Peter Goldscheider was not interested in mass production. What he wanted was to continue the artistic and technical quality standards set by the manufactory: "Goldscheider figures are made by hand by skilled craftsmen. They are works of art, not just in design but also in execution."[662] The first series of models include a selection of some fifteen of the most popular Goldscheider models made between 1885 and 1938.

Reproducing sophisticated Goldscheider figures of such high quality required specialist skills mastered by only a very few crafts-

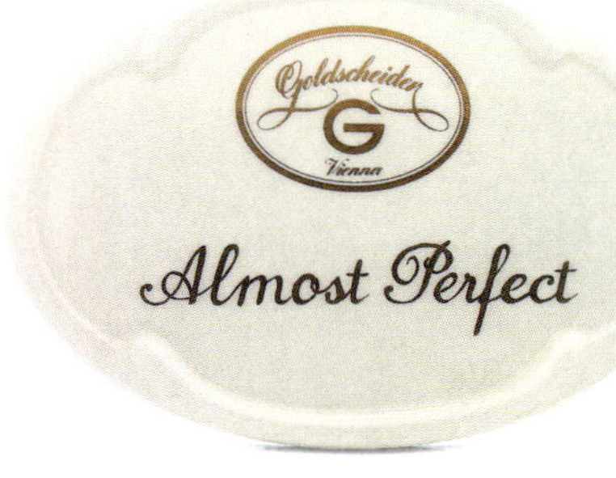

Ill. 196

men. Since there were no longer enough well-trained skilled craftsmen available in Austria – only the Augarten Porcelain Factory could afford a division manned just by modelers – the plaster of Paris models were made by skilled ceramicists from abroad.[663] Making Goldscheider figures was particularly demanding because they were "made from separate pieces, luted, fired, painted and then put back in the kiln."[664] Peter Goldscheider has described the work process in much more detail: Making Goldscheider ceramics on the basis of modeling originals was by then a far more elaborate process than creating a comparable figure from scratch. Numerous detailed steps were necessary to complete a figure. The modeler made the original sculpture. The model was then worked over by an expert, ie, it was taken apart depending on the technical requirements and the plaster of Paris piece molds were made.[665] Piece molds were filled with faience paste to preserve the individual intricate parts such as hands, hair decoration, etc. After luting, that is, assembling the pieces to produce a whole figure, the first firing took place at approx 900 degrees Celsius, to lend the body a certain firmness (cheese-hard state). The body, however,

Die Reproduktion dieser komplexen und hoch qualitativen Goldscheider Figuren benötigt ein Fachwissen, das nur noch wenige Facharbeiter beherrschen. Da es in Österreich nicht mehr genügend gut ausgebildete Fachkräfte gibt – nur die Porzellanmanufaktur Augarten leistet sich eine eigene Abteilung mit Modelleuren – werden die Gipsmodelle von internationalen Keramikfachleuten modelliert.[663] Die Herstellung der Goldscheider-Figuren ist besonders anspruchsvoll, da sie aus mehreren „Einzelteilen zusammengesetzt, garniert, gebrannt, bemalt, und dann nochmals in den Ofen gesteckt" werden.[664] Peter Goldscheider beschreibt den Arbeitsprozess wie folgt: „Die Herstellung der Goldscheider Keramik auf Basis der Originalmodellierung ist heute wesentlich aufwendiger als die Neugestaltung einer vergleichbaren Figur. Bis zur Vollendung sind zahlreiche Detailarbeitsschritte erforderlich. Der Modelleur stellt die Original-Plastik her. Das Modell wird von einem Experten eingerichtet, d.h. es wird entsprechend den technischen Gegebenheiten zerlegt und die Gipsarbeitsformen gestaltet."[665] Diese Arbeitsformen werden mit Fayencemasse gefüllt, um die einzelnen feinen Teile, wie Hände, Haarschmuck etc., zu erhalten. Nach dem Garnieren, also dem

Ill. 197

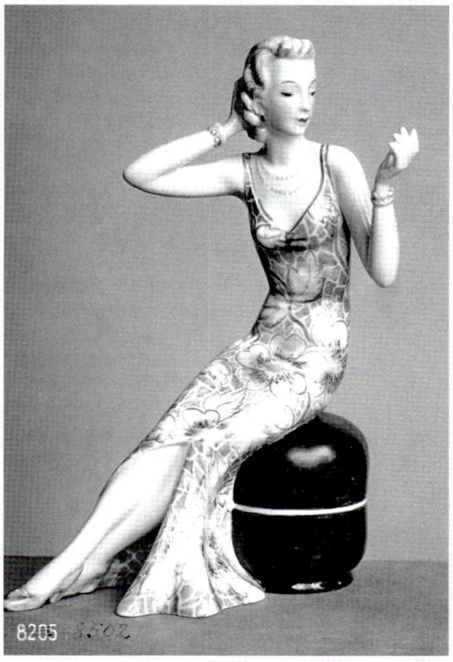

Ill. 198

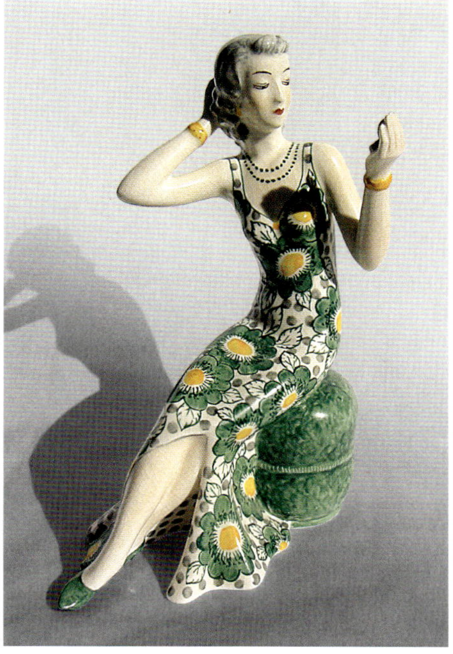

Ill. 196–198
Erfolgreiche Goldscheider Modelle aus der gesamten Geschichte der traditionsreichen Manufaktur werden in Stoob reproduziert. So wird das Modell 8502 unter dem Titel „Almost Perfect" neu aufgelegt. (Quelle: Katalogfoto aus Fotoarchiv Karl Täuber; Stoober Modell aus Sammlung Alexandra Herrmann-Goldscheider; Werbeschild aus Sammlung a@goldscheider.net)

Successful Goldscheider Models from the entire history of the venerable manufactory were reproduced at Stoob. Model 8502, for instance, was reproduced and given the title "Almost Perfect." (Source: catalogue photograph from Fotoarchiv Karl Täuber; Stoob model from the Alexandra Herrmann-Goldscheider Collection; advertising sign from Collection a@goldscheider.net)

Zusammenfügen der Teile zur Gesamtfigur, erfolgt der erste Brand bei ca. 900 Grad Celsius, um der Masse eine gewisse Festigkeit zu geben. Der Ton bleibt jedoch porös und hat eine sehr schmutzempfindliche Oberfläche. Daher wird er in einen überwiegend aus „Metalloxyden bestehenden, mit Wasser angerührten Glasurbrei getaucht. Infolge der Durchlässigkeit des Scherbens wird das Wasser von diesem aufgesogen und das Zinnemail bleibt als feuchter Niederschlag auf der Oberfläche haften. Auf diesem Untergrund werden nun mit dem Pinsel die Malfarben, die aus Metalloxyden bestehen, angebracht. Es leuchtet von selbst ein, dass der Fayencemaler bei der Saugfähigkeit des Grundes nicht in der Lage ist, an seiner Arbeit nachträglich noch Veränderungen vorzunehmen."[666] Jeder Pinselstrich muss sitzen, denn nach dem Trocknen können nur noch Schattierungen oder Glanzstellen hinzugefügt werden. Im darauf folgenden Garbrand werden die Farben eingebrannt, da aber nur verhältnismäßig wenige Metalloxyde die hohe Temperatur aushalten, ist normaler-

remained porous and had a very vulnerable surface. That is why it was "plunged into a frit mainly composed of metal oxides mixed with water. Due to the porosity of the body, the water is absorbed by it and the tin enamel clings as a damp precipitate to the surface. Now the paint colors, which are composed of metal oxides, are applied on this ground with a brush. It goes without saying that the faience painter, considering the absorptive properties of the ground, cannot make any later corrections to his work."[666] Every brush stroke had to be right because after drying only shading or highlights might be added. The glazes were then glost-fired but, since only relatively few metal oxides can withstand high temperatures, the palette was limited. At Goldscheider, however, the color range was considerably widened by the use of special procedures and blends.

This high-quality craftsmanship was to guarantee the venerable tradition of the Goldscheider quality and aesthetic and provide the Goldscheider clientele with objects

Ill. 199

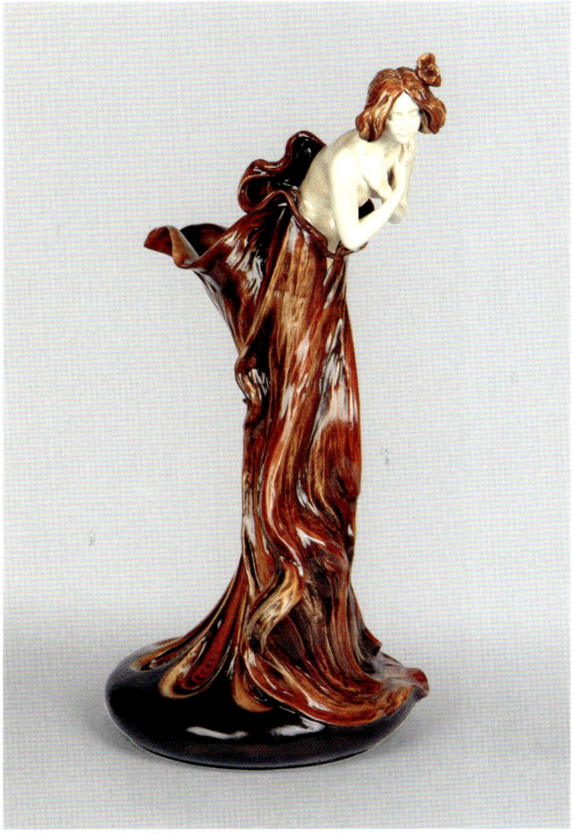

Ill. 200

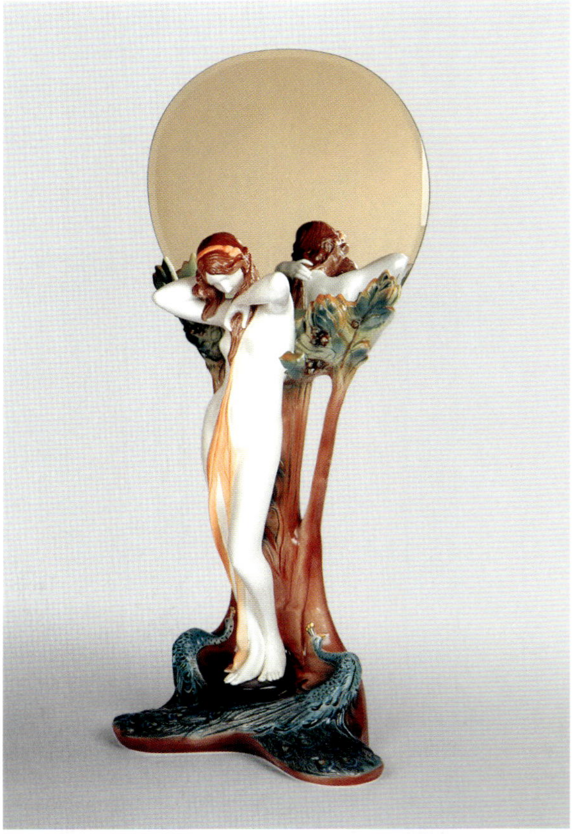

Ill. 199–202
Die Stoober Modelle tragen einen eigenen Stempel sowie den Hinweis „Worldwide Limited Edition" mit der Angabe der limitierten Stückzahl sowie der Nummer des jeweiligen Objekts. Reproduktions-Beispiele der Modelle 841, 2543, 4387 und 5230. (Quelle: Archiv Goldscheider)

Stoob models bear a stamp of their own plus "Worldwide Limited Edition," the number of pieces in the limited edition as well as the object number. Reproductions of models 841, 2543, 4387 and 5230. (Source: Goldscheider Archives)

of lasting value. The high-priced reproductions of the old Goldscheider models are virtually identical with the original models. They were produced in limited editions of 1000 to 3000. The pieces were individually numbered in sequence.[667] The stamp used by the Goldscheider factory at Stoob consists of "Goldscheider" in lettering above a large "G" and "Vienna" ("GWM" was no longer in use), usually also with a slightly modernized reproduction of the sculpture mark. (Ill. 196–202)

The finished art objects were, like their precursors, sold worldwide and met with particularly keen interest in the US and Japan. The following models from this series were especially popular: "La Tendresse," a Jugendstil vamp in a golden dress; an amphora-shaped vase after A. Bertrand; an exquisitely wrought faience figure "Muse des Tanzes" ["The Muse of Dance"] clad in reticently hued yet colorfully decorated dance costume; the celebrated Lorenzl "Gefangener Vogel" ["Captured Bird"] and "Perpetuum mobile," a faience candelabrum

weise die Farbauswahl beschränkt. Bei Goldscheider wird durch besondere Verfahren und Beimischungen die Farbpalette stark erweitert.

Diese hochwertige Handarbeit soll den traditionellen Anspruch von Goldscheider an Qualität und Ästhetik gewährleisten und den Kunden ein wertbeständiges Objekt bieten. Die hochpreisigen Reproduktionen der alten Goldscheider Modelle sind fast identisch mit den ursprünglichen Originalmodellen und erscheinen in limitierter Edition von 1.000 bis 3.000 Exemplaren, jedes einzelne Stück ist mit fortlaufender Nummer versehen.[667] Der Stempel der Stoober Goldscheider Manufaktur zeigt den Schriftzug Goldscheider sowie darunter ein großes G und Vienna (also nicht mehr GWM), meistens auch eine leicht modernisierte Abbildung der Sculpture-Marke (Ill. 196–202).

Die fertigen Kunstgegenstände werden genauso wie ihre Vorgängermodelle weltweit vertrieben und stoßen vor allem in den USA und Japan auf Interesse. Besonders beliebt sind aus dieser Reihe die Figur „La

Ill. 201

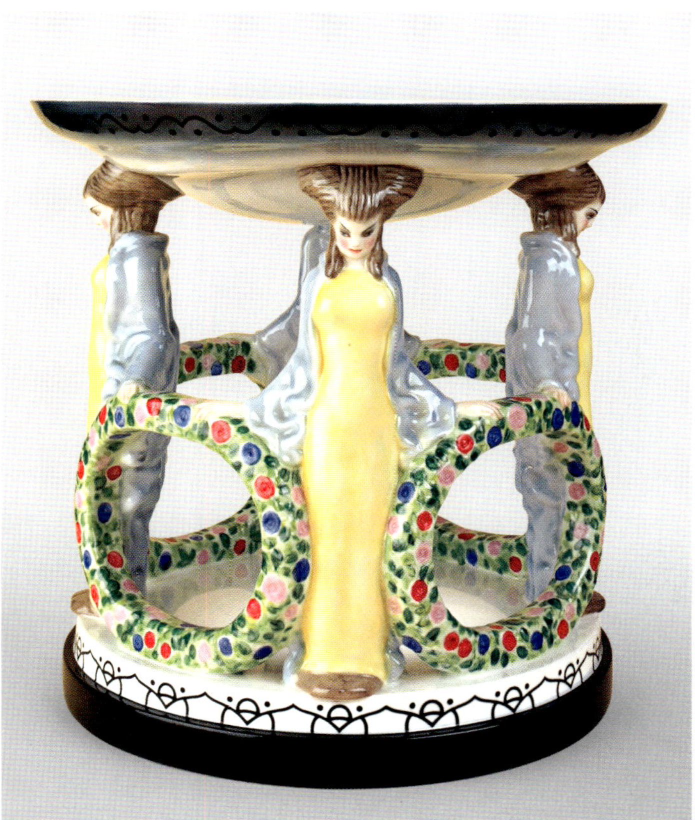

Ill. 202

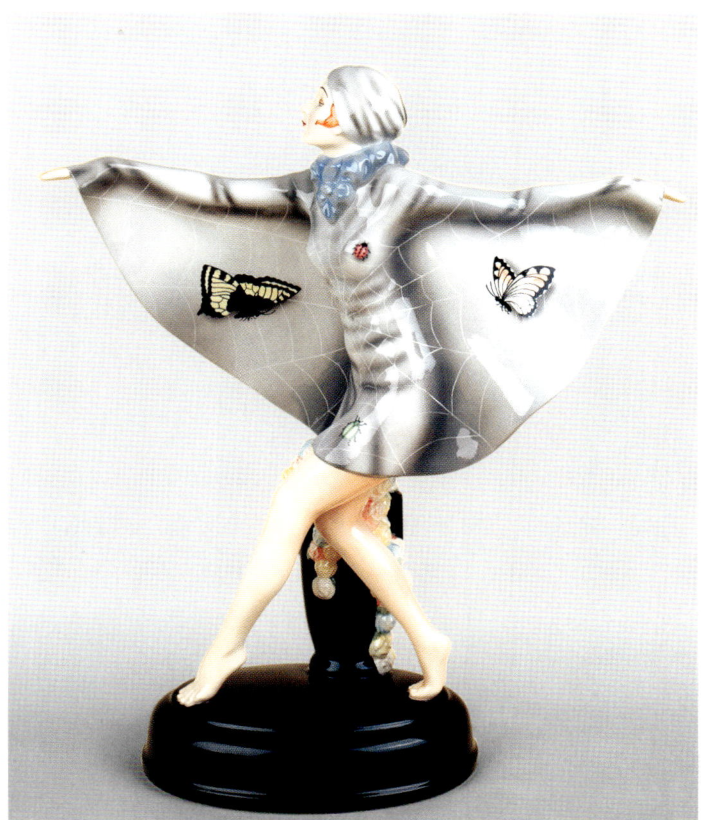

Ill. 203, 204
Der Vasenkopf „Konzert für ein goldenes Telefon" vom Surrealisten Leherb ist eine handausgeformte, handbemalte Keramik (40 x 25 x 25 cm), die auf 140 Exemplare limitiert ist. (Quelle: Archiv Goldscheider)

The head vase "Konzert für ein goldenes Telefon" ["Concert for a Golden Telephone"] by the surrealist Leherb is a hand-cast, hand-painted piece (40 x 25 x 25 cm), a limited edition of 140. (Source: Goldscheider Archives)

Tendresse", eine Jugendstil-Femme im goldenen Kleid, und eine Amphora-Vase nach A. Bertrand, die fein gearbeitete Fayence-Figur „Muse des Tanzes" mit einer dezent bunten Dekoration des Tanzkleides, der berühmte „Gefangene Vogel" von Lorenzl oder ein Fayenceleuchter mit drei Kerzentüllen „Perpetuum mobile" von Thomasch, die in einem Ring eine weiß glasierte Tänzerin umranden.

Im Jahr 1990 werden bereits 20 Mitarbeiter beschäftigt, die an drei Produktionslinien arbeiten. Neben den klassischen Reproduktionen werden bei Goldscheider auch Objekte zeitgenössischer Künstler produziert. Diese moderne Keramikkunst entsteht „getreu dem Grundsatz, immer auch die großen zeitgenössischen Künstler einzubinden."[668] Diese Objekte werden von Facharbeitern nach detailtreuen Entwürfen in limitierter Auflage hergestellt. Jedes dieser hochwertigen Exemplare wird nummeriert, handsigniert und mit einem Authentizitätszertifikat versehen.[669] Zusätzlich erhalten einige Modelle das österreichische Gütezei-

by Thomasch featuring three branches encircling a white-glazed dancer.

By 1990 the factory employed a workforce of 20, who worked on three product lines. Besides the classic reproductions, objects by contemporary artists were also produced at Goldscheider. This modern art ceramics was made "faithfully following the principle of always engaging great contemporary artists as well."[668] These objects were made after detailed designs in limited editions by skilled workers. Each of these high-quality pieces was numbered, signed by hand and issued with a certificate of authenticity.[669] In addition, some models were awarded the Austrian quality seal, consisting of a red-and-white striped "A" and "Austria" in lettering. One of the contemporary artists employed was Leherb, the Viennese surrealist, who created a monumental work, "Tor für ein imaginäres Museum" ["Entrance to an Imaginary Museum"] – an outsize bronze and ceramics door – for the Stoob factory between 1989 and 1991.[670] The head vases "Konzert für ein goldenes Telefon" ["Concert

Ill. 203

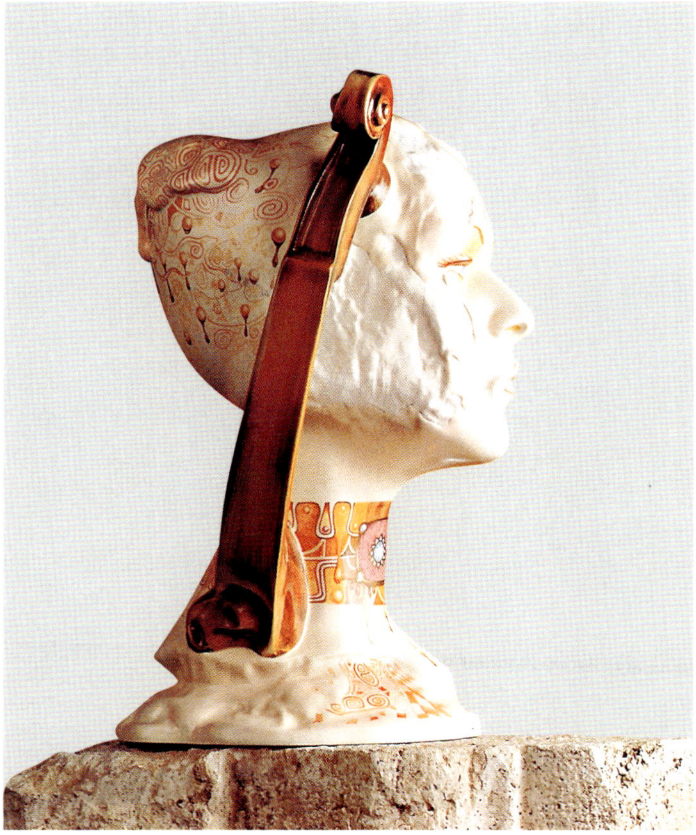

Ill. 204

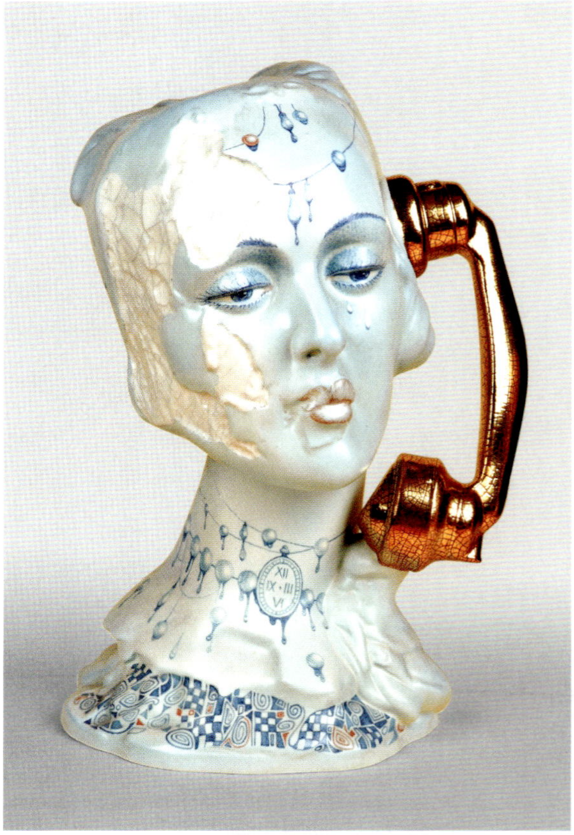

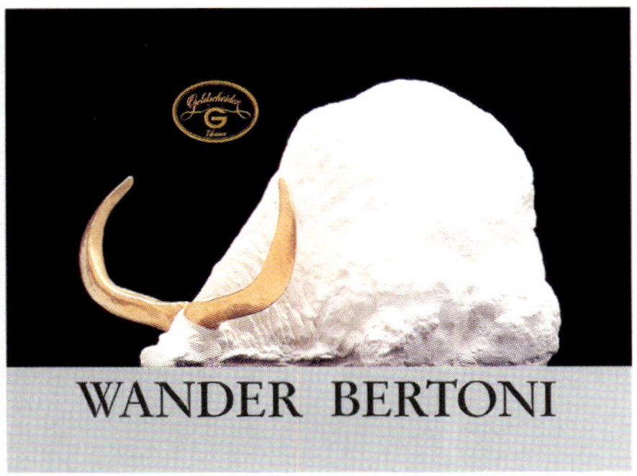

Ill. 205

for a Golden Telephone"] and "Sonate für eine Träumerin" ["Sonata for a Dreamer"], both produced in limited editions of 140, were also by Leherb (Ill. 203 and 204). The Italian sculptor Wander Bertoni also contributed three faience models from his "Indisches Tagebuch" ["Indian Diary"] around 1992 to this second product line in high-quality modern ceramics (Ill. 205–207).

The third product line was to be affordable for everyone. Unlike the art series, utilitarian ceramics (tiles, vases, etc) were not produced in limited editions but were sold as mass-produced series.[671] This line included utilitarian objects such as ashtrays, vases and candlesticks, marketed under the "GEO-ART" label. They were designed by A. Stepan, the porcelain designer, following sophisticated geometric patterns (Ill. 208). A specialty was the GEO-ART Signs of the Zodiac collection, representing the signs of

chen für geprüfte Qualität, das aus einem rotweiß-gestrichenen A und dem Schriftzug Austria besteht. Einer der zeitgenössischen Künstler ist der Wiener Surrealist Leherb, der zwischen 1989 und 1991 in der Stoober Manufaktur das Monumentalwerk „Tor für ein imaginäres Museum" erstellt – ein überdimensionales Tor aus Bronze und Keramik.[670] Die auf 140 Exemplare limitierten Vasenköpfe „Konzert für ein goldenes Telefon" und „Sonate für eine Träumerin" sind ebenfalls von Leherb (Ill. 203 und 204). Auch der aus Italien gebürtige Bildhauer Wander Bertoni beteiligt sich um 1992 mit drei Fayence-Modellen aus seinem „Indischen Tagebuch" an dieser zweiten Produktionsreihe hochwertiger moderner Keramik (Ill. 205–207).

Die dritte Produktionsreihe soll für jeden preislich erschwinglich sein. Im Gegensatz zu den Kunstserien ist die produzierte Gebrauchskeramik (Fließen, Vasen etc.) nicht limitiert und wird als Großserie angeboten.[671] Darunter fallen auch die unter dem Label „GEO-ART" vertriebenen Gebrauchsgegenstände wie Aschenbecher, Vasen und Kerzen-

Ill. 205–207
Der italienische Bildhauer Wander Bertoni steuert die auf Indienreisen beruhenden Modelle „Der Büffel", „Der reiche Bettler" und „Lingam" zur modernen Keramikkunst bei. (Quelle: Archiv Goldscheider)

The Italian sculptor Wander Bertoni contributed "Der Büffel" ["Water Buffalo"], "Der reiche Bettler" ["Rich Beggar"] and "Lingam" to the art of modern ceramics. (Source: Goldscheider Archives)

Ill. 206

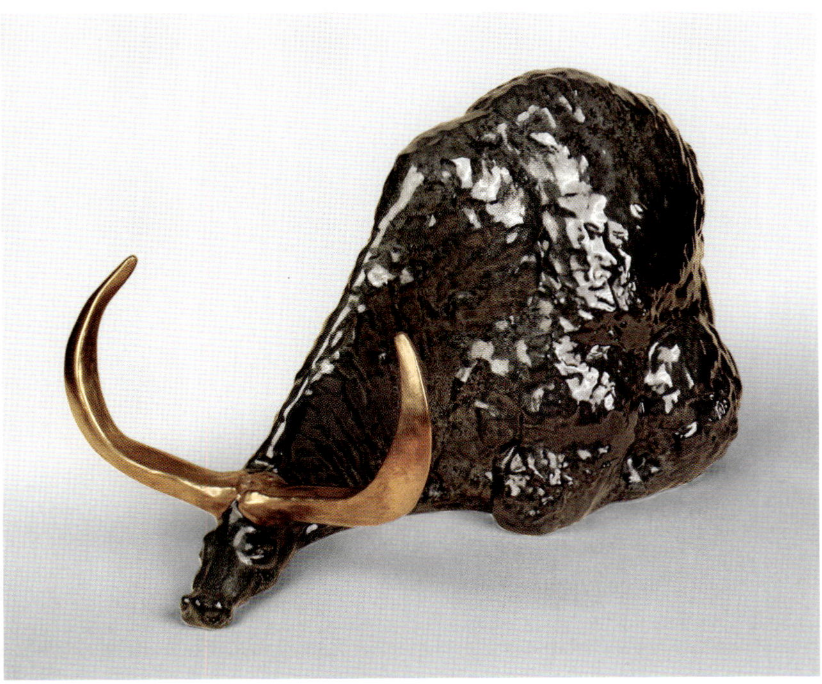

Ill. 207

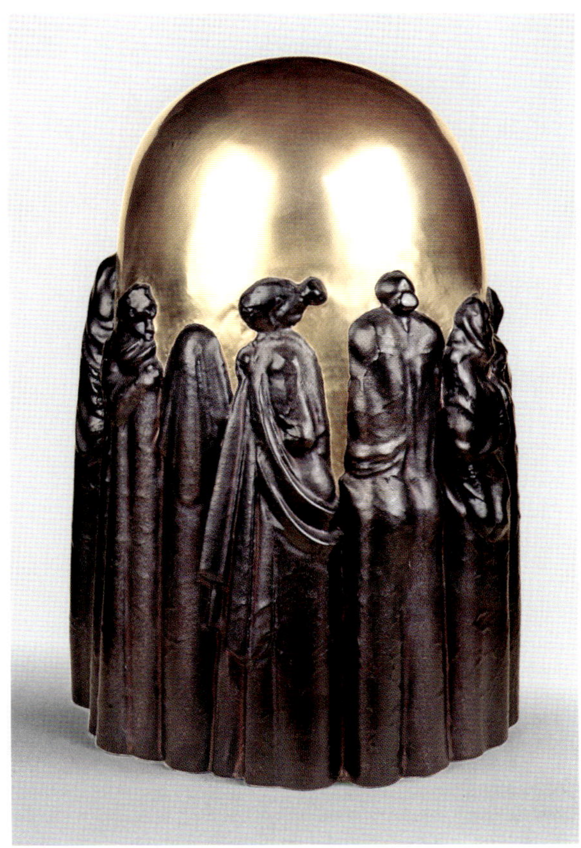

halter, die vom Porzellan-Designer A. Stepan nach durchdachten, geometrischen Mustern entworfen werden (Ill. 208). Eine Besonderheit ist die GEO-ART Sternzeichen-Kollektion, die Sternzeichen in ungewöhnlicher Vasenform darstellt. Stepan sieht in diesen Objekten „geometrische Magie" entstehen und empfiehlt, „das geometrische Objekt an einem Schnittpunkt der Geometrie des Wohnraumes" aufzustellen.[672]

Diese Großserie von Gebrauchsgegenständen ist auch die profitabelste der Produktionsreihen. Aufgrund der hohen Produktionskosten und des nicht ausreichenden Absatzes auf den ausländischen Märkten wird die Produktion der klassischen und modernen Goldscheider Keramik eingestellt. Im Juli 1994 kauft Peter Goldscheider, der inzwischen seine Karriere in der Finanzbranche erfolgreich fortsetzt, die Anteile seiner Mitgesellschafter auf und führt eigenständig als Geschäftsführer das Unternehmen weiter.[673] Am 20. Oktober 1994 wird der Sitz des Unternehmens mit Plößlgasse 8, Wien angemeldet.[674] Mit der Standortverlegung wird die eigene Produktion eingestellt, da die Produktionskosten aufgrund der hohen Ansprüche an die handwerkliche Ausführung zu hoch waren und die Lebensverhältnisse und der Geschmack der Menschen sich geändert haben.[675] Ein Bereich der Goldscheider-Keramik wird jedoch weiter hergestellt und bei anderen Keramikproduzenten in Auftrag gegeben: Urnen. Diese sind aufgrund ihrer Form relativ kostengünstig herzustellen und finden dank eines größeren Rahmenvertrags stete Abnahmemengen.[676] Neben der Produktion der Urnen konzentriert sich die Goldscheider Keramik GmbH von nun an in reduziertem Maße nur noch auf den Handel.[677] Die Geschichte der weltberühmten Goldscheider Manufaktur kehrt somit zu den Anfängen von Friedrich Goldscheider zurück, als dieser um 1880 sich vor allem dem Handel mit Porzellan und Ton-Erzeugnissen widmete.

the Zodiac in an unusual vase form. Stepan saw "the magic of geometry" invoked in these objects and recommended setting up "the geometrical object at a point of intersection in the geometry of the living space."[672]

This mass-produced series of utilitarian objects was also the most lucrative of the product lines. High production overheads and insufficient sales forced production of the classic and modern Goldscheider ceramics to close down. In July 1994, Peter Goldscheider, who by then was again successfully pursuing a career in finance, bought out his partners' shares and ran the business on his own as managing director.[673] On 20 October 1994 the business was registered as based at 8 Plößlgasse, Vienna.[674] The change in location was accompanied by cessation of production of proprietary designs since the cost of having so much done by hand was too high. Changes in lifestyle and tastes were a further contributing factor.[675] One type of Goldscheider ceramics, however, continued in production on commission from other ceramics producers: urns. Their form made them relatively cheap to produce and, thanks to a wider-reaching agreement, sales volume was guaranteed.[676] Apart from producing urns, Goldscheider Keramik GmbH now concentrated solely on marketing and distribution.[677] Thus the history of the world-famous Goldscheider factory returned to its beginnings, when Friedrich Goldscheider devoted himself primarily to trading in porcelain and terracotta products around 1880.

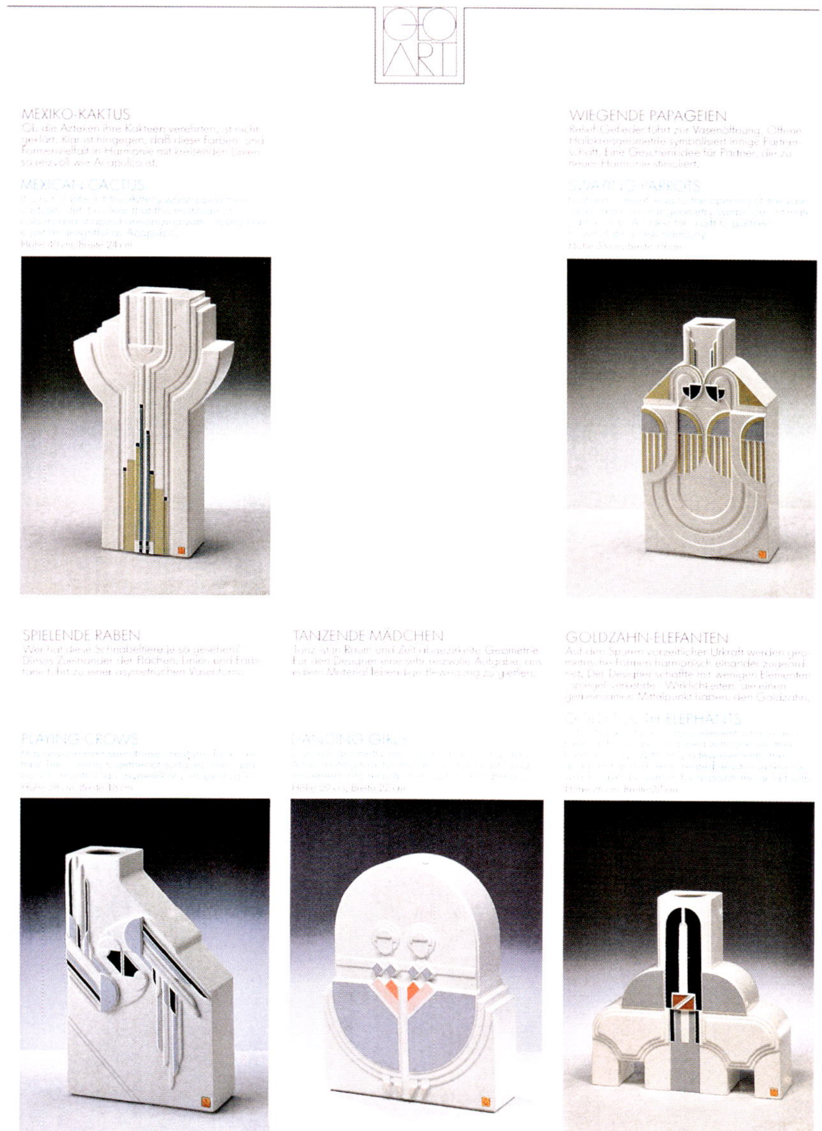

Ill. 208
Moderne Gebrauchskeramik wird in unlimitierter Großserie unter dem Label „GEO-ART" produziert. (Quelle: Archiv Goldscheider)

Modern utilitarian ceramics were mass produced under the "GEO-ART" label. (Source: Goldscheider Archives)

Notes

[661] Gewerbeschein [trade licence] of the firm of "Goldscheider Keramik Gesellschaft m.b.H," granted by the Bezirkshauptmannschaft Oberpullendorf on 19 Nov. 1987. The business was based at 24 Neugasse.
[662] M. Kattinger: Goldscheider Keramiken, in: Parnass, 6/87, p. 44
[663] W. Pohl: Schönheiten aus Porzellan, in: AZ/Tagblatt, November 1987
[664] Ibid
[665] Catalogue "The Royal Family of Porcelain Since 1885" of the firm of Goldscheider, ca 1988 (GSN)
[666] Goldscheider factory firm prospectus on "Leherb," ca 1991
[667] M. Kattinger: Goldscheider Keramiken, in: Parnass, 6/87, p. 45
[668] Darstellung der Firmengeschichte durch die Goldscheider Keramik Ges.m.b.H, Stoob, 1992
[669] Goldscheider Keramik Ges. m.b.H firm prospectus "Wander Bertoni" (1992)
[670] Darstellung der Firmengeschichte durch die Goldscheider Keramik Ges.m.b.H, Stoob, 1992
[671] H. A. Gansterer: Erfolg: Die kunstvolle Wiederauferstehung der Goldscheiders, in: Trend, 21.1990, p. 149
[672] Goldscheider firm prospectus "GEO-ART" (1992)
[673] Notariatsakt zum Abtretungsvertrag zwischen [Certified instrument of conveyance between] Michael Sommer Metall-Systeme, Mag.Dr. Walter St. Weinhandl & Engelbert Katt, Vermögensverwaltungsgesellschaft and Dr. Peter Goldscheider, of 12 July 1994
[674] Handelregister Österreich, Jupiter Verlag, Vienna 1995, p. 590
[675] Letter ("Verständigung" ["Notification"]) from the Magistracy of the City of Vienna to the Goldscheider Keramik Gesellschaft m.b.H., 27 Jan. 1995
[676] Interview with Dr. Peter Goldscheider, Vienna, 16 Nov. 2004
[677] Gewerbeschein of the firm of "Goldscheider Keramik Gesellschaft m.b.H," granted by the Magistracy of the City of Vienna on 14 March 1995. Trade registration had taken place on 5 Jan. 1995.

Anmerkungen

[661] Gewerbeschein der Firma „Goldscheider Keramik Gesellschaft m.b.H", ausgestellt von der Bezirkshauptmannschaft Oberpullendorf, am 19.11.1987. Das Unternehmen ist in der Neugasse 24 angesiedelt.
[662] Kattinger, M.: Goldscheider Keramiken, in: Parnass, 6/87, S. 44
[663] Pohl, W.: Schönheiten aus Porzellan, in: AZ/Tagblatt, November 1987
[664] Ebenda
[665] Katalog „The Royal Family of Porcelain Since 1885" der Firma Goldscheider, um 1988 (GSN)
[666] Firmenprospekt der Manufaktur Goldscheider zu „Leherb", um 1991
[667] Kattinger, M.: Goldscheider Keramiken, in: Parnass, 6/87, S. 45
[668] Darstellung der Firmengeschichte durch die Goldscheider Keramik Ges.m.b.H, Stoob, aus dem Jahr 1992
[669] Firmenprospekt „Wander Bertoni" der Goldscheider Keramik Ges.m.b.H, aus dem Jahr 1992
[670] Darstellung der Firmengeschichte durch die Goldscheider Keramik Ges.m.b.H, Stoob, aus dem Jahr 1992
[671] Gansterer, H.A.: Erfolg: Die kunstvolle Wiederauferstehung der Goldscheiders, in: Trend, 21.1990, S. 149
[672] Firmenprospekt „GEO-ART" der Firma Goldscheider, aus dem Jahr 1992
[673] Notariatsakt zum Abtretungsvertrag zwischen Michael Sommer Metall-Systeme, Mag.Dr. Walter St. Weinhandl & Engelbert Katt, Vermögensverwaltungsgesellschaft und Dr. Peter Goldscheider, vom 12.7.1994
[674] Handelregister Österreich, Jupiter Verlag, Wien 1995, S. 590
[675] Brief („Verständigung") des Magistrats der Stadt Wien an die Goldscheider Keramik Gesellschaft m.b.H., 27.1.1995
[676] Interview mit Dr. Peter Goldscheider, Wien, 16.11.2004
[677] Gewerbeschein der Firma „Goldscheider Keramik Gesellschaft m.b.H", ausgestellt vom Magistrat der Stadt Wien, am 14.3.1995. Die Gewerbeanmeldung für den Handel erfolgte bereits am 5.1.1995.

Werkverzeichnis I

Catalogue of Works

Robert E. Dechant

Anmerkungen zum Werkverzeichnis

Die Produktion der Goldscheider-Manufaktur umfasste ab 1885 mehrere hunderttausend Stück, die auf über 10.000 individuellen Modellvorlagen basierten. Denn die Modelle wurden nicht nur in diversen Materialien und für verschiedene Verwendungsmöglichkeiten (z.B. Erweiterung von Figuren zu Lampen) hergestellt, sie weisen auch mannigfaltige und abwechslungsreiche Dekore auf. Hätte man das utopische Ziel gehabt, alle einzelnen Modelle und Variationen ermitteln sowie allen Objekten die Künstler zuweisen zu wollen (viele Objekte sind zudem unsigniert), wäre das vorliegende Werkverzeichnis nie abgeschlossen worden. Dieses Werkverzeichnis bietet jedoch mit Abstand die bislang umfangreichste Übersicht zu Goldscheider-Objekten mit mehreren tausend Modellen. Es dürfte jedem Sammler, Händler und Kunstinteressierten neue und interessante Einblicke in die österreichische und französische Produktion des Historismus, des Jugendstils und des Art Décos gewähren. Zudem zeigt es Beispiele der amerikanischen bzw. englischen Produktion von Walter und Marcel Goldscheider und alle in Deutschland (Fredelsloh) zwischen 1953–1960 erzeugten Modelle.

Bei der Aufnahme von Modellen ins Werkverzeichnis wurde möglichst detailliert vorgegangen. Soweit Informationen zu einem Modell vorlagen, wurden diese übernommen. Dies musste in manchen Fällen ungeprüft geschehen, da beispielsweise Informationen aus Auktionen schon länger zurücklagen und Rückfragen aus Datenschutzgründen nicht mehr möglich waren. Zum anderen konnte nicht jedes Angebot auf seine Richtigkeit geprüft werden. Oft waren die Fotos unzureichend und die Angaben (Firmenmarken, Datierungen, Signaturen und Modellnummern) sehr spärlich bzw. falsch. Da das Ziel eine möglichst breite Erfassung der Goldscheider-Modelle war, wurde bewusst in Kauf genommen, dass nicht alle Modelle exakt und in allen Details beschrieben werden konnten. Wären nur Objekte in das Verzeichnis aufgenommen worden, die die Autoren selber in Augenschein genommen haben, wäre nur ein Bruchteil des Materials und der Informationen möglich gewesen.

Dass in diesem Werkverzeichnis nicht – wie sonst in der Porzellan- und Keramikliteratur üblich – je nur ein Nachweis für ein Modell aufgelistet wird, hat einen wichtigen Grund. Durch die bei Goldscheider übliche Praxis, die Objekte neben einer Modellnummer auch mit der Stückzahl und der Zahlenkombination für den ausführenden Modelleur zu versehen, kann dargelegt werden, wie häufig ein Modell produziert wurde und welchen Weg die Figuren genommen haben (Provenienz). Anhand der unterschiedlichen Firmenstempel kann festgestellt werden, zu welcher Zeit es von der Firma bzw. deren Nachfolgern (Josef Schuster in Wien, die amerikanische Produktion in Trenton oder die deutsche Weiterführung in Fredelsloh durch die Firma Carstens) reproduziert wurde.

Aufbau des Verzeichnis-Eintrags:

Nacheinander sind im Werkverzeichnis folgende Angaben aufgeführt: gesicherte, von Goldscheider vergebene **Modellnummer** (grau hinterlegt), gesicherter **Künstlername** (fett), die Objektart (in Versalien), der **Modelltitel** (dahinter in Klammern nähere Erläuterungen zum Titel/Objekt) sowie die **Datierung**. In Sonderfällen steht vor der Datierung eine Angabe zum **Material**. Hierzu im Einzelnen auch weiter unten nähere Informationen.

Unter diesem Grundeintrag können nach dem gleichen Schema spezielle Stücke des Modells aufgeführt sein, die im Markt gehandelt wurden. Hier wird dann die meist dreigliedrige **Nummerierung** angegeben sowie die Beschreibung der Marke (**Firmenstempel/Zusätze/Pickerl**), ob und wie das Stück **signiert** ist. Danach folgen Maßangaben; falls die Provenienz eines Stücks bekannt ist, wird diese ebenfalls aufgeführt. Als Bemerkungen („Bem.:") werden nähere Informationen zu einer Darstellung, zum historischen/kulturellen Hintergrund, Ausstellungen oder Literatur angegeben (s.a. **Bibliografie**). Weiter gibt es Verweise zu ähnlichen/gleichen Modellnummern, Pendants. Ein Eintrag „–> vgl. .... (H. 47 cm)" bedeutet zum Beispiel, dass das gleiche Modell auch mit einer anderen Höhe nachweisbar ist; die zugehörige Modellnummer konnte jedoch (noch) nicht verifiziert werden.

Derselben Systematik des Werkverzeichnisses für die Wiener Manufaktur FG folgen auch das Verzeichnis der „Modernen Keramik" und Gebrauchskeramik, das Verzeichnis der Modelle „ohne Nummerierung" sowie die Verzeichnisse für Goldscheider West Germany, Arthur Goldscheider Paris (La Stèle) sowie Marcell Goldscheider, Wien. Die Angaben für die Verzeichnisse von Goldscheider USA sowie Marcell Goldscheider, Stoke-on-Trent sind verkürzt mit einfacher Nennung der Modellnummer, Objektart, Titel und Maßangaben. Die Einträge werden hier sowohl nach Nummern wie auch nach Künstlern geordnet aufgeführt.

Alle Abkürzungen sind im **Abkürzungsverzeichnis** aufgelöst.

Modellnummer / Nummerierung

Bei Goldscheider findet sich meist eine dreigliedrige Zahlenfolge, wobei die erste Zahl die Modellnummer darstellt. Die zweite Ziffernfolge gibt Aufschluss über die jeweils produzierte Stückzahl (also unter welcher Produktionsnummer das vorliegende Stück geführt wird, nicht – wie häufig angenommen – die gesamte Anzahl der produzierten Stücke). Die dritte Zahl wurde als Kürzel für den Modelleur/Ausformer vergeben.

In der Regel erfolgte die Vergabe der Modellnummern bei Goldscheider chronologisch. Identische Modelle in verschiedenen Größen bekamen jedoch keine direkt aufeinander folgenden Zahlen, da sie erst bei eindeutigem Markterfolg in anderen Größen und dementsprechend später produziert wurden.

In der Auflistung bedeutet zum Beispiel:
„5230 (ohne Ang.)": in der Quelle ist ausschließlich die Modellnummer ohne nähere Angaben zu Stückzahl, Modelleur, Malerzeichen etc. vermerkt.
„ohne Ang.": in der Quelle sind keinerlei Angaben zur Modellnummer bzw. Firmenmarke, diese wurden alleine durch Modell- und Größenvergleich ermittelt.

Die zusätzlichen Bezeichnungen hinter den Modellnummern (die etwa ab 1930 eingeführt werden) „F" für Fayence und „M" für Moderne (Kunst)-Keramik sind dem Firmenkatalog „Wiener Manufaktur Friedrich Goldscheider Nach(folger)" entnommen. Für einige Modelle ab 1900 findet sich ebenfalls ein „F" (direkt in den Scherben vor der Modellnummer eingepresst), was ebenfalls die Ausführung in Fayence bzw. farbig glasiert bezeichnet.

Modellnummern, die im Werkverzeichnis in Klammern stehen, bezeichnen Objekte, deren Urheberschaft für Goldscheider fragwürdig ist. Sie wurden hier auch nur in Ausnahmefällen aufgenommen als Hinweis oder wenn es sich nach Qualität und Sujet um ein Stück ohne Firmenstempel handeln könnte; dies wird dann unter „Bem." näher erläutert.

Modelle mit dem Zusatz „(GSB)" belegen Objekte aus der Pariser Produktion und sind im Firmenkatalog Frédéric Goldscheider „Fabrique de Bronzes" (ca. 1895/1900) nachgewiesen.

Notes on the Catalogue of Works

From 1885 the output of the Goldscheider Manufactory comprised several hundred thousand pieces, which were based on more than ten thousand individual models. After all, the models were not just manufactured in all sorts of different materials and for various uses (for instance, figures were enlarged into lamps); they also reveal manifold and diverse decoration schemes. If the authors had had the utopian goal of ascertaining all individual models and variations as well as attributing all objects to specific artists (many objects are, moreover, unsigned), the present catalogue of works would never have been finished. Nevertheless, this catalogue of works provides by far the most comprehensive survey of Goldscheider objects available since it includes several thousand models. It should give all collectors, dealers and those interested in art interesting new insights into the Austrian and French products in the Historicizing, Jugendstil/Art Nouveau and Art Déco period styles. In addition, it shows examples of the American and English products made by Walter and Marcel Goldscheider, respectively, as well as all models made in Germany (Fredelsloh) in the years 1953–1960.

Recording models in the catalogue of works necessitated processing and including as many details as possible. All items of information available on a particular model were taken over into the catalogue. This had to be done in some cases without checking since information from auctions, to take one example, was quite far back in the past and data protection regulations made it impossible to inquire in more detail. On the other hand, it was not possible to verify the accuracy of each offer. Photographs were often insufficient and the data provided (firm marks, dating, signatures and model numbers) were very scant indeed or even erroneous. Since the aim was to provide as broad a record of the Goldscheider models as possible, it was accepted from the outset that not all models could be precisely described in details. If only objects that the authors had personally seen had been taken into the catalogue, only a fraction of the material and information could have been covered in here.

There is an important reason why not just a single entry for each model – which is the usual practice in publications on porcelain and pottery – has been listed in this catalogue of works. The usual practice at Goldscheider of giving objects not only a model number but also marking them by piece, that is, with the product number under which a particular piece was listed and a combination of numerals for the modeler who executed them makes it possible to ascertain how often a model was produced and where the figures went (provenance). The various firm stamps make it possible to ascertain when a particular model was reproduced by the firm or its successors (Josef Schuster in Vienna, the American factory in Trenton, New Jersey, or the continued production in Germany at Fredelsloh by the firm of Carstens).

How catalogue entries are structured:

The following items of information are presented in the catalogue of works in the sequence given as follows: the verified **model number** (shaded in grey) given by Goldscheider; the verified **artist name** (bold-face); the type of object (in capital initials); the **model title** (behind it in parentheses more precise explanations on the title/object) and the **date**. In special cases, the **material** is noted before the date. See below for more detailed information on the individual points.

Under the basic entry as outlined above, special pieces made up from a model and marketed as that model may be listed according to the same scheme. Here the **numbering**, which usually has three parts, is given with a description of the marks (**firm stamp / additional marks / labels**), and whether the piece is signed, and if so, how it **is signed**. The dimensions of the object then follow. If the provenance of a piece is known, it, too, is given. More precise items of information on a representation, such as historical/cultural background, exhibitions or publications (also see the **bibliography**) are added under the heading "Bem.:" (note). Further, there are cross-references to similar/identical model numbers and/or companion pieces ("Pendant"). An entry that looks like "–> vgl. .... (H. 47 cm)" means, for instance, that the same model is known to have been produced in a different size (height); however, the model number that would belong to this variant has (not yet) been verified.

The same structuring system that has been used for the pieces from the Wiener Manufaktur FG also applies for the catalogues of "Modern Ceramics" and utilitarian ceramics, the list of "unnumbered" models and the Goldscheider West Germany, Arthur Goldscheider Paris (La Stèle) and Marcell Goldscheider, Vienna, catalogues. The data given for the Goldscheider USA and Marcell Goldscheider Stoke-on-Trent catalogues are abbreviated to the following: the simple listing of the model number, object type, title and dimensions. Here the entries are arranged both by number and by artist.

All abbreviations are explained in the **list of abbreviations**.

Model number / Numbering

Goldscheider usually used three numerical sequences, with the first number representing the model number. The second numerical sequence imparts information on the number of pieces produced (that means under what production number the piece concerned is listed, not – as is often assumed – the entire number of pieces produced). The third number given is the code for the modeler/caster.

As a rule, numbers were assigned chronologically at Goldscheider. Models that were identical except in size, however, were not numbered in sequence since they were not produced in different sizes until later, after they had definitely proved successful on the market.

The following explanations clarify the above:
"5230 (ohne Ang.)": = "no data," in the source, only the model number is given without more precise data on the product number, the modeler, a painter sign, etc.
"ohne Ang.": = "no data,"the source provides no data on the model number or firm mark: in this case, they were ascertained solely by comparison of models and sizes.

The additional designations following the model numbers (which were introduced around 1930), such as "F" for faience and "M" for Modern (Art) Ceramics are taken from the "Wiener Manufaktur Friedrich Goldscheider Nach(folger)" firm catalogue. Some models from 1900 also bear an "F" (impressed directly into the clay body before the model number), which also denotes that it was made in faience and/or glazed in color.

Model numbers given in parentheses in the catalogue of works designate objects that may not have been made for or by Goldscheider, objects that are in any case dubious, if not spurious. They have only been included here in exceptional cases for reference or if the piece concerned might plausibly be a Goldscheider without a firm stamp because the quality is good enough and the

Künstlername / Signatur

Ist im Werkverzeichnis bei einem Modell ein Künstlername (fettgedruckt) angegeben, so wurde in der Auflistung der einzelnen Stücke, falls diese nur mit dem Familiennamen des Künstlers signiert wurden, auf die erneute Nennung seines Namens verzichtet. Differierende Namensformen wie z.B. bei Lorenzl/Laurenzl, Petri/Petrides usw. wurden, soweit in der Quelle (Auktionskataloge, Kunsthandel, private Sammlungen etc.) angegeben, in ihrer entsprechenden Namensschreibweise berücksichtigt und angeführt. Vollständige Signaturen mit Vornamen oder Monogramme (z.B. Claire Herczeg oder E. Tell) wurden explizit vermerkt.

Bei der Bearbeitung des Werkverzeichnisses verstärkte sich der Verdacht, dass man es bei Goldscheider mit den Künstlersignaturen nicht sehr genau nahm. Besonders auffällig ist, dass z.B. Figuren wie etwa die Windhunde (–> vgl. Modellnummern 3859, 4187 und 4687) fast identisch sind und einmal mit Gemignani, ein anderes Mal mit Latour signiert sind. Bei anderen Modellen wurden die Namen angepasst („Fischer" = „Pecheur", „Bormann" = „Beaume", „Schörk" = „Cherc"), damit sie französisch klangen und für den internationalen, insbesondere französischen Markt attraktiver wurden. Es scheint auch, dass etliche Künstlersignaturen wie „Simon" oder „Podola" einfach nur erfunden sind, da sich diese Namen in der einschlägigen Literatur nicht nachweisen lassen.

Eines der Hauptprobleme bei „industriell" erzeugter Keramik ist, dass nur ein Teil der Objekte vom Künstler (bzw. von der Firma) signiert wurden. Als besonders problematisch erweist es sich bei Goldscheider, dass Modelle von Josef Lorenzl und Stephan Dakon oftmals kaum eindeutige stilistische Unterscheidungsmerkmale aufweisen. Auch die Wandmasken von Rudolf Knörlein und Adolf Prischl weisen häufig Ähnlichkeiten auf, so dass nur die wenigen, meist mit „Pickerln" signierten Objekte eindeutigen Aufschluss über den Künstler geben. Bei Goldscheider kommt hinzu, dass fest angestellte Mitarbeiter wie Viktor Matula ihre Modelle nicht signiert haben, da die Firma alle Rechte durch die Mitarbeiterschaft des Künstlers besaß. Ähnlich war es auch bei Adolf Prischl. Wie im Werkverzeichnis belegt werden kann, gibt es aber auch hier Ausnahmen.

Modelltitel

Es wurden nur die Originaltitel, die sich in der zeitgenössischen Literatur, in Ausstellungs-Katalogen bzw. in Zeitschriften nachweisen ließen, übernommen und in Anführungszeichen gesetzt. Vereinzelt neu vergebene, beschreibende Titel, wie z.B. „Der Kuss" (wenn sich ein Paar küsst), wurden ohne Anführungszeichen ins Werkverzeichnis eingebracht.

Datierungen

Die Datierung verläuft in der Regel nach der Reihenfolge der Modellnummern. Abgewichen wurde nur durch die Bezeichnung „(für den Entwurf)", was bedeutet, dass hier der ursprüngliche Entwurf bzw. die Erstausformung einer anderen Größe schon früher existierte.

Material

Das Material bzw. der Scherben der Objekte wurde nur dort angegeben, wo es unbedingt notwendig erschien (z.B. bei Bronze und anderen Metallen, Marmor und Alabaster sowie Elfenbein). Die Erzeugnisse vor 1900 wurden meist in Terrakotta, teils mit Kaltbemalung ausgeführt, ab 1904 wurde auch in Fayence (glasierte Keramik) gearbeitet. In den 1930er Jahren gewann dann wieder der terrakottafarbene Scherben verstärkt an Bedeutung, vor allem in der Majolika der angewandten modernen Keramik.

Firmenstempel / Zusätze / Pickerl (Sticker)

Bei der Manufaktur war es üblich, die Produkte durch einen im Laufe der Jahrzehnte sich verändernden Firmenstempel oder eine Prägung eindeutig als Erzeugnis von Goldscheider kenntlich zu machen (vgl. Markenverzeichnis, S. 604ff). Sehr selten unterblieb das Anbringen von Marken oder Stempel.

Es gab natürlich Firmen, die im Erfolgssog von Goldscheider ähnlich produzierten oder nummerierten und den Stempel „Made in Austria" oder „Austria" bzw. ähnliche Nummernfolgen als Modellnummer auf den Objekten anbrachten, um vom Anschein her Goldscheider zu imitieren (vgl. hierzu das Kapitel „Tipps für Sammler", S. 610ff). Dies waren teils Wiener Konkurrenten (vgl. hierzu: Annonce von Pilar, Leipz. Messe, in: Keram. RS) oder noch häufiger böhmische oder auch französische Firmen. Aus der Produktion der Kölner Firma König & Lengsfeld, die in ihrem Eintrag in Greven's Adressbuch für Köln und Umgebung (1904; Teil II, S. 327) wie folgt wirbt: „Kunstgewerbliche Ateliers, Galvanoplastische Kunstanstalt, Figuren, Vasen, Büsten etc. für Zimmerdekoration ... Ausführungen in Terrakotta, Marmor ...", lässt sich eine Jugendstilbüste von Affortunato Gory in der Sammlung John Nash nachweisen. Das kupferfarbene Sigel, das der Goldscheider-Marke „Sculpture" sehr ähnlich sieht (s.a. Ill. S. 611 oben) und ebenfalls von Maurice Bouval signiert ist, könnte darauf hindeuten, dass König & Lengsfeld ihr Firmenlogo bewusst an Goldscheider anlehnen wollten, zumal auch die Modellnummer 3311 dem Duktus von Goldscheider entspricht. Selbst Ora Pinhas bezeichnet in ihrem Buch („Goldscheider, A Catalogue of Selected Models", S. 200 m. Abb.) diese Marke mit dem ligierten Monogram KL als eine „späte ungewöhnliche Plakette" von Goldscheider. Als weiteres Beispiel für die Imitation einer Goldscheider-Firmenmarke sei noch Wilhelm Thomasch genannt. Dieser Künstler, der auch für Goldscheider arbeitete, lehnte sich bei seinem eigenen Manufakturstempel (Thomasch-Sierndorf) an die Goldscheider-Marke der 1920er Jahre an (ähnlich dem ovalen Schriftzug GWM).

Bei Stücken, die keinen Goldscheider-Stempel tragen, sei immer Vorsicht geboten! Erfahrene Sammler können „Konkurrenzfirmen" aber mit einiger Sicherheit als solche entlarven. Man kann grundsätzlich sagen, dass deren Modelle in den seltensten Fällen die Qualität von Goldscheider erreichen.

Der Zusatz „Fabriqué en Autriche" (hergestellt in Österreich), bedeutet, dass diese Objekte speziell für den frankophonen Markt produziert wurden.

Größenangaben

Größenangaben wurden, sofern sie in Sammlungen und Museen selbst nachgemessen werden konnten, direkt ermittelt. Bei Objekten, die nicht selbst in Augenschein zu nehmen waren, wurden offensichtlich falsche Höhenangaben aus der Literatur berichtigt.

Zuschreibungen

Bei Zuschreibungen wurde in aller Regel sehr vorsichtig und nicht spekulativ vorgegangen. Es soll dem kundigen und ernsthaften Sammler überlassen bleiben, seine eigenen Zuschreibungen zu tätigen bzw. die Goldscheider-Stücke seiner Sammlung auf Signaturen zu überprüfen und mit dem Werkverzeichnis zu vergleichen.

Dieses Goldscheider Werkverzeichnis wird sicherlich vielen Sammlern, dem Kunsthandel und Museen neue Aspekte und Zusammenhänge eröffnen. Es wird natürlich auch zahlreiche Modelle von Goldscheider geben, die hier fehlen und in einer eventuellen Neuauflage aufgenommen und berücksichtigt werden sollten. Wenden Sie sich bitte diesbezüglich an den Verlag oder die Autoren, am besten per E-Mail an info@goldscheider.de.

motif one that Goldscheider produced; all this is explained then in more detail under "Bem.:" (= note).

Models with the additional mark "(GSB)" were made in Paris and have been verified from the Frédéric Goldscheider "Fabrique de Bronzes" (ca 1895/1900) firm catalogue.

Artist name/Signature

If an artist name is given (in bold-face) in the catalogue of works for a particular model, the artist name is not given again when individual models by the same artist are listed if they are signed only with the artist's surname. Varying spellings for, and forms of, names, for instance Lorenzl/Laurenzl, Petri/Petrides, etc, as given in the sources (auction catalogues, art market, private collections, etc), are reproduced as in the source and the variations in spelling are duly listed. Complete signatures with first names or monograms (for instance, Claire Herczeg or E. Tell) are explicitly noted.

While we were drawing up the catalogue of works, our suspicions were confirmed that at Goldscheider no one was too particular about artist signatures and attribution. It is especially noticeable that figures such as the whippets (–> cf model numbers 3859, 4187 and 4687) are almost identical yet in one instance the signature is Gemignani, in another Latour. Other models bore artist names that had been Gallicized, either by being literally translated from German into French or by being spelled or phonologically transcribed so as to sound French ("Fischer" = "Pecheur," "Bormann" = "Beaume," "Schörk" = "Cherc") to make them more attractive on the international, specifically the French, market. It also seems as if quite a few artist signatures, such as "Simon" or "Podola," were simply invented since these names cannot be verified in the relevant sources and specialist publications.

One of the chief problems with "industrially" produced ceramics is that only some but not all objects are signed by the artist (or the firm). A particularly knotty problem with Goldscheider is that many models by Josef Lorenzl and Stephan Dakon do not reveal unequivocally diagnostic stylistic features. Further, wall masks by Rudolf Knörlein and Adolf Prischl are often so similar that only a few objects, most of them "signed" with "stickers", can be attributed with assurance to an artist. Moreover, at Goldscheider, employees on the payroll such as Viktor Matula did not sign their models because the firm reserved all rights to such models and the artist had agreed to the conditions imposed when he entered on permanent employment. Adolf Prischl's case is similar. However, as the catalogue of works shows, there were exceptions to this practice, too.

Model titles

Only the original titles that have been verified in contemporaneous publications, in exhibition catalogues or journals and magazines have been taken over and are given here in double quotation marks. In a few cases, modern titles have been used in the present catalogue of works, such as "The Kiss" (when a couple is quite obviously kissing), without being placed in quotation marks.

Dating

Dating is done by checking the sequence of model numbers. The only deviation from this standard practice has been indicated by "(for the design)," which means that the original design or even a first cast in a different size previously existed.

Material

The material and the body of which the objects are made, has only been indicated when doing so seemed absolutely necessary (for instance, in the case of bronze and other metals as well as marble, alabaster and ivory). Most Goldscheider products made before 1900 were in terracotta, some of them with cold painting, and from 1904 the firm also worked in faience (glazed pottery). In the 1930s terracotta body gained importance again, especially majolica for the modern ceramics.

Firm stamps/Additional marks/Labels (Stickers)

It was usual at the Goldscheider Manufactory to mark objects clearly as Goldscheider products with a firm stamp that changed over decades of use or an impressed stamp mark (cf Marks Table, p. 604f). Only very rarely was a Goldscheider object not designated as such by marks or a stamp on it.

There were, of course, copycat firms that wanted to benefit from the Goldwater success story by producing similar objects, numbering them according to a similar system, stamping them "Made in Austria" or simply "Austria" and employing similar number sequences as model numbers to make them look like Goldscheider pieces (cf on this the chapter "Pointers for Collectors", p. 610f). Some of those imitators were the competition in Vienna (cf on this: an advertisement placed by Pilar, Leipz. Messe [Leipzig Trade Fair], in: Keram. RS) or, more frequently, Bohemian or even French firms. A Jugendstil bust of Affortunato Gory in the John Nash Collection can be identified from the following entry in Grevens Adressbuch für Köln und Umgebung (1904; Part II, p. 327) on the product line marketed by the Cologne firm of König & Lengsfeld, who advertised as follows: "Crafts studios, galvanoplastic art institute, figures, vases, busts, etc, for interior decoration [...] Executed in terracotta, marble [...]." The copper-colored seal, which looks very similar to the Goldscheider "Sculpture" mark (see also ill. on p. 611, above), might be taken as an indication that König & Lengsfeld deliberately sought to borrow from the Goldscheider firm logo, especially since the model number, 3311, also corresponds to the Goldscheider numbering system. Even Ora Pinhas in her book ("Goldscheider, A Catalogue of Selected Models," p. 200 with illustration) calls this mark, which is a monogram because the letters "KL" are in ligature, an "unusual late plaquette" from Goldscheider. Mention should be made here of Wilhelm Thomasch as another example of imitating a Goldscheider firm mark. An artist who also worked for Goldscheider, Thomasch borrowed stylistically from the 1920s Goldscheider mark for his own factory stamp (Thomasch-Sierndorf), which is similar to the oval GWM lettering.

Always be careful with pieces that bear no Goldscheider stamp! Experienced collectors can, of course, expose "rival firms" as such with some assurance. Basically, one can say that their models only very rarely attained Goldscheider quality.

The addition "Fabriqué en Autriche" [Made in Austria] means that objects bearing that mark were made specifically for the Francophone market.

Dimensions

Dimensions, if they could be checked in collections and museums, were ascertained directly from the object. Even in cases when it has not been possible to examine objects personally, heights given in publications which quite obviously seemed erroneous have been corrected.

Attribution

When attribution was concerned, extreme caution has been the order of the day and conjecture has not been relied on. It should be left up to knowledgeable, serious collectors to do their own attributing or to check the Goldscheider pieces in their collections for signatures and compare them with the catalogue of works.

This catalogue of Goldscheider works is bound to open up new aspects and links for many collectors, antiques dealers and museums. Of course there will also be numerous Goldscheider models that are missing here, which should be taken into consideration in a possible revised new edition. If you have anything to add in this respect, please contact the publishers or the authors, preferably by e-mail to info@goldscheider.de.

Wiener Manufaktur Friedrich Goldscheider

Viennese Manufactory Friedrich Goldscheider

Die mit der Notiz „Überschneidung" gelisteten Modelle sind damit zu erklären, dass man in der Pariser Produktion von Friedrich (nachgewiesen im GSB Katalog um 1900) neu mit den Modellnummern begann und sich dadurch diese teils mit den Wiener Modellen überschneiden. Weitere Doppelbelegungen sind, wegen der umfangreichen und vielfältigen Herstellung, meist auf Versehen der ausführenden Modelleure zurückzuführen.

The explanation for models being listed with the remark "Überschneidung" [overlap] is that in Friedrich's Paris product range (as shown in the GSB catalogue ca 1900), model number series began from the beginning again so that some of these numbers overlap with those on Viennese models. Other occurrences of the same number being used for two different models are usually due to the modelers executing them mistakenly re-using numbers because so many and so very different models were made.

XIII
PLATTE (Wandrelief? aus Terrakotta)
1895
XIII ohne Ang.; KeramRS, 3.1895, S. 345
Bem.: s. Modell-Nr. 6

XIV
PLATTE (Wandrelief? aus Terrakotta)
1895
XIV ohne Ang.; KeramRS, 3.1895, S. 345
Bem.: s. Modell-Nr. 6

4
BÜSTE
1895
4 ohne Ang.; KeramRS, 3.1895, S. 545
Bem.: Es wurden im Juni 1895 „drei Büsten mit eigenartiger Bemalung und Farbenzusammenstellung (Dekorationsmuster)" zum Musterschutz eingereicht. Warum alle drei Objekte dieselbe Nummer erhielten kann nicht mehr nachvollzogen werden. Diese Praxis bedeutet jedoch, dass die Manufaktur zumindest in den Anfängen die Modellnummernvergabe nicht als so wichtig erachtete und diese auch nicht exakt in chronologischer Reihenfolge vornahm. Der deutlichste Beleg für dieses Faktum zeigt die gleichzeitige Einreichung der Modell-Nrn. 211, 468, 884 sowie 1162, die alle am 17.9.1896 in der KeramRS veröffentlicht wurden. In der Folge kommt es daher immer wieder zu „Überschneidungen" bei den Nummern, welche auch im Werkverzeichnis als solche gekennzeichnet sind. Weitere Bem. s. Modell-Nr. 6.

6
WANDTELLER (Rokokostil)
1893
6 ohne Ang.; KeramRS, 1.1893, S. 321
Bem.: Hier die Einreichung in Leipzig, um den „Musterschutz" zu erwirken. Die Muster wurden durch „photographische Abb." angemeldet, um sie für jeweils drei Jahre vor Nachbildung durch andere Hersteller gesetzlich schützen zu lassen.

7
WANDTELLER (Rokokostil)
1893
7 ohne Ang.; KeramRS, 1.1893, S. 321
Bem.: s. Modell-Nr. 6

8
WANDTELLER (mit Spiegeleinsatz, vergoldet)
1893
8 ohne Ang.; KeramRS, 1.1893, S. 321
Bem.: s. Modell-Nr. 6

12
RAHMEN
1893
12 ohne Ang.; KeramRS, 1.1893, S. 531
Bem.: s. Modell-Nr. 6

13
RAHMEN
1893
13 ohne Ang.; KeramRS, 1.1893, S. 531
Bem.: s. Modell-Nr. 6

14
RAHMEN
1893
14 ohne Ang.; KeramRS, 1.1893, S. 531
Bem.: s. Modell-Nr. 6

14A
Houssin, Edouard Charles Marie
BÜSTE Mädchen
BRONZE/Marmorsockel
ca. 1900
14,A FG Wien oder Arthur GS Paris (Firmenmarke?), sign. E. Houssin/Paris, Gesamth. 28 cm; Quit. 1
-> vgl. .... (H. 47 cm); Soth. L. 20.5.1994

15
RAHMEN
1893
15 ohne Ang.; KeramRS, 1.1893, S. 531
Bem.: s. Modell-Nr. 6

16
RAHMEN
1893
16 ohne Ang.; KeramRS, 1.1893, S. 531
Bem.: s. Modell-Nr. 6

17
RAHMEN
1893
17 ohne Ang.; KeramRS, 1.1893, S. 531
Bem.: s. Modell-Nr. 6

17A
Maurel, E(mile)
VASE mit Heuschrecke
BRONZE (poliert)
ca. 1900
17,A Firmenmarke Sculpture Bronze 1, H. 29 cm; PB Wien 19

18
RAHMEN
1893
18 ohne Ang.; KeramRS, 1.1893, S. 531
Bem.: s. Modell-Nr. 6

19
SCHALE
1893
19 ohne Ang.; KeramRS, 1.1893, S. 531
Bem.: s. Modell-Nr. 6

20
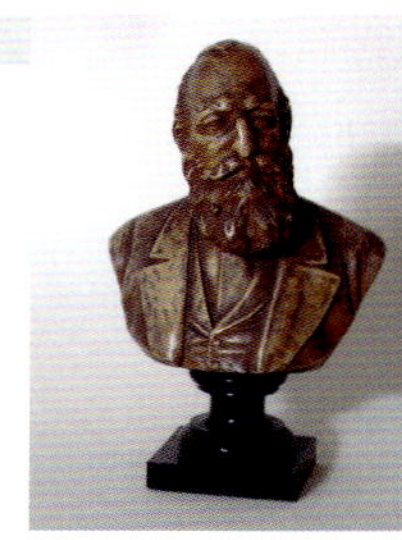
BÜSTE „Theodor Billroth" (auf Holzsockel)
ca. 1889
20/12 ger., Firmenmarke 3, gepr. „I"= Z17, H. 16,5 cm (ohne Sockel); PB Deutschland
Bem.: Dargestellt ist der Chirurg und Mediziner Theodor Billroth (1829–1894).

22
Stehender Page (in mittelalterlicher Tracht), ein Buch haltend
BRONZE/Marmorsockel
ca. 1900
22 Plak. A, Gesamth. 48,3 cm; eBay # 64751608

23
Delagrange, Léon Noel
Stehender Deutschordensritter mit Schwert
BRONZE/ELFENBEIN/Marmorsockel
ca. 1895/1900
23 Bronze AG Paris (wohl Arthur GS Paris= Firmenmarke Bronze 4), H. 33 cm; Soth. L. 9.10.1992
Bem.: An der Plinthe sek. bezeichnet „1914–1919".
Lit.: zur Marke -> vgl. Berman, Vol. 2, S. 486 (Marke 42, m. Abb.)

30
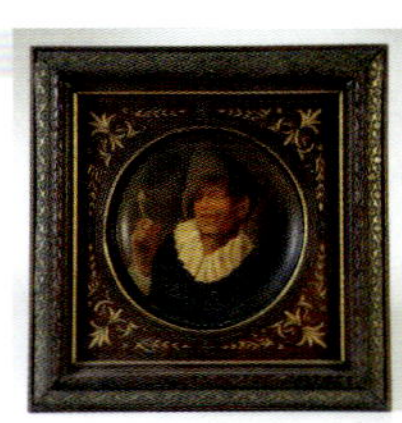
WANDTELLER Rauchender Bauernbursche (im Holzrahmen)
ca. 1897
30,8 ger., handschriftl. datiert 1897, Firmenmarke 3, Vervielf. vorbeh., Ø 47 cm; Wien Museum Inv. 219.143
Bem.: Der Bilderrahmen ist aus der Zeit, könnte jedoch sek. für einen der Vorbesitzer angefertigt worden sein.

35
Chalon, Louis
Walküre (La Walkyrie) (Kämpferin mit Speer auf springendem Pferd)
BRONZE/ELFENBEIN/Onyxsockel
ca. 1897/98
35,JD (Firmenmarke Bronze 1 und/oder 2?), Vrai Bronze, H. (mit Sockel) 63,5 cm; Christie's NY 25.10.2005
Bem.: Für seinen Entwurf der Walküre erhielt Chalon 1898 eine „Mention Honorable".
-> vgl. 1199
Lit.: Forrest, M.: Art Bronzes (1988), S. 454; Berman, Vol. 3, Nr. 2568 (m. Abb.)

38
Kleiner Bär
ca. 1889
38,3 Firmenmarke 3, H. 14 cm; PB Wien 19

40
Wind, Josef
„Schlangenbändigerin"/„Charmeuse des Serpents (nue)" (Stehende nackte Orientalin mit Turban, Flöte und Schlange)
BRONZE
ca. 1897/98
40 „GS Paris" Vraie Bronze, H. 54 cm; Soth. L. 30.3.1990
Bem.: Vermutlich ist die von Sotheby's angegebene Modell-Nr. 40 nicht korrekt, oder es handelt sich um die Stückzahl und nicht um die Produktionsnummer. In der Lit. GSB finden sich nur 3-stellige Nummern ab etwa 500 aufwärts. Eine Ausführung in der H. 85 cm jedenfalls läuft unter der Modell-Nr. 644.
Ausst.: ÖMKI (1898/99) ausgest. war die Ausführung in Terrakotta, hier mit Ang. des Originaltitels „Schlangenbändigerin".
-> vgl. 1399 TERRAKOTTA
-> vgl. .... (H. 55 cm); Richemond (1995), S. 62
-> vgl. 644 (GSB)
Lit.: GSB, Pl. 56, hier ist die Figur mit dem franz. Originaltitel „Charmeuse des Serpents (nue)" unter Nr. 644 (H. 85 cm) abgebildet. Ang. zum Material (Terrakotta/Bronze) fehlen.

42 -> ÜBERSCHNEIDUNG
Chalon, Louis
Walküre (La Walkyrie) (Kämpferin mit Speer auf springendem Pferd)
BRONZE/Marmorsockel
ca. 1897/98
42,1199,LX (Firmenmarke Bronze 1 und/oder 2?), H. (mit Sockel) 120 cm; Soth. L. 23.11.1990
Bem.: Für seinen Entwurf der Walküre erhielt Chalon 1898 eine „Mention Honorable".
-> vgl. 35
Lit.: Forrest, M.: Art Bronzes (1988), S. 454; Berman, Vol. 3, Nr. 2568 (m. Abb.)42

42 -> ÜBERSCHNEIDUNG
Cardona, José
Stehender, Zigarette rauchender Junge mit Schiebermütze
BRONZE
ca. 1900
42,L,X FirmenmarkFirmenmarke Bronze 1), sign. J. Cardona, H. 37 cm; eBay # 861649232

44(?)
Delagrange, Léon Noel
BÜSTE Venezianischer Edelmann
BRONZE/ELFENBEIN?/Achatsockel
ca. 1900
44(?) Plak. A, H. 16,5 cm; KH

61
Clarà, Juan (Joan)
Mädchen mit Katze spielend (auf Rechtecksockel)
BRONZE
ca. 1900
61,A,PR Firmenmarke Sculpture Bronze 1, sign. Juan Clarà, H. 27 cm; Wien Museum Inv. 190.096
Lit.: Catley (1987), S. 105 (m. Abb.).

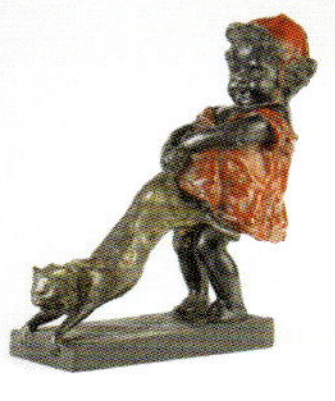

107
Feixender, sich verbeugender Hofnarr (Zwerg)
ca. 1887/88
107 (Firmenmarke ohne Ang.), H. 49,5 cm; Pinhas, S. 34 (m. Abb.)

109
VASE Stehender Minnesänger mit Mandoline (am Vasenkorpus)
ca. 1897
0109 (Firmenmarke ohne Ang.), H. 66 cm; Pinhas, S. 52 (m. Abb.)

111
Philippe, Paul
„Le Réveil" (Stehender Damenakt mit ausgestrecktem Arm) (auf Rundsockel)
Inschrift: LE RÉVEIL
BRONZE/Marmorsockel
ca. 1900 (für den Entwurf)
111,D,1413 ger., (Firmenmarke ohne Ang.), sign. P. Philippe, H. 71 cm; Christie's Sy. 15./16.5.2004
Text der Inschrift: Das Erwachen (Die Erwachende)
Bem.: Eine Version in Marmor durch Arthur Goldscheider, Paris verlegt wurde 1907 im Pariser Salon ausgestellt (vgl. Lit.)
Lit.: N.N.: Bronze-Elfenbein Statuetten der Neuzeit. Künstlerkat. 7. Forts., in: AZ, 1979, S. 581

119
GRUPPE Drei Negerknaben auf einer Mauer sitzend
ca. 1890
0119,030,10 FG Wien (Firmenmarke?), H. 56,5 cm; Acatos 1991, S. 262
119 (BRONZE) vgl. 1308 (BRONZE)

184
Jacques, E.
JARDINIERE Zwei Frauen, die sich an den Händen halten und deren Kleider die Schale bilden
ca. 1897/98
0184 (Firmenmarke ohne Ang.), H. 38 cm; Pinhas, S. 49 (m. Abb.)

196
LAMPE Stehendes junges Mädchen mit herabhängendem Haar
ca. 1897/98
0196 (Firmenmarke ohne Ang.), H. 63,5 cm; Pinhas, S. 51 (m. Abb.)

203
BÜSTE „Schiller"
ca. 1887/88
203 (Firmenmarke ohne Ang.), H. 24 cm; GSF, S. 3
Bem.: Dargestellt ist der Dichter Friedrich Schiller (1759–1805).
Lit.: GSF, S. 3, hier mit der Ang. des Originaltitels „Schiller".

204
BÜSTE „Goethe"
ca. 1887/88
204 (Firmenmarke ohne Ang.), H. 24 cm; GSF, S. 3
Bem.: Dargestellt ist der Dichter Johann Wolfgang von Goethe (1749–1832).
Lit.: GSF, S. 3, hier mit der Ang. des Originaltitels „Goethe".

206
GRUPPE Drei, auf einem Besen reitende Hexen
ca. 1890
206 (Firmenmarke ohne Ang.), H. 57 cm; Pinhas, S. 34 (m. Abb.)

211
Strasser, Arthur (zugeschr.)
BÜSTE Orientale mit Vollbart, Turban und Fransenschal
ca. 1895/96
211,12 (Firmenmarke ohne Ang.), Repr. Res., unsign., H. 20 cm; Schloss Hag. 37
211,58 (Firmenmarke ohne Ang.), Repr. Res., unsign., H. 18,5 cm; Doro 23.5.1991
Bem.: s. Modell-Nr. 6
Lit.: KeramRS, 4.1896, S. 863, hier die Einreichung zum „Musterschutz".

212
Strasser, Arthur (zugeschr.)
BÜSTE Orientale mit Kopfputz (Fez)
ca. 1895/96
212 Firmenmarke 10, unsign., H. 18 cm; WKH

213
BÜSTE Henry VIII.
ca. 1890
213,2 ger., Firmenmarke 4a, Ges. gesch. (ger.)= Z8/a, H. 14 cm; Sammlung f@goldscheider.de; Bem.: Sockel fehlt
Bem.: Dargestellt ist Henry VIII., König von England (1491–1547).
Pendant -> vgl. 214

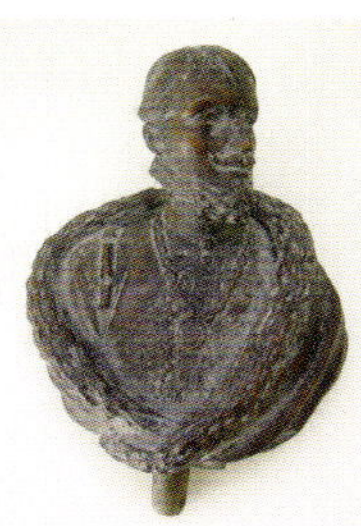

214
BÜSTE Anne Boleyn
ca. 1890
214, No. 3 ger., Firmenmarke 4a, Ges. gesch. (ger.) = Z8/a, H. 15 cm; Sammlung f@goldscheider.de;
Bem.: Sockel fehlt
Bem.: Dargestellt ist Anne Boleyn (1502–1536), die zweite Frau des engl. Königs Henry VIII.
Pendant -> vgl. 213

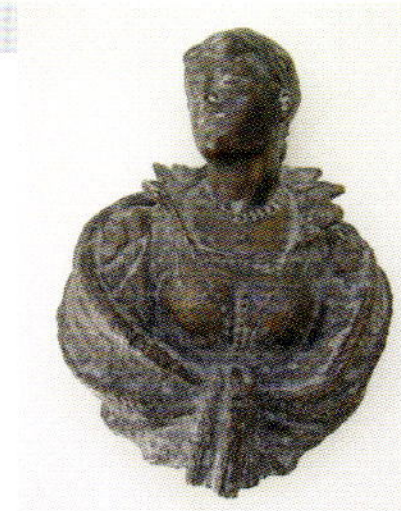

217
Strasser, Arthur (zugeschr.)
BÜSTE Orientale mit Kopfputz (Fez)
ca. 1888
217 Firmenmarke 3, Ges. gesch., unsign., H. 44 cm; Schmidt 65

228
BÜSTE Junger Mann mit Krawatte und Schürze
ca. 1888
228 ger., Firmenmarke 3, Ges. gesch., H. 11,4 cm; eBay # 1255297312

231
BÜSTE Joseph Haydn
ca. 1888
231 ger., Firmenmarke 3, Ges. gesch., H. fehlt; eBay # 73887670
Bem.: Dargestellt ist der Komponist Joseph Haydn (1732–1809).

236
Stehende Zigeunerin mit Pfeife und Korb (auf Landschaftssockel mit Mauer)
ca. 1888
236 ger., Firmenmarke 3, Vervielf. vorbeh., H. 35 cm; Wien Museum Inv. 219.137
236 Firmenmarke 3, Ges. gesch., H. 37 cm; eBay # 2216715429
236 Firmenmarke 3, Ges. gesch., H. 37 cm; eBay # 2573229236
Bem.: Pfeife evtl. sek.

237
BÜSTE Offizier (auf Holzsockel)
ca. 1888
No. 2,237 ger., Firmenmarke 3, Ges. gesch., H. (ohne Holzsockel) 43,7 cm, Gesamth. 55,7 cm;
MAK Wien, Inv. Ke 10521-7

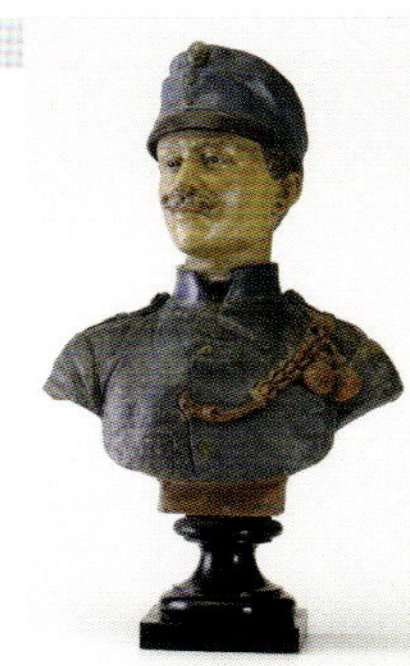

242
Conquistador (Stehender Ritter/Eroberer in Rüstung)
ca. 1890
242,38 (Firmenmarke ohne Ang.), H. 101 cm; Ribeyre-Baron 14.3.1995
250 (BRONZE) vgl. 1372 (BRONZE)

252
BÜSTE King Henry VIII (auf Holzsockel)
ca. 1887/88
252 (Firmenmarke ohne Ang.), H. 18 cm; Pinhas, S. 35 (m. Abb.)
Pendant -> vgl. 271

254
Stehender Mann mit Kappe (sog. Wiener Type)
ca. 1887/88
254 FG Wien (Firmenmarke ohne Ang.), Ges. gesch., H. 39,9 cm; Doro 13.9.1990
Bem.: zu den sog. „Wiener Typen" -> vgl. Schmidt, Otto: Wiener Typen und Strassenbilder, Wien 2000

259
Stehender Mann mit Hut und großer Schleife (Mascherl) (sog. Wiener Type)
ca. 1887/88
259,H ger. Firmenmarke 3, Ges. gesch., gepr. „P"= Z3, H. 58,5 cm; GK, S. 10, 49 (m. Abb.); Wien Museum Inv. 128.052
Bem.: zu den sog. „Wiener Typen" –> vgl. Schmidt, Otto: Wiener Typen und Strassenbilder, Wien 2000
Ausst.: HMW (jetzt Wien Museum) 1985/86.
Lit.: GK, S. 10, 49 (m. Abb.)

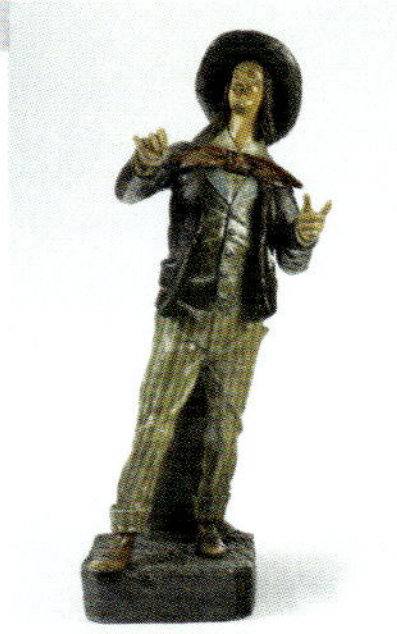

262
BÜSTE Bauer in Tracht mit Hut (auf Holzsockel)
ca. 1890
262,26 ger., Firmenmarke 3, Ges. gesch., H. (ohne Holzsockel) 16 cm, Gesamth. 21,5 cm; KH

271
BÜSTE Anne Boleyn (auf Holzsockel)
ca. 1887/88
271 (Firmenmarke ohne Ang.), H. 18 cm; Pinhas, S. 35 (m. Abb.)
Pendant –> vgl. 252

300
Stehender Bauernbursche (Winzer?), Kirschen (Trauben?) essend (auf Natursockel)
ca. 1890
300,144 (Firmenmarke ohne Ang.), H. 54,6 cm; eBay # 26832292
Bem.: Da der Arm der Figur stark restauriert wurde, kann nicht eindeutig festgestellt werden, ob die Kirschen evtl. eine sek. Beigabe sind und es sich beim Original nicht eher um Trauben handelte.

311
BÜSTE „Italienischer Knabe Alvalore"
ca. 1890
311 (Firmenmarke ohne Ang.), H. 59 cm; GSF, S. 3
Pendant –> vgl. 312
Lit.: GSF, S. 3, hier mit der Ang. des Originaltitels „Alvalore"; Pinhas, S. 71 (m. Abb.).

312
BÜSTE „Italienischer Knabe Con amore"
ca. 1890
312,V ger., Firmenmarke 11, Repr. Res., H. 54,6 cm; eBay # 148174536
312 (Firmenmarke ohne Ang.), H. 55 cm; GSF, S. 3
Pendant –> vgl. 311
Lit.: GSF, S. 3, hier mit der Ang. des Originaltitels „Con amore"; Pinhas, S. 71 (m. Abb.).

(314)
Le Guluche, J(oseph)
Wasserträger (Stehender Afrikaner mit Krug auf der Schulter)
ca. 1890
314 (Firmenmarke Hanne), H. 102 cm; PB Wien 19
Bem.: Es handelt sich hier nicht um Erzeugnis der Manufaktur GS, denn mehrere Quellen belegen für dieses Objekt den Hersteller „A. Hanne", vgl. hierzu: Richemond (1995), S. 61, hier identische Figur (H. 100 cm); vgl. Richemond (1995), S. 61 (H. 51,5 cm); vgl. Weichmann 18.3.1988 (H. 52 cm), jeweils gefertigt v. der Ed. „A. Hanne". Dass dieses Objekt hier Aufnahme findet, hat seinen Grund: häufig sind die Modelle von Hanne ungemarkt und werden ebenso oft wegen themat. Parallelen und ähnl. Ausführung mit GS verwechselt bzw. im Handel als solche angeboten. Die franz. Firma „Alphonse Hanne" (L' Isle Adam) war, besonders das Genre der Orientalen (Büsten und Figuren) betreffend, eine wichtige Konkurrenz zu GS. Hanne gewann ebenso wie GS zahlreiche Medaillen auf Ausst. und unterhielt in Paris eine Niederlassung in der Rue d' Hauteville 66.

318
BÜSTE Nordafrikaner mit Bart (nach rechts blickend) (auf Holzsockel)
ca. 1888/90
318 ger., Firmenmarke 3, Ges. gesch., H. (m. Holzsockel) 32,5 cm; GK, S. 10, 50 (m. Abb.); Wien Museum Inv. 182.053
Ausst.: HMW (jetzt Wien Museum) 1985/86.
Pendant –> vgl. 319
Lit.: GK, S. 10, 50 (m. Abb.)

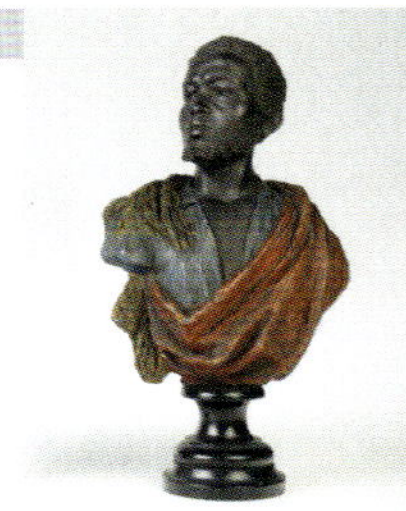

319
BÜSTE Nordafrikanerin mit Münzkette als Halsschmuck (nach links blickend) (auf Holzsockel)
ca. 1888/90
319, J.K. ger., gem. IV, Firmenmarke 3, H. (mit Holzsockel) 32 cm; GK, S. 10, 51 (m. Abb.); Wien Museum Inv. 182.054
Ausst.: HMW (jetzt Wien Museum) 1985/86.
Pendant –> vgl. 318
Lit.: GK, S. 10, 51 (m. Abb.)

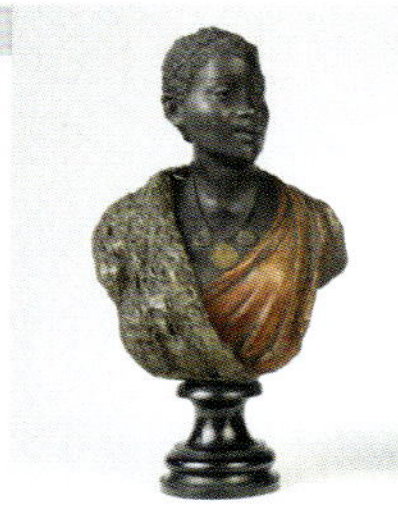

321
BÜSTE „Onkel Nolte" (Lachender alter Mann mit Zipfelmütze)
ca. 1895
321 ger., Firmenmarke 3, Ges. gesch., H. (m. Holzsockel) 27 cm; Sammlung f@goldscheider.de
Bem.: Die Vorlage für diese Büste geht zurück auf die Figur des Onkel Nolte aus Wilhelm Buschs „Die fromme Helene".

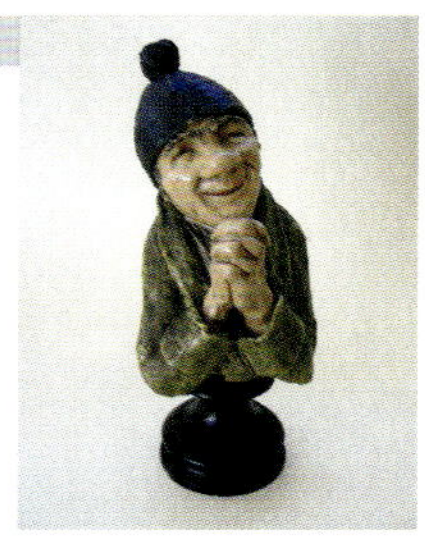

323
BÜSTE Nordafrikanerin mit Stirnband und Münzkette als Halsschmuck
ca. 1888/90
323 ger., Firmenmarke 3, Vervielf. vorbeh., H. 26,7 cm; eBay # 263088157
323 ger.,2, Firmenmarke 3, Vervielf. vorbeh., H. 26,5 cm; eBay # 893942034
323,16 Firmenmarke 11, Repr. Res., gepr. „P"= Z3, H. 27 cm; Doro 20.4.1993
–> vgl. 1044

324
BÜSTE Spanischer Eroberer mit Helm
ca. 1890/92
324,26 FG Firmenmarke 8a, H. 33 cm; eBay # 3163470369

329
Stehende nordafrikanische Wasserverkäuferin mit Krug auf der rechten Schulter (auf Landschaftssockel)
ca. 1889
329 (Firmenmarke ohne Ang.), H. 36 cm; Pinhas, S. 29 (m. Abb.)
Bem.: Hier lässt sich das Jahr der Edition ziemlich genau mit 1889 festlegen. Findet sich doch auf der Figur ein Aufkleber „Exposition Paris 1889".

330
Capèque (d.i. Czapek, Stanislaus)
FIGUR „Melonenverkäufer" (Kniender Araber mit ausgestrecktem Arm) (auf Rechtecksockel)
ca. 1890
330,III ger., Firmenmarke 8b, unsign.?, gepr. „P"= Z3, H. 42,5 cm; WKA 44
ohne Ang. GS (Firmenmarke ohne Ang.), sign. Capèque, H. 38 cm; Soth. L. 6.11.1986
330 (Firmenmarke ohne Ang.), H. 43 cm; GSF, S. 3
Bem.: Der nicht unerhebliche Unterschied bei der Größenang. resultiert aus unterschiedlichen Sockelhöhen.
Lit.: GSF, S. 3, hier mit der Ang. des Originaltitels „Melonenverkäufer".

331a
Strasser. Arthur (zugeschr.)
FIGUR „Schlangenbändiger" (Stehender Araber mit Turban, Schlange und Flöte)
ca. 1890
331 ger. Firmenmarke 8, Vervielf. vorbeh., unsign., H. 76 cm; PB Wien 19
331,V,a ger. Firmenmarke 8b, unsign., H. 72 cm; Wien Museum Inv. 186.602;
Bem.: Flöte fehlt hier.

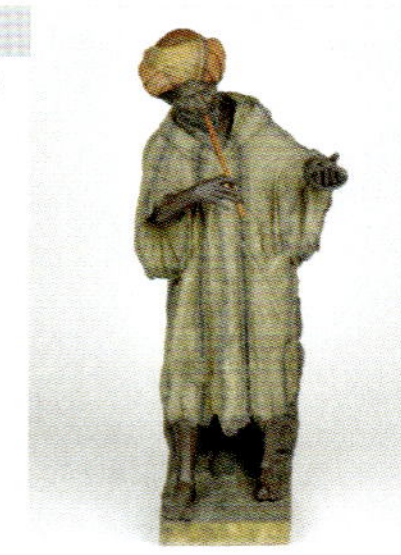

331b
Strasser. Arthur (zugeschr.)
FIGUR „Wasserträger"
ca. 1890
331b (Firmenmarke ohne Ang.), H. 71 cm; GSF, S. 3
Lit.: GSF, S. 3, hier mit der Ang. des Originaltitels „Wasserträger".

333
Schreitender Bauernjunge mit Hut und Schürze, eine Gans tragend (sog. Wiener Type)
ca. 1890/92
333,H ger., Firmenmarke 3, Vervielf. vorbeh., H. 34 cm; Wien Museum Inv. 219.131
333,(5,4,2,5 vertikal), Firmenmarke 7, H. 35 cm; Peege 93
Bem.: zu den sog. „Wiener Typen" –> vgl. Schmidt, Otto: Wiener Typen und Strassenbilder, Wien 2000
Pendant –> vgl. 1333

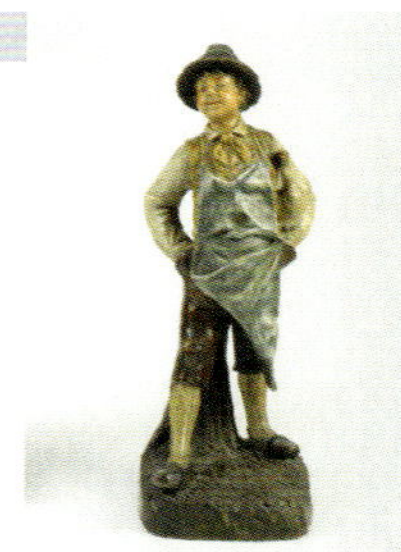

334
Schreitender Schusterjunge mit Kappe und Schürze, einen Schuh tragend (sog. Wiener Type)
ca. 1890/92
334 ger., Firmenmarke 3, Vervielf. vorbeh., H. 34 cm; Wien Museum Inv. 219.136
Bem.: zu den sog. „Wiener Typen" –> vgl. Schmidt, Otto: Wiener Typen und Strassenbilder, Wien 2000

336
Strasser, Arthur (zugeschr.)
BÜSTE Araberin mit Kopftuch (nach rechts blickend)
ca. 1890/92
336 ohne Firmenmarke, unsign., H. 23,5 cm; Wien Museum Inv. 186.636
Ausst.: SSH 1992

341
BÜSTE Araberin mit Kopftuch und Halsschmuck (nach links blickend)
ca. 1890/92
341,C ger., Firmenmarke 3, Vervielf. vorbeh., gepr. „P"= Z3, unsign., H. 31,5 cm; Wien Museum Inv. 186.635

357
BÜSTE Fiaker (Herr mit Schnauzbart und Zylinder) (sog. Wiener Type)
ca. 1890/92
357 ger., Firmenmarke 3, Vervielf. vorbeh., H. 22,8 cm, (mit Holzsockel 30,5 cm;); eBay # 929134382
Bem.: Zu den sog. „Wiener Typen" –> vgl. Schmidt, Otto: Wiener Typen und Strassenbilder, Wien 2000

374 –> ÜBERSCHNEIDUNG
Gracien, G./J.?
Wasserverkäufer (Stehender Araber mit Turban, Wasserkrug, Kanne und Schöpfkelle)
ca. 1890/92
374,80,43 (Firmenmarke ohne Ang.), sign. G. Gracien, H. 39,8 cm; Christie's SK 4.11.1999
Bem.: An der Plinthe betitelt mit arab. Schriftzeichen.
–> vgl. 857 (Strasser zugeschr.)

374 -> ÜBERSCHNEIDUNG
FIGUR „Incroyable Puvis"
ca. 1896/97
347 ohne Ang.; KeramRS, 5.1897, S. 347
Bem.: s. Modell-Nr. 6

380
FIGUR „Schusterjunge"
ca. 1890/92
380 (Firmenmarke ohne Ang.), H. 22 cm; GSF, S. 3
Lit.: GSF, S. 3, hier mit der Ang. des Originaltitels „Schusterjunge".

385
BÜSTE Junger Bursche mit Berét
ca. 1890/92
385 ger., Firmenmarke 3, H. 41,9 cm; eBay # 6134420232

389
BÜSTE Dame mit Halskette (auf Holzsockel)
ca. 1890/92
389 ger., Firmenmarke 3, Vervielf. vorbeh., H. fehlt; KH

390
TORSO Orientalin eine Wasserschale (Muschel) haltend
ca. 1890/92
390 (Firmenmarke ohne Ang.), H. 63 cm; Pinhas, S. 18 (m. Abb.)

411
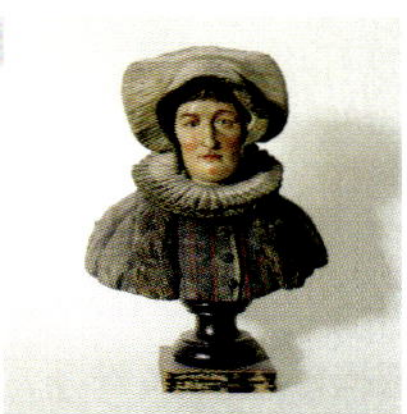
BÜSTE Edelmann? (auf Holzsockel)
ca. 1890/92
411 ger., Firmenmarke 3, Vervielf. vorbeh., gepr. „P"= Z3, H. (ohne Holzsockel) 17,8 cm; PB Deutschland
Bem.: Trotz intensiver Recherche ließ sich nicht ermitteln was oder wen genau diese besonders seltene Büste darstellt.

418
BÜSTE „Singhalese"
ca. 1890/92
418 (Firmenmarke ohne Ang.), H. 59 cm; GSF, S. 3
-> vgl. 669; 768
Pendant -> vgl. 419
Lit.: GSF, S. 3, hier mit der Ang. des Originaltitels „Singhalese".

419
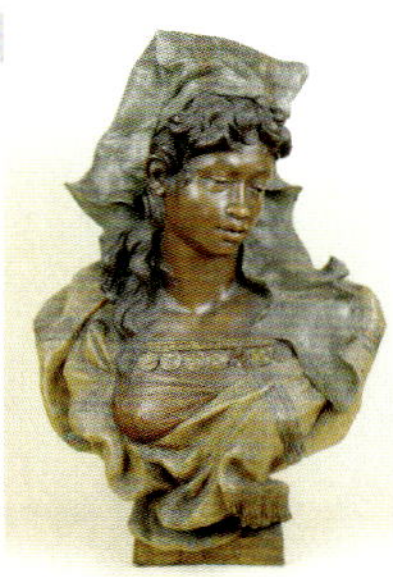
BÜSTE „Singhalesin" (mit Kopftuch und Münzkette als Halsschmuck)
ca. 1890/92
419,11 ger., FG Wien (Firmenmarke?), H. 57,6 cm; Doro 26.11.2002
419,XI ger., Firmenmarke 12, Vervielf. vorbeh.= Z10, H. 54 cm; Sammlung Renée und Otto Schenk, Wien
419,14,47 Firmenmarke 7, undeutl. sign. B....ey, H. 53 cm; PB Wien 19
419 (Firmenmarke ohne Ang.), H. 59 cm; GSF, S. 3
-> vgl. 670; 769
Pendant -> vgl. 418
Lit.: GSF, S. 3, hier mit der Ang. des Originaltitels „Singhalesin".

425
BÜSTE Araber mit Vollbart, Turban und breiter Schärpe (auf Holzsockel)
ca. 1890/92
425,48 ger., Firmenmarke 8b, H. 22,4 cm; PB Wien 19
425 ger., GS Wien (Firmenmarke?), H. (mit Holzsockel) 24,3 cm; Doro 6.4.1999
Pendant -> vgl. 426

426
BÜSTE Araberin mit Kopf- und Halsschmuck
ca. 1890/92
426,12 Firmenmarke 8b, gepr. „P"= Z3, H. 23,4 cm; PB Wien 19
Pendant -> vgl. 425

437
BÜSTE „Landsknecht"
ca. 1890/92
437,0 ger., Firmenmarke 8, Repr. Res., H. 43 cm; Schuler 72
437 (Firmenmarke ohne Ang.), H. 41 cm; GSF, S. 3
Pendant -> vgl. 438
Lit.: GSF, S. 3, hier mit der Ang. des Originaltitels „Landsknecht".

(437)
BÜSTE Faun
ca. 1890
437 (ohne Firmenmarke), Made in Austria, Repr. Res., H. 33,5 cm; GK, S. 10, 52 (m. Abb.); Wien Museum 182.055
Bem.: Keramik mit Kupferpatina; dieses Objekt kein GS, es stammt von einer der zahlreichen nicht verifizierbaren Konkurrenzfirmen. Diese Büste soll an dieser Stelle nur aufgeführt werden da sie ein zu GS ähnliches Erzeugnis darstellt und bereits in der großen Ausst. im heutigen Wien Museum 1985/86 als Objekt von GS fälschlicherweise ausgestellt wurde.
Ausst.: HMW (jetzt Wien Museum) 1985/86.
Pendant -> vgl. 438
Lit.: GK, S. 10, 52 (m. Abb.)

438
BÜSTE „Landsknecht"
ca. 1890/92
438 (Firmenmarke ohne Ang.), H. 42 cm; GSF, S. 3
Pendant -> vgl. 437
Lit.: GSF, S. 3, hier mit der Ang. des Originaltitels „Landsknecht".

(438)
BÜSTE Nymphe
ca. 1890
438 (ohne Firmenmarke), Made in Austria, Repr. Res., H. 31,5 cm; GK, S. 10, 53 (m. Abb.); Wien Museum Inv. 82.056
Bem.: Keramik mit Kupferpatina; dieses Objekt kein GS, es stammt von einer der zahlreichen nicht verifizierbaren Konkurrenzfirmen Diese Büste soll an dieser Stelle nur aufgeführt werden da sie ein zu GS ähnliches Erzeugnis darstellt und bereits in der großen Ausst. im heutigen Wien Museum 1985/86 als Objekt von GS fälschlicherweise ausgestellt wurde.
Ausst.: HMW (jetzt Wien Museum) 1985/86.
Pendant -> vgl. 437
Lit.: GK, S. 10, 53 (m. Abb.)

447
FIGUR Othello (auf Rechtecksockel)
ca. 1890/92
447 GS Wien (Firmenmarke 4/5?), Vervielf. vorbeh., H. 88,9 cm; Cheffins 23.4.1997
-> vgl. 473
Bem.: Dargestellt ist die gleichnamige Titelfigur nach William Shakespeares Trauerspiel „Othello, der Mohr von Venedig" (entstanden ca. 1604).

454
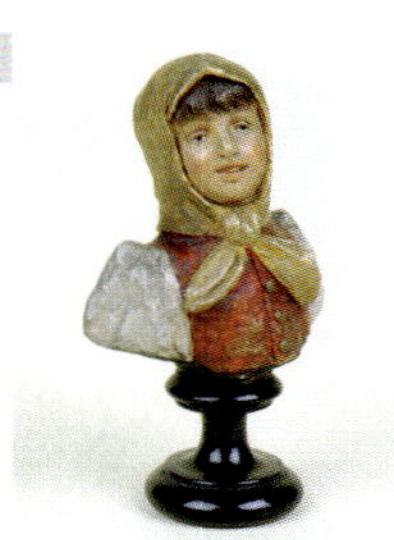
BÜSTE Bauernmädchen mit Kopftuch (auf Holzsockel)
ca. 1890/92
454,453 ger., Firmenmarke 3, H. (mit Holzsockel) 12,4 cm; PB Wien 19

458
BÜSTE Mann in historischer Tracht (Spanier?) mit großem federverziertem Hut (auf Holzsockel)
ca. 1890/92
458 (Firmenmarke ohne Ang.), H. (mit Holzsockel) 38 cm; KH
Pendant -> vgl. 459

459
BÜSTE Frau in historischer Tracht (Spanierin?) mit großem federverziertem Hut (auf Holzsockel)
ca. 1890/92
458 (Firmenmarke ohne Ang.), H. (mit Holzsockel) 38 cm; KH
Pendant -> vgl. 458

460
BÜSTE Alter Araber mit Vollbart und Turban (auf Holzsockel)
ca. 1890/92
460,KW ger., Firmenmarke 3, Vervielf. vorbeh., H. 24,1 cm; eBay # 2559218578
460 Firmenmarke 3, H. (mit Holzsockel) 21,5 cm; GK, S. 10, 54 (m. Abb.); Wien Museum Inv. 223.742
Ausst.: HMW (jetzt Wien Museum) 1985/86.
Pendant -> vgl. 461
Lit.: GK, S. 10, 54 (m. Abb.)

461
BÜSTE Araberin mit Kopftuch (auf Holzsockel)
ca. 1890/92
461 (ohne Ang.), H. 21 cm; eBay # 25592185780
Pendant -> vgl. 460

468
BÜSTE mit eigenartiger Bemalung
ca. 1895/96
468 ohne Ang.; KeramRS, 4.1896, S. 863
Bem.: s. Modell-Nr. 6

473
FIGUR „Othello" (auf Rechtecksockel)
ca. 1895/96
473 (Firmenmarke ohne Ang.), H. 88 cm; Pinhas, S. 26 (m. Abb.)
473 (Firmenmarke ohne Ang.), H. 84 cm; GSF, S. 3
-> vgl. 447
Lit.: GSF, S. 3, hier mit der Ang. des Originaltitels „Othello".

486
BÜSTE „Othello"
ca. 1895/96
486 (Firmenmarke ohne Ang.), H. 39 cm; GSF, S. 3
Lit.: GSF, S. 3, hier mit der Ang. des Originaltitels „Othello".

498
Stehender Ritter mit Schwert (auf quadratischem Sockel)
ca. 1895/96
498 ger., FG Wien (Firmenmarke?), H. 84 cm; Doro 28.9.1993
Bem.: Schwert aus Metall gefertigt.
Lit.: Pinhas, S. 66 (m. Abb.)

501 (GSB)
STATUE „Faune"
BRONZE oder ZINKGUSS
ca. 1892/1900
501 H. 130 cm; GSB, Pl. 1 (m. Abb.)
501 H. 62 cm; GSB, Pl. 1 (m. Abb.)
501 H. 30 cm; GSB, Pl. 1 (m. Abb.)
Bem.: Diese imposante Figur wurde in Bronze und Zinkguss, sowie in den Höhen 130 cm; 62 cm sowie 30 cm Friedrich Goldscheiders Pariser Produktion gefertigt.
Pendant -> vgl. 502 (GSB)

502 (GSB)
STATUE „Bacchante"
BRONZE oder ZINKGUSS
ca. 1892/1900
502 H. 130 cm; GSB, Pl. 1 (m. Abb.)
502 H. 62 cm; GSB, Pl. 1 (m. Abb.)
502 H. 30 cm; GSB, Pl. 1 (m. Abb.)
Bem.: Diese imposante Figur wurde in Bronze und Zinkguss, sowie in den Höhen 130 cm; 62 cm sowie 30 cm Friedrich Goldscheiders Pariser Produktion gefertigt.
Pendant -> vgl. 501 (GSB)

503 (GSB)
Anglis, J.
SCHALE Plateau „Amour Chasseur"
ca. 1892/1900
503 L. 45 cm; GSB, Pl. 3 (m. Abb.)

505 -> ÜBERSCHNEIDUNG
VASE
1895
505 ohne Ang.; KeramRS, 3.1895, S. 545
Bem.: s. Modell-Nr. 6

505 -> ÜBERSCHNEIDUNG
Stehender Ritter
ca. 1893/95
505 FG Wien (Firmenmarke ohne Ang.), H. 36 cm; Doro 8.3.2004

505 -> ÜBERSCHNEIDUNG
Stehender Haremswächter mit Turban
ca. 1893/95
505 Firmenmarke 11, Repr. Res., H. ca. 70 cm; KH

505 (GSB)
Baron, René
STATUE „La Chanson"
ZINKGUSS
ca. 1892/1900
505 H. 104 cm; GSB, Pl. 5 (m. Abb.)
-> vgl. 509 (GSB); 952; 954; 972

506
VASE
1895
506 ohne Ang.; KeramRS, 3.1895, S. 153
Bem.: s. Modell-Nr. 6

507
VASE
1895
507 ohne Ang.; KeramRS, 3.1895, S. 153
Bem.: s. Modell-Nr. 6

509 (GSB)
Baron, René
STATUE „La Chanson"
BRONZE
ca. 1892/1900
509 H. 22 od. 23 cm; GSB, Pl. 5, 10 (m. Abb.)
-> vgl. 505 (GSB); 952; 954; 972

510
Stoltenberg-Lerche, Hans Stephan
VASE „Chasse aux Papillons" (Schmetterlingsjagd) (Zwei Putti, einer plastisch, der andere reliefiert, auf Schmetterlingsjagd)
ca. 1893
510,41,13 GS Wien (Firmenmarke?), Repr. Res., sign. Lerche, H. 29,7 cm; Doro 21.5.1997
Bem.: Diese Vase, eigentl. eher ein Henkelkrug (mit plastischem und floralem Reliefdekor), findet sich unter der Nr. 561 im Kat. GSB wieder. Sie wurde in der Pariser Produktion in einer Höhe von 36 cm aus Zinn (Etain) und „Kunstguss" (Fonte d'Art, d.i. wohl Zinkguss, patiniert) gefertigt.
-> vgl. 561 (GSB)
Lit.: GSB, Pl. 26, hier mit Ang. des Originaltitels „Chasse aux Papillons".

510 (GSB)
STATUE „Bouquetière" (No. 1)
ca. 1892/1900
510 H. 90 cm; GSB, Pl. 6 (m. Abb.)
-> vgl. 558 (GSB)

511 (GSB)
STATUE „Amour aux Castagnettes"
ca. 1892/1900
511 H. 38 cm; GSB, Pl. 12 (m. Abb.)

512 -> ÜBERSCHNEIDUNG
FIGUR „Suis je assez beau" (Stehender, eleganter Knabe mit Zylinder)
Inschrift: SUIS JE ASSEZ BEAU
1893
512/60 ohne Ang.; KeramRS, 1.1893, S. 531, hier mit der Ang.: Figur (mit eigenartiger Bemalung und Farbenzusammenstellung).
512 (Firmenmarke ohne Ang.), H. 22 cm; GSF, S. 3
Bem.: s. Modell-Nr. 6
Text der Inschrift: Bin ich nicht schön?
-> vgl. 1524; 1832
Lit.: GSF, S. 3, hier mit der Ang. des Originaltitels „Suis je assez beau".

512 -> ÜBERSCHNEIDUNG
VASE
1895
512 ohne Ang.; KeramRS, 3.1895, S. 545
Bem.: s. Modell-Nr. 6

513
VASE
1895
513 ohne Ang.; KeramRS, 3.1895, S. 153
Bem.: s. Modell-Nr. 6

514
VASE
1895
514 ohne Ang.; KeramRS, 3.1895, S. 153
Bem.: s. Modell-Nr. 6

516 -> ÜBERSCHNEIDUNG
WANDTELLER
1895
516 ohne Ang.; KeramRS, 3.1895, S. 153
Bem.: s. Modell-Nr. 6

516 -> ÜBERSCHNEIDUNG
Stehende Japanerin (im Kimono)
ca. 1893/95
516 FG Wien (Firmenmarke?), H. 68 cm; Doro 12.12.1991

516 (GSB)
VASE „Renaissance"
ZINN
ca. 1892/1900
516 H. 36 cm, B. 21 cm; GSB, Pl. 22 (m. Abb.)

518
Stoltenberg-Lerche, Hans Stephan
VASE Zwei Putti auf der Vasenwandung
ca. 1895/1900
518 GS Wien (Firmenmarke?), Ges. gesch., sign. Lerche (od. H. St. Lerche, Paris?), H. 29,7 cm; Rieber 21.9.1999

519 (GSB)
Stoltenberg-Lerche, Hans Stephan (zugeschr.)
VASE „Enfants jouant"
ZINN
ca. 1892/1900
519 H. 25 cm, B. 40 cm; GSB, Pl. 15 (m. Abb.)

521
VASE
1895
521 ohne Ang.; KeramRS, 3.1895, S. 545
Bem.: s. Modell-Nr. 6

521 (GSB)
Stoltenberg-Lerche, Hans Stephan (zugeschr.)
VASE „Enfant au Tambourin"
ZINN
ca. 1892/1900
521 H. 21 cm, B. 16 cm; GSB, Pl. 10, 82 (m. Abb.)
Pendant -> vgl. 679 (GSB)

522
LAMPE Elfe in Blütenkelch kletternd
ca. 1895
522 (ohne Firmenmarke?), Made in Austria, Repr. Res., H. ca. 59 cm; Henry's 28.12.1995
Bem.: Vermutlich GS (Zuschreibung unter Vorbehalt, jedoch sprechen, trotz der fehlenden Firmenmarke, Nummerierung und Qualität für GS).

522 (GSB)
VASE „Sirène"
ZINN
ca. 1892/1900
522 H. 45 cm, B. 26 cm; GSB, Pl. 22 (m. Abb.)

523
VASE
1895
523 ohne Ang.; KeramRS, 3.1895, S. 153
Bem.: s. Modell-Nr. 6

523 (GSB)
Stoltenberg-Lerche, Hans Stephan (zugeschr.)
VASE „Enfant Négrillon"
ZINN
ca. 1892/1900
523 H. 40 cm, B. 22 cm; GSB, Pl. 17 (m. Abb.)

524
Strasser, Arthur
BÜSTE Araber mit Vollbart und Turban
ca. 1895
524,40 Firmenmarke 11, Repr. Res., H. 11,5 cm; Neuwirth (1974), S. 27 (m. Abb.); Bem.: Sockel fehlt.
524 Firmenmarke 3, unsign., H. 16,5 cm; Wien Museum Inv. 219.124
524 FG Wien (Firmenmarke?), Repr. Res., H. (mit Terrakottasockel) 17 cm; Mehlis 33
524 GS (Firmenmarke ohne Ang.), H. (mit Terrakottasockel) 17,5 cm; Doro 10.–18.12.1985
524 FG Wien (Firmenmarke?), H. (mit Terrakottasockel) 17,5 cm; Spik 532

524 (GSB)
Stoltenberg-Lerche, Hans Stephan (zugeschr.)
VASE „Les deux Amis"
ZINN
ca. 1892/1900
524 H. 18 cm, B. 18 cm; GSB, Pl. 13 (m. Abb.)

525 -> ÜBERSCHNEIDUNG
Strasser, Arthur
BÜSTE Jüngling mit Kappe in Renaissancetracht
ca. 1895
525 GS (Firmenmarke ohne Ang.), H. 17,5 cm; Doro 10.–18.12.1985
525,VIII ger., (ohne Firmenmarke), H. 18 cm; Wien Museum Inv. 219.125
525 FG Wien (Firmenmarke?), Repr. Res., H. (mit Terrakottasockel) 17 cm; Mehlis 33

525 -> ÜBERSCHNEIDUNG
VASE
1895
525 ohne Ang.; KeramRS, 3.1895, S. 153
Bem.: s. Modell-Nr. 6

525 (GSB)
Stoltenberg-Lerche, Hans Stephan (zugeschr.)
VASE „Enfants Grimpant"
ZINN
ca. 1892/1900
525 H. 28 cm, B. 22 cm; GSB, Pl. 17 (m. Abb.)

526 (GSB)
Stoltenberg-Lerche, Hans Stephan (zugeschr.)
VASE „Ibsen"
ca. 1892/1900
526 H. 42 cm, B. 65 cm; GSB, Pl. 11 (m. Abb.)

527 (GSB)
Stoltenberg-Lerche, Hans Stephan (zugeschr.)
VASE „Saut de Gamin"
ZINN
ca. 1892/1900
527 H. 28 cm; GSB, Pl. 12 (m. Abb.)

528 (GSB)
Stoltenberg-Lerche, Hans Stephan
PLATEAU „Turc"
ZINN
ca. 1892/1900
528 H. 40 cm, B. 30 cm; GSB, Pl. 10 (m. Abb.)
Pendant -> vgl. 564 (GSB)

529 (GSB)
Stoltenberg-Lerche, Hans Stephan
PLATEAU „Nayade"
ZINN
ca. 1892/1900
529 sign. H. S. Lerche, Paris, H. 15 cm; GSB, Pl. 37 (m. Abb.)

530
PLATTE (Wandrelief?)
1895
530 ohne Ang.; KeramRS, 3.1895, S. 153
Bem.: s. Modell-Nr. 6

530 (GSB)
PLATEAU „Le petit Paradis"
ZINN
ca. 1892/1900
530 H. 25 cm, L. 39 cm; GSB, Pl. 17 (m. Abb.)

531
VASE
1895
531 ohne Ang.; KeramRS, 3.1895, S. 545
Bem.: s. Modell-Nr. 6

532
SPIEGEL Stehendes Mädchen vor Lorbeerbaum
ca. 1895
532 ger., FG Wien? (Firmenmarke ohne Ang.), Made in Austria, Repr. Res., H. ca. 67 cm; Doro 18.4.1994
Bem.: Vermutlich GS (Zuschreibung unter Vorbehalt, jedoch sprechen, trotz der fehlenden Firmenmarke, Nummerierung und Qualität für GS).
-> vgl. 2040 (ähnl. Modell)

532 (GSB)
Stoltenberg-Lerche, Hans Stephan
PLATEAU „Printemps"
ZINN/ZINKGUSS
ca. 1892/1900
532 sign. H. S. Lerche, Paris, H. 45 cm; GSB, Pl. 24 (m. Abb.)

533
VASE
1895
533 ohne Ang.; KeramRS, 3.1895, S. 803
Bem.: s. Modell-Nr. 6

534
VASE
1895
534 ohne Ang.; KeramRS, 3.1895, S. 803
Bem.: s. Modell-Nr. 6
Lit.: KeramRS, 6.1898, S. 421, hier, im Mai 1898, die Verlängerung der Schutzfrist auf sechs Jahre.

535
VASE
1895
535 ohne Ang.; KeramRS, 3.1895, S. 545; 805
Bem.: s. Modell-Nr. 6

537 (GSB)
Strasser, Arthur
BUSTE „Nègre"
ca. 1892/1900
BRONZE
ohne Sockel
537 H. 31 cm; GSB, Pl. 16 (m. Abb.)
ZINKGUSS
mit Sockel
537 H. 45 cm; GSB, Pl. 52 (m. Abb.)

538 (GSB)
BUSTE „Cynghalais"
ca. 1892/1900
538 H. 43 cm; GSB, Pl. 17 (m. Abb.)

539 (GSB)
Strasser, Arthur (zugeschr.)
BUSTE „Fumeur" (Téte d'étude)
ca. 1892/1900
539 H. 31 cm; GSB, Pl. 15 (m. Abb.)
-> vgl. 1219

540 (GSB)
PLATEAU „Déjeuner"
ca. 1892/1900
540 H. fehlt; GSB, Pl. 14 (m. Abb.)
Pendant, bzw. rechter Teil -> vgl. 541 (GSB)

541 (GSB)
PLATEAU „Déjeuner"
ca. 1892/1900
541 H. fehlt; GSB, Pl. 14 (m. Abb.)
Pendant, bzw. linker Teil -> vgl. 541 (GSB)

542 (GSB)
BUSTE „Jeannette"
ca. 1892/1900
542 H. 31 cm; GSB, Pl. 13 (m. Abb.)
Pendant -> vgl. 543 (GSB)

543 (GSB)
BUSTE „Henriette"
ca. 1892/1900
543 H. 29 cm; GSB, Pl. 13 (m. Abb.)
Pendant -> vgl. 542 (GSB)

544 (GSB)
STATUETTE „Après le Bain"
Vraie BRONZE
ca. 1892/1900
544 H. 20 cm; GSB, Pl. 23 (m. Abb.)

545 (GSB)
Strasser, Arthur
BUSTE „Antique"
ca. 1892/1900
ohne Sockel
545 H. 34 cm; GSB, Pl. 16 (m. Abb.)
-> vgl. 545 (GSB) mit Sockel „Femme Fellah"

545 (GSB)
Strasser, Arthur
BUSTE „Femme Fellah"
ca. 1892/1900
mit Sockel
545 H. 49 cm; GSB, Pl. 58 (m. Abb.)
-> vgl. 545 (GSB) ohne Sockel „Antique"

546 (GSB)
PLATEAU „Fumeur"
ca. 1892/1900
546 H. fehlt; GSB, Pl. 14 (m. Abb.)
Pendant -> vgl. 547 (GSB)

547 (GSB)
PLATEAU „Fumeur"
ca. 1892/1900
547 H. fehlt; GSB, Pl. 14 (m. Abb.)
Pendant -> vgl. 546 (GSB)

548 (GSB)
BUSTE „Moqueuse"
Inschrift: MOQUEUSE
ca. 1892/1900
548 H. 50 cm; GSB, Pl. 16 (m. Abb.)

549 (GSB)
Strasser, Arthur (zugeschr.)
BUSTE „Soudanais"
ca. 1892/1900
ohne Sockel
549 H. 49 cm; GSB, Pl. 15 (m. Abb.)
mit Sockel
549 H. 67 cm; GSB, Pl. 25 (m. Abb.)

550 (GSB)
STATUE „Amour Content"
ca. 1892/1900
550 H. 52 cm; GSB, Pl. 12 (m. Abb.)

551 (GSB)
Strasser, Arthur
STATUE „Retour de la Fontaine" No. 1
ca. 1898
551 H. 90 cm; GSB, Pl. 18 (m. Abb.)
-> vgl. 576 (GSB); 579 (GSB); 7915; 8727

553 (GSB)
VASE „Plaisirs D'Automne"
ZINN
ca. 1892/1900
553 H. 40 cm; GSB, Pl. 12 (m. Abb.)

554 (GSB)
Stoltenberg-Lerche, Hans Stephan (zugeschr.)
VASE „Précoce Vigneron"
ZINN/ZINKGUSS
ca. 1892/1900
554 H. 33 cm; GSB, Pl. 26 (m. Abb.)

555 (GSB)
BUSTE „Femme" (Téte d'étude)
ca. 1892/1900
555 H. 10 cm; GSB, Pl. 10 (m. Abb.)

556 (GSB)
VASE „Enfants Pècheurs"
ca. 1892/1900
556 H. 75 cm; GSB, Pl. 19 (m. Abb.)

557
FIGUR „Gutenberg"
ca. 1893
557 (Firmenmarke ohne Ang.), H. 67 cm; GSF, S. 3
Bem.: Dargestellt ist der Erfinder des Buchdrucks Johannes Gensfleisch, genannt Gutenberg (ca. 1400–1468).
Lit.: GSF, S. 3, hier mit der Ang. des Originaltitels „Gutenberg".

557 (GSB)
Stoltenberg-Lerche, Hans Stephan (zugeschr.)
VASE „Enfant a Califourchon"
ZINN/ZINKGUSS
ca. 1892/1900
557 H. 39 cm; GSB, Pl. 26 (m. Abb.)

558 (GSB)
STATUE „Bouquetière" No. 2
Vraie BRONZE
ca. 1892/1900
558 H. 34 cm; GSB, Pl. 26 (m. Abb.)
-> vgl. 510 (GSB)

560
Mann in spanischer Tracht des 16. Jahrhunderts (auf quadratischem Sockel mit Säule)
ca. 1895
560,18,(5,9,90 vertikal?), FG Wien (Firmenmarke 7?), Vervielf. vorbeh., H. 26,2 cm; Kett. 174; Ruef 457

560 (GSB)
Stoltenberg-Lerche, Hans Stephan (zugeschr.)
VASE „Enfants Sauveteurs"
ZINKGUSS
ca. 1892/1900
560 H. 55 cm; GSB, Pl. 35 (m. Abb.)

561 (GSB)
Stoltenberg-Lerche, Hans Stephan
VASE „Chasse aux Papillons"
ZINN/ZINKGUSS
ca. 1892/1900
561 H. 36 cm; GSB, Pl. 26 (m. Abb.)
-> vgl. 510

562 (GSB)
PANNEAU „Tète D' Ėtude"
ca. 1892/1900
562 H. 47 cm, B. 32 cm; GSB, Pl. 22 (m. Abb.)

563 (GSB)
STATUETTE „Falstaff"
Inschrift: FALSTAFF
ZINKGUSS
ca. 1892/1900
563 H. 55 cm; GSB, Pl. 23 (m. Abb.)

564 (GSB)
Stoltenberg-Lerche, Hans Stephan
PANNEAU „Turc"
ZINN/ZINKGUSS
ca. 1892/1900
564 H. 38 cm; GSB, Pl. 35 (m. Abb.)
Pendant -> vgl. 528 (GSB)

566 (GSB)
Palma?
BUSTE „Jeanne D' Arc"
ZINN / ZINKGUSS / BRONZE
ca. 1892/1900
BRONZE / TERRACOTTA
566,37,AG (od. RG) Firmenmarke Sculpture Bronze 1, unles. sign. Palma?, H. 48,3 cm; eBay # 363752673
566 H. 50 cm; GSB, Pl. 25; 53 (m. Abb.)
Bem.: Diese Büste wurde sowohl in Zinn, als Zinkguss und in Bronze, kombiniert mit Terracotta ausgeführt.

568 (GSB)
BUSTE „Bouquetière" (No. 2)
Vraie BRONZE
ca. 1892/1900
568 H. 15 cm; GSB, Pl. 47 (m. Abb.)
-> vgl. 710 (GSB)

569 (GSB)
CRUCHE „Antique"
ZINN / ZINKGUSS
ca. 1892/1900
569 H. 37 cm; GSB, Pl. 24 (m. Abb.)

570 (GSB)
PLATEAU „Danse des Roses"
ZINN / ZINKGUSS
ca. 1892/1900
570 H. 40 cm; GSB, Pl. 24 (m. Abb.)

571
Cherc
FIGUR „Die Saat" (Säender Bauer mit Hut)
ca. 1895
571,53,17 Firmenmarke 19, Ausformung 20er Jahre, H. 71 cm; Doro 20.4.1993
571 (Firmenmarke ohne Ang.), H. 69 cm; GSF, S. 3
Ausst.: ÖMKI 1899/1900, hier mit Ang. des Originaltitels „Die Saat".
Pendant -> vgl. 1935; 2220
Lit.: GSF, S. 3, hier mit der Ang. des abweichenden Originaltitels „Saemann".

571 (GSB)
Lorieux, Julien
PLATEAU „Le Cauchemar"
ZINN / ZINKGUSS
ca. 1892/1900
571 H. 40 cm; GSB, Pl. 38 (m. Abb.)

572 (GSB)
PLATEAU „Oriental ou Croissant"
ca. 1892/1900
572 H. 40 cm; GSB, Pl. 30 (m. Abb.)

574 (GSB)
STATUE „Faune assoiffé"
ZINKGUSS
ca. 1892/1900
574 H. 70 cm; GSB, Pl. 32 (m. Abb.)
-> vgl. 1255

575 (GSB)
STATUETTE „Souffleuse de Bulles"
Vraie BRONZE
ca. 1892/1900
575 H. 15 cm; GSB, Pl. 23 (m. Abb.)

576
BÜSTE Südamerikaner (Indianer?) mit Halsschmuck, in leidender Pose
ca. 1895
576 (Firmenmarke ohne Ang.), H. 50 cm; Soth. Mailand 29.–31.5.1989
576 (Firmenmarke ohne Ang.), H. 40 cm; Pinhas, S. 17 (m. Abb.)

576 (GSB)
Strasser, Arthur
STATUE „Retour de la Fontaine" No. 3
ca. 1898
576 H. 32 cm; GSB, Pl. 18 (m. Abb.)
-> vgl. 551 (GSB); 579 (GSB); 7915; 8727

577 (GSB)
BUSTE „Rieuse"
ZINKGUSS
ca. 1892/1900
577 H. 45 cm; GSB, Pl. 35 (m. Abb.)

579 (GSB)
Strasser, Arthur
STATUE „Retour de la Fontaine" No. 2
ca. 1898
576 H. 61 cm; GSB, Pl. 18 (m. Abb.)
-> vgl. 551 (GSB); 576 (GSB); 7915; 8727

581 (GSB)
PLATEAU „Amour & Psyché"
ca. 1892/1900
581 H. 32 cm; GSB, Pl. 30 (m. Abb.)

582 (GSB)
PLATEAU „Dryades"
ZINN
ca. 1892/1900
582 H. 40 cm, L. 32 cm; GSB, Pl. 37 (m. Abb.)

583 (GSB)
Lorieux, Julien
PLATEAU „Offrande a Paris"
ZINN / ZINKGUSS
ca. 1892/1900
583 H. 45 cm; GSB, Pl. 38 (m. Abb.)

584 (GSB)
STATUE „Judith nue" No. 1
BRONZE / ZINKGUSS
ca. 1892/1900
584 H. 65 cm; GSB, Pl. 27 (m. Abb.)
Bem.: Die Figur der Judith zeigt sich hier mit entblößtem Oberkörper. Mit derselben Modell-Nr. wurde sie auch „angezogen" ausgeführt.
-> vgl. 602 (GSB); 604 (GSB); 734

584 (GSB)
STATUE „Judith habillée" No. 1
BRONZE / ZINKGUSS
ca. 1892/1900
584 H. 65 cm; GSB, Pl. 27 (m. Abb.)
Bem.: Die Figur der Judith zeigt sich hier mit „angezogenem" Oberkörper. Mit derselben Modell-Nr. wurde sie auch mit entblößtem Oberkörper ausgeführt.
-> vgl. 602 (GSB); 604 (GSB); 734

585 (GSB)
VASE „Baigneuse"
ZINN
ca. 1892/1900
585 H. 38 cm; GSB, Pl. 34 (m. Abb.)

586 (GSB)
BUSTE „Napoléon"
Vraie BRONZE
ca. 1892/1900
586 H. 25 cm; GSB, Pl. 33 (m. Abb.)
-> vgl. 1535

587 (GSB)
STATUETTE „Napoléon"
ca. 1892/1900
587 H. 30 cm; GSB, Pl. 30 (m. Abb.)

588 (GSB)
STATUETTE „Berger Grec"
ca. 1892/1900
588 H. 32 cm; GSB, Pl. 29 (m. Abb.)

589 (GSB)
STATUETTE „Dans L' Enclos"
ca. 1892/1900
589 H. 68 cm; GSB, Pl. 29 (m. Abb.)
-> vgl. 594 (GSB)

590
FIGUR „Todtenkopf (natürl. Größe)"
ca. 1895
590 (Firmenmarke ohne Ang.), H. 16 cm; GSF, S. 3
Lit.: GSF, S. 3, hier mit der Ang. des Originaltitels „Todtenkopf (natürl. Größe)".

590 (GSB)
STATUE „Sapho"
Vraie BRONZE
ca. 1892/1900
590 H. 33 cm; GSB, Pl. 34 (m. Abb.)

591 (GSB)
STATUE „Sphinx"
Vraie BRONZE
ca. 1892/1900
591 H. 31 cm; GSB, Pl. 34 (m. Abb.)

592
BÜSTE Tirolerin mit Hut und Zither
ca. 1895
592,VIII ger., Firmenmarke 8, H. 24,1 cm; eBay #531198788; Bem.: Da dieses Exemplar als völlig unbemalte Terrakotta ausgeführt ist und auch der Firmenstempel in dieser Form nur hier einzig nachgewiesen werden kann, könnte es sich um ein „Muster" bzw. einen Entwurf handeln.

592 (GSB)
STATUETTE „Porteuse D' Amphores"
ca. 1892/1900
592 H. 32 cm; GSB, Pl. 29 (m. Abb.)

593 (GSB)
STATUETTE „Baigneuse Léda"
ca. 1892/1900
593 H. 25 cm; GSB, Pl. 30 (m. Abb.)

594 (GSB)
STATUETTE „Dans L' Enclos"
ca. 1892/1900
594 H. 31 cm; GSB, Pl. 29 (m. Abb.)
-> vgl. 589 (GSB)

595 (GSB)
STATUETTE „Réverie"
ZINKGUSS
ca. 1892/1900
595 H. 56 cm; GSB, Pl. 51 (m. Abb.)

596 (GSB)
Lemoine, B.(?)
„Premiére Flèche"
KUNSTGUSS / METALLGUSS
ca. 1892/1900
596 H. 57 cm; GSB, Pl. 36 (m. Abb.)
-> vgl. 598 (GSB)
-> vgl. 1067; 1818

597 (GSB)
STATUETTE „Salambo"
ca. 1892/1900
597 H. 59 cm; GSB, Pl. 57 (m. Abb.)

598 (GSB)
Lemoine, B.(?)
„Premiére Flèche"
BRONZE
ca. 1892/1900
598 H. 32 cm; GSB, Pl. 36 (m. Abb.)
-> vgl. 596 (GSB)
-> vgl. 1067; 1818

599 (GSB)
STATUE „Charmeuse de Serpents (drapée)" No. 1
BRONZE
ca. 1892/1900
599 H. 86 cm; GSB, Pl. 54 (m. Abb.)
-> vgl. 647 (GSB)

600 (GSB)
GROUPE „Baiser D' Amour"
ca. 1892/1900
600 H. 65 cm; GSB, Pl. 28 (m. Abb.)
-> vgl. 605 (GSB)
-> vgl. 1292; 1520

601
FIGUR „Kegelschieber"
ca. 1895
601 (Firmenmarke ohne Ang.), H. 20 cm; GSF, S. 3

Pendant –> vgl. 602
Lit.: GSF, S. 3, hier mit der Ang. des Originaltitels „Kegelschieber".

601 (GSB)
PLATEAU „Nénuphar"
ZINN / ZINKGUSS
ca. 1892/1900
601 L. 34 cm; GSB, Pl. 50 (m. Abb.)

602
FIGUR „Kegeljunge"
ca. 1895
602 (Firmenmarke ohne Ang.), H. 20 cm; GSF, S. 3
Pendant –> vgl. 601
Lit.: GSF, S. 3, hier mit der Ang. des Originaltitels „Kegeljunge".

602 (GSB)
STATUETTE „Judith nue" No. 2
BRONZE / ZINKGUSS
ca. 1892/1900
602 H. 30 cm; GSB, Pl. 27 (m. Abb.)
Bem.: Die Figur der Judith zeigt sich hier mit entblößtem Oberkörper. Mit der Modell-Nr. 584 (H. 65 cm) wurde sie auch „angezogen" ausgeführt.
–> vgl. 584 (GSB); 604 (GSB); 734

603 (GSB)
STATUETTE „Mignon"
ZINKGUSS
ca. 1892/1900
603 H. 50 cm; GSB, Pl. 51 (m. Abb.)

604 (GSB)
STATUETTE „Judith nue" No. 3
BRONZE / ZINKGUSS
ca. 1892/1900
604 H. 22 cm; GSB, Pl. 27 (m. Abb.)
Bem.: Die Figur der Judith zeigt sich hier mit entblößtem Oberkörper. Mit der Modell-Nr. 584 (H. 65 cm) wurde sie auch „angezogen" ausgeführt.
–> vgl. 584 (GSB); 602 (GSB); 734

605 (GSB)
GROUPE „Baiser D' Amour"
ca. 1892/1900
605 H. 40 cm; GSB, Pl. 28 (m. Abb.)
–> vgl. 600 (GSB)
–> vgl. 1292; 1520

606 (GSB)
Cherc
PLATEAU „Paris et Hélène"
BRONZE / Fonte d'Art
ca. 1892/1900
606 H. 46 cm; GSB, Pl. 47 (m. Abb.)

607
TABAKSDOSE (FIGUR) „Todtenkopf (klein)"
Form: Totenkopf (vollplastisch)
Knauf: Knochen
ca. 1895
607,5 ger., Firmenmarke 3, H. 14,8 cm; Sammlung.schmutz@chello.at 83
607 (Firmenmarke ohne Ang.), H. ca. 15 cm; GSF, S. 3
Bem.: Ein besonderes Stück der frühen Goldscheiderproduktion. Realistische und dadurch zugleich makabre Darstellung eines Totenschädels. Als Knauf des Deckels dieser Tabakdose dient ein Knochen.
–> vgl. 607a
Lit.: GSF, S. 3, hier mit der Ang. des Originaltitels „Todtenkopf (klein)".

607 (GSB)
BUSTE „Stella"
ca. 1892/1900
607 H. 46 cm; GSB, Pl. 58 (m. Abb.)

607a
FIGUR „Todtenkopf (kleinst)"
ca. 1895
607a (Firmenmarke ohne Ang.), H. 6 cm; GSF, S. 3
–> vgl. 607
Lit.: GSF, S. 3, hier mit der Ang. des Originaltitels „Todtenkopf (kleinst)".

608 (GSB)
Strasser, Arthur
STATUE „Porteur D'Eau" No. 1
ZINKGUSS
ca. 1898
608 H. 65 cm; GSB, Pl. 52 (m. Abb.)
–> vgl. 609 (GSB); 1174; 1600

609 (GSB)
Strasser, Arthur
STATUE „Porteur D'Eau" No. 2
BRONZE
ca. 1898
609 H. 40 cm; GSB, Pl. 52 (m. Abb.)
–> vgl. 608 (GSB); 1174; 1600

610 (GSB)
Strasser, Arthur
STATUE „Marchand D'Eau" No. 1
ZINKGUSS
ca. 1892/1900
610 H. 65 cm; GSB, Pl. 18 (m. Abb.)
Bem.: Evtl. handelt sich um den im ÖMKI (1898/99) ausgestellten „Wasserverkäufer".
–> vgl. 611 (GSB)

611 (GSB)
Strasser, Arthur
STATUE „Marchand D'Eau" No. 2
BRONZE
ca. 1892/1900
611 H. 40 cm; GSB, Pl. 18 (m. Abb.)
Bem.: Evtl. handelt sich um den im ÖMKI (1898/99) ausgestellten „Wasserverkäufer".
–> vgl. 610 (GSB)

613
Prémont, J. du
WANDMASKE Nordafrikanerin mit Turban
ca. 1892/95
613,81,44 Firmenmarke 7, Repr. Res., gepr. „P"= Z3, H. 26,5 cm; Wien Museum Inv. 186.639
–> vgl. 1613

614 (GSB)
BOUGEOIR „Gothique"
ca. 1892/1900
614 H. 14 cm; GSB, Pl. 63 (m. Abb.)

615 (GSB)
Waderé, Heinrich
BUSTE „Grecque"
BRONZE
ca. 1892/1900
615 H. 45 cm; GSB, Pl. 53 (m. Abb.)
–> vgl. 1268; 2692

616
Della Robbia (nach / in der Art)
RELIEFPLATTE Madonna mit Kind
ca. 1895/97
646,30,28 weiß gem. III., Firmenmarke 4, Ges. gesch., H. 70 cm, B. 47,5 cm; Soth.com #IWQ

616 (GSB)
BAS-RELIEF „Aurore"
ca. 1892/1900
616 H. 22 cm; GSB, Pl. 65 (m. Abb.)

617 (GSB)
Callot, Jacques H.
VIDE-POCHE „Médaillon"
BRONZE oder ZINN
ca. 1892/1900
617 H. 15 cm; GSB, Pl. 48 (m. Abb.)
Pendant –> vgl. 660 (GSB)

618
BÜSTE „Richard Wagner"
ca. 1895
618 (Firmenmarke ohne Ang.), H. 26 cm; GSF, S. 3
Bem.: Dargestellt ist der Komponist Richard Wagner (1813–1883).
–> vgl. 981; 1030; 1897
Lit.: GSF, S. 3, hier mit der Ang. des Originaltitels „Richard Wagner"; Pinhas, S. 67 (m. Abb.).

618 (GSB)
Barabino, Armando
PLAQUE „Vierge et enfant Jésus"
ca. 1892/1900
618 H. 42 cm; GSB, Pl. 53 (m. Abb.)

619
BÜSTE „Franz Liszt"
ca. 1895
619 (Firmenmarke ohne Ang.), H. 26 cm; Pinhas, S. 67 (m. Abb.)
619 (Firmenmarke ohne Ang.), H. 26 cm; GSF, S. 3
Bem.: Dargestellt ist der Komponist und Pianist Franz Liszt (1811–1886).
–> vgl. 992; 1031
Lit.: GSF, S. 3, hier mit der Ang. des Originaltitels „Franz Liszt".

619 (GSB)
STATUETTE „Dois-je le Croire?" No. 1
ca. 1892/1900
619 H. 90 cm; GSB, Pl. 31 (m. Abb.)
–> vgl. 648 (GSB)

620 (GSB)
COQUILLE „Naïade du Rhin"
ca. 1892/1900
620 H. 25 cm; GSB, Pl. 63 (m. Abb.)

621
Chinard, Joseph (nach)
BÜSTE „Madame Récamier" (mit hochgestecktem Haar und Haarband)
ca. 1895
621, 46 (oder 45) Firmenmarke 11, Repr. Res., unsign., H. 40 cm; Weidler 300; 320; Sebök 35; 40
Bem.: Diese Büste wurde nach dem Original Joseph Chinard's (ca. 1795/96) gefertigt.

622 (GSB)
Ceccarelli, Ezio
BUSTE „Automne"
ca. 1898
622 H. 52 cm; GSB, Pl. 71 (m. Abb.)
–> vgl. 734 (GSB)
–> vgl. 1297; 1619

623
BÜSTE „Neger mit Laute"
ca. 1895
623 (Firmenmarke ohne Ang.), H. 71 cm; GSF, S. 3
Lit.: GSF, S. 3, hier mit der Ang. des Originaltitels „Neger mit Laute".

623 (GSB)
Rosier, (David?)
VIDE-POCHE Série III
ZINN / ZINKGUSS
ca. 1892/1900
623 L. 21 cm; GSB, Pl. 49 (m. Abb.)
Pendant –> vgl. 624 (GSB)

624 (GSB)
Rosier, (David?)
VIDE-POCHE Série IV
ZINN / ZINKGUSS
ca. 1892/1900
624 L. 21 cm; GSB, Pl. 49 (m. Abb.)
Bem.: Bei GSB wohl Druckfehler der Modell-Nr., die mit 524 angegeben wird.
Pendant –> vgl. 623 (GSB)

(624)
LAMPE Schreitendes Mädchen, einen Schleier über dem Kopf haltend (halbkugelförmiger Sockel mit Sternzeichensymbolen)
ca. 1895/97
624 ger. (ohne Firmenmarke), Repr. Res., Made in Austria, H. 88 cm; Schmidt 111
624 (Firmenmarke ohne Ang.), Repr. Res., Made in Austria, H. 88 cm; Doro 9.11.1993
Bem.: Evtl. GS (Zuschreibung unter großem Vorbehalt, da neben der fehlenden Firmenmarke auch die Qualität von GS nicht erreicht wird, vermutlich eine der zahlreichen Konkurrenzfirmen).

626 (GSB)
VIDE-POCHE „Soleil"
ZINN
ca. 1892/1900
626 Ø 22 cm; GSB, Pl. 50 (m. Abb.)

627 (GSB)
VIDE-POCHE „Nénuphar"
ZINN
ca. 1892/1900
627 Ø 19 cm; GSB, Pl. 50 (m. Abb.)

628
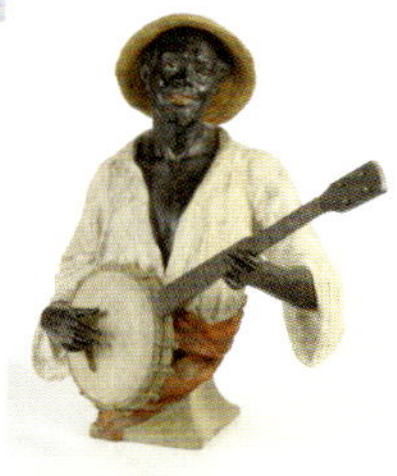
BÜSTE Schwarzer Banjospieler mit Strohhut
ca. 1895/97
623 (vielmehr 628), 21, (4,0,5,8 vertikal ger.), Firmenmarke 7, Firmenmarke 11, Repr. Res., gepr. „P"= Z3, H. 70,8 cm; Wien Museum Inv. 190.056; SS, S. 50 (m. Abb.), hier Modell-Nr. 623 angegeben
628 (Firmenmarke ohne Ang.), H. 65 cm; Pinhas, S. 31 (m. Abb.)
–> vgl. 1734 (ähnl. Modell)
Bem.: Banjo aus Holz gefertigt.
Ausst.: SSH 1992

628 (GSB)
VIDE-POCHE „Pavot"
ZINN
ca. 1892/1900
628 Ø 22 cm; GSB, Pl. 50 (m. Abb.)

629 (GSB)
Stoltenberg-Lerche, Hans Stephan
VIDE-POCHE „Grenouille"
ZINN
ca. 1892/1900
629 sign. H. S. Lerche, H. 15 cm; GSB, Pl. 48 (m. Abb.)

630 (GSB)
SCHALE Vide-Poche
ca. 1892/1900
630 L. 21 cm; GSB, Pl. 48 (m. Abb.)

631 (GSB)
SCHALE Vide-Poche
ca. 1892/1900
631 L. 21 cm; GSB, Pl. 48 (m. Abb.)

632 (GSB)
VASE „Scarabée"
ZINN
ca. 1892/1900
632 H. 14 cm; GSB, Pl. 47 (m. Abb.)

633 (GSB)
VASE „Fleurs"
ca. 1892/1900
633 H. 18 cm; GSB, Pl. 63 (m. Abb.)

634 (GSB)
VASE „Fleurs Fantaisie"
ca. 1892/1900
634 H. 18 cm; GSB, Pl. 63 (m. Abb.)

635
Stehender, junger Fürst in prächtigem Gewand (auf rechteckigem Sockel)
ca. 1895
11,4,1890,XV,635 Firmenmarke 5, Vervielf. vorbeh., H. 52 cm; Wien Museum Inv. 186.605

635 (GSB)
POT „Aux Chats"
ca. 1892/1900
635 H. 18 cm; GSB, Pl. 55 (m. Abb.)

636 (GSB)
PLUMIER
BRONZE
ca. 1892/1900
636 L. 25 cm; GSB, Pl. 49 (m. Abb.)

638 (GSB)
ENCRIER
BRONZE
ca. 1892/1900
638 H. 10 cm, L. 15 cm; GSB, Pl. 49 (m. Abb.)

639 (GSB)
BOUGEOIR
BRONZE
ca. 1892/1900
639 H. 6 cm, L. 25 cm; GSB, Pl. 49 (m. Abb.)

642
Bouval, Maurice
RELIEF „Sculpture" (umrankt von Blütenzweigen) (Wandrelief)
TERRAKOTTA
ca. 1898
0642 (Firmenmarke ohne Ang.), H. 27 cm; Pinhas, S. 50 (m. Abb.)
Bem.: Dieses Relief, bzw. dessen Motiv der sitzenden töpfernden, vielmehr an einer Skulptur arbeitenden nackten Frau, wurde dann ab 1900 die Firmenmarke der Manufaktur. Die Reklameschilder die für die Anbringung an der Aussenwand eines Geschäftes welches Goldscheidererzeugnisse führte, gedacht waren, bzw. für die Schaufensterdekoration bestimmt waren, wurden in Bronze oder Zinn ausgeführt (–> vgl. Maurice Bouval, Künstlerliste)

644 (GSB)
Wind, Josef
STATUETTE „Charmeuse de Serpents (nue)"
ca. 1898
644 H. 85 cm; GSB, Pl. 56 (m. Abb.)
–> vgl. 40 BRONZE; 1399

646 (GSB)
BONBONNIÈRE „A L'Enfant"
ca. 1892/1900
646 H. 26 cm; GSB, Pl. 66 (m. Abb.)

647 (GSB)
STATUE „Charmeuse de Serpents (drapée)" No. 2
BRONZE
ca. 1892/1900
647 H. 40 cm; GSB, Pl. 54 (m. Abb.)
–> vgl. 599 (GSB)

648 (GSB)
STATUETTE „Dois-je le Croire?" No. 2
ca. 1892/1900
648 H. 40 cm; GSB, Pl. 31 (m. Abb.)
–> vgl. 619 (GSB)

649
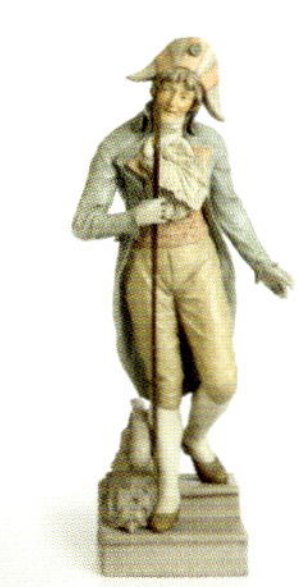
Schreitender Herr mit Zweispitz und Spazierstock (auf getrepptem Sockel)
ca. 1892/94
649 ger., Firmenmarke 3, Vervielf. vorbeh., H. 64,2 cm; GK, S. 10, 55 (m. Abb.); Wien Museum Inv. 182.057
Ausst.: HMW (jetzt Wien Museum) 1985/86.
Pendant –> vgl. 660
Lit.: GK, S. 10, 55 (m. Abb.)

650 (GSB)
VASE „Libellule"
ca. 1892/1900
650 H. 24 cm; GSB, Pl. 78 (m. Abb.)

651 (GSB)
COQUILLE „Atlas"
ca. 1892/1900
651 H. 32 cm; GSB, Pl. 78 (m. Abb.)

652
Haniroff
VASE Kleines Mädchen mit Frosch an der Vasenwandung stehend
ca. 1898/1900
0652 (Firmenmarke ohne Ang.), H. 56 cm; Pinhas, S. 49 (m. Abb.)

652 (GSB)
Kloster, M.
PLATEAU „Lézards"
ZINN / ZINKGUSS
ca. 1892/1900
652 H. 19 cm; GSB, Pl. 48 (m. Abb.)

653 (GSB)
„Enfant aux Paniers"
Vraie BRONZE
ca. 1892/1900
653 H. 23 cm; GSB, Pl. 51 (m. Abb.)

654 (GSB)
„Enfant au Polchinelle"
ca. 1892/1900
654 H. 22 cm; GSB, Pl. 55 (m. Abb.)
Pendant –> vgl. 655 (GSB)

655 vgl. 4866

655 (GSB)
„Enfant a la Lecture"
ca. 1892/1900
655 H. 22 cm; GSB, Pl. 55 (m. Abb.)
Pendant –> vgl. 654 (GSB)

656
SÄULE / POSTAMENT (reliefiert mit Figuren aus der griechischen Mythologie)
ca. 1895
0656,117,7 Firmenmarke Sculpture 13, Firmenmarke Adler 17, Fabriqué en Autriche, Ausformung ca. 1900, H. 107,3 cm; PB Wien 19
Bem.: Dargestellt sind Zeus, Demeter, Poseidon und Hermes.

656 (GSB)
VASE „Le Néant"
ca. 1892/1900
656 H. 40 cm; GSB, Pl. 57 (m. Abb.)

657 (GSB)
VASE „Joyeux Ébats"
ca. 1892/1900
657 H. 40 cm; GSB, Pl. 57 (m. Abb.)

658 (GSB)
BOUGEOIR „Louix XV"
Vraie BRONZE
ca. 1892/1900
658 H. 7 cm; GSB, Pl. 47 (m. Abb.)

659
Tell, E.
VASE Stehender Mädchenakt (am Vasenkorpus)
ca. 1898/1900
0659 (Firmenmarke ohne Ang.), unsign., H. 63 cm; Pinhas, S. 51 (m. Abb.)
–> vgl. 853
Pendant –> vgl. 675

659 (GSB)
PORTE-BOUQUET „Antique"
ca. 1892/1900
659 H. 22 cm; GSB, Pl. 66 (m. Abb.)

660
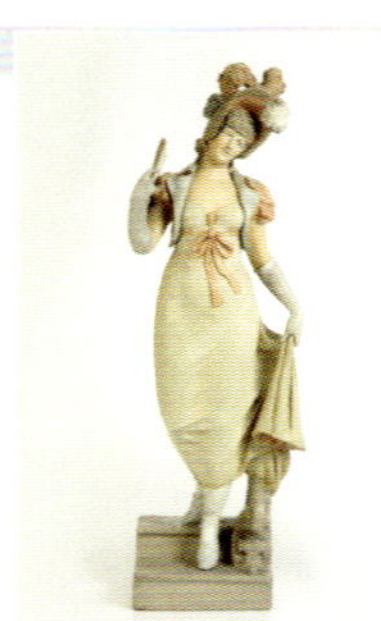
Schreitende Dame im Empirekleid (auf getrepptem Sockel)
ca. 1892/94
660 ger.,8 gepr., Firmenmarke 3, Vervielf. vorbeh., H. 68 cm; GK, S. 11, 56 (m. Abb.); Wien Museum Inv. 182.058
Ausst.: HMW (jetzt Wien Museum) 1985/86.
Pendant –> vgl. 649
Lit.: GK, S. 10, 56 (m. Abb.)

660 (GSB)
Callot, Jacques H.
VIDE-POCHE „Fantaisie"
BRONZE oder ZINN

ca. 1892/1900
660 H. 15 cm; GSB, Pl. 48 (m. Abb.)
Pendant –> vgl. 617 (GSB)

661
FIGUR „Grüß Gott"
ca. 1895
661 (Firmenmarke ohne Ang.), H. 53 cm; GSF, S. 3
Lit.: GSF, S. 3, hier mit der Ang. des Originaltitels „Grüß Gott".

661 (GSB)
ENCRIER „Lecture"
ca. 1892/1900
661 H. 19 cm; GSB, Pl. 62 (m. Abb.)

662 (GSB)
ENCRIER „Watteau"
ca. 1892/1900
662 H. 10 cm; GSB, Pl. 62 (m. Abb.)

663 (GSB)
Kinsburger, Sylvain
VIDE-POCHE „Curieux"
ca. 1892/1900
663 H. 17 cm; GSB, Pl. 51 (m. Abb.)

665 (GSB)
Coudray, Georges-Charles (zugeschr.)
VASE „Enfant Prisonnier"
ca. 1892/1900
665 H. 51 cm; GSB, Pl. 73 (m. Abb.)

666 (GSB)
Coudray, Georges-Charles
VASE „Nymphe a la Fontaine"
ca. 1892/1900
666 H. 29 cm; GSB, Pl. 58 (m. Abb.)
–> vgl. 682

667
BÜSTE Bärtiger Mann mit Hut
ca. 1895
667,20 ger. Firmenmarke 8, Repr. Res., H. 52 cm; eBay # 160033374952

667 (GSB)
VASE „Baroque"
ZINN / ZINKGUSS
ca. 1892/1900
667 H. 20 cm; GSB, Pl. 49 (m. Abb.)

668
Stoltenberg-Lerche, Hans Stephan
VASE An der Vasenwandung sitzender lachender Knabe
ca. 1895
0668,175,69 FG Wien (Firmenmarke 7?), sign. H. St. Lerche, H. 35,5 cm; WKH

668 (GSB)
PORTE-CORNET „Soleil"
ca. 1892/1900
668 H. 28 cm; GSB, Pl. 77 (m. Abb.)

669
BÜSTE „Singhalese"
ca. 1890/92 (für den Entwurf)
669 (Firmenmarke ohne Ang.), H. 26 cm; GSF, S. 3
–> vgl. 418; 768
Pendant –> vgl. 670
Lit.: GSF, S. 3, hier mit der Ang. des Originaltitels "Singhalese".

669 (GSB)
PORTE-CORNET „Pavot"
ca. 1892/1900
669 H. 31 cm; GSB, Pl. 78 (m. Abb.)

670
BÜSTE „Singhalesin" (mit Kopftuch und Münzkette als Halsschmuck)
ca. 1890/92 (für den Entwurf)
670,26 FG (Firmenmarke ohne Ang.), Repr. Res., H. 26 cm; Doro 9.11.1989
670 (Firmenmarke ohne Ang.), H. 26 cm; GSF, S. 3
–> vgl. 419; 769
Pendant –> vgl. 669
Lit.: GSF, S. 3, hier mit der Ang. des Originaltitels „Singhalesin".

670 (GSB)
PORTE-CORNET „Nénuphar"
ca. 1892/1900
670 H. 28 cm; GSB, Pl. 77 (m. Abb.)

671
FIGUR „Elephant"
ca. 1895
671 (Firmenmarke ohne Ang.), H. 22 cm; GSF, S. 3
Lit.: GSF, S. 3, hier mit der Ang. des Originaltitels „Elephant".

671 (GSB)
VASE „Adam et Ève"
ca. 1892/1900
671 H. 37 cm; GSB, Pl. 77 (m. Abb.)

672
Bouval, Maurice
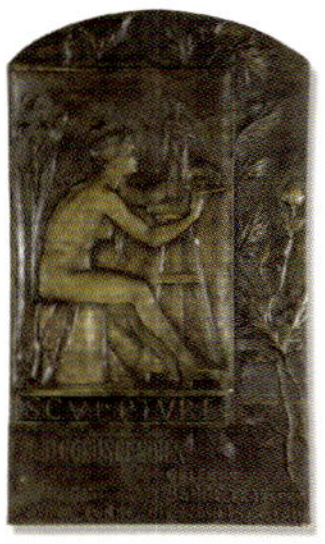
REKLAMESCHILD „Sculpture" Fréd. Goldscheider, Editeur, Paris, Vienne, Florence, Bronze, Terre cuite, Marbre, Etain
TERRAKOTTA
1898
672,36,28 Firmenmarke 4, Ges. gesch., sign. M. Bouval-98., H. 95,6 cm, B. 56,5 cm; MAK Wien, Inv. Ke 10521-1
Bem.: Dieses übergroße Reklameschild (für die Werbung innerhalb eines Verkaufslokales gedacht), welches Maurice Bouval für Goldscheider als Allegorie der Bildhauerei entwarf, ist nicht nur wegen seiner Größe ungewöhnlich. Es zeigt auch das Selbstbewusstsein und die Marktführerschaft der Manufaktur um die Jahrhundertwende.
Ausst.: PSK 1995
–> vgl. o. Nr. Version in Zinn und Bronze
Lit.: ADGS, Nr. 1 (m. Abb.)

672 (GSB)
COQUILLE „Baigneuse"
ca. 1892/1900
672 H. 8 cm; GSB, Pl. 62 (m. Abb.)

673
Tell, E.
SCHALE Nymphe am Teich
ca. 1895
0673,189,3 Firmenmarke 4, Ges. gesch., H. 6,5 cm, B. 29 cm; Wien Museum Inv. 220.784

675
Tell, E.
VASE Stehender Mädchenakt (am Vasenkorpus)
ca. 1898/1900
0675 (Firmenmarke ohne Ang.), unsign., H. 63 cm; Pinhas, S. 51 (m. Abb.)
Pendant –> vgl. 659

675 (GSB)
„Œuf aux Chats"
ca. 1892/1900
675 H. 14 cm; GSB, Pl. 58 (m. Abb.)

676
Tell, E.
SCHALE Nymphe in eine Muschel kletternd
ca. 1895
0676,180,10 Firmenmarke 7, Repr. Res., sign. E. Tell, H. 40 cm; Sammlung a@goldscheider.net

676 (GSB)
COQUILLE „Trésor de la Mer"
ca. 1892/1900
676 H. 17 cm; GSB, Pl. 66 (m. Abb.)

677 (GSB)
Kinsburger, Sylvain
PLATEAU „Danse du Printemps"
ca. 1892/1900
677 H. 32 cm; GSB, Pl. 66 (m. Abb.)

678 (GSB)
Stoltenberg-Lerche, Hans Stephan (zugeschr.)
JARDINIÈRE „Rocaille"
ca. 1892/1900
678 H. 30 cm; GSB, Pl. 82 (m. Abb.)

679 (GSB)
Stoltenberg-Lerche, Hans Stephan (zugeschr.)
VASE „Enfant joueur de Flute"
ZINN?
ca. 1892/1900
679 H. 21 cm; GSB, Pl. 82 (m. Abb.)
Pendant –> vgl. 521 (GSB)

680
FIGUR „Münchner Kind'l"
ca. 1895
680 (Firmenmarke ohne Ang.), H. 61 cm; GSF, S. 3
–> vgl. 738; 741
Lit.: GSF, S. 3, hier mit der Ang. des Originaltitels „Münchner Kind'l".

680 (GSB)
COQUILLE „Sirène"
ca. 1892/1900
680 H. 13 cm; GSB, Pl. 60 (m. Abb.)

681 (GSB)
PLATEAU „Plaisirs de Pêche"
ca. 1892/1900
681 H. 35 cm; GSB, Pl. 61 (m. Abb.)
Pendant –> vgl. 688 (GSB)

682
Coudray, Georges-Charles
VASE „Nymphe a la Fontaine" (Nymphe am Brunnen) (Frauengestalt neben wasserspeiendem Gesicht sitzend)
ca. 1895
F682,6,6 Firmenmarke 9, Repr. Res., unsign., H. 26 cm; GK, S. 11, 57 (m. Abb.); Wien Museum Inv. 182.059
Bem.: Sign. Exemplar in Weißmetall (Zinkguss, patiniert) aus der Pariser Produktion, H. 28 cm –> vgl. Ruef 481; diese Vase, eigentl. eher ein Henkelkrug (plastische Nymphe an der Vasenmündung neben wasserspeiendem Gesicht), findet sich unter der Nr. 666 im Kat. GSB wieder. Sie wurde in der Pariser Niederlassung in einer Höhe von 29 cm aus Zinn (Etain) und „Kunstguss" (Fonte d'Art, d.i. wohl Zinkguss, patiniert) gefertigt.
Ausst.: HMW (jetzt Wien Museum) 1985/86.
–> vgl. 666 (GSB)
Lit.: GSB, Pl. 58, hier mit Ang. des Originaltitels „Nymphe a la Fontaine"; GK, S. 11, 57 (m. Abb.)

682 (GSB)
STATUE „Fillette au Parapluie"
ca. 1892/1900
682 H. 37 cm (mit Sockel); GSB, Pl. 61 (m. Abb.)

683 (GSB)
ENCRIER „Stalactite"
ca. 1892/1900
683 H. 7 cm; GSB, Pl. 20 (m. Abb.)

684
Kinsburger, Sylvain
VASE „Adam" (Adam an einem Baumstamm sitzend)
ca. 1895
0684,266,35 FG Wien (Firmenmarke unleserlich), H. 34,5 cm; Doro 9.11.1993
Pendant –> vgl. 685
Lit.: Pinhas, S. 51 (m.Abb.), hier die Künstlerangabe „Lemoine".

684 (GSB)
Lévèque, Auguste Maurice / Thiele, (P.?)
BUSTE „Esclave" No. 1
ca. 1897/1900
684 H. 62 cm; GSB, Pl. 59 (m. Abb.)
Bem.: Die Büsten der Wiener Produktion wurden mit dem Originaltitel „Bedouine" bezeichnet.
–> vgl. 733 (GSB)
–> vgl. 1101; 1877

685
Kinsburger, Sylvain
VASE „Eva" (Eva an einem Baumstamm stehend)
ca. 1895
0685,271,27 FG Wien (Firmenmarke unleserlich), H. 34,5 cm; Doro 9.11.1993
Pendant –> vgl. 684
Lit.: Pinhas, S. 51 (m.Abb.), hier die Künstlerangabe „Lemoine".

685 (GSB)
STATUE „Pauline Borghèse"
ca. 1892/1900
685 H. 36 cm; GSB, Pl. 21 (m. Abb.)

686 (GSB)
Callot, Jacques H.
VASE Etrusque „Eglantine"
ca. 1892/1900
686 H. 23 cm; GSB, Pl. 40 (m. Abb.)
Pendant –> vgl. 687 (GSB)
–> vgl. 692

687 (GSB)
Callot, Jacques H.
VASE Etrusque „Chrysanthéme"
ca. 1892/1900
687 H. 23 cm; GSB, Pl. 40 (m. Abb.)
Pendant –> vgl. 686 (GSB)
–> vgl. 693

688 (GSB)
PLATEAU „Péche a L'Écrevisse"
ca. 1892/1900
688 H. 35 cm; GSB, Pl. 39 (m. Abb.)
Pendant –> vgl. 686 (GSB)

689 (GSB)
VASE „Caiman"
ca. 1892/1900
689 H. 20 cm; GSB, Pl. 62 (m. Abb.)

690
Stoltenberg-Lerche, Hans Stephan
VASE An der Vasenwandung sitzender Knabe (in die Mündung blickend)
ca. 1895
0690,140,25 Firmenmarke 7, Repr. Res., H. 17 cm; Sammlung Gruber, S. 104, 105 (m. Abb.)

690 (GSB)
VIDE-POCHE „Fantaisie"
ca. 1892/1900
690 H. 23 cm; GSB, Pl. 60 (m. Abb.)

691 (GSB)
VASE „Antique"
ca. 1892/1900
691 H. 26 cm; GSB, Pl. 20 (m. Abb.)

692
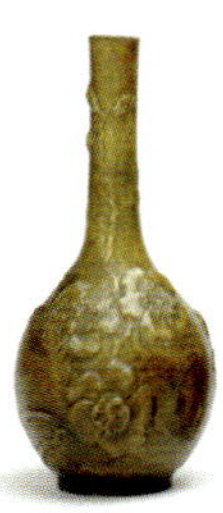
Callot, J(acques H.)
VASE Etrusque „Eglantine" (reliefiertes Mädchenbildnis, nach rechts blickend)
ca. 1895
0692,28,69 Firmenmarke 4a, Repr. Res., gepr. „P"= Z3, H. 22 cm; PB Wien 19
–> vgl. 686 (GSB)
Pendant –> vgl. 693

693
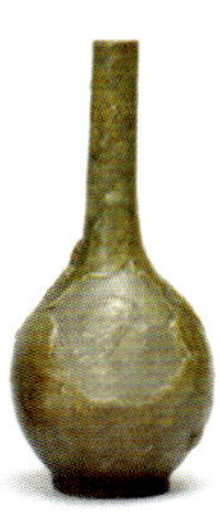
Callot, J(acques H.)
VASE Etrusque „Chrysanthème" (reliefiertes Mädchenbildnis, nach links blickend)
ca. 1895
0693,23,69 Firmenmarke 4a, Repr. Res., H. 22 cm; PB Wien 19
693,94,37 (Firmenmarke ohne Ang.), H. 20 cm; KaM 89
–> vgl. 687 (GSB)
Pendant –> vgl. 692

693 (GSB)
ENCRIER „Empire"
ca. 1892/1900
693 H. 9 cm; GSB, Pl. 40 (m. Abb.)

694 (GSB)
PLUMIER „Empire"
ca. 1892/1900
694 H. 8 cm; GSB, Pl. 40 (m. Abb.)

695 (GSB)
CENDRIER „Empire"
ca. 1892/1900
695 H. 12 cm; GSB, Pl. 40 (m. Abb.)

696 (GSB)
BOUGEOIR „Empire"
ca. 1892/1900
696 H. 8 cm; GSB, Pl. 40 (m. Abb.)

697
BÜSTE „Tyrolerknabe, reseda" (Bursche in tirolischer Tracht mit Hut) (auf Holzsockel)
ca. 1896/97
B 697 ger., Firmenmarke 5, H. ca. 7 cm, H. (mit Holzsockel) ca. 10 cm; KH
Bem.: s. Modell-Nr. 6
Lit.: KeramRS, 5.1897, S. 347, hier die Einreichung zum „Musterschutz" und mit Ang. des Originaltitels BÜSTE „Tyrolerknabe, reseda".

697 (GSB)
Schuller, L(?)
GARNITURE ANTIQUE „Encrier" No. 1
ca. 1892/1900
697 H. 10 cm; GSB, Pl. 46 (m. Abb.)
Weitere Teile dieser Garnitur –> vgl. 698–702 (GSB)

698 (GSB)
Schuller, L(?)
GARNITURE ANTIQUE „Encrier" No. 2
ca. 1892/1900
698 H. 7 cm; GSB, Pl. 46 (m. Abb.)
Weitere Teile dieser Garnitur –> vgl. 697 (GSB); 699–702 (GSB)

699 (GSB)
Schuller, L(?)
GARNITURE ANTIQUE „Plumier"
ca. 1892/1900
699 H. 9 cm; GSB, Pl. 46 (m. Abb.)
Weitere Teile dieser Garnitur –> vgl. 697 (GSB); 698 (GSB); 700–702 (GSB)

700 (GSB)
Schuller, L(?)
GARNITURE ANTIQUE „Cendrier"
ca. 1892/1900
700 H. 7 cm; GSB, Pl. 46 (m. Abb.)
Weitere Teile dieser Garnitur –> vgl. 697–699 (GSB); 700 (GSB); 702 (GSB)

701 (GSB)
Schuller, L(?)
GARNITURE ANTIQUE „Flambeau"
ca. 1892/1900
701 H. 10 cm; GSB, Pl. 46 (m. Abb.)
Weitere Teile dieser Garnitur –> vgl. 697–700 (GSB); 702 (GSB)

702 (GSB)
Schuller, L(?)
GARNITURE ANTIQUE „Bougeoir"
ca. 1892/1900
702 H. 8 cm; GSB, Pl. 46 (m. Abb.)
Weitere Teile dieser Garnitur –> vgl. 697–701 (GSB)

704 (GSB)
VASE „Crabe"
ca. 1892/1900
704 H. 15 cm; GSB, Pl. 39 (m. Abb.)

710 (GSB)
BUSTE „Bouquetière" No. 1
ca. 1892/1900
710 H. 35 cm; GSB, Pl. 62 (m. Abb.)
–> vgl. 568 (GSB)

711 (GSB)
VASE „Houx"
ca. 1892/1900
711 H. 16 cm; GSB, Pl. 39 (m. Abb.)

712
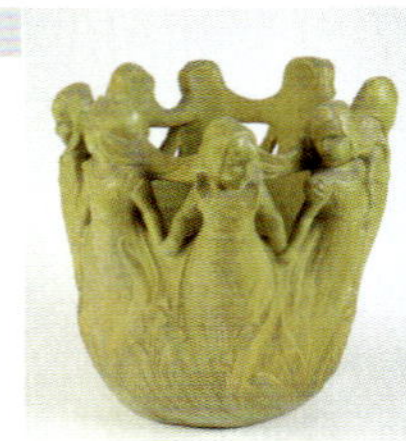
Simon
VASE Acht Mädchen, Ringelreigen tanzend
ca. 1898/1900
0712,150,10 Firmenmarke 7, Repr. Res., monogr. S, H. 37 cm, Ø 30 cm; Wien Museum Inv. 219.128
Bem.: Evtl. handelt es sich um das im ÖMKI 1901/02 ausgestellte Objekt „Frühling".
Ausst.: ÖMKI 1901/02?
Lit.: R., F.: Wiener Kunstkeramik, in: KM, 2.1902, S. 89, 90 (m. Abb.), „Eine von Simon modellierte Vase stellt Ringelreigen spielende kleine Mädchen dar, deren unten zusammen fließende Kleider die Wände der Vase bilden. Die Füßchen verschwinden in dem hohen Grase der Wiese, auf welcher die Kinder spielend gedacht sind. Jedes Köpfchen der singenden Kleinen ist liebevoll und sorgsam durchgeführt, die ganze Gruppe von frappierender Charakteristik".

712 (GSB)
STATUE „Captive" No. 1
ca. 1892/1900
712 H. 50 cm; GSB, Pl. 67 (m. Abb.)
–> vgl. 744 (GSB)
–> vgl. 1304

715 (GSB)
ENCRIER „Louis XVI"
ca. 1892/1900
715 H. 19 cm; GSB, Pl. 40 (m. Abb.)

716 (GSB)
BUSTE „Beethoven"
ca. 1892/1900
716 H. 25 cm (mit Sockel); GSB, Pl. 61 (m. Abb.)
Bem.: Dargestellt ist der Komponist Ludwig van Beethoven (1770–1827).

717 (GSB)
Mercier
SERVICE FUMEUR „Narghileh"
ca. 1892/1900
PLATEAU
717 H. 23 cm; GSB, Pl. 42 (m. Abb.)
weitere Teile der Garnitur ebenfalls mit der Modell-Nr. 717
Pot a Tabac H. 20 cm; GSB, Pl. 42 (m. Abb.)
Porte-Allumettes H. 8 cm; GSB, Pl. 42 (m. Abb.)
Cendrier H. 11 cm; GSB, Pl. 42 (m. Abb.)

719 (GSB)
PLATEAU „Oasis"
ca. 1892/1900
719 H. 30 cm; GSB, Pl. 20 (m. Abb.)
Pendant –> vgl. 722 (GSB)

722 (GSB)
PLATEAU „Simoun"
ca. 1892/1900
719 H. 30 cm; GSB, Pl. 20 (m. Abb.)
Pendant –> vgl. 719 (GSB)

724
Pecheur (d.i. vermutl. Friedrich Fischer)
VASE Zwei Frauen (Wassernixen?) (an der Vasenwandung)
ca. 1898/1900
724 (Firmenmarke ohne Ang.), H. 87 cm; Pinhas, S. 51 (m. Abb.)
Bem.: Evtl. handelt es sich um das im ÖMKI 1900/01 ausgestellte Objekt VASE „Wassernixen".
Ausst.: ÖMKI 1900/01?
–> vgl. 795

726
Simon
SCHALE Liegender Frauenakt
ca. 1898/1900
0726,46,40 Firmenmarke
Sculpture 13, Ausformung
ca. 1900, H. 28,2 cm, L. 60 cm; PB Wien 19

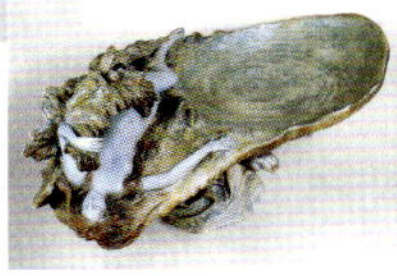

726 (GSB)
Jonchery, Charles Émile
„Feuille Almée"
ca. 1892/1900
726 H. 51 cm; GSB, Pl. 39 (m. Abb.)

727
FIGUR „Mohr" (Stehender afrikanischer Fruchtverkäufer mit Korb) (auf Rundsockel)
ca. 1895
727 (Firmenmarke ohne Ang.), H. 97 cm; Pinhas, S. 26 (m. Abb.)
727 (Firmenmarke ohne Ang.), H. 95 cm; GSF, S. 3
Pendant -> vgl. 794
Lit.: GSF, S. 3, hier mit der Ang. des Originaltitels „Mohr".

727 (GSB)
STATUETTE „Premier Baisers"
ca. 1892/1900
727 H. 28 cm; GSB, Pl. 60 (m. Abb.)

728
BÜSTE Posierende junge Dame mit langem Haar
ca. 1898/1900
728,4 ger., Firmenmarke 11a, Repr. Res., gepr. „P"= Z3, H. 26,2 cm; Wien Museum Inv. 186.638
728,14 ger., Firmenmarke 8, H. 27 cm;
eBay # 320024512462
728,27 ger., Firmenmarke 8a, H. 26,7 cm;
eBay # 1419209269; 3720534480

728 (GSB)
SERVICE FUMEUR „Médailles"
ca. 1892/1900
PLATEAU
728 H. 26 cm; GSB, Pl. 41 (m. Abb.)
weitere Teile der Garnitur ebenfalls mit der Modell-Nr. 728
Pot a Tabac H. 19 cm; GSB, Pl. 41 (m. Abb.)
Porte-Cigares H. 10 cm; GSB, Pl. 41 (m. Abb.)
Porte-Cigarettes H. 7 cm; GSB, Pl. 41 (m. Abb.)
Porte-Allumettes H. 6 cm; GSB, Pl. 41 (m. Abb.)
Cendrier H. 7 cm; GSB, Pl. 41 (m. Abb.)
Frottoir H. 7 cm; GSB, Pl. 41 (m. Abb.)

729 (GSB)
COQUILLE „Baigneuse surprise"
ca. 1892/1900
729 H. 52 cm; GSB, Pl. 64 (m. Abb.)

732 (GSB)
Maison, Rudolf
GROUPE „Le bon Chien" („Gute Freunde")
ca. 1892/1900
732 H. 28 cm; GSB, Pl. 67 (m. Abb.)
ohne Ang., H. 28 cm; Spik 572
Bem.: Ob dieses Modell in der Wiener Produktion (Fayence) auch die Nummer 732 hatte, ist fraglich und durch die fehlenden Ang. bei der Auktion Spik 572 leider nicht mehr nachprüfbar. Auch lässt sich für die Pariser Produktion nichts endgültiges über das Material aussagen, es ist aber zu vermuten dass es sich um Bronze oder Zinkguss handelt.
Lit.: Pudor (1906), o. S. (m. Abb.), hier mit Ang. des Künstlers R. Maison u. des deutschen Originaltitels "Gute Freunde". Pudor vermerkt als Material „Fayence".

733 (GSB)
Lévêque, Auguste Maurice/Thiele, (P.?)
BUSTE „Esclave" No. 2
ca. 1897/1900
733 H. 30 cm; GSB, Pl. 59 (m. Abb.)
Bem.: Die Büsten der Wiener Produktion wurden mit dem Originaltitel „Bedouine" bezeichnet.
-> vgl. 684 (GSB)
-> vgl. 1101; 1877

734
FIGUR „Judith" (Stehende Frauengestalt mit Schwert, Stirnband und Gürtel)
ca. 1897/98
734,14 Firmenmarke 11, H. 61 cm; KH; Bem.: An der Plinthe betitelt (Metallschild).
734,?,3 Firmenmarke 11?, H. 61,5 cm; Doro 1.12.1998; Bem.: Schwert fehlt.
Lit.: GSB, Pl. 27, hier ist die Figur mit dem Originaltitel „Judith (Habillée)" unter Nr. 584 (H. 65 cm) abgebildet. Eine identische Figur, jedoch mit entblößtem Oberkörper „Judith (Nue)" in drei verschiedenen Größen (H. 65, 30, 22 cm) findet sich an gleicher Stelle. Ausgeformt wurden diese Objekte in der Pariser Niederlassung von Friedrich Goldscheider jeweils in „Vrai Bronze" und „Fonte d'Art" unter den Nrn. 584; 602; 604.
-> vgl. 584 (GSB); 602 (GSB); 604 (GSB)
-> vgl. 1012; 1012a; 1012b
-> vgl. 5105= ähnl. Modell in einer Ausformung der 20er Jahre
Lit.: GSF, S. 3; Pinhas, S. 71 (m. Abb.)

734a
FIGUR „Judith (nackt)"
ca. 1897/98
734a (Firmenmarke ohne Ang.), H. 63 cm; GSF, S. 3
-> vgl. 734; 1012; 1012a
Lit.: GSF, S. 3, hier mit der Ang. des Originaltitels „Judith (nackt)".

734b
FIGUR „Judith (bekleidet)"
ca. 1897/98
734b (Firmenmarke ohne Ang.), H. 63 cm; GSF, S. 3
-> vgl. 734; 1012; 1012b
Lit.: GSF, S. 3, hier mit der Ang. des Originaltitels „Judith (bekleidet)".

734 (GSB)
Ceccarelli, Ezio
BUSTE „Automne"
ca. 1898
734 H. 30 cm; GSB, Pl. 71 (m. Abb.)
-> vgl. 622 (GSB)
-> vgl. 1297; 1619

735a
GRUPPE „Knabenspiel (3 Knaben)"
ca. 1898
735a (Firmenmarke ohne Ang.), H. 27 cm; GSF, S. 3
-> vgl. 735b; 743
Lit.: GSF, S. 3 hier mit der Ang. des Originaltitels „Knabenspiel (3 Knaben)".

735b
GRUPPE „Knabenspiel (2 Knaben)"
ca. 1898
735b (Firmenmarke ohne Ang.), H. 21 cm; GSF, S. 3
-> vgl. 735b; 743
Lit.: GSF, S. 3, hier mit der Ang. des Originaltitels „Knabenspiel (2 Knaben)".

736
SCHALE / MUSCHEL Zwei stehende Nymphen
ca. 1898/1900
0736 (Firmenmarke ohne Ang.), H. 44 cm; Pinhas, S. 48 (m. Abb.)

738
FIGUR „Münchner Kind'l (groß)"
ca. 1895 (für den Entwurf)
738 (Firmenmarke ohne Ang.), H. 94 cm; GSF, S. 3
Bem.: Das sog. „Münchner Kindl", ist eine figürliche Darstellung des Stadtwappens von München.
-> vgl. 680; 741
Lit.: GSF, S. 3, hier mit der Ang. des Originaltitels „Münchner Kind'l (groß)".

738 (GSB)
Reyam, (David?)
STATUETTE „Libellule"
ca. 1892/1900
738 H. 49 cm; GSB, Pl. 79 (m. Abb.)

739
FIGUR „Falstaff"
ca. 1898/1900
739,X GS Wien (Firmenmarke?), H. 53 cm;
eBay # 2559472658
739,25,16 Firmenmarke 7, Repr. Res., gepr. „P"= Z3, undeutl. sign., H. 53 cm; KH Steinigeweg, D-Tecklenburg
Lit.: GSF, S. 3, hier mit der Ang. des Originaltitels „Falstaff".

739 (GSB)
Debienne, Noémi(e)
„Triomphe de Bacchante"
ca. 1892/1900
739 H. 92 cm; GSB, Pl. 75 (m. Abb.)
Bem.: In der Wiener Produktion trägt diese Figur ein langes Gewand und wird unter der Modell-Nr. 2019 (H. 81 cm) geführt.
-> vgl. 2019

739
FIGUR „Kellermeister"
ca. 1898/1900
739 (Firmenmarke ohne Ang.), H. 51 cm; GSF, S. 3
Lit.: GSF, S. 3, hier mit der Ang. des Originaltitels „Kellermeister".

740 (GSB)
Masse, (Charles?)
„Enfant Sculpteur"
ca. 1892/1900
740 H. 27 cm; GSB, Pl. 76 (m. Abb.)
Pendant -> vgl. 741 (GSB)

741
FIGUR „Münchner Kind'l (kleinst)"
ca. 1895 (für den Entwurf)
741 (Firmenmarke ohne Ang.), H. 19 cm; GSF, S. 4
Bem.: Das sog. „Münchner Kindl", ist eine figürliche Darstellung des Stadtwappens von München.
-> vgl. 680; 741
Lit.: GSF, S. 3, hier mit der Ang. des Originaltitels „Münchner Kind'l (kleinst)".

741 (GSB)
Masse, (Charles?)
„Enfant Peinture"
ca. 1892/1900
741 H. 27 cm; GSB, Pl. 76 (m. Abb.)
Pendant -> vgl. 740 (GSB)

742 (GSB)
BUSTE „Olympe"
Inschrift: OLYMPE
ca. 1892/1900
742 H. 39 cm; GSB, Pl. 68 (m. Abb.)
Pendant -> vgl. 743 (GSB)

743 -> ÜBERSCHNEIDUNG
GRUPPE „Knabenspiel"
ca. 1898
743 (Firmenmarke ohne Ang.), H. 21 cm; GSF, S. 4
-> vgl. 735a; 735b
Lit.: GSF, S. 4, hier mit der Ang. des Originaltitels „Knabenspiel".

743 -> ÜBERSCHNEIDUNG
Stehende junge Frau mit entblößten Brüsten und Lyra (auf Rechtecksockel)
ca. 1898/1900
743 (Firmenmarke ohne Ang.), H. 97 cm; Pinhas, S. 42 (m. Abb.)

743 (GSB)
BUSTE „Madelon"
Inschrift: MADELON
ca. 1892/1900
743 H. 39 cm; GSB, Pl. 68 (m. Abb.)
Pendant -> vgl. 742 (GSB)

744 (GSB)
STATUE „Captive" No. 2
ca. 1892/1900
744 H. 40 cm; GSB, Pl. 67 (m. Abb.)
-> vgl. 712 (GSB)
-> vgl. 1304

745
Delagrange, Léon Noel
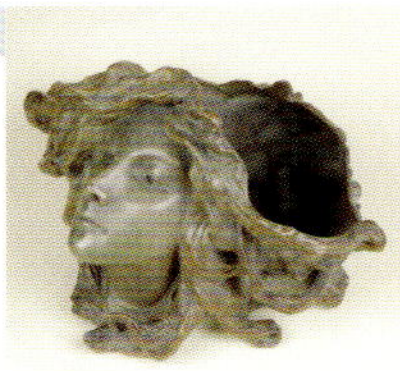
MASKE/JARDINIERE Frauengesicht mit wallendem Haar (Gefäß für Blumen)
ca. 1900
0745,17,13 Firmenmarke 7, Repr. Res., H. 24 cm; eBay # 2038065651
0745,151,69 Firmenmarke 7, Repr. Res., H. 24 cm; Wien Museum Inv. 219.127
0745,245,35 gem. XI, ohne Firmenmarke, Repr. Res., unsign., H. 24 cm; PB Wien 19
Bem.: Evtl. handelt es sich um das im ÖMKI 1900/01 ausgestellte Objekt „Jardiniere".
Ausst.: ÖMKI 1900/01? „Jardiniere"?
-> vgl. 868

746
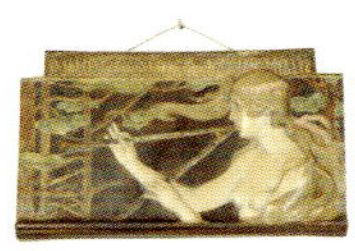
RELIEF Junge Frau im Profil, eine Doppelflöte spielend (Frontfigur eines Bacchantenzuges?)
ca. 1898/1900
0746,288,47 Firmenmarke 4, H. 38,1 cm, L. 63,5 cm; eBay # 2628578376
0746 ohne Ang., H. 38 cm; PB Wien 19

747
VASE An der Vasenwandung stehende Frau mit Wasserlilien
ca. 1898/1900
0747 (Firmenmarke ohne Ang.), H. 48 cm; Pinhas, S. 48 (m. Abb.)

747 (GSB)
BUSTE „Fiancée"
ca. 1892/1900
747 H. 53 cm; GSB, Pl. 68 (m. Abb.)

752 (GSB)
STATUE „Petit Lever"
ca. 1892/1900
752 H. 82 cm; GSB, Pl. 43 (m. Abb.)
-> vgl. 1804

753 (GSB)
Fontini (d.i. Ceccarelli, Ezio)
BÜSTE „Coquette" (Lachende, junge Dame mit Haarschleife) (nach rechts oben blickend)
ca. 1892/1900
753 H. 40 cm; GSB, Pl. 76 (m. Abb.)
Bem.: In der Wiener Produktion wird diese Büste, die von dem italienischen Bildhauer Ezio Ceccarelli stammt, unter der Modell-Nr. 1784 (H. ca. 60 cm) geführt.
-> vgl. 1784

755 (GSB)
Caussé, Julien
CHOPE „Antique"
ca. 1892/1900
755 H. 10 cm; GSB, Pl. 83 (m. Abb.)

756 (GSB)
Bozzi
RELIEF Plateau „Mater Purissima"
ca. 1892/1900
756 H. 45 cm; GSB, Pl. 76 (m. Abb.)

757 (GSB)
Cherc
„Charmeur de Serpents"
ca. 1892/1900
757 H. 66 cm; GSB, Pl. 70 (m. Abb.)

758 (GSB)
Savine, Léopold Pierre Antoine
„Moissonneur"
ca. 1892/1900
758 H. 40 cm; GSB, Pl. 80 (m. Abb.)
Pendant -> vgl. 759 (GSB)

759
BÜSTE Dante
Inschrift: DANTES
ca. 1898/1900
759,123,27 gem. weiß S.K., Firmenmarke 7, Repr. Res., gepr. „P"= Z3, H. 20,3 cm, B. 27,9 cm; eBay # 3772063078; 7375865619

759 (GSB)
Savine, Léopold Pierre Antoine
„Moissonneuse"
ca. 1892/1900
759 H. 40 cm; GSB, Pl. 80 (m. Abb.)
Pendant -> vgl. 758 (GSB)

761 (GSB)
Mercier
JARDINIÈRE „Sphynx"
ca. 1892/1900
761 H. 12 cm, L. 49 cm; GSB, Pl. 69 (m. Abb.)

764 (GSB)
GROUPE „Bacchante et Silène"
ca. 1892/1900
764 H. 40 cm; GSB, Pl. 67 (m. Abb.)
-> vgl. 1464

765 (GSB)
Borsi
VASE „Nymphe aux Nénuphars"
ca. 1892/1900
765 H. 48 cm; GSB, Pl. 83 (m. Abb.)

766 -> ÜBERSCHNEIDUNG
Cherc
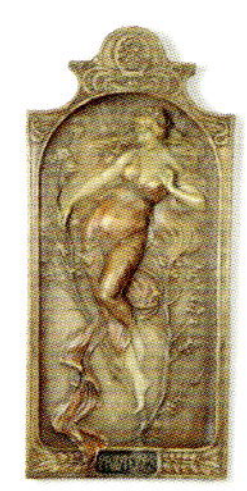
RELIEF „Printemps"
Inschrift: PRINTEMPS
ca. 1895/97
0766,32,17 Firmenmarke 4a, Repr. Res., H. 73 cm; PB Wien 19
Bem.: Diese Darstellung, wie auch die Pendants dazu, sind in der Art von Alphonse (eigentl. Alfons Maria) Mucha gestaltet.
Text der Inschrift: Frühling
Pendants -> vgl. 768; 770; 772

766 -> ÜBERSCHNEIDUNG
Lévèque, Auguste-Maurice (zugeschr.)
BÜSTE Junger Afrikaner mit entblößtem Oberkörper und Halsschmuck
ca. 1893
766,44 (Firmenmarke ohne Ang.), Repr. Res.,unsign., H. 46,2 cm; Christie's SK 16.11.2006
766 (ohne Ang.), Firmenmarke 11, Vervielf. vorbeh., unsign., H. 45 cm; Ruef 486
766 (ohne Ang.), FG Wien (Firmenmarke?), Vervielf. vorbeh., unsign., H. 46 cm; Christie's L. 1.11.2001
766 (ohne Ang.), H. 43 cm; Richemond (1999), S. 77 (m. Abb.), hier mit der Ang. datiert „Février 1893"; Richemond schreibt diese Büste dem Künstler (Auguste-Maurice) Lévèque zu.
Pendant -> vgl. 767

767
Lévèque, Auguste-Maurice (zugeschr.)
BÜSTE Junge Afrikanerin mit entblößten Brüsten und Halsschmuck
ca. 1893
767,V FG Wien (Firmenmarke?), Vervielf. vorbeh., unsign., H. 45 cm; Ruef 486
767,8 Firmenmarke 8a, unsign., H. 43 cm; Wien Museum Inv. 219.131
767 (ohne Ang.), Firmenmarke 11?, Vervielf. vorbeh., unsign., H. 46 cm; Christie's L. 1.11.2001
767 (ohne Ang.), unsign., H. 43 cm; Richemond (1999), S. 77 (m. Abb.), hier mit der Ang. datiert „Février 1893"; Richemond schreibt diese Büste dem Künstler (Auguste-Maurice) Lévèque zu.
Pendant -> vgl. 766

768 -> ÜBERSCHNEIDUNG
BÜSTE „Singhalese" (mit Bart und hochgebundenem Haar)
ca. 1890/95
768,44 FG (Firmenmarke ohne Ang.), Repr. Res., H. 45 cm; Soth. L. 20.3.1986
768 ger., FG (Firmenmarke ohne Ang.), Repr. Res., H. 46,5 cm; Ahlden 102
768 GS Wien (Firmenmarke?), Vervielf. vorbeh., H. 46,3 cm; Christie's SK 27.9.2000
ohne Nr., (Firmenmarke unleserlich), H. 48 cm; Schuler 68
768 (Firmenmarke ohne Ang.), H. 45 cm; GSF, S. 4
-> vgl. 418; 669
Pendant -> vgl. 769
Lit.: GSF, S. 4, hier mit der Ang. des Originaltitels „Singhalese".

768 -> ÜBERSCHNEIDUNG
Cherc
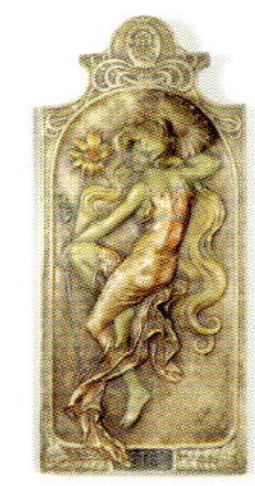
RELIEF „Été"
Inschrift: ÉTÉ
ca. 1895/97
0768,12,3 Firmenmarke 4a, Ges. gesch., H. 73 cm; PB Wien 19
ohne Ang., FG Wien (Firmenmarke ohne Ang.), Repr. Res., H. 72,4 cm; Soth. N.Y. 26.11.1993
Bem.: Diese Darstellung, wie auch die Pendants dazu, sind in der Art von Alphonse (eigentl. Alfons Maria) Mucha gestaltet.
Text der Inschrift: Sommer
Pendants -> vgl. 766; 770; 772

769
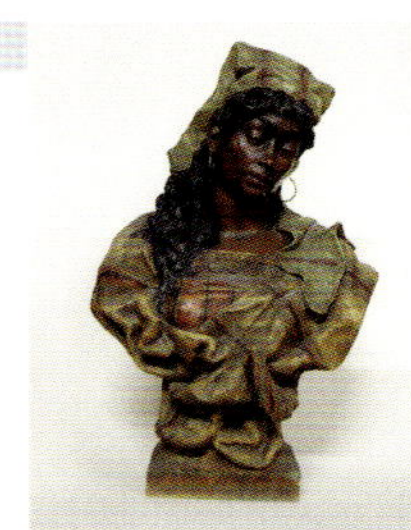
BÜSTE „Singhalesin" (mit Kopftuch und Münzkette als Halsschmuck)
ca. 1890/95
769,XII ger., (Firmenmarke unleserlich), H. 45 cm; Schuler 68
769,33 Firmenmarke 8a, gepr. „P"= Z3, H. 45,4 cm; MAK Wien, Inv. Ke 10602-108
769,49 FG (Firmenmarke?), Repr. Res., gepr. „P"= Z3, H. 45 cm; Doro 20.4.1993; Soth. L. 20.3.1986; Bem.: Soth. gibt falsche Nr. 759 an.
769 GS Wien (Firmenmarke?), Vervielf. vorbeh., H. 46,3 cm; Christie's SK 27.9.2000
769 (Firmenmarke ohne Ang.), H. 45 cm; GSF, S. 4
-> vgl. 419; 670
Pendant -> vgl. 768
Lit.: GSF, S. 4, hier mit der Ang. des Originaltitels „Singhalesin".

770
Cherc
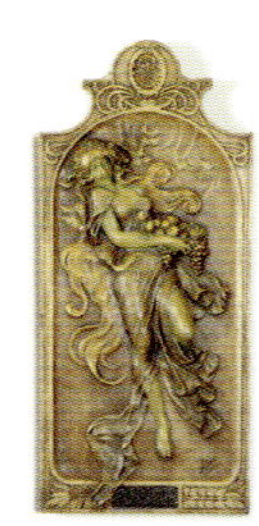
RELIEF „Autumne"
Inschrift: AUTUMNE
ca. 1895/97
0770,16,3 Firmenmarke 4a, Ges. gesch., H. 73 cm; PB Wien 19
Bem.: Diese Darstellung, wie auch die Pendants dazu, sind in der Art von Alphonse (eigentl. Alfons Maria) Mucha gestaltet.
Text der Inschrift: Herbst
Pendants -> vgl. 766; 768; 772

771
BÜSTE Mädchen mit Mandoline
ca. 1895/97
771,82 Firmenmarke 8a, gepr. „P"= Z3, H. 43 cm; Wien Museum Inv. 219.130

772
Cherc
RELIEF „Hiver"
Inschrift: HIVER
ca. 1895/97
0772,30,7 Firmenmarke 4a, Fabriqué en Autriche, H. 73 cm; PB Wien 19
ohne Ang., FG Wien (Firmenmarke?), Repr. Res., H. 72,4 cm; Soth. N.Y. 26.11.1993
Bem.: Diese Darstellung, wie auch die Pendants dazu, sind in der Art von Alphonse (eigentl. Alfons Maria) Mucha gestaltet.
Text der Inschrift: Winter
Pendants -> vgl. 766; 768; 770

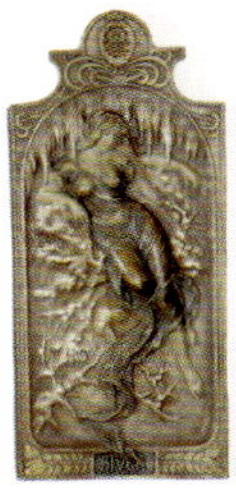

772 (GSB)
Arnoux
„Caresses"
ca. 1892/1900
772 H. 75 cm; GSB, Pl. 72 (m. Abb.)

774
DOSE Schmuckschatulle
Form: quaderförmig
Dekor: reliefiert mit Medaillon (Frauenporträt im Profil nach links) sowie Blumen- und Blattwerk
ca. 1900
0774 (Firmenmarke ohne Ang.), H. 13 cm, B. 29 cm; Pinhas, S. 50 (m. Abb.)

774 (GSB)
Halou, Alfred Jean
PLATEAU „Chasse"
ca. 1892/1900
774 H. 28 cm, L. 35 cm; GSB, Pl. 74 (m. Abb.)

775 (GSB)
Nelson, (Alphonse-Henri?)
BUSTE „Femme Fantaisie"
ca. 1892/1900
775 H. 48 cm; GSB, Pl. 79 (m. Abb.)

776
Dornauszieher (Sitzender Knabe, der sich einen Dorn aus dem Fuß zieht)
ca. 1895/97 (für die Ausführung)
776,98,19 gepr., ger. II, gem. weiß XXI, Firmenmarke 7, gepr. „P"= Z3, H. 32 cm; eBay # 3351674372
Bem.: Dieses in Terrakotta ausgeführte Objekt geht zurück auf einen frühhellenistischen Figurentypus (3. Jh. v. Chr.), vor allem aber auf den kapitolinischen Dornauszieher (Rom, 1. Jh. v. Chr.), der in Bronze überliefert ist. Zahlreiche Nachbildungen, auch in nachantiker Zeit, besonders der ital. Renaissance zeugen von der Beliebtheit der Darstellung.
Lit.: Lexikon der Antike: Dornauszieher, Digitale Bibliothek Band 18: Lexikon der Antike, S. 1445 (vgl. LDA, S. 145)

776 (GSB)
Mercié, Antonin
BUSTE „Faunesse"
ca. 1892/1900
776 H. 32 cm; GSB, Pl. 83 (m. Abb.)

777 (GSB)
Huneker (d.i. Bracken, Clio Hinton)
„Cendrier Libellule"
ca. 1892/1900
777 H. 12 cm; GSB, Pl. 69 (m. Abb.)

778
Lesca
RELIEF „Victoribus" (Junge Frau im Profil, die einen Lorbeerzweig an einen Jüngling reicht) Inschrift: VICTORIBUS
ca. 1898/1899
0778,10,3 gem. gold III, GS Wien (Firmenmarke?), Ges. gesch., sign. Leska, zwei Originalpick.: 1. Kunstsalon Banger, Wiesbaden (Motiv: Lesende junge Frau, sign. Ludwig Hohlwein), 2. Originalpreisauszeichnung M 55.– H. 39,5 cm, L. 83 cm; Zezschwitz 20
Text der Inschrift: Den Siegern.
Ausst.: ÖMKI (1900/01)

778 (GSB)
Huneker (d.i. Bracken, Clio Hinton)
„Coque Marine"
ca. 1892/1900
778 H. 9 cm, L. 21 cm; GSB, Pl. 69 (m. Abb.)

780 (GSB)
Bruyneel, (Victor?)
STATUE „Le Piège"
ca. 1892/1900
780 H. 58 cm; GSB, Pl. 44 (m. Abb.)
-> vgl. 832 (GSB)

782 (GSB)
Paris, Auguste
STATUE „Éternel Moqueur"
ca. 1892/1900
782 H. 51 cm; GSB, Pl. 80 (m. Abb.)
-> vgl. 783 (GSB)= BÜSTE

783 (GSB)
Paris, Auguste
BUSTE „Éternel Moqueur" No. 1
ca. 1892/1900
783 H. fehlt; GSB, Pl. 4 (m. Abb.)
Ausst.: Paris, Salon 1898
-> vgl. 842 (GSB)
-> vgl. 782 (GSB)= FIGUR
-> vgl. 1991

790
BÜSTE Frau mit Dutt (Haarschmuck)
ca. 1900
790 Firmenmarke 8a, Repr. Res., H. 29 cm; eBay # 6599185238

791 (GSB)
Croisy, Aristide
STATUETTE „Baigneuse"
ca. 1892/1900
791 H. 61 cm; GSB, Pl. 79 (m. Abb.)
Ausst.: Paris, Salon 1898
Lit.: Neuwirth (1974), S. 61

792 (GSB)
Steuer, Bernard Adrien
PLATEAU „Femme Moyen Age"
ca. 1892/1900
792 H. 37 cm, B. 28 cm; GSB, Pl. 69 (m. Abb.)

793 (GSB)
Huneker (d.i. Bracken, Clio Hinton)
„Femme au Lotus"
ca. 1892/1900
793 H. 14 cm, L. 22 cm; GSB, Pl. 100 (m. Abb.)

794
FIGUR „Mohrin"
ca. 1895
794 (Firmenmarke ohne Ang.), H. 94 cm; GSF, S. 4
Pendant -> vgl. 727
Lit.: GSF, S. 4, hier mit der Ang. des Originaltitels „Mohrin".

794 (GSB)
Huneker (d.i. Bracken, Clio Hinton)
„Cendrier Femme a la Rose sauvage"
ca. 1892/1900
794 H. 8 cm, L. 12 cm; GSB, Pl. 83 (m. Abb.)

795
Pecheur (d.i. vermutl. Friedrich Fischer)
VASE Zwei Frauen (Wassernixen?) (an der Vasenwandung)
ca. 1900
0795,46,35 Firmenmarke Sculpture 13, Repr. Res., H. 48,2 cm; eBay # 901267110
Bem.: Evtl. handelt es sich um das im ÖMKI 1900/01 ausgestellte Objekt VASE „Wassernixen".
Ausst.: ÖMKI 1900/01?
-> vgl. 724

795 (GSB)
Huneker (d.i. Bracken, Clio Hinton)
„Cendrier Femme couchée sur feuille"
ca. 1892/1900
795 H. 5 cm, L. 11 cm; GSB, Pl. 83 (m. Abb.)

796
Tell, E.
VASE
Form: gebaucht, hochgezogene Mündung
Dekor: an der Vasenwandung liegender Frauenakt (vollplastische Figur), Vasenkorpus mit Libellen(dekor)
ca. 1898
0796,14,6 Firmenmarke 9, Repr. Res., H. 32 cm; Sammlung Renée und Otto Schenk, Wien
Lit.: Neuwirth (1974), S. 34, 35 (Abb.), hier Ang. zu Modell-Nr. und H. falsch.

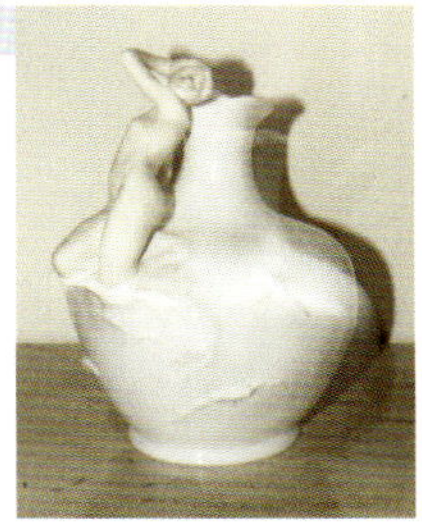

796 (GSB)
Huneker (d.i. Bracken, Clio Hinton)
BAS-RELIEF „Naiade au Dauphin"
ca. 1892/1900
796 H. 20 cm, L. 32 cm; GSB, Pl. 74 (m. Abb.)
Pendant -> vgl. 797 (GSB)

797
Cherc
VASE „Lustiger Tanz" (Tanzendes Paar)
ca. 1898/1900
0797,31,35 Firmenmarke Sculpture 13, H. 59,7 cm; eBay # 7769481557
0797,59,(1,8,5,8 vertikal) (Firmenmarke ohne Ang.), H. 60 cm, B. 48 cm; Henry's 21.4.1993; 27.4.1995
Ausst.: ÖMKI 1900/01, hier mit Ang. des Originaltitels „Lustiger Tanz".
Lit.: Neuwirth (1974), S. 61

797 (GSB)
Huneker (d.i. Bracken, Clio Hinton)
BAS-RELIEF „Naiade au Cheval Marin"
ca. 1892/1900
797 H. 20 cm, L. 32 cm; GSB, Pl. 74 (m. Abb.)
Pendant -> vgl. 796

802 (GSB)
Huneker (d.i. Bracken, Clio Hinton)
„Femme au Coquillage"
ca. 1892/1900
802 H. 16 cm; GSB, Pl. 83 (m. Abb.)

804
Carrièr-Belleuse, (Albert-Ernest)
Stehender Frauenakt mit Tuch und Katze (auf Rundsockel)
ca. 1898
804,4,1 Firmenmarke 7, Repr. Res., sign. Carrier-Belléuse, H. 78 cm; Wien Museum Inv. 219.141

806
Jacques, E.
LAMPE Liebespaar
ca. 1898
0806 (Firmenmarke ohne Ang.), H. 52 cm; Pinhas, S. 49 (m. Abb.)

811
Cherc
VASE Sterbender (toter?) Jüngling in der Meeresbrandung (von Frauengestalt mit Fackel und Möwe beobachtet)
ca. 1898/1900
0811,45,55 Firmenmarke Sculpture 13, Repr. Res., H. 86,5 cm; Quit. 14

812
Hermes (als stehende Figur, nach historischem Vorbild) (auf rundem Landschaftssockel)
ca. 1898
812 Firmenmarke 7, H. 59 cm; Sammlung mad
Lit.: Matschiner (2006), S. 6,7 (m. Abb.).

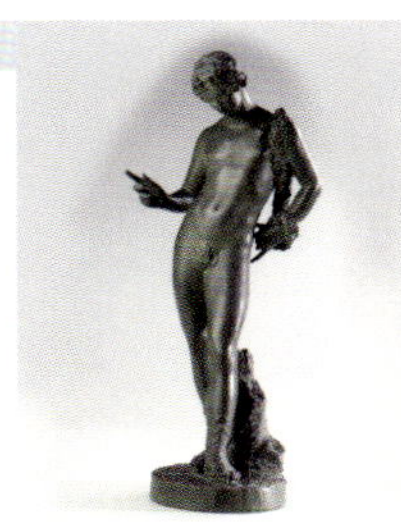

815 (GSB)
Lemaire, Hector
„Musique Profane"
ca. 1892/1900
815 H. 42 cm; GSB, Pl. 4 (m. Abb.)

816 (GSB)
Lemaire, Hector
„Musique Sacrée"
ca. 1892/1900
816 H. 46 cm; GSB, Pl. 4 (m. Abb.)

818 (GSB)
Carlier, Émile Joseph
GROUPE „Sur la Gréve"
ca. 1892/1900
818 H. 71 cm; GSB, Pl. 84 (m. Abb.)

819 (GSB)
Huneker (d.i. Bracken, Clio Hinton)
„Cendrier Femme a la Draperie"
ca. 1892/1900
819 H. 19 cm; GSB, Pl. 83 (m. Abb.)

820 (GSB)
Huneker (d.i. Bracken, Clio Hinton)
VIDE-POCHE „Femme a la Coquille"
ca. 1892/1900
820 H. 9 cm; GSB, Pl. 100 (m. Abb.)

821 (GSB)
Injalbert, Jean Antoine
VASE „Les Flots"
ca. 1892/1900
821 H. 29 cm; GSB, Pl. 95 (m. Abb.)

822 (GSB)
Injalbert, Jean Antoine
BUSTE „Hippomène"
ca. 1892/1900
822 H. 52 cm; GSB, Pl. 87 (m. Abb.)
Bem.: Ein Exemplar dieser Gruppe wurde seinerzeit vom Musée du Luxembourg angekauft.
Lit.: Catley (1987), S. 180 (m. Abb.).

823 (GSB)
Injalbert, Jean Antoine
BUSTE „Fierté"
ca. 1892/1900
823 H. 60 cm; GSB, Pl. 87 (m. Abb.)

824
Nordafrikanischer Straßenverkäufer mit Turban, einen Wasserkrug und Sirupkanne
ca. 1892
824 (Firmenmarke ohne Ang.), H. 40 cm; Pinhas, S. 23 (m. Abb.)

824 (GSB)
Injalbert, Jean Antoine
BUSTE „Le Dédain"
ca. 1892/1900
824 H. 37 cm; GSB, Pl. 87 (m. Abb.)
Lit.: Catley (1987), S. 180 (m. Abb.).

825 (GSB)
Lemaire, Hector
„La Vague"
ca. 1892/1900
825 H. 67 cm; GSB, Pl. 88 (m. Abb.)

826
FIGUR „Ruth" (Stehende junge Frau Haarband und langem Gewand, ein Bündel Ähren in ihrer Rechten haltend) (auf Landschaftssockel)
Inschrift: RUTH
ca. 1895/97
826,(9,0,9,7 vertikal) Firmenmarke 7, Repr. Res., H. 62 cm; KH
826 (Firmenmarke ohne Ang.), H. 64 cm; GSF, S. 4
Lit.: GSF, S. 4, hier mit der Ang. des Originaltitels „Ruth"; Pinhas, S. 64 (m. Abb.).

827
FIGUR „Egypter mit Stab/Gewehr"
Inschrift: arab./egypt. Schriftzeichen
ca. 1895/97
ohne Nr., (827) Firmenmarke 11, Gesamth. 73 cm; KH
827 (Firmenmarke ohne Ang.), H. 73 cm; GSF, S. 4
Bem.: Am Sockel arab., nicht näher zu verifizierende Inschrift.
Pendant –> vgl. 828
Lit.: GSF, S. 4, hier mit der Ang. des Originaltitels „Egypter mit Stab/Gewehr".

828
FIGUR „Egypterin" (Stehende Wasserträgerin)
Inschrift: arab. Schriftzeichen
ca. 1895/97
828,V ger., Firmenmarke 11, H. 75 cm; KH
828 (Firmenmarke ohne Ang.), H. 75 cm; GSF, S. 4
Bem.: Am Sockel arab., nicht näher zu verifizierende Inschrift.
Pendant –> vgl. 827
Lit.: GSF, S. 4, hier mit der Ang. des Originaltitels „Egypterin".

828 (GSB)
Ledru, Auguste
PLAT „Femme aux Chardons"
ca. 1892/1900
828 Ø 35 cm; GSB, Pl. 94 (m. Abb.)
Ausst.: Salon des Beaux-Arts (ohne Jahresangabe)
Pendant –> vgl. 929 (GSB)

829 (GSB)
Ledru, Auguste
VIDE-POCHE „Nymphe aux Coquillages"
ca. 1892/1900
829 H. 9 cm, L. 17 cm; GSB, Pl. 90 (m. Abb.)
Ausst.: Salon des Beaux-Arts (ohne Jahresangabe)

830
BÜSTE „Falstaff"
ca. 1895/97
830,V ger., Firmenmarke 3, Vervielf. vorbeh., H. 30,5 cm; eBay # 142822496
Bem.: Diese Büste, mehr noch ein Halbtorso, geht auf die Figur des Falstaff in William Shakespeares Bühnenstück „Heinrich IV" zurück.

830 (GSB)
Ledru, Auguste
VIDE-POCHE „Ondine"
ca. 1892/1900
830 H. 9 cm, L. 15 cm; GSB, Pl. 90 (m. Abb.)
Ausst.: Salon des Beaux-Arts (ohne Jahresangabe)

831 (GSB)
Ledru, Auguste
VIDE-POCHE „Grenouilles"
ca. 1892/1900
831 H. 9 cm, L. 18 cm; GSB, Pl. 90 (m. Abb.)
Ausst.: Salon des Beaux-Arts (ohne Jahresangabe)

832 (GSB)
Bruyneel, (Victor?)
STATUE „Le Piège"
ca. 1892/1900
832 H. 42 cm; GSB, Pl. 44 (m. Abb.)
–> vgl. 780 (GSB)

833 (GSB)
Huneker (d.i. Bracken, Clio Hinton)
ENCRIER „Repos"
ca. 1892/1900
833 H. 17 cm; GSB, Pl. 100 (m. Abb.)

834 (GSB)
Huneker (d.i. Bracken, Clio Hinton)
ENCRIER „Vigilante"
ca. 1892/1900
834 H. 17 cm; GSB, Pl. 100 (m. Abb.)

835 (GSB)
Levasseur, Henri-Louis
„Diane" No. 1; No. 2; No. 3
ca. 1892/1900
835 H. 88 cm; GSB, Pl. 86 (m. Abb.)
Bem.: Im Firmenkat. GSB werden 3 verschiedene Größen offeriert, wobei nur von einer Ausführung die Höhe vermerkt wird. Ob diese Figur, eine schreitende halbnackte Diana mit Bogen, wirklich in den weiteren Größen in Produktion ging, lässt sich nicht mit letzter Sicherheit sagen, da sich keine dieser Statuen im Handel oder in einer Sammlung nachweisen lässt.

836 (GSB)
Bouval, Maurice
LAMPE „Flambeau électrique" (Blütenstängel in der rechten Hand)
ca. 1892/1900
836 H. 44 cm; GSB, Pl. 101 (m. Abb.)
Pendant –> vgl. 853 (GSB)

837 (GSB)
Moreau-Vauthier, Paul
VASE „Curieuse"
ca. 1892/1900
837 H. 12 cm; GSB, Pl. 94 (m. Abb.)
Ausst.: Salon des Beaux-Arts (ohne Jahresangabe)

838 (GSB)
Moreau-Vauthier, Paul
BOUGEOIR „Sommeil"
ca. 1892/1900
838 H. 9 cm; GSB, Pl. 94 (m. Abb.)
Ausst.: Salon des Beaux-Arts (ohne Jahresangabe)

839 (GSB)
Moreau-Vauthier, Paul
VASE „Triangulaire"
ca. 1892/1900
839 H. 13 cm; GSB, Pl. 94 (m. Abb.)
Ausst.: Salon des Beaux-Arts (ohne Jahresangabe)

840 (GSB)
Levasseur, Henri-Louis
„Déesse de L' Électricité" No. 1
ca. 1892/1900
840 H. 100 cm; GSB, Pl. 96 (m. Abb.)
–> vgl. 921 (GSB)
–> vgl. 1943= ähnl. Mod. (Simon)

841
Tell, E.
VASE Stehende, nach vorne gebeugte junge Frau (deren Kleid sich zur Vase formt)
ca. 1895
0841,53,65 Firmenmarke 7, Repr. Res., gepr. „P"= Z3, H. 72,5 cm; GK, S. 11, 58 (m. Abb.); Wien Museum Inv. 182.060
Ausst.: HMW (jetzt Wien Museum) 1985/86.
Lit.: GSN, S. 10, 11 (m. Abb.); GK, S. 11, 58 (m. Abb.)
GS neu: Modell 1 „Melodie" (H. 72 cm), Aufl. 1000 Stück

841 (GSB)
Lemaire, Hector
BUSTE „Fantaisie"
ca. 1892/1900
841 H. 16 cm; GSB, Pl. 95 (m. Abb.)

842
Moulin, Eugène Emile
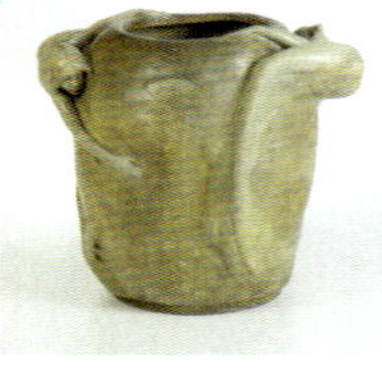
VASE „Sirenenspiel" (Zwei schwimmende, plastische Nixen, reliefierte Fische)
ca. 1898/1900
0842,50,35 ger. S, (ohne Firmenmarke), Repr. Res., H. 42 cm; GK, S. 11, 59 (m. Abb.); Wien Museum Inv. 223.898
Ausst.: HMW (jetzt Wien Museum) 1985/86.
Lit.: Pudor (1902), S. 70 (Abb.); VKMH, 16.1901/02, S. 592 (m. Abb.), hier mit Ang. des Originaltitels "Sirenenspiel"; GK, S. 11, 59 (m. Abb.)

842 (GSB)
Paris, Auguste
BUSTE „Éternel Moqueur" No. 2
ca. 1892/1900
842 H. 22 cm; GSB, Pl. 95 (m. Abb.)
Ausst.: Paris, Salon 1898
-> vgl. 783 (GSB)
-> vgl. 782 (GSB)= FIGUR

843 (GSB)
Moreau-Vauthier, Paul
STATUE „Peinture"
ca. 1892/1900
843 H. 33 cm; GSB, Pl. 94 (m. Abb.)

844
FIGUR „Mohr" (Stehender Afrikaner mit rechteckigem Korb)
ca. 1895
844,16 ger., Firmenmarke 3, Vervielf. vorbeh., H. 53,3 cm; KH
ohne Ang., H. 55 cm; Richemond (1999), S. 81
844 (Firmenmarke ohne Ang.), H. 61 cm; GSF, S. 4
Pendant -> vgl. 845
Lit.: GSF, S. 4, hier mit der Ang. des Originaltitels „Mohr"; Richemond (1999), S. 81 (m. Abb.).

844 (GSB)
Moreau-Vauthier, Paul
CENDRIER „Perle"
BRONZE
ca. 1892/1900
ohne Ang., L. 15 cm; Soth. L. 18.2.1983
844 L. 15 cm; GSB, Pl. 94 (m. Abb.)
Ausst.: Salon des Beaux-Arts (ohne Jahresangabe)

845
FIGUR „Mohrin" (Stehende Afrikanerin mit entblößtem Oberkörper und rechteckigem Korb)
ca. 1895
ohne Ang., H. 55 cm; Richemond (1999), S. 81 (m. Abb.)
845 (Firmenmarke ohne Ang.), H. 61 cm; GSF, S. 4
Pendant -> vgl. 844
Lit.: GSF, S. 4, hier mit der Ang. des Originaltitels „Mohrin"; Richemond (1999), S. 81 (m. Abb.).

845 (GSB)
Potet, Louis (Loys)
„Au Champs"
ca. 1892/1900
845 H. 29 cm; GSB, Pl. 100 (m. Abb.)

848 -> ÜBERSCHNEIDUNG
Richer, Paul Marie Louis Pierre
Schnitter (Le Faucheur) (Mähender Bauer mit Hut auf Landschaftssockel)
BRONZE
ca. 1895/97
G5,848 (Firmenmarke o. Ang.), H. 57 cm; KH
Lit.: Berman, Vol. 4, Nr. 4500

848 -> ÜBERSCHNEIDUNG
FIGUR „Falstaff" (auf mauerartigem Sockel)
ca. 1895/97
848 (Firmenmarke ohne Ang.), H. fehlt; Pinhas, S. 65 (m. Abb.)
848 (Firmenmarke ohne Ang.), H. 36 cm; GSF, S. 4
Lit.: GSF, S. 4, hier mit der Ang. des Originaltitels „Falstaff".

848 (GSB)
Ledru, Auguste
PLAT „Naiades"
ca. 1892/1900
848 H. 29 cm, L. 49 cm; GSB, Pl. 90 (m. Abb.)
Ausst.: Salon des Beaux-Arts (ohne Jahresangabe)

849
FIGUR „Kellermeister"
ca. 1895/97
849 (Firmenmarke ohne Ang.), H. fehlt; Pinhas, S. 65 (m. Abb.)
849 (Firmenmarke ohne Ang.), H. 37 cm; GSF, S. 4
Lit.: GSF, S. 4, hier mit der Ang. des Originaltitels „Kellermeister".

849 (GSB)
Noël, Tony (Edme Antony Paul)
„Querelle D'Amour" No. 2
ca. 1892/1900
849 H. 43 cm; GSB, Pl. 93 (m. Abb.)
-> vgl. 880 (GSB)

850 -> ÜBERSCHNEIDUNG
Griser

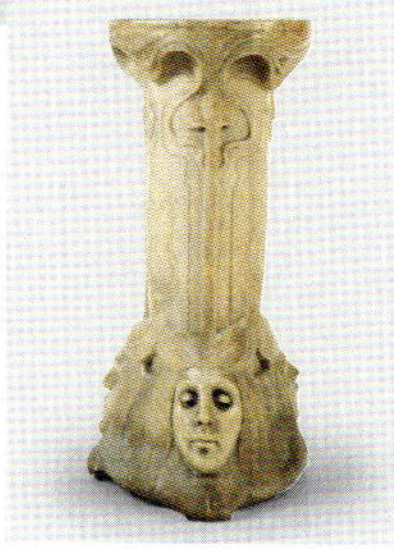

SÄULE / POSTAMENT Drei Frauengesichter mit verflochtenen Haaren
ca. 1895/97
0850,8,3 Firmenmarke 4, Ges. gesch., sign. Grißer, H. 93 cm; GK, S. 11 (hier Ang. „Srisser?"), 60 (m. Abb.); Wien Museum Inv. 223.899
Bem.: Sicher handelt es sich um den Künstler „Griser", der im ÖMKI (1902/03) unter Nr. 346 das Objekt „Perle" ausgestellt hatte; Pudor (1902) erwähnt ebenfalls einen Künstler namens „Griser".
Ausst.: HMW (jetzt Wien Museum) 1985/86; SSH 1992.
Lit.: GK, S. 11, 60 (m. Abb.)

850 -> ÜBERSCHNEIDUNG
BÜSTE Araber (nach links blickend)
ca. 1895/97
850,XII ger., Firmenmarke 3, Vervielf. vorbeh., H. 23 cm; Wien Museum Inv. 186.633; Bem.: Holzsockel sekundär.
Ausst.: SSH 1992
Pendant -> vgl. 851

850 (GSB)
Huneker (d.i. Bracken, Clio Hinton)
JARDINIÈRE
ca. 1892/1900
850 H. 20 cm, B. 15 cm; GSB, Pl. 100 (m. Abb.)

851
BÜSTE Araberin (nach rechts blickend)
ca. 1895/97
851,XII ger., Firmenmarke 3, Vervielf. vorbeh., H. 22,2 cm; Wien Museum Inv. 186.634; Bem.: Holzsockel sekundär.
Ausst.: SSH 1992
Pendant -> vgl. 850

851 (GSB)
Marquet, Alix
BOUGEOIR „Femme sur Feuille"
ca. 1892/1900
851 H. 17 cm; GSB, Pl. 91 (m. Abb.)
Ausst.: Salon des Beaux-Arts (ohne Jahresangabe)

852
BÜSTE Araberin (nach links blickend)
ca. 1895/97
852,19 Firmenmarke 8a, Repr. Res., H. 27 cm; eBay # 2551373533

852 (GSB)
Noël, Tony (Edme Antony Paul)
„Rétaire" No. 1
ca. 1892/1900
852 H. 56 cm, L. 49 cm; GSB, Pl. 92 (m. Abb.)
-> vgl. 920 (GSB)

853
Tell, E.
VASE Stehender Mädchenakt (am Vasenkorpus)
ca. 1898/1900
0853 (Firmenmarke ohne Ang.), H. 44,5 cm; Pinhas, S. 51 (m. Abb.)
-> vgl. 659
-> vgl. 675= ähnl. Mod.

853 (GSB)
Bouval, Maurice
LAMPE „Flambeau électrique" (Blütenstängel in der linken Hand)
ca. 1892/1900
853 H. 44 cm; GSB, Pl. 101 (m. Abb.)
Pendant -> vgl. 836 (GSB)

855 (GSB)
Nelson, (Alphonse-Henri?)
BUSTE „Ondine" No. 1
ca. 1892/1900
855 H. 53 cm; GSB, Pl. 99 (m. Abb.)
-> vgl. 918 (GSB); 919(GSB)
-> vgl. 2091; 2212

856a
FIGUR „Vasenträgerin (nackt)" (Stehendes nacktes junges Mädchen mit Krug und Handtuch) (auf mit Löwenkopfmaske und Muschel verziertem Sockel)
ca. 1895/97
856 ger., GS (Firmenmarke ohne Ang.), H. 74,9 cm; Christie's SK 11.7.2001
856a (Firmenmarke ohne Ang.), H. 74 cm; GSF, S. 4
-> vgl. 856b; 1215
Lit.: GSF, S. 4, hier mit der Ang. des Originaltitels „Vasenträgerin (nackt)".

856b
FIGUR „Vasenträgerin (bekleidet)"
ca. 1895/97
856b (Firmenmarke ohne Ang.), H. 74 cm; GSF, S. 4
-> vgl. 856a; 1215
Lit.: GSF, S. 4, hier mit der Ang. des Originaltitels „Vasenträgerin (bekleidet)".

857
Strasser, Arthur (zugeschr.)
Wasserverkäufer (Stehender Araber mit Turban, Wasserkrug, Gefäß und kleiner Schale)
ca. 1897/98
857 Firmenmarke 8, Repr. Res., unsign., H. 66 cm; PB Wien 19
857 FG (Firmenmarke?), Repr. Res., unsign., H. 62 cm; Tajan 13.3.1995
ohne Ang., unsign. H. 62 cm; Zeller 33
Bem.: An der Plinthe betitelt mit arab. Schriftzeichen (Metallschild); evtl. handelt es sich um das im ÖMKI 1898/99 ausgestellte Objekt „Wasserverkäufer".
Ausst.: ÖMKI 1898/99?
-> vgl. 374 (Gracien); 1175; 1601

857 (GSB)
Braecke, Pieter Jan (Pierre)
„Désespoir"
ca. 1892/1900
857 H. 89 cm; GSB, Pl. 97 (m. Abb.)

859
Meißner, (Max?)
RELIEF Mädchen zwischen zwei Bäumen, an einer Blüte riechend
ca. 1896/97
0859,23,3 gem. weiß III, Firmenmarke 4, Ges. gesch., sign. Meissner, H. 20 cm, L. 38 cm; eBay # 74161522253

860
FIGUR „Araberin mit Krug"
ca. 1896/97
860 (Firmenmarke ohne Ang.), H. 114 cm; GSF, S. 4
-> vgl. 1138
Lit.: GSF, S. 4, hier mit der Ang. des (von der Mod.-Nr. 1138 abweichenden) Originaltitels „Araberin mit Krug".

860 (GSB)
Bouval, Maurice
LAMPE „Femme au Lys, électricité" (Blütenstängel in der rechten Hand)
ca. 1892/1900
860 H. 65 cm; GSB, Pl. 99 (m. Abb.)
Pendant -> vgl. 870 (GSB)
Lit.: Magyar Iparmüvészet, 3.1900, S. 293 (m. Abb.)

861 (GSB)
Loiseau-Rousseau, Paul Louis Émile
VIDE-POCHE „Repos du Coquillage“
ca. 1892/1900
861 H. 19 cm; GSB, Pl. 95 (m. Abb.)

862 (GSB)
Ledru, Auguste
Surtout „Femme au Poisson“
ca. 1892/1900
862 H. 19 cm, L. 53 cm; GSB, Pl. 101 (m. Abb.)
Ausst.: Salon des Beaux-Arts (ohne Jahresangabe)

863 (GSB)
Levasseur, Henri-Louis
„Le Rameaux“ No. 1; No. 2; No. 3
ca. 1892/1900
863 H. 82 cm; GSB, Pl. 89 (m. Abb.)
Bem.: Im Firmenkat. GSB werden 3 verschiedene Größen offeriert, wobei nur von einer Ausführung die Höhe vermerkt wird. Ob diese Figur, eine schreitende halbnackte Diana mit Bogen, wirklich in den weiteren Größen in Produktion ging, lässt sich nicht mit letzter Sicherheit sagen, da sich keine dieser Statuen im Handel oder in einer Sammlung nachweisen lässt.

868
Delagrange, Léon Noel
MASKE/JARDINIERE Frauengesicht mit wallendem Haar (Gefäß für Blumen)
ca. 1900
0868 (Firmenmarke ohne Ang.), H. fehlt, B. 32 cm; Pinhas, S. 49 (m. Abb.)
Bem.: Evtl. handelt es sich um das im ÖMKI 1900/01 ausgestellte Objekt „Jardiniere“.
Ausst.: ÖMKI 1900/01? „Jardiniere“?
–> vgl. 745

868 (GSB)
Mérite, Édouard Paul
„Martin Pécheur“
ca. 1892/1900
868 H. 11 cm; GSB, Pl. 90 (m. Abb.)

868b
FIGUR „Judith pleinair chromé“
ca. 1896/97
868b ohne Ang.; KeramRS, 5.1897, S. 347
Bem.: s. Modell-Nr. 6

869 (GSB)
Mérite, Édouard Paul
„Souris“
ca. 1892/1900
869 H. 6 cm; GSB, Pl. 90 (m. Abb.)

870 (GSB)
Bouval, Maurice
LAMPE „Femme au Pavot, électricité“ (Blütenstängel in der linken Hand)
ca. 1892/1900
870 H. 65 cm; GSB, Pl. 99 (m. Abb.)
Pendant –> vgl. 860 (GSB)
Lit.: Magyar Iparmüvészet, 3.1900, S. 293 (m. Abb.)

871
FIGUR „Hirtenknabe“
ca. 1896/97
871 (Firmenmarke ohne Ang.), H. 75 cm; GSF, S. 4
–> vgl. 1214
Lit.: GSF, S. 4, hier mit der Ang. des Originaltitels "Hirtenknabe".

871 (GSB)
Levasseur, Henri-Louis
„Légende des Ruines“
ca. 1892/1900
871 H. 40 cm; GSB, Pl. 95 (m. Abb.)
Ausst.: Salon des Beaux-Arts (ohne Jahresangabe)

872
Strasser, Arthur (zugeschr.)
FIGUR „Egyptischer Krieger“ (auf passigem Sockel mit arab. Schriftzeichen)
ca. 1895/96
872,19 Firmenmarke 3, Vervielf. vorbeh., unsign., H. 41,9 cm; eBay # 3704204714
872 ger., FG (Firmenmarke?), Vervielf. vorbeh.= Z10, unsign., H. 41 cm; Nagel 378
872 ger.?, FG (Firmenmarke?), unsign., H. 39,5 cm; Ahlden 115
ohne Modell-Nr., FG (Firmenmarke?), Vervielf. vorbeh., gepr. „P“= Z3, unsign., H. ca. 35 cm; KH
872 (Firmenmarke ohne Ang.), H. 35 cm; GSF, S. 4
–> vgl. 884 BÜSTE
Pendant –> vgl. 873
Lit.: GSF, S. 4, hier mit der Ang. des Originaltitels „Egyptischer Krieger“.

873
Strasser, Arthur (zugeschr.)
FIGUR „Vasenträgerin“ (auf passigem Sockel mit arab. Schriftzeichen)
ca. 1897
873,0 ger., FG (Firmenmarke?), Vervielf. vorbeh., gepr. „P“= Z3, unsign., H. ca. 43 cm; KH
873 ger., FG (Firmenmarke?), Vervielf. vorbeh., unsign., H. 43 cm; Nagel 378
873 ger.?, FG Wien (Firmenmarke?), unsign., H. 42 cm; Ahlden 115
873 (Firmenmarke ohne Ang.), H. 35 cm; GSF, S. 4
–> vgl. 885 BÜSTE
Pendant –> vgl. 872
Lit.: GSF, S. 4, hier mit der Ang. des Originaltitels „Vasenträgerin“.

874
FIGUR „Sorbetverkäufer“ (mit Turban, einen Krug tragend)
ca. 1897
874 (Firmenmarke ohne Ang.), H. 38 cm; Pinhas, S. 23 (m. Abb.)
874 (Firmenmarke ohne Ang.), H. 38 cm; GSF, S. 4
Lit.: GSF, S. 4, hier mit der Ang. des Originaltitels „Sorbetverkäufer“.

874 (GSB)
Loiseau-Rousseau, Paul Louis Émile
VASE „Aquatique“
ca. 1892/1900
874 H. 38 cm; GSB, Pl. 95 (m. Abb.)

875
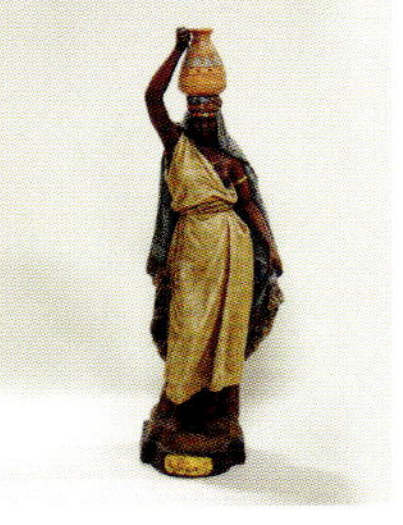
FIGUR „Krugträgerin“ (einen länglichen Krug auf dem Kopf tragend) (auf Sockel mit arab. Schriftzeichen)
ca. 1897
875,(4,2,9,3 vertikal),7 ohne Firmenmarke, Repr. Res., H. 43 cm; MAK Wien, Inv. Ke 10602-90
875 (Firmenmarke ohne Ang.), H. 43 cm; GSF, S. 4
–> vgl. 1183?; .... (H. 108 cm); Richemond (1999), S. 75 (m. Abb.)
Lit.: GSF, S. 4, hier mit der Ang. des Originaltitels „Krugträgerin“; Richemond (1999), S. 75 (m. Abb.) Krugvariante und für die H. 108 cm (bezeichnet in arab. Schriftzeichen, „Sakka-ha“, Metallschild).

875 (GSB)
Loiseau-Rousseau, Paul Louis Émile
BUSTE „Libellule“
ca. 1892/1900
875 H. 13 cm; GSB, Pl. 95 (m. Abb.)

878
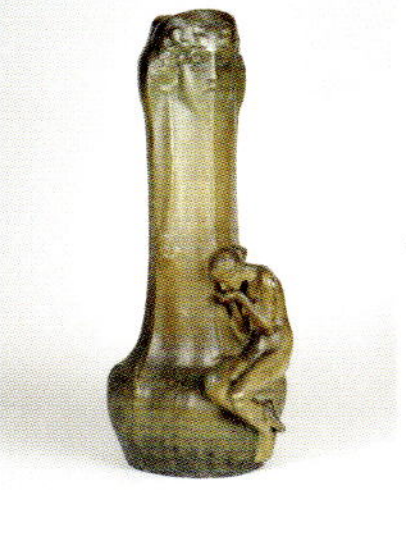
Simon
VASE „Am Brunnen“ (Kniende, trinkende Frauengestalt, Vasenmündung als Wasserspeier)
ca. 1898
0878,263,35 ohne Firmenmarke, H. 86 cm; PB Wien 19
0878,281,35 Firmenmarke 7, Repr. Res., gepr. „C“= Z4, H. 83,8 cm; eBay # 2665696164
Ausst.: Turin 1902
Lit.: Minkus, F.: Die Erste Internationale Ausstellung für moderne dekorative Kunst in Turin (1902), in: KuKHW, 5.1902, S. 407 (Abb.); R., F.: Keramische Wanderungen durch den Kunstsalon von Keller & Reiner in Berlin, in: KM, 3.1903, S. 127 (m. Abb.), hier mit Ang. des Originaltitels „Am Brunnen“, „Simons' pikantem Meißel verdanken die Uhr „Frühling“ und die „Am Brunnen“ betitelte Vase ihr Entstehen. Dieser Künstler versteht es gut zu stilisieren und seine jedesmaligen Motive geschmackvoll dem gewählten Sujet anzupassen“.

879 (GSB)
Bouval, Maurice
„Le Secret“
BRONZE
ca. 1892/1900
879 H. 38 cm; GSB, Pl. 91 (m. Abb.)
Lit.: Thomas, A.: Petit Bronzes d'Art, in: L'Art Dec., 3.1900/01, S. 187.

880
FIGUR „Schlangenbändiger“
ca. 1897
880 (Firmenmarke ohne Ang.), H. 41 cm; GSF, S. 4
Lit.: GSF, S. 4, hier mit der Ang. des Originaltitels „Schlangenbändiger“.

880 (GSB)
Noël, Tony (Edme Antony Paul)
„Querelle D'Amour“ No. 1
ca. 1892/1900
880 H. 74 cm; GSB, Pl. 93 (m. Abb.)
–> vgl. 849 (GSB)

881
FIGUR „Wasserträger“ (mit Fez, einen großen Wasserbehälter tragend)
ca. 1897
881 (Firmenmarke ohne Ang.), H. 40 cm; Pinhas, S. 23 (m. Abb.)
881 (Firmenmarke ohne Ang.), H. 41 cm; GSF, S. 4
Lit.: GSF, S. 4, hier mit der Ang. des Originaltitels „Wasserträger“.

881 (GSB)
Arnoux
„Éléphant“
ca. 1892/1900
881 H. 18 cm, L. 27 cm; GSB, Pl. 100 (m. Abb.)

882 (GSB)
Croisy, Aristide
GROUPE „Le Nid“
ca. 1892/1900
882 H. 46 cm, B. 43 cm; GSB, Pl. 98 (m. Abb.)
Bem.: Ein Exemplar dieser Gruppe wurde seinerzeit vom Musée du Luxembourg angekauft.
–> vgl. 883 (GSB)

883 (GSB)
Croisy, Aristide
GROUPE „Le Nid“
ca. 1892/1900
883 H. 31 cm, B. 25 cm; GSB, Pl. 98 (m. Abb.)
Bem.: Ein Exemplar dieser Gruppe wurde seinerzeit vom Musée du Luxembourg angekauft.
–> vgl. 882 (GSB)

884
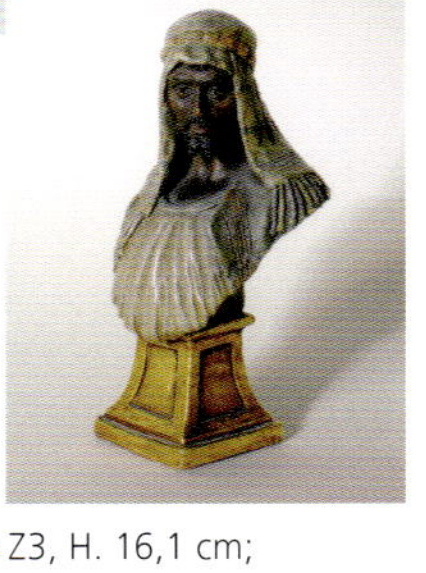
Strasser, Arthur (zugeschr.)
BÜSTE Ägypter mit Bart (nach links blickend)
ca. 1895/96
884,19 ger., Firmenmarke 8a, Repr. Res., gepr. „P“= Z3, unsign., H. 17,9 cm; Schopmann 17.12.1993; Stahl 6/2003
0884,23 Firmenmarke 8, Repr. Res., unsign., gepr. „P“= Z3, H. 16,1 cm; PB Deutschland
Bem.: s. Modell-Nr. 6

-> vgl. .... (H. 25 cm); Richemond (1999), S. 76 (m. Abb.) (Wien Museum?)
-> vgl. 872 FIGUR
Pendant -> vgl. 885
Lit.: KeramRS, 4.1896, S. 863, hier Einreichung zum „Musterschutz"; Richemond (1999), S. 76 (m. Abb.), (Wien Museum?).

884 (GSB)
Bouval, Maurice
ENCRIER „Pivoine"
ca. 1892/1900
884 H. 11 cm; GSB, Pl. 91 (m. Abb.)
Ausst.: Salon des Beaux-Arts, Paris 1900.

885
Strasser, Arthur (zugeschr.)
BÜSTE Ägypterin (nach rechts blickend)
ca. 1895/96
885,19 ger., Firmenmarke 8a, Repr. Res., unsign., H. 17,9 cm; Schopmann 17.12.1993; Stahl 6/2003
-> vgl. .... (H. 25 cm); Richemond (1999), S. 76 (m. Abb.) (Wien Museum?)
-> vgl. 873 FIGUR
Pendant -> vgl. 884
Lit.: Richemond (1999), S. 76 (m. Abb.), (Wien Museum?)

885 (GSB)
Charpentier, Alexandre-Louis-Marie
„Christ au Linceul"
ca. 1892/1900
885 H. 5 cm, L. 18 cm; GSB, Pl. 91 (m. Abb.)

886 (GSB)
Charpentier, Alexandre-Louis-Marie
„Femme au Bain"
ca. 1892/1900
886 H. 17 cm, L. 13 cm; GSB, Pl. 91 (m. Abb.)

887 (GSB)
Charpentier, Alexandre-Louis-Marie
„La Lutte"
ca. 1892/1900
887 H. 17 cm, L. 12 cm; GSB, Pl. 91 (m. Abb.)

888 (GSB)
Charpentier, Alexandre-Louis-Marie
„Enfant Musicien"
ca. 1892/1900
888 H. 12 cm; GSB, Pl. 91 (m. Abb.)
Pendant -> vgl. 889 (GSB)

889 (GSB)
Charpentier, Alexandre-Louis-Marie
„Enfant Chantant"
ca. 1892/1900
889 H. 12 cm; GSB, Pl. 91 (m. Abb.)
Pendant -> vgl. 888 (GSB)

890
BÜSTE „Hirtenknabe"
ca. 1895/96
890,A Firmenmarke 8a, Repr. Res., H. 39 cm; Wien Museum Inv. 186.631
890 (Firmenmarke ohne Ang.), H. 39 cm; GSF, S. 4
Bem.: Keramik bronzefarben staffiert.
Pendant -> vgl. 891
Lit.: GSF, S. 4, hier mit der Ang. des Originaltitels "Hirtenknabe".

891
BÜSTE „Vasenträgerin" (Halbakt)
ca. 1895/96
891,10,A Firmenmarke 8a, Repr. Res., H. 37,5 cm; Wien Museum Inv. 186.632
891 (Firmenmarke ohne Ang.), H. 36 cm; GSF, S. 4
Bem.: Keramik bronzefarben staffiert.
Pendant -> vgl. 890
Lit.: GSF, S. 4, hier mit der Ang. des Originaltitels „Vasenträgerin".

891 (GSB)
Ledru, Auguste
„Coquille" No. 1
ca. 1892/1900
891 H. 18 cm, L. 19 cm; GSB, Pl. 90 (m. Abb.)

892
BÜSTE „Todtenkopf (mittel)"
ca. 1895/96
892 (Firmenmarke ohne Ang.), H. 8 cm; GSF, S. 4
Lit.: GSF, S. 4, hier mit der Ang. des Originaltitels „Todtenkopf (mittel)".

892 (GSB)
Ledru, Auguste
„Coquille" No. 2
ca. 1892/1900
892 H. 15 cm, L. 13 cm; GSB, Pl. 100 (m. Abb.)

895
BÜSTE „Mohr mit Muschel"
ca. 1897
895,37 ger., Firmenmarke 8b, Repr. Res., gepr. „P"= Z3, H. 56,8 cm, B. 37,5 cm; Wien Museum Inv. 219. 114
895 (Firmenmarke ohne Ang.), H. 55 cm; GSF, S. 4
Pendant -> vgl. 896
Lit.: GSF, S. 4, hier mit der Ang. des Originaltitels „Mohr mit Muschel"; Pinhas, S. 66 (m. Abb.).

896 -> ÜBERSCHNEIDUNG
BÜSTE „Mohrin mit Muschel"
ca. 1897
Pendant -> vgl. 895
896 (Firmenmarke ohne Ang.), H. 55 cm; GSF, S. 4
Lit.: GSF, S. 4, hier mit der Ang. des Originaltitels „Mohrin mit Muschel"; Pinhas, S. 66 (m. Abb.).

896 -> ÜBERSCHNEIDUNG
Louger, J.
VASE Kind und Frosch im Schilf mit Libelle
ca. 1897
0896,8,35 gem. weiß XXI, Firmenmarke 4a, Repr. Res., H. 69,7 cm; PB Wien 19
Pendant -> vgl. 897

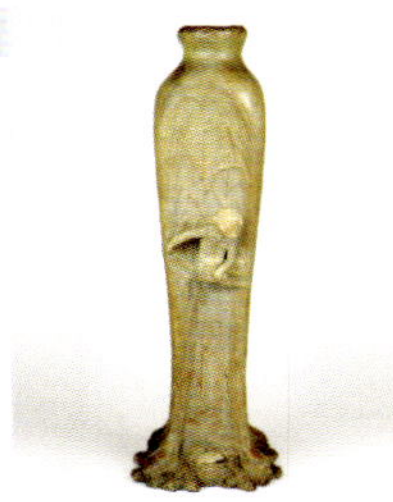

897
Louger, J.
VASE Kind und Frosch im Schilf mit Libelle
ca. 1897
0897 (Firmenmarke ohne Ang.), H. 70 cm; Pinhas, S. 49 (m. Abb.)
Pendant -> vgl. 896

898 (GSB)
Bruyneel, (Victor?)
BUSTE „Le Piège"
ca. 1892/1900
898 H. 26 cm; GSB, Pl. 100 (m. Abb.)

899
FIGUR „Flora"
ca. 1897
899 (Firmenmarke ohne Ang.), H. 76 cm; GSF, S. 4
-> vgl. 1823
Lit.: GSF, S. 4, hier mit der Ang. des Originaltitels „Flora".

901
BÜSTE „Neger mit Korb"
ca. 1897
901 (Firmenmarke ohne Ang.), H. 67 cm; GSF, S. 4
-> vgl. 1677
Pendant -> vgl. 902
Lit.: GSF, S. 4, hier mit der Ang. des Originaltitels „Neger mit Korb".

902
BÜSTE „Mohrin mit Korb" (mit Tuch und Halsschmuck)
ca. 1897
902,5 Firmenmarke 8a, gepr. „P"= Z3, H. 75 cm; Wien Museum Inv. 219.119
902 (Firmenmarke ohne Ang.), H. 73 cm; GSF, S. 4
-> vgl. 1678
Pendant -> vgl. 901
Lit.: GSF, S. 4, hier mit der Ang. des Originaltitels „Mohrin mit Korb".

903
Strasser, Arthur
FIGUR „Kabyle" („Antar")
ca. 1895
903 (Firmenmarke ohne Ang.), H. 100 cm; GSF, S. 4
-> vgl. 961; 1055
Pendant -> vgl. 904; 962; 1056
Lit.: GSF, S. 4, hier mit der Ang. des Originaltitels „Kabyle"; Richemond (1995), S. 57, hier ohne Ang. der Modell-Nr., jedoch m. Ang. der Sign.: A. Strasser.

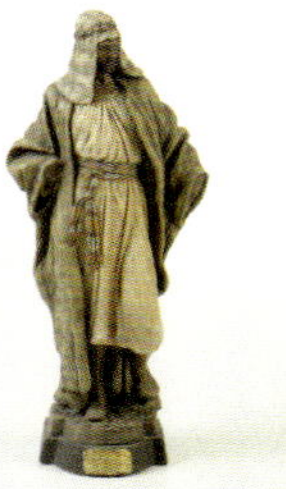

904
Désurmont, (Gil-D.?)
FIGUR „Kabylin" („Abla")
Inschrift: ABLA (in arab. Schriftzeichen)
ca. 1895
904,5 Firmenmarke 8a, Repr. Res., unsign., H. 97 cm; PB Wien 19
ohne Ang., FG (Firmenmarke?), H. 98 cm; Richemond (1995), S. 56 (m. Abb.)
ohne Ang., H. 100 cm; Richemond (1999), S. 85 (m. Abb.)
904 (Firmenmarke ohne Ang.), H. 100 cm; GSF, S. 4
-> vgl. 962; 1056
-> vgl. 927 BÜSTE
Pendant -> vgl. 903; 961; 1055
Lit.: GSF, S. 4, hier mit der Ang. des Originaltitels „Kabylin".

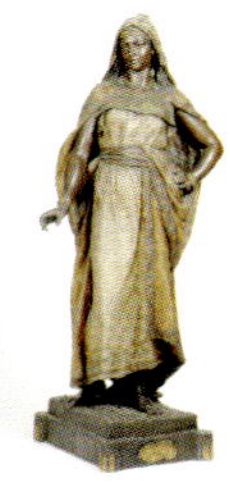

906
FIGUR / LAMPE „Nachtwächter"
ca. 1897
906 (Firmenmarke ohne Ang.), H. 56 cm; GSF, S. 4
Bem.: Diese Figur wurde lt. Katalogbeschreibung auch als Lampe geliefert.
-> vgl. 908
Lit.: GSF, S. 4, hier mit der Ang. des Originaltitels „Nachtwächter" und der Bem. „detto elektr. montirt".

907
FIGUR „Mignon"
ca. 1897
907 (Firmenmarke ohne Ang.), H. 78 cm; GSF, S. 4
-> vgl. 1822
Lit.: GSF, S. 4, hier mit der Ang. des Originaltitels „Mignon".

908
FIGUR / LAMPE „Nachtwächter"
ca. 1897
908 (Firmenmarke ohne Ang.), H. 38 cm; GSF, S. 5
Bem.: Diese Figur wurde lt. Katalogbeschreibung auch als Lampe geliefert.
-> vgl. 906
Lit.: GSF, S. 5, hier mit der Ang. des Originaltitels „Nachtwächter" und der Bem. „detto elektr. montirt".

908 (GSB)
Larroux, Antonin
BUSTE „Méditation"
ca. 1892/1900
908 H. 44 cm; GSB, Pl. 101 (m. Abb.)

909
Renard, (Léopold)
VASE Drei stehende Frauenfiguren
ca. 1897
0909,166 Firmenmarke 4, Repr. Res., H. 68 cm; Sammlung Gruber, S. 104; Peretz 62
Bem.: Diese Vase wurde wohl von Léopold Renard (geb. 3.3.1868) und nicht von dessen Sohn Marcel (geb. 1893) entworfen, wie bei Neuwirth (1974), S. 72 angegeben.
Lit.: Artis, 1977, Nr. 7, S. 43 (m. Abb.)

917
FIGUR „Mönch (trinkend, groß)" (Stehender, aus Kugelflasche trinkender Mönch mit Schürze)
ca. 1895/97
917,89,(3,1,8,1 vertikal), Firmenmarke 11, H. 33 cm; eBay # 2576333347
917 (Firmenmarke ohne Ang.), H. 34 cm; GSF, S. 5
Lit.: GSF, S. 5, hier mit der Ang. des Originaltitels „Mönch (trinkend, groß)".

918
FIGUR „Mönch (schnupfend, groß)" (Stehender Mönch mit Pfeife und Tabaksdose)
ca. 1895/97
918 (Firmenmarke ohne Ang.), H. 32,8 cm; Fischer 151
918 (Firmenmarke ohne Ang.), H. 34 cm; GSF, S. 5
-> vgl. 948a
Lit.: GSF, S. 5, hier mit der Ang. des Originaltitels „Mönch (schnupfend, groß)".

918 (GSB)
Nelson, (Alphonse-Henri?)
BUSTE „Ondine" No. 2
ca. 1892/1900
918 H. 40 cm; GSB, Pl. 99 (m. Abb.)
-> vgl. 855 (GSB); 919(GSB)
-> vgl. 2091; 2212

919
FIGUR „Mönch (Glas haltend, groß)"
ca. 1895/97
919 (Firmenmarke ohne Ang.), H. 34 cm; GSF, S. 5
-> vgl. 921
Lit.: GSF, S. 5, hier mit der Ang. des Originaltitels „Mönch (Glas haltend, groß)".

919 (GSB)
Nelson, (Alphonse-Henri?)
BUSTE „Ondine" No. 3
ca. 1892/1900
919 H. 12 cm; GSB, Pl. 99 (m. Abb.)
-> vgl. 855 (GSB); 918(GSB)
-> vgl. 2091; 2212

920
FIGUR „Mönch (einschenkend, groß)"
ca. 1895/97
920 (Firmenmarke ohne Ang.), H. 34 cm; GSF, S. 5
-> vgl. 949
Lit.: GSF, S. 5, hier mit der Ang. des Originaltitels „Mönch (einschenkend, groß)".

920 (GSB)
Noël, Tony (Edme Antony Paul)
„Rétaire" No. 2
ca. 1892/1900
920 H. 27 cm, L. 24 cm; GSB, Pl. 92 (m. Abb.)
-> vgl. 852 (GSB)

921
FIGUR „Mönch (Glas haltend, klein)"
ca. 1895/97
921 (Firmenmarke ohne Ang.), H. 21 cm; GSF, S. 5
-> vgl. 919
Lit.: GSF, S. 5, hier mit der Ang. des Originaltitels „Mönch (Glas haltend, klein)".

921 (GSB)
Levasseur, Henri-Louis
„Déesse de L' Électricité" No. 2
BRONZE (auf Marmorsockel)/BRONZE/ELFENBEIN
ca. 1892/1900
BRONZE (auf Marmorsockel)
921,1540 (Firmenmarke ohne Ang.), H. 59 cm; Soth. L. 29.11.1991
BRONZE/ELFENBEIN
921,134,B (Firmenmarke ohne Ang.), H. 60,2 cm; Christie's SK 11.11.2004
921 H. fehlt; GSB, Pl. 96 (m. Abb.)
Bem.: Der Firmenkat. GSB gibt keinen Aufschluss über das Material der Figur. Jedoch zeigt das bei Christie's SK angebotene Exemplar, dass neben der Ausführung in Bronze (und Zinkguss?), auch eine Ausführung mit Hautpartien aus Elfenbein erzeugt wurde.
-> vgl. 840 (GSB)
-> vgl. 1943= ähnl. Mod. (Simon)

922
FIGUR „Mönch (trinkend)" (Stehender Mönch mit Bierhumpen) (auf getrepptem Sockel)
ca. 1895/97
922,31 ger., GS Wien (Firmenmarke ohne Ang.), H. 19,8 cm; Doro 10.3.2003
922 (Firmenmarke ohne Ang.), H. 21 cm; GSF, S. 5
-> vgl. 963?
Lit.: GSF, S. 5, hier mit der Ang. des Originaltitels „Mönch (trinkend)".

923 -> ÜBERSCHNEIDUNG
Langer, (Leopold?)
VASE Frauenakt mit Kind und Wassergefäß auf Ast- und Blattwerk
ca. 1895/97
0923,72,35 gem. XIX, Firmenmarke Sculpture 13, Repr. Res., Ausformung nach 1900, H. 86,4 cm; eBay # 483641590

923 -> ÜBERSCHNEIDUNG
BÜSTE „Münchner Kind'l"
ca. 1895/97
923 (Firmenmarke ohne Ang.), H. 41 cm; GSF, S. 5
Lit.: GSF, S. 5, hier mit der Ang. des Originaltitels "Münchner Kind'l".

924
BÜSTE „Marocca ner art nouvadouré" (offen)
ca. 1898/99
924 ohne Ang.; KeramRS, 7.1899, S. 881
Bem.: s. Modell-Nr. 6

925
Langer, (Leopold?)
BÜSTE Araberin mit Kopftuch und Halsschmuck
ca. 1895/97
925,3,7 FG (Firmenmarke ohne Ang.), H. 24 cm; Schuler 63
925,20 ger., ohne Firmenmarke, unsign., H. 29 cm; eBay # 3775831363
-> vgl. 935

926
Strasser, Arthur
BÜSTE „Kabyle" („Antar") (nach links blickend)
ca. 1895
926,(7,5,8,– vertikal), Firmenmarke 7, Repr. Res., gepr. „P"= Z3, H. 43 cm; Wien Museum Inv. 219.108
926,51 Firmenmarke 8a?, Repr. Res., gepr. „P"= Z3, unsign., H. ca. 45 cm; WKA 9
926,68 (Firmenmarke ohne Ang.), unsign., H. 48,2 cm; eBay # 2215203202 (Jenack Auctioneers, Chester N.Y. USA)
926 (ohne Ang.), unsign., H. 45 cm; Arnold 28.5.1983
926 (Firmenmarke ohne Ang.), H. 45 cm; GSF, S. 5
Bem.: Zu „Antar ben Cheddad" vgl. Marcel Devic (Übers. aus dem Arab.): Roman d'Antar (1864).
-> vgl. 934
-> vgl. 903; 961; 1055 FIGUR
Pendant -> vgl. 927
Lit.: GSF, S. 5, hier mit der Ang. des Originaltitels "Kabyle"; Richemond (1999), S. 223 (m. Abb.).

927
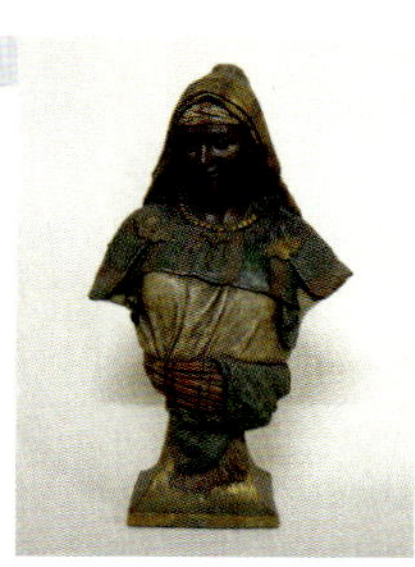
Désurmont, (Gil-D.?)
BÜSTE „Kabylin" („Abla") (nach rechts blickend)
ca. 1895
927,33 ger., Firmenmarke 8a, Firmenmarke 12, unsign., gepr. „P"= Z3, H. 43,3 cm; MAK Wien, Inv. Ke 10602-110
927,38 Firmenmarke 8a, Repr. Res., H. 44 cm; eBay # 2654520571
927,55 Firmenmarke 8a, Repr. Res., H. ca. 44 cm; eBay # 2573430174
927 Firmenmarke 11?, Repr. Res., gepr. „P"= Z3, H. ca. 45 cm; WKA 9
927 (Firmenmarke ohne Ang.), H. 45 cm; GSF, S. 5
-> vgl. 935
-> vgl. 904; 962; 1056 jeweils FIGUR
Pendant -> vgl. 926
Lit.: GSF, S. 5, hier mit der Ang. des Originaltitels „Kabylin".

928a
FIGUR „Baigneuse"
ca. 1897/98
928a (Firmenmarke ohne Ang.), H. 89 cm; GSF, S. 5
-> vgl. 928b
Lit.: GSF, S. 5, hier mit der Ang. des Originaltitels „Baigneuse".

928b
FIGUR „Baigneuse (mit Spiegel)"
ca. 1897/98
928b (Firmenmarke ohne Ang.), H. 89 cm; GSF, S. 5
-> vgl. 928a
Lit.: GSF, S. 5, hier mit der Ang. des Originaltitels „Baigneuse".

929
BÜSTE „Gabrielle" (Frau mit Schal/Stola) (nach rechts blickend)
ca. 1897/98
929,1261 Firmenmarke 11, H. 47 cm; eBay # 101231043; 121514594
929 (Firmenmarke ohne Ang.), H. 47 cm; GSF, S. 5
-> vgl. 1646
Lit.: GSF, S. 5, hier mit der Ang. des Originaltitels „Gabrielle".

929 (GSB)
Ledru, Auguste
PLAT „Femme aux Chardons"
ca. 1892/1900
929 Ø 35 cm; GSB, Pl. 94 (m. Abb.)
Ausst.: Salon des Beaux-Arts (ohne Jahresangabe)
Pendant -> vgl. 828 (GSB)

930
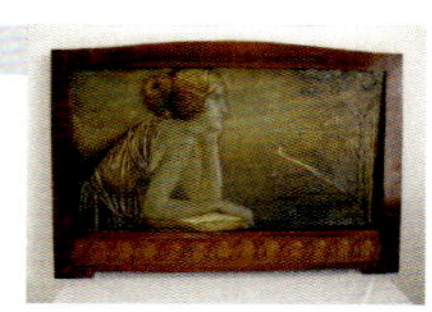
Montenave
RELIEF Frau mit Buch vor Landschaft (Wandrelief)
ca. 1897/98
930 ger., 19,28 gem. III, Firmenmarke 4a, Ges. gesch., H. 50 cm, L. 93 cm (mit Rahmen H. 68 cm, L. 102 cm; Sammlung f@goldscheider.de

930 (GSB)
Ledru, Auguste
BAS-RELIEF „Effroi"
ca. 1892/1900
930 H. 24 cm, L. 40 cm; GSB, Pl. 90 (m. Abb.)

931
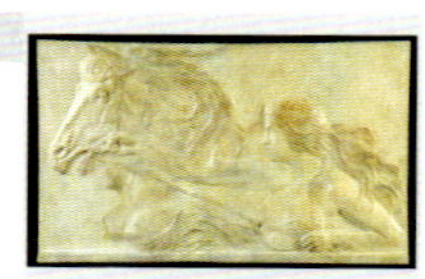
Montenave
RELIEF Amazone mit Pferd (nach links) (Wandrelief)
ca. 1897/98
0931,51,10 Firmenmarke 7, Repr. Res., unsign., H. 63,3 cm, L. 104 cm; Wien Museum Inv. 223.900
-> vgl. 1035
Pendant -> vgl. RELIEF Walküre mit Pferd (nach rechts),
Lit.: H. H.: Internationale Kunstgewerbliche Ausstellung in Petersburg 1908, in: KM, 8.1908, S. 166 (m. Abb.).
Lit.: Pudor (1902), S. 72 (m. Abb.)

933
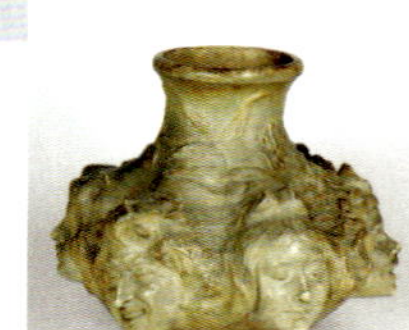
Larroux, (Antonin)
VASE „Masques" (Sechs plastische Gesichter: Nymphen/Faun)
ca. 1897/98
0933,31,21 gem. XIX, Firmenmarke 7, sign. Laroux, H. 47 cm; PB Wien 19

Ausst.: Turin 1902; Paris Salon 1902.
–> vgl. 1097; WANDMASKEN 2287; 2288; 2289; 2290; 2291
Lit.: Gmelin, L.: Die 1. intern. Ausst. für moderne dekorative Kunst in Turin 1902, in: KuHW, 52.1901/02, S. 325–342 (S. 334, hier als Künstler „Larone" und Material „Steinzeug" angegeben); L'Art décoratif aux Expositions des Beaux-Arts 1902, Vol. 1, Paris (1902), Pl. 95 (Abb.); Vol. 2, Pl. 200 (Abb.).

934
Strasser, Arthur
BÜSTE „Kabyle" („Antar") (nach links blickend)
ca. 1895
934 (Firmenmarke ohne Ang.), H. 35 cm; Pinhas, S. 20 (m. Abb.)
934 (Firmenmarke ohne Ang.), H. 34 cm; GSF, S. 5
Bem.: Zu „Antar ben Cheddad" vgl. Marcel Devic (Übers. aus dem Arab.): Roman d'Antar (1864).
–> vgl. 926
–> vgl. 903; 961; 1055 FIGUR
Pendant –> vgl. 935
Lit.: GSF, S. 5, hier mit der Ang. des Originaltitels „Kabyle"; Richemond (1999), S. 223 (m. Abb.).

935 –> ÜBERSCHNEIDUNG
Désurmont, (Gil-D.?)
BÜSTE „Kabylin" („Abla") (nach rechts blickend)
ca. 1895
935 (Firmenmarke ohne Ang.), sign. Désurmont, H. 35 cm; Pinhas, S. 20 (m. Abb.)
935 (Firmenmarke ohne Ang.), H. 34 cm; GSF, S. 5
–> vgl. 927
–> vgl. 904; 962; 1056 jeweils FIGUR
Pendant –> vgl. 926
Lit.: GSF, S. 5, hier mit der Ang. des Originaltitels „Kabylin".

935 –> ÜBERSCHNEIDUNG
Langer, (Leopold?)
BÜSTE Araberin mit Kopftuch und Halsschmuck
ca. 1895/97
935 ger., Firmenmarke 8, Vervielf. vorbeh., unsign., H. 38 cm; eBay # 6505142835
–> vgl. 925

943
BÜSTE Münchner Kindl (mit Bierhumpen „HB" und Radi)
ca. 1899
943 (Firmenmarke ohne Ang.), H. fehlt; Pinhas, S. 65 (m. Abb.)
Bem.: Die Darstellung dieses „Münchner Kindls" erinnert an eine dralle Bedienung des Münchener Hofbräuhauses (HB) welche mit einem Maßkrug und dem obligatorischen Rettich ausgestattet ist.

944
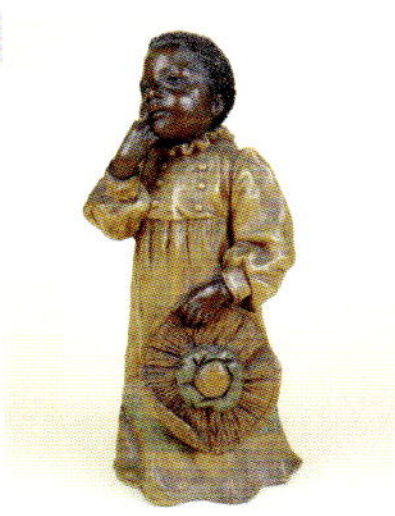
Stehendes Negerkind mit großem Hut in der Hand
ca. 1895
944,10 ohne Firmenmarke, H. 25 cm; PB Wien 19
944,A,41 ger., Firmenmarke 11, H. 25,2 cm; Doro 29.10.1996; 26.5.1998
ohne Ang., Firmenmarke 8b, Repr. Res., H. 25,4 cm; eBay # 1419618609
ohne Ang., FG (Firmenmarke?), H. 24 cm; Soth. L. 4.11.1988
Pendants –> vgl. 945; 946; 947

945 –> ÜBERSCHNEIDUNG
Stehendes Negerkind
ca. 1895
945,31 Firmenmarke 11, Repr. Res., H. 25 cm; Neuwirth (1974), S. 29 (m. Abb.)
Pendants –> vgl. 944; 946; 947

945 –> ÜBERSCHNEIDUNG
VASE Stehendes Paar, mit ihren großen Tüchern die Vase bildend
ca. 1895
0945 (Firmenmarke ohne Ang.), H. 56 cm; Pinhas, S. 52 (m. Abb.)

946 –> ÜBERSCHNEIDUNG
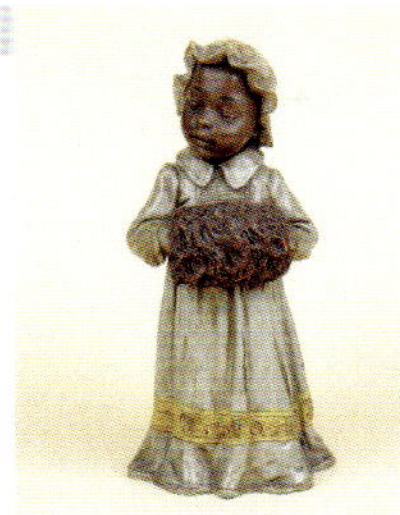
Stehendes Negerkind mit Muff
ca. 1895
946,24 ger., Firmenmarke 8b, Repr. Res., H. 24 cm; PB Wien 19
946,25 Firmenmarke 11, Vervielf. vorbeh., H. 23,3 cm; Neuwirth (1974), S. 29 (m. Abb.)
946,28 ger., Firmenmarke 11, H. 25,2 cm; Doro 29.10.1996; 26.5.1998
Pendants –> vgl. 944; 945; 947

946 –> ÜBERSCHNEIDUNG (Nr. wohl falsch)
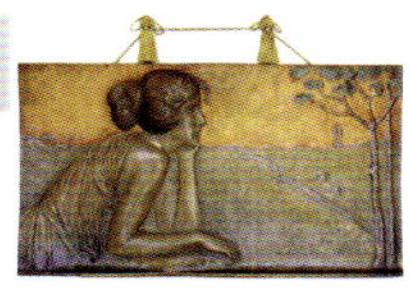
RELIEF Frauengestalt in weite Landschaft blickend (Wandrelief)
ca. 1897/98
0946,259,17 Firmenmarke 4, Fabriqué en Autriche, H. 34,3 cm, L. 62,2 cm; eBay # 449005647; Sammlung Robert Senft
Lit.: Pudor (1902), S. 73 (m. Abb.)

947
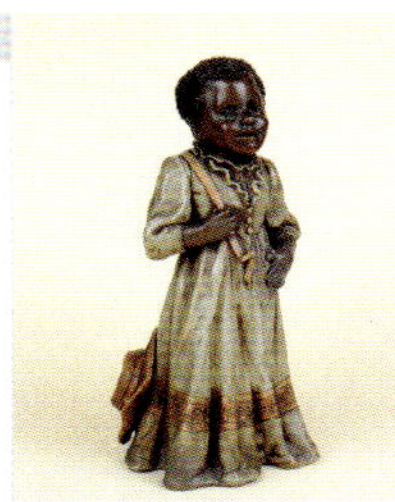
Stehendes Negerkind mit großem Hut
ca. 1895
947,13 Firmenmarke 8b, Repr. Res., H. 23 cm; PB Wien 19
947,21 Firmenmarke 8a, Repr. Res., H. 24,1 cm; KH
ohne Ang., Firmenmarke 11a, Vervielf. vorbeh., H. 22 cm; eBay # 1419618609
ohne Ang., FG (Firmenmarke?), H. 24 cm; Soth. L. 4.11.1988
Pendants –> vgl. 944; 945; 946

948a
FIGUR „Mönch (schnupfend, klein)"
ca. 1895/97
948a (Firmenmarke ohne Ang.), H. 21 cm; GSF, S. 5
–> vgl. 918
Lit.: GSF, S. 5, hier mit der Ang. des Originaltitels „Mönch (schnupfend, klein)".

948b
FIGUR „Mönch (niessend, klein)"
ca. 1895/97
948b (Firmenmarke ohne Ang.), H. 21 cm; GSF, S. 5
Lit.: GSF, S. 5, hier mit der Ang. des Originaltitels „Mönch (niessend, klein)".

949 –> ÜBERSCHNEIDUNG
FIGUR „Mönch (einschenkend, klein)"
ca. 1895/97
949 (Firmenmarke ohne Ang.), H. 21 cm; GSF, S. 5
–> vgl. 920
Lit.: GSF, S. 5, hier mit der Ang. des Originaltitels „Mönch (einschenkend, klein)".

949 –> ÜBERSCHNEIDUNG
Pecheur
RELIEF Junges Paar mit Buch und Blütenzweig (Wandrelief)
ca. 1897/98
0949,16,28 (Firmenmarke ohne Ang.), unsign., H. 18,5 cm, L. 39 cm; Henry's 29.11.1995
Pendant –> vgl. 950
Lit.: N., N., Friedrich Goldscheider, Kunstgewerbliche Etablissements, Kunst-Terrakotta-Fabrik, Atelier künstlerischer Fayencen, in: Jahrbuch der Gesellschaft österreichischer Architekten, 1909, 1 Bl. o. Pag. (m. Abb.), hier eine Größenvariante (H. 43 cm, L. 96 cm) des Motivs m. der Ang. der Inschrift „POESIA" die jedoch auf dem nachgewiesenen Ex. der Auktion von Henry's fehlt.

950
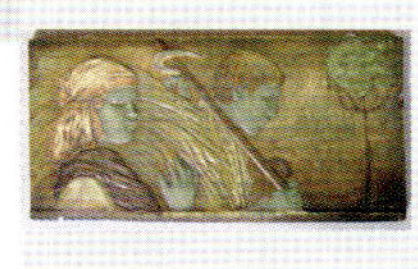
Pecheur
RELIEF Schnitter (Junges Paar bei der Feldarbeit) (Wandrelief)
ca. 1897/98
0950,68,69 Firmenmarke 4, Ges. gesch., H. 29,6 cm; L. 61,6 cm; PB Deutschland
Pendant –> vgl. 949
Lit.: N., N., Friedrich Goldscheider, Kunstgewerbliche Etablissements, Kunst-Terrakotta-Fabrik, Atelier künstlerischer Fayencen, in: Jahrbuch der Gesellschaft österreichischer Architekten, 1909, 1 Bl. o. Pag. (m. Abb.), hier zwei Größenvarianten (H. 43 cm, L. 96 cm; H. 18 cm, L. 38 cm) des Motivs m. der Ang. der Inschrift „PROSA" die jedoch auf dem nachgewiesenen Ex. der PB Deutschland fehlt.

952 –> ÜBERSCHNEIDUNG
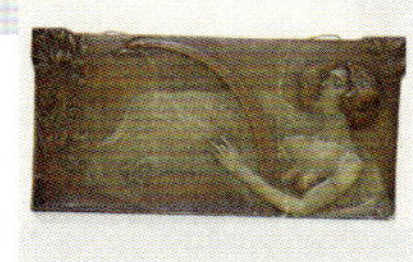
Lesca
RELIEF Junge Frau mit Harfe und Blumenschmuck (Wandrelief)
ca. 1897/98
0952,31,28 Firmenmarke 4, Ges. gesch., sign. Leska (!), H. 31,6 cm, B. 64,3 cm; MAK Wien, Inv. Ke 10602-112
Lit.: Pinhas, S. 50 (m. Abb.), auf S. 23 wird eine stehende orientalische Früchteverkäuferin ebenfalls unter Mod.-Nr. 952 gezeigt.

952 –> ÜBERSCHNEIDUNG
Baron, René
FIGUR „Lied"
ca. 1897/98
952 (Firmenmarke ohne Ang.), H. 100 cm; GSF, S. 5
–> vgl. 954; 972
–> vgl. 505 (GSB); 509 (GSB)
Lit.: GSF, S. 5, hier mit der Ang. des Originaltitels „Lied".

954 –> ÜBERSCHNEIDUNG
VASE / JARDINIERE Stehendes Paar, ein großes Tuch haltend
ca. 1897/98
0954,37,10 (Firmenmarke ohne Ang.), Repr. Res., H. 56 cm; Soth. L. 28.10.1998

954 –> ÜBERSCHNEIDUNG
Baron, René
FIGUR „Lied"/„La Chanson" (Stehende Sängerin mit Notenheft) (auf getrepptem Sockel)
ca. 1897/98 (für den Entwurf)
954 (Firmenmarke ohne Ang.), unsign., H. 100 cm; Pinhas, S. 44 (m. Abb.); Bem.: Pinhas gibt als Mod.-Nr. 1954 an. Diese dürfte aber eher als 954 zu verifizieren sein.
Bem.: An der Plinthe betitelt (Metallschild).
–> vgl. 952; 972
–> vgl. 505 (GSB); 509 (GSB)
Lit.: GSB, Pl. 5, 10, hier ist die Figur (jeweils ohne Ang. des Entwerfers) mit dem Originaltitel „La Chanson" abgebildet; Nr. 505 (H. 104 cm, „Fonte d'Art"/Zinkguss) und Nr. 509 (H. 23 cm, Bronze); Berman, Vol. 2, Nr. 1303 (m. Abb.), hier mit Ang. des Entwerfers „René Baron".

957
FIGUR „Sorbetverkäufer"
ca. 1897/98
957 (Firmenmarke ohne Ang.), H. 63 cm; GSF, S. 5
Lit.: GSF, S. 5, hier mit der Ang. des Originaltitels „Sorbetverkäufer".

958
FIGUR „Obstverkäuferin"
ca. 1897/98
958 (Firmenmarke ohne Ang.), H. 63 cm; GSF, S. 5
Lit.: GSF, S. 5, hier mit der Ang. des Originaltitels „Obstverkäuferin".

959
Strasser, Arthur (zugeschr.)
BÜSTE Junger Orientale (Araber) mit Vollbart und Turban
ca. 1890/95
959,44 ger., Firmenmarke 8a, Repr. Res., unsign., H. 58,5 cm; Wien Museum Inv. 186.608
959,52 ger., (ohne Firmenmarke), unsign., H. 58 cm; Kaisersl. 1987, S. 69

961
Strasser, Arthur

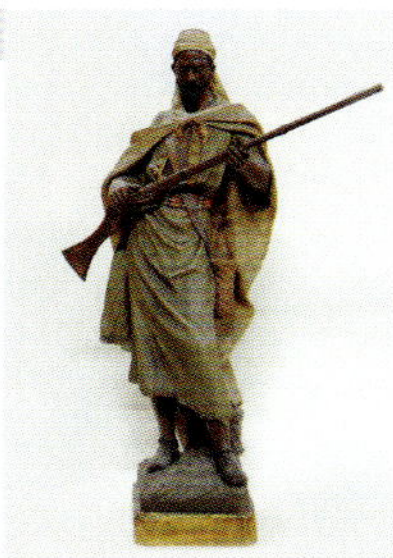

FIGUR „Kabyle" („Antar")
Inschrift: ANTAR
ca. 1895
961,10,83 gem. weiß VIII, Firmenmarke 7, Repr. Res., sign. Desurmont, gepr. „P"= Z3, H. 69,3 cm; MAK Wien, Inv. Ke 10602-27
961,(1,5,1,6,6 vertikal) Firmenmarke 11, Repr. Res., unsign., gepr. „P"= Z3, H. 72 cm; GK, S. 11, 61 (m. Abb.); Wien Museum Inv. 220.865
961,(8,6,7,7 vertikal),53 (Firmenmarke ohne Ang.), unsign., H. 74 cm; Christie's SK 4.11.1999; Bem.: Sockel ohne Inschrift.
961 (Firmenmarke ohne Ang.), unsign., H. 69 cm; Doro 1.6.2005; 31.5.2006
ohne Ang., H. 72 cm; Gros/Del. 11.12.1995
961 (Firmenmarke ohne Ang.), H. 71 cm; GSF, S. 5
Bem.: Vgl. Richemond (1995), S. 57, hier sign. A. Strasser (Modell-Nr. 903, H. 98 cm); zu „Antar ben Cheddad" vgl. Marcel Devic (Übers. aus dem Arab.): Roman d'Antar (1864).
Ausst.: HMW (jetzt Wien Museum) 1985/86; Orientalische Reise, Wien Museum 2003.
-> vgl. 903; 1055
-> vgl. 926 BÜSTE
Pendant -> vgl. 904; 962; 1056
Lit.: GSF, S. 5, hier mit der Ang. des Originaltitels „Kabyle"; GK, S. 11, 61 (m. Abb.); Mayr-Oehring, E. u. E. Doppler: Orientalische Reise, Wien 2003, S. 252 (m. Abb.).

962
Désurmont, (Gil-D.?)
FIGUR „Kabylin" („Abla")
Inschrift: ABLA (in arab. Schriftzeichen)
ca. 1895
962,6,(8,2,7,8,6 vertikal) (ohne Firmenmarke), unsign.?, gepr. „P"= Z3, H. 70 cm; GK, S. 11, 62 (m. Abb.); Wien Museum Inv. 220.866
962,52,5 Firmenmarke 7, unsign., Repr. Res., H. 70 cm; PB Wien 19
962,152,55 (Firmenmarke ohne Ang.), unsign., H. ca. 72 cm; Christie's SK 4.11.1999
962 (Firmenmarke ohne Ang.), H. 70 cm; GSF, S. 5
ohne Ang., H. 70 cm; Gros/Del. 11.12.1995
Ausst.: HMW (jetzt Wien Museum) 1985/86; Orientalische Reise, Wien Museum 2003.
-> vgl. 904; 1056
-> vgl. 927 BÜSTE
Pendant -> vgl. 903; 961; 1055
Lit.: GSF, S. 5, hier mit der Ang. des Originaltitels „Kabylin"; Mayr-Oehring, E. u. E. Doppler: Orientalische Reise, Wien 2003, S. 252 (m. Abb.); GK, S. 11, 62 (m. Abb.).

963?
Stehender Mönch mit Bierhumpen (auf getrepptem Sockel)
ca. 1895/97
963? (Firmenmarke ohne Ang.), H. 36,5 cm; Pinhas, S. 34 (m. Abb.)
Bem.: Pinhas gibt als Mod.-Nr. 1963 an, diese ist vermutlich falsch. Es könnte sich um die Mod. Nr. 963 handeln.
-> vgl. 922

964
BÜSTE Junge Dame nach links blickend (mit Rose am Dekolleté)
ca. 1897/98
964,32 Firmenmarke 8a, Repr. Res.,, H. 61 cm; eBay # 7314956112

965
FIGUR „Baer"
ca. 1897/98
965 (Firmenmarke ohne Ang.), H. 13 cm; GSF, S. 5
Lit.: GSF, S. 5, hier mit der Ang. des Originaltitels „Baer".

966 -> ÜBERSCHNEIDUNG
Cherc
VASE Zwei stehende Frauen, die eine mit Baum- und Blattwerk reliefierte Kugel in die Höhe halten
ca. 1897/98
0966 (Firmenmarke ohne Ang.), H. 70 cm; Pinhas, S. 48 (m. Abb.)

966 -> ÜBERSCHNEIDUNG
FIGUR „Löwe"
ca. 1897/98
966 (Firmenmarke ohne Ang.), H. 24 cm; GSF, S. 5
Lit.: GSF, S. 5, hier mit der Ang. des Originaltitels „Löwe".

967 -> ÜBERSCHNEIDUNG
(Marchand, Harry?)
VASE Zwei an der Vasenwandung stehende Frauen mit Mohnblumen
ca. 1897/98
0967 (Firmenmarke ohne Ang.), H. 66 cm; Pinhas, S. 48 (m. Abb.); Bem.: Der Künstler Harry Marchand lässt sich in der einschlägigen Literatur nicht nachweisen. Wahrscheinlich handelt es sich um ein Pseudonym, bzw. um einen Lesefehler.

967 -> ÜBERSCHNEIDUNG
FIGUR „Baer"
ca. 1897/98
967 (Firmenmarke ohne Ang.), H. 69 cm; GSF, S. 5
Lit.: GSF, S. 5, hier mit der Ang. des Originaltitels „Baer".

969 -> ÜBERSCHNEIDUNG
Lesca
RELIEF Junge Bäuerin mit erlegtem Fasan unter Apfelbaumzweig (Wandrelief)
ca. 1897/98
0969,57,69 (Firmenmarke ohne Ang.), H. 72,8 cm, B. 51,8 cm; Doro 19.9.2000

969 -> ÜBERSCHNEIDUNG
BÜSTE „Rieuse" (Junge Dame) (nach rechts blickend)
ca. 1897/98
969B (Firmenmarke ohne Ang.), H. ca. 72 cm; Pinhas, S. 71 (m. Abb.)
969 (Firmenmarke ohne Ang.), H. 72 cm; GSF, S. 5
Lit.: GSF, S. 5, hier mit der Ang. des Originaltitels „Rieuse".

972
Baron, René
FIGUR „Lied (mittel)"/"La Chanson" (Stehende Sängerin mit Notenheft) (auf getrepptem Sockel)
ca. 1897/98
972,X ger., FG (Firmenmarke?), Repr. Res., unsign., H. 73 cm; Doro 3.5.1995
972 (Firmenmarke ohne Ang.), H. 50 cm; GSF, S. 5
Bem.: An der Plinthe betitelt (Metallschild).
-> vgl. 952; 954
-> vgl. 505 (GSB); 509 (GSB)
Lit.: GSF, S. 5, hier mit der Ang. des Originaltitels „Lied (mittel)"; GSB, Pl. 5, 10, hier ist die Figur (jeweils ohne Ang. des Entwerfers) mit dem Originaltitel „La Chanson" abgebildet; Nr. 505 (H. 104 cm, „Fonte d'Art"/Zinkguss) und Nr. 509 (H. 23 cm, Bronze); Berman, Vol. 2, Nr. 1303 (m. Abb.), hier mit Ang. des Entwerfers „René Baron".

974
FIGUR „Schenkmädchen"
ca. 1897/98
974 (Firmenmarke ohne Ang.), H. fehlt; GSF, S. 5
Lit.: GSF, S. 5, hier mit der Ang. des Originaltitels „Schenkmädchen".

975
Weiher
FIGUR „Tiger" (liegend) (auf Rechtecksockel)
ca. 1897/98
975,17 (Firmenmarke ohne Ang.), gepr. „P"= Z3, (H. 18,5 cm, L. 34,5 cm; eBay # 2560069782
975 (Firmenmarke ohne Ang.), H. fehlt; GSF, S. 5
Lit.: GSF, S. 5, hier mit der Ang. des Originaltitels „Tiger".

976
RELIEF Frauenporträt im Profil nach links (Wandrelief)
ca. 1897/98
0976 (Firmenmarke ohne Ang.), H. 51 cm; Pinhas, S. 50 (m. Abb.)

978
FIGUR „Phyrne (nackt)"
ca. 1897/98
978 (Firmenmarke ohne Ang.), H. 90 cm; GSF, S. 5
Bem.: Phyrne, eine griech. Hetäre (leichtes Mädchen) des 4. Jahrhunderts, war die Geliebte des Praxiteles.
-> vgl. 979 (Variante); 1125
Lit.: GSF, S. 5, hier mit der Ang. des Originaltitels „Phyrne (nackt)".

979
FIGUR „Phyrne (bekleidet)"
ca. 1897/98
979 (Firmenmarke ohne Ang.), H. 90 cm; GSF, S. 5
Bem.: s. 978
-> vgl. 978; 1159
Lit.: GSF, S. 5, hier mit der Ang. des Originaltitels „Phyrne (bekleidet)".

980
BÜSTE „Bouquetèrie" (Junge Dame mit großer Schleife im Dékolleté)
ca. 1897/98
980,0 Firmenmarke 11?, Vervielf. vorbeh., gepr. „P"= Z3, H. 33 cm; eBay # 179155964
980 (Firmenmarke ohne Ang.), H. 32 cm; GSF, S. 5
Lit.: GSF, S. 5, hier mit der Ang. des Originaltitels „Bouquetèrie".

981
BÜSTE „Richard Wagner"
ca. 1895 (für den Entwurf)
981 (Firmenmarke ohne Ang.), H. 63 cm; Pinhas, S. 67 (m. Abb.)
981 (Firmenmarke ohne Ang.), H. 63 cm; GSF, S. 5
Bem.: Dargestellt ist der Komponist Richard Wagner (1813–1883).
-> vgl. 618; 1030; 1897
Lit.: GSF, S. 5, hier mit der Ang. des Originaltitels „Richard Wagner".

982a
FIGUR „Wahrheit (nackt)"
ca. 1897/98
982a (Firmenmarke ohne Ang.), H. 126 cm; GSF, S. 5
-> vgl. 982b
Lit.: GSF, S. 5, hier mit der Ang. des Originaltitels „Wahrheit".

982b
FIGUR „Wahrheit (bekleidet)"
ca. 1897/98
982b (Firmenmarke ohne Ang.), H. 126 cm; GSF, S. 5
-> vgl. 982a; 1422
Lit.: GSF, S. 5, hier mit der Ang. des Originaltitels „Wahrheit (bekleidet)".

985
FIGUR „Wallone"
ca. 1897/98
985 (Firmenmarke ohne Ang.), H. 43 cm; GSF, S. 5
-> vgl. 986= BÜSTE
Lit.: GSF, S. 5, hier mit der Ang. des Originaltitels „Wallone".

986 –> ÜBERSCHNEIDUNG
BÜSTE „Wallone"
ca. 1897/98
986 (Firmenmarke ohne Ang.), H. 44 cm; GSF, S. 6
–> vgl. 985= FIGUR
Lit.: GSF, S. 6, hier mit der Ang. des Originaltitels „Wallone".

986 –> ÜBERSCHNEIDUNG
Mondi
RELIEF Mädchen mit Blume (Wandrelief)
ca. 1897/98
0986,62,69 GS Wien (Firmenmarke ohne Ang.), Ges. gesch., H. 49,8 cm; Soth. N.Y. 26.11.1993
0986 ohne Ang., H. 50 cm; PB Wien 19

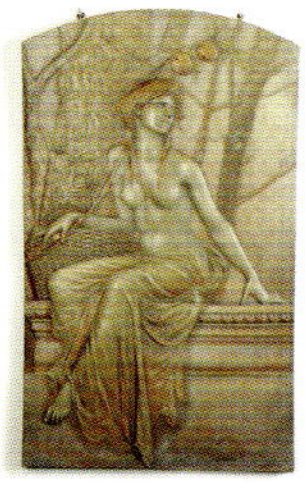

989
BÜSTE „Landsknecht"
ca. 1897/98
989 (Firmenmarke ohne Ang.), H. 52 cm; GSF, S. 6
Lit.: GSF, S. 6, hier mit der Ang. des Originaltitels „Landsknecht".

990 –> ÜBERSCHNEIDUNG
Suchan(ek), Vacslav
VASE Zwei Chamäleons
ca. 1897/98
0990,3,35 (ohne Firmenmarke), Repr. Res., sign. Suchanek, H. 36,5 cm; Sammlung Bon/Venedig
0990,33,27 Firmenmarke 7, Repr. Res., sign. „Suchan", H. 37 cm; Sammlung Gruber, S. 104, 105 (m. Abb.)

990 –> ÜBERSCHNEIDUNG
BÜSTE „Junger Wein" (Junges Bauernmädchen mit Kappe) (nach links blickend)
ca. 1897/98
990,51 ger., „Wortmarke" (Firmenmarke ohne Ang.), H. 48,3 cm; Doro 7.3.2001
990 (Firmenmarke ohne Ang.), H. 50 cm; GSF, S. 6
Lit.: GSF, S. 6, hier mit der Ang. des Originaltitels „Junger Wein".

992
BÜSTE „Franz Liszt"
ca. 1895 (für den Entwurf)
992 (Firmenmarke ohne Ang.), H. 60 cm; Pinhas, S. 67 (m. Abb.)
992 (Firmenmarke ohne Ang.), H. 60 cm; GSF, S. 6
Bem.: Dargestellt ist der Komponist und Pianist Franz Liszt (1811–1886).
–> vgl. 619; 1031
Lit.: GSF, S. 6, hier mit der Ang. des Originaltitels „Franz Liszt".

1000 –> ÜBERSCHNEIDUNG
Cherc
„Salome" (Stehende Frauengestalt mit lyraähnlichem Instrument)
Inschrift: SALOME
ca. 1897/98
1000,50,53 Firmenmarke Sculpture 13, unsign.?, H. 106 cm; Millon & Robert 14.11.1995
ohne Ang. GS Wien (Firmenmarke ohne Ang.), Repr. Res., H. 105 cm; Soth. L. 30.3.1990
ohne Ang. FG (Firmenmarke ohne Ang.), Repr. Res., H. 105 cm; Soth. Arcade 14.6.1989
–> vgl. 1900; 2085; 2530

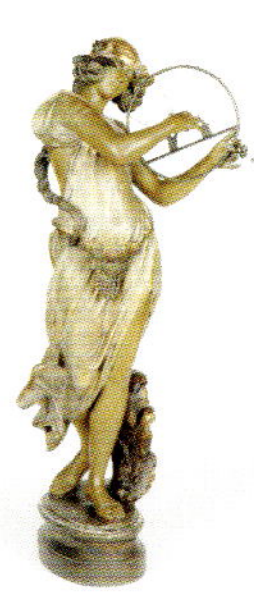

1000 –> ÜBERSCHNEIDUNG
GRUPPE „Liebeserwachen"
ca. 1897/98
1000 (Firmenmarke ohne Ang.), H. 54 cm; GSF, S. 6
Lit.: GSF, S. 6, hier mit der Ang. des Originaltitels „Liebeserwachen".

1001
BÜSTE „Beethoven"
Inschrift: BEETHOVEN
ca. 1897
1001 (Firmenmarke ohne Ang.), H. 72 cm; Pinhas, S. 67 (m. Abb.)
1001 (Firmenmarke ohne Ang.), H. 72 cm; GSF, S. 6
Bem.: Dargestellt ist der Komponist Ludwig van Beethoven (1770–1827).
–> vgl. 1028; 1053
Lit.: GSF, S. 6, hier mit der Ang. des Originaltitels „Beethoven".

1002
BÜSTE „Mozart"
Inschrift: MOZART
ca. 1897
1002 (Firmenmarke ohne Ang.), H. 61 cm; Pinhas, S. 67 (m. Abb.)
1002 (Firmenmarke ohne Ang.), H. 61 cm; GSF, S. 6
Bem.: Dargestellt ist der Komponist Wolfgang Amadeus Mozart (1756–1791).
–> vgl. 1029; 1054
Lit.: GSF, S. 6, hier mit der Ang. des Originaltitels „Mozart".

1003
FIGUR „Schöne Helena"
ca. 1897/98
1003 (Firmenmarke ohne Ang.), H. 94 cm; GSF, S. 6
–> vgl. 1124
Lit.: GSF, S. 6, hier mit der Ang. des Originaltitels „Schöne Helena".

1005
Meier, Emil
SCHREIBZEUG / TINTENFASS
Mädchenkopf mit ausgebreiteten Armen
ca. 1897/98
1005,33,40 ohne Firmenmarke, sign. E. Meier, gepr. „S"= Z2, H. 11,7 cm; PB Wien 19

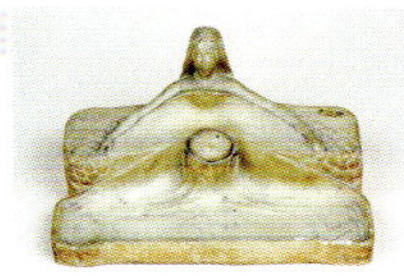

1006 –> ÜBERSCHNEIDUNG
Simon
VASE Mädchen mit Vogel spielend (reliefiert)
ca. 1897/98
01006,13,69 Firmenmarke 4, Ges. gesch., H. 43 cm; Wien Museum Inv. 186.630
01006,16,69 Firmenmarke 7, Repr. Res., H. 42,5 cm; www.wishfulthings.com/items/6860

1006 –> ÜBERSCHNEIDUNG
FIGUR „Kellermeister"
ca. 1897/98
1006 (Firmenmarke ohne Ang.), H. 98 cm; GSF, S. 6
Lit.: GSF, S. 6, hier mit der Ang. des Originaltitels „Kellermeister".

1009
Strasser, Arthur (zugeschr.)
BÜSTE Araber mit Vollbart und Turban
ca. 1896/97
1009,25 ger., GS (Firmenmarke?), Repr. Res., unsign., H. 28,5 cm; Doro 29.10.1996

1010
Strasser, Arthur
BÜSTE Junger Araber mit Bart und Turban
ca. 1896/97
1010,(1,3,2,6 vertikal),44 Firmenmarke 11, Repr. Res., unsign., gepr. „P"= Z3, H. ca. 49 cm; Neuwirth (1974), Farbtaf. nach S. 24; Doro 7.3.2001; eBay # 2573441798;
Bem.: Bei allen Quellen unterschiedliche Höhenangabe.

1012
FIGUR „Judith" (Stehende Frauengestalt mit Schwert, Stirnband und Gürtel)
ca. 1898/1900
1012 (Firmenmarke ohne Ang.), H. 92 cm; Pinhas, S. 71 (m. Abb.); Bem.: An der Plinthe betitelt (Metallschild).
Lit.: GSB, Pl. 27, hier ist die Figur mit dem Originaltitel „Judith (Habillée)" unter Nr. 584 (H. 65 cm) abgebildet. Eine identische Figur, jedoch mit entblößtem Oberkörper „Judith (Nue)" in drei verschiedenen Größen (H. 65, 30, 22 cm) findet sich an gleicher Stelle. Ausgeformt wurden diese Objekte in der Pariser Niederlassung von Friedrich Goldscheider jeweils in „Vrai Bronze" und „Fonte d'Art" unter den Nrn. 584; 602; 604.
–> vgl. 584 (GSB); 602 (GSB); 604 (GSB)
–> vgl. 734
–> vgl. 5105= ähnl. Modell in einer Ausformung der 20er Jahre

1012a
FIGUR „Judith (nackt)"
ca. 1897/98
1012a (Firmenmarke ohne Ang.), H. 92 cm; GSF, S. 6
–> vgl. 734; 1012
Lit.: GSF, S. 6, hier mit der Ang. des Originaltitels „Judith (nackt)".

1012b
FIGUR „Judith (bekleidet)"
ca. 1897/98
1012b (Firmenmarke ohne Ang.), H. 92 cm; GSF, S. 6
–> vgl. 734; 1012
Lit.: GSF, S. 6, hier mit der Ang. des Originaltitels „Judith (bekleidet)".

1013
VASE Stehendes Mädchen mit Vogel (an der Vasenwandung)
ca. 1897/98
01013 (Firmenmarke ohne Ang.), H. 81 cm; Pinhas, S. 49 (m. Abb.)

1014
BÜSTE „Lied"
ca. 1897/98
1014 (Firmenmarke ohne Ang.), H. 46 cm; GSF, S. 6
Lit.: GSF, S. 6, hier mit der Ang. des Originaltitels „Lied".

1016
WANDMASKE Araberin mit Kopftuch
ca. 1896/97
1016,88,7 (Firmenmarke ohne Ang.), H. ca. 27 cm; Christie's SK 4.11.1999

1018a
FIGUR „Mignon (nackt)" (Stehende Sängerin mit Lyra) (auf Rechtecksockel)
ca. 1897/98
1018a (Firmenmarke ohne Ang.), H. 46 cm; GSF, S. 6
–> vgl. 1018b; 1235
Lit.: GSF, S. 6, hier mit der Ang. des Originaltitels „Mignon (nackt)".

1018b
FIGUR „Mignon (bekleidet)" (Stehende Sängerin mit Lyra) (auf Rechtecksockel)
ca. 1897/98
1018B (Firmenmarke ohne Ang.), H. fehlt; Pinhas, S. 71 (m. Abb.)
1018b (Firmenmarke ohne Ang.), H. 46 cm; GSF, S. 6
–> vgl. 1018a; 1235
Lit.: GSF, S. 6, hier mit der Ang. des Originaltitels „Mignon (bekleidet)".

1019
Stehender Araber mit Turban und Dolch (auf verziertem Natursockel mit Baumstumpf)
ca. 1896/97
1019,56 Firmenmarke 8a, H. 63 cm; Wien Museum Inv. 219.120

1025
RELIEF Ein junger Mann umgeben von zwei junge Frauen (Wandrelief)
ca. 1896/97
01025 (Firmenmarke ohne Ang.), H. fehlt, B. 45,5 cm; Pinhas, S. 50 (m. Abb.)

1026
Strasser, Arthur (zugeschr.)
Stehender Araber (auf Quadratsockel)
ca. 1896/97
1026,6 Firmenmarke 8a, unsign., H. 75 cm; Wien Museum Inv. 219.116

1027
LAMPE Adam und Eva am Baum der Erkenntnis stehend
ca. 1896/97
01027 (Firmenmarke ohne Ang.), H. 110 cm; Pinhas, S. 48 (m. Abb.)

1028
BÜSTE „Beethoven"
Inschrift: BEETHOVEN
ca. 1897
1028 (Firmenmarke ohne Ang.), H. 50 cm; Pinhas, S. 67 (m. Abb.)
1028 (Firmenmarke ohne Ang.), H. 50 cm; GSF, S. 6
Bem.: Dargestellt ist der Komponist Ludwig van Beethoven (1770–1827).
-> vgl. 1001; 1053
Lit.: GSF, S. 6, hier mit der Ang. des Originaltitels „Beethoven".

1029
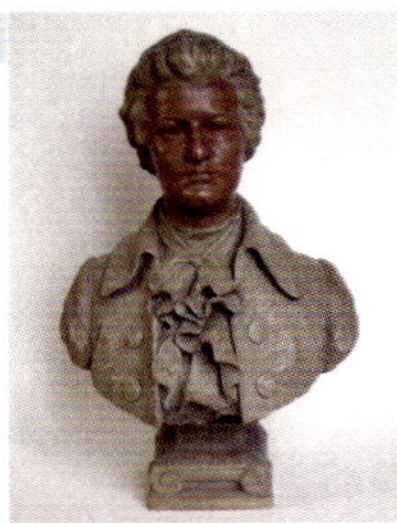
BÜSTE „Mozart"
Inschrift: MOZART
ca. 1897
1029 (ohne Nr.), Firmenmarke 8a, H. 50 cm; Sammlung Weilinger, Wien
1029 (Firmenmarke ohne Ang.), H. 50 cm; GSF, S. 6
Bem.: Dargestellt ist der Komponist Wolfgang Amadeus Mozart (1756–1791).
-> vgl. 1002; 1054
Lit.: GSF, S. 6, hier mit der Ang. des Originaltitels „Mozart"; Pinhas, S. 67 (m. Abb.).

1030
BÜSTE „Richard Wagner"
ca. 1895 (für den Entwurf)
1030 (Firmenmarke ohne Ang.), H. 45 cm; Pinhas, S. 67 (m. Abb.)
1030 (Firmenmarke ohne Ang.), H. 45 cm; GSF, S. 6
Bem.: Dargestellt ist der Komponist Richard Wagner (1813–1883).
-> vgl. 618; 981; 1897
Lit.: GSF, S. 6, hier mit der Ang. des Originaltitels „Richard Wagner".

1031
BÜSTE „Franz Liszt"
ca. 1895 (für den Entwurf)
1031 (Firmenmarke ohne Ang.), H. 48 cm; Pinhas, S. 67 (m. Abb.)
1031 (Firmenmarke ohne Ang.), H. 48 cm; GSF, S. 6
Bem.: Dargestellt ist der Komponist und Pianist Franz Liszt (1811–1886).
-> vgl. 619; 992
Lit.: GSF, S. 6, hier mit der Ang. des Originaltitels „Franz Liszt".

1032
TORSO Afrikaner (mit Tuch um die Hüfte)
ca. 1896/97
1032 (Firmenmarke ohne Ang.), H. 58 cm; Pinhas, S. 18 (m. Abb.)

1033
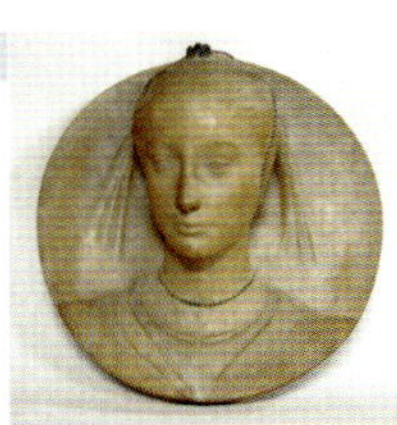
TONDO Frauenkopf mit Schleier, Halskette und Kopfschmuck (Renaissance)
ca. 1896/97
01033,27,69 Firmenmarke 7, Repr. Res., Ø 39,2 cm; MAK Wien, Inv. Ke 10602–96
Bem.: Die Manufaktur edierte diesen Frauenkopf nach einem historischen Vorbild.

1034
BÜSTE Zigeunerin
ca. 1896/97
1034 (Firmenmarke ohne Ang.), H. 50 cm; Pinhas, S. 20 (m. Abb.)

1035 -> ÜBERSCHNEIDUNG
Montenave
RELIEF Amazone mit Pferd (nach links) (Wandrelief)
ca. 1897/98
01035,3,28 Firmenmarke 4, Ges. gesch., H. 33 cm, L. 55 cm; Bergmann 50; Hampel 18./19.9.1998
Bem.: Modell-Nr. musste ungeprüft aus dem Auktionskat. übernommen werden.
-> vgl. 931
Pendant -> vgl. RELIEF Walküre mit Pferd (nach rechts),
Lit.: H. H.: Internationale Kunstgewerbliche Ausstellung in Petersburg 1908, in: KM, 1908, S.166 (m. Abb.)
Lit.: Pudor (1902), S. 72 (m. Abb.)

1035a
Gracien, J./G.?
FIGUR „Vasenträgerin (halb bekleidet)" (Stehende junge Frau mit Krug auf der Schulter) (auf Rechtecksockel)
ca. 1897/98
1035a (Firmenmarke ohne Ang.), H. 62 cm; GSF, S. 6
-> vgl. 1035b; 1440
Lit.: GSF, S. 6, hier mit der Ang. des Originaltitels „Vasenträgerin (halb bekleidet)"

1035b
Gracien, J./G.?
FIGUR „Vasenträgerin (ganz bekleidet)" (Stehende junge Frau mit Krug auf der Schulter) (auf Rechtecksockel)
ca. 1897/98
1035,186,19 gem. weiss X, Firmenmarke 7, Repr. Res., sign. J. Gracien, gepr. „P"= Z3, H. 63 cm; GK, S.11, 63 (m. Abb.); Wien Museum Inv. 182.062
1035b (Firmenmarke ohne Ang.), H. 62 cm; GSF, S. 6
Ausst.: HMW (jetzt Wien Museum) 1985/86.
-> vgl. 1035a; 1440
Lit.: GSF, S. 6, hier mit der Ang. des Originaltitels „Vasenträgerin (ganz bekleidet)"; GK, S. 11, 63 (m. Abb.); Pinhas, S. 66 (m. Abb.), hier mit der Mod.-Nrn. Ang. „B"; S. 73 (m. Abb.), hier mit der Mod.-Nrn. Ang. „C".

1041 -> ÜBERSCHNEIDUNG
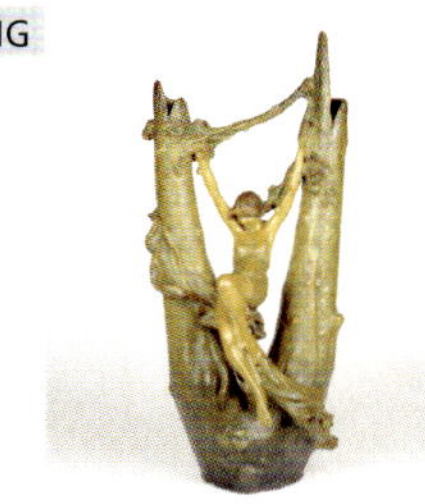
Pecheur
„Schaukel" (Doppel-VASE Junge Frau, zwischen einer Astgabel schwingend)
ca. 1897/98
1041,20,35 (Marke ohne Ang.), H. 111 cm; Herr 32
ohne Ang., H. 111 cm; Soth. L. 4.12.1985
-> vgl. 1072
Lit.: Pudor (1902), S. 36 (m. Abb.), hier mit Ang. des Originaltitels „Schaukel" und mit Ang. des Künstlers Pecheur; KPrJb, 53.1998, Bd. 3, S. 405; Weltkunst 1997, S. 2810.

1041 -> ÜBERSCHNEIDUNG
BÜSTE „Judith (nackt, mittel)"
ca. 1897/98
1041 (Firmenmarke ohne Ang.), H. 29 cm; GSF, S. 6
-> vgl. 1042a; 1042b
Lit.: GSF, S. 6, hier mit der Ang. des Originaltitels „Judith (nackt, mittel)".

1042 -> ÜBERSCHNEIDUNG
BÜSTE Junger Nordafrikaner mit halb entblößtem Oberkörper
ca. 1897/98
1042,56 Firmenmarke 8a?), H. 57 cm; Nagel 376; Hampel 36

1042a -> ÜBERSCHNEIDUNG
BÜSTE „Judith (nackt, groß)"
ca. 1897/98
1042a (Firmenmarke ohne Ang.), H. 44 cm; GSF, S. 6
-> vgl. 1041; 1042b
Lit.: GSF, S. 6, hier mit der Ang. des Originaltitels „Judith (nackt, groß)".

1042b -> ÜBERSCHNEIDUNG
BÜSTE „Judith (bekleidet, groß)"
ca. 1897/98
1042b (Firmenmarke ohne Ang.), H. 44 cm; GSF, S. 6
-> vgl. 1041; 1042a
Lit.: GSF, S. 6, hier mit der Ang. des Originaltitels „Judith (bekleidet, groß)".

1043 -> ÜBERSCHNEIDUNG
Gornik, Friedrich
JARDINIERE „Kakadu"
ca. 1898/99
01043,8,10 gem. X, Firmenmarke 7, gepr. „C"= Z4, H. 60,5 cm, B. 88 cm; MAK Inv. W.I. 395
Ausst.: ÖMKI 1903/04, hier mit Ang. des Originaltitels „Kakadu"
Lit.: KuKHW, 7.1904, S. 26 (m. Abb.), hier als Jardiniere bezeichnet; Neuwirth (1974a), S. 291, 292 (m. Abb.).

1043 -> ÜBERSCHNEIDUNG
BÜSTE Nordafrikaner mit Fez
ca. 1892 (für den Entwurf)
1043 (Firmenmarke ohne Ang.), H. 38 cm; Pinhas, S. 20 (m. Abb.)
Pendant -> vgl. 1044

1044
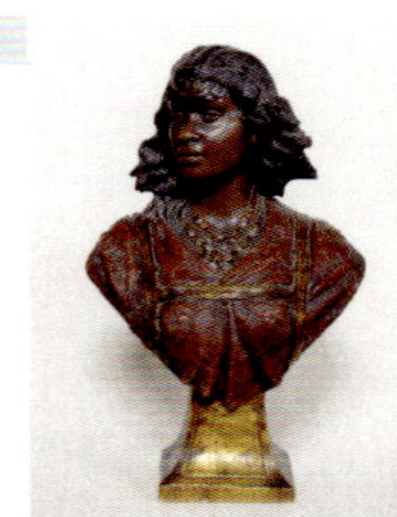
BÜSTE Nordafrikanerin mit Stirnband und Münzkette als Halsschmuck
ca. 1892 (für den Entwurf)
1044,42,(1,8,6,7 vertikal) Firmenmarke 11, Repr. Res., gepr. „P"= Z3, H. 39,9 cm; MAK Wien, Inv. Ke 10602-88
-> vgl. 323
Pendant -> vgl. 1043

1047
Huneker (d.i. Bracken, Clio Hinton)
VASE Frau auf der Vasenmündung sitzend
ca. 1897/98
01047 (Firmenmarke ohne Ang.), sign. „Huncken", H. 53 cm; Pinhas, S. 50 (m. Abb.); Bem.: Die Sign. „Huncken" muss Huneker heißen, eindeutig ein Lese- oder Druckfehler bei Pinhas.

1048
FIGUR „Mohr mit Flöte" (auf getrepptem Sockel)
ca. 1897/98
1048,3 Firmenmarke 8a, H. 109 cm; eBay # 5351544196
1048 (Firmenmarke ohne Ang.), H. 110 cm; GSF, S. 6
Lit.: GSF, S. 6, hier mit der Ang. des Originaltitels „Mohr mit Flöte".

1053
BÜSTE „Beethoven"
Inschrift: BEETHOVEN
ca. 1897
1053,323,14 Firmenmarke Sculpture 13, Fabriqué en Autriche, Ausformung nach 1900, H. ca. 25 cm; WKH
1053 (Firmenmarke ohne Ang.), H. 25 cm; GSF, S. 6
Bem.: Dargestellt ist der Komponist Ludwig van Beethoven (1770–1827).
-> vgl. 1001; 1028
Lit.: GSF, S. 6, hier mit der Ang. des Originaltitels „Beethoven".

1054
BÜSTE „Mozart"
Inschrift: MOZART
ca. 1897
1054 (Firmenmarke ohne Ang.), H. 25 cm; Pinhas, S. 67 (m. Abb.)
1054 (Firmenmarke ohne Ang.), H. 25 cm; GSF, S. 6
Bem.: Dargestellt ist der Komponist Wolfgang Amadeus Mozart (1756–1791).
-> vgl. 1002; 1029
Lit.: GSF, S. 6, hier mit der Ang. des Originaltitels „Mozart".

1055
Strasser, Arthur
FIGUR „Kabyle" („Antar")
Inschrift: ANTAR (in arab. Schriftzeichen)
ca. 1895
1055,79,69 FG (Firmenmarken?), unsign., H. 37 cm;

Doro 12.6.1995; Bem.: Hier mit „Speer“ anstatt Gewehr.
1055 (Firmenmarke ohne Ang.), H. 37 cm; GSF, S.6
Bem.: Vgl. Richemond (1995), S. 57, dort sign. A. Strasser (Modell-Nr. 903, H. 98 cm); zu „Antar ben Cheddad“ vgl. Marcel Devic (Übers. aus dem Arab.): Roman d' Antar (1864).
-> vgl. 903; 961
-> vgl. 926 BÜSTE
Pendant -> vgl. 904; 962; 1056
Lit.: GSF, S. 5, hier mit der Ang. des Originaltitels „Kabyle“.

1056
Désurmont, (Gil-D.?)
FIGUR „Kabylin“ („Abla“)
Inschrift: ABLA (in arab. Schriftzeichen)
ca. 1895
1056,37 Firmenmarke 8a, unsign., H. 35,8 cm; Wien Museum Inv. 219.133
1056,82,64 FG (Firmenmarken?), unsign., H. 36,3 cm; Doro 12.6.1995
1056 (Firmenmarke ohne Ang.), H. 37 cm; GSF, S. 6
-> vgl. 904; 962
-> vgl. 927 BÜSTE
Pendant -> vgl. 903; 961; 1055
Lit.: GSF, S. 6, hier mit der Ang. des Originaltitels „Kabylin“.

1059
BÜSTE Schwarzer Araber mit Fez
ca. 1897
1059 (Firmenmarke ohne Ang.), H. 47 cm; Pinhas, S. 17 (m. Abb.)

1060
Strasser, Arthur (zugeschr.)
BÜSTE Schwarzer Araber mit Turban
ca. 1897
1060,12 Firmenmarke 8, Repr. Res., unsign., H. 47 cm; PB Wien 19

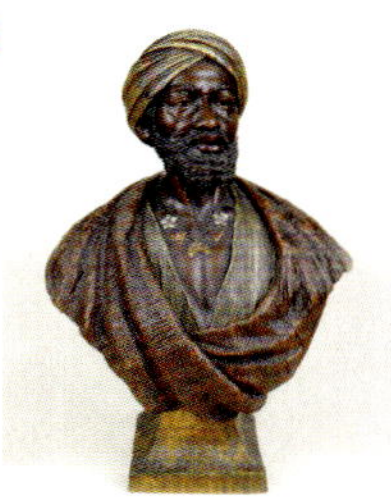

1061
BÜSTE „Shakespeare“
ca. 1897/98
1061 (Firmenmarke ohne Ang.), H. 69 cm; GSF, S. 6
Bem.: Dargestellt ist der Schauspieler, Bühnenautor und Dramatiker William Shakespeare (1564–1616).
-> vgl. 1144
Lit.: GSF, S. 6, hier mit der Ang. des Originaltitels „Shakespeare“.

1062
Rozel
BÜSTE „Schubert“
ca. 1897/98
1062 (Firmenmarke ohne Ang.), H. 68 cm; GSF, S. 6;
Bem.: Hier ohne Ang. der Künstlersign.
-> vgl. 1145; 1157
Bem.: Dargestellt ist der Komponist Franz Schubert (1797–1828), die Büste wurde vermutlich zum 100. Geburtstag des Komponisten aufgelegt.
Lit.: GSF, S. 6, hier mit der Ang. des Originaltitels „Schubert“.

1063
BÜSTE „Gutenberg“
ca. 1897/98
1063 (Firmenmarke ohne Ang.), H. 72 cm; GSF, S. 6
Bem.: Dargestellt ist der Erfinder des Buchdrucks Johannes Gensfleisch, genannt Gutenberg (ca. 1400–1468).
Lit.: GSF, S. 6, hier mit der Ang. des Originaltitels „Gutenberg“.

1064
BÜSTE „Napoleon I.“
ca. 1897/98
1064 (Firmenmarke ohne Ang.), H. 70 cm; GSF, S. 6
Bem.: Dargestellt ist franz. Kaiser Napoleon I. (1769–1821).
-> vgl. 1126; 1154
-> vgl. 1723; 1726= jew. WANDMASKE
Lit.: GSF, S. 6, hier mit der Ang. des Originaltitels "Napoleon I.".

1065
FIGUR „Gefesselte Amazone“ (auf Natursockel mit Baumstamm)
ca. 1897/98
32,1065,(7,2,7,2 vertikal) (Firmenmarke ohne Ang.), Repr. Res., H. 74 cm; Christie's SK 12.5.2000
1065 (Firmenmarke ohne Ang.), H. 73 cm; GSF, S. 6
Lit.: GSF, S. 6, hier mit der Ang. des Originaltitels „Gefesselte Amazone“; Pinhas, S. 71 (m. Abb.).

1066
Brenner, (Victor David?)
GRUPPE „Adam and Eva“ (nach dem Sündenfall auf Stein sitzend)
ca. 1897/98
1066,50 (Firmenmarke ohne Ang.), sign. Adam and Eve par Brenner (Metallschild), H. 75,5 cm; Doro 1.12.2000
ohne Ang., unleserlich sign., „MUSTER“ ger., H. 78 cm; Doro 20.4.1993
1066 (Firmenmarke ohne Ang.), H. 76 cm; GSF, S. 6
Lit.: GSF, S. 6, hier mit der Ang. des Originaltitels „Adam und Eva“.

1067
Lemoine, B.(?)
FIGUR „Pfeilschnitzer“/ „Prémiere Flèche“ (Sitzender Araberknabe mit Kopftuch, an einem Pfeil schnitzend) (auf getrepptem Steinsockel)
ca. 1898
1067,27 Firmenmarke 11, unsign., H. 76 cm; Wien Museum Inv. 186.625

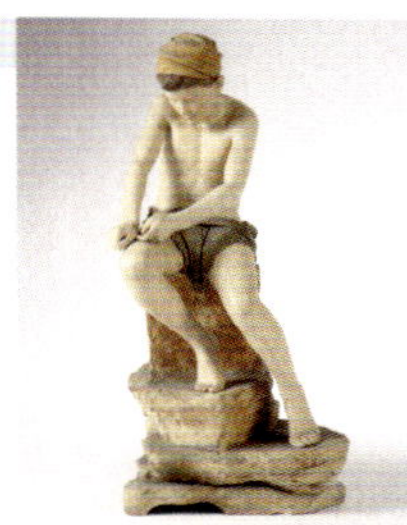

ohne Ang., sign. B. Lemoine, H. 78 cm; Richemond (1999), S. 103 (m. Abb.)
1067 (Firmenmarke ohne Ang.), H. 77 cm; GSF, S. 6
Bem.: Diese Statue findet sich unter den Nrn. 596/598 im Kat. GSB wieder. Sie wurde in der Pariser Produktion in einer Höhe von 32 cm aus Bronze (Vrai Bronze) und in „Kunstguss“ (Fonte d'Art, d.i. wohl Terrakotta, patiniert) in der H. 57 cm gefertigt.
-> vgl. 1818
-> vgl. GSB 596; 598
Lit.: GSF, S. 6, hier mit der Ang. des Originaltitels „Pfeilschnitzer“; GSB, Pl. 36, hier mit Ang. des Originaltitels „Prémiere Flèche“; Richemond (1999), S. 103 (m. Abb.).

1069
FIGUR „Pierrot“ (Stehender Pierrot mit erhobenem Zeigefinger) (auf Rundsockel)
ca. 1898
1069 (Firmenmarke ohne Ang.), H. 36 cm; Pinhas, S. 40 (m. Abb.)
1069 (Firmenmarke ohne Ang.), H. 38 cm; GSF, S. 6
Lit.: GSF, S. 6, hier mit der Ang. des Originaltitels „Pierrot“.

1070 -> vgl. 1828

1072
Pecheur / Simon
„Schaukel“ (Doppel-VASE Junge Frau, zwischen einer Astgabel schwingend)
ca. 1897/98
01072,44,27 Firmenmarke 7, Repr. Res., sign. „Simon“, H. 80 cm; PB Wien 19
-> vgl. 1041
Lit.: Pudor (1902), S. 36

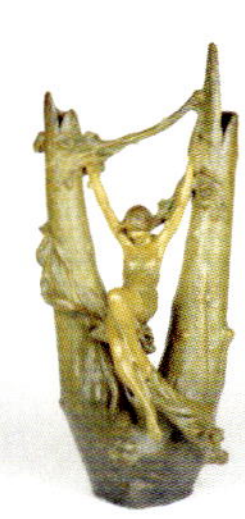

(m. Abb.), hier mit Ang. des Originaltitels „Schaukel“ und mit Ang. des Künstlers Pecheur.

1074
FIGUR „Napoleon I.“ (Stehend, mit verschränkten Armen) (auf Rechtecksockel)
ca. 1898
1074 (Firmenmarke ohne Ang.), H. 102 cm; Pinhas, S. 64 (m. Abb.)
1074 (Firmenmarke ohne Ang.), H. 102 cm; GSF, S. 6
Bem.: Dargestellt ist franz. Kaiser Napoleon I. Bonaparte (1769–1821).
-> vgl. 1198
Lit.: GSF, S. 6, hier mit der Ang. des Originaltitels „Napoleon I.“.

1075 -> ÜBERSCHNEIDUNG
Louger, J.
VASE „Musik“ (Damenhalbakt mit Lyra vollplastisch an der Vasenwandung sitzend, Totenmaske Beethovens an der Vasenmündung)
ca. 1897/98
01075 ohne Ang., GS Wien (Firmenmarke?), Ges. gesch., unsign.?, H. 84 cm; Schuler 92
-> vgl. 1171; 1172
Lit.: KM, 4.1904, S. 112 (m. Abb.), hier mit Ang. des Originaltitels „Musik“ und mit Ang. des Künstlers „J. Louger“, „Ein prächtiges Stück ist auch die Vase, an deren Fuß eine träumerische Gestalt sitzt. Das leichte Gewand umschließt den Körper von der Büste abwärts und schmiegt sich dem Unterkörper und Beinen dicht an. Während sich die Linke des Mädchens auf die Rücklehne der Bank stützt, auf welcher sie sitzt, greift die Rechte in die Saiten der an der anderen Seite der Bank ruhenden Leier, derselben harmonische Töne entlockend. Vielleicht entstammen sie einer Melodie, welche jener Künstler komponiert hat, dessen lorbeerumranktes Gesicht am oberen Rande der Vase angebracht ist“.

1075 -> ÜBERSCHNEIDUNG
BÜSTE „Epicuräer“
ca. 1897/98
1075,8,24 Firmenmarke 7, Repr. Res., unleserlich sign. A... R...? H. 57 cm; eBay # 2659317313
1075 (Firmenmarke ohne Ang.), H. 56 cm; GSF, S. 7
Lit.: GSF, S. 7, hier mit der Ang. des Originaltitels „Epicuräer“.

1077
FIGUR „Mönch mit Todtenkopf“ (Sitzender Mönch mit Totenkopf und Bücherstapel)
ca. 1897/98
1077,25 ger., Firmenmarke 8a, H. 34 cm; eBay # 1452317868
1077,39 Firmenmarke 8a, gepr. „P“= Z3, H. 34 cm; eBay # 6538749734
1077 (Firmenmarke ohne Ang.), H. 35 cm; GSF, S. 7
Lit.: GSF, S. 7, hier mit der Ang. des Originaltitels „Mönch mit Todtenkopf“.

1078
BÜSTE „Clown“ (Pierrot mit Zylinder, Zigarette und Spielkarte) (auf konischem Sockel)
ca. 1897/98
1078 (Firmenmarke ohne Ang.), H. 25 cm; Pinhas, S. 40, 67; 96 (m. Abb.)
1078 (Firmenmarke ohne Ang.), H. 23 cm; GSF, S. 7
Lit.: GSF, S. 7, hier mit der Ang. des Originaltitels „Clown“.

1079
BÜSTE Ägypterin (mit Kopfputz, Ureusschlange und Halskette) (auf Sockel mit Palmblättern)
ca. 1897/98
1079,12 FG (Firmenmarke?), Repr. Res., H. 40 cm; Doro 20.4.1993
1079,43 Firmenmarke 8a, Repr. Res., H. 40 cm; eBay # 2220503659

1080
FIGUR „Schreihals“
ca. 1897/98
1080 (Firmenmarke ohne Ang.), H. 26 cm; GSF, S. 7
Bem.: Die Modell-Nr. 1080 B könnte eine Variante, bzw. eine Ausführung in Bronze dieser Figur sein. Da sich aber keine Stücke nachweisen lassen, bleibt dies vorerst eine Vermutung.
-> vgl. 1080 B
Lit.: GSF, S. 7, hier mit der Ang. des Originaltitels „Schreihals“.

1080 B
SCHALE Kind
BRONZE
ca. 1897/98
1080 B (ohne Ang.), H. fehlt; Neuwirth (1974), S. 58;
Bem.: Hier keine nähere Beschreibung des Objektes.
Bem.: Die Modell-Nr. 1080 könnte eine Variante, bzw. eine Ausführung in Terrakotta dieser Schale sein. Da sich aber keine Stücke nachweisen lassen, bleibt dies vorerst eine Vermutung.
-> vgl. 1080

1082
RELIEF Holländerin, ein Tablett mit Fischen haltend (im Hintergrund Nordseelandschaft mit Windmühlen) (Wandrelief als Mosaikimitation)
ca. 1897/98
01082,18,27 Firmenmarke 7, Repr. Res., Breveté S.G.D.G. DRGM, H. 55,9 cm, L. 101,6 cm; eBay # 514709580 (KH: www.tias.com/stores/frozenbeauty)

1085
Klimt, Gustav (nach)
RELIEF Göttin Juno (Allegorie des Juni) (Wandrelief als Mosaikimitation)
ca. 1897/98
01085,47,69 GS Wien (Firmenmarke?), Ges. gesch., bezeichnet: nach Klimt; Breveté S.G.D.G., Patent ang. D.R.G.M., H. 68 cm, B. 55 cm; GK, S. 11, 33 (m. Abb.); PB ohne Ang., GS Wien (Firmenmarke?), Ges. gesch., bezeichnet: nach Klimt; Breveté S.G.D.G., Patent ang., H. 68 cm, B. 55 cm; Doro 12.12.1991
Bem.: Die Vorlage zu diesem Wandrelief bildet das Blatt „Junius" aus der Mappe „Allegorien, Neue Folge" Wien, Gerlach & Schenk (ca. 1896).
Ausst.: HMW (jetzt Wien Museum) 1985/86.
Lit.: GK, S. 11, 33 (m. Abb.)

1087
BÜSTE Adelige mit Kopftuch und Halsschmuck (auf reliefiertem mit Blumen- und Vogelmotiven geschmücktem konischen Sockel)
ca. 1897/98
1087 (Firmenmarke ohne Ang.), H. 44 cm; Pinhas, S. 35 (m. Abb.)

1088
BÜSTE Junges Mädchen (nach rechts blickend) mit Kopftuch/Schal
ca. 1897/98
1088 (Firmenmarke ohne Ang.), H. fehlt; Pinhas, S. 78 (m. Abb.)

1093
BÜSTE „Holländerin"
ca. 1897/98
1093 (Firmenmarke ohne Ang.), H. 66 cm; GSF, S. 7
-> vgl. 1402
Lit.: GSF, S. 7, hier mit der Ang. des Originaltitels „Holländerin".

1096
Jacques, E.
GRUPPE „Römerin im Bade" (Nackte Schönheit mit Kopfputz, die von ihrem Sklaven nach dem Bade abgetrocknet wird) (auf getrepptem Sockel)
ca. 1897/98
1096 (Firmenmarke ohne Ang.), H. 89 cm; Pinhas, S. 27, 70 (m. Abb.)
1096 (Firmenmarke ohne Ang.), H. 88 cm; GSF, S. 7
-> vgl. 1579
Lit.: GSF, S. 7, hier mit der Ang. des Originaltitels „Römerin im Bade".

1097 -> ÜBERSCHNEIDUNG
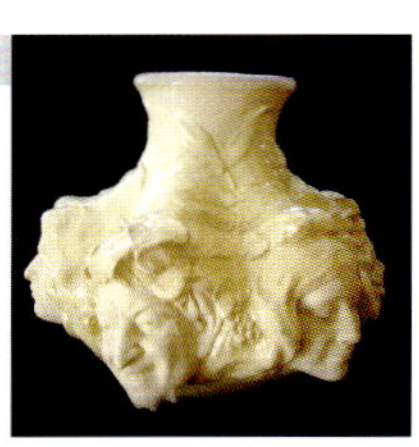
Larroux, (Antonin)
VASE „Masques"
(Sechs plastische Gesichter: Nymphen/Faun)
ca. 1897/98
01097,1,10 FG Wien (Firmenmarke?), Repr. Res., sign. Laroux (!), H. 36 cm, Ø 43 cm; KH
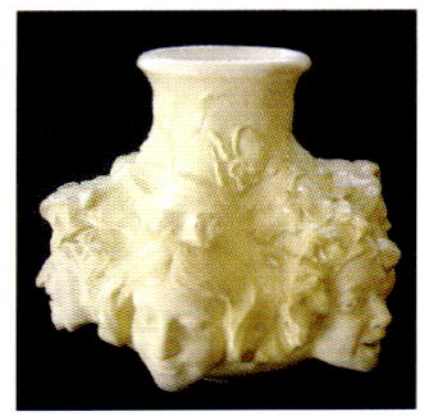
F1097,3,6 Firmenmarke 9, unsign., H. 38 cm, B. 40 cm, Mündungs-Ø 17 cm; Sammlung f@goldscheider.de
Ausst.: Turin 1902; Paris Salon 1902.
-> vgl. 933; WANDMASKEN 2287; 2288; 2289; 2290; 2291
Lit.: Gmelin, L.: Die 1. internat. Ausst. für moderne dekorative Kunst in Turin 1902, in: KuHW, 52.1901/02, S. 325–342 (S. 334, hier als Künstler „Larone" angegeben); L´Art décoratif aux Expositions des Beaux-Arts 1902, Vol. 1, Paris (1902) Pl. 95; Vol. 2, Pl. 200.

1097 -> ÜBERSCHNEIDUNG
BÜSTE oder FIGUR „Zigeunerknabe"
-> vgl. 1735

1098
Prémont, F. du
BÜSTE Junge Frau mit Spitzenhaube
ca. 1897
1098,24 Firmenmarke 8a, unsign., H. 68 cm; GK, S. 11, 64 (m. Abb.); Wien Museum Inv. 182.063
1098,24,47 Firmenmarke 7, Repr. Res., H. ca. 68 cm; KH ohne Ang., Firmenmarke 11?, unsign.?, H. 66 cm; Christie's SK 7.2.2001
Ausst.: HMW (jetzt Wien Museum) 1985/86.
-> vgl. 1402
Lit.: GK, S. 11, 64 (m. Abb.); Pinhas, S. 97 (m. Abb.).

1101
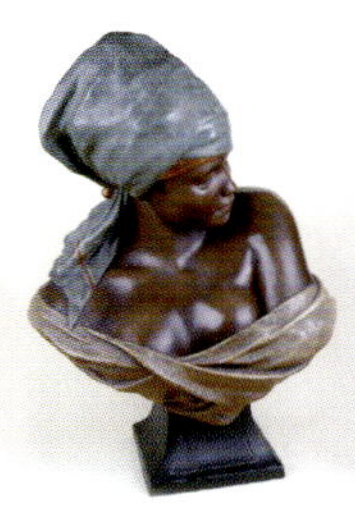
Lévèque, Auguste Maurice / Thiele, (P.?)
BÜSTE „Sklavin"/„Bedouine" (Sklavin, Araberin mit großem Turban) (nach links blickend)
ca. 1897
1101,(8,1,4,8,10 vertikal) Firmenmarke 11, Repr. Res., sign. Bedouine par Lévèque (Metallschild), H. 66 cm; GK, S. 11, 65 (m. Abb.); Wien Museum Inv. 223.917
1101,253,9 ger. IV, Firmenmarke 7, Repr. Res., sign. Thiele, H. 67 cm; PB Wien 19
1101,92,53 gem. IV, innen gepr. IV., Firmenmarke 7, (unsign.?), H. 66 cm; KH: www.trocadero.com/eclectic/items/9585
1101 (Firmenmarke ohne Ang.), H. 60 cm; GSF, S. 7
Ausst.: HMW (jetzt Wien Museum) 1985/86.
-> vgl. (H. 66 cm) (sign. „P. Thiele") ausgeführt in MARMOR; Soth. L. 9.10.1992 und 9.12.1993
-> vgl. 1338; 1877
-> vgl. 684 (GSB); 733 (GSB)
-> vgl. 1616= WANDMASKE
Pendant -> vgl. 1105
Lit.: GSF, S. 7, hier mit der Ang. des Originaltitels „Sklavin"; GSB, Pl. 59, hier ist die BÜSTE mit dem Originaltitel „Esclave" unter Nr. 684= No.1 (H. 62 cm) und unter Nr. 733= No. 2 (H. 30 cm) abgebildet. Ang. zum Material (Terrakotta/Bronze fehlen); Richemond (1999), S. 78 (m. Abb.); GK, S. 11, 65 (m. Abb.); Pinhas, S. 67; 68 (m. Abb.).

1102
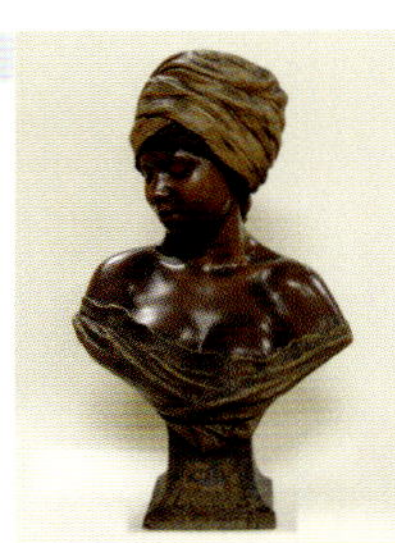
Prémont, F. de
BÜSTE „Mohrin"/„Négresse au Papillon" (Afrikanerin / Mulattin mit Turban)
ca. 1897
1102,88,(1,8,1,2 vertikal) Firmenmarke 11, sign. F. de Prémont (Metallschild), H. ca. 69 cm; KH
1102,115,52 gem. weiß IV, Firmenmarke 7, Repr. Res., unsign., H. 69,5 cm; Doro 15.3.2005
1102,(5,0,7 vertikal),1,53 Firmenmarke 7, Repr. Res., gepr. „P"= Z3, unsign., H. 67,7 cm; MAK Wien, Inv. Ke 10521-8
1102 (1162? lt. AH),21,(7,1,3,8 vertikal), Firmenmarke 11?, sign. T. de Prémont (Metallschild), H. 69 cm; Millon & Robert 14.11.1995
1102 (1002?, lt. AH) (Firmenmarke ohne Ang.), sign. T. du Prémont, H. 72 cm; Soth. P. 18.11.2003
1102 (Firmenmarke ohne Ang.), H. 66 cm; GSF, S. 7
-> vgl. 1878
-> vgl. 1341; 1613= WANDMASKE
Lit.: GSF, S. 7, hier mit der Ang. des Originaltitels „Mohrin"; Richemond (1999), S. 103 (m. Abb.).

1105
Lévèque, Auguste Maurice
BÜSTE „Araber" (mit Turban) (nach rechts blickend)
ca. 1897
1105,(6,6,7,4 vertikal?),58 Firmenmarke 8a?, unsign., H. 63,5 cm; Christie's SK 18.2.2004
1105 (Firmenmarke ohne Ang.), H. 65 cm; GSF, S. 7
-> vgl. 1339; 1899
-> vgl. 1612= WANDMASKE
Pendant -> vgl. 1101
Lit.: GSF, S. 7, hier mit der Ang. des Originaltitels „Araber"; Richemond (1999), S. 78 (m. Abb.); Pinhas, S. 67, 69 (m. Abb.).

1106
FIGUR „Falstaff" (auf Rechtecksockel)
ca. 1897
1106 (Firmenmarke ohne Ang.), H. 38 cm; Pinhas, S. 34 (m. Abb.)
1106 (Firmenmarke ohne Ang.), H. 40 cm; GSF, S. 7
Bem.: Die Figur des Falstaff geht auf William Shakespeares Bühnenstück „Heinrich IV" zurück.
Lit.: GSF, S. 7, hier mit der Ang. des Originaltitels „Falstaff".

1109
Thiele, (P.?)
BÜSTE „Bulgarin" (mit langem Kopftuch) (nach links blickend)
ca. 1897
1109,79,8 FG (Firmenmarke?), sign. „Rohely"(?), H. 66 cm; WKA 2
1109,51 Firmenmarke 7, Repr. Res., unsign., gepr. „P"= Z3, H. 65 cm; Nagel 38. Varia
1109 (Firmenmarke ohne Ang.), H. 66 cm; GSF, S. 7
-> vgl. 1340; 1615= WANDMASKE
-> vgl. 1409 (sign. Rochella)
-> vgl. 1898 (sign. Thiele)
Lit.: GSF, S. 7, hier mit der Ang. des Originaltitels „Bulgarin".

1111
Stehende Nordafrikanerin mit Wasserkrug
ca. 1897
1111,(6,8,4,6 vertikal?) FG Wien (Firmenmarke?), Repr. Res., gepr. „P"= Z3, H. 59,5 cm; Doro 20.4.1993

1114
Petri (d.i. Petrides, János) (Johann)
VASE Holländisches Mädchen, ihre Schürze hochhebend
ca. 1897
01114,41,69 Firmenmarke 4, Ges. gesch., sign. Petri, H. 57,3 cm; Wien Museum Inv. 190.059
Pendant -> vgl. 1115
Lit.: GSN, S. 14, 15 (m. Abb.)
GS neu: Modell-Nr. 12 „Anmut"

1115
Petri (d.i. Petrides, János) (Johann)
VASE Holländischer (Fischer-) Knabe mit Korb
ca. 1897
01115,32,69 gem. weiß VI, Firmenmarke 4, Ges. gesch., sign. Petri, H. 61 cm; Weidler 520; Bergmann 67
Pendant -> vgl. 1114

1117 -> ÜBERSCHNEIDUNG
BÜSTE „Incroyable" (Lachender Mann mit Zweispitz)
ca. 1897/98
1117 (Firmenmarke ohne Ang.), H. 47 cm; GSF, S. 7
-> vgl. 1939= WANDMASKE
Lit.: GSF, S. 7, hier mit der Ang. des Originaltitels „Incroyable".

1117 -> ÜBERSCHNEIDUNG
Bläsche
BÜSTE Mädchen mit Lorbeerkranz (Sockel reliefiert mit Sonne und Baum)
ca. 1897/98
1117,3,212 (Firmenmarke?), H. 52,5 cm; Doro 24.10.1995
Bem.: Trotz der doppelten Nummernbelegung vermutlich ebenfalls ein Erzeugnis von GS (Zuschreibung unter Vorbehalt, jedoch sprechen, trotz der fehlenden Firmenmarke, Qualität und Nummerierung für GS). Möglicherweise ist aber auch die vom Dorotheum angegebene Mod.-Nr. falsch, bzw. ein Druckfehler im Kat.

1118
FIGUR „Odaliske"
ca. 1897/98
1118 (Firmenmarke ohne Ang.), H. 87 cm; GSF, S. 7
-> vgl. 1414
Lit.: GSF, S. 7, hier mit der Ang. des Originaltitels „Odaliske".

1119
Wind, Josef (zugeschr.)
FIGUR „Gefesselte Sklavin" (Stehende, an einen Baumstamm mit Handfesseln gebundene nackte Araberin mit Turban)
ca. 1897/98
1119 ohne Ang., Firmenmarke 8a, unsign., H. 76 cm; Z 66
1119 (Firmenmarke ohne Ang.), H. 76 cm; GSF, S. 7
Bem.: Bildhauerische Arbeit, Figur und Turban sprechen eindeutig für einen Entwurf von Josef Wind.
-> vgl. 1393
Lit.: GSF, S. 7, hier mit der Ang. des Originaltitels „Gefesselte Sklavin".

1121
FIGUR „Stoffverkäuferin" (Kniender Halbakt einer jungen Nordafrikanerin) (auf rechteckigem Landschaftssockel)
ca. 1897/98
1121 (Firmenmarke ohne Ang.), H. fehlt; Pinhas, S. 66 (m. Abb.)
1121 (Firmenmarke ohne Ang.), H. 47 cm; GSF, S. 7
Lit.: GSF, S. 7, hier mit der Ang. des Originaltitels „Stoffverkäuferin".

1123
BÜSTE „Königin Louise"
ca. 1897/98
1123 (Firmenmarke ohne Ang.), H. 83 cm; GSF, S. 7
Bem.: Dargestellt ist die preuss. Königin Luise, Herzogin zu Mecklenburg (1776–1810).
-> vgl. 1229; 1794
-> vgl. 1843= WANDMASKE
Lit.: GSF, S. 7, hier mit der Ang. des Originaltitels „Königin Louise".

1124 -> ÜBERSCHNEIDUNG
VASE mit Blumen und drei Handhaben
ca. 1897/98
F01124,6,7 Firmenmarke 9, Repr. Res., H. 17,9 cm; Wien Museum Inv. 220.778
01124,10,45 gepr. H, Firmenmarke 9, Repr. Res., H. 17,9 cm; Wien Museum Inv. 220.779

1124 -> ÜBERSCHNEIDUNG
FIGUR „Schöne Helena"
ca. 1897/98
1124 (Firmenmarke ohne Ang.), H. 71 cm; GSF, S. 7
-> vgl. 1003
Lit.: GSF, S. 7, hier mit der Ang. des Originaltitels „Schöne Helena".

1125
FIGUR „Phyrne"
ca. 1897/98
1125 (Firmenmarke ohne Ang.), H. 72 cm; GSF, S. 7
Bem.: s. 978
-> vgl. 978; 979
Lit.: GSF, S. 7, hier mit der Ang. des Originaltitels „Phyrne".

1126
BÜSTE „Napoleon (mittel)"
ca. 1897/98
1126 (Firmenmarke ohne Ang.), H. 54 cm; GSF, S. 7
Bem.: Dargestellt ist franz. Kaiser Napoleon I. (1769–1821).
-> vgl. 1064; 1154
-> vgl. 1723; 1726= jew. WANDMASKE
Lit.: GSF, S. 7, hier mit der Ang. des Originaltitels „Napoleon (mittel)".

1127
FIGUR „Mohr" (Sitzender junger Araber mit Flöte)
ca. 1897/98
1127 (Firmenmarke ohne Ang.), H. 110 cm; Pinhas, S. 64 (m. Abb.)
1127 (Firmenmarke ohne Ang.), H. 110 cm; GSF, S. 7
-> vgl. 1366; 1668
Lit.: GSF, S. 7, hier mit der Ang. des Originaltitels „Mohr".

1134
GRUPPE „Meerkönigin"
ca. 1897/98
1134 (Firmenmarke ohne Ang.), H. 45 cm; GSF, S. 7
Lit.: GSF, S. 7, hier mit der Ang. des Originaltitels „Meerkönigin".

1136
FIGUR „Fischermädchen"
ca. 1897/98
1136 (Firmenmarke ohne Ang.), H. 74 cm; GSF, S. 7
Lit.: GSF, S. 7, hier mit der Ang. des Originaltitels „Fischermädchen".

1137
FIGUR „Fischerknabe" (Stehender Junge mit Fisch in der linken Hand) (auf Natursockel)
ca. 1897/98
1137,27,31 Firmenmarke 7, H. 75 cm; Wien Museum Inv. 186.603
1137 (ohne Ang.) FG Wien (Firmenmarke?), Repr. Res., H. 73,5 cm; Nagel 369; 371; Bem.: Das AH gibt als Modell-Nr. 1136 an.
1137 (Firmenmarke ohne Ang.), H. 76 cm; GSF, S. 7
-> vgl. 1387
Lit.: GSF, S. 7, hier mit der Ang. des Originaltitels „Fischerknabe".

1138
FIGUR „Krugträgerin"
ca. 1897/98
1138 (Firmenmarke ohne Ang.), H. 76 cm; GSF, S. 7
-> vgl. 860
Lit.: GSF, S. 7, hier mit der Ang. des (von der Mod.-Nr. 860 abweichenden) Originaltitels „Krugträgerin".

1139a
RELIEFKOPF „Bacchant"
ca. 1897/98
1139a (Firmenmarke ohne Ang.), H. 47 cm; GSF, S. 7
Pendant -> vgl. 1139b
Lit.: GSF, S. 7, hier mit der Ang. des Originaltitels „Bacchant".

1139b
RELIEFKOPF „Bacchant"
ca. 1897/98
1139b (Firmenmarke ohne Ang.), H. 47 cm; GSF, S. 7
Pendant -> vgl. 1139a
Lit.: GSF, S. 7, hier mit der Ang. des Originaltitels „Bacchant".

1140
Posierende junge Dame mit Buch (auf Landschaftssockel)
ca. 1897/98
1140 (Firmenmarke ohne Ang.), H. fehlt; Pinhas, S. 71 (m. Abb.)

1142
Strasser, Arthur
FIGUR „Haruspex" (Stehender römischer Wahrsager) (auf Rechtecksockel)
Inschrift: CAES. HARUSPEX
ca. 1897/98
1142 (Firmenmarke ohne Ang.), unsign., H. ca. 100 cm; Pinhas, S. 67 (m. Abb.)
1142 (Firmenmarke ohne Ang.), H. 100 cm; GSF, S. 7
Bem.: Dargestellt ist die Figur eines römischen (bzw. etruskischen) Wahrsagers, der aus den Eingeweiden von Opfertieren weissagte. Die Zuordnung an Arthur Strasser konnte durch eine sign. Figur (H. 98 cm), welche im Doro Wien (637. Kunstauktion, 14.–20.9.1982) angeboten wurde, vorgenommen werden. Ob es sich bei diesem Ex. um ein Erzeugnis von GS handelte oder um eine Originalausformung Strassers, konnte nicht mehr nachvollzogen werden.
-> vgl. 1262
Lit.: GSF, S. 7, hier mit der Ang. des Originaltitels „Haruspex".

1143
FIGUR „Rubens"
ca. 1897/98
1143 (Firmenmarke ohne Ang.), H. 98 cm; GSF, S. 7
Bem.: Dargestellt ist der flämische Maler Peter Paul Rubens (1577–1640).
-> vgl. 1261
Lit.: GSF, S. 7, hier mit der Ang. des Originaltitels „Rubens".

1144
BÜSTE „Shakespeare (klein)"
ca. 1897/98
1144 (Firmenmarke ohne Ang.), H. 98 cm; GSF, S. 7
Bem.: Dargestellt ist der Schauspieler, Bühnenautor und Dramatiker William Shakespeare (1564–1616).
-> vgl. 1061
Lit.: GSF, S. 7, hier mit der Ang. des Originaltitels „Shakespeare (klein)".

1145
Rozel
BÜSTE „Schubert (klein)"
ca. 1897/98
1145,JR,5 Plak. E, Ausführung 20er Jahre, H. 26,1 cm; PB Wien 19
1145 (Firmenmarke ohne Ang.), H. 25 cm; GSF, S. 7
Bem.: Dargestellt ist der Komponist Franz Schubert (1797–1828), die Büste wurde vermutlich zum 100. Geburtstag des Komponisten aufgelegt.
-> vgl. 1062; 1157
Lit.: GSF, S. 7, hier mit der Ang. des Originaltitels „Schubert (klein)".

1150
Haniroff
FIGUR „Fanchon"
ca. 1897/98
1150 (Firmenmarke ohne Ang.), H. 100 cm; GSF, S. 7
-> vgl. 2325= BÜSTE
Lit.: GSF, S. 7, hier mit der Ang. des Originaltitels „Fanchon".

1154
BÜSTE „Napoleon I. (mittel)"
ca. 1897/98
1154 (Firmenmarke ohne Ang.), H. 39 cm; GSF, S. 7
-> vgl. 1064; 1126
-> vgl. 1723; 1726= jew. WANDMASKE
Lit.: GSF, S. 7, hier mit der Ang. des Originaltitels „Napoleon (mittel)".

1156
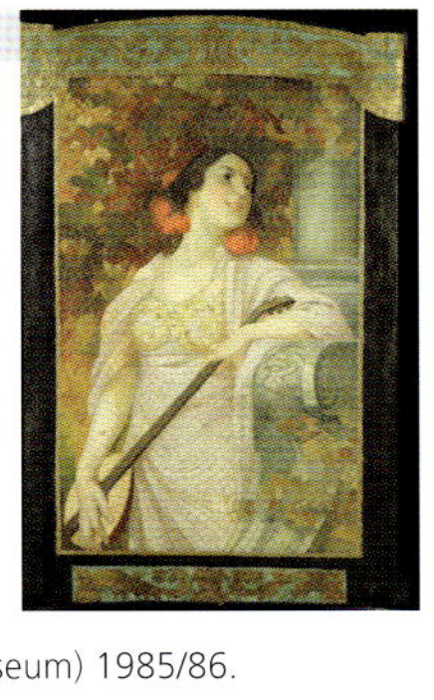
Schram, Alois Hans
RELIEF Lautenspielerin (Wandrelief als Mosaikimitation)
ca. 1897/98
AHS,01156,12,69 Firmenmarke 4, Ges. gesch., Breveté S.G.D.G., Patent ang., sign. lig. AHS, H. 85,5 cm, B. 59 cm; GK, S. 11, 34 (m. Abb.); Wien Museum Inv. 223.902
Ausst.: HMW (jetzt Wien Museum) 1985/86.
Lit.: GK, S. 11, 34 (m. Abb.)

1157
Rozel
BÜSTE „Schubert"
ca. 1897/98
1157 (Firmenmarke ohne Ang.), unsign., H. 49 cm; Doro 24.10.1995
1157 (Firmenmarke ohne Ang.), H. 48 cm; GSF, S. 7; Bem.: Hier ohne Ang. der Künstlersign.
Bem.: Dargestellt ist der Komponist Franz Schubert (1797–1828), die Büste wurde vermutlich zum 100. Geburtstag des Komponisten aufgelegt.
-> vgl. 1062; 1145
Lit.: GSF, S. 7, hier mit der Ang. des Originaltitels „Schubert".

1159
FIGUR „Phyrne (bekleidet)"
ca. 1897/98
1159 (Firmenmarke ohne Ang.), H. 74 cm; GSF, S. 7
Bem.: s. 978
-> vgl. 978; 979
Lit.: GSF, S. 7, hier mit der Ang. des Originaltitels „Phyrne (bekleidet)".

1162
BÜSTE mit eigenartiger Bemalung
ca. 1895/96
1162 ohne Ang.; KeramRS, 4.1896, S. 863
Bem.: s. Modell-Nr. 6

1170
FIGUR „Shocking"
ca. 1897/98
1170 (Firmenmarke ohne Ang.), H. 69 cm; GSF, S. 7
-> vgl. 2275
Lit.: GSF, S. 7, hier mit der Ang. des Originaltitels „Shocking".

1171
Louger, J.
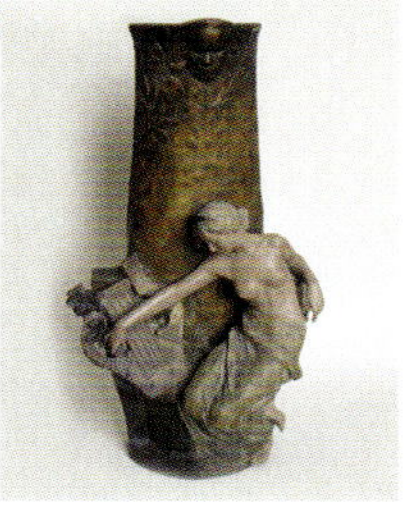
VASE „Musik" (Damenhalbakt mit Lyra vollplastisch an der Vasenwandung sitzend, Totenmaske Beethovens an der Vasenmündung)
ca. 1897/98
01171,65,17 Firmenmarke Sculpture 13, unsign.?, H. 63,5 cm; eBay # 1312431123
-> vgl. 1075; 1172
Lit.: KM, 4.1904, S. 112 (m. Abb.), hier mit Ang. des Originaltitels „Musik" und mit Ang. des Künstlers „J. Louger".

1172 -> ÜBERSCHNEIDUNG
Louger, J.
VASE „Musik" (Damenhalbakt mit Lyra vollplastisch an der Vasenwandung sitzend, Totenmaske Beethovens an der Vasenmündung)
ca. 1897/98
01172,13,27 (Firmenmarke ohne Ang.), unsign.?, H. 47 cm; Henry's 31.10.2003
-> vgl. 1075; 1171
Lit.: KM, 4.1904, S. 112 (m. Abb.), hier mit Ang. des Originaltitels „Musik" und mit Ang. des Künstlers „J. Louger".

1172 -> ÜBERSCHNEIDUNG
FIGUR „Niederländer" (Stehender bärtiger Mann mit Hut) (auf Landschaftssockel)
ca. 1897/98
1172,20 Firmenmarke 8a, H. 98,3 cm; Wien Museum Inv. 219.140
1172 (Firmenmarke ohne Ang.), H. 100 cm; GSF, S. 7
Lit.: GSF, S. 7, hier mit der Ang. des Originaltitels „Niederländer".

1173 -> ÜBERSCHNEIDUNG
Lefont
VASE „Der Handkuss" (Amphore mit zwei Henkeln, reliefierte Wandung mit Dekor: Handkuss und Puttenreigen)
ca. 1897/98
01173,2,17 Firmenmarke 7, Repr. Res., gepr. „C"= Z4, H. 75,5 cm; Wien Museum Inv. 219.166

1173 -> ÜBERSCHNEIDUNG
FIGUR „Kosake"
ca. 1897/98
1173 (Firmenmarke ohne Ang.), H. 105 cm; GSF, S. 7
Lit.: GSF, S. 7, hier mit der Ang. des Originaltitels „Kosake".

1174
Strasser, Arthur
FIGUR „Wasserträger" (Stehender Nordafrikaner mit großem Wasserschlauch/tierischer Wasserbehälter) (auf Natursockel)
ca. 1897/98
ohne Ang., H. 98 cm; Richemond (1995), S. 227
1174 (Firmenmarke ohne Ang.), H. 94 cm; GSF, S. 7
Ausst.: ÖMKI 1898/99, hier mit Ang. des Originaltitels „Wasserträger".
-> vgl. 1600; 608-GSB; 609-GSB
Lit.: GSF, S. 7, hier mit der Ang. des Originaltitels „Wasserträger".

1175
Strasser, Arthur
FIGUR „Wasserverkäufer"
ca. 1897/98
1175 (Firmenmarke ohne Ang.), H. 100 cm; GSF, S. 8
Ausst.: ÖMKI 1898/99, hier mit Ang. des Originaltitels „Wasserverkäufer"
-> vgl. 1601
Lit.: GSF, S. 8, hier mit der Ang. des Originaltitels „Wasserverkäufer".

1176 -> ÜBERSCHNEIDUNG
Capèque (d.i. Czapek, Stanislaus)
VASE „Die Quelle" (aus einer Quelle, als wasserspeiendes Gesicht an der Vasenmündung, trinkendes Figurenpaar)
ca. 1897/98
01176,15,35 (Firmenmarke ohne Ang.), undeutl. sign., H. 70 cm; Christie's SK 18.9.1998
1176,21,2 GS Wien (Firmenmarke?), sign. Capèque, H. 71,1 cm; G 68 Auctions Toronto 11.9.2000

1176 -> ÜBERSCHNEIDUNG
Strasser, Arthur
FIGUR „Afrikanischer Sprecher"
ca. 1897/98
1176 (Firmenmarke ohne Ang.), H. 94 cm; GSF, S. 8
Ausst.: ÖMKI 1899/1900, hier mit Ang. des Originaltitels „Afrikanischer Sprecher".
Lit.: GSF, S. 8, hier mit der Ang. des Originaltitels „Afrikanischer Sprecher"; Neuwirth (1974), S. 74.

1177
Blumen überreichende junge Dame (Halbakt) mit Blumenkorb (auf Landschaftssockel)
ca. 1897/98
1177 (Firmenmarke ohne Ang.), H. fehlt; Pinhas, S. 71 (m. Abb.)

1178
BÜSTE „Mohr"
ca. 1897/98
1178 (Firmenmarke ohne Ang.), H. 50 cm; GSF, S. 8
Lit.: GSF, S. 8, hier mit der Ang. des Originaltitels „Mohr".

1179
GRUPPE Alter Araber mit Odaliske
ca. 1897
1179,27 Firmenmarke 8a, Repr. Res., gepr. „P"= Z3, H. 41,9 cm; MAK Wien, Inv. Ke 10602-14
1179,80 Firmenmarke 8a, Repr. Res., gepr. „P"= Z3, H. 44 cm; PB Wien 19

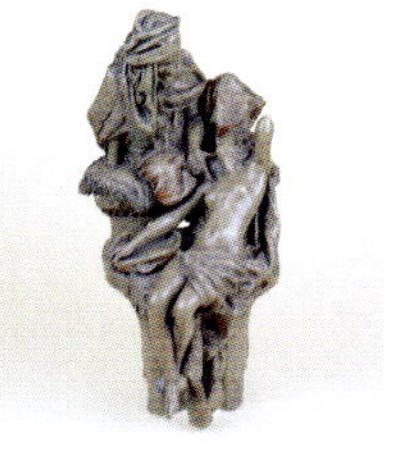

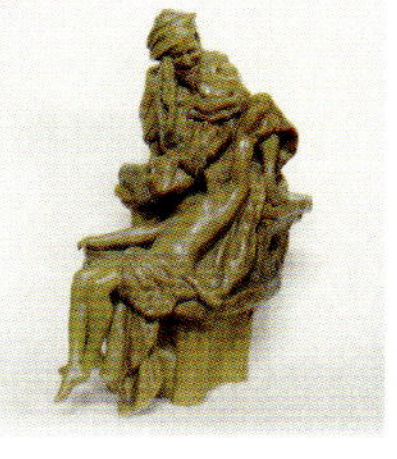

1180
Mutter mit Kind, das sie auf dem linken Arm trägt (in antikisierendem Gewand)
ca. 1897/98
1180,20,(1,2,8,0 vertikal), Firmenmarke 11?, Repr. Res., H. 121 cm; Schloss Rick. 105

1182
VASE Jeanne d'Arcs Ritterschlag (durch Blanche de Castille?)
ca. 1897/98
01182 (Firmenmarke ohne Ang.), H. 69 cm; Pinhas, S. 49 (m. Abb.)

1183?
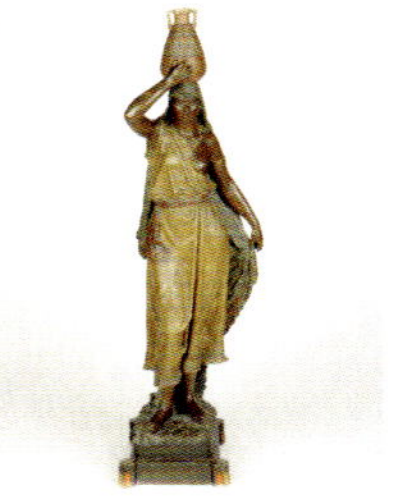
FIGUR Krugträgerin (Araberin, einen länglichen Krug auf dem Kopf tragend)
ca. 1897
1183? Firmenmarke 8a, H. 87 cm; PB Wien 19
-> vgl. 875; .... (H. 108 cm); Richemond (1999), S. 75 (m. Abb.)
Lit.: Richemond (1999), S. 75 (m. Abb.) Krugvariante und für die H. 108 cm (bezeichnet in arab. Schriftzeichen, „Sakka-ha", Metallschild).

1187
GRUPPE „Pièta"
ca. 1897/98
1187 (Firmenmarke ohne Ang.), H. 117 cm; GSF, S. 8
Lit.: GSF, S. 8, hier mit der Ang. des Originaltitels „Pièta".

1189
GRUPPE „Liebesgeflüster"
ca. 1897/98
1189 (Firmenmarke ohne Ang.), H. 117 cm; GSF, S. 8
Lit.: GSF, S. 8, hier mit der Ang. des Originaltitels „Liebesgeflüster".

1190
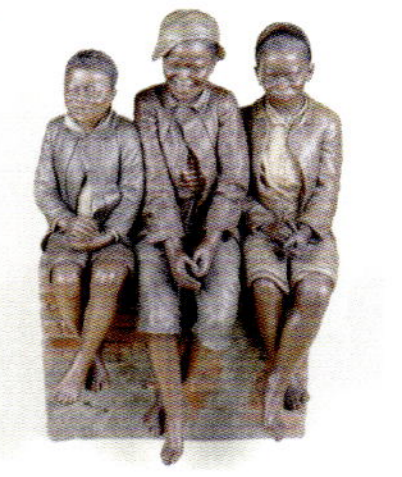
GRUPPE „Beim Photographen" (Drei lachende Negerjungen) (auf einer Mauer sitzend)
ca. 1897
1190,26 Firmenmarke 8a, gepr. „P"= Z3, H. 60 cm; PB Wien 19
1190,34 Firmenmarke 8a, H. 60 cm; PB Wien 19
1190 (Firmenmarke ohne Ang.), H. 58 cm; GSF, S. 8
-> vgl. 1224= linke EinzelFIGUR
-> vgl. 1236 und 1225= jeweils mittlere EinzelFIGUR
-> vgl. 1617= WANDMASKE Gesicht der mittleren FIGUR
-> vgl. 1695= BÜSTE (Einzelfigur)
Lit.: GSF, S. 8, hier mit der Ang. des Originaltitels „Beim Photographen".

1191
FIGUR „Mohrin"
ca. 1897/98
1191 (Firmenmarke ohne Ang.), H. 95 cm; GSF, S. 8
-> vgl. 1356; 1573
Pendant -> vgl. 1196
Lit.: GSF, S. 8, hier mit der Ang. des Originaltitels „Mohrin".

1192
BÜSTE „Hassan"
ca. 1897/98
1192 (Firmenmarke ohne Ang.), H. 43 cm; GSF, S. 8
Lit.: GSF, S. 8, hier mit der Ang. des Originaltitels „Hassan".

1193
Strasser, Arthur
BÜSTE „Berber"
ca. 1897/98
1193 (Firmenmarke ohne Ang.), H. 38 cm; GSF, S. 8
Ausst.: ÖMKI 1899/1900, hier mit Ang. des Originaltitels „Berber".
Lit.: GSF, S. 8, hier mit der Ang. des Originaltitels „Berber"; Neuwirth (1974), S. 74.

1194 -> ÜBERSCHNEIDUNG
VASE Landschaft mit Bäumen
ca. 1897
01194,19,37 Firmenmarke 6, Firmenmarke 4, Ges. gesch., H. 31,5 cm; eBay # 6585237585

1194 -> ÜBERSCHNEIDUNG
FIGUR „Knabe (rauchend)"
ca. 1897/98
1194 (Firmenmarke ohne Ang.), H. 76 cm; GSF, S. 8
Lit.: GSF, S. 8, hier mit der Ang. des Originaltitels „Knabe (rauchend)".

1195 -> ÜBERSCHNEIDUNG
POKAL / VASE (oder Deckelvase) mit Medaillon „Napoleon" (auf der Schauseite)
ca. 1896/97
01195,3,37 Firmenmarke 6, Ges. gesch., H. 30 cm; WKH
F1195,22,50 FG Wien (Firmenmarke?), Repr. Res., H. 30,5 cm; Mehlis 21
Bem.: Dargestellt ist der franz. Kaiser Napoleon I. Bonaparte (1769–1821), dieses Objekt wurde vermutlich zum 75. Todestag des Monarchen aufgelegt. Möglicherweise handelte es sich ursprünglich um eine Deckelvase, deren Deckel dann bei diesen beiden, hier einzig nachgewiesenen Exemplaren fehlen würde.

1195 -> ÜBERSCHNEIDUNG
FIGUR „Knabe, mit zerbr(ochenem) Krug" (Stehender Bub mit hochgekrempelter Hose) (auf Landschaftssockel)
ca. 1897/98
1195 (Firmenmarke ohne Ang.), H. 76 cm; Pinhas, S. 36 (m. Abb.)
1195 (Firmenmarke ohne Ang.), H. 76 cm; GSF, S. 8
-> vgl. 1400
Lit.: GSF, S. 8, hier mit der Ang. des Originaltitels „Knabe, mit zerbr(ochenem) Krug".

1196
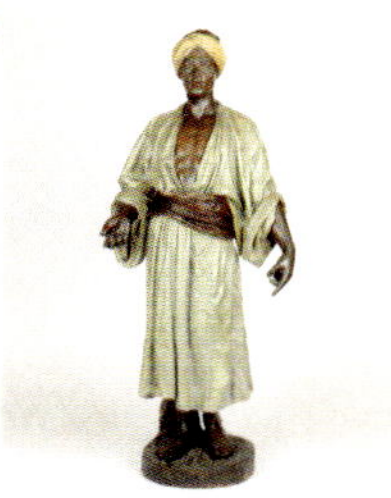
FIGUR „Mohr (bekleidet)"
ca. 1897/98
1196 (Firmenmarke ohne Ang.), H. 96 cm; Pinhas, S. 25 (m. Abb.)
1196 ohne Ang., H. 96 cm; PB Wien 19
1196 (Firmenmarke ohne Ang.), H. 93 cm; GSF, S. 8
-> vgl. 1355; 1572
Pendant -> vgl. 1191
Lit.: GSF, S. 8, hier mit der Ang. des Originaltitels „Mohr (bekleidet)".

1198
FIGUR „Napoleon I. (mittel)" (Stehend, mit verschränkten Armen) (auf Rechtecksockel)
ca. 1898
1198 (Firmenmarke ohne Ang.), H. 71 cm; Pinhas, S. 64 (m. Abb.)
1198 (Firmenmarke ohne Ang.), H. 71 cm; GSF, S. 8
Bem.: Dargestellt ist der franz. Kaiser Napoleon I. Bonaparte (1769–1821).
-> vgl. 1074
Lit.: GSF, S. 8, hier mit der Ang. des Originaltitels „Napoleon I. (mittel)".

1199
BÜSTE Junge Bäuerin mit Kopftuch
ca. 1897/98
1199 (Firmenmarke ohne Ang.), H. 39 cm; Pinhas, S. 35 (m. Abb.)

1201
BÜSTE „Kinderkopf"
ca. 1897/98
1201 (Firmenmarke ohne Ang.), H. 29 cm; GSF, S. 8
Lit.: GSF, S. 8, hier mit der Ang. des Originaltitels „Kinderkopf".

1202
SCHREIBZEUG Stehendes, schreibendes Biedermeiermädchen
ca. 1899/1900
F1202 GS (Firmenmarke?), H. 20 cm; BM 17
Lit.: AZ 1979, S. 514 (m. Abb.)

1204
BÜSTE Junge Afrikanerin
ca. 1897/98
1204,(2,7,5,6 vertikal) FG (Firmenmarke?), Repr. Res., H. 33,5 cm; Doro 20.4.1993

(1205)
LAMPE Mädchenpaar mit Weidenkorb
ca. 1898/99
1205,5 (ohne Firmenmarke?), Repr. Res., Made in Austria, H. 66 cm; Doro 4.11.1998
Bem.: Evtl. GS (Zuschreibung unter großem Vorbehalt, da neben der fehlenden Firmenmarke auch die Qualität von GS nicht erreicht wird, vermutlich eine der zahlreichen Konkurrenzfirmen).

1208
FIGUR „Poesie"
ca. 1897/98
1208 (Firmenmarke ohne Ang.), H. 72 cm; GSF, S. 8
Lit.: GSF, S. 8, hier mit der Ang. des Originaltitels „Poesie".

1209
VASE mit Laufglasur
ca. 1898/99
01209, 35,7 Firmenmarke 4, Ges. gesch., H. 24 cm; WKH

1210
Lavergne, Adolphe-Jean
VASE „Orpheus" mit Lyra, von drei Löwen und zwei Vögeln begleitet (vollplastisch an der Vasenwandung stehend)
ca. 1898/99
1210,(33?),17 (Firmenmarke ohne Ang.), H. 123 cm; KH
Ausst.: Leipzig 1905 „Michaelismesse".
-> vgl. 1251
Lit.: N.N.: Michaelismesse 1905, in: KeramRS, 13.1905, S. 916 „Als neu sahen wir eine im vornehmen Geschmack gehaltene Vase mit Orpheus".

1213
FIGUR „Wildschütz"
ca. 1897/98
1213 (Firmenmarke ohne Ang.), H. 74 cm; GSF, S. 8
-> vgl. 1750
Lit.: GSF, S. 8, hier mit der Ang. des Originaltitels "Wildschütz".

1214
FIGUR „Hirtenknabe"
ca. 1896/97 (für den Entwurf)
1214 (Firmenmarke ohne Ang.), H. 57 cm; GSF, S. 8
-> vgl. 871
Lit.: GSF, S. 8, hier mit der Ang. des Originaltitels „Hirtenknabe".

1215
FIGUR „Vasenträgerin (nackt)" (Stehendes nacktes junges Mädchen mit Krug und Handtuch) (auf mit Löwenkopfmaske und Muschel verziertem Sockel)
ca. 1895/97 (für den Entwurf)
856 ger., GS (Firmenmarke ohne Ang.), H. 74,9 cm; Christie's SK 11.7.2001
1215 (Firmenmarke ohne Ang.), H. 57 cm; GSF, S. 8
-> vgl. 856a; 856b
Lit.: GSF, S. 8, hier mit der Ang. des Originaltitels „Vasenträgerin (nackt)".

1219
Strasser, Arthur (zugeschr.)
BÜSTE „Raucher"/„Fumeur" (Téte d'étude) (Rauchender Bursche)
ca. 1898/99
1219 (Firmenmarke ohne Ang.), H. fehlt; Pinhas, S. 97 (m. Abb.)
1219 (Firmenmarke ohne Ang.), H. 52 cm; GSF, S. 8
-> vgl. 539 (GSB), hier mit Ang. des franz. Originaltitels „Fumeur".
Lit.: GSF, S. 8, hier mit der Ang. des Originaltitels „Raucher".

1221
GRUPPE „Centauer"
ca. 1897/98
1221 (Firmenmarke ohne Ang.), H. 118 cm; GSF, S. 8
Lit.: GSF, S. 8, hier mit der Ang. des Originaltitels „Centauer".

1222 -> ÜBERSCHNEIDUNG
GRUPPE „Der Durst"
ca. 1897/98
1222 (Firmenmarke ohne Ang.), H. 132 cm; GSF, S. 8
Lit.: GSF, S. 8, hier mit der Ang. des Originaltitels „Der Durst".

1222 -> ÜBERSCHNEIDUNG
Colombier, Amelie
„Carmencita"
BRONZE / Marmorsockel
ca. 1899/1900
1222,75 Sculpture (Firmenmarke Bronze 1), H. 53 cm, (Sockel 2,1 cm); Weidler 520; 530; Sigalas 63; Doro 26.4.2004
Bem.: Am Marmorsockel sign. und betitelt.
Lit.: Berman, Vol. 3, Nr. 1877 (m. Abb.), hier H. 21,4 cm, (Firmenmarke Bronze 4) (wohl Arthur GS, Paris); Catley (1987), S. 120 (m. Abb.).

1224
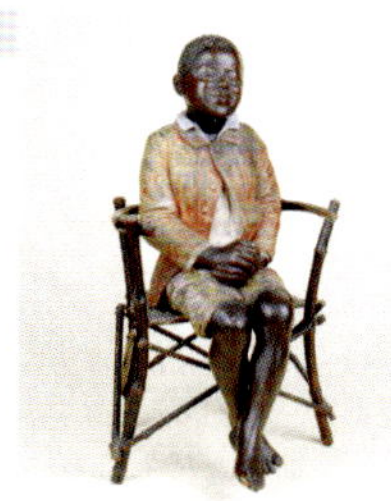
FIGUR „Beim Photographen" (Sitzender Negerjunge) (auf Sessel)
ca. 1897/98
1224,4 Firmenmarke 8a, gepr. „P"= Z3, H. 42 cm, PB Wien 19
1224 (Firmenmarke ohne Ang.), H. 40 cm; GSF, S. 8
Bem.: Der Sessel, der separat aber über Goldscheider zu kaufen war, ist aus einem anderen Material (Bambus?) gefertigt. Es handelt sich hier um die linke Figur der Gruppe „Drei lachende Negerjungen" (auf einer Mauer sitzend).
-> vgl. 1190= DreierGRUPPE
Lit.: GSF, S. 8, hier mit der Ang. des Originaltitels „Beim Photographen".

1225
FIGUR „Beim Photographen" (Sitzender, lachender Negerjunge mit Kappe) (auf Sessel)
ca. 1897/98
1225,38,(6,7,4,2, vertikal) Firmenmarke 7, H. 53 cm (mit Sessel); GK, S. 14, 66 (m. Abb.), Buchdeckel= Farbabb.; Wien Museum Inv. 223.918
1225 (Firmenmarke ohne Ang.), H. 49 cm; GSF, S. 8
Bem.: Der Sessel ist aus Bambus gefertigt und konnte bei Goldscheider separat dazu bezogen werden. Es handelt sich hier um die mittlere Figur der Gruppe „Drei lachende Negerjungen" (auf einer Mauer sitzend).
Ausst.: HMW (jetzt Wien Museum) 1985/86.
-> vgl. 1236; 1805
-> vgl. 1617 WANDMASKE (Variante Schiebermütze)
-> vgl. 1695= BÜSTE
-> vgl. 1190= DreierGRUPPE
Lit.: GSF, S. 8, hier mit der Ang. des Originaltitels „Beim Photographen"; GSF, S. 8, hier mit der Ang. des Originaltitels „Beim Photographen"; GK, S. 11, 66 (m. Abb.); GSN, S. 12, 13 (m. Abb.).
GS neu Modell-Nr. 8 „Sunnyboy" (H. 50 cm), Aufl. ohne Ang.

1226
FIGUR „Beim Photographen" (Sitzender, lachender Negerjunge mit Schiebermütze) (auf Sessel)
ca. 1897/98
1226 (Firmenmarke ohne Ang.), H. 57 (?) cm; Pinhas, S. 32 (m. Abb.)
1226 (Firmenmarke ohne Ang.), H. 45 cm; GSF, S. 8
Bem.: Der Sessel ist aus Bambus gefertigt und konnte bei Goldscheider separat dazu bezogen werden. Es handelt sich hier um die rechte Figur der Gruppe „Drei lachende Negerjungen" (auf einer Mauer sitzend).
-> vgl. 1190= DreierGRUPPE
Lit.: GSF, S. 8, hier mit der Ang. des Originaltitels „Beim Photographen".

1228
FIGUR „Vasenträgerin (bekleidet)"
ca. 1895/97 (für den Entwurf)
1228 (Firmenmarke ohne Ang.), H. 57 cm; GSF, S. 8
-> vgl. 856a; 856b
Lit.: GSF, S. 8, hier mit der Ang. des Originaltitels „Vasenträgerin (bekleidet)".

1229
BÜSTE „Königin Louise"
ca. 1897/98
1229 (Firmenmarke ohne Ang.), H. 63 cm; GSF, S. 8
Bem.: Dargestellt ist die preuss. Königin Luise, Herzogin zu Mecklenburg (1776–1810).
-> vgl. 1123; 1794
Lit.: GSF, S. 8, hier mit der Ang. des Originaltitels „Königin Louise".

1231
Perigot
DECKELDOSE Mädchen mit Weihnachtsbaum
ca. 1898/99
01231,33,10 Firmenmarke 9, Firmenmarke 6, Repr. Res., H. 32 cm; Wien Museum Inv. 190.099/1+2

1232
Strasser, Arthur
FIGUR „Othello"
ca. 1898/99
1232 (Firmenmarke ohne Ang.), H. 98 cm; GSF, S. 8
Bem.: Dargestellt ist die gleichnamige Titelfigur nach William Shakespeares Trauerspiel „Othello, der Mohr von Venedig" (entstanden ca. 1604).
Ausst.: ÖMKI 1899/1900, hier mit Ang. des Originaltitels „Othello".
Lit.: GSF, S. 8, hier mit der Ang. des Originaltitels „Othello"; Neuwirth (1974), S. 74.

1233
FIGUR „Goliath" (Auf Felsen stehender und mit Speer und Schwert bewaffneter Goliath)
Inschrift: GOLIATH
ca. 1898/99
1233 (Firmenmarke ohne Ang.), H. 85 cm; Pinhas, S. 70 (m. Abb.)
1233 (Firmenmarke ohne Ang.), H. 150 cm; GSF, S. 8
Bem.: Dargestellt ist der aus der Bibel (Altes Testament) bekannte riesenhafte Krieger Goliath.
-> vgl. 1721
Lit.: GSF, S. 8, hier mit der Ang. des Originaltitels „Goliath".

1235
FIGUR „Mignon (bekleidet)" (Stehende Sängerin mit Lyra) (auf Rechtecksockel)
ca. 1897/98
1235 (Firmenmarke ohne Ang.), H. 73 cm; Pinhas, S. 71 (m. Abb.)
1235 (Firmenmarke ohne Ang.), H. 73 cm; GSF, S. 8
-> vgl. 1018a; 1018b
Lit.: GSF, S. 8, hier mit der Ang. des Originaltitels „Mignon (bekleidet)".

1236
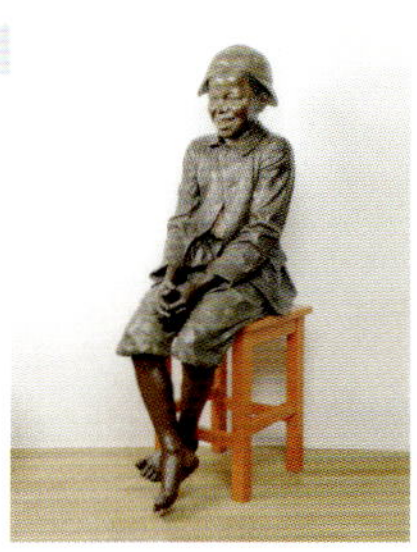
FIGUR „Beim Photographen" (Sitzender lachender Negerjunge mit Kappe) (auf Sessel)
ca. 1897/98
22,1236,(6,2,6,7 vertikal) ohne Firmenmarke, H. 120 cm; PB Wien 19
1236 (Firmenmarke ohne Ang.), H. 123 cm; GSF, S. 8
Bem.: Es handelt sich hier um die mittlere Figur der „Drei lachende Negerjungen" (auf einer Mauer sitzend), Modell-Nr. 1190. Der Sessel ist aus Bambus gefertigt und konnte durch Goldscheider separat dazu bezogen werden.
-> vgl. 1225; 1805
-> vgl. 1617 WANDMASKE (Variante: Schiebermütze)
-> vgl. 1695= BÜSTE
-> vgl. 1190= DreierGRUPPE
Lit.: GSF, S. 8, hier mit der Ang. des Originaltitels „Beim Photographen"; Pinhas, S. 96 (m. Abb.).

1238
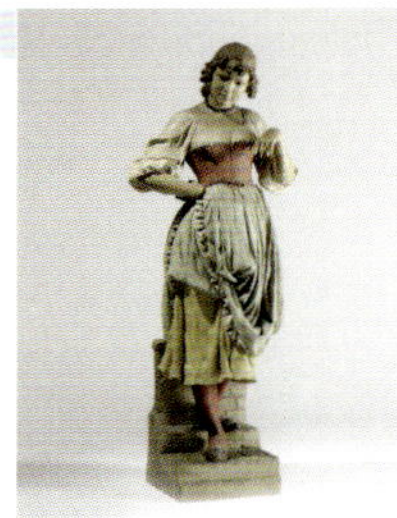
Stehendes Bauernmädchen in Pose (auf getrepptem Sockel)
ca. 1896/97
1238,(2,9,1,5 vertikal),24, Firmenmarke 11, Repr. Res., gepr. „P"= Z3, H. 104 cm; MAK Wien, Inv. Ke 10521-5

1241 -> ÜBERSCHNEIDUNG
Petri (d.i. Petrides, János) (Johann)
VASE Jugendlicher Ballwerfer in antikisierendem Gewand (große Bodenvase)
ca. 1900
01241,204,17 gem. Z, Firmenmarke Adler 15 (gepr.), Firmenmarke Sculpture 13 (gepr.), Fabriqué en Autriche, gepr., „S"= Z2, sign. Petri, H. 61 cm; Kett. 214

1241 -> ÜBERSCHNEIDUNG
FIGUR „Judith (nackt)"
ca. 1897/98
1241 (Firmenmarke ohne Ang.), H. 72 cm; GSF, S. 8
-> vgl. 734; 1012
Lit.: GSF, S. 8, hier mit der Ang. des Originaltitels „Judith (nackt)".

1243
Lemassi
FIGUR/LAMPE „Negerknabe mit Leuchter" (Stehender Negerjunge mit Kerzenhalter) (Nordafrikaner, vor Hocker stehend) (auf Rundsockel)
ca. 1897/98
1234 (vielmehr 1243),51 Firmenmarke 8a, Repr. Res., unsign., H. 75 cm; GK, S. 14, 67 (m. Abb.); Wien Museum Inv. 224.301
1243 (Firmenmarke ohne Ang.), sign. Lemassi, H. 73 cm; Pinhas, S. 29, 65, 68 (m. Abb.)
1243 (Firmenmarke ohne Ang.), H. 75 cm; GSF, S. 8
Bem.: Dieses Modell wurde auch mit einer Glühbirne elektrifiziert geliefert.
Ausst.: HMW (jetzt Wien Museum) 1985/86.
-> vgl. 1291; 1863
Lit.: GSF, S. 8, hier mit der Ang. des Originaltitels "Negerknabe mit Leuchter", „detto elektr. montiert mit 1 Lampe"; GK, S. 14, 67 (m. Abb.); Pinhas, S. 29, 65, 68 (m. Abb.).

1245
BÜSTE Lachender Mönch (Sockel reliefiert mit Trauben, Weinblättern und Flasche)
ca. 1898
1245,54 FG (Firmenmarke?), Repr. Res., H. 30,5 cm; eBay # 1326916408
Pendants -> vgl. 1246; 1247

1246
BÜSTE Schmunzelnder Mönch
ca. 1898
1246,(5,9,4,8 vertikal), Firmenmarke 11, Repr. Res., H. 30 cm; eBay # 160017432617
Pendants -> vgl. 1245; 1247

1247
BÜSTE Niesender Mönch (Sockel reliefiert mit Trauben)
Inschrift: HA-ZI
ca. 1898
1247,13 FG (Firmenmarke?), Repr. Res., H. 30,5 cm; eBay # 1326916408
1247 Firmenmarke 11, H. 30,5 cm; eBay # 272779316
Pendants -> vgl. 1245; 1246

1251 -> ÜBERSCHNEIDUNG
Lavergne, Adolphe-Jean
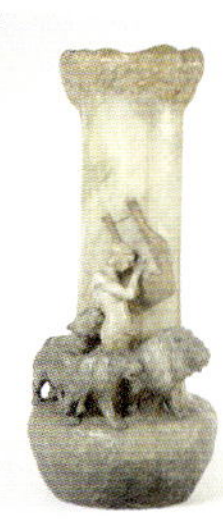
VASE „Orpheus" mit Lyra, von drei Löwen und zwei Vögeln begleitet (vollplastisch an der Vasenwandung stehend)
ca. 1898/99
01251,85,17 Firmenmarke Sculpture 13, Repr. Res., H. 79 cm; Wien Museum Inv. 219.138
Ausst.: Leipzig 1905 „Michaelismesse".
-> vgl. 1210
Lit.: N.N.: Michaelismesse 1905, in: KeramRS, 13.1905, S. 916 „Als neu sahen wir eine im vornehmen Geschmack gehaltene Vase mit Orpheus".

1251 -> ÜBERSCHNEIDUNG
FIGUR „Lautenschläger (nackt)"
ca. 1898/99
1251 (Firmenmarke ohne Ang.), H. 80 cm; GSF, S. 8
-> vgl. 1406 (bekleidet); 1551 (bekleidet)
Lit.: GSF, S. 8, hier mit der Ang. des Originaltitels „Lautenschläger (nackt)".

1253
BÜSTE „Italienischer Knabe" (nach rechts blickend)
ca. 1898/99
1253 (Firmenmarke ohne Ang.), H. 63 cm; Pinhas, S. 83 (m. Abb.)
1253 (Firmenmarke ohne Ang.), H. 63 cm; GSF, S. 8
-> vgl. 1779
Lit.: GSF, S. 8, hier mit der Ang. des Originaltitels „Italienischer Knabe".

1255
Telsch (d.i. Telcs, Ede/Eduard)
GRUPPE „Der kleine Durstige"/„Faune assoiffé" (Kleiner bocksbeiniger Faun, von Erwachsenem zum Trinken verführt) (auf Landschaftssockel)
ca. 1898/99
1255 (Firmenmarke ohne Ang.), H. fehlt; Pinhas, S. 85 (m. Abb.)
1255 (Firmenmarke ohne Ang.), H. 64 cm; GSF, S. 8
Ausst.: ÖMKI 1899/1900, hier jeweils mit Ang. des Originaltitels „Der kleine Durstige" und des Künstlernamens „Telsch".
-> vgl. 574 (GSB), hier mit Ang. des franz. Originaltitels.
Lit.: GSF, S. 8, hier mit der Ang. des Originaltitels „Der kleine Durstige"; Neuwirth (1974), S. 75.

1256 -> ÜBERSCHNEIDUNG
FIGUR „Judith (bekleidet)"
ca. 1897/98
1256 (Firmenmarke ohne Ang.), H. 72 cm; GSF, S. 9
-> vgl. 734; 1012a; 1012b
Lit.: GSF, S. 9, hier mit der Ang. des Originaltitels „Judith (bekleidet)".

1256 -> ÜBERSCHNEIDUNG
Falguière, Alexandre
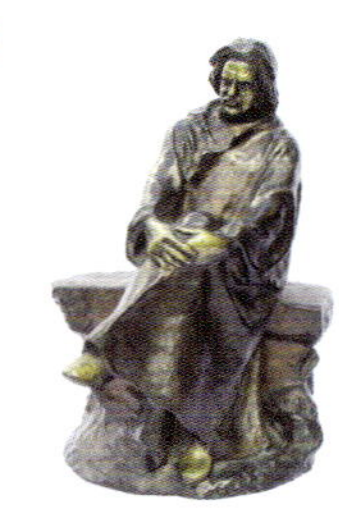
Balzac (auf einer Bank sitzend)
BRONZE
ca. 1899/1900
1256,67,4V ger., Sculpture (Firmenmarke Bronze 1), H. 25,4 cm; Sammlung a@goldscheider.net
Bem.: Dargestellt ist der Dichter Honoré de Balzac (1799–1850)

1258
Portou (d.i. Bortone, Ipazzio Antonio)
BÜSTE Dame mit Hut
ca. 1898/99
1258,(10?),39 Firmenmarke 7, undeutlich sign. („Portou"?), gepr. „P"= Z3, H. 31,7 cm; eBay # 9014499135

1259
Araber auf Kamel sitzend (vor Steinmauer) (auf Rechtecksockel)
ca. 1898/99
1259 (Firmenmarke ohne Ang.), H. 66 cm; Pinhas, S. 31 (m. Abb.)

1260
FIGUR „Gefesselte Amazone"
ca. 1898/99
1256 (Firmenmarke ohne Ang.), H. 46 cm; GSF, S. 9
Lit.: GSF, S. 9, hier mit der Ang. des Originaltitels „Gefesselte Amazone".

1261
FIGUR „Rubens"
ca. 1897/98
1261 (Firmenmarke ohne Ang.), H. 54 cm; GSF, S. 9
Bem.: Dargestellt ist der flämische Maler Peter Paul Rubens (1577–1640).
-> vgl. 1143
Lit.: GSF, S. 9, hier mit der Ang. des Originaltitels „Rubens".

1262
Strasser, Arthur
FIGUR „Haruspex" (Stehender römischer Wahrsager) (auf Rechtecksockel)
Inschrift: CAES. HARUSPEX
ca. 1897/98
1262 (Firmenmarke ohne Ang.), unsign., H. 58 cm; Pinhas, S. 67 (m. Abb.)
1262 (Firmenmarke ohne Ang.), H. 54 cm; GSF, S. 9
Bem.: Dargestellt ist die Figur eines römischen (bzw. etruskischen) Wahrsagers, der aus den Eingeweiden von Opfertieren weissagte. Die Zuordnung an Arthur Strasser konnte durch eine sign. Figur (H. 98 cm), welche im Doro Wien (637. Kunstauktion, 14.–20.9.1982) angeboten wurde, vorgenommen werden. Ob es sich bei diesem Ex. um ein Erzeugnis von GS handelte oder um eine Originalausformung Strassers, konnte nicht mehr nachvollzogen werden.
-> vgl. 1142
Lit.: GSF, S. 9, hier mit der Ang. des Originaltitels „Haruspex".

1264
BÜSTE Junge Frau mit Haube (auf Blumen verziertem Sockel)
ca. 1899/1900
1264 (Firmenmarke ohne Ang.), H. fehlt; Pinhas, S. 65 (m. Abb.)

1265
FIGUR „Bacchantin" (Stehende Sängerin mit halb entblößtem Busen) (auf getrepptem Sockel mit Maskaron und Musikinstrumenten)
ca. 1898
1265,(3,1,2,7 vertikal),50 Firmenmarke 7, Repr. Res., H. 117 cm; Wien Museum Inv. 219.104
1265 (Firmenmarke ohne Ang.), H. 100 cm; GSF, S. 9
Bem.: An diesem Modell zeigt sich der Variantenreichtum der GS-Entwürfe. Ein und dieselbe Figur wurde als Sängerin bzw. als Blumenträgerin, in diesem Fall mit einem langen Stab, der von einem Maiskolben bekrönt wird, angeboten. Zudem wird der entblößte Busen durch das Kleid bedeckt.
-> vgl. 1558
Lit.: GSF, S. 9, hier mit der Ang. des Originaltitels „Bacchantin"; Pinhas, S. 65 (m. Abb.), hier als Variante mit Bekleidung und Blumenstab (Maiskolben).

1268
Waderé, Heinrich
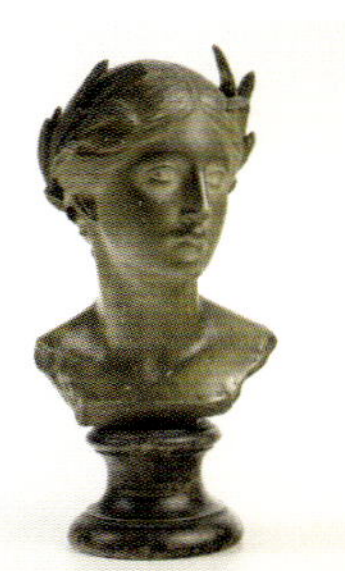
BÜSTE „Venuskopf"/„Vénus Grecque"
ca. 1898
1268,109,10 GS (Firmenmarke?), Firmenmarke Sculpture 13, unsign., H. 41 cm; WKA 23
1268,155,11 Firmenmarke 7, Repr. Res., H. 40 cm; WKH
1268,187,19 Firmenmarke 7, H. 40 cm; Wien Museum Inv. 186.610
1268 (Firmenmarke ohne Ang.), H. 40 cm; GSF, S. 9
Bem.: An der Plinthe betitelt und sign. (Metallschild). In der Pariser Produktion läuft diese Büste (H. 45 cm) unter der Modell-Nr. 615 und wird dort in Bronze ausgeführt.
Ausst.: ÖMKI 1898/99; 1899/1900; 1902/03; 1906/07, hier jeweils mit Ang. des Originaltitels „Venus", im Ausst.-Kat. 1906/07 Ang. des Künstlernames Prof. „Vadère".
-> vgl. 615 (GSB)
-> vgl. 2692
Lit.: GSF, S. 9, hier mit der Ang. des Originaltitels „Venuskopf"; GSB, Pl. 53 (m. Abb.); Neuwirth (1974), S.75, 76; Pinhas, S. 95 (m. Abb.).

1271
BÜSTE Nordafrikaner mit Turban und Halsschmuck
ca. 1898
1271,3,2 FG (Firmenmarke?), Repr. Res., H. 45 cm; Doro 20.4.1993
Pendant -> vgl. 1272

1272
BÜSTE Nordafrikanerin mit Münzketten als Kopf- und Halsschmuck
ca. 1898
1272,29,6 FG Wien (Firmenmarke?), bezeichnet „R"?, H. 43,5 cm; Doro 4.11.1998
Pendant -> vgl. 1271

1277
FIGUR „Paris"
ca. 1898
1277 (Firmenmarke ohne Ang.), H. 125 cm; GSF, S. 9
-> vgl. 1362
Lit.: GSF, S. 9, hier mit der Ang. des Originaltitels „Paris".

1279
FIGUR „Seifenbläser"
ca. 1898
1279 (Firmenmarke ohne Ang.), H. 75 cm; GSF, S. 9
-> vgl. 1817
Lit.: GSF, S. 9, hier mit der Ang. des Originaltitels „Seifenbläser".

1280
RELIEF „Niederländische Dame"
ca. 1898
1280 (Firmenmarke ohne Ang.), H. 78 cm; GSF, S. 9
-> vgl. 1281
Bem.: Für dieses Relief wurde ein Rahmen angeboten, der separat durch Goldscheider bezogen werden konnte.
Lit.: GSF, S. 9, hier mit der Ang. des Originaltitels „Niederländische Dame".

1281
RELIEF „Niederländischer Knabe"
ca. 1898
1281 (Firmenmarke ohne Ang.), H. 78 cm; GSF, S. 9
Bem.: Für dieses Relief wurde ein Rahmen angeboten, der separat durch Goldscheider bezogen werden konnte.
-> vgl. 1280
Lit.: GSF, S. 9, hier mit der Ang. des Originaltitels „Niederländischer Knabe".

1284
Sitzender Araber mit Turban, ein Musikinstrument (Zither?) spielend (auf kissenartigem Sockel)
ca. 1898
1284 (Firmenmarke ohne Ang.), H. 66 cm; Pinhas, S. 25 (m. Abb.)

1285
Wind, Josef
FIGUR „Krugträgerin"/„Krughändlerin" (Schreitende Frau mit Turban und zwei/drei Krügen) (auf Rundsockel)
ca. 1898
1285 (Firmenmarke ohne Ang.), unsign., H. 78 cm; Pinhas, S. 75, 82 (m. Abb.)
1285 (Firmenmarke ohne Ang.), H. 78 cm; GSF, S. 9
Bem.: Diese Größe wurde in der Variante mit zwei, oder auch drei Krügen angeboten.
Ausst.: ÖMKI 1898/99, hier mit Ang. des Originaltitels „Krughändlerin".
-> vgl. 1727; 1997
Lit.: GSF, S. 9, hier mit der Ang. des Originaltitels „Krugträgerin".

1289
FIGUR „Kellermeister (groß)" mit großem Humpen (einen getreppten Sockel mit Mauer hinab schreitend)
ca. 1898
1289 (Firmenmarke ohne Ang.), H. 78 cm; GSF, S. 9
-> vgl. 1290
Lit.: GSF, S. 9, hier mit der Ang. des Originaltitels „Kellermeister (groß)".

1290
FIGUR „Kellermeister (mittel)" mit großem Humpen (einen getreppten Sockel mit Mauer hinab schreitend)
ca. 1898
1290,(9,1,7,3 vertikal) Firmenmarke 11, H. 58 cm; eBay # 6527007435
1290 (Firmenmarke ohne Ang.), H. 58 cm; GSF, S. 9
-> vgl. 1289
Lit.: GSF, S. 9, hier mit der Ang. des Originaltitels „Kellermeister (mittel)".

1291
Lemassi
FIGUR „Negerknabe mit Leuchter" (Stehender Negerjunge mit Kerzenhalter) (Nordafrikaner, vor Hocker stehend) (auf Rundsockel)
ca. 1897/98
1791 (vielmehr 1291) GS (Firmenmarke Sculpture 13?), Repr. Res., H. 152,5 cm; Soth. L. 9.10.1992
1291 (Firmenmarke ohne Ang.), H. 143 cm; Pinhas, S. 29, 65, 68 (m. Abb.)
ohne Ang., (ohne Firmenmarke?), H. 148 cm; Soth. L. 17.3.1989
1291 (Firmenmarke ohne Ang.), H. 143 cm; GSF, S. 9
-> vgl. 1243; 1863
-> vgl. 2000= WANDMASKE
Lit.: GSF, S. 9, hier mit der Ang. des Originaltitels „Negerknabe mit Leuchter"; GK, S. 14, 67 (m. Abb.); Pinhas, S. 29, 65, 68 (m. Abb.).

1292
GRUPPE „Süsses Geheimnis"/„Baiser D' Amour" (Amor, eine junge Frau neckend) (auf Rundsockel)
ca. 1898
1292 (Firmenmarke ohne Ang.), H. 77 cm; Pinhas, S. 69 (m. Abb.)
1292 (Firmenmarke ohne Ang.), H. 77 cm; GSF, S. 9
-> vgl. 600; 605 (GSB), hier mit Ang. des franz. Originaltitels.
-> vgl. 1520
Lit.: GSF, S. 9, hier mit der Ang. des Originaltitels „Süsses Geheimnis".

1293
BÜSTE „Tuareg" (mit Turban) (Kopf nach links gedreht)
ca. 1898
6,1293 (1298?),(9,0,8 vertikal) Firmenmarke 11, Repr. Res., gepr. „P"= Z3, H. 64,5 cm; Quit. 26.6.1998; Doro 18.5.1999
1293 (Firmenmarke ohne Ang.), H. 68 cm; GSF, S. 9
Lit.: GSF, S. 9, hier mit der Ang. des Originaltitels „Tuareg".

1294
Stehende junge Frau mit zwei Körben (auf Landschaftssockel)
ca. 1898/99
1294,6,4 (Marke ohne Ang.), undeutlich sign., H. 60,5 cm; Doro 4.3.2003

1295
Wind, Josef
FIGUR „Schlangenbändigerin" (Stehender Akt einer Orientalin mit Tuch, Turban, Flöte und Schlange)
ca. 1898
1295 (Firmenmarke ohne Ang.), H. ca. 55 cm; Pinhas, S. 66 (m. Abb.)
ohne Ang. H. 55 cm; Richemond (1995), S. 62
1295 (Firmenmarke ohne Ang.), H. 77 cm; GSF, S. 9
Ausst.: ÖMKI 1898/99, hier mit Ang. des Originaltitels „Schlangenbändigerin".
-> vgl. 40 BRONZE; 644 (GSB)
-> vgl. 1399 (Variante)
Lit.: GSF, S. 9, hier mit der Ang. des Originaltitels „Schlangenbändigerin"; GSB, Pl. 56, hier ist die Figur mit dem franz. Originaltitel „Charmeuse des Serpents (nue)" unter Nr. 644 (H. 85 cm) abgebildet. Ang. zum Material (Terrakotta/Bronze) fehlen. Die Differenz bei der Größenangabe dürfte von unterschiedlichen Sockelhöhen herrühren; Richemond (1999), S. 235 (m. Abb.).

1297
Ceccarelli, Ezio
BÜSTE „Winzerin"/„Automne" (Herbst) (Frau mit großem Hut und Trauben)
ca. 1898
1297,34,24 Firmenmarke 7, Ausformung ca. 1898, H. 72 cm; eBay # 7344452694
1297,238,24 Firmenmarke Sculpture 13, Repr. Res., Ausformung ca. 1900, H. 72,8 cm; Doro 4.6.1996
1297 (Firmenmarke ohne Ang.), H. 72 cm; GSF, S. 9
–> vgl. 1619; 1961
–> vgl. 622 (GSB); 734 (GSB)
Lit.: GSF, S. 9, hier mit der Ang. des Originaltitels „Winzerin"; GSB, Pl. 71, hier ist die BÜSTE mit dem Originaltitel „Automne" unter Nr. 622= No.1 (H. 52 cm) und unter Nr. 734= No. 2 (H. 30 cm) abgebildet. Ang. zum Material (Terrakotta/Bronze) fehlen.
1298 s. 1293

1299 –> ÜBERSCHNEIDUNG
Lesca
VASE Mädchen in antikisierendem Gewand, neben einer Quelle sitzend (Vasenhals als antike Säule geformt) (große Bodenvase)
ca. 1899/1900
01299,22,17 Firmenmarke Sculpture 13, Originalpick.: Fa. E. Canezin, Toulouse, H. 85 cm; Neumeister 270

1299 –> ÜBERSCHNEIDUNG
Hiller, (Max)
„Porteuse d'Eau Aveugle" (Schreitende Wasserträgerin mit Krug und Blindenstab) (auf Holzsockel)
ca. 1899
1299,(3,1,2,1 vertikal),50 Firmenmarke 7, Repr. Res., sign. Porteuse d'Eau Aveugle par Hiller (Metallschild), H. 85,1 cm, H. mit Holzsockel 90,2 cm; eBay # 2539695318
Bem.: An der Plinthe betitelt und sign. (Metallschild).

1300
FIGUR „Lustiger Zecher"
ca. 1898/99
1300 (Firmenmarke ohne Ang.), H. 80 cm; GSF, S. 9
Lit.: GSF, S. 9, hier mit der Ang. des Originaltitels „Lustiger Zecher".

1301
BÜSTE Robinson Crusoe
Inschrift: ROBINSON
ca. 1898/99
1301,3,4,49 Firmenmarke 11, Repr. Res., H. 30,5 cm; Sammlung Weilinger, Wien
Bem.: Dargestellt ist Robinson Crusoe, die Hauptfigur eines Abenteuerromans (1719) von Daniel Defoe.

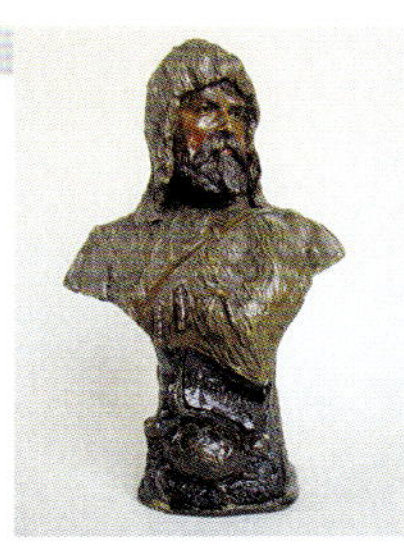

1303
FIGUR „Pfeilschnitzer (orientalisch)" (auf mauerartigem Sockel)
ca. 1898/99
1303 (Firmenmarke ohne Ang.), H. fehlt; Pinhas, S. 65 (m. Abb.)
1303 (Firmenmarke ohne Ang.), H. 77 cm; GSF, S. 9
Lit.: GSF, S. 9, hier mit der Ang. des Originaltitels „Pfeilschnitzer (orientalisch)".

1304
FIGUR „Gefesselte Amazone (orientalisch)"/„Captive" (Afrikanische Sklavin als Halbakt) (an einen Baumstamm gefesselt)
ca. 1898
1304 (Firmenmarke ohne Ang.), H. fehlt; Pinhas, S. 82 (m. Abb.)
1304 (Firmenmarke ohne Ang.), H. 77 cm; GSF, S. 9
–> vgl. 712; 744 (GSB), hier mit Ang. des franz. Originaltitels.
Lit.: GSF, S. 9, hier mit der Ang. des Originaltitels „Gefesselte Amazone (orientalisch)".

1308 –> ÜBERSCHNEIDUNG
BÜSTE „Mahdi-Krieger"
ca. 1898
1308 (Firmenmarke ohne Ang.), H. 70 cm; GSF, S. 9
–> vgl. 1423; 1614= WANDMASKE
Lit.: GSF, S. 9, hier mit der Ang. des Originaltitels „Mahdi-Krieger".

1308 –> ÜBERSCHNEIDUNG
Clarà, Juan (Joan)
Zappelndes Kind auf Stuhl BRONZE
ca. 1898/99
119,1308 (Firmenmarke ohne Ang., wohl Arthur GS, Paris), H. 23 cm; Soth. L. 20.3.1992

1310
FIGUR „Ist es wahr?" (Stehende junge Dame in langem Kleid) (auf Landschaftssockel mit Pflanzen)
ca. 1898/99
1310 (Firmenmarke ohne Ang.), H. 77 cm; Pinhas, S. 77 (m. Abb.)
1310 (Firmenmarke ohne Ang.), H. 80 cm; GSF, S. 9
–> vgl. 1660
Lit.: GSF, S. 9, hier mit der Ang. des Originaltitels „Ist es wahr?".

1312
Stehendes Bauernmädchen mit Wassereimer (auf Rechtecksockel)
ca. 1898/99
1312,(6,8,6,8 vertikal) (Firmenmarke ohne Ang.), H. 77,5 cm; Doro 20.3.2006

1313
Stehende junge Frau mit Haube und langem Kleid (auf rechteckigem Landschaftssockel)
ca. 1898/99
1313 (Firmenmarke ohne Ang.), H. fehlt; Pinhas, S. 66 (m. Abb.)

1314
Strasser, Arthur
FIGUR „Stoffverkäufer" (Stehender Nordafrikaner, Stoffballen auf Kopf und Schulter tragend) (auf Rechtecksockel)
ca. 1898
1314 (Firmenmarke ohne Ang.), unsign., H. 110 cm; Pinhas, S. 75 (m. Abb.)
1314 (Firmenmarke ohne Ang.), H. 110 cm; GSF, S. 9
Ausst.: ÖMKI 1898/99, hier mit Ang. des Originaltitels „Stoffverkäufer".
–> vgl. 1609
Lit.: GSF, S. 9, hier mit der Ang. des Originaltitels „Stoffverkäufer".

1316
BÜSTE Junge Frau mit Haube (auf Blumen verziertem Sockel)
ca. 1899/1900
1316 (Firmenmarke ohne Ang.), H. fehlt; Pinhas, S. 65 (m. Abb.)

1317
FIGUR „Fischerknabe"
ca. 1898/99
1317 (Firmenmarke ohne Ang.), H. 43 cm; GSF, S. 9
–> vgl. 1950
Lit.: GSF, S. 9, hier mit der Ang. des Originaltitels „Fischerknabe".

1318
GRUPPE „Amor und Psyche"
ca. 1898/99
1318 (Firmenmarke ohne Ang.), H. 124 cm; GSF, S. 9
Lit.: GSF, S. 9, hier mit der Ang. des Originaltitels „Amor und Psyche".

1324
DECKELDOSE Kleines Biedermeiermädchen mit Spitztüte (Schultüte)
Form: rund, gebaucht
Knauf/Bekrönung: stehendes Mädchen (vollplastische Figur)
Dekor: Blumengirlanden
ca. 1900
01324,10,36 FG Wien (Firmenmarke?), Repr. Res., Gesamth. 25 cm; eBay # 891582086

1328
BÜSTE (chamois)
ca. 1897/98
1328 ohne Ang.; KeramRS, 6.1898, S. 353
Bem.: s. Modell-Nr. 6

1329
Chéret, Joseph
BÜSTE „Modern"
ca. 1896/97
1329 ohne Ang.; KeramRS, 5.1897, S. 347
Bem.: s. Modell-Nr. 6
Lit.: KeramRS, 5.1897, S. 347, hier die Einreichung zum „Musterschutz" und mit Ang. des Originaltitels BÜSTE „Modern", sowie des Künstlers „Cheret"; KeramRS, 6.1898, S. 353, hier, am 1.3.1898, die erneute Einreichung in „Terracotta" ohne Titel und Künstlerang.

1330
Schreitende Holländerin mit Wasserkrug und Brotkorb
ca. 1898/99
1330,641,? (Firmenmarke ohne Ang.), Repr. Res., Made in Austria H. 69,5 cm; Doro 21.5.1997

1333
Schreitender Schusterjunge mit Kappe und Schürze, ein Paar Stiefel haltend (sog. Wiener Type)
ca. 1890/95
1333,(8,0,3,0 vertikal) Firmenmarke 7, H. 37 cm; Peege 93
Bem.: Zu den sog. „Wiener Typen" –> vgl. Schmidt, Otto: Wiener Typen und Strassenbilder, Wien 2000.
Pendant –> vgl. 333

1334
Strasser, Arthur
FIGUR „Wollverkäufer" (Stehender Nordafrikaner) (auf Natursockel)
ca. 1898
1334,(8,1,9,4 vertikal) Firmenmarke 11, H. 107 cm; PB Wien 19
1334 (Firmenmarke ohne Ang.), H. 102 cm; GSF, S. 9

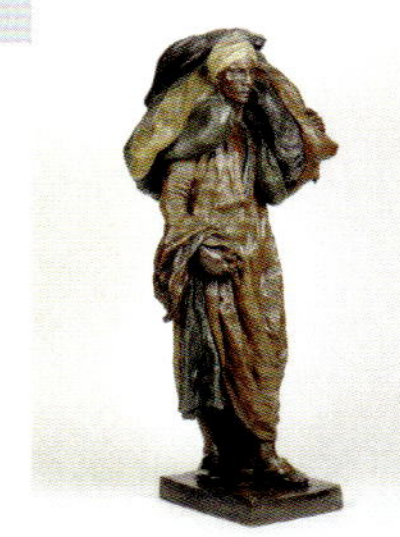

Ausst.: ÖMKI 1898/99, hier mit Ang. des Originaltitels „Wollverkäufer"
–> vgl. 1608
Lit.: GSF, S. 9, hier mit der Ang. des Originaltitels „Wollverkäufer"; Neuwirth (1973), in: Keramos 59.1973, S. 54–63, hier S. 55 (m. Abb.) (Samml. F. Witton, St. Albans/England); Neuwirth (1974), S. 58, 307 (Abb.); Pinhas, S. 75 (m. Abb.), hier mit Ang.: H. 102 cm.

1335
WANDMASKE Faun?
ca. 1898
1335 (Firmenmarke ohne Ang.), H. fehlt; Pinhas, S. 96 (m. Abb.)

1338 –> ÜBERSCHNEIDUNG
Lévèque, Auguste Maurice/Thiele, (P.?)
BÜSTE „Sklavin"/„Bedouine" (Sklavin, Araberin mit großem Turban) (nach links blickend)
ca. 1897
1338,280,21 ger. VIII., gem. weiß IV, Firmenmarke 7, sign. Thiele, H. 47 cm; Wien Museum Inv. 223.901 ohne Ang., sign. Lévèque, H. 47 cm; Richemond (1999), S. 78 (m. Abb.)
1338 (Firmenmarke ohne Ang.), H. 60 cm; GSF, S. 9
–> vgl. (H. 66 cm) (sign. „P. Thiele") ausgeführt in MARMOR; Soth. L. 9.10.1992 und 9.12.1993
–> vgl. 1101; 1877
–> vgl. 684 (GSB); 733 (GSB)
Lit.: GSF, S. 9, hier mit der Ang. des Originaltitels „Sklavin"; GSB, Pl. 59, hier ist die BÜSTE mit dem Originaltitel „Esclave" unter Nr. 684= No.1 (H. 62 cm) und unter Nr. 733 = No. 2 (H. 30 cm) abgebildet. Ang. zum Material (Terrakotta/Bronze fehlen); Richemond (1999), S. 78 (m. Abb.).

1338 -> ÜBERSCHNEIDUNG

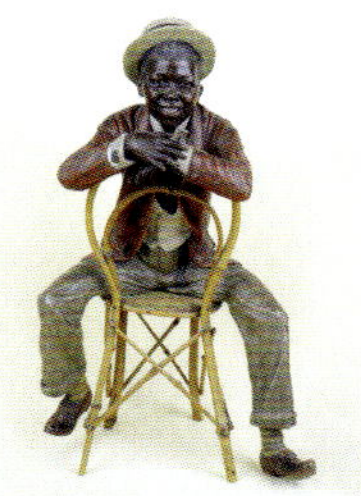

Sitzender, lachender Negerjunge (mit Anzug und Strohhut, rittlings auf Bambussessel)
ca. 1898
1338,(3,8,9,3 vertikal),52
Firmenmarke 7, Repr. Res., gepr. „P"= Z3, H. 52 cm; PB Wien 19

1339
Lévèque, Auguste Maurice
BÜSTE „Araber" (mit Turban) (nach rechts blickend)
ca. 1897
1339 (Firmenmarke ohne Ang.), unsign., H. 48 cm; Pinhas, S. 68 (m. Abb.)
ohne Ang. H. 47 cm; Richemond (1999), S. 79 (m. Abb.)
1339 (Firmenmarke ohne Ang.), H. 48 cm; GSF, S. 9
-> vgl. 1105; 1899
Pendant -> vgl. 1101
Lit.: GSF, S. 9, hier mit der Ang. des Originaltitels „Araber"; Richemond (1999), S. 78 (m. Abb.); Pinhas, S. 67, 69 (m. Abb.).

1340 -> ÜBERSCHNEIDUNG
BÜSTE „Bulgarin" (mit langem Kopftuch) (nach links blickend)
ca. 1897
1340 (Firmenmarke ohne Ang.), H. 49 cm; Pinhas, S. 67 (m. Abb.)
1340 (Firmenmarke ohne Ang.), H. 49 cm; GSF, S. 9
-> vgl. 1109
-> vgl. 1409 (sign. D. Rochella)
-> vgl. 1898 (sign. Thiele)
Lit.: GSF, S. 9, hier mit der Ang. des Originaltitels „Bulgarin".

1340 -> ÜBERSCHNEIDUNG
Simon
VASE Stehendes Paar (sie reicht ihm eine Schale mit frischem Wasser)
ca. 1898
01340 (Firmenmarke ohne Ang.), H. 61 cm; Pinhas, S. 52 (m. Abb.)

1341
Lévèque, (Auguste Maurice)/Thiele, (P.?)

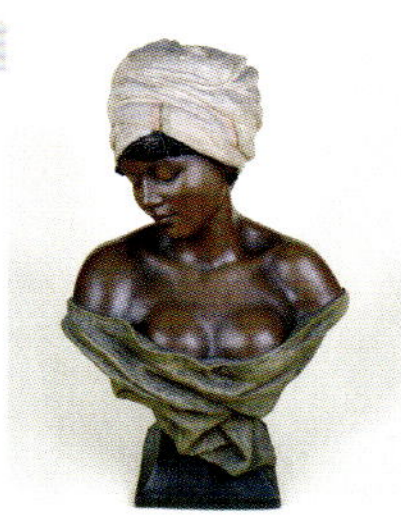

BÜSTE „Mohrin" (mit Turban) (nach rechts blickend)
ca. 1898
1341,122,7 ger., 1341,122,7 gepr., ger. X, Firmenmarke 7, Repr. Res., sign. Lévèque, gepr. „P"= Z3, H. 48,7 cm; MAK Wien, Inv. Ke 10602-109
1341,189,8 ger. VIII, gem. IV, FG Wien (Firmenmarke?), Repr. Res., sign. Lévèque, gepr. „P"= Z3, H. 49 cm; Christie's SK 15.11.2001; 9.10.2002
1341,202,8, 1341 ger., gem. weiß IV, Firmenmarke 7, sign. Lévèque, H. 48 cm; PB Wien 19
1341,361,40 Firmenmarke 7, Repr. Res., sign. Thiele, gepr. „P"= Z3, H. 50 cm; PB Wien 19
1341 (Firmenmarke ohne Ang.), H. 48 cm; GSF, S. 9
-> vgl. 1102
Lit.: GSF, S. 9, hier mit der Ang. des Originaltitels „Mohrin"; Pinhas, S. 69 (m. Abb.).

1342
FIGUR „(Junger) Raucher"
ca. 1897/98
1342 (Firmenmarke ohne Ang.), H. 58 cm; GSF, S. 9
-> vgl. 1194
Lit.: GSF, S. 9, hier mit der Ang. des Originaltitels "(Junger) Raucher".

1345
FIGUR „Mohrenbub"
ca. 1898/99
1345 (Firmenmarke ohne Ang.), H. 106 cm; GSF, S. 9
Bem.: Ein Sessel, bzw. lt. Katalog ein Stuhl, konnte bei Goldscheider separat dazu bezogen werden.
Lit.: GSF, S. 9, hier mit der Ang. des Originaltitels „Mohrenbub".

1346
FIGUR „Mohrenbub"
ca. 1898/99
1346 (Firmenmarke ohne Ang.), H. 116 cm; GSF, S. 9
Bem.: Ein Sessel, bzw. lt. Katalog ein Stuhl, konnte bei Goldscheider separat dazu bezogen werden.
Lit.: GSF, S. 9, hier mit der Ang. des Originaltitels „Mohrenbub".

1348
Schreitende Wasserträgerin (Ägypterin, einen Krug auf dem Kopf tragend)
ca. 1898
1348 Firmenmarke 11?, H. 74 cm; Doro 29.10.1996

1349
Schreitende Römerin (auf Landschaftssockel)
ca. 1898
1349 (Firmenmarke ohne Ang.), H. fehlt; Pinhas, S. 68 (m. Abb.)

1350
FIGUR „Feldwache" Stehender Araber mit Turban (auf Rechtecksockel)
ca. 1898/99
1350 (Firmenmarke ohne Ang.), H. 92 cm; Pinhas, S. 27 (m. Abb.)
1350 (Firmenmarke ohne Ang.), H. 87 cm; GSF, S. 10
-> vgl. 1513
Lit.: GSF, S. 10, hier mit der Ang. des Originaltitels „Feldwache".

1353
Dupré, J.
FIGUR „Lustiger Zecher"/„Le joyeux Buveur" (Stehender Zecher mit Hut, Humpen und Korbweinflasche)
ca. 1898
1353,82,(9,3,9,6 vertikal) Firmenmarke 11, H. 57 cm; WKA 14
Bem.: An der Plinthe mit dem franz. Originaltitel bezeichnet und sign. (Metallschild).
Text der Inschrift: Der fröhliche Zecher
Bem.: Vom gleichen Künstler wurde im ÖMKI 1902/03 ein Objekt mit dem Originaltitel „Heuernte" ausgestellt, im Ausst.-Kat. wird „J. Dupri" angegeben.
-> vgl. 1583
Lit.: GSF, S. 10, hier mit der Ang. des Originaltitels „Lustiger Zecher".

1355
FIGUR „Mohr" (Stehender Nordafrikaner mit Turban und Bambusstab)
ca. 1898
1355 (Firmenmarke ohne Ang.), H. 74 cm; Pinhas, S. 75 (m. Abb.)
-> vgl. 1196; 1572
Pendant -> vgl. 1356
Lit.: GSF, S. 10, hier mit der Ang. des Originaltitels „Mohr".

1356

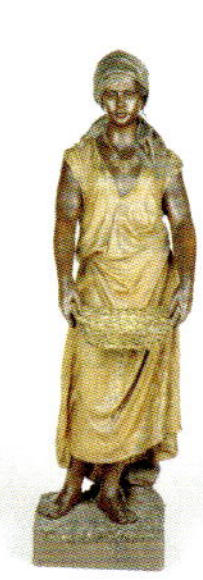

FIGUR „Mohrin" (Stehende Nordafrikanerin mit Turban, einen Korb haltend)
ca. 1898
1356,(3,4,9,9 vertikal),52
Firmenmarke 11, Repr. Res., H. 76 cm; PB Wien 19
1356 (Firmenmarke ohne Ang.), H. 77 cm; GSF, S. 10
-> vgl. 1191; 1573
Pendant -> vgl. 1355
Lit.: GSF, S. 10, hier mit der Ang. des Originaltitels „Mohrin"; Pinhas, S. 75 (m. Abb.).

1357
BÜSTE „Nacht"
ca. 1898/99
1357 (Firmenmarke ohne Ang.), H. 59 cm; GSF, S. 10
-> vgl. 1702
Lit.: GSF, S. 10, hier mit der Ang. des Originaltitels „Nacht".

1358
Langer, (Leopold?)
BÜSTE „Orientalin"
ca. 1898/99
1358 (Firmenmarke ohne Ang.), H. 69 cm; GSF, S. 10
-> vgl. 1424
Lit.: GSF, S. 10, hier mit der Ang. des Originaltitels „Orientalin".

1360
Langer, (Leopold?)

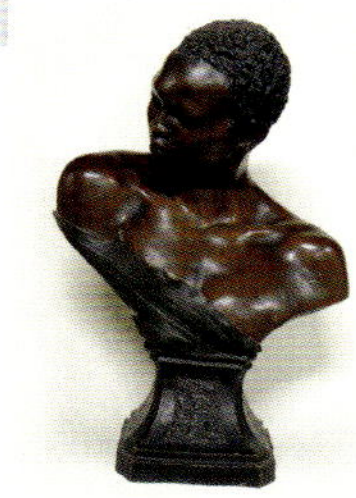

BÜSTE Afrikaner (nach rechts blickend)
ca. 1898/99
1360,(1,3,5,3 vertikal),42
Firmenmarke 11, unsign., gepr. „P"= Z3, H. 60,1 cm; MAK Wien, Inv. Ke 10602-29
Lit.: Pinhas, S. 28 (m. Abb.), hier ein mit „Langer" sign. Ex. abgebildet.

1362
FIGUR „Paris"
ca. 1898
1362 (Firmenmarke ohne Ang.), H. 76 cm; GSF, S. 10
-> vgl. 1277
Lit.: GSF, S. 10, hier mit der Ang. des Originaltitels „Paris".

1363
Sitzendes Mädchen (in historischem Kostüm) mit Laute (auf Thonet-Hocker)
ca. 1897/98
1363 (Firmenmarke ohne Ang.), H. fehlt; Pinhas, S. 71 (m. Abb.)

1364
FIGUR „Arab(isches) Mädchen (ohne elektr. Montage)"
ca. 1898/99
1364 (Firmenmarke ohne Ang.), H. 112 cm; GSF, S. 10
Bem.: Wie die Beschreibung im Firmenkatalog verdeutlicht, dürfte auch eine Ausführung als elektrifizierte Lampe produziert worden sein. Die dazugehörige Mod.-Nr. ließ sich jedoch bis Redaktionsschluss nicht nachweisen.
Lit.: GSF, S. 10, hier mit der Ang. des Originaltitels „Arab(isches) Mädchen (ohne elektr. Montage)".

1365
GRUPPE „Neckerei"
ca. 1898
1365 (Firmenmarke ohne Ang.), H. 87 cm; GSF, S. 10
-> vgl. 2008
Lit.: GSF, S. 10, hier mit der Ang. des Originaltitels „Neckerei".

1366
FIGUR „Mohr, Flöte spielend"
ca. 1897/98
1366,(6,9,8,6,5 vertikal) Firmenmarke 7, gepr. „P"= Z3, H. 81 cm; GK, S. 14, 35 (m. Abb.); Wien Museum Inv. 220.864
1366 (Firmenmarke ohne Ang.), H. 73 cm; GSF, S. 10
Ausst.: HMW (jetzt Wien Museum) 1985/86; Orientalische Reise, Wien Museum 2003.
-> vgl. 1127; 1668
Lit.: GSF, S. 10, hier mit der Ang. des Originaltitels „Mohr, Flöte spielend"; GK, S. 14, 35 (m. Abb.); Mayr-Oehring, E. u. E. Doppler: Orientalische Reise, Wien 2003, S. 252, 253 (m. Abb.); Pinhas, S. 64, 82 (m. Abb.).

1367
FIGUR „Mohrengigerl" (Sitzender rauchender Neger mit Strohhut) (auf Sessel)
ca. 1898
1367 (Firmenmarke ohne Ang.), H. 49 cm; Pinhas, S. 32, 69 (m. Abb.)
1367 (Firmenmarke ohne Ang.), H. 52 cm; GSF, S. 10
Bem.: Als „Gigerl" wird in Wien ein Modegeck bzw. ein Mann bezeichnet, der sich von einer Frau aushalten lässt. Lt. Katalog war einen Stuhl oder ein Schaukelstuhl bei Goldscheider separat zu diesem Modell zu beziehen.
-> vgl. 1806
Lit.: GSF, S. 10, hier mit der Ang. des Originaltitels „Mohrengigerl".

1368
FIGUR „Mohrengigerl" (Sitzender Neger mit Strohhut) (rittlings auf Sessel sitzend)
ca. 1898
1368 (Firmenmarke ohne Ang.), H. 49 cm; Pinhas, S. 33; 69; 71 (m. Abb.)
1368 (Firmenmarke ohne Ang.), H. 54 cm; GSF, S. 10
Bem.: Als „Gigerl" wird in Wien ein Modegeck bzw. ein Mann bezeichnet, der sich von einer Frau aushalten lässt. Lt. Katalog war ein Stuhl bei Goldscheider separat zu diesem Modell zu beziehen.
-> vgl. 1419; 1807
Lit.: GSF, S. 10, hier mit der Ang. des Originaltitels „Mohrengigerl".

1372
BÜSTE Mädchen mit großer Haube
ca. 1897/98
1372,47,(43,7,7 vertikal) Firmenmarke 11, H. 53 cm; eBay # 5397327531

1372?
Clarà, Juan (Joan)
Kind auf Hocker, sich einen Schuh anziehend
BRONZE
ca. 1900
250.1372, A. Goldscheider, Paris Vrai Bronze 8 (Firmenmarke?), H. 25 cm; Soth. L. 21.6.1991
ohne Ang. GS (Firmenmarke ohne Ang., wohl Arthur GS, Paris), H. 25 cm; Soth. L. 20.6.1985

1379
Prunier, (D./G.?)
Stehender Araber mit Gewehr (auf Landschaftssockel)
ca. 1898
1379 (Firmenmarke ohne Ang.), sign. Prunier, H. 105 cm; Pinhas, S. 29 (m. Abb.)
-> vgl. 1397; 1421; 1574
-> vgl. 1596= BÜSTE

1379
Prunier, (D./G.?)
FIGUR „Haremswächter" (Stehender Araber mit Gewehr) (auf Landschaftssockel)
ca. 1898
1379 (Firmenmarke ohne Ang.), sign. Prunier, H. 105 cm; Pinhas, S. 29 (m. Abb.)
1379 (Firmenmarke ohne Ang.), H. 104 cm; GSF, S. 10
-> vgl. 1397; 1421; 1574
Lit.: GSF, S. 10, hier mit der Ang. des Originaltitels „Haremswächter".

1382
Stehender Knabe mit Kopftuch und Schildkröte (auf Landschaftssockel mit Zaun)
ca. 1898
1382 (Firmenmarke ohne Ang.), H. fehlt; Pinhas, S. 68 (m. Abb.)

1383
BÜSTE Mädchen
ca. 1898
1383 (Firmenmarke ohne Ang.), H. 50 cm; Weichmann 24.6.1988
Bem.: Im Auktionskat. weder nähere Angaben zum Objekt, noch eine Abb.

1387
FIGUR „Fischerknabe" (Stehender Junge mit Fisch in der linken Hand) (auf Natursockel)
ca. 1897/98
1387,(16,2,3,4,0 vertikal) Firmenmarke 11, H. 57,7 cm; Henry's 26.8.1998; Doro 21.9.1999 (Fisch fehlt)
1387 (Firmenmarke ohne Ang.), H. 58 cm; GSF, S. 10
-> vgl. 1137
Lit.: GSF, S. 10, hier mit der Ang. des Originaltitels „Fischerknabe".

1389
Sitzender lachender Negerjunge mit Hut und Schal (auf Sessel)
ca. 1897/98
1389 (Firmenmarke ohne Ang.), H. 50 cm; Pinhas, S. 32; 69 (m. Abb.)
Bem.: Der Sessel ist aus Bambus gefertigt.

1390
Mercadier, J.
BÜSTE „Beppo" (mit gelockten Haaren)
ca. 1898
1390,72,37 Firmenmarke 7, Repr. Res., sign. J. Mercadier, gepr. „P"= Z3, H. 54 cm; Ruef 478; 480
1390 (Firmenmarke ohne Ang.), H. 57 cm; GSF, S. 10
Lit.: GSF, S. 10, hier mit der Ang. des Originaltitels „Beppo".

1393
Wind, Josef (zugeschr.)
FIGUR „Gefesselte Sklavin" (Stehende, an einen Baumstamm mit Handfesseln gebundene nackte Araberin mit Turban)
ca. 1898
1393,49,31 gem. X, FG Wien (Firmenmarke?), unleserlich sign., gepr. „P"= Z3, H. 57,2 cm; Soth. com. # 1BXQ
1393,141,59 Firmenmarke Sculpture 13, Ausformung um/kurz nach 1900, unleserlich sign. „Y. Kaye"?, H. 57 cm; Soth. L. 18.3.1988
1393 (Firmenmarke ohne Ang.), H. 57 cm; GSF, S. 10
Bem.: Bildhauerische Arbeit, Figur und Turban sprechen eindeutig für einen Entwurf von Josef Wind.
-> vgl. .1119
(-> vgl. Modell 40 BRONZE; zum Künstler Josef Wind -> vgl. 1399)
Lit.: GSF, S. 10, hier mit der Ang. des Originaltitels „Gefesselte Sklavin".

1395
FIGUR „Bachelse"
ca. 1898/99
1395 (Firmenmarke ohne Ang.), H. 60 cm; GSF, S. 10
Lit.: GSF, S. 10, hier mit der Ang. des Originaltitels „Bachelse".

1397
Prunier, (D./G.?)
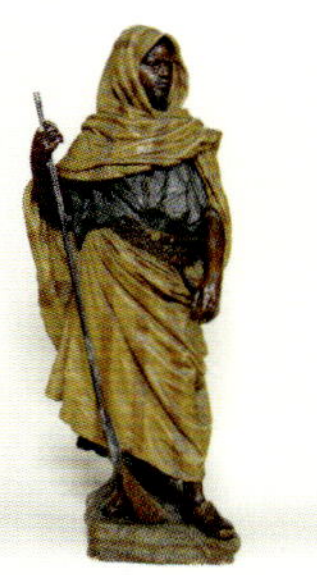
FIGUR „Haremswächter" (Stehender Araber mit Gewehr) (auf Landschaftssockel)
ca. 1898
1397,30,22 Firmenmarke 7, unsign., gepr. „P"= Z3, H. 77 cm; MAK Wien, Inv. Ke 10602-16
-> vgl. 1379; 1421; 1574
-> vgl. 1596= BÜSTE

1399
Wind, Josef
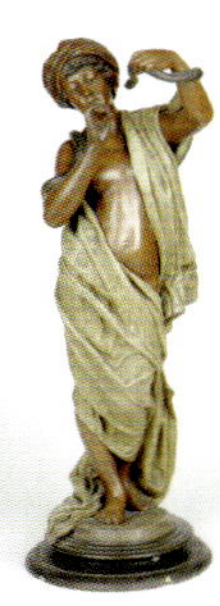
FIGUR „Schlangenzauberin"/ „Schlangenbändigerin" (Stehender Halbakt einer Orientalin mit Tuch, Turban, Flöte und Schlange)
ca. 1898
1399,714,3 Firmenmarke Sculpture 13, H. 78 cm; Wien Museum Inv. 219.110
(1399, ohne Nr.), ohne Firmenmarke, H. 76 cm; PB Wien 19; Bem.: Durch Beschädigung fehlt sowohl die vollständige Modellnummer sowie der Firmenstempel.
ohne Ang. H. 80 cm; Richemond (1996), S. 70
1399 (Firmenmarke ohne Ang.), H. 78 cm; GSF, S. 10
Ausst.: ÖMKI 1898/99, hier mit Ang. des Originaltitels „Schlangenbändigerin".
-> vgl. 40 BRONZE; 644 (GSB)
-> vgl. 1295 (Variante)
Lit.: GSF, S. 10, hier mit der Ang. des Originaltitels „Schlangenzauberin"; GSB, Pl. 56, hier ist die Figur mit dem franz. Originaltitel „Charmeuse des Serpents (nue)" unter Nr. 644 (H. 85 cm) abgebildet. Ang. zum Material (Terrakotta/Bronze) fehlen. Die Differenz bei der Größenangabe dürfte von unterschiedlichen Sockelhöhen herrühren; Richemond (1999), S. 235 (m. Abb.); Pinhas, S. 68 (m. Abb.).

1400
FIGUR „Knabe, mit zerbr(ochenem) Krug"
ca. 1897/98
1400 (Firmenmarke ohne Ang.), H. 59 cm; GSF, S. 10
-> vgl. 1195
Lit.: GSF, S. 10, hier mit der Ang. des Originaltitels „Knabe, mit zerbr(ochenem) Krug".

1402
Prémont, F. du
BÜSTE „Holländerin" (Junge Frau mit Spitzenhaube)
ca. 1897/98
1402,5,29 Firmenmarke 7, unsign., H. 47 cm; eBay # 2829559627
1402 (Firmenmarke ohne Ang.), H. 48 cm; GSF, S. 10
-> vgl. 1098
Lit.: GSF, S. 10, hier mit der Ang. des Originaltitels „Holländerin"; Pinhas, S. 97 (m. Abb.).

1403
Clarà, Juan (Joan)
„Alé" (Sitzendes, lachendes Kind)
BRONZE
Inschrift: ALÉ
ca. 1900
1403,76 (Firmenmarke Bronze1 und/oder 2?), H. 22,5 cm; Soth. L. 20.5.1994
Bem.: Es handelt sich hier um die rechte Figur der „Drei, auf einer Bank sitzenden Kinder" Modell-Nr. 1485.
-> vgl. 1485 GRUPPE

1406
FIGUR „Lautenschläger (bekleidet)"
ca. 1898/99
1406 (Firmenmarke ohne Ang.), H. 76 cm; GSF, S. 10
-> vgl. 1251 (nackt); 1551 (bekleidet)
Lit.: GSF, S. 10, hier mit der Ang. des Originaltitels „Lautenschläger (bekleidet)".

1409
Rochella, D.
BÜSTE Araberin mit langem Kopftuch (nach links blickend) (ohne/mit Halskette)
ca. 1897
OHNE HALSKETTE
1409,(6,5,3,1,5 vertikal?) FG Wien (Firmenmarke ohne Ang.), Repr. Res., sign. Rochella, H. 65,5 cm; Doro 3.5.1995
MIT HALSKETTE
1409 (ohne Ang.), FG Wien (Firmenmarke ohne Ang.), sign. D. Rochella, H. 65,5 cm; Doro 24.5.2000
-> vgl. 1109 (sign. Thiele)
-> vgl. 1340
-> vgl. 1898 (sign. Thiele)

1410
Telsch (d.i. Telcs, Ede/Eduard)
BÜSTE „Raucher"
ca. 1898/99
1410 (Firmenmarke ohne Ang.), H. 52 cm; GSF, S. 10
Ausst.: ÖMKI 1899/1900, hier jeweils mit Ang. des Originaltitels „Raucher" und des Künstlernamens „Telsch". Ausgestellt war die Bronzevariante dieser Büste.
Lit.: GSF, S. 10, hier mit der Ang. des Originaltitels „Raucher; Neuwirth (1974), S. 75.

1412
BÜSTE „Shocking"
ca. 1898/99
1412 (Firmenmarke ohne Ang.), H. 32 cm; GSF, S. 10
Lit.: GSF, S. 10, hier mit der Ang. des Originaltitels „Shocking".

1413
Langer, (Leopold?)
FIGUR „Korbflechter" (Sitzender Knabe, einen Korb flechtend) (Sockel als Steinmauer)
ca. 1898
1413,60,16 ger. 14, Firmenmarke 7, gepr. „P"= Z3, H. 83 cm; Wien Museum Inv. 219.139
ohne Ang. GS (Firmenmarke ohne Ang.), H. 80,5 cm; Soth. L. 9.6.1989
1413 (Firmenmarke ohne Ang.), H. 83 cm; GSF, S. 10

-> vgl. 1650
Lit.: GSF, S. 10, hier mit der Ang. des Originaltitels „Korbflechter"; Pinhas, S. 69 (m. Abb.).

1414
FIGUR „Odaliske"
ca. 1897/98 (für den Entwurf)
1414 (Firmenmarke ohne Ang.), H. 62 cm; GSF, S. 10
-> vgl. 1118
Lit.: GSF, S. 10, hier mit der Ang. des Originaltitels „Odaliske".

1417
Stehender Negerjunge mit großem Tuch (auf Rechtecksockel)
ca. 1898
1417 (Firmenmarke ohne Ang.), H. 128 cm; Pinhas, S. 29 (m. Abb.)

1419
FIGUR „Gigerl" (Sitzender Neger mit Strohhut) (rittlings auf Sessel sitzend)
ca. 1898
1419 (Firmenmarke ohne Ang.), H. fehlt; Pinhas, S. 33; 69; 71 (m. Abb.)
1419 (Firmenmarke ohne Ang.), H. 122 cm; GSF, S. 10
Bem.: Als „Gigerl" wird in Wien ein Modegeck bzw. ein Mann bezeichnet, der sich von einer Frau aushalten lässt. Lt. Katalog war einen Stuhl oder ein Schaukelstuhl bei Goldscheider separat zu diesem Modell zu beziehen.
-> vgl. 1368; 1807
-> vgl. 1666= WANDMASKE
Lit.: GSF, S. 10, hier mit der Ang. des Originaltitels „Gigerl".

1421
Prunier, G.
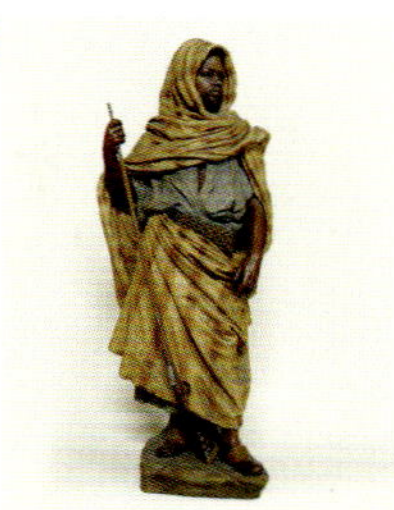
FIGUR „Haremswächter" (Stehender Araber mit Gewehr) (auf Landschaftssockel)
ca. 1898
1421,98,70 ger. 14, FG Wien (Firmenmarke 7, zwei mal), Repr. Res., unsign., H. cm; MAK Wien, Inv. Ke 10602-28
ohne Ang. GS (Firmenmarke Sculpture 13?), sign. G. Prunier, H. 81 cm; Soth. L. 12.6.1986; 5.11.1987
1421 (Firmenmarke ohne Ang.), H. 80 cm; GSF, S. 10
-> vgl. 1379; 1397; 1574
-> vgl. 1596= BÜSTE
Lit.: GSF, S. 10, hier mit der Ang. des Originaltitels „Haremswächter"; Pinhas, S. 68 (m. Abb.).

1422
FIGUR „Wahrheit"
ca. 1897/98 (für den Entwurf)
1422 (Firmenmarke ohne Ang.), H. 90 cm; GSF, S. 10
-> vgl. 982b
Lit.: GSF, S. 10, hier mit der Ang. des Originaltitels „Wahrheit".

1423
BÜSTE „Mahdi-Krieger"
ca. 1898
1423 (Firmenmarke ohne Ang.), H. 48 cm; GSF, S. 10
-> vgl. 1308; 1614= WANDMASKE
Lit.: GSF, S. 10, hier mit der Ang. des Originaltitels „Mahdi-Krieger".

1424
Langer, (Leopold?)
BÜSTE „Orientalin"
ca. 1898/99
1424,97,19 gem. II, FG Wien (Firmenmarke ohne Ang.), Repr. Res., sign. Lanogé (!), H. 49 cm; Herr 34
1424,158,5 Firmenmarke 7, Repr. Res., sign. Langer, H. 48,3 cm; eBay # 2850514394
1424 (Firmenmarke ohne Ang.), H. 50 cm; GSF, S. 10
Bem.: Es dürfte sich wohl um den Künstler Leopold Langer handeln.
-> vgl. 1358
Lit.: GSF, S. 10, hier mit der Ang. des Originaltitels „Orientalin".

1427 -> ÜBERSCHNEIDUNG
Weihwasserbecken (vierpassig, auf gedrehten Säulen)
ca. 1898/99
1427,6,27 Firmenmarke 9, Repr. Res., H. 53 cm; eBay # 378727693; WKA 27
Bem.: Diese vierpassige „Kanzel" (lt. WKA), meiner Ansicht jedoch stilistisch eher ein „Weihwasserbecken", ist in seiner Art einmalig für GS. Es werden in diesem Pasticcio romanische und gotische Stilelemente vermischt. Die Reliefs stellen u. a. neben dem Kirchenvater Ambrosius auch Maria mit Kind und zwei Anbetungsengel dar. Die Inschriften, die sich um das ganze Objekt ziehen, lassen sich leider anhand der vorliegenden Fotos nicht eindeutig verifizieren. Die Säulen gehen auf einen Typus des 12. Jahrhunderts zurück. Dieses ungewöhnliche Modell könnte auch einem realen Vorbild, einem Weihwasserbecken (Mailand?), nachgebildet sein.

1427 -> ÜBERSCHNEIDUNG
BÜSTE Pfeife rauchender Knabe mit Kappe
ca. 1898
1427 (Firmenmarke ohne Ang.), H. fehlt; Pinhas, S. 68 (m. Abb.)

1428
Rust, L.
BÜSTE „Triumphator" (Römer mit Lorbeerkranz)
Inschrift: IO TRIVMPHE
ca. 1898
1428,202,21 gem. weiß XIX, Firmenmarke Sculpture 13, Repr. Res., sign. L. Rust, H. 52,5 cm; Doro 13.9.2004
1428 (Firmenmarke ohne Ang.), H. 53 cm; GSF, S. 10
-> vgl. 2123
Ausst.: ÖMKI 1898/99; 1902/03; 1903/04, hier jeweils mit Ang. des Originaltitels „Triumphator".
Lit.: GSF, S. 10, hier mit der Ang. des Originaltitels „Triumphator"; Neuwirth (1974), S. 72; Pinhas, S. 68 (m. Abb.).

1431
Sitzende Afrikanerin (Halbakt) (auf Ovalsockel)
ca. 1898
1431 (Firmenmarke ohne Ang.), H. 64 cm; Pinhas, S. 27 (m. Abb.)

1433
FIGUR „La lune coissante (Mondgöttin)"
ca. 1898/99
1433 (Firmenmarke ohne Ang.), H. 69 cm; GSF, S. 10
Bem.: Goldscheider führte beide, sowohl den franz. als auch den deutschen, Originaltitel für diese Figur.
-> vgl. 1798; 1937
Lit.: GSF, S. 10, hier mit der Ang. des Originaltitels „La lune coissante (Mondgöttin)".

1435
FIGUR „Am Strande"
ca. 1898/99
1435 (Firmenmarke ohne Ang.), H. 59 cm; GSF, S. 10
Lit.: GSF, S. 10, hier mit der Ang. des Originaltitels „Am Strande.

1440
Gracien, J./G.?
FIGUR „Vasenträgerin" (Stehende junge Frau mit Krug auf der Schulter) (auf Rechtecksockel)
ca. 1898
1440 (Firmenmarke ohne Ang.), unsign., H. fehlt; Pinhas, S. 73 (m. Abb.)
1440 (Firmenmarke ohne Ang.), H. 102 cm; GSF, S. 10
-> vgl. 1035a; 1035b
Lit.: GSF, S. 10, hier mit der Ang. des Originaltitels "Vasenträgerin".

1441
Allouard, Henri Emile
BÜSTE „Rococo" (Junge Dame nach rechts blickend)
ca. 1898
1441 (Firmenmarke ohne Ang.), H. fehlt; Pinhas, S. 68 (m. Abb.)
1441 (Firmenmarke ohne Ang.), H. 59 cm; GSF, S. 11
Ausst.: ÖMKI 1898/99, hier mit Ang. des Originaltitels „Rococo". Da die Büste in drei verschiedenen Höhen produziert wurde und die Mod.-Nr. im Ausst.-Kat. nicht angegeben ist, lässt sich nicht sagen welche Größe ausgestellt war.
-> vgl. 2023; 2079
Lit.: GSF, S. 11, hier mit der Ang. des Originaltitels „Rococo".

1442
BÜSTE „Ophelie"
Inschrift: OPHELIE
ca. 1898/99
1442,67,34 FG Wien (Firmenmarke ohne Ang.), Repr. Res., H. 38 cm; Henry's 28.12.1994
1442 (Firmenmarke ohne Ang.), H. 38 cm; GSF, S. 11
Bem.: Der Name Ophelie (bzw. Ophelia) bezeichnet die weibliche Hauptfigur aus William Shakespeares "Hamlet".
Lit.: GSF, S. 11, hier mit der Ang. des Originaltitels „Ophelie".

1443
FIGUR „Graziella" (Stehender Mädchenakt mit Haarband) (auf Landschaftssockel)
ca. 1898/99
1443 (Firmenmarke ohne Ang.), H. fehlt; Pinhas, S. 70 (m. Abb.)
1443 (Firmenmarke ohne Ang.), H. 90 cm; GSF, S. 11
Bem.: Die Mod.-Nr. 1477 zeigt dieselbe Figur in ident. Höhe, jedoch mit einem Tuch um die Hüften.
-> vgl. 1477 (Variante)
Lit.: GSF, S. 11, hier mit der Ang. des Originaltitels „Graziella".

1444
GRUPPE „Vertrauliche Mitteilung"
ca. 1898/99
1444 (Firmenmarke ohne Ang.), H. 50 cm; GSF, S. 11
-> vgl. 1484= FIGUR Knabe; 1485= FIGUR Mädchen
Lit.: GSF, S. 11, hier mit der Ang. des Originaltitels „Vertrauliche Mitteilung".

1445
Etoile, A.
GRUPPE „Erste Liebe" (Schäfer und Schäferin in antikisierendem Gewand) (auf Landschaftssockel)
ca. 1898/99
1445,9,50 Firmenmarke 7, sign. A. Etoile, H. 89 cm; GK, S. 14, 70 (m. Abb.); Wien Museum Inv. 182.066
1445 (Firmenmarke ohne Ang.), H. 88 cm; GSF, S. 11
Ausst.: HMW (jetzt Wien Museum) 1985/86.
-> vgl. 1800
Lit.: GSF, S. 11, hier mit der Ang. des Originaltitels „Erste Liebe"; GK, S. 14, 70 (m. Abb.); Pinhas, S. 71 (m. Abb.).

1446
Maison, Rudolf
FIGUR „Neger" (Stehender Neger mit Schild und Speer) (auf Rundsockel)
ca. 1898
1446 (Firmenmarke ohne Ang.), unsign., H. fehlt; Pinhas, S. 69 (m. Abb.)
1446 (Firmenmarke ohne Ang.), H. 74 cm; GSF, S. 11
Bem.: Die Figur dieses Afrikaners lässt sich durch die hier angegebene Lit. eindeutig R. Maison zuordnen.
Lit.: GSF, S. 11, hier mit der Ang. des Originaltitels „Neger"; Schmid, M.: Rudolf Maison, in: VKMH, 1.1898/99, S. 4 (m. Abb.), hier eine vergleichbare Statuette.

1447
Maison, Rudolf
FIGUR „Augur"
ca. 1898
1447,144,13 GS Wien (Firmenmarke ohne Ang.), H. 47,2 cm; Doro 14.9.2001
1447 (Firmenmarke ohne Ang.), H. 49 cm; GSF, S. 11
Bem.: Die Figur des „Augur" (altrömischer Priester) wurde 1890 entworfen, dies ließ sich durch ein sign. und datiertes Exemplar „R. Maison 1890" (Gipsausformung, H. 53 cm) im KH nachweisen.
Ausst.: ÖMKI 1898/99, hier mit Ang. des Originaltitels „Augur".
-> vgl. 2147 BÜSTE
Lit.: Schmid, M.: Rudolf Maison, in: VKMH, 1.1898/99,

S. 8 (m. Abb.), hier mit dem Titel „Römischer Augur"; GSF, S. 11, hier mit der Ang. des Originaltitels „Augur"; Ostini, F. v.: Rudolf Maison in DK, 3.1901, S. 137 (m. Abb.), hier bezeichnet „Römischer Augur"; N.N.: Wiener Kleinplastik einst und jetzt, in: MW, 2.1920/21, Nr. 12, S. 29 (m. Abb.) hier mit Ang. des Originaltitels „Augur"; Neuwirth (1974), S. 67; Pinhas, S. 69 (m. Abb.).

1448
GRUPPE „Eselreiter"
ca. 1898/99
1448 (Firmenmarke ohne Ang.), H. 96 cm; GSF, S. 11
-> vgl. 1835
Lit.: GSF, S. 11, hier mit der Ang. des Originaltitels „Eselreiter".

1449
BÜSTE Frau mit großer Schleife und Spitzenbesatz am Kleid
ca. 1898
1449,129,18 Firmenmarke 7, Repr. Res., H. 31,5 cm; eBay # 2821184951
Lit.: Pinhas, S. 70 (m. Abb.)

1450
BÜSTE Frau mit zwei Rosen am Kleid
ca. 1898
1450,157,25 gem. weiß IV, Firmenmarke 7, Repr. Res., H. 31,1 cm; eBay # 280003629703
Lit.: Pinhas, S. 70 (m. Abb.)

1452
Stehende Orientalin mit verschränkten Armen (auf Rechtecksockel)
ca. 1898
1452 (Firmenmarke ohne Ang.), H. fehlt; Pinhas, S. 73 (m. Abb.)

1453
BÜSTE „Madame Lebrun"
ca. 1898/99
1453 (Firmenmarke ohne Ang.), H. 98 cm; GSF, S. 11
-> vgl. 1918
Bem.: Dargestellt ist die franz. Malerin Marie Louise Élisabeth Vigée-Lebrun (1755–1842).
Lit.: GSF, S. 11, hier mit der Ang. des Originaltitels „Madame Lebrun".

1455
BÜSTE Frau (nach links blickend) mit großem Hut
ca. 1898
1455 (Firmenmarke ohne Ang.), H. fehlt; Pinhas, S. 70 (m. Abb.)

1456
BÜSTE Frau (mit leicht gesenktem Blick) mit großem Hut mit Schleife
ca. 1898
1456 (Firmenmarke ohne Ang.), H. fehlt; Pinhas, S. 70 (m. Abb.)

1457
BÜSTE Frau (leicht gesenkter Blick nach rechts) mit kleinem Hut mit Schleife
ca. 1898
1457 (Firmenmarke ohne Ang.), H. fehlt; Pinhas, S. 70 (m. Abb.)

1458
Poirier, B.
Stehende Frau mit langem Kleid und entblößten Brüsten
ca. 1898
1458 (Firmenmarke ohne Ang.), H. 100 cm; Pinhas, S. 23 (m. Abb.)

1459
Cherc
BÜSTE Frau mit altägyptischem Kopfputz
ca. 1898
1459,28,10 Firmenmarke 7, H. 61 cm; GK, S. 14, 71 (m. Abb.); Wien Museum Inv. 182.067
Ausst.: HMW (jetzt Wien Museum) 1985/86.
Lit.: GK, S. 14, 71 (m. Abb.)

1461
Petri (d.i. Petrides, János) (Johann)
VASE Diana als Bogenschützin (vollplastisch an der Vasenwandung stehend, säulenartiger Vasenschaft mit Löwenmaskaron) (große Bodenvase)
ca. 1899/1900
01461,34,17 Firmenmarke Sculpture 13, Fabriqué en Autriche, sign. Petri, H. 61 cm; eBay # 2003463539; 6505365746
Bem.: Der Bogen ist lt. Anbieter aus Bronze gefertigt.

1462
Maison, Rudolf
FIGUR „Philosoph"
ca. 1898
1462 (Firmenmarke ohne Ang.), H. 70 cm; Pinhas, S. 69 (m. Abb.).
1462 (Firmenmarke ohne Ang.), H. 70 cm; GSF, S. 11
Ausst.: ÖMKI 1898/99, hier mit Ang. des Originaltitels „Philosoph".
-> vgl. 1888
Lit.: GSF, S. 11, hier mit der Ang. des Originaltitels „Philosoph"; Schmid, M.: Rudolf Maison, in: VKMH, 1.1898/99, S. 4 (m. Abb.), hier mit dem Titel „Der Philosoph", S. 5 „Aber Maison weiß seine Studien der menschlichen Figur stets interessant mit geistigem Inhalt zu füllen, ihnen ein Motiv zu geben, ernst oder heiter ohne dabei banaler Neugierde und Schaulust Konzessionen zu machen. So in dem wundervollen, klassisch geformten Körper des Philosophen, mit den strengen, ausdrucksvoll geprägten Zügen, der sinnend dem Liebesgirren der Tauben zu seinen Füßen zusieht"; Ostini, F. v.: Rudolf Maison in DK, 3.1901, S. 132–133, 138 (m. Abb.), hier bezeichnet „Ein Philosoph"; Neuwirth (1974), S. 67.

1463 -> ÜBERSCHNEIDUNG
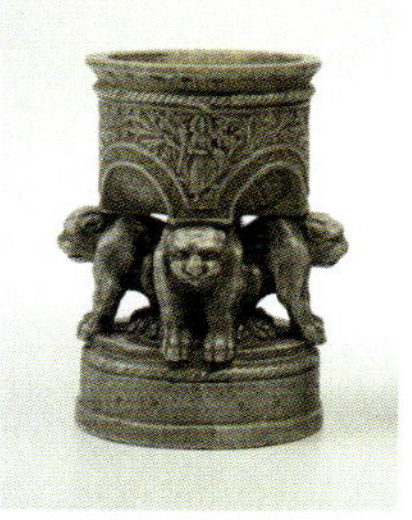
TAFELAUFSATZ mit vier löwenartigen Tieren als Schalenträger
ca. 1898/1899
01463,7,7 Firmenmarke 4, Fabriqué en Autriche, H. 14,8 cm, Ø 11 cm; Wien Museum Inv. 220.781
01463,47,55 Firmenmarke 4, Firmenmarke Adler 17, Fabriqué en Autriche, H. 14,8 cm, Ø 11 cm; Wien Museum Inv. 220.780

1463 -> ÜBERSCHNEIDUNG
BÜSTE „Mohrin"
ca. 1898/99
1463 (Firmenmarke ohne Ang.), H. 43 cm; GSF, S. 11
Lit.: GSF, S. 11, hier mit der Ang. des Originaltitels „Mohrin".

1464
GRUPPE „Pan und Bacchantin"/„Bacchante et Silène" (Stehende nackte Bacchantin, eine Bacchusbüste umarmend) (auf Rundsockel)
ca. 1898/99
1464 (Firmenmarke ohne Ang.), H. fehlt; Pinhas, S. 73 (m. Abb.)
1464 (Firmenmarke ohne Ang.), H. 76 cm; GSF, S. 11
-> vgl. 764 (GSB), hier mit Ang. des franz. Originaltitels.
Lit.: GSF, S. 11, hier mit der Ang. des Originaltitels „Pan und Bacchantin".

1470
Cherc
FIGUR „Negerknabe (fischend)" (Stehender Negerjunge mit Hut, Angel und Fisch)
ca. 1898/99
1470,117,8 Firmenmarke 8, Repr. Res., H. 60 cm; Doro 1573 (5.–14.12.1989)
1470,170,37 Firmenmarke 7, Repr. Res., H. 59 cm; Wien Museum Inv. 219.122; Bem.: Hier fehlt der Fisch.
1470 (Firmenmarke ohne Ang.), H. 60 cm; GSF, S. 11
Bem.: Dieses Modell ist von der Figur her identisch mit Modell-Nr. 1471. Die Attribute Angel und Fisch wurden bei GS als Variante zu Zündholz und Zündholzschachtel produziert.
-> vgl. 2134
-> vgl. 1471 (Variante)
-> vgl. 1472 (ähnl. Mod.)
Lit.: GSF, S. 11, hier mit der Ang. des Originaltitels „Negerknabe (fischend)"; Pinhas, S. 72 (m. Abb.).

1471
Cherc
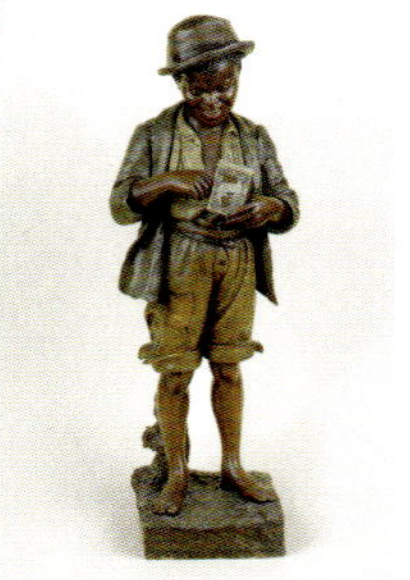
FIGUR „Negerknabe (rauchend)" (Stehender Negerjunge mit Hut) (und Zündholzschachtel)
ca. 1898/99
1471,65,8 Firmenmarke 7, Repr. Res., unsign.?, H. 59,7 cm; eBay # 859226090; Bem.: Zündholzschachtel scheint hier Original.
1471,253,8 Firmenmarke 7, Repr. Res., H. 58 cm; PB Wien 19; Bem.: Zündholzschachtel hier sek.
?,143,? FG (Firmenmarke ohne Ang.), unsign., H. 58,5 cm Ribeyre-Baron 14.3.1995; Bem.: Hier fehlt(en) Attribut(e).
1471 (Firmenmarke ohne Ang.), H. 60 cm; GSF, S. 11
Bem.: Dieses Modell ist von der Figur her identisch mit Modell-Nr. 1470. Die Attribute Zündholz und Zündholzschachtel wurden bei GS als Variante zu Angel und Fisch produziert.
-> vgl. 2135
-> vgl. 1470 (Variante)
-> vgl. 1472 (ähnl. Mod.)
Lit.: GSF, S. 11, hier mit der Ang. des Originaltitels „Negerknabe (rauchend)"; Richemond (1999), S. 81 (m. Abb.); Pinhas, S. 72 (m. Abb.).

1472
Cherc
FIGUR „Negerknabe mit zerbrochenem Krug" (Stehender Negerjunge mit Hut und zerbrochener Vase)
ca. 1898/99
1472 (Firmenmarke ohne Ang.), unsign., H. ca. 60 cm; Pinhas, S. 72 (m. Abb.).
1472 (Firmenmarke ohne Ang.), H. 60 cm; GSF, S. 11
Bem.: Dieses Modell ist von der Figur her ähnl. der Mod.-Nrn. 1470 und 1471. Die zerbrochene Vase am Sockel wurde bei GS als Variante, bzw. als Pendant zu Zündholz und Zündholzschachtel, sowie Angel und Fisch produziert.
-> vgl. 2136
-> vgl. 1470; 1471 (jew. ähnl. Mod.)
Lit.: GSF, S. 11, hier mit der Ang. des Originaltitels „Negerknabe mit zerbrochenem Krug".

1473 -> vgl. 1816

1475
FIGUR „Jungfrau von Orleans"
ca. 1898/99
1475 (Firmenmarke ohne Ang.), H. 119 cm; GSF, S. 11
-> vgl. 1836
Lit.: GSF, S. 11, hier mit der Ang. des Originaltitels „Jungfrau von Orleans".

1476
Stehende Nordafrikanerin mit Wassergefäß und Korb (auf konischem reliefiertem Sockel)
ca. 1898/99
1476 (Firmenmarke ohne Ang.), H. fehlt; Pinhas, S. 73 (m. Abb.)

1477
FIGUR „Graziella" (Stehender Mädchenhalbakt mit Hüfttuch und Haarband) (auf Landschaftssockel)
ca. 1898/99
1477 (Firmenmarke ohne Ang.), H. fehlt; Pinhas, S. 70 (m. Abb.)
1477 (Firmenmarke ohne Ang.), H. 90 cm; GSF, S. 11
Bem.: Die Mod.-Nr. 1443 zeigt dieselbe Figur in ident. Höhe, jedoch ohne Tuch um die Hüften.
-> vgl. 1443 (Variante)
Lit.: GSF, S. 11, hier mit der Ang. des Originaltitels „Graziella".

1478 -> ÜBERSCHNEIDUNG
Chavanne, (Jean Maurice?)
FIGUR „Vasenmaler" (Sitzender Römerjüngling mit Krug)
ca. 1898/99
1478 ger. IV, (Firmenmarke ohne Ang.), sign. Cha(r?)anne, H. 76 cm; Mehlis 28; Bem.: Der Krug, also das Attribut fehlt hier.
1478 (Firmenmarke ohne Ang.), H. 71 cm; GSF, S. 11
-> vgl. 1973
Lit.: GSF, S. 11, hier mit der Ang. des Originaltitels „Vasenmaler".

1478 -> ÜBERSCHNEIDUNG
Stehende, junge Frau mit Strohhut und Früchtekorb (auf getrepptem Sockel)
ca. 1898/99
1478,26,3 Firmenmarke 7, H. 118 cm; Wien Museum Inv. 219.106

1479
BÜSTE „Circassierin"
ca. 1898/99
1479 (Firmenmarke ohne Ang.), H. 67 cm; GSF, S. 11
Bem.: Dargestellt ist eine Frau der Tscherkessen (auch Zirkassier genannt, früheste Bewohner des Kaukasus), (Quelle: Wikipedia).
Lit.: GSF, S. 11, hier mit der Ang. des Originaltitels „Circassierin".

1484 -> ÜBERSCHNEIDUNG
Bormann, Wilhelm
VASE Putti
Form: quadratisch/becherförmig
Dekor: reliefierte Wandung mit Putti und Blütengirlanden
ca. 1899/1900
01484,47,10 Firmenmarke 10, sign. Bormann, H. 14,5 cm; eBay # 3353181993
Pendant -> vgl. 1486

1484 -> ÜBERSCHNEIDUNG
FIGUR „Vertrauliche Mitteilung (Knabe)"
ca. 1898/99
1484 (Firmenmarke ohne Ang.), H. 50 cm; GSF, S. 11
-> vgl. 1444= GRUPPE; 1485= FIGUR Mädchen
Lit.: GSF, S. 11, hier mit der Ang. des Originaltitels „Vertrauliche Mitteilung (Knabe)".

1485 -> ÜBERSCHNEIDUNG
FIGUR „Vertrauliche Mitteilung (Mädchen)"
ca. 1898/99
1485 (Firmenmarke ohne Ang.), H. 50 cm; GSF, S. 11
-> vgl. 1444= GRUPPE; 1484= FIGUR Knabe
Lit.: GSF, S. 11, hier mit der Ang. des Originaltitels „Vertrauliche Mitteilung (Mädchen)".

1485 -> ÜBERSCHNEIDUNG
Clarà, Juan (Joan)
Drei, auf einer Bank sitzende Kinder
BRONZE
ca. 1899/1900
1485,75 (Firmenmarke Bronze 1 und/oder 2?), H. 25 cm; Soth. L. 17.3.1989
-> vgl. 1403= rechte EinzelFIGUR
Lit.: Catley (1987), S. 105 (m. Abb.), hier bez. „Children watching Punch and Judy Show".

1486 -> ÜBERSCHNEIDUNG
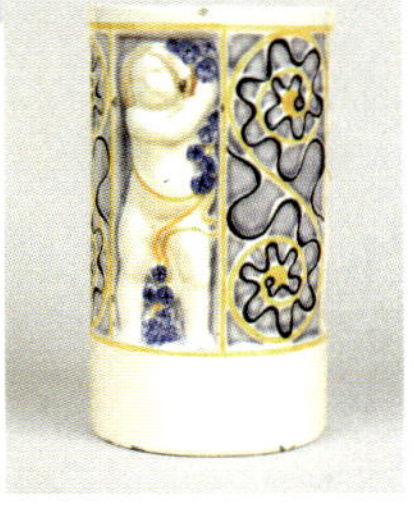
Bormann, Wilhelm / Beaume (d.i. Bormann, Wilhelm)
VASE Putti
Form: zylinder/becherförmig
Dekor: reliefierte Wandung mit Putti und floralen Elementen
ca. 1899/1900
01486,91,10 Firmenmarke 10, sign. Bormann, H. 14,5 cm; PB Wien 19
ohne Ang. sign. Beaume, H. 14,2 cm; Kett. 100
Bem.: Die beiden aufgelisteten Stücke belegen beide Signaturvarianten: Bormann / Beaume.
Pendant -> vgl. 1484

1486 -> ÜBERSCHNEIDUNG
Schmid
Stehender nackter Knabe
BRONZE / Marmorsockel (mit Rosengirlande)
ca. 1900
II.1486 Sculpture (Firmenmarke Bronze 1), H. der Figur 36,8 cm, Gesamth. 50,8 cm; eBay # 1448044328

1487
FIGUR „Knabe mit Cigarre" (Stehender rauchender Bauernbub mit Hut) (auf Rechtecksockel)
ca. 1898/99
1487 (Firmenmarke ohne Ang.), H. 97 cm; Pinhas, S. 36 (m. Abb.)
1487 (Firmenmarke ohne Ang.), H. 98 cm; GSF, S. 11
Bem.: Bei diesem Modell verwendete Goldscheider für die beiden produzierten Größen zwei unterschiedliche Titel: Knabe mit Cigarre und Raucher (Mod.-Nr. 1927).
-> vgl. 1927
Lit.: GSF, S. 11, hier mit der Ang. des Originaltitels „Knabe mit Cigarre"; Pinhas, S. 73 (m. Abb.).

1492
FIGUR „Chansonette"
ca. 1898/99
1492 (Firmenmarke ohne Ang.), H. 103 cm; GSF, S. 11
Lit.: GSF, S. 11, hier mit der Ang. des Originaltitels „Chansonette".

1493
BÜSTE Winzerin (Junge Frau mit großer Haube, nach links blickend) (auf mit Weintrauben geschmücktem Sockel)
ca. 1899/1900
1493 (Firmenmarke ohne Ang.), H. fehlt; Pinhas, S. 74 (m. Abb.)
Pendant -> vgl. 1494

1494
BÜSTE Bäuerin (Junge Frau mit Hut, nach rechts blickend) (auf mit Ähren und Sichel geschmücktem Sockel)
ca. 1899/1900
1494 (Firmenmarke ohne Ang.), H. fehlt; Pinhas, S. 74 (m. Abb.)
Pendant -> vgl. 1493

1497
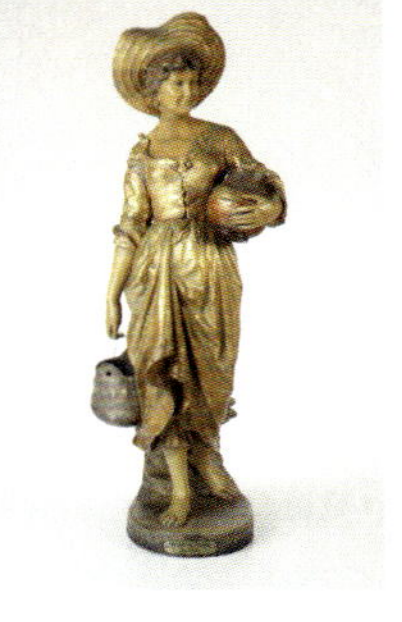
Sardon / Tell, E.
FIGUR „Caprieserin"/„Fille de Capri" (Stehende Wasserträgerin mit Strohhut) (auf Natursockel)
ca. 1899/1900
1497,1671,25 Firmenmarke Sculpture 13, Fabriqué en Autriche, sign. E. Tell, H. 65 cm; Wien Museum Inv. 219.126
1497 (Firmenmarke ohne Ang.), H. 67 cm; GSF, S. 11
Bem.: An der Plinthe betitelt („Fille de Capri") und sign. „Sardon" (Metallschild).
-> vgl. 1913
-> vgl. 1761= BÜSTE
Pendant -> vgl. 1565
Lit.: GSF, S. 11, hier mit der Ang. des Originaltitels „Caprieserin"; Pinhas, S. 73 (m. Abb.).

1498
GRUPPE „Neger und Leopard"
ca. 1898/99
1498 (Firmenmarke ohne Ang.), H. 128 cm; GSF, S. 11
Lit.: GSF, S. 11, hier mit der Ang. des Originaltitels „Neger und Leopard".

1499 -> ÜBERSCHNEIDUNG
Cherc
VASE Disteln
Form: balusterförmig
Dekor: reliefierte Disteln und Insekten
ca. 1900
01499,4,7 Firmenmarke Sculpture 13, H. 40 cm; Doro 4.3.2002

1499 -> ÜBERSCHNEIDUNG
Maison, Rudolf
GRUPPE „Gänsemädchen"
ca. 1898/99
1499 (Firmenmarke ohne Ang.), H. 57 cm; GSF, S. 11
Lit.: GSF, S. 11, hier mit der Ang. des Originaltitels „Gänsemädchen"; DK, 3.1901, S. 138 (m. Abb.); Neuwirth (1974), S. 67.

1500 -> ÜBERSCHNEIDUNG
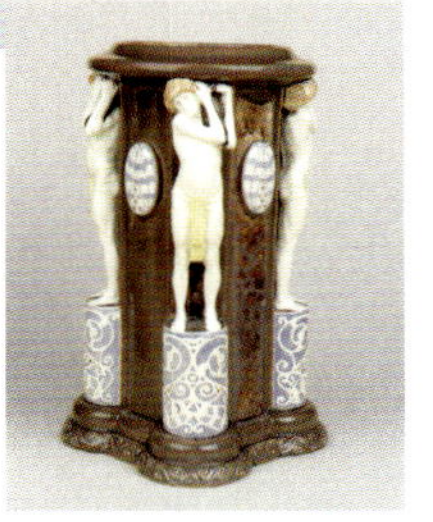
VASE Vier Atlanten
Form: achtpassig
Dekor: vier stehende, plastische Jünglinge als Atlanten, ovale, reliefierte Medaillons
ca. 1900
01500,3,10 Firmenmarke 9, Repr. Res., unleserlich sign., H. 33,3 cm; PB Wien 19

1500 -> ÜBERSCHNEIDUNG
FIGUR „Venet(ianische) Wasserträgerin" (Stehendes, nach vorne gebeugtes Bauernmädchen mit zwei Wassergefäßen, die durch einen Stab verbunden sind) (auf getrepptem Sockel)
ca. 1898/99
1500 (Firmenmarke ohne Ang.), H. 93 cm; Pinhas, S. 70 (m. Abb.)
1500 (Firmenmarke ohne Ang.), H. 93 cm; GSF, S. 11
-> vgl. 1947
Lit.: GSF, S. 11, hier mit der Ang. des Originaltitels „Venet(ianische) Wasserträgerin".

1502
Bortone, Ipazzio Antonio
BÜSTE „Annita" (Lachendes Bauernmädchen mit Weintrauben)
ca. 1898/99
1502 (Firmenmarke ohne Ang.), H. fehlt; Pinhas, S. 74 (m. Abb.)
1502 (Firmenmarke ohne Ang.), H. 49 cm; GSF, S. 11
Bem.: Im ÖMKI war 1900/01 die Maske „Annita" (Mod.-Nr. 2029) von Bortone ausgestellt. Es ist daher anzunehmen, dass die hier vorliegende Büste ebenfalls von diesem Künstler entworfen wurde. Ein sign. Ex. der Maske bzw. der Büste, lässt sich aber derzeit nicht belegen.
-> vgl. 1740; 2130= jew. Variante
-> vgl. 2029= WANDMASKE
Lit.: GSF, S. 11, hier mit der Ang. des Originaltitels „Annita".

1505
BÜSTE Lachendes Mädchen mit Kirschbaumzweig, einen darauf sitzenden Vogel beobachtend
ca. 1899/1900
1505 (Firmenmarke ohne Ang.), H. fehlt; Pinhas, S. 74 (m. Abb.)

1506 -> ÜBERSCHNEIDUNG
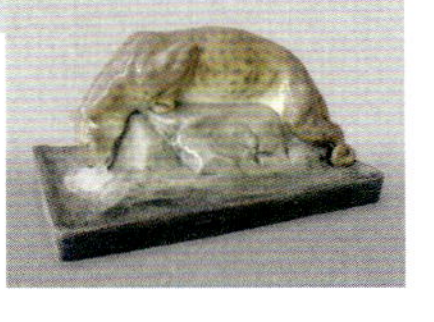
Gesa, Jos(ep)
Wassertrinkender Leopard auf Felsen (auf Rechtecksockel)
ca. 1899/1900
01506,5,69 Firmenmarke 9, Repr. Res., sign. Gesa Jos., H. 14 cm, L. 28 cm; Sammlung Stressless, Portugal

1506 -> ÜBERSCHNEIDUNG
Cherc
BÜSTE „Epheu" (Junge Frau mit Efeuschmuck)
ca. 1899/1900
1506 (Firmenmarke ohne Ang.), unsign., H. 75 cm; Pinhas, S. 73 (m. Abb.)
1506 (Firmenmarke ohne Ang.), H. 75 cm; GSF, S. 11
-> vgl. 1949; 2082
Lit.: GSF, S. 11, hier mit der Ang. des Originaltitels „Epheu".

1507
GRUPPE „Avant" (Biedermeierpaar, sie flehend) (auf gekehltem Sockel)
ca. 1899/1900

1507 (Firmenmarke ohne Ang.), H. fehlt; Pinhas, S. 73 (m. Abb.)
1507 (Firmenmarke ohne Ang.), H. 47 cm; GSF, S. 11
-> vgl. 1508
Lit.: GSF, S. 11, hier mit der Ang. des Originaltitels „Avant".

1508
GRUPPE „Après" (Biedermeierpaar, er sie belehrend) (auf gekehltem Sockel)
ca. 1899/1900
1508 (Firmenmarke ohne Ang.), H. fehlt; Pinhas, S. 73 (m. Abb.)
1508 (Firmenmarke ohne Ang.), H. 47 cm; GSF, S. 11
-> vgl. 1507
Lit.: GSF, S. 11, hier mit der Ang. des Originaltitels „Après".

1509
FIGUR „Mohrengigerl"
ca. 1898
1509 (Firmenmarke ohne Ang.), H. 150 cm; GSF, S. 11
Bem.: Als „Gigerl" wird in Wien ein Modegeck bzw. ein Mann bezeichnet, der sich von einer Frau aushalten lässt. Lt. Katalog war einen Bambusstuhl oder ein Schaukelstuhl bei Goldscheider separat zu diesem Modell zu beziehen.
-> vgl. 1667= WANDMASKE
Lit.: GSF, S. 11, hier mit der Ang. des Originaltitels „Mohrengigerl".

1512
BÜSTE Junge Dame mit Federhut (in historischem Kostüm)
ca. 1899/1900
1512 (Firmenmarke ohne Ang.), H. fehlt; Pinhas, S. 74 (m. Abb.); Bem.: Am Sockel der Büste befindet sich ein Schriftband dessen Text sich aber nicht entziffern lässt.

1513
FIGUR „Feldwache" (Stehender Araber mit Turban) (auf Rechtecksockel)
ca. 1898/99
1513 (Firmenmarke ohne Ang.), H. 57 cm; GSF, S. 11
-> vgl. 1350
Lit.: GSF, S. 11, hier mit der Ang. des Originaltitels „Feldwache".

1514
FIGUR „Münchner Kind'l"
ca. 1898/99
1514 (Firmenmarke ohne Ang.), H. 72 cm; GSF, S. 11
Bem.: Das sog. „Münchner Kindl", ist eine figürliche Darstellung des Stadtwappens von München.
-> vgl. 1581
Lit.: GSF, S. 11, hier mit der Ang. des Originaltitels „Münchner Kind'l".

1516
FIGUR „Rückkehr vom Felde"
ca. 1898/99
1516 (Firmenmarke ohne Ang.), H. 25 cm; GSF, S. 11
Lit.: GSF, S. 11, hier mit der Ang. des Originaltitels „Rückkehr vom Felde".

1517
BÜSTE Lachende junge Dame mit Hut
ca. 1899/1900
1517 (Firmenmarke ohne Ang.), H. fehlt; Pinhas, S. 74 (m. Abb.)
Pendant -> vgl. 1518

1518
BÜSTE Junge Dame mit großer Blüte im Haar
ca. 1899/1900
1518 (Firmenmarke ohne Ang.), H. fehlt; Pinhas, S. 74 (m. Abb.)
Pendant -> vgl. 1517

1519
FIGUR „Chloë" (Stehendes junges Mädchen mit zwei Flöten)
Inschrift: CHLOË
ca. 1898/99
1519 (Firmenmarke ohne Ang.), H. fehlt; Pinhas, S. 68 (m. Abb.)
1519 (Firmenmarke ohne Ang.), H. 96 cm; GSF, S. 11
Bem.: Chloë ist der Beiname von Demeter, einer Figur der griech. Mythologie. Es lässt sich zwar ein RELIEF „Chloë" (Terrakotta) ÖMKI (1899/1900) des Künstlers Simon nachweisen, ob jedoch zu dieser Figur ein Zusammenhang besteht, ist zu bezweifeln.
-> vgl. 1948
-> vgl. 1737= BÜSTE
Lit.: GSF, S. 11, hier mit der Ang. des Originaltitels „Chloë".

1520
GRUPPE „Süsses Geheimnis"/„Baiser D' Amour" (Amor, eine junge Frau neckend) (auf Rundsockel)
ca. 1898
1520 (Firmenmarke ohne Ang.), H. 67 cm; Pinhas, S. 69 (m. Abb.)
1520 (Firmenmarke ohne Ang.), H. 67 cm; GSF, S. 12
-> vgl. 600; 605 (GSB), hier mit Ang. des franz. Originaltitels.
-> vgl. 1292
Lit.: GSF, S. 12, hier mit der Ang. des Originaltitels „Süsses Geheimnis".

1521
Stehende Orientalin mit Turban und Halsschmuck (auf Rechtecksockel mit Mauer)
ca. 1899/1900
1521 (Firmenmarke ohne Ang.), H. fehlt; Pinhas, S. 74 (m. Abb.)

1524
FIGUR „Suie je assez beau" (Stehender, eleganter Knabe mit Zylinder)
Inschrift: SUIE JE ASSEZ BEAU
1893
1524 (Firmenmarke ohne Ang.), H. 86; GSF, S. 12
Text der Inschrift: Bin ich nicht schön?
-> vgl. 512; 1832
Lit.: GSF, S. 12, hier mit der Ang. des Originaltitels „Suis je assez beau"; Pinhas, S. 74 (m. Abb.).

1530
Carvin, Louis Albert
Lauernder Luchs auf Baumstamm
BRONZE
ca. 1899/1900
1530,45,A Sculpture (Firmenmarke Bronze 1), sign. L. Carvin, H. 27,3 cm; GK, S. 14, 72 (m. Abb.); Wien Museum Inv. 224.301
Ausst.: HMW (jetzt Wien Museum) 1985/86.
Lit.: GK, S. 14, 72 (m. Abb.)

1535
BÜSTE „Napoleon"
BRONZE
ca. 1898/99
MT.7.1535 Sculpture (Firmenmarke Bronze1), H. 10,2 cm; eBay # 893785805
Bem.: Dargestellt ist der franz. Kaiser Napoleon I. Bonaparte (1769–1821).
-> vgl. 586 (GSB)
Lit.: GSB, Pl. 33, hier ist die BÜSTE mit dem Originaltitel „Buste Napoléon" unter Nr. 586 (H. 25 cm) in Vrai Bronze abgebildet.

1547
FIGUR „Andromeda" (an einen Baumstamm gefesselt)
ca. 1898/99
1547 (Firmenmarke ohne Ang.), H. 145 cm; Pinhas, S. 74 (m. Abb.)
1547 (Firmenmarke ohne Ang.), H. 145 cm; GSF, S. 12
Bem.: Andromeda, eine Figur der griech. Sage, war die Tochter von Kassiopeia und Kepheus.
-> vgl. 1871
Lit.: GSF, S. 12, hier mit der Ang. des Originaltitels „Andromeda".

1551
FIGUR „Lautenschläger (bekleidet)"
ca. 1898/99
1551,88,25 gem. weiß II, Firmenmarke 7, H. 44 cm; eBay # 3718336617; Bem.: Laute fehlt.
1551 (Firmenmarke ohne Ang.), H. 44 cm; GSF, S. 12
-> vgl. 1251 (nackt); 1406 (bekleidet)
Lit.: GSF, S. 12, hier mit der Ang. des Originaltitels „Lautenschläger (bekleidet)".

1552
BÜSTE „Le Rebell"
ca. 1898/99
1552 (Firmenmarke ohne Ang.), H. 91 cm; GSF, S. 12
-> vgl. 1701
Lit.: GSF, S. 12, hier mit der Ang. des Originaltitels „Le Rebell".

1553
BÜSTE „Diana, Marmor chromé" (offen)
MARMOR
ca. 1898/99
1553 ohne Ang.; KeramRS, 7.1899, S. 881
Bem.: s. Modell-Nr. 6

1554
JARDINIERE mit Cherubs
ca. 1898/99
01554 (Firmenmarke ohne Ang.), H. 126 cm; Pinhas, S. 50 (m. Abb.)

1555
FIGUR „Medea"
ca. 1898/99
1555 (Firmenmarke ohne Ang.), H. 85 cm; GSF, S. 12
Lit.: GSF, S. 12, hier mit der Ang. des Originaltitels „Medea".

1558
FIGUR „Bacchantin" (Stehende Sängerin mit halb entblößtem Busen) (auf getrepptem Sockel mit Maskaron und Musikinstrumenten)
ca. 1898
1558 (Firmenmarke ohne Ang.), H. 72 cm; Pinhas, S. 65 (m. Abb.)
1558 (Firmenmarke ohne Ang.), H. 72 cm; GSF, S. 12
Bem.: An diesem Modell zeigt sich der Variantenreichtum der GS-Entwürfe. Ein und dieselbe Figur wurde als Sängerin bzw. als Blumenträgerin, in diesem Fall mit einem langen Stab, der von einem Maiskolben bekrönt wird, angeboten. Zudem wird der entblößte Busen durch das Kleid bedeckt.
-> vgl. 1265
Lit.: GSF, S. 12, hier mit der Ang. des Originaltitels „Bacchantin".

1559
Cherc
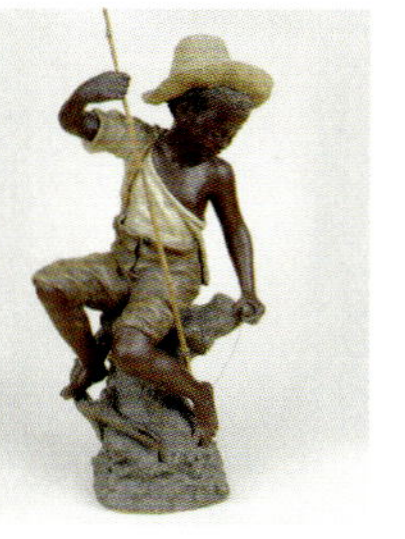
FIGUR „Fischerknabe (Neger)" (Sitzender Negerjunge mit Strohhut und Angel) (auf Natursockel)
ca. 1899/1900
1559,108,54 Firmenmarke 7, Repr. Res., H. 45 cm; PB Wien 19
1559 (Firmenmarke ohne Ang.), H. 45 cm; GSF, S. 12
Lit.: GSF, S. 12, hier mit der Ang. des Originaltitels „Fischerknabe (Neger)".

1560
MUSCHEL (SCHALE) „Echo des Meeres" (Nacktes Mädchen auf einer Muschel liegend)
ca. 1899/1900
1560 (Firmenmarke ohne Ang.), H. fehlt; Pinhas, S. 74, 79 (m. Abb.)
1560 (Firmenmarke ohne Ang.), H. 72 cm; GSF, S. 12
Lit.: GSF, S. 12, hier mit der Ang. des Originaltitels MUSCHEL „Echo des Meeres".

1561
Waderé, Heinrich
BÜSTE „Elsässerin" (Lachendes Mädchen mit großer Haarschleife)
ca. 1899/1900
1561 (Firmenmarke ohne Ang.), unsign., H. fehlt; Pinhas, S. 74 (m. Abb.)
1561 (Firmenmarke ohne Ang.), H. 27 cm; GSF, S. 12
Ausst.: ÖMKI 1900/01, hier mit Ang. des Originaltitels „Elsässerin". In der Ausst. 1906/07 dann ebenfalls ein Objekt von Waderé mit dem Titel „Elsässerin". Ob es sich dabei auch um die Büste (evtl. in einer anderen Größe), oder um eine Ausführung als Figur handelt, kann wegen fehlender nachweisbarer Stücke, nicht eruiert werden.
Lit.: GSF, S. 12, hier mit der Ang. des Originaltitels „Elsässerin"; Neuwirth (1974), S. 75.

1564
Strasser, Arthur
FIGUR „Araber mit Säbel" (auf felsigem Landschaftssockel)
ca. 1898/99
1564 (Firmenmarke ohne Ang.), unsign., H. 105 cm; Pinhas, S. 77 (m. Abb.)
1564 (Firmenmarke ohne Ang.), H. 145 cm; GSF, S. 12
-> vgl. 1824
Lit.: GSF, S. 12, hier mit der Ang. des Originaltitels „Araber mit Säbel".

1565
Tell, E.
FIGUR „Caprieser"
ca. 1898/99
1565 (Firmenmarke ohne Ang.), H. 65 cm; GSF, S. 12
-> vgl. 1655
-> vgl. 1760= BÜSTE
Pendant -> vgl. 1497
Lit.: GSF, S. 12, hier mit der Ang. des Originaltitels „Caprieser".

1569
FIGUR „Jason"
ca. 1898/99
1569 (Firmenmarke ohne Ang.), H. 100 cm; GSF, S. 12
Bem.: Dargestellt ist Jason (Iason), einer der Argonauten und Heldengestalt der griech. Sage.
Lit.: GSF, S. 12, hier mit der Ang. des Originaltitels „Jason".

1570
FIGUR „Jungfrau von Orleans"/„Jeanne d'Arc" (als Kriegerin mit Rüstung und Schwert) (auf Rechtecksockel)
Inschrift: JEANNE D'ARC
ca. 1899/1900
1570 (Firmenmarke ohne Ang.), H. fehlt; Pinhas, S. 74 (m. Abb.)
1570 (Firmenmarke ohne Ang.), H. 32 cm; GSF, S. 12
Bem.: Dargestellt ist die franz. Nationalheldin Johanna von Orléans, bzw. Jeanne d'Arc (1412–1431).
Lit.: GSF, S. 12, hier mit der Ang. des Originaltitels „Jungfrau von Orleans".

1572
FIGUR „Mohr" (Stehender Nordafrikaner mit Turban und Bambusstab)
ca. 1898 (für den Entwurf)
1572 (Firmenmarke ohne Ang.), H. 39 cm; Pinhas, S. 75 (m. Abb.)
1572 (Firmenmarke ohne Ang.), H. 39 cm; GSF, S. 12
Pendant -> vgl. 1573
-> vgl. 1196; 1355
Lit.: GSF, S. 12, hier mit der Ang. des Originaltitels „Mohr".

1573
FIGUR „Mohrin" (Stehende Nordafrikanerin mit Turban, einen Korb haltend)
ca. 1898 (für den Entwurf)
1573 (Firmenmarke ohne Ang.), H. 40 cm; Pinhas, S. 75 (m. Abb.)
1573 (Firmenmarke ohne Ang.), H. 40 cm; GSF, S. 12
-> vgl. 1191; 1356
Pendant -> vgl. 1572
Lit.: GSF, S. 12, hier mit der Ang. des Originaltitels „Mohrin".

1574
Prunier
FIGUR „Haremswächter" (Stehender Araber mit Gewehr) (auf Landschaftssockel)
ca. 1898 (für den Entwurf)
1574 (Firmenmarke ohne Ang.), unsign., H. 44 cm; Pinhas, S. 68 (m. Abb.)
1574 (Firmenmarke ohne Ang.), H. 44 cm; GSF, S. 12
-> vgl. 1379; 1397; 1421
-> vgl. 1596= BÜSTE
Lit.: GSF, S. 12, hier mit der Ang. des Originaltitels „Haremswächter"; Pinhas, S. 75 (m. Abb.).

1575
Simon
FIGUR „Aschenbrödel"
ca. 1898/99
1575 (Firmenmarke ohne Ang.), H. 87 cm; GSF, S. 12
Bem.: Dargestellt ist Ludwig Bechsteins Märchenfigur Aschenbrödel (Aschenputtel, Cinderella).
-> vgl. 1936
Lit.: GSF, S. 12, hier mit der Ang. des Originaltitels „Aschenbrödel".

1578
BÜSTE „Elisabeth"
ca. 1898/99
1578 (Firmenmarke ohne Ang.), H. 33 cm; GSF, S. 12
Lit.: GSF, S. 12, hier mit der Ang. des Originaltitels „Elisabeth".

1579
Jacques, E.
GRUPPE „Römerin im Bade" (Nackte Schönheit mit Kopfputz, die von ihrem Sklaven nach dem Bade abgetrocknet wird) (auf getrepptem Sockel)
ca. 1897/98 (für den Entwurf)
1579 (Firmenmarke ohne Ang.), H. 62 cm; Pinhas, S. 27, 70 (m. Abb.)
1579 (Firmenmarke ohne Ang.), H. 62 cm; GSF, S. 12
-> vgl. 1096
Lit.: GSF, S. 12, hier mit der Ang. des Originaltitels „Römerin im Bade".

1581
FIGUR „Münchner Kind'l"
ca. 1898/99
1581 (Firmenmarke ohne Ang.), H. 25 cm; GSF, S. 12
Bem.: Das sog. „Münchner Kindl", ist eine figürliche Darstellung des Stadtwappens von München.
-> vgl. 1514
Lit.: GSF, S. 12, hier mit der Ang. des Originaltitels „Münchner Kind'l".

1582
BÜSTE Nordafrikanischer Junge mit Turban
ca. 1898
1582 (Firmenmarke ohne Ang.), H. fehlt; Pinhas, S. 74, 75 (m. Abb.)

1583
Dupré, J.
FIGUR „Lustiger Zecher" (Stehender Zecher mit Hut, Humpen und Korbweinflasche)
ca. 1898 (für den Entwurf)
1583 (Firmenmarke ohne Ang.), H. 34 cm; GSF, S. 12
-> vgl. 1353
Lit.: GSF, S. 12, hier mit der Ang. des Originaltitels „Lustiger Zecher".

1585
BÜSTE (offen)
ca. 1898
1585 ohne Ang.; KeramRS, 6.1898, S. 757
Bem.: s. Modell-Nr. 6

1588
Lavergne, Adolphe-Jean
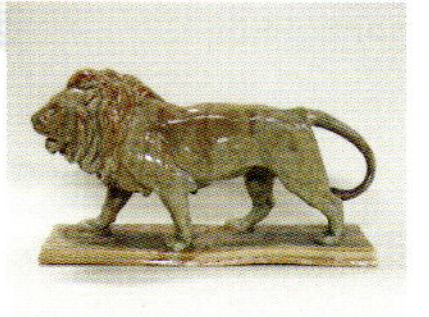
Schreitender Löwe (auf Rechtecksockel)
ca. 1899/1900
1588,5,69 Firmenmarke 9, Repr. Res., unsign., H. 19,9 cm, L. 36,3 cm; MAK Wien, Inv. Ke 10602-43
1588,35,1 ger. K, gem. lig. TS. DI, GWM Wien (rot), Made in Austria (rot), H. 20,5 cm, L. 37 cm; Wien Museum Inv. 220.752

1589
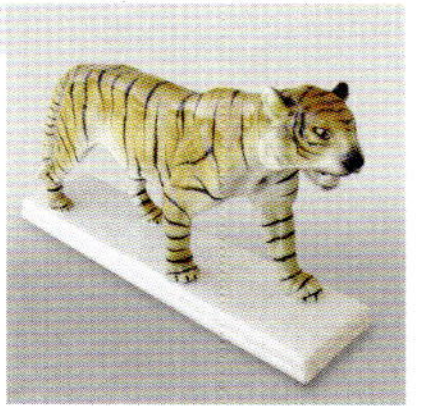
Schreitender Tiger (auf profiliertem Rechtecksockel)
ca. 1899/1900
1589,31,10 gem. lig. AS, D1, Firmenmarke 19, Made in Austria, Ausformung 20er Jahre, H. 18 cm, L. 36,5 cm; Sammlung Stressless, Portugal

1594
BÜSTE (offen)
ca. 1898
1594 ohne Ang.; KeramRS, 6.1898, S. 757
Bem.: s. Modell-Nr. 6

1595
BÜSTE Pharao
ca. 1899/1900
1595,6,59 FG Wien (Firmenmarke?), H. 34,5 cm, B. 35 cm; KH

1596
Prunier, D.
BÜSTE „Haremswächter"
ca. 1898/99
1596,22,5 Firmenmarke 7, Repr. Res., sign. D. Prunier, H. 51,5 cm; eBay # 2529408525
1596 (Firmenmarke ohne Ang.), H. 52 cm; GSF, S. 12
-> vgl. 1379; 1397; 1421; 1574= jew. FIGUR
Lit.: GSF, S. 12, hier mit der Ang. des Originaltitels „Haremswächter"; Pinhas, S. 75 (m. Abb.).

1597
BÜSTE „Feldwache"
ca. 1898/99
1597 (Firmenmarke ohne Ang.), H. 42 cm; GSF, S. 12
Lit.: GSF, S. 12, hier mit der Ang. des Originaltitels „Feldwache".

1600
Strasser, Arthur
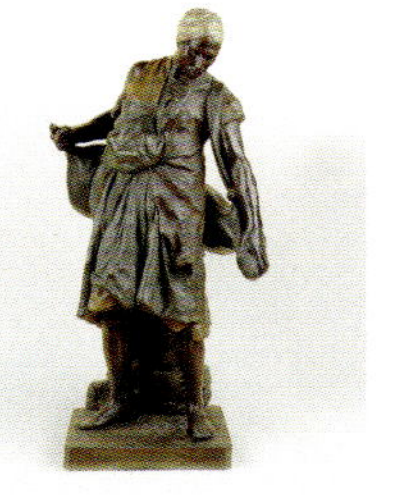
FIGUR „Wasserträger"/ „Porteur D'Eau" (Stehender Nordafrikaner mit großem Wasserschlauch/tierischer Wasserbehälter) (auf Natursockel)
ca. 1898
1600,7,26 ger., IV. Firmenmarke 7, Repr. Res., H. 60,2 cm; PB Deutschland
1600,18,34 Firmenmarke 7, Repr. Res., (Metallschild: A Mrs. Joseph Tenneson, ses Ouvriers, ses Employés, Janvier 1901), H. 63 cm; PB Wien 19
1600 (Firmenmarke ohne Ang.), H. 60 cm; GSF, S. 12
Ausst.: ÖMKI 1898/99, hier mit Ang. des Originaltitels „Wasserträger".
-> vgl. 1174; 608-GSB; 609-GSB
Lit.: GSF, S. 12, hier mit der Ang. des Originaltitels „Wasserträger"; GSB, Pl. 52, hier ist die Figur mit dem franz. Originaltitel „Porteur D'Eau" (Metallschild), unter Nr. 609 (H. 65 cm) Material: „Fonte d'Art" (Zinkguss?) und Nr. 608 (H. 40 cm) Material: Bronze abgebildet; Richemond (1999), S. 227 (m. Abb.).

1601
Strasser, Arthur
FIGUR „Wasserverkäufer"
ca. 1897/98
1601 (Firmenmarke ohne Ang.), H. 60 cm; GSF, S. 12
Ausst.: ÖMKI 1898/99, hier mit Ang. des Originaltitels „Wasserverkäufer".
-> vgl. 857; 1175
Lit.: GSF, S. 12, hier mit der Ang. des Originaltitels „Wasserverkäufer"; Neuwirth (1974), S. 74.

1602
FIGUR (LAMPE) „Illumination" (Stehende junge Dame, eine Laterne in der ausgestreckten rechten Hand haltend) (auf Rundsockel)
ca. 1898/99
1602 (Firmenmarke ohne Ang.), H. fehlt; Pinhas, S. 76 (m. Abb.); Bem.: Hier die Ausführung als Lampe.
1602 (Firmenmarke ohne Ang.), H. 98 cm; GSF, S. 12
Bem.: Dieses Modell wurde auch mit einer Glühbirne elektrifiziert geliefert.
-> vgl. 2070
Lit.: GSF, S. 12, hier mit der Ang. des Originaltitels „Illumination".

1604
Strasser, Arthur
Stehender Nordafrikaner mit Turban, breitem Gürtel und langem Stab (auf oktogonalem Sockel)
ca. 1898
1604 (Firmenmarke ohne Ang.), unsign., H. fehlt; Pinhas, S. 76 (m. Abb.)

1608
Strasser, Arthur
FIGUR „Wollverkäufer" (Stehender Nordafrikaner, einen großen Sack Wolle auf der Schulter tragend) (auf Natursockel)
ca. 1898
1608 (Firmenmarke ohne Ang.), H. 60 cm; Pinhas, S. 75 (m. Abb.)
1608 (Firmenmarke ohne Ang.), H. 60 cm; GSF, S. 12
Ausst.: ÖMKI 1898/99, hier mit Ang. des Originaltitels „Wollverkäufer".
-> vgl. 1334
Lit.: GSF, S. 12, hier mit der Ang. des Originaltitels „Wollverkäufer"; Neuwirth (1973), in: Keramos 59.1973, S. 54–63, hier S. 55 (m. Abb.) (Samml. F. Witton, St. Albans/England); Neuwirth (1974), S. 58, 307 (Abb.); Pinhas, S. 75 (m. Abb.), hier mit Ang.: H. 102 cm.

1609
Strasser, Arthur
FIGUR „Stoffverkäufer" (Stehender Nordafrikaner, Stoffballen auf Kopf und Schulter tragend) (auf Rechtecksockel)
ca. 1898
1609 (Firmenmarke ohne Ang.), unsign., H. 60 cm; Pinhas, S. 75 (m. Abb.)
1609 (Firmenmarke ohne Ang.), H. 60 cm; GSF, S. 12
Ausst.: ÖMKI 1898/99, hier mit Ang. des Originaltitels „Stoffverkäufer".
-> vgl. 1314
Lit.: GSF, S. 12, hier mit der Ang. des Originaltitels „Stoffverkäufer"; Neuwirth (1974), S. 74.

1610
Carrièr-Belleuse, A(lbert-Ernest)
FIGUR „Harmonie"/„Mélodie" (Sitzende junge Dame mit Lyra) (auf Rechtecksockel)
ca. 1898/99
1610,59,12 GS Wien (Firmenmarke?), sign. A. Carrier-Belleuse (Metallschild), H. 71,1 cm; G 68 Auctions, Toronto 11.9.2000
1610 (Firmenmarke ohne Ang.), H. 73 cm; GSF, S. 12
-> vgl. 2216; 1925= BÜSTE
Lit.: , GSF, S. 12, hier mit der Ang. des Originaltitels "Harmonie"; Berman, Vol. 2, S. 391(m. Abb.), Nr. 1447 (Bronze, H. 78,7 cm), hier mit Ang. des franz. Originaltitels „Mélodie" (Metallschild) und der Datierung 1875 für den Entwurf; Pinhas, S: 75 (m. Abb.).

1611
Cherc
GRUPPE „Kunst bringt Gunst" (Strickender Junge mit Mütze, daneben zwei Mädchen, die seine Künste bewundern) (alle auf einem Mauersockel sitzend)
ca. 1898/99
1611,27,63 FG Wien (Firmenmarke ohne Ang.), Repr. Res., H. 58 cm, B. 54 cm; eBay # 2500921454
1611 (Firmenmarke ohne Ang.), H. 54 cm; GSF, S. 12
Lit.: GSF, S. 12, hier mit der Ang. des Originaltitels „Kunst bringt Gunst".

1612
Haniroff
WANDMASKE „Araber" (mit Vollbart und Turban)
ca. 1898
1612,79,38 (Firmenmarke ohne Ang.), unsign., H. ca. 27 cm; Christie's SK 4.11.1999
1612 (Firmenmarke ohne Ang.), H. 26 cm; GSF, S. 12
-> vgl. 1105 FIGUR
Pendant -> vgl. 1613
Lit.: GSF, S. 12, hier mit der Ang. des Originaltitels „Araber".

1613
Piémont, T. du
WANDMASKE „Mohrin" (mit Turban)
ca. 1898
1613 (Firmenmarke ohne Ang.), sign. T. du Piémont, H. 27 cm; Pinhas, S. 30 (m. Abb.)
1613 (Firmenmarke ohne Ang.), H. 29 cm; GSF, S. 12
-> vgl. 1102 FIGUR
Pendant -> vgl. 1612
Lit.: GSF, S. 12, hier mit der Ang. des Originaltitels „Mohrin".

1614
Haniroff
WANDMASKE „Mahdi"
ca. 1898
1614,67,64 Firmenmarke 7, Repr. Res., unsign., H. 26,5 cm; Bonhams, Chester/UK 27.4.2004
1614 (Firmenmarke ohne Ang.), H. 30 cm; Pinhas, S. 30 (m. Abb.)
1614 (Firmenmarke ohne Ang.), H. 29 cm; GSF, S. 12
-> vgl. 1308; 1423= jew. BÜSTE
Lit.: GSF, S. 12, hier mit der Ang. des Originaltitels „Mahdi".

1615
Thiele, (P.?)
WANDMASKE „Bulgarin"
ca. 1898
1615 (Firmenmarke ohne Ang.), H. 22 cm; GSF, S. 12
-> vgl. 1109; 1340; 1898= jew. BÜSTE
Lit.: GSF, S. 12, hier mit der Ang. des Originaltitels „Bulgarin".

1616
WANDMASKE „Sclavin" (mit Kopftuch)
ca. 1898
1616 (Firmenmarke ohne Ang.), unsign., H. 27 cm; Pinhas, S. 30 (m. Abb.)
1616 (Firmenmarke ohne Ang.), H. 23 cm; GSF, S. 12
-> vgl. 1101= BÜSTE
Lit.: GSF, S. 12, hier mit der Ang. des Originaltitels „Sclavin".

1617
Haniroff
WANDMASKE „Negerknabe" (Lachender Negerjunge mit Schiebermütze)
ca. 1898
1617 ger.,1614,4,15 FG Wien (Firmenmarke 7, zwei mal), Repr. Res., gepr. „P"= Z3, unsign., H. 25,5 cm; eBay # 2664921765; Sammlung John Nash

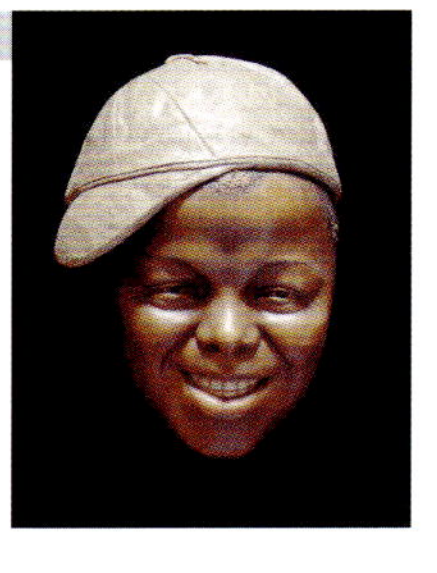

1617,104,21 Firmenmarke 7, H. 24 cm; eBay # 2515701187
1617 (Firmenmarke ohne Ang.), H. 25 cm; GSF, S. 12
-> vgl. 1236 und 1225= mittlere EinzelFIGUR (aus dieser entstammt der Entwurf für diese Maske)
-> vgl. 1190= Dreier GRUPPE
-> vgl. 1695= BÜSTE
Lit.: GSF, S. 12, hier mit der Ang. des Originaltitels „Negerknabe".

1618
Sitzender Bub mit Angel (auf mauerartigem Sockel)
ca. 1898
1618 (Firmenmarke ohne Ang.), H. 80 cm; Pinhas, S. 36 (m. Abb.)

1619
Ceccarelli, Ezio
BÜSTE „Winzerin"/„Automne" (Herbst) (Frau mit großem Hut und Trauben)
ca. 1898
1619,105,13 Firmenmarke 7, Repr. Res., H. 48,3 cm; eBay # 347652312
1619 (Firmenmarke ohne Ang.), H. 48 cm; GSF, S. 13
-> vgl. 1297; 1961
-> vgl. 622 (GSB); 734 (GSB)
Lit.: GSF, S. 13, hier mit der Ang. des Originaltitels „Winzerin"; GSB, Pl. 71, hier ist die BÜSTE mit dem Originaltitel „Automne" unter Nr. 622= No.1 (H. 52 cm) und unter Nr. 734= No. 2 (H. 30 cm) abgebildet. Ang. zum Material (Terrakotta/Bronze) fehlen.

1620
Cherc
FIGUR „Raucher" (Stehender lachender Negerjunge mit Zündhölzern)
ca. 1898/99
1620 (Firmenmarke ohne Ang.), H. 136 cm; Pinhas, S. 29 (m. Abb.)
1620 (Firmenmarke ohne Ang.), H. 137 cm; GSF, S. 13
Lit.: GSF, S. 13, hier mit der Ang. des Originaltitels „Raucher".

1621
FIGUR „Pierrot mit Mandoline"
ca. 1898/99
1621 (Firmenmarke ohne Ang.), H. 54 cm; GSF, S. 13
Bem.: Lt. Katalog war ein Stuhl separat bei Goldscheider zu diesem Modell zu beziehen.
-> vgl. 2006
Lit.: GSF, S. 13, hier mit der Ang. des Originaltitels „Pierrot mit Mandoline".

1622
FIGUR „Pierrot, Ruhepause"
ca. 1898/99
1622 (Firmenmarke ohne Ang.), H. 55 cm; GSF, S. 13
Bem.: Lt. Katalog war einen Stuhl bei Goldscheider separat zu diesem Modell zu beziehen.
Lit.: GSF, S. 13, hier mit der Ang. des Originaltitels „Pierrot, Ruhepause".

1623
FIGUR „Pierrot (demaskiert)"
ca. 1898/99
1623 (Firmenmarke ohne Ang.), H. 50 cm; GSF, S. 13
Bem.: Lt. Katalog war einen Stuhl bei Goldscheider separat zu diesem Modell zu beziehen.
-> vgl. 2007
Lit.: GSF, S. 13, hier mit der Ang. des Originaltitels „Pierrot (demaskiert)".

1624
BÜSTE „Madame Sans Gêne" (mit großem Hut/Haube) (nach rechts blickend)
ca. 1898/99
1624 (Firmenmarke ohne Ang.), H. fehlt; Pinhas, S. 78 (m. Abb.)
1624 (Firmenmarke ohne Ang.), H. 71 cm; GSF, S. 13
Bem.: Dargestellt ist die Hauptfigur Catherine (Madame Sans Gêne) aus dem Lustspiel in 4 Akten von Victorien Sardou. (Quelle: Wikipedia)
-> vgl. 2038; 2052
Lit.: GSF, S. 13, hier mit der Ang. des Originaltitels „Madame Sans Gêne".

1626
BÜSTE „Corsicanerin" (Junges Mädchen mit Haarband und Halsschmuck) (nach links blickend)
ca. 1898/99
1626 (Firmenmarke ohne Ang.), H. fehlt; Pinhas, S. 78 (m. Abb.)
1626 (Firmenmarke ohne Ang.), H. 69 cm; GSF, S. 13
Lit.: GSF, S. 13, hier mit der Ang. des Originaltitels „Corsicanerin".

1627
FIGUR „Musik"
ca. 1898/99
1627 (Firmenmarke ohne Ang.), H. 91 cm; GSF, S. 13
Lit.: GSF, S.13, hier mit der Ang. des Originaltitels „Musik".

1628
Cherc
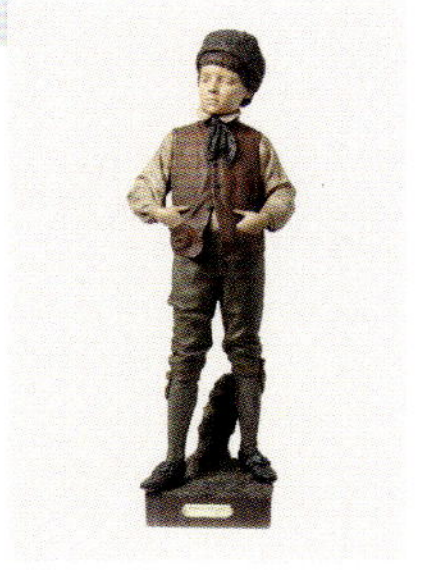
FIGUR „Dorfprinz" (Stehender rauchender Bursche) (auf quadratischem Landschaftssockel)
ca. 1898
1628,69,55 gem. weiß XV., Firmenmarke 7, Repr. Res., H. 88 cm; Kunsthaus Wiesinger, Wels; Bem.: Zigarette bzw. Pfeife fehlt hier.
1628,108,47 (Firmenmarke ohne Ang.), H. 89 cm; Christie's SK 13.5.1999; Bem.: Zigarette bzw. Pfeife fehlt hier.
1628 (Firmenmarke ohne Ang.), H. 88 cm; GSF, S. 13
Bem.: An der Plinthe betitelt und sign. „COQ du VILLAGE, Cherc" (Metallschild).
-> vgl. 1928
-> vgl. 3823=BÜSTE
Lit.: GSF, S. 13, hier mit der Ang. des Originaltitels „Dorfprinz"; Pinhas, S. 96 (m. Abb.).

1629
BÜSTE „Jane"
ca. 1898/99
1629 (Firmenmarke ohne Ang.), H. 35 cm; GSF, S. 13
Lit.: GSF, S. 13, hier mit der Ang. des Originaltitels „Jane".

1630
BÜSTE „Paulette"
ca. 1898/99
1630 (Firmenmarke ohne Ang.), H. 37 cm; GSF, S. 13
Lit.: GSF, S. 13, hier mit der Ang. des Originaltitels „Paulette".

1631
BÜSTE „Tuareg"
ca. 1898/99
1631 (Firmenmarke ohne Ang.), H. 46 cm; GSF, S. 13
Lit.: GSF, S. 13, hier mit der Ang. des Originaltitels „Tuareg".

1632
FIGUR „Fischermädchen" (Sitzendes Mädchen mit Strohhut und Fisch) (auf Felssockel)
ca. 1898/99
1632,14,20 FG (Firmenmarke ohne Ang.), Repr. Res., H. ca. 87 cm; Burchard 30.8.1998
1632 (Firmenmarke ohne Ang.), H. 84 cm; GSF, S. 13
-> vgl. 1901
Lit.: GSF, S. 13, hier mit der Ang. des Originaltitels „Fischermädchen".

1642
GRUPPE „Ein lustig Lied"
ca. 1898/99
1642 (Firmenmarke ohne Ang.), H. 60 cm; GSF, S. 13
Lit.: GSF, S. 13, hier mit der Ang. des Originaltitels „Ein lustig Lied".

1643
Cherc
FIGUR „Mignon" (Stehende junge Frau mit langem Haar) (nach rechts blickend)
ca. 1898/99
1643,233,70 (Firmenmarke ohne Ang.), H. 85 cm; Christie's SK 27.6.1997
1643,361,17 Firmenmarke 23, Ausformung 20er Jahre, H. 87 cm; eBay # 926322951
1643 (Firmenmarke ohne Ang.), H. 82 cm; GSF, S. 13
-> vgl. 2072; 2436
Lit.: GSF, S. 13, hier mit der Ang. des Originaltitels „Mignon".

1644
FIGUR „Sicilianerin"
ca. 1898/99
1644 (Firmenmarke ohne Ang.), H. 65 cm; GSF, S. 13
Lit.: GSF, S. 13, hier mit der Ang. des Originaltitels „Sicilianerin".

1645
Mayre, G.
FIGUR „Ein altes Lied" (Sitzender Flöte spielender Junge als Halbakt) (auf Podest mit sitzendem Löwen)
ca. 1898/99
1645 (Firmenmarke ohne Ang.), H. fehlt; Pinhas, S. 78 (m. Abb.)
1645 (Firmenmarke ohne Ang.), H. 109 cm; GSF, S. 13
Ausst.: ÖMKI 1898/99, hier mit Ang. des Originaltitels „Ein altes Lied" und des Künstlernamens „Mayre".
Lit.: GSF, S. 13, hier mit der Ang. des Originaltitels „Ein altes Lied".

1646
BÜSTE „Gabrielle" (Frau mit Schal/Stola) (nach rechts blickend)
ca. 1897/98
1646 (Firmenmarke ohne Ang.), H. 33 cm; Pinhas, S. 96 (m. Abb.)
1646 (Firmenmarke ohne Ang.), H. 33 cm; GSF, S. 13
-> vgl. 929
Lit.: GSF, S. 13, hier mit der Ang. des Originaltitels „Gabrielle".

1647
Carrièr-Belleuse, (Albert-Ernest)
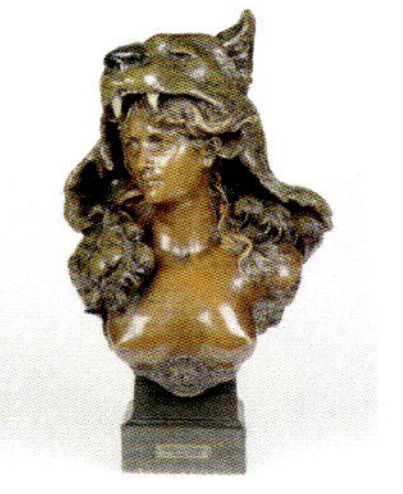
BÜSTE „Chriemhilde"/ „Frédègonde" (Barbusige Frau mit Löwenkopf und -fell als Kopfbedeckung)
ca. 1898/99
1647,35,12 Firmenmarke 7, Repr. Res., H. 73 cm; PB Wien 19
1647,67,9 FG (Firmenmarke?), Repr. Res., H. 73,9 cm; Doro 18.5.1999
1647 (Firmenmarke ohne Ang.), H. 68 cm; GSF, S. 13
Bem.: An der Plinthe betitelt und sign. (Metallschild). Diese Büste findet sich nicht bei Hargrove. Nur eine ähnl. gestaltete, barbusige „Diana" kann als Vergleichsstück herangezogen werden (-> vgl. Hargrove, S. X, Taf. 81). Dargestellt ist Kriemhild, eine Hauptfigur des Nibelungenliedes.
-> vgl. 1979
-> vgl. 1772 (bekleidet)
-> vgl. 2106= WANDMASKE
Pendant -> vgl. 1648
Lit.: GSF, S. 13, hier mit der Ang. des Originaltitels „Chriemhilde".

1648 -> ÜBERSCHNEIDUNG
Carrièr-Belleuse, (Albert-Ernest)
BÜSTE „Brunhilde"/(„Brunhaut") (Frau mit Adlerkopf und -schwingen als Kopfbedeckung)
ca. 1899/1900
1648 (Firmenmarke ohne Ang.), H. 66 cm; Pinhas, S. 80 (m. Abb.)
1648 (Firmenmarke ohne Ang.), H. 64 cm; GSF, S. 13
Bem.: Dargestellt ist Brunhild, die sagenhafte Königin aus der Nibelungensage.
-> vgl. 1773; 1980= jew. Variante (nackt)
-> vgl. 2107= WANDMASKE
Pendant -> vgl. 1647
Lit.: GSF, S. 13, hier mit der Ang. des Originaltitels „Brunhilde".

1648 -> ÜBERSCHNEIDUNG
BÜSTE Othello
ca. 1898/99
1648 (Firmenmarke ohne Ang.), H. 40 cm; Pinhas, S. 28 (m. Abb.)

1649
FIGUR „Zeitungsleser" (Sitzender rauchender und lesender Neger mit Strohhut) (auf Sessel)
ca. 1898/99
1649 (Firmenmarke ohne Ang.), H. 71 cm; Pinhas, S. 76 (m. Abb.)
1649 (Firmenmarke ohne Ang.), H. 71 cm; GSF, S. 13
Bem.: Der Sessel, bzw. Stuhl ist aus Bambus gefertigt und konnte bei Goldscheider separat zu dem Modell bezogen werden.
-> vgl. 1663; 1890
Lit.: GSF, S. 13, hier mit der Ang. des Originaltitels „Zeitungsleser".

1650
Langer, (Leopold?)
FIGUR „Korbflechter" (Sitzender Knabe, einen Korb flechtend) (Sockel als Steinmauer)
ca. 1898/99
1650 (Firmenmarke ohne Ang.), H. 50 cm; GSF, S. 13
-> vgl. 1413
Lit.: GSF, S. 13, hier mit der Ang. des Originaltitels „Korbflechter".

1651
GRUPPE „Bitte um Feuer" (Zwei rauchende Burschen) (auf Landschaftssockel)
ca. 1898/99
1651 (Firmenmarke ohne Ang.), H. fehlt; Pinhas, S. 77 (m. Abb.)
1651 (Firmenmarke ohne Ang.), H. 65 cm; GSF, S. 13
Lit.: GSF, S. 13, hier mit der Ang. des Originaltitels „Bitte um Feuer".

1652
Strasser, Arthur
FIGUR „Janitschare" (mit Turban, Schild, Säbel und langem Stab) (auf Landschaftssockel)
ca. 1898/99
1652 (Firmenmarke ohne Ang.), unsign., H. fehlt; Pinhas, S. 77 (m. Abb.)
1652 (Firmenmarke ohne Ang.), H. 102 cm; GSF, S. 13
Bem.: Ein „Janitschar" ist ein Krieger (Infantrie) der Elitetruppen des osmanischen Reichs, bzw. ein Leibwächter des Sultans (Quelle: Wikipedia).
-> vgl. 2043; 2225
-> vgl. 1921= BÜSTE
Lit.: GSF, S. 13, hier mit der Ang. des Originaltitels „Janitschare".

1654
Wilhelm Tell (Stehender Tell mit Armbrust und Pfeil) (auf Felssockel)
ca. 1898/99
1654 (Firmenmarke ohne Ang.), H. fehlt; Pinhas, S. 76 (m. Abb.)
Bem.: Dargestellt ist hier der Schweizer Nationalheld Wilhelm Tell in typischer Pose.

1655
Tell, E.
FIGUR „Caprieser"
ca. 1898/99
1655 (Firmenmarke ohne Ang.), H. 45 cm; GSF, S. 13
-> vgl. 1565
-> vgl. 1760= BÜSTE
Pendant -> vgl. 1656
Lit.: GSF, S. 13, hier mit der Ang. des Originaltitels „Caprieser".

1656
Tell, E.
FIGUR „Capriesierin"
ca. 1898/99
1656 (Firmenmarke ohne Ang.), H. 45 cm; GSF, S. 13
-> vgl. 1761= BÜSTE
Pendant -> vgl. 1655
Lit.: GSF, S. 13, hier mit der Ang. des Originaltitels „Capriesierin".

1657
BÜSTE „Eté" (Junge Bäuerin mit Kopftuch und Ähren über der Schulter)
ca. 1898/99
1657 (Firmenmarke ohne Ang.), H. fehlt; Pinhas, S. 77 (m. Abb.)
1657 (Firmenmarke ohne Ang.), H. fehlt; GSF, S. 13
Bem.: Bei dieser Büste handelt es sich um eine Allegorie des Sommers.
-> vgl. 2073
Lit.: GSF, S. 13, hier mit der Ang. des Originaltitels „Eté".

1658
BÜSTE „Premier Present" (Mädchen mit Haube und Blumenschmuck)
ca. 1898/99
1658 (Firmenmarke ohne Ang.), H. fehlt; Pinhas, S. 77 (m. Abb.)
1658 (Firmenmarke ohne Ang.), H. fehlt; GSF, S. 13
-> vgl. 2073
Lit.: GSF, S. 13, hier mit der Ang. des Originaltitels „Premier Present".

1659
FIGUR „Zuckerbäcker"
ca. 1898/99
1659 (Firmenmarke ohne Ang.), H. 76 cm; GSF, S. 13
-> vgl. 1738 (ähnl. Thematik)
Lit.: GSF, S. 13, hier mit der Ang. des Originaltitels „Zuckerbäcker".

1660
FIGUR „Ist es wahr?" (Stehende junge Dame in langem Kleid) (auf Landschaftssockel mit Pflanzen)
ca. 1898/99
1660 (Firmenmarke ohne Ang.), H. 62 cm; Pinhas, S. 77 (m. Abb.)
1660 (Firmenmarke ohne Ang.), H. 62 cm; GSF, S. 13
-> vgl. 1310
Lit.: GSF, S. 13, hier mit der Ang. des Originaltitels „Ist es wahr?".

1663
FIGUR „Zeitungsleser" (Sitzender rauchender und lesender Neger mit Strohhut) (auf Sessel)
ca. 1898/99
1663 (Firmenmarke ohne Ang.), H. 150 cm; Pinhas, S. 76 (m. Abb.)
1663 (Firmenmarke ohne Ang.), H. 150 cm; GSF, S. 13
Bem.: Der Sessel, bzw. Stuhl ist aus Bambus gefertigt und konnte bei Goldscheider separat zu dem Modell bezogen werden.
-> vgl. 1649; 1890
-> vgl. 1757= WANDMASKE
Lit.: GSF, S. 13, hier mit der Ang. des Originaltitels „Zeitungsleser".

1664
BÜSTE Biedermeiermädchen (nach rechts blickend) mit großem Hut
ca. 1898/99
1664 (Firmenmarke ohne Ang.), H. fehlt; Pinhas, S. 77 (m. Abb.)

1665
Stehende junge Dame mit Blumen (auf Rechtecksockel)
ca. 1898/99
1665 (Firmenmarke ohne Ang.), H. fehlt; Pinhas, S. 79 (m. Abb.)

1666
Haniroff
WANDMASKE „Gigerl" (Negerjunge mit Strohhut)
ca. 1898/99
1666 (Firmenmarke ohne Ang.), H. 27 cm; Pinhas, S. 30; 96 (m. Abb.)
1666 (Firmenmarke ohne Ang.), H. 25 cm; GSF, S. 13
Bem.: Als „Gigerl" wird in Wien ein Modegeck bzw. ein Mann bezeichnet, der sich von einer Frau aushalten lässt.
-> vgl. 1368; 1419; 1807= jew. FIGUR
Pendant -> vgl. 1667
Lit.: GSF, S. 13, hier mit der Ang. des Originaltitels „Gigerl".

1667
Haniroff (zugeschr.)
WANDMASKE „Gigerl" (Neger mit Strohhut, Monokel und Zigarette)
ca. 1898/99
1667 (Firmenmarke ohne Ang.), unsign., H. fehlt; Pinhas, S. 96 (m. Abb.)
1666 (Firmenmarke ohne Ang.), H. 25 cm; GSF, S. 13
Bem.: Als „Gigerl" wird in Wien ein Modegeck bzw. ein Mann bezeichnet, der sich von einer Frau aushalten lässt.
-> vgl. 1509= FIGUR
Pendant -> vgl. 1666
Lit.: GSF, S. 13, hier mit der Ang. des Originaltitels „Gigerl".

1668
FIGUR „Mohr mit Flöte"
ca. 1897/98
1668,55 ger., (Firmenmarke ohne Ang. oder evtl. ohne Firmenmarke?), „MUSTER" ger., H. 49 cm; Doro 20.4.1993; Bem.: Flöte fehlt hier.
1668 (Firmenmarke ohne Ang.), H. 49 cm; GSF, S. 13
-> vgl. 1127; 1366
Lit: GSF, S. 13, hier mit der Ang. des Originaltitels „Mohr, Flöte"; Pinhas, S. 82 (m. Abb.).

1669
BÜSTE „Phyrne" (Junge Dame mit entblößter rechter Brust)
ca. 1898/99
1669 (Firmenmarke ohne Ang.), H. fehlt; Pinhas, S. 77 (m. Abb.)
1669 (Firmenmarke ohne Ang.), H. 50 cm; GSF, S. 14
Bem.: Phyrne, eine griech. Hetäre (leichtes Mädchen) des 4. Jahrhunderts, war die Geliebte des Praxiteles.
Lit: GSF, S. 14, hier mit der Ang. des Originaltitels „Phyrne".

1670
FIGUR „Harfenspielerin" (Stehende Nordafrikanerin mit fünfsaitigem Instrument und Fidelbogen)
ca. 1898/99
1670 (Firmenmarke ohne Ang.), H. 102 cm; Pinhas, S. 78 (m. Abb.)
1670 (Firmenmarke ohne Ang.), H. 103 cm; GSF, S. 14
-> vgl. 1859
Lit: GSF, S. 14, hier mit der Ang. des Originaltitels „Harfenspielerin".

1671
BÜSTE Pierrot
ca. 1898/99
1671 (Firmenmarke ohne Ang.), H. fehlt; Pinhas, S. 79 (m. Abb.)
1671 (Firmenmarke ohne Ang.), H. 51 cm; GSF, S. 14
Lit: GSF, S. 14, hier mit der Ang. des Originaltitels „Pierrot".

1676
Stehendes Bauernmädchen mit Wasserkrug
ca. 1898/99
1676 (Firmenmarke ohne Ang.), H. fehlt; Pinhas, S. 78 (m. Abb.)

1677
BÜSTE „Mohr mit Korb"
ca. 1897 (für den Entwurf)
1677 (Firmenmarke ohne Ang.), H. 45 cm; GSF, S. 14
-> vgl. 901
Pendant -> vgl. 1678
Lit.: GSF, S. 14, hier mit der Ang. des Originaltitels „Mohr mit Korb".

1678
BÜSTE „Mohrin mit Korb"
ca. 1897 (für den Entwurf)
1678 (Firmenmarke ohne Ang.), H. 40 cm; GSF, S. 14
-> vgl. 902
Pendant -> vgl. 1677
Lit.: GSF, S. 14, hier mit der Ang. des Originaltitels „Mohrin mit Korb".

1680
FIGUR „Knabe (Zeitung lesend)" (Stehender lesender Bursche mit Brille und Hut) (auf Landschaftssockel mit Baumstumpf)
ca. 1898/99
1680 (Firmenmarke ohne Ang.), H. 95 cm; Pinhas, S. 76 (m. Abb.)
1680 (Firmenmarke ohne Ang.), H. 95 cm; GSF, S. 14
-> vgl. 1929
Lit.: GSF, S. 14, hier mit der Ang. des Originaltitels „Knabe (Zeitung lesend)".

1681
FIGUR „Knabe mit Kürbis" (Stehender Bursche mit Strohhut und Kürbis) (auf Landschaftssockel)
ca. 1898/99
1681 (Firmenmarke ohne Ang.), H. fehlt; Pinhas, S. 76 (m. Abb.)
1681 (Firmenmarke ohne Ang.), H. 98 cm; GSF, S. 14
Lit.: GSF, S. 14, hier mit der Ang. des Originaltitels „Knabe mit Kürbis".

1682
Dupré, J.
BÜSTE „Lustiger Zecher" (Mann mit Hut)
ca. 1898/99
1682 (Firmenmarke ohne Ang.), sign. J. Dupré, H. 52 cm; Christie's SK 14.11.2006
1682 (Firmenmarke ohne Ang.), H. 52 cm; GSF, S. 14
Lit.: GSF, S. 14, hier mit der Ang. des Originaltitels „Lustiger Zecher".

1683
BÜSTE „Kellermeister"
ca. 1898/99
1683 (Firmenmarke ohne Ang.), H. 52 cm; GSF, S. 14
Lit.: GSF, S. 14, hier mit der Ang. des Originaltitels „Kellermeister".

1685
FIGUR „Bitte um Feuer"
ca. 1898/99
1685 (Firmenmarke ohne Ang.), H. 59 cm; GSF, S. 14
-> vgl. 1651
Lit.: GSF, S. 14, hier mit der Ang. des Originaltitels „Bitte um Feuer".

1686
FIGUR „Zu dienen!"
ca. 1898/99
1686 (Firmenmarke ohne Ang.), H. 61 cm; GSF, S. 14
Lit.: GSF, S. 14, hier mit der Ang. des Originaltitels „Zu dienen!".

1687
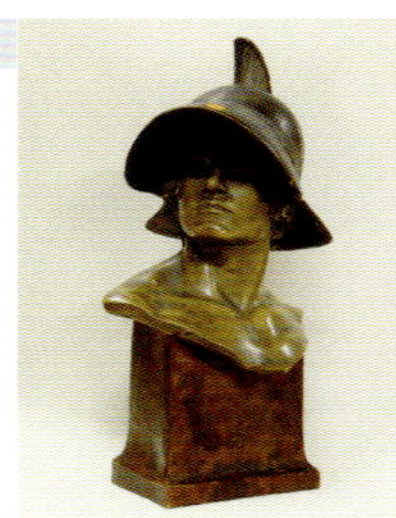
Rozel
BÜSTE „Gladiator"
Inschrift: „AVE CAESARE ! MORITURI TE SALVTANT"
ca. 1898
1687,2,11 Firmenmarke 7, Repr. Res., sign. Rozel, H. 56,5 cm; KH Ritschka, Wien; Doro 24.5.2000
1687,60,14 Firmenmarke 7, Repr. Res., sign. Rozel, H. 56,5 cm; MAK Wien, Inv. Ke 10521-6
1687,237,3 Firmenmarke 7, gepr. „P"= Z3, unsign.?, H. ca. 55 cm; WKH
1687,334,2 gem. weiß VIII., Firmenmarke Sculpture 13, Repr. Res., sign. Rozel, H. 56 cm; Doro 13.9.2004
1687 (Firmenmarke ohne Ang.), H. 57 cm; GSF, S. 14
Text der Inschrift: Ave Caesar! Die Sterbenden grüßen dich.
Ausst.: ÖMKI 1898/99, hier mit Ang. des Originaltitels Büste „Gladiator" und des Künstlers „Rozel".
-> vgl. 2184; 2294
-> vgl. GRUPPE „Animus et vis"= Modell 3680
Lit.: GSF, S. 14, hier mit der Ang. des Originaltitels „Gladiator"; Neuwirth (1974), S. 72; Pinhas, S. 79 (m. Abb.).

1689
Mayre, G.
FIGUR „Flötenspieler" (Sitzender nackter Junge, eine Flöte spielend) (auf getrepptem Sockel)
ca. 1898/99
1689 (Firmenmarke ohne Ang.), H. fehlt; Pinhas, S. 85 (m. Abb.)
1689 (Firmenmarke ohne Ang.), H. 70 cm; GSF, S. 14
Ausst.: ÖMKI 1899/1900, hier mit Ang. des Originaltitels „Flötenspieler" und des Künstlers „Mayre".
Lit.: GSF, S. 14, hier mit der Ang. des Originaltitels „Flötenspieler"; Neuwirth (1974), S. 68.

1690
BÜSTE „Darde d'amore" (Junge Mutter mit Kind)
ca. 1898/99
1690 (Firmenmarke ohne Ang.), H. fehlt; Pinhas, S. 81 (m. Abb.)
1690 (Firmenmarke ohne Ang.), H. 65 cm; GSF, S. 14
Lit.: GSF, S. 14, hier mit der Ang. des Originaltitels „Darde d'amore".

1695
Haniroff
BÜSTE Lachender Negerjunge mit Kappe
ca. 1898
1695 (Firmenmarke ohne Ang.), H. 66 cm; Pinhas, S. 29 (m. Abb.)
-> vgl. 1617 (Variante: Schiebermütze)

1696
BÜSTE Junges Mädchen in mittelalterlicher Tracht
ca. 1898
1696 (Firmenmarke ohne Ang.), H. fehlt; Pinhas, S. 79 (m. Abb.)

1697
BÜSTE „Schiller"
Inschrift: SCHILLER
ca. 1898
1697,4,13 Firmenmarke o. Ang., Repr. Res., H. 60 cm; Kunst- u. Auktionshaus Wiesbaden 39
1697 (Firmenmarke ohne Ang.), H. 61 cm; GSF, S. 14
Bem.: Dargestellt ist der Dichter Friedrich Schiller (1759–1805).
Lit.: GSF, S. 14, hier mit der Ang. des Originaltitels „Schiller"; Pinhas, S. 79 (m. Abb.).

1698
BÜSTE „Goethe"
ca. 1898
1698,110,11 gem. weiß X, Firmenmarke Sculpture 13, Firmenmarke Adler 17, Fabriqué en Autriche, Ausformung ca. 1900, H. 61 cm; Wien Museum Inv. 219.112
1698 (Firmenmarke ohne Ang.), H. 63 cm; GSF, S. 14
Bem.: Dargestellt ist der Dichter Johann Wolfgang von Goethe (1749–1832), die Büste wurde vermutlich zum 250. Geburtstag des Schriftstellers aufgelegt.
Lit.: GSF, S. 14, hier mit der Ang. des Originaltitels „Goethe"; Pinhas, S. 79 (m. Abb.).

1699
BÜSTE „Gluck"
ca. 1898/99
1699 (Firmenmarke ohne Ang.), H. fehlt; Pinhas, S. 79 (m. Abb.)
1699 (Firmenmarke ohne Ang.), H. 70 cm; GSF, S. 14
Bem.: Dargestellt ist der Komponist Christoph Willibald Ritter von Gluck (1714–1887).
Lit.: GSF, S. 14, hier mit der Ang. des Originaltitels „Gluck".

1700
FIGUR „Musik (bekleidet)" (Stehende junge Dame mit Violine) (auf Rundsockel)
ca. 1898/99
1700 (Firmenmarke ohne Ang.), H. fehlt; Pinhas, S. 79 (m. Abb.)
1700 (Firmenmarke ohne Ang.), H. 91 cm; GSF, S. 14
Lit.: GSF, S. 14, hier mit der Ang. des Originaltitels „Musik (bekleidet)".

1701
BÜSTE „Le Rebell"
ca. 1898/99
1701 (Firmenmarke ohne Ang.), H. 58 cm; GSF, S. 14
-> vgl. 1552
Lit.: GSF, S. 14, hier mit der Ang. des Originaltitels „Le Rebell".

1702
BÜSTE „Nacht"
ca. 1898/99
1702 (Firmenmarke ohne Ang.), H. 44 cm; GSF, S. 14
-> vgl. 1357
Lit.: GSF, S. 14, hier mit der Ang. des Originaltitels „Nacht".

1719
BÜSTE Lachendes Mädchen
ca. 1898/99
1719 (Firmenmarke ohne Ang.), H. fehlt; Pinhas, S. 79 (m. Abb.)

1720
BÜSTE Junge Frau mit geschlossenen Augen und Kopftuch
ca. 1898
1720 (Firmenmarke ohne Ang.), H. fehlt; Pinhas, S. 79 (m. Abb.)

1721
FIGUR „Goliath" (Auf Felsen stehender und mit Speer und Schwert bewaffneter Goliath)
Inschrift: GOLIATH
ca. 1898/99
1721 (Firmenmarke ohne Ang.), H. fehlt; Pinhas, S. 70 (m. Abb.)
1721 (Firmenmarke ohne Ang.), H. 85 cm; GSF, S. 14
Bem.: Dargestellt ist der aus der Bibel (Altes Testament) bekannte riesenhafte Krieger Goliath.
-> vgl. 1233
Lit.: GSF, S. 14, hier mit der Ang. des Originaltitels „Goliath".

1722
FIGUR „Blumenmädchen" (Sich verneigende, Blumen geschmückte junge Dame mit großem Strohhut)
ca. 1898/99
1722 (Firmenmarke ohne Ang.), H. fehlt; Pinhas, S. 78 (m. Abb.)
1722 (Firmenmarke ohne Ang.), H. 87 cm; GSF, S. 14
-> vgl. 2098
-> vgl. 1926= BÜSTE
Lit.: GSF, S. 14, hier mit der Ang. des Originaltitels „Blumenmädchen".

1723
WANDMASKE „Napoleon" (mit Zweispitz)
ca. 1898/99
1723 (Firmenmarke ohne Ang.), H. 29 cm; Pinhas, S. 96 (m. Abb.)
1723 (Firmenmarke ohne Ang.), H. 31 cm; GSF, S. 14
Bem.: Dargestellt ist der franz. Kaiser Napoleon I. Bonaparte (1769–1821).
-> vgl. 1726
-> vgl. 1064; 1126; 1154= jew. BÜSTE
Lit.: GSF, S. 14, hier mit der Ang. des Originaltitels „Napoleon".

1724
WANDMASKE „Epicuräer" (geschmückt mit Weintrauben)
ca. 1898/99
1724 (Firmenmarke ohne Ang.), H. fehlt; Pinhas, S. 96 (m. Abb.)
1724 (Firmenmarke ohne Ang.), H. 20 cm; GSF, S. 14
Lit.: GSF, S. 14, hier mit der Ang. des Originaltitels „Epicuräer".

1725
BÜSTE „Klosterblume" (Junges Mädchen mit großer Kapuze)
ca. 1898/99
1725 (Firmenmarke ohne Ang.), H. fehlt; Pinhas, S. 84 (m. Abb.)
1725 (Firmenmarke ohne Ang.), H. 43 cm; GSF, S. 14
Lit.: GSF, S. 14, hier mit der Ang. des Originaltitels „Klosterblume".

1726
WANDMASKE „Napoleon" (mit Zweispitz)
ca. 1898/99
1726,19,40 Firmenmarke 7, Repr. Res., unles. sign., H. 20 cm; KH
1726 (Firmenmarke ohne Ang.), H. 20 cm; GSF, S. 14
Bem.: Dargestellt ist der franz. Kaiser Napoleon I. Bonaparte (1769–1821).
-> vgl. 1723
-> vgl. 1064; 1126; 1154= jew. BÜSTE
Lit.: GSF, S. 14, hier mit der Ang. des Originaltitels „Napoleon"; Pinhas, S. 96 (m. Abb.)

1727
Wind, Josef
FIGUR „Krugträgerin"/„Krughändlerin" (Schreitende Frau mit Turban und zwei/drei Krügen) (auf Rundsockel)
ca. 1898
1727 (Firmenmarke ohne Ang.), H. 58 cm; Pinhas, S. 26, 82 (m. Abb.)
1727 (Firmenmarke ohne Ang.), H. 55 cm; GSF, S. 14
Bem.: Diese Größe wurde in der Variante mit zwei, oder auch drei Krügen angeboten.
Ausst.: ÖMKI 1898/99, hier mit Ang. des Originaltitels „Krughändlerin".
-> vgl. 1285; 1997
Lit.: GSF, S. 14, hier mit der Ang. des Originaltitels „Krugträgerin"; Neuwirth (1974), S. 76.

1728
BÜSTE Orientalin (mit gesenktem Blick) und Kopfputz
ca. 1898/99
1728 (Firmenmarke ohne Ang.), H. fehlt; Pinhas, S. 80 (m. Abb.)

1730
BÜSTE „Madame Recamier"
ca. 1898/99
1730 (Firmenmarke ohne Ang.), H. fehlt; Pinhas, S. 80 (m. Abb.)
1730 (Firmenmarke ohne Ang.), H. 62 cm; GSF, S. 14
Bem.: Dargestellt ist Julie Récamier (1777–1849), Gattin eines Pariser Bankiers und Vertraute Chateaubriands.
Lit.: GSF, S. 14, hier mit der Ang. des Originaltitels „Madame Recamier".

1731
FIGUR „Lola" (Stehendes lachendes Bauermädchen mit Wasserkrug) (auf Landschaftssockel)
ca. 1898/99
1731 (Firmenmarke ohne Ang.), H. fehlt; Pinhas, S. 79 (m. Abb.)
1731 (Firmenmarke ohne Ang.), H. 62 cm; GSF, S. 14
Pendant -> vgl. 1732
Lit.: GSF, S. 14, hier mit der Ang. des Originaltitels „Lola".

1732
FIGUR „Addio" (Stehendes lachendes und mit Taschentuch winkendes Bauermädchen mit Weinkrug) (auf Landschaftssockel)
ca. 1898/99
1732 (Firmenmarke ohne Ang.), H. fehlt; Pinhas, S. 79 (m. Abb.)
1732 (Firmenmarke ohne Ang.), H. 59 cm; GSF, S. 14
Pendant -> vgl. 1731
Lit.: GSF, S. 14, hier mit der Ang. des Originaltitels „Addio".

1733
BÜSTE „Fahtuma" (Orienatlin mit Kopftuch und Halsschmuck)
ca. 1898/99
1733 (Firmenmarke ohne Ang.), H. fehlt; Pinhas, S. 82 (m. Abb.)
1733 (Firmenmarke ohne Ang.), H. 68 cm; GSF, S. 14
Lit.: GSF, S. 14, hier mit der Ang. des Originaltitels „Fahtuma".

1734
BÜSTE „Neger mit Laute" (Schwarzer Banjospieler mit Strohhut)
ca. 1898/99
1734 (Firmenmarke ohne Ang.), (Gesamth. 65 cm?); Pinhas, S. 30 (m. Abb.)
1734 (Firmenmarke ohne Ang.), H. 47 cm; GSF, S. 14
Bem.: Banjo (welches im Firmenkatalog als „Laute" bezeichnet wird, ist aus Holz gefertigt.
-> vgl. 628 (ähnl. Modell)
Lit.: GSF, S. 14, hier mit der Ang. des Originaltitels „Neger mit Laute".

1735
WANDMASKE „Zigeunerknabe"
ca. 1898/99
1735 (Firmenmarke ohne Ang.), H. 26 cm; GSF, S. 14
Bem.: Im Firmenkatalog wird auf die Mod.-Nr. 1097

verwiesen. Hier wäre die aktuell jedoch noch nicht nachweisbare BÜSTE oder FIGUR dieser Maske zu vermuten.
Lit.: GSF, S. 14, hier mit der Ang. des Originaltitels „Zigeunerknabe".

1737 -> ÜBERSCHNEIDUNG
BÜSTE Chloë (Junges Mädchen mit zwei Flöten)
ca. 1898/99
1737 (Firmenmarke ohne Ang.), H. fehlt; Pinhas, S. 79 (m. Abb.)
1737 (Firmenmarke ohne Ang.), H. 45 cm; GSF, S. 14
Bem.: Dargestellt ist Chloë, eine Figur der griech. Mythologie.
-> vgl. 1519= FIGUR
Lit.: GSF, S. 14, hier mit der Ang. des Originaltitels „Chloë".

1737 -> ÜBERSCHNEIDUNG
Morice, Léon
„Enfant au Lezard" (Erschreckter Knabe mit Eidechse) (auf Natursockel)
BRONZE
Inschrift: ENFANT AU LEZARD
ca. 1898/1900
6,1737,G GS Edit A (Firmenmarke?), sign. Léon Morice, H. 55,3 cm; Christie's SK 9.10.2002
Text der Inschrift: Kind mit Eidechse

1738
FIGUR „Zuckerbäcker (Neger)"
ca. 1898/99
1738 (Firmenmarke ohne Ang.), H. 72 cm; GSF, S. 14
-> vgl. 1659 (ähnl. Thematik)
Lit.: GSF, S. 14, hier mit der Ang. des Originaltitels „Zuckerbäcker (Neger)".

1739
Bortone, Ipazzio Antonio
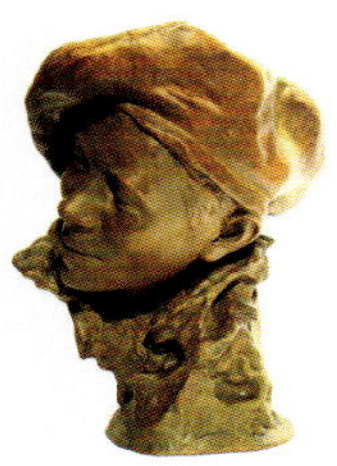
BÜSTE „Bummler" (Junger Mann mit Barett)
ca. 1898/99
1739,18,14 gem. weiß P, Firmenmarke 7, undeutlich sign. Borthone? (gemeint ist wohl Bortone), H. 32,5 cm; Sammlung a@goldscheider.net
1739 (Firmenmarke ohne Ang.), H. 47 cm; GSF, S. 14
-> vgl. 1940= WANDMASKE
Lit.: GSF, S. 14, hier mit der Ang. des Originaltitels „Bummler"; Pinhas, S. 81 (m. Abb.).

1740
Bortone, Ipazzio Antonio
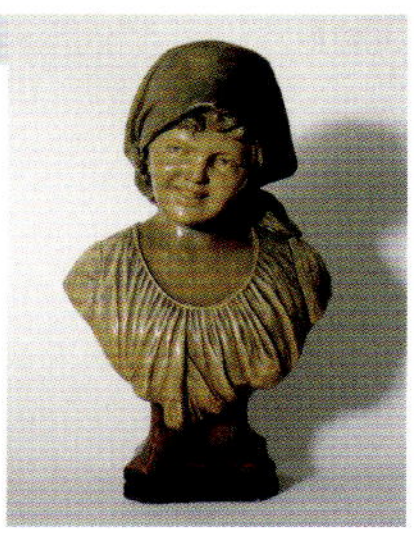
BÜSTE „Annita"
ca. 1898/99
1740,98,57 gem. weiß III, Firmenmarke 7, Repr. Res., sign. Bortone, H. 48 cm; eBay # 1508393181
1740,369,9 Firmenmarke Sculpture 13, Repr. Res., sign. Bertrand, H. 48 cm; KH
1740 (Firmenmarke ohne Ang.), H. 48 cm; GSF, S. 14
-> vgl. 2130
-> vgl. 1502 (Variante)
-> vgl. 2029= WANDMASKE
Lit.: GSF, S. 14, hier mit der Ang. des Originaltitels „Annita"; Pinhas, S. 80 (m. Abb.).

1741
Cherc
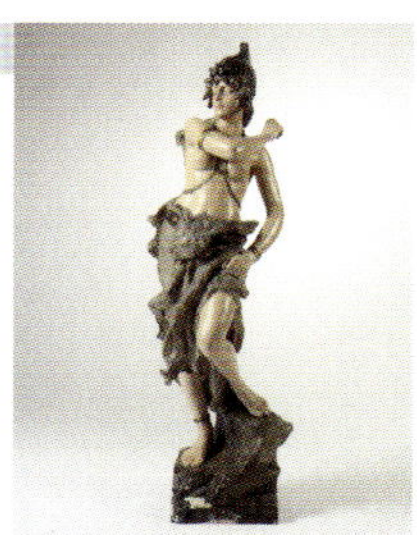
FIGUR „Amazonenkönigin" (Stehende Kriegerin, Halbakt auf Felssockel)
ca. 1898
1741,35,34 Firmenmarke 7, Repr. Res., H. 113 cm; Wien Museum Inv. 186.619
1741 (Firmenmarke ohne Ang.), H. 112 cm; GSF, S. 14
Bem.: An der Plinthe betitelt „La Reine des Amazones" (Metallschild).
-> vgl. 1880
-> vgl. 1923= BÜSTE
Ausst.: ÖMKI 1898/99, hier mit Ang. des Originaltitels „Amazonenkönigin".
Lit.: GSF, S. 14, hier mit der Ang. des Originaltitels „Amazonenkönigin"; Neuwirth (1974), S. 61; Pinhas, S. 80 (m. Abb.).

1742
FIGUR „Cyrano de Bergerac" (als stehende Figur in Pose mit Degen) (auf Rechtecksockel)
Inschrift: CYRANO DE BERGERAC
ca. 1898/99
1742 (Firmenmarke ohne Ang.), H. fehlt; Pinhas, S. 83 (m. Abb.)
1742 (Firmenmarke ohne Ang.), H. 92 cm; GSF, S. 15
Bem.: Dargestellt ist der franz. Schriftsteller Cyrano de Bergerac (1619–1655).
-> vgl. 1802 (ähnl. Mod.)
Lit.: GSF, S. 15, hier mit der Ang. des Originaltitels „Cyrano de Bergerac".

1744
WANDMASKE „Styl Louis XIV" (Lachendes Mädchen mit hochgestecktem Haar)
ca. 1898/99
1744 (Firmenmarke ohne Ang.), H. fehlt; Pinhas, S. 81 (m. Abb.)
1744 (Firmenmarke ohne Ang.), H. 31 cm; GSF, S. 15
Lit.: GSF, S. 15, hier mit der Ang. des Originaltitels „Styl Louis XIV".

1745
WANDMASKE „Secession" (Lachendes Kind mit großer Haube und gerüschtem Kragen)
ca. 1898/99
1745 (Firmenmarke ohne Ang.), H. fehlt; Pinhas, S. 81 (m. Abb.)
1745 (Firmenmarke ohne Ang.), H. 38 cm; GSF, S. 15
-> vgl. 1954 (Büste mit gleichem Titel)
Lit.: GSF, S. 15, hier mit der Ang. des Originaltitels „Secession".

1746
Schuhputzer (Stehender lachender Negerbursche mit Putzzeug) (auf Rechtecksockel)
ca. 1898/99
1746 (Firmenmarke ohne Ang.), H. fehlt; Pinhas, S. 79 (m. Abb.)

1747
BÜSTE Mädchen (nach links blickend)
ca. 1899
1747 (Firmenmarke ohne Ang.), H. fehlt; Pinhas, S. 84 (m. Abb.)

1748
FIGUR „Knabe (strickend)" (auf Bambussessel)
ca. 1898/99
1748 (Firmenmarke ohne Ang.), H. fehlt; Pinhas, S. 81 (m. Abb.)
1748 (Firmenmarke ohne Ang.), H. 45 cm; GSF, S. 15
Bem.: Der Stuhl, bzw. Bambussessel gehörte zum Lieferumfang dieses Modells.
Lit.: GSF, S. 15, hier mit der Ang. des Originaltitels „Knabe (strickend)".

1750
FIGUR „Wildschütz"
ca. 1897/98 (für den Entwurf)
1750 (Firmenmarke ohne Ang.), H. 50 cm; GSF, S. 15
-> vgl. 1213
Lit.: GSF, S. 15, hier mit der Ang. des Originaltitels „Wildschütz".

1751
FIGUR „Flösser" (Jüngling, mit Fellschurz bekleidet und mit langem Stab) (auf Sockel mit Mauer)
ca. 1898/99
1751 (Firmenmarke ohne Ang.), H. fehlt; Pinhas, S. 84 (m. Abb.)
1751 (Firmenmarke ohne Ang.), H. 98 cm; GSF, S. 15
Lit.: GSF, S. 15, hier mit der Ang. des Originaltitels „Flösser".

1752
Lesca
BÜSTE „Bébé" (Junger Pierrot, einen Apfel haltend)
ca. 1899
1752,42,9 „Wortmarke" (Firmenmarke ohne Ang.), H. 33 cm; Doro 14.9.2001
1752 (Firmenmarke ohne Ang.), H. 33 cm; GSF, S. 15
Lit.: GSF, S. 15, hier mit der Ang. des Originaltitels „Bébé"; Pinhas, S. 81 (m. Abb.).

1754
FIGUR „Pfiffikus" (Stehender verschmitzt lachender Negerbub mit Strohhut) (auf Rechtecksockel mit Korb)
ca. 1899
1754 (Firmenmarke ohne Ang.), H. fehlt; Pinhas, S. 85 (m. Abb.)
1754 (Firmenmarke ohne Ang.), H. 67 cm; GSF, S. 15
Lit.: GSF, S. 15, hier mit der Ang. des Originaltitels „Pfiffikus".

1755
BÜSTE „Salambo" (Junge Dame mit Hochsteckfrisur und Medaillon um den Hals)
Inschrift: SALAMBO
ca. 1899
1755 (Firmenmarke ohne Ang.), H. fehlt; Pinhas, S. 81 (m. Abb.)
1755 (Firmenmarke ohne Ang.), H. 60 cm; GSF, S. 15
Bem.: Dargestellt ist die Priesterin Salambo, die Tochter des punischen Heerführers Hamilkars und Schwester Hannibals. Gustave Flaubert (1821–1864) widmete ihr auch einen gleichnamigen Roman, der 1863 erschien und dann 1904 ins dt. übersetzt wurde.
-> vgl. 2183; 2430
Lit.: GSF, S. 15, hier mit der Ang. des Originaltitels „Salambo".

1756
Stehender, an einen Baumstamm gelehnter Mädchenakt (in Pose) mit Hüfttuch, Pfeilen und Köcher (auf Rundsockel)
ca. 1899
1756 (Firmenmarke ohne Ang.), H. fehlt; Pinhas, S. 81 (m. Abb.)
Bem.: Die Darstellung erinnert an eine „Eva", nur dass der Apfel durch einen Pfeil ersetzt wurde und die Figur um den Bauch einen Gürtel mit Köcher trägt.

1757
WANDMASKE „Gigerl" (Lachender Neger mit Strohhut)
ca. 1899
1757 (Firmenmarke ohne Ang.), H. fehlt; Pinhas, S. 81 (m. Abb.)
1757 (Firmenmarke ohne Ang.), H. 27 cm; GSF, S. 15
-> vgl. 1663= FIGUR
Lit.: GSF, S. 15, hier mit der Ang. des Originaltitels „Gigerl".

1758
FIGUR „Wasserträger" (Junger Bursche, ein großes Wassergefäß tragend) (auf Landschaftssockel)
ca. 1899
1758 (Firmenmarke ohne Ang.), H. fehlt; Pinhas, S. 84 (m. Abb.)
1758 (Firmenmarke ohne Ang.), H. 84 cm; GSF, S. 15
-> vgl. 2157
Lit.: GSF, S. 15, hier mit der Ang. des Originaltitels „Wasserträger".

1759
Donatello (nach)
BÜSTE „Nobel-Dame" (Junges Mädchen mit gesenktem Haupt) (auf verziertem Sockel)
ca. 1899
1759,13,34 FG Wien (Firmenmarke 7?), Repr. Res., sign. Donatello?, H. 45 cm; KH
1759 (Firmenmarke ohne Ang.), H. 45 cm; GSF, S. 15
Ausst.: ÖMKI 1899/1900, hier mit Ang. des Originaltitels „Nobel-Dame".
Lit.: GSF, S. 15, hier mit der Ang. des Originaltitels „Nobel-Dame"; Neuwirth (1974), S. 62.

1760
Tell, E.
BÜSTE „Caprieser" (Junger Mann mit Strohhut) (auf ährenverziertem Sockel)
ca. 1899
1760 (Firmenmarke ohne Ang.), unsign., H. ca. 38 cm; Pinhas, S. 83 (m. Abb.)
1760 (Firmenmarke ohne Ang.), H. 38 cm; GSF, S. 15
Pendant –> vgl. 1761
Lit.: GSF, S. 15, hier mit der Ang. des Originaltitels „Caprieser".

1761
Tell, E.
BÜSTE „Capriesérin" (Junge Frau mit Strohhut) (auf ährenverziertem Sockel)
ca. 1899
1761,167,2 Firmenmarke 7, Repr. Res., H. 38 cm; Zengen 398
1761 (Firmenmarke ohne Ang.), H. 37 cm; GSF, S. 15
Pendant –> vgl. 1760
Lit.: GSF, S. 15, hier mit der Ang. des Originaltitels „Capriesérin"; Pinhas, S. 83 (m. Abb.).

1762
Donatello (nach)
BÜSTE „San Lorenzo" (Junger Mann mit Kappe)
ca. 1898/99
1762 (Firmenmarke ohne Ang.), H. fehlt; Pinhas, S. 97 (m. Abb.)
1762 (Firmenmarke ohne Ang.), H. 46 cm; GSF, S. 15
Ausst.: ÖMKI 1898/99, hier mit Ang. des Originaltitels „San Lorenzo".
Lit.: GSF, S. 15, hier mit der Ang. des Originaltitels „San Lorenzo"; Neuwirth (1974), S. 62.

1763
FIGUR „Schmetterling" (Sitzender Mädchenhalbakt) (auf oktogonalem, mit Blumen geschmücktem Sockel)
ca. 1899
1763 (Firmenmarke ohne Ang.), H. fehlt; Pinhas, S. 85 (m. Abb.)
1763 (Firmenmarke ohne Ang.), H. 113 cm; GSF, S. 15
–> vgl. 1974, hier mit dem abweichenden Originaltitel „Printemps".
Lit.: GSF, S. 15, hier mit der Ang. des Originaltitels „Schmetterling".

1764
FIGUR „Chianti" (Stehendes Bauernmädchen mit Haube und Schürze, eine Chiantiflasche haltend) (auf mit Weintrauben und Laub verziertem Rundsockel)
Inschrift: CHIANTI
ca. 1899
1764 (Firmenmarke ohne Ang.), H. fehlt; Pinhas, S. 80 (m. Abb.)
1764 (Firmenmarke ohne Ang.), H. 106 cm; GSF, S. 15
–> vgl. 2027
Lit.: GSF, S. 15, hier mit der Ang. des Originaltitels „Chianti".

1765
Mayre, G.
BÜSTE „Jason"
Inschrift: JASON
ca. 1899
1765 (Firmenmarke ohne Ang.), H. 45 cm; Pinhas, S. 39, 84 (m. Abb.)
1765 (Firmenmarke ohne Ang.), H. 44 cm; GSF, S. 15
Bem.: Dargestellt ist Jason, eine Figur der griech. Sagenwelt.
Pendant –> vgl. 1766
Lit.: GSF, S. 15, hier mit der Ang. des Originaltitels „Jason".

1766
Tell, E.
BÜSTE „Medea"/„Medee"
Inschrift: MEDEE
ca. 1899
1766 FG (Firmenmarke ohne Ang.), H. 40 cm; Sigalas 47
1766 (Firmenmarke ohne Ang.), H. 39 cm; GSF, S. 15
Bem.: Dargestellt ist Medea, eine Figur der griech. Sagenwelt. Medea verhalf Jason zum Goldenen Vlies.
Pendant –> vgl. 1765
Lit.: GSF, S. 15, hier mit der Ang. des Originaltitels „Medea"; Pinhas, S. 84 (m. Abb.).

1767
Tell, E.
GRUPPE „Geschwisterkuss" (Stehendes sich küssendes Kinderpaar) (auf Ovalsockel)
ca. 1899
1767,43,35 Firmenmarke 7, Repr. Res., H. 51 cm; eBay # 6521971052
1767 (Firmenmarke ohne Ang.), H. 52 cm; GSF, S. 15
Lit.: GSF, S. 15, hier mit der Ang. des Originaltitels „Geschwisterkuss"; Pinhas, S. 84 (m. Abb.).

1768
Tell, E. (zugeschr.)
GRUPPE „Der kleine Doktor" (Biedermeierkinderpaar) (auf Rechtecksockel mit Schemel)
ca. 1899
1768 (Firmenmarke ohne Ang.), unsign., H. fehlt; Pinhas, S. 83 (m. Abb.)
1768 (Firmenmarke ohne Ang.), H. 45 cm; GSF, S. 15
Lit.: GSF, S. 15, hier mit der Ang. des Originaltitels „Der kleine Doktor".

1769
Tell, E.
FIGUR „Krugträgerin" (Wasserträgerin mit Gefäß auf dem Kopf)
ca. 1899
1769,48,34 Firmenmarke 7, Repr. Res., gepr. „P"= Z3, H. 94 cm; Wien Museum Inv. 186.624
1769 (Firmenmarke ohne Ang.), H. 96 cm; GSF, S. 15
–> vgl. 2274
Lit.: GSF, S. 15, hier mit der Ang. des Originaltitels „Krugträgerin"; Neuwirth (1974), S. 75; Pinhas, S. 85 (m. Abb.).

1771
BÜSTE „Soubrette"(Junges, mit Blumen geschmücktes Mädchen, stolz schauend)
ca. 1899
1771 (Firmenmarke ohne Ang.), H. 65 cm; GSF, S. 15
Pendant –> vgl. 1761
Lit.: GSF, S. 15, hier mit der Ang. des Originaltitels „Soubrette".

1772
Carrièr-Belleuse, (Albert-Ernest)

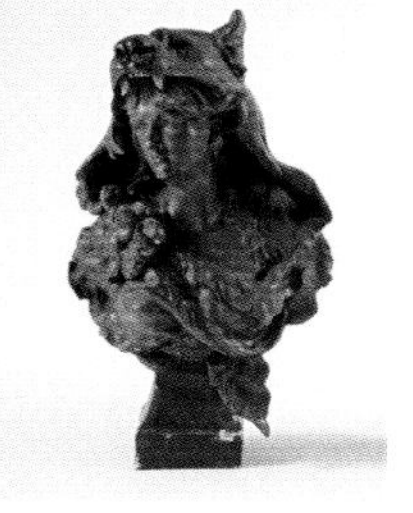

BÜSTE „Chriemhild (bekleidet)"/(„Frédégonde") (Bekleidete Frau mit Löwenkopf und -fell als Kopfbedeckung)
ca. 1899/1900
1772,28,70 ohne Firmenmarke, unsign., gepr. „P"= Z3, H. 77 cm; Wien Museum Inv. 219.111
1772,117,34 ger. I, Firmenmarke Sculpture 13, Repr. Res., unsign., gepr. „S"= Z2, H. 75 cm; GK, S. 14, 73 (m. Abb.); Wien Museum Inv. 223.903
1772,135,2 Firmenmarke Sculpture 13, Repr. Res., H. 74 cm; Soth. L. 15.3.1991; Phillips L. 23.9.1997
1772 (Firmenmarke ohne Ang.), H. 68 cm; GSF, S. 15
Ausst.: HMW (jetzt Wien Museum) 1985/86.
–> vgl. 1647 (nackt)
Pendant –> vgl. 1773; 1980
Lit.: GSF, S. 15, hier mit der Ang. des Originaltitels „Chriemhilde (bekleidet)"; GK, S. 14, 73 (m. Abb.); Pinhas, S. 89 (m. Abb.).

1773
Carrièr-Belleuse, (Albert-Ernest)
BÜSTE „Brunhilde (bekleidet)"/(„Brunhaut") (Frau mit Adlerkopf und -schwingen als Kopfbedeckung)
ca. 1899/1900
1773,70,12 Firmenmarke Sculpture 13, Repr. Res., H. 71 cm; Soth. L. 15.3.1991; Phillips L. 23.9.1997
1773 (Firmenmarke ohne Ang.), H. 64 cm; GSF, S. 15
–> vgl. 1648; 1980= jew. Variante (nackt)
Pendant –> vgl. 1647; 1772
Lit.: GSF, S. 15, hier mit der Ang. des Originaltitels „Brundhilde (bekleidet)"; Pinhas, S. 89 (m. Abb.).

1774
BÜSTE „La Cigarett" (Junge rauchende Dame)
Inschrift: LA CIGARETT
ca. 1898/99
1774 (Firmenmarke ohne Ang.), H. fehlt; Pinhas, S. 81 (m. Abb.)

1775
BÜSTE Junge Dame mit großer Spitzenhaube
ca. 1898/99
1775 (Firmenmarke ohne Ang.), H. fehlt; Pinhas, S. 92 (m. Abb.)

1776
BÜSTE Junge Dame mit Feder geschmücktem Hut
ca. 1898/99
1776 (Firmenmarke ohne Ang.), H. fehlt; Pinhas, S. 92 (m. Abb.)

1777
FIGUR „Schusterlehrling" (Schreitender rauchender Schusterbub, bepackt mit Stiefeln) (auf Rundsockel)
ca. 1899
1777 (Firmenmarke ohne Ang.), H. fehlt; Pinhas, S. 92 (m. Abb.)
1777 (Firmenmarke ohne Ang.), H. 33 cm; GSF, S. 15
Lit.: GSF, S. 15, hier mit der Ang. des Originaltitels „Schusterlehrling".

1778
FIGUR „Zeitungsträgerin" (Stehende alte Frau, einen Stapel mit Zeitungen in Händen haltend) (auf Rundsockel)
ca. 1899
1778 (Firmenmarke ohne Ang.), H. fehlt; Pinhas, S. 92 (m. Abb.)
1778 (Firmenmarke ohne Ang.), H. 44 cm; GSF, S. 15
Lit.: GSF, S. 15, hier mit der Ang. des Originaltitels „Zeitungsträgerin".

1779
BÜSTE „Italienischer Knabe" (nach rechts blickend)
ca. 1898/99
1779 (Firmenmarke ohne Ang.), H. 42 cm; Pinhas, S. 83 (m. Abb.)
1779 (Firmenmarke ohne Ang.), H. 44 cm; GSF, S. 15
–> vgl. 1253
Lit.: GSF, S. 15, hier mit der Ang. des Originaltitels „Italienischer Knabe".

1780
BÜSTE „Curiosité" (Neugieriger Bub über einen Holzzaun (Wand) schauend)
ca. 1899
1780 (Firmenmarke ohne Ang.), H. fehlt; Pinhas, S. 83 (m. Abb.)
1780 (Firmenmarke ohne Ang.), H. 42 cm; GSF, S. 15
Lit.: GSF, S. 15, hier mit der Ang. des Originaltitels „Curiosité".

1781 –> ÜBERSCHNEIDUNG
BÜSTE „Au Bal" (Junge Dame mit Kopfputz) (nach rechts blickend) (auf konischem Sockel mit Theater/Ball-Maske)
ca. 1899/1900
1781 (Firmenmarke ohne Ang.), H. fehlt; Pinhas, S. 81 (m. Abb.)
1781 (Firmenmarke ohne Ang.), H. 56 cm; GSF, S. 15
Lit.: GSF, S. 15, hier mit der Ang. des Originaltitels „Au Bal".

1781 –> ÜBERSCHNEIDUNG
BÜSTE Mädchen mit Blumen geschmücktem Haar und Kleid (Flora)
ca. 1899/1900
1781,8,39 gem. weiß VII., Firmenmarke 7, Repr. Res., gepr. „P"= Z3, H. 32 cm; eBay # 2269005084

1783
BÜSTE „Bismarck“
ca. 1899/1900
1783 (Firmenmarke ohne Ang.), H. 70 cm; GSF, S. 15
Bem.: Dargestellt ist der deutsche Staatsmann und Reichskanzler Otto von Bismarck (1815–1898).
-> vgl. 1914
Lit.: GSF, S. 15, hier mit der Ang. des Originaltitels „Bismarck“.

1784 -> ÜBERSCHNEIDUNG
Ceccarelli, Ezio
BÜSTE „Coquette“ (Lachende, junge Dame mit Haarschleife) (nach rechts oben blickend)
ca. 1899/1900
1784,174,34 Firmenmarke Sculpture 13, Repr. Res., H. ca. 60 cm; eBay # 896993118; 906496026
Lit.: GSB, Pl. 76, hier ist eine identische Büste mit dem franz. Originaltitel „Coquette“ unter Nr. 753 (H. 40 cm) abgebildet. Ang. zum Material (Terrakotta/Bronze fehlen), als Künstler wird „Fontini“ (Pseud. für Ceccarelli) genannt.
-> vgl. 753 (GSB)

1784 -> ÜBERSCHNEIDUNG
BÜSTE Junge Dame mit Blumen geschmücktem Haar (nach rechts blickend)
ca. 1899
1784 (Firmenmarke ohne Ang.), H. fehlt; Pinhas, S. 84 (m. Abb.)

1785
BÜSTE Junge Dame mit geschlossenen Augen
ca. 1899
1785 (Firmenmarke ohne Ang.), H. fehlt; Pinhas, S. 84 (m. Abb.)

1787
Stehender Mädchenakt (in Pose) mit großem Tuch (auf Rundsockel mit Schriftband/Banner)
Inschrift: NEME REGARD...
ca. 1899/1900
1787 (Firmenmarke ohne Ang.), H. fehlt; Pinhas, S. 81 (m. Abb.)
Bem.: Der vollständige Text auf dem Banner, lässt sich auf der Abb. dieses zeitgenössischen (ca. 1900) Katalogs der Firma, nicht entziffern.

1788
BÜSTE „Beatrice“
ca. 1899/1900
1788 (Firmenmarke ohne Ang.), H. 64 cm; GSF, S. 15
Lit.: GSF, S. 15, hier mit der Ang. des Originaltitels „Beatrice“.

1789
Waderé, Heinrich
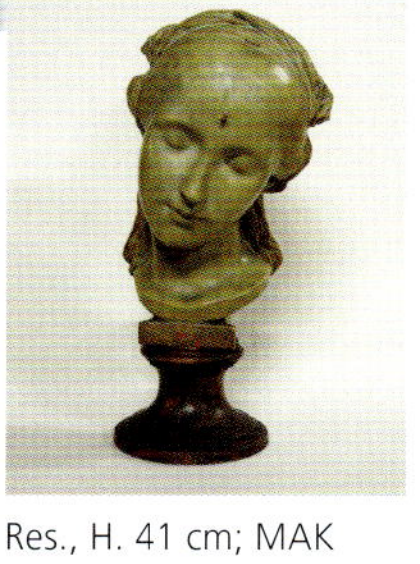
BÜSTE „Madonnenkopf“ (Frau mit geneigtem Haupt) (auf Rundsockel)
ca. 1898/99
1789,76,13 Firmenmarke 7, Repr. Res., gepr. „P“= Z3, H. (mit Sockel, lt. Anbieter) 40 cm; eBay # 3349429427
1789,219,13 gem. 8, Firmenmarke Sculpture 13, Repr. Res., H. 41 cm; MAK Wien, Inv. Ke 10602-93
1789 (Firmenmarke ohne Ang.), H. 39 cm; GSF, S. 15
Ausst.: ÖMKI 1900/01, hier mit Ang. des Originaltitels Büste „Madonnenkopf“.
Lit.: GSF, S. 15, hier mit der Ang. des Originaltitels „Madonnenkopf“; Neuwirth (1974), S. 76; Pinhas, S. 83 (m. Abb.).

1790
FIGUR „Schwerttänzerin“ (Stehende orientalische Schönheit mit zwei kurzen Schwertern)
ca. 1899/1900
1790 (Firmenmarke ohne Ang.), H. fehlt; Pinhas, S. 83 (m. Abb.)
1790 (Firmenmarke ohne Ang.), H. 39 cm; GSF, S. 15
-> vgl. 1883
-> vgl. 2139= BÜSTE
Lit.: GSF, S. 15, hier mit der Ang. des Originaltitels „Schwerttänzerin“.

1792
Stehende Dame (im Renaissancekostüm) eine Schale hochhebend (auf getrepptem Sockel)
ca. 1898/99
1792 (Firmenmarke ohne Ang.), H. fehlt; Pinhas, S. 85 (m. Abb.)

1793
FIGUR „Castagnettentänzerin (nackt)“ (auf Rundsockel)
ca. 1899/1900
1793 (Firmenmarke ohne Ang.), H. fehlt; Pinhas, S. 83 (m. Abb.)
1793 (Firmenmarke ohne Ang.), H. 39 cm; GSF, S. 15
-> vgl. 1872 (bekleidet)
Lit.: GSF, S. 15, hier mit der Ang. des Originaltitels „Castagnettentänzerin (nackt)“.

1794
BÜSTE „Königin Louise“
ca. 1897/98 (für den Entwurf)
1794 (Firmenmarke ohne Ang.), H. 42 cm; GSF, S. 15
Bem.: Dargestellt ist die preuss. Königin Luise, Herzogin zu Mecklenburg (1776–1810).
-> vgl. 1123; 1229
Lit.: GSF, S. 15, hier mit der Ang. des Originaltitels „Königin Louise“.

1795
BÜSTE „En avant“ (Arabischer Krieger mit Speer)
ca. 1898/99
1795 (Firmenmarke ohne Ang.), H. fehlt; Pinhas, S. 82 (m. Abb.)
1795 (Firmenmarke ohne Ang.), H. 89 cm; GSF, S. 15
Pendant -> vgl. 1796
Lit.: GSF, S. 15, hier mit der Ang. des Originaltitels „En avant“.

1796
BÜSTE „Au marché“ (Araberin, einen Korb auf dem Kopf tragend)
ca. 1898/99
1796 (Firmenmarke ohne Ang.), H. fehlt; Pinhas, S. 82 (m. Abb.)
1796 (Firmenmarke ohne Ang.), H. 82 cm; GSF, S. 15
Pendant -> vgl. 1795
Lit.: GSF, S. 15, hier mit der Ang. des Originaltitels „Au marché“.

1797
FIGUR „Melancholie“ (Sitzende sinnierende junge Frau)
Inschrift: MELANCHOLIE
ca. 1899/1900
1797 (Firmenmarke ohne Ang.), H. fehlt; Pinhas, S. 92 (m. Abb.)
1797 (Firmenmarke ohne Ang.), H. 29 cm; GSF, S. 16
Lit.: GSF, S. 16, hier mit der Ang. des Originaltitels „Melancholie“.

1798
FIGUR „Mondgöttin (La lune coissante)“
ca. 1898/99
1798 (Firmenmarke ohne Ang.), H. 46 cm; GSF, S. 16
Bem.: Goldscheider führte beide, sowohl den franz. als auch deutschen, Originaltitel für diese Figur.
-> vgl. 1433; 1937
Lit.: GSF, S. 16, hier mit der Ang. des Originaltitels „Mondgöttin (La lune coissante)“.

1799
FIGUR „Kugelspielerin (bekleidet)“ (Mädchenhalbakt, eine Kugel werfend) (auf Rundsockel)
ca. 1898/99
1799 (Firmenmarke ohne Ang.), H. fehlt; Pinhas, S. 85 (m. Abb.)
1799 (Firmenmarke ohne Ang.), H. 82 cm; GSF, S. 16
Lit.: GSF, S. 16, hier mit der Ang. des Originaltitels „Kugelspielerin (bekleidet)“.

1799
Kugelspielerin (Mädchenhalbakt eine Kugel werfend) (auf Rundsockel)
ca. 1898/99
1799 (Firmenmarke ohne Ang.), H. fehlt; Pinhas, S. 85 (m. Abb.)

1800
Etoile, A.
GRUPPE „Erste Liebe“ (Schäfer und Schäferin in antikisierendem Gewand) (auf Landschaftssockel)
ca. 1898/99
1800 (Firmenmarke ohne Ang.), H. 64 cm; GSF, S. 16
-> vgl. 1445
Lit.: GSF, S. 11, hier mit der Ang. des Originaltitels „Erste Liebe“; GK, S. 14, 70 (m. Abb.); Pinhas, S. 71 (m. Abb.).

1801
GRUPPE „Gondoliens“ (Spazierendes Biedermeierkinderpaar) (auf Ovalsockel mit Rosenstrauch)
ca. 1899/1900
1801 (Firmenmarke ohne Ang.), H. fehlt; Pinhas, S. 84 (m. Abb.)
1801 (Firmenmarke ohne Ang.), H. 54 cm; GSF, S. 16
Lit.: GSF, S. 16, hier mit der Ang. des Originaltitels „Gondoliens“.

1802
FIGUR „Cyrano de Bergerac“ (als stehende Figur in Pose mit langem Stock) (auf Rechtecksockel)
Inschrift: CYRANO DE BERGERAC
ca. 1899/1900
1802 (Firmenmarke ohne Ang.), H. fehlt; Pinhas, S. 83 (m. Abb.)
1802 (Firmenmarke ohne Ang.), H. 96 cm; GSF, S. 16
Bem.: Dargestellt ist der franz. Schriftsteller Cyrano de Bergerac (1619–1655).
-> vgl. 1889
-> vgl. 1742 (ähnl. Mod.)
Lit.: GSF, S. 16, hier mit der Ang. des Originaltitels „Cyrano de Bergerac“.

1803
BÜSTE Junge lachende Dame mit großem Strohhut und Blumenschmuck
ca. 1898/99
1803 (Firmenmarke ohne Ang.), H. fehlt; Pinhas, S. 82 (m. Abb.)

1804 -> ÜBERSCHNEIDUNG
FIGUR „Petit Lever“ (Stehender Mädchenakt mit großem Tuch) (auf Rundsockel mit Katze)
ca. 1899/1900
1804 (Firmenmarke ohne Ang.), H. fehlt; Pinhas, S. 83 (m. Abb.)
1804 (Firmenmarke ohne Ang.), H. 76 cm; GSF, S. 16
-> vgl. 752 (GSB)
Lit.: GSF, S. 16, hier mit der Ang. des Originaltitels „Petit Lever“.

1804 -> ÜBERSCHNEIDUNG
BÜSTE „Printemps“
ca. 1899/1900
1804 (Firmenmarke ohne Ang.), H. 74 cm; GSF, S. 16
vgl. 1959; 2003; 2065
Lit.: GSF, S. 16, hier mit der Ang. des Originaltitels „Printemps“.

1805
FIGUR „Mohrenknabe“ (Sitzender lachender Negerjunge mit Kappe) (auf Sessel)
ca. 1897/98 (für den Entwurf)
1805 (Firmenmarke ohne Ang.), H. 31 cm; Pinhas, S. 96 (m. Abb.)
1805 (Firmenmarke ohne Ang.), H. 34 cm; GSF, S. 16
Bem.: Es handelt sich hier um die mittlere Figur der „Drei lachende Negerjungen“ (auf einer Mauer sitzend), Modell-Nr. 1190. Der Sessel ist aus Bambus gefertigt und gehörte zum Lieferumfang dieses Modells.
-> vgl. 1225; 1236
-> vgl. 1617 WANDMASKE (Variante: Schiebermütze)
-> vgl. 1695= BÜSTE
-> vgl. 1190= DreierGRUPPE
Lit.: GSF, S. 16, hier mit der Ang. des Originaltitels „Mohrenknabe“.

1806
FIGUR „Mohrengigerl" Sitzender rauchender Neger mit Strohhut) (auf Sessel)
ca. 1898
1806 (Firmenmarke ohne Ang.), H. 33 cm; Pinhas, S. 32, 69 (m. Abb.)
1806 (Firmenmarke ohne Ang.), H. 33 cm; GSF, S. 16
Bem.: Als „Gigerl" wird in Wien ein Modegeck bzw. ein Mann bezeichnet, der sich von einer Frau aushalten lässt. Lt. Katalog war einen Stuhl bei Goldscheider separat zu diesem Modell zu beziehen.
-> vgl. 1367
Lit.: GSF, S. 16, hier mit der Ang. des Originaltitels „Mohrengigerl".

1807
FIGUR „Mohrengigerl" (Sitzender Neger mit Strohhut) (rittlings auf Sessel sitzend)
ca. 1898
1807 (Firmenmarke ohne Ang.), H. 33 cm; GSF, S. 16
1807 (Firmenmarke ohne Ang.), H. 33 cm; Pinhas, S. 33, 69; 71 (m. Abb.)
Bem.: Als „Gigerl" wird in Wien ein Modegeck bzw. ein Mann bezeichnet, der sich von einer Frau aushalten lässt. Lt. Katalog war ein Stuhl bei Goldscheider separat zu diesem Modell zu beziehen.
-> vgl. 1368; 1419
Lit.: GSF, S. 16, hier mit der Ang. des Originaltitels „Mohrengigerl".

1808
FIGUR „Karrenschieber" (Arbeiter mit schwerem Schubkarren) (auf Rechtecksockel mit Felsen)
ca. 1899/1900
1808 (Firmenmarke ohne Ang.), H. fehlt; Pinhas, S. 86 (m. Abb.)
1808 (Firmenmarke ohne Ang.), H. 48 cm; GSF, S. 16
Lit.: GSF, S. 16, hier mit der Ang. des Originaltitels „Karrenschieber".

1811
Haniroff
FIGUR „Säemann" (Säender Bauer) (auf Landschaftssockel)
ca. 1899/1900
1811 (Firmenmarke ohne Ang.), unsign., H. fehlt; Pinhas, S. 81; 89 (m. Abb.)
1811 (Firmenmarke ohne Ang.), H. 84 cm; GSF, S. 16
Bem.: Die Zuordnung an Haniroff erfolgt aufgrund des im ÖMKI durch Künstlernamen belegten Pendants.
-> vgl. 1838
Pendant -> vgl. 1812
Lit.: GSF, S. 16, hier mit der Ang. des Originaltitels „Säemann".

1812
Haniroff
FIGUR „Schnitterin" (Mit Sense mähende Bäuerin) (auf Landschaftssockel)
ca. 1899/1900
1812 (Firmenmarke ohne Ang.), H. fehlt; Pinhas, S. 81; 89 (m. Abb.)
1812 (Firmenmarke ohne Ang.), H. 84 cm; GSF, S. 16
Ausst.: ÖMKI 1899/1900, hier mit Ang. des Künstlers „Xaniroff" und des Originaltitels „Schnitterin".
-> vgl. 1837
Pendant -> vgl. 1811
Lit.: GSF, S. 16, hier mit der Ang. des Originaltitels „Schnitterin"; Neuwirth (1974), S. 76.

1813
Haniroff
FIGUR „Marmor-Arbeiter" (Stehender Arbeiter mit großem Hammer/Schlegel) (auf Landschaftssockel)
ca. 1899/1900
1813,42,15 gem. weiß AIII, Firmenmarke Sculpture 13, Repr. Res., H. 107 cm; eBay # 6600121653
1813 (Firmenmarke ohne Ang.), H. 98 cm; GSF, S. 16
Lit.: GSF, S. 16, hier mit der Ang. des Originaltitels „Marmor-Arbeiter"; Pinhas, S. 87 (m. Abb.).

1814
FIGUR „Mädchen (sitzend)" (Nacktes kleines Mädchen) (auf Felssockel)
ca. 1899/1900
1814 (Firmenmarke ohne Ang.), H. fehlt; Pinhas, S. 92 (m. Abb.)
1814 (Firmenmarke ohne Ang.), H. 39 cm; GSF, S. 16
Lit.: GSF, S. 16, hier mit der Ang. des Originaltitels „Mädchen (sitzend)".

1815
GRUPPE „Amor und Psyche" (auf Felssockel)
ca. 1899/1900
1815 (Firmenmarke ohne Ang.), H. fehlt; Pinhas, S. 89 (m. Abb.)
1815 (Firmenmarke ohne Ang.), H. 80 cm; GSF, S. 16
Lit.: GSF, S. 16, hier mit der Ang. des Originaltitels „Amor und Psyche".

1816
FIGUR „Bettina"
ca. 1899/1900
1816 (Firmenmarke ohne Ang.), H. 70 cm; GSF, S. 16
-> vgl. 1473?
Lit.: GSF, S. 16, hier mit der Ang. des Originaltitels „Bettina" sowie dem Verweis auf eine größere Ausformung (Mod.-Nr. 1473), die sich aber weder in der Quelle findet, noch anderweitig als Stück nachweisen lässt.

1817
FIGUR „Seifenbläser"
ca. 1898
1817 (Firmenmarke ohne Ang.), H. 45 cm; GSF, S. 16
-> vgl. 1279
Lit.: GSF, S. 16, hier mit der Ang. des Originaltitels „Seifenbläser".

1818
Lemoine, B.(?)
FIGUR „Pfeilschnitzer"/„Prémiere Flèche" (Sitzender Araberknabe mit Kopftuch, an einem Pfeil schnitzend) (auf getrepptem Steinsockel)
ca. 1898
1818,56,25 Firmenmarke 7, unsign.?, H. 53 cm; KH
1818 (Firmenmarke ohne Ang.), H. 53 cm; GSF, S. 16
Bem.: Diese Statue findet sich unter den Nrn. 596/598 im Kat. GSB wieder. Sie wurde in der Pariser Produktion in einer Höhe von 32 cm aus Bronze (Vrai Bronze) und in „Kunstguss" (Fonte d'Art, d.i. wohl Terrakotta, patiniert) in der H. 57 cm gefertigt.
-> vgl. 1067
-> vgl. GSB 596; 598
Lit.: GSF, S. 16, hier mit der Ang. des Originaltitels „Pfeilschnitzer"; GSB, Pl. 36, hier mit Ang. des Originaltitels „Prémiere Flèche"; Richemond (1999), S. 103 (m. Abb.).

1819
Schreitender Junge mit großer Amphore (auf Rechtecksockel mit Gefäß)
ca. 1899/1900
1819 (Firmenmarke ohne Ang.), H. fehlt; Pinhas, S. 86 (m. Abb.)

1820
Sitzende junge Dame mit halb entbößter Brust (auf Rechtecksockel mit Säule)
ca. 1898/1900
1820 (Firmenmarke ohne Ang.), H. fehlt; Pinhas, S. 86 (m. Abb.)
Bem.: Die weiteren beigegebenen Attribute sind auf der Abb. nicht zu erkennen.

1821
FIGUR „Träumerei" (Sitzende junge Biedermeierdame) (auf Landschaftssockel)
ca. 1899/1900
1821 (Firmenmarke ohne Ang.), H. fehlt; Pinhas, S. 86 (m. Abb.)
1821 (Firmenmarke ohne Ang.), H. 50 cm; GSF, S. 16
Lit.: GSF, S. 16, hier mit der Ang. des Originaltitels „Träumerei".

1822
FIGUR „Mignon"
ca. 1897 (für den Entwurf)
1822 (Firmenmarke ohne Ang.), H. 44 cm; GSF, S. 16
-> vgl. 907
Lit.: GSF, S. 16, hier mit der Ang. des Originaltitels „Mignon".

1823
FIGUR „Flora"
ca. 1897
1823 (Firmenmarke ohne Ang.), H. 53 cm; GSF, S. 16
-> vgl. 899
Lit.: GSF, S. 16, hier mit der Ang. des Originaltitels „Flora".

1824
Strasser, Arthur
FIGUR „Araber mit Säbel" (auf felsigem Landschaftssockel)
ca. 1898/99
1824 (Firmenmarke ohne Ang.), unsign., H. 76 cm; Pinhas, S. 77 (m. Abb.)
1824 (Firmenmarke ohne Ang.), H. 89 cm; GSF, S. 16
-> vgl. 1564
Lit.: GSF, S. 16, hier mit der Ang. des Originaltitels „Araber mit Säbel".

1825
BÜSTE „Doux mots" (Junges Mädchen mit gesenktem Haupt) (auf Rundsockel)
ca. 1899/1900
1825 (Firmenmarke ohne Ang.), H. fehlt; Pinhas, S. 88 (m. Abb.)
1825 (Firmenmarke ohne Ang.), H. 62 cm; GSF, S. 16
Lit.: GSF, S. 16, hier mit der Ang. des Originaltitels „Doux mots".

1826
FIGUR „Scherzando" (Stehender Bub, ein Notenblatt in der linken Hand betrachtend) (auf Rechtecksockel)
Inschrift: ? (und Notenzeile)
ca. 1899/1900
1826 (Firmenmarke ohne Ang.), H. fehlt; Pinhas, S. 89 (m. Abb.)
1826 (Firmenmarke ohne Ang.), H. 84 cm; GSF, S. 16
Bem.: Die Inschrift, bzw. die Notenzeile am Sockel, ist auf der Abb. nicht zu entziffern.
Pendant -> vgl. 1827
Lit.: GSF, S. 16, hier mit der Ang. des Originaltitels „Scherzando".

1827
FIGUR „Lamentoso" (Stehendes weinendes Mädchen mit zerbrochenem Krug) (auf Rechtecksockel mit Mauer)
Inschrift: ? (und Notenzeile)
ca. 1899/1900
1827 (Firmenmarke ohne Ang.), H. fehlt; Pinhas, S. 89 (m. Abb.)
1827 (Firmenmarke ohne Ang.), H. 83 cm; GSF, S. 16
Bem.: Die Inschrift, bzw. die Notenzeile am Sockel, ist auf der Abb. nicht zu entziffern.
Pendant -> vgl. 1826
Lit.: GSF, S. 16, hier mit der Ang. des Originaltitels „Lamentoso".

1828
FIGUR „Floreal"
ca. 1898 (für den Entwurf)
1828 (Firmenmarke ohne Ang.), H. 67 cm; GSF, S. 16
-> vgl. 1070
Lit.: GSF, S. 16, hier mit der Ang. des Originaltitels „Floreal" sowie dem Verweis auf eine größere Ausformung (Mod.-Nr. 1070), die sich aber weder in der Quelle findet, noch anderweitig als Stück nachweisen lässt.

1830
FIGUR „Pierrot" (Stehender Harlekin mit Musikinstrument)
ca. 1899/1900
1830,21,27 Firmenmarke 9, Firmenmarke 6, Repr. Res., H. 29,5 cm; Wien Museum Inv. 190.167
1830 (Firmenmarke ohne Ang.), H. 28 cm; GSF, S. 16
Lit.: GSF, S. 16, hier mit der Ang. des Originaltitels „Pierrot"; Pinhas, S. 92 (m. Abb.).

1832
FIGUR „Suie je assez beau" (Stehender, eleganter Knabe mit Zylinder)
Inschrift: SUIE JE ASSEZ BEAU
1893 (für den Entwurf)
1832,60,2 Firmenmarke Sculpture 13, Fabriqué en Autriche, H. 61 cm; eBay # 86084485
1832 (Firmenmarke ohne Ang.), H. 58 cm; GSF, S. 16
Text der Inschrift: Bin ich nicht schön?
-> vgl. 512; 1524
Lit.: GSF, S. 16, hier mit der Ang. des Originaltitels „Suis je assez beau"; Pinhas, S. 74 (m. Abb.).

1833
Cherc
FIGUR „Gladiator" (Römischer Krieger mit Helm, Schild und Schwert) (auf Ovalsockel)
ca. 1899/1900
1833 (Firmenmarke ohne Ang.), H. fehlt; Pinhas, S. 89 (m. Abb.)
1833 (Firmenmarke ohne Ang.), H. 87 cm; GSF, S. 16
-> vgl. 2137
-> vgl. 2033= BÜSTE
Lit.: GSF, S. 16, hier mit der Ang. des Originaltitels „Gladiator".

1834
Cherc
FIGUR „Bogenspanner" (Römischer Bogenschütze in Rüstung) (auf Ovalsockel)
ca. 1899/1900
1834 (Firmenmarke ohne Ang.), unsign., H. fehlt; Pinhas, S. 88 (m. Abb.)
1834 (Firmenmarke ohne Ang.), H. 87 cm; GSF, S. 16
-> vgl. 2138
-> vgl. 2034= BÜSTE
Lit.: GSF, S. 16, hier mit der Ang. des Originaltitels „Bogenspanner".

1835
GRUPPE „Eselreiter"
ca. 1898/99
1835 (Firmenmarke ohne Ang.), H. 66 cm; GSF, S. 16
-> vgl. 1448
Lit.: GSF, S. 16, hier mit der Ang. des Originaltitels „Eselreiter".

1836
FIGUR „Jungfrau von Orleans"
ca. 1898/99
1836 (Firmenmarke ohne Ang.), H. 75 cm; GSF, S. 16
-> vgl. 1475
Lit.: GSF, S. 16, hier mit der Ang. des Originaltitels „Jungfrau von Orleans".

1837
Haniroff
FIGUR „Schnitterin" (Stehende Bäuerin mit Kopftuch) (Ähren bindend) (auf Landschaftssockel)
ca. 1899/1900
1837,46,65 Firmenmarke Sculpture 13, Repr. Res., H. 54,5 cm; Kaupp 26.–28.10.2006; eBay # 250053098589
1837 (Firmenmarke ohne Ang.), H. 53 cm; GSF, S. 16
Ausst.: ÖMKI 1899/1900, hier mit Ang. des Künstlers „Xaniroff" und des Originaltitels „Schnitterin".
> vgl. 1812
Pendant -> vgl. 1838
Lit.: GSF, S. 16, hier mit der Ang. des Originaltitels „Schnitterin".

1838
Haniroff
FIGUR „Säemann" (Säender Bauer) (auf Landschaftssockel)
ca. 1899/1900
1838 (Firmenmarke ohne Ang.), H. 52 cm; GSF, S. 16
Bem.: Die Zuordnung an Haniroff erfolgt aufgrund des im ÖMKI durch Künstlernamen belegten Pendants.
-> vgl. 1811
Pendant -> vgl. 1837
Lit.: GSF, S. 16, hier mit der Ang. des Originaltitels „Säemann".

1839
FIGUR „Mohr mit Korb" (Schreitender Afrikaner, einen großen Korb auf der Schulter tragend) (auf Landschaftssockel)
ca. 1898/99
1839 (Firmenmarke ohne Ang.), H. fehlt; Pinhas, S. 91 (m. Abb.)
1839 (Firmenmarke ohne Ang.), H. 112 cm; GSF, S. 16
Pendant -> vgl. 1840
Lit.: GSF, S. 16, hier mit der Ang. des Originaltitels „Mohr mit Korb".

1840
FIGUR „Mohrin mit Korb" (Schreitende Afrikanerin mit Wassergefäß, einen Korb auf dem Kopf tragend) (auf Landschaftssockel)
ca. 1898/99
1840 (Firmenmarke ohne Ang.), H. fehlt; Pinhas, S. 91 (m. Abb.)
1840 (Firmenmarke ohne Ang.), H. 113 cm; GSF, S. 16
Pendant -> vgl. 1839
Lit.: GSF, S. 16, hier mit der Ang. des Originaltitels „Mohrin mit Korb".

1841
Stehende junge Dame in prächtigem Blumenkleid (eine Blume in der linken Hand haltend) (auf Rundsockel)
ca. 1898/99
1841 (Firmenmarke ohne Ang.), H. fehlt; Pinhas, S. 89 (m. Abb.)

1843
WANDMASKE „Königin Louise"
ca. 1897/98 (für den Entwurf)
1843 (Firmenmarke ohne Ang.), H. 42 cm; GSF, S. 17
Bem.: Dargestellt ist die preuss. Königin Luise, Herzogin zu Mecklenburg (1776–1810).
-> vgl. 1123; 1229; 1794= jew. BÜSTE
Lit.: GSF, S. 17, hier mit der Ang. des Originaltitels „Königin Louise"; Pinhas, S. 96 (m. Abb.).

1844
Larroux, Antonin
BÜSTE „Modern" (Mädchen mit Hochsteckfrisur)
ca. 1899/1900
1844 (Firmenmarke ohne Ang.), H. fehlt; Pinhas, S. 86 (m. Abb.)
1844 (Firmenmarke ohne Ang.), H. 41 cm; GSF, S. 17
Ausst.: ÖMKI 1902/03, hier mit Ang. des Originaltitels „Modern" sowie der Ang. des Künstlers „Laroux".
-> vgl. 2292= WANDMASKE
Lit.: GSF, S. 17, hier mit der Ang. des Originaltitels „Modern"; Neuwirth (1974), S. 66.

1845
BÜSTE „Olmype" (Junge Dame mit Haarschmuck) (stolz nach links blickend)
ca. 1899/1900
1845 (Firmenmarke ohne Ang.), H. fehlt; Pinhas, S. 92 (m. Abb.)
1845 (Firmenmarke ohne Ang.), H. 34 cm; GSF, S. 17
Lit.: GSF, S. 17, hier mit der Ang. des Originaltitels „Olympe".

1846
BÜSTE „Madelon" (Lachende junge Dame mit Hut und Schnürmieder) (nach rechts blickend)
ca. 1899/1900
1846 (Firmenmarke ohne Ang.), H. fehlt; Pinhas, S. 92 (m. Abb.)
1846 (Firmenmarke ohne Ang.), H. 33 cm; GSF, S. 17
Lit.: GSF, S. 17, hier mit der Ang. des Originaltitels „Madelon".

1847
FIGUR „Sensenschleifer" (auf Landschaftssockel mit Ähren)
ca. 1899/1900
1847 (Firmenmarke ohne Ang.), H. fehlt; Pinhas, S. 87 (m. Abb.)
1847 (Firmenmarke ohne Ang.), H. 87 cm; GSF, S. 17
Lit.: GSF, S. 17, hier mit der Ang. des Originaltitels „Sensenschleifer".

1848
FIGUR „Malerei" (Sitzende junge Frau mit Malerpalette und Pinsel)
ca. 1899/1900
1848 (Firmenmarke ohne Ang.), H. fehlt; Pinhas, S. 87 (m. Abb.)
1848 (Firmenmarke ohne Ang.), H. 78 cm; GSF, S. 17
-> vgl. 2104
-> vgl. 1941= BÜSTE
Lit.: GSF, S. 17, hier mit der Ang. des Originaltitels „Malerei".

1849
Wind, Josef
BÜSTE „Heilige Cäcilie" (auf Rechtecksockel mit sechs reliefierten Porträts und einer Orgel spielenden Mädchenfigur)
ca. 1899/1900
1849 (Firmenmarke ohne Ang.), H. fehlt; Pinhas, S. 89 (m. Abb.)
1849 (Firmenmarke ohne Ang.), H. 54 cm; GSF, S. 17
Bem.: Dargestellt ist die Hl. Cäcilia, die Patronin der Kirchenmusik. Sie war Märtyrerin und lebte um 200 in Rom. Im ÖMKI 1899/1900 und 1900/01 wurde jew. ein weiteres Modell von Wind, ebenfalls mit dem Titel „Hl. Cäcilie, ausgestellt. Ob es sich um Größenvarianten der Büste handelte, oder um eine Ausformung als Figur kann derzeit nicht näher bestimmt werden (vgl. auch hierzu Neuwirth, 1974, S. 76).
Ausst.: ÖMKI 1900/01, hier mit Ang. des Originaltitels „Heilige Cäcilie".
Lit.: GSF, S. 17, hier mit der Ang. des Originaltitels „Heilige Cäcilie"; Neuwirth (1974), S. 76.

1850
Houdon, Jean-Antoine (nach)
BÜSTE „Molière"
ca. 1899/1900
1850 (Firmenmarke ohne Ang.), H. fehlt; Pinhas, S. 89 (m. Abb.)
1850 (Firmenmarke ohne Ang.), H. 55 cm; GSF, S. 17
Bem.: Dargestellt ist der franz. Dramatiker Molière (1622–1673).
Lit.: GSF, S. 17, hier mit der Ang. des Originaltitels „Molière".

1851
Houdon, Jean-Antoine (nach)
BÜSTE Voltaire
ca. 1899/1900
1851 (Firmenmarke ohne Ang.), H. fehlt; Pinhas, S. 89 (m. Abb.)
1851 (Firmenmarke ohne Ang.), H. 52 cm; GSF, S. 17
Bem.: Dargestellt ist der franz. Schriftsteller François-Marie Arouet, besser bekannt unter dem Namen Voltaire (1694–1778). Die Büste ist dem Original von 1781, welche Jean-Antoine Houdon schuf, nachempfunden.
Lit.: GSF, S. 17, hier mit der Ang. des Originaltitels „Voltaire".

1852
FIGUR „Parze" (Stehendes junges Mädchen mit entblößten Brüsten und Spindel) (auf rundem Landschaftssockel)
ca. 1899/1900
1852 (Firmenmarke ohne Ang.), H. fehlt; Pinhas, S. 88 (m. Abb.)
1852 (Firmenmarke ohne Ang.), H. 77 cm; GSF, S. 17
Bem.: Dargestellt ist die Parze Nona, eine der drei Schicksalsgöttinnen der röm Mythologie. (Quelle: Wikipedia).
Lit.: GSF, S. 17, hier mit der Ang. des Originaltitels „Parze".

1853
FIGUR „Sandalenbinderin" (Stehendes junges Mädchen mit entblößten Brüsten, sich die rechte Sandale schnürend (auf getrepptem Sockel)
ca. 1899/1900
1853 (Firmenmarke ohne Ang.), H. fehlt; Pinhas, S. 88 (m. Abb.)
1853 (Firmenmarke ohne Ang.), H. 93 cm; GSF, S. 17
Lit.: GSF, S. 17, hier mit der Ang. des Originaltitels „Sandalenbinderin".

1854
FIGUR „Falconière" (Stehende Falknerin mit Falke) (auf rundem Felssockel)
ca. 1899/1900
1854 (Firmenmarke ohne Ang.), H. fehlt; Pinhas, S. 88 (m. Abb.)
1854 (Firmenmarke ohne Ang.), H. 86 cm; GSF, S. 17
Lit.: GSF, S. 17, hier mit der Ang. des Originaltitels „Falconière".

1855 -> ÜBERSCHNEIDUNG
FIGUR „Gigerl am Telefon"
ca. 1899/1900
1855 (Firmenmarke ohne Ang.), H. 74 cm; GSF, S. 17
Bem.: Als „Gigerl" wird in Wien ein Modegeck bzw. ein Mann bezeichnet, der sich von einer Frau aushalten lässt.
Lit.: GSF, S. 17, hier mit der Ang. des Originaltitels „Gigerl am Telefon".

1855 -> ÜBERSCHNEIDUNG
Fotograf (Stehender lachender Neger mit Strohhut beim Fotografieren) (auf hexagonalem Sockel mit Säule)
ca. 1899/1900
1855 (Firmenmarke ohne Ang.), H. fehlt; Pinhas, S. 86 (m. Abb.)

1856
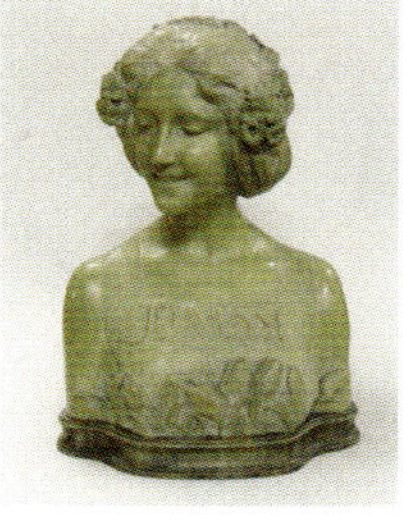
Ceccarelli, Ezio
BÜSTE „Jugend"/„Jeunesse" (Lachendes nach rechts blickendes Mädchen mit Tulpenschmuck)
Inschrift: JEUNESSE
ca. 1899
1856,341,1 ger. II, Firmenmarke Sculpture 13, Repr. Res., H. 50,1 cm; MAK Wien, Inv. Ke 10521-10
1856 (Firmenmarke ohne Ang.), H. 50 cm; GSF, S. 17
Ausst.: ÖMKI 1899/1900, hier mit Ang. des Originaltitels „Jeunesse".
-> vgl. 2120; 2412
Lit.: GSF, S. 17, hier mit der Ang. des Originaltitels „Jugend"; Neuwirth (1974), S. 61.

1857
FIGUR „Cleopatra" (Stehende Frau mit entblößten Brüsten und Kopfputz, in Form eines Schwans) (auf reliefiertem Sockel mit Säule und Blumenschmuck)
ca. 1899/1900
1857 (Firmenmarke ohne Ang.), H. fehlt; Pinhas, S. 85 (m. Abb.)
1857 (Firmenmarke ohne Ang.), H. 105 cm; GSF, S. 17
-> vgl. 1922= BÜSTE
Lit.: GSF, S. 17, hier mit der Ang. des Originaltitels „Cleopatra".

1858
FIGUR „Wasserspiegel" (Stehendes nacktes Mädchen) (auf Rundsockel mit Blume)
ca. 1899/1900
1858 (Firmenmarke ohne Ang.), H. fehlt; Pinhas, S. 92 (m. Abb.)
1858 (Firmenmarke ohne Ang.), H. 68 cm; GSF, S. 17
Lit.: GSF, S. 17, hier mit der Ang. des Originaltitels „Wasserspiegel".

1859
FIGUR „Harfenspielerin" (Stehende Nordafrikanerin mit fünfsaitigem Instrument und Fidelbogen)
ca. 1898/99
1859 (Firmenmarke ohne Ang.), H. 75 cm; Pinhas, S. 78 (m. Abb.)
1859 (Firmenmarke ohne Ang.), H. 75 cm; GSF, S. 17
-> vgl. 1670
Lit: GSF, S. 17, hier mit der Ang. des Originaltitels „Harfenspielerin".

1860
Waderé, Heinrich
BÜSTE „Psyche" (Junges Mädchen mit Schmetterlingsflügeln) (mit Blumen und Perlen verziertem Kleid) (auf Rundsockel)
ca. 1899/1900
1860 (Firmenmarke ohne Ang.), H. fehlt; Pinhas, S. 88 (m. Abb.)
1860 (Firmenmarke ohne Ang.), H. 66 cm; GSF, S. 17
Ausst.: ÖMKI 1906/07, hier mit Ang. des Originaltitels „Psyche".
-> vgl. 2055; 2081
Lit.: GSF, S. 17, hier mit der Ang. des Originaltitels „Psyche".

1861
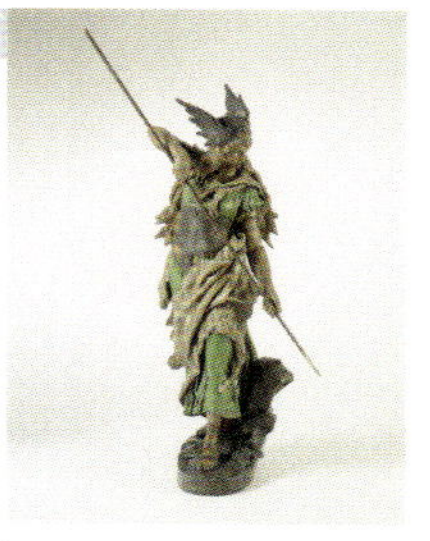
Cherc
FIGUR „Walküre"/„Le Walkyrie" (Stehende Kriegerin mit Lanze) (auf Felssockel)
ca. 1899
1861,12,34 Firmenmarke 7, Repr. Res., H. 110 cm; Wien Museum Inv. 223.916;
Bem.: An der Plinthe betitelt und sign. (Metallschild).
1861 (Firmenmarke ohne Ang.), H. 110 cm; GSF, S. 17
Bem.: Eine Walküre, auch Schlacht- oder Schildjungfer, ist in der nordischen Mythologie ein weibliches Geisterwesen aus dem Gefolge des (auch so genannten) Göttervaters Wotan bzw. Odin. (Quelle: Wikipedia).
Lit.: GSF, S. 17, hier mit der Ang. des Originaltitels „Walküre"; Pinhas, S. 87 (m. Abb.).

1862
Thom, A.
BÜSTE „Römischer Krieger" (Römer mit Lorbeerkranz und Brustpanzer)
Inschrift: CIVIS ROMANUS SUM
ca. 1899/1900
1862 (Firmenmarke ohne Ang.), H. fehlt; Pinhas, S. 87 (m. Abb.)
1862 (Firmenmarke ohne Ang.), H. 86 cm; GSF, S. 17
Bem.: Mit dem Ausspruch: „Ich bin ein römischer Bürger", berief man sich auf sein römisches Bürgerrecht, genoss dadurch Vorteile bzw. entzog man sich der Gerichtsbarkeit im Ausland. Dargestellt könnte auch der römische Kaiser Augustus (63 v. Chr.–14 n. Chr.) sein.
Ausst.: ÖMKI 1903/1904, hier mit Ang. des Originaltitels „Römischer Krieger" und des nicht näher verifizierbaren Künstlers A. Thom.
Lit.: GSF, S. 17, hier mit der Ang. des Originaltitels „Römischer Krieger"; Neuwirth (1974), S. 75.

1863
Lemassi
FIGUR „Negerknabe mit Leuchter" (Stehender Negerjunge mit Kerzenhalter) (Nordafrikaner, vor Hocker stehend) (auf Rundsockel)
ca. 1897/98 (für den Entwurf)
1863 (Firmenmarke ohne Ang.), H. 43 cm; Pinhas, S. 29, 65, 68 (m. Abb.)
1863 (Firmenmarke ohne Ang.), H. 43 cm; GSF, S. 17
-> vgl. 1243; 1291
Lit.: GSF, S. 17, hier mit der Ang. des Originaltitels „Negerknabe mit Leuchter".

1864
Stehende junge Dame in einem Buch lesend (auf Rundsockel)
ca. 1898/99
1864 (Firmenmarke ohne Ang.), H. fehlt; Pinhas, S. 89 (m. Abb.)

1865
BÜSTE „Dubarry" (nach links blickend)
ca. 1899/1900
1865 (Firmenmarke ohne Ang.), H. fehlt; Pinhas, S. 92 (m. Abb.)
1862 (Firmenmarke ohne Ang.), H. 86 cm; GSF, S. 17
Bem.: Dargestellt ist Madame Dubarry (Marie-Jeanne Bécu, Comtesse du Barry 1743–1793), die Mätresse Ludwigs XV. (Quelle: Wikipedia).
Lit.: GSF, S. 17, hier mit der Ang. des Originaltitels „Dubarry".

1866
BÜSTE „Ophelie" (Mädchen mit Blumen)
ca. 1899/1900
1866 (Firmenmarke ohne Ang.), H. fehlt; Pinhas, S. 89 (m. Abb.)
1866 (Firmenmarke ohne Ang.), H. 44 cm; GSF, S. 17
Bem.: Die Bezeichnung „cadre extra" im orig. Firmenkat. der Jahrhundertwende, auf den sich Pinhas hier stützt, belegt, dass zu dieser Büste auch ein separater Rahmen/Wandsockel geliefert wurde, der die Hängung an der Wand ermöglicht. Ophelia ist eine der Hauptfiguren aus William Shakespeares Tragödie Hamlet.
Lit.: GSF, S. 17, hier mit der Ang. des Originaltitels „Ophelie".

1867
FIGUR „Wiener Walzer" (Tanzende junge Dame, ihr langes Kleid mit beiden Armen anhebend) (auf Ovalsockel)
Inschrift: WIENER WALZER
ca. 1899/1900
1867 (Firmenmarke ohne Ang.), H. fehlt; Pinhas, S. 87 (m. Abb.)
1867 (Firmenmarke ohne Ang.), H. 82 cm; GSF, S. 17
-> vgl. 2112
Lit.: GSF, S. 17, hier mit der Ang. des Originaltitels „Wiener Walzer".

1869
Simon
FIGUR „Fischer" (mit Korb und Käscher?) (auf Landschaftssockel)
ca. 1899
1869 (Firmenmarke ohne Ang.), H. fehlt; Pinhas, S. 91 (m. Abb.)
1869 (Firmenmarke ohne Ang.), H. 70 cm; GSF, S. 17
Ausst.: ÖMKI 1899/1900, hier mit Ang. des Originaltitels „Fischer".
Pendant -> vgl. 1870
Lit.: GSF, S. 17, hier mit der Ang. des Originaltitels „Fischer"; Neuwirth (1974), S. 74.

1870
Simon
FIGUR „Fischerin"(sich auf ein Paddel stütztend) (auf Landschaftssockel)
ca. 1899
1870 (Firmenmarke ohne Ang.), H. fehlt; Pinhas, S. 91 (m. Abb.)
1870 (Firmenmarke ohne Ang.), H. 70 cm; GSF, S. 17
Ausst.: ÖMKI 1899/1900, hier mit Ang. des Originaltitels „Fischerin".
Pendant -> vgl. 1869
Lit.: GSF, S. 17, hier mit der Ang. des Originaltitels „Fischerin"; Neuwirth (1974), S. 74.

1871
FIGUR „Andromeda" (an einen Baumstamm gefesselt)
ca. 1898/99 (für den Entwurf)
1871 (Firmenmarke ohne Ang.), H. 88 cm; Pinhas, S. 74 (m. Abb.)
1871 (Firmenmarke ohne Ang.), H. 89 cm; GSF, S. 17
Bem.: Andromeda, eine Figur der griech. Sage, war die Tochter von Kassiopeia und Kepheus.
-> vgl. 1547
Lit.: GSF, S. 17, hier mit der Ang. des Originaltitels „Andromeda".

1872
FIGUR „Castagnettentänzerin (bekleidet)" (auf Rundsockel)
ca. 1899/1900
1872 (Firmenmarke ohne Ang.), H. fehlt; Pinhas, S. 86 (m. Abb.)
1872 (Firmenmarke ohne Ang.), H. 82 cm; GSF, S. 17
-> vgl. 1793 (nackt)
Lit.: GSF, S. 17, hier mit der Ang. des Originaltitels „Castagnettentänzerin (bekleidet)".

1873
FIGUR „Mädchen am Zaun"
ca. 1899/1900
1873 (Firmenmarke ohne Ang.), H. 61 cm; GSF, S. 17
Bem.: Im Firmenkatalog wird auf die Mod.-Nr. 1047

verwiesen. Hier wäre eine aktuell jedoch noch nicht nachweisbare BÜSTE oder FIGUR in anderer Größe dieses Modells zu vermuten.
Lit.: GSF, S. 17, hier mit der Ang. des Originaltitels „Mädchen am Zaun".

1874
Simon
FIGUR „Eva (ohne/mit elektr. Broncezweig)"
ca. 1899/1900
1874 (Firmenmarke ohne Ang.), H. 106 cm; GSF, S. 17
Bem.: Der Firmenkatalog macht darauf aufmerksam, dass es zwei Varianten gab. Eine mit elektrifiziertem Broncezweig und eine weitere Ausführung ohne diesen.
Ausst.: ÖMKI 1902/03, hier mit Ang. des Originaltitels „Eva".
Lit.: GSF, S. 17, hier mit der Ang. des Originaltitels „Eva"; Neuwirth (1974), S. 74.

1875
Barabino, Armando (nach)
GRUPPE „Madonna mit Kind" (Sitzende Gottesmutter mit Jesuskind, auf getrepptem Steinsockel)
ca. 1899
1875,7,9 Firmenmarke 7, Repr. Res., sign. „d'apres Barabino", H. fehlt; KH
1875 (Firmenmarke ohne Ang.), H. 73 cm; GSF, S. 17
Bem.: Diese Gruppe wurde nach einem Gemälde von Armando Barabino gestaltet.
-> vgl. 1879 BÜSTE
Lit.: GSF, S. 17, hier mit der Ang. des Originaltitels „Madonna mit Kind"; Pinhas, S. 89 (m. Abb.).

1876
Strasser, Arthur (zugeschr.)
FIGUR „Araber mit Krügen" (mit Turban und zwei Wassergefäßen) (auf Landschaftssockel)
ca. 1899/1900
1876 (Firmenmarke ohne Ang.), unsign., H. fehlt; Pinhas, S. 91 (m. Abb.)
1876 (Firmenmarke ohne Ang.), H. 91 cm; GSF, S. 17
Lit.: GSF, S. 17, hier mit der Ang. des Originaltitels „Araber mit Krügen".

1877
Thiele, (P.?)/Lévèque, Auguste Maurice
BÜSTE „Sklavin"/„Bedouine" (Sklavin, Araberin mit großem Turban) (nach links blickend)
ca. 1897 (für den Entwurf)
1877 (Firmenmarke ohne Ang.), sign. Thiele, H. 33 cm; Pinhas, S. 21 (m. Abb.)
1877 (Firmenmarke ohne Ang.), H. 33 cm; GSF, S. 17
Ausst.: HMW (jetzt Wien Museum) 1985/86.
-> vgl. (H. 66 cm) (sign. „P. Thiele") ausgeführt in MARMOR; Soth. L. 9.10.1992 und 9.12.1993
-> vgl. 1101; 1338
-> vgl. 684 (GSB); 733 (GSB)
Pendant -> vgl. 1105
Lit.: GSF, S. 17, hier mit der Ang. des Originaltitels „Sklavin"; GSB, Pl. 59, hier ist die BÜSTE mit dem Originaltitel „Esclave" unter Nr. 684= No.1 (H. 62 cm) und unter Nr. 733= No. 2 (H. 30 cm) abgebildet. Ang. zum Material (Terrakotta/Bronze fehlen); Richemond (1999), S. 78 (m. Abb.); GK, S. 11, 65 (m. Abb.).

1878
Thiele, (P.?)/Prémont, F. de
BÜSTE „Mohrin mit Schmetterling"/„Négresse au Papillon" (Afrikanerin/Mulattin mit Turban)
ca. 1897 (für den Entwurf)
1878 (Firmenmarke ohne Ang.), sign. Thiele, H. 35 cm; Pinhas, S. 21 (m. Abb.)
1878 (Firmenmarke ohne Ang.), H. 33 cm; GSF, S. 17
-> vgl. 1102
-> vgl. .... (H. 47 cm) bis auf den unteren Faltenwurf des Tuches identisch; A/T 20.4.1994
-> vgl. .... (H. 48 cm) (sign. T. du Piémont, wohl Prémont?); Richemond (1999), S. 103 (m. Abb.)
Lit.: GSF, S. 17, hier mit der Ang. des Originaltitels „Mohrin mit Schmetterling".

1879
Barabino, Armando (nach)
BÜSTE „Madonna (auf Sockel)" (Rechtecksockel reliefiert mit Krippenszene)
ca. 1899
1879,55,53 FG (Firmenmarke?), Repr. Res., sign. „d'apres Barabino", gepr. „P"= Z3, H. 40,7 cm; eBay # 61956736
1879,83,13 Firmenmarke Sculpture 13, Repr. Res., sign. „d' apres Barabino", Ausformung um/kurz nach 1900, H. 40 cm; Doro 24.10.1994
1879,86,39 Firmenmarke Sculpture 13, Repr. Res., sign. Barabino, Ausformung um/kurz nach 1900, H. 41 cm; WKH; eBay # 2560840217
1879 (Firmenmarke ohne Ang.), H. 41 cm; GSF, S. 17
Bem.: Diese Büste wurde nach einem Gemälde von Armando Barabino gestaltet.
Ausst.: ÖMKI 1899/1900; 1900/01, hier jeweils mit Ang. des Originaltitels „Madonna".
-> vgl. 1875 GRUPPE
Lit.: GSF, S. 17, hier mit der Ang. des Originaltitels „Madonna mit Kind"; Neuwirth (1974), S. 59; Pinhas, S. 89 (m. Abb.).

1880
Cherc
FIGUR „Amazonenkönigin" (Stehende Kriegerin, Halbakt auf Felssockel)
ca. 1898 (für den Entwurf)
1880 (Firmenmarke ohne Ang.), H. 79 cm; Pinhas, S. 37; 80 (m. Abb.)
1880 (Firmenmarke ohne Ang.), H. 75 cm; GSF, S. 17
Ausst.: ÖMKI 1898/99, hier mit Ang. des Originaltitels „Amazonenkönigin".
-> vgl. 1741
-> vgl. 1923= BÜSTE
Lit.: GSF, S. 17, hier mit der Ang. des Originaltitels „Amazonenkönigin".

1881
FIGUR „Hirtin" (Schreitendes blindes Mädchen mit Stock und Wasserkrug) (auf Landschaftssockel)
ca. 1899/1900
1881 (Firmenmarke ohne Ang.), H. fehlt; Pinhas, S. 86 (m. Abb.)
1881 (Firmenmarke ohne Ang.), H. 66 cm; GSF, S. 17
Lit.: GSF, S. 17, hier mit der Ang. des Originaltitels „Hirtin".

1882
„Medee" (Stehendes Mädchen mit Dolch und aufwändigem Kopfputz) (auf Rundsockel)
Inschrift: MEDEE
ca. 1899/1900
1882 (Firmenmarke ohne Ang.), H. fehlt; Pinhas, S. 91 (m. Abb.)
Bem.: Dargestellt ist Medea, , eine Figur der griech. Sagenwelt. Medea verhalf Jason zum Goldenen Vlies. Der Kopfputz mit seinen Blumen und Spitzen erinnert an die Freihheitsstatue in New York, die 1886 aufgestellt wurde.

1883
FIGUR „Schwerttänzerin" (Stehende orientalische Schönheit mit zwei kurzen Schwertern)
ca. 1899/1900
1883 (Firmenmarke ohne Ang.), H. 70 cm; GSF, S. 18
-> vgl. 1790
-> vgl. 2139= BÜSTE
Lit.: GSF, S. 18, hier mit der Ang. des Originaltitels „Schwerttänzerin".

1884
Amy, Jean Barnabé
BÜSTE „Defi" (Kleines Mädchen mit Kapuze)
ca. 1899
1884 (Firmenmarke ohne Ang.), H. fehlt; Pinhas, S. 86 (m. Abb.)
1884 (Firmenmarke ohne Ang.), H. 34 cm; GSF, S. 18
Ausst.: ÖMKI 1899/1900; 1900/01, hier jew. mit Ang. des Originaltitels „Defi".
Lit.: GSF, S. 18, hier mit der Ang. des Originaltitels „Defi"; Neuwirth (1974), S. 59.

1886
Stehender, römischer Kämpfer (Herkules?) (auf rechteckigem Landschaftssockel mit Baumstumpf)
ca. 1899
1886,50,24 gem. weiß V, Firmenmarke Sculpture 13, Repr. Res., H. 64 cm; eBay # 1431634729
Bem.: Da dieses Genre (auch als Büste) in den Jahren 1895–1905 sehr beliebt war, gibt es zahlreiche von GS ausgeführte ähnliche Modelle. Diese lassen sich jedoch in der Regel keiner bestimmten dargestellten Person zuordnen.

1887
Villieu
BÜSTE „Beatrice" (Junge Dame mit Diadem und Kopftuch)
ca. 1899/1900
1887 (Firmenmarke ohne Ang.)., Sign. ohne Ang., H. fehlt; Pinhas, S. 89 (m. Abb.)
1887 (Firmenmarke ohne Ang.), H. 50 cm; GSF, S. 18
Bem.: Dargestellt ist Beatrice Portinari (1266–1290), die Jugendliebe des Dante Alighieri (1265–1321). Die Kartusche, wo sich die Inschrift „Beatrice" befinden sollte, ist in der Abb. aus dem Originalfirmenkat. der Jahrhundertwende leer (vgl. Pinhas, S. 89). Zur Inschrift -> vgl. 2593.
-> vgl. 2593
-> vgl. 2122= ähnl. BÜSTE
-> vgl. 2200 WANDMASKE
Lit.: GSF, S. 18, hier mit der Ang. des Originaltitels „Beatrice".

1888
Maison, Rudolf
FIGUR „Philosoph"
ca. 1898
1888,21,65 FG Wien (Firmenmarke ohne Ang.), Repr. Res., Ausformung kurz vor 1900, H. 48 cm; Doro 20.4.1993
1888 (Firmenmarke ohne Ang.), H. 46 cm; GSF, S. 18
-> vgl. 1462
Lit.: GSF, S. 18, hier mit der Ang. des Originaltitels „Philosoph".

1889
FIGUR „Cyrano de Bergerac" (als stehende Figur in Pose mit langem Stock) (auf Rechtecksockel)
Inschrift: CYRANO DE BERGERAC
ca. 1899/1900
1889 (Firmenmarke ohne Ang.), H. 71 cm; GSF, S. 18
Bem.: Dargestellt ist der franz. Schriftsteller Cyrano de Bergerac (1619–1655).
-> vgl. 1802
-> vgl. 1742 (ähnl. Mod.)
Lit.: GSF, S. 18, hier mit der Ang. des Originaltitels „Cyrano de Bergerac".

1890
FIGUR „Mohrengigerl" (Sitzender rauchender und lesender Neger mit Strohhut) (auf Sessel)
ca. 1898/99
1890 (Firmenmarke ohne Ang.), H. 30 cm; Pinhas, S. 76 (m. Abb.)
1890 (Firmenmarke ohne Ang.), H. 30 cm; GSF, S. 18
Bem.: Warum für die kleinste der drei Ausformungen dieses sitzenden Rauchers der Originaltitel von „Zeitungsleser" auf „Mohrengigerl" umbenannt wurde, geht aus dem Firmenkatalog nicht hervor. Der Sessel, bzw. Stuhl ist aus Bambus gefertigt und konnte bei Goldscheider separat zu dem Modell bezogen werden.
-> vgl. 1649; 1663
Lit.: GSF, S. 18, hier mit der Ang. des Originaltitels „Mohrengigerl".

1891
GRUPPE „Süsse Last" (Junger Mann seine Geliebte auf Händen tragend) (auf Landschaftssockel)
ca. 1899/1900
1891 (Firmenmarke ohne Ang.), H. 80 cm; Pinhas, S. 37, 88 (m. Abb.)
1891 (Firmenmarke ohne Ang.), H. 84 cm; GSF, S. 18
Lit.: GSF, S. 18, hier mit der Ang. des Originaltitels „Süsse Last".

1892
BÜSTE „Gallischer Krieger" (mit Helm)
Inschrift: A......US?
ca. 1899/1900
1892 (Firmenmarke ohne Ang.), H. fehlt; Pinhas, S. 87 (m. Abb.)
1892 (Firmenmarke ohne Ang.), H. 83 cm; GSF, S. 18
Bem.: Der Text der Inschrift ist auf der Abb. nicht lesbar.
Lit.: GSF, S. 18, hier mit der Ang. des Originaltitels „Gallischer Krieger".

1893
FIGUR „Flötenspielerin" (Stehendes junges Mädchen, eine große Flöte spielend) (auf Ovalsockel)
ca. 1899/1900
1893 (Firmenmarke ohne Ang.), H. fehlt; Pinhas, S. 92 (m. Abb.)
1893 (Firmenmarke ohne Ang.), H. 87 cm; GSF, S. 18
-> vgl. 2153
Lit.: GSF, S. 18, hier mit der Ang. des Originaltitels „Flötenspielerin".

1897
BÜSTE „Richard Wagner"
ca. 1895 (für den Entwurf)
1897 (Firmenmarke ohne Ang.), H. 26 cm; GSF, S. 18
Bem.: Dargestellt ist der Komponist Richard Wagner (1813–1883).
-> vgl. 618; 981; 1030
Lit.: GSF, S. 18, hier mit der Ang. des Originaltitels „Richard Wagner".

1898
Thiele, (P.?)
BÜSTE „Bulgarin" (mit langem Kopftuch) (nach links blickend)
ca. 1897 (für den Entwurf)
1898,24,23 ger. K, gem. IV, Firmenmarke 7, Repr. Res., Made in Austria= Z13, H. 35,5 cm; eBay # 89188916
1898 (Firmenmarke ohne Ang.), H. 36 cm; GSF, S. 18
-> vgl. 1109
-> vgl. 1340
-> vgl. 1409 (sign. D. Rochella)
Pendant -> vgl. 1899
Lit.: GSF, S. 18, hier mit der Ang. des Originaltitels „Bulgarin".

1899
Lévèque, Auguste Maurice
BÜSTE „Beduine" (Araber mit Turban) (nach rechts blickend)
ca. 1897 (für den Entwurf)
1899 (Firmenmarke ohne Ang.), H. 34 cm; GSF, S. 18
-> vgl. 1105; 1339
-> vgl. 1612= WANDMASKE
Pendant -> vgl. 1898
Lit.: GSF, S. 18, hier mit der Ang. des Originaltitels „Beduine".

1900
Cherc
FIGUR „Salome" (Stehende Frauengestalt mit lyraähnlichem Instrument)
Inschrift: SALOME
ca. 1897/98 (für den Entwurf)
1900 (Firmenmarke ohne Ang.), H. fehlt; Pinhas, S. 91 (m. Abb.)
1900 (Firmenmarke ohne Ang.), H. 100 cm; GSF, S. 18
-> vgl. 1000; 2085; 2530
Lit.: GSF, S. 18, hier mit der Ang. des Originaltitels „Salome".

1901
FIGUR „Fischermädchen" (Sitzendes Mädchen mit Strohhut und Fisch) (auf Felssockel)
ca. 1898/99
1901 (Firmenmarke ohne Ang.), H. 43 cm; GSF, S. 18
-> vgl. 1632
Lit.: GSF, S. 18, hier mit der Ang. des Originaltitels „Fischermädchen".

1902
Cherc
BÜSTE/LAMPE „Sumpfblume" (Blumengeschmückte junge Frau nach rechts blickend)
ca. 1899/1900
1902,176,13 Firmenmarke Sculpture 13, Repr. Res., H. 60 cm; Doro 12.12.1991; Bem.: Staffierte Bronze-Imitation.
1902 (Firmenmarke ohne Ang.), H. 56 cm; GSF, S. 18
Bem.: Im Firmenkat. ist vermerkt, dass auch eine Ausführung als Lampe (elektrifiziert mit vier Glühbirnen) angeboten wurde.
-> vgl. 2265
Lit.: GSF, S. 18, hier mit der Ang. des Originaltitels „Sumpfblume"; Pinhas, S. 92 (m. Abb.).

1903
Cherc
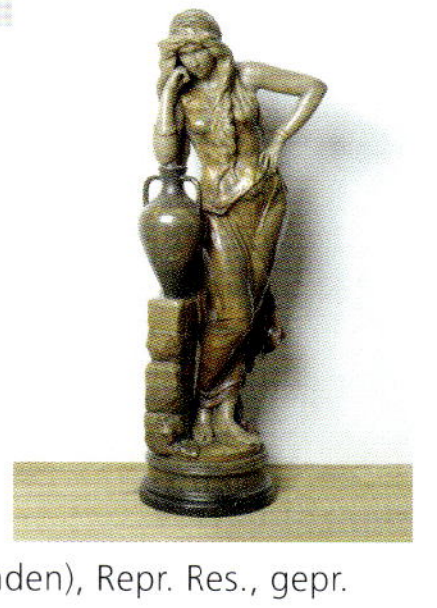
FIGUR „Rebecca" (Stehende Frauengestalt neben einer Mauer, ihren Ellenbogen auf eine Amphore stützend)
Inschrift: REBECCA
ca. 1899/1900
1903,110,55 Firmenmarke 7, Firmenmarke Sculpture 13 (offenbar nach Restaurierung durch eine „ähnliche" Marke, „sitzender Faun" nachempfunden), Repr. Res., gepr. „P"= Z3, H. 93 cm; Wien Museum Inv. 219.105
1903,146,47 Firmenmarke Sculpture 13, Repr. Res., H. 95 cm; PB Wien 19
1903,231,14 Firmenmarke 11, Repr. Res., H. 92 cm; G 68 Auctions 15.10.2001
1903,408,55 Firmenmarke Sculpture 13, Repr. Res., H. 92 cm; PB Wien 19
1903 (Firmenmarke ohne Ang.), H. 87 cm; GSF, S. 18
-> vgl. 2171; 2537
Lit.: GSF, S. 18, hier mit der Ang. des Originaltitels „Rebecca"; Pinhas, S. 92 (m. Abb.).

1904
BÜSTE Kriegerin/Amazone mit Helm und Brustpanzer, ein Gefäß in der rechten Hand haltend (Athene?) (auf konischer Säule)
ca. 1899/1900
1904 (Firmenmarke ohne Ang.), H. fehlt; Pinhas, S. 93 (m. Abb.)
Pendant -> vgl. 1905

1905
BÜSTE Blumen geschmücktes Mädchen mit Kopfputz, ein Gefäß in beiden Händen haltend (auf konischer Säule)
ca. 1899/1900
1905 (Firmenmarke ohne Ang.), H. fehlt; Pinhas, S. 93 (m. Abb.)
Pendant -> vgl. 1904

1906
Mondi
BÜSTE „Kleeblatt" (Junge Dame mit Schmetterlingsflügeln) (nach rechts blickend)
ca. 1899/1900
1906 (Firmenmarke ohne Ang.), H. fehlt; Pinhas, S. 92 (m. Abb.)
1906 (Firmenmarke ohne Ang.), H. 31 cm; GSF, S. 18
Ausst.: ÖMKI 1902/03, hier mit Ang. des Originaltitels „Kleeblatt" und des Künstlers „Mondi".
Pendant -> vgl. 1907
-> vgl. 2328
Lit.: GSF, S. 18, hier mit der Ang. des Originaltitels „Kleeblatt"; Neuwirth (1974), S. 68.

1907 -> ÜBERSCHNEIDUNG
BÜSTE „Schmetterling" (Junge Dame mit Schmetterlingsflügeln) (nach links blickend)
ca. 1899/1900
1907 (Firmenmarke ohne Ang.), H. fehlt; Pinhas, S. 92 (m. Abb.)
1907 (Firmenmarke ohne Ang.), H. 31 cm; GSF, S. 18
-> vgl. 2327
Pendant -> vgl. 1906
Lit.: GSF, S. 18, hier mit der Ang. des Originaltitels „Schmetterling".

1907 -> ÜBERSCHNEIDUNG
Tell, E.
Stehende junge Frau mit Strohhut und zwei Wassergefäßen (auf Landschaftssockel)
ca. 1899/1900
1907,147,6 gem. weiß J. (oder lig. JT.), Firmenmarke Sculpture 13, H. 66 cm; eBay # 517889603

1908
BÜSTE „Barroque"
ca. 1900
1908 (Firmenmarke ohne Ang.), H. 78 cm; GSF, S. 18
Lit.: GSF, S. 18, hier mit der Ang. des Originaltitels „Barroque":

1910
FIGUR „Knabe (schreibend)" (Sitzender Jüngling mit Haarband) (auf Landschaftssockel)
ca. 1899/1900
1910 (Firmenmarke ohne Ang.), H. fehlt; Pinhas, S. 94 (m. Abb.)
1910 (Firmenmarke ohne Ang.), H. 65 cm; GSF, S. 18
Lit.: GSF, S. 18, hier mit der Ang. des Originaltitels „Knabe (schreibend)".

1911
BÜSTE „Bianca" (Junges Mädchen mit Blumen geschmücktem Haar)
Inschrift: BIANCA
ca. 1899/1900
1911 (Firmenmarke ohne Ang.), H. fehlt; Pinhas, S. 94 (m. Abb.)
1911 (Firmenmarke ohne Ang.), H. 47 cm; GSF, S. 18
Lit.: GSF, S. 18, hier mit der Ang. des Originaltitels „Bianca".

1912
WANDMASKE „Clown" (Lachender Narr mit Narrenkappe)
ca. 1899/1900
1912 (Firmenmarke ohne Ang.), H. fehlt; Pinhas, S. 94 (m. Abb.)
1912 (Firmenmarke ohne Ang.), H. 29 cm; GSF, S. 18
Lit.: GSF, S. 18, hier mit der Ang. des Originaltitels „Clown".

1913
Sardon/Tell, E.
FIGUR „Caprieserin (groß)" (Stehende Wasserträgerin mit Strohhut) (auf Natursockel)
ca. 1899/1900
1913 (Firmenmarke ohne Ang.), H. 100 cm; GSF, S. 18
-> vgl. 1497
-> vgl. 1761= BÜSTE
Lit.: GSF, S. 18, hier mit der Ang. des Originaltitels „Caprieserin (groß)".

1914
BÜSTE „Bismarck"
ca. 1899/1900
1914 (Firmenmarke ohne Ang.), H. 45 cm; GSF, S. 18
Bem.: Dargestellt ist der deutsche Staatsmann und Reichskanzler Otto von Bismarck (1815–1898).
-> vgl. 1783
Lit.: GSF, S. 18, hier mit der Ang. des Originaltitels „Bismarck".

1915
Zardier?
BÜSTE „Angelica" (Junge Dame mit Haarschmuck) (nach links blickend) (auf Ovalsockel)
Inschrift: ANGELICA
ca. 1899/1900
1915,40,13 ger. VIII, gem. V, Firmenmarke 7, Repr. Res., undeutl. sign. Zardier?, H. 31,5 cm; Wien Museum Inv. 220.787
1915 (Firmenmarke ohne Ang.), H. 31 cm; GSF, S. 18
Lit.: GSF, S. 18, hier mit der Ang. des Originaltitels „Angelica".

1916
FIGUR „Versuchet das Glück" (Losverkäuferin mit Bauchladen) (auf Landschaftssockel)
ca. 1899/1900
1916 (Firmenmarke ohne Ang.), H. fehlt; Pinhas, S. 94 (m. Abb.)
1916 (Firmenmarke ohne Ang.), H. 58 cm; GSF, S. 18
Lit.: GSF, S. 18, hier mit der Ang. des Originaltitels „Versuchet das Glück".

1917
FIGUR „Scheikh" (Stehender Scheich mit Fez/Kopfputz und Säbel) (auf mit Reliefs verziertem Rechtecksockel)
ca. 1899/1900
1917 (Firmenmarke ohne Ang.), H. fehlt; Pinhas, S. 93 (m. Abb.)
1917 (Firmenmarke ohne Ang.), H. 110 cm; GSF, S. 18
Bem.: Dargestellt ist ein sog. Scheich, ein Stammesoberhaupt der Beduinen. (Quelle: Wikipedia).
–> vgl. 2071; 2233
Lit.: GSF, S. 18, hier mit der Ang. des Originaltitels „Scheikh".

1918
BÜSTE „Madame Lebrun"
ca. 1898/99 (für den Entwurf)
1918 (Firmenmarke ohne Ang.), H. 49 cm; GSF, S. 18
–> vgl. 1453
Bem.: Dargestellt ist die franz. Malerin Marie Louise Élisabeth Vigée-Lebrun (1755–1842).
Lit.: GSF, S. 18, hier mit der Ang. des Originaltitels „Madame Lebrun".

1919
GRUPPE „Corrigez la fortune!" (Zwei Karten spielende Buben) (auf Landschaftssockel mit Baumstumpf)
ca. 1899/1900
1919 (Firmenmarke ohne Ang.), H. fehlt; Pinhas, S. 93 (m. Abb.)
1919 (Firmenmarke ohne Ang.), H. 52 cm; GSF, S. 18
–> vgl. 2042 (hier unter dem Titel „Kartenspieler")
Lit.: GSF, S. 18, hier mit der Ang. des Originaltitels „Corrigez la fortune!".

1920
Cherc
FIGUR „Der kleine Meisterschütze" (Stehender Bauernjunge mit Armbrust)
ca. 1899/1900
1920,23,70 gem. weiß II, Firmenmarke 7, H. ca. 80 cm; KH C. Scalini/Firenze; Bem.: Die Armbrust ist aus Holz gefertigt.
1920 (Firmenmarke ohne Ang.), H. 78 cm; GSF, S. 18
Lit.: GSF, S. 18, hier mit der Ang. des Originaltitels „Der kleine Meisterschütze"; Pinhas, S. 93 (m. Abb.).

1921
Strasser, Arthur
BÜSTE „Janitschare"
ca. 1898/99
1921 (Firmenmarke ohne Ang.), H. 57 cm; GSF, S. 18
Bem.: Ein „Janitschar" ist ein Krieger (Infantrie) der Elitetruppen des osmanischen Reichs, bzw. ein Leibwächter des Sultans (Quelle: Wikipedia).
–> vgl. 1652; 2043; 2225
Lit.: GSF, S. 18, hier mit der Ang. des Originaltitels „Janitschare".

1922
BÜSTE „Cleopatra" (mit Kopfputz/Diadem und breitem Gürtel mit großer Schnalle) (Kopf nach links, geschlossene Augen)
ca. 1899/1900
1922 (Firmenmarke ohne Ang.), H. fehlt; Pinhas, S. 95 (m. Abb.)
1922 (Firmenmarke ohne Ang.), H. 53 cm; GSF, S. 18
–> vgl. 1857= FIGUR
Lit.: GSF, S. 18, hier mit der Ang. des Originaltitels „Cleopatra".

1923
Cherc
BÜSTE „Amazonenkönigin" (mit Kopfputz und Perlenkette) (Kopf nach rechts, geschlossene Augen)
ca. 1898/99 (für den Entwurf)
1923 (Firmenmarke ohne Ang.), H. fehlt; Pinhas, S. 95 (m. Abb.)
1923 (Firmenmarke ohne Ang.), H. 54 cm; GSF, S. 18
Bem.: Die der Büste entsprechende Figur wurde im ÖMKI 1898/99 ausgestellt, hier findet sich auch die Ang. des Künstlernamens „Cherc".
–> vgl. 1741; 1880= jew. FIGUR
Lit.: GSF, S. 18, hier mit der Ang. des Originaltitels „Amazonenkönigin".

1924
Simon
WANDMASKE „Poppaea"
ca. 1898/99
1924 (Firmenmarke ohne Ang.), H. 44 cm; GSF, S. 18
Bem.: Dargestellt ist römische Kaiserin Poppaea (ca. 30/32–65 n. Chr.), die Frau Kaisers Nero (37–68 n. Chr.).
Ausst.: ÖMKI 1899/1900, hier mit Ang. des Originaltitels „Poppaea".
Pendant –> vgl. 1952
Lit.: GSF, S. 18, hier mit der Ang. des Originaltitels „Poppaea"; Neuwirth (1974), S. 74.

1925
Carrièr-Belleuse, A(lbert-Ernest)
BÜSTE „Harmonie" (Junge Dame mit Hochsteckfrisur) (nach rechts blickend)
ca. 1898/99
1925 (Firmenmarke ohne Ang.), H. fehlt; Pinhas, S. 95 (m. Abb.)
1925 (Firmenmarke ohne Ang.), H. 45 cm; GSF, S. 18
–> vgl. 1610; 2216= jew. FIGUR
Pendant –> vgl. 1941
Lit.: GSF, S. 18, hier mit der Ang. des Originaltitels „Harmonie".

1926
BÜSTE „Fleuron" (Junge Dame mit großem Strohhut)
ca. 1898/99
1926 (Firmenmarke ohne Ang.), H. fehlt; Pinhas, S. 95 (m. Abb.)
1926 (Firmenmarke ohne Ang.), H. 44 cm; GSF, S. 18
–> vgl. 1722; 2098= jew. FIGUR
Lit.: GSF, S. 18, hier mit der Ang. des Originaltitels „Fleuron".

1927
FIGUR „Raucher" (Stehender rauchender Bauernbub mit Hut) (auf Rechtecksockel)
ca. 1898/99
1927 (Firmenmarke ohne Ang.), H. 58 cm; GSF, S. 18
Bem.: Bei diesem Modell verwendete Goldscheider für die beiden produzierten Größen zwei unterschiedliche Titel: Raucher und Knabe mit Cigarre (Mod.-Nr. 1487).
–> vgl. 1487
Lit.: GSF, S. 18, hier mit der Ang. des Originaltitels „Raucher".

1928
Cherc
FIGUR „Dorfprinz" (Stehender rauchender Bursche) (auf quadratischem Landschaftssockel)
ca. 1898/99
1928,44,69 gem. weiß II., Firmenmarke 7, Repr. Res., H. 60 cm; eBay # 3346545155; Bem.: Zigarette bzw. Pfeife fehlt hier.

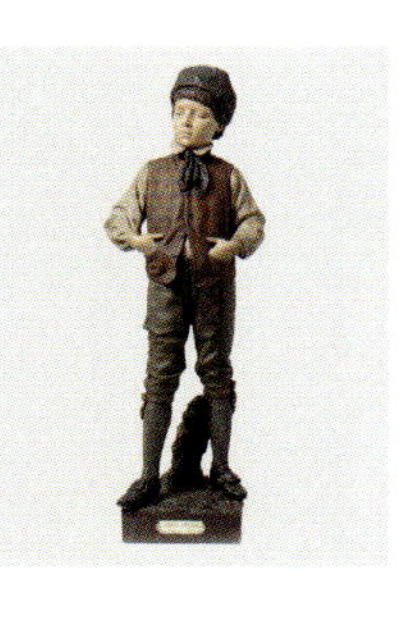

ohne Ang., (ohne Firmenmarke), unsign.?, H. 59,7 cm; eBay # 3702875408: Bem.: Dieses Stück mit einer alten, offenbar original zum Modell gehörigen Pfeife.
1928 (Firmenmarke ohne Ang.), H. 60 cm; GSF, S. 19
–> vgl. 1628
–> vgl. 3823 BÜSTE
Lit.: GSF, S. 19, hier mit der Ang. des Originaltitels „Dorfprinz"; Pinhas, S. 96 (m. Abb.).

1929
FIGUR „Zeitungsleser" (Stehender lesender Bursche mit Brille und Hut) (auf Landschaftssockel mit Baumstumpf)
ca. 1898/99
1929 (Firmenmarke ohne Ang.), H. 58 cm; Pinhas, S. 76 (m. Abb.)
1929 (Firmenmarke ohne Ang.), H. 58 cm; GSF, S. 19
–> vgl. 1680
Lit.: GSF, S. 19, hier mit der Ang. des Originaltitels „Zeitungsleser".

1931
BÜSTE „Jeanne d' Arc" (als Kriegerin mit Rüstung und Lorbeerkranz) (auf Sockel mit Schwert und Bourbonenlilien)
ca. 1899/1900
1931 (Firmenmarke ohne Ang.), H. fehlt; Pinhas, S. 95 (m. Abb.)
1931 (Firmenmarke ohne Ang.), H. 44 cm; GSF, S. 19
Bem.: Bei der dargestellten Person handelt es sich um Jeanne d' Arc.
Lit.: GSF, S. 19, hier mit der Ang. des Originaltitels „Jeanne d' Arc.

1932
Simon
HERME „Kind mit Schmetterling" (Kleines Mädchen mit Blumen geschmücktem Haar und Schmetterling) (auf hoher Blumen verzierter Säule)
ca. 1898/99
1932 (Firmenmarke ohne Ang.), H. fehlt; Pinhas, S. 95 (m. Abb.)
1932 (Firmenmarke ohne Ang.), H. 127 cm; GSF, S. 19
Bem.: Die hier dokumentierte Herme (hoher Pfeilerschaft, eine Art Sockel, mit aufgesetzem Kopf), wird auch im Firmenkat. als solche bezeichnet. Geliefert wurde dieses Objekt auch als Lampe (3 Glühbirnen).
Ausst.: ÖMKI 1899/1900, hier mit Ang. des Originaltitels „Kind mit Schmetterling".
Lit.: GSF, S. 19, hier mit der Ang. des Originaltitels „Kind mit Schmetterling"; Neuwirth (1974), S. 74.

1935
Cherc
FIGUR „Saat" (Säende Bäuerin)
ca. 1895 (für den Entwurf)
1935 (Firmenmarke ohne Ang.), H. 83 cm; GSF, S. 19
1935 (Firmenmarke ohne Ang.), unsign., H. fehlt; Pinhas, S. 95 (m. Abb.)
Pendant –> vgl. 571; 2220
Lit.: GSF, S. 19, hier mit der Ang. des Originaltitels „Saat".

1936
Simon
FIGUR „Aschenbrödel" (Tauben fütternd)
ca. 1898/99
1936,157,12 gem. Z, Firmenmarke Sculpture 13, Repr. Res., H. 62 cm; Soth. com. # 1KJM (Hickmet Fine Art); eBay # 1495485969
1936 (Firmenmarke ohne Ang.), H. 62 cm; GSF, S. 19
Bem.: Dargestellt ist Ludwig Bechsteins Märchenfigur Aschenbrödel (Aschenputtel, Cinderella).
–> vgl. 1575
Lit.: GSF, S. 19, hier mit der Ang. des Originaltitels „Aschenbrödel".

1937
FIGUR „Mondgöttin (La lune coissante)"
ca. 1898/99
1937 (Firmenmarke ohne Ang.), H. 50 cm; GSF, S. 19
Bem.: Goldscheider führte beide, sowohl den franz. als auch deutschen, Originaltitel für diese Figur.
–> vgl. 1433; 1798
Lit.: GSF, S. 19, hier mit der Ang. des Originaltitels „Mondgöttin (La lune coissante)".

1939
WANDMASKE „Incroyable" (Lachender Mann mit Zweispitz)
ca. 1897/98 (für den Entwurf)
1939 (Firmenmarke ohne Ang.), H. fehlt; Pinhas, S. 94 (m. Abb.)
1939 (Firmenmarke ohne Ang.), H. 40 cm; GSF, S. 19

–> vgl. 1117= BÜSTE
Lit.: GSF, S. 19, hier mit der Ang. des Originaltitels „Incroyable".

1940
Bortone, Ipazzio Antonio
WANDMASKE „Bummler" (Junger Mann mit Barett)
ca. 1898/99
1940 (Firmenmarke ohne Ang.), unsign., H. fehlt; Pinhas, S. 94 (m. Abb.)
1940 (Firmenmarke ohne Ang.), H. 32 cm; GSF, S. 19
–> vgl. 1739= BÜSTE
Lit.: GSF, S. 19, hier mit der Ang. des Originaltitels „Bummler".

1941
Carrièr-Belleuse, A(lbert-Ernest)
BÜSTE „Malerei" (Junge Dame mit Hochsteckfrisur und wallendem Gewand) (nach links blickend)
ca. 1898/99
1941 (Firmenmarke ohne Ang.), H. fehlt; Pinhas, S. 95 (m. Abb.)
1941 (Firmenmarke ohne Ang.), H. 46 cm; GSF, S. 19
–> vgl.. 1848; 2104= jew. FIGUR
Pendant –> vgl. 1925
Lit.: GSF, S. 19, hier mit der Ang. des Originaltitels „Malerei".

1942
BÜSTE „Flamländerin" (Mädchen mit Haube, die mit einer Schleife / Mascherl bekrönt ist)
ca. 1899/1900
1942 (Firmenmarke ohne Ang.), H. fehlt; Pinhas, S. 95 (m. Abb.)
1942 (Firmenmarke ohne Ang.), H. 56 cm; GSF, S. 19
Lit.: GSF, S. 19, hier mit der Ang. des Originaltitels „Flamländerin".

1943
Simon
GRUPPE „Die Phantasie" (Nackte Frauengestalt mit Harfe) (auf Adler thronend)
ca. 1899
1943 (Firmenmarke ohne Ang.), H. 91 cm; Pinhas, S. 37, 96 (m. Abb.)
1943 (Firmenmarke ohne Ang.), H. 98 cm; GSF, S. 19
Bem.: Geliefert wurde dieses Objekt auch als Lampe (9 Glühbirnen).
Ausst.: ÖMKI 1899/1900, hier mit Ang. des Originaltitels „Die Phantasie".
–> vgl. 2180
Lit.: GSF, S. 19, hier mit der Ang. des Originaltitels „Die Phantasie"; Neuwirth (1974), S. 74.

1944
FIGUR „Troubadour" (Stehender junger Violinspieler mit Kappe) (auf Rundsockel)
ca. 1899/1900
1944 (Firmenmarke ohne Ang.), H. fehlt; Pinhas, S. 95 (m. Abb.)
1944 (Firmenmarke ohne Ang.), H. 82 cm; GSF, S. 19
Lit.: GSF, S. 19, hier mit der Ang. des Originaltitels „Troubadour".

1945
Lenoir, Pierre Charles
FIGUR „Muschelträgerin" (Stehendes Blumen geschmücktes junges Mädchen, eine große Muschel haltend) (auf Ovalsockel)
ca. 1899/1900
1945 (Firmenmarke ohne Ang.), H. fehlt; Pinhas, S. 95 (m. Abb.)
1945 (Firmenmarke ohne Ang.), H. 83 cm; GSF, S. 19
Bem.: Geliefert wurde dieses Objekt auch als Lampe (5 Glühbirnen).
–> vgl. 2188; 2221
–> vgl. 2063= BÜSTE
Lit.: GSF, S. 19, hier mit der Ang. des Originaltitels „Muschelträgerin"; N.N.: Illustrierte Rundschau, Terrakotten von Fr. Goldscheider, Wien, in: VKMH, 17.1902/03, S. 717 (m. Abb.), hier mit Ang. des Originaltitels und des Künstlernamens Lenoir.

1946
Cherc
BÜSTE „Die Ranke" (Blumengeschmückte junge Frau, nach rechts blickend)
ca. 1899/1900
1946,369,55 Firmenmarke Sculpture 13, Repr. Res., gepr. „S"= Z2, H. 65 cm; Nagel 381
1946,349,14 Firmenmarke Sculpture 13, Repr. Res., H. 66 cm; Soth. Olympia 7.11.2002
1946 (Firmenmarke ohne Ang.), H. 67 cm; GSF, S. 19
Ausst.: ÖMKI 1901/02, hier mit Ang. des Originaltitels „Die Ranke".
–> vgl. 2195
–> vgl. 2159= WANDMASKE
Lit.: GSF, S. 19, hier mit der Ang. des Originaltitels „Die Ranke"; Neuwirth (1974), S. 61; Pinhas, S. 95 (m. Abb.).

1947
FIGUR „Venet(ianische) Wasserträgerin" (Stehendes, nach vorne gebeugtes Bauernmädchen mit zwei Wassergefäßen, die durch einen Stab verbunden sind) (auf getrepptem Sockel)
ca. 1898/99
1947 (Firmenmarke ohne Ang.), H. 66 cm; Pinhas, S. 70 (m. Abb.)
1947 (Firmenmarke ohne Ang.), H. 66 cm; GSF, S. 19
–> vgl. 1500
Lit.: GSF, S. 19, hier mit der Ang. des Originaltitels „Venet(ianische) Wasserträgerin".

1948
FIGUR „Chloë" (Stehendes junges Mädchen mit zwei Flöten)
ca. 1898/99
1948 (Firmenmarke ohne Ang.), H. 66 cm; GSF, S. 19
Bem.: Es lässt sich zwar ein RELIEF „Chloé" (Terrakotta) ÖMKI (1899/1900) des Künstlers Simon nachweisen, ob jedoch zu dieser Figur ein Zusammenhang besteht, ist zu bezweifeln.
–> vgl. 1519
Lit.: GSF, S. 19, hier mit der Ang. des Originaltitels „Chloë".

1949
Cherc
BÜSTE „Epheu" (Junge Frau mit Efeuschmuck)
ca. 1899/1900
1949 (Firmenmarke ohne Ang.), unsign., H. 50 cm; Pinhas, S. 73 (m. Abb.)
1949 (Firmenmarke ohne Ang.), H. 50 cm; GSF, S. 19
–> vgl. 1506; 2082
Lit.: GSF, S. 19, hier mit der Ang. des Originaltitels „Epheu".

1950
FIGUR „Fischerknabe"
ca. 1898/99
1950 (Firmenmarke ohne Ang.), H. 87 cm; GSF, S. 19
–> vgl. 1317
Lit.: GSF, S. 19, hier mit der Ang. des Originaltitels „Fischerknabe".

1951
Cherc
FIGUR / LAMPE „Ex Oriente Lux" (Stehende mit Blüten geschmückte Frauengestalt / mit Schlange) (auf Rundsockel)
Inschrift: EX ORIENTE LUX
ca. 1899/1900
1951,26,22 (Firmenmarke ohne Ang.), Repr. Res., H. 124 cm; Christie's SK 11.5.2001; Bem.: Hier in elektrifizierter Ausführung.
1951 (Firmenmarke ohne Ang.), H. 128 cm; GSF, S. 19
Text der Inschrift: Aus dem Osten kommt das Licht.
Lit.: GSF, S. 19, hier mit der Ang. des Originaltitels „Ex oriente lux"; Pinhas, S. 94 (m. Abb.).

1952
Simon
WANDMASKE / STEHKOPF „Nero" (auf quaderförmigem Sockel)
Inschrift: NERO
ca. 1899
1952 (Firmenmarke ohne Ang.), H. fehlt; Pinhas, S. 94 (m. Abb.); Bem.: Hier als Stehkopf abgebildet.
1952 (Firmenmarke ohne Ang.), H. 42 cm; GSF, S. 19
Bem.: Dargestellt ist der römische Kaiser Nero (37–68 n. Chr.).
Ausst.: ÖMKI 1899/1900, hier mit Ang. des Originaltitels „Nero".
Pendant –> vgl. 1924
Lit.: GSF, S. 19, hier mit der Ang. des Originaltitels „Nero"; Neuwirth (1974), S. 74.

1953
FIGUR „Der Kampf"/„La Guerre" (Stehender römischer Krieger in Rüstung mit Schild, Schwert und Axt) (auf Landschaftssockel)
Inschrift: LA GUERRE
ca. 1899/1900
1953 (Firmenmarke ohne Ang.), H. fehlt; Pinhas, S. 94 (m. Abb.); Bem: Hier mit der franz. Inschrift „La Guerre", welche aber nicht wie der deutsche Originaltitel besagt „Der Kampf" heißt.
1953 (Firmenmarke ohne Ang.), H. 100 cm; GSF, S. 19
Text der Inschrift: Der Krieg.
–> vgl. 2182; 2307
–> vgl. 2140= BÜSTE
Lit.: GSF, S. 19, hier mit der Ang. des Originaltitels „Der Kampf".

1954
Cherc (zugeschr.)
BÜSTE „Secession" (Junges Blumen geschmücktes Mädchen) (nach rechts blickend)
ca. 1899/1900
1954 (Firmenmarke ohne Ang.), H. fehlt; Pinhas, S. 95 (m. Abb.)
1954 (Firmenmarke ohne Ang.), H. 50 cm; GSF, S. 19
Bem.: Geliefert wurde dieses Objekt auch als Lampe (5 Glühbirnen).
–> vgl. 1754 (Wandmaske mit gleichem Titel)
Lit.: GSF, S. 19, hier mit der Ang. des Originaltitels „Secession".

1955
Amy, Jean Barnabé
WANDMASKE „Defi" (Kleinkind (Mädchen) mit Kapuze)
ca. 1899
1955,86,25 Firmenmarke 7, Repr. Res., unsign.?, H. 25 cm; eBay # 2526071882
1955,144,2 gem. weiß VIII, Firmenmarke 7, Repr. Res., sign. Cherc, H. 25,4 cm; eBay # 885741987
1955 (Firmenmarke ohne Ang.), H. 24 cm; GSF, S. 19
Ausst.: ÖMKI 1899/1900; 1900/01, hier jew. mit Ang. des Originaltitels „Defi".
–> vgl. 1884= BÜSTE
Lit.: GSF, S. 19, hier mit der Ang. des Originaltitels „Defi"; Pinhas, S. 94 (m. Abb.).

1956
Simon
BÜSTE „Lorely" (Junge Frau mit Adler als Kopfputz)
Inschrift: LORELY
ca. 1899
1956 gepr., 291,55 ger., Firmenmarke Sculpture 13, H. 62 cm; Doro 5.6.1997
Bem.: Die Ausführung mit der Inschrift „Lorely" wurde auch, wie der folgende Eintrag belegt, mit dem Schriftzug „Harmonia" angeboten. Der Adler als Kopfputz ist dort dann ein Schwan.
–> vgl. 1956 (Variante)
–> vgl. 2226

1956
Simon
BÜSTE „Harmonia" (Junge Frau mit Schwan als Kopfputz)
Inschrift: HARMONIA
ca. 1899
1956 (Firmenmarke ohne Ang.), H. fehlt; Pinhas, S. 95 (m. Abb.)
1956 (Firmenmarke ohne Ang.), H. 63 cm; GSF, S. 19
Ausst.: ÖMKI 1899/1900; hier mit Ang. des Originaltitels „Harmonia".
Bem.: Die Ausführung mit der Inschrift „Harmonia"

wurde auch, wie der vorher gehende Eintrag belegt, mit dem Schriftzug „Lorely" angeboten. Der Schwan als Kopfputz ist dort dann ein Adler.
-> vgl. 1956 (Variante)
-> vgl. 2119
Lit.: GSF, S. 19, hier mit der Ang. des Originaltitels „Harmonia"; Neuwirth (1974), S. 74.

1957
FIGUR „Knabe (tanzend)" (auf Rundsockel mit Gebüsch)
ca. 1899/1900
1957 (Firmenmarke ohne Ang.), H. fehlt; Pinhas, S. 95 (m. Abb.)
1957 (Firmenmarke ohne Ang.), H. 54 cm; GSF, S. 19
Pendant -> vgl. 1958
Lit.: GSF, S. 19, hier mit der Ang. des Originaltitels „Knabe (tanzend)".

1958 -> ÜBERSCHNEIDUNG
FIGUR „Mädchen (tanzend)" (auf Rundsockel mit Gebüsch)
ca. 1899/1900
1958 (Firmenmarke ohne Ang.), H. fehlt; Pinhas, S. 95 (m. Abb.)
1958 (Firmenmarke ohne Ang.), H. 54 cm; GSF, S. 19
Pendant -> vgl. 1957
Lit.: GSF, S. 19, hier mit der Ang. des Originaltitels „Mädchen (tanzend)".

1958 -> ÜBERSCHNEIDUNG
Stehende Araberin
ca. 1900
1958,37 Firmenmarke 8a, H. 34 cm; Wien Museum Inv. 219.132

1959
BÜSTE „Printemps"
ca. 1899/1900
1959 (Firmenmarke ohne Ang.), H. 43 cm; GSF, S. 19
-> vgl. 1804; 2003; 2065
Lit.: GSF, S. 19, hier mit der Ang. des Originaltitels „Printemps".

1960
BÜSTE „Mädchen von Lille" (nach rechts blickend) (nach klassischem Vorbild)
ca. 1899/1900
1960 (Firmenmarke ohne Ang.), H. fehlt; Pinhas, S. 95 (m. Abb.)
1960 (Firmenmarke ohne Ang.), H. 43 cm; GSF, S. 19
Lit.: GSF, S. 19, hier mit der Ang. des Originaltitels „Mädchen von Lille".

1961
Ceccarelli, Ezio
BÜSTE „Winzerin"/„Automne" (Herbst) (Frau mit großem Hut und Trauben)
ca. 1898
1961 (Firmenmarke ohne Ang.), H. 28 cm; GSF, S. 19
-> vgl. 1297; 1619
-> vgl. 622 (GSB); 734 (GSB)
Lit.: GSF, S. 19, hier mit der Ang. des Originaltitels „Winzerin"; GSB, Pl. 71, hier ist die BÜSTE mit dem Originaltitel „Automne" unter Nr. 622= No.1 (H. 52 cm) und unter Nr. 734= No. 2 (H. 30 cm) abgebildet. Ang. zum Material (Terrakotta/Bronze) fehlen.

1962
BÜSTE „Relief" (Junge Frau im Profil nach rechts) (auf Felssockel)
ca. 1899/1900
1962 (Firmenmarke ohne Ang.), H. fehlt; Pinhas, S. 95 (m. Abb.)
1962 (Firmenmarke ohne Ang.), H. 33 cm; GSF, S. 19
Pendant -> vgl. 1963
Lit.: GSF, S. 19, hier mit der Ang. des Originaltitels „Relief".

1963
BÜSTE „Relief" (Junge Frau im Profil nach links) (auf Felssockel)
ca. 1899/1900
1963 (Firmenmarke ohne Ang.), H. fehlt; Pinhas, S. 95 (m. Abb.)
1963 (Firmenmarke ohne Ang.), H. 33 cm; GSF, S. 19
Pendant -> vgl. 1962
Lit.: GSF, S. 19, hier mit der Ang. des Originaltitels „Relief (Pendant)".

1964
Schram, Alois Hans
BÜSTE „Ilka" (Junge Frau mit Blumenschmuck)
ca. 1899/1900
1964 (Firmenmarke ohne Ang.), H. fehlt; Pinhas, S. 97 (m. Abb.)
1964 (Firmenmarke ohne Ang.), H. 60 cm; GSF, S. 19
Ausst.: ÖMKI 1900/01; hier mit Ang. des Originaltitels „Ilka".
-> vgl. 1968
Lit.: GSF, S. 19, hier mit der Ang. des Originaltitels „Ilka"; Neuwirth (1974), S. 74.

1966
BÜSTE Junge Frau mit Blumen geschmücktem Haar
ca. 1899/1900
1966 (Firmenmarke ohne Ang.), H. fehlt; Pinhas, S. 97 (m. Abb.)

1967
BÜSTE „Egypterin" (Junge Dame mit Kopfputz/Diadem? und Schal)
ca. 1899/1900
1967 (Firmenmarke ohne Ang.), H. fehlt; Pinhas, S. 96 (m. Abb.)
1967 (Firmenmarke ohne Ang.), H. 60 cm; GSF, S. 19
-> vgl. 1972= FIGUR
Lit.: GSF, S. 19, hier mit der Ang. des Originaltitels „Egypterin".

1968
Schram, Alois Hans
BÜSTE „Ilka" (Junge Frau mit Blumenschmuck) (auf hohem Sockel mit Blumenranken)
ca. 1899/1900
1968 (Firmenmarke ohne Ang.), H. fehlt; Pinhas, S. 97 (m. Abb.)
1968 (Firmenmarke ohne Ang.), H. 72 cm; GSF, S. 20
Ausst.: ÖMKI 1900/01; hier mit Ang. des Originaltitels „Ilka".
-> vgl. 1964
Lit.: GSF, S. 20, hier mit der Ang. des Originaltitels „Ilka".

1969
Haniroff/Xaniroff
FIGUR „Wasserträgerin"
ca. 1899/1900
1969 (Firmenmarke ohne Ang.), H. 168 cm; GSF, S. 20
Ausst.: ÖMKI 1900/01, im Ausst.-Kat. wird der falsche Künstlername „Xaniroff" genannt, hier mit Ang. des Originaltitels „Wasserträgerin".
Lit.: GSF, S. 20, hier mit der Ang. des Originaltitels „Wasserträgerin"; Neuwirth (1974), S. 76.

1970
BÜSTE „Fiancée" (Junge Dame mit halb entblößter Brust) (verträumt nach rechts blickend)
ca. 1899/1900
1970 (Firmenmarke ohne Ang.), H. fehlt; Pinhas, S. 96 (m. Abb.)
1970 (Firmenmarke ohne Ang.), H. 49 cm; GSF, S. 20
Lit.: GSF, S. 20, hier mit der Ang. des Originaltitels „Fiancée".

1971
FIGUR „Schlangenzauberin"
ca. 1899/1900
1971 (Firmenmarke ohne Ang.), H. 67 cm; GSF, S. 20
Lit.: GSF, S. 20, hier mit der Ang. des Originaltitels „Schlangenzauberin".

1972
FIGUR „Egypterin"
ca. 1899/1900
1972 (Firmenmarke ohne Ang.), H. 57 cm; GSF, S. 20
-> vgl. 1967= BÜSTE
Lit.: GSF, S. 20, hier mit der Ang. des Originaltitels „Egypterin".

1973
Chavanne, (Jean Maurice?)
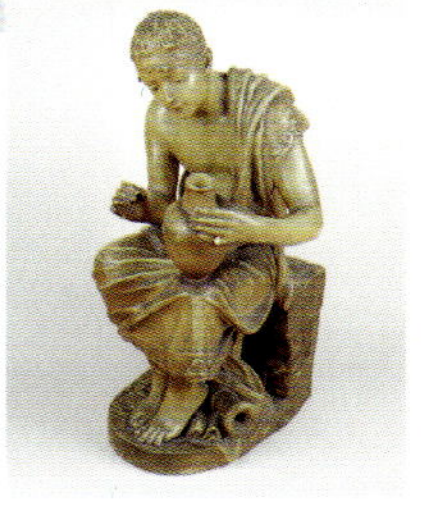
FIGUR „Vasenmaler" (Sitzender Römerjüngling mit Krug)
ca. 1898/99
1973,177,67 Firmenmarke Sculpture 13, Repr. Res., H. 48 cm; PB Wien 19
1973 (Firmenmarke ohne Ang.), H. 48 cm; GSF, S. 20
-> vgl. 1478
Lit.: GSF, S. 20, hier mit der Ang. des Originaltitels „Vasenmaler".

1974
FIGUR „Printemps" (Sitzender Mädchenhalbakt) (auf oktogonalem, mit Blumen geschmücktem Sockel)
ca. 1899
1974 (Firmenmarke ohne Ang.), H. 92 cm; GSF, S. 20
-> vgl. 1763, hier mit dem abweichenden Originaltitel „Schmetterling".
Lit.: GSF, S. 20, hier mit der Ang. des Originaltitels „Printemps".

1976
FIGUR „Le Piège (klein)" (Sich verneigende junge Dame mit langem Kleid) (auf Rundsockel)
ca. 1899/1900
1976 (Firmenmarke ohne Ang.), H. fehlt; Pinhas, S. 97 (m. Abb.)
1976 (Firmenmarke ohne Ang.), H. 57 cm; GSF, S. 20
-> vgl. 2036= BÜSTE
Lit.: GSF, S. 20, hier mit der Ang. des Originaltitels „La Piège (klein)".

1977
Rubelle, A.
BÜSTE „Heilige Elisabeth"
mit/ohne Inschrift: HL. ELISABETH
ca. 1899/1900
1977,17,8 FG Wien (Firmenmarke?), Repr. Res., sign. A. Rubelle, (mit Inschrift), H. 50 cm, B. 51 cm; KH
1977,47,14 Firmenmarke Sculpture 13, Repr. Res., unsign.?, (ohne Inschrift), H. 50 cm; KH Dirk Balduyck Antiques, Belgien/Belgium
1977 (Firmenmarke ohne Ang.), H. 50 cm; GSF, S. 20
Ausst.: ÖMKI 1900/01, hier mit Ang. des Originaltitels „Heilige Elisabeth".
Lit.: GSF, S. 20, hier mit der Ang. des Originaltitels „Elisabeth (mit Sockel)"; Pinhas, S. 96 (m. Abb.).

1978
FIGUR „Sandalenbinderin" (stehend)
ca. 1899/1900
1978,29,1 Firmenmarke Sculpture 13, H. 97 cm; Doro 31.5.1990
1978 (Firmenmarke ohne Ang.), H. 98 cm; GSF, S. 20
-> vgl. 2156
Lit.: GSF, S. 20, hier mit der Ang. des Originaltitels „Sandalenbinderin"; Pinhas, S. 97 (m. Abb.).

1979
Carrièr-Belleuse, (Albert-Ernest)
BÜSTE „Chriemhilde" (Barbusige Frau mit Löwenkopf und -fell als Kopfbedeckung)
ca. 1898/99
1979 (Firmenmarke?), unsign., H. 47 cm; Pinhas, S. 80 (m. Abb.)
1979 (Firmenmarke ohne Ang.), H. 47 cm; GSF, S. 20
Bem.: Dargestellt ist Kriemhild, eine Hauptfigur der Nibelungensage.
-> vgl. 1647
-> vgl. 1772 (bekleidet)
Lit.: GSF, S. 20, hier mit der Ang. des Originaltitels „Chriemhilde".

1980
Carrièr-Belleuse, (Albert-Ernest)
BÜSTE „Brunhilde"/„Brunhaut" (Frau mit Adlerkopf und -schwingen als Kopfbedeckung)
ca. 1899/1900
1980 (Firmenmarke ohne Ang.), H. 43 cm; Pinhas, S. 39, 80 (m. Abb.); Bem.: An der Plinthe betitelt und sign. (Metallschild).

1980 (Firmenmarke ohne Ang.), H. 43 cm; GSF, S. 20
Bem.: Dargestellt ist Brunhild, die sagenhafte Königin aus der Nibelungensage.
-> vgl. 1648; 1773= jew. Variante (nackt)
Pendant -> vgl. 1772
Lit.: GSF, S. 20, hier mit der Ang. des Originaltitels „Brunhilde".

1981
Villieu
BÜSTE „Lilie" (Junge Frau in antikisierendem Gewand mit Blumenkranz im Haar, Dekolleté mit Lilienblüten geschmückt)
ca. 1899/1900
1981,307,68 gem. V, Firmenmarke Sculpture 13, Repr. Res., gepr. „P"= Z3, H. 52 cm; PB; Bem.: Vermutl. handelt sich bei diesem Ex. um die Ausführung mit „niedrigem Sockel".
1981 (Firmenmarke ohne Ang.), H. 58 cm; GSF, S. 20
Bem.: Wie der Firmenkat. belegt, gab es auch eine Ausführung mit „niedrigem Sockel". Die entsprechenden Maßangaben (H. 52 cm?) fehlen jedoch in der Quelle. (Ausst.: ÖMKI 1900/01, hier mit dem franz. Originaltitel „Pureté" ausgestellt.)
-> vgl. 2213; 2411
Lit.: GSF, S. 20, hier mit der Ang. des Originaltitels „Lilie".

1982
FIGUR „Hamlet"
ca. 1899/1900
1982 (Firmenmarke ohne Ang.), H. 89 cm; GSF, S. 20
Pendant -> vgl. 1983
Lit.: GSF, S. 20, hier mit der Ang. des Originaltitels „Hamlet".

1983
FIGUR „Ophelie"
ca. 1899/1900
1983 (Firmenmarke ohne Ang.), H. 87 cm; GSF, S. 20
Bem.: Der Name Ophelie (bzw. Ophelia) bezeichnet die weibliche Hauptfigur aus William Shakespeares „Hamlet".
-> vgl. 2121; 2281
Pendant -> vgl. 1982
Lit.: GSF, S. 20, hier mit der Ang. des Originaltitels „Ophelie".

1986
FIGUR „Knappe" (Stehender Jüngling mit Kappe und Schwert) (auf Ovalsockel)
ca. 1899/1900
1986 (Firmenmarke ohne Ang.), H. fehlt; Pinhas, S. 97 (m. Abb.)
1986 (Firmenmarke ohne Ang.), H. 82 cm; GSF, S. 20
Lit.: GSF, S. 20, hier mit der Ang. des Originaltitels „Knappe".

1987
FIGUR „Araber, Harfe spielend"
ca. 1899/1900
1987 (Firmenmarke ohne Ang.), H. 55 cm; GSF, S. 20
Lit.: GSF, S. 20, hier mit der Ang. des Originaltitels „Araber, Harfe spielend".

1988
GRUPPE „Ländliche Liebe"
ca. 1899/1900
1988 (Firmenmarke ohne Ang.), H. 85 cm; GSF, S. 20
Lit.: GSF, S. 20, hier mit der Ang. des Originaltitels „Ländliche Liebe".

1989
WANDAPPLIK Beethoven mit Medaillon (Monogramm „B") (montiert auf wappenartigem Holzschild)
ca. 1899/1900
1989 (Firmenmarke ohne Ang.), H. 54 cm; Pinhas, S. 35 (m. Abb.)
Bem.: Ob das Holzschild zu der Beethovendarstellung ursprünglich schon gehörte, lässt sich nicht sagen, mit hoher Wahrscheinlichkeit wurde die Montage nachträglich vorgenommen. Dargestellt ist der Komponist Ludwig van Beethoven (1770–1827).

1991 -> ÜBERSCHNEIDUNG
Cherc
BÜSTE „Boa" (Frau mit Schlange und großem extravagantem Hut)
Inschrift: BOA
ca. 1899/1900
01991 (Firmenmarke ohne Ang.), H. 50 cm; Pinhas, S. 53 (m. Abb.)

1991 -> ÜBERSCHNEIDUNG
Paris, Auguste
BÜSTE „Spottvogel"/„Éternel Moqueur" (Lachender Bursche)
ca. 1898/99 (für den Entwurf)
1991 (Firmenmarke ohne Ang.), unsign., H. fehlt; Pinhas, S. 97 (m. Abb.)
1991 (Firmenmarke ohne Ang.), H. 47 cm; GSF, S. 20
Ausst.: Paris, Salon 1898
-> vgl. 783; 842 (GSB), hier mit Ang. des franz. Originaltitels.
-> vgl. 782 (GSB)= FIGUR
Lit.: GSF, S. 20, hier mit der Ang. des Originaltitels „Spottvogel".

1993
Cherc
BÜSTE „Athenais" (Blumengeschmückte junge Frau, nach links blickend)
ca. 1899/1900
1993,99,34 ger. III., Firmenmarke Scupture, Repr. Res., H. 65,5 cm; Wien Museum Inv. 190.058
1993,277,? Firmenmarke Sculpture 13, H. 66 cm; Soth. Olympia 29.5.2003
1993,279,46 Firmenmarke Sculpture 13, Repr. Res., H. 63,5 cm; Pickford 8.4.1999
1993 (Firmenmarke ohne Ang.), H. 67 cm; GSF, S. 20
Bem.: Im Firmenkat. ist vermerkt, dass auch eine Ausführung als Lampe (elektrifiziert mit fünf Glühbirnen) angeboten wurde.
Ausst.: ÖMKI 1901/02, hier mit Ang. des Originaltitels „Athenais".
-> vgl. 2254
-> vgl. 2087= HERME
-> vgl. 2158= WANDMASKE
Lit.: GSF, S. 20, hier mit der Ang. des Originaltitels „Athenais"; Pinhas, S. 96 (m. Abb.).

1994
Cherc
BÜSTE „Theodora" (mit Helm)
ca. 1899/1900
1994,52,15 Firmenmarke 7, Repr. Res., gepr." P"= Z3, H. 60 cm; Keup 67
1994 (Firmenmarke ohne Ang.), H. 67 cm; GSF, S. 20
-> vgl. 2253
-> vgl. 2088= HERME
Lit.: GSF, S. 20, hier mit der Ang. des Originaltitels „Theodora"; Pinhas, S. 96 (m. Abb.).

1995
GRUPPE „Schönheit und Kraft" (Liegender Damenakt mit Löwen) (auf mit Löwenköpfen verziertem Sockel)
ca. 1899/1900
1995 (Firmenmarke ohne Ang.), Originalpick.: Bentejac, Bordeaux, H. 51 cm; Pinhas, S. 38 (m. Abb.)
1995 (Firmenmarke ohne Ang.), H. 49 cm; GSF, S. 20
Lit.: GSF, S. 20, hier mit der Ang. des Originaltitels „Schönheit und Kraft".

1996 -> ÜBERSCHNEIDUNG
FIGUR „Cigaretten-Raucherin" (Stehende rauchende Dame in langem Kleid) (auf getrepptem Sockel)
ca. 1899/1900
1996 (Firmenmarke ohne Ang.), H. fehlt; Pinhas, S. 97 (m. Abb.)
1996 (Firmenmarke ohne Ang.), H. 87 cm; GSF, S. 20
Lit.: GSF, S. 20, hier mit der Ang. des Originaltitels „Cigaretten-Raucherin".

1996 -> ÜBERSCHNEIDUNG
Cherc
Muschelträgerin (Stehende, blumengeschmückte Frauengestalt, eine große Muschel haltend) (auf Rundsockel)
ca. 1899/1900
1996 (Firmenmarke ohne Ang.), H. 80 cm; Pinhas, S. 63 (m. Abb.)
Bem.: Das Dekor „Heliotrop", wie in den KM beschrieben, ist eine helle pastellfarbene Lila Staffierung der ansonsten in Brauntönen ausgeführten Figur.
-> vgl. 2111 (Variante)

1997
Wind, Josef
FIGUR „Krugträgerin"/„Krughändlerin" (Schreitende Frau mit Turban und zwei/drei Krügen) (auf Rundsockel)
ca. 1898 (für den Entwurf)
1997 (Firmenmarke ohne Ang.), H. 102 cm; GSF, S. 20
-> vgl. 1285; 1727
Lit.: GSF, S. 20, hier mit der Ang. des Originaltitels „Krugträgerin".

1998
WANDMASKE „Hera"
ca. 1899/1900
1998 (Firmenmarke ohne Ang.), H. 55 cm; GSF, S. 20
Bem.: Im Firmenkat. ist vermerkt, dass auch eine Ausführung als Lampe (elektrifiziert mit fünf Glühbirnen) angeboten wurde.
Lit.: GSF, S. 20, hier mit der Ang. des Originaltitels „Hera".

1999
WANDMASKE „Gassenjunge"
ca. 1899/1900
1999 (Firmenmarke ohne Ang.), H. 33 cm; GSF, S. 20
Lit.: GSF, S. 20, hier mit der Ang. des Originaltitels „Gassenjunge".

2000
Haniroff
WANDMASKE „Maske von Knabe mit Leuchter" (Negerjunge mit Fez)
ca. 1899/1900
2000 (Firmenmarke ohne Ang.), H. 28 cm; Pinhas, S. 30, 96 (m. Abb.)
2000 (Firmenmarke ohne Ang.), H. 27 cm; GSF, S. 20
-> vgl. 1291= FIGUR
Lit.: GSF, S. 20, hier mit der Ang. des Originaltitels „Maske von Knabe mit Leuchter".

2002
FIGUR „Melonenverkäufer"
ca. 1899/1900
2002 (Firmenmarke ohne Ang.), H. 28 cm; GSF, S. 20
Lit.: GSF, S. 20, hier mit der Ang. des Originaltitels „Melonenverkäufer".

2003
BÜSTE „Printemps"
ca. 1899 (für den Entwurf)
2003 (Firmenmarke ohne Ang.), H. 26 cm; GSF, S. 20
-> vgl. 1804; 1959; 2065
Lit.: GSF, S. 20, hier mit der Ang. des Originaltitels „Printemps".

2005
BÜSTE „Rahel"
ca. 1899/1900
2005 (Firmenmarke ohne Ang.), H. 34 cm; GSF, S. 20
Lit.: GSF, S. 20, hier mit der Ang. des Originaltitels „Rahel".

2006
FIGUR „Pierrot mit Mandoline"
ca. 1899
2006 (Firmenmarke ohne Ang.), H. 30 cm; GSF, S. 20
-> vgl. 1621
Lit.: GSF, S. 20, hier mit der Ang. des Originaltitels „Pierrot mit Mandoline".

2007
FIGUR „Pierrot (demaskiert)"
ca. 1899
2007 (Firmenmarke ohne Ang.), H. 30 cm; GSF, S. 20
-> vgl. 1623
Lit.: GSF, S. 20, hier mit der Ang. des Originaltitels „Pierrot (demaskiert)".

2008
GRUPPE „Neckerei (bekleidet)" (Amor, einer jungen Frau ein Auge zuhaltend) (auf Rundsockel)
ca. 1898 (für den Entwurf)
2008 (Firmenmarke ohne Ang.), H. fehlt; Pinhas, S. 96 (m. Abb.)
2008 (Firmenmarke ohne Ang.), H. 63 cm; GSF, S. 20
-> vgl. 1365
Lit.: GSF, S. 20, hier mit der Ang. des Originaltitels „Neckerei (bekleidet)".

2009
Vidal, Henri
FIGUR „Diana"
ca. 1899/1900
2009 (Firmenmarke ohne Ang.), H. 150 cm; GSF, S. 20
Ausst.: ÖMKI 1900/01, hier mit Ang. des Originaltitels „Diana".
-> vgl. 2114
Lit.: GSF, S. 20, hier mit der Ang. des Originaltitels „Diana"; Neuwirth (1974), S. 75.

2010
FIGUR „Blumenorakel" (Stehende junge Dame mit großem Strohhut) (auf Rundsockel)
ca. 1899/1900
2010 (Firmenmarke ohne Ang.), H. fehlt; Pinhas, S. 97 (m. Abb.)
2010 (Firmenmarke ohne Ang.), H. 60 cm; GSF, S. 20
Lit.: GSF, S. 20, hier mit der Ang. des Originaltitels „Blumenorakel".

2012
BÜSTE „Nürnberger Madonna"
ca. 1899
2012,42,5 Firmenmarke 7, Repr. Res., H. fehlt; eBay # 552688053
Bem.: Als Vorbild für diese Büste diente die Figur der sog. „Nürnberger Madonna", welche Teil einer Kreuzigungsgruppe ist (Original um 1510/20 von einem unbekannten Meister geschnitzt aus Lindenholz, Germanisches Nationalmuseum Nürnberg). Lit. hierzu: Stadtlexikon Nürnberg, 2. verb. Aufl., Nürnberg 2000; Anzeiger des German. Nationalmuseums 1976, S. 63–71; Berman, Vol. 3, Nr. 2521 (m. Abb.).

2014
WANDMASKE „Frauenkopf" („Die unbekannte Tote aus der Seine") (mit Halsansatz)
ca. 1899/1900
2014 (Firmenmarke ohne Ang.), H. fehlt; Pinhas, S. 96 (m. Abb.)
2014 (Firmenmarke ohne Ang.), H. 28 cm; GSF, S. 21
Bem.: Diese beliebte und oft reproduzierte Maske wurde auch noch zw. 1953–1960 bei Carstens (GWM West-Germany) produziert.
-> vgl. 6816; 6817
-> vgl. 6818 (ohne Halsansatz)
Lit.: GSF, S. 21, hier mit der Ang. des Originaltitels „Frauenkopf".

2017
Simon
FIGUR „Ruth" (Stehende Frau mit Kornähren) (auf Natursockel)
Inschrift: RUTH
ca. 1899/1900
2017,103,15 (Firmenmarke ohne Ang.), Fabriqué en Autriche, H. 84,5 cm; Christie's SK 13.9.2005
2017 (Firmenmarke ohne Ang.), H. 82 cm; GSF, S. 21
-> vgl. 2110 (Variante)
Lit.: GSF, S. 21, hier mit der Ang. des Originaltitels „Ruth".

2018
Polair
LAMPE „Berggeist" (Bergsteigender Bursche, einen Blumenstrauß haltend) (auf Felssockel)
ca. 1899/1900
2018 (Firmenmarke ohne Ang.), H. fehlt; Pinhas, S. 96 (m. Abb.)
2018 (Firmenmarke ohne Ang.), H. 54 cm; GSF, S. 21
Bem.: Als Lampenfassung dient der Blumenstrauß, der mit einer Glühbirne elektrifiziert ist.
Ausst.: ÖMKI 1900/01, hier mit Ang. des Originaltitels „Berggeist".
Lit.: GSF, S. 21, hier mit der Ang. des Originaltitels „Berggeist"; Neuwirth (1974), S. 70.

2019
Debienne, Noémi(e)
Stehende, posierende junge Frau in langem Gewand
ca. 1899/1900
2019,3,47 FG (Firmenmarke ohne Ang.), unsign., H. 81 cm; Doro 20.4.1993
-> vgl. 739 (GSB)
Lit.: GSB, Pl. 75, hier ist dieses Modell als Variante abgebildet: Akt vor einem Rebstock stehend, Originaltitel „Triomphe de Bacchante", Modell-Nr. 739, (H. 92 cm). Als Künstlerin wird (Noémie) Debienne genannt. Ang. zum Material (Terrakotta/Bronze) fehlen.

2022
FIGUR „Der Hieb"
ca. 1899/1900
2022 (Firmenmarke ohne Ang.), H. 118 cm; GSF, S. 21
Lit.: GSF, S. 21, hier mit der Ang. des Originaltitels „Der Hieb".

2023
Allouard, Henri Emile
BÜSTE „Rococo" (Junge Dame nach rechts blickend)
ca. 1898 (für den Entwurf)
2023,27,25 gem. weiß IV, Firmenmarke 7, Repr. Res., H. 42 cm; eBay # 2502316366
2023 (Firmenmarke ohne Ang.), H. 42 cm; GSF, S. 21
Ausst.: ÖMKI 1898/99, hier mit Ang. des Originaltitels „Rococo". Da die Büste in drei verschiedenen Höhen produziert wurde und die Mod.-Nr. im Ausst.-Kat. nicht angegeben ist, lässt sich nicht sagen welche Größe ausgestellt war.
-> vgl. 1441; 2079
Lit.: GSF, S. 21, hier mit der Ang. des Originaltitels „Rococo"; Neuwirth (1974), S. 59.

2026
FIGUR „Harfenspielerin"
ca. 1899/1900
2026 (Firmenmarke ohne Ang.), H. 109 cm; GSF, S. 21
Lit.: GSF, S. 21, hier mit der Ang. des Originaltitels „Harfenspielerin".

2027
FIGUR „Chianti" (Stehendes Bauernmädchen mit Haube und Schürze, eine Chiantiflasche haltend)
(auf mit Weintrauben und Laub verziertem Rundsockel)
Inschrift: CHIANTI
ca. 1899
2027 (Firmenmarke ohne Ang.), H. 79 cm; GSF, S. 21
-> vgl. 1764
Lit.: GSF, S. 21, hier mit der Ang. des Originaltitels „Chianti".

2028
FIGUR (LAMPE) „Astarte"
ca. 1899/1900
2028 (Firmenmarke ohne Ang.), H. 78 cm; GSF, S. 21
Bem.: Dargestellt ist „Astarte", die Fruchtbarkeitsgöttin der westsemitischen Länder. In Ägypten wurde sie als Liebes- und Kriegsgöttin verehrt (Quelle: Wikipedia). Dieses Modell wurde lt. Firmenkat. auch mit einem „elektr. Zweig mit oder ohne Strahl" geliefert, die jew. durch Glühbirnen beleuchtet waren.
Lit.: GSF, S. 21, hier mit der Ang. des Originaltitels „Astarte".

2029
Bortone, Ipazzio Antonio/Portou
WANDMASKE „Annita"
ca. 1898/99 (für den Entwurf)
2029 (Firmenmarke ohne Ang.), H. 22 cm; GSF, S. 21
Ausst.: ÖMKI 1900/01, hier mit Ang. des Originaltitels „Annita".
-> vgl. 1740; 2130= jew. BÜSTE
Lit.: GSF, S. 21, hier mit der Ang. des Originaltitels „Annita"; Neuwirth (1974), S. 60.

2030
Cherc
FIGUR „Zeitungsständer mit Sockel" (Stehende junge Frau (mit eulengeschmücktem Diadem) auf Bücherstapel, zwei große Papierbögen auffächernd) (auf getrepptem obeliskartigem mit Eulen und Maskarons verziertem Sockel)
ca. 1899/1900
2030 (Firmenmarke ohne Ang.), H. 150 cm; Pinhas, S. 44 (m. Abb.)
2030 (Firmenmarke ohne Ang.), H. 145 cm; GSF, S. 21
Lit.: GSF, S. 21, hier mit der Ang. des Originaltitels „Zeitungsständer mit Sockel".

2032
BÜSTE „Loreley"
ca. 1899/1900
2032 (Firmenmarke ohne Ang.), H. 43 cm; GSF, S. 21
-> vgl. 2314
Lit.: GSF, S. 21, hier mit der Ang. des Originaltitels „Loreley".

2033
Cherc
BÜSTE „Gladiator"
ca. 1899/1900
2033 (Firmenmarke ohne Ang.), H. 40 cm; GSF, S. 21
-> vgl. 1833; 2137= jew. FIGUR
Lit.: GSF, S. 21, hier mit der Ang. des Originaltitels „Gladiator".

2034
Cherc
BÜSTE „Bogenspanner"
ca. 1899/1900
2034 (Firmenmarke ohne Ang.), H. 38 cm; GSF, S. 21
-> vgl. 1834; 2138= jew. FIGUR
Lit.: GSF, S. 21, hier mit der Ang. des Originaltitels „Bogenspanner".

2035
Lemassi
BÜSTE „Ernte"
ca. 1899/1900
2035 (Firmenmarke ohne Ang.), H. 60 cm; GSF, S. 21
Ausst.: ÖMKI 1900/01, hier mit Ang. des Originaltitels „Ernte".
-> vgl. 2315
Lit.: GSF, S. 21, hier mit der Ang. des Originaltitels „Ernte"; Neuwirth (1974), S. 66.

2036
BÜSTE „Piège"
ca. 1899/1900
2036 (Firmenmarke ohne Ang.), H. 33 cm; GSF, S. 21
-> vgl. 1976= FIGUR
Lit.: GSF, S. 21, hier mit der Ang. des Originaltitels „Piège".

2037
FIGUR „Assyrer"
ca. 1899/1900
2037 (Firmenmarke ohne Ang.), H. 99 cm; GSF, S. 21
Lit.: GSF, S. 21, hier mit der Ang. des Originaltitels „Assyrer".

2038
BÜSTE „Madame Sans Gêne" (mit großem Hut/Haube) (nach rechts blickend)
ca. 1898/99 (für den Entwurf)
2038 (Firmenmarke ohne Ang.), H. 51 cm; GSF, S. 21
Bem.: Dargestellt ist die Hauptfigur Catherine (Madame Sans Gêne) aus dem Lustspiel in 4 Akten von Victorien Sardou. (Quelle: Wikipedia)
-> vgl. 1624; 2052
Lit.: GSF, S. 21, hier mit der Ang. des Originaltitels „Madame Sans Gêne".

2040
Tell, E.
FIGUR (SPIEGEL) „Spiegelträgerin" (Stehender Frauenakt, einen Spiegel haltend und in diesen schauend)
ca. 1899/1900
2040,583,3 gem. weiß 23, Firmenmarke Sculpture 13, Repr. Res., gepr. „P"= Z3, H. 74 cm; PB Wien 19

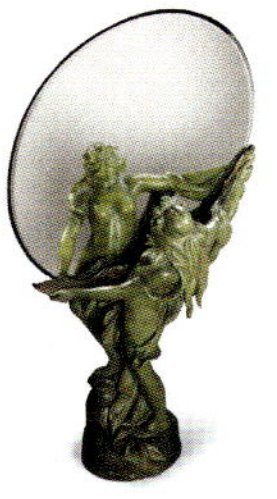

2040,627,14 Firmenmarke Sculpture 13, Repr. Res., H. 78,5 cm; Doro 25.4.2000
2040,669,1 Firmenmarke Sculpture 13, H. 75 cm; Christie' s SK 16.7.2006
2040 (Firmenmarke ohne Ang.), H. 101 cm; GSF, S. 21
–> vgl. 532 (ähnl. Modell)
–> vgl. 2109 (Variante, bekleidet); 2359
Lit.: GSF, S. 21, hier mit der Ang. des Originaltitels „Spiegelträgerin".

2041
BÜSTE „Heldenmuth"
ca. 1899/1900
2041 (Firmenmarke ohne Ang.), H. 61 cm; GSF, S. 21
Lit.: GSF, S. 21, hier mit der Ang. des Originaltitels „Heldenmuth".

2042
GRUPPE „Kartenspieler" (Zwei Karten spielende Buben) (auf Landschaftssockel mit Baumstumpf)
ca. 1899/1900
2042 (Firmenmarke ohne Ang.), H. 40 cm; GSF, S. 21
–> vgl. 1919 (hier unter dem Titel „Corrigez la fortune!")
Lit.: GSF, S. 21, hier mit der Ang. des Originaltitels „Kartenspieler".

2043
Strasser, Arthur
FIGUR „Janitschare" (mit Turban, Schild, Säbel und langem Stab) (auf Landschaftssockel)
ca. 1898/99 (für den Entwurf)
2043 (Firmenmarke ohne Ang.), H. 72 cm; GSF, S. 21
Bem.: Ein „Janitschar" ist ein Krieger (Infantrie) der Elitetruppen des osmanischen Reichs, bzw. ein Leibwächter des Sultans (Quelle: Wikipedia).
–> vgl. 1652; 2225
–> vgl. 1921= BÜSTE
Lit.: GSF, S. 21, hier mit der Ang. des Originaltitels „Janitschare".

2044
FIGUR „Mode"
ca. 1900
2044 (Firmenmarke ohne Ang.), H. 144 cm; GSF, S. 21
–> vgl. 2086
Lit.: GSF, S. 21, hier mit der Ang. des Originaltitels „Mode".

2046
Tell, E.
FIGUR „Muschelträgerin" (Stehende, nach vorne gebeugte nackte Frauengestalt, eine große Muschel haltend)
ca. 1900
2046 (Firmenmarke ohne Ang.), H. 60 cm; Pinhas, S. 63 (m. Abb.)
2046 (Firmenmarke ohne Ang.), H. 57 cm; GSF, S. 21
Bem.: Das Dekor „Heliotrop", wie in den KM beschrieben, ist eine helle pastellfarbene Lila Staffierung der ansonsten in Brauntönen ausgeführten Figur.
–> vgl. 2111 (Variante)
Lit.: GSF, S. 21, hier mit der Ang. des Originaltitels „Muschelträgerin"; R., F.: Wiener Kunstkeramik, in: KM, 2.1902, S. 90 (m. Abb.), hier auch mit Ang. des Originaltitels „Muschelträgerin", S. 91: „Zarteste lila Farbenvariation gibt das Dekor „Heliotrop", in welchem Tells „Muschelträgerin" ausgeführt ist. Die feinen lichten Tönungen dieser duftigen Nuancen ergeben Farbakkorde von ausserordentlicher Reinheit".

2047
BÜSTE „Scheikh"
ca. 1900
2047 (Firmenmarke ohne Ang.), H. 53 cm; GSF, S. 21
Lit.: GSF, S. 21, hier mit der Ang. des Originaltitels „Scheikh".

2048
BÜSTE „Krugträgerin"
ca. 1900
2048 (Firmenmarke ohne Ang.), H. 46 cm; GSF, S. 21
Lit.: GSF, S. 21, hier mit der Ang. des Originaltitels „Krugträgerin".

2049
WANDMASKE „Baroque"
ca. 1900
2049 (Firmenmarke ohne Ang.), H. 38 cm; GSF, S. 21
Lit.: GSF, S. 21, hier mit der Ang. des Originaltitels „Baroque".

2050
WANDMASKE „Griechisch"
ca. 1900
2050 (Firmenmarke ohne Ang.), H. 34 cm; GSF, S. 21
Lit.: GSF, S. 21, hier mit der Ang. des Originaltitels „Griechisch".

2051
Lesca
WANDMASKE „Chrisis"
ca. 1900
2051 (Firmenmarke ohne Ang.), H. 29 cm; GSF, S. 21
Ausst.: ÖMKI 1900/01, hier mit Ang. des Originaltitels „Chrisis".
Lit.: GSF, S. 21, hier mit der Ang. des Originaltitels „Chrisis"; Neuwirth (1974), S. 66.

2052
BÜSTE „Madame Sans Gêne" (mit großem Hut/Haube) (nach rechts blickend)
ca. 1898/99 (für den Entwurf)
2052 (Firmenmarke ohne Ang.), H. 34 cm; GSF, S. 21
Bem.: Dargestellt ist die Hauptfigur Catherine (Madame Sans Gêne) aus dem Lustspiel in 4 Akten von Victorien Sardou. (Quelle: Wikipedia)
–> vgl. 1624; 2038
Lit.: GSF, S. 21, hier mit der Ang. des Originaltitels „Madame Sans Gêne".

2053
Cherc
FIGUR „Das tägliche Brod"
ca. 1900
2053 (Firmenmarke ohne Ang.), H. 81 cm; GSF, S. 21
Bem.: Warum die Manufaktur für dieses Modell zwei fast ident. aber eben doch abweichende Titel (vgl. 2217) vergab, bleibt unklar. In den Ausst.-Katalogen des ÖMKI erscheinen beide Ausformungen, die in den beiden aufeinander folgenden Jahren gezeigt wurden ebenfalls mit diesen Titeln.
Ausst.: ÖMKI 1900/01, hier mit Ang. des Originaltitels „Das tägliche Brod".
–> vgl. 2217
Lit.: GSF, S. 21, hier mit der Ang. des Originaltitels „Das tägliche Brod"; Neuwirth (1974), S. 61.

2054
FIGUR „Nordpol"
ca. 1900
2054 (Firmenmarke ohne Ang.), H. 74 cm; GSF, S. 21
Lit.: GSF, S. 21, hier mit der Ang. des Originaltitels „Nordpol".

2055
Waderé, Heinrich
BÜSTE „Psyche" (Junges Mädchen mit Schmetterlingsflügeln) (mit Blumen und Perlen verziertem Kleid) (auf Rundsockel)
ca. 1899/1900
2055 (Firmenmarke ohne Ang.), H. 48 cm; GSF, S. 21
Ausst.: ÖMKI 1906/07, hier mit Ang. des Originaltitels „Psyche".
–> vgl. 1860; 2081
Lit.: GSF, S. 21, hier mit der Ang. des Originaltitels „Psyche".

2056
Delagrange, Léon Noël
FIGUR „Brise"
ca. 1900
2056 (Firmenmarke ohne Ang.), H. 88 cm; GSF, S. 21
Ausst.: ÖMKI 1900/01, hier mit Ang. des Originaltitels „Brise".
Lit.: GSF, S. 21, hier mit der Ang. des Originaltitels „Brise"; Neuwirth (1974), S: 62.

2057
FIGUR „L' Histoire" (Stehendes Mädchen an einem Felsen)
ca. 1900
2057 (Firmenmarke ohne Ang.), H. 75 cm; Pinhas, S. 62 (m. Abb.)
2057 (Firmenmarke ohne Ang.), H. 76 cm; GSF, S. 21
Lit.: GSF, S. 21, hier mit der Ang. des Originaltitels „L' Histoire".

2058
FIGUR „Parze"
ca. 1900
2058 (Firmenmarke ohne Ang.), H. 54 cm; GSF, S. 21
Bem.: Dargestellt ist eine Parze, eine Schicksalsgöttin der röm Mythologie. (Quelle: Wikipedia).
Lit.: GSF, S. 21, hier mit der Ang. des Originaltitels „Parze".

2059
Delagrange, Léon Noël
FIGUR „Le Secret"
ca. 1900
2059 (Firmenmarke ohne Ang.), H. 59 cm; GSF, S. 21
Ausst.: ÖMKI 1900/01, hier mit Ang. des Originaltitels „Le Secret".
Lit.: GSF, S. 21, hier mit der Ang. des Originaltitels „Le Secret"; Neuwirth (1974), S: 62.

2060
Mercier
FIGUR (LAMPE) „Uhu"
ca. 1900
ohne Ang., (Firmenmarke ohne Ang.), sign. Mercier, H. 54 cm; Nagel „D2" (21.10.1995)
2060 (Firmenmarke ohne Ang.), H. 48 cm; GSF, S. 22
Bem.: Dieses Modell wurde auch mit einer Glühbirne elektrifiziert geliefert.
Lit.: GSF, S. 22, hier mit der Ang. des Originaltitels „Uhu".

2061
BÜSTE „Editha"
Inschrift: EDITH
ca. 1900
2061,125,31 Firmenmarke Sculpture 13, Repr. Res., H. 39 cm; Doro 20.3.1992; Bem.: Hier mit falscher Mod.-Nr. 2081.
2061 (Firmenmarke ohne Ang.), H. 36 cm; GSF, S. 22
Lit.: GSF, S. 22, hier mit der Ang. des Originaltitels „Editha".

2062
BÜSTE „Salomé"
ca. 1900
2062 (Firmenmarke ohne Ang.), H. 48 cm; GSF, S. 22
Lit.: GSF, S. 22, hier mit der Ang. des Originaltitels „Salomé".

2063
Lenoir, Pierre Charles
BÜSTE „Felicia"
ca. 1900
2063 (Firmenmarke ohne Ang.), H. 42 cm; GSF, S. 22
–> vgl. 1945; 2221= jew. FIGUR
Lit.: GSF, S. 22, hier mit der Ang. des Originaltitels „Felicia"; N.N.: Illustrierte Rundschau, Terrakotten von Fr. Goldscheider, Wien, in: VKMH, 17.1902/03, S. 717 (m. Abb.), hier die Abb. als Figur.

2064
Tell, E.
FIGUR „Serpentintänzerin" (Loie Fuller) (auf Natursockel)
ca. 1900
2064,139,47 (Firmenmarke ohne Ang.), H. 72 cm; Privatsammlung
2064,347,53 (Firmenmarke Sculpture 13?), Repr. Res., H. 71 cm; Soth. L. 27.3.1997
ohne Ang., (Firmenmarke ohne Ang.), H. 72 cm; Christie's SK 13.5.1999
2064 (Firmenmarke ohne Ang.), H. 68 cm; GSF, S. 22
Bem.: Dargestellt ist die berühmte Tanzkünstlerin Loie Fuller (1862–1928) bei einem ihrer Tänze. Im Firmenkat. ist vermerkt, dass auch eine Ausführung als Lampe (elektrifiziert mit drei Glühbirnen) angeboten wurde.
-> vgl. 2428
Lit.: GSF, S. 22, hier mit der Ang. des Originaltitels „Serpentintänzerin"; Pinhas, S. 61 (m. Abb.), hier mit der Mod.-Nr. 3784, diese aber offenbar falsch, da die Höhe 72 cm identisch mit Mod.-Nr. 2064.

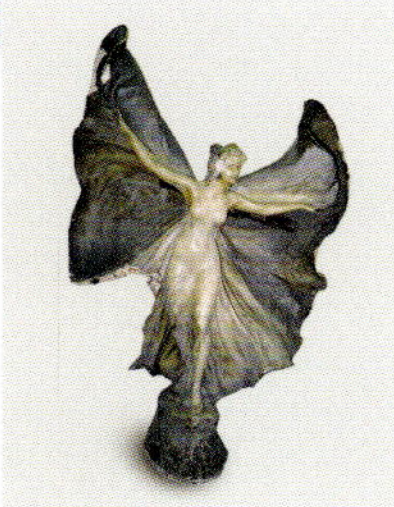

2065
BÜSTE „Printemps"
ca. 1899/1900
2065 (Firmenmarke ohne Ang.), H. 29 cm; GSF, S. 22
-> vgl. 1804; 1959; 2003
Lit.: GSF, S. 22, hier mit der Ang. des Originaltitels „Printemps".

2066
Oury, Louis
BÜSTE „L'Aiglon" (Napoleon II.)
ca. 1900
2066 (Firmenmarke ohne Ang.), unsign.?, H. 49 cm; Pinhas, S. 35 (m. Abb.)
2066 (Firmenmarke ohne Ang.), sign. Louis Oury, H. 65 cm (mit Sockel); www.desouzy.com/images/g5207.jpg (14.08.2006)
2066 (Firmenmarke ohne Ang.), H. 64 cm (mit Sockel); GSF, S. 22
Bem.: Die hier dargestellte Person ist der Sohn Napoleons I. und nicht wie in der Lit. angegeben der Sohn Napoleons III. Napoleon II. (1811–1832) war König von Rom.
-> vgl. 2133
Lit.: GSF, S. 22, hier mit der Ang. des Originaltitels „L'Aiglon".

2067
BÜSTE „Cyrano"
ca. 1900
2067 (Firmenmarke ohne Ang.), H. 42 cm; GSF, S. 22
Bem.: Dargestellt ist der franz. Schriftsteller Cyrano de Bergerac (1619–1655).
Lit.: GSF, S. 22, hier mit der Ang. des Originaltitels „Cyrano de Bergerac".

2068
WANDMASKE „Frauenkopf"
ca. 1900
2068 (Firmenmarke ohne Ang.), H. 41 cm; GSF, S. 22
Lit.: GSF, S. 22, hier mit der Ang. des Originaltitels „Frauenkopf".

2069
FIGUR „Eisfee"
ca. 1900
2069 (Firmenmarke ohne Ang.), H. 81 cm; GSF, S. 22
Bem.: Dieses Modell wurde auch mit einer Glühbirne elektrifiziert geliefert.
Lit.: GSF, S. 22, hier mit der Ang. des Originaltitels „Eisfee".

2070
FIGUR (LAMPE) „Illumination" (Stehende junge Dame, eine Laterne in der ausgestreckten rechten Hand haltend) (auf Rundsockel)
ca. 1898/99
2070 (Firmenmarke ohne Ang.), H. 68 cm; GSF, S. 22
Bem.: Dieses Modell wurde auch mit einer Glühbirne elektrifiziert geliefert.
-> vgl. 1602
Lit.: GSF, S. 22, hier mit der Ang. des Originaltitels „Illumination".

2071
FIGUR „Scheikh"
ca. 1900
2071 (Firmenmarke ohne Ang.), H. 77 cm; GSF, S. 22
Bem.: Dargestellt ist ein sog. Scheich, ein Stammesoberhaupt der Beduinen. (Quelle: Wikipedia).
-> vgl. 1917; 2233
Lit.: GSF, S. 22, hier mit der Ang. des Originaltitels „Scheikh".

2072
Cherc/Mercadier, J.
FIGUR „Mignon" (Stehende junge Frau mit langem Haar) (nach rechts blickend)
ca. 1898/99 (für den Entwurf)
2072,404,19 gem. Z, Firmenmarke Sculpture 13, Fabriqué en Autriche, gepr. „S"= Z2, sign. J. Mercadier, H. 59 cm; Wien Museum Inv. 186.611
ohne Ang., Firmenmarke Sculpture 13, Repr. Res., sign. „Macadier", H. 60 cm; Soth. L. 10.11.1983
2072 (Firmenmarke ohne Ang.), H. 60 cm; GSF, S. 22
-> vgl. 1643; 2436
Lit.: GSF, S. 22, hier mit der Ang. des Originaltitels „Mignon".

2073
BÜSTE „Eté" (Junge Bäuerin mit Kopftuch und Ähren über der Schulter)
ca. 1898/99 (für den Entwurf)
2073 (Firmenmarke ohne Ang.), H. 24 cm; GSF, S. 22
Bem.: Bei dieser Büste handelt es sich um eine Allegorie des Sommers.
-> vgl. 1657
Lit.: GSF, S. 22, hier mit der Ang. des Originaltitels „Eté".

2076
WANDMASKE „Beethoven"
ca. 1900
2076 (Firmenmarke ohne Ang.), H. 32 cm; GSF, S. 22
Bem.: Dargestellt ist der Komponist Ludwig van Beethoven (1770–1827).
Lit.: GSF, S. 22, hier mit der Ang. des Originaltitels „Beethoven".

2077
WANDMASKE „Mozart"
ca. 1900
2077 (Firmenmarke ohne Ang.), H. 31 cm; GSF, S. 22
Bem.: Dargestellt ist der Komponist Wolfgang Amadeus Mozart (1756–1791).
-> vgl. 2330
Lit.: GSF, S. 22, hier mit der Ang. des Originaltitels „Mozart".

2078
FIGUR „Wasserträgerin"
ca. 1900
2078 (Firmenmarke ohne Ang.), H. 98 cm; GSF, S. 22
-> vgl. 2206
Lit.: GSF, S. 22, hier mit der Ang. des Originaltitels „Wasserträgerin".

2079
Allouard, Henri Emile
BÜSTE „Rococo" (Junge Dame nach rechts blickend)
ca. 1898 (für den Entwurf)
2079 (Firmenmarke ohne Ang.), H. 26 cm; GSF, S. 22
Ausst.: ÖMKI 1898/99, hier mit Ang. des Originaltitels „Rococo". Da die Büste in drei verschiedenen Höhen produziert wurde und die Mod.-Nr. im Ausst.-Kat. nicht angegeben ist, lässt sich nicht sagen welche Größe ausgestellt war.
-> vgl. 1441; 2023
Lit.: GSF, S. 22, hier mit der Ang. des Originaltitels „Rococo".

2080
BÜSTE „Doux mots"
ca. 1900
2080 (Firmenmarke ohne Ang.), H. 25 cm; GSF, S. 22
Lit.: GSF, S. 22, hier mit der Ang. des Originaltitels „Doux mots".

2081
Waderé, Heinrich
BÜSTE „Psyche" (Junges Mädchen mit Schmetterlingsflügeln) (mit Blumen und Perlen verziertem Kleid) (auf Rundsockel)
ca. 1899/1900
2081 (Firmenmarke ohne Ang.), H. 25 cm; GSF, S. 22
Ausst.: ÖMKI 1906/07, hier mit Ang. des Originaltitels „Psyche".
-> vgl. 1860; 2055
Lit.: GSF, S. 22, hier mit der Ang. des Originaltitels „Psyche".

2082
Cherc
BÜSTE „Epheu" (Junge Frau mit Efeuschmuck)
ca. 1899/1900
2082 (Firmenmarke ohne Ang.), H. 35 cm; GSF, S. 22
-> vgl. 1506; 2082
Lit.: GSF, S. 22, hier mit der Ang. des Originaltitels „Epheu".

2083
Simon
FIGUR „Orpheus"
ca. 1900
2083 (Firmenmarke ohne Ang.), H. 97 cm; GSF, S. 22
Bem.: Orpheus, ein Sänger und Saitenspieler der griech. Sage, war Sohn von Apollo und Kalliope.
Ausst.: ÖMKI 1900/01, hier mit Ang. des Originaltitels „Orpheus".
Lit.: GSF, S. 22, hier mit der Ang. des Originaltitels „Orpheus"; Neuwirth (1974), S. 74.

2084
Bate, Firmin
FIGUR „Daphne"
ca. 1900
2084 (Firmenmarke ohne Ang.), H. 106 cm; GSF, S. 22
Bem.: Dargestellt ist Daphne, eine Nymphe der griech. Sage. Wie der Firmenkat. belegt, wurde dieses Modell auch mit zwei Glühbirnen elektrifiziert geliefert.
Ausst.: Paris, Salon 1902
Lit.: GSF, S. 22, hier mit der Ang. des Originaltitels „Daphne"; Neuwirth (1974), S. 59.

2085
Cherc
FIGUR „Salome" (Stehende Frauengestalt mit lyraähnlichem Instrument)
Inschrift: SALOME
ca. 1897/98 (für den Entwurf)
2085 (gepr.),51 (ger.),10 (ger.), Firmenmarke Sculpture 13, Repr. Res., H. 75 cm; PB Wien 19
2085 (Firmenmarke ohne Ang.), H. 69 cm; GSF, S. 22
-> vgl. 1000; 1900; 2530
Lit.: GSF, S. 22, hier mit der Ang. des Originaltitels „Salome".

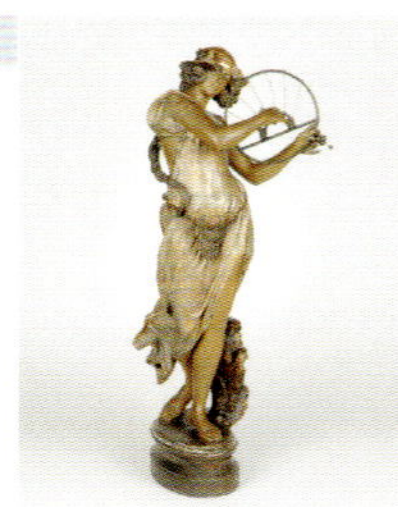

2086
FIGUR „Mode"
ca. 1900
2086 (Firmenmarke ohne Ang.), H. 90 cm; GSF, S. 22
-> vgl. 2044
Lit.: GSF, S. 22, hier mit der Ang. des Originaltitels „Mode".

2087
Cherc
HERME „Athenais"
ca. 1900
2087 (Firmenmarke ohne Ang.), H. 150 cm; GSF, S. 22
Bem.: Die hier dokumentierte Herme (hoher Pfeilerschaft, eine Art Sockel, mit aufgesetzem Kopf), wird auch im

Firmenkat. als solche bezeichnet.
-> vgl. 1993; 2254= jew. BÜSTE
-> vgl. 2158= WANDMASKE
Lit.: GSF, S. 22, hier mit der Ang. des Originaltitels „Athenais".

2088
Cherc
HERME „Theodora"
ca. 1900
2088 (Firmenmarke ohne Ang.), H. 150 cm; GSF, S. 22
Bem.: Die hier dokumentierte Herme (hoher Pfeilerschaft, eine Art Sockel, mit aufgesetzem Kopf), wird auch im Firmenkat. als solche bezeichnet.
-> vgl. 1994; 2254= jew. BÜSTE
Lit.: GSF, S. 22, hier mit der Ang. des Originaltitels „Theodora".

2089
FIGUR „Spiegelträgerin"
ca. 1900
2089 (Firmenmarke ohne Ang.), H. 80 cm; GSF, S. 22
Bem.: Lt. Firmenkatalog wurde dieses Modell auch mit 24 Lampen (Glühbirnen) montiert angeboten.
-> vgl. 2144
Lit.: GSF, S. 22, hier mit der Ang. des Originaltitels „Spiegelträgerin".

2090
Donatello, F(l)agolli A.
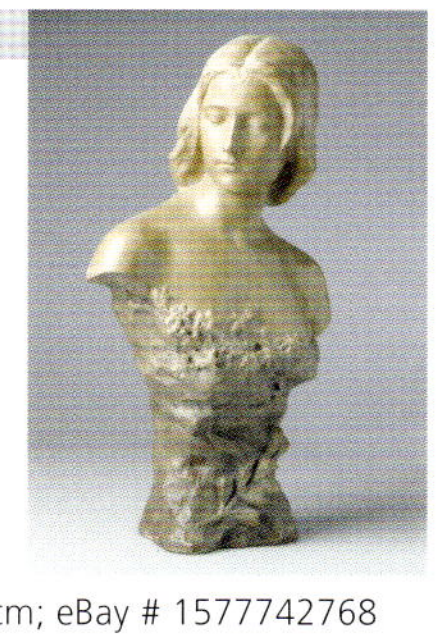
BÜSTE „Edelweiss"
(Frau nach rechts blickend, auf Natursockel mit Blattwerk und Edelweiß)
ca. 1900
2090,117,32 Firmenmarke Sculpture 13, Repr. Res., unsign.?, H. 58,5 cm; Christie's L. 17.7.1991
2090,125,32 (Firmenmarke ohne Ang.), unsign.?, H. 60 cm; eBay # 1577742768
2090,232,68 ger. IV., gem. II, Firmenmarke Sculpture 13, Repr. Res., gepr. „P"= Z3, sign. Fagolli Donatello, H. 57 cm, B. 36 cm; Sammlung mad
2090 (Firmenmarke ohne Ang.), H. 57 cm; GSF, S. 22
Ausst.: ÖMKI 1902/03, hier mit Ang. des Originaltitels „Edelweiß" und des Künstlernamens Flagolli Donatello (welcher sich jedoch in dieser Form in keinem Künstlerlexikon nachweisen lässt).
-> vgl. 2268
Lit.: GSF, S. 22, hier mit der Ang. des Originaltitels „Edelweiss"; Matschiner (2006), S. 81 (m. Abb.).

2091
Nelson, (Alphonse-Henri?)
BÜSTE „Ondine"/„Oudine" (Junge Dame mit hochgebundenem Haar)
ca. 1900
2091,122,21 Firmenmarke Sculpture 13, Repr. Res., H. 60 cm; eBay # 3708413368
2091,336,5 gem. weiß 16, Firmenmarke Sculpture 13, H. 62 cm; KH
2091 (Firmenmarke ohne Ang.), H. 60 cm; GSF, S. 22
Bem.: Diese Büste findet sich unter den Nrn. 855/918/919 im Kat. GSB wieder. Sie wurde in der Pariser Produktion in den Höhen 53/40/12 cm gefertigt. Ang. zum Material (Terrakotta/Bronze) fehlen.
Ausst.: ÖMKI 1900/01; 1902/03, hier jeweils mit Ang. des Originaltitels „Ondine".
-> vgl. 2212; 2432
-> vgl. GSB 855; 918; 919
Lit.: GSF, S. 22, hier mit der Ang. des Originaltitels „Oudine"; GSB, Pl. 99; Neuwirth (1974), S. 68.

2092
Haniroff/Xaniroff
FIGUR „Triumph"
ca. 1900
2092 (Firmenmarke ohne Ang.), H. 98 cm; GSF, S. 23
Bem.: Lt. Firmenkatalog wurde dieses Modell auch mit 12 Lampen (Glühbirnen) montiert angeboten.
Ausst.: ÖMKI 1900/01, im Ausst.-Kat. wird „Xaniroff" als Künstler genannt, hier mit Ang. des Originaltitels „Triumph".
-> vgl. 2201
Lit.: GSF, S. 23, hier mit der Ang. des Originaltitels „Triumph".

2093
Haniroff
FIGUR „Orchidee" (Stehende Frauengestalt mit Blütenschmuck) (auf Blumen geschmücktem Rundsockel)
ca. 1900
2093 (Firmenmarke ohne Ang.), H. 96 cm; GSF, S. 23
Bem.: Lt. Firmenkatalog wurde dieses Modell auch mit sechs Lampen (Glühbirnen) montiert angeboten.
-> vgl. 2277
Lit.: GSF, S. 23, hier mit der Ang. des Originaltitels „Orchidee".

2094
Cherc
BÜSTE „Flieder"
ca. 1900
2094 (Firmenmarke ohne Ang.), H. 64 cm; GSF, S. 23
Bem.: Lt. Firmenkatalog wurde dieses Modell auch mit sechs Lampen (Glühbirnen) montiert angeboten.
-> vgl. 2369; 2370
Lit.: GSF, S. 23, hier mit der Ang. des Originaltitels „Flieder"; Neuwirth (1974), S. 61.

2095
BÜSTE „Holländerin"
ca. 1900
2095 (Firmenmarke ohne Ang.), H. 52 cm; GSF, S. 23
Lit.: GSF, S. 23, hier mit der Ang. des Originaltitels „Holländerin".

2096
FIGUR „Lilie"
ca. 1900
2096 (Firmenmarke ohne Ang.), H. 94 cm; GSF, S. 23
Bem.: Lt. Firmenkatalog wurde dieses Modell auch mit sechs Lampen (Glühbirnen) montiert angeboten.
-> vgl. 2345
Lit.: GSF, S. 23, hier mit der Ang. des Originaltitels „Lilie".

2097
Donatello, F(l)agolli A.
BÜSTE „Je meurs ou je m'attache"/„Lierre" (Junge Dame mit Efeu auf Rechtecksockel)
Inschrift: JE MEURS OU JE M'ATTACHE/LIERRE
ca. 1900
2097,84,2 (Firmenmarke ohne Ang.), Repr. Res., sign. Fagoli, H. 61 cm; Soth. Olympia 1.6.2005; hier am Sockel die Inschrift: Lierre.
2097 (Firmenmarke ohne Ang.), H. fehlt; Pinhas, S. 55 (m. Abb.), hier am Sockel die Inschrift: Je meurs ou je M'attache.
2097 (Firmenmarke ohne Ang.), H. 59 cm; GSF, S. 23
Text der Inschrift: Efeu, bzw. Ich sterbe wo ich mich festhalte.
Lit.: GSF, S. 23, hier mit der Ang. des Originaltitels „Je meurs ou je m'attache".

2098
FIGUR „Blumenmädchen" (Sich verneigende, Blumen geschmückte junge Dame mit großem Strohhut)
ca. 1898/99 (für den Entwurf)
2098 (Firmenmarke ohne Ang.), H. 63 cm; GSF, S. 23
-> vgl. 1722
-> vgl. 1926= BÜSTE
Lit.: GSF, S. 23, hier mit der Ang. des Originaltitels „Blumenmädchen".

2099
FIGUR „Diana auf der Jagd"
ca. 1900
2099 (Firmenmarke ohne Ang.), H. 73 cm; GSF, S. 23
Lit.: GSF, S. 23, hier mit der Ang. des Originaltitels „Diana auf der Jagd".

2100
BÜSTE „Aschenbrödel"
ca. 1900
2100 (Firmenmarke ohne Ang.), H. 44 cm; GSF, S. 23
Bem.: Dargestellt ist Ludwig Bechsteins Märchenfigur Aschenbrödel (Aschenputtel, Cinderella).
Lit.: GSF, S. 23, hier mit der Ang. des Originaltitels „Aschenbrödel".

2101
BÜSTE „Rebecca"
ca. 1900
2101 (Firmenmarke ohne Ang.), H. 45 cm; GSF, S. 23
Lit.: GSF, S. 23, hier mit der Ang. des Originaltitels „Rebecca".

2102
FIGUR „Furcht vor Überraschung"
ca. 1900
2102 (Firmenmarke ohne Ang.), H. 62 cm; GSF, S. 23
Lit.: GSF, S. 23, hier mit der Ang. des Originaltitels „Furcht vor Überraschung".

2103
Montenave
BÜSTE „Lucrezia" (nach links blickend)
(auf konischem Sockel)
ca. 1900
2103,138,7 (Firmenmarke Sculpture 13?), Repr. Res., H. 53,7 cm; Christie's SK 18.9.2002
2103 (Firmenmarke ohne Ang.), H. 53 cm; GSF, S. 23
Bem.: Dargestellt ist Lukrezia (Lucrezia) Borgia (1480–1519), eine ital. Fürstin span. Abstammung. (Quelle: Wikipedia)
-> vgl. 2224; 2434
Lit.: GSF, S. 23, hier mit der Ang. des Originaltitels „Lucrezia".

2104
FIGUR „Malerei" (Sitzende junge Frau mit Malerpalette und Pinsel)
ca. 1899/1900
2104 (Firmenmarke ohne Ang.), H. 51 cm; GSF, S. 23
-> vgl. 1848
-> vgl. 1941= BÜSTE
Lit.: GSF, S. 23, hier mit der Ang. des Originaltitels „Malerei".

2105
FIGUR „Lautenspielerin"
ca. 1900
2105 (Firmenmarke ohne Ang.), H. 77 cm; GSF, S. 23
Lit.: GSF, S. 23, hier mit der Ang. des Originaltitels „Lautenspielerin".

2106
Carrièr-Belleuse, (Albert-Ernest)
WANDMASKE „Chriemhilde"
ca. 1900
2106 (Firmenmarke ohne Ang.), H. 44 cm; GSF, S. 23
-> vgl. 1772; 1979
Pendant -> vgl. 2107
Lit.: GSF, S. 23, hier mit der Ang. des Originaltitels „Chriemhilde".

2107
Carrièr-Belleuse, (Albert-Ernest)
WANDMASKE „Brunhilde"
ca. 1900
2107 (Firmenmarke ohne Ang.), H. 40 cm; GSF, S. 23
-> vgl. 1773; 1980
Pendant -> vgl. 2106
Lit.: GSF, S. 23, hier mit der Ang. des Originaltitels „Brunhilde".

2108
Bouval, Maurice
BÜSTE „Jvy" (Mädchen mit efeugeschmücktem Haar)
ca. 1900
2108,153,69 ger. III, Firmenmarke Sculpture 13, Repr. Res., H. 39.5 cm; KH
2108 (Firmenmarke ohne Ang.), H. 38 cm; GSF, S. 23
Ausst.: ÖMKI 1900/01, hier mit Ang. des abweichenden Originaltitels „Ivy".
Lit.: GSF, S. 23, hier mit der Ang. des Originaltitels „Jvy"; Neuwirth (1974), S. 60.

2109
Tell, E.
FIGUR (SPIEGEL) „Spiegelträgerin (bekleidet)" (Stehende Frau, einen Spiegel haltend und in diesen schauend)
ca. 1899/1900

2109 (Firmenmarke ohne Ang.), H. 71 cm; GSF, S. 23
Bem.: Diese Variante und Größe wurde lt. Firmenkatalog mit und ohne einen „elektrifizierten Zweig" angeboten.
-> vgl. 532 (ähnl. Modell)
-> vgl. 2040; 2359 (= jew. Variante, nackt)
Lit.: GSF, S. 23, hier mit der Ang. des Originaltitels „Spiegelträgerin (bekleidet)".

2110
Simon
FIGUR „Ruth" (Stehendes, sich bückendes Bauernmädchen mit Kornähren) (auf Natursockel)
ca. 1899/1900
2110,149,7 Firmenmarke Sculpture 13, Repr. Res., H. 62 cm; Mehlis 26
2110 (Firmenmarke ohne Ang.), H. 61 cm; GSF, S. 23
Ausst.: ÖMKI 1902/03, hier mit Ang. des Originaltitels „Ruth".
-> vgl. 2017 (Variante)
Lit.: GSF, S. 23, hier mit der Ang. des Originaltitels „Ruth".

2111
Tell, E.
FIGUR „Muschelträgerin (bekleidet)" (Stehende, nach vorne gebeugte bekleidete Frauengestalt, eine große Muschel haltend)
ca. 1900
2111 (Firmenmarke ohne Ang.), H. 60 cm; Pinhas, S. 63 (m. Abb.)
2111 (Firmenmarke ohne Ang.), H. 57 cm; GSF, S. 23
-> vgl. 2046 (nackt)
Lit.: GSF, S. 23, hier mit der Ang. des Originaltitels „Muschelträgerin (bekleidet)".

2112
FIGUR „Wiener Walzer" (Tanzende junge Dame, ihr langes Kleid mit beiden Armen anhebend) (auf Ovalsockel)
ca. 1899/1900
2112 (Firmenmarke ohne Ang.), H. 62 cm; GSF, S. 23
-> vgl. 1867
Lit.: GSF, S. 23, hier mit der Ang. des Originaltitels „Wiener Walzer".

2113
BÜSTE „Printemps"
ca. 1900
2113 (Firmenmarke ohne Ang.), H. 67 cm; GSF, S. 23
Lit.: GSF, S. 23, hier mit der Ang. des Originaltitels „Printemps".

2114
Vidal, Henri
FIGUR „Diana"
ca. 1899/1900
2114 (Firmenmarke ohne Ang.), H. 100 cm; GSF, S. 23
Ausst.: ÖMKI 1900/01, hier mit Ang. des Originaltitels „Diana".
-> vgl. 2009
Lit.: GSF, S. 23, hier mit der Ang. des Originaltitels „Diana"; Neuwirth (1974), S. 75.

2115
Bouval, Maurice
BÜSTE „Meditation" (Frau mit Kopftuch, auf Sockel)
Inschrift: MEDITATION
ca. 1900
2115,3,13 Firmenmarke Sculpture 13, Repr. Res., H. 43 cm; KH
2115 (Firmenmarke ohne Ang.), H. 41 cm; GSF, S. 23
Lit.: GSF, S. 23, hier mit der Ang. des Originaltitels „Meditation".

2116
FIGUR „Jeunesse"
ca. 1900
2116 (Firmenmarke ohne Ang.), H. 87 cm; GSF, S. 23
Bem.: Lt. Firmenkatalog wurde dieses Modell auch mit sechs Lampen (Glühbirnen) montiert angeboten.
-> vgl. 2264
Lit.: GSF, S. 23, hier mit der Ang. des Originaltitels „Jeunesse".

2117
FIGUR „Les Lys"
ca. 1900
2117 (Firmenmarke ohne Ang.), H. 67 cm; GSF, S. 23
Bem.: Lt. Firmenkatalog wurde dieses Modell auch mit einer Lampe (Glühbirne) montiert angeboten.
Lit.: GSF, S. 23, hier mit der Ang. des Originaltitels „Les Lys".

2118
FIGUR „Les Pavots"
ca. 1900
2118 (Firmenmarke ohne Ang.), H. 71 cm; GSF, S. 23
Bem.: Lt. Firmenkatalog wurde dieses Modell auch mit einer Lampe (Glühbirne) montiert angeboten.
Lit.: GSF, S. 23, hier mit der Ang. des Originaltitels „Les Pavots".

2119
Simon
BÜSTE „Harmonia" (Junge Frau mit Schwan als Kopfputz)
Inschrift: HARMONIA
ca. 1899 (für den Entwurf)
2119 (Firmenmarke ohne Ang.), H. 40 cm; GSF, S. 23
Ausst.: ÖMKI 1899/1900; hier mit Ang. des Originaltitels „Harmonia".
-> vgl. 1956
Lit.: GSF, S. 23, hier mit der Ang. des Originaltitels „Harmonia".

2120
Ceccarelli, Ezio
BÜSTE „Jugend"/„Jeunesse" (Lachendes nach rechts blickendes Mädchen mit Tulpenschmuck)
Inschrift: JEUNESSE
ca. 1899 (für den Entwurf)
2120 (Firmenmarke ohne Ang.), H. 29 cm; GSF, S. 23
Ausst.: ÖMKI 1899/1900, hier mit Ang. des Originaltitels „Jeunesse".
-> vgl. 1856; 2412
Lit.: GSF, S. 23, hier mit der Ang. des Originaltitels „Jeunesse".

2121
FIGUR „Ophelia" (Ophelie)
ca. 1899/1900
2121 (Firmenmarke ohne Ang.), H. 40 cm; GSF, S. 23
Bem.: Der Name Ophelie (bzw. Ophelia) bezeichnet die weibliche Hauptfigur aus William Shakespeares „Hamlet".
-> vgl. 1983; 2281
Lit.: GSF, S. 23, hier mit der Ang. des zu den beiden anderen angebotenen Größen abweichenden Originaltitels „Ophelia".

2122
Villieu
BÜSTE „Beatrice" (Junge Dame mit Diadem und Kopftuch) (auf konischem Sockel)
ca. 1899/1900
2122,135,33 (ohne Firmenmarke, bzw. ohne Ang.), Repr. Res., H. 45,5 cm; Christie's SK 4.11.1999
2122 (Firmenmarke ohne Ang.), H. 28 cm; GSF, S. 24
Bem.: Dargestellt ist Beatrice Portinari (1266–1290), die Jugendliebe des Dante Alighieri (1265–1321).
-> vgl. 1887; 2593= ähnl. BÜSTE
-> vgl. 2200 WANDMASKE
Lit.: GSF, S. 24, hier mit der Ang. des Originaltitels „Beatrice".

2123
Rust, L.
BÜSTE „Triumphator" (Römer mit Lorbeerkranz)
ca. 1898/99 (für den Entwurf)
2123 (Firmenmarke ohne Ang.), H. 28 cm; GSF, S. 24
-> vgl. 1428
Lit.: GSF, S. 24, hier mit der Ang. des Originaltitels „Triumphator".

2126
FIGUR „Austernverkäuferin"
ca. 1900
2126 (Firmenmarke ohne Ang.), H. 50 cm; GSF, S. 24
Lit.: GSF, S. 24, hier mit der Ang. des Originaltitels „Austernverkäuferin".

2127
FIGUR „Drache"
ca. 1900
2127 (Firmenmarke ohne Ang.), H. 23 cm; GSF, S. 24
Bem.: Lt. Firmenkatalog wurde dieses Modell auch mit drei Lampen (Glühbirnen) montiert angeboten.
Lit.: GSF, S. 24, hier mit der Ang. des Originaltitels „Drache".

2128
WANDMASKE „Parisienne"
ca. 1900
2128 (Firmenmarke ohne Ang.), H. 27 cm; GSF, S. 24
Lit.: GSF, S. 24, hier mit der Ang. des Originaltitels „Parisienne".

2129
FIGUR „Stier"
ca. 1900
2129 (Firmenmarke ohne Ang.), H. 35 cm; GSF, S. 24
Lit.: GSF, S. 24, hier mit der Ang. des Originaltitels „Stier".

2130
Bortone, Ipazzio Antonio / Portou
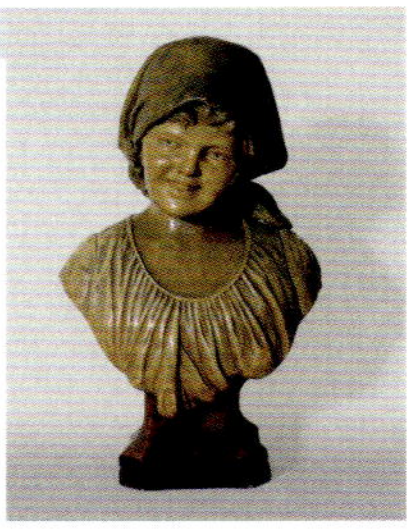
BÜSTE „Annita" (Mädchen mit Kopftuch)
ca. 1898/99 (für den Entwurf)
2130,156,11 gem. XXVI, Firmenmarke 7, Repr. Res., sign. Portou, gepr. „C"= Z4, H. 33,3 cm; PB Deutschland
2130 (Firmenmarke ohne Ang.), H. 33 cm; GSF, S. 24
-> vgl. 1740
-> vgl. 1502 (Variante)
-> vgl. 2029= WANDMASKE
Lit.: GSF, S. 24, hier mit der Ang. des Originaltitels „Annita".

2131
Donatello (nach)
BÜSTE Mädchen (Renaissancestil)
ca. 1900 (für die Ausführung)
2131,12,67 Firmenmarke Sculpture 13, Repr. Res., bezeichnet: Donatello, H. 48 cm; Doro 23.5.1991; 12.12.1991
Bem.: Vermutlich eine der im ÖMKI ausgestellten Büsten nach Donatello (1898/99 „Vornehme Dame" oder 1899/1900 „Nobel Dame").

2133 -> ÜBERSCHNEIDUNG
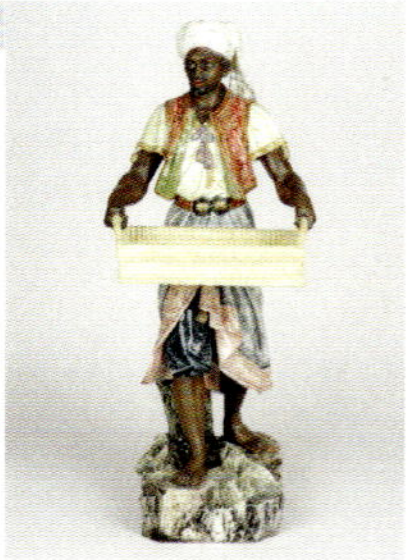
Stehender Araber mit Turban, einen Korb haltend
ca. 1900
2133 FG Wien (Firmenmarke?), H. 75 cm; PB Wien 19

2133 -> ÜBERSCHNEIDUNG
Oury, Louis
BÜSTE „L'Aiglon" (Napoleon II.)
ca. 1900
2133 (Firmenmarke ohne Ang.), H. 37 cm; GSF, S. 24
-> vgl. 2066
Bem.: Die hier dargestellte Person ist Napoleon II. (1811–1832).
Lit.: GSF, S. 24, hier mit der Ang. des Originaltitels „L'Aiglon".

2134
Cherc
FIGUR „Negerknabe (fischend)"
ca. 1898/99 (für den Entwurf)
2134 (Firmenmarke ohne Ang.), H. 43 cm; GSF, S. 24
-> vgl. 1470
Lit.: GSF, S. 24, hier mit der Ang. des Originaltitels „Negerknabe (fischend)".

2135
Cherc
FIGUR „Raucher" (Stehender Negerjunge mit Hut und Zündholzschachtel)
ca. 1898/99 (für den Entwurf)
2135 (Firmenmarke ohne Ang.), H. 42 cm; GSF, S. 24
-> vgl. 1471
Lit.: GSF, S. 24, hier mit der Ang. des Originaltitels „Raucher".

2136
Cherc
FIGUR „Zerbrochener Krug" (Stehender Negerjunge mit Hut und zerbrochener Vase)
ca. 1898/99 (für den Entwurf)
2136 (Firmenmarke ohne Ang.), H. 42 cm; GSF, S. 24
-> vgl. 1472
Lit.: GSF, S. 24, hier mit der Ang. des abweichenden Originaltitels „Zerbrochener Krug".

2137
Cherc
FIGUR „Gladiator" (Römischer Krieger mit Helm, Schild und Schwert) (auf Ovalsockel)
ca. 1899/1900
2137,4,9 Firmenmarke Sculpture 13, Repr. Res., H. 68,6 cm; eBay # 21769698
2137 (Firmenmarke ohne Ang.), H. 66; GSF, S. 24
-> vgl. 1833
-> vgl. 2033= BÜSTE
Lit.: GSF, S. 24, hier mit der Ang. des Originaltitels „Gladiator".

2138
Cherc
FIGUR „Bogenspanner" (Römischer Bogenschütze in Rüstung) (auf Ovalsockel)
ca. 1899/1900
2138 (Firmenmarke ohne Ang.), H. 65; GSF, S. 24
-> vgl. 1834
-> vgl. 2034= BÜSTE
Lit.: GSF, S. 24, hier mit der Ang. des Originaltitels „Bogenspanner".

2139
BÜSTE „Schwerttänzerin"
ca. 1900
2139 (Firmenmarke ohne Ang.), H. 50 cm; GSF, S. 24
-> vgl. 1790; 1883= jew. FIGUR
Lit.: GSF, S. 24, hier mit der Ang. des Originaltitels „Schwerttänzerin".

2140
BÜSTE „La Guerre"
ca. 1900
2140 (Firmenmarke ohne Ang.), H. 49 cm; GSF, S. 24
-> vgl. 1953; 2182; 2307= jew. FIGUR
Lit.: GSF, S. 24, hier mit der Ang. des Originaltitels „La Guerre".

2141
BÜSTE „Walküre"
ca. 1900
2141 (Firmenmarke ohne Ang.), H. 52 cm; GSF, S. 24
Lit.: GSF, S. 24, hier mit der Ang. des Originaltitels „Walküre".

2142
BÜSTE „Todtenkopf auf Buch"
ca. 1900
2142 (Firmenmarke ohne Ang.), H. 11 cm; GSF, S. 24
Lit.: GSF, S. 24, hier mit der Ang. des Originaltitels „Todtenkopf auf Buch".

2144
FIGUR „Spiegelträgerin"
ca. 1900
2144 (Firmenmarke ohne Ang.), H. 63 cm; GSF, S. 24
-> vgl. 2089
Lit.: GSF, S. 24, hier mit der Ang. des Originaltitels „Spiegelträgerin".

2145
FIGUR „Muschelträgerin"
ca. 1900
2145 (Firmenmarke ohne Ang.), H. 71 cm; GSF, S. 24
-> vgl. 2145a
Lit.: GSF, S. 24, hier mit der Ang. des Originaltitels „Muschelträgerin".

2145a
LAMPE „Muschelträgerin"
ca. 1900
2145a (Firmenmarke ohne Ang.), H. 71 cm; GSF, S. 24
-> vgl. 2145
Lit.: GSF, S. 24, hier mit der Ang. des Originaltitels „Muschelträgerin", sowie dem Vermerk: mit „überhängenden Zapfen, elektr. montiert mit drei Lampen".

2146
FIGUR „Spiegelfigur, knieend"
ca. 1900
2146 (Firmenmarke ohne Ang.), H. 100 cm; GSF, S. 24
Lit.: GSF, S. 24, hier mit der Ang. des Originaltitels „Spiegelfigur, knieend".

2147
Maison, Rudolf
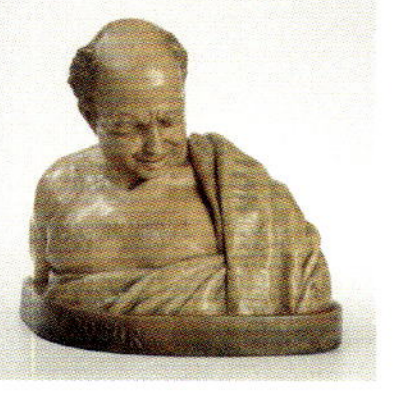
BÜSTE „Augur"
Inschrift: AUGUR
ca. 1900
2147,24,19 gem. XXI, innen ger. II, Firmenmarke Sculpture 13, Repr. Res., H. 38 cm, B. 41,1 cm; KH (Antiquitäten Baumann, Regensburg)
2147,55,13 ger. III, Firmenmarke Sculpture 13, Repr. Res., gepr. „C"= Z5, Ausformung ca. 1900, H. 37 cm; Wien Museum Inv. 186.609
2147,70,13 Firmenmarke Sculpture 13, H. 37,1 cm; Doro 15.9.2003
2147 (Firmenmarke ohne Ang.), H. 17 (sic!) cm; GSF, S. 24; Bem.: Die hier angegebene Höhe 17 cm dürfte ein Druckfehler sein da die nachgewiesenen Stücke ca. 37 cm hoch sind.
Bem.: Die Figur des „Augur" (altrömischer Priester) wurde 1890 entworfen. Dies ließ sich durch ein sign. und datiertes Exemplar „R. Maison 1890" (Gipsausformung, H. 53 cm) im KH nachweisen. Die Version als Büste dürfte Maison speziell für GS entworfen haben und muss um 1900 datiert werden.
Ausst.: ÖMKI 1898/99, hier mit Ang. des Originaltitels „Augur".
-> vgl. 1447= FIGUR
Lit.: GSF, S. 24, hier mit der Ang. des Originaltitels „Augur"; Schmid, M.: Rudolf Maison, in: VKMH, 1.1898/99, S. 8 (m. Abb.), hier mit dem Titel „Römischer Augur"; N.N.: Wiener Kleinplastik einst und jetzt, in: MW, 2.1920/21 Nr. 12, S. 29 (m. Abb.), hier mit Ang. des Originaltitels „Augur"; Neuwirth (1974), S. 67.

2148
Corbeille
FIGUR „Der Traum"
ca. 1900
2148 (Firmenmarke ohne Ang.), H. 74 cm; GSF, S. 24
Bem.: Lt. Firmenkatalog wurde dieses Modell auch mit einer Lampe (Glühbirne) montiert angeboten.
Ausst.: ÖMKI 1901/02 hier mit Ang. des Originaltitels „Der Traum".
-> vgl. 2214= BÜSTE
Lit.: GSF, S. 24, hier mit der Ang. des Originaltitels „Der Traum"; Neuwirth (1974), S. 61.

2149
Tell, E.
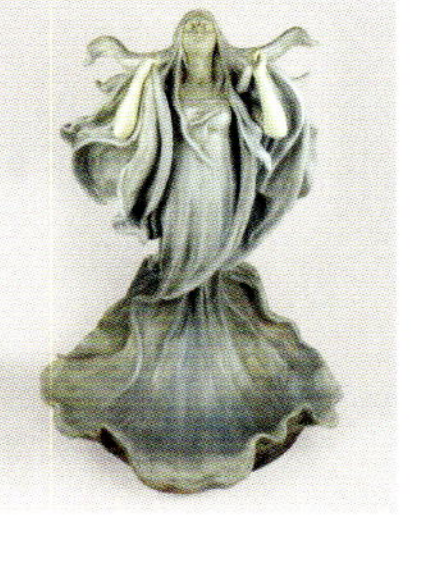
FIGUR „Die Nacht" (Junge Frau in wallendem Gewand das sich zu einer Schale formt)
ca. 1900
2149,22,15 Firmenmarke Sculpture 13, Repr. Res., gepr. „P"= Z3, H. 67 cm; PB Wien 19
2149 (Firmenmarke ohne Ang.), H. 68 cm; GSF, S. 24
Lit.: GSF, S. 24, hier mit der Ang. des Originaltitels „Die Nacht".

2150
Lemassi
WANDMASKE „Römischer Krieger"
ca. 1900
2150,32,20 GS (Firmenmarke ohne Ang.), H. 25,5 cm; Soth. Olympia 29.3.2003
2150,66,16 gepr. H, Firmenmarke 7, Repr. Res., gepr. „P"= Z3, H. 25 cm; Wien Museum Inv. 220.785
2150 (Firmenmarke ohne Ang.), H. 22 cm; GSF, S. 24
Ausst.: ÖMKI 1901/02 hier mit Ang. des Originaltitels „Römischer Krieger".
Bem.: Dargestellt ist eigentlich der römische Kaiser Augustus (63 v. Chr.–14 n. Chr.).
-> vgl. 2152= BÜSTE
Lit.: GSF, S. 24, hier mit der Ang. des Originaltitels „Römischer Krieger"; Neuwirth (1974), S. 66.

2151
BÜSTE „Antoinette"
ca. 1900
2151 (Firmenmarke ohne Ang.), H. 75 cm; GSF, S. 24
Lit.: GSF, S. 24, hier mit der Ang. des Originaltitels „Antoinette".

2152
Lemassi
BÜSTE „Römischer Krieger"
Inschrift: CIVIS ROMANUS SUM
ca. 1900
ohne Ang., (Firmenmarke ohne Ang.), H. 51 cm; Zeller 41
2152 (Firmenmarke ohne Ang.), H. 57 cm; GSF, S. 24
Ausst.: ÖMKI 1901/02 hier mit Ang. des Originaltitels „Römischer Krieger".
Bem.: Dargestellt ist eigentlich der römische Kaiser Augustus (63 v. Chr.–14 n. Chr.).
-> vgl. 2150= WANDMASKE
Lit.: GSF, S. 24, hier mit der Ang. des Originaltitels „Römischer Krieger".

2153
FIGUR „Flötenspielerin" (Stehendes junges Mädchen, eine große Flöte spielend) (auf Ovalsockel)
ca. 1899/1900
2153 (Firmenmarke ohne Ang.), H. 63 cm; GSF, S. 24
-> vgl. 1893
Lit.: GSF, S. 24, hier mit der Ang. des Originaltitels „Flötenspielerin".

2154
Waderé, Heinrich
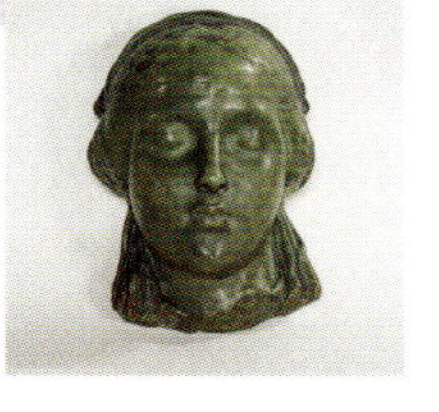
WANDMASKE „Madonnenkopf" (Frauenkopf mit Kopftuch)
ca. 1898
2154,37,23 gem. II, Firmenmarke 7, Repr. Res., gepr. „P"= Z3, H. 26 cm; Wien Museum Inv. 220.786
2154,57,23 Firmenmarke 7, Repr. Res., gepr. „P"= Z3, H. 25,3 cm; MAK Wien, Inv. Ke 10602-103
2154 (Firmenmarke ohne Ang.), H. 24 cm; GSF, S. 24
Bem.: Vermutlich gleiches Modell auch als BÜSTE Madonnenkopf (vgl. ÖMKI 1900/01) bzw. BÜSTE Madonna (ÖMKI 1906/07).
Ausst.: ÖMKI 1898/99, hier mit Ang. des Originaltitels „Madonnenkopf" und des falschen Künstlernamens „Wadré"; ÖMKI 1901/02.
Lit.: GSF, S. 24, hier mit der Ang. des Originaltitels „Madonnenkopf"; Neuwirth (1974), S. 75, 76.

2155
FIGUR „Ikarus"
ca. 1900
2155 (Firmenmarke ohne Ang.), H. 78 cm; GSF, S. 24
Bem.: Lt. Firmenkatalog wurde dieses Modell auch mit vier Lampen (Glühbirnen) montiert angeboten. Dargestellt ist die nackte Figur des geflügelten Ikarus (griech. Ikaros), der Sohn des Dädalus aus der griech. Sagenwelt.
Lit.: GSF, S. 24, hier mit der Ang. des Originaltitels „Ikarus".

2156
FIGUR „Sandalenbinderin" (stehend)
ca. 1899/1900
2156 (Firmenmarke ohne Ang.), H. 73 cm; GSF, S. 24
-> vgl. 1978
Lit.: GSF, S. 24, hier mit der Ang. des Originaltitels „Sandalenbinderin"; Pinhas, S. 97 (m. Abb.).

2157
FIGUR „Wasserträger" (Junger Bursche, ein großes Wassergefäß tragend) (auf Landschaftssockel)
ca. 1899
2157 (Firmenmarke ohne Ang.), H. 58 cm; GSF, S. 24
-> vgl. 1758
Lit.: GSF, S. 24, hier mit der Ang. des Originaltitels „Wasserträger".

2158
Cherc
WANDMASKE „Athenais" (Frauenkopf mit Blumenschmuck im Haar)
ca. 1900
2158,92,23 Firmenmarke 7, Repr. Res., gepr. „P"= Z3, H. 35 cm; eBay # 663936150
2158,123,23 GS (Firmenmarke ohne Ang.), Repr. Res., H. 35 cm; Doro L 13.5.2004; Soth. Olympia 1.6.2005
2158 (Firmenmarke ohne Ang.), H. 34 cm; GSF, S. 25
-> vgl. 1993; 2254= jew. BÜSTE
-> vgl. 2087= HERME
Pendant -> vgl. 2159
Lit.: GSF, S. 25, hier mit der Ang. des Originaltitels „Athenais".

2159
Cherc
WANDMASKE „Ranke" (Frauenkopf mit Dutt und Blumenschmuck im Haar)
ca. 1900
2159,74,18 Firmenmarke Sculpture 13, Repr. Res., H. 34 cm; eBay # 6639359654
2159,100,67 GS (Firmenmarke ohne Ang.), Repr. Res., H. 35 cm; Doro L 13.5.2004; Soth. Olympia 1.6.2005
2159 (Firmenmarke ohne Ang.), H. 33 cm; GSF, S. 25
-> vgl. 1946; 2195= jew. BÜSTE
Pendant -> vgl. 2158
Lit.: GSF, S. 25, hier mit der Ang. des Originaltitels „Ranke".

2160
FIGUR „Jagdbeute"
ca. 1900/01
2160 (Firmenmarke ohne Ang.), H. 57 cm; GSF, S. 25
Lit.: GSF, S. 25, hier mit der Ang. des Originaltitels „Jagdbeute".

2161
Baudry, R. E.
BÜSTE „Silence"
ca. 1900/01
2161 (Firmenmarke ohne Ang.), H. 26 cm; GSF, S. 25
Ausst.: ÖMKI 1901/02 hier mit Ang. des Originaltitels „Silence".
Lit.: GSF, S. 25, hier mit der Ang. des Originaltitels „Silence"; Neuwirth (1974), S. 59.

2162
BÜSTE „Mignon"
ca. 1900/01
2162 (Firmenmarke ohne Ang.), H. 48 cm; GSF, S. 25
Lit.: GSF, S. 25, hier mit der Ang. des Originaltitels „Mignon".

2163
FIGUR „Pythia"
ca. 1900/01
2163 (Firmenmarke ohne Ang.), H. 93 cm; GSF, S. 25
Bem.: Dargestellt ist „Pythia", eine Priesterin, die im Tempel von Delphi den Ratsuchenden weissagte. (Quelle: Wikipedia).
Lit.: GSF, S. 25, hier mit der Ang. des Originaltitels „Pythia".

2164
BÜSTE „Marguerite"
ca. 1900/01
2164 (Firmenmarke ohne Ang.), H. 51 cm; GSF, S. 25
Lit.: GSF, S. 25, hier mit der Ang. des Originaltitels „Marguerite".

2166
Della Robbia (nach)
BÜSTE Mädchen mit Haarband
ca. 1900 (für die Ausführung)
2166,18,25 Firmenmarke 7, Repr. Res., bezeichnet: Della Robbia, H. 33 cm; eBay # 300019498528

2170
FIGUR „Abendstern"
ca. 1900/01
2170 (Firmenmarke ohne Ang.), H. 90 cm; GSF, S. 25
Bem.: Lt. Firmenkatalog wurde dieses Modell auch mit einer Lampe (Glühbirne) montiert angeboten.
Lit.: GSF, S. 25, hier mit der Ang. des Originaltitels „Abendstern".

2171
Cherc
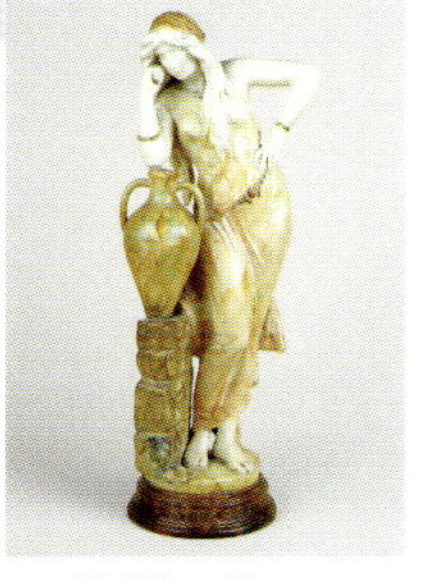
FIGUR „Rebecca" (Stehende Frauengestalt neben einer Mauer, ihren Ellenbogen auf eine Amphore stützend)
Inschrift: REBECCA
ca. 1900/01
2171,132,70 Firmenmarke Sculpture 13, H. 63,3 cm; Wien Museum Inv. 219.115
2171,565,24 Firmenmarke Sculpture 13, Repr. Res., H. 63 cm; PB Wien 19
2171,873,14 ger., ohne Firmenmarke, H. 62 cm; Wien Museum Inv. 219.125; Bem.: Ausführung in Terrakotta.
2171 (Firmenmarke ohne Ang.), H. 62 cm; GSF, S. 25
-> vgl. 1903; 2537
Lit.: GSF, S. 25, hier mit der Ang. des Originaltitels „Rebecca".

2172
FIGUR „Maske mit Muschel und Spiegel"
ca. 1900/01
2172 (Firmenmarke ohne Ang.), H. 65 cm; GSF, S. 25
Lit.: GSF, S. 25, hier mit der Ang. des Originaltitels „Maske mit Muschel und Spiegel".

2173
BÜSTE „Sommeil"
ca. 1900/01
2173 (Firmenmarke ohne Ang.), H. 49 cm; GSF, S. 25
Bem.: Lt. Firmenkatalog wurde dieses Modell auch mit einer Lampe (Glühbirne) montiert angeboten.
Lit.: GSF, S. 25, hier mit der Ang. des Originaltitels „Sommeil".

2175
FIGUR „Spiegelträgerin"
ca. 1900/01
2175 (Firmenmarke ohne Ang.), H. 82 cm; GSF, S. 25
Lit.: GSF, S. 25, hier mit der handschriftlichen Ang. des Originaltitels „Spiegelträgerin".

2176
Sydan / Tell, E.
FIGUR „Le Parfum" (Stehende Frauengestalt mit wallendem Tuch und großer (Mohn?-) Blume)
ca. 1900/01
2176,26,34 Firmenmarke Sculpture 13, sign. E. Tell, H. 73,5 cm; Doro 12.9.2006; Bem.: Lampe.
2176 (Firmenmarke ohne Ang.), H. 73 cm; GSF, S. 25
Bem.: Lt. Firmenkatalog wurde dieses Modell auch mit einer Lampe (Glühbirne) montiert angeboten.
Ausst.: ÖMKI 1901/02, hier mit Ang. des Originaltitels „Le Parfum", sowie des Künstlernamens „Sydan".
Lit.: GSF, S. 25, hier mit der Ang. des Originaltitels „Le Parfum"; Neuwirth (1974), S. 75.

2177
Bertrand, A.
BÜSTE „Sonnenblume" (Junge Frau, nach links blickend, geschmückt mit Sonnenblumen)
ca. 1900/01
2177 (Firmenmarke ohne Ang.), unsign., H. 72,5 cm; Pinhas, S. 53 (m. Abb.)
2177 (Firmenmarke ohne Ang.), H. 73 cm; GSF, S. 25
Bem.: Lt. Firmenkatalog wurde dieses Modell auch mit vier Lampen (Glühbirnen) montiert angeboten.
-> vgl. 2296; 2458
Lit.: GSF, S. 25, hier mit der Ang. des Originaltitels „Sonnenblume".

2178
Haniroff / Xaniroff
BÜSTE „La Fierté"
Inschrift: LA FIERTÉ
ca. 1900/01
2178,23,13 (Firmenmarke ohne Ang.), sign. Haniroff, H. 65 cm; Rammel 65
2178,79,21 ger. 16 Firmenmarke Sculpture 13, Repr. Res., sign. Haniroff, H. 63,5 cm; Soth. com # 6VP4; Soth. com # 3543N
2178,220,67 Firmenmarke Sculpture 13, Repr. Res., sign. Haniroff, H. 68,6 cm; eBay # 523880239
2178,293,2 Firmenmarke Sculpture 13, Repr. Res., sign. Haniroff, H. 65 cm; eBay # 2837144289
2178 (Firmenmarke ohne Ang.), H. 62 cm; GSF, S. 25
Text der Inschrift: Der Stolz
Ausst.: ÖMKI 1901/02, hier mit Ang. des Originaltitels „La Fierté" und des falschen Künstlernamens „Xaniroff".
-> vgl. 2371; 2435
Lit.: GSF, S. 25, hier mit der Ang. des Originaltitels „La Fierté".

2179
FIGUR „Loreley"
ca. 1900/01
2179 (Firmenmarke ohne Ang.), H. 78 cm; GSF, S. 25
Lit.: GSF, S. 25, hier mit der Ang. des Originaltitels „Loreley".

2180
Simon
GRUPPE „Phantasie" (Nackte Frauengestalt (mit Harfe) auf Adler thronend)
ca. 1899 (für den Entwurf)
2180 (Firmenmarke ohne Ang.), H. 63 cm; GSF, S. 25
Ausst.: ÖMKI 1899/1900, hier mit Ang. des Originaltitels „Die Phantasie".
-> vgl. 1943
Lit.: GSF, S. 25, hier mit der Ang. des Originaltitels „Phantasie".

2181
FIGUR „Schalenträgerin"
ca. 1900/01
2181 (Firmenmarke ohne Ang.), H. 65 cm; GSF, S. 25
Lit.: GSF, S. 25, hier mit der Ang. des Originaltitels „Schalenträgerin".

2182
FIGUR „Der Kampf" (Stehender römischer Krieger in Rüstung mit Schild, Schwert und Axt) (auf Landschaftssockel)
ca. 1900
2182 (Firmenmarke ohne Ang.), H. 71 cm; GSF, S. 25
-> vgl. 1953; 2307
-> vgl. 2140= BÜSTE
Lit.: GSF, S. 25, hier mit der Ang. des Originaltitels „Der Kampf".

2183
BÜSTE „Salambo" (Junge Dame mit Hochsteckfrisur und Medaillon um den Hals)
Inschrift: SALAMBO
ca. 1899 (für den Entwurf)
2183 (Firmenmarke ohne Ang.), H. 43 cm; GSF, S. 25
Bem.: Dargestellt ist die Priesterin Salambo, die Tochter des punischen Heerführers Hamilkars und Schwester Hannibals. Gustave Flaubert (1821–1864) widmete ihr auch einen gleichnamigen Roman, der 1863 erschien und dann 1904 ins dt. übersetzt wurde.

-> vgl. 1755; 2430
Lit.: GSF, S. 25, hier mit der Ang. des Originaltitels „Salambo".

2184
Rozel
BÜSTE „Gladiator"
ca. 1898 (für den Entwurf)
2184 (Firmenmarke ohne Ang.), H. 38 cm; GSF, S. 25
Ausst.: ÖMKI 1898/99, hier mit Ang. des Originaltitels Büste „Gladiator".
-> vgl. 1687; 2294
-> vgl. GRUPPE „Animus et vis"= Modell 3680
Lit.: GSF, S. 25, hier mit der Ang. des Originaltitels „Gladiator"; Neuwirth (1974), S. 72.

2185
WANDMASKE „Beethoven, lebende Maske"
ca. 1900/01
2185 (Firmenmarke ohne Ang.), H. 21 cm; GSF, S. 25
-> vgl. 2186
Lit.: GSF, S. 25, hier mit der Ang. des Originaltitels „Beethoven, lebende Maske".

2186
WANDMASKE „Beethoven, Todtenmaske"
ca. 1900/01
2186 (Firmenmarke ohne Ang.), H. 23 cm; GSF, S. 25
-> vgl. 2185
Lit.: GSF, S. 25, hier mit der Ang. des Originaltitels „Beethoven, Todtenmaske".

2187
Simon
BÜSTE „Pax" (Junge Frau mit Diadem aus Lorbeer)
Inschrift: PAX
ca. 1900/01
2187 (Firmenmarke ohne Ang.), H. 61 cm; Pinhas, S. 55 (m. Abb.)
2187 (Firmenmarke ohne Ang.), H. 60 cm; GSF, S. 25
Ausst.: ÖMKI 1901/02, hier mit Ang. des Originaltitels „Pax".
-> vgl. 2280; 2433
Lit.: GSF, S. 25, hier mit der Ang. des Originaltitels „Pax"; Neuwirth (1974), S. 74.

2188
Lenoir, Pierre Charles
FIGUR „Muschelträgerin" (Stehendes Blumen geschmücktes junges Mädchen, eine große Muschel haltend (auf Ovalsockel)
ca. 1900
2188 (Firmenmarke ohne Ang.), H. 90 cm; GSF, S. 25
-> vgl. 1945; 2221
-> vgl. 2063= BÜSTE
Lit.: GSF, S. 25, hier mit der Ang. des Originaltitels „Muschelträgerin"; N.N.: Illustrierte Rundschau, Terrakotten von Fr. Goldscheider, Wien, in: VKMH, 17.1902/03, S. 717 (m. Abb.), hier mit Ang. des Originaltitels und des Künstlernamens Lenoir.

2189
BÜSTE „Le vin nouveau"
ca. 1900/01
2189 (Firmenmarke ohne Ang.), H. 66 cm; GSF, S. 25
Lit.: GSF, S. 25, hier mit der Ang. des Originaltitels „Le vin nouveau".

2190
BÜSTE Junge Dame (Renaissancestil)
ca. 1900/01
2190,63,36 Firmenmarke 9, H. 21,6 cm; eBay # 2630742168

2191
Cherc
FIGUR „Die Ernte"
ca. 1900/01
2191 (Firmenmarke ohne Ang.), H. 88 cm; GSF, S. 25
Ausst.: ÖMKI 1902/03, hier mit Ang. des Originaltitels „Ernte".
-> vgl. 2278
-> vgl. 2303= BÜSTE
Lit.: GSF, S. 25, hier mit der Ang. des Originaltitels „Die Ernte"; Neuwirth (1974), S. 61.

2192
Ceccarelli, Ezio
BÜSTE „Ave virgo"
ca. 1900/01
2192 (Firmenmarke ohne Ang.), H. 58 cm; GSF, S. 25
Ausst.: ÖMKI 1901/02, hier mit Ang. des Originaltitels „Ave virgo".
Lit.: GSF, S. 25, hier mit der Ang. des Originaltitels „Ave virgo"; Neuwirth (1974), S. 61.

2193
FIGUR „Wasserblume"
ca. 1900/01
2193 (Firmenmarke ohne Ang.), H. 70 cm; GSF, S. 25
Bem.: Lt. Firmenkatalog wurde dieses Modell auch mit einer Lampe (Glühbirne) montiert angeboten.
Lit.: GSF, S. 25, hier mit der Ang. des Originaltitels „Wasserblume".

2194
FIGUR „Libelle"
ca. 1900/01
2194 (Firmenmarke ohne Ang.), H. 68 cm; GSF, S. 25
Bem.: Lt. Firmenkatalog wurde dieses Modell auch mit drei Lampen (Glühbirnen) montiert angeboten.
Lit.: GSF, S. 25, hier mit der Ang. des Originaltitels „Libelle".

2195
Cherc / Thiele
BÜSTE „Ranke" (Blumengeschmückte junge Frau, nach rechts blickend)
ca. 1900
2195 (Firmenmarke ohne Ang.), sign. Thiele, H. 47 cm; Pinhas, S. 53 (m. Abb.)
2195 (Firmenmarke ohne Ang.), H. 49 cm; GSF, S. 25
Ausst.: ÖMKI 1901/02, hier mit Ang. des Originaltitels „Ranke".
-> vgl. 1946
-> vgl. 2159= WANDMASKE
Lit.: GSF, S. 25, hier mit der Ang. des Originaltitels „Ranke".

2196
Waderé, Heinrich
BÜSTE „Madonna"
ca. 1900/01
2196,6,39 (Firmenmarke ohne Ang.), unsign.?, H. 31,5 cm; KH
2196 (Firmenmarke ohne Ang.), H. 31 cm; GSF, S. 25
Ausst.: ÖMKI 1906/07, hier mit Ang. des Originaltitels „Madonna" und des Künstlers Waderé.
Lit.: GSF, S. 25, hier mit der Ang. des Originaltitels „Madonna"; Neuwirth (1974), S. 76.

2197
BÜSTE „Doppelmaske"
ca. 1900/01
2197 (Firmenmarke ohne Ang.), H. fehlt; GSF, S. 26
Lit.: GSF, S. 26, hier mit der Ang. des Originaltitels „Doppelmaske".

2198
GRUPPE „Leopard und Krokodil"
ca. 1900/01
2198 (Firmenmarke ohne Ang.), H. 18/39 cm; GSF, S. 26
Bem.: Lt. Firmenkatalog wurden wohl zwei Ausführungen offeriert, eine davon „naturalistisch bemalt".
Lit.: GSF, S. 26, hier mit der Ang. des Originaltitels „Leopard und Krokodil".

2199
FIGUR „Sonnenaufgang" (Stehende junge Frau auf Felssockel)
ca. 1900/01
2199 (Firmenmarke ohne Ang.), H. 104 cm; GSF, S. 26
Bem.: Lt. Firmenkatalog wurde dieses Modell auch mit zwei Lampen (Glühbirnen) montiert angeboten.
-> vgl. 2199b (Variante bekleidet)
Lit.: GSF, S. 26, hier mit der Ang. des Originaltitels „Sonnenaufgang".

2199b
FIGUR „Sonnenaufgang (bekleidet)" (Stehende junge Frau mit wallendem Tuch auf Felssockel)
ca. 1900/01
2199b (Firmenmarke ohne Ang.), H. 104 cm; GSF, S. 26
Bem.: Lt. Firmenkatalog wurde dieses Modell auch mit zwei Lampen (Glühbirnen) montiert angeboten. Der Zusatz „bekleidet" hier lässt vermuten, dass es sich bei der die Mod.-Nr. 2199 um die „nackte" Variante handeln müsste.
-> vgl. 2199 (Variante nackt?)
-> vgl. 2368
Lit.: GSF, S. 26, hier mit der Ang. des Originaltitels „Sonnenaufgang (bekleidet)".

2200
Villieu
WANDMASKE „Beatrice" (Frauenkopf mit Diadem und Kopftuch)
ca. 1900/01
2200,32,23 Firmenmarke 7, Repr. Res., gepr. „P"= Z3, H. 22 cm; GK, S. 14, 74 (m. Abb.); Wien Museum Inv. 182.070
2200,41,27 Firmenmarke 7, Repr. Res., gepr. „P"= Z3, unsign.? H. 22,8 cm; eBay # 2040796851
2200 (Firmenmarke ohne Ang.), H. 22 cm; GSF, S. 26
Bem.: Dargestellt ist Beatrice Portinari (1266–1290), die Jugendliebe des Dante Alighieri (1265–1321).
Ausst.: ÖMKI 1901/02; 1902/03 jeweils mit Ang. des Originaltitels „Beatrice"; HMW (jetzt Wien Museum) 1985/86.
-> vgl. 1887; 2122; 2593= jeweils BÜSTE
Lit.: GSF, S. 26, hier mit der Ang. des Originaltitels „Beatrice"; GK, S. 14, 74 (m. Abb.).

2201
Haniroff / Xaniroff
FIGUR „Triumph"
ca. 1900
2201 (Firmenmarke ohne Ang.), H. 87 cm; GSF, S. 26
Ausst.: ÖMKI 1900/01, im Ausst.-Kat. wird „Xaniroff" als Künstler genannt, hier mit Ang. des Originaltitels „Triumph".
-> vgl. 2092
Lit.: GSF, S. 26, hier mit der Ang. des Originaltitels „Triumph".

2202
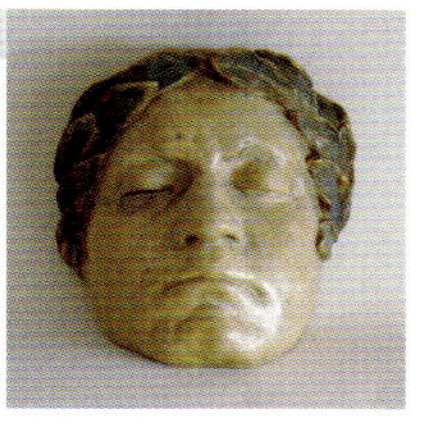
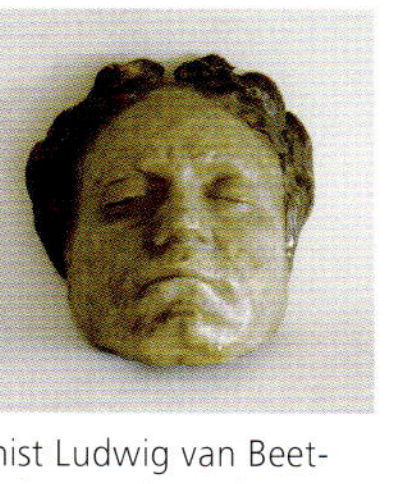
WANDMASKE „Beethoven mit Lorberkranz" (Totenmaske mit Lorbeerkranz)
ca. 1900/01
2202,27,10 Firmenmarke 10, H. 20,5 cm; Sammlung John Nash
2202,123,39 FG Wien (Firmenmarke?), Repr. Res., H. 24 cm; eBay # 3323774394
2202,231,8 Firmenmarke Sculpture 13, Ges. gesch., H. 24 cm; Sammlung John Nash
2202 (Firmenmarke ohne Ang.), H. 20 cm; GSF, S. 26
Bem.: Dargestellt ist der Komponist Ludwig van Beethoven (1770–1827) nach der durch Joseph Danhauser abgenommenen Totenmaske. Diese Vorlage wurde durch einen Lorbeerkranz ergänzt (vgl. auch das Bild von Franz von Stuck „Beethoven"). Wie die bis in die 50er Jahre ausgeformten Modelle belegen, war diese Maske beim Publikum sehr gefragt und wurde daher zwischen 1900–1960 von GS produziert. Da dieses Modell vor allem am Lorbeerkranz, aber auch an der Hautstruktur individuell gestaltet und staffiert wurde, ergeben sich die teils erheblichen Größenunterschiede und variantenreichen Ausführungen.
Ausst.: Turin 1902
-> vgl. 2285
Lit.: GSF, S. 26, hier mit der Ang. des Originaltitels „Beethoven mit Lorbeerkranz"; Pudor (1902a), S. 119; KGSF1, S. (16) (m. Abb.); KGSF2, S. (102) (m. Abb.); Könecke, S. 56 (m. Abb.) 76, 79 (mit Künstlerang. Lorenzl, dieser

Hinweis kann sich jedoch nur auf die Ausformung bei GS Tönnieshof beziehen, wo die Grundform der Maske evtl. von Josef Lorenzl minimal überarbeitet wurde; der eigentliche Entwurf stammt eindeutig nicht von Lorenzl).

2203
FIGUR „Schneeglöckchen"
ca. 1900/01
2203 (Firmenmarke ohne Ang.), H. 70 cm; GSF, S. 26
Lit.: GSF, S. 26, hier mit der Ang. des Originaltitels „Schneeglöckchen".

2204
FIGUR „Pferd"
ca. 1900/01
2204 (Firmenmarke ohne Ang.), H. 27 cm; GSF, S. 26
Lit.: GSF, S. 26, hier mit der Ang. des Originaltitels „Pferd".

2205
Bertrand, A.
FIGUR „Paradieswächter"
ca. 1900/01
2205 (Firmenmarke ohne Ang.), H. 103 cm; GSF, S. 26
Ausst.: ÖMKI 1901/02, hier mit Ang. des Originaltitels „Paradieswächter".
Lit.: GSF, S. 26, hier mit der Ang. des Originaltitels „Paradieswächter"; Neuwirth (1974), S. 59.

2206
FIGUR „Wasserträgerin"
ca. 1900
2206 (Firmenmarke ohne Ang.), H. 80 cm; GSF, S. 26
-> vgl. 2078
Lit.: GSF, S. 26, hier mit der Ang. des Originaltitels „Wasserträgerin".

2207
Cherc
BÜSTE „Tulpe"
ca. 1900/01
2207 (Firmenmarke ohne Ang.), H. 71 cm; GSF, S. 26
Ausst.: ÖMKI 1901/02; 1902/03, hier mit Ang. des Originaltitels „Tulpe".
-> vgl. 2363; 2367
Lit.: GSF, S. 26, hier mit der Ang. des Originaltitels „Tulpe"; Neuwirth (1974), S. 61.

2208
Laurana, Francesco (nach)
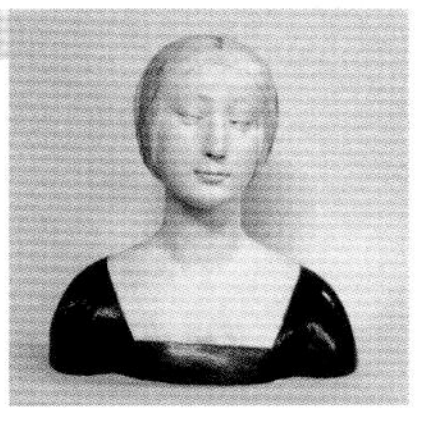
BÜSTE „Isabella de Aragon"
ca. 1900 (für die Ausführung)
2208,5,69 Firmenmarke 9, Repr. Res., H. 43 cm; MAK Wien, Inv. Ke 10602-94
Bem.: Dargestellt ist Isabella de Aragon (1430–1502?). GS übernimmt die im KHM Wien befindliche und in Marmor ausgeformte Büste Lauranas, deren H. 44 cm beträgt (vgl. John Pope-Hennessy: An introduction to Italian sculpture, Vol. 2, London 1996).

2209
Montenave
BÜSTE „Fides"
ca. 1900/01
2209 (Firmenmarke ohne Ang.), H. 60 cm; GSF, S. 26
Ausst.: ÖMKI 1901/02, hier mit Ang. des Originaltitels „Fides".
Lit.: GSF, S. 26, hier mit der Ang. des Originaltitels „Fides"; Neuwirth (1974), S. 68.

2210
GRUPPE „Der Blitz"
ca. 1900/01
2210 (Firmenmarke ohne Ang.), H. 106 cm; GSF, S. 26
Bem.: Lt. Firmenkatalog wurde dieses Modell auch mit einer Lampe (Glühbirne) montiert angeboten.
Lit.: GSF, S. 26, hier mit der Ang. des Originaltitels „Der Blitz".

2211
Tell, E.
Stehender Frauenhalbakt (umgeben von wucherndem Blattwerk und Sonnenblume)
ca. 1900/01
2211? (Firmenmarke ohne Ang.), H. 85 cm; Soth. L. 19.12.1986

2212
Nelson, (Alphonse-Henri?)
BÜSTE „Ondine"/„Oudine" (Junge Dame mit hochgebundenem Haar)
ca. 1900
2212,261,27 Firmenmarke Sculpture 13, Repr. Res., H. 42 cm; Christie's SK 15.10.2003; WKA 52
2212 (Firmenmarke ohne Ang.), H. 42 cm; GSF, S. 26
Bem.: Diese Büste findet sich unter den Nrn. 855/918/919 im Kat. GSB wieder. Sie wurde in der Pariser Produktion in den Höhen 53/40/12 cm gefertigt. Ang. zum Material (Terrakotta/Bronze) fehlen.
Ausst.: ÖMKI 1900/01; 1902/03, hier mit Ang. des Originaltitels „Ondine".
-> vgl. 2091; 2432
-> vgl. GSB 855; 918; 919
Lit.: GSF, S. 26, hier mit der Ang. des Originaltitels „Oudine"; GSB, Pl. 99; Neuwirth (1974), S. 68.

2213
Villieu
BÜSTE „Lilie"/„Pureté" (Junge Frau in antikisierendem Gewand mit Blumenkranz im Haar, Dekolleté mit Lilienblüten geschmückt)
Inschrift: PURETÉ
ca. 1899/1900 (für den Entwurf)
2213,62,32 Firmenmarke Sculpture 13, Repr. Res., H. 45 cm; Nagel 381
2213 (Firmenmarke ohne Ang.), H. 45 cm; GSF, S. 26
Text der Inschrift: Reinheit
Ausst.: ÖMKI 1900/01, hier im Ausst.-Kat. unter dem Titel „Bureté" (!) aufgeführt.
-> vgl. 1981; 2411
Lit.: GSF, S. 26, hier mit der Ang. des Originaltitels „Lilie".

2214
Corbeille
BÜSTE „Der Traum"
ca. 1900/01
2214 (Firmenmarke ohne Ang.), H. 72 cm; GSF, S. 26
-> vgl. 2148= FIGUR
Lit.: GSF, S. 26, hier mit der Ang. des Originaltitels „Traum".

2215
Tell, E.
FIGUR „Schmetterling" (Nackte Frau mit Schmetterlingsflügeln, an einem Blütenkelch stehend)
ca. 1900/01
2215,102,33 (Firmenmarke ohne Ang.), H. 79 cm; Soth. L. 17.6.1988
2215 (Firmenmarke ohne Ang.), H. 75 cm; GSF, S. 26
Bem.: Lt. Firmenkatalog wurde dieses Modell auch mit einer Lampe (Glühbirne) montiert angeboten, bzw. auch mit vier Glasschirmen (lt. handschriftl. Vermerk, GSF, S. 26).
Lit.: GSF, S. 26, hier mit der Ang. des Originaltitels „Schmetterling"; B., F.: Keramische Wanderungen durch den Kunstsalon Keller & Reiner Berlin, in: KM, 3.1903, S. 125–128 (m. Abb.), hier mit Ang. des Titels „Schmetterlingsweibchen", Der graziöse Entwurf zu dem Schmetterlingsweibchen, das neugierig in eine große Callablüte schaut stammt von dem Bildhauer Tell. Im Innern der Blüte befindet sich eine elektrische Glühlampe, die aufflammend das feine Körperchen des Schmetterlings mit leuchtenden Reflexen bestrahlt".

2216
Carrièr-Belleuse, A(lbert-Ernest)
FIGUR „Harmonie" (Sitzende junge Dame mit Lyra) (auf Rechtecksockel)
ca. 1898/99 (für den Entwurf)
2216 (Firmenmarke ohne Ang.), H. 52 cm; GSF, S. 26
-> vgl. 1610; 1925= BÜSTE
Lit.: GSF, S. 26, hier mit der Ang. des Originaltitels „Harmonie".

2217
Cherc
FIGUR „Tägliches Brod"
ca. 1900 (für den Entwurf)
2217 (Firmenmarke ohne Ang.), H. 60 cm; GSF, S. 26
Bem.: Warum die Manufaktur für dieses Modell zwei fast ident. aber eben doch abweichende Titel (vgl. 2053) vergab, bleibt unklar. In den Ausst.-Katalogen des ÖMKI erscheinen beide Ausformungen, die in den beiden aufeinander folgenden Jahren gezeigt wurden ebenfalls mit diesen Titeln.
Ausst.: ÖMKI 1901/02, hier mit Ang. des Originaltitels „Tägliches Brod".
-> vgl. 2053
Lit.: GSF, S. 26, hier mit der Ang. des Originaltitels „Tägliches Brod"; Neuwirth (1974), S. 61.

2218
Stellmacher, (Eduard?)
BÜSTE „Erwachen"
ca. 1900/01
2218 (Firmenmarke ohne Ang.), H. 76 cm; GSF, S. 26
Ausst.: ÖMKI 1901/02, hier mit Ang. des Originaltitels „Erwachen".
-> vgl. 2343
Lit.: GSF, S. 26, hier mit der Ang. des Originaltitels „Erwachen"; Neuwirth (1974), S. 74.

2219
FIGUR „La Lumière"
ca. 1900/01
2219 (Firmenmarke ohne Ang.), H. 65 cm; GSF, S. 26
Lit.: GSF, S. 26, hier mit der Ang. des Originaltitels „La Lumière".

2220
Cherc
„Die Saat" (Säende Bäuerin)
ca. 1895 (für den Entwurf)
2220,885,25 Firmenmarke Sculpture 13, Ausformung nach 1900, H. 58 cm; Kaisersl. 1987, S. 69
2220 (Firmenmarke ohne Ang.), H. 57 cm; GSF, S. 26
Ausst.: ÖMKI 1899/1900, hier mit Ang. des Originaltitels „Die Saat".
Pendant -> vgl. 571; 1935
Lit.: GSF, S. 26, hier mit der Ang. des Originaltitels „Saat".

2221
Lenoir, Pierre Charles
FIGUR „Muschelträgerin" (Stehendes Blumen geschmücktes junges Mädchen, eine große Muschel haltend (auf Ovalsockel)
ca. 1900
2221 (Firmenmarke ohne Ang.), H. 62 cm; GSF, S. 26
-> vgl. 1945; 2188
-> vgl. 2063= BÜSTE
Lit.: GSF, S. 26, hier mit der Ang. des Originaltitels „Muschelträgerin"; N.N.: Illustrierte Rundschau, Terrakotten von Fr. Goldscheider, Wien, in: VKMH, 17.1902/03, S. 717 (m. Abb.), hier mit Ang. des Originaltitels und des Künstlernamens Lenoir.

2222
Langer, (Leopold?)/Longer
FIGUR (SPIEGEL) „Bei der Toilette" (Sitzender Mädchenakt in einen Spiegel schauend) (auf Landschaftssockel mit Lilien/Callas)
ca. 1900/01
2222 (Firmenmarke ohne Ang.), Gesamth. 66 cm; Pinhas, S. 60 (m. Abb.)
2222 (Firmenmarke ohne Ang.), H. 57 cm; GSF, S. 26
Bem.: Der in der Lit. angegebene Artikel weist auf den nicht näher identifizierbaren Künstler „Longer" hin.
Lit.: GSF, S. 26, hier mit der Ang. des Originaltitels „Bei der Toilette"; B., F.: Keramische Wanderungen durch den Kunstsalon Keller & Reiner Berlin, in: KM, 3.1903, S. 128, „Vor dem Spiegel, der all ihre Reize zurückgibt, steckt sie sich Blüten, die sie von dem Rahmen aufspriessenden Gerank abgebrochen hat, in ihr üppiges Haargewoge. Ihre hingeschmiegten Glieder sind von plastischer Schönheit. Ihre siegesbewußte Haltung verrät den Triumph, den sie darüber empfindet".

2223
Haniroff
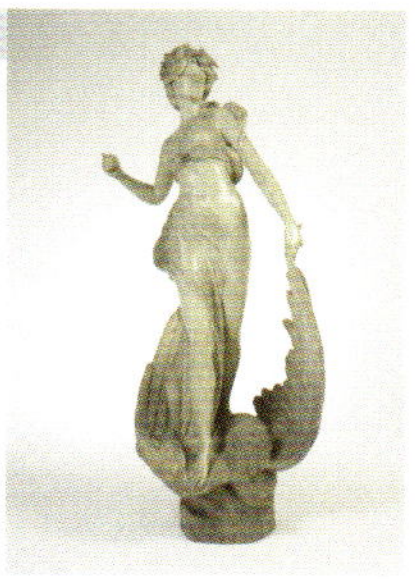
FIGUR „Jagdtrophäe"
(Leda besiegt den Schwan)
ca. 1900/01
2223,45,8 gem. weiß 17, Firmenmarke Sculpture 13, Repr. Res., sign. Haniroff, gepr. „P"= Z3, Originalmetalletikette: Bentejac, Cours de Tourny, Bordeaux, H. 114 cm; GK, S.14, 76 (m. Abb.); Wien Museum Inv. 223.904; Bem.: Bei diesem Exemplar fehlt der Speer, den Leda in der Hand hält.
2223 (Firmenmarke ohne Ang.), H. 112 cm; GSF, S. 26
Ausst.: HMW (jetzt Wien Museum) 1985/86.
–> vgl. 2444
Lit.: GSF, S. 26, hier mit der Ang. des Originaltitels „Jagdtrophäe"; GK, S. 14, 76 (m. Abb.).

2224
Montenave
BÜSTE „Lucrezia"
Inschrift: LUCRECE
ca. 1900
2224,28,2 Firmenmarke Sculpture 13, Repr. Res., undeutl. sign. „Montenave", H. 38 cm; KH; Bem.: Diese bislang einzig nachgewiesene Büste ist mit dem franz. Titel „Lucrece" bezeichnet.
2224 (Firmenmarke ohne Ang.), H. 38 cm; GSF, S. 26
Bem.: Dargestellt ist Lukrezia (Lucrezia) Borgia (1480–1519), eine ital. Fürstin span. Abstammung.
–> vgl. 2103; 2434
Lit.: GSF, S. 26, hier mit der Ang. des Originaltitels „Lucrezia".

2225 –> ÜBERSCHNEIDUNG
Strasser, Arthur
FIGUR „Janitschare" (mit Turban, Schild, Säbel und langem Stab) (auf Landschaftssockel)
ca. 1898/99 (für den Entwurf)
2225 (Firmenmarke ohne Ang.), H. 43 cm; GSF, S. 26
Bem.: Ein „Janitschar" ist ein Krieger (Infantrie) der Elitetruppen des osmanischen Reichs, bzw. ein Leibwächter des Sultans (Quelle: Wikipedia).
–> vgl. 1652; 2043
–> vgl. 1921= BÜSTE
Lit.: GSF, S. 26, hier mit der Ang. des Originaltitels „Janitschare".

2225 –> ÜBERSCHNEIDUNG
Cherc
BÜSTE Frau mit Kastanien(?) und Blattwerk (Flora?)
ca. 1900/01
2225,23,34 gem. XXI, Firmenmarke Sculpture 13, Repr. Res., H. 67,3 cm; eBay # 472789705

2226
Simon
BÜSTE „Loreley" (Junge Frau mit Adler als Kopfputz)
ca. 1899 (für den Entwurf)
2226 (Firmenmarke ohne Ang.), H. 33 cm; GSF, S. 26
–> vgl. 1956
Lit.: GSF, S. 26, hier mit der Ang. des Originaltitels „Loreley".

2227
FIGUR „Fleur de Pensee"
ca. 1900/01
2227 (Firmenmarke ohne Ang.), H. 117 cm; GSF, S. 26
Lit.: GSF, S. 26, hier mit der Ang. des Originaltitels „Fleur de Pensee".

2228
GRUPPE „Najaden"
ca. 1900/01
2228 (Firmenmarke ohne Ang.), H. 55 cm; GSF, S. 27
Bem.: Najaden sind Nymhen der griech. Mythologie und Töchter des Zeus oder des Okeanos. (Quelle: Wikipedia).
Lit.: GSF, S. 27, hier mit der Ang. des Originaltitels „Najaden".

2229
Simon
UHR „Frühling" (Stehende Frauengestalt vor stilisiertem Kirschbaum)
ca. 1901
2229 (Firmenmarke ohne Ang.), H. 95 cm; Pinhas, S. 58 (m. Abb.)
2229 (Firmenmarke ohne Ang.), H. 95 cm; GSF, S. 27
Bem.: Lt. Firmenkatalog wurde dieses Modell auch „elektr. montiert" also mit Lampe angeboten, bzw. war es auch ohne das Uhrwerk lieferbar. Dies erklärt auch, warum in den diversen Goldscheideruhren sich gelegentlich auch Uhrwerke von anderen Firmen als Lenzkirch finden.
Ausst.: ÖMKI 1901/02, hier mit Ang. des Originaltitels „Frühling".
–> vgl. 2299
Lit.: GSF, S. 27, hier mit der Ang. des Originaltitels „Frühling"; Pudor (1902), S. 71 (m. Abb.); B., F.: Keramische Wanderungen durch den Kunstsalon Keller & Reiner Berlin, in: KM, 3.1903, S. 128, „Dieser Künstler versteht es gut zu stilisieren und seine jedesmaligen Motive geschmackvoll dem gewählten Sujet anzupassen"; Neuwirth (1974), S. 74; Weltkunst, 50.1980, S. 1504 (m. Abb.).

2230
FIGUR „Am Bachesrande"
ca. 1901
2230 (Firmenmarke ohne Ang.), H. 62 cm; GSF, S. 27
Bem.: Lt. Firmenkatalog wurde dieses Modell auch „elektr. montiert" mit einer, zwei oder drei Lampen angeboten.
Lit.: GSF, S. 27, hier mit der Ang. des Originaltitels „Am Bachesrande".

2231
Simon
FIGUR „Carmen"
ca. 1901
2231 (Firmenmarke ohne Ang.), H. 71 cm; GSF, S. 27
Ausst.: ÖMKI 1901/02, hier mit Ang. des Originaltitels „Carmen".
Lit.: GSF, S. 27, hier mit der Ang. des Originaltitels „Carmen"; Neuwirth (1974), S. 74.

2232
SPIEGEL „Träumerei"
ca. 1901
2232 (Firmenmarke ohne Ang.), H. 53 cm; GSF, S. 27
Bem.: Lt. Firmenkatalog wurde dieses Modell auch ohne den eigentl. dazugehörigen Spiegel angeboten.
Lit.: GSF, S. 27, hier mit der Ang. des Originaltitels „Träumerei".

2233
FIGUR „Scheikh"
ca. 1900
2233 (Firmenmarke ohne Ang.), H. 51 cm; GSF, S. 27
Bem.: Dargestellt ist ein sog. Scheich, ein Stammesoberhaupt der Beduinen. (Quelle: Wik pedia).
–> vgl. 1917; 2071
Lit.: GSF, S. 27, hier mit der Ang. des Originaltitels „Scheikh".

2234
Haniroff / Xaniroff
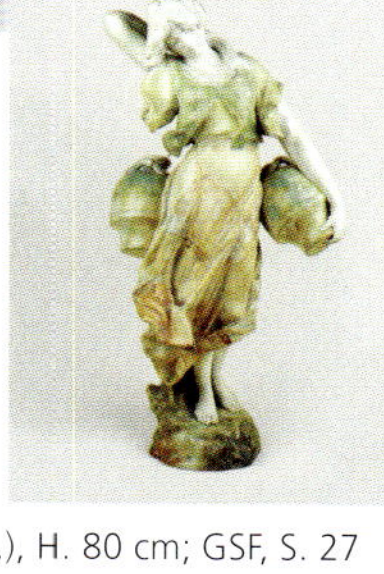
FIGUR „Der Wind"
(Stehendes Mädchen mit zwei Krügen)
ca. 1901
2234,494,16 Firmenmarke Sculpture 13, Repr. Res., sign. Haniroff, H. 85 cm; GK, S. 14, 75 (m. Abb.); Wien Museum Inv. 182.071
2234 (Firmenmarke ohne Ang.), H. 80 cm; GSF, S. 27
Ausst.: ÖMKI 1901/02, hier mit Ang. des Originaltitels „Der Wind"; HMW (jetzt Wien Museum) 1985/86.
–> vgl. 2452
Lit.: GSF, S. 27, hier mit der Ang. des Originaltitels „Der Wind"; KuKHW, 1902, S. 27; Neuwirth (1974), S. 76; GK, S. 14, 75 (m. Abb.).

2235
Fischer, Friedrich
BÜSTE „Renaissance"
ca. 1901
2235 (Firmenmarke ohne Ang.), H. 40 cm; GSF, S. 27
Ausst.: ÖMKI 1903/04, hier mit Ang. des Originaltitels „Renaissance".
Lit.: GSF, S. 27, hier mit der Ang. des Originaltitels „Renaissance"; Neuwirth (1974), S. 63.

2236
Montenave
BÜSTE „Sphinx"
ca. 1901
2236 (Firmenmarke ohne Ang.), H. 51 cm; GSF, S. 27
Ausst.: ÖMKI 1901/02, hier mit Ang. des Originaltitels „Sphinx".
Lit.: GSF, S. 27, hier mit der Ang. des Originaltitels „Sphinx"; Neuwirth (1974), S. 68.

2237
Cárdenas
WANDMASKE „Odorat" (Frauenkopf mit langem geschwungenem Haar)
ca. 1901
2237,34,23 Firmenmarke 7, Repr. Res., H. 53,3 cm; eBay # 5359668952
2237 (Firmenmarke ohne Ang.), H. 51 cm; GSF, S. 27
Ausst.: ÖMKI 1901/02, hier mit Ang. des Originaltitels „Odorat".
Lit.: GSF, S. 27, hier mit der Ang. des Originaltitels „Odorat"; Neuwirth (1974), S. 60.

2238
Haniroff
FIGUR „Vasenträgerin" (Stehende junge Frau eine Amphore auf der rechten Schulter tragend) (auf Landschaftssockel)
ca. 1901
2238 (Firmenmarke ohne Ang.), H. 122 cm; Pinhas, S. 43 (m. Abb.)
2238 (Firmenmarke ohne Ang.), H. 118 cm; GSF, S. 27
–> vgl. 2409; 2583
Lit.: GSF, S. 27, hier mit der Ang. des Originaltitels „Vasenträgerin".

2239
Powolny, Michael
UHR „Die Zeit" (Stehende Frauengestalt mit Sanduhr)
ca. 1901
2239 (Firmenmarke ohne Ang.), H. 70 cm; GSF, S. 27
Bem.: Lt. Firmenkatalog war dieses Modell auch ohne das Uhrwerk lieferbar. Dies erklärt auch, warum in den diversen Goldscheideruhren sich gelegentlich auch Uhrwerke von anderen Firmen als Lenzkirch finden.
Ausst.: Paris, Salon (1902), hier mit Ang. des franz. Originaltitels „Le Temps".
Lit.: GSF, S. 27, hier mit der Ang. des Originaltitels „Die Zeit"; KM, 1903, S. 126, 128 (m. Abb.), „Powoley(!) betitelt seine sehr große, schöne Frauengestalt, die starr und unerbittlich auf die Uhr in ihrer Hand blickt, „Die Zeit". Das grausam Unaufhaltsame der rinnenden Stunden hat der bildende Künstler als tiefen Ernst über seine Arbeit ausgegossen. Als Symbol der flüchtigen Lebensfreuden legte er schnell ein paar schnellverblühende Rosen in die Hände, welche unerbittlich die Zeitenuhr (Sanduhr, Anm. des Autors) wenden"; Neuwirth (1974), S. 71.

2240
Müllner, Josef
FIGUR „Rautendelein" (Mädchen, an einem Teich kniend)
ca. 1901
2240 (ohne Ang.), Firmenmarke Sculpture 13, Repr. Res., unsign.?, H. fehlt; L. 41 cm; Nagel 361
2240 (Firmenmarke ohne Ang.), H. 31 cm; GSF, S. 27
Bem.: Die Märchenfigur Rautendelein ist ein elfenhaftes Wesen aus Gerhart Hauptmanns Märchendrama „Die versunkene Glocke" (1896). Lt. Firmenkatalog wurde dieses Modell auch „elektr. montiert" also mit Lampe angeboten, bzw. war es auch mit und ohne einen Spiegel lieferbar.
Ausst.: ÖMKI 1902/03, hier mit Ang. des Originaltitels „Rautendelein".
Lit.: GSF, S. 27, hier mit der Ang. des Originaltitels „Rautendelein"; Neuwirth (1974), S. 38.

2241
FIGUR „Katze"
ca. 1901
2241 (Firmenmarke ohne Ang.), H. 52 cm; GSF, S. 27
Bem.: Lt. Firmenkatalog wurde dieses Modell auch „elektr. montiert" angeboten.
Lit.: GSF, S. 27, hier mit der Ang. des Originaltitels „Katze".

2242
Simon
UHR „Psyche"
Inschrift: PSYCHE
ca. 1901
2242,11,69 (Firmenmarke ohne Ang.), H. 48 cm; Rammel 65
2242,50,25 Firmenmarke Sculpture 13, Repr. Res., H. 47 cm; KHW um 1900, Berlin (1986), S. 184
2242 (Firmenmarke ohne Ang.), H. 45 cm; GSF, S. 27
Bem.: Lt. Firmenkatalog war dieses Modell auch ohne das Uhrwerk lieferbar. Dies erklärt auch, warum in den diversen Goldscheideruhren sich gelegentlich auch Uhrwerke von anderen Firmen als Lenzkirch finden.
Lit.: GSF, S. 27, hier mit der Ang. des Originaltitels „Psyche"; Pudor (1902), S. 70 (m. Abb.).

2243
BÜSTE „Kirsche"
ca. 1901
2243 (Firmenmarke ohne Ang.), H. 72 cm; GSF, S. 27
Lit.: GSF, S. 27, hier mit der Ang. des Originaltitels „Kirsche".

2244
FIGUR „Zephir"
ca. 1901
2244 (Firmenmarke ohne Ang.), H. 76 cm; GSF, S. 27
Bem.: Wie der Firmenkat. belegt, wurde dieses Modell auch mit „zwei Lampen elektr. montirt" geliefert.
Lit.: GSF, S. 27, hier mit der Ang. des Originaltitels „Zephir".

2245
BÜSTE „Innocence"
ca. 1901
2245 (Firmenmarke ohne Ang.), H. 38 cm; GSF, S. 27
Lit.: GSF, S. 27, hier mit der Ang. des Originaltitels „Innocence".

2246
SPIEGEL „Träumerei"
ca. 1901
2246 (Firmenmarke ohne Ang.), H. 58 cm; GSF, S. 27
Bem.: Lt. Firmenkatalog war dieses Modell auch ohne Spiegel lieferbar.
Lit.: GSF, S. 27, hier mit der Ang. des Originaltitels „Träumerei".

2247
FIGUR „Fischerin"
ca. 1901
2247 (Firmenmarke ohne Ang.), H. 75 cm; GSF, S. 27
Lit.: GSF, S. 27, hier mit der Ang. des Originaltitels „Fischerin".

2248
FIGUR „Weinlese"
ca. 1901
2248 (Firmenmarke ohne Ang.), H. 108 cm; GSF, S. 27
Bem.: Wie der Firmenkat. belegt, wurde dieses Modell auch „mit Glastraube elektr. montirt" geliefert.
Lit.: GSF, S. 27, hier mit der Ang. des Originaltitels „Weinlese".

2249
BÜSTE „Tulpe"
ca. 1901
2249 (Firmenmarke ohne Ang.), H. 49 cm; GSF, S. 27
Lit.: GSF, S. 27, hier mit der Ang. des Originaltitels „Tulpe".

2250
Griser
FIGUR „Perle"
ca. 1901
2250 (Firmenmarke ohne Ang.), H. 64 cm; GSF, S. 28
Bem.: Wie der Firmenkat. belegt, wurde dieses Modell auch „elektr. montirt mit nat(ürlicher) Muschel" geliefert.
Ausst.: ÖMKI 1902/03, hier mit Ang. des Originaltitels „Perle".
Lit.: GSF, S. 28, hier mit der Ang. des Originaltitels „Perle"; Neuwirth (1974), S. 63.

2251
FIGUR „Tänzerin"
ca. 1901
2251 (Firmenmarke ohne Ang.), H. 91 cm; GSF, S. 28
Lit.: GSF, S. 28, hier mit der Ang. des Originaltitels „Tänzerin".

2252
BÜSTE „Orangenblüthe"
ca. 1901
2252 (Firmenmarke ohne Ang.), H. 60 cm; GSF, S. 28
-> vgl. 2399
Lit.: GSF, S. 28, hier mit der Ang. des Originaltitels „Orangenblüthe".

2253
Cherc
BÜSTE „Theodora" (mit Helm)
ca. 1899/1900
2253 (Firmenmarke ohne Ang.), H. 44 cm; GSF, S. 28
-> vgl. 1994
-> vgl. 2088= HERME
Lit.: GSF, S. 28, hier mit der Ang. des Originaltitels „Theodora".

2254
Cherc
BÜSTE „Athenais" (Blumengeschmückte junge Frau, nach links blickend)
ca. 1899/1900
2254 (Firmenmarke ohne Ang.), H. 44 cm; GSF, S. 28
Ausst.: ÖMKI 1901/02, hier mit Ang. des Originaltitels „Athenais".
-> vgl. 1993
-> vgl. 2087= HERME
-> vgl. 2158= WANDMASKE
Lit.: GSF, S. 28, hier mit der Ang. des Originaltitels „Athenais".

2255
Cherc
BÜSTE „Kastanie" (Junge Frau mit Kastanienzweigen und Blattwerk)
ca. 1901
2255 (Firmenmarke ohne Ang.), H. 67 cm; GSF, S. 28
-> vgl. 2410
Lit.: GSF, S. 28, hier mit der Ang. des Originaltitels „Kastanie".

2256
BÜSTE „Mercedes"
ca. 1901
2256 (Firmenmarke ohne Ang.), H. 62 cm; GSF, S. 28
Lit.: GSF, S. 28, hier mit der Ang. des Originaltitels „Mercedes".

2257
FIGUR „Gesang"
ca. 1901
2257 (Firmenmarke ohne Ang.), H. 76 cm; GSF, S. 28
Lit.: GSF, S. 28, hier mit der Ang. des Originaltitels „Gesang".

2258
FIGUR „Der Morgen"
ca. 1901
2258 (Firmenmarke ohne Ang.), H. 89 cm; GSF, S. 28
Bem.: Wie der Firmenkat. belegt, wurde dieses Modell auch „mit Porzellanblume", bzw. mit „einer Lampe elektr. montirt" geliefert.
Lit.: GSF, S. 28, hier mit der Ang. des Originaltitels „Der Morgen".

2259
Tell, E.
FIGUR (SPIEGEL) „Beim Spiegel" (Stehende, sich an einen Spiegel lehnende Frau, umgeben von Blattwerk)
ca. 1901
2259,58,– Firmenmarke Sculpture 13, gepr. „P"= Z3, H. 55,9 cm; eBay # 2588839528
2259 (Firmenmarke ohne Ang.), H. 58 cm; GSF, S. 28
Bem.: Wie der Firmenkat. belegt, wurde dieses Modell auch als Figur „ohne Spiegel" geliefert.
Lit.: GSF, S. 28, hier mit der Ang. des Originaltitels „Beim Spiegel".

2260
GRUPPE „Sternschnuppe"
ca. 1901
2260 (Firmenmarke ohne Ang.), H. 88 cm; GSF, S. 28
Bem.: Wie der Firmenkat. belegt, wurde dieses Modell auch „elektr. montirt mit einer Lampe" geliefert.
Lit.: GSF, S. 28, hier mit der Ang. des Originaltitels „Sternschnuppe".

2261
FIGUR „Wandspiegel"
ca. 1901
2261 (Firmenmarke ohne Ang.), H. 34 cm; GSF, S. 28
Bem.: Wie der Firmenkat. belegt, wurde dieses Modell auch als Figur „ohne Spiegel" geliefert.
Lit.: GSF, S. 28, hier mit der Ang. des Originaltitels „Wandspiegel".

2262
BÜSTE „Juanita"
ca. 1901
2262 (Firmenmarke ohne Ang.), H. 55 cm; GSF, S. 28
Lit.: GSF, S. 28, hier mit der Ang. des Originaltitels „Juanita".

2263
FIGUR „Comödie"
ca. 1901
2263 (Firmenmarke ohne Ang.), H. 76 cm; GSF, S. 28
Bem.: Wie der Firmenkat. belegt, wurde dieses Modell auch mit einer „elektr. montirten Terracottamaske" geliefert.
Lit.: GSF, S. 28, hier mit der Ang. des Originaltitels „Comödie".

2264
FIGUR „Jeunesse"
ca. 1900
2264 (Firmenmarke ohne Ang.), H. 65 cm; GSF, S. 28
-> vgl. 2116
Lit.: GSF, S. 28, hier mit der Ang. des Originaltitels „Jeunesse".

2265
Cherc
BÜSTE „Sumpfblume" (Blumengeschmückte junge Frau nach rechts blickend)
ca. 1899/1900
2265 (Firmenmarke ohne Ang.), H. 44 cm; GSF, S. 28
-> vgl. 1902
Lit.: GSF, S. 28, hier mit der Ang. des Originaltitels „Sumpfblume".

2266
FIGUR „Fischerin"
ca. 1901
2266 (Firmenmarke ohne Ang.), H. 75 cm; GSF, S. 28
Lit.: GSF, S. 28, hier mit der Ang. des Originaltitels „Fischerin".

2267
FIGUR „Schleiertanz"
ca. 1901
2267 (Firmenmarke ohne Ang.), H. 78 cm; GSF, S. 28
Bem.: Wie der Firmenkat. belegt, wurde dieses Modell auch „elektr. montirt mit einer Lampe" geliefert.
Lit.: GSF, S. 28, hier mit der Ang. des Originaltitels „Schleiertanz".

2268
Donatello, F(l)agolli A.
BÜSTE „Edelweiss" (Frau nach rechts blickend, auf Natursockel mit Blattwerk und Edelweiß)
ca. 1900/01
2268,80,26 gem. weiß XIX, Firmenmarke Sculpture 13, Repr. Res., gepr. „P"= Z3, sign. A. Fagolli Donatello (!), H. 39,2 cm; Doro 10.3.2003
2268 (Firmenmarke ohne Ang.), H. 38 cm; GSF, S. 28
Ausst.: ÖMKI 1902/03, hier mit Ang. des Originaltitels „Edelweiss" und des Künstlernamens Flagolli, Donatello (welcher sich jedoch in dieser Form in keinem Künstlerlexikon nachweisen lässt).
-> vgl. 2090
Lit.: GSF, S. 28, hier mit der Ang. des Originaltitels „Edelweiss".

2269
BÜSTE „Tänzerin"
ca. 1901/02
2269 (Firmenmarke ohne Ang.), H. 70 cm; GSF, S. 28
Lit.: GSF, S. 28, hier mit der Ang. des Originaltitels „Tänzerin".

2270
BÜSTE „Goldelse"
ca. 1901/02
2270 (Firmenmarke ohne Ang.), H. 69 cm; GSF, S. 28
Lit.: GSF, S. 28, hier mit der Ang. des Originaltitels „Goldelse".

2271
BÜSTE „Dedain"
ca. 1901/02
2271 (Firmenmarke ohne Ang.), H. 58 cm; GSF, S. 28
Lit.: GSF, S. 28, hier mit der Ang. des Originaltitels „Dedain".

2272
Bertrand, A.

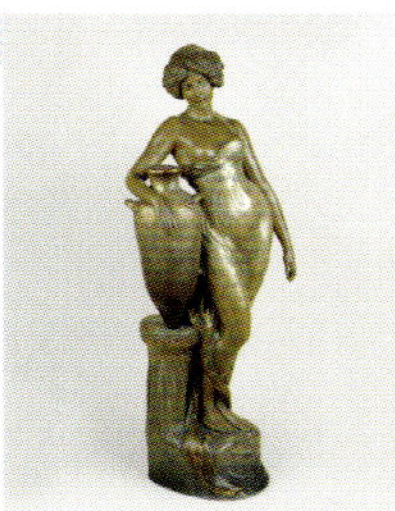

FIGUR „Träumerei" (Stehende Frau mit Amphore)
ca. 1901/02
2272 ger., (2272,101,19 gepr.) ger. II, Firmenmarke 7, Repr. Res., H. 92,5 cm; GK, S. 15, 36 (m. Farbabb.); Wien Museum Inv. 223.905
2272 (Firmenmarke ohne Ang.), H. 92 cm; GSF, S. 28
Ausst.: HMW (jetzt Wien Museum) 1985/86.
-> vgl. 2443
Lit.: GSF, S. 28, hier mit der Ang. des Originaltitels „Träumerei"; GK, S. 15, 36 (m. Abb.); GSN, S. 8, 9 (m. Abb.).
GS neu: Modell 63 „La Tendresse" (H. 82 cm), Aufl. 1000 Stück

2273
FIGUR „Affe im Topf"
ca. 1901/02
2273 (Firmenmarke ohne Ang.), H. 23 cm; GSF, S. 28
Lit.: GSF, S. 28, hier mit der Ang. des Originaltitels „Affe im Topf".

2274
Tell, E.
FIGUR „Krugträgerin" (Wasserträgerin mit Gefäß auf dem Kopf)
ca. 1899 (für den Entwurf)
2274 (Firmenmarke ohne Ang.), H. 71 cm; GSF, S. 28
-> vgl. 1769
Lit.: GSF, S. 28, hier mit der Ang. des Originaltitels „Krugträgerin".

2275
FIGUR „Shocking"
ca. 1897/98 (für den Entwurf)
2275 (Firmenmarke ohne Ang.), H. 47 cm; GSF, S. 28
-> vgl. 1170
Lit.: GSF, S. 28, hier mit der Ang. des Originaltitels „Sho(c)king".

2276
FIGUR „Cellospielerin"
ca. 1901/02
2276 (Firmenmarke ohne Ang.), H. 67 cm; GSF, S. 28
Lit.: GSF, S. 28, hier mit der Ang. des Originaltitels „Cellospielerin".

2277
Haniroff
FIGUR „Orchidee" (Stehende Frauengestalt mit Blütenschmuck) (auf Blumen geschmücktem Rundsockel)
ca. 1900/01
2277 (Firmenmarke ohne Ang.), H. 75 cm; Pinhas, S. 62 (m. Abb.)
2277 (Firmenmarke ohne Ang.), H. 73 cm; GSF, S. 28
Bem.: Lt. Firmenkatalog wurde dieses Modell auch mit fünf Lampen (Glühbirnen) montiert angeboten.
-> vgl. 2093
Lit.: GSF, S. 28, hier mit der Ang. des Originaltitels „Orchidee".

2278
Cherc
FIGUR „Die Ernte"
ca. 1900/01
2278 (Firmenmarke ohne Ang.), H. 62 cm; GSF, S. 28
Ausst.: ÖMKI 1902/03, hier mit Ang. des Originaltitels „Ernte".
-> vgl. 2191
-> vgl. 2303= BÜSTE
Lit.: GSF, S. 28, hier mit der Ang. des Originaltitels „Die Ernte".

2279
FIGUR „Kartoffelernte"
ca. 1901/02
2279 (Firmenmarke ohne Ang.), H. 85 cm; GSF, S. 29
Lit.: GSF, S. 29, hier mit der Ang. des Originaltitels „Kartoffelernte".

2280
Simon
BÜSTE „Pax" (Junge Frau mit Diadem aus Lorbeer)
ca. 1900/01
2280 (Firmenmarke ohne Ang.), H. 37 cm; GSF, S. 29
Ausst.: ÖMKI 1901/02, hier mit Ang. des Originaltitels „Pax".
-> vgl. 2187; 2433
Lit.: GSF, S. 29, hier mit der Ang. des Originaltitels „Pax".

2281
FIGUR „Ophelie" (Ophelia)
ca. 1899/1900 (für den Entwurf)
2281 (Firmenmarke ohne Ang.), H. 60 cm; GSF, S. 29
Bem.: Der Name Ophelie (bzw. Ophelia) bezeichnet die weibliche Hauptfigur aus William Shakespeares „Hamlet". Im Firmenkat. ist vermerkt, dass auch Ausführungen als Lampe (elektrifiziert mit drei, bzw. mit sechs Glühbirnen) angeboten wurden.
-> vgl. 1983; 2121
Lit.: GSF, S. 29, hier mit Ang. des Originaltitels „Ophelie".

2282
FIGUR „Guitarrespielerin"
ca. 1901/02
2282 (Firmenmarke ohne Ang.), H. 86 cm; GSF, S. 29
-> vgl. 2445
Lit.: GSF, S. 29, hier mit der Ang. des Originaltitels „Guitarrespielerin".

2283
Tell, E.
FIGUR „Kugelträgerin"
ca. 1901/02
2283 (Firmenmarke ohne Ang.), H. 71 cm; GSF, S. 29
Ausst.: ÖMKI 1902/03, hier mit Ang. des Originaltitels „Kugelträgerin".
Lit.: GSF, S. 29, hier mit der Ang. des Originaltitels „Kugelträgerin"; Neuwirth (1974), S. 75.

2284
Lesca
BÜSTE „Mispel"
ca. 1901/02
2284 (Firmenmarke ohne Ang.), H. 71 cm; GSF, S. 29
Ausst.: ÖMKI 1902/03, hier mit Ang. des Originaltitels „Mispel".
Lit.: GSF, S. 29, hier mit der Ang. des Originaltitels „Mispel"; Neuwirth (1974), S. 66.

2285

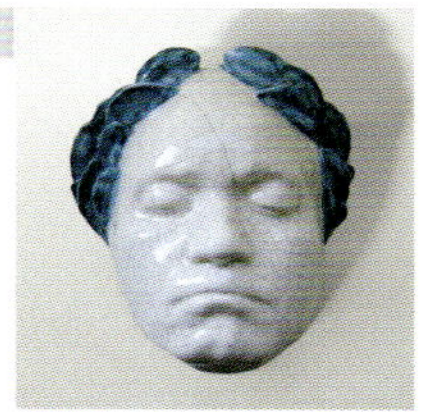

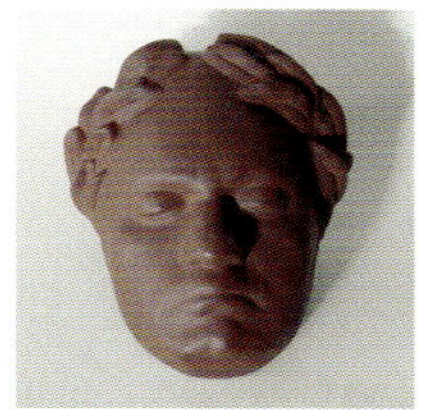

WANDMASKE „Beethoven" (Totenmaske mit Lorbeerkranz)
ca. 1900/01
2285,22,33 Firmenmarke 20, Hand dec., Ausformung 20er Jahre, H. 24 cm; eBay # 2664488750
2285,200,65 Firmenmarke 19, Made in Austria, Ausformung 20er Jahre, H. 24,5 cm; Sammlung f@goldscheider.de
2285,22/II,411 GS (Firmenstempel WG 1), Ausformung 1953–1960, H. 24 cm; PB Deutschland
2285 (Firmenmarke ohne Ang.), H. 39 cm (wohl Druckfehler); GSF, S. 29
Bem.: Dargestellt ist der Komponist Ludwig van Beethoven (1770–1827) nach der durch Joseph Danhauser abgenommenen Totenmaske. Diese Vorlage wurde durch einen Lorbeerkranz ergänzt (vgl. auch das Bild von Franz von Stuck „Beethoven"). Wie die bis in die 50er Jahre ausgeformten Modelle belegen, war diese Maske beim Publikum sehr gefragt und wurde daher zwischen 1900–1960 von GS produziert.
Ausst.: Turin 1902
-> vgl. 2202
Lit.: GSF, S. 29, hier mit der Ang. des Originaltitels „Beethoven"; Pudor (1902a), S. 119; KGSF1, S. (16) (m. Abb.); KGSF2, S. (102) (m. Abb.); Neuwirth (1974), S. 58; Könecke, S. 56 (m. Abb.) 76, 79 (mit Künstlerang. Lorenzl), dieser Hinweis ist jedoch eindeutig falsch, denn zum Einen wurde die Maske bereits um 1900 produziert und zum Anderen verstarb J. Lorenzl bereits 1950, also lange vor der Verlegung der Manufaktur an den Tönnieshof).
Quelle: AGM (Bestellung der Firma Wahliss vom 27.2.1946 an die „Arbeitsgemeinschaft der Firma Friedrich Goldscheider" in Ausführung „Terrakotta").

2286
WANDMASKE „Venuskopf"
ca. 1901/02
2286 (Firmenmarke ohne Ang.), H. 19 cm; GSF, S. 29
-> vgl. 2329
Lit.: GSF, S. 29, hier mit der Ang. des Originaltitels „Venuskopf".

2287
Larroux, Antonin
WANDMASKE „Junger Faun" (mit Blüten im Haar)
ca. 1901/02
2287,27,26 Firmenmarke 7, gepr. „P"= Z3, unsign., H. fehlt; eBay # 903944743
2287 (Firmenmarke ohne Ang.), H. 24 cm; GSF, S. 29
Bem.: Diese und mindestens drei weitere Wandmasken entstammen dem Entwurf VASE „Masques" Modell-Nrn. 933 und 1097. Eine dieser als Pendants angeführten Faun-Masken wurde auch 1903 im Kunstsalon Keller & Reiner in Berlin gezeigt.
Ausst.: ÖMKI 1902/03, hier mit Ang. des Originaltitels „Junger Faun", Künstlerang. „Laroux".
Pendants -> vgl. 2288; 2289; 2290; 2291
Lit.: GSF, S. 29, hier mit der Ang. des Originaltitels „Junger Faun"; R., F.: Keramische Wanderungen durch den Kunstsalon von Keller & Reiner, in: KM, 1903, S. 125f.; Neuwirth (1974), S. 66.

2288
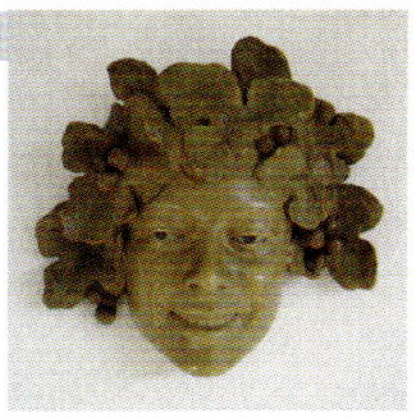
Larroux, Antonin
WANDMASKE „Bacchantin" (mit Weinblättern im Haar)
ca. 1901/02
2288,55,27,XXX,(6,3,102 vertikal), Firmenmarke 7, Repr. Res., sign. Larroux, gepr. „P"= Z3, H. 20 cm; Sammlung f@goldscheider.de
2288 (Firmenmarke ohne Ang.), H. 22 cm; GSF, S. 29
Bem.: Diese und mindestens drei weitere Wandmasken entstammen dem Entwurf VASE „Masques" Modell-Nrn. 933 und 1097.
Ausst.: ÖMKI 1902/03, hier mit Ang. des Originaltitels „Bacchantin", Künstlerang. „Laroux".
Pendants -> vgl. 2287; 2289; 2290; 2291
Lit.: GSF, S. 29, hier mit der Ang. des Originaltitels „Bacchantin"; Neuwirth (1974), S. 66.

2289
Larroux, Antonin
WANDMASKE „Faun"
ca. 1901/02
2289 (Firmenmarke ohne Ang.), H. 26 cm; GSF, S. 29
Bem.: Diese und mindestens drei weitere Wandmasken entstammen dem Entwurf VASE „Masques" Modell-Nrn. 933 und 1097. Eine dieser als Pendants angeführten Faun-Masken wurde auch 1903 im Kunstsalon Keller & Reiner in Berlin gezeigt.
Ausst.: ÖMKI 1902/03, im Ausst.-Kat. werden insgesamt drei „Masken" ohne nähere Bezeichnung/Titel aufgeführt, Künstlerang. „Laroux".
Pendants -> vgl. 2287; 2288; 2290; 2291
Lit.: GSF, S. 29, hier mit der Ang. des Originaltitels „Faun"; R., F.: Keramische Wanderungen durch den Kunstsalon von Keller & Reiner, in: KM, 1903, S. 125f.

2290
Larroux, Antonin
WANDMASKE „Klatschblume" (Frauenkopf mit langem Haar und Mohnblüten)
ca. 1901/02
2290,4,27 Firmenmarke 7, Repr. Res., gepr. „P"= Z3 sign. „Leroux", H. 24 cm, B. 24 cm; eBay # 1465953964
2290,14,27 gem. X, Firmenmarke 7, Repr. Res., sign. „Leroux", H. 23 cm; eBay # 3702331016
2290 (Firmenmarke ohne Ang.), H. 24 cm; GSF, S. 29
Bem.: Diese und mindestens drei weitere Wandmasken entstammen dem Entwurf VASE „Masques" Modell-Nrn. 933 und 1097.
Ausst.: ÖMKI 1902/03, hier mit Ang. des Originaltitels „Klatschblume"; Künstlerang. „Laroux".
Pendants -> vgl. 2287; 2288; 2289; 2291
Lit.: GSF, S. 29, hier mit der Ang. des Originaltitels „Klatschblume"; Neuwirth (1974), S. 66.

2291
Larroux, Antonin
WANDMASKE „Coquette"/„Kokette" (Frauenkopf mit wallendem Haar)
ca. 1901/02
2291,48,26 FG Wien (Firmenmarke?), Repr. Res., sign. Laroux, H. 28,5 cm; Soth. L. 20.10.1989
2291 (Firmenmarke ohne Ang.), H. 29 cm; GSF, S. 29
Bem.: Diese und mindestens zwei weitere Wandmasken entstammen dem Entwurf VASE „Masques" Modell-Nrn. 933 und 1097.
Ausst.: ÖMKI 1902/03, hier mit Ang. des deutschen Originaltitels „Kokette", Künstlerang. „Laroux".
Pendants -> vgl. 2287; 2288; 2289; 2290
Lit.: GSF, S. 29, hier mit der Ang. des franz. Originaltitels „Coquette"; Neuwirth (1974), S. 66.

2292
Larroux, Antonin
WANDMASKE „Modern"
ca. 1901/02
2292 (Firmenmarke ohne Ang.), H. 30 cm; GSF, S. 29
-> vgl. 1844= BÜSTE
Lit.: GSF, S. 29, hier mit der Ang. des Originaltitels „Modern".

2293
BÜSTE „Althaea"
ca. 1901/02
2293 (Firmenmarke ohne Ang.), H. 63 cm; GSF, S. 29
Lit.: GSF, S. 29, hier mit der Ang. des Originaltitels „Althaea".

2294
Rozel
BÜSTE „Gladiator"
ca. 1898 (für den Entwurf)
2294 (Firmenmarke ohne Ang.), H. 29 cm; GSF, S. 29
Ausst.: ÖMKI 1898/99, hier mit Ang. des Originaltitels Büste „Gladiator".
-> vgl. 1687; 2184
-> vgl. GRUPPE „Animus et vis"= Modell 3680
Lit.: GSF, S. 29, hier mit der Ang. des Originaltitels „Gladiator".

2295
FIGUR „Thermometerträger"
ca. 1901/02
2295 (Firmenmarke ohne Ang.), H. 22 cm; GSF, S. 29
Lit.: GSF, S. 29, hier mit der Ang. des Originaltitels „Thermometerträger".

2296
Bertrand, A.
BÜSTE „Sonnenblume" (Junge Frau, nach links blickend)
ca. 1900/01
2296 (Firmenmarke ohne Ang.), H. 44 cm; Pinhas, S. 53 (m. Abb.)
2296 (Firmenmarke ohne Ang.), H. 47 cm; GSF, S. 29
-> vgl. 2177; 2458
Lit.: GSF, S. 29, hier mit der Ang. des Originaltitels „Sonnenblume".

2297
BÜSTE „Räthsel"
ca. 1901/02
2297 (Firmenmarke ohne Ang.), H. 58 cm; GSF, S. 29
Lit.: GSF, S. 29, hier mit der Ang. des Originaltitels „Räthsel".

2299
Simon
UHR „Frühling" (Stehende Frauengestalt vor stilisiertem Kirschbaum)
ca. 1901
2299,29,25 (Firmenmarke ohne Ang.), H. 62 cm; Soth. L. 3.3.1989; Bem.: Aufl. lt. Soth. „90 Stück"!?
2229 (Firmenmarke ohne Ang.), H. 64 cm; GSF, S. 29
Bem.: Lt. Firmenkatalog war dieses Modell auch ohne das Uhrwerk lieferbar. Dies erklärt auch, warum in den diversen Goldscheideruhren sich gelegentlich auch Uhrwerke von anderen Firmen als Lenzkirch finden.
Ausst.: ÖMKI 1901/02, hier mit Ang. des Originaltitels „Frühling".
-> vgl. 2229
Lit.: GSF, S. 29, hier mit der Ang. des Originaltitels „Frühling"; Pudor (1902), S. 71 (m. Abb.); B., F.: Keramische Wanderungen durch den Kunstsalon Keller & Reiner Berlin, in: KM, 3.1903, S. 128, „Dieser Künstler versteht es gut zu stilisieren und seine jedesmaligen Motive geschmackvoll dem gewählten Sujet anzupassen"; Neuwirth (1974), S. 74; Weltkunst, 50.1980, S. 1504 (m. Abb.).

2300
FIGUR „Das Glück"
ca. 1901/02
2300 (Firmenmarke ohne Ang.), H. 103 cm; GSF, S. 29
Lit.: GSF, S. 29, hier mit der Ang. des Originaltitels „Das Glück".

2301
Powolny, Michael
FIGUR „Die Zeit"
ca. 1901/02
2301 (Firmenmarke ohne Ang.), H. 88 cm; GSF, S. 29
Bem.: Lt. Firmenkatalog war dieses Modell auch „elektr. montirt" lieferbar. Die Anzahl der Beleuchtungskörper geht jedoch nicht aus der Quelle hervor. Ob es sich bei dem vorliegenden Modell wirklich um die vergrößerte Ausformung der Uhr „Die Zeit" handelt, kann derzeit nur vermutet werden. Weder nachweisbare Stücke noch die Abbildungstafeln der Quelle lagen vor.
Ausst.: Paris, Salon (1902), hier mit Ang. des franz. Originaltitels „Le Temps".
-> vgl. 2239= UHR?
Lit.: GSF, S. 29, hier mit der Ang. des Originaltitels „Die Zeit".

2302
Cherc
BÜSTE „Saat" (Holländerin mit Haube) (nach links blickend)
ca. 1901/02
2302,188,2 Firmenmarke 7, Repr. Res., H. 21 cm; eBay # 280026178938
2302,215,7 Firmenmarke Sculpture 13, unsign., H. 21 cm; Henry's 28.7.2000
2302 (Firmenmarke ohne Ang.), H. 19 cm; GSF, S. 29
Lit.: GSF, S. 29, hier mit der Ang. des Originaltitels „Saat".

2303
Cherc
BÜSTE „Ernte"
ca. 1901/02
2303 (Firmenmarke ohne Ang.), H. 20 cm; GSF, S. 29
-> vgl. 2191; 2278= jew. FIGUR
Lit.: GSF, S. 29, hier mit der Ang. des Originaltitels „Ernte".

2304
Bertrand, A.
BÜSTE „Gebet"
ca. 1901
2304 (Firmenmarke ohne Ang.), H. 21 cm; GSF, S. 29
Ausst.: ÖMKI 1901/02, hier mit Ang. des Originaltitels „Gebet".
Lit.: GSF, S. 29, hier mit der Ang. des Originaltitels „Gebet"; Neuwirth (1974), S. 59.

2305
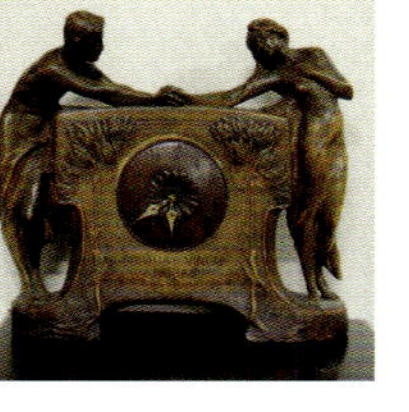
Simon
UHR „Amicitia" (Stehendes Liebespaar reicht sich die Hände)
Inschrift: AMICITA VINCIT HORAS
ca. 1901/02
2305,198,3 Firmenmarke Sculpture 13, Repr. Res., L. 62 cm; GK, S. 14, 77 (m. Abb.); Wien Museum Inv. 223.906
2305,281,17 Firmenmarke Sculpture 13, Repr. Res., H. 53 cm; Sammlung Gruber, S. 104
2305,281,S17 Firmenmarke Sculpture 13, Repr. Res., H. 53 cm; Venzmer, S. 341
2305,419,17 Firmenmarke Sculpture 13, Repr. Res., H. 54,3 cm; MAK Wien, Inv. Ke 10602-107
2305,524,10 Firmenmarke Sculpture 13, H. 53,8 cm; Doro 30.11.2001; 10.3.2003
2305 (Firmenmarke ohne Ang.), H. 51 cm, B. 60 cm; GSF, S. 29
Bem.: Lt. Firmenkatalog war dieses Modell auch ohne das Uhrwerk lieferbar. Dies erklärt auch, warum in den diversen Goldscheideruhren sich gelegentlich auch Uhrwerke von anderen Firmen als Lenzkirch finden.
Text der Inschrift: Die Freundschaft besiegt die Stunden.
Ausst.: Turin 1902; HMW (jetzt Wien Museum) 1985/86.
-> vgl. 2523; 2736
Lit.: GSF, S. 29, hier mit der Ang. des Originaltitels „Amicitia"; N.N.: Von der Internationalen Ausstellung für dekorative Kunst in Turin 1902, in: KM, 2.1902, S. 162–163 (m. Abb.), „Die in einfachster Umrahmung gehaltene Standuhr wird auf der Vorderseite durch zwei baumartige Gewächse flankiert, welche aus dem Fuß des Gestelles herauswachsen und oben Zweige über die Uhrscheibe hinweg einander zusenden. Zwei jugendliche Gestalten, welche seitlich des Gehäuses stehen, die den untenstehenden Spruch: Amicitia vincit horas (die Freundschaft besiegt die Stunden)"; GK, S. 14, 77 (m. Abb.).

2306
Jaray, Sándor (Alexander)
GRUPPE „Adam und Eva" (als sitzende nackte Kinder)
Inschrift: ADAM UND EVA
ca. 1901/02
2306,107,68 Firmenmarke 4, Ges. Gesch., H. 33,6 cm; eBay # 730724427
2306 (ohne Ang.), (Firmenmarke Sculpture 13?), undeutlich sign., H. 34 cm; RM 32
2306 (Firmenmarke ohne Ang.), H. 32 cm; GSF, S. 29
Lit.: GSF, S. 29, hier mit der Ang. des Originaltitels „Adam und Eva".

2307
FIGUR „Der Kampf" (Stehender römischer Krieger in Rüstung mit Schild, Schwert und Axt) (auf Landschaftssockel)
ca. 1900 (für den Entwurf)
2307 (Firmenmarke ohne Ang.), H. 53 cm; GSF, S. 29
-> vgl. 1953; 2182
-> vgl. 2140= BÜSTE
Lit.: GSF, S. 29, hier mit der Ang. des Originaltitels „Der Kampf".

2308
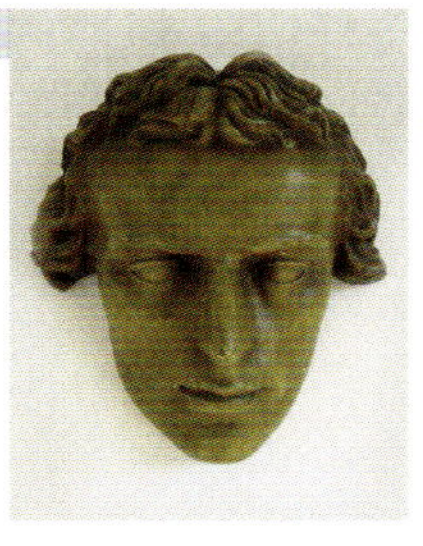
WANDMASKE „Schiller"
ca. 1901/02
2308,22,7, gem. 13, (ohne Firmenmarke), Repr. Res., H. 26 cm; Sammlung f@goldscheider.de
2308 (Firmenmarke ohne Ang.), H. 26 cm; GSF, S. 29
Bem.: Dargestellt ist der Dichter Friedrich Schiller (1759–1805).
Lit.: GSF, S. 29, hier mit der Ang. des Originaltitels „Schiller".

2309
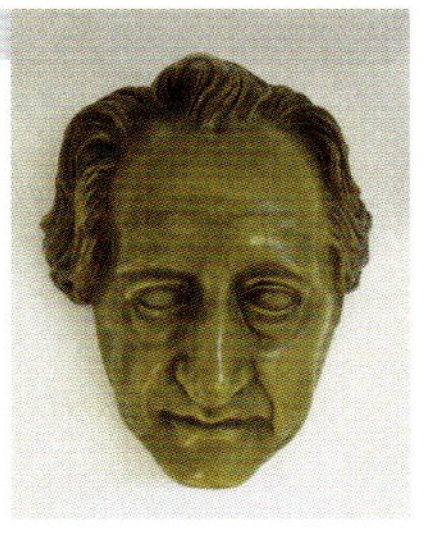
WANDMASKE „Goethe"
ca. 1901/02
2309,5,46 FG (Firmenmarke?), Repr. Res., gepr. „C"= Z4, H. ca. 26 cm; WKH
2309,35,31 Firmenmarke 7, Repr. Res., gepr. „C"= Z4, H. 26 cm; Sammlung f@goldscheider.de Wien (Firmenmarke),
2309,51,33 GS Wien (handschriftl. Firmenmarke?), Repr. Res., H. 26 cm; eBay # 7412590242
2309 (Firmenmarke ohne Ang.), H. 26 cm; GSF, S. 29
Bem.: Dargestellt ist der Dichter Johann Wolfgang von Goethe (1749–1832). Die Ausführung der Maske ist etwas angelehnt an die Goethe-Büste von Johann Gottfried Schadow (1764–1850) deren Original sich in der Nationalgalerie Berlin befindet.
Lit.: GSF, S. 29, hier mit der Ang. des Originaltitels „Goethe".

2310
BÜSTE „Gute Nacht"
ca. 1901/02
2310 (Firmenmarke ohne Ang.), H. 67 cm; GSF, S. 29
Lit.: GSF, S. 29, hier mit der Ang. des Originaltitels „Gute Nacht".

2311
Lehmann, Ida
FIGUR „Eitelkeit"
ca. 1901/02
2311 (Firmenmarke ohne Ang.), H. 90 cm; GSF, S. 30
Bem.: Wie der Firmenkat. belegt, wurde der Pfau dieses Modells „mit oder ohne Glasaugen", bzw. mit „fünf Lampen elektr. montirt" geliefert. Zusätzlich mit handschr. Vermerk „detto mit Reflection".
Ausst.: ÖMKI 1912, hier mit Ang. des Originaltitels „Eitelkeit" sowie der Künstlerin Ida Lehmann.
Lit.: GSF, S. 30, hier mit der Ang. des Originaltitels „Eitelkeit"; Neuwirth (1974), S. 74.

2312
FIGUR „Tafelaufsatz"
ca. 1901/02
2312 (Firmenmarke ohne Ang.), H. 56 cm; GSF, S. 30
Lit.: GSF, S. 30, hier mit der Ang. des Originaltitels „Tafelaufsatz".

2313
UHR „Klio"
ca. 1901/02
2313 (Firmenmarke ohne Ang.), H. 52 cm; GSF, S. 30
Bem.: Lt. Firmenkatalog war dieses Modell auch ohne das Uhrwerk lieferbar. Dies erklärt auch, warum in den diversen Goldscheideruhren sich gelegentlich auch Uhrwerke von anderen Firmen als Lenzkirch finden.
Lit.: GSF, S. 30, hier mit der Ang. des Originaltitels „Klio".

2314
BÜSTE „Loreley (ohne Sockel)"
ca. 1899/1900 (für den Entwurf)
2314 (Firmenmarke ohne Ang.), H. 23 cm; GSF, S. 30
-> vgl. 2032
Lit.: GSF, S. 30, hier mit der Ang. des Originaltitels „Loreley (ohne Sockel)".

2315
Lemassi
BÜSTE „Ernte"
ca. 1899/1900 (für den Entwurf)
2315 (Firmenmarke ohne Ang.), H. 19 cm; GSF, S. 30
Ausst.: ÖMKI 1900/01, hier mit Ang. des Originaltitels „Ernte".
-> vgl. 2035
Lit.: GSF, S. 30, hier mit der Ang. des Originaltitels „Ernte".

2316
Lesca
FIGUR „Labor" (Frau mit Spaten)
Inschrift: LABOR
ca. 1902
2316,15,9 FG Wien (Firmenmarke?), H. 86 cm; Schloss Hag. 37
2316,41,19 (Firmenmarke ohne Ang.), H. 91 cm; Arnold 26.11.1988
2316 (Firmenmarke ohne Ang.), H. 85 cm; GSF, S. 30
Bem.: Mit dieser Figur ist der Sockel der Büste „Labor" (Mod.-Nr. 2387) reliefiert.
Text der Inschrift: Arbeit
Ausst.: ÖMKI 1902/03, hier mit Ang. des Originaltitels „Labor".
-> vgl. 2573
-> vgl. 2387= BÜSTE
Lit.: GSF, S. 30, hier mit der Ang. des Originaltitels „Labor"; Neuwirth (1974), S. 66.

2317
Sidan
FIGUR „Das Gehör" (Stehendes Mädchen an Muschel hörend)
ca. 1902
2317 (Firmenmarke ohne Ang.), H. 66 cm; Pinhas, S. 43 (m. Abb.)
2317 (Firmenmarke ohne Ang.), H. 62 cm; GSF, S. 30
Lit.: GSF, S. 30, hier mit der Ang. des Originaltitels „Das Gehör"; N.N.: Illustrierte Rundschau, Terrakotten von Fr. Goldscheider, Wien, in: VKMH, 17.1902/03, S. 717 (m. Abb.), hier mit der Ang. des Originaltitels „Das Gehör" und des Künstlers „Sidan".

2318
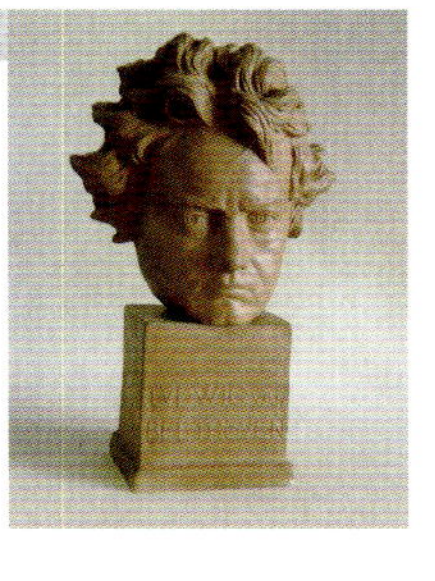
Skall (d.i. vermutlich Skála, Jaroslav)
BÜSTE „Beethoven" (halbplastischer Kopf auf quaderförmigem Sockel)
Inschrift: LUDWIG VAN BEETHOVEN
ca. 1902
2318,49,11 Firmenmarke Sculpture 13, Repr. Res., gepr. „C"= Z4 undeutlich sign. „Skall", H. 40,8 cm; PB Deutschland
2318 (Firmenmarke ohne Ang.), H. 38 cm; GSF, S. 30
Bem.: Dargestellt ist der Komponist Ludwig van Beethoven (1770–1827); vermutlich handelt es sich bei dem Entwerfer dieser Büste um den Bildhauer Jaroslav Skála.
Ausst.: ÖMKI 1902/03, im Ausst.-Kat. wird „Skal" als Künstler genannt, hier auch mit Ang. des Originaltitels „Beethoven".
Lit.: GSF, S. 30, hier mit der Ang. des Originaltitels „Beethoven", (handschriftl. Vermerk: auf Sockel); N.N.: Illustrierte Rundschau, Terrakotten von Fr. Goldscheider, Wien, in: VKMH, 17.1902/03, S. 717 (m. Abb.); Neuwirth (1974), S. 74.

2319
Bertrand, A.
BÜSTE „Constance" (Junge Dame verträumt nach rechts blickend)
ca. 1902
2319 (Firmenmarke ohne Ang.), H. 36 cm; GSF, S. 30
Lit.: GSF, S. 30, hier mit der Ang. des Originaltitels „Constance"; N.N.: Illustrierte Rundschau, Terrakotten von Fr. Goldscheider, Wien, in: VKMH, 17.1902/03, S. 717 (m. Abb.), hier mit Ang. des Titels „Constanze" und des Künstlers Bertrand.

2320
Meier, Emil
FIGUR „Heimkehr"
ca. 1902
2320 (Firmenmarke ohne Ang.), H. 80 cm (handschr. verbessert: 93 cm); GSF, S. 30
Ausst.: ÖMKI 1912, hier mit Ang. des Originaltitels „Heimkehr".
Lit.: GSF, S. 30, hier mit der Ang. des Originaltitels „Heimkehr"; Neuwirth (1974), S. 68.

2321
FIGUR „Holländerin mit Terracottamuschel/Naturmuschel"
ca. 1902
2321 (Firmenmarke ohne Ang.), H. 73 cm; GSF, S. 30
Bem.: Wie der Firmenkat. belegt, wurde der dieses Modell „mit Terracottamuschel" oder „Naturmuschel", bzw. mit „elektr. montirt" geliefert. Die Anzahl der Beleuchtungskörper geht jedoch nicht aus der Quelle hervor.
Lit.: GSF, S. 30, hier mit der Ang. des Originaltitels „Holländerin mit Terracottamuschel/Naturmuschel".

2322
FIGUR „Gesang"
ca. 1902
2322 (Firmenmarke ohne Ang.), H. 79 cm; GSF, S. 30
Lit.: GSF, S. 30, hier mit der Ang. des Originaltitels „Gesang".

2323
UHR „Am Brunnen"
ca. 1902
2323 (Firmenmarke ohne Ang.), H. 60 cm; GSF, S. 30
Bem.: Lt. Firmenkatalog war dieses Modell auch ohne das Uhrwerk lieferbar. Zusätzlich wurde eine „mit Uhrwerk elektr. montirte" Variante angeboten.
Lit.: GSF, S. 30, hier mit der Ang. des Originaltitels „Am Brunnen".

2324
BÜSTE „Ruth"
ca. 1902
2324 (Firmenmarke ohne Ang.), H. 38 cm; GSF, S. 30
Lit.: GSF, S. 30, hier mit der Ang. des Originaltitels „Ruth".

2325
Haniroff
BÜSTE „Fanchon" (Junge Dame nach links blickend)
ca. 1902
2325 (Firmenmarke ohne Ang.), H. 28 cm; GSF, S. 30
-> vgl. 1150= FIGUR
Lit.: GSF, S. 30, hier mit der Ang. des Originaltitels „Fanchon"; N.N.: Illustrierte Rundschau, Terrakotten von Fr. Goldscheider, Wien, in: VKMH, 17.1902/03, S. 717 (m. Abb.), hier mit Ang. des Künstlers Haniroff.

2326
BÜSTE „Gloria"
ca. 1902
2326 (Firmenmarke ohne Ang.), H. 47 cm; GSF, S. 30
Lit.: GSF, S. 30, hier mit der Ang. des Originaltitels „Gloria".

2327
BÜSTE „Schmetterling" (Junge Dame mit Schmetterlingsflügeln) (nach links blickend)
ca. 1900 (für den Entwurf)
2327 (Firmenmarke ohne Ang.), H. 55 cm; GSF, S. 30
-> vgl. 1907
Pendant -> vgl. 2328
Lit.: GSF, S. 30, hier mit der Ang. des Originaltitels „Schmetterling".

2328
Mondi
BÜSTE „Kleeblatt" (Junge Dame mit Schmetterlingsflügeln) (nach rechts blickend)
ca. 1900 (für den Entwurf)
2328 (Firmenmarke ohne Ang.), H. 50 cm; GSF, S. 30
Ausst.: ÖMKI 1902/03, hier mit Ang. des Originaltitels „Kleeblatt" und des Künstlers „Mondi".
-> vgl. 1906
Pendant -> vgl. 2327
Lit.: GSF, S. 30, hier mit der Ang. des Originaltitels „Kleeblatt".

2329
WANDMASKE „Venuskopf"
ca. 1901/02
2329 (Firmenmarke ohne Ang.), H. 21 cm; GSF, S. 30
-> vgl. 2286
Lit.: GSF, S. 30, hier mit der Ang. des Originaltitels „Venuskopf".

2330
WANDMASKE „Mozart"
ca. 1900 (für den Entwurf)
2330 (Firmenmarke ohne Ang.), H. 37 cm; GSF, S. 30
Bem.: Dargestellt ist der Komponist Wolfgang Amadeus Mozart (1756–1791).
-> vgl. 2077
Lit.: GSF, S. 30, hier mit der Ang. des Originaltitels „Mozart".

2331
Waldmann, (Oscar?)
FIGUR „Löwe"
ca. 1902
2331 (Firmenmarke ohne Ang.), H. 25 cm, L. 38 cm; GSF, S. 30
Ausst.: ÖMKI 1902/03, hier mit Ang. des Originaltitels „Löwe" und des Künstlers Waldmann.
Lit.: GSF, S. 30, hier mit der Ang. des Originaltitels „Löwe"; Newirth (1974), S. 76.

2332
Meunier, Constantin Emile
FIGUR „Fischer" (mit Netz) (auf Natursockel)
ca. 1900/01
2332,29,10 Firmenmarke 9, H. 28 cm; Doro 25.4.2000
2332,33,10 FG Wien (Firmenmarke 9?), H. 27,5 cm; Quit. 29
2332 (Firmenmarke ohne Ang.), H. 26 cm; GSF, S. 30
Lit.: GSF, S. 30, hier mit der Ang. des Originaltitels „Fischer".

2333
BÜSTE „Seerose"
ca. 1902
2333 (Firmenmarke ohne Ang.), H. 28 cm; GSF, S. 30
Lit.: GSF, S. 30, hier mit der Ang. des Originaltitels „Seerose".

2334
BÜSTE „Schwalbe"
ca. 1902
2334 (Firmenmarke ohne Ang.), H. 45 cm; GSF, S. 30
Lit.: GSF, S. 30, hier mit der Ang. des Originaltitels „Schwalbe".

2335
Tell, E.
UHR „Patientia"
ca. 1902
2335 (Firmenmarke ohne Ang.), H. 58 cm; GSF, S. 30
Bem.: Lt. Firmenkatalog war dieses Modell auch ohne das Uhrwerk lieferbar. Dies erklärt auch, warum in den diversen Goldscheideruhren sich gelegentlich auch Uhrwerke von anderen Firmen als Lenzkirch finden.
Ausst.: ÖMKI 1902/03, hier mit Ang. des Originaltitels „Patientia".
Lit.: GSF, S. 30, hier mit der Ang. des Originaltitels „Patientia"; Newirth (1974), S. 78.

2336
BÜSTE „La Chasse" (Junge Frau ihren Kopf nach links neigend)
Inschrift: LA CHASSE
ca. 1902
2336,81,26 (Firmenmarke ohne Ang.), Repr. Res., H. 34 cm; eBay # 3723298362
2336 (Firmenmarke ohne Ang.), H. 31 cm; GSF, S. 30
Lit.: GSF, S. 30, hier mit der Ang. des Originaltitels „La Chasse".

2337
BÜSTE „Sappho"
ca. 1902
2337 (Firmenmarke ohne Ang.), H. 23 cm; GSF, S. 30
Lit.: GSF, S. 30, hier mit der Ang. des Originaltitels „Sappho".

2338
BÜSTE „Narcisse" (Junge Frau mit Blumen im Haar)
Inschrift: NARCISSE
ca. 1899/1900
2338,84,8 Firmenmarke 7, Repr. Res., H. 24 cm, B. 21,7 cm; Wien Museum Inv. 186.637
2338,134,14 FG Wien (Firmenmarke?), Repr. Res., H. 23,5 cm; Wendl 51
ohne Ang., FG Wien (Firmenmarke?), H. 24,4 cm, B. 21,6 cm; eBay # 456626811
2338 (Firmenmarke ohne Ang.), H. 23 cm; GSF, S. 30
Lit.: GSF, S. 30, hier mit der Ang. des Originaltitels „Narcisse".

2339
BÜSTE „Gantalée"
ca. 1902
2339 (Firmenmarke ohne Ang.), H. 44 cm; GSF, S. 30
Lit.: GSF, S. 30, hier mit der Ang. des Originaltitels „Gantalée".

2340
Werner, Franz
FIGUR „Affe als Philosoph" (sitzend auf Buch)
Inschrift: PHILOSOPHIE
ca. 1899/1900 (für den Entwurf)
2340,49,39 Firmenmarke 7, Repr. Res., H. 28,6 cm; Sammlung John Nash

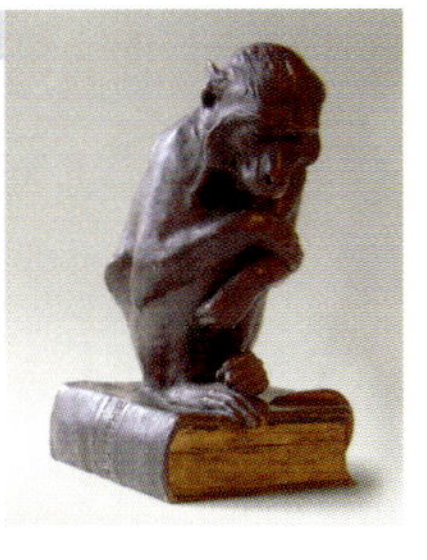

2340 (Firmenmarke ohne Ang.), H. 28 cm; GSF, S. 30
Bem.: Evtl. identisch mit „Affe" (ÖMKI 1902/03)?, vgl. hierzu auch Neuwirth (1974), S. 76.
Lit.: GSF, S. 30, hier mit der Ang. des Originaltitels „Affe als Philosoph".

2341
BÜSTE „La vertu" (Junge Frau mit Lilien im Haar)
Inschrift: LA VERTU
ca. 1902
2341,13,55 Firmenmarke Sculpture 13, Repr. Res., H. 26 cm; eBay # 6561204354
2341 (Firmenmarke ohne Ang.), H. 24 cm; GSF, S. 31
Text der Inschrift: Die Tugend
Lit.: GSF, S. 31, hier mit der Ang. des Originaltitels „La vertu".

2342
FIGUR „Aphrodite mit Naturmuschel"
ca. 1902
2342 (Firmenmarke ohne Ang.), H. 89 cm; GSF, S. 31
Bem.: Wie der Firmenkat. belegt, wurde der dieses Modell auch „elektr. montirt" geliefert. Die Anzahl der Beleuchtungskörper geht jedoch nicht aus der Quelle hervor. Dargestellt ist Aphrodite, die in der griech. Mythologie Göttin der Liebe, Schönheit und sinnlichen Begierde ist. (Quelle: Wikipedia).
-> vgl. 2390
Lit.: GSF, S. 31, hier mit der Ang. des Originaltitels „Aphrodite mit Naturmuschel".

2343
Stellmacher, (Eduard?)
BÜSTE „Erwachen"
ca. 1900/01
2343 (Firmenmarke ohne Ang.), H. 56 cm; GSF, S. 31
Ausst.: ÖMKI 1901/02, hier mit Ang. des Originaltitels „Erwachen".
-> vgl. 2218
Lit.: GSF, S. 31, hier mit der Ang. des Originaltitels „Erwachen"; Neuwirth (1974), S. 74.

2344
Dupré, J.
FIGUR „Heuernte"
ca. 1902
2344 (Firmenmarke ohne Ang.), H. 92 cm; GSF, S. 31
Ausst.: ÖMKI 1902/03, hier mit Ang. des Originaltitels „Heuernte" und des Künstlers „J. Dupri".
Lit.: GSF, S. 31, hier mit der Ang. des Originaltitels „Heuernte"; Neuwirth (1974), S. 62.

2345 -> ÜBERSCHNEIDUNG
FIGUR „Lilie"
ca. 1900/01
2345 (Firmenmarke ohne Ang.), H. 74 cm; GSF, S. 31
Bem.: Lt. Firmenkatalog wurde dieses Modell auch mit vier Lampen (Glühbirnen) montiert angeboten.
-> vgl. 2096
Lit.: GSF, S. 31, hier mit der Ang. des Originaltitels „Lilie".

2345 -> ÜBERSCHNEIDUNG
Petri (d.i. Petrides, János) (Johann)
UHR Stehendes betendes Mädchen (auf abgerundeter Plinthe, stelen/quaderförmiger Hintergrund/Aufbau)
ca. 1901/02
2345,32,10 gem. F, (Firmenmarke ohne Ang.), Repr. Res., sign. Petri, H. 25 cm; Herr 47

2346
Gross, F. (Karl?)
FIGUR „Orangenpflückerin"
ca. 1902
2346 (Firmenmarke ohne Ang.), H. 90 cm; GSF, S. 31
Bem.: Lt. Firmenkatalog wurde dieses Modell auch mit zwei Lampen (Glühbirnen) montiert angeboten.
Ausst.: ÖMKI 1902/03, hier mit Ang. des Originaltitels „Orangenpflückerin" und des Künstlers „Gross".
Lit.: GSF, S. 31, hier mit der Ang. des Originaltitels „Orangenpflückerin"; Neuwirth (1974), S. 63.

2347
FIGUR „Kind mit Blumentopf"
ca. 1902
2347 (Firmenmarke ohne Ang.), H. 84 cm; GSF, S. 31
Bem.: Lt. Firmenkatalog wurde dieses Modell auch mit zwei Lampen (Glühbirnen) montiert angeboten.
Lit.: GSF, S. 31, hier mit der Ang. des Originaltitels „Kind mit Blumentopf".

2348
FIGUR „Vasenmalerin"
ca. 1902
2348 (Firmenmarke ohne Ang.), H. 84 cm; GSF, S. 31
Lit.: GSF, S. 31, hier mit der Ang. des Originaltitels „Vasenmalerin".

2349
FIGUR „Die Neugier"
ca. 1902
2349 (Firmenmarke ohne Ang.), H. 68 cm; GSF, S. 31
Bem.: Lt. Firmenkatalog wurde dieses Modell auch mit einer Lampe (Glühbirne) montiert angeboten.
Lit.: GSF, S. 31, hier mit der Ang. des Originaltitels „Die Neugier".

2350
FIGUR „Sirene"
ca. 1902
2350 (Firmenmarke ohne Ang.), H. 50 cm; GSF, S. 31
Lit.: GSF, S. 31, hier mit der Ang. des Originaltitels „Sirene".

2351
FIGUR „Nacht"
ca. 1902
2351 (Firmenmarke ohne Ang.), H. 80 cm; GSF, S. 31
Bem.: Lt. Firmenkatalog wurde dieses Modell auch mit einer Lampe (Glühbirne) montiert angeboten.
Lit.: GSF, S. 31, hier mit der Ang. des Originaltitels „Nacht".

2352
Sydan / Tell, E.
FIGUR „Edelstein" (Stehende Frauengestalt mit Kopfputz und gesenktem Haupt, in ihren Händen Blumenblüten und einen Kristall haltend)
ca. 1901/02
2352 (Firmenmarke ohne Ang.), sign. E. Tell, H. 70 cm; Pinhas, S. 62 (m. Abb.)
2352 (Firmenmarke ohne Ang.), H. 69 cm; GSF, S. 31
Bem.: Lt. Firmenkatalog wurde dieses Modell auch mit einer Lampe (Glühbirne) montiert angeboten, die den Kristall illuminiert.
Ausst.: ÖMKI 1902/03, hier mit Ang. des Originaltitels „Edelstein" und des Künstlers „Sydan".
Lit.: GSF, S. 31, hier mit der Ang. des Originaltitels „Edelstein"; Pudor (1902), S. 108; Neuwirth (1974), S. 75.

2353
FIGUR „Stolz"
ca. 1902
2353 (Firmenmarke ohne Ang.), H. 70 cm; GSF, S. 31
Lit.: GSF, S. 31, hier mit der Ang. des Originaltitels „Stolz".

2354
Bertrand, A.
BÜSTE „Ziniae"
ca. 1901/02
2354 (Firmenmarke ohne Ang.), H. 69 cm; GSF, S. 31
Lit.: GSF, S. 31, hier mit der Ang. des Originaltitels „Ziniae"; Pudor (1902), S. 108; Neuwirth (1974), S. 59.

2355
Suchanek, Vacslav
FIGUR „Falke" (auf Natursockel)
ca. 1901/02
2355,226,10 Firmenmarke 19, Ausformung 20er Jahre, H. 29 cm; Doro L. 24.11.2000
2355 ohne Ang., (Firmenmarke ohne Ang.), H. 27,5 cm; Nagel 324
2355 ohne Ang., (Firmenmarke ohne Ang.), H. fehlt, L. 55 cm; Mehlis 37
2355 ohne Ang., Firmenmarke 19, Ausformung 20er Jahre, H. 28 cm; Ahlden 66
2355 (Firmenmarke ohne Ang.), H. 33 cm; GSF, S. 31
Ausst.: ÖMKI 1911/12, hier mit Ang. des Originaltitels „Falke".
Lit.: GSF, S. 31, hier mit der Ang. des Originaltitels „Falke"; Neuwirth (1974), S. 75.

2356
FIGUR „Lampe Chrysantheme"
ca. 1902
2356 (Firmenmarke ohne Ang.), H. 40 cm; GSF, S. 31
Bem.: Lt. Firmenkatalog wurde dieses Modell auch mit einer Lampe (Glühbirne) montiert angeboten.
Lit.: GSF, S. 31, hier mit der Ang. des Originaltitels „Lampe Chrysantheme".

2357
Tell, E.

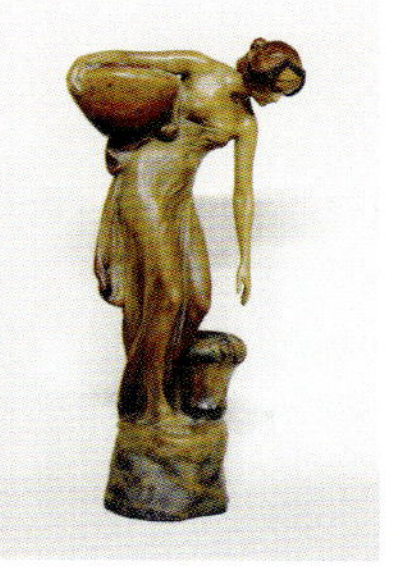

FIGUR „Rückkehr vom Brunnen" (Frau mit Gefäß, sich nach Krug bückend)
ca. 1901/02
2357,7,65 Firmenmarke 9, Repr. Res., H. 67 cm; KH
2357,747,19 Firmenmarke Sculpture 13, Fabriqué en Autriche, gepr. „S"= Z2, H. 64 cm; MAK Wien, Inv. Ke 10602-47
ohne Ang., H. 65,7 cm; Soth. Arcade 1486
2357 (Firmenmarke ohne Ang.), H. 62 cm; GSF, S. 31
Ausst.: ÖMKI 1902/03; 1903/04, hier jeweils mit Ang. des Originaltitels „Rückkehr vom Brunnen".
-> vgl. 2447; 2876
Lit.: GSF, S. 31, hier mit der Ang. des Originaltitels „Rückkehr vom Brunnen"; Pudor (1902), S. 107 (m. Abb.), hier mit Ang. des Originaltitels „Rückkehr vom Brunnen".

2358
Tell, E.
FIGUR „Kugelträgerin mit Glaskugel"
ca. 1902
2358 (Firmenmarke ohne Ang.), H. 70 cm; GSF, S. 31
Bem.: Lt. Firmenkatalog wurde die Glaskugel mit einer Glühbirne elektrifiziert.
Lit.: GSF, S. 31, hier mit der Ang. des Originaltitels „Kugelträgerin mit Glaskugel".

2359
Tell, E.
FIGUR (SPIEGEL) „Spiegelträgerin" (Stehender Frauenakt, einen Spiegel haltend und in diesen schauend)
ca. 1899/1900 (für den Entwurf)
2359 (Firmenmarke ohne Ang.), H. 80 cm; GSF, S. 31
Bem.: Diese Größe wurde lt. Firmenkatalog mit und ohne Spiegel angeboten.
-> vgl. 532 (ähnl. Modell)
-> vgl. 2040; 2109 (Variante, bekleidet)
Lit.: GSF, S. 31, hier mit der Ang. des Originaltitels „Spiegelträgerin".

2360
BÜSTE „Orchidée"
ca. 1902
2360 (Firmenmarke ohne Ang.), H. 70 cm; GSF, S. 31
Lit.: GSF, S. 31, hier mit der Ang. des Originaltitels „Orchidée".

2361
Cherc
FIGUR „Die Fütterung"
ca. 1902
2361 (Firmenmarke ohne Ang.), H. 90 cm; GSF, S. 31
Ausst.: ÖMKI 1902/03, hier jeweils mit Ang. des Titels „Fütterung".
Lit.: GSF, S. 31, hier mit der Ang. des Originaltitels „Die Fütterung"; Neuwirth (1974), S. 61.

2362
FIGUR „Fischerin"
ca. 1902
2362 (Firmenmarke ohne Ang.), H. 110 cm; GSF, S. 31
Lit.: GSF, S. 31, hier mit der Ang. des Originaltitels „Fischerin".

2363
Cherc
BÜSTE „Tulpe, II. Gr(öße)"
ca. 1900/01 (für den Entwurf)
2363 (Firmenmarke ohne Ang.), H. 54 cm; GSF, S. 31
Ausst.: ÖMKI 1901/02; 1902/03, hier mit Ang. des Originaltitels „Tulpe".
-> vgl. 2207; 2367
Lit.: GSF, S. 31, hier mit der Ang. des Originaltitels „Tulpe, II. Gr(öße)".

2364
Gornik, Friedrich
FIGUR „Eule" (mit Schreibzeug, auf zwei Büchern sitzend)
ca. 1901/02
2364,9,13 Firmenmarke Sculpture 13, Repr. Res., H. 48 cm; Wendl 26; Bem.: „mit elektrifizierten. Glasaugen".
2364 (Firmenmarke ohne Ang.), H. 50 cm; GSF, S. 31
Bem.: Die Glasaugen mit denen das Tier auch ausgestattet war, wurden durch eine Glühbirne elektrifiziert.
-> vgl. 3853; 3854
Lit.: GSF, S. 31, hier mit der Ang. des Originaltitels „Eule"; Pudor (1902), S. 109 (m. Abb.) und Ang. des Originaltitels „Eule"; KuKHW, 6.1903, S. 526 (m. Abb.); Neuwirth (1974), S. 63.

2365
Suchanek, Vacslav
FIGUR „Adler" (mit Beute) (auf Natursockel)
ca. 1902
2365,507,17 (Firmenmarke ohne Ang.), H. 70,5 cm; Doro 20.10.1997
2365 (Firmenmarke ohne Ang.), H. 68 cm; GSF, S. 31
Ausst.: ÖMKI 1902/03, hier mit Ang. des Originaltitels „Adler".
Lit.: GSF, S. 31, hier mit der Ang. des Originaltitels „Adler"; Neuwirth (1974), S. 75.

2366
Suchanek, Vacslav
FIGUR „Auerhahn"
ca. 1902
2366 (Firmenmarke ohne Ang.), H. 58 cm; GSF, S. 31
Ausst.: ÖMKI 1902/03, 1911/12, hier jew. mit Ang. des Originaltitels „Auerhahn".
Lit.: GSF, S. 31, hier mit der Ang. des Originaltitels „Auerhahn"; Neuwirth (1974), S. 75.

2367
Cherc
BÜSTE „Tulpe, III. Gr(öße)"
ca. 1900/01 (für den Entwurf)
2367 (Firmenmarke ohne Ang.), H. 40 cm; Pinhas, S. 54 (m. Abb.)
2367 (Firmenmarke ohne Ang.), H. 53 cm (vermutl. Druckfehler); GSF, S. 31
Ausst.: ÖMKI 1901/02; 1902/03, hier mit Ang. des Originaltitels „Tulpe".
-> vgl. 2207; 2363
Lit.: GSF, S. 31, hier mit der Ang. des Originaltitels „Tulpe, III. Gr(öße)".

2368
Sydan

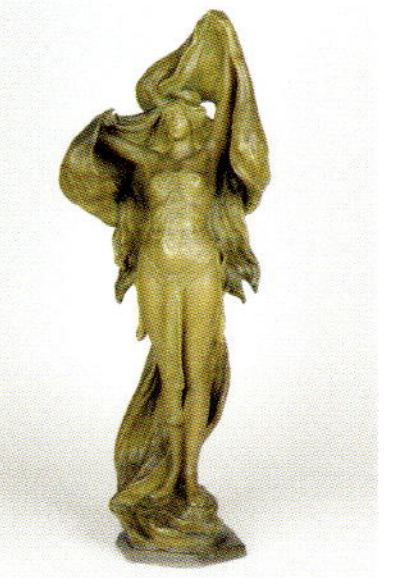

FIGUR „Sonnenaufgang, II. Gr(öße)" (Stehende junge Frau mit wallendem Tuch auf Felssockel) (auf sechseckiger Plinthe)
ca. 1900/01 (für den Entwurf)
2368,13,58 gem. II, Firmenmarke Sculpture 13, H. 76 cm; PB Wien 19
2368 (Firmenmarke ohne Ang.), H. 78 cm; GSF, S. 32
Bem.: Lt. Firmenkatalog wurde dieses Modell auch elektrifiziert angeboten.
-> vgl. ohne Nr.
-> vgl. 2199b
Lit.: GSF, S. 32, hier mit der Ang. des Originaltitels „Sonnenaufgang, II. Gr(öße)".

2369
Cherc
BÜSTE „Flieder"
ca. 1900 (für den Entwurf)
2369 (Firmenmarke ohne Ang.), H. 47 cm; GSF, S. 32
-> vgl. 2094; 2370
Lit.: GSF, S. 32, hier mit der Ang. des Originaltitels „Flieder, II. Gr(öße)".

2370
Cherc
BÜSTE „Flieder"
ca. 1900 (für den Entwurf)
2370 (Firmenmarke ohne Ang.), H. 36 cm; GSF, S. 32
-> vgl. 2094; 2369
Lit.: GSF, S. 32, hier mit der Ang. des Originaltitels „Flieder, III. Gr(öße)".

2371
Haniroff/Xaniroff
BÜSTE „La Fierté"
Inschrift: LA FIERTÉ
ca. 1900/01 (für den Entwurf)
2371 (Firmenmarke ohne Ang.), H. 44 cm; GSF, S. 32
Text der Inschrift: Der Stolz
Ausst.: ÖMKI 1901/02, hier mit Ang. des Originaltitels „La Fierté" und des falschen Künstlernamens „Xaniroff".
-> vgl. 2178; 2435
Lit.: GSF, S. 32, hier mit der Ang. des Originaltitels „La Fierté".

2372
Pecheur
FIGUR „Poesie"
ca. 1900/01 (für den Entwurf)
2372 (Firmenmarke ohne Ang.), H. 69 cm; GSF, S. 32
Ausst.: ÖMKI 1901/02, hier mit Ang. des Originaltitels „Poesie".
Lit.: GSF, S. 32, hier mit der Ang. des Originaltitels „Poesie"; Neuwirth (1974), S. 69.

2373
FIGUR „Jongleuse"
ca. 1902
2373 (Firmenmarke ohne Ang.), H. 78 cm; GSF, S. 32
Bem.: Lt. Firmenkatalog wurde dieses Modell auch elektrifiziert angeboten.
Lit.: GSF, S. 32, hier mit der Ang. des Originaltitels „Jongleuse".

2374
BÜSTE „Weißheit" (Frau mit Eule als Kopfputz) (auf Rechtecksockel aus Büchern)
ca. 1902
2374,9,13 ger. S, gem. weiß 32, Firmenmarke 7, Repr. Res., H. 67 cm; Wien Museum Inv. 219.121
2374 (Firmenmarke ohne Ang.), H. 62 cm; GSF, S. 32
Lit.: GSF, S. 32, hier mit der Ang. des Originaltitels „Weißheit".

2375
WANDMASKE „Beethoven, ohne Sockel"
ca. 1902
2375 (Firmenmarke ohne Ang.), H. 27 cm; GSF, S. 32
Bem.: Dargestellt ist der Komponist Ludwig van Beethoven (1770–1827).
Lit.: GSF, S. 32, hier mit der Ang. des Originaltitels „Beethoven, ohne Sockel".

2376
FIGUR „Elefant"
ca. 1902
2376 (Firmenmarke ohne Ang.), H. 38 cm; GSF, S. 32
Lit.: GSF, S. 32, hier mit der Ang. des Originaltitels „Elefant".

2377
FIGUR „Korbträgerin"
ca. 1902
2377 (Firmenmarke ohne Ang.), H. 72 cm; GSF, S. 32
Lit.: GSF, S. 32, hier mit der Ang. des Originaltitels „Korbträgerin".

2378
FIGUR „Naive"
ca. 1902
2378 (Firmenmarke ohne Ang.), H. 50 cm; GSF, S. 32
-> vgl. 2403= BÜSTE
Lit.: GSF, S. 32, hier mit der Ang. des Originaltitels „Naive".

2379
FIGUR „Mignon"
ca. 1902
2379 (Firmenmarke ohne Ang.), H. 76 cm; GSF, S. 32
Lit.: GSF, S. 32, hier mit der Ang. des Originaltitels „Mignon".

2380
FIGUR „Fischerknabe"
ca. 1902
2380 (Firmenmarke ohne Ang.), H. 104 cm; GSF, S. 32
Lit.: GSF, S. 32, hier mit der Ang. des Originaltitels „Fischerknabe".

2381
Cherc
BÜSTE „Dictator" (Kaiser Caligula? mit Lorbeerkranz)
Inschrift: ODERINT DUM METUANT
ca. 1902
2381,71,19 Firmenmarke Sculpture 13, Repr. Res., H. 64 cm; Wendl 43
2381,73,1(9?) gem. weiß 19., Firmenmarke Sculpture 13, Repr. Res., H. 64,1 cm; eBay # 1288044855
2381 (Firmenmarke ohne Ang.), H. 63 cm; GSF, S. 32
Text der Inschrift: „Mögen sie (mich) nur hassen, wenn sie (mich) nur fürchten" (nach Caligula 12–41)
Bem.: Dieses Zitat (Inschrift) entstammt der Tragödie „Atreus" von Accius (170–104 v. Chr.) und war ein Lieblingsspruch Caligulas (vgl. hierzu: Büchmann, Geflügelte Worte, Berlin 1928, S. 202). Ob es sich bei dem Dargestellten um ein Porträt Caligulas handelt, erscheint zweifelhaft.
Lit.: GSF, S. 32, hier mit der Ang. des Originaltitels „Dictator".

2382
BÜSTE „Herbstzeitlose"
ca. 1902
2382 (Firmenmarke ohne Ang.), H. 68 cm; GSF, S. 32
Lit.: GSF, S. 32, hier mit der Ang. des Originaltitels „Herbstzeitlose".

2383
Gornik, Friedrich
FIGUR „Hahn"
ca. 1902
2383,7,34 Firmenmarke Sculpture 13, Repr. Res., sign. Gornic (!), H. 75 cm; Wien Museum Inv. 190.054
2383 (Firmenmarke ohne Ang.), H. 75 cm; GSF, S. 32
Bem.: Der Entwurf dieses Modells ist zwischen 1900–1902 zu datieren.
-> vgl. 4168
Ausst.: ÖMKI 1912, hier mit Ang. des Originaltitels „Hahn".
Lit.: GSF, S. 32, hier mit der Ang. des Originaltitels „Hahn"; KuKHW, 6.1903, S. 528 (m. Abb.).

2384
FIGUR „Truthahn"
ca. 1902
2384 (Firmenmarke ohne Ang.), H. 61 cm; GSF, S. 32
Lit.: GSF, S. 32, hier mit der Ang. des Originaltitels „Truthahn".

2385 -> ÜBERSCHNEIDUNG
Gory, A(ffortunato?)
BÜSTE „Arôme" (Frau mit Callablütenstengeln)
Inschrift: ARÔME
ca. 1902
2385,84,19 (Firmenmarke ohne Ang.), undeutlich sign., H. 59 cm; Christie's SK 4./5.12.1996; SK 14.3.1997; 4.11.1999; Bem.: Diese Büste ohne die Inschrift.
2385,103,15 gem. XXX, ger. 4, Firmenmarke Sculpture 13, Repr. Res., sign. „F. Gori", H. 60,5 cm; Quit. 14
2385,108,19 (Firmenmarke ohne Ang.), Repr. Res., unsign.?, H. 61 cm; Christie's SK 21.4.2004
2385 (Firmenmarke ohne Ang.), H. 60 cm; GSF, S. 32
Text der Inschrift: Aroma
-> vgl. 2885 (ähnl. Modell)
Lit.: GSF, S. 32, hier mit der Ang. des Originaltitels „Arôme"; Pinhas, S. 53 (m. Abb.), hier mit der falschen Künstlerang. „F. Gornik".

2385 -> ÜBERSCHNEIDUNG
Lavergne, Adolphe-Jean
JARDINIERE Mädchen mit zwei Körben
ca. 1902
2385 (ohne Ang.), Neuwirth (1974), S. 58, 66

2386
BÜSTE „Yvette"
ca. 1902
2386 (Firmenmarke ohne Ang.), H. 44 cm; GSF, S. 32
Lit.: GSF, S. 32, hier mit der Ang. des Originaltitels „Yvette".

2387
Lesca
BÜSTE „Labor" (Frau mit langem Haar und Kopftuch) (Sockelrelief: Frau mit Spaten)
ca. 1902
2387 (ohne Ang.), H. 47 cm; Henry's 26.6.1996
2387 (Firmenmarke ohne Ang.), H. 47 cm; GSF, S. 32
Bem.: Das Relief am Sockel dieser Büste, ist die Wiedergabe der Figur „Labor" (Mod.-Nr. 2316; 2573).
-> vgl. 2316; 2573= jew. FIGUR
Lit.: GSF, S. 32, hier mit der Ang. des Originaltitels „Labor".

2388
Puchegger, Anton
FIGUR „Bär"
ca. 1902/03
2388 (Firmenmarke ohne Ang.), H. 25 cm; GSF, S. 32
Ausst.: ÖMKI 1911/12, hier mit Ang. des Originaltitels „Bär".
Lit.: GSF, S. 32, hier mit der Ang. des Originaltitels „Bär"; Neuwirth (1974), S. 72.

2389
FIGUR „Kätzchen"
ca. 1902/03
2389 (Firmenmarke ohne Ang.), H. 21 cm; GSF, S. 32
Lit.: GSF, S. 32, hier mit der Ang. des Originaltitels „Kätzchen".

2390
FIGUR „Aphrodite mit Terracottamuschel"
ca. 1902/03
2390 (Firmenmarke ohne Ang.), H. 86 cm; GSF, S. 32
Bem.: Wie der Firmenkat. belegt, wurde der dieses Modell auch „elektr. mit Glasmuschel (Lampe)" geliefert. Die Anzahl der Beleuchtungskörper geht jedoch nicht aus der Quelle hervor. Dargestellt ist Aphrodite, die in der griech. Mythologie Göttin der Liebe, Schönheit und sinnlichen Begierde ist. (Quelle: Wikipedia).
-> vgl. 2342
Lit.: GSF, S. 32, hier mit der Ang. des Originaltitels „Aphrodite mit Terracottamuschel".

2391
FIGUR „Sackträgerin"
ca. 1902/03
2391 (Firmenmarke ohne Ang.), H. 40 cm; GSF, S. 32
Lit.: GSF, S. 32, hier mit der Ang. des Originaltitels „Sackträgerin".

2392
FIGUR „Parze"
ca. 1902/03
2392 (Firmenmarke ohne Ang.), H. 50 cm; GSF, S. 32
Bem.: Dargestellt ist eine der drei Schicksalsgöttinnen der röm Mythologie. (Quelle: Wikipedia).
Lit.: GSF, S. 32, hier mit der Ang. des Originaltitels „Parze".

2393
BÜSTE „Les Cerises"
ca. 1902/03
2393 (Firmenmarke ohne Ang.), H. 60 cm; GSF, S. 32
Lit.: GSF, S. 32, hier mit der Ang. des Originaltitels „Les Cerises".

2394
Piquard
BÜSTE „Mais" (Bauernmädchen mit Kopftuch)
(Sockel reliefiert mit Maiskolben)
ca. 1902/03
2394,(107,83?) Firmenmarke Sculpture 13, Repr. Res., H. 36,5 cm; eBay # 1400152015
2394 (Firmenmarke ohne Ang.), H. 34 cm; GSF, S. 32
Lit.: GSF, S. 32, hier mit der Ang. des Originaltitels „Mais".

2395
BÜSTE „Venetianerin"
ca. 1902/03
2395 (Firmenmarke ohne Ang.), H. 25 cm; GSF, S. 32
Lit.: GSF, S. 32, hier mit der Ang. des Originaltitels „Venetianerin".

2396
BÜSTE „Traumverloren"
ca. 1902/03
2396 (Firmenmarke ohne Ang.), H. 21 cm; GSF, S. 32
Lit.: GSF, S. 32, hier mit der Ang. des Originaltitels „Traumverloren".

2397
BÜSTE „Immortelle"
ca. 1902/03
2397 (Firmenmarke ohne Ang.), H. 38 cm; GSF, S. 32
Lit.: GSF, S. 32, hier mit der Ang. des Originaltitels „Immortelle".

2398
Vansteen
BÜSTE „Holländisches Kind" (Mädchen mit Spitzenhaube) (nach rechts blickend)
ca. 1902/03
FAYENCE
F2398,59,45 FG (Firmenmarke?), H. 24 cm; KaM 110
F2398,60,68 Firmenmarke 9, H. 24 cm; KH
TERRAKOTTA
2398,200,41,G gem. weiß VII, Firmenmarke 4, Ges. gesch., H. 24,1 cm; eBay # 2643164307
2398 (Firmenmarke ohne Ang.), H. 23 cm; GSF, S. 32
MARMOR
2398,H,12 Plak. A, H. 24,5 cm; Fischer 71
Bem.: Dieses Modell wurde in Fayence, Terrakotta und Marmor (Alabaster/Kunststein?) ausgeführt.
–> vgl. 2635
Lit.: GSF, S. 32, hier mit der Ang. des Originaltitels „Holländisches Kind".

2399
BÜSTE „Orangenblüte"
ca. 1901 (für den Entwurf)
2399 (Firmenmarke ohne Ang.), H. 38 cm; GSF, S. 32
–> vgl. 2252
Lit.: GSF, S. 32, hier mit der Ang. des Originaltitels „Orangenblüte".

2400
BÜSTE „Vision"
ca. 1902/03
2400 (Firmenmarke ohne Ang.), H. 38 cm; GSF, S. 32
Lit.: GSF, S. 32, hier mit der Ang. des Originaltitels „Vision".

2401
Lexa, H. (V.?)
FIGUR „Erdarbeiter"
ca. 1902/03
2401 (Firmenmarke ohne Ang.), H. 28 cm; GSF, S. 32
Ausst.: ÖMKI 1910/11, hier mit Ang. des Originaltitels „Erdarbeiter".
Lit.: GSF, S. 32, hier mit der Ang. des Originaltitels „Erdarbeiter"; Neuwirth (1974), S. 66.

2402
FIGUR „Gastfreundschaft"
ca. 1902/03
2402 (Firmenmarke ohne Ang.), H. 72 cm; GSF, S. 33
Bem.: Wie der Firmenkat. belegt, wurde der dieses Modell auch „elektr. mit einer Lampe mit Bronzeaufsatz" geliefert.
Lit.: GSF, S. 33, hier mit der Ang. des Originaltitels „Gastfreundschaft".

2403
BÜSTE „Naive" (Junge Frau mit geflochtenem Zopf)
ca. 1902/03
2403 (Firmenmarke ohne Ang.), H. 65 cm; Pinhas, S. 54 (m. Abb.)
2403 (Firmenmarke ohne Ang.), H. 61 cm; GSF, S. 33
–> vgl. 2378= FIGUR
Lit.: GSF, S. 33, hier mit der Ang. des Originaltitels „Naive".

2404
Cherc
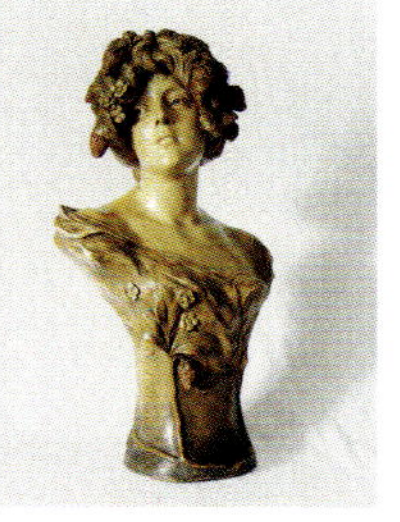
BÜSTE „Erdbeere" (Junge Frau nach rechts blickend) (Haare und Sockel geschmückt mit Erdbeeren)
ca. 1902/03
2404,1,30,7 gem. XXX, Firmenmarke Sculpture 13, Repr. Res., gepr. „S"= Z2, unsign., H. 64 cm; PB Deutschland
2404,61,13 gem. XXX, Firmenmarke Sculpture 13, Repr. Res., unsign., H. 62 cm; eBay # 7346161718
2404 (Firmenmarke ohne Ang.), H. 61 cm; GSF, S. 33
–> vgl. 2621
Lit.: GSF, S. 33, hier mit der Ang. des Originaltitels „Erdbeere".

2405
Haniroff
FIGUR „Harfenspielerin" (Stehende junge Frau mit Harfe) (auf Landschaftssockel)
ca. 1902/03
2405 (Firmenmarke ohne Ang.), H. 102,5 cm; Pinhas, S. 38 (m. Abb.)
2405 (Firmenmarke ohne Ang.), H. 90 cm; GSF, S. 33
Lit.: GSF, S. 33, hier mit der Ang. des Originaltitels „Harfenspielerin".

2406
FIGUR „Holländische Vasenträgerin" (Stehende, junge Holländerin, einen großen Henkelkrug tragend)
ca. 1902/03
2406 (ohne Ang.), Firmenmarke Sculpture 13, gepr. „C"= Z5, H. 95 cm; RM 34
2406 (Firmenmarke ohne Ang.), H. 97 cm; GSF, S. 33
Lit.: GSF, S. 33, hier mit der Ang. des Originaltitels „Holländische Vasenträgerin".

2407
FIGUR „Katze (am Dach)"
ca. 1902/03
2407 (Firmenmarke ohne Ang.), H. 46 cm; GSF, S. 33
Bem.: Wie der Firmenkat. belegt, wurde der dieses Modell auch „elektr." geliefert.
Lit.: GSF, S. 33, hier mit der Ang. des Originaltitels „Katze (am Dach)".

2408
GRUPPE „Wüstenfüchse"
ca. 1902/03
2408 (Firmenmarke ohne Ang.), H. 38 cm, L. 61 cm; GSF, S. 33
Bem.: Möglicherweise stammen diese Wüstenfüchse von Sigismund Wernekinck, einem Tierbildhauer, der für Goldscheider etliche Figuren lieferte und dieses Sujet in der Mod.-Nr. 4383 ebenfalls aufgriff. Da aber bis Redaktionsschluss weder ein sign. Exemplar dieser Gruppe, noch eine Abb. vorlag, muss dies eine vage Vermutung bleiben.
Lit.: GSF, S. 33, hier mit der Ang. des Originaltitels „Wüstenfüchse".

2409
Haniroff
FIGUR „Vasenträgerin" (Stehende junge Frau eine Amphore auf der rechten Schulter tragend) (auf Landschaftssockel)
ca. 1901 (für den Entwurf)
2409 (Firmenmarke ohne Ang.), H. 86 cm; GSF, S. 33
–> vgl. 2238; 2583
Lit.: GSF, S. 33, hier mit der Ang. des Originaltitels „Vasenträgerin".

2410
Cherc
BÜSTE „Kastanie" (Junge Frau mit Kastanienzweigen und Blattwerk)
ca. 1901 (für den Entwurf)
2410,188,67 FG Wien (Firmenmarke?), Repr. Res., H. 53,5 cm; Soth. Olympia 1.6.2005
2410,254,11 FG Wien (Firmenmarke?), Repr. Res., H. 53 cm; eBay # 1498761508
2410 (Firmenmarke ohne Ang.), H. 52 cm; GSF, S. 33
–> vgl. 2255
Lit.: GSF, S. 33, hier mit der Ang. des Originaltitels „Kastanie".

2411
Villieu
BÜSTE „Lilie" (Junge Frau in antikisierendem Gewand mit Blumenkranz im Haar, Dekolleté mit Lilienblüten geschmückt)
ca. 1899/1900 (für den Entwurf)
2411 (Firmenmarke ohne Ang.), H. 33 cm; GSF, S. 33
Ausst.: ÖMKI 1900/01, hier mit dem franz. Originaltitel „Pureté" ausgestellt.
–> vgl. 1981; 2213
Lit.: GSF, S. 33, hier mit der Ang. des Originaltitels „Lilie".

2412
Ceccarelli, Ezio
BÜSTE „Jeunesse" (Lachendes nach rechts blickendes Mädchen mit Tulpenschmuck)
Inschrift: JEUNESSE
ca. 1899 (für den Entwurf)
2412 (Firmenmarke ohne Ang.), H. 26 cm; Pinhas, S. 58 (m. Abb.)
2412 (Firmenmarke ohne Ang.), H. 25 cm; GSF, S. 33
Ausst.: ÖMKI 1899/1900, hier mit Ang. des Originaltitels „Jeunesse".
–> vgl. 1856; 2120
Lit.: GSF, S. 33, hier mit der Ang. des Originaltitels „Jugend"; Neuwirth (1974), S. 61.

2413
SPIEGEL „Vanitas"
ca. 1902/03
2413 (Firmenmarke ohne Ang.), H. 126 cm; GSF, S. 33
Lit.: GSF, S. 33, hier mit der Ang. des Originaltitels „Vanitas".

2414
FIGUR „Mond"
ca. 1902/03
2414 (Firmenmarke ohne Ang.), H. 65 cm; GSF, S. 33
Bem.: Wie der Firmenkat. belegt, wurde der dieses Modell auch „elektr." geliefert.
Lit.: GSF, S. 33, hier mit der Ang. des Originaltitels „Mond".

2415
FIGUR „Ampelträgerin"
ca. 1902/03
2415 (Firmenmarke ohne Ang.), H. 76 cm; GSF, S. 33
Bem.: Wie der Firmenkat. belegt, wurde der dieses Modell auch „elektr." geliefert.
Lit.: GSF, S. 33, hier mit der Ang. des Originaltitels „Ampelträgerin".

2416
BÜSTE „Freiheit"
ca. 1902/03
2416 (Firmenmarke ohne Ang.), H. 63 cm; GSF, S. 33
Lit.: GSF, S. 33, hier mit der Ang. des Originaltitels „Freiheit".

2419
BÜSTE „Winter"
ca. 1902/03
2419 (Firmenmarke ohne Ang.), H. 72 cm; GSF, S. 33
Lit.: GSF, S. 33, hier mit der Ang. des Originaltitels „Winter".

2420
BÜSTE „Francesca"
ca. 1902/03
2420 (Firmenmarke ohne Ang.), H. 49 cm; GSF, S. 33
Lit.: GSF, S. 33, hier mit der Ang. des Originaltitels „Francesca".

2421
GRUPPE „Spielkatzen"
ca. 1902/03
2421 (Firmenmarke ohne Ang.), H. 29 cm; GSF, S. 33
Bem.: Wie der Firmenkat. belegt, wurde der dieses Modell auch „elektr." geliefert.
Lit.: GSF, S. 33, hier mit der Ang. des Originaltitels „Spielkatzen".

2422
BÜSTE „Pensée"
ca. 1902/03
2422 (Firmenmarke ohne Ang.), H. 47 cm; GSF, S. 33
Lit.: GSF, S. 33, hier mit der Ang. des Originaltitels „Pensée".

2423
Gornik, Friedrich
FIGUR „Kakadu"
ca. 1902/03
2423 (Firmenmarke ohne Ang.), H. 73 cm; GSF, S. 33
Ausst.: ÖMKI 1903/04, hier mit Ang. des Originaltitels „Kakadu".
Lit.: GSF, S. 33, hier mit der Ang. des Originaltitels „Kakadu"; Neuwirth (1974), S. 63.

2424
TAFELAUFSATZ „Sturmflut"
ca. 1902/03
2424 (Firmenmarke ohne Ang.), H. 52 cm, L. 65 cm; GSF, S. 33
Bem.: Wie der Firmenkat. belegt, wurde der dieses Modell auch „elektr." geliefert.
Lit.: GSF, S. 33, hier mit der Ang. des Originaltitels „Sturmflut".

2425
FIGUR „Gärtnerin"
ca. 1902/03
2425 (Firmenmarke ohne Ang.), H. 78 cm; GSF, S. 33
Lit.: GSF, S. 33, hier mit der Ang. des Originaltitels „Gärtnerin".

2426
Gambeauche
FIGUR „Lygia"
ca. 1902/03
2426 (Firmenmarke ohne Ang.), H. 74 cm; GSF, S. 33
Bem.: Dargestellt ist die Sklavin „Lygia", eine Figur des alten Roms aus der Zeit Neros. Die Zuordnung an den nicht näher verifizierbaren Künstler „Gambeauche", konnte durch die sign. Büste „Lygia" (Mod.-Nr. 2639) vorgenommen werden.
–> vgl. 2639= BÜSTE
Lit.: GSF, S. 33, hier mit der Ang. des Originaltitels „Lygia".

2427
FIGUR „Schmied"
ca. 1902/03
2427 (Firmenmarke ohne Ang.), H. 61 cm; GSF, S. 33
Bem.: Möglicherweise handelt sich hier um die auf der Leipziger Michaelimesse 1904 vorgestellte Figur „Schmied mit Schurzfell" von Haniroff.
Lit.: GSF, S. 33, hier mit der Ang. des Originaltitels „Schmied".

2428
Tell, E.
FIGUR „Serpentintanz" (Loie Fuller) (auf Natursockel)
ca. 1900 (für den Entwurf)
2428 (Firmenmarke ohne Ang.), H. 68 cm; GSF, S. 33
Bem.: Dargestellt ist die berühmte Tanzkünstlerin Loie Fuller (1862–1928) bei einem ihrer Tänze. Im Firmenkat. ist vermerkt, dass auch eine Ausführung als Lampe (elektrifiziert mit zwei Glühbirnen) angeboten wurde.
–> vgl. 2064
Lit.: GSF, S. 33, hier mit der Ang. des abweichenden Originaltitels „Serpentintanz".

2429
BÜSTE „Friede"
ca. 1902/03
2429 (Firmenmarke ohne Ang.), H. 54 cm; GSF, S. 33
Lit.: GSF, S. 33, hier mit der Ang. des Originaltitels „Friede".

2430
BÜSTE „Salambo" (Junge Dame mit Hochsteckfrisur und Medaillon um den Hals)
Inschrift: SALAMBO
ca. 1899 (für den Entwurf)
2430 (Firmenmarke ohne Ang.), H. 30 cm; GSF, S. 33
Bem.: Dargestellt ist die Priesterin Salambo, die Tochter des punischen Heerführers Hamilkars und Schwester Hannibals. Gustave Flaubert (1821–1864) widmete ihr auch einen gleichnamigen Roman, der 1863 erschien und dann 1904 ins dt. übersetzt wurde.
–> vgl. 1755; 2183
Lit.: GSF, S. 33, hier mit der Ang. des Originaltitels „Salambo".

2431
BÜSTE „Adrienne"
ca. 1902/03
2431 (Firmenmarke ohne Ang.), H. 54 cm; GSF, S. 33
Lit.: GSF, S. 33, hier mit der Ang. des Originaltitels „Adrienne".

2432
Nelson, (Alphonse-Henri?)
BÜSTE „Oudine"
ca. 1900 (für den Entwurf)
2432 (Firmenmarke ohne Ang.), H. 30 cm; GSF, S. 33
Bem.: Diese Büste findet sich unter den Nrn. 855/918/919 im Kat. GSB wieder. Sie wurde in der Pariser Produktion in den Höhen 53/40/12 cm gefertigt. Ang. zum Material (Terrakotta/Bronze) fehlen.
Ausst.: ÖMKI 1900/01; 1902/03, hier mit Ang. des abweichenden Originaltitels „Ondine".
–> vgl. 2212; 2432
–> vgl. GSB 855; 918; 919
Lit.: GSF, S. 33, hier mit der Ang. des Originaltitels „Oudine".

2433
Simon
BÜSTE „Pax" (Junge Frau mit Diadem aus Lorbeer)
ca. 1900/01 (für den Entwurf)
2433 (Firmenmarke ohne Ang.), H. 29 cm; GSF, S. 33
Ausst.: ÖMKI 1901/02, hier mit Ang. des Originaltitels „Pax".
–> vgl. 2187; 2280
Lit.: GSF, S. 33, hier mit der Ang. des Originaltitels „Pax".

2434
Montenave
BÜSTE „Lucrezia" (nach links blickend) (auf konischem Sockel)
ca. 1900 (für den Entwurf)
2434 (Firmenmarke ohne Ang.), H. 26 cm; GSF, S. 33
Bem.: Dargestellt ist Lukrezia (Lucrezia) Borgia (1480–1519), eine ital. Fürstin span. Abstammung. (Quelle: Wikipedia)
–> vgl. 2103; 2224
Lit.: GSF, S. 33, hier mit der Ang. des Originaltitels „Lucrezia".

2435
Haniroff / Xaniroff
BÜSTE „Stolz"
ca. 1900/01 (für den Entwurf)
2435 (Firmenmarke ohne Ang.), H. 31 cm; GSF, S. 34
–> vgl. 2178; 2371
Lit.: GSF, S. 34, hier mit der Ang. des Originaltitels „Stolz".

2436
Cherc
FIGUR „Mignon" (Stehende junge Frau mit langem Haar) (nach rechts blickend)
ca. 1898/99 (für den Entwurf)
2436 (Firmenmarke ohne Ang.), H. 39 cm; GSF, S. 34
–> vgl. 1643; 2072
Lit.: GSF, S. 34, hier mit der Ang. des Originaltitels „Mignon".

2437
BÜSTE „Schilf"
ca. 1903
2437 (Firmenmarke ohne Ang.), H. 56 cm; GSF, S. 34
Lit.: GSF, S. 34, hier mit der Ang. des Originaltitels „Schilf".

2438
SPIEGEL „Neugierige"
ca. 1903
2438 (Firmenmarke ohne Ang.), H. 105 cm; GSF, S. 34
Lit.: GSF, S. 34, hier mit der Ang. des Originaltitels „Neugierige".

2439
FIGUR „Muschelträgerin"
ca. 1903
2439 (Firmenmarke ohne Ang.), H. 65 cm; GSF, S. 34
Lit.: GSF, S. 34, hier mit der Ang. des Originaltitels „Muschelträgerin".

2440
Fischer, Friedrich / Haniroff
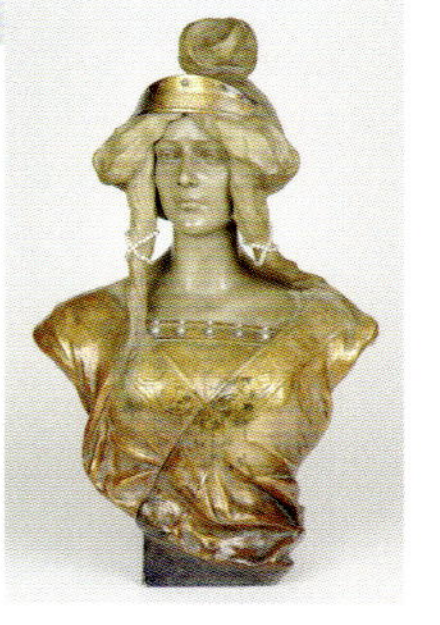
BÜSTE „Byzanz" (Frau mit Zöpfen und Diadem)
ca. 1903
2440,25,34 gem. weiß 8, Firmenmarke Sculpture 13, Repr. Res., sign. Haniroff, H. 74,1 cm; PB Wien 19
2440 (Firmenmarke ohne Ang.), H. 73 cm; GSF, S. 34
Bem.: Besetzt mit Perlen und Glassteinen.
Ausst.: ÖMKI 1903/04, hier mit Ang. des Originaltitels „Byzanz", sowie der Ang. des Künstlers Friedrich Fischer.
Lit.: GSF, S. 34, hier mit der Ang. des Originaltitels „Byzanz"; Neuwirth (1974), S. 63.

2441
BÜSTE „Unschuld"
ca. 1903
2441 (Firmenmarke ohne Ang.), H. 58 cm; GSF, S. 34
Lit.: GSF, S. 34, hier mit der Ang. des Originaltitels „Unschuld".

2442
Gornik, Friedrich
FIGUR „Eisbär"
ca. 1903
ohne Ang., (Firmenmarke ohne Ang.), H. 41 cm; Acatos, 1991; Bem.: Hier mit der Künstlerang. „Lornic", gemeint ist aber Gornik.
2442 (Firmenmarke ohne Ang.), H. 40 cm; GSF, S. 34
Bem.: Wie der Firmenkat. belegt, wurde der dieses Modell auch „elektr. mit Glassockel" geliefert.
Lit.: GSF, S. 34, hier mit der Ang. des Originaltitels „Eisbär".

2443
Bertrand, A.
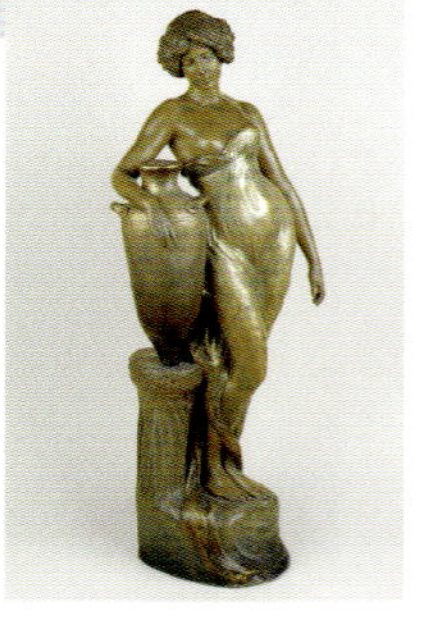
FIGUR „Träumerei" (Stehende Frau mit Amphore)
ca. 1901/02 (für den Entwurf)
2443,148,12 ohne Firmenmarke, H. 67 cm; PB Wien 19
2443 (Firmenmarke ohne Ang.), H. 65 cm; GSF, S. 34
–> vgl. 2272
Lit.: GSF, S. 34, hier mit der Ang. des Originaltitels „Träumerei"; GSN, S. 8, 9 (m. Abb.); Pinhas, S. 45 (m. Abb.), hier sign. Cherc.
GS neu: Modell 63 „La Tendresse" (H. 82cm), Aufl. 1000 Stück

2444
Haniroff
FIGUR „Jagdtrophäe" (Leda besiegt den Schwan)
ca. 1900/01 (für den Entwurf)
2444,77,24 Firmenmarke Sculpture 13, Repr. Res., sign. Haniroff, H. 83 cm; Christie's SK 4.11.1999; Bem.: Vollständiges Exemplar mit Speer, den Leda in der Hand hält. Christie's gibt die Sign. „Caniroff" an, die jedoch Haniroff bzw. Xaniroff zu lesen wäre.
2444 (Firmenmarke ohne Ang.), H. 86 cm; GSF, S. 34
Ausst.: HMW (jetzt Wien Museum) 1985/86, hier aus-

gestellt in der Größe 114 cm (Mod.-Nr. 2223).
-> vgl. 2223
Lit.: GSF, S. 34, hier mit der Ang. des Originaltitels „Jagdtrophäe".

2445
FIGUR „Guitarrespielerin"
ca. 1901/02 (für den Entwurf)
2445 (Firmenmarke ohne Ang.), H. 71 cm; GSF, S. 34
-> vgl. 2282
Lit.: GSF, S. 34, hier mit der Ang. des Originaltitels „Guitarrespielerin".

2446
BÜSTE „Cleo"
ca. 1903
2446 (Firmenmarke ohne Ang.), H. 61 cm; GSF, S. 34
Lit.: GSF, S. 34, hier mit der Ang. des Originaltitels „Cleo".

2447
Tell, E.
FIGUR „Rückkehr vom Brunnen" (Frau mit Gefäß, sich nach Krug bückend)
ca. 1901/02 (für den Entwurf)
2447,184,3 Firmenmarke Sculpture 13, unsign., gepr. „S"= Z2, H. 100 cm; Wien Museum Inv. 186.621
2447,205,16 Firmenmarke Sculpture 13, Repr. Res., sign. E. Tell, H. 99 cm; Christie's SK 7.2.2001
2447 (Firmenmarke ohne Ang.), H. 100 cm; GSF, S. 34
Ausst.: ÖMKI 1902/03; 1903/04, hier jeweils mit Ang. des Originaltitels „Rückkehr vom Brunnen".
-> vgl. 2357; 2876
Lit.: GSF, S. 34, hier mit der Ang. des Originaltitels „Rückkehr vom Brunnen"; Pudor (1902), S. 107 (m. Abb.), hier mit Ang. des Originaltitels „Rückkehr vom Brunnen".

2448
Montenave
BÜSTE „Egypterin" (Sockel reliefiert: Sphinx vor aufgehender Sonne u.a.)
ca. 1903
2448,2,67 gem. F, Firmenmarke 7, unsign., gepr. „C"= Z5, H. 60 cm; Wien Museum Inv. 186.600
2448 (Firmenmarke ohne Ang.), H. 61 cm; GSF, S. 34
-> vgl. 2595
Lit.: GSF, S. 34, hier mit der Ang. des Originaltitels „Egypterin"; Berman, Vol. 2, Nr. 1797 (m. Abb.); Hillenbrand, S. 486; Kashey, S. 272; Bem.: Berman, Hillenbrand und Kashey geben als Künstlernamen jeweils „Monterowe" an.

2449
Escoula, J(ean/Jacques?)
FIGUR „Angelus"
ca. 1902
2449 (Firmenmarke ohne Ang.), H. 93 cm; GSF, S. 34
Ausst.: Paris, Salon 1902, hier mit Ang. des franz. Originaltitels „Angélus" und des Künstlers Escoula.
Lit.: GSF, S. 34, hier mit der Ang. des Originaltitels „Angelus"; Neuwirth (1974), S. 63.

2450
FIGUR „Feldarbeiter"
ca. 1903
2450 (Firmenmarke ohne Ang.), H. 65 cm; GSF, S. 34
Lit.: GSF, S. 34, hier mit der Ang. des Originaltitels „Feldarbeiter".

2451
FIGUR „Tanz"
ca. 1903
2451 (Firmenmarke ohne Ang.), H. fehlt; GSF, S. 34
Lit.: GSF, S. 34, hier mit der Ang. des Originaltitels „Tanz".

2452
Haniroff/Xaniroff
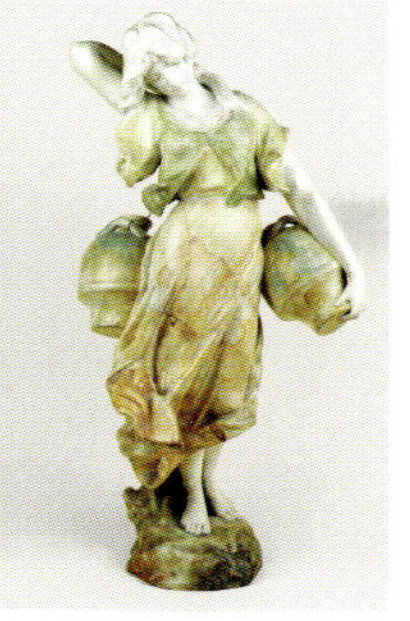
FIGUR „Der Windstoß" (Stehendes Mädchen mit zwei Krügen)
ca. 1901 (für den Entwurf)
2452,156,55 gem. weiß III, Firmenmarke Sculpture 13, Repr. Res., unsign., H. 62 cm; PB Wien 19
2452,424,11 Firmenmarke Sculpture 13, Ges. geschützt., sign. Haniroff, H. 61 cm; PB Wien 19
2452 (Firmenmarke ohne Ang.), H. fehlt; GSF, S. 27
-> vgl. 2234
Lit.: GSF, S. 27, hier mit der Ang. des abweichenden Originaltitels „Der Windstoß"; KuKHW, 1902, S. 27.

2453
UHR „Die Schöpfung"
ca. 1903
2453 (Firmenmarke ohne Ang.), H. 87 cm; GSF, S. 34
Bem.: Dieses Modell belegt die Angebotsvielfalt der Manufaktur. Die Uhr erscheint im Firmenkat. „mit Uhr- und Schlagwerk, Pendelgang", sowie in der gleichen Variante „elektr.". Zusätzlich war sie mit „Schlagwerk, Cylinder", sowie dieses ebenfalls „elektr." und als „gewöhnliche Uhr", diese wiederum auch „elektr." angeboten. Die Preise steigerten sich vom reinen Uhrgehäuse für 85 Mark bis 149 Mark für die Luxusvariante „elektr. Schlagwerk, Cylinder".
Lit.: GSF, S. 34, hier mit der Ang. des Originaltitels „Die Schöpfung".

2454
Cherc
FIGUR „Aus der Bretagne" (Bäuerin mit Krug und Schale)
ca. 1903
2454 (Firmenmarke ohne Ang.), H. 83,2 cm; Pinhas, S. 34 (m. Abb.)
2454 (Firmenmarke ohne Ang.), H. 84 cm; GSF, S. 34
-> vgl. 2662
Pendants -> vgl. 2917; 3840
Lit.: GSF, S. 34, hier mit der Ang. des Originaltitels „Aus der Bretagne".

2455
FIGUR „Venusopfer"
ca. 1903
2455 (Firmenmarke ohne Ang.), H. 97 cm; GSF, S. 34
Bem.: Wie der Firmenkat. belegt, wurde der dieses Modell auch „elektr." und „mit Bronze Gestell" (handschr. Vermerk) geliefert.
Lit.: GSF, S. 34, hier mit der Ang. des Originaltitels „Venusopfer".

2456
BÜSTE „Cyclame"
ca. 1903
2456 (Firmenmarke ohne Ang.), H. 64 cm; GSF, S. 34
Lit.: GSF, S. 34, hier mit der Ang. des Originaltitels „Cyclame".

2457
Pecheur
BÜSTE „Empire" (Junge Dame mit Hochsteckfrisur) (nach rechts blickend)
ca. 1903
2457,9,19 Firmenmarke Sculpture 13, Repr. Res., unsign., H. 58,4 cm; eBay # 180027540320
2457 (Firmenmarke ohne Ang.), H. 58 cm; GSF, S. 34
Ausst.: ÖMKI 1906/07, hier mit Ang. des Originaltitels „Empire" und der Ang. des Künstlers „Pecheur".
Lit.: GSF, S. 34, hier mit der Ang. des Originaltitels „Empire".

2458
Bertrand, A.
BÜSTE „Sonnenblume" (Junge Frau, nach links blickend)
ca. 1900/01 (für den Entwurf)
2458 (Firmenmarke ohne Ang.), H. 35 cm; GSF, S. 34
-> vgl. 2177; 2296
Lit.: GSF, S. 34, hier mit der Ang. des Originaltitels „Sonnenblume".

2459
BÜSTE „Nymphe"
ca. 1903
2459 (Firmenmarke ohne Ang.), H. fehlt; GSF, S. 34
Lit.: GSF, S. 34, hier mit der Ang. des Originaltitels „Nymphe".

2460
FIGUR „Reifenspielerin"
ca. 1903
2460 (Firmenmarke ohne Ang.), H. 74 cm; GSF, S. 34
Bem.: Wie der Firmenkat. belegt, wurde der dieses Modell auch „elektr." geliefert.
Lit.: GSF, S. 34, hier mit der Ang. des Originaltitels „Reifenspielerin".

2466
Montenave
BÜSTE Junge Frau (nach links blickend) geschmückt mit großen Blättern
ca. 1902/03
2466 (Firmenmarke ohne Ang.), H. 65 cm; Pinhas, S. 57 (m. Abb.)
Pendant -> vgl. 2511

2474
Gory, A(ffortunato?)
BÜSTE Junge Frau (nach links blickend)
ca. 1903
2474 (Firmenmarke ohne Ang.), sign. A. Gory, H. 62 cm; Pinhas, S. 58 (m. Abb.)

2482
Cherc
Badende mit Tuch (auf Rundsockel)
ca. 1903
2482 (Firmenmarke ohne Ang.), H. 39 cm; Pinhas, S. 47 (m. Abb.)

2485
VASE mit ägyptischer Landschaft, auf der Vasenmündung die (plastische) Figur einer sitzenden Ägypterin, die eine Obstschale hält (gem. Dekor: Reiter und Palmen)
ca. 1903
2485 ger., GS Wien (Firmenmarke ohne Ang.), H. 46 cm; Doro 20.10.1997

2486 -> ÜBERSCHNEIDUNG
Tell, E.
UHR Der Kuss
Inschrift: FUGIAT AETAS MANEAT AMOR
ca. 1903
2486,39,34 Firmenmarke Sculpture 13, Repr. Res., H. ca.75 cm; WKH
Text der Inschrift: Mag die Zeit auch vergehen, die Liebe wird bleiben.

2486 -> ÜBERSCHNEIDUNG
Lapigre
Mädchen, an heckenumwundenem Mauerdurchbruch stehend
ca. 1903
2486 (Firmenmarke), unsign., H. 83 cm; Pinhas, S. 60 (m. Abb.)
-> vgl. 2763

2490
Petri (d.i. Petrides, János) (Johann)
BÜSTE Junge Frau mit Kappe (Holländerin?)
ca. 1903
2490 (Firmenmarke ohne Ang.), sign. Petri, H. 35 cm; Pinhas, S. 54 (m. Abb.)

2491
Louger, (J.)
BÜSTE Holländerin mit Haube (Sockelrelief: Landschaft mit Windmühlen)
ca. 1903
2491,78,48 Firmenmarke 7, Repr. Res., H. 41,9 cm; eBay # 3765880120
2491,237,40 (ohne Firmenmarke), Repr. Res., gepr. „S"= Z2, H. 41 cm; eBay # 901445088
2491,264,1 Firmenmarke Sculpture 13, Repr. Res., H. 42 cm; Henry's 31.7.1999

2492
Sydan
SPIEGEL Stehender Mädchenhalbakt in einen Spiegel schauend (auf Landschaftssockel mit Baum)
ca. 1903
2492 (Firmenmarke ohne Ang.), H. 108 cm (wohl Gesamth.); Pinhas, S. 60 (m. Abb.)

2494
Capèque (d.i. Czapek, Stanislaus)
UHR Weinbergschnecke
ca. 1903
2494,21,15 Firmenmarke Sculpture 13, Repr. Res., sign. Capéque, H. 33,5 cm, L. 51 cm; Doro 12.6.1995

2497
Schulz
Stehende Elsässerin mit Blumenkorb, Blumengirlande und Gans (auf oktogonalem Sockel)
ca. 1903
2497,129,– Firmenmarke Sculpture 13, Repr. Res., H. 45,5 cm; eBay # 260002737113

2504
Louger, J.
„Venetienne" (Junge, am Ruder stehende Gondel fahrende Frau)
Inschrift: VENETIENNE
ca. 1903
2504 (Firmenmarke ohne Ang.), H. 89 cm; Pinhas, S. 44 (m. Abb.); Bem.: Pinhas gibt als Künstler zwar „Sauger" an, es handelt sich aber sicher um J. Louger, der für Goldscheider etliche weitere Modelle schuf.

2506
Pecheur
UHR Stehendes, Seifenblasen pustendes Holländermädchen (Uhrgehäuse reliefiert mit zwei Blumentöpfen/Tulpen)
ca. 1903
2506,35,69 gem. 16, Firmenmarke 7, Repr. Res., H. 71,1 cm; eBay # 1225801611; Bem.: Lenzkirch Uhrwerk.
Bem.: Besetzt mit Farbsteinen/Edelsteinimitation.

2510
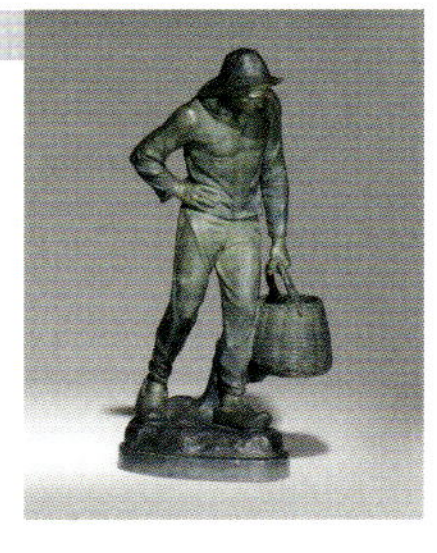
Petri (d.i. Petrides, János) (Johann)
Stehender Fischer mit Sturmhaube und Weidenkorb (auf Landschaftssockel)
ca. 1903
2510,15,25 Firmenmarke 7, Repr. Res., sign. Petri, H. 56 cm (Ang. wohl falsch); eBay # 2501627965; 3351653612; 5349546551
2510,70,25 weiß gem. 18, Firmenmarke Sculpture 13, gepr. „C"= Z4, sign. Petri, H. 63,9 cm; MAK Wien, Inv. Ke 10521-2
Pendant –> vgl. 2535

2511
Montenave
BÜSTE Junge Frau (nach rechts blickend) geschmückt mit großen Blättern
ca. 1902/03
2511 (Firmenmarke ohne Ang.), H. 70 cm; Pinhas, S. 57 (m. Abb.)
Pendant –> vgl. 2466

2516
BÜSTE/LAMPE Römischer Krieger mit Helm und Schild
ca. 1903
2516,78(76?),3 Firmenmarke Sculpture 13, Repr. Res., H. 70 cm; eBay # 3353317965
Bem.: Das Schild, welches die Lampenfassungen verdeckt, ist besetzt mit vier Farbsteinen/Edelsteinimitation.

2517
Louger, J.
UHR Sitzendes Mädchen am Felsen
ca. 1903
2517,30,15 gem. X, Firmenmarke Sculpture 13, H. ca. 54 cm; Henry's 29.1.1997

2519
Cherc
BÜSTE Junge Frau mit orientalischem (byzantinischem?) Kopfputz
ca. 1903
2519 (Firmenmarke ohne Ang.), H. 33 cm; Pinhas, S. 57 (m. Abb.)

2520
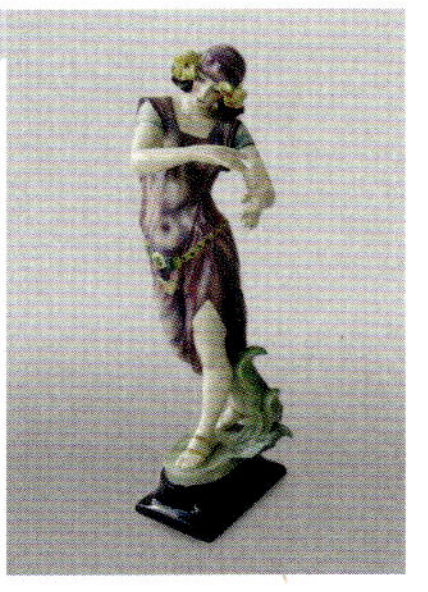
Lorenzl, Josef (?)
Orientalische Tänzerin mit Kappe und Blumenschmuck im Haar (auf profiliertem Rechtecksockel)
ca. 1903
2520,32,65 Firmenmarke 19, Made in Austria, Ausformung 20er Jahre, H. 47 cm; Sammlung Stressless, Portugal
Bem.: Die Sign. „Lorenzl", ist definitiv falsch vom Modelleur eingeritzt worden, da Modelle von Lorenzl erst ab 1920 nachgewiesen werden können. Die Ausformung des vorliegenden Stückes ist aufgrund des Firmenstempels jedoch in die 20er Jahre zu datieren.
2520 s.a. 2523

2523
Simon
UHR mit zwei Figuren (Stehendes Liebespaar reicht sich die Hände)
Inschrift: AMICITA VINCIT HORAS
ca. 1901/02
F2523,11,6 FG Wien (Firmenmarke?), Firmenmarke Sculpture 13, H. 42 cm; PB (Melk, NÖ)
2523,52,45 FG Wien (Firmenmarke?), H. 42,5 cm; Nagel 303
2523,66,24 Firmenmarke Sculpture 13, gepr. „S"= Z2, H. ca. 42 cm; Nagel 362
2523,88,20 Firmenmarke Sculpture 13, H. 39,5 cm; WKA 18
2523,121,45 Firmenmarke Sculpture 13, H. 40 cm; Quit. 35; Doro 25.11.2003
2523,167,17 Firmenmarke 9, H. 39,2 cm; Sammlung Dr. Kurt Berger, Wien; Neuwirth (1974), S. 74, 77 (m. Abb.)
2523,247,10,IV Firmenmarke 4, Firmenmarke Sculpture 13, Ges. gesch., H. 40 cm; Sebök 46 (Lit.: AZ, 1997, S. 971); KNS 10; KNS 17
2523,307,7 GS (Firmenmarke ohne Ang.), H. 40,8 cm; Doro 14.9.2001
Bem.: Diese Uhr wurde (ohne weitere Ang.) noch bei mehreren kleineren Auktionshäusern angeboten, jeweils mit Lenzkirch Uhrwerk.
Text der Inschrift: Die Freundschaft besiegt die Stunden.
–> vgl. 2305; 2736
Lit.: N.N.: Von der Internationalen Ausstellung für dekorative Kunst in Turin 1902, in: KM, 2.1902, S. 162–163 (m. Abb.); Neuwirth (1974), S. 58, 74, hier Modell-Nr. 2520 angegeben, vermutlich Lesefehler/Druckfehler, richtig ist wohl Modell 2523.

2524
Klein, (Max)
GRUPPE Hagar und Ismael
ca. 1902/03
2524,2,34 Firmenmarke Sculpture 13, Repr. Res., H. 65 cm; Doro 18.5.1999
ohne Ang., Firmenmarke Sculpture 13, H. 64 cm; Henry's 9.12.1993
Bem.: Dargestellt sind Hagar (Nebenfrau Abrahams) und deren gemeinsamer Sohn Ismael.
Ausst.: ÖMKI 1903/04, im Ausst.-Kat. wird fälschlicherweise P. Klein als Künstler genannt, hier mit Ang. des Originaltitels „Hagar und Ismael".
Lit.: Schulz Ser. 1, o. Pag., Text zum Künstler, hier bezeichnet als „Hauptwerk"; Schulz Ser. 3, Taf. 9 (Abb.); Neuwirth (1974), S. 64

2530
Cherc
FIGUR „Salome" (Stehende Frauengestalt mit lyraähnlichem Instrument)
Inschrift: SALOME
ca. 1897/98 (für den Entwurf)
2530 (Firmenmarke ohne Ang.), H. 59 cm; Pinhas, S. 62 (m. Abb.)
2530 Firmenmarke 19, H. 59 cm; Pinhas, S. 62 (m. Abb.)
Bem.: Dieses erfolgreiche Modell, wurde auch noch in den 20er Jahren produziert, wie ein Ex. bei Pinhas belegt, welches den entsprechenden Firmenstempel der Zeit trägt.
–> vgl. 1000; 1900; 2085

2535
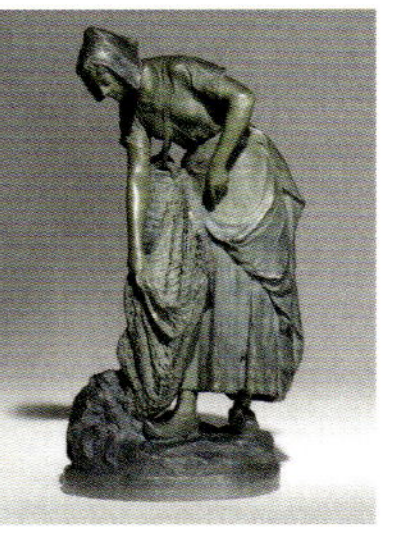
Petri (d.i. Petrides, János) (Johann)
Stehende Fischerin (Holländerin) mit Haube, Netz und Korb (auf Landschaftssockel)
ca. 1903
2535,76,15 Firmenmarke Sculpture 13, Repr. Res., unsign. H. 63,2 cm; MAK Wien, Inv. Ke 10521-3
Pendant –> vgl. 2510

2537
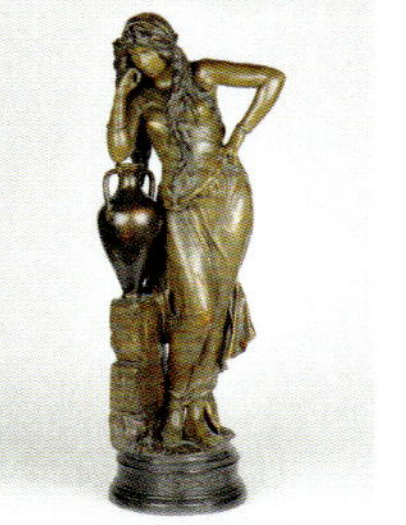
Cherc
FIGUR „Rebecca" (Stehende Frauengestalt neben einer Mauer, ihren Ellenbogen auf eine Amphore stützend)
Inschrift: REBECCA
ca. 1902
2537 (2387?),649,15 Firmenmarke Sculpture 13, Fabriqué en Autriche, H. 44,5 cm; eBay # 2535499438; Bem.: Der Anbieter gibt als Mod.-Nr. 2387 an.
2537,734,35 Firmenmarke Sculpture 13, Fabriqué en Autriche, H. 46 cm; PB Wien 19
–> vgl. 1903; 2171

2543
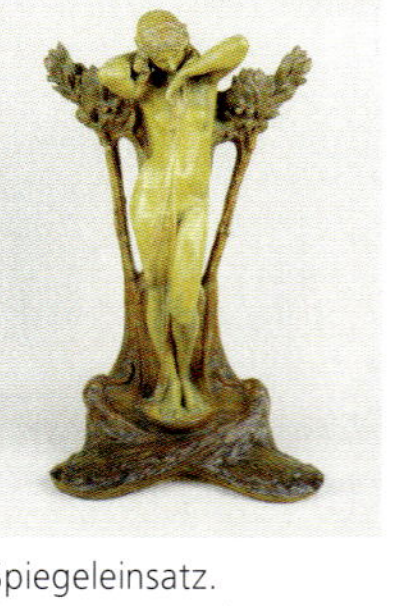
SPIEGEL „Iris" (Stehender Frauenakt mit Tuch vor zwei Bäumen/Blüten?)
ca. 1903
2543,4,17 (Firmenmarke ohne Ang.), H. 62 cm; Schöner Wohnen 1979 (August)
2543,26,24 Firmenmarke Sculpture 13, Repr. Res., gepr. „S"= Z2, H. 60,5 cm; PB Wien 19
Bem.: Höhenang. jeweils ohne Spiegeleinsatz.
Lit.: KeramRS, 13.1905, S. 916, hier mit Ang. des Originaltels „Iris"; GSN, S. 22, 23 (m. Abb.)
GS neu: Modell 88 „Schönheit und Leben" (H. 60 cm), Aufl. 1000 Stück

2551
Zago, (Emil/Emile/Emilio)
„Dante" (Schreitender Dante Alighieri, auf Felssockel)
ca. 1903
2551,17,16 (Firmenmarke ohne Ang.), sign. Emile(io) Zago, H. 45,3 cm; WKA 33; Doro 10.3.2003
Bem.: Dargestellt ist der italienische Dichter Dante Alighieri (1265–1321).
Ausst.: ÖMKI 1903/04; 1906/07, hier jeweils mit Ang. des Originaltitels FIGUR „Dante" und des Künstlernamens „Emil Zago".
Lit.: Neuwirth (1974), S. 76

2560
Capèque (d.i. Czapek, Stanislaus)
Diskuswerferin (in antikisierendem Gewand)
ca. 1903
2560,40(oder 49?),55 Firmenmarke Sculpture 13, Repr. Res., sign. Capèque, gepr. „S"= Z2, H. 66 cm; WKA 34; 38
Bem.: Diese Figur wurde evtl. anlässlich der olympischen Sommerspiele 1904 in St. Louis geschaffen.
-> vgl. 2793

2572
Cherc
BÜSTE Junge Frau mit Fuchsienblüten
ca. 1903
2572 (Firmenmarke ohne Ang.), H. 62 cm; Pinhas, S. 56 (m. Abb.)

2573
Lesca
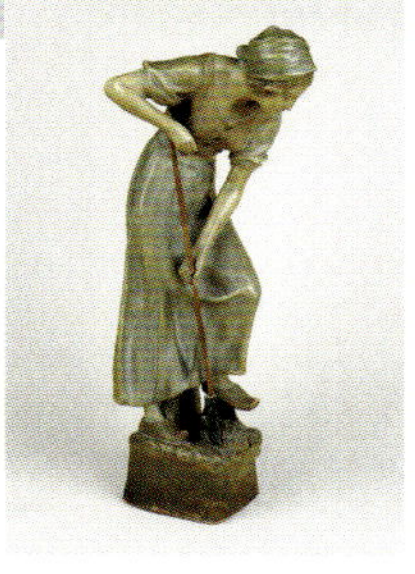
FIGUR „Labor" (Stehende Frau mit Spaten)
Inschrift: LABOR
ca. 1902/03
2573,274,18 gem. weiß XXX, ohne Firmenmarke, Repr. Res., H. 43,3 cm; PB Wien 19
Bem.: Mit dieser Figur ist der Sockel der Büste „Labor" (Mod.-Nr. 2387) reliefiert.
Text der Inschrift: Arbeit
Ausst.: ÖMKI 1902/03, hier mit Ang. des Originaltitels „Labor".
-> vgl. 2316
-> vgl. 2387= BÜSTE

2575
Simon
Stehendes Holländermädchen mit Korb und Krug (auf Landschaftssockel)
ca. 1903
2575,60,24 Firmenmarke Sculpture 13, Repr. Res., H. 62 cm; eBay # 6503619502

2576
Montenave
BÜSTE Junges Mädchen mit beidseitigem Haarschmuck (Bänder oder Blumenblüten?)
ca. 1903
2576 (Firmenmarke ohne Ang.), H. 60 cm; Pinhas, S. 58 (m. Abb.)

2579
Petri (d.i. Petrides, János) (Johann)
BÜSTE Römischer Krieger mit Helm (bekrönt mit Adler) (auf antiker Säule/Kapitell)
ca. 1903
25(79),45,2 (Firmenmarke ohne Ang.), sign. Petri, H. 66 cm; KH
2579,260,19 Firmenmarke Sculpture 13, Fabriqué en Autriche, sign. Petri, H. ca. 65 cm; KH
2579 (ohne Ang.), sign. Petri, H. 66 cm; Herr 25

2580
Montenave
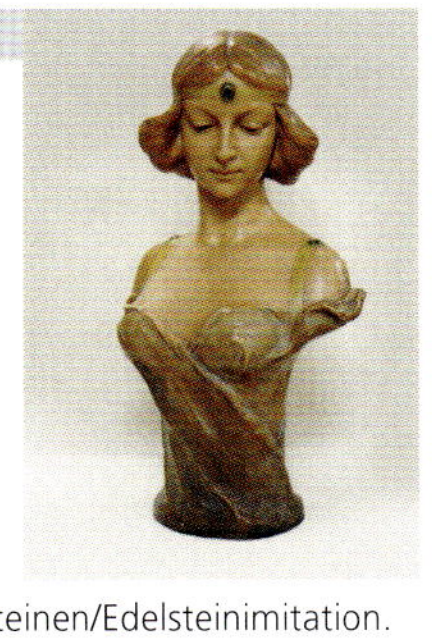
BÜSTE Junge Frau mit gesenktem Haupt
ca. 1903
2580,2,47 Firmenmarke Sculpture 13, H. 66,9 cm; MAK Wien, Inv. Ke 10521-4
2580,(38,25?) gem. VII, Firmenmarke Sculpture 13, Repr. Res., gepr. „L"= Z1, H. ca. 60 cm; KH
Bem.: Kleid besetzt mit Farbsteinen/Edelsteinimitation.

2581
Fecler?
BÜSTE Junge Frau mit Kopfputz in Form eines Pfaus
ca. 1903
2581 (Firmenmarke ohne Ang.), sign. Fecler, H. 46 cm; Pinhas, S. 56 (m. Abb.); Bem.: Die bei Pinhas angegebene Sign. „Fecler" ist anzuzweifeln, vermutl. ein Lesefehler.

2583
Haniroff
FIGUR „Vasenträgerin" (Stehende junge Frau eine Amphore auf der rechten Schulter tragend) (auf Landschaftssockel)
ca. 1901 (für den Entwurf)
2583 (Firmenmarke ohne Ang.), H. 65 cm; Pinhas, S. 43 (m. Abb.)
-> vgl. 2238; 2409

2585
Czapek, Stanislaus
UHR/LAMPE „Feenwagen" (Stehende Fee in einem römischen Streitwagen, der von zwei Panthern gezogen wird)
ca. 1903
ohne Ang., (Firmenmarke ohne Ang.), sign. Capèque, H. 68 cm; Soth. L. 20.3.1986; 6.11.1986
2585 (Firmenmarke ohne Ang.), H. 65 cm; Pinhas, S. 59 (m. Abb.)
Lit.: N.N., Figürliche Terrakotten, in: KM, 4.1904, S. 111 (m. Abb.), „Zwei Panther ziehen den zweirädrigen, niedrigen Wagen, auf dem die Fee steht. Das im Winde flatternde Gewand wird nach hinten getrieben und lässt die Formen des jugendlich schönen, weiblichen Körpers erkennen, der mit einem schmiegsamen Rock bedeckt ist. Auf diesem Wagen steht eine Uhr, über welche sich die Fee mit weit vorgestrecktem Oberkörper beugt, ihr linker Arm hält einen Stab, auf dessen Ende ein elektrisches Glühlicht aufgesetzt ist. Modelliert ist dieses schöne Stück von Capèque".

2592
Montenave
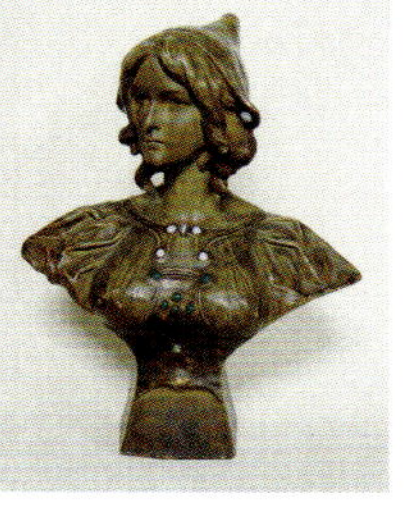
BÜSTE Mädchen mit gelockten Haaren und Kappe
ca. 1903
2592,35,12 Firmenmarke Sculpture 13, Repr. Res., H. 48,8 cm; Franzke, S. 149; Franzke 1, S. 165; MAK Wien, Inv. Ke 10602-48
2592,95,7 Firmenmarke Sculpture 13, Repr. Res., H. 49,5 cm; eBay # 2521678831
2592,188,48 (Firmenmarke ohne Ang.), H. 50 cm; www.profibid.com (Auktion 3922, Nov. 2001)

2593
Villieu
BÜSTE „Beatrice" (Junge Dame mit Diadem und Kopftuch)
Inschrift: BEATRICE
ca. 1899/1900 (für den Entwurf)
2593,64,32 gem. weiß 22., Firmenmarke Sculpture 13, Repr. Res., H. 23 cm; eBay # 1456242576; 920031848
Bem.: Dargestellt ist Beatrice Portinari (1266–1290), die Jugendliebe des Dante Alighieri (1265–1321).
-> vgl. 1887; 2122
-> vgl. 2200 WANDMASKE

2595
Montenave
BÜSTE „Egypterin" (Sockel reliefiert: Sphinx vor aufgehender Sonne u.a.)
ca. 1902/03
2595 (Firmenmarke ohne Ang.), H. 45 cm; Pinhas, S. 58 (m. Abb.)
-> vgl. 2448
Lit.: GSF, S. 34, hier mit der Ang. des Originaltitels „Egypterin" (für die Größenvariante Mod.-Nr. 2448).

2597
Sydan
Wasserträgerin (Junge Bäuerin mit geschultertem Krug) (auf Landschaftssockel)
ca. 1903
2597 (Firmenmarke ohne Ang.), H. cm; Pinhas, S. 34 (m. Abb.)

2600
Capèque (d.i. Czapek, Stanislaus)
UHR Männliche Figur an Glocke (Gehäuseaufbau mit „byzantinischem Wandmosaik")
Inschrift am Gehäuse: VIVOS VOCO MORTVOS PLANGO FVLGVRA FRANGO
Inschrift an der Glocke: ANNO MDCCCII
ca. 1902
2600,64,34 GS Wien (Firmenmarke ohne Ang.), unsign.?, H. 57 cm; Fischer 88
2600,64,35 GS Wien (Firmenmarke ohne Ang.), sign. Capèque, H. 60,7 cm, L. 57 cm; Doro 29.10.1996
Bem.: Diese Uhr wurde in drei Größen gefertigt, auch in der Gestaltung des Gehäusedekors kommen unterschiedliche Entwürfe zur Ausführung (vgl. hierzu 2678; 2718). Die Inschrift ist ein Zitat des Leitspruchs von Friedrich Schillers Dichtung „Das Lied von der Glocke" (1799).
Text der Inschrift am Gehäuse: Lebende rufe ich, Tote beklage ich, Blitze breche ich.
Text der Inschrift an der Glocke: 1902
-> vgl. 2678; 2718

2605
Simon
BÜSTE Junge Frau mit orientalischer Kopfbedeckung und Halsschmuck
ca. 1903
2605 (Firmenmarke ohne Ang.), H. 64 cm; Pinhas, S. 57 (m. Abb.)

2606
BÜSTE Holländerin mit Haube
ca. 1903
2606,140,40 ger. J (Firmenmarke ohne Ang.), H. 53,3 cm; Soth. Chester 6.–8.9.1988

2611
BÜSTE Frau mit Diadem und Perlenschmuck
ca. 1903
2611 (Firmenmarke ohne Ang.), H. 67 cm; Pinhas, S. 57 (m. Abb.); Bem.: Diadem mit Opalen (Farbsteinen) besetzt.

2616
Gross, F.
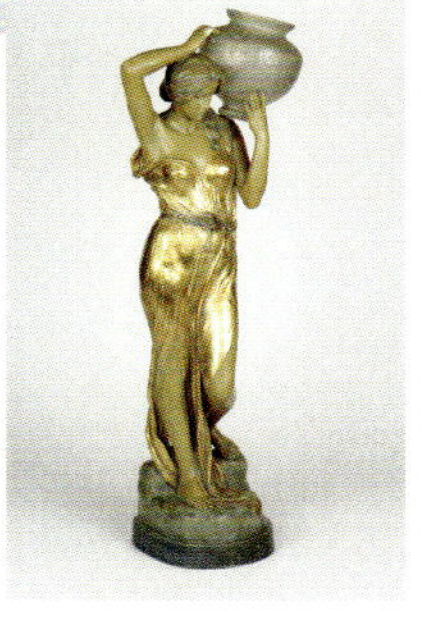
FIGUR/LAMPE Wasserträgerin (einen Krug/Henkeltopf auf der Schulter tragend)
ca. 1903
FIGUR
2616,378,17 (Firmenmarke ohne Ang.), unsign., H. 105,5 cm; Phillips L. 25.1.2001
LAMPE
ohne Ang. Gesamth. 137 cm; Phillips E. 29.10.1999
Bem.: Gürtel besetzt mit Opale (Farbsteine).
-> vgl. 2741; 2770
-> vgl. .... (H. 39 cm); Zeller 29
Lit.: Neuwirth (1974), S. 58

2621
Cherc
BÜSTE „Erdbeere" (Junge Frau nach rechts blickend) (Haare und Sockel geschmückt mit Erdbeeren)
ca. 1902/03
2621,75,40 Firmenmarke Sculpture 13, H. 46 cm; Soth. Olympia 29.5.2003
-> vgl. 2404

2627
Brünelle/Brunelle
Stehende junge Frau mit Lyra (in antikisierendem Kleid) (Sappho?)
ca. 1903/05
2627,64,24 gem. Z, Firmenmarke Sculpture 13, Repr. Res., undeutl. sign. Brü(u)nelle, gepr. „S"= Z2, H. 76 cm; Wien Museum Inv. 186.627
Bem.: Bei diesem Modell handelt es sich vermutlich um eine Darstellung der griech. Lyrikerin Sappho (um 660 v. Chr.).

2633
Simon
„Neugierde“ (Stehendes Holländermädchen neben einer Standuhr)
ca. 1903/05
2633,234,25 Firmenmarke Sculpture 13, Repr. Res., H. 66,5 cm; Bergmann 64; Fischer 137 (Sammlung Werner Jahn, Porzellan des Jugendstils und Art Déco).
2633 (ohne Ang.), Firmenmarke Sculpture 13, Repr. Res., H. 67 cm; Spik 503
-> vgl. 2693
Lit.: Pudor (1906), hier mit Abb. und Ang. des Originaltitels „Neugierde“.

2635
Vansteen
BÜSTE „Holländisches Kind“ (Mädchen mit Spitzenhaube) (nach rechts blickend)
ca. 1902/03
2635,78,19 gem. weiß H., Firmenmarke Sculpture 13, H. 40,6 cm; eBay # 2559912396
-> vgl. 2398

2639
Gambeauche
BÜSTE „Lygia“ (auf eckigem Sockel mit Fischsymbolen)
Inschrift: LYGIA
ca. 1903
2639 (o. Ang.) Firmenmarke o. Ang., H. 26,5 cm; KH
Bem.: Dargestellt ist die Sklavin „Lygia“, eine Figur des alten Roms aus der Zeit Neros.
-> vgl. 2426= FIGUR

2643
Haniroff
BÜSTE Römischer Krieger mit Helm (auf mit Opferszene reliefiertem Sockel)
ca. 1904
2643,64,4 Firmenmarke Sculpture 13, Repr. Res., H. 40 cm; Nagel 381; Dawo 117

2647
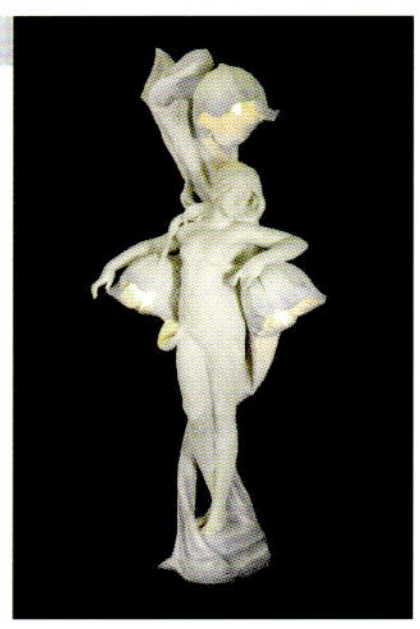
LAMPE Stehende junge Frau, von drei übergroßen Maiglöckchen umgeben
ca. 1903
2647,12,10 Firmenmarke Sculpture 13, H. 107 cm; Brigantine 1900, München
Bem.: Bei dieser Darstellung des Frühlings dienen die überdimensionierten Maiglöckchenblüten als Lampenschirme.
Lit.: Weltkunst 70.2000, S. 387 (m. Abb.)

2655
Gornik, Friedrich
„Eisbär“
ca. 1903
2655,117,1,2 gem. lig. AD, Firmenmarke 19, Made in Austria, sign. „Gornick“ (!), Ausformung 20er Jahre, H. 31,2 cm; GK, S. 17, 95 (m. Abb.); Wien Museum Inv. 223.734
Ausst.: ÖMKI 1906/07, hier mit Ang. des Originaltitels „Eisbär“; HMW (jetzt Wien Museum) 1985/86.
Lit.: KuKHW, 6.1903, S. 527 (m. Abb.); Neuwirth (1974), S. 63; GK, S. 17, 95 (m. Abb.).

2662
Cherc
FIGUR „Aus der Bretagne“ (Bäuerin mit Krug und Schale)
ca. 1903
2662,224,7 gem. weiß 8, Firmenmarke Sculpture 13, Fabriqué en Autriche, H. 64 cm; Wien Museum Inv. 186.601
-> vgl. 2454
Pendants -> vgl. 2917; 3840

2664
Montenave
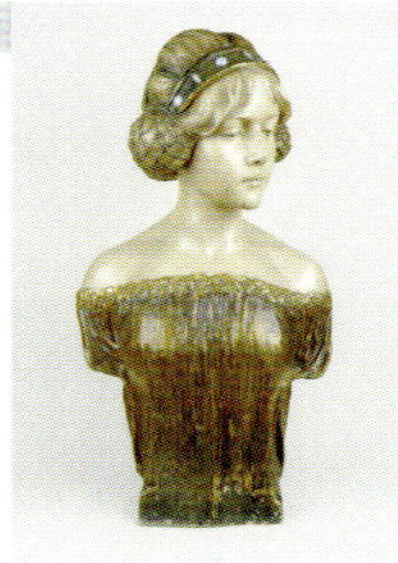
BÜSTE Mädchen mit seitlich geknotetem Haar und Diadem
ca. 1903
2664,220,35 Firmenmarke Sculpture 13, Fabriqué en Autriche , H. 67 cm; PB Wien 19
2664,233,19 FG (Firmenmarke?), Firmenmarke Sculpture 13, Fabriqué en Autriche, H. 70 cm; PB
-> vgl. 2768; 2842

2666
Horsetzky, Melanie von
JARDINIERE Stehende Holländerin mit Korb
ca. 1903/05
2666,440,27 gem. 5, Firmenmarke 9, Repr. Res., H. 34,5 cm; PB Deutschland
2666,532,3 Firmenmarke Adler 16, H. 35,5 cm; (KH www.tias.com/stores/frozenbeauty)
2666,537,3 Firmenmarke 13a, H. 35 cm; WKH
2666 (ohne Ang.), H. 33 cm; Schlüter 2659
Bem.: Diese Figur kann sowohl als Jardiniere wie auch als Visitenkartenschale dienen.
-> vgl. 3591

2668
Montanave
BÜSTE Junge Frau mit Schilf (und Lilienblüten?)
ca. 1903
2581 (Firmenmarke ohne Ang.), H. 47 cm; Pinhas, S. 56 (m. Abb.)

2671
Cherc
BÜSTE Frau mit Blumengirlande
ca. 1903
2671,32,67 gem. Z, Firmenmarke Sculpture 13, Repr. Res., H. 63,5 cm; eBay # 7384788871

2678
Capèque
(d.i. Czapek, Stanislaus)
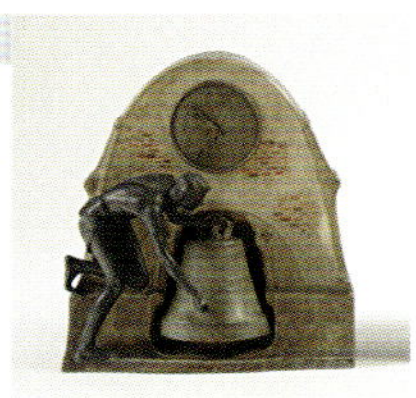
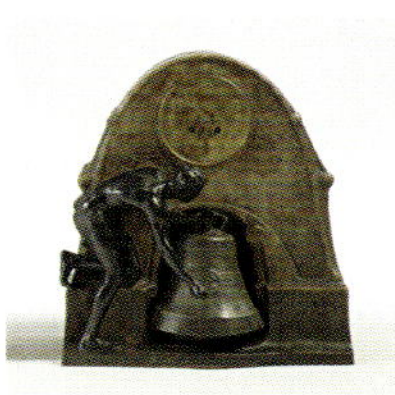
UHR Männliche Figur an Glocke (Gehäuseaufbau mit „abblätternder Mauer“)
Inschrift: VIVOS VOCO MORTVOS PLANGO FVLGVRA FRANGO
Inschrift an der Glocke: ANNO MDCCCII
ca. 1902
2678,80,17 Firmenmarke 7, Repr. Res., sign. Capeque, L. 46 cm; GK, S. 14, 15, 78 (m. Abb.); Wien Museum Inv. 182.072
3678(= 2678), 327,17 gem. XI, Firmenmarke 4, Repr. Res., sign. Capeque, H. 46,8 cm; Wien Museum Inv. 219.144;
Bem.: Modell-Nr. 3678 vom ausführenden Modelleur falsch eingestempelt, richtig ist 2678.
Bem.: Diese Uhr wurde in drei Größen gefertigt, auch in der Gestaltung des Gehäusedekors kommen unterschiedliche Entwürfe zur Ausführung (vgl. hierzu 2600; 2718). Die Inschrift ist ein Zitat des Leitspruchs von Friedrich Schillers Dichtung „Das Lied von der Glocke“ (1799). Text der Inschrift am Gehäuse: Lebende rufe ich, Tote beklage ich, Blitze breche ich.
Text der Inschrift an der Glocke: 1902
Ausst.: HMW (jetzt Wien Museum) 1985/86.
-> vgl. 2600; 2718
Lit.: GK, S. 14, 15, 78 (m. Abb.)

2690
Rosé
UHR „Théodora“ (Sarah Bernhardt als „Théodora“ auf einem Thron sitzend)
ca. 1903
2690 (Firmenmarke ohne Ang.), H. 61 cm; Pinhas, S. 59 (m. Abb.); Bem.: Besetzt mit Glasperlen und Cabochons.
Bem.: Dargestellt ist die Schauspielerin Sarah Bernhardt (1844–1923) in der Titelrolle der „Théodora“ des gleichnamigen Stückes (Urauff. Paris 1885) von Victorien Sardou (vgl. hierzu: Le Théatre, 5.1902, Nr. 75, S. 4–20).
-> vgl. 2786
Lit.: N.N.: Ausstellung des Kunstgewerbevereins in Wien, in: Sprechsaal, 24.1891, S. 63, „Stets steigende Erfolge hat auch FG, Wien-Währing mit seinen bemalten und unbemalten Terracottawerken zu verzeichnen. Wir finden wieder zahlreiche neue Figuren und Gruppen von aufstrebenden jungen Plastikern, die sich gerne der Industrie zuwenden. So das „Tirolerpaar“, das von der Kronprinzessin-Witwe angekauft wurde, Sarah Bernhard als Theodora...“; N.N.: Die keramische Ausstellung in Dresden 1891, in: Sprechsaal, 24.1891, S. 816, „Man möchte fast annehmen, dass einige Künstler von Goldscheider ihre Studien in Paris gemacht haben, denn die in gelblicher Terrakotta gearbeitete Figur der Sarah Bernard entspricht der Auffassung, in der Stellung und freien Durchbildung der Gewandtheile ganz den Pariser Terracotten.“ Bem.: Ob es sich hier bei diesen beiden zeitgenössischen Berichten um exakt dieses Modell handelt, kann nicht mit letzter Sicherheit gesagt werden, möglicherweise wurde die berühmte Sarah Bernhardt auch öfter durch die Manufaktur zur Ausführung gebracht.

2692
Waderé, Heinrich
BÜSTE „Venus“
ca. 1898
F2692,181,15 Firmenmarke 6, H. 15 cm; Wien Museum Inv. 186.676
Bem.: In der Pariser Produktion läuft diese Büste (H. 45 cm) unter der Modell-Nr. 615 und wird dort in Bronze ausgeführt.
Ausst.: ÖMKI 1898/99; 1899/1900; 1902/03; 1906/07, hier jeweils mit Ang. des Originaltitels „Venus“, im Ausst.-Kat. 1906/07 Künstlername Prof. „Vadère“.
-> vgl. 615 (GSB)
-> vgl. 1268
Lit.: Neuwirth (1974), S. 58, hier als „Frauenbüste“ bezeichnet.

2693
Simon
„Neugierde“ (Stehendes Holländermädchen neben einer Standuhr)
ca. 1903/05
F2693,44,6 Firmenmarke 9, Repr. Res., H. 36 cm; eBay # 539437261
-> vgl. 2633
Lit.: Pudor (1906), hier mit Abb. und Ang. des Originaltitels „Neugierde“.

2698
Simon
BÜSTE Holländerin (in der Sockelkartusche Landschaftsdarstellung)
ca. 1903
F2698,19,49,3,C,+ Firmenmarke 9, unsign., H. 28,2 cm; Wien Museum Inv. 219.789
F2698,33,6,12 GS Wien (Firmenmarke?), (Repr.) Res., H. 28 cm; KH

2699
Sydan
BÜSTE Holländermädchen (ihre Schürze zu einer Vase / Vide Poche(?) geformt)
ca. 1903
F2699,12,68 Firmenmarke 9, Repr. Res., H. 37 cm; Soth. Ama # 961508248/058; eBay # 1221730426
Bem.: Der Kopf dieses Modells ist identisch mit Simon „Topfguckerin“ Modell Nr. 2702; 2807. Aufgrund dieser Gegebenheit ist anzunehmen, dass die Künstler Sydan und Simon identisch sind.

2700
Montenave
BÜSTE Mädchen mit zwei großen Schleifen im Haar
ca. 1903
F2700,35,65 Firmenmarke 9, Firmenmarke 6, Repr. Res., H. 35 cm; Wien Museum Inv. 186.686

2702
Simon
FIGUR/LAMPE „Topfguckerin" (Kleines Holländermädchen in einen großen Topf schauend)
ca. 1903
FIGUR (Fayence)
2702,220,2 Firmenmarke 19 (rosa), Made in Austria, Ausformung 20er Jahre, H. 31,5 cm; Doro 21.9.1999
LAMPE (Fayence)
2702,79,45,2 Firmenmarke 6, Firmenmarke 9, Repr. Res., unsign., H. 31,5 cm; Wien Museum Inv. 190.157
LAMPE (Bronze)
ohne Nummer, CA, 17 Sculpture (Firmenmarke Bronze 1), H. 31 cm; Wien Museum Inv. 219.146
Bem.: vgl. Sydan 2699. Dieses Modell wurde in Fayence und Bronze ausgeführt.
Lit.: N.N.: Leipziger Michaelismesse 1904, in: KeramRS, 12.1904, S. 945, „Salonplastiken par excellence stellte die Firma FG in Wien aus. Novität: „Topfguckerin" mit Beleuchtungseffekten; Pazaurek (1904) in: Mitt. des Nordböhm. Gewerbemus., 22.1904., S. 91 (m. Abb.), „Friedrich Goldscheider in Wien, dessen graziöse und liebliche Terrakotta-Figuren manche originellen Effekte hervorbringen".

2706
Gross, F.
BÜSTE Kriegerin mit Helm
ca. 1903
2706,11,32 gem. weiß VII., Firmenmarke 7, Repr. Res., sign. F. Gross, H. 45 cm; eBay # 2273795392

2711
Tell, E.
Tanzendes Biedermeier-Paar
ca. 1903/05
F2711,14,8 Firmenmarke 9, Firmenmarke 6, unsign., H. 40 cm; Wien Museum Inv. 190.072
F2711,49,45 FG Wien (Firmenmarke ohne Ang.), Firmenmarke 6, Repr. Res., H. 42 cm; Spik 573
F2711,67,68 FG Wien (Firmenmarke ohne Ang.), Firmenmarke 6, Repr. Res., H. 42 cm; Neumeister 293
Lit.: Neuwirth (1974), S. 58, 75

2713
BÜSTE Der Kuss (Sich küssendes Paar, auf Landschaftssockel)
ca. 1903
F2713,9,6 FG Wien (Firmenmarke ohne Ang.), Repr. Res., H. 40 cm; Henry's 31.7.1999

2718
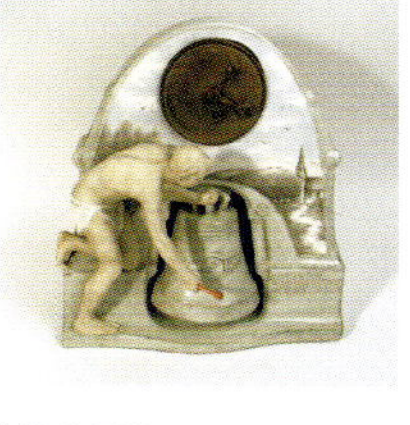
Czapek, Stanislaus
UHR Männliche Figur an Glocke (Gehäuseaufbau mit „Kirche in weiter Landschaft")
Inschrift am Gehäuse: VIVOS VOCO MORTVOS PLANGO FVLGVRA FRANGO
Inschrift an der Glocke: ANNO MDCCCII
ca. 1902
F2718,2,6 Firmenmarke 9, Repr. Res., unsign., H. 35 cm, B. 32,5 cm; city-antik
Bem.: Diese Uhr wurde in drei Größen gefertigt, auch in der Gestaltung des Gehäusedekors kommen unterschiedliche Entwürfe zur Ausführung (vgl. hierzu 2600; 2718). Die Inschrift ist ein Zitat des Leitspruchs von Friedrich Schillers Dichtung „Das Lied von der Glocke" (1799).
Text der Inschrift am Gehäuse: Lebende rufe ich, Tote beklage ich, Blitze breche ich.
Text der Inschrift an der Glocke: 1902
-> vgl. 2600; 2678

2719
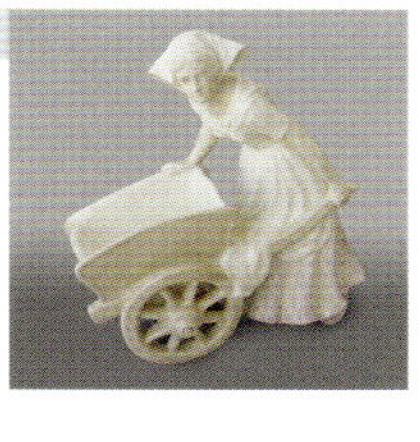
Clarà, Juan (Joan)/Sydan
Mädchen mit Schubkarren
ca. 1902
FAYENCE
F2719,89,27 gem. X, Firmenmarke 9, Firmenmarke 6, Repr. Res., unsign., H. 26 cm; Sammlung Stressless, Portugal
-> vgl. 2812

2727
Sydan
Stehendes Bauernmädchen mit Korb auf dem Rücken und Schale (auf Rundsockel)
ca. 1902/03
2727,30,85 ohne Firmenmarke, H. 49 cm; Wien Museum Inv. 219.168

2736
Simon
UHR „Amicitia" (Stehendes Liebespaar reicht sich die Hände)
Inschrift AMICITA VINCIT HORAS
ca. 1901/02
F2736,3,6 ger., gepr. H, Firmenmarke Sculpture 13, Firmenmarke 6, unsign., H. 26 cm; Wien Museum Inv. 220.772
Text der Inschrift: Die Freundschaft besiegt die Stunden.
-> vgl. 2305; 2523
Lit.: N.N.: Von der Internationalen Ausstellung für dekorative Kunst in Turin 1902, in: KM, 2.1902, S. 162–163 (m. Abb.)

2737
Cherc
Stehender rauchender Gaucho mit Hut (auf Rechtecksockel)
ca. 1902
2737 (Firmenmarke ohne Ang.), H. 68 cm; Pinhas, S. 34 (m. Abb.)

2741
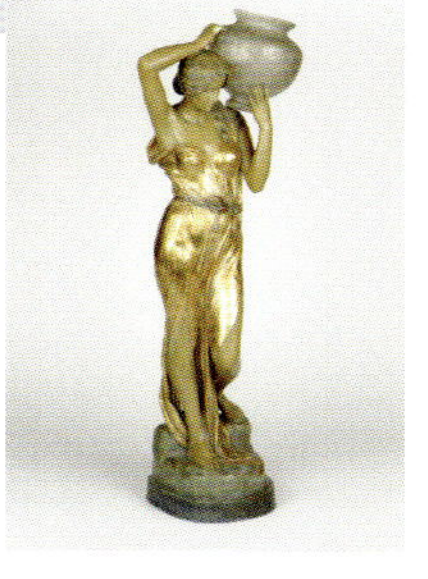
Gross, F.
Wasserträgerin (einen Krug/Henkeltopf auf der Schulter tragend)
ca. 1902
2741,114,10 Firmenmarke 4, Firmenmarke Sculpture 13, Ges. gesch., H. 61 cm; eBay # 4143665998
2741 (ohne Ang.), Firmenmarke Sculpture 13, Repr. Res., H. 55 cm; PB Wien 19
Bem.: Gürtel besetzt mit Opale (Farbsteine).
-> vgl. 2616; 2770
-> vgl. .... (H. 39 cm); Zeller 29
Lit.: Neuwirth (1974), S. 58

2749
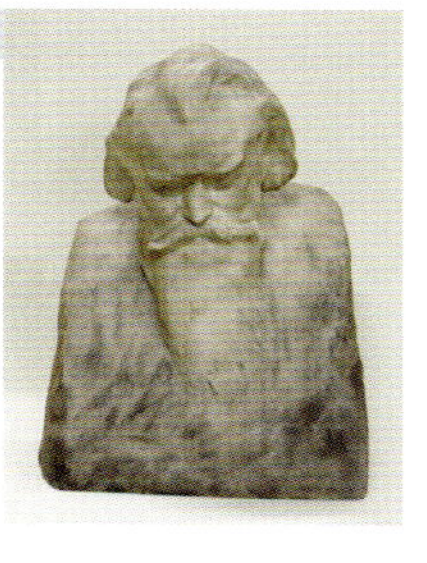
Conrat, Ilse
BÜSTE „Brahms"
ca. 1902/04
F2749,8,6 Firmenmarke 9, sign. Ilse Conrat, H. 32,1 cm; MAK Wien, Inv. Ke 10521-11
Bem.: Dargestellt ist der Komponist Johannes Brahms (1833–1897).
Ausst.: ÖMKI 1906/07
Lit.: Neuwirth (1974), S. 61

2758
Lavergne, Adolphe-Jean
UHR Am Brunnen (Paar auf säulenartigem Sockel mit Baum)
ca. 1902/04
2758,4,35 Firmenmarke 4, Ges. gesch., H. 70 cm; Soth. N.Y. 10.3.1983

2759
Haniroff
Schreitendes Bauernmädchen, einen Korb tragend (auf Landschaftssockel)
ca. 1902/04
2759,32,65 (Firmenmarke ohne Ang.), H. 42 cm; Doro 7.3.2001

2760
Ciseau
Kniende betende Frau mit Urne/Vase (auf baumartigem Sockel)
ca. 1902/04
2760 (Firmenmarke ohne Ang.), H. 71 cm; Pinhas, S. 45 (m. Abb.)

2763
Lapigre
Mädchen, an heckenumwundenem Mauerdurchbruch stehend
ca. 1903 (für den Entwurf)
FIGUR
F2763,10,63 Firmenmarke 9, Repr. Res., H. 47 cm; Neuwirth (1974), S. 58
LAMPE
F2763,65,10 Firmenmarke 9, Firmenmarke 6, H. ca. 47 cm; WKH
Lit.: Neuwirth (1973), in: Keramos, 59.1973, S. 54–63, hier S. 58 (m. Abb.) (PB Wien); Neuwirth (1974), S. 58, 307 (Abb.).
-> vgl. 2486

2768
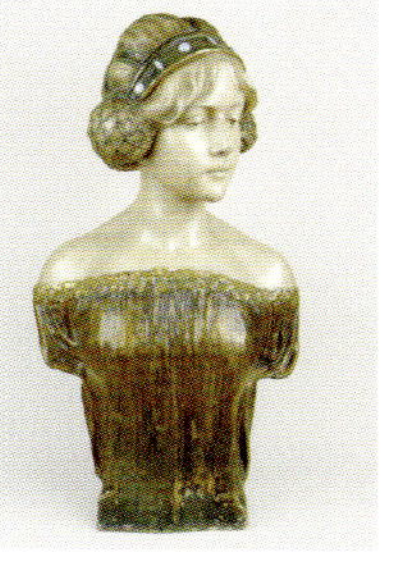
Montenave
BÜSTE Mädchen mit seitlich geknotetem Haar und Diadem
ca. 1903 (für den Entwurf)
2768,260,IV,1 Firmenmarke Sculpture 13, Repr. Res., Monogr. IE(?), (unsign.?), H. 49,5 cm; Neumeister 4.6.1994
-> vgl. 2664; 2842

2770
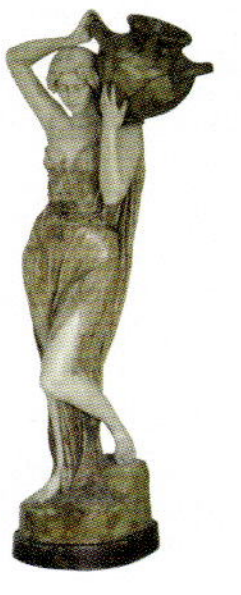
Gross, F.
LAMPE Wasserträgerin (eine(n) Krug/Amphore auf der Schulter tragend)
ca. 1902/04
2770,155,16 Firmenmarke Sculpture 13, Repr. Res., sign. F. Gross, H. 78 cm; Peege 88
2770,409,2 Firmenmarke Sculpture 13, sign. F. Gross, H. 80,5 cm; Doro 31.5.2006
2770,568,13 Firmenmarke Sculpture 13, Fabriqué en Autriche, sign. F. Gross, H. 81 cm; PB Wien 19; Bem.: Der Krug, der als Lampenschirm dient, besteht aus böhm. Glas.
2770,651,17 Firmenmarke Sculpture 13, Firmenmarke Adler 17, Fabriqué en Autriche, gepr. „S"= Z2, sign. F. Gross, H. 84 cm; Sammlung a@goldscheider.net
Bem.: vgl. Cherc 2770
-> vgl. 2616; 2741
-> vgl. .... (H. 39 cm); Zeller 29
Lit.: Pinhas, S. 45 (m. Abb.)

2770
Cherc
Wasserträgerin (eine Amphore auf der Schulter tragend)
ca. 1902/04
2770,(7,0,5,1,1 vertikal?), ohne Firmenmarke?, Made in Austria, H. 80 cm; Soth. L. 9.6.1989
Bem.: Bis auf die Amphore identisch mit Gross Modell-Nr. 2770.

2771
Sydan
LAMPE Stehendes Biedermeiermädchen eine Schatulle öffnend (auf Rundsockel)
ca. 1902/04
2771 (Firmenmarke ohne Ang.), H. 68 cm; Pinhas, S. 45 (m. Abb.)

2773
Leclancher
LAMPE Wasserträgerin (Schale in der Rechten)
ca. 1903/05
2773,91,15 Firmenmarke Sculpture 13, Repr. Res., H. 75 cm; Wien Museum Inv. 186.622

2775
LAMPE/oder BRUNNENGEHÄUSE von Taube bekrönt
ca. 1903/05
2775,5,10 Firmenmarke 7, Repr. Res., H. 73 cm, B. 40 cm; eBay # 6608553125

2781
SPIEGEL Stehende junge Dame in einen, mit großem Tuch, drapierten Spiegel schauend
ca. 1903
2781 (Firmenmarke ohne Ang.), H. 40 cm; Pinhas, S. 60 (m. Abb.)

2786
Rosé
UHR „Théodora" (Sarah Bernhardt als „Théodora" auf einem Thron sitzend)
ca. 1903
2786 gepr.,50,10 ger., Firmenmarke Sculpture 13, Repr. Res., H. 58 cm; Doro 5.6.1997; Bem.: Besetzt mit Glasperlen und Cabochons, Kugelbekrönung aus böhm. Farbglas; Lit.: AZ, 1997, S. 452, 453.
Bem.: Dargestellt ist die Schauspielerin Sarah Bernhardt (1844–1923) in der Titelrolle der „Théodora" des gleichnamigen Stückes (Urauff. Paris 1885) von Victorien Sardou (vgl. hierzu: Le Théatre, 5.1902, Nr. 75, S. 4–20).
-> vgl. 2690
Lit.: N.N.: Ausstellung des Kunstgewerbevereins in Wien, in: Sprechsaal, 24.1891, S. 63, „Stets steigende Erfolge hat auch FG, Wien-Währing mit seinen bemalten und unbemalten Terracottawerken zu verzeichnen. Wir finden wieder zahlreiche neue Figuren und Gruppen von aufstrebenden jungen Plastikern, die sich gerne der Industrie zuwenden. So das „Tirolerpaar", das von der Kronprinzessin-Witwe angekauft wurde, Sarah Bernhard als Theodora..."; N.N.: Die keramische Ausstellung in Dresden 1891, in: Sprechsaal, 24.1891, S. 816, „Man möchte fast annehmen, dass einige Künstler von Goldscheider ihre Studien in Paris gemacht haben, denn die in gelblicher Terrakotta gearbeitete Figur der Sarah Bernard entspricht der Auffassung, in der Stellung und freien Durchbildung der Gewandtheile ganz den Pariser Terracotten". Bem.: Ob es sich hier bei diesen beiden zeitgenössischen Berichten um exakt dieses Modell handelt, kann nicht mit letzter Sicherheit gesagt werden, möglicherweise wurde die berühmte Sarah Bernhardt auch öfter durch die Manufaktur zur Ausführung gebracht.

2789
Ciseau
BÜSTE Junge Dame mit Kopftuch und Pelzkragen
ca. 1903/04
2789,33,19 Firmenmarke Sculpture 13, Repr. Res., H. 58 cm; eBay # 7303442732

2790
Cherc
Krugträgerin (Stehende junge Holländerin, ein Gefäß tragend) (auf Landschaftssockel)
ca. 1903
2790 (Firmenmarke ohne Ang.), H. 88 cm; Pinhas, S. 43 (m. Abb.)

2791
JARDINIERE Frau mit zwei Wasserbehältern am Brunnen
ca. 1903/04
F2791,1,65 Firmenmarke 9, Firmenmarke 6, Repr. Res., H. 27 cm, L. 33 cm; Wien Museum Inv. 219.169

2793
Czapek, Stanislaus
Diskuswerferin (in antikisierendem Gewand)
ca. 1903/04
2793,55,3 (Firmenmarke ohne Ang.), unleserlich sign. (Czapek?), H. 51,5 cm; Doro 4.11.1998
Bem.: Diese Figur wurde evtl. anlässlich der olympischen Sommerspiele 1904 in St. Louis geschaffen.
-> vgl. 2560

2800
Ricard
Biedermeiertänzerin mit langem Kleid (ein großes Tuch über dem Kopf schwingend) (auf Blumen verziertem Rundsockel)
ca. 1903/04
2800,47,10 FG Wien (Firmenmarke ohne Ang.), Repr. Res., H. 52,7 cm; BM 70
Bem.: Evtl. identisch mit „Tänzerin" (ÖMKI 1906/07)?, vgl. hierzu auch Neuwirth (1974), S. 72.
Lit.: Pinhas, S. 63 (m. Abb.), hier H. 63 cm, ohne Künstlersign.

2807
Simon
„Topfguckerin" (Kleines Holländermädchen in einen großen Topf schauend)
ca. 1903
2807,208,13 gem. XXX, Firmenmarke Sculpture 13, Repr. Res., H. fehlt; WKH Bem.: Beschädigter Topf ersetzt durch Lampen-Glas.
Bem.: vgl. Sydan 2699
-> vgl. 2702

2812
Clarà, Juan (Joan) / Sydan
Mädchen mit Schubkarren
ca. 1902 (für den Entwurf)
TERRAKOTTA
2812,5,2 gem. VI, Firmenmarke Sculpture 13, sign. Sydan, H. 35 cm; Doro 21.9.1999
FAYENCE
2812 (ohne Ang.), sign. Clara, H. fehlt; Neuwirth (1974), S. 58, 61
Bem.: Dieses Modell wurde in Terrakotta und Fayence ausgeführt.
-> vgl. 2719

2820
Couturier, (Léon-Philibert)
UHR „Napoleon" (mit verschränkten Armen) vor Schlachtenszene (Mosaikimitation) stehend
ca. 1904
2820,43,67 Firmenmarke Sculpture 13, Repr. Res., H. 75,4 cm; Doro 18.5.1999
Bem.: Diese Uhr, sowie die Einzelfigur des Napoleon wurde unter Verwendung eines Gemäldes (Schlachtenszene) des Malers Léon-Philibert Couturier gestaltet und daher auch von der Manufaktur mit diesem Künstlernamen signiert. Lit.: zum Künstler vgl. Martin, S. 117. Die Figur findet sich unter der Mod. Nr. 2855.
-> vgl. 2855= FIGUR

2824
BÜSTE Mädchen mit Ähren
ca. 1904
2824,206,8 (Firmenmarke unleserlich), Repr. Res., H. 41 cm; Doro 9.11.1989

2826
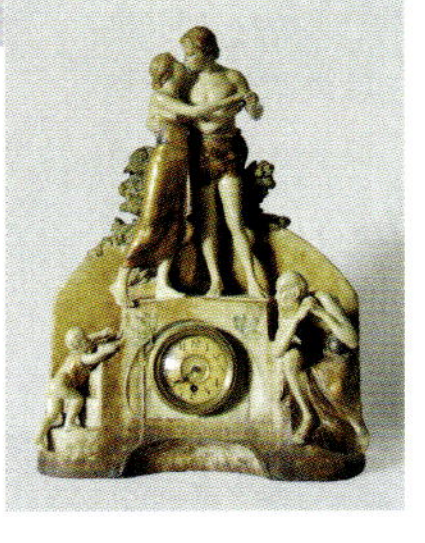
Capèque (d.i. Czapek, Stanislaus)
Lavergne, Adolphe-Jean
UHR Allegorie des Lebens
ca. 1904
2826,57,10 gem. 16, Firmenmarke Sculpture 13, sign. Capeque, H. 61 cm, B. 44 cm; Neumeister 270
2826,143,25 Firmenmarke Sculpture 13, Fabriqué en Autriche, sign. Lavergne, H. 59 cm, B. 43 cm; PB Deutschland
Bem.: Diese Uhr, wie auch die zahlreichen anderen mit „Lavergne" sign. Modelle, dürften von Stanislaus Czapek für GS geschaffen worden sein, da sich die Mitarbeiterschaft (trotz zahlreicher mit „Lavergne" sign. Modelle) für Jean-Adolphe Lavergne nicht sicher belegen lässt.

2833
Bulier
BÜSTE Betende Frau (mit großem Kopftuch)
ca. 1904/05
2833,26,7,VIII gem., GS (Firmenmarke?), Made in Austria gepr., „P"= Z3, H. 19,7 cm, L. 33,7 cm; eBay # 165562457
2833,30,1 Firmenmarke 19, Ausformung 20er Jahre, H. 20 cm; Doro 19.9.2000

2839
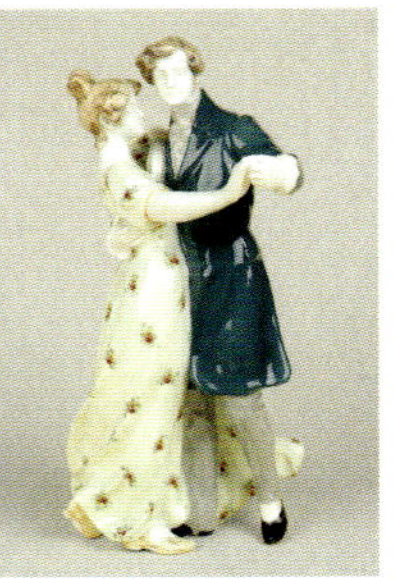
Tanzendes Biedermeierpaar
ca. 1904/05
2839,156,10 Firmenmarke 18, Ausformung ca. 1920, H. 28 cm; PB Wien 19

2840
Gross, F.
UHR Stehendes Holländermädchen an einem Schiffssteuer drehend (Uhrgehäuse als Felsstück)
ca. 1904/05
2840,(14?),17 (Firmenmarke ohne Ang.), Repr. Res., sign. F. Gross, H. 55,5 cm; Schuler 92

2842
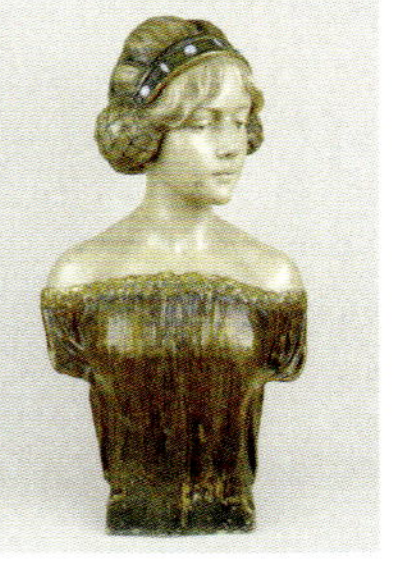
Montenave
BÜSTE Mädchen mit seitlich geknotetem Haar und Diadem
ca. 1903 (für den Entwurf)
2842,58,1 (Firmenmarke Sculpture 13?), H. 35,1 cm; Doro 15.5.2002
-> vgl. 2664; 2768

2850
Petri (d.i. Petrides, János) (Johann)
Stehendes Holländermädchen, in einem Teich angelnd (Fishbowl?)
ca. 1905
2850,559,17 Firmenmarke 22a, sign. Petri, Ausformung 20er Jahre, H. 66 cm, Ø 36,8 cm; eBay # 337380223; # 491151791
2850,594,17 gem. II, Firmenmarke 23, unsign.?, H. 66 cm; eBay # 3272719602

2853
Gross, F.
BÜSTE Mädchenkopf mit schulterlangem Haar (in der Sockelkartusche die Darstellung von Rotkäppchen mit dem Wolf)
ca. 1905
2853,5,25 gem. weiß XI, Firmenmarke Sculpture 13, Repr. Res., sign. F. Gross, gepr. „S"= Z2, H. 49 cm; Wien Museum Inv. 219.118

2854
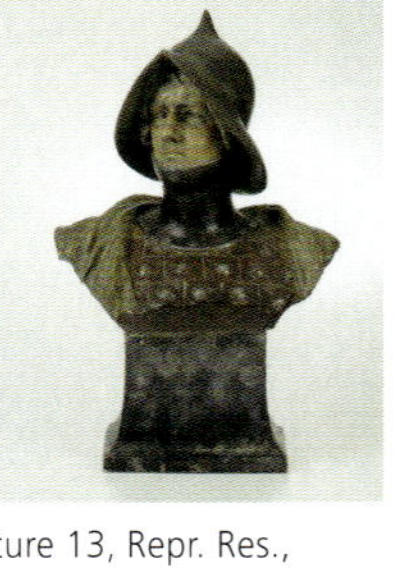
Ciseau
BÜSTE Jeanne d'Arc (als Kriegerin mit Rüstung und Helm)
ca. 1905
2854,6,12 FG Wien (Firmenmarke ohne Ang.), Repr. Res., Patent ang., D.R.G.M., gepr. „C"= Z4, H. 62 cm; Zezschwitz 22
2854,47,13 Firmenmarke Sculpture 13, Repr. Res., H. 60 cm; eBay # 5366968475
2854,57,19 ger. XI, Firmenmarke Sculpture 13, Repr. Res., H. 60 cm; Wien Museum Inv. 219.113
Bem.: Bei der dargestellten Person handelt es sich um Jeanne d'Arc.

2855
Couturier, (Léon-Philibert)
„Napoleon" (mit verschränkten Armen) stehend
ca. 1904
2855,38,2 Firmenmarke Sculpture 13, Repr. Res., Patent ang. D.R.G.M., H. 50 cm; KH
Bem.: Diese Figur, sowie die Uhr wurde unter Verwendung eines Gemäldes (Schlachtenszene) des Malers Léon-Philibert Couturier gestaltet und daher auch von der Manufaktur mit diesem Künstlernamen signiert.

Lit.: zum Künstler vgl. Martin, S. 117. Die stehende Figur des Napoleon findet sich unter der Mod. Nr. 2820.
-> vgl. 2820= UHR

2857
Simon
UHR „Das Rendezvouz" (Biedermeierkavalier mit Taschenuhr, auf seine Geliebte wartend, diese versteckt sich hinter einer Mauer/Uhrgehäuse)
ca. 1905
2857,24,45 Firmenmarke 6, (und weitere Firmenmarke ohne Ang.), H. 65 cm; Nagel 346

2859
Leclancher
LAMPE Bergmann (mit Pickel am Bergkristall)
ca. 1904
2859,27,67 gem. weiß V., Firmenmarke Sculpture 13, Repr. Res., gepr. „C"= Z4, H. 76,6 cm; GK, S. 15, 79 (m. Abb.); Wien Museum Inv. 182.073
Ausst.: HMW (jetzt Wien Museum) 1985/86.
Lit.: GK, S. 15, 79 (m. Abb.)

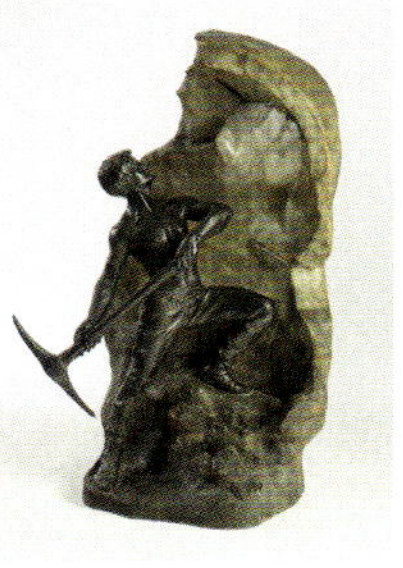

2868
Schram, Alois Hans
BÜSTE Frau mit Kopftuch
ca. 1904
2868 (ohne Ang.), GS (Firmenmarke?), sign.: A.H. Schram, H. 35 cm; RM 21
Bem.: Evtl. identisch mit Büste „Ilka" (ÖMKI 1900/01)?, vgl. hierzu auch Neuwirth (1974), S. 74.

2875
Petri (d.i. Petrides, János) (Johann)
BÜSTE Ritter mit Helm (bekrönt mit Bourbonenlilie) (auf mit drei Bourbonenlilien verziertem Sockel)
ca. 1904
2875 (Firmenmarke ohne Ang.), H. 41 cm; Pinhas, S. 38 (m. Abb.)

2876
Tell, E.
FIGUR „Rückkehr vom Brunnen" (Frau mit Gefäß, sich nach Krug bückend)
ca. 1901/02 (für den Entwurf)
2876 (ohne Ang.), FG Wien (Firmenmarke ohne Ang.), H. 40,5 cm; BM 17; Lit.: AZ 1979, S. 514
Ausst.: ÖMKI 1902/03; 1903/04, hier jeweils mit Ang. des Originaltitels „Rückkehr vom Brunnen".
-> vgl. 2357; 2447
Lit.: Pudor (1902), S. 107 (m. Abb.), hier mit Ang. des Originaltitels „Rückkehr vom Brunnen".

2884
BÜSTE Frau mit Spitzenhaube (auf reliefiertem Sockel)
ca. 1905
2884 (Firmenmarke ohne Ang.), H. 71 cm; Pinhas, S. 35 (m. Abb.)

2885
Gory, A(ffortunato?)
BÜSTE Flora (Frau mit Callablüten)
ca. 1904
2885,146,19 Firmenmarke Sculpture 13, Repr. Res., gepr. „L"= Z1, sign.: T. Gori, H. ca. 50,2 cm; KH
-> vgl. 2385 (ähnl. Modell)

2888
Skall (d.i. vermutlich Skála, Jaroslav)
BÜSTE „Mignon" (Junge Dame (Burgunderin?) mit Haube) (auf Kartuschensockel mit Landschaftsdarstellungen)
ca. 1901
F2888,3,6 Firmenmarke 6, Repr. Res., sign. Skall, H. 26,5 cm; Wendl 47
Ausst.: ÖMKI 1901/02, im Ausst.-Kat. mit der Künstlerang. „Skall", hier auch mit Ang. des Originaltitels „Mignon".
Lit.: Neuwirth (1974), S. 74

2891
Pecheur
BÜSTE Junge Dame mit Diadem (Sockelrelief: lyraspielender Jüngling)
ca. 1904
2891,5,19 Firmenmarke Sculpture 13, H. 46 cm; eBay # 2656504611

2893
Latour, (Louis-Marie-Blaise?)
LAMPE Stehendes Mädchen mit Blasebalg (auf Rundsockel)
ca. 1904
2893,9,45 (Firmenmarke Sculpture 13?), Firmenmarke 6, H. 29,3 cm; Quit. 6

2895
Gemignani, Valmore
BÜSTE Junge Frau mit Haube
ca. 1904
2895,K,19,33 (Firmenmarke ohne Ang.), sign. V. Gemignani, H. 27,4 cm; Quit. 29

2896
Lavergne, Adolphe-Jean
GRUPPE Reitende indische Schönheit auf einem von Sklaven geführten Elefanten (auf Rechtecksockel)
ca. 1904
2896 (Firmenmarke ohne Ang.), H. 85 cm; Pinhas, S. 31 (m. Abb.)
Bem.: Offenbar handelt es bei dem bei Pinhas nachgewiesenem Ex. um eine Ausformung der 20er Jahre. Da die Autorin in ihrem Werk keinerlei Nachweise auf Stempel oder Zeichen gibt, kann dies aufgrund des Sockels, der in der glasurtypisch für diese Zeit ist bzw. des Hinweises auf das Editionsjahr „1919" jedoch nur vermutet werden.

2897
Tell, E.
SPIEGEL „Malerei" (Vor einem Spiegel sitzende Malerin mit Palette)
ca. 1904
2897,62,24 gem. VI, Firmenmarke Sculpture 13, Repr. Res., H. 50 cm; Wien Museum Inv. 186.626
Lit.: KM, 4.1904, S. 138, 139 (m. Abb.), hier mit Ang. des Originaltitels „Malerei", als Künstler wird „Sydan" angegeben.

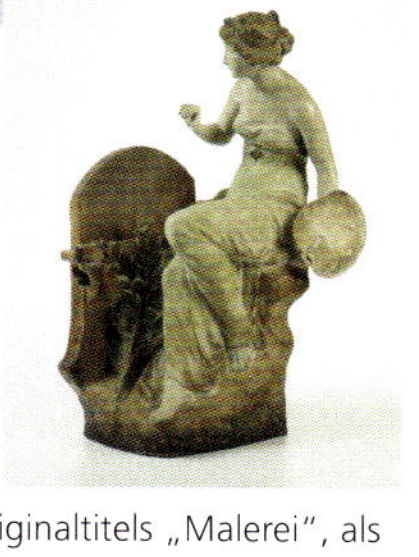

2901
Gross, F.
Wasserträgerin (ein Gefäß in die Höhe haltend)
ca. 1903/05
2901,65,16 gem. XIX, Firmenmarke Sculpture 13, Repr. Res., H. 119,5 cm; eBay # 928936939
-> vgl. 3031

2903
Roy, P. (oder F.?)
„Allegorie der Kunst" (Stehende junge Frau in antikisierendem Gewand, eine kleine Statuette betrachtend)
Inschrift: ARS LONGA VITA BREVIS
ca. 1905
2903,4,35 (Firmenmarke ohne Ang.), H. 86,5 cm; Doro 1.12.1998
Text der Inschrift: Die Kunst besteht, das Leben vergeht.

2905
Sydan
Stehendes Mädchen mit großem Weidenkorb
ca. 1905
2905,17,27 Firmenmarke 9, Firmenmarke 6, Repr. Res., H. 42 cm; GK, S. 15, 80 (m. Abb.); Wien Museum Inv. 190.074
Ausst.: HMW (jetzt Wien Museum) 1985/86.
Lit.: GK, S. 15, 80 (m. Abb.)

2907
Rosé
UHR „Der Morgen" (Hahn und drei Hennen)
ca. 1905
2907,35,27 Firmenmarke 9, Repr. Res., H. 21 cm; Wien Museum Inv. 190.101
Lit.: Pudor (1906), hier mit Ang. des Originaltitels „Der Morgen".

2909
Schreitendes Mädchen (Bäuerin), einen Weidenkorb auf der Schulter tragend
ca. 1905
2909,66,16 gem. III, Firmenmarke Sculpture 13 (gepr.), gepr. „S"= Z2, H. 86 cm; Weidler 530

2912
Leclancher
Stehende Frau mit zwei Amphoren (auf Natursockel)
ca. 1905
2912,(31?),(4?) (Firmenmarke ohne Ang.), Repr. Res., H. 52 cm; eBay # 6638871508

2917
Stehende Holländerin (Bäuerin) mit Pflug (auf Landschaftssockel)
ca. 1905
2917,241,17 Firmenmarke 19, Ausformung 20er Jahre, H. 70 cm; Wien Museum Inv. 186.623
Pendants -> vgl. 2454; 2662; 3840

2920
Lefont
Stehendes Biedermeiermädchen mit großem Hut (einen übergroßen Korb tragend)
ca. 1905
2920,360,69 FG Wien (Firmenmarke ohne Ang.), Repr. Res., unsign., H. 27 cm; Bödiger 240
2920,399,10 gem. 2, Firmenmarke Adler 15, Firmenmarke Sculpture 13, H. 27 cm; Henry's 23.9.1998

2934
Rosé
Mädchen mit Schirm
ca. 1905
2934,56,2 (2 Firmenmarken ohne Ang.), H. 20 cm; Kett. 165; 177; 181

2948
Ceccarelli, Ezio
„Cicero" (Stehende Figur) (auf quadratischer Plinthe)
Inschrift: CICERO
ca. 1905
2948,57,1 (ohne Firmenmarke), sign. „Prof. E. Ceccarelli", H. 61 cm; eBay # 2212428457
Bem.: Vermutlich GS (Zuschreibung unter Vorbehalt, jedoch sprechen, trotz der fehlenden Firmenmarke, Qualität und Nummerierung für GS), Objekt konnte leider nicht selbst geprüft werden.

2955
Rosé
UHR-LAMPE Frau an verschneitem Glockenturm, die Glocke läutend
ca. 1905
2955,42,69 Firmenmarke 9, Repr. Res., H. 44 cm; KH
Bem.: Eines der wenigen Objekte mit Doppelfunktion. Uhr und Lampe wurden hier bei einem Werk vereint.

2960
Perigot
Paar an einem Tisch stehend (sich mit Gläsern zuprostend)
ca. 1905
2960,22,10 FG Wien (Firmenmarke ohne Ang.), Firmenmarke 6, H. 26,8 cm; Doro 12.6.1995
F2960,34,6 Firmenmarke 9, Repr. Res., H. 27 cm, B. 25 cm; Wien Museum Inv. 190.118

2962
Rosé
UHR „Beethoven“ (mit einem Notenbuch) auf einem Thron sitzend
ca. 1905
2962,20,32 Firmenmarke Sculpture 13, Repr. Res., H. 65 cm; Doro 21.9.1999
Bem.: Dargestellt ist der Komponist Ludwig van Beethoven (1770–1827).

2975
Ciseau
BÜSTE Junge Frau mit gesenktem Blick
ca. 1905
2975 (Firmenmarke ohne Ang.), H. 65 cm; Pinhas, S. 56 (m. Abb.)
–> vgl. 2976

2976
Ciseau
BÜSTE Junge Frau mit gesenktem Blick
ca. 1905
2976 (Firmenmarke ohne Ang.), H. 41 cm; Pinhas, S. 56 (m. Abb.)
–> vgl. 2975

2980
Haniroff
BÜSTE Biedermeierdame mit Schutenhut (Sockelrelief: Taube auf zwei Rosengirlanden sitzend)
ca. 1905
2980,2,38 gem. Z, Firmenmarke Sculpture 13, Repr. Res., H. 33 cm; eBay # 2593923035

2982
Tell, E.
Stehendes Mädchen, einen Hasen haltend (auf Landschaftssockel)
STEINGUSS?
ca. 1905
2982,4,69 FG Wien (Firmenmarke ohne Ang.), Repr. Res., sign. E. Tell, Goldscheider, H. 38 cm; WKA 27; Bem.: Im Auktionskat. als „Steinguss“ bezeichnet, dies konnte am Original nicht überprüft werden.

2985
Petri (d.i. Petrides, János) (Johann)
Bauernmädchen, einen Wagen mit Korb ziehend
ca. 1905
2985,57,19 (Firmenmarke ohne Ang.), sign. Petri, H. 43 cm; Soth. Monrepos Lot 1878
–> vgl. 3163

2987
Ciseau
BÜSTE Junge Dame mit Kopftuch
ca. 1905
2987,26,33 (Firmenmarke ohne Ang.), Repr. Res., H. 39 cm; eBay # 6558055870

2991
Lefont
Stehende junge Frau, eine Lyra spielend (auf Rundsockel)
ca. 1905
2991 (Firmenmarke ohne Ang.), H. 76 cm; Pinhas, S. 34 (m. Abb.)

2998
Rosé
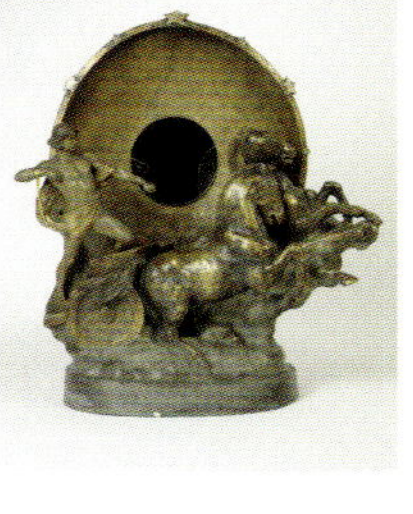
LAMPE Sonnenwagen (Römer im zweispännigen Streitwagen)
ca. 1905
2998,35,3 Firmenmarke Sculpture 13, Repr. Res., H. 83 cm; GK, S. 15, 37 (m. Farbabb.); Wien Museum Inv. 182.075
Bem.: Der Titel „Sonnenwagen“ wird nach dem Ausst.-Kat. GK, S. 15 zit.
Ausst.: HMW (jetzt Wien Museum) 1985/86.
Lit.: GK, S. 15, 37 (m. Abb.)

3009
Tell, E.
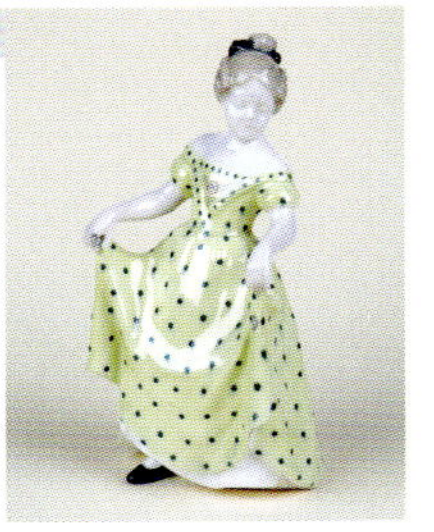
Stehendes Biedermeiermädchen (ihr Kleid etwas hochhebend) (nach links blickend)
ca. 1904/05
FAYENCE
3009,301,27 (Firmenmarke ohne Ang.), H. 18,3 cm; Zeller 47
3009,523,13 gem. IV., H., Firmenmarke Adler 15, H. 20 cm; PB Wien 19
3009,776,16 Firmenmarke Adler 15, H. 19 cm; Fischer Luz. 328
3009,1381,10 (Firmenmarke ohne Ang.), unsign?, H. 18 cm; Wendl 31
3009 (ohne Ang.), H. 19 cm; RM 26
ALABASTER
3009,72,H1 ger., Plak. A, H. 21,2 cm; PB Wien 19
3009, lig. HM, 243 ger., Plak. C, Ausformung 20er Jahre, H. ca. 22 cm; KH
Bem.: Dieses Modell wurde in Fayence und Alabaster ausgeführt.
–> vgl. 4656
Pendant –> vgl. 3067

3013
BÜSTE Mädchen mit großem breitkrempigem Hut
ca. 1904/05
3013,(316,50?) Firmenmarke Sculpture 13, Repr. Res., H. fehlt; eBay # 7332272258

3018
Latour, (Louis-Marie-Blaise?)
BÜSTE Frau mit Diadem und Halsband
ca. 1904/05
3018,31,19 Firmenmarke Sculpture 13, Repr. Res., H. 59 cm; Doro 10.4.1997; Bem.: Diadem mit Opalen (Farbsteinen) besetzt.

3022
Lefont
BÜSTE Junge Frau mit Kopftuch
ca. 1904/05
3022,–,35 gem. VII., Firmenmarke Sculpture 13, Ges. gesch., H. 29,2 cm; eBay # 247093288

3024
Schmidt-Kestner, Erich
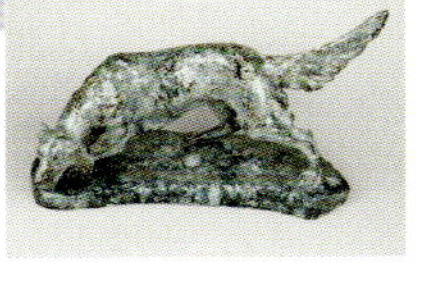
„Setter“ (mit/ohne Landschaftssockel)
ca. 1904/05
mit Landschaftssockel
3024,55,65 (Firmenmarke ohne Ang.), sign. Schmidt-Kestner, H. 28,4 cm; eBay # 2058808280
3024,96,17 Firmenmarke 19, Ausformung 20er Jahre, sign. Strobl, H. 27 cm; PB Wien 19
ohne Landschaftssockel
3024,138,15 (Firmenmarke ohne Ang.), wohl Ausformung 20er Jahre, sign. Strobl, H. 24,5 cm, L. 50 cm; Zezschwitz 13
3024F, H. 31 cm, B. 48 cm; GSNachf. 5,9d
Bem.: Dieses Modell kann eindeutig dem Tierbildhauer Erich Schmidt-Kestner zugeordnet werden. Der Name „Strobl“ diente wohl als Pseudonym für den in der Zeit bekannten Bildhauer Schmidt-Kestner. Möglicherweise sollte die Zusammenarbeit mit Goldscheider dadurch verschleiert werden.
–> vgl. 6135
–> vgl. 6659; 7100= jeweils GRUPPE
–> vgl. GRUPPE Knabe, Setter am Halsband führend
Lit.: GSNachf. 5,9 (m. Abb.), hier mit Ang. des Originaltitels „Setter“ (vorne); Neuwirth (1974), S. 58 (für die Modell-Nr. 3024); Pinhas, S. 100 (m. Abb.).

3025
Lefont
GRUPPE „Reigen“ (Junges Paar einen Reigen tanzend) (auf blumenverziertem mehrpassigem Sockel)
ca. 1905
3025,16,10 Firmenmarke 9, Firmenmarke 6, Repr. Res., H. 43,8 cm; eBay # 1299674036
Ausst.: ÖMKI 1906/07, hier mit Ang. des Originaltitels „Reigen“.
Lit.: Neuwirth (1974), S. 66

3027
Simon
GRUPPE Stehendes Biedermeiermädchen, ein Kleinkind im Arm haltend
ca. 1904/05
3027,22,12 Firmenmarke Sculpture 13, Repr. Res., gepr. „S“= Z2, H. 45,7 cm; eBay # 2884735161

3031
Gross, F.
Wasserträgerin (ein Gefäß in die Höhe haltend)
ca. 1904/05
3031,139,35 Firmenmarke Sculpture 13, Fabriqué en Autriche, H. 85 cm; Wien Museum Inv. 186.618
–> vgl. 2901

3042
Bofill, Antoine
Stehender Holländerjunge mit Zipfelhaube und Laterne (auf Landschaftssockel)
ca. 1904/05
3042,18,1 ohne Firmenmarke, Repr. Res., sign. Antoine Bofils (!), H. 43 cm; Soth. Monrepos Lot 1870
F3042,48 (Firmenmarke Sculpture 13?), undeutlich sign., H. 41 cm; Doro 3.2.1998 (ohne Kat.)
Bem.: Der Künstler wurde möglicherweise durch das Märchen „Die 7 Zwerge“ zu dieser Darstellung inspiriert.

3044
(Leisek, Georg)
Kleines Mädchen mit Puppe am Waschtrog
ca. 1905
3044,20,65 (Firmenmarke Sculpture 13?), undeutlich sign. „Leiser“, H. 31 cm; Henry's 26.8.1998; 25.3.1999

3058
Cherc
GRUPPE „Musique“ (Stehendes Kinderpaar, sich verschämt einen Apfel zureichend) (auf Landschaftssockel)
Inschrift: MUSIQUE
ca. 1904/05
3058,167,2 Firmenmarke Sculpture 13, Repr. Res., H. 48,9 cm; eBay # 3156525375
3058,274,35 Firmenmarke 23, Ausformung 20er Jahre, H. 50,8 cm; eBay # 3711071458
Text der Inschrift: Musik

3063
UHR Stehendes, sich küssendes Biedermeierpaar (auf oktogonalem Sockel)
ca. 1904/05
F3063,10,6 FG (Firmenmarke o. Ang.), Repr. Res., H. ca. 51 cm; KH, Bem.: Lenzkirch Uhrwerk.

3067
Tell, E.
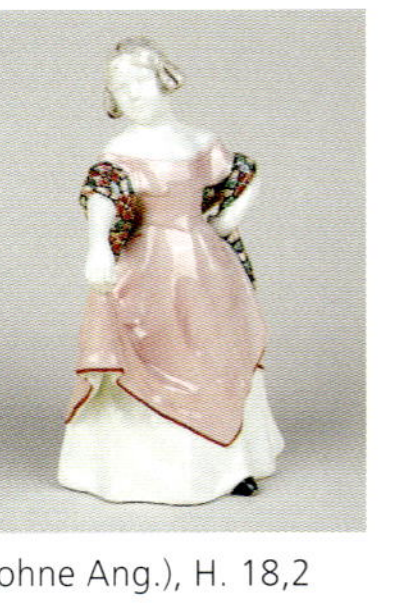
FIGUR/LAMPE Stehendes Biedermeiermädchen (ihr Kleid etwas hochhebend) (nach rechts blickend)
ca. 1904/05
FIGUR
3067,219,11 (Firmenmarke Adler ohne Ang.), H. 18,5 cm; Wendl 43
3067,300,6 (Firmenmarke Adler ohne Ang.), H. 18,2 cm; Wendl 30
3067,357,35 gem. B, XXIX, Firmenmarke 24, Ausformung 20er Jahre, H. 19 cm; PB Wien 19
3067,439,11 Firmenmarke 19, Ausformung 20er Jahre, H. 17,5 cm; Herr 31; Quit. 29; Metz 23.10.2004
3067,563,65 Firmenmarke Adler 16, H. 17,5 cm; KaM 84; 87
3067,629,13 (Firmenmarke Adler ohne Ang.), H. 17,4 cm; Doro 10.4.1997
3067,632,10 ohne Firmenmarke, H. 18 cm; Wien Museum Inv. 219.183
3067,638,65,4 gem. 4., Firmenmarke Adler 16, unsign.,

H. 18 cm; Wien Museum Inv. 190.165
3067,822,65 gem. X,V, Firmenmarke 18, Ausformung ca. 1920, H. ca. 18 cm; eBay # 1219687937; 2624484833
LAMPE
3067,887,5 Firmenmarke 19, Ausformung 20er Jahre, H. 24,4 cm; Doro 25.4.2000
–> vgl. 4657
Pendant –> vgl. 3009
Lit.: Neuwirth (1974), S. 58, 75

3069
Labatut, Jules Jacques
Stehende Frau, eine Mandoline um ihre Schulter tragend (auf Landschaftssockel)
ca. 1904/05
3069 (Firmenmarke ohne Ang.), H. 87 cm; Pinhas, S. 44 (m. Abb.)

3072
Kleines Biedermeiermädchen mit übergroßem Korb
ca. 1904/05
3072,31,36 Firmenmarke 9, Firmenmarke 6, Repr. Res., H. 17,5 cm; Wien Museum Inv. 186.682

3075
Meier, (Emil oder Meier-Michel, Johanna)
LAMPE Stehende Biedermeierdame mit Schutenhut und Blumenkorb (auf Rundsockel)
ca. 1905
3075,7,63 gem. T., Firmenmarke 3, Firmenmarke Sculpture 13 (gepr.), sign. Meier, H. 46 cm; PB

3077
Rosé
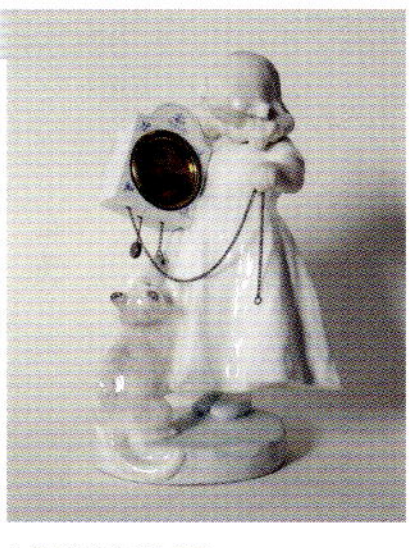
UHR Biedermeiermädchen mit Uhr und spielender Katze
ca. 1905
F3077,48,6 gem. 1, Firmenmarke 9, Firmenmarke 6, Repr. Res., H. 27,5 cm; PB Deutschland
3077,52,36 gem. F., Firmenmarke 9, Firmenmarke 6, Repr. Res., H. 27,5 cm; eBay # 1478008265
3077,165,36 Firmenmarke 6, Firmenmarke 9, Repr. Res., H. 27,5 cm; Wien Museum Inv. 190.100

3078
Hauf
Sitzender Windhund
ca. 1905
3078,146,3 gem G., GS Wien (Firmenmarke?), H. 21,2 cm, L. 19,5 cm; Kett. 28
Lit.: Neuwirth (1974), S. 64

3084
Cherc
Stehender Junge mit Hut und Schulranzen (auf Felssockel)
ca. 1905
3084,196,3 gem. weiß 15,VII., Firmenmarke Sculpture 13, Repr. Res., gepr. „S"= Z2, H. 69 cm; eBay # 6639084471

3085
Gambeauche
Stehende Römerin mit Fisch (auf Rundsockel)
ca. 1905
3085,9,11 gem. XII.D.1., Firmenmarke 19, Made in Austria, Ausformung 20er Jahre, H. 55,9 cm; eBay # 150392786

3088
Leclancher
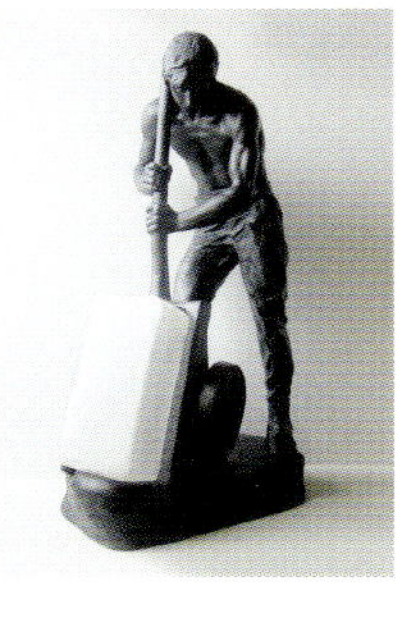
LAMPE Arbeiter, einen Eisblock auf Karren schiebend
ca. 1905
3088,23,24 Firmenmarke Sculpture 13, Repr. Res., undeutl. sign. Le(clancher?), H. 72 cm, B. 43 cm; Sammlung.schmutz@chello.at 111

3089
Leclancher
BÜSTE Kinderkopf
ca. 1900/05
3089,22,36 Firmenmarke 9, Repr. Res., H. 30 cm; Wien Museum Inv. 220.788

3091
Persan
Gänseliesel (Kleines Mädchen mit Kükenschar) (auf Natursockel)
ca. 1905
F3091,150,6 Firmenmarke 8c, Repr. Res., H. 34 cm; Wien Museum Inv. 219.161

3095
Petri (d.i. Petrides, János) (Johann)
UHR Sitzender Knabe
ca. 1905
3095,26,65 Firmenmarke 6, Firmenmarke 9, Repr. Res., sign. Petri, H. 30 cm; WKH

3100
BÜSTE Madame Dubarry (nach links blickend) (auf rechteckiger Plinthe)
ca. 1905
3100,37,19 (ohne Firmenmarke), gepr. „L"=Z1, H. 76,5 cm; Ahlden 112
Bem.: Die Dargestellte soll laut AH „Madame Dubarry" (1743–1793), die Geliebte Ludwigs XV., sein. Ähnliche Büsten gibt es von Augustin Pajou (1730–1809) und Jean-Baptiste Lemoyne (1704–1778).
Pendant –> vgl. 3111

3110
Philippe, Paul
Auf Hocker sitzender Mädchenhalbakt mit Katze
ca. 1905
3110 (Firmenmarke ohne Ang.), H. 32 cm; Pinhas, S. 47 (m. Abb.); Bem.: Laut Pinhas auch in Bronze ausgeführt.

3111
BÜSTE Höfische Dame der Zeit Ludwigs XV. (nach rechts blickend) (auf rechteckiger Plinthe)
ca. 1905
3111,16,13 Firmenmarke Sculpture 13, Fabriqué en Autriche, H. 76 cm; Ahlden 112; 117; Rieber 16.–18.9.2002
Bem.: Die Dargestellte soll laut AH eine nicht näher zu verifizierende „Hofdame" der „Madame Dubarry" sein.
Pendant –> vgl. 3100

3116
Rosé
UHR Drei Mädchen (durch eine geöffnete Tür laufend)
ca. 1905
3116,8,10 gem. W., Firmenmarke 9, Repr. Res., H. 51 cm; Wien Museum Inv. 186.640

3117
Pfeffer, L.
LAMPE Stehendes Mädchen mit zwei Mohnblüten
ca. 1905
3117,40,15 (Firmenmarke ohne Ang.), H. 67 cm; Phillips L. 25.1.2001

3120
Rosé
„Heine" (Heinrich Heine mit Buch, auf einer Bank mit antikisierender Kratervase sitzend)
Inschrift: HEINE
ca. 1905
3120,95,17 Firmenmarke Sculpture 13, Fabriqué en Autriche, sign. Rose (!), H. 55,5 cm; GK, S. 15, 82 (m. Abb.); Wien Museum Inv. 182.077
Bem.: Dargestellt ist der Dichter Heinrich Heine (1797–1856) in der Art eines Miniaturdenkmales.
Ausst.: HMW (jetzt Wien Museum) 1985/86.
Lit.: GK, S. 15, 82 (m. Abb.)

(3129)
Auber
Stehende Holländerin mit zwei Körben
ca. 1905
A 3129,169 X, Deposé (!), H. 31 cm; Zeller 41
Bem.: Meiner Ansicht nach handelt es sich hier eindeutig um ein Erzeugnis des Wiener Emailfarbenwerkes „Schauer". Die fehlende Firmenmarke sowie das Wort „Deposé", welches bei GS nicht verwendet wurde, deuten darauf hin. Die Zuordnung zur Manufaktur GS erfolgte durch das AH Zeller.

3131
Haniroff
FIGUR/LAMPE Schreitender Harlekin mit Schirm
ca. 1905
FIGUR
3131,14,15 gepr. K, gem. N, GS Wien (Firmenmarke ohne Ang.), unleserlich sign., H. 54 cm; Kastern 79
LAMPE
3131,8,11 gem. 34 L, Firmenmarke 19, Made in Austria, sign. Petri, Ausformung 20er Jahre, H. 54,6cm; eBay # 3707184302 (KH www.tias.com/stores/frozenbeauty);
Bem.: Bei der Sign. „Petri" dürfte es sich um ein Versehen des Modelleurs handeln.
3131,15,14 Firmenmarke 19, Made in Austria, sign. Haniroff, Ausformung 20er Jahre, H. ca. 54 cm; KH
–> vgl. 3133
–> vgl. ohne Nr. BRONZE

3133
Haniroff
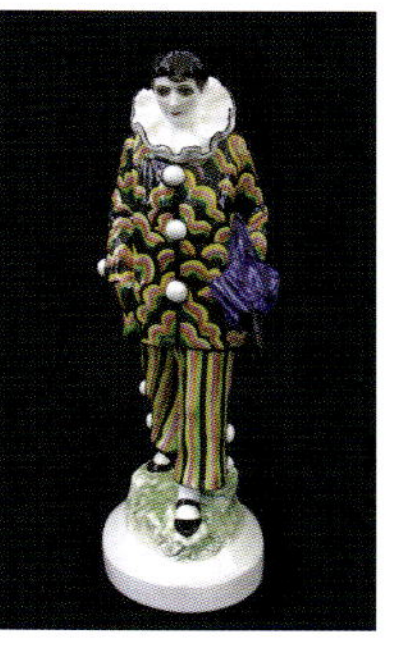
Schreitender Harlekin mit Schirm
ca. 1905
3133,13,27 (Firmenmarken ohne Ang.), sign. Haniroff, H. 29 cm; Doro 20.11.1996
3133,98,1 (Firmenmarke ohne Ang.), sign. Haniroff, H. 30,5 cm; Fischer 127
3133,151,10 gem. XII., Firmenmarke 19, sign. Haniroff, Ausformung 20er Jahre, H. 29,9 cm; Sammlung Robert Senft
3133,156,8 ger., gem. XII., gem. lig. AS, Firmenmarke 19, sign. Haniroff, Ausformung 20er Jahre, H. 30 cm; Wien Museum Inv. 190.114
3133,167,7 gem. VF., Firmenmarke 19?, sign. Haniroff, H. 29,8 cm; Quit. 3
3133,198,65 gem. F XII, Firmenmarke 19, Made in Austria, sign. Haniroff, Ausformung 20er Jahre, H. 30,5 cm; Soth. com. # 31HD; eBay # 1202012721 (KH www.tias.com/stores/frozenbeauty)
ohne Ang., H. 29 cm; Zacke 6, S. 67
Ausst.: SSH 1992
–> vgl. 3131
–> vgl. ohne Nr. BRONZE

3150
LAMPE/SPIEGEL Liegender Mädchenhalbakt auf Felsen, nach einer Wasserlilie greifend (auf rechteckigem Sockel/ Wasser-Uferzone)
ca. 1905
3150,15,19 (Firmenmarke fehlt, wohl ehem. Firmenmarke Sculpture 13?), H. fehlt, Ø 50 cm; Christie's SK 18.9.1998
3150 (Firmenmarke ohne Ang.), H. 50 cm; Pinhas, S. 60 (m. Abb.)
Bem.: Das Lampenglas hat die Form einer Blumenblüte, die Wasseroberfläche wird durch einen Spiegel dargestellt.

3156
Leclancher
LAMPE Stehender, an der Feuerstelle (Esse) arbeitender Schmied
ca. 1905
3156,33,24 Firmenmarke Sculpture 13, Repr. Res., H. 80 cm; eBay # 3507160501
Bem.: Die Feuerstelle ist von unten her durch mehrere Lämpchen beleuchtet, so dass der Eindruck lodernden Feuers entsteht.

3163
Petri (d.i. Petrides, János) (Johann)
Bauernmädchen, einen Wagen mit Korb ziehend
ca. 1905
3163,5,10 Firmenmarke 9, Repr. Res., sign. Petri, H. 25 cm, L. 28 cm; Wien Museum Inv. 186.685
-> vgl. 2985

3169
Coudray, Georges-Charles
„Aida" (Stehende junge Frau mit Harfe)
Inschrift: AIDA
ca. 1905
3169 (Firmenmarke ohne Ang.), H. 64 cm; Pinhas, S. 37 (m. Abb.)

3170
Rosé
LAMPE „Goethe" (lesender Johann Wolfgang von Goethe, an einem Tischchen mit Lampe sitzend)
Inschrift: GOETHE
ca. 1905
3170 (ohne Ang.), (Firmenmarke ohne Ang.), Gesamth. 68 cm; Zof. 15; Lit.: AZ 1996, S. 339; TuS 1996 Nr. 199, S. 156; Weltkunst 66.1999, S. 1175
Bem.: Dargestellt ist der Dichter Johann Wolfgang von Goethe (1749–1832).

3191
Schmidt, Henri
Stehendes Mädchen mit fünf Katzen spielend (auf Rundsockel)
ca. 1905
3191,21,65 gem. X, Firmenmarke 9, Repr. Res., sign. Henri Schmidt, H. 26 cm; Sammlung Stressless, Portugal

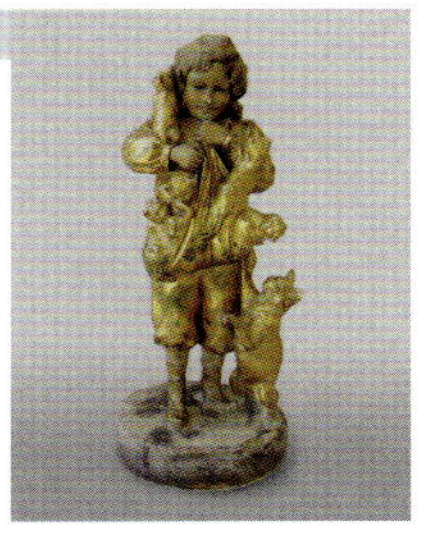

3200
Haniroff
Stehende junge Frau mit großem Hammer und Amboss (auf hohem Rundsockel)
ca. 1905
3200 (Firmenmarke ohne Ang.), H. 88 cm; Pinhas, S. 45 (m. Abb.)

3208
Villanis, Emmanuel
„Captive" (Stehende Sklavin mit Handfesseln) (auf Mauersockel)
ca. 1905/07
3208 (Firmenmarke ohne Ang.), H. 100 cm; Pinhas, S. 62 (m. Abb.)
Lit.: Berman, Vol. 4, S. 981 (m. Abb.), hier die Ausführung als Büste mit dem Originaltit. „Captive".

3215
Biedermeiermädchen mit Schale, Spitztüte und Brief in Händen
ca. 1905
3215,29,– FG (Firmenmarke ohne Ang.), Firmenmarke 6, H. 61 cm?; BM 65; Bem.: H. 61 cm wohl Druckfehler, richtige H. vermutlich ca. 40 cm.

3207
Lavergne, Adolphe-Jean
Stehende junge Frau eine Schale in die Höhe haltend (auf Rundsockel)
ca. 1905
3207 (Firmenmarke ohne Ang.), H. 118,5 cm; Pinhas, S. 38 (m. Abb.)
Bem.: Die Schale kann sowohl als Visitenkartenschale als auch als Lampensockel dienen.
-> vgl. 3590

3217
Rosé
BÜSTE „Goethe" (mit Lorbeerzweig)
Inschrift: GOETHE
ca. 1905
3217,39,11 gem. weiß B, (Firmenmarke ohne Ang.), H. 30,5 cm; eBay # 3776049249
Bem.: Dargestellt ist der Dichter Johann Wolfgang von Goethe (1749–1832).

3234
Rosé
BÜSTE „Schiller"
Inschrift: SCHILLER
ca. 1905
3234,73,4 Firmenmarke 23, gem. AUSTRIA, Ausformung 20er Jahre, H. 30 cm; eBay # 180026398489
3234,78,14 Firmenmarke 23, gem. AUSTRIA, Ausformung 20er Jahre, H. 30,5 cm; eBay # 7354276598
Bem.: Dargestellt ist der Dichter Friedrich Schiller (1759–1805). Die Büste wurde zum 100. Todestag aufgelegt.

3244
Leclancher
Kniender Gärtnerjunge (einen Baum/Rebstock pflanzend?)
ca. 1905
3244,3,46 Firmenmarke Sculpture 13, Repr. Res., H. 42 cm; WKH (Doro FVK)

3246
Tanzendes Paar in historisierendem Kostüm (auf getrepptem Sockel)
ca. 1899/1905
3246,16,10 Firmenmarke 9, Repr. Res., H. 33 cm; Wien Museum Inv. 190.120

3248
Rosé
„Richard Wagner" (sitzende, lesende Figur)
ca. 1905/06
3248,138,2 Firmenmarke Sculpture 13, H. 48 cm, L. 51 cm; Wendl 29
Bem.: Dargestellt ist der Komponist Richard Wagner (1813–1883).
-> vgl. 3921

3251
Abschlag
UHR Der Abschied (Junger Ritter in Rüstung auf Pferd, sich von seinem Mädchen verabschiedend)
ca. 1899/1905
3251,5,69 Firmenmarke 9, Firmenmarke 6, Repr. Res., H. 55,9 cm; eBay # 253371483; 2615210956

3256
Lavergne, (Adolphe-Jean)
UHR/LAMPE „Dans Le Arene" (In der Arena) (Römischer Sklave lässt zwei Löwen aus dem Käfig) (Uhrgehäuse als Löwenkäfig, mit zwei auf Säulen stehenden Urnen, welche als Lampenfassungen dienen)
ca. 1905/06
3256,62,10 Firmenmarke Sculpture 13, Repr. Res., sign. Lavergne, H. 78 cm (mit Lampenfassung); Walldorf 14
Bem.: Diese Skulptur ist eines der wenigen Stücke, bei dem die Manufaktur GS Uhr und Lampenfunktion in einem Objekt zur Ausführung bringt. Ang. zum Originaltitel -> vgl. 3444.
-> vgl. 3444

3265
Rosé
BÜSTE Lächelnde Frau mit Kopftuch
ca. 1899/1905
3265,17,1 Firmenmarke 7, Repr. Res., H. 37,5 cm; Sammlung Holzer, Regensburg

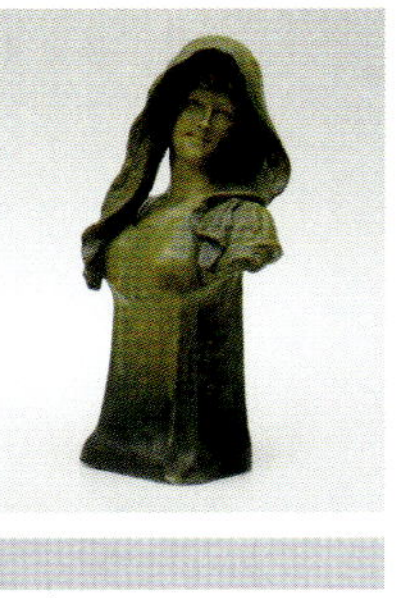

3283
Capèque (d.i. Czapek, Stanislaus)
Spähende Amazone (mit Schild und Speer) auf Pferd (auf Landschaftssockel)
ca. 1905
3283,63,24 Firmenmarke Sculpture 13, H. 84 cm; Doro 10.3.2003
Bem.: Der Speer ist aus Metall gefertigt.
Pendant -> vgl. 3288

3285
Rosé
„Mozart" (mit Geige, auf Sessellehne sitzend)
ca. 1905
3285,24,67 gem. weiß XXX, Firmenmarke Sculpture 13, Repr. Res., gepr. „S"= Z2, H. 71 cm; Wien Museum Inv. 186.604
Bem.: Dargestellt ist der Komponist Wolfgang Amadeus Mozart (1756–1791).
-> vgl. 3522; 5483

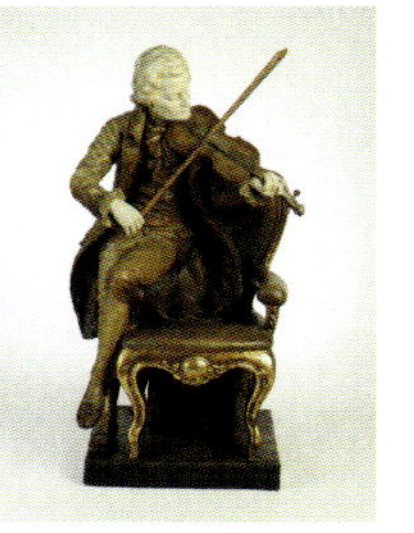

3288
Capèque (d.i. Czapek, Stanislaus)
Spähende Amazone (mit Schild und Dolch) auf Pferd (auf Landschaftssockel)
ca. 1905
3288,75,55 (Firmenmarke ohne Ang.), H. 76 cm; Doro 4.3.2002; 8.3.2004; 15.3.2005
Pendant -> vgl. 3283

3291
LAMPE Stehendes Biedermeiermädchen, ein Kleinkind haltend (auf Ovalsockel)
ca. 1905
3291 (Firmenmarke ohne Ang.), H. 67 cm; Pinhas, S. 46 (m. Abb.)
Bem.: Lampenkörper aus böhmischem Glas.

3294
LAMPE Sarah Bernhardt als Kleopatra? (Stehende Frau in antikisierendem Gewand)
ca. 1905
3294,3,67 Firmenmarke Sculpture 13, Repr. Res., H. ca. 97,5 cm, Gesamth. (mit Lampe) 116 cm; Doro 1.12.1998; Bem.: Zwei Lampen auf goldgefassten Holzsäulen, diese auf separatem Sockel; Lit.: Weltkunst, 66.1999, S.100.
Bem.: Dargestellt ist vermutlich die Schauspielerin Sarah Bernhardt (1844–1923) in der Rolle der Kleopatra, im gleichnamigen Stück „Cléopâtre" von Victorien Sardou (Urauff. Paris 1890).

3295
Rosé
UHR als Denkmal mit Jüngling und Adler
Inschrift: UT ABSIN(?) CURAE NE OBSIM(?)INJURIOS VIGILO
ca. 1905
3295,66,17 GS Wien (Firmenmarke?), H. ca. 60 cm; Henry's 26.2.1997; 26.3.1998
Text der Inschrift: Ich wache, dass deine Sorgen vergehen und dass kein Unrecht dir widerfährt.

3298
Rosé
BÜSTE „Wagner" (auf quaderförmigem Sockel)
Inschrift: WAGNER
ca. 1905
3298,66,55 weiß gem. XXX, rot gem. 1194, (Firmenmarke ohne Ang.), H. 30,5 cm; eBay # 3703887896
3298,183,14 Firmenmarke 19, H. 31 cm; Kett. 74
Bem.: Dargestellt ist der Komponist Richard Wagner (1813–1883).

3301
Daubling (d.i. Doblinger, Ferdinand)
Stehende Biedermeierdame mit Schultertuch
ca. 1905/06
3301 (ohne Ang.), GS Wien (Firmenmarke?), sign. Daubling, H. 35 cm; Henry's 28.5.1998

3312
Sydan
Bäuerin mit Futtertrog und Kuh (auf Natursockel)
ca. 1905/06
3312,29,12 gem. weiß B8, Firmenmarke Sculpture 13, Repr. Res., H. 42,5 cm; Wien Museum Inv. 219.134

3319
Lavergne, Adolphe-Jean
UHR „Historia" (Sitzende Frau auf Sockel mit Buch, in das ein stehender Jüngling (mit Lorbeer und Kugel) schreibt)
Inschrift: CAVE NE INDIGNUM
ca. 1905/06
3319,147,14 Firmenmarke Sculpture 13, unsign., H. ca. 70 cm, B. ca. 55 cm; KH
Text der Inschrift: Hüte Dich vor dem Unwürdigen.
-> vgl. 3396; 3763

3322
Baudry, R. E.
BÜSTE Junge Frau mit Haarband
ca. 1905/06
3322 (Firmenmarke ohne Ang.), sign. R. Baudry, H. 50 cm; Pinhas, S. 58 (m. Abb.)

3323
Stehender Bauernbub (auf Landschaftssockel)
ca. 1905/06
3323 (Firmenmarke ohne Ang.), H. 44,5 cm; Pinhas, S. 36 (m. Abb.)
Bem.: Möglicherweise handelt es sich bei der Nummer 3323 um einen Lese- oder Druckfehler. Zu vermuten ist eher die Mod.-Nr. 3327, denn normalerweise folgen eindeutige Pendants, wie in diesem Fall aufeinander. Bei Pinhas wird auch eine GRUPPE dieser beiden Figuren abgebildet, die Mod.-Nr. fehlt jedoch.
Pendant -> vgl. 3328

3328
Stehendes Bauernmädchen mit Korb (auf Landschaftssockel)
ca. 1905/06
3328 (Firmenmarke ohne Ang.), H. 44,5 cm; Pinhas, S. 36 (m. Abb.)
Pendant -> vgl. 3323

3330
Rosé
„Friedrich der Große" (Stehende Figur in Uniform) (auf Landschaftssockel)
ca. 1905/06
3330,17,1(5?) ger. M VIII, gem. XIV (A?), Firmenmarke 19, Made in Austria, Hand dec., unsign., Ausformung 20er Jahre, H. 60 cm; eBay # 1474632635
Bem.: Dargestellt ist Friedrich der Große (der II.) (gen. „Der alte Fritz", 1712–1786); Lit.: Venohr, W.: Der große König, Friedrich II., Bergisch-Gladbach 1997, Umschlag (m. Abb.).
-> vgl. 4020

(3333)
Ballfigur (Frau in Ballkleid)
ca. 1905
3333,45,10 H. 31,1 cm; Brö 2,1, S. 400; u.a. Nachweise bei Auktionshäusern, KH etc.
Bem.: Dass dieses Objekt hier Aufnahme findet hat seinen Grund: immer wieder wird dieses Modell im KH wegen der ähnl. Ausführung und des Genres als GS-Figur angeboten. Es handelt sich jedoch eindeutig um ein Erzeugnis der Firma Bernhard Bloch, Eichwald. Die meisten der angebotenen Figuren sind leicht von GS zu unterscheiden. Zum einen sind sie mit „BB" (für Bernhard Bloch) gemarkt, zum anderen fehlen jegliche Marken und Bezeichnungen für GS. Die böhm. Firma „Bernhard Bloch" kann zwar als Konkurrenzfirma angesehen werden, erreichte aber weder Bekanntheit noch die Qualität von GS.

3346
Lesca
LAMPE Stehendes Mädchen (Tänzerin) in Pose
ca. 1905/06
3346,3,24 (15) Firmenmarke Sculpture 13, Repr. Res., H. 82 cm; Wien Museum Inv. 186.617

3347
Petri (d.i. Petrides, János) (Johann)
Schreitende Frau, einen flachen Korb haltend (auf Ovalsockel)
ca. 1905/06
3347 (ohne Ang.), Firmenmarke Sculpture 13, sign. Petri, H. 67 cm; Doro 10.3.2003

3367
Schreitende Frau mit Krug und Eimer (auf Landschaftssockel)
ca. 1905/06
3367 (ohne Ang.), (Firmenmarke Sculpture 13?), Repr. Res., H. 59 cm; Henry's 30.6.2000

3368
Cherc
Stehende Frau (in antikisierendem Gewand), eine Amphore auf eine Säule stellend
ca. 1905/06
3368,182,– gem. 21, Firmenmarke Sculpture 13 (gepr.), „P"= Z3, gepr. „Doppeladler"= Z7, unsign., H. 75 cm; eBay # 3652004882
3368,210,25 Firmenmarke 19a, sign. Cherc, Ausformung 20er Jahre, H. 73 cm; eBay # 7417496018

3369
Lavergne, Adolphe-Jean
„Le Grand Condé" (Louis, Prince de Condé zu Pferde)
ca. 1905/06
3369,32,15 GS Wien (Firmenmarke?), Ausformung nach 1900, H. 84 cm, L. 80 cm; Zof. 19
Bem.: Dargestellt ist Louis, Prince de Condé (1621–1686). Diese Figur wurde nach dem Original, welches um 1875 zu datieren ist, gefertigt.
Ausst.: Salon Paris 1876 (für den Originalentwurf, zit. nach Ang. des AH).

3373
Rosé
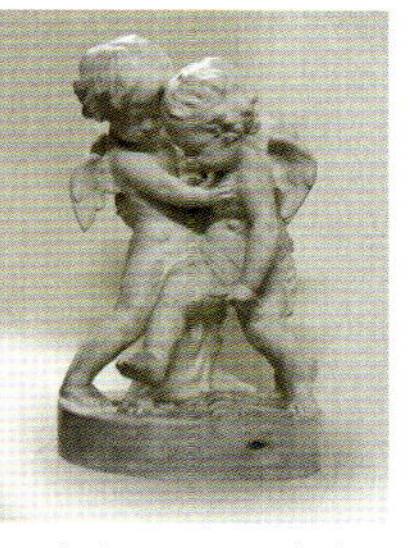
Zwei Putti um ein Herz kämpfend (auf Ovalsockel mit Herz, Rosen, Köcher und Bogen)
ca. 1905/06
3373,5,10 Firmenmarke 9, Repr. Res., H. 56,5 cm; MAK Wien, Inv. Ke 10521-09
Bem.: Diese Gruppe geht zurück auf eine Marmorskulptur von François-Joseph LeClercq, die 1780 entstand. Das Original befindet sich im Philbrook Museum of Art, Tulsa OK/USA (www.philbrook.org).
-> vgl. 3762

3377
Rosé
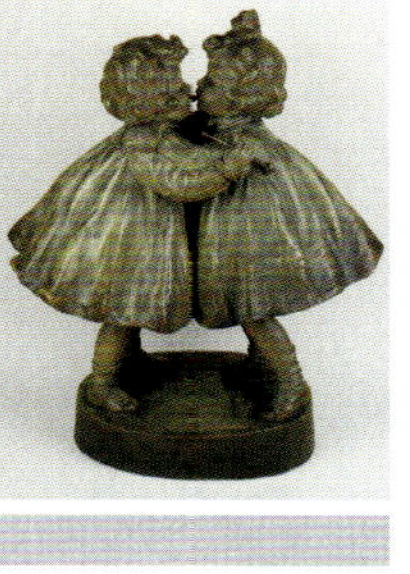
Zwillinge (Zwei, sich umarmende und küssende kleine Mädchen)
ca. 1905/06
3377,56,15 Firmenmarke Sculpture 13, H. 58 cm; PB Wien 19

3389
Rosé
UHR Stehende Schönheit, auf einem von vier Hengsten gezogenen Wagen, der aus der Sonne kommt und über den Wolken schwebt (bekrönt von Adler)
ca. 1905/06
3389 (Firmenmarke ohne Ang.), H. 80 cm; Pinhas, S. 59 (m. Abb.)

3396
Lavergne, Adolphe-Jean
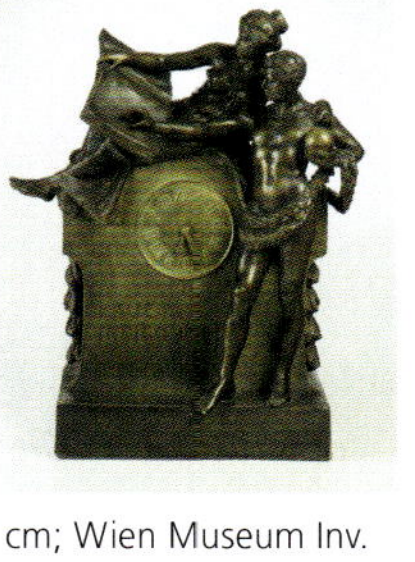
UHR „Historia" (Sitzende Frau auf Sockel mit Buch, in das ein stehender Jüngling (mit Lorbeer und Kugel) schreibt)
Inschrift: „CAVE NE INDIGNUM"
ca. 1905/06
3396,46,9 gem. X, ohne Firmenmarke, Repr. Res., H. 51 cm; Wien Museum Inv. 186.628
Bem.: Keramik bronzefarben staffiert..
Text der Inschrift: Hüte Dich vor dem Unwürdigen.
-> vgl. 3319; 3763

3401
Montenave
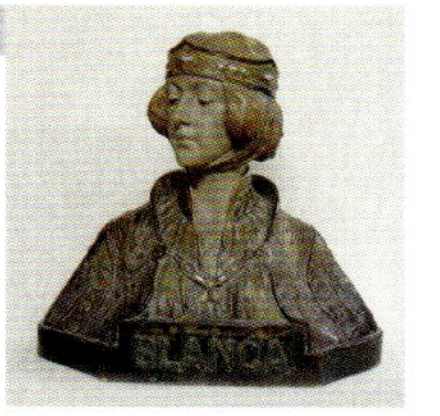
BÜSTE „Blanca" (Frau mit Diadem und Halskette)
Inschrift: BLANCA
ca. 1905
3401,2,25 ger. IX, gem. 25, Firmenmarke Sculpture 13, Repr. Res., H. 48 cm; MAK Wien, Inv. Ke 10602-66

3416
Beaume
(d.i. Bormann, Wilhelm)
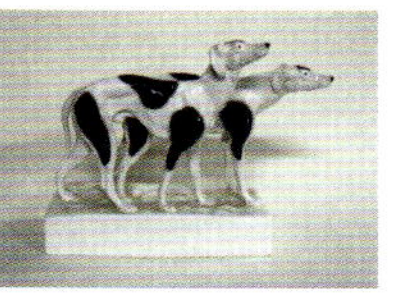
Zwei stehende Windhunde (auf Rechtecksockel)
ca. 1905/06
3416,199,25 gem. HA, Firmenmarke 13a, sign. Lavergne (!), Originalpick.: P. Raddatz & Co., Hoflieferanten, Berlin, H. ca. 25.5 cm; Nagel 379
3416,232,2 gem. F., Firmenmarke 22, sign. Lavernge (!), H. 25,4 cm; MAK Wien, Inv. Ke 10602-56
3416,278,65 Firmenmarke 19, unsign., Ausformung 20er Jahre, H. 25,4 cm; eBay # 378066973; 1220306921
3416,283,6 gem. lig. AH, Firmenmarke 19, Ausformung 20er Jahre, H. ca. 25,5 cm; eBay # 2065236590
3416 (ohne Ang.), gem. lig. AS D.1., Firmenmarke 19, Made in Austria, undeutlich sign. „Beaume", Ausformung 20er Jahre, H. 24,8 cm L. 34,3 cm; eBay # 35230713
3416,310,18 Firmenmarke 19, Made in Austria, sign. Lavergne (!), H. 25,4 cm; eBay # 3702029401
Bem.: Trotz der zwei mit „Lavergne" sign. Exemplare ist diese Gruppe eindeutig Wilhelm Bormann zuzuschreiben.
Ähnl. Modell -> vgl. 3878; 4407
-> vgl. 4406= linke Einzelfigur

3424
Capèque (d.i. Czapek, Stanislaus)
LAMPE Fackelträgerin (Stehende Frau auf säulenartigem Sockel, zwei Fackeln in die Höhe haltend)
ca. 1905/06
3424,21,65 gem. FIII., Firmenmarke 22, Ausformung 20er Jahre, sign. Capeque, H. 63 cm; KH

3427
Tell, E.
Sitzende Dame und stehender Kavalier mit Dreispitz in Rokokokostümen
ca. 1906
3427,31,4 Firmenmarke 19, Made in Austria, Ausformung 20er Jahre, H. 32,5 cm; Wien Museum Inv. 190.102

3437
Haniroff
Der erste Kuss (Mädchen mit Weidenkorb, von Jungen geküsst)
ca. 1906
3437,57,35 gem. weiß VI., XXX, Firmenmarke Sculpture 13, Fabriqué en Autriche, sign. Haniroff, gepr. „S"= Z2, H. 49 cm; Wien Museum Inv. 219.135

3444
Lavergne, (Adolphe-Jean)
UHR/LAMPE „Dans Le Arene" (In der Arena) (Römischer Sklave lässt zwei Löwen aus dem Käfig) (Uhrgehäuse als Löwenkäfig, mit zwei auf Säulen stehenden Urnen, welche als Lampenfassungen dienen)
ca. 1905/06
3444,18,7 gem. XXVI., Firmenmarke Sculpture 13, Repr. Res., H. 43 cm; eBay # 969440041; Bem.: An der Plinthe sign. und betitelt „Dans L'Arene par Lavergue Vienne" (Metallschild); Lenzkirch Uhrwerk.
3444,120,17 Firmenmarke Sculpture 13, Fabriqué en Autriche, gepr. „Doppeladler"= Z7, sign. Lavergne, Gesamth. 52 cm (mit Lampenfassung); eBay # 740852204; Bem.: Lenzkirch Uhrwerk.
Bem.: Diese Skulptur ist eines der wenigen Stücke, bei dem die Manufaktur GS Uhr und Lampenfunktion in einem Objekt zur Ausführung bringt.
-> vgl. 3256

3447
Hegenbarth, Ernst
Stehende Frau mit Rosengirlande
(auf vorne abgeschrägtem Rechtecksockel)
ca. 1906
3447,1,16 Firmenmarke Sculpture 13, H. 110 cm;
GK, S. 15, 83 (m. Abb.); Wien Museum Inv. 182.078
Ausst.: HMW (jetzt Wien Museum) 1985/86.
Lit.: GK, S. 15, 83 (m. Abb.).

3454
Lesca
„Amerika" (Junge Frau mit Fackel, vor Bison stehend)
ca. 1906/07
3454,28,55 Firmenmarke Sculpture 13, Repr. Res.,
H. 61 cm; Neumeister 270; Lit.: ARGOS 1995/96, S. 677.

3455
Lesca
Stehende Frau mit Kopftuch und Krug (auf Ovalsockel)
ca. 1906/07
3455,17,65 Firmenmarke 9, Repr. Res., H. 36 cm;
Wien Museum Inv. 190.073

3473
Haniroff
LAMPE Stehender Nachtwächter mit Helm
(eine eiserne Lampe haltend)
ca. 1906/07
3473,4 Firmenmarke 19?, Made in Austria, (Ausformung 20er Jahre?), H. 41 cm;
Soth. Ama # 954925933-2

3485
Rosé
BÜSTE „Mozart"
Inschrift: MOZART
ca. 1906
3485 (ohne Ang.), H. 38 cm; Henry's 25.3.2000
Bem.: Dargestellt ist der Komponist Wolfgang Amadeus Mozart (1756–1791). Die Büste wurde zum 150. Geburtstag aufgelegt.

3487
Sydan
JARDINIERE Ovaler Korb mit darauf sitzendem Mädchen und Knaben
ca. 1906/07
3487,2,10 (Firmenmarke ohne Ang.), unsign., H. 30 cm; Metz 21.3.1992
3487 (ohne Ang.), FG Wien (gepr. Firmenmarke?), Firmenmarke 6, H. 28 cm; RM 37; Bem.: hier als Modell-Nr. 3437 angegeben.

3493
Junger Page mit Schale
ca. 1907
3493,5,9 Firmenmarke Sculpture 13, Repr. Res.,
H. 113 cm; KH

3497
Tell, E.
Sitzende junge Frau (Akt) mit Lyra (Sappho?)
ca. 1907
3497,19,15 FG Wien (Firmenmarke ohne Ang.),
Repr. Res., H. 62 cm; eBay # 882188886
–> vgl. 3576; 3577; 4625

3501
Daubling (d.i. Doblinger, Ferdinand)
Schubertlied
ca. 1905/07
3501 (ohne Ang.), sign. Daubling, H. fehlt; Neuwirth (1974), S. 58, 61 (WKH 1972); Bem.: Hier keine Beschreibung des Objektes oder nähere Ang. zum Titel „Schubertlied" (evtl. Inschrift?).

3511
Tell, E.
Sitzender, sich bückender Frauenakt (auf Felssockel)
ca. 1905
3511,37,10,O Firmenmarke Adler 16, Firmenmarke Sculpture 13, H. 48,3 cm; MAK Wien, Inv. Ke 10602-38
–> vgl. 3512

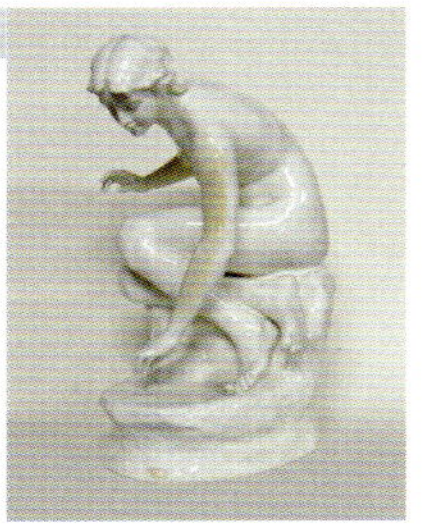

3512
Tell, E.
Sitzender, sich bückender Frauenakt (auf Felssockel)
ca. 1905
FAYENCE
3512,55,65 (Firmenmarke ohne Ang.), H. 36,8 cm;
Doro 8.3.2004
ALABASTER / Marmorsockel
ohne Ang. (Plak. ohne Ang.), H. 41 cm; Doro 23.5.1991
Bem.: Dieses Modell wurde in Fayence und Alabaster (Kunststein?) ausgeführt.
–> vgl. 3511

3513
Daubling (d.i. Doblinger, Ferdinand)
Tänzerin
ca. 1905/07
3513 (ohne Ang.), sign. Daubling, H. fehlt; Neuwirth (1974), S. 61; Bem.: Hier keine nähere Beschreibung des Objektes.

3520
BÜSTE Mädchen (Renaissancestil) (Prinzessin Borghese?)
ca. 1906/07
3520,13,7 Firmenmarke Sculpture 13, gepr. „S" (=Z 2?), H. 47 cm; Doro 12.12.1991; Bem.: Die Bezeichnung „Prinzessin Borghese" wurde aus den Ang. des AH übernommen.

3522
Rosé / Perigot
FIGUR / LAMPE „Mozart"
(mit Geige, auf Sessellehne sitzend)
(Inschrift: MOZART)
ca. 1905
FIGUR (mit Inschrift)
3522,255,10 gem. 7., Firmenmarke Adler 16, Firmenmarke Sculpture 13, sign. Rosé, H. ca. 36 cm; KH
FIGUR (ohne Inschrift)
3522,451,2 GS (Firmenmarke ohne Ang.), sign. Rosé, H. 37 cm; Doro S 29.6.2000
3522,533,10 Firmenmarke 19?, unsign.?, (Ausformung 20er Jahre,?), H. 36,5 cm; Zeller 45
3522,660,2 Firmenmarke 19?, unsign.?, (Ausformung 20er Jahre?), H. 36 cm; Mehlis 18
3522,792,65 gem. N2, Firmenmarke 19, Hand dec., unsign.?, Ausformung 20er Jahre, H. 36,5 cm; KH
3522,825,12 gem. S.G.3., ger. K3, Firmenmarke 19, Made in Austria, Hand dec., Ausformung 20er Jahre, sign. „Perigot" (!), H. ca. 36 cm; eBay # 501053292
3522 ohne Ang., sign.?, H. 36 cm; PB Wien 19
ohne Ang., gepr. K, gem. N, Firmenmarke 18, Ausformung ca. 1920, sign. Rosé, H. 36 cm; Henry's 26.2.1997; 28.5.1997
LAMPE
ohne Ang., H. fehlt; Neuwirth (1974), S. 58
Bem.: Dargestellt ist der Komponist Wolfgang Amadeus Mozart (1756–1791).
–> vgl. 3285; 5483

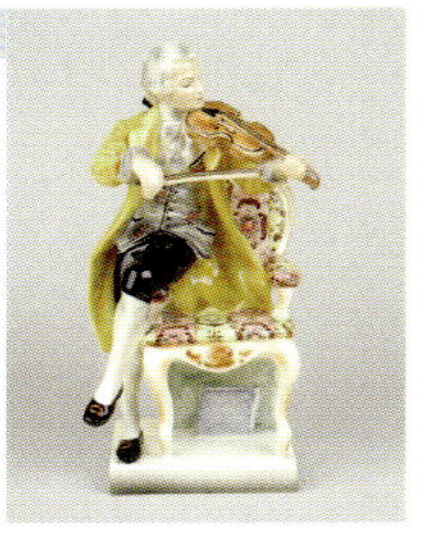

3525
Rosé
Europa auf dem Stier reitend (auf Landschaftssockel)
ca. 1905
3525 (Firmenmarke ohne Ang.), H. 71,5 cm; Pinhas, S. 41 (m. Abb.)

3526
Lavergne, Adolphe-Jean
„Animus et vis" (Stehender römischer Krieger mit Helm, Speer und Tiger) (auf passigem Rechtecksockel)
Inschrift: ANIMUS ET VIS
ca. 1905/08
3526 (Firmenmarke ohne Ang.), unsign., H. 107 cm;
Pinhas, S. 41 (m. Abb.)
Bem.: Der Kopf des Kriegers mit Helm wurde, bis auf eine leichte Abwandlung der Haartracht, als Büste unter Modell-Nr. 1687 produziert, Künstlersign. hier „Rozel". Text der Inschrift: Mut und Kraft.
–> vgl. 3680
–> vgl. auch 3536; 3988= Tiger einzeln

3531
Czapeque (d.i. Czapek, Stanislaus)
LAMPE „Aquarium" (Zwei Mädchen auf Felsen, in einen Brunnen schauend)
ca. 1906/07
3531,16,15 gem. 20, Firmenmarke Sculpture 13, Repr. Res., sign. Czapeque, H. 60 cm; Wien Museum Inv. 186.606
3531,116,14 (Firmenmarke ohne Ang.), sign. Petri, H. 64 cm; Soth. Olympia 3.5.2002; Bem.: Die Sign. „Petri" dürfte wohl als Versehen des Modelleurs angesehen werden.
Lit.: Neuwirth (1974), S. 58, hier mit Ang. des Originaltitels (?) „Aquarium".

3534
Lavergne, Adolphe-Jean
„Wiener Walzer" (Tanzendes Paar im Biedermeierkostüm)
ca. 1907
3534,4,12 gem. XX. 919, F 23, Firmenmarke Sculpture 13, H. 32 cm; Wien Museum Inv. 190.117
Lit.: (vgl. evtl.) N.N.: Leipziger Michaelismesse 1904, in: KeramRS, 12.1904, S. 945, „Salonplastiken par excellence stellte die Firma FG in Wien aus. Novität: Tanzendes Paar, Wiener Walzer".
Lit.: Alt-Wien, S. 463 (m. Abb.).

3536
Lavergne, Adolphe-Jean
Tiger (auf Rechtecksockel)
ca. 1907
3536,5,25 gem. X, Firmenmarke Sculpture 13, Repr. Res., H. 42,7 cm, L. 55 cm;
GK, S. 15, 84 (m. Abb.); Wien Museum Inv. 223.907;
Neuwirth (1974), S. 58.
3536,147,10 Firmenmarke 19, Made in Austria, Ausformung 20er Jahre, unsign., H. 30 cm, L. 62 cm;
eBay # 2235013215
3536,214,16 weiß gem. IV, Firmenmarke Sculpture 13, Fabriqué en Autriche, H. 35,1 cm, L. 59 cm, T. 18,5 cm;
MAK Wien, Inv. Ke 10602-95
3536,289,25 Firmenmarke 22a, Ausformung 20er Jahre, unsign., H. 34 cm, L. 60 cm, T. 19 cm; Sammlung f@goldscheider.de; Bem.: Figur bronzefarben staffiert, Sockel in Marmoroptik bemalt.
3536 (ohne Ang.), Firmenmarke 19, Ausformung 20er Jahre, H. fehlt; Nagel 312
Ausst.: HMW (jetzt Wien Museum) 1985/86.
–> vgl. 3988
–> vgl. GRUPPE „Animus et vis" Modell-Nr. 3680
Lit.: Neuwirth (1974), S. 58; GK, S. 15, 84 (m. Abb.).

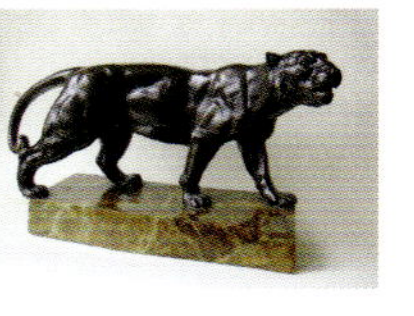

3538
Daubling (d.i. Doblinger, Ferdinand)
Biedermeierdame mit Hut und zwei Weidenkörben (auf getrepptem Sockel)
ca. 1907/08
3538,1,10 Firmenmarke 9, Repr. Res., sign. Daubling, H. 41 cm; Wien Museum Inv. 219.162

3545
Daubling / Dobling(er) (d.i. Doblinger, Ferdinand)
Herr mit Zylinder (Biedermeierfigur) (auf Rechtecksockel)
ca. 1907/08
3545,48,10 Firmenmarke Adler 15, Firmenmarke Sculpture 13, sign. Daubling, H. 25,7 cm; Nagel Leipzig 3; Ahlden 117

3545,59,10 gem. FK, XIX, H19, Firmenmarke Adler 16, Firmenmarke Sculpture 13 (gepr.), sign. Dobling, H. 25 cm; Wien Museum Inv. 190.151; Neuwirth (1974), S. 58, 61 (m. Abb.), 62.
3545,58,10 Firmenmarke Adler 15, Firmenmarke Sculpture 13, sign. Daubling, H. 25 cm; BröPK, S. 146; Brö 2,1, S. 400
3545 (ohne Ang.), sign. Daubling, H. 26 cm; Kett. 196; Ahlden 91
-> vgl. 4085 (Rundsockel)
Pendant -> vgl. 3546

3546
Daubling (d.i. Doblinger, Ferdinand)
Dame mit Hut (Biedermeierfigur) (auf Rechtecksockel)
ca. 1907/08
3546,54,65 Firmenmarke Adler 15, Firmenmarke Sculpture 13, sign. Daubling, H. 25,7 cm; Nagel Leipzig 3; Ahlden 117
3546 (ohne Ang.), Firmenmarke Adler 15, Firmenmarke Sculpture 13, sign. Daubling, H. 26 cm; Kett. 196; Ahlden 91
Pendant -> vgl. 3545

3551
Daubling (d.i. Doblinger, Ferdinand)
Sitzende Biedermeierdame mit lesender Kinderschar (Putti) (auf getrepptem Sockel)
ca. 1907/08
3551,80,27 Firmenmarke 9, Repr. Res., sign. Daubling, H. 46 cm; GK, S. 15, 85 (m. Abb.); Wien Museum Inv. 182.079
Ausst.: HMW (jetzt Wien Museum) 1985/86.
Lit.: GK, S. 15, 85 (m. Abb.)

3563
Tell, E.
Stehende Wäscherin mit Trog (auf Rechtecksockel)
ca. 1907/08
3563,5,27 Firmenmarke 9, Repr. Res., H. 44 cm; Wien Museum Inv. 219.163

3565
Daubling (d.i. Doblinger, Ferdinand)
UHR Rokokopaar
ca. 1907/08
3565,10,10 FG Wien (Firmenmarke ohne Ang.), sign. Daubling, H. 51,5 cm; KNS 12; Bem.: Lenzkirch Uhrwerk.

3576
Tell, E.
Sitzende junge Frau (Akt) mit Lyra (Sappho?)
ca. 1907
3576,137,17 FG (Firmenmarke ohne Ang.), H. 47 cm; Doro 28.5.1991
3576,143,10 Firmenmarke Sculpture 13, Fabriqué en Autriche, H. 45,5 cm; Wien Museum Inv. 186.612
3576,144,15 gem. 22, (Firmenmarke Adler ohne Ang.), Firmenmarke Sculpture 13, Fabriqué en Autriche, H. ca. 46 cm; KH
-> vgl. 3497; 3577; 4625; 4697
Lit.: Pinhas, S. 99 (m. Abb.)

3577
Tell, E.
Sitzende junge Frau (Akt) mit Lyra (Sappho?)
ca. 1907
3577,189,65 Firmenmarke 19, Ausformung 20er Jahre, unsign.?, H. ca. 33 cm; eBay # 3703249677
-> vgl. 3576; 3577; 4625; 4697
Lit.: Pinhas, S. 99 (m. Abb.)

3581
Petri (d.i. Petrides, János) (Johann)
Stehendes Biedermeiermädchen mit Korb und Handtasche (Beutel)
ca. 1907
3581,17,65 Firmenmarke 9, Repr. Res., sign. Petri, H. 31,7 cm; eBay # 3735251193

3590
Lavergne, Adolphe-Jean
LAMPE Stehende junge Frau eine Schale in die Höhe haltend (auf Rundsockel)
ca. 1905
3590 (Firmenmarke ohne Ang.), H. 69 cm; Pinhas, S. 38 (m. Abb.)
Bem.: Die Schale kann sowohl als Visitenkartenschale als auch, wie bei diesem Ex. als Lampensockel dienen.
-> vgl. 3207

3591
Horsetzky, Melanie von
JARDINIERE Stehende Holländerin mit Korb
ca. 1903/05
3591,221,10 Firmenmarke Adler 14, sign. Horsetzky, H. 20 cm; Mehlis 6.3.2004 (Sonderaukt. Sammlung Langer)
3591,235,16 gem. VI, Firmenmarke Adler 14, unsign.?, H. 21 cm; city-antik
Bem.: Diese Figur kann sowohl als Jardiniere, wie auch als Visitenkartenschale dienen.
-> vgl. 2666

3592
Petri (d.i. Petrides, János) (Johann)
Spielendes Kinderpaar (Das Mädchen hält dem Jungen die Augen zu) (auf Ovalsockel)
ca. 1909/10 (für den Entwurf)
3592,138,14 Firmenmarke 23, sign. Petri, Ausformung 20er Jahre, H. 47 cm; eBay # 7415525072

3596
LAMPE Stehende Frauengestalt mit nach oben gestrecktem Arm
ca. 1907/08
3596,24,9 (Firmenmarke ohne Ang.), H. 69 cm; Christie's SK 11.5.2001

3599
Valentin, Max
„Badende" (Stehender Damenakt mit Tuch) (auf Felssockel)
ca. 1907/08
3599,29,10 Firmenmarke Adler 15, Firmenmarke Sculpture 13 (gepr.), H. 56 cm; WKA 35
-> vgl. 3681= ähnl. Modell
Lit.: N.N.: Aus Berliner Musterlagern 1909, in: Sprechsaal, 42.1909, S. 308, hier mit Ang. des Originaltitels „Badende" (möglicherweise ist diese Fig. identisch mit der 1913 im Münchner Glaspalast auf der XI. Internationalen Kunstausst. präsentierten Bronze „Badendes Mädchen"). Denkbar für die „Badende" wäre auch das Modell 3681, welches das gleiche Genre in ähnl. Auffassung wiedergibt; Neuwirth (1974), S. 75.

3606
Rosé
„Beethoven" (als stehende Figur)
ca. 1907/08
3606,11,65 FG Wien (Firmenmarke?), Repr. Res., H. 65 cm; Doro 24.11.1992
Bem.: Dargestellt ist der Komponist Ludwig van Beethoven (1770–1827).
(-> vgl. auch 4645?)

3610
Leclancher
Stehender Arbeiter mit Spaten (auf Rechtecksockel)
ca. 1907/08
3610,21,10 FG (Firmenmarke?), H. 36 cm; eBay # 903623039
3610,91,10 FG Wien (Firmenmarke?), H. 35 cm; KH

3615
Schmidt-Kestner, Erich
Brunst-Hirsch (Röhrender Hirsch)
ca. 1908/09
3615,29,65 gem. 2, Firmenmarke 3, Firmenmarke Sculpture 13, H. 45 cm, L. 45 cm; Weidler 370
3615,58,2 Firmenmarke Adler 15, Firmenmarke Sculpture 13 (gepr.), H. 39 cm, L. 45 cm; Kühling 6.11.1998
3615,371,10 Firmenmarke 19, Made in Austria, Ausformung 20er Jahre, H. 40 cm; Doro L 24.11.2000
Lit.: Vermutlich identisch mit „Brunst-Hirsch", in: Sprechsaal, 1909, S. 308; Neuwirth (1974), S. 74.

3620
Leclancher
Mutter mit Kind (auf Felssockel)
ca. 1908/09
3620 (Firmenmarke ohne Ang.), H. 66 cm; Pinhas, S. 47 (m. Abb.)

3622
Bormann, Wilhelm
„Kind mit Hund" (auf Rundsockel)
ca. 1908/09
3622,3,10 FG Wien (Firmenmarke ohne Ang.), Repr. Res., H. 37,7 cm; Doro 18.4.1994
Ausst.: ÖMKI 1910/11, hier mit Ang. des Originaltitels „Kind mit Hund".
Lit.: Neuwirth (1974), S. 60

3625
Capèque
(d.i. Czapek, Stanislaus)
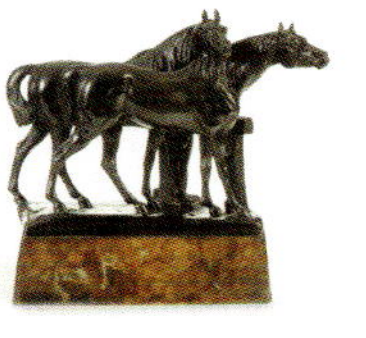
„Pferde" (Zwei Pferde, bronziert, auf marmoriertem Sockel)
ca. 1908/09
3625,198,15 Firmenmarke Adler 17, Firmenmarke Sculpture 13, Fabriqué en Autriche, sign. Capèque, gepr. „S"= Z2, H. 43 cm; Wien Museum Inv. 190.053
Bem.: Keramik bronzefarben staffiert, Sockel in Marmoroptik.
Ausst.: ÖMKI 1912, hier mit Ang. des Originaltitels „Pferde".
Lit.: Neuwirth (1974), S. 61

3629
Petri (d.i. Petrides, János) (Johann)
Putto vor (Grab?)-Denkmal sitzend (Keramik/Holz)
Inschrift: 1885–1910
ca. 1909/10
mit Inschrift
3629,51,7 Firmenmarke Sculpture 13, Fabriqué en Autriche, sign. Petri, H. 55 cm; Wien Museum Inv. 186.641
ohne Inschrift
3629 (Firmenmarke ohne Ang.), H. 55,5 cm; Pinhas, S. 47 (m. Abb.)
Bem.: Dieses mit der Inschrift 1885–1910 (auch beleuchtbare) Modell könnte zum 25-jährigen Firmenjubiläum editiert worden sein.

3630
Schmidt-Kestner, Erich
Tänzerin in Bewegung (Halbakt) (auf Rechtecksockel)
ca. 1908/09
3630,97,12 Firmenmarke 19, Made in Austria, Ausformung 20er Jahre, H. 53,3 cm; eBay # 1204329875
ohne Ang., GS (Firmenmarke ohne Ang.), unsign.?, H. 52,7 cm; Soth. N.Y. 11.6.1997

3643
Augmüller, Ludwig
JARDINIERE
ca. 1908/09
3643 (ohne Ang.), H. fehlt; Neuwirth (1974), S. 59;
Bem.: Hier keine nähere Beschreibung des Objektes.

3649
Perigot
JARDINIERE / KONFEKTSCHALE Biedermeiermädchen mit übergroßem quadratischen Korb
ca. 1908/09
3649,1,10 Firmenmarke 9, Repr. Res., H. 27 cm, L. 31 cm; Wien Museum Inv. 219.175
3650,15,25 gem. weiß V., Firmenmarke Sculpture 13, H. 25 cm, L. 30 cm; KH

3655
Lesca
„Pierrot und Pierrette" (auf einer Bank sitzend) (auf Ovalsockel)
ca. 1908/09
3655,210,14 gem. X1., Firmenmarke 18, Ausformung ca. 1920, H. 35,5 cm; KH www.nightingaleantiques.com/pottery/3146.html
3655,225,5 ger. K, gem. B, Firmenmarke 18, Ausformung ca. 1920, H. 34 cm; Wendl 13; Mehlis 5
3655,333,9,8 gem. lig. AS, gem. rot XI., Firmenmarke 22, Ausformung ab 1922, H. 34,9 cm; eBay # 376173809
-> vgl. 5016; 5327
Lit.: MW, 2.1920/21, Nr. 12, S. 2 (m. Abb.); DD, 50.1922/23, Nr. 10, S. 30 (m. Abb.), hier jeweils bezeichnet mit dem Künstlernamen „Lexa" (!) und mit Ang. des Originaltitels „Pierrot und Pierrette".

3657
Polter
Biedermeiermädchen einen großen Korb auf dem Rücken tragend
ca. 1908/09 (für den Entwurf)
3657,78,8 Firmenmarke 19, Ausformung ca. 1920, H. 23 cm; eBay # 2264220624; 6508166303

3662
Polter
Biedermeierdame mit Hut und zwei quaderförmigen Gefäßen (als Blumenbehälter?)
ca. 1908/09
3662,12,10 gem. 27 R, 27 G, 27 H, III, Firmenmarke 9, Firmenmarke Adler 16, Repr. Res., H. 33,5 cm, L. 23 cm; Wien Museum Inv. 219.158

3667
GRUPPE Zwei scherzende Putti (auf Landschaftssockel)
ca. 1908/09
3667,8,7 Firmenmarke Sculpture 13, Fabriqué en Autriche, H. 45 cm; eBay # 6605874912
Bem.: Diese Figurengruppe geht wohl auf ein klassisches (franz.) Vorbild zurück.

3673
Leclancher
Stehender Arbeiter mit Spitzhacke (auf Rechtecksockel)
ca. 1908/09
3673,45,11 gem. 4, Firmenmarke Sculpture 13, Firmenmarke Adler 16, H. 35 cm; Sammlung Helmut Riener, St. Florian, Austria

3679
Sydan
LAMPE Scherzendes Rokokopaar (auf Ovalsockel)
ca. 1908/09
3679,23,27 gem. 4, Firmenmarke 9, H. 44,5 cm; Wien Museum Inv. 219.147

3680
Lavergne, Adolphe-Jean / Cherc
„Animus et vis" (Stehender römischer Krieger mit Helm, Speer und Tiger) (auf passigem Rechtecksockel)
Inschrift: ANIMUS ET VIS
ca. 1905/08
3680,61,25 Firmenmarke Sculpture 13, H. 68 cm; Nusser 196
3680,170,35 GS Wien (Firmenmarke ohne Ang.), sign. Cherc, H. 76,8 cm; Doro 14.9.2001; Bem.: Dieses Exemplar ohne Inschrift, Speer aus Holz/Metall gefertigt.
Bem.: Der Kopf des Kriegers mit Helm wurde, bis auf eine leichte Abwandlung der Haartracht, als Büste unter Modell-Nr. 1687 produziert, Künstlersign. hier „Rozel". Text der Inschrift: Mut und Kraft.
-> vgl. 3526
-> vgl. auch 3536; 3988= Tiger einzeln

3681
Valentin, Max
„Badende" (Stehender Damenakt mit Tuch) (auf achteckigem Sockel)
ca. 1908/09
3681,17,65 (Firmenmarke ohne Ang.), H. 47,5 cm; Doro 14.9.2001
-> vgl. 3599= ähnl. Modell
Lit.: N.N.: Aus Berliner Musterlagern 1909, in: Sprechsaal, 42.1909, S. 308, hier mit Ang. des Originaltitels „Badende" (möglicherweise ist diese Fig. identisch mit der 1913 im Münchner Glaspalast auf der XI. Internationalen Kunstausst. präsentierten Bronze „Badendes Mädchen"). Denkbar für die „Badende" wäre auch das Modell 3599, welches das gleiche Genre in ähnl. Auffassung wiedergibt; Neuwirth (1974), S. 75.

3682
Rosé
GRUPPE Kleines Mädchen mit Hund, der einen Korb trägt
ca. 1908/09
3682,10,65 (Firmenmarke ohne Ang.), Repr. Res., H. 24 cm; Mehlis 32
3682,–,– Firmenmarke 9, H. 25 cm; Nitschareuth 5.6.2004
-> vgl. 3683

3683
Rosé
GRUPPE Kleines Mädchen mit Hund, der einen Korb trägt
ca. 1908/09
3683,21,27 gem. 2, Firmenmarke 9, Repr. Res., H. 17 cm; eBay # 3355152538
-> vgl. 3682

3684
Latour, (Louis-Marie-Blaise?)
„Pro rude" (Zwei Gladiatoren, gegen eine Löwin kämpfend)
Inschrift: PRO RUDE
ca. 1909
3684,44,15 ger., Firmenmarke Sculpture 13, Fabriqué en Autriche, gepr. „L"= Z1, H. 51 cm, B. 54 cm; KAHS 23.7.1998; Lit.: AZ, 1998, S. 557

3697
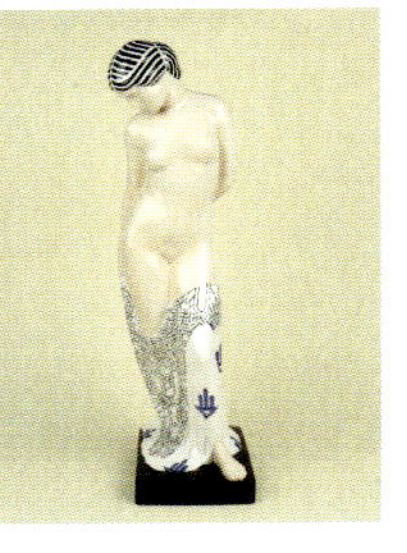
Seger, Ernst
„Helena"/„Helene" (Stehender Frauenakt mit Tuch)
ca. 1909
3697,40,2 Firmenmarke 19, sign. Seeger (!), Ausformung 20er Jahre, H. 56 cm; Kett. 70
3697,54,5 gem. lig. AH, XI, Firmenmarke 19, unsign., Ausformung 20er Jahre, H. 56 cm; GK, S. 17, 96 (m. Abb.); Wien Museum Inv. 223.908; SS, S. 6, 7, 12 (m. 2 Abb.)
3697,55,11 gem. X, AH, Firmenmarke 19, Ausformung 20er Jahre, unsign.?, H. 56 cm; Neuwirth (1974), S. 43 (m. Abb.)
3697,64,8 gem. VIII, Firmenmarke 19, Ausformung 20er Jahre, sign. Seeger (!), H. 56,5 cm; Doro 1.12.2000
3697 (ohne Ang.), Firmenmarke 19, sign. Seger, Ausformung 20er Jahre, H. 58 cm; Spik 535
Ausst.: HMW (jetzt Wien Museum) 1985/86; SSH 1992
-> vgl. 3698; .... (H. 80 cm) sign. Seeger (!); Spik 506
Lit.: N.N.: Aus Berliner Musterlagern, in: Sprechsaal, 42.1909, S. 308, hier mit Ang. des Originaltitels „Helena", „...Alles hervorragende Arbeiten. Bei allen diesen Plastiken werden die Körperteile, soweit sie nicht vom Gewand bedeckt sind, ganz matt-glänzend gehalten, während das Gewand sind in glänzender Frische abhebt"; DD, 50.1922/23, Nr. 5, S. 26 (m. Abb.), hier mit Ang. des (Original)titels „Helene"; GK, S. 17, 96 (m. Abb.); GSN, S. 18, 19 (m. Abb.); SS, S. 6, 7, 12 (m. 2 Abb.)
GS neu: Modell-Nr. 7 „Schönheit" (H. 54 cm), Aufl. 2000

3698
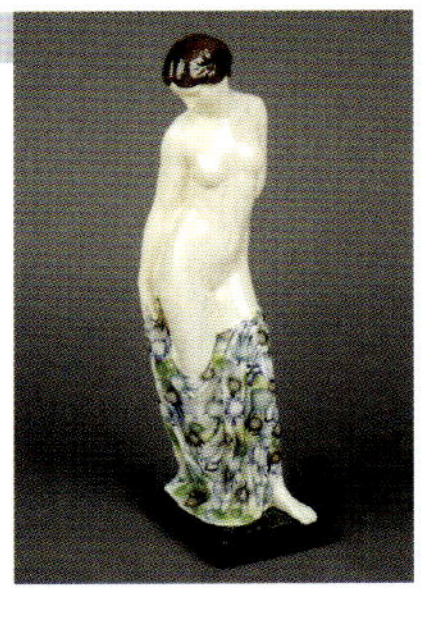
Seger, Ernst
„Helena"/„Helene" (Stehender Frauenakt mit Tuch)
ca. 1909
3698,37,5 Firmenmarke 13a, sign. Seeger (!), H. 39,4 cm; eBay # 7378945197
3698,79,8 Firmenmarke 19, Made in Austria, Hand dec., unsign., Ausformung 20er Jahre, H. 39 cm; PB Wien 19
3698,87,75 Firmenmarke 19, Made in Austria, Hand dec., sign. Seeger (!), Ausformung 20er Jahre, H. 41 cm; Soth. L. 4.6.1993
Ausst.: SSH 1992
-> vgl. 3697; .... (H. 80 cm) sign. Seeger (!); Spik 506
Lit.: N.N.: Aus Berliner Musterlagern, in: Sprechsaal, 42.1909, S. 308, hier mit Ang. des Originaltitels „Helena"; DD, 50.1922/23, Nr. 5, S. 26 (m. Abb.), hier mit Ang. des (Original)titels „Helene"; GSN, S. 18, 19 (m. Abb.); SS, S. 6, 7, 12 (m. 2 Abb.)
GS neu: Modell-Nr. 7 „Schönheit" (H. 54 cm), Aufl. 2000

3714
Bormann, Wilhelm / Beaume (d.i. Wilhelm Bormann)
GRUPPE „Freundschaft" (Kind mit Hund= Neufundländer)
ca. 1910
FIGUR
3714,0 ger. S, gem. F, Firmenmarke 19, sign. Beaume, Ausformung 20er Jahre, H. 20,5 cm; eBay # 225462641
3714,4,? Firmenmarke 19, Made in Austria, unsign.?, Ausformung 20er Jahre, H. 19 cm eBay # 2003141214
3714,127,11 (Firmenmarke ohne Ang.), unsign.?, H. 23 cm, L. 28 cm; Bloss 13.–15.4.2000
3714,193,13 Firmenmarke Adler 16?, Firmenmarke Sculpture 13 (gepr.), sign. Beaume, H. fehlt; eBay # 1216347029
3714,418,8 (Firmenmarke ohne Ang.), sign. Beaume, H. 23 cm, L. 32 (!) cm; Doro S 11.4.2001; eBay # 1165197592
3714,441,6 gem. lig. AH, IV., Firmenmarke 19, sign. Beaume, Ausformung 20er Jahre, H. 23,5 cm, L. 27,5 cm; city-antik
3714,449,1 gem. VIII., T, Firmenmarke 19, sign. Beaume, H. 23,3 cm; Wien Museum Inv. 190.086
3714,1070,10 (Firmenmarke ohne Ang.), unsign.?, Ausformung vermutlich 20er Jahre, H. fehlt; Doro G 18.5.1993
DECKELDOSE
3714,96,65 Firmenmarke 3, unleserlich sign., Gesamth. 26 cm, L. 28 cm; Doro 9.11.1989; 21.5.1997
-> vgl. 5908
-> vgl. auch 4014= Hund einzeln
Lit.: MW, 11. 1929/30, Nr. 13, S. VI (m. Abb.), hier mit Ang. des Originaltitels „Freundschaft"; Schilling-Schletter, A.: Hunde-Plastiken, in: MW, 11.1929/30, Nr. 6, S. 12 (m. Abb.); Neuwirth (1974), S. 58, 60.

3716
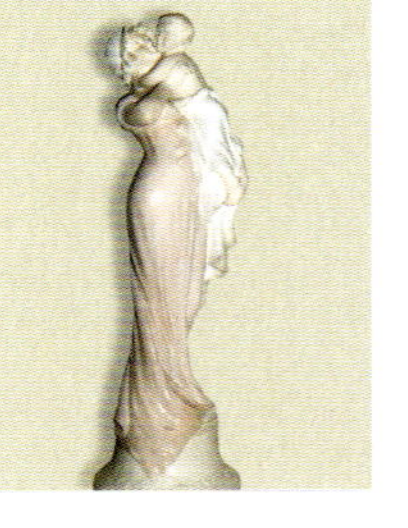
Tell, E.
GRUPPE „Mutterglück" (Mutter mit Kind auf reliefiertem Sockel)
ca. 1905/06
3716,5,65 Firmenmarke 9, zus. (sek.?) innen im Sockel gem. Firmenmarke 19, (Ausformung 20er Jahre?), H. 53,5 cm; Sammlung Renée und Otto Schenk, Wien
Ausst.: ÖMKI 1906/07, hier mit Ang. des Originaltitels GRUPPE „Mutterglück".
Lit.: Neuwirth (1974), S. 75

3718
Lavergne, Adolphe-Jean
Stehende Frauengestalt, eine flache Schale in die Höhe haltend (auf zylindrischem, mit Blumengirlanden verziertem Sockel)
ca. 1909/10
3718,3,25 FG (Firmenmarke ohne Ang.), Fabriqué en Autriche, H. 88 cm; Ruef 475
-> vgl. Tell 3810 (bis auf die Schale identisches Modell)

3721
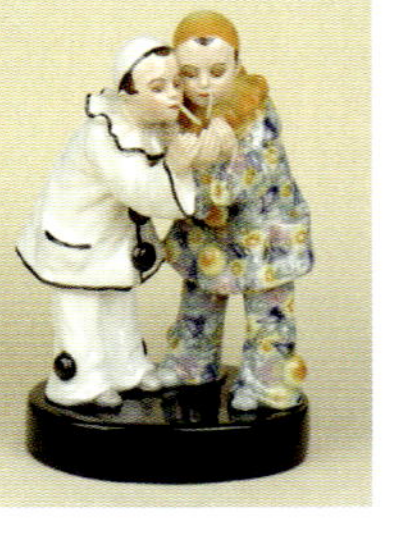
Lavergne, Adolphe-Jean
GRUPPE Zwei Zigaretten rauchende Harlekine
ca. 1909/10
FIGUREN
3721,141,10,05 ger. M, gem. S.G.5., gem. lig. A7, Firmenmarke 19, Made in Austria, Hand dec., Ausformung 20er Jahre,

H. 37,1 cm; PB Wien 19
LAMPE
3721,20,69 FG Wien (Firmenmarke ohne Ang.), H. 40 cm; Phillips E. 25.1.2001; Bem.: Keramik alabasterfarben staffiert.
-> vgl. 3766= rechte Einzelfigur
-> vgl. 3735= LAMPE

3724
Petri (d.i. Petrides, János (Johann))
UHR Stehendes Biedermeiermädchen (auf konischem, mit Blumengirlanden verziertem Sockel)
ca. 1909/10
3724,25,10 Firmenmarke Sculpture 13, sign. Petri, H. 45 cm; eBay # 3346931717

3726
Latour, (Louis-Marie-Blaise?)
Lucretia (Stehender Frauenhalbakt mit Dolch) (auf Rechtecksockel)
ca. 1909/10
3726,62,17 ohne Firmenmarke, unsign., H. 76 cm; Sammlung mad
3726,66,35 Firmenmarke 22, Ausformung 20er Jahre, H. 78 cm; Ahlden 126
Lit.: Matschiner (2006), S. 16 (m. Abb.)

3729
Rosé
Stehender Knabe mit Büchern (auf Rundsockel)
ca. 1909/10
3729,3,65 Firmenmarke 9, Repr. Res., H. 43,5 cm; GK, S. 15, 86 (m. Abb.); Wien Museum Inv. 182.080
Ausst.: HMW (jetzt Wien Museum) 1985/86.
Lit.: GK, S. 15, 86 (m. Abb.)

3735
Lavergne, Adolphe-Jean
LAMPE Zwei Zigaretten rauchende Harlekine
ca. 1909/10
3735,13,65 FG Wien (Firmenmarke 9?), Gesamth. 67 cm; Doro 19.9.2000
3735,49,12 Firmenmarke Sculpture 13, Gesamth. 71 cm; Doro 4.6.1996
3735,57,65 Firmenmarke Adler 16, H. 45,7 cm, B. 25,4 cm; eBay # 498719848
3735,62,15 ger. D, KV?, Firmenmarke 13a, Gesamth. 68 cm; Mehlis 23; 24
-> vgl. 3721 GRUPPE
-> vgl. 3766= rechte Einzelfigur

3762
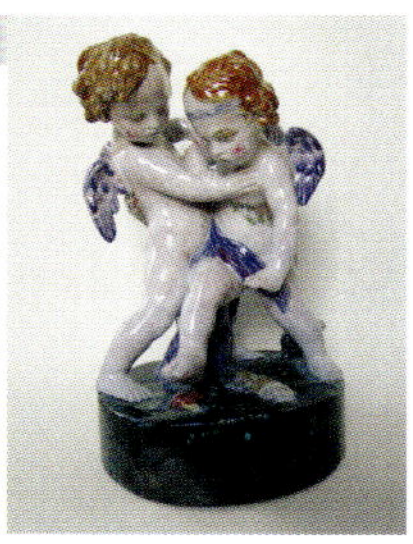
Zwei Putti um ein Herz kämpfend (auf Ovalsockel mit Herz, Rosen, Köcher und Bogen)
ca. 1909
3762,22,7 gem. 2.L., Firmenmarke 19, Made in Austria, Hand dec., Ausformung 20er Jahre, H. 40 cm; Sammlung Weilinger, Wien
Bem.: Diese Gruppe geht zurück auf eine Marmorskulptur von François-Joseph LeClercq, die 1780 entstand. Das Original befindet sich im Philbrook Museum of Art, Tulsa, OK/USA (www.philbrook.org).
-> vgl. 3373

3763
Lavergne, Adolphe-Jean
UHR „Historia" (Sitzende Frau auf Sockel mit Buch, in das ein stehender Jüngling (mit Lorbeer und Kugel) schreibt)
Inschrift: CAVE NE INDIGNUM
ca. 1905/06 (für den Entwurf)
3763,27,10 Firmenmarke 10, Firmenmarke Sculpture 13, H. 38,1 cm; eBay # 220003153699
Text der Inschrift: Hüte Dich vor dem Unwürdigen.
-> vgl. 3319; 3396

3766
Lavergne, Adolphe-Jean
Zigarette rauchender Harlekin
ca. 1909/10
3766,48,8 (Firmenmarke ohne Ang.), H. 39 cm; WKH
3766,120,10,19 Firmenmarke 20, Made in Austria, Hand dec., unsign.?, Ausformung 20er Jahre, H. 36,5 cm; eBay # 2221512624; Doro 15.3.2005
-> vgl. 3721= GRUPPE
-> vgl. 3735= LAMPE

3768
Cherc
Stehende Frau (daneben große Amphore, reliefiert mit mythologischen Szenen)
ca. 1910
3768,53,17 Firmenmarke 23, Ausformung 20er Jahre, H. 94 cm; KH Mercantile Antiques Auburn, CA. (www.//tace.com/items/2487); eBay # 220353929; Bem.: Hier mit Lampenfassung angeboten.
-> vgl. 3923

3769
Sitzendes Mädchen mit Buch und kleinem Hund
ca. 1910
3769,81,10 gem. XXIX, Firmenmarke 18, Ausformung ca. 1920, H. 12 cm; eBay # 2291619644

3778
Stehender Putto mit Gans (auf profiliertem Rundsockel)
ca. 1910
3778,36,10 FG Wien (Firmenmarke 7?), Repr. Res., undeutl. sign., H. 39 cm; Zezschwitz 29
3780 s. 3789

3782
Petri (d.i. Petrides, János (Johann))
Schreitender Knabe in Tracht (mit Ledertasche und erlegtem Wild) (auf Landschaftssockel)
ca. 1910
3782,278,16 Firmenmarke Sculpture 13, sign. Petri, H. 53,5 cm; Doro 4.3.2002

3784
Haniroff
Stehender nackter Römerjüngling mit Lorbeerkranz und Degen (auf Landschaftssockel)
ca. 1910
3784,118,15 Firmenmarke Sculpture 13, Fabriqué en Autriche, gepr. „S"= Z2, H. fehlt; eBay # 2550766910

3785
Schmidt-Kestner, Erich
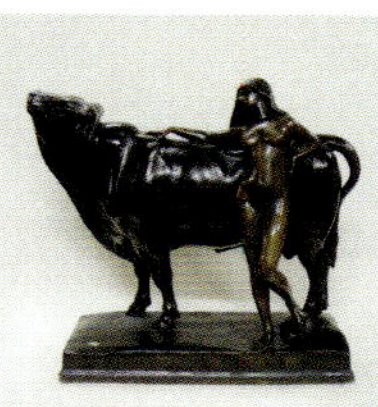
Europa mit dem Stier (auf Rechtecksockel)
ca. 1910
3785,23,2 Firmenmarke Sculpture 13, Fabriqué en Autriche, gepr. „S"= Z2, H. 35,8 cm; MAK Wien, Inv. Ke 10602-4; Bem.: Bronzefarben staffiert.
3785,36,1 (Firmenmarke unleserlich), Made in Austria, H. 35,5 cm, L. 45,7 cm; eBay # 454102226

3789
Abschlag
Stehender Ziegenbock (auf Rechtecksockel)
ca. 1910
3789(od. 3780?),30(od. 39?),65 (Firmenmarke ohne Ang.), Repr. Res., H. 24,5 cm, L. 37,5 cm; eBay # 3733380260; Bem.: Hier zur Modellnummer und zur Firmenmarke unklare Angaben.

3793
Pferd (auf rechteckigem Landschaftssockel)
ca. 1910
3793,20,10 Firmenmarke Adler 14, Firmenmarke Sculpture 13 (gepr.), H. 17,2 cm L. 24 cm; GK, S. 15, 87 (m. Abb.);Wien Museum Inv. 223.738
Ausst.: HMW (jetzt Wien Museum) 1985/86.
Lit.: GK, S. 15, 87 (m. Abb.)

3796
Pferd (auf rechteckigem Grassockel)
ca. 1910
3796,12,10 FG (Firmenmarke ohne Ang.), Repr. Res., L. 33 cm; Nagel 311
3796,87,11 Firmenmarke Adler 15, Firmenmarke Sculpture 13, H. 27, L. 33 cm; eBay # 2501077346
3796,107,16 FG Wien (Firmenmarke?), H. 27 cm, L. 32,5 cm; BM 73; Henry's 26.5.1993

3800
Butensky, Jules Léon
Nathan der Weise (Stehender Gelehrter mit Turban und weitem Mantel)
ca. 1910
3800,25,40 Firmenmarke Sculpture 13, Ges. gesch., Ausformung ca. 1900/05, H. 60 cm; Wien Museum Inv. 186.615
-> vgl. 3839

3804
Rosé
Stehendes Mädchen mit Rosenstrauß (an einer Rose riechend)
ca. 1910
3804,9,10 Firmenmarke 8c, Repr. Res., H. 49 cm; Wien Museum Inv. 219.152

3810 -> ÜBERSCHNEIDUNG
Tell, E.
LAMPE Stehende Frauengestalt, einen geschmückten Kranz (?) in die Höhe haltend (auf zylindrischem, mit Blumengirlanden verziertem Sockel)
ca. 1910
3810,1,3 Firmenmarke Sculpture 13, Repr. Res., H. 96,5 cm; KH www.nightingaleantiques.com/ pottery/ 7795.html
-> vgl. Lavergne 3718 (bis auf den Kranz (?) identisches Modell)

3810 -> ÜBERSCHNEIDUNG
Haniroff
BÜSTE Frau mit Diadem und Kopftuch (nach links blickend)
ca. 1910
3810,17,55 gem. weiß IV, Firmenmarke Sculpture 13, Fabriqué en Autriche, H. 26,7 cm; eBay # 519179096

3815
Lehmann, Ida
Stehende Dame mit Blumenbukett (auf quaderförmigem blumenverziertem Sockel)
ca. 1910/11
3815,35,27 Firmenmarke 9, H. 37,5 cm; Hennig 2; Bem.: Höhenang. falsch?
3815,148,11 (Firmenmarke ohne Ang.), H. 27 cm; Nagel 376; Bem.: Höhenang. falsch?
Bem.: Evtl. identisch mit dem Modell „Vom Garten", ÖMKI 1912?

3822
Capèque (d.i. Czapek, Stanislaus)
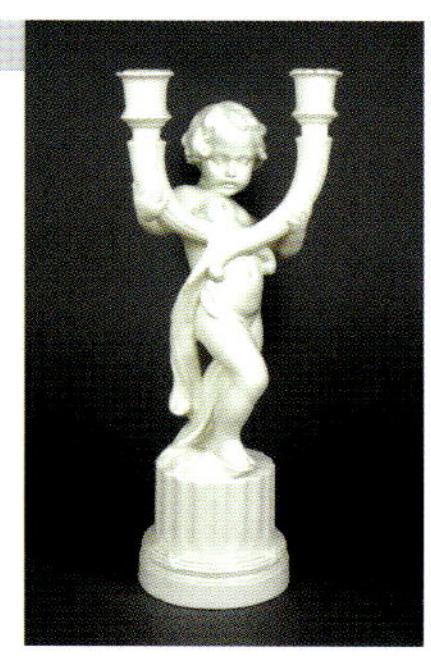
KERZENHALTER (zweiflammig) Stehender Putto (auf säulenartigem Sockel)
ca. 1910
3822,51,65 Firmenmarke 19, Made in Austria, Hand dec., sign. Capèque, Ausformung 20er Jahre, H. 39 cm; PB Wien 19
3822,52,7 ger., ger. M14, Firmenmarke 19, Made in Austria, sign. Capèque, Ausformung 20er Jahre, H. 39 cm; PB Wien 19
Lit.: ÖBWK, 1.1924/25, Sept.-Heft, S. III (m. Abb.)

3823
Cherc
BÜSTE Rauchender Bursche (auf konischem Sockel)
ca. 1899 (für den Entwurf)
3823,3,11 gem. Z, gem. weiß 31., Firmenmarke Sculpture 13, Ges. gesch., Ausformung ca. 1905/10, H. 40,6 cm; eBay # 3702874428; Bem.: Zigarette, bzw. Pfeife fehlt hier.
-> vgl. 1928 FIGUR

3831
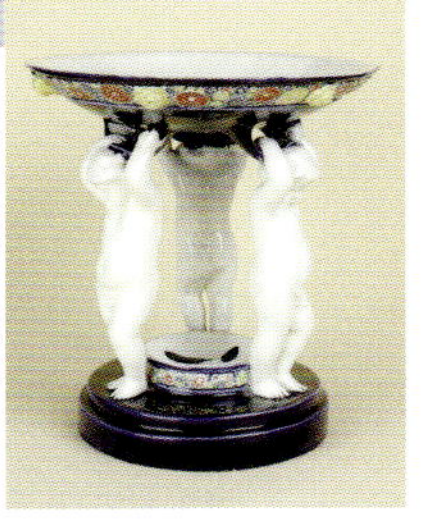
SCHALE Drei stehende Putti als Schalenträger (auf Rundsockel)
ca. 1910
3831,21,65,7 Firmenmarke Adler 16, Firmenmarke Sculpture 13 (gepr.), H. 27 cm; PB Wien 19
3831,174,16 (Firmenmarke Adler ohne Ang.), Firmenmarke Sculpture 13, undeutl. sign. „Holu"?, H. 28 cm; Stahl 8.12.2001
Bem.: Dieses Modell lässt in seiner Ausführung an die Putti (Schalenträger) von Michael Powolny denken.

3834
Latour, (Louis-Marie-Blaise?)
Judith (Stehende bekleidete Frauengestalt mit Schwert)
ca. 1910
3834,46,25 ger. II, Firmenmarke Sculpture 13, Fabriqué en Autriche, gepr. „S"= Z2, H. 54,3 cm; Wien Museum Inv. 219.129
-> vgl. 3835, hier als Halbakt (und Größenvariante)

3835
(Latour, (Louis-Marie-Blaise?))
Judith (Stehender Frauenhalbakt mit Schwert)
ca. 1910
3835,57,18 Firmenmarke 19, unleserlich sign., Ausformung 20er Jahre, H. 39 cm; Doro 7.11.1995
-> vgl. 3834, hier als bekleidete Figur (und Größenvariante)

3836
BÜSTE Mädchen mit Haarband
ca. 1910
3836,18,10 gem. TD, Firmenmarke Sculpture 13, Firmenmarke 10, H. 20,3 cm; eBay # 7303756039

3839
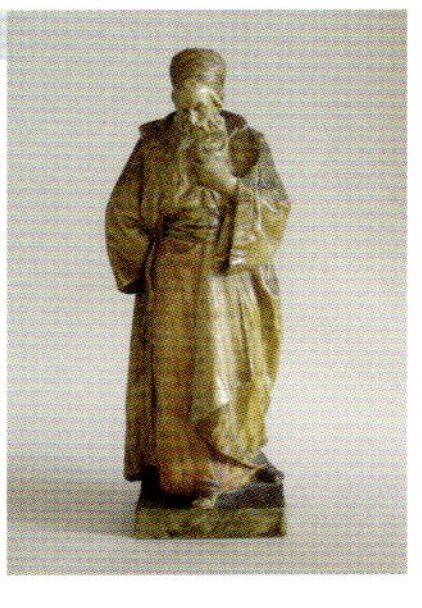
Butensky, Jules Léon
Nathan der Weise (Stehender Gelehrter mit Turban und weitem Mantel)
ca. 1910
3839,2,32 gem. lig. TK, Firmenmarke 4, Fabriqué en Autriche, Ausformung kurz vor 1900, gepr. „S"= Z2, H. 39,5 cm; Bode 20; PB Deutschland
-> vgl. 3800

3840
Persan
Holländische Gärtnerin mit Korb und Holzrechen
ca. 1910
3840,5,11 gem. XII, gepr. J (oder S.), Firmenmarke Sculpture 13, Fabriqué en Autriche, H. 44,7 cm; PB Deutschland
Pendants -> vgl. 2454; 2662; 2917

3841
Lesca
Stehendes Mädchen mit Harke
ca. 1910
3841,15,27 (Firmenmarke ohne Ang.), H. 38 cm; Doro 19.10.1992 (ohne Kat.)
3841 (ohne Ang.), H. 35 cm; Ahlden 38; Bem.: Hier „8841" als Modell-Nr. angegeben.
Bem.: Die Figur konnte nicht selbst geprüft werden. Falls die vom AH angegebene Modell-Nr. 8841 stimmt bzw. keine GS Firmenmarken vorhanden sind, so ist dieses Modell vermutlich ein Erzeugnis der Firma Bernhard Bloch, Eichwald.
Pendant -> vgl. 3842

3842
Lesca
Stehendes Mädchen eine Garbe tragend
ca. 1910
3842 (ohne Ang.), H. 35 cm; Ahlden 38; Bem.: Hier „8842" als Modell-Nr. angegeben.
Bem.: Die Figur konnte nicht selbst geprüft werden. Falls die vom AH angegebene Modell-Nr. 8841 stimmt bzw. keine GS Firmenmarken vorhanden sind, so ist dieses Modell vermutlich ein Erzeugnis der Firma Bernhard Bloch, Eichwald.
Pendant -> vgl. 3841

3843
Rosé
Stehendes kleines Mädchen mit Buch (auf Quadratsockel mit Stuhl)
ca. 1910
3843,189,3 ger. L, gem. V, Firmenmarke 22, H. 31,5 cm; Wien Museum Inv. 220.729

3844
Perigot
UHR Stehendes, sich umarmendes Rokokopaar (auf mit Blumengirlanden verziertem Sockel/Gehäuse)
ca. 1911
3844,28,10 Firmenmarke 15, H. 32 cm; Wendl 37

3846
Schmidt-Kestner, Erich
„Hirschkuh mit Kalb" (auf Landschaftssockel)
ca. 1911
3846,91,55 (Firmenmarke ohne Ang.), H. 36 cm; Kett. 181; Kett. 192
3846,186,25 Firmenmarke 22, Ausformung 20er Jahre, H. 35,5 cm; eBay # 921476252
3846,267,11 Firmenmarke 19, Ausformung 20er Jahre, H. 35 cm; Fischer 80
Ausst.: ÖMKI 1912, hier mit Ang. des Originaltitels „Hirschkuh mit Kalb".
Lit.: Neuwirth (1974), S. 74

3849
Stehendes lesendes Mädchen (auf Stuhl mit (zwei?) Büchern)
ca. 1911
3849,189,3 gem. V., Firmenmarke 22, Ausformung 20er Jahre, H. 33 cm; Wien Museum Inv. 220.729
-> vgl. 4534= ähnl. Modell (sign. Polter/Rosé)

3853
Gornik, Friedrich
FIGUR „Eule" (mit Schreibzeug, auf zwei Büchern sitzend)
ca. 1901/02 (für den Entwurf)
3853,61,15 Firmenmarke 19, Made in Austria, sign. Gornick (!), Ausformung 20er Jahre, H. 36,2 cm; eBay # 2233361186
-> vgl. 3853; 3854
Lit.: Pudor (1902), S. 109 (m. Abb.) und Ang. des Originaltitels „Eule"; KuKHW, 6.1903, S. 526 (m. Abb.); Neuwirth (1974), S. 58.

3854
Gornik, Friedrich
FIGUR „Eule" (mit Schreibzeug, auf zwei Büchern sitzend)
ca. 1901/02 (für den Entwurf)
3854,154,14 Firmenmarke 19, Made in Austria, Ausformung 20er Jahre, H. 25 cm; WKA 23
3854,156,8 (gem./ger. M99?), Firmenmarke 19, Made in Austria, Ausformung 20er Jahre, H. 24,9 cm; eBay # 3160816034
-> vgl. 2364; 3853
Lit.: Pudor (1902), S. 109 (m. Abb.) und Ang. des Originaltitels „Eule"; KuKHW, 6.1903, S. 526 (m. Abb.); Neuwirth (1974), S. 58.

3855
Zwei schleichende Pumas (auf rechteckigem Landschaftssockel)
ca. 1911/12
3855,6,65 FG Wien (Firmenmarke ohne Ang.), H. fehlt; Tradera # 54222; (Antik Batiran, Berlin)

3856
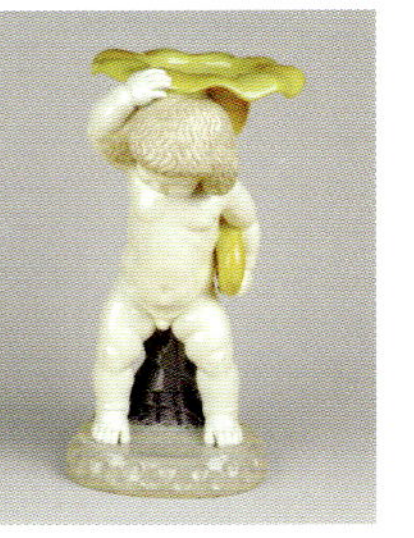
Müller, Armin
Putto, einen großen Pfifferling tragend
ca. 1911/12
3856,9,10 Firmenmarke 9, Repr. Res., sign. Armin Müller, H. 44,2 cm; Doro 25.11.2003
3856,10,27 Firmenmarke Adler 16, Firmenmarke 10, Repr. Res., H. 44 cm; PB Wien 19
3856,16,10 gem. VIII, Firmenmarke Adler 15, Firmenmarke Sculpture 13 (gepr.), sign. Armin Müller, H. 43,4 cm; eBay # 161348463; Doro 26.11.2002; Doro 8.3.2004
Pendant(?) -> vgl. 4643
Lit.: Pinhas, S. 99 (m. Abb.)

3858
Gemignani, Valmore
Sitzender Windhund
ca. 1911/12
3858,150,13 Firmenmarke Adler 15, Firmenmarke Sculpture 13, H. 27,5 cm; Scheffler, S. 224 (m. Abb.), KGWMus. Berlin Inv. WA 145; Neuwirth (1973), S. 59 (m. Abb.); Neuwirth (1974), S. 58, 63
3858,172,12 gem. W., Firmenmarke Adler 16, Firmenmarke Sculpture 13 (gepr.), H. 27 cm; eBay # 2226296383
Lit.: Neuwirth (1973), in: Keramos, 59.1973, S. 54–63, hier S. 59 (m. Abb.)

3859
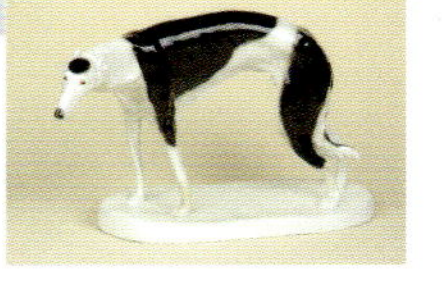
Gemignani, Valmore
Stehender Windhund
ca. 1911/12
3859,455,35 Firmenmarke Adler 15, Firmenmarke Sculpture 13, sign. Gemignani, H. 19 cm; eBay # 923664634
3859,483,59 Firmenmarke Adler 15, Firmenmarke Sculpture 13 (gepr.), sign. Gemignani, H. ca. 19 cm; KH
3859,763,6 Firmenmarke Adler 15, Firmenmarke Sculpture 13 (gepr.), sign. Gemignani, H. 19,8 cm, L. 28 cm; eBay # 1416084418
3859,1041,10 (Firmenmarke ohne Ang.), H. 19,5 cm; KH
3859,1023,11 Firmenmarke Adler 15, Firmenmarke Sculpture 13 (gepr.), sign. Gemignani, H. 20 cm; eBay # 8795578137
3859,1161,10 Firmenmarke Adler 14, Firmenmarke Sculpture 13 (gepr.), sign. Gemignani, H. 19,3 cm, L. 29,3 cm; MAK Wien, Inv. Ke 10602-57
3859,1182,10 Firmenmarke Adler 14, Firmenmarke Sculpture 13 (gepr.), sign. Gemignani, H. 19,5 cm, L. 30 cm; BröPK, S. 146; Brö 2,1, S. 401
3859,1834,? Firmenmarke 19, sign. Gemignani, Ausformung 20er Jahre, H. 20 cm; Wilbois 65
3859,2072,10 Firmenmarke 19, sign. Gemignani, Ausformung 20er Jahre, H. ca. 19,5 cm; KH
3859,2290,1 ger. K, gem. braun B., Firmenmarke 22, sign. Gemignani, Ausformung 20er Jahre, H. 18,5 cm, L. 28,5 cm; GK, S. 17, 97 (m. Abb.); Wien Museum Inv. 182.085
3859,2414,2 gem. B, (Firmenmarke ohne Ang.), sign. „Lavergne" (!), H. 20 cm, L. 26 cm; Karrenbauer 176; Bem.: Die Sign. „Lavergne" dürfte wohl als Versehen des Modelleurs angesehen werden.
3859,2573,24 Firmenmarke 19, sign. Gemignani, Ausformung 20er Jahre, H. 19 cm; Wendl 42
3859,2735,(?) Firmenmarke 19, sign. Gemignani, Ausformung 20er Jahre, L. 27 cm; RM 28
3859,2935,11 gem. lig. AC, Firmenmarke 19, sign. Gemignani, Ausformung 20er Jahre, H. 20 cm; eBay # 2074918961

3859,2983,12 Firmenmarke 19, Made in Austria, sign. Gemignani, Ausformung 20er Jahre, H. ca. 19,5 cm; KH
3859,3196,10 Firmenmarke 19, sign. Gemignani, Ausformung 20er Jahre, H. 18,1 cm; Doro 21.5.1997
3859 (ohne Ang.), Firmenmarke 19, sign. Gemignani, Ausformung 20er Jahre, L. 28 cm; Nagel 313; 317; 325
3859 (ohne Ang.), sign. „Latour" (!), H. 19 cm; Sworders 30.5.2001; Bem.: Die Sign. „Latour" dürfte wohl als Versehen des Modelleurs angesehen werden.
Bem.: Wenn auch hier mit „Latour" und „Lavergne" sign. Exemplare nachgewiesen werden können, so ist doch eindeutig der Entwurf dieses Windhundes dem Künstler Valmore Gemignani zuzuordnen, der für GS etliche dieser Tierfiguren entworfen hat. Dies wird u.a. auch durch die Höhenvariante Mod. 4687 belegt.
Ausst.: HMW (jetzt Wien Museum) 1985/86.
-> vgl. 4687
Lit.: Neuwirth (1974), S. 58, 63; GK, S. 17, 97 (m. Abb.).

3860
Perigot
Stehende Rokokodame mit Hut, sich auf großen ovalen Korb stützend
ca. 1910/11
3860,22,27 FG Wien (Firmenmarke 9?), H. 35 cm
Wendl 38; eBay # 1468066665
3860 (ohne Ang.), H. 36 cm; Fischer 30
3860 (ohne Ang.), H. 35,5 cm; BM 34

3861
Meier, Emil
Stehende Rokokodame mit Blumenkorb (Sommer)
ca. 1910/11
3861,30,27 Firmenmarke Sculpture 13, H. 30 cm; WKA 12
-> vgl. 4244
Lit.: Studio YB, 1911, S. 250 (m. Abb.), hier mit Ang. des Entwerfers Emil Meier.

3862
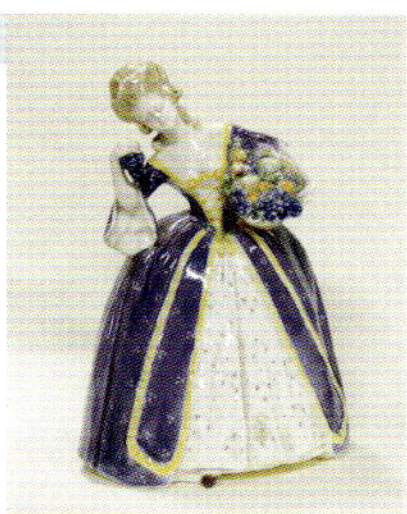
Meier, Emil
Stehende Rokokodame mit Früchtekorb (Herbst)
ca. 1910/11
3862,21,69 FG Wien (Firmenmarke?), unsign.?, H. 28 cm; Kett. 14
3862,68,69 Firmenmarke 10, Firmenmarke Sculpture 13, H. 27,7 cm; MAK Wien, Inv. Ke 10521-42
3862,71,65 (Firmenmarke ohne Ang.), H. 28,5 cm; Herr 53
3862,155,8 GS Wien (Firmenmarke ohne Ang.), unsign.?, H. 27 cm; Soth. N.Y. 10.6.1993; Soth. Arcade 1462
-> vgl. 4052; 4245

3868
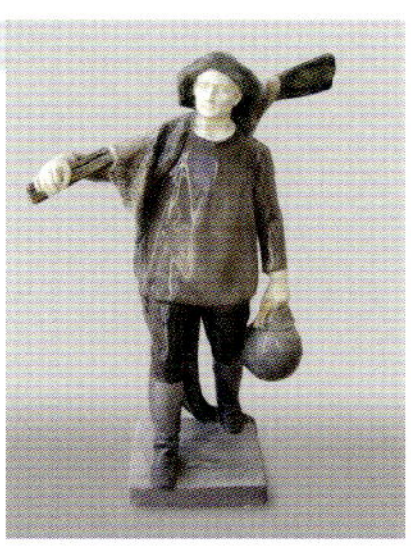
Lesca
Fischer mit Netz, Paddel und Krug (auf Rechtecksockel)
ca. 1910/11
3868,36,10 ger. B, gem. XXVIII, gem. lig. JF, Firmenmarke 18, Ausformung ca. 1920, H. 36 cm; Sammlung Stressless, Portugal
3868,43,16 gem. V., Firmenmarke 19, Made in Austria, Ausformung 20er Jahre, H. 34,3 cm; eBay # 2500907547

3870
Rosé
„Ruth" (Tänzerin auf dreipassigem Sockel mit zwei Rosenkörben)
ca. 1911/12
3870,1 FG Wien (Firmenmarke ohne Ang.), Repr. Res., H. 75 cm, L. 53 cm; BM 55
3870,10,? (Firmenmarke ohne Ang.), H. 75 cm; Schloß Rick 61
Bem.: Dargestellt ist die Tänzerin Ruth Saint Denis (1877/79–1968, eigentl. Ruth Dennis, gen. „Miss Ruth"), in ihrem Tanz „Radha" (Tanz der fünf Sinne, Urauff. 1906), -> vgl. hierzu Ludwigg, H. (Hg.): 40 Jahre Wintergarten, Fs. Berlin 1928, S. 68 (m. Abb.); Berühmte Tänzerinnen (Zigarettenbildersammelalbum) Berlin, Garbaty (ca. 1930) Bild 84; Ochaim, B.: Varieté-Tänzerinnen ca. 1900, Frankfurt/M., Basel 1998, S. 69 (m. Abb.).
-> vgl. 3992; 4141; 5171; 5513
Lit.: MW, 11.1929/30, Nr. 11, S. VII (m. Abb.), hier mit Ang. des Originaltitels „Ruth" und des falschen Künstlernamens „Bosi"; BT, o. Pag., Bild 84.

3871
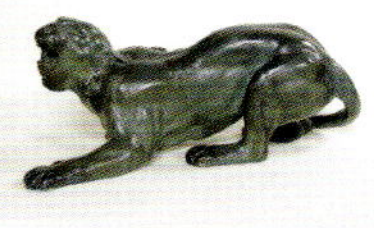
Fauchender Berglöwe (in Abwehrhaltung)
ca. 1910/11
3871,2,32 gem. L, VIII, Firmenmarke 4, Fabriqué en Autriche, H. 12 cm, L. 30,8 cm; MAK Wien, Inv. Ke 10602-5; Bem.: Bronzefarben patiniert.

3873
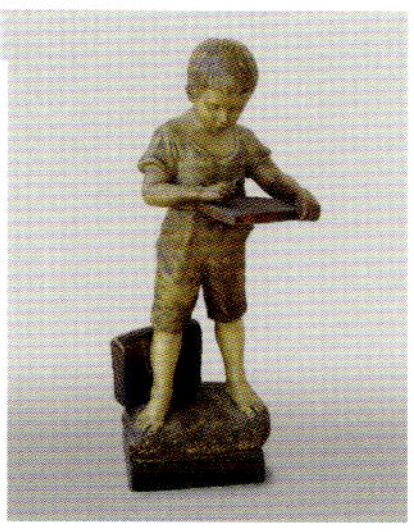
Stehender Bub auf einer Schiefertafel schreibend (auf rechteckigem Landschaftssockel mit Schultasche)
ca. 1910
3873,31,25 Firmenmarke Sculpture 13, Fabriqué en Autriche, H. 50,5 cm; Sammlung Stressless, Portugal

3876
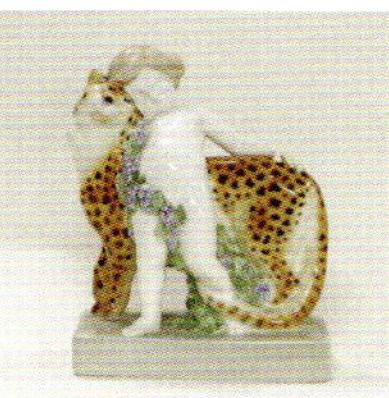
Meier-Michel, Johanna
„Kind mit Löwen"
(Putto mit (Raub)katze)
ca. 1910
3876,11,10 (Firmenmarke ohne Ang.), sign. Meier, H. 20 cm; Doro 20.10.1997
3876,73,65 Firmenmarke Adler 15a, sign. Meier, H. 20 cm; MAK Wien, Inv. Ke 10521-18
ohne Ang., gem. F.S., (Firmenmarke ohne Ang.), unsign., H. 19 cm; RM 34
Bem.: Diese Gruppe ist vermutlich mit dem im ÖMKI 1910/11 ausgestellten „Kind mit Löwen" identisch.
Ausst.: ÖMKI 1910/11?
Lit.: Neuwirth (1974), S. 68; Pinhas, S. 99 (m. Abb.).

3878
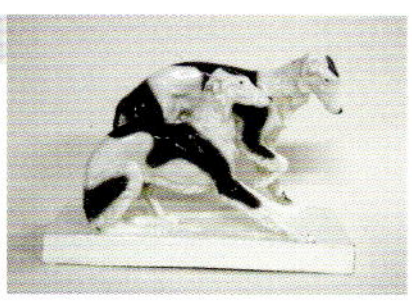
Beaume
(d.i. Bormann, Wilhelm)
GRUPPE / LAMPE Zwei stehende Windhunde (Barsois) (auf Rechtecksockel)
ca. 1910/12
GRUPPE
3878,30,27 Firmenmarke 3, Firmenmarke 7, Firmenmarke Sculpture 13, H. 21 cm; Wien Museum Inv. 190.023
3878,283,65 gem. W, II/G, Firmenmarke Adler 16, Firmenmarke Sculpture 13 (gepr.), H. 20,5 cm, L. 29,9 cm; MAK Wien, Inv. Ke 10602-44
3878,94,10 Firmenmarke Sculpture 13 (gepr.), H. 22 cm, L. 29,5 cm; WKH
3878,185,10 Firmenmarke Sculpture 13 (gepr.), (H. ca. 22 cm), L. 30 cm; Z 74
3878,592,6 Firmenmarke 19, Made in Austria, Ausformung 20er Jahre, H. 19,5 cm, L. 28,6 cm; eBay # 127557451
3878,709,12 gem. 260.Z/B, Firmenmarke 19, Made in Austria, Ausformung 20er Jahre, H. 19 cm, L. 28 cm; Soth. com. # 3J6X7
3878,773,11 gem. lig. SB, Firmenmarke 19, Ausformung 20er Jahre, H. 20 cm, L. 27,5 cm; Doro 9.11.1989
3878,1196,16 Firmenmarke 19?, Ausformung 20er Jahre, H. 21 cm; Ahlden 42
LAMPE
3878,695,10 (Firmenmarke ohne Ang.), Gesamth. 89 cm; eBay # 7378231534
-> vgl. 4407= rechte Einzelfigur
-> vgl. 4637= linke Einzelfigur

3879
Lehmann, Ida
Stehende Rokokodame mit Fächer (auf Rundsockel)
ca. 1911
3879,108,65 Firmenmarke Adler 15, Firmenmarke Sculpture 13, sign. Ida Lehmann, H. 29 cm; Zezschwitz 10; Fischer 141
3879 (ohne Ang.), (zwei Firmenmarken ohne Ang.), H. 28,7 cm; Hennig 15

3885
Czapeque (d.i. Czapek, Stanislaus)
Stehendes Biedermeiermädchen mit Rose
ca. 1911
3885,8,10 FG (Firmenmarke?), Repr. Res., unsign., H. 29 cm; Rieber 30.11.–12.12.2000
3885,82,35 (Firmenmarke ohne Ang.), sign. Czapeque, H. 29 cm; Doro 15.9.2003
Bem.: Ähnl. Modell vgl. Modell-Nr. 3913 Telzec (d.i. Telcs, Ede / Eduard).
-> vgl. 3948

3887
Cherc
LAMPE „Muschelträgerin" (Stehender Damenakt, eine Muschel über den Kopf haltend)
ca. 1899
3887,39,14 GS Wien (Firmenmarke ohne Ang.), Ausformung ca. 1900, H. 79 cm; Soth. L. 25.10.1991
Bem.: Evtl. identisch mit „Muschelträgerin" ÖMKI 1899/1900?
Ausst.: ÖMKI 1899/1900?

3890
Meier, Emil
Stehende Rokokodame mit Rosen
ca. 1911
3890,61,65 (Firmenmarke ohne Ang.), H. 29,2 cm; Doro 3.5.1995
-> vgl. 4247
Bem.: Evtl. handelt es sich um die im ÖMKI 1913/14 ausgestellte Figur „Herbst".
Ausst.: ÖMKI 1913/14?
Lit.: Studio YB, 1911, S. 250 (m. Abb.), hier mit Ang. des Entwerfers Emil Meier.

3894
Czapek, Stanislaus
Amazone mit Schild, vor Pferd stehend (auf gewölbtem Rechtecksockel)
ca. 1911
3894 (Firmenmarke ohne Ang.), sign. Czapèque, H. 56 cm; Pinhas, S. 41 (m. Abb.)
-> vgl. 4155

3895
Petri (d.i. Petrides, János) (Johann)
LAMPE Blumenmädchen, eine Kugel aus Rosen tragend
ca. 1910/12
3895,6,19 Firmenmarke Sculpture 13, Fabriqué en Autriche, sign. Petri, H. 67,5 cm; Wien Museum Inv.186.629
Bem.: Die Lampe ist durch vier Glühbirnenfassungen elektrifiziert.

3900
Cherc
Wasserträgerin (ein Henkelgefäß über dem Kopf haltend) (auf Rundsockel)
ca. 1910/12
3900 (Firmenmarke ohne Ang.), H. 88 cm; Pinhas, S. 63 (m. Abb.)

3913
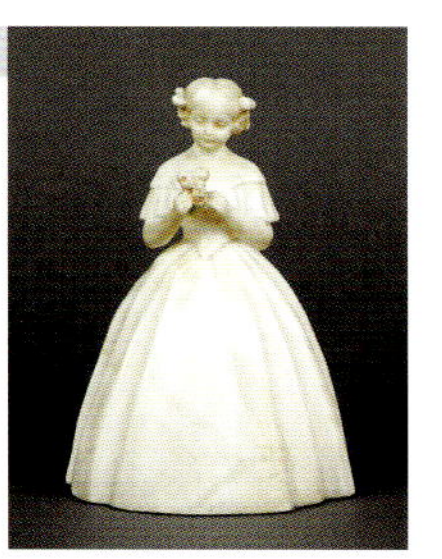
Telcs, Ede / Eduard
Stehendes Mädchen mit Rosen und Korb
ca. 1911
FAYENCE
3913,4,65 FG Wien (Firmenmarke ohne Ang.), Repr. Res., sign. Telzer, H. 17,5 cm; WKA 18; 23
3913,10,65,1 Firmenmarke

Sculpture 13 (gepr.), sign. Telzer, H. 17,8 cm; Wien Museum Inv. 190.169
3913,36,9 Firmenmarke Adler 14, unsign., H. ca. 17,5 cm; WKH
3913,659,? Firmenmarke Sculpture 13, sign. Telsch, H. 17 cm; BM 74
ALABASTER
Stehendes Mädchen mit Rosen (ohne Korb!)
T3913, F37 Plak. A, sign. Telzer, H. 19 cm; PB Wien 19
3913, lig. JF (oder TF),16 (Firmenmarke ohne Ang.), sign. Telzec, H. 19,5 cm; Sammler-Journal 24.1995, S. 1822
Bem.: Dieses Modell wurde in Fayence und Alabaster (Kunststein?) ausgeführt. Bei der Ausführung in Alabaster wurde jeweils der Korb weg gelassen. Die vorliegenden unterschiedlichen Signaturvarianten werden sogar in den offiziellen Ausst.-Kat. des ÖMKI aufgeführt. Es handelt sich aber eindeutig um den ungar. Bildhauer Ede/Eduard Telcs, der in Wien unter K. Zumbusch studierte.

3921
Rosé
„Richard Wagner" (sitzende, lesende Figur)
ca. 1905/06 (für den Entwurf)
3921,20,01 FG Wien (Firmenmarke ohne Ang.), Firmenmarke Adler 14, H. 30 cm; Henry's 24.8.1994
3921,36,14 Firmenmarke Adler 14, Firmenmarke Sculpture 13 (gepr.), H. 30 cm; Weidler 460
Bem.: Dargestellt ist der Komponist Richard Wagner (1813–1883). Wahrscheinlich wurde die Figur zum 100. Geburtstag (1913) noch einmal „neu" aufgelegt.
-> vgl. 3248

3923
Polter
Stehende Frau (daneben große Amphore, reliefiert mit mythologischen Szenen)
ca. 1910 (für den Entwurf)
3923 (Firmenmarke ohne Ang.), H. 80 cm; Pinhas, S. 47 (m. Abb.)
-> vgl. 3768
3934 vgl. 2934

3938
Gornik, Friedrich
Balzender Auerhahn
ca. 1911/12
3938,1,10 (Firmenmarke undeutl.), H. 26 cm, L. 35 cm; B. 26 cm; Kett. 167
Bem.: Vermutlich GS (Zuschreibung unter Vorbehalt, jedoch sprechen, trotz der undeutlichen Firmenmarke, Nummerierung, Qualität und nicht zuletzt die Mitarbeiterschaft Gorniks für GS).

3941
Lefont
Stehendes Bauernpaar (auf abfallendem Landschaftssockel)
ca. 1911/12
3941,12,10 Firmenmarke 10, Firmenmarke Sculpture 13, H. 43 cm; Wien Museum Inv. 190.076

3943
Rosé
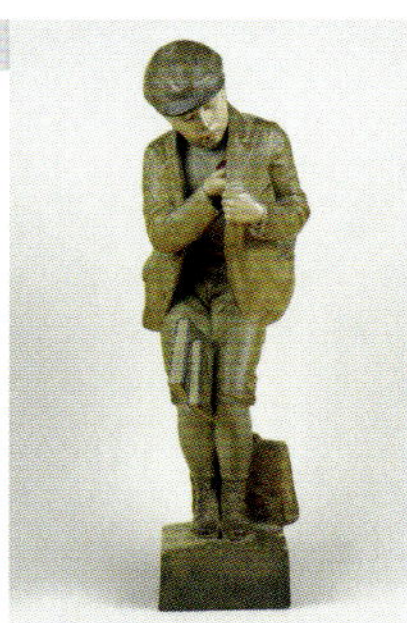
Knabe mit Büchern und Schultasche
ca. 1911/12
3943,4,16 Firmenmarke Sculpture 13, Fabriqué en Autriche, H. 50 cm; Stahl 16.10.1993
3943,81,25 Firmenmarke Sculpture 13, Fabriqué en Autriche, H. 50 cm; PB Wien 19

3945
Simon
Sitzendes Biedermeiermädchen (auf Scherenstuhl) mit geigenartigem Instrument
ca. 1911/12
3945(Firmenmarke ohne Ang.), H. 56 cm; Pinhas, S. 103 (m. Abb.)

3947
Czapek, Stanislaus
Stehendes Biedermädchen mit Blume
ca. 1911/12
FAYENCE
3974 (ohne Ang.), H. fehlt; Neuwirth (1974), S. 58, 61; Bem.: Hier keine nähere Beschreibung des Objektes.
ALABASTER
3974,+,15 ger. Plak. A, sign. Capeque, H. 27,2 cm; PB Wien 19

3948
Capèque / Czapeque (d.i. Czapek, Stanislaus)
Stehendes Biedermeiermädchen mit Rose
ca. 1911
FAYENCE
3948,94,65 Firmenmarke Sculpture 13, sign. Czapeque, H. 19,5 cm; Doro 7.3.2001
3948,192,65 (Firmenmarke ohne Ang.), unsign.?, H. 20 cm; RM 39
3948,314,16 (Firmenmarke ohne Ang.), sign. Czapeque, H. 20 cm; Henry's 26.2.1999; Doro 21.9.1999
3948,374,76 (Firmenmarke ohne Ang.), unsign.?, H. 18,5 cm; Kastern 87
ohne Ang., sign. Capèque, H. 19,5 cm; Doro 21.5.1997
ALABASTER
3948,W,53 ger., Plak. E, Made in Austria ger., Ausformung 20er Jahre, H. 20,3 cm; eBay # 5390409221
Bem.: Dieses Modell wurde in Fayence und Marmor (Alabaster/Kunststein?) ausgeführt. Ähnl. Modell vgl. Telcs 3913.
-> vgl. 3885

3951
Schwetz-Lehmann, Ida
BÜSTE Mädchen mit Blumen geschmücktem Haar und Vogerl auf der Schulter
ca. 1911
3951,2,6 Firmenmarke Sculpture 13, Sign. o. Ang., H. 33 cm; Doro 31.5.2006

3952
Dackel
ca. 1911/12
3952 (ohne Ang.), H. fehlt; Neuwirth (1974), S. 58; Bem.: Hier keine nähere Beschreibung des Objektes.

3963
Meier, (Emil oder Meier-Michel, Johanna)
LAMPE Sitzende Biedermeierdame mit großem Hut und Blumen
ca. 1911/12
3963,16,10 Firmenmarke 10, Firmenmarke Sculpture 13, sign. Meier. H. 40,6 cm; eBay # 45350781; Doro 18.5.1999
3963,30,10 Firmenmarke 10, Firmenmarke Sculpture 13, sign. Meier, H. ca. 40,5 cm; Doro 3.5.1995

3969
Leclancher
UHR „Eisengießer" (Zwei Arbeiter mit Handwagen vor dem Ofen)
ca. 1906
3969,47,7 Firmenmarke Sculpture 13, Repr. Res., H. 77,5 cm, B. 62 cm; Doro 8.11.1994
Ausst.: ÖMKI 1906/07, hier mit Ang. des Originaltitels „Eisengießer".
Lit.: Neuwirth (1974), S. 66

3977
Meier-Michel, Johanna
„Velazquez" (Infantin Margarita)
ca. 1911/12
3977,1,35 (ohne Firmenmarke), Fabriqué en Autriche, unsign., gepr. „L"= Z1, H. 29,5 cm; SLW 176
3977,11,6 (Firmenmarke ohne Ang.), sign. Meier, H. 30,4 cm; Doro 13.9.2004
3977,30,11 (Firmenmarke ohne Ang.), sign. Meier, H. 31 cm; WKA 38
3977,56,14 (Firmenmarke ohne Ang.), unsign., H. 29,7 cm; Doro 26.11.2002; 8.3.2004
3977,74,14 gem. B XXIII, Firmenmarke 22, sign. Meier, Ausformung 20er Jahre, H. 30,5 cm; eBay # 1218644987
Ausst.: Deutsche Gewerbeschau München 1922.
-> vgl. 5520
Lit.: Burger, W.: Die Deutsche Gewerbeschau in München, Keramik und Glas, in: Dekor. Kunst, 25.1921/22, S. 265 (m. Abb.), hier mit Ang. des Originaltitels „Velasquez" und der Künstlerin „Johanna Meier-Michel"; Neuwirth (1974), S. 68.

3980
Rosé
Stehendes Mädchen mit Cape und großer Schleife (Mascherl) (einen Koffer tragend)
ca. 1911/12
3980,5,65 Firmenmarke 9, Firmenmarke Sculpture 13, H. 27,5 cm; Wien Museum Inv. 220.732

3982
Latour, (Louis-Marie-Blaise?)
Tänzerin (sich nach links drehend)
ca. 1911/12
3982,109,35 Firmenmarke Sculpture 13?, H. 46 cm; RM 5
3982,120,13 Firmenmarke Adler 15, Firmenmarke Sculpture 13, H. 45 cm; Brö 2,1, S. 401; H. 44,5 cm; BröPK, S. 146, Farbtaf. 51
3982,125,13 Firmenmarke Adler 15, Firmenmarke Sculpture 13 (gepr.), H. 45 cm; Wien Museum Inv. 219.188
3982,126,65 Firmenmarke Sculpture 13, H. 45 cm; Doro 9.11.1989
3982,180,8 gem. F, III, Firmenmarke Sculpture 13, H. 46 cm; eBay # 7318220677
Lit.: Neuwirth (1974), S. 58, 66

3983
Sitzende Bäuerin mit Korb
ca. 1911/12
3983,24,4 (Firmenmarke ohne Ang.), H. 13,5 cm, Doro 7.3.2001

3988
Lavergne, Adolphe-Jean
Tiger (auf Rechtecksockel)
ca. 1905/08
3988,38,27 Firmenmarke Sculpture 13?, unsign., Ausformung wohl um 1910, H. 27 cm; Doro S 7/1998 (ohne Kat.)
3988 (ohne Ang.), Firmenmarke Sculpture 13?, Ausformung wohl um 1910, L. 43,5 cm; BM 101
-> vgl. 3536
-> vgl. GRUPPE „Animus et vis"= Modell-Nr. 3680

3990
Rosé
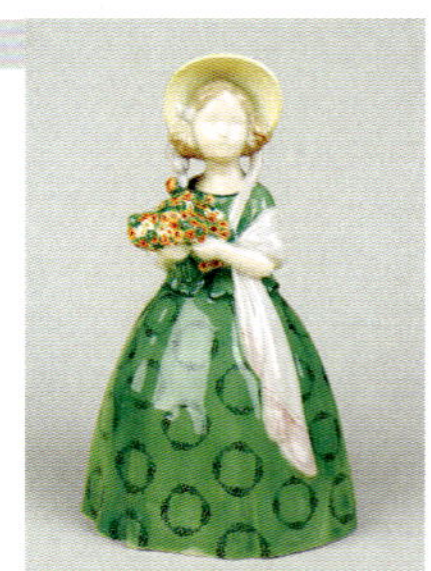
Stehendes Biedermeiermädchen mit Schutenhut und Blumenstrauß
ca. 1911
3990,29,6 Firmenmarke Sculpture 13, H. 39,8 cm; Brö 2,1, S. 402; H. 41 cm; BröPK, S. 146 (m. Abb.)
3990,37,10 Firmenmarke 10, Firmenmarke Sculpture 13 (gepr.), H. 40,5 cm; Quit. 17
3990,45,27 (Firmenmarke ohne Ang.), unsign., H. 39,8 cm; Christie's L. 16.3.1982
3990,77,6 Firmenmarke Adler 14, Firmenmarke Sculpture 13, H. 40,2 cm; BM 71; 73
3990 (ohne Ang.), H. 39,8 cm; Weichmann 18.10.1985
-> vgl. 4083; 4103
Lit.: Neuwirth (1974), S. 58, 71; Pinhas, S. 101 (m. Abb.).

3991
Perigot
UHR Sitzende Lautespielerin (Rokokodame)
ca. 1911/12
3991,43,24 ger. H, F, IV, Firmenmarke 22, H. 35,5 cm; eBay # 3703243025; Sammlung Wade Sanders

3992
Rosé
„Ruth" (Tänzerin auf dreipassigem Sockel mit zwei Rosenkörben)
ca. 1911/12
FAYENCE
3992,20,10 (Firmenmarke ohne Ang.), H. 55,9 cm; eBay # 1270520122
3992,33,6 FG (Firmenmarke ohne Ang.), Firmenmarke Sculpture 13, H. 55,6 cm; Doro 18.5.1999
3992,59,10 gem. 2019, gem. braun VI, (Firmenmarke ohne Ang.), H. 55 cm; Herr 53; Christie's SK 13.10.2004
3992,77,65 ger. 2, gem. F III, gem. grün H 19, Firmenmarke Adler 16, Firmenmarke Sculpture 13 (gepr.), H. 55 cm; Koller Z A114
3992,90,32, gem. IA, Firmenmarke 13a, gepr. „S"= Z2, H. 56 cm; Sammlung f@goldscheider.de
3992,132,25 ger. B, gem. M III, Firmenmarke 19, Ausformung 20er Jahre, H. 53,5 cm; BM 84
3992,135,2 gem. X.I N., Firmenmarke 18, Ausformung ca. 1920, H. 55 cm; Sammlung Robert Senft
3992,136,1 gem. NB XXVII, GS (Firmenmarke ohne Ang.), H. 54,5 cm; Soth. L. 31.10.1997
3992,204,10 Firmenmarke 19, Ausformung 20er Jahre, H. 55,8 cm; Soth. com. # 1BX8
3992,281,65 gem. L., gem. lig. AS, Firmenmarke 19, Made in Austria, Hand dec., Ausformung 20er Jahre, H. 52 cm; KH Ruby Lane (Jane Perlow / Lacremeantik, Great Neck N.Y.)
3992,298,10 ger. BJ., gem. lig. SB, Firmenmarke 19, Hand dec. (2x), Ausformung 20er Jahre, H. 53 cm; eBay # 340907312; # 1204263044
3992 (ohne Ang.), Firmenmarke 22?, Ausformung 20er Jahre, H. 55 cm; Spik 551; Battenberg 1991, S. 225
ALABASTER/MARMOR/KUNSTSTEIN?
3992,VR 3 ger., (Firmenmarke ohne Ang.), unsign.?, H. 54 cm; BM 17
Bem.: Dargestellt ist die Tänzerin Ruth Saint Denis (1877/79–1968, eigentl. Ruth Dennis, gen. „Miss Ruth"), in ihrem Tanz „Radha" (Tanz der fünf Sinne, Urauff. 1906), vgl. hierzu Ludwig, H. (Hg.): 40 Jahre Wintergarten, Fs. Berlin 1928, S. 68 (m. Abb.); Berühmte Tänzerinnen (Zigarettenbildersammelalbum) Berlin, Garbaty (ca. 1930) Bild 84; Ochaim, B.: Varieté-Tänzerinnen ca. 1900, Frankfurt/M., Basel 1998, S. 69 (m. Abb.).
Bem.: Dieses Modell wurde in Fayence und Marmor (Alabaster/Kunststein?) ausgeführt.
-> vgl. 3870; 4141; 5171; 5513
Lit.: MW, 11.1929/30, Nr. 11, S. VII (m. Abb.), hier mit Ang. des Originaltitels „Ruth" und des falschen Künstlernamens „Bosi".

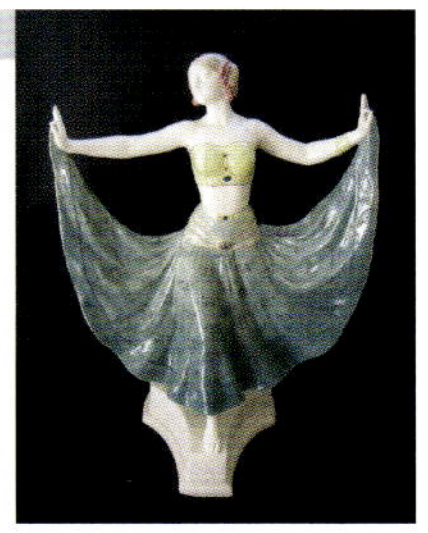

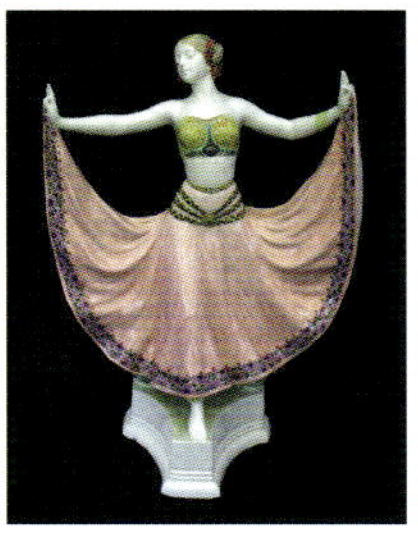

4000
Polter
Stehendes kleines Mädchen mit großem Schirm und Handtasche (auf profiliertem Rundsockel)
ca. 1912
4000,144,10 Firmenmarke 16?, Firmenmarke Sculpture 13, H. 22 cm; Wendl 28
4000,521,32 gem. XX, Firmenmarke Adler 16, unsign.?, H. 22,5 cm; Wendl 41
4000,548,3 ger. 1, gem. 20. 4 F, Firmenmarke Sculpture 13, unsign., H. 22 cm; Brö 2,1, S. 402, 403 (m. Abb.); BröPK, S. 146
4000,615,11 gem. D2, (III?), Firmenmarke 19, Made in Austria, Hand dec., H. 21,5 cm; city-antik
-> vgl. 4016
Lit.: Neuwirth (1974), S. 58, 71 (m. Abb. der Modell-Nr. 4200 von Petrides).

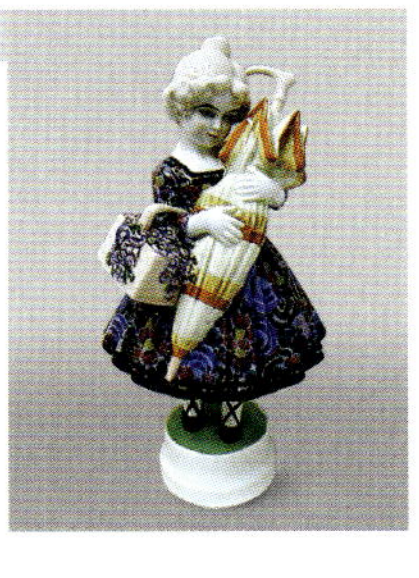

4002
Meier, (Emil oder Meier-Michel, Johanna)
Stehende Biedermeierdame mit Schutenhut und Blumenkorb (Sommer)
ca. 1912
4002,15,65 Firmenmarke Sculpture 13, sign. Meier, H. 28 cm; Doro 9.11.1993

4007
Montini, (Tullio?) / Massolini
Dante (stehende Figur)
ca. 1912
ohne Nr., Firmenmarke Sculpture 13, sign. Montini, H. 50 cm; Wien Museum Inv. 190.063
4007,41,11 (Firmenmarke ohne Ang.), sign. Massolini, H. 49 cm; Henry's 27.9.1995
4007,46,6, gem. VIII, Firmenmarke 22, unleserlich sign. „Hourtoni"? (wohl eher Montini), Ausformung 20er Jahre, H. 50,8 cm; eBay # 161345977
4007,60,16 gem. VIII, Firmenmarke 19 (zwei mal), Made in Austria, unsign.?, Ausformung 20er Jahre, H. 46 cm; eBay # 49698102
Bem.: Dargestellt ist der italienische Dichter Dante Alighieri (1265–1321).

4014
Bormann, Wilhelm / Beaume (d.i. Wilhelm Bormann)
Stehender Neufundländer (auf Ovalsockel)
ca. 1910
4014,110,? (Firmenmarke ohne Ang.), sign. Beaume, Ausformung 20er Jahre, H. 20 cm, L. 34 cm; KaM 180
4014,119,7 ger. B, Firmenmarke 22, sign. Beaume, Ausformung 20er Jahre, H. 20 cm, L. 28 cm; Keup Varia Juli 2000
4014,719,7 (Firmenmarke ohne Ang.), unsign.?, H. 20 cm; Sebök 30
-> vgl. 3714 (GRUPPE / Deckeldose); 5908 (GRUPPE)

4016
Polter
UHR Stehendes kleines Mädchen mit großem Schirm und Handtasche (auf winkelförmigem Rechtecksockel)
ca. 1912
4016,2,10 Firmenmarke ohne Ang., unsign., H. 25 cm; KH
-> vgl. 4000

4017
Perigot
UHR Dame mit Kavalier
ca. 1912
4017,305,– (Firmenmarke ohne Ang.), H. 52 cm; Doro WA 12.12.1992

4020
Rosé
„Friedrich der Große" (Stehende Figur in Uniform) (auf Landschaftssockel)
ca. 1905/06 (für den Entwurf)
4020,23,16 gem. III. F., Firmenmarke 19, Made in Austria, Ausformung 20er Jahre, H. 34 cm; eBay # 859901795
Bem.: Dargestellt ist Friedrich der Große (d. II.) (gen. „Der alte Fritz", 1712–1786); Lit.: Venohr, W.: Der große König, Friedrich II., Bergisch-Gladbach 1997, Umschlag (m. Abb.).
-> vgl. 3330

4026
Petri (d.i. Petrides, János) (Johann)
POSTKARTENHALTER Stehendes Biedermeiermädchen mit Schirm (auf getrepptem Sockel)
ca. 1911/12
4026,2,20 Firmenmarke Sculpture 13, sign. Petri, H. 16 cm; KH

4028
Raum, Alfred
„Dachshund" (Sitzender Dackel)
ca. 1911/12
4028,2 gem. L., Firmenmarke 19, Made in Austria, Ausformung zw. 1945–1953, H. 16,2 cm; Sammlung a@goldscheider.net
4028,16,54 (Firmenmarke ohne Ang.), H. 15 cm; WKH
4028,21 gem. L, Firmenmarke 19, Made in Austria, Ausformung zw. 1945–1953, unsign., Pick. 5/a, H. 15,2 cm; eBay # 1078432894
4028,208,10 (Firmenmarken ohne Ang.), H. 15,2 cm; Doro 29.6.1993
4028,489,11 gem. 1, gem. lig. AS, Firmenmarke 19, Made in Austria, Ausformung 20er Jahre, H. 16,2 cm; Sammlung a@goldscheider.net
4028,566,3 Firmenmarke 19, Made in Austria, Ausformung 20er Jahre, H. 15 cm; Doro 21.9.1999
4028 (ohne Ang.), unsign.?, H. 14 cm; Henry's 2.6.2001
Ausst.: ÖMKI 1912, hier mit Ang. des Originaltitels „Dachshund".
-> vgl. 4080 (ähnl. Modell)
Lit.: Neuwirth (1974), S. 58, 72

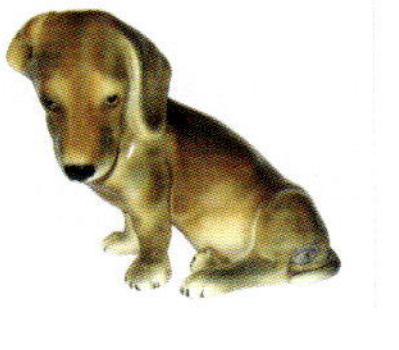

4031
Polter
Gänseliesel (Kleines Mädchen mit zwei Gänsen) (auf rechteckigem Landschaftssockel)
ca. 1911/12
4031,110,10 gem. 13, Firmenmarke Adler 16, Firmenmarke Sculpture 13 (gepr.), H. 23,5 cm, L. 28 cm; Sammlung Stressless, Portugal

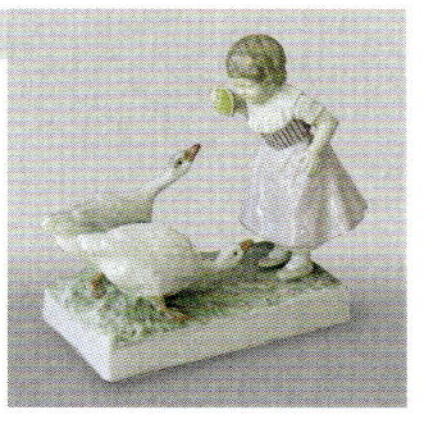

4036
Rosé
LAMPE Stehendes kleines Mädchen an einem kleinen Ofen (sog. „Kanonenofen") kochend (auf Rechtecksockel)
ca. 1911/12
4036, (8?),6 Firmenmarke Sculpture 13, H. 35,5 cm; eBay # 27449774

4038 s. 4938

4040
Petri (d.i. Petrides, János) (Johann)
Stehender Bauernbub in Tracht mit zwei Vögeln und Umhängetasche (auf Rechtecksockel)
ca. 1911/12 (für den Entwurf)
4040,2,8 Firmenmarke 19, Made in Austria, sign. Petri, Ausformung 20er Jahre, H. 43 cm; eBay # 7401420247

4041
Rosé
Stehendes Mädchen mit Puppe und Teddybär (auf Rundsockel)
ca. 1911/12
4041,42,15 Firmenmarke 13a, H. 29 cm; eBay # 110050123565

4042
Stehender Putto mit Blume und Blumenkorb (auf Rechtecksockel)
ca. 1911/12
4042,2,35 Firmenmarke Sculpture 13, Fabriqué en Autriche, gepr. „S"= Z2?, H. 27 cm; Doro 20.4.1993

4043
Perigot
Rokokodame mit großem Korb
ca. 1911/12
4043,11,10 (Firmenmarken ohne Ang.), H. 30,5 cm; BM 83

4045
Rosé
Stehender Mädchenhalbakt
ca. 1911/12
4045,7,65 (Firmenmarke ohne Ang.), H. 29 cm; Doro 20.5.1996

4049
Kugelträgerin (Stehender Mädchenhalbakt, eine Eisenkugel tragend) (auf Rundsockel)
ca. 1911/12
4049 (Firmenmarke ohne Ang.), H. 74 cm; Pinhas, S. 100 (m. Abb.)

4052
Meier, Emil
LAMPE Stehende Rokokodame mit Früchtekorb (Herbst)
ca. 1910/11 (für den Entwurf)
4052,27,3 sign. Meier, H. 64 cm; Walldorf 10
4052,–,– Firmenmarke Sculpture 13, Gesamth. 65 cm; Mehlis 20
–> vgl. 3862; 4245

4059
Perigot
Stehendes Biedermeiermädchen mit zwei großen Körben
ca. 1911/12
4059,38,6 (Firmenmarke Sculpture 13?), H. 30 cm; Wendl 24; Mehlis 8
4059,75,65 Firmenmarke 22, Ausformung 20er Jahre, H. 30,5 cm; eBay # 2623609613

4062
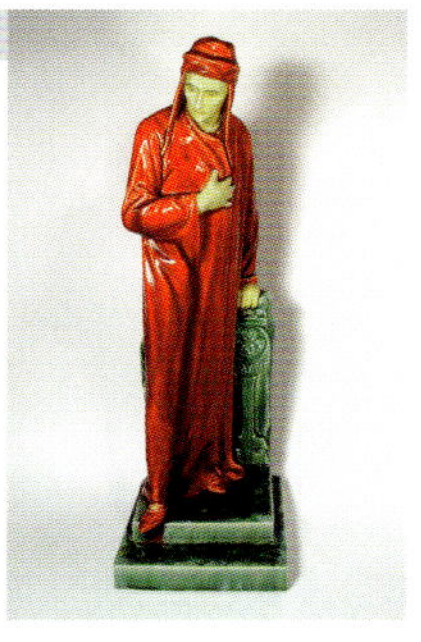
Dante (stehende Figur) (auf getrepptem Sockel mit Balustrade)
ca. 1911/12
4062,62,8 gem. G, SG, CR, Firmenmarke 19, Made in Austria, Ausformung 20er Jahre, H. 48,3 cm; Sammlung a@goldscheider.net
Bem.: Dargestellt ist der italienische Dichter Dante Alighieri (1265–1321).

4068
Meier, (Emil oder Meier-Michel, Johanna)
LAMPE Stehende Biedermeierdame mit federverziertem Hut
ca. 1911/12
4068,7,6 Firmenmarke Sculpture 13 (gepr.), sign. Meier, H. 40 cm; Doro 25.4.2000
–> vgl. .... FIGUR (H. 19 cm); RM 25

4079
Perigot
Krinolinendame mit Blumenvase
ca. 1911/12
4079 (ohne Ang.), H. fehlt; Neuwirth (1974), S. 58, 59; Bem.: Hier keine nähere Beschreibung des Objektes.

4080
Postl, Hugo
Sitzender, kurzhaariger Dachshund
ca. 1911/12
4080,6 ger., gem. T, Firmenmarke 19, unsign.?, Ausformung 20er Jahre, H. 8,9 cm; eBay # 1265105899
4080,470,4 gem. F., Firmenmarke 19, Made in Austria, Ausformung 20er Jahre, H. 7,9 cm; eBay # 1421534908
Lit.: Neuwirth (1974), S. 58 (ohne nähere Ang.)

4083
Rosé
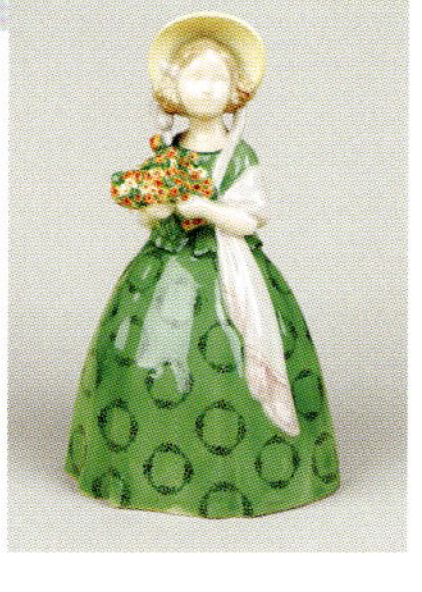
Stehendes Biedermeiermädchen mit Schutenhut und Blumenstrauß
FAYENCE
ca. 1911/12
4083,29,65 (Firmenmarke ohne Ang.), unsign.?, H. 21 cm; Rehm 166
4083,190,35 Firmenmarke Adler 15, Firmenmarke Sculpture 13 (gepr.), H. 21 cm; eBay # 2230480512
4083,241,10 Firmenmarke Sculpture 13, H. 20 cm; Doro 18.4.1994
4083,324,6 Firmenmarke Sculpture 13 (gepr.), H. 20,5 cm; BM 68
4083,536,6 Firmenmarke Adler 15, Firmenmarke Sculpture 13 (gepr.), H. 20,7 cm; city-antik
4083,716,16 Firmenmarke Adler 14, H. 22 cm; PB Wien 19
4083,777,10 (oder 19) gem. AV, Firmenmarke Adler 14, Firmenmarke Sculpture 13 (gepr.), H. 20 cm; Quit. 3; Ahlden 102
4083,851,12 Firmenmarke Sculpture 13, H. 20,5 cm; Doro WA 11.12.1993; Doro 8.11.1994
ALABASTER
4083,MR,80 Plak. A, H. 23 cm, B. 8 cm; Sammlung mad
Bem.: Dieses Modell wurde in Fayence und Alabaster/ (Kunststein?) ausgeführt.
–> vgl. 3990; 4103

4084
Haniroff
Jüngling mit Adler kämpfend (auf Felssockel)
ca. 1908
4084,17,3 Firmenmarke Sculpture 13, Firmenmarke 16?, H. 54 cm; Sammlung Helmut Riener, St. Florian, Austria
Bem.: Keramik bronzefarben staffiert.

4085
Daubling (d.i. Doblinger, Ferdinand)
Herr mit Zylinder (Biedermeierfigur) (auf verziertem Rundsockel)
ca. 1908
4085,20,65 Firmenmarke Sculpture 13, sign. Daubling, H. 32 cm; Doro 9.11.1993
–> vgl. 3545 (auf Rechtecksockel)
Pendant –> vgl. 4086

4086
Daubling (d.i. Doblinger, Ferdinand)
Dame mit Hut (Biedermeierfigur) (auf verziertem Rundsockel)
ca. 1911/12
4086,50,11 Firmenmarke Sculpture 13, Firmenmarke Adler 15, sign. Daubling, H. 32 cm; Zezschwitz 24; 26

4087
Mörtel (d.i. Mörtl, Michael)
Drei sitzende Eulen (auf Rundsockel mit vier Büchern)
ca. 1911/12
4087,2,65 Firmenmarke Sculpture 13 (gepr.), sign. Mörtel (!), H. 21,5 cm; eBay # 2276913870

4089
Stehendes Biedermeiermädchen (mit Strohhut und Blumenkorb) und kleiner Amor, der ihr ins Ohr flüstert (auf Landschaftssockel)
ca. 1912
4089 (ohne Ang.), H. 29,2 cm; BM 34

4090
Lindner, Helmut
Tänzerin
ca. 1912
4090 (ohne Ang.), H. fehlt; Neuwirth (1974), S. 58, 67; Bem.: Hier keine nähere Beschreibung des Objektes.

4094
Latour, (Louis-Marie-Blaise?)
Sitzender Cellist (im Biedermeiergewand) mit stehendem Biedermeiermädchen (mit zwei Rosen am Dekolleté) (auf mit Blumengirlanden verziertem, kanneliertem Rundsockel)
ca. 1911/12
4094,21,10 Firmenmarke Sculpture 13 (gepr.), H. 36 cm; city-antik
Bem.: Die Einzelfiguren dieser Biedermeiergruppe wurden unter den Modell-Nrn. 4108 u. 4109 produziert.
–> vgl. Cellist 4108; Biedermeiermädchen 4109

4095
Latour, (Louis-Marie-Blaise?)
Die Liebeserklärung (Sitzendes, händchenhaltendes Rokokopaar an einem Tisch) (auf mit Blumengirlanden verziertem, kanneliertem Rundsockel)
ca. 1912
4095,112,16 GS Wien (Firmenmarke ohne Ang.), H. 35,5 cm; eBay # 870916613

4101
Perigot
UHR Sitzende Biedermeierdame mit Hut, ihren Schuh anziehend
ca. 1912
4101,31,2 gem. X.D.I., Firmenmarke Adler 15, H. fehlt; KH

4103
Rosé
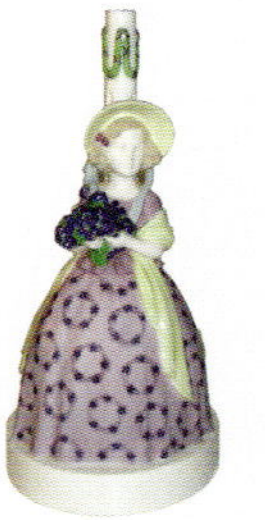
FIGUR/LAMPE Stehendes Biedermeiermädchen mit Schutenhut und Blumenstrauß
ca. 1911/12
FIGUR
4103,211,10 gem. blau P., gem. grün VII, (Firmenmarke ohne Ang.), H. 27,5 cm; Kett.157
ohne Ang., (Firmenmarke ohne Ang.), H. 28 cm; Henry's 29.6.1993
ohne Ang., (Firmenmarke ohne Ang.), H. ca. 29cm; Doro 20.3.1992
LAMPE
4103,54,6 Firmenmarke Adler 15, Firmenmarke Sculpture 13 (gepr.), H. 38 cm; Sammlung a@goldscheider.net
–> vgl. 3990; 4083

4105
BÜSTE Mädchen mit Spitzenhaube (auf konischem Rechtecksockel)
ALABASTER
ca. 1911/12
ger. lig. PW(?),4105,5 Plak. A, H. 45 cm; Fischer 56

4108
Latour, (Louis-Marie-Blaise?)
Sitzender Cellist (im Biedermeiergewand)
ca. 1911/12
4108,50,05 gem. VI, Firmenmarke Adler 16, H. ca. 30 cm; CuxAH 25.7.1998
4108,82,32 (Firmenmarke ohne Ang.), H. ca. 30 cm; KH
–> vgl. GRUPPE 4094

4109
Latour, (Louis-Marie-Blaise?)
Stehendes Biedermeiermädchen (mit zwei Rosen am Dekolleté)
ca. 1911/12
4109,188,4 gem. F.V., Firmenmarke 19, Ausformung 20er Jahre, H. 25 cm; Wien Museum Inv. 190.112
–> vgl. GRUPPE 4094

4111
Rosé
BÜSTE „Liszt"
Inschrift: LISZT
ca. 1911
4111,11,16 Firmenmarke Sculpture 13, Fabriqué en Autriche, H. 30,5 cm; eBay # 3703884202
4111,41,14 Firmenmarke Sculpture 13, H. 30,5 cm; Kett. 181; BM 83; Ruef 477
Bem.: Dargestellt ist der Komponist und Pianist Franz Liszt (1811–1886). Die Büste wurde zum 100. Geburtstag aufgelegt.

4114
Horsetzky, Melanie von
„Lautenspiel" (Sitzende Lautenspielerin)
ca. 1911/12
4114,1,10 gem. T, Firmenmarke Sculpture 13 (gepr.), H. 25,6 cm; Wien Museum Inv. 223.732
4114,48,65 Firmenmarke Adler 15, Firmenmarke Sculpture 13 (gepr.), H. 25,7 cm; Wien Museum Inv. 219.184
4114,62,20,15 gem. F., H19, IV, Firmenmarke 13a, H. 25 cm; Wien Museum Inv. 219.186
4114,74,10 Firmenmarke Sculpture 13, unsign.?, H. 25 cm; Doro G 7.11.1995
4114,(108,6?), Firmenmarke Sculpture 13 (gepr.), H. 25,4 cm; eBay # 498067994
4114,171,10 gem. D.2, VIII, Firmenmarke Adler 15, Firmenmarke Sculpture 13 (gepr.), H. 25,8 cm; GK, S. 15, 88 (m. Abb.); Wien Museum Inv. 219.185
4114,302,14 Firmenmarke 13a, H. 25,4 cm; eBay # 257600046
4114,321,2 gem. FF II, Firmenmarke 22, Ausformung 20er Jahre, H. 25,3 cm; eBay # 2062853121
Bem.: Die im Originaltitel angeführte „Laute" ist im eigentlichen Sinn eine Gitarre.
Ausst.: ÖMKI 1912, hier mit Ang. des Originaltitels

„Lautenspiel"; HMW (jetzt Wien Museum) 1985/86; SSH 1992.
-> vgl. 4580
Lit.: Neuwirth (1974), S. 64; GK, S. 15, 88 (m. Abb.).

4115
Cherc
Löwe (auf rechteckigem Felssockel)
ca. 1911/12
4115,74,14 Firmenmarke 22 (gepr.), Ausformung 20er Jahre, H. 57 cm, L. 60 cm; eBay # 20598001139; 2065820876

4120
Wernekinck, Sigismund
Ballspielende Katze (auf Ovalsockel)
ca. 1911/12
4120,88,3 Firmenmarke 19, Ausformung 20er Jahre, H. 19 cm, L. 22 cm, B. 10 cm; eBay # 3729740898

4123
Lavergne, Adolphe-Jean
BÜSTE Chopin
Inschrift: CHOPIN
ca. 1911/12
4123,60,4 Firmenmarke 22a, Ausformung 20er Jahre, H. 30 cm; eBay # 190056189243
Bem.: Dargestellt ist der Komponist Frédéric Chopin (1810–1849).

4124
Petri (d.i. Petrides, János) (Johann)
BÜSTE Holländermädchen (auf oktogonalem Längssockel)
ALABASTER
ca. 1911/12
4124,W,200 ger. GS Wien (Firmenmarke Plak.?), sign. Petri, H. 14 cm; Bergmann 45

4130
Polter
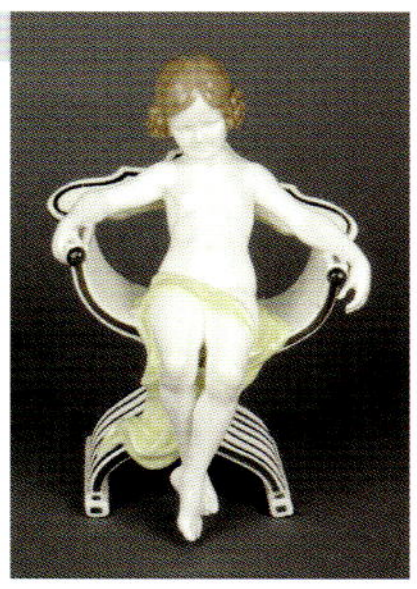
Sitzendes nacktes Mädchen auf Scherenstuhl
ca. 1911/12
4130,112,11 Firmenmarke Adler 15, H. 23 cm; eBay # 25822967604
4130,382,10 (Firmenmarke ohne Ang.), H. 23 cm; Doro 21.5.1991
4130,508,11 Firmenmarke 13a, H. 22,5 cm; eBay # 2578983894
4130,543,10 Firmenmarke 13a, H. 22,5 cm; PB Wien 19
4130,554,25 gem. XXVI, Firmenmarke 13a, H. 22,5 cm; eBay # 1283635317
4130,579,17 Firmenmarke 19, Made in Austria, Hand dec., unsign.?, Ausformung 20er Jahre, H. ca. 20 cm; eBay # 1445950245
4130,612,5 Firmenmarke 22, H. 24 cm; Wien Museum Inv. 220.731
Lit.: Pinhas, S. 99 (m. Abb.)

4131
Lehmann, Ida
Stehendes Biedermeiermädchen mit Blumenstrauß
ca. 1911/12
4131,269,65 gem. K, Firmenmarke Adler 15, Firmenmarke Sculpture 13 (gepr.), H. 18 cm; Wien Museum Inv. 219.171

4132
Müller, Leo
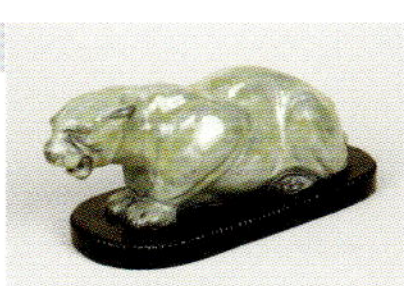
Panther (auf abgerundetem Rechtecksockel)
ca. 1911/12
4132,25,6 Firmenmarke 19, sign. Müller, Ausformung 20er Jahre, H. 11 cm; PB Wien 19
Ausst.: ÖMKI 1912, hier mit Ang. des Originaltitels „Panther" und mit Ang. des Entwerfers „Leo" Müller.
Lit.: Neuwirth (1974), S. 68

4134 ?
Perigot
Stehendes lesendes Mädchen (auf Rundsockel)
ca. 1911/12
4134? (ohne Ang.), H. 20,5 cm; Ahlden 66
ohne Ang., H. 20 cm; Schloß Rick. 55

4141
Rosé
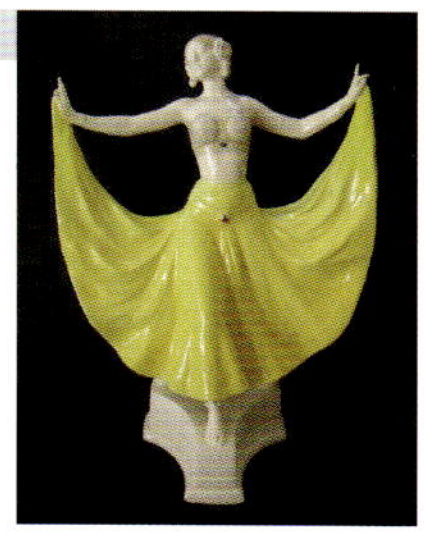
„Ruth" (Tänzerin auf dreipassigem Sockel mit zwei Rosenkörben)
ca. 1911/12
4141,31,10 Firmenmarke Adler 15, Firmenmarke Sculpture 13, H. 38 cm; Sammlung f@goldscheider.de
4141,124,12 Firmenmarke Adler 16?, Firmenmarke Sculpture 13, H. 33 cm; Quit. H. 28.2.2004
4141,126,12 Firmenmarke Sculpture 13, Firmenmarke Adler 16, H. 38 cm; eBay # 6505752795
4141,243,14 gem. lig. HA, XXVII, Firmenmarke 13a, H. 36,8 cm; (bzw. 39,3 cm); eBay # 1204263042; # 467003160
4141,273,35 (Firmenmarke ohne Ang.), H. 37,5 cm; WKA 35
4141,402,65 Firmenmarke 22?, unsign.?, Ausformung 20er Jahre, H. 37 cm; Schlüter 2678
4141,424,8 Firmenmarke 19, Ausformung 20er Jahre, H. 37,4 cm; Doro 1.12.1998
4141,612,7 Firmenmarke 19, Made in Austria, Hand dec., Ausformung 20er Jahre, H. 35,5 cm; Doro 7.3.2001; WKA 35; WKA 38; Doro 10.3.2003
4141,649,9 (Firmenmarke ohne Ang.), H. 35,5 cm; Fischer 130
4141,681,11 (Firmenmarke ohne Ang.), H. ca. 36 cm; KH
4141 (ohne Ang.), unsign.?, H. 38 cm; Rehm 137
Bem. und Lit.: vgl. bei 3870, hier mit Ang. des Originaltitels „Ruth" und des falschen Künstlernamens „Bosi".
-> vgl. 3870; 3992; 5171; 5513

4144
Wernekinck, Sigismund
„Eisbär"
ca. 1911/12
4144,6,69 Firmenmarke Adler 15, unsign.?, H. 22 cm, L. 42 cm; eBay # 923029494
4144,49,6 gem. 1, Firmenmarke Adler 14, Firmenmarke Sculpture 13 (gepr.), H. ca. 19 cm; KH
Ausst.: ÖMKI 1912, hier mit Ang. des Originaltitels „Eisbär" und der falschen Ang. „A. Wernekink".
Lit.: Neuwirth (1974), S. 58, 76

4145
Wernekinck, Sigismund
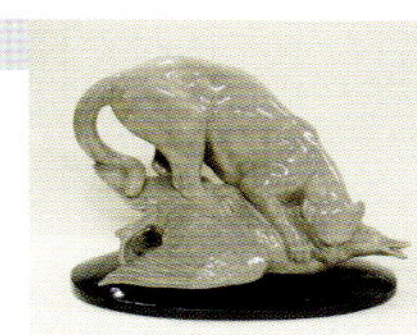
„Ente und Wildkatze" (auf Ovalsockel)
ca. 1911/12
4145,24,35 Firmenmarke 19, Ausformung 20er Jahre, H. 23 cm, L. 35,4 cm; BM 62; Fischer 67; Stahl 26.3.1994
4145,27,6 (4145,27,6 gem.), gem. VIII., Firmenmarke 22, sign. S. Wernekinck, H. 23,8 cm, L. 35,7 cm; MAK Wien, Inv. Ke 10602-64
4145 (ohne Ang.), Firmenmarke 19, Ausformung 20er Jahre, H. 25 cm, L. 34 cm; Spik 565
Ausst.: ÖMKI 1912, hier mit Ang. des Originaltitels „Ente und Wildkatze" und der falschen Ang. „A. Wernekink".
Lit.: Neuwirth (1974), S. 76

4155
Czapek, Stanislaus
Amazone mit Schild, vor Pferd stehend (auf gewölbtem Rechtecksockel)
ca. 1911
TERRAKOTTA
4155,69,5 (Firmenmarke ohne Ang.), unleserlich sign. (Czapèque?), H. 38,5 cm; Doro 15.9.2003
BRONZE / Marmorsockel
4155,1,25 (Firmenmarke ohne Ang.), unleserlich sign., H. 48,5 cm; Nagel 346
Bem.: Dieses Modell wurde in Terrakotta und Bronze ausgeführt.
-> vgl. 3894

4156
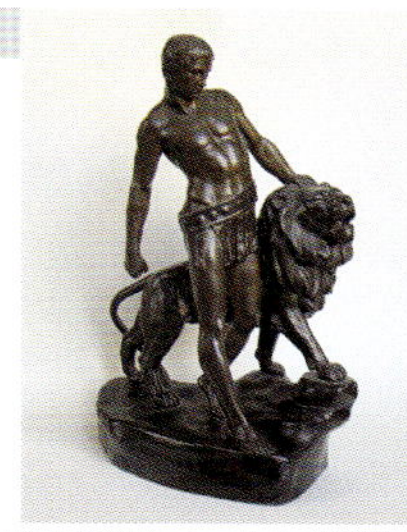
Gladiator, einen Löwen führend (auf Landschaftssockel)
ca. 1911
4156,19,3 Firmenmarke Sculpture 13, Fabriqué en Autriche, undeutl. sign., gepr. „S" = Z2, H. 56,5 cm; Sammlung Weilinger, Wien

4158
Latour, (Louis-Marie-Blaise?)
Biedermeiertänzerin (auf Rundsockel)
ca. 1911/12
4158,21,6 Firmenmarke Adler 15, Firmenmarke Sculpture 13, H. 47,5 cm; BM 57; 58
4158,51,65 Firmenmarke Sculpture 13, H. 47,3 cm; BM 43
4158,57,10 (Firmenmarke ohne Ang.), Firmenmarke Sculpture 13, H. 59 (!) cm; Nagel 379; Bem.: Höhenang. sicher falsch, bzw. Druckfehler im Kat.
4158,87,11 (grüne Firmenmarke ohne Ang.), H. 47,5 cm; WKA 29
4158,112,12 gem. W 22/3, Firmenmarke Adler 16, Firmenmarke Sculpture 13 (gepr.), gepr. „P" = Z3, H. 47 cm; Brö 2,1, S. 403 (m. Abb.); BröPK, S. 146
4158,181,11 (Firmenmarke ohne Ang.), H. 48,5 cm; Stahl 8.12.2001
-> vgl. 4252; 4253
Lit.: Neuwirth (1974), S. 58, 66

4163
Perigot
TAFELAUFSATZ Zwei Putti mit Trauben, eine Schale tragend (Tafelaufsatz / Obstschale)
ca. 1912
4163 (ohne Ang.), Firmenmarke Sculpture 13, H. 30 cm; WKH

4164
Latour, (Louis-Marie-Blaise?)
Diana mit Pfeil und Bogen neben Windhund stehend (auf Vierkantsockel)
ca. 1912
4164,65,16 (Firmenmarke ohne Ang.), H. 70 cm; Mehlis 32
4164,145,15,S Firmenmarke 13a, H. 61 cm; eBay # 2217235927
4164,168,25 (Firmenmarke ohne Ang.), H. 59 cm; eBay # 2195811437
Bem.: Der Bogen der Diana ist aus Metall (Messing?) und Leder gefertigt. Die stark differierenden Höhenangaben resultieren von den unterschiedlich gefertigten Bögen.
-> vgl. 4310

4168
Gornik, Friedrich
FIGUR „Hahn"
ca. 1911/12
4168,1,7,Firmenmarke Sculpture 13, Fabriqué en Autriche, sign. Gornic (!), H. 51 cm; Doro 20.4.1993
Bem.: Der Entwurf dieses Modells ist zwischen 1900–1902 zu datieren.
-> vgl. 2383
Ausst.: ÖMKI 1912, hier mit Ang. des Originaltitels „Hahn".
Lit.: KuKHW, 6.1903, S. 528 (m. Abb.); Neuwirth (1974), S. 63

4170
Mallener, Theodor
„Spitz" (liegender Hund)
ca. 1911/12
4170,28,2 (Firmenmarken ohne Ang.), sign. Malthner (!), H. 12,3 cm, L. 24,4 cm; BM 64; 73
ohne Ang, sign. „Malthner", H. 12 cm; Henry's 26.5.1994
Ausst.: ÖMKI 1912, hier mit Ang. des Originaltitels „Spitz".

4182
Lehmann, Ida
Stehende junge Dame mit Hut und Rosenstrauß
(auf reliefiertem Ovalsockel)
ca. 1911/12
4182,66,13 gem. X D4, gem. grün H4, Firmenmarke Adler 15, Firmenmarke Sculpture 13 (gepr.), sign. Lehmann, H. 36 cm; Sammlung Eva Stümpfl/Wien

4187
Gemignani, Valmore
Stehender Windhund
ca. 1911/12
4187,90,11 Firmenmarke Sculpture 13 (gepr.), sign. Gemignani/Latour, H. 26 cm; Doro S 7/1998 ohne Kat., hier Sign. Ang. Gemignani; Doro 4.3.2002, hier Sign. Ang. Latour (!).
4187,307,8 (Firmenmarke ohne Ang.), sign. Latour, H. 25,4 cm; eBay # 1215777476
Bem.: Wenn auch hier zwei mit „Latour" sign. Exemplare nachgewiesen werden können, so ist doch eindeutig der Entwurf dieses Windhundes dem Künstler Valmore Gemignani zuzuordnen, der für GS etliche dieser Tierfiguren entworfen hat. Dies gilt trotz mehrfach mit „Latour" sign. Stücke bei der Höhenvariante Modell-Nr. 4374; sowie die ähnl. Modelle 3859; 4687.
–> vgl. Latour 4374
Lit.: Pinhas, S. 100 (m. Abb.).

4192
Lehmann, Ida
Stehende lesende Biedermeierdame
ca. 1911/12
4192,141,6 (Firmenmarken ohne Ang.), H. 18 cm; WKA 38
4192,282,65 Firmenmarke Adler 15, Originalpick.: „Kgl. Hofl. Anhäuser, Dresden", H. 17 cm; Mehlis 6; Ahlden 102; Henry's 29.4.1998
4192,355,10 (Firmenmarke Adler ohne Ang.), Firmenmarke Sculpture 13, H. 17 cm; WKA 33
4192 (ohne Ang.), Firmenmarke Adler 14, unsign.?, H. 17 cm; RM 21
4192 (ohne Ang.), Firmenmarke Sculpture 13, H. 19 cm; Spik 569
Bem.: Vermutlich identisch mit dem Modell „Lektüre" (ÖMKI 1912; 1913/14).
Ausst.: ÖMKI 1912; 1913/14?

4194
Cherc
Römischer Krieger mit Schild (Medusenhaupt) und Speer (auf Rechtecksockel)
Inschrift: ET MIHI SUNT VIRES ET MEA TELA NOCENT
ca. 1911/12
4194,61,14 Firmenmarke 23, gem. Made in Austria, Ausformung 20er Jahre, Fig. H. 48,3 cm, Sockel 8,9 cm, L. 25,4 cm; eBay # 1447526548 Bem.: Speer fehlt
Bem.: Terrakotta bronzefarben staffiert.
Text der Inschrift: Ich habe Kräfte (Stärke) und meine Wurfgeschosse fügen Schaden zu.

4195
Bormann, Wilhelm
„Kaninchen"
ca. 1911/12
4195,152,16,3 (Firmenmarke ohne Ang.), H. 8,9 cm; KH
4195,329,3 Firmenmarke 19, sign. Borman (!), Ausformung 20er Jahre, H. 8,4 cm; Wien Museum Inv. 190.041
Ausst.: ÖMKI 1912; 1913/14, hier jeweils mit Ang. des Originaltitels „Kaninchen".
Lit.: Neuwirth (1974), S. 60

4200 –> ÜBERSCHNEIDUNG
Petri (d.i. Petrides, János) (Johann)
Kleines Biedermeiermädchen mit Hut und Schirm
ca. 1912/13
4200,6,11 gem. D II, Firmenmarke Sculpture 13, sign. Petri, H. 28,5 cm; Brö 2,1, S. 404 (m. Abb.)
Lit.: Neuwirth (1974), S. 71 (m. Abb.), hier falsche Ang. zu Modell-Nr. etc. des Exemplars der Sammlung Bröhan/ Berlin.

4200 –> ÜBERSCHNEIDUNG
Lavergne, Adolphe-Jean
LAMPE Salome (Tanzende junge Frau)
(auf Rundsockel mit drei Muscheln)
ca. 1912/13
4200,9,6 Firmenmarke Sculpture 13, H. 79 cm; Doro 1.12.2000; Bem.: Dieses Objekt konnte im Original nicht geprüft werden, möglicherweise handelt es sich bei der Modell-Nr. um einen Druckfehler im Kat.
Bem.: Dargestellt ist Salome bei ihrem Tanz. Die drei Lampenfassungen befinden sich jeweils hinter einer Muschel, die den Sockel umsäumen.

4202
Petri (d.i. Petrides, János) (Johann)
Stehender Jockey (auf quaderförmigem Landschaftssockel)
ca. 1912/13
FAYENCE
4202,30,16 gem. S., II, Firmenmarke 19, sign. Petri, Ausformung 20er Jahre, Originalpick.: Sammlung Skalitzki, Wien, H. 21 cm; GK, S. 17, 98 (m. Abb.); Wien Museum Inv. 182.081
ALABASTER
4202,L,2 Plak. A, sign. Petri, H. 24 cm; eBay # 5355571851
Bem.: Dieses Modell wurde in Fayence und Alabaster/ (Kunststein?) ausgeführt.
Ausst.: HMW (jetzt Wien Museum) 1985/86.
Lit.: GK, S. 17, 98 (m. Abb.)

4203
Rosé
Schreitende Bacchantin (Damenhalbakt mit Kelch, einen Leoparden an blumengeschmücktem Halsband führend) (auf mit Blumengirlanden verziertem Ovalsockel)
ca. 1912/13
4203,3,35 (Firmenmarke ohne Ang.), unsign., H. 58,5 cm; Doro 6.4.1999
4203,26,10 Firmenmarke 19, Made in Austria, Ausformung 20er Jahre, H. 58,5 cm; eBay # 2052212939 (KH www.tias.com/stores/frozenbeauty)
ohne Ang., unsign., H. 58 cm; Aukcni Sin Nuselska 40 (21.10.1999, Lot 76)
Lit.: Pinhas, S. 99 (m. Abb.)

4209
Podany, Rudolf
LAMPE Stehender Pierrot mit zwei Pierretten
(auf Rundsockel mit Blumenstrauß)
ca. 1912/13
318,4209,5 gem. ID.2/8, Firmenmarke 19, Made in Austria, Ausformung 20er Jahre, Pick. 1, H. 53,3 cm; eBay # 459946247; 1232387747
(KH www.tias.com/stores/frozenbeauty)

4210
Stehendes Biedermeiermädchen mit großem Hut, Rosenstrauß und Beuteltasche
ca. 1912/13
4210,1,65 Firmenmarke Sculpture 13 (gepr.), H. 43 cm; Mehlis 11

4219
Petri (d.i. Petrides, János) (Johann)
Stehendes Biedermeiermädchen mit Korb, Krug und Blumentopf
ca. 1912/13
4219,31,11 (Firmenmarke ohne Ang.), sign. Petri, H. 16,5 cm; Quit. 31
4219,42,65 gem. 4, Firmenmarke Adler 15, Firmenmarke Sculpture 13 (gepr.), sign. Petri, H. 16 cm; Wien Museum Inv. 219.177
4219,67,11 Firmenmarke Adler 15, Firmenmarke Sculpture 13 (gepr.), sign. Petri, H. 17 cm; Quit. 17
4219,68,6 Firmenmarke 19?, sign. Petri, Ausformung 20er Jahre, H. 16,5 cm; Doro 21.6.1991
4219,95,10 (Firmenmarke ohne Ang.), sign. Petri, H. 17 cm; Herr 46
4219 (ohne Ang.), gem. H., sign. Petri, H. 17 cm; Zeller 39
Lit.: Pinhas, S. 103 (m. Abb.)

4221
Lavergne, Adolphe-Jean
Stehendes Biedermeiermeiermädchen mit Chrysanthemenstrauß
ca. 1911/12
4221,12,10 Firmenmarke Sculpture 13 (gepr.), H. 28 cm; PB

4222
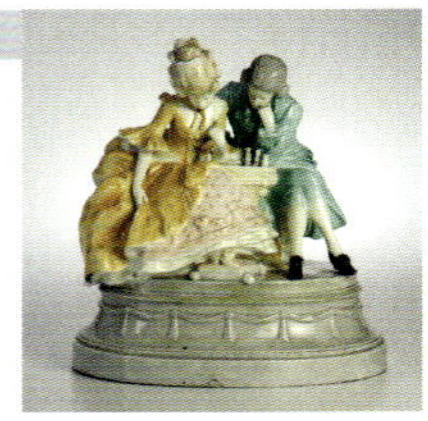
Podola
(d.i. Czapek, Stanislaus)
„Schachmatt" (Sitzendes Rokokopaar beim Schachspiel) (auf girlandenverziertem Ovalsockel)
ca. 1911/12
4222,34,65 Firmenmarke Adler 16, Firmenmarke Sculpture 13, sign. Podola, H. 38,5 cm; Wien Museum Inv. 190.078
Ausst.: Herbstmesse Leipzig 1912
Lit.: N.N.: Die Herbstmesse Leipzig 1912, in: DER 12.1912/13, S. 674, hier mit Ang. des Originaltitels „Schachmatt".

4227
Perigot
Stehendes Mädchen mit Schokoladentasse
(auf Rundsockel)
ca. 1911/12
4227,25,65 Firmenmarke Sculpture 13, H. 29,3 cm; Doro 10.3.2003

4232
Lehmann, Ida
Stehende Dame mit Hut, zwei Blumentöpfe haltend
ca. 1911/12
4232 (Firmenmarke ohne Ang.), H. 27 cm; BM 101
Bem.: Evtl. identisch mit „Vom Garten", ÖMKI 1912.
Ausst.: ÖMKI 1912?

4234
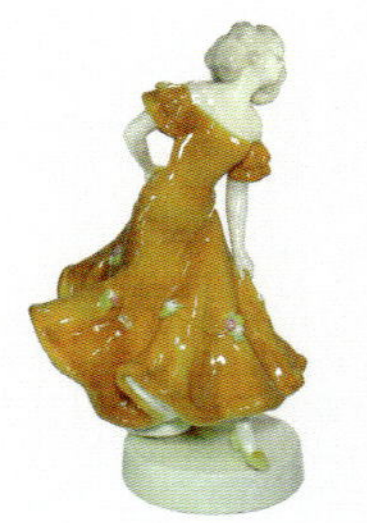
Latour,
(Louis-Marie-Blaise?)/Rosé
Tänzerin (nach links drehend)
ca. 1911/12
FAYENCE
4234,26,11 Firmenmarke Adler 15a, H. 31,5 cm; Sammlung a@goldscheider.net
4234,31,11 Firmenmarke Adler 15, Firmenmarke Sculpture 13, sign. Latour, H. 31,5 cm; Bödiger 236; Zezschwitz 10; Fischer 141
4234,187,25 Firmenmarke 18, sign. Latour, Ausformung ca. 1920, H. 31 cm; Quit. 17; Herr 46; Blank 102; Wendl 47
ALABASTER
4234,M,12 ger., Plak. E, sign. Rosé, Ausführung 20er Jahre, H. 30 cm; Stahl 13.6.1998
Bem.: Dieses Modell wurde in Fayence und Alabaster/ (Kunststein?) ausgeführt.
–> vgl. 4235 (Latour); 4236
–> vgl. .... LAMPE (ohne Nr.), ger. K, gem. F. XXXVII, Firmenmarke 22, unsign., Ausformung 20er Jahre, Gesamth. 62 cm; eBay # 1425611131

4235
Latour, (Louis-Marie-Blaise?)
Tänzerin (nach links drehend)
ca. 1911/12
4235,12,6 gem. 7, Firmenmarke Adler 15, töpf. (gepr.), sign. Latour, H. 23 cm; Wien Museum Inv. 219.172
4235,152,2 gem. III, Firmenmarke Adler 15, Firmenmarke Sculpture 13, sign. Latour, H. 23 cm; eBay # 3309591569
4235,202,76 (2 Firmenmarken ohne Ang.), sign. Latour, H. 24,5 cm; Schloss Hag. 20.9.2003; Bem.: Das AH gibt die Modell-Nr. 4233 an, was wohl ein Lese-/Druckfehler sein dürfte.
4235,240,12 Firmenmarke Adler 14, Firmenmarke Sculpture 13, sign. Latour, H. 23 cm; Wien Museum Inv. 190.142
4235,243,16 (Firmenmarken ohne Ang.), sign. Latour,

H. 23 cm; Stahl 5.9.1992
-> vgl. 4234 (Latour/Rosé); 4236
-> vgl. .... LAMPE (ohne Nr.), ger. K, gem. F. XXXVII, Firmenmarke 22, unsign., Ausformung 20er Jahre, Gesamth. 62 cm; eBay # 1425611131

4236
Latour, (Louis-Marie-Blaise?)
Tänzerin (nach links drehend)
ca. 1911/12
4236,240,12 Firmenmarke Adler 16, Firmenmarke Sculpture 13, H. 22 cm; Wien Museum Inv. 190.142

4237
Schmidt-Kestner, Erich
Reh
ca. 1911/12
4237,25,10 (Firmenmarke ohne Ang.), H. 35 cm; Kaupp 30.11.1996

4241
Pollak, Fritz (Friedrich)
Dame mit Pagen
ca. 1911/12
4241 (ohne Ang.), H. fehlt; Neuwirth (1974), S. 58, 71

4242
Heinemann, Fritz
Schreitende Holländerin mit zwei Milchkannen (auf Rundsockel)
ca. 1911/12
4242,50,6 (Firmenmarken ohne Ang.), H. 37,2 cm; BM 105
Bem.: Diese Figur könnte mit dem „Milchmädchen" (Museum Dessau) identisch sein, welches in TB, Bd. 19, S. 294 expliziert aufgeführt ist.
-> vgl. 4642

4243
Meier, Emil
Stehende Biedermeierdame mit großem Hut
ca. 1911/12
4243,79,? gem. F, Firmenmarke 18, sign. Meier, Ausformung ca. 1920, H. ca. 20 cm; KH

4244
Meier, Emil
FIGUR/LAMPE Stehende Rokokodame mit Blumenkorb (Sommer)
ca. 1910/11
FIGUR
4244,132,65 gem. FS, Firmenmarke 19, Made in Austria, Ausformung 20er Jahre, H. 20,3 cm; eBay # 68394511 ohne Ang., Firmenmarke Adler 15, H. 21 cm; eBay # 195782944
LAMPE
4244,50,10 Firmenmarke Adler 14, Firmenmarke Sculpture 13 (gepr.), sign. Petri (!), Gesamth. 32 cm; Wien Museum Inv. 219.148
4244,181,18 gem. XIV. D 2, Firmenmarke 19, Made in Austria, Ausformung 20er Jahre, H. 21,6 cm, Gesamth. 30,5 cm; eBay # 252234709
-> vgl. 3861
Lit.: Studio YB, 1911, S. 250 (m. Abb.), hier mit Ang. des Entwerfers Emil Meier.

4245
Meier, Emil
Stehende Rokokodame mit Früchtekorb (Herbst)
ca. 1910/11
4245,61,12 (Firmenmarke ohne Ang.), H. 20 cm; Doro 31.5.1990
-> vgl. 3862; 4052

4246
Meier, Emil
Stehende gitarrespielende Rokokodame (Krinoline)
ca. 1911
4246,66,10 Firmenmarke Sculpture 13, Firmenmarke Adler 15, H. 19,5 cm; eBay # 6506821073
4246,94,10 Firmenmarke Sculpture 13, Firmenmarke Adler 16?, H. 19,5 cm; Wendl 40; Ruef 481 (hier unvollständige und ungenaue Ang.)

4247
Meier, Emil

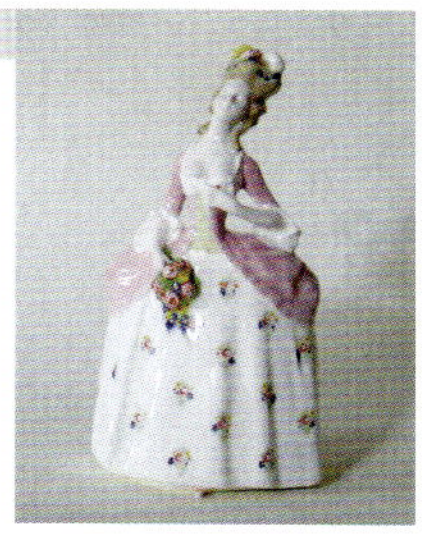

Stehende Rokokodame mit Rosen
ca. 1911
4247,17,11 Firmenmarke Adler 15, Firmenmarke Sculpture 13, sign. Meier, H. 20,1 cm; Sammlung K. Dechant
4247,31,65 Firmenmarke Sculpture 13, sign. Meier, H. 19,5 cm; Doro 15.9.2003
4247,226,6 Firmenmarke 19, Made in Austria, unsign.?, Ausformung 20er Jahre, H. 20 cm; eBay # 3358188180
4247 (ohne Ang.), Firmenmarke Adler 14?, sign. Lehmann (!), H. 19 cm; Kett. 177; Bem.: Die Sign. „Lehmann" ist eindeutig ein Versehen des Modelleurs.
-> vgl. 3890
Lit.: Studio YB, 1911, S. 250 (m. Abb.), hier mit Ang. des Entwerfers Emil Meier; Pinhas, S. 101 (m. Abb.).

4250
Paltier/Paltir/Polter
Stehende Biedermeierdame (daneben kleines, rundes Tischchen mit Blumenstrauß)
ca. 1911/12
4250,10,35 (Firmenmarke ohne Ang.), sign. „Paltier" (wohl Polter), H. 34,3 cm; Quit. 29
4250,82,6 Firmenmarke Sculpture 13, sign. „Paltir" (wohl Polter), H. 34 cm; WKA 14
Bem.: Die zwei bislang mit „Paltier/Paltir" nachweisbaren Schreibformen, können eindeutig den Figuren zugeordnet werden, die mit Polter sign. sind.

4252
Latour, (Louis-Marie-Blaise?)
Biedermeiertänzerin (auf Rundsockel)
ca. 1911/12
4252,78,13 Firmenmarke Adler 16?, H. 31 cm; KaM 83
4252,105,10 Firmenmarke Sculpture 13, H. 31 cm; eBay # 3157381708; 3161977846
4252,197,10 Firmenmarke Adler 16, Firmenmarke Sculpture 13, H. 32,5 cm; GK, S. 15, 89 (m. Abb.); Wien Museum Inv. 182.082
4252,262,5 Firmenmarke 13a, H. 32 cm; WKA 22
4252,297,35 gem. F (Firmenmarke ohne Ang.), H. 31,8 cm; Soth. Chester 7.–9.6.1988
4252,350,12 (Firmenmarke ohne Ang.), H. 30,4 cm; Doro 15.9.2003
Ausst.: HMW (jetzt Wien Museum) 1985/86.
-> vgl. 4158; 4253
Lit.: GK, S. 15, 89 (m. Abb.); Alt-Wien, S. 463.

4253
Latour, (Louis-Marie-Blaise?)
Biedermeiertänzerin (auf Rundsockel)
ca. 1911/12
FAYENCE
4253,135,11 (Firmenmarke ohne Ang.), sign. Latour, H. 23 cm; Henry's 27.10.2001
4253,192,16 (Firmenmarke ohne Ang.), sign. Latour, H. 23 cm; Mehlis 6.3.2004 (Sonderaukt. Sammlung Langer)
ALABASTER
4253,W,11 ger., (Firmenmarke Plak. ohne Ang.), H. 23,8 cm; Doro 15.3.2005
4253,W,40 ger., (Plak. A?), unsign.?, H. 24,2 cm; Kett. 64
Bem.: Dieses Modell wurde in Fayence und Alabaster ausgeführt.
-> vgl. 4158; 4252

4256
Meier, (Emil oder Meier-Michel, Johanna)
Sitzende Dame auf Sofa (mit Blumenvase)
ca. 1912/13
4256,65,11 Firmenmarke Adler 14?, Firmenmarke Sculpture 13, H. 21,5 cm; Doro 18.11.1997, Bem.: Laut Ang. des AH sign. mit „Künstlermonogramm".
4256,85,8 Firmenmarke 19, sign. Meier, Ausformung 20er Jahre, H. ca. 21,5 cm, L. 25 cm; GK, S. 17, 99 (m. Abb.); Samml. Eva Stümpfl/Wien
Ausst.: HMW (jetzt Wien Museum) 1985/86.
Lit.: GK, S. 17, 99 (m. Abb.)

4262
Schimmelpfennig, (Oswald)
Sitzender Frauenakt auf Stuhl (mit jungem Löwen)
ca. 1912/13
4262,27,13 (Firmenmarke ohne Ang.), H. 21,2 cm; BM 62

4264
Petri (d.i. Petrides, János) (Johann)

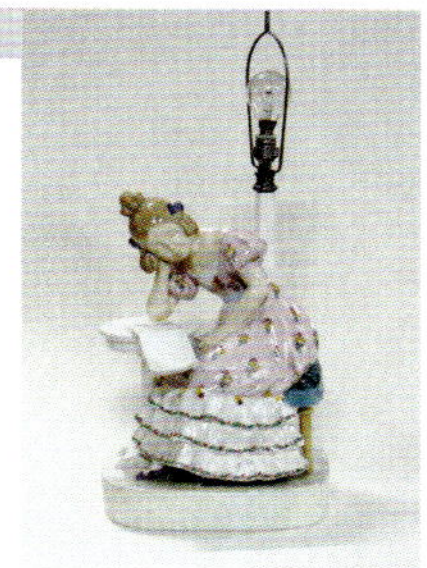

LAMPE Sitzendes lesendes Biedermeiermädchen
ca. 1912/13
4264,12,10 Firmenmarke 15, Firmenmarke Sculpture 13, sign. Petri, Gesamth. 47,4 cm, H. 30,2 cm; MAK Wien, Inv. Ke 10521-26

4267
Polter
Sitzende Tänzerin, sich einen Schuh ausziehend (auf Hocker/Stuhl)
ca. 1912/13
Hocker
4267,147,9 gem. XXIX, Firmenmarke Sculpture 13, H. 30,2 cm; Neuwirth (1974), S. 58, 71, 73 (m. Abb.) ohne Ang., gem. IX., Firmenmarke 18, unsign.?, Ausformung ca. 1920, H. 30,5 cm; eBay # 2002702668
Stuhl
4267,82,14 (Firmenmarke ohne Ang.), H. 30 cm; WKH
Bem.: Dieses Modell wurde in zwei verschiedenen Varianten unter der gleichen Modell-Nr. angeboten, einmal mit Hocker bzw. mit Stuhl.
-> vgl. 4895/4897? (auf Hocker)

4268
Pfister, (Émile?)
Stehende, sich kämmende junge Frau (Hera?) mit Pfau (auf profiliertem Rundsockel)
ca. 1912/13
4268,3,7 Firmenmarke Sculpture 13, H. 90,5 cm; Doro 15.5.2003; 31.5.2006
Bem.: Die Darstellung lässt an Hera, die oberste griech. Göttin, denken. Ihr Attribut war der Pfau. Vgl. hierzu Lit.: Lexikon der Antike, Hg.: Johannes Irmscher, Berlin 1999 (Digitale Bibliothek; Bd. 18); Goldmann Lexikon, CD-ROM Ausg., München 1998; Ebenso gilt der Pfau auch als „Symboltier für Eitelkeit, Luxus und Hochmut (Superbia)", zit. nach: Lexikon der Symbole: Pfau, S. 2. (Digitale Bibliothek; Bd. 16): Knaurs Lexikon der Symbole, Berlin 2000, S. 817.

4271
Polter
LAMPE Tanzendes Rokokopaar (Dame und Kavalier) (auf mit Blumengirlanden verziertem Rundsockel)
ca. 1912/13
4271,15,65 (Firmenmarke ohne Ang.), sign. „Paltier", H. 55 cm; (ohne Elektrifizierung); Doro 7.3.2001
-> vgl. 4915
-> vgl. 4305= ähnl. Modell
-> vgl. 4539= ähnl. Modell

4278
Petrides, János (Johann) (zugeschr.)
BÜSTE Holländermädchen mit Haube
ALABASTER/Steinsockel
ca. 1912/13
4278,F,21 ger., Plak. C, unsign., Ausformung 20er Jahre, H. 19,4 cm; Doro 25.4.2000
-> vgl. 4664= ähnl. Modell

4283
Petri (d.i. Petrides, János) (Johann)
BÜSTE Bub mit Hut
ALABASTER
ca. 1912/13
4283,M,6 ger., (Plak. E?), sign. Petri, Ausformung 20er Jahre? H. ca. 18 cm; WKH

4286
Polter
Kaffee trinkendes Mädchen, an einem Tisch sitzend (auf blütenverziertem Rechtecksockel)
ca. 1912/13
4286,80,10 (Firmenmarke Adler ohne Ang.), Firmenmarke Sculpture 13, H. 28 cm; Nagel 379
4286,221,10 ger. H, gem. I, Firmenmarke 19, Ausformung 20er Jahre, H. 28 cm; PB Wien 19
4286,222,5 Firmenmarke 19, Ausformung 20er Jahre, H. 26 cm; WKA 40
4286,311,15 Firmenmarke 19, Made in Austria, Hand dec., Ausformung 20er Jahre, H. 26 cm; eBay # 3721151070

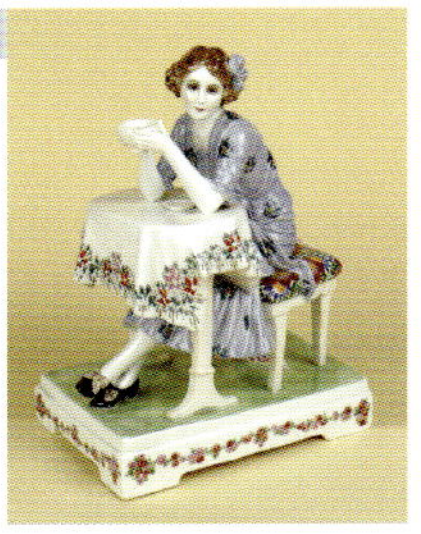

4287
Brynski
Stehender Knabe mit zwei Enten, die er in den Armen hält (auf Ovalsockel)
ca. 1912/13
4287,14,10 Firmenmarke Sculpture 13, sign. Brynski, H. 34,3 cm; Doro 10.3.2003
4287,36,14 Firmenmarke Adler 16, Firmenmarke Sculpture 13, sign. Brynski, H. 33,5 cm; Zeller 74; city-antik
4287 (ohne Ang.), Firmenmarke Sculpture 13, sign. Brynski, H. 34 cm; Ahlden 33

4295
Polter
Stehender Knabe (Putto) mit Blumenstrauß und Korb auf dem Rücken (auf Rundsockel)
ca. 1912/13
4295,11,9 ger. K, gem. XXXIX, Firmenmarke 22, Ausformung 20er Jahre, H. 21 cm; Wien Museum Inv. 186.673

4301
Schoerk, Hans
BÜSTE Frau mit Kopftuch (nach links blickend)
ALABASTER
ca. 1912/13
4301,JH,53 Plak. E, sign. Scherk (!), Ausformung 20er Jahre, H. 20,3 cm; eBay # 31217127; 78156090
4301,W,23 Plak. A, sign. Scherk (!), H. 24 cm; Wien Museum Inv. 220.792

4302
Schoerk, Hans
BÜSTE Frau mit Kopftuch (nach rechts blickend)
ALABASTER
ca. 1912/13
4302,Sch.,52 (Firmenmarke Plak. ohne Ang.), sign. Scherk (!), H. 25,1 cm; Doro 14.9.2001

4304
Monginot, Charlotte
FIGUR/LAMPE Schlafendes Kind auf Bücherstapel (Buchstütze?)
ca. 1912/13
FIGUR
4304,80,8 Firmenmarke 24, Ausformung 20er Jahre, H. 17,5 cm; eBay # 1467215453
4304,115,24 gepr./ger. „A", Firmenmarke 19, Ausformung 20er Jahre, H. 18 cm; Wien Museum Inv. 186.674
LAMPE
4304,150,65 gem. X.D.1., Firmenmarke 19, Made in Austria, Hand dec., Ausformung 20er Jahre, Gesamth. 40,6 cm; eBay # 54040617
4304,152,11 (Firmenmarke ohne Ang.), H. fehlt; Doro 16.6.1992; WKH
Lit.: Neuwirth (1974), S. 58, hier nur Ang. „Buchstütze" (Kind).

4305
LAMPE Scherzendes Rokokotrio (Zwei Damen und ein Herr an mit Blumengirlanden verzierter Säule) (auf volutenartigem Sockel)
ca. 1912/13
4305,62,65 gem. 2, (VI?), Firmenmarke Sculpture 13, H. 60 cm; WKA 12
-> vgl. 4539= ähnl. Modell
-> vgl. 4915= ähnl. Modell

4310
Latour, (Louis-Marie-Blaise?)
Diana mit Pfeil und Bogen neben Windhund stehend (auf Vierkantsockel)
ca. 1912
4310,27,13 gem. 2, (Firmenmarke ohne Ang.), H. 38 cm; Quit. 14; Bem.: Bogen fehlt
4310,199,14 (Firmenmarke ohne Ang.), H. 43,8 cm; Soth. Chester 23.–25.5.1989
4310,273,1 Firmenmarke 19, Ausformung 20er Jahre, H. 43 cm; KaM 116
ohne Ang., Firmenmarke Sculpture 13, unsign.?, H. 44 cm; BM 64; 65
Bem.: Der Bogen der Diana ist aus Metall (Messing?) gefertigt.
-> vgl. 4164

4313
Fleury
Stehende Dame (Krinoline) mit Fächer und Kakadu
ca. 1912/13
4313,24,25 gem. blau VIII, Firmenmarke 24, Firmenmarke 19, Ausformung 20er Jahre, H. 26 cm; Wien Museum Inv. 190.109

4315
Brynski / Brynsky
Zwei Putti mit störrischem Widder (auf Rechtecksockel)
ca. 1912/13
4315,1,10 Firmenmarke Adler 15, Firmenmarke Sculpture 13 (gepr.), sign. Brysnki, H. 24,1 cm; Soth. com. # 32 AI
4315,42,10 (Firmenmarke ohne Ang.), unsign.?, L. 24 cm; Nagel 303
4315,54,10 gem. 1, Firmenmarke Adler 15, Firmenmarke Sculpture 13 (gepr.), sign. Brysnky (!), H. 23,8 cm, L. 23,5 cm; Sammlung K. Dechant
ohne Ang., Firmenmarke Adler 15, Firmenmarke Sculpture 13 (gepr.), sign. Brynsky (!), H. 24,1 cm, L. 25,4 cm; eBay # 1202011585 (KH www.tias.com/stores/frozen-beauty)

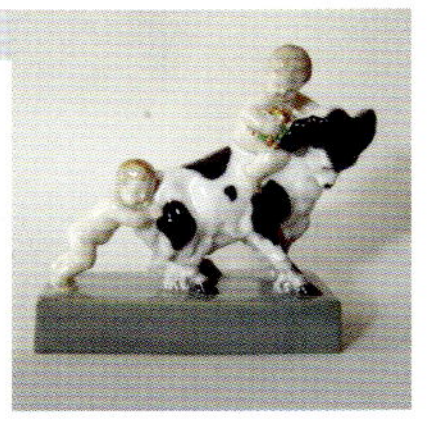

4317
Polter
Biedermeiermädchen auf Stuhl stehend, mit Dackel
ca. 1912/13
4317,36,11 Firmenmarke Adler 15, Firmenmarke Sculpture 13, H. 30 cm; KH
Bem.: -> vgl. Petri .... Biedermeiermädchen auf Stuhl kniend, mit Katze, = ähnl. Modell.

4318
Mallener, Theodor
Stehender Knabe in Tracht mit Hut, zwei Blumenkörbe tragend
ca. 1912/13
4318,45,2 Firmenmarke Adler 15, Firmenmarke Sculpture 13, H. 20,5 cm; Fischer 151
4318,60,6 Firmenmarke Adler 15, Firmenmarke Sculpture 13, H. 19,5 cm; eBay # 25221212362; Zezschwitz 16
4318,67,27 (Firmenmarke ohne Ang.), H. 19,4 cm; Doro 8.3.2004
4318,156,55 gem. N. 12, Firmenmarke Adler 16, Firmenmarke Sculpture 13, H. 18,5 cm; eBay # 3343168274
4318,180,10 gem. MJ., Firmenmarke Adler 16, Firmenmarke Sculpture 13, H. 19,6 cm; Wien Museum Inv. 219.182
4318,233,5 Firmenmarke Sculpture 13, H. 19,4 cm; Doro 4.3.2003
-> vgl. SCHALE 5437
Lit.: Studio YB, 1914, S. 191 (m. Abb.) , hier mit Ang. des Entwerfers Theodor Mallener.

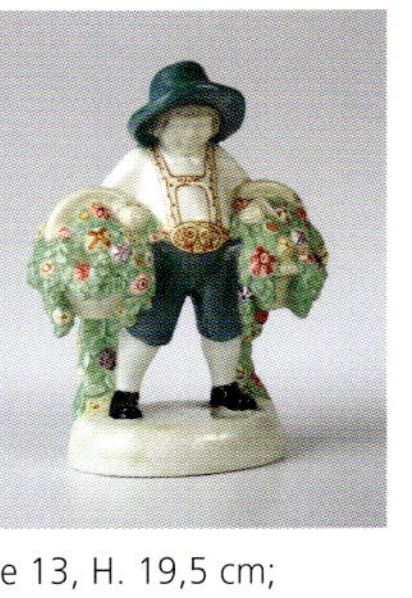

4319
Mallener, Theodor
Stehender Knabe in Tracht mit Hut, einen Hasen beim Ohr ziehend (auf Quadratsockel)
ca. 1912/13
4319,99,36 gem. IV, Dl., Firmenmarke Adler 15, Firmenmarke Sculpture 13 (gepr.), H. 17 cm; PB Wien 19
4319,187,15 Firmenmarke Adler 15, Firmenmarke Sculpture 13 (gepr.), H. 17 cm; WKA 33

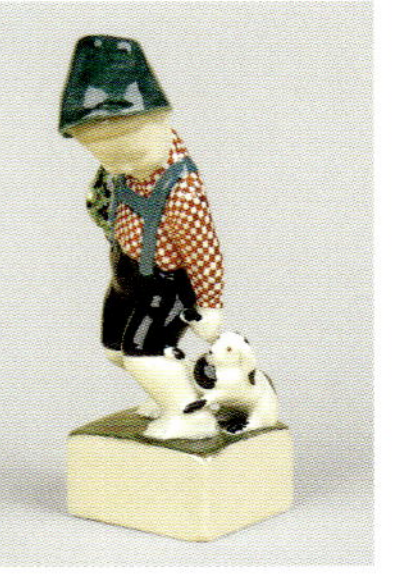

4324 s. 4624

4332
Schimmelpfennig, (Oswald)
UHR „Herzensfrage"
(Liebespaar der Antike als Akt/Halbakt)
ALABASTER / KUNSTSTEIN
Inschrift: HERZENSFRAGE
ca. 1912/13
4332,W,5 Plak. A, sign. Schimelpfenig (!), H. 37 cm; Sammlung Holzer, Regensburg; Bem.: Lenzkirch Uhrwerk.
Bem.: Eine mit „O. Schimmelpfennig / Berlin" sign. Uhr, Inschrift „QUESTIO" (ähnl. sowohl hinsichtlich des Genres, als auch in der Ausführung, H. 59,1 cm) wurde am 17.3.1995 von Soth. NY angeboten. Soth. gibt als Material Marmor an und nennt einen falschen Vornamen (Oscar) des Künstlers. Es handelt sich jedoch eindeutig um den Berliner Bildhauer Oswald Schimmelpfennig, der 1902 und 1904 an der Großen Berliner Kunst-Ausst. beteiligt war und von dem heute noch im öffentlichen Raum eine Brunnenplastik „Gänsejunge" (Bronze, datiert 1939) in Berlin (Eberswalderstrasse, Krankenhaus) steht.

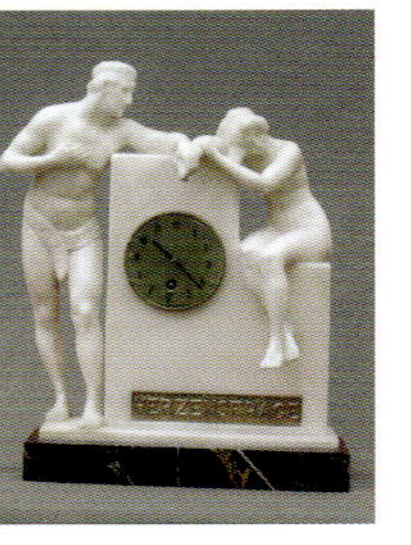

4341
Meier, (Emil oder Meier-Michel, Johanna)
Biedermeiertänzerin im Rosenblütenkleid (auf Rundsockel) (Anna Pawlowa?)
ca. 1912/13
4341,36,15 Firmenmarke Sculpture 13, (und Firmenmarke ohne Ang.), sign. Meier, H. 46 cm; Doro 24.10.1995
4341,38,5 (Firmenmarke ohne Ang.), sign. Meier, H. 46,3 cm; Doro 8.3.2004
Bem.: Dargestellt ist vermutlich die Tänzerin Anna Pawlowa (1882–1931), vgl. hierzu: WeltM, 3.1927, Nr. 3, S. 72 (m. Abb.); DKT, S. 1, Bild 5 „Anna Pawlowa in einer Biedermeierszene".
-> vgl. 4616; 4619
Lit.: Pinhas, S. 101 (m. Abb.).

4343
Pecheur
BÜSTE Biedermeiermädchen mit Schutenhut
ALABASTER
ca. 1912/13
4343,WR,22 Plak. A, H. 25,5 cm; Wien Museum Inv. 186.675

4347
Latour, (Louis-Marie-Blaise?)
BÜSTE Biedermeiermädchen mit Schutenhut
ALABASTER
ca. 1912/13
4347,F,83 ger., (Plak. C/E?), Ausformung 20er Jahre, H. 15,5 cm; Doro 20.4.1993; 21.5.1997

4355
Petrides, János (Johann)
„Die kleine Strickerin" (Stehendes, strickendes Biedermeiermädchen mit Hut)
ca. 1913
4355,258,10 Firmenmarke Adler 16, unsign.?, H. 18 cm; Wendl 41
4355,269,10 Firmenmarke Adler 14, Firmenmarke Sculpture 13, sign. Petri, H. 18,6 cm; Quit. 17
4355,271,12 gem. lig. hellbraun HT, Firmenmarke Adler 16, Firmenmarke Sculpture 13, sign. Petri, H. 18 cm; Brö 2,1, S. 404, 405 (m. Abb.); BröPK, S. 147; Neuwirth (1974), S. 70 (m. Abb.); Himmelheber, G.: Kunst des Biedermeier 1815–1835. München 1998, S. 288 (m. Abb.)
4355,313,65 Firmenmarke Adler 16, unsign.?, H. fehlt; eBay # 881275757
4355,371,2 Firmenmarke 19, sign. Petri, Ausformung 20er Jahre, H. 17,5 cm; Heickmann 18
Lit.: Studio YB, 1914, S. 190 (m. Abb.), hier mit Ang. des Originaltitels „Die kleine Strickerin" und mit Ang. des Entwerfers „Joseph Petrides". Bem.: Hiermit kann u.a. belegt werden, dass es sich bei den mit „Petri" sign. Modellen um den, auch für Zsolnay tätigen, ungar. Bildhauer und Keramiker János Petrides handelt, und nicht um den Berliner Künstler Otto Petri, der gerne und oft von Auktionshäusern als Entwerfer für GS genannt wird; Neuwirth (1974), S. 58, 70 (m. Abb.), hier auch mit Hinweis auf (Otto?) Petri.

4359
Mallener, Theodor
„Jeunesse" (Stehendes Biedermeiermädchen mit Rosenstrauß)
ca. 1913
4359,87,(7?) gem. S. F. II. I., Firmenmarke 19, Ausformung 20er Jahre, H. 19 cm; eBay # 55832767
Ausst.: ÖMKI 1913/14, hier mit Ang. des Originaltitels „Jeunesse".
Lit.: Studio YB, 1914, S. 190 (m. Abb.), hier mit Ang. des Originaltitels „Jeunesse"; Neuwirth (1974), S. 67.

4361
Petrides, János (Johann)
„Vorbereitung zum Tanz" (Stehendes Biedermeiermädchen mit Handspiegel)
ca. 1913
4361 (ohne Ang.), gem. H19, (Firmenmarke ohne Ang.), sign. Petri, H. 21 cm; Doro 31.5.1990
Lit.: Studio YB, 1914, S. 190 (m. Abb.), hier mit Ang. des Originaltitels „Vorbereitung zum Tanz" und mit Ang. des Entwerfers „Joseph Petrides". -> vgl. auch Lit. bei Modell-Nr. 4355.

4362
Hentschel (d.i. Handzel, August)
Sitzende Biedermeierdame mit blumenverziertem Schutenhut
ca. 1913
4362,17,10 gem. IX. K, Firmenmarke Adler 15, Firmenmarke Sculpture 13 (gepr.), sign. Hentschel, H. 30 cm; eBay # 1216680000; Doro 14.9.2001
4362,21,6 Firmenmarke Adler 15, Firmenmarke Sculpture 13, sign. Hentschel, H. 29,8 cm; Doro 20.12.1994; WKH
4362,22,13 Firmenmarke Adler 15, Firmenmarke Sculpture 13, sign. Hentschel, H. 30,2 cm; Doro 8.3.2004
Bem.: Zum Künstler vgl. Bem. bei Modell-Nr. 4390.
Lit.: Pinhas, S. 100 (m. Abb.).

4364
Wernekinck, Sigismund
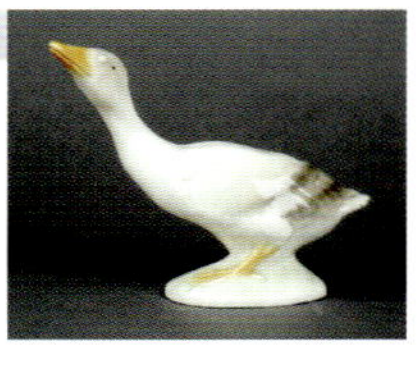
Gans (auf Ovalsockel)
ca. 1910
4364,20,65 Firmenmarke 15, sign. sign. Wernekinck, H. 14 cm, L. 19 cm; eBay # 6532961961
4364,59,10 Firmenmarke 10, Firmenmarke Adler 14, sign. Wernekinck, H. 14 cm; PB Wien 19
4364,111,25 Firmenmarke 6, sign. Wernekink (!), H. 14 cm; Wien Museum Inv. 190.044
4364,125,7 gem. lig. AS, Firmenmarke 22, sign. Wernekink (!), Ausformung 20er Jahre, H. 14 cm, L. 18 cm; Brö 2,1, S. 404; BröPK, S. 146, hier mit abweichenden Maßen H. 13,5 cm, L. 19 cm
Lit.: DK, 22.1910, S. 450 (m. Abb.), hier die Ang., dass dieses Modell auch von der Firma Metzler & Ortloff (Ilmenau) ausgeführt wurde; KeramRS, 19.1911, S. 422; Neuwirth (1974), S. 58, 76.

4365
Meier-Michel, Johanna
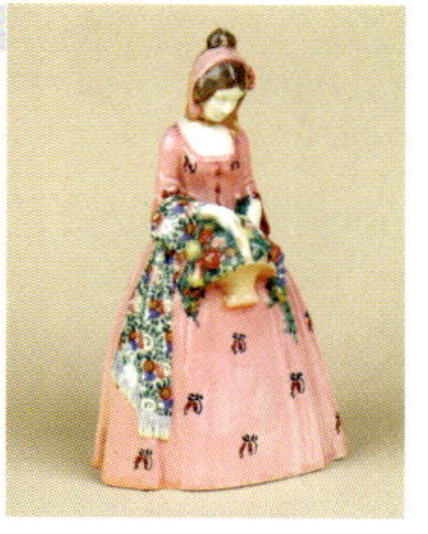
„Frühling" (Stehende Biedermeierdame mit Blumenkorb)
ca. 1913
4365,24,6 Firmenmarke Sculpture 13, sign. Meier, H. 32 cm; Doro 21.5.2003
4365,26,2 (Firmenmarke ohne Ang.), sign. Meier, H. 31,8 cm; Doro 14.9.2001
4365,47,65 (Firmenmarke ohne Ang.), unsign.?, H. 31,5 cm; Schuler 64
4365,84,15 Firmenmarke Adler 16, Firmenmarke Sculpture 13, sign. Meier, H. 31 cm; Doro 3.5.1995
-> vgl. 4563
Pendant -> vgl. 4366
Lit.: Studio YB, 1914, S. 193 (m. Abb.), hier mit Ang. des Originaltitels „Frühling" und mit Ang. der Entwerferin Johanna Meier-Michel.

4366
Meier-Michel, Johanna
„Sommer" (Stehende Biedermeierdame mit Hündchen)
ca. 1913
4366,17,11 Firmenmarke Adler 15, Firmenmarke Sculpture 13, sign. Meier, H. 30,5 cm; Brö 2,1, S. 404 (m. Abb.); BröPK, S. 147; Neuwirth (1974), S. 70 (m. Abb.)
4366,45,2 Firmenmarke Adler 16?, Firmenmarke Sculpture 13, unsign.?, H. 30 cm; WKA 22
4366,103,8 (Firmenmarke ohne Ang.), sign. Meier, H. 28,6 cm; Doro 15.9.2003
4366 (ohne Ang.), Firmenmarke Adler 15, sign. Meier, H. 30 cm; Ahlden 104
-> vgl. .... (H. ca. 21 cm); Nagel 347; Ahlden 79
Pendant -> vgl. 4365
Lit.: Studio YB, 1914, S. 193 (m. Abb.), hier mit Ang. des Originaltitels „Sommer" und mit Ang. der Entwerferin Johanna Meier-Michel; Neuwirth (1974), S. 58, 68, 70 (m. Abb.)

4372
Meier, Emil
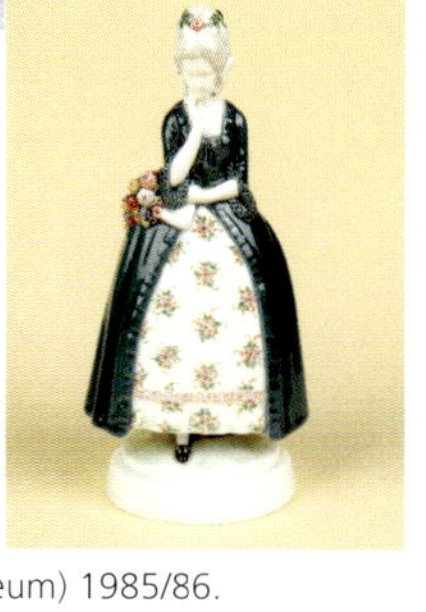
„Pensée" (Stehende Rokokodame in Gedanken)
ca. 1913
4372,106,1 gem. F., Firmenmarke 19, Ausformung 20er Jahre, H. 32 cm; GK, S. 17 (hier falsche Zuschreibung an J. Meier-Michel), 100 (m. Abb.); Wien Museum Inv. 223.733
Ausst.: HMW (jetzt Wien Museum) 1985/86.
-> vgl. 4622
Lit.: Studio YB, 1914, S. 189 (m. Abb.), hier mit Ang. des Originaltitels „Pensée" und mit Ang. des Entwerfers Emil Meier; Lit.: GK, S. 17, 100 (m. Abb.).

4374
Gemignani, Valmore / Latour, (Louis-Marie-Blaise?)
Stehender Windhund
ca. 1911/12
4374,171,10 Firmenmarke Sculpture 13 (gepr.), sign. „LaTour", H. 19,7 cm; SLW 82
4374,372,25 ger. B, (Firmenmarke ohne Ang.), sign. Latour, H. 19 cm, L. 39 cm; Kett. 39
4374 (ohne Ang.), sign. Latour, H. 20 cm; KaM 171 ohne Ang., sign. Latour, H. 20 cm; Schlüter 2676
Bem.: Wenn auch hier mehrere mit „Latour" sign. Exemplare nachgewiesen werden können, so ist doch eindeutig der Entwurf dieses Windhundes dem Künstler Valmore Gemignani zuzuordnen, der für GS etliche dieser Tierfiguren entworfen hat. Dies belegen die Höhenvariante Modell-Nr. 4187, sowie die ähnl. Modelle 3859; 4687.
-> vgl. Gemignani / Latour 4187

4376
Hentschel (d.i. Handzel, August)
LAMPE Stehendes Chinesenmädchen einen Topf haltend (auf Rechtecksockel)
ca. 1913
4376 (Firmenmarke ohne Ang.), sign. Hentschel, H. 55 cm; Pinhas, S. 102 (m. Abb.)
-> vgl. 5587
Pendant -> vgl. 4377

4377
Hentschel (d.i. Handzel, August)
LAMPE Stehendes Chinesenmädchen mit Fächer (auf Rechtecksockel)
ca. 1913
4377,143,74 Firmenmarke 19, unsign.?, Ausformung 20er Jahre, H. 58,4 cm; eBay # 250014678312
-> vgl. 5588
Pendant -> vgl. 4376

4380
Podola (d.i. Czapek, Stanislaus)
Tanzende junge Frau mit Leopard (Bacchantin)
ca. 1913
4380,153,10 gem. IV, F. D. 2., Firmenmarke 19, Made in Austria, Ausformung 20er Jahre, sign. Podola, H. 59 cm; Soth. com. # 338E

4383
Wernekinck, Sigismund
Zwei Wüstenfüchse
ca. 1913
4383,42,13 Firmenmarke Adler 15, H. 16,1 cm; Scheffler S. 198, KGW Mus. Berlin Inv. WA 144; Neuwirth (1974), 37 (m. Abb.), 76

4386
Wernekinck, Sigismund
Haubentaucher mit zwei Jungen
ca. 1913
4386,56,13 FG Wien (Firmenmarke ohne Ang.), H. 20,5 cm; Scheffler S. 198; KGW Mus. Berlin Inv. WA 143
4386,131,10 gem. FH, Firmenmarke 19, Ausformung 20er Jahre, H. 20,2 cm; WKA 16
4386,207,65 ger. K, gem. lig. AH, Firmenmarke 22, H. 20,3 cm; eBay # 3276419794
4386,353,8 Firmenmarke 19, Made in Austria, Hand dec., Ausformung 20er Jahre, H. 19,5 cm; eBay # 1284478884
-> vgl. 5081 DECKELDOSE
Lit.: N.N.: Keramik und Glas auf der Intern. Baufach-Ausst. in Leipzig, in: KeramRS, 21.1913, S. 473, „FG, Wien, führt eine Fülle von künstlerischer Kleinplastik vor, darunter charakteristische Trachtenfiguren und vorzügliche Tierfiguren, von denen der „Haubentaucher" von Wernekinck besonders erwähnt sei"; Neuwirth (1974), S. 58, 76.

4387
Pollak, Fritz
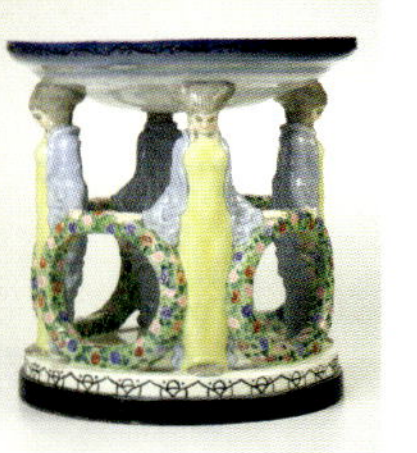
SCHALE Vier stehende Frauengestalten mit Blumenkränzen (Tafelaufsatz)
ca. 1913
4387,10,8 Firmenmarke 19, Ausformung 20er Jahre, H. 29,5 cm, Ø 29,3 cm; GK, S. 18, 38 (m. Abb.); Wien Museum Inv. 222.553
Ausst.: HMW (jetzt Wien Museum) 1985/86; SSH 1992.
Lit.: GK, S. 18, 38 (m. Abb.); GSN, S. 24, 25 (m. Abb.).
GS neu: Modell-Nr. 42 „Schale des Frühlings" (H. 27 cm), Aufl. 1000 Stück

4390

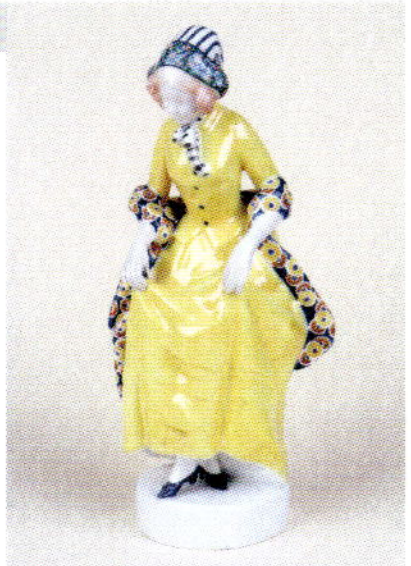

Hentschel
(d.i. Handzel, August)
„Im Regen" (Biedermeierdame mit Hut, den Rock anhebend)
ca. 1913
4390,15,2 (Firmenmarke ohne Ang.), unsign.?, H. 29 cm; Stahl 20.4.1991
4390,20,11 gem. DI, *, Firmenmarke Adler 15, Firmenmarke Sculpture 13, sign. Hentschel, H. 27,5 cm; Wien Museum Inv. 219.170
4390,76,2 Firmenmarke Sculpture 13 (gepr.), (und Firmenmarke ohne Ang.), sign. Hentschel, H. 27,5 cm; Doro 23.5.1991
4390,78,13 gem. F., Firmenmarke Adler 15, Firmenmarke Sculpture 13 (gepr.), sign. Hentschel, H. 28 cm; Soth. com. # 1ANP; # 30DA
4390,118,2 gem. III, Firmenmarke Adler 15, Firmenmarke Sculpture 13 (gepr.), sign. Hentschel, H. 27,9 cm; city-antik
4390,134,2 gem. K, Firmenmarke Adler 15, Firmenmarke Sculpture 13, sign. Hentschel, H. 29 cm; PB Wien 19
4390,152,2 Firmenmarke Sculpture 13, sign. Hentschel, H. 27,2 cm; BM 62
4390,277,? gepr. M, gem. HAT, (Firmenmarke ohne Ang.), sign. Hentschel, H. 28 cm; Z 39; Z 40
4390,302,10 (Firmenmarken ohne Ang.), sign. Hentschel, H. 27 cm; Doro 10.4.1997
4390,488,6 gem. IV., gem. lig. AH, Firmenmarke 19, sign. Hentschel, Ausformung 20er Jahre, H. 27 cm; eBay # 3716090421
Ausst.: ÖMKI 1913/14, hier mit Ang. des Originaltitels und der „falschen" Künstlerang. „Hentschel".
Bem.: Wie die Auflistung der einzelnen Exemplare zeigt, wurde bei GS immer die Sign. „Hentschel" verwendet. Vielleicht um mit dem für Meissen tätigen und bekannteren Entwerfer Konrad Hentschel (berühmt für seine „Hentschel-Kinder") eine Verbindung herzustellen. Eindeutig kommt aber hier der Künstler August Handzel in Frage, lässt sich dieses Modell doch durch die angegebene Lit. korrekt zuweisen.
Lit.: Studio YB, 1914, S. 191 (m. Abb.), hier mit Ang. des Entwerfers August Handzel.

4393

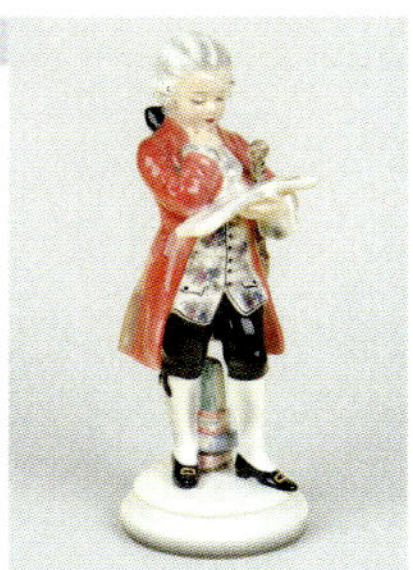

Perigot
„Mozart" (Stehende Figur als Knabe mit Geige, Notenheft und ohne/mit Bücherstapel)
ca. 1913
ohne Bücherstapel
4393,89,6 Firmenmarke Sculpture 13, (und Firmenmarke ohne Ang.), H. 42,8 cm; Doro 1.12.1998
4393,122,65 (Firmenmarke Adler ohne Ang.), Firmenmarke Sculpture 13, H. 43,1 cm; WKA 9
mit Bücherstapel
4393,521,25 (Firmenmarke ohne Ang.), H. 43 cm; Nagel 381
4393,753,8 Firmenmarke 19, Ausformung 20er Jahre, H. 42 cm; Heickmann 8; Doro 16.4.1996
4393,808,8 Firmenmarke 19, Ausformung 20er Jahre, H. 40 cm; Karrenbauer 176
Bem.: Ähnl. Darstellungen (Mozart mit Geige) finden sich auch bei L.-E. Barrias und M. Debut, vgl. hierzu: Berman, Vol. 3, (Nr. 2619: 2622 jeweils mit Abb.).
-> vgl. 4610; 5519
Lit.: MW, 11.1929/30, Nr. 8, S. V (mit Abb. „mit Bücherstapel"), hier mit Ang. des Originaltitels „Mozart"; GSN, Erg. Bl. (o. Pag., m. Abb.) und Preisliste 1992. GS neu ohne Modell-Nr. „Mozart" (H. 25 cm), Aufl. 300 Stück

4401
Polter
Stehendes Biedermeiermädchen mit Haube und Blumen (auf mit Blumengirlanden verziertem Rundsockel)
ca. 1913
4401,24,10 (Firmenmarke Adler ohne Ang.), Firmenmarke Sculpture 13, unsign., H. 21,5 cm; Schloss Rick. 65
4401,43,11 (Firmenmarke ohne Ang.), H. 21,1 cm; Doro 14.9.2001

4406
Beaume (d.i. Bormann, Wilhelm) (zugeschr.)
Stehender Windhund (auf Ovalsockel)
ca. 1905/06
4406,40,10 FG (Firmenmarke ohne Ang.), unsign., H. 24 cm, L. 25,4 cm; eBay # 36891712
Bem.: Bei dieser Figur handelt es sich um die linke Einzelfigur von Modell-Nr. 3416.
-> vgl. 3416= zweier GRUPPE

4407
Beaume (d.i. Bormann, Wilhelm)
Stehender Windhund (Barsoi/Greyhound) (auf Ovalsockel)
ca. 1910/12
4407,166,10 (Firmenmarke ohne Ang.), sign. Beaume, H. 21 cm, L. 27 cm; Metz 34
4407,186,20 Firmenmarke 15, sign. Beaume, H. 20,5 cm; Doro S FVK
4407,204,65 (Firmenmarke Sculpture 13?), sign. Beaume, H. 20 cm, L. 26,5 cm; BM 69
4407,508,16 Firmenmarke 19, Made in Austria, sign. Beaume, Ausformung 20er Jahre, H. 20 cm, L. 25 cm; Wien Museum Inv. 190.021
4407,514,11 gem. F., Firmenmarke 19, Made in Austria
4407,680,19 Firmenmarke 19, unsign.?, Ausformung 20er Jahre, H. 18,5 cm; Ilko
4407,806,15,7,K gem. lig. SC, Firmenmarke 19, Made in Austria, Ausformung 20er Jahre, sign. Beaume, H. 18,4 cm, L. 25,6 cm; MAK Wien, Inv. Ke 10521-21
4407,1009,7 Firmenmarke 19, Made in Austria, sign. Beaume, H. 19,5 cm; eBay # 91482952
Bem.: Bei dieser Figur handelt es sich um die rechte Einzelfigur von Modell-Nr. 3878.
-> vgl. 6011
-> vgl. 3878= zweier GRUPPE
Lit.: Studio YB, 1914, S. 197 (m. Abb.), hier bezeichnet „Greyhound" und mit Künstlerang. „Borrmann" (!).

4408
Meier, (Emil oder Meier-Michel, Johanna)
BÜSTE Mädchen mit leicht geneigtem Kopf
ALABASTER
ca. 1913
4408,?,14 (Firmenmarke Plak. ohne Ang.), sign. Meier, H. 26,3 cm; Doro 7.3.2001
4408 (ohne Ang.), H. fehlt; Neuwirth (1974), S. 58, 68, hier keine nähere Ang. zu Material, Sign. oder Künstler.

4412
Rosé
BÜSTE Beethoven
ALABASTER
ca. 1913
4412 Plak. A, H. 32 cm; eBay # 5354954713
Bem.: Dargestellt ist der Komponist Ludwig van Beethoven (1770–1827).

4442
Stehendes Biedermeiermädchen mit Blumenstrauß (auf mit Blumengirlanden verziertem Rundsockel)
ca. 1913
4442,64,12 Firmenmarke Adler 15, H. 27,5 cm; Schopmann 7.5.1998

4445
Zeiller, Paul
Geier (auf Felssockel)
ALABASTER
ca. 1913
4445,T,6 ger., Plak. A, H. 32,3 cm; Wien Museum Inv. 190.049

4446
Hentschel (d. i. Handzel, August)
Stehende Rokokodame, ihren Reifrock anhebend
ca. 1913/14
4446,13,65 F Firmenmarke Adler 16?, Firmenmarke Sculpture 13, H. 45,5 cm; Quit. 31; 39; Schloss Hag. 40; Ahlden 129
Lit.: Pinhas, S. 101 (m. Abb.).

4450
Hentschel (d. i. Handzel, August)
UHR Zwei stehende Putti (vor klassizistischem, urnenbekröntem Denkmal)
ca. 1913/14
4450,21,15 Firmenmarke 19, Made in Austria, sign. Hentschel, Ausformung 20er Jahre, H. 61 cm; Quit. 14; eBay # 910621212
Lit.: ÖBWK 1.1924/25, Heft Sept., S. III (m. Abb.); Pinhas, S. 100 (m. Abb.).

4451

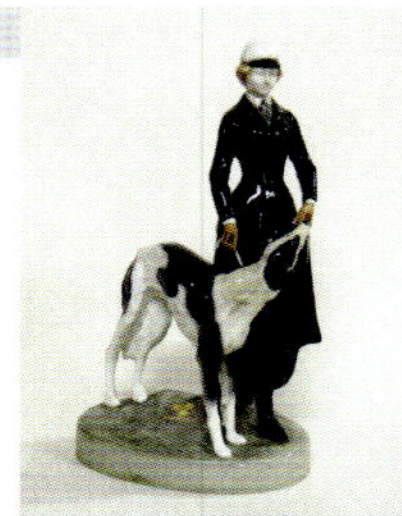

Perigot
Stehende Dame im Jagdoutfit mit Barsoi (auf Landschaftssockel)
ca. 1913/14
4451,56,06 gem. VIII,8,F,9, Firmenmarke Sculpture 13, H. 28,5 cm; MAK Wien, Inv. Ke 10602-20

4458

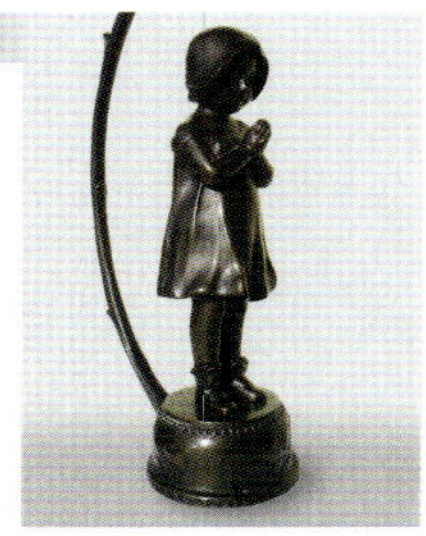

Petri (d.i. Petrides, János) (Johann)
FIGUR/LAMPE Stehendes Mädchen mit Haube und gefalteten Händen (auf zylindrischem Rundsockel)
ca. 1913/14
FAYENCE FIGUR
4458,274,75 gem. Blau XXVIII, Firmenmarke 13a, sign. Petri, H. 26 cm; PB Wien 19
FAYENCE LAMPE
4458,336,2 gem. III, Firmenmarke 18, sign. Polter, Ausformung ca. 1920, H. 25 cm; Wien Museum Inv. 220.730
ALABASTER
4458,JR,73 ger., Plak. A, sign. Petri, H. 24,5 cm; GK, S. 15, 90 (m. Abb.) ; Wien Museum Inv. 182.083
4458,247,K ger., Plak. A, sign. Petri, H. 25 cm; Nagel, G. in: TuS Nr. 185, 1995, S. 6 (m. Abb.)
BRONZE
als LAMPE
ohne Nr., ger., F.G. (Firmenmarke Bronze 3), sign. Petri, H. 26 cm, Gesamth. 52 cm; Sammlung HR Dr. K. Berger, Wien
Ausst.: HMW (jetzt Wien Museum) 1985/86.
Bem.: Dieses Modell wurde in Fayence, Alabaster und Bronze (als Lampe) ausgeführt.
Lit.: GK, S. 15, 90 (m. Abb.)

4463
Rosé
BÜSTE Mädchen (nach links blickend)
ALABASTER
ca. 1913/14
4463,102,NM ger., Plak. A, H. 12,9 cm; eBay # 871617311
4463,350,K ger., (Firmenmarke/Plak. ohne Ang.), H. 12,9 cm; Doro 21.5.1997
4463,R,130 ger., Plak. A, H. 12,5 cm; eBay # 926322987; 2875310071
4463,491,HM ger., Plak. A, H. 12 cm; Stahl 13.6.1998; WKA 23; eBay # 282920098

4465
Rosé
BÜSTE Junges Mädchen mit Schutenhut (nach links blickend)
ALABASTER
ca. 1913/14
K4465,R4,16 ger., Plak. A, H. 12,7 cm; eBay # 1484764255
4465,MR,269 ger., Plak. A, H. 12,7 cm; PB Wien 19
4465,W,316 ger., Plak. A, H. 12 cm; Wien Museum Inv. 220.790

4472
Polter
Stehende Rokokodame mit Blumenstrauß
ca. 1913/14
4472,22,11 FG (Firmenmarke ohne Ang.), Firmenmarke Sculpture 13, H. ca. 30 cm; Doro 20.4.1993

4475
Brynsky
Stehende Dame mit Fächer (auf verziertem Rundsockel)
ca. 1913/14
4475,19,10 gem. B., XV, ohne Firmenmarke, H. 28 cm; Wien Museum Inv. 219.173

4476
Padola
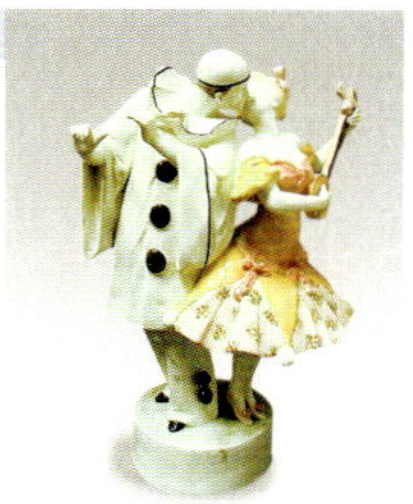
FIGUR/LAMPE Pierrot und Pierrette mit Musikinstrumenten (auf Rundsockel)
ca. 1913/14
FIGUR
4476,3,10 (Firmenmarke ohne Ang.), unsign.?, H. 48 cm; Ruef 481
4476,282,6 (Firmenmarke ohne Ang.), H. 47 cm; Zeller 72
4476,303,24 gem. F. XXIII, Firmenmarke 18, Ausformung ca. 1920, H. 47 cm; Sammlung Robert Senft
LAMPE
4476,14,10 Firmenmarke Sculpture 13, H. 57 cm; Soth. R. 16.10.1993
4476,129,12 (Firmenmarke ohne Ang.), H. 58,5 cm; Doro 10.3.2003
4476,308,14 Firmenmarke 19, Ausformung 20er Jahre, Gesamth. 85,5 cm; Doro 1.12.1998
-> vgl. 4773

4477
Wernekinck, Sigismund
BÜSTE Mädchen
ca. 1913/14
4477 (ohne Ang.), H. fehlt; Neuwirth (1974), S. 58, 76;
Bem.: Hier keine nähere Beschreibung des Objektes.

4481
Kind mit Katze
ca. 1913/14
4481,56,10 gem. 2, Firmenmarke Adler 15, Firmenmarke Sculpture 13, H. fehlt; KH

4482
Hentschel (d. i. Handzel, August)
Stehende Dame (in langem Kleid) mit Blumenstrauß und Buch (Kirchgängerin?)
ca. 1913/14
4482,25,25 gem. F.XIX, Firmenmarke 19, sign. Hentschel, Ausformung 20er Jahre, H. 36,8 cm; WKA 2

4483
Ruß, (Rudolf?/Willibald?)
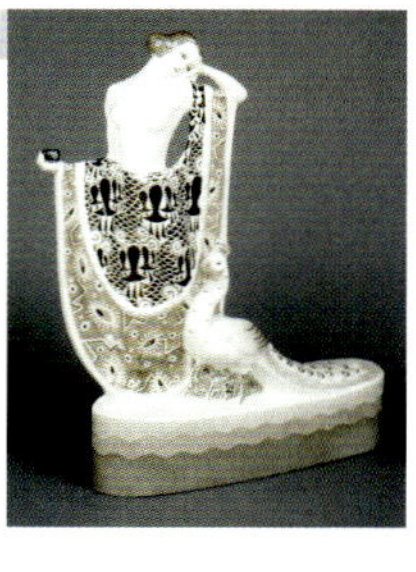
FIGUR/LAMPE Juno mit Pfau (Stehender Damenhalbakt mit großen Tuch) (auf längsovalem Sockel)
ca. 1913/14
FIGUR
4483,12,35 Firmenmarke 22, sign. Ruß, Ausformung 20er Jahre, H. 40,2 cm; eBay # 1450930613
4483,13,10 gem. XXVII, gem. lig. AH, Firmenmarke 22, sign. Ruß, Ausformung 20er Jahre, H. 40 cm, L. 32 cm; PB Wien 19
4483,15,8 ger. H, gem. XXVII, Firmenmarke 24, sign. Ruß, H. 39,9 cm; MAK Wien, Inv. Ke 10602-1
LAMPE
4438,45,65 (Firmenmarke ohne Ang.), sign. (N.?) Russ, H. 46,5 cm; Doro 31.5.2006
Ausst.: HMW (jetzt Wien Museum) 1985/86; SSH 1992.
Bem.: Als Künstler kommen der Keramiker Willibald (Willy) Ruß oder der Wiener Bildhauer Rudolf Ruß (der im Wiener Adressbuch von den 20er bis 40er Jahren nachgewiesen werden kann) in Frage. Für N. Ruß, der in GSN S. 16 genannt wird, konnte kein eigener Nachweis gefunden werden.
Lit.: GK, S. 18, 19 (m. Abb.); GSN S. 16, 17 (m. Abb.), hier als Künstler N. Ruß angegeben.
GS neu Modell-Nr. 74 „Die Göttin" (H. 40 cm), Aufl. 2000 Stück

4484
Petri (d.i. Petrides, János) (Johann)
Lesende Dame auf Récamiere (Sofa)
ca. 1913/14
4484,29,11 Firmenmarke Adler 15, sign. Petri, H. 24 cm, L. 30 cm; PB

4485
Hentschel (d.i. Handzel, August)
Tänzerin in Pose (auf Rundsockel)
ca. 1913/14
4485(Firmenmarke ohne Ang.), sign. Hentschel, H. 31,5 Pinhas, S. 103 (m. Abb.)

4490
LAMPE Fackelträgerin (Stehender Mädchenhalbakt, zwei Fackeln tragend) (auf Ovalsockel)
ca. 1913/14
4490 (Firmenmarke ohne Ang.), H. 120 cm; Pinhas, S. 100 (m. Abb.)

4491
Hentschel (d.i. Handzel, August)
Stehendes Mädchen mit zwei Blumensträußen (auf Rundsockel)
ca. 1913/14
4491,8,2 gem. H., IV, F, Firmenmarke Adler 15, Firmenmarke Sculpture 13, H. 31,5 cm; Herr 42
Bem.: Zum Künstler vgl. Bem. bei Modell-Nr. 4390.

4495
Podola
(d.i. Czapek, Stanislaus)/ Rosé
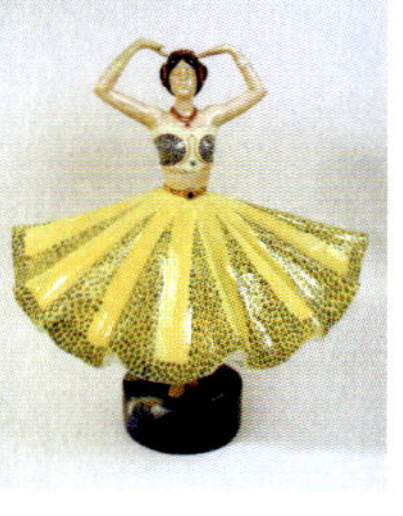
„Indische Tänzerin" (auf Rundsockel)
ca. 1913/14
4495,4,11 Firmenmarke 19, sign. Rosé, Ausformung 20er Jahre, H. 50,8 cm; eBay # 2052216829
4495,10,2 Firmenmarke Adler 15, Firmenmarke Sculpture 13 (gepr.), sign. Podola, H. 52 cm; city-antik
Bem.: Dargestellt ist die Tänzerin Ruth Saint Denis (1877/79–1968, eigentl. Ruth Dennis, gen. „Miss Ruth") in ihrem Tanz „Radha" (Tanz der fünf Sinne, Urauff. 1906), -> vgl. hierzu Modell-Nr. 3870.
Lit.: Heimatkalender für Thüringen und das Osterland 1914, S. 217 (m. Abb.), hier mit Ang. des Originaltitels „Indische Tänzerin" und des Künstlernamens „Podola" (Pseud. für Stanislaus Czapek).

4496
Valentin, Max
Schreitende Tänzerin, ihr langes Kleid seitlich hochhebend (auf getrepptem Ovalsockel)
ca. 1913/14
4496,137,65 Firmenmarke 22, Ausformung 20er Jahre, H. 50,8 cm; eBay # 1326008106
-> vgl. 4794

4497
Eule (auf Natursockel)
ca. 1913/14
4497,183,12 Firmenmarke 19, Made in Austria, Ausformung 20er Jahre, H. 22,5 cm; GK, S: 18, 101 (m. Abb.); Wien Museum Inv. 182.084
Ausst.: HMW (jetzt Wien Museum) 1985/86.
Lit.: GK, S. 18, 101 (m. Abb.)

4507
Gemignani, Valmore
Zwei Windhunde
ca. 1913/14
4507,266,16 ger. 5, gem. lig. SB, Firmenmarke 19, Made in Austria, Ausformung 20er Jahre, H. ca. 16 cm; KH

4511
Lindner, Helmut
FIGUR/LAMPE Junge, sich verneigende Biedermeierdame mit drei Rosen
ca. 1913/14
FIGUR
4511,74,65 gem. AV, Firmenmarke Adler 16, Firmenmarke Sculpture 13, unsign., H. 32,5 cm; Wien Museum Inv. 219.155
4511,319,14 gem. lig. SO, N 5, Firmenmarke 19, Made in Austria, Hand dec., Ausformung 20er Jahre, H. 32 cm; Wien Museum Inv. 190.105
4511 (ohne Ang.), Firmenmarke Sculpture 13, H. 35 cm; Spik 540
LAMPE
4511,313,20 ger. F, gem. F. SE 3., Firmenmarke 19, Made in Austria, Hand dec., Ausformung 20er Jahre, H. fehlt; eBay # 2075163302
Bem.: Dieses Modell wurde auch noch in den USA unter der GS-Everlast Corp. produziert (Modell-Nr. 844)
-> vgl. eBay # 1296010535
-> vgl. 4712; 5491

4518
Kucharzyk, Franz
Stehendes Bauernmädchen
ca. 1913/14
4518,2,10 Firmenmarke Adler 15, Firmenmarke Sculpture 13 (gepr.), H. ca. 30 cm; KH

4522
Nelson, (Alphonse Henri?)
BÜSTE Frau, nach links blickend
MARMOR
ca. 1913/14
4522,K,3 ger., Plak. A, H. fehlt; PB Austria
Bem.: Der Entwurf der Büste geht sicher auf die Jahre zw. 1880–1900 zurück.

4527
Petri (d.i. Petrides, János) (Johann)
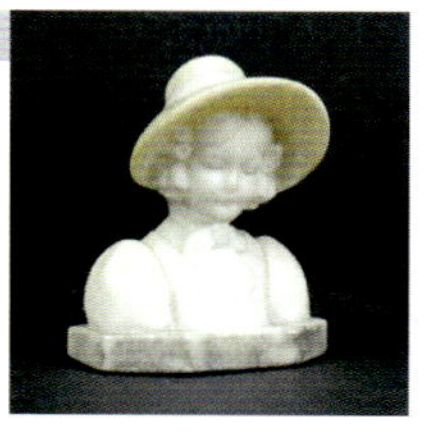
BÜSTE Biedermeiermädchen mit Schutenhut
ALABASTER
ca. 1913/14
4527,W,12 ger., (ohne Firmenmarke?), sign. Petri, H. 8 cm; PB Wien 19

4533
Puchegger, Anton
Adler (auf Felssockel)
ca. 1913/14
4533,218,24 Firmenmarke 22, Ausformung 20er Jahre, H. 31 cm; Wendl 44; 45
-> vgl. 4781; 4782

4534
Rosé
Stehendes lesendes Mädchen (auf Landschaftssockel mit 4 Büchern)
ca. 1913/14
4534,14,10 Firmenmarke Adler 15, Firmenmarke Sculpture 13, sign. Rosé, H. 24,4 cm; Nagel L. 3
4534,142,65 gem. L.B., (Firmenmarke ohne Ang.), sign. „Golter/Polter?", H. 24 cm; Heickmann 17; Wendl 24
4534 (ohne Ang.), sign. Rosé, H. 24,3 cm; Ahlden 91
-> vgl. 3849= ähnl. Modell

4535
Rosé
Liegender Barsoi
ca. 1913/14
4535,140,55 Firmenmarke Adler 14, H. 13,6 cm, L. 34 cm; BM 45
4535,170,12 Firmenmarke Adler 16, unsign., H. 13,5 cm; KH
4535,868,12 Firmenmarke 19, Made in Austria, Ausformung 20er Jahre, H. fehlt, L. 32,4 cm; eBay # 4097894053
4535,875,15 gem. F D 1, Firmenmarke 19, Made in Austria, Ausformung 20er Jahre, H. 11,4 cm, L. 33 cm; eBay # 71620555
4535,1026,65 Firmenmarke 19, Made in Austria, unsign., gepr. „2"= Z18, Ausformung 20er Jahre, H. 12,7 cm, L. 30,5 cm; eBay # 193202378
-> vgl. 5796

4539
Padola / Podola
(d.i. Czapek, Stanislaus)
LAMPE Scherzendes Biedermeierpaar (Dame und Kavalier) (auf mit Blumengirlanden verziertem Rundsockel)
ca. 1913/14
4539,24,65 gem. XV, gem. blau L., Firmenmarke 13a, sign. Padola, H. 58 cm; MAK Wien, Inv. Ke 10521-23
4539 (ohne Ang.), (Firmenmarke Sculpture 13?), sign. Podola, H. 62,7 cm; BM 37
4539 (ohne Ang.), (Firmenmarke Sculpture 13?), sign. Padola, Gesamth. 87 cm; Ahlden 28
-> vgl. 4305= ähnl. Modell
-> vgl. 4915= ähnl. Modell

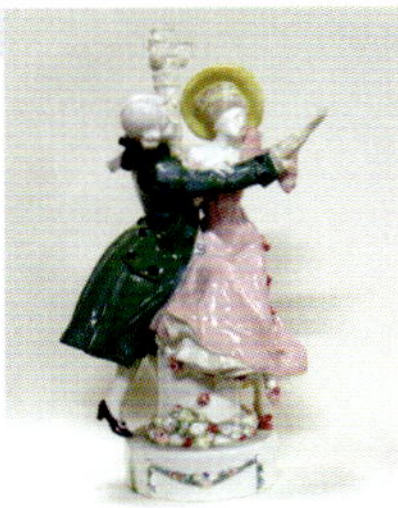

4544
Polter
Sitzender Knabe beim Schreiben (als „Tisch" dient ein hoher Barockstuhl)
ca. 1913/14
4544 (ohne Ang.), Firmenmarke Sculpture 13, (und weitere Firmenmarke ohne Ang.), H. 24,5 cm; Spik 530

4546
Petri (d.i. Petrides, János) (Johann)
Stehendes Mädchen mit Blumenstrauß
ALABASTER / STEINGUSS?
ca. 1913/14
4546,F,56 ger., Firmenmarke Sculpture 13 (Firmenmarke wohl Plak. B), sign. Petri, H. 28,2 cm; Neuwirth (1974), S. 33 (m. Abb.), 70
4546,R,48 ger., Firmenmarke Sculpture 13 (Firmenmarke wohl Plak. B), sign. Petri, H. 29 cm; Pforzheimer Kunst- und AH 49

4550
Polter
Stehendes Mädchen mit Haube und Blumenstrauß, daneben Hund (Dalmatiner?) (auf ovalem Natursockel)
ca. 1913/14
4550,30,10 gem. D. I. IX., (Firmenmarke Adler ohne Ang.), H. 28 cm; Quit. 14
4550,77,12 (Firmenmarke ohne Ang.), H. 27 cm; Quit. 3
4550 (ohne Ang.), H. 28 cm; Doro G 16.5.1995

4551
Petri (d.i. Petrides, János) (Johann)
Stehendes Mädchen (Halbakt) mit Gitarre (auf hohem Rundsockel)
ca. 1913/14
FAYENCE
4551 (ohne Ang.), sign. Petri, H. 27,5 cm; Kett. 198
4551 (ohne Ang.), sign. Petri, H. 27 cm; Ahlden 87
Lit.: Pinhas, S. 123 (m. Abb.), hier mit der falschen Mod.-Nr. 4557, in einer ALABASTER Version.

4554
Sitzender Mädchenhalbakt mit Laute
ca. 1913/14
4554 (Firmenmarke ohne Ang.), H. 64 cm; Pinhas, S. 99 (m. Abb.)

4557
Perigot
Stehendes Mädchen in Pumphosen (auf Ovalsockel)
ca. 1913/14
4557,334,6 gem. VI I, Firmenmarke 19, gepr. „P"= Z3, Ausformung 20er Jahre, H. ca. 22 cm; KH
4557,357,11 gem. VIII, Firmenmarke 19, H. 22 cm; eBay # 1414142288

4558
Perigot
Stehendes Mädchen mit Hut (auf Rundsockel)
ca. 1913/14
4558,78,10 gem. VI I, DDO, Firmenmarke Adler 15, Firmenmarke Sculpture 13 (gepr.), H. 22 cm; PB Wien 19
ohne Ang., Firmenmarke Sculpture 13, unsign.?, H. 21,5 cm; Doro 24.11.1992

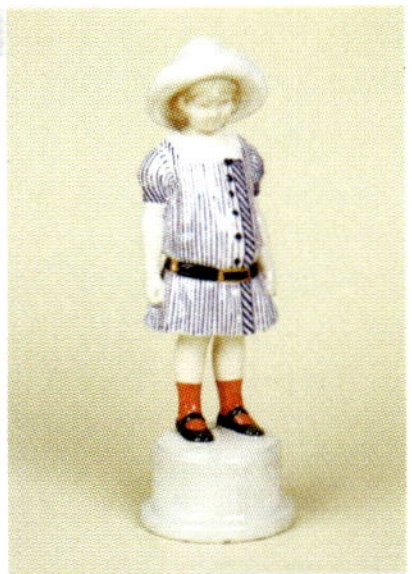

4563
Meier-Michel, Johanna
„Frühling" (Stehende Biedermeierdame mit Blumenkorb)
ca. 1913
4563,111,1 gem. V, gem. rosa 19 P, Firmenmarke 19, sign. Meier, Ausformung 20er Jahre, H. 19 cm; PB Wien 19
-> vgl. 4365
Lit.: Studio YB, 1914, S. 193 (m. Abb.), hier mit Ang. des Originaltitels „Frühling" und mit Ang. der Entwerferin Johanna Meier-Michel.

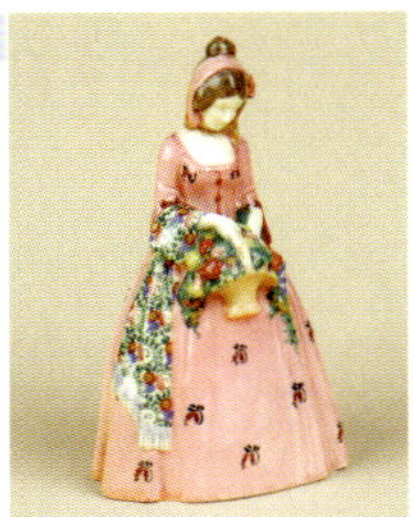

4565
Padola
Junge, sich drehende Biedermeiertänzerin (auf Rundsockel)
ca. 1913/14
4565,15,65 (Firmenmarke ohne Ang.), unsign.?, H. 28 cm; eBay # 61277783
4565,28,10 (Firmenmarke Adler ohne Ang.), Firmenmarke Sculpture 13, sign. Padola, H. 30,5 cm; Stahl 8.12.2001

4568
Meier, (Emil oder Meier-Michel, Johanna) / Polter
Stehendes Biedermeiermädchen, einen großen, blumengeschmückten Hut haltend
ca. 1913/14
4568,19,2 gem. III 1., Firmenmarke Adler 15, Firmenmarke Sculpture 13 (gepr.), sign. Polter, H. 30,5 cm; eBay # 422073650
4568,163,3 gem. XXVII, (Firmenmarke ohne Ang.), sign. Polter, (Ausformung 20er Jahre?), H. 37 cm; Soth. A. 31.3.1993; Bem.: Die Sign. „Polter" dürfte wohl als Versehen des Modelleurs angesehen werden.
4568,171,1 Firmenmarke 19 (braun), sign. Meier, Ausformung 20er Jahre, H. 36 cm; Hampel 26.10.1990
Bem.: Als Entwerfer(in) kommt wohl am ehesten Johanna Meier-Michel in Frage.

4569
Meier, (Emil oder Meier-Michel, Johanna)
Mädchen mit Blumenstrauß
ca. 1913/14
4569 (ohne Ang.), H. fehlt; Neuwirth (1974), S. 58, 68;
Bem.: Hier keine nähere Beschreibung des Objektes.

4570
Meier, (Emil oder Meier-Michel, Johanna)
„Herbst" (Stehendes Biedermeiermädchen, einen Früchtekorb auf der linken Schulter tragend)
ca. 1913/14
4570,1,10 Firmenmarke Sculpture 13, (und Firmenmarke ohne Ang.), sign. Meier, H. 31 cm; Doro 8.11.1994
4570,6,13 gem. lig. HK, gem. VIII/P D I, Firmenmarke Adler 15, Firmenmarke Sculpture 13, sign. Meier, H. 31 cm; Brö 2,1, S. 406 (m. Abb.); BröPK, S. 147
4570,37,1 GS (Firmenmarke?), sign. Meier, H. 30,5 cm; Doro 30.11.2001; 26.11.2002
Lit.: Neuwirth (1974), S. 58, 68; Pinhas, S. 101 (m. Abb.).

4573
Lindner, Helmut
Diana mit Hund (Stehender Akt mit Köcher, Ausschau haltend) (auf hohem Ovalsockel)
ca. 1913/14
4573,15,17 Firmenmarke 19a, Ausformung 20er Jahre, H. 90 cm; Wien Museum Inv. 186.620

4575
Petri (d.i. Petrides, János) (Johann)
BÜSTE Mädchen mit Haube und gefalteten Händen
ALABASTER
ca. 1913/14
4575,V,260 ger., Plak. E, sign. Petri, Ausformung 20er Jahre, H. 11,4 cm; eBay # 7408417301
Lit.: Neuwirth (1974), S. 58 (hier ohne näheren Ang.)

4577
Polter
FIGUR / LAMPE Sitzende Lautenspielerin mit Hut (Biedermeierfigur)
ca. 1913/14
FAYENCE (Figur)
4577,22,11 Firmenmarke Sculpture 13 (gepr.), H. 27,7 cm; Doro 16.4.1996
4577,25,11 gem. X D. I., (Firmenmarke ohne Ang.), H. 28 cm; Kett. 165; 177
4577,26,2 gem. IV, D.I., Firmenmarke Adler 15, Firmenmarke Sculpture 13, H. 29 cm; Wien Museum Inv. 219.174
4577,122,10 (Firmenmarken ohne Ang.), H. 27 cm; Mehlis 24
4577,360,25 Firmenmarke Sculpture 13 (gepr.), H. 28 cm; Kett. 39; 160
4577 (ohne Ang.), H. 28 cm; Ruef 463
FAYENCE (Lampe)
4577,484,8 gem. XXVII, Firmenmarke 22, Ausformung 20er Jahre, H. 49,8 cm; Wien Museum Inv. 219.149
ALABASTER (Figur)
4577,W,4 Plak. A, H. 30 cm, Gesamth. mit Sockel 39 cm; eBay # 931993380; 5350248893
Bem.: Dieses Modell wurde in Fayence und Alabaster ausgeführt.
Lit.: Alt-Wien, S. 463; Pinhas, S. 100 (m. Abb.).

4580
Horsetzky, Melanie von
„Lautenspiel" (Sitzende Lautenspielerin)
ca. 1911/12
4580,125,65 FG (Firmenmarke ohne Ang.), H. 17,8 cm; BM 62
4580,158,3 Firmenmarke 13a, H. 18 cm; Mehlis 33
Bem.: Die im Originaltitel angeführte „Laute" ist im eigentlichen Sinn eine Gitarre.
Ausst.: ÖMKI 1912, hier mit Ang. des Originaltitels „Lautenspiel"; SSH 1992.
-> vgl. 4114
Lit.: Neuwirth (1974), S. 64

4587
Rosé
BÜSTE Mädchen mit Schutenhut
ALABASTER
ca. 1913/14
4587,lig. JK,68 ger., Plak. A, H. ca. 17 cm; KH

4588
Rosé
BÜSTE Mädchen mit Biedermeierfrisur (nach rechts blickend)
ALABASTER / Marmorsockel
ca. 1913/14
4588 (Plak. A?), H. 15,2 cm; Doro 13.9.2004

4593
Petri (d.i. Petrides, János) (Johann)
BÜSTE Mädchen mit Haube
ca. 1913/14
ALABASTER
4593,TK,19 ger., Plak. A, sign. Petri, H. 17,5 cm; eBay # 6527141969

4594
Petri (d.i. Petrides, János) (Johann) / Perigot
Stehendes weinendes Mädchen (auf Rundsockel)
ca. 1913/14
FAYENCE
4594,144,65 gem. VII., Firmenmarke 19, sign. Perigot, Ausformung 20er Jahre, H. 25,5 cm; eBay # 2506964909
ALABASTER
4594,?,176 (Firmenmarke Plak. ohne Ang.), sign. Petri, H. 28 cm; WKA 33
Bem.: Dieses Modell wurde in Fayence und Alabaster ausgeführt.
Lit.: Neuwirth (1974), S. 58, hier nur Hinweis auf die Modell-Nr., sonst ohne weitere Ang.

4597
BÜSTE Mädchen mit Haube
ALABASTER
ca. 1913/14
4597,lig. HM,1, Plak. A, unleserlich sign., H. 20,3 cm; eBay # 1456051384; Bem.: Laut Anbieter unleserlich sign.
Bem.: Ähnl. Entwürfe gibt es von Petri (d.i. Petrides) oder Rosé.

4603
Rudolff
Mädchenakt mit Hündchen (King Charles Spaniel) im Sessel sitzend
ca. 1913/14
4603,29,19 Firmenmarke Sculpture 13, H. 23,7 cm; Doro 15.9.2003
4603,81,? Firmenmarke 19?, Ausformung 20er Jahre, H. 23 cm, B. 18 cm; Schlüter 2677; Bem.: Das AH gibt als Entwerfer „Kurt Rudolff" an, ein Künstler dieses Namens lässt sich jedoch nicht nachweisen.

4609
Lindner, Helmut

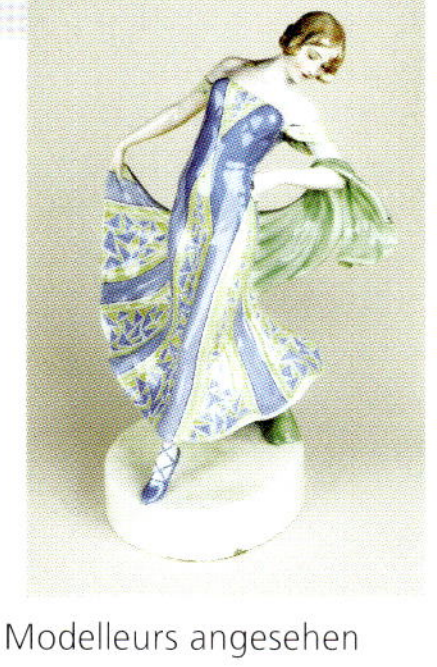

Tänzerin in drehender Bewegung (auf Rundsockel)
ca. 1913/14
4609,15,13 gem. X.D.I, Firmenmarke Adler 15, Firmenmarke Sculpture 13 (gepr.), H. 50 cm; Dawo 142
4609,43,2 (Firmenmarke Adler ohne Ang.), sign. Latour, H. 49,8 cm; BM 32; Bem.: Die Sign. „Latour" dürfte wohl als Versehen des Modelleurs angesehen werden.
4609,61,10 gem. grün 3., Firmenmarke 9, Firmenmarke Adler 14, Firmenmarke Sculpture 13, sign. Lindner, H. 49,6 cm; Wien Museum Inv. 219.189
4609,67,55 Firmenmarke Sculpture 13, sign. Lindner, H. 49 cm; BM 72; 73
4609,77,10 (Firmenmarke ohne Ang.), sign. Lindner, H. 49,5 cm; Doro 8.3.2004
4609,146,65 (Firmenmarke ohne Ang.), sign. H. Lindner, H. 49,5 cm; Sammlung Robert Senft
4609,135,10 Firmenmarke Sculpture 13 (gepr.), unsign.?, H. 49 cm; eBay # 2043868658
4609,310,35 ger. K, gem. Bo II., Firmenmarke 20?, sign. Lindner, Ausformung 20er Jahre?, H. 50 cm; Nagel 375; Lit.: TuSA 9, S. 185 (m. Abb.)
4609,330,1 Firmenmarke 20, Made in Austria, Ausformung 20er Jahre, sign. Lindner, H. 50 cm; PB Wien 19 (ehem.)
ohne Ang., Firmenmarke Adler 16, sign. H. Lindner, H. 48,3 cm; eBay # 1204328514 (KH www.tias.com/stores/frozenbeauty)
ohne Ang., sign. Lindner, H. 49,5 cm; Zeller 54; Lit.: TuS, 1995, Nr. 193, S. 172 (m. Abb.)
ohne Ang., (Firmenmarke ohne Ang.), Repr. Res., unsign., H. 50 cm; Christie's L. 29.4.1992
-> vgl. 4690; 4749
Lit.: Pinhas, S. 103 (m. Abb.).

4610
Perigot

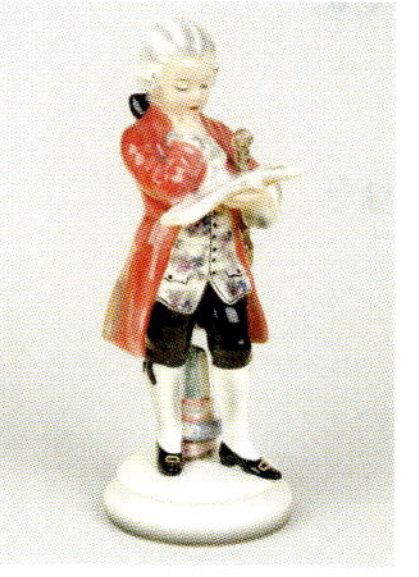

„Mozart" (Stehende Figur als Knabe mit Geige, Notenheft und Bücherstapel)
ca. 1913
FAYENCE
4610,662,14 gem. F. VIII, Firmenmarke 16?, Firmenmarke 22, Ausformung 20er Jahre, H. ca. 30 cm; eBay # 496633515
4610,668,65 Adler Firmenmarke 24, Ausformung 20er Jahre, H. 33 cm; Henry's 21.12.2001; city-antik
4610,961,14 ger. M03, gem. G., G., 2., Firmenmarke 19b, Made in Austria, Hand dec., Ausformung 20er Jahre, H. 30 cm; Wien Museum Inv. 219.179
4610,1049,14 (Firmenmarke ohne Ang.), H. 32 cm; Schloss Rick. 70
ALABASTER
4610,KS,8 Plak. E, H. 30,5 cm; eBay # 5386255419
ohne Ang., H. 30 cm; Spik 552
Bem.: Dieses Modell wurde in Fayence und Alabaster ausgeführt. Ähnl. Darstellungen (Mozart mit Geige) finden sich auch bei L.-E. Barrias und M. Debut, vgl. hierzu: Berman, Vol. 3, (Nr. 2619: 2622 jeweils m. Abb.).
-> vgl. 4393; 5519
Lit.: MW, 11.1929/30, Nr. 8, S. V (m. Abb.), hier mit Ang. des Originaltitels „Mozart"; Neuwirth (1974), S. 58, hier ohne Ang. zum Künstler; GSN, Erg. Bl. (o. Pag., m. Abb.) und Preisliste 1992.
GS neu (ohne Modell-Nr.) „Mozart" (H. 25 cm), Aufl. 300 Stück

4612
Hentschel (d. i. Handzel, August)
Stehendes Bauernmädchen in Tracht (Böhmen / Ungarn?) (auf profiliertem Rundsockel)
ca. 1913/14
4612,8,11 Firmenmarke Sculpture 13, (und Firmenmarke ohne Ang.), sign. Hentschel, H. 33,2 cm; Doro 24.10.1994
4612,14,10 Firmenmarke Adler 16, sign. Hentschel, H. 34 cm; Doro S FVK VII/1999
-> vgl. 4701= ähnl. Modell (Pendant?)

4614
Petri (d.i. Petrides, János) (Johann)
Stehender Mädchenhalbakt mit Tuch (auf Rundsockel)
ca. 1913/14
FAYENCE
4614,346,65 Firmenmarke 19, sign. Petri, Ausformung 20er Jahre, H. 28,6 cm; Doro 14.9.2001
ALABASTER / Marmorsockel
4614,HM,6(49?) Made in Austria, ger., Plak. E, sign. Petri, Ausformung 20er Jahre, eBay # 2300144
4614,MA3,6 ger., Plak. E, unsign.?, Ausformung 20er Jahre, H. 29,2 cm; eBay # 1413387186
Bem.: Dieses Modell wurde in Fayence und Alabaster / (Kunststein?) ausgeführt.

4616
Meier, (Emil oder Meier-Michel, Johanna)

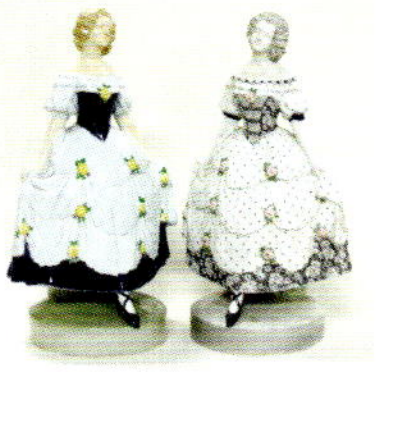

Biedermeiertänzerin im Rosenblütenkleid (auf Rundsockel) (Anna Pawlowa?)
ca. 1912/13
4616,1,11 gem. lig. WT/DB?, XI, Firmenmarke Adler 15, Firmenmarke Sculpture 13, sign. Meier, H. 31,6 cm; MAK Wien, Inv. Ke 10521-38
4616,34,7 Firmenmarke Adler 16, Firmenmarke Sculpture 13, sign. Meier, H. 31 cm; KaM 99
4616,43,? Firmenmarke Adler 16, sign. Meier, H. 30,8 cm; BM 57
4616,70,(4?) Firmenmarke Sculpture 13 (gepr.), Firmenmarke 13a, sign. Meier, H. 30,8 cm; Doro 21.9.1999
4616,102,65 Firmenmarke 18, Ausformung ca. 1920, sign. Meier, H. 31,3 cm; BM 62
4616,109,86 Firmenmarke 19, sign. Meier, H. 32 cm; Pforzheimer KAH 42
4616,123,1 gem. II., Firmenmarke 22, sign. Meier, Ausformung 20er Jahre, H. 31,2 cm; MAK Wien, Inv. Ke 10521-40
4616,154,8 Firmenmarke 19, sign. Meier, H. 32 cm; eBay # 913059656
4616 (ohne Ang.), unsign.?, H. 31 cm; Henry's 28.5.1998
Bem.: s. Bem. bei 4341
-> vgl. 4341; 4619

4619
Meier, (Emil oder Meier-Michel, Johanna)

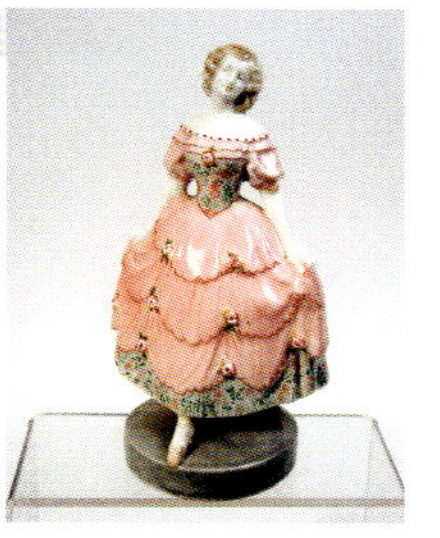

Biedermeiertänzerin im Rosenblütenkleid (auf Rundsockel) (Anna Pawlowa?)
ca. 1912/13
461(9),9,10 Firmenmarke Sculpture 13, sign. Meier, H. 23,3 cm; Doro 7.3.2001; Bem.: Im Auktionskat. wohl Druckfehler, dort Modell-Nr. 4612 angegeben.
4619,35,7 gem. VI,Firmenmarke Adler 14?, sign. Meier, H. 23 cm; Kett. 39
4619,58,8 gem. XXIII, Firmenmarke Sculpture 13, sign. Meier, H. 23 cm; Wendl 29; city-antik
4619,136,5 Firmenmarke 19, sign. „Lorenzl", Ausformung 20er Jahre, H. 23,5 cm; WKH; Bem.: Die Sign. „Lorenzl" dürfte wohl als Versehen des Modelleurs angesehen werden.
4619,146,1 Firmenmarke 19, sign. Meier, Ausformung 20er Jahre, H. 23,5 cm; Doro 15.3.2005
4619 (ohne Ang.), (Firmenmarke Adler ohne Ang.), Firmenmarke Sculpture 13, sign. Meier, H. 23,5 cm; Wendl 21
Bem.: s. Bem. bei 4341
-> vgl. 4341; 4616

4622
Meier, Emil

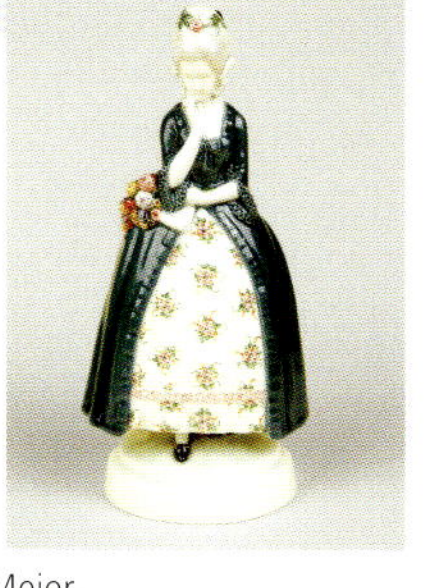

„Pensée" (Stehende Rokokodame in Gedanken)
ca. 1913
4622,101,65 Firmenmarke 19, Ausformung 20er Jahre, H. 22 cm; PB Wien 19
-> vgl. 4372
Lit.: Studio YB, 1914, S. 189 (m. Abb.), hier mit Ang. des Originaltitels „Pensée" und mit Ang. des Entwerfers Emil Meier.

4624
Rudolff
Stehendes Biedermeiermädchen mit Schutenhut und Blumenstrauß (Krinoline)
ca. 1913/14
4624,4,10 (Firmenmarke Adler ohne Ang.), Firmenmarke Sculpture 13, sign. Rudolf, H. 34,4 cm; WKA 18
4624,164,14 Firmenmarke Adler 16, sign. Rudolff, H. 34,3 cm; eBay # 1243691650
4624,322,1 (Firmenmarke ohne Ang.), unsign.?, Ausformung 20er Jahre, H. 33 cm; KaM 99
4624,323,16 gem. V, Firmenmarke 19, sign. Rudolf, Ausformung 20er Jahre, H. 34 cm; Wien Museum Inv. 190.111
ohne Ang., sign. Rudolff, H. 35 cm; Fischer 30; Weichmann 26.4.1985
-> vgl. 4819; 4820

4625
Tell, E.
Sitzende junge Frau (Akt) mit Lyra (Sappho?)
ca. 1908
FAYENCE
4625,183,65 (Firmenmarke ohne Ang.), H. 37 cm; Henry's 29.4.1998
4625,207,65 Firmenmarke 19, Made in Austria, unsign.?, Ausformung 20er Jahre, H. 37 cm; Zacke 15, S. 49
ALABASTER
ohne Ang., (ohne Firmenmarke?), H. 39 cm; Soth. L. 5.11.1987
Bem.: Dieses Modell wurde in Fayence und Alabaster/ (Kunststein?) ausgeführt.
-> vgl. 3497; 3576; 3577; 4697

4629
Polter
Stehende Elsässerin in Tracht (mit Kopfputz/Haube und langem Kleid) (auf quaderförmigem Sockel)
ca. 1913/14
4629,26,10 (Firmenmarke Adler ohne Ang.), Firmenmarke Sculpture 13, H. 34 cm; Henry's 23.2.1994

4632
Petri (d.i. Petrides, János) (Johann)
Stehendes kleines Mädchen mit Schiefertafel und Buch (auf Landschaftssockel)
ca. 1913/14
4632 (ohne Ang.), sign. Petri, H. 23 cm; BM 28
4632 (ohne Ang.), sign. Petri, H. 23 cm; Koller Z 61

4634
Tacke

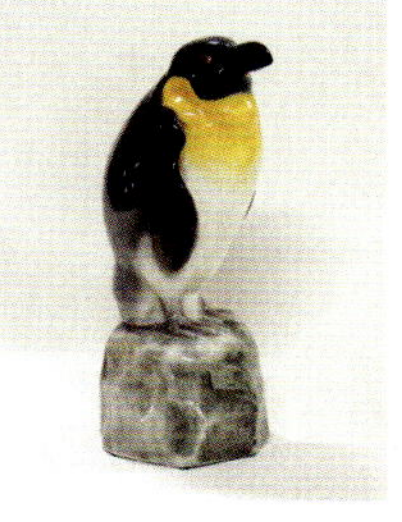

Stehender Königspinguin (auf Felssockel)
ca. 1913/14
FAYENCE
4634,4,10 Firmenmarke Adler 15, Firmenmarke Sculpture 13, sign. Tacke, H. 26,8 cm; MAK Wien, Inv. Ke 10521-20
4634,14,10 (Firmenmarke ohne Ang.), sign. Tacke, H. 27 cm; Schloss Hag. 34
ALABASTER
4634,JF,6 ger., Plak. A, sign. Taake (!), H. 28 cm; eBay # 5458522286
Bem.: Dieses Modell wurde in Fayence und Alabaster/ (Kunststein?) ausgeführt.

4635
Kopp, (Josef?)

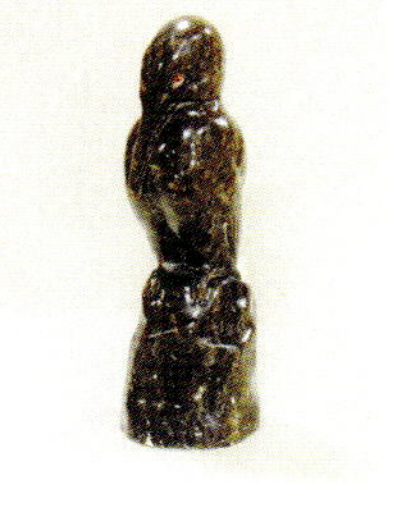

Marabu (auf Felssockel)
MARMOR
ca. 1914/15
4635.K.8 ger., Plak. A, H. 25,9 cm; eBay # 1424058130
4635.TK.6 ger., Plak. A, H. 25,7 cm; MAK Wien, Inv. Ke 10602-85
Bem.: Mit eingesetzten Glasaugen.

4637
Beaume (d.i. Bormann, Wilhelm)
Stehender Windhund (Barsoi/Greyhound) (auf Rechtecksockel)
ca. 1910/12
4637,39,1 gepr. K., gem. F. F., Firmenmarke 22, sign. Beaume, Ausformung 20er Jahre, H. ca. 20 cm; eBay # 548238265
Bem.: Bei dieser Figur handelt es sich um die linke Einzelfigur von Modell-Nr. 3878.
-> vgl. 3878= zweier GRUPPE

4642
Heinemann, Fritz
Schreitende Holländerin mit zwei Milchkannen (auf Rundsockel)
ca. 1911/12
4642,84,8 gem. B., XIX, Firmenmarke 22, gepr. „P"= Z3, H. 29 cm; Wien Museum Inv. 219.159
Bem.: Diese Figur könnte mit dem „Milchmädchen" (Museum Dessau) identisch sein, welches in TB, Bd. 19, S. 294 explizit aufgeführt ist.
-> vgl. 4242

4643
Müller, Armin
Stehender Putto mit Blumengirlande (auf Rundsockel)
ca. 1914/15
4643,4,65 gem., VII,1,2 Firmenmarke Adler 15, sign. Armin Müller, H. 37 cm; Quit. 58
4643,40,10 Firmenmarke Adler 16, Firmenmarke Sculpture 13, sign. Müller, H. 38 cm; Wendl 26
4643,29,1 Firmenmarke Sculpture 13, sign. Müller?, H. 36,2 cm; Doro 15.9.2003
4643,42,10 gem. grün 7, Firmenmarke Sculpture 13, Firmenmarke 21?, sign. Müller, Ausformung 20er Jahre?, H. ca. 37 cm; Kett. 70; 74; 191
4643,145,11 ger., gem. H. XXIII, Firmenmarke 19, sign. Müller, Ausformung 20er Jahre, H. 36,5 cm; Brö 2,1, S. 406 (m. Abb.); BröPK, S. 147
4643 (ohne Ang.), (Firmenmarke ohne Ang.), sign. Armin Müller, H. 38 cm; Mehlis 27; 29
Pendant (?) -> vgl. 3856
Lit.: Neuwirth (1974), S. 58, 68

4644
Petri (d.i. Petrides, János) (Johann)
Stehendes kleines Mädchen (auf Landschaftssockel)
ALABASTER
ca. 1914/15
4644,ST (lig.),2 Plak. A, sign. Petri, H. 30,5 cm; eBay # 1508602757

4645
Rosé
Beethoven
ca. 1914/15
4645 (ohne Ang.), H. fehlt; Neuwirth (1974), S. 58, 72;
Bem.: Hier keine nähere Beschreibung des Objektes.
-> vgl. auch 3606?

4646
Carlus, Jean

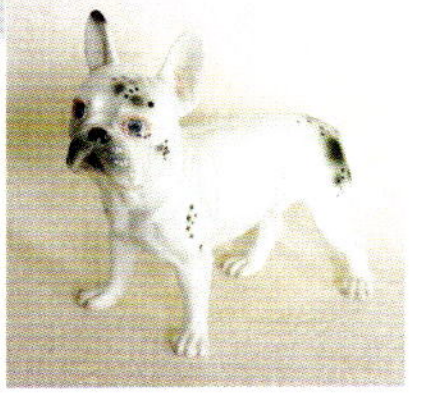

Stehende französische Bulldogge
ca. 1914/15
4646,31,55 ohne Firmenmarke, H. 21 cm; eBay # 729999053
4646,312,65 gem. R., Firmenmarke 19, Ausformung 20erJahre, H. 20 cm, L. 26 cm; Sammlung F. & J.P. Girard/Frankreich

4650
Petri (d.i. Petrides, Janos) (Johann)/Polter
Stehendes Mädchen mit Haube und Schiefertafel (auf Rundsockel)
ca. 1915/16
FAYENCE
4650,125,14 Firmenmarke Sculpture 13, sign. Polter, H. 28 cm; Doro 18.11.1997
ALABASTER
4650,P.,59 ger., (Firmenmarke Plak. A, fehlt), sign. Petri, H. 29 cm; Wien Museum Inv. 219.157
Bem.: Dieses Modell wurde in Fayence sowie Alabaster ausgeführt.

4651
Petri (d.i. Petrides, János) (Johann)/Polter
FIGUR/LAMPE Stehender Knabe mit Haube und Apfel (auf Rundsockel)
ca. 1915/16
FIGUR (Fayence)
4651,220,65 gem. AN, Firmenmarke 19, sign. „Polter", Ausformung 20er Jahre, H. 29,5 cm; WKH; Die Sign. „Polter" dürfte wohl als Versehen des Modelleurs angesehen werden.
LAMPE (Fayence)
4651,233,8 gem. VII., Firmenmarke 19, sign. „Polter", Ausformung 20er Jahre, H. 29 cm, Gesamth. 38 cm; eBay # 930595018; Bem.: Die Sign. „Polter" dürfte wohl als Versehen des Modelleurs angesehen werden.
ALABASTER/STEINGUSS (?)
K4651,F,75 (vermutlich ger.), Plak. A, sign. Petri, H. 28,3 cm; Neuwirth (1974), S. 33 (m. Abb.), 70 (Samml. Christine Moser, Wien)
4651 (ohne Ang.), Plak. A, sign. Petri, H. 28,3 cm; Neuwirth (1973), in: Keramos, 59.1973, S. 54–63, hier auf S. 61 (m. Abb.) (Sammlung Tadler, Wien).
Bem.: Dieses Modell wurde in Fayence (als Figur und Lampe) sowie Alabaster/Steinguss (?) ausgeführt. Die Figur ist mit ziemlicher Sicherheit Petrides zuzuordnen, da die Sign. „Petri" auf den zum Entwurf gehörigen, zeitnahen Stücken aufgebracht wurde. Weitere Indizien sind zahlreiche von Petrides ausgeführte ähnl. Modelle. Beide mit „Polter" sign. Stücke stammen zudem aus späterer Zeit.
Lit.: Neuwirth (1974) 1, S. 61 (m. Abb.); Pinhas, S. 99 (m. Abb.).

4653
Schörk, Hans
BÜSTE Mädchen mit Haube (nach links blickend)
ALABASTER
ca. 1915/16
4653,F,48 ger., Plak. A, H. 17,1 cm; eBay # 905102233
4653,J,2,W ger., Plak. A, unsign., H. 13,5 cm; eBay # 5382358131

4654
BÜSTE Holländermädchen
ALABASTER
ca. 1915/16
4654,W,187 ger., Plak. C, Ausformung 20er Jahre, H. 15,5 cm; WKH

4655
Petri (d.i. Petrides, János) (Johann)
BÜSTE Kleiner Bub
ALABASTER/Marmorsockel
ca. 1915/16
4655,HM,28 ger., Plak. A, sign. Petri, H. 16,5 cm, B. 15,2 cm; eBay # 471624099

4656
Tell, E.

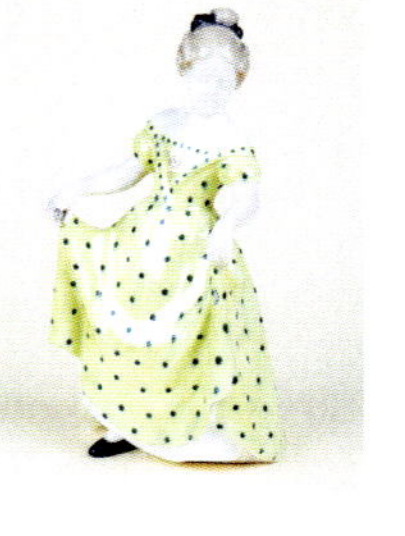

Stehendes Biedermeiermädchen (ihr Kleid etwas hochhebend) (nach links blickend)
ca. 1904/05 (für den Entwurf)
4665,16,10 (Firmenmarken ohne Ang.), H. 23,2 cm; Doro 16.4.1996
-> vgl. 3009
Pendant -> vgl. 4657

4657
Tell, E.

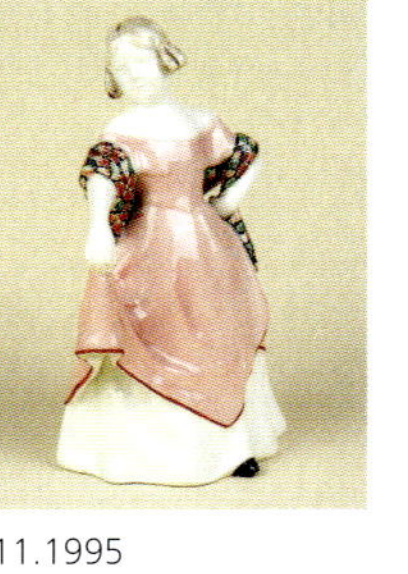

Stehendes Biedermeiermädchen (ihr Kleid etwas hochhebend) (nach rechts blickend)
ca. 1904/05 (für den Entwurf)
4657,8,65 FG Wien (Firmenmarke?), H. 23 cm; eBay # 3317177038
4657,88,5 (Firmenmarken ohne Ang.), H. 22,5 cm; Doro 7.11.1995
-> vgl. 3067
Pendant -> vgl. 4656

(4657) ?
Hund
ca. 1915/16?
4657 (ohne Ang.), H. fehlt; Neuwirth (1974), S. 58;
Bem.: Hier keine nähere Beschreibung des Objektes.
Bem.: Da sich kein „Hund" unter dieser Modell-Nr. nachweisen lässt, sondern das o.g. Biedermeiermeiermädchen von E. Tell, ist davon auszugehen, dass es sich um einen Druckfehler bei Neuwirth handelt, bzw. die Tierfigur von einer anderen Firma stammt.

4659
Perigot
Stehendes Biedermeiermädchen mit Blumenstrauß und Schleife im Haar (auf Rundsockel)
ca. 1915/16
4659,53,12 gem. F23, (H4?), Firmenmarke Adler 15, H. 30 cm; eBay # 3336228561

4660
Tacke
Stehender Collie (auf Rechtecksockel)
ca. 1915/16
4660,104,8 gem. lig. JT, D1., Firmenmarke 19, Made in Austria, H. 26,6 cm, L. 36,8 cm; eBay # 3667136051

4662
Kienast
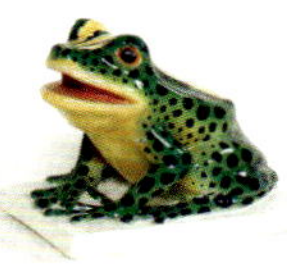
Sitzende Kröte (auf Rechtecksockel)
ca. 1915/16
4662,44,2 gem. lig. AH, Firmenmarke 19, Made in Austria, Ausformung 20er Jahre, H. 8 cm; MAK Wien, Inv. Ke 10602-58

4664
Petri (d.i. Petrides, János) (Johann)
BÜSTE Holländermädchen mit Haube
ALABASTER
ca. 1915/16
4664,A,136 ger., Plak. A, sign. Petri, H. 13,5 cm; Wien Museum Inv. 186.681
4664,MR,13 ger., Plak. A, sign. Petri, H. 13 cm; Wien Museum Inv. 220.791
-> vgl. 4287= ähnl. Modell

4674
Stehendes Biedermeiermädchen mit gefalteten Händen
ca. 1915/16
4674,73,9 (Firmenmarke Adler ohne Ang.), Firmenmarke Sculpture 13, H. 20,5 cm; Fischer 63

4676
Padola / Podola (d.i. Czapek, Stanislaus)/ Podolak
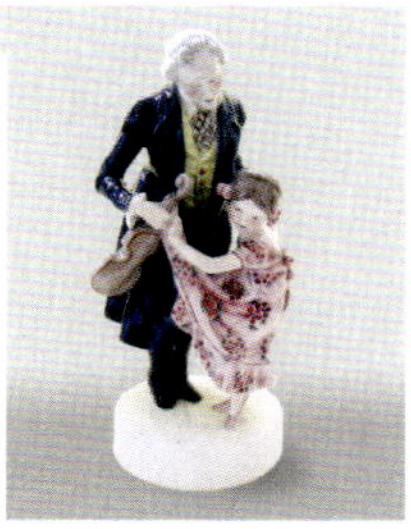
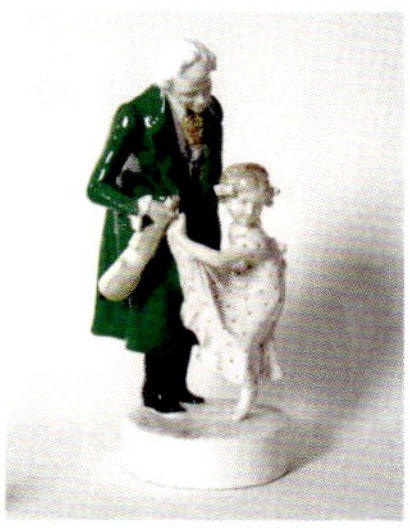
Die Tanzstunde (Violinspieler im Biedermeiergewand mit einem Mädchen tanzend) (auf Rundsockel)
ca. 1915/16
(4676),88,10 (Firmenmarke ohne Ang.), sign. Padola, H. 35(?) cm; Metz 16.3.1991; Bem.: Im Kat. falsche Modell-Nr. 4846 und wohl auch falsche H. angegeben.
4676,191,10 Firmenmarke 16?, Firmenmarke Sculpture 13 (gepr.), sign. Padola, H. fehlt; Quit. H. 28.2.2004
4676,237,12 gem. IX, 2878 Fadrich?, „Dreieck“, gem. blau X, Firmenmarke Adler 16, Firmenmarke Sculpture 13 (gepr.), sign. Podola, H. 31,2 cm; PB Deutschland
4676,527,1 (Firmenmarke ohne Ang.), sign. Padola, H. 31 cm; WKH
4676,566,11 gem. F., (Firmenmarke ohne Ang.), unsign.?, H. 31 cm; Kastern 75
4676,667,10 gem. F.D. 2.II, Firmenmarke 19, sign. Podolak, Ausformung 20er Jahre, H. 31 cm; eBay # 3367828831
4676,794,7 ger. M I, gem. R, Firmenmarke 19, Made in Austria, Hand dec., sign. Podolak, Ausformung 20er Jahre, H. 29,5 cm; Zezschwitz 13
4676,834,16 (Firmenmarke ohne Ang.), sign. Podolak, Ausformung 20er Jahre, H. 28 cm; Doro 15.3.2005
4676,854,15,04,K, gem. AE, MR,3, Firmenmarke 19, Made in Austria, Hand dec., sign. Podolak, Ausformung 20er Jahre, H. 30 cm; Sammlung f@goldscheider.de
4676 (ohne Ang.), Firmenmarke Sculpture 13, sign. Podola, H. 32 cm; RM 5

4677
Polter
Stehender Knabe, einen Brief lesend (in knielangen Hosen, mit Hut und Blumenstrauß) (auf Rundsockel)
ca. 1913/15
4677,58,16 GS Wien (Firmenmarke ohne Ang.), H. 53,3 cm; eBay # 46121319; Bem.: Brief fehlt hier.
-> vgl. ....= Nr. unleserlich durch Restauration, gem. lig. HT, gem. rot VIII, Firmenmarke Adler 16, H. 36 cm; Wien Museum Inv. 219.156

4678
Petri (d.i. Petrides, János) (Johann)
Stehendes, sich verneigendes Biedermeiermädchen (ihr Kleid etwas hochhebend) (auf Rundsockel)
ca. 1913/15
4678,1,11 Firmenmarke Adler 14, unsign.?, H. 22 cm; eBay # 2578928757
ohne Ang., Firmenmarke 19, sign. Petri, Ausformung 20er Jahre, H. 21,5 cm; Herr 22

4680
Haniroff
Stehender Knabe, einen Brief lesend (in knielangen Hosen und Strohhut) (auf rechteckigem Landschaftssockel)
ca. 1913/15
4680,22,34 Firmenmarke Sculpture 13, H. 95 cm; Millon & Robert 14.11.1995; Bem.: Brief fehlt hier.

4684
Podola(k) (d.i. Czapek, Stanislaus)
„Ballpause“ (Sitzende Krinoline, eine Maske haltend)
ca. 1915/16
FAYENCE
4684,74,14 (Firmenmarke ohne Ang.), sign. Podola, H. 28,5 cm; Christie's SK 5.12.1997: Bem.: Hier als Modell-Nr. 4864 angegeben.
4684,349,65 gem. H.H., gem. grün F 23, Firmenmarke Adler 16, sign. Podola, Ausformung 20er Jahre, H. 27,7 cm; Schloss Rick. 11
4684,397,83 (Firmenmarke ohne Ang.), sign. A. Podolak, H. 30 cm; Doro 31.5.1990; Bem.: Hier als Modell-Nr. 4694 angegeben.
4684,570,1 (Firmenmarke ohne Ang.), sign. Podola, (Ausformung 20er Jahre?), H. 27 cm; Wendl 36; Dawo 123
4684,593,65 gem. XXVII, Firmenmarke 19, sign. Podola, Ausformung 20er Jahre, H. 27,3 bzw. 28,5 cm; BM 44; 54; Kett. 148
4684 (ohne Ang.), sign. „Padala“, H. ca. 27,5 cm; Henry's 29.11.1995
ALABASTER
4684,W.62 (Firmenmarke Plak. ohne Ang.), sign. Podola, H. 32 cm; Doro 15.9.2003
Bem.: Dieses Modell wurde in Fayence und Alabaster ausgeführt.
-> vgl. 4883
-> vgl. 4884 FIGUR / LAMPE
Lit.: N.N.: Wiener Kleinplastik einst und jetzt, in: MW, 2.1920/21, Nr. 12, S. 28 (m. Abb.), hier mit Künstlerang. Stanislaus Czapek; MW, 3.1921/22, Nr. 5, S. 37 (m. Abb.), hier mit Künstlerang. Stanislaus Czapek.

4685
Wernekinck, Sigismund
Fuchs
ca. 1915
4685,118,12 ger. M, Firmenmarke 19, Made in Austria, Ausformung 20er Jahre, H. 10,8 cm, L. 23 cm; Wien Museum Inv. 190.039

4686
Padola
Sitzendes Mädchen, einen blumengeschmückten Hut haltend (auf Rechtecksockel)
ca. 1915/16
4686,47,15 (Firmenmarke ohne Ang.), H. fehlt; WKH

4687
Gemignani, Valmore
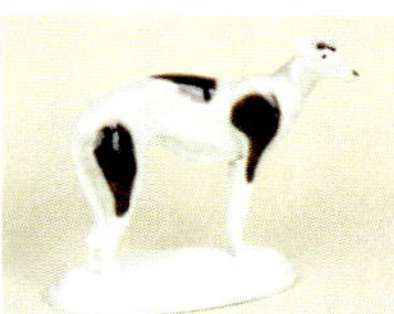
Stehender Windhund
ca. 1911/12
4687,269,25 ger. B, gem. FF, Firmenmarke 19, sign. Gemignani, Ausformung 20er Jahre, H. 26 cm; PB Wien 19
4687,306,11 Firmenmarke 19, Made in Austria, Ausformung 20er Jahre, unsign?, H. 25,4 cm; eBay # 2195811442
4687,358,? (Firmenmarke ohne Ang.), sign. Latour (!), (Ausformung 20er Jahre?), H. 23,5 cm; KaM 91; Bem.: Die Sign. „Latour“ dürfte wohl als Versehen des Modelleurs angesehen werden.
Bem.: Wenn auch hier ein mit „Latour“ sign. Exemplar nachgewiesen werden kann, so ist doch eindeutig der Entwurf dieses Windhundes dem Künstler Valmore Gemignani zuzuordnen, der für GS etliche dieser Tierfiguren entworfen hat. Dies wird u.a. auch durch die Höhenvariante Modell-Nr. 3859 belegt.
-> vgl. 3859

4688?
Zeiller, Paul
Hahn
ca. 1915/16
4688(oder 4888?), 6,10 Firmenmarke Adler 15, Firmenmarke Sculpture 13 (gepr.), sign. „Zailler“, H. 22 cm; KH
ohne Ang., unsign.?, H. 22 cm; SJ 26.1997, Nr. 11, S. 208 (m. Abb.)
Bem.: Wenn auch bei der Sign. „Zailler“ eine andere Namensschreibweise vorliegt, so kann dieser Entwurf dem Tierbildhauer Paul Zeiller zugeordnet werden.

4689
Tacke
Perlhuhn
ca. 1916
4689,21,10 Firmenmarke Adler 16, Firmenmarke Sculpture 13, H. 19,5 cm, L. 30 cm; Doro 4.6.1996
4689,22,7 Firmenmarke Adler 16, Firmenmarke Sculpture 13, unsign., H. 19,5 cm, L. 29 cm; Wien Museum Inv. 190.046
-> vgl. 4744

4690
Lindner, Helmut
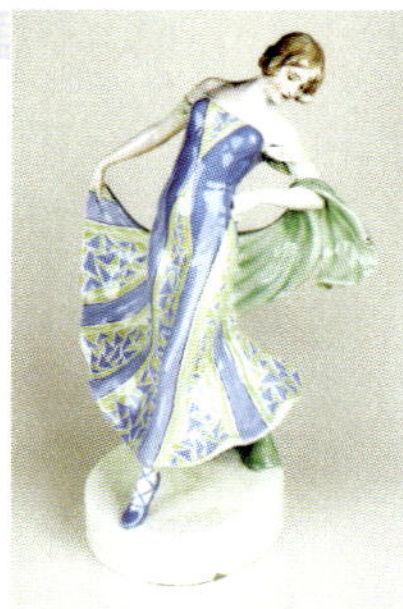
Tänzerin in drehender Bewegung (auf Rundsockel)
ca. 1913/14 (für den Entwurf)
4690,33,10 Firmenmarke 14, Firmenmarke Sculpture 13, H. 34,3 cm; eBay # 1302860824
4690,34,2 gem. A, VIII, Firmenmarke 19, Ausformung 20er Jahre, H. 33 cm; Mehlis 6.3.2004 (Sonderaukt. Sammlung Langer)
-> vgl. 4609; 4749

4695
Schmidt-Kestner, Erich
Zwei Pudel (auf Ovalsockel)
ca. 1916
4695,2,11 Firmenmarke Adler 14?, Firmenmarke Sculpture 13, H. ca. 20 cm; KH

4697
Tell, E.
Sitzende junge Frau (Akt) mit Lyra (Sappho?)
ca. 1907 (für den Entwurf)
4697,?,? (Firmenmarke ohne Ang.), sign. E. Tell, H. 23 cm; KH; Bem.: Die weiteren Nummern nach der Modell-Nr. waren unleserlich.
-> vgl. 3577; 3576; 3577; 4625

4699
Schoerk, Hans / Haniroff
Sitzender Pan mit Panflöte
ca. 1916/17
4699,16,24 (Firmenmarke ohne Ang.), sign. „Schörck“, H. 50 cm; Metz 13.3.1993

4700
Haniroff
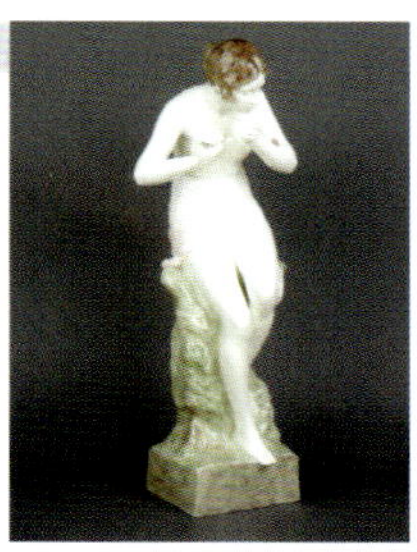
Sitzende Nymphe mit Flöte
ca. 1916/17
4700,47,5 gem. L., Firmenmarke 19, Ausformung 20er Jahre, H. 52 cm; PB Wien 19
4700,70,65 gem. F. gem. lig. JH, Firmenmarke 19, Made in Austria, Hand dec., Ausformung 20er Jahre, H. 50,3 cm; GK, S. 18, 102 (m. Abb.); Wien Museum Inv. 223.909
Ausst.: HMW (jetzt Wien Museum) 1985/86.
Pendant -> vgl. 4699 (Schoerk/Haniroff)
Lit.: GK, S. 18, 102 (m. Abb.)

4701
Hentschel (d. i. Handzel, August)
Stehendes Bauernmädchen in Tracht (Österreich/Böhmen?) (auf profiliertem Rundsockel)
ca. 1916/17
4701,7,10 Firmenmarke Adler 14, Firmenmarke Sculpture 13 (gepr.), sign. Hentschel, H. 39 cm; GK, S. 16, 91 (m. Abb.); Wien Museum Inv. 182.068
4701,95,10 gem. XXIII, Firmenmarke 19, Made in Austria, sign. Hentschel, Ausformung 20er Jahre, H. 38 cm; Wien Museum Inv. 219.164
Ausst.: HMW (jetzt Wien Museum) 1985/86.
–> vgl. 4612= ähnl. Modell (Pendant?)
Lit.: GK, S. 16, 91 (m. Abb.)

4708
Meier, (Emil oder Meier-Michel, Johanna)
Stehendes Mädchen mit Fächer
ca. 1916/17
4708,9,10 (Firmenmarke ohne Ang.), sign. „Padola", H. 30,6 cm; Doro 10.3.2003; Bem.: Die Sign. „Padola" kann hier nur als Versehen des Modelleurs angesehen werden. Das Modell ist Emil oder Johanna Meier(-Michel) zuzuordnen.
4708,35,? (Firmenmarke ohne Ang.), unsign., H. 31 cm; Fischer Luz. 388
4708,54,11 gem. XII A., Firmenmarke 18?, sign. Meier, Ausformung 20er Jahre, H. 30,5 cm; Doro 24.11.1992

4709
Kinderpaar im Rokokogewand, das Mädchen auf einer Bank sitzend
ca. 1916/17
4709,101,4 (Firmenmarke ohne Ang.), unsign.?, H. fehlt (ca. 20 cm?); Doro G 7.11.1995

4712
Lindner, Helmut
FIGUR/LAMPE Junge, sich verneigende Biedermeierdame mit drei Rosen
ca. 1913/14
FIGUR
4712,32,16,4 Firmenmarke Sculpture 13 (gepr.), H. 23 cm; WKA 14
4712,124,3,3 (Firmenmarke ohne Ang.), H. 22 cm; Fischer Luz. 388
4712,185,20 gem. lig. JF XXVI., Firmenmarke 13a, unsign., H. 23,3 cm; eBay # 121501819
4712,286,14 (Firmenmarke ohne Ang.), H. 22,5 cm; WKA 14
4712,358,1 gem. F.R.II, Firmenmarke 19, unsign.?, Ausformung 20er Jahre, H. 22 cm; Quit. 9
4712,409,16 gem. F.VIII, Firmenmarke 19, Ausformung 20er Jahre, H. 21,5 cm; Wien Museum Inv. 190.108
4712,410,5 gem. F.IX, Firmenmarke 19, Ausformung 20er Jahre, H. 21,5 cm; Kett. 192; BM 83
4712,420,2 Firmenmarke 19, Ausformung 20er Jahre, H. 22 cm; KH
4712,456,11 (Firmenmarke ohne Ang.), H. 22 cm; Doro WN 20.10.2004
LAMPE
4712,364,10 Firmenmarke 19, Ausformung 20er Jahre, Gesamth. 46 cm; Nagel 349
Bem.: Dieses Modell wurde auch noch in den USA unter der GS-Everlast Corp. produziert (Modell-Nr. 844)
–> vgl. 4511; 5491

4713
Petri (d.i. Petrides, János) (Johann)
„Rast" (Sitzendes Bauernmädchen mit Kopftuch)
ca. 1916/17
FAYENCE
4713,12,? Firmenmarke 9, Firmenmarke Adler 14, unsign.?, H. 14,4 cm; BM 55; 56
4713,111,9 (Firmenmarke ohne Ang.), sign. Petri, H. 14,5 cm; Doro 20.11.1996; 15.3.2005
Bem.: Dieses Modell wurde in Fayence und farbigem Marmor/(Alabaster?) ausgeführt.
Lit.: Trenkwald, H.: Keramik, Glas, Textilien, in: MW, 2.1920/21, Nr. 12, S. 17 (m. Abb.), hier Ausführung in „farbigem Marmor" und mit Ang. des Künstlers „J. Petrides"; Neuwirth (1974), S. 70.

4714
Petri (d.i. Petrides, János) (Johann)
Sitzendes Mädchen
ALABASTER
ca. 1916/17
4714,W,12 ger., Plak. A, sign. Petri, H. 13,5 cm; Wien Museum Inv. 186.678
ohne Ang., unsign.?, H. 14 cm; eBay # 25150516

4720
Meier-Michel, Johanna

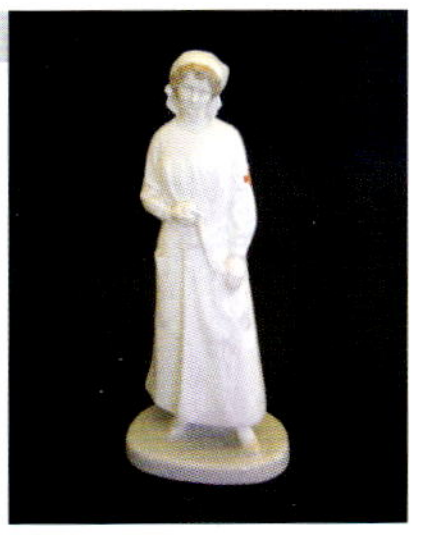

Stehende Krankenschwester (Rotkreuzschwester)
ca. 1916/17
4720,20,10 Firmenmarke Adler 15, Firmenmarke Sculpture 13 (gepr.), sign. Johanna Meier-Michel, Originalpick: 40 Kronen, H. 29 cm; Sammlung.schmutz@chello at 110
4720,47,10 Firmenmarke Adler 14, Firmenmarke Sculpture 13, sign. Johanna Meier-Michel, H. 28,5 cm; Wien Museum Inv. 190.133
ohne Ang., sign. Johanna Meier-Michel, H. 29 cm; Herr 29
Bem.: Hier zwei von den wenigen Stücken, die mit dem ausgeschriebenen Namen der Künstlerin sign. sind.
–> Vgl. auch hierzu die Bem. bei Modell-Nr. 4763.
Ausst.: SSH 1992
–> vgl. 4725

4725
Meier-Michel, Johanna
Stehende Krankenschwester (Rotkreuzschwester)
ca. 1916/17
4725,40,35 Firmenmarke Sculpture 13, Firmenmarke Adler 14, sign. Meier, H. 20,4 cm; Doro 15.9.2003
–> vgl. 4720

4726
Petri (d.i. Petrides, János) (Johann)
BÜSTE Kleines Mädchen (nach links blickend)
ALABASTER/Marmorsockel
ca. 1916/17
4726,JR lig.,13 ger., Plak. A, sign. Petri, H. 9,5 cm; eBay # 879862808

4727
Petri (d.i. Petrides, János) (Johann)
BÜSTE Biedermeiermädchen mit Schutenhut
ALABASTER
ca. 1916/17
4727,V,129 ger., Plak. E, Ausführung 20er Jahre, H. 8,6 cm; PB Wien 19

4730
Petri (d.i. Petrides, János) (Johann)
BÜSTE Kleines Mädchen mit Hut
ALABASTER
ca. 1916/17
4730,W,114 ger., Plak. E, Made in Austria ger., sign. Petri, Ausformung 20er Jahre, H. 8,3 cm; eBay # 1458838797

4739
Petri (d.i. Petrides, János) (Johann)
BÜSTE Bub mit Kappe
ALABASTER
ca. 1916/17
4739,HM,385 ger., Plak. E, sign. Petri, Ausformung 20er Jahre, H. 9,8 cm; Doro 9.6.1998
Pendant –> vgl. 4740

4740
Petri (d.i. Petrides, János) (Johann)
BÜSTE Mädchen mit Kappe
ALABASTER
ca. 1916/17
4740,HM,488 ger., Plak E, sign. Petri, Ausformung 20er Jahre, H. 10 cm; KH
4740,HM,493 ger., Plak E, sign. Petri, Ausformung 20er Jahre, H. 10 cm; Doro 9.6.1998
Pendant –> vgl. 4739

4741
Hoyos
Sitzendes Mädchen mit Hasen (Kaninchen) (auf Rechtecksockel)
ca. 1916/17
4741,3,10 Firmenmarke Adler 14, Firmenmarke Sculpture 13 (gepr.), unsign.?, H. 14 cm, L. 16 cm; eBay # 921962287
Lit.: Neuwirth (1974), S. 58, 64, hier nur mit Ang. des Künstlers „Hoyos", der sich jedoch in der Lit. nicht nachweisen lässt. Möglicherweise auch Lesefehler, bzw. undeutlich sign.

4744
Tacke
Perlhuhn
ca. 1916
4744,106,65 gem. I.D, Firmenmarke 19, Ausformung 20er Jahre, H. 8 cm; WKA 16
–> vgl. 4689

4745
Schörck (d.i. Hans Schoerk)
Stehender Knabe mit großen Augen (Google Eyes) (auf Rundsockel)
ca. 1916/17
4745,28,4 gem. XXVII, Firmenmarke 25, sign. Schörck, gepr. „P"= Z3, Ausformung 20er Jahre, H. 25 cm; Wien Museum Inv. 220.742

4747
Cherc
Stehende Frau in Tracht (mit Haube und Blumenstrauß) (auf Rundsockel)
ca. 1916/17
4747,15,9 gem. F., Firmenmarke 19, Made in Austria, Ausformung 20er Jahre, H. 34 cm; eBay # 917535794; 931476961; 3349398534
Bem.: Bei der Tracht dürfte es sich um eine balkanische handeln (Ungarn, Rumänien etc.). Endgültig konnte dies jedoch nicht geklärt werden.

4749
Lindner, Helmut

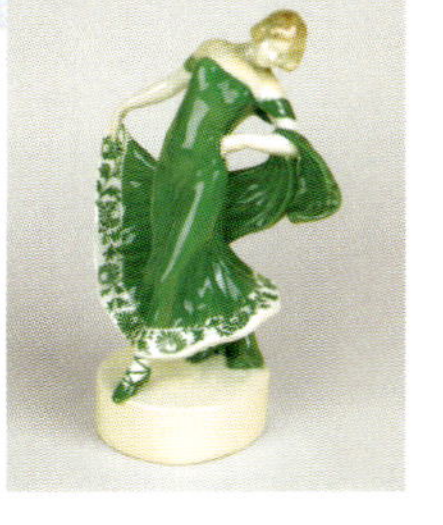

Tänzerin in drehender Bewegung (auf Rundsockel)
ca. 1912/14
4749,3,22 (Firmenmarke ohne Ang.), H. 24,5 cm; Schloss Hag. 40; Ahlden 129
4749,49,10 ger. X, gem. 14., Firmenmarke Adler 14, Firmenmarke Sculpture 13 (gepr.), H. 26 cm; PB Wien 19
4749,192,3 (Firmenmarke ohne Ang.), Ausformung 20er Jahre, H. 24 cm; Stahl 5.9.1992
–> vgl. 4609; 4690
Lit.: Pinhas, S. 103 (m. Abb.).

4753
Tell, E.
BÜSTE Dame mit entblößter Schulter
ALABASTER
ca. 1916/17
4753,HM,14 ger., Plak. A, H. 17,4 cm, B. 18,3 cm; KNS 6; Keup 146
4753,W,1 ger., (Firmenmarke ohne Ang.), H. 18 cm; Mehlis 31

4757
Hentschel (d.i. Handzel, August)
„Erwartung" (Sitzendes, kleines Mädchen in Tracht, mit Regenschirm)
ca. 1916/17
4757,105,8 gem. III, Firmenmarke 25, sign. Hentschel, Ausformung 20er Jahre, H. 25,4 cm; eBay # 2544843559
4757,206,15 (Firmenmarke ohne Ang.), sign. Hentschel, H. 23,2 cm; Doro 20.3.2006
Bem.: Zum Künstler vgl. Bem. bei Modell-Nr. 4390.
Lit.: Trenkwald, H.: Keramik, Glas, Textilien, in: MW, 2.1920/21, Nr. 12, S. 17 (m. Abb.), hier mit Ang. des Originaltitels „Erwartung" und mit Ang. des Künstlers „A. Handzel".

4761
Schwetz-(Lehmann, Ida)
Flora (Tanzendes Mädchen mit blumengeschmücktem Haar) (auf oktogonalem Sockel)
ca. 1916/17
4761,32,8 gem. lig. JF, Firmenmarke Adler 16, Firmenmarke Sculpture 13, sign. Schwetz, H. 26,4 cm; MAK Wien, Inv. Ke 10521-39
4761,35,4 gem. III 1, Firmenmarke Sculpture 13, handschriftl. „GS Wien 295,20.H", sign. Schwetz, H. 26,5 cm; Wien Museum Inv. 190.123

4763
Meier-Michel, Johanna

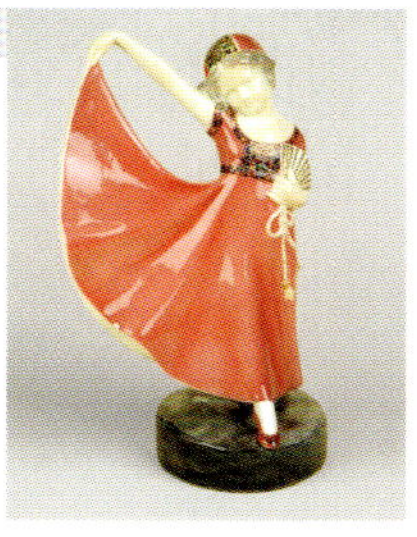

Stehendes Mädchen mit Fächer (den Rock hochhebend) (auf Rundsockel)
ca. 1916/17
FAYENCE
4763,11,65 Firmenmarke Adler 16, Firmenmarke Sculpture 13 (gepr.), sign. Meier, H. 27, 9 cm; eBay # 1215192648
4763,38,14 gem. 20, gem. lig. AH, Firmenmarke Adler 16, Firmenmarke Sculpture 13 (gepr.), sign. J. Meier-Michel, H. 28 cm; PB Wien 19; Bem.: Hier wieder eines der wenigen Stücke, das mit dem ausgeschriebenen Namen der Künstlerin sign. ist.
4763,189,14 (Firmenmarke ohne Ang.), sign. Meier, H. 27 cm; Doro 13.9.2004
ALABASTER
4763,F,18 ger., Plak. E, sign. Meier, H. 25,4 cm; eBay # 861779786
Bem.: Dieses Modell belegt die Schwierigkeit der Unterscheidung von mit „Meier" und „Meier-Michel" sign. Werken. Die Sign. „Meier" wurde von Emil Meier und zugleich mehrfach von seiner Frau, der Künstlerin Johanna Meier-Michel, verwendet. Da beide Entwerfer auch noch stilistisch ähnlich arbeiteten, lassen sich die Modelle eindeutig nur die durch die in der zeitgenössischen Literatur mit unterschiedlichen Angaben versehenen Reklamen bzw. anhand der Ang. in Ausst.-Katalogen sowie der mit „Meier-Michel" sign. Werke auseinander halten. Dieses Modell wurde in Fayence und Alabaster/(Kunststein?) ausgeführt.

4767 s. 4757

4769
Schoerk, Hans
Stehender Putto mit Blumengirlande (auf Rundsockel)
ca. 1916/17
4769,98,3 gem. IV., Firmenmarke Sculpture 13 (gepr.), unsign.?, H. 34,8 cm; eBay # 7340570647
4769,126,65 Firmenmarke 13a?, sign. Schörck, H. 35 cm; Doro 24.10.1995; Bem.: Hier Ang. „Schärek", es müsste sich aber um „Schörck" und somit um den Künstler Hans Schoerk handeln.
4769,158,10 (Firmenmarke ohne Ang.), sign. Schörck, H. 35,3 cm; Doro 12.9.2006
ohne Ang., sign. Schörck, H. 35 cm; BM 18

4770
Stoltenberg-Lerche, Hans
Sitzender, Flöte spielender Pan
ca. 1900 (für den Entwurf)
4770,86,10 (Firmenmarke ohne Ang.), sign. H.S. Lerche, Ausformung ca. 1915, H. 15,5 cm; Quit. 6; Bem.: Da im Kat. die Firmenmarke nicht eindeutig bezeichnet ist, lässt sich auch die Zeit der Ausformung für dieses Exemplar nicht exakter angegeben.
4770,135,35 Firmenmarke 25, sign. Lerche, Ausformung 20er Jahre, H. 16 cm, B. 14 cm; eBay # 225075813
–> vgl. 5032 DECKELDOSE

4773
Padola
Pierrot und Pierrette mit Musikinstrumenten (auf Rundsockel)
ca. 1913/14
4773,230,2 ger. K, Firmenmarke 18, Ausformung ca. 1920, H. 30 cm; Wendl 44
–> vgl. 4476
Lit.: Neuwirth (1974), S. 58, 69

4774
Padola/Podany, Rudolf
Sitzende Pierrette mit Laute
ca. 1916/17
4774,58,35 ger. lig. KW, Firmenmarke Sculpture 13, Firmenmarke 13a, sign. Padola, H. 39 cm (gemessen mit Laute!); GK, S. 16, 92 (m. Abb.); Wien Museum Inv. 182.086
4774,90,65 gem. X, GS (Firmenmarke ohne Ang.), sign. Podany, H. 33 cm; Soth. L. 3.5.1991
4774,116,1(oder 4?) gem. XXIII, Firmenmarke 22, Ausformung 20er Jahre, sign. Podany, H. 33 cm; eBay # 1232158541; # 1280260939
4774,157,1 (Firmenmarke ohne Ang.), sign. Podany, H. 33,5 cm; WKA 22
Bem.: Dieses Modell belegt die Schwierigkeit der Unterscheidung von mit „Padola/Podola..." und „Czapek/Podany" sign. Werken. Die Signaturen Padola etc. fanden häufig bei GS Verwendung. Da die Entwürfe und Genres dieser Figuren ähnlich sind, lassen sich die Modelle eindeutig nur durch die in der zeitgenössischen Literatur mit unterschiedlichen Ang. versehenen Reklamen bzw. anhand der Ang. in Ausst.- Katalogen sowie der mit Czapek/Podany sign. Werke auseinander halten. Es ist davon auszugehen, dass die Namensformen „Padola/Podola..." nur Pseud. sind. Sie lassen sich in der einschlägigen Lit. nicht gesondert nachweisen.
Ausst.: HMW (jetzt Wien Museum) 1985/86.
–> vgl. 4996
Lit.: GK, S. 16, 92 (m. Abb.)

4781
Puchegger, Anton
Adler (auf Felssockel)
ca. 1913/14 (für den Entwurf)
4781,112,6 gem. S, D I, Firmenmarke 19, Made in Austria, Ausformung 20er Jahre, H. 19 cm, L. 25 cm; eBay # 897779321
4781 (ohne Ang.), Firmenmarke 19, Ausformung 20er Jahre, H. 20 cm; Spik 540; Bem.: Das AH gibt als Modell-Nr. 4731 an. Eindeutig richtig ist aber 4781, dies konnte an dem zweiten hier nachgewiesenen Exemplar selbst geprüft werden.
–> vgl. 4533; 4782

4782
Puchegger, Anton
Adler (auf Felssockel)
ca. 1913/14 (für den Entwurf)
4782,38,65 Firmenmarke Adler 17?, Firmenmarke Sculpture 13, H. 15 cm; Schuler 68
4782,72,2 (Firmenmarke ohne Ang.), H. 15 cm; Doro 6.4.1999
–> vgl. 4533; 4781

4788
Petri (d.i. Petrides, János) (Johann)
BÜSTE Bub (etwas nach rechts blickend)
ca. 1913/14
ALABASTER
4788,TK,11 ger., Plak. A, sign. Petri, H. 12,5 cm; eBay # 5329414409

4793
Schoerk, Hans
BÜSTE Frau mit Haube/Hut
ALABASTER
ca. 1916/17
4793,F,40 ger., Plak. E, sign. „Scherk", Ausformung 20er Jahre, H. 15,5 cm; Wien Museum Inv. 186.680

4794
Valentin, Max
Schreitende Tänzerin, ihr langes Kleid seitlich hochhebend (auf getrepptem Ovalsockel)
ca. 1913/14 (für den Entwurf)
4794,40,9 (Firmenmarke Adler ohne Ang.), Firmenmarke Sculpture 13, H. fehlt; Quit. H. 28.2.2004
4794,65,10 Firmenmarke Sculpture 13, (und Firmenmarke ohne Ang.), H. 35,2 cm; BM 62
4794,130,10 gem. N, Firmenmarke 22, Ausformung 20er Jahre, H. 34 cm; eBay # 3779586133
–> vgl. 4496

4797
Jarl, Karin
Stehender Wolfs(Schäfer?)hund (auf abgerundetem Längssockel)
ca. 1917
4797,44,5 Firmenmarke 19, Ausformung 20er Jahre, H. ca. 15 cm; Doro ohne Kat. 16.4.1998

4798
Schwetz-Lehmann, Ida
„Rottraut" (Kniendes Mädchen mit erhobenen Armen) (auf oktogonalem Sockel)
ca. 1917
4798,10,14 GS Wien (Firmenmarke ohne Ang.), sign. Schwetz, H. 42 cm; Soth. com # 6 LLT
Lit.: N.N.: Wiener Kleinplastik einst und jetzt, in: MW, 2.1920/21, Nr. 12, S. 29 (m. Abb.), hier mit Ang. des Originaltitels „Rottraut" und der Künstlerin „I. Schwetz-Lehmann"; MW, 3. 1921/22, Nr. 6, S. 37 (m. Abb.), identische Reklame auch in MW, 3.1921/22, Nr. 8, 9, 11 (jew. m. Abb.).

4799
Podany, Rudolf
Tanzendes Kinderpaar (im Biedermeiergewand) (auf getrepptem Rundsockel)
ca. 1917
4799,209,35 Firmenmarke 13a, H. 28,5 cm; Soth. com. # 1LUX
4799,233,4 (Firmenmarke ohne Ang.), H. 26,5 cm; Doro 4.11.1998

4807
Maison, Rudolf

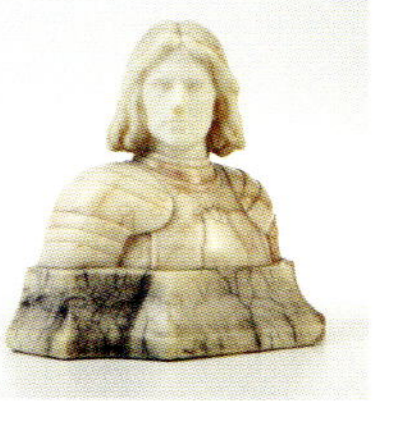

BÜSTE Jeanne D' Arc
MARMOR
ca. 1900 (für den Entwurf)
4807,W,2 ger., Plak. A, Ausformung ca. 1917, H. 23,3 cm, B. 24 cm; Wien Museum Inv. 186.684

4810
Tacke
Adler (auf Felssockel)
ca. 1917
4810,77,8 Firmenmarke 19, Made in Austria, sign. Tacke, Ausformung 20er Jahre, H. 26,5 cm; BM 42
ohne Ang., sign. „Take", H. 26 cm; Doro 13.9.1990; Bem.: Bei der Sign. könnte es sich auch um einen Druckfehler im Kat. handeln.

4814
Podany, Rudolf

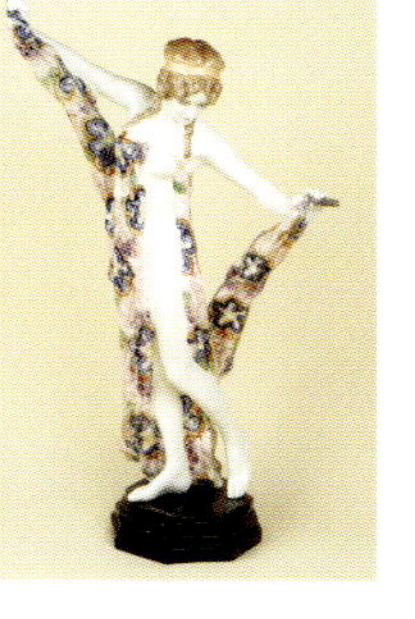

Adorée-Via Villany („Tanzspiel mit Bändern") (auf oktogonalem Sockel)
ca. 1917
4814,13,14 (Firmenmarke ohne Ang.), H. 58 cm; KaM 93
4814,40,8 gem. lig. AM, Firmenmarke 19, Ausformung 20er Jahre, H. 59 cm; PB Wien 19
4814,52,65 gem. lig. AS, XII, Firmenmarke 19, Made in Austria, unsign., Ausformung 20er Jahre, H. 59 cm; city-antik
4814 (ohne Ang.), H. 58 cm; BM 17; Bem.: Das AH gibt als Sign. „Padany" an, dies ist sicher ein Lese- oder Druckfehler im Kat.
Bem.: Dargestellt ist seinerzeit berühmt-berüchtigte (Nackt)-Tänzerin Adorée-Via Villany (eigentl. Erna Reich), in ihrer Tanz-Caprice „Tanzspiel mit Bändern" (–> vgl. Villany, S. 70–72 m. 3 Abb.). Da die Bänder für eine Keramikfigur zu fragil wären, wurden sie durch einen Schal ersetzt und geben so den ausgestreckten Armen der Tänzerin noch zusätzliche Stabilität. Wegen ihrer Tänze wurde Via-Villany laut Polizeibericht vom 16.3.1912 aus Bayern ausgewiesen. Die Bayerische Regierung warf ihr ein Vergehen gegen die Sittlichkeit (Unzucht) vor. Diese und zahlreiche weitere Anekdoten

sowie 154 Fotos aus dem Leben der Villany finden sich in ihrem Buch „Tanzreform und Pseudo-Moral", Paris 1912, 360 S. (m. Abb.).
Ausst.: SSH 1992
–> vgl. 4908
Lit.: GSN, S. 20, 21 (m. Abb.)
GS neu: Modell-Nr. 71 „Muse des Tanzes" (H. 50 cm), Aufl. 3000 Stück

4818
Podany, Rudolf
Stehende Dame mit entblößter Schulter und langem Kleid (auf Rechtecksockel)
ca. 1917
4818,15,11 (Firmenmarke ohne Ang.), H. 47 cm; Kett. 50

4819
Rudolff
Stehendes Biedermeiermädchen mit Schutenhut und Blumenstrauß (Krinoline)
ca. 1913/14 (für den Entwurf)
4819,137,8 gem. VIII., Firmenmarke 19, unsign.?, Ausformung 20er Jahre, H. fehlt; eBay # 71861617
4819,146,1 Firmenmarke 19, unsign.?, Ausformung 20er Jahre, H. 22,5 cm; WKH
ohne Ang., sign. Rudolf, H. 21 cm; Nagel 328
–> vgl. 4624; 4820
Lit.: Neuwirth (1974), S. 58, hier keine näheren Ang. zu Objekt oder Künstler.

4820
Rudolff
Stehendes Biedermeiermädchen mit Schutenhut und Blumenstrauß (Krinoline)
ca. 1913/14 (für den Entwurf)
4820,113,1 gem. XIX, Firmenmarke 22, sign. Rudolff, Ausformung 20er Jahre, H. 17 cm; eBay # 3352996061
–> vgl. 4624; 4819

4823
Jaray, Sándor (Alexander)
Sitzendes, nacktes Figurenpaar (Adam und Eva?)
ca. 1917
FAYENCE
4823,16,35 gepr. T, gem. XXIV, Firmenmarke 18, sign. Jaray, Ausformung ca. 1920, H. 17,6 cm; eBay # 2284217717
ALABASTER / Steinsockel
4823,R,7 ger., (Firmenmarke Plak. ohne Ang.), sign. „Jarey", H. 38,5 cm; Doro 18.4.1994; 3.5.1995
Bem.: Dieses Modell wurde in Fayence und Alabaster ausgeführt.

4828
Wernekinck, Sigismund
Eule (auf Felssockel)
ca. 1917
4828,65,14 Firmenmarke Adler 16, Firmenmarke Sculpture 13, H. 40,3 cm; Wien Museum Inv. 190.051

4831
Belleuse
Stehender Damenhalbakt (Badende mit großem Tuch) (auf Rundsockel)
ca. 1917
4831,41,16 Firmenmarke 19, Ausformung 20er Jahre, H. ca. 39 cm; eBay # 26258525 (Bem.: Laut Anbieter „Bellaruse" sign.); Christie's SK 14.11.2002 (Bem.: Hier mit der Ang. „Belleuse")
Bem.: Die undeutliche Sign. konnte nicht selbst geprüft werden, ein Künstler namens „Bellaruse" lässt sich in der einschlägigen Lit. nicht nachweisen, wahrscheinlicher ist die Variante „Belleuse", die bei Christie's angegeben wird.
Lit.: Pinhas, S. 103 (m. Abb.); hier mit der Ang. „Carrier-Belleuse".

4834
Podany, Rudolf
FIGUR/LAMPE „Bei der Toilette" (Sitzende Balldame im Rokokokleid) (auf Ovalsockel)
ca. 1917/18

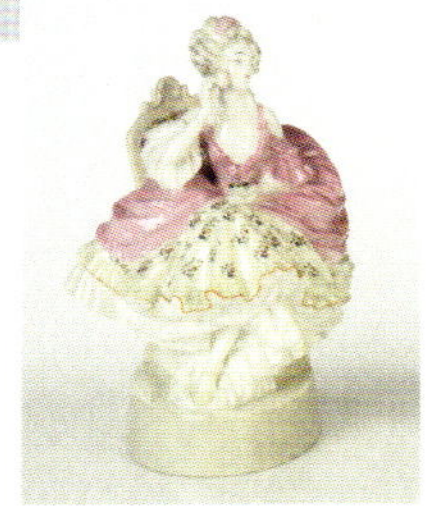

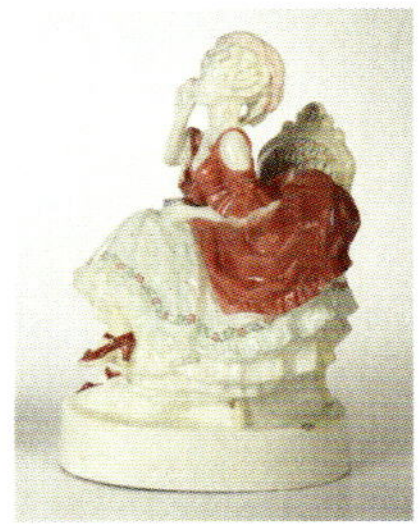

FIGUR
4834,5,35 Firmenmarke Sculpture 13, (und Firmenmarke ohne Ang.), H. 40 cm; Doro 23.5.1991
4834,33,10 gem. ST, 20.H., Firmenmarke Sculpture 13, Firmenmarke Adler 16, H. 39 cm; Wien Museum Inv. 219.153
4834,98,14 (Firmenmarke ohne Ang.), H. 38 cm; eBay # 2269648487
4831,175,? gem. lig. AH, XXVII D2, Firmenmarke 19, Made in Austria, Ausformung 20er Jahre, H. 38 cm; eBay # 1227991127
4834,228,10 gem. F VIII, Firmenmarke 22, Ausformung 20er Jahre, H. 37,5 cm; Wien Museum Inv. 190.106
4834,424,11 Firmenmarke 19, Made in Austria, Ausformung 20er Jahre, H. 38 cm; Fischer 71
4834,445,18 gem. FH XXVII D2, Firmenmarke 19, Made in Austria, Ausformung 20er Jahre, H. 35 cm; eBay # 63394748
4834,452,15 (Firmenmarke ohne Ang.), Ausformung 20er Jahre, H. 36,5 cm; Christie's SK 11.11.2004
4834,475,4 gem. lig. AH XXVII D2, Firmenmarke 19, Made in Austria, Ausformung 20er Jahre, H. 40 cm; eBay # 1284611586; 2070823006; 3705678865
4834,521,8 gem. M29, gem. lig. AH XXVII D2, Firmenmarke 19, Made in Austria, Hand dec., Ausformung 20er Jahre, H. 38 cm; Keup 62
4834,553,8 Firmenmarke 19, Ausformung 20er Jahre, H. fehlt; BM 78
4834,576,10 gem. N 2 AE, Firmenmarke 19, Made in Austria, Hand dec., Ausformung 20er Jahre, H. 38,1 cm; eBay # 105332619
4834,612,7 Firmenmarke 19, unsign.?, Ausformung 20er Jahre, H. 37 cm; Ruef 473
LAMPE
4834,436,11 (Firmenmarke ohne Ang.), Gesamth. (m. Lampenschirm) 67 cm; Christie's SK 14.3.1997
4834,473,11 gem. XXVIII, D2, gem. lig. AH, Firmenmarke 19, Made in Austria, Gesamth. (m. Lampenschirm) 68 cm, H. 44,8 cm; MAK Wien, Inv. Ke 10521-27
ohne Ang., H. (mit Fassung) 43,2 cm; eBay # 1214214327
Bem.: Wie die Auflistung der zahlreichen Stücke belegt, war dieses Modell eines der Beliebtesten in den ersten Jahren nach dem 1. Weltkrieg und Anfang der 20er Jahre.
–> vgl. 5169; 5329
Lit.: DD 50.1922/23, Nr. 8, S. 38 (m. Abb.), Nr. 9, S. 39 (m. Abb.), hier jeweils mit Ang. des Originaltitels „Bei der Toilette" und mit Ang. des Entwerfers Rudolf Podany.

4835
Gruber, (Franz?)
Stehende Biedermeierdame mit Schutenhut und Blumenbukett (auf oktogonalem Sockel)
ca. 1917/18
4835,76,3 (Firmenmarke ohne Ang.), H. 36,8 cm; eBay # 188939107

4837
Gruber, (Franz?)
Tanzendes Paar (Biedermeierdame mit Mädchen) (auf Rundsockel)
ca. 1917/18
4837,84,65 ger. M, FG Wien (Firmenmarke ohne Ang.), Firmenmarke Sculpture 13, H. 30,8 cm; BM 75; 76
4837,123,1 gem. F. III, (Firmenmarke ohne Ang.), H. 31 cm; eBay # 335089745
4837,138,7 gem. III, gem. lig. FG?, Firmenmarke 19, Ausformung 20er Jahre, H. 31 cm; KH
4837,185,5 gem. T XVII, FG Wien (Firmenmarke ohne Ang.), Firmenmarke 13a, H. 31 cm; Keup 62
4837,209,25 Firmenmarke 19, Ausformung 20er Jahre, H. 31,5 cm; Doro 19.9.2000
4837,221,8 (Firmenmarke ohne Ang.), H. 31,2 cm; Lemp. 653
Lit.: Neuwirth (1974), S. 58

4846
Meier, (Emil oder Meier-Michel, Johanna)
BÜSTE Knabe mit Schleife / Mascherl (Kronprinz Otto von Habsburg 1916)
ALABASTER
ca. 1917/18
4846,MR,3 ger., Plak. A, H. 12,6 cm; PB Wien 19
Bem.: Laut altem Zettelvermerk ist die dargestellte Person Kronprinz Otto von Habsburg 1916.

4850
Petri (d.i. Petrides, János) (Johann)
BÜSTE Bub mit Hut
ALABASTER
ca. 1917/18
4850,St.,5 ger., Plak. E, sign. Petri, Ausformung 20er Jahre, H. ca. 10 cm; KH

4854
Petri (d.i. Petrides, János) (Johann)
BÜSTE Kleines Mädchen
ALABASTER
ca. 1917/18
4854,H1,47 ger., Plak. E, sign. Petri, Ausformung 20er Jahre, H. 9,8 cm; Wien Museum Inv. 186.677

4860
Podany, Rudolf
BÜSTE Pierrot
ALABASTER
ca. 1917/18
4860,0,23 ger., Plak. E, sign. „Podoni", Ausformung 20er Jahre, H. 13 cm; Stahl 13.6.1998; Bem.: Ob es sich bei der Sign. „Podoni" um einen Lesefehler des AH oder um ein Versehen des Ausformers handelt, konnte am Original nicht nachgeprüft werden.

4864
Czapek, Stanislaus
BÜSTE Mädchen
ALABASTER / MARMOR?
ca. 1917/18
4864,(1?),K,2 ger., Plak. E, Ausformung 20er Jahre, H. 9 cm; / 9,5 cm; Stahl 13.6.1998; WKA 23; eBay # 2830064290
Lit.: Pinhas, S. 101 (m. Abb.).

4865
Padola
Sich verbeugende Tänzerin (mit Schal) (auf Rundsockel)
ca. 1917/18
4865,25,10 (Firmenmarke ohne Ang.), H. 31 cm; Karrenbauer 176

4866
Podany, Rudolf
BÜSTE Junge Dame mit Haarband
ALABASTER
ca. 1917/18
4866.BJ.8 ger., (Firmenmarke Plak. ohne Ang.), H. 16,5 cm; Doro 18.4.1994

4873
Wernekinck, Sigismund
Eule (auf Felssockel)
ca. 1917/18
4873,60,1 (Firmenmarke ohne Ang.), sign. „Wernekink", H. 27 cm; Doro WA 12.12.1992

4880
Horsetzky, Melanie von
Kind mit Rabe (auf Fels sitzend)
ca. 1917/18
4880,22,7 gem. II, Firmenmarke 18, Ausformung ca. 1920, H. 17,6 cm; Ahlden 85; Quit. 14

4882
Schoerk, Hans
BÜSTE Knabe
ALABASTER
ca. 1914/15
4882,BJ,23 ger., Plak. A, sign. „Scholz?" (laut Anbieter), H. 15,5 cm; eBay # 4143797205
Bem.: Es ist bei der angegebenen Sign. „Scholz" ein Lesefehler des Anbieters, bzw. ein Fehler des Modelleurs anzunehmen. Infrage kommt eindeutig der für Goldscheider vielfach tätige Künstler Hans Schoerk.

4883
Podola(k) (d.i. Czapek, Stanislaus)
FIGUR/LAMPE „Ballpause" (Sitzende Krinoline, eine Maske haltend)
ca. 1914/15
FIGUR
4883,169,5 Firmenmarke 19, Made in Austria, sign. Podolak, Ausformung 20er Jahre, H. 14,5 cm; eBay # 1460414425
LAMPE
4883,156,8 gem. IV., Firmenmarke 19, Made in Austria, sign. Podola, Ausformung 20er Jahre, Gesamth. 30,5 cm; eBay # 170645981
–> vgl. 4684 FIGUR
–> vgl. 4884 FIGUR/LAMPE
Lit.: vgl. bei 4684

4884
Podola (d.i. Czapek, Stanislaus)
FIGUR/LAMPE „Ballpause" (Sitzende Krinoline, eine Maske haltend)
ca. 1914/15
FIGUR
4884,99,10 Firmenmarke Adler 14, unsign.?, H. 30 cm; eBay # 2226717705
LAMPE
4884,198,20 Firmenmarke 19, unsign., Ausformung 20er Jahre, Gesamth. 46 cm; Wien Museum Inv. 219.151
–> vgl. 4684 FIGUR
–> vgl. 4883 LAMPE
Lit.: vgl. bei 4684

4888 s. 4688

4890
Lindner, Helmut
Tänzerin
ca. 1917/18
4890,196,35 (Firmenmarke ohne Ang.), H. 34 cm; Schuler 72

4892
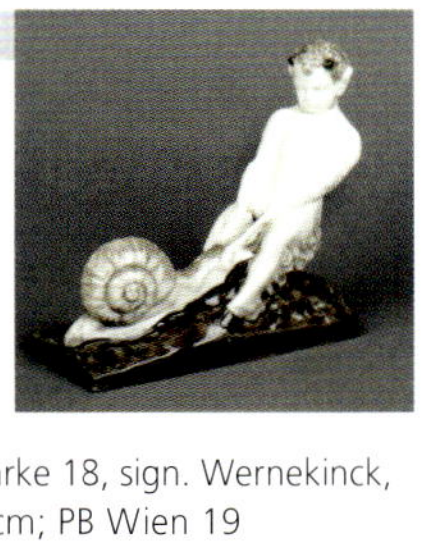
Kostial, Josef / Wernekinck, Sigismund
Faun mit Schnecke
ca. 1917/18
4892,5,5 Firmenmarke 13a, Firmenmarke Sculpture 13 (gepr.), sign. Kostial, H. 18 cm; PB Wien 19
4892,27,35 gem. XI, Firmenmarke 18, sign. Wernekinck, Ausformung ca. 1920, H. 18 cm; PB Wien 19
4892,37,2 gem. VI, GS (Firmenmarke ohne Ang.), Firmenmarke Sculpture 13, unsign.?, H. 16,5 cm; Kett. 60
4892,37,3 gem. VI, Firmenmarke Sculpture 13, unsign., H. 17,2 cm; Neuwirth (1974), S. 38, 39 (m. Abb.); WKA 9
4892 (ohne Ang.), sign. Wernekinck, L. 23 cm; RM 25
Bem.: Von welchem der beiden Künstler diese Gruppe stammt, lässt sich nicht mit letzter Sicherheit sagen. Von Kostial lässt sich in diesem Genre eine Gruppe „Duett" (Putto auf Vogel sitzend) in: Keramos, 2.1923, S. 309 (m. Abb.) nachweisen. Wernekincks Domäne war die Tierskulptur.

4893
Perigot
Stehendes Mädchen mit Haube und Buch, einen Blumenstrauß hinter dem Rücken versteckend (auf Rundsockel)
ca. 1912/13
4893,17,2 (Firmenmarke ohne Ang.), H. 20,2 cm; Doro 12.9.2006

4895/4897 ?
Polter
Sitzende Tänzerin, sich einen Schuh anziehend (auf Hocker)
ca. 1912/13
4895,15,9 gem. IX, Firmenmarke 22, Ausformung 20er Jahre, H. 22,9 cm; eBay # 1461692191

4897
4897 (ohne Ang.), H. 22 cm; Ahlden 54/55; 60/61
Bem.: Bei diesem Modell werden bei zwei Anbietern differierende Modell-Nrn. angegeben. Da beide Objekte nicht selbst in Augenschein genommen werden konnten, werden beide Nummern im WVZ geführt.
–> vgl. 4267 (auf Hocker)
–> vgl. 4267 (auf Stuhl)

4900
Krinolinendame mit Schutenhut
ca. 1918/19
4900,65,1 gem. S. IX, Firmenmarke 19, Ausformung 20er Jahre, H. ca. 20 cm; WKH
4900,615,11 gem. D. 2, III, Firmenmarke 19, Made in Austria, Hand dec., unsign.?, Ausformung 20er Jahre, H. ca. 20 cm; city-antik

4902
Stehender Putto mit Trauben (auf Rundsockel)
ca. 1918/19
4902,30,7 gem. VIII, Firmenmarke 19, Ausformung 20er Jahre, H. 40,6 cm; eBay # 2615980921; Doro 15.9.2003

4903
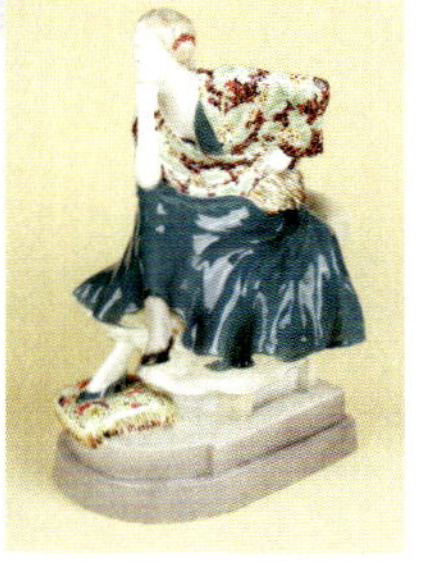
Kostial, Josef
FIGUR/LAMPE „Carmen" (Sitzende Dame mit Fußkissen)
ca. 1918/19
FIGUR
4903,120,8 Firmenmarke Sculpture 13, H. 37 cm; PB Wien 19
T(oder F?)4903,140,10 Firmenmarke 19, Ausformung 20er Jahre, H. 38 cm; Doro 22.11.1990
4903,441,20 gem. XIII F.N., Firmenmarke 19, Ausformung 20er Jahre, H. 36 cm; eBay # 316031990
4903,547,5 gem. D2, F, XXIII, Firmenmarke 19, Made in Austria, Ausformung 20er Jahre, H. 32 cm; Sammlung Stressless, Portugal
4903,585,11 (Firmenmarke 19, geschwärzt/übermalt), Made in Austria, Ausformung 20er Jahre, H. 34,3 cm; eBay # 3703251466
4903,591,65 gem. XXIII. D2., Firmenmarke 19, Made in Austria, Hand dec., Ausformung 20er Jahre, H. 35,5 cm; eBay # 181640203
LAMPE
4903,510,1 Firmenmarke 19, Ausformung 20er Jahre, H. 64 cm; Doro 4.3.2002
Lit.: N.N.: Wiener Kleinplastik einst und jetzt, in: MW, 2.1920/21, Nr. 12, S. 28 (m. Abb.); MW, 3.1921/22, Nr. 3, S. 39 (m. Abb.), hier jeweils mit Ang. des Originaltitels „Carmen".
–> vgl. 5235; 5326

4906
Stehendes Mädchen mit Blumenkranz im Haar und ausgestrecktem Arm (auf Blumengirlanden verziertem Rundsockel)
ca. 1918/19
4906,23,24 Firmenmarke 19, Ausformung 20er Jahre, H. 49 cm; WKA 44

4908
Podany, Rudolf
Adorée-Via Villany („Tanzspiel mit Bändern") (auf oktogonalem Sockel)
ca. 1917 (für den Entwurf)
4908,65,4 gem. AV, Firmenmarke 19, H. 32 cm; KH
Bem.: s. bei 4814
–> vgl. 4814
Lit.: GSN, S. 20, 21 (m. Abb.)

GS neu: Modell-Nr. 71 „Muse des Tanzes" (H. 50 cm), Aufl. 3000 Stück

4914
Gruber, (Franz?)
Tänzerin in Bewegung, ihr Kleid nach hinten hochhaltend (sechseckiger Sockel mit Blumenstrauß)
ca. 1918/19
4914,9,02 Firmenmarke 14?, H. 56 cm, B. 35 cm; Kett. 39

4915
Podola/Polter (d.i. Czapek, Stanislaus)
LAMPE Tanzendes Rokokopaar (Dame und Kavalier) (auf mit Blumengirlanden verziertem Rundsockel)
ca. 1912/13 (für den Entwurf)
4915,26,10 (Firmenmarke ohne Ang.), sign. Polter, Gesamth. 75 cm; KH
4915,43,11 gem. lig. AS, XI, gem. rot 105, Firmenmarke 22, Made in Austria, sign. Podola, Ausformung 20er Jahre, H. 53,3 cm; eBay # 376848208
–> vgl. 4271
–> vgl. 4305= ähnl. Modell
–> vgl. 4539= ähnl. Modell

4920
Postl, Hugo
Stehender Schäferhund (auf Ovalsockel)
ca. 1919
4920,230,16 gem. lig. AC., Firmenmarke 19, Made in Austria, H. 26,6 cm; eBay # 3738003699
4920,241,5 (Firmenmarke ohne Ang.), H. 26,6 cm, L. 30,5 cm; eBay # 406637761
4920,264,16 gem. AL., Firmenmarke 19, Made in Austria, Ausformung 20er Jahre, H. 25,4 cm; eBay # 1254907843
4920,367,15 gem. J. D1, Firmenmarke 19, Made in Austria, Ausformung 20er Jahre, H. 25,4 cm; eBay # 2642061754

4922
Podola (d.i. Czapek, Stanislaus)
Stehender Harlekin mit Laute und Colombine (auf oktogonalem Sockel)
ca. 1919
4922,82,1 ger. K, gem. VI, (ohne Firmenmarke), H. 49,7 cm; GK, S. 16, 94 (m. Abb.); Wien Museum Inv. 223.915
Ausst.: HMW (jetzt Wien Museum) 1985/86.
–> vgl. 4935; 5262; 5263 jeweils sign. „Podolak"= Harlekin einzeln
Lit.: GK, S. 16, 94 (m. Abb.)

4925
Putto, auf Blumenstrauß sitzend
ca. 1919
4925,7,32 Firmenmarke 13a, undeutl. sign., H. fehlt; Doro 20.10.1997
–> vgl. 4939= stilistisch ähnl. Modell
–> vgl. 4946= stilistisch ähnl. Modell

4935
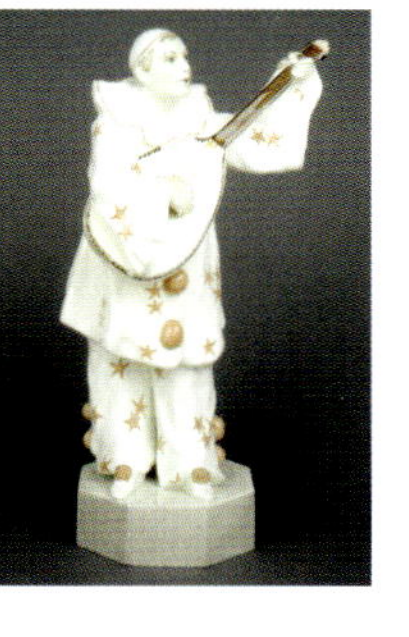
Padola/Podolak, A./Karl? (d.i. Czapek, Stanislaus?)
Stehender Harlekin mit Laute (auf Rundsockel/auf oktogonalem Sockel)
ca. 1919
Rundsockel
4935,10,10 Firmenmarke Sculpture 13, sign. Padola, H. 39 cm; Nagel 385
oktogonaler Sockel
4935,304,3 gem. RH, XIX, Firmenmarke 19, sign. Podolak, Ausformung 20er Jahre, H. 39,6 cm; PB Wien 19
4935,365,– (Firmenmarke ohne Ang.), sign. Podolak, H. 38,5 cm; Christie's SK 27.6.1997; 12.9.1997
–> vgl. 5262; 5263
–> vgl. 4922 (Padola) GRUPPE

4937
Kostial, Josef
„Duett" (Flöte spielender Putto auf Meise sitzend) (auf Natursockel)
ca. 1919
4937,4,– (Firmenmarken ohne genaue Ang.), unleserlich sign., H. 27 cm; Mehlis 36
Lit.: Meisenbach, J. A.: Putten mit Tieren, in: Keramos, 2.1923, S. 309 (m. Abb.), hier mit der Ang. des Originaltitels „Duett" und des Künstlers J. Kostial.

4938
Kostial, Josef
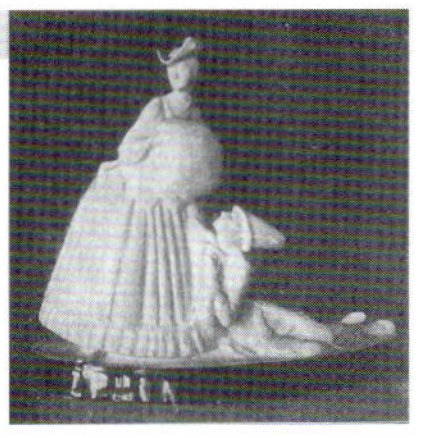
„Maskerade" (Lo Hesse und Joachim von Seewitz)
ca. 1919
4938,16,20 Firmenmarke 19, Made in Austria, sign. Kostial, Ausformung 20er Jahre, H. 34 cm;
eBay # 3714745633
4938,52,14 gem. lig. JF. XIX., Firmenmarke 18, sign. Kostial, Ausformung ca. 1920, H. 35,5 cm;
eBay # 2048139908; Bem.: Der Verkäufer gibt als Modell-Nr. 4038 an (KH www.tias.com/stores/frozenbeauty).
4938,107,35 (Firmenmarke ohne Ang.), sign. „Louger" (!), H. 35 cm; Doro 29.5.2001 Bem.: Die Sign. „Louger" ist eindeutig ein Versehen des Modelleurs.
4938,153,65 (Firmenmarke ohne Ang.), sign. Kostial, H. 34 cm, L. 36 cm; WKA 50
4938,160,90 (Firmenmarke ohne Ang.), sign. Kostial, H. ca. 35 cm; KH
Bem.: Dargestellt sind die beiden Tanzkünstler Lo Hesse (1889–1983?) und Joachim von Seewitz (1891–1966) mit ihrem Tanz „Maskerade". Wann und wo die Uraufführung dieser Tanzschöpfung stattfand, ließ sich schlussendlich nicht ermitteln. Das Deutsche Tanzarchiv in Köln bewahrt jedoch einen Programmzettel vom 21.7.1917 des Theaters an der Königgrätzerstrasse (Berlin) auf, der den Komponisten Moszkowski (Moritz Moszkowski, 1854–1925) nennt. Dieser Hinweis deckt sich auch mit dem bei Niecol, Bd. 2, S. 205 zit. Ausst.-Kat. der Deutschen Gewerbeschau 1922: „Die Bühnentänzerin Lo Hesse in der Tanzpantomime Maskerade (1917)". Für die Kostümentwürfe zeichnet Walter Schnackenberg (1880–1961) verantwortlich. –> vgl. Lit. Rochowanski (1923), S. 22 (m. Abb.); zu Schnackenberg –> vgl. TB, Bd. 30, S. 386; Vollmer, Bd. 4, S. 205; Bd. 6, S. 402; Die Alpträume und Visionen des Walter Schnackenberg, München 1984; Der Künstler Constantin Holzer-Defanti nahm sich desselben Themas an. Er schuf 1918 für die Firma Rosenthal eine Figur, die Lo Hesse („Dame mit Muff") in der gleichen Pose wiedergibt. –> vgl. hierzu: Niecol, Bd. 1, S. 79 (m. Abb.); Bd. 2, S. 205 (m. Abb.). Für den Hinweis auf den Entwerfer der Kostüme und zahlreiche weitere wertvolle Tipps zum Tanz und seinen Protagonisten danke ich Herrn Dr. Frank-Manuel Peter vom Deutschen Tanzarchiv in Köln recht herzlich.
Ausst.: Deutsche Gewerbeschau München 1922.
–> vgl. 5251= Lo Hesse (als Einzel-Figur)
Lit.: MW, 2.1920/21, Nr.12, S. 28 (m. Abb.); MW, 3.1921/22, Nr. 1, S. 39 (m. Abb.); DD, 50.1922/23, Nr. 7, S. 38 (m. Abb.); Burger, W.: Die Deutsche Gewerbeschau in München, Keramik und Glas, in: Dekor. Kunst, 25.1921, S. 265 (m. Abb.), hier bezeichnet als „Schwarz-Weiss Keramik"; bei allen Lit.-Stellen jeweils mit Ang. des Originaltitels „Maskerade".

4939
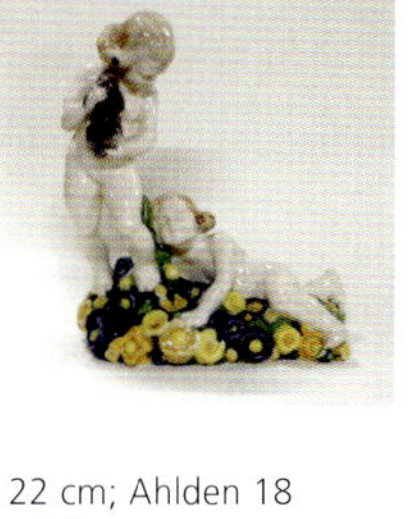
Zwei Putti mit einem Kaninchen (auf Blumensockel)
ca. 1919
4939,7,35 gem. IV, Firmenmarke 13a, H. 28,4 cm; MAK Wien, Inv. Ke 10521-19
4939,26,12 (Firmenmarke ohne Ang.), H. 26,5 cm; Doro 20.5.1996
4939 (ohne Ang.), H. 28 cm, L. 22 cm; Ahlden 18
–> vgl. 4925= stilistisch ähnl. Modell
–> vgl. 4946= stilistisch ähnl. Modell

4944
Postl, Hugo
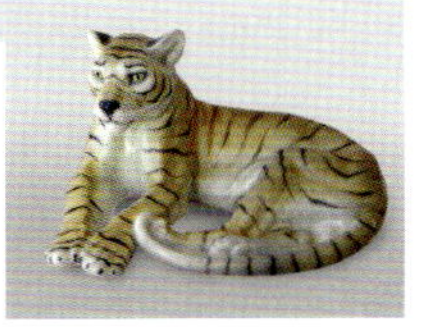
Liegender Tiger
ca. 1919
4944,93,14 gem. lig. TJ. X. D1, Firmenmarke 19, Made in Austria, Ausformung 20er Jahre, unsign., H. 12 cm, L. 21 cm; Sammlung Stressless, Portugal
–> vgl. 6322= ähnl. Modell

4946
Putto mit Schmetterling (auf Blumensockel)
ca. 1919
4946,30,65 gem. IX, Firmenmarke 19, Ausformung 20er Jahre, H. 19 cm; Neumeisters Moderne 26
4946,31,8 Firmenmarke 22, Ausformung 20er Jahre, H. 19,7 cm; eBay # 7302061095
4946,41 gem. K. 1., Firmenmarke 19, Made in Austria, Hand dec., Ausformung 20er Jahre, H. 18 cm; Doro 20.5.1996
–> vgl. 4925= stilistisch ähnl. Modell
–> vgl. 4939= stilistisch ähnl. Modell

4951
Kostial, Josef
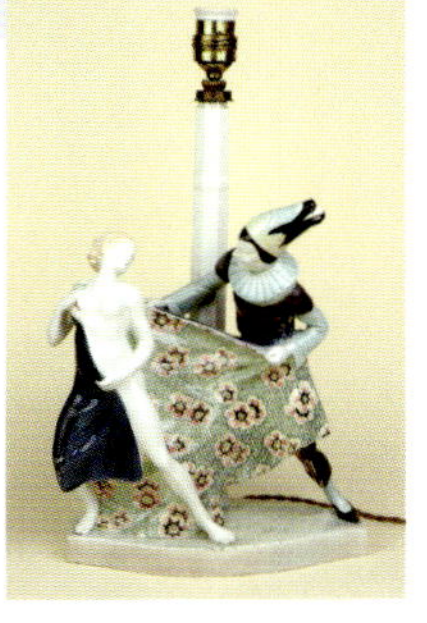
FIGUR / LAMPE Pierrot und Colombine (Halbakt) (auf passigem Sockel)
ca. 1919/20
FIGUREN (Gruppe)
4951,63,8 Firmenmarke 19, unsign.?, Ausformung 20er Jahre, H. 32/33 cm; L. 26 cm; Weichmann 7.11.1986; BM 54; RM 33 (unterschiedliche Höhenang. der drei AH).
LAMPE
4951,25,5 ger. K, gem. XV, Firmenmarke 22, Ausformung 20er Jahre, H. 48,6 cm; PB Wien 19
4951,30,1 gem. F, V (ohne Firmenmarke), sign. Lavergne, Ausformung 20er Jahre, H. der Figuren 40 cm; eBay # 2233360923; Bem.: Die Sign. „Lavergne" dürfte wohl als Versehen des Modelleurs angesehen werden.
4951,68,1 ger. K, gem. F., VI, Firmenmarke 22, sign. J. Kostial, Ausformung 20er Jahre, H. 40,5 cm; Sammlung Robert Senft
Bem.: Eine ähnl. Szene, ebenfalls von Josef Kostial gestaltet, kann unter dem Originaltitel „Maskerade" (Modell-Nr. 4938, –> vgl. dort auch Lit.) nachgewiesen werden. Dargestellt in dieser stilistisch verwandten Gruppe sind die Tänzerin Lo Hesse und der Tanzkünstler Joachim von Seewitz.

4958
Krebs?
DECKELDOSE
Form: länglich, hochoval
Knauf: längsoval
Dekor: Blattranken, Zackenfries
ca. 1920
4958,8,35 gem. XI., Firmenmarke 18, undeutl. sign. „Krebs"?, Ausformung ca. 1920, H. 10,8 cm; WKA 14

4969
Gult, Johann
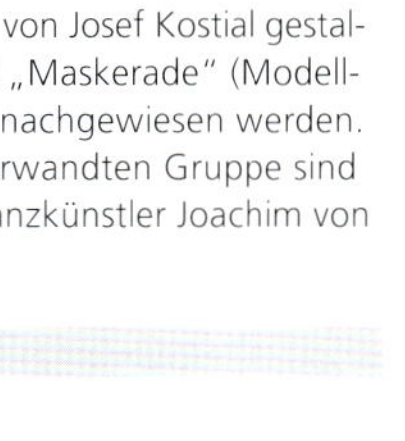
DECKELDOSE Taube
Form: rechteckig / balusterförmig
Knauf: Taube (vollplastisch)
Dekor: Ast- und Blattwerk mit Blüten
ca. 1915
4969,111,14 ger. lig. MI, gem. XII.2., Firmenmarke 19, Made in Austria, Hand dec., Ausformung 20er Jahre, H. 13 cm, L. 12,5 cm, B. 10 cm; Sammlung Irsigler, Wien

4975
Gult, Johann
DECKELDOSE Chinese
Form: quaderförmig
Knauf: sitzender Chinese (vollplastische Fig.)
Dekor: vielfarbig, tunkpapierähnl.
ca. 1920
4975,23,9 gem. XXVIII, Firmenmarke 19, H. 12 cm; WKA 22; Bem.: Hier Ang. der Künstlersign. „Galt" (vgl. Zühlsdorff T. 2, S. 409, der diesen Fehler bei Neuwirth (1974), S. 63, 471 übernommen hat). Richtig ist der Wiener Bildhauer Johann Gult (vgl. Künstlerliste).
4975,48,65 gem. lig. AT/AC?, Firmenmarke 19, sign. Gult, H. 12 cm, L. 9 cm, B. 9 cm; PB Stuttgart

4976
Gult, Johann (zugeschr.)
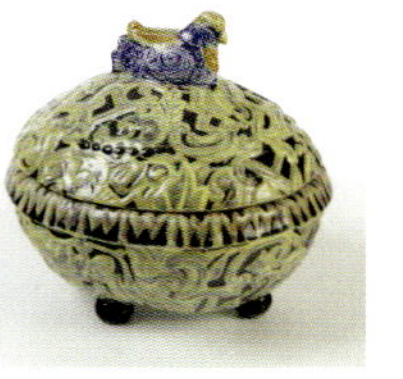
DECKELDOSE Vogel
Form: eiförmig, vier Kugelfüße
Knauf: Liegende Henne
Dekor: durchbrochene Metallarbeit suggerierend
ca. 1920
4976,11,35 gem. G, Firmenmarke 18, unsign., Ausformung ca. 1920, H. 13 cm; Wien Museum Inv. 190.005/1+2

4978
Thomasch, Wilhelm
Putto mit Vase auf der Schulter (sich auf einen Blumentopf kniend) (profilierter Rundsockel)
ca. 1920/21
4978,16,2 Firmenmarke 19, undeutl. sign. „Thoma..."?, H. 48,8 cm; Doro 21.5.1997; Bem.: Durch die schlecht leserliche Sign. kann der Entwurf Wilhelm Thomasch zugeordnet werden.

4981
Krebs
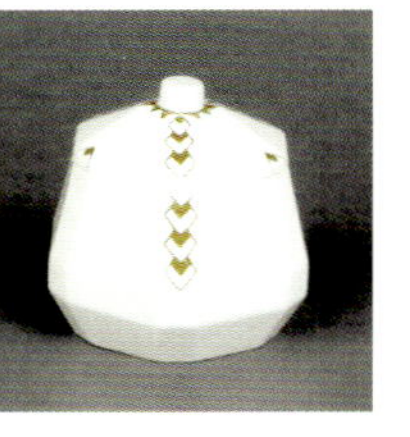
DECKELDOSE
Form: zwölfpassig (facettiert), gebaucht, zu Stand und Mündung konisch zulaufend
Dekor: vertikal lig. „Herzen"/ Rankenmuster, bzw. stilisierte Blumen
ca. 1920/21
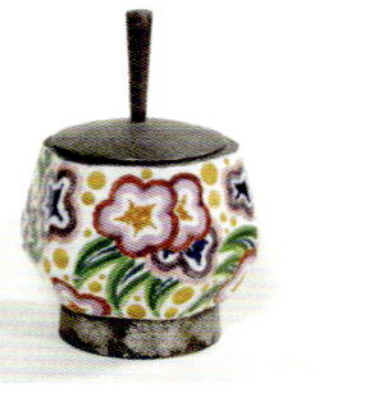
4981,52,4 gem. VIII, Firmenmarke 19, sign. Krebs, H. 10,7 cm; Zezschwitz 20
4981,53,5 gem. WG, gem. gold II, Firmenmarke 19, sign. Krebs, H. 10,5 cm; PB Wien 19
4981,71,3 gem. D.1.II, Firmenmarke 19, Made in Austria, sign. Krebs, H. 9,3 cm, Gesamth. mit Deckel 15,2 cm; MAK Wien, Inv. Ke 10602-73; Bem.: Dose mit Metallmontierung, Deckel aus gehämmertem Weißblech.

4982
Gult, Johann
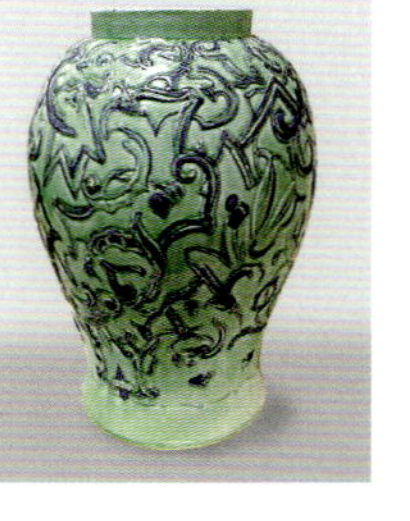
VASE
Form: balusterförmig
Dekor: Arabesken, reliefartig aufgelegt
ca. 1920/21
4982,6,1 Firmenmarke 18, unsign.?, H. ca. 33 cm; city-antik
4982,15,8 Firmenmarke 19, H. 33,2 cm; Doro 27.10.1993

4983
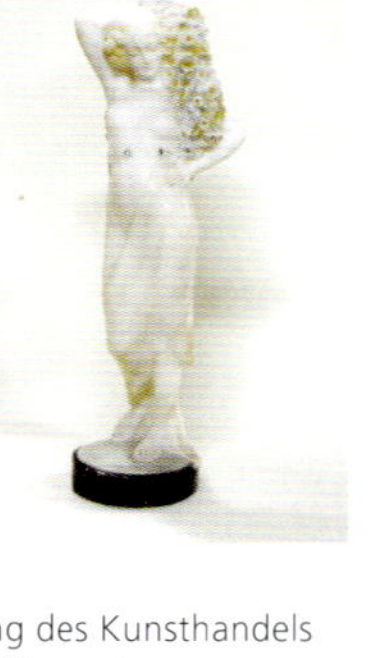
Flora (Stehende junge Dame mit Blumengirlande um die Schulter) (auf Rundsockel)
ALABASTER
ca. 1920/21
4983,FK./3 Plak. A, H. 39,5 cm; MAK Wien, Inv. Ke 10602-45
Bem.: Nach der Modell-Nr. zu schließen, könnte das Material Alabaster sein. Es handelt sich bei dieser Angabe um eine mündliche Mitteilung des Kunsthandels und konnte nicht selbst nachgeprüft werden.

4984
Schoerk, Hans
Damenakt auf einer Sessellehne sitzend (auf rechteckigem Podest)
ca. 1921
4984,21,35 gem. F. IX. SO(G?), Firmenmarke 22, sign. Schörk, H. 26 cm; (Sockel L. 17 cm, B. 12,5 cm); eBay # 1241234755

4987
Müller
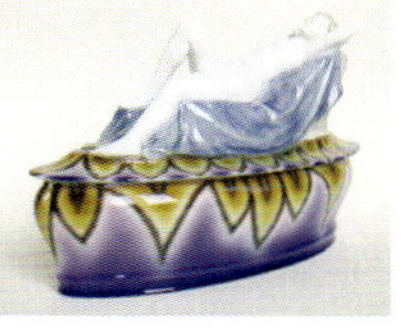
DECKELDOSE Mädchenakt
Form: oval
Bekrönung: liegender Mädchenakt (vollplastische Fig.)
Dekor: mehrfarbiger Spritzdekor, Zackenmuster
ca. 1921
4987,5471,93,10 gem. lig. AC, XXIII. D.1., Firmenmarke 19, Made in Austria, gepr. „L"= Z1, H. 20 cm, L. 27,5 cm; Sammlung Weilinger, Wien
Bem.: Die bekrönende Figur dieser Deckeldose wurde auch als eigenständiges Modell, mit selber Mod.-Nr. von Goldscheider editiert.
-> vgl. auch folgenden Eintrag
Lit.: Neuwirth (1974), S. 58, 68

4987
Müller
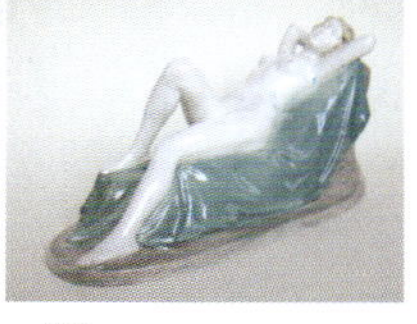
FIGUR Mädchenakt
(Liegender Mädchenakt)
ca. 1921
4987,9,65 gepr. RP, Firmenmarke 22, H. 12,2 cm, L. 23,4 cm; Sammlung Weilinger, Wien
Bem.: Diese Figur wurde auch als Deckeldose mit selber Mod.-Nr. von Goldscheider editiert.
-> vgl. auch vorherigen Eintrag

4988
Sitzender Affe (auf Rundsockel)
ca. 1921
4822,34,1 gem. lig. MM, AH, Firmenmarke 19, H. 17 cm; eBay # 268936825
4822,46,65 Firmenmarke 19, Made in Austria, H. 17 cm; Doro 20.4.1993
4822,48,5 gem. lig. AT, Firmenmarke 19, H. 17,8 cm; eBay # 396850088

4996
Podany, Rudolf
Sitzende Pierrette mit Laute
ca. 1916/17 (für den Entwurf)
4996,13,1 (Firmenmarke ohne Ang.), sign. Podany, H. 23,5 cm; Doro 8.3.2004
-> vgl. 4774

4997
Herkner, Friedrich
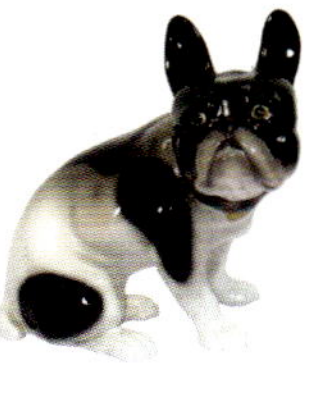
Sitzende französische Bulldogge(n) (oder Boston Terrier)
ca. 1921
EINZELFIGUR
4997,108,8 gem. F, D.1., Firmenmarke 19, Made in Austria, H. 15,2 cm; eBay # 1270686938
4997,111,10 gem. lig. AC, Firmenmarke 19, Made in Austria, gepr. „S"= Z2, H. 15,9 cm; Sammlung a@goldscheider.net
GRUPPE (PAAR)
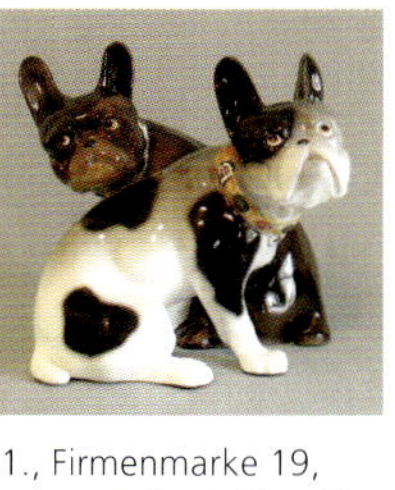
4997,70,2 gem. F., Firmenmarke 19, H. 15,9 cm, L. 17,8 cm; Sammlung Gary Bachman, USA
4997,92,11 gem. F., Firmenmarke 19, Made in Austria, Ausformung 20er Jahre, H. 16 cm; eBay # 7418242970
4997,157,11 ger. M14, gem. D.1., Firmenmarke 19, Made in Austria, H. 16 cm; Sammlung F. & J.P. Girard/ Frankreich
Bem.: Diese Hunde wurden unter derselben Modell-Nr. in zwei verschiedenen Ausführungen angeboten. Einmal das linke Einzeltier als separate Figur, dann das Paar als Gruppe. Dies ist ungewöhnlich für GS, da in fast allen ähnl. gelagerten Fällen für jede Variante eine eigene Modellnummer vergeben wurde.

5001
Perl, Karl
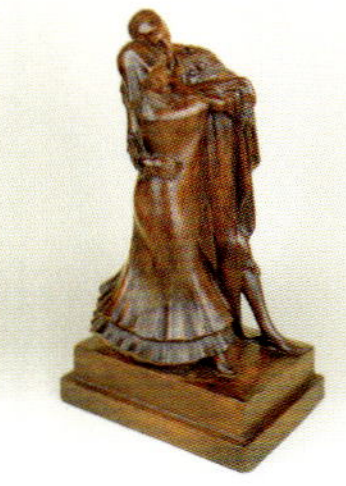
Spanisches Paar beim Tanz (auf Rechtecksockel)
ca. 1922
FAYENCE
5001,26,65 Firmenmarke 19, unsign.?, H. 38 cm; Schlüter 2675
5001,30,10 gem. B, 22, unsign.?, H. 39,6 cm; city-antik
5001,32,2 GS Wien (Firmenmarke?), H. 38,5 cm; Doro 14.9.2001
5001,62,2 Firmenmarke Adler 16?, unsign.?, H. 43 cm; KH
5001,107,? (Firmenmarke ohne Ang.), H. 41 cm; Fischer 30
5001 ohne Ang., (Firmenmarke ohne Ang.), H. 41 cm; PB Wien 19
ALABASTER
5001,X,4 Plak. A, H. 43 cm; eBay # 2248955911
Bem.: Diese Figur wurde in Fayence und Alabaster ausgeführt.
-> vgl. 5260

5004
Müller
Sitzendes Mädchen mit Blumenstrauß (auf einer Steinmauer)
ca. 1922
5004 (ohne Ang.), gem. AT, Firmenmarke 19, H. 25 cm; Ahlden 68/69

5005
Perl, Karl
Göttin der Fruchtbarkeit (Ceres/Demeter?) (Stehender Halbakt mit Feldfrüchten, Trauben, Obst und Getreideähren) (auf getrepptem Quadratsockel)
ca. 1922
5005,2,35 Firmenmarke 23, H. 54,5 cm; Wien Museum Inv. 186.614
(5005,7,6 gem.!), gem. VIII., 22, H. 55,8 cm; Soth. com. # 325Z; eBay # 1202183942 (KH www.tias.com/stores/frozenbeauty)
5005,16,10 Firmenmarke 18, sign. K. Perl, H. 54,5 cm; MAK Wien, Inv. Ke 10602-49

5009
Perl, Karl
Stehender Frauenakt mit Tuch (auf getrepptem Quadratsockel)
ALABASTER / Marmorsockel
ca. 1922
5009,WK,8 ger., Plak. E, H. 33 cm; GK, S. 18, 103 (m. Abb.); Wien Museum Inv. 223.737
Ausst.: HMW (jetzt Wien Museum) 1985/86.
Lit.: GK, S. 18, 103 (m. Abb.)

5014
Lorenzl, Josef
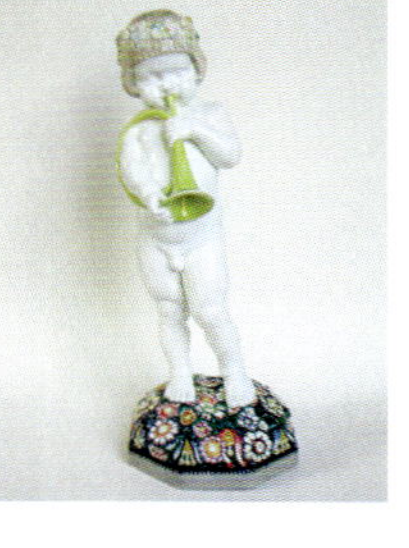
Stehender Putto, in ein Horn blasend (auf oktogonalem Blumensockel)
ca. 1922
5014,13,5 ger. K., gem. V, Firmenmarke 25, sign. Lorenzl, H. 51,8 cm; Sammlung Weilinger, Wien
5014,25,10 Firmenmarke 19, sign. Laurenzl, H. 51 cm; KH München

5015
Lorenzl, Josef
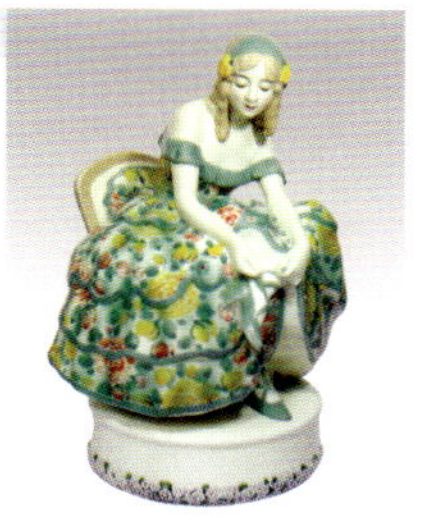
FIGUR / LAMPE Sitzende Tänzerin beim Schuhe schnüren (auf blumenverziertem Rundsockel)
ca. 1922
FIGUR
5015,143,16 gem. XXVII, N, (Firmenmarke o. Ang.), sign. Laurenzl, H. 25,5 cm; Kett. 39
5015,207,16 Firmenmarke 19, Made in Austria, Hand dec., sign. Laurenzl, H. 27 cm; Doro 24.10.1995
5015,310,20 gem. XXVII D2, gem. lig. AH, Firmenmarke 19, Made in Austria, sign. Laurenzl, H. 26/27 cm; BM 48; Weichmann 26.4.1985; RM 32; Kett. 136
5015,384,16 ger. IR, gem. H. N8, Firmenmarke 19, sign. Lorenzl, H. 27,5 cm; Sammlung a@goldscheider.net
5015,447,19 gem. N8, Firmenmarke 19, Made in Austria, Hand dec., sign. Lorenz(l?), H. ca. 27 cm; Henry's 29.11.1995
5015,457,11 Firmenmarke 19, Made in Austria, Hand dec., sign. Lorenzl, H. 27 cm; WKA 35
LAMPE
5015,276,65 gem. XXVII D2, Firmenmarke 19, Made in Austria, sign. Laurenzl, H. 34,3 cm, Gesamth. 55,9 cm; eBay # 3724026739

5016
Lesca
GRUPPE / LAMPE Pierrot und Pierrette (auf einer Bank sitzend) (auf Ovalsockel)
ca. 1910/12 (für den Entwurf)
FIGURENGRUPPE
5016,77,8 gem. F. L., gem. lig. JR, Firmenmarke 19, Ausformung 20er Jahre, H. 20,3 cm; eBay # 1248322607
LAMPE
5016,42,35 gem. F., 22, H. 40,6 cm; eBay # 3712081454
-> vgl. 3655; 5327
Lit.: MW, 2.1920/21, Nr. 12, S. 2 (m. Abb.); DD, 50.1922/23, Nr. 10, S. 30 (m. Abb.), hier jeweils bezeichnet „Lexa" (!) und mit Ang. des Originaltitels „Pierrot und Pierrette".

5018
Lorenzl, Josef
Odaliske (Haremstänzerin mit langem Tuch) (auf facettiertem Ovalsockel)
ca. 1922
5018,260,11,1,P gem. MR 1, Firmenmarke 19, Made in Austria, Hand dec., unsign.?, H. 48 cm; Soth. L. 31.10.1997
-> vgl. 5613; 6063

5022
Lorenzl, Josef
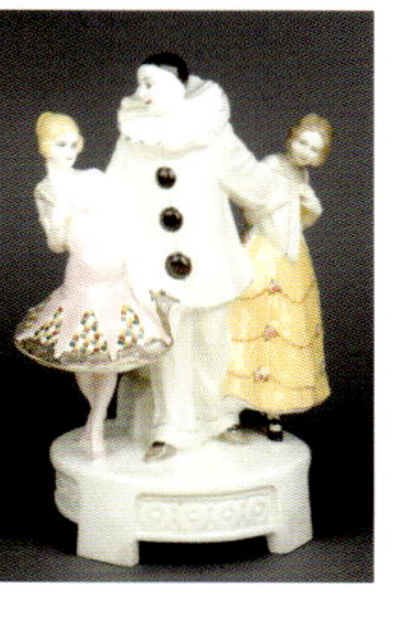
GRUPPE / LAMPE Pierrot mit zwei Pierretten (auf vierfüßigem Rundsockel)
ca. 1922
FIGURENGRUPPE
5022,29,10 ger. K, Firmenmarke 24, gem. VI, gem. lig. AS, IX., Firmenmarke 19, H. 41,8 cm; PB Wien
LAMPE
5022,16,10 gem. XI, Firmenmarke 19, H. 54 cm; Quit. 14; Fischer 151
5022,48,8 Firmenmarke 19, Gesamth. 83 cm; Ahlden 97
-> vgl. 5508= ähnl. Modell

5024
Perl, Karl
Sitzender Araber mit Turban und langer Pfeife (auf Rechtecksockel)
ca. 1922
5024,7,2 Firmenmarke 19, sign. K(arl?) Perl, H. 33,8 cm; Doro 10.3.2003
5024,70,65 Firmenmarke 19, sign. K. Perl, H. 34 cm; WKA 14

5024,92,10 Firmenmarke 19, sign. K. Perl, H. 34,5 cm; WKH
5024,101,4 Firmenmarke 19, sign. Perl, H. 33 cm; Doro 13.9.2004
ohne Ang., sign. Karl Perl, H. 35 cm; Nagel 332
ohne Ang., sign. K. Perl, H. fehlt; Zacke 3, S.166, 167
ohne Ang., Firmenmarke 19 (zwei mal), (unsign.?), H. 34,5 cm; eBay # 86173945

5025
Perl, Karl
Sitzender Damenakt mit Leoparden scherzend (auf Landschaftssockel)
ca. 1922
5025,71,11 Firmenmarke 19, Made in Austria, H. 20,5 cm; Sammlung Stressless, Portugal

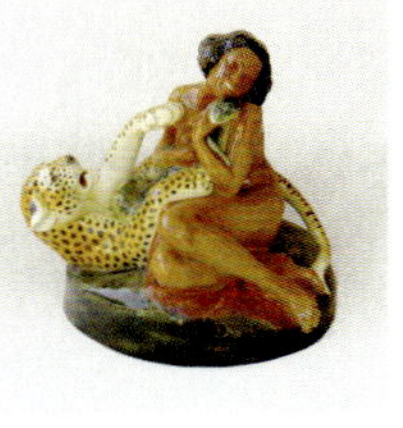

5027
Geiger, Benno
BUCHSTÜTZE Sitzende Frau (Japanerin)
ca. 1922
5027,49,4 gem. SG, Firmenmarke 19, Made in Austria, H. 20,6 cm; PB Wien 19

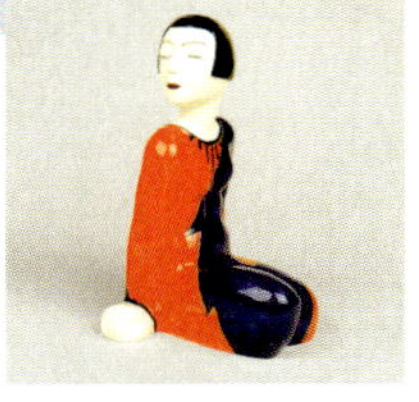

5030
Thomasch, Wilhelm
FIGUR / LAMPE Sitzende Pierrette mit Laute
ca. 1922
FIGUR (Fayence)
5030,48,10 (Firmenmarke ohne Ang.), H. 34 cm; Christie's SK 12.9.1997
5030,64,2 ger. K, 22, H. 34 cm; Sammlung Robert Senft
5030,78,8 gepr. G, ger. R, gem. F XXIX, 22, H. 34 cm; KH
5030,79,2 gem. F.K, XXIX, Firmenmarke 22, H. 34 cm, B. 18 cm; Wien Museum Inv. 190.103
5030,243,4 Firmenmarke 19, Made in Austria, H. 32 cm; WKA 38
LAMPE (Fayence)
5030,40,5 (Firmenmarke ohne Ang.), Gesamth. 49 cm; Doro 21.12.1993
5030,132,24 Firmenmarke 22, Gesamth. 50,8 cm; eBay # 1279697744
5030,276,8 Firmenmarke 19, Made in Austria, Hand dec., Gesamth. 48,3 cm; eBay # 176788420
ohne Ang., gem. XXVII. N, gem. lig. AH, Firmenmarke 19, H. 34,2 cm, (Gesamth. ca. 50 cm); Soth. com # 6NVL
FIGUR (Alabaster)
5030,B7 (Firmenmarke ohne Ang.), unsign.?, H. 33,5 cm; Wendl 47
Bem.: Diese Figur wurde in Fayence und Alabaster sowie als Lampe (Fayence) ausgeführt.

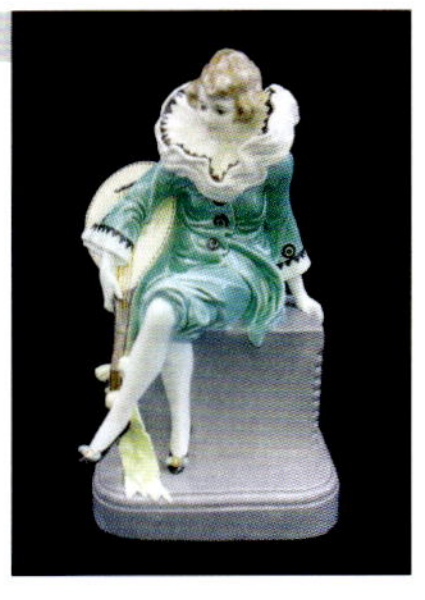

5032
Stoltenberg-Lerche, Hans
DECKELDOSE Sitzender, Flöte spielender Pan
Form: oval
Knauf / Bekrönung: sitzender Pan (vollplastische Figur)
Dekor: schwarz / grau
ca. 1900 (für den Entwurf)
5032,7,1 Firmenmarke 19, sign. Lerche, Ausformung 20er Jahre, H. 22,4 cm; Doro 30.11.2001
-> vgl. 4770 FIGUR

5042
Perl, Karl
König (Stehende posierende Figur mit Krone) (auf quaderförmigem Sockel)
ca. 1922
5042,22,85 22, sign. K. Perl, H. 40,6 cm; KH
Bem.: Bei dieser Figur handelt sich um eine Darstellung eines Königs aus einem Theaterstück, bzw. einer Oper. Genaueres lässt sich aufgrund einer zu undeutlichen vorliegenden Abb. nicht sagen.

5043
Perl, Karl
Stehender Chinese mit Kopfbedeckung und zwei erhobenen Zeigefingern (auf zylindrischem Sockel)
ca. 1922
5043,7,24,K gem. F., XV, 22, unsign., H. 26,5 cm; MAK Wien, Inv. Ke 10602-62
5043,77,8 (Firmenmarke ohne Ang.), H. 26,5 cm; Doro 9.11.1993; 8.3.1994
Bem.: Bei dieser Figur handelt sich um eine Darstellung des „Prinzen Sou-Chong" aus der Operette „Land des Lächelns" von Franz Lehar (1870–1948).

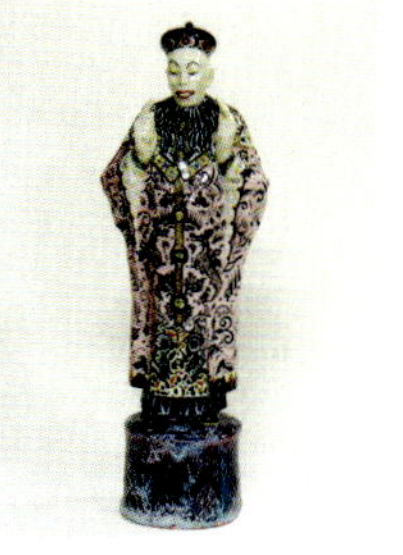

5048
Perl, Karl
Hänsel und Gretel
ca. 1922
5048 (ohne Ang.), H. 26 cm; Doro 16.12.1989

5050
Podany, Rudolf
Sitzende Pierrette (auf Rechtecksockel)
ca. 1922
5050,13,65 22, H. 20,3 cm; eBay # 2078518037
5050,73,8 Firmenmarke 19, H. 20,5 cm; BM 74; Herr 36
5050,118,11 gem. lig. AH, XXIII., Firmenmarke 19, H. 20 cm; eBay # 559953960

5052
Postl, Hugo
Stehender Pinguin (auf getrepptem Quadratsockel)
ca. 1922
5052,34,3 gem. B., Firmenmarke 24, H. 25,5 cm; Wien Museum Inv. 220.765
-> vgl. 5053= GRUPPE

5053
Postl, Hugo
GRUPPE Zwei stehende Pinguine (auf Rechtecksockel)
ca. 1922
5053 (Firmenmarke ohne Ang.) H. 26 cm; Pinhas, S. 104 (m. Abb.)
-> vgl. 5052= Einzelfigur

5054
Lorenzl, Josef
„Gefangener Vogel" (Der Gefangene Vogel, Niddy Impekoven)
BRONZE / Marmorsockel
ca. 1922
5054,10 (Firmenmarke Plak. ohne Ang.), sign. Lorenzl, H. 26 cm; Karrenbauer 122
5054,24 Plak. E, sign. Lorenzl, H. 26 cm; Doro 22.5.1989; PB Wien
5054,112 o. Firmenmarke, Made in Austria, unsign., H. 26 cm; eBay # 3732243251
ohne Nr., ohne Firmenmarke, H. 27 cm; PB Wien 19
-> vgl. FAYENCE 5230; 5960; 5964= Pendant zu 5960; 6204; 6600; 8176
-> vgl. 5530 GRUPPE (Trio)
Bem.: s. bei 5230
Lit.: s. bei 5230; DD, 50.1922/23, Nr. 6, S. 29 (m. Abb. für die BRONZE-Ausführung).

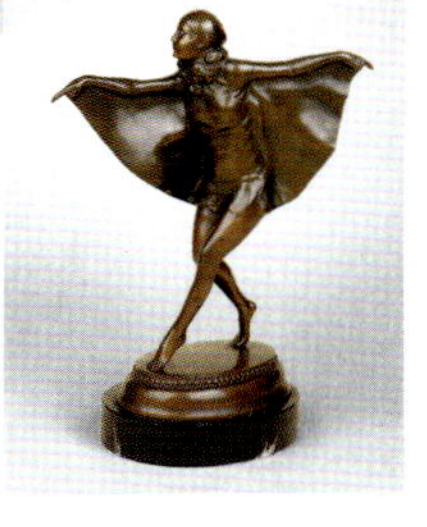

5055
Sitzender Dudelsackspieler
ca. 1922
5055,17,8 Firmenmarke 19, Made in Austria, Hand dec., H. 25,4 cm; eBay # 3754905765

5057
Lorenzl, Josef
GRUPPE / LAMPE Stehender Pierrot mit Laute und Pierrette (als Kinder) (auf Rundsockel)
ca. 1922
GRUPPE
5057 o. Ang., Firmenmarke 19, Made in Austria, sign. J. Lorenzl, H. 29,5 cm; Ahlden 129
LAMPE
5057,12,10 ger.; ger. RP, Firmenmarke 19, Made in Austria, sign. J. Lorenzl, H. 47 cm (Gesamth. 64 cm); Henry's/Hess 42 (2.6.2001)
5057,56,8 gem. F, IV, Firmenmarke 19, Made in Austria, sign. JLorenzl, Gesamth. 68 cm; PB Wien 19

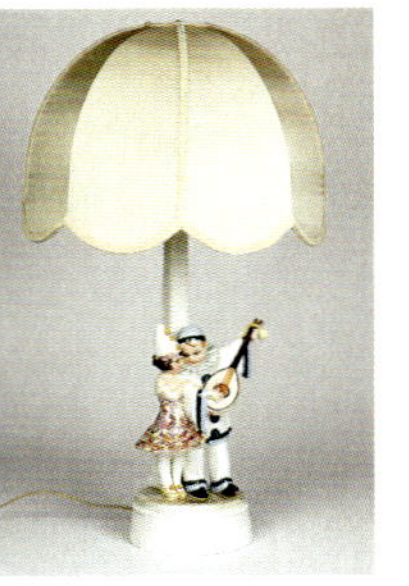

5058
Lorenzl, Josef
„Erna Pinner-Puppe" (mit Laute) (auf facettiertem, mit Blumengirlanden verziertem Rundsockel) (Niddy Impekoven)
ca. 1922
5058,98,8 Firmenmarke 19, Made in Austria, sign. Laurenzl, H. 30,5 cm; WKA 9
Bem.: Dargestellt ist die Tänzerin Niddy Impekoven (1904–2002) in ihrem Tanz „Erna Pinner-Puppe", Moment musical, Musik: Franz Schubert (Urauff. Frankfurt/M. 4.12.1918) -> vgl. hierzu Lit.-Auswahl (m. Abb.): Impekoven, L.: Niddy Impekoven, Berlin 1920; Frentz, H.: Niddy Impekoven und ihre Tänze, Freiburg/Br. 1929; Frentz, H.: Weg und Entfaltung Niddy Impekovens, Leipzig 1933; Stöckemann, P.: Niddy Impekoven, Geburtstag eines Wunderkindes, in: Tanzdrama, Nr.11., 1990, S. 26–28, die hier wiedergegebene zeitgenössische Fotografie von Anny Eberth (Berlin) diente als Vorlage für die Ausführung dieses Modells. Hinzugefügt wurde eine Laute, die am Boden (Sockel) steht und für die zusätzliche Stabilität der Figur sorgt. Erna Pinner (1890–1987) war Künstlerin und Publizistin, sie schuf durch den Entwurf ihrer etwa lebensgroßen (H. 150 cm) Puppe „Pierrot" die Vorlage für die tänzerische Umsetzung Niddy Impekovens, -> vgl. hierzu Lit.-Auswahl (m. Abb.): Pinner, E.: Das Puppenbuch, Berlin 1921; Pinner, E. und K. Edschmid: Wir wollen nicht mehr darüber reden, München 1999; Weidle, B.: Unser eigenes Gespenst neben uns: Die Pinner-Puppen, in: Ich reise durch die Welt – Die Zeichnerin und Publizistin Erna Pinner, Bonn 1997.

5060 -> ÜBERSCHNEIDUNG
Thomasch, Wilhelm
„Faszination" (Stehender weiblicher Halbakt mit großem Tuch und Kopfputz) (auf facettiertem Rundsockel)
ca. 1922
5060,1,10 Firmenmarke 19, H. 46,9 cm; Doro 30.11.2001
5060,1,35 ger. B., gem. B., Malersign.: R. Spuller, 22, H. 48 cm; Privatsammlung, Wien; Bem.: Einziges bisher nachgewiesenes Exemplar, welches mit dem vollen Namenszug des (Dekor)-Malers, sign. ist. Zu R. Spuller -> vgl. Künstlerliste.
5060,54?,65 (Firmenmarke ohne Ang.), H. 47 cm; WKA 35
5060,71,10 ger. K, gem. lig. AS, Firmenmarke 24, H. 47 cm; eBay # 7314217386
5060,88,9 Firmenmarke 24, H. 50 cm; PB Wien 19
5060,149,3 (Firmenmarke ohne Ang.), H. 46,5 cm; Neumeister Varia 54 (15.3.1990); Acatos, 1991, S.262
5060,155,2 gem. RH N XXVII, (Firmenmarke ohne Ang.), H. 47 cm; Christie's SK 15.10.2003

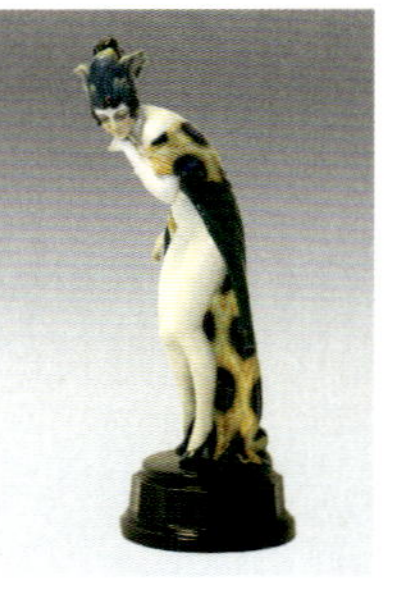

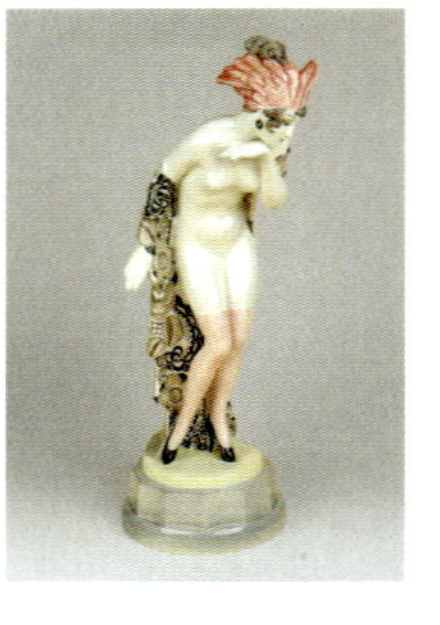

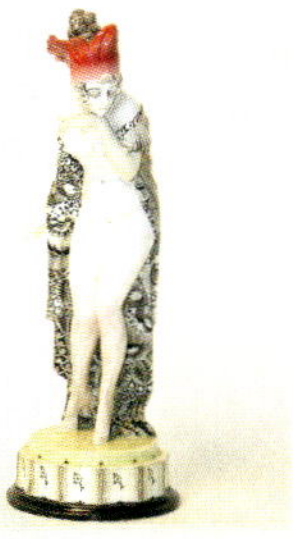

5060,163,4 (Firmenmarke ohne Ang.), H. 46,3 cm; Doro 24.5.2000; Lit.: O. Kronsteiner: Ergebnisse aus Wien, in: AZ, 2000, S. 474 (m. Abb.); Kurier (Wien) vom 26.5.2000 (m. Abb.)
5060,219,11 (Firmenmarke ohne Ang.), H. 45,7 cm; eBay # 2195811439
5060,242,65 (Firmenmarke ohne Ang.), Made in Austria, H. 46,7 cm; WKA 38
5060,277,8 gem. D1 XXVII, gem. lig. AH, Firmenmarke 19, Made in Austria, H. 47,5 cm; Doro WA 12.12.1992; Privatsammlung, Wien
5060,291,11 (Firmenmarke ohne Ang.), H. 47 cm; Christie's East 12.6.1990; Acatos, 1991, S.262
5060,304,12 gem. TS D. 3., Firmenmarke 19, Made in Austria, Hand dec., H. 47 cm; Soth. com. # 6LMT
5060 (ohne Ang.), H. 48 cm; Stahl 27.11.1993; Lit.: KPJb, 49.1994, Bd. 3, S.420
Lit.: DD, 50.1922/23, Nr. 4, S. 32 (m. Abb.), hier mit Ang. des Originaltitels „Faszination"; GSN, S. 26, 27 (m. Abb.); GSNF, o. Pag. (m. Abb.)
GS neu: Modell-Nr. 89 „Das Versprechen" (H. 50 cm), Aufl. 2000 Stück

5060 –> ÜBERSCHNEIDUNG

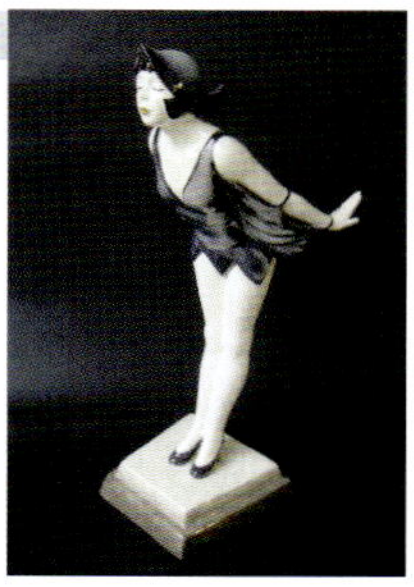

Dakon, Stephan
Stehende Tänzerin, nach vorne zum Kuss gebeugt (auf trapezförmigem Sockel)
ca. 1928
5060,304,18 Firmenmarke 19, Made in Austria, Hand dec., H. 42 cm; Sammlung Stressless, Portugal
5060,304,22 Firmenmarke 19, Made in Austria, Hand dec., H. 42 cm; Sammlung f@goldscheider.de
Bem.: Die Modell-Nr. 5060 ist, auch wenn sich hier zwei Ex. (mit der selben Stücknr.) nachweisen lassen sicher falsch, da Dakons Zusammenarbeit mit Goldscheider ca. 1924 beginnt und die Modell-Nr. 5060 sicher schon um 1922 entstanden sein müsste. Bei der dargestellten Tänzerin handelt sich aller Wahrscheinlichkeit nach um eine der beiden Dolly Sisters.
–> vgl. auch 5712, 5781 (ähnl. Modelle der Dolly Sisters)

5061

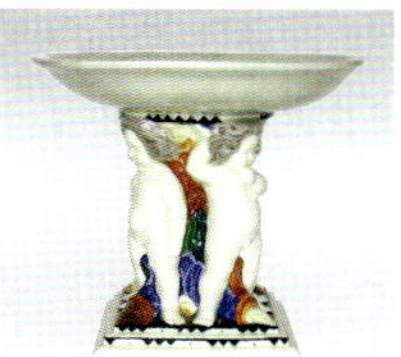

Wiegand, (Martin?)
TAFELAUFSATZ / OBST-SCHALE Vier, eine runde Schale tragende Putti (auf getrepptem Quadratsockel)
ca. 1922
5061,15,35 gem. V, gem. lig. AH, Firmenmarke 24, H. fehlt; eBay # 3753944716
5061,17,5 gem. FH, Firmenmarke 19, H. 18,5 cm; Ø 20,5 cm; Herr 40
5061,36,4 gem. V, S, gem. lig. AH, Firmenmarke 19, Made in Austria, H. 17,6 cm; Ø 19,5 cm; Sammlung a@goldscheider.net
5061,38,16 gem. WG, Firmenmarke 19, H. 18,3 cm; Ø 20,3 cm; city-antik
Bem.: In der Art Michael Powolnys, der ähnl. Schalen für die Gmundner Keramik und die Wiener kunstkeramische Werkstätte schuf.

5066
VASE
Form: birnförmig, an der Mündung mit vier durchbrochenen Tüllen
Handhabe(n): seitlich ein(?) durchbrochener Henkel
Dekor: Craqueléglasur weiß / braun
ca. 1922/23
5066 (ohne Ang.), H. 31 cm; Metz 21.3.1992
Bem.: Die Beschreibung musste ungeprüft vom Auktionskat. übernommen werden, dort ohne Abb.

5077
Perl, Karl
Stehender Seemann (auf quaderförmigem Sockel)
ca. 1922/23
5077,15,34 Firmenmarke 19, H. 29,8 cm; Doro 12.6.1995

5081

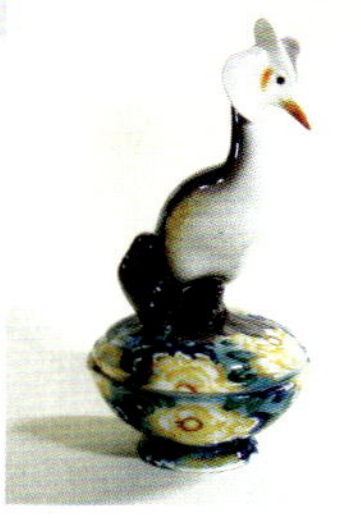

Wernekinck, Sigismund
DECKELDOSE Haubentaucher mit zwei Jungen
Form: eiförmig
Knauf/Bekrönung: Haubentaucher mit zwei Jungtieren
Dekor: stilisierte Blütenmalerei
ca. 1909/10 (für den Entwurf)
5081,22,1 gem. lig. AH, AS, Firmenmarke 19, H. 26,9 cm; MAK Wien, Inv. Ke 10602-19-1+2
5081 ger., gem. R.H., (Firmenmarke ohne Ang.), Ausformung 20er Jahre, H. 27 cm; RM 22
–> vgl. 4386 FIGUR

5085
Lorenzl, Josef
„Wilder Tanz" (Tänzerin auf einem Bein) (auf oktogonal-facettiertem Sockel)
BRONZE / Marmorsockel
ca. 1923
5085,60 Firmenmarke 20? oder Plak. F?, sign. J. Lorenzl, H. 30 cm, Gesamth. 40 cm; Nagel Leipzig 5; Soth. L. 1.11.1996
Lit.: Ros, L.: Wiener Fayencen und Bronzen, Begleitworte zu den Abb. über neue Werke von FG Wien, in: KuKGW, 4.1923/24, S. 16, 17 (m. Abb.), hier mit Ang. des Originaltitels „Wilder Tanz", „In seinem wilden Tanz hat Lorenzl echte Bronze zu meisterlicher, in Stil und Form ganz ursprünglicher Wirkung gebracht. Der Ausdruck des Gesichtes hat etwas Ekstatisches, Übersinnliches. Jedem noch so eintönigem Raum wird diese Figur Leben, Lust und Freude bringen".

5087
Lorenzl, Josef
Stehende Rokokodame mit Handspiegel (auf Rundsockel)
ca. 1923
5087,1,10 ger. K, gem. F.IX, Firmenmarke 22, sign. Lorenzl, H. 53,5 cm; Wien Museum Inv. 190.077
5087,40,8 Firmenmarke 19, sign. Laurenzl, H. 53,5 cm; Doro 20.12.1994
5087,55,11 Firmenmarke 19, sign. Lorenzl, H. 52,5 cm; Nagel 393
5087,56,– gem. XXVII. F., Firmenmarke 19, Made in Austria, sign. Lorenzl, H. 55 cm; eBay # 854235280
5087 (ohne Ang.), sign. Lorenzl, H. 54 cm; Spik 570

5088
Lorenzl, Josef
FIGUR / LAMPE „Carmen" (Stehende Spanierin mit Fächer) (auf Rundsockel)
ca. 1923
FIGUR
5088,12,1 Firmenmarke 19, sign. Lorenzl, H. 49 cm; KaM 99
5088,63,16 Firmenmarke 19, sign. Lorenzl, H. 50,5 cm; Herr 38; Christie's SK 4.11.1999; 12.5.2000
5088,137,11 Firmenmarke 19, Made in Austria, sign. Laurenzl, H. 50 cm; eBay # 154022545
ohne Ang., Firmenmarke 19, sign. Lorenzl, H. 49,5 cm; Soth. Arcade 1282 (14.6.1982); Christie's East 9.12.1982; Ahlden 38
LAMPE
5088,7,11 (Firmenmarke ohne Ang.), sign. Laurenzl, Gesamth. 74,5 cm; KaM 85
ohne Nr., gem. 04, III, Firmenmarke 19, Made in Austria, unsign.?, Gesamth. 72,4 cm; eBay # 256508890
–> vgl. 5232; 5277
–> vgl. 5534= sitzende Figur
Lit.: Ros, L.: Wiener Fayencen und Bronzen, Begleitworte zu den Abb. über neue Werke von FG Wien, in: KuKGW, 4.1923/24, S. 15–16 (m. Abb.), hier mit Ang. des Originaltitels „Carmen", „Lorenzl's Carmen ist eine anmutig bewegte, feingliedrige Statue, biegsam und schmiegsam. Ein gefälliger Rhythmus liegt in dem gelb und braun reich durchwirktem Gewande, in der feinen Formung der Hüften, mit denen sie sich wie ein Pfau zu wiegen scheint. Das schmale lächelnde Gesicht mit den dunklen Augen und die Grazie der Fächerhaltung lassen sie als würdiges Dokument des spanischen Frauentyps erscheinen".

5092
Lorenzl, Josef
Stehende Pierrette mit Laute (auf oktogonal facettiertem Sockel)
ca. 1923
5092,79,65 gem. II., Firmenmarke 19, sign. Laurenzl, H. 34,3 cm; Wien Museum Inv. 219.180
5092,111,2 Firmenmarke 19, sign. Laurenzl, H. 34,6 cm; Doro 1.6.2005
5092,127,2 gem. F.X., Firmenmarke 19, Made in Austria, sign. Laurenzl, H. 35 cm; Soth. com. # 31H5
5092,134 (oder 124?),65 gem. lig. HZ?, gem. grün III, Firmenmarke 19, Made in Austria, sign. Laurenzl, H. 33,6 cm; eBay # 22229425; 1237646444

5094

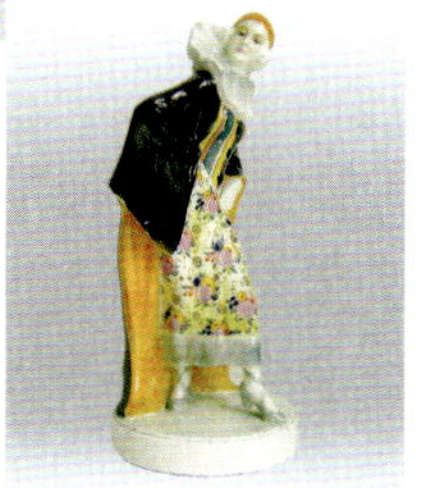

Kostial, Josef
„Pierrette" (Stehende junge Dame im Harlekinskostüm) (auf Rundsockel)
ca. 1923
5094,5,65 gem. lig. AH, XXVII, (Firmenmarke ohne Ang.), sign. J. Kostial, H. 44,5 cm; Soth. L. 3.5.1991
5094,7,9 gem. B.I.M., Firmenmarke 24, sign. J. Kostial, H. 44,4 cm; Soth. com. # 6LOA; Soth. com. # 346 YC
5094,10,8 gepr. P, gem. FH, VIII, (Firmenmarke ohne Ang.), sign. J. Kostial, H. 44,2 cm; Kett. 148
5094,19,10 gepr./ger. K, gem. AS, XXVII, Firmenmarke Sculpture 13, sign. J. Kostial, H. 44,3 cm; Neuwirth (1974), S. 65 (m. Abb.)
5094,59,65 gem. XXVII, (Firmenmarke o. Ang.), sign. J. Kostial, H. 44,3 cm; Kett. 14
5094,76,16 gem. F, XXVII, Firmenmarke 19, sign. Kostial, H. 45 cm; Privatsammlung, Wien
–> vgl. 5485
Lit.: Ros, L.: Wiener Fayencen und Bronzen, Begleitworte zu den Abb. über neue Werke von FG Wien, in: KuKGW, 4.1923/24, S.15–16 (m. Abb.), hier mit Ang. des Originaltitels „Pierrette", „Die leicht vornüber gebeugte Figur mit der tänzelnden Schrittstellung und das schwarzrotbraune Gewand mit dem Überwurf, wie auch die Neigung des Kopfes, geben dem Ganzen etwas Verheißungsvolles, Lockendes".

5101
Meier, (Emil oder Meier-Michel, Johanna)
GRUPPE / LAMPE Rokokopaar
ca. 1923
GRUPPE
5101,39,6 gem. lig. AM, gem. T., XXIII, Firmenmarke 19, unsign., H. 18 cm; Wien Museum Inv. 219.178
LAMPE
5101,93,19 Firmenmarke 19, Gesamth. 42 cm; Doro 7.3.2001

5102
Kostial, Josef
„Harlekin" (Stehender Till Eulenspiegel mit Flöte) (auf halbovalem Sockel)
ca. 1923
5102,113,10 Firmenmarke 19, unsign.?, H. 30,2 cm; Doro 26.5.1998
5102,7,10 gem. B. XXIII, Firmenmarke 19, sign. J. Kostial, H. 31 cm; KH
Lit.: Ros, L.: Wiener Fayencen und Bronzen, Begleitworte zu den Abb. über neue Werke von FG Wien, in: KuKGW, 4.1923/24, S. 16 (m. Abb.), hier mit Ang. des Originaltitels „Harlekin" und des Künstlers „Kostial", „Mit rotbraun blaukariertem Gewande und einem seelenvollen Gesichtsausdruck lässt der Künstler durch die Schalmeiden den kleinen Harlekin zu uns sprechen in einer neuen Form von dem alten Liede, das verdammt ist, soviel schmerzvolle Lächerlichkeit auszulösen und das doch so reizvoll ist: von dem Bajazzoschicksal einer unglücklichen Liebe".

5105
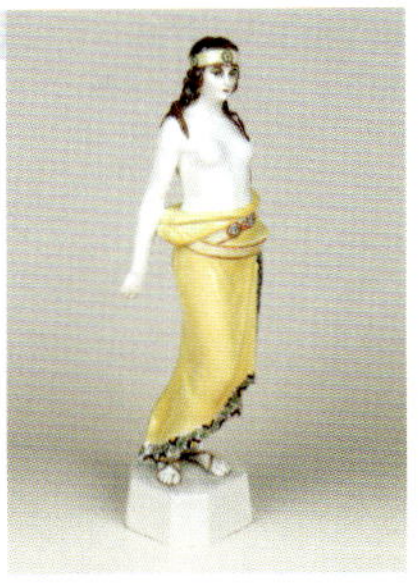
Tell, E.
„Judith" (Stehender Frauenhalbakt mit Stirnband und Gürtel) (auf oktogonal facettiertem Sockel)
ca. 1923
5105,28,11 Firmenmarke 19, Made in Austria, sign. E. Tell, H. 41 cm; PB Wien 19
-> vgl. 734= ähnl. Modell in einer Ausformung ca. 1898.

5108
Gult, Johann
VASE
Form: stangenförmig, Mündung etwas geweitet, mit umlaufendem Relief, Fuß mit zwei floralen Applikationen
Dekor: Rustika, kobalt, grün
ca. 1923
5108 (ohne Ang.), H. fehlt; Ros, S. 17, 19
Lit.: Ros, L.: Wiener Fayencen und Bronzen, Begleitworte zu den Abb. über neue Werke von FG Wien, in: KuKGW, 4.1923/24, S.17, 19 (m. Abb.), hier mit Originalbezeichn. des Dekors „Rustika, kobalt, grün", „Die letzte Vase Nr. 5108 möchten wir Wien's Königin unter den Vasen nennen, in ihrer Form und in ihrer Farbe, dem russisch Kobalt und grün. Das Breiterwerden ihres Fußes verstärkt ihr dekoratives Element. Nur schönste, königliche Blumen sind ihrer würdig, vielleicht der Blütentraum der Chrysanthemen".

5109
Gult, Johann
VASE
Form: becherförmig, leicht geschwungen, Fuß zweifach gewellt
Dekor: Korpus vertikale Blütenranken, Fuß horizontale Blütenranken
ca. 1923
5109 (ohne Ang.), H. fehlt; Ros, S. 17, 18
Lit.: Ros; L.: Wiener Fayencen und Bronzen, Begleitworte zu den Abb. über neue Werke von FG Wien, in: KuKGW, 4.1923/24, S.17, 18 (m. Abb.), „Der Blumenbecher No. 5109 hat eine schöne, leicht geschwungene Form. Seine Rankenkomposition verdichtet sich am Fuße zu einem lebendigen Akkord in grün, grau und schwarz-weiß".

5110
Gult, Johann
VASE
Form: Mündung und Fuß etwas geweitet, mittig gebaucht
Dekor: florale und Phantasieelemente
ca. 1923
5110 (ohne Ang.), H. fehlt; Ros, S. 17, 18
Lit.: Ros, L.: Wiener Fayencen und Bronzen, Begleitworte zu den Abb. über neue Werke von FG Wien, in: KuKGW, 4.1923/24, S.17, 18 (m. Abb.), „In dem Blumenbehälter Nr. 5110 sind die kleinen Phantasien mit freiem, kecken Pinsel auf das Gefäß gebannt, voll lebendigen Schwunges, gelb und bunt".
Dekor-> vgl. 5111

5111
Gult, Johann
DECKELVASE (balusterförmig)
Knauf: blütenförmig
Dekor: floral und Phantasie-Elemente
ca. 1923
5111,38,1 gem. XX, II., Firmenmarke 19, Made in Austria, H. 24,6 cm; GK, S. 18, 104 (m. Abb.); Wien Museum Inv. 182.088/1+2
Ausst.: HMW (jetzt Wien Museum) 1985/86.
Dekor-> vgl. 5110
Lit.: GK, S. 18, 104 (m. Abb.)

5115
Kostial, Josef
„Tanz der Eitelkeit" (Lucie Kieselhausen) (Stehende Pierrette mit Kopfbedeckung) (auf Rundsockel)
ca. 1923
5115,120,8 gem. lig. AC?, XII, Firmenmarke 19, unsign., H. 45,8 cm; KH
5115,240,4 gem. lig. SG, Firmenmarke 19, Made in Austria, Hand dec., unsign., H. 49 cm; eBay # 1208847942; # 1483681029
Bem.: Dargestellt ist die Tänzerin Lucie Kieselhausen (1897–1927) in ihrem Tanz „Tanz der Eitelkeit". Eine zeitgenössische Fotografie des Ateliers (Anny) Eberth (Berlin) diente als Vorlage für die Ausführung dieses Modells. Die Figur wurde zwar an den Schuppungen des Kostüms (Entwurf von Ludwig Kainer) und des Kopfputzes leicht abgewandelt und vereinfacht, wohl vorwiegend wegen der technischen Machbarkeit. Ergänzt wurde auch eine Haarlocke, die unter der Kopfbedeckung seitlich hervorlugt. Was aber die Haltung sowie den Gesichtsausdruck betrifft, so ist unverkennbar L. K. dargestellt.
Lit.: Ros, L.: Wiener Fayencen und Bronzen, Begleitworte zu den Abb. über neue Werke von FG Wien, in: KuKGW, 4.1923/24, S. 16 (m. Abb.), hier mit Ang. des Originaltitels „Tanz der Eitelkeit" und des Künstlers „Kostial", „Der ganze Farbenakkord in Rotbraun, Orange und Gelb überrascht durch Eleganz und Geschmeidigkeit der Linien. Es liegt eine zarte Erotik über dem Ganzen, eine kleine Sensation, die die weibliche Eitelkeit erlebt"; Oberzaucher-Schüller, G.: Der freie Tanz in Wien bis 1938, in: Amort, A. (Hrsg.): Österreich tanzt, Wien, Köln, Weimar 2001, S. 58 (m. Abb.), Fotografie: Atelier Eberth; Koegler, 1972, S. 305.

5116
Jarl, Karin
Liegende Katze
ca. 1923
5116,6,1 gem. lig. AH, Firmenmarke 24, H. 16 cm, L. 27 cm; Wien Museum Inv. 190.040
5116,47,10 gem. H., Firmenmarke 19, Made in Austria, H. 15,9 cm, L. 26 cm; eBay # 290170839
Bem.: Im Bericht von Johanna Grell: Die Pariser Kunstgewerbeausstellung 1925 Sonderdr. (KeramRS 1925, Nr. 38, S. 617–656) werden zwar auf S. 33 „Tierplastiken von Karin Jarl, Firma Goldscheider" genannt, jedoch lässt sich nicht nachvollziehen, ob auch diese Katze der bekannten Tierbildhauerin ausgestellt war.

5117
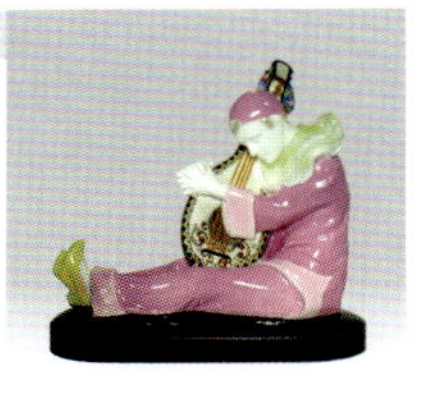
Thomasch, Wilhelm
Sitzender Pierrot mit Laute (auf abgerundetem Längssockel)
ca. 1923
5117,4,10 gem. lig. FH XXVII, 22, unsign., H. 24,5 cm; Sammlung Robert Senft
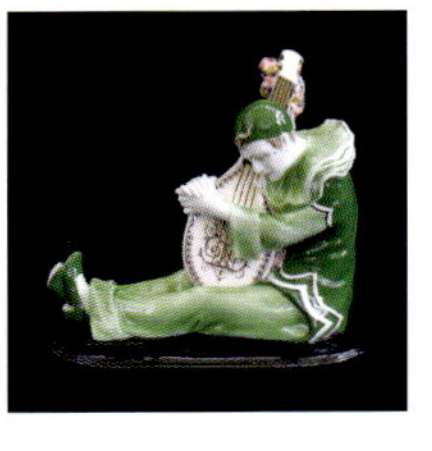
5117,5,8 (Firmenmarke ohne Ang.), unsign.?, H. 22 cm; KaM 92
5117,20,35 ger. H., gem. AHXXVII, gem. lig. AM (oder VW), Firmenmarke 25, unsign., H. 25 cm; eBay # 2501300028
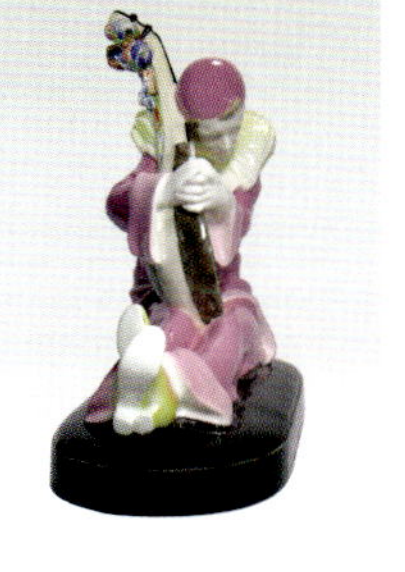
5117,69,65 gem. F, PPEBEE, III, Firmenmarke 19, unsign., H. 24,5 cm; Sammlung a@goldscheider.net
5117,71,24 gem. F, unsign.?, H. 25 cm; Henry's 31.7.1996; 25.6.1997; Sammler-Markt 1997, Nr. 2
5117,108,11 gem. F VI, Firmenmarke 19, unsign., H. 24,2 cm; Doro 10.4.1997
5117,114,5 gem. F., Firmenmarke 19, sign. Thomasch, H. 24 cm; eBay # 435523699
5117,119,2 Firmenmarke 19, sign. Thomasch, H. 24 cm; Doro 18.4.1994
5117,159,7 Firmenmarke 19, sign. Thomasch, H. 22,5 cm; Doro 5.6.1997
5117,166,4 gem. AH, N 2, Firmenmarke 19, Made in Austria, Hand dec., sign. „Tomasch", H. 22,4 cm; Art 1900; Quit 3

5119
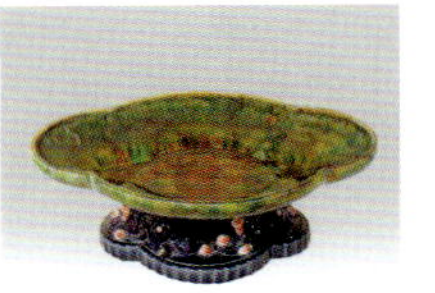
TAFELAUFSATZ / OBSTSCHALE (auf vertikal geriffeltem Vierpasssockel)
Form: vierpassig, oval
Dekor: mehrfarbige Laufglasur
ca. 1923
5119,90,14 gem. F.1, Firmenmarke 19, Made in Austria, Hand dec., H. 16,2 cm, L. 42,8 cm, B. 30,7 cm; Sammlung a@goldscheider.net
5119,137,15 Firmenmarke 19, H. 14,5 cm, L. 41,5 cm, B. 30,5 cm; Doro 3,5,1995; 29.10.1996

5120
Jarl, Karin
Reh (auf rechteckigem Landschaftssockel)
ca. 1923
5120,7,17 Firmenmarke 23, H. 30,5 cm, L. 38 cm; Doro 24.10.1995; eBay # 2581450168
5120,10,7 gepr. M 14, (Firmenmarke ohne Ang.), unsign.?, H. 31 cm, L. 34 cm; Schloss Hag. 28

5129
Hauf
Hund
ca. 1923
5129,30,11 gem. lig. GW, F, (Firmenmarke o. Ang.), H. fehlt, L. 19 cm; Kett. 10
Bem.: Hier keine nähere Beschreibung der Hunderasse möglich, da im Auktionskat. weder eine Abb. sich befindet, bzw. keine weiteren Ang. zur Art des Hundes gemacht werden.

5131
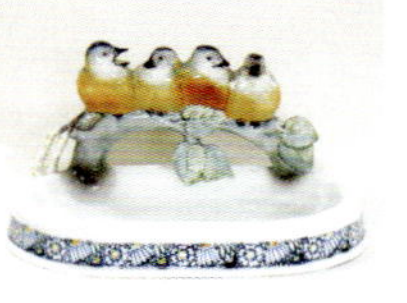
Jarl, Karin
Vogeltränke (Vier sitzende Spatzen auf einem Birkenast)
ca. 1923
5131,18,5 gem. V, Firmenmarke 22, Made in Austria, H. 15,8 cm, L. 31,6 cm; MAK Wien, Inv. Ke 10521-17
5131,18,9,2 gem. F V., Firmenmarke 22, H. 15,2 cm, L. 30,8 cm; eBay # 3712032374

5133
Jarl, Karin
Tiger
ca. 1923
5133 (ohne Ang.), (Firmenmarke ohne Ang.), H. 11 cm; KaM 171

5135
Rosé
FIGUR / LAMPE Stehende Rokokodame mit Rundfächer (auf Rundsockel)
ca. 1911/12 (für den Entwurf)
FIGUR
5135,25,11 gem. F.,VIII, Firmenmarke 25, H. 27,9 cm; eBay # 3773682421
5135,97,65 gem. F. IV., Firmenmarke 19, Ausformung 20er Jahre, H. 28 cm; city-antik
LAMPE
5135,180,18 Firmenmarke 19, Ausformung 20er Jahre, H. 33 cm; Doro 20.11.1996

5136
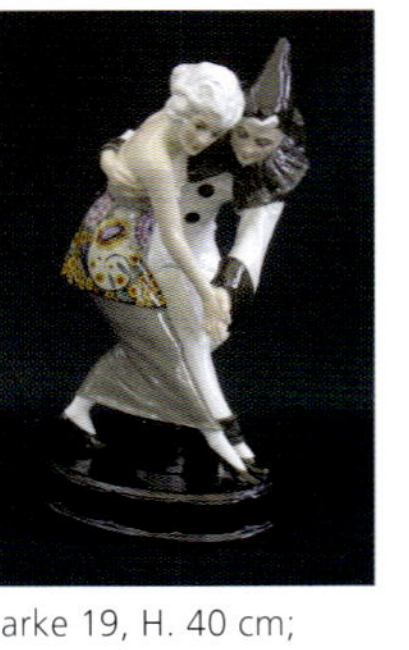
Thomasch, Wilhelm
Tanzender Pierrot mit Dame (auf facettiertem Ovalsockel)
ca. 1923
5136,79,24 gem. F. IX., MM S, Firmenmarke 19, H. 40,6 cm; Soth. com # 6LNP
5136,85,10 gem. F., IX., gem. lig. AM, Firmenmarke 19, H. 39 cm; Wien Museum Inv. 190.070
5136,120,8 gem. F. IV., Firmenmarke 19, H. 40 cm; Doro 1.12.2000; Sammlung Robert Senft
ohne Ang., (Firmenmarke ohne Ang.), H. 39,5 cm; Christie's SK 20.9.2000
Bem.: Wenn auch gelegentlich behauptet wird, die

Schauspielerin Jean Harlow (1911–1937) sei eine der beiden dargestellten Figuren, so lässt sich das sicher ausschließen. Sie wäre bei Entstehung der Gruppe ca. zwölf Jahre alt gewesen.

5144
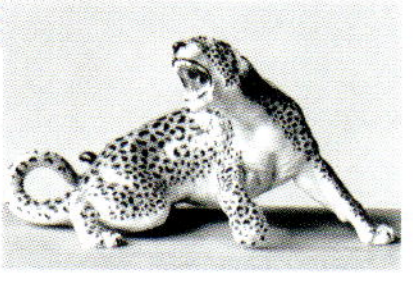
Jarl, Karin (zugeschr.)
Fauchender Leopard
ca. 1923
5144,43,2 Firmenmarke 19, Made in Germany, unsign., Ausformung nach 1938, H. 14,5 cm, L. 25,5 cm; Wien Museum Inv. 190.029
5144 H. 14 cm, L. 28 cm; Fotoarchiv K. Täuber
Quelle: Fotoarchiv K. Täuber

5147
Jarl, Karin
DECKELDOSE Schwan
Form: rund, gebaucht
Knauf: Schwan (vollplastisch)
Dekor: Seerosenteich
ca. 1923
5147,49,5 gem. F V, Firmenmarke 19, H. 11,4 cm, Ø 15,2 cm; Soth. com # 1LJB

5148
Jarl, Karin
Stehender Windhund (auf Längssockel)
ca. 1923
5148,37,4 Firmenmarke 19, H. 22 cm; PB Wien 19

5149
Jarl, Karin
Liegender Windhund
ca. 1923
5149,74,8 gem. lig. A.L.I.D., Firmenmarke 19, Made in Austria, H. 11,4 cm, L. 27,9 cm; eBay # 26992724
-> vgl. 5299= ähnl. Modell

5151
Thomasch, Wilhelm
DECKELDOSE / RAUMLUFTERFRISCHER Elefant
Form: kugelförmig
Knauf / Bekrönung: Elefant (vollplastische Figur)
Dekor: Mündung und Deckel länglich oval durchbrochen
ca. 1923
5151,2,5 Firmenmarke 19, H. 20; WKH
Bem.: Sog. „Rauchverzehrer" waren in den zwanziger Jahren beliebt. Somit war es auch für Goldscheider ein „Muss" solche Objekte im Programm zu haben. Ob die Elektrifizierung, die der zusätzlichen Beleuchtung dient nachträglich angebracht wurde, oder schon so von Goldscheider so konzipiert war, ließ sich nicht mehr feststellen.

5166
Gult, Johann
DECKELDOSE
ca. 1923
5166 (ohne Ang.), H. fehlt; Neuwirth (1974), S. 58, 63, hier keine nähere Ang. zu Form/Dekor oder Sign.

5167
Gult, Johann
SCHALE
ca. 1923
5167 (ohne Ang.), H. fehlt; Neuwirth (1974), S. 58, hier keine nähere Ang. zu Form/Dekor oder Sign.

5169
Podany, Rudolf
„Bei der Toilette" (Sitzende Balldame im Rokokokleid) (auf Ovalsockel)
ca. 1917/18 (für den Entwurf)
FIGUR
5169,288,12,0 gem. XII. D.1., Firmenmarke 19, Made in Austria, Ausformung 20er Jahre, H. ca. 21 cm; Henry's 29.6.2001
5169,340,19 Firmenmarke 19, Ausformung 20er Jahre, H. 19,5 cm; BM 124
ohne Ang., H. 21 cm; Schuler 49
-> vgl. 4834; 5329
Lit.: DD, 50.1922/23, Nr. 8, S. 38 (m. Abb.), Nr. 9, S. 39 (m. Abb.), hier jeweils mit Ang. des Originaltitels „Bei der Toilette" und des Entwerfers Rudolf Podany.

5171
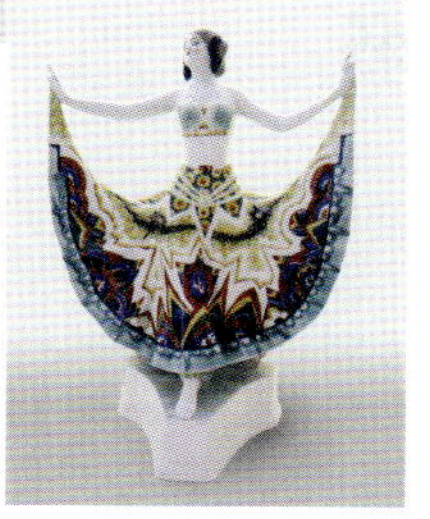
Rosé
„Ruth" (Tänzerin auf dreipassigem Sockel mit zwei Rosenkörben)
ca. 1911/12 (für den Entwurf)
5171,37,7,1 Firmenmarke 19 (Wien schwarz übermalt), Made in Austria, Hand dec., Ausformung 20er Jahre, H. 25,4 cm; Soth. com # AOD6
5171,50,8 gem. XIX, SA, Firmenmarke 19, Ausformung 20er Jahre, H. 27 cm; PB München
5171,350,8,03 ger. P, gem. AD, Firmenmarke 19, Made in Austria, Hand dec., unsign., Ausf. 20erJahre, H. 27 cm; Sammlung f@goldscheider.de
5171,375,9 (Firmenmarke ohne Ang.), H. 25,5 cm; Christie's SK 4.11.1999
5171,446,10 ger. VI T, gem. lig. MR, R 7, R, Firmenmarke 19, Made in Austria, Hand dec., Ausformung 20er Jahre, H. 28 cm; eBay # 497887859; WKA 33; Sammlung Alain Bernard
5171,480,7 ger. P4, gem. lig. MR 7, R., Firmenmarke 19, Made in Austria, Hand dec., Ausformung 20er Jahre, H. 28 cm; city-antik
5171 (ohne Ang.), Firmenmarke 18 od. 19?, Ausformung 20er Jahre, H. 26 cm; Spik 533
Bem. und Lit.: -> vgl. bei 3870, hier mit Ang. des Originaltitels „Ruth" und des falschen Künstlernamens „Bosi"; BT, o. Pag., Bild 84.
-> vgl. 3870; 3992; 4141; 5513
Lit.: Witkiewicz, S. 124–125 (m. Abb.).

5174
Jarl, Karin
Papagei (auf glockenförmigem Sockel)
ca. 1923
5174,42,11 gem. lig. CC, Firmenmarke 19, Made in Austria, Hand dec., H. 35 cm; KNS 17 (Schloss Hellbrunn, Salzburg 3.10.2001)
5174 (ohne Ang.), H. 35 cm; Schuler 50

5178
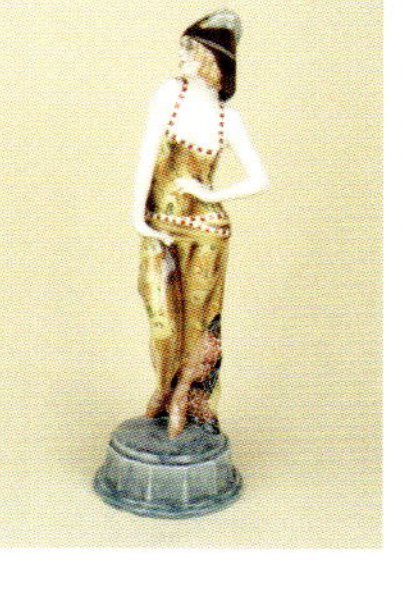
Lorenzl, Josef
FIGUR / LAMPE Tänzerin in Pose (mit federverziertem Hut) (auf facettiertem Rundsockel)
ca. 1923
FIGUR
5178,25,8 Firmenmarke 19, H. 48,3 cm; Doro 15.5.2002
5178,46,2 gem. O, XXIII, Firmenmarke 19, unsign., H. 50 cm; PB Wien 19
5178,100,8 Firmenmarke 19, Made in Austria, sign. Lorenzl, H. 48,3 cm; eBay # 921130669
5178,115,8 Firmenmarke 19, Made in Austria, unsign., H. 48 cm; Phillips L. 25.1.2001
5178 (ohne Ang.), sign. Lorenzl, H. 48,6 cm; Doro 24.5.2000
LAMPE
ohne Ang., Firmenmarke 19, unsign.?, H. 49,5 cm; Miller's A 16.1995 (m. Abb.)

5187
Thomasch, Wilhelm
Tänzerin (Halbakt auf dem rechten Bein tanzend) (auf passigem Sockel mit umlaufendem Perlstab) BRONZE
ca. 1923/24
5187,5 (ohne Firmenmarke), H. 38,2 cm; Doro 18.5.1999
Bem.: Vermutlich GS (Zuschreibung unter Vorbehalt, jedoch sprechen, trotz der fehlenden Firmenmarke, Qualität und Nummerierung für GS, auch ist Wilhelm Thomasch einer der bedeutenden Entwerfer für die Manufaktur GS).

5190
Lorenzl, Josef
Kniende Geisha (sich schminkend) (auf quaderförmigem Sockel)
ca. 1923/24
5190,18,16 (Firmenmarke ohne Ang.), sign. Lorenzl, H. 35 cm; Christie's SK 18.9.1998
5190,31,15 Firmenmarke 19, sign. Laurenzl, H. 33,1 cm; Doro 7.3.2001

5192
Männchenmachender Pekinese (auf profiliertem Rundsockel)
ca. 1923/24
5192,12,2 Firmenmarke 19, Made in Austria, H. 8,5 cm; eBay # 7355282510

5193
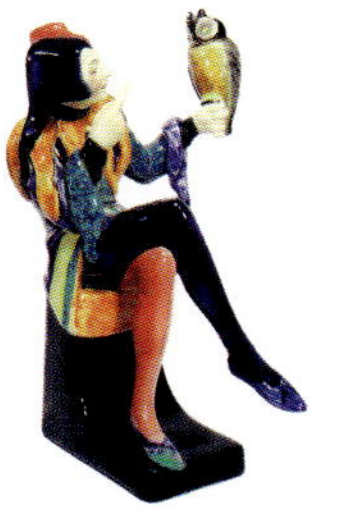
Lorenzl, Josef
Sitzender Till Eulenspiegel mit Eule
ca. 1923/24
5193,101,8 gem. F XIII, gem. lig. WW, Firmenmarke 19, unsign.?, H. 23 cm; eBay # 1497707409
5193,134,? gem. lig. AM, (Firmenmarke ohne Ang.), sign. Laurenzl, H. 24 cm; Hennig 17
5193,469,6,4,+ gem. lig. MR, (AvD?), Firmenmarke 19, Made in Austria, Hand dec., sign. Lorenzl, H. 24 cm; Sammlung a@goldscheider.net
-> vgl. 5509

5196
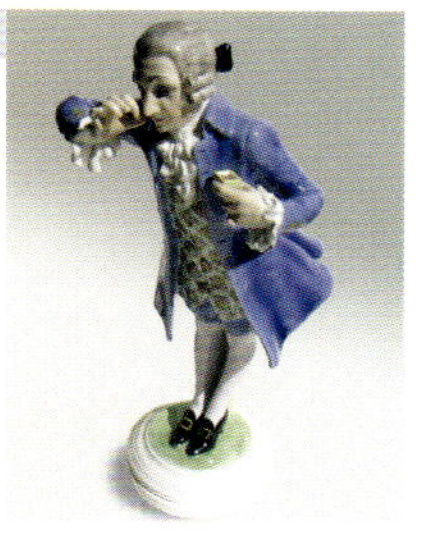
Lorenzl, Josef
„Eine kleine Prise" (Schnupfender Kavalier im Rokokokostüm) (auf Rundsockel)
ca. 1923/24
5196,22,65 (Firmenmarke ohne Ang.), H. 35 cm; Doro 30.5.1994
5196,40,65 gem. F, Firmenmarke 19, unsign.?, H. 35 cm; eBay # 1418014225
5196,48,65 Firmenmarke 19, sign. Laurenzl, H. 35 cm; Sammlung f@goldscheider.de
5196,53,3 Firmenmarke 19, Made in Austria, H. 34 cm; eBay # 928052222
5196 (ohne Ang.), Firmenmarke 19, Made in Austria, unsign.?, H. 34 cm; Schloss Rick. 38
5196 (ohne Ang.), H. 35 cm; Ahlden 42
Lit.: Alt-Wien, S. 464.

5202
Jarl, Karin
Liegender Rehpinscher (auf Rundsockel)
ca. 1923/24
5202,16 gem. S, D1, Firmenmarke 19, Made in Austria, H. 5,5 cm; AH Historia 77

5211
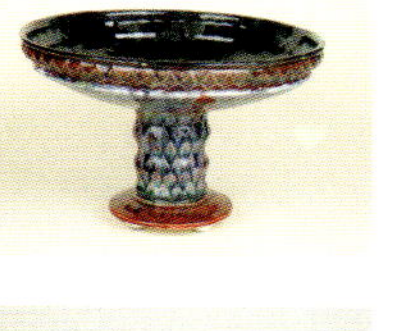
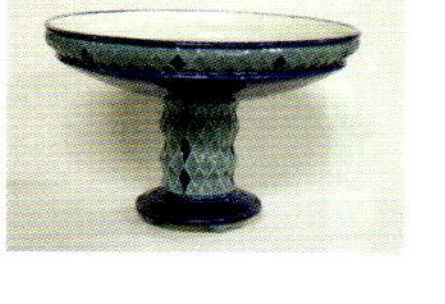
Thomasch, Wilhelm
TAFELAUFSATZ
Form: rund, flache Schale über zylinderförmigem, rautenreliefiertem Stand, perlenstabverzierte Halbkugelfüße
Dekor: mehrfarbige Laufglasur
ca. 1923/24
5211,14,5 gem. VIII, Firmenmarke 19, sign. Thomasch, H. 17,6 cm; Ø 28,1 cm; MAK Wien, Inv. Ke 10602-11
5211,22,12 ger. MI, gem. XIV, Firmenmarke 19, Made in Austria, Hand dec., unsign., H. 16,8 cm, Ø 27 cm; PB Wien 19

5214
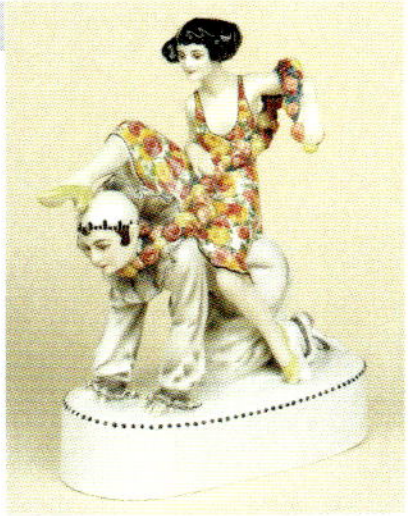
Thomasch, Wilhelm
GRUPPE Pierrot und Pierrette (Pierrette mit Blumengirlande, rittlings auf Pierrot) (div. Sockelformen)
ca. 1923/24
Sockel (Rocailleornamente)
5214,12,3 gem. lig. AM, F. XXIII, Firmenmarke 19, H. 34,3 cm; Quit. 17
Sockel (oval, Perlenstabverzierung)
5214,13,8 (Firmenmarke ohne Ang.), H. 35,5 cm; eBay # 2048140702
5214,70,12 gem. XXVII., D2., Firmenmarke 19, Made in Austria, H. 34,5 cm; PB Wien 19
5214,72,65 gem. D 2 XXVII, (Firmenmarke ohne Ang.), H. 35,5 cm; Soth. L. 29.11.1984

5218
Kuhn, Dina (Bernhardine)
LAMPENFUSS
Form: kegelförmig, darauf eiförmige Bekrönung
Dekor: abstraktes Muster, mehrfarbig
ca. 1923/24
5218,39,8 (Firmenmarke ohne Ang.), sign. Dina Kuhn, Gesamth. 64,2 cm; Christie's SK 13.10.2004
Bem.: Höchstwahrscheinlich gehört dieser Lampenfuß zu jenen, die 1925 von Dina Kuhn auf der Kunstgewerbeausstellung in Paris ausgestellt waren.

5227
Jarl, Karin
Steigendes Pferd (auf Ovalsockel)
ca. 1923/24
5227,57,18 Firmenmarke 19, H. 35,7 cm; Doro 4.3.2002
5227,59,11 Firmenmarke 19, Made in Austria, Hand dec., H. 36 cm; eBay # 870027895

5230
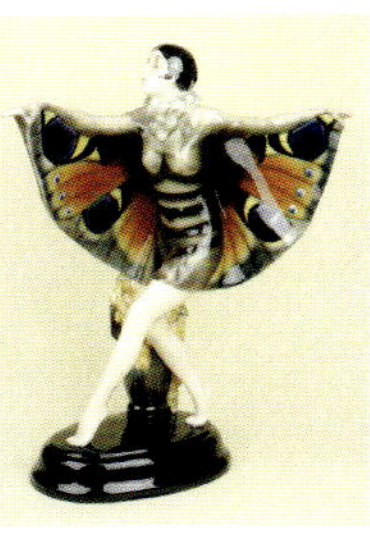
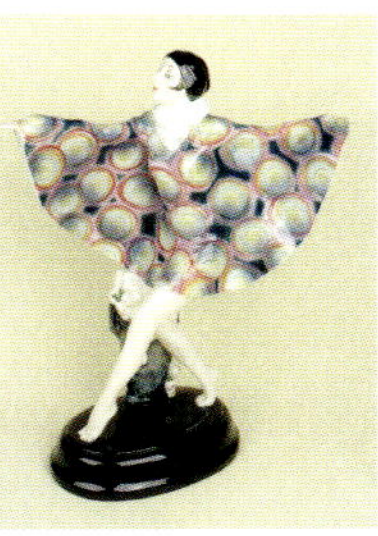
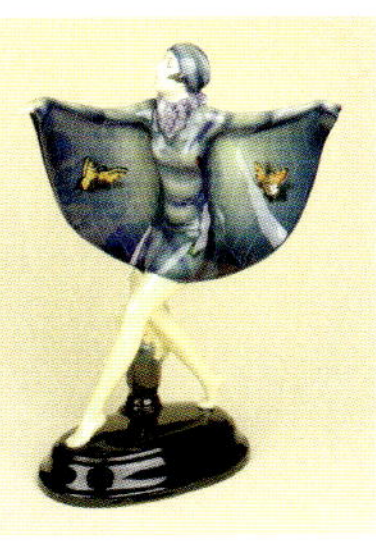
Lorenzl, Josef
FIGUR/LAMPE „Gefangener Vogel" (Der Gefangene Vogel, Niddy Impekoven) (auf Ovalsockel)
ca. 1922
FIGUR
5230,1,5 Firmenmarke 19, Made in Austria, unsign., H. 50 cm; KH
5230,11,? Firmenmarke 19, Made in Austria, unsign., H. 49,5 cm; Christie's SK 9.2.2000
5230,50,10 Firmenmarke 19, Made in Austria, sign. Lorenzl, H. 49 cm; PB Wien 19
5230,30,11 gem. AM, XXVII, Firmenmarke 19, sign. Laurenzl, H. 47 cm; Sammlung f@goldscheider.de
5230,41,11 gem. XO.J.7, gem. lig. AS, Firmenmarke 19, Made in Austria, sign. Lorenzl, H. 47,5 cm; eBay # 230016556462
5230,62,11 Firmenmarke 19, Made in Austria, sign. Lorenzl, H. 46 cm; Soth. Chester 6.–8.9.1988
5230,64,65 gem. lig. AM, AH, XXVII, Firmenmarke 19, unsign., H. 46,8 cm; PB
5230,88,8 gem. lig. MA, AV, Firmenmarke 19, sign. Laurenzl, H. 49 cm; PB Wien 19
5230,140,16 Firmenmarke 19, unsign.?, H. 49 cm; Soth. L. 25.9.1979; Lit.: AZ 1979, S. 502, 503 (m. Abb.)
5230,163,1 gem. lig. AH, XXVII, Firmenmarke 19, unsign., H. 49 cm; Sammlung Weilinger, Wien
5230,164,7 gem. lig. AH, gem. grün XXVII, Firmenmarke 19, unsign., H. 48,2 cm; Sammlung.schmutz@chello.at 33
5230,171,7 gem. XXVII, I, I, Firmenmarke 19, unsign.?, H. 49 cm; Soth. com. # 1BV4
5230,296,4,A,I,X, Firmenmarke 19, Made in Austria, sign. Laurenzl, H. 46,5 cm; PB
5230,366,4 gem. XI., gem. grün III., Firmenmarke 19, Made in Austria, sign. Laurenzl, H. 47,5 cm; eBay # 1229766477
5230,422,16 Firmenmarke 19, Made in Austria, sign. Laurenzl, H. 47,5 cm; Christie's SK 27.4.2005; 21.2.2006
5230,456,65 Firmenmarke 19, Made in Austria, sign. Laurenzl, H. 47 cm; BM 62
5230,458,4 Firmenmarke 19, Made in Austria, sign. Lorenzl, H. 48 cm; Ilko
5230,537,65 gem. lig. AS, gem. D2a, XII, Firmenmarke 19, Made in Austria, unsign. H. 47 cm; Sammlung.schmutz@chello.at 60
5230,545,11 Firmenmarke 19, Made in Austria, sign. Laurenzl, H. 47 cm; BM 103
5230,724,8 gem. lig. AH, M.L., XVII, Firmenmarke 19, Made in Austria, gepr. „L"= Z1, unsign., H. 47 cm; Sammlung Weilinger, Wien
5230,750,4 gem. D.4., XI, XII., Firmenmarke 19, Made in Austria, sign. Lorenzl, gepr. „L"= Z1, H. 47 cm; Sammlung.schmutz@chello.at 59
5230,753,18 gem. PH, D7, XXVI, Firmenmarke 19, Made in Austria, sign. Laurenzl, H. 47,5 cm; Soth. L. 31.10.1997; Soth. com # 1860110198; eBay # 2636234050; Bem.: Mit Originalpick. „F. Hengstmann Berlin/Leipzig".
5230,770,20 gem. D.2., XXIII., Firmenmarke 19, Made in Austria, sign. Laurenzl, H. 48 cm; PB Berlin
5230,807,8 gem. lig. AS., D.2., XII, Firmenmarke 19, sign. Laurenzl, Extra-Stempel 1, H. 48,3 cm; eBay # 410733704; Sammlung Joan Ferguson
5230,886,? gem. PH, Firmenmarke 19, Made in Austria, sign. Laurenzl, H. 46,5 cm; Quit. 25
5230,1014,18 Firmenmarke 19, sign. Laurenzl, H. 47,2 cm; Doro 5.6.1997
5230,1018,8 gem. lig. AH, XXVII, D7., Firmenmarke 19, Made in Austria, sign. Laurenzl, H. 48 cm; Sammlung Weilinger, Wien
5230,1050,56 gem. XXIV?, Firmenmarke 19, Made in Austria, Hand dec., sign. Laurenzl, H. 47 cm; Soth. com # 322D
5230,1068,10 gem. (lig.?) AS, 04, Firmenmarke 19, Made in Austria, sign. Laurenzl, H. 48,3 cm; Butterfields 20.11.2000/ zugl. eBay # 49994742; 532480362
5230,1232,11 gem. lig. HB(?), gem. XXVII, Firmenmarke 19, Made in Austria, sign. Lorenzl, H. 48,3 cm; eBay # 421787070 (KH www.tias.com/stores/ frozenbeauty)
5230,1246,11 gem. XXI., D.7, gem. lig. AS, Firmenmarke 19, Made in Austria, sign. Laurenzl, H. 47 cm; eBay # 7318588031
5230,1265,11 Firmenmarke 19, Made in Austria, Hand dec., sign. Laurenzl, H. 48,5 cm; Soth. L. 15.2.1985
5230,1488,7,0 ger. M8, gem. XII.D7, Firmenmarke 19, Made in Austria, Hand dec., sign. Lorenzl, H. 47 cm; eBay # 3769662173
5230,1499,7 ger. lig. M18, Firmenmarke 19, Made in Austria, Hand dec., sign. Laurenzl, H. 48,2 cm; eBay # 3734533196
5230,1592,4 Firmenmarke 19, sign. Laurenzl, H. 46 cm; KaM 87
5230,1604,6 ger. M29, gem. XXVII, D2, Firmenmarke 19, Made in Austria, sign. Laurenzl, H. 47 cm; Wendl 46
5230,1605,7 Firmenmarke 19, sign. Laurenzl, H. 46 cm; Soth. South, Billingshurst 11.10.2000
5230,1638,65 ger. M1, gem. lig. AH, HA, XXVII, 7, Firmenmarke 19, Made in Austria, Hand dec., sign. Laurenzl, H. 47 cm; Sammlung Weilinger, Wien
5230,1640,10 gem. lig. AH, Firmenmarke 19, Made in Austria, Hand dec., sign. Lorenzl, H. 47 cm; eBay # 1481689918
5230,1790,4,03 gem. XXVII, 13, gem. lig. AH, Firmenmarke 19, Made in Austria, sign. Laurenzl, H. 46,5 cm; Sammlung Weilinger, Wien
5230,1794,7 Firmenmarke 19, Made in Austria, Hand dec., sign. Lorenzl, H. 46 cm; Neumeister 278
5230,1800,8,04 gem. lig. AH, XXVII, 13, Firmenmarke 19, unsign.?, H. 46 cm; Soth. L. 19.10.1990
5230,1811,12,03 gem. F., XXVII, 12, Firmenmarke 19, Made in Austria, Hand dec., sign. Laurenzl, H. 46 cm; Sammlung Weilinger, Wien
5230,1815,11,03 gem. lig. AH, XXVI, 13, Firmenmarke 19, Made in Austria, Hand dec., sign. Laurenzl, H. 48 cm; city-antik
5230,1825,4,04 s. 5613
5230,1827,65 Firmenmarke 19, unsign.?, H. 46,5 cm; Zeller 44
5230,1843,7,04 gem. XII, D3, III, Firmenmarke 19, Made in Austria, Hand dec., unsign., H. 45,7 cm; eBay # 1485375825
5230,1860,8,03 gem. GD I, Firmenmarke 19, Made in Austria, Hand dec., sign. Laurenzl, H. 46 cm; Sammlung f@goldscheider.de
5230,1864,7,M03, Firmenmarke 19, Made in Austria, Hand dec., sign. Laurenzl, H. 44,5 cm; Soth. N.Y. 23.3.1996; Soth. Arcade 1556 (5.12.1996)
5230,1870,7 gem. N12, F, Firmenmarke 19, unsign.?, H. 45 cm; Schloss Hag. 14
5230,1916,4,8 gem. N4, F, Firmenmarke 19, Made in Austria, Hand dec., sign. Laurenzl, H. 46,5 cm; Sammlung.schmutz@chello.at 103
5230,1953,12,03 gem. N12, F, Firmenmarke 19, Made in Austria, Hand dec., sign. Laurenzl, H. 46,5 cm; Soth. L. 31.10.1997
5230,1976,65 ger. M03, gem. LN, 12, Firmenmarke 19, Made in Austria, Hand dec., sign. Lorenzl/Laurenzl, H. 46 cm; BM 44; Sammlung.schmutz@chello.at 34
5230,1981,18,O,3 gem. N13, Firmenmarke 19, Made in Austria, Hand dec., sign. Laurenzl, H. 45,7 cm; eBay # 3734533325
5230,1998,12,3,O gem. F.N, 11, Firmenmarke 19, Made in Austria, Hand dec., sign. Lorenzl, H. 46,5 cm; eBay # 1208573705
5230,2008,16,03, gem. 20, gem. lig. AH, JH, Firmenmarke 19, Made in Austria, Hand dec., sign. Laurenzl, H. 45,5 cm; Sammlung Weilinger, Wien
5230,2064,16,03 gem. N7, Firmenmarke 19, Made in Austria, Hand dec., sign. Laurenzl, H. 45 cm; Sammlung Weilinger, Wien
5230,2131,14 ger. 03, P, gem. H6, I, II, Firmenmarke 19, sign. Laurenzl, H. 47 cm; Sammlung f@goldscheider.de; Foto www.artwerk-online.de
5230,2158,14,03,P gem. lig. IO, N13, Firmenmarke 19, Made in Austria, Hand dec., sign. Laurenzl, H. 45,7 cm; eBay # 1213584339
5230,2261,8 Firmenmarke 19, Hand painted, sign. Lorenzl, H. 46 cm; Soth. L. 23.10.1987; Bem.: Die Ang. „Hand painted" ist sonst bei keinem weiteren GS-Stück bekannt, es handelt sich wohl um einen Druckfehler im Auktionskat.
5230,2327,4 gem. N11, Firmenmarke 19, Made in Austria, sign. Lorenzl, H. 46 cm; Henry's (Hess) 28.8.1999
5230,2342,10 gem. N13, gem. lig. AH, Firmenmarke 19, Made in Austria, Hand dec., sign. Lorenzl, H. 46 cm; eBay # 7312157484
5230,2381,6 Firmenmarke 19, sign. Lorenzl, H. 46 cm; Christie's SK 11.5.2001
5230,2476,6,1 gem. N11, F, Firmenmarke 19, Made in Austria, Hand dec., sign. Lorenzl, H. 46 cm; GK, S. 18, 40 (m. Abb.); Wien Museum Inv. 182.089
5230,2503,8,3 gem. F.N12, Firmenmarke 19, Made in Austria, Hand dec., sign. Lorenzl, H. 46 cm; eBay # 7346317808
5230,2519,19 gem. N22, F., Firmenmarke 19, Made in Austria, Hand dec., unsign.?, H. 46 cm; eBay # 7346089955
5230,2563,12 Firmenmarke 19, sign. Lorenzl, H. 45 cm; BM 70
5230,2595,65 (Firmenmarke ohne Ang.), sign. Lorenzl, H. 45 cm; Christie's SK 10.11.2000
5230,2622,12 gem. lig. AH, NM, Firmenmarke 19, Made in Austria, sign. Lorenzl, H. 47 cm; Phillips S. 31.7.2001
5230,2625,11 Firmenmarke 19, Made in Austria, Hand dec., sign. Lorenzl, H. 46 cm; PB Wien 19
5230,2668,75 gem. N22, Firmenmarke 19, Made in Austria, Hand dec., sign. Lorenzl, H. 47 cm; Sammlung Joan Ferguson
5230,2725,65 gepr. VI/p, gem. SO, Firmenmarke 19, Made in Austria, Hand dec., sign. Lorenzl, H. 49 cm; Henry's 25.10.1997
5230,2737,19 Firmenmarke 19, Made in Austria, Hand dec., sign. Lorenzl, H. 45,6 cm; city-antik
5230,2809,10,IO gem. N7, I, II, gem. blau O, Firmenmarke 19, Made in Austria, Hand dec., unsign.?, H. 49 cm;

Soth. com # 338J; Soth.. com # 346 WW
5230,2928,7,04 P, gem. N7, F., Firmenmarke 19, Made in Austria, Hand dec., sign. Lorenzl, H. 47,5 cm; Soth. L. 3.11.1995
5230,2971,19 gem. FN 12, (Firmenmarke ohne Ang.), sign. Lorenzl, H. fehlt; Soth. Olympia 3.5.2002
5230,2974,65 gepr. 1 P, gem. N28, F, Firmenmarke 19, sign. Lorenzl, H. 46 cm; Doro 12.12.1991
5230,3015,7 gem. lig. AS, 22, N, Firmenmarke 19, Made in Austria, Hand dec., sign. Lorenzl, H.48,3 cm; Sammlung.schmutz@chello.at 84
5230,3023,? gem. AS, 22, Firmenmarke 19, Made in Austria, sign. Lorenzl, H. 47 cm; Karrenbauer 146; Lit.: AZ 1997, S.485
5230,3037,10 ger. 16P, gem. S., gem. lig. AH7., P, X, Firmenmarke 19, Made in Austria, Hand dec., sign. Lorenzl, H. 47 cm; PB Deutschland
5230,3070,11,20, gem. lig. AH, N3, P, Firmenmarke 19, Made in Austria, Hand dec., sign. Laurenzl, H. 47,5 cm; Sammlung f@goldscheider.de
5230,3078,15,19 gem. lig. AH, Firmenmarke 19, Made in Austria, Hand dec., sign. Lorenzl, H. 47 cm; PB Wien
5230 (ohne Ang.) Firmenmarke 19, sign. Laurenz, H. 46 cm; RM 20
ohne Ang., Firmenmarke 19, Made in Austria, sign. Laurenzl, H. 46 cm; Soth. Mon. 17.4.1988
ohne Ang., Firmenmarke 19, Made in Austria, Hand dec., sign. Lorenzl, H. 46 cm; Christie's N.Y. 15.6.1985
ohne Ang., gem. lig. HJ, NB?, Firmenmarke 19, Made in Austria, Hand dec., sign. Lorenzl, H. ca. 46 cm; eBay # 76555917
LAMPE

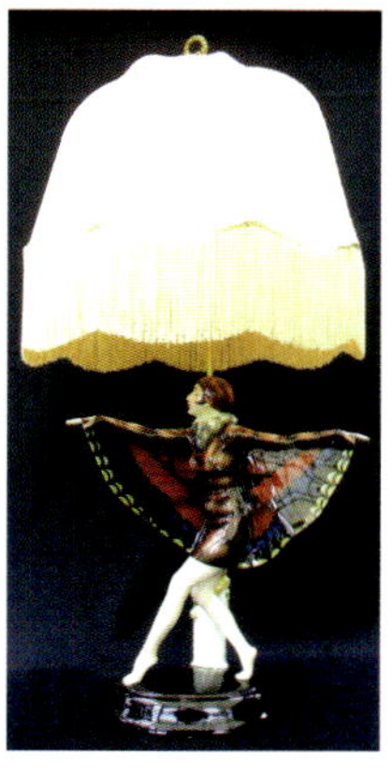

5230,648,65 Firmenmarke 19, Made in Austria, unsign., H. 53 cm; PB Wien 19
5230,914,65 gem. 4160, Firmenmarke 19, sign. Laurenzl, H. 53,2 cm; Soth. L. 2.5.1990
5230,1560,7 (Firmenmarke ohne Ang.), sign. Lorenzl, H. ca. 53 cm, Gesamth. 85 cm; Doro 15.3.2005
5230,2960,8 gem. 1P, SA 22 N, Firmenmarke 19, Made in Austria, Hand dec., sign. Lorenzl, H. 47,5 cm; Sammlung Robert Senft

Bem.: Dargestellt ist die Ausdruckstänzerin Niddy Impekoven (1904–2002), bzw. eine Pose aus ihrem Tanz „Der Gefangene Vogel", Musik: Bruno Hartl (Urauff. Frankfurt/M. 4.12.1918) –> vgl. hierzu Lit.-Auswahl: Impekoven, L.: Niddy Impekoven, Berlin 1920; Niddy Impekoven, Werdegang, Dresden 1920; Niddy Impekoven: Die Geschichte eines Wunderkindes, Zürich 1955 ; Aubel, 1928, S. 76, 77; Frentz, H.: Niddy Impekoven und ihre Tänze, Freiburg/Br. 1929; Frentz, H.: Weg und Entfaltung Niddy Impekovens, Leipzig 1933; Koegler, 1972, S. 275; Stöckemann, P.: Niddy Impekoven, Geburtstag eines Wunderkindes, in: Tanzdrama, Nr.11, 1990, S. 26–28. Die dargestellte Person kann alleine wegen des Körperbaues sicher nicht Niddy Impekoven selber sein, denn sie war 1918, als die Uraufführung stattfand, erst 14 Jahre alt und hat, wie eine Fotoserie aus dieser Zeit und vor allem das für den Entwurf ausschlaggebende Foto von Anny Eberth, Berlin belegen, nicht der von Lorenzl nachgebildeten, voll entwickelten Frau entsprochen. Auch konnte sie sich, als ich sie zwei Jahre vor ihrem Tod per Brief befragte, weder an die Figur, noch an die Firma GS erinnern. Möglicherweise verblasste auch durch ihr hohes Alter die Erinnerung. Die Haarlocke, die unter der Kappe jeweils seitlich hervorlugt, wurde ebenso hinzugefügt, wie auch die für die Stabilität der Figur sorgende rosengefüllte Blumenurne. Eine Besonderheit stellt der als Trio gefertigte Reigen von drei sich an den Händen anfassenden Figuren dar (Modell-Nr. 5530). Der mittig platzierte Säulenschaft, der die Elektrifizierung und die Glühbirnenfassung aufnimmt, ist mit Rosengirlanden verziert. Zwei weitere Varianten, die man durch ihre nach rechts gewandte Richtung, im Gegensatz zu allen anderen Modellen, die nach links tanzen nennen kann, sind die Modell-Nr. 5964 (H. 40 cm), also das direkte Pendant zur Nr. 5960, sowie die Nr. 5434 (H. 45 cm), die durch ihre Haartracht mit Haarband weitere Unterschiede aufweist. Marcell Goldscheider, versuchte sich in England auch mit einem Entwurf der dem „Gefangenen Vogel" angelehnt war. Dieser erreicht aber weder die Qualität noch konnte damit der Erfolg der dem Original beschieden wiederholt werden. Die Modell-Nr. für die englische Produktion lautet 893, die Höhe beträgt ca. 27,5 cm. Die Darstellung der Niddy Impekoven als „Gefangener Vogel" inspirierte zahlreiche Künstler und Porzellanfirmen. Genannt seien hier nur als Beispiele, die in Bronze ausgeführten Modelle von Barner und Zametzer (–> vgl. Christie's SK 15.11.2001), Otto Poertzel (zwei unterschiedliche Höhen –> vgl. Christie's SK 13.5.1999 und 20.9.2000), sowie u. a. die von der Porzellanfirma Hutschenreuther, Kunstabt. Selb von Karl Tutter entworfene Version.
Anm. zur Sign. Lorenzl / Laurenzl / Laurenz etc.: Wie bei GS oft, so verwendete man auch hier verschiedene Signaturen bzw. Namensformen. Die Variante „Laurenzl" wurde offensichtlich für den „weltweiten Markt" benutzt um dem Namen einen „internationalen Klang" zu geben und somit das Modell weltweit besser anbieten und vermarkten zu können.
Ausst.: HMW (jetzt Wien Museum) 1985/86; SSH 1992.
–> vgl. 5054 (BRONZE); 5960; 5964= Pendant zu 5960; 6204; 6600; 8176
–> vgl. 5530 GRUPPE (Trio)
Pendant –> vgl. 5434
Lit.: GSNachf. 1,50 (m. Abb. als Figur), hier mit Ang. des Originaltitels „Gefangener Vogel";
MW, 10.1928/29, Nr. 21, S. IV (m. Abb.); N.N.: Farbe und Bewegung, Die Fayencen der Wiener Manufaktur Friedrich GS Wien-Währing, in: DB, 5.1928, Nr. 213, S. 38 und 40 (m. Abb. als LAMPE), „Künstlerinnen von Rang haben den Tanzfiguren, die ihren Liebreiz im spröden Material festhalten, als Modell gedient. Man sieht Niddy Impekovens „Gefangenen Vogel"... und viele andere"; DD, 50.1922/23, Nr. 6, S. 29 (m. Abb. als BRONZE-Figur); Neuwirth (1974), S. 45, 59, 67; GK, S. 18, 40 (m. Abb.); GSN, S. 28, 29 (m. Abb.); Witkiewicz, S. 132–133 (m. Abb.).
GS neu: Modell-Nr. 82 „Gefangener Vogel" (H. 40 cm), Aufl. 3000 Stück

5232
Lorenzl, Josef
„Carmen" (Stehende Spanierin mit Fächer) (auf Rundsockel)
ca. 1923
5232,47,4 Firmenmarke 19, Made in Austria, sign. Lorenzl, H. 36 cm; Sammlung Dr. Peter Hieke, Wien
5232,58,11 Firmenmarke 19, Made in Austria, sign. Laurenzl, H. 35,7 cm; WKA 38
5232,79,18 gem. D.F. IV, Firmenmarke 19, Made in Austria, sign. Laurenzl, H. 35,5 cm; eBay # 94244247
5232,133,14 Firmenmarke 19, Made in Austria, Hand dec., sign. Laurenzl, H. 34 cm; Kett. 39
5232,141,12 Firmenmarke 19, Made in Austria, Hand dec., sign. Laurenzl, H. 34,9 cm; eBay # 37083227656
5232,84,65 gem. XXVII, D3, F, (Firmenmarke ohne Ang.), sign. Lorenzl, H. 34,3 cm; eBay # 2233361190
5232,171,8 ger. M26, Firmenmarke 19, Made in Austria, Hand dec., sign. Laurenzl, H. 34,5 cm; Doro 8.11.1994
5232,182,65 gem. X.9, Firmenmarke 19, Made in Austria, Hand dec., sign. Laurenzl, H. 33 cm; eBay # 2073536677
5232,281,4,8 Firmenmarke 19, Made in Austria, Hand dec., sign. Lorenzl, H. 38 cm; eBay # 2222654854
5232,29?,1 (Firmenmarke ohne Ang.), unsign., H. 34,1 cm; Doro 8.3.2004; 14.3.2005
5232,345,15 (Firmenmarke ohne Ang.), sign. Lorenzl, H. 34,5 cm; Christie's SK 12.5.2000
5232,365,13,12 (Firmenmarke ohne Ang.), sign. Lorenzl, H. 35 cm; Christie's SK 18.9.1998
5232,366,6 gem. 5.N, gem. lig. SG, Firmenmarke 19, Made in Austria, sign. Lorenzl, H. 34,5 cm; WKA 40; Doro 10.3.2003; Schloss Hag. 33
–> vgl. 5088; 5277
–> vgl. 5534= sitzende Figur
Lit.: Ros, L.: Wiener Fayencen und Bronzen, Begleitworte zu den Abb. über neue Werke von FG Wien, in: KuKGW, 4.1923/24, S. 15–16 (m. Abb.), hier mit Ang. des Originaltitels „Carmen".

5235
Kostial, Josef

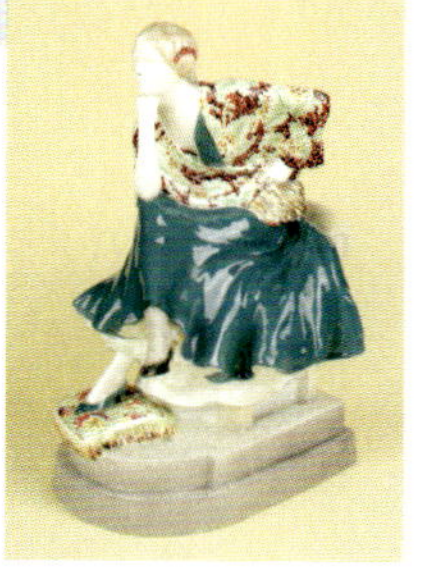

„Carmen" (Sitzende Dame mit Fußkissen)
ca. 1918/19 (für den Entwurf)
FAYENCE
5235,3,10 (Firmenmarke ohne Ang.), H. 23,9 cm; Doro 29.5.2001
5235,110,15 Firmenmarke 19, Made in Austria, Hand dec., Ausformung 20er Jahre, H. 23 cm; Stahl 14.9.1996
ALABASTER
5235,W,5 ger., Plak. E, Made in Austria (ger.), Ausformung 20er Jahre, H. 24,1 cm; eBay # 193495424
Bem.: Dieses Modell wurde in Fayence und Alabaster/Kunststein ausgeführt.
–> vgl. 4903; 5326
Lit.: N.N.: Wiener Kleinplastik einst und jetzt, in: MW, 2.1920/21, Nr. 12, S. 28 (m. Abb.); MW, 3. 1921/22, Nr. 3, S. 39 (m. Abb.), hier jeweils mit Ang. des Originaltitels „Carmen".

5236
Perl, Karl
„Die sieben Schwaben" (Siebenköpfige Figurengruppe auf rechteckigem Landschaftssockel)
ca. 1923
5236,20,20 Firmenmarke 19, Made in Austria, unsign., H. 27,3 cm, L. 40,4 cm; PB Wien 19
5236 (ohne Ang.), Firmenmarke 19, Made in Austria, unsign.?, H. 27 cm, L. 40 cm; WKH
Lit.: Meyer, K.: Arbeitsstoff und Arbeitswert, eine Betrachtung über Wiener Fayenceplastiken, in: KuKGW, 5.1925, SH. 3, S. 11 (m. Abb.), hier mit Ang. des Originaltitels „Die sieben Schwaben" und des Künstlers „Karl Perl", „Das gelungenste Stück unserer Zusammenstellung scheint mir aber das letzte zu sein: „Die sieben Schwaben", die Karl Perl entworfen hat. Die Aufgabe, diese sieben Hasenhelden aus dem Volksmärchen in eine einzige Gruppe zusammenzubringen und den sogenannten „fruchtbaren Moment" zu finden, war wohl nicht ganz einfach. Die Komposition ist überaus glücklich. Die nach oben gewölbte Grundplatte, vielleicht eine leichte Erdwelle im Gelände andeutend, ermöglichte in natürlicher, ungezwungener Weise die Anordnung des Ganzen zur Gestalt eines unregelmäßigen Dreiecks. Wie sich in den Gesichtern der Helden das Gegenteil von Intelligenz und Mut ausprägt, ist mit köstlichem Humor dargestellt. Und dabei ist jedes Gesicht individuell behandelt, in jedem drückt sich Blödigkeit sowie Schrecken über den plötzlich erschienen Hasen, der natürlich vom Künstler mit vollem Recht weggelassen wurde, verschieden aus; dazu das Gewusel, das Durcheinander der Beine, die Verschiedenheit der Gewänder und Kopfbedeckungen, und alles das in leuchtender Buntheit wie eine bizarre Farbensymphonie vor uns hingestellt: wahrlich, wer angesichts dieser überwältigend komischen Gruppe noch der Fayence scharf einzuhaltende Grenzen zuweisen wollte, innerhalb deren sie sich mit ihrer Formengebung und Farbengebung, mit Umfang und Wahl des künstlerischen Vorwurfes usw. zu beschränken habe, damit sie ja nicht etwa freibeuterische Streifzüge ins Gebiet des Porzellans hinüber mache, dem wäre nicht zu helfen".
Bem.: Die Darstellung geht auf einen mittelalterlichen Schwank zurück, der um 1545 von Hans Sachs bearbeitet wurde.

5238
Ingels, Domien

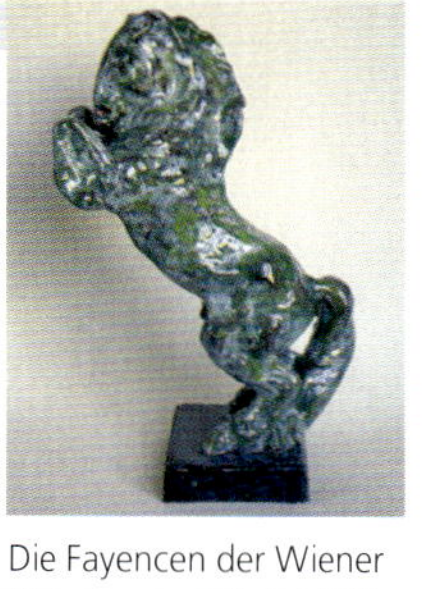

„Wallonisches Pferd" (Steigendes Pferd auf Rechtecksockel)
ca. 1923
5238,47,14 gem. rot 1L., Firmenmarke 19a, Firmenmarke 19, Made in Austria, Hand dec., unsign., H. 45 cm; Sammlung Weilinger, Wien
Lit.: N.N.: Farbe und Bewegung, Die Fayencen der Wiener Manufaktur Friedrich GS Wien-Währing, in: DB, 5.1928, Nr. 213, S. 38 und 40 (m. Abb.), hier mit Ang. des Originaltitels „Wallonisches Pferd" und des Künstlers Domien Ingels.

5240
Ingels, Domien
Eisbär
ca. 1923
5240,2,8 gem. B., Firmenmarke 19, Made in Austria, H. fehlt; PB

5242
Verbanck, Georges
Stehender Putto mit Früchten (auf Rundsockel)
ca. 1923
5242,6,? (Firmenmarke ohne Ang.) Made in Austria, H. 42 cm; BM 86; Bem.: Hier ist der Künstler S. Wernekinck als „schwer lesbare Sign." angegeben, es dürfte sich aber eindeutig um einen Lesefehler handeln, die Sign. muss wohl G. Verbanck lauten.
5242 (ohne Ang.) H. 42 cm; Ahlden 56; 66; Bem.: Hier keine exakte Ang. zur Sign.
5242 (ohne Ang.), Made in Austria, sign. Geo. Verbanck, H. 42,5 cm; Nagel 332
Lit.: Neuwirth (1974), S. 75, hier ohne Ang. zur Modell-Nr., jedoch mit Hinweis auf den Salon Paris 1928, wo ein Putto ausgestellt war. Möglicherweise handelt es sich hierbei um eine Ausformung der Manufaktur GS.

5245
Jarl, Karin
Liegendes Lama
ca. 1923
5242,6,? Firmenmarke 19, H. 14 cm, L. 20 cm; Doro 19.9.2000

5246 s. 5248

5248
Kuhn, Dina (Bernhardine)
„Adam und Eva" (als vollplastische Figuren mit Baum und Schlange)
ca. 1923
5248,3,8 (ohne Firmenmarke), unsign?, H. 31 cm; Zacke, S. 54, Nr. 26
5248(5246?),11,11 gem. L., Firmenmarke 19, Made in Austria, Hand dec., unsign.?, H. 37 cm; eBay # 2214367185; Bem.: Der Anbieter nennt die Modell-Nr. 5246, auch die Höhe differiert zum zweiten nachgewiesenen Exemplar. Genaueres konnte nicht geklärt werden.
Ausst.: Int. Exh. CA 1928/29, Abt. „Austria," Nr. 17: Group „Adam and Eve", polychrome Pottery, hier mit Ang. des Originaltitels und der Künstlerin Dina Kuhn.

5249
Kuhn, Dina (Bernhardine)
Jäger mit zwei springenden Hirschen
ca. 1923
5249,10,12 ger. M26, Firmenmarke 19, H. 24,1 cm; Doro 8.11.1994

5251
Kostial, Josef
„Maskerade" Stehende Dame mit Muff (auf Ovalsockel) (Lo Hesse)
ca. 1919 (für den Entwurf)
5251,1,10 gem. lig. AH II., Firmenmarke 19, H. 25,4 cm; Zezschwitz 16; 20; 24
-> vgl. 4938 GRUPPE „Maskerade"
Bem.: Dargestellt ist die Tänzerin Lo Hesse (1889–1983?) in ihrem Kostüm zur Tanzschöpfung „Maskerade". Lit. und Bem. -> vgl. bei Modell-Nr. 4938.
Lit.: DM 3.1926/27, Nr. 30, S. 576 (m. Abb.), hier eine zeitgenössische Fotografie, die wohl als Vorlage für die Ausführung diente. Ohne Ang. des Fotografen bzw. des Fotostudios.

5255
Rottenberg, Ena (Emma Helena)
Liegende weibliche Figur (Halbakt)
MARMOR
ca. 1923
5255,HM.,2 Plak. E, unsign, H. 23,5 cm, L. 40 cm; Doro 8.3.2004
Lit.: N.N.: Österreich in Paris, in: ÖBWK, 1.1924/25, S. 341 (m. Abb.); Hefte Mai, Juni jeweils S. III (m. Abb.) (Reklame).

5258
Meier-Michel, Johanna
Stehende Dame mit Fächer (auf Rundsockel)
ca. 1923
5258,8,11 gem. VIII. Firmenmarke 19, Made in Austria, sign. J. Meier-Michel, H. 32,3 cm; Doro 1.12.2000; eBay # 7389090031

5260
Perl, Karl
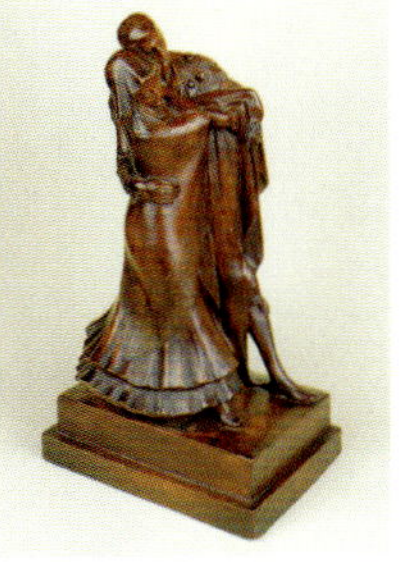
Spanisches Paar beim Tanz (auf Rechtecksockel)
ca. 1922 (für den Entwurf)
5260,13,65 gem. XIV, D2, 22, Made in Austria, unsign. H. 28 cm; Wien Museum Inv. 190.119
-> vgl. 5001

5262
Podolak, (A./Karl?)
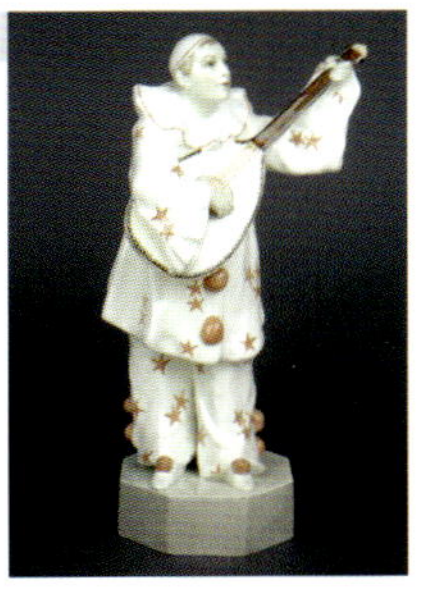
Stehender Harlekin mit Laute (auf oktogonalem Sockel)
ca. 1919 (für den Entwurf)
5262,22,5 Firmenmarke 19, Made in Austria, unsign.?, H. 30,5 cm; eBay # 2048139919
-> vgl. 4935; 5263
-> vgl. 4922 GRUPPE

5263
Podolak, (A./Karl?)
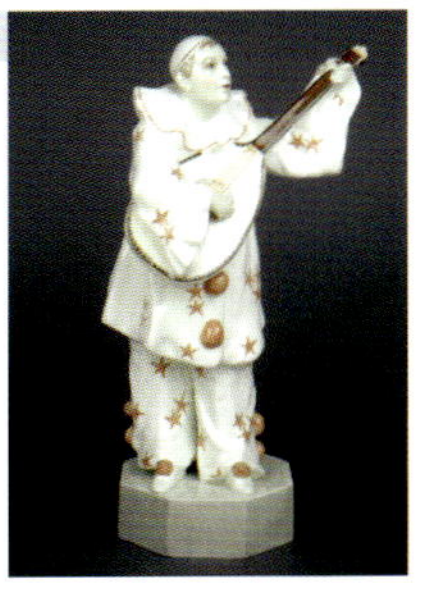
Stehender Harlekin mit Laute (auf oktogonalem Sockel)
ca. 1919 (für den Entwurf)
5263,22,20 (Firmenmarke ohne Ang.), H. 21,7 cm; Doro 19.9.2000
5263,23,10 (Firmenmarke ohne Ang.), H. 21,5 cm; Schopmann 23.4.2002
-> vgl. 4935; 5262
-> vgl. 4922 GRUPPE

5266
Lorenzl, Josef
Kartenlegerin (Sitzende Zigeunerin mit Spielkarten)
ca. 1923
5266,7,13 Firmenmarke 19, Made in Austria, sign. Laurenzl, H. 32 cm; Kett. 136
5266,25,11 gem. A1, D1, XXIII, Firmenmarke 19, Made in Austria, sign. Laurenzl, H. 32 cm; eBay # 295771353
5266,33,8 Firmenmarke 19, Made in Austria, Hand dec., sign. Laurenzl, H. 34,3 cm; eBay # 398890841
5266,54,10 (Firmenmarke ohne Ang.), unsign.?, H. 34,5 cm; Soth. L. 23.10.1987

5267
Lorenzl, Josef
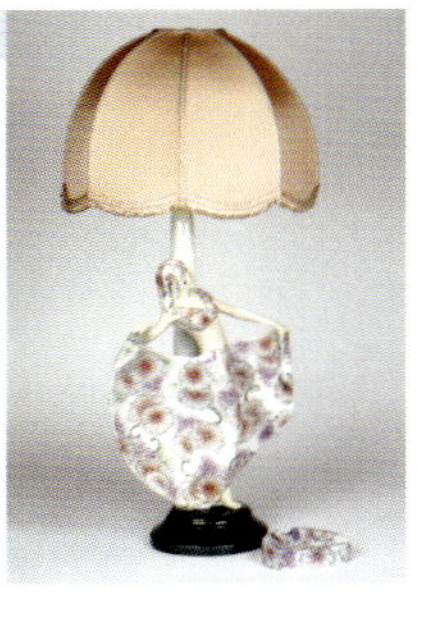
FIGUR/LAMPE Orientalische Tänzerin in Pose (auf Rundsockel)
ca. 1923
FIGUR
5267,40,65 Firmenmarke 19, sign. Lorenzl, H. 44,5 cm; Doro 30.11.2001
5267,53,10 Firmenmarke 19, Made in Austria, sign. Laurenzl, H. 44,4 cm; eBay # 1212682165
5267,60,11 Firmenmarke 19, Made in Austria, sign. Laurenzl, gepr. „L"= Z1, H. 45,7 cm; KH www.tias.com/stores/frozenbeauty
5267,125,6 Firmenmarke 19, Made in Austria, Hand dec., unsign.?, H. 45 cm; WKA 12
LAMPE
5267,32,11 Firmenmarke 19, Made in Austria, unsign., H. 46 cm; PB Wien 19; Bem.: Zu dieser Lampe wurde ein Aschenbecher (ohne Modellnummer, Firmenmarke 19, Made in Austria) im gleichen Dekor gefertigt.
-> vgl. 6023
Lit.: Pinhas, S. 146 (m. Abb.), hier mit der falschen Mod.-Nr. 5237.

5275
Kuhn, Dina (Bernhardine)
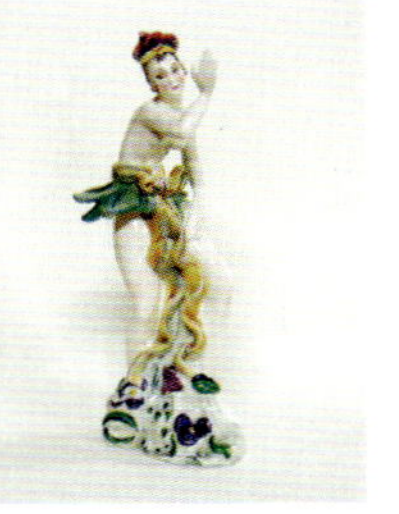
FIGUR/LAMPE „Papagena" (Tänzerin mit exotischem Kostüm und Kopfputz)
ca. 1923/24
FIGUR
5275,23,14 gem. lig. AZ?, gem. grün XIV.D, Firmenmarke 19, Made in Austria, H. 32,3 cm; MAK Wien, Inv. Ke 10201
5275,31,4 Firmenmarke 19 (zwei mal), Made in Austria, Hand dec., Z 18, H. 31,5 cm; Wien Museum Inv. 190.130
LAMPE
5275,12,10 Firmenmarke 19, Made in Austria, H. 30 cm, Gesamth. 59,7 cm; eBay # 7306593161
Bem.: Die Bezeichnung „Papagena" geht zurück auf das Inventarbuch des MAK (Wien). Ob die Firma Goldscheider wirklich diesen Titel verwandt hat, lässt sich nicht mehr nachvollziehen.
Ausst.: Enthüllt, ein Jahrhundert Akte österr. Künstlerinnen, Frauenbad, Baden 1998/99; Galerie im Stadthaus Klagenfurt 1999.
Lit.: Neuwirth (1974), S. 58

5276
Moroder, C.
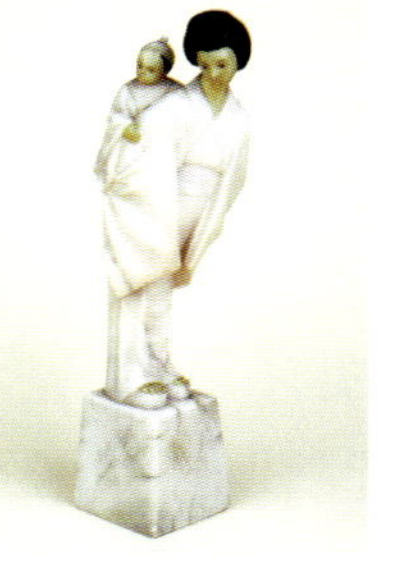
„Madame Butterfly" (Stehende Japanerin mit Kind)
ca. 1923/24
KERAMIK/FAYENCE
5276,61,10 Firmenmarke 19, sign. „Monder", H. 34,5 cm; Doro 30.5.1994
SPECKSTEIN
5276,W,2 Plak. E, sign. „Monder", H. 34,6 cm; PB Wien 19
Bem.: Dieses Modell wurde in Fayence und Speckstein (Alabaster?) ausgeführt.
Lit.: Meyer, K.: Arbeitsstoff und Arbeitswert, eine Betrachtung über Wiener Fayenceplastiken, in: KuKGW, 5.1925, SH. 3, S. 11 (m. Abb.), hier mit Ang. des Originaltitels „Madame Butterfly" und des Künstlers „C. Moroder", „Es ist bei aller herben Geschlossenheit doch ein ungemein zartes, inniges Bildchen, das hier der verstorbene Südtiroler Künstler Moroder geschaffen hat".

5277
Lorenzl, Josef
„Carmen" (Stehende Spanierin mit Fächer) (auf Rundsockel)
ca. 1923
5277,12,15 gem. lig. AD, gem. N, Firmenmarke 19, Made in Austria, Hand dec., H. 26,6 cm; Soth. com. # 32AN
-> vgl. 5088; 5232
-> vgl. 5534= sitzende Figur
Lit.: Ros, L.: Wiener Fayencen und Bronzen, Begleitworte zu den Abb. über neue Werke von FG Wien, in: KuKGW, 4.1923/24, S. 15–16 (m. Abb.), hier mit Ang. des Originaltitels „Carmen".

5281
Lorenzl, Josef
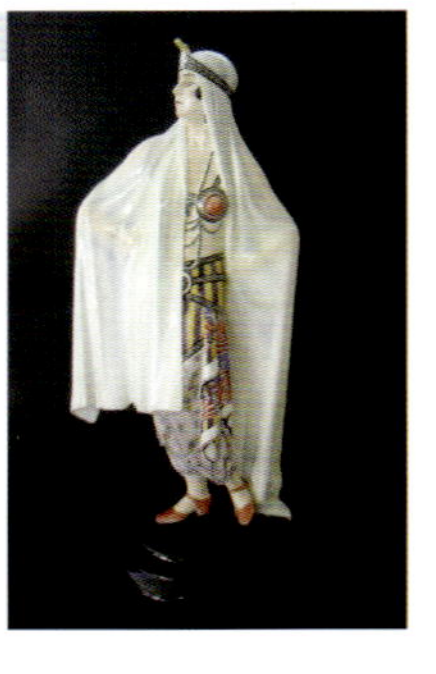
FIGUR/LAMPE „Aida" (Stehende Odaliske) (auf Ovalsockel)
ca. 1923/24
FIGUR
5281,38,1 gem. lig. AS. XII., Firmenmarke 19, unsign.?, H. 47,5 cm; KH Art 1900
5281,80,11 gem. XXIII Firmenmarke 19, Made in Austria, sign. Laurenzl, H. 46 cm;

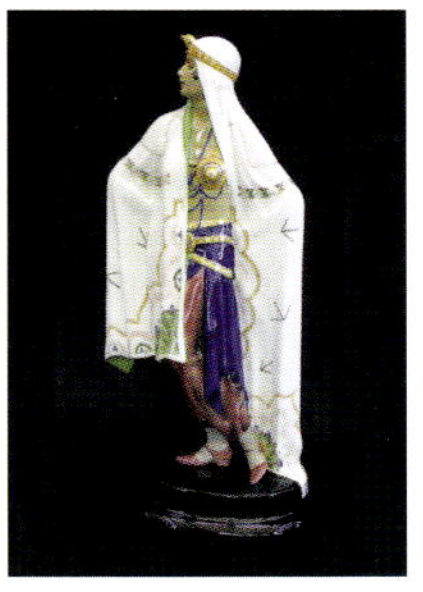

Soth. L. 3.5.1991
5281,107,? gem. F XIII, Firmenmarke 19, Made in Austria, sign.?, H. 47,8 cm; Sammlung Joan Ferguson
5281,168,5 gem. IX., Firmenmarke 19, Made in Austria, sign. Lorenzl, H. 46,5 cm; eBay # 406659001; Sammlung Robert Senft
5281,192,10 gem. F., XXIII, D1., Firmenmarke 19, Made in Austria, sign. Laurenzl, H. 47 cm; eBay # 3762354454
5281,218,65 (Firmenmarke ohne Ang.), sign. Lorenzl, H. 46 cm; Christie's A. 15.6.2004
5281,332,11 gem. F XXIII D1, Firmenmarke 19, Made in Austria, sign. Laurenzl, H. 48,2 cm; eBay # 12716406
5281,358,8 gem. F., XXIII., D1., Firmenmarke 19, Made in Austria, sign. Laurenzl, gepr. „L"= Z1, H. 47,5 cm; Sammlung f@goldscheider.de
5281,381,8 gem. F. XXIII D1, Firmenmarke 19, Made in Austria, sign. Lorenzl, H. 47 cm; Quit. H. 9
5281,512,8 gem. XXIII, Firmenmarke 19, Made in Austria, Hand dec., sign. Laurenzl, H. 47 cm; eBay # 1236677828
5281,518,12 Firmenmarke 19, Made in Austria, Hand dec., sign. Laurenzl, H. 46 cm; eBay # 7306786945
5281,534,7 ger. M5, Firmenmarke 19, unsign.?, H. 44,5 cm; BM 34
5281,635,7 gem. O2, R.E, X4, Firmenmarke 19, Made in Austria, sign. Laurenzl, H. 45,8 cm; Sammlung Joan Ferguson
5281,659,65 (Firmenmarke ohne Ang.), sign. Laurenzl, H. 46 cm; Christie's SK 27.4.2005
5281,667,12 Firmenmarke 19, sign. Lorenzl, H. 46 cm; Doro 1.12.2005
5281,734,16 ger. MO3, gem. N7 D, Firmenmarke 19, Made in Austria, Hand dec., sign. Lorenzl, H. 45,7 cm; eBay # 2562916531
5281,745,8 gem. 2.1.1. F., (Firmenmarke ohne Ang.), sign. Lorenzl, H. 46,3 cm; Sammlung Robert Senft
5281,757,10 (Firmenmarke ohne Ang.), unsign.?, H. 46 cm; KH
5281,795,7,04P Firmenmarke 19, Made in Austria, Hand dec., sign. Lorenzl, H. 46 cm; Quit. 6
5281,846,16 gem. F.J.R. 13A, gem. rot N, Firmenmarke 19, Made in Austria, Hand dec., sign. Lorenzl, H. 47 cm; eBay # 1205855101
5281,905,7 gem. MR13, Firmenmarke 19, Made in Austria, Hand dec., sign. Lorenzl?, H. 47,7 cm; Soth. L. 31.10.1997
5281,928,8 gem. lig. MR, AS, 2Jp, Firmenmarke 19, Made in Austria, Hand dec., Soth. com # 32 US; Sammlung Joan Ferguson
5281,942,11 (Firmenmarke ohne Ang.), sign.?, H. 47 cm; KH
LAMPE
5281,422,8 Firmenmarke 19, Made in Austria, unsign., H. (mit Lampenschirm) 82 cm; Nagel 352; Quit. 9
5281,561,1,12 ger. M27, Firmenmarke 19, Made in Austria, Hand dec., unsign., H. 53,3 cm; Soth. L. 22.4.1994
5281,834,15 Firmenmarke 19, Made in Austria, Hand dec., sign. Lorenzl, H. 49,5 cm; WKA 35
5281,915,12 gem. F., Firmenmarke 19, Made in Austria, Hand dec., unsign., Gesamth. 77 cm; eBay # 3222821547
5281,944,15 gem. F, gem. blau S.13a, Firmenmarke 19, Made in Austria, sign. Lorenzl, Extra-Stempel 2, H. 48,2 cm; Soth. com. # 1BWP
Bem.: Das in den Kopfschmuck eingearbeitete Tier ist die sog. Ureusschlange. Bei der Figur handelt es sich allenfalls um eine Tempeltänzerin aus der Verdi Oper „Aida". Ich danke Herrn F. M. Peter (Deutsches Theatermuseum, Köln) für seine freundliche Auskunft.
-> vgl. 6021
Lit.: Meyer, K.: Arbeitsstoff und Arbeitswert, eine Betrachtung über Wiener Fayenceplastiken, in: KuKGW, 5.1925, SH. 3, S. 10 (m. Abb.), hier mit Ang. des Originaltitels „Aida" und des Künstlers „Josef Lorenzl", „Ich habe den Eindruck, als müsste diese, einen halben Meter hohe Plastik – eine harmonische Vereinigung vornehmer Ruhe und anmutiger, aber bestimmter Bewegung, wobei der bunte Mantel eine wirksame Folie zu dem Ebenmaß der Glieder bildet, beim ersten Anblick wie polychrom behandelter Marmor wirken"; MW, 11.1929/30, Nr. 20, Umschlaginnenseite (m. Abb.); Witkiewicz, S. 126–127 (m. Abb.).

5282
Lorenzl, Josef
Stehende junge Tänzerin mit Trommel (auf Rundsockel)
ca. 1923/24
5282,192,12 gem. XIV, D4, gem. lig. AS, Firmenmarke 19, Made in Austria, Hand dec., sign. Laurenzl, Originalpick.: „R. Morath, Arensburg", H. 30 cm; Quit. 14; eBay # 3705113040

5288
Bormann, Willy
„Englischer Greyhound" (auf Ovalsockel)
ca. 1923/24
5288,9,16 gem. lig. JT, Firmenmarke 19, unleserlich sign., H. 27,3 cm; eBay # 73616583
Lit.: Meyer, K.: Arbeitsstoff und Arbeitswert, eine Betrachtung über Wiener Fayenceplastiken, in: KuKGW, 5.1925, SH. 3, S. 9, 10 (m. Abb.), hier mit Ang. des Originaltitels „Englischer Greyhound" und des Künstlers „Willy Bormann", „Der von Willy Bormann entworfene „Englische Greyhound" gibt die ganze Geschmeidigkeit, man möchte fast sagen: Eleganz, dieser Hunderasse wieder. Da die Bemalung bei diesem 27 cm hohen Stück nur ganz gering ist, so wirkt der Werkstoff an sich, ohne weitere Unterstützung. Wer möchte nun als Käufer so enghorizontig sein und fragen, ob dieses Stück aus Porzellan oder „bloß" aus Fayence ist, und, wenn er die entsprechende Auskunft erhält, das Stück dann vielleicht als minderwertig ablehnen und nach einem – vielleicht recht nichtssagenden, kitschigen – Motive aus Porzellan greifen, nur dem pretiösen Werkstoff zuliebe?".

5289
Perl, Karl
Das letzte Abendmahl (auf Rechtecksockel)
ca. 1923/24
5289,5,65 gem. VIII., Firmenmarke 19, H. 20 cm, L. 45,7 cm; eBay # 7315756330
5289,16,10 (Firmenmarke ohne Ang.), H. 18,7 cm, L. 44,2 cm; Doro 7.3.2001
Bem.: Nach historischem Vorbild gearbeitete Figurengruppe, Jesus mit seinen Jüngern um den Abendmahlstisch versammelt.

5296
Lorenzl, Josef

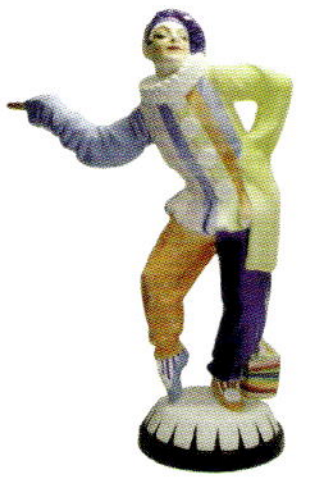

„Herein spaziert!" (Posierender Harlekin mit Akkordeon) (auf Rundsockel)
ca. 1923/24
5296,2,10 gem. F.III, gem. rot P, Firmenmarke 19, sign. Laurenzl, H. 30,3cm; Sammlung a@goldscheider.net
5296,11,7 gem. F., gem. lig. AM, Firmenmarke 19, H. 30,5 cm; eBay # 1422584731
5296,20,85 Firmenmarke 19, sign. Lorenzl, H. 29,7 cm; Doro 29.10.1996
Lit.: Meyer, K.: Arbeitsstoff und Arbeitswert, eine Betrachtung über Wiener Fayenceplastiken, in: KuKGW, 5.1925, SH. 3, S. 9, 10 (m. Abb.), hier mit Ang. des Originaltitels „Herein spaziert!" und des Künstlers „Laurenz" (!), „Das alte, aber scheinbar unerschöpfliche Motiv des Pierrot ist hier in frischer Weise von Laurenz neu behandelt worden. „Herein spaziert" nannte der Künstler diesen seinen Entwurf (Höhe 30 cm). Das vielfarbige Kostüm, der keck zurückgeworfene Kopf mit dem schelmischen Gesicht, die einladende Geste des rechten Arms: alles das ist quicklebendig geschaut und zeigt überwiegend südländisches Gepräge".

5299
Jarl, Karin
Liegender Windhund
ca. 1924
FAYENCE
5299,540,11 gem. XI, 22, sign. Jarl, H. 6 cm, L. 14 cm; Henry's 30.3.2000
5299,628,7 ger. M14, Firmenmarke 19, Made in Austria, sign. Jarl, H. ca. 6 cm; KH
BRONZE
5299,15 (Firmenmarke ohne Ang.), sign./bezeichnet Czapek (!), H. 5,7 cm, L. 14 cm; Doro 19.9.2000
Bem.: Dieses Modell wurde in Fayence und Bronze ausgeführt. Wie die Sign. „Czapek" zustande kommt, lässt sich nicht nachvollziehen, das Modell stammt eindeutig von Karin Jarl, was nicht nur die beiden nachgewiesenen, mit Jarl sign. Fayencen und das ebenso Jarl sign. Ex. der Modell-Nr. 5461 belegen.
-> vgl. 5461
-> vgl. 5149= ähnl. Modell

5305
Sitzender Putto mit Kaktus (auf Landschaftssockel)
ca. 1924
5305,17,14,F gem. 4.L., Firmenmarke 19, Made in Austria, Hand dec., H. 29 cm, L. 28 cm; Sammlung Weilinger, Wien

5314
Thomasch, Wilhelm

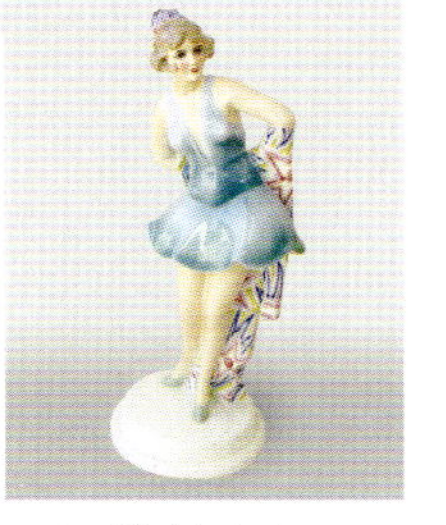

Stehende Tänzerin mit Kopfputz (einen langen Schal hinter sich haltend) (auf Rundsockel)
ca. 1924
5314,06,11 gem. D.3, IV, A, Firmenmarke 19, Made in Austria, H. 35,5 cm; eBay # 2531936272
5314,16, 3 gem. W, IV, Firmenmarke 19, Made in Austria, H. 35 cm; BM 55
5314,38 Firmenmarke 19, Made in Austria, Originalpick.: „El Trust Joyero, Buenos Aires", H. 35 cm; Sammlung Stressless, Portugal
5314,61,18 gem. 3823, Firmenmarke 19, Made in Austria, H. 35,5 cm; eBay # 344285984
5314,83,18 (Firmenmarke ohne Ang.), H. 35cm; Doro 31.5.1990
5314,101,? (Firmenmarke ohne Ang.), unsign.?, H. 35 cm; Herr11

5316
Lorenzl, Josef
Stehende, nach vorne gebeugte Pierrette (auf Rundsockel)
ca. 1924
5316,11,11 gem. lig. AH, XVII., Firmenmarke 19, sign. Laurenzl, H. 30 cm; eBay # 6511568147
5316,294,4 gem. FK, Firmenmarke 19, Made in Austria, Hand dec., sign. Laurenzl, H. 28,5 cm; Doro 9.11.1993
5316,298,8,0,3 Firmenmarke 19, Made in Austria, Hand dec., sign. Lorenzl, H. 29,2 cm; eBay # 936712454; Doro 15.9.2003
5316,359,(10),8,4 K gem. lig. MR8, Firmenmarke 19, Made in Austria, sign. Lorenzl, H. 29,2 cm; Soth. com # 1BVP; KH www.tias.com/stores/frozenbeauty

5326
Kostial, Josef

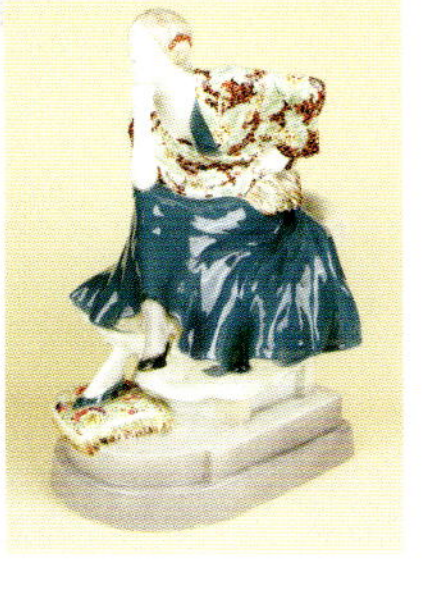

„Carmen" (Sitzende Dame mit Fußkissen)
ca. 1918/19 (für den Entwurf)
5326,140,5 gem. XIV, Firmenmarke 19, Made in Austria, Ausformung 20er Jahre, H. 10 cm; eBay # 339959994; Fischer 118;
Bem.: Dieses Exemplar wurde zusätzlich (sek.?) mit dem gem. Schriftzug „GOLDSCHEIDER-WIEN" bezeichnet.
5326,248,4 gem. MDXXIII, Firmenmarke 19, Made in Austria, unsign.?, Ausformung 20er Jahre, H. 9,5 cm; Kett. 57
-> vgl. 4903; 5235
Lit.: N.N.: Wiener Kleinplastik einst und jetzt, in: MW, 2.1920/21, Nr. 12, S. 28 (m. Abb.); MW, 3.1921/22, Nr. 3, S. 39 (m. Abb.), hier jeweils mit Ang. des Originaltitels „Carmen".

5327
Lesca
GRUPPE „Pierrot und Pierrette" (auf einer Bank sitzend) (auf Ovalsockel)
ca. 1910/12 (für den Entwurf)
5327 (ohne Ang.), Firmenmarke 19, Made in Austria, Ausformung 20er Jahre, H. 10 cm; RM 19
5327 (ohne Ang.), H. 10 cm; Fischer 61
-> vgl. 3655; 5016
Lit.: MW, 2.1920/21, Nr. 12, S. 2 (m. Abb.); DD, 50.1922/23, Nr. 10, S. 30 (m. Abb.), hier jeweils bezeichnet „Lexa" (!) und mit Ang. des Originaltitels „Pierrot und Pierrette".

5329
Podany, Rudolf
„Bei der Toilette" (Sitzende Balldame im Rokokokleid) (auf Ovalsockel)
ca. 1917/18 (für den Entwurf)
5329,97,4 Firmenmarke 19, Made in Austria, unsign., Ausformung 20er Jahre, H. 12 cm; WKH
5329,194,65 gem. XII, Firmenmarke 19, Made in Austria, Ausformung 20er Jahre, H. 11,5 cm; eBay # 2066023798
5329,206,11 Firmenmarke 19, Made in Austria, Ausformung 20er Jahre, H. 12 cm; Doro 21.5.1991
5329,298,4 gem. 1., III., Firmenmarke 19, Made in Austria, Ausformung 20er Jahre, H. 11 cm; eBay # 2004626294
5329,310,12 gem. D1., XXIII., Firmenmarke 19, Made in Austria, Ausformung 20er Jahre, H. 11 cm; eBay # 2004626294
-> vgl. 4834; 5169
Lit.: DD, 50.1922/23, Nr. 8, S. 38 (m. Abb.), Nr. 9, S. 39 (m. Abb.), hier jeweils mit Ang. des Originaltitels „Bei der Toilette" und des Entwerfers Rudolf Podany.

5340
SPIEGELRAHMEN Frauengestalt, zwei Affen und Fledermaus
ca. 1924
5340,4 ger., Firmenmarke 19, Made in Austria, H. 36 cm; Soth. L. 23.6.1989

5342
Kuhn, Dina (Bernhardine)
VASE
Form: gebaucht, seitlich verschoben, mit einer Handhabe
Dekor: abstraktes Muster, mehrfarbig
ca. 1924
5342,42,10 gem. N2, Firmenmarke 19, Made in Austria, H. 21 cm; eBay # 3714047720

5343
Kuhn, Dina (Bernhardine)
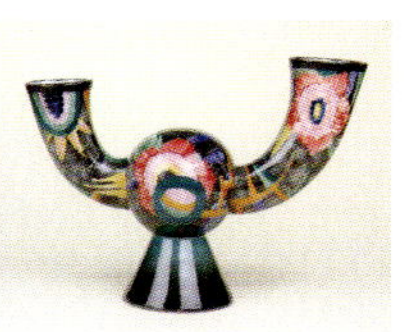
KERZENHALTER (zweiflammig)
ca. 1924
5343,36,7 gem. R., II, 2, GX, Firmenmarke 19, Made in Austria, sign. Dina Kuhn, H. 16,3 cm, B. 23 cm; PB Wien 19

5345
Bucher, Hertha
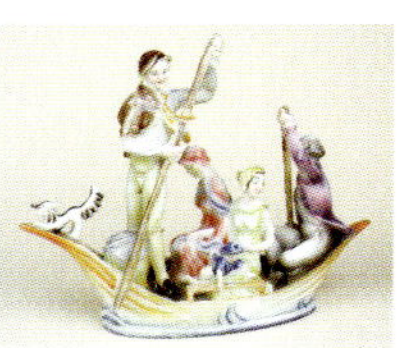
„Kahnfahrt" (Figurengruppe mit Hund in einem Kahn)
ca. 1924
5345,7, 65 gem. F. Firmenmarke 19, sign. Pucher (!), H. 30 cm, L. 39 cm; PB Wien 19
5345,12,18 gem. 3, III, Firmenmarke 19, Made in Austria, sign. Bucher, H. 29,5 cm, L. 38 cm; Brö 2,1, S. 407; Brö PK, S. 147
Ausst.: Int. Exh. CA 1928/29; HMW (jetzt Wien Museum) 1985/86.
Lit.: ÖBWK, 1.1924/25, Juli/Aug., S. III (Reklame); KeramRS, 33.1925, S. 32; Grell, S. 32 (m. Abb.), 33 „Bei der Firma GS finden wir als keramische Mitarbeiterinnen Dina Kuhn und Hertha Bucher, von welch letzterer besonders die sehr gelungene „Kahnfahrt" in Feinsteingut zu nennen ist..."; KuKGW, 6.1925/26, o. Pag. (Reklame m. Abb.), hier mit Ang. des Originaltitels und des Künstlernamens „Pucher" (!); L., A. S.: Vienna, in: Studio, 95.1928, S. 367 (m. Abb.), „Nothing could serve to convey more grace, charm and sense of beauty than the ‚Kahnfahrt' by Hertha Pucher. It is deliciously composed, all is in movement, pleasure is revealed everywhere in the features of those enjoying the lovely still waters, for one may imagine such a scene on Lake Atter, Lake Mond or one of the many lakes of the Salzkammergut ..."; Neuwirth (1974), S. 42, 43, 60 (m. Abb.); Neuwirth (1973), S. 60 (m. Abb.) (Bröhan Museum, Berlin); GK, S. 18, 41 (m. Abb.).

5351
Wieselthier, Vally
TEEGARNITUR Zuckerdose, Sahnekännchen, Tasse
ca. 1924/25
Zuckerdose: 5351,2,3 Firmenmarke 19, gem. sign. Wieselthier, H. (geschätzt) ca. 8 cm; Doro 20.11.1996
Sahnekännchen: unbezeichnet, H. (geschätzt) ca. 6,5 cm; Doro 20.11.1996
Tasse: unbezeichnet, H. 5 cm; Doro 20.11.1996
Lit.: Hörmann S. 273, hier ist das im Doro angebotene Teilservice unter der Nr. WV-K 473 aufgeführt. Diese Teegarnitur gab es in zweifacher Farbgebung und der Entwurf ist um das Jahr 1921 zu datieren (-> vgl. Hörmann, S. 59, 60 Abb. Nr. 24). Offenbar ist von diesen Serviceteilen kaum etwas erhalten, da im Auktions- und Antiquitätenhandel bislang keine weiteren Stücke nachgewiesen werden konnten.

5354
Thomasch, Wilhelm
DECKELDOSE Kugel
Form: kugelförmig mit zwei Handhaben
Knauf: drei Zacken
Dekor: Blüten- und Blattwerk
ca. 1924/25
5354,42,4 Firmenmarke 19, H. 11 cm; eBay # 1502238502

5356
Thomasch, Wilhelm (zugeschr.)
DECKELDOSE Kugel
Form: kugelförmig mit drei Kugelfüßen
Knauf: Kugel
Dekor: Blüten
ca. 1924/25
5356,147,6 gem. 6.II, Firmenmarke 19, Made in Austria, Hand dec., H. 17 cm; eBay # 6627538436

5366
Thomasch, Wilhelm
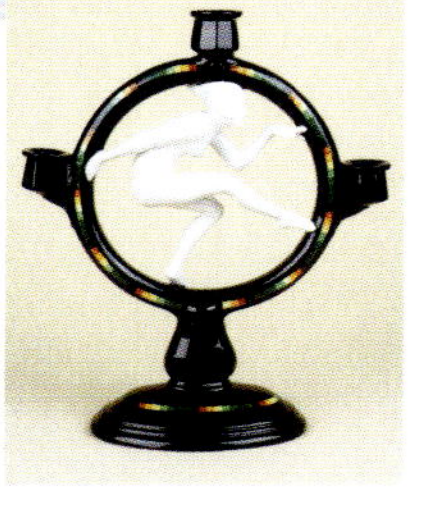
KERZENHALTER (dreiflammig) Tanzender Frauenakt in rundem Reifen (auf Ovalsockel)
ca. 1924/25
5366,12,4 Firmenmarke 19, Made in Austria, Hand dec., unsign.?, H. 37,5 cm; Bukowskis 27.–29.4.1998; TuSA 1998, H. 2, S. 128
5366,59,11 gem. F, VV, Firmenmarke 19, Made in Austria, H. 37,2 cm; eBay # 297616819; Bem.: Durch sekundäre Elektrifizierung als Lampe umgebaut.
5366,148,11 gem. D5, F, XXVIII, Firmenmarke 19, Made in Austria, H. 38,3 cm; Soth. N.Y. 19.3.1993
5366,156,18 gem. D5, IV, Firmenmarke 19, Made in Austria, H. 38 cm; Soth. N.Y. 19.3.1993
5366,172,12 Firmenmarke 19, Made in Austria, unsign.?, H. 38,5 cm; Bonhams 25.2.2003
5366,209,10 gem. F., XII D.G., Firmenmarke 19, Made in Austria, H. 39,4 cm; eBay # 1254939167; Quit. 21
5366,259,10 ger. lig. M18, gem. XIV., D., Firmenmarke 19, Made in Austria, Hand dec., H. 39 cm; GK, S. 18, 42 (m. Abb.); Wien Museum Inv. 222.548
Bem.: Der Leuchter erinnert in seiner Darstellung an den indischen Schiwa.
Lit.: GK, S. 18, 42 (m. Abb.); GSN, S. 36, 37 (m. Abb.); GSNF, o. Pag. (m. Abb.)
GS neu: Modell-Nr. 43 „Perpetuum mobile" (H. 40 cm), Aufl. 3000 Stück
No. 43-31/3000, GS Stoob 2, GS Stoob 4, Worldwide limited Edition 3000 pieces, Ausformung ca. 1988, H. 35,2 cm; Sammlung.schmutz@chello.at 71; Bem.: Dieses Exemplar stammt aus der Produktion Stoob/Burgenland, die in den achtziger Jahren des 20. Jh. versuchte, an die früheren Firmenerfolge anzuknüpfen -> vgl. Firmengeschichte Stoob. Der Größenunterschied beruht darauf, dass für Neuausformung, ein Original-Leuchter der Produktion der 20er Jahre, verwendet wurde und durch den Brand bzw. Schwund des Scherbens dann das neu ausgefertigte Stück kleiner ausfällt.

5376
Schoerk, Hans
BÜSTE Mädchen
MARMOR
ca. 1924/25
5376,R,10 ger., (Firmenmarke ohne Ang.), sign. Scherk, H. 10 cm; KNS 10
Bem.: Im Auktionskat. keine Ang. zur Darstellung. Ob es sich bei dieser Büste wirklich um Marmor handelt oder sie evtl. in Alabaster ausgeführt ist, konnte nicht geprüft werden.

5382
Petri (d.i. Petrides, János (Johann))
BÜSTE Knabe mit Kappe
ALABASTER
ca. 1924/25
5382,W,41 ger., Plak. E, H. 9,5 cm, B. 9 cm; eBay # 277142878

5385
Lorenzl, Josef
BÜSTE Mädchen mit langen Haaren
ALABASTER
ca. 1924/25
5385,JR (lig.),1 (Plak. ohne Ang.), H. 11,5 cm; Doro 20.4.1993
Lit.: Neuwirth (1974), S. 58, 67

5386
Petri (d.i. Petrides, János (Johann))
BÜSTE Kleines Mädchen mit Kappe
ALABASTER
ca. 1924/25
5386,JR (lig.),157 Plak. E, sign. Petri, H. 9,5 cm; eBay # 1486203491
5386 (Plak. ohne Ang.), unsign., H. 9 cm; Doro 8.3.2004

5388
BÜSTE Frau mit Schleier
ALABASTER
ca. 1924/25
5388,F,8 ger., Plak. E, H. 11,5 cm; Schuler 56

5391
Montini, (Tullio?)
BÜSTE Dante (nach rechts blickend)
ALABASTER
ca. 1924/25
5391,WK,4 ger., Plak. F, Made in Austria (ger.), H. 13,3 cm; eBay # 2899398370
Bem.: Dargestellt ist der italienische Dichter Dante Alighieri (1265–1321).

5393
Latour, (Louis-Marie-Blaise?)
BÜSTE Junge Biedermeierdame (nach rechts blickend)
ALABASTER
ca. 1924/25
5393,WF,2 Plak. E, H. 8,9 cm; eBay # 1510640424

5396
Kauerndes Mädchen mit französischer Bulldogge (auf Ovalsockel)
ca. 1924/25
5396,7 (Firmenmarke ohne Ang.), H. 23 cm, L. 28 cm; Zof. 19

5397
Lorenzl, Josef
FIGUR / LAMPE Stehende Tänzerin in russischer Tracht (Mantel und Pelzkappe) (auf facettiertem Rundsockel)
ca. 1924/25
FIGUR
5397,44,4 (Modell-Nr. zweimal gepr.), gem. F., Firmenmarke 19, Made in Austria, sign. Laurenzl, H. 31 cm; Wien Museum Inv. 220.750
LAMPE
5397,15,(11?) gem. F., Firmenmarke 19, Made in Austria, sign. Lorenzl, Gesamth. 42 cm; eBay # 6165946610

5403
Sitzende Columbine (Junge Dame in Karnevalskostüm)
ca. 1924/25
5403,12,5 gem. F., gem. grün III., Firmenmarke 19, Made in Austria, H. 24,5 cm; Wien Museum Inv. 190.145

5405
Philippe, Paul
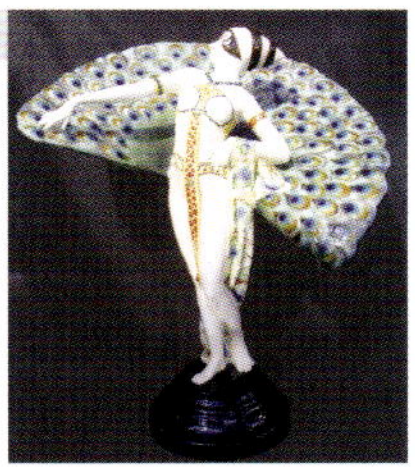
Tänzerin in Pfauenkostüm (auf Rundsockel)
ca. 1924/25
5405,14,3 (Firmenmarke ohne Ang.), sign. „Phillippe", H. 48,5 cm; Christie's SK 17.10.2001
5405,30,10 Firmenmarke 19, Made in Austria, unsign., H. 48,3 cm; eBay # 1225758194 (KH www.tias.com/stores/frozenbeauty)
5405,41,10 gem. XII, gem. lig. AS, D 4, Firmenmarke 19, Made in Austria, sign. „Phillippe", H. 48,5 cm; Sammlung Robert Senft
o. Ang., gem. XII.II, gem. lig. AS, Firmenmarke 19, Made in Austria, sign. Philippe, H. 47,5 cm; Soth. com. # 338D

5406
Lorenzl, Josef
Stehende Sängerin mit Kopftuch und langen Handschuhen (auf Ovalsockel)
ca. 1924/25
5406,22,18 Firmenmarke 19, Made in Austria, Hand dec., sign. Laurenzl, H. 23,8 cm; Soth. L. 23.10.1988
5406,27,4 Firmenmarke 19, Made in Austria, sign. Laurenzl, H. 24 cm; eBay # 7385879409

5407
Meier, Emil
Stehende Rokokodame mit Rose und Puderquaste
ca. 1924/25
5407,19,12 gem. F, XXVII, D2., Firmenmarke 19, Made in Austria, sign. E. Meier, H. 30 cm; Wien Museum Inv. 190.110

5425
Petri (d.i. Petrides, János (Johann))
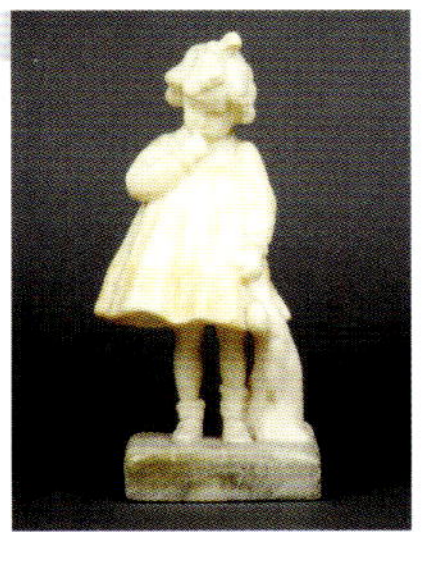
Stehendes kleines Mädchen mit Puppe (auf Rechtecksockel)
ca. 1924/25
ALABASTER
5425,Sch,13 ger., Plak. E, sign. Petri, H. 25,3 cm; PB Wien 19

5427
Thomasch, Wilhelm
Stehender Knabe mit Kappe und Tragebeutel (auf oktogonalem Sockel)
ca. 1924/25
5427,28,3 gem. IV., D.1, F., Firmenmarke 19, Made in Austria, H. 27 cm; eBay # 230056208838

5431
Jarl, Karin
SCHALE Vier Spatzen (an der Tränke sitzend)
ca. 1924/25
5431,18,5 Firmenmarke 19, Made in Austria, H. 17 cm; PB Wien 19 (ehem.)
Bem.: Diese Schale in Form einer Vogeltränke kann als Schreibzeugablage oder Visitenkartenschale Verwendung finden.

5433
Herlinger-Schwarz, Ida / Zlata
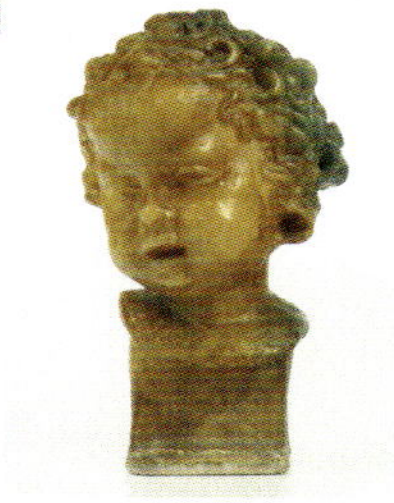
BÜSTE „Kinderkopf" (nach links blickend)
ca. 1925 (für den Entwurf)
FAYENCE
5433,5,8 Firmenmarke 19, Made in Austria, sign. Z.HSch/1928 (ger.), H. 27,5 cm; WKA 20
SPECKSTEIN
5433,SCH,3 ger., Plak. C, H. 27 cm; Wien Museum Inv. 220.793
Ausst.: Paris, Salon 1928, hier unter dem Originaltitel „Tête d'enfant" ausgestellt.
Bem.: Da diese Büste bereits 1926 in ÖBWK (–> vgl. Lit.) abgebildet war, kann der Entwurf etwa um 1925 datiert werden. Die Jahreszahl 1928 (beim ger. Monogramm) kann sich daher nur auf die Ausformung bei GS beziehen, nicht auf die Entstehungszeit.
Lit.: ÖBWK, 3.1926/27, S. 104 (m. Abb.), hier mit Ang. des Originaltitels „Kinderkopf" und mit Ang. der Entwerferin „Marie" Herlinger-Schwarz; Neuwirth (1974), S. 64.

5434
Lorenzl, Josef
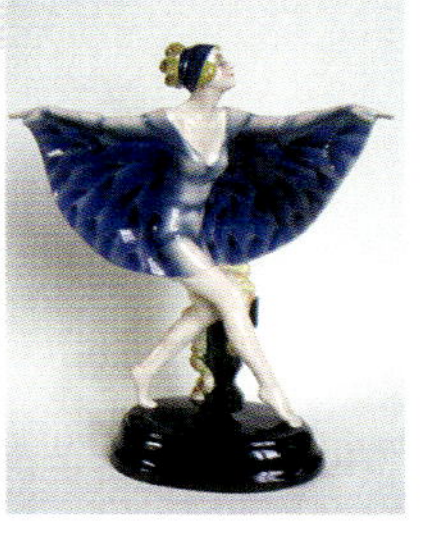
Tänzerin (mit breitem Haarband) im Schmetterlingskostüm (nach rechts schreitend) (auf Ovalsockel)
ca. 1924/25
5434,43,65 gem. XXIII, D2., Firmenmarke 19, Made in Austria, sign. Laurenzl, H. 44,5 cm; Sammlung Weilinger, Wien
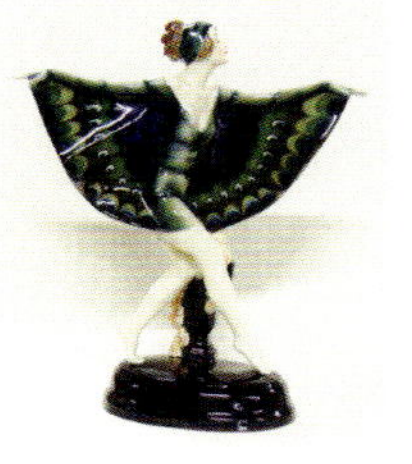
5434,87,18 Firmenmarke 19, Made in Austria, unsign., H. 45,4 cm; Soth. Arcade 1339 (29.11.1990)
5434,250,10 ger. lig. M22, gem. lig. HA, Firmenmarke 19, Made in Austria, Hand dec., sign. Laurenzl, H. 44,9 cm; MAK Wien, Inv. Ke 10602-25
5434,335,16 (Firmenmarke ohne Ang.), sign. Laurenzl, H. 44 cm; Weichmann 30.5.1987
5434,349,4 Firmenmarke 19, Made in Austria, Hand dec., unsign.?, H. 44 cm; Lempertz 702
5434,383,4,4 ger. P, gem. SB, N22, Firmenmarke 19, Made in Austria, Hand dec., sign. Lorenzl, H. 44 cm; WKA 29
Bem.: Mit dieser Figur wurde bis auf den Kopf (Gesicht, Frisur, Haarfarbe und das Haarband) ein Pendant zur weitaus bekannteren, beliebteren und ungleich häufiger produzierten Figur des „Gefangenen Vogel" (nach links schreitend) geschaffen. Sie steht seitenverkehrt zum Originalentwurf und bietet daher die Möglichkeit beide Figuren einander gegenüberzustellen um somit in dieser Größe (H. 45 cm) ein komplettes Ensemble zu erstellen.
Pendant –> vgl. 5230

5435
Lorenzl, Josef
Tänzerin in Fantasiekostüm mit Kopfputz (auf facettiertem Rundsockel)
ca. 1924/25
5435,41,12 gem. XXVII, D2., F., Firmenmarke 19, Made in Austria, sign. Laurenzl, Extra-Stempel 1, H. 34,3 cm; eBay # 7390544170

5437
Mallener, Theodor
SCHALE Stehender Knabe in Tracht mit Hut, zwei Blumenkörbe tragend
ca. 1912/13 (für den Entwurf)
5437,15,4, zusätzlich 4318 seitl. gepr., gem. JS, I, Firmenmarke 19, Made in Austria, Ausformung 20er Jahre, H. 22,2 cm; Doro 15.9.2003; eBay # 2626096421; Doro 13.9.2004
–> vgl. FIGUR 4318
Lit.: Studio YB, 1914, S. 191 (m. Abb.) , hier als Figur abgebildet, mit Ang. des Entwerfers Theodor Mallener.
Bem.: Diese kleine Schale belegt, dass bei GS oft aus einem Modell durch Erweiterung des Originalentwurfes mehrere Varianten und Funktionen geschaffen wurden. Hier z. B. wurde das ursprünglich als Figur konzipierte Modell zur Visitenkartenschale umfunktioniert. In anderen Fällen wurden den Figuren Tiere beigegeben oder Lampen, Schreibzeug und Aschenbecher geschaffen.

5443
Jarl, Karin (zugeschr.)
Paradiesvogel (auf Ast sitzend) (auf Rundsockel)
ca. 1925
5443,8 gem. II, S, Firmenmarke 19, Made in Austria, Hand dec., unsign., Pick. 1, H. 40 cm; Wien Museum Inv. 190.050
5443,11,I Firmenmarke 20, Made in Austria, unsign., H. 40 cm; eBay # 1472758925
5443,17,12 Firmenmarke 19, Made in Austria, Sign. ohne Ang., H. 39 cm; Henry's 27.1.2000

5447
Lorenzl, Josef
Tänzerin (auf einem Bein) im Schmetterlingskostüm
BRONZE / Marmorsockel
ca. 1925
5447,60 Firmenmarke 20? oder Plak. F?, Made in Austria, H. 16,2 cm; Doro 25.4.2000; Bem.: Farbig patiniert.

5457
Bormann, Wilhelm
Stehender Schäferhund (auf Rechtecksockel)
ca. 1925
5457,16,10 (Firmenmarke ohne Ang.), H. 27,3 cm, L. 28 cm; eBay # 3710555211
Bem.: Möglicherweise handelt es sich bei dem hier einzig nachgewiesenen Exemplar um einen Entwurf von Hugo Postl. Dieser ist bekannt für seine naturgetreuen Hundedarstellungen. Ein weiteres Indiz könnte die auch an Postl vergebene Modell-Nr. 5458 (ebenfalls ein Schäferhund) sein. Die Sign. „Bormann" ist möglicherweise ein Versehen des Modelleurs.

5458F
Postl, Hugo
„Wolfshund" (Liegender Schäferhund)
ca. 1925
5458,5,72 Firmenmarke 19, Made in Austria, Hand dec., unsign., H. 15,2 cm, L. 35,5 cm; eBay # 1206244284
5458,790,10,3 gem. F., Firmenmarke 19, Made in Austria, unsign.?, H. fehlt, L. 34 cm; Kastern 85
5458,825,4,12 gem. lig. NO, gem. blau X, Firmenmarke 21a, H. 15 cm, L. 38 cm; Wien Museum Inv. 220.755
5458,930,3 gem. F., Firmenmarke 19, Made in Austria, H. 15,4 cm; Schmidt 66; 84
5458,1224,7 gepr. G, Firmenmarke 19, Made in Germany, unsign., Ausformung zw. 1938–1941, H. fehlt, L. 36,5 cm; eBay # 487064866
5458,1304,68 Firmenmarke 19, Made in Germany, gepr. „G"= Z19, Ausformung zw. 1938–1941, H. 16 cm, L. 36 cm; eBay # 914192710
5458,1363,3 gem. XX30, Firmenmarke Schuster 1, Ausformung zw. 1941–1944, H. 16 cm, L. 36 cm; eBay # 3778215352
5458F, H. 15 cm, L. 37 cm; GSNachf. 5,13b
–> vgl. 5541; 5542; 5883
Lit.: GSNachf. 5,13 (m. Abb.), hier mit Ang. des Originaltitels „Wolfshund" (liegend); Schilling-Schletter, A.: Hunde-Plastiken, in: MW, 11.1929/30, Nr. 6, S. 12–13 (m. Abb.)

5461
Jarl, Karin
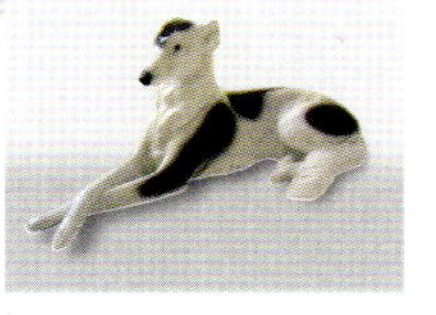
Liegender Windhund
ca. 1924 (für den Entwurf)
5461,71,18, gem. AT., D1., Firmenmarke 19, Made in Austria, gepr. „L"= Z1, H. 9 cm, B. 21 cm; Sammlung f@goldscheider.de
–> vgl. 5299
–> vgl. 5149= ähnl. Modell

5464
Rottenberg, Ena (Emma Helena)
Sitzender männlicher Akt
JADE / SPECKSTEIN?
ca. 1925
5464,JR,3 ger., Plak. F, sign. Rottenberg, H. 11 cm; Wien Museum Inv. 182.090

5468
Doblinger, Ferdinand
Kamelreitender Afrikanerjunge (auf Landschaftssockel)
ca. 1925
5468,8,12 gem. S, DI, X., Firmenmarke 19, Made in Austria, H. 35 cm; Christie's SK 14.3.1997; eBay # 1471858049

5483F
Rosé
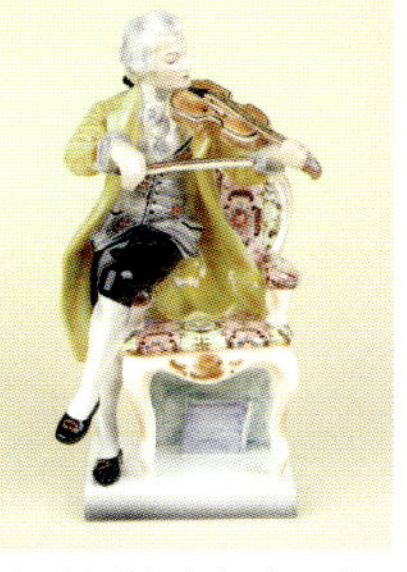
„Mozart" (mit Geige, auf Sessellehne sitzend)
ca. 1905 (für den Entwurf)
5483,1,83 Firmenmarke 19, Made in Austria, Hand dec., Ausformung 20er Jahre, unsign., H. 29 cm; city-antik
5483,18,20 (Firmenmarke ohne Ang.), H. 26,5 cm; Fischer 136
5483,37,66 gem. S., Firmenmarke 19, Made in Austria, unsign., H. 27 cm; eBay # 449399174
5483,277,7 gem. A.E, Firmenmarke 19, Made in Austria, Ausformung 20er Jahre, H. 27 cm; eBay # 2286841868
5483F, H. 27,5 cm, B. 15,5 cm; GSNachf. 1,52
Bem.: Dargestellt ist der Komponist Wolfgang Amadeus Mozart (1756–1791). Dieses beliebte Modell wurde auch noch in der Wiederaufbauphase nach dem Krieg in Wien produziert (–> vgl. Quelle).
–> vgl. 3285; 3522
Lit.: GSNachf. (2. Ausg., bzw. Erg. Bl.) 1,52 (m. Abb.)
Quelle: AGM (Rechnung an Maurice Mizrahi, Alexandria, Ägypten, vom 22. April 1953), hier mit Ang. des Originaltitels „Mozart".

5485
Kostial, Josef
„Pierrette" (Stehende junge Dame im Harlekinskostüm) (auf Rundsockel)
ca. 1923 (für den Entwurf)
5485,5,11 gem. grün D1, XXVII, Firmenmarke 19, Made in Austria, H. 21,8 cm; GK, S. 18, 106 (m. Abb.); Wien Museum Inv. 182.091
Ausst.: HMW (jetzt Wien Museum) 1985/86.
–> vgl. 5094
Lit.: Ros, L.: Wiener Fayencen und Bronzen, Begleitworte zu den Abb. über neue Werke von FG Wien, in: KuKGW, 4.1923/24, S. 15–16 (m. Abb.), hier mit Ang. des Originaltitels „Pierrette"; GK, S. 18, 106 (m. Abb.).

5486
Lorenzl, Josef
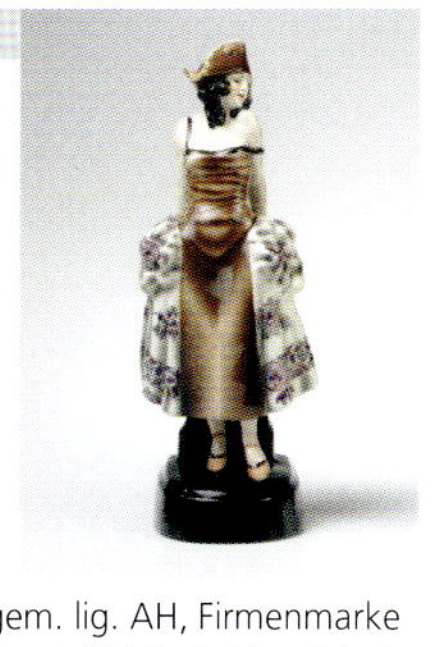
FIGUR / LAMPE Stehende junge Dame in Karnevalskostüm (mit Dreispitz)
ca. 1925
FIGUR
5486,88,14 gem. lig. HR, Firmenmarke 19, Made in Austria, Hand dec., sign. Laurenzl, H. 35 cm; Doro 21.5.1991
5486,(51?),(18?) gem. D.1.II, gem. lig. AH, Firmenmarke 19, Made in Austria, sign. Laurenzl, Originalpick. „Schellenberg", H. 34,8 cm; Wien Museum Inv. 170.149
5486,109,4 gem. G, J., gem. lig. JH.1, Firmenmarke 19, Made in Austria, Hand dec., H. 33,2 cm; Wien Museum Inv. 220.703
5486 (weitere Zahlen durch Pick. verdeckt), gem. lig. AH, gem. D.1.II, Firmenmarke 19, Made in Austria, sign. Laurenzl, H. 34,5 cm; Wien Museum Inv. 190.149
LAMPE
5486,70,15 gem. lig AZ, Firmenmarke 19, Made in Austria, Hand dec., sign. Laurenzl, H. 34,3 cm; Soth. com. # 6NWK; Bem.: Dieses Objekt war ursprünglich eine Lampe, die Elektrifizierung und der Lampenfuß wurden jedoch entfernt.
5486,98,7,O3 Firmenmarke 19, Made in Austria, Hand dec., sign. Lorenzl, H. 34,3 cm, Gesamth. 73,6 cm; eBay # 1203286169

5491
Lindner, Helmut
Junge, sich verneigende Biedermeierdame mit drei Rosen
ca. 1913/14 (für den Entwurf)
5491,90,4 gem. G1.III., Firmenmarke 19, Made in Austria, Ausformung 20er Jahre, H. 11,5 cm; eBay # 6505630139
Bem.: Dieses Modell wurde auch noch in den USA (in größerer Ausführung) unter der GS-Everlast Corp. produziert (Modell-Nr. 844) –> vgl. eBay # 1296010535 –> vgl. 4511; 4712

5492
LAMPE Papagei (dreiflammig)
ca. 1925
5492,13,8 Firmenmarke 19, Gesamth. 53 cm, B. 39 cm; WKH

5493
Thomasch, Wilhelm
TELLER
Form: rund
Dekor an der Fahne: Blumen und Blattwerk
ca. 1925
5493,13,14 gem. MR, Firmenmarke 19, Made in Austria, Hand dec., Ø 21,2 cm; Wien Museum Inv. 190.003

5500
Lorenzl, Josef
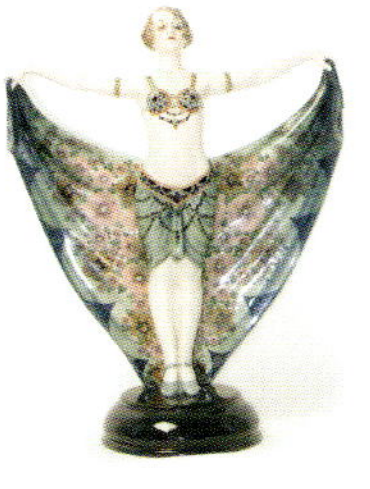
FIGUR / LAMPE Stehende Haremstänzerin mit/ohne großem Kopfputz (mit ausgestreckten Armen, ihr Kleid auseinanderhaltend) (Dolores del Rio) (auf Rundsockel)
ca. 1925
FIGUR
mit Kopfputz
5500,31,18 gem. XXVII, D75, Firmenmarke 19, Made in Austria, sign. Laurenzl, H. 50,8 cm; eBay # 194588877
5500,127,10 ger. lig. M29, gem. XXVII, D5, Firmenmarke 19, Made in Austria, Hand dec., sign. Laurenzl, H. 50,8 cm; Soth. com. # 6LMW
FIGUR
ohne Kopfputz
5500,117,10 ger. lig. M27, gem. D5 XXVII, Firmenmarke 19, Made in Austria, Hand dec., sign. Laurenzl, H. 48 cm; Privatsammlung, Wien
LAMPE (mit Kopfputz)
5500,89,12 gem. XXVII, D5, gem. lig. AH, Firmenmarke 19, Made in Austria, Hand dec., sign. Lorenzl, H. 52,5 cm, Gesamth. 94 cm; eBay # 7321322960
5500,160,7 (Firmenmarke ohne Ang.), sign. Laurenzl, Gesamth. 80 cm; Christie's SK 13.5.2004
Bem.: Laut Ang. des AH Sotheby's (jedoch ohne Quellenang.) soll die Dargestellte „Theda Bara" sein. Theda Bara, ihr richtiger Name war Theodosia Burr Goodman (1885–1955), war ein Stummfilmstar der 20er Jahre. Leider ist kein Film zu nennen, in dem sie die dargestellte Rolle verkörpert haben sollte. Wahrscheinlicher ist, dass es sich um die Schauspielerin Dolores del Rio handelt (s. Abb. Postkarte).
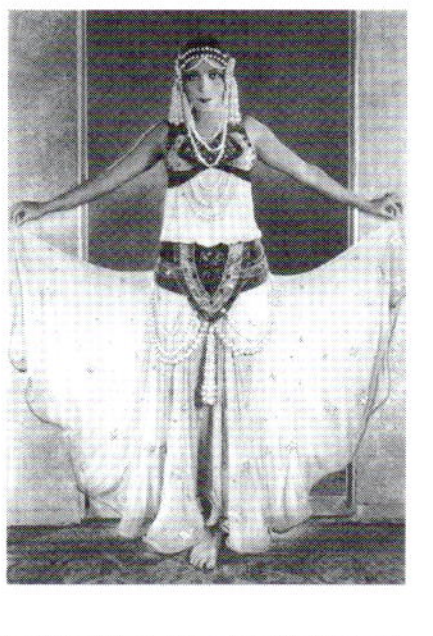

5501
Lorenzl, Josef
FIGUR / LAMPE Sitzende Rokokodame mit Buch und Papagei
ca. 1925
FIGUR
5501,2,11 Firmenmarke 19, H. 34,7 cm; WKH
LAMPE
5501,25,12 (Firmenmarke ohne Ang.), sign. Laurenzl, H. 44 cm; Metz 30.3.1996; eBay # 6553699964

5504
Lorenzl, Josef
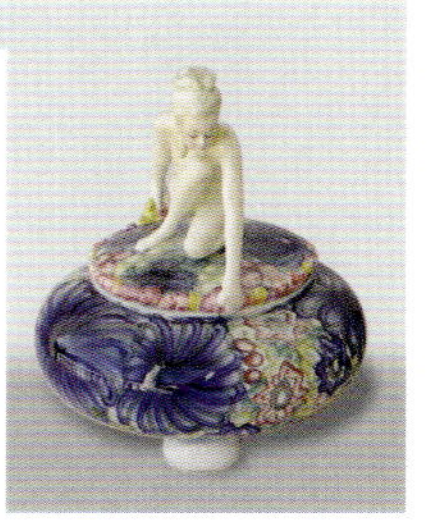
DECKELDOSE „Akt auf einer Dose" (Kniender Damenakt)
ca. 1925
5504,326,65 gem. lig. JH., 3., Firmenmarke 19, Made in Austria, Hand dec., unsign., H. 16 cm, Ø 14 cm; eBay # 6627288874
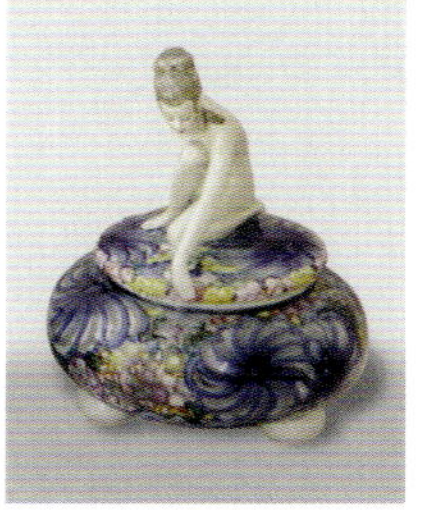
5504,344,65 gem. lig. HC, gem. lig. AS.3., Firmenmarke 19, Made in Austria, Hand dec., unsign., H. ohne Deckel 6,5 cm, Ø 13,8 cm; Sammlung f@goldscheider.de; Bem.: Deckel fehlt.
5504,352,11 gem. S.G., R3., Firmenmarke 19, Made in Austria, Hand dec., unsign., H. ohne Deckel 6,4 cm, Ø 14 cm; Sammlung a@goldscheider.net; Bem.: Deckel fehlt.
5504,483,14 gem. DE, Firmenmarke 19, Made in Austria, Hand dec., H. 16 cm, Ø 14 cm; Sammlung Stressless, Portugal
5504,542,4 gem. P1, gem. lig. JR, Firmenmarke 19, Made in Austria, Hand dec., H. 16 cm, Ø 14 cm; Sammlung Stressless, Portugal
Lit.: MW, 11.1929/30, Nr. 9, S. II (m. Abb.), hier mit Ang. des Originaltitels „Akt auf einer Dose".

5508
Lorenzl, Josef
LAMPE Zwei Pierrots mit einer Pierrette (auf facettiertem Rundsockel)
ca. 1925
5508,11,65,L gem. FAC, D.2., Firmenmarke 19, Made in Austria, sign. Laurenzl, H. 57,1 cm; Soth. com. # 6LPL; eBay # 20481444565 (KH www.tias.com/stores/frozenbeauty)
–> vgl. 5022= ähnl. Modell

5509
Lorenzl, Josef
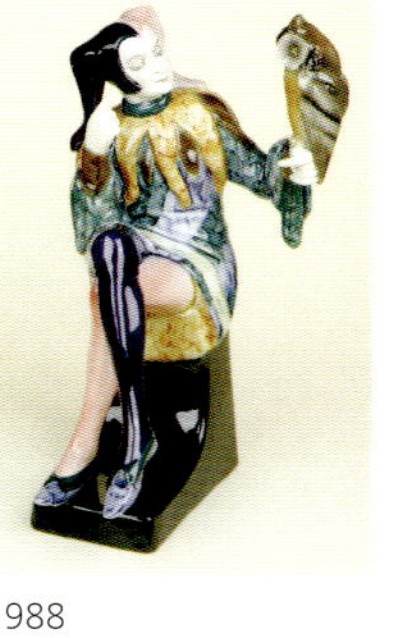
FIGUR / LAMPE Sitzender Till Eulenspiegel mit Eule
ca. 1923/24 (für den Entwurf)
FIGUR
5509,25,12 gem. F., XII., D1, Firmenmarke 19, Made in Austria, sign. Laurenzl, H. 35 cm; PB Wien 19
5509,75,8 gem. F., XII., D.1, Firmenmarke 19, Made in Austria, sign. Lorenzl, H. 35 cm; Soth. Chester 7.–9.6.1988
5509,268,12 Firmenmarke 19, Made in Austria, Hand dec., sign. Lorenzl, H. 33 cm; Sammlung Dr. Peter Hieke, Wien
5509,373,11 (Firmenmarke ohne Ang.), sign. Lorenzl, H. 35 cm; Doro F. 10.12.1999; Bem.: Im Kat. falsche Modell-Nr. 7509 angegeben.
LAMPE
5509,90,12 ger. M12, gem. R. XI, Firmenmarke 19, Made in Austria, Hand dec., unsign., H. 41 cm; Sammlung Robert Senft
–> vgl. 5193
Lit.: Neuwirth (1974), S. 58

5512
Calm, Lotte (Charlotte Alice)
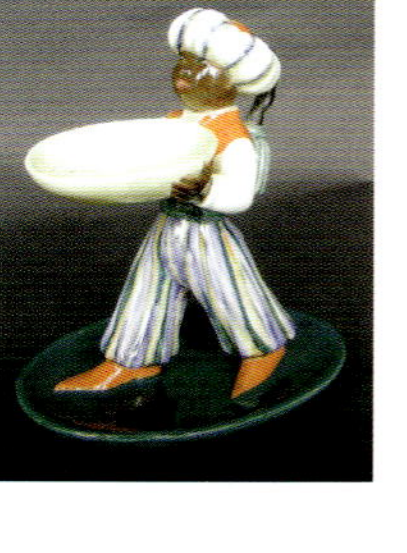
Schreitender Orientale mit Turban als Schalenträger (auf Ovalsockel / Schale)
ca. 1925
5512,51,9 gem. H., Firmenmarke 19, Made in Austria, Hand dec., H. 12,2 cm, B. 11 cm; PB Wien

5513
Rosé
„Ruth" (Tänzerin auf dreipassigem Sockel mit zwei Rosenkörben)
ca. 1911/12 (für den Entwurf)
5513,162,14 Firmenmarke 19, Made in Austria, unsign., Ausformung 20er Jahre, H. 14 cm; eBay # 39588984
Bem.: Dargestellt ist die Tänzerin Ruth Saint Denis (1877/79–1968, eigentl. Ruth Dennis, gen. „Miss Ruth") in ihrem Tanz „Radha" (Tanz der fünf Sinne, Urauff. 1906), -> vgl. hierzu Ludwigg, H. (Hg.): 40 Jahre Wintergarten, Fs. Berlin 1928, S. 68 (m. Abb.); Berühmte Tänzerinnen (Zigarettenbildersammelalbum) Berlin, Garbaty (ca. 1930) Bild 84; Ochaim, B.: Varieté-Tänzerinnen um 1900, Frankfurt/M., Basel 1998, S. 69 (m. Abb.).
-> vgl. 3870; 3992; 4141; 5171
Lit.: MW, 11.1929/30, Nr. 11, S. VII (m. Abb.), hier mit Ang. des Originaltitels „Ruth" und des falschen Künstlernamens „Bosi"; BT, o. Pag., Bild 84.

5516
Jarl, Karin
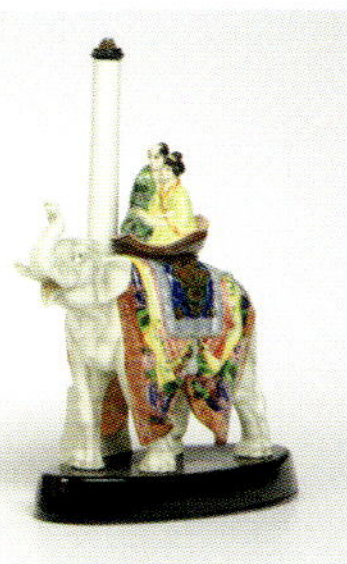
GRUPPE/LAMPE Elefant mit reitendem Chinesenpaar (auf Ovalsockel)
ca. 1925
GRUPPE
5516,56,16 gem. 1.2., ger. KO3, gem. lig. AT, Firmenmarke 19, Made in Austria, Hand dec., gepr. „M"= Z20, H. 30,5 cm, L. 28,5 cm; Wien Museum Inv. 220.756
ohne Ang., H. 32 cm; Weichmann 30.11.1984
LAMPE
5516,50,10,30 gem. M, III, gem. lig. AC, Firmenmarke 19, Made in Austria, Hand dec., H. 38 cm, L. 28,5 cm; Wien Museum Inv. 182.092

5517
Kostial, Josef
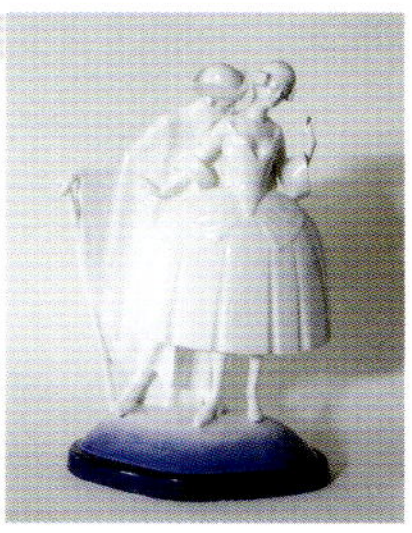
FIGUR/LAMPE Galante Rokokoszene (Kavalier mit Dreispitz küsst sie auf die Schulter) (auf abgerundetem, trapezförmigem Sockel)
ca. 1925
FIGUR
5517,16,11 gem. 75, gem. lig. AT, D.1, gepr. „S"= Z2, H. 36,5 cm; Sammlung K. Dechant
5517,18,15 Firmenmarke 19, Made in Austria, gepr. „L"= Z1, H. 35,6 cm; Doro 14.9.2001;13.9.2004
5517,38,12 Firmenmarke 19, Made in Austria, gepr. „S"= Z2, H. 36,8 cm; eBay # 22508189
5517,26,65 gem. lig. AS, 2, III., Firmenmarke 19, Made in Austria, H. 36,8 cm; Soth. com. # A0DD
5517,49,4 gem. XI, Firmenmarke 19, Made in Austria, gepr. „L"= Z1, H. 36,8 cm; eBay # 504945118; Quit. 21
5517,86,8 gem. lig. AH, Firmenmarke 19, Made in Austria, H. 36,9 cm; MAK Wien, Inv. Ke 10964
5517,193,10 Firmenmarke 19, Made in Austria, Hand dec., H. 38 cm; Doro S 15.10.2003
5517,338,10,03 Firmenmarke 19, Made in Austria, Hand dec., unsign., H. 35,5 cm; Soth. L. 4.6.1993
5517,352,11 gem. lig. R(Z?), Firmenmarke 19, Made in Austria, Hand dec., H. 35,5 cm; eBay # 2049836053
LAMPE
5517,6,4 gem. lig. AS, D.2, (Firmenmarke ohne Ang.), H. 45,1 cm; eBay # 3773959107
Bem.: Dieses Modell wurde in diversen Dekoren angeboten. Von der polychromen, überreichen Bemalung über die Variante (weiße Figuren, kräftig blauer Sockel in Sprühtechnik) bis hin zur ganz weißen, porzellanartig scheinenden Glasur.

5519F
Perigot
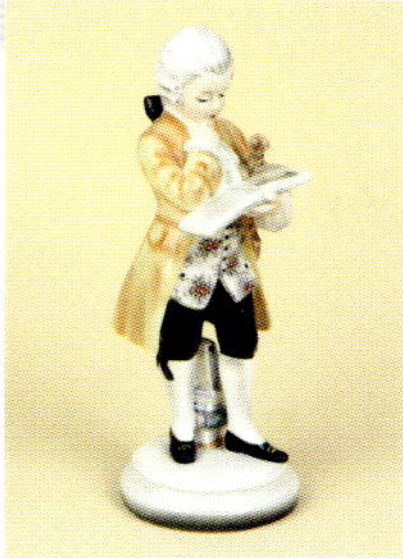
„Mozart" (Stehende Figur als Knabe mit Geige, Notenheft und Bücherstapel)
ca. 1913 (für den Entwurf)
5519,5,21 Firmenmarke 19, Made in Austria, Hand dec. (alles blau gestempelt), unsign., Ausformung 20er Jahre, H. 26 cm; WKH
5519,22,28 Firmenmarke 19, unsign.?, Ausformung 20er Jahre, H. 25 cm; Doro 14.9.2001
5519,14,52 Firmenmarke 19, unsign., Ausformung 20er Jahre, H. 27 cm; eBay # 39562555
5519,22,28 Firmenmarke 20, Made in Austria, Hand dec., unsign., H. 26,5 cm; WKA 27
5519,33,? gem. L.S., Firmenmarke 19, Made in Austria, unsign., Pick. 5/a, H. 26,5 cm; eBay # 926322968
5519,66,40,60 gem. HD, Firmenmarke 19, Made in Austria, Ausformung 20er Jahre, H. ca. 26 cm; eBay # 294124573
5519,66,89 gem. G., Firmenmarke 19, Made in Austria, Hand dec., unsign., Ausformung 20er Jahre, H. 26 cm; eBay # 3326473017
5519,100,7 ger. M26, gem. XII, D.2, Firmenmarke 19, Made in Austria, Hand dec., unsign.?, Ausformung 20er Jahre, H. 24 cm; Wilbois 61
5519 ohne Ang., unsign.?, H. 26 cm; PB Wien 19
5519F, H. 26 cm, B. 10 cm; GSNachf. 1,53
Bem.: Wie beliebt dieses Modell des jungen Mozart war, belegt die in der Quelle angegebene Rechnung aus dem Jahr 1953. Das heißt, dass diese Figur von der Entstehungszeit etwa 40 Jahre lang im Programm der Firma zu finden und somit erfolgreich war. Ähnl. Darstellungen (Mozart mit Geige) finden sich auch bei L.-E. Barrias und M. Debut, -> vgl. hierzu: Berman, Vol. 3, Nrn. 2619; 2622 jeweils m. Abb.
-> vgl. 4393; 4610
Lit.: MW, 11.1929/30, Nr. 8, S. V (m. Abb. „mit Bücherstapel"), hier mit Ang. des Originaltitels „Mozart". GSNachf. (2. Ausg., bzw. Erg. Bl.) 1,53 ; GSN, Erg. Bl. (o. Pag., m. Abb.) und Preisliste 1992.
Quelle: AGM (Rechnung an Maurice Mizrahi, Alexandria, Ägypten, vom 22. April 1953), hier mit Ang. des Originaltitels „Mozart".
GS neu ohne Modell-Nr. „Mozart" (H. 25 cm), Aufl. 300 Stück

5520
Meier-Michel, Johanna
„Velazquez" (Infantin Margarita)
ca. 1911/12 (für den Entwurf)
5520,15,10 (Firmenmarke ohne Ang.), sign. Meier, Ausformung 20er Jahre, H. 22 cm; KaM 79
-> vgl. 3977
Lit.: Dekorative Kunst, 25.1921/22, S. 265 (m. Abb.), „Die Deutsche Gewerbeschau", hier mit Ang. des Originaltitels „Velasquez" und die Künstlerin „Johanna Meier-Michel".

5523
Lorenzl, Josef
Stehende Varietetänzerin mit Zylinder (auf facettiertem Rundsockel)
ca. 1925
5523,8,20 (Firmenmarke ohne Ang.), sign. Laurenzl, H. 31 cm; Christie's SK 4.11.1999
5523,79,5 ger. X, Firmenmarke 19, sign. Lorenzl, H. 33 cm; eBay # 1280446442
5523,160,12,0,4 gem. 1.II, Firmenmarke 19, Made in Austria, sign. Laurenzl, H. 30,5 cm; eBay # 140458903
ohne Ang., Firmenmarke 19, sign. Laurenzl, H. 32 cm; Herr 19
Lit.: Pinhas, S. 146 (m. Abb.)

5525
Lorenzl, Josef
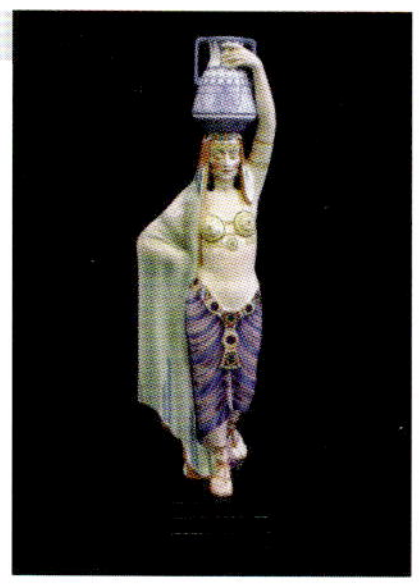
Schreitende Sklavin, eine Amphore/Henkelkrug auf dem Kopf tragend (auf getrepptem Rechtecksockel)
ca. 1925
5525,2,8 gem. rot XXIII., Firmenmarke 19, Made in Austria, sign. Laurenzl, H. 61 cm; Sammlung Robert Senft
5525,23,12 ger. lig. M29, Firmenmarke 19, Made in Austria, Hand dec., sign. Lorenzl, H. 61 cm; KH

5528
Lorenzl, Josef
KERZENHALTER (dreiflammig) Putto mit Blumengirlanden (auf dreifach getrepptem Quadratsockel)
ca. 1925
5528,4,68 (od. 65?) gem. 169, F.D.2, Firmenmarke 19, Made in Austria, sign. Laurenzl, H. 35 cm, B. 34 cm; Soth. com. # 4BWE
5528,32,7 gem. M4, 5, Firmenmarke 19, Made in Austria, Hand dec., sign. Laurenzl, H. 40,6/41 cm; eBay # 164562970; Christie's SK 20.9.2000; Bem.: unterschiedliche Größenangaben der Anbieter.

5530
Lorenzl, Josef
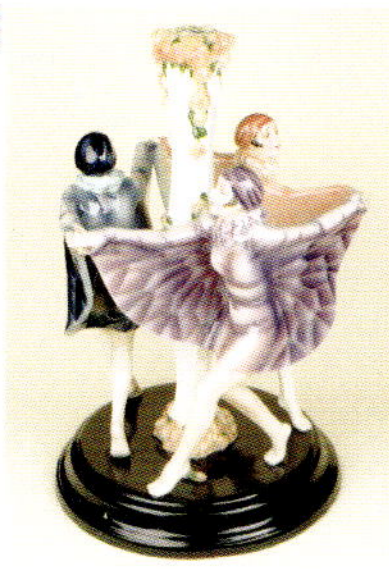
GRUPPE/LAMPE „Gefangener Vogel" (Der Gefangene Vogel, Niddy Impekoven) (als Trio auf Rundsockel)
ca. 1922 (für den Entwurf)
GRUPPE
ohne Ang., H. fehlt; ehem. Sammlung R. Schmutz/Wien;
Bem.: Dieses Trio, das sich ehem. in der Sammlung Rudolf Schmutz, Wien befand, ist verschollen. Jedoch war es ein Beleg dafür, dass es nicht nur die Ausführung als Lampe gab, sondern die Figuren auch als Dreier-Gruppe produziert wurden.
LAMPE
5530,14,11 Firmenmarke 19, Made in Austria, sign. Lorenzl, H. 60 cm (mit Elektrifizierung); Herr 51
5530,60,8,03 gem. AZ, Firmenmarke 19, Made in Austria, Hand dec., sign. Laurenzl, H. 46 cm; PB Wien 19
5530,104,65 gem. A1, ZN, Firmenmarke 19, Made in Austria, Hand dec., sign. Lorenzl, H. 49,5 cm; Sammlung Stressless, Portugal
-> vgl. 5054 (BRONZE); 5230; 5960; 5964 = Pendant zu 5960; 6204; 6600; 8176
Pendant, bzw. ähnl. Modell -> vgl. 5434
Bem.: s. bei 5230
Lit.: s. bei 5230

5532
Dakon, Stephan / Lorenzl, Josef
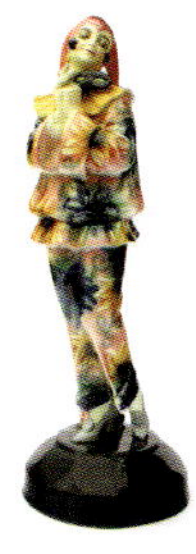
FIGUR/LAMPE Stehende Pierrette mit Maske (auf Rundsockel)
ca. 1925
FAYENCE/FIGUR
5532,168,8 Firmenmarke 19, Made in Austria, Hand dec., unsign., H. 34 cm; eBay # 1256004331
5532,298,7,3 gem. . lig. AD, N4, Firmenmarke 19, Made in Austria, sign. Lorenzl, H. 31 cm; Hennig 11
5532,438,16 Firmenmarke 19, unsign., H. 33,2 cm; Doro 26.11.2002
5532,445,6 gem. MR4, Firmenmarke 19, unsign.?, H. 34 cm; Henry's 27.5.1999
5532,608,19 ger. 1283, ger. lig. TP, gem. R, Firmenmarke 19, Made in Austria, Hand dec., unsign.?, H. 32,5 cm; WKA 44; city-antik
LAMPE
5532,257,12,03, gem. SA.3.K, Firmenmarke 19, sign. Dakon, H. 37,6 cm; Sammlung Robert Senft

ALABASTER / FIGUR
5532,K,4 ger., Plak. F, sign. Dakon, H. 30 cm; Christie's L. 25.9.1987
5532,W,7 ger., (Firmenmarke Plak. ohne Ang.), sign. Dakon, H. 30,6 cm; Doro 15.3.2005
Bem.: Welchem der beiden Künstler dieses Modell endgültig zuzuordnen ist, lässt sich nicht eindeutig sagen. Jedoch erscheint, nicht nur wegen der zwei mit „Dakon" sign. Stücke, die Tendenz eher zu diesem zu gehen. Dieses Modell wurde in Fayence und Alabaster, sowie als Lampe ausgeführt, was ebenso wie die verhältnismäßig hohe Stückzahl für die Beliebtheit dieser Figur spricht.

5534
Lorenzl, Josef
Carmen (Sitzende Spanierin mit Fächer) (auf halbrundem Sockel)
ca. 1925
5534,42,11,0,3 gem. lig. JH., 2, BS, Firmenmarke 19, Made in Austria, Hand dec., H. 17 cm; city-antik
5534,51,6 Firmenmarke 19, H. 18,2 cm; BM 68
-> vgl. 5555; 6666
-> vgl. 5088; 5232; 5277= stehende Figur

5535
Lorenzl, Josef
Stehende Haremsdame mit federverziertem Turban (auf Ovalsockel)
ca. 1925
5535,1,10 gem. blau XI., D.7., Firmenmarke 19, Made in Austria, sign. Laurenzl, H. 52 cm; Sammlung Robert Senft

5536
Lorenzl, Josef
Sitzende Haremsdame mit reich geschmückter Kopfbedeckung / Turban (auf podestartigem Sockel, im Schneidersitz)
ca. 1925/26
5536,38,11 Firmenmarke 19, sign. Laurenzl, H. 38 cm; Doro 29.5.2001
5536,42,8 gem. F.0.3, XII, Firmenmarke 19, Made in Austria, Hand dec., H. 38 cm; Sammlung Wade Sanders

5537
Lorenzl, Josef
FIGUR / LAMPE Stehender „Mephisto"
ca. 1925/26

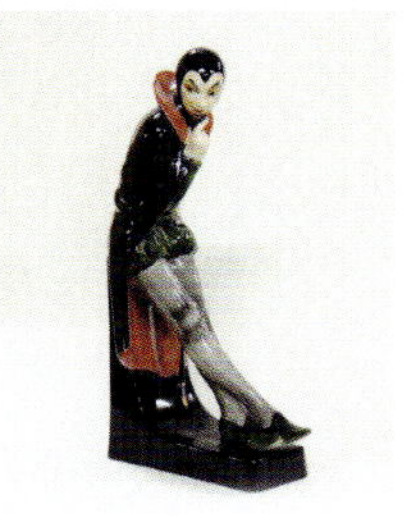

FIGUR
5537,88,65 (Firmenmarke ohne Ang.), sign. Laurenzl, H. 36,5 cm; Christie's SK 14.11.2002; 15.5.2003
5537,154,65,O,4 Firmenmarke 19, Made in Austria, Hand dec., sign. Laurenzl, H. 36,3 cm; GK, S. 21, 108 (m. Abb.); Wien Museum Inv. 223.735

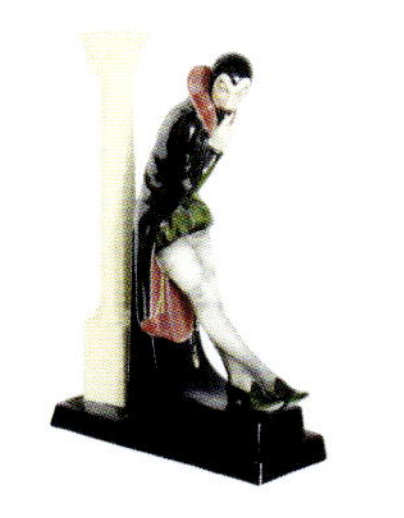

5537,186,6 gem. lig. LN, F, Firmenmarke 19, Made in Austria, Hand dec., sign. Laurenzl, H. 35,6 cm; Sammlung Joan Ferguson
LAMPE
5537,178,6,3 gem. N1, gem. lig. SB, Firmenmarke 19, Made in Austria, Hand dec., sign. Laurenzl, H. 39,8 cm; Sammlung a@goldscheider.net
5537,199,11 gem. F., N1., Firmenmarke 19, Made in Austria, Hand dec., sign. Lorenzl, H. 36,3 cm; MAK Wien, Inv. Ke 10602-63
Ausst.: HMW (jetzt Wien Museum) 1985/86; SSH 1992.
Lit.: KuKGW, 8.1927/28, o. Pag. (Figur „Mephisto" Nr. 5537, Fayence, Entwurf Bildhauer Josef Lorenzl, Dek. rot, grün... Wert: Mk. 76,–), hier mit Ang. des Originaltitels „Mephisto"; GK, S. 21, 108 (m. Abb.).

5541F
Postl, Hugo
FIGUR „Wolfshund" (Liegender Schäferhund)
ca. 1925
5541,440,6 gem. F., Firmenmarke 19, H. 11,5 cm, L. 28 cm; eBay # 3354818244
5541F, H. 10 cm, L. 26 cm; GSNachf. 5,13c
-> vgl. 5458; 5542; 5883
Lit.: GSNachf. 5,13 (m. Abb.), hier mit Ang. des Originaltitels „Wolfshund" (liegend); Schilling-Schletter, A.: Hunde-Plastiken, in: MW, 11.1929/30, Nr. 6, S. 12–13 (m. Abb.)

5542F
Postl, Hugo
FIGUR „Wolfshund" (Liegender Schäferhund)
ca. 1925
5542F, H. 7 cm, L. 19 cm; GSNachf. 5,13d
-> vgl. 5458; 5541; 5883
Lit.: GSNachf. 5,13 (m. Abb.), hier mit Ang. des Originaltitels „Wolfshund" (liegend); Schilling-Schletter, A.: Hunde-Plastiken, in: MW, 11.1929/30, Nr. 6, S. 12–13 (m. Abb.)

5543
Lorenzl, Josef
FIGUR / LAMPE Tänzerin mit Kopfputz und großem Schal (auf Ovalsockel)
ca. 1925

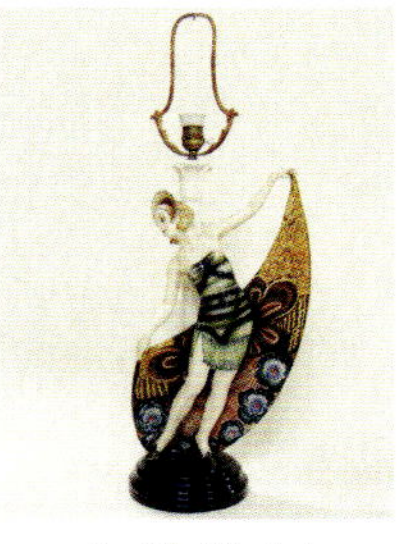

FIGUR
o. Ang., H. fehlt; McClinton, K.M.: Art Deco, A guide for collectors, 1972
LAMPE
5543,46,8 gem. lig. M12, Firmenmarke 19, Made in Austria, Hand dec., sign. Laurenzl, H. 53,5 cm; MAK Wien, Inv. Ke 10521-25

5547
Lorenzl, Josef
Stehender Mädchenakt mit Tuch (auf quaderförmigem Sockel)
ALABASTER / Marmorsockel
ca. 1925/26
5547,51 ger., Plak. E, Made in Austria (ger.), H. 32 cm; eBay # 890697050

5548
Lorenzl, Josef
LAMPE Sitzende Spanierin mit Hut und Gitarre (auf halbrundem Sockel)
ca. 1925/26
5548,45,12 Firmenmarke 19, sign. Laurenzl, H. 57 cm; Doro 24.5.2000

5549
Kostial, Josef
FIGUR/LAMPE Tänzerin mit großem Turban (ein Bein hebend) (auf Ovalsockel)
ca. 1925/26

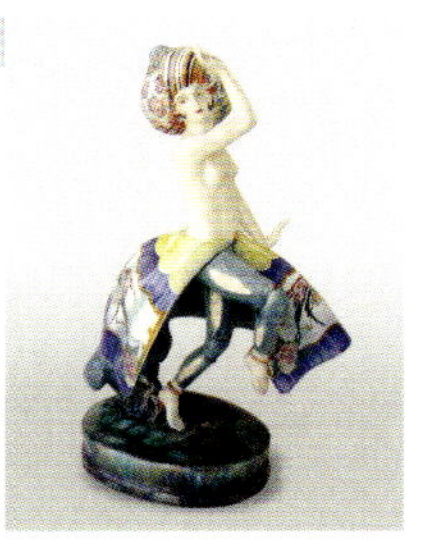

FIGUR
5549,31,7 ger. M18, gem. X.D.1., Firmenmarke 19, Made in Austria, Hand dec., unsign., H. 42,5 cm; Sammlung Stressless, Portugal
5549,81,12 gem. FO,3, Firmenmarke 19, Made in Austria, Hand dec., H. 44,8 cm; Soth. L. 3.11.1995
LAMPE
5549,121,7 gem. 3P, Firmenmarke 19, Made in Austria, Hand dec., unsign., H. 45,4 cm; Soth. N.Y. 10.6.1993; Soth. Arcade 1462 (18.3.1994)

5555
Lorenzl, Josef
Carmen (Sitzende Spanierin mit Fächer) (auf halbrundem Sockel)
ca. 1925
5555,42,12 Firmenmarke 19, sign. Laurenzl, H. 38 cm; Doro 1.12.2000
-> vgl. 5534; 6666
-> vgl. 5088; 5232; 5277= stehende Figur

5556
Lorenzl, Josef
Stehender singender Harlekin mit Akkordeon (auf Rundsockel)
ca. 1925/26

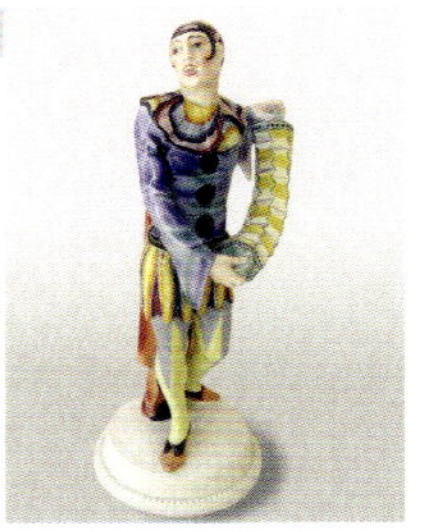

5556,1,8 gem. F.II, Firmenmarke 19, Made in Austria, Hand dec., H. 34 cm; Sammlung Stressless, Portugal

5558
Lorenzl, Josef
Stehender Harlekin mit Handpuppe (auf Rechtecksockel)
ca. 1925/26

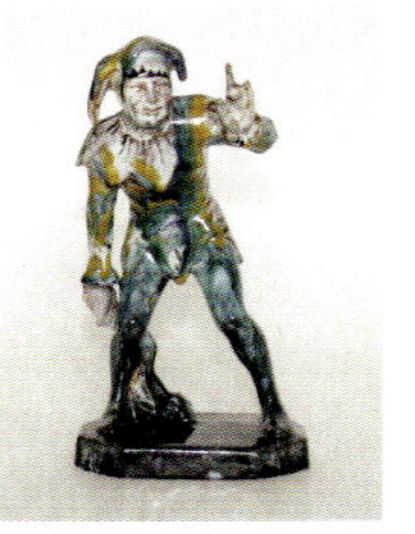

5558,4,5 Firmenmarke 19, Made in Austria, Hand dec., H. 24 cm; Sammlung Dr. Peter Hieke, Wien
5558,25,4 ger. lig. M26, Firmenmarke 19, Made in Austria, Hand dec., unsign., H. 23 cm; Sammlung Irsigler, Wien

5568
Lorenzl, Josef
Boxer
ca. 1925/26
5568,6,15 (Firmenmarke ohne Ang.), sign. Laurenzl, H. 30 cm; Doro WA 11.12.1993
Bem.: Da sich im Auktionskat. weder eine nähere Beschreibung noch eine Abb. findet, kann nicht eindeutig verifiziert werden, ob es sich bei dieser Figur um einen Sportler oder die Hunderasse „Boxer" handelt.

5569
Lorenzl, Josef
GRUPPE / LAMPE „Katta Sterna und Maria Solveg" (hintereinander stehend, mit ausgestreckten Armen) (auf oktogonalem, gekehltem Sockel)
ca. 1926
FIGURENGRUPPE
5569,93,10 (Firmenmarke ohne Ang.), sign. Laurenzl, H. 44 cm; WKA 44
5569,113,6 gepr. 8K, gem. lig. MR, 8, Firmenmarke 19, Made in Austria, Hand dec., sign. Lorenzl, H. 44 cm; GK, S. 21, 109 (m. Abb.); Wien Museum Inv. 222.550
LAMPE
5569,120,19 (Firmenmarke ohne Ang.), sign. Lorenzl, H. 51,5 cm; Phillips L. 25.1.2001
Bem.: Die beiden Tänzerinnen Katta Sterna und Maria Solveg von der Matray-Pantomimen-Gesellschaft könnten hier leicht mit den „Dolly Sisters" verwechselt werden, was bei GK, S. 21 auch geschehen ist. Eine zeitgenössische Abb. gibt jedoch näheren Aufschluss über dieses Modell. Die Tänzerinnen gehörten der Truppe um Ernst Matray (Berlin) an, der selbst Tänzer war und mit Katta Sterna auch in Wien gastierte. Der Tanz, den die beiden Tänzerinnen darstellen, ist nicht näher bezeichnet (DB 3.1926, Nr. 109, S. 14 m. Abb.)
Ausst.: HMW (jetzt Wien Museum) 1985/86.
> vgl. 5600= Einzelfigur
Lit.: GK, S. 21, 109 (m. Abb.)

5570F
Dakon, Stephan
FIGUR „Schleiertanz"
ca. 1926
Quelle: Archiv Mimi Dakon (Tantiemenabrechnung 1.7.–31.12.1937; Modellliste Dakon, undatiert, ca. 1940; Wiener Manufaktur FG, Modellliste Dakon, 7.11.1946), hier jeweils mit Ang. des Originaltitels „Schleiertanz".

5573
Lorenzl, Josef
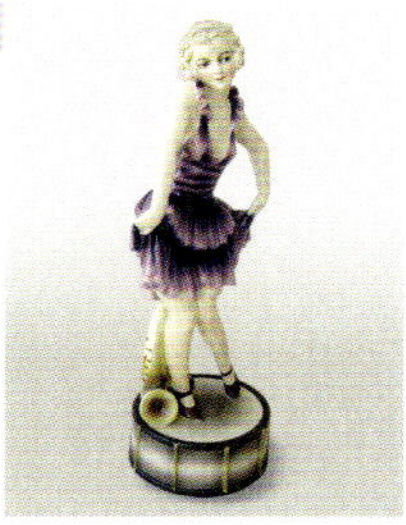
Posierende Varietékünstlerin (auf Sockel: Trommel, Saxophon und Tamburin)
ca. 1926
5573,49,7 gem. S, gem. lig. AH, XII, Firmenmarke 19, Made in Austria, Hand dec., sign. Laurenzl, H. 35,5 cm; Sammlung Stressless, Portugal
5573,120,7 ger. GO, W, gem. R., Firmenmarke 19, Made in Austria, Hand dec., H. 37 cm; Kastern 87; Art 1900

5574
Jarl, Karin
Stehender Boxer (Hund) (auf Landschaftssockel)
ca. 1926
5574,82,65 gem. lig. TV od. AZ?, Firmenmarke 19, Made in Austria, H. 17,8 cm, L. 21,6 cm; eBay # 39355444

5575
Jarl, Karin
Sitzende Bulldogge (auf Landschaftssockel)
ca. 1926
5575,4,2 ger. M1, Firmenmarke 19, H. 18,5 cm; Henry's 23.4.1997; 26.3 1999
5575,43,14 ger. M1, Firmenmarke 19, H. 18,5 cm; WKH
5575,67,8 ger. MII, Firmenmarke 19, H. 18 cm; Wien Museum Inv. 220.757

5580
Lorenzl, Josef
Stehende Tänzerin (mit Biedermeierfrisur), ihr langes Kleid anhebend
ca. 1926
5580,96,7 gem. MR4, Firmenmarke 19, Made in Austria, Hand dec., H. ca. 32 cm; WKH

5581
Lorenzl, Josef
Kniender Damenakt (Trauben essend) (auf Ovalsockel)
ca. 1926
5581,9,– Firmenmarke 19, Made in Austria, Hand dec., unsign., H. 36,8 cm; eBay # 6628397509

5582
Lorenzl, Josef
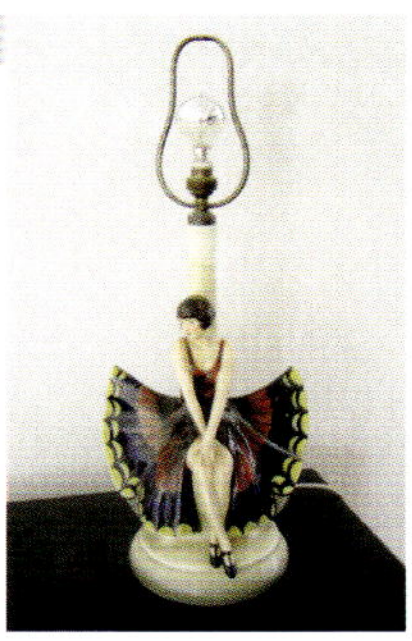
FIGUR / LAMPE Sitzende Tänzerin im Schmetterlingskleid (auf Rundsockel)
ca. 1926
FIGUR
5582,23,7,0,2 gem. XXVII,1, Firmenmarke 19, Made in Austria, Hand dec., sign. Laurenzl, H. 24,3 cm; eBay # 3645235992
5582,87,8 (Firmenmarke ohne Ang.), unsign.?, H. 23 cm; Christie's SK 19.3.1993
5582,196,12 Firmenmarke 19, Made in Austria, Hand dec., sign. Laurenzl, H. 23,5 cm; Quit. H. 28.2.2004
5582,232,8, gem. FB, JH 2, Firmenmarke 19, Made in Austria, Hand dec., sign. Laurenzl, H. 23,5 cm; Wien Museum Inv. 190.141
5582,379,17 ger. VP, gem. MR 22, Firmenmarke 19, Made in Austria, Hand dec., H. 25 cm; Doro S 2./3.12.1998
5582,415,12 (Firmenmarke ohne Ang.), sign. Lorenzl, H. 25,2 cm; Doro 30.11.1999
5582,484,8 Firmenmarke 19, Made in Austria, Hand dec., sign. Lorenzl, H. 26 cm; WKA 52
5582,491,4 gem. HB, Firmenmarke 19, Made in Austria, Hand dec., sign. Lorenzl, H. 25 cm; Mehlis 28
LAMPE
5582,336,12 Firmenmarke 19, Made in Austria, Hand dec., sign. Lorenzl, H. 30 cm; Sammlung Dr. Peter Hieke, Wien
5582,441,15 gem. MR 22, F., Firmenmarke 19, Made in Austria, Hand dec., sign. Lorenzl, H. 30,5 cm, Gesamth. 47,5 cm; Privatsammlung, Wien

5583
Lorenzl, Josef
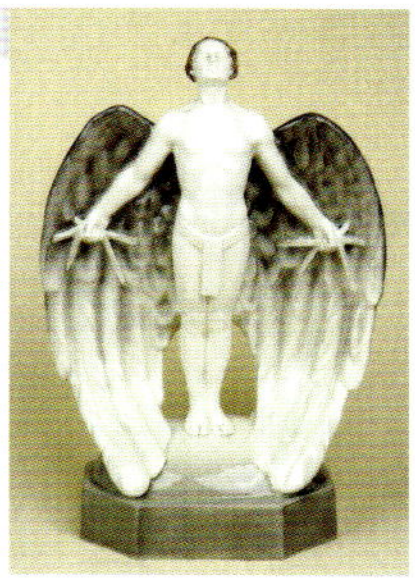
Stehender Ikarus (auf sechseckigem Sockel)
ca. 1926
5583,8,14 ger. M1, gem. III, MHA, Firmenmarke 19, Made in Austria, Hand dec., sign. Laurenzl, H. 37,1 cm; PB Wien 19
Bem.: Dargestellt ist die nackte Figur des geflügelten Ikarus (griech. Ikaros), der Sohn des Dädalus aus der griech. Sagenwelt.

5584
Dakon, Stephan
„Würde" (Stehende junge Dame in halbtransparentem Kleid und großem Umhang) (auf oktogonalem, abgerundetem Sockel)
ca. 1926
5584,47,8 Firmenmarke 19, Made in Austria, H. 45 cm; eBay # 1479347123
Bem.: Typische Art Deco Figur, laszive Pose, halbtransparentes Outfit.
Quelle: Archiv Mimi Dakon (Wiener Manufaktur FG, Modellliste Dakon, 7.11.1946), hier mit Ang. des Originaltitels „Würde".

5586
Boschull
DOPPELBÜSTE Lachendes Kinderpaar
ca. 1926
5586,1,11 Firmenmarke 19, Made in Austria, Hand dec., H. 12 cm; eBay # 910553899

5587
Hentschel (d.i. Handzel, August)
Stehendes Chinesenmädchen einen Topf haltend (auf Rundsockel)
ca. 1926
FAYENCE
5587,2,8 gem. XXVII, 1, M31, Firmenmarke 19, unsign.?, H. 19,8 cm; Neuwirth (1974), S. 76 (m. Abb.)
ALABASTER / SPECKSTEIN?
5587,W,2 ger., Plak. F, sign. Hentschel, H. 19,6 cm; Doro 9.6.1998
–> vgl. 4376
Pendant –> vgl. 5588

5588
Hentschel (d.i. Handzel, August)
Stehendes Chinesenmädchen mit Fächer (auf Rundsockel)
ca. 1926
FAYENCE
5588,6,14 gem. XXVII, 1, AX, Firmenmarke 19, unsign.?, H. 19,8 cm; Neuwirth (1974), S. 76 (m. Abb.)
5588,19,16 Firmenmarke 19, sign. (Laurenzl, wohl Lesefehler), H. 19,8 cm; Doro 10.4.1997
–> vgl. 4377
Pendant –> vgl. 5587

5592
Dakon, Stephan
„Kugelwerferin"
ca. 1926/27
Quelle: Archiv Mimi Dakon (Wiener Manufaktur FG, Modellliste Dakon, 7.11.1946), hier mit Ang. des Originaltitels „Kugelwerferin".

5595
Dakon, Stephan
Sitzende junge Dame mit Fächer (auf Rechtecksockel, mit abgeschrägten Ecken an der Schauseite)
ca. 1926/27
5595,26,4 gem. B1, gem. lig. SB, (Firmenmarke ohne Ang.), unsign.?, H. 33 cm; PB
5595,27,10 gem. M1, (Firmenmarke ohne Ang.), sign. Dakon, H. 33 cm; KaM 81

5596
Dakon, Stephan
„Pyjama" (Emmy Sturm stehend im Pyjama) (auf querovalem Sockel)
ca. 1926/27
5596,170,7 ger. M5, gem. KA, Firmenmarke 19, Made in Austria, Hand dec., H. 36 cm; BM 60
5596,296,10 Firmenmarke 19, Made in Austria, H. 37,3 cm; Doro 18.5.1999
Lit.: DD, 52.1924/25, H. 13, S. 60, hier Reklame der Wäschefabrik F. v. Grünfeld, die wohl als Vorlage des Entwurfs von S. Dakon diente. Modell: Emmy Sturm, –> vgl.: „Emmi Sturm in einem herrenmäßig geschnittenen Pyjama aus bunt gestreiftem Chinakrepp" (Nagy, L. v.: Die Frau im Pyjama, in: Uhu, 4.1925/26, Nr. 3, S. 51 m. Abb.).
Quelle: Archiv Mimi Dakon (Modellliste Dakon, undatiert, ca. 1940; Wiener Manufaktur FG, Modellliste Dakon, 7.11.1946), hier jeweils mit Ang. des Originaltitels „Pyjama".

5600
Lorenzl, Josef
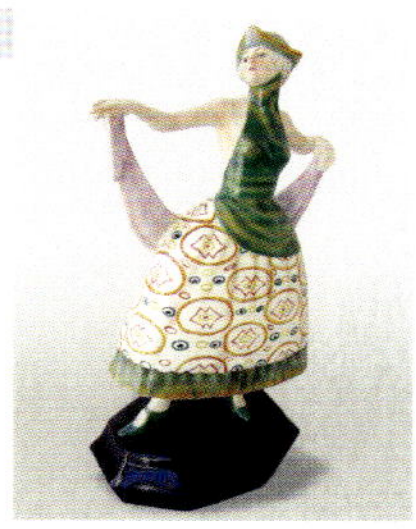
„Katta Sterna" (Einzelfigur, stehend, mit ausgestrecktem Arm) (auf oktogonalem, gekehltem Sockel)
ca. 1926/27
5600,23,6,03 Firmenmarke 19, Made in Austria, Hand dec., Originalpick. „Farmers, Sydney", H. 39,5 cm; eBay # 2008850594
5600,44,11 gem. 7E, Firmenmarke 19, Made in Austria, Hand dec., sign. Laurenzl, H. 41,5 cm; Sammlung Stressless, Portugal
Bem.: –> vgl. bei 5569. Ob bei der Einzelfigur wirklich die Tänzerin „Katta Sterna" dargestellt ist, ließ sich nicht mit allerletzter Sicherheit verifizieren.
–> vgl. 5569= GRUPPE

5604
Jarl, Karin
Liegendes Reh
BRONZE
ca. 1926/27
5604,1 GS Vienna (Firmenmarke Bronze 4), H. 6,9 cm; Herr 36

5609
Jarl, Karin
Zwei stehende Terrier (auf Ovalsockel)
ca. 1926/27
5609,64,11 gem. F., Firmenmarke 19, Made in Austria, H. 23 cm; Henry's 2.6.2001

5612
Dakon, Stephan
GRUPPE „Revue" (Die „Dolly Sisters" nebeneinander stehend, ihre kurzen Röcke hochhebend) (auf Rechtecksockel)
ca. 1926/27
5612,3,11 Firmenmarke 19, H. 37 cm; Doro 15.5.2002
5612,20(od. 29?),7 gem. S.4, Firmenmarke 19, Made in Austria, Hand dec., unsign.?, H. 40,6 cm; eBay # 520934413
5612,41,6 gem. F., Firmenmarke 19, Made in Austria, Hand dec., H. 38,1 cm; (KH www.tias.com/stores/frozenbeauty)
5612,108,12 Firmenmarke 19, Made in Austria, Hand dec., H. 37 cm; Phillips L. 25.1.2001
5612,201,7 gem. D4, Firmenmarke 19, unsign.?, H. 37,5 cm; Soth. South, Billingshurst 11.10.2000
Bem.: –> vgl. bei 5569.
–> vgl. 5634; 5649= jeweils Einzelfigur
–> vgl. 5819= ViererGRUPPE
Quelle: Archiv Mimi Dakon (Wiener Manufaktur FG, Modellliste Dakon, 7.11.1946), hier mit Ang. des Originaltitels „Revue".

5613

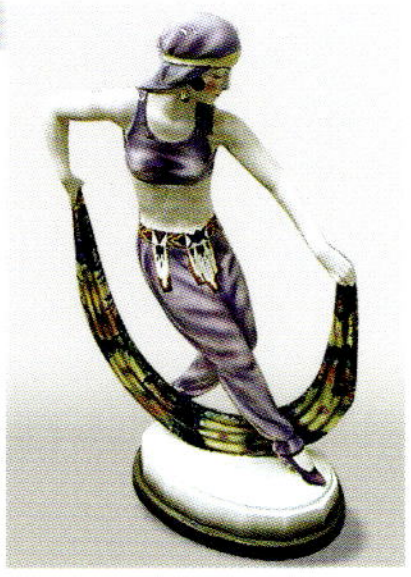

Lorenzl, Josef
Odaliske (Haremstänzerin mit langem Tuch) (auf facettiertem Ovalsockel)
ca. 1922 (für den Entwurf)
5613,70,65,03 gem. lig. AS1, Firmenmarke 19, Made in Austria, Hand dec., sign. Laurenzl, H. 46 cm; Doro 25.11.2003
5613,71,10 Firmenmarke 19, Made in Austria, sign. Laurenzl, H. 46 cm; Karrenbauer 149; Lit.:. AZ 1997, S. 767, 972 (m. Abb.)
5613,119,4,M03 ger., gem. K.1., Firmenmarke 19, Made in Austria, Hand dec., sign. Lorenzl, H. 45,4 cm; Doro 1.12.2000
5613,136,8 Firmenmarke 19, Made in Austria, Hand dec., sign. Laurenzl, H. 45,8 cm; BM 124
5613,213,7 gem. MR, AS, Firmenmarke 19, Made in Austria, Hand dec., sign. Lorenzl, H. 46,5 cm; KH
5613,235,7 Firmenmarke 19, sign. Lorenzl, H. 49 cm; Doro 15.5.2002
5613,258,7 (Firmenmarke ohne Ang.), sign. Lorenzl?, H. 47,5 cm; Doro 1.6.2005
5613,281,8 gem. XII P, gem. lig. MR4, Firmenmarke 19, Made in Austria, sign. Lorenzl, H. 45,7 cm; Sammlung Joan Ferguson/UK (www.paddycliff.com)
5613,282,10 (Firmenmarke ohne Ang.), unsign.?, H. 47 cm; Christie's SK 27.4.2005
5613,289,4 Firmenmarke 19, sign. Lorenzl, H. 46 cm; Soth. L. 4.6.1987
5613 gem. (!), (gepr. 5230,1825,4,04) gem. XXIII. 1, Firmenmarke 19, Made in Austria, Hand dec., sign. Laurenzl, H. 45,6 cm; Soth. com. # 1BSQ; Bem.: Die versehentlich falsch gepr. Modell-Nr. 5230 wurde vom Staffierer durch die gem. Nr. 5613 berichtigt.
5613 (ohne Ang.), Firmenmarke 19, sign. Laurenzl, H. 45,5 cm; Nagel 324
-> vgl. 5018; 6063

5616

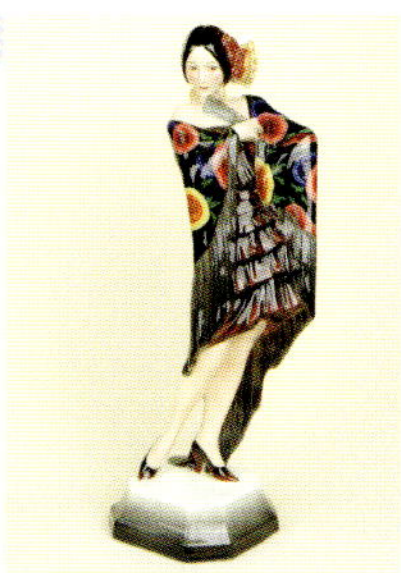

Lorenzl, Josef
„Dolores del Rio als Carmen" (Stehende Spanierin mit Fächer) (auf oktogonalem, gekehltem Sockel)
Inschrift: DOLORES DEL RIO (mit/ohne Inschrift)
ca. 1927
ohne Inschrift
5616,47,12,03 gem. XXIII.1., Firmenmarke 19, Made in Austria, Hand dec., sign. Laurenzl, H. 45,7 cm; Soth. com. # 6NV4
5616,66,14 Firmenmarke 19, Made in Austria, Hand dec., sign. Laurenzl, H. 46 cm; Kett. 184
5616,84,4,03 Firmenmarke 19, Made in Austria, Hand dec., sign. Laurenzl, H. 46 cm; KH Ruby Lane (Jane Perlow/Lacremeantik, Great Neck N.Y.)
5616,133,6 gem. lig. AH, 2, Firmenmarke 19, Made in Austria, Hand dec., unsign.?, H. 45,2 cm; Soth. com., Lot 2525 (London, Sept. 2002)
5616,173,4,03 gem. HH, Firmenmarke 19, Made in Austria, Hand dec., unsign.?, H. 45,4 cm; Soth. L. 25.10.1991
5616,220,11 Firmenmarke 19, Made in Austria, Hand dec., sign. Lorenzl, H. 45 cm; KH
5616,237,65 (Firmenmarke ohne Ang.), unsign.?, H. 45,7 cm; Christie's SK 27.4.2005
5616,255,6 (Firmenmarke ohne Ang.), sign. Lorenzl, H. 49 cm; WKA 38
5616,276,6 Firmenmarke 19, Made in Austria, Hand dec., sign. Lorenzl, H. 50 cm; PB Wien 19
mit Inschrift: DOLORES DEL RIO
5616,202,8 Firmenmarke 19, sign. Laurenzl, H. 46 cm; Sammlung Joan Ferguson
5616,207,10,K5 gem. MR1, Firmenmarke 19, Made in Austria, Hand dec., sign. Laurenzl, H. 46,4 cm; Soth. com. # A0D9
Bem.: Dargestellt ist die mexikan. Filmschauspielerin Dolores del Rio (eigentl. Dolores Martinez Aúnsolo y López Negrete, 3.8.1905–11.4.1983) in einer ihrer bedeutendsten Rolle als „Carmen" (The loves of Carmen, 1927. Regie: Raoul Walsh).
Ausst.: SSH 1992
-> vgl. 6208
Lit.: N.N.: Berühmte Frauen und moderne Keramik, in: MW, 10.1928/29, Nr. 6, S. 29 (m. Abb.), hier mit Ang. des Originaltitels „Dolores del Rio als Carmen" und der falschen Künstlerang. „Helmut Lindner", „Dolores del Rio, die glutäugige Mexikanerin mit dem interessanten exotischen Gesichtsausdruck, die von ihrem Manager und Regisseur trotz des Widerstrebens ihrer Eltern für den Film gewonnen wurde und binnen weniger Jahre zu einem der berühmtesten Filmstars aufstieg, zeigt sich hier in einer ihrer bedeutendsten Rollen, der Carmen".

5625

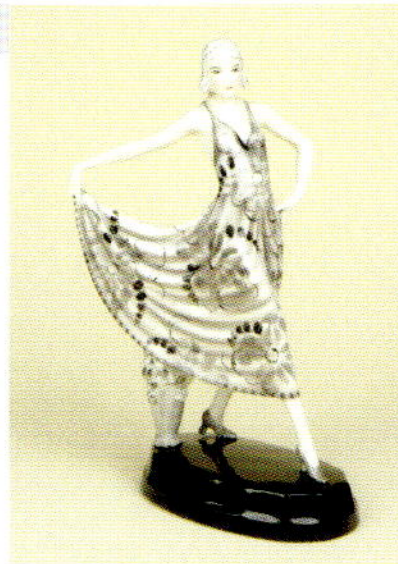

Dakon, Stephan
FIGUR „Harvey" (Lilian Harvey als schreitende Tänzerin, ihren Rock mit der rechten Hand anhebend) (auf facettiertem Ovalsockel)
ca. 1926/27
5625,30,4 gem. lig. M.H.3., Firmenmarke 19, Made in Austria, Hand dec., H. 47,7 cm; Soth. com. # A0DA
5625,50,8 Firmenmarke 19, H. 47,5 cm; Doro 18.11.1997
-> vgl. 6009; 6211
Lit./Bem.: Becker, K.: Lilian, das Mädchen mit den 100 Gesichtern, in: Uhu 4.1927/28, Nr. 10, S. 54 (m. Abb.), hier eine zeitgenössische Fotografie, die wohl als Inspiration für die Ausführung diente. Dargestellt ist die Schauspielerin Lilian Harvey (eigentl. Lilian Pape, 19.1.1906–27.7.1968) hier mit dunkler Perücke, betitelt „Spanischer Tanz". Fotografie: Binder.

5629
Dakon, Stephan
DOSE „Mädchen"
ca. 1926/27
Quelle: Archiv Mimi Dakon (Wiener Manufaktur FG, Modellliste Dakon, 7.11.1946), hier mit Ang. des Originaltitels DOSE „Mädchen".

5634
Dakon, Stephan
„Revue"
ca. 1926/27
5643 H. fehlt
-> vgl. 5649
-> vgl. 5612= ZweierGRUPPE
-> vgl. 5819= ViererGRUPPE
Quelle: Archiv Mimi Dakon (Wiener Manufaktur FG, Modellliste Dakon, 7.11.1946), hier mit Ang. des Originaltitels „Revue".

5649

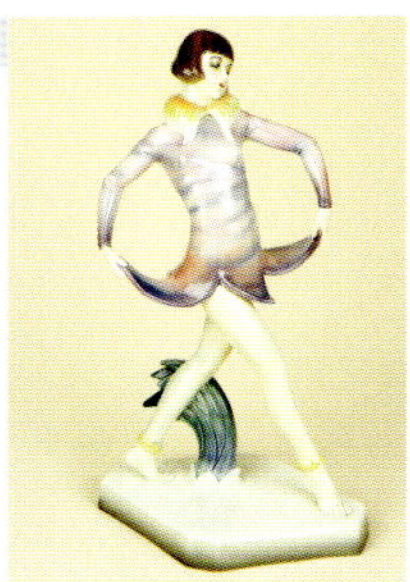

Dakon, Stephan
FIGUR/LAMPE „Revue" (Schreitende Einzelfigur der „Dolly Sisters", ihren kurzen Rock hochhebend) (auf Rechtecksockel)
ca. 1926/27
FIGUR
5649,44,6 (Firmenmarke - ohne Ang.), H. 37 cm; Christie's SK 12.5.2000
5659 (5649),139,15 gem. MK, 1F, Firmenmarke 19, Made in Austria, Originalpick.: „Holstein & Düren, Köln", H. 37,5 cm; PB Wien 19
5649,144,11 gem. lig. MR1, Firmenmarke 19, Made in Austria, Hand dec., H. 37 cm; eBay # 917206029
LAMPE (als Straßenlaterne)
5649,164,16 Firmenmarke 19, Made in Austria, Gesamth. 74 cm; Doro 18.5.1999
5649,177,6 gem. lig. TR, Firmenmarke 19, Made in Austria, Hand dec., Gesamth. 64 cm; GK, S. 21, 110 (m. Abb.); Wien Museum Inv. 223.910
Bem.: -> vgl. bei 5569.
Ausst.: HMW (jetzt Wien Museum) 1985/86.
-> vgl. 5634
-> vgl. 5612= ZweierGRUPPE
-> vgl. 5819= ViererGRUPPE
Lit.: GK, S. 21, 110 (m. Abb.); GSN, S. 38, 39 (m. Abb.); GSNF, o. Pag. (m. Abb.)
Quelle: Archiv Mimi Dakon (Wiener Manufaktur FG, Modellliste Dakon, 7.11.1946), hier mit Ang. des Originaltitels „Revue".
GS neu: Modell-Nr. 6 „Lilli Marlen" (H. 55 cm), Aufl. 2000 Stück
Bem.: Denkt man an das Lied „Lilli Marlen", so ist der für die neue Produktion in Stoob/Bgld. vergebene Titel recht schlüssig. Tatsächlich handelt es sich bei der dargestellten Person aber um eine der beiden Dolly Sisters.

5650
Fuchs
GEFÄSS Zwei sitzende Affen (jeder mit einer Schale, als Salz- und Pfeffergarnitur)
ca. 1927
5650,3,14 gem. L., Firmenmarke 19, Made in Austria, Hand dec., H. 8,5 cm, L. 18,2 cm; Wien Museum Inv. 186.679

5657
Hafenruckler
Ägyptische Tempeltänzerin (auf abgeschrägtem, mit umlaufendem Reliefdekor (Elefanten, Palmen) verziertem Rechtecksockel)
ca. 1927
5657,2,8 Firmenmarke 19, Made in Austria, Hand dec., H. 47 cm; WKA 38

5658
Dakon, Stephan
„Dreispitz" (Stehende junge Dame in Karnevalskostüm und Dreispitz) (Lissi Arna?) (auf Dreiecksockel)
ca. 1927
FAYENCE
5658,137,8 gem. H3, (Firmenmarke ohne Ang.), H. 32,5 cm; Kett. 157
5658,375,15 gem. B1P, Firmenmarke 19, Made in Austria, Hand dec., H. 32,5 cm; Rehm 127
5658,464,16 (Firmenmarke ohne Ang.), H. 35 cm; Christie's SK 20.9.2000
ALABASTER
5658,W,8 ger., (Firmenplak. ohne Ang.), H. 34 cm; WKA 50
Bem.: Dieses Modell wurde in Fayence und Alabaster ausgeführt. Die Variante in Alabaster birgt als besondere Zugabe einen kleinen sitzenden Pierrot am Sockel.
Lit.: DM 4.1927/28, Nr. 42, S. 1866 (m. Abb.), hier eine zeitgenössische Fotografie, die wohl als Inspiration für die Ausführung diente. Dargestellt ist die Schauspielerin Lissi Arna (eigentl. Elisabeth Arndt, 20.12.1904–23.1.1964). Fotografie: Feldscharck, Wien.
Quelle: Archiv Mimi Dakon (Wiener Manufaktur FG, Modellliste Dakon, 7.11.1946), hier mit Ang. des Originaltitels „Dreispitz".

5659
Dakon, Stephan
„Banjospiel" (Zerline Balten als stehende Banjospielerin) (auf Rundsockel mit kleiner sitzender Pierrotfigur)
ca. 1927
5659,3,11 gem. lig. AH, Firmenmarke 19, Made in Austria, Hand dec., H. 34,5 cm; Wien Museum Inv. 190.146
5659,54,8 (Firmenmarke ohne Ang.), H. 34,6 cm; Doro 10.4.1997
5659,164,16 (Firmenmarke ohne Ang.), Originalpick.: „Corthy, Nantes", H. 36,2 cm; Christie's SK 27.4.2005
Bem.: Dargestellt ist Zerline Balten, eine Wiener Tänzerin und Schauspielerin die sich bei ihrem „Phantasietanz" auf einem Banjo begleitet" (Hirschfeld, H.: Von der Laute zum Banjo, in: EW, 14.1925, Nr. 16, S. 18 (m. Abb.), hier eine Fotografie des Fotostudios d'Ora, Wien. Wenngleich auch das Kostüm etwas abgewandelt ist, so sprechen Physiognomie und Pose eindeutig für die Darstellung dieser Tänzerin). Die Firma Rosenthal (Kunstabt. Bahnhof-Selb) brachte um 1923 ebenfalls eine Figur von Constantin Holzer-Defanti zur Ausführung (Niecol, Bd. 1, S. 173 m. Abb.; Bd. 4, S. 160 m. Abb.).
Quelle: Archiv Mimi Dakon (Wiener Manufaktur FG, Modellliste Dakon, 7.11.1946), hier mit Ang. des Originaltitels „Banjospiel".

5660
Lorenzl, Josef
Stehende Rokokodame mit Handspiegel (auf Ovalsockel)
ca. 1927
5660,9,10 Firmenmarke 19, Made in Austria, Hand dec., H. 40 cm; Sammlung Dr. Peter Hieke, Wien

5663
Watzka, Friedrich Roland
Papagei, auf Kugel sitzend
ca. 1927
5663,4,6 ger. O3M, gem. L. D1, Firmenmarke 19, Made in Austria, Hand dec., sign. Watzga (!), H. 19,8 cm; Wien Museum Inv. 190.082

5670
Dakon, Stephan
„Bananentanz"
ca. 1927
Quelle: Archiv Mimi Dakon (Wiener Manufaktur FG, Modellliste Dakon, 7.11.1946), hier mit Ang. des Originaltitels „Bananentanz".

5672
Hahn
Chinesischer Mönch (auf Ovalsockel)
ca. 1927
5672,2,10 Firmenmarke 19, Made in Austria, Hand dec., H. 21 cm; eBay # 6599337720

5677
Schwarz, K.
BÜSTE Schubert
ca. 1927
5677,4,12 gem. AH, Firmenmarke 19, Made in Austria, sign. K. Schwarz, H. fehlt, L. 26 cm; Mehlis 31
Bem.: Dargestellt ist der Komponist Franz Schubert (1797–1828)

5678
Lorenzl, Josef
ASCHER/RAUCHERGARNITUR „Aschenboy" (Page mit zylindrischem Gefäß für Zigaretten, auf einem Aschenbecher mit vier Ablagen stehend)
ca. 1927
5678,30,8 (Firmenmarke ohne Ang.), H. 21,6 cm; Christie's SK 15.10.2003
5678,420,4 (Firmenmarke ohne Ang.), unsign.?, H. 22 cm; Schuler 73
5678,501,11 gem. lig. SO, N, gem. blau M, Firmenmarke 19, Made in Austria, H. 22,5 cm, Ø 15 cm; eBay # 2506964949
Lit.: MW, 11.1929/30, Nr. 10, S. V (m. Abb.), hier mit Ang. des Originaltitels „Aschenboy" und des Entwerfers Josef Lorenzl; Neuwirth (1974), S. 67.

5681
Geiger, Benno (zugeschr.)
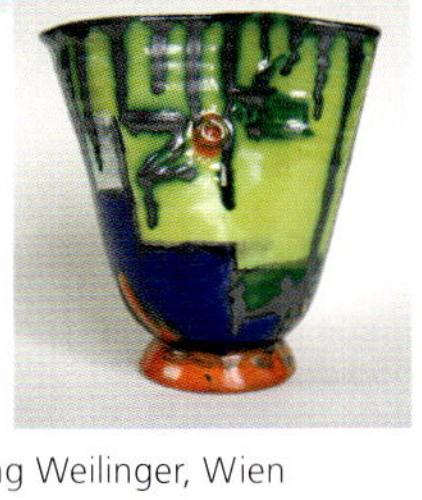
VASE
Form: becherförmig, Mündung mittig gedrückt
Dekor: geometrische Formen, mehrfarbige Laufglasur
ca. 1927
5681,127,1 gem. G., Firmenmarke 19, Made in Austria, unsign., H. 15,9 cm; Sammlung Weilinger, Wien

5691
Pejic, (R.?)
BÜSTE Araber mit Turban (auf quaderförmigem Sockel)
MARMOR
ca. 1927
5691,HM,2 ger., Plak. E, Made in Austria (ger.), sign. Pejic, H. 20,3 cm; eBay # 2866573241

5693
Pejic, (R.?)
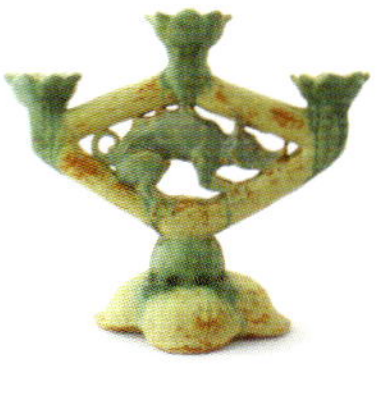
KERZENHALTER (dreiflammig)
„Leuchter (mit) Fabeltier"
ca. 1927
5693,24,6 gem. ML, Firmenmarke 19, Made in Austria, sign. Pejic, H. 29 cm, L. 33 cm; Sammlung a@goldscheider.net
5693,26,6 gem. L.L., Firmenmarke 19, Made in Austria, unsign., H. 29 cm; Sammlung f@goldscheider.de
5693,27,6 Firmenmarke 19, Made in Austria, unsign., H. 29 cm; MAK Wien, Inv. Ke 10521-24
Lit.: N.N.: Farbe und Bewegung, Die Fayencen der Wiener Manufaktur Friedrich GS Wien-Währing, in: DB, 5.1928, Nr. 213, S. 38 und 40 (m. Abb.), hier mit Ang. des Originaltitels „Leuchter mit Fabeltier"; Lehmann, G.: Neuheiten, in: KuKGW, 9.1928/29, S. 126 (m. Abb.), „gute töpferische Handarbeit"; MW, 11.1929/30, Nr. 15, S. VI (m. Abb.), hier mit Ang. des Originaltitels „Leuchter-Fabeltier" und des Entwerfers „Pejic"; Neuwirth (1974), S. 69.

5705
Meisinger, Ida
Sitzender Yorkshire Terrier (Malteser?)
ca. 1927
5705,72,15 gem. 1.L, Firmenmarke 19, Made in Austria, H. 21,5 cm; eBay # 2660779800
5705,103,16 gem. L, Firmenmarke 19, H. 20 cm; Herr 45
-> vgl. 5784= ähnl. Modell.

5706
Meisinger, Ida
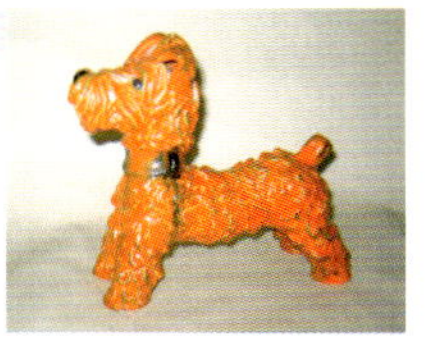
Stehender Airedaleterrier
ca. 1927
5706,186,8 ohne Firmenmarke, unsign., H. 18 cm, L. 21 cm; PB Wien

5708
SCHALE mit zwei Handhaben
Form: rund, trichterförmig
Handhaben: zwei Doppelösen
Dekor (appliziert): Blütenzweige
ca. 1927
5708,64,19 Firmenmarke 19, Made in Austria, H. 11 cm, Ø 24 cm; Kaupp 25.9.1999

5710
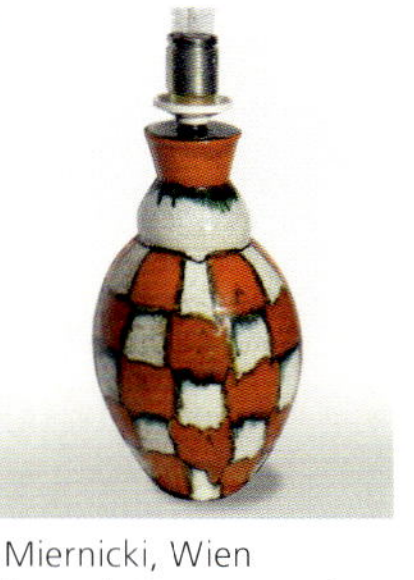
LAMPE
Form: gebaucht
Dekor: Schachbrettmuster, orange-weiß, abgesetzt mit schwarz-grüner Laufglasur
ca. 1927/28
5710,21,6 gem. N, Firmenmarke 19, Made in Austria, H. 29 cm; Gesamth. 36 cm; Soth. L. 30.11.1993; Doro 26.5.1998; Sammlung H. Miernicki, Wien
Bem.: Zu diesem Lampenfuß gehört als Ergänzung eine vielseitig verwendbare Schale, die auch als Vase oder Blumentopf dienen kann. Sie wurde unter der Modell-Nr. 564 in der neuen Modellnummernreihe (Angewandte Kunst) produziert.
Pendant -> vgl. 564= VASE/CACHEPOT/TOPF

5712
Dakon, Stephan
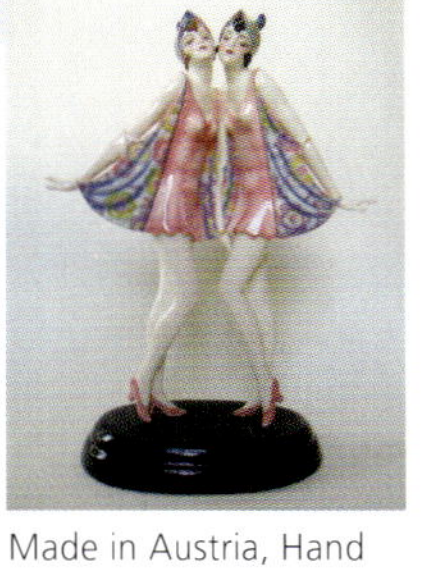
GRUPPE/LAMPE „Turteltauben" (Die Dolly Sisters, Cheek to cheek, zueinander stehend, ihre kurzen Kostüme anhebend) (auf Ovalsockel)
ca. 1927/28
FIGURENGRUPPE
5712,49,65,03 gem. 2N, gem. lig. AS, Firmenmarke 19, Made in Austria, Hand dec., unsign., H. 39,5 cm; Sammlung Weilinger, Wien
5712,82,6 gem. R.2.L., Firmenmarke 19, Made in Austria, Hand dec., H. 39,5 cm; eBay # 6555504386
5712,91,16 gem. 2.D.L., gem. lig. SB, Firmenmarke 19, Made in Austria, Hand dec., unsign., H. 39,5 cm; eBay # 1496594480
5712,208,16 (Firmenmarke ohne Ang.), H. 38 cm; Christie's SK 20.9.2000
5712,528,11 (Firmenmarke ohne Ang.), H. ca. 39 cm; Doro 29.5.2001; Christie's SK 14.11.2002
5712,546,15,12 gem. AS2, gem. rot N2, Firmenmarke 19, Made in Austria, Hand dec., unsign.?, H. 41,2 cm; Soth. com. # A0DF
LAMPE
5712,6,10,0,3 gem. K, gem. lig. SA Firmenmarke 19, Made in Austria, Hand dec., H. 64 cm; Soth. L. 28.10.1998
Bem.: -> vgl. bei 5569.
-> vgl. 5781= ähnl. Modell (Einzelfigur)
Lit.: N.N.: Berühmte Frauen und moderne Keramik, in: MW, 10.1928/29, Nr. 6, S. 28 (m. Abb. als LAMPE), hier mit Ang. „Die Dolly Sisters" und der falschen Künstlerang. „A. Stefan" (d.i. Stephan Dakon), S. 29 „Aber auch die Dolly Sisters mag man wahrhaft beneiden, denn Rosi Dolly hat es ja vor etwa einem Jahr verstanden, die Bank von Monte Carlo zu sprengen. Die beiden Schwestern sind in einer typischen Pose wiedergegeben, wie wir sie in deren Revuetänzen finden. Der eigenartige Reiz, der gerade in dem gleichsam Mechanisierten dieser Bewegungen liegt, hier vor allem durch die Symmetrie der Stellungen zum Ausdruck kommend, steckt in diesen Figuren. Die Gesichter haben den Charme, die reizvolle Koketterie von Revuegirls, die süßen Plappermäulchen sind wie zum Sprechen oder Singen geöffnet"; MW, 11.1929/30, Nr. 22, S. IV (m. Abb. als FIGURENGRUPPE), hier mit Ang. des Originaltitels „Turteltauben" und der falschen Künstlerang. „Josef Lorenzl"; Neuwirth (1974), S. 74, hier falsch übernommene Ang. aus MW „A. Stefan"; Pinhas, S. 149 (m. Abb.).

5713
Dakon, Stephan
„Zylinder"
ca. 1927/28
Quelle: Archiv Mimi Dakon (Wiener Manufaktur FG, Modellliste Dakon, 7.11.1946), hier mit Ang. des Originaltitels „Zylinder".

5714
Dakon, Stephan
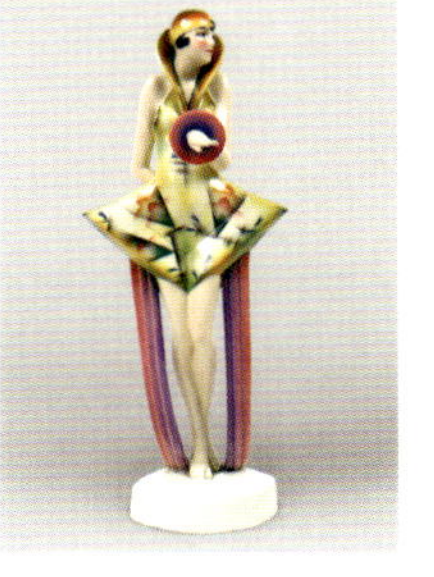
„Modern" (Stehende Tänzerin in expressivem Kostüm) (auf Rundsockel)
ca. 1928
5714,9,6,O3,J gem. S, 1., gem. lig. SA.1., Firmenmarke 19, Made in Austria, Hand dec., H. 43,5 cm; Sammlung Weilinger, Wien
Bem.: Wer in diesem wirklich außergewöhnlichen und seltenen Modell dargestellt ist, ließ sich bis Redaktionsschluss nicht eruieren. Es ist jedoch sicher davon auszugehen, dass es sich hier um eine (bekannte) Tänzerin oder Schauspielerin handelt.
Lit.: N.N.: Berühmte Frauen und moderne Keramik, in: MW, 10.1928/29, Nr. 6, S. 29 (m. Abb.) .), hier mit Ang. des Originaltitels „Modern" und der falschen Künstlerang. „A. Stefan", „Die letzte Figur schließlich wirkt besonders apart durch das eigenartige Tanzkostüm und durch die beiden, bis zum Boden reichenden Stoffstreifen, die im Gegensatz der einfachen Kurve des fallenden Stoffs zur feinen Modellierung der Schenkel den Ausdruck dieser schlanken Beine so sehr zu heben verstehen. Das Köpfchen selbst hat die ganze Anmut einer modernen, schönen, verwöhnten Frau"; Neuwirth (1974), S. 74.

5715
Lorenzl, Josef
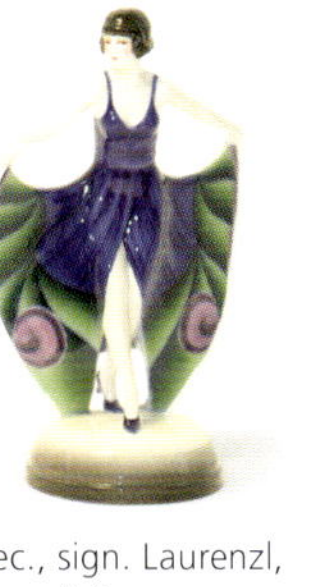
Stehende Tänzerin im Schmetterlingskleid (auf Rundsockel)
ca. 1928
5715,6,4,03 (Firmenmarke o. Ang.), unsign., H. 39,5 cm; city-antik
5715,49,8 ger. lig. M5, gem. S.2., gem. lig. SB, Firmenmarke 19, Made in Austria, Hand dec., sign. Laurenzl, H. 41,2 cm, B. 26 cm; Privatsammlung, Wien
5715,73,9 gem. S.2, gem. lig. SO3, Firmenmarke 19, Made in Austria, Hand dec., H. 39 cm; eBay # 3716833384
5715,73,9 gem. S.2, gem. lig. SB, Firmenmarke 19, Made in Austria, unsign.?, H. 39,5 cm; Schloss Hag. 33
5715,108,9 gem. F., gem. lig. MR, Firmenmarke 19, Made in Austria, Hand dec., sign. Lorenzl, H. 39 cm; Sammlung Stressless, Portugal
5715,147,10,8 gem. N2, Firmenmarke 19, H. 39 cm; Kastern 68; Soth. Olympia 3.5.2002
5715,264,16 gepr. 1P, gem. N7, Firmenmarke 19, Hand dec., H. 40,6 cm; Sammlung Joan Ferguson

5719
Postl, Hugo
„Polizeihund (sitzend)" (Sitzender Schäferhund/Deutscher Schäfer)
ca. 1928
5719,371,12,5 gem. lig. RZ, Firmenmarke 19, Made in Austria, unsign., H. 23 cm; eBay # 3659373079
-> vgl. 6069= GRUPPE
-> vgl. 6069= GRUPPE Stehendes Kind, einen sitzenden Schäferhund umarmend (Modell: Lorenzl/Postl)
Lit.: MW, 10.1928/29, Nr. 23, S. V (m. Abb.), hier mit Ang. des Originaltitels „Polizeihund (sitzend)" und des Entwerfers Hugo Postl, Salzburg; Schilling-Schletter, A.: Hunde-Plastiken, in: MW, 11.1929/30, Nr. 6, S. 12 (m. Abb.), hier betitelt als „Deutscher Schäfer".
Quelle: AGM (Bestellungen der Firma Wahliss (Wien), vom 9. und 24.11.1945 an die „Arbeitsgemeinschaft der Firma Friedrich Goldscheider", hier jeweils mit Ang. des Titels „Wolfshund (sitzend)"; Bestellung der Firma Wahliss (Wien), vom 27.2.1946 in Ausführung „Terrakotta").

5720
Postl, Hugo
Sitzender Fox-Terrier
ca. 1928
5720,25,10 (Firmenmarke ohne Ang.), H. 17,8 cm; eBay # 423713882
5720,117,13 Firmenmarke 19, Made in Austria, H. 19 cm; eBay # 1440144415

5724
Lorenzl, Josef
Schreitender Amor (auf Natursockel)
ca. 1928
5724,30,10 Firmenmarke 19, Made in Austria, unsign., H. 25 cm; eBay # 7316523072
5724 ger., (Firmenmarke ohne Ang.), monogr. JL, H. 24,2 cm; Doro 26.5.2004; 15.3.2005

5726
Lorenzl, Josef
Stehendes Rokokopaar (auf Ovalsockel)
ca. 1928
5726,15,? gem. 582,708,NF, Firmenmarke 19, Made in Austria, H. 35,5 cm; eBay # 59423279

5728
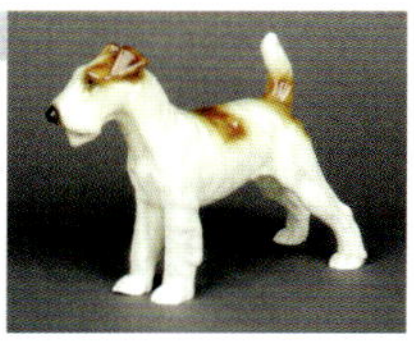
Postl, Hugo
Stehender Fox-Terrier
ca. 1928
5728,84,10 ger. lig. M5, gem. H, Firmenmarke 19, Made in Austria, H. 19 cm; eBay # 6526784902
5728,351,10 gem. lig. SO Firmenmarke 19, Made in Austria, H. 19 cm; eBay # 342660112
5728 ohne Ang., Firmenmarke 19, Made in Austria, H. 19 cm; PB Wien 19
Lit.: N.N.: Farbe und Bewegung, Die Fayencen der Wiener Manufaktur Friedrich GS Wien-Währing, in: DB 5.1928, Nr. 213, S. 38 und 40 (m. Abb.).

5729
Meisinger, Ida
DECKELDOSE Klarinetten- bzw. Flötenspieler
Form: quaderförmig
Knauf: würfelartig
Dekor (appliziert): Klarinetten- bzw. Flötenspieler (Rattenfänger von Hameln?), Blumendekor
ca. 1928
5729,19,16 gem. 1.L., gem. lig. AH, Firmenmarke 19, Made in Austria, Hand dec., H. 17,8 cm, B. 12,7 cm; eBay # 128241342
5729,22,8,B gem. G., Firmenmarke 19, Made in Austria, Hand dec., unsign.?, H. 17 cm; eBay # 3653487430

5732
Lorenzl, Josef
„Clotilde von Derp" (Schreitende Tänzerin in Kniehosen) (auf Ovalsockel)
ca. 1928
5732,3,7 Firmenmarke 19, sign. Laurenzl, H. 33,8 cm; Doro 24.5.2000
5732,4,10 Firmenmarke 19, Made in Austria, Hand dec., sign. Lorenzl, H. 35 cm; Nagel 379
5732,68,12 gem. J.N2, Firmenmarke 19, Made in Austria, sign. Lorenzl, H. 35,4 cm; Mehlis 22; 23; city-antik
5732,281,6 Firmenmarke 19, sign. Lorenzl, H. 37,1 cm; BM 68
5732,296,65,04,K, Firmenmarke 19, Made in Austria, Hand dec., sign. Lorenzl, H. 36 cm; Wien Museum Inv. 190.113
-> vgl. 6248
Bem.: Dargestellt ist die mit Alexander Sacharoff verheiratete Tänzerin Clotilde von Derp (5.11.1892, Berlin – 11.1.1974, Rom) in ihrem Tanz zu „Papillons", Musik: Robert Schumann, 1912. Als Vorlage für den Entwurf der Figur diente eine Fotografie (die aus einer ganzen Serie von Bildern stammt) von Hugo Erfurth, abgebildet bei Peter (2002), S. 111; -> vgl. auch Peter (1992), S. 45–52 (m. Abb.); Aubel, S. 25, hier eine Fotografie in fast identischer Pose von Stephanie Held.

5737
Waldmüller
FIGUR Katze
ca. 1928
5737 Firmenmarke 19, Made in Austria, H. fehlt; eBay # 7403974953
-> vgl. 5769= BUCHSTÜTZE

5739
Geiger, Benno (zugeschr.)
VASE Vier Jahreszeiten
Form: becherförmig, geweitete Mündung
Dekor: vier Figuren (Skifahrerin, Jäger, Mädchen mit Gitarre, Dame mit Tuch)
ca. 1928
5739,503,132 Firmenmarke 19, Made in Austria, unsign., H. 27 cm, Ø 17,4 cm; eBay # 3329114033; PB
-> vgl. 583

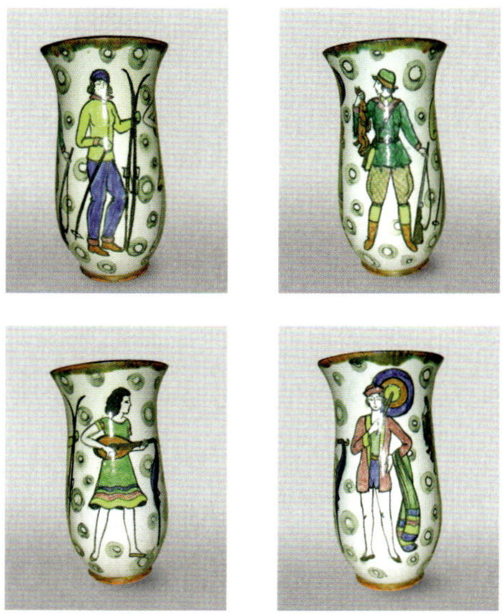

5742
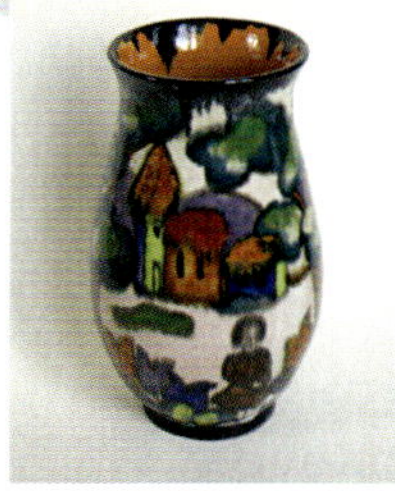
Geiger, Benno (zugeschr.)
VASE
Form: balusterförmig
ca. 1928
Dekor: Landschaft, Kirche und Brücke mit Staffagen
5742,19 gem. D.2.G., Firmenmarke 19, Made in Austria, unsign., H. 23,4 cm; Sammlung Weilinger, Wien
Dekor: Reiter, stilisierte Bäume
5742,42 Firmenmarke 19, Made in Austria, unsign., H. 23,5 cm; Doro 3.5.1995
Bem.: Die Zuschreibung an Geiger erfolgt aufgrund der Dekore, denn Geiger hat etliche dieser Landschaften mit Staffagen entworfen (-> vgl. Künstlerverzeichnis).

5744
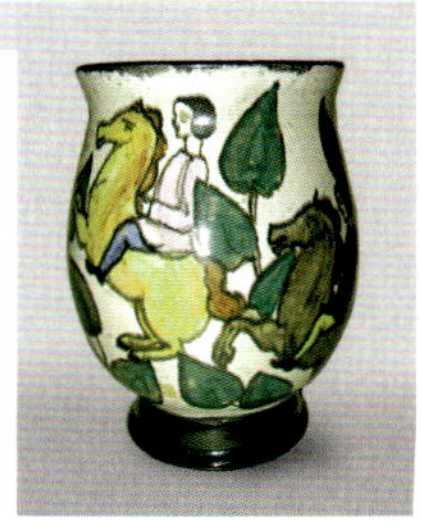
Geiger, Benno (zugeschr.)
VASE Fünf Reiter
Form: becherförmig, gewulsteter Standring
Dekor: Fünf Reiter auf steigenden Pferden, stilisiertes Blattwerk
ca. 1928
5744,38,1 gem. VI, B, Firmenmarke 19, unsign., Pick. 9, H. 21,4 cm, Ø 17 cm; PB Wien

5746
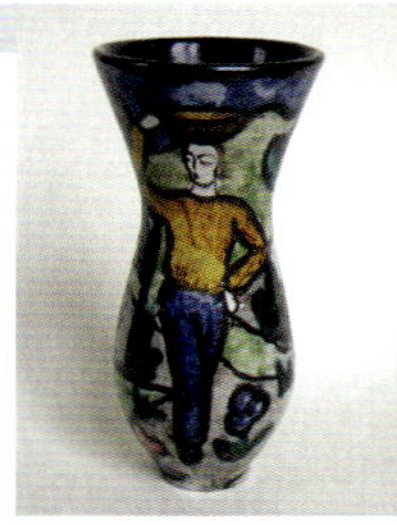
Geiger, Benno (zugeschr.)
VASE Mann und Frau im Garten
Form: balusterförmig, geweitete Mündung
Dekor: Paar bei der Gartenarbeit
ca. 1928
5746,5 gem. B, Firmenmarke 19, unsign., H. 16 cm; Sammlung Weilinger, Wien

5767 -> ÜBERSCHNEIDUNG
Geiger, Benno
SCHALE Frauengesicht
Form: rund, tief gemuldete Form, zylindrischer Stand
Dekor: Frauengesicht vor Stadtlandschaft
ca. 1928
5767,2 gem. G., Firmenmarke 19, Ø 13 cm; Doro 12.12.1991
5767 gem. DVIII., G, Firmenmarke 19, Hand dec. (zweimal), Pick. 11a (Entwurf Benno Geiger); H. 4 cm, Ø 13 cm; eBay # 3329114033; Zezschwitz 13

5767 -> ÜBERSCHNEIDUNG
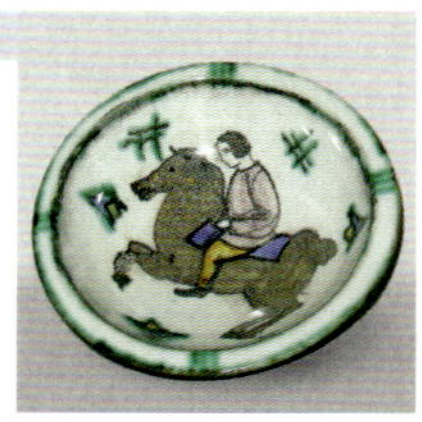
Geiger, Benno
SCHALE Reiter
Form: rund, tief gemuldete Form, zylindrischer Stand
Dekor: Reiter auf springendem Pferd
ca. 1928
5767,102 gem. B., Firmenmarke 19, Made in Austria, unsign., H. 5,2 cm, Ø 17,3 cm; PB Wien

5769
Waldmüller
BUCHSTÜTZE Katze
ca. 1928
5769,39,9 Firmenmarke 19, Made in Austria, H. 14 cm, L. 20 cm; eBay # 6627527892
-> vgl. 5737= FIGUR

5773
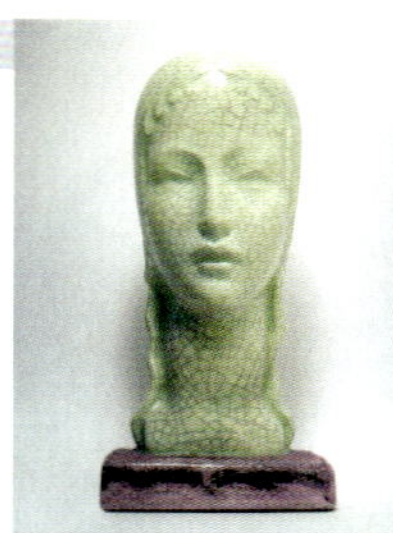
Motto, (Joseph C.)
STEHMASKE Frauenkopf mit langem Haar (auf Rechtecksockel)
ca. 1928
FAYENCE
5773,5,4 Firmenmarke 19, Made in Austria, unsign., H. 23,7 cm; Sammlung a@goldscheider.net
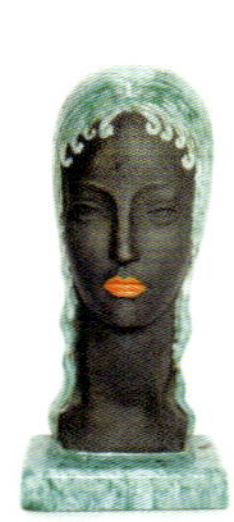
5773,5,14 ger., gem. GI, Firmenmarke 19, Made in Austria, unsign.?, H. 23,5 cm; eBay # 437668034; Herr 51
5773,8,5 Firmenmarke 19, Made in Austria, ger., H. 25 cm; eBay # 3161903721
5773,20 ohne Firmenmarke, unsign., H. 23,7 cm; Sammlung a@goldscheider.net
5773,22,8 ger. 4, XK, Firmenmarke 19, Made in Austria, H. 23,5 cm; eBay # 1289963790

5773,41,14 ger. K., gem. SG., Firmenmarke 19, Made in Austria, H. 23,6 cm; Sammlung HR Dr. K. Berger/Wien
5773,65 (Firmenmarke ohne Ang.), H. 23,8 cm; Doro 8.3.2004
BRONZE
ohne Ang., H. 22,8 cm; Soth. L. 28.9.1984; Bem.: Da das AH keinerlei Ang. zu Firmenstempel, Nummern etc. macht, könnte es sich auch um eine Ausführung einer anderen Gießerei, bzw. Firma handeln.
Lit.: Pinhas, S. 164 (m. Abb.)

5774
Lorenzl, Josef
FIGUR / LAMPE „Maria Jeritza" (Kammersängerin Maria Jeritza als stehende Figur auf reliefiertem Rundsockel)
ca. 1928
FIGUR
ohne Ang., H. fehlt; MW, 10.1929/30, Nr. 6, S. 28 (m. Abb.)
LAMPE
5774 (Firmenmarke ohne Ang.), Gesamth. 79 cm; Pinhas, S. 125 (m. Abb.)
Bem.: Dargestellt ist die Wiener Kammersängerin Maria Jeritza, (1887–1982).
Lit.: N.N.: Berühmte Frauen und Keramik, in: MW, 10.1929/30, Nr. 6, S. 28 (m. Abb.), hier mit falscher Künstlerang. „Josef Althan", S. 29 „Die Manufaktur Goldscheider, die für ihre Arbeiten die bedeutendsten Wiener Keramiker beschäftigt – Wien besitzt durch diese Künstler auf keramischem Gebiet einen internationalen Ruf – hat nun für diese Figuren eine sehr gute Auswahl getroffen. Da ist zunächst die berühmte Jeritza, der vergötterte Liebling zweier Kontinente, sie, die in sehr kurzer Zeit groß geworden istund nun in Amerika Triumphe feiert, wie kaum eine Sängerin vor ihr, sie, die vielleicht als die bedeutendste Sängerin der Welt zu bezeichnen ist; Schaulade, 6.1930, S. 797, ...Figurenserie „Berühmte Frauen"..., die u.a. auch die Kammersängerin Maria Jeritza naturalistisch wiedergibt".

5775
Lorenzl, Josef
GRUPPE „Spanischer Tanz" (Tanzendes Paar in spanischen Kostümen) (auf Ovalsockel)
ca. 1928
5775,68,13 (Firmenmarke ohne Ang.), H. 43,2 cm; Christie's SK 13.10.2004
5775,77,11 gem. N2, Firmenmarke 19, Made in Austria, Hand dec., H. 44,5 cm; eBay # 1521001729
5775,168,7 gem. F.L.N., Firmenmarke 19, H. 45 cm; Kastern 68
Lit.: MW, 11.1929/30, Nr. 5, S. XI (m. Abb.), hier mit Ang. des Originaltitels „Spanischer Tanz" und des Entwerfers Josef Lorenzl; Neuwirth (1974), S. 67
–> vgl. 5888= weibliche Einzelfigur

5776
Dakon, Stephan
GRUPPE „Matrosentanz" (Maria Solveg und Katta Sterna im Matrosenkostüm, ZweierGRUPPE) (auf Ovalsockel)
ca. 1928
5776,12,4 Firmenmarke 19, Made in Austria, Hand dec., H. 39,8 cm; Doro 12.12.1991
5776,92,12 (Firmenmarke ohne Ang.), H. 38,5 cm; Christie's SK 15.2.2001
5776,165,17 gem. VI, K, Firmenmarke 19, Made in Austria, Hand dec., H. 42,7 cm; Soth. L. 30.11.1993
5776,229,11 Firmenmarke 19, Made in Austria, unsign.?, H. 40,6 cm; eBay # 2195811445
–> vgl. 5818= Einzelfigur
–> vgl. 5961
–> vgl. .... DreierGRUPPE; Soth. N.Y. 14.6.1991
Lit.: SM, 4.1928, Nr. 9 (Sept.), S. 985 (m. Abb.), hier eine zeitgenössische Fotografie, die wohl als Vorlage für die Ausführung diente. Dargestellt sind die Varietétänzerinnen Maria Solveg und Katta Sterna vom Matray-Ballett / Scala, Berlin bei ihrem „Matrosentanz". Fotografie: Badekow; MW, 10.1928/29, Nr. 26, S. V (m. Abb.), hier mit Ang. des Originaltitels „Matrosentanz" und des Entwerfers Stefan Dakon.
Quelle: Archiv Mimi Dakon (Wiener Manufaktur FG, Modellliste Dakon, 7.11.1946), hier mit Ang. des Originaltitels „Matrosentanz".

5780
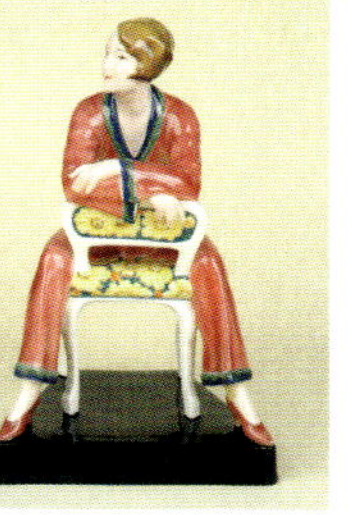
Lorenzl, Josef
„Pyjama" (Junge Dame, rittlings auf einem Stuhl sitzend) (auf Rechtecksockel)
ca. 1928
5780,216,15,3 gem. FN, Firmenmarke 19, unsign., H. 29 cm; PB Wien 19
Lit.: MW, 11.1929/30, Nr. 4, S. IV (m. Abb.), hier mit Ang. des Originaltitels „Pyjama" und des Entwerfers Josef Lorenzl; Neuwirth (1974), S. 67.

5781
Dakon, Stephan
„Dolly Sister(s)" (Stehend, zum Kuss nach vorne gebeugt, ihr kurzes Kostüme nach hinten weg haltend) (auf getrepptem Trapezsockel)
ca. 1928
5781,61,75,1 gem. Z, Firmenmarke 19, Made in Austria, Hand dec., H. 41,2 cm; Soth. com. # 1BVO
5781,162,12 Firmenmarke 19, Made in Austria, H. 41 cm; Zof. 36
5781,385,19 (Firmenmarke ohne Ang.), H. 44,2 cm; Soth. L. 16.2.1984
Bem.: –> vgl. bei 5569. Dieses Modell, in Physiognomie, Pose und Schnitt des Kostüms sehr nahe an der Gruppe 5712 stehenden Figur, deutet auf eine der beiden „Dolly Sisters" hin.
–> vgl. 5712= ähnl. Modell (GRUPPE)
Lit.: Pinhas, S. 141 (m. Abb.)

5783
Ingels, Domien
Elefant
ca. 1928
5783,5,12 Firmenmarke 19, Made in Austria, H. 32 cm, L. 52 cm; Wien Museum Inv. 220.767

5784
Meisinger, Ida
Liegender Yorkshire Terrier (Malteser?)
ca. 1928
5784,67,11 gem. E, Firmenmarke 19, Made in Austria, H. 17,5 cm, L. 30 cm; eBay # 2034227461; Zezschwitz 10
–> vgl. 5705= ähnl. Modell

5787
Meisinger, Ida
„Elefant"
ca. 1928
FIGUR
5787,1,10 Firmenmarke 19, sign. Meisinger, H. 13,8 cm; Hennig 6
BUCHSTÜTZE
5787,61,32 gem. lig. JP, Firmenmarke 19, unsign., H. 15,8 cm, L. 16,5 cm; Wien Museum Inv. 220.782
Lit.: MW, 11.1929/30, Nr. 23, S. 20 (m. Abb.), hier mit Ang. des Originaltitels „Elefant" und der Entwerferin Ida Meisinger.

5790
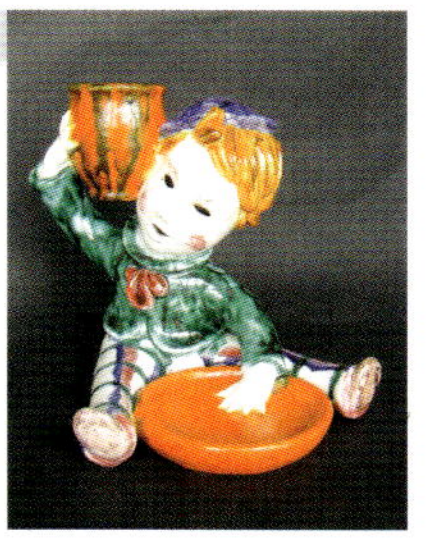
Weber, Ernst
„Aschenkindl" (Sitzender Junge mit Aschenschale und Streichholztopf)
ca. 1928
5790,23,15 Firmenmarke 19, Made in Austria, H. 17,6 cm; Doro 16.4.1996
5790,35,14 gem. L., Firmenmarke 19, Made in Austria, H. 17,4 cm; Doro 18.11.1997; PB
5790,38,? gem. L, Firmenmarke 19, Made in Austria, H. 19 cm; Zacke 3, S. 28; Zacke Sonderauktion 30.3.1996
5790,57,15 ger., Firmenmarke 19, Made in Austria, H. 18 cm, B. 20 cm; Sammlung mad
Bem.: Die Darstellung, die entfernt an Bosses Grotesken erinnert oder an sog. „Kakteenträger" der Wiener Werkstätte (–> vgl. ExpKWW, S, 86, 101), dient als Rauchergarnitur. Wenngleich auch das Dekor mit der teilweise aufgebrachten Laufglasur den o. g. Objekten nahe steht, so ist in diesem „Aschenkindl" dennoch ein eigener künstlerischer Ausdruck in der Behandlung des Themas, der Form und Bemalung zu erkennen.
Lit.: MW, 11.1929/30, Nr. 23, S. 20 (m. Abb.), hier mit Ang. des Originaltitels „Aschenkindl" und des Entwerfers Ernst Weber.

5794
Putto (Amor mit Bogen)
ca. 1928
5794,(26?),10 (Firmenmarke ohne Ang.), H. 25 cm; KH
Bem.: Leider können zu diesem Modell keine näheren Ang. gemacht werden, da es sich nur um eine mündliche Mitteilung des KH handelt.

5791
Waldmüller
FIGUR Strauß (auf Rechtecksockel)
ca. 1928
5791,41,11 gem. F., Firmenmarke 19, , H. 12,2 cm; Zezschwitz 24

5796
Rosé
Liegender Barsoi
ca. 1913/14 (für den Entwurf)
5796,24,14 Firmenmarke 19, Made in Austria, H. 20,3 cm, L. 52 cm; eBay # 114951527
5796,202,14 gem. R, Firmenmarke 19, Made in Austria, unsign.?, H. 22 cm, L. 55 cm; eBay # 868249903
–> vgl. 4535

5797
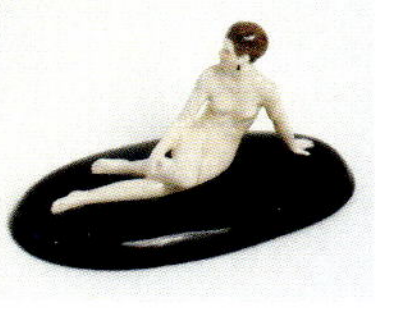
Lorenzl, Josef
Sitzender Damenakt (auf Ovalsockel)
ca. 1928/29
5797,93,4 gem. N, Firmenmarke 19, Made in Austria, H. 10,5 cm, B. 17 cm; city-antik

5802
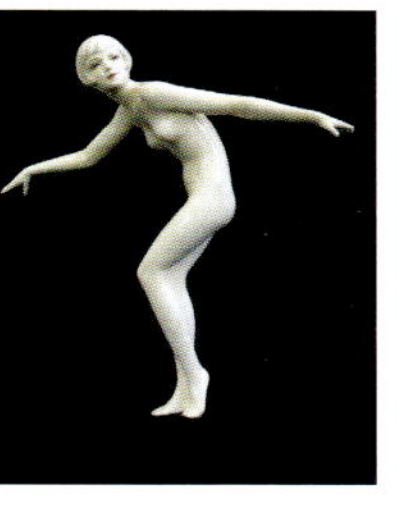
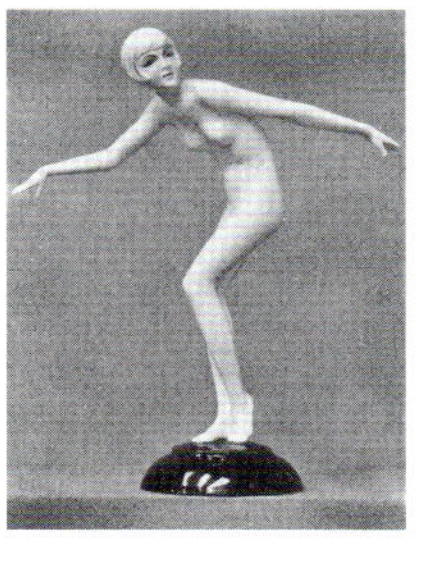
Lorenzl, Josef
FIGUR „Kokettchen" (Stehender Damenakt mit ausgestreckten Armen) (auf Rundsockel)
ca. 1928/29
5802,68,2 gem. N, Firmenmarke 19, Made in Austria, H. 28 cm; city-antik
5802,70,10,6 gem. N, Firmenmarke 19, Made in Austria, Pick. 6, Originalpick.: „J.K. Klammerth, Graz", H. 27,3 cm; PB
5802,464,? Firmenmarke 19, Made in Austria, H. 27,9 cm; Soth. Arcade 1432; 1449
5802,482,14,8 gem. N, gem. lig. ON?, Firmenmarke 19, Made in Austria, H. 28,5 cm; Sammlung Robert Senft
5802,489,4,8 Firmenmarke 19, Made in Austria, H. 28 cm; Soth. Olympia 3.6.2004
5802,520,11,8 gem. N 11, Firmenmarke 19, Made in Austria, H. 27 cm; eBay # 12142206828
5802,593,31 ger. XII,3, gem. X, Firmenmarke 19, Made in Austria, H. 28,6 cm; Soth. com. # 1BTO
5802,596,7 Firmenmarke 19, Made in Austria, H. 30,5 cm; eBay # 4964679201 (KH www.tias.com/stores/frozenbeauty)
5802F, H. 29 cm, B. 25 cm; GSNachf. 1,30a
–> vgl. 7227
–> vgl. Dakon 5907= fast identisches Modell.
Lit.: GSNachf. 1,30 (m. Abb.), hier mit Ang. des Originaltitels „Kokettchen".

5808F
Dakon, Stephan
FIGUR „Clown mit Ball"
ca. 1928/29
Quelle: Archiv Mimi Dakon (Tantiemenabrechnung 1.7.–31.12.1937; Wiener Manufaktur FG, Modellliste Dakon, 7.11.1946, hier mit Ang. des Originaltitels „Clown").

5811
Meisinger, Ida
BUCHSTÜTZE Sitzender Chinese mit Buch
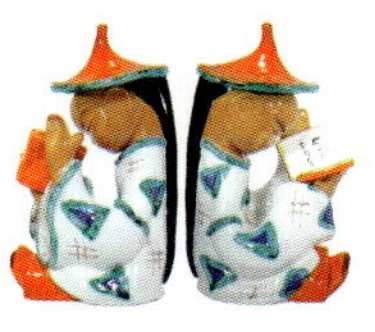
ca. 1928/29
5811,35,4 Firmenmarke 19, Made in Austria, Hand dec., H. 19,8 cm; PB
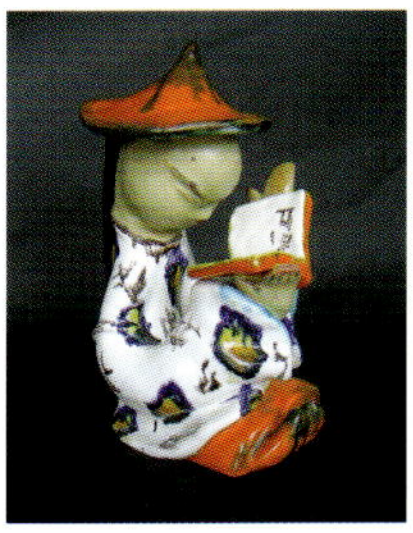
5811,132,11 gem. lig. JT, Firmenmarke 19, Made in Austria, unsign., H. 20,2 cm; Sammlung a@goldscheider.net
5811,138,19 gem. S.G, Firmenmarke 19, Made in Austria, unsign., H. 21 cm; eBay # 2589867400
5811,148,65 Firmenmarke 19, Made in Austria, unsign., H. 20,8 cm; Soth. L. 4.3.1988
5811,165,12 Firmenmarke 19, Made in Austria, unsign., H. 20,8 cm; Soth. L. 4.3.1988

5814
Postl, Hugo
Sitzender Pointer (nach links blickend)
ca. 1928/29
5814,147,10 gem. H., Firmenmarke 19, Made in Austria, unsign., H. 26,7 cm; eBay # 2077850923
–> vgl. 5861= ähnl. Modell

5815
Dakon, Stephan
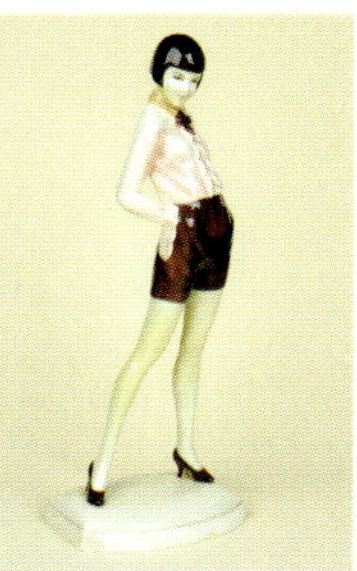
„Bubi" (Louise Brooks als stehende junge Dame in kurzen Hosen und Bubikopf) (auf abgerundetem Rechtecksockel)
ca. 1928/29
5815,22,11 gem. lig. SB, N1, Firmenmarke 19, Made in Austria, Hand dec., sign. St. Dakon, H. 36,9 cm; eBay # 77782709
5815,27,4 (Firmenmarke ohne Ang.), H. 37 cm; Christie's SK 13.5.1999
5815,201,4 ger. VI, K, gem. AKZ, gem. lig. DH, Firmenmarke 19, Made in Austria, Hand dec., H. 42 cm; PB Wien 19
5815,220,16 gem. VI, K, Firmenmarke 19, Made in Austria, Hand dec., H. 40 cm; eBay # 882439289; Christie's SK 15.11.2001
5815,371,12 ger. G, Ie, gem. lig. MR, 2L, Firmenmarke 19, Made in Austria, Hand dec., H. 39,1 cm; Sammlung.schmutz@chello.at 58
5815,364,11 (Firmenmarke ohne Ang.), H. 38 cm; Christie's SK 17.10.2001
Ausst.: SSH 1992
Lit.: SM, 4.1928, Nr. 11, Umschlag (Titelbild), hier eine zeitgenössische Fotografie der Tänzerin und Schauspielerin Louise Brooks. Sie war Revuetänzerin bei den „Ziegfeld Follies" und den „George White's Scandals" (–> vgl. Gabor, M.: Die Geschichte des Pin-up, München 1995, S. 48 (m. Abb.); MW, 11.1929/30, Nr. 6, S. VI (m. Abb.), hier mit Ang. des Originaltitels „Bubi" und des Entwerfers St. Dakon.
–> vgl. 5959
–> vgl. 6210= Frisurvariante
Quelle: Archiv Mimi Dakon (Tantiemenabrechnung 1.7.–31.12.1937; Modellliste Dakon, undatiert, ca. 1940; Wiener Manufaktur FG, Modellliste Dakon, 7.11.1946), hier jeweils mit Ang. des Originaltitels „Bubi".

5818
Dakon, Stephan
„Matrosentanz" (Maria Solveg im Matrosenkostüm) (auf Ovalsockel)
ca. 1928
FAYENCE
5818,47,11 Firmenmarke 19, Made in Austria, Hand dec., H. 36 cm; Kett. 123
ALABASTER
5818,W,1 ger., Plak. E, H. 35 cm, (Höhe ohne Sockel 29 cm); eBay # 3710104534; Christie's SK 27.4.2005
Bem.: Dieses Modell wurde in Fayence und Alabaster/Kunststein ausgeführt. Bei der Ausführung in Alabaster wurde am Sockel zusätzlich noch ein Segelboot angebracht, welches der Figur zusätzliche Stabilität verleiht.
–> vgl. 5776= ZweierGRUPPE
–> vgl. .... DreierGRUPPE; Soth. N.Y. 14.6.1991
Lit.: RDM 2.1927/28, S. 1187, (m. Abb.), hier eine zeitgenössische Fotografie der Varietétänzerin Maria Solveg vom Matray-Ballett/Scala, Berlin bei ihrem „Matrosentanz". Fotografie: Badekow, Titel „Der Leichtmatrose"; SM, 4.1928, Nr. 9 (Sept.), S. 985 (m. Abb.); MW, 10.1928/29, Nr. 26, S. V (m. Abb. der ZweierGRUPPE), hier mit Ang. des Originaltitels „Matrosentanz" und des Entwerfers Stefan Dakon.
Quelle: Archiv Mimi Dakon (Wiener Manufaktur FG, Modellliste Dakon, 7.11.1946), hier mit Ang. des Originaltitels „Matrosentanz".

5819
Dakon, Stephan
„Revue zu 4" (Die „Dolly Sisters" als „Vierer Gruppe" nebeneinander stehend, ihre kurzen Röcke hochhebend) (auf Rechtecksockel)
ca. 1926/27 (für den Entwurf)
ohne Ang., Firmenmarke 19, H. 38,1 cm; Soth. N.Y. 1.12.1990
Bem.: Diese Gruppe mit vier Figuren ist natürlich nur eine Schöpfung der Firma GS, da bekanntermaßen die Dolly Sisters nur zwei Tänzerinnen waren.
–> vgl. 5612 ZweierGRUPPE
–> vgl. 5634; 5649= jeweils Einzelfigur
Quelle: Archiv Mimi Dakon (Wiener Manufaktur FG, Modellliste Dakon, 7.11.1946), hier mit Ang. des Originaltitels „Revue zu 4".

5820
Weber, Ernst
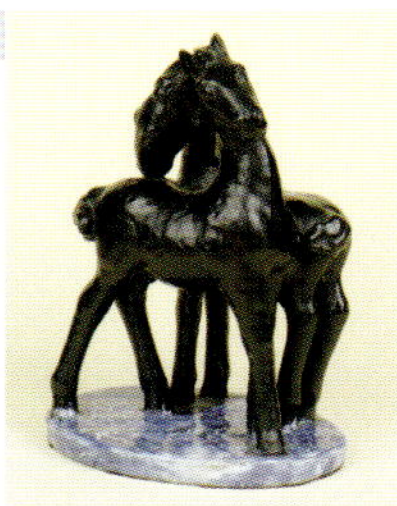
GRUPPE Zwei Pferde (auf Ovalsockel)
ca. 1928/29
5820,36,10 Firmenmarke 19, Made in Austria, unsign., H. 21 cm; PB Wien 19
5820,43,10 gem. L, Firmenmarke 19, Made in Austria, H. 20,4 cm; eBay # 211156133
5820,107,65 ger., Firmenmarke 19, Made in Austria, H. 21 cm; eBay # 3774330537

5821
Pferd (auf Ovalsockel?)
ca. 1928/29
5821,46,7 (Firmenmarke ohne Ang.), H. 20,7 cm; Doro 4.6.1996
Bem.: Möglicherweise handelt sich hier um eines der beiden Pferde der Gruppe (Modell-Nr. 5820). Da sich im Auktionskat. weder eine Abb. findet, noch eine Sign. angeben ist, wurde von einer möglichen Zuschreibung an Ernst Weber abgesehen.

5822
Lorenzl, Josef
„Spanischer Shawl" (Stehende Tänzerin (Dolores del Rio?) in Pose, mit großem Schal) (auf Rundsockel)
ca. 1928/29
5822,3,8 gem. 04, B, E1, Firmenmarke 19, Made in Austria, H. fehlt; eBay # 2226755255
5822,37,8,P ger., gem. E1., Firmenmarke 19, Made in Austria, Hand dec., H. 42 cm; eBay # 223440491
5822,75,75 gem. lig. SG, S.O.T., Firmenmarke 19, Made in Austria, Hand dec., H. 44 cm; eBay # 2003079420
5822,208,6 gem. N3, Firmenmarke 19, Made in Austria, Hand dec., sign. Lorenzl, H. 46,4 cm; Sammlung.schmutz@chello.at 104
5822,337,15 (Firmenmarke ohne Ang.), H. 46 cm; Doro 26.5.2004
5822,568,11,16,P gem. MR, Firmenmarke 19, Made in Austria, Hand dec., H. 45 cm; Soth. L. 31.10.1997
5822, 587,1,1 (Firmenmarke ohne Ang.), H. 45 cm; PB
Bem.: Möglicherweise handelt es sich bei der dargestellten Person um die mexikan. Filmschauspielerin Dolores del Rio (3.8.1905–11.4.1983). Dies konnte jedoch bis Redaktionsschluss nicht eindeutig verifiziert werden. Die verhältnismäßig hohe Stückzahl, sowie zwei weitere ausgeführte Größen belegen, dass es sich hier um eines der beliebteren Modelle der Firma handelte.
–> vgl. 6062; 6345
Lit.: MW, 10.1928/29, Nr. 25, S. IV (m. Abb.), hier mit Ang. des Originaltitels „Spanischer Shawl" und des Entwerfers Josef Lorenzl.

5824
Meisinger, Ida
Sitzende Perserkatze
ca. 1928/29
5824,?,? (Firmenmarke ohne Ang.), H. 25,4 cm; eBay # 15538290

5828
Stehendes kleines Mädchen mit drei Geschenkpäckchen (auf Rundsockel)
ca. 1928/29
5828,157,35 gem. lig. ST(?), Firmenmarke 19, Made in Austria, H. 18,4 cm; eBay # 1218356431

5836
Goebel, Kurt (zugeschr.)
STEHKOPF Frauenkopf mit geschlossenen Augen, Locken (Curls) und Kragenansatz)
ca. 1928/29
5836 (Firmenmarke), unsign., H. 24,5 cm; Pinhas, S. 164 (m. Abb.)

5837
Meisinger, Ida
GRUPPE „Scotchterrier" (Zwei sitzende Scotchterrier)
ca. 1928/29
5837,177,14 gem. S.C, O, Firmenmarke 19, Made in Austria, H. 22 cm; eBay # 3774320172
–> vgl. 5846
–> vgl. ohne Nr.
Lit.: MW, 11.1929/30, Nr. 1, S. IV (m. Abb.), hier mit Ang. des Originaltitels „Scotchterrier" und der Künstlerin „Ida Meisinger".

5838
Damenakt auf Delphin
ca. 1928/29
5838,115,19 gem. XX, 30, Firmenmarke Schuster 1, Ausformung zw. 1941–1944, H. 18,6 cm; PB Wien 19

5839
Lorenzl, Josef
Stehendes Mädchen in Hosenanzug (auf Rechtecksockel)
ALABASTER
ca. 1928/29
5839,W,6 Plak. E, H. 29,4 cm; Doro 24.11.2004; 1.12.2005

5840
Sailer, Karl
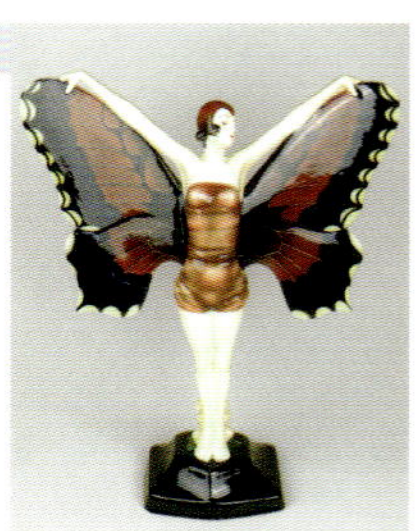
FIGUR/LAMPE Stehende Tänzerin im Schmetterlingskostüm (auf Rundsockel)
ca. 1928/29
FIGUR
5840,112,4,20 gem. N, gem. lig. AS 22, Firmenmarke 19, Made in Austria, Hand dec., H. 48,7 cm; Herr 49
5840 ohne Ang., Firmenmarke 19, Made in Austria, H. 48 cm; PB Wien 19
LAMPE
5840,89,16 ger. SC, gem. N22, Firmenmarke 19, Made in Austria, Hand dec., H. 47,2 cm; Soth. com. # 6LKL
–> vgl. 6366

5846
Meisinger, Ida
GRUPPE „Scotchterrier" (Zwei sitzende Scotchterrier)
ca. 1928/29
5846,250,20 gem. S.C, Firmenmarke 19, Made in Austria, H. 13 cm, B. 14 cm; eBay # 65013855935; # 2278897210
–> vgl. 5837
–> vgl. ohne Nr.
Lit.: MW, 11.1929/30, Nr. 1, S. IV (m. Abb.), hier mit Ang. des Originaltitels „Scotchterrier" und der Künstlerin „Ida Meisinger".

5850
Jaray, Sándor (Alexander)
Josef Kainz als Hamlet (Denkmal)
ca. 1928/29
5850,3,11 Firmenmarke 19, Made in Austria, Hand dec., H. 34,8 cm; Doro 9.11.1993
5850,11,15 ger. TK, gem. IN, J., Firmenmarke 19, Made in Austria, Hand dec., H. 37,5 cm; Wien Museum Inv. 190.064
Bem.: Dargestellt ist der österreich. Schauspieler Josef Kainz (*2.1.1858 †20.9.1910) nach dem Wiener Josef-Kainz-Denkmal (–> vgl. Große Berliner Kunstausst., 1911). Dieses Modell wurde bereits 1912 von der Berliner Firma Rosenthal & Maeder ausgeführt und auf der Leipziger Messe gezeigt (–> vgl. Die Leipziger Messe, Nr. 9, Ostervormesse 1912, S. 7 (m. Abb.).
Lit.: Das Zeitalter Kaiser Franz Josephs, Bd. 2, S. 242 (für die Bronzefassung, Kat. Nr. 18/27), Wien 1987; zu J. Kainz: www.aeiou.at (Österreich Lexikon).

5854
Meisinger, Ida
„Der Modehund"
(Sitzender Scotch-Terrier)
ca. 1928/29

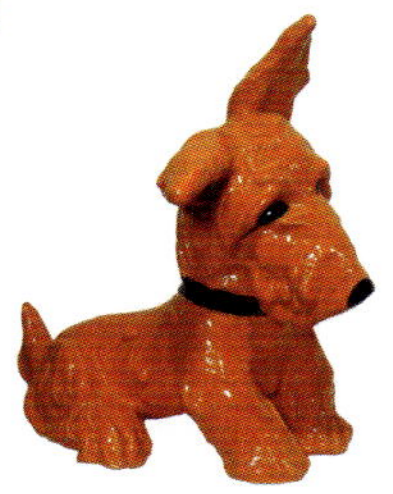

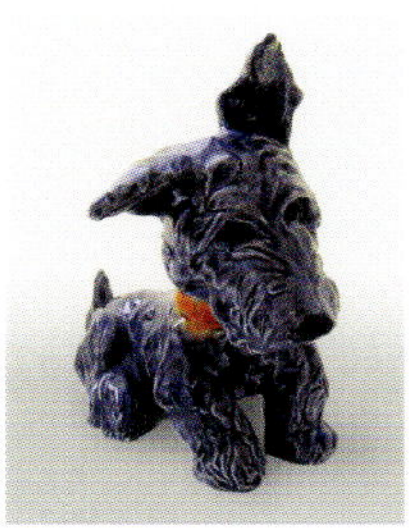

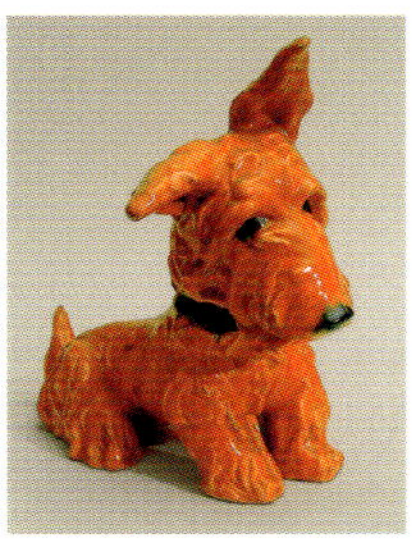

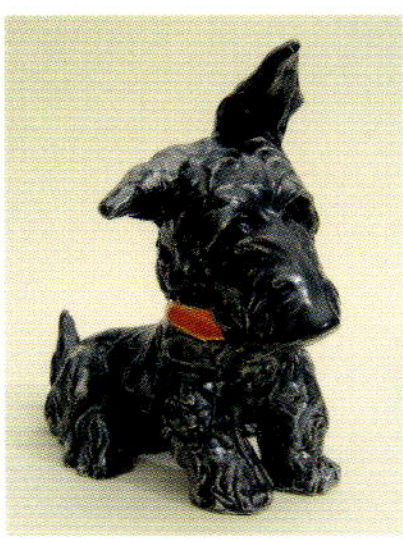

5854,540,65 Firmenmarke 19, H. 22 cm; Kett. 184; eBay # 2529999609
5854,560,19 Firmenmarke 19, Made in Austria, H. 21,3 cm; PB Deutschland
5854,623,20 gem. LM, Firmenmarke 19, Made in Austria, H. 21,3 cm; PB Deutschland
5854,690,17 gem. F5, Firmenmarke 19, Made in Austria, H. 21,5 cm; PB Deutschland
5854,706,12,K4 Firmenmarke 19, Made in Austria, H. 22 cm; Henry's 28.10.1998
5854,(867 ger.),4,1P, gem. WG, Firmenmarke 19, Made in Austria, H. 22 cm; Sammlung f@goldscheider.de
5854,891,7 ger. K, Firmenmarke 19, Made in Austria, H. 21 cm; eBay # 46289257
5854,893,12 ger. K, gem. SG, Firmenmarke 19, Made in Austria, H. 21,5 cm; PB Wien
5854,943,15,4 ger. P, Firmenmarke 19, Made in Austria, H. 23 cm; KNS 11
5854,1053,4 Firmenmarke 19, Made in Austria, H. 21 cm; Henry's 23.11.1994
5854,2016,– Firmenmarke 19, H. 21 cm; PB Wien 19
5854,2027,10 gem. SG, Firmenmarke 19, Made in Austria, H. 21,5 cm; KH
5854,2429,20 Firmenmarke 19, Made in Austria, H. 22 cm; BM 105
5854,2498,20 (Firmenmarke ohne Ang.), H. fehlt; eBay # 2219770550
5854,2684,20 gem. SO, Firmenmarke 19, Made in Austria, H. 22 cm; Henry's 25.11.1998; Doro 21.9.1999
5854,2751,21 Firmenmarke 19, H. 21,7 cm; Hennig 43
5854,2780,20 Firmenmarke 19, Made in Austria, H. 21,5 cm, L. 23 cm; GK, S. 21, 112 (m. Abb.); Wien Museum Inv. 182.093
5854,2810,20 Firmenmarke 19, Made in Austria, H. 22,1 cm; PB Wien
5854,3016,10 gem. lig. JP, Firmenmarke 19, Made in Austria, H. 21,1 cm; PB Wien 19
5854,3148,20 gem. lig. JP, Firmenmarke 19, Made in Austria, unsign., H. 22,3 cm; Sammlung a@goldscheider.net
5854,8780,20 Firmenmarke 19, H. ca. 22 cm; WKH
–> vgl. 5879; 6026; ohne Nr.
–> vgl. .... (H. 39 cm); Doro 9.11.1989
–> vgl. 6043= ASCHER
–> vgl. 6309= KERZENHALTER
Bem.: Dieses Modell gehört eindeutig zu den beliebtesten der ganzen Produktionspalette bei Goldscheider. Es wurde nicht nur in hohen Stückzahlen (besonders in der mittleren Ausführung, Modell-Nr. 5879) hergestellt, sondern auch in fünf verschiedenen Größen (H. 5,4–6,2 cm, 12 cm, 16,5 cm, 22 cm, 39 cm). Eine Auswertung der Glasurvarianten ergab, dass die Farben Orange und Weiß am häufigsten produziert wurden, aber auch Schwarz, Braun und sogar Türkis (sehr selten) im Programm waren. Der Scotch-Terrier im Besonderen, aber auch andere Terrierrassen (z.B. Airedale-, Skye- oder Cairn-Terrier) waren am Ende der 20er Jahre vorwiegend in der Damenwelt in Mode, was zahlreiche Abb. in den illustrierten zeitgenössischen Magazinen belegen. Somit ist der Originaltitel „Der Modehund", den Goldscheider verwendete, nicht verwunderlich, sondern sehr schlüssig.
Ausst.: HMW (jetzt Wien Museum) 1985/86.
Lit.: Schilling-Schletter, A.: Hunde-Plastiken, in: MW, 11.1929/30, Nr. 6, S. 12 (m. Abb.); MW, 11.1929/30, Nr. 19, S. 20 (m. Abb.), hier mit Ang. des Originaltitels „Der Modehund" und des Entwerferin Ida Meisinger; GK, S. 21, 112 (m. Abb.).

5856
Meisinger, Ida
BUCHSTÜTZE Pferd (kniend)
ca. 1928/29

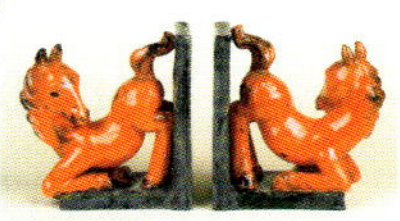

5856,63,9 ger., 1B. ger., Firmenmarke 19, Made in Austria, unsign., H. 19,5 cm; PB Wien 19
5856,64,9 ger., Firmenmarke 19, Made in Austria, unsign., H. 19,5 cm; PB Wien 19
5856,67,10 ger. 2.G., gem. L., Firmenmarke 19, Made in Austria, H. 20 cm, B. 16 cm; PB

5858
Postl, Hugo
Pekinese
ca. 1928/29
5858,667,19 gem. F.1, Firmenmarke 19, Made in Austria, H. fehlt, L. 21 cm; eBay # 3355551523
–> vgl. 6136; 7320
Lit.: MW, 11.1929/30, Nr. 3, S. IX (m. Abb.), hier mit Ang. des Originaltitels „Pekinese" und des Entwerfers Hugo Postl; Schilling-Schletter, A.: Hunde-Plastiken, in: MW, 11.1929/30, Nr. 6, S. 12–13 (m. Abb.), hier bez. „Pekingeser Palasthündchen"; Neuwirth (1974), S. 71.

5861
Postl, Hugo
Sitzender Pointer (nach oben blickend)
ca. 1928/29
5861,32,12 gem. G, I., Firmenmarke 19, Made in Austria, H. 23,5 cm, B. 16 cm; eBay # 30996198; KH
–> vgl. 5814= ähnl. Modell

5863
Dakon, Stephan
„Matrosin"
ca. 1928/29
Quelle: Archiv Mimi Dakon (Wiener Manufaktur FG, Modellliste Dakon, 7.11.1946), hier mit Ang. des Originaltitels „Matrosin".

5864
Dakon, Stephan
„Maskentanz" (Rita Zabekow in einem Kostüm mit Kapuze, eine Maske haltend) (auf profiliertem Sockel mit Maske)
ca. 1928/29
5864,10,65 gem. K7 Firmenmarke 19, Made in Austria, Hand dec., sign. St. Dakon, H. 39,5 cm; Soth. L. 2.11.1999
Lit.: MW, 11.1929/30, Nr. 7, S. VIII (m. Abb.), hier mit Ang. des Originaltitels „Maskentanz" und des Künstlers Stefan Dakon; „...jeder Mensch ist ein Tänzer", Ausdruckstanz in Deutschland zw. 1900 und 1945, (1993), S. 129 (m. Abb.), das zeitgenössische Foto belegt, dass es sich bei der dargestellten Person um die Tänzerin Rita Zabekow handelt; Pinhas, S. 141 (m. Abb.).
Quelle: Archiv Mimi Dakon (Wiener Manufaktur FG, Modellliste Dakon, 7.11.1946), hier mit Ang. des Originaltitels „Maskentanz".

5865
Dakon, Stephan
„Jagdfee"
ca. 1928/29
Quelle: Archiv Mimi Dakon (Wiener Manufaktur FG, Modellliste Dakon, 7.11.1946), hier mit Ang. des Originaltitels „Jagdfee".

5867
Stehende, schlanke Madonna (auf Rundsockel)
ca. 1928/29
5867,4,12 gem. F, Firmenmarke 19, Made in Austria, Hand dec., H. 58 cm; Soth. com. # 4VKZ

5870
Lorenzl, Josef
„Fächerdame" (Stehende junge Dame in japanischem Kostüm, mit erhobener Hand und großem Fächer) (auf Rundsockel)
ca. 1929

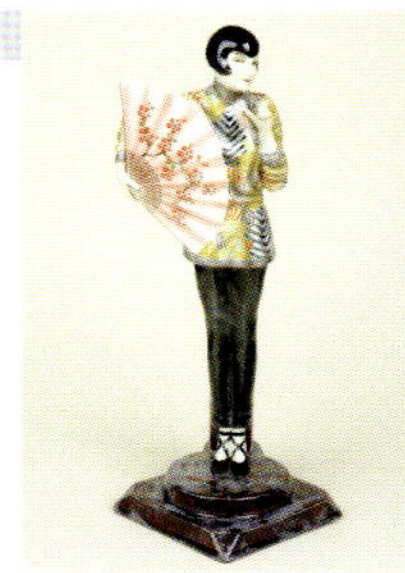

5870,65,11,4,K gem. 2.N, gem. lig. TO, Firmenmarke 19, Made in Austria, Hand dec., H. 33,9 cm; MAK Wien, Inv. Ke 10521-44
5870,93,12,4,K gem. lig. ML, Firmenmarke 19, Made in Austria, Hand dec., H. 33,7 cm; MAK Wien, Inv. Ke 10602-59
5870 ohne Ang., Firmenmarke 19, Made in Austria, H. 33,5 cm; PB Wien 19
Lit.: MW, 11.1929/30, Nr. 12, S. VI (m. Abb.), hier mit Ang. des Originaltitels „Fächerdame" und des Entwerfers Josef Lorenzl.

5871
Lorenzl, Josef
GRUPPE Stehende Tänzerin in fächerartig ausgestelltem Kleid, hinter ihr ein Mann mit Geige (auf Rundsockel)
ca. 1929
5871,40,7 ger. JISI, Firmenmarke 19, Made in Austria, Hand dec., H. 48,2 cm; BM 45; 54
5871,99,11,16,P gem. lig. MR 1, Firmenmarke 19, Made in Austria, Hand dec., H. 43,2cm; Soth. com. # 6LOK; 3479L
5871,103,6,12 gem. S1, Firmenmarke 21a, Made in Austria, H. 44 cm; Wien Museum Inv. 219.190
Bem.: Josef Lorenzl verwendete bei diesem Modell und bei Nr. 5875 identische Figuren. Ist hier ein elegantes Tänzerpaar mit einem Geigenspieler dargestellt, so wurde durch die Kostümierung der Frau (Spanierin), sowie der Änderung der Haartracht des Mannes und seines Instrumentes (Gitarre) ein „neues" Modell kreiert.
–> vgl. 5875= ähnl. Modell

5872
Lorenzl, Josef
Stehende Tänzerin, ihren langen Rock schmetterlingsflügelartig auseinander haltend (auf Rundsockel)
ca. 1929
5872,23,6 Firmenmarke 19, Made in Austria, Hand dec., H. 49 cm; Soth. L. 1.11.1996
–> vgl. 6173
–> vgl. 6399= fast identisches Modell

5875
Lorenzl, Josef
GRUPPE Stehende Spanierin in fächerartig ausgestelltem Kleid, hinter ihr ein Spanier mit Gitarre (auf Rundsockel)
ca. 1929
5875 ohne Ang., Firmenmarke 19, Made in Austria, H. 44 cm; Schloss Rick. 38
Bem.: Josef Lorenzl verwendete bei diesem Modell und bei Nr. 5871 identische Figuren. Ist hier ein spanisches Tänzerpaar mit einem Gitarrespieler dargestellt, so wurde durch die geänderte Kostümierung der Frau (elegante Tänzerin), sowie der Änderung der Haartracht des Mannes und seines Instrumentes (Geige) ein „neues" Modell kreiert.
-> vgl. 5871= ähnl. Modell

5876F
Dakon, Stephan
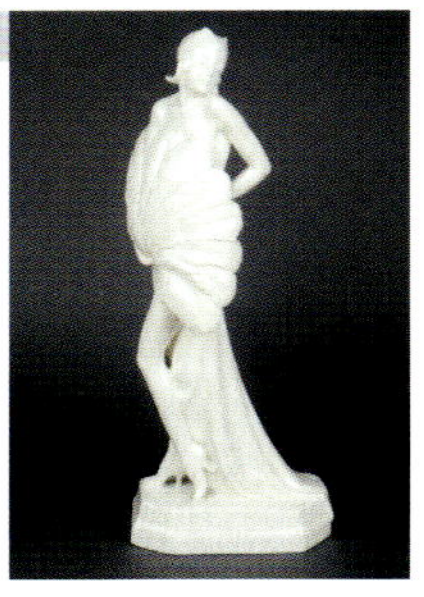
FIGUR „Fächer(dame)" (Stehende Varietétänzerin mit großem Feder-Fächer) (auf an den Ecken abgeschrägtem, gekehltem Rechtecksockel)
ca. 1928/29
5876,16,8 Firmenmarke 19, Made in Austria, Hand dec., H. 46 cm; PB Wien 19
5876,112,11 gem. NF, X11 S, Firmenmarke 19, Made in Austria, H. 45,7 cm; Sammlung Joan Ferguson / UK
-> vgl. 6214
Quelle: Archiv Mimi Dakon (Tantiemenabrechnung 1.7.–31.12.1937), hier mit Ang. des Originaltitels FIGUR „Fächer"; (Modellliste Dakon, undatiert, ca. 1940; Wiener Manufaktur FG, Modellliste Dakon, 7.11.1946), hier jeweils mit Ang. des Originaltitels „Fächerdame".

5878
BUCHSTÜTZE Springendes Reh
ca. 1928/29
5878,5,5,5 Firmenmarke 19, Made in Austria, Hand dec., H. 20 cm, B. 7,5 cm, L. 19 cm; Wien Museum Inv. 190.036,1
5878,17,57,B Firmenmarke 19, Made in Austria, Hand dec., H. 20 cm, B. 7,5 cm, L. 19 cm; Wien Museum Inv. 190.036,2

5879
Meisinger, Ida
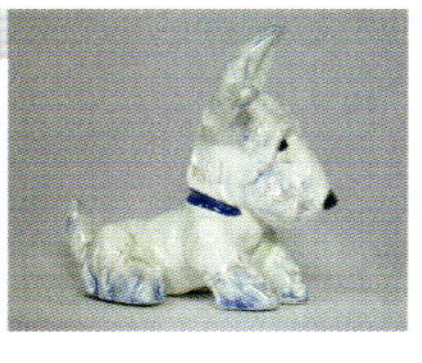
„Der Modehund" (Sitzender Scotch-Terrier)
ca. 1928/29
FIGUR
5879,1 ger., Firmenmarke 19, Made in Austria, unsign., H. 11,5 cm; Wien Museum Inv. 182.094
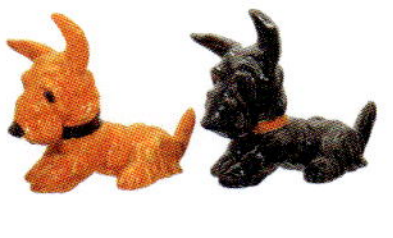
5879,14,19 ger. J, Firmenmarke 19, Made in Austria, H. 13cm; Sammlung a@goldscheider.net
5879,31,1 ger. M, gem. O, Firmenmarke 19, unsign., H. 13cm; Sammlung a@goldscheider.net
5879,66,4 Firmenmarke 19, Made in Austria, H. 11 cm; eBay # 3357129692
5879,168,1 gem. E, Firmenmarke 19, Made in Austria, H. 12 cm; PB Wien
5879,482,26 gem. SC, Firmenmarke 19, Made in Austria, H. 11,5 cm; Wien Museum Inv. 220.715
5879,538,20 Firmenmarke 19, Made in Austria, H. 11,5 cm; eBay # 94938320
5879,0633,21 gem. lig. JP, Firmenmarke 19, Made in Austria, H. 11,5 cm; Wien Museum Inv. 220.713
5879,643,20 Firmenmarke 19, Made in Austria, H. 11,5 cm; Fischer 63
5879,683,6 Firmenmarke 19, H. 12,5 cm; eBay # 3370316027
5879,1427,2 Firmenmarke 19, Made in Austria, H. 12 cm, L. 12 cm; PB Deutschland
5879,1840,49 Firmenmarke 19, Made in Austria, H. 11,5 cm, L. 11,2 cm; BM 55; 65

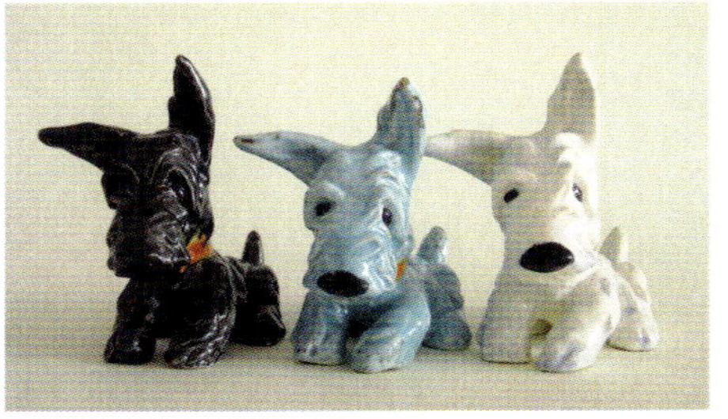

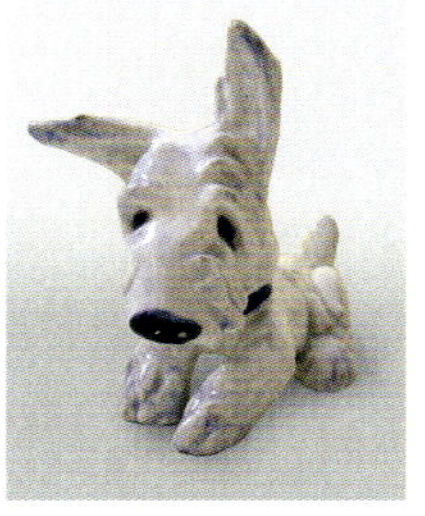

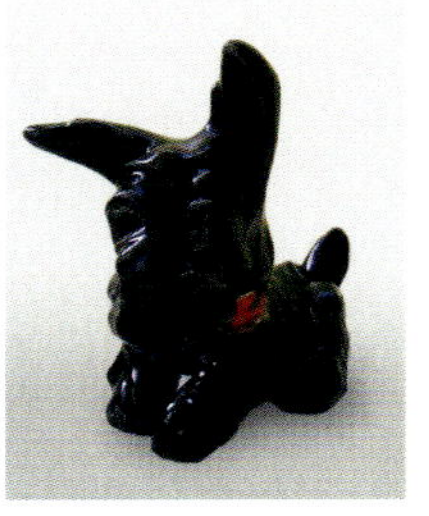

5879,2165,4,K5 Firmenmarke 19, Made in Austria, H. 14,6 cm(!); eBay # 2006592187
5879,2595,14,4 gem. G, Firmenmarke 19, Made in Austria, H. 11 cm; eBay # 1266166606
5879,2792,12 gem. S.G., Firmenmarke 19, Made in Austria, H. 15 cm(!); eBay # 421926091
5879,2920,11 Firmenmarke 19, Made in Austria, H. 11,8 cm; Wendl 43
5879,3003,6 (Firmenmarke ohne Ang.), H. 11,5 cm; Christie's SK 21.6.1996
5879,3008,65 Firmenmarke 19, Made in Austria, unsign., H. 11,9 cm; Wien Museum Inv. 220.712
5879,3181,20 gem. lig. JP, Firmenmarke 19, Made in Austria, H. 11,5 cm; eBay # 3717656782
5879,3209,6 Firmenmarke 19, Made in Austria, H. 12 cm; eBay # 3664444774
5879,3248,20 gem. lig. JP, Firmenmarke 19, Made in Austria, H. 11,5 cm; eBay # 2615444636
5879,3267,12 Firmenmarke 19, Made in Austria, H. 11,5 cm; Henry's 28.10.1998
5879,3351,24 Firmenmarke 19, Made in Austria, H. 11,4 cm; eBay # 1258949842
5879,3432,20 gem. SG, Firmenmarke 19, Made in Austria, H. 11,5 cm; PB Wien
5879,3665,1,81, gepr. M, gem. P, Firmenmarke 19, Made in Austria, H. 12 cm; Sammlung f@goldscheider.de
5879,4768,20 gem. lig. JP, Firmenmarke 19, Made in Austria, H. 11,7 cm; Sammlung mad
5879,5265,2 Firmenmarke 19, Made in Austria, H. 11,9 cm; PB Wien
5879,6768,32 gem. AR, Firmenmarke 19, Made in Austria, H. 12 cm; Wendl 31
5879,7150,24 Firmenmarke 19, H. 13 cm; PB Wien 19
5879,7171,24 Firmenmarke 19, H. 13 cm; PB Wien 19
5879,7594,21 gem. SG, Firmenmarke 19, (Made in) Austria, H. 11,2 cm; PB Deutschland
5879,7661,20 gem. SI, Firmenmarke 19, Made in Austria, H. 11,5 cm; eBay # 197443773
5879,7907,20 Firmenmarke 19, Made in Austria, H. 11,5 cm; eBay # 229559517
5879,9923,20 Firmenmarke 19, Made in Austria, H. 12 cm; eBay # 50720408
5879,10064,20 Firmenmarke 19, Made in Austria, H. 12 cm; Doro 14.6.1993 (ohne Auktionskat.)
5879,10669,29 gem. lig. JP, Firmenmarke 19, Made in Austria, H. 12 cm; eBay # 2065072571
5879,10791,? Firmenmarke 19, Made in Austria, H. 12 cm; KAHS 17.1.2000
5879,11186,20 ohne Firmenmarke, Made in Austria, H. 13 cm; Sammlung a@goldscheider.net
5879,11666(od. 11888?),20 Firmenmarke 19, Made in Austria, H. 12 cm; eBay # 1475644641
5879,11744,20 gem. SG, Firmenmarke 19, H. 13cm; Sammlung a@goldscheider.net
5879,11871,31, gem. SG, Firmenmarke 19, Made in Austria, H. 12,5 cm; Sammlung f@goldscheider.de
05879,12107,6 gem. lig. JP, Firmenmarke 19, Made in Austria, H. 12,2 cm; PB Deutschland
5879,12771,14 gem. SG, Firmenmarke 19, Made in Austria, H. 12,5 cm; PB Wien
5879,12869,14 gem. R, Firmenmarke 19, H. 12,5 cm; PB Wien 19
5879,13205,20 Firmenmarke 19, Made in Austria, H. 12 cm; KH
5879,13288,2 Firmenmarke 19, Made in Austria, H. 12 cm; KH
5879,13665,31 Firmenmarke 19, Made in Austria, H. 13,5 cm; eBay # 915782916
5879,13729,31 Firmenmarke 19, Made in Austria, H. 12,7 cm; eBay # 1434566214
5879,13966,31 ger. M, gem. lig. JP, Firmenmarke 19, Made in Austria, H. 11,7 cm; Sammlung mad
5879,14126,20 Firmenmarke 19, H. 12 cm; WKH
5879,14258,31 gepr. M, Firmenmarke 19, Made in Austria, unsign., H. 13 cm; KH
BUCHSTÜTZE
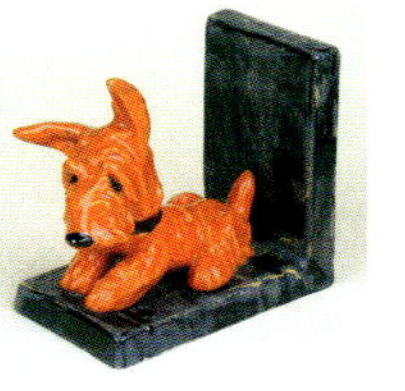
5879,2799,12 gem. G, Firmenmarke 19, Made in Austria, H. 15,2 cm; eBay # 25285641
5879,6469,32 gem. lig. JP, Firmenmarke 19, Made in Austria, H. 13,5 cm, B. 14,5 cm; PB Wien
5879,6482,24 gem. lig. JP, Firmenmarke 19, Made in Austria, H. 13,5 cm, B. 15 cm; PB Wien
5879,7150,24 gem. R, Firmenmarke 19, Made in Austria, H. 13,5 cm; PB Wien 19
5879,7171,24 gem. R, Firmenmarke 19, Made in Austria, H. 13,5 cm; PB Wien 19
-> vgl. 5854; 6026; ohne Nr.
-> vgl. .... (H. 39 cm); Doro 9.11.1989
-> vgl. 6043= ASCHER
-> vgl. 6309= KERZENHALTER
Bem. und Lit. vgl. bei 5854

5881
Dakon, Stephan
„Frühlingstanz"
ca. 1929
Quelle: Archiv Mimi Dakon (Wiener Manufaktur FG, Modellliste Dakon, 7.11.1946), hier mit Ang. des Originaltitels „Frühlingstanz".

5882
Lorenzl, Josef
Stehendes Kind mit Ball und Schubkarren
ALABASTER
ca. 1929
5882,Sch.3 ger., Plak. E, H. 35 cm; Wien Museum Inv. 186.683

5883F
Postl, Hugo
FIGUR „Wolfshund" (Liegender Schäferhund)
ca. 1925 (für den Entwurf)
5883,139,65 Firmenmarke 19, Made in Austria, H. ca. 25 cm, L. 60 cm; eBay # 1246725894
5883,671,6 Firmenmarke 19, unsign., H. 26 cm; Doro L 24.11.2000
5883,?,? gepr. P, (Firmenmarke ohne Ang.), unsign., H. fehlt, L. 61 cm; Spik 538
5883F, H. 25 cm, L. 61 cm; GSNachf. 5,13a
-> vgl. 5458; 5541; 5542
Lit.: GSNachf. 5,13 (m. Abb.), hier mit Ang. des Originaltitels „Wolfshund" (liegend); Schilling-Schletter, A.: Hunde-Plastiken, in: MW, 11.1929/30, Nr. 6, S. 12–13 (m. Abb.)
Quelle: AGM (Rechnung an Maurice Mizrahi, Alexandria, Ägypten, vom 22. April 1953), hier mit Ang. des Originaltitels „Wolfshund".

5888
Lorenzl, Josef
„Spanischer Tanz" (Spanierin in Tanzpose) (auf Ovalsockel)
ca. 1928
5888,43,7 gem. lig. AVD, MR1, Firmenmarke 19, Made in Austria, Hand dec., H. 43,5 cm; ; eBay # 3717051821
Lit.: MW, 11.1929/30, Nr. 5, S. XI (m. Abb.), hier mit Ang. des Originaltitels „Spanischer Tanz" und des Entwerfers Josef Lorenzl, für die GRUPPE.
-> vgl. 5775= GRUPPE

5891
Meisinger, Ida
BUCHSTÜTZE Elefant
ca. 1928

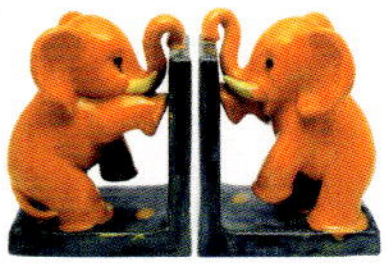

5891,2 ger. B Firmenmarke 19, Made in Austria, unsign., H. 16 cm, B. 13 cm; eBay # 2269692584
5891,10,92 Firmenmarke 19 (blau), Made in Austria (blau), Hand dec. (blau), unsign., H. 18 cm, B. 19 cm; Sammlung mad
5891,21,14 ger., Firmenmarke 19 (blau), Made in Austria (blau), Hand dec. (blau), unsign., H. 17 cm, B. 19 cm; Sammlung mad
5891,98,65 gem. L, Firmenmarke 19, Made in Austria, H. 15,5 cm; Sammlung a@goldscheider.net
5891 ohne Ang., H. fehlt; Könecke (2002), S. 76
Bem.: Diese Figur wurde auch noch zw. 1953–1960 bei Carstens (GWM West-Germany) produziert.
Lit.: Könecke (2002), S. 76, hier ohne weitere Ang.

5892
Meisinger, Ida
DECKELDOSE (Zigaretten-/Zigarrendose) Rauchende Dame

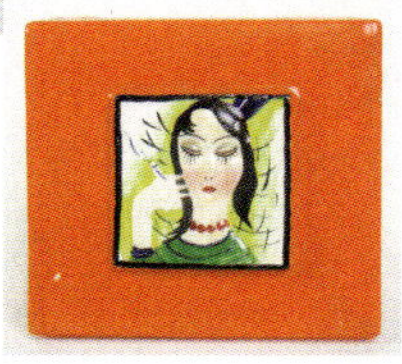

Form: viereckig (fast quadratisch)
Dekor: rauchende Dame, leicht reliefiert
ca. 1929
5892,312,1 gem. 1P, Firmenmarke 19, Made in Austria, unsign., H. 2,54 cm, B. 12,7 cm, L. 14 cm; eBay # 12161822494
5892,343,8 ger. XII, gem. SG, Deckel ger. 291,20 Firmenmarke 19, Made in Austria, H. 4,5 cm, B. 12,3 cm, L. 14 cm; Sammlung Weilinger, Wien
5892,440,8 ohne Marke, H. fehlt, B. 12,3 cm, L. 13,8 cm; Doro 9.6.1998
5892,479,6 Firmenmarke 19, unsign., H. 4,5 cm, B. 12,5 cm, L. 13,5 cm; Doro 25.10.1997
Bem.: Neben dem mittig (in einer Art Kartusche) gemalten Dekor der rauchenden Dame wurde als „Einrahmung" dem Maler quasi freie Hand gelassen. Die verschiedenen Ausführungen der nachgewiesenen Dosen belegen dies. Das Stück mit der Nr. 440 ist verziert mit Blattranken und geometrischen Linien, die beiden anderen wurden einfach in der Fondfarbe (in diesem Fall orange) belassen. Sicher gibt es weitere Farb- und Dekorvarianten.

5900
Lorenzl, Josef
Tänzerin im Schmetterlingskostüm (sich drehend) (auf Rundsockel)
ca. 1929

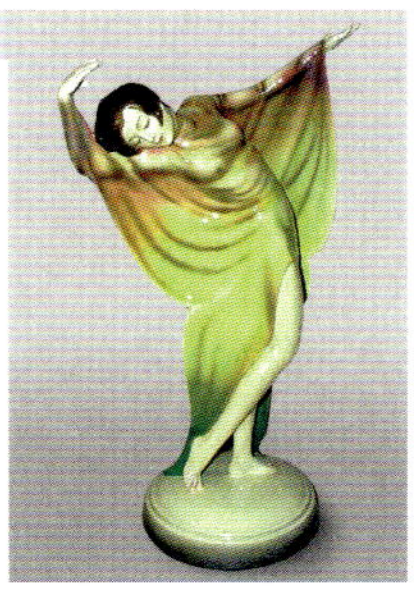

5900,30,8 ger. ISIP, gem. F., N2, Firmenmarke 19, Made in Austria, Hand dec., H. 48 cm; city-antik
5900,72,7 ger. VI+5, gem. W, NF2, Firmenmarke 19, Made in Austria, Hand dec., H. 46,6 cm; MAK Wien, Inv. Ke 10521-47
5900,73,15 Firmenmarke 19, H. 46,5 cm; Zeller 42
5900,87,10 ger. 01P, gem. F., N2, W, Firmenmarke 19, Made in Austria, Hand dec., H. 45,8 cm; Soth. com. # 6LNA; 346XX
5900,179,7 ger. 4K, gem. S5, gem. lig. AS, Firmenmarke 19, Made in Austria, Hand dec., H. 47 cm; eBay # 2531934933
5900,184,4 Firmenmarke 19, H. 44,2 cm; Doro 1.6.2005
5900,187,19 gem. F4, Firmenmarke 19, Made in Austria, Hand dec., H. 48,2 cm; Dreaweatt Neate 1.11.2001; eBay # 3652482355
Bem.: Ähnliches Kostüm wie bei Modell-Nr. 6394.
-> vgl. 6206

5902
Lorenzl, Josef
Posierende (tanzende) junge Dame in knielanger Hose, Rüschenbluse und Dreispitz (auf Ovalsockel)
ca. 1929
5902 (Firmenmarke ohne Ang.), H. 36 cm; Pinhas, S. 129 (m. Abb.)

5903
Dakon, Stephan
„Aschermittwoch" (Stehende Pierrette mit Gitarre) (auf getreppten rautenförmigem Sockel)
ca. 1929

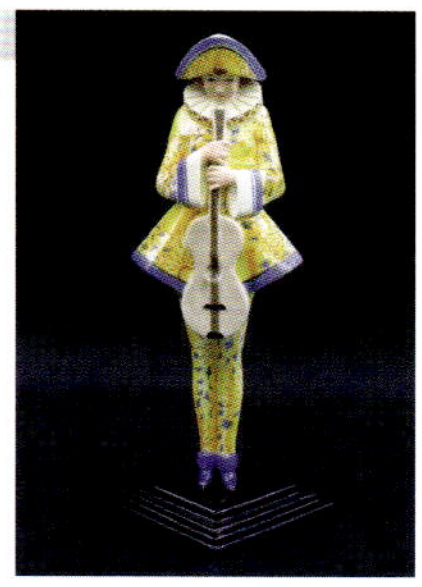

5903,78,19 ger. P, gem. S.G. D.2 gem. lig. MR, Firmenmarke 19, Made in Austria, Hand dec., H. 34 cm; Herr 42; Lit.: WK, 2000, S. 508–509.
5903,105,15,16,P rot gem. S., Firmenmarke 19, Made in Austria, Hand dec., unsign., H. 34,5 cm; Sammlung Robert Senft
Quelle: Archiv Mimi Dakon (Wiener Manufaktur FG, Modellliste Dakon, 7.11.1946), hier mit Ang. des Originaltitels „Aschermittwoch".

5907F
Dakon, Stephan
FIGUR „Akt" („Kugelspiel") (Stehender Damenakt mit ausgestreckten Armen) (auf Rundsockel)
ca. 1929
5907,(2,94?),5 gem. lig. JK II., Firmenmarke 19, Made in Austria, H. 27,9 cm; eBay # 1222584144
5907,60,40 gem. lig. HS?, Firmenmarke 19, Made in Germany, unsign., Ausformung zw. 1938–1941, Pick. 6 Ang., H. 30 cm; WKA 38; 44
-> vgl. Lorenzl 5802= fast identisches Modell.
Lit.: Pinhas, S. 143 (m. Abb.)
Quelle: Archiv Mimi Dakon (Tantiemenabrechnung 1.7.–31.12.1937), hier mit Ang. des Originaltitels FIGUR „Akt"; (Wiener Manufaktur FG, Modellliste Dakon, 7.11.1946), hier mit Ang. des (Original)-Titels „Kugelspiel".

5908
Beaume
(d.i. Wilhelm Bormann)
GRUPPE „Freundschaft" (Kind mit Hund= Neufundländer)

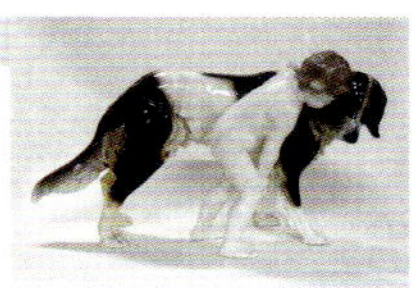

ca. 1910 (für den Entwurf)
5908,99,10,K,5 Firmenmarke 19, Made in Austria, unsign., Ausformung 20er Jahre, H. 28 cm, L. 48,6 cm; MAK Wien, Inv. Ke 10521-13
5908,139,6 (Firmenmarke ohne Ang.), unsign.?, Ausformung wohl 20er Jahre, H. 27 cm, L. 50 cm; Heickmann 8
5908 ohne Ang., (Firmenmarke ohne Ang.), sign. Beaume, Ausformung wohl 20er Jahre, H. 27 cm, L. 48 cm; Berlinghof 32
-> vgl. 3714
-> vgl. auch 4014= Hund einzeln
Lit.: MW, 11. 1929/30, Nr. 13, S. VI (m. Abb.), hier mit Ang. des Originaltitels „Freundschaft"; Schilling-Schletter, A.: Hunde-Plastiken, in: MW, 11.1929/30, Nr. 6, S. 12 (m. Abb.); Neuwirth (1974), S. 58, 60.

5911
Lorenzl, Josef
Sitzende spanische Tänzerin, ihren Kopf auf die linke Hand aufgestützt (Carmen?) (auf oktogonalem Sockel)
ca. 1929
5911,77,19 Firmenmarke 19, H. 35,5 cm; Christie's SK 17.10.2001; Soth. Olympia 3.5.2002
5911,131,11 (Firmenmarke ohne Ang.), H. 37 cm; Soth. L. 23.10.1987
5911,151,11 ger. 1P, gem. AJ, Firmenmarke 19, Made in Austria, Hand dec., H. 34,5 cm; Kett. 35
5911,165,10 Firmenmarke 19, Made in Austria, Hand dec., H. 37 cm; Sammlung Dr. Peter Hieke, Wien
5911,172,6 (Firmenmarke ohne Ang.), H. 35 cm; Kett. 196

5912
Lorenzl, Josef
Stehende Tänzerin in fächerartig ausgestelltem Kleid, Gitarre spielend (auf Rundsockel)
ca. 1929
5912,42,15 (Firmenmarke ohne Ang.), H. 37,6 cm; Doro 24.5.2000
5912,60,19 (Firmenmarke ohne Ang.), H. 37,5 cm; Christie's SK 9.4.1999

5914
Dakon, Stephan
„Schwestern" (Die Vigo-Drillinge, drei Varietétänzerinnen, als Chorusline hintereinander stehend) (auf abgerundetem Längssockel)
ca. 1929

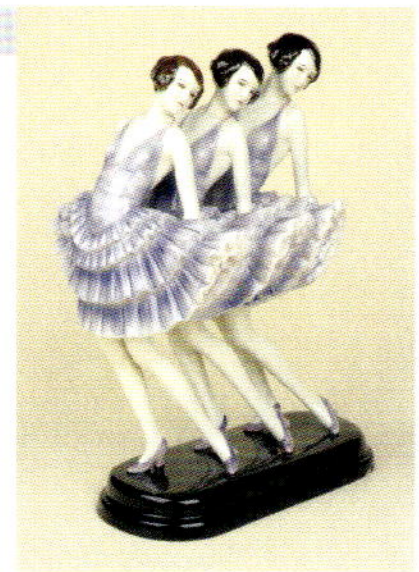

5914,43,11 (Firmenmarke ohne Ang.), H. 35,5 cm; Christie's SK 20.9.2000
5914,47,4 Firmenmarke 19, Made in Austria, Hand dec., H. 36 cm; PB Wien 19
5914,55,7 Firmenmarke 19, Made in Austria, H. 38,1 cm; eBay # 3160380811
5914,68,8 (gem.? FN), Firmenmarke 19, Made in Austria, H. 36,5 cm; Soth. L. 31.10.1997
Lit.: DM 5.1928/29, Nr. 57, S. 3487 (m. Abb.), hier eine zeitgenössische Fotografie, die wohl als Vorlage für die Ausführung diente. Dargestellt sind die Varietétänzerinnen und Drillingsschwestern „Vigo". Fotografie: d'Ora-Benda, Wien.
Quelle: Archiv Mimi Dakon (Wiener Manufaktur FG, Modellliste Dakon, 7.11.1946), hier mit Ang. des Originaltitels „Schwestern".

5915
Dakon, Stephan
„Sportgirl"
ca. 1929
Quelle: Archiv Mimi Dakon (Wiener Manufaktur FG, Modellliste Dakon, 7.11.1946), hier mit Ang. des Originaltitels „Sportgirl".

5916
Dakon, Stephan
„Tennisspielerin"
ca. 1929
Quelle: Archiv Mimi Dakon (Modellliste Dakon, undatiert, ca. 1940; Wiener Manufaktur FG, Modellliste Dakon, 7.11.1946), hier jeweils mit Ang. des Originaltitels „Tennisspielerin".

5917
Lorenzl, Josef
FIGUR/LAMPE Schreitende Tänzerin im Schmetterlingskostüm (ihren Kopf auf die rechte Schulter gelegt) (auf sechseckigem Sockel)
ca. 1929

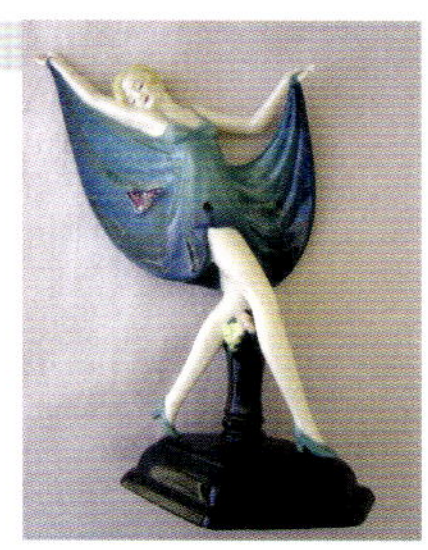

FIGUR
5917,155,8 gem. J.N 4, Firmenmarke 19, Made in Austria, Hand dec., H. ca. 26 cm; eBay # 2651888849
5917,359,4,19 gem. lig. AS, JR, Firmenmarke 19, Made in Austria, Hand dec., H. 26,7 cm; eBay # 1203974098 (KH www.tias.com/stores/frozenbeauty)
5917,413,19 (Firmenmarke ohne Ang.), unsign.?, H. 25,2 cm; Christie's SK 4./5.12.1996
5917,515,4 gem. FZ, Firmenmarke 19, Made in Austria, H. 25,5 cm; Sammlung Stressless, Portugal;
Lit.: Miller's A, Vol. 13.1992, S. 617 (m. Abb.)
5917 ohne Ang., gem. AJ, Firmenmarke 19, Made in Austria, Hand dec., H. 25 cm; eBay # 44811692
LAMPE
5917,329,4 Firmenmarke 19, Made in Austria, Hand dec., H. 28 cm; Sammlung Dr. Peter Hieke, Wien
Quelle: AGM (Tantiemenabrechnung Lorenzl 10. März 1953), hier unter dem Titel „Tänzerin" geführt.

5918
Dakon, Stephan
„Vogelgezwitscher"
ca. 1929
Quelle: Archiv Mimi Dakon (Wiener Manufaktur FG, Modellliste Dakon, 7.11.1946), hier mit Ang. des Originaltitels „Vogelgezwitscher".

5921
Lorenzl, Josef
Stehender Till Eulenspiegel mit Laute (auf oktogonalem, getrepptem Sockel)
ca. 1929
5921,29,11,K gem. AE, ND1, Firmenmarke 19, Made in Austria, Hand dec., H. 37 cm; eBay # 3718476408

5922
Dakon, Stephan
„Coolen more"
ca. 1929
Quelle: Archiv Mimi Dakon (Wiener Manufaktur FG, Modellliste Dakon, 7.11.1946), hier mit Ang. des Originaltitels „Coolen more".

5923
Lorenzl, Josef
Stehende kleine Spanierin (Betty Boop) (auf Rundsockel)
ca. 1929
5923,3,8 ger., Firmenmarke 19, H. 17,5 cm; Doro 15.3.2005
Bem.: Dieser Entwurf geht zurück auf die, nicht nur in den USA, bekannte Comicfigur Betty Boop von Max Fleischer (Fleischer Studios, New York). Die Adaption durch Lorenzl erfolgte bereits sehr früh, denn der Erfolg im Kino stellte sich Anfangs der 30er Jahre unmittelbar ein.

5927
Lorenzl, Josef

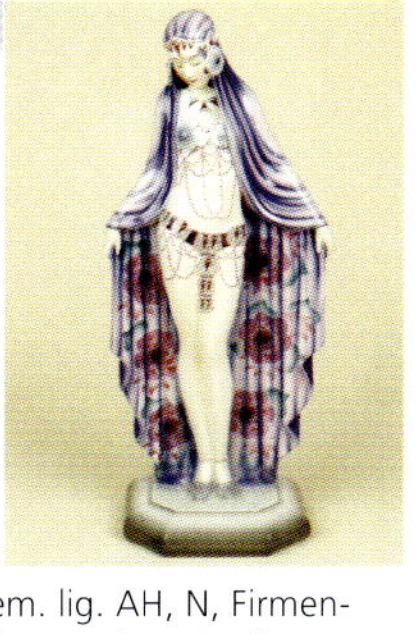

Stehende Haremstänzerin (mit gesenktem Kopf) (auf an den Ecken abgeschrägtem Quadratsockel)
ca. 1929
5927,9,11 ger. 04 F, gem. F.S.2, Firmenmarke 19, Made in Austria, Hand dec., H. 44,5 cm; eBay # 2622736782
5927,74,15,20 gem. S.C.3., gem. lig. AH, N, Firmenmarke 21a, H. 45,7 cm; Sammlung Robert Senft
(5927 Nrn. unleserlich) Firmenmarke 19, Made in Austria, Hand dec., unsign., H. 45 cm; PB Wien 19
–> vgl. 6245

5930
Dakon, Stephan

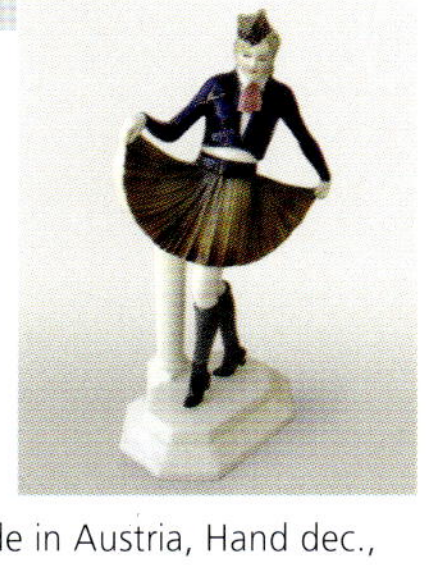

„Schottländerin" (Stehende junge Dame mit Hut (Schiffchen) und Kniestrümpfen, ihren (Schotten)-Plisseerock fächerartig auseinander haltend) (auf getrepptem, oktogonalem Sockel)
ca. 1929
5930,21,65 gem. lig. MR, AS, M., Firmenmarke 19, Made in Austria, Hand dec., H. 34 cm; Sammlung Stressless, Portugal; Bem.: Diese Figur war ehem. elektrifiziert und fand somit als Lampe Verwendung.
Quelle: Archiv Mimi Dakon (Wiener Manufaktur FG, Modellliste Dakon, 7.11.1946), hier mit Ang. des Originaltitels „Schottländerin".

5931
Lorenzl, Josef

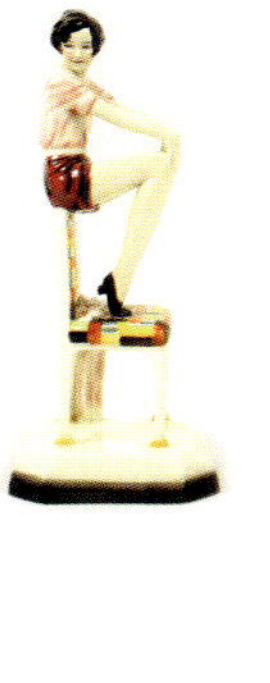

Auf Stuhl sitzende junge Dame in kurzen Hosen, ihre Hände keck auf den Knien abstützend (auf abgeschrägtem Quadratsockel)
ca. 1929
5931 (Firmenmarke ohne Ang.), H. 35 cm; Pinhas, S. 127 (m. Abb.)
5931,33,11 Firmenmarke 19, Made in Austria, Hand dec., sign. Lorenzl, H. 34 cm; Yves Siebers Auktionen, Stuttgart

5932
Dakon, Stephan
„Matrosengirl"
ca. 1929
Quelle: Archiv Mimi Dakon (Wiener Manufaktur FG, Modellliste Dakon, 7.11.1946), hier mit Ang. des Originaltitels „Matrosengirl".

5937
Meisinger, Ida (zugeschr.)
Stehender Terrier mit Halsband (große Öse)
ca. 1929
5937,715,31 gem. SG, Firmenmarke 19, Made in Austria, unsign., H. ca. 9 cm; KH

5938
Dakon, Stephan
„Autogirl" (Stehende junge Dame im enganliegendem Overall mit Kappe und Sportbrille) (auf getrepptem Quadratsockel)
ca. 1929
5938,78,4 (Firmenmarke ohne Ang.), H. 37,5 cm; Christie's SK 17.10.2001
Quelle: Archiv Mimi Dakon (Wiener Manufaktur FG, Modellliste Dakon, 7.11.1946), hier mit Ang. des Originaltitels „Autogirl".
Lit.: Pinhas, S. 116 (m. Abb.), hier mit der Ang., dass es sich um die bekannte Fliegerin Amelia Erhart (vielmehr Earhart) handeln solle. Das kann eindeutig durch die Quelle (Archiv Mimi Dakon) widerlegt werden, außerdem zeigt die Figur weder die Gesichtszüge Earharts, noch das typische Fliegeroutfit (mit etwas weiterer Hose und Cargotaschen). Wollte Dakon wirklich die Fliegerin nachbilden wollen, so hätte er und vor allem auch die Manufaktur sicher den Titel anders gewählt um das Stück besser vermarkten zu können.

5939
Lorenzl, Josef
GRUPPE Harlekin, seiner Columbine die Hand küssend (auf bogenartig facettiertem Rundsockel)
ca. 1929
5939,1,8 Firmenmarke 19, Made in Austria, Hand dec., H. 39 cm; WKA 33; Doro 10.3.2003
5939,8,4 (Firmenmarke ohne Ang.), unsign.?, H. 39 cm; Christie's SK 11.11.2004
5939,30,1 gem. lig. AN, N2, Firmenmarke 19, Made in Austria, Hand dec., H. 38,1 cm; eBay # 2523908314
5939,37,12 (Firmenmarke ohne Ang.), H. 35 cm; Soth. L. 23.10.1987; Bem.: Die etwas geringere Höhe resultiert auf einem kürzeren Sockel (ohne die bogenartige Facettierung).
5939,42,S1 Firmenmarke 19, Made in Austria, H. 39 cm; A/P/T 6.3.1989
5939,53,6 gem. F.I.S., Firmenmarke 19, H. 38,7 cm; Doro 1.12.2000
5939,78,19 (Firmenmarke ohne Ang.), H. 39 cm; Christie's SK 27.6.1997
5939,82,10 gem. F., Firmenmarke 19, Made in Austria, H. 39 cm; GK, S. 21, 115 (m. Abb.); Wien Museum Inv. 182.095
Ausst.: HMW (jetzt Wien Museum) 1985/86.
Lit.: DM, 3.1926/27, Nr. 30, S. 614 (m. Abb.), hier die fotografische Vorlage der weiblichen Figur aus dieser Karnevalsgruppe, „Frau Castiglione auf einem Wiener Kostümfest"; John, T.: Revue-Toiletten, in: RDM, 4.1929/30, Nr. 1, S. 82 (m. Abb.), „Venezianerin aus dem Karneval von Venedig", Fotografie: Willinger, Wien; GK, S. 21, 115 (m. Abb.).

5940
Lorenzl, Josef

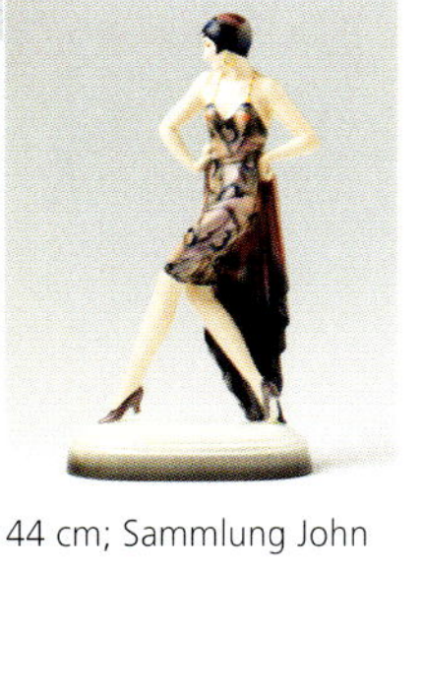

Schreitende Dame in kurzem Kleid mit Schleppe (auf Ovalsockel)
ca. 1929
5940,4,1,9 gem. lig. MR, F., Firmenmarke 19, Made in Austria, Hand dec., H. 44,1 cm; Doro 1.12.2000
5940,21,4 Firmenmarke 19, Made in Austria, Hand dec., H. 44 cm; Sammlung John Jesse, London

5946
Meisinger, Ida

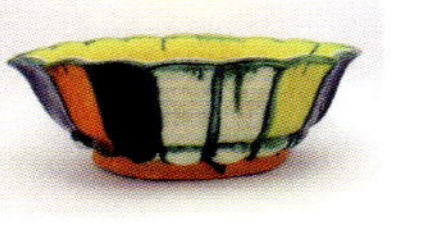

SCHALE
Form: rund, sechzehnpassig, vertikale Segmente
Dekor: mehrfarbig, Laufglasur
ca. 1929
5946,2 gem. B, Firmenmarke 19, Made in Austria, H. 10,5 cm, Ø 32,3 cm; Sammlung a@goldscheider.net

5949
Dakon, Stephan
„Donauwalzer"
ca. 1929
Quelle: Archiv Mimi Dakon (Wiener Manufaktur FG, Modellliste Dakon, 7.11.1946), hier mit Ang. des Originaltitels „Donauwalzer".

5950
Dakon, Stephan
„Tänzerin"
ca. 1929
Quelle: Archiv Mimi Dakon (Wiener Manufaktur FG, Modellliste Dakon, 7.11.1946), hier mit Ang. des Originaltitels „Tänzerin".

5953
Kremser, Lizzi

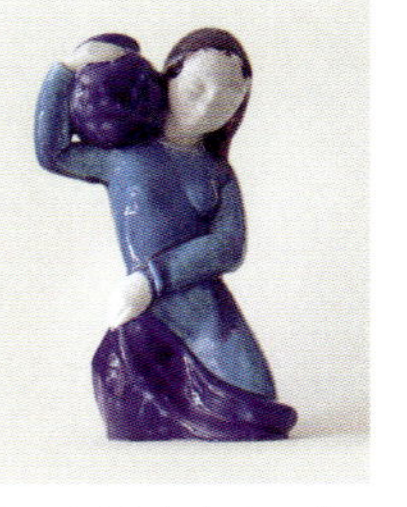

„Vasenfigur" (Kniendes Mädchen mit balusterförmigem Topf auf der rechten Schulter)
ca. 1929
5953,6,19,1 ger. K, gem. Z, Firmenmarke 19, Made in Austria, unsign.?, H. 29 cm; Sammlung Fritz Meyer / Hannover
5953,11,12 Firmenmarke 19 (zwei mal), Made in Austria, sign. Kremser, H. 29,2 cm; Sammlung John Nash
5953 ohne Ang., Firmenmarke 19, sign. Kremser, H. 29 cm; Ahlden 83
Lit.: MW, 11.1929/30, Nr. 25, S. III (m. Abb.), hier mit Ang. des Originaltitels „Vasenfigur" und der Entwerferin Lizzi Kremser; ÖK 3.1932, Nr. 25, S. 35 (m. Abb.); Neuwirth (1974), S. 65.

5954
RELIEF Madonna mit Kind
ca. 1929
5954,11,19 Firmenmarke 19, Made in Austria, H. 46,5 cm; BM 74

5957
Postl, Hugo (zugeschr.)
Stehender Fox-Terrier
BRONZE
ca. 1929
5957,10 ger., GS Vienna (Firmenmarke Bronze 4), unsign., H. 12 cm, L. 15,2 cm; eBay # 457139978
–> vgl. 5728= ähnl. Modell

5959
Dakon, Stephan
„Bubi" („Verkl.") (Louise Brooks als stehende junge Dame in kurzen Hosen und Bubikopf)
ca. 1928/29 (für den Entwurf)
Bem.: Wie der Zusatz „verkl.", also verkleinert zum Titel besagt, handelt es sich hier um die kleinere Version von Modell-Nr. 5815. Leider wird in der Quelle weder die Sockelform noch die Höhe aufgeführt.
–> vgl. 5815
–> vgl. 6210= Frisurvariante
Quelle: Archiv Mimi Dakon (Wiener Manufaktur FG, Modellliste Dakon, 7.11.1946), hier mit Ang. des Originaltitels „Bubi" („Verkl.").

5960
Lorenzl, Josef

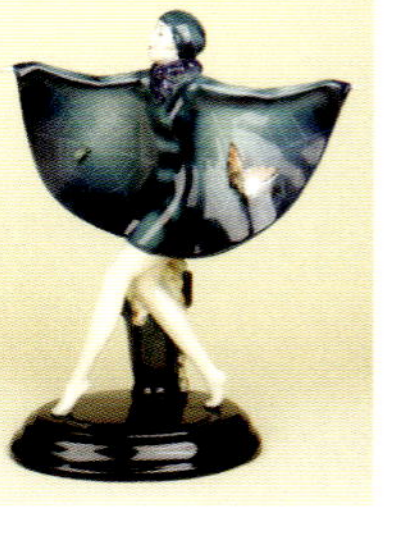

FIGUR / LAMPE „Gefangener Vogel" (Der Gefangene Vogel, Niddy Impekoven, nach links gewandt) (auf Ovalsockel)
ca. 1922 (für den Entwurf)
FIGUR
5960,41,4 gepr. K, gem. 1, Firmenmarke 19, Made in

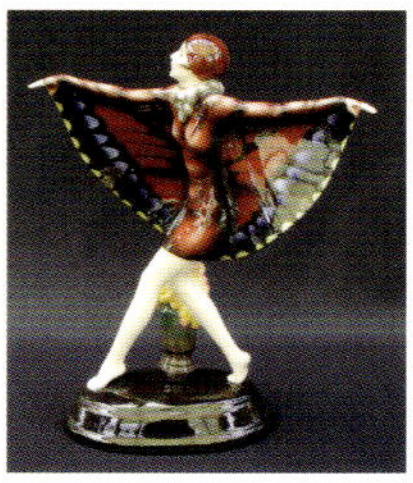

Austria, Hand dec., H. 38 cm; city-antik
5960,43,7 Firmenmarke 19, Made in Austria, Hand dec., unsign., H. 37,9 cm; Doro 19.9.2000
5960,47,10 ger. 1,K, Firmenmarke 19, Made in Austria, Hand dec., sign. Lorenzl, H. 38 cm; Doro G 28.5.2004; 3.12.2004
5960,61,10 (Firmenmarke ohne Ang.), sign. Lorenzl, H. 39 cm; Christie's SK 16.9.1999
5960,81,65,1,P gem. lig. MR4, Firmenmarke 19, Made in Austria, unsign.?, H. 39,6 cm; eBay # 1300732658
5960,155,7 gem. lig. MR, gem. F22, Firmenmarke 19, Made in Austria, Hand dec., unsign., H. 40,5 cm; Sammlung Joan Ferguson
5960,165,11, XII gem. lig. AS, gem. S., Firmenmarke 19, Made in Austria, Hand dec., H. 39 cm; Sammlung.schmutz@chello.at 35
5960,(191?),15 ger., gem. lig. MR, NO.7, Firmenmarke 19, Made in Austria, Pick. 13 (Gefangener Vogel, Bildhauer Josef Lorenzl), H. 39 cm; eBay # 1453634613
5960,193,4 gem. lig. MR, AS22, Firmenmarke 19, Made in Austria, Hand dec., H. 39,2 cm; Soth. L. 3.5.1991
5960,346,25 Firmenmarke 19, Made in Austria, Hand dec., H. 40 cm; PB Wien 19
5960,371,4 Firmenmarke 19, H. 40 cm; Doro 18.5.1999
5960,384,10 ger. 16P, gem. lig. AH, JP, N22, Firmenmarke 19, Made in Austria, sign. Laurenzl, H. 39,6 cm; Sammlung Weilinger, Wien
5960,457,11 Firmenmarke 19, unsign.?, H. 40 cm; Doro G 19.5.1992
5960,480,11,15 gem. lig. JP, AS 22N, Firmenmarke 19, Made in Austria, Hand dec., unsign., H. 40,4 cm; Sammlung Robert Senft
5960,490,19,20 gem. N, F.12, Firmenmarke 19, Made in Austria, Hand dec., H. 40,6 cm; Sammlung Joan Ferguson
5960,492,6 gem. N, F.12, Firmenmarke 19, H. 41 cm; Schloss Hag. 18; Soth. Olympia 3.5.2002
5960,508,19 Firmenmarke 19, Made in Austria, Hand dec., H. 41 cm; Soth. L. 16.2.1984
5960,512,4,19 Firmenmarke 19, H. 40 cm; Christie's SK 17.10.2001; Soth. Olympia 3.5.2002
5960,516,6,19 Firmenmarke 19, Made in Austria, Hand dec., unsign.?, H. 40,5 cm; Christie's L. 16.2.1994, Miller's A, Vol. 15.1995
5960,520,4,3,A, gem. 22N, gem. lig. OS, Firmenmarke 19, Made in Austria, Hand dec., H. 41 cm; Kett. 131
5960,585,10 Firmenmarke 19, H. 41,5 cm; Soth. L. 29.11.1984; Bem.: Mit Originalpick. „Butterfly by J. Lorenzl".
5960,645,6,5C gem. B, Firmenmarke 19, Made in Germany, Ausformung zw. 1938–1941, H. 43 cm; PB
5960,684,37,G gem. F., gem. lig. HK, Firmenmarke 19, Made in Germany, Ausformung zw. 1938–1941, H. 42 cm; city-antik
5960,685,1,10 gem. F, Firmenmarke 19, Made in Austria, unsign., H. 42,5 cm; Sammlung Alain Bernard
5960,686,37, ger. G, Firmenstempel WG 1, Ausformung 50er Jahre, unsign.?, H. 42 cm; Fischer 83
LAMPE
5960,237,11 (Firmenmarke ohne Ang.), unsign.?, H. 39 cm; Doro WA 12.12.1992
Ausst.: SSH 1992
-> vgl. 5054 (BRONZE); 5230; 6204; 6600; 8176
-> vgl. 5530 GRUPPE (Trio)
Pendant -> vgl. 5964
Pendant, bzw. ähnl. Modell -> vgl. 5434
Bem.: s. bei 5230 und bei 5964.
Lit.: s. bei 5230
GS neu: Modell-Nr. 82 „Gefangener Vogel" (H. 40 cm), Aufl. 3000 Stück

5961
Dakon, Stephan
„Matrosentanz"
ca. 1929
Bem.: Ob es sich hier um eine Größenvariante der Modell-Nr. 5776 oder um ein eigenständiges Modell handelt muss offen bleiben. Die Quelle macht hierzu keine näheren Angaben.
-> vgl. 5776
Quelle: Archiv Mimi Dakon (Wiener Manufaktur FG, Modellliste Dakon, 7.11.1946), hier mit Ang. des Originaltitels „Matrosentanz".

5962 -> ÜBERSCHNEIDUNG
Dakon, Stephan
„Küsschen"
ca. 1929
-> vgl. 6209
Quelle: Archiv Mimi Dakon (Modellliste Dakon, undatiert, ca. 1940; Wiener Manufaktur FG, Modellliste Dakon, 7.11.1946), hier jeweils mit Ang. des Originaltitels „Küsschen".

5962 -> ÜBERSCHNEIDUNG
KERZENHALTER (einflammig) Reiter auf springendem Pferd (auf Ovalsockel)
ca. 1929
5962,28,15 Firmenmarke 19, H. 27,5 cm; Herr 45
Bem.: Offenbar liegt hier ein Druck- bzw. Lesefehler des Auktionshauses vor, denn die Nachweise aus den Originalunterlagen bezeichnen mit der Modell-Nr. 5962 ein Modell von Dakon.

5963
Lorenzl, Josef

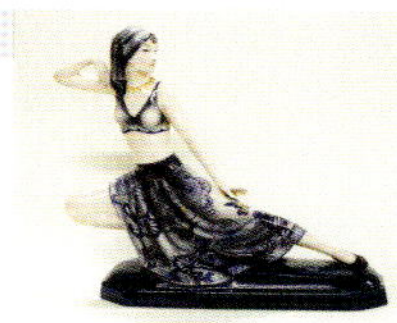

Tänzerin in Haremskostüm (hockend, ihr linkes Bein nach hinten gestreckt) (auf Rechtecksockel, mit abgeschrägten Ecken)
ca. 1929
5963,77,4 gem. lig. OL, gem. grün S.3, Firmenmarke 19, Made in Austria, Hand dec., sign. Lorenzl, H. 32,5 cm; MAK Wien, Inv. Ke 10521-45
5963,?,? Firmenmarke 19, Made in Austria, sign. Lorenz (!), H. 32 cm; RM 25

5964
Lorenzl, Josef

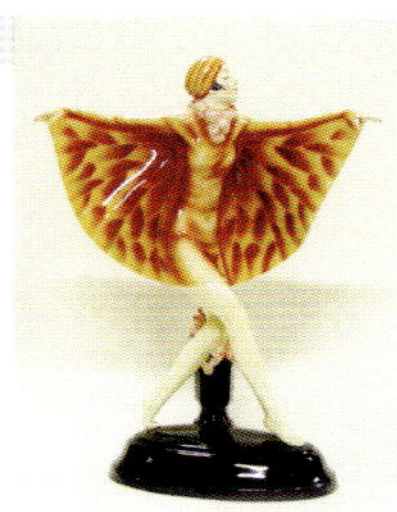

FIGUR „Gefangener Vogel" (Der Gefangene Vogel, Niddy Impekoven, nach rechts gewandt) (auf Ovalsockel)
ca. 1922 (für den Entwurf)
5964,30,4 ger. GI, gem. S.G., gem. lig. AH, Firmenmarke 19, Made in Austria, sign. Lorenzl, H. 38,8 cm; MAK Wien, Inv. Ke 10602-24
5964,36,15 gem. lig. SO, 22N, Firmenmarke 19, Made in Austria, Hand dec., sign. Lorenzl, H. 40 cm; eBay # 6153788089
-> vgl. 5054 (BRONZE); 5230; 6204; 6600; 8176
-> vgl. 5530 GRUPPE (Trio)
Pendant -> vgl. 5960
Pendant, bzw. ähnl. Modell -> vgl. 5434
Bem.: Diese seitenverkehrte Ausführung ist äußerst rar. Sie ist das direkte Pendant, weil auch in derselben Größe, zur Modell-Nr. 5960. Wie der Vergleich der beiden Modelle zeigt, sind bei der links gewandten Stellung zahlreiche Exemplare nachweisbar. Von der hier vorliegenden, nach rechts schreitenden Tänzerin bislang jedoch nur ein Stück. Weitere Bem. zum Modell s. bei 5230.
Lit.: s. bei 5230

5974F
Postl, Hugo
FIGUR „Windhund" (Liegender Barsoi)
ca. 1929
5974,613,15 Firmenmarke 19, unsign.?, H. 16 cm; Doro L 24.11.2000
5974,987,19 gem. S, Firmenmarke 19, Made in Austria, sign. Postl, H. 15,5 cm, B. 40 cm; Wien Museum Inv. 190.019
5974F, H. 16 cm, B. 40 cm; GSNachf. 5,2a
-> vgl. 6422F; 7143F
Lit.: MW, 11.1929/30, Nr. 17, S. III (m. Abb.), hier mit Ang. des Originaltitels „Windhund" und des Entwerfers Hugo Postl; GSNachf. 5,2 (m. Abb.), hier mit Ang. des Originaltitels „Windhund".

5975
Dakon, Stephan

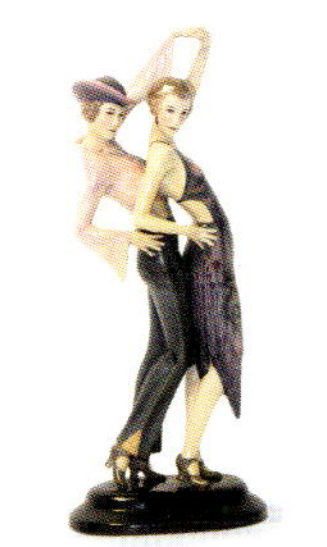

GRUPPE „Tanz" (Zwei Tänzerinnen, eine davon mit Hut, im Tanzschritt hintereinander posierend (auf Ovalsockel)
ca. 1929
5975,73,10 gem. F. IN., Firmenmarke 19, Made in Austria, Hand dec., H. 39,5 cm, B. 16,3 cm; Privatsammlung, Wien
Quelle: Archiv Mimi Dakon (Wiener Manufaktur FG, Modellliste Dakon, 7.11.1946), hier mit Ang. des Originaltitels GRUPPE „Tanz".

5984
Dakon, Stephan
„Kind im Sessel"
ca. 1929
Quelle: Archiv Mimi Dakon (Wiener Manufaktur FG, Modellliste Dakon, 7.11.1946), hier mit Ang. des Originaltitels „Kind im Sessel".

5993
Kremser, Lizzi (zugeschr.)

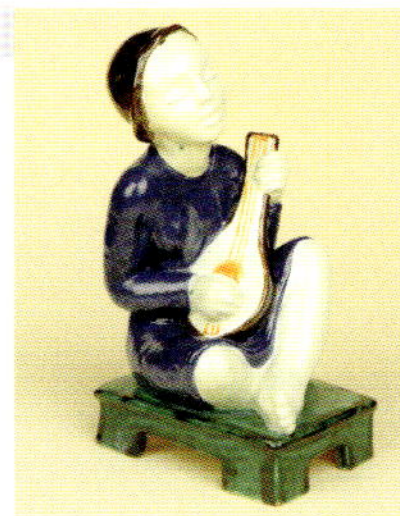

Sitzender Lautenspieler
ca. 1929
5993,8,6 Firmenmarke 19, unsign., H. 24,8 cm; Doro 20.10.1997
5993,11,11,0 gem. lig. AH, Firmenmarke 19, Made in Austria, unsign., H. 25 cm; PB Wien 19

5997
Meisinger, Ida

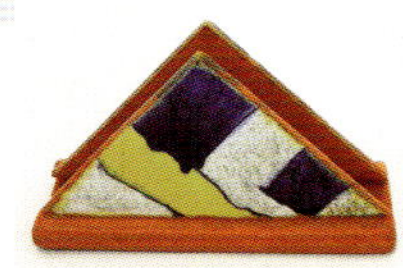

SERVIETTENHALTER Dreieck
Form: Dreieck
Dekor: geometrische Formen, bunte Laufglasur
ca. 1929
05997,24,1 gem. lig. HH., Firmenmarke 19, Made in Austria, H. 11,4 cm, B. 18 cm; Sammlung a@goldscheider.net

5998
Meisinger, Ida (zugeschr.)
SERVIETTENHALTER Halbrund
Form: halbrunde Scheibe
Dekor: geometrische Formen, bunte Laufglasur
ca. 1929
05998,20,27 Firmenmarke 19, Made in Austria, unsign., Pick. ohne Ang., H. 11 cm, B. 15 cm; Herr 48

6003
Lorenzl, Josef

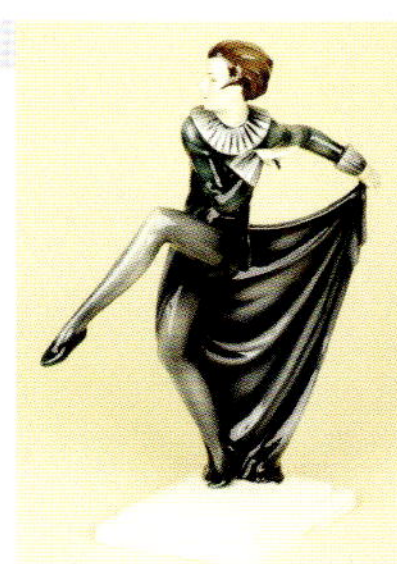

Tänzerin, auf dem rechten Bein tänzelnd, das linke Bein nach vorne schwingend (auf Rautensockel)
ca. 1929/30
6003,1,10 ger. GIR, gem. 119MR, Firmenmarke 19, Made in Austria, Hand dec., H. ca. 38 cm; Soth. com. # 1BSP; eBay # 399488088 (KH www.tias.com/stores/frozenbeauty)
6003,7,6 Firmenmarke 19, Made in Austria, Hand dec., H. 39 cm; PB Wien 19
6003,47,4 gem. N2., Firmenmarke 19, Made in Austria, Hand dec., H. 38 cm; eBay # 7328712217
6003,86,4,15 gem. D1, N1, Firmenmarke 19, Made in Austria, Hand dec., gem. „II. Sorte"= Z21, H. 38 cm; eBay # 190007322342
6003,134,6 Firmenmarke 19, Made in Austria, H. 38 cm; WKH
6003,150,7 gem. FN, Firmenmarke 19, Made in Austria, H. 38 cm; Christie's 13.12.1989
-> vgl. 6323; 6935

6004
Lorenzl, Josef
Schreitender Till Eulenspiegel, die Eule auf dem ausgestreckten Arm haltend (auf getrepptem, hexagonalem Sockel)
ca. 1929/30
6004,15,4,16P ger., Firmenmarke 19, Made in Austria, Hand dec., H. 39 cm; Nagel 368; Lit.: TuS, 1999, Nr. 238, S. 52 (m. Abb.)
Bem.: Die Figur des Till Eulenspiegel (Ulenspegel, ein mittelalterlicher Schalksnarr, hat ca. von 1300–1350 gelebt), geht zurück auf ein anonym erschienenes Volksbuch des 15. Jahrhunderts, das auf einem niederdeutschen Text beruht, der aus altem Schwankgut und erfundenen Geschichten zusammengetragen wurde (-> vgl. hierzu: www.eulenspiegel-museum.de). Lorenzl nahm sich des Themas „Till Eulenspiegel" mehrfach an, -> vgl. hierzu die Modelle 5193; 5509 und 5921.

6005
Lorenzl, Josef
Tänzerin (mit Bikini und großem Tuch) auf dem linken Bein tänzelnd, das rechte Bein angewinkelt (auf Rautensockel)
ca. 1929/30
6005,25,4 (Firmenmarke ohne Ang.), H. 41,5 cm; Doro 1.6.2005

6006
Lorenzl, Josef
Stehende Japanerin, ein Lampion in die Höhe haltend (auf Quadratsockel)
ca. 1929/30
6006,5,5 gem. N2, Firmenmarke 19, Made in Austria, Hand dec., H. 40,7 cm; eBay # 3710766848
6006,35,4 gem. 1FS, Firmenmarke 19, Made in Austria, Hand dec., H. 42 cm; Soth. South, Billingshurst 11.10.2000; eBay # 1320620519

6009
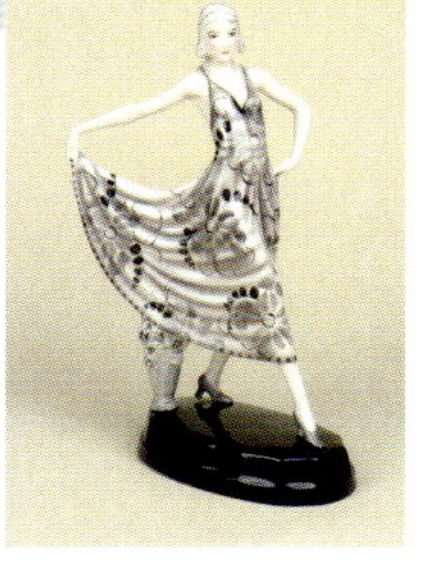
Dakon, Stephan
FIGUR „Harvey" (Lilian Harvey als schreitende Tänzerin, ihren Rock mit der rechten Hand anhebend) (auf facettiertem Ovalsockel)
ca. 1926/27 (für den Entwurf)
6009,272,15,15 gem. F., gem. lig. NB, Firmenmarke 19, Made in Austria, Hand dec., unsign.?, H. 26,5 cm; eBay # 3701651620; 2278063124
6009,282,10 gem. lig. AS 19, JR 3, Firmenmarke 19, sign. Dakon, H. 27 cm; PB Wien 19
6009,489,4 gem. N4, gem. lig. AS 4, Firmenmarke 19, Made in Austria, Hand dec., H. 26,5 cm; eBay # 2263867261
6009,540,15 Firmenmarke 19, sign. Dakon, H. 26 cm; Doro 10.3.2003
6009,616,4 (Firmenmarke ohne Ang.), unsign.?, H. 26 cm; Christie's SK 12.9.1997
6009,675,7 gem. AD, Firmenmarke 19, Made in Austria, sign. „Lorenzl"!, H. 26,5 cm; WKA 27; 33; Bem.: Die Sign. „Lorenzl" dürfte wohl als Versehen des Modelleurs angesehen werden.
-> vgl. 5625; 6211
Lit./Bem.: Becker, K.: Lilian, das Mädchen mit den 100 Gesichtern, in: Uhu, 4.1927/28, Nr. 10, S. 54 (m. Abb.), hier eine zeitgenössische Fotografie, die wohl als Inspiration für die Ausführung diente. Dargestellt ist die Schauspielerin Lilian Harvey (eigentl. Lilian Pape, 19.1.1906–27.7.1968), hier mit dunkler Perücke, betitelt „Spanischer Tanz". Fotografie: Binder.
Quelle: Archiv Mimi Dakon (Tantiemenabrechnung 1.7.–31.12.1937), hier mit Ang. des Originaltitels FIGUR „Harvey"; (Modellliste Dakon, undatiert, ca. 1940; Wiener Manufaktur FG, Modellliste Dakon, 7.11.1946), hier jeweils mit Ang. des Originaltitels „Harvey" („Verkl.").

6011F
Beaume (d.i. Bormann, Wilhelm)
Stehender Windhund (Barsoi / Greyhound) (auf Ovalsockel)
ca. 1910/12 (für den Entwurf)
6011,298,6 ger. A, Firmenmarke 19, Made in Austria, sign. Beaume, Ausformung 20er Jahre, H. 31 cm, L. 49 cm; Zeller 72; eBay # 2080115875
6011,403,6 ger. A, gem. F., Firmenmarke 19, Made in Austria, sign. Beaume, Ausformung 20er Jahre, H. 33 cm, L. 48,2 cm; eBay # 1437999022
Bem.: Bei dieser Figur handelt es sich um die rechte Einzelfigur von Modell-Nr. 3878.
-> vgl. 4407
-> vgl. 3878= ZweierGRUPPE
Lit.: Studio YB, 1914, S. 197 (m. Abb.), hier bezeichnet „Greyhound" und mit Künstlerang. „Borrmann" (!).
Quelle: AGM (Rechnung an Maurice Mizrahi, Alexandria, Ägypten, vom 22. April 1953), hier unter dem Titel „Windhund" geführt.

6012
Meisinger, Ida
Stehender Scotchterrier
ca. 1929/30
6012,40,20 Firmenmarke 19, Made in Austria, H. 23 cm; Schuler 75

6013
Meisinger, Ida
Sitzender West Highland Terrier (mit heraushängender Zunge)
ca. 1929/30
6013,48,7 ger. K5, gem. lig. JP, Firmenmarke 19, Made in Austria, H. ca. 20 cm, L. ca. 21 cm; eBay # 922205152
6013,275,20 Firmenmarke 19, Made in Austria, H. 21,5 cm; WKA 23
6013,519,31 gem. S, Firmenmarke 19, Made in Austria, unsign., H. 22,2 cm; eBay # 288673105
6013,617,65 Firmenmarke 19, Made in Austria, unsign., H. 21,6 cm; eBay # 1295269398; 1230008430
-> vgl. 6025; 6500
-> vgl. 6466= KERZENHALTER

6017
Geiger, Benno
ASCHER (mit zwölf Ablagen)
Form: rund, zwölf kreisförmig angeordnete Zigarettenablagen
Dekor: mehrfarbig, Laufglasur
ca. 1929/30
6017,164,16 gem. lig. AH, Firmenmarke 19, Made in Austria, H. 3 cm, Ø 15,8 cm; Sammlung John Nash

6018
Dakon, Stephan
„Skiläuferin"
ca. 1929/30
Quelle: Archiv Mimi Dakon (Wiener Manufaktur FG, Modellliste Dakon, 7.11.1946), hier mit Ang. des Originaltitels „Skiläuferin".

6019
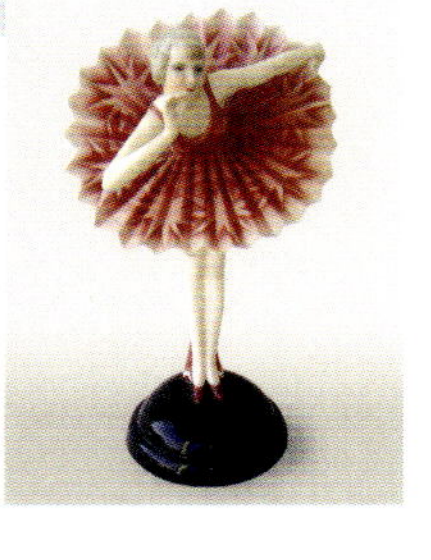
Dakon, Stephan
„Wiener Walzer" („Stern") (Stehende, sich verneigende Pierrette auf Rundsockel mit Spitzhut)
ca. 1929/30
6019,215,10 gem. 3, Firmenmarke 19, Made in Austria, Hand dec., H. 31 cm; Sammlung Stressless, Portugal
-> vgl. 6647
Quelle: Archiv Mimi Dakon (Modellliste Dakon, undatiert, ca. 1940), hier mit Ang. des (Original)-Titels „Wiener Walzer"; Wiener Manufaktur FG, Modellliste Dakon, 7.11.1946), hier mit Ang. des Titels „Stern".

6020
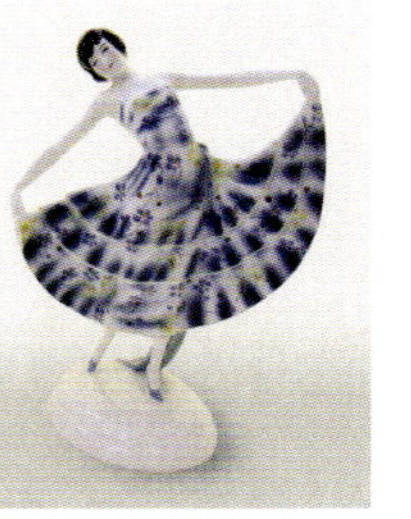
Dakon, Stephan
FIGUR „Wiener Walzer (Stehende Tänzerin, ihr Kleid mit langer schmaler Schleppe, fächerartig ausbreitend) (auf Ovalsockel)
ca. 1929/30
6020,117,11, ger. 16,P, gem. AD, NZ, Firmenmarke 19, Made in Austria, Hand dec., H. 29 cm; Sammlung @goldscheider.de
6020,201,15 gem. IN, gem. lig. AS, Firmenmarke 20, Made in Austria, Hand dec., H. 29,2 cm; eBay # 2580102653
6020,550,6,3 gem. S., Firmenmarke 19, Made in Austria, Hand dec., H. 29 cm; Wien Museum Inv. 220.706
6020,582,4 Firmenmarke 19, Made in Austria, H. 31 cm; Soth. com. # 8LL7
6020 ohne Ang., Firmenmarke 19, H. 30 cm; Nagel 313
Quelle: Archiv Mimi Dakon (Tantiemenabrechnung 1.1.–30.6.1940; Modellliste Dakon, undatiert, ca. 1940; 2. Halbjahr 1949; 1. Halbjahr 1950; 2. Halbjahr 1950; Wiener Manufaktur FG, Modellliste Dakon, 7.11.1946), hier jeweils mit Ang. des Originaltitels FIGUR „Wiener Walzer".

6021
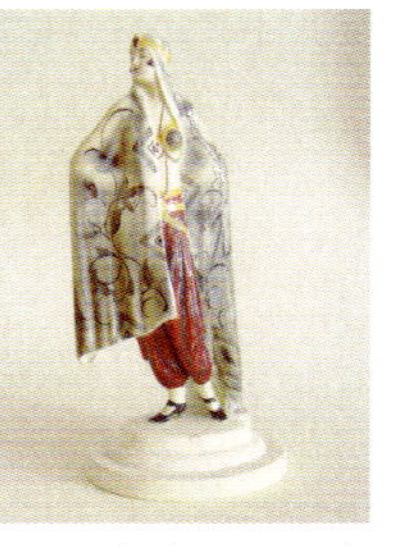
Lorenzl, Josef
FIGUR / LAMPE „Aida" (Stehende Odaliske) (auf Ovalsockel)
ca. 1923/24 (für den Entwurf)
FIGUR
6021,66,16 Firmenmarke 19, Made in Austria, Hand dec., sign. Lorenzl, H. 28,5 cm; Sammlung Alain Bernard
6021,222,15,3 (Firmenmarke ohne Ang.), sign. Lorenzl, H. 28 cm; Christie's SK 15.11.2001
LAMPE
6021,8 ger. XIX, Firmenmarke 19, Made in Austria, Hand dec., sign. Lorenzl, H. 28 cm; PB Deutschland
Lit. und Bem. s. bei Modell-Nr. 5281

6023
Lorenzl, Josef
Orientalische Tänzerin in Pose (auf Rundsockel)
ca. 1923 (für den Entwurf)
6023,21,8 (Firmenmarke ohne Ang.), sign. Lorenzl, H. 26 cm; Christie's SK 22.4.2004
6023,195,5,5 gem. N7 Firmenmarke 19, Made in Austria, sign. Lorenzl, H. 26,5 cm; Soth. L. 2.11.1999
-> vgl. 5267
Lit.: Pinhas, S. 146 (m. Abb.)

6025
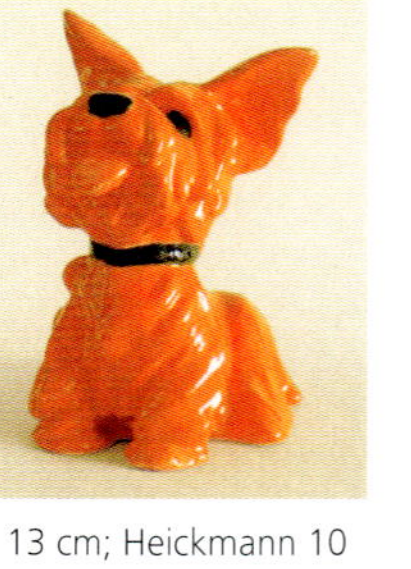
Meisinger, Ida
Sitzender West Highland Terrier (mit heraushängender Zunge)
ca. 1929/30
FIGUR
6025,251,24 gem. N.U., Firmenmarke 19, Made in Austria, H. 13 cm; Schloss Hag. 34
6025,862,24 Firmenmarke 19, H. 13 cm; Heickmann 10
6025,1006,32 Firmenmarke 19, Made in Austria, H. 13 cm; SS, S. 46 (m. Abb.); Wien Museum Inv. 190.016
6025,1620,20 gem. PX, (Firmenmarke ohne Ang.), H. 13 cm; eBay # 1434563529
6025,1725,21 Firmenmarke 19, Made in Austria, H. ca. 14 cm; eBay # 2519695638
6025,1836,20 Firmenmarke 19, Made in Austria, H. 14 cm; WKH
6025,1936,20 Firmenmarke 19, Made in Austria, H. 14,2 cm; PB Wien
6025,2209,31 Firmenmarke 19, Made in Austria, H. 14 cm; WKH
6025,2253,20 gem. lig. JP, Firmenmarke 19, Made in Austria, H. 13,2 cm; PB Wien
6025,2259,31 Firmenmarke 19, Made in Austria,

H. 13 cm; Wien Museum Inv. 190.017
6025,2383,2 gem. SG, X, Firmenmarke 19, Made in Austria, H. 13,4 cm; PB Wien
6025,2581,34 gem. lig. JP, Firmenmarke 19, Made in Austria, H. ca. 15 cm; eBay # 3759075090
6025,2865,20 gem. lig. BJ, Firmenmarke 19, Made in Austria, H. 14 cm; eBay # 2050840419
6025,8109,31,M gem. D, Firmenmarke 19, Made in Austria, unsign., H. 14 cm; PB Deutschland
BUCHSTÜTZE

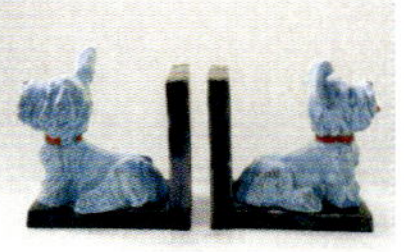

6025, 23, gem. P., Firmenmarke 19, Made in Austria, H. 13,5 cm; Sammlung f@goldscheider.de
6025, gem. P., Firmenmarke 19, Made in Austria, H. 13,5 cm; Sammlung f@goldscheider.de
6025 gem., gem. lig. JP, Firmenmarke 19, Made in Austria, H. 15 cm, L. 14,5 cm; eBay # 1317415028
Bem.: Das hier vorläufig schon durch die hohe Stückzahl 8109 nachgewiesene Modell kann als äußerst erfolgreich bezeichnet werden und wurde in den Farben schwarz, weiß und orange ausgeführt.
Ausst.: SSH 1992
-> vgl. 6013; 6500
-> vgl. 6466= KERZENHALTER
Lit.: SS, S. 46 (m. Abb.)

6026

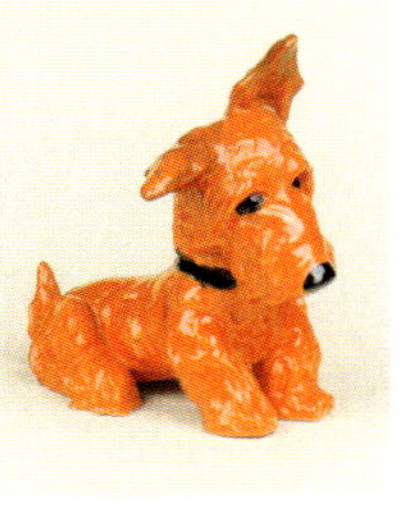

Meisinger, Ida
„Der Modehund"
(Sitzender Scotchterrier)
ca. 1928/29 (für den Entwurf)
6026,73,21 Firmenmarke 19, H. 16,7 cm; BM 68
6026,377,20 ger., Firmenmarke 19, Made in Austria, unsign.?, H. 16,5 cm; eBay # 2216616083
6026,406,31 Firmenmarke 19, H. 17 cm, B. 19 cm; Sammlung mad
6026,546,65 gem. lig. JP, Firmenmarke 19, Made in Austria, H. ca. 17 cm; eBay # 2223417111
-> vgl. 5854; 5879; ohne Nr.
-> vgl. .... (H. 39 cm); Doro 9.11.1989
-> vgl. 6043= ASCHER
-> vgl. 6309= KERZENHALTER
Bem. und Lit. vgl. bei 5854

6030

GRUPPE Entenmutter mit drei Küken
(auf Rechtecksockel)
ca. 1929/30
6030 ohne Ang., (Firmenmarke ohne Ang.), H. 11 cm, L. 14 cm; Herr 11

6043

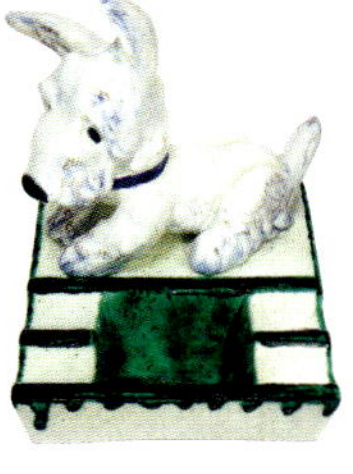

Meisinger, Ida
ASCHER „Der Modehund"
(Sitzender Scotchterrier)
(mit vier Ablagen)
Form: rechteckig, seitlich jeweils zwei Zigarettenablagen
Dekor: abstraktes Dekor, grüne Laufglasur
ca. 1928/29 (für den Entwurf)
06043,1 gem. TT, Firmenmarke 19, Made in Austria, unsign., H. 15 cm, B. 14 cm, T. 12 cm; Sammlung f@goldscheider.de
06043,1 gem. lig. AH, Firmenmarke 19, Made in Austria, unsign., H. 15,5 cm; eBay # 6543806700
6043,311,1 gem. B, Firmenmarke 19, Made in Austria, unsign., H. 15 cm, B. 14 cm, T. 12 cm; Sammlung a@goldscheider.net
-> vgl. 5854; 5879; 6026
-> vgl. o. Modell-Nr. (H. 6,2 cm); PB Deutschland
-> vgl. .... (H. 39 cm); Doro 9.11.1989
-> vgl. 6309= KERZENHALTER
Bem. und Lit. vgl. bei 5854

6044

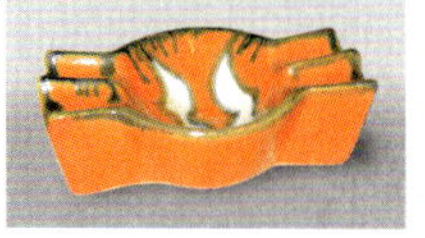

ASCHER (mit vier Ablagen)
Form: rund, seitlich jeweils zwei Zigarettenablagen
Dekor: abstraktes Dekor, Laufglasur
ca. 1929/30
6044,187,19 gem. lig. AH, Firmenmarke 19, Made in Austria, H. 4,5 cm, B. 14,5 cm; PB Wien

6052

Dakon, Stephan
„Libelle"
ca. 1929/30
Quelle: Archiv Mimi Dakon (Wiener Manufaktur FG, Modellliste Dakon, 7.11.1946), hier mit Ang. des Originaltitels „Libelle".

6062

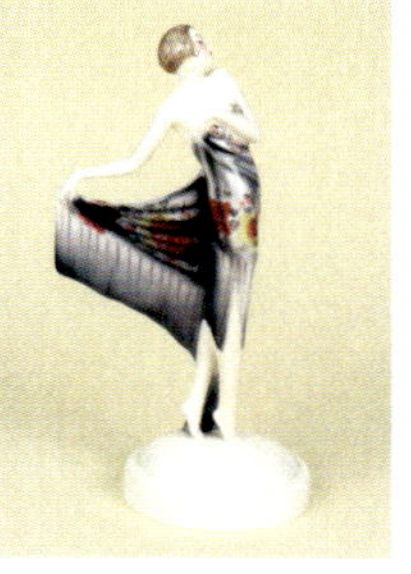

Lorenzl, Josef
„Spanischer Shawl" (Stehende Tänzerin (Dolores del Rio?) in Pose, mit großem Schal) (auf Rundsockel)
ca. 1928/29 (für den Entwurf)
6062,154,10 (Firmenmarke ohne Ang.), unsign.?, H. 29 cm; Christie's SK 14.3.1997
6062,185,10,20 Firmenmarke 19, Made in Austria, Hand dec., unsign., H. 28,8 cm; Christie's L. 19.4.1989

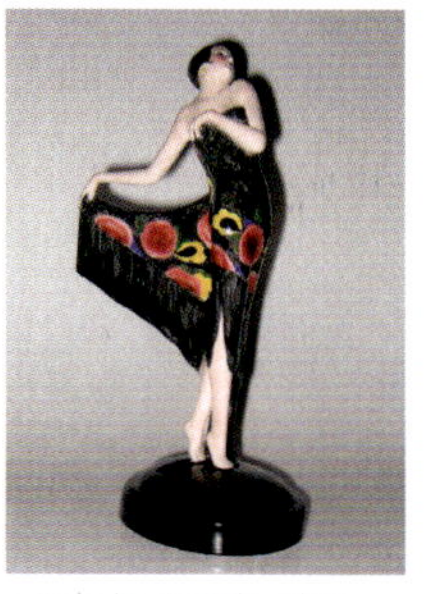

6062,276,15 Firmenmarke 19, sign. Lorenzl, H. 30 cm; PB Wien 19
6062,315,5,(3 ger.), gem. F., JR, V, Firmenmarke 19, Made in Austria, Hand dec., sign. Lorenzl, H. 28,2 cm; Sammlung.schmutz@chello.at 105
6062,405,31 Firmenmarke 19, Made in Austria, sign. Lorenzl, H. 28,1 cm; GK, S. 21, 44 (m. Abb.); Wien Museum Inv. 182.096
Bem. und Lit. vgl. bei 5822
Ausst.: HMW (jetzt Wien Museum) 1985/86.
-> vgl. 5822; 6345
Lit.: GK, S. 21, 44 (m. Abb.)

6063

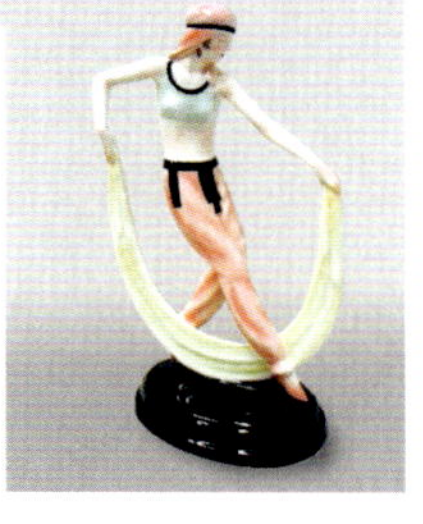

Lorenzl, Josef
Odaliske (Haremstänzerin mit langem Tuch) (auf facettiertem Ovalsockel)
ca. 1922 (für den Entwurf)
6063,69,15 ger. XIP, gem. lig. AS, NS.G, Firmenmarke 19, Made in Austria, Hand dec., unsign., H. 27,5 cm; Dawo 138; eBay # 2256700254
6063,78,11 ger. 16 P, Firmenmarke 19, unsign.?, H. 26,8 cm; Doro 12.6.1995
6063,110,10 Firmenmarke 19, unsign.?, H. 27,5 cm; Bonhams 25.2.2003
6063,226,15,19 Firmenmarke 19, Made in Austria, Hand dec., sign. Lorenzl, H. 29 cm; city-antik
-> vgl. 5018; 5613

6066

Stehende elegante Dame in Pose (auf achtfach gekehltem und abgerundetem Quadratsockel)
ca. 1929/30
6066,1,11 (Firmenmarke ohne Ang.), H. 40,5 cm; Christie's SK 12.5.2000

6067F

Dakon, Stephan
FIGUR „Hund" („Figur mit Hund")
ca. 1929/30
Quelle: Archiv Mimi Dakon (Tantiemenabrechnung 1.7.–31.12.1937; 1.1.–30.6.1940; Wiener Manufaktur FG, Modellliste Dakon, 7.11.1946), hier jeweils mit Ang. des Originaltitels FIGUR „Hund"; (Modellliste Dakon, undatiert, ca. 1940), hier unter dem Titel „Figur mit Hund" geführt.

6068B

Dakon, Stephan
FIGUR „Ramona" („Tanzfreude")
ca. 1929/30
Quelle: Archiv Mimi Dakon (Tantiemenabrechnung 1.7.–31.12.1937), hier mit Ang. des Originaltitels FIGUR „Ramona"; (Modellliste Dakon, undatiert, ca. 1940), hier unter dem Titel „Tanz" geführt; (Wiener Manufaktur FG, Modellliste Dakon, 7.11.1946), hier mit Ang. des (Original-) Titels „Tanzfreude".

6069

Lorenzl, Josef und Postl, Hugo
GRUPPE Stehendes Kind, einen sitzenden Schäferhund umarmend
ca. 1929/30
6069,17,19 gem. SO, Firmenmarke 19, Made in Austria, H. 28,3 cm; Herr 58
6069,270,10 gem. lig. LC(N od. Z?), Firmenmarke 19, Made in Austria, sign. Postl, H. 28,4 cm; city-antik
6069,332,10 gem. F, gem. rot S, Firmenmarke 19, Made in Austria, sign. Lorenzl, H. 29,2 cm; eBay # 40119669
Bem.: Bei dieser Gruppe wurden Entwürfe bzw. Elemente zweier Künstler verwendet. Für das stehende Kind zeichnet Josef Lorenzl verantwortlich, der sitzende Hund ist von Hugo Postl. Eine Nachfrage bei beiden Anbietern brachte das hier bislang einmalig vorliegende Ergebnis: jede der beiden Figuren wurde nur mit einer Signatur der jeweils entwerfenden Künstler versehen. Postl war bei der Ausformung des Modells durch die Manufaktur vermutlich schon verstorben (-> vgl. Biografie). Ähnliche Kombinationen kommen auch bei Claire (Klára) Herczeg vor. Auch bei ihr wurden teils Tänzerinnen mit eleganten Hunden zu einem Modell kombiniert. Laut freundlicher Auskunft von Herrn Heinz Hoever, dem Herczeg Spezialisten schlechthin, war sie mit dieser Praxis der Manufaktur nicht einverstanden. Zumal sie offenbar nicht einmal gefragt wurde, bzw. keine Autorisierung dieser Modelle bei ihr eingeholt wurde.
-> vgl. 5719= Einzelfigur Hund
-> vgl. 7041= Damenakt (Lorenzl) mit Hund (Postl)

6070

Dakon, Stephan
„Wind"
ca. 1929/30
Quelle: Archiv Mimi Dakon (Wiener Manufaktur FG, Modellliste Dakon, 7.11.1946), hier mit Ang. des Originaltitels „Wind".

6071F

Dakon, Stephan
FIGUR „Cimono" („Kimono")
ca. 1929/30
Quelle: Archiv Mimi Dakon (Tantiemenabrechnung 1.7.–31.12.1937), hier mit Ang. des (Original)-Titels FIGUR „Cimono"; (Modellliste Dakon, undatiert, ca. 1940; Wiener Manufaktur FG, Modellliste Dakon, 7.11.1946; Tantiemenabrechnung 1. Halbjahr 1949), hier jeweils mit Ang. des (Original)-Titels FIGUR „Kimono".

6072

Meisinger, Ida
Sitzender Scotchterrier (mit Halsband/Schleife)
ca. 1930
6072,61,21 gem. lig. JP, Firmenmarke 19, Made in Austria, H. 19 cm; eBay # 446363688; 2236200439
6072,71,20 gem. SG., Firmenmarke 19, Made in Austria, H. 19 cm; Wien Museum Inv. 220.718
-> vgl. 6101
Lit.: Bernd, G.: Heitere Wiener Tierkunst, Ida Meisinger, in: Schaulade, 8.1932, S. 480 (m. Abb.)

6074F
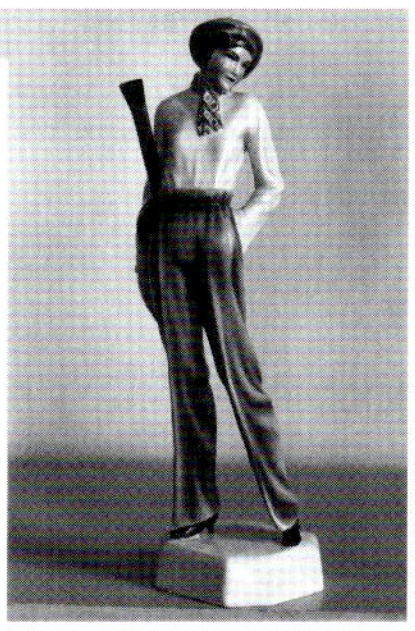
Dakon, Stephan
FIGUR „Daisy" (Stehende junge Dame mit Béret/Pullmannkappe und langen Hosen, eine Gitarre haltend) (auf oktogonalem Sockel)
ca. 1930
6074,4,4 Firmenmarke 19, Made in Austria, Hand dec., H. 40 cm; eBay # 1514453142
Bem.: Diese Figur spiegelt die so genannte Garconne Mode wider, die in den Jahren ab 1924 bis nach 1928 eine beliebte, im männlichen Stil gehaltene Damenmode darstellte. Kostümjacken in Herrenfasson sowie Paletot-Mäntel und der typische Bubikopf als Frisur waren charakteristisch für diese Strömung (-> vgl. Hartmann).
-> vgl. 6629
Quelle: Archiv Mimi Dakon (Originalfoto); (Wiener Manufaktur FG, Modellliste Dakon, 7.11.1946), hier mit Ang. des Originaltitels „Daisy".

6075
Dakon, Stephan
„Schiffsgirl"
ca. 1930
Quelle: Archiv Mimi Dakon (Wiener Manufaktur FG, Modellliste Dakon, 7.11.1946), hier mit Ang. des Originaltitels „Schiffsgirl".

6076
Stehendes Reh (auf Rechtecksockel)
ca. 1930
6076,5,8 gem. G., Firmenmarke 19, Made in Austria, H. 22,8 cm; eBay # 2001363630

6077F
Dakon, Stephan
FIGUR „Page im Sessel" (Sitzende junge Dame mit Bubikopf/Pagenschnitt) (auf Rechtecksockel)
ca. 1930
6077,6,6 gem. lig. AS, Firmenmarke 19, Made in Austria, Hand dec., wohl unsign.?, H. fehlt; Rieber 23.4.2002;
Bem.: Das AH gibt zwar den Künstler Josef Lorenzl an, jedoch sprechen sowohl der Entwurf selbst als auch die Belege aus dem Archiv Mimi Dakon für die Autorschaft Dakons.
Quelle: Archiv Mimi Dakon (Tantiemenabrechnung 1.1.–30.6.1940, Modellliste Dakon, undatiert, ca. 1940; Wiener Manufaktur FG, Modellliste Dakon, 7.11.1946), hier jeweils mit der Ang. des Originaltitels FIGUR „Page im Sessel".

6079
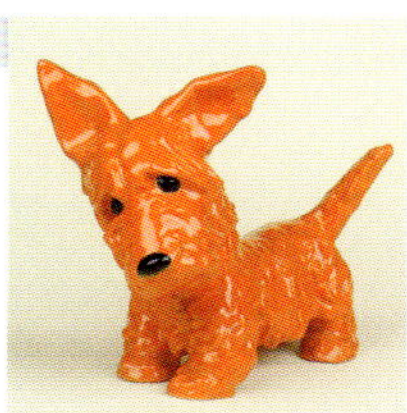
Meisinger, Ida
Stehender Scotchterrier (Kopf nach links drehend)
ca. 1930
6079,39,20 (Firmenmarke ohne Ang.), H. 19,5 cm; Christie's SK 14.3.1997
6079,136,31 gepr. M, gem. O, Firmenmarke 19, unsign., H. 20,4 cm; PB Wien 19
Lit.: Bernd, G.: Heitere Wiener Tierkunst, Ida Meisinger, in: Schaulade, 8.1932, S. 480 (m. Abb.)

6083
Dakon, Stephan
„Ryth(mischer) Tanz"
ca. 1930
Quelle: Archiv Mimi Dakon (Wiener Manufaktur FG, Modellliste Dakon, 7.11.1946), hier mit Ang. des Originaltitels „Ryth(mischer) Tanz".

6085
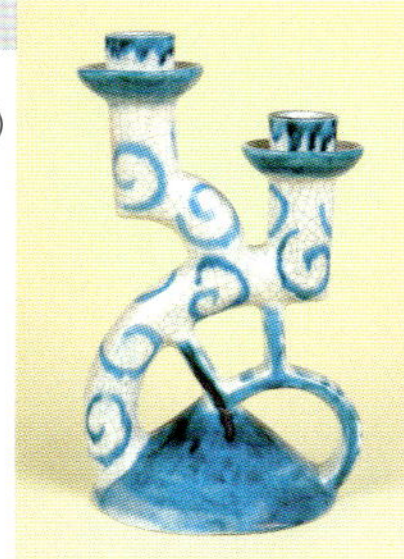
Piesch
KERZENHALTER (zweiflammig)
Form: zwei Kerzentüllen auf rohrartigem Unterbau, über kegelförmigem Sockel
Handhabe: halbkreisförmig
Dekor: gemalte, offene Ringe auf Craquelé/Laufglasur
ca. 1930
6085,2 (Firmenmarke ohne Ang.), unsign.?, H. 24,5 cm; Doro 4.3.2002
06085,108,19 gem. lig. MK, Firmenmarke 19, Made in Austria, H. 24,7 cm; PB Wien 19
Ausst.: PSK 1995
Lit.: ADGS, Nr. 37 (m. Abb.)

6086
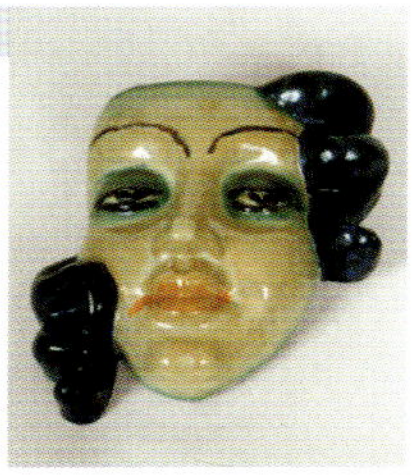
Weber, Ernst
WANDMASKE/WANDVASE
Mädchenkopf mit Locken
ca. 1930
WANDMASKE
6086,54,1 Firmenmarke 19, H. 12 cm; PB Wien 19
6086,58,31 Firmenmarke 19, H. 12 cm; PB Wien 19
6086,97,20 Firmenmarke 19, H. 13,8 cm; Doro 24.10.1995
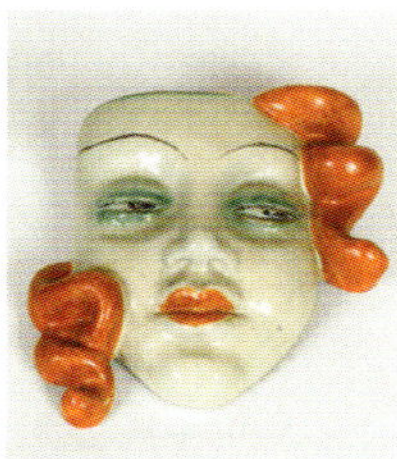
6086,189,69 gem. H., Firmenmarke 19, Made in Austria, unsign.?, H. 12,7 cm; eBay # 42855304
6086,277,31 gem. G, Firmenmarke 19, Made in Austria, unsign., H. 14 cm; Sammlung.schmutz@chello.at 3
WANDVASE
6086,A ger., Firmenmarke 19, unsign., H. 14 cm; Doro 16.6.1992; eBay # 2042946681
6086,72,34 gem. lig. AH, Firmenmarke 19, Made in Austria, H. 14 cm; eBay # 2005778108
6086,75,33 Firmenmarke 19, Made in Austria, H. 13,7 cm; Sammlung.schmutz@chello.at 4
Bem.: Dieses Modell wurde in zwei Versionen produziert. Zum einen die gängige Ausführung als Wandmaske, zum anderen die als Wandvase. Bei Goldscheider übrigens eine durchaus übliche Praxis, durch Erweiterung der Funktion Neues zu kreieren und zugleich die Modellpalette zu erweitern. Diese Vase zum Aufhängen zeigt dazu, wie bereits Anfang der dreißiger Jahre dieses dann in den Fünfzigern so beliebte Genre Eingang in die angewandte Kunst fand.

6099
Meisinger, Ida
Stehende Chinesin mit zwei kleinen Schalen (Kakteentöpfchen/Gewürzschalen?)
ca. 1930
6099,3,2 gem. lig. JP, Firmenmarke 19, Made in Austria, unsign., H. 19,5 cm; eBay # 2526387929; 2559552177
Pendant -> vgl. 6100

6100
Meisinger, Ida
Stehender Chinese mit zwei kleinen Schalen (Rauchzeug/Gewürzschalen?)
ca. 1930
6100,6,2 Firmenmarke 19, unsign.?, H. 19,5 cm; Neumeister 278
6100,16,20 Firmenmarke 19, Made in Austria, H. 20 cm; Zacke 3, S. 93 (m. Abb.); WKH
Pendant -> vgl. 6099

6101
Meisinger, Ida
Sitzender Scotchterrier (mit Halsband/Schleife)
ca. 1930
6101,145,33 gem. SG, Firmenmarke 19, Made in Austria, H. 13,5 cm; eBay # 7405397327
6101,456,31 ger. Z, gem. S.G, Firmenmarke 19, Made in Austria, H. 13,5 cm; eBay # 2240324835
-> vgl. 6072
Lit.: Bernd, G.: Heitere Wiener Tierkunst, Ida Meisinger, in: Schaulade, 8.1932, S. 480 (m. Abb.)

6103
Meisinger, Ida
BUCHSTÜTZE Springender Scotchterrier (Kopf nach rechts drehend)
ca. 1930
6103,18,21 Firmenmarke 19, Made in Austria, unsign., H. 18 cm; Quit. 14
6103,201,20 ger., gepr. M, Firmenmarke 19, Made in Austria, unsign., H. 18 cm; eBay # 1227637689
Pendant -> vgl. 6104
Lit.: Bernd, G.: Heitere Wiener Tierkunst, Ida Meisinger, in: Schaulade, 8.1932, S. 480 (m. Abb.)

6104
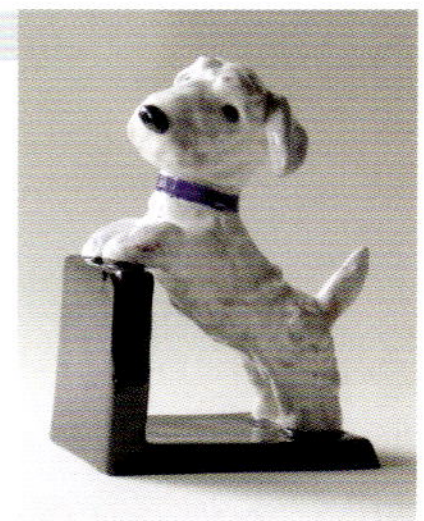
Meisinger, Ida
BUCHSTÜTZE Springender Scotchterrier (Kopf nach halb links drehend)
ca. 1930
6104,17,21 Firmenmarke 19, Made in Austria, unsign., H. 18 cm; Quit. 14
6104,249,31 ger. M, gem. SG, Firmenmarke 19, Made in Austria, unsign., H. 17,9 cm; Sammlung John Nash
Pendant -> vgl. 6103
Lit.: Bernd, G.: Heitere Wiener Tierkunst, Ida Meisinger, in: Schaulade, 8.1932, S. 480 (m. Abb.)

6113F
FIGUR „Pferd" (springend) (auf Landschaftssockel mit „Hürde")
ca. 1930
6113F, H. 33 cm, B. 46 cm; GSNachf. 5,18
Lit.: GSNachf. 5,18 (m. Abb.), hier mit Ang. des Originaltitels FIGUR „Pferd" (springend).

6116F
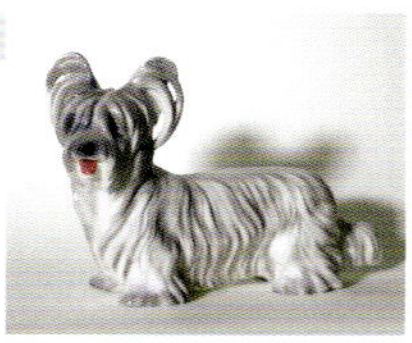
Postl, Hugo
FIGUR „Skyeterrier"
um/ab 1930 (für die Ausführung)
6116,41,19 gem. lig. AS, Firmenmarke 19, H. 17,3 cm, B. 9 cm; Sammlung mad
6116,87,6,20 gem. lig. LG, Firmenmarke 19, Made in Austria, H. 17,2 cm; PB Deutschland
6116,125,4,19 gem. lig. SH, Firmenmarke 19, Made in Austria, H. 17 cm; KH
6116F, H. 17 cm, B. 27 cm; GSNachf. 5,11a
-> vgl. 6503; 6802; 7300; 7519
-> vgl. 6975= Variante (auf Ovalsockel)
Lit.: GSNachf. 5,11 (m. Abb.), hier mit Ang. des Originaltitels „Skyeterrier".

6119
Motto, (Joseph C.)
Stehende Madonna mit Kind (auf Rundsockel)
ca. 1930
6119,340,4 Firmenmarke 19, Made in Austria, H. 45 cm; eBay # 929523735

6128F
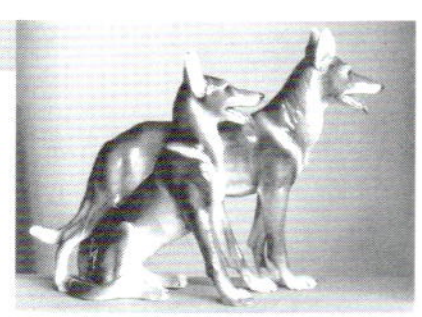
GRUPPE „Wolfshunde" (Schäferhunde)
ca. 1930
6128,124,7 gem. rot XX5, Firmenmarke Schuster 1, Made in Germany, Ausformung zw. 1941–1944, H. 35 cm, B. 46 cm; Wien Museum Inv. 190.024
6128F, H. 33 cm, B. 49 cm; GSNachf. 5,28a
-> vgl. 8822
Lit.: GSNachf. 5,28 (m. Abb.)
Quelle: AGM (Bestellungen der Firma Wahliss (Wien), vom 9. und 24.11.1945 an die „Arbeitsgemeinschaft der Firma Friedrich Goldscheider", hier jeweils mit Ang.

des (Original?-)Titels GRUPPE „Wolfshunde"; Bestellung der Firma Wahliss (Wien), vom 27.2.1946 in Ausführung „Terrakotta" und „Fayence").

6135F
Schmidt-Kestner, Erich
FIGUR „Setter" (vorne)
ca. 1904/05 (für den Entwurf)
6135,(6)77,19 Firmenmarke 19, Made in Austria, unsign., H. 12 cm, L. 35 cm; eBay # 2018635342
6135F, H. 14 cm, B. 36 cm; GSNachf. 5,9c
-> vgl. 3024
-> vgl. 6659; 7100= jeweils GRUPPE
Lit.: GSNachf. 5,9 (m. Abb.), hier mit Ang. des Originaltitels „Setter" (vorne).
Quelle: AGM (Rechnung an Maurice Mizrahi, Alexandria, Ägypten, vom 22. April 1953), hier mit Ang. des Originaltitels „Setter".

6136
Postl, Hugo

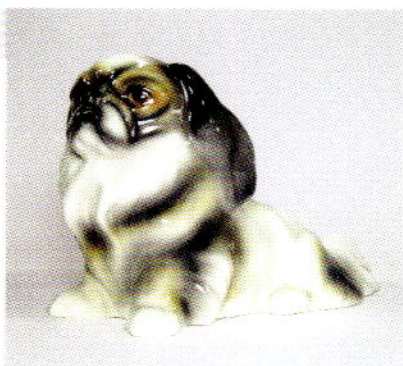

Pekinese
ca. 1928/29 (für den Entwurf)
6136,189,3,1 gem. F., Firmenmarke 19, Made in Austria, H. 10,1 cm, L. 17,8 cm; eBay # 46174435
6163,534,15 gem. F., Firmenmarke 19, Made in Austria, H. 10,2 cm, L. 17,5 cm; Sammlung a@goldscheider.net
6136,553,15 gem. F., Firmenmarke 19, Made in Austria, H. 10 cm; Doro FVK, Dez. 1999
6136,826,14 gem. F., Firmenmarke 19, Made in Austria, H. 10 cm, L. 17,8 cm; eBay # 1277919229
-> vgl. 5858; 7320
Lit.: s. 5858

6137
Postl, Hugo
Stehender Dachshund (Dackel)
ca. 1928/29 (für den Entwurf)
6137 Firmenmarke 19, H. 14 cm, L. 31 cm; eBay # 6614819943

6143
Ingels, Domien (zugeschr.)
Pferd (auf Rechtecksockel)
ca. 1930
6143,6,20 Firmenmarke 19, Made in Austria, H. 21,6 cm, B. 34,3 cm; city-antik

6150F
Dakon, Stephan
FIGUR „Mary" (Stehende Tänzerin, ihr Kleid mit beiden Händen anhebend) (auf Rundsockel mit 3 kleinen Würfeln)
ca. 1929/30
6150,24,20 Firmenmarke 19, Made in Austria, Hand dec., H. 36,5 cm; eBay # 6520721440
Quelle: Archiv Mimi Dakon (Tantiemenabrechnung 1.7.–31.12.1937; Modellliste Dakon, undatiert, ca. 1940; Wiener Manufaktur FG, Modellliste Dakon, 7.11.1946), hier jeweils mit Ang. des Originaltitels FIGUR „Mary".

6154
Lorenzl, Josef
Sitzende Dame mit zwei Hunden
ca. 1930
6154,35,15 (Firmenmarke ohne Ang.), H. 28,5 cm; Doro 30.11.1999
6154,55,15 Firmenmarke 20, Made in Austria, H. 28,4 cm; Doro 7.11.1995
6154,57,10 gem. F.N., Firmenmarke 19, Made in Austria, Hand dec., H. 28,8 cm; eBay # 7300553746
6154,99,5 gem. F., Firmenmarke 19, Made in Austria, unsign., H. 28,6 cm; Soth. com. # 74CB

6156
Stehender Mädchenakt mit hinter dem Kopf verschränkten Armen (auf Rundsockel)
ca. 1930
6156,219,19 Firmenmarke 19, Made in Germany, Ausformung zw. 1938–1941, H. 37,5 cm; WKA 38

6158
KERZENHALTER / GIRANDOLE (fünfflammig)
Form: fünf Kerzentüllen, rohrartig, V- und W-förmig, über halbkugelförmigem Sockel
Dekor: mehrfarbig, teilweise gesprenkelte Glasur
ca. 1930
6158,W (Firmenmarke ohne Ang.), H. 23 cm; Doro 12.12.1991
ohne Nr., gem. lig. AH, Firmenmarke 19, Made in Austria, H. 22,4 cm; B. 24 cm; Wien Museum Inv. 190.001
ohne Ang. (Firmenmarke ohne Ang.), H. 22,5 cm; Soth. L. 28.9.1984
Ausst.: SSH 1992
Lit.: SS, S. 32, 33 (m. Abb.).

6163
BUCHSTÜTZE

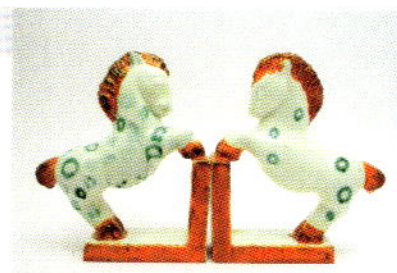

Springendes Pferd
ca. 1930
6163,1,21 gem. G., Firmenmarke 19, H. 14 cm; eBay # 1240921358
6163,2,21 gem. G., Firmenmarke 19, H. 14 cm; eBay # 1240921358
6163,4,20 gem. B, Firmenmarke 19, Made in Austria, H. 17,3 cm, L. 14,8 cm; Sammlung a@goldscheider.net
6163,5,20 gem. B, Firmenmarke 19, Made in Austria, H. 17,3 cm, L. 14,8 cm; Sammlung a@goldscheider.net
-> vgl. 6612= ähnl. Modell

6164
Piesch
BUCHSTÜTZE Max (als stehende Figur)
ca. 1930
6164,10,20 Firmenmarke 19, Made in Austria, H. ca. 19 cm; Doro FVK
6164,39,20 ger. lig. MK, Firmenmarke 19, Made in Austria, unsign., H. 19,7 cm; Doro 1.12.1998
Pendant -> vgl. 6165
Bem.: Dieses Modell, sowie das folgende Pendant, gehen auf Wilhelm Buschs Verserzählung mit Bildern aus „Max und Moritz" (1858) zurück.

6165
Piesch
BUCHSTÜTZE Moritz (als stehende Figur)
ca. 1930
6165,10,21 Firmenmarke 19, Made in Austria, H. ca. 19 cm; Doro FVK
6165,35,31 ger. lig. MK, Firmenmarke 19, Made in Austria, unsign., H. 18,8 cm; Doro 1.12.1998
Pendant -> vgl. 6164

6170
Dakon, Stephan
„Schwimmerin" (Posierende junge Dame im Badedress, ihre Kappe mit der Rechten in die Höhe haltend (auf Rundsockel)
ca. 1930
6170 (Firmenmarke ohne Ang.), H. 43 cm; Pinhas, S. 142 (m. Abb.)
Quelle: Archiv Mimi Dakon (Modellliste Dakon, undatiert, ca. 1940; Wiener Manufaktur FG, Modellliste Dakon, 7.11.1946), hier jeweils mit Ang. des Originaltitels „Schwimmerin".

6173
Lorenzl, Josef
Stehende Tänzerin, ihren langen Rock schmetterlingsflügelartig auseinander haltend (auf Rundsockel)
ca. 1929
6173,10,10,15 gem. lig. AH, gem. blau N, Firmenmarke 19, Made in Austria, Hand dec., unsign., H. 50,2 cm; Wien Museum Inv. 190.081
-> vgl. 5872
-> vgl. 6399= fast identisches Modell

6201F
GRUPPE Windhunde
ca. 1930
6201,66 (Firmenmarke ohne Ang.), H. fehlt, L. 58 cm; Doro 30.5.1994
Quelle: AGM (Rechnung an Maurice Mizrahi, Alexandria, Ägypten, vom 22. April 1953), hier mit Ang. des Originaltitels „Windhund(e)".

6204
Lorenzl, Josef

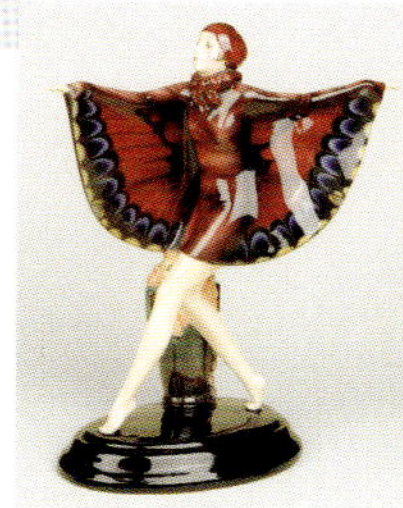

„Gefangener Vogel"
(Der Gefangene Vogel, Niddy Impekoven) (auf Ovalsockel)
ca. 1922 (für den Entwurf)
6204,3,2(oder 7?) gem. S., gem. rot II, Firmenmarke 19, Made in Austria, Hand dec., unsign., H. 31,5 cm; Herr 38; KH Art 1900
6204,4,5 ger. A, Firmenmarke 19, Made in Austria, sign. Lorenzl, H. 30,5 cm; Soth. NY 19.3.1993
6204,5,57 gem. X, S., gem. lig. HS, Firmenmarke 19, Made in Austria, unsign., H. 31 cm; Sammlung Weilinger, Wien
6204,46,19,15 gem. NS22, Firmenmarke 19, Made in Austria, Hand dec., sign. Lorenzl, H. 30 cm; SS, S. 18, 19 (m. Abb.); Wien Museum Inv. 190.122
6204,77,19,20 gem. NF11, Firmenmarke 19, Made in Austria, Hand dec., sign. Lorenzl, H. 31 cm; WKA 38
6204,102,10 gepr. L, (Firmenmarke ohne Ang.), unsign.?, H. 31 cm; Christie's SK 13.5.2004
6204,168,15,18 gem. N22, Firmenmarke 19, Made in Austria, sign. Lorenzl, H. 30,5 cm; eBay # 2663362840
6204,186,10 (Firmenmarke ohne Ang.), sign. Lorenzl, H. 30 cm; Doro 8.3.2004
6204,288,10 gem. F., Firmenmarke 19, Made in Austria, Hand dec., sign. Lorenzl, H. 30,5 cm; WKA 35
6204,347,6,19 gem. grün N, gem. lig. AS, Firmenmarke 19, Made in Austria, Hand dec., sign. Lorenzl, H. 29,8 cm; MAK Wien, Inv. Ke 10521-46
6204,348,19 Firmenmarke 19, sign. Lorenzl, H. 29,5 cm; Bonhams & Brooks L. 23.5.2001
6204,367,6 Firmenmarke 19, Made in Austria, Hand dec., sign. Lorenzl, H. 30 cm; PB Wien 19
6204,372,19 Firmenmarke 19, sign. Lorenzl, H. 30 cm; Kett. 74; Bem.: Das AH gibt zwar die Nr. 6294 an, dies kann jedoch nur ein Lese/Schreibfehler sein. Richtig ist die Modell-Nr. 6204.
6204,387,10 Firmenmarke 19, Made in Austria, sign. Lorenzl, H. 30,5 cm; Herr 30
6204,579,7 Firmenmarke 19, Made in Austria, sign. Lorenzl, H. 29 cm; PB Wien 19
6204,580,4,4 gem. N, Firmenmarke 19, Made in Austria, sign. Lorenzl, H. 29,6 cm; Sammlung.schmutz@chello.at 101
6204,681,16 gem. N., Firmenmarke 19, Made in Austria, sign. Lorenzl, Pick. 2, H. 31 cm; Nagel 373
6204,718,31 gem. N, Firmenmarke 19, Made in Austria, sign. Lorenzl, Originalpick.: „J.C. Herold, Hannover", H. 29,6 cm; Sammlung.schmutz@chello.at 102
6204,738,31 gem. II., X,E, gem. lig. NO4, Firmenmarke 19, Made in Austria, sign. Lorenzl, H. 29,3 cm; Wien Museum Inv. 220.704
6204,809,14,8,A gem. N, gem. lig. AS, Firmenmarke 19, Made in Austria, sign. Lorenzl, H. 31 cm; eBay # 871659924
6204,818,6 ger. VII, gem. N, Firmenmarke 19, Made in Austria, sign. Lorenzl, H. 29,9 cm; eBay # 2671560752
6204,833,11 gem. SB, Firmenmarke 19, Made in Austria, sign. Lorenzl, H. 31 cm; Lyon / T. 24.4.2001
6204,891,11 gem. NG, Firmenmarke 19, Made in Austria, sign. Lorenzl, H. 33 cm; eBay # 2572075107
6204,906,16 ger. A, gem. lig. NO, X, Firmenmarke 19, Made in Austria, sign. Lorenzl, Pick. 8, Originalpick.: „Kunstgewerbehaus Sticher, Wuppertal", H. 30,6 cm; Sammlung Weilinger, Wien
6204,970,14 gem. N6, Firmenmarke 19, Made in Austria, sign.?, H. 31,8 cm; eBay # 194239044
6204,979,14 ger. A, XII, gem. SG, Firmenmarke 19, Made in Austria, unsign., H. 30,5 cm; eBay # 120892497
6204,1091,4 Firmenmarke 19, Made in Germany, unsign., Ausformung zw. 1938–1941, H. 32 cm; Wendl 31
6204,1109,33 Firmenmarke 19, (vermutlich Made in Germany), Ausformung zw. 1938–1941, H. 32 cm; Doro 30.5.1994
6204,1129,19,G gem. rot 3, Firmenmarke 19, Made in Germany, unsign., Ausformung zw. 1938–1941, H. 31 cm; Sammlung a@goldscheider.net
Ausst.: SSH 1992
-> vgl. 5054 (BRONZE); 5230; 5960; 5964= Pendant zu 5960; 6600; 8176

-> vgl. 5530 GRUPPE (Trio)
Pendant, bzw. ähnl. Modell -> vgl. 5434
Bem.: s. bei 5230
Lit.: s. bei 5230
Quelle: AGM (Tantiemenliste Lorenzl, ohne Datierung (ca. 1940).
GS neu: Modell-Nr. 82 „Gefangener Vogel" (H. 40 cm), Aufl. 3000 Stück

6206
Lorenzl, Josef
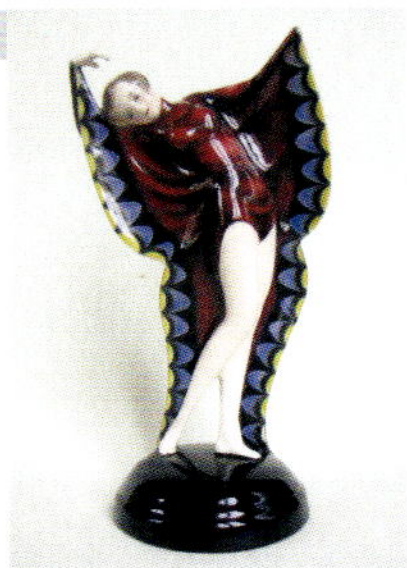
Tänzerin im Schmetterlings-kostüm (sich drehend)
(auf Rundsockel)
ca. 1929 (für den Entwurf)
6206,90,4,20 gem. lig. AH, +, Firmenmarke 19, Made in Austria, Hand dec., H. 30,5 cm; eBay # 2053076228; 3727146923
Bem.: Ähnliches Kostüm wie bei Modell-Nr. 6394
-> vgl. 5900

6207
Dakon, Stephan
„Tanz" (Schreitende Tänzerin, ihr Kleid fächerartig auseinander haltend) (auf Ovalsockel)
ca. 1930
6207,15,6 (Firmenmarke ohne Ang.), H. 30 cm; Christie's SK 13.5.2004
Quelle: Archiv Mimi Dakon (Wiener Manufaktur FG, Modellliste Dakon, 7.11.1946), hier mit Ang. des Originaltitels „Tanz".

6208
Lorenzl, Josef
„Dolores del Rio als Carmen" (Stehende Spanierin mit Fächer) (auf oktogonalem, gekehltem Sockel)
ca. 1927 (für den Entwurf)
6208,49,15 Firmenmarke 19, H. 31 cm; Doro 30.11.2001
Bem.: Dargestellt ist die mexikan. Filmschauspielerin Dolores del Rio (eigentl. Dolores Martinez Aúnsolo y López Negrete, 3.8.1905–11.4.1983) in einer ihrer bedeutendsten Rolle als „Carmen" (The loves of Carmen, 1927. Regie: Raoul Walsh).
-> vgl. 5616
Lit.: N.N.: Berühmte Frauen und moderne Keramik, in: MW, 10.1928/29, Nr. 6, S. 29 (m. Abb.), hier mit Ang. des Originaltitels „Dolores del Rio als Carmen" und der falschen Künstlerang. „Helmut Lindner".

6209
Dakon, Stephan
„Küsschen"
ca. 1929 (für den Entwurf)
-> vgl. 5962
Quelle: Archiv Mimi Dakon (Modellliste Dakon, undatiert, ca. 1940; Wiener Manufaktur FG, Modellliste Dakon, 7.11.1946), hier jeweils mit Ang. des Originaltitels „Küsschen".

6210F
Dakon, Stephan
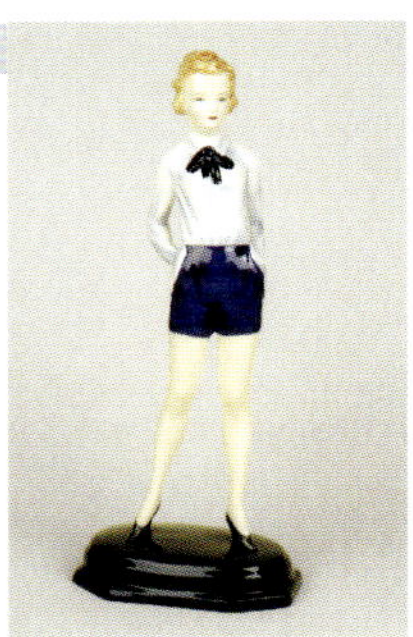
FIGUR „Bubi" (Louise Brooks als stehende junge Dame mit Bubikopf bzw. hellem, lockigem Haar) (auf abgerundetem Rechtecksockel)
ca. 1930
Variante: Bubikopf
6210,109,4,18 gem. F., gem. lig. TJ, S2, Firmenmarke 19, Made in Austria, Hand dec., H. 33 cm; eBay # 501582221 (KH www.tias.com/stores/frozenbeauty)
Variante: helles, lockiges Haar
6210,439,2 Firmenmarke 19, Made in Austria, H. 33 cm; PB Wien 19
6210F, H. 33 cm; GSNachf. 1,7
Bem.: Dieses Modell wurde, was die Frisur betrifft, in zwei Versionen produziert. Dem sog. „Bubikopf" (dunkles, kurz geschnittenes Haar) wie bei Modell 5815 wird nun in dieser Größe noch die Variante mit hellem, lockigem, etwas längerem Haar gegenübergestellt. Das „bravere", nicht mehr so „kesse" Aussehen entsprach eher dem Geschmack der späten 30er Jahre. Zur dargestellten Person Louise Brooks -> vgl. Modell-Nr. 5815.
-> vgl. 5815; 5959= jeweils Bubikopf als Frisurvariante
Lit.: GSNachf. 1,7 (m. Abb.), hier mit Ang. des Originaltitels „Bubi"; SM 4.1928, Nr. 11, Umschlag (Titelbild)
Quelle: Archiv Mimi Dakon (Tantiemenabrechnung 1.7.–31.12.1937; 1.1.–30.6.1940; Modellliste Dakon, undatiert, ca. 1940; Wiener Manufaktur FG, Modellliste Dakon, 7.11.1946).

6211
Dakon, Stephan
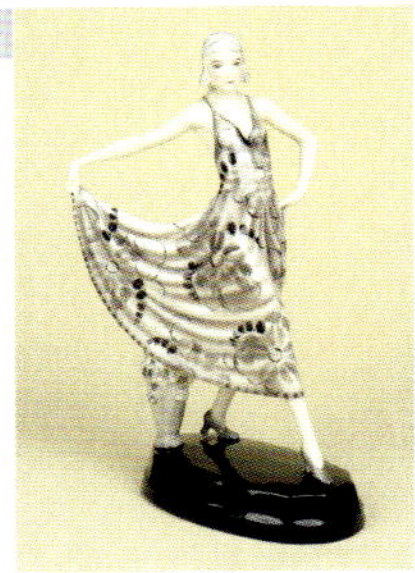
FIGUR „Harvey" (Lilian Harvey schreitende Tänzerin, ihren Rock mit der rechten Hand anhebend) (auf facettiertem Ovalsockel)
ca. 1926/27 (für den Entwurf)
6211,45,15,20 gem. NI, gem. lig. AH, Firmenmarke 19, Made in Austria, Hand dec., H. 40,6 cm; MAK Wien, Inv. Ke 10602-36
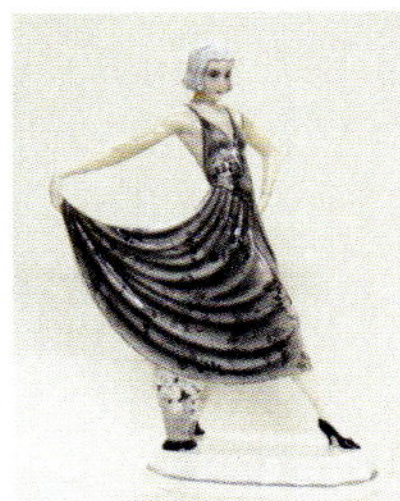
6211,231,19,3 gem. F., gem. blau NI, Firmenmarke 19, Made in Austria, Hand dec., sign. „Lorenzl"!, H. 40 cm; eBay # 1247114552;
Bem.: Die Sign. „Lorenzl" ist ein Versehen des Modelleurs.
6211,245,11 (Firmenmarke ohne Ang.), sign. „Lorenzl"!, H. 40 cm; Christie's SK 18.9.1998; Bem.: Die Sign. „Lorenzl" ist ein Versehen des Modelleurs.
-> vgl. 5625; 6009
Lit./Bem.: Becker, K.: Lilian, das Mädchen mit den 100 Gesichtern, in: Uhu 4.1927/28, Nr. 10, S. 54 (m. Abb.), hier eine zeitgenössische Fotografie, die wohl als Inspiration für die Ausführung diente. Dargestellt ist die Schauspielerin Lilian Harvey (eigentl. Lilian Pape, 19.1.1906–27.7.1968), hier mit dunkler Perücke, betitelt „Spanischer Tanz". Fotografie: Binder; Pinhas, S. 156 (m. Abb.).
Quelle: Archiv Mimi Dakon (Tantiemenabrechnung 1.1.–30.6.1940, Modellliste Dakon, undatiert, ca. 1940; Wiener Manufaktur FG, Modellliste Dakon, 7.11.1946), hier jeweils mit Ang. des Originaltitels FIGUR „Harvey".

6214
Dakon, Stephan
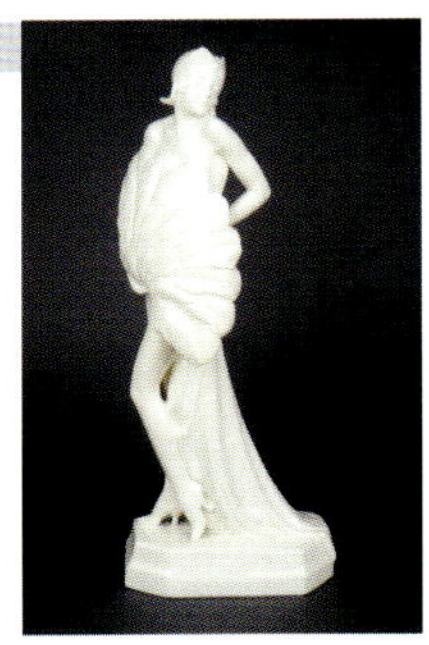
FIGUR „Fächer(dame)" (Stehende Varietétänzerin mit großem Fächer) (auf an den Ecken abgeschrägtem, gekehltem Rechtecksockel)
ca. 1928/29 (für den Entwurf)
6214,17,15,15 (Firmenmarke ohne Ang.), unsign., H. 33 cm; Christie's SK 15.2.2001
6214,54,10,19 gem. S., Firmenmarke 19, Made in Austria, Hand dec., unsign., H. 33,3 cm; Wien Museum Inv. 219.192
6214,140,4 (Firmenmarke ohne Ang.), H. 33 cm; Miller's A, Vol. 16.1995, (m. Abb.), hier Sockelvariante mit abgerundeten Ecken.
-> vgl. 5876
Quelle: Archiv Mimi Dakon (Tantiemenabrechnung 1.7.–31.12.1937), hier mit Ang. des Originaltitels FIGUR „Fächer"; (Modellliste Dakon, undatiert, ca. 1940; Wiener Manufaktur FG, Modellliste Dakon, 7.11.1946), hier jeweils mit Ang. des Originaltitels „Fächerdame".

6218
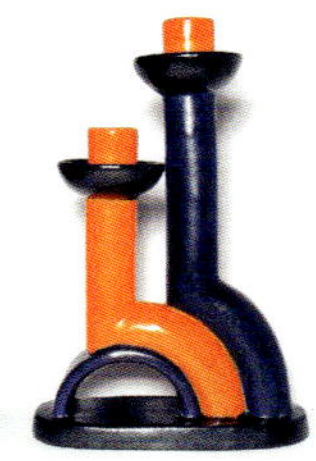
KERZENHALTER
(zweiflammig)
Form: zwei Kerzentüllen auf rohrartigem Unterbau, (auf Ovalsockel)
Dekor: zweifarbig
ca. 1930
06218,19 gem. W.H, Firmenmarke 19, H. 28,5 cm; PB Wien
6218 Firmenmarke 19, Made in Austria, H. 28,7 cm; Sammlung a@goldscheider.net

6227
Lorenzl, Josef
Schreitende Tänzerin (etwas nach vorne gebückt), ihr Kleid schmetterlingsflügelartig auseinander haltend (auf Ovalsockel)
ca. 1930
6227,22,14 Firmenmarke 19, Made in Austria, Hand dec., H. 34,3 cm; Quit. 21
6227,25,10 (Firmenmarke ohne Ang.), H. 35 cm; Christie's SK 4.11.1999
6227,82,11,19 (Firmenmarke ohne Ang.), sign.?, H. 35 cm; Doro L 25.–27.11.1996
6227,135,4,3 gem. F., N2, Firmenmarke 19, Made in Austria, Hand dec., H. 34 cm; WKA 38
6227,191,8,4 gem. lig. CO, Firmenmarke 19, Made in Austria, H. 34,5 cm; WKA 35
6227,220,2 Firmenmarke 19, H. 34,5 cm; SOTH. 1987, S. 292
-> vgl. 6693

6229
Lorenzl, Josef
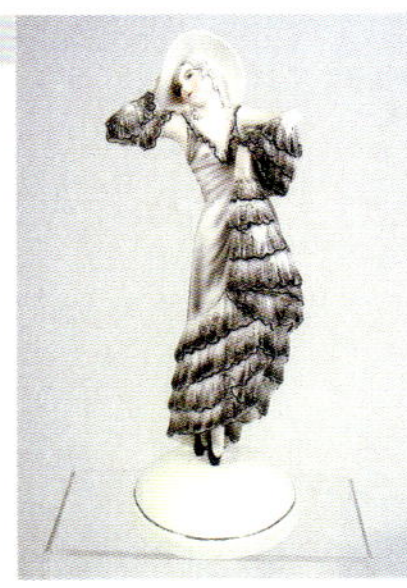
Stehende junge Dame mit großem Hut (in Pose)
(auf Rundsockel)
ca. 1930
6229,19,4 gem. NIF, Firmenmarke 19, Made in Austria, Hand dec., Pick. 8, H. 32,5 cm; WKA 35
6229,116,19,18 (Firmenmarke ohne Ang.), unsign., H. 33 cm; WKA 33
6229,133,15,19 Firmenmarke 19, Made in Austria, Hand dec., H. 33 cm; city-antik
6229,139,15 (Firmenmarke ohne Ang.), H. 32 cm; Doro 15.3.2005
6229,189,4 Firmenmarke 19, H. 31,5 cm; Doro 10.3.2003
6229,197,15 ger., Firmenmarke 19, H. 32 cm; Herr 54
6229,199,11 (Firmenmarke ohne Ang.), unsign.?, H. 32 cm; Christie's SK 15.11.2001
6229,201,19 Firmenmarke 19, H. 32 cm; Doro 8.3.2004
6229,256,10 gem. lig. AS, grün gem. 2,S., Firmenmarke 19, Made in Austria, Hand dec., H. 32,1 cm; MAK Wien, Inv. Ke 10521-43
6229,544,10 Firmenmarke 19, Made in Austria, Fabriqué en Autriche, H. 32 cm; WKA 38
6229,552,– (Firmenmarke ohne Ang.), H. 31,7 cm; Christie's SK 9.4.1999
6229,592,16 gem. lig. JK, Firmenmarke 19, Made in Austria, H. 34,4 cm; eBay # 1302967466 (KH www.tias.com/stores/frozenbeauty)

6233
Bosse, Walter
BUCHSTÜTZE Max (als sitzende Figur)
ca. 1930
6233,12,82 gem. lig. MK, Firmenmarke 19, Made in Austria, H. ca. 14 cm; eBay # 181232950
BUCHSTÜTZE Moritz (als sitzende Figur)
6233,21,31 gem. lig. AH, Firmenmarke 19, Made in Austria, H. 15,2 cm; eBay # 181232950
Bem.: Diese beiden Grotesken von Walter Bosse gehen auf Wilhelm Buschs Verserzählung mit Bildern aus „Max und Moritz" (1858) zurück. GS weicht hier von seiner an sich üblichen Modellnummernvergabe (für jedes Objekt eine eigene Nummer) ab. Beide Figuren erhielten dieselbe Modell-Nr. 6233. -> vgl. zu dieser Problematik die Nummern 6164 sowie 6165.

6234
Sitzende französische Bulldogge
ca. 1930
6234,7,31 Firmenmarke 19, Made in Austria, H. 18 cm, L. 21 cm; Sammlung F. & J.P. Girard/Frankreich

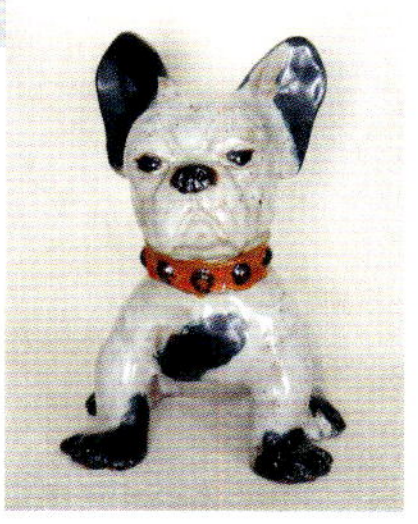

6236
Bosse, Walter (zugeschr.)
BUCHSTÜTZE Blumen
Form: rechtwinkelig, angesetzter Viertelkreisbogen
Dekor: zwei stilisierte Blumen, Laufglasur
ca. 1930
6236,3,32 gem. lig. HH., Firmenmarke 19, Made in Austria, unsign., H. 10,4 cm; Wien Museum Inv. 190.176/2
6236,6,32 gem. lig. HH., Firmenmarke 19, Made in Austria, unsign., H. 10,4 cm; Wien Museum Inv. 190.176/1
Ausst.: SSH 1992

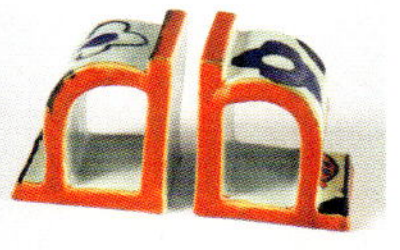

6239
Meisinger, Ida (zugeschr.)
GRUPPE Hochzeitspaar (stehende Braut mit Bräutigam) (auf Rechtecksockel)
ca. 1930
6239,13,7 gem. lig. AH, Firmenmarke 19, Made in Austria, unsign., H. 16 cm; eBay # 3721478080
6239,68,65 Firmenmarke 19, unsign., H. 15,6 cm; Doro 9.6.1998; Lit.: AZ, 1998, S. 409 (m. Abb.)
6239,83,20 gem. lig. TK, Firmenmarke 19, Made in Austria, unsign., H. 16 cm; Sammlung a@goldscheider.net
6239,98,65 Firmenmarke 19, Made in Austria, unsign., H. 15,8 cm; BM 76; Fischer 151
6239 ger., gem. A, Firmenmarke 19, Made in Austria, unsign., H. 15,6 cm; PB

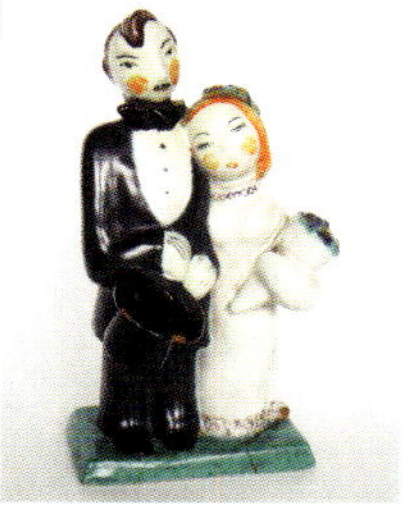

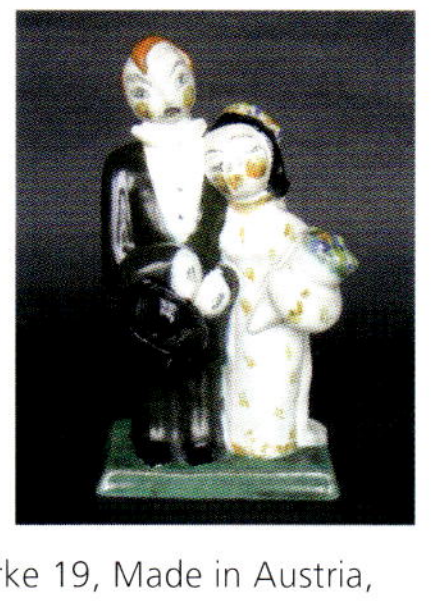

6245
Lorenzl, Josef
Stehende Haremstänzerin (mit gesenktem Kopf) (auf an den Ecken abgeschrägtem Quadratsockel)
ca. 1929 (für den Entwurf)
6245,30,4 Firmenmarke 19, Made in Austria, Hand dec., unsign.?, H. 30,5 cm; Soth. L. 17.10.1985
–> vgl. 5927

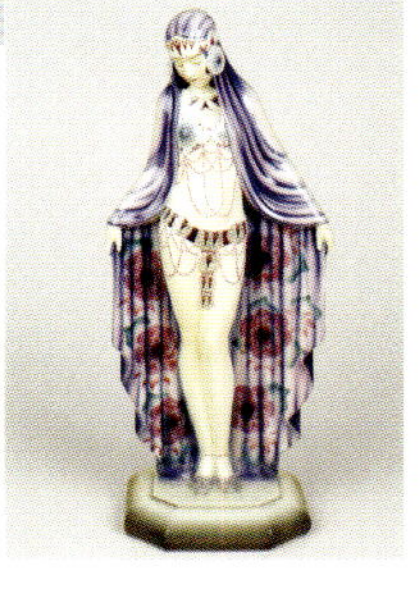

6248
Lorenzl, Josef
„Clotilde von Derp" (Schreitende Tänzerin in Kniehosen) (auf Ovalsockel)
ca. 1928 (für den Entwurf)
6248,20,11 Firmenmarke 19, Made in Austria, Hand dec., sign. Lorenzl, H. 27 cm; KH
6248,45,11,20 gem. F.2, Firmenmarke 19, Made in Austria, Hand dec., unsign., H. 28 cm; WKA 40; Doro 10.3.2003
6248,117,6,18 gem. FN Ru 2, Firmenmarke 19, Made in Austria, Hand dec., sign. Lorenzl, H. 27,9 cm; eBay # 1234348060
6258 (vielmehr 6248),146 ger., 19 gepr., gem. NF, gem. lig. JR2, Firmenmarke 19, Made in Austria, Hand dec., sign. Lorenzl, H. 27 cm; eBay # 1250720026, Bem.: Durch ein Versehen des Modelleurs wurde die Nr. 6258 eingeritzt, richtig muss es aber 6248 heißen.
6248,256,4,3 gem. NIF, Firmenmarke 21a, Made in Austria, Hand dec., sign. Lorenzl, H. 27,8 cm; Sammlung a@goldscheider.net
6248,478,4 Firmenmarke 19, Made in Austria, Hand dec., sign. Lorenzl, H. 27,8 cm; Doro 1.12.2000
Bem.: Dargestellt ist die mit Alexander Sacharoff verheiratete Tänzerin Clotilde von Derp (5.11.1892, Berlin – 11.1.1974, Rom) in ihrem Tanz zu „Papillons", Musik: Robert Schumann, 1912. (–> vgl. Bem. bei 5732).
–> vgl. 5732

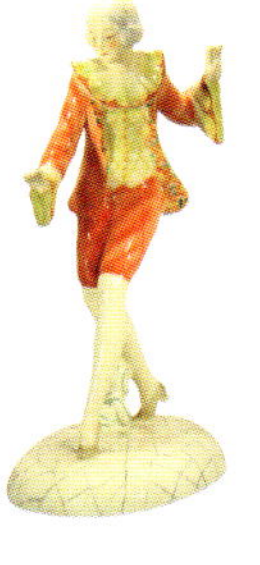

6252
Meisinger, Ida
DECKELDOSE Springendes Reh
Form: rechteckig
Dekor: springendes Reh (reliefiert)
ca. 1930
6252,JP gem., (06253,17,1, gepr.), Firmenmarke 19, Made in Austria, sign. Meisinger, Originalpick.: „Holstein & Düren, Köln", H. 3,8 cm, B. 9,6 cm, L. 11,4 cm; PB
6252 gem., gem. lig. AH, Firmenmarke 19, Made in Austria, unsign., H. 3,7 cm, B. 9,5 cm, L. 11,2 cm; Sammlung a@goldscheider.net
6252 gem., gem. lig. JP, Firmenmarke 19, Made in Austria, sign. Meisinger, H. 3,5 cm, B. 9,5 cm, L. 11 cm; eBay # 1468015856
Lit.: GSN, S. 54, 55 (m. Abb.)
GS neu: Modell-Nr. 31 „Schwarzer Wald, rotes Reh" (L./B. je 15 cm), Aufl. ohne Ang.

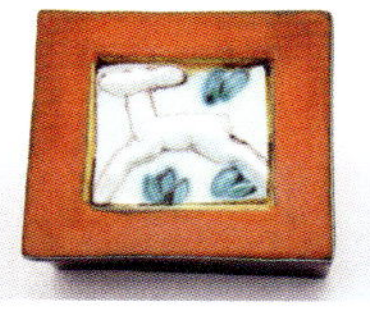

6258
Thronender Buddha
ca. 1930
6258,20,7 gem. XXVII, D1, Firmenmarke 19, Made in Austria, Hand dec., H. 34,3 cm, B. 20,3 cm; eBay # 1215362607

6263
Knörlein, Rudolf (zugeschr.)
KERZENHALTER (zweiflammig)
Form: U-förmig, vier leicht gebogene Standfüße
Dekor: gemalte Linien, Laufglasur
ca. 1930
6263,28,14 Firmenmarke 19, unsign., H. 15,3 cm; WKA 9
6263,37,20 gem. H, Firmenmarke 19, Made in Austria, unsign., H. 15,3 cm; Doro 6.4.1999
Bem.: Für die Zuschreibung –> vgl. ähnl. Kerzenhalter ohne Nr. (Originalfoto, Privatarchiv).

6269
Knörlein, Rudolf (zugeschr.)
KERZENHALTER (fünfflammig)
Form: fünf Kerzentüllen auf rohrartigem Unterbau, über halbkugelförmigem Sockel
Dekor: runde Farbflecken, Laufglasur
ca. 1930
6269,53,20 gem. SCH, Firmenmarke 19, Made in Austria, unsign., Originalpick.: „J.K. Klammert, Graz", H. 16,6 cm, L. 32,2 cm; PB Wien 19
6269,106,31 gem. lig. AH, Firmenmarke 19, Made in Austria, unsign., H. 16,5 cm, L. 32,2 cm; Sammlung a@goldscheider.net
Bem.: Für die Zuschreibung –> vgl. ähnl. Kerzenhalter ohne Nr. (Originalfoto, Privatarchiv).
Ausst.: PSK 1995
Lit.: ADGS, Nr. 36 (m. Abb.)

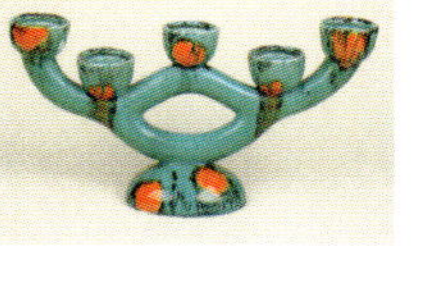

6270
Meisinger, Ida
Bellender Scotchterrier (in spielerischer Pose)
ca. 1930
6270,129,31 Firmenmarke 19, H. 12 cm; eBay # 65474838834
6270,167,20 Firmenmarke 19, Made in Austria, H. 14,2 cm; PB, Wien
ohne Ang., H. fehlt, L. 20,3 cm; Soth. Arcade 1576

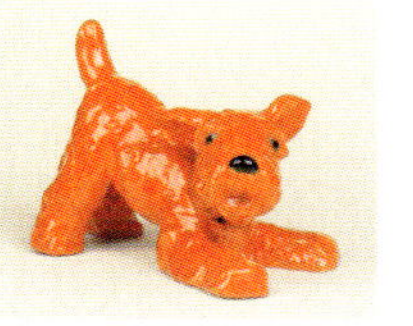

6277
Dakon, Stephan
Ballerina beim Spitzentanz (auf Rundsockel mit Hut)
ca. 1930
6277,50,4,19 gem. S., Firmenmarke 19, Made in Austria, Hand dec., H. 34 cm; Wien Museum Inv. 190.135

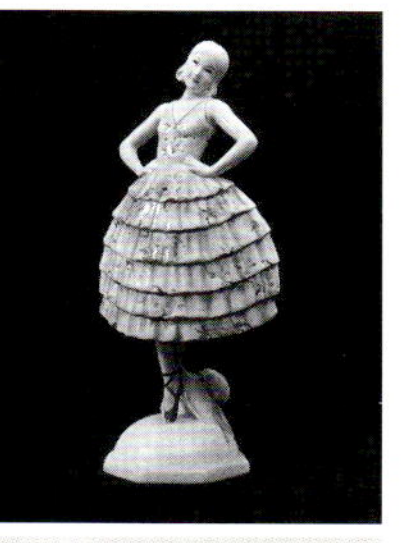

6282
Geiger, Benno
SCHALE
Form: oval
Dekor: mehrfarbig, Laufglasur
ca. 1930
6282 Firmenmarke 19, Made in Austria, H. 8,5 cm, L 31 cm; eBay # 6603243387
6282 gem. lig. AH, Firmenmarke 19, Made in Austria, H. 9 cm, L 31 cm; Sammlung Weilinger, Wien
–> vgl. 677= ähnl. Modell.

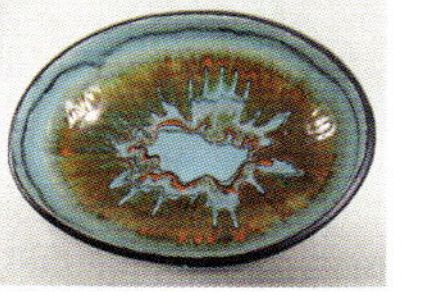

6283
Geiger, Benno (zugeschr.)
SCHALE
Form: oval
Handhaben: zwei M-förmige Griffe mit Öse
Dekor: mehrfarbig, Laufglasur
ca. 1930
06283 gem. lig. AH, Firmenmarke 19, Made in Austria, unsign., H. 10 cm, L 36,5 cm, B. 21 cm; eBay # 300056387750

6284
Meisinger, Ida
Sitzendes Mädchen mit Trompete
ca. 1930
6284,10,7 ger., gem. AH, Firmenmarke 19, Made in Austria, H. 13,5 cm; Sammlung Stressless, Portugal
6284,24,20 gem. GI (oder CI), Firmenmarke 19 (zwei mal), Made in Austria, unsign., H. 13,5 cm, B. 11 cm; eBay # 914797778
Bem.: Dieses Modell belegt, wie ähnl. sich manche grotesken Entwürfe von Ida Meisinger und Walter Bosse sind. Gerne werden, u. a. vom Kunsthandel, Meisingers (teils, wie bei Bosse üblich, unsignierten) Arbeiten aus Gründen der „Preissteigerung" Walter Bosse zugeordnet (–> vgl. Modell-Nr. 6239).

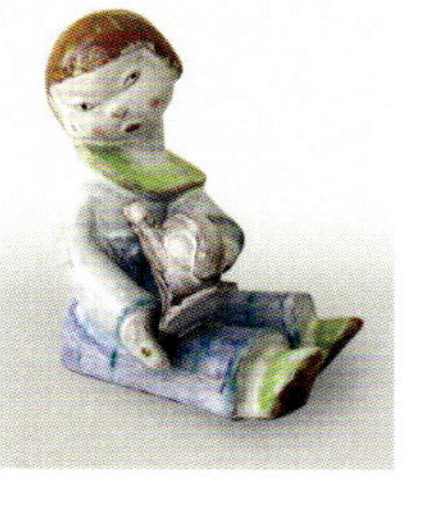

6285
Meisinger, Ida
RELIEF Sitzender Scotchterrier (Wandfliese)
ca. 1930
6285,64,33 gem. W.H., Firmenmarke 19, Made in Austria, H. 10,1 cm, B. 8,8 cm; eBay # 3650875001
6285,72,65 gem. W.H., Firmenmarke 19, Made in Austria, H. 10 cm, B. 9,6 cm; Sammlung a@goldscheider.net
–> vgl. 6316= DECKELDOSE

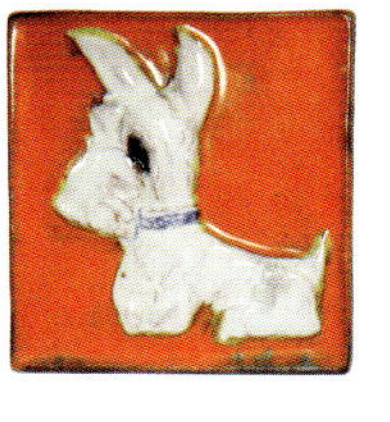

6288
Weber, Ernst
WANDMASKE Frauenkopf mit Hand, eine Maske haltend
ca. 1930

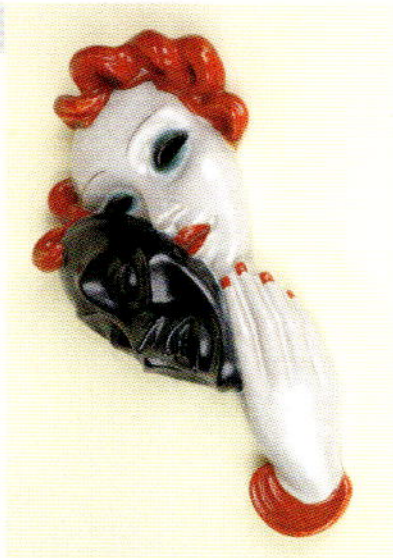

6288,8 (Firmenmarke ohne Ang.), unsign., H. 35 cm; Christie's SK 4./5.12.1996
6288,12 Firmenmarke 19, unsign.?, H. 36 cm; Soth. L. 17.10.1985
6288,63,14 gem. A, Firmenmarke 19, Made in Austria, H. 35 cm; Sammlung.schmutz@chello.at 5

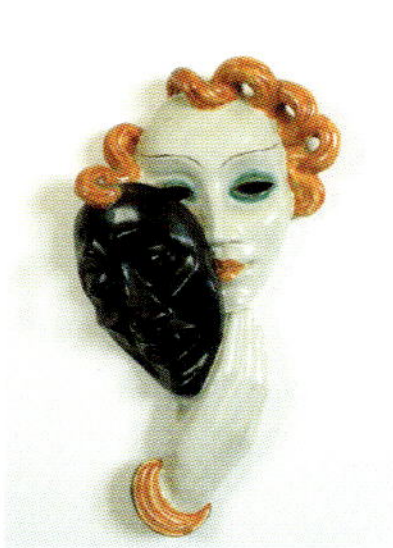

(62)88,74 ger., Firmenmarke 19, Made in Austria, H. 36 cm; Christie's L. 25.9.1987
6288,99,65 gem. A, Firmenmarke 19, Made in Austria, unsign.?, H. 35 cm; KH
6288,107,65 (Firmenmarke ohne Ang.), unsign., H. 35 cm; Doro 15.9.2003
6288,S,+,R gem., Firmenmarke 19, Made in Austria, unsign.?, H. 35,5 cm; BM 40

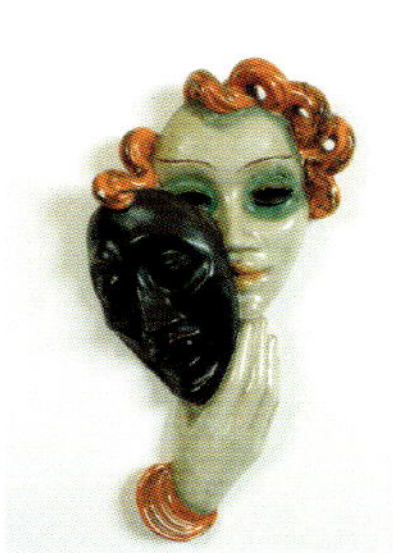

6288 ger., Firmenmarke 19, Made in Austria, unsign.?, H. 34,7 cm; Sammlung.schmutz@chello.at 36
6288 gem. lig. AH, Firmenmarke 19, Made in Austria, unsign., H. 36 cm; GK, S. 21, 116 (m. Abb.); Wien Museum Inv. 182.097
6288 Firmenmarke 19, unsign., H. 34 cm; PB Wien 19
Ausst.: HMW (jetzt Wien Museum) 1985/86.
–> vgl. 6787
Lit.: GK, S. 21, 116 (m. Abb.); Arwas (1987), S. 292 (m. Abb.).

6295
Lorenzl, Josef
Kniender Mädchenakt mit trinkendem Rehkitz (auf ovalem Landschaftssockel)
ca. 1930
6295,68,19 Firmenmarke 19, Made in Austria, Hand dec., H. 14 cm, B. 20 cm; Kett. 148
6295,123,11,4 Firmenmarke 19 (blau), Made in Austria (blau), H. 13,5 cm, B. 20 cm; Wien Museum Inv. 190.095; Bem.: Ausführung in Terrakotta.
6295 (Firmenmarke ohne Ang.), H. 14 cm, B. 20,5 cm; Fischer 110

6305
Knörlein, Rudolf
STEHMASKE / WANDMASKE Frauenkopf mit Haarschmuck mit/ohne Blumenblüte
ca. 1930/31

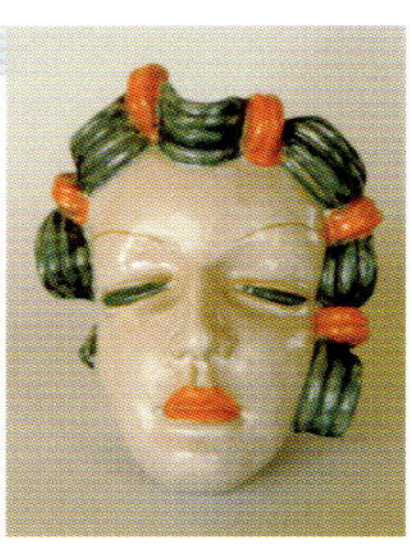

STEHMASKE mit Blumenblüte
6305,3 ger., gem. grün G., Firmenmarke 19, Made in Austria, H. 23 cm; MAK Wien, Inv. Ke 10602-81
6305,31 ger., gem. grün G., Firmenmarke 19, Made in Austria, sign. Monogr. RK, H. 23,5 cm; GK, S. 21, 121 (m. Abb.); Wien Museum Inv. 223.911
6305,67,20 gem. lig. AH, Firmenmarke 19, Made in Austria, unsign., H. 28 cm (Messfehler?); eBay # 355523456
6305,69,31 Firmenmarke 19, Made in Austria, unsign., H. 22 cm; Soth. L. 23.6.1989
6305,A,5i, (Firmenmarke ohne Ang.), unsign., H. 23,4 cm; Christie's SK 14.3.1997
6305 ger., gem. S., Firmenmarke 19, Made in Austria, unsign., H. 22,6 cm; Sammlung.schmutz@chello.at 78
ohne Ang., gem. B, Firmenmarke 19, Made in Austria, unsign., H. 23 cm; Christie's L. 18.7.1985
WANDMASKE ohne Blumenblüte
6305,A ger., gem. S., Firmenmarke 19 (zwei mal), Made in Austria, unsign., H. 18,7 cm; Sammlung.schmutz@chello.at 61
6305 ger., gem. AB, Firmenmarke 19, Made in Austria, H. 18 cm; Soth. L. 20.10.1989
Ausst.: HMW (jetzt Wien Museum) 1985/86; SSH 1992.
Lit.: GK, S. 21, 121 (m. Abb.); GSN, S. 56, 57 (m. Abb.).
GS neu: Modell-Nr. 10 „Meine Blüte" (H. 20 cm), Aufl. ohne Ang.

6306
Knörlein, Rudolf (zugeschr.)
WANDMASKE Frauenkopf mit zurückgekämmten Haaren
ca. 1930/31

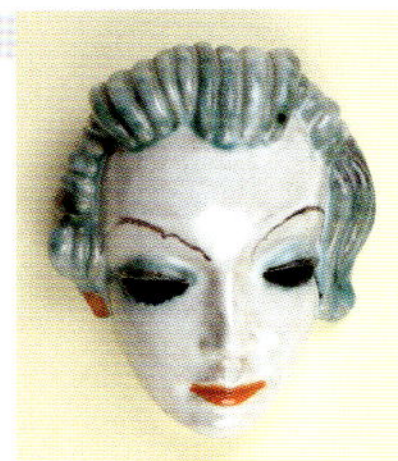

6306,39,20 gem. lig. TK, Firmenmarke 19, Made in Austria, unsign., H. 16,5 cm; Soth. com. # 3J5HD
6306,72,33 gem. H., Firmenmarke 19, Made in Austria, unsign., H. 15 cm; PB

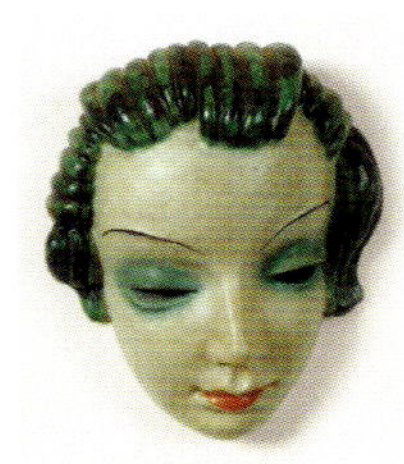

6306,74,33 Firmenmarke 19, Made in Austria, unsign., H. 15 cm; PB Wien 19
6306,92,20 gem. G, Firmenmarke 19, unsign., H. 16,5 cm; Wien Museum Inv. 186.666
6306 ger., gem. G., Firmenmarke 19 (zwei mal), Made in Austria, unsign., H. 16,8 cm; Sammlung.schmutz@chello.at 37

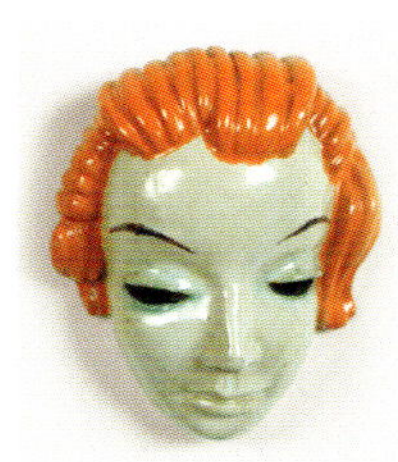

6306 ger., Firmenmarke 19, Made in Austria, unsign., H. 15 cm; Sammlung f@goldscheider.de
ohne Nr. Firmenmarke 19, Made in Austria, unsign., H. 16,9 cm; Sammlung.schmutz@chello.at 55

6309
Meisinger, Ida
KERZENHALTER (einflammig) „Der Modehund" (Sitzender Scotchterrier auf runder Schale)
ca. 1930/31

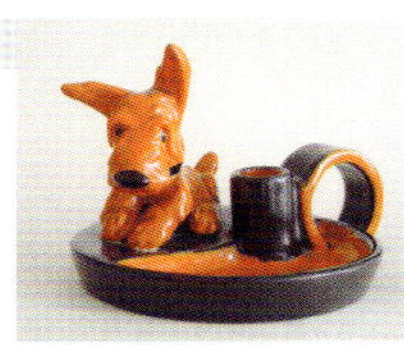

6309,3 Firmenmarke 19, Pick. 15 (Entwurf Ida Meisinger, Wien), H. 16 cm; Doro 9.11.1989
06309,24,1,20 (Firmenmarke ohne Ang.), unsign., H. fehlt, Ø 22,5 cm; Christie's SK 4./5.12.1996
6309 gem. M, Firmenmarke 19, Made in Austria, Pick. 15 (Entwurf Ida Meisinger, Wien), H. 14,2 cm, Ø 22 cm; PB Deutschland
–> vgl. 5854; 5879; 6026= jeweils FIGUR

6315
Meisinger, Ida
Sitzende Bulldogge (oder Boston Terrier) (mit Halsband)
ca. 1930/31
6315,16,34 Firmenmarke 19, Made in Austria, H. 13,5 cm; Privatsammlung

6316
Meisinger, Ida
DECKELDOSE Sitzender Scotchterrier
Form: rechteckig
Dekor: sitzender Scotchterrier (reliefiert)
ca. 1930/31

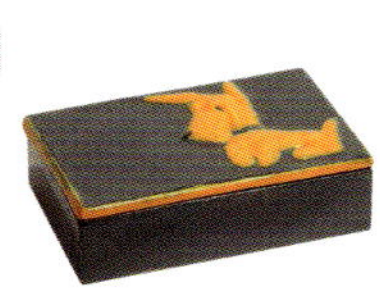

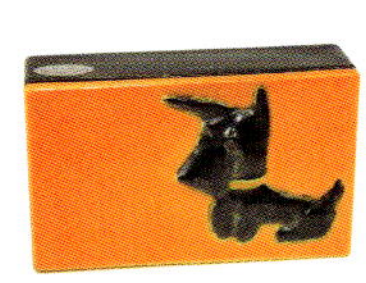

6316,3 gem. G., Firmenmarke 19, Made in Austria, Pick. 6, H. 4 cm, B. 8 cm, L. 12,8 cm; Sammlung a@goldscheider.net
6316,22 ger., gem. lig. CI, Firmenmarke 19, Made in Austria, H. 3,5 cm, B. 7,7 cm, L. 12,5 cm; Sammlung John Nash
6316 ger., gem. B, Firmenmarke 19, Made in Austria, unsign., H. 4 cm, B. 8 cm, L. 12,8 cm; Sammlung a@goldscheider.net
–> vgl. 6285= RELIEF (Wandfliese)

6322
Postl, Hugo
Liegender Tiger
ca. 1930/31 (für die Ausführung)

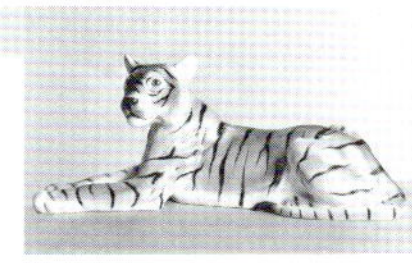

6322,82,10 gem. lig. HH, Firmenmarke 19, H. 17 cm, L. 40 cm; Ölschläger 14
6322,159,19 Firmenmarke 19, Made in Austria, unsign., H. 16,5 cm, L. 38 cm; Wien Museum Inv. 190.030
–> vgl. 4944= ähnl. Modell
6322 H. 17 cm, L. 42 cm; Fotoarchiv K. Täuber
Quelle: Fotoarchiv K. Täuber

6323
Lorenzl, Josef
Tänzerin, auf dem rechten Bein tänzelnd, das linke Bein nach vorne schwingend (auf Rautensockel)
ca. 1929/30 (für den Entwurf)
6323,143,19 Firmenmarke 19, Made in Austria, sign. Lorenzl, H. 32 cm; WKA 38
6323,252,6,9 (Firmenmarke ohne Ang.), sign. Lorenzl, H. 29,5 cm; Christie's SK 20.9.2000
6323,598,37,G (Malerzeichen ohne Ang.), Firmenmarke 19, unsign., H. 31,5 cm; Zeller 45
6323 ohne Ang., Firmenmarke 19, unsign., H. 30 cm; RM 18
–> vgl. 6003; 6935

6325F
„Primavera"
ca. 1930/31
Quelle: AGM (Rechnung an Maurice Mizrahi, Alexandria, Ägypten, vom 22. April 1953), hier mit Ang. des Originaltitels „Primavera".

6328
Weiss, Claire (Klára)
(d.i. Herczeg, Claire (Klára)
„Muff" (Stehende Dame mit Muff und großem Hut) (auf Ovalsockel)
ca. 1930/31

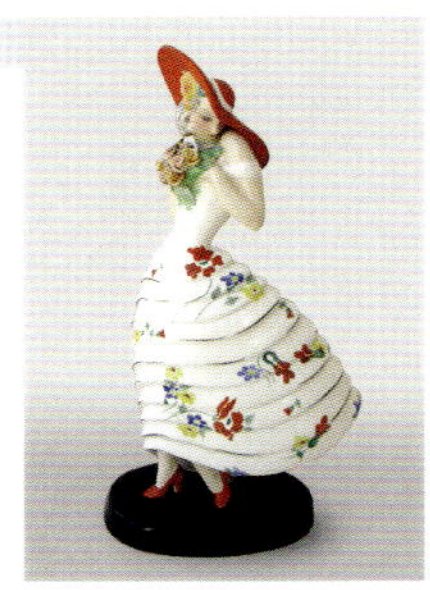

6328,27,4,19 gem. AH, Firmenmarke 19, Made in Austria, Hand dec., sign. Claire Weis, H. 30 cm; eBay # 2075105325
6328,50,19 Firmenmarke 19, Made in Austria, Hand dec., sign. Claire Weiss, H. 31,4 cm; Wendl 44
6328,149,6,19 gem. E., Firmenmarke 19, Made in Austria, sign. Claire Weiss, H. 31,5 cm; Schloss Hag. 34; Christie's SK 13.10.2004; Doro 1.12.2005
6328,245,11,3 Firmenmarke 19, Made in Austria, sign. Claire Weis, H. 31,7 cm; eBay # 1201992664; (KH www.tias.com/stores/frozenbeauty)
6328,335,10 ohne Firmenmarke, Made in Austria, sign. Claire Weiss, H. 30,5 cm; eBay # 401531586
6328,484,4 gem. BC, XII, Firmenmarke 19, Made in Austria, sign. Claire Weiss, H. 32,5 cm; Sammlung Stressless, Portugal; Bem.: Die hier vorliegende Figur wurde im Gegensatz zu allen anderen nachweisbaren Stücken mit einem Rosenstrauß anstatt des Muffs ausgestattet. Somit darf man dieses Exemplar als absolute Besonderheit bezeichnen, wenn es nicht sogar in dieser speziellen Ausführung ein Unikat ist.
–> vgl. 7044
–> vgl. 7720= GRUPPE

6330F
Dakon, Stephan
FIGUR „Kongress" (Lilian Harvey) (auf Rundsockel)
ca. 1931

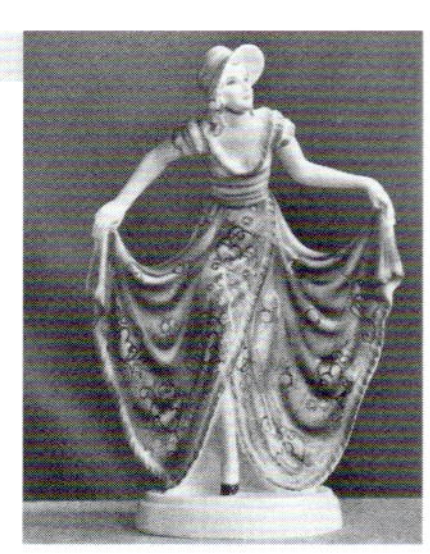

6330,2,33 gem. MR, Firmenmarke 19, Made in Austria, unsign., H. 33,5 cm; WKA 29
6330,3,28 Firmenmarke 19, Made in Austria, Hand dec., unsign., H. 33 cm; WKA 38;

Doro 31.5.2006
6330,42,6 gem. XX, gem. blau H, gem. rot D., Firmenmarke 19, Made in Austria, unsign., H. 32,5 cm; Soth. com. # 30DI; Soth. com. # 6TKB
6330,44,15,19 Firmenmarke 19, Made in Austria, H. 31 cm; Schloss Hag. 16
6330,68,19 Firmenmarke 19, Made in Austria, Hand dec., H. 31 cm; WKA 40
6330,164,4,3 gem. N2, F, Firmenmarke 19, Made in Austria, Hand dec., H. 30,8 cm; eBay # 1860110240 (Soth.)
6330,214,4,6 gem. JR II, Firmenmarke 19, Made in Austria, H. 31 cm; KNS 7; WKA 33; Doro 15.5.2002; 10.3.2003
6330,241,– Firmenmarke 19, Made in Austria, H. ca. 30,5 cm; Zengen 402
6330,244,19 (Firmenmarke ohne Ang.), H. 31 cm; Christie's SK 9.4.1999
6330,286,14 gem. SG, Firmenmarke 19, Made in Austria, H. 30,5 cm; eBay # 3728419213
6330,335,10 Firmenmarke 19, Made in Austria, H. 30,5 cm; eBay # 189187957
6330,364,19 (Firmenmarke ohne Ang.), H. 31 cm; Christie's SK 4.11.1999
6330,404,16 (Firmenmarke ohne Ang.), H. 32 cm; Christie's SK 18.9.1998
6330F, H. 32 cm, B. 22 cm; GSNachf. 1,8a
Bem.: Die Figur geht zurück auf eine Szene des Ufa-Films „Der Kongress tanzt" (Uraufführung 29.9.1931, Wien), in dem Lilian Harvey die weibliche Hauptrolle verkörperte (-> vgl. www.deutscher-tonfilm.de).
-> vgl. 6646
Lit.: GSNachf. 1,8 (m. Abb.), hier mit Ang. des Originaltitels „Kongress".
Quelle: AGM (Bestellungen der Firma Wahliss (Wien), vom 10.10.1945 und 27.2.1946 an die „Arbeitsgemeinschaft der Firma Friedrich Goldscheider", hier jeweils mit Ang. des Originaltitels „Kongress").

6335
Knörlein, Rudolf
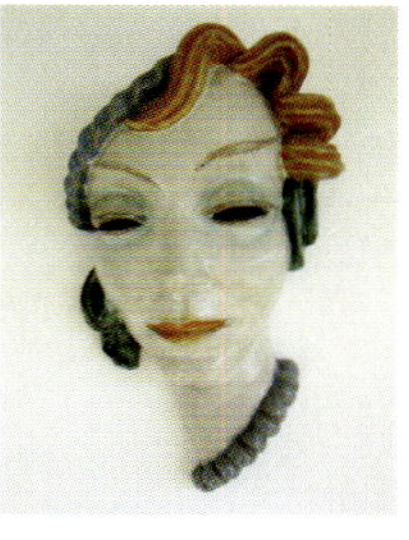
WANDMASKE Frauenkopf mit Halskette und Haarband
ca. 1931
ohne Nr.,20 gem. A, Firmenmarke 19, Made in Austria, Hand dec., unsign., H. 25,5 cm; Sammlung f@goldscheider.de
6335 Firmenmarke 19, Made in Austria, unsign., H. 29 cm; eBay # 3756777036

6345
Lorenzl, Josef
„Spanischer Shawl" (Stehende Tänzerin (Dolores del Rio?) in Pose, mit großem Schal) (auf Rundsockel)
ca. 1928/29 (für den Entwurf)
6345,102,15,3 Firmenmarke 19, Made in Austria, sign. Laurenzl, H. 38 cm; KH Ruby Lane (Jane Perlow/Lacreme-antik, Great Neck N.Y.)
6345,146,10,4 gem. S., Firmenmarke 19, Made in Austria, sign. Lorenzl, H. 38,5 cm; WKA 38
6345,166,14 Firmenmarke 19, Made in Austria, sign. Lorenzl, H. 37,2 cm; Soth. L. 3.5.1990
Bem. und Lit. vgl. bei 5822
-> vgl. 5822; 6062

6346
Lorenzl, Josef
Schreitende Tänzerin, ihr Kleid nach vorne hoch haltend (auf Ovalsockel)
ca. 1931
6346,16,19,20 gem. NF.11, Firmenmarke 19, Made in Austria, Hand dec., sign. Lorenzl, H. 31 cm; WKA 40
6346,24,4,3 Firmenmarke 19, Made in Austria, Hand dec., sign. Lorenzl, H. 32 cm; eBay # 2000795460
6346,50,11 gem. NF, Firmenmarke 19, sign. Lorenzl, H. 30,5 cm; Franzisi 10
6346 ohne Ang., sign. Lorenzl, H. 30 cm; Herr 18

6352F
Dakon, Stephan
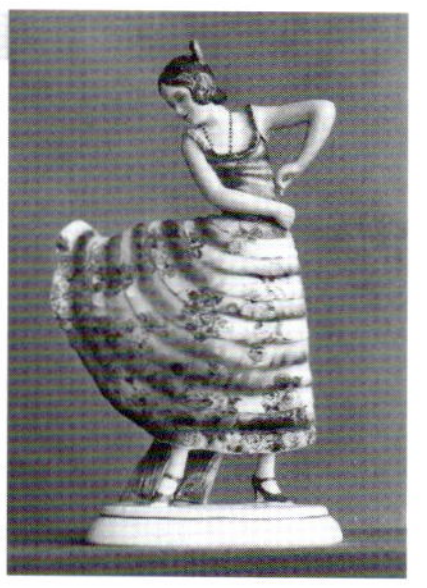
FIGUR „Castanet" (Tanzende Spanierin) (auf Rundsockel)
ca. 1931
6352F, H. 32 cm, B. 19 cm
Quelle: Archiv Mimi Dakon (Originalfoto)

6353
Matula, Viktor
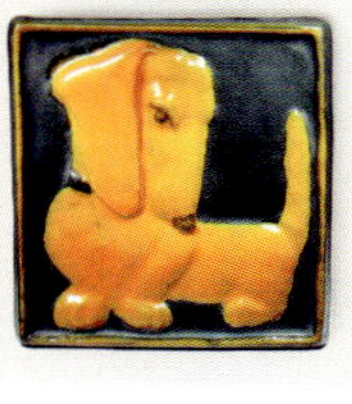
RELIEF Dackel (Wandrelief)
ca. 1934
ohne Nr. (6353), gem. lig. TK, Firmenmarke 19, Made in Austria, unsign., H. 10,1 cm, B. 10,2 cm; Sammlung a@goldscheider.net
Quelle: Nachlass Matula, Wien

6354
Matula, Viktor
RELIEF Springendes Reh (Wandrelief)
ca. 1931
6354 gem., gem. lig. TK, Firmenmarke 19, Made in Austria, unsign., H. ca. 13,5 cm, B. 10,2 cm; eBay # 3741884227
Quelle: Nachlass Matula, Wien

6356
Lorenzl, Josef
Stehende junge Dame in japanischem Kostüm, mit Blumen geschmücktem Haar und großem Fächer) (auf Rundsockel)
ca. 1931
6356 (Firmenmarke ohne Ang.), H. 33 cm; Pinhas, S. 125 (m. Abb.)

6360F
Dakon, Stephan
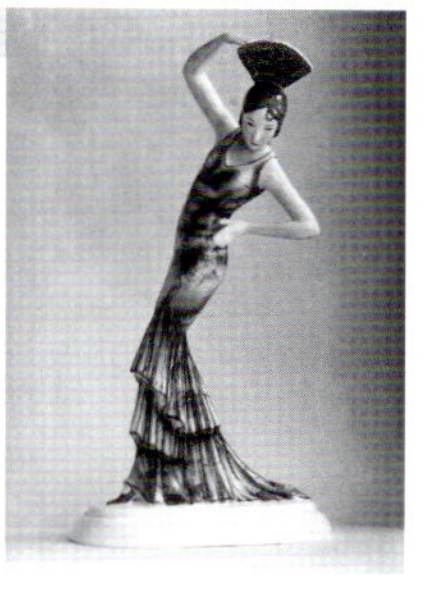
Spanierin (Carmen) in Pose (auf profiliertem länglichem Sockel)
ca. 1931
6360F, H. 34 cm, B. 18 cm
Quelle: Archiv Mimi Dakon (Originalfoto)

6361F
Dakon, Stephan
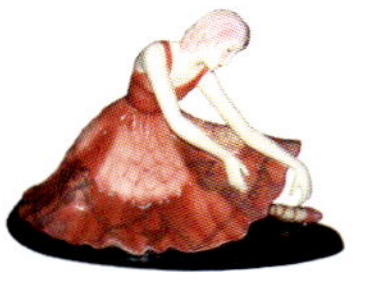
FIGUR „Sterbender Schwan" (Anna Pawlowa?) mit Federputz im Haar (auf Ovalsockel)
ca. 1931
6361,1 gem. GS, Firmenmarke 19, Made in Austria, unsign.?, H. 20,3 cm; eBay # 2015137389
6361,8 gem. lig. HS., E, Firmenmarke 19, Made in Austria, H. 18,5 cm, L. 27,5cm; Sammlung Weilinger, Wien
6361,8 Firmenmarke 19, Made in Austria, unsign., H. 18,5 cm, L. 28 cm; Wien Museum Inv. 190.127
6361,22,15 (Firmenmarke ohne Ang.), H. 18 cm; Christie's SK 12.5.2000
6361,34,11,19 Firmenmarke 19, Made in Austria, Hand dec., H. 18,5 cm; Quit. 25; WKA 38; Sammlung Alain Bernard
6361,111,26,3 gem. S., gem. lig. NO, Firmenmarke 19, Made in Austria, H. 18 cm, L. 27 cm; Sammlung a@goldscheider.net
6361,208,12 Firmenmarke 19, Made in Austria, H. 17,5 cm, L. 26,7 cm; WKA 33
6361,210,19 Firmenmarke 19, H. 18 cm; PB Wien 19
6361,249,6,F Firmenmarke 19, Made in Austria, H. 19 cm, L. 27,9 cm; eBay # 20906373
6361,279,? Firmenmarke 19, Made in Austria, L. 27,5 cm; RM 29
6361,273,12,G gem. N (od. Z?), Firmenmarke 19, Made in Austria, Pick. 1, H. 18 cm; 3362589404
6361,287,2 Firmenmarke 19, Made in Austria, H. 18 cm; Doro 8.3.2004; 15.3.2005
6361,409,19 Firmenmarke 19, Made in Germany, unsign., Ausformung zw. 1938–1941, H. 18 cm; eBay # 1280156623
Bem.: Dargestellt ist vermutlich die Ballerina Anna Pawlowa (12.2.1881–23.1.1931) -> vgl. Lit.: Dictionnaire du Ballet moderne, Paris 1957, S. 265 (m. Abb.). Diese Figur wurde auch noch zw. 1953–1960 bei Carstens (GWM West-Germany) produziert, dort jedoch mit leicht abgewandeltem Federputz.
-> vgl. 7426= ähnl. Modell (mit/ohne Federputz im Haar, mit/ohne Schwanenflügel am Kleid)
Lit.: KGSF1; KGSF2; Könecke (2002), S. 34, S. 76 (m. Abb.); Witkiewicz, S. 130–131 (m. Abb.).
Quelle: AGM (Bestellung der Firma Wahliss (Wien), vom 30.8.1945 an die „Arbeitsgemeinschaft der Firma Friedrich Goldscheider"; Bestellung der Firma Wahliss (Wien), vom 27.2.1946, hier mit Ang. des Originaltitels „Sterbender Schwan" in Ausführung „Terrakotta" und „Fayence"; Rechnung an Maurice Mizrahi, Alexandria, Ägypten, vom 22. April 1953).

6362
Meisinger, Ida
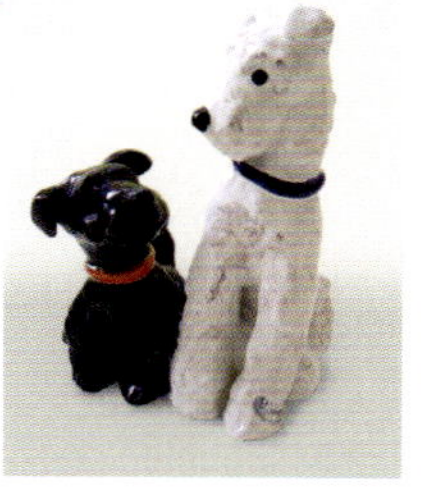
„Pat und Patachon" (zwei sitzende Airedaleterrier)
ca. 1931
6362,33 Firmenmarke 19, Made in Austria, unsign., Pick. 6, H. 16,5 cm; Sammlung f@goldscheider.de
6362,634,65 Firmenmarke 19, Made in Austria, H. fehlt; Zengen 386
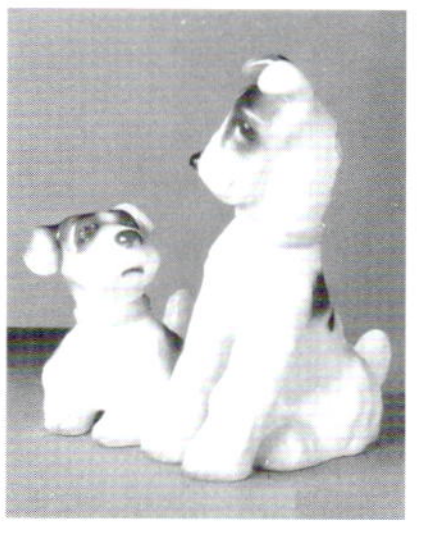
6362,209,31 gem. lig. JP, Firmenmarke 19, Made in Austria, unsign., H. 17 cm; eBay # 2894046389
6362,240,31 gem. lig. JP, Firmenmarke 19, Made in Austria, unsign., H. 16,5 cm; eBay # 1298561695
6362,257,20 ger., gem. lig. JF, Firmenmarke 19, Made in Austria, unsign., H. 17 cm; eBay # 914442488
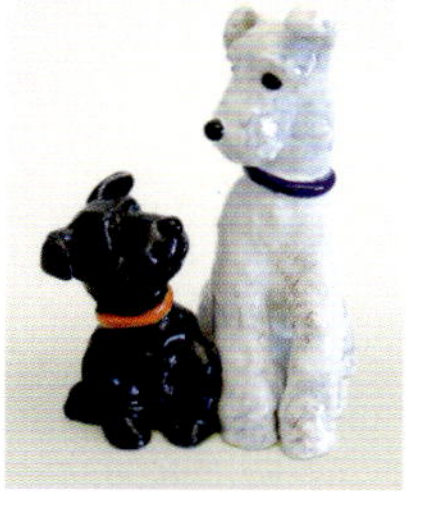
6362,634,65 ger. M, gem. S.G, Firmenmarke 19, Made in Austria, unsign., H. 17 cm; Sammlung John Nash
6362,1341,20 Firmenmarke 19, Made in Austria, unsign., H. 16,3 cm; PB Deutschland
6362,–,– gem. lig. ET?, Firmenmarke 19, Made in Austria, unsign., H. 16,5 cm; Wien Museum Inv. 220.721
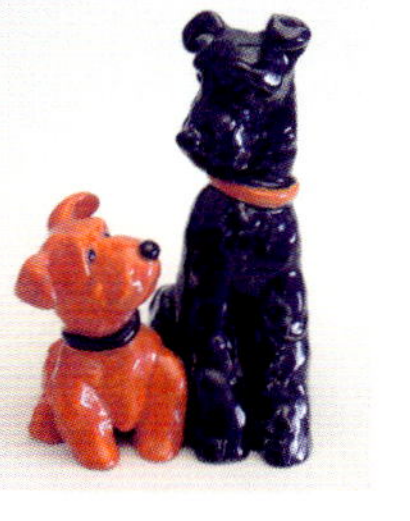
Lit.: Bernd, G.: Heitere Wiener Tierkunst, Ida Meisinger, in: Schaulade, 8.1932, S. 480 (m. Abb.), hier mit Ang. des Originaltitels „Pat und Patachon", „Die groteske Tierfigur ist eigentlich im Österreichischen zuhause. Hier ist der Sinn für die Fragwürdigkeit jeder Kreatur mit Einschluss der menschlichen besonders lebendig, für die kurze Spanne, die zwischen Ernsthaftigkeit und lässiger Selbstpreisgabe, zwischen Heroismus und Lächerlichkeit besteht. Da ist beispielsweise Ida Meisinger. Sie hat den Wiener Blick für die ernsthafte Drolligkeit dieser kleinen und großen Hundeviecher. Mögen sie auch nicht sämtlich rasse- und stubenrein sein – „gar so herzig" sind sie in jeder Bewegung. Sie schnuppern, machen Männchen, sehen trist philosophisch drein – ein positiver Pessimismus, der zum Lachen reizt! – gefallen sich, unvereinbar und doch mit der Pose von Denkmälern, etwa nach dem

klassischen Vorbild Pat und Patachons, in ernsthaftem tête à tête, springen auch mal possierlich hoch und markieren so nebenbei Bücherstütze. Ida Meisinger charakterisiert durch die Bewegung, sozusagen durch die Bewegung der keramischen Masse, und durch die Akzente, die sie bestimmten Details, der Schnauze, den Ohren, dem Schwanz, dem tappigen Gehwerk gibt. Diese erlaubten, ja erwünschten Übertreibungen vermitteln erst das, worauf es der Künstlerin ankommt, den liebenswürdigen grotesken Charme. Die Farben sind leuchtend, aber nicht schreiend. Ein emailähnliches Weiß wechselt mit feinem, mattem Schwarz, mit intensivem Korallrot. Die „Widernatürlichkeit" dieser Farben entspricht der Natur des Werkstoffes vollkommen".
-> vgl. 6996
-> vgl. 6463= DreierGRUPPE
6362 H. 17 cm, B. 10 cm; Fotoarchiv K. Täuber
Quelle: Fotoarchiv K. Täuber

6364
Knörlein, Rudolf
STEHKOPF Frauenkopf mit Kragen (offene Augen) (auf flachem Rundsockel)
ca. 1931
farbig glasiert
6364,1 (Firmenmarke ohne Ang., evtl. ohne Firmenmarke?), unsign., H. 11,4 cm; Christie's SK 14.3.1997
6364 Firmenmarke 19, Made in Austria, unsign., H. 11,1 cm; Sammlung a@goldscheider.net
6364 ger., Firmenmarke 19, unsign., H. 11 cm; PB
6364 ger., gem. AUSTRIA, ohne Firmenmarke, unsign., H. 10,4 cm; MAK Wien, Inv. Ke 10602-31
terrakottafarben
6364 ger., ohne Firmenmarke, unsign., H. 11 cm; PB
Pendant -> vgl. 6446
-> vgl. 6393= ähnl. Modell.

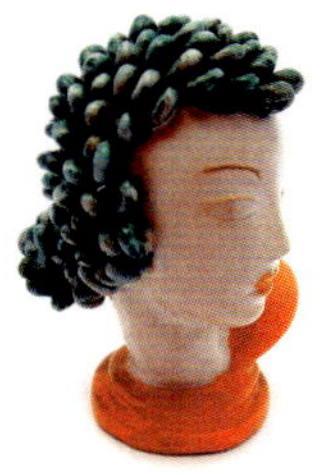

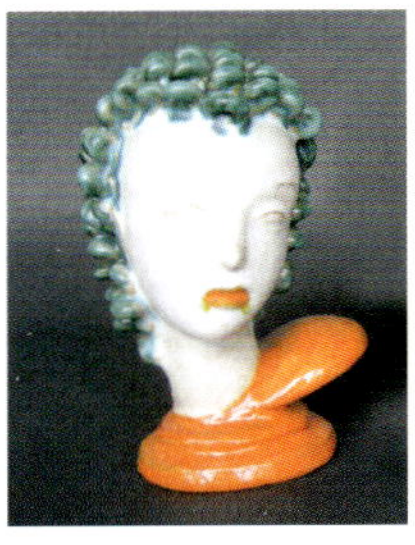

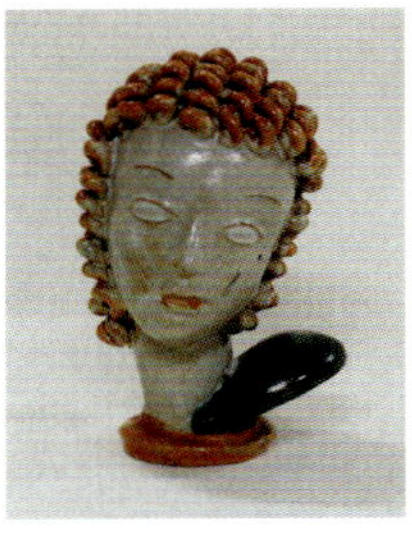

6366
Sailer, Karl
Stehende Tänzerin im Schmetterlingskostüm (auf Rundsockel)
ca. 1928/29 (für den Entwurf)
6366,7,11 Firmenmarke 19, Made in Austria, Hand dec., unsign., H. 33 cm; city-antik
-> vgl. 5840

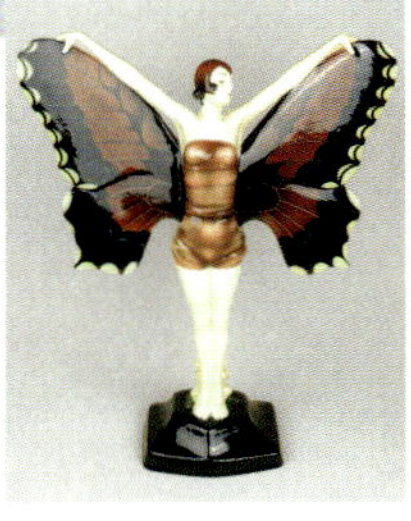

6375
Geiger, Benno (zugeschr.)
BUCHSTÜTZE
Form: dreieckig, mit zwei Erhebungen
Dekor: streifenartig (türkis, orange, schwarz)
ca. 1931
6375,108 ger., gepr. PL?, gem. H., Firmenmarke 19, Made in Austria, unsign., H. 11,9 cm; MAK Wien, Inv. Ke 10602-9-1
6375,109 ger., gepr. PL?, gem. H., Firmenmarke 19, Made in Austria, unsign., H. 11,9 cm; MAK Wien, Inv. Ke 106029-2
Ausst.: PSK 1995
Lit.: ADGS, Nr. 19 (m. Abb.)

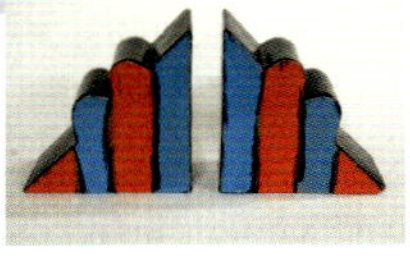

6379
DECKELDOSE
Form: hochrechteckig
Handhaben: zwei halbkreisförmige Ösen am Deckel, vier ebensolche an den beiden Schmalseiten
Dekor: geometrische Formen, Laufglasur
ca. 1931
6379,8 gem. K?, Firmenmarke 19, Made in Austria, H. 13 cm, L. 15 cm, B. 9 cm; PB Wien 19
Ausst.: PSK 1995
Lit.: ADGS, Nr. 38 (m. Abb.)

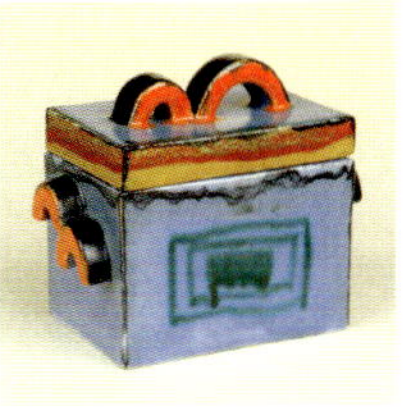

6392
Fried, Nellie
Stehende junge Dame „im Wind" (auf Rundsockel, mit Baumstamm)
ca. 1931
6392,40,10 gem. 19, Firmenmarke 19, Made in Austria, Hand dec., H. 30 cm; Sammlung Stressless, Portugal
6392 ohne Ang., Firmenmarke 19, H. 31 cm; Fischer 50

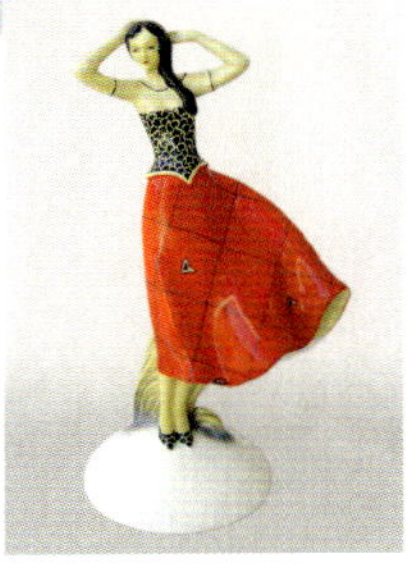

6393
Knörlein, Rudolf
STEHKOPF Frauenkopf mit Kragen (geschlossene Augen) (auf flachem Quadratsockel)
ca. 1930/31
6393,105 gepr., ger. 2, gem. T, Firmenmarke 19, unsign., H. 10,5 cm; PB Wien 19
6393 ohne Ang., (Firmenmarke ohne Ang.), unsign., H. 11 cm; Doro 6.4.1999
-> vgl. 6364; 6446= jeweils ähnl. Modell.

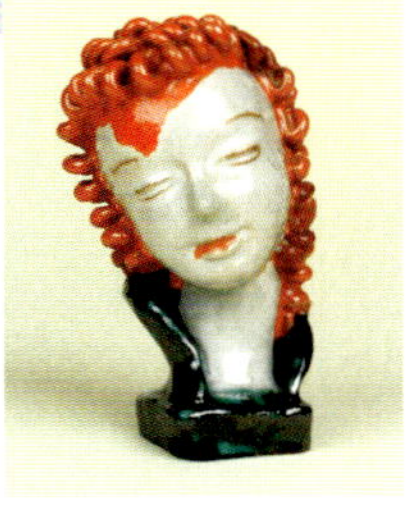

6394
Lorenzl, Josef
Tänzerin im Schmetterlingskostüm (in Pose) (auf Ovalsockel)
ca. 1932
6394,15,11 gem. F, gem. rot N, Firmenmarke 19, Made in Austria, H. 31,5 cm; Sammlung Stressless, Portugal
Bem.: Ähnliches Kostüm wie bei Modell-Nr. 5900; 6206.

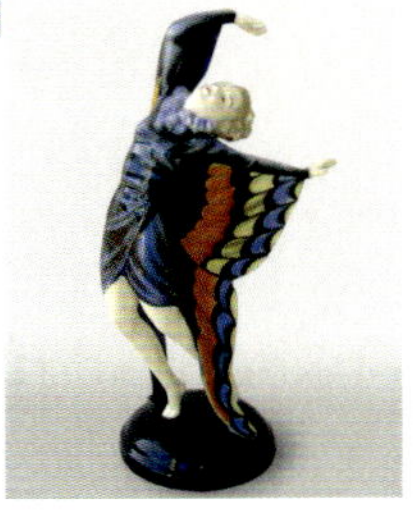

6399
Lorenzl, Josef
Stehende Tänzerin, ihren langen Rock schmetterlingsflügelartig auseinander haltend (auf Rundsockel)
ca. 1931
6399,35,10 Firmenmarke 19, Made in Austria, Hand dec., sign. Lorenzl, H. 29,5 cm; Doro 24.11.1992
6399,65,11,19 gem. N, gem. lig. AS, XFe, Firmenmarke 19, Made in Austria, Hand dec., unsign., Originalpick.: „Carl Deesz, Saarbrücken", H. 30 cm; Dawo 136
6399,129,11 ger. 4X, gem. N, Firmenmarke 19, Made in Austria, sign. „Dakon", H. 29 cm; Sammlung Stressless, Portugal; Bem.: Bei der Sign. Dakon handelt es sich um ein Versehen des Modelleurs.
6399,136,19 ger. XI, Firmenmarke 19, Made in Austria, sign. Lorenzl, H. 29,2 cm; eBay # 2609631582
6399,149,6 (Firmenmarke ohne Ang.), sign. Lorenzl, H. 28,5 cm; Soth. L. 23.10.1987
6399,193,4 (Firmenmarke ohne Ang.), sign. Lorenzl, H. 31 cm; Soth. Olympia 7.11.2002
-> vgl. 5872; 6173= jeweils fast identisches Modell

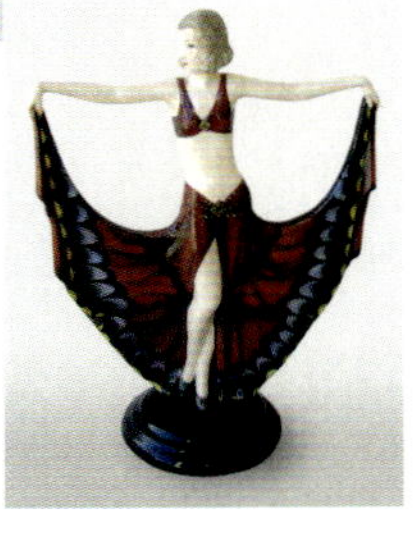

6401
DECKELDOSE Springendes Reh
Form: quadratisch
Dekor: springendes Reh, reliefiert
ca. 1931
06401,1 gem. lig. AH, Firmenmarke 19, Made in Austria, H. 3,5 cm, B. 10 cm, L. 10 cm; GK, S. 22, 118 (m. Abb.); Wien Museum Inv. 222.543/1+2
Ausst.: HMW (jetzt Wien Museum) 1985/86.
Lit.: GK, S. 22, 118 (m. Abb.)

6410
SCHALE Schachbrett
Form: rechteckig
Handhaben: zwei seitlich, an der kürzeren Seiten angesetzte quadratische Griffe
Dekor: Schachbrettmuster
ca. 1931
06410,26 gem. M oder W, Firmenmarke 19, Made in Austria, H. 1,6 cm, B. 12,7 cm, L. 18,5 cm; Sammlung a@goldscheider.net

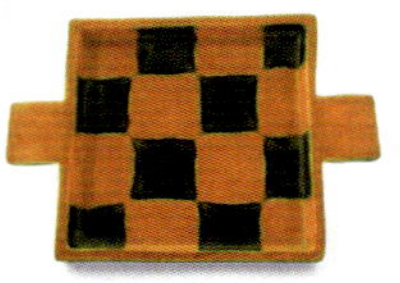

6412
Knörlein, Rudolf (zugeschr.)
BÜSTE Madonna mit Kind
ca. 1931
6412,22,65 gem. lig. AH, Firmenmarke 19, Made in Austria, H. 14 cm; Wien Museum Inv. 220.794
6412,65 Firmenmarke 19, Made in Austria, unsign., Pick. 2, H. 14,5 cm; eBay # 2005786537; 2037761222; 3759307512

6417
Meisinger, Ida
Stehender Airedaleterrier
ca. 1931
6417,15,20 gem. 0, Firmenmarke 19, unsign., H. 18,4 cm; Sammlung John Nash

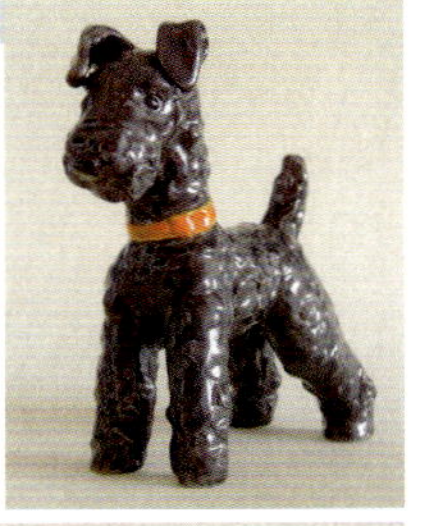

6422F
Postl, Hugo
FIGUR „Windhund" (Liegender Barsoi)
ca. 1929 (für den Entwurf)
6422,6 ger. G, Firmenmarke 19, Made in Austria, sign.?, H. 10 cm; Doro (o. Kat. 6.6.1999)
6422,471,19 gem. lig. AZ, Firmenmarke 19, Made in Austria, H. 9,5 cm, L. 25 cm; Wien Museum Inv. 220.753
6422,775,3 Firmenmarke 19, Made in Germany, unsign., Ausformung zw. 1938–1941, H. ca. 10 cm, L. 27 cm; Wendl 47
6422,803,19 gem. AN, Firmenmarke 19, Made in Germany, unsign., Ausformung zw. 1938–1941, H. 10 cm, L. 27 cm; Sigalas 56
6422F, H. 10 cm, B. 26 cm; GSNachf. 5,2b
-> vgl. 5974F; 7143F
Lit.: MW, 11.1929/30, Nr. 17, S. III (m. Abb.), hier mit Ang. des Originaltitels „Windhund" und des Entwerfers Hugo Postl; GSNachf. 5,2 (m. Abb.), hier mit Ang. des Originaltitels „Windhund".

6427M
Knörlein, Rudolf
WANDMASKE „Turandot" (Frauenkopf mit Schulteransatz)
ca. 1932
6427,1,20 gem. lig. MG., ohne Firmenmarke, unsign., H. 25,4 cm; eBay # 2512962641
6427,3 Firmenmarke 19, Made in Austria, unsign., H. 27 cm; Doro 21.9.1999
6427,8 gem. G, Firmenmarke 19, Made in Austria, unsign., H. 27 cm; PB Deutschland
6427,(f?),16 gem. SC, Firmenmarke 19, unsign., H. 28 cm; Sammlung a@goldscheider.net
6427 gepr. H, gem. S, Firmenmarke 19, unsign., H. 27 cm; Sammlung.schmutz@chello.at 73

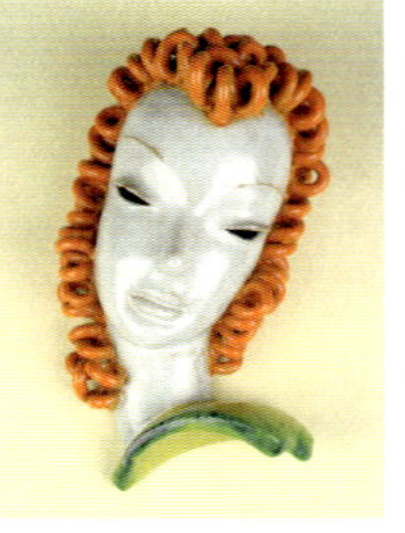

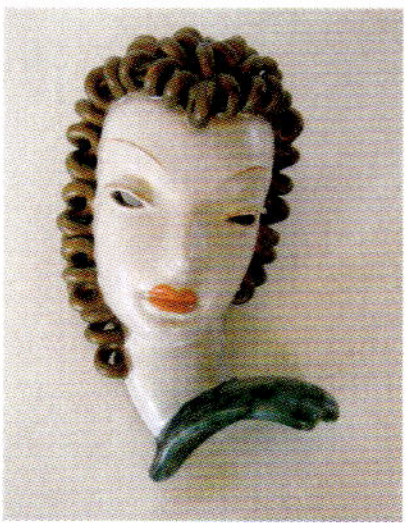
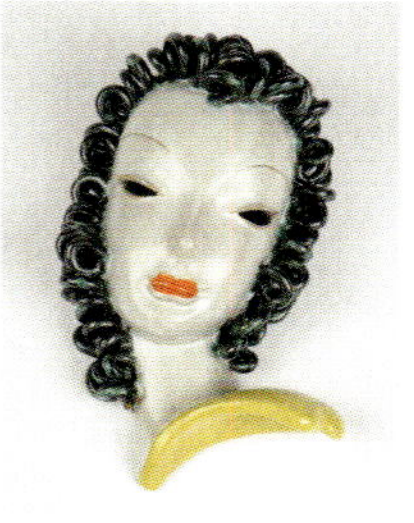
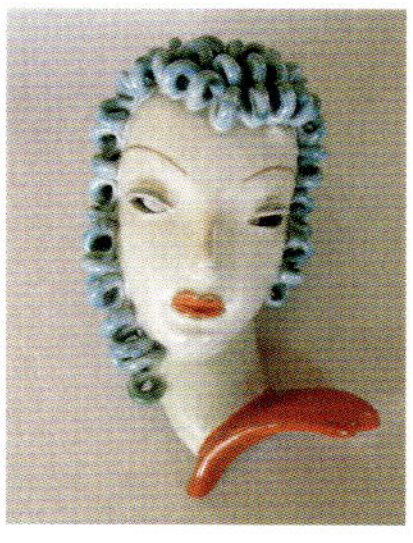
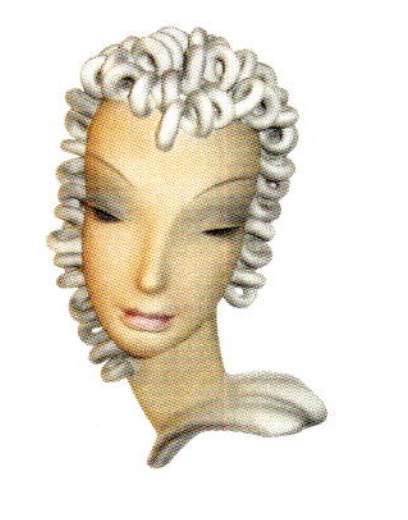

6427 ger., gem. G, Firmenmarke 19, unsign., H. 24,1 cm; eBay # 431619228
6427 Firmenmarke 19, unsign., H. 27,4 cm; Sammlung.schmutz@chello.at 7
6427 Firmenmarke 19, unsign., H. 28 cm; PB Wien 19
6427 ger., (Firmenmarke ohne Ang.), unsign., H. 26,5 cm; Christie's SK 14.3.1997; 12.9.1997
o. Nr., gem. R, Firmenmarke 19, Made in Austria, unsign., H. 26,5 cm; Sammlung.schmutz@chello.at 69
6427,535/2,6 Firmenstempel WG 1, unsign., Pick. WG 1, Ausformung 50er Jahre, H. 28 cm; Sammlung.schmutz@chello.at 80
6427M, H. 28 cm, B. 17 cm; GSNachf. 2,1a

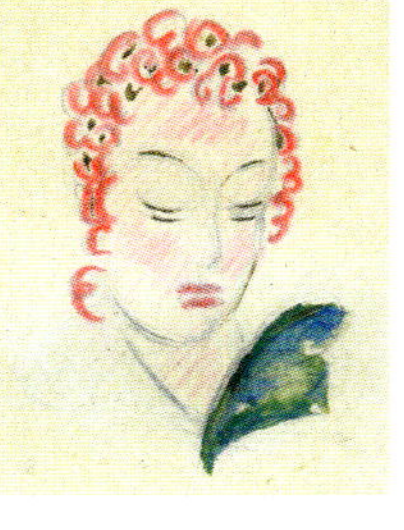

Bem.: Wenngleich keine der hier nachgewiesenen Masken sign. ist, so fiele eine Zuschreibung an Rudolf Knörlein leicht, denn nicht nur die gedrehten Haare (Curls) waren ein Markenzeichen von ihm, sondern auch der typische Gesichtsausdruck spricht eindeutig für diesen Künstler. Kurz vor Redaktionsschluss fand sich dann als endgültiger Beleg auch die originale Entwurfszeichnung (Aquarell über Graphit) von der Hand Knörleins. Durch die eigenhändige Datierung (1932) Knörleins auf dem Trägerblatt eines Originalfotos konnte das Jahr 1932 exakt bestimmt werden. Dieses Modell war höchst erfolgreich wie zahlreiche weitere nachweisbare unsign. Exemplare im KH zeigen. Alle mir bekannten Stücke sind, bis auf das in der Sammlung Schmutz/Wien befindliche Exemplar ohne Stückzahl, jedoch nur mit der Modellnummer versehen, so dass eine Auflistung an dieser Stelle keinen Sinn machen würde. Ein Schreiben Walter Goldscheiders bestätigt zusätzlich „Von Ihren Modellen in diesem Genre haben sich die Maske 6427 und die Madonna 7188 am Besten verkauft" (Brief von Walter Goldscheider an Rudolf Knörlein, vom 4. Juli 1952). Wie das bei Quit. 17 angebotene Exemplar sowie Könecke (2002) auf S. 76 belegen kann, wurde diese beliebte Wandmaske auch noch in den 50er Jahren in Deutschland produziert. Der Teint ist in dem typischen Braun gehalten, wie es bei Carstens üblich war. Ausformungen dieser Zeit tauchen im Handel kaum auf.
-> vgl. 6623; 7453
-> vgl. 7354= ASCHER
Lit.: GSNachf. 2,1 (m. Abb.), hier mit Ang. des Originaltitels MASKE „Turandot"; Könecke (2002), S. 76, 85 (m. Abb.).
Quelle: Privatsammlung D (Originalzeichnung); Privatarchiv (Originalfoto).

6428
Knörlein, Rudolf
WANDMASKE Frauenkopf mit flachem Hut
ca. 1931
6428,3 gem. R, Firmenmarke 19, Made in Austria, unsign., H. 22,9 cm; MAK Wien, Inv. Ke 10521-30
6428,8 Firmenmarke 19, unsign., Pick. ohne Ang., H. 23,2 cm; Doro 1.6.2005
6428,31 ger., gem. lig. AH, Firmenmarke 19, Made in Austria, unsign., H. 22,9 cm; eBay # 1425339317
6428 ger. gem. lig. AB, Firmenmarke 19, Made in Austria, unsign., H. 23 cm; Sammlung.schmutz@chello.at 8
Bem.: Durch eine Originalzeichnung (Aquarell über Graphit), kann diese Maske R. Knörlein zugeordnet werden.
Quelle: Privatsammlung D (Originalzeichnung).

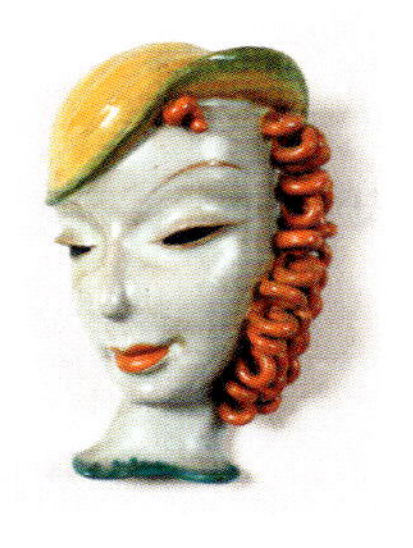
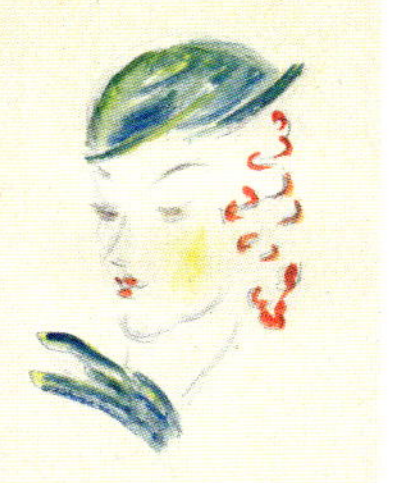

6442
Heindl, Hermann
BÜSTE Madonna mit Kind
ca. 1931
6442,46,20 Firmenmarke 19, Made in Austria, H. 30 cm; Franzisi 21

6444F
Heindl, Hermann
FIGUR „Madonna mit Kind"
(Stehende Madonna mit Kind)
ca. 1931
6444,21,2 gem. lig. AH, Firmenmarke 19, Made in Austria, unsign., H. fehlt; eBay # 898180334
6444,46,31 Firmenmarke 19, Made in Austria, unsign., H. 39 cm; Fischer 151
6444,154,65 Firmenmarke 19, Made in Austria, Pick. 10 (Entwurf Hermann Heindl Wien), H. 39 cm; KH
6444F, H. 39 cm, B. 18 cm; GSNachf. 6,10a
6444 H. 39 cm, B. 18 cm; Fotoarchiv K. Täuber
-> vgl. 7730
Lit.: GSNachf. 6,10 (m. Abb.), hier mit Ang. des Originaltitels FIGUR „Madonna mit Kind".
Quelle: Fotoarchiv K. Täuber

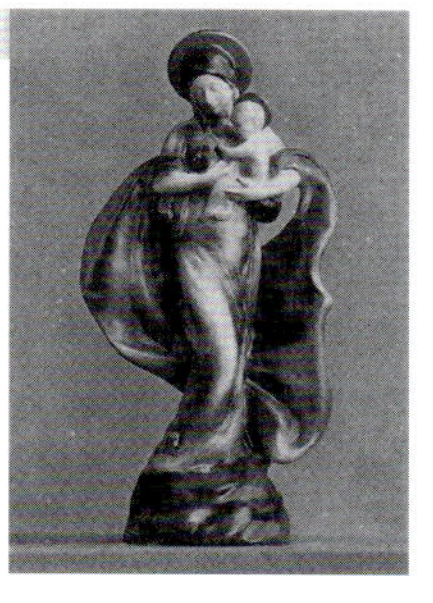

6445
Knörlein, Rudolf
STEHKOPF Mädchenkopf mit leicht gesenktem Haupt (Lockenfrisur, Curls) (auf Keramik / Holzsockel)
ca. 1931
Keramiksockel
6445,1 gem. B, Firmenmarke 19 (zwei mal), unsign., H. 12 cm; eBay # 7347471340
Holzsockel
6445 Plak. C, unsign., H. 15 cm; Mehlis 18
6445 (Firmenmarke ohne Ang.), unsign., H. 11 cm; Schuler 71
Bem.: Eindeutig, wenngleich auch nicht signiert, handelt es sich hier um einen Entwurf Rudolf Knörleins. Vergleiche ähnliche Stehköpfe mit den Modell-Nrn. 6364, 6393 und 6446.

6446
Knörlein, Rudolf
STEHKOPF Frauenkopf mit Kragen (geschlossene Augen) (auf flachem Rundsockel)
ca. 1931
6446,17,65 gem. G, ohne Firmenmarke, unsign., H. 10,2 cm; MAK Wien, Inv. Ke 10602-22

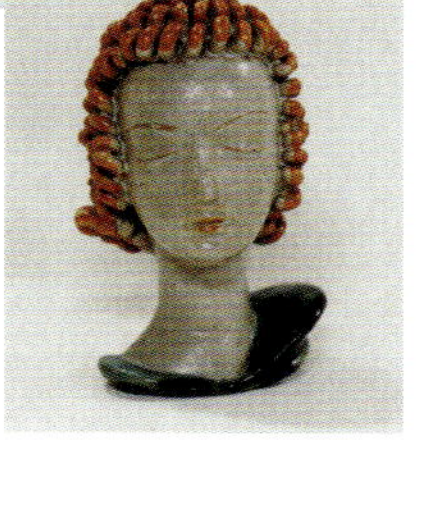

6446 ger., ohne Firmenmarke, Austria, unsign., H. 10,2 cm; PB
Pendant -> vgl. 6364
-> vgl. 6393= ähnl. Modell.

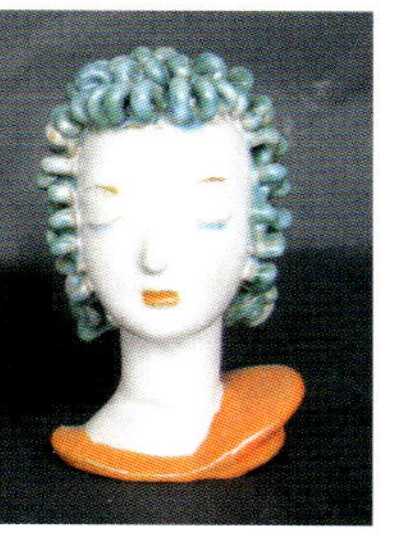

6447
Weiss, Claire (Klára) (d.i. Herczeg, Claire (Klára)
Posierende junge Dame mit Hut (einen zusammengefalteten Fächer in ihrer rechten Hand haltend) (auf Ovalsockel)
ca. 1931
6447 (Firmenmarke ohne Ang.), sign. Claire Weiss, H. 27 cm; Pinhas, S. 130 (m. Abb.)

6449
Meisinger, Ida
FIGUR „Hundewetter"
(Sitzender Terrier mit Regenschirm) (mit/ohne Ovalsockel)
ca. 1931
mit Sockel
6449 Firmenmarke 19, Made in Austria, unsign., H. 11,5 cm, L. 16,5 cm; Sammlung a@goldscheider.net
ohne Sockel
6449,3 Firmenmarke 19, Made in Austria, unsign., H. 10,5 cm; WKA 33
6449,5 ger. K, Firmenmarke 19, Made in Austria, unsign., H. 12 cm; GK, S. 22, 119 (m. Abb.); Wien Museum Inv. 222.542
6449,6,14 gem. lig. WC, Firmenmarke 19 (blau), Made in Austria (blau), Hand dec. (blau), Pick. 1, H. 10 cm; PB Deutschland
6449,816,65 Firmenmarke 19, Made in Austria, Pick. 8, Pick. 15 (Entwurf Ida Meisinger, Wien), H. 10,2 cm; eBay # 418968234
6449M, H. 10,5 cm, B. 14 cm; GSNachf. 3,25a
Bem.: Dieser Hund gehört zu einer ganzen Reihe von Hundegrotesken, die in ihrem Variantenreichtum kaum zu überbieten sind. Vom Hund mit Zahnschmerzen (Modell-Nr. 7070) über den „Paddler" (Modell-Nr. 7321), den „Skifahrer" (Modell-Nr. 7700), die Gruppe „Naschkätzchen" (Modell-Nr. 7701) bis hin zum „Hund mit Bassgeige" (Modell-Nr. 6977; 7878), um hier nur einige wenige zu nennen. Der Phantasie der Entwerferin Ida Meisinger, die diese „Nippesfiguren" entwarf, waren keine Grenzen gesetzt. Glücklicherweise haben sich nun auf diesem nachweisbaren Stück zwei Aufkleber erhalten. Ersterer (Pick. 15) belegt den Entwurf durch Ida Meisinger, der zweite (Pick 8) dokumentiert den damaligen Ladenpreis von RM 9,–. Die Entwürfe, die ohne diese Aufkleber nicht eindeutig hätten verifiziert werden können, hätten auch von Wilhelm Thomasch stammen können. In seinem Betrieb in Sierndorf hat auch er ähnliche dieser Figuren produziert. Dadurch, dass sich die Entwürfe Meisingers der Grenze des Kitsches nähern, wird klar, dass diese, ganz im Gegensatz zu ihren Terriern, wie der „Modehund" (Modell-Nr. 5854 etc.), alle im Scherben unsign. geblieben sind. Diesen Entwürfen fehlt die „freiere" Auffassung und die expressive Handschrift, wenngleich sie lustig grotesk sind und durchaus ihre Berechtigung haben und auch beim Publikum erfolgreich waren.
Ausst.: HMW (jetzt Wien Museum) 1985/86; SSH 1992.
-> vgl. 6997
Lit.: GSNachf. 3,25 (m. Abb.), hier mit Ang. des Originaltitels FIGUR „Hundewetter"; SS, S. 44, 45 (m. Abb.); GK, S. 22, 119 (m. Abb.).

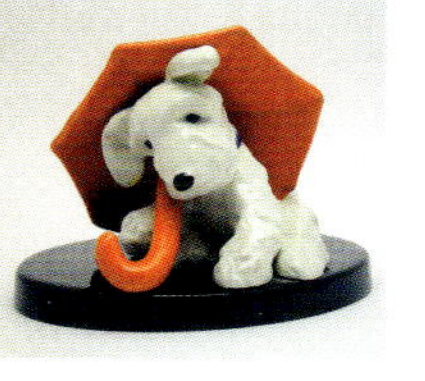
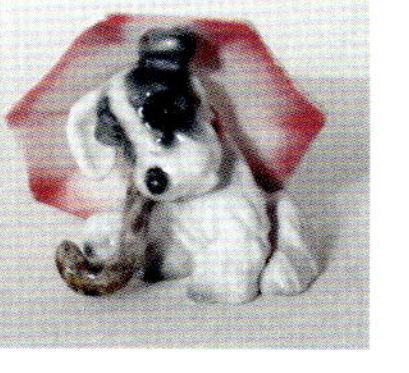

6451M
Matula, Viktor
FIGUR „Reh liegend"
ca. 1931
6451M, H. 7 cm, B. 7 cm; GSNachf. 3,12b
Lit.: GSNachf. 3,12 (m. Abb.), hier mit Ang. des Originaltitels „Reh liegend".
Quelle: Nachlass Matula, Wien

6453
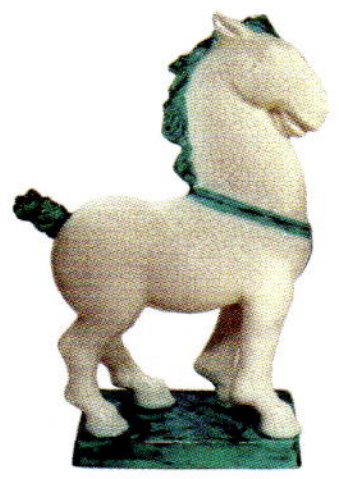
Stehendes Fohlen
(auf Rechtecksockel)
ca. 1931
6453,20,K Firmenmarke 19, Made in Austria, H. 25,5 cm; eBay # 1200540543
6453,25,20 ger., Firmenmarke 19, Made in Austria, undeutlich sign., H. 25,4 cm; eBay # 52662871
6453,55 Firmenmarke 19, Made in Austria, H. 26 cm; Sammlung a@goldscheider.net
-> vgl. 6896
Lit.: GSN, S. 50, 51 (m. Abb.)
(GS neu: Modell-Nr. 15 „Mein kleines Pferd" (H. 15 cm), Aufl. ohne Ang.

6454
Weiss, Claire (Klára) (d.i. Herczeg, Claire (Klára)
FIGUR „Komtesse" (Stehende Dame mit Hut)
(auf Ovalsockel)
ca. 1931
6454,49,6 Firmenmarke 19, Made in Austria, sign. Claire Weis, H. 31 cm; Schuler 66
6454,57,11 Firmenmarke 19, Made in Austria, Hand dec., sign. Claire Weis, H. ca. 31 cm; Herr 40
6454,97,10 Firmenmarke 19, Made in Austria, Hand dec., sign. Claire Weis, H. 31 cm; eBay # 1282204399
6454,211,11 gem. N, Firmenmarke 19, Made in Austria, sign.?, H. 32,5 cm; eBay # 540026979
6454,361,6 gem. lig. CWS?, Firmenmarke 19, Made in Austria, sign. Claire, H. 33 cm; Nagel 370
6454,513,14 gem. F., gem. lig. JR, Firmenmarke 19, Made in Austria, sign. Claire Weiss, H. 33 cm; eBay # 35555292
-> vgl. 6951
-> vgl. 7327; 8145= jeweils GRUPPE „Komtesse mit Hund".

6456
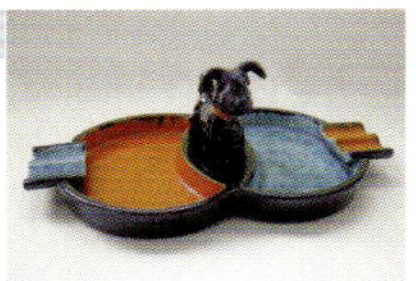
Meisinger, Ida
ASCHER Foxterrier
Form: zweifach gemuldet, als sich überschneidende Kreise, vier Zigarettenablagen
Dekor: mittig stehender Foxterrier (vollplastische Fig.)
ca. 1931
6456 gem. lig. TF, Firmenmarke 19, Made in Austria, Pick. 15 (Entwurf Ida Meisinger, Wien), H. 8,5 cm, B. 24 cm; Sammlung f@goldscheider.de

6458
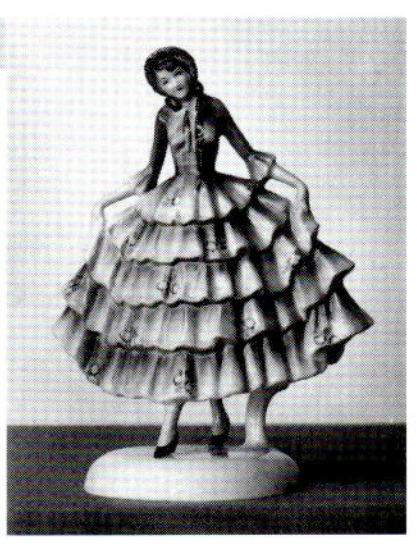
Dakon, Stephan
FIGUR „Alt-Wien"
(Schreitende junge Dame mit Schutenhut und gerüschtem Kleid (auf Rundsockel)
ca. 1931
6458,74,15 Firmenmarke 19, Made in Austria, Hand dec., H. 18,5 cm; Wilbois 61
6458,192,Firmenmarke 19, Made in Austria, Hand dec., unsign.?, H. 18 cm; eBay # 891234794
6458,470,11 gem. rot N, Firmenmarke 19, Made in Austria, H. 19 cm; eBay # 1421574602
6458,675,15,3 gem. N, Firmenmarke 19, Made in Austria, H. 18,8 cm; eBay # 1247885661
6458,840,7,4 gem. lig. JR III, Firmenmarke 19, Made in Austria, H. 18 cm; eBay # 1464944653
Bem.: Dieses Modell wurde noch in Trenton (USA) unter der GS Everlast Corp. (Modell-Nr. 195) in der H. 20,3 cm weiter produziert (eBay # 190011220196).
-> vgl. 7268
Lit.: GSNachf. 1,9 (m. Abb.), hier mit Ang. des Originaltitels „Alt-Wien".

6459
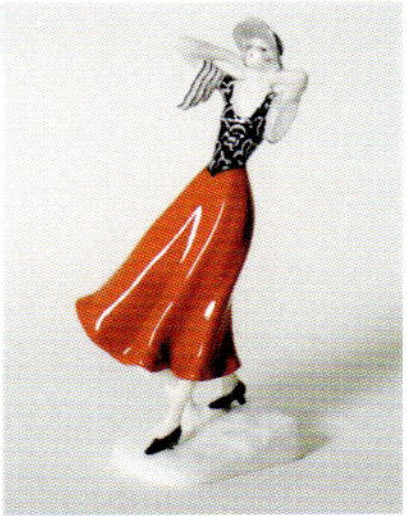
Dakon, Stephan
Schreitende junge Dame mit Hut „im Wind" (auf abgerundetem Rechtecksockel)
ca. 1930 (für den Entwurf)
6459,529,19 gem. lig. AS., Firmenmarke 19, Made in Austria, unsign., H. 18,5 cm; eBay # 913940968; # 3777927881
6459,544,4,3 Firmenmarke 19, Made in Austria, H. ca. 19 cm; Weidler 259
6459,600,15,5 gem. JR, gem. lig. SH, Firmenmarke 19, Made in Austria, H. ca. 18,5 cm; Wien Museum Inv. 220.751
6459,673,10,4,A gem. JR III, Firmenmarke 19, Made in Austria, H. 18,7 cm; PB Deutschland
6459,1071,2 Firmenmarke 19, H. 19,7 cm; BM 56

6460F
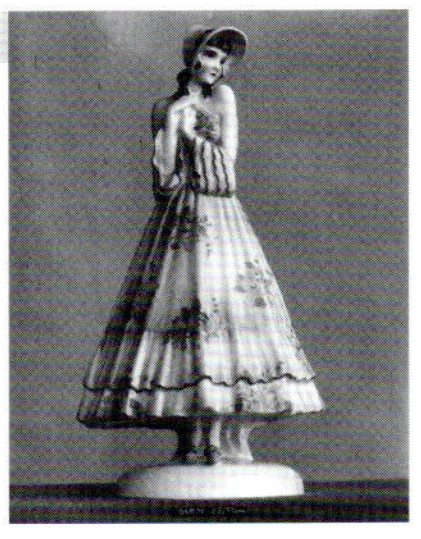
Dakon, Stephan
FIGUR „Innocense" (Stehende junge Dame mit Schutenhut) (auf Rundsockel)
ca. 1931
6460,18,10 gem. lig. AS, gem. grün N2, Firmenmarke 19, Made in Austria, Hand dec., H. 32 cm; eBay # 230007744883
6460F, H. 33 cm, B. 17 cm
Quelle: Archiv Mimi Dakon (Originalfoto), hier mit Ang. des Originaltitels „Innocense".

6461
Dakon, Stephan
GRUPPE „Tanz" („Tanzgruppe")
ca. 1931
Quelle: Archiv Mimi Dakon (Modellliste Dakon, undatiert, ca. 1940), hier mit Ang. des Originaltitels GRUPPE „Tanz"; (Wiener Manufaktur FG, Modellliste Dakon, 7.11.1946), hier unter dem Titel „Tanzgruppe" geführt.

6462
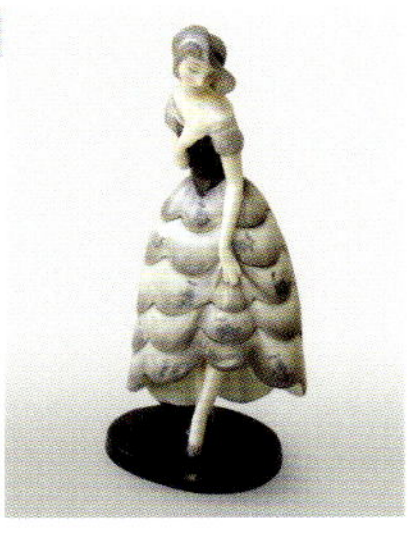
Weiss, Claire (Klára)
(d.i. Herczeg, Claire (Klára)
Stehende Dame mit Hut, ihr Kleid etwas hochhebend (auf Rundsockel)
ca. 1931
6462,0,19 gem. H13, N1, gem. lig. AH, Firmenmarke 19, Made in Austria, Hand dec., sign. Claire Weiss, H. 26 cm; Sammlung Stressless, Portugal
6462,30,4,3 gem. F.S., Firmenmarke 19, Made in Austria, Hand dec., sign. Claire Weis, H. 26 cm; Wien Museum Inv. 190.129

6463
Meisinger, Ida
Drei sitzende Airedaleterrier
ca. 1931
6463,96,65 Firmenmarke 19, Made in Austria, unsign.?, H. 16,5 cm; eBay # 1235325126
-> vgl. 6362; 6996= jew. ZweierGRUPPE „Pat und Patachon".

6464
Meisinger, Ida
Drei sitzende junge Terrier
ca. 1931
6464,184,65 Firmenmarke 19, unsign., H. 9,5 cm; Doro 4.3.2002

6466
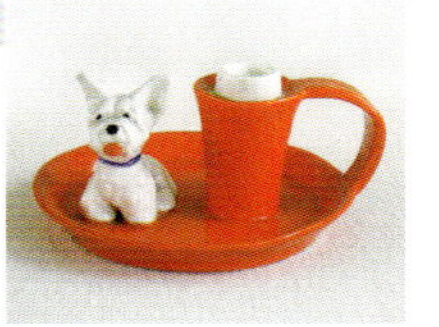
Meisinger, Ida
KERZENHALTER (einflammig)
Sitzender West Highland Terrier (mit heraushängender Zunge)
ca. 1929/30 (für den Entwurf des Hundes)
6466,7 ger., gem. S., Firmenmarke 19, Made in Austria, unsign., H. 7,2 cm, Ø 15,5 cm; Privatsammlung, Wien
Bem.: Dieses Modell wurde wohl um 1931 ins Programm genommen, der Entwurf des Terriers stammt etwa aus den Jahren 1929/30.
-> vgl. 6013; 6025; 6500= jeweils FIGUR

6478
Herczeg, Claire (Klára) (zugeschr.)
Stehendes Mädchen mit Hut, einem Terrier eine große Schleife (Mascherl) um den Hals bindend (auf Ovalsockel)
ca. 1931
6478,132,2 gem. grün S.1., Firmenmarke 19, Made in Austria, Hand dec., unsign., H. 13 cm; eBay # 230011053763
6478,335,20 Firmenmarke 19, Made in Austria, unsign., H. 13,9 cm; eBay # 7392680599
6478,482,31 gem. S, gem. lig. SC, Firmenmarke 19, Made in Austria, unsign., H. 13,5 cm; eBay # 7412220076
6478,663,20 gem. MR2, Firmenmarke 19, Made in Austria, unsign., H. 13 cm; Wien Museum Inv. 190.091

6482
WANDMASKE Kopf eines Chinesenmädchens
ca. 1931/32
6482 ger., Firmenmarke 19, Made in Austria, H. 11,5 cm; eBay # 3161490895
Lit.: Pinhas, S. 174 (m. Abb.)

6491
Lorenzl, Josef
Stehende junge Tänzerin in Pose (auf Rundsockel)
ca. 1931/32
6491,3,11 Firmenmarke 19, H. 19 cm; Fischer 136
6491,82,? Firmenmarke 19, Made in Austria, „handgemalt" (gemeint ist wohl Hand dec.?), H. 18 cm; RM 31; 32
6491,104,19,3 gem. F, Firmenmarke 19, Made in Austria, KH

6492
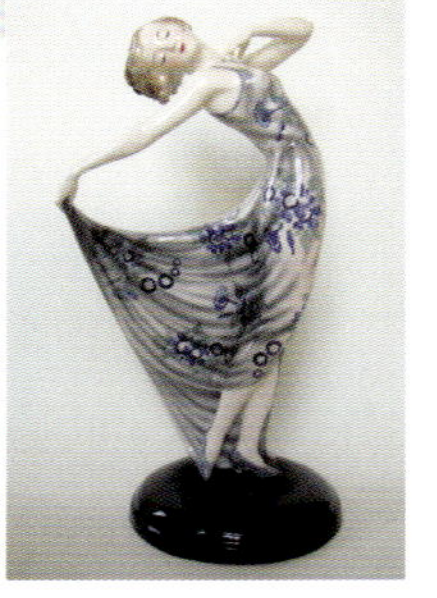
Lorenzl, Josef
Stehende Tänzerin in Pose, ihr Kleid mit der rechten Hand zurückhaltend (auf Ovalsockel)
ca. 1931/32
6492,14,6,3 ger. B, gem. F, Firmenmarke 19, Made in Austria, Hand dec., unsign., H. 39,4 cm; eBay # 2559479295
6492,55,4,3 gem. NF., Firmenmarke 19, Made in Austria, Hand dec., sign. Lorenzl, H. 40,2 cm; Sammlung Weilinger, Wien
6492,59,10,3 Firmenmarke 19, Made in Austria, Hand dec., sign. Lorenzl, H. 40 cm; Wien Museum Inv. 190.071
6492,77,6 gem. S., Firmenmarke 19, Made in Austria, Extra-Stempel 2, unsign., H. 39,4 cm; eBay # 25095115369
6492,80,4 (Firmenmarke ohne Ang.), unsign.?, H. 40,5 cm; Christie's SK 13.5.2004; 11.11.2004
-> vgl. 7732
-> vgl. 7193= ähnl. Modell.

6496
Meisinger, Ida
ASCHER Terrier
Form: rund, eine Zigarettenablage
Dekor: sitzender Terrier (vollplastische Fig.)
ca. 1931/32
06496,1 Firmenmarke 19, Made in Austria, unsign., H. 9 cm, Ø 15 cm; PB Wien
6496,14 ger., gem. lig. TK, Firmenmarke 19, Made in Austria, unsign., H. 9,4 cm, Ø 15,7 cm, L. 17,5 cm; Sammlung a@goldscheider.net

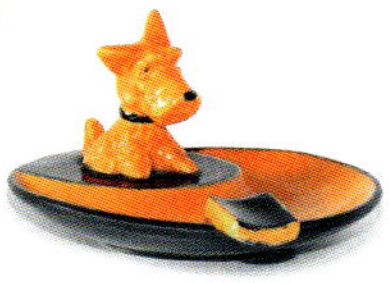

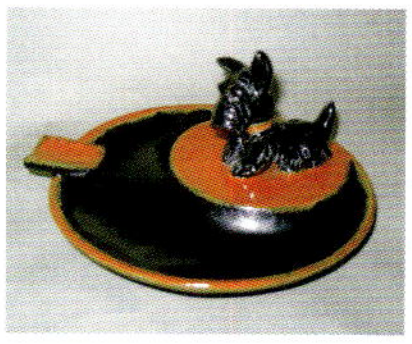

6500
Meisinger, Ida
FIGUR / BUCHSTÜTZE Sitzender West Highland Terrier (mit heraushängender Zunge)
ca. 1929/30 (für den Entwurf)
FIGUR
6500,76,31 gem. SG, Firmenmarke 19, Made in Austria, unsign., H. 10 cm; Wien Museum Inv. 220.719
6500 Firmenmarke 19, Made in Austria, H. 10,2 cm; Privatsammlung, Wien
BUCHSTÜTZE
6500 Firmenmarke 19, Made in Austria, unsign., H. 13,4 cm, L. 15,2 cm; Sammlung a@goldscheider.net
(6500 fehlt),31 Firmenmarke 19, Made in Austria, unsign., H. 13,4 cm, L. 15,2 cm; Sammlung a@goldscheider.net
-> vgl. 6013; 6025
-> vgl. 6466= KERZENHALTER

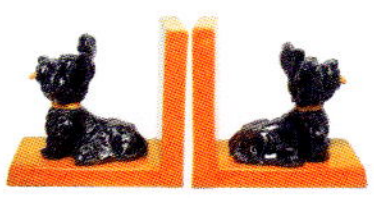

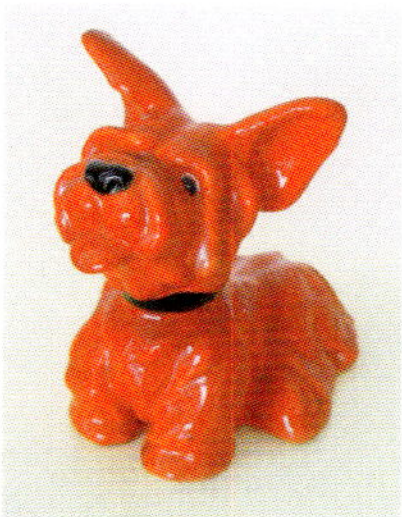

6503F
Postl, Hugo
FIGUR „Skyeterrier"
um/ab 1930 (für die Ausführung)
6503,114,11 Firmenmarke 19, Made in Austria, unsign., H. 13 cm; PB Wien 19
6503F, H. 12 cm, B. 22 cm; GSNachf. 5,11b
-> vgl. 6116; 6802; 7300; 7519
-> vgl. 6975= Variante (auf Ovalsockel)
Lit.: GSNachf. 5,11 (m. Abb.), hier mit Ang. des Originaltitels „Skyeterrier".

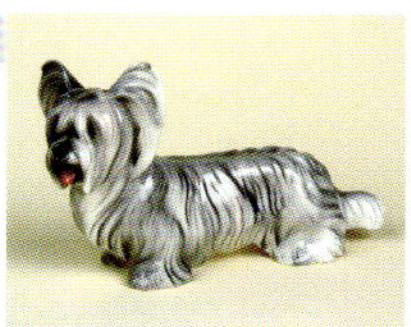

6504/2F
Dakon, Stephan
„Rosenkönigin" (Tänzerin mit Reifrock) (auf Ovalsockel)
ca. 1931/32
6504/2F, H. 26 cm, B. 16 cm
Quelle: Archiv Mimi Dakon (Originalfoto); Wiener Manufaktur FG, Modellliste Dakon, 7.11.1946), hier mit Ang. des Originaltitels „Rosenkönigin".

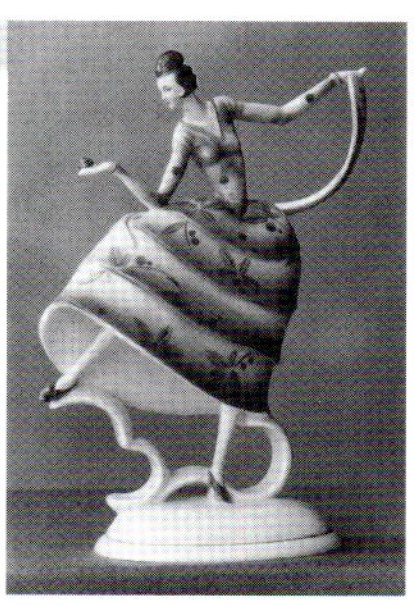

6507
Weiss, Claire (Klára) (d.i. Herczeg, Claire (Klára)
Stehende Dame mit Kappe und Muff (auf Ovalsockel)
ca. 1931/32
6507,16,19 Firmenmarke 19, sign.?, H. 33 cm; Herr 26
6507,30,10 gem. S.I., Firmenmarke 19, Made in Austria, Hand dec., sign. Claire Weis, H. 33 cm; WKA 38; eBay # 1255481823; Doro 26.5.2004
6507,75,7 Firmenmarke 19, Made in Austria, sign. Claire Weis, H. 33 cm; eBay # 119070190; Fischer 144
6507,81,11 gem. F3, gem. rot S.1., Firmenmarke 19, Made in Austria, sign. Claire Weis, H. 33 cm; PB Berlin
6507,111,15 (Firmenmarke ohne Ang.), unsign.?, H. 33 cm; Christie's SK 20.9.2000
6507,126,31,8 Firmenmarke 19, Made in Austria, unsign., H. 32,5 cm; Kastern 79; KH Art 1900 / Berlin

6508
Dakon, Stephan
„Fusspitzentanz"
ca. 1931/32
Quelle: Archiv Mimi Dakon (Wiener Manufaktur FG, Modellliste Dakon, 7.11.1946), hier mit Ang. des Originaltitels „Fusspitzentanz".

6518
Sitzende Mutter mit Kind
ca. 1931/32
6518,20 gem. A, Firmenmarke 19, Made in Austria, unleserlich sign., H. 18 cm; Rammel 54; Sebök 26; Quit. 14
6518 ohne Ang., Firmenmarke 19, Made in Austria, H. 18 cm; RM 31; 32

6520F
Dakon, Stephan
Stehende junge Dame in weiten Hosen (sog. Schlaghosen) mit großem Hut und Ball (auf getrepptem Rechtecksockel?)
ca. 1931/32
6520F, H. 29 cm
Quelle: Archiv Mimi Dakon (Originalfoto)

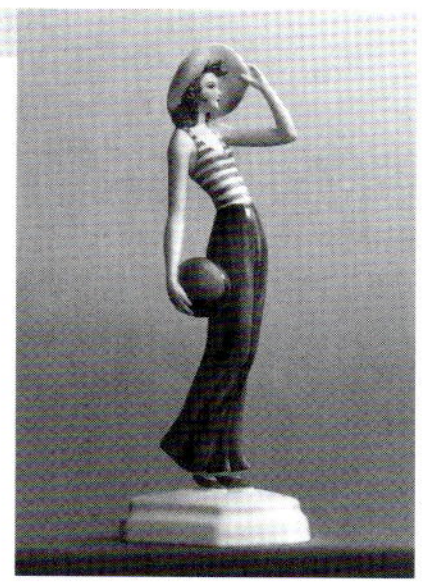

6521
Dakon, Stephan
Schreitende Tänzerin mit Hut (ihr Kleid mit beiden Händen anhebend) (auf abgerundetem Rechtecksockel)
ca. 1931/32
6521,15,15,3 Firmenmarke 19, Made in Austria, H. 20,3 cm; eBay # 3726143537

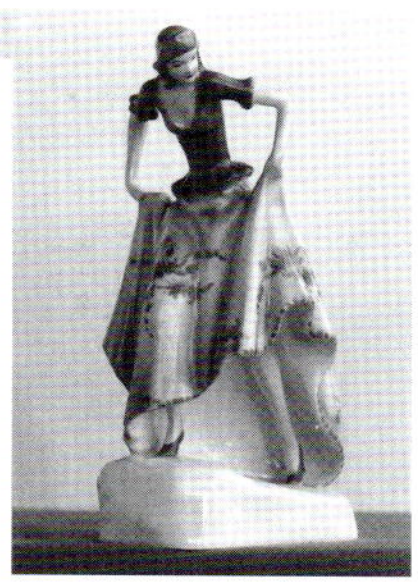

6525
Postl, Hugo
Stehender Pekinese
ca. 1931/32
6525,132,11 Firmenmarke 19, H. 8,4 cm, L. 12 cm; Sammlung a@goldscheider.net
6525,478,14 gem. F, Firmenmarke 19, unsign., H. 9 cm, L. 12,5 cm; Henry's 26.2.1998; 27.1.2000

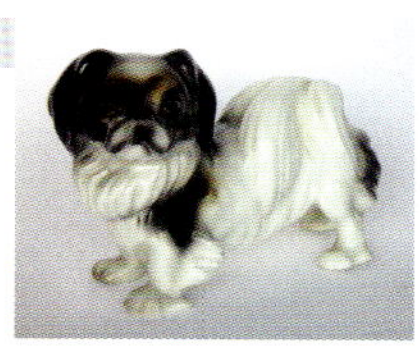

6531
Elefant
ca. 1931/32
6531,8,20 gem. H, Firmenmarke 19, H. 11,3 cm; Wien Museum Inv. 190.028
Lit.: GSN, S. 48, 49 (m. Abb.); GSN, Erg. Bl. (o. Pag., m. Abb.) und Preisliste 1992.
GS neu: Modell-Nr. 14 „Gefleckter Elefant" (H. 15 cm), Aufl. 500 Stück

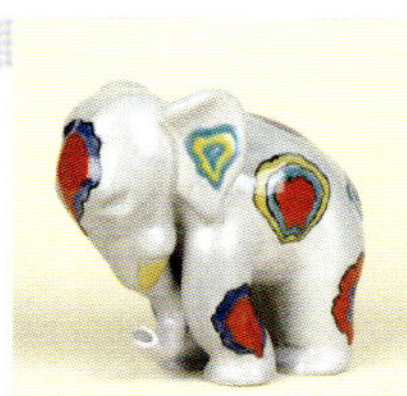

6533
Meisinger, Ida
Zwei stehende, sich beschnuppernde Scotchterrier (auf Rechtecksockel)
ca. 1931/32
6533,34,31 ger. lig. HLP, Firmenmarke 19, unsign., H. 8,4 cm, L. 13,5 cm; PB Wien

6535F
Lorenzl, Josef
„Rosenkavalier" (Octavian als schreitende Figur mit Rose in der rechten Hand) (diverse ovale Sockelformen)
ca. 1931/32
6535,5,66 gem. K., Firmenmarke 19, Made in Austria, Hand dec., unsign.?, H. ca. 21 cm; city-antik
6535,66,8 Firmenmarke 19, Made in Austria, Hand dec., sign. Lorenzl, H. 21,5 cm; PB Wien 19
6535,113,15 gem. S, Firmenmarke 19, Made in Austria, unsign., H. 19,3 cm; Sammlung a@goldscheider.net
6535,251,10 gem. S.1., Firmenmarke 19, Made in Austria, sign. Lorenzl und Dakon, H. 19,5 cm; eBay # 375840439; Bem.: Die zusätzliche Sign. „Dakon" ist eindeutig ein Versehen des Modelleurs.
6535,293,? (Firmenmarke ohne Ang.), sign. Lorenzl, H. 20 cm; Henry´s 31.7.2004
6535,532,4 (Firmenmarke ohne Ang.), sign. Lorenzl, H. 20,1 cm; Doro 12.9.2006
6535,589,10 Firmenmarke 19, Made in Germany, sign. Lorenzl, Ausformung zw. 1938–1941, H. 20 cm; Doro S (FVK VII/1999); eBay # 518311852; 501050333
6535,592,4 Firmenmarke 19, Made in Germany, sign. Lorenzl, Ausformung zw. 1938–1941, H. 20,5 cm; Wendl 45; Dawo 134
Bem.: Dargestellt ist die Figur des Octavian (gen. Quinquin), eine Hosenrolle aus der Komödie für Musik „Der Rosenkavalier" (Urauff. Dresden 1911) von Richard Strauss (-> vgl. Westerman, S. 367–372; Leopold, S. 242–243). Die Figur wurde mit verschiedenen ovalen Sockeln produziert. Vorwiegend wurde der mit Rocaillen verzierte Ovalsockel verwendet, es gibt aber auch die einfache, ohne Verzierung versehene ovale Form.
-> vgl. 6555 (falsche Ang.); 6602
Lit.: GSNachf. (2. Ausg., bzw. Erg. Bl.) 1,54 (m. Abb.); Bem.: Hier mit falscher Modell-Nr. 6555.
Quelle: AGM (Bestellung von Ferd. Lundquist (Göteborg, Schweden), vom 23. August 1950; Kommissionsfaktura der „Österreichischen Gruppe-Austria" für die Messe in Utrecht vom 26.2.1953; Rechnung an Maurice Mizrahi, Alexandria, Ägypten, vom 22. April 1953), hier jeweils mit Ang. des Originaltitels „Rosenkavalier".

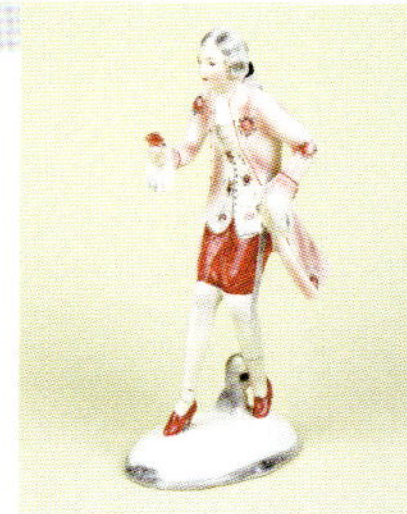

6536
Lorenzl, Josef
Sitzende elegante junge Dame mit Hut und ausgestrecktem Bein (auf Ovalsockel)
ca. 1931/32
6536,124,15,3 gem. lig. AS, gem. rot N, Firmenmarke 19, Made in Austria, H. 13,5 cm, L. 20.3 cm; city-antik
6536,331,11,5 Firmenmarke 19, Made in Austria, H. 12,5 cm, L. 20,5 cm; Sammlung Stressless, Portugal

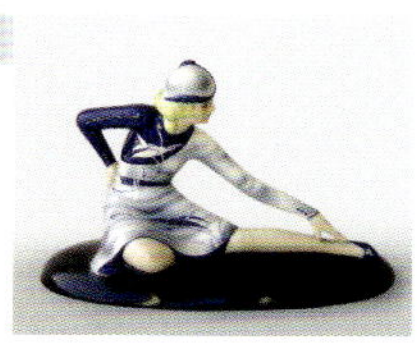

6537
Knörlein, Rudolf
WANDMASKE Frauenkopf mit Hut (Kappe)
ca. 1931/32
6537 ger., gem. T, Firmenmarke 19, Made in Austria, Pick. 12 (Entwurf Rudolf Knörlein, Wien), H. 17 cm; PB Deutschland

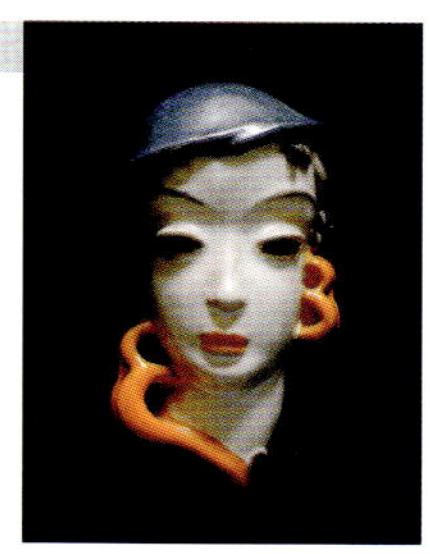

6542
Knörlein, Rudolf (zugeschr.)
WANDMASKE Frauenkopf mit Hand und Fächer
ca. 1931/32
6542,12,K gem. Z, Firmenmarke 19, Made in Austria, unsign., H. ca. 26 cm; eBay # 18371013 (Soth.)
6542,20 ger., gem. H, Firmenmarke 19, Made in Austria, unsign., H. 26,4 cm;

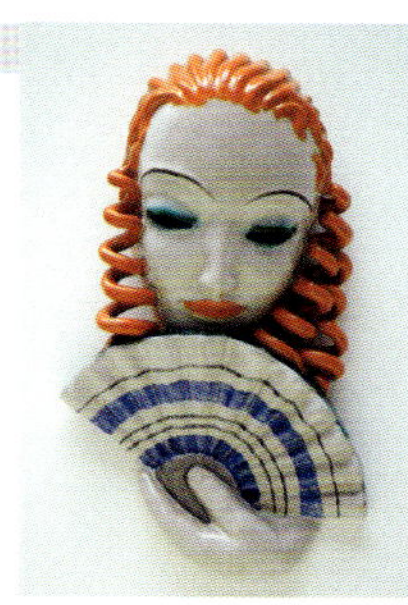

Sammlung.schmutz@chello.at 95
6542,31 gem. W, Firmenmarke 19, Made in Austria, unsign., H. 26,5 cm; Sammlung.schmutz@chello.at 9
6542,65 (Firmenmarke ohne Ang.), unsign., H. 26,5 cm; Christie's SK 12.9.1997; 2.4.1998
6542 Firmenmarke 19, Made in Austria, unsign., H. 26 cm; PB Wien 19
6542 Firmenmarke 19, Made in Austria, unsign., H. 26,2 cm; Soth. L. 20.10.1989; Soth. South, Billingshurst 11.10.2000

6543
Meisinger, Ida
WANDMASKE Terrierkopf mit Schleife / Mascherl
ca. 1931/32
6543,85,20 Firmenmarke 19, Made in Austria, unsign., H. 25 cm, B. 12 cm; eBay # 3766851220
-> vgl. 6611
-> vgl. 6610= ZweierGRUPPE

6545
Meisinger, Ida
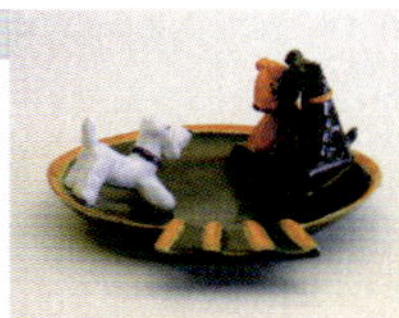
ASCHER Drei Scotchterrier
Form: rund, drei Zigarettenablagen
Dekor: ein stehender bellender und zwei sitzende Scotchterrier
ca. 1931/32
6545,E gem., Firmenmarke 19, Made in Austria, unsign., H. 7,5 cm, Ø 19,5 cm; PB Deutschland

6547
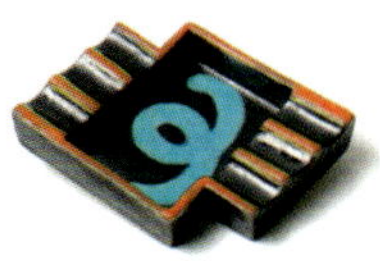
ASCHER (mit sechs Zigarettenablagen)
Form: rechteckig
Dekor: Spiegel mit gemaltem Kringel
ca. 1931/32
6547,2 gem. M, Firmenmarke 19, Made in Austria, H. 2,5 cm, B. 11 cm, L. 13,8 cm; Sammlung R. Krenn, Wien

6548
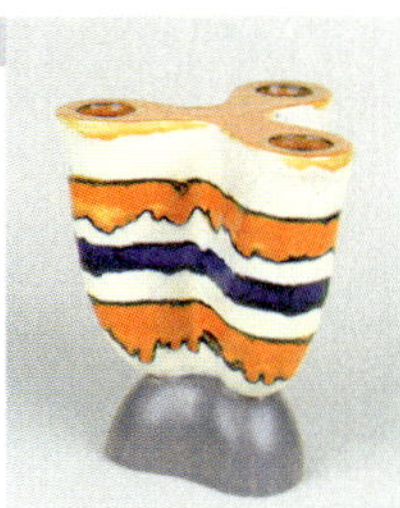
KERZENHALTER (dreiflammig)
Form: drei röhrenartig geformte, zusammenhängende Kerzentüllen (auf hexagonalem gewölbtem Sockel)
Dekor: quergestreift, orange, weiß, blau,
ca. 1932
6548,3,22 gem. H., Firmenmarke 19, Made in Austria, H. 16 cm, B. 13,8 cm; PB Wien 19
Ausst.: PSK 1995
Lit.: ADGS, Nr. 5 (m. Abb.)

6551F
Dakon, Stephan
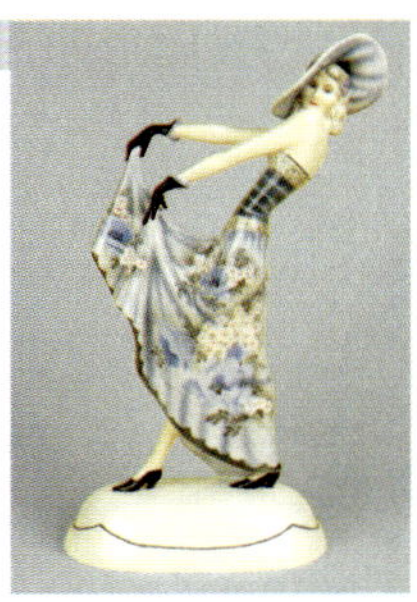
FIGUR „Blonder Traum" (Lilian Harvey als schreitende Dame mit großem Hut und Handschuhen, ihr Kleid mit beiden Händen anhebend) (auf Ovalsockel)
ca. 1932
6551,32,11 (Firmenmarke ohne Ang.), H. 40 cm; Gorringes 12.2.2002
6551,85,15 (Firmenmarke ohne Ang.), H. 39,5 cm; Christie's SK 4./5.12.1996
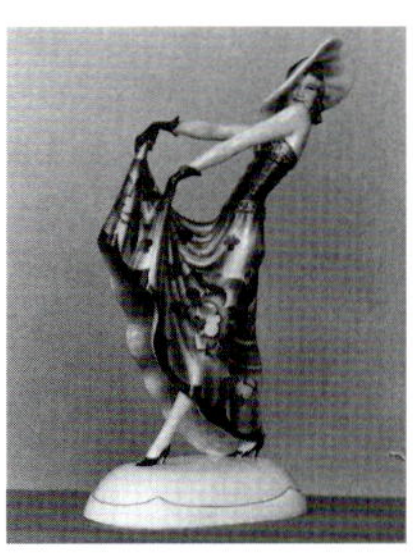
6551,159,8 Firmenmarke 19, Made in Austria, H. 40 cm; PB Wien 19
6551,187,6 Firmenmarke 19, H. 39,5 cm; KaM 85
6551,342,11 Firmenmarke 19, Made in Austria, H. 39,5 cm; WKA 33
6551F, H. 39 cm
Bem.: Lebensdaten zu Lilian Harvey s. 6009.
-> vgl. 6606; 7208
Lit.: GSN, S. 44, 45 (m. Abb.) ; GSNF, o. Pag. (m. Abb.)
Quelle: Archiv Mimi Dakon (Originalfoto; Tantiemenabrechnung 1.7.–31.12.1937; 1.1.–30.6.1940; Modellliste Dakon, undatiert, ca. 1940; Wiener Manufaktur FG, Modellliste Dakon, 7.11.1946), hier mit jeweils Ang. des Originaltitels FIGUR „Blonder Traum".
GS neu: Modell-Nr. 29 „Noblesse" (H. 20 cm), Aufl. 3000 Stück

6552F
Dakon, Stephan
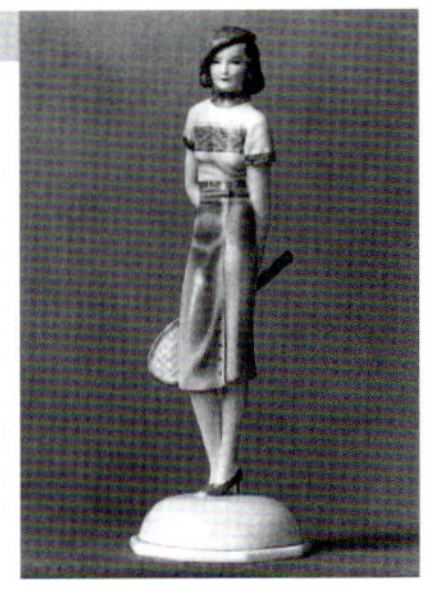
Stehende Tennisspielerin (mit Kappe und Schläger) (auf Rundsockel)
ca. 1932
6552,388,19 (Firmenmarke ohne Ang.), H. 28 cm; Christie's SK 12.5.2000
6552F, H. 28 cm
Quelle: Archiv Mimi Dakon (Originalfoto)

6553F
Dakon, Stephan
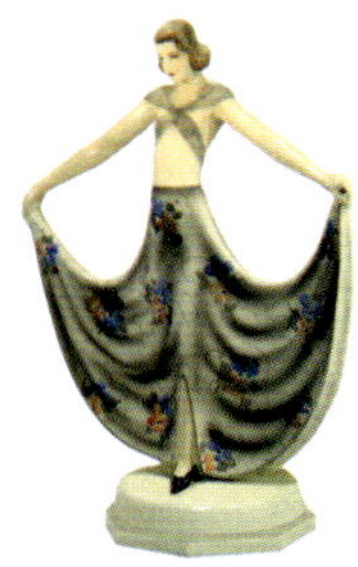
Stehende Tänzerin, ihr Kleid auseinander haltend (auf oktogonal facettiertem Rundsockel)
ca. 1932
6553,283,4,4 ger. A, gem. lig. JR, AMD, Firmenmarke 19, Made in Austria, H. 20,1 cm; Sammlung a@goldscheider.net
6553F, H. 19 cm, B. 15 cm
Quelle: Archiv Mimi Dakon (Originalfoto)
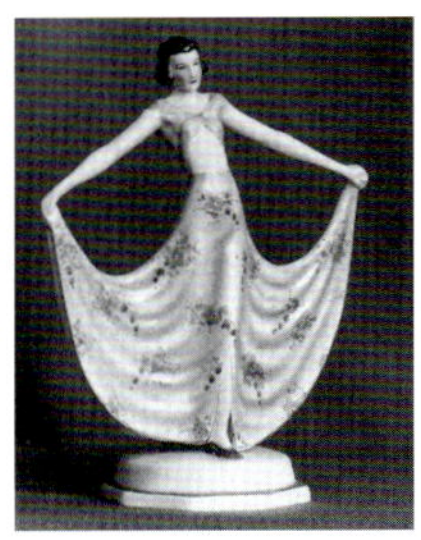

6554
Dakon, Stephan
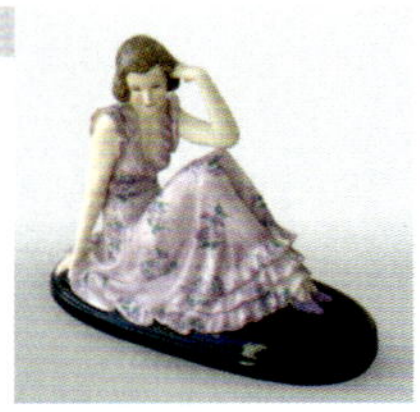
Sitzende junge Dame im langen Kleid (auf Ovalsockel)
ca. 1932
6554,40,6 gem. FN1, Firmenmarke 19, Made in Austria, Originalpick.: „Holstein & Düren, Köln", H. 22 cm; Sammlung Stressless, Portugal
6554,53,10 (Firmenmarke ohne Ang.), H. 22,5 cm; Doro 12.9.2006

6555F
Dakon, Stephan
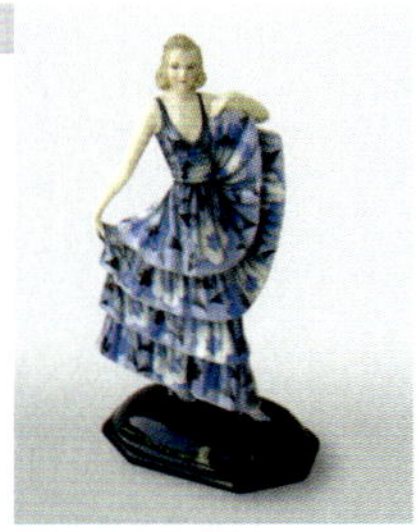
„Jane" (Schreitende Tänzerin in Pose, ihr Kleid mit der linken Hand anhebend) (auf Ovalsockel)
ca. 1932
6555,49,15 gem. Z, Firmenmarke 19, Made in Austria, unsign., H. 29 cm; Sammlung Stressless, Portugal
6555 ohne Ang., Firmenmarke 19, H. 30 cm; RM 18
Bem.: Der Firmenkat. GSNachf. (2. Ausg., bzw. Erg. Bl.) 1,54 (m. Abb.) weist an dieser Stelle einen Druckfehler auf. Die dort angegebene Modell-Nr. 6555 muss eigentlich 6535 heißen. Eindeutig handelt es sich also nicht um das dort aufgeführte Modell von Dakon, sondern um den Rosenkavalier von Lorenzl (-> vgl. 6535). Ein Originalfoto mit dem handschriftlichen Vermerk „Jane" und der korrekten Modell-Nr. 6555F, welches sich im Archiv Mimi Dakon befindet, belegt dies zusätzlich.
Quelle: Archiv Mimi Dakon (Originalfoto)
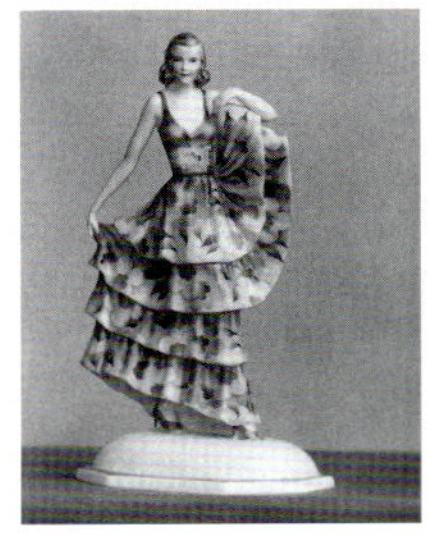

6557
Lorenzl, Josef
Stehende Dame in venezianischem Kostüm mit sitzendem Windhund (auf Ovalsockel)
ca. 1932
6557,15,? Firmenmarke 19, H. 30,6 cm; Doro 30.11.2001
6557,148,? Firmenmarke 19, Made in Austria, H. 32 cm; RM 29
6557 ohne Ang., Firmenmarke 19, H. 31 cm; Ahlden 56

6560
Knörlein, Rudolf
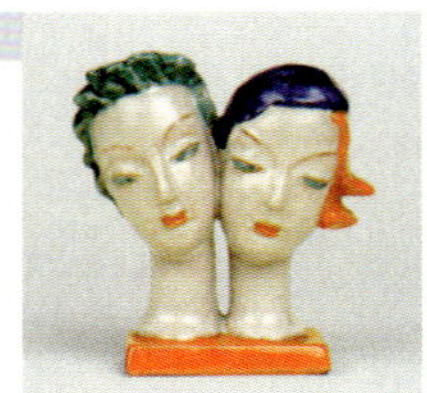
DOPPELSTEHKOPF Zwei Frauenköpfe (einer mit Kappe) (auf flachem Rechtecksockel)
ca. 1932
6560 ger., Firmenmarke 19, Made in Austria, Pick. 12 (Entwurf Rudolf Knörlein, Wien), H. 11 cm; Fischer 127
6560 ger., (ohne Firmenmarke), Pick. 12 (Entwurf Rudolf Knörlein, Wien), H. 11 cm; Sammlung R. Krenn, Wien
6560 ger., gem. S., Firmenmarke 19, Made in Austria, unsign., H. 10,6 cm; PB Wien 19

6562
KERZENHALTER (zweiflammig)
Form: zwei Kerzentüllen auf U-förmiger Röhre, über ovalem Sockel
Dekor: Punkte, Kreise, Striche, mehrfarbig
ca. 1932
6562,7 gem. lig. TK, Firmenmarke 19, H. 18 cm, B. 23,9 cm; eBay # 6529595344

6577
WANDMASKE Frauenkopf mit Hut und Terrier
ca. 1932
6577 gem. B, Firmenmarke 19, Made in Austria, H. 23,3 cm; Wien Museum Inv. 190.007

6586F
Dakon, Stephan
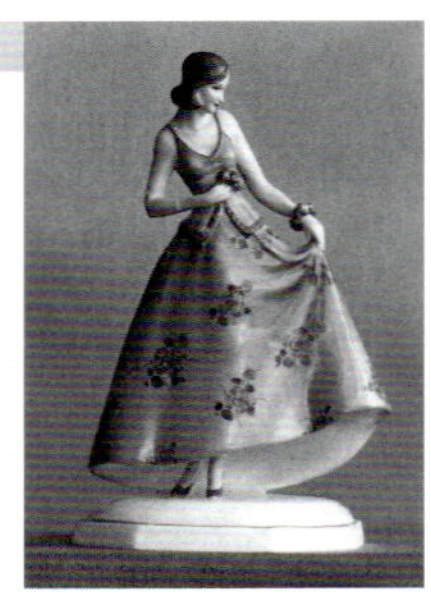
Stehende junge Dame mit Blumen (ihr langes Kleid etwas anhebend) (auf Ovalsockel)
ca. 1932
6586,62,11 (Firmenmarke ohne Ang.), H. 27,7 cm; Doro 20.3.2006
6586,115,2 Firmenmarke 19, Made in Austria, H. 28 cm; WKA 38
6586F, H. 27 cm
Quelle: Archiv Mimi Dakon (Originalfoto)

6590
Lorenzl, Josef
Sitzendes Mädchen in langem Kleid mit liegendem Barsoi (auf Ovalsockel)
ca. 1932
6590 (Firmenmarke), H. 27 cm; Pinhas, S. 137 (m. Abb.)

6591F

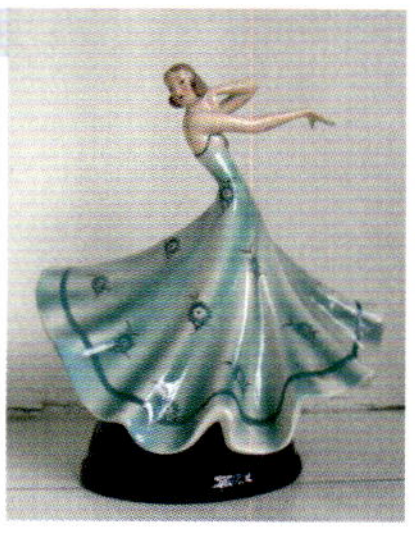

Dakon, Stephan
FIGUR „Kitty" (Tänzerin mit langem Kleid, in Pose) (auf Ovalsockel)
ca. 1932
6591,138,14,4 gem. S.1.F, Firmenmarke 19, Made in Austria, unsign., H. 19,6 cm, B. 17,5 cm; Privatsammlung, Wien
6591,161,14,4 gem. S, Firmenmarke 19, Made in Austria, H. 20 cm; Christie's SK 15.11.2001; city-antik
Quelle: Archiv Mimi Dakon (Originalfoto), hier mit Ang. des Originaltitels „Kitty".

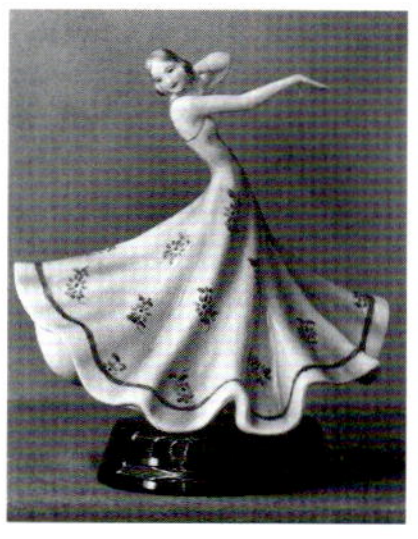

6596

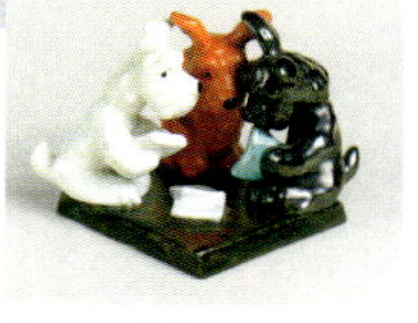

Meisinger, Ida
Drei sitzende Terrier beim Kartenspiel (auf flachem Rechtecksockel)
ca. 1932
6596,17,31 gem. rot G, Firmenmarke 19, Made in Austria, Hand dec., unsign., H. 10,5 cm; Wien Museum Inv. 220.716
6596,61,11 gem. HE, Firmenmarke 19, Made in Austria, unsign., H. 10,6 cm; Wien Museum Inv. 220.717
6596,81,1 ger. Firmenmarke 19, Made in Germany, Ausformung zw. 1938–1941, H. 11,1 cm; PB Wien 19

6599
Knörlein, Rudolf (zugeschr.)
BÜSTE Frauenkopf (nach rechts blickend) mit Schulteransatz
ca. 1932
6599 gem. AB, Firmenmarke 19, Made in Austria, unsign., H. 25 cm; Kett. 136; 206; Zezschwitz 22
Bem.: Die beiden Münchener AH Kett. und Zezschwitz schreiben dieses Modell dem Künstler Armel Beaufils zu. Diese Zuordnung ist aber weder belegbar noch haltbar, da der Kopf mit den für Rudolf Knörlein typischen Locken (Curls) garniert ist. Auch der Gesichtsausdruck zeigt eindeutige Züge seiner Entwürfe. Bei Soth. L. wurde diese Büste am 6.7.1984 unter der Modell-Nr. 6654 angeboten.

6600
Lorenzl, Josef
„Gefangener Vogel" (Der Gefangene Vogel, Niddy Impekoven) (auf Ovalsockel)
ca. 1922 (für den Entwurf)
6600,51,10 gem. N, (Firmenmarke ohne Ang.), H. 21 cm; AZ, 1979, S. 502, 503 (m. Abb.)
6600,213,4 gem. N, Firmenmarke 19, Made in Austria, H. 21 cm; eBay # 2259146861; 7319565081
6600,591,19 Firmenmarke 19, Made in Austria, H. 21 cm; Kett. 184
Ausst.: SSH 1992
–> vgl. 5054 (BRONZE); 5230; 5960; 5964= Pendant zu 5960; 6204; 8176
–> vgl. 5530 GRUPPE (Trio)
Pendant, bzw. ähnl. Modell –> vgl. 5434
Bem.: s. bei 5230
Lit.: s. bei 5230
GS neu: Modell-Nr. 82 „Gefangener Vogel" (H. 40 cm), Aufl. 3000 Stück

6602

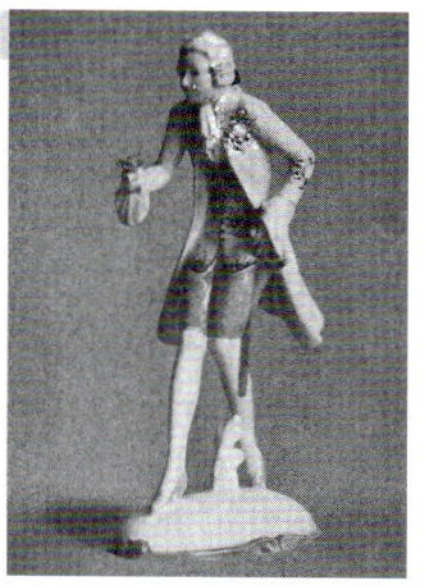

Lorenzl, Josef
„Rosenkavalier" (Octavian als schreitende Figur mit Rose in der rechten Hand) (diverse ovale Sockelformen)
ca. 1931/32
6602F, H. 35 cm, B. 14,5 cm; GSNachf. 1,54b
Bem.: s. 6535
–> vgl. 6535; 6555 (falsche Ang.)
Lit.: GSNachf. (2. Ausg., bzw. Erg. Bl.) 1,54 (m. Abb.).

6603

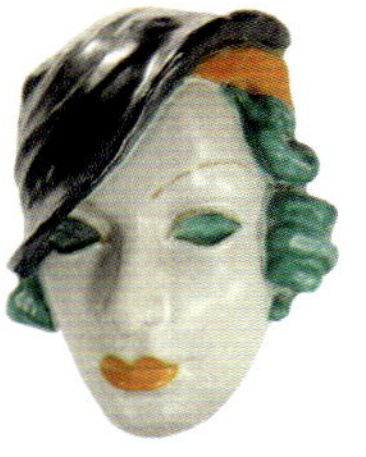

WANDMASKE Frauenkopf mit Hut (Marlene Dietrich?)
ca. 1931/32
6603 ger., gem. lig. HW, ohne Firmenmarke, H. 18,5 cm; Sammlung a@goldscheider.net

6605
Meisinger, Ida (oder Bosse, Walter) (zugeschr.)
Stehender Elefant
ca. 1932
6605 ger., Firmenmarke 19, Made in Austria, unsign., H. 12,5 cm; Fischer 144
Bem.: Die Zuschreibung an Bosse durch das AH scheint zwar auf den ersten Blick durchaus nachvollziehbar, jedoch dürfte eher Ida Meisinger als Künstlerin infrage kommen, die einen ähnlichen Dickhäuter unter der Modell-Nr. 5787 für GS schuf.

6606

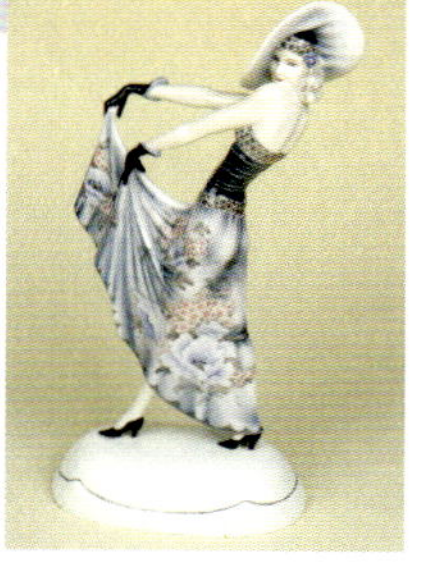

Dakon, Stephan
FIGUR „Blonder Traum" (Lilian Harvey als schreitende Dame mit großem Hut und Handschuhen, ihr Kleid mit beiden Händen anhebend) (auf Ovalsockel)
ca. 1932
6606,17,10 Firmenmarke 19, Made in Austria, H. 32 cm; PB Wien 19
6606,115,19,6 gem. N, Firmenmarke 19, Made in Austria, H. 32,3 cm; KNS 7
6606,297,19 gem. S, gem. lig. AS, Firmenmarke 19, Made in Austria, H. 32 cm; city-antik
6606,300,4 Firmenmarke 19, Made in Austria, H. 32 cm; Christie's SK 27.4.2005
6606,371,16 Firmenmarke 19, H. 32,5 cm; Soth. South, Billingshurst 11.10.2000

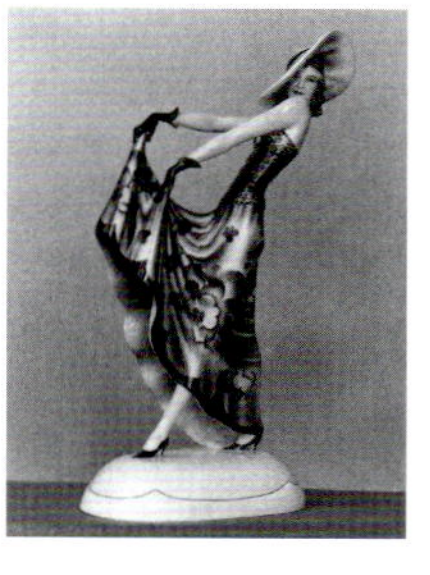

6606,377,6,6 Firmenmarke 19, Made in Austria, H. 31,3 cm; Christie's L. 27.9.1989
6606,472,12 gem. Z, Firmenmarke 19, Made in Austria, H. 32 cm; WKA 29
6606,575,16 gem. N, Firmenmarke 19, Made in Austria, H. 32 cm; eBay # 2607425582
6606,585,6 Firmenmarke 19, H. 32 cm; eBay # 39688643
6606,587,19 gem. E.F., Firmenmarke 19, Made in Austria, Pick. 4, H. 38,5 cm; WKA 35
6606 ohne Ang., Firmenmarke 19, Made in Germany, Ausformung zw. 1938–1941, H. 33,5 cm; Doro 18.11.1997
Bem.: Lebensdaten zu Lilian Harvey s. 6009.
–> vgl. 6551; 7208
Lit.: GSN, S. 44, 45 (m. Abb.) ; GSNF, o. Pag. (m. Abb.)
Quelle: Archiv Mimi Dakon (Tantiemenabrechnung 1.7.–31.12.1937; 1.1.–30.6.1940; Modellliste Dakon, undatiert, ca. 1940; Wiener Manufaktur FG, Modellliste Dakon, 7.11.1946), hier jeweils mit Ang. des Originaltitels FIGUR „Blonder Traum".
GS neu: Modell-Nr. 29 „Noblesse" (H. 20 cm), Aufl. 3000 Stück

6607
WANDMASKE Frauenkopf mit Kappe (Greta Garbo?)
ca. 1932/33
6607,4,33 Firmenmarke 19, Made in Austria, H. 17 cm; Fischer 151

6610

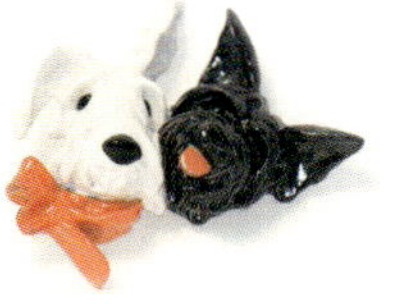

Meisinger, Ida
WANDMASKE Zwei Terrierköpfe mit Schleife / Mascherl
ca. 1932/33
6610,15,20 Firmenmarke 19, Made in Austria, Pick. 15 (Entwurf Ida Meisinger, Wien), H. 25,8 cm, B. 24 cm; Privatsammlung, Wien
–> vgl. 6611= linker Kopf

6611

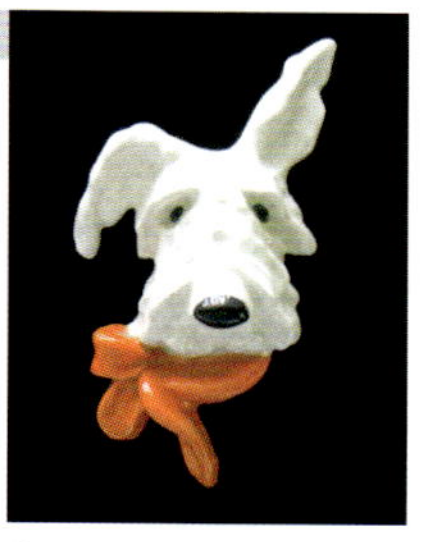

Meisinger, Ida
WANDMASKE Terrierkopf mit Schleife / Mascherl
ca. 1931/32 (für den Entwurf)
6611,20 ger., gem. lig. JP, Firmenmarke 19, Made in Austria, unsign., H. 17,5 cm; PB Deutschland
6611,65 gem. S, Firmenmarke 19, Made in Austria, unsign., H. 17,5 cm; Privatsammlung, Wien
–> vgl. 6543
–> vgl. 6610= ZweierGRUPPE

6612

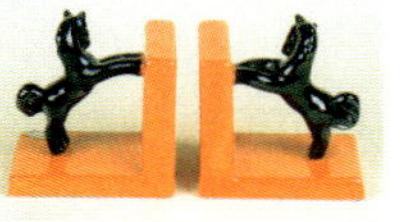

BUCHSTÜTZE Springendes Pferd
ca. 1932/33
6612,B,20 ger. gem. S, Firmenmarke 19, Made in Austria, Pick. 6, H. 10,3 cm; PB Wien 19
6612,65 ger., gem. S, Firmenmarke 19, Made in Austria, H. 10,3 cm; PB Wien 19
–> vgl. 6163= ähnl. Modell

6615
WANDMASKE Frau mit Kappe (nach rechts blickend)
ca. 1932/33
6615,15,20 Firmenmarke 19, Made in Austria, H. 19,8 cm; Quit. 29
Bem.: Diese Maske könnte evtl. von Rudolf Podany stammen. Ähnliche Entwürfe seiner Hand wurden von Keramos produziert. Für eine schlussendliche Zuschreibung erscheint mir der Vergleich und die Nähe zu den Keramoserzeugnissen nicht ausreichend.

6623M

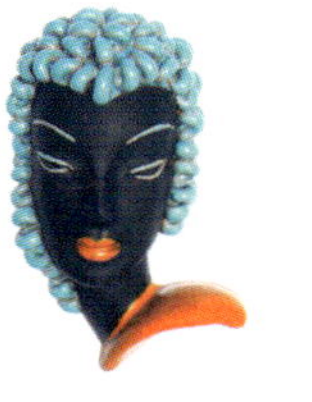

Knörlein, Rudolf
WANDMASKE „Turandot" (Frauenkopf mit Schulteransatz)
ca. 1932 (für den Entwurf)
6623,67 gem. Si, Firmenmarke 19, unsign., H. 13 cm; Sammlung Fritz Meyer/Hannover
6623 ger., gem. G, Firmenmarke 19, unsign., Pick. 12 (Entwurf Rudolf Knörlein, Wien), H. 10 cm; Sammlung.schmutz@chello.at 93
6623,S ger., gem. H, Firmenmarke 19, Made in Austria, H. 12,7 cm; Sammlung a@goldscheider.net
6623M, H. 12 cm, B. 8 cm; GSNachf. 2,1c
Bem. s. 6427
–> vgl. 6427; 7453
Lit.: GSNachf. 2,1 (m. Abb.), hier mit Ang. des Originaltitels MASKE „Turandot".
Quelle: Privatarchiv (Originalfoto)

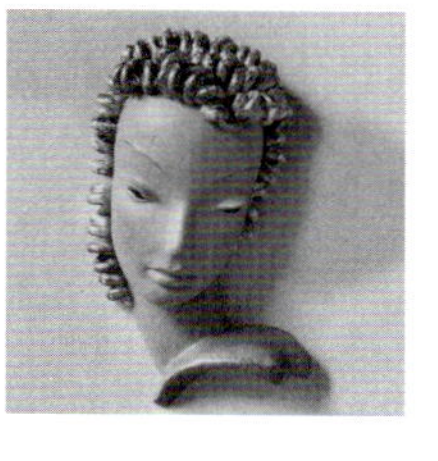

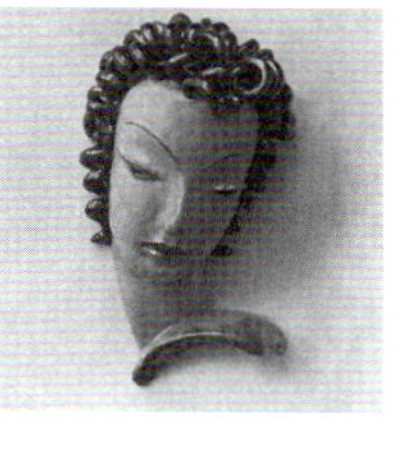

6625
WANDMASKE Frauenkopf mit Berét und Schal
ca. 1932/33
6625,20 gem. lig. MA, Firmenmarke 19, Made in Austria, H. 22 cm; eBay # 73277857797
6625,65 (Firmenmarke ohne Ang.), H. 21,2 cm; Christie's SK 2.4.1998
Lit.: Pinhas, S. 168 (m. Abb.)

6628F
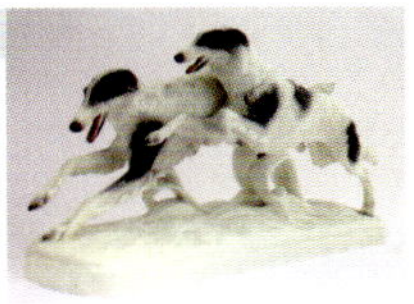
Postl, Hugo
„Windhunde"
(Zwei laufende Windhunde) (auf Landschaftssockel)
ca. 1932/33
6628,65,15 gem. lig. NO, II., Firmenmarke 19, Made in Austria, sign. H. Postl, H. 18 cm., L. 40 cm; Sammlung a@goldscheider.net
Quelle: AGM (Rechnung an Maurice Mizrahi, Alexandria, Ägypten, vom 22. April 1953), hier mit Ang. des Originaltitels „Windhunde".

6629F
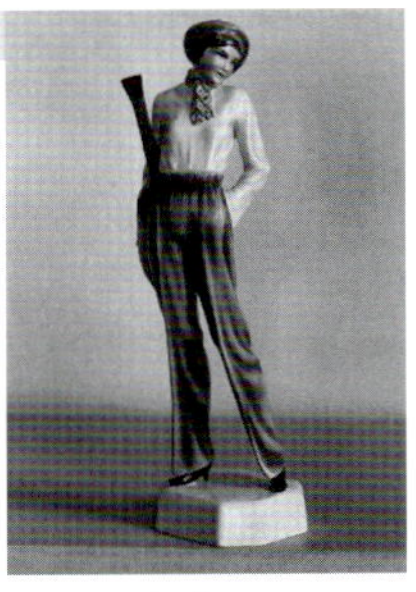
Dakon, Stephan
FIGUR „Daisy" (Stehende junge Dame mit Béret / Pullmannkappe und langen Hosen, eine Gitarre haltend) (auf oktogonalem Sockel)
ca. 1930 (für den Entwurf)
6629,2,4 (Firmenmarke ohne Ang.), unsign.?, H. 31 cm; Christie's SK 4.11.1999
6629F, H. 31 cm
Bem.: Diese Figur spiegelt die sogenannte Garconne Mode wider, die in den Jahren ab 1924 bis nach 1928 eine beliebte, im männlichen Stil gehaltene Damenmode darstellte. Kostümjacken in Herrenfasson sowie Paletot-Mäntel und der typische Bubikopf als Frisur waren charakteristisch für diese Strömung (–> vgl. Hartmann).
–> vgl. 6074
Quelle: Archiv Mimi Dakon (Originalfoto)

6630F
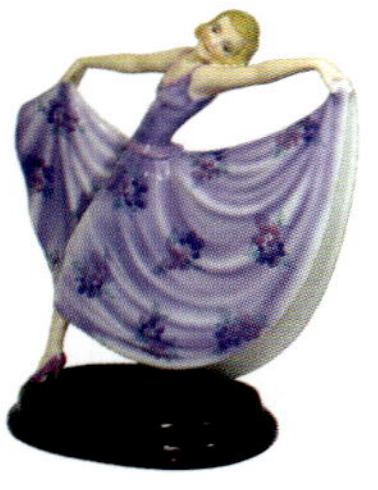
Dakon, Stephan
FIGUR „Walzer" (Schreitende Tänzerin, ihr Kleid schmetterlingsflügelartig auseinander haltend) (auf Ovalsockel)
ca. 1932/33
6630,74,11 ger. 4, gem. N, gem. lig. HS, Firmenmarke 19, Made in Austria, H. 24,8 cm; Sammlung a@goldscheider.net
6630,179,16,8 Firmenmarke 19, Made in Austria, H. 27,9 cm; eBay # 1421919686
6630,206,19 Firmenmarke 19, H. 27,1 cm; Doro 7.3.2001
–> vgl. 7285F
Lit.: GSNachf. 1,9 (m. Abb.), hier in der H. 16 cm; mit Ang. des Originaltitels „Walzer".

6631F
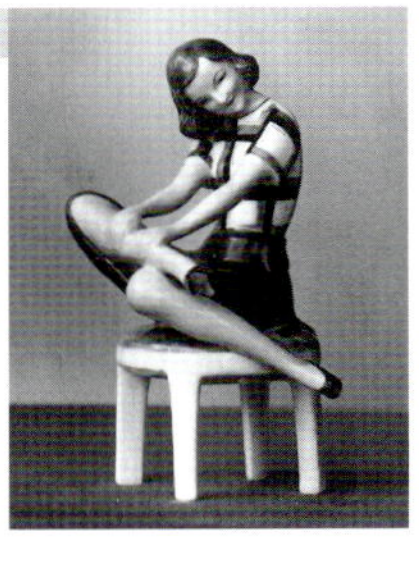
Dakon, Stephan
Sitzendes Mädchen in koketter Pose (auf vierfüßigem Hocker)
ca. 1933
6631F, H. 15 cm
Quelle: Archiv Mimi Dakon (Originalfoto)

6635
Ski fahrendes kleines Mädchen mit ausgebreiteten Armen
ca. 1932/33
6635,24,31 Firmenmarke 19, Made in Austria, H. 10 cm, B. 7,5 cm, L. 9 cm; eBay # 256387934

6636
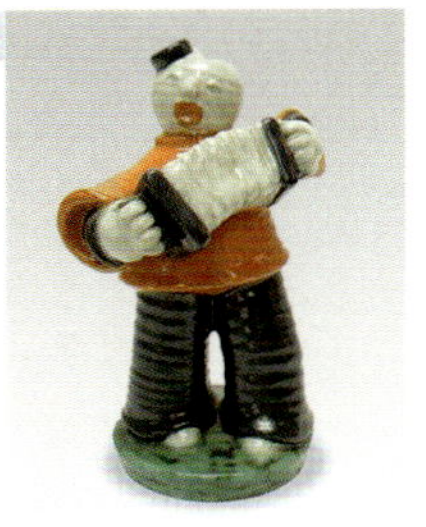
Bosse, Walter (zugeschr.)
Stehender Clown mit Bandoneon (auf Rundsockel)
ca. 1932/33
6636,4,66,5 gem. lig. AH, Firmenmarke 19, Made in Austria, unsign., H. 20 cm; eBay # 906393034
6636,13,20 Firmenmarke 19, Made in Austria, unsign., H. 20,3 cm; eBay # 7380145943

6640
WANDMASKE Grock (Lachender Clown)
ca. 1932/33
6640 (Firmenmarke ohne Ang.), H. 19,5 cm; Pinhas, S. 174 (m. Abb.)
Bem.: Dargestellt ist der weltbekannte Schweizer Clown Grock (Charles Adrien Wettach, *10.1.1880 †14.7.1959). Näheres zu seiner Biografie –> vgl.: www.clown-grock.ch.

6642
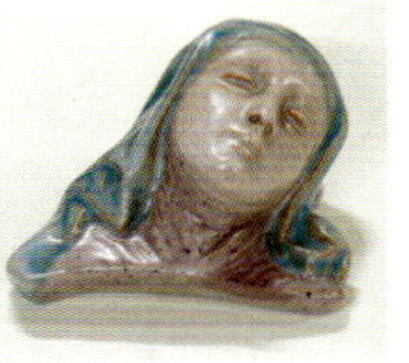
WANDMASKE Madonna (nach links blickend)
ca. 1932/33
6642,65 gem. lig. AH, Firmenmarke 19, Made in Austria, H. 10,4 cm; MAK Wien, Inv. Ke 10521-37

6645F
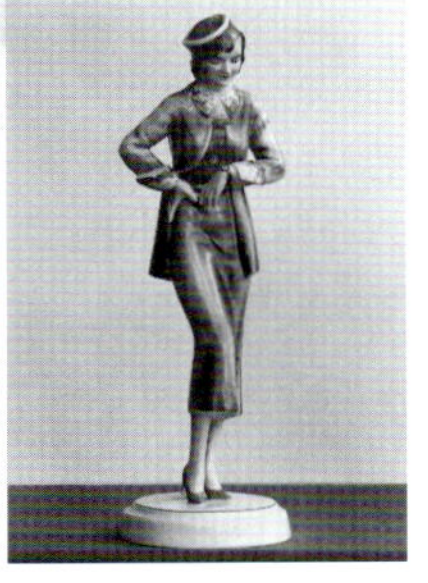
Dakon, Stephan
FIGUR „Rendezvous" (Stehende Dame im Kostüm mit Hut) (auf Rundsockel)
ca. 1932/33
6645,50,7 (Firmenmarke ohne Ang.), H. 29 cm; Christie's SK 12.5.2000
6645F, H. 29 cm
Quelle: Archiv Mimi Dakon (Originalfoto)

6646F
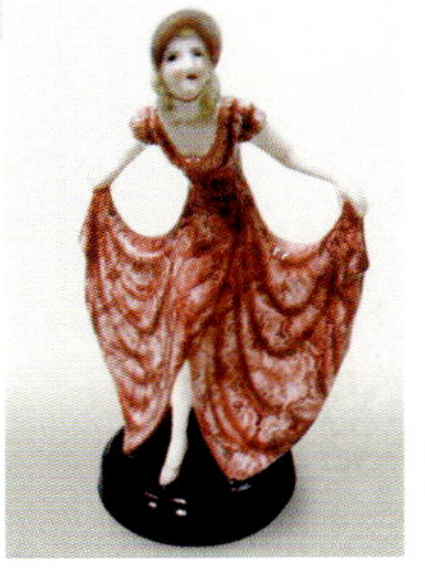
Dakon, Stephan
FIGUR „Kongress" (Lilian Harvey) (auf Rundsockel)
ca. 1931 (für den Entwurf)
6646,14,52 gem. MR, Firmenmarke 19, Made in Austria, Hand dec., H. 20 cm; GK, S. 22, 120 (m. Abb.); Wien Museum Inv. 182.098
6646,132,4 gem. MR1, Firmenmarke 19, Made in Austria, H. 19, cm; eBay # 1501140980
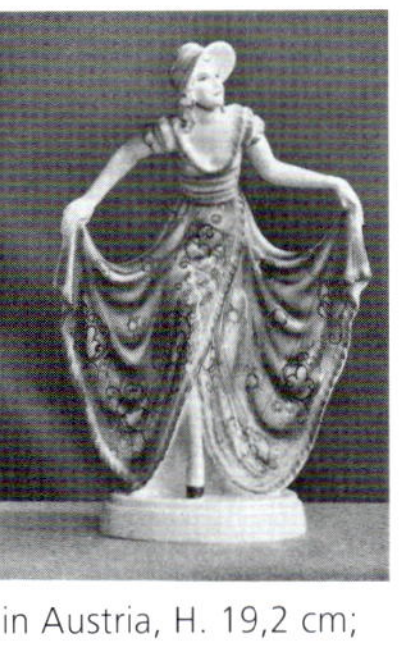
6646,134,11 ger. 5A, D, Firmenmarke 19, Made in Germany, unsign., Ausformung zw. 1938–1941, H. 20,5 cm; eBay # 1239896850
6646,148,0 Firmenmarke 19, H. 20,3 cm; BM 68
6646,290,4 Firmenmarke 19, Made in Austria, H. 18,8 cm; BM 62
6646,401,16,7 gem. S, gem. lig. AS, Firmenmarke 19, Made in Austria, H. 19,2 cm; Doro 21.9.1999; city-antik
6646,579,10 Firmenmarke 19, Made in Austria, H. ca. 19 cm; WKA 38
6646F, H. 19 cm, B. 13 cm; GSNachf. 1,8b
Bem.: Die Figur geht zurück auf eine Szene des Ufa-Films „Der Kongress tanzt" (Uraufführung 29.9.1931, Wien), in dem Lilian Harvey die weibliche Hauptrolle verkörperte (–> vgl. www.deutscher-tonfilm.de).
Ausst.: HMW (jetzt Wien Museum) 1985/86.
–> vgl. 6330F
Lit.: GSNachf. 1,8 (m. Abb.), hier mit Ang. des Originaltitels „Kongress"; GK, S. 22, 120 (m. Abb.).

6647
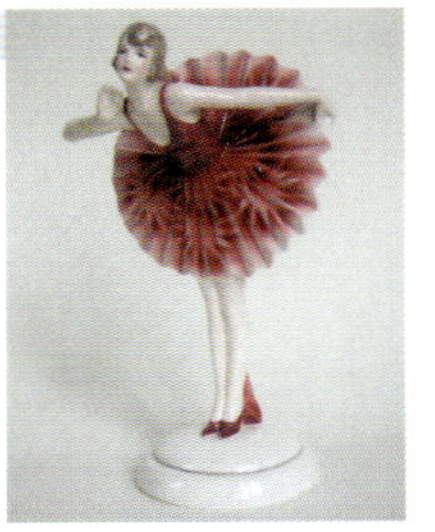
Dakon, Stephan
„Wiener Walzer" („Stern") (Stehende, sich verneigende Pierrette auf Rundsockel mit Spitzhut)
ca. 1929/30 (für den Entwurf)
6647,12,10,5 gem. S.1.F, Firmenmarke 19, Made in Austria, H. 19,3 cm; PB Deutschland
6647,48,12,9 gem. lig. JR., Firmenmarke 19, Made in Austria, H. 18,9 cm; Sammlung Weilinger, Wien
Bem.: Der Spitzhut, der am Sockel liegt, dient zugleich als Stütze für die Figur.
–> vgl. 6019

6648F
Schmidt-Kestner, Erich
FIGUR „Setter" (rückwärts)
ca. 1904/05 (für den Entwurf)
6648F, H. 14,5 cm, B. 35 cm; GSNachf. 5,9e
–> vgl. 7066
–> vgl. 6659; 7100= jeweils GRUPPE
Lit.: GSNachf. 5,9 (m. Abb.), hier mit Ang. des Originaltitels „Setter" (rückwärts).
Quelle: AGM (Rechnung der Wiener Manufaktur Josef Schuster an Laszlo Bösze (Szeged, Ungarn), vom 5. August 1944), hier mit Ang. des Originaltitels „Setter".
6654 s. 6599

6658
Knörlein, Rudolf
RELIEF Frauenkopf mit Kappe (Wandrelief)
ca. 1932/33
6658,20 Firmenmarke 19, monogr. RK, H. 20 cm, B. 15 cm; PB Wien
6658,31 gem. B, Firmenmarke 19, Made in Austria, monogr. RK, H. 19,8 cm, B. 15 cm; KH Art 1900/Berlin

6659F
Schmidt-Kestner, Erich
GRUPPE „Setter"
ca. 1904/05 (für den Entwurf)
6659F, H. 15 cm, B. 36 cm; GSNachf. 5,9a
–> vgl. 7100
–> vgl. 3024; 6135= jeweils EinzelFIGUR (vorne)
–> vgl. 6648; 7066= jeweils EinzelFIGUR (rückwärts)
Lit.: GSNachf. 5,9 (m. Abb.), hier mit Ang. des Originaltitels „GRUPPE Setter".
Quelle: AGM (Rechnung an Maurice Mizrahi, Alexandria, Ägypten, vom 22. April 1953), hier mit Ang. des Originaltitels GRUPPE „Setter".

6660
Knörlein, Rudolf
WANDMASKE Frauenkopf mit Kragen
ca. 1932/33
6660,12 ger. KR, Firmenmarke 19, Made in Austria, unsign., H. 27 cm; eBay # 2561223246
6660 ger., Firmenmarke 19, Made in Austria, Pick. 12 (Entwurf Rudolf Knörlein, Wien), H. 27 cm; Wien Museum Inv. 190.009
6660 ger., gem. R Firmenmarke 19, Made in Austria, unsign., H. 27 cm; Kett. 191

6661
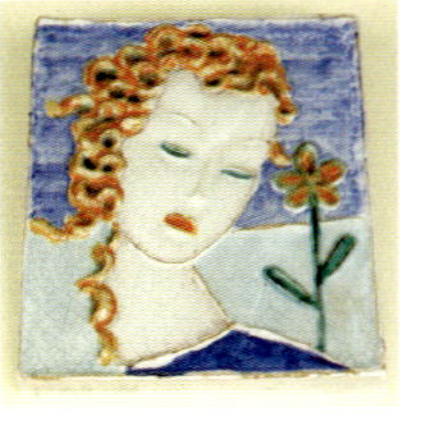
Knörlein, Rudolf
RELIEF Frauenkopf mit Blume (Wandrelief)
ca. 1932/33
6661,20 Firmenmarke 19, Made in Austria, monogr. RK, H. 27 cm; PB Wien
6661,20 ger., gem. B, Firmenmarke 19, Made in Austria, unsign., H. 27 cm; PB Wien 19
6661,20 Firmenmarke 19, Made in Austria, unsign., H. 27 cm, B. 23 cm; Doro 22.11.1990
Lit.: Pinhas, S. 178 (m. Abb.), hier mit Ang. der falschen Mod.-Nr. 6611.

6662
Knörlein, Rudolf
RELIEF Frauenkopf im Profil (nach links) (Wandrelief)
ca. 1932/33
6662,31 Firmenmarke 19, Made in Austria, unsign., H. 18 cm; PB Wien 19

6663
Knörlein, Rudolf
RELIEF Zwei Frauenköpfe vor Fenster (Wandrelief)
ca. 1932/33
6663,31 Firmenmarke 19, Made in Austria, Pick. 12 (Entwurf Rudolf Knörlein, Wien), H. 24 cm, B. 20 cm; eBay # 2211311903

6664
Knörlein, Rudolf
WANDMASKE Zwei Frauenköpfe mit Schal
ca. 1932/33
6664 gem. L., Firmenmarke 19, Made in Austria, unsign., H. 18 cm; eBay # 7407728059

6665
Knörlein, Rudolf
STEHKOPF Frauenkopf mit Hut und Halskette
ca. 1932/33
6665 ger., 415 gem., Firmenmarke 19, Made in Austria, Pick. 12 (Entwurf Rudolf Knörlein, Wien), H. 10,8 cm; eBay # 292170372

6666
Lorenzl, Josef
Carmen (Sitzende Spanierin mit Fächer)
(auf halbrundem Sockel)
ca. 1925 (für den Entwurf)
6666,38,7 gem. F, (Firmenmarke ohne Ang.), H. fehlt; eBay # 230664515
–> vgl. 5534; 5555
–> vgl. 5088; 5232; 5277= stehende Figur

6667
Matula, Viktor
FIGUR Foxterrier
ca. 1932/33
6667 H. fehlt
Quelle: Nachlass Matula, Wien

6668
Meisinger, Ida (zugeschr.)
GRUPPE Zwei männchenmachende Terrier
ca. 1932/33
6668,6 Firmenmarke 19, Made in Austria, unsign., H. 15 cm; PB Deutschland
6668,6,16 Firmenmarke 19, Made in Austria, Hand dec., unsign., H. 15,4 cm; Sammlung a@goldscheider.net
6668,7,31 Firmenmarke 19, Made in Austria, unsign., H. 15 cm; Wien Museum Inv. 190.015
6668 H. 14 cm, B. 9 cm; Fotoarchiv K. Täuber
Ausst.: SSH 1992
Lit.: SS, S. 44, 45 (m. Abb.)
Quelle: Fotoarchiv K. Täuber

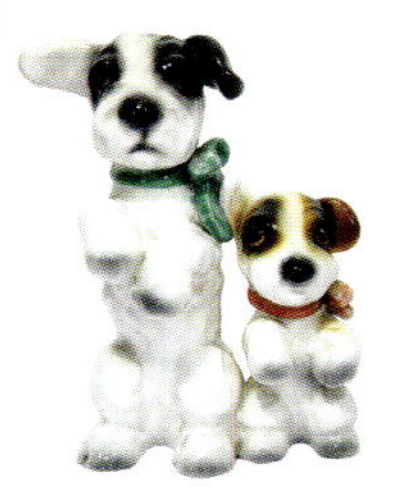

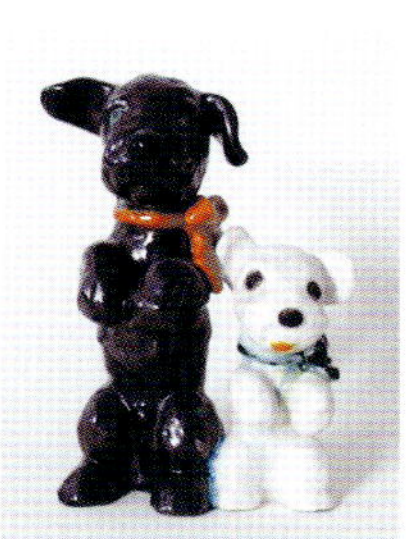

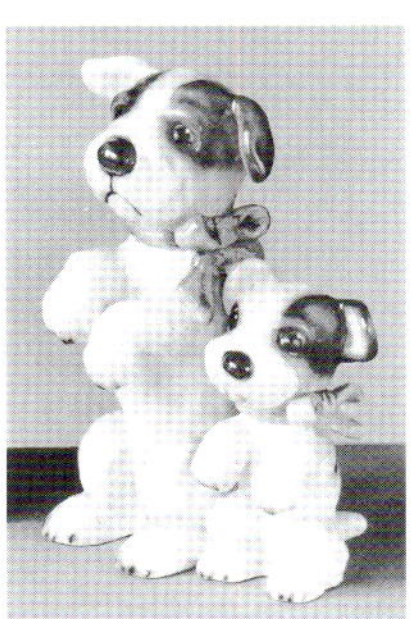

6672
Lorenzl, Josef
Stehende, nach vorne gebeugte Rokokodame
(auf Ovalsockel)
ca. 1932/33
6672,39,10 (Firmenmarke ohne Ang.), H. 18,5 cm; Christie's SK 13.5.2004
6672,67,19,4 gem. lig. MR, Firmenmarke 19, Made in Austria, H. 18,5 cm; Sammlung a@goldscheider.net
6672,78 ,11 Firmenmarke 19, Made in Austria, H. 18 cm; Fischer 71

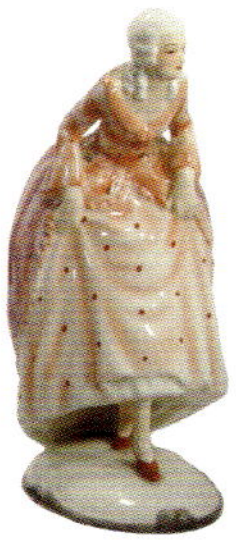

6675
WANDAPPLIK Madonna (Maria als Ordensfrau mit Aureole)
ca. 1932/33
6675,65 Firmenmarke 19, Made in Austria, Ausformung zw. 1945–1953, H. 14 cm; KH

6676
Bosse, Walter (zugeschr.)
Stehender kleiner Engel in Trachtenkostüm
(auf Rundsockel)
ca. 1932/33
6676,65 gem. AB, Firmenmarke 19, unsign., H. 9,2 cm; Wien Museum Inv. 190.163

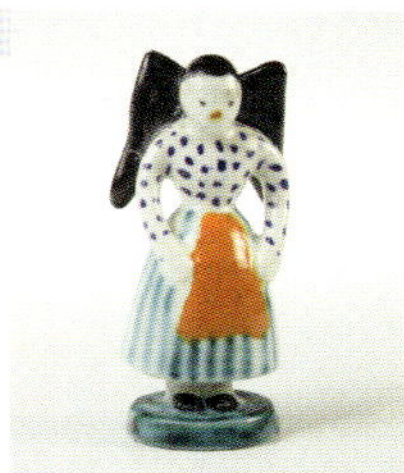

6679?
Sitzende Katze mit Schleife / Mascherl um den Hals
ca. 1932/33
6679?,10 Firmenmarke 19, Made in Austria, Pick. 5/a, H. 20,3 cm; eBay # 146731045

6681
RELIEF Madonna (nach links blickend) (Wandrelief)
ca. 1932/33
6681,20 ger., Firmenmarke 19, Made in Austria, H. fehlt; eBay # 3365859870
Bem.: Der Rand des Reliefs ist als „Bilderrahmen" gearbeitet.
Pendant –> vgl. 6689

6689
RELIEF Madonna (nach rechts blickend) (Wandrelief)
ca. 1932/33
6689,31 ger., Firmenmarke 19, Made in Austria, H. fehlt; eBay # 3365859870
6689 gem., gem. lig. AH, Firmenmarke 19, H. 10,5 cm; WKH
Bem.: Der Rand des Reliefs ist als „Bilderrahmen" gearbeitet.
Pendant –> vgl. 6681

6690
Lorenzl, Josef
GRUPPE Biedermeierpaar
ca. 1932/33
6690,33,4,4 gem. JR, Firmenmarke 19, Made in Austria, H. ca. 20 cm; WKH
Bem.: Mündliche Mitteilung aus dem WKH, daher leider keine nähere Beschreibung oder Abb. des Objektes möglich.

6692
Dakon, Stephan
Flamencotänzerin in langem Rock (Halbakt) mit Tamburin
(auf Ovalsockel)
ca. 1933
6692,11,29 Firmenmarke 19, H. 40,8 cm; Doro 20.4.1993
6692 ger., Firmenmarke 19, Made in Germany, Ausformung zw. 1938–1941, H. 41,2 cm; Doro 29.5.2001
–> vgl. 7699
Lit.: RDM, 5.1930/31, S. 488 (m. Abb.), hier eine zeitgenössische Fotografie, die wohl als Vorlage für die durch das Tamburin abgewandelte Ausführung diente. Dargestellt ist hier eine nicht näher bezeichnete Tänzerin. Fotografie: d'Ora-Benda, Wien, hier betitelt „Wiener Blut".

6693
Lorenzl, Josef
Schreitende Tänzerin (etwas nach vorne gebückt), ihr Kleid schmetterlingsflügelartig auseinander haltend
(auf Ovalsockel)
ca. 1930 (für den Entwurf)
6693,22,10,4 gem. lig. NO, JR, Firmenmarke 19, Made in Austria, H. 18,8 cm; Wien Museum Inv. 220.707
6693,49,7 (Firmenmarke ohne Ang.), H. 19 cm; Christie's SK 13.5.2004
6693,74,14 (Firmenmarke ohne Ang.), H. 19,5 cm; WKA 38; 44; Doro 15.3.2005
6693,420,6 gem. GEHE, Firmenmarke 19, Made in Austria, H. 19 cm; Herr 34
–> vgl. 6227

6694F
„Madonna" (Stehende Madonna mit Kind)
(auf gewölbtem Ovalsockel)
ca. 1933
6694,31 gem. lig. AH, Firmenmarke 19, Ausformung zw. 1945–1953, H. 28,5 cm; eBay # 6504783319
6694,118,3 Firmenmarke 19, Made in Austria, H. 30 cm; eBay # 1480709009
6694 H. 30 cm, B. 10 cm; Fotoarchiv K. Täuber
Quelle: AGM (Kontoblatt der Wiener Manufaktur Josef Schuster für die Firma Porslins Ericsson-Lund, Stockholm?, Schweden vom 7. Januar 1943), hier mit Ang. des Originaltitels „Madonna"; Fotoarchiv K. Täuber.

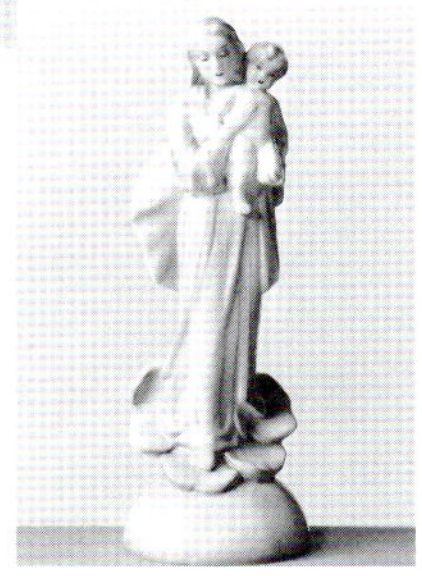

6696
Bosse, Walter (zugeschr.)
Stehender vom Notenblatt singender Engel
(auf Rundsockel)
ca. 1932/33
6696,16 gem. lig. AH, Firmenmarke 19, Made in Austria, unsign., H. 18 cm; eBay # 6605218495

6698
Goebel, Kurt (zugeschr.)
GRUPPE Tanzendes Spanierpaar (auf Rundsockel)
ca. 1933
6698,3,65 gem. lig. AH, Firmenmarke 19, Made in Austria, unsign., H. 30,8 cm; PB Wien
6698,7,31 gem. A, Firmenmarke 19, Made in Austria, unsign., H. 30,8 cm; RM 11; 13

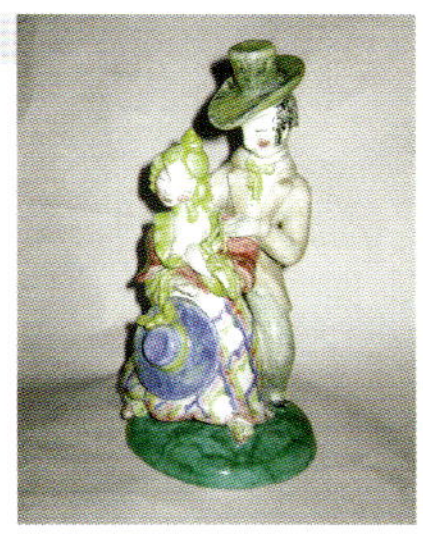

6699
Goebel, Kurt (zugeschr.)
Skifahrer
ca. 1933
6699,45,8 Firmenmarke 19, Made in Austria, unsign., H. 19 cm; Herr 38

6702F
Dakon, Stephan
Stehende junge Dame, an eine Reling gelehnt
(auf Segmentalsockel)
ca. 1933
6702,13,4 (Firmenmarke ohne Ang.), H. 29,1 cm; Doro 26.11.2002
ohne Ang. Firmenmarke 19, Made in Austria, H. 29,9 cm; Soth. Arcade 1593
ohne Ang. (Firmenmarke ohne Ang.), unsign.?, H. 29,2 cm; Soth. Arcade 1358
6702F, H. 30 cm
Quelle: Archiv Mimi Dakon (Originalfoto)

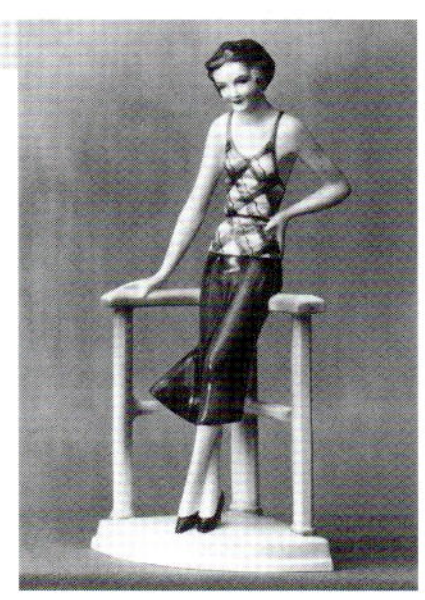

6703
Meisinger, Ida
Fressender West Highland Terrier mit Schleife / Mascherl (an der Futterschale) (auf flachem Rechtecksockel)
ca. 1933
6703,292,65 gem. S, Firmenmarke 19, Made in Austria, unsign., H. 8,3 cm, L. 16,3 cm, B. 7,5 cm; Sammlung John Nash

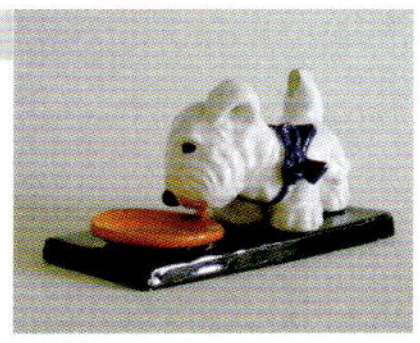

6704
Dakon, Stephan
FIGUR „Madelaine" („Madeleine"/„Tanz") (Tänzerin auf dem linken Bein tänzelnd, das rechte Bein nach vorne schwingend) (auf Ovalsockel)
ca. 1933
6704,49,6,9 gem. MR 1, Firmenmarke 19, Made in Austria, H. 30,5 cm; Fischer Luz. 377
6704,132,19 gem. FX, gem. blau N, Firmenmarke 19, Made in Austria, H. 31 cm; eBay # 1431042065
6704,446,2 (Firmenmarke ohne Ang.), H. 32 cm; Christie's SK 17.10.2001
6704,471,2 gem. blau H, gem. grün M, gem. rot EW, Firmenmarke 19, Made in Germany, Ausformung zw. 1938–1941, H. 31,9 cm; Nagel 365; Kühling 10.7.1998
Lit.: Pinhas, S. 154 (m. Abb.)
Quelle: Archiv Mimi Dakon (Tantiemenabrechnung 1.7.–31.12.1937; Modellliste Dakon, undatiert, ca. 1940), hier jeweils mit Ang. des (Original)-Titels FIGUR „Madelaine"; (Tantiemenabrechnung 1. Halbjahr 1950; 2. Halbjahr 1950 (1. Halbjahr 1951, AGM), hier jeweils mit Ang. des (Original)-Titels FIGUR „Madeleine"; (Wiener Manufaktur FG, Modellliste Dakon, 7.11.1946), hier mit Ang. des (Original)-Titels „Tanz".

6705
Dakon, Stephan
FIGUR „Quadrille" (Stehende Tänzerin mit Hut und Handschuhen) (auf Rundsockel)
ca. 1933
6705,72,7 gem. S.G.S.F., Firmenmarke 19, Made in Austria, H. 20,3 cm; eBay # 197327693
Quelle: Archiv Mimi Dakon (Tantiemenabrechnung 1.7.–31.12.1937, Modellliste Dakon, undatiert, ca. 1940; Wiener Manufaktur FG, Modellliste Dakon, 7.11.1946), hier jeweils mit Ang. des Originaltitels FIGUR „Quadrille".

6706
Stehender Mädchenakt mit Blume (auf Rundsockel)
ca. 1933
6706 (ohne Ang.), Firmenmarke 19, Made in Austria, H. 28 cm; Mehlis 16; 17

6707F
Dakon, Stephan
Schreitende junge Dame (auf Ovalsockel mit Tamburin)
ca. 1933
6707,63,31,4 gem. S.1, Firmenmarke 19, Made in Austria, H. 30 cm; eBay # 7391418624
6707F, H. 30 cm, B. 23 cm
Quelle: Archiv Mimi Dakon (Originalfoto)

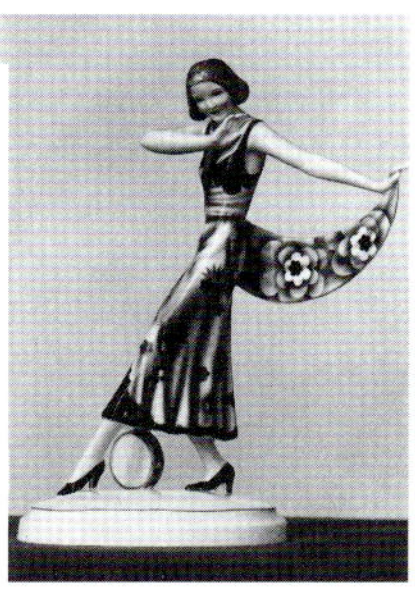

6708
Meisinger, Ida
„Herzensdieb" (Stehender West Highland Terrier, ein Herz mit dem Maul haltend)
ca. 1933
6708,57,(84?) ger., Firmenmarke 19, Pick. 15 (Entwurf Ida Meisinger, Wien), und Originalpick.: „J.F. Maerklin, Stuttgart, Preis RM 8,50, Modell HERZENSDIEB 126/815", H. 13 cm, L. 16 cm; Rieber 22.4.–5.5. 2004
Bem.: Besonderer Glücksfall: an einem Objekt befinden sich sowohl das Entwerfer-Pick. für Ida Meisinger/Wien als auch ein weiteres Verkaufs-Etikett der Firma Maerklin (Stuttgart) mit Ang. des Originaltitels „Herzensdieb" und zusätzlich auf einem weiteren Aufkleber der damalige Ladenpreis von RM 8,50.

6712
Knörlein, Rudolf
Stehendes Mädchen mit Reh (auf Spitzovalsockel)
ca. 1933
6712 ger., gem. lig. A.B, Firmenmarke 19, Made in Austria, unsign., H. fehlt; KH
Lit.: Pinhas, S. 178 (m. Abb.), hier ebenfalls unsign.

6713
DECKELDOSE
Form: dreipassig, gewölbt
Dekor: geometrische Ornamente
Knauf: Dreizylindermotiv
ca. 1933
6713 ger., Firmenmarke 19, Made in Austria, H. 11 cm; Zezschwitz 26

6715
Postl, Hugo
Sitzender Spaniel
ca. 1933 (für die Ausführung)
6715,131,11 gem. lig. AS, Firmenmarke 19, Made in Austria, Pick. 2, H. 19,5 cm; eBay # 3702737539
6715,170,19 gem. F, Firmenmarke 19, Made in Austria, H. 20 cm; eBay # 850819499
-> vgl. 7144

6716
Dakon, Stephan
WANDMASKE „Maske Sissy"
ca. 1933
Quelle: Archiv Mimi Dakon (Wiener Manufaktur FG, Modellliste Dakon, 7.11.1946), hier mit Ang. des Originaltitels „Maske Sissy".

6717
Dakon, Stephan
„Rel. Blonder Traum"
ca. 1933
Quelle: Archiv Mimi Dakon (Wiener Manufaktur FG, Modellliste Dakon, 7.11.1946), hier mit Ang. des Originaltitels „Rel. Blonder Traum".

6726
Lorenzl, Josef
GRUPPE Galantes Paar (auf Ovalsockel)
ca. 1933
6726,15,19,Y gem. N, 582/108, Firmenmarke 19, Made in Austria, H. 35,5 cm; eBay # 882861808; Christie's SK 14.11.2002

6730
Liegender Barsoi
ca. 1933
6730 H. 17 cm, L. 45 cm
Quelle: Fotoarchiv K. Täuber

6731F
FIGUR „Windhund" (Stehender Barsoi)
ca. 1933
6731F, H. 22 cm, B. 27 cm; GSNachf. 5,3a
-> vgl. 7541
Lit.: GSNachf. 5,3 (m. Abb.), hier mit Ang. des Originaltitels „Windhund".
Quelle: AGM (Rechnung der Wiener Manufaktur Josef Schuster an Laszlo Bösze (Szeged, Ungarn), vom 5. August 1944); Fotoarchiv K. Täuber.

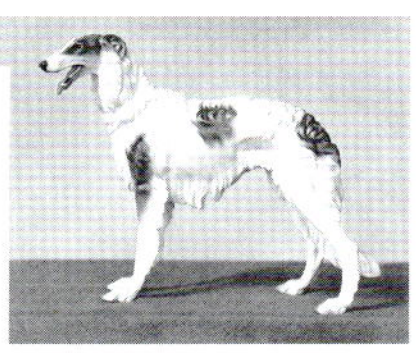

6735
Knörlein, Rudolf
WANDMASKE Frauenkopf mit Halstuch
ca. 1933
6735,20 ger., gem. lig. AH, Firmenmarke 19, Made in Austria, unsign., H. 26,2 cm; Sammlung.schmutz@chello.at 38
Lit.: Pinhas, S. 170 (m. Abb.), hier mit Ang. der falschen Mod.-Nr. 6775.

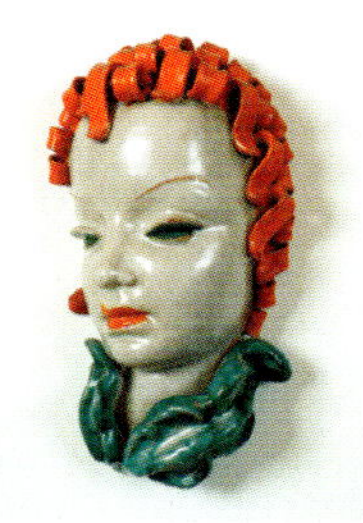

6737F
Dakon, Stephan
Ballerina mit ausgestelltem Kleid beim Spitzentanz (auf Rundsockel)
ca. 1933
6737F, H. 29 cm
Quelle: Archiv Mimi Dakon (Originalfoto)

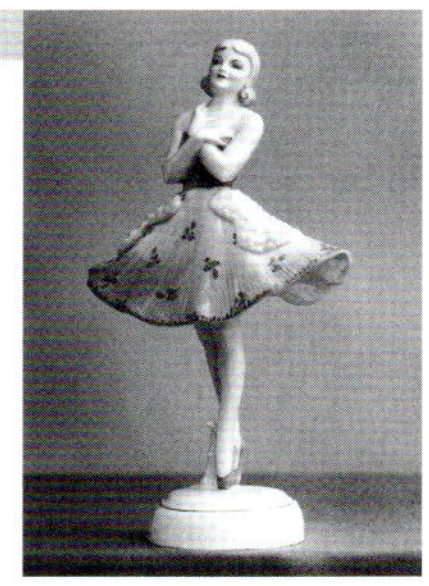

6738F
Dakon, Stephan
Stehendes Biedermeiermädchen mit Schutenhut und Blumenstrauß (auf Rundsockel)
ca. 1933
6738F, H. 18 cm
Quelle: Archiv Mimi Dakon (Originalfoto)

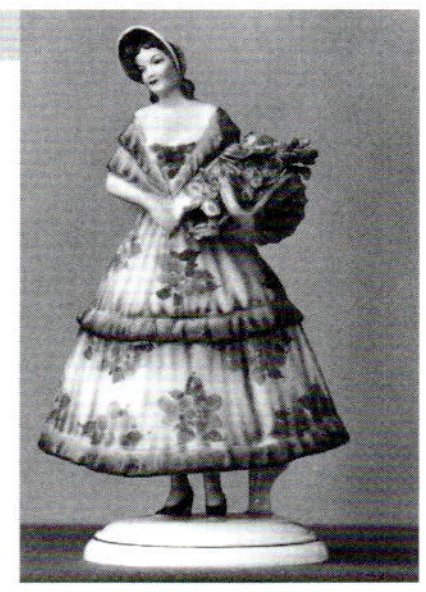

6739
Knörlein, Rudolf
DOPPELSTEHKOPF Zwei Frauenköpfe (Cheek to cheek) (auf flachem Rechtecksockel)
ca. 1933
6739,391,4 ger., gem. G., Firmenmarke 19, Made in Austria, unsign., H. 14,6 cm; PB Wien 19
6739 ger., gem. WS, Firmenmarke 19, Made in Austria, unsign., H. 14,8 cm; Sammlung.schmutz@chello.at 82
Bem.: Derartige Doppelköpfe wurden von Knörlein öfter entworfen. Eine aus dem Jahre 1929 stammende originale Entwurfszeichnung (Aquarell über Graphit) zeigt eine Variante bei der der Sockel identisch mit der Modell-Nr. 6739 ist.
Quelle: Privatsammlung D.

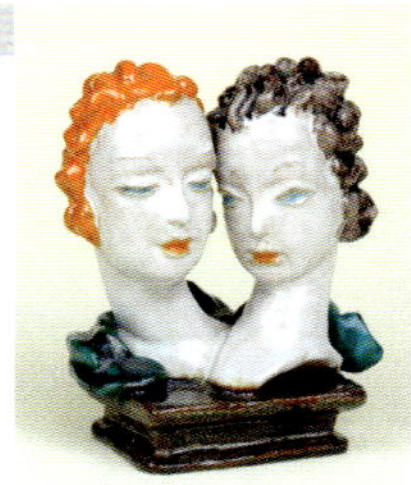

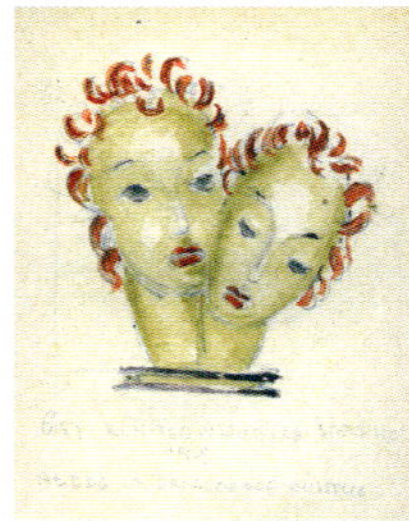

6745
Lorenzl, Josef
BÜSTE Madonna mit gefalteten Händen
ca. 1933
6745,6 gem. N, Firmenmarke 19, Made in Austria, sign. Lorenzl, H. 28,6 cm; Sammlung a@goldscheider.net
6745,7 ger., gem. G, Firmenmarke 19, Made in Austria, unsign., H. 30 cm; Wien Museum Inv. 190.075
6745,8 Firmenmarke 19, Made in Austria, unsign., H. 29,8 cm; Doro 25.4.2000
6745,(26 oder 46?),4 gem. A, Firmenmarke 19, Made in Austria, unsign., H. 28 cm; eBay # 432365558
-> vgl. 7052; 7736
-> vgl. 6923= WANDMASKE

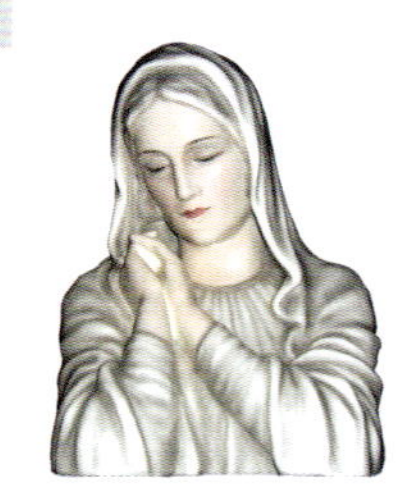

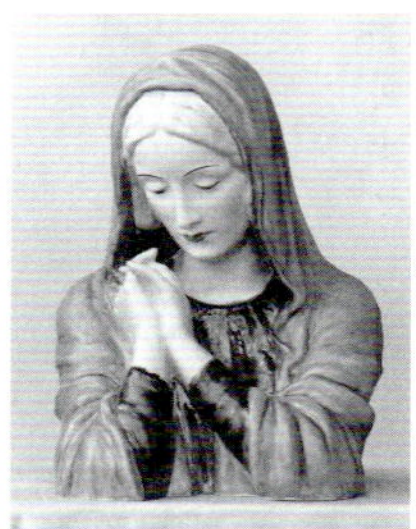

6746
Lorenzl, Josef
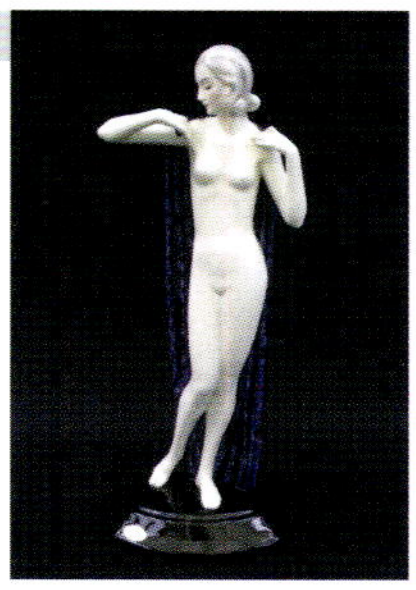
Stehender Damenakt mit großem Tuch (auf gekehltem, geschwungenem Rechtecksockel)
ca. 1933
6746,51,4 gem. lig. AS, Firmenmarke 19, Made in Austria, H. 43,2 cm; eBay # 3738668483 (KH www.tias.com/stores/frozenbeauty)
6746,74,4,8 ger. A, gem. H, Firmenmarke 19, Made in Austria, Pick. 5, H. 44 cm; Sammlung Robert Senft

6747
Herczeg, Claire (Klára) (zugeschr.)
Stehendes Mädchen mit Hut, ein Schaf betrachtend (auf Ovalsockel)
ca. 1933
6747,44,20,P Firmenmarke 19, unsign., H. 18 cm; Doro 9.11.1989
6747,153,31 gem. lig. AS, Firmenmarke 19, Made in Austria, unsign., H. 18 cm; eBay # 6637512868

6753
Lorenzl, Josef
Tänzerin mit Tamburin (La Jana) (auf Ovalsockel)
ca. 1933
6753,65,19,5 Firmenmarke 19, Made in Austria, H. 37,5 cm; WKA 38
6753,96,11 Firmenmarke 19, Made in Austria, H. 37 cm; WKA 33
-> vgl. 6759
Lit.: RDM, 5.1930/31, S. 80 (m. Abb.), hier eine zeitgenössische Fotografie, die wohl als Vorlage für die Ausführung diente. Dargestellt ist die Tänzerin „La Jana" (eigentl. Henriette Margarethe Hiebel, 24.2.1905–13.3.1940). Fotografie: Manassé, Wien, hier betitelt „Tarantella". Da die Fotografien des Ateliers Manassé in Zeitschriften und Magazinen oft anders betitelt sind, kann hier nicht von einem Originaltitel ausgegangen werden (-> vgl. Faber, S. 153); Pinhas, S. 155 (m. Abb.).

6755F
Dakon, Stephan
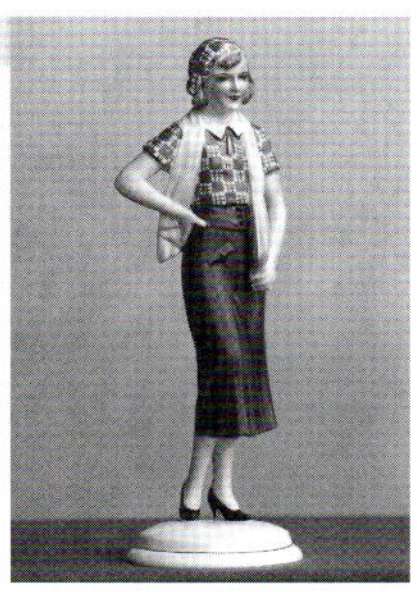
Stehende junge Dame mit Hut und Schal (auf Rundsockel)
ca. 1933
6755F, H. 28 cm
Quelle: Archiv Mimi Dakon (Originalfoto)

6756
Dakon, Stephan
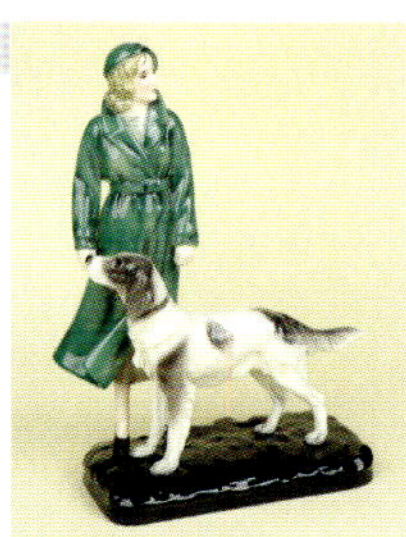
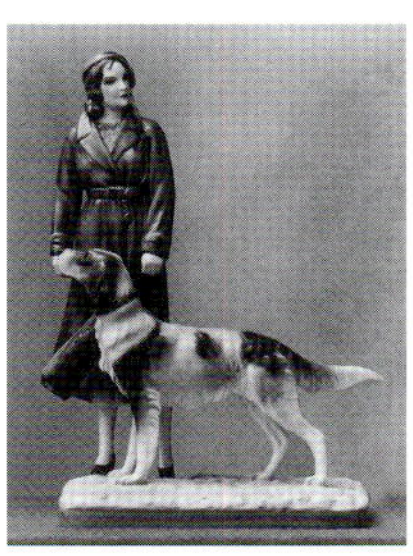
Stehende Dame (Marlene Dietrich?) mit Kappe, Mantel und einem Setter (auf abgerundetem Rechtecksockel)
ca. 1933
6756,12,5 ger. lig. JT, Firmenmarke 19, Made in Austria, H. 32 cm; PB Wien 19
Bem.: Bei der dargestellten Dame ist eine gewisse Ähnlichkeit zur berühmten Filmschauspielerin und Sängerin Marlene Dietrich (27.12.1901–6.5.1992) nicht zu verleugnen (-> vgl. www.marlene.com). Der Hund (Modell-Nr. 6804) dieser Gruppe ist vermutlich kein Entwurf Dakons, sondern wurde der stehenden Frauengestalt durch die Manufaktur „beigefügt". Näheres hierzu -> vgl. 6069.
Quelle: Archiv Mimi Dakon (Originalfoto)

6757
Dakon, Stephan
Stehende Tänzerin mit großem Hut und Handschuhen (ihren langen Rock mit beiden Armen auseinander haltend) (auf Ovalsockel)
ca. 1933
6757,4,65 gem. S., Firmenmarke 19, Made in Austria, H. 32 cm; Soth. com. # 5YX0
Lit.: Pinhas, S. 143 (m. Abb.)

6758F
Dakon, Stephan
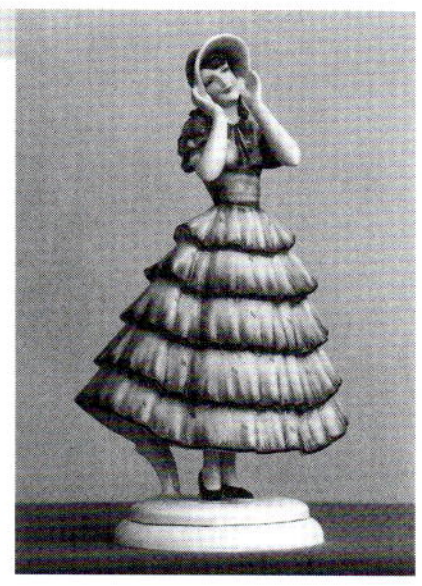
Stehende junge Dame ihren Schutenhut mit beiden Händen haltend (auf Rundsockel)
ca. 1933
6758F, H. 29 cm
Quelle: Archiv Mimi Dakon (Originalfoto)

6759
Lorenzl, Josef
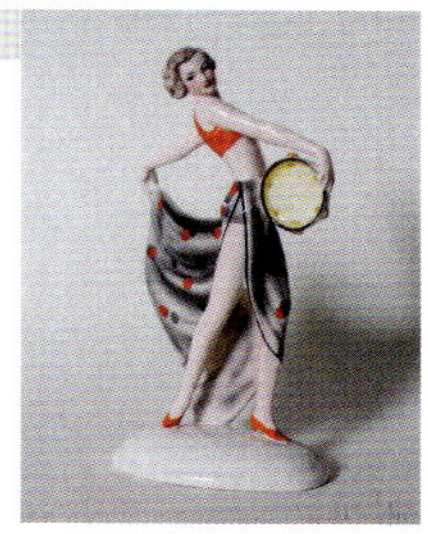
Tänzerin mit Tamburin (La Jana) (auf Ovalsockel)
ca. 1933
6759,45,7,4 Firmenmarke 19, Made in Austria, H. 20 cm; PB Deutschland
6759,128,11 (Firmenmarke ohne Ang.), H. 19 cm; Christie's SK 18.9.1998
-> vgl. 6753
Lit.: RDM, 5.1930/31, S. 80 (m. Abb.), hier eine zeitgenössische Fotografie, die wohl als Vorlage für die Ausführung diente. Dargestellt ist die Tänzerin „La Jana" (eigentl. Henriette Margarethe Hiebel, 24.2.1905–13.3.1940). Fotografie: Manassé, Wien, hier betitelt „Tarantella". Da die Fotografien des Ateliers Manassé in Zeitschriften und Magazinen oft anders betitelt sind, kann hier nicht von einem Originaltitel ausgegangen werden (-> vgl. Faber, S. 153).

6760
Lorenzl, Josef
Stehender Damenakt mit großem Tuch (auf Rundsockel)
ca. 1933
6760,23,14,4 gem. N, Firmenmarke 19, Made in Austria, H. 34,3 cm; eBay # 3772506385
6760,93,4,4 Firmenmarke 19, Made in Austria, H. 33,6 cm; Christie's L. 19.4.1989

6761
Postl, Hugo
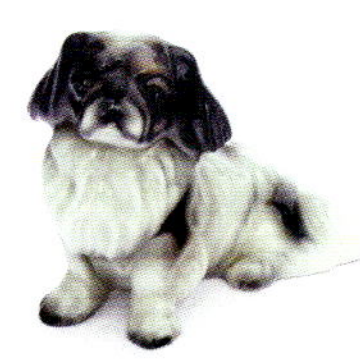
GRUPPE / FIGUR Pekinesen (Zwei sitzende Pekinesen)
ca. 1933 (für die Ausführung)
GRUPPE
6761,12,14 Firmenmarke 19, Made in Austria, Hand dec., unsign., H. 12,2 cm; Doro 29.6.1993
6761,101,4 (Firmenmarke ohne Ang.), H. 12 cm; Schuler 70
6761,233,11 Firmenmarke 19, H. 13 cm, L. 22 cm; eBay # 2262269077
6761,645,19 gem. F., Firmenmarke 19, Made in Austria, H. 12 cm; KH
linke EINZELFIGUR
6761,244,31 gem. F., Firmenmarke 19, Made in Germany, unsign., Ausformung zw. 1938–1941, H. 12 cm, L. 21 cm; Schloss Hag. 31; 34
Bem.: Hier eines der wenigen Beispiele, bei dem eine identische Modell-Nr. Verwendung für mehrere Modellvarianten findet. Zur gleichen Problematik -> vgl. 7039.

6763
Knörlein, Rudolf (zugeschr.)
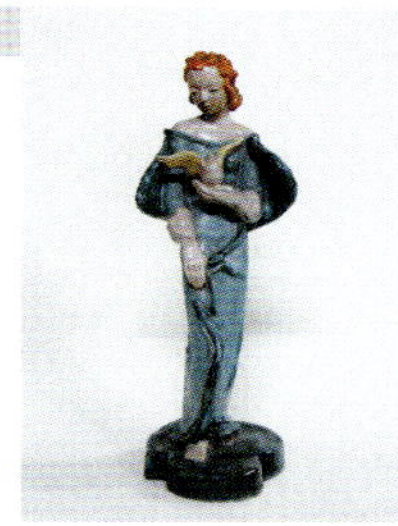
Stehendes Mädchen mit Taube (auf unregelmäßig geschwungenem Sockel)
ca. 1933
6763,27,? ger., gem. lig. grün AH, Firmenmarke 19, Made in Austria, unsign., H. 31,5 cm; eBay # 7416154857
6763 (ohne Nr.), gem. lig. grün AH, Firmenmarke 19, Made in Austria, unsign., H. 31,8 cm; MAK Wien, Inv. Ke 10602-91; Bem.: Der Kopf der Taube fehlt.

6764
Wieselthier, Vally
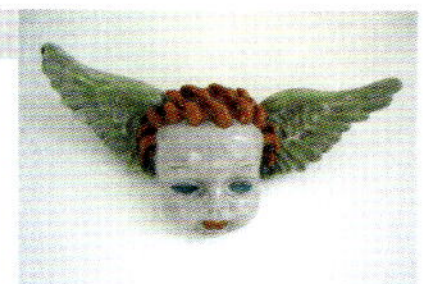
WANDMASKE Cherub (Engelskopf mit Flügeln)
ca. 1933
6764,31 gem. lig. TK, Firmenmarke 19, Made in Austria, unsign., H. 10,2 cm, B. 25,4 cm; eBay # 3745909127
6764 ger., gem. A, Firmenmarke 19, Made in Austria, unsign., H. 11 cm, B. 26 cm; Sammlung f@goldscheider.de
6764 gem. M, Firmenmarke 19, Made in Austria, unsign., H. 10,2 cm, B. 25,4 cm; eBay # 902657273
Bem.: Eine sichere Zuweisung an Wieselthier wäre durch die stilistische Nähe zu Entwürfen Walter Bosses schwierig gewesen. Die vorliegende groteske Darstellung hätte man im ersten Moment sicher Bosse zugeschr., wenn es nicht einen für die GS Produktion in Trenton, auch in der Größe, fast identischen und signierten Entwurf von Vally Wieselthier gäbe. Vergleicht man bei Hörmann (WV-K 812), so bestätigt der Gesichtsausdruck, besonders die Augen, diese sichere Zuschreibung.

6765
Lorenzl, Josef
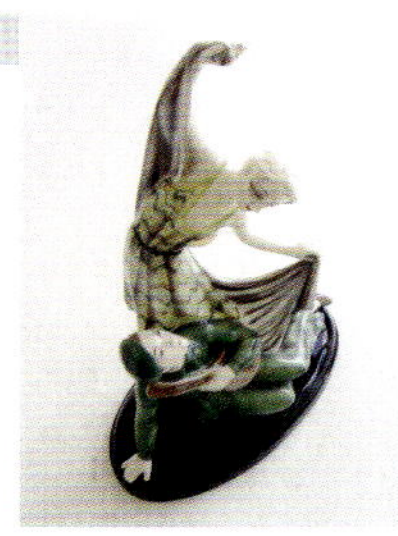
FIGUR / GRUPPE Tänzerin in Pose und liegender Pierrot mit Laute (auf Ovalsockel)
ca. 1933
FIGUR (Tänzerin)
6765,10,10 gem. S., Firmenmarke 19, Made in Austria, sign. Lorenzl, H. 31,8 cm; eBay # 3779503856
GRUPPE
6765,16,7,4 gem. S.2., GH, Firmenmarke 19, Made in Austria, sign. Dakon, H. 32 cm; Sammlung Stressless, Portugal; Bem.: Die Sign. „Dakon" muss als Versehen des Modelleurs angesehen werden, denn das Modell stammt eindeutig von Lorenzls Hand.
6765,26,10 Firmenmarke 19, Made in Austria, sign. Lorenzl, H. 31,8 cm, B. 21,6 cm; eBay # 859156718

6766
WANDMASKE Frauenkopf mit Hut
ca. 1933
6766 ger., gem. G, Firmenmarke 19, H. 7 cm; eBay # 2275426256

6769
Weiss, Claire (Klára)
(d.i. Herczeg, Claire (Klára)
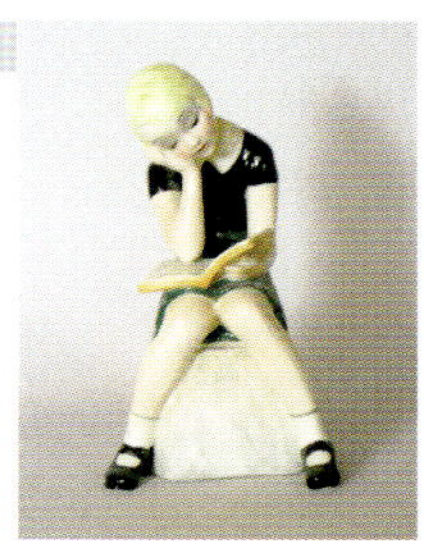
Sitzendes, lesendes Mädchen
ca. 1933
6769,83,11 gem. AD, Firmenmarke 19, Made in Austria, sign. Claire Weiss, H. 19 cm; Doro 24.11.1992
6769,136,10 gem. AMD, gem. lig. JR, Firmenmarke 19, Made in Austria, sign. Claire Weiss, H. 19,8 cm; Sammlung.schmutz@chello.at 10
6769,231,16 Firmenmarke 19, Made in Austria, sign. Weiss, H. 20 cm; eBay # 290003854299
6769,332,1 ger., Firmenmarke 19, unsign., Pick. 6, H. 22 cm; Keup 124

6770
Knörlein, Rudolf
STEHKOPF Kinderkopf mit Kragen (auf Quadratsockel)
ca. 1931 (für den Entwurf)
6770,3 ger., Firmenmarke 19, Made in Austria, sign. Prischl und Pick. 12 (Entwurf Rudolf Knörlein Wien), H. 13,3 cm; Wien Museum Inv. 186.670
6770 ger., gem. R, Firmenmarke 19, Made in Austria, sign. Dakon (!), H. 13,5 cm; KH Art 1900/Berlin
Bem.: Wie es zu diesem Wirrwarr von falschen Signaturen und dem einzig richtigen Pick. kommt, bleibt ungeklärt. Der Entwurf ist jedenfalls eindeutig Rudolf Knörlein zuzuordnen.
Ausst.: SSH 1992
-> vgl. 6770 Dakon / Knörlein / Prischl

6773
Weiss, Claire (Klára)
(d.i. Herczeg, Claire (Klára)
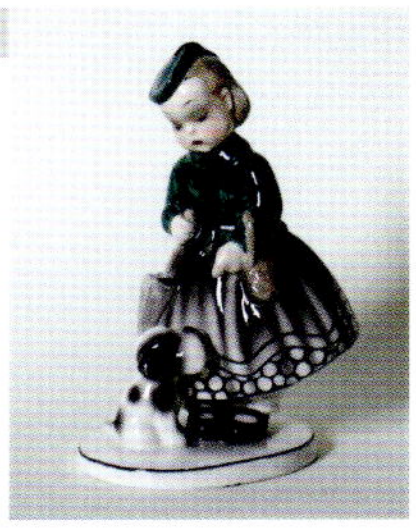
Stehendes Mädchen mit Hut, Schirm und Tasche, einen kleinen Hund belehrend (auf Ovalsockel)
ca. 1933
6773,42,19 gem. G.H., Firmenmarke 19, Made in Austria, sign. Claire Weiss, H. 17,2 cm; PB Deutschland
6773,72,14 Firmenmarke 19, unsign., H. ca. 17 cm; eBay # 2029998720

6774
Knörlein, Rudolf
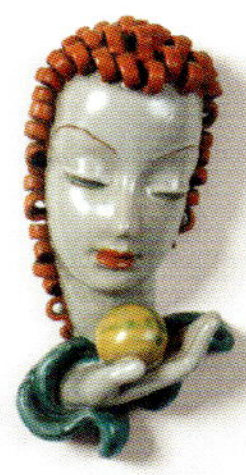
WANDMASKE / STEHMASKE „Eva" (Frauenkopf mit Hand und Apfel)
ca. 1933
WANDMASKE
(H. ca. 20/24 cm)
6774,3,1 Firmenmarke 19, Made in Austria, Hand dec., unsign., H. 24 cm; eBay # 2232704263
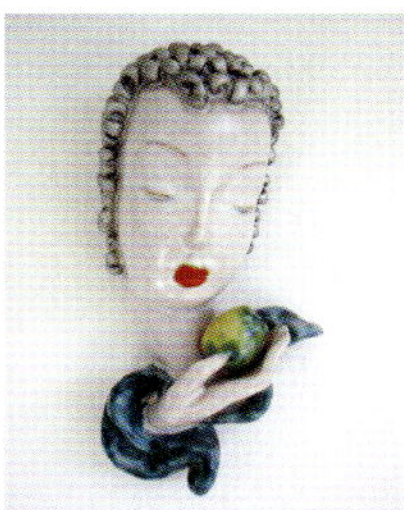
6774,4 ger. XVE, gem. N, gem. lig. AS, Firmenmarke 19, H. 22 cm; Sammlung a@goldscheider.net; Bem.: Dieser Kopf ist nicht nur einer der ganz wenigen Stücke, der mit dem Namen Knörlein sign. ist, er hat auch die Besonderheit als Variante zum Apfel, eine Rose in der Hand zu halten.
6774,12,K (Firmenmarke ohne Ang.), unsign., H. 20 cm; Christie's SK 10.11.2000
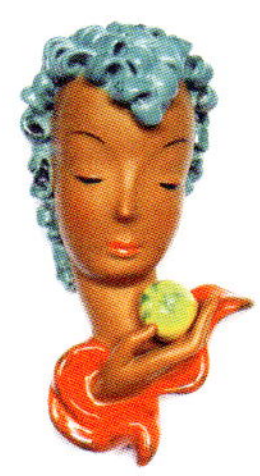
6774,18,63 Firmenmarke 19 (blau), unsign., H. 22,5 cm; Sammlung a@goldscheider.net
6774,20 Firmenmarke 19, Made in Austria, unsign., H. 24 cm; eBay # 37187533355
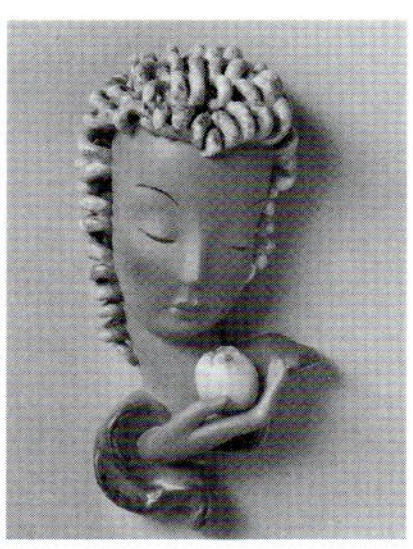
6774,26 gem. G1, Firmenmarke 19, Made in Austria, unsign., H. 20 cm; Sammlung f@goldscheider.de
6774,34,16 ger., gem. G, Firmenmarke 19, unsign., H. 23 cm; eBay # 1481310804
6774 gem. N, G., Firmenmarke 19, Made in Austria, unsign., H. 20,8 cm; eBay # 328468121
6774 gem. O, Firmenmarke 19, Made in Austria, unsign., H. 20,6 cm; Soth. com. # 3J5HD
6774 Firmenmarke 19, Made in Austria, unsign., H. 20,5 cm; PB Deutschland
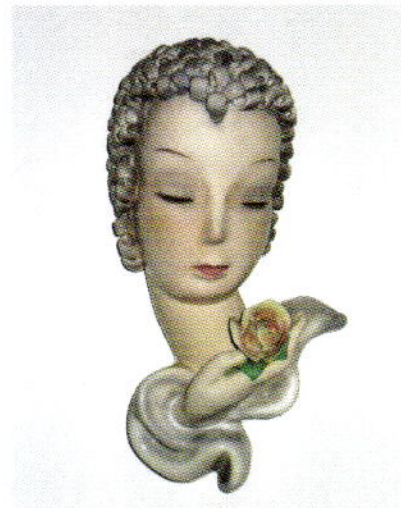
6774,44 Firmenmarke 19, Made in Germany, unsign., Ausformung zw. 1938–1941, H. 22,5 cm; eBay # 879559142
6774M, H. 24,5 cm, B. 11 cm; GSNachf. 2,12a; Fotoarchiv K. Täuber (hier mit der Ang.: H. 21 cm, B. 10 cm)
WANDMASKE
(H. ca. 34/40 cm)
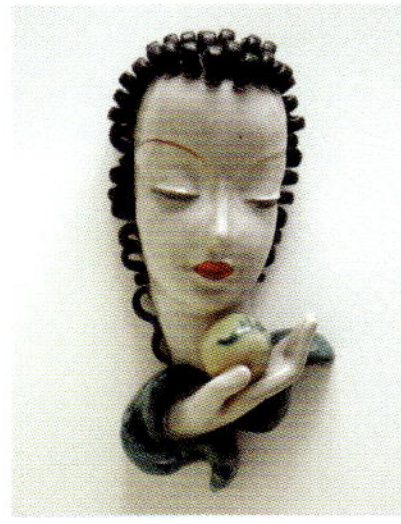
6774,3,13 Firmenmarke 19, Made in Austria, unsign., H. 34 cm; PB Wien 19
6774,44 gem. grün H, Firmenmarke 19, Made in Austria, Ausformung zw. 1945–1953, H. 34,9 cm; MAK Wien, Inv. Ke 10602-104
6774,I,S ger., Firmenmarke 19, Made in Austria, unsign., H. 34,5 cm; Soth. L. 30.11.1983
6774 gem. lig. TK, Firmenmarke 19, Made in Austria, unsign., H. 34,9 cm; Sammlung.schmutz@chello.at 67
6774,A,3 (Firmenmarke ohne Ang.), unsign., H. 33 cm; Christie's SK 11.5.2001
6774,A ger., Firmenmarke 19, Made in Germany, unsign., Ausformung zw. 1938–1941, H. 34 cm; Soth. L. 21.10.1988; Mehlis 32
6774aM, H. 40 cm, B. 19 cm; GSNachf. 2,12b
STEHMASKE (auf Holzsockel)
H. ca. 24 cm
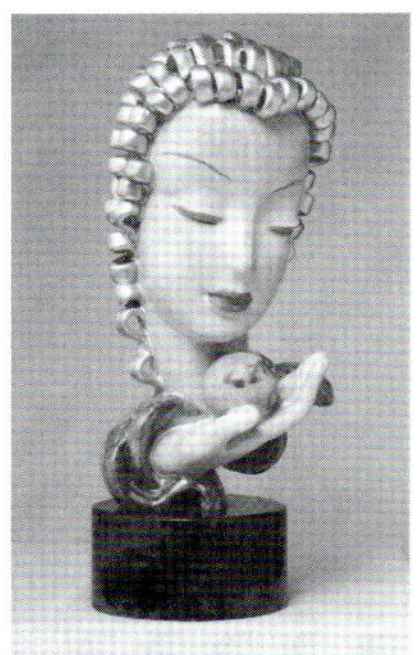
6774,3,13 Firmenmarke 19 (zwei mal), Made in Austria, Pick. 1, H. 22 cm, (Sockel 4,5 cm); Wien Museum Inv. 220.795
6774M, H. 24,5 cm, B. 11 cm; GSNachf. 2,12ab
STEHMASKE (auf Holzsockel)
H. ca. 34 cm
6774,8 ger. A, gem. lig. AH, Firmenmarke 19, Made in Austria, H. 32,7cm, (Sockel 6,8 cm); Wien Museum Inv. 190.173
6774aM, H. 40 cm, B. 19 cm; GSNachf. 2,12bb
Bem.: Die Datierung (1935) Knörleins auf dem Trägerblatt des Originalfotos muss auf 1932 korrigiert werden. Es fand sich nämlich kurz vor Redaktionsschluss die originale Entwurfszeichnung (Aquarell über Graphit). Diese ist von Knörlein eigenhändig mit 1932 datiert. Die Maske wurde wohl ab 1933 bei Goldscheider produziert, im Mai 1934 publiziert (-> vgl. Lit.) und war im gleichen Jahr auf der Ausst. „Österreichische Kunst in London" (Dorland-Hall) ausgestellt. Hierzu schreibt Grete Donau: „Mit einigen Frauenköpfen von weichpikantem Ausdruck und lässiger Anmut hat R.K., einer der wichtigsten keramischen Künstler dieser Erzeugung, Stücke geschaffen, die in London den Wiener Charm mir Glück vertreten". Besonderheit bei diesem Modell: es wurde eine Modell-Nr. für zwei verschiedene Größen vergeben. Beide Größen wurden auch als Stehmaske auf Holzsockel angeboten. Eine weitere Variante auf rechteckiger Plinthe (H. 22 cm) wurde unter der Modell-Nr. 7947 produziert. -> vgl. 7947
Ausst.: Österreichische Kunst in London 1934; SSH 1992.
Lit.: Donau, in: ÖK, 5.1934, Nr. 5, S. 17 (m. Abb.); GSNachf. 2,12 (m. Abb.), hier mit Ang. des Originaltitels MASKE „Eva"; GSN, S. 30, 31 (m. Abb.)
Quelle: AGM (Rechnung an Maurice Mizrahi, Alexandria, Ägypten, vom 22. April 1953, geliefert in Ausführung „matt"); Privatarchiv (Originalfoto); Privatsammlung D (Originalzeichnung); Fotoarchiv K. Täuber.
GS neu: Modell-Nr. 9 „Die Auserwählte" (H. 40 cm), Aufl. ohne Ang.

6775 s. 6735

6779
Meisinger, Ida
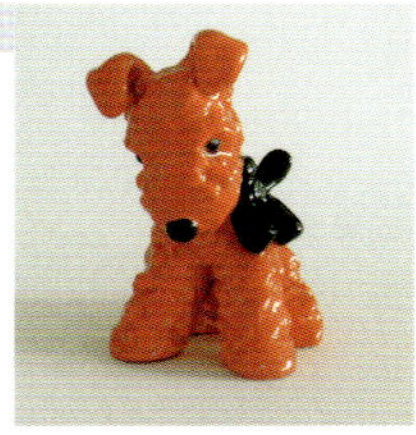
Sitzender West Highland Terrier mit Schleife / Mascherl
ca. 1933
6779,87,65 gem. lig. JP, Firmenmarke 19, Made in Austria, unsign., H. 11,2 cm; PB Deutschland

6780
Meisinger, Ida
Sitzender Foxterrier mit Schleife / Mascherl (auf Sockel als Buch / mit kleinem Hufeisen)
ca. 1933
6780,6 ger., gem. lig. KS, gem. grün W.H., Firmenmarke 19, Made in Austria, Hand dec., unsign., Pick. 1, H. 12,3 cm; Wien Museum Inv. 190.013
6780,25,20 Firmenmarke 19, Made in Austria, unsign., H. 12,5 cm; eBay # 6540664296
6780,31 (Firmenmarke ohne Ang.), H. 12 cm; Doro WA 12.12.1992
Bem.: Das Hufeisen bei dem Exemplar des Wien Museums ist möglicherweise sek. angebracht worden.

6781M
Matula, Viktor
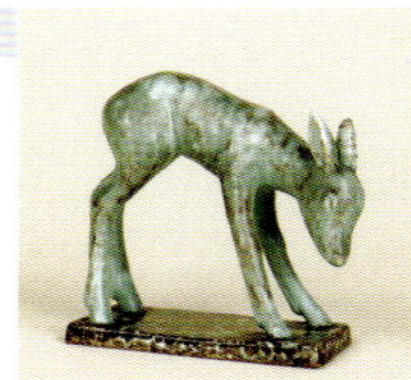
FIGUR „Reh" (auf flachem Rechtecksockel)
ca. 1933
6781,17,76 Firmenmarke 19, Made in Austria, Hand dec., unsign., H. 10,6 cm; PB Deutschland
6781,99,11 Firmenmarke 19, Made in Austria, unsign., H. 10,5 cm; Sammlung a@goldscheider.net
6781,293,20 ger., gem. G., Firmenmarke 19, Made in Austria, unsign., H. 10 cm; Sammlung f@goldscheider.de
6781,981,1 Firmenmarke 19, Made in Austria, unsign., H. 11 cm; eBay # 890805335
6781,1071,? Firmenmarke 19, unsign., H. 10 cm; RM 33
6781 ohne Ang., Firmenmarke 19, unsign., H. 10 cm; PB Wien 19
6781M, H. 10,5 cm, B. 11 cm; GSNachf. 3,15e
Lit.: GSNachf. 3,15 (m. Abb.), hier mit Ang. des Originaltitels „Reh".
Quellen: Nachlass Matula, Wien; AGM (Bestellung der Firma Wahliss (Wien), vom 24.11.1945 an die „Arbeitsgemeinschaft der Firma Friedrich Goldscheider"; Bestellung der Firma Wahliss (Wien), vom 27.2.1946 in Ausführung „Terrakotta").

6783
Meisinger, Ida
Sitzender, jaulender West Highland Terrier mit Schleife / Mascherl
ca. 1933
6783,20 gem. M, Firmenmarke 19, Made in Austria, unsign., Ausformung zw. 1945–1953, H. 16 cm; eBay # 150029408378
6783,47,65 Firmenmarke 19, Made in Austria, unsign., H. 16 cm; Schuler 66
6783,61,21 gem. XS, Firmenmarke 19, Made in Austria, unsign., H. 15,5 cm; Ruef 489; eBay # 3352526804

6784F
Dakon, Stephan
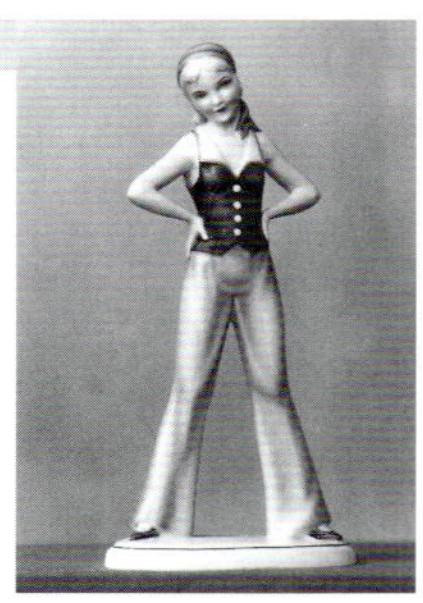
Stehendes junges Mädchen in weiten Hosen (sog. Schlaghosen) und Kopftuch (auf Ovalsockel)
ca. 1933
6784,90,6 gem. Z, Firmenmarke 19, Made in Austria, H. 31,1 cm; eBay # 164575318
6784,123,11 (Firmenmarke ohne Ang.), H. 30,5 cm; Christie's SK 27.4.2005
6784,140,2 gem. F, gem. lig. JR, Firmenmarke 19, Made in Austria, H. 31,1 cm; eBay # 102299856
6784F, H. 31 cm, B. 14 cm
Quelle: Archiv Mimi Dakon (Originalfoto)

6786
WANDPLASTIK Madonna mit Kind
ca. 1933
6786,5,65 (Firmenmarke ohne Ang.), H. 30,3 cm; Doro 20.10.1997

6787
Weber, Ernst
WANDMASKE Frauenkopf mit Hand, eine Maske haltend
ca. 1930 (für den Entwurf)
6787 Firmenmarke 19, unsign., H. 22 cm; Doro 4.3.2002
6787 ger., gem. lig. TK, Firmenmarke 19, Made in Austria, unsign., H. 22 cm; Doro 26.5.2004
-> vgl. 6288

6788
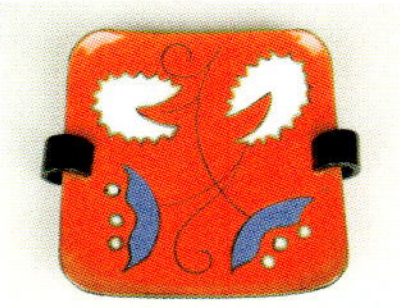
SCHALE
Form: rechteckig/quadratisch mit abgerundeten Ecken
Handhaben: zwei runde Ösen jeweils an der (kurzen) Seite
Dekor: zwei stilisierte Blumen mit Blättern
ca. 1933
6788 ger., gem. B, Firmenmarke 19, Made in Austria, H. 4,8 cm, L. 28 cm, B. 24,5 cm; PB Wien 19
6788 ger., gem. B, Firmenmarke 19, Made in Austria, H. 3,5 cm, L. 24 cm, B. 24 cm; WKA 27; eBay # 405257546
Bem.: Der hier angegebene Größenunterschied könnte auch davon herrühren, dass einmal die Handhaben mit gerechnet wurden.
Ausst.: PSK 1995
Lit.: ADGS, Nr. 11 (m. Abb.)

6792
Bosse, Walter (zugeschr.)
Stehender Matrose mit Bandoneon
ca. 1933
6792,31 gem. lig. AH, Firmenmarke 19, Made in Austria, unsign., Pick. 3, H. 21 cm; eBay # 180118236

6793
Sitzender Bub mit kleinem Terrier (auf getrepptem Sockel)
ca. 1933
6793,8,20 gem. lig. AS, Firmenmarke 19, Made in Germany, Ausformung zw. 1938–1941, H. 15,2 cm; eBay # KH

6796
Meisinger, Ida (zugeschr.)
„Pferd" (Springendes Fohlen)
ca. 1933
6796 Firmenmarke 19, unsign., H. 11,4 cm, L. 17,8 cm; eBay # 3713938783
-> vgl. 6844= DECKELDOSE
Quelle: AGM (Bestellungen der Firma Wahliss (Wien), vom 9. und 24.11.1945 an die „Arbeitsgemeinschaft der Firma Friedrich Goldscheider", hier jeweils mit Ang. des (Original?-)Titels „Pferd"; Bestellung der Firma Wahliss (Wien), vom 27.2.1946 in Ausführung „Terrakotta").

6797M
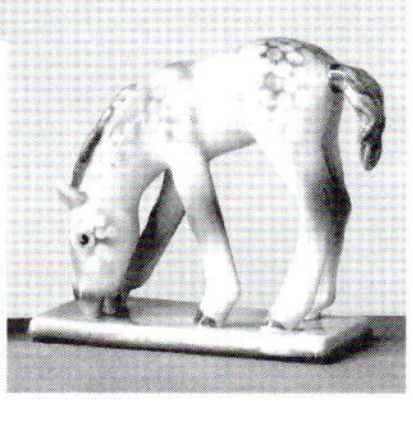
Matula, Viktor
FIGUR „Pferd" (Stehendes fressendes Pferd) (auf flachem Rechtecksockel)
ca. 1933
6797,891,65 Firmenmarke 19, Made in Austria, unsign., H. 10,5 cm, B. 12 cm; Sammlung a@goldscheider.net
6797,H ger., gem. Z, Firmenmarke 19, Made in Austria, unsign., H. 10,5 cm, B. 12 cm; Sammlung a@goldscheider.net
6797 ger., gem. lig. TK, Firmenmarke 19, Made in Austria, unsign., H. 10 cm, B. 10 cm; eBay # 1429138332
6797M, H. 10 cm, B. 11 cm; GSNachf. 3,18c; 3,21f (hier mit der Ang.: H. 10,5 cm, B. 12 cm); Fotoarchiv K. Täuber (hier mit der Ang.: H. 11 cm, B. 12 cm).
-> vgl. 7386; 7737
-> vgl. 6986= GRUPPE
-> vgl. 7939= GRUPPE (als Höhenvariante).
Lit.: GSNachf. 3,18 und 3,21 (m. Abb.), hier mit Ang. des Originaltitels „Pferd".
Quelle: Fotoarchiv K. Täuber

6801
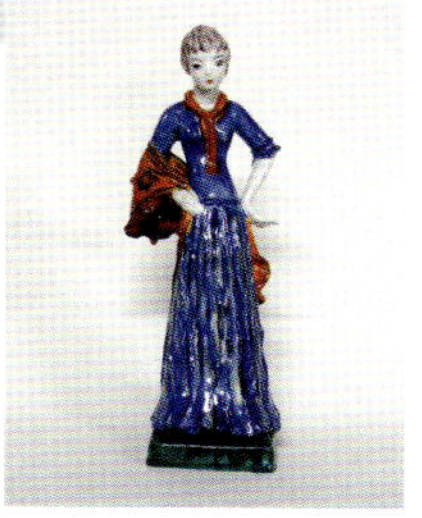
Stehende Dame mit Tuch (auf Quadratsockel)
ca. 1933
6801,4,65 gem. lig. AH, Firmenmarke 19, Made in Austria, H. 30,9 cm; MAK Wien, Inv. Ke 10602-6
6801,65 gem. lig. AH, Firmenmarke 19, Made in Austria, H. 30,8 cm; PB

6802F
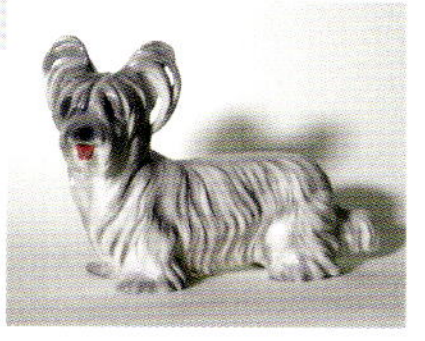
Postl, Hugo
FIGUR „Skyeterrier"
um/ab 1930 (für die Ausführung)
6802F, H. 8 cm, B. 12 cm; GSNachf. 5,11c
-> vgl. 6116; 6503; 7300; 7519
-> vgl. 6975= Variante (auf Ovalsockel)
Lit.: GSNachf. 5,11 (m. Abb.), hier mit Ang. des Originaltitels „Skyeterrier".

6803F
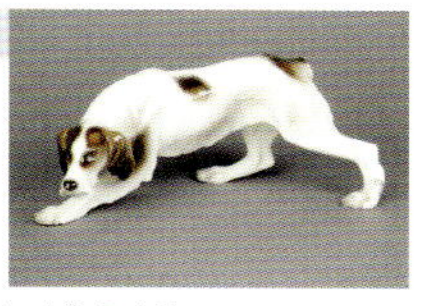
FIGUR „Setter" (Witternder Setter)
ca. 1933
6803 ohne Ang., Firmenmarke 19, H. 8 cm; PB Wien 19
6803F, H. 8 cm, B. 23 cm; GSNachf. 5,10b
Pendant -> vgl. 6804
-> vgl. ohne Nr.
-> vgl. 6818= GRUPPE
Lit.: GSNachf. 5,10 (m. Abb.), hier mit Ang. des Originaltitels FIGUR „Setter".

6804F
FIGUR „Setter" (Mit erhobenem Kopf)
ca. 1933
6804F, H. 11 cm, B. 22 cm; GSNachf. 5,10c
Pendant -> vgl. 6803
-> vgl. 6818= GRUPPE
-> vgl. 6756= GRUPPE mit Dame
Lit.: GSNachf. 5,10 (m. Abb.), hier mit Ang. des Originaltitels FIGUR „Setter".

6811
Matula, Viktor
FIGUR/BUCHSTÜTZE Rehe
ca. 1933
FIGUR
6811 H. fehlt; Nachlass Matula, Wien
BUCHSTÜTZE
6811 gem. B, Firmenmarke 19, H. 16 cm; Spik 560
Quelle: Nachlass Matula, Wien

6812
Lorenzl, Josef
Stehende junge Dame in langem Kleid, Handschuhen und Pekinese (auf Ovalsockel)
ca. 1933
6812 (Firmenmarke ohne Ang.), H. 28,5 cm; Pinhas, S. 139 (m. Abb.)

6816M
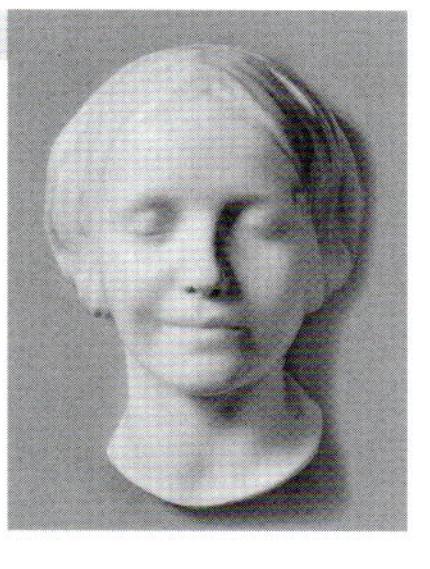
WANDMASKE „Die unbekannte Tote aus der Seine" (mit Halsansatz)
ca. 1953–1960
6816,3 ger., ohne Firmenmarke, Ausformung 50er Jahre, Pick. WG 1, H. 27,1 cm, B. 18,5 cm; Sammlung John Nash
6816 ohne Ang., H. 27 cm, B. 19 cm; Könecke (2002), S. 55, 76, 85 (m. Abb.)
6816 H. 29 cm, B. 19 cm; Fotoarchiv K. Täuber
Bem.: Diese Maske wurde zw. 1953–1960 bei Carstens (GWM West-Germany) produziert.
-> vgl. 2014; 6817
-> vgl. 6818 (ohne Halsansatz)
Lit.: KGSF1; Könecke (2002), S. 55, 76, 86 (m. Abb.)
Quelle: AGM (Rechnung an Maurice Mizrahi, Alexandria, Ägypten, vom 22. April 1953), hier in der Ausführung „matt" bezogen; Fotoarchiv K. Täuber.

6817
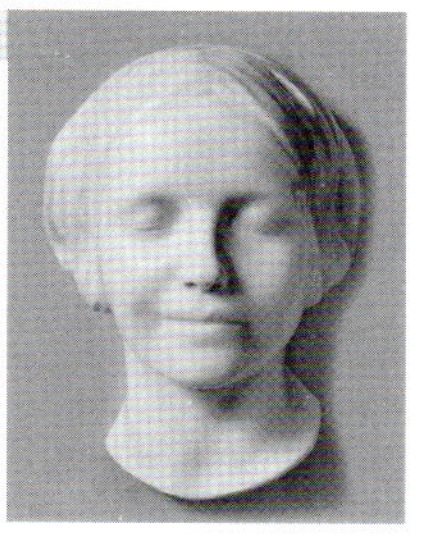
WANDMASKE „Die unbekannte Tote aus der Seine" (mit Halsansatz)
ca. 1953–1960
6817 ohne Ang., H. 13 cm, B. 9,5 cm; Könecke (2002), S. 55, 76, 85 (m. Abb.)
Bem.: Diese Maske wurde zw. 1953–1960 bei Carstens (GWM West-Germany) produziert.
-> vgl. 2014; 6816
-> vgl. 6818 (ohne Halsansatz)
Lit.: KGSF1; KGSF2; Könecke (2002), S. 55, 76, 86 (m. Abb.)

6818 -> ÜBERSCHNEIDUNG
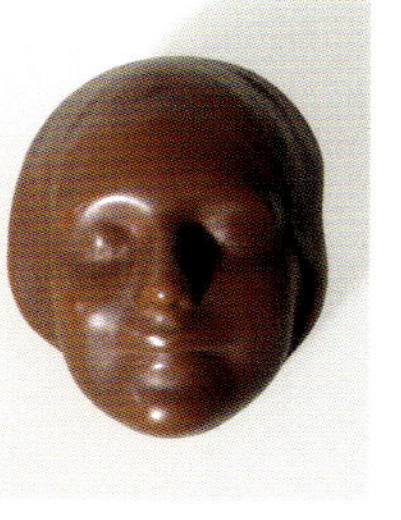
WANDMASKE „Die unbekannte Tote aus der Seine" (ohne Halsansatz)
ca. 1953–1960
6818 ohne Ang., H. 17 cm, B. 13,5 cm; Könecke (2002), S. 55, 76, 85 (m. Abb.)
Bem.: Diese Maske wurde zw. 1953–1960 bei Carstens (GWM West-Germany) produziert.
-> vgl. 6817; 6818 (= jeweils ohne Halsansatz)
Lit.: KGSF1; KGSF2; Könecke (2002), S. 55, 76, 86 (m. Abb.)

6818F -> ÜBERSCHNEIDUNG
GRUPPE „Setter" (Zwei laufende Setter)
ca. 1933
6818F, H. 11 cm, B. 23 cm; GSNachf. 5,10a
-> vgl. 6803; ohne Nr.= jeweils EinzelFIGUR (vorne)
-> vgl. 6804= EinzelFIGUR (rückwärtig)
Lit.: GSNachf. 5,10 (m. Abb.), hier mit Ang. des Originaltitels GRUPPE „Setter".

6821
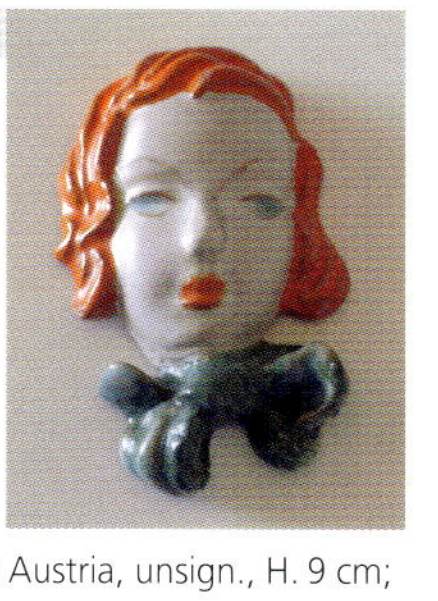
Knörlein, Rudolf (zugeschr.)
WANDMASKE Frauenkopf mit halblangen Haaren und Schleife/Mascherl
ca. 1933/34
6821 gem. S/H, (ohne Firmenmarke), unsign., H. 9,9 cm; Sammlung.schmutz@chello.at 15
6821 Firmenmarke 19, Made in Austria, unsign., H. 9 cm; eBay # 1426180089
-> vgl. 6881= TONDO

6822
Matula, Viktor
ASCHER Foxterrier
ca. 1933/34
6822, H. fehlt, B. fehlt
Quelle: Nachlass Matula, Wien

6824
Dakon, Stephan
Stehende, lesende junge Dame mit Berét (auf Rundsockel)
ca. 1933/34
6824,132,4 gem. N., Firmenmarke 19, Made in Austria, Fabriqué en Autriche, Pick. 3, H. 31 cm; Wien Museum Inv. 220.743

6825

Dakon, Stephan (zugeschr.)
Schreitende junge Dame im Wind, ihren Hut haltend (auf Ovalsockel)
ca. 1933
6825 (Firmenmarke ohne Ang.), unsign., H. 28 cm; Pinhas, S. 160 (m. Abb.)

6829

Weiss, Claire (Klára) (d.i. Herczeg, Claire (Klára) (zugeschr.)
Stehendes kleines Mädchen mit Haarschleife und Teddybär (auf Rundsockel)
ca. 1933/34
6829,74,4 gem. blau M, Firmenmarke 19, Made in Austria, unsign., H. 17,8 cm; Soth. com. # 1LJH

6830

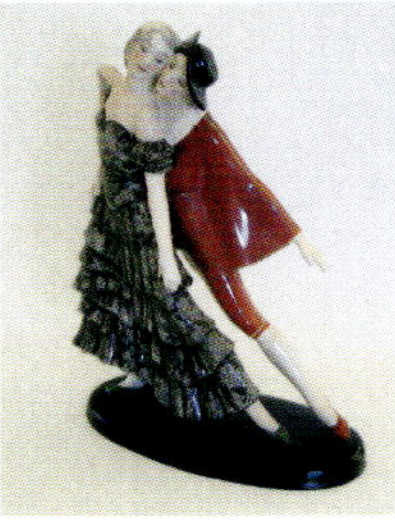

Weiss, Claire (Klára) (d.i. Herczeg, Claire (Klára)
GRUPPE Tanzendes Paar (Russisches Ballett) (auf Ovalsockel)
ca. 1933/34
6830,30,4 (Firmenmarke ohne Ang.), sign. Claire Weiss, H. 35 cm; Christie's SK 20.9.2000
6830,86,10,8A, gem. N7, Firmenmarke 19, Made in Austria, sign. Claire Weiss, H. 35 cm; Sammlung.schmutz@chello.at 11
Lit.: RDM, 8.1933/34, S. 145 (m. Abb.), hier eine zeitgenössische Fotografie, die wohl als Vorlage für das Modell diente, bei dem lediglich der Hut der Tänzerin weggelassen wurde. Dargestellt ist hier ein nicht näher bezeichnetes Tänzerpaar. Fotografie: Harlip, hier betitelt „Russisches Ballett".

6832

Knörlein, Rudolf (zugeschr.)
WANDMASKE Frauenkopf mit halblangen Haaren und Kragen
ca. 1933/34
6832,27 gem. MT, Firmenmarke 19, H. 10,1 cm; eBay # 2001626542
6832,65 Firmenmarke 19, Made in Austria, unsign., H. 9,5 cm; eBay # 3317373993
-> vgl. 6880= ABREISSKALENDER
-> vgl. 6998= DECKELDOSE

6835

Cancantänzerin (auf Rundsockel)
ca. 1933/34
6835,27,31 (Firmenmarke ohne Ang.), H. 19,4 cm; Doro 21.5.1997

6836

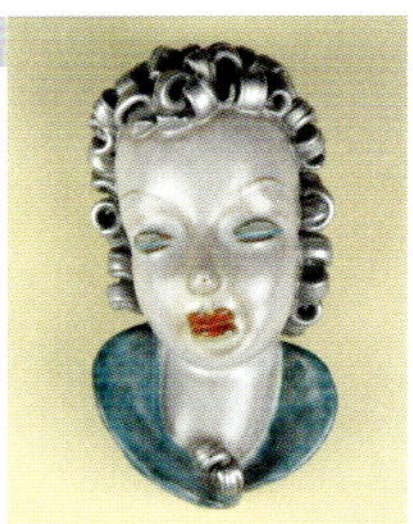

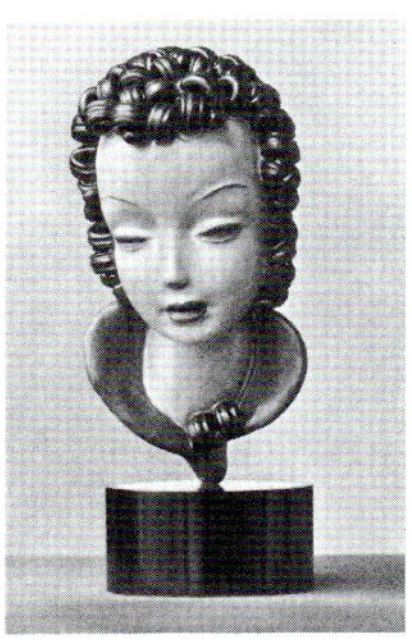

Prischl, Adolf
WANDMASKE/STEHMASKE Frauenkopf mit Kragen
ca. 1933/34
WANDMASKE
6836,8 Firmenmarke 19, Made in Austria, H. 23 cm; eBay # 1480092755
6836,12,K Firmenmarke 19, Made in Austria, Pick. 1, H. 20 cm; Quit. 21
6836 Firmenmarke 19, Made in Austria, Pick. 2, H. 20,3 cm; eBay # 455277155
6836 Firmenmarke 19, H. 19,5 cm; Stahl 4.2.1995
6836 Firmenmarke 20, Made in Austria, unsign., H. 19 cm; PB Wien 19
6836 gem. M, (ohne Firmenmarke), H. 20,5 cm; Sammlung.schmutz@chello.at 39
6836 gem. B, Firmenmarke 19, Made in Germany, unsign., Ausformung zw. 1938–1941, H. 20 cm; Wien Museum Inv. 186.659
6836M, H. 20 cm, B. 12 cm; GSNachf. 2,13a
STEHMASKE (auf Holzsockel)
6836 (Firmenmarke ohne Ang.), unsign., Gesamth. 25 cm; Fischer 59
6836 Firmenmarke 19, Made in Austria, unsign., Gesamth. 25 cm; Soth. L. 17.10.1985
6836 (Firmenmarke Plak. ohne Ang.), Gesamth. 24,5 cm; Soth. L. 23.6.1989
6836M, H. 20 cm, B. 12 cm, Gesamth. 25 cm; GSNachf. 2,13a
-> vgl. 7549
-> vgl. 7340= DECKELDOSE
Lit.: GSNachf. 2,13 (m. Abb.), hier mit Ang. des Originaltitels „Maske" und „Maske auf Sockel".

6839

Matula, Viktor
FIGUR Pferd
ca. 1933/34
6839 H. fehlt
Quelle: Nachlass Matula, Wien

6842

Dakon, Stephan
Stehende junge Dame mit Berét und Handtasche (auf Rundsockel)
ca. 1933/34
6842,17,6 Firmenmarke 19, Made in Austria, H. 30 cm; eBay # 1241654024

6844

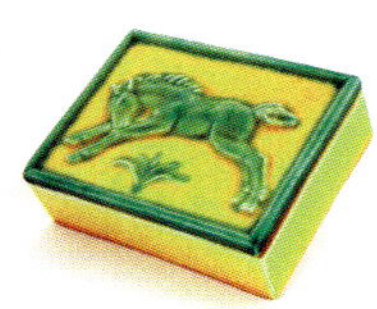

Meisinger, Ida (zugeschr.)
DECKELDOSE „Pferd" (Springendes Fohlen)
Form: rechteckig
Dekor: springendes Fohlen mit Pflanzenbüschel (reliefiert)
ca. 1933/34
6844 gem. A, Firmenmarke 19, Made in Austria, unsign., H. 4 cm, B. 9,8 cm, L. 12,8 cm; Sammlung a@goldscheider.net
6844 gem. TK, Firmenmarke 19, Made in Austria, unsign., H. 4 cm, B. 9,7 cm, L. 13 cm; eBay # 3321032605
-> vgl. 6796= FIGUR
Bem.: Diese Deckeldose zeigt die unter der Modell-Nr. 6796 ausgeführte Figur des Fohlens als Relief.

6853

Weiss, Claire (Klára) (d.i. Herczeg, Claire (Klára) (zugeschr.)
Schreitendes Mädchen mit kleinem Terrier, welcher an ihrem Kleid zerrt (auf Ovalsockel)
ca. 1933/34
6853,130,26 gem. EF, Firmenmarke 19, Made in Austria, unsign., H. 16,2 cm; BM 50
6853,165,6 gem. B, Firmenmarke 19, Made in Austria, unsign., H. 15,8 cm; eBay # 7400534910
6853,230,20 ger., (Firmenmarke ohne Ang.), unsign., H. 16 cm; Christie's SK 12.9.1997
Lit.: Pinhas, S. 136 (m. Abb.)

6859

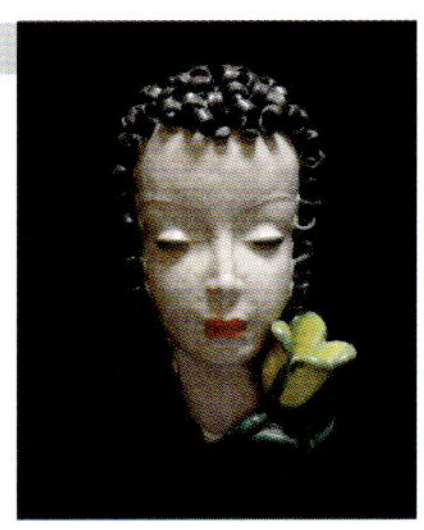

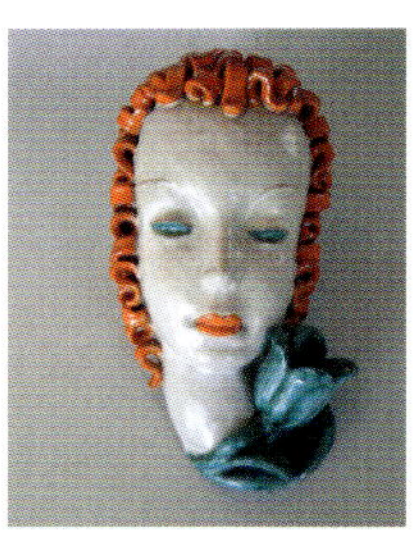

Knörlein, Rudolf
WANDMASKE Frauenkopf mit/ohne Tulpe
ca. 1933/34
mit Tulpe
6859,3 gem. lig. AH, Firmenmarke 19, Made in Austria, unsign., H. 24,5 cm; eBay # 105722645
6859,8 Firmenmarke 19, unsign., H. 25,5 cm; Soth. L. 30.11.1983
6859,12 gem. G., Firmenmarke 19, Made in Austria, H. 26,3 cm; PB
6859 gem. B, Firmenmarke 19, Made in Austria, Pick. 12 (Entwurf Rudolf Knörlein, Wien), Originalpick.: „Holstein & Düren/Köln", H. 26,3 cm; PB Deutschland
6859 ger. K, (Firmenmarke ohne Ang.), unsign., H. 26,5 cm; Christie's SK 11.11.2004
6859 Firmenmarke 19, Made in Austria, unsign., H. 26,6 cm; Sammlung.schmutz@chello.at 40
6859 Firmenmarke 19, unsign., H. 25,5 cm; Lyon/T. 7.11.2001
6859 (Firmenmarke ohne Ang.), unsign., H. 26 cm; Christie's SK 2.4.1998
ohne Tulpe
6859 (Firmenmarke ohne Ang.), unsign., H. 25 cm; Christie's SK 14.3.1997
Bem.: Wie die Auflistung der einzelnen Stücke zeigt, wurde die Maske in der Regel mit der Tulpe ausgeformt. Die Version ohne die Blume ist somit wesentlich rarer (-> vgl. Pinhas. S. 170 (m. Abb.), deren Abb. aus dem Kat. von Christie's SK 14.3.1997 stammt).
-> vgl. 7548

6861

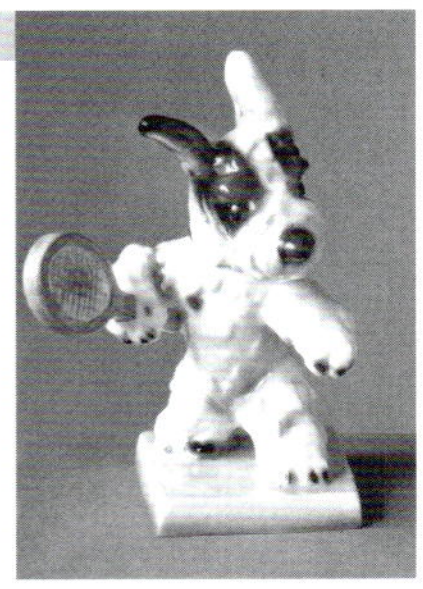

Meisinger, Ida
Foxterrier mit Tennisschläger (auf Quadratsockel)
ca. 1933/34
6861,6 gem. W, gem. rot X, Firmenmarke 19, Made in Austria, unsign., H. 18 cm; Wien Museum Inv. 190.012
6861 H. 19 cm, B. 10 cm; Fotoarchiv K. Täuber
Bem.: s. 6449
Ausst.: SSH 1992
Lit.: SS, S. 44, 45 (m. Abb.)
Quelle: Fotoarchiv K. Täuber.

6862

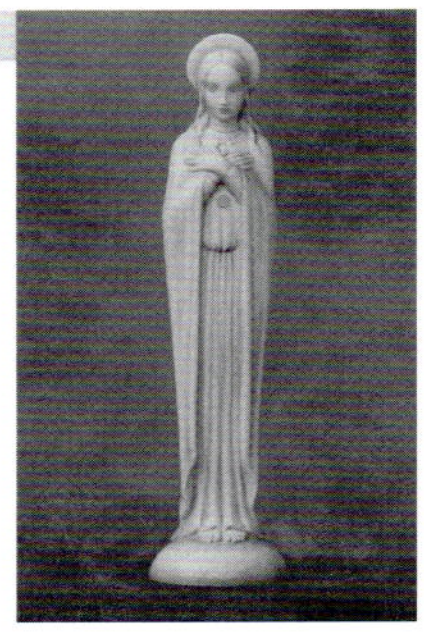

Lorenzl, Josef
Stehende Madonna mit gekreuzten Händen (auf Rundsockel)
ca. 1934
6862,39,6 Firmenmarke 19, sign. Lorenzl, H. 42,2 cm; Doro 15.3.2005; 20.3.2006
6862,182,19,L Firmenmarke 19, Made in Germany, unsign., Ausformung zw. 1938–1941, H. 44,5 cm; eBay # 141095284
6862 H. 43 cm, B. fehlt; Fotoarchiv K. Täuber
-> vgl. 7735
Quelle: Fotoarchiv K. Täuber

6863

STEHKOPF Frauenkopf mit Blume (Tulpe?)
ca. 1934
6863,20 ger., Firmenmarke 19, Made in Austria, H. 23 cm; eBay # 7345364636
6863 (Firmenmarke Plak. ohne Ang.), H. 19 cm; WKA 29

6866

Matula, Viktor
RELIEF Pferd (Wandrelief)
ca. 1934
6866 H. fehlt, B. fehlt
Quelle: Nachlass Matula, Wien

6871

Knörlein, Rudolf/ Prischl, Adolf
WANDMASKE Frauenkopf mit oben glattem, unten seitlich gelocktem Haar und Halstuch
ca. 1934

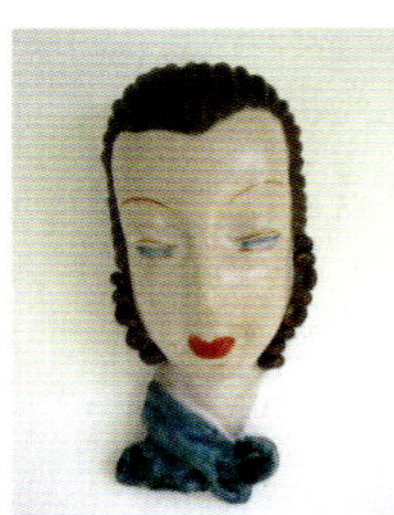

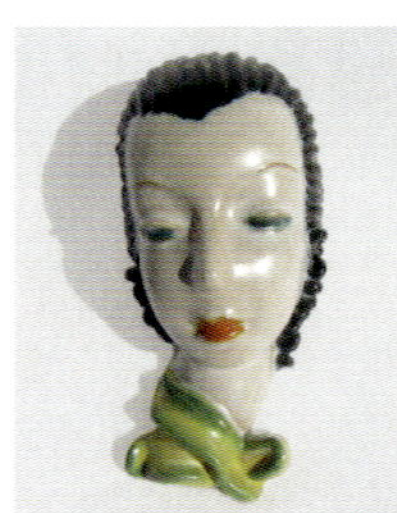

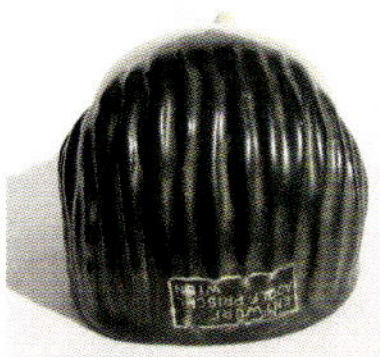

6871,26, gem. R, Firmenmarke 19, Made in Austria, unsign., H. 16 cm; Sammlung f@goldscheider.de
6871 gem. G, Firmenmarke 19, Made in Austria, Pick. 12 (Entwurf Rudolf Knörlein, Wien), H. 16 cm; BM 40
6871 gem. BN1, Firmenmarke 19, Made in Austria, Pick. 17 (Entwurf Adolf Prischl, Wien), H. 15 cm; Sammlung Alexandra Herrmann-Goldscheider

6873
Meisinger, Ida
BUCHSTÜTZE Sitzender Terrier (auf rechtwinkelig aufgeschlagenem Buch)
ca. 1934
6873,74,20 Firmenmarke 19, Pick. 15 (Entwurf Ida Meisinger, Wien), H. 10 cm; eBay # 6547473087
6873,146,31 ger., (Firmenmarke ohne Ang.), unsign., H. 10 cm; Schuler 59

6874
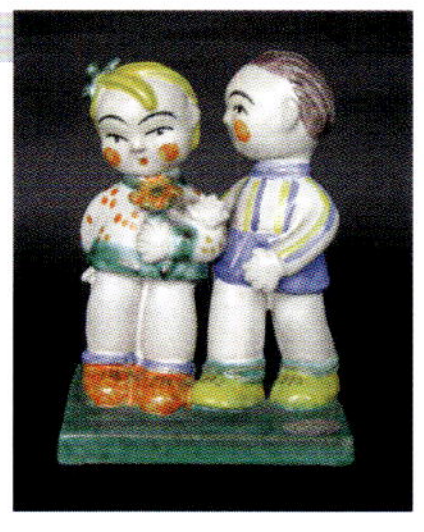
Meisinger, Ida (zugeschr.)
GRUPPE Stehendes Kinderpaar (auf Rechtecksockel)
ca. 1934
6874,5,4 gem. lig. AH, Firmenmarke 19, Made in Austria, unsign., Pick. 2, H. 15,2 cm, B. 11,2 cm; PB

6876
Meisinger, Ida (zugeschr.)
GRUPPE Stehendes Mädchen mit Puppe und Terrier (auf Ovalsockel)
ca. 1934
6876,4(2,? Nummern überklebt), Firmenmarke 19, Made in Austria, unsign., H. 20 cm, B. 15,5 cm; PB

6880
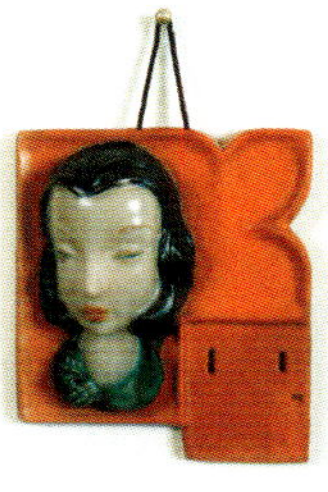
Knörlein, Rudolf (zugeschr.)
ABREISSKALENDER Frauenkopf mit halblangen Haaren und Kragen
ca. 1933/34
6880,T Firmenmarke 19, Made in Austria, unsign., H. 12 cm, B. 10,7 cm; Sammlung.schmutz@chello.at 109
6880 Firmenmarke 19, Made in Austria, unsign., H. 11,9 cm, B. 10,4 cm; eBay # 1424059362
–> vgl. 6832= WANDMASKE

6881
Knörlein, Rudolf (zugeschr.)
TONDO Frauenkopf mit halblangen Haaren und Schleife / Mascherl
ca. 1933/34
6881 ger., gem. S/H, Firmenmarke 19, Made in Austria, unsign., Ø 11 cm; KH Art 1900/Berlin
–> vgl. 6821= WANDMASKE

6882
Knörlein, Rudolf (zugeschr.)
ABREISSKALENDER Frauenkopf mit Hut
ca. 1933/34
6882 Firmenmarke 19, unsign., H. 10 cm; PB Wien 19

6883
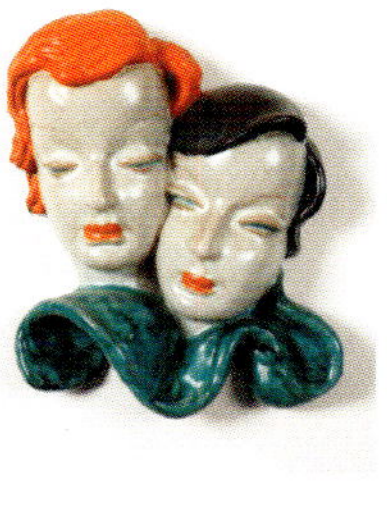
Knörlein, Rudolf
WANDMASKE Zwei Frauenköpfe (Cheek to cheek) mit Halstuch
ca. 1934
6883,5 gem. S/H, Firmenmarke 19, Made in Austria, unsign., H. 30,8 cm; Sammlung.schmutz@chello.at 13
Bem.: Durch eine Originalzeichnung (Aquarell über Graphit) konnte diese Doppelwandmaske eindeutig R. Knörlein zugeordnet werden.
Quelle: Privatsammlung D.

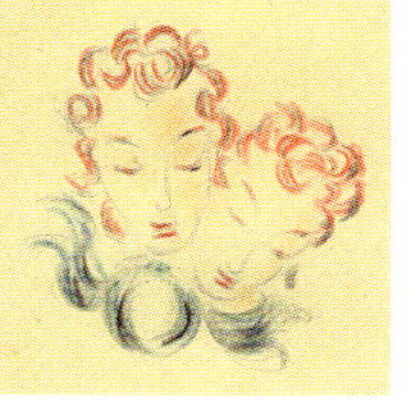

6888
Matula, Viktor
ASCHER
ca. 1934
6888 H. fehlt, B. fehlt
Quelle: Nachlass Matula, Wien

6890
Matula, Viktor
ASCHER
ca. 1934
6890 H. fehlt, B. fehlt
Quelle: Nachlass Matula, Wien

6891
WANDMASKE Frauenkopf mit Hut
ca. 1934
6891,461/2,6 Firmenmarke 19, Pick. ohne Ang., H. 21,4 cm; BM 47
Bem.: Da sich im Auktionskat. weder eine genauere Beschreibung noch eine Abb. findet, kann diese Maske nicht näher beschrieben werden. Eine Zuschreibung an Rudolf Knörlein läge vermutlich auf der Hand, jedoch sind die Angaben des AH zum Modell zu vage.

6894
Matula, Viktor
ASCHER
ca. 1934
6894 H. fehlt, B. fehlt
Quelle: Nachlass Matula, Wien

6896
Stehendes Fohlen (auf Rechtecksockel)
ca. 1931 (für den Entwurf)
6896,20 ger., gem. L., Firmenmarke 19, Made in Austria, H. 17,4 cm; Wien Museum Inv. 190.032
6896,26 Firmenmarke 19, Made in Austria, H. 17,8 cm; eBay # 1298301292
–> vgl. 6453
Lit.: GSN, S. 50, 51 (m. Abb.)
GS neu: Modell-Nr. 15 „Mein kleines Pferd" (H. 15 cm), Aufl. ohne Ang.

6900
Weiss, Claire (Klára) (d.i. Herczeg, Claire (Klára)
Tänzerin in Bewegung (auf Ovalsockel)
ca. 1934
6900,140,2,7 (Firmenmarke ohne Ang.), sign. Claire Weiss, H. 29 cm; Christie's SK 9.4.1999; 16.9.1999
6900,221,10 ger. L, gem. S., Firmenmarke 19, Made in Austria, sign. Claire Weiss, Pick. 2, H. 29,2 cm; eBay # 3728353965
Lit.: Pinhas, S. 160 (m. Abb.)

6901F
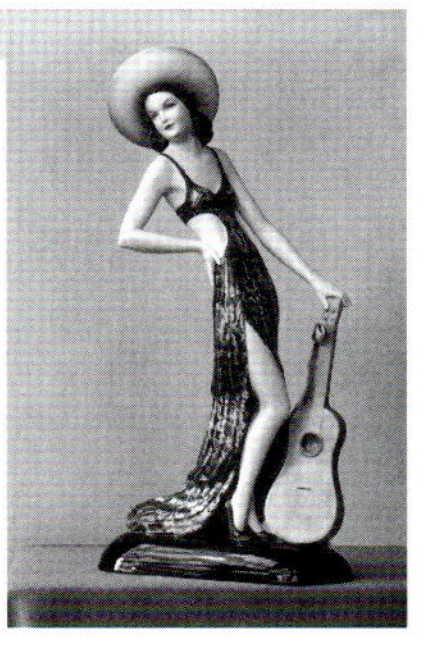
Dakon, Stephan
Stehende elegante Dame in geschlitztem langen Kleid mit großem Hut und Gitarre (auf Ovalsockel)
ca. 1934
ohne Ang. Firmenmarke 19, Made in Austria, H. 44,5 cm; Soth. NY 23.11.1991
6901F, H. 43 cm, B. 21 cm
Quelle: Archiv Mimi Dakon (Originalfoto)

6903
Sitzender Dackel
ca. 1934
6903 H. 12 cm, L. 17 cm
Quelle: Fotoarchiv K. Täuber

6905F
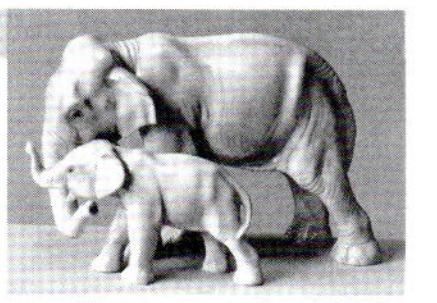
Jarl, Karin
GRUPPE „Elefanten" (Zwei stehende Elefanten, groß / klein)
ca. 1934
6905,2 Firmenmarke 19, Made in Austria, unsign., H. 21 cm, L. 31 cm; Wien Museum Inv. 190.027
6905 ohne Ang., Firmenmarke 19, Made in Austria, unsign., H. 22,9 cm, L. 33 cm; Bergmann 42; eBay # 202669554
6905F, H. 20 cm, B. 30 cm; GSNachf. 5,17a; Fotoarchiv K. Täuber, hier mit der Ang.: H.21 cm, B. 30 cm.
–> vgl. 6906= EinzelFIGUR (groß)
–> vgl. 6907= EinzelFIGUR (klein)
Lit.: GSNachf. 5,17 (m. Abb.), hier mit Ang. des Originaltitels GRUPPE „Elefanten"; Könecke (2000), S. 76, 85 (m. Abb.).
Quelle: AGM (Bestellungen der Firma Wahliss (Wien), vom 9. und 24.11.1945 an die „Arbeitsgemeinschaft der Firma Friedrich Goldscheider"; Bestellung der Firma Wahliss (Wien), vom 27.2.1946 in den Ausführungen „Terrakotta" je 56 Stück, und „Fayence" je 3 Stück); Fotoarchiv K. Täuber.

6906F
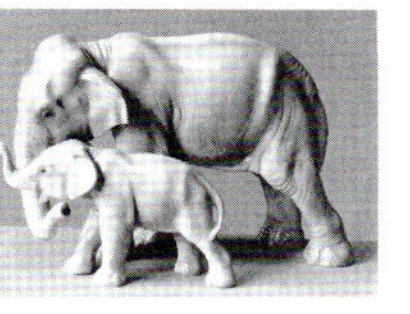
Jarl, Karin
FIGUR „Elefant" (Stehender Elefant, groß)
ca. 1934
6906F, H. 20 cm, B. 30 cm; GSNachf. 5,17b
–> vgl. 6905= GRUPPE
Pendant –> vgl. 6907
Lit.: GSNachf. 5,17 (m. Abb.), hier mit Ang. des Originaltitels FIGUR „Elefant" (groß); Könecke (2000), S. 76, 85 (m. Abb.).

6907F
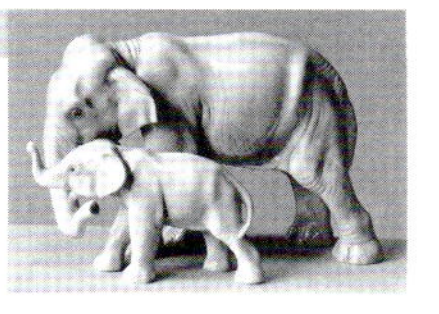
Jarl, Karin
FIGUR „Elefant" (Stehender Elefant, klein)
ca. 1934
6907F, H. 12 cm, B. 20 cm; GSNachf. 5,17c
–> vgl. 6905= GRUPPE
Pendant –> vgl. 6906
Lit.: GSNachf. 5,17 (m. Abb.), hier mit Ang. des Originaltitels FIGUR „Elefant" (klein); KGSF1; Könecke (2000), S. 76, 85 (m. Abb.).

6908
Knörlein, Rudolf
SCHALE
Form: rund
Dekor: vier Streifen
ca. 1934
6908,7 (Firmenmarke ohne Ang.), Pick. 12 (Entwurf Rudolf Knörlein, Wien), Ø 23,5 cm; Doro 7.3.2001

6910
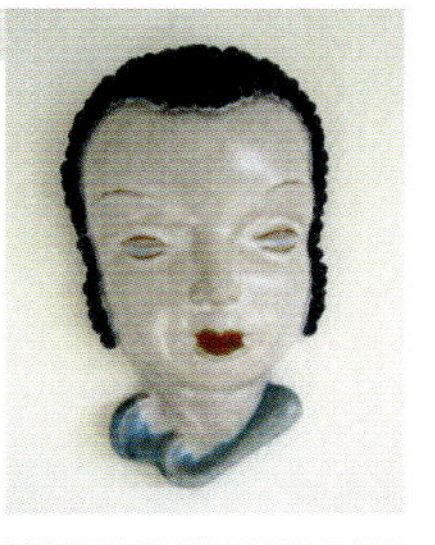
WANDMASKE Mädchenkopf mit Locken (Curls) und Kragen
ca. 1934
6910 ger., gem. G, Z, Firmenmarke 19, Made in Austria, H. 16 cm; Sammlung f@goldscheider.de

6911
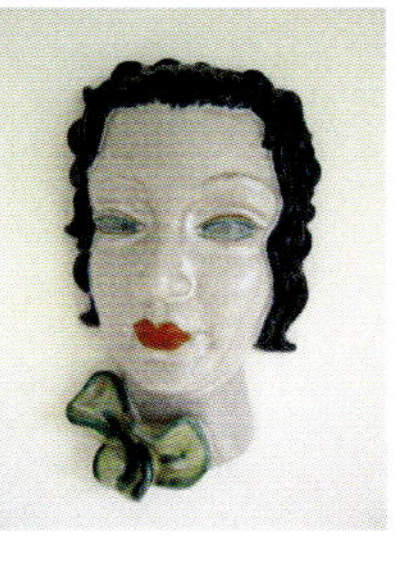
WANDMASKE Frauenkopf mit Schleife / Mascherl am Hals
ca. 1934
6911,0 (Firmenmarke ohne Ang.), H. 26,5 cm; Christie's SK 14.3.1997
6911,3 (Firmenmarke ohne Ang.), H. 26 cm; Doro 14.9.2001

6911,12 Firmenmarke 19, Made in Austria, H. 26 cm; eBay # 3761254200
6911 gem. B, Firmenmarke 19, Made in Austria, H. 26 cm; Sammlung f@goldscheider.de
6911 (Firmenmarke ohne Ang.), H. 25,5 cm; Christie's SK 16.9.1999

6912F
Dakon, Stephan
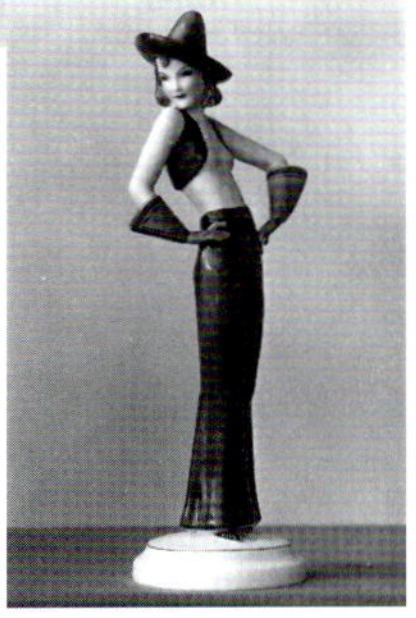
FIGUR/LAMPE Stehende junge Dame mit Hut (sombreroartig) und langen Handschuhen (auf profiliertem Rundsockel)
ca. 1934
FIGUR
6912,107,10 Firmenmarke 19, Made in Austria, H. 32,8 cm; Doro 9.11.1993
LAMPE
6912,7,1 gm. X.S., gem. grün S., Firmenmarke 19, Made in Austria, H. 30,5 cm; eBay # 1432812157
6912F, H. 31 cm, B. 12 cm
Quelle: Archiv Mimi Dakon (Originalfoto)

6917
Matula, Viktor
GRUPPE Enten
ca. 1934
6917 H. fehlt
Quelle: Nachlass Matula, Wien

6918
Meisinger, Ida
Sitzender West Highland Terrier mit Schottenmütze
ca. 1934
6918,139,20 Firmenmarke 19, Made in Austria, unsign., Pick. 4, H. ca. 15 cm; KH

6921
Matula, Viktor
ASCHER Fisch
ca. 1934
6921, H. fehlt, B. fehlt
Quelle: Nachlass Matula, Wien

6922
Goebel, Kurt
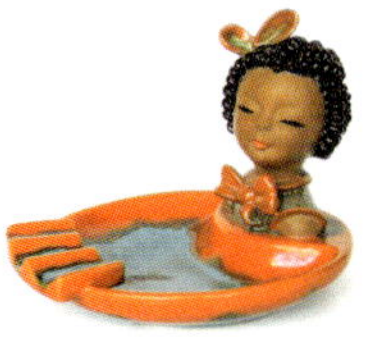
ASCHER Ascher mit Kopf
Form: rund (mit Standfläche für den Kopf), drei Zigarettenablagen
Dekor: Mädchenkopf mit zwei Mascherln/Schleifen (vollplastisch)
ca. 1934
6922 ger. 8, gem. lig. SI, Firmenmarke 19, Made in Austria, unsign., H. 14 cm, Ø 18,5 cm; Sammlung a@goldscheider.net

6923
Lorenzl, Josef
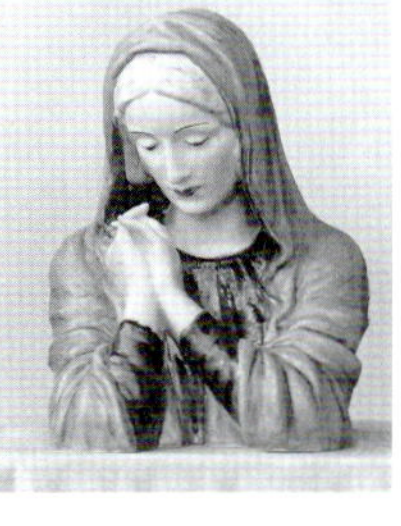
WANDMASKE Madonna
ca. 1933 (für den Entwurf)
6923,1 ger. XVD, gem. N, Firmenmarke 19, Made in Austria, Pick. 6, H. 21,6 cm; eBay # 423252341; 1302978938
6923,23 Firmenmarke 19, unsign.?, H. 18 cm; BM 75
–> vgl. 6745; 7052; 7736= jeweils BÜSTE

6929
Jarl, Karin
Stehender Sealyham Terrier
ca. 1934
6929,87,14 gem. S., Firmenmarke 19, Made in Austria, H. 12,5 cm; eBay # 339974690

6930
Dakon, Stephan
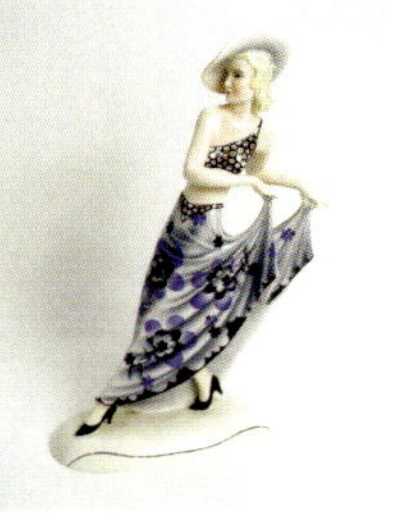
FIGUR „Blonder Traum" (Schreitende Tänzerin mit Hut, ihr Kleid mit beiden Händen seitlich weg haltend) (auf Ovalsockel)
ca. 1934
6930,1,10 gem. N, Firmenmarke 19, Made in Austria, Originalpick.: „Carl Deesz, Saarbrücken", H. 40,5 cm; Dawo 121; city-antik
6930,25,16 gem. S., Firmenmarke 19, Made in Austria, H. 41 cm; eBay # 1419884510
Lit.: Pinhas, S. 153 (m. Abb.)
Quelle: Archiv Mimi Dakon (Tantiemenabrechnung 1.7.–31.12.1937, Wiener Manufaktur FG, Modellliste Dakon, 7.11.1946), hier mit Ang. des Originaltitels FIGUR „Blonder Traum".

6931
Lorenzl, Josef
Liegender Damenakt (auf Ovalsockel)
ca. 1934
6931,34,11 (Firmenmarke ohne Ang.), H. ca. 20 cm, L. 42 cm; Christie's SK 13.5.1999
6931,36,11 (Firmenmarke ohne Ang.), H. ca. 20 cm, L. 43,5 cm; Soth. L. 4.6.1987
6931,55,16 Firmenmarke 19, Made in Austria, H. 20 cm, L. 44 cm; Herr 36
Lit.: Pinhas, S. 134 (m. Abb.)

6932
Lorenzl, Josef
Stehender Damenakt mit ausgestrecktem Arm und großem Tuch (auf Rechteck/Rundsockel)
ca. 1934
Rechtecksockel
6932,69,7,5 gem. N, Firmenmarke 19, Made in Austria, H. 35 cm; Wien Museum Inv. 219.200
Rundsockel
6932 (Firmenmarke ohne Ang.), H. 37 cm; Pinhas, S. 144 (m. Abb.)
–> vgl. 7525= Figur ohne Tuch

6933
Lorenzl, Josef
Schreitende Dame mit Cape, Hut und Muff (auf Ovalsockel)
ca. 1934
6933,4,11 (Firmenmarke ohne Ang.), H. 29,6 cm; Doro 19.12.1995
6933,127,2 gem. lig. AD, Firmenmarke 19, Made in Austria, H. 30 cm; eBay # 2573558674
6933,423,14 (Firmenmarke ohne Ang.), Pick. 2 oder 3, H. 29,5 cm; Doro 29.5.2001
Lit.: Neuwirth (1973), S. 61 (m. Abb.) (PB Wien); ExpKWW, S. 32 (m. Abb.).

6935
Lorenzl, Josef
Tänzerin auf dem rechten Bein tänzelnd, das linke Bein nach vorne schwingend (auf Rautensockel)
ca. 1929/30 (für den Entwurf)
6935,119,11 gem. H, Firmenmarke 19, Made in Austria, H. 21 cm; eBay # 6598150609
–> vgl. 6003; 6323

6940
Dakon, Stephan
Schreitende junge Dame (einen großen Hut in der Hand haltend) (auf Natursockel)
ca. 1934
6940,10,10 ger. 4, Firmenmarke 19, Made in Austria, H. 29 cm; Quit. 1; Doro 1.12.2000

6942F
Dakon, Stephan
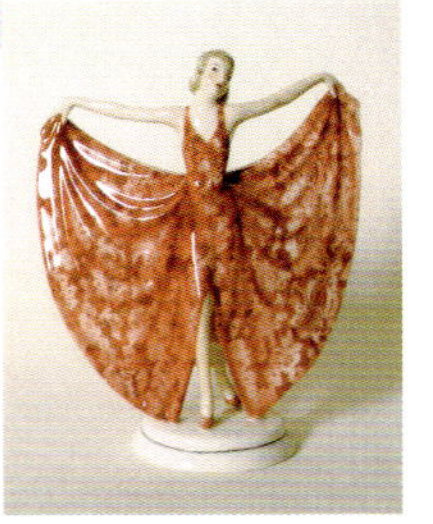
Stehende Tänzerin (ihr Kleid schmetterlingsflügelartig auseinander haltend) (auf Rundsockel)
ca. 1934
6942,80,6 gem. F, Firmenmarke 19, Made in Austria, H. 20 cm; city-antik
6942,91,4 Firmenmarke 19, H. 19,6 cm; BM 41
6942,113,11,7 gem. N, Firmenmarke 19, Made in Austria, H. 20 cm; PB Deutschland
Quelle: AGM (Rechnung der Wiener Manufaktur Josef Schuster an Laszlo Bösze (Szeged, Ungarn), vom 5. August 1944), hier unter dem Titel „Tanz" geführt.

6945
Matula, Viktor
ASCHER Schiff
ca. 1934
6945, H. fehlt, B. fehlt
Quelle: Nachlass Matula, Wien

6946
Matula, Viktor
RELIEF Schiff (Wandrelief)
ca. 1934
6946, H. fehlt, B. fehlt
Quelle: Nachlass Matula, Wien

6947 s. 7947

6948
STEHKOPF Frauenkopf mit Halstuch/Schal
ca. 1934
6948 Firmenmarke 19, H. 12,6 cm; Doro 14.9.2001

6950
Meisinger, Ida
GRUPPE Zwei sitzende West Highland Terrier
ca. 1934
6950,40,81 gem. lig. JP, Firmenmarke 19, Made in Austria, unsign., H. 10 cm; eBay # 7316963377

6951
Weiss, Claire (Klára) (d.i. Herczeg, Claire (Klára)
FIGUR „Komtesse" (Stehende Dame mit Hut) (auf Ovalsockel)
ca. 1931 (für den Entwurf)
6951,6,5 gem. lig. JR, gem. F, Firmenmarke 19, Made in Austria, sign. Claire Weis(s?), Pick. 6, H. 20,5 cm; eBay # 1279479277
6951,15,1 (Firmenmarke ohne Ang.), sign. Claire Weihs, H. 20,4 cm; Doro 15.9.2003
6951,96,10,4 gem. SG, Firmenmarke 19, Made in Austria, Fabriqué en Autriche, sign. Claire Weiss, H. 20,5 cm; eBay # 200046599543
–> vgl. 6454
–> vgl. 7327; 8145= jeweils GRUPPE „Komtesse mit Hund".

6955
Weiss, Claire (Klára) (d.i. Herczeg, Claire (Klára) (zugeschr.)
Gehendes Kinderpaar (auf Ovalsockel)
ca. 1934
6955,30,20 Firmenmarke 19, Made in Austria, unsign., H. 18,5 cm; Zezschwitz 29

6958
WANDMASKE Mädchenkopf mit Zöpfen und Kasperle (Pinocchio?)
ca. 1934
6958 (Firmenmarke ohne Ang.), H. 18 cm; Pinhas, S. 173 (m. Abb.)

6959
Dakon, Stephan
Sitzende Dame mit großem Hut (auf zwei Reisekoffern mit Zeitung)
ca. 1934

6959,58,10 Firmenmarke 19, Made in Austria, H. 25 cm; Doro 9.11.1993
-> vgl. 7364= ähnl. Modell, hier wird die auf dem Koffer liegende Zeitung durch einen Pekinesen ersetzt.

6960F
Dakon, Stephan
Stehende junge Dame (im Wind) ihren Hut fest haltend (auf profiliertem Rundsockel)
ca. 1934
6960F, H. 28 cm
Quelle: Archiv Mimi Dakon (Originalfoto)

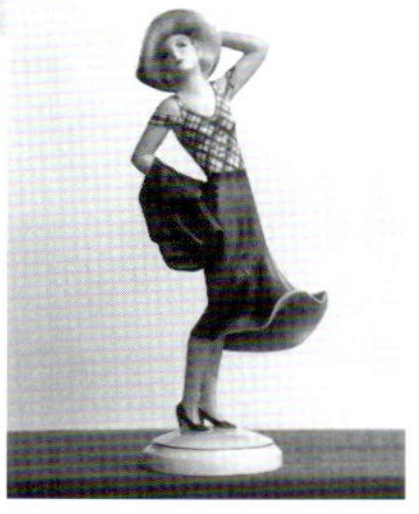

6975
Postl, Hugo
Skyeterrier (auf Ovalsockel)
um/ab 1930 (für die Ausführung)
6975,20 Firmenmarke 19, Made in Austria, unsign., H. ca. 19 cm; WKH
Bem.: Diese Variante mit Ovalsockel könnte als Schreibtischablage dienen bzw. so von der Firma gedacht gewesen sein. Die Figur des Hundes ist nicht mittig auf dem Sockel aufgebracht, sondern nach hinten versetzt. Somit bleibt im vorderen Bereich Platz für Schreibzeug etc.
-> vgl. 6503; 6802; 7300; 7519= jeweils FIGUR (ohne Sockel)
Lit.: GSNachf. 5,11 (m. Abb. als Figur), hier mit Ang. des Originaltitels „Skyeterrier".

6977
Meisinger, Ida
FIGUR „Hund mit Bassgeige" (Stehender Terrier mit Kontrabass)
ca. 1934
6977,7,25 Firmenmarke 19, unsign., H. 13,6 cm; PB Deutschland
Bem.: s. 6449
-> vgl. 7878
Lit.: GSNachf. 3,28 (m. Abb.) für die Modell-Nr. 7878.

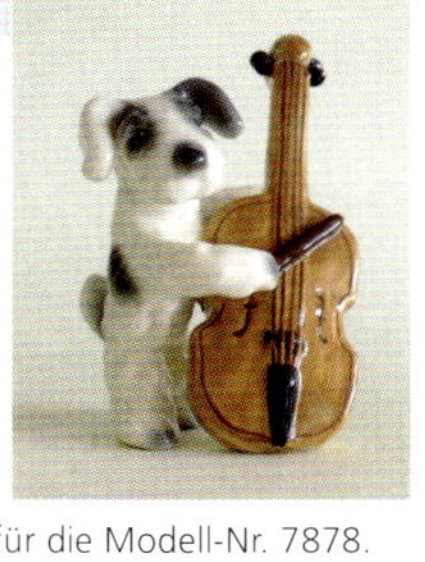

6979
Matula, Viktor
GRUPPE Rehe
ca. 1934
6979 H. fehlt
Quelle: Nachlass Matula, Wien

6982
Matula, Viktor
RELIEF Schiff (Wandrelief)
ca. 1934
6982, H. fehlt, B. fehlt
Quelle: Nachlass Matula, Wien

6983
Matula, Viktor
BUCHSTÜTZE Reh
ca. 1934
6983,20 Firmenmarke 19, Made in Austria, unsign., Ausformung zw. 1945–1953, H. 13 cm, L. 20,5 cm; eBay # 6565468333

6984
Meisinger, Ida
GRUPPE Zwei an einem Halsband zerrende Foxterrier (auf Ovalsockel)
ca. 1934
6984,94,10 Firmenmarke 19, Made in Austria, unsign., H. 12 cm, L. 16 cm; eBay # 3756049905
Bem.: s. 6449

6985
Matula, Viktor
ASCHER
ca. 1934
6985 H. fehlt, B. fehlt
Quelle: Nachlass Matula, Wien

6986
Matula, Viktor
GRUPPE Zwei Pferde (auf flachem Rechtecksockel)
ca. 1934
6986,9 (Firmenmarke ohne Ang.), unsign., H. 10 cm; Christie's SK 14.3.1997
6986,40 gem. Z, Firmenmarke 19, Made in Austria, unsign., H. 11,4 cm, B. 15,8 cm; eBay # 1456707618
-> vgl. 6797= Pferd (vorne)
-> vgl. 7804= Pferd (hinten)
Lit.: GSNachf. 3,18 und 3,21 (m. Abb.) für die Einzelfiguren, hier jeweils mit Ang. des Originaltitels „Pferd".

6989
Neuwalder-Breuer, Grete (zugeschr.)
STEHKOPF Mädchenkopf mit Blumenkranz im Haar
ca. 1934
6989,20 ger., gem. lig. AH, ohne Firmenmarke, Fabriqué en Autriche, unsign., H. 21 cm; Sammlung John Nash

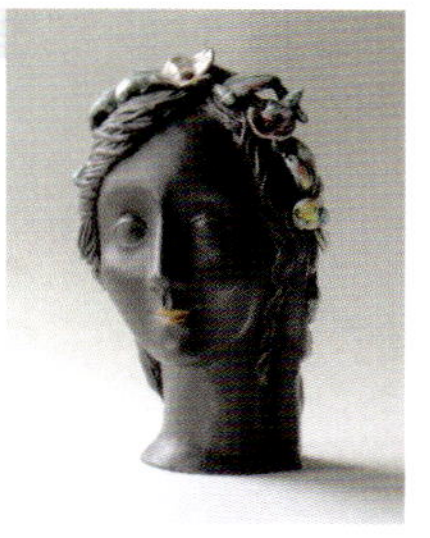

Bem.: Für die Zuschreibung wurde der Stehkopf, abgebildet in: ÖBWK, 1.1924/25, S. 340, herangezogen. Die zurückgekämmten, strähnigen Haare sowie der Mund und die schmale elegante Nase sprechen eindeutig für einen Entwurf von Neuwalder-Breuer.
Lit.: N.N.: Österreich in Paris, in: ÖBWK, 1.1924/25, S. 340 (m. Abb.), für die Zuschreibung.

6990
Stehender Frauenakt (Torso) (auf Quadratsockel)
ca. 1934
6990 Firmenmarke 19, Made in Austria, Originalpick.: „Küster Perry", H. 20 cm; Sammlung.schmutz@chello.at 54

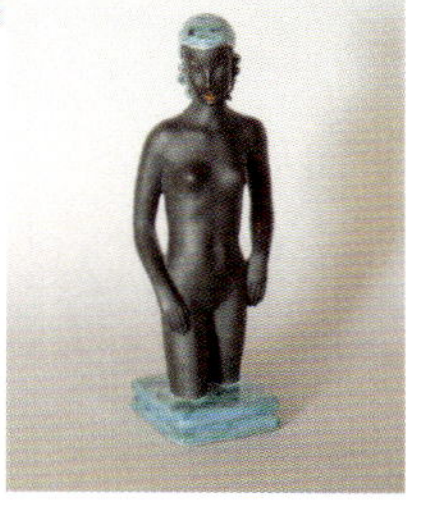

6992
Matula, Viktor
ASCHER Reh
ca. 1934
6992, H. fehlt, B. fehlt
Quelle: Nachlass Matula, Wien

6994
Matula, Viktor
ASCHER Elefant
ca. 1934
6994, H. fehlt, B. fehlt
Quelle: Nachlass Matula, Wien

6996
Meisinger, Ida
„Pat und Patachon" (zwei sitzende Airedaleterrier)
ca. 1931 (für den Entwurf)
6996,9,65 gem. S, Firmenmarke 19, Made in Austria, unsign., H. 8 cm; eBay # 120024258796
Lit.: Bernd, G.: Heitere Wiener Tierkunst, Ida Meisinger, in: Schaulade, 8.1932, S. 480 (m. Abb.), hier mit Ang. des Originaltitels „Pat und Patachon", „Die groteske Tierfigur ist eigentlich im Österreichischen zuhause. Hier ist der Sinn für die Fragwürdigkeit jeder Kreatur mit Einschluss der menschlichen besonders lebendig, für die kurze Spanne, die zwischen Ernsthaftigkeit und lässiger Selbstpreisgabe, zwischen Heroismus und Lächerlichkeit besteht. Da ist beispielsweise Ida Meisinger. Sie hat den Wiener Blick für die ernsthafte Drolligkeit dieser kleinen und großen Hundeviecher. Mögen sie auch nicht sämtlich rasse- und stubenrein sein – „gar so herzig" sind sie in jeder Bewegung. Sie schnuppern, machen Männchen, sehen trist philosophisch drein – ein positiver Pessimismus, der zum Lachen reizt! – gefallen sich, unvereinbar und doch mit der Pose von Denkmälern, etwa nach dem klassischen Vorbild Pat und Patachons, in ernsthaftem tête à tête, springen auch mal possierlich hoch und markieren so nebenbei Bücherstütze. Ida Meisinger charakterisiert durch die Bewegung, sozusagen durch die Bewegung der keramischen Masse, und durch die Akzente, die sie bestimmten Details, der Schnauze, den Ohren, dem Schwanz, dem tappigen Gehwerk gibt. Diese erlaubten, ja erwünschten Übertreibungen vermitteln erst das, worauf es der Künstlerin ankommt, den liebenswürdigen grotesken Charme. Die Farben sind leuchtend, aber nicht schreiend. Ein emailähnliches Weiß wechselt mit feinem, mattem Schwarz, mit intensivem Korallrot. Die „Widernatürlichkeit" dieser Farben entspricht der Natur des Werkstoffes vollkommen".
-> vgl. 6362
-> vgl. 6463= DreierGRUPPE

6997M
Meisinger, Ida
FIGUR/BUCHSTÜTZE „Hundewetter" (Sitzender Terrier mit Regenschirm)
ca. 1931 (für den Entwurf)
FIGUR
6997,10,R Firmenmarke 19, Made in Austria, Hand dec., unsign., H. 8 cm; eBay # 3714562675

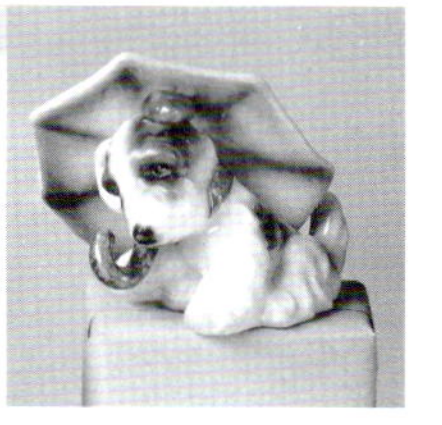

6997,20 ger., Firmenmarke 19, Made in Austria, Ausformung zw. 1945–1953, H. 8 cm; eBay # 160013166513
6997,158,31 gem. W, Firmenmarke 19, Made in Austria, unsign., Pick. 2, H. 8 cm; eBay # 2279470186
6997M, H. 8 cm, B. 9 cm; GSNachf. 3,25b; Fotoarchiv K. Täuber, hier mit der Ang.: H. 8 cm, B. 10 cm
BUCHSTÜTZE
6997 ger., (ohne Firmenmarke), unsign., H. ca. 14 cm; WKH
Bem.: s. 6449
-> vgl. 6449
Lit.: GSNachf. 3,25 (m. Abb.), hier mit Ang. des Originaltitels FIGUR „Hundewetter".
Quelle: Fotoarchiv K. Täuber

6998
Knörlein, Rudolf (zugeschr.)
DECKELDOSE Frauenkopf mit halblangen Haaren und Kragen
Form: rechteckig
Dekor: Frauenkopf (reliefiert)
ca. 1933/34
6998,3 gem. lig. TK, Firmenmarke 19, Made in Austria, unsign., H. fehlt; eBay # 3714842735
-> vgl. 6832= WANDMASKE
-> vgl. 6880= ABREISSKALENDER

7015F
Thomasch, Wilhelm
GRUPPE „Spaniels" (Zwei laufende Spaniels)
ca. 1934

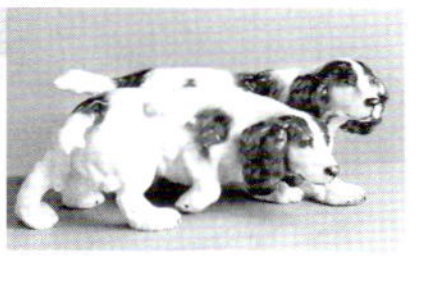

7015,8 Firmenmarke 19, Made in Austria, unsign., Ausformung zw. 1945–1953, H. 14 cm; WKH
7015F, H. 14 cm, B. 35,5 cm; GSNachf. 5,8a; Fotoarchiv K. Täuber, hier mit der Ang.: H. 14 cm, B. 33 cm
Bem.: Für die Zuordnung dieser Gruppe und der damit verbundenen Einzeltiere, danke ich Herrn Herwig Thomasch, dem Sohn von Wilhelm Thomasch.
-> vgl. 7281F; 8342F
-> vgl. 7061F; 7329F; 8344F= jeweils EinzelFIGUR (vorderer Hund)
-> vgl. 7062F; 7330F; 8343F= jeweils EinzelFIGUR (rückwärtiger Hund)
Lit.: GSNachf. 5,8 (m. Abb.), hier mit Ang. des Originaltitels GRUPPE „Spaniels".
Quelle: Fotoarchiv K. Täuber

7019
RELIEF Madonna (Wandrelief)
ca. 1934
7019,65 Firmenmarke 19, Made in Austria, Ausformung zw. 1945–1953, H. 28 cm, L. 11,5 cm; Fischer 67; BM 62
Bem.: Da sich in keinem der Auktionskataloge eine nähere Beschreibung oder eine Abb. findet, kann dieses Relief nicht exakter beschrieben bzw. keinem Künstler zugeordnet werden.

7021
Knörlein, Rudolf
KERZENHALTER (dreiflammig)
Form: astartig, blütenartige Kerzentüllen, sternförmig gezackter Stand (Sockel)
Dekor: vermutlich diverse Dekore
ca. 1934
7021, H. 29 cm, B. 28 cm
Bem.: Durch die eigenhändige Datierung Knörleins (1934) auf dem Trägerblatt des Originalfotos kann das Jahr 1934 exakt bestimmt werden.
Quelle: Privatarchiv (Originalfoto)

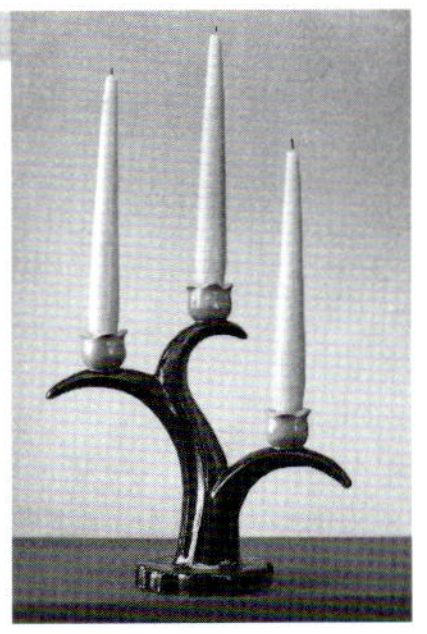

7022
Kniende junge Frau
ca. 1934
7022,14,65 Firmenmarke 19, Made in Austria, H. 23 cm; WKA 35 (Lot 1071, außer Kat.)

7025M
Knörlein, Rudolf
RELIEF „Madonna mit Kind" (Madonna mit Schleier) (Wandrelief)
ca. 1934
7025 Firmenmarke 19, Made in Germany, unsign., Ausformung zw. 1938–1941, H. 15 cm, B. 12 cm; eBay # 2223685795
7025M, H. 14,5 cm, B. 12 cm; GSNachf. 6,1b; Fotoarchiv K. Täuber, hier mit der Ang.: H. 15 cm, B. 12 cm
Bem.: Durch die eigenhändige Datierung Knörleins (1934) auf dem Trägerblatt des Originalfotos kann das Jahr 1934 exakt bestimmt werden. Dieses Relief wurde auch noch zw. 1953–1960 bei Carstens (GWM West-Germany) produziert.
Lit.: GSNachf. 6,1 (m. Abb.), hier mit Ang. des Originaltitels RELIEF „Madonna mit Kind"; KGSF1; Könecke (2002), S. 76, die Ang. der Künstlerin K. Jarl hier ist falsch, dies konnte durch ein Originalfoto aus dem Nachlass R. Knörleins (Privatarchiv) belegt werden.
Quelle: Fotoarchiv K. Täuber

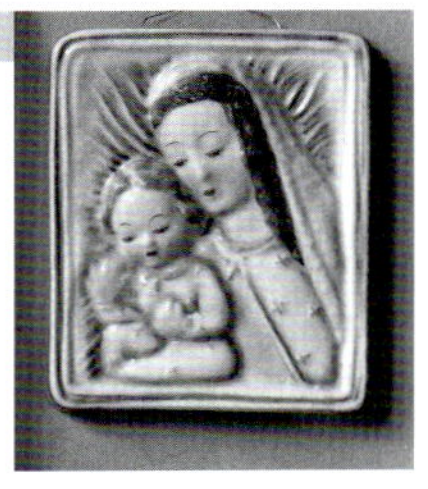

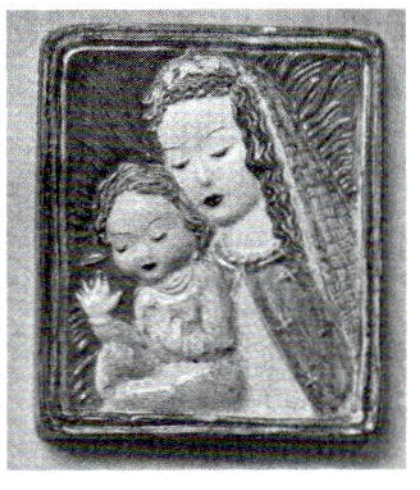

7027
Weiss, Claire (Klára) (d.i. Herczeg, Claire (Klára)
GRUPPE Rotkäppchen (mit Korb) und der Wolf (auf Ovalsockel)
ca. 1934/35
7027,174,20 gem. IE, Firmenmarke 19, Made in Austria, unsign., H. 19 cm; eBay # 390094616
7027,201,9 gem. N, Firmenmarke 19, Made in Austria, unsign., H. fehlt, L. 27 cm; Mehlis 36
7027 gepr.,239 ger.,5 gepr., ger. XV, gem. F., N, Firmenmarke 19, Made in Austria, H. 19 cm; Wien Museum Inv. 220.728
7027,289,9 ger. G, gem. F., Firmenmarke 19, Made in Austria, unsign., H. 20 cm; eBay # 2288775400
Bem.: Die Darstellung geht auf das Märchen „Rotkäppchen und der Wolf", welches die Gebrüder Grimm in ihrer Märchensammlung publizierten, zurück. Wiedergegeben ist hier die Szene, in der das kleine Mädchen mit der roten Kappe dem Wolf begegnet. Die Figur des Mädchens wurde auch bei den Modellen 7847, 7861 sowie 8003 verwendet. Dort jeweils als Darstellung einer „Gänseliesel".

7028
Weiss, Claire (Klára) (d.i. Herczeg, Claire (Klára) (zugeschr.)
Schreitender Junge in Matrosenkostüm (ein Modellsegelschiff hinter sich herziehend) (auf flachem Rechtecksockel)
ca. 1934/35
7028,16,35 Firmenmarke 19, Made in Austria, unsign., H. 18 cm; Phillips L. 25.1.2001

7029M
Knörlein, Rudolf
RELIEF „Madonna mit Kind" (Madonna und Kind mit Aureole) (Wandrelief)
ca. 1934
7029 gem. lig. AH, Firmenmarke 19, Made in Germany, Ausformung zw. 1938–1941, H. 17 cm; B. 14 cm; Sammlung John Nash
7029M, H. 16,5 cm, B. 13 cm; GSNachf. 6,1a; Fotoarchiv K. Täuber, hier mit der Ang.: H. 17 cm, B. 14 cm
Bem.: Durch die eigenhändige Datierung Knörleins (1934) auf dem Trägerblatt des Originalfotos kann das Jahr 1934 exakt bestimmt werden. Dieses Relief wurde auch noch zw. 1953–1960 bei Carstens (GWM West-Germany) produziert.
Lit.: GSNachf. 6,1 (m. Abb.), hier mit Ang. des Originaltitels RELIEF „Madonna mit Kind"; KGSF1; Könecke (2002), S. 76, die Ang. der Künstlerin K. Jarl hier ist falsch, dies konnte durch ein Originalfoto aus dem Nachlass R. Knörleins (Privatarchiv) belegt werden.
Quelle: Fotoarchiv K. Täuber

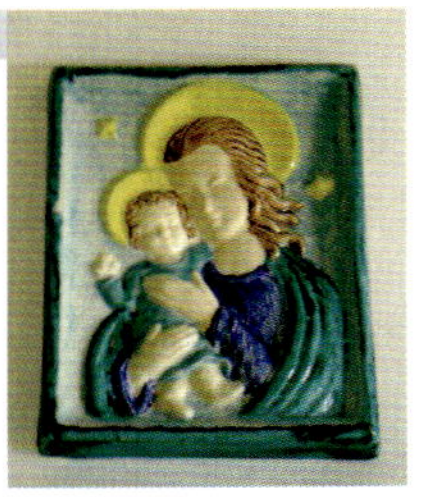

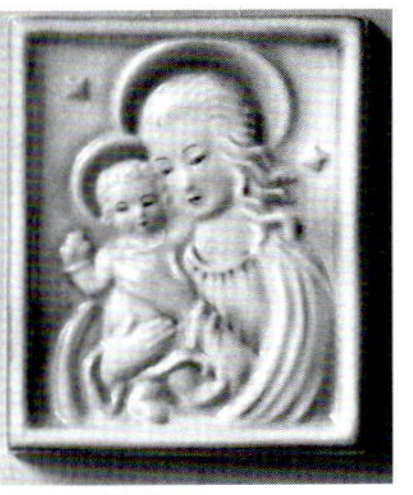

7035
Weiss, Claire (Klára) (d.i. Herczeg, Claire (Klára)
Stehende Tennisspielerin im Wind (auf Ovalsockel)
ca. 1934/35
7035,8,14 gem. S., gem. rot lig. TS, Firmenmarke 19, Made in Austria, sign. Claire Weiss, H. 30 cm; eBay # 2269142480

7037
Lorenzl, Josef
„Eifersucht" (Stehende Dame mit Hut und zwei Pekinesen) (auf Rundsockel)
ca. 1934/35
7037,22,16 (Firmenmarke ohne Ang.), H. 38 cm; Christie's SK 14.11.2002
7037,29,1(1?),7 gem. lig. ON, gem. lig. rot JR, Firmenmarke 19, Made in Austria, Pick. 2, H. 37,5 cm; eBay # 1439881822
7037,193,10 Firmenmarke 19, H. 37 cm; Soth. South, Billingshurst 11.10.2000
-> vgl. 7731
Quelle: AGM (Bestellungen der Firma Wahliss (Wien) vom 17.10.1945 und 27.2.1946 an die „Arbeitsgemeinschaft der Firma Friedrich Goldscheider"), hier jeweils mit Ang. des (Original?-)Titels „Eifersucht".

7039
Bouret, G(ermaine)
DreierGRUPPE/ZweierGRUPPE/EinzelFIGUR/SCHREIBTISCHABLAGE
Drei schreitende Kinder (zwei Jungen mit einem Mädchen, Arm in Arm gehend) (auf abgerundetem Rechtecksockel)
ca. 1934/35
DreierGRUPPE
7039,20 Firmenmarke 19, Made in Germany, Ausformung zw. 1938–1941, H. 22 cm; PB
7039,20 Firmenmarke 19, Made in Austria, unsign., Ausformung zw. 1945–1953, H. 22 cm, B. 16 cm; eBay # 2571601333
7039,218,20 ger. III, Firmenmarke 19, Made in Austria, unsign., Pick. 3, H. ca. 22 cm; KH
7039,293,1 Firmenmarke 19, Made in Austria, unsign., H. 20,5 cm; Sammlung Bernard Amigon, Frankreich
7039,447,16 gem. M.W., Firmenmarke 19, Made in Austria, unsign., H. 22,8 cm; eBay # 173768836
7039,570,9 (ger./gem.? T.W.), Firmenmarke 19, Made in Austria, sign. G. Bouret, Pick. ohne Ang., H. 21,6 cm; eBay # 32516972
ZweierGRUPPE (mittlere und rechte Figur)
7039A,282,36 ger. XVE, Firmenmarke 19, Made in Austria, sign. G. Bouret, H. 21,5 cm; eBay # 3780717173
7039,374,65 ger. A, gem. lig. blau GH., JR, Firmenmarke 19, Made in Austria, unsign., H. 20,5 cm; Sammlung Bernard Amigon, Frankreich
EINZELFIGUR (linke Figur)
7039,7,20 gem. lig. JR, Firmenmarke 19, Made in Austria, unsign., H. 21 cm; eBay # 880881760
SCHREIBTISCHABLAGE (als DreierGRUPPE)
7039,65 gem. GH., gem. lig. blau JK, Firmenmarke 19, Made in Austria, unsign., H. ca. 24 cm; WKH
Bem.: Hier eines der wenigen Beispiele für die Vergabe einer Modell-Nr. für mehrere Modellvarianten. Zur gleichen Problematik -> vgl. 6761. Warum hier von der bei GS üblichen Praxis, für jedes eigene Modell eine eigene Nummer zu vergeben abgewichen wurde, lässt sich nicht sagen. Bei der Schreibtischablage könnte es sich sogar um eine Einzelanfertigung handeln bzw. sind ziemlich sicher von dieser Variante nur sehr wenige Stücke produziert worden. Durch ein Foto aus dem Nachlass Dakon kann eindeutig nachgewiesen werden, dass die Figuren, die auf die Entwürfe von Germaine Bouret zurückgehen, von Stephan Dakon ausmodelliert wurden.
Quelle: Archiv Mimi Dakon

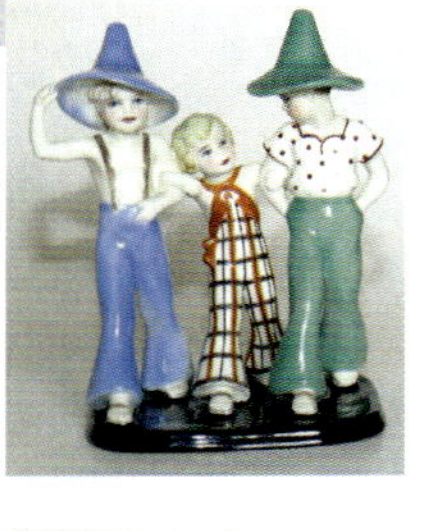

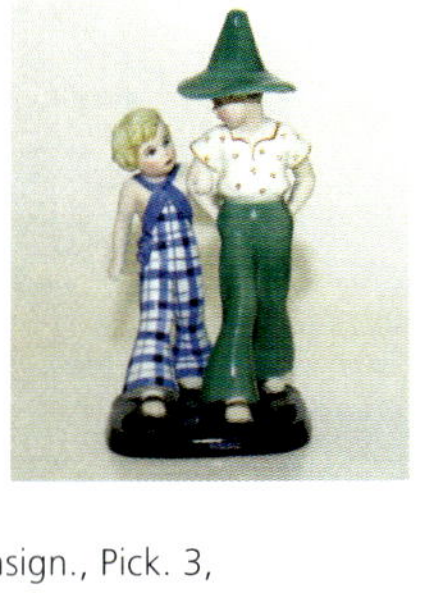

7040
Weiss, Claire (Klára) (d.i. Herczeg, Claire (Klára)
Tänzerin in langem Kleid (auf Ovalsockel)
ca. 1934/35
7040,116,11 (Firmenmarke ohne Ang.), sign. Claire Weiss, H. 28,5 cm; Christie's SK 10.11.2000
Lit.: Pinhas, S. 160 (m. Abb.)

7041
Lorenzl, Josef und Postl, Hugo
GRUPPE Stehender Damenakt mit sitzendem Schäferhund (auf Rundsockel)
ca. 1934/35
7041,74,6 gem. F., Firmenmarke 19, Made in Austria, H. 40,6 cm; eBay # 3734760926
> vgl. 5719= EinzelFIGUR Hund
-> vgl. 6069= Kind (Lorenzl) mit Hund (Postl)
Lit.: Pinhas, S. 142 (m. Abb.)

7042
Lorenzl, Josef
Stehende Dame mit Hut, Fuchsstola, Handtasche und Foxterrier (auf Rundsockel)
ca. 1934/35
7042,26,11 Firmenmarke 19, Made in Austria, H. 28,5 cm; eBay # 2635510515
Lit.: Pinhas, S. 139 (m. Abb.)

7044F
Weiss, Claire (Klára) (d.i. Herczeg, Claire (Klára)
„Muff" (Stehende Dame mit Muff und großem Hut) (auf Ovalsockel)
ca. 1930/31 (für den Entwurf)
7044,20,14 Firmenmarke 19, Made in Austria, sign. Claire Weiss, H. 20,5 cm; Henry's 27.4.2001
7044 ohne Ang., (Firmenmarke ohne Ang.), sign. Claire Weiss, H. 20 cm; KaM 146
-> vgl. 6328
-> vgl. 7720= GRUPPE
Quelle: AGM (Rechnung an Maurice Mizrahi, Alexandria, Ägypten, vom 22. April 1953), hier mit Ang. des (Original)-Titels „Muff".

7048
KERZENHALTER (zweiflammig) Posthorn
Form: röhrenartig zu einer Öse geformt (Posthorn), zwei geweitete Kerzentüllen (auf gewölbtem Rechtecksockel)
Dekor: Laufglasur mit Craquelé
ca. 1934/35

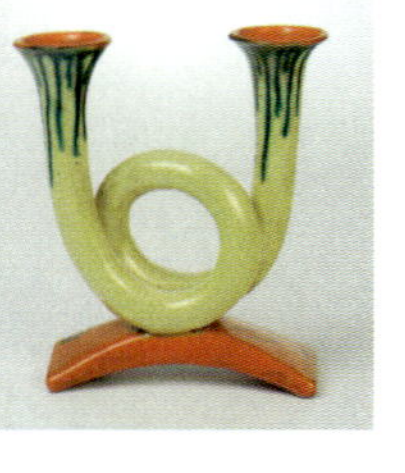

7048,47,4 gem. lig. TK, Firmenmarke 19, Pick. 2, H. 25 cm, B. 19 cm; Wien Museum Inv. 186.700
Ausst.: SSH 1992
Lit.: GSN, S. 52, 53 (m. Abb.)
GS neu: Modell-Nr. 47 „Das andere Posthorn" (H. 22 cm), Aufl. ohne Ang.

7050F
Dakon, Stephan
Stehende junge Dame in einteiligem Hosenanzug mit großem Hut und Ball (auf Ovalsockel)
ca. 1934/35
7050F, H. 30 cm
Quelle: Archiv Mimi Dakon (Originalfoto)

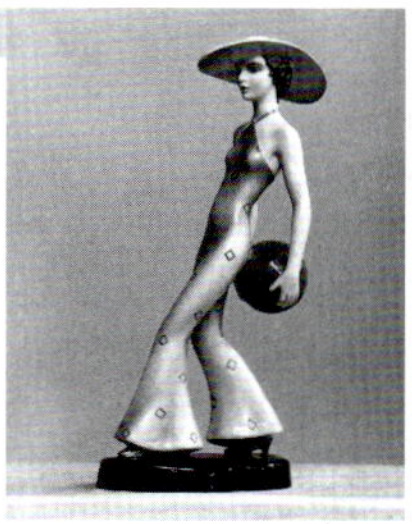

7051
Lorenzl, Josef
BÜSTE Flehende Frau mit gefalteten Händen
ca. 1934/35
7051,23,14 Firmenmarke 19, Made in Austria, H. 16,5 cm, B. 19 cm; eBay # 2201779679
7051,32,6 gem. lig. AS, Firmenmarke 19, Made in Austria, H. 17 cm, B. 20 cm; eBay # 7319558220

7052
Lorenzl, Josef
BÜSTE Madonna mit gefalteten Händen
ca. 1933 (für den Entwurf)
7052,221,31 gem. N, Firmenmarke 19, Made in Austria, H. 17,8 cm; eBay # 1205471172
7052 H. 18 cm, B. 11 cm; Fotoarchiv K. Täuber
-> vgl. 6745; 7736
-> vgl. 6923= WANDMASKE
Quelle: Fotoarchiv K. Täuber

7053
Lorenzl, Josef
Tänzerin, das linke Bein abgewinkelt nach oben schwingend (ihren langen Rock mit beiden Händen auseinander haltend) (auf Rautensockel)
ca. 1934/35
7053,107,31 gem. A.S. Firmenmarke 19, Made in Austria, H. 38,1 cm; Soth. Chester 7.–9.6.1988
7053,145,16,8A Firmenmarke 19, Made in Austria, H. 38 cm; Soth. L. 2.11.1999
7053,152,6 (Firmenmarke ohne Ang.), H. 38,4 cm; Christie's SK 15.10.2003
7053,169,31,8 gem. lig. AS, Firmenmarke 19, Made in Austria, H. 39,8 cm; PB Wien
7053 ohne Ang., Firmenmarke 19, Pick. 2, H. 38 cm; Weichmann 24.6.1988

7054F
Ullmann, Robert
GRUPPE „Windhunde"
(Ein stehender und ein liegender Barsoi)
ca. 1934/35

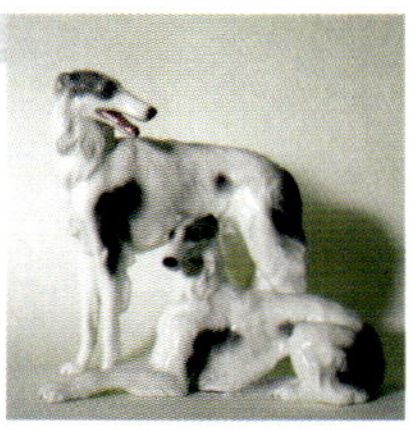

7054,1 Firmenmarke 19, Made in Austria, unsign., Ausformung zw. 1945–1953, H. 39 cm, B. 44 cm; Doro 9.6.1998
7054,8 Firmenmarke 19, Made in Austria, Hand dec., unsign., Ausformung zw. 1945–1953, H. 39 cm; WKH
7054,66,33 gem. G, Firmenmarke 19, Made in Austria, unsign., H. 40,6 cm, B. 46,7 cm; city-antik
7054,407,10 Firmenmarke 19, H. 37 cm; KaM 89
7054,626,11 Firmenmarke 19, Made in Austria, unsign., H. 39 cm; Fischer 132
7054,1027,19 Firmenmarke 19, Made in Germany, unsign., Ausformung zw. 1938–1941, H. 38,5 cm; Doro 6.4.1999
7054,1144,31 gem. AN, Firmenmarke 19, unsign., Ausformung zw. 1938–1941, H. 38,5 cm; Schloss Hag. 16
7054,1590,66 Firmenmarke 19, Made in Germany, unsign., Ausformung zw. 1938–1941, H. 38,5 cm; Doro 29.6.1993
7054,1690,3 Firmenmarke Schuster 1, Ausformung zw. 1941–1944, H. ca. 39 cm, B. 45 cm; Mehlis 26
7054,1698,3 Firmenmarke Schuster 1, Ausformung zw. 1941–1944, H. 38 cm, Christie's SK 18.9.1998
7054F, H. 36 cm, B. 41 cm; GSNachf. 5,4a
Bem.: zum Künstler s. 7054AF; zur Modellnummernvergabe s. 6761.
-> vgl. 7521F; 7972F
-> vgl. 7054AF; 7571AF; 7972AF= jeweils stehende EinzelFIGUR
-> vgl. 7054BF; 7571BF; 7972BF= jeweils liegende EinzelFIGUR
Lit.: GSNachf. 5,4 (m. Abb.), hier mit Ang. des Originaltitels GRUPPE „Windhunde".
Quelle: AGM (Rechnung der Wiener Manufaktur Josef Schuster an Laszlo Bösze (Szeged, Ungarn), vom 5. August 1944; Bestellungen der Firma Wahliss (Wien) vom 9. und 24.11.1945 an die „Arbeitsgemeinschaft der Firma Friedrich Goldscheider"), hier jeweils mit Ang. des Titels GRUPPE „Barsoyes"; Bestellung der Firma Wahliss (Wien) vom 27.2.1946 in Ausführung „Terrakotta" je 22 Stück, 1 Stück in „Fayence"; Rechnung an Maurice Mizrahi, Alexandria, Ägypten, vom 22. April 1953), hier unter dem Titel „Windhunde" geführt.

7054AF
Ullmann, Robert
FIGUR „Windhund" (Stehender Barsoi)
ca. 1934/35
7054,128,? Firmenmarke 19, Made in Austria, sign. Ullmaier (!), H. 35 cm; RM 47; Bem.: Hier handelt es sich um einen Lesefehler des AH. Zwar sind fast alle bislang nachgewiesenen Stücke unsigniert, jedoch konnte ich ein mit dem Namen Ullmann signiertes selbst in Augenschein nehmen (Modell-Nr. 7571).
7054,148,7 (Firmenmarke ohne Ang.), unsign., H. 38 cm; Metz 13.3.1993
7054AF, H. 37 cm, B. 36 cm; GSNachf. 5,4d
Bem.: Trotz des Lesefehlers „Ullmaier" bei RM 47 und der fast ausschließlich unsignierten Exemplare stammen die Barsoigruppe sowie die Einzelfiguren natürlich von Robert Ullmann, der u.a. auch für Keramos und die Porzellanmanufaktur Augarten (beide Wien) Modelle entwarf (-> vgl. Neuwirth (1974), S. 176, 404–407).
Zur Modellnummernvergabe s. 6761.
-> vgl. 7571AF; 7972AF
-> vgl. 7054F; 7571F= jeweils GRUPPE
Lit.: GSNachf. 5,4 (m. Abb.), hier mit Ang. des Originaltitels FIGUR „Windhund" (stehend).
Quelle: AGM (Bestellung der Firma Wahliss (Wien) vom 27.2.1946 an die „Arbeitsgemeinschaft der Firma Friedrich Goldscheider", hier mit dem Titel „Windhund (groß)", in Ausführung „Terrakotta").

7054BF
Ullmann, Robert
FIGUR „Windhund" (Liegender Barsoi)
ca. 1934/35
7054,9 ger. B, gem. rot G., Firmenmarke 19, Made in Austria, unsign., Ausformung zw. 1945–1953, H. 22,8 cm; city-antik
7054BF, H. 21 cm, B. 41 cm; GSNachf. 5,4g
Bem.: zum Künstler s. 7054AF; zur Modellnummernvergabe s. 6761.
-> vgl. 7571BF; 7972BF
-> vgl. 7054F; 7571F= jeweils GRUPPE
Lit.: GSNachf. 5,4 (m. Abb.), hier mit Ang. des Originaltitels FIGUR „Windhund" (liegend).
Quelle: AGM (Rechnung der Wiener Manufaktur Josef Schuster an Laszlo Bösze (Szeged, Ungarn), vom 5. August 1944), hier mit Ang. des Originaltitels „Windhund".

7058
Lorenzl, Josef
Schreitende Tänzerin in Pose (mit Bikinioberteil und langem gerüschtem Rock) (auf Ovalsockel)
ca. 1934/35
7058,153,19 (Firmenmarke ohne Ang.), sign. Lorenzl, H. 31 cm; Christie's SK 4.11.1999
7058,159,6,5 (Firmenmarke ohne Ang.), sign. Lorenzl, H. 35,4 cm; Christie's SK 13.5.2004
7058,332,6 gem. SG, Firmenmarke 19, Made in Austria, sign. Lorenzl, H. 35,4 cm; eBay # 7308270147
Lit.: Pinhas, S. 146 (m. Abb.)

7059F
Dakon, Stephan
Tango tanzendes Paar (auf Ovalsockel)
ca. 1934/35
7059 (Firmenmarke ohne Ang.), H. 31 cm; Pinhas, S. 140 (m. Abb.)
7059F, H. 30 cm, B. 16 cm
Quelle: Archiv Mimi Dakon (Originalfoto)

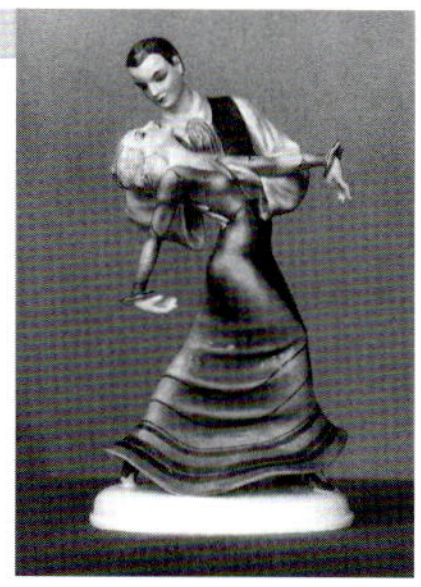

7061F
Thomasch, Wilhelm
FIGUR „Spaniel"
(Laufender Spaniel)
ca. 1934

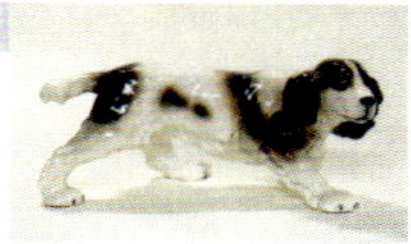

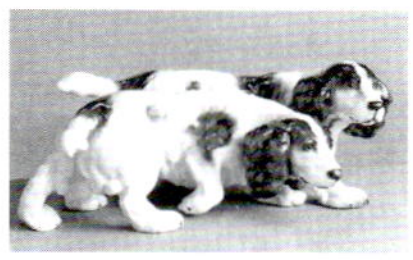

7061,1 ger. I, Firmenmarke 19, Made in Austria, unsign., Ausformung zw. 1945–1953, H. 12,5 cm, B. 30 cm; eBay # 879262816
7061,14,4 ger., Firmenmarke 20, Made in Austria, Hand dec., H. 14,4 cm, B. 32,2 cm; MAK Wien, Inv. Ke 10521-22
7061F, H. 15 cm, B. 30 cm; GSNachf. 5,8d
Bem.: Für die Zuordnung dieser Figur und der damit verbundenen Gruppe, danke ich Herrn Herwig Thomasch, dem Sohn von Wilhelm Thomasch.
-> vgl. 7329F; 8344F; ohne Nr.
-> vgl. 7015F; 7281F; 8342F= jeweils GRUPPE
Pendant -> vgl. 7062
Lit.: GSNachf. 5,8 (m. Abb.), hier mit Ang. des Originaltitels FIGUR „Spaniel" (Vorderer Hund aus der GRUPPE 7015).

7062F
Thomasch, Wilhelm
FIGUR „Spaniel"
(Laufender Spaniel)
ca. 1934

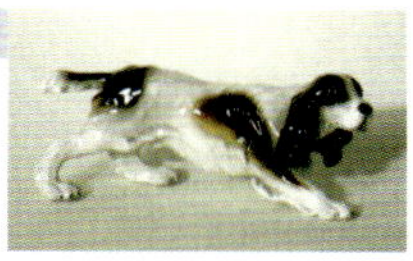

7062F, H. 14 cm, B. 30 cm; GSNachf. 5,8g; Fotoarchiv K. Täuber

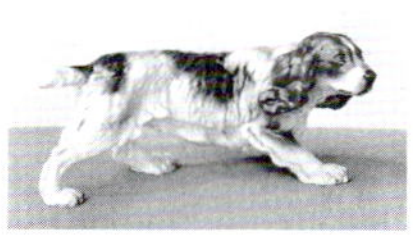

Bem.: Für die Zuordnung dieser Figur und der damit verbundenen Gruppe, danke ich Herrn Herwig Thomasch, dem Sohn von Wilhelm Thomasch.
-> vgl. 7299F; 7330F; 8343F
-> vgl. 7015F; 7281F; 8342F= jeweils GRUPPE
Pendant -> vgl. 7061
Lit.: GSNachf. 5,8 (m. Abb.), hier mit Ang. des Originaltitels FIGUR „Spaniel" (Rückwärtiger Hund aus der GRUPPE 7015).
Quelle: Fotoarchiv K. Täuber

7063AF
Herczeg, Claire (Klára)
FIGUR „Furchtsame" (Stehendes kleines Mädchen mit Schirm) (auf Ovalsockel)
ca. 1935
7063,386,4 gem. SG, Firmenmarke 19, Made in Austria, unsign., H. 20 cm; eBay # 2292786165
7063,389,4 gem. lig. JR, Firmenmarke 19, Made in Austria, H. 19 cm; Wien Museum Inv. 220.725
7063AF, H. 20 cm, B. 8 cm; GSNachf. 1,1a; Fotoarchiv K. Täuber, hier mit der Ang.: H. 22 cm, B. 11 cm
Bem.: Diese Figur gelangte 1935 erstmalig in den Verkauf (Brief von GS an die Wiener Firma Keramos, vom 25.2.1953).
-> vgl. 7063F (Mädchen mit Hund)
Lit.: GSNachf. 1,1 (m. Abb.), hier mit Ang. des Originaltitels „Furchtsame"; Hoever (1996).
Quelle: AGM (Bestellung von Ferd. Lundquist (Göteborg, Schweden), vom 23. August 1950, hier mit Ang. des Originaltitels „Furchtsame"; Brief von der Wiener Manufaktur Goldscheider an die Firma Keramos, vom 25.2.1953); Fotoarchiv K. Täuber.

7063F
Herczeg, Claire (Klára)

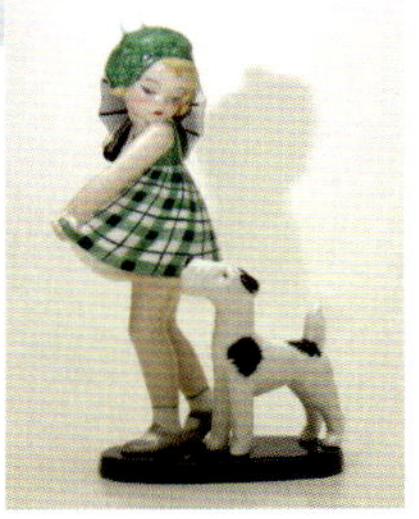

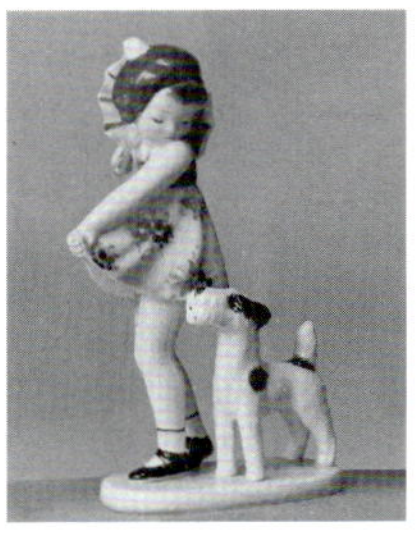

FIGUR (bzw. GRUPPE)
„Furchtsame" (Stehendes kleines Mädchen mit Schirm und Foxterrier) (auf Ovalsockel)
ca. 1935
7063,17 gem. HK, Firmenmarke 19, Made in Germany, unsign., Ausformung zw. 1938–1941, H. 20,5 cm; eBay # 3328933231
7063,66 gem. E, gem. rot X, Firmenmarke 19, Made in Austria, unsign., Ausformung zw. 1945–1953, H. 21,6 cm; eBay # 502138893
7063,342,? Firmenmarke 19, Made in Austria, unsign.?, H. 20 cm; Bonhams 25.2.2003
7063,389,4 gem. lig. JR, Firmenmarke 19, Made in Austria, unsign., H. 19,5 cm; Wien Museum Inv. 220.725
7063,867,5 gem. AN, Firmenmarke 19, Made in Germany, sign. G. Bouret, Ausformung zw. 1938–1941, H. 20 cm; Sammlung Bernard Amigon, Frankreich
7063F, H. 20 cm, B. 14 cm; GSNachf. 1,1b
-> vgl. 7063AF (Mädchen ohne Hund)
Lit.: GSNachf. 1,1 (m. Abb.), hier mit Ang. des Originaltitels „Furchtsame".

7064
Lorenzl, Josef
Reisende (Stehende junge Dame mit Hut, zwei Koffern und Hutschachtel)
ca. 1935
7064,74,14 gem. E, Firmenmarke 19, Made in Austria, H. 27,5 cm; eBay # 320020887268
Lit.: Pinhas, S. 131 (m. Abb.)

7065
Goebel, Kurt
Julia (Stehendes schüchternes Mädchen) (auf Rundsockel)
ca. 1935
7065 (Firmenmarke ohne Ang.), unsign., H. 27,5 cm; Pinhas, S. 177 (m. Abb.)
Bem.: Der Titel „Julia" liegt auf der Hand. Ist diese Figur doch als Gegenstück zum „Romeo" (Modell-Nr. 7656) zu sehen.
-> vgl. 7657

7066F
Schmidt-Kestner, Erich
FIGUR „Setter" (rückwärts)
ca. 1904/05 (für den Entwurf)
7066,29 ger., Firmenmarke 19, unsign., Ausformung zw. 1945–1953, H. 22,8 cm, L. 50 cm; eBay # 191554447
7066F, H. 21 cm, B. 47 cm; GSNachf. 5,9f
-> vgl. 6648
-> vgl. 6659; 7100= jeweils GRUPPE
Lit.: GSNachf. 5,9 (m. Abb.), hier mit Ang. des Originaltitels „Setter" (rückwärts).
Quelle: AGM (Rechnung an Maurice Mizrahi, Alexandria, Ägypten, vom 22. April 1953), hier mit Ang. des Originaltitels „Setter".

7069
Meisinger, Ida
Sitzender Terrier mit Damenschuh
ca. 1935
7069 H. 10 cm, L. 9 cm
Quelle: Fotoarchiv K. Täuber

7070
Meisinger, Ida

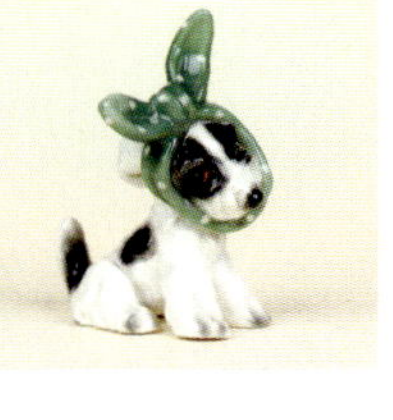

„Zahnweh" (Sitzender Terrier mit Kopftuch)
ca. 1935
7070,I,4,5 Firmenmarke 19, unsign., H. 13,5 cm; Schloss Hag. 37
7070,10,51 Firmenmarke 19, Made in Austria, unsign., H. 13,3 cm; eBay # 2012158119
7070,13 Firmenmarke 19, Made in Austria, Hand dec., unsign., Ausformung zw. 1945–1953, H. 13 cm; PB Wien 19
Bem.: s. 6449
Quelle: AGM (Kommissionsfaktura der „Österreichischen Gruppe-Austria" für die Messe in Utrecht vom 26.2.1953), hier mit Ang. des Originaltitels „Zahnweh".

7074
Matula, Viktor
FIGUR Pferd
ca. 1935
7074 H. fehlt
Quelle: Nachlass Matula, Wien

7075
Matula, Viktor (zugeschr.)
Sitzender Foxterrier
ca. 1935
7075,(18,3?) Firmenmarke 19, unsign., H. 15,2 cm; eBay # 293326141

7077
Matula, Viktor (zugeschr.)
Liegender Foxterrier
ca. 1935
7077,66,19 gem. N, Firmenmarke 19, Made in Austria, unsign., H. 8,9 cm; eBay # 109114520

7078F
Dakon, Stephan

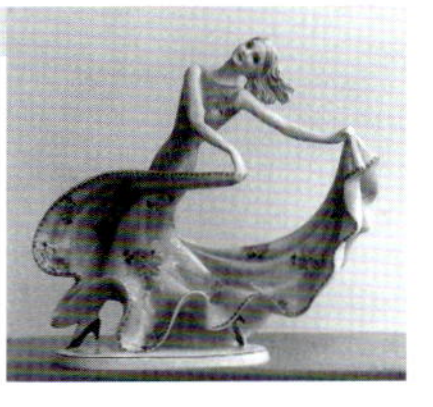

Tänzerin in schwungvoller drehender Bewegung (auf Ovalsockel)
ca. 1935
7078,66,52 Firmenmarke 19, Made in Austria, unsign., H. ca. 25 cm; eBay # 39525991
7078,210,11 gem. lig. GH., Firmenmarke 19, Made in Austria, sign. Dakon, Pick. 2, H. 25 cm; eBay # 2009162777
7078,216,31 Firmenmarke 19, Made in Austria, sign. Dakon, H. 24,4 cm; Quit. 21
7078,272,2 (Firmenmarke ohne Ang.), sign. Dakon, H. 25,2 cm; Christie's SK 15.5.2003
7078F, H. 25 cm, B. 26 cm
-> vgl. 8136
Lit.: Pinhas, S. 159 (m. Abb.)
Quelle: Archiv Mimi Dakon (Originalfoto)

7081
Knörlein, Rudolf (zugeschr.)

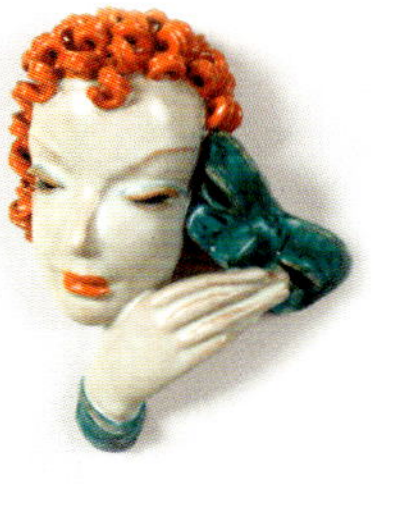

WANDMASKE/STEHMASKE Frauenkopf mit Hand und Maske
ca. 1935
WANDMASKE
7081,3 gem. lig. AH, Firmenmarke 19, Made in Germany, unsign., Ausformung zw. 1938–1941, H. 28,8 cm; Sammlung.schmutz@chello.at 94
7081,7 gem. S41, Firmenmarke 19, Made in Austria, unsign., Ausformung zw. 1945–1953, H. 27,5 cm; Sammlung.schmutz@chello.at 68
7081,S ger., gem. Z, Firmenmarke 19, Made in Austria, unsign., Ausformung zw. 1945–1953, H. 28,2 cm; Sammlung.schmutz@chello.at 41
7081 (Firmenmarke ohne Ang.), unsign., H. 28 cm; Christie's SK 4./5.12.1996
STEHMASKE (auf Holzsockel)
7081,1 gem. FO, Plak. C, unsign., Ausformung zw. 1945–1953, Gesamth. 32,8 cm; Doro 21.9.1999

7087
WANDMASKE Frauenkopf (gewelltes Haar) mit Hand und Fächer
ca. 1935
7087,31 Firmenmarke 19, Made in Austria, Ausformung zw. 1945–1953, H. 27 cm; Soth. L. 20.10.1989

7088
STEHKOPF Frauenkopf mit zurück gekämmtem Haar (auf Ovalsockel)
ca. 1935
7088 Firmenmarke 19, H. 12,7 cm; eBay # 250005839174

7089

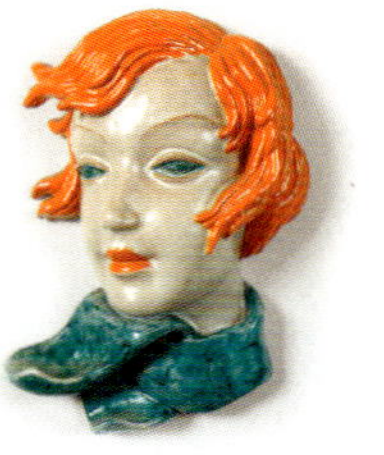

WANDMASKE Frauenkopf mit wehendem Haar und geknotetem Schal
ca. 1935
7089,12 Firmenmarke 19, H. 28,3 cm; Doro 15.9.2003
7089,69 Firmenmarke 19, H. 29 cm; KH
7089 gem. RK, Firmenmarke 19, Made in Austria, H. 30,5 cm; eBay # 115779717
7089 gem. S.P., Firmenmarke 19, Made in Austria, H. 28,5 cm; Sammlung.schmutz@chello.at 14
7089 gem. lig. TK, Firmenmarke 19, Made in Austria, H. 28,5 cm; Wien Museum Inv. 186.664
Bem.: Diese Maske wurde ebenfalls zw. 1953–1960 bei Carstens (GWM West-Germany) produziert.
Lit.: Könecke (2002), S. 76, 86 (m. Abb.).

7091
Matula, Viktor
FIGUR Reh
ca. 1935
7091 H. fehlt
Quelle: Nachlass Matula, Wien

7093
Matula, Viktor
FIGUR Katze
ca. 1935
7093 H. fehlt
Quelle: Nachlass Matula, Wien

7094F

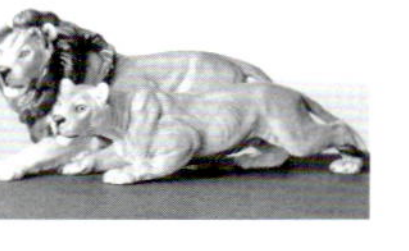

GRUPPE Löwenpaar
ca. 1935
7094,1,92 gem. F., Firmenmarke 19, Made in Austria, Hand dec., H. 24,6 cm, L. 54,5 cm; Wien Museum Inv. 190.031
7094,6 Firmenmarke 19, H. 25 cm, L. 55 cm; Wien Museum Inv. 220.764
7094,64,9 Firmenmarke 19, H. 25 cm, L. 55 cm; eBay # 2058532139
7094,163,7 Firmenmarke 19, Made in Germany, Ausformung zw. 1938–1941, H. 25 cm, L. 55 cm; KH
7094F, H. 51 cm, B. 26 cm; GSNachf. 5,21a
-> vgl. 8533F
-> vgl. 7094= jeweils EinzelFIGUR Löwe/Löwin
Lit.: GSNachf. (2. Ausg., bzw. Erg. Bl.) 5,21 (m. Abb.)
Quelle: AGM (Bestellungen der Firma Wahliss (Wien) vom 9. und 24.11.1945 an die „Arbeitsgemeinschaft der Firma Friedrich Goldscheider"), hier jeweils mit Ang. des Titels „Löwengruppe"; Bestellung der Firma Wahliss (Wien) vom 27.2.1946 in Ausführung „Terrakotta"; Fotoarchiv K. Täuber.

7094
FIGUR Löwe
ca. 1935
7094,8 ger. XX, gem. LS, Firmenmarke 19, Made in Austria, Hand dec., Ausformung zw. 1945–1953, H. 25 cm, L. 48 cm; Wien Museum Inv. 220.754
-> vgl. 7094F= GRUPPE

7094
FIGUR Löwin
ca. 1935
7094,8,74 Firmenmarke 19, Made in Austria, Hand dec., H. 15 cm, L. 45 cm; Wien Museum Inv. 220.762
7094,255,7 ger., Firmenmarke 19, Made in Germany, Ausformung zw. 1938–1941, H. 14,6 cm, L. 47,2 cm; KNS 12

7095
Matula, Viktor
FIGUR Kamel
ca. 1935
7095 H. fehlt
Quelle: Nachlass Matula, Wien

7098
Dakon, Stephan
Stehende junge Dame mit großem Hut, ihr Kleid mit der linken Hand anhebend (auf Ovalsockel)
ca. 1935
7098 (Firmenmarke ohne Ang.), H. 40,5 cm; Pinhas, S. 150 (m. Abb.)

7100F
Schmidt-Kestner, Erich
GRUPPE „Setter"
ca. 1904/05 (für den Entwurf)
7100F, H. 21,5 cm, B. 46 cm; GSNachf. 5,9b
-> vgl. 6659
-> vgl. 3024; 6135= jeweils EinzelFIGUR (vorne)
-> vgl. 6648; 7066= jeweils EinzelFIGUR (rückwärts)
Lit.: GSNachf. 5,9 (m. Abb.), hier mit Ang. des Originaltitels „GRUPPE Setter".

7101
GRUPPE Drei sitzende Terrier
ca. 1935
7101,129,6 Firmenmarke 19, Made in Austria, H. 15,2 cm; eBay # 7410468973

7102
WANDMASKE Frauenkopf mit Hand und kleinem Vogel (Kohlmeise)
ca. 1935
7102 Firmenmarke 19, Made in Austria, H. 32 cm; Doro 20.4.1993
7102 Firmenmarke 19, Made in Austria, H. 31 cm; Sammlung.schmutz@chello.at 42
7102 Firmenmarke 19, Pick. 3, H. 31,5 cm; Soth. L. 30.11.1983
Lit.: Pinhas, S. 164 (m. Abb.)

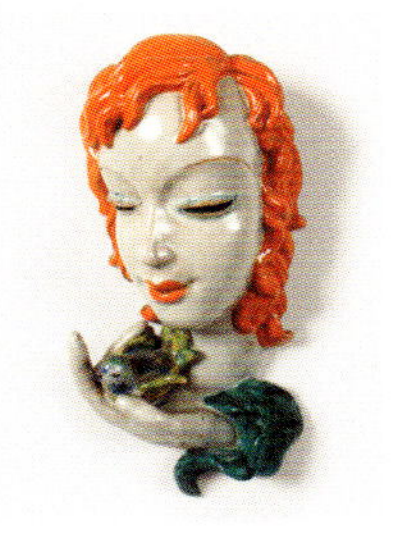

7104
SCHALE
Form: länglich, gemuldete Segmente
Dekor: Streifen, rot/weiß/schwarz
ca. 1935
7104 gem. lig. TK, Firmenmarke 19, Made in Austria, L. 31,2 cm, B. 23,8 cm; PB
7104,26 gem. lig. AH, Firmenmarke 19, Made in Austria, Ausformung zw. 1945–1953, L. 30 cm, B. 23,2 cm; GK, S. 22, 45 (m. Abb.); Wien Museum Inv. 223.912
7104,26 gem. B, Firmenmarke 19, Made in Austria, Ausformung zw. 1945–1953, H. 7,5 cm, L. 32 cm, B. 23,6 cm; MAK Wien, Inv. Ke 10602-74
Ausst.: HMW (jetzt Wien Museum) 1985/86; PSK 1995.
Lit.: GK, S. 22, 45 (m. Abb.); ADGS, Nr. 14 (m. Abb.); GSN, S. 58, 59 (m. Abb.)
GS neu: Modell-Nr. 86 „Nur eine Schale" (L. 35 cm), Aufl. ohne Ang.

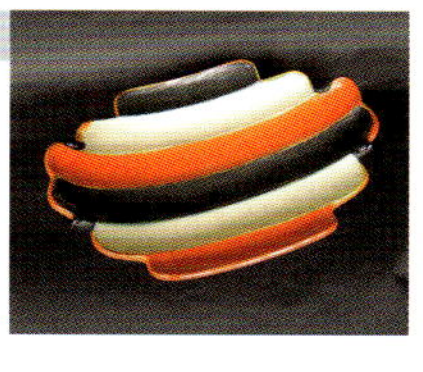

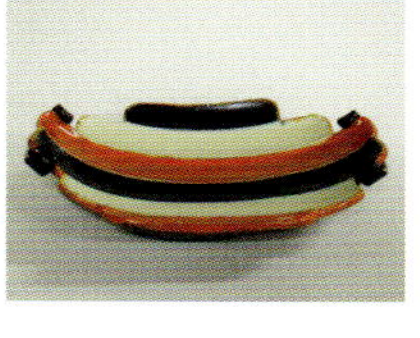

7105
SCHALE
Form: fast quadratisch, mehrpassig, sternförmig gemuldete Segmente
Dekor: Segmente, rot/weiß/schwarz
ca. 1935
7105,13 gem. P, Firmenmarke 19, Made in Germany, Ausformung zw. 1938–1941, H. 5,5 cm, L. 29 cm, B. 28 cm; WKA 27

7106
Matula, Viktor
FIGUR Springbock
ca. 1935
7106 H. fehlt
Quelle: Nachlass Matula, Wien

7107M
Matula, Viktor
FIGUR „Esel"
ca. 1935
7107M, H. 6 cm, B. 6 cm; GSNachf. 3,12a
Lit.: GSNachf. 3,12 (m. Abb.), hier mit Ang. des Originaltitels „Esel".
Quellen: Nachlass Matula, Wien; AGM (Kontoblatt der Wiener Manufaktur Josef Schuster für die Firma Porslins Ericsson-Lund, Stockholm?, Schweden vom 7. Januar 1943), hier mit Ang. 12 „Kleintiere".

7109M
Matula, Viktor
FIGUR „Pferd"
ca. 1935
7109M, H. 6 cm, B. 8 cm; GSNachf. 3,12c
Lit.: GSNachf. 3,12 (m. Abb.), hier mit Ang. des Originaltitels „Pferd".
Quelle: Nachlass Matula, Wien

7110
Matula, Viktor
Stehendes Reh (auf flachem Rechtecksockel)
ca. 1935
7110, 5 ger., Firmenmarke 19, Made in Germany, Ausformung zw. 1938–1941, H. 13 cm; eBay # 6532975837
Quelle: Nachlass Matula, Wien

7114
WANDMASKE Frauenkopf mit Hundekopf (Skyeterrier)
ca. 1935
7114 gem. R, Firmenmarke 19, Made in Austria, H. 24,3 cm, Ø 17,7 cm; Sammlung.schmutz@chello.at 92

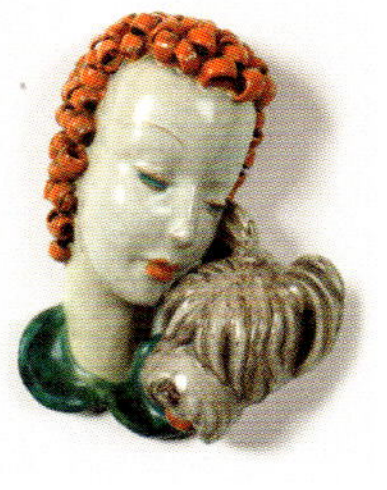

7119
WANDMASKE Frauenkopf mit Hand und Rosenblüte
ca. 1935
7119,8 gem. B, Firmenmarke 19, Made in Austria, Ausformung zw. 1945–1953, H. 28 cm; eBay # 2060439029
7119 Firmenmarke 19, Made in Austria, Pick. 2, H. 27 cm; Sammlung Stressless, Portugal
7119 Firmenmarke 19, Made in Austria, Pick. 2, H. 28 cm; Soth. L. 20.10.1989

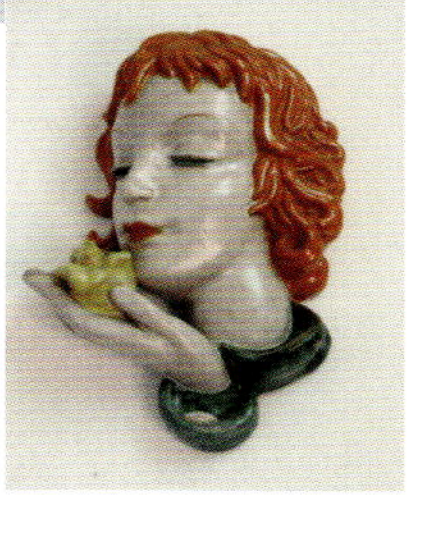

7120
WANDMASKE Frauenkopf mit kurzem Kragenansatz
ca. 1935
7120 gem. 15, Firmenmarke 19, Made in Austria, H. 19,7 cm; Soth. com. # 3J5HF

7121M
Knörlein, Rudolf
WANDMASKE „Maske" (Frauenkopf mit Hand und Halskette)
ca. 1935/36
7121,5 ger., Firmenmarke 19, Made in Austria, unsign., Ausformung zw. 1945–1953, H. 32 cm; Sammlung Stressless, Portugal
7121,33 gem. lig. AH, Firmenmarke 19, Made in Austria, unsign., Ausformung zw. 1945–1953, H. 32 cm; GK, S. 22, 117 (m. Abb.); Wien Museum Inv. 223.913
7121,96 Firmenmarke 19, unsign., Ausformung zw. 1945–1953, H. 32 cm; Sammlung a@goldscheider.net
7121 (Firmenmarke ohne Ang.), unsign., H. 32 cm; Christie's SK 12.9.1997; 2.4.1998
7121M, H. 32 cm, B. 18 cm; GSNachf. 2,2
Bem.: Durch die eigenhändige Datierung Knörleins (1936) auf dem Trägerblatt des Originalfotos kann die Entstehung um 1935/36 ziemlich exakt bestimmt werden.
Ausst.: HMW (jetzt Wien Museum) 1985/86.
Lit.: GSNachf. 2,2 (m. Abb.), hier mit Ang. des Originaltitels „Maske"; GK, S. 22, 117 (m. Abb.).
Quelle: Privatarchiv (Originalfoto)

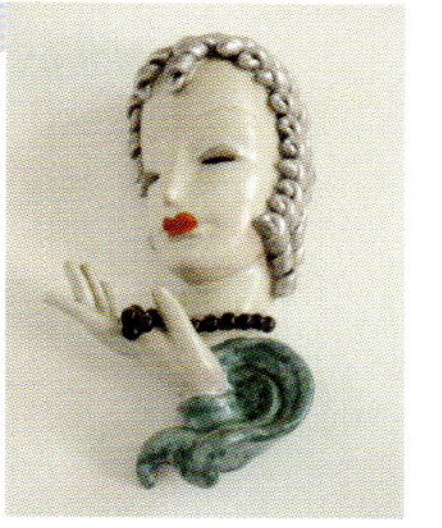

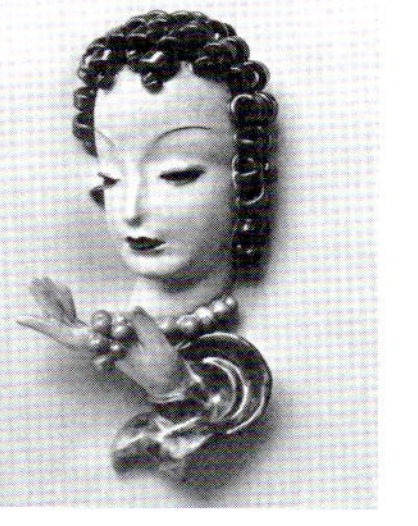

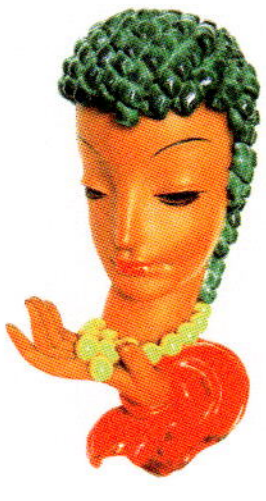

7122
Matula, Viktor
FIGUR Pferd
ca. 1935
7122 H. fehlt
Quelle: Nachlass Matula, Wien

7123F
GRUPPE „Hirschkuh mit Jungem" (auf Ovalsockel)
ca. 1935
7123F, H. 27 cm, B. 29 cm; GSNachf. 5,16
Lit.: GSNachf. 5,16 (m. Abb.), hier mit Ang. des Originaltitels GRUPPE „Hirschkuh mit Jungem".

7128
Matula, Viktor
ASCHER Elefanten
ca. 1935
7128 H. fehlt, B. fehlt
Quelle: Nachlass Matula, Wien

7143F
Postl, Hugo
FIGUR „Windhund" (Liegender Barsoi)
ca. 1929 (für den Entwurf)
7143F, H. 7 cm, B. 20 cm; GSNachf. 5,2c
-> vgl. 5974F; 6422F
Lit.: MW, 11.1929/30, Nr. 17, S. III (m. Abb.), hier mit Ang. des Originaltitels „Windhund" und des Entwerfers Hugo Postl; GSNachf. 5,2 (m. Abb.), hier mit Ang. des Originaltitels „Windhund".

7144
Postl, Hugo
Sitzender Spaniel
ca. 1933 (für die Ausführung)
7144,22 gem. 94, Firmenmarke 19, unsign., Pick. 1, H. 11,4 cm; eBay # 3702737539
-> vgl. 6715

7149F
Dakon, Stephan
FIGUR „Skiläuferin" (Stehende Skiläuferin mit Skistock) (auf Ovalsockel)
ca. 1935
7149,2,34 gem. W., Firmenmarke 19, Made in Austria, unsign., H. 29,2 cm; eBay # 1242415092
7149,9 ger, I, gem. rot G, Firmenmarke 19, Made in Austria, unsign., Ausformung zw. 1945–1953, H. 29,8 cm; MAK Wien, Inv. Ke 10602-60
7149,11,? Firmenmarke 19, H. 30 cm; Doro 7.3.2001
7149,36,11 Firmenmarke 19, unsign.?, H. 27,5 cm; Doro 21.1.1998 (o. Kat.)
7149,92,19 Firmenmarke 19, Made in Austria, H. 28 cm; Soth. Chester 23.–25.5.1989
7149,183,2 gem. H., SG 2, Firmenmarke 19, Made in Austria, Originalpick.: „Ewald Frey, Mainz", H. 28 cm; eBay # 1410883326
7149,195,14 Firmenmarke 19, Made in Austria, Pick. 2, H. 29 cm; Kett. 192
7149,263,16 gem. F, N, H, Firmenmarke 19, Made in Austria, H. 28 cm; Wien Museum Inv. 220.702
7149,618,31 gem. W, gem. lig. JR, Firmenmarke 19, Made in Austria, H. 30 cm; eBay # 6522625981
7149,646,19 Firmenmarke 19, Made in Germany, Ausformung zw. 1938–1941, H. 30,5 cm; eBay # 5400262352
7149F, H. 29 cm; GSNachf. 1,10a
-> vgl. 8157
Lit.: GSNachf. 1,10 (m. Abb.), hier mit Ang. des Originaltitels „Skiläuferin".

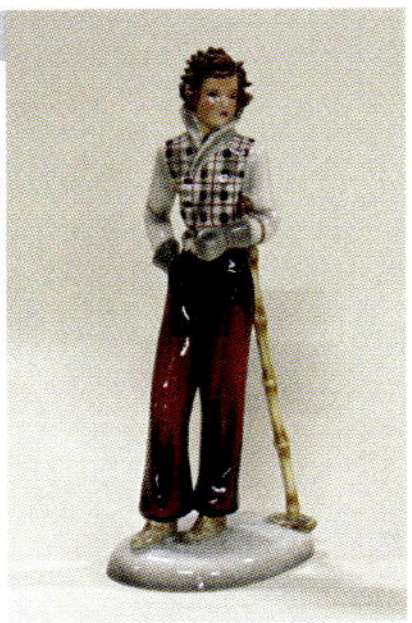

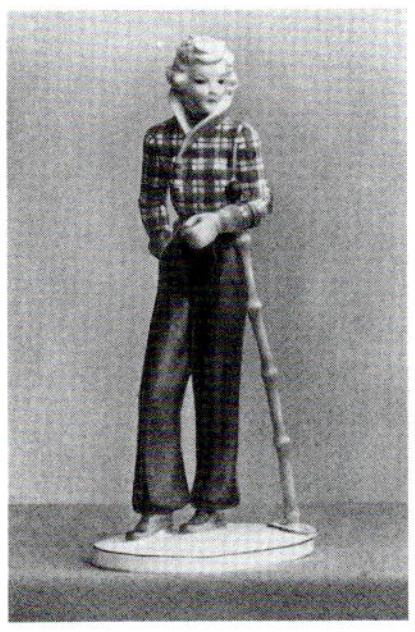

7154M
Matula, Viktor
FIGUR „Rehbock"
ca. 1935
7154M, H. 9 cm, B. 7 cm; GSNachf. 3,12e
Lit.: GSNachf. 3,12 (m. Abb.), hier mit Ang. des Originaltitels „Rehbock".
Quelle: Nachlass Matula, Wien

7155
Prischl, Adolf
WANDMASKE Frauenkopf mit Hut
ca. 1935
7155 Firmenmarke 19, unsign., H. 22 cm, B. 15 cm; Sammlung mad

7157
Matula, Viktor
FIGUR Pferd
ca. 1935
7157 H. fehlt
Quelle: Nachlass Matula, Wien

7158M
Matula, Viktor
GRUPPE „Pferde" (Zwei stehende Pferde)
ca. 1935
7158M, H. 12 cm, B. 13 cm; GSNachf. 3,19c
Lit.: GSNachf. 3,19 (m. Abb.), hier mit Ang. des Originaltitels GRUPPE „Pferde".
Quelle: Nachlass Matula, Wien

7165
WANDMASKE Frauenkopf mit Huf und Schleife/ Mascherl
ca. 1935
7165,20 gem. lig. AH, Firmenmarke 19, Made in Austria, Ausformung zw. 1945–1953, H. 22 cm; PB Wien 19

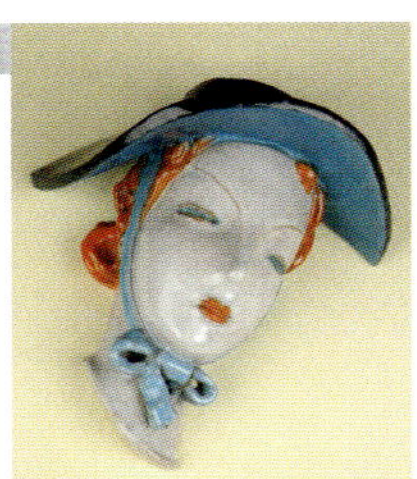

7166
Knörlein, Rudolf
RELIEF Stehendes Mädchen mit zwei Vögeln
ca. 1935
7166,65 Firmenmarke 19, Made in Austria, Monogr. RK, Ausformung zw. 1945–1953, H. 36,3 cm, B. 15 cm; Sammlung John Nash
Pendant -> vgl. 7167

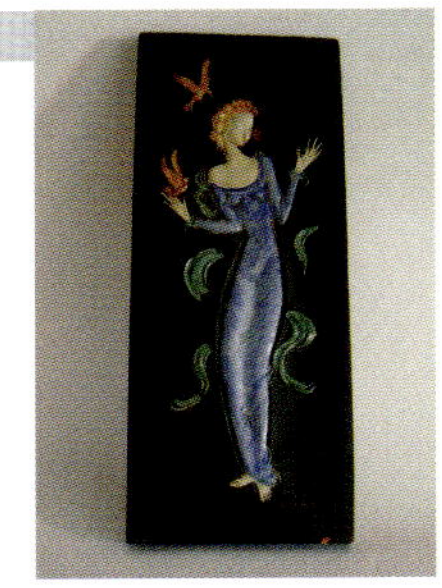

7167
Knörlein, Rudolf
RELIEF Stehender Mädchenakt
ca. 1935
7167,20 gem. lig. TK, Firmenmarke 19, Made in Austria, Monogr. RK, Ausformung zw. 1945–1953, H. 36,2 cm, B. 15,2 cm; eBay # 858716164
Pendant -> vgl. 7166

7171
Stehender Mädchenakt
ca. 1935
7171 (ohne Ang.), Firmenmarke 19, Made in Austria, H. 21,3 cm; Christie's Genf 12.11.1989
Bem.: Hier keine nähere Ang. zum Objekt möglich, da im Auktionskat. weder nähere Beschreibung, bzw. keine Abb. enthalten ist.

7172F
Dakon, Stephan
Sitzende junge Dame mit kleinem Hut (auf zwei Reisekoffern)
ca. 1935
7172,33,31 gem. SA Firmenmarke 19, Made in Austria, H. 25,4 cm; Soth. Chester 23.–25.5.1989
7172,47,19 Firmenmarke 19, Made in Austria, H. 27 cm; PB Wien 19
7172F, H. 26 cm, B. 10 cm
Quelle: Archiv Mimi Dakon (Originalfoto)

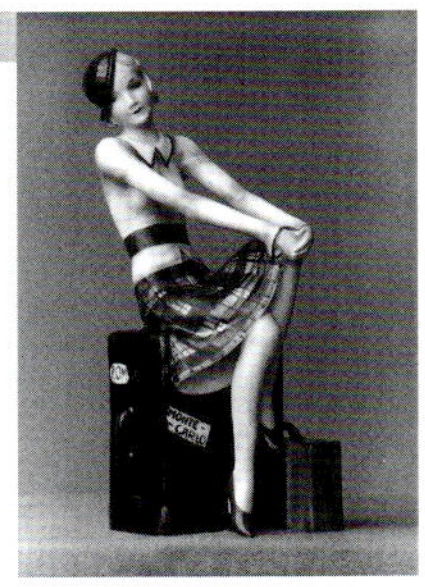

7176
WANDMASKE Frauenkopf mit Hut und Halstuch
ca. 1935
7176,8 ger., Firmenmarke 19, Made in Austria, Ausformung zw. 1945–1953, H. 31 cm; Sammlung Stressless, Portugal
7176,65 (Firmenmarke ohne Ang.), Ausformung zw. 1945–1953, H. 30 cm; Doro 30.11.1999
7176 (Firmenmarke ohne Ang.), H. 30,5 cm, B. 27 cm; Soth. L. 30.9.1983

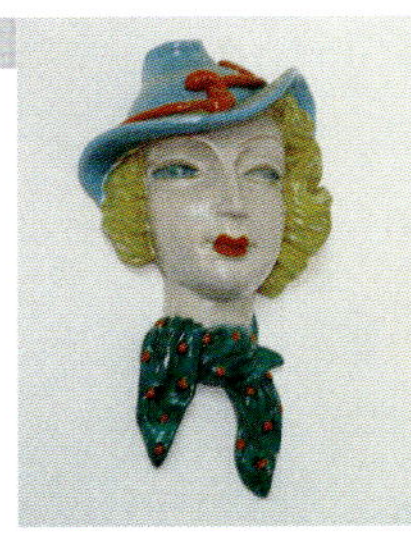

7182
Matula, Viktor
ASCHER
ca. 1935
7182 H. fehlt, B. fehlt
Quelle: Nachlass Matula, Wien

7183
RELIEF Madonna mit Kind
ca. 1935
7183,20 ger., gem. T, Firmenmarke 19, Made in Austria, Ausformung zw. 1945–1953, H. ca. 27 cm, B. ca. 21 cm; KH

7185M
Matula, Viktor
RELIEF „Pferd" (Springendes Pferd nach rechts)
ca. 1935
7185,20 gem. H, Firmenmarke 19, Made in Austria, unsign., Ausformung zw. 1945–1953, L. 30,5 cm; eBay # 2049377083
7185M, H. 16 cm, B. 30 cm; GSNachf. 3,19a
Lit.: GSNachf. 3,19 (m. Abb.), hier mit Ang. des Originaltitels RELIEF „Pferd".
Quelle: Nachlass Matula, Wien

7186
Meisinger, Ida
„Zeitungsleser" (Sitzender, in Zeitung lesender Terrier mit Brille)
ca. 1935
7186,5,14 gem. M, Firmenmarke 19, Made in Austria, Hand dec., unsign., H. 14 cm; eBay # 1205571285
7186,5,G gem. M, Firmenmarke 19, Made in Austria, Hand dec., unsign., H. 14,1 cm; Sammlung a@goldscheider.net
7186 ger., Firmenmarke 19, Made in Germany, unsign., Ausformung zw. 1938–1941, H. 14,5 cm; eBay # 1206139709
Bem.: s. 6449
Quelle: AGM (Bestellung der Firma Wahliss (Wien) hier mit Ang. des (Original?-)Titels „Zeitungsleser").

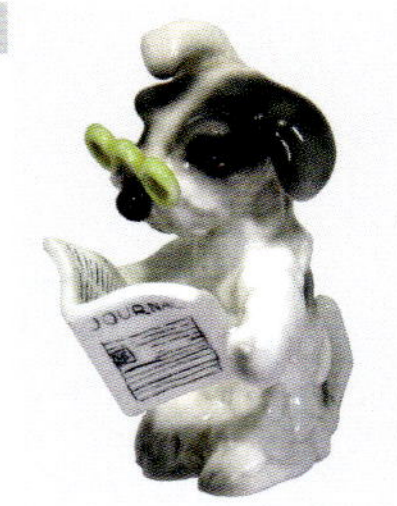

7187
Thomasch, Wilhelm (zugeschr.)
GRUPPE Zwei Musikanten (Akkordeonspieler und Sängerin als bekleidetes Terrierpärchen) (auf flachem Rechtecksockel)
ca. 1935
7187,0 gem. lig. HS, Firmenmarke 19, Made in Austria, unsign., Ausformung zw. 1945–1953, H. 15,5 cm; eBay # 3714561053
7187,2,17 Firmenmarke 19, Made in Austria, H. 15,5 cm, B. 12,5 cm; Sammlung mad
7187,?6 ger. I, Firmenmarke 19, Made in Austria, unsign., Ausformung zw. 1945–1953, Originalpick. „Rasper & Söhne, Salzburg-Wien", H. 15 cm; eBay # 2280831452; Bem.: Die Zahl(en) vor der „6" sind durch das Originalpick. überklebt.
7187 H. 15 cm, B. 11 cm; Fotoarchiv K. Täuber
Bem.: Für die Zuschreibung an Wilhelm Thomasch danke ich Frau Mag. Uta Matschiner.
Quelle: Fotoarchiv K. Täuber

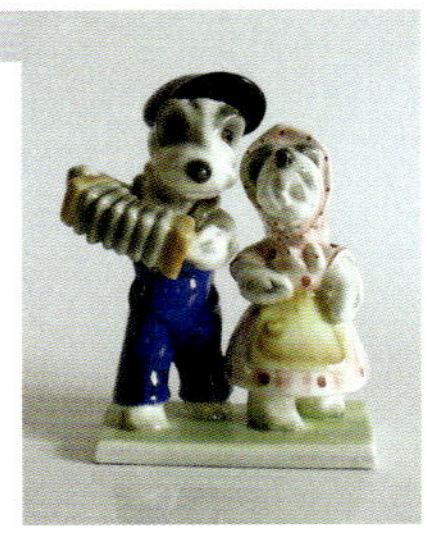

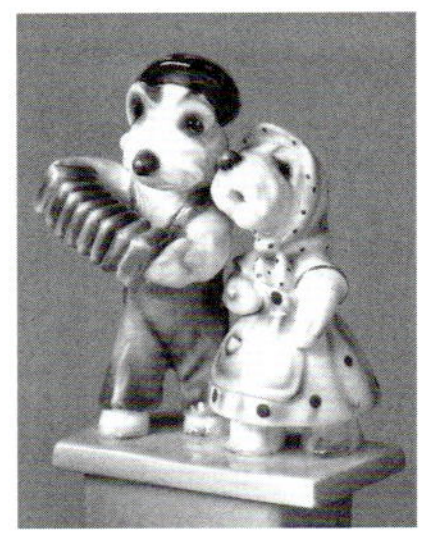

7188F
Knörlein, Rudolf
BÜSTE „Ave Maria" (Madonna mit gefalteten Händen)
ca. 1934/35
7188,9,65 gem. lig. AH, Firmenmarke 19 (zwei mal), Made in Austria (zwei mal), unsign., H. 25,5 cm; MAK Wien, Inv. Ke 10602-50
7188,72,65 Firmenmarke 19, Made in Austria, unsign., H. 26 cm, B. 12 cm; eBay # 146494670
7188,152,(65?) Firmenmarke 19, Pick. 2, Pick. 8, H. 25,4 cm; eBay # 1204779825
7188,441,6 Firmenmarke 19, Made in Austria, H. 26,3 cm; Doro 20.10.1997; WKA 38
7188,462,20 gem. lig. AS, Firmenmarke 19, Made in Austria, H. 25,7 cm; MAK Wien, Inv. Ke 10602-53
7188,824,7 (Firmenmarke ohne Ang.), unsign., H. 29,5 cm; Christie's SK 21.6.1996

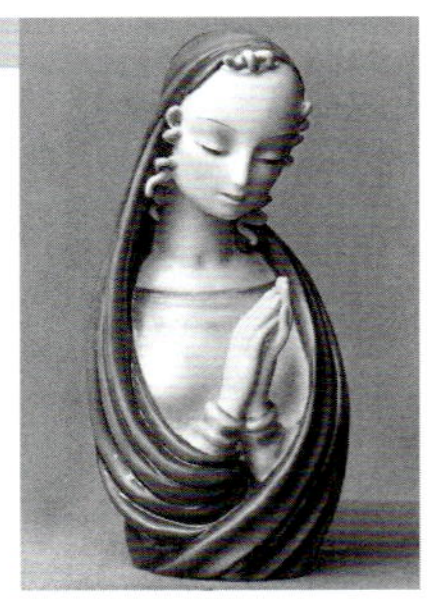

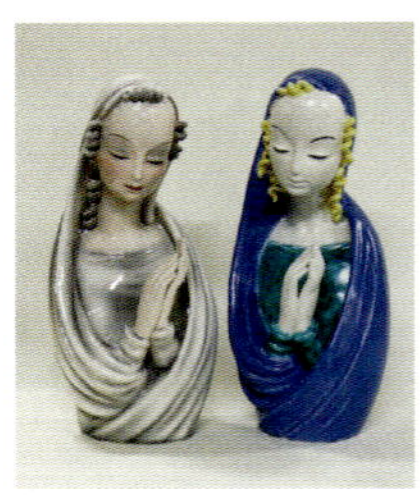

7188,1015,7 gem. K, Firmenmarke 19, Made in Germany, unsign., Ausformung zw. 1938–1941, H. 26,4 cm; eBay # 923147396
7188F, H. 26 cm, B. 12 cm; GSNachf. 6,5a
Bem.: Dieses Modell war sehr erfolgreich, wie alleine das nachgewiesene Exemplar mit der Stücknr. 1015 zeigt. Ein Schreiben Walter Goldscheiders bestätigt „...von Ihren Modellen in diesem Genre haben sich die Maske 6427 und die Madonna 7188 am Besten verkauft" (Brief von Walter Goldscheider an Rudolf Knörlein, vom 4. Juli 1952). Durch die eigenhändige Datierung Knörleins (1934) auf dem Trägerblatt des Originalfotos sowie der Originalzeichnung (Aquarell über Graphit), kann das Jahr 1934 exakt bestimmt werden. Diese Büste wurde auch von GS USA (Everlast Corp.) unter der Modell-Nr. 819 in der H. 21,6 cm produziert.

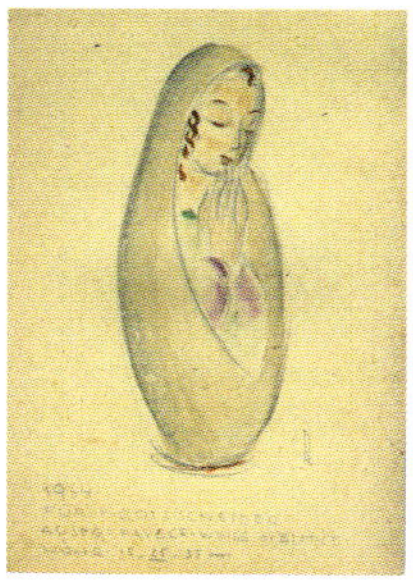

-> vgl. 7263; 7858; 8743
Lit.: GSNachf. 6,5 (m. Abb.), hier mit Ang. des Originaltitels BÜSTE „Ave Maria".
Quelle: Privatarchiv (Originalfoto); Privatsammlung D (Originalzeichnung).

7190

WANDMASKE Frauenkopf mit Kappe / Hut und Schal / Halstuch
ca. 1935
7190 Firmenmarke 19, Made in Austria, H. 29 cm; eBay # 160040499935
Lit.: Pinhas, S. 174 (m. Abb.)

7191

Weiss, Claire (Klára) (d.i. Herczeg, Claire (Klára)
Stehender Junge mit Golfsack
ca. 1935
7191,7 (Firmenmarke ohne Ang.), Ausformung zw. 1945–1953, H. 23,2 cm; Doro 20.3.2006
7191,(8?),(9?) ger., Firmenmarke 19, Made in Austria, unsign.?, H. 23 cm; eBay # 334559300
7191,154,3 gem. H, Firmenmarke 19, Made in Austria, sign. Claire Weiss, H. 23,5 cm; Schloss Hag. 34; Herr 53
ohne Ang. gem. W4, Firmenmarke 19, Made in Austria, sign. Claire Weiss, H. 22,9 cm; eBay # 60854974; 1302704241
Pendant -> vgl. 7564

7192F

Lorenzl, Josef
FIGUR „Revue" (Stehende Tänzerin, ihr Kleid auseinanderhaltend) (auf Ovalsockel)
ca. 1935

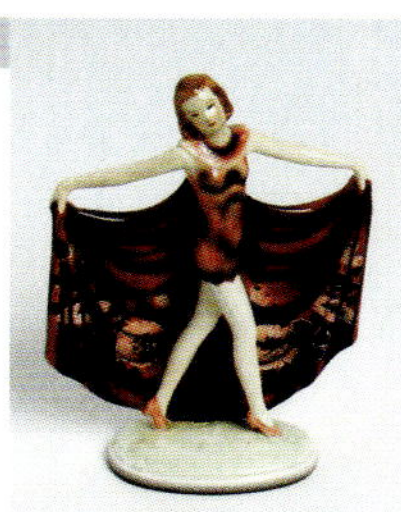

7192,2,30 Firmenmarke 19, Made in Austria, Hand dec., unsign., H. 21,3 cm, B. 16 cm; city-antik
7192,16,16 Firmenmarke 19, Made in Austria, Hand dec., unsign., Pick. 3, Originalpick.: „Karl Skala, Wien XX., Wallensteinstr. 7", H. 22 cm; city-antik

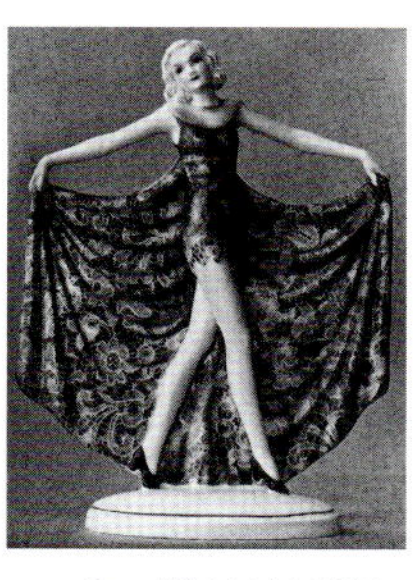

7192,55,10 gem. H, Firmenmarke 19, Made in Austria, sign. Lorenzl, Pick. 3, H. 20,3 cm; eBay # 2632258654
7192,248,6 gem. blau H, SG, Firmenmarke 19, Made in Austria, sign. Lorenzl, H. ca. 21 cm; eBay # 2014114608
7192,303,19 (Firmenmarke ohne Ang.), sign. Lorenzl, H. 21,5 cm; Doro 12.9.2006
7192,336,31 gem. GH., Firmenmarke 19, Made in Austria, sign. Lorenzl, Pick. 3, H. ca. 21 cm; eBay # 2623657997
7192F, H. 20 cm, B. 16 cm; GSNachf. 1,11a
-> vgl. 7396
Lit.: GSNachf. 1,11 (m. Abb.), hier mit Ang. des Originaltitels „Revue".

7193

Lorenzl, Josef
Stehende Tänzerin in Pose, ihr Kleid mit der rechten Hand zurückhaltend (auf Ovalsockel)
ca. 1935
7193,62,16 Firmenmarke 19, Made in Austria, H. 33 cm; eBay # 194211516
7193,90,31 Firmenmarke 19, Made in Austria, H. 33,5 cm; Doro 18.11.1997
7193,185,14,8 Firmenmarke 19, Made in Austria, H. 35 cm; eBay # 3362367222; Christie's SK 13.10.2004
7193,206,11 ger. XII,A, Firmenmarke 19, Made in Austria, H. ca. 33 cm; eBay # 1500088974
7193,247,16 gem. N, Firmenmarke 19, Made in Austria, gepr. „L"= Z6, Pick. 6, H. 35 cm; eBay # 2071619574
7193,258,10 Firmenmarke 19, Made in Austria, H. 35,4 cm; Christie's L. 19.4.1989
-> vgl. 6492; 7732= jeweils ähnl. Modell.
Lit.: Pinhas, S. 155 (m. Abb.)

7194

Dakon, Stephan
Stehendes Mädchen mit Hut und Foxterrier (auf Ovalsockel)
ca. 1935
7194,2,11 ger. 7, gem. rot S, Firmenmarke 19, Made in Austria, H. 31,7 cm; eBay # 299581032
7194,66,27 gem. G, gem. rot H., Firmenmarke 19, Made in Austria, unsign., H. 32,2 cm; Wien Museum Inv. 190.089
Lit.: GSN, S. 42, 43 (m. Abb.)
GS neu: Modell-Nr. 50 „So bin ich" (H. 27 cm), Aufl. 3000 Stück

7195F

Dakon, Stephan
FIGUR / LAMPE „Beauty" (Tänzerin, nach rechts geneigt, ihr Kleid flügelartig zurückhaltend) (auf Ovalsockel)
ca. 1935

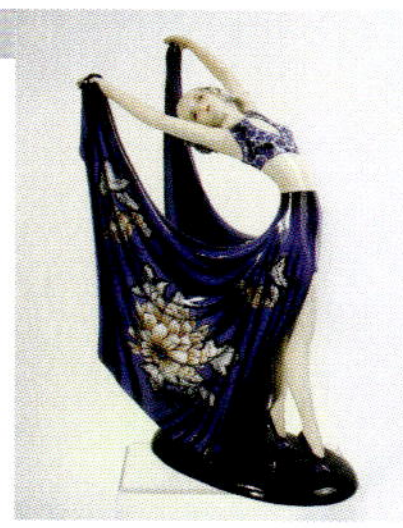

FIGUR
7195,0 ger., gem. blau G, Firmenmarke 19, Made in Austria, unsign., Ausformung zw. 1945–1953, H. 38,1 cm; Wien Museum Inv. 190.068

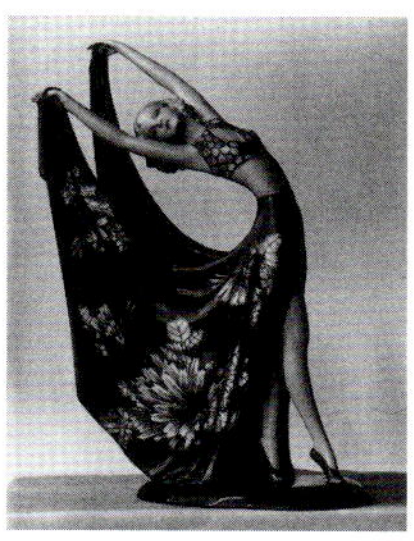

7195,7 (Firmenmarke ohne Ang.), Ausformung zw. 1945–1953, H. ca. 40 cm; Doro S 13.10.2004
7195,52,4 (Firmenmarke ohne Ang.), H. 34,5 cm; Christie's SK 4./5.12.1996
7195,66 gem. B.W., Firmenmarke 19, Made in Austria, unsign., Ausformung zw. 1945–1953, H. ca. 40 cm; eBay # 915124036
7195,108,11 (Firmenmarke ohne Ang.), H. 35,8 cm; Doro 30.11.1999
7195,175,31 Firmenmarke 19, Made in Austria, unsign.?, H. 36,7 cm; Soth. L. 23.10.1987
7195,182,16 Firmenmarke 19, Made in Austria, unsign.?, Pick. 5, H. 37,2 cm; eBay # 1860110216
7195,193,31 Firmenmarke 19, Made in Austria, H. 35 cm; Phillips L. 25.1.2001
7195,231,4 Firmenmarke 19, Made in Austria, H. 36 cm; eBay # 320018814035
7195,284,10 (Firmenmarke ohne Ang.), H. 37,5 cm; Soth. Chester 7.–9.6.1988
7195,306,31 ger. lig. (L./XII?), Firmenmarke 19, Made in Austria, H. 40 cm; WKA 38
7195,424,10,8A gem. S., gem. rot ON, Firmenmarke 19, Made in Austria, H. 36,8 cm; eBay # 1239226952
7195,435,6 Firmenmarke 19, unsign.?, H. 36 cm; eBay # 1443512491
7195,448,4 Firmenmarke 19, Made in Austria, H. 36,2 cm; eBay # 193675182
7195,514,6,8 ger. B., gem. HI, Firmenmarke 19, Made in Austria, H. 36,8 cm; city-antik
7195,582,10 gem. CV, Firmenmarke 19, Made in Austria, H. 36,5 cm; KH
7195,615,14 gem. F., Firmenmarke 19, Made in Austria, H. 36 cm; Quit. 14
7195,697,16 gem. F., Firmenmarke 19, Made in Austria, H. 38,5 cm; Soth. L. 28.9.1984; WKA 35; Doro 15.5.2002
7195,783,6,8 gem. lig. AS, E, S, Firmenmarke 19, Made in Austria, H. 37,5 cm; Sammlung Robert Senft
7195,784,? gem. F.S.#, Firmenmarke 19, Made in Austria, H. 37 cm; eBay # 2588265783
7195,788,2 (Firmenmarke ohne Ang.), H. 39 cm; Soth. Olympia 24.2.2004
7195,911,31 Firmenmarke 19, Made in Austria, H. 38 cm; Doro 1.12.1998
7195,992,10 Firmenmarke 19, Made in Austria, H. 38,6 cm; eBay # 16282215
7195,1023,9 Firmenmarke 19, Made in Austria, H. 40,5 cm; Doro 15.3.2005
7195,1056,16 Firmenmarke 19, Made in Austria, H. 40 cm; WKA 33
7195,1095,65 ger. XII A, gem. SG., Firmenmarke 19, Made in Austria, H. 40 cm; city-antik; Wendl 36
7195,1147,11 Firmenmarke 19, (Made in Austria?), H. 39,5 cm; Herr 35
7195,1210,8 ger., gem. F., Firmenmarke 19, Made in Austria, H. 40 cm; eBay # 2229482935
7195,1213,2 (Firmenmarke ohne Ang.), H. 40 cm; Soth. Olympia 3.5.2002
7195,1303,65 (Firmenmarke ohne Ang.), H. 39,3 cm; Doro 1.6.2005
7195,1364,4 Firmenmarke 19, Made in Austria, H. 39,3 cm; city-antik
7195,1371,4 Firmenmarke 19, Made in Austria, H. 39,4 cm; Doro 21.9.1999
7195,1397,31 gepr. L, Firmenmarke 19, (Made in Austria?), Pick. 1, H. 39,5 cm; BM 28
7195,1798,31 Firmenmarke 19, Made in Germany, Ausformung zw. 1938–1941, H. 39,5 cm; Wendl 41
7195,1812,6 ger. A, Firmenmarke 19, Made in Germany, Ausformung zw. 1938–1941, Originalpick.: „J.C. Herhold, Hannover", H. 40,2 cm; eBay # 1456653552
7195,1835,2 gem. D, Firmenmarke 19, Made in Germany, Ausformung zw. 1938–1941, H. 40,5 cm; city-antik
7195,1863,7 gepr. D, gem. F., Firmenmarke 19, Made in Germany, Ausformung zw. 1938–1941, Pick. 8 (R.M. 60,–), H. 40,3 cm; PB Wien; Bem.: Eines der wenigen Exemplare, bei dem das Originalpreisschild noch erhalten ist. Dieses belegt, dass die Figur um 1940 für 60,– Reichsmark verkauft wurde.
7195,(19476?),19 Firmenmarke 19, unsign.?, H. 38,5 cm; Wendl 17
7195F, H. 36 cm, B. 22 cm; GSNachf. 1,12a; 1,50a
LAMPE

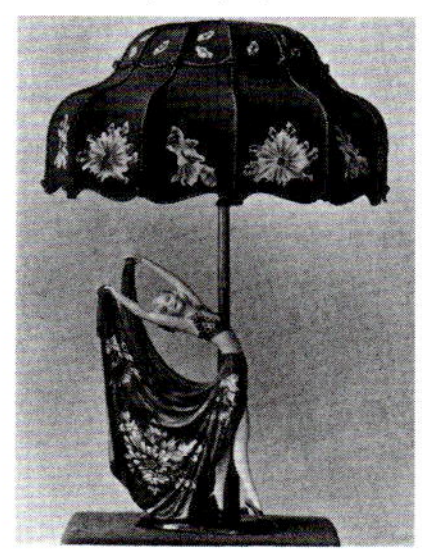

7195,1424,19 Firmenmarke 19, (Made in Austria?), H. (Figur) 38,5 cm; Doro 26.5.1998
7195,1756,0 (Firmenmarke ohne Ang.), Gesamth. 56 cm; Doro 8.3.2004: 15.3.2005
7195F, H. 69 cm, B. 47 cm; GSNachf. 1,12f; 1,50f
Bem.: Dieses Modell gibt es in zwei Varianten, einmal neigt sich der Körper der Tänzerin nach rechts wie in diesem Fall, das andere Mal nach links.
-> vgl. 7897; 7996; 8746
Lit.: GSNachf. 1,12 (m. Abb. als Figur), hier mit Ang. des Originaltitels „Beauty"; GSNachf. 1,50 (m. Abb. als Lampe montiert, komplett, mit Seidenschirm), hier mit Ang. des Originaltitels „Beauty"; Neuwirth (1974), S. 67, hier falsche Ang. des Entwerfers „Lorenzl".
Quelle: AGM (Rechnung an Maurice Mizrahi, Alexandria, Ägypten, vom 22. April 1953), hier mit Ang. des Originaltitels „Beauty"; Archiv Mimi Dakon (Originalfoto).

7199

Dakon, Stephan (zugeschr.)
Tanzende Spanierin mit Kastagnetten (auf Rundsockel)
ca. 1935
7199,11 ger. X, gem. SG, Firmenmarke 19, Made in Austria, unsign., Ausformung zw. 1945–1953, H. 33 cm; GK, S. 22, 122 (m. Abb.) Wien Museum Inv. 222.555
Ausst.: HMW (jetzt Wien Museum) 1985/86.
Lit.: GK, S. 22, 122 (m. Abb.)

7200
Dakon, Stephan
Surprise (Sitzendes Mädchen mit Hut in einer Dose)
ca. 1935
7200,3 Firmenmarke 19, Made in Austria, Hand dec., unsign., Ausformung zw. 1945–1953, H. 16 cm, B. 20,5 cm; city-antik

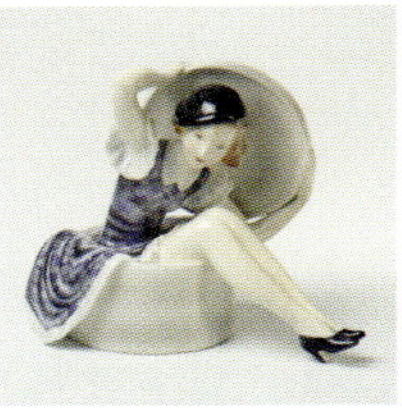

7200,11,96 gem. X., Firmenmarke 19, Made in Austria, Hand dec., unsign., H. 17 cm, B. 20 cm; GK, S. 22, 123 (m. Abb.); Wien Museum Inv. 222.546
Bem.: Diese Darstellung spielt sicher auf die Art „Überraschung" zum Junggesellengeburtstag oder anderen Anlässen an, wo leicht bekleidete Mädchen aus tortenähnlichen Behältnissen (hier eine Hutschachtel nachahmend) steigen und erotische Tänze (Striptease) aufführen.
Ausst.: HMW (jetzt Wien Museum) 1985/86; SSH 1992.
Lit.: GK, S. 22, 123 (m. Abb.); SS, S. 28, 29 (m. Abb.); GSN, S. 34, 35 (m. Abb.).
GS neu: Modell-Nr. 53 „Die Überraschung" (H. 15 cm), Aufl. ohne Ang.

7201
Weiss, Claire (Klára)
(d.i. Herczeg, Claire (Klára)
Stehendes Mädchen, ihren Hut haltend (auf Rundsockel mit Kaninchen)
ca. 1935
7201,3,33 gem. G, E, Firmenmarke 20, Made in Austria, Hand dec., unsign., H. 21 cm; Sammlung Stressless, Portugal

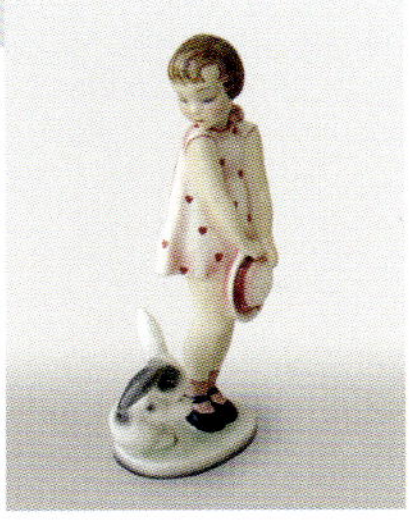

7201,112,20 Firmenmarke 19, Made in Austria, sign. Claire Weiss, H. ca. 22,5 cm; KH
7201,160,4 (Firmenmarke ohne Ang.), unsign.?, H. 21 cm; Christie's SK 4./5.12.1996
7201,211,35 ger. XVI, gem. lig. JR, Firmenmarke 19, Made in Austria, sign. Claire Weiss, H. fehlt; eBay # 1207929572
7201,1902,4 (Firmenmarke ohne Ang.), unsign., H. 27 cm; Christie's SK 14.11.2002
Bem.: Sollte es sich bei Christie's im Auktionskat. vom 14.11.2002 nicht um einen Druckfehler handeln, so gehört dieses gefällige Modell aufgrund der hohen Stückzahl (Exemplarnr. 1902) auch zu den Erfolgsgaranten der Manufaktur.

7205
BUCHSTÜTZE Blätter
Form: quaderförmig
Dekor: verschiedene vegetabile Blattformen (geritzt und farbig liniert)
ca. 1935

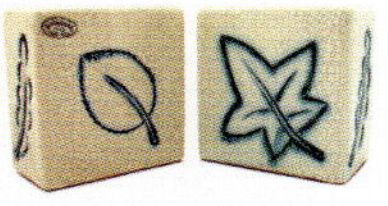

7205,20 gem. B, Firmenmarke 19, Made in Austria, Pick. 8, H. 11,3 cm; Sammlung a@goldscheider.net

7208F
Dakon, Stephan
FIGUR „Blonder Traum" (Lilian Harvey als schreitende Dame mit großem Hut und Handschuhen, ihr Kleid mit beiden Händen anhebend) (auf Ovalsockel)
ca. 1932 (für den Entwurf)
7208,47,7 gem. lig. ON, Firmenmarke 19, Made in Austria, Pick. 2/a, H. 21,8 cm; eBay # 3335056975

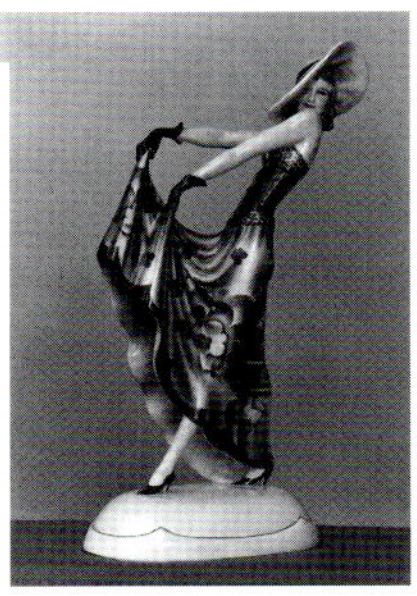

7208,98,14 gem. F, Firmenmarke 19, Made in Austria, H. 20,2 cm; Wien Museum Inv. 190.140
7208,174,16 gem. SG., Firmenmarke 19, Made in Austria, H. 21,2 cm; eBay # 941623068
Bem.: Lebensdaten zu Lilian Harvey s. 6009.
–> vgl. 6551; 6606
Lit.: GSN, S. 44, 45 (m. Abb.) ; GSNF, o. Pag. (m. Abb.)
Quelle: Archiv Mimi Dakon (Tantiemenabrechnung 1.7.–31.12.1937; Modellliste Dakon, undatiert, ca. 1940; Wiener Manufaktur FG, Modellliste Dakon, 7.11.1946), hier jeweils mit Ang. des Originaltitels FIGUR „Blonder Traum".
GS neu: Modell-Nr. 29 „Noblesse" (H. 20 cm), Aufl. 3000 Stück

7211
Stuart-Willfort, Irma
STEHKOPF Frauenkopf mit gelockten Haaren und Kopftuch (halbplastisch)
ca. 1935
7211 gem. grün lig. AH, Firmenmarke 19, Made in Austria, sign. I. Willfort, H. 24,5 cm; Sammlung.schmutz@chello.at 57

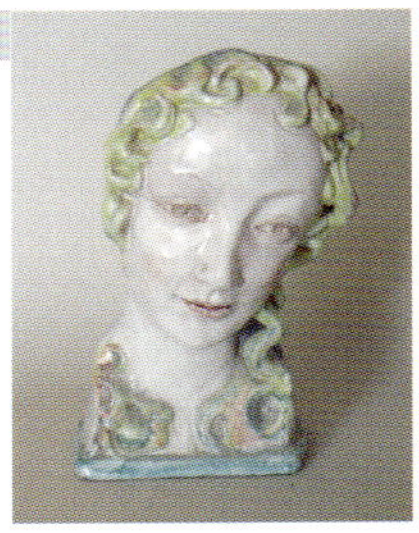

7211 Firmenmarke 19, Made in Austria, sign. I. Willfort, H. 24,5 cm; Nagel, 2. Auktion „Moderne Kunst und KGW des 20. Jh. (14.11.1992)

7214F
FIGUR „Tiger" (schleichend)
ca. 1935
7214,5,22 (Firmenmarke ohne Ang.), H. ca. 18 cm; Doro G 7.11.1995
7214F, H. 18 cm, B. 42 cm; GSNachf. 5,20
Lit.: GSNachf. 5,20 (m. Abb.), hier mit Ang. des Originaltitels FIGUR „Tiger".
Quelle: AGM (Kommissionsfaktura der „Österreichischen Gruppe-Austria" für die Messe in Utrecht vom 26.2.1953; Rechnung an Maurice Mizrahi, Alexandria, Ägypten, vom 22. April 1953), hier jeweils mit Ang. des Originaltitels „Tiger".

7217
RELIEF Pferdekopf (nach links) (Wandrelief)
ca. 1935
7217,65 gem. lig. AH, Firmenmarke 19, Made in Austria, Ausformung zw. 1945–1953, H. 12 cm, B. 12 cm; eBay # 6627582904

7216F
Matula, Viktor
ASCHER „Ascher, groß"
Dekor: Vasen
ca. 1935
Quelle: AGM (Rechnung an Maurice Mizrahi, Alexandria, Ägypten, vom 22. April 1953), hier mit Ang. des Originaltitels „Ascher, groß".

7218F
Herczeg, Claire (Klára)
GRUPPE „Jugend" (Nacktes, kniendes Mädchen mit Reh) (auf Ovalsockel)
ca. 1935

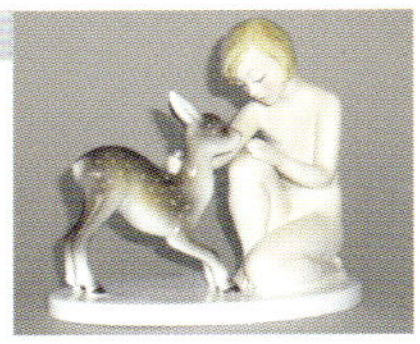

7218,3 Firmenmarke 19, Made in Austria, unsign., Ausformung zw. 1945–1953, H. 14,8 cm; Wien Museum Inv. 190.093

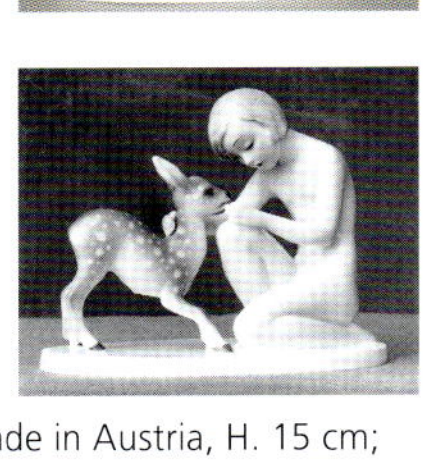

7218,10,5 gem. G, Firmenmarke 19, Made in Austria, Hand dec., H. 14,8 cm; eBay # 1269452986
7218,17 Firmenmarke 19, Made in Austria, H. 15 cm; Wendl 49
7218,38 ger., gem. LS., Firmenmarke 19, Made in Austria, Hand dec., Ausformung zw. 1945–1953, H. 15,5 cm, B. 18,4 cm; eBay # 1241199258
7218,263,7 gem. lig. HK., Firmenmarke 19, Made in Germany, sign. Claire Herczeg, Ausformung zw. 1938–1941, H. 15,4 cm; Sammlung Möller, Deutschland
7218,281,44 (Firmenmarke ohne Ang.), H. 15 cm; Schuler 64
7218F, H. 15 cm, B. 17 cm; GSNachf. 1,2; Fotoarchiv K. Täuber
Lit.: GSNachf. 1,2 (m. Abb.), hier mit Ang. des Originaltitels „Jugend".
Quelle: Fotoarchiv K. Täuber

7222
Herczeg, Claire (Klára)
Stehendes Mädchen mit Haarband und Männchen machendem Pekinesen (auf Ovalsockel)
ca. 1935
7222,67 gem. II, Firmenmarke 19, Made in Germany, unsign., Ausformung zw. 1938–1941, H. 23 cm; eBay # 6524722692
7222,86,2 (Firmenmarke ohne Ang.), unsign., H. 21 cm; Christie's SK 4./5.12.1996
7222,113,20 gem. EV, F, Firmenmarke 19, Made in Austria, unsign., H. 21,6 cm; eBay # 7303419446

7224F
Lorenzl, Josef
„Abendkleid" (Stehende elegante Dame mit Fächer)
ca. 1935
7224,123,19 Firmenmarke 19, Made in Austria, H. 29 cm; WKA 33
7224,286,19 ger. 8A, gem. rot GH, Firmenmarke 19, Made in Austria, H. 29 cm; eBay # 7317300006
7224,295,31 Firmenmarke 19, H. 29 cm; Metz 18.10.1997
7224,497,4 Firmenmarke 19, Made in Austria, H. 30 cm; Sammlung Dr. Peter Hieke, Wien
Quelle: AGM (Tantiemenliste Lorenzl, ohne Datierung (ca. 1940); Rechnung an Maurice Mizrahi, Alexandria, Ägypten, vom 22. April 1953), hier mit Ang. des Originaltitels „Abendkleid", geliefert in Ausführung „gold".

7225F
Lorenzl, Josef (zugeschr.)
„Gaby" (Sich verbeugende Rokokodame) (auf Ovalsockel)
ca. 1935
7225,8,85 gem. lig. SP., Firmenmarke 19, Made in Austria, Hand dec., unsign., H. 20,3 cm; eBay # 462416139

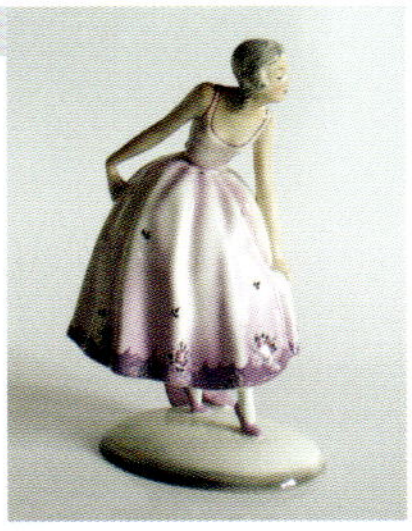

7225,11 gem. 4, Firmenmarke 19, Made in Austria, unsign., Ausformung zw. 1945–1953, H. 18,8 cm; eBay # 1418022056
7225,22,15 ger., Firmenmarke 19, Made in Austria, Hand dec., unsign., H. 19 cm; Sammlung mad
7225F, H. 19,5 cm, B. 10 cm; GSNachf. 1,55
Lit.: GSNachf. (2. Ausg., bzw. Erg. Bl.) 1,55 (m. Abb.)

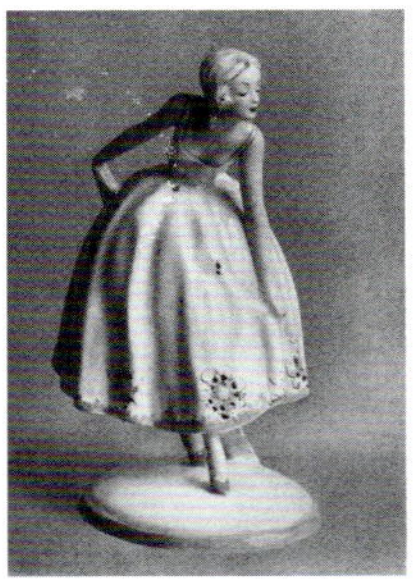

Quelle: AGM (Bestellung von Ferd. Lundquist (Göteborg, Schweden), vom 23. August 1950; Rechnung an Maurice Mizrahi, Alexandria, Ägypten, vom 22. April 1953), hier jeweils mit Ang. des Originaltitels „Gaby".

7226F
Dakon, Stephan
Sitzendes Mädchen mit Buch (auf Bücherregal/Sockel)
ca. 1935
7226,34,16 gem. lig. JR, Firmenmarke 19, Made in Austria, Hand dec., H. 23 cm; Sammlung Stressless, Portugal
7226F, H. 24,5 cm, B. 15 cm
Quelle: Archiv Mimi Dakon (Originalfoto)

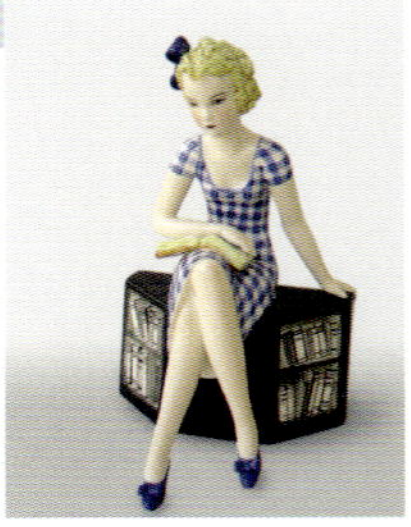

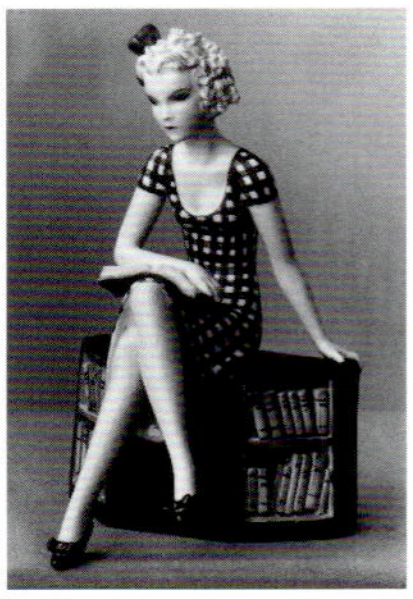

7227F
Lorenzl, Josef
FIGUR „Kokettchen"
(Stehender Damenakt mit ausgestreckten Armen)
(auf Rundsockel)
ca. 1928/29 (für den Entwurf)
7227F, H. 19 cm, B. 15 cm; GSNachf. 1,30b
-> vgl. 5802
-> vgl. Dakon 5907= fast identisches Modell.
Lit.: GSNachf. 1,30 (m. Abb.), hier mit Ang. des Originaltitels „Kokettchen".

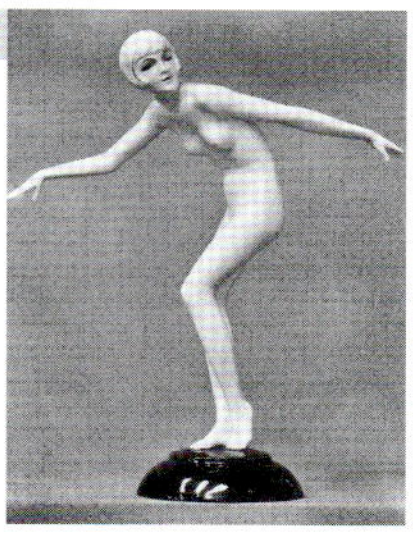

7230
Lorenzl, Josef
Tänzerin in Bewegung (Halbakt mit langem Kleid)
(auf Ovalsockel)
ca. 1935
7230 (Firmenmarke ohne Ang.), H. 29,5 cm; Pinhas, S. 130 (m. Abb.)

7234
Stehende Madonna mit Kind (auf glockenartigem Sockel)
ca. 1935
7234 H. 36 cm, B. 14 cm
Quelle: Fotoarchiv K. Täuber

7235
Matula, Viktor
ASCHER
ca. 1935
7235 H. fehlt, B. fehlt
Quelle: Nachlass Matula, Wien

7236
RELIEF Dahlienstrauß in bäuerlichem Krug / Vase
ca. 1935
7236,20 gem. lig. TK, Firmenmarke 19, Made in Austria, H. 27,3 cm, B. 21,3 cm; Sammlung John Nash

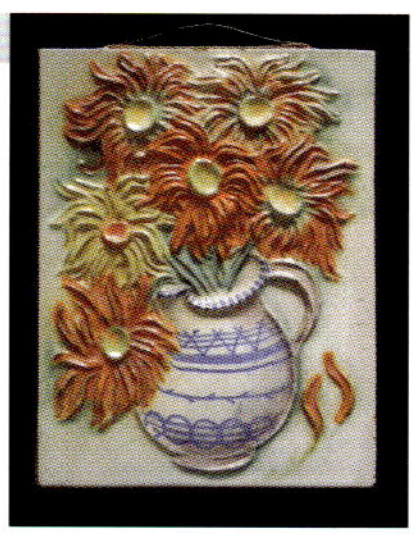

7237
Meisinger, Ida (zugeschr.)
BUCHSTÜTZE Stehender Pinguin mit Zipfelhaube und Muff
ca. 1935
7237,6,32 gem. S., Firmenmarke 19, Made in Austria, unsign., H. 16,4 cm, B. 10,3 cm; PB Wien

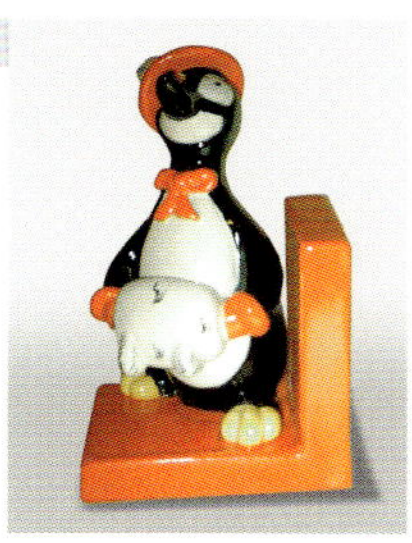

7243
Matula, Viktor
GRUPPE Rehe
ca. 1935
7243 H. fehlt
Quelle: Nachlass Matula, Wien

7246
Jarl, Karin
„Zicklein" (liegend)
ca. 1935
7246,87,31 Firmenmarke 19, Made in Germany, Ausformung zw. 1938–1941, H. 19 cm; PB Wien 19
7246 H. 18 cm, B. 29 cm; Fotoarchiv K. Täuber
Quelle: AGM (Bestellungen der Firma Wahliss (Wien) vom 9. und 24.11.1945 an die „Arbeitsgemeinschaft der Firma Friedrich Goldscheider", hier jeweils mit Ang. des (Original?-)Titels „Zicklein"; Bestellung der Firma Wahliss (Wien) vom 27.2.1946 in Ausführung „Terrakotta"); Fotoarchiv K. Täuber.

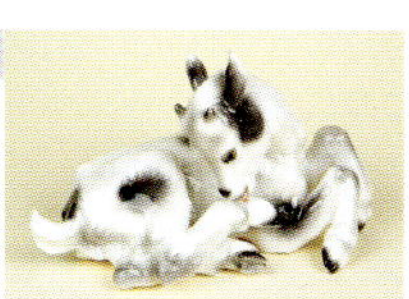

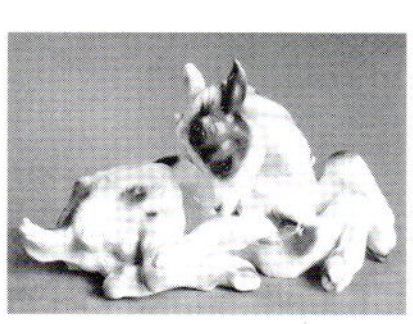

7247
Jarl, Karin
GRUPPE Zwei Barsois (Einer sitzend, einer liegend)
ca. 1935
7247,101,19 Firmenmarke 19, Made in Austria, gepr. „B"= Z6/a, L. 35 cm; Weidler 600; 610; 630; eBay # 123979197

7248F
Jarl, Karin
FIGUR „Spaniel" (sitzend)
ca. 1935
7248,2,15 Firmenmarke 19, Made in Austria, unsign., H. 21,5 cm; Wien Museum Inv. 190.020
7248,98,66 (Firmenmarke ohne Ang.), unsign., H. fehlt; Doro G 7.11.1995
7248F, H. 20 cm, B. 30 cm; GSNachf. 5,6
Pendant -> vgl. 7578F
Lit.: GSNachf. 5,6 (m. Abb.), hier mit Ang. des Originaltitels FIGUR „Spaniel".

7253F
Dakon, Stephan
Lesendes Mädchen am Stehpult (auf Rundsockel)
ca. 1935
7253F, H. 22 cm
Quelle: Archiv Mimi Dakon (Originalfoto)

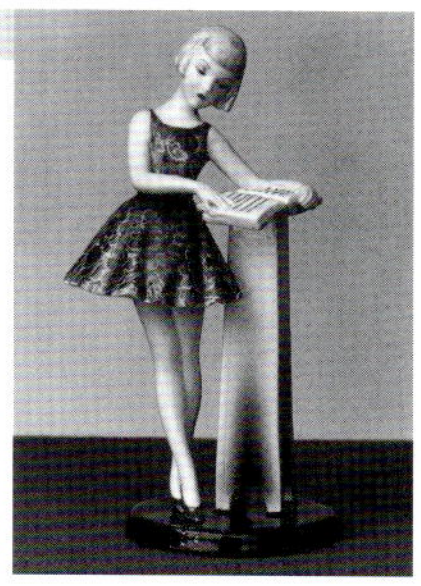

7254F
Dakon, Stephan
Posierende junge Dame mit Hut (auf Treppe / Sockel)
ca. 1935
7254F, H. 21 cm, B. 16 cm
Quelle: Archiv Mimi Dakon (Originalfoto)

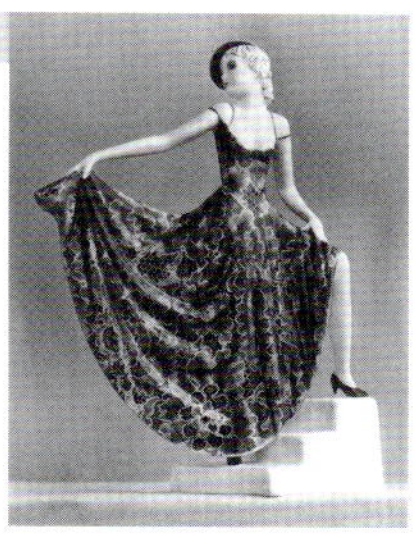

7256
Dakon, Stephan
Elegante Dame mit großem Hut und Foxterrier (in einem Boot sitzend)
ca. 1935
7256,23,19 gem. AS, gem. blau L7, Firmenmarke 19, Made in Austria, unsign., H. 17 cm, L. 38 cm; Sammlung Stressless, Portugal
7256,26,16 gem. AS, Firmenmarke 19, Made in Austria, Hand dec., H. 17 cm, L. 38 cm; Sammlung Stressless, Portugal
7256,28,31 (Firmenmarke ohne Ang.), L. 34,3 cm; Soth. Chester 7.–9.6.1988
7256,112,4 (Firmenmarke ohne Ang.), H. ca. 17,5 cm, L. 37 cm; Christie's SK 12.9.1997
7256,237,14 gem. EM, Firmenmarke 19, Made in Austria, H. 17,8 cm, L. 39,4 cm; Soth. South, Billingshurst 11.10.2000; eBay # 1233646528
Quelle: Archiv Mimi Dakon (Originalfoto)

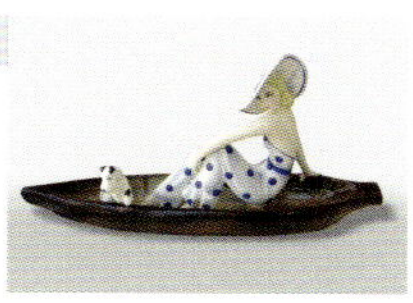

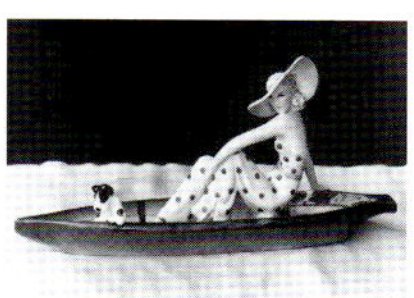

7257F
Dakon, Stephan
FIGUR / LAMPE „Gavotte"
(Jessie Matthews als stehende Tänzerin mit Hut, ihr Kleid schmetterlingsflügelartig auseinander haltend)
(auf Ovalsockel)
ca. 1935
FIGUR
7257,52,14 gem. F.N., Firmenmarke 19, Made in Austria, sign. Dakon, H. 31,5 cm; WKA 38; Doro 1.6.2005; 12.9.2006

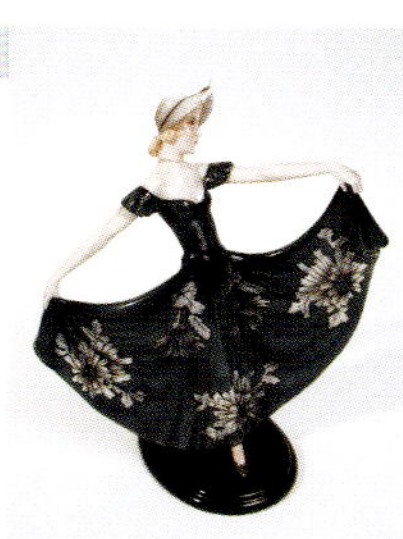

7257,61,6 Firmenmarke 19, Made in Austria, H. 32 cm; city-antik
7257,96,31 (Firmenmarke ohne Ang.), sign. Lorenzl, H. 32 cm; Christie's SK 10.11.2000; Bem.: Die Sign. „Lorenzl" ist ein Versehen des Modelleurs.
FIGUR / LAMPE
7257,34,31 gem. lig. NO, Firmenmarke 19, Made in Austria, H. (Figur) 33 cm, Gesamth. mit neuem Schirm 52,3 cm; Doro 7.3.2001; eBay # 1239802584;
Bem.: Dieses Objekt wurde wohl, da zuerst als Figur im Doro versteigert und kurz darauf bei eBay als Lampe angeboten, vom Besitzer „umgebaut".
Bem.: Dargestellt ist die engl. Sängerin und Schauspielerin Jessie Matthews (11.3.1907–19.8.1981).
-> vgl. 8318
-> vgl. Lorenzl 7581= ähnl. Modell.
Lit.: DTW, S. 35 (m. Abb.), die hier wiedergegebene zeitgenössische Fotografie von Sascha, diente als Vorlage für die Ausführung dieses Modells.
Quelle: Archiv Mimi Dakon (Tantiemenabrechnung 1.7.–31.12.1937; Modellliste Dakon, undatiert, ca. 1940; Wiener Manufaktur FG, Modellliste Dakon, 7.11.1946; Tantiemenabrechnung 2. Halbjahr 1949; 1. Halbjahr 1950), hier jeweils mit Ang. des Originaltitels FIGUR „Gavotte"; AGM.

7261
WANDMASKE „Maske"
(Frauenkopf mit Hand und Schmetterling)
ca. 1935
7261,(3?) Firmenmarke 19, Made in Germany, Ausformung zw. 1938–1941, H. 23 cm; Sammlung Rudolf Hansl, St. Florian
7261,8 Firmenmarke 19, Made in Austria, Ausformung zw. 1945–1953, H. 22 cm; Kett. 160
7261 Firmenmarke 19, Made in Austria, Ausformung zw. 1945–1953, H. 22,4 cm; Sammlung a@goldscheider.net; Bem.: Dieses Ex. wurde ohne den auf der Hand sitzenden Schmetterling ausgeführt.
Lit.: Pinhas, S. 173 (m. Abb.), Farbvariante.
Quelle: AGM (Kontoblatt der Wiener Manufaktur Josef Schuster für die Firma Porslins Ericsson-Lund, Stockholm?, Schweden vom 7. Januar 1943), hier mit Ang. des Originaltitels „Maske".

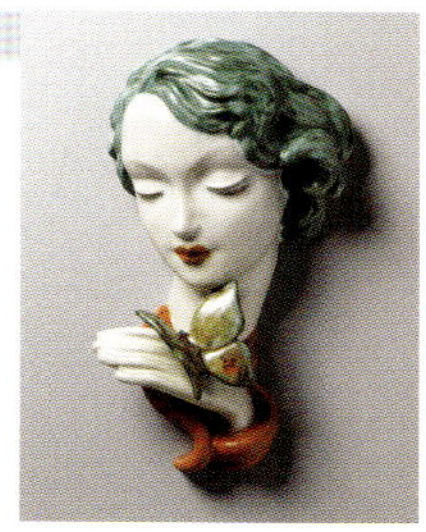

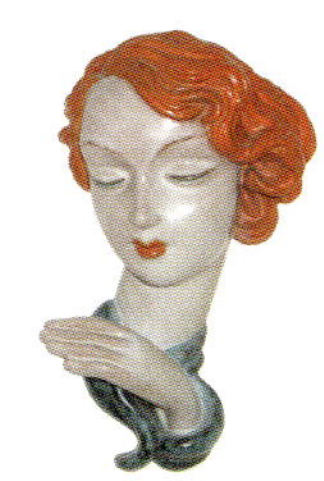

7263F
Knörlein, Rudolf
BÜSTE „Ave Maria" (Madonna mit gefalteten Händen)
ca. 1934
7263,90,66 Firmenmarke 19, Made in Austria, Hand dec., unsign., H. 41,5 cm; Bonhams Chester 2.9.2003
7263,119,16 ger. XII, E, Firmenmarke 19, Made in Austria, Monogr. RK, H. 42 cm; GK, S. 22, 124 (m. Abb.); Wien Museum Inv. 222.549
7263,135,31 gem. AN, Firmenmarke 19, Made in Germany, Ausformung zw. 1938–1941, unsign., Pick. 6, H. 42 cm; Kastern 70; Schloss Rick. 82
ohne Ang., Pick. 1, H. 40,5 cm; Schlüter 2658
7263F, H. 39,5 cm, B. 17,5 cm; GSNachf. 6,5b
Bem.: Durch die eigenhändige Datierung Knörleins (1934) auf dem Trägerblatt des Originalfotos sowie der Originalzeichnung (Aquarell über Graphit), kann das Jahr

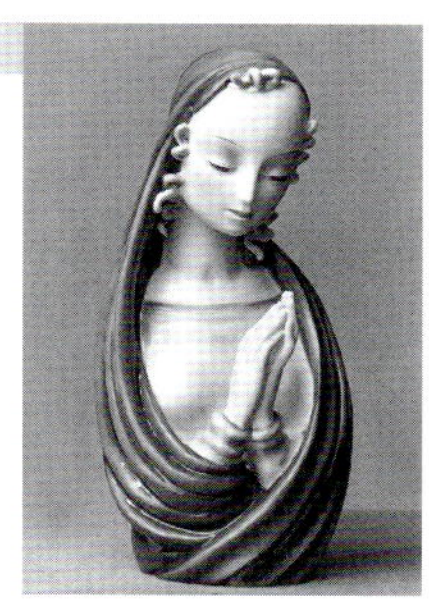

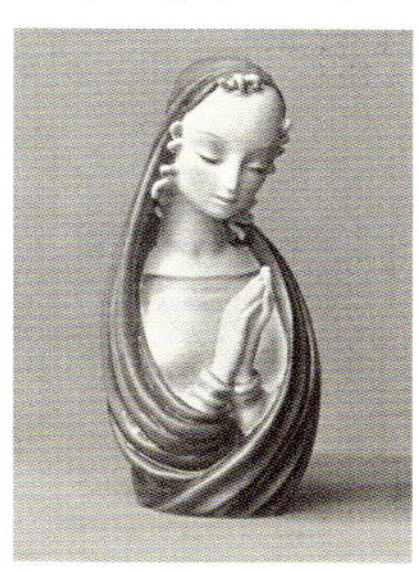

1934 exakt bestimmt werden. Diese Büste wurde auch von GS USA (Everlast Corp.) unter der Modell-Nr. 819 in der H. 21,6 cm produziert.
Ausst.: HMW (jetzt Wien Museum) 1985/86.
–> vgl. 7188; 7858; 8743
Lit.: GSNachf. 6,5 (m. Abb.), hier mit Ang. des Originaltitels BÜSTE „Ave Maria"; GK, S. 22, 124 (m. Abb.).
Quelle: Privatarchiv (Originalfoto); Privatsammlung D (Originalzeichnung).

7273
Bouret, Germaine

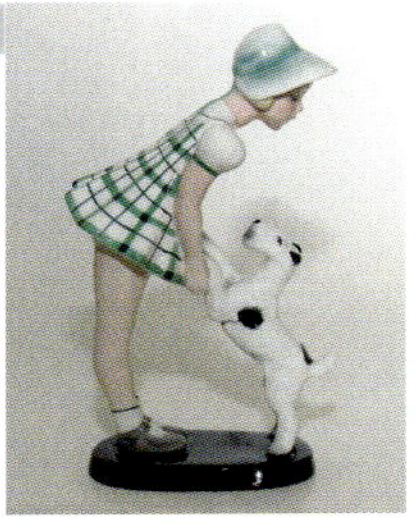

Stehendes Mädchen mit Hut mit aufspringendem Foxterrier (auf Ovalsockel)
ca. 1935
7273,44,1 gem. EV, gem. lig. AS, Firmenmarke 19, Made in Austria, unsign., H. 19 cm; Sammlung Bernard Amigon, Frankreich
Bem.: Dieses Modell wurde auch in Japan, wenngleich in einigermaßen guter Qualität, nachgebildet. Es kann somit als Fälschung bezeichnet werden (–> vgl. Sammlung Bernard Amigon, Frankreich).
Lit.: GSN, S. 46, 47 (m. Abb.)
GS neu: Modell-Nr. 30 „Mein Hund und ich" (H. 20 cm), Aufl. ohne Ang.

7274
GRUPPE Stehendes Mädchen mit Berét und kurzem Kleid und Foxterrier (auf Ovalsockel)
ca. 1935
7274,107,0 (Firmenmarke ohne Ang.), H. 24 cm; Christie's SK 27.6.1997

7275
Stehende Dame mit langem Kleid und Hut (auf Rund/Ovalsockel)
ca. 1935
7275,85,14 gem. F., Firmenmarke 19, Made in Austria, H. 33 cm; KH
Lit.: Pinhas, S. 147 (m. Abb.)

7276M
Knörlein, Rudolf

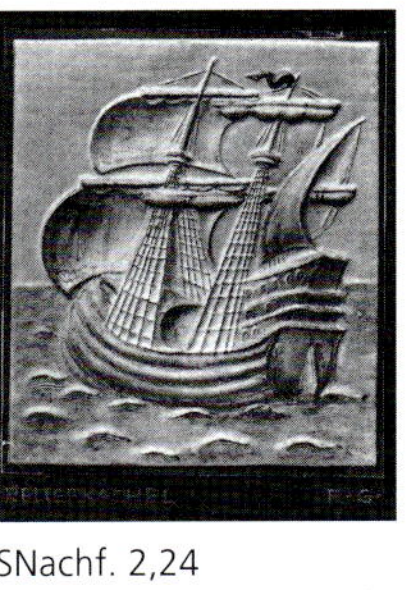

RELIEF „Segler" (Segelschiff nach links) (Wandrelief)
ca. 1935
7276,65 gem. lig. AH, Firmenmarke 19, Made in Austria, Ausformung zw. 1945–1953, H. 29,5 cm, B. 25,5 cm; Doro FVK Sept. 2004
7276M, H. 29 cm, B. 25 cm; GSNachf. 2,24
Lit.: GSNachf. 2,24 (m. Abb.), hier mit Ang. des Originaltitels „Relief Segler".
Quelle: AGM (Kontoblatt der Wiener Manufaktur Josef Schuster für die Firma Porslins Ericsson-Lund, Stockholm?, Schweden vom 7. Januar 1943), hier mit Ang. des Titels „Relief"; Privatarchiv (Originalfoto).

7277
Dakon, Stephan
FIGUR „Schaukelpferd" („Kutschpferd")
ca. 1935
7277, H. fehlt
Quelle: Archiv Mimi Dakon (Tantiemenabrechnung 1.7.–31.12.1937), hier mit Ang. des Originaltitels FIGUR „Schaukelpferd"; Wiener Manufaktur FG, Modellliste Dakon, 7.11.1946), hier mit Ang. des (Original)-Titels „Kutschpferd".

7278
Mädchen mit Hund spielend (auf Ovalsockel)
ca. 1935
7278,2,10 gem. W., Firmenmarke 19, Made in Austria, H. 20 cm; Wien Museum Inv. 220.724
7278,17,3 (Firmenmarke ohne Ang.), H. 20 cm; KaM 88
7278,197,65 gem. lig. LS, Firmenmarke 19, Made in Austria, H. 19,8 cm; Wien Museum Inv. 190.090

7279
Weiss, Claire (Klára)
(d.i. Herczeg, Claire (Klára)

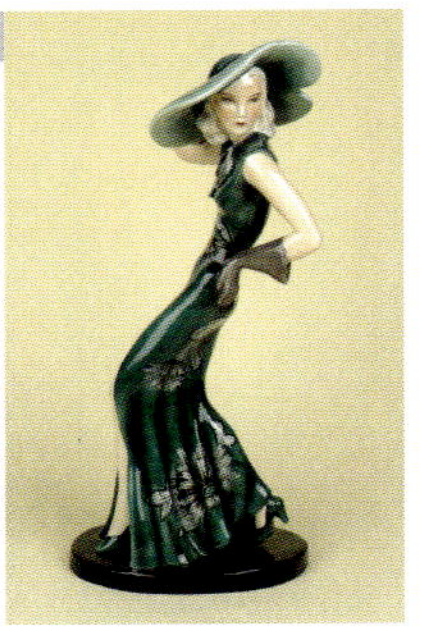

Stehende elegante Dame mit großem Hut und langen Handschuhen (auf Ovalsockel)
ca. 1935
7279,95,31 Firmenmarke 19, Made in Austria, unsign., H. 39,5 cm; WKA 35
7279,169,4 Firmenmarke 19, Made in Austria, unsign., H. 40 cm; PB Wien 19
7279,285,19 gem. EM., AH., Firmenmarke 19, Made in Austria, sign. Claire Weiss, gepr. „L"= Z6, H. 41 cm; eBay # 3733172230
7279,299,19 gem. AH, B, Firmenmarke 19, Made in Germany, sign. Claire Herczeg, Ausformung zw. 1938–1941, H. 40 cm; Wien Museum Inv. 190.080
Ausst.: SSH 1992
–> vgl. 8151
–> vgl. 7367; 8155= jeweils GRUPPE „Maskerade" (mit Barsoi).

7280F
Herczeg, Claire (Klára)

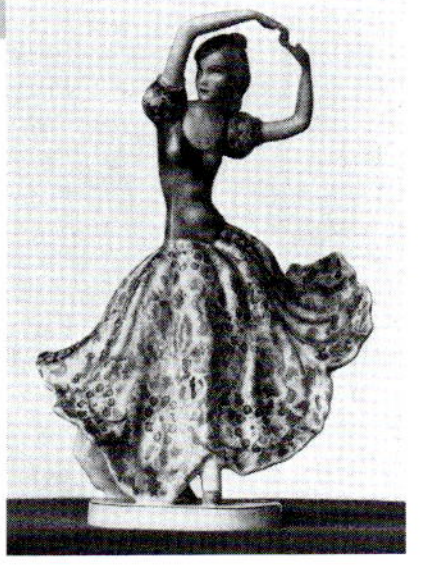

FIGUR „Ballnacht" (Tänzerin mit erhobenen Armen in Bewegung) (auf Ovalsockel)
ca. 1935
7280,230,2 Firmenmarke 19, Made in Germany, sign. Claire Herczeg, Ausformung zw. 1938–1941, H. 32,5 cm; Fischer Luz. 322
7280F, H. 31 cm, B. 12 cm; GSNachf. 1,13
Lit.: GSNachf. 1,13 (m. Abb.), hier mit Ang. des Originaltitels „Ballnacht".

7281F
Thomasch, Wilhelm

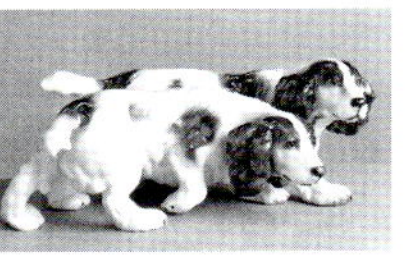

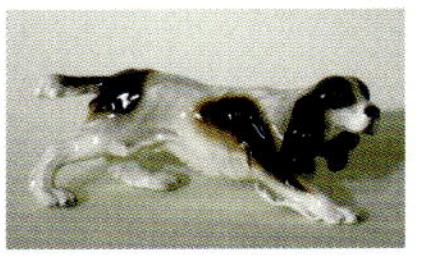

GRUPPE „Spaniels" (Zwei laufende Spaniel)
ca. 1934 (für den Entwurf)
7281,8 gem. G, Firmenmarke 19, Made in Austria, unsign., Ausformung zw. 1945–1953, H. 20 cm, B. 57 cm; Wien Museum Inv. 220.763
7281,66,35 gem. G, Firmenmarke 19, Made in Austria, unsign., H. 20,5 cm, B. 55 cm; Wien Museum Inv. 190.026
7281F, H. 18 cm, B. 55 cm; GSNachf. 5,8b
Bem.: Für die Zuordnung dieser Gruppe und der damit verbundenen Einzeltiere, danke ich Herrn Herwig Thomasch, dem Sohn von Wilhelm Thomasch.
–> vgl. 7015F; 8342F
–> vgl. 7061F; 7329F; 8344F= jeweils EinzelFIGUR (vorderer Hund)
–> vgl. 7062F; 7330F; 8343F= jeweils EinzelFIGUR (rückwärtiger Hund)
Lit.: GSNachf. 5,8 (m. Abb.), hier mit Ang. des Originaltitels GRUPPE „Spaniels".
Quelle: Fotoarchiv K. Täuber

7282
Meisinger, Ida
Zwei Fohlen (auf Landschaftssockel)
ca. 1935
7282,20 Firmenmarke 19, Made in Austria, sign. Meis(i)n(g)er, H. 15 cm, B. 21,5 cm; WKA 2; Bem.: Das AH gibt als Sign. „Meisner" an. Es handelt es sich aber, wenn nicht ein Lese-/Druckfehler um eine Figur der Tierbildhauerin Ida Meisinger.

7285F
Dakon, Stephan

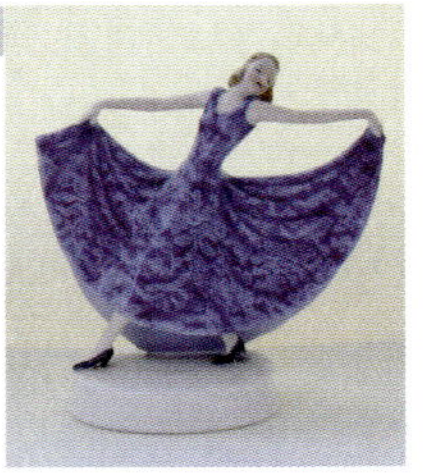

FIGUR „Walzer" (Schreitende Tänzerin, ihr Kleid schmetterlingsflügelartig auseinander haltend) (auf Ovalsockel)
ca. 1932/33 (für den Entwurf)
7285,896,4 Firmenmarke 19, H. 16 cm; KaM 124
7285,913,6 Firmenmarke 19, Made in Austria, H. 16 cm; Sammlung f@goldscheider.de
7285F, H. 16 cm, B. 16 cm; GSNachf. 1,9b
–> vgl. 6630
Lit.: GSNachf. 1,9 (m. Abb.), hier mit Ang. des Originaltitels „Walzer".

7286F
Dakon, Stephan

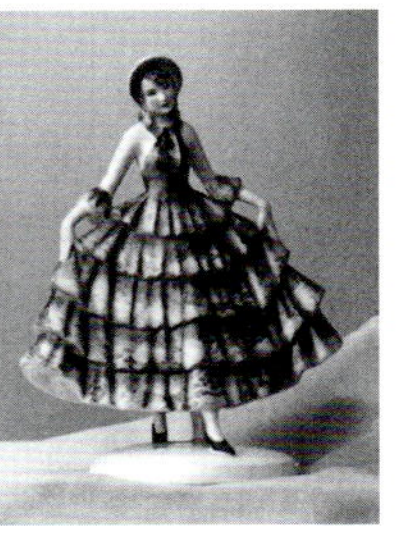

FIGUR „Alt-Wien" (Schreitende junge Dame mit Schutenhut und gerüschtem Kleid (Lilian Harvey?) (auf Rundsockel/Ovalsockel)
ca. 1931 (für den Entwurf)
7286,2,33 gem. TV, Firmenmarke 19, Made in Austria, Hand dec., unsign.?, Pick. 1, H. 14,5 cm;
eBay # 2223436401
7286,78,6 (Firmenmarke ohne Ang.), Pick. 2, H. 14,5 cm; Schuler 63; Lit.: TuS 1996, Nr. 205, S. 149 (m. Abb.)
7286,112,19 gem. lig. rot GH, Firmenmarke 19, Made in Austria, H. 14 cm; Wien Museum Inv. 190.131
7286F, H. 17 (!) cm, B. 12,5 cm; GSNachf. 1,9c; Bem.: Hier mit der falschen Höhenang. 17 cm, richtig sind 14 cm, wie das Originalfoto aus dem Archiv Mimi Dakon und die nachgewiesenen Exemplare belegen.
Bem.: Dieses Modell wurde noch in Trenton (USA) unter der GS Everlast Corp. (Modell-Nr. 195) weiter in der H. 20,3 cm produziert (eBay # 190011220196).
–> vgl. 6458
Lit.: GSNachf. 1,9 (m. Abb.), hier mit Ang. des Originaltitels „Alt-Wien".
Quelle: Archiv Mimi Dakon (Originalfoto)

7287F
Dakon, Stephan

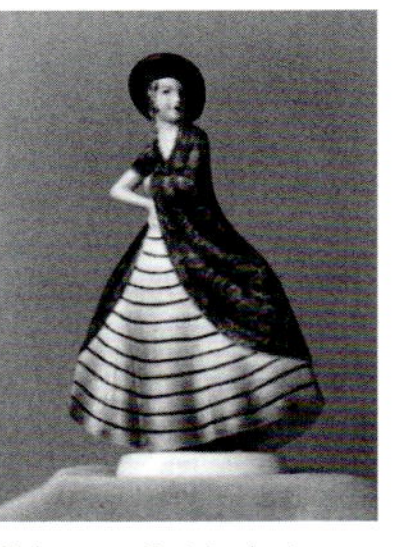

Stehende Dame mit Hut, langem Kleid und Spitzenumhang (auf Ovalsockel)
ca. 1935
7287,97,11,8 gem. rot B., Firmenmarke 19, Made in Austria, unsign., H. ca. 15 cm;
eBay # 2640638193
7287,1973,XX11, Firmenmarke Schuster 2, Made in Germany, unsign., Ausformung ca. 1944, H. 15,5 cm; eBay # 1435452397
7287F, H. 15 cm, B. 10 cm
Quelle: AGM (Rechnung der Wiener Manufaktur Josef Schuster an Laszlo Bösze (Szeged, Ungarn), vom 5. August 1944), hier unter dem Titel „Tanz" geführt; Archiv Mimi Dakon (Originalfoto)

7288F
Dakon, Stephan (zugeschr.)
„Ninon" (Stehende junge Dame mit Hut und gestuftem Rüschenkleid) (auf Ovalsockel)
ca. 1935
7288,176,31 gem. G., E.M., Firmenmarke 19, Made in Austria, unsign., Pick. 5, H. 16,5 cm; eBay # 741635993; # 3772531895; 658128350
Quelle: AGM (Rechnung der Wiener Manufaktur Josef Schuster an Laszlo Bösze (Szeged, Ungarn), vom 5. August 1944), hier mit Ang. des Originaltitels „Ninon".

7289F
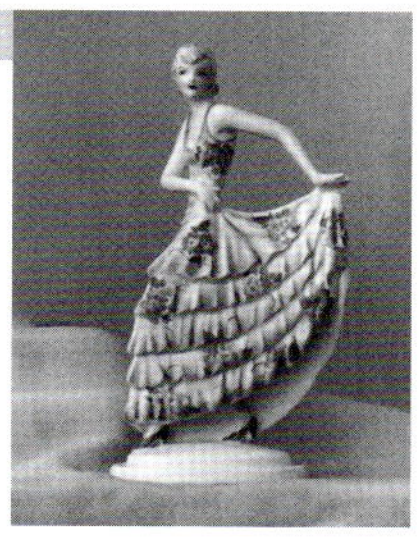
Dakon, Stephan
Schreitende Tänzerin, ihr Kleid seitlich weg haltend (auf Ovalsockel)
ca. 1935
7289,?,10 gem. blau H., Firmenmarke 19, Made in Austria, H. 17,8 cm; eBay # 2009577662
7289F, H. 15 cm, B. 9 cm
Quelle: Archiv Mimi Dakon (Originalfoto)

7290F
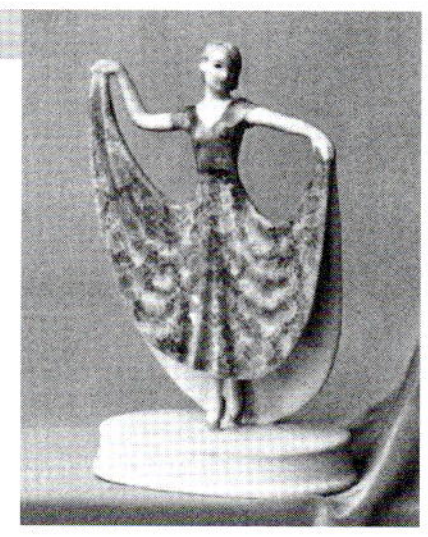
Dakon, Stephan
FIGUR „Nora" (Stehende Tänzerin, ihr Kleid schmetterlingsflügelartig auseinander haltend) (auf Ovalsockel)
ca. 1935
7290,182,4 ger. A, gem. EM, Firmenmarke 19, Made in Austria, H. 15,3 cm; eBay # 1236937239
7290,286,16 Firmenmarke 19, Made in Austria, H. 17,1 cm; eBay # 186872618
7290,365,6 Firmenmarke 19, Made in Austria, H. 17,1 cm; eBay # 186872618
7290F, H. 17,5 cm, B. 11,5 cm; GSNachf. 1,9a
Lit.: GSNachf. 1,9 (m. Abb.), hier mit Ang. des Originaltitels „Nora".

7291
Dakon, Stephan
WANDMASKE Kinderkopf (Mädchen mit Hut und Schleife/Mascherl)
ca. 1935
7291,1 gem. rot N, Firmenmarke 19, Made in Austria, H. 24 cm; eBay # 7342375389
7291,5 gem. AN, Firmenmarke 19, unsign., H. 25 cm; Mehlis 34
7291 (Firmenmarke ohne Ang.), unsign., H. 23 cm; eBay # 1577742575
-> vgl. 7413

7292
WANDMASKE Frauenkopf mit Hut und Kragen
ca. 1935
7292,8 Firmenmarke 19, Made in Austria, Ausformung zw. 1945–1953, H. 25 cm; eBay # 7423650989
7292 gem. 2, Firmenmarke 19, Made in Austria, H. 26 cm; eBay # 3357380180

7293
Matula, Viktor
ASCHER Foxterrier
ca. 1935
7293 H. fehlt, B. fehlt
Quelle: Nachlass Matula, Wien

7294
Matula, Viktor
ASCHER Esel
ca. 1935
7294 H. fehlt, B. fehlt
Quelle: Nachlass Matula, Wien

7295F
ASCHER „Ascher, groß"
ca. 1935
Quelle: AGM (Rechnung an Maurice Mizrahi, Alexandria, Ägypten, vom 22. April 1953), hier mit Ang. des Originaltitels „Ascher, groß".

7299F
Thomasch, Wilhelm
FIGUR „Spaniel" (Laufender Spaniel)
ca. 1934 (für den Entwurf)
7299F, H. 4 cm, B. 10 cm
Bem.: Für die Zuordnung dieser Figur und der damit verbundenen Gruppe, danke ich Herrn Herwig Thomasch, dem Sohn von Wilhelm Thomasch.
-> vgl. 7062F; 7330F; 8343F
-> vgl. 7015F; 7281F; 8342F= jeweils GRUPPE
Pendant -> vgl. 7061
Lit.: GSNachf. 5,8 (m. Abb.), hier mit Ang. des Originaltitels FIGUR „Spaniel" (Rückwärtiger Hund aus der GRUPPE 7015).
Quelle: Fotoarchiv K. Täuber

7300F
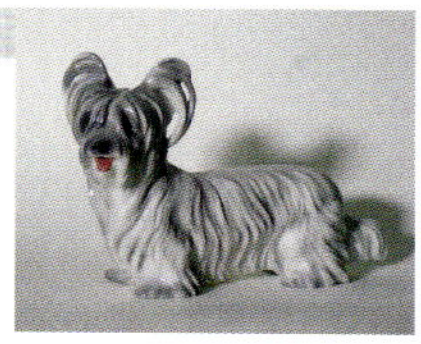
Postl, Hugo
FIGUR „Skyeterrier"
um/ab 1930 (für die Ausführung)
7300F, H. 6 cm, B. 9 cm; GSNachf. 5,11d
-> vgl. 6116; 6503; 6802; 7519
-> vgl. 6975= Variante (auf Ovalsockel)
Lit.: GSNachf. 5,11 (m. Abb.), hier mit Ang. des Originaltitels „Skyeterrier".

7308M
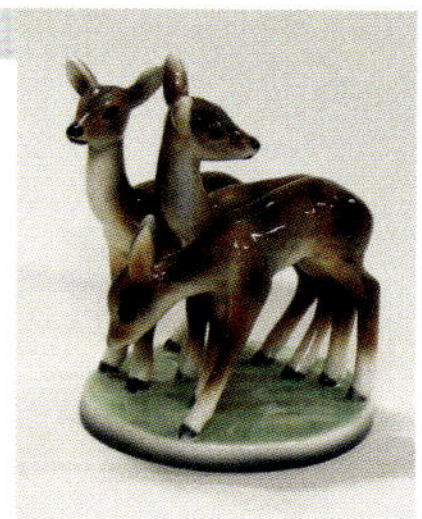
Matula, Viktor
GRUPPE „Rehe" (Drei stehende Rehe) (auf Ovalsockel)
ca. 1935
7308,10,26 gem. W., Firmenmarke 20, Made in Austria, Hand dec., unsign., H. 15 cm; eBay # 932822818
7308,20 Firmenmarke 19, Made in Austria, unsign., Ausformung zw. 1945–1953, H. 14,6 cm; eBay # 6510152128
7308,93,20 Firmenmarke 19, Made in Austria, Hand dec., unsign., H. 15,1 cm; MAK Wien, Inv. Ke 10521-15
7308M, H. 14 cm, B. 16 cm; GSNachf. 3,19b
Lit.: GSNachf. 3,19 (m. Abb.), hier mit Ang. des Originaltitels GRUPPE „Rehe".
Quelle: Nachlass Matula, Wien

7312
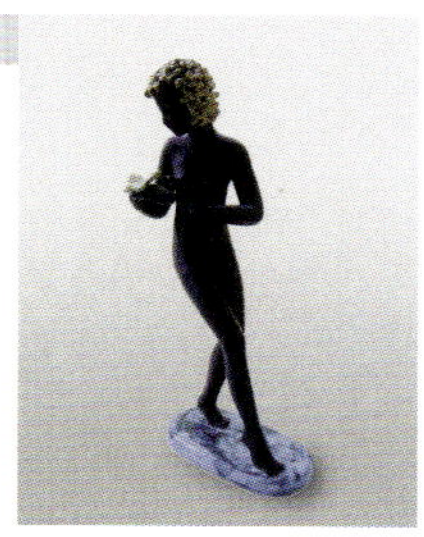
Prischl, Adolf (zugeschr.)
Schreitender Mädchenakt mit Schmetterling (auf Ovalsockel)
ca. 1935
7312,102,0 gem. A, Firmenmarke 19, Made in Austria, unsign., H. 22,5 cm; Sammlung f@goldscheider.de
Bem.: Die Zuschreibung an A. Prischl und den Hinweis, dass eine Variante mit Vogel anstatt des Schmetterlings, sich in einer österr. Privatsamml. befindet, verdanke ich Frau Mag. Uta Matschiner.

7313
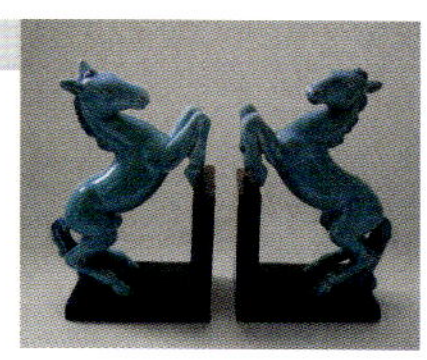
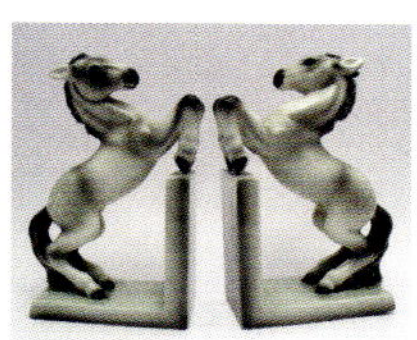
Prischl, Adolf
BUCHSTÜTZE Pferd (links/rechts)
ca. 1935
Pferd nach links
7313,10 ger. B, gem. W, Firmenmarke 19, Made in Austria, Pick. 6, Ausformung zw. 1945–1953, H. 20 cm; eBay # 1411313159
7313,18 ger. B, gepr. S, Firmenmarke 19, Made in Austria, Ausformung zw. 1945–1953, H. 19,6 cm, B. 12,3 cm; Wien Museum Inv. 182.099/1
7313,44 gem. A, Firmenmarke 19, Made in Germany, Ausformung zw. 1938–1941, H. 21 cm, B. 12 cm; Sammlung f@goldscheider.de
Pferd nach rechts
7313,9 Firmenmarke 19, Made in Austria, Ausformung zw. 1945–1953, H. 20,5 cm, B. 13 cm; Sammlung a@goldscheider.net
7313,18 ger. B, gepr. S, Firmenmarke 19, Made in Austria, Ausformung zw. 1945–1953, H. 20,3 cm, B. 13 cm; GK, S. 22, 125 (m. Abb.) Wien Museum Inv. 182.099/2
ohne Nr., gem. A, Firmenmarke 19, Made in Germany, Ausformung zw. 1938–1941, H. 20,5 cm, B. 12 cm; Sammlung f@goldscheider.de
Ausst.: HMW (jetzt Wien Museum) 1985/86.
Lit.: Neuwirth (1973), S. 62 (m. Abb.) (PB Wien); GK, S. 22, 125 (m. Abb.).

7314F
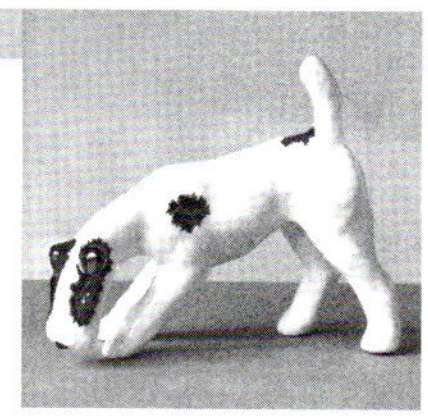
FIGUR „Foxterrier" (Witterung aufnehmend)
ca. 1935
7314F, H. 8 cm, B. 10cm; GSNachf. 5,1f; Fotoarchiv K. Täuber, hier mit der Ang.: H. 9 cm, B. 10 cm
Lit.: GSNachf. 5,1 (m. Abb.), hier mit Ang. des Originaltitels „Foxterrier".
Quelle: Fotoarchiv K. Täuber

7320
Postl, Hugo
Pekinese
ca. 1928/29 (für den Entwurf)
7320,7 Firmenmarke 19, Made in Austria, unsign.?, Ausformung zw. 1945–1953, H. fehlt; WKH
-> vgl. 5858; 6136
Lit.: s. bei Modell-Nr. 5858

7321F
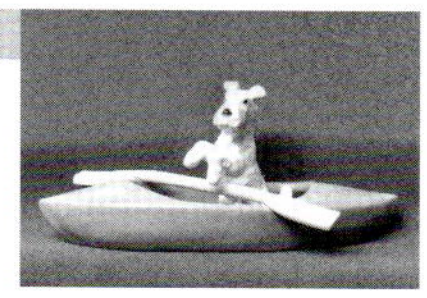
Meisinger, Ida
FIGUR Hund „Paddler" (In einem Ruderboot sitzender Terrier)
ca. 1935
7321F, H. 9 cm, B. 19 cm; GSNachf. 3,22c; Fotoarchiv K. Täuber
Bem.: s. 6449
Lit.: GSNachf. 3,22 (m. Abb.), hier mit Ang. des Originaltitels FIGUR Hund „Paddler".
Quelle: Fotoarchiv K. Täuber

7327F
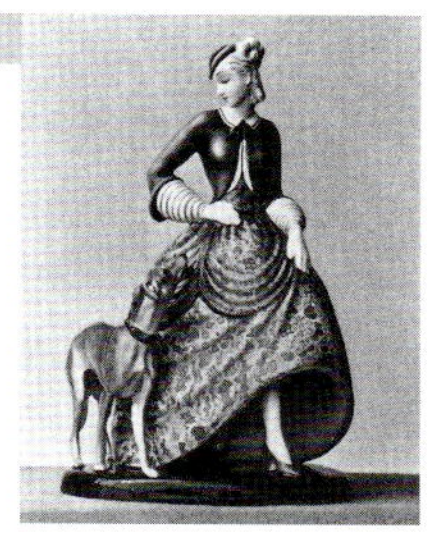
Weiss, Claire (Klára)
(d.i. Herczeg, Claire (Klára)
GRUPPE „Komtesse mit Hund" (Schreitende Tänzerin mit Windhund) (auf Ovalsockel)
ca. 1931 (für den Entwurf)
7327,1,50 gem. LS, Firmenmarke 19, Made in Austria, unsign., H. 35,5 cm; eBay # 399488110 (KH www.tias.com/stores/frozenbeauty)
7327,4 gem. G, Firmenmarke 19, Made in Austria, Hand dec., unsign., Ausformung zw. 1945–1953, H. 34,3 cm; eBay # 2600955650
7327,122,31 Firmenmarke 19, Made in Austria, unsign., H. 35 cm; eBay # 2061809843
7327,126,10 (Firmenmarke ohne Ang.), unsign., H. 33 cm; Christie's SK 17.10.2001
7327,145,81 gem. N, Firmenmarke 19, Made in Austria, Hand dec., sign. C. Weiss, H. 33 cm; eBay # 900504332
7327 ohne Ang., Firmenmarke 19, unsign., H. fehlt; Fischer 151
7327F, H. 33 cm, B. 20 cm; GSNachf. 1,14a
Bem.: Ob es sich bei dem beigestellten Windhund wirklich um einen Entwurf von Klára Herczeg handelt, ist sehr zu bezweifeln, da ihren Tänzerinnen oder Aktdarstellungen durch die Manufaktur teils ungefragt Hunde beigefügt wurden, welche ziemlich sicher nicht von ihrer Hand stammen. Auch sie stand dieser Praxis sehr kritisch gegenüber. Diese Information verdanke ich Herrn Heinz Hoever, der die Künstlerin vor ihrem Tod persönlich befragen konnte und mit ihr längere Zeit in Kontakt stand.
-> vgl. 8145
-> vgl. 6454; 6951= jeweils „Komtesse" als EinzelFIGUR.
Lit.: GSNachf. 1,14 (m. Abb.), hier mit Ang. des Originaltitels „Komtesse mit Hund".
Quelle: AGM (Rechnung an Maurice Mizrahi, Alexandria, Ägypten, vom 22. April 1953), hier mit Ang. des Originaltitels „Komtesse/Hund".

7328
Matula, Viktor
ASCHER
ca. 1935
7328 H. fehlt, B. fehlt
Quelle: Nachlass Matula, Wien

7329F
Thomasch, Wilhelm
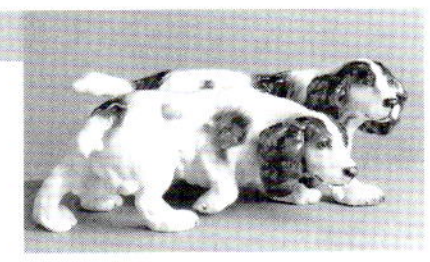
FIGUR „Spaniel"
(Laufender Spaniel)
ca. 1934 (für den Entwurf)
7329,11,44 Firmenmarke 19, Made in Austria, unsign., Pick. 1, Originalpick.: „Mimi Wodak, Edelhausrat, Wien XV, Lerchenfeldergürtel 31", H. 19 cm, B. 45 cm; Wien Museum Inv. 190.022
7329 ger., (Firmenmarke ohne Ang.), unsign., H. 19,4 cm, B. 43 cm; Doro 9.6.1998
7329F, H. 17 cm, B. 42 cm; GSNachf. 5,8e
Bem.: Für die Zuordnung dieser Figur und der damit verbundenen Gruppe, danke ich Herrn Herwig Thomasch, dem Sohn von Wilhelm Thomasch.
-> vgl. 7061F; 8344F
-> vgl. 7015F; 7281F; 8342F= jeweils GRUPPE
Pendant -> vgl. 7330F
Lit.: GSNachf. 5,8 (m. Abb.), hier mit Ang. des Originaltitels FIGUR „Spaniel" (Vorderer Hund aus der GRUPPE 7281).

7330F
Thomasch, Wilhelm
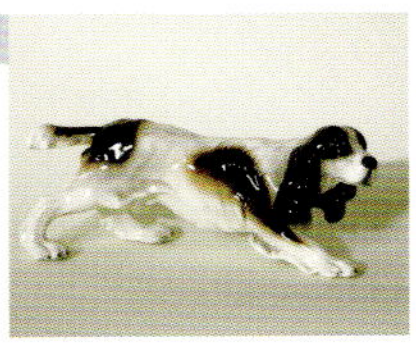
FIGUR „Spaniel"
(Laufender Spaniel)
ca. 1934 (für den Entwurf)
7330F, H. 18 cm, B. 44 cm; GSNachf. 5,8h; Fotoarchiv K. Täuber
Bem.: Für die Zuordnung dieser Figur und der damit verbundenen Gruppe, danke ich Herrn Herwig Thomasch, dem Sohn von Wilhelm Thomasch.
-> vgl. 7062F; 8343F
-> vgl. 7015F; 7281F; 8342F= jeweils GRUPPE
Pendant -> vgl. 7329F
Lit.: GSNachf. 5,8 (m. Abb.), hier mit Ang. des Originaltitels FIGUR „Spaniel" (Rückwärtiger Hund aus der GRUPPE 7281).
Quelle: Fotoarchiv K. Täuber

7338
WANDMASKE Frauenkopf mit Blumen im Haar und Schleife/Mascherl um den Hals
ca. 1935
7338,3 Firmenmarke 19, H. 24 cm; Doro 20.12.1994
7338 (Firmenmarke ohne Ang.), H. 24 cm; Christie's SK 21.6.1996
Lit.: Pinhas, S. 172 (m. Abb.)

7340
Prischl, Adolf
DECKELDOSE Frauenkopf mit Kragen
Form: rechteckig
Dekor: Frauenkopf, reliefiert
ca. 1935
7340,25 Firmenmarke 19, Made in Austria, unsign., Ausformung zw. 1945–1953, H. 4,5 cm, L. 11 cm; Schuler 58
Bem.: Das Motiv des Frauenkopfes für diese Deckeldose geht auf den Entwurf Prischls (ca. 1933/34) für eine Wandmaske zurück. Bekannt ist auch die Variante als Abreißkalender für die Wand. Dort wurde ebenfalls das Motiv des Frauenkopfes verwendet und auf eine rechteckige Platte appliziert. Leider lässt sich die Modell-Nr. beim hier einzig nachweisbaren Exemplar nicht korrekt ablesen.
-> vgl. 6836; 7549= jeweils WANDMASKE

7344M
FIGUR/GRUPPE „Reh(e)" (auf länglichem Natursockel)
ca. 1935/36
FIGUR
7344,246,9 Firmenmarke 19, Made in Austria, H. 22,8 cm, B. 36,9 cm; eBay # 515897689
7344,647,? Firmenmarke 19, H. fehlt; Schuler 67
GRUPPE
7344,44,65 Firmenmarke 19, Made in Austria, H. 23 cm, L. 33 cm; eBay # 7334498798
Quelle: AGM (Kontoblatt der Wiener Manufaktur Josef Schuster für die Firma Porslins Ericsson-Lund, Stockholm?, Schweden vom 7. Januar 1943), hier mit Ang. des Originaltitels GRUPPE „Rehe".

7346
„Rehbock"
ca. 1935/36
Quelle: AGM (Bestellungen der Firma Wahliss (Wien) vom 17.10.1945 und 27.2.1946 an die „Arbeitsgemeinschaft der Firma Friedrich Goldscheider", hier jeweils mit Ang. des (Original?-)Titels „Rehbock", in Ausführung „Terrakotta").

7350F
ASCHER „Ascher, klein"
ca. 1935/36
Quelle: AGM (Rechnung an Maurice Mizrahi, Alexandria, Ägypten, vom 22. April 1953), hier mit Ang. des Originaltitels „Ascher, klein".

7353
Matula, Viktor
ASCHER Kopf und Vase
ca. 1935/36
7353 H. fehlt, B. fehlt
Quelle: Nachlass Matula, Wien

7354
Knörlein, Rudolf (zugeschr.)
ASCHER
Form: quadratisch, eine Zigarettenablage
Dekor: Frauenkopf mit Schulteransatz, reliefiert
ca. 1935/36
7354,26 Firmenmarke 19, Made in Austria, Ausformung zw. 1945–1953, B. 12,7 cm; eBay # 2602014060
Bem.: Das Dekor im Spiegel des Aschers geht auf eine Wandmaske, die Rudolf Knörlein zuzuschreiben ist, zurück -> vgl. 6427.

7355
Matula, Viktor
ASCHER Reh
ca. 1935/36
7355 H. fehlt, B. fehlt
Quelle: Nachlass Matula, Wien

7356
Matula, Viktor
ASCHER Fisch
ca. 1935/36
7356 H. fehlt, B. fehlt
Quelle: Nachlass Matula, Wien

7364
Dakon, Stephan
Sitzende Dame mit großem Hut (auf zwei Reisekoffern mit Pekinesen)
ca. 1934 (für den Entwurf)
7364,42,19 gem. H, Firmenmarke 19, Made in Austria, H. 25,5 cm; Doro 24.11.2004
-> vgl. 6959= ähnl. Modell, hier wird der auf dem Koffer sitzende Pekinese durch eine Zeitung ersetzt.

7367F
Weiss, Claire (Klára)
(d.i. Herczeg, Claire (Klára)
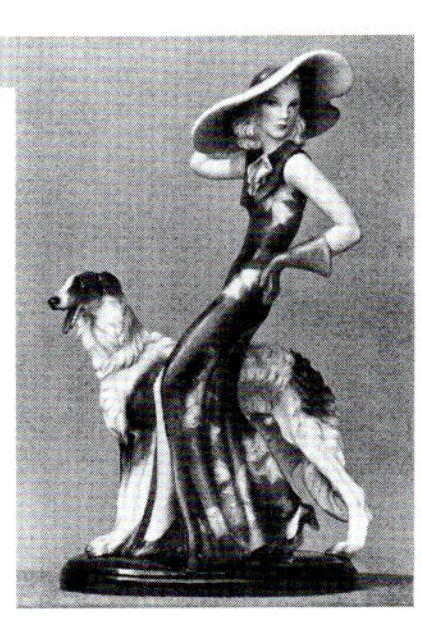
GRUPPE „Maskerade"
(Schreitende elegante Dame mit Barsoi) (auf Ovalsockel)
ca. 1935/36
7367,11 gem. GE, Firmenmarke 19, Made in Austria, unsign., Ausformung zw. 1945–1953, H. 40,6 cm; eBay # 399488106 (KH www.tias.com/stores/frozenbeauty)
7367,37,11 (Firmenmarke ohne Ang.), sign. Claire Weiss, H. 38,2 cm; Christie's SK 9.2.2000
7367,41,19 (Firmenmarke ohne Ang.), unsign.?, Pick. 6, H. 39,2 cm; Christie's SK 13.5.1999
7367,172,10 (Firmenmarke ohne Ang.), sign. Claire Weiss, H. 40,6 cm; Christie's SK 27.6.1997
7367,194,10 Firmenmarke 19, Made in Austria, unsign., H. 42,4 cm; eBay # 1860110203 (Soth.)
7367,393,31,G gem. F(K?), Firmenmarke 19, Made in Germany, Ausformung zw. 1938–1941, unsign., H. 43 cm; Heickmann 18
7367F, H. 40 cm, B. 32 cm; GSNachf. 1,15a
-> vgl. 8155
-> vgl. 7279 (Dame ohne Hund)
Lit.: GSNachf. 1,15 (m. Abb.), hier mit Ang. des Originaltitels „Maskerade".
Quelle: AGM (Rechnung an Maurice Mizrahi, Alexandria, Ägypten, vom 22. April 1953), hier mit Ang. des Originaltitels „Maskerade/Hund".

7374F
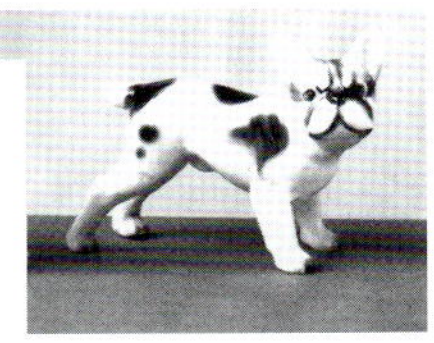
FIGUR „Bully"
(Stehender Bullterrier)
ca. 1935/36
7374F, H. 8 cm, B. 9,5 cm; GSNachf. 5,1a; Fotoarchiv K. Täuber, hier mit der Ang.: H. 8 cm, B. 10 cm
Lit.: GSNachf. 5,1 (m. Abb.), hier mit Ang. des Originaltitels „Bully".
Quelle: Fotoarchiv K. Täuber

7375F
FIGUR „Foxterrier" (liegend)
ca. 1935/36
7375F, H. 6 cm, L. 14 cm
Quelle: Fotoarchiv K. Täuber

7376F
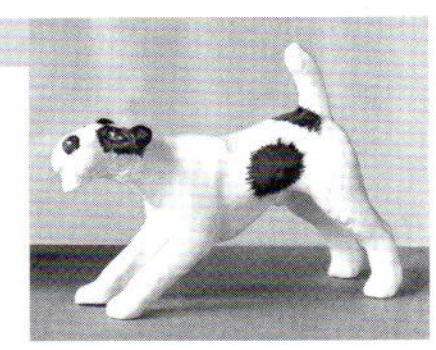
FIGUR „Foxterrier"
(Bellender Foxterrier)
ca. 1935/36
7376F, H. 8,5 cm, B. 11 cm; GSNachf. 5,1c; Fotoarchiv K. Täuber, hier mit der Ang.: H. 8 cm, B. 13 cm
Lit.: GSNachf. 5,1 (m. Abb.), hier mit Ang. des Originaltitels „Foxterrier".
Quelle: Fotoarchiv K. Täuber

7377F
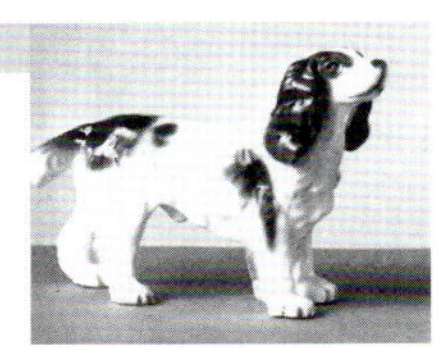
FIGUR „Spaniel" (Stehender aufmerksamer Spaniel)
ca. 1935/36
7377F, H. 8 cm, B. 11 cm; GSNachf. 5,1d; Fotoarchiv K. Täuber, hier mit der falschen Mod.-Nr. 7378 bezeichnet.
Lit.: GSNachf. 5,1 (m. Abb.), hier mit Ang. des Originaltitels „Spaniel".
Quelle: Fotoarchiv K. Täuber

7378F
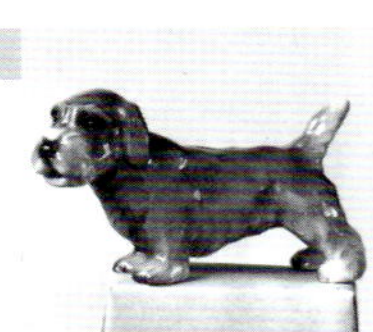
FIGUR „Cairnterrier"
(Stehender Cairnterrier)
ca. 1935/36
7378F, H. 5,5 cm, B. 10,5 cm; GSNachf. 5,1b; Fotoarchiv K. Täuber, hier mit der Ang.: H. 6 cm, B. 10 cm und der falschen Mod.-Nr. 7377 bezeichnet.
Lit.: GSNachf. 5,1 (m. Abb.), hier mit Ang. des Originaltitels „Cairnterrier".
Quelle: Fotoarchiv K. Täuber

7385
Knörlein, Rudolf
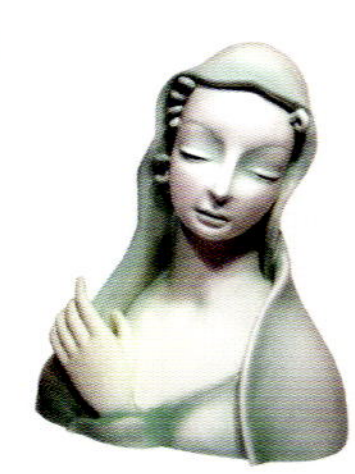
BÜSTE Madonna mit einer Hand
ca. 1935/36
7385,8,65 Firmenmarke 19, Made in Austria, H. 27,1 cm; Sammlung a@goldscheider.net

7385,20 gem. lig. NO, Firmenmarke 19, Made in Austria, Ausformung zw. 1945–1953, H. 25,9 cm; Sammlung Witzel
7385,31,4 gem. N, (Firmenmarke ohne Ang.), H. 26,5 cm; Schuler 70

7386M
Matula, Viktor
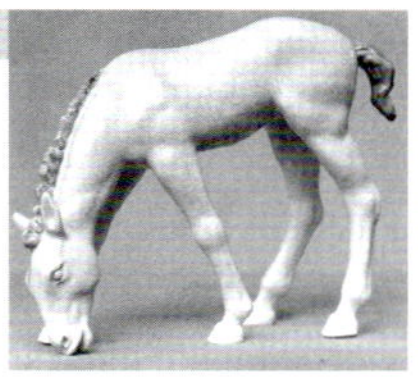
FIGUR „Pferd" (Stehendes fressendes Pferd)
ca. 1933 (für den Entwurf)
7386M, H. 28 cm, B. 33 cm; GSNachf. 3,18a
-> vgl. 6797; 7737
-> vgl. 6986
-> vgl. 7939= GRUPPE (als Höhenvariante).
Lit.: GSNachf. 3,18 (m. Abb.), hier mit Ang. des Originaltitels „Pferd".
Quelle: AGM (Bestellung der Firma Wahliss (Wien) vom 27.2.1946 an die „Arbeitsgemeinschaft der Firma Friedrich Goldscheider", in Ausführung „Terrakotta"); Fotoarchiv K. Täuber.

7390F
Weiss, Claire (Klára)
(d.i. Herczeg, Claire (Klára)
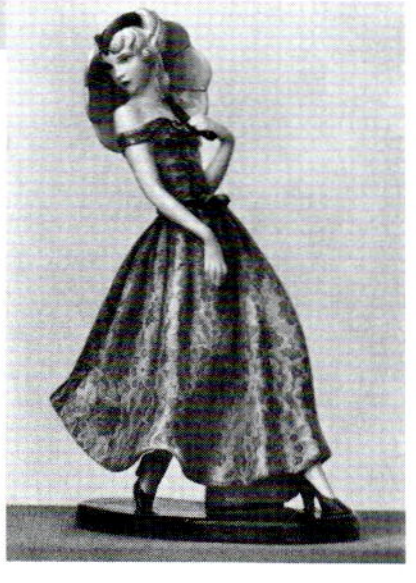
FIGUR/LAMPE „Parisienne" (Schreitence elegante Dame, ihren Schirm in der linken Hand haltend) (auf Ovalsockel)
ca. 1935/36
FIGUR
7390,39,31 gem. F., Firmenmarke 19, Made in Austria, sign. Claire Weiss, H. 31,5 cm; Müller 93
7390,65,66 (Firmenmarke ohne Ang.), unsign., H. 35 cm; Doro 6.4.1999
7390,86,8 gem. lig. ST, Firmenmarke 19, Made in Austria, Hand dec., unsign., H. 36 cm (wohl Messfehler); eBay # 7321407557
7390,124,31,8 gem. 94, gem. lig. STH, Firmenmarke 19, Made in Austria, unsign., H. 33 cm; eBay # 159450552
7390,258,4 gem. lig. blau OS, Firmenmarke 19, Made in Austria, sign. Claire Weiss, H. 34 cm; KH
7390F, H. 32 cm, B. 19 cm; GSNachf. 1,16a
LAMPE
7390,16,6 (Firmenmarke ohne Ang.), unsign.?, Gesamth. 57 cm; Henry's 26.2.1998
-> vgl. 7723; 8138
-> vgl. 8361= seitenverkehrtes Modell
Lit.: GSNachf. 1,16 (m. Abb.), hier mit Ang. des Originaltitels „Parisienne". Der Hinweis auf die Ausführung als Lampe fehlt jedoch.
Quelle: AGM (Rechnung an Maurice Mizrahi, Alexandria, Ägypten, vom 22. April 1953), hier mit Ang. des (Original)-Titels „Parisienne".

7393 -> ÜBERSCHNEIDUNG
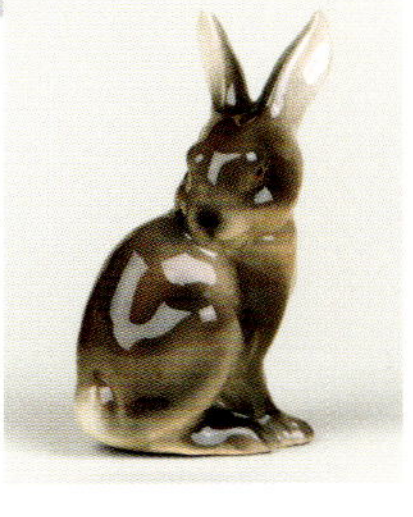
Sitzender Hase
ca. 1935/36
7393,4 Firmenmarke 19, H. 11 cm; PB Wien 19
7393,6 (od. 9?) Firmenmarke 19, Pick. 1, L. 10,1 cm; eBay # 1264794020
7393 H. 10 cm, B. 6 cm; Fotoarchiv K. Täuber
Quelle: Fotoarchiv K. Täuber

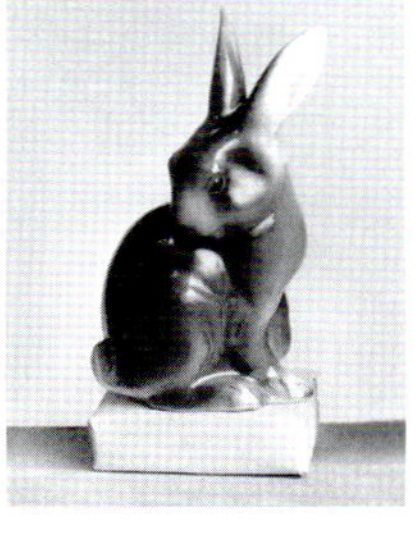

7393 -> ÜBERSCHNEIDUNG
Dakon, Stephan (zugeschr.)
Tennisspielerin (Stehendes Mädchen mit Tennisracket) (auf Ovalsockel)
ca. 1935/36
7393 (Firmenmarke ohne Ang.), H. 25 cm; Pinhas, S. 135 (m. Abb.)

7394
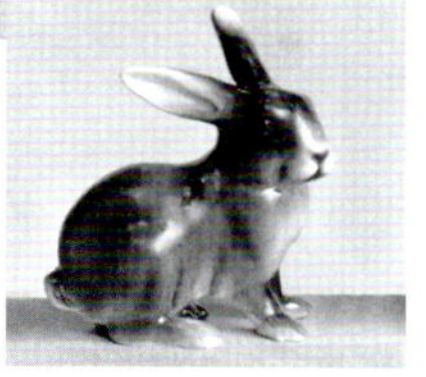
Sitzender Hase
ca. 1935/36
7394 H. 10 cm, L. 7 cm
Quelle: Fotoarchiv K. Täuber

7395
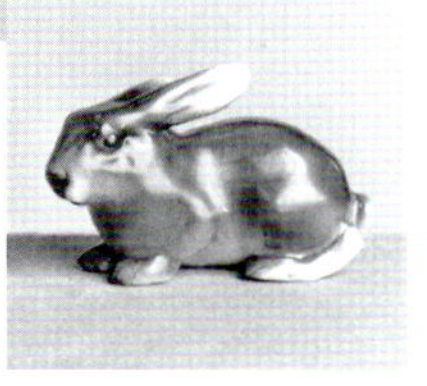
Liegender Hase
ca. 1935/36
7395,4 Firmenmarke 19, Ausformung zw. 1945–1953, H. 7 cm; eBay # 6615350983
7395 H. 7 cm, L. 10 cm
Quelle: Fotoarchiv K. Täuber

7396F
Lorenzl, Josef
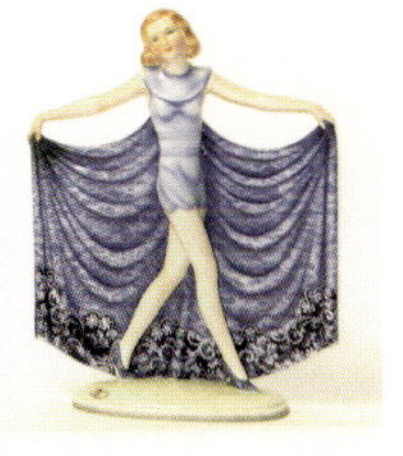
FIGUR „Revue" (Stehende Tänzerin, ihr Kleid auseinander haltend) (auf Ovalsockel)
ca. 1935
7396,11 gem. S., gem. lig. HS, Firmenmarke 19, Made in Austria, Pick. 1, Ausformung zw. 1945–1953, H. 39,0 cm, B. 30 cm; Privatsammlung, Wien
7396,26 Firmenmarke 19, unsign., Ausformung zw. 1945–1953, H. 39 cm; KH
7396 ohne Ang., Firmenmarke 19, sign. Lorenzl, H. 37 cm; Ahlden 129
7396F, H. 38 cm, B. 28 cm; GSNachf. 1,11b
-> vgl. 7192
Lit.: GSNachf. 1,11 (m. Abb.), hier mit Ang. des Originaltitels „Revue".

7398
Matula, Viktor
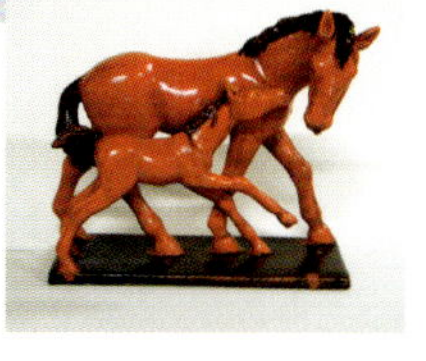
Zwei laufende Pferde (auf Rechecksockel)
ca. 1935/36
7398,20 ger., Firmenmarke 19, Made in Austria, unsign., Ausformung zw. 1945–1953, H. ca. 20 cm; KH
7398,113,7 ger. 1, gem. lig. AH, Firmenmarke 19, Made in Austria, unsign., H. 20,9 cm, L. 24,5 cm; MAK Wien, Inv. Ke 10602-23
Quelle: Nachlass Matula, Wien

7399
Matula, Viktor
FIGUR Pferd
ca. 1935/36
7399 H. fehlt
Quelle: Nachlass Matula, Wien

7401M
Matula, Viktor
FIGUR „Reh"
ca. 1935/36
7401M, H. 6 cm, B. 8,5 cm; GSNachf. 3,12d
Lit.: GSNachf. 3,12 (m. Abb.), hier mit Ang. des Originaltitels „Reh".
Quelle: Nachlass Matula, Wien

7402M
Matula, Viktor
FIGUR „Pferd"
ca. 1935/36
7402M, H. 6 cm, B. 8 cm; GSNachf. 3,12f
Lit.: GSNachf. 3,12 (m. Abb.), hier mit Ang. des Originaltitels „Pferd".
Quelle: Nachlass Matula, Wien

7405
Dakon, Stephan (zugeschr.)
GRUPPE Stehender Knabe mit Brezel und Dackel
ca. 1935/36
7405,10,11 Firmenmarke 19, unsign., H. 20,3 cm; eBay # 921138223

7406
Dakon, Stephan
FIGUR „Triton" (Stehendes Mädchen mit Kappe und Tretroller) (auf abgerundetem Längssockel)
ca. 1935/36
7406,8,9 Firmenmarke 19, Made in Austria, unsign., H. 26,7 cm; Wien Museum Inv. 190.159
7406,213,16 gem. 5173, Firmenmarke 19, Made in Austria, H. 26,7 cm; eBay # 189540819
7406,238,31 unsign., H. 25 cm; Christie's SK 12.9.1997
7406,319,31 ger. D, 5A, gem. AN, Firmenmarke 19, Made in Germany, Ausformung zw. 1938–1941, H. 27 cm; eBay # 1284478882
7406,425,19,50,D gem. lig. AW, Firmenmarke 19, Made in Germany, Ausformung zw. 1938–1941, H. 26,8 cm; Quit. 43
Lit.: GSN, S. 66, 67 (m. Abb.)
Quelle: Archiv Mimi Dakon (Tantiemenabrechnung 1.7.–31.12.1937; Modellliste Dakon, undatiert, ca. 1940; Wiener Manufaktur FG, Modellliste Dakon, 7.11.1946; Tantiemenabrechnung 1. Halbjahr 1949; 2. Halbjahr 1949), hier jeweils mit Ang. des Originaltitels FIGUR „Triton".
GS neu: Modell-Nr. 58 „Rotkäppchen" (H. 22 cm), Aufl. ohne Ang.

7407
Dakon, Stephan
Sitzendes Mädchen mit Hut und Schulranzen
ca. 1935/36
7407,30,19 gem. lig. JR, Firmenmarke 19, Made in Austria, H. ca. 24 cm; KH

7409
Dakon, Stephan
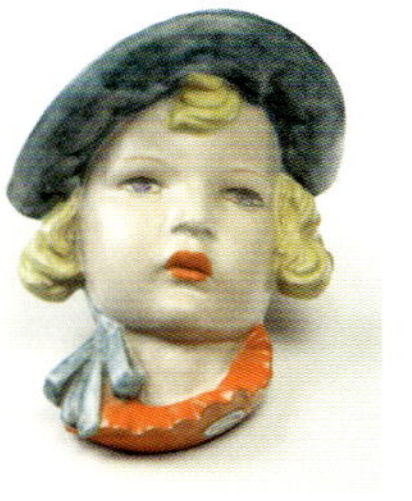
WANDMASKE Mädchenkopf mit Hut, Schleife/Mascherl und Kragen
ca. 1935/36
7409,2 Firmenmarke 19, Made in Austria, unsign., Pick. 6, Ausformung zw. 1945–1953, H. 22 cm; eBay # 2523691999
7409,4 gem. blau H., Firmenmarke 19, Made in Austria, H. 22,8 cm; eBay # 2536659132; Soth. com. # 3J5HG
7409,8 gem. lig. AH, Firmenmarke 19, Made in Austria, unsign., Pick. 5, Ausformung zw. 1945–1953, H. 22 cm; Wien Museum Inv. 220.184
7409,20,5 Firmenmarke 19, Made in Austria, unsign., H. 22,8 cm; eBay # 2515046563
7409,20 ger. XVB, gem. W.B, Firmenmarke 19, Made in Austria, Ausformung zw. 1945–1953, H. 22,8 cm; eBay # 3719961130
7409,20 Firmenmarke 19, Made in Germany, Ausformung zw. 1938–1941, H. 22,8 cm; KH
Bem.: Bei der dargestellten Person könnte es sich um den Kinderstar Shirley Temple handeln.

7412
Dakon, Stephan
WANDMASKE Mädchenkopf mit Béret und Schal
ca. 1935/36
7412,4 Firmenmarke 19, Made in Austria, unsign., Ausformung zw. 1945–1953, H. 25 cm; eBay # 3351445189
7412,35 gem. lig. HG, Firmenmarke 19, Made in Ger-

many, Ausformung zw. 1938–1941, unsign., H. 25 cm; eBay # 2540435950
7412 (Firmenmarke), H. 18 cm; Pinhas, S. 167, 174 (m. Abb.), in zwei Farbvarianten.

7413
Dakon, Stephan
WANDMASKE Mädchenkopf mit Kappe und Schleife / Mascherl
ca. 1935/36
7413,20 Firmenmarke 19, Made in Austria, Ausformung zw. 1945–1953, H. 25,4 cm; eBay # 3725737474

7415
Dakon, Stephan
Schreitende Spanierin mit großem Fächer
(auf Ovalsockel)
ca. 1935/36

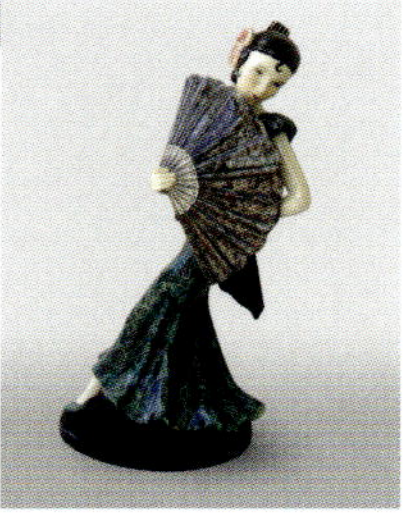

7415,22,6 Firmenmarke 19, H. 40 cm; eBay # 2247324497; Doro 13.9.2004
7415,29,6 gem. F., gem. lig. JR, Firmenmarke 19, Made in Austria, H. 40 cm (bzw. 45,7 cm); Christie's SK 4./5.12.1996; eBay # 1509820950; Bem.: Sehr abweichende Größenang. bei beiden Auktionen.

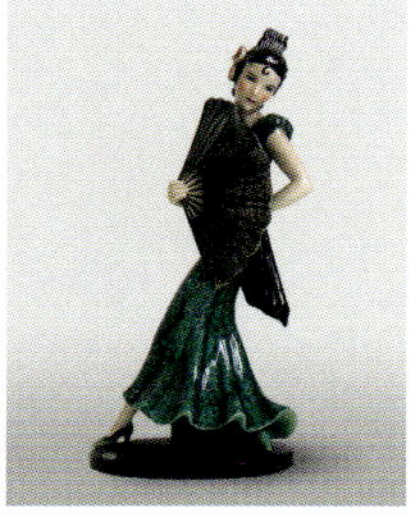

7415,51,19 gem. blau F., Firmenmarke 19, Made in Austria, H. 42 cm; eBay # 233600337
7415,70,31 ger. PV, gem. F, gem. lig. JR, Firmenmarke 19, Made in Austria, H. 41 cm; Sammlung Stressless, Portugal
7415,87,11 ger. XII, D, gem. F., Firmenmarke 19, Made in Austria, Pick. 5/b, H. 45,4 cm, B. 22 cm; Privatsammlung, Wien
7415,108,58 gepr. G, Firmenmarke 19, Made in Germany, Ausformung zw. 1938–1941, unsign., H. 43 cm; Soth. L. 2.11.1999

7416F
Dakon, Stephan
FIGUR „Golf" (Golfspielerin mit Hut und Schläger) (auf Rundsockel)
ca. 1935/36

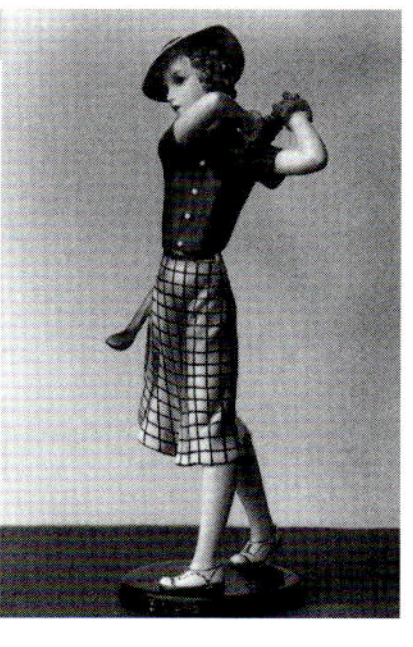

7416,76,66 gem. lig. SP, Firmenmarke 20, Made in Austria, Hand dec., unsign., H. 29 cm; Wien Museum Inv. 190.136
7416F, H. 28 cm; GSNachf. 1,17a
Ausst.: SSH 1992
-> vgl. 8140
Lit.: GSNachf. 1,17 (m. Abb.), hier mit Ang. des Originaltitels „Golf".
Quelle: Archiv Mimi Dakon (Originalfoto)

7418
Knörlein, Rudolf
KERZENHALTER (dreiflammig)
Form: röhrenartig, scheibenartige Kerzentüllen (auf Rund- / Ovalsockel)
Dekor: vermutlich diverse Dekore
ca. 1935/36

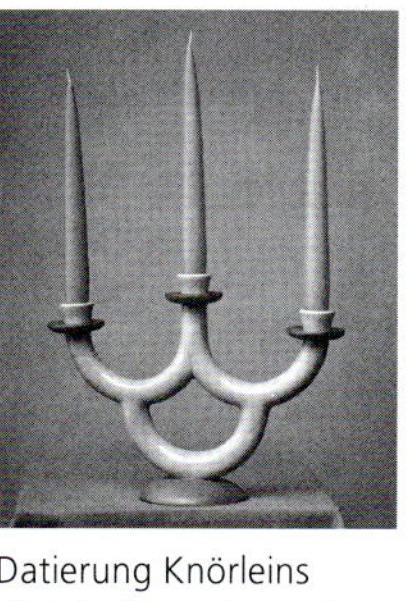

7418, H. 25,5 cm, B. 34 cm
Bem.: Durch die eigenhändige Datierung Knörleins (1936) auf dem Trägerblatt des Originalfotos kann die Entstehung um 1935/36 ziemlich exakt bestimmt werden.
Quelle: Privatarchiv (Originalfoto)

7419
WANDMASKE Frauenkopf mit Hand und Buch
ca. 1935/36

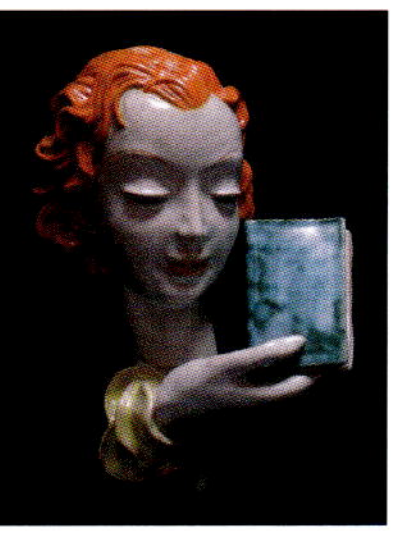

7419,3 Firmenmarke 19, Originalpick.: „Carl Deesz, Saarbrücken", H. 28 cm; PB Deutschland
7419,3 gem. G, Firmenmarke 19, Made in Austria, Ausformung zw. 1945–1953, H. 28 cm; eBay # 1837101013 (Soth., Harriman Judd Coll. of British Art Pottery)

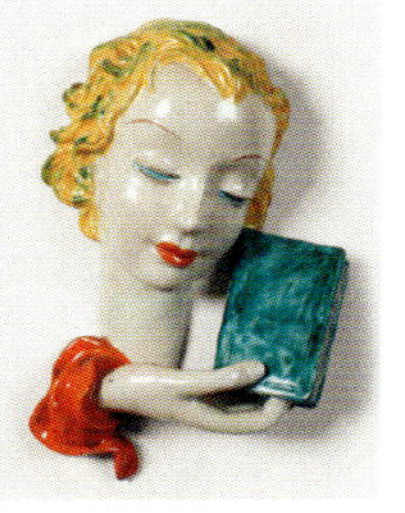

7419,8 gem. lig. AH, Firmenmarke 19, Made in Austria, Ausformung zw. 1945–1953, H. 28 cm; KH
7419,9 ger., Firmenmarke 19, H. 28 cm; KH Art 1900/Berlin
7419,12 gem. lig. AH, Firmenmarke 19, Made in Austria, Ausformung zw. 1945–1953, H. 28 cm; Sammlung.schmutz@chello.at 44

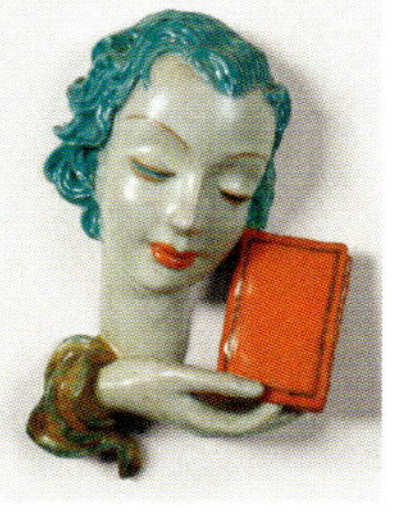

7419,S ger., gem. HX, Firmenmarke 19, Made in Austria, H. 28 cm; Sammlung.schmutz@chello.at 43

7420M
Prischl, Adolf
WANDMASKE „Maske" (Frauenkopf mit wehendem Haar)
ca. 1935/36

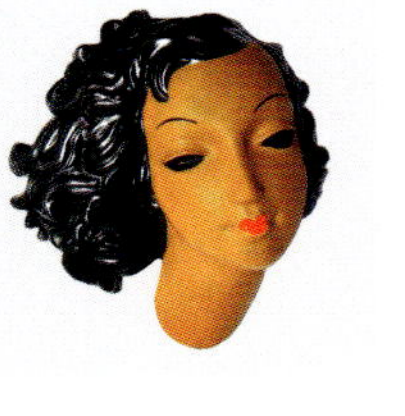

7420,2,33 Firmenmarke 19, Made in Austria, unsign., H. 23,5 cm, B. 22 cm; Sammlung a@goldscheider.net
7420,3 Firmenmarke 19, unsign., H. 24 cm; Soth. L. 30.11.1983

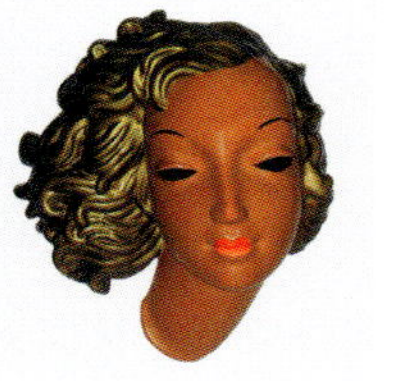

7420,4 ger., Firmenmarke 19, Made in Austria, unsign., Ausformung zw. 1945–1953, H. 24 cm, B. 21 cm; eBay # 873507541
7420,4,28 Firmenmarke 19, unsign., H. 23,5 cm, B. 22 cm; Sammlung a@goldscheider.net
7420,5,77 Firmenmarke 19, Made in Austria, unsign., H. 24,5 cm, B. 22,5 cm; eBay # 887945579
7420,8 gem. lig. AH, Firmenmarke 19, Made in Austria, unsign., Ausformung zw. 1945–1953, H. 22 cm; Wien Museum Inv. 220.801
7420,12 (Firmenmarke ohne Ang.), unsign., H. 22,3 cm; Christie's SK 5.12.1997
7420,24,28 Firmenmarke 19, unsign., H. 24 cm; city-antik
7420,83 Firmenmarke 19, unsign., H. 25 cm; Herr 34
7420 gem.,2669,6 gepr., (Firmenmarke ohne Ang.), unsign., H. 24,2 cm; Christie's SK 18.9.1998
7420 ger., (ohne Firmenmarke), unsign., Ausformung 50er Jahre, H. 24,5 cm; Sammlung John Nash
7420M, H. 24 cm, B. 21 cm; GSNachf. 2,3a; Fotoarchiv K. Täuber, hier mit der Ang.: H. 25 cm, B. 23 cm
Bem.: Diese Maske wurde auch noch zw. 1953–1960 bei Carstens (GWM West-Germany) produziert.
-> vgl. 7750; 8207
Lit.: GSNachf. 2,3 (m. Abb.), hier mit Ang. des Originaltitels „Maske"; KGSF1; KGSF2; Könecke (2002), S. 77, hier mit Ang. des Künstlers A. Prischl (für die Modell-Nr. 8207), S. 88 (m. Abb.).
Quelle: AGM (Bestellung der Firma Elisabeth Salle (Wiesbaden), vom 8.1.1951); Fotoarchiv K. Täuber.

7423F
Dakon, Stephan
Sitzende Dame mit Hut und zwei Foxterriern (auf Backsteinmauer/Sockel)
ca. 1935/36

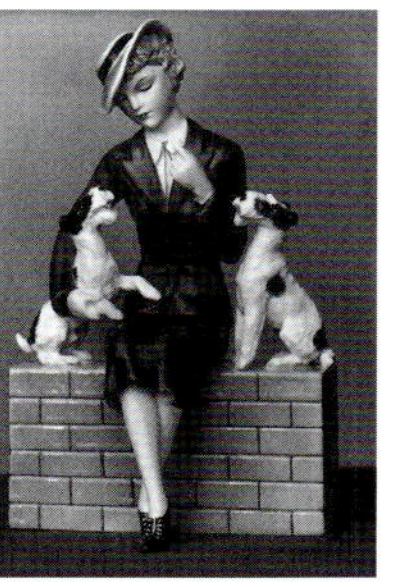

7423F, H. 26 cm, B. 16 cm
Quelle: Archiv Mimi Dakon (Originalfoto)

7424F
Dakon, Stephan / Lorenzl, Josef
Stehende Dame mit Hut, Cape (großer Fransenschal), einen kleinen Hund (King Charles?) haltend
(auf Rundsockel)
ca. 1935/36

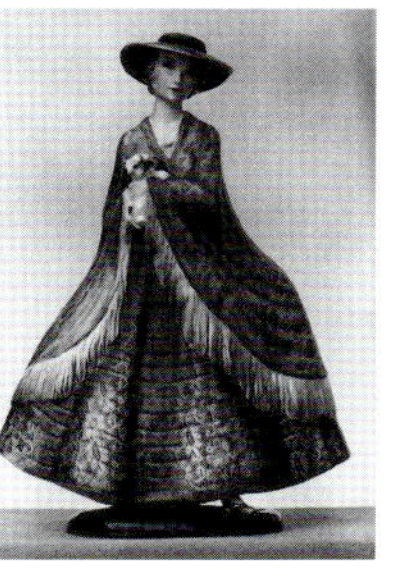

7424,11,10 gem. lig. ON, Firmenmarke 19, sign. Lorenzl, H. 33 cm; eBay # 1227349696; Bem.: Die Sign. „Lorenzl" ist ein Versehen des Modelleurs, denn durch das Originalfoto aus dem Archiv Mimi Dakon lässt sich diese Figur eindeutig Stephan Dakon zuweisen.
7424F, H. 33 cm
Quelle: Archiv Mimi Dakon (Originalfoto)

7426F
Dakon, Stephan
FIGUR „Sterbender Schwan" (mit/ohne Schwanenflügel am Kleid) (auf Ovalsockel)
ca. 1935/36

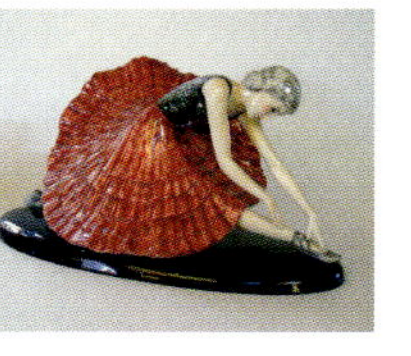

ohne Schwanenflügel
7426,4,4 gem. .X., Firmenmarke 19, Made in Austria, Hand dec., unsign., H. 21,4 cm, L. 42,4 cm; MAK Wien, Inv. Ke 10602-54

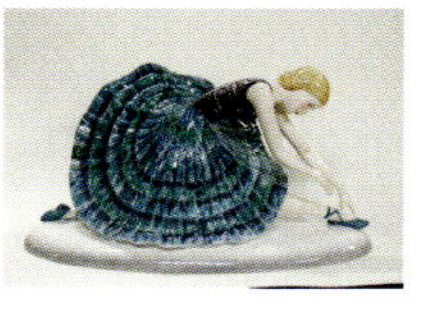

7426,8 gem. W, Firmenmarke 19, Made in Austria, unsign., Ausformung zw. 1945–1953, H. 22 cm, L. 42 cm; WKA 35; WKA 40; Doro 10.3.2003; Quit. 43; Doro 15.3.2005

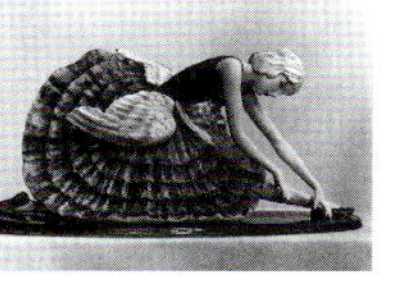

7426,11 Firmenmarke 19, Made in Austria, unsign., Ausformung zw. 1945–1953, H. 23 cm, L. 42 cm; eBay # 110267164
7426,66,44 gem. lig. HS, Firmenmarke 19, Made in Austria, Hand dec., unsign., Pick. 1 (zwei mal), H. 21 cm, L. 42 cm; Wien Museum Inv. 190.137
7426,137,19 Firmenmarke 19, Made in Austria, H. 20 cm, L. 41 cm; Falkkloo's, Malmö 21.4.2001
7426,297,19,6 gem. AN, Firmenmarke 19, unsign., H. 22,3 cm, L. 41 cm; Sammlung.schmutz@chello.at 100
7426,322,17 Firmenmarke 19, unsign., H. 23 cm, L. 44 cm; Sammlung Alain Bernard
7426,345,7 gem. F., Firmenmarke 19, Made in Germany, unsign., Ausformung zw. 1938–1941, H. 23 cm; eBay # 2239362571
7426,348,7 Firmenmarke 19, Made in Germany, unsign., Ausformung zw. 1938–1941, H. 23 cm, L. 42 cm; eBay # 6506613791
7426,363,? Firmenmarke 19, Made in Germany, unsign., Ausformung zw. 1938–1941, L. 44 cm; RM 28
mit Schwanenflügel
7426,71,11,8A gem. NO gem. lig. rot JR, Firmenmarke 19, Made in Austria, H. 23 cm, L. 40 cm; eBay # 3778303872
7426F, H. 18 cm, L. 40 cm; GSNachf. 1,18
-> vgl. 6361= ähnl. Modell (mit Federputz im Haar, ohne Schwanenflügel)
Lit.: GSNachf. 1,18 (m. Abb.), hier mit Ang. des Originaltitels „Sterbender Schwan"; Witkiewicz, S. 128–129 (m. Abb.).
Quelle: AGM (Rechnung an Maurice Mizrahi, Alexandria, Ägypten, vom 22. April 1953); Archiv Mimi Dakon (Originalfoto).

7428
„Elefant"
ca. 1935/36
Quelle: AGM (Kontoblatt der Wiener Manufaktur Josef Schuster für die Firma Porslins Ericsson-Lund, Stockholm?, Schweden vom 7. Januar 1943); (Bestellungen der Firma Wahliss (Wien) vom 9. und 24.11.1945 an die „Arbeitsgemeinschaft der Firma Friedrich Goldscheider", hier jeweils mit Ang. des Originaltitels „Elefant"; Bestellung der Firma Wahliss (Wien) vom 24.11.1945 und 27.2.1946).

7430
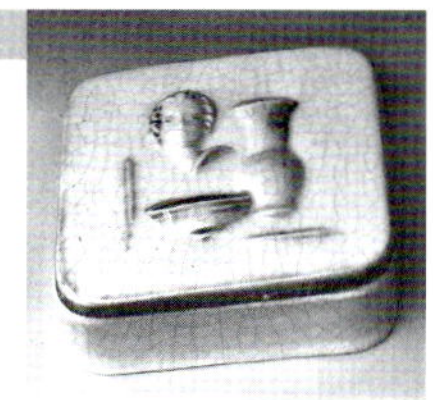
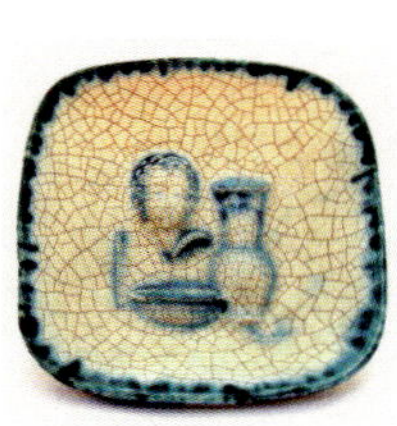
Knörlein, Rudolf
DECKELDOSE
Form: quadratisch, abgerundete Ecken
Dekor: Frauengesicht, Vase, Schale
ca. 1934/35
7430, H. 4 cm, B. 11 cm
Bem.: Das Dekor dieser Dose findet sich auch auf einer Werbeschale der Manufaktur, die zum 50-jährigen Firmenjubiläum 1935 editiert wurde. Durch die eigenhändige Datierung Knörleins (1934) auf dem Trägerblatt des Originalfotos kann die Entstehung um 1934 bestimmt werden.
Quelle: Privatarchiv (Originalfoto)

7431F
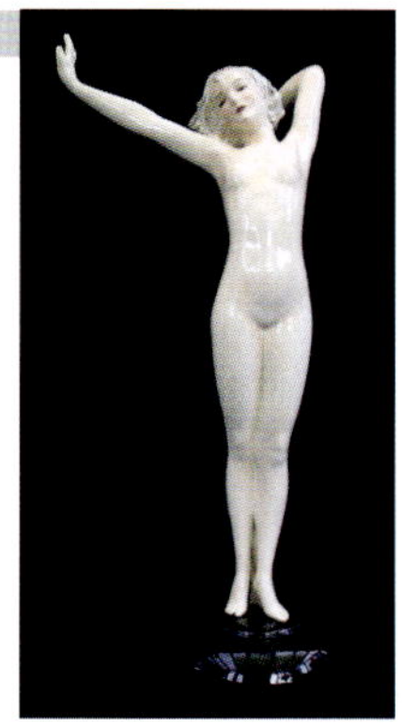
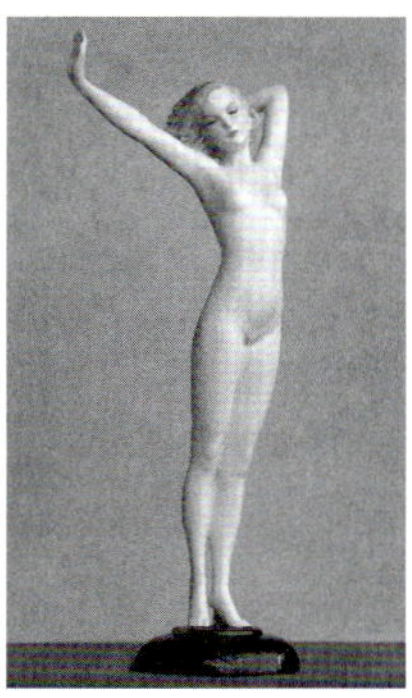
FIGUR „Erwachen" (Stehender Damenakt mit ausgestrecktem Arm) (auf Rundsockel)
ca. 1935/36
7431,6,14 ger., Firmenmarke 19, Made in Austria, H. 35,5 cm; eBay # 2654999609
7431,72,6 gem. F., gem. rot N, Firmenmarke 19, Made in Austria, H. 36,8 cm; Sammlung Robert Senft
7431,775,10 gem. lig. FD, Firmenmarke 19, Made in Germany, Ausformung zw. 1938–1941, H. 36,2 cm; eBay # 2586494706
7431F, H. 38 cm; GSNachf. 1,31a
Bem.: Diese Figur wurde auch noch zw. 1953–1960 bei Carstens (GWM West-Germany) produziert.
-> vgl. 8137
-> vgl. 7525= ähnl. Modell
Lit.: GSNachf. 1,31 (m. Abb.), hier mit Ang. des Originaltitels „Erwachen"; Könecke (2002), S. 76, 86 (m. Abb.); Fotoarchiv K. Täuber.
Quelle: AGM (Rechnung an Maurice Mizrahi, Alexandria, Ägypten, vom 22. April 1953), hier unter dem Titel „Akt" geführt, geliefert in Ausführung „matt".

7432
Stehende Fechterin mit Dreispitz und Degen
ca. 1936
7432,31,10 (Firmenmarke ohne Ang.), Pick. 6, H. 32 cm; Doro 29.5.2001; Bem.: Degenklinge aus Metall gefertigt.

7433
Matula, Viktor
LEUCHTER
ca. 1936
7433 H. fehlt
Quelle: Nachlass Matula, Wien

7435M / F
BÜSTE „Büste mit Blume" (Mädchenkopf leicht nach rechts blickend, Blumenblüte auf der rechten Schulter)
ca. 1936
7435,S Firmenmarke 19, Made in Austria, H. 20,3 cm, Holzsockel ca. 5 cm; eBay # 2837611184
7435M, H. 19 cm, B. 14 cm; GSNachf. 2,17
Bem.: Dieser Entwurf wurde unter der Mod. Nr. 594 auch in Trenton (USA) produziert (H. 20,3 cm) -> vgl. Ex. aus dem WKH.
Lit.: GSNachf. 2,17 (m. Abb.), hier mit Ang. des Originaltitels „Büste mit Blume".
Quelle: AGM (Rechnung an Maurice Mizrahi, Alexandria, Ägypten, vom 22. April 1953), geliefert in Ausführung „matt".

7442
Matula, Viktor
GRUPPE Pferde
ca. 1936
7442 H. fehlt
Quelle: Nachlass Matula, Wien

7444
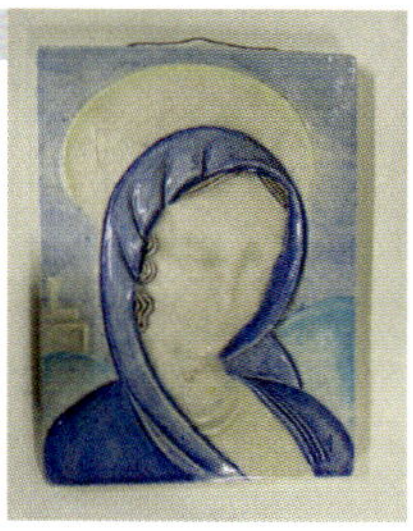
Knörlein, Rudolf
RELIEF Madonna (mit Aureole) (Wandrelief)
ca. 1936
7444,20 ger., Firmenmarke 19, Made in Austria, Monogr. RK, H. 26,9 cm, B. 20,1 cm; MAK Wien, Inv. Ke 10602-106

7446
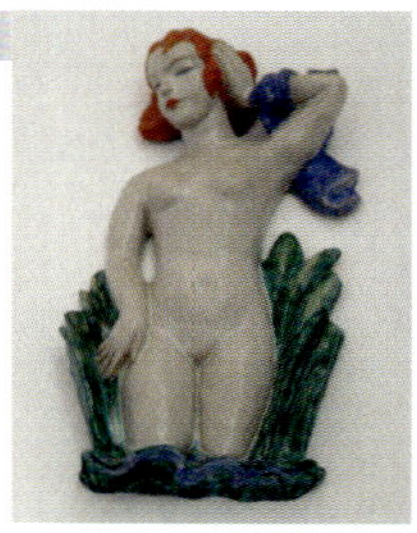
WANDAPPLIK Badendes Mädchen (Torso) mit Handtuch im Schilf stehend
ca. 1936
7446,0 Firmenmarke 19, Made in Austria, H. 38 cm; Sammlung Stressless, Portugal

7447
RELIEF Skifahrerin (Wandrelief)
ca. 1936
7447,20 gem. B, Firmenmarke 19, Pick. 6, Ausformung zw. 1945–1953, H. 17,8 cm, B. 19 cm; eBay # 7379820115
Lit.: Pinhas, S. 178 (m. Abb.)

7448
Goebel, Kurt
Sitzendes posierendes Mädchen mit Hut (auf Rundsockel mit Kaktus)
ca. 1936
7448 Firmenmarke 19, Made in Austria, unsign., H. 19 cm; eBay # 7305387180

7449M
Goebel, Kurt
(STEH)KOPF „Kopf ohne / mit Sockel" (Frauenkopf mit geschlossenen Augen, Haarschmuck und Kragen)
ca. 1936
7449M, H. 28,5 cm, B. 17 cm; GSNachf. 2,19a
Lit.: GSNachf. 2,19 (m. Abb.), hier mit Ang. des Originaltitels „Kopf ohne / mit Sockel".
Quelle: AGM (Tantiemenliste Goebel, ohne Datierung (ca. 1940); (Bestellung der Firma Wahliss (Wien) vom 27.2.1946 an die „Arbeitsgemeinschaft der Firma Friedrich Goldscheider", in Ausführung „Terrakotta"; Rechnung an Maurice Mizrahi, Alexandria, Ägypten, vom 22. April 1953).

7453M / F
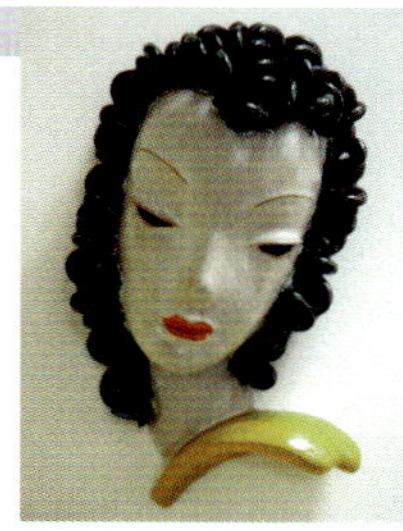
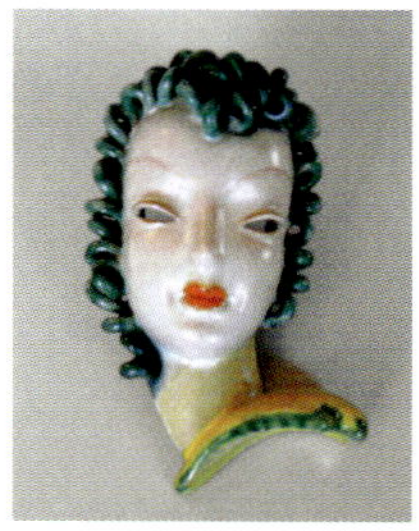
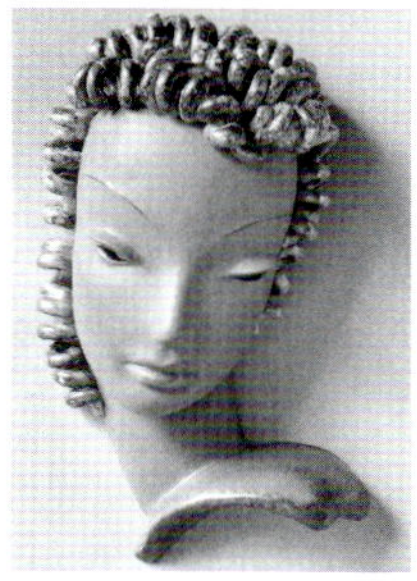
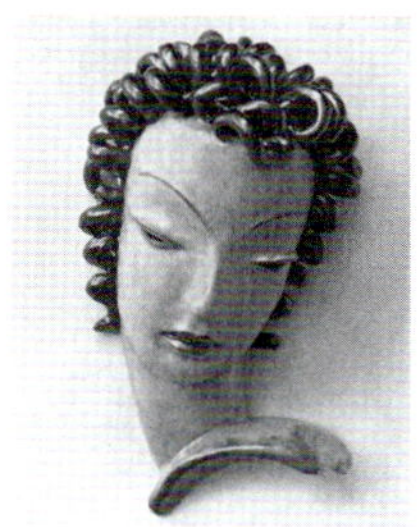
Knörlein, Rudolf
WANDMASKE/STEHMASKE „Turandot" (Frauenkopf mit Schulteransatz)
ca. 1932 (für den Entwurf)
WANDMASKE
7453,3 gem. G, Firmenmarke 19 (blau), Made in Austria (blau), unsign., Ausformung zw. 1945–1953, H. 20,3 cm; Sammlung.schmutz@chello.at 6
7453,3 ger. Firmenmarke 19, Made in Austria, unsign., Ausformung zw. 1945–1953, H. 20 cm; Sammlung Stressless, Portugal
7453,9 gem. X, gem. lig. AH, Firmenmarke 19, Made in Austria, unsign., Ausformung zw. 1945–1953, H. 21,5 cm; Doro 24.11.2004
7453,13,17 Firmenmarke 19, unsign., H. 20,8 cm; eBay # 1418018793
7453,33 (Firmenmarke ohne Ang.), unsign., H. 21 cm; Christie's SK 27.6.1997
7453,44 Firmenmarke 19, Made in Austria, unsign., Ausformung zw. 1945–1953, H. 20,3 cm; eBay # 3735953632
7453M, H. 20 cm, B. 10 cm; GSNachf. 2,1b
STEHMASKE (auf Holzsockel)
7453 (Firmenmarke ohne Ang.), unsign., Gesamth. 26,5 cm; eBay #.357954015
Bem. s. 6427
-> vgl. 6427; 6623
Lit.: GSNachf. 2,1 (m. Abb.), hier mit Ang. des Originaltitels MASKE „Turandot".
Quelle: AGM (Rechnung an Maurice Mizrahi, Alexandria, Ägypten, vom 22. April 1953), geliefert in Ausführung „matt"; Privatarchiv (Originalfoto).

7454
Lorenzl, Josef
Schreitende Dame, ihr Kleid mit der linken Hand zurückhaltend (auf Ovalsockel)
ca. 1936
7454,3,14 Firmenmarke 19, Made in Austria, H. 31,7 cm; eBay # 1223171808
7454,153,19 ger. III, gem. lig. AS, Firmenmarke 19, Made in Austria, H. 31,7 cm; eBay # 260023521800

7458
WANDMASKE Frauenkopf mit geschlossenen Augen, Hut und Kragen
ca. 1936
7458,8 Firmenmarke 19, H. 28 cm; Doro 9.11.1993; 8.3.1994
Quelle: AGM (Bestellung der Firma Elisabeth Salle (Wiesbaden), vom 8.1.1951), hier ohne nähere Ang. zur Maske.

7460F
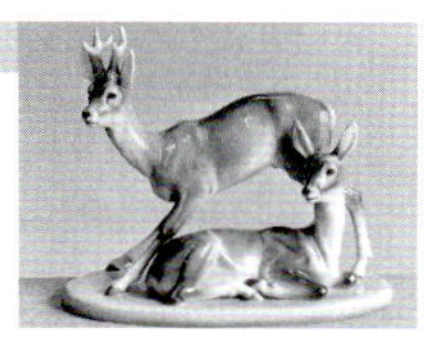
Jarl, Karin
GRUPPE „Rehe" (Zwei Rehe auf Ovalsockel)
ca. 1936
7460,11,97 Firmenmarke 19, Made in Austria, Hand dec., unsign., H. 30 cm, B. 35,5 cm; KH
7460,51,4 Firmenmarke 19, Made in Austria, H. 30 cm; Sammlung Helmut Riener, St. Florian, Austria
7460 H. 29 cm, B. 35 cm; Fotoarchiv K. Täuber
-> vgl. 7462= EinzelFIGUR
Quelle: AGM (Kontoblatt der Wiener Manufaktur Josef

Schuster für die Firma Porslins Ericsson-Lund, Stockholm?, Schweden vom 7. Januar 1943), hier mit Ang. des Originaltitels GRUPPE „Rehe"; Fotoarchiv K. Täuber.

7462F

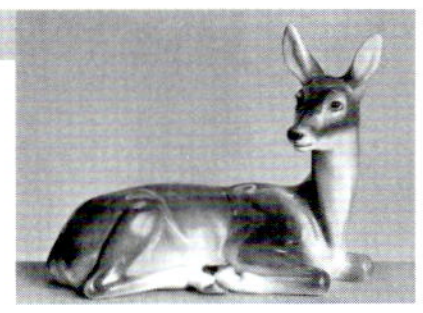

Jarl, Karin
FIGUR „Reh" (liegend)
ca. 1936
7462,3,27 Firmenmarke 19, Made in Austria, unsign., H. 18 cm, B. 22,2 cm; eBay # 925854310
7462,56 Firmenmarke 19, Made in Austria, unsign., Ausformung zw. 1945–1953, H. 18 cm, B. 23 cm; eBay # 3735260590
7462,84,16 gem. 8001, Firmenmarke 19, Made in Austria, unsign., H. 17,8 cm, B. 22 cm; eBay # 142805267
7462,172,7 ger. J., Firmenmarke 19, Made in Germany, unsign., Ausformung zw. 1938–1941, H. 18 cm; Wien Museum Inv. 190.035
7462F, H. 17,5 cm, B. 22 cm; GSNachf. 5,15; Fotoarchiv K. Täuber, hier mit der Ang.: H. 18 cm, B. 23 cm
-> vgl. 7460= GRUPPE
Lit.: GSNachf. 5,15 (m. Abb.), hier mit Ang. des Originaltitels FIGUR „Reh" (liegend).
Quelle: AGM (Kontoblatt der Wiener Manufaktur Josef Schuster für die Firma Porslins Ericsson-Lund, Stockholm?, Schweden vom 7. Januar 1943), hier mit Ang. des Originaltitels „Reh"; (Rechnung an Maurice Mizrahi, Alexandria, Ägypten, vom 22. April 1953); Fotoarchiv K. Täuber.

7463
Lorenzl, Josef
Stehender Damenakt mit Barsoi (Marussia und Iwan) (auf Ovalsockel)
ca. 1936
7463,117,14 Firmenmarke 19, H. 39 cm; Doro 21.5.1997
7463,121,19 Firmenmarke 19, Made in Germany, Ausformung zw. 1938–1941, H. 39 cm; Soth. L. 29.3.1996
Lit.: DM, 5.1928/29, Nr. 53, S. 3078 (m. Abb.), hier eine zeitgenössische Fotografie, die wohl als Vorlage bzw. Inspiration für die Ausführung diente. Fotografie: Zielke; Pinhas, S. 156 (m. Abb.).

7464

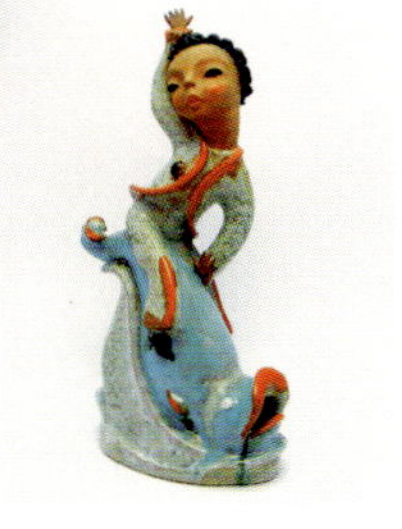

Goebel, Kurt
Knabe auf Fisch („reitend")
ca. 1936
7464 ger., Firmenmarke 19, Made in Austria, ger. sign. GOEBEL, H. 33 cm; Wien Museum Inv. 190.168
7464 ger., Firmenmarke 19, Made in Austria, ger. sign. GOEBEL, H. 34 cm; Sammlung a@goldscheider.net
Bem.: Die Figuren und Büsten des Bildhauers Kurt Goebel können, wenngleich auch die allermeisten unsigniert sind, an ihrer eigenständigen Darstellung und Behandlung der Oberfläche bzw. der Glasur leicht erkannt werden. Ihre autarken Gesichtszüge und Posen sind nicht ganz so grotesk wie die eines Walter Bosse, können jedoch jederzeit mit dessen Qualität konkurrieren. Sie werden im KH auch öfter als Erzeugnisse von Bosse angeboten. Dies geschieht teils aus Unwissenheit, vorwiegend aber wegen der doch höheren Bewertung der Bosse-Figuren und somit aus nicht ganz lauteren pekuniären Gründen. Weitere Modelle, die mit dem hier beschriebenen in einer stilistischen Linie stehen, sind zu finden unter den Modell-Nrn. 7465; 7466; 7476; 7618; 7622; 7655 usw.
Ausst.: SSH 1992
Pendant -> vgl. 7467
Lit.: SS, S. 24, 25 (m. Abb.)

7465
Goebel, Kurt
Stehendes Mädchen im Badeanzug (auf Natursockel mit Baumstamm)
ca. 1936
7465,35 gem. L., Firmenmarke 19, Made in Austria, unsign., Pick. 5, Ausformung zw. 1945–1953, H. 32,3 cm; GK, S. 22, 46 (m. Abb.) Wien Museum Inv. 223.734
Bem.: s. 7464
Ausst.: HMW (jetzt Wien Museum) 1985/86; SSH 1992.
Lit.: GK, S. 22, 46 (m. Abb.); SS, S. 24, 25 (m. Abb.).

7466

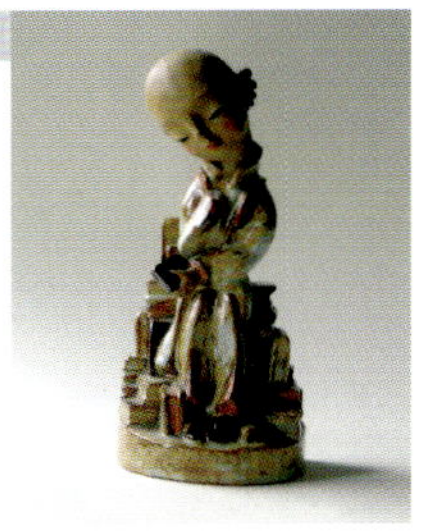

Goebel, Kurt
Der Bücherwurm (Stehender, fast glatzköpfiger Mann, ein Buch haltend) (auf Rundsockel mit Büchern)
ca. 1936
7466,0,P ger., gem. L., Firmenmarke 19, Made in Austria, Ausformung zw. 1945–1953, H. 24,5 cm; Sammlung John Nash
Bem.: s. 7464

7467
Goebel, Kurt
Stehendes Mädchen mit Fisch (auf Rundsockel)
ca. 1936
7467 ger., Firmenmarke 19, sign. Goebel, H. 19 cm; eBay # 264716921
Bem.: s. 7464
Pendant -> vgl. 7464
7469 s. 7496

7470
Stehendes Reh (auf flachem Rechtecksockel)
ca. 1936
7470,20 ger., Firmenmarke 19, H. 14 cm; eBay # 3343638626

7471
BÜSTE Mädchen mit Wellensittich auf der Hand
ca. 1936
7471,65 (ohne Firmenmarke), Ausformung zw. 1945–1953, H. 21,5 cm; Wien Museum Inv. 190.010
7471 ger., gem. lig. AH, Firmenmarke 19, Made in Austria, H. 21,5 cm; Wien Museum Inv. 220.796

7484
Matula, Viktor
ZIGARETTENDOSE
ca. 1936
7484 H. fehlt, B. fehlt
Quelle: Nachlass Matula, Wien

7486
Matula, Viktor
ZIGARETTENDOSE
ca. 1936
7486 H. fehlt, B. fehlt
Quelle: Nachlass Matula, Wien

7488
KERZENHALTER Engel (nach rechts blickend) mit zwei Kerzentüllen (Wandapplik)
ca. 1936
7488,11,65 Firmenmarke 19, Made in Austria, H. 25,4 cm; eBay # 112721979
Pendant -> vgl. 7489

7489
KERZENHALTER Engel (nach links blickend) mit zwei Kerzentüllen (Wandapplik)
ca. 1936
7489,11 (ohne Firmenmarke), Ausformung zw. 1945–1953, H. 25,4 cm; eBay # 112721979
7489,65,12 Firmenmarke 19, Made in Austria, H. 25,5 cm; KH
Pendant -> vgl. 7488

7492
Jarl, Karin
GRUPPE Zwei Schäferhunde (Einer stehend, einer liegend)
ca. 1936
7492,3,14 gem. F., Firmenmarke 19, Made in Austria, H. 24,4 cm, L. 33,5 cm; eBay # 854388473; 2031992873
7492,28,6,8,A gem. F., Firmenmarke 19, Made in Austria/oder Germany, H. 24,5 cm, L. 32,5 cm; Wien Museum Inv. 190.025

7494F

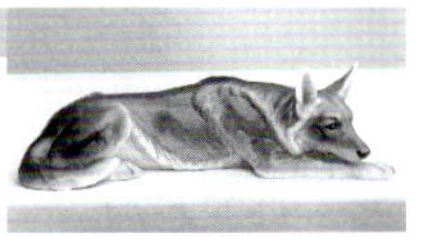

Jarl, Karin
„Wolfshund" (Liegender Schäferhund)
ca. 1936
7494,7 Firmenmarke 19, Made in Austria, unsign., Ausformung zw. 1945–1953, H. ca. 8 cm, L. ca. 28 cm; WKH
7494 H. 10 cm; B. 31 cm; Fotoarchiv K. Täuber
Quelle: AGM (Bestellung der Firma Wahliss (Wien) vom 24.11.1945 an die „Arbeitsgemeinschaft der Firma Friedrich Goldscheider"), hier mit Ang. des Originaltitels „Wolfshund"; Bestellung der Firma Wahliss (Wien) vom 27.2.1946 in Ausführung „Terrakotta"; Bestellung von Ferd. Lundquist (Göteborg, Schweden), vom 23. August 1950), in Ausführung „Fayence"; Fotoarchiv K. Täuber.

7495M

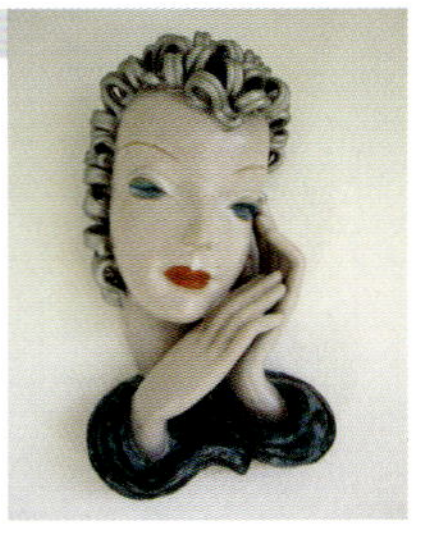

Knörlein, Rudolf
WANDMASKE „Maske" (Frauenkopf mit zwei Händen und Halstuch)
ca. 1935/36
7495,8,H gem. lig. AH, Firmenmarke 19, Made in Austria, Ausformung zw. 1945–1953, H. 30 cm; Herr 35
7495,R Firmenmarke 19, H. 29,5 cm; Soth. L. 28.4.1983

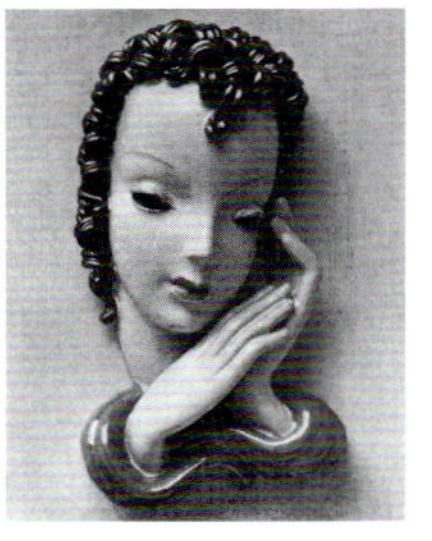

7495 gem. A, Firmenmarke 19, Made in Austria, unsign., H. 30 cm; Sammlung f@goldscheider.de
7495 Firmenmarke 19, Made in Austria, H. 30,5 cm; eBay # 343268684
7495M, H. 30 cm, B. 17 cm; GSNachf. 2,4
STEHMASKE
(auf Holzsockel)

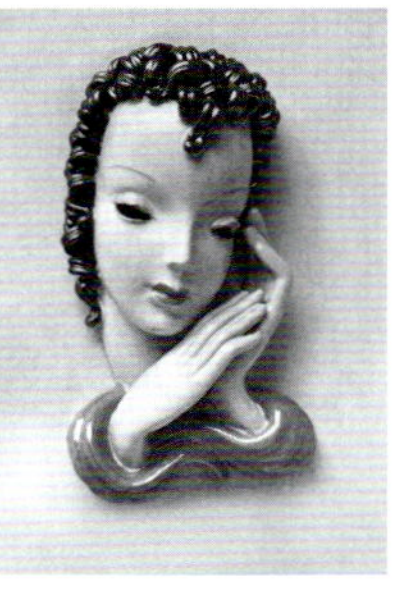

7495,12 gem. G, Plak. C, H. 29,2 cm (ohne Sockel), Gesamth. 36 cm; PB Wien 19
Bem.: Durch die eigenhändige Datierung Knörleins (1935) auf dem Trägerblatt des Originalfotos kann die Entstehung um 1935/36 ziemlich exakt bestimmt werden.
Lit.: GSNachf. 2,4 (m. Abb.), hier mit Ang. des Originaltitels „Maske".
Quelle: Privatarchiv (Originalfoto)

7496M

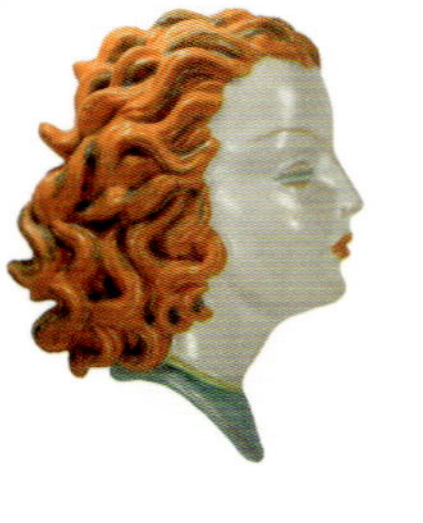

WANDMASKE „Maske" (Frauenkopf mit Kragen, im Profil nach rechts)
ca. 1936
7496,7,S ger., Firmenmarke 19, Made in Austria, H. 29 cm; BM 40; 43
7496,8 Firmenmarke 19, Made in Austria, Ausformung zw. 1945–1953, H. 25,3 cm; Sammlung a@goldscheider.net

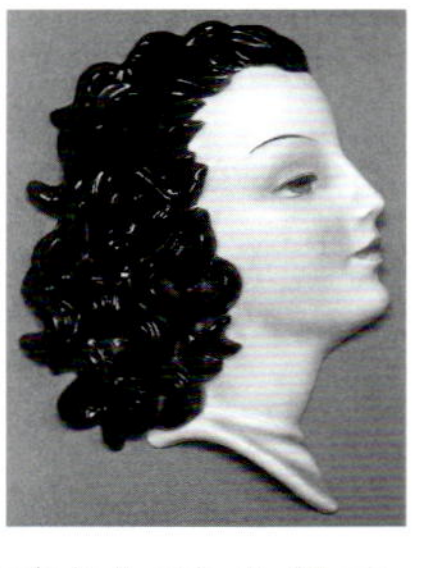

7496,8 ger., Firmenmarke 19, Made in Austria, Ausformung zw. 1945–1953, H. 25,3 cm; Herr 34
7496,12 gem. AH, Firmenmarke 19, Made in Austria, Ausformung zw. 1945–1953, H. 28 cm; Wien Museum Inv. 220.802
7496 handschriftlich,
7469,6,55 Firmenmarke 19, Made in Austria, H. 28 cm; Wien Museum Inv. 186.665
7496 Firmenmarke 19, Made in Austria, H. 28,5 cm; Soth. L. 21.10.1988

7496,3,A gem. lig. AH, Firmenmarke 19, Made in Germany, Ausformung zw. 1938–1941, H. 29 cm; eBay # 2650628424
7496M, H. 28 cm, B. 20 cm; GSNachf. 2,5a; Fotoarchiv K. Täuber, hier mit der Ang.: H. 29,5 cm, B. 22 cm
-> vgl. 7771; 7815
Lit.: GSNachf. 2,5 (m. Abb.), hier mit Ang. des Originaltitels „Maske".
Quelle: AGM (Bestellung der Firma Elisabeth Salle (Wiesbaden), vom 8.1.1951), hier ohne nähere Ang. zur Maske; Fotoarchiv K. Täuber.

7497
WANDMASKE Frauenkopf mit mehrpassigem Kragen
ca. 1936
7497,3,B Firmenmarke 19, H. 28 cm; eBay # 1411627320
7497,8,R ger., Firmenmarke 19, Made in Austria, Ausformung zw. 1945–1953, H. 28 cm; KH Art 1900/Berlin
7497 Firmenmarke 19, Made in Austria, Pick. 1, H. 27,5 cm; Soth. L. 17.10.1985

7500
Dakon, Stephan
Schreitende Dame in sportlichem Kostüm und Kappe, zwei Pointer ausführend (auf abgerundetem Rechtecksockel)
ca. 1936
7500,88,11,8,A gem. N., Firmenmarke 19, Made in Austria, H. 32,5 cm; Wien Museum Inv. 190.092
7500,250,31 gem. rot H, Firmenmarke 19, Made in Germany, unsign., Ausformung zw. 1938–1941, H. 33 cm; eBay # 345722527
Ausst.: SSH 1992

7501
Dakon, Stephan
Stehender Mädchenakt auf einer Wasserlilienblüte (einen Schmetterling betrachtend)
ca. 1936
7501,50,11 ger. 123, gem. N., Firmenmarke 26, H. 29,2 cm; Zezschwitz 20

7502
Lorenzl, Josef
FIGUR/LAMPE Stehende schlanke Dame mit Puderdose
ca. 1936
FIGUR
7502,131,4 gem. lig. LS, SG Firmenmarke 19, Made in Austria, H. 30,5 cm; eBay # 511127673; (KH www.tias.com/stores/frozenbeauty)
7502,165,11 gem. R, gem. blau EM., Firmenmarke 19, Made in Austria, H. 32 cm; Wien Museum Inv. 190.139
LAMPE
7502,22,6 gem. 5203, Firmenmarke 19, Made in Austria, H. 30,5 cm, Gesamth. 56 cm; eBay # 1034446172

7505
Herczeg, Claire (Klára)
Stehendes Mädchen mit großer Blume (auf Rundsockel)
ca. 1936
7505,63,44 ger. 16, gem. SG, H., Firmenmarke 19, Made in Austria, sign. Claire Herczeg, H. 24 cm; Wien Museum Inv. 220.739

7506 -> ÜBERSCHNEIDUNG
VASE
Form: becherförmig
Dekor: Laufglasur
ca. 1936
7506 K, Firmenmarke 19, Made in Austria, H. 14 cm; Wien Museum Inv. 186.694
Ausst.: SSH 1992

7506 -> ÜBERSCHNEIDUNG
Stehender Mädchenakt mit großem Tuch (auf Ovalsockel)
ca. 1936
7506 (Firmenmarke ohne Ang.), H. 33 cm; Pinhas, S. 134 (m. Abb.)
Bem.: Wahrscheinlich ist die Mod.-Nr. 7506 in der Quelle (Wolley & Wallis, Auctioneers) falsch, da die Vase mit der gleichen Nr. im Wien Museum selbst geprüft werden konnte.

7509
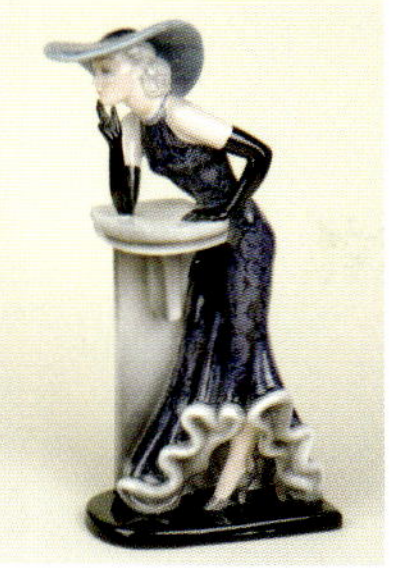
Stehende elegante Dame mit großem Hut und langen Handschuhen (sich auf ein Tischchen stützend)
ca. 1936
7509,6,14 gem. lig. HS, Firmenmarke 19, Made in Austria, Pick. 4, H. 30 cm; WKA 35
7509,10,19 gem. SG, Firmenmarke 19, Made in Austria, H. 30,5 cm; eBay # 7308454588
7509,24,10 gem. F., Firmenmarke 19, Made in Austria, H. 30,6 cm; PB Wien 19
Lit.: GSN, S. 62, 63 (m. Abb.); Pinhas, S. 158 (m. Abb.).
GS neu: Modell-Nr. 39 „Blue Hour" (H. 25 cm), Aufl. 3000 Stück

7511F
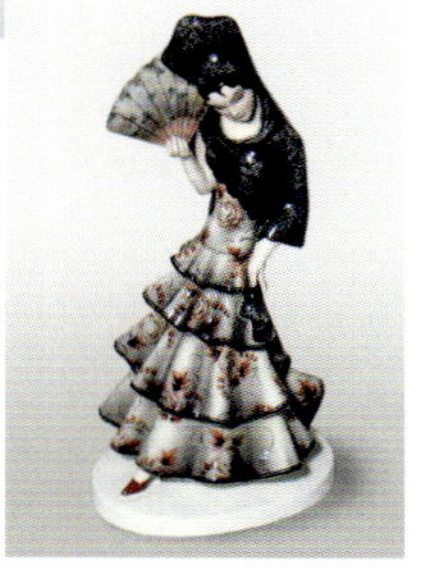
Weiss, Claire (Klára) (d.i. Herczeg, Claire (Klára)
„Spanierin" (Stehende Spanierin mit Fächer) (auf Ovalsockel)
ca. 1936
7511,3,6 gem. lig. rot NC, Firmenmarke 19, Made in Austria, sign. Claire Weiss (an der Oberseite der Plinthe), unter dem Boden sign. Dakon, H. 38,8 cm; eBay # 3703452822; Bem.: Die Sign. „Dakon" ist ein Versehen des Modelleurs. Wie die anderen durchgehend mit Weiss sign. Modelle belegen, kommt Stephan Dakon als Entwerfer nicht infrage.
7511,8,A Firmenmarke 19, Made in Austria, unsign., Ausformung zw. 1945–1953, H. 38 cm; city-antik
7511,11 ger. NT, A, gem. S., Firmenmarke 19, Made in Austria, unsign.?, Ausformung zw. 1945–1953, H. 40,6 cm; eBay # 365396140
7511,19 ger. XII, AS, gem. IF, Firmenmarke 19, Made in Austria, sign. Claire Weiss, Ausformung zw. 1945–1953, H. 40,6 cm; eBay # 303157913
7511,40,2 ger. XII, A, gem. NF, Firmenmarke 19, Made in Austria, sign. Claire Weiss, H. 40 cm; eBay # 1860110207 (Soth.); eBay # 2632251003
Bem.: Dieses Modell wurde auch in Trenton (USA) produziert (H. 37 cm) -> vgl. Fischer 144.
Quelle: AGM (Rechnung an Maurice Mizrahi, Alexandria, Ägypten vom 22. April 1953), hier mit Ang. des (Original)-Titels „Spanierin".

7516F
„Windhund"
ca. 1936
Quelle: AGM (Rechnung der Wiener Manufaktur Josef Schuster an Laszlo Bösze (Szeged, Ungarn) vom 5. August 1944), hier mit Ang. des Originaltitels „Windhund".

7518
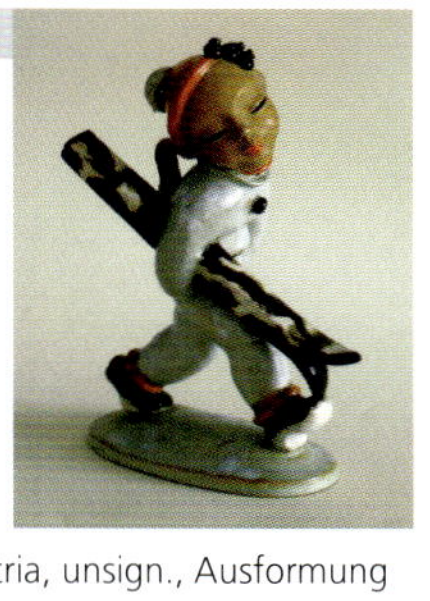
Goebel, Kurt
Skifahrer (Schreitender Skifahrer, seine Skier unter dem rechten Arm tragend) (auf Ovalsockel)
ca. 1936
7518,H ger., Firmenmarke 19, unsign., H. 23 cm; PB Deutschland
7518,35 ger., gem. B, Firmenmarke 19, Made in Austria, unsign., Ausformung zw. 1945–1953, H. 24,1 cm; eBay # 2513039058
7518,35 gem. M, Firmenmarke 19, unsign., Ausformung zw. 1945–1953, H. 23 cm; Doro 25.11.2003
-> vgl. 7655M; 7935M= jeweils ähnl. GRUPPE
Pendants -> vgl. 7622; 7887

7519F
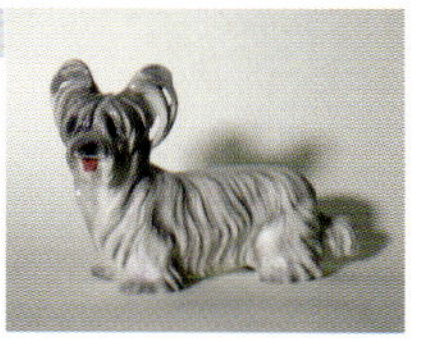
Postl, Hugo
FIGUR „Skyeterrier" um/ab 1930 (für die Ausführung)
7519,17,9 gem. grün G, Firmenmarke 19, Made in Austria, Hand dec., unsign., H. 33 cm, B. 50 cm; eBay # 2583335323; Zezschwitz 29
7519,85,31 Firmenmarke 19, unsign.?, H. 33 cm, B. 51 cm; Doro 21.10.1998 (ohne Auktionskat.)
7519F, H. 30 cm, B. 50 cm; GSNachf. 5,11e
-> vgl. 6116; 6503; 6802; 7300
-> vgl. 6975= Variante (auf Ovalsockel)
Lit.: GSNachf. 5,11 (m. Abb.), hier mit Ang. des Originaltitels „Skyeterrier".

7521
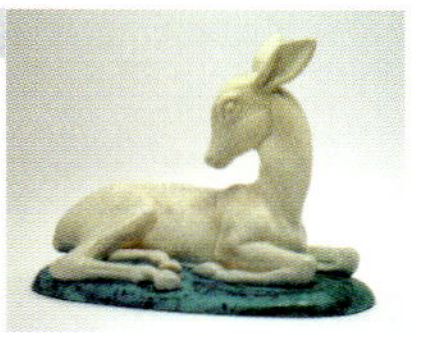
Liegendes Reh (auf Ovalsockel)
ca. 1936
7521,55,4 gem. grün A, Firmenmarke 19, Made in Austria, H. 19,8 cm, L. 34,5 cm; Sammlung a@goldscheider.net

7522
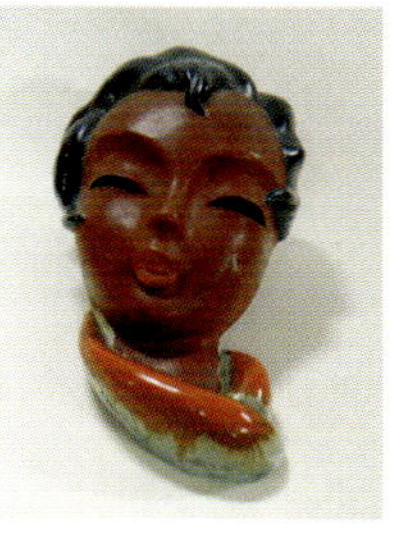
Goebel, Kurt
WANDMASKE Frauenkopf (dunkelhäutig) mit Halstuch/Schal
ca. 1936
7522,12 gem. L., Firmenmarke 19, Made in Austria, Ausformung zw. 1945–1953, H. 28,9 cm; MAK Wien, Inv. Ke 10521-34

7523
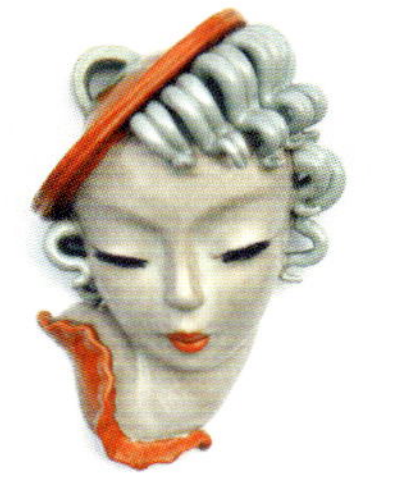
Goebel, Kurt
WANDMASKE Frauenkopf mit Béret/Kappe und Kragen
ca. 1936
7523,8 gem. L, Firmenmarke 19, unsign., Ausformung zw. 1945–1953, H. 24 cm; eBay # 882987417; Christie's SK 16.9.1999

7525
Lorenzl, Josef
Stehender Damenakt mit ausgestrecktem Arm (auf abgeschrägtem Quadratsockel bzw. Rundsockel)
ca. 1936
Quadratsockel
7525,14,11 (Firmenmarke ohne Ang.), Pick. 6, H. 36 cm; Doro 14.9.2001
Rundsockel
7525,46,31 (Firmenmarke ohne Ang.), H. 35 cm; Christie's SK 10.11.2000
Bem.: Diese Figur wurde auch noch zw. 1953–1960 bei Carstens (GWM West-Germany) produziert.
-> vgl. 7431; 8137= jeweils ähnl. Modell
-> vgl. 6932= Figur mit Tuch
-> vgl. 7724= GRUPPE
Lit.: Könecke (2002), S. 76, S. 86 (m. Abb.).

7527
Knörlein, Rudolf (zugeschr.)
WANDMASKE Frauenkopf mit Hand und Weintrauben
ca. 1936
7527,17 Firmenmarke 19, Made in Austria, unsign., Ausformung zw. 1945–1953, H. 27 cm; eBay # 2525098895
Lit.: Pinhas, S. 173 (m. Abb.), hier ebenfalls unsign., H. 28 cm.

7528
Knörlein, Rudolf
Stehender Damenakt (Negerin) (auf Rundsockel)
ca. 1936
7528,5,65 gem. B, Firmenmarke 19, Made in Austria, unsign., H. 30,5 cm; eBay # 1223145690
(KH www.tias.com/stores/frozenbeauty)

7541F

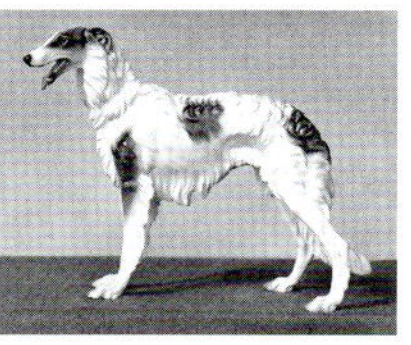

FIGUR „Windhund“
(Stehender Barsoi)
ca. 1933 (für den Entwurf)
7541F, H. 16 cm, B. 22 cm; GSNachf. 5,3b
-> vgl. 6731
-> vgl. 8117= GRUPPE mit Mädchen im Badeanzug.
Lit.: GSNachf. 5,3 (m. Abb.), hier mit Ang. des Original'-titels „Windhund“.
Quelle: Fotoarchiv K. Täuber

7542

Knörlein, Rudolf
BÜSTE Damenakt mit Lockenkopf (die Hände an die Brust gelegt)
ca. 1936
7542,11,20 (Firmenmarke ohne Ang.), unsign., H. 37 cm; city-antik
-> vgl. 7543= Ausführung mit Tulpe

7543M

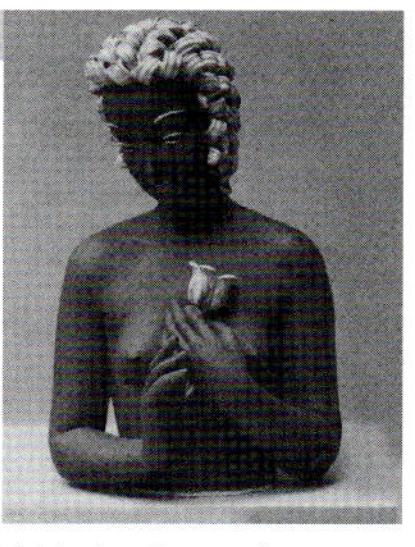

Knörlein, Rudolf
BÜSTE „Frühling“
(Damenakt mit Lockenkopf, eine Tulpe haltend)
ca. 1936
7543,8 gem. A, Firmenmarke 19, Made in Austria, unsign., Ausformung zw. 1945–1953, H. ca. 38 cm; eBay # 1837400151
(Soth., Harriman Judd Coll. of British Art Pottery)
7543,71 Firmenmarke 19, unsign., Ausformung zw. 1945–1953, H. 38 cm; Phillips Z. 13.10.1997
7543 ohne Ang., Firmenmarke 19, unsign., H. 37 cm; Soth. Olympia 7.11.2002
7543M, H. 38 cm, B. 20 cm; GSNachf. 3,6
Bem.: In der Nachkriegsproduktion unter dem Titel KOPF „Jugend“ geführt (Brief von Walter Goldscheider an Rudolf Knörlein vom 10. Januar 1953). In dem selbem Schreiben wird deutlich, wie Walter Goldscheider für ein weiteres Modell auf Grundlage dieser Büste das Aussehen bzw. die Umgestaltung des Entwurfes lenken wollte. Er schlug Knörlein vor „dem Gesicht einen etwas anderen Ausdruck zu geben..., soll der Oberkörper nicht nackt, sondern mit einer modernen Draperie, aber mit Armen oder Händen allein versehen sein, dieselben könnten entweder einen Schmetterling halten oder eine Orange, evtl. könnte der Schmetterling auch auf der Schulter sitzen oder evtl. ein kleiner Vogel“.
-> vgl. 7542= Ausführung ohne Tulpe
Lit.: GSNachf. 3,6 (m. Abb.), hier mit Ang. des Originaltitels BÜSTE „Frühling“; Pinhas, S. 164 (m. Abb.), hier mit Ang. der falschen Mod.-Nr. 7534.
Quelle: AGM (Rechnung an Maurice Mizrahi, Alexandria, Ägypten vom 22. April 1953); Fotoarchiv K. Täuber.

7544F

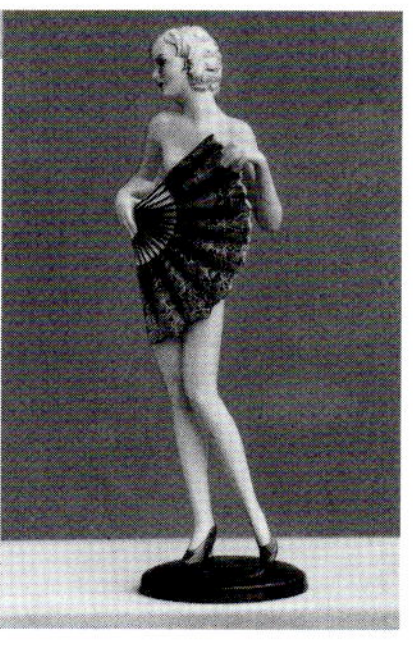

Lorenzl, Josef
FIGUR „Akt mit Fächer“
(Stehender Damenakt, einen großen Fächer haltend)
(auf Rundsockel)
ca. 1936
7544,22 Firmenmarke 19, Made in Austria, Hand dec., unsign., Ausformung zw. 1945–1953, Pick. 1, H. 36,8 cm; Soth. com. # 5K9Y
7544,185,4 (Firmenmarke ohne Ang.), H. 36 cm; Soth. L. 2.5.1985
7544F, H. 34,5 cm; GSNachf. 1,32
Bem.: Dieses Modell wurde noch in Trenton (USA) unter der GS Everlast Corp. (Modell-Nr. 899) weiter produziert (eBay # 2640187532).
-> vgl. 7688= ähnl. Modell, hier die gleiche Tänzerin mit zwei Fächern.
Lit.: GSNachf. 1,32 (m. Abb.), hier mit Ang. des Originaltitels „Akt mit Fächer“.
Quelle: AGM (Rechnung an Maurice Mizrahi, Alexandria, Ägypten vom 22. April 1953); Fotoarchiv K. Täuber.

7545

„Wisent“
ca. 1936
Quelle: AGM (Bestellungen der Firma Wahliss (Wien) vom 9. und 24.11.1945 an die „Arbeitsgemeinschaft der Firma Friedrich Goldscheider“, hier jeweils mit Ang. des (Original?-)Titels „Wisent“; Bestellung der Firma Wahliss (Wien) vom 27.2.1946 in Ausführung „Terrakotta“).

7545 s.a. 7546

7546

Knörlein, Rudolf
BÜSTE Frauenkopf mit gelockten Haaren und Kleid/ Blusenansatz
ca. 1936
7546,8 (Firmenmarke ohne Ang.), unsign., Ausformung zw. 1945–1953, H. 20 cm; Phillips L. 25.1.2001
7546,20 (XVB ger./gem.?), Firmenmarke 19, Made in Austria, Ausformung zw. 1945–1953, H. 19,5 cm; Christie's East 15.2.1983
7546 Firmenmarke 19, Made in Austria, unsign., H. 19,5 cm; Soth. L. 5.7.1985
Bem.: Der Kleid- bzw. Blusenansatz wurde offenbar in zwei Varianten ausgeführt. Zum Einen gibt es ihn in einer rechtwinkeligen und zum Anderen in spitz zulaufender Form. Auch die Haarlocken variieren in ihrer Länge.
-> vgl. 7550
Lit.: Pinhas, S. 166 (m. Abb.), hier mit Ang. der falschen Mod.-Nr. 7545. Richtig ist 7546, diese wurde durch direkten Kontakt mit Phillips geprüft und wird belegt durch die Ang. von Christie's.

7547

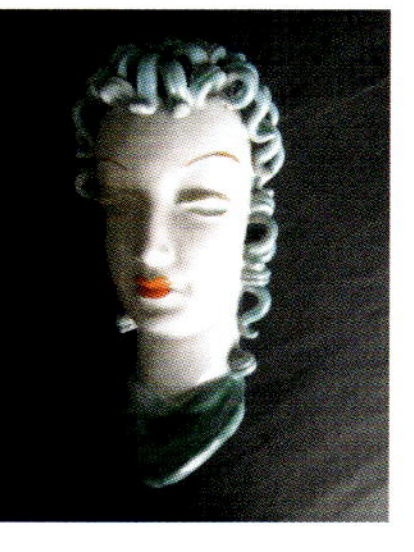

Knörlein, Rudolf (zugeschr.)
WANDMASKE Frauenkopf mit Kragen
ca. 1936
7547,33 Firmenmarke 19, unsign., Ausformung zw. 1945–1953, H. 20 cm; PB

7548

Knörlein, Rudolf
WANDMASKE Frauenkopf mit Tulpe
ca. 1933/34 (für den Entwurf)
7548,33 ger., Firmenmarke 19, unsign., Ausformung zw. 1945–1953, Pick. 6, H. 21 cm; Schuler 74; Doro 24.5.2000; eBay # 1268009411
-> vgl. 6859

7549M

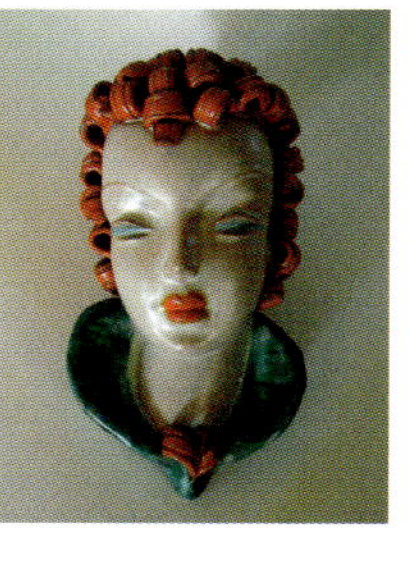

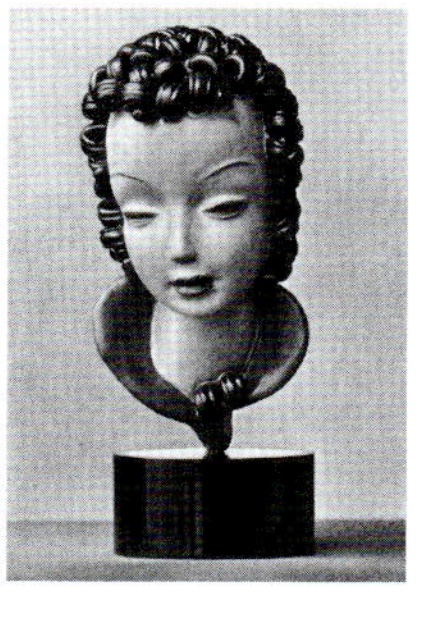

Prischl, Adolf
WANDMASKE Frauenkopf mit Kragen
ca. 1933/34 (für den Entwurf)
WANDMASKE
7549,12 ger. K, gem. G., Firmenmarke 19, Made in Austria, unsign., Ausformung zw. 1945–1953, H. 28 cm; Sammlung.schmutz@chello.at 45
7549,?5 Firmenmarke 19, Made in Austria, unsign., Ausformung zw. 1945–1953, H. 25 cm; WKA 33
7549M, H. 31 cm, B. 14,5 cm; GSNachf. 2,13b
Bem.: Die im Firmenkatalog abgebildete Stehmaske (auf schwarzem Holzsockel) konnte bislang noch in keinem Exemplar nachgewiesen werden.
-> vgl. 6836
-> vgl. 7340= DECKELDOSE
Lit.: GSNachf. 2,13 (m. Abb. als Stehmaske), hier mit Ang. des Originaltitels „Maske“ und „Maske auf Sockel“.

7550

Knörlein, Rudolf
BÜSTE Frauenkopf mit gelockten Haaren und Kleid/ Blusenansatz
ca. 1936
7550,S ger., Firmenmarke 19, Made in Austria, unsign., H. 19,5 cm; Wien Museum Inv. 186.668
-> vgl. 7546

7551F

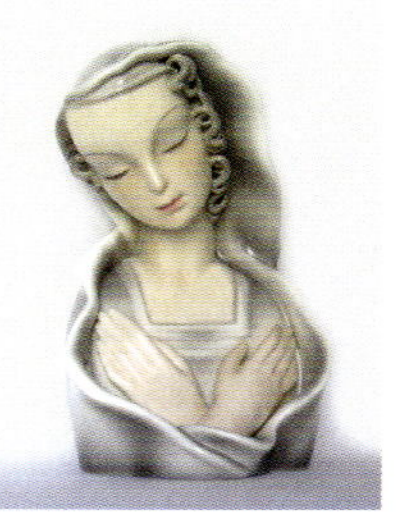

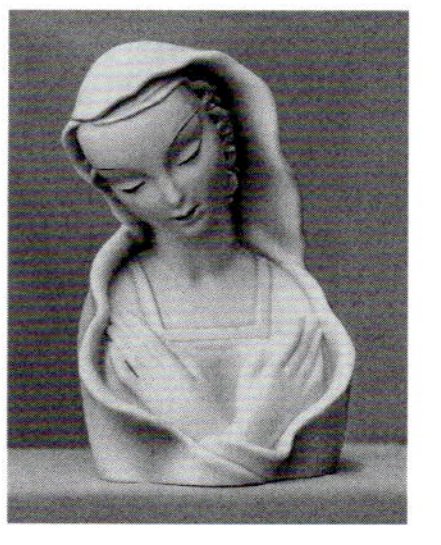

Prischl, Adolf
BÜSTE „Madonna“
(Madonna mit gekreuzten Händen)
ca. 1936
7551,3,65 Firmenmarke 19, Made in Austria, unsign., H. 26 cm; KH
7551,58,9 ger., gem. S., Firmenmarke 19, Made in Austria, H. 26 cm; Wien Museum Inv. 220.797
7551,246,65 Firmenmarke 19, Made in Austria, unsign., H. 26,5 cm; KH
7551,270,3 ger. XIIA, N, Firmenmarke 19, Made in Austria, H. 27,5 cm; Sammlung a@goldscheider.net
7551,N ger., Firmenmarke 19, Made in Austria, H. 26 cm; eBay # 922154522
7551,387,1 gem. SG, Firmenmarke 19, Made in Germany, Ausformung zw. 1938–1941, H. 28 cm; eBay # 403770972
7551,586,3 Firmenmarke Schuster 1, unsign., Ausformung zw. 1941–1944, H. 27 cm; Sammlung mad
7551F, H. 26 cm, B. 16 cm; GSNachf. 6,4a; Fotoarchiv K. Täuber, hier mit der Ang.: H. 25 cm.
-> vgl. 8245F
-> vgl. 7985F= RELIEF
Lit.: GSNachf. 6,4 (m. Abb.), hier mit Ang. des Originaltitels BÜSTE „Madonna“.
Quelle: Fotoarchiv K. Täuber

7552

Fohlen
ca. 1936
7552 ohne Ang., H. fehlt; Könecke (2002), S. 76
Bem.: Diese Figur wurde auch noch zw. 1953–1960 bei Carstens (GWM West-Germany) produziert.
Lit.: Könecke (2002), S. 76, hier ohne weitere Ang.

7553M/T

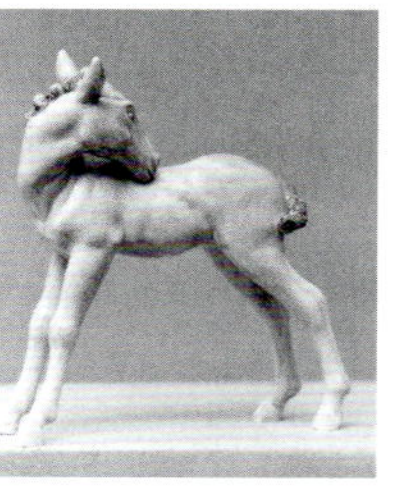

Matula, Viktor
FIGUR „Pferd“ (Stehendes Fohlen mit nach hinten gewandtem Kopf)
ca. 1936
7553M, H. 21,5 cm, B. 21 cm; GSNachf. 3,20a
Bem.: Diese Figur wurde auch noch zw. 1953–1960 bei Carstens (GWM West-Germany) produziert.
-> vgl. 8062M
Lit.: GSNachf. 3,22 (m. Abb.), hier mit Ang. des Originaltitels FIGUR „Pferd“; KGSF1; KGSF2; Könecke (2002), S. 76, hier mit Ang. des Künstlers Matula, S. 86 (m. Abb.).
Quelle: AGM (Kontoblatt der Wiener Manufaktur Josef Schuster für die Firma Porslins Ericsson-Lund, Stockholm?, Schweden vom 7. Januar 1943); (Bestellung der Firma Wahliss (Wien) vom 27.2.1946 an die „Arbeitsgemeinschaft der Firma Friedrich Goldscheider“, in Ausführung „Terrakotta“); (Kommissionsfaktura der „Österreichischen Gruppe-Austria“ für die Messe in Utrecht vom 26.2.1953; Rechnung an Maurice Mizrahi, Alexandria, Ägypten vom 22. April 1953), hier jeweils mit Ang. des Originaltitels „Pferd“; Fotoarchiv K. Täuber.

7555M
Matula, Viktor
RELIEF „Eichhörnchen" (Eichhörnchen auf Ast sitzend) (Wandrelief)
ca. 1936

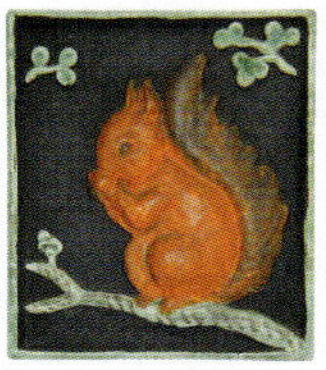

7555 gem. MT, Firmenmarke 19, unsign., H. 13,4 cm, B. 12 cm; eBay # 136761356

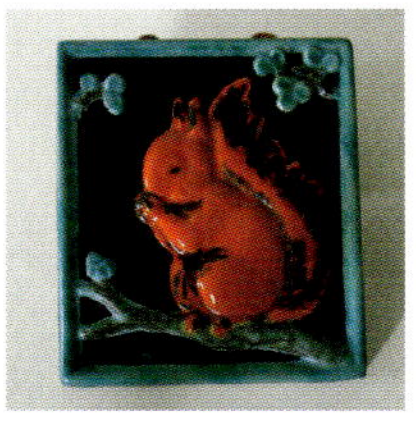

7555,20 Firmenmarke 19, Made in Germany, unsign., Ausformung zw. 1938–1941, H. 13,9 cm, B. 12,2 cm; eBay # 2229073515
7555 gem. L., Firmenmarke 19, Made in Germany, unsign., Ausformung zw. 1938–1941, H. 13,7 cm, B. 12,2 cm; PB Deutschland
7555,20 gem. A, Firmenmarke 19, Made in Austria, unsign., Ausformung 50er Jahre, H. 13,4 cm, B. 11,8 cm; Sammlung a@goldscheider.net
7555,20 ger., gem. G, Firmenmarke 19, Made in Austria, unsign., Ausformung 50er Jahre, H. 13,4 cm, B. 11,6 cm; Wien Museum Inv. 220.777
7555M, H. 11,5 cm, B. 13 cm; GSNachf. 2,23c
Lit.: GSNachf. 2,23 (m. Abb.), hier mit Ang. des Originaltitels „Relief Eichhörnchen".
Quelle: Nachlass Matula, Wien

7557
Weiss, Claire (Klára) (d.i. Herczeg, Claire (Klára)
Schreitende junge Dame mit Hut, zwei Hutschachteln tragend (auf Ovalsockel)
ca. 1936
7557 (Firmenmarke ohne Ang.), sign. Claire Weiss, H. 30 cm; Pinhas, S. 131 (m. Abb.)

7558F
Ungarische Tänzerin in Tracht
ca. 1936
7558,11 Firmenmarke 19, Made in Austria, H. ca. 22 cm; WKH
Quelle: AGM (Rechnung an Maurice Mizrahi, Alexandria, Ägypten vom 22. April 1953), hier unter dem Titel „Tanz" geführt.

7559
Matula, Viktor
FIGUR Pferd
ca. 1936
7559 H. fehlt
Quelle: Nachlass Matula, Wien

7560
Dakon, Stephan
Junge Golfspielerin beim Abschlag (auf Rundsockel)
ca. 1936
7560,99,15 (Firmenmarke ohne Ang.), H. 26,8 cm; Christie's SK 4.11.1999
Lit.: Pinhas, S. 128 (m. Abb.), hier in der Variante auf Ovalsockel.

7561F
Dakon, Stephan
Sitzende lesende junge Dame (auf Büchersockel)
ca. 1936
7561,4,11,8,A Firmenmarke 19, Made in Austria, Hand dec., H. 20,5 cm; Sammlung Stressless, Portugal
7561F, H. 20,5 cm, B. 20 cm
Quelle: Archiv Mimi Dakon (Originalfoto)

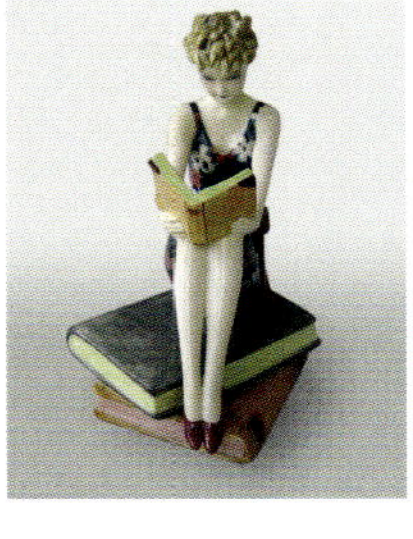

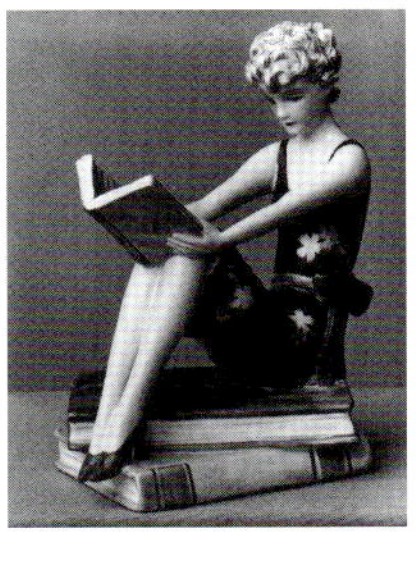

7562
Dakon, Stephan
Schreitender Skibub, seine Skier tragend (auf Ovalsockel)
ca. 1936
7562,39,4 Firmenmarke 19, Made in Austria, H. 25 cm; eBay # 6250948151
7562,186,2,8 gem. SW., Firmenmarke 19, Made in Austria, H. 25,4 cm; eBay # 1279999443

7563F
Dakon, Stephan
Stehendes Skifahrerpaar (auf Rundsockel)
ca. 1936/37
7563F, H. 28 cm
Quelle: Archiv Mimi Dakon (Originalfoto)

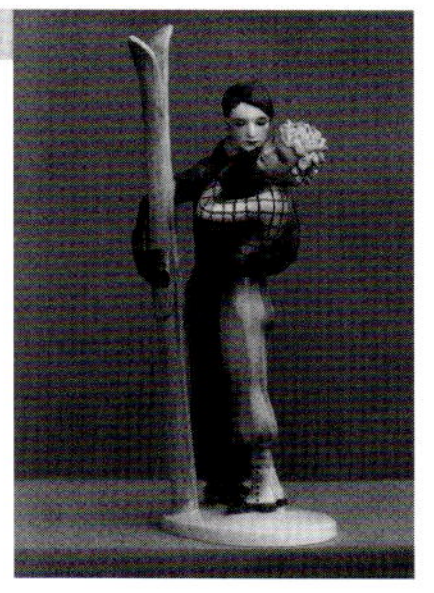

7564F
Dakon, Stephan
Stehendes Mädchen mit Golfsack (auf Rundsockel)
ca. 1936/37
7564,14,11,8 gem. SG, gem. blau H., Firmenmarke 19, Made in Austria, H. 23,4 cm; Wien Museum Inv. 220.740
7564F, H. 23,5 cm
Quelle: Archiv Mimi Dakon (Originalfoto)
Pendant –> vgl. 7191

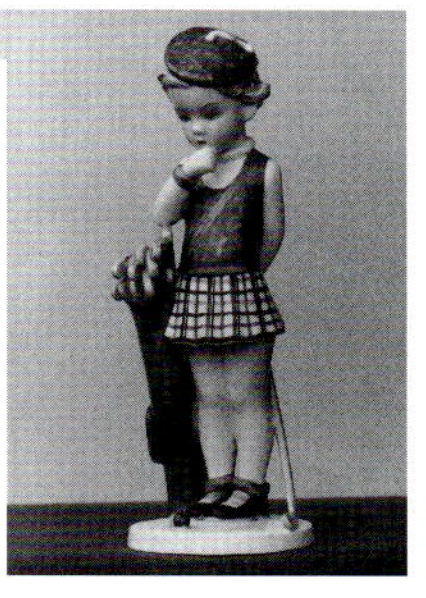

7565
Matula, Viktor
Stehendes Mädchen mit Rehkitz (auf flachem Rechtecksockel)
ca. 1936
7565,59,31 Firmenmarke 19, Made in Austria, H. 21,6 cm; eBay # 375710055

7566
Matula, Viktor
FIGUR Pferd
ca. 1936
7566 H. fehlt
Quelle: Nachlass Matula, Wien

7567
Matula, Viktor
FIGUR Pferd
ca. 1936
7567 H. fehlt
Quelle: Nachlass Matula, Wien

7568
Matula, Viktor
„Zicklein" (stoßend)
ca. 1936
7568,2 Firmenmarke 19, unsign., H. 10,5 cm; Wien Museum Inv. 190.038
Quellen: Nachlass Matula, Wien; AGM (Bestellungen der Firma Wahliss (Wien) vom 9. und 24.11.1945 an die „Arbeitsgemeinschaft der Firma Friedrich Goldscheider"), hier mit Ang. des (Original-?)Titels „Zicklein (stoßend)".

7569M/T
Matula, Viktor
GRUPPE „Rehe" (Ricke mit säugendem Kitz) (auf flachem Rechtecksockel)
ca. 1936

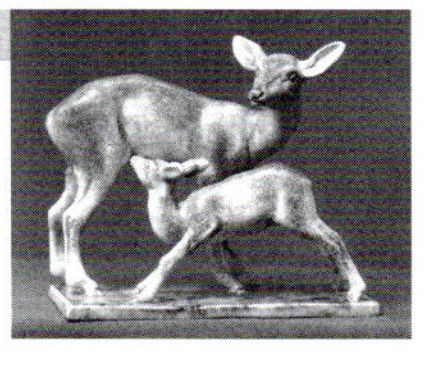

7569 gepr., 68B ger., Firmenmarke 19, Made in Austria, unsign., H. 18 cm, L. 19 cm; eBay # 3723293891
7569,9 Firmenmarke 19, Made in Austria, unsign., H. 18 cm; WKH
7569,20 Firmenmarke 19, Made in Germany, unsign., Ausformung zw. 1938–1941, H. 19 cm; eBay # 2223216319
7569,37 ger. G, Firmenmarke 19, Made in Germany, unsign., Ausformung zw. 1938–1941, H. 18 cm; Doro 20.10.1996
7569,30 ger., ger. 49C, gem. lig. AH, Firmenmarke 19, Made in Germany, unsign., Ausformung zw. 1938–1941, H. 18,2 cm; Wien Museum Inv. 190.037
7569M, H. 17 cm, B. 19 cm; GSNachf. 3,14
Lit.: GSNachf. 3,14 (m. Abb.), hier mit Ang. des Originaltitels GRUPPE „Rehe".
Quellen: Nachlass Matula, Wien; AGM (Bestellungen der Firma Wahliss (Wien) vom 9. und 24.11.1945 an die „Arbeitsgemeinschaft der Firma Friedrich Goldscheider", hier jeweils mit Ang. des Titels „Rehgruppe"; Bestellung der Firma Wahliss (Wien) vom 27.2.1946 (Ausführung Terrakotta); Rechnung an Maurice Mizrahi, Alexandria, Ägypten vom 22. April 1953), hier mit Ang. des Originaltitels GRUPPE „Rehe"; Fotoarchiv K. Täuber.

7571F
Ullmann, Robert
GRUPPE „Windhunde" (Ein stehender und ein liegender Barsoi)
ca. 1934/35 (für den Entwurf)

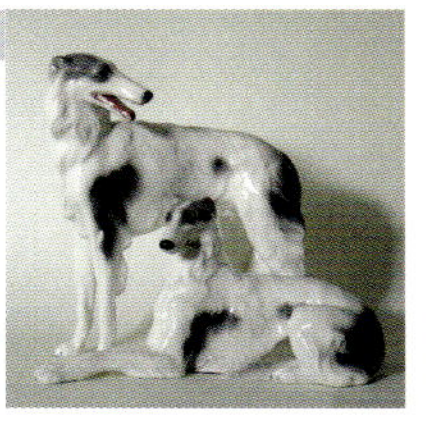

7571,4 Firmenmarke 19, Made in Austria, Hand dec., unsign., Ausformung zw. 1945–1953, H. 30 cm; eBay # 332410740
7571,11 gem. G, Firmenmarke 19, Made in Austria, unsign., Ausformung zw. 1945–1953, H. 30,5 cm; WKH
7571,66,64 Firmenmarke 19, Made in Austria, unsign., H. 30,5 cm; PB Deutschland
7571,88,31,5 Firmenmarke 19, Made in Austria, sign. Ullmann, H. ca. 29 cm; KH
7571F, H. 28 cm, B. 30 cm; GSNachf. 5,4b
Bem.: Zum Künstler s. 7054AF; zur Modellnummernvergabe s. 6761. Diese Figurengruppe wurde auch noch zw. 1953–1960 bei Carstens (GWM West-Germany) produziert.
–> vgl. 7054F; 7972F
–> vgl. 7054AF; 7571AF; 7972AF= jeweils stehende EinzelFIGUR
–> vgl. 7054BF; 7571BF; 7972BF= jeweils liegende EinzelFIGUR
Lit.: GSNachf. 5,4 (m. Abb.), hier mit Ang. des Originaltitels GRUPPE „Windhunde"; Könecke (2002), S. 76, S. 86 (m. Abb.).
Quelle: AGM (Rechnung der Wiener Manufaktur Josef Schuster an Laszlo Bösze (Szeged, Ungarn) vom 5. August 1944; Bestellungen der Firma Wahliss (Wien) vom 9. und 24.11.1945 an die „Arbeitsgemeinschaft der Firma Friedrich Goldscheider"; Bestellung der Firma Wahliss (Wien) vom 27.2.1946 an die „Arbeitsgemeinschaft der Firma Friedrich Goldscheider", in den Ausführungen „Terrakotta" und „Fayence").

7571AF
Ullmann, Robert
FIGUR „Windhund" (Stehender Barsoi)
ca. 1934/35 (für den Entwurf)
7571AF, H. 28 cm, B. 28 cm; GSNachf. 5,4e
Bem.: Zum Künstler s. 7054AF; zur Modellnummernvergabe s. 6761.
–> vgl. 7054AF; 7972AF
–> vgl. 7054F; 7571F= jeweils GRUPPE
Lit.: GSNachf. 5,4 (m. Abb.), hier mit Ang. des Originaltitels FIGUR „Windhund" (stehend).
Quelle: AGM (Rechnung der Wiener Manufaktur Josef Schuster an Laszlo Bösze (Szeged, Ungarn) vom 5. August 1944); Bestellung der Firma Wahliss (Wien) vom 24.11.1945 an die „Arbeitsgemeinschaft der Firma Friedrich Goldscheider", hier jeweils mit Ang. des Originaltitels „Windhund"; Bestellung der Firma Wahliss (Wien) vom 27.2.1946 in Ausführung „Terrakotta").

7571BF
Ullmann, Robert
FIGUR „Windhund" (Liegender Barsoi)
ca. 1934/35 (für den Entwurf)
7571BF, H. 16 cm, B. 31 cm; GSNachf. 5,4h
Bem.: Zum Künstler s. 7054AF; zur Modellnummernvergabe s. 6761.

-> vgl. 7054BF; 7972BF
-> vgl. 7054F; 7571F= jeweils GRUPPE
Lit.: GSNachf. 5,4 (m. Abb.), hier mit Ang. des Originaltitels FIGUR „Windhund" (liegend).
Quelle: AGM (Kommissionsfaktura der „Österreichischen Gruppe-Austria" für die Messe in Utrecht vom 26.2.1953), hier mit Ang. des Originaltitels „Windhund".

7572
Matula, Viktor
FIGUR Hahn
ca. 1936/37
7572 H. fehlt
Quelle: Nachlass Matula, Wien

7578F
Jarl, Karin
FIGUR „Spaniel" (sitzend)
ca. 1936/37
7578,8 Firmenmarke 19, unsign., Ausformung zw. 1945–1953, H. 23 cm; WKH
7578,81,19 Firmenmarke 19, Made in Austria, sign. Jarl, Pick. 6, H. 24 cm; eBay # 3739107591
7578 Firmenmarke 19, Made in Austria, unsign., Ausformung zw. 1945–1953, Pick. 1, H. 24 cm; Mehlis 24; 29
7578F, H. 23 cm, B. 23 cm; GSNachf. 5,7
Pendant -> vgl. 7248F
Lit.: GSNachf. 5,7 (m. Abb.), hier mit Ang. des Originaltitels FIGUR „Spaniel".

7579F
Prischl, Adolf
WANDMASKE „Madonna" (Madonna mit gefalteten Händen)
ca. 1936

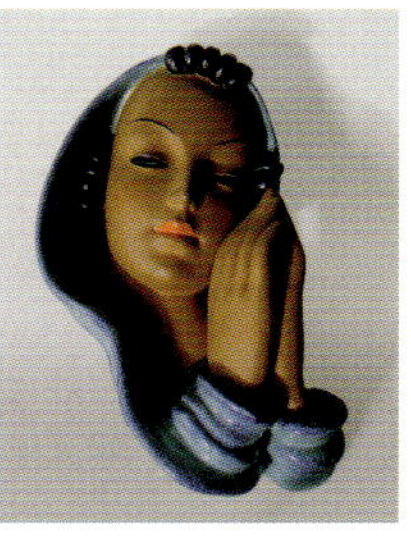

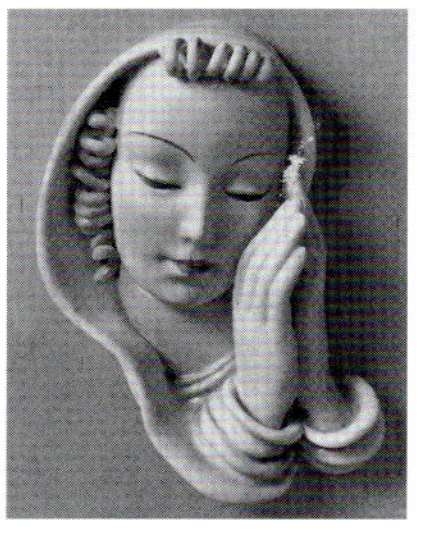

7579,8,12 Firmenmarke 19, H. 33 cm; WKH
7579,10,42 Firmenmarke 19, unsign., H. 32 cm; eBay # 1227326731
(7579ohne Nr.),13 Firmenmarke 19, Made in Austria, unsign., Pick. 1, Ausformung zw. 1945–1953, H. 32,5 cm; MAK Wien, Inv. Ke 10602-105
7579,14,37 Firmenmarke 19, Made in Austria, unsign., H. 33 cm; eBay # 358609977
7579,18,16 Firmenmarke 19, Made in Austria, H. 31 cm; eBay # 2063279192
7579,62,16 Firmenmarke 19, unsign., H. 32 cm; Soth. L. 30.9.1983
7579,94,66 Firmenmarke 19, Made in Austria, H. 33 cm; PB Deutschland
7579,420/2,6 Firmenstempel WG 1, unsign., Ausformung 50er Jahre, Pick. WG 1, H. 31,7 cm; KH
7579F, H. 32 cm, B. 19 cm; GSNachf. 6,8a; Fotoarchiv K. Täuber, hier mit der Ang.: H. 29 cm.
Bem.: Diese Maske wurde auch noch zw. 1953–1960 bei Carstens (GWM West-Germany) produziert.
-> vgl. 7865F
-> vgl. 7794= ähnl. Modell
Lit.: GSNachf. 6,8 (m. Abb.), hier mit Ang. des Originaltitels MASKE „Madonna"; Könecke (2002), S. 76, S. 87 (m. Abb.).
Quelle: Fotoarchiv K. Täuber

7581F
Lorenzl, Josef
„Liane" (Stehende Tänzerin mit Hut, ihr Kleid schmetterlingsflügelartig auseinander haltend) (auf abgerundetem rautenförmigem Sockel)
ca. 1936
7581,7,16 gem. lig. rot ON, X, Firmenmarke 19, Made in Austria, H. 34,3 cm; eBay # 1204263015; (KH www.tias.com/stores/frozenbeauty)
7581,48,31 Firmenmarke 19, Made in Austria, H. 34 cm; eBay # 1210566822; Doro 29.5.2001
7581,71,7 Firmenmarke 19, Made in Austria, H. 34 cm; eBay # 858954012
7581,279,0 Firmenmarke 19, Made in Germany, Ausformung zw. 1938–1941, H. 37 cm; Sammlung Dr. Peter Hieke, Wien
7581,291,10 gem. lig. SG., Firmenmarke 19, Made in Germany, Ausformung zw. 1938–1941, gepr. „L"= Z6, H. 37 cm; eBay # 876163497; WKA 38
7581,334,31,G (Firmenmarke ohne Ang.), unsign.?, H. 36 cm; Christie's SK 10.11.2000
-> vgl. 8321
-> vgl. Dakon 7257= ähnl. Modell.
Quelle: AGM (Rechnung der Wiener Manufaktur Josef Schuster an Laszlo Bösze (Szeged, Ungarn) vom 5. August 1944), hier mit Ang. des Originaltitels „Liane".

7582F
Lorenzl, Josef
„Sonate" (Schreitende Tänzerin mit Bikinioberteil, ihren Rock schmetterlingsflügelartig auseinanderhaltend) (auf Dreiecksockel)
ca. 1936
7582,24,11 Firmenmarke 19, unsign., H. 34 cm; eBay # 2244412336
7582,66 gem. S., Firmenmarke 19, Made in Austria, unsign., Ausformung zw. 1945–1953, H. 34,5 cm; eBay # 3355303689
7582,149,7 (Firmenmarke ohne Ang.), H. 37 cm; Doro 12.12.1991
7582,150,0 (Firmenmarke ohne Ang.), H. 35 cm; Schloss Hag. 35
-> vgl. 8459
Lit.: Pinhas, S. 153, 155 (m. Abb.)
Quelle: AGM (Bestellungen der Firma Wahliss (Wien) vom 30.8.1945 und 27.2.1946 an die „Arbeitsgemeinschaft der Firma Friedrich Goldscheider"; Rechnung an Maurice Mizrahi, Alexandria, Ägypten vom 22. April 1953), hier jeweils mit Ang. des (Original-?)Titels „Sonate".

7583
Goebel, Kurt
RELIEF Madonna mit Kind und Sternen (Wandrelief)
ca. 1936
7583,26 ger., gem. M, Firmenmarke 19, Made in Austria, unsign., Ausformung zw. 1945–1953, H. 25 cm, B. 20,5 cm; BM 68; Henry's 29.1.1998; 25.11.1998

7584
Goebel, Kurt
WANDMASKE Frauenkopf mit Locken (Curls) und Halstuch
ca. 1936

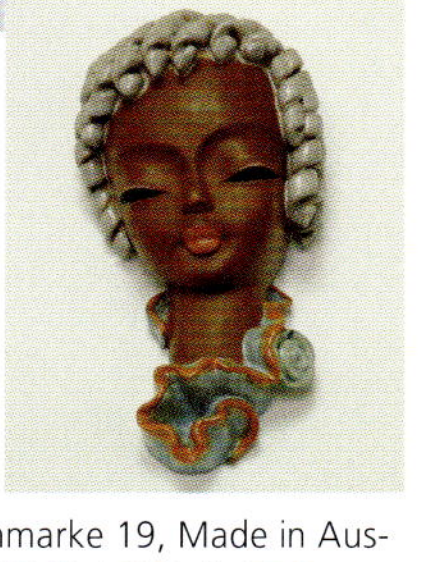

7584,8 ger., Firmenmarke 19, Made in Austria, Ausformung zw. 1945–1953, H. 26,5 cm; Sammlung Stressless, Portugal
7584,S, H ger., gem. L., Firmenmarke 19, Made in Austria, unsign., Ausformung zw. 1945–1953, H. 28,2 cm; Sammlung a@goldscheider.net

7585
Goebel, Kurt
WANDMASKE Frauenkopf mit Hut und Hand
ca. 1936

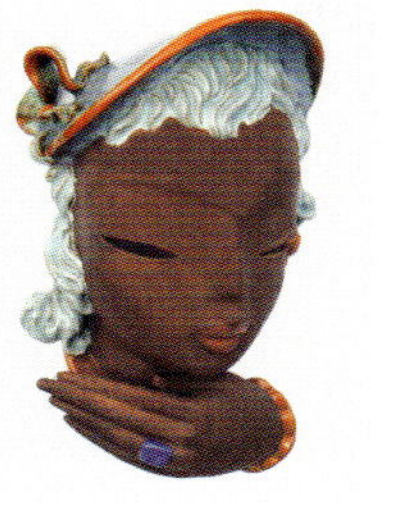

7585,8 gem. O, Firmenmarke 19, Made in Austria, unsign., Ausformung zw. 1945–1953, H. 23 cm; Sammlung a@goldscheider.net
7585 Firmenmarke 19, Made in Austria, unsign., H. 22,9 cm; eBay # 3761212268

7586
Goebel, Kurt
STEHKOPF Frauenkopf mit Kragen
ca. 1936
7586,0 ger. H, G., Firmenmarke 19, Made in Austria, unsign., H. 25,4 cm; eBay # 250420627
7586,26 ger., gem. L., Firmenmarke 19, Made in Austria, sign. GOEBEL, Ausformung zw. 1945–1953, H. 25 cm; Wien Museum Inv. 190.011
7586 Firmenmarke 19a, H. 27 cm; Soth. L. 6.7.1984
7586 Firmenmarke 19, H. 26,5 cm; Soth. L. 15.2.1985; SOTH. 1986, S. 297
Ausst.: SSH 1992

7587
Goebel, Kurt
BÜSTE Frauenkopf mit Hand und Halskette
ca. 1936
7587,35 ger. R, gem. L., Firmenmarke 19, unsign., Ausformung zw. 1945–1953, H. 16 cm; eBay # 2541569433; 3782029725
7587 Firmenmarke 19, Made in Austria, unsign., H. 16,3 cm; eBay # 2534110255

7588
RELIEF Madonna mit Kind (Wandrelief)
ca. 1936
7588,4 gem. G., Firmenmarke 19, Made in Austria, H. 30 cm, B. 24,9 cm; eBay # 1424058789

7589
Goebel, Kurt
Stehender Knabe mit Hut und Schmetterling (auf Rundsockel)
ca. 1936
7589 (Firmenmarke ohne Ang.), unsign., H. 20 cm; Pinhas, S. 177 (m. Abb.)

7594F
Bouret, Germaine / Dakon, Stephan
FIGUR „Joujou" (Stehendes Mädchen mit Ball) (auf Rundsockel)
ca. 1936

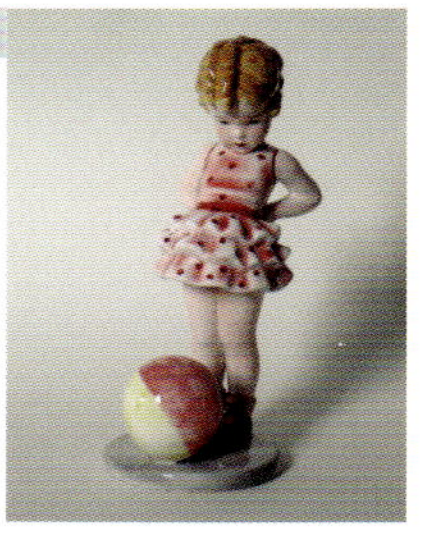

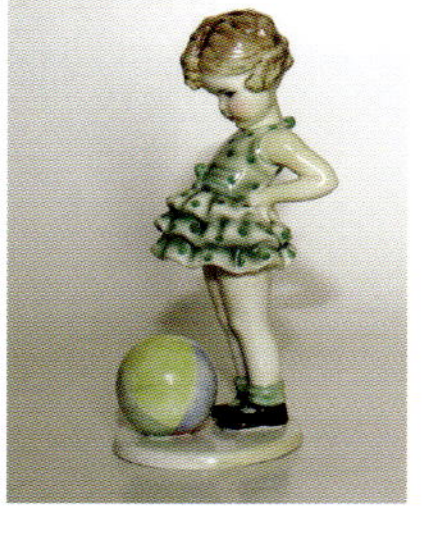

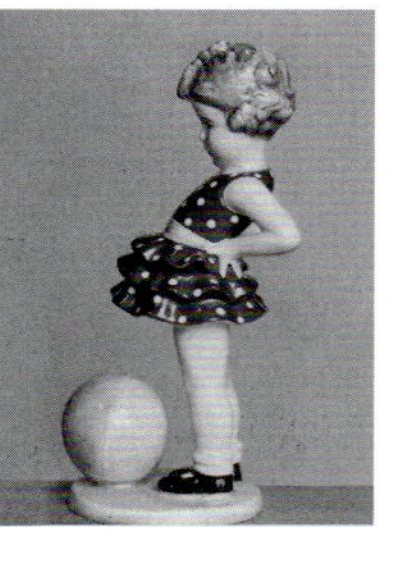

7594,18,14 ger., Firmenmarke 19, Made in Austria, Hand dec., unsign., H. 22,7 cm; city-antik
7594,96,16 gem. E, !, gem. rot H., Firmenmarke 19, Made in Austria, sign. G. Bouret, H. 22 cm; Sammlung Bernard Amigon, Frankreich
7594,S,7,G ger., Firmenmarke 19, Made in Germany, Ausformung zw. 1938–1941, unsign., H. 22,5 cm; eBay # 858834875
7594,70 (od. 76?) Firmenmarke 19, Made in Germany, Ausformung zw. 1938–1941, unsign., H. 21,8 cm; PB Deutschland
7594F, H. 21 cm; GSNachf. 1,3; Fotoarchiv K. Täuber, hier mit der Ang.: H. 22 cm, B. 9 cm
Bem.: Durch ein Foto aus dem Nachlass Dakon, welches dieses Modell zeigt, kann eindeutig nachgewiesen werden, dass die Figuren, die auf die Entwürfe von Germaine Bouret zurückgehen, von Stephan Dakon ausmodelliert wurden. Diese Figur gelangte 1936 erstmalig in den Verkauf (Brief von GS an die Wiener Firma Keramos vom 25.2.1953).
Lit.: GSNachf. 1,3 (m. Abb.), hier mit Ang. des Originaltitels „Joujou".
Quelle: AGM (Brief von der Wiener Manufaktur Goldscheider an die Firma Keramos vom 25.2.1953); Archiv Mimi Dakon (Foto); Fotoarchiv K. Täuber.

7595
Dakon, Stephan
Schreitender Dudelsackspieler (auf Ovalsockel)
ca. 1936
7595,34,31 Firmenmarke 19, Made in Austria, H. 28 cm; Doro L 23.11.2001
7595,63,11 ger. D, Firmenmarke 19, Made in Germany, Ausformung zw. 1938–1941, H. 28 cm; Keup 68

Bem.: Diese Figur wurde auch noch zw. 1953–1960 bei Carstens (GWM West-Germany) produziert.
Lit.: Könecke (2002), S. 76, S. 87 (m. Abb.).

7596
Dakon, Stephan
Schreitendes Mädchen in blauer Schuluniform mit Kappe und Schultasche (auf Ovalsockel)
ca. 1936
7596,22,11 Firmenmarke 19, Made in Austria, H. 31,7 cm; eBay # 2059846423
7596,111,7 gem. HK, Firmenmarke 19, Made in Germany, unsign., Ausformung zw. 1938–1941, H. 35,5 cm; eBay # 920920827

7604M
Goebel, Kurt

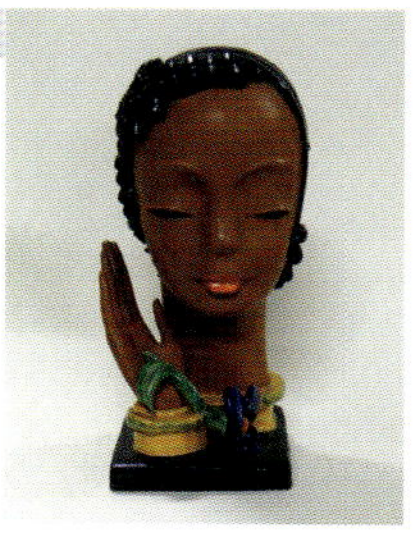

(STEH)KOPF „Kopf ohne/mit Sockel" (Frauenkopf mit Hand und Blumenblüte)
ca. 1936
7604,8,6 Firmenmarke 19, unsign., H. 31,1 cm; eBay # 123100901
7604,8,12 Firmenmarke 19 (blau), Made in Austria (blau), Hand dec. (blau), unsign., H. 29,6 cm; MAK Wien, Inv. Ke 10602-82
7604,14,29 Firmenmarke 19, Made in Austria, unsign., H. 30,5 cm; eBay # 1234150687
7604,22 Firmenmarke 19, Made in Austria, unsign., Ausformung zw. 1945–1953, H. 30,5 cm; eBay # 1273447387
7604M, H. 33 cm, B. 14,5 cm; GSNachf. 2,20
Lit.: GSNachf. 2,20 (m. Abb.), hier mit Ang. des Originaltitels „Kopf ohne/mit Sockel".
Quelle: AGM (Rechnung an Maurice Mizrahi, Alexandria, Ägypten vom 22. April 1953), hier mit Ang. des (Original)-Titels „Kopf".

7605
Goebel, Kurt
STEHKOPF Frauenkopf mit Hand und Buch
ca. 1936
7605 (Firmenmarke ohne Ang.), H. 16,5 cm; Pinhas, S. 166 (m. Abb.)

7607
Goebel, Kurt
WANDMASKE Madonna mit Kind
ca. 1936
7607,S Firmenmarke 19, Made in Austria, unsign., Ausformung zw. 1945–1953, H. 26 cm; eBay # 7417303840

7608
Goebel, Kurt

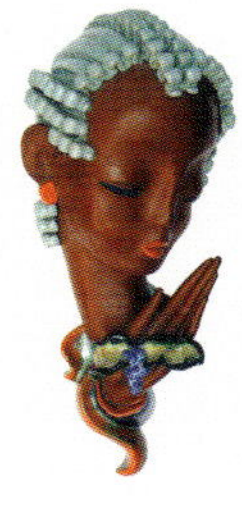

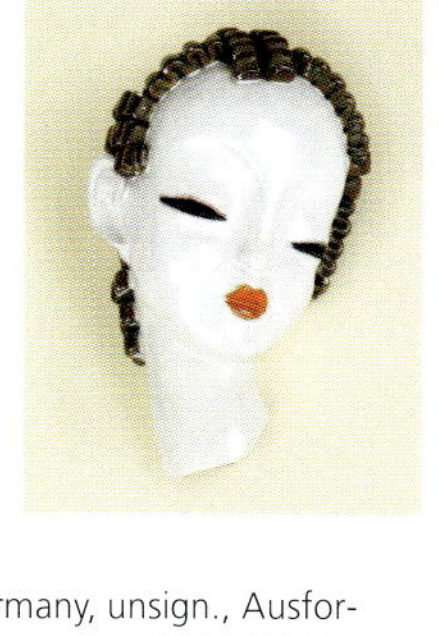

WANDMASKE Frauenkopf mit Hand und Blumenblüte/Trauben
ca. 1936
7608,8 gem. R, Firmenmarke 19, Made in Austria, unsign., Ausformung zw. 1945–1953, H. 43,2 cm; eBay # 2037758499
7608,12 Firmenmarke 19, unsign., Ausformung zw. 1945–1953, H. 27 cm; PB Wien 19; Bem.: Die geringe Höhe von 27 cm rührt daher, dass dieses Exemplar ohne Hand und Blumenblüte ausgefertigt wurde.
7608,35 ger., Firmenmarke 19, Made in Austria, unsign., H. 38,8 cm; Sammlung a@goldscheider.net
7608,II ger., gem. 6, S, Firmenmarke 19, Made in Germany, unsign., Ausformung zw. 1938–1941, H. 36 cm; eBay # 3357380201
Lit.: Pinhas, S. 172 (m. Abb.), hier auch ohne Blumenblüte/Trauben.

7614
Goebel, Kurt

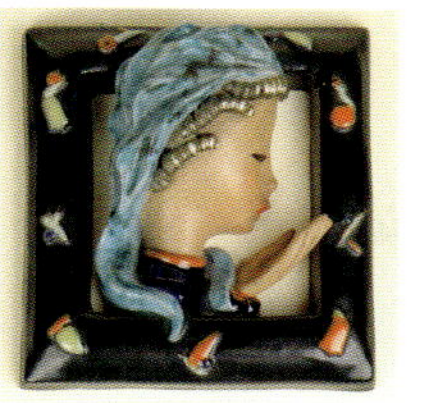

WANDBEHANG Frauenkopf im Profil (gerahmt)
ca. 1936
7614,26 Firmenmarke 19, Made in Austria, sign. Goebel, Ausformung zw. 1945–1953, H. 19 cm; PB Wien 19

7616
Goebel, Kurt

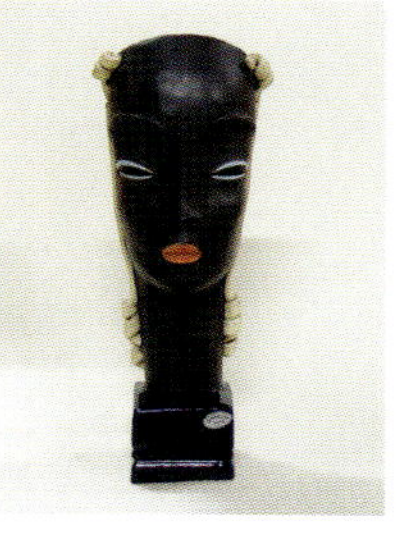

STEHMASKE/WANDMASKE/BUCHSTÜTZE Frauenkopf (auf Rechtecksockel)
ca. 1936
STEHMASKE
7616,3 gem. K, Firmenmarke 19, Made in Austria, unsign., Ausformung zw. 1945–1953, H. 36,5 cm; eBay # 3741558475
7616,8 gem. M., Firmenmarke 19, Made in Austria, unsign., Pick. 5, Ausformung zw. 1945–1953, H. 36,7 cm; MAK Wien, Inv. Ke 10602-97
7616,8 ohne Firmenmarke, unsign., Ausformung zw. 1945–1953, H. 37,5 cm; Wien Museum Inv. 190.172
7616,8 Firmenmarke 19, unsign., Ausformung zw. 1945–1953, H. 37,5 cm; Soth. L. 23.10.1987
7616 (ohne Firmenmarke?), unsign., H. 37,5 cm; Doro 20.11.1996
WANDMASKE
s. Quelle
BUCHSTÜTZE
s. Quelle
Bem.: Wie die Bestellung der Firma Wahliss (Wien) vom 24.11.1945 belegt, wurde diese Stehmaske auf Wunsch auch als Buchstütze und Wandmaske angefertigt.
Ausst.: Wiener Herbstmesse 1936
-> vgl. 7879
Lit.: N.N.: Neue Keramik, in: ÖK, 7.1936, Nr. 10, S. 25 (m. Abb.) „Man sah auf der Wiener Herbstmesse eine Reihe keramischer Arbeiten, die Originalität und richtiges Erfassen der lebendigen Strömungen in der Entwicklung moderner Keramik zeigten. Besonders die Arbeiten des Keramikers Kurt Goebel, die die Wiener Manufaktur Friedrich Goldscheider mit viel Erfolg herausgebracht hat, fielen vorteilhaft auf".
Quelle: AGM (Bestellungen der Firma Wahliss (Wien) vom 9. und 24.11.1945 an die „Arbeitsgemeinschaft der Firma Friedrich Goldscheider", hier in der Ausführung als BUCHSTÜTZE „Maskenkopf" und „Maske ohne Sockel, zum Aufhängen" in Auftrag gegeben; Bestellung der Firma Wahliss (Wien) vom 27.2.1946 als BUCHSTÜTZE „Maskenkopf" und als „Maske ohne Sockel" jeweils in Ausführung „Terrakotta").

7617
Goebel, Kurt

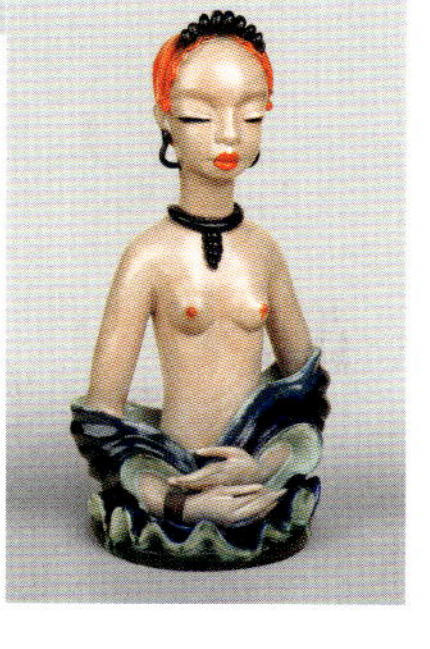

TORSO Mädchen mit Halsschmuck, Ohrringen und Haarband
ca. 1936
7617,8 gem. L., D III, Firmenmarke 19, Made in Austria, unsign., Ausformung zw. 1945–1953, H. 42,3 cm; PB Wien 19

7618
Goebel, Kurt
Golfspieler beim Abschlag (auf Rundsockel)
ca. 1936
7618,35,M,12 ger., Firmenmarke 19, Made in Austria, unsign., H. 30,5 cm; eBay # 265285884
7618 ger., gem. K, Firmenmarke 19, Made in Austria, Pick. 5, H. 29,3 cm; Wien Museum Inv. 190.166
Ausst.: SSH 1992
Lit.: SS, S. 24, 25 (m. Abb.)

7619
Goebel, Kurt
WANDBEHANG Madonna (Frauenkopf) im Profil (gerahmt)
7619,? Firmenmarke 19, Made in Austria, unsign., H. 19 cm, B. 19 cm; Sammlung Rudolf Hansl, St. Florian
Lit.: Pinhas, S. 178 (m. Abb.)

7622M
Goebel, Kurt
FIGUR „Skiläufer" (Schreitende Skiläuferin mit geschulterten Skiern) (auf Ovalsockel)
ca. 1936
7622,0 ger., gem. L., DI, Firmenmarke 19, Made in Austria, sign. Goebel, H. 22,5 cm; GK, S. 22, 126 (m. Abb.); Wien Museum Inv. 223.914
7622,4 gem. O, Firmenmarke 19, Made in Austria, unsign., Ausformung zw. 1945–1953, H. ca. 23 cm; eBay # 354444256
7622,67 Firmenmarke 19, unsign., Ausformung zw. 1945–1953, H. 23,8 cm; Doro 12.9.2006; Bem.: Hier mit der falschen Zuschreibung an Walter Bosse.
7622M, H. 23 cm, B. 20 cm; GSNachf. 3,8a
Bem.: Diese Figur gelangte 1936 erstmalig in den Verkauf (Brief von GS an die Wiener Firma Keramos vom 25.2.1953). Die von der Firma als „Skiläufer" bezeichnete Figur ist eindeutig eine „Skiläuferin".
Ausst.: HMW (jetzt Wien Museum) 1985/86; SSH 1992.
-> vgl. 7887M
-> vgl. 7655M; 7935M= jeweils ähnl. GRUPPE
Pendants -> vgl. 7518; 7887
Lit.: GSNachf. 3,8 (m. Abb.), hier mit Ang. des Originaltitels FIGUR „Skiläufer"; SS, S. 24, 25 (m. Abb.); GK, S. 22, 126 (m. Abb.).
Quelle: AGM (Tantiemenliste Goebel, ohne Datierung (ca. 1940); Brief von der Wiener Manufaktur Goldscheider an die Firma Keramos vom 25.2.1953).

7625
Goebel, Kurt
Stehendes Mädchen (an einem Kaktus oder exotischem Baumstamm mit Blüten) (auf Natursockel mit Schnecke)
ca. 1936
7625,3 Firmenmarke 19, sign. Goebel, Ausformung zw. 1945–1953, H. 30 cm; WKA 20; 29
ohne Ang., Firmenmarke 19, Made in Austria, unsign.?, H. 31 cm; Soth. L. 8.7.1986

7629F
FIGUR „Pekinese"
ca. 1936
7629,16 gem. lig. HS, Firmenmarke 19, Made in Austria, Ausformung zw. 1945–1953, H. 12,9 cm; eBay # 1416085886
7629F, H. 12 cm, B. 17 cm; GSNachf. 5,12b
-> vgl. 7630= EinzelFIGUR (rückwärtig)
-> vgl. 7631= GRUPPE
Lit.: GSNachf. 5,12 (m. Abb.), hier mit Ang. des Originaltitels FIGUR „Pekinese" (Vorderer Hund aus der GRUPPE 7631).

7630F
FIGUR „Pekinese"
ca. 1936
7630F, H. 13 cm, B. 20 cm; GSNachf. 5,12c
-> vgl. 7629= EinzelFIGUR (vorne)
-> vgl. 7631= GRUPPE
Lit.: GSNachf. 5,12 (m. Abb.), hier mit Ang. des Originaltitels FIGUR „Pekinese" (Rückwärtiger Hund aus der GRUPPE 7631).

7631F

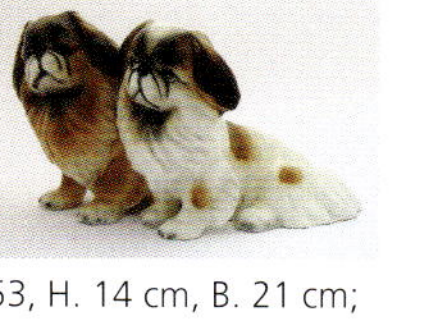

GRUPPE „Pekinesen" (Zwei Pekinesen)
ca. 1936
7631,10 ger., Firmenmarke 19, Made in Austria, Hand dec., Ausformung zw. 1945–1953, H. 14 cm, B. 21 cm; Sammlung a@goldscheider.net
7631,15,14 Firmenmarke 19, Made in Austria, H. 11,4 cm, B. 20,3 cm; eBay # 39441491
7631,16,14 gem. F, Firmenmarke 19, Made in Austria, H. 12,5 cm, B. 20,3 cm; eBay # 237052942

7631F, H. 12,5 cm, B. 20 cm; GSNachf. 5,12a
-> vgl. 7629= EinzelFIGUR (vorne)
-> vgl. 7630= EinzelFIGUR (rückwärtig)
Lit.: GSNachf. 5,12 (m. Abb.), hier mit Ang. des Originaltitels GRUPPE „Pekinesen".

7638
Lorenzl, Josef
Stehender Mädchenakt mit Tuch / Schal (auf Rundsockel)
ca. 1936
7638 (Firmenmarke ohne Ang.), H. 37 cm;
Pinhas, S. 142 (m. Abb.)

7639
Lorenzl, Josef

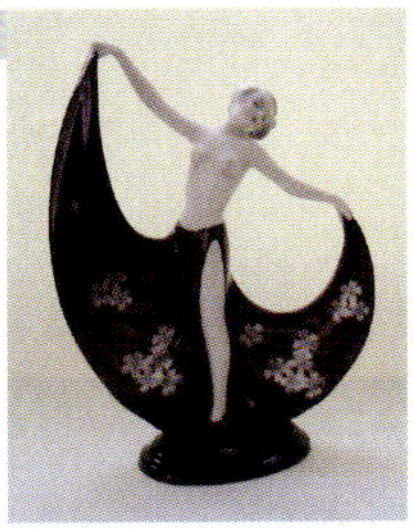

„Sibille" (Stehende Tänzerin in Pose, mit nacktem Oberkörper, ihren langen Rock auseinander haltend) (auf Ovalsockel)
ca. 1936
7639,138,4 ger. A, gem. !EM, Firmenmarke 19, Made in Austria, sign. Lorenzl, H. 25 cm;
eBay # 1268666484
7639,222,? Firmenmarke 19, sign. Lorenzl, Pick. ohne Ang., H. 24 cm; RM 29
7639,234,31 ger., gem. AN, Firmenmarke 19, Made in Germany, sign. Lorenzl, Ausformung zw. 1938–1941, H. 24,5 cm; Sammlung f@goldscheider.de
7639,265,4 gem. GB, Firmenmarke 19, Made in Germany, unsign., Ausformung zw. 1938–1941, H. 25 cm; Henry's 25.10.1997
7639,426 Firmenmarke Schuster 2, Made in Germany, unsign., Ausformung ca. 1944, H. 25 cm;
Doro 16.6.1992
Quelle: AGM (Bestellungen der Firma Wahliss (Wien) vom 9. und 24.11.1945 an die „Arbeitsgemeinschaft der Firma Friedrich Goldscheider", hier jeweils mit Ang. des (Original-?)Titels „Sibille"; Bestellung der Firma Wahliss (Wien) vom 27.2.1946 in Ausführung „Terrakotta").

7640F
Lorenzl, Josef

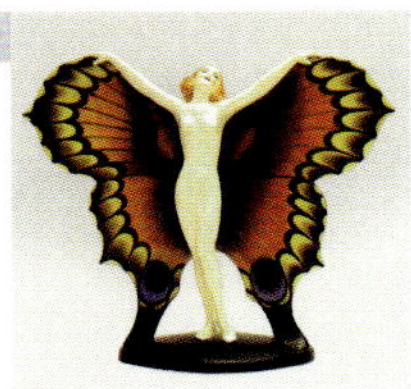

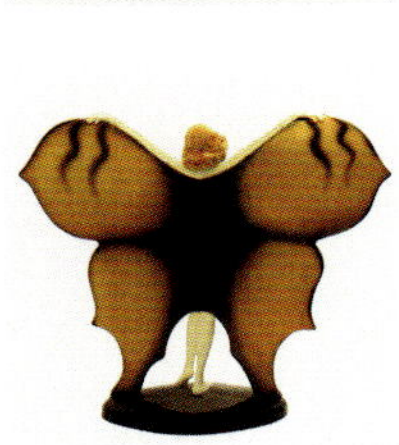

„Falter" (Stehende nackte Tänzerin mit Schmetterlingskostüm) (auf abgerundetem Rautensockel)
ca. 1936
7640,54,2,8 gem. F.S., Firmenmarke 19, Made in Austria, sign. Lorenzl, Pick. 5, H. 26,5 cm, B. 31,5 cm;
Sammlung Weilinger, Wien
7640,74,9 ger. XIIA, gem. AS, Firmenmarke 19, Made in Austria, sign. Lorenzl, Pick. 6, H. 27 cm; Sammlung Stressless, Portugal
7640,83,4 Firmenmarke 19, Made in Austria, sign. Lorenzl, H. 28 cm; eBay # 2195811436
7640,129,19 gem. F., X, Firmenmarke 19, Made in Germany, unsign., Ausformung zw. 1938–1941, H. 27,9 cm, B. 32,7 cm; Sammlung a@goldscheider.net
Quelle: AGM (Rechnung an Maurice Mizrahi, Alexandria, Ägypten vom 22. April 1953), hier mit Ang. des Originaltitels „Falter".

7641
Lorenzl, Josef
Sitzender Damenakt (auf Ovalsockel)
ca. 1936
7641,39,2,8 Firmenmarke 19, Made in Austria, sign. Lorenzl, H. 20,5 cm; Sammlung Robert Senft
7641,89,19 Firmenmarke 19, sign. Lorenzl, H. 20,5 cm;
Doro 14.9.2001
7641,162,9 (Firmenmarke ohne Ang.), unsign.?, H. fehlt, B. 35 cm; Christie's SK 14.11.2002

7642
Lorenzl, Josef

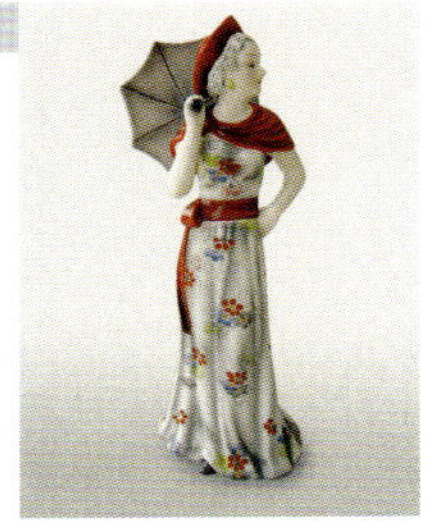

Stehende junge Dame mit Hut und Sonnenschirm
ca. 1936
7642,65,31 ger. XI, gem. AS, Firmenmarke 19, Made in Austria, sign. Lorenzl, H. 34 cm; Sammlung Stressless, Portugal
Lit.: Pinhas, S. 148 (m. Abb.)

7644M
„Akt mit Spiegel"
ca. 1936
Quelle: AGM (Kontoblatt der Wiener Manufaktur Josef Schuster für die Firma Porslins Ericsson-Lund, Stockholm?, Schweden vom 7. Januar 1943), hier mit Ang. des Originaltitels „Akt mit Spiegel".

7648
Weiss, Claire (Klára) (d.i. Herczeg, Claire (Klára)
LAMPE Stehende Dame mit großem Hut und Fächer (auf flachem Rechtecksockel)
ca. 1936
7648,12,4 (Firmenmarke ohne Ang.), sign. Claire Weiss, Gesamth. 65 cm; Doro 4.3.2002; 8.3.2004; 15.3.2005
Lit.: Pinhas, S. 150 (m. Abb.)

7653M
Goebel, Kurt

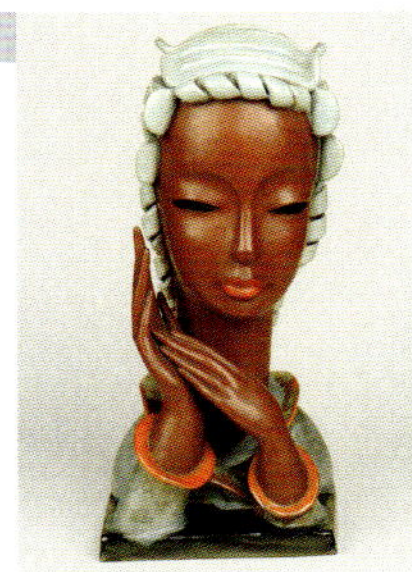

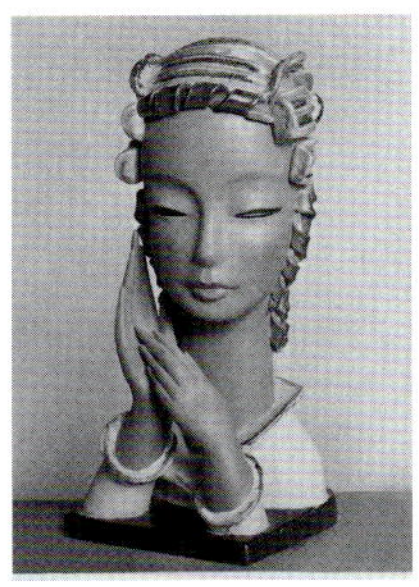

(STEH)KOPF „Kopf ohne/mit Sockel" (Frauenkopf mit zwei Händen)
ca. 1936
7653,1,51 Firmenmarke 19, Made in Austria, unsign., original Papieretikette „Madame Turandot", H. 37,9 cm; MAK Wien, Inv. Ke 10602-2
7653,3 Firmenmarke 19 (blau), unsign., Ausformung zw. 1945–1953, H. 38,6 cm; eBay # 1233327331
7653,3,25 Firmenmarke 19, unsign., H. 37,5 cm;
Doro 10.3.2003
7653,4 Firmenmarke 19, Made in Austria, unsign., Ausformung zw. 1945–1953, H. 38 cm;
eBay # 2511269983
7653,8,31 gem. rot G, Firmenmarke 19, Made in Austria, Hand dec., unsign., H. 38,1 cm;
eBay # 1245483332
7653,8,97 Firmenmarke 19, unsign., Pick. ohne Ang., H. 38 cm; Doro 24.5.2000
7653,19 ger., Firmenmarke 19, Made in Austria, Hand dec., unsign., Ausformung zw. 1945–1953, Pick. 5/a, H. 38,1 cm; eBay # 2032567773
7653,26 gem. L, Firmenmarke 19, Made in Austria, sign. Goebel, Ausformung zw. 1945–1953, H. 38 cm; Soth. L. 17.10.1985; 23.10.1987; eBay # 1208401932
7653,26 gem. L., Firmenmarke 19, Made in Austria, unsign., Ausformung zw. 1945–1953, H. 37,7 cm; MAK Wien, Inv. Ke 10602-84
7653,35 gem. L., Firmenmarke 19, Made in Austria, unsign., Pick. 6, Ausformung zw. 1945–1953, H. 37,5 cm; MAK Wien, Inv. Ke 10602-83
7653,65 Firmenmarke 19, Made in Austria (zwei mal), unsign., Ausformung zw. 1945–1953, H. 37,4 cm; PB Wien 19
7653M, H. 41 cm, B. 19,5 cm; GSNachf. 2,21; Fotoarchiv K. Täuber, hier mit der Ang.: H. 37 cm, B. 28 cm
Lit.: GSNachf. 2,21 (m. Abb.), hier mit Ang. des Originaltitels „Kopf ohne/mit Sockel".
Quelle: AGM (Rechnung an Maurice Mizrahi, Alexandria, Ägypten vom 22. April 1953), hier mit Ang. des (Original)-Titels „Kopf"; Fotoarchiv K. Täuber.

7654
Goebel, Kurt

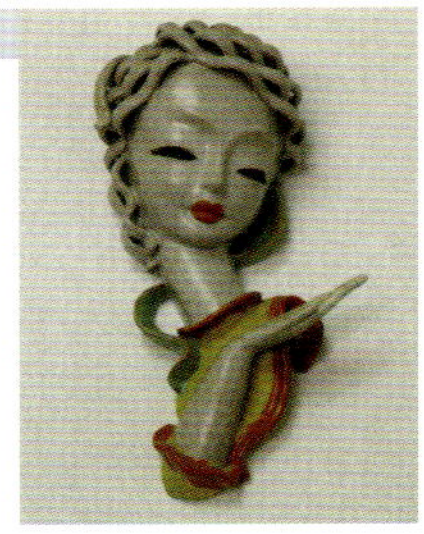

WANDMASKE Frauenkopf mit Hand und kleiner Vase
ca. 1936
7654,6 Firmenmarke 19, Made in Germany, unsign., Ausformung zw. 1938–1941, H. 39 cm; Soth. L. 23.6.1989
7654 Firmenmarke 19, Made in Austria, unsign., H. 40,5 cm; Sammlung Stressless, Portugal
Lit.: Pinhas, S. 167, 173 (m. Abb.)

7655M
Goebel, Kurt

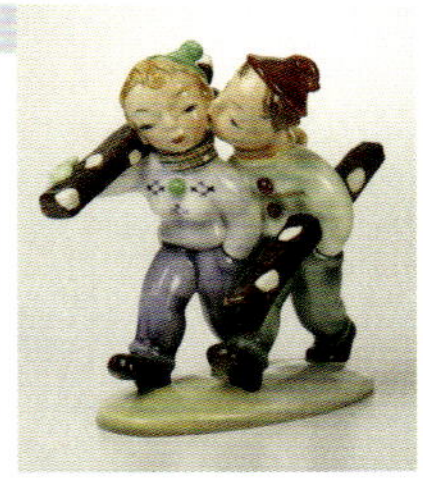

GRUPPE „Skifahrer" (Schreitendes Skifahrerpaar) (auf Ovalsockel)
ca. 1936
7655,2,31 gem. lig. HS, gem. braun H, Firmenmarke 19, Made in Austria, unsign., H. 22,7 cm; Wien Museum Inv. 190.162
7655,26 gem. L., Firmenmarke 19, Made in Austria, sign. Goebel, Ausformung zw. 1945–1953, Pick. 5, H. 22,8 cm; eBay # 3653156207
7655,65,0 Firmenmarke 19, Made in Austria, unsign., H. 22,9 cm; eBay # 512908071
7655M, H. 22 cm, B. 24 cm; GSNachf. 3,7a
-> vgl. 7935M
-> vgl. 7518= Skifahrer, ähnl. FIGUR
-> vgl. 7622; 7887= jeweils Skifahrerin, ähnl. FIGUR
Ausst.: Wiener Herbstmesse 1936
Lit.: N.N.: Neue Keramik, in: ÖK, 7.1936, Nr. 10, S. 25 (m. Abb.) „Kleinplastiken, Gruppen, voll frischer Lebendigkeit im Umriss und in der Oberflächenbehandlung, voll frischer Farbigkeit in Glasur und Dekor, gewannen insbesondere durch den Humor und Charme der Darstellung auch das anspruchsvolle Publikum"; GSNachf. 3,7 (m. Abb.), hier mit Ang. des Originaltitels GRUPPE „Skifahrer".

7656M
Goebel, Kurt
FIGUR „Romeo" (Stehender junger Mann mit Hut und Kaktus) (auf Rundsockel)
ca. 1936
7656,4 ger., Firmenmarke 19, Made in Austria, unsign., Ausformung zw. 1945–1953, H. 40,6 cm; eBay # 1201994425; (KH www.tias.com/stores/frozenbeauty)
7656,8,34 gem. G, Firmenmarke 19, Made in Austria, unsign., H. 41 cm; eBay # 894718792
7656,9 ger., Firmenmarke 19, Made in Austria, unsign., Ausformung zw. 1945–1953, H. 40,6 cm;
eBay # 1236587671
7656M, H. 38 cm, B. fehlt; GSNachf. 3,9a
-> vgl. 7912M; 7978M
Pendant -> vgl. 7657
Lit.: GSNachf. 3,9 (m. Abb.), hier mit Ang. des Originaltitels FIGUR „Romeo".

7657
Goebel, Kurt
Julia (Stehendes schüchternes Mädchen) (auf Rundsockel)
ca. 1936
7657,4 ger., (Firmenmarke ohne Ang.), unsign., Ausformung zw. 1945–1953, H. 37 cm; Soth. L. 18.2.1983
Bem.: Der Titel „Julia" liegt auf der Hand. Ist diese Figur doch als Gegenstück zum „Romeo" (Modell-Nr. 7656) zu sehen.
-> vgl. 7065
Pendant -> vgl. 7656

7658
Prischl, Adolf
Stehende Madonna mit Krone und erhobenen Armen (auf gekehltem Rundsockel)
ca. 1936
7658,28,14 ger. XII, A, gem. N, Firmenmarke 19, Made in Austria, H. 33 cm; eBay # 885881182

7659
Dakon, Stephan (zugeschr.)
Stehender pausbäckiger Holländerjunge mit Kappe und Clogs, ein Modellsegelboot haltend (auf Rundsockel)
ca. 1936
7659,3,16 (Firmenmarke ohne Ang.), H. 28 cm; Christie's SK 4./5.12.1996

7662F
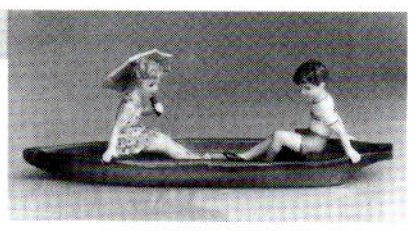
Dakon, Stephan
Kinderpaar im Boot
ca. 1936
7662F, H. 14,5 cm, L. 38 cm
Quelle: Archiv Mimi Dakon (Originalfoto)

7663F
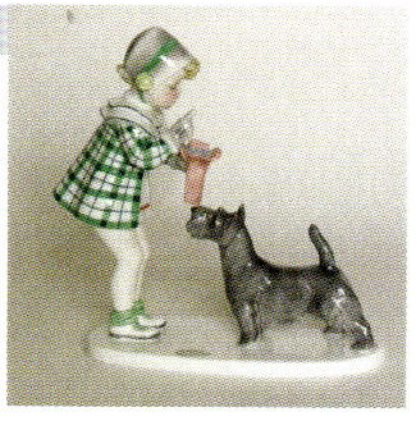
Bouret, Germaine
GRUPPE „Naschkätzchen" (Stehendes Mädchen, einen Terrier fütternd) (auf Ovalsockel)
ca. 1936
7663,0 ger., gem. G, Pfeil, Firmenmarke 19, unsign., Ausformung zw. 1945–1953, H. 22,5 cm, B. 22,5 cm; Wien Museum Inv. 220.726
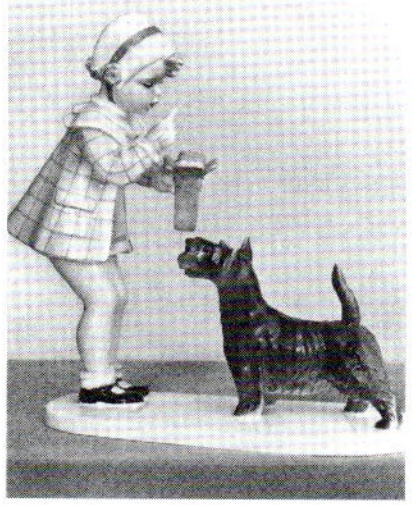
7663,6 (Firmenmarke ohne Ang.), unsign., Ausformung zw. 1945–1953, H. 23 cm; WKA 38
7663,14 ger., gem. F.W., Firmenmarke 19, Made in Austria, Hand dec. unsign., Ausformung zw. 1945–1953, Pick. 1, H. 24 cm; Sammlung Bernard Amigon, Frankreich
7663,20,16 ger. 52, Firmenmarke 19, Made in Austria, sign. G. Bouret, Pick. 6, H. 22,5 cm; Sammlung Bernard Amigon, Frankreich
7663,206,4 Firmenmarke 19, Made in Germany, sign. G. Bouret, Ausformung zw. 1938–1941, H. 22,5 cm; Sammlung Bernard Amigon, Frankreich
7663,474,19 ger. X, 5A, Firmenmarke 19, Made in Germany, sign. G. Bouret, Ausformung zw. 1938–1941, H. 24 cm; eBay # 2056192158; 3367535244
7663,530,40 Firmenmarke 19, Made in Germany, Ausformung zw. 1938–1941, H. 23,3 cm, B. 24 cm; Wien Museum Inv. 220.727
7663F, H. 21,5 B. 22 cm; GSNachf. 1,4
Bem.: Die Entwürfe von Germaine Bouret wurden von Stephan Dakon ausmodelliert. Diese Figur gelangte 1936 erstmalig in den Verkauf (Brief von GS an die Wiener Firma Keramos vom 25.2.1953).
Lit.: GSNachf. 1,4 (m. Abb.), hier mit Ang. des Originaltitels „Naschkätzchen".
Quelle: AGM (Bestellungen der Firma Wahliss (Wien) vom 9. und 24.11.1945 an die „Arbeitsgemeinschaft der Firma Friedrich Goldscheider"; Bestellung der Firma Wahliss (Wien) vom 27.2.1946, in Ausführung „Terrakotta" je 55 Stück, in Fayence je 2 Stück; Brief von der Wiener Manufaktur Goldscheider an die Firma Keramos vom 25.2.1953).

7664F
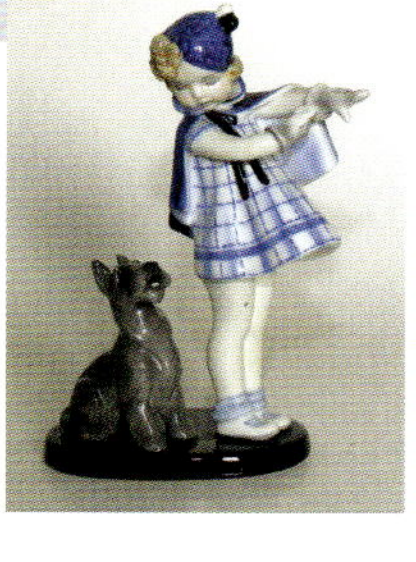
Bouret, Germaine
GRUPPE „Komm mit!" (Stehendes Mädchen mit sitzendem Terrier) (auf Ovalsockel)
ca. 1936
7664,1 gem. G, Firmenmarke 19, Made in Austria, unsign., Ausformung zw. 1945–1953, H. 23,5 cm; eBay # 1301346541
7664,4,12 Firmenmarke 19, Made in Austria, Hand dec., unsign., H. 22,5 cm; KH
7664,40,20 ger. XV(II?). gem. W., Firmenmarke 19, Made in Austria, sign. G. Bouret, H. 22,5 cm; Sammlung Bernard Amigon, Frankreich
7664,98,4 gem. SG, gem. lig. LS, Firmenmarke 19, Made in Austria, sign. Dakon (!), H. ca. 21 cm; eBay # 544652225;
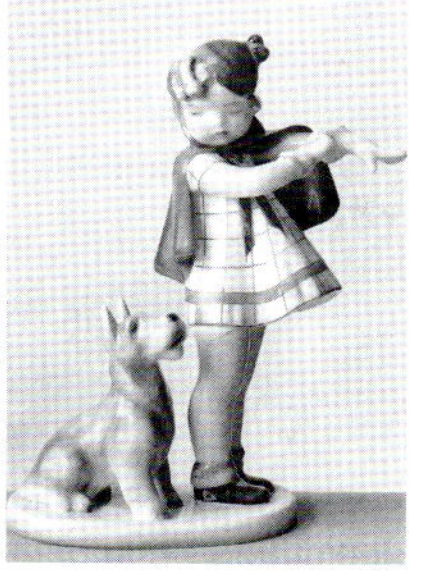
Bem.: Die Signatur „Dakon" ist dem Umstand zuzuschreiben, dass Dakon die Modelle von Germaine Bouret für die Manufaktur ausgeformt und retouchiert hat.
7664,369,14 Firmenmarke 19, Made in Austria, sign. G. Bouret, H. 23 cm, B. 18 cm; Sammlung mad
7664,442,9 ger. XVI, Firmenmarke 19, Made in Austria, sign. G. Bouret, H. 22,8 cm; eBay # 3740009336
7664,459,21 Firmenmarke 19, Made in Austria, sign. G. Bouret, H. 22,5 cm; Sammlung Bernard Amigon, Frankreich
7664,537,77,47,M gem. AN, Firmenmarke 19, Made in Germany, Ausformung zw. 1938–1941, Originalpick.: „McLellan Est. 1893, San Francisco", H. ca. 22 cm; eBay # 1238103081
7664,756,19 gem. AN, Firmenmarke 19, Made in Germany, unsign., Ausformung zw. 1938–1941, H. 23,5 cm; eBay # 6503411666
7664F, H. 21 cm, B. 13 cm; GSNachf. 1,5; Fotoarchiv K. Täuber, hier mit der Ang.: H. 23 cm, B. 16 cm
Bem.: Die Entwürfe von Germaine Bouret wurden von Stephan Dakon ausmodelliert. Diese Figur gelangte 1936 erstmalig in den Verkauf (Brief von GS an die Wiener Firma Keramos vom 25.2.1953). Sie wurde ebenso, wenngleich minimal abgewandelt durch die Sitzposition des Hundes, unter Marcel Goldscheider in England (Staffordshire) produziert. Je eine Version mit dem sitzenden Terrier sowie eine ohne diesen befinden sich in der Sammlung Bernard Amigon, Frankreich. Die Größe dieser Ausformungen ist jeweils nur 21 cm.
Lit.: GSNachf. 1,5 (m. Abb.), hier mit Ang. des Originaltitels „Komm mit!".
Quelle: AGM (Bestellungen der Firma Wahliss (Wien) vom 9.11.1945 und 27.2.1946 an die „Arbeitsgemeinschaft der Firma Friedrich Goldscheider" in Ausführung „Terrakotta"; Brief von der Wiener Manufaktur Goldscheider an die Firma Keramos vom 25.2.1953); Fotoarchiv K. Täuber.

7665F
„Pyjama"
ca. 1936
Bem.: Es ist zu vermuten, dass es sich hierbei um einen Entwurf von Stephan Dakon handelt. Da aber weder ein Stück oder eine Abbildung nachweisbar ist, noch der Rechnungsbeleg den Künstler nennt, wäre eine Zuschreibung an Dakon zu unsicher, zumal auch von Josef Lorenzl ein Modell mit dem Titel „Pyjama" gestaltet wurde. Es könnte sich auch um eine Größenvariante (Modell-Nr. 5596 oder 5780) handeln.
Quelle: AGM (Rechnung an Maurice Mizrahi, Alexandria, Ägypten vom 22. April 1953), hier mit Ang. des Originaltitels „Pyjama".

7667F
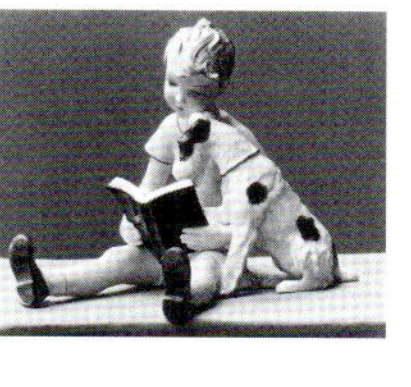
Dakon, Stephan
Sitzendes lesendes Mädchen mit Foxterrier
ca. 1936/37
7667,38,16 gem. B.i., Firmenmarke 19, Made in Austria, H. 19 cm; PB
7667F, H. 28 cm, B. 19 cm
Quelle: Archiv Mimi Dakon (Originalfoto)

7669
Dakon, Stephan
„Badende" (Sitzendes Mädchen im Badedress)
ca. 1936/37
7669,6,2 gem. 94II, gem. lig. NC, Firmenmarke 19, Made in Austria, H. 28 cm; WKH
Lit.: Pinhas, S. 144 (m. Abb.)
Quelle: AGM (Bestellungen der Firma Wahliss (Wien) vom 30.8.1945, 17.10.1945 und 27.2.1946 an die „Arbeitsgemeinschaft der Firma Friedrich Goldscheider"), hier jeweils mit Ang. des (Original-?)Titels „Badende".

7670
Matula, Viktor
GRUPPE Rehe
ca. 1936/37
7670 H. fehlt
Quelle: Nachlass Matula, Wien

7671
Matula, Viktor
GRUPPE „Rehgruppe" (Zwei Rehe, eines stehend, das andere liegend) (auf Ovalsockel)
ca. 1936/37
7671,20 Firmenmarke 19, unsign., H. 30 cm, B. 30 cm; eBay # 2230522466
7671,108,9 Firmenmarke 19, unsign., H. 28 cm, B. 28 cm; eBay # 1487985815
Quellen: Nachlass Matula, Wien; AGM (Bestellungen der Firma Wahliss (Wien) vom 9. und 24.11.1945 an die „Arbeitsgemeinschaft der Firma Friedrich Goldscheider", hier jeweils mit Ang. des (Original?-)Titels „Rehgruppe"; Bestellung der Firma Wahliss (Wien) vom 27.2.1946 in Ausführung „Terrakotta" und „Fayence").

7672M
Prischl, Adolf
FIGUR „Akt mit Perlenkette" (auf Rundsockel)
ca. 1936/37
7672,4 ger., gem. Z (oder N?), Firmenmarke 19, Made in Austria, Ausformung zw. 1945–1953, H. 30,5 cm; eBay # 415481918; (KH www.tias.com/stores/frozen-beauty); eBay # 7416903028
7672,(26?) ger., gem. G., Firmenmarke 19, Made in Austria, unsign., Ausformung zw. 1945–1953, H. 28 cm; eBay # 911665704
7672M, H. 30 cm, B. 14 cm; GSNachf. 3,1; Fotoarchiv K. Täuber, hier mit der Ang.: H. 31 cm, B. 12 cm
Lit.: GSNachf. 3,1 (m. Abb.), hier mit Ang. des Originaltitels „Akt mit Perlenkette".
Quelle: Fotoarchiv K. Täuber

7676M
Matula, Viktor
RELIEF „Segler" (Segelschiff nach rechts) (Wandrelief)
ca. 1936/37
7676 Firmenmarke 19, Made in Germany, unsign., Ausformung zw. 1938–1941, H. 10,3 cm, B. 13,3 cm; eBay # 159563568
7676M, H. 10 cm, B. 13 cm; GSNachf. 2,23a
Lit.: GSNachf. 2,23 (m. Abb.), hier mit Ang. des Originaltitels „Relief Segler".
Quelle: Nachlass Matula, Wien

7679
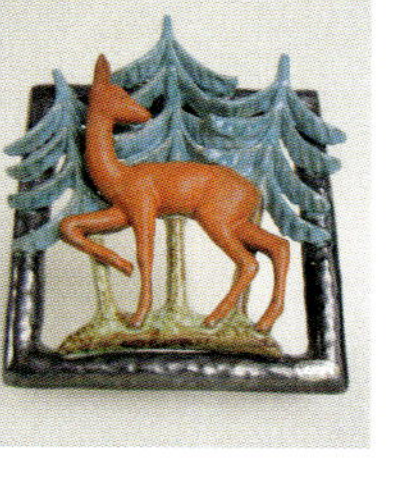
RELIEF Reh vor drei Tannen stehend (durchbrochenes Wandrelief)
ca. 1936/37
7679 (Bleistift), gem. lig. AH, Firmenmarke 19, Made in Austria, Pick. 6; H. 22,5 cm, B. 19,5 cm; Sammlung Weilinger, Wien

7680M
GRUPPE „Pferde" (Zwei Fohlen, eines steigend) (auf flachem Rechtecksockel)
ca. 1936/37
7680,5 gem. XVI, Firmenmarke 19, Made in Austria, Ausformung zw. 1945–1953, H. 20 cm; eBay # 260001038825
7680,7 gem. lig. grün TK, Firmenmarke 19, Made in Austria, Ausformung zw. 1945–1953, H. 18 cm, L. 18 cm; eBay # 2228663200; 7319589640
Quelle: AGM (Kontoblatt der Wiener Manufaktur Josef Schuster für die Firma Porslins Ericsson-Lund, Stockholm?, Schweden vom 7. Januar 1943), hier mit Ang. des Originaltitels GRUPPE „Pferde".

7684F
Dakon, Stephan
Schreitende Tänzerin in mexikanischem Kostüm mit Hut (auf Ovalsockel)
um
7684F, H. 35 cm, B. 16 cm
Quelle: Archiv Mimi Dakon (Originalfoto)

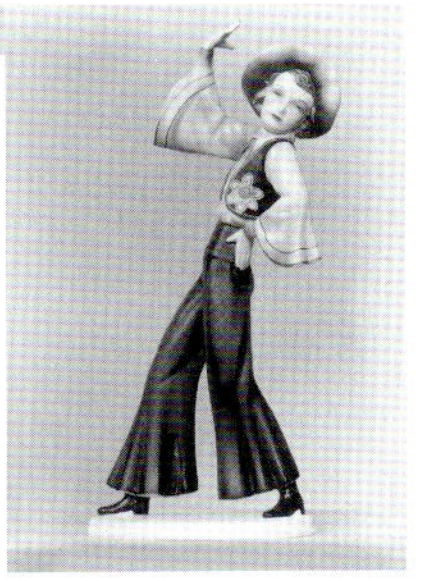

7685
Dakon, Stephan
Schreitende Tänzerin, ihr langes Kleid auseinander haltend (auf Ovalsockel)
ca. 1936/37
7685,13,– Firmenmarke 19, Made in Austria, H. 36,3 cm; Soth. L. 23.6.1989
7685,35,2 Firmenmarke 19, H. 38,3 cm; Doro 24.5.2000
Lit.: Pinhas, S. 145, 156 (m. Abb.)

7686
Goebel, Kurt
BUCHSTÜTZEN
Beim Fensterln
ca. 1936/37
7686,0 gem. L., K, Firmenmarke 19, Made in Austria, H. 18 cm; PB Wien 19

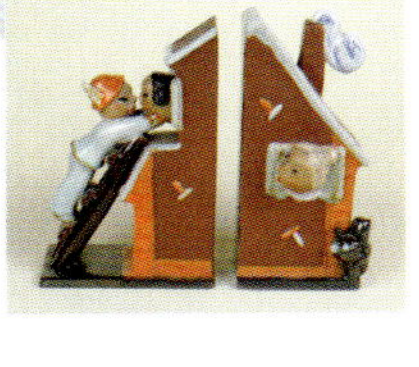

7688 –> ÜBERSCHNEIDUNG
Lorenzl, Josef
Stehender Damenakt, mit zwei großen Fächer posierend (auf Ovalsockel)
ca. 1936/37
7688,30,7 Firmenmarke 19, Made in Austria, H. 47 cm; PB Wien 19
–> vgl. 7544= ähnl. Modell, hier die gleiche Tänzerin mit nur einem Fächer.

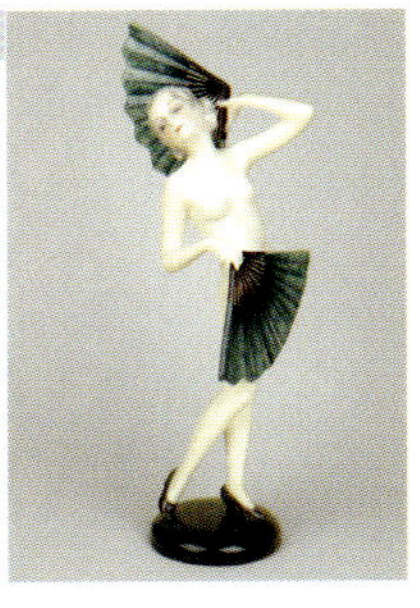

7688 –> ÜBERSCHNEIDUNG
Dakon, Stephan
Stehende junge Dame (im Jagd / Reitoutfit) mit Jagdhund (auf Ovalsockel)
ca. 1936/37
7688 (Firmenmarke ohne Ang.), H. 32,8 cm; Pinhas, S. 142 (m. Abb.)

7689
Lorenzl, Josef
Stehende Tänzerin in Pose, mit erhobenem Arm (auf Rundsockel)
ca. 1936/37
7689,8,16 Firmenmarke 19, Made in Austria, H. ca. 33 cm; KH
7689,22,31 Firmenmarke 19, Made in Austria, unsign., H. 33 cm; eBay # 1426853434

7690
Lorenzl, Josef
RELIEF Betende Madonna (im Profil nach rechts) (Wandrelief)
ca. 1936/37
7690,83,31 ger. XII, BG, gem. lig. blau SN, Firmenmarke 19, Made in Austria, H. 35,5 cm; eBay # 914440840

7691F
Weiss, Claire (Klára)
(d.i. Herczeg, Claire (Klára)
FIGUR „Demaskierung"
(Stehende junge Dame, ihre Maske abnehmend) (auf Ovalsockel)
ca. 1936/37
7691,22,4 gem. I., Firmenmarke 20, Made in Austria, Hand dec., unsign., H. 38 cm; eBay # 1281680680

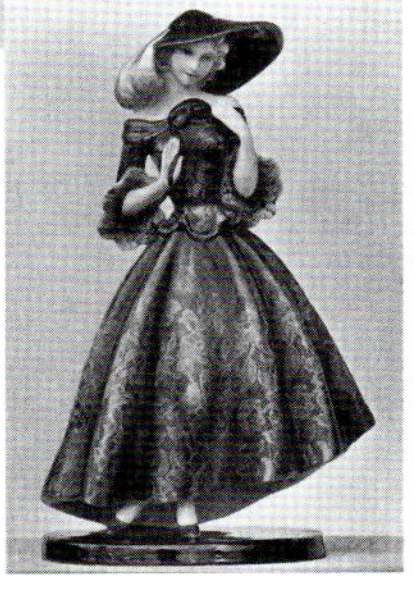

7691,103,4 Firmenmarke 19, Made in Austria, sign. C. Weiss, H. 35 cm; eBay # 899584440
7691,171,4 Firmenmarke 19, Made in Austria, sign. C. Weiss, H. 38 cm; eBay # 3718552890
7691,210,4 Firmenmarke 19, Made in Germany, sign. Claire Herczeg, Ausformung zw. 1938–1941, H. 38 cm; Christie's SK 27.4.2005
7691,224,0 Firmenmarke 19, Made in Germany, sign. Claire Herczeg, Ausformung zw. 1938–1941, H. 38 cm; Christie's SK 13.10.2004
7691,274,7 Firmenmarke Schuster 1, Sign. ohne Ang., Ausformung zw. 1941–1944, H. 38 cm; Herr 26
7691F, H. 35 cm, B. 23 cm; GSNachf. 1,19a
Bem.: Dieses Modell ist eines der wenigen, das auch noch in Trenton (USA) erfolgreich produziert wurde.
–> vgl. 8183
Lit.: GSNachf. 1,19 (m. Abb.), hier mit Ang. des Originaltitels „Demaskierung"; Hoever (1996).

7693
WANDMASKE Frauenkopf mit Hand, sich durch die Haare fahrend
ca. 1936/37
7693,33 gem. G, Firmenmarke 19, Made in Austria, H. 24 cm; PB Wien 19

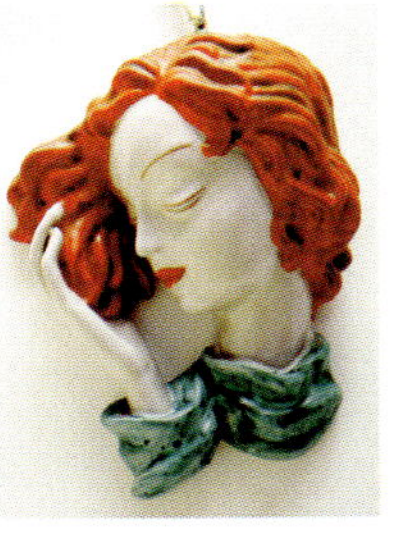

7695
WANDMASKE Frau mit langem Haar (an einer Seite lockig herabhängend)
ca. 1936/37
7695,3 Firmenmarke 19, Made in Germany, Ausf. nach 1938, H. 25 cm; Wien Museum Inv. 186.649
7695,3 gem. A, Firmenmarke 19, Made in Germany, Ausf. nach 1938, H. 25 cm; Sammlung.schmutz@chello.at 46
7695,8 Firmenmarke 19, Made in Austria, Pick. 6, H. 24,4 cm; MAK Wien, Inv. Ke 10521-31
Bem.: Trotzdem die beiden nachgewiesenen Stücke identische Nummernfolgen aufweisen, sind sie in der Staffierung unterschiedlich. Die Haare des Exemplars des Wien Museums sind in dem typischen uranrot gehalten, die Maske aus der Sammlung Rudolf Schmutz/Wien hat eine brünette Haarfarbe.
–> vgl. 7698

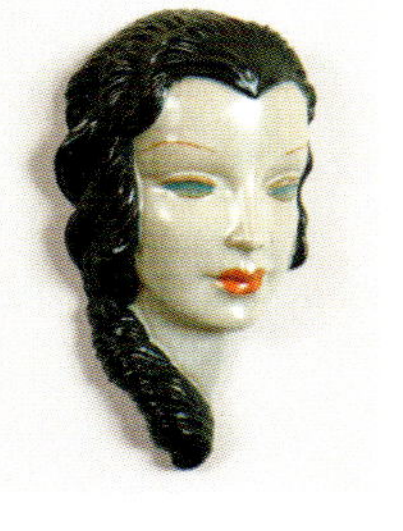

7696
WANDMASKE Frauenkopf mit klatschenden Händen (im Profil nach links)
ca. 1936/37
7696,3 gem. G, Firmenmarke 19, Made in Germany, Ausformung zw. 1938–1941, Pick. 6, H. 33 cm; eBay # 2227071958
7696,33 gem. lig. AH, Firmenmarke 19, Made in Austria, Ausformung zw. 1945–1953, H. 32 cm; eBay # 200014521142
Lit.: Pinhas, S. 172 (m. Abb.), hier mit Ang. der falschen Mod.-Nr. 7676.

7698
WANDMASKE Frau mit langem Haar (an einer Seite lockig herabhängend)
ca. 1936/37
7698 gem. Z, Firmenmarke 19, Made in Austria, H. 24,5 cm; Sammlung f@goldscheider.de
Bem.: Diese Maske ist mit der Pressnummer 7698 gestempelt. Es kann sich hier nur um ein Versehen des Modelleurs handeln, da die beiden anderen nachgewiesenen Stücke, die in Aussehen und Größe identisch sind, die Modell-Nr. 7695 geprägt haben.
–> vgl. 7695

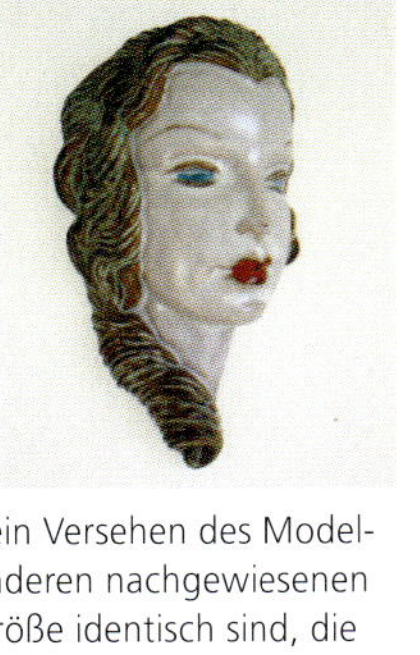

7699F
Dakon, Stephan
Flamencotänzerin in langem Rock (Halbakt) mit Tamburin (auf Ovalsockel)
ca. 1933 (für den Entwurf)
7699,11 gem. G, Firmenmarke 19, Made in Austria, unsign., Ausformung zw. 1945–1953, H. 42 cm; Wien Museum Inv. 190.065
7699,79,11 (Firmenmarke ohne Ang.), H. 43,7 cm; Doro 20.3.2006
7699,88,31 gem. F., Firmenmarke 19, Made in Germany, unsign., Ausformung zw. 1938–1941, H. 41 cm; WKA 35
7699,93,19 Firmenmarke 19, Made in Germany, unsign., Ausformung zw. 1938–1941, H. 41,5 cm; WKA 52
7699,108,2 Firmenmarke 19, Ausformung wohl ab 1938–1941, unsign., H. 43,5 cm; KaM 91
7699 ger., Firmenmarke 19, Made in Germany, unsign., Ausformung zw. 1938–1941, H. 41,2 cm; Doro 29.5.2001
7699,8 ger. K5, Firmenmarke 19, Made in Austria, unsign., Ausformung 50er Jahre, H. 41 cm; Wien Museum Inv. 219.197
7699F, H. 41 cm, B. 25 cm
–> vgl. 6692
Lit.: RDM, 5.1930/31, S. 488 (m. Abb.), hier eine zeitgenössische Fotografie, die wohl als Vorlage für die durch das Tamburin abgewandelte Ausführung diente. Dargestellt ist hier eine nicht näher bezeichnete Tänzerin. Fotografie: D'Ora-Benda, Wien, hier betitelt „Wiener Blut".
Quelle: Archiv Mimi Dakon (Originalfoto)

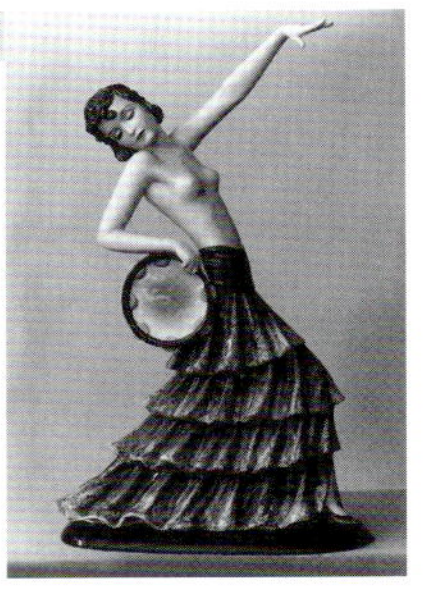

7700F
Meisinger, Ida
FIGUR Hund „Skifahrer"
(Skifahrender Terrier)
ca. 1936/37
7700,3,9 Firmenmarke 19, Made in Austria, Hand dec., unsign., H. 12,7 cm, L. 11,9 cm; Sammlung a@goldscheider.net
7700,20 ger. NSW, 1733, GZ, Firmenmarke 19, Made in Austria, Hand dec., unsign., Ausformung zw. 1945–1953, H. 11,4 cm; KH
7700F, H. 11 cm, B. 12 cm; GSNachf. 3,22b; Fotoarchiv K. Täuber
Bem.: s. 6449
Lit.: GSNachf. 3,22 (m. Abb.), hier mit Ang. des Originaltitels FIGUR Hund „Skifahrer".
Quelle: Fotoarchiv K. Täuber

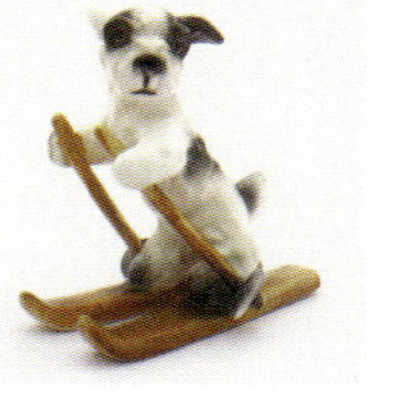

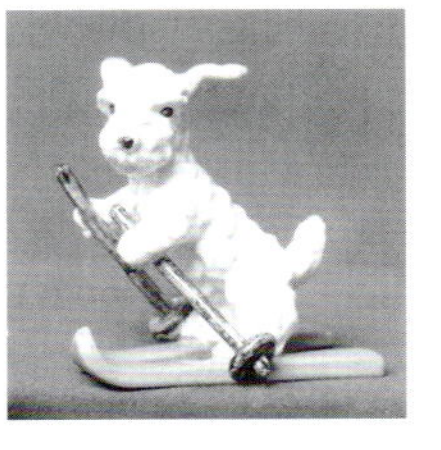

7701F
Meisinger, Ida
GRUPPE Hunde „Naschkätzchen" (Zwei in einem Topf sitzende Terrier)
ca. 1936/37
7701,3,13 gem. grün G, Firmenmarke 19, Made in Austria, unsign., H. 13,9 cm; eBay # 1282948283
7701,6 ger., Firmenmarke 19, Made in Austria, Ausformung zw. 1945–1953, H. 13 cm; PB Deutschland
7701,6,22 Firmenmarke 19, Made in Austria, unsign., H. 12,7 cm; eBay # 2148849398
7701,70,13 Firmenmarke 19, Made in Austria, unsign., H. 13,5 cm; Wien Museum Inv. 190.014
7701F, H. 13,5 cm, B. 11 cm; GSNachf. 3,22a; Fotoarchiv K. Täuber
Bem.: s. 6449
Lit.: GSNachf. 3,22 (m. Abb.), hier mit Ang. des Originaltitels GRUPPE Hunde „Naschkätzchen".
Quelle: Fotoarchiv K. Täuber

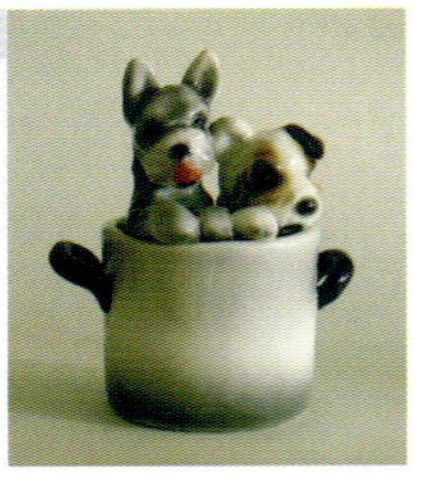

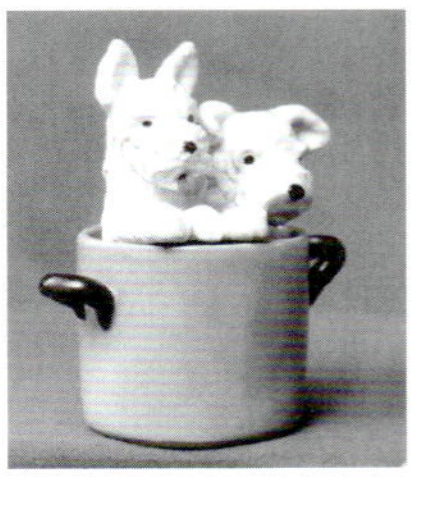

7704
Sitzendes Eichhörnchen (auf Tannenzweig)
ca. 1936/37
7704 Firmenmarke 19, Made in Austria, Hand dec., Pick. 1, H. 19 cm; eBay # 3200081454

7705
Dakon, Stephan (zugeschr.)
WANDMASKE Mädchenkopf mit gerüschtem Kragen
ca. 1936/37
7705,11,4 Firmenmarke 19, Made in Austria, H. 25,5 cm; Sammlung Stressless, Portugal

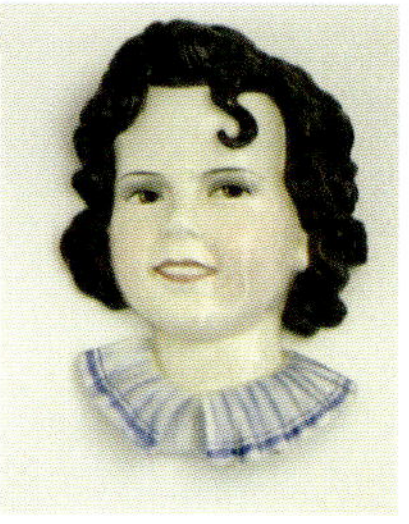

7706
Dakon, Stephan
WANDMASKE Kinderkopf (Bub mit Kragen und Mütze)
ca. 1936/37
7706 gem. W., N, Firmenmarke 19, Made in Austria, Pick. 6, H. 24 cm; PB Österreich

7707
WANDMASKE Mädchenkopf mit großen Locken und großer Schleife
ca. 1936/37
7707,27 gem. B, Firmenmarke 19, Made in Austria, Ausformung zw. 1945–1953, H. 27 cm; eBay # 6594461818

7713M
Matula, Viktor
KERZENHALTER „Leuchter" (vierflammig)
Form: zweifach, u-förmig geschwungene Röhren, auf wellen-förmigem Sockel
Dekor: Laufglasur
ca. 1936/37
7713,2 Firmenmarke 19, Made in Austria, unsign., H. 19 cm, L. 34 cm; GK, S. 22, 127 (m. Abb.); Wien Museum Inv. 222.554
7713M, H. 18 cm, L. 33 cm; GSNachf. 4,9b
Ausst.: HMW (jetzt Wien Museum) 1985/86; SSH 1992.
Lit.: GSNachf. 4,9b (m. Abb.), hier mit Ang. des Originaltitels „Leuchter"; GK, S. 22, 127 (m. Abb.); GSN, S. 64, 65 (m. Abb.).
Quelle: Nachlass Matula, Wien
GS neu: Modell-Nr. 80 „Gemeinsam" (L. 30 cm), Aufl. ohne Ang.

7717M
Knörlein, Rudolf (zugeschr.)
FIGUR „Akt liegend"
(mit/ohne Ovalsockel)
ca. 1936/37
ohne Sockel
7717,65 ohne Firmenmarke, unsign., Ausformung zw. 1945–1953, H. 13,5 cm, L. 29 cm; eBay # 3760702921
7717M, H. 13,5 cm, B. 28 cm; GSNachf. 3,2a
mit Sockel
7717M, H. 18 cm, B. 32,5 cm; GSNachf. 3,2b; Fotoarchiv K. Täuber, hier mit der Ang.: H. 18 cm, B. 32 cm
Lit.: GSNachf. 3,2 (m. Abb.), hier mit Ang. des Originaltitels „Akt liegend".
Quelle: Fotoarchiv K. Täuber

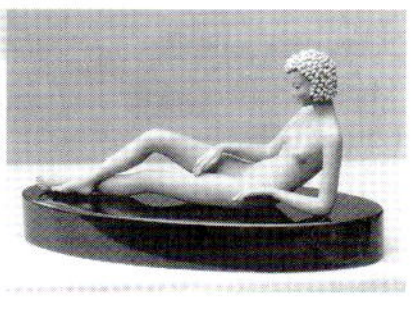

7719
BECHER/ZIGARETTEN- bzw. BLEISTIFTHALTER
Form: zylindrisch
Dekor: Frauenkopf
ca. 1936/37
7719,99 Firmenmarke 19, Made in Germany, Ausformung zw. 1938–1941, H. 5,1 cm, Ø 5,8 cm; Sammlung a@goldscheider.net
-> vgl. 718= ähnl. Modell.

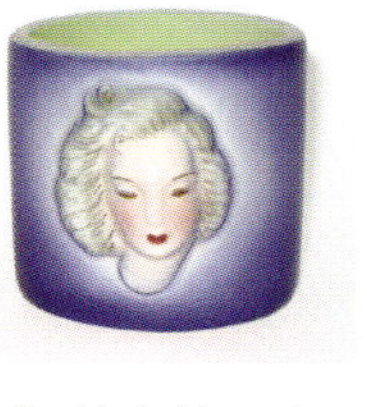

7720
Weiss, Claire (Klára) (d.i. Herczeg, Claire (Klára)
GRUPPE Stehende Dame mit Muff, großem Hut und Windhund (Windspiel) (auf Ovalsockel)
ca. 1936/37
7720,18,4 Firmenmarke 19, Made in Austria, sign. Claire Weiss, Pick. 5, H. 31 cm; Soth. L. 3.11.1995; Christie's SK 13.5.1999; eBay # 1503208625
Bem.: Der Entwurf der Damenfigur ist in die Jahre 1930/31 zu datieren. Der beigegebene Hund dieser Gruppe stammt eindeutig nicht aus der Hand von Claire Weiss, sondern wurde, wie an anderer Stelle bereits dokumentiert, von einem Modelleur der Firma „hinzukomponiert".
-> vgl. 6328; 7044= jeweils FIGUR Dame
Lit.: Pinhas, S. 134 (m. Abb.)

7721
Lorenzl, Josef
GRUPPE Stehende Rokokodame mit Federfächer und Windhund (Windspiel) (auf Ovalsockel)
ca. 1936/37
7721,5,11 Firmenmarke 19, Made in Austria, H. 31 cm; Doro 20.4.1993
7721,12,14 gem. lig. AS, S., Firmenmarke 19, Made in Austria, H. 31,3 cm, B. 21,5 cm; Privatsammlung, Wien

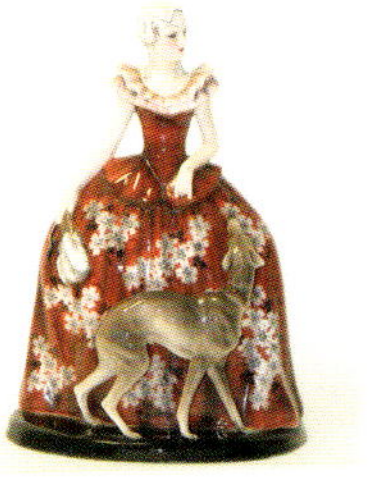

7722
Lorenzl, Josef
GRUPPE Stehender Damenakt mit großem Fächer, Hut und Barsoi (auf Ovalsockel)
ca. 1936/37
7722,85,19 ger., gem. A, Firmenmarke 19, Made in Austria, H. 37 cm; Soth. L. 4.11.1994
Lit.: Pinhas, S. 132 (m. Abb.)

7723
Weiss, Claire (Klára) (d.i. Herczeg, Claire (Klára)
FIGUR „Parisienne" (Schreitende elegante Dame, ihren Schirm in der linken Hand haltend) (auf Ovalsockel)
ca. 1935/36 (für den Entwurf)
7723,143,7 gem. G, gem. rot XX36, Firmenmarke Schuster 1, Made in Germany, unsign., Ausformung zw. 1941–1944, H. 35 cm; Wien Museum Inv. 219.193
-> vgl. 8138
-> vgl. 8361= seitenverkehrtes Modell
Lit.: GSNachf. 1,16 (m. Abb.), hier mit Ang. des Originaltitels „Parisienne". Hier für die Höhe 32 cm.

7724F
Lorenzl, Josef
GRUPPE Stehender Damenakt mit ausgestrecktem Arm und Barsoi (auf Ovalsockel)
ca. 1936/37
7724,28,14 Firmenmarke 19, Made in Austria, sign. Lorenzl, Pick. 6, H. 38,5 cm; eBay # 1200818244
7724,39,31 Firmenmarke 19, Made in Austria, sign. Lorenzl, H. 41 cm; KH Ruby Lane (Jane Perlow/ Lacremeantik, Great Neck N.Y.)
7724,48,31 ger. 47C, (Firmenmarke ohne Ang.), unsign., Pick. ohne Ang., H. 37,5 cm; Christie's SK 22.4.2004
-> vgl. 7525= FIGUR Damenakt
Quelle: AGM (Rechnung der Wiener Manufaktur Josef Schuster an Laszlo Bösze (Szeged, Ungarn) vom 5. August 1944), hier unter dem Titel „Akt/Windhund" geführt.

7726M
Matula, Viktor
RELIEF „Rehe" (Zwei springende Rehe) (Wandrelief)
ca. 1936/37
7726 Firmenmarke 19, unsign., Pick. 6, H. 15 cm, B. 22,5 cm; eBay # 913483023
7726M, H. 14,5 cm, B. 22 cm; GSNachf. 2,23b
Lit.: GSNachf. 2,23 (m. Abb.), hier mit Ang. des Originaltitels „Relief Rehe".
Quelle: Nachlass Matula, Wien

7730F
Heindl, Hermann
FIGUR „Madonna mit Kind" (Stehende Madonna mit Kind)
ca. 1931 (für den Entwurf)
7730F, H. 31 cm, B. 15 cm; GSNachf. 6,10b
7730 H. 32 cm, B. 18 cm; Fotoarchiv K. Täuber
Bem.: Diese Figur wurde auch noch zw. 1953–1960 bei Carstens (GWM West-Germany) produziert.
-> vgl. 6444
Lit.: GSNachf. 6,10 (m. Abb.), hier mit Ang. des Originaltitels FIGUR „Madonna mit Kind"; Könecke (2002), S. 76, S. 87 (m. Abb.).
Quelle: Fotoarchiv K. Täuber

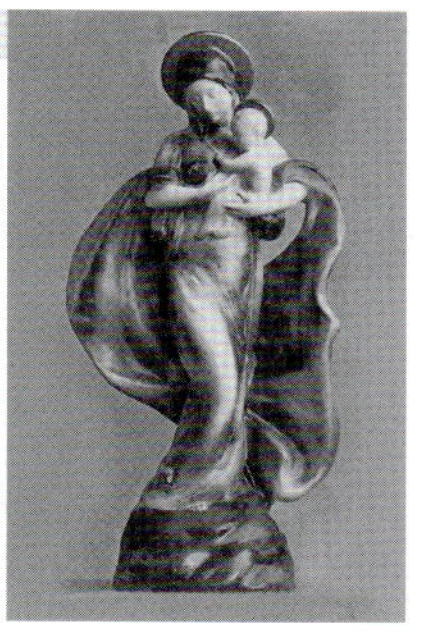

7731
Lorenzl, Josef
„Eifersucht" (Stehende Dame mit Hut und zwei Pekinesen) (auf Rundsockel)
ca. 1934/35 (für den Entwurf)
7731,26,19 (Firmenmarke ohne Ang.), H. 24,2 cm; Christie's SK 13.5.2004
7731,32,14 gem. lig. LS, gem. rot B., Firmenmarke 19, H. 24,8 cm; eBay # 1283809869
-> vgl. 7037

7732
Lorenzl, Josef
Stehende Tänzerin in Pose, ihr Kleid mit der rechten Hand zurückhaltend (auf Ovalsockel)
ca. 1931/32 (für den Entwurf)
7732,45,6 (Firmenmarke ohne Ang.), H. 23 cm; Soth. Olympia 3.5.2002
-> vgl. 6492
-> vgl. 7193= ähnl. Modell.
Lit.: Pinhas, S. 148 (m. Abb.)

7733
Stehende Revuetänzerin (mit gerüschtem langem Rock) in Pose (auf Ovalsockel)
ca. 1936/37
7733,164,68,2,G, gem. lig. HK, gem. rot M, Firmenmarke 19, Made in Germany, Ausformung zw. 1938–1941, Pick. 6, H. 25,3 cm; Christie's SK 15.11.2001; WKA 38

7735
Lorenzl, Josef
Stehende Madonna mit gekreuzten Händen (auf Rundsockel)
ca. 1934 (für den Entwurf)
7735 H. 30 cm
-> vgl. 6862
Quelle: Fotoarchiv K. Täuber

7736
Lorenzl, Josef
BÜSTE Madonna mit gefalteten Händen
ca. 1933 (für den Entwurf)
7736 H. 11 cm, B. 7 cm; Fotoarchiv K. Täuber
-> vgl. 6745; 7052
-> vgl. 6923= WANDMASKE
Quelle: Fotoarchiv K. Täuber

7737M
Matula, Viktor
FIGUR „Pferd" (Stehendes fressendes Pferd)
ca. 1933 (für den Entwurf)
7737M, H. 23 cm, B. 26 cm; GSNachf. 3,18b
-> vgl. 6797; 7386
-> vgl. 6986= GRUPPE
-> vgl. 7939= GRUPPE (als Höhenvariante).
Lit.: GSNachf. 3,18 (m. Abb.), hier mit Ang. des Originaltitels „Pferd".
Quelle: AGM (Kontoblatt der Wiener Manufaktur Josef Schuster für die Firma Porslins Ericsson-Lund, Stockholm?, Schweden vom 7. Januar 1943), hier mit Ang. des Originaltitels „Pferd".

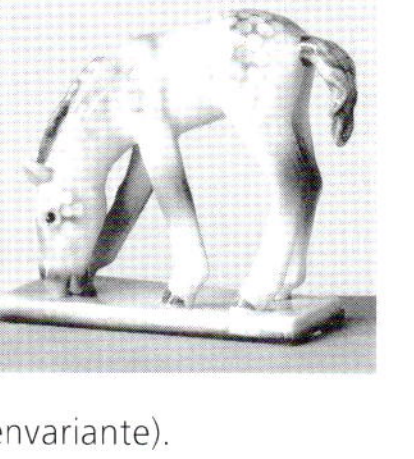

7739M
Matula, Viktor
FIGUR „Reh" (Stehendes Reh)
ca. 1936/37
7739M, H. 39 cm, B. 32 cm; GSNachf. 3,16; Fotoarchiv K. Täuber, hier mit der Ang.: H. 42 cm, B. 33 cm
Bem.: Diese Figur wurde auch noch zw. 1953–1960 bei Carstens (GWM West-Germany) produziert.
Lit.: GSNachf. 3,16 (m. Abb.), hier mit Ang. des Originaltitels „Reh"; KGSF1; KGSF2; Könecke (2002), S. 76, hier mit Ang. des Künstlers Matula, S. 59 (m. Abb.).
Quelle: Fotoarchiv K. Täuber

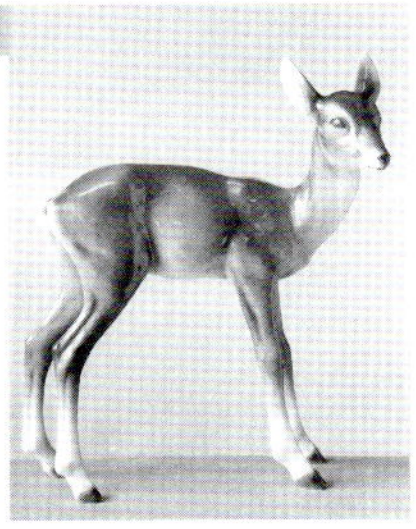

7747
RELIEF Mistelzweig (Wandrelief)
ca. 1937
7747,3 Firmenmarke 19, H. 10 cm, B. 10 cm; PB Wien 19

7750M
Prischl, Adolf
WANDMASKE „Maske" (Frauenkopf mit wehendem Haar)
ca. 1935/36 (für den Entwurf)
7750,5,80 Firmenmarke 19, Made in Austria, unsign., H. 20 cm, B. 19,5 cm; PB Deutschland
7750,10,67 gem. RS, Firmenmarke 19, Made in Austria, unsign., H. 20 cm; Herr 48
7750,17,11 Firmenmarke 19, Made in Austria, unsign., H. 20 cm; city-antik
7750,78,10 Firmenmarke 19, Made in Austria, unsign., H. 20 cm; PB Wien 19
7750M, H. 19,5 cm, B. 19,5 cm; GSNachf. 2,3b; Fotoarchiv K. Täuber, hier mit der Ang.: H. 22 cm, B. 20 cm und der falschen Mod.-Nr. 7756 bezeichnet.
-> vgl. 7420; 8207
Lit.: GSNachf. 2,3 (m. Abb.), hier mit Ang. des Originaltitels „Maske"; Könecke (2002), S. 77, hier mit Ang. des Künstlers A. Prischl (für die Modell-Nr. 8207), S. 88 (m. Abb.).
Quelle: Fotoarchiv K. Täuber

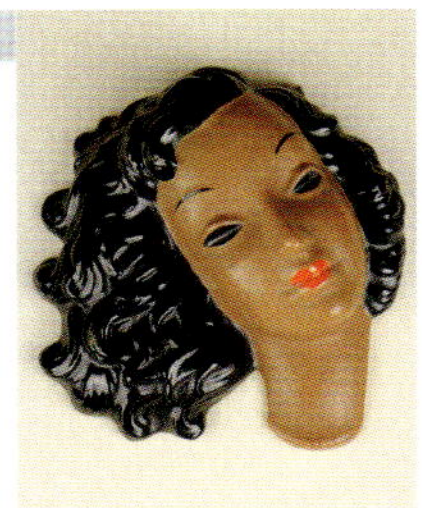

7752
Knörlein, Rudolf (zugeschr.)
BÜSTE Damenakt mit erhobenen Armen
ca. 1937
7752 Firmenmarke 19, Made in Austria, unsign., H. 23 cm; Schuler 58
Lit.: Pinhas, S. 165 (m. Abb.)

7756
Knörlein, Rudolf (zugeschr.)
STEHKOPF Frauenkopf mit Hut und Kragen
ca. 1937
7756,65 gem. S, Firmenmarke 19, Made in Austria, unsign., Ausformung zw. 1945–1953, Pick. 6, H. 13,5 cm, B. 28 cm; eBay # 292170387

7766
BÜSTE Madonna mit gefalteten Händen
ca. 1937
7766,14,9 ger., Firmenmarke 20, Made in Austria, Hand dec., H. 17 cm; eBay # 2015794774

7771M
WANDMASKE „Maske" (Frauenkopf mit Kragen, im Profil nach rechts)
ca. 1936 (für den Entwurf)
7771,3 gem. A, Firmenmarke 19, Made in Germany, Ausformung zw. 1938–1941, H. 20 cm; eBay # 6523079687
7771,S ger., Firmenmarke 19, H. 21,5 cm; Soth. com. # 3J5HF
7771 (Firmenmarke ohne Ang.), H. 21 cm; Doro 4.3.2002
7771M, H. 21 cm, B. 16,5 cm; GSNachf. 2,5b
-> vgl. 7496; 7815
Lit.: GSNachf. 2,5 (m. Abb.), hier mit Ang. des Originaltitels „Maske".
Quelle: Fotoarchiv K. Täuber

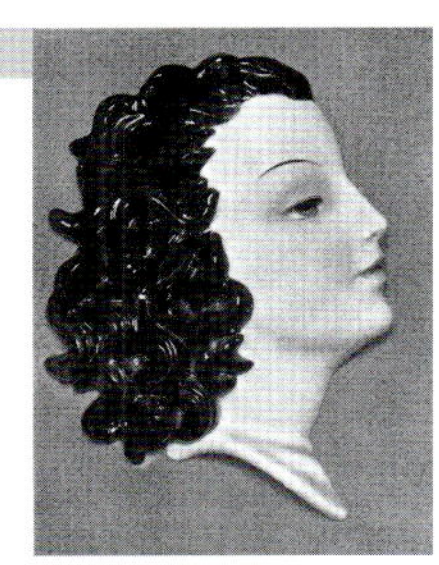

7772
WANDMASKE
ca. 1937
7772 ohne Ang., H. ca. 29 cm, B. ca. 22 cm; Könecke (2002), S. 76
Bem.: Diese Maske wurde auch noch zw. 1953–1960 bei Carstens (GWM West-Germany) produziert.
Lit.: Könecke (2002), S. 76, hier ohne weitere Ang.

7774
Dakon, Stephan (zugeschr.)
Kleines Mädchen, das verzweifelt versucht ihren großen Koffer zu schließen
ca. 1937
7774 (Firmenmarke ohne Ang.), unsign., H. 18 cm; Pinhas, S. 134 (m. Abb.)
Bem.: Der Koffer, der Aufkleber von Wien und Paris trägt, lässt sich deswegen nicht schließen, weil in sich in der Öffnung eine Puppe eingeklemmt hat.

7776F
Dakon, Stephan
FIGUR „Arabella" (Stehende junge Dame, ihr langes Kleid schmetterlingsflügelartig auseinander haltend) (auf Ovalsockel)
ca. 1937
7776,11,N gem. HS, Firmenmarke 19, Made in Austria, unsign., Ausformung zw. 1945–1953, H. 41,5 cm; PB Deutschland
7776,11 Firmenmarke 19, Made in Austria, unsign., Ausformung zw. 1945–1953, H. 41,5 cm; Soth. Olympia 3.6.2004
7776,28,2 Firmenmarke 19, Made in Austria, H. 41,9 cm; eBay # 1204263041; (KH www.tias.com/stores/frozenbeauty)
7776,41,31 Firmenmarke 19, H. 42,5 cm; Doro 24.5.2000
7776,51,6 Firmenmarke 19, Made in Austria, unsign., H. 43 cm; eBay # 3701227246
7776,67,31 Firmenmarke 19, Made in Germany, unsign., Ausformung zw. 1938–1941, H. 42 cm; Soth./eBay # 1860110222
7776 (ohne Ang.), Firmenmarke 19, Made in Germany, unsign., Ausformung zw. 1938–1941, H. 41 cm; RM 21
7776F, H. 42 cm, B. 37 cm; GSNachf. 1,20a
-> vgl. 7961; 8317
Lit.: GSNachf. 1,20 (m. Abb.), hier mit Ang. des Originaltitels „Arabella".
Quelle: Archiv Mimi Dakon (Tantiemenabrechnung 1.7.–31.12.1937), hier wird die Figur unter dem Titel „Tanz" geführt; (Tantiemenabrechnung 1.1.–30.6.1940; Modellliste Dakon, undatiert, ca. 1940; Wiener Manufaktur FG, Modellliste Dakon, 7.11.1946), hier wird die Figur jeweils unter dem Titel „Tänzerin" geführt; (Tantiemenabrechnung 1. Halbjahr 1949; 2. Halbjahr 1949; 1. Halbjahr 1950; 2. Halbjahr 1950; 1. Halbjahr 1951, AGM); 2. Halbjahr 1951; 1. Halbjahr 1952), hier wird die Figur jeweils unter dem Originaltitel „Arabella" geführt.

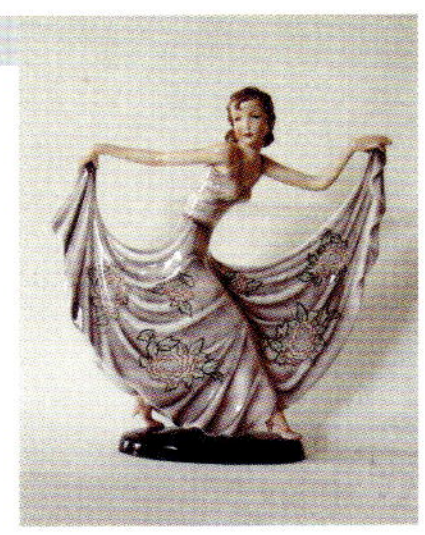

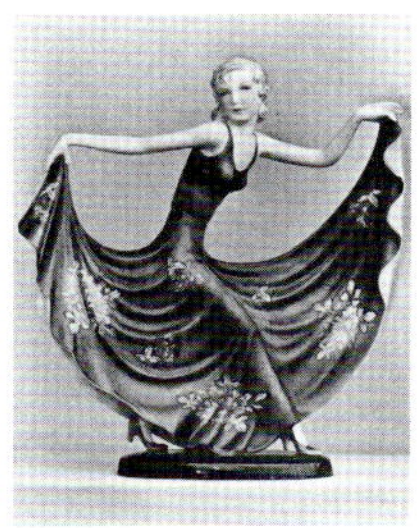

7777F
Dakon, Stephan
Schreitende Tänzerin, ihr Kleid mit beiden Armen auseinander haltend (auf Ovalsockel)
ca. 1937
7777,8,39,9 Firmenmarke 19, Made in Austria, H. 29,8 cm; eBay # 34729923
7777,171,31 gem. rot H, Firmenmarke 19, Made in Germany, Ausformung zw. 1938–1941, H. 30 cm; *eBay # 2249637710
7777,199,7,G gem. F., Firmenmarke 19, Made in Germany, unsign., Ausformung zw. 1938–1941, H. 30 cm; Wien Museum Inv. 190.154
7777F, H. 30 cm, B. 24 cm
Ausst.: SSH 1992
Lit.: SS, S. 22, 23 (m. Abb.)
Quelle: Archiv Mimi Dakon (Originalfoto)

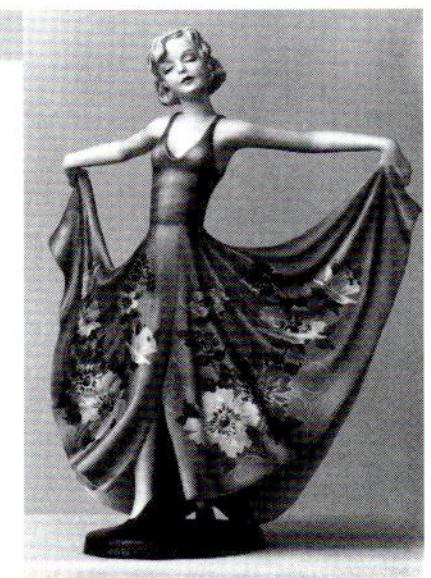

7780
Matula, Viktor
Liegendes Pferd
ca. 1937
7780,7,20 gem. lig. grün AH, Firmenmarke 19, Made in Austria, unsign., H. 19 cm, L. 24 cm; eBay # 7334797814
7780,68,G Firmenmarke 19, Made in Germany, unsign., Ausformung zw. 1938–1941, H. 18 cm, L. 24cm; Wien Museum Inv. 220.768
Quelle: Nachlass Matula, Wien

7781
Matula, Viktor
Stehendes Reh (mit nach hinten gewandtem Kopf)
ca. 1937
7781 ohne Ang., H. 30 cm, B. 22 cm; Könecke (2002), S. 76, 59 (m. Abb.)
Bem.: Diese Figur wurde auch noch zw. 1953–1960 bei Carstens (GWM West-Germany) produziert.
Lit.: KGSF1; KGSF2; Könecke (2002), S. 76, hier mit Ang. des Künstlers Matula, S. 59 (m. Abb.); Pinhas, S. 162 (m. Abb.), hier ohne Künstlerang., H. 18 cm.

7784M
Prischl, Adolf
WANDMASKE „Maske" (Frauenkopf mit Schal)
ca. 1937
7784,3,90 Firmenmarke 19, Made in Austria, unsign., H. 30 cm; eBay # 439118157
7784,3 gem. lig. AH, Firmenmarke 19, Made in Germany, Ausformung zw. 1938–1941, H. 29,2 cm; MAK Wien, Inv. Ke 10602-98
ohne Nr., Firmenmarke 19, Made in Austria, unsign., H. 29 cm; Sammlung f@goldscheider.de
7784,9 gem. AF, Firmenmarke 19, Made in Germany, unsign., Ausformung zw. 1938–1941, H. 29,2 cm; eBay # 168595378
7784,9 gem. M, Firmenmarke 19, Made in Austria, unsign., Ausformung zw. 1945–1953, H. 29 cm; Sammlung a@goldscheider.net
7784,15 Firmenmarke 19, Made in Austria, unsign., Ausformung zw. 1945–1953, H. 29,3 cm; Doro 25.4.2000
7784,33 gem. lig. AH, Firmenmarke 19, Made in Austria, unsign., Ausformung zw. 1945–1953, Originalpick.: „Kieser, Basel", H. 29 cm; Wien Museum Inv. 190.008

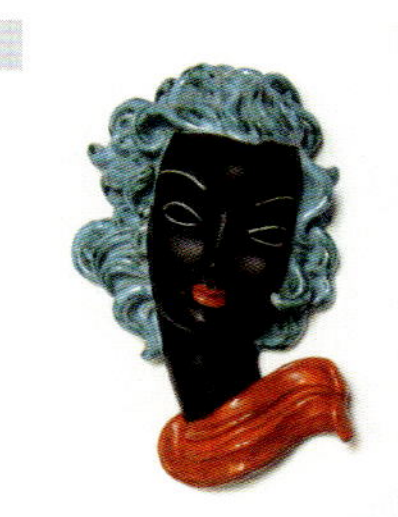

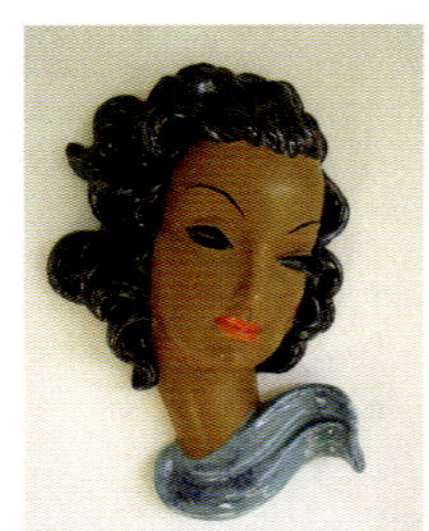

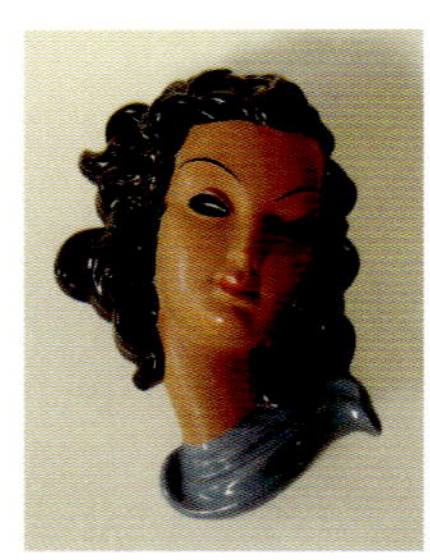

7784 ger., gem. TS, gem. lig. AH, Firmenmarke 19, Made in Austria, unsign., Ausformung zw. 1945–1953, Pick. 6, H. 29,2 cm; eBay # 1290816528
7784,185,(930?),9 gem. AF, Firmenmarke 19, Made in Germany, unsign., Ausformung zw. 1938–1941, H. 29,2 cm; eBay # 168595378
7784,91,VII Firmenstempel WG 1, unsign., Ausformung 50er Jahre, H. 29 cm; PB Deutschland
7784,529,1,6 Firmenstempel WG 1, unsign., Ausformung 50er Jahre, H. 29,5 cm; Stahl 7.5.1994
7784 gem. G., Firmenmarke 19 (blau), Made in Austria (blau), H. 32 cm, B. 21 cm; Sammlung mad
7784M, H. 28,5 cm, B. 20 cm; GSNachf. 2,6a

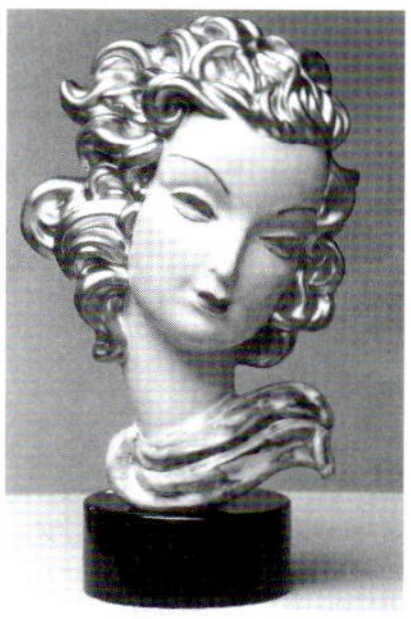

Bem.: Dieses überaus erfolgreiche Modell wurde auch als Stehmaske auf Holzsockel angeboten (-> vgl. Originalfoto, Fotoarchiv K. Täuber) sowie zw. 1953–1960 weiter bei Carstens (GWM West-Germany) produziert. Ebenso auch in Trenton (USA), dort jedoch wurde es unter der Modell-Nr. 598 auf den Markt gebracht.
-> vgl. 8266
Lit.: GSNachf. 2,6 (m. Abb.), hier mit Ang. des Originaltitels „Maske"; KGSF1; Könecke (2002), S. 76, hier mit Ang. des Künstlers A. Prischl, S. 54, 87 jeweils (m. Abb.).
Quelle: AGM (Bestellungen der Firma Wahliss (Wien) vom 9. und 24.11.1945 an die „Arbeitsgemeinschaft der Firma Friedrich Goldscheider", hier jeweils mit Ang. des Titels „Maske mit Schal"; Bestellung der Firma Wahliss (Wien) vom 27.2.1946 in Ausführung „Terrakotta"); (Bestellung der Firma Elisabeth Salle, Wiesbaden vom 8.1.1951); Fotoarchiv K. Täuber.

7785
STEHMASKE / WANDMASKE Frauenkopf mit kurzem Kragenansatz / mit Hand und Kerzenhalter
ca. 1937
STEHMASKE
7785 ger., gem. 7998, Firmenmarke 19, H. 35,5 cm; Soth. L. 23.10.1987
WANDMASKE
7785,3 (Firmenmarke ohne Ang.), H. 39,5 cm; Doro 4.3.2002
7785,55,16 Firmenmarke 19, H. 39 cm; WKA 33; Doro 10.3.2003
7785 ger., gem. RS, Firmenmarke 19, H. 39 cm; WKH
Bem.: Diese Maske wurde abgewandelt in zwei Varianten angeboten, einmal als Stehmaske ohne die Hand und den Kerzenhalter, sowie als Wandmaske mit den genannten Attributen.

7786
Motto, (Joseph C.)
WANDMASKE Madonna (Frauengesicht)
ca. 1937
77(8?)6,4,20 gem. SG, Firmenmarke 19, Made in Austria, H. 22 cm; eBay # 3745904488

7787
Prischl, Adolf

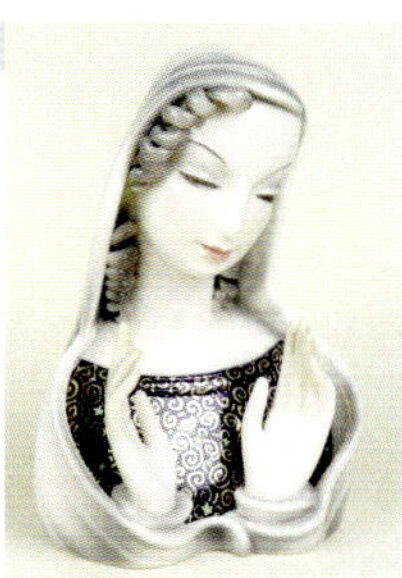

BÜSTE Madonna mit erhobenen Händen
ca. 1937
7787,39,0 ger. III, Firmenmarke 19, Made in Austria, unsign., H. 27,9 cm; eBay # 72591025
7787,43,4 Firmenmarke 19, Made in Austria, H. 29,2 cm; eBay # 1238682462
7787,63,31 gem. lig. AS, Firmenmarke 19, Made in Austria, H. 29,2 cm; eBay # 1304107351
7787,93,19 Firmenmarke 19, H. 27 cm; PB Wien 19

7788
STEHKOPF Mädchenkopf mit Zöpfen (auf mit Kugeln verziertem Ovalsockel)
ca. 1937
7788,0 ger. S, Firmenmarke 19, Made in Austria, H. 25,5 cm; Wien Museum Inv. 220.799

7789
Wieselthier, Vally

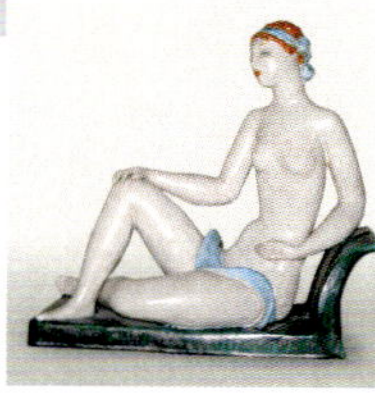

Sitzender Frauenakt
ca. 1937
7789,0,S gem. lig. AH, Firmenmarke 19, Made in Austria, unsign., H. 37 cm, B. 37 cm; PB
7789,2 ger., Firmenmarke 19, Made in Austria, unsign., Ausformung zw. 1945–1953, H. 37 cm, B. 38,5 cm; Doro 1.12.2000
Lit.: N.N.: Neue Keramik, in: ÖK, 8.1937, Nr. 7, S. 22 (m. Abb.) „Vally Wieselthier, die kapriziös begabte Wiener Keramikerin, die seit mehreren Jahren erfolgreich in Amerika wirkt und ausstellt, wurde nun wieder für Österreich beschäftigt. Wir begrüßen diesen Entschluss der Manufaktur. Die Verbindung der Terrakotta mit farbigen Partien gibt den Arbeiten besonderen Reiz. Die weiche Grazie, die Wieselthier ihrem Frauenakt gibt, wirkt mit leicht exotischer Anmut"; Neuwirth (1974), S. 76; Hörmann, S. 302 (m. Abb.), hier ist diese Figur unter der Nr. WV-K 788 aufgeführt. Neben der Ausführung in weißer Glasur mit orangefarben glasierten Haaren gab es auch eine „Variante in schwarz brennendem Manganton und Akzenten in Goldfarbe". Unter der Nr. WV-K 754 findet sich noch eine weitere Version mit einem stehenden „Amor". Datiert 1930, wird dieser nicht von GS produzierte Entwurf „Venus und Amor" betitelt (-> vgl. Hörmann, S. 73 Abb. 41, sowie S. 300; DKuD, 1930/31, Bd. 67, S. 397).

7790?
WANDMASKE Frauenkopf mit Spitzenschal und Kragen
ca. 1937
7790(oder 7796?),8,32 Firmenmarke 20, Made in Austria, Hand dec., H. 33 cm; city-antik

7794
Prischl, Adolf
WANDMASKE Madonna mit Hand
ca. 1937
7794,20 Firmenmarke 19, Made in Austria, unsign., Ausformung zw. 1945–1953, H. 34 cm; Doro FVK
7794,101,20 (Firmenmarke ohne Ang.), unsign., H. 34 cm; Christie's SK 2.4.1998
-> vgl. 7579= ähnl. Modell

7795
Matula, Viktor
GRUPPE Rehe (Ricke mit Kitz) (auf Landschaftssockel)
ca. 1937
7795,32,9 gem. lig. grün TK, Firmenmarke 19, Made in Austria, unsign., H. 14 cm, L. 22 cm; eBay # 7400929616
Quelle: Nachlass Matula, Wien

7796 s. 7790

7797F
Prischl, Adolf

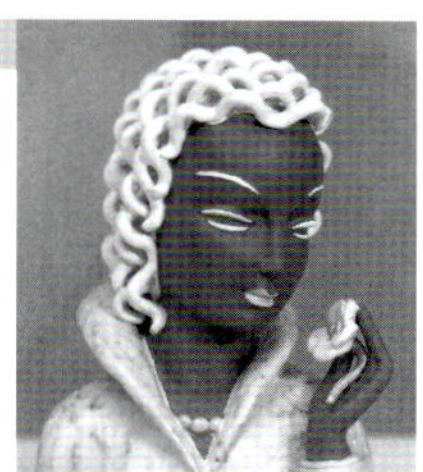

BÜSTE Frau mit Lockenkopf einen Vogel haltend
ca. 1937
7797,9 Firmenmarke 19, Ausformung zw. 1945–1953, H. 23 cm; Christie's East 15.2.1983
7797,37 ger. KG, gem. lig. AH, Firmenmarke 19, unsign., Ausformung zw. 1945–1953, H. 21,6 cm; eBay # 20622167362
7797 H. 22 cm, B. 16 cm
Quelle: AGM (Rechnung an Maurice Mizrahi, Alexandria, Ägypten vom 22. April 1953, geliefert in Ausführung „matt"); Fotoarchiv K. Täuber.

7799F / M
Prischl, Adolf

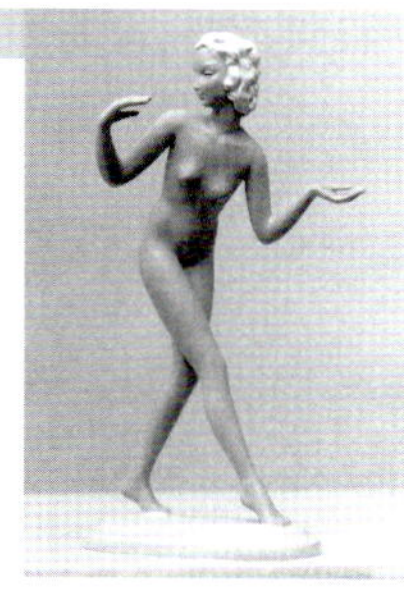

„Akt" (Schreitender Damenakt) (auf Ovalsockel)
ca. 1937
7799 H. 27 cm, B. 13 cm
Quelle: AGM (Rechnung an Maurice Mizrahi, Alexandria, Ägypten vom 22. April 1953), hier mit Ang. des Originaltitels „Akt", geliefert in Ausführung „matt"; Fotoarchiv K. Täuber.

7800M
Matula, Viktor

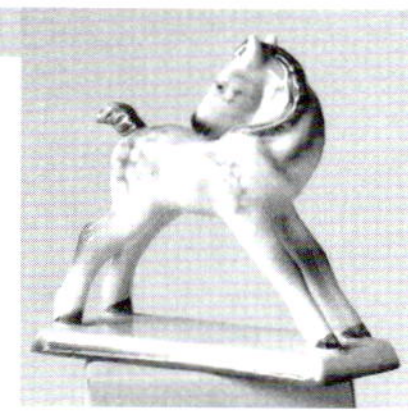

FIGUR/BUCHSTÜTZE „Pferd" (Stehendes Pferd/Fohlen mit nach hinten gewandtem Kopf) (auf flachem Rechtecksockel)
ca. 1937
FIGUR
7800,10,94 Firmenmarke 19, Made in Austria, Hand dec., unsign., H. 10,4 cm; Wien Museum Inv. 220.758
7800,41 ger., gem. Z, Firmenmarke 19, Made in Austria, unsign., Ausformung zw. 1945–1953, H. 11,4 cm; eBay # 1238675020
7800,65,1 ger. XVB, Firmenmarke 19, Made in Austria, unsign., H. 11,4 cm; eBay # 1236012716
7800M, H. 11 cm, B. 11,5 cm; GSNachf. 3,21e; Fotoarchiv K. Täuber, hier mit der Ang.: H. 11 cm, B. 11 cm
BUCHSTÜTZE
s. Quellen
Lit.: GSNachf. 3,21 (m. Abb.), hier mit Ang. des Originaltitels „Pferd".
Quelle: Nachlass Matula, Wien; AGM (Bestellungen der Firma Wahliss (Wien) vom 9. und 24.11.1945 an die „Arbeitsgemeinschaft der Firma Friedrich Goldscheider", bei diesem Aufträgen wurden sowohl Figuren als auch die Ausführung als Buchstütze bestellt; Bestellung der Firma Wahliss (Wien) vom 7.12.1945 als Figur je 28 Stück, als Buchstützen je 32 Stück jeweils in der Ausführung „Terrakotta"; Bestellung der Firma Wahliss (Wien) vom 27.2.1946 als Figur je 51 Stück, als Buchstützen je 64 Paar jeweils in der Ausführung „Terrakotta"); Fotoarchiv K. Täuber.

7804M
Matula, Viktor

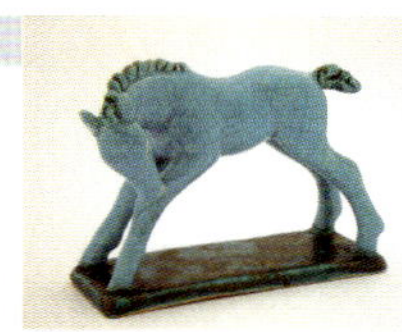

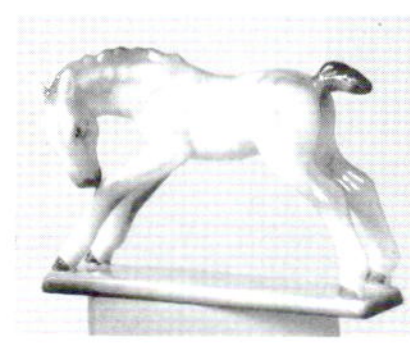

FIGUR „Pferd" (auf flachem Rechtecksockel)
ca. 1934 (für den Entwurf)
7804,27,99 Firmenmarke 19, Made in Austria, unsign., H. 8,5 cm, L. 12 cm; Sammlung a@goldscheider.net
7804M, H. 9 cm, B. 12 cm; GSNachf. 3,21d
-> vgl. 6986= GRUPPE
Lit.: GSNachf. 3,21 (m. Abb.), hier mit Ang. des Originaltitels „Pferd".
Quelle: AGM (Kontoblatt der Wiener Manufaktur Josef Schuster für die Firma Porslins Ericsson-Lund, Stockholm?, Schweden vom 7. Januar 1943), hier mit Ang. des Originaltitels „Pferd"; Fotoarchiv K. Täuber.

7805M
Matula, Viktor

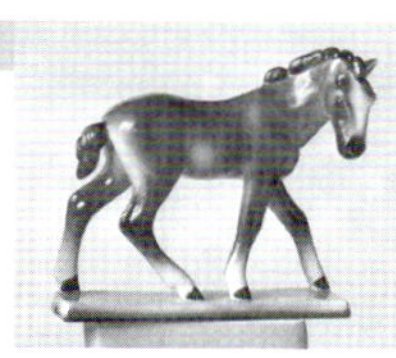

FIGUR „Pferd" (Schreitendes Pferd/Fohlen) (auf flachem Rechtecksockel)
ca. 1937
7805,17,69 Firmenmarke 19, Made in Austria, unsign., Pick. 5/a, H. 11,4 cm, B. 12,7 cm; eBay # 1220557560
7805,68 gepr. G, Firmenmarke 19, Made in Germany, unsign., Ausformung zw. 1938–1941, H. 12 cm, B. 13 cm; eBay # 1423561682; # 6562294506
7805M, H. 11 cm, B. 12 cm; GSNachf. 3,21b
Lit.: GSNachf. 3,21 (m. Abb.), hier mit Ang. des Originaltitels „Pferd".
Quelle: AGM (Bestellung von Ferd. Lundquist (Göteborg, Schweden) vom 23. August 1950); Fotoarchiv K. Täuber.

7810M
Matula, Viktor
GRUPPE „Pferde"
ca. 1937
7810M H. fehlt
Quellen: AGM (Kontoblatt der Wiener Manufaktur Josef Schuster für die Firma Porslins Ericsson-Lund, Stockholm?, Schweden vom 7. Januar 1943), hier mit Ang. des Originaltitels GRUPPE „Pferde".

7813M
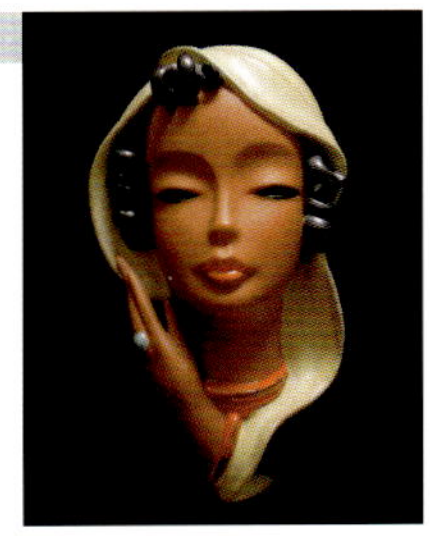
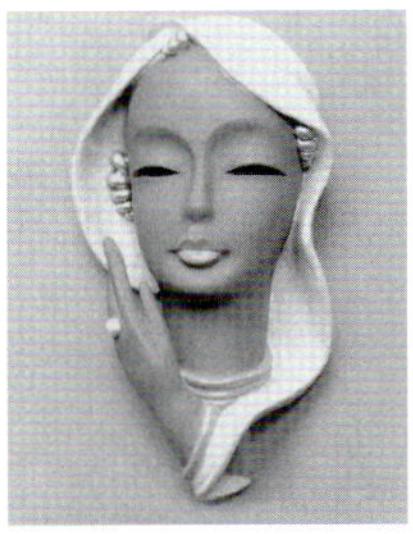
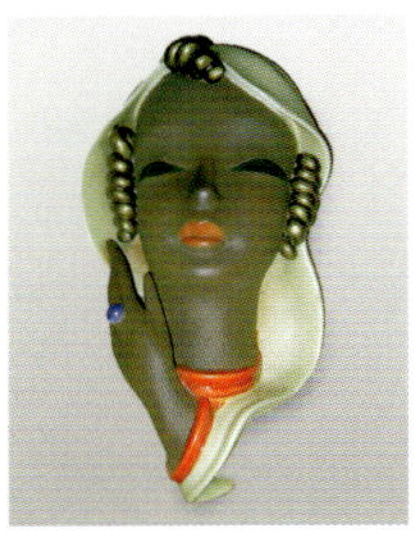
Goebel, Kurt
WANDMASKE/STEHMASKE „Maske Araberin" (Frauenkopf mit Hand, Ring und Kopftuch)
ca. 1937
WANDMASKE
7813,1,9 Firmenmarke 19, Made in Austria, unsign., H. 31,7 cm; eBay # 342883860
7813,3,15 Firmenmarke 20 (blau), Made in Austria, Hand dec., unsign., Pick. 6, H. 31,6 cm; PB Deutschland
7813,4,29 Firmenmarke 19, Made in Austria, unsign., H. 32 cm; Sammlung Rudolf Hansl, St. Florian
7813,7 (Firmenmarke ohne Ang.), unsign., Ausformung zw. 1945–1953, H. 32 cm; Christie's SK 16.9.1999
7813,9 (Firmenmarke ohne Ang.), unsign., Ausformung zw. 1945–1953, H. 30,2 cm; Christie's SK 4./5.12.1996
7813,11,82 Firmenmarke 19, Made in Austria, unsign., H. 31,3 cm; Soth. L. 21.10.1988
7813,17,56 gem. G, Firmenmarke 19, Made in Austria, unsign., H. 32 cm; eBay # 439120007
7813,20 gem. M., Firmenmarke 19, Made in Germany, unsign., Ausformung zw. 1938–1941, H. 31,6 cm; PB Wien
7813,28,10 Firmenmarke 19, unsign., Pick. ohne Ang., H. 31,5 cm; Doro 19.9.2000
7813,33 Firmenmarke 19, Made in Austria, unsign., Ausformung zw. 1945–1953, H. 30,5 cm; Soth. L. 23.6.1989; eBay # 410533393
7813,50 Firmenmarke 19, Made in Austria, unsign., Ausformung zw. 1945–1953, H. 31 cm; KH
7813,3 ger., Firmenstempel WG 1, Ausformung 50er Jahre, H. 31 cm; Sammlung John Nash
7813M, H. 31 cm, B. 17 cm; GSNachf. 2,7a
STEHMASKE
7813,5 Firmenmarke 19, Made in Austria, Ausformung zw. 1945–1953, H. 36 cm; Soth. L. 2.11.1999
Bem.: Diese Maske wurde auch noch zw. 1953–1960 bei Carstens (GWM West-Germany) produziert.
-> vgl. 8247
Lit.: GSNachf. 2,7 (m. Abb.), hier mit Ang. des Originaltitels „Maske Araberin"; KGSF1; KGSF2; Könecke (2002), S. 76, hier mit Ang. des Künstlers Goebel, S. 89 (m. Abb.).
Quelle: AGM (Bestellung der Firma Wahliss (Wien) vom 17.10.1945 an die „Arbeitsgemeinschaft der Firma Friedrich Goldscheider", hier unter dem Titel „Maske mit Hand", in Ausführung „Fayence" bestellt; Rechnung an Maurice Mizrahi, Alexandria, Ägypten vom 22. April 1953, geliefert in Ausführung „matt"); Fotoarchiv K. Täuber.

7814
Goebel, Kurt
WANDMASKE Frauenkopf mit gelocktem Haar und seitlichem Kragenansatz
ca. 1937
7814,8 gem. L., Firmenmarke 19, Made in Austria, unsign., Ausformung zw. 1945–1953, Pick. 6, H. 28,5 cm; eBay # 332263707

7815M
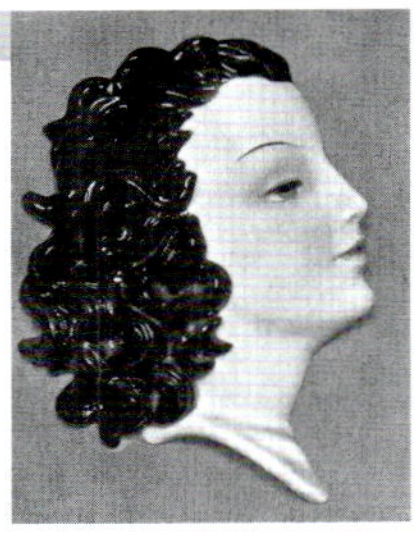
WANDMASKE „Maske" (Frauenkopf mit Kragen, im Profil nach rechts)
ca. 1936 (für den Entwurf)
7815M, H. 16,5 cm, B. 12 cm; GSNachf. 2,5c
-> vgl. 7496; 7771
Lit.: GSNachf. 2,5 (m. Abb.), hier mit Ang. des Originaltitels „Maske".
Quelle: Fotoarchiv K. Täuber

7816
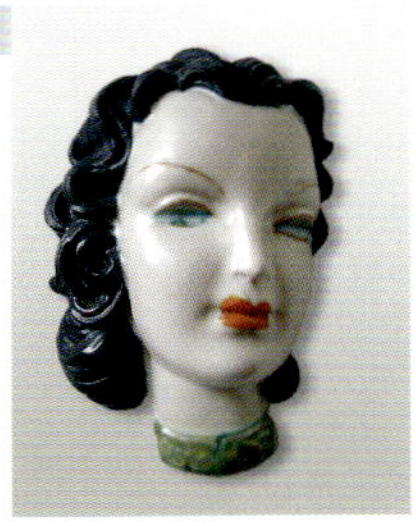
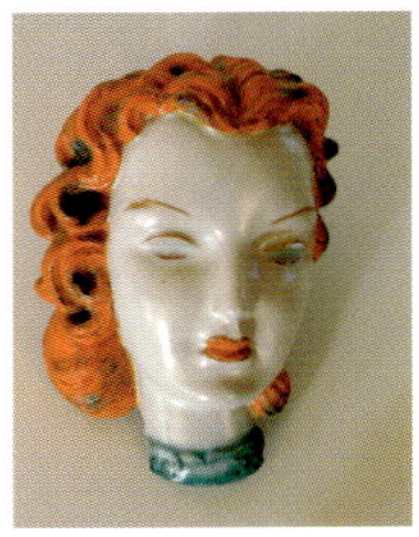
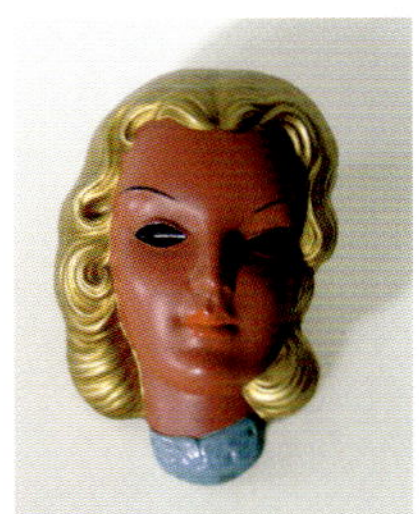
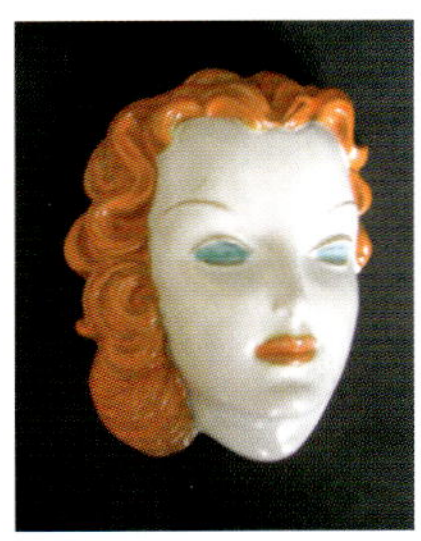
Prischl, Adolf
WANDMASKE „Maske" (Frauenkopf mit kurzem Kragenansatz)
ca. 1937
7816,3 gem. R., Firmenmarke 19, Made in Austria, sign. Prischl, H. 15 cm; PB
7816,33 ger. lig. V11, gem. 8, Firmenmarke 19, Made in Austria, unsign., Ausformung zw. 1945–1953, H. 15,8 cm; eBay # 1323556915
7816,S,III ger., gem. lig. AH, Firmenmarke 19, Made in Austria, unsign., Ausformung zw. 1945–1953, H. 16,3 cm; Sammlung.schmutz@chello.at 96
7816,III ger., Firmenmarke 19, Made in Austria, unsign., H. 16 cm; Sammlung Stressless, Portugal
7816 Firmenstempel WG 1, unsign., Ausformung 50er Jahre, H. 17,8 cm; Sammlung John Nash
Bem.: Diese Maske wurde auch noch zw. 1953–1960 bei Carstens (GWM West-Germany) produziert.
-> vgl. 7931; 8248
Lit.: KGSF1; KGSF2; Könecke (2002), S. 76, hier mit Ang. des Künstlers A. Prischl, S. 51 (m. Abb.).

7817M
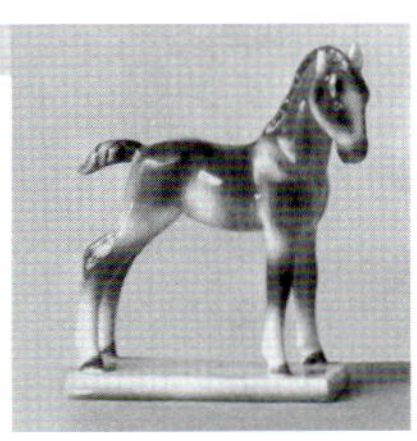
Matula, Viktor
FIGUR „Pferd" (Stehendes Pferd/Fohlen) (auf flachem Rechtecksockel)
ca. 1937
7817,40,9 gem. 8032, G., Firmenmarke 19, Made in Austria, unsign., H. 12,7 cm; eBay # 3645172835
7817,149,55 gem./(ger.?) XLT, Firmenmarke 19, Made in Austria, unsign., H. 12,7 cm; eBay # 1268548546
7817M, H. 12,5 cm, B. 11 cm; GSNachf. 3,21a; Fotoarchiv K. Täuber, hier mit der Ang.: H. 12 cm, B. 10 cm
Lit.: GSNachf. 3,21 (m. Abb.), hier mit Ang. des Originaltitels „Pferd".
Quellen: Nachlass Matula, Wien; AGM (Kontoblatt der Wiener Manufaktur Josef Schuster für die Firma Porslins Ericsson-Lund, Stockholm?, Schweden vom 7. Januar 1943), hier mit Ang. des Originaltitels „Pferd"; Fotoarchiv K. Täuber.

7818M
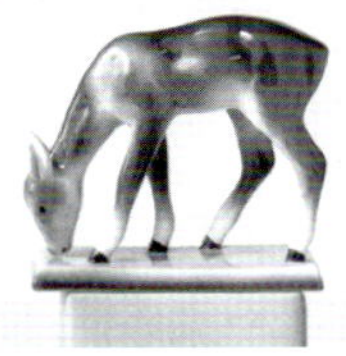
Matula, Viktor
FIGUR „Reh" (Äsendes Reh) (auf flachem Rechtecksockel)
ca. 1937
7818,7,75 Firmenmarke 19, Made in Austria, Hand dec., unsign., H. 10 cm; Doro FVK
7818M, H. 10 cm, B. 11,5 cm; GSNachf. 3,15b; Fotoarchiv K. Täuber, hier mit der Ang.: H. 9 cm, B. 10 cm
-> vgl. 7863
Lit.: GSNachf. 3,15 (m. Abb.), hier mit Ang. des Originaltitels „Reh".
Quellen: Nachlass Matula, Wien; AGM (Kontoblatt der Wiener Manufaktur Josef Schuster für die Firma Porslins Ericsson-Lund, Stockholm?, Schweden vom 7. Januar 1943), hier mit Ang. des Originaltitels „Reh"; Fotoarchiv K. Täuber.

7819M
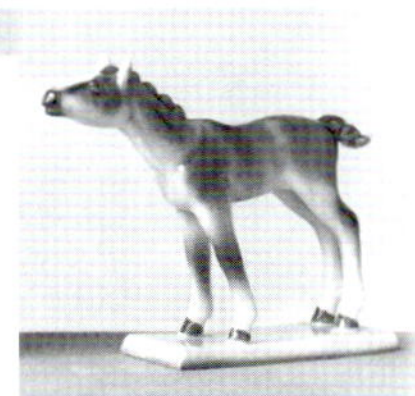
Matula, Viktor
FIGUR „Pferd" (Stehendes wieherndes Pferd/Fohlen) (auf flachem Rechtecksockel)
ca. 1937
7819M, H. 11,5 cm, B. 12 cm; GSNachf. 3,21c; Fotoarchiv K. Täuber, hier mit der Ang.: H. 11 cm, B. 14 cm
Lit.: GSNachf. 3,21 (m. Abb.), hier mit Ang. des Originaltitels „Pferd".
Quellen: Nachlass Matula, Wien; AGM (Kontoblatt der Wiener Manufaktur Josef Schuster für die Firma Porslins Ericsson-Lund, Stockholm?, Schweden vom 7. Januar 1943), hier mit Ang. des Originaltitels „Pferd"; Fotoarchiv K. Täuber.

7820M
Matula, Viktor
FIGUR „Reh" (Stehendes Reh mit nach vorne gerecktem Kopf) (auf flachem Rechtecksockel)
ca. 1937
7820M, H. 10,5 cm, B. 15 cm; GSNachf. 3,15f
Lit.: GSNachf. 3,15 (m. Abb.), hier mit Ang. des Originaltitels „Reh".
Quellen: Nachlass Matula, Wien; AGM (Bestellungen der Firma Wahliss (Wien) vom 9. und 24.11.1945 an die „Arbeitsgemeinschaft der Firma Friedrich Goldscheider"; Bestellung der Firma Wahliss (Wien) vom 27.2.1946 in Ausführung „Terrakotta"; Bestellung von Ferd. Lundquist (Göteborg, Schweden) vom 23. August 1950, in Ausführung „Fayence").

7822 s. 7874

7823
Schreitende Dame mit Hut und (Schirm?) (auf Rundsockel)
ca. 1937
7823,11 Firmenmarke 19, H. 30,8 cm; Doro 4.6.1996

7825
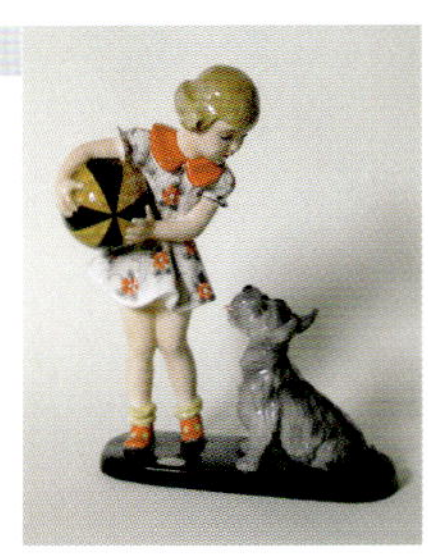
Bouret, Germaine
GRUPPE Stehendes Mädchen mit Ball und Terrier (auf abgerundetem Rechteckssockel)
ca. 1937
7825,67,11 ger. III, gem. W.SG, Firmenmarke 19, Made in Austria, sign. G. Bouret, H. 26,3 cm; PB Deutschland
7825,419,10 Firmenmarke 19, sign. G. Bouret, H. 27,3 cm; Phillips L. 15.2.1990; Actatos, 1991, S. 262
Bem.: Die Entwürfe von Germaine Bouret wurden von Stephan Dakon ausmodelliert.

7826F
Dakon, Stephan
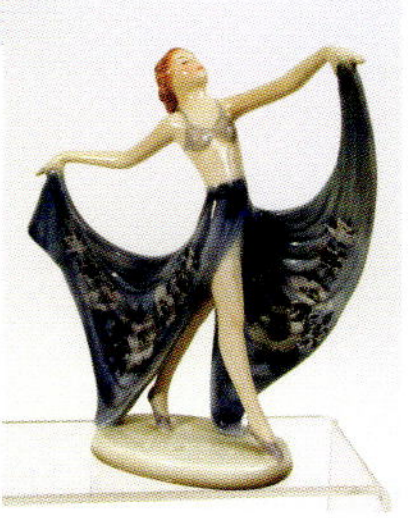
FIGUR „Champagner" (Nach rechts schreitende junge Dame mit Bikinioberteil, ihren langen Rock schmetterlingsflügelartig auseinander haltend) (auf Ovalsockel)
ca. 1937
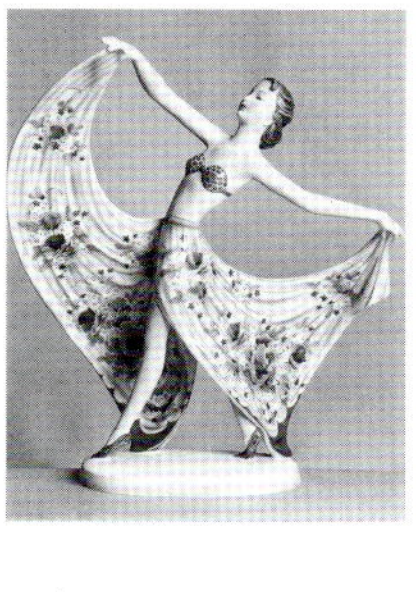
7826,3,69 Firmenmarke 19, Made in Austria, Hand dec., unsign., H. 22,9 cm; eBay # 423617694
7826,8 Firmenmarke 19, Made in Austria, unsign., Ausformung zw. 1945–1953, H. 21,7 cm; Wien Museum Inv. 220.708
7826,11 ger. II.G, Firmenmarke 19, Made in Austria, unsign., Ausformung zw. 1945–1953, H. 21 cm; city-antik
7826,25,14 Firmenmarke 19, Made in Austria, Hand dec., unsign.?, H. 21,5 cm; city-antik
7826,66,61 Firmenmarke 19, Made in Austria, Hand dec., unsign., H. 21 cm; city-antik
7826,113,31 gem. H, Firmenmarke 19, Made in Austria, H. 22 cm; WKH
7826,144,65 Firmenmarke 19, Made in Austria, Pick. 4, H. 22 cm; eBay # 170281150
7826,691,4 gem. HS., Firmenmarke 19, Made in Germany, Pick. 6, Ausformung zw. 1938–1941, H. 22 cm; eBay # 6507216774
7826,K,2 ger., gepr. 8, Firmenmarke 19, Made in Austria, unsign., Ausformung zw. 1945–1953, H. 22 cm; city-antik
7826F, H. 22 cm, B. 20,5 cm; GSNachf. 1,21a; Fotoarchiv K. Täuber, hier mit der Ang.: H. 22 cm, B. 18 cm
Bem.: s. 7857
-> vgl. 8319
Pendant -> vgl. 7857= seitenverkehrtes Modell.
Lit.: GSNachf. 1,21 (m. Abb.), hier mit Ang. des Originaltitels „Champagner".
Quelle: AGM (Bestellung von Ferd. Lundquist (Göteborg, Schweden) vom 23. August 1950; Rechnung an Maurice Mizrahi, Alexandria, Ägypten vom 22. April 1953), hier jeweils unter dem Titel „Sekt" geführt; Fotoarchiv K. Täuber.

7827F
Lorenzl, Josef (zugeschr.)
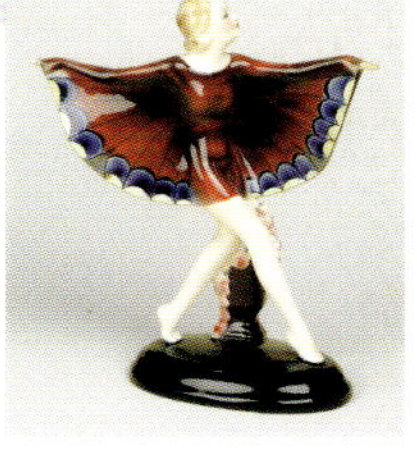
„Gefangener Vogel" (Nach rechts schreitende Tänzerin im Schmetterlingskostüm) (auf Ovalsockel)
ca. 1937
7827,30,16,A gem. lig. CH?, Firmenmarke 19, Made in Austria, unsign., H. 21,5 cm; city-antik
7827,166,7 Firmenmarke 19, Made in Germany, unsign., Pick. 6, Ausformung zw. 1938–1941, H. 23 cm; eBay # 1476913332
7827,179,31 Firmenmarke 19, unsign., H. 23 cm; PB Wien 19
7827,283,11 (Firmenmarke ohne Ang.), unsign., H. 22,5 cm; Christie's SK 11.5.2001
7827,503,40 Firmenmarke 19, Made in Germany, unsign., Ausformung zw. 1938–1941, H. 23,5 cm; eBay # 6626064571
7827 ohne Ang., Firmenmarke 19, unsign., H. 23 cm; Ahlden 129
Bem.: Diese Figur geht auf den „Gefangenen Vogel" (-> vgl. Modell-Nr. 5230), der die Tänzerin Niddy Impekoven darstellt, zurück. Abgewandelt wurde diese kleine Ausführung sowohl in der Haartracht, als auch in der Richtung, in welche die Tänzerin schreitet. Da sich der ursprüngliche Entwurf und die Ausführung hier ziemlich ähnlich sind, kann wohl auch dieses Modell Josef Lorenzl zugeordnet werden.
Quelle: AGM (Rechnung an Maurice Mizrahi, Alexandria, Ägypten vom 22. April 1953), hier mit Ang. des Originaltitels „Gefangener Vogel".

7829
Matula, Viktor
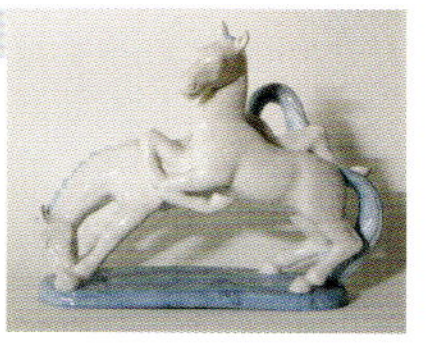
Zwei springende Pferde (auf längsovalem Landschaftssockel)
ca. 1935/36 (für den Entwurf)
7829,16,9 Firmenmarke 19, Made in Austria, unsign., H. 35,3 cm, L. 42,5 cm; Soth. Chester 23.–25.5.1989; Doro 21.5.1997
7829,20,49,C ger., gem. grün L., Firmenmarke 19, Made in Germany, unsign., Ausformung zw. 1938–1941, H. 34 cm, L. 44 cm; PB Deutschland
Lit.: Pinhas, S. 178 (m. Abb.), hier H. 36 cm.

7831F
Prischl, Adolf
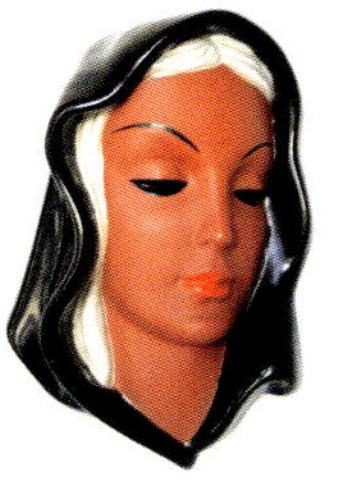
(WAND)MASKE „Madonna" (Madonna mit Schleier/Kopftuch)
ca. 1937
7831 Firmenmarke 19, Made in Germany, Ausformung zw. 1938–1941, H. 20 cm; WKA 33
7831,0 ger., gem. G, Firmenmarke 19, unsign., Ausformung zw. 1945–1953, H. 19 cm; Wien Museum Inv. 186.643
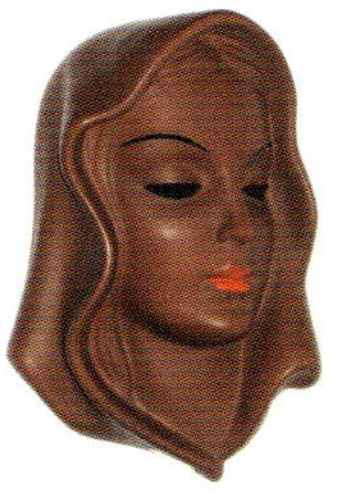
7831,7 Firmenmarke 19, unsign., Ausformung zw. 1945–1953, H. 20,5 cm; WKH
7831,20 Firmenstempel WG 1, unsign., Ausformung 50er Jahre, H. 21 cm; Stahl 5.2.1994; 7.5.1994
7831 Firmenstempel WG 1, unsign., Ausformung 50er Jahre, H. 20 cm; Sammlung a@goldscheider.net
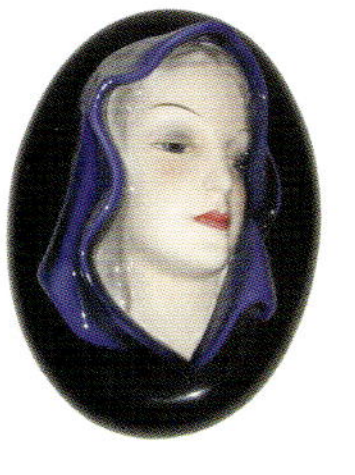
7831 ohne Firmenmarke, Pick. WG 1, H. 20 cm; Sammlung a@goldscheider.net
7831 (Firmenmarke u. Modell-Nr. nicht sichtbar), H. 20 cm; Gesamth. 26,8 cm; Sammlung a@goldscheider.net; Bem.: Diese Maske ist auf schwarzen ovalen Holzbasis montiert, daher sind weder Modell-Nr. noch Marken sichtbar. Ob es bei diesem Stück um eine Eigenanfertigung handelt, oder ob Goldscheider diese Ausführung auf Wunsch lieferte, lässt sich nicht sagen.
7831 Firmenstempel WG 1, unsign., Ausformung 50er Jahre, H. 21 cm; Sammlung John Nash
7831F, H. 20 cm, B. 13 cm; GSNachf. 6,9
Bem.: Diese Wandmaske war durchgängig ab 1937 bis in die späten 50er Jahre beliebt und wurde sehr häufig hergestellt. Die meisten Nachweise (die an dieser Stelle nicht einzeln aufgelistet werden können) betreffen die Produktion bei Carstens (GWM West-Germany) in Fredelsloh.
Lit.: GSNachf. 6,9 (m. Abb.), hier mit Ang. des Originaltitels MASKE „Madonna"; KGSF1; KGSF2; Könecke (2000), S. 55 (m. Abb.), 76.
Quelle: AGM (Kontoblatt der Wiener Manufaktur Josef Schuster für die Firma Porslins Ericsson-Lund, Stockholm?, Schweden vom 7. Januar 1943), hier mit Ang. des Originaltitels MASKE „Madonna"; Archiv Matschiner (Originalfoto).

7833
Stehender Mädchenakt mit Tier (Reh, Hund oder ähnliches)
ca. 1937
7833,65 gem. G, Firmenmarke 19, Made in Austria, H. 28 cm; Müller 83
Bem.: Da bei dem bislang einzig nachweisbaren Exemplar das dem Mädchenakt beigestellte Tier fehlt (abgebrochen und verschollen), lässt sich nur vermuten, dass es sich entweder um ein Reh oder einen Hund handeln könnte.

7834
Matula, Viktor
GRUPPE Pferde
ca. 1937
7834 H. fehlt
Quelle: Nachlass Matula, Wien

7835
Bouret, Germaine
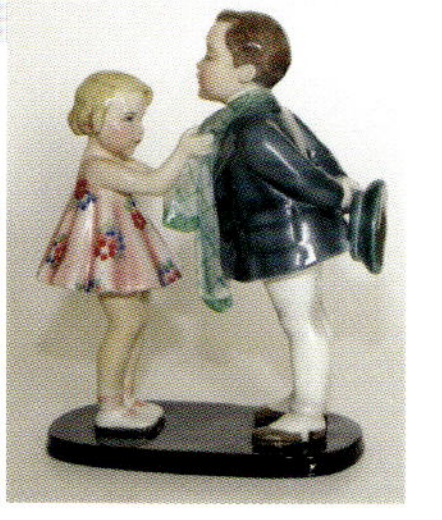
GRUPPE Kleines Mädchen, einem Jungen den Schal bindend (auf abgerundetem Rechtecksockel)
ca. 1937
7835,56,9 gem. E, LS.I, Firmenmarke 19, Made in Austria, sign. G. Bouret, H. 25,4 cm; eBay # 3723277898
7835,98,16 ger. 6, Firmenmarke 19, Made in Austria, sign. G. Bouret, H. 24 cm; Sammlung Bernard Amigon, Frankreich
Bem.: Die Entwürfe von Germaine Bouret wurden von Stephan Dakon ausmodelliert.

7836
Dakon, Stephan
Stehendes kleines Mädchen mit drei Tauben (auf Ovalsockel)
ca. 1937
7836,31,6,8,A Firmenmarke 19, Made in Austria, H. 32 cm; eBay # 907549800
Lit.: Pinhas, S. 135 (m. Abb.), hier mit der Künstlerang. Claire Weiss. Sollte dieses Ex. aus einer Auktion bei Christie' s wirklich mit Weiss sign. sein, so wäre das ein Versehen des Modelleurs. Das Mädchen mit den Tauben ist eindeutig Stephan Dakon zuzuweisen, was nicht nur ein hier nachgewiesenes sign. Ex. (aus eBay) belegt, sondern der Entwurf an sich zeigt.

7838 -> ÜBERSCHNEIDUNG
Bosse, Walter (zugeschr.)
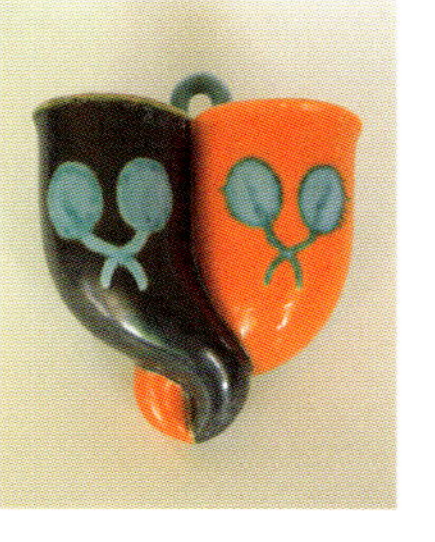
WANDVASE
Form: zwei becherförmige, miteinander verbundene Gefäße
Aufhänger: eine flache Öse
Dekor: vier stilisierte Blütenkapseln oder Blätter, rot-schwarz-türkis
7838,6 Firmenmarke 19, Made in Austria, unsign., Ausformung zw. 1945–1953, H. 17,7 cm, B. 14,8 cm; Sammlung.schmutz@chello.at 90
Bem.: Für die Zuschreibung findet sich eine vergleichbare Wandvase mit Blume bei WKA 33 (Lot 1019).

7838 -> ÜBERSCHNEIDUNG
Dakon, Stephan
Stehende junge Gitarrespielerin mit Schottenrock und Kappe (auf Ovalsockel)
ca. 1935
7838 (Firmenmarke ohne Ang.), H. 32 cm; Pinhas, S. 151 (m. Abb.)

7839
Goebel, Kurt
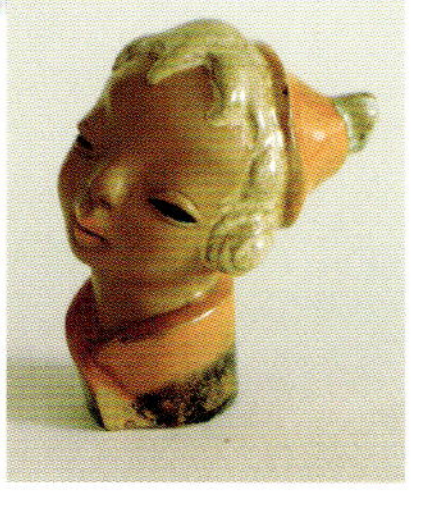
STEHKOPF Frauenkopf mit Mütze und Schal
ca. 1937
7839,1 ohne Firmenmarke, unsign., Ausformung zw. 1945–1953, H. 9 cm; Sammlung John Nash

7843
Goebel, Kurt
GRUPPE Wandersmann (Bursche) mit springendem Hund (auf Landschaftssockel)
ca. 1937
7843,35 ger. M1, gem. L., Firmenmarke 19, Made in Austria, unsign., Ausformung zw. 1945–1953, H. 28 cm; eBay # 12205329942

7844M
Goebel, Kurt
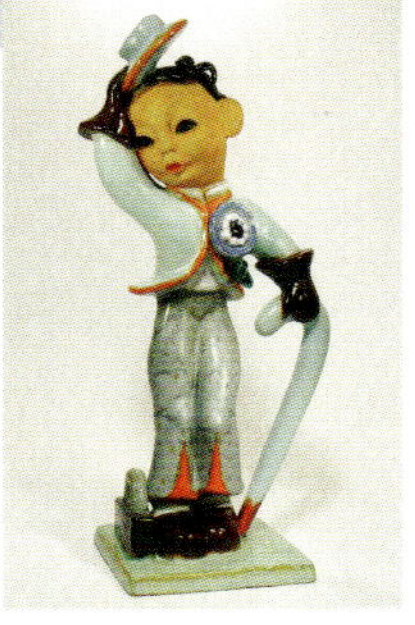
FIGUR „Rendez-vous" (Stehender Bursche mit Stock und Tasche, mit Hut grüßend) (auf flachem Rechtecksockel)
ca. 1937
7844,35 gem. B., Firmenmarke 19, Made in Austria, unsign., H. 26,8 cm; Sammlung a@goldscheider.net
7844M, H. 26 cm, B. 13 cm; GSNachf. 3,11b
Pendant –> vgl. 7845
–> vgl. 7962= GRUPPE
Lit.: GSNachf. 3,11 (m. Abb.), hier mit Ang. des Originaltitels „Rendez-vous".

7845M
Goebel, Kurt
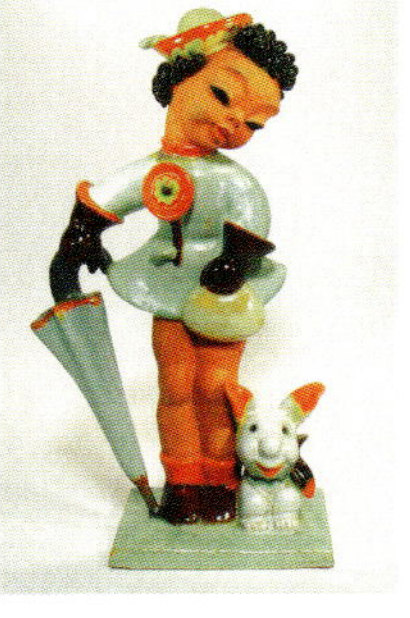
FIGUR „Rendez-vous" (Stehendes Mädchen mit Regenschirm und Tasche) (auf flachem Rechtecksockel mit Hund)
ca. 1937
7845,0 ger., gem. B., Firmenmarke 19, Made in Austria, unsign., H. 25,6 cm; Sammlung a@goldscheider.net
7845M, H. 25 cm, B. 15 cm; GSNachf. 3,11a
Pendant –> vgl. 7844
–> vgl. 7962= GRUPPE
Lit.: GSNachf. 3,11 (m. Abb.), hier mit Ang. des Originaltitels „Rendez-vous".

7847
Weiss, Claire (Klára) (d.i. Herczeg, Claire (Klára)
GRUPPE Gänseliesel (Stehendes kleines Mädchen mit Gänsen) (auf Ovalsockel)
ca. 1937
7847,3,11 gem. SG., Firmenmarke 19, Made in Austria, sign. Claire Weiss, H. 18 cm; Wien Museum Inv. 190.094
Bem.: Die Figur des Mädchens wurde auch beim Modell 7027 verwendet. Dort als „Rotkäppchen".
–> vgl. 7861; 8003=SCHALE

7849
Dakon, Stephan
Stehende Schlittschuhläuferin, ihren Rock fächerartig auseinander haltend (auf Ovalsockel)
ca. 1937
7849,32,6 Firmenmarke 19, Made in Austria, H. 28 cm; Soth. A. 30.6.1992; 22.3.1995
7849,98,31 Firmenmarke 19, Made in Germany, Ausformung zw. 1938–1941, H. 29 cm; Christie's SK 10.11.2000; WKA 25; Doro 7.3.2001
Lit.: Pinhas, S. 128 (m. Abb.)

7851M
Matula, Viktor
GRUPPE „Zicklein"
ca. 1937
7851M H. fehlt
Quellen: Nachlass Matula, Wien; AGM (Kontoblatt der Wiener Manufaktur Josef Schuster für die Firma Porslins Ericsson-Lund, Stockholm?, Schweden vom 7. Januar 1943), hier mit Ang. des Originaltitels GRUPPE „Zicklein".

7854
Knörlein, Rudolf
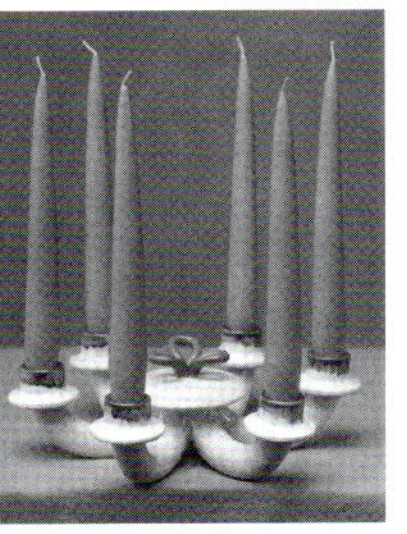
KERZENHALTER (sechsflammig)
Form: u-förmig geschwungene Röhren, mittig runde Bekrönung mit drei Ösen
Dekor: Craquelé- und Laufglasur
ca. 1937
7854, H. 9 cm, Ø 30 cm
Bem.: Durch die eigenhändige Datierung Knörleins (1937) auf dem Trägerblatt des Originalfotos kann das Jahr 1937 exakt bestimmt werden.
Quelle: Privatarchiv (Originalfoto)

7855F
Weiss, Claire / Herczeg, Claire (Klára)
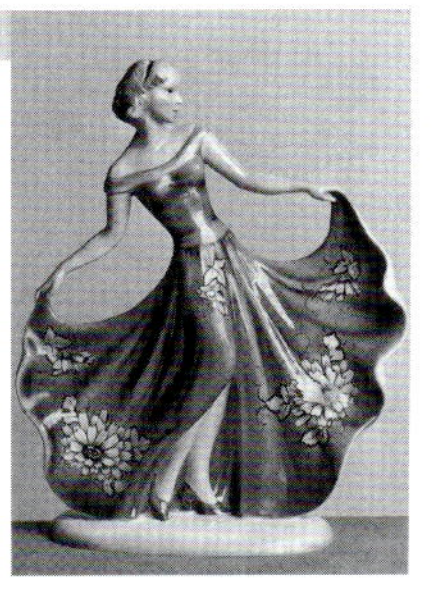
FIGUR / LAMPE „Lydia" (Schreitende junge Tänzerin, ihr langes Kleid auseinander haltend) (auf Ovalsockel)
ca. 1937
FIGUR
7855,3,76 gem. lig. ST, Firmenmarke 19, Made in Austria, Hand dec., unsign., H. 38 cm; eBay # 7319564207
7855,11(0?),14 gem. lig. ON, Firmenmarke 19, Made in Austria, sign. Claire Weiss, H. 39,4 cm; eBay # 1300668925
7855,136,0 Firmenmarke 19, Made in Germany, sign. Claire Herczeg, Ausformung zw. 1938–1941, H. 39 cm; Doro 7.3.2001
7855,137,11 Firmenmarke 19, Made in Germany, sign. Claire Herczeg, Ausformung zw. 1938–1941, H. 39 cm; Doro 19.9.2000
7855F, H. 39 cm, B. 31 cm; GSNachf. 1,22a; Fotoarchiv K. Täuber, hier mit der Ang.: H. 40 cm, B. 29 cm
LAMPE
7855 ohne Ang., Firmenmarke 19, sign. C(laire?) Weiss, H. 38 cm; Herr 22; Bem.: Elektrifizierung fehlt.
–> vgl. 7963; 8322
Lit.: GSNachf. 1,22 (m. Abb.), hier mit Ang. des Originaltitels „Lydia".
Quelle: AGM (Bestellungen der Firma Wahliss (Wien) vom 9. und 24.11.1945 an die „Arbeitsgemeinschaft der Firma Friedrich Goldscheider"), hier jeweils mit dem Titel „Tänzerin"; Bestellung der Firma Wahliss (Wien) vom 27.2.1946 in Ausführung „Terrakotta"; Rechnung an Maurice Mizrahi, Alexandria, Ägypten vom 22. April 1953), hier unter dem Titel „Tanz" geführt; Fotoarchiv K. Täuber.

7856
Tirolerin in Tracht mit Hut
ca. 1937
7856 ohne Ang., Firmenmarke 19, Made in Germany, Ausformung zw. 1938–1941, H. ca. 30 cm; KH; Bem.: Mündliche Mitteilung, daher keine exakten Ang. über etwaige Signatur oder Sockelform möglich.

7857F
Dakon, Stephan
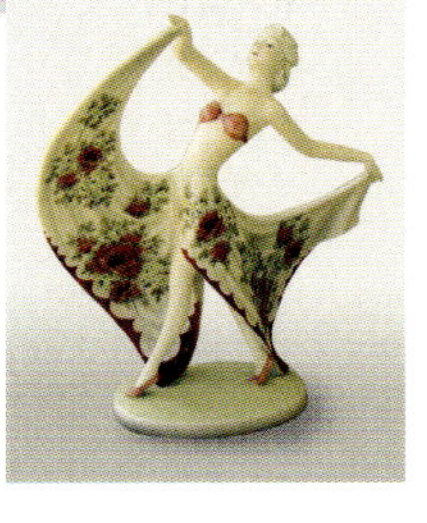
FIGUR „Champagner" (Nach links schreitende junge Dame mit Bikinioberteil, ihren langen Rock schmetterlingsflügelartig auseinander haltend) (auf Ovalsockel)
ca. 1937
7857,7 gem. G, Firmenmarke 19, Made in Austria, Ausformung zw. 1945–1953, H. 39 cm; Sammlung Stressless, Portugal
7857,11 gem. lig. HS, Firmenmarke 19 (blau), Made in Austria (blau), Hand dec. (blau), Ausformung zw. 1945–1953, H. 40 cm; WKH
7857,74,6 gem. F., gem. lig. JR, Firmenmarke 19, Made in Austria, H. 40 cm; eBay # 370981975
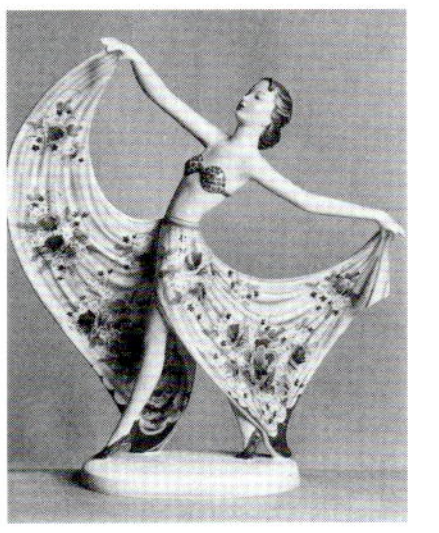
7857,1(0?)5,14 Firmenmarke 19, Made in Austria, gepr. „L"= Z6, H. 40,6 cm; eBay # 1204328515; (KH www.tias.com/stores/frozenbeauty)
7857,148,19 Firmenmarke 19, Made in Germany, unsign.?, Ausformung zw. 1938–1941, H. 38 cm; RM 27; 29
7857,173,9 (od. 6?),D Firmenmarke 19, Made in Germany, unsign.?, Ausformung zw. 1938–1941, H. 40 cm; Dawo 141
7857,S gem. S., gem. lig. HS, Firmenmarke 19, Made in Austria, Hand dec., unsign., Originalpick.: „C. Bachl, Wien", Ausformung zw. 1945–1953, H. 39,4 cm; Gorringes 12.2.2002; Bay # 731620212
Bem.: Die in der Quelle angegebene Bestellung der Firma Wahliss (Wien) belegt, dass auch das „seitenverkehrte" Modell den Originaltitel „Champagner" trug. Diese Figur wurde auch noch in den 50er Jahren in Wien und später dann bei Carstens (GWM West-Germany) produziert. Hier wurde es unter dem Titel „Sekt" geführt (–> vgl. Könecke 2002, S. 76; s.a. Quelle).
–> vgl. 8360
Pendant –> vgl. 7826; 8319= jeweils seitenverkehrtes Modell.
Lit.: GSNachf. 1,21 (m. Abb.), hier das seitenverkehrte Pendant mit Ang. des Originaltitels „Champagner"; KGSF1; KGSF2; Könecke (2002), S. 31, 76, hier unter dem Titel „Sekt" (m. Abb.).
Quelle: AGM (Bestellung der Firma Wahliss (Wien) vom 27.2.1946 an die „Arbeitsgemeinschaft der Firma Friedrich Goldscheider", in Ausführung „Terrakotta" sowie mit Ang. des Originaltitels „Champagner"; Rechnung an Maurice Mizrahi, Alexandria, Ägypten vom 22. April 1953), hier unter dem Titel „Sekt" geführt; Fotoarchiv K. Täuber.

7858F
Knörlein, Rudolf
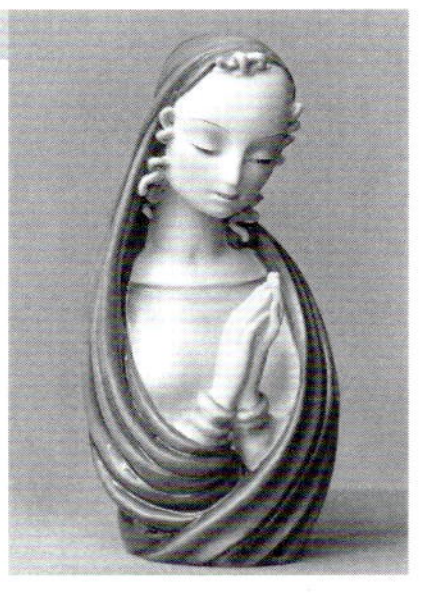
BÜSTE „Ave Maria" (Madonna mit gefalteten Händen)
ca. 1934 (für den Entwurf)
7858,20 Firmenmarke 19, unsign., Ausformung zw. 1945–1953, H. 12 cm; Schopmann 25.3.1994
7858,25 Firmenmarke 19, Ausformung zw. 1945–1953, H. 11,5 cm; Schuler 59
7858F, H. 11 cm, B. 5/6 cm; GSNachf. 6,5c
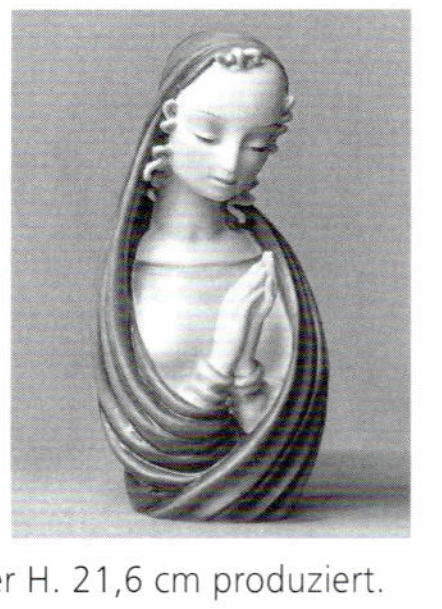
Bem.: Durch die eigenhändige Datierung Knörleins (1934) auf dem Trägerblatt des Originalfotos sowie der Originalzeichnung (Aquarell über Graphit), kann das Jahr 1934 exakt bestimmt werden. Diese Büste wurde auch von GS USA (Everlast Corp.) unter der Modell-Nr. 819 in der H. 21,6 cm produziert.
–> vgl. 7188; 7263; 8743
Lit.: GSNachf. 6,5 (m. Abb.), hier mit Ang. des Originaltitels BÜSTE „Ave Maria".
Quelle: Privatarchiv (Originalfoto); Privatsammlung D (Originalzeichnung).

7859
Goebel, Kurt
Stehender Matrose (auf Rundsockel)
ca. 1937
7859,0 ger. G, gem. O, Firmenmarke 19, Made in Austria, unsign., Ausformung zw. 1945–1953, H. 24 cm; eBay # 2629174376
7859,0 gem. B, Firmenmarke 19, Made in Austria, unsign., Ausformung zw. 1945–1953, H. 24 cm; eBay # 6174186133
Pendant –> vgl. 7977
–> vgl. 7960= GRUPPE; Bem.: Bis auf die Armstellung und den Kragen des Hemdes identische Figur zur Gruppe.

7860
Goebel, Kurt
GRUPPE Stehender Matrose mit Matrosenbraut (auf Ovalsockel)
ca. 1937
7860,65 gem. M., Firmenmarke 19, Made in Austria, unsign., Ausformung zw. 1945–1953, Pick. 6, H. 21 cm; Wien Museum Inv. 220.747
-> vgl. 7977= EinzelFIGUR (Matrosenbraut)
-> vgl. 7858= ähnl. FIGUR (Matrose); Bem.: Bis auf die Armstellung und den Kragen des Hemdes identische Figur zur Gruppe.

7861
Weiss, Claire (Klára) (d.i. Herczeg, Claire (Klára)
GRUPPE Gänseliesel (Stehendes kleines Mädchen mit Gans) (auf Ovalsockel)
ca. 1937
7861,20 (Firmenmarke ohne Ang.), unsign., Ausformung zw. 1945–1953, H. 20 cm; Doro S 15.4.1992
7861 ohne Ang., Firmenmarke 19, sign. Claire Weiss, H. 19,5 cm; KaM 71
Bem.: Die Figur des Mädchens wurde auch beim Modell 7027 verwendet. Dort als „Rotkäppchen".
-> vgl. 7847; 8003=SCHALE

7862
Prischl, Adolf
Stehende Madonna mit gekreuzten Händen (auf Rundsockel)
ca. 1937
7862,50,31 Firmenmarke 19, Made in Germany, Ausformung zw. 1938–1941, H. 39,4 cm; eBay # 13351589

7863F
Matula, Viktor
FIGUR „Reh" (Äsendes Reh) (auf länglichem Ovalsockel)
ca. 1937
7863F, H. 24 cm, B. 33,5 cm; Fotoarchiv K. Täuber, hier mit der Ang. des Originalpreises RM 39,75
-> vgl. 7818
Lit.: GSNachf. 3,15 (m. Abb.), hier mit Ang. des Originaltitels „Reh" in der H. 10 cm auf flachem Rechtecksockel.
Quelle: Fotoarchiv K. Täuber

7864M
Matula, Viktor
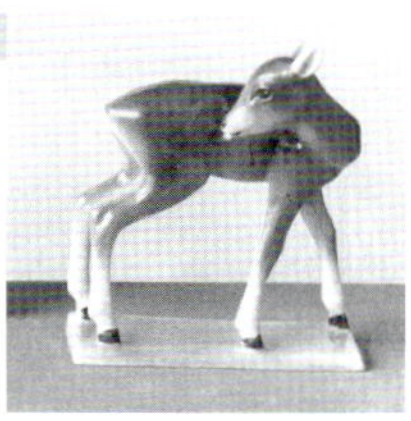
FIGUR „Reh" (Stehendes Reh mit nach hinten gewandtem Kopf) (auf flachem Rechtecksockel)
ca. 1937
7864,16 Firmenmarke 19, unsign., H. 12 cm; eBay # 2225687050
7864,20,16 ger., Firmenmarke 19, Made in Austria, unsign., H. 12 cm; Wien Museum Inv. 190.034
7864M, H. 12 cm, B. 10 cm; GSNachf. 3,15d; Fotoarchiv K. Täuber, hier mit der Ang.: H. 11 cm, B. 10 cm
Lit.: GSNachf. 3,15 (m. Abb.), hier mit Ang. des Originaltitels „Reh".
Quellen: Nachlass Matula, Wien; AGM (Bestellungen der Firma Wahliss (Wien) vom 9. und 24.11.1945 an die „Arbeitsgemeinschaft der Firma Friedrich Goldscheider"; Bestellung der Firma Wahliss (Wien) vom 27.2.1946 in Ausführung „Terrakotta"); Fotoarchiv K. Täuber.

7865F
Prischl, Adolf
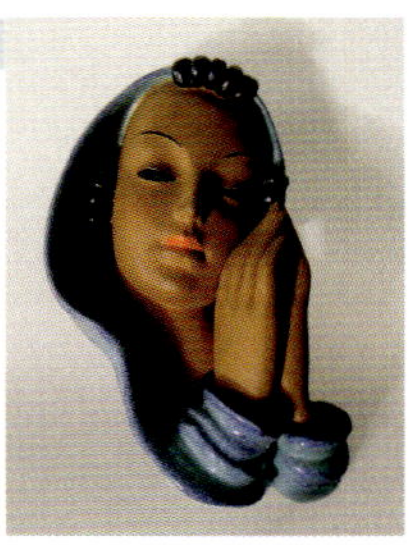
WANDMASKE „Madonna" (Madonna mit gefalteten Händen)
ca. 1936/37
7865,14,57 gem. G., Firmenmarke 19, Made in Austria, unsign., H. 23,3 cm; MAK Wien, Inv. Ke 10602-100
7865,20 ger. XVI, Firmenmarke 19, Made in Austria, unsign., Ausformung zw. 1945–1953, H. 23,5 cm; eBay # 2232728431
7865,66,29 Firmenmarke 20, Made in Austria, Hand dec., unsign., H. 23 cm; city-antik
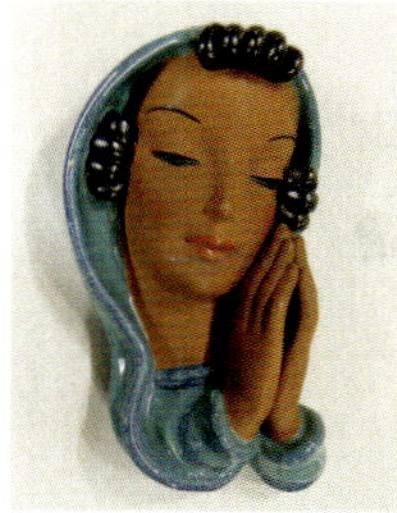
7865,94,10 (Firmenmarke ohne Ang.), unsign., H. 23 cm; Christie's SK 27.6.1997
7865F, H. 24 cm, B. 13 cm; GSNachf. 6,8b
-> vgl. 7579F
Lit.: GSNachf. 6,8 (m. Abb.), hier mit Ang. des Originaltitels MASKE „Madonna".
Quelle: Fotoarchiv K. Täuber

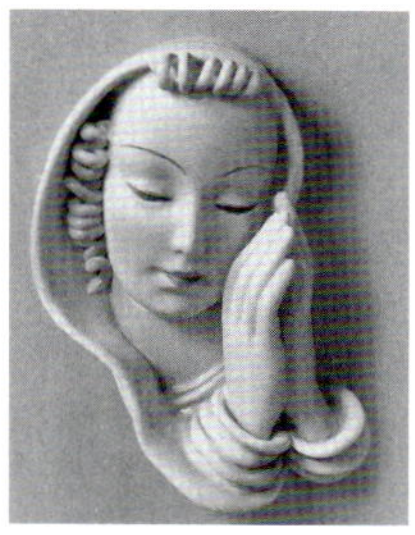

7866
„Kitty" (Stehendes kleines Mädchen mit Hut und bellendem, aufspringendem Foxterrier) (auf Ovalsockel)
ca. 1937
7866,5,6 ger., Firmenmarke 19, Made in Austria, Hand dec., H. 18,5 cm; Henry's 25.11.1998; Quit. 14
7866,10,13 gem. H, ohne Firmenmarke, Pick. 1, H. 19 cm; Wien Museum Inv. 220.723
7866,30,20,A,9 gem. G., Firmenmarke 19, Made in Austria, H. 19 cm; eBay # 14462406
Quelle: AGM (Bestellung von Ferd. Lundquist (Göteborg, Schweden) vom 23. August 1950), hier mit Ang. des Originaltitels „Kitty".

7867M
Matula, Viktor
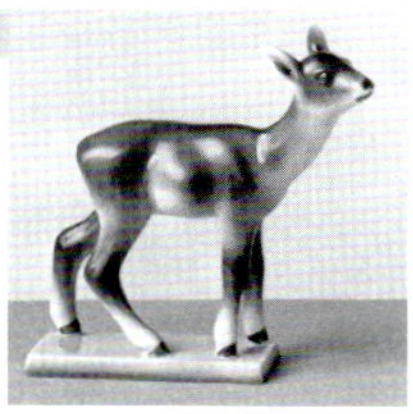
FIGUR „Reh" (auf flachem Rechtecksockel)
ca. 1937
7867M, H. 13 cm, B. 11,5 cm; GSNachf. 3,15a; Fotoarchiv K. Täuber, hier mit der Ang.: H. 12 cm, B. 9 cm
Lit.: GSNachf. 3,15 (m. Abb.), hier mit Ang. des Originaltitels „Reh".
Quellen: Nachlass Matula, Wien; AGM (Kontoblatt der Wiener Manufaktur Josef Schuster für die Firma Porslins Ericsson-Lund, Stockholm?, Schweden vom 7. Januar 1943), hier mit Ang. des Originaltitels „Reh"; Fotoarchiv K. Täuber.

7868F
Dakon, Stephan
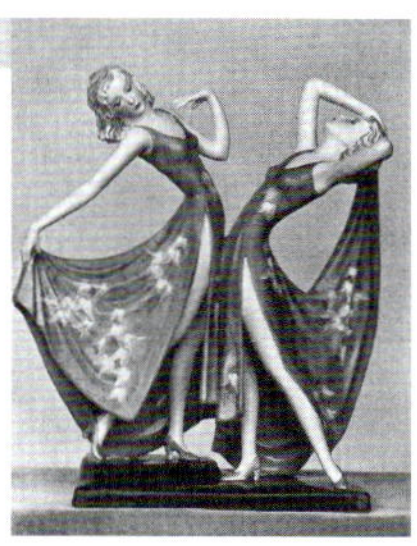
GRUPPE „Dolly Sisters" (Stehend, mit langen Kleidern, in Tanzpose) (auf Ovalsockel/Rechtecksockel)
ca. 1937
auf Ovalsockel
7868,68,16 Firmenmarke 19, H. 34,3 cm; eBay # 2013746225
auf Rechtecksockel
7868,88,31 ger. XII,(H?), gem. SG, Firmenmarke 19, Made in Austria, H. 33 cm; Soth. com. # 329H
7868,115,19 (Firmenmarke ohne Ang.), H. 34 cm; Doro 26.5.2004
7868F, H. 35 cm, B. 29 cm; GSNachf. 1,23a
-> vgl. 7991= GRUPPE
-> vgl. 7957; 7992= rechte EinzelFIGUR
-> vgl. 7958; 7993= linke EinzelFIGUR
Lit.: GSNachf. 1,23 (m. Abb. der Gruppe auf Ovalsockel), hier mit Ang. des Originaltitels „Dolly Sisters".
Quelle: Archiv Mimi Dakon (Originalfoto)

7869M
Matula, Viktor
FIGUR „Reh" (Stehendes Reh mit erhobenem Kopf) (auf flachem Rechtecksockel)
ca. 1937
7869M, H. 12,5 cm, B. 10 cm; GSNachf. 3,15c
Lit.: GSNachf. 3,15 (m. Abb.), hier mit Ang. des Originaltitels „Reh".
Quelle: Nachlass Matula, Wien

7871
Liegender Windhund
ca. 1937
7871,765,– gem. AN, Firmenmarke 19, H. fehlt, L. 33 cm; Karrenbauer 171

7874
Goebel, Kurt
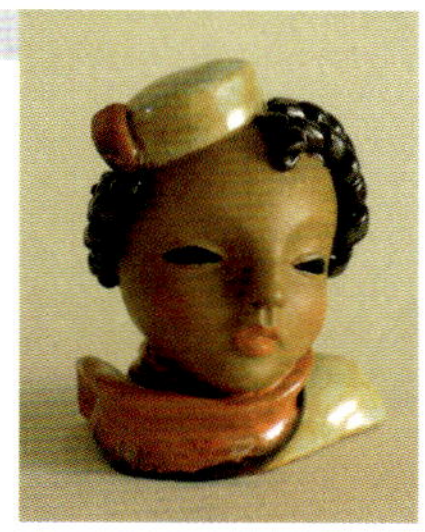
STEHKOPF Frauenkopf mit modischer Kappe und Schal
ca. 1937
7874,65 gem. 7822, ohne Firmenmarke, unsign., Ausformung zw. 1945–1953, H. 9 cm; PB Deutschland

7875
Matula, Viktor
GRUPPE Zicklein
ca. 1937
7875 H. fehlt
Quelle: Nachlass Matula, Wien

7877M
Meisinger, Ida
FIGUR „Grippe" (Im Bett liegender Terrier mit Schal)
ca. 1934 (für den Entwurf)
7877M, H. 13,5 cm, B. 10 cm; GSNachf. 3,28a
Bem.: s. 6449
Lit.: GSNachf. 3,28 (m. Abb.), hier mit Ang. des Originaltitels FIGUR „Grippe".

7878M
Meisinger, Ida
FIGUR „Hund mit Bassgeige" (Stehender Terrier mit Kontrabass)
ca. 1934 (für den Entwurf)
7878M, H. 10,5 cm, B. 13 cm; GSNachf. 3,28b
Bem.: s. 6449
-> vgl. 6977
Lit.: GSNachf. 3,28 (m. Abb.), hier mit Ang. des Originaltitels FIGUR „ Hund mit Bassgeige".

7879
Goebel, Kurt
BUCHSTÜTZE Frauenkopf (auf halbkreisförmigem Sockel)
ca. 1936 (für den Entwurf)
7879,65,5 gem. M (oder W?), Firmenmarke 19, Made in Austria, unsign., H. 21,6 cm; eBay # 1505007100
Ausst.: Wiener Herbstmesse 1936
-> vgl. 7616
Lit. u. Quelle s. 7616

7883M
Prischl, Adolf
WANDMASKE Frauenkopf mit Ohrring und Schal
ca. 1937
7883,33 Firmenmarke 19, Made in Austria, unsign., Originalpick.: „T. Waldenborgs Möbelaffär, Göteborg", Ausformung zw. 1945–1953, H. 30 cm; eBay # 7304468520
7883,S,M16 ger., gem. lig. TK, Firmenmarke 19, Made in Austria, unsign., Ausformung zw. 1945–1953, H. 31,3 cm; Wien Museum Inv. 186.645
-> vgl. 8198
Lit.: Könecke (2002), S. 77, hier mit Ang. des Künstlers A. Prischl (für die Modell-Nr. 8198), S. 88 (m. Abb.).
Quelle: AGM (Kontoblatt der Wiener Manufaktur Josef Schuster für die Firma Porslins Ericsson-Lund, Stockholm?, Schweden vom 7. Januar 1943), hier mit Ang. des Titels „Maske/Goldband".

7885F
GRUPPE „Windhunde"
ca. 1937
Quelle: AGM (Rechnung der Wiener Manufaktur Josef Schuster an Laszlo Bösze (Szeged, Ungarn) vom 5. August 1944).

7887M / F
Goebel, Kurt
FIGUR „Skiläufer" (Schreitende Skiläuferin mit geschulterten Skiern) (auf Ovalsockel)
ca. 1936 (für den Entwurf)
7887,6,18 gem. lig. rot WB, Firmenmarke 19, Made in Austria, Hand dec., unsign., H. 14 cm; eBay # 2655126536
7887,20,14 ger., gem. braun LS, gem. lig. LR, Firmenmarke 19, Made in Austria, Hand dec., unsign., Original-pick.: „Ludwig Fuchs, Kitzbühel Stadttor", H. 14 cm; eBay # 28167121
7887,55,2 gem. W, Firmenmarke 19, H. 14,3 cm; Wien Museum Inv. 220.737
7887M, H. 13,5 cm, B. 11 cm; GSNachf. 3,8b
Bem.: Die von der Firma als „Skiläufer" bezeichnete Figur ist eindeutig eine „Skiläuferin".
-> vgl. 7622M
-> vgl. 7655M; 7935M= jeweils ähnl. GRUPPE
Pendant -> vgl. 7518
Lit.: GSNachf. 3,8 (m. Abb.), hier mit Ang. des Originaltitels FIGUR „Skiläufer".
Quelle: AGM (Bestellung von Ferd. Lundquist (Göteborg, Schweden) vom 23. August 1950), hier unter dem Titel „Skiläuferin" bestellt.

7889
Goebel, Kurt
Stehendes Mädchen mit zwei Hasen (auf Rundsockel)
ca. 1937
7889,25 Firmenmarke 19, Made in Austria, unsign., Ausformung zw. 1945–1953, H. 26,7 cm; eBay # 32519131

7890
Goebel, Kurt (zugeschr.)
Stehendes Mädchen mit Puppenwiege (auf abgerundetem Rechtecksockel)
ca. 1937
7890,26 Firmenmarke 19, Made in Austria, unsign., Ausformung zw. 1945–1953, H. 26 cm; RM 42; Fischer 67; 74; 83

7891
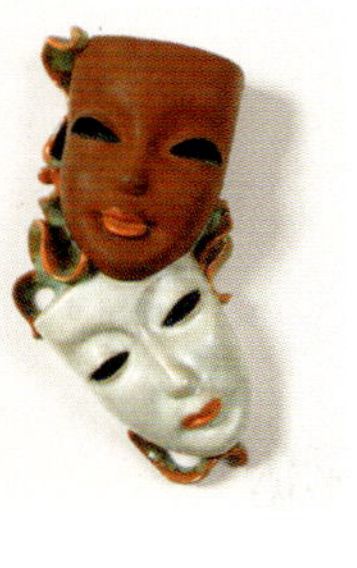
WANDMASKE / DOPPELMASKE Zwei (Theater)-Masken mit Girlanden
ca. 1937
7891 ger. S, gem. U?, Firmenmarke 19, Made in Austria, unsign., H. 30,5 cm; Sammlung.schmutz@chello.at 12
7891,33 ger. VII, gem. L., Firmenmarke 19, Made in Austria, unsign., Ausformung zw. 1945–1953, H. 25 cm (Mess- oder Druckfehler?); eBay # 2070804075
7891 ger., Firmenmarke 19, Made in Austria, unsign., H. 30 cm; eBay # 6519441464

7894
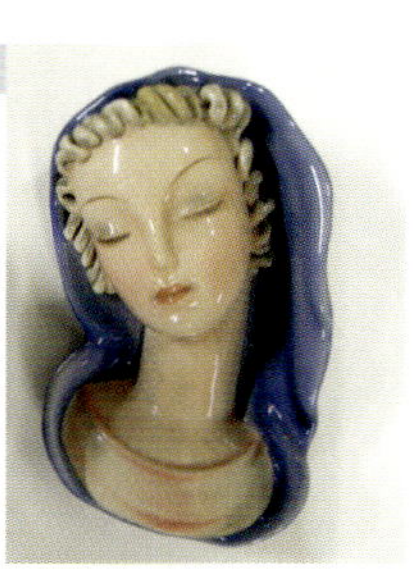
Prischl, Adolf
BÜSTE Madonna mit Schleier/Kopftuch
ca. 1937
7894,92 gem. NF,S., Firmenmarke 19, Made in Austria, unsign., Ausformung zw. 1945–1953, H. 21,8 cm; MAK Wien, Inv. Ke 10602-99

7897F
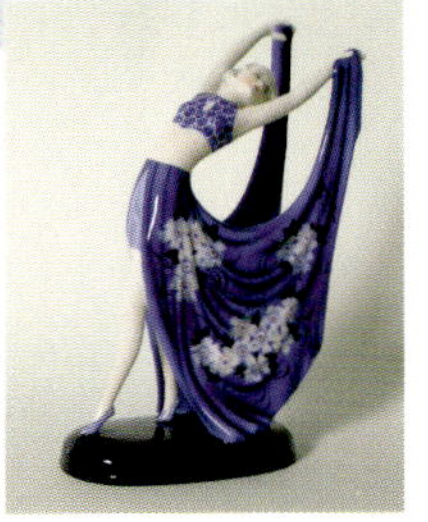
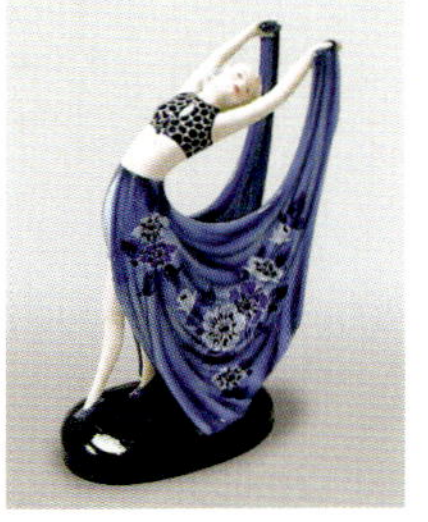
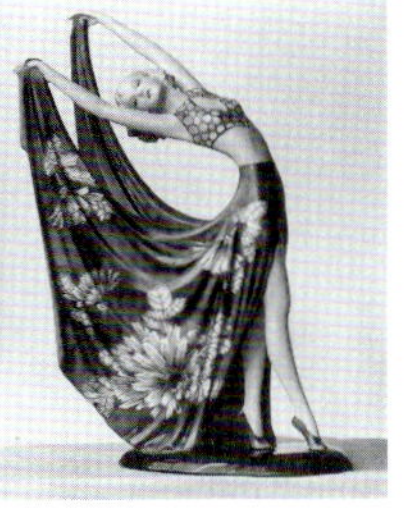
Dakon, Stephan
FIGUR / LAMPE „Beauty" (Tänzerin, nach links geneigt, ihr Kleid flügelartig zurückhaltend) (auf Ovalsockel)
ca. 1935 (für den Entwurf)
FIGUR
7897,89,16 gem. rot N, Firmenmarke 19, Made in Austria, H. 24 cm; city-antik
7897,252,0 gem. H., Firmenmarke 19, Made in Germany, Ausformung zw. 1938–1941, H. 23,5 cm; eBay # 2258790289; 3360682470
7897,271,0 ger. M, gem. B, Firmenmarke 19, Made in Germany, Ausformung zw. 1938–1941, H. ca. 24 cm; city-antik
7897,282,? Firmenmarke 19, Made in Germany, Ausformung zw. 1938–1941, H. 24,5 cm; eBay # 3318531214
7897,331,10 gem. lig. FD, X, gem. rot EW., Firmenmarke 19, Made in Germany, Ausformung zw. 1938–1941, H. 24,5 cm; Soth. com. # 1BT7
7897,363,19 gem. B, Firmenmarke 19, Made in Germany, Ausformung zw. 1938–1941, unsign., H. 24,1 cm; Christie's L. 19.4.1989
7897,399,7 ger. SB, Firmenmarke 19, Made in Germany, Ausformung zw. 1938–1941, Pick. 6, H. 25 cm; Mehlis 32
7897,559,2 Firmenmarke 19, Made in Germany, Ausformung zw. 1938–1941, unsign., Pick. 5c, H. 24 cm; GK, S. 22, 128 (m. Abb.); Wien Museum Inv. 222.547
7897,722,7 gem. rot XX32, Firmenmarke Schuster 1, Made in Germany, unsign., Ausformung zw. 1941–1944, H. 24 cm; Wien Museum Inv. 190.138
7897F, H. 23 cm, B. 16 cm; GSNachf. 1,12c; 1,50c
LAMPE
7897F, H. 40 cm, B. 39 cm; GSNachf. 1,12e; 1,50e
Bem.: Dieses Modell gibt es in zwei Varianten, einmal neigt sich der Körper der Tänzerin nach links, wie in diesem Fall, das andere Mal nach rechts.
Ausst.: HMW (jetzt Wien Museum) 1985/86.
-> vgl. 7195; 7996; 8746
Lit.: GSNachf. 1,12 (m. Abb. als Figur), hier mit Ang. des Originaltitels „Beauty"; GSNachf. 1,50 (m. Abb. als Lampe montiert, komplett, mit Seidenschirm), hier mit Ang. des Originaltitels „Beauty"; GK, S. 22, 128 (m. Abb.).
Quelle: AGM (Bestellung der Firma Wahliss (Wien) vom 17.10.1945 und 27.2.1946 an die „Arbeitsgemeinschaft der Firma Friedrich Goldscheider", in der Ausführung „Terrakotta"); Archiv Mimi Dakon (Originalfoto).

7900M
Matula, Viktor
ASCHER „Ascher mit Schildkröte"
Form: rund (mit Standfläche für die Schildkröte), zwei Zigarettenablagen
Dekor: Schildkröte (vollplastische Fig.)
ca. 1937
7900,16,20 Firmenmarke 19, Made in Austria, Hand dec., unsign., H. 7,5 cm, B. 20 cm; eBay # 6610418034
7900,89,20 Firmenmarke 19, Made in Austria, Hand dec., unsign., H. 8 cm, B. 20 cm; eBay # 7372050409
7900M, H. 8 cm, B. 19,5 cm; GSNachf. 4,3a
Lit.: GSNachf. 4,3a (m. Abb.), hier mit Ang. des Originaltitels „Ascher mit Schildkröte".
Quelle: Nachlass Matula, Wien

7909
Meisinger, Ida (zugeschr.)
GRUPPE Mutter und Kind (Stehende Terrierhündin mit Welpe in der Wiege)
ca. 1937
7909,15,65 Firmenmarke 19, Made in Austria, unsign., H. 11,5 cm; eBay # 2529031440
7909,43,65 Firmenmarke 19, Made in Austria, unsign., H. ca. 11,5 cm; eBay # 920095613
Bem.: s. 6449

7910
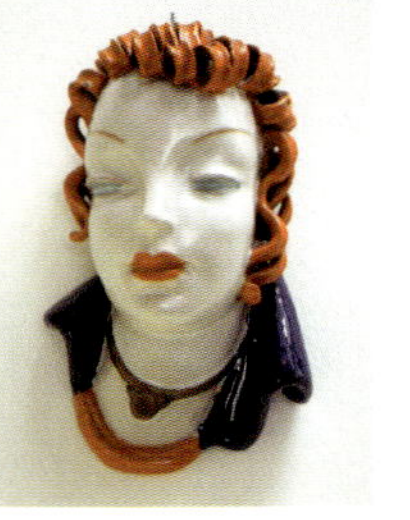
Prischl, Adolf (zugeschr.)
WANDMASKE Frauenkopf mit Kragen und Halsschmuck
ca. 1937
7910,33 gem. lig. AH, Firmenmarke 19, Made in Austria, unsign., Ausformung zw. 1945–1953, H. 22 cm; PB Wien 19

7912M
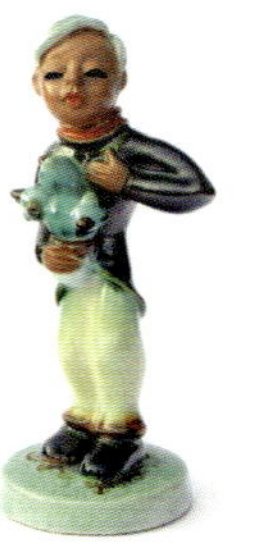
Goebel, Kurt
FIGUR „Romeo" (Stehender junger Mann mit Hut und Kaktus) (auf Rundsockel)
ca. 1936/37
7912,2 gem. G, Firmenmarke 19, Made in Austria, H. 14 cm; Wien Museum Inv. 220.738
7912M, H. 14 cm, B. 5,5 cm; GSNachf. 3,9c
-> vgl. 7656M; 7978M
Lit.: GSNachf. 3,9 (m. Abb.), hier mit Ang. des Originaltitels FIGUR „Romeo".

7915M
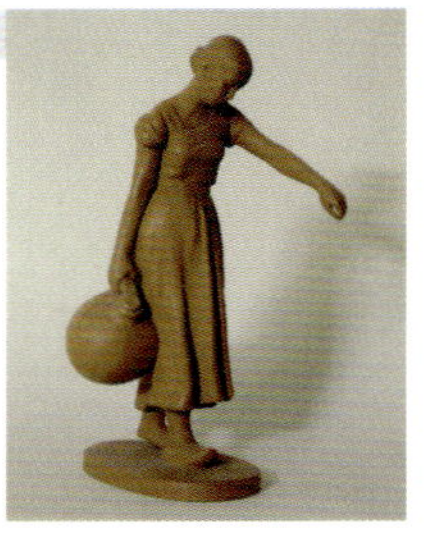
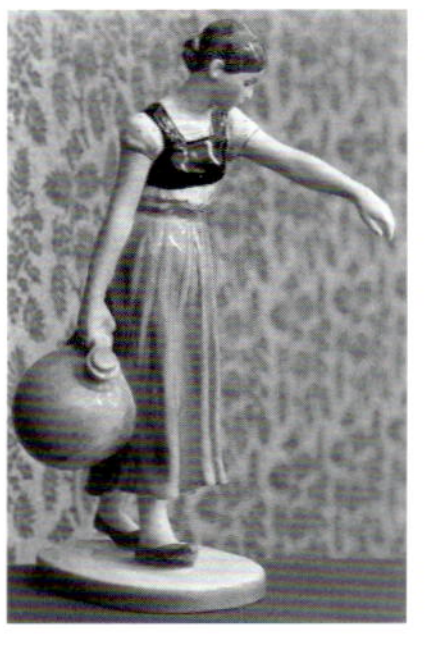
Strasser, Arthur (nach)
„Das Krüglein"/„Krugträgerin" (Schreitendes Bauernmädchen mit Krug) (auf Ovalsockel)
ca. 1898 (für den Entwurf)
7915,6 Firmenmarke 19a, unsign., Ausformung zw. 1945–1953, H. 30 cm; Wien Museum Inv. 219.160
7915,8 Firmenmarke 19a, unsign., Ausformung zw. 1945–1953, H. 30 cm; Doro 7.3.2001
7915,8,6,8 gem. lig. ON, S, Firmenmarke 19, Made in Austria, unsign., Pick. 6, H. ca. 30 cm; eBay # 2042258248
7915,84,110 Firmenmarke 19, Made in Germany, unsign., Ausformung zw. 1938–1941, H. 31 cm; Kastern 87
7915,93,31,G, gem. l., Firmenmarke 19, Made in Germany, unsign., Ausformung zw. 1938–1941, H. 31 cm; Weidler 370; Mehlis 33
7915,307,7 Firmenmarke 19, unsign., H. 31 cm; KH
7915 ger., ohne Firmenmarke, unsign., H. 30,5 cm; PB Deutschland
Bem.: Diese Figur stammt im ursprünglichen Entwurf von Arthur Strasser, datiert um das Jahr 1898, und war mehrfach im ÖMKI ausgestellt (-> vgl. Ausst.). Die zeitgenössische Pariser Produktion nahm dieses Modell ebenfalls auf und stellte es in drei Größen her (-> vgl. Lit.).
Ausst.: ÖMKI 1906/07, hier mit dem Originaltitel „Das Krüglein"; ÖMKI 1910/11, hier mit dem Titel „Krüglein".
-> vgl. 8727
-> vgl. 551 (GSB); 576 (GSB); 579 (GSB)
Lit.: GSB, Pl. 18, hier ist die Figur mit dem franz. Originaltitel „Retour de la Fontaine" unter Nr. 551 (H. 90 cm) abgebildet. Zwei weitere Größen mit der Nr. 579 (H. 61 cm) sowie der Nr. 576 (H. 32 cm) finden sich an gleicher

Stelle. Ausgeformt wurden diese Objekte in der Pariser Niederlassung. Angaben zum Material werden im Firmenkat. nicht gemacht; KuKHW, 1.1898, S. 354, 362; Neuwirth (1974), S. 74.
Quelle: AGM (Kontoblatt der Wiener Manufaktur Josef Schuster für die Firma Ohlson jr., Malmö, Schweden vom 16. September 1942); (Kontoblatt der Wiener Manufaktur Josef Schuster für die Firma Porslins Ericsson-Lund, Stockholm?, Schweden vom 7. Januar 1943), hier mit Ang. des Titels „Krugträgerin"; (Bestellungen der Firma Wahliss (Wien) vom 9. und 24.11.1945 an die „Arbeitsgemeinschaft der Firma Friedrich Goldscheider"), hier jeweils mit Ang. des Titels „Krugträgerin"; Bestellung der Firma Wahliss (Wien) vom 27.2.1946 in Ausführung „Terrakotta"; Fotoarchiv K. Täuber.

7916
Lorenzl, Josef
Sitzender lesender Mädchenakt mit Pumps
(auf Polstersessel)
ca. 1937
7916 (Firmenmarke ohne Ang.), H. 21 cm;
Pinhas, S. 142 (m. Abb.)

7920
Dame mit Hund
ca. 1937
7920,50,9 (Firmenmarke ohne Ang.), H. 17,5 cm;
Doro 20.3.2006

7924
Jarl, Karin
Pekinese (stehend)
ca. 1937
7924,56,3 gem. F., Firmenmarke 19, Made in Germany, Ausformung zw. 1938–1941, H. 12 cm;
eBay # 1426916519

7926
Jarl, Karin
GRUPPE Vier Pekinesen
ca. 1937
7926,15,14 gem. F., Firmenmarke 19, Made in Austria, H. 12 cm, L. 22 cm; eBay #20807931610
7926,60,37 ger. X, Firmenmarke 19, Made in Germany, unsign., Ausformung zw. 1938–1941, H. 12 cm; Quit. 17

7930
Matula, Viktor
LEUCHTER
ca. 1937
7930 H. fehlt
Quelle: Nachlass Matula, Wien

7931
Prischl, Adolf
WANDMASKE „Maske" (Frauenkopf mit kurzem Kragenansatz)
ca. 1937
7931,20 Firmenmarke Schuster 1, Made in Germany, unsign., Ausformung zw. 1941–1944, H. 26,7 cm;
city-antik
–> vgl. 7816; 8248
Lit.: Könecke (2002), S. 76, hier mit Ang. des Künstlers A. Prischl (für die Modell-Nr. 7816).

7934
WANDMASKE Frauenkopf mit Kappe und Rollkragenpullover
ca. 1937
7934 (Firmenmarke ohne Ang.), H. 27,5 cm;
Pinhas, S. 172 (m. Abb.)

7935M
Goebel, Kurt
GRUPPE „Skifahrer" (Schreitendes Skifahrerpaar)
(auf Ovalsockel)
ca. 1936
7935,44 Firmenmarke 19, Made in Germany, unsign., Ausformung zw. 1938–1941, H. 15 cm;
eBay # 2245112637
7935M, H. 14 cm, B. 14 cm; GSNachf. 3,7b
–> vgl. 7655M
–> vgl. 7518= Skifahrer, ähnl. FIGUR
–> vgl. 7622; 7887= jeweils Skifahrerin, ähnl. FIGUR
Lit.: N.N.: Neue Keramik, in: ÖK, 7.1936, Nr. 10, S. 25 (m. Abb.) „Man sah auf der Wiener Herbstmesse eine Reihe keramischer Arbeiten, die Originalität und richtiges Erfassen der lebendigen Strömungen in der Entwicklung moderner Keramik zeigten. Besonders die Arbeiten des Keramikers Kurt Goebel..."; GSNachf. 3,7 (m. Abb.), hier mit Ang. des Originaltitels GRUPPE „Skifahrer".
Quelle: AGM (Tantiemenliste Goebel, ohne Datierung (ca. 1940).

7936
Prischl, Adolf
Frauenakt als Torso
(auf Rundsockel)
ca. 1937

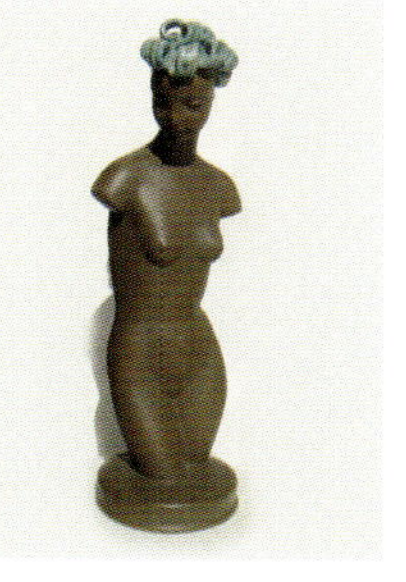

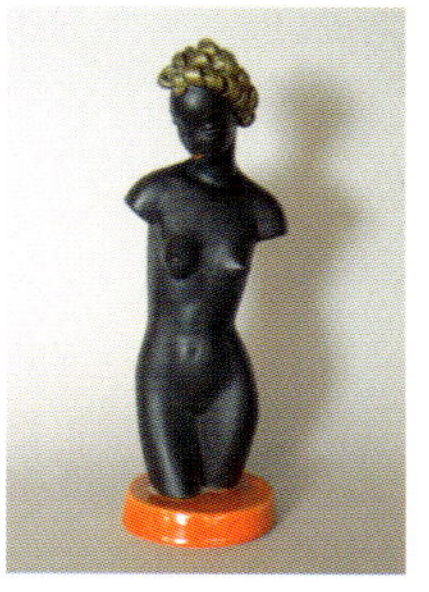

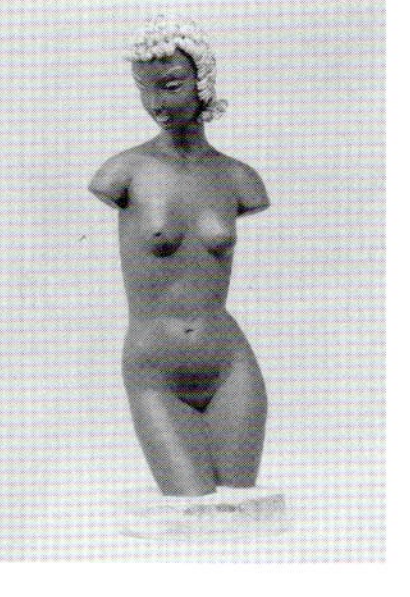

7936,0 ger., Firmenmarke 19, unsign., Ausformung zw. 1945–1953, H. 29,2 cm;
Quit. 52
7936 ger., Firmenmarke 19 (blau), unsign., Originalpick.: „Edelhausrat Mimi Wodak, Wien XVI, Lerchenfeldergürtel", H. 31 cm; Sammlung.schmutz@chello.at 74
7936,4 (Firmenmarke ohne Ang.), unsign., Ausformung zw. 1945–1953, H. 29,5 cm;
Christie's SK 4./5.12.1996
7936,8 Firmenmarke 19, Made in Austria, Pick. 0, H. 30 cm;
eBay # 290014506895
7936,9 (Firmenmarke ohne Ang.), unsign., Ausformung zw. 1945–1953, H. 30 cm;
Christie's SK 4.11.1999
7936,18,X ger., Firmenmarke 19 (blau), unsign., Ausformung zw. 1945–1953, H. 31 cm; Doro 6.4.1999
7936,800/VIII, Firmenstempel WG 1, unsign., Ausformung 50er Jahre, H. 31 cm;
Sammlung Stressless, Portugal
7936,1394/VI, Firmenstempel WG 1, Ausformung 50er Jahre, H. 31 cm;
eBay # 3350358756
7936 ger., Firmenstempel WG 1, Ausformung 50er Jahre, H. 31 cm; Christie's SK 14.11.2002
7936 (Firmenmarke ohne Ang.), unsign., H. 32 cm; Christie's SK 10.11.2000
7936 H. 29,5 cm, B. 10 cm; Fotoarchiv K. Täuber
Bem.: Diese Figur wurde auch noch zw. 1953–1960 bei Carstens (GWM West-Germany) produziert und weist unterschiedliche Frisuren auf. Dies liegt zum einen am Zeitgeschmack, zum anderen wurden die Haare (Curls) einzeln angarniert und differieren deshalb zusätzlich etwas.
Lit.: Könecke (2002), S. 76, 87 (m. Abb.).
Quelle: AGM (Auftragszettel für die Firma „Haus der Geschenke", Aachen vom 16. Januar 1953), hier „mit Gold" (staffiert, Anm. des Autors); Fotoarchiv K. Täuber.

7939M
Matula, Viktor
GRUPPE „Pferde"
(Zwei Pferde) (auf flachem Rechtecksockel)
ca. 1933 (für den Entwurf)

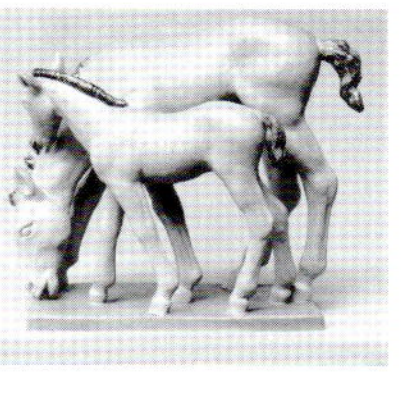

7939,18,9 ger. ND, gem. H, Firmenmarke 19, Made in Austria, sign. Matula, H. 31,5 cm; Wien Museum Inv. 190.033
7939,20 Firmenmarke Schuster 1, Made in Germany, unsign., Ausformung zw. 1941–1944, H. 31 cm; KH
7939M, H. 31,5 cm, B. 35 cm; GSNachf. 3,13; Fotoarchiv K. Täuber, hier mit der Ang. des Originalpreises RM 66.–
–> vgl. 6797; 7386; 7737= jeweils EinzelFIGUR (rückwärts), als Höhenvarianten.
Lit.: GSNachf. 3,13 (m. Abb.), hier mit Ang. des Originaltitels GRUPPE „Pferde".
Quelle: AGM (Kontoblatt der Wiener Manufaktur Josef Schuster für die Firma Porslins Ericsson-Lund, Stockholm?, Schweden vom 7. Januar 1943), hier mit Ang. des Originaltitels GRUPPE „Pferde"; Bestellungen der Firma Wahliss (Wien) vom 9. und 24.11.1945 an die „Arbeitsgemeinschaft der Firma Friedrich Goldscheider"; Bestellung der Firma Wahliss (Wien) vom 27.2.1946 in Ausführung „Terrakotta"; Fotoarchiv K. Täuber.

7941
Skifahrerin
ca. 1937
7941,254,14 (Firmenmarke ohne Ang.), H. 27,5 cm;
Doro 28.11.1989
Bem.: Bei diesem Modell könnte es sich um eine Höhenvariante zu den Skifahrerinnen Goebels (oder Dakons) handeln. Da sich im Auktionskat. weder eine nähere Beschreibung noch eine Abb. findet, lässt sich keine schlussendliche Aussage treffen.

7943M
Goebel, Kurt
ASCHER „Ascher mit Kopf"
Form: rund (mit Standfläche für den Kopf), vier Zigarettenablagen
Dekor: Frauenkopf (vollplastisch)
ca. 1937
7943M, H. 13,5 cm, B. 19 cm; GSNachf. 4,4c
–> vgl. 7944; 7945; 7946= jeweils ähnl. Modell.
Lit.: GSNachf. 4,4c (m. Abb.), hier mit Ang. des Originaltitels „Ascher mit Kopf".
Quelle: AGM (Rechnung an Maurice Mizrahi, Alexandria, Ägypten vom 22. April 1953).

7944M
Goebel, Kurt
ASCHER „Ascher mit Kopf"
Form: rund (mit Standfläche für den Kopf), drei Zigarettenablagen
Dekor: Knabenkopf (vollplastisch)
ca. 1937
7944M, H. 14 cm, B. 17,5 cm; GSNachf. 4,4a
–> vgl. 7943; 7945; 7946= jeweils ähnl. Modell.
Lit.: GSNachf. 4,4a (m. Abb.), hier mit Ang. des Originaltitels „Ascher mit Kopf".

7945M
Goebel, Kurt
ASCHER „Ascher mit Kopf"
Form: rund (mit Standfläche für den Kopf), vier Zigarettenablagen
Dekor: Mädchenkopf (vollplastisch)
ca. 1937

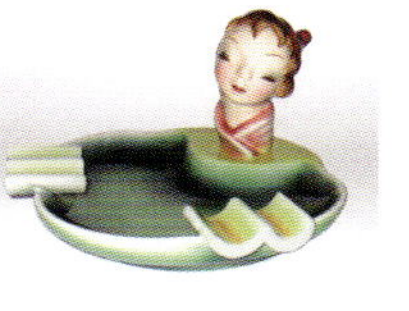

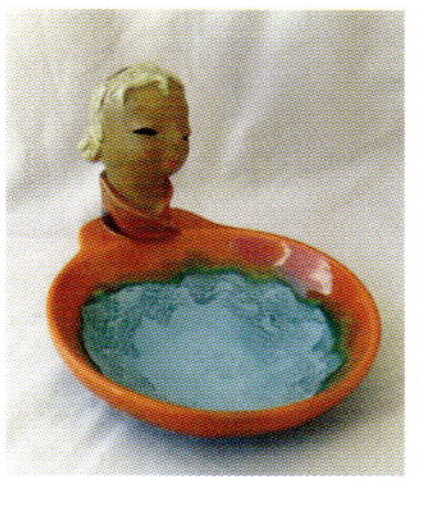

7945,17 ger. X, gem. lig. HS, gem. rot 6., Firmenmarke 19, Made in Austria, unsign., Ausformung zw. 1945–1953, H. 12 cm, Ø 23 cm; Sammlung a@goldscheider.net
7945,37 Firmenmarke 19, Made in Austria, unsign., Ausformung zw. 1945–1953, H. 12,5 cm, Ø 16 cm; Sammlung Stressless, Portugal;
Bem.: Dieses Exemplar wurde ohne die Zigarettenablagen ausgeführt und weicht somit von der im GSNachf. gezeigten Variante als Ascher ab.
7945M, H. 13 cm, B. 19 cm; GSNachf. 4,4b
–> vgl. 7943; 7944; 7946= jeweils ähnl. Modell.
Lit.: GSNachf. 4,4b (m. Abb.), hier mit Ang. des Originaltitels „Ascher mit Kopf".

7946
Goebel, Kurt
ASCHER Mädchenkopf
Form: rund, drei Zigarettenablagen
Dekor: BÜSTE Mädchenkopf mit Schleife/Mascherl im Haar (vollplastisch)
ca. 1937
7946,8 gem. lig. TK, Firmenmarke 19, Made in Austria, unsign., Ausformung zw. 1945–1953, H. 13,5 cm, Ø 15 cm; eBay # 2033908260

7946,37 gem. lig. TK, Firmenmarke 19, Made in Austria, unsign., Ausformung zw. 1945–1953, H. 14 cm, Ø 16,5 cm; eBay # 208125987, Doro 10.3.2003
Bem.: Der Mädchenkopf wurde, quasi als Mehrfachverwertung eines Entwurfes, auch als Büste auf Holzsockel angeboten (–> vgl. Soth. L. 15.2.1985).
–> vgl. 7943; 7944; 7945= jeweils ähnl. Modell.

7947
Knörlein, Rudolf

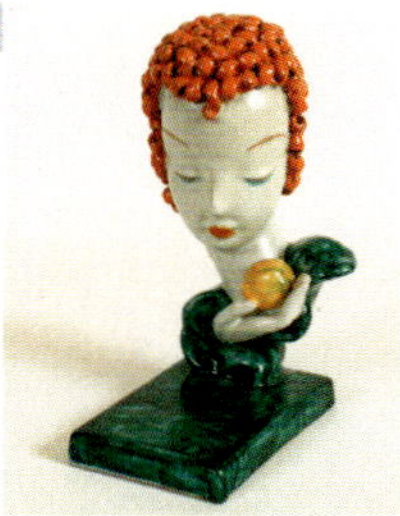

STEHMASKE „Eva" (Frauenkopf mit Hand und Apfel) (auf Rechtecksockel)
ca. 1933 (für den Entwurf)
7947,S ger., Firmenmarke 19, Made in Austria, unsign., Ausformung zw. 1945–1953, H. 22 cm; Sammlung.schmutz@chello.at 79
Bem.: Ausst., Lit. s. 6774
–> vgl. 6774
Lit.: Pinhas, S: 164 (m. Abb.), hier mit Ang. der falschen Mod.-Nr. 6947 (H. 25 cm)
Quelle: Privatarchiv (Originalfoto).
GS neu: Modell-Nr. 9 „Die Auserwählte" (H. 40 cm), Aufl. ohne Ang.

7949M
Matula, Viktor
ASCHER „Ascher mit Gänsen"
Form: rund, eine Zigarettenablage
Dekor: zwei Gänse (vollplastische Figuren)
ca. 1937
7949M, H. 10 cm, B. 22 cm; GSNachf. 4,3b
Lit.: GSNachf. 4,3b (m. Abb.), hier mit Ang. des Originaltitels „Ascher mit Gänsen".
Quelle: Nachlass Matula, Wien

7950
WANDMASKE Frauenkopf mit Perlenkette
ca. 1937
7950,33,17 Firmenmarke 19, Made in Austria, H. 23,2 cm; Wien Museum Inv. 220.803

7957F
Dakon, Stephan

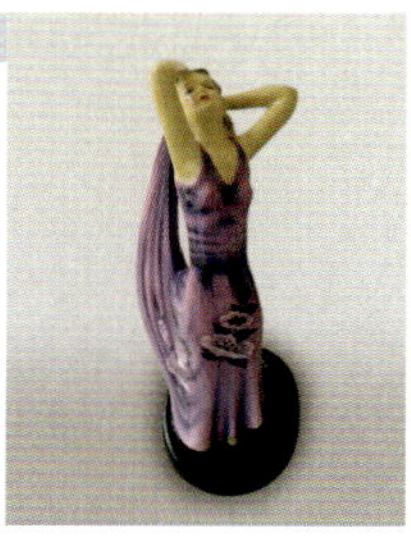

FIGUR „Dolly" (Stehend, mit langem Kleid, in Tanzpose, rechte Figur) (auf Ovalsockel)
ca. 1937
7957,21,7 ger. XIIA, gem. AL, SG, Firmenmarke 19, Made in Austria, H. 33 cm; Sammlung Stressless, Portugal
7957F, H. 34,5 cm, B. 16 cm; GSNachf. 1,23e

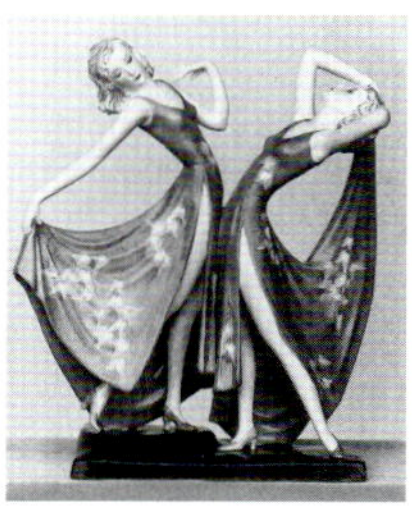

–> vgl. 7992
–> vgl. 7868; 7991= jeweils GRUPPE
–> vgl. 7958; 7993= jeweils linke EinzelFIGUR/Pendant
Lit.: GSNachf. 1,23 (m. Abb.), hier mit Ang. des Originaltitels „Dolly" (rechte Figur).

7958F
Dakon, Stephan

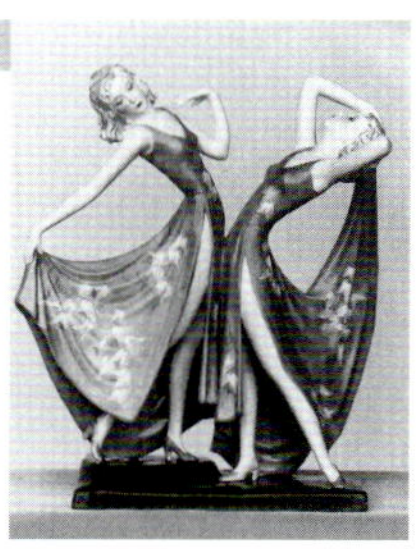

FIGUR „Dolly" (Stehend, mit langem Kleid, in Tanzpose, linke Figur) (auf Ovalsockel)
ca. 1937
7958,33,4 ger. XII, gem. lig. AS, Firmenmarke 19, Made in Austria, H. 32 cm; eBay # 6504484364;
Bem.: Besonderheit an diesem Exemplar: ein Rosenstrauß in der linken Hand.
7958,69,2 gem. lig. JK (oder JT?), AS, Firmenmarke 19, Made in Austria, H. 33 cm; eBay # 513379469; (KH www.tias.com/stores/frozenbeauty)
7958,128,14 Firmenmarke 19, Made in Austria, H. 33 cm; WKA 38
7958F, H. 33 cm, B. 17,5 cm; GSNachf. 1,23c
–> vgl. 7993
–> vgl. 7868; 7991= jeweils GRUPPE
–> vgl. 7957; 7992= jeweils rechte EinzelFIGUR/Pendant
Lit.: GSNachf. 1,23 (m. Abb.), hier mit Ang. des Originaltitels „Dolly" (linke Figur).

7960F
Dakon, Stephan
Sich verbeugende Tänzerin (mit ausgestreckten Armen) (auf Ovalsockel)
ca. 1937
7960, H. 33 cm, B. 28,5 cm
Quelle: AGM (Rechnung der Wiener Manufaktur Josef Schuster an Laszlo Bösze (Szeged, Ungarn) vom 5. August 1944), hier unter dem Titel „Tanz" geführt; Archiv Mimi Dakon (Originalfoto)

7961F
Dakon, Stephan

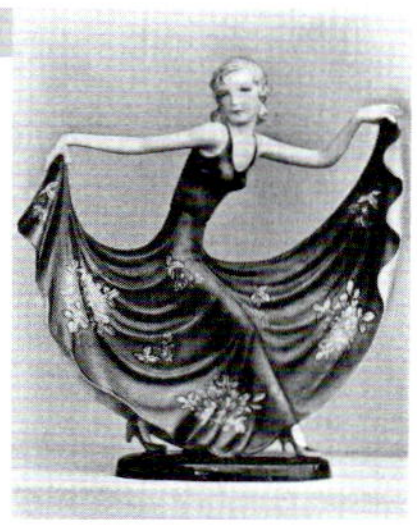

FIGUR „Arabella" (Stehende junge Dame, ihr langes Kleid schmetterlingsflügelartig auseinander haltend) (auf Ovalsockel)
ca. 1937
7961,45,31 gem. F., Firmenmarke 19, Made in Germany, Ausformung zw. 1938–1941, H. 34,5 cm; WKA 35; Doro 15.5.2002
7961,55,2 Firmenmarke 19, Made in Germany, unsign., Ausformung zw. 1938–1941, H. 34 cm; Doro 8.3.1994
7961F, H. 33 cm, B. 29 cm; GSNachf. 1,20b
–> vgl. 7776; 8317
Lit.: GSNachf. 1,20 (m. Abb.), hier mit Ang. des Originaltitels „Arabella".
Quelle: Archiv Mimi Dakon (Tantiemenabrechnung 1.7.–31.12.1937), hier wird die Figur unter dem Titel „Tanz" geführt; (Tantiemenabrechnung 1.1.–30.6.1940), hier wird die Figur unter dem Titel „Tänzerin" geführt.

7962
Goebel, Kurt
GRUPPE „Rendez-vous" (Stehender Bursche mit Mädchen) (auf flachem Rechtecksockel mit Hund)
ca. 1937
7962 (Firmenmarke ohne Ang.), unsign., H. 28 cm; Pinhas, S. 177 (m. Abb.)
–> vgl. 7844; 7845= jew. EinzelFIGUREN

7963F
Herczeg, Claire (Klára)

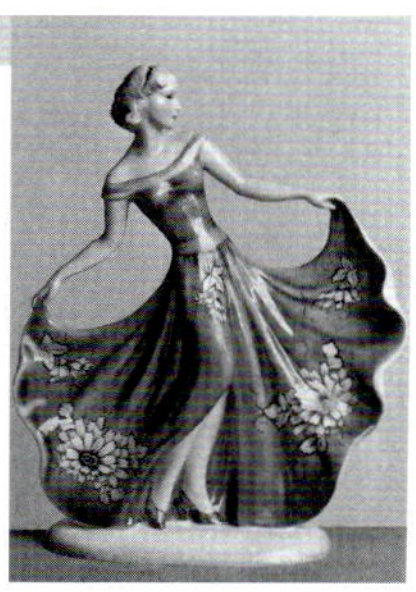

FIGUR „Lydia" (Schreitende junge Tänzerin, ihr langes Kleid auseinander haltend) (auf Ovalsockel)
ca. 1937
7963,7,11 gem. SG., Firmenmarke 19, Made in Austria, unsign.?, H. 32,4 cm; eBay # 1226585015
7963,65,11,47 gem. H., SG, Firmenmarke 19, Made in Germany, sign. Claire Herczeg, Ausformung zw. 1938–1941, H. 34 cm; Soth. com. # 42OT
7963F, H. 32 cm, B. 24 cm; GSNachf. 1,22b
–> vgl. 7855; 8322
Lit.: GSNachf. 1,22 (m. Abb.), hier mit Ang. des Originaltitels „Lydia".
Quelle: Fotoarchiv K. Täuber

7965
Dakon, Stephan

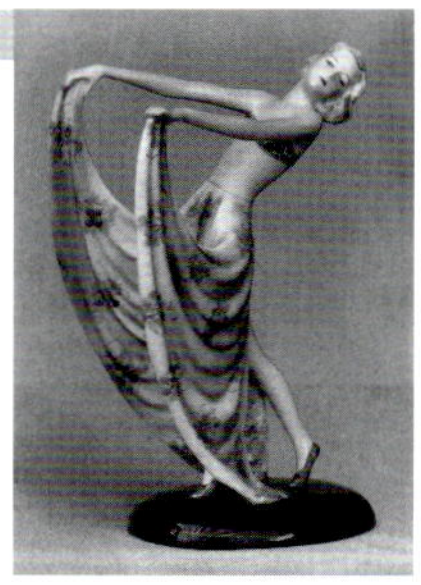

Schreitende Tänzerin mit Bikinioberteil, ihren langen Rock nach vorne weg haltend (auf Ovalsockel)
ca. 1937
7965, H. 20 cm, B. 14 cm
Quelle: AGM (Rechnung der Wiener Manufaktur Josef Schuster an Laszlo Bösze (Szeged, Ungarn) vom 5. August 1944), hier unter dem Titel „Tanz" geführt; Archiv Mimi Dakon (Originalfoto)

7966F
Weiss, Claire (Klára)
(d.i. Herczeg, Claire (Klára)

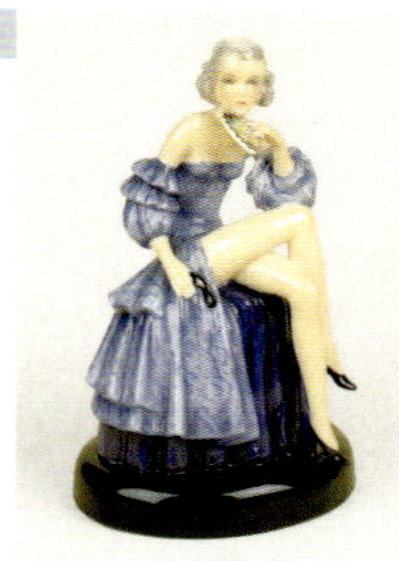

„Ballpause" (Sitzende Tänzerin mit Ballmaske und Fächer) (auf Rundsockel)
ca. 1937
7966,10,6 gem. SG, gem. lig. NO, Firmenmarke 19, Made in Austria, sign. Claire Weiss, H. 28 cm; PB Wien 19
7966,78,4 Firmenmarke 19, Made in Germany, sign. Dakon, Ausformung zw. 1938–1941, H. 27,5 cm; Doro 30.11.1999; Bem.: Entweder ist die Ang. der Sign. „Dakon" im Auktionskat. falsch, oder es handelt sich um ein Versehen des Modelleurs, der bei der Ausarbeitung die falsche Sign. eingeprägt hat.
7966,A Firmenmarke 19, Made in Austria, unsign., Ausformung zw. 1945–1953, Pick. 1, H. 27 cm; KH Ruby Lane (Jane Perlow/Lacremeantik, Great Neck N.Y.)
Quelle: AGM (Rechnung an Maurice Mizrahi, Alexandria, Ägypten vom 22. April 1953), hier mit Ang. des Originaltitels „Ballpause".

7971
„Akt mit Krug"
ca. 1937
7971 ohne Ang., H. fehlt; AGM
Quelle: AGM (Bestellung der Firma Wahliss (Wien) vom 27.2.1946 an die „Arbeitsgemeinschaft der Firma Friedrich Goldscheider", hier mit Ang. des (Original?-)Titels „Akt mit Krug", in Ausführung „Terrakotta").

7972F
Ullmann, Robert

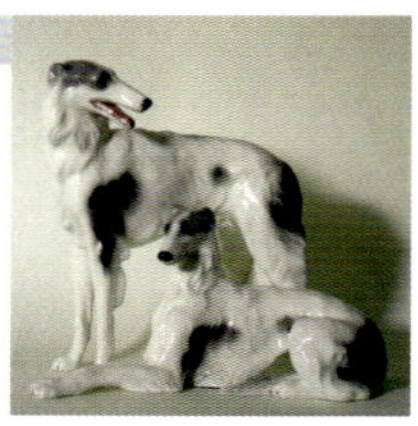

GRUPPE „Windhunde" (Ein stehender und ein liegender Barsoi)
ca. 1934/35 (für den Entwurf)
7972,3 (Firmenmarke ohne Ang.), unsign., Ausformung zw. 1945–1953, L. 25 cm; eBay # 2500863709
7972,44,71 ger., gem. lig. HS, Firmenmarke 19, Made in Austria, unsign., Pick. 5/a, H. 11 cm, L. 24 cm; Sammlung Alexandra Herrmann-Goldscheider
7972F, H. 17,5 cm, B. 23 cm; GSNachf. 5,4c
Bem.: zum Künstler s. 7054AF; zur Modellnummernvergabe s. 6761.
–> vgl. 7054F; 7571F
–> vgl. 7054AF; 7571AF; 7972AF= jeweils stehende EinzelFIGUR
–> vgl. 7054BF; 7571BF; 7972BF= jeweils liegende EinzelFIGUR
Lit.: GSNachf. 5,4 (m. Abb.), hier mit Ang. des Originaltitels GRUPPE „Windhunde".
Quelle: AGM (Bestellungen der Firma Wahliss (Wien) vom 9. und 24.11.1945 an die „Arbeitsgemeinschaft der Firma Friedrich Goldscheider"; Bestellung der Firma Wahliss (Wien) vom 27.2.1946 in Ausführung „Terrakotta").

7972AF
Ullmann, Robert
FIGUR „Windhund" (Stehender Barsoi)
ca. 1934/35 (für den Entwurf)
7972AF, H. 17 cm, B. 17 cm; GSNachf. 5,4f
Bem.: zum Künstler s. 7054AF; zur Modellnummern-

vergabe s. 6761.
-> vgl. 7054AF; 7571AF
-> vgl. 7054F; 7571F= jeweils GRUPPE
Lit.: GSNachf. 5,4 (m. Abb.), hier mit Ang. des Originaltitels FIGUR „Windhund" (stehend).
Quelle: AGM (Rechnung der Wiener Manufaktur Josef Schuster an Laszlo Bösze (Szeged, Ungarn) vom 5. August 1944), hier mit Ang. des Originaltitels „Windhund".

7972BF
Ullmann, Robert
FIGUR „Windhund" (Liegender Barsoi)
ca. 1934/35 (für den Entwurf)

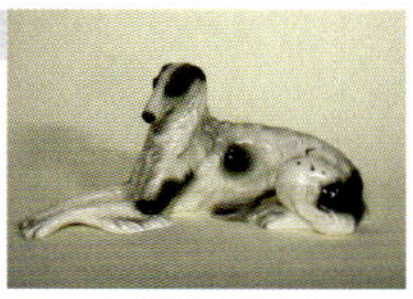

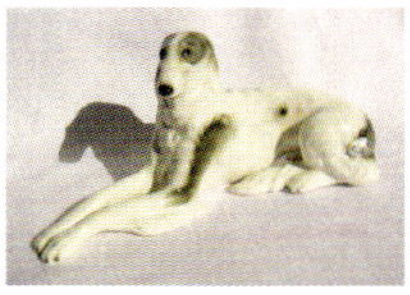

7972,3 ger. B, Firmenmarke 19, Made in Austria, unsign., H. 10,1 cm, B. 22,9 cm; MAK Wien, Inv. Ke 10602-21
7972BF, H. 10 cm, B. 23 cm; GSNachf. 5,4i
Bem.: zum Künstler s. 7054AF; zur Modellnummernvergabe s. 6761.
-> vgl. 7054BF; 7571BF
-> vgl. 7054F; 7571F= jeweils GRUPPE
Lit.: GSNachf. 5,4 (m. Abb.), hier mit Ang. des Originaltitels FIGUR „Windhund" (liegend).

7974M
Matula, Viktor
RELIEF „Fisch" (Wandrelief)
ca. 1937
7974M, H. 10 cm, B. 10 cm; GSNachf. 2,22c
Lit.: GSNachf. 2,22 (m. Abb.), hier mit Ang. des Originaltitels „Relief Fisch".
Quelle: Nachlass Matula, Wien

7975/M
Prischl, Adolf
„Akt mit Draperie" (Stehender Damenakt, ein langes Tuch mit den Armen hochhaltend) (auf Rundsockel)
ca. 1937

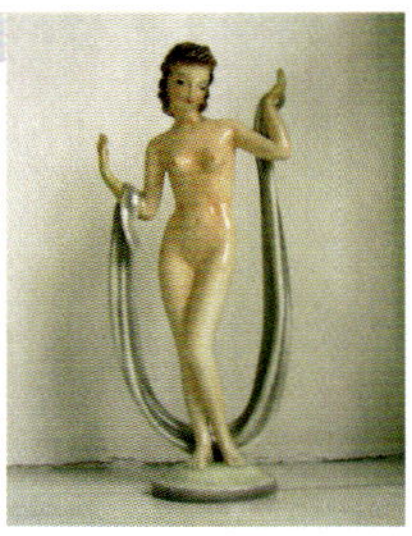

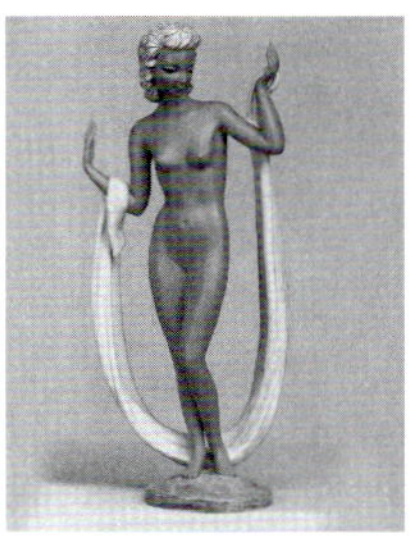

7975,6 gem. G, Firmenmarke 19, unsign., Ausformung zw. 1945–1953, H. 36,6 cm; Privatsammlung, Wien
7975, H. 35,5 cm, B. 15 cm
Quelle: AGM (Bestellung der Firma Wahliss (Wien) vom 27.2.1946 an die „Arbeitsgemeinschaft der Firma Friedrich Goldscheider", hier mit Ang. des (Original?-) Titels „Akt mit Draperie", in Ausführung „Terrakotta"; Rechnung an Maurice Mizrahi, Alexandria, Ägypten vom 22. April 1953, geliefert in Ausführung „matt"); Archiv Matschiner (Originalfoto); Fotoarchiv K. Täuber.

7977
Goebel, Kurt
Stehendes Mädchen mit Schürze und Schal (Matrosenbraut) (auf Rundsockel)
ca. 1937
7977,0 gem. O, Firmenmarke 19, Made in Austria, unsign., H. 22 cm; eBay # 6174196605
7977,3 Firmenmarke 19, Made in Austria, unsign., Ausformung zw. 1945–1953, H. 21,5 cm; eBay # 2629173522
7977,4 Firmenmarke 19, Made in Austria, unsign., Ausformung zw. 1945–1953, H. 20,3 cm; eBay # 343431210
7977,25 ger., gem. L., Firmenmarke 19, Made in Austria, unsign., Ausformung zw. 1945–1953, H. 20,3 cm; eBay # 343431210
-> vgl. 7960= GRUPPE
Pendant -> vgl. 7859

7978M
Goebel, Kurt
FIGUR „Romeo" (Stehender junger Mann mit Hut und Kaktus) (auf Rundsockel)
ca. 1936/37 (für den Entwurf)
7978M, H. 19,5 cm, B. 8 cm; GSNachf. 3,9b
-> vgl. 7656M; 7912M
Lit.: GSNachf. 3,9 (m. Abb.), hier mit Ang. des Originaltitels FIGUR „Romeo".

7980M
Matula, Viktor
RELIEF „Pferd" (Wandrelief)
ca. 1937
7980,9 gem. R, Firmenmarke 19, Made in Germany, unsign., Ausformung zw. 1938–1941, H. 14,5 cm, B. 12 cm; eBay # 913479973
7980M, H. 14 cm, B. 12 cm; GSNachf. 2,22b
Lit.: GSNachf. 2,22 (m. Abb.), hier mit Ang. des Originaltitels „Relief Pferd".
Quelle: Nachlass Matula, Wien

7981
Matula, Viktor
FIGUR Pelikan
ca. 1937
7981 H. fehlt
Quelle: Nachlass Matula, Wien

7982
Prischl, Adolf
STEHKOPF Mädchenkopf mit aufrecht stehender Hand (auf Ovalsockel)
ca. 1937

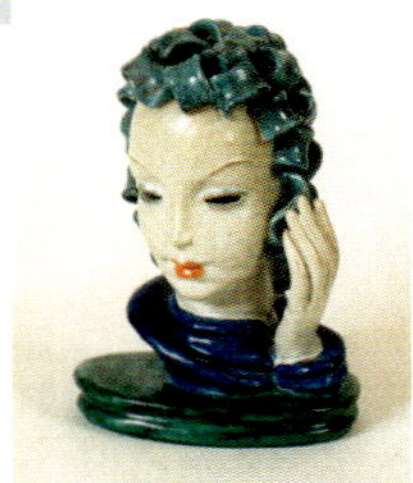

7982,3,N/16 ger., gem. grün N, Firmenmarke 19, Made in Austria, unsign., H. 26 cm; Sammlung.schmutz@chello.at 66
Lit.: N.N.: Neue Keramik, in: ÖK, 8.1937, Nr. 7, S. 22 (m. Abb.) „Auch der Keramiker Adolf Prischl, den wir zum ersten Male zur Veröffentlichung bringen, ist eine Begabung, die entwicklungsfähig ist und eigene Wege finden wird. Die Verbindung der Terrakotta mit farbigen Partien gibt den Arbeiten besonderen Reiz".

7983
Prischl, Adolf (zugeschr.)
Stehender Mädchenakt mit Tuch (auf Ovalsockel)
ca. 1937
7983 ger., Firmenmarke 19, unsign., H. 31,1 cm; Doro 4.3.2002

7985F
Prischl, Adolf
RELIEF „Madonna" (Madonna mit gekreuzten Händen) (Wandrelief)
ca. 1937/38
ohne Ang., gem. grün N, Firmenmarke 19, Made in Austria, unsign., H. 27,5 cm, B. 21 cm; Fischer Luz. 328
7985F, H. 28 cm, B. 21,5 cm; GSNachf. 6,2
-> vgl. 7551; 8245= jeweils BÜSTE
Bem.: Der Entwurf der Büste und somit der Grundgedanke zu dieser Madonnendarstellung geht auf das Jahr 1936 zurück, das Relief wurde wohl erst ab ca. 1937/38 bei Goldscheider produziert.
Lit.: GSNachf. 6,2 (m. Abb.), hier mit Ang. des Originaltitels RELIEF „Madonna".

7987
Jarl, Karin
GRUPPE Zwei Barsois (auf Ovalsockel)
ca. 1937/38
7987,46,14 gem. T., Firmenmarke 19, Made in Austria, H. ca. 20 cm, L. 26 cm; eBay # 1472143028
7987,22 Firmenmarke 19, Made in Germany, unsign., Ausformung zw. 1938–1941, Pick. 1, H. 19,5 cm; Doro FVK

7988 s. 7998

7991F
Dakon, Stephan
GRUPPE „Dolly Sisters" (Stehend, mit langen Kleidern, in Tanzpose) (auf Ovalsockel)
ca. 1937

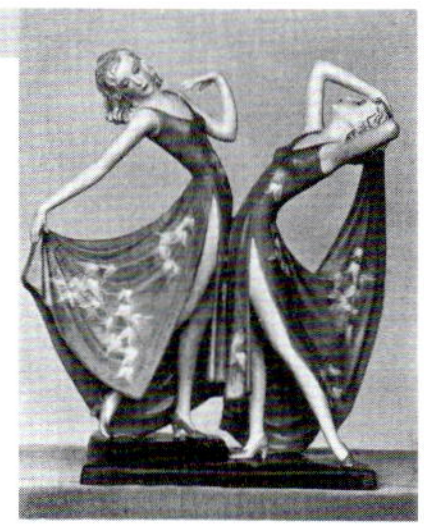

7991F, H. 30 cm, B. 30 cm; GSNachf. 1,23b
-> vgl. 7868= GRUPPE
-> vgl. 7957; 7992= jeweils rechte EinzelFIGUR
-> vgl. 7958; 7993= jeweils linke EinzelFIGUR
Lit.: GSNachf. 1,23 (m. Abb.), hier mit Ang. des Originaltitels „Dolly Sisters".

7992F
Dakon, Stephan
FIGUR „Dolly" (Stehend, mit langem Kleid, in Tanzpose, rechte Figur) (auf Ovalsockel)
ca. 1937

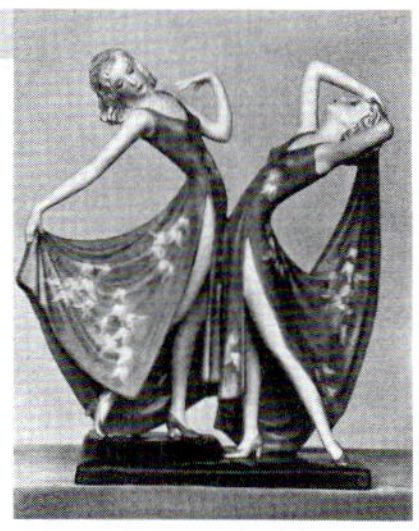

7992F, H. 30 cm, B. 14 cm; GSNachf. 1,23f
-> vgl. 7957
-> vgl. 7868; 7991= jeweils GRUPPE
-> vgl. 7958; 7993= jeweils linke EinzelFIGUR/Pendant
Lit.: GSNachf. 1,23 (m. Abb.), hier mit Ang. des Originaltitels „Dolly" (rechte Figur).

7993F
Dakon, Stephan
FIGUR „Dolly" (Stehend, mit langem Kleid, in Tanzpose, linke Figur) (auf Ovalsockel)
ca. 1937

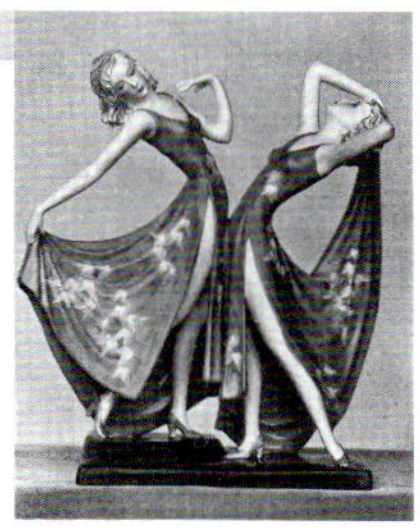

7993F, H. 29 cm, B. 17 cm; GSNachf. 1,23d
-> vgl. 7958
-> vgl. 7868; 7991= jeweils GRUPPE
-> vgl. 7957; 7992= jeweils rechte EinzelFIGUR/Pendant
Lit.: GSNachf. 1,23 (m. Abb.), hier mit Ang. des Originaltitels „Dolly" (linke Figur).

7996F
Dakon, Stephan
FIGUR/LAMPE „Beauty" (Tänzerin, nach links geneigt, ihr Kleid flügelartig zurückhaltend) (auf Ovalsockel)
ca. 1935 (für den Entwurf)

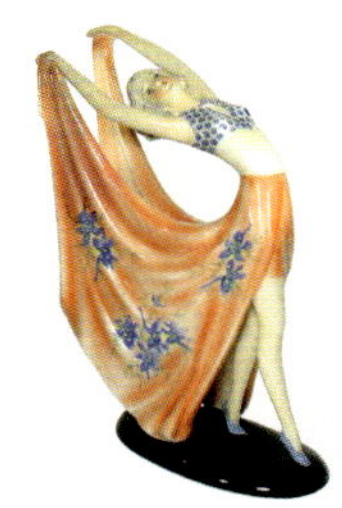

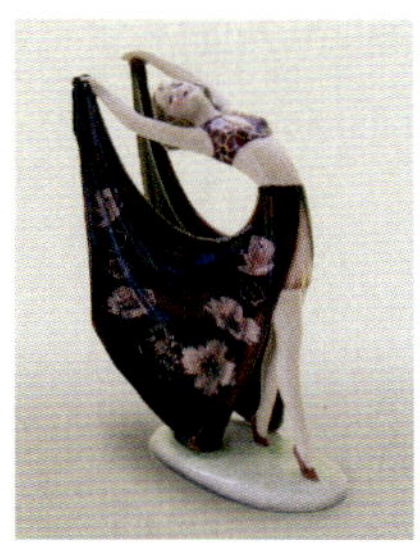

FIGUR
7996,3,7 gem. W, FP, (ohne Firmenmarke), Made in Austria, Hand dec., H. 27 cm; eBay # 293124854
7996,6 Firmenmarke 19, H. 26,7 cm; Doro 20.4.1993; WKA 38
7996,69,6 ger. XII, gem. lig. AS, Firmenmarke 19, Made in Austria (zwei mal), H. ca. 27 cm; city-antik
7996,170,7 Firmenmarke 19, H. 26,5 cm; BM 68
7996,317,13 gem. S., 4.1, Firmenmarke 19, Made in Austria, H. 27 cm; Sammlung a@goldscheider.net
7996,S gem. E, W., Z, Firmenmarke 19, Made in Austria, unsign., Ausformung zw. 1945–1953, H. 25 cm; Sammlung f@goldscheider.de
7996F, H. 27 cm, B. 19,5 cm; GSNachf. 1,12b; 1,50b
LAMPE
(7996, Nummern durch aufmontierten Sockel verdeckt), unsign., Pick. ohne Ang., H. der Figur 28 cm (Elektrifizierung fehlt); eBay # 52510583
7996F, H. 49 cm, B. 39 cm; GSNachf. 1,12d; 1,50d
Bem.: Dieses Modell gibt es in zwei Varianten, einmal neigt sich der Körper der Tänzerin wie in diesem Fall nach links, das andere Mal nach rechts.

-> vgl. 7195; 7897; 8746
Lit.: GSNachf. 1,12 (m. Abb. als Figur), hier mit Ang. des Originaltitels „Beauty"; GSNachf. 1,50 (m. Abb. als Lampe montiert, komplett, mit Seidenschirm), hier mit Ang. des Originaltitels „Beauty".
Quelle: AGM (Rechnung der Wiener Manufaktur Josef Schuster an Laszlo Bösze (Szeged, Ungarn) vom 5. August 1944; Rechnung an Maurice Mizrahi, Alexandria, Ägypten vom 22. April 1953), hier jeweils mit Ang. des Originaltitels „Beauty".

7997
Knörlein, Rudolf (zugeschr.)
WANDMASKE Frauenkopf mit kurzem Kragenansatz (nach rechts blickend)
ca. 1937
7997,3 gem. R, Firmenmarke 19, Made in Germany, Ausformung zw. 1938–1941, unsign., H. 30,5 cm; eBay # 3769471218
7997,71 gem. MT, Firmenmarke 19a, unsign., Ausformung zw. 1945–1953, H. 30,5 cm; eBay # 3775005608

7998
Knörlein, Rudolf (zugeschr.)
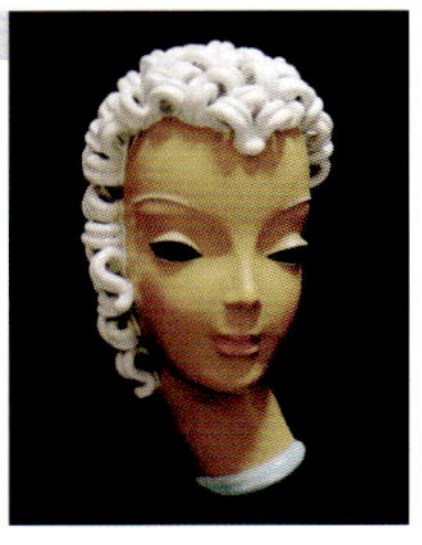
WANDMASKE Frauenkopf mit kurzem Kragenansatz (nach links blickend)
ca. 1937
7998,3 Firmenmarke 19, Made in Austria, unsign., Ausformung zw. 1945–1953, H. 25,2 cm, B. 14,1 cm; PB Deutschland
Bem.: Diese Maske wurde auch noch zw. 1953–1960 bei Carstens (GWM West-Germany) produziert.
Lit.: Könecke (2002), S. 76 (hier falsche Ang. der Modell-Nr. 7988), 87 (m. Abb.).
Quelle: AGM (Rechnung an Maurice Mizrahi, Alexandria, Ägypten vom 22. April 1953).

7999
Knörlein, Rudolf (zugeschr.)
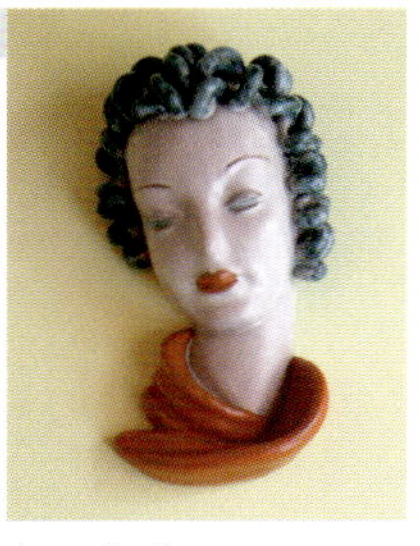
WANDMASKE Frauenkopf mit Halstuch / Schal
ca. 1937
7999,3 gem. B, Firmenmarke 19, Made in Germany, unsign., Ausformung zw. 1938–1941, H. 23 cm; Sammlung Irsigler, Wien
7999,3 gem. B, Firmenmarke 19, Made in Austria, unsign., Ausformung zw. 1945–1953, H. 23,5 cm; BM 50

8003
Weiss, Claire (Klára) (d.i. Herczeg, Claire (Klára)
SCHALE Gänseliesel (Stehendes kleines Mädchen mit Gans)
Form: rund
Dekor: Mädchen und Gans (vollplastische Fig.)
ca. 1937
8003,44,15 ger. B., Firmenmarke 19, Made in Austria, unsign., H. 18,5 cm; eBay # 7393005756
Bem.: Die Figur des Mädchens wurde auch beim Modell 7027 verwendet. Dort als „Rotkäppchen".
-> vgl. 7847; 7861

8006F
Dakon, Stephan
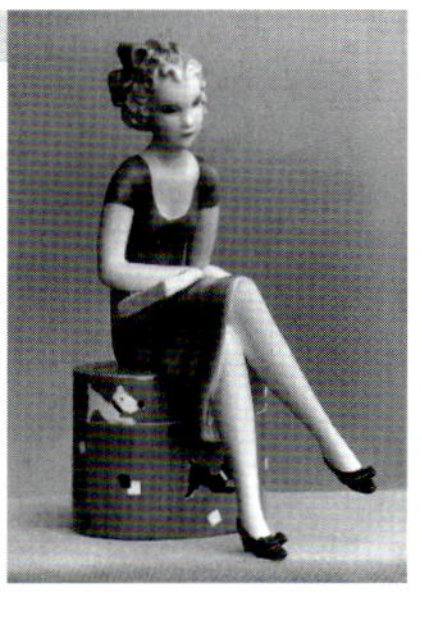
FIGUR „Lektüre" (Sitzende junge Dame mit Buch) (auf ovalem Sockel, Dekor: Fox-Terrier)
ca. 1937
8006,4,19 gem. F., gem. blau B., Firmenmarke 19, Made in Austria, Pick. 5, H. ca. 25 cm; eBay # 905615416
8006,10,31 Firmenmarke 19, Made in Germany, Ausformung zw. 1938–1941, H. 24,8 cm; eBay # 2565633313
8006F, H. 24,5 cm, B. 18 cm; GSNachf. 1,24
Lit.: GSNachf. 1,24 (m. Abb.), hier mit Ang. des Originaltitels „Lektüre".
Quelle: Archiv Mimi Dakon (Originalfoto)

8019F
GRUPPE „Rehe" (Rehkuh mit Kitz) (auf Ovalsockel)
ca. 1937
8019,32 Firmenmarke 19, Made in Austria, Ausformung zw. 1945–1953, H. 31,7 cm; Doro 26.8.1998
Quelle: AGM (Rechnung an Maurice Mizrahi, Alexandria, Ägypten, vom 22. April 1953), hier mit Ang. des Originaltitels GRUPPE „Rehe".

8020F / M
Matula, Viktor
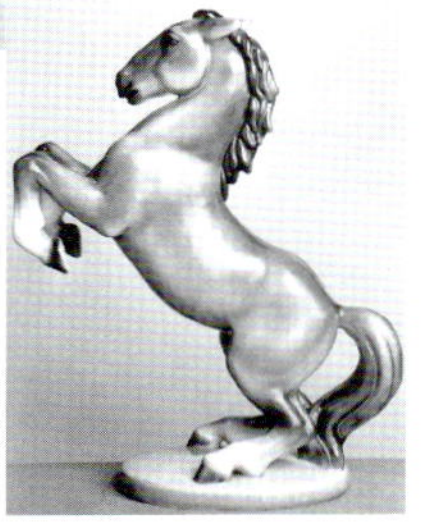
FIGUR „Pferd", aufbäumend (Steigendes Pferd) (auf Ovalsockel)
ca. 1937
8020,20 gem. blau N, Firmenmarke 19, Made in Germany, unsign., Ausformung zw. 1938–1941, H. 41 cm; Wien Museum Inv. 190.052
8020,20 gem. T, Firmenmarke 19, Made in Germany, unsign., Ausformung zw. 1938–1941, H. 41 cm; Wendl 47
8020,65 gem. blau N, Firmenmarke 19, Made in Austria, unsign., Ausformung zw. 1945–1953, H. 40 cm; eBay # 1446419529
8020,12 Firmenstempel WG 1, unsign., Ausformung 50er Jahre, H. 41 cm; eBay # 2222725811
8020M, H. 41,5 cm, B. 32 cm; GSNachf. 3,23; Fotoarchiv K. Täuber, hier mit der Ang.: H. 40 cm, B. 31 cm
Bem.: Diese Figur wurde auch noch zw. 1953–1960 bei Carstens (GWM West-Germany) produziert.
Lit.: GSNachf. 3,23 (m. Abb.), hier mit Ang. des Originaltitels FIGUR „Pferd", aufbäumend; Könecke (2002), S. 77, S. 88 (m. Abb.).
Quellen: AGM (Rechnung an Maurice Mizrahi, Alexandria, Ägypten, vom 22. April 1953), hier mit Ang. des Originaltitels „Pferd"; Nachlass Matula, Wien; Fotoarchiv K. Täuber.

8021F
„Wolfshund"
ca. 1937
Quelle: AGM (Kontoblatt der Wiener Manufaktur Josef Schuster für die Firma Porslins Ericsson-Lund, Stockholm?, Schweden vom 7. Januar 1943), hier mit Ang. des Originaltitels „Wolfshund".

8025
GRUPPE Zwei Kinder (auf Ovalsockel?)
ca. 1937
8025 ohne Ang., H. fehlt; Könecke (2002), S. 77, 88 (m. Abb.)
Bem.: Diese Figur wurde auch noch zw. 1953–1960 bei Carstens (GWM West-Germany) produziert.
Lit.: Könecke (2002), S. 76, 88 (m. Abb.), hier ohne weitere Ang.

8026
Dakon, Stephan
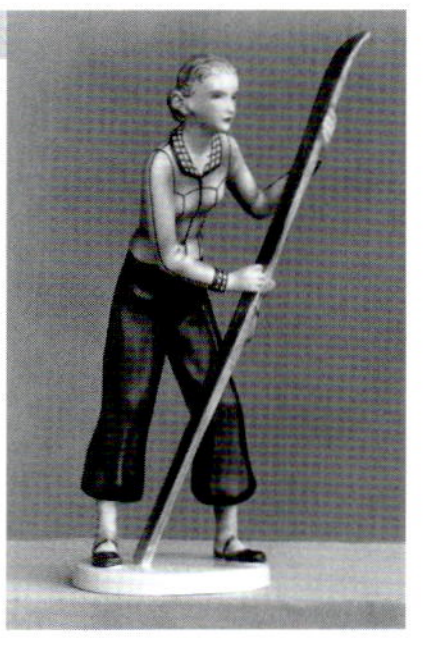
Junge Dame beim Skiwachsen (auf Ovalsockel)
ca. 1937
8026, H. 32,5 cm, B. 14 cm
Quelle: Archiv Mimi Dakon (Originalfoto)

8029
Weiss, Claire (Klára) (d.i. Herczeg, Claire (Klára)
Stehendes Mädchen mit Kappe, Schirm und Skyeterrier (auf Ovalsockel)
ca. 1937
8029,0 ger. 42, gem. B, T, Firmenmarke 19, Made in Austria, sign. Claire Weiss, H. 24,5 cm; GK, S. 23, 129 (m. Abb.); Wien Museum Inv. 222.556
Bem.: Der kleine Hund, der neben der Mädchenfigur steht, wurde von Hugo Postl für GS entworfen (-> vgl. 6116; 6503; 6802; 6975; 7300; 7519). Somit kann an dieser Stelle abermals belegt werden, dass GS bei den Zusammenstellungen von Gruppen die Entwürfe mehrerer Künstler kombinierte. Hugo Postl war zum Zeitpunkt der Edition durch GS lange tot, Klára Herczeg mit dieser Praxis durch die Manufaktur nicht immer einverstanden (-> vgl. Bem. bei 7327).
Ausst.: HMW (jetzt Wien Museum) 1985/86.
Lit.: GK, S. 23, 129 (m. Abb.)

8030F
Herczeg, Claire (Klára)
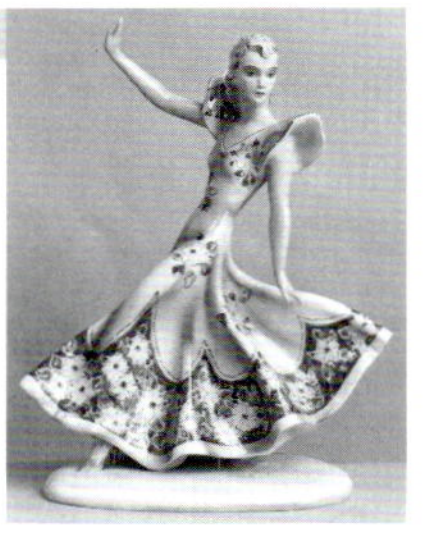
FIGUR „Allegretto" (Tänzerin in Pose) (auf Ovalsockel)
ca. 1937
8030,S,67 ger. HS, (Firmenmarke ohne Ang.), unsign., H. 35,5 cm; WKA 33
8030,1,6 Firmenmarke 19, Made in Austria, Hand dec., unsign., H. 34 cm; WKA 40
8030,8,67 gem. lig. HS, Firmenmarke 19, Made in Austria, unsign., H. 35 cm; WKH
8030,21,7 (Firmenmarke ohne Ang.), sign. Claire Weiss, H. 35 cm; Doro 31.5.2006
8030,22,21 Firmenmarke 19, Made in Austria, Hand dec., unsign., H. 36 cm; WKA 38
8030,90,2 ger. L, gem. lig. AS, Firmenmarke 26, sign. Claire Herczeg, H. ca. 35,5 cm; eBay # 1297802844
8030,297,2,1,12 Firmenstempel WG 1, unsign., Ausformung 50er Jahre, H. 35,5 cm; KNS 17
8030F, H. 35 cm, B. 29 cm; GSNachf. 1,25; Fotoarchiv K. Täuber, hier mit der Ang.: H. 33 cm, B. 27 cm
Lit.: GSNachf. 1,25 (m. Abb.), hier mit Ang. des Originaltitels „Allegretto".
Quelle: AGM (Rechnung an Maurice Mizrahi, Alexandria, Ägypten, vom 22. April 1953), hier unter dem Titel „Tanz" geführt; Fotoarchiv K. Täuber.

8032
Goebel, Kurt (zugeschr.)
WANDMASKE Frauenkopf mit langem Zopf und Kragen
ca. 1937
8032,9 gem. O, Firmenmarke 19, Made in Austria, unsign., Ausformung zw. 1945–1953, H. 30 cm; eBay # 2228841783

8033
Goebel, Kurt (zugeschr.)
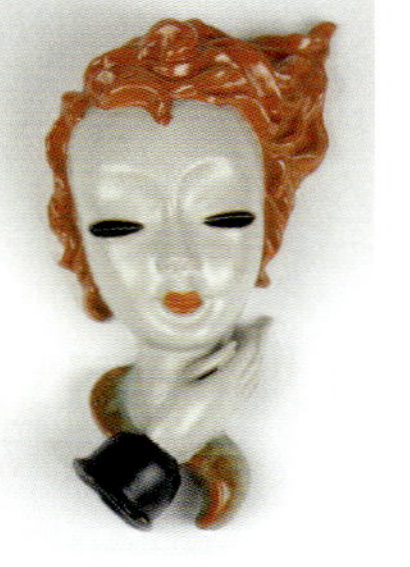
WANDMASKE Frauenkopf mit wehendem Haar und Hand
ca. 1937
8033,33 gem. O, Firmenmarke 19, Made in Austria, unsign., Ausformung zw. 1945–1953, H. 30 cm; PB Wien 19

8035
Goebel, Kurt (zugeschr.)
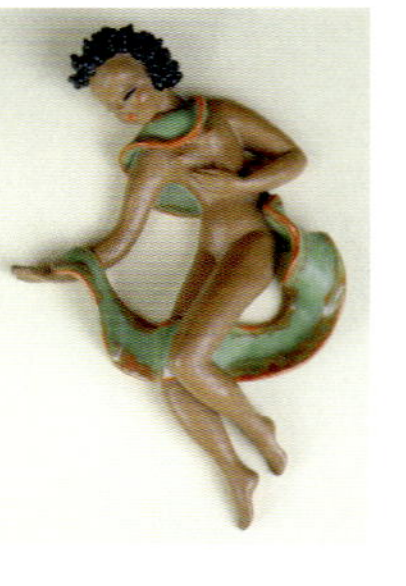
WANDRELIEF Mädchenakt mit langem Schal (Tuch)
ca. 1937
8035,26 Firmenmarke 19, Made in Austria, Ausformung zw. 1945–1953, H. 26 cm; PB Wien 19

8036
Goebel, Kurt
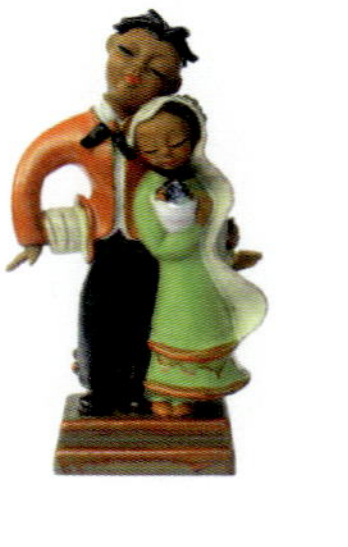
GRUPPE Brautpaar (Stehendes Paar auf Quadratsockel)
ca. 1937
8036,26 gem. L., Firmenmarke 19, Made in Austria, unsign., Ausformung zw. 1945–1953, H. 25 cm; eBay # 7385632080

8038M
Matula, Viktor
RELIEF „Reh" (Wandrelief)
ca. 1937
8038,7 Firmenmarke 19, Made in Austria, unsign., Ausformung zw. 1945–1953, H. 27 cm, L. 31 cm; eBay # 7332564035
Bem.: Wie das bislang einzig nachgewiesene Ex. zeigt, sind auf diesem Wandrelief zwei Rehe dargestellt. Ob auch eine Ausführung mit nur einem Reh hergestellt wurde, wie es der Originaltitel „Reh" vermuten lässt, konnte bis Redaktionsschluß nicht verifiziert werden.
Quellen: AGM (Kontoblatt der Wiener Manufaktur Josef Schuster für die Firma Porslins Ericsson-Lund, Stockholm?, Schweden vom 7. Januar 1943), hier mit Ang. des Originaltitels RELIEF „Reh"; Nachlass Matula, Wien.

8039M
Matula, Viktor
RELIEF „Klee" (Wandrelief)
ca. 1937
8039M, H. 10,5 cm, B. 10,5 cm; GSNacht. 2,22a
Lit.: GSNachf. 2,22 (m. Abb.), hier mit Ang. des Originaltitels „Relief Klee".
Quelle: Nachlass Matula, Wien

8040M
RELIEF „Akt liegend" (Liegender Damenakt mit Tuch und Blume) (Wandrelief)
ca. 1937
8040M, H. 17,5 cm, B. 33,5 cm; GSNachf. 2,25
Lit.: GSNachf. 2,25 (m. Abb.), hier mit Ang. des Originaltitels „Relief Akt liegend".

8041M
Prischl, Adolf
WANDMASKE
„Maske / Goldband"
(Frauenkopf mit Haarband und Halstuch)
ca. 1937

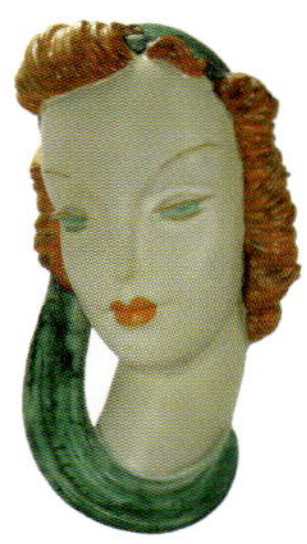

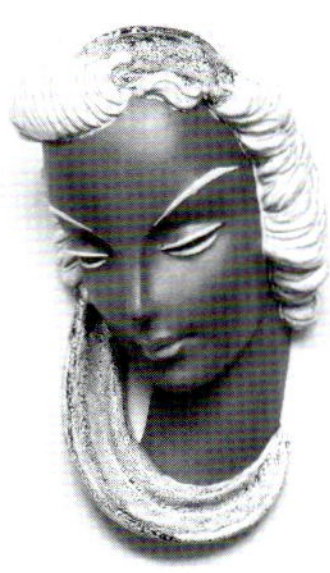

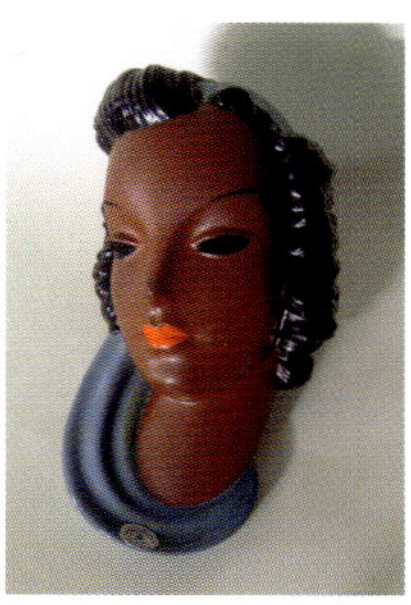

8041,1 Firmenmarke 19, Made in Austria, unsign., Ausformung zw. 1945–1953, H. 27 cm; city-antik
8041,3,20 Firmenmarke 19 (blau), Made in Austria (blau), unsign., H. 27 cm; eBay # 7302033968
8041,3 (Firmenmarke ohne Ang.), unsign., Ausformung zw. 1945–1953, H. 27 cm; Fischer Luz. 324
8041,4 Firmenmarke 19, Made in Austria, unsign., Ausformung zw. 1945–1953, H. 26,9 cm; eBay # 1414731883
8041,5 Firmenmarke 19, Made in Austria, unsign., Ausformung zw. 1945–1953, H. 30,4 cm; Sammlung a@goldscheider.net
8041,9 ger. M19, Firmenmarke 19, Made in Austria, unsign., Ausformung zw. 1945–1953, H. 28 cm; Wien Museum Inv. 186.663
8041,15 gem. SG, Firmenmarke 19, Firmenmarke 19a, unsign., Made in Germany, Ausformung zw. 1938–1941, H. 26,7 cm; Wien Museum Inv. 186.661
8041,17 (Firmenmarke ohne Ang.), vermutl. Ausformung zw. 1945–1953, H. 27 cm; Christie's SK 14.3.1997
8041,1920,94 Firmenmarke 19, Made in Austria, unsign., Ausformung ca. 1937/38, H. 27,4 cm; eBay # 1457628266
8041M, H. 27 cm, B. 15 cm; GSNachf. 2,8a; Fotoarchiv K. Täuber, hier mit der Ang.: H. 32 cm, B. 17 cm
Bem.: Diese Maske wurde auch noch zw. 1953–1960 bei Carstens (GWM West-Germany) und in Trenton, USA (Modell-Nr. 732) weiter produziert.
-> vgl. 8249
Lit.: GSNachf. 2,8 (m. Abb.), hier mit Ang. des Originaltitels „Maske"; KGSF1; KGSF2; Könecke (2002), S. 77, hier mit Ang. des Künstlers A. Prischl, S. 54 (m. Abb.).
Quelle: AGM (Kontoblatt der Wiener Manufaktur Josef Schuster für die Firma Porslins Ericsson-Lund, Stockholm?, Schweden vom 7. Januar 1943), hier mit Ang. des Originaltitels „Maske / Goldband"; (Bestellung der Firma Elisabeth Salle (Wiesbaden) vom 8.1.1951); Archiv Matschiner (Originalfoto); Fotoarchiv K. Täuber.

8052
Matula, Viktor
LEUCHTER
ca. 1937
8052 H. fehlt, B. fehlt
Quelle: Nachlass Matula, Wien

8053
Matula, Viktor
DECKELDOSE

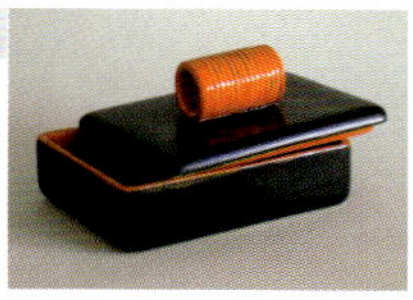

Form: rechteckig
Knauf: röhrenartige, geriffelte Handhabe
Dekor: zweifarbig (schwarz / orange)
ca. 1937
8053 gem. lig. AH, Firmenmarke 19, Made in Germany, unsign., Ausformung zw. 1938–1941, H. 8,7 cm, L. 16,2 cm, B. 11,2 cm; PB Deutschland

8055
Matula, Viktor
LEUCHTER
ca. 1937
8055 H. fehlt, B. fehlt
Quelle: Nachlass Matula, Wien

8062M
Matula, Viktor
FIGUR „Pferd" (Stehendes Fohlen mit nach hinten gewandtem Kopf)

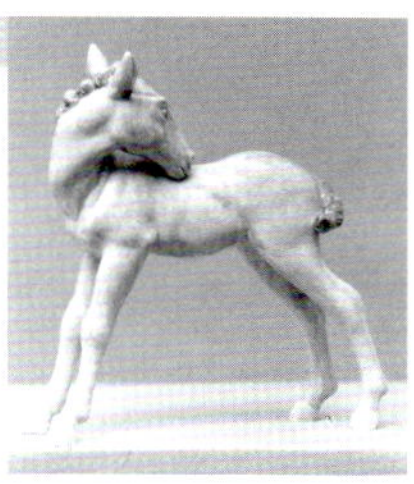

ca. 1936 (für den Entwurf)
8062M, H. 7 cm, B. 7 cm; GSNachf. 3,20b
Bem.: Diese Figur wurde auch noch zw. 1953–1960 bei Carstens (GWM West-Germany) produziert.
-> vgl. 7553M
Lit.: GSNachf. 3,22 (m. Abb.), hier mit Ang. des Originaltitels FIGUR „Pferd"; Könecke (2002), S. 76, hier mit Ang. des Künstlers Matula (für die Modell-Nr. 7553), S. 86 (m. Abb.).
Quelle: Fotoarchiv K. Täuber

8063
Knörlein, Rudolf (zugeschr.)
BÜSTE Damenakt mit Lockenkopf
ca. 1937
8063,9 Firmenmarke 19, unsign., Ausformung zw. 1945–1953, H. 24 cm; Wien Museum Inv. 186.669
8063,65 Firmenmarke 19, unsign., Ausformung zw. 1945–1953, H. 23,5 cm; Doro 15.9.2003
Ausst.: SSH 1992

8069M
Dakon, Stephan
FIGUR „Großer Akt"
ca. 1937
Quelle: Archiv Mimi Dakon (Tantiemenabrechnung 1.1.–30.6.1940; Wiener Manufaktur FG, Modellliste Dakon, 7.11.1946), hier wird die Figur jeweils unter dem Titel „Großer Akt" geführt.

8070M / F
Herczeg, Claire (Klára)
FIGUR „Akt" (Stehender Damenakt) (auf Rundsockel)
ca. 1937

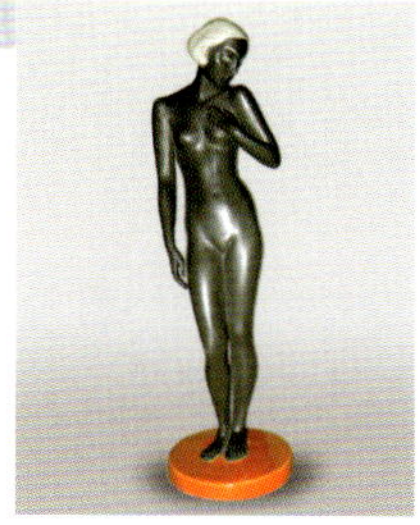

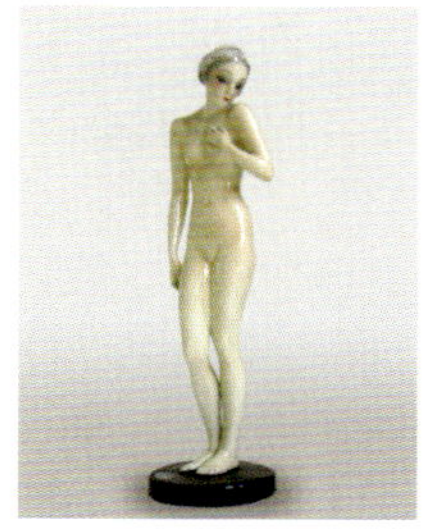

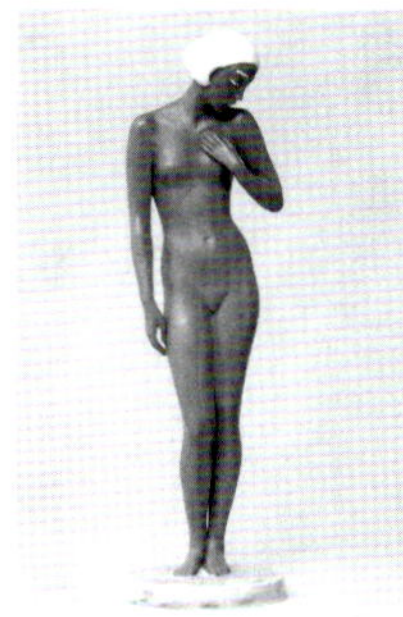

8070,4,44 gem. qw1, Firmenmarke 19, Made in Austria, sign. C. Herczeg, Originalpick.: „JCH, Herhold, Hannover", H. 31 cm; Privatsammlung, Wien
8070,8 Firmenmarke 19, Made in Austria, unsign, Pick. 1, Ausformung zw. 1945–1953, H. 31,3 cm; Wien Museum Inv. 190.132
8070,68,20 ger. 16a, gem. SG, Firmenmarke 19, Made in Germany, Pick. 6, Ausformung zw. 1938–1941, H. 31 cm; Privatsammlung, Wien
8070M, H. 31 cm, B. 9 cm; GSNachf. 3,4
Ausst.: SSH 1992
Lit.: GSNachf. 3,2 (m. Abb.), hier mit Ang. des Originaltitels „Akt liegend".
Quelle: AGM (Rechnung an Maurice Mizrahi, Alexandria, Ägypten, vom 22. April 1953), geliefert in Ausführung „matt";
Fotoarchiv K. Täuber.

8072
WANDMASKE Frauenkopf mit Kragen
ca. 1937
8072,65 (Firmenmarke ohne Ang.), vermutl. Ausformung zw. 1945–1953, H. 22 cm; Christie's SK 11.11.2004

8073M
Prischl, Adolf
FIGUR „Pferd" (Stehendes Fohlen)
ca. 1937
8073M, H. 42 cm, B. 35 cm; GSNachf. 3,24
Bem.: Diese Figur wurde auch noch zw. 1953–1960 bei Carstens (GWM West-Germany) produziert.
Lit.: N.N.: Neue Keramik, in: ÖK, 8.1937, Nr. 7, S. 22 (m. Abb.), „Auch der Keramiker Adolf Prischl, den wir zum ersten Male zur Veröffentlichung bringen, ist eine Begabung, die entwicklungsfähig ist und eigene Wege finden wird"; GSNachf. 3,24 (m. Abb.), hier mit Ang. des Originaltitels FIGUR „Pferd"; Könecke (2002), S. 77.

8075
Goebel, Kurt (zugeschr.)
STEHKOPF Frauenkopf mit Locken und Kopftuch
ca. 1937
8075,31 Firmenmarke 19, unsign., Ausformung zw. 1945–1953, H. 30 cm; RM 9; 10

8076M
Goebel, Kurt
WANDMASKE „Maske" (Frauenkopf mit Hand und Blumenschmuck im Haar)
ca. 1937

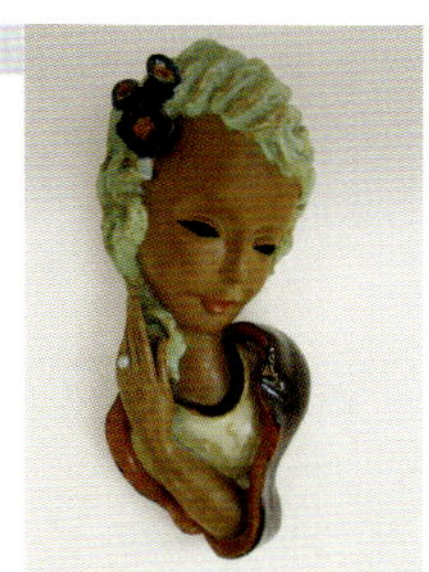

8076,0, Firmenmarke 19, Made in Austria, unsign., Ausformung zw. 1945–1953, H. 34 cm; Sammlung f@goldscheider.de
8076,S ger., Firmenmarke 19, Made in Austria, unsign., Pick. 6, Ausformung zw. 1945–1953, H. 34,4 cm; MAK Wien, Inv. Ke 10602-101

8076 Firmenmarke 19, Made in Austria, unsign., Ausformung zw. 1945–1953, H. 35 cm; Doro L. 24.11.2000
8076M, H. 34 cm, B. 15 cm; GSNachf. 2,11c
Lit.: GSNachf. 2,11 (m. Abb.), hier mit Ang. des Originaltitels „Maske".

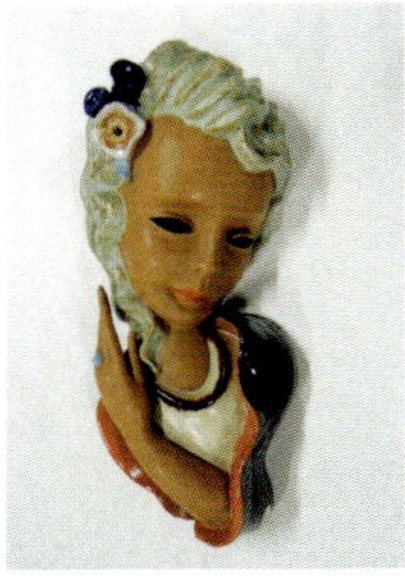

8077F
ASCHER „Ascher, groß"
ca. 1937
Quelle: AGM (Rechnung an Maurice Mizrahi, Alexandria, Ägypten, vom 22. April 1953), hier mit Ang. des Originaltitels „Ascher, groß".

8080
Matula, Viktor
FIGUR Pferd
ca. 1937
8080 H. fehlt
Quelle: Nachlass Matula, Wien

8081M
Matula, Viktor
BUCHSTÜTZE „Elefant" (Stehender, den Rüssel erhebender indischer Elefant)
ca. 1937
8081M, H. 12 cm, B. 16,5 cm; GSNachf. 4,6a
Lit.: GSNachf. 4,6a (m. Abb.), hier mit Ang. des Originaltitels „Buchstütze Elefant".
Quelle: Nachlass Matula, Wien

8083F
Dakon, Stephan
FIGUR „Susette" (Lilian Harvey, seitlich schreitend) (auf Ovalsockel)
ca. 1937

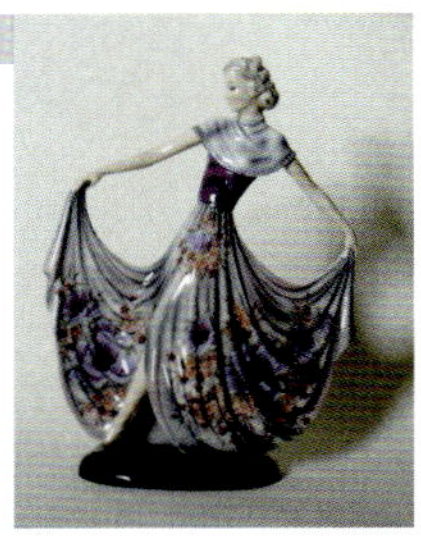

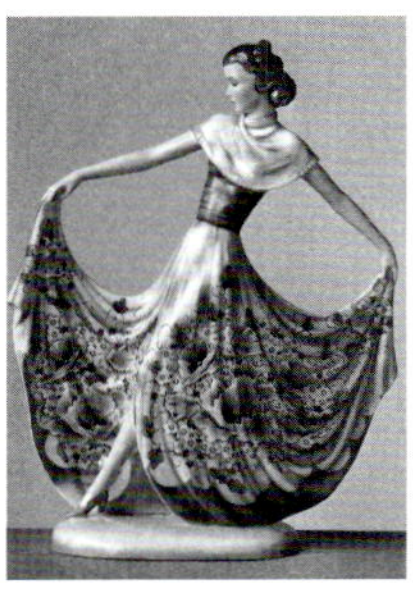

8083,0 gem. lig. HZ., Firmenmarke 19, unsign., Ausformung zw. 1945–1953, H. 34,5 cm; eBay # 1223978702
8083,1,7 gem. lig. UH., AS, Firmenmarke 19, Made in Austria, Hand dec., unsign., Ausformung zw. 1937/38, H. 34,9 cm; eBay # 2052646759
8083,3,60 Firmenmarke 19, Made in Austria, unsign., Ausformung zw. 1937/38, H. 34 cm; WKA 29
8083,6 Firmenmarke 19, unsign., Ausformung zw. 1945–1953, H. 33,5 cm; Doro 30.11.1999
8083,11 (Firmenmarke ohne Ang.), unsign., vermutl. Ausformung zw. 1945–1953, H. 33,3 cm; Doro 13.9.2004
8083,117,19 gepr. L, (Firmenmarke ohne Ang.), unsign.?, Ausformung zw. 1937/38, H. 34 cm; Christie's SK 13.5.2004
8083,182,19 Firmenmarke 19, Made in Germany, Ausformung zw. 1938–1941, unsign.?, H. 34 cm; Doro 10.3.2003
8083,192,68,G Firmenmarke 19, Made in Germany, unsign., Ausformung zw. 1938–1941, H. 34 cm; KH
8083,366,31 Firmenmarke 19, Made in Germany, Ausformung zw. 1938–1941, H. 34 cm; WKA 29
8083,460,19 Firmenmarke 19, unsign., Ausformung zw. 1938–1941, H. 33,8 cm; BM 71
8083,528,19,G Firmenmarke 19, Made in Germany, Ausformung zw. 1938–1941, unsign., H. 34 cm; Nagel 373
8083,573,VIII ger., Firmenstempel WG 1, unsign., Ausformung 50er Jahre, H. 34 cm; Wien Museum Inv. 190.148
8083,577,11 Firmenmarke 19, unsign. (oder wie fälschlicherweise im Auktionskat. angegeben „Lorenzl"?), Ausformung zw. 1938–1941, H. 35 cm; Kastern 68
8083,625,68,G gem. B., gem. blau H, Firmenmarke 19, Made in Germany, unsign., Ausformung zw. 1938–1941, H. 33,3 cm; PB Deutschland
8083,800,19 Firmenmarke 19, Ausformung zw. 1938–1941, H. 33,7 cm; BM 36
8083,803,10 (Firmenmarke ohne Ang.), Made in Germany, Ausformung zw. 1938–1941, H. 34 cm; Christie's SK 10.11.2000
8083,831,7 gem. SG., Firmenmarke 19, Made in Germany, Ausformung zw. 1938–1941, H. 34,3 cm; eBay # 1233339778
8083F, H. 34 cm, B. 26 cm; GSNachf. 1,26a; Fotoarchiv K. Täuber, hier mit der Ang.: H. 35 cm, B. 26 cm
Bem.: Diese Figur wurde auch noch zw. 1953–1960 bei Carstens (GWM West-Germany) produziert.
Ausst.: SSH 1992
-> vgl. 8320
Lit.: GSNachf. 1,26 (m. Abb.), hier mit Ang. des Originaltitels „Susette"; DM, 13.1936/37, Nr. 151, S. 98 (m. Abb.), hier eine zeitgenössische Fotografie vom März 1937, die wohl als Vorlage für die Ausführung diente. Dargestellt ist die Schauspielerin Lilan Harvey (eigentl. Lilian Pape, 19.1.1906–27.7.1968), Fotografie: Hämmerer (Ufa); SSH, S. 22, 23 (m. Abb.), Könecke (2002), S. 77, 88 (m. Abb.).
Quelle: AGM (Rechnung der Wiener Manufaktur Josef Schuster an Laszlo Bösze, Szeged, Ungarn, vom 5. August 1944), hier mit Ang. des Originaltitels „Susette"; Fotoarchiv K. Täuber.

8086M
WANDMASKE „Maske" (Frauenkopf mit Hand und Zigarette)
ca. 1937
8086 gepr., 8086 gem., ger. S, gem. lig. TK, Firmenmarke 19, Made in Austria, Pick. 5, Ausformung zw. 1945–1953, H. 33,5 cm; eBay # 3771225567
8086M, H. 34 cm, B. 18 cm; GSNachf. 2,11a
Lit.: GSNachf. 2,11 (m. Abb.), hier mit Ang. des Originaltitels „Maske".

8087M
WANDMASKE „Maske" (Frauenkopf mit Halstuch bzw. Kragen)
ca. 1937
8087 Firmenmarke 19, Made in Austria, Pick. 6, Ausformung zw. 1945–1953, H. 23 cm; Doro 24.11.1992
8087M, H. 25 cm, B. 15 cm; GSNachf. 2,11b
Lit.: GSNachf. 2,11 (m. Abb.), hier mit Ang. des Originaltitels „Maske".

8089
Dakon, Stephan
Stehendes nacktes Mädchen mit Rehkitz (auf Ovalsockel?)
ca. 1937
8089, H. 25 cm, B. 20,5 cm
Quelle: Archiv Mimi Dakon (Originalfoto)

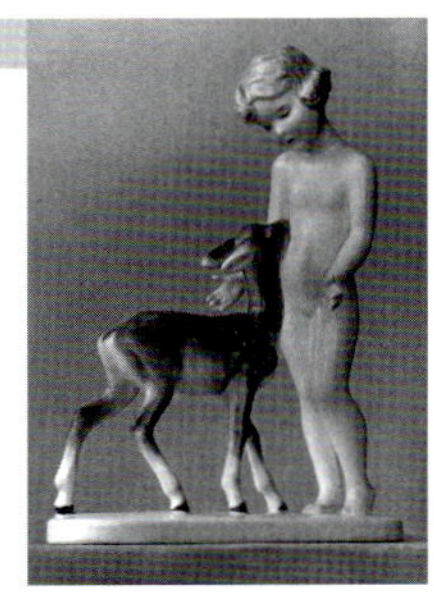

8090
WANDBEHANG (oder WANDRELIEF?) Madonna mit Kind
ca. 1937
8090,23,9 Firmenmarke 19, Made in Austria, Ausformung zw. 1937/38, H. 51 cm; Schuler 72
Bem.: Das AH bezeichnet dieses Objekt als „Relieffigur". Ob es sich nicht evtl. um ein Wandrelief handelt, kann, da keine Abb. vorliegt, abschließend nicht gesagt werden.

8091
Matula, Viktor
GRUPPE Zwei „Lämmchen" (Craqueléglasur)
ca. 1937
8091,6,9 Firmenmarke 19, Made in Austria, unsign., Ausformung zw. 1937/38, H. 14,5 cm; PB Deutschland

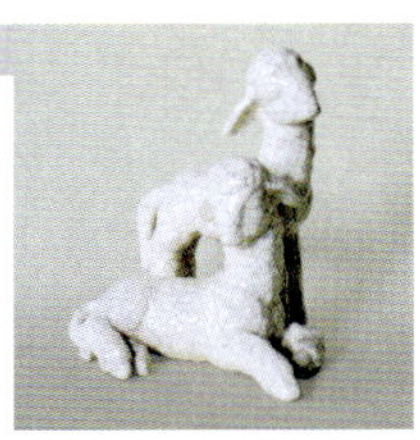

Bem.: Da im Kat. GSNachf. nur mehr die Einzelfiguren gezeigt werden, ist davon auszugehen, dass die Gruppe ab 1940 nicht mehr in Produktion war.
-> vgl. 8092= liegende EinzelFIGUR
-> vgl. 8093= stehende EinzelFIGUR

8092M
Matula, Viktor
FIGUR „Lämmchen" (liegend)
ca. 1937
8092M, H. 9,5 cm, B. 12,5 cm; GSNachf. 3,27c
-> vgl. 8091= GRUPPE
-> vgl. 8093= stehende EinzelFIGUR
Lit.: GSNachf. 3,27 (m. Abb.), hier mit Ang. des Originaltitels FIGUR „Lämmchen".
Quelle: Nachlass Matula, Wien

8093M
Matula, Viktor
FIGUR „Lämmchen" (stehend)
ca. 1937
8093M, H. 14 cm, B. 11 cm; GSNachf. 3,27b
-> vgl. 8091= GRUPPE
-> vgl. 8092= liegende EinzelFIGUR
Lit.: GSNachf. 3,27 (m. Abb.), hier mit Ang. des Originaltitels FIGUR „Lämmchen".
Quelle: Nachlass Matula, Wien

8094F
GRUPPE „Doggen" (Eine stehende und ein sitzende Dogge)
ca. 1937
8094F, H. 20 cm, B. 17,5 cm; GSNachf. 5,14a
-> vgl. 8095= sitzende EinzelFIGUR
-> vgl. 8096= stehende EinzelFIGUR
Lit.: GSNachf. 5,14 (m. Abb.), hier mit Ang. des Originaltitels GRUPPE „Doggen".

8095F
FIGUR „Dogge" (Sitzende Dogge)
ca. 1937
8095F, H. 19 cm, B. 13 cm; GSNachf. 5,14b; Fotoarchiv K. Täuber, hier mit der Ang.: H. 18 cm, B. 12 cm

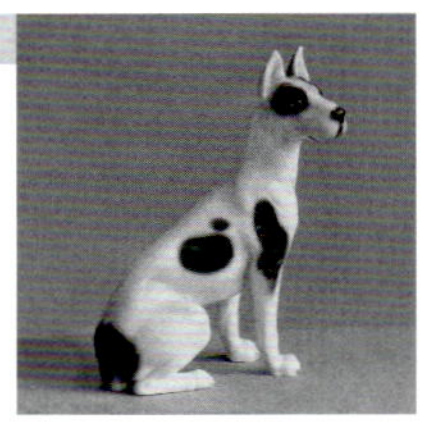

-> vgl. 8094= GRUPPE
-> vgl. 8096= stehende EinzelFIGUR
Lit.: GSNachf. 5,14 (m. Abb.), hier mit Ang. des Originaltitels FIGUR „Dogge".
Quelle: Fotoarchiv K. Täuber

8096F
FIGUR „Dogge" (Stehende Dogge)
ca. 1937
8096F, H. 20 cm, B. 17,5 cm; GSNachf. 5,14c; Fotoarchiv K. Täuber, hier mit der Ang.: H. 20 cm, B. 18 cm

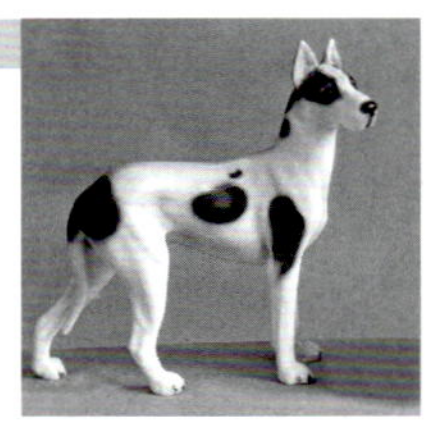

-> vgl. 8094= GRUPPE
-> vgl. 8095= sitzende EinzelFIGUR
Lit.: GSNachf. 5,14 (m. Abb.), hier mit Ang. des Originaltitels FIGUR „Dogge".
Quelle: Fotoarchiv K. Täuber

8102
Goebel, Kurt
ASCHER Mandolinenspieler
Form: rund
Zigarettenablagen: vier
Dekor: stehender Mandolinenspieler (vollplastisch)
ca. 1937
8102 gem. U, Firmenmarke 19, Made in Germany, unsign., Ausformung zw. 1938–1941, H. 14,6 cm, Ø 16,5 cm; eBay # 474474494

8106
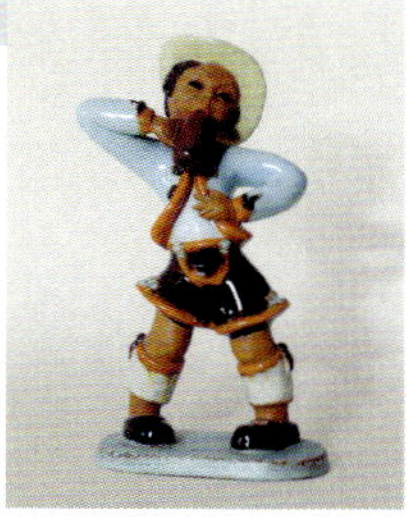

Goebel, Kurt
Stehender, aus einem Bierkrug trinkender Bursche in Tracht (auf Ovalsockel)
ca. 1937
8106,44 ger. 17, gem. O, Firmenmarke 19, Made in Austria, unsign., Ausformung zw. 1945–1953, H. 24 cm; PB Deutschland

8108F
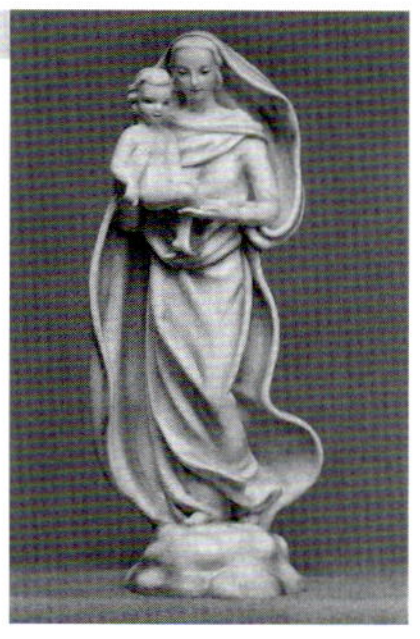

FIGUR „Madonna mit Kind" (Stehende Madonna mit Kind) (auf wolkenartigem Sockel)
ca. 1937
8108F, H. 43 cm, B. 18 cm; GSNachf. 6,11; Fotoarchiv K. Täuber, hier mit der Ang. des Originalpreises RM 30.–
Lit.: GSNachf. 6,11 (m. Abb.), hier mit Ang. des Originaltitels FIGUR „Madonna mit Kind".
Quelle: AGM (Rechnung an Maurice Mizrahi, Alexandria, Ägypten, vom 22. April 1953), geliefert in Ausführung „matt"; Fotoarchiv K. Täuber.

8111
Matula, Viktor
GRUPPE Zwei stehende Rehe (auf Ovalsockel)
ca. 1937
8111,9 ger. XVD, gem. lig. AH, Firmenmarke 19, Made in Austria, unsign., Ausformung zw. 1945–1953, H. 12,7 cm, L. 19 cm; eBay # 290163581
Quelle: Nachlass Matula, Wien

8112M
Matula, Viktor
GRUPPE „Pferde"
ca. 1937
Quellen: Nachlass Matula, Wien; AGM (Kontoblatt der Wiener Manufaktur Josef Schuster für die Firma Porslins Ericsson-Lund, Stockholm?, Schweden vom 7. Januar 1943), hier mit Ang. des Originaltitels GRUPPE „Pferde".

8113
Matula, Viktor
FIGUR Zicklein
ca. 1937
8113 H. fehlt
Quelle: Nachlass Matula, Wien

8116
Matula, Viktor
BUCHSTÜTZE Lämmer
ca. 1937
8116 H. fehlt
Quelle: Nachlass Matula, Wien

8117
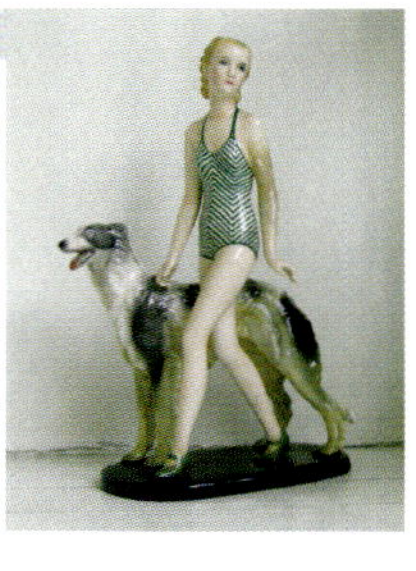

Dakon, Stephan
Schreitendes Mädchen im Badeanzug mit einem Barsoi (auf Ovalsockel)
ca. 1937
8117,34,31 gem. grün HE, Firmenmarke 19, Made in Germany, unsign., Ausformung zw. 1938–1941, H. 33 cm, B. 22,5 cm; Privatsammlung, Wien
Bem.: Der Barsoi, welcher dem Mädchen beigestellte wurde ist jener, der als Einzelfigur die Modell-Nr. 7541 trägt.
-> vgl. 7541= EinzelFIGUR Barsoi
Lit.: Pinhas, S. 132 (m. Abb.), hier sign. Dakon.

8120
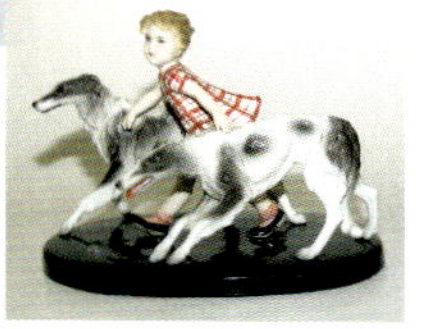

Bouret, Germaine
GRUPPE Laufendes Mädchen mit zwei Barsois (auf Ovalsockel)
ca. 1937
8120,70,33 gem. AN, B., Firmenmarke 19, Made in Germany, sign. G. Bouret, Ausformung zw. 1938–1941, H. 22 cm; Sammlung Bernard Amigon
8120,96,6,50 gem. AN, Firmenmarke 19, Made in Germany, unsign., Ausformung zw. 1938–1941, H. 22,5 cm, L. 29,5 cm; Wien Museum Inv. 190.087
ohne Ang., (Firmenmarke ohne Ang.), sign. G. Bouret, H. fehlt, L. 30,5 cm; Christie's SK 15.11.2001
Bem.: Die Entwürfe von Germaine Bouret wurden von Stephan Dakon ausmodelliert.

8121F
GRUPPE „Windhunde" (Zwei laufende Barsois)
ca. 1937
8121F, H. 13,5 cm, B. 25,5 cm; GSNachf. 5,5a
-> vgl. 8123= EinzelFIGUR (vorne)
-> vgl. 8122= EinzelFIGUR (rückwärts)
Lit.: GSNachf. 5,5 (m. Abb.), hier mit Ang. des Originaltitels GRUPPE „Windhunde".

8122F
FIGUR „Windhund" (Laufender Barsoi)
ca. 1937
8122F, H. 14 cm, B. 24 cm; GSNachf. 5,5b
-> vgl. 8121= GRUPPE
Lit.: GSNachf. 5,5 (m. Abb.), hier mit Ang. des Originaltitels FIGUR „Windhund".

8123F
FIGUR „Windhund" (Laufender Barsoi)
ca. 1937
8123,22,16 Firmenmarke 19, Ausformung zw. 1937/38, H. 11 cm; Wien Museum Inv. 190.018
8123F, H. 11,5 cm, B. 23 cm; GSNachf. 5,5a
-> vgl. 8121= GRUPPE
Lit.: GSNachf. 5,5 (m. Abb.), hier mit Ang. des Originaltitels FIGUR „Windhund".

8125
Matula, Viktor
BUCHSTÜTZE Pferd
ca. 1937
8125 H. fehlt
Quelle: Nachlass Matula, Wien

8126F
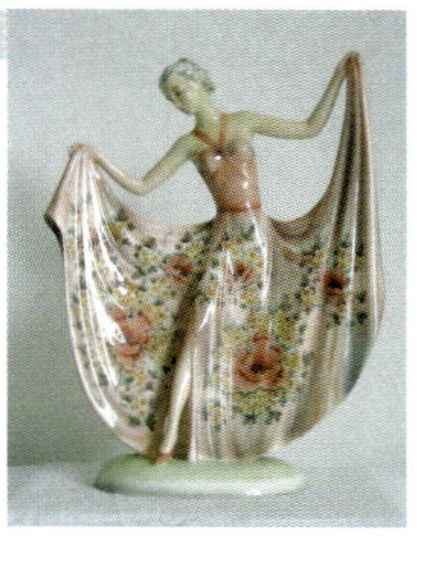

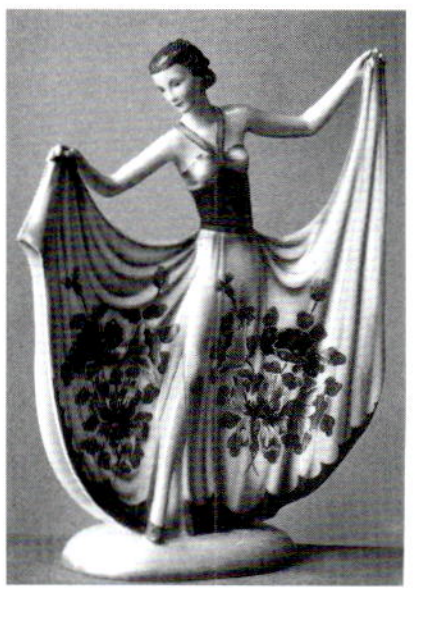

Dakon, Stephan
FIGUR „Tanzstudie" (Tänzerin mit geneigtem Kopf, ihr langes Kleid mit beiden Händen hochhebend) (auf Ovalsockel)
ca. 1937
8126,1,20, gem. S, Firmenmarke 19, Made in Austria, Hand dec., unsign., Ausformung zw. 1937/38, H. 39,4 cm; Wien Museum Inv. 190.066
8126,1,39 (Firmenmarke ohne Ang.), unsign., Ausformung zw. 1937/38, H. 40 cm; Doro 29.5.2001
8126,8,20 Firmenmarke 19 (blau), Made in Austria (blau), unsign., Ausformung zw. 1937/38, H. 40 cm; WKA 29
8126,11 Firmenmarke 19, Made in Germany, unsign., Ausformung zw. 1945–1953, H. 39 cm; Doro 20.10.1997; 13.9.2004
8126,22,10 gem. N, Firmenmarke 19, Made in Austria, H. 40 cm; eBay # 6516504134
8126,31,11 (Firmenmarke ohne Ang.), Ausformung zw. 1937/38, H. 39 cm; Christie's SK 12.5.2000
8126,66 gem. lig. HS, gem. rot X., Firmenmarke 19, Made in Austria, unsign., Ausformung zw. 1945–1953, H. 40,6 cm; eBay # 3710056457
8126,69,31 Firmenmarke 19, Made in Germany, Ausformung zw. 1938–1941, H. 38,5 cm; Doro 10.3.2003
8126,93,11 Firmenmarke 19, Made in Germany, Ausformung zw. 1938–1941, H. 39,5 cm; eBay # 2258151873
8126,96,4 Firmenmarke 19, Made in Germany, Ausformung zw. 1938–1941, H. 39,5 cm; Christie's SK 27.4.2005
8126,111,1 Firmenmarke 19, Made in Germany, Ausformung zw. 1938–1941, H. 39,5 cm; BM 61
8126,163,7 gem. AN, (Firmenmarke ohne Ang.), unsign.?, Ausformung zw. 1938–1941, H. 40 cm; Schloss Rick. 61
8126,176,19 Firmenmarke 19, Made in Germany, unsign., Ausformung zw. 1938–1941, H. 39,5 cm; WKA 35
8126,245,37 gepr. G, Firmenmarke 19, Made in Germany, Ausformung zw. 1938–1941, H. 39 cm; Christie's SK 13.10.2004
8126,346,7 gem. lig. SP., gem. rot XX, 34, Firmenmarke Schuster 1, Made in Germany, unsign., Ausformung zw. 1941–1944, H. 39,5 cm; Sammlung f@goldscheider.de
8126F, H. 40 cm, B. 30 cm; GSNachf. 1,27; Fotoarchiv K. Täuber, hier mit der Ang.: H. 40 cm, B. 34 cm
Lit.: GSNachf. 1,27 (m. Abb.), hier mit Ang. des Originaltitels „Tanzstudie".
Quelle: Fotoarchiv K. Täuber

8127F
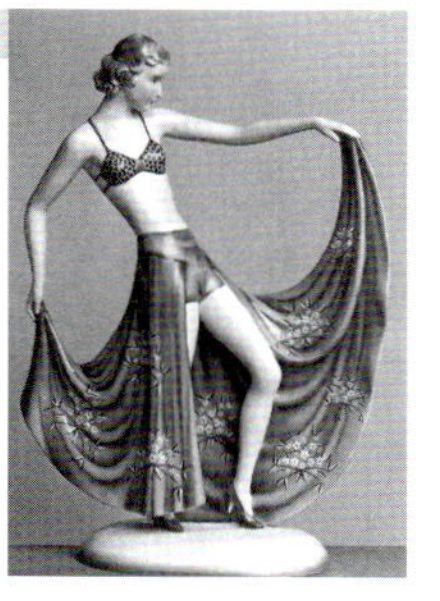

Dakon, Stephan
FIGUR „Frasquita" (Tänzerin mit Bikini und langem Rock, seitlich, in Pose) (auf Ovalsockel)
ca. 1937
8127,1,81 Firmenmarke 19, unsign., vermutl. Ausformung zw. 1938–1941, H. 38,5 cm; Doro 24.5.2000
8127,7 Firmenmarke 19, Made in Austria, Ausformung zw. 1945–1953, H. 39 cm; Z 41
8127,8 gem. S., gem. lig. HS, Firmenmarke 19, Made in Austria, unsign., Ausformung zw. 1945–1953, H. 38,1 cm; eBay # 112761297
8127,42,2 Firmenmarke 19, H. 39 cm; WKH
8127,154,19 Firmenmarke 19, Made in Germany, unsign., Ausformung zw. 1938–1941, H. 38 cm; Christie's L. 29.4.1992
8127F, H. 39 cm, B. 29,5 cm; GSNachf. 1,28; Fotoarchiv K. Täuber, hier mit der Ang.: H. 39 cm, B. 27 cm
Bem.: Das Badekostüm „Bikini" war zwar schon um die Jahrhundertwende durchaus gebräuchlich, wenn auch anfangs verpönt. Diesem Outfit wurde erst nach dem 2. Weltkrieg durch Jacques Heim und Bart Luis zum Durchbruch verholfen. Sie brachten den Zweiteiler zuerst 1946 unter dem Namen „Atom", dann, nach dem Angriff auf Hiroshima, mit dem geändertem Namen „Bikini" auf den Markt. (-> vgl.: www.bikinihome.de)
Lit.: GSNachf. 1,28 (m. Abb.), hier mit Ang. des Originaltitels „Frasquita"; DM, 1938, Nr. 111, (m. Abb.), hier eine zeitgenössische Fotografie die wohl als Vorlage für die Ausführung diente. Dargestellt ist die Schauspielerin Lida Baarova (eigentl. Ludmilla Babkova, 7.9.1914–28.10.2000). Bekannt wurde sie in Deutschland durch ihren Film „Barcarole" (1935) und durch eine zweijährige, heftige Liebesaffäre mit Joseph Goebbels, die zu einem Skandal geriet, in den sich sogar Hitler einschaltete. 1938 wurde sie ausgewiesen und ging wieder in die Tschechoslowakei. (-> vgl.: www.kinokessel.de/baarova.htm; Baarova, L.: Die süße Bitterkeit meines Lebens, Koblenz 2001, 304 S.). Bei der Ausführung durch die Manufaktur wurde die an sich dunkelhaarige Schauspielerin mit blondem Haar ausgestattet. Dies entsprach eher dem „deutschen" Zeitgeschmack.
Quelle: AGM (Rechnung an Maurice Mizrahi, Alexandria, Ägypten, vom 22. April 1953), hier unter dem Titel „Tanz" geführt; Fotoarchiv K. Täuber.

8129

Dakon, Stephan

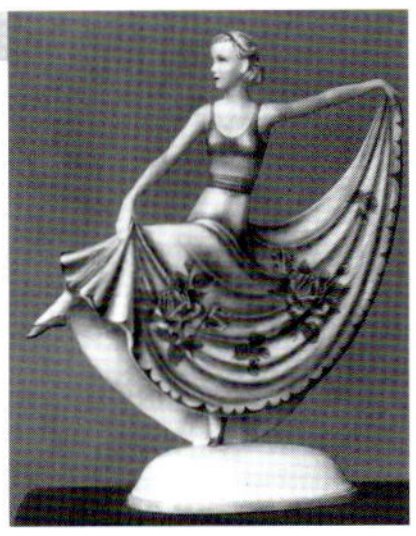

FIGUR / LAMPE Tänzerin, ihr linkes Bein nach vorne schwingend (auf Ovalsockel)
ca. 1937
FIGUR
8129, H. 34 cm, B. 30 cm
LAMPE
8129 ohne Ang., Firmenmarke 19, Made in Austria, vermutl. Ausformung zw. 1945–1953, H. fehlt; WKH
Lit.: Pinhas, S. 150 (m. Abb.), hier als Lampe.
Quelle: Archiv Mimi Dakon (Originalfoto), hier als Figur abgebildet.

8132F & F / Kr

Schwetz-Lehmann, Ida

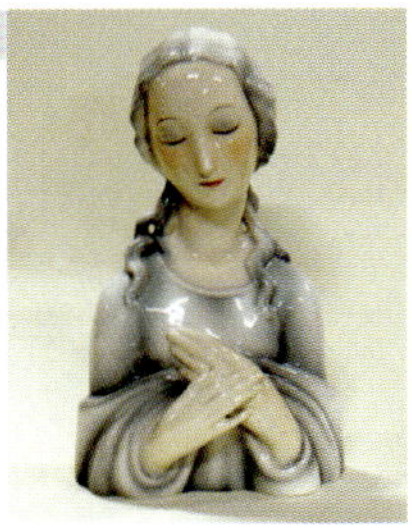

(BÜSTE) „Himmelskönigin" (Madonna mit gekreuzten Händen, mit/ohne Krone)
ca. 1937
ohne Krone
8132,5 ger., gem. rot D, Firmenmarke 19, Made in Austria, unsign., Ausformung zw. 1945–1953, H. 28,7 cm; city-antik

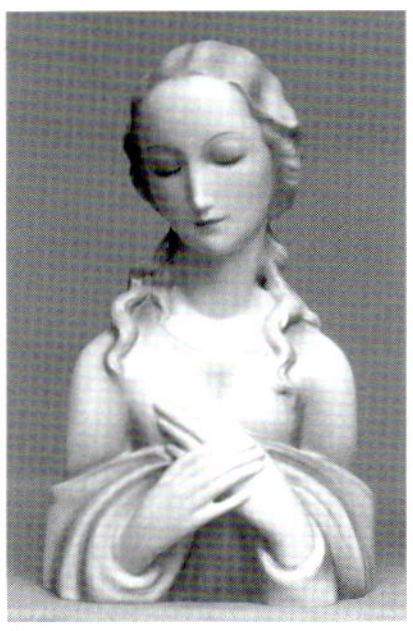

8132,27,14 ger. XII,B, gem. lig. NO, Firmenmarke 19, Made in Austria, sign. Schwetz-Lehmann, Ausformung zw. 1937/38, H. 28 cm; MAK Wien, Inv. Ke 10602-52
8132,200,4 gem. SG, Firmenmarke 19, Made in Germany, sign. S. Lehmann, Ausformung zw. 1938–1941, H. 28 cm; eBay # 250002613105
8132 (Firmenmarke o. Ang.), unsign., H. 28 cm, B. 17 cm; Sammlung mad

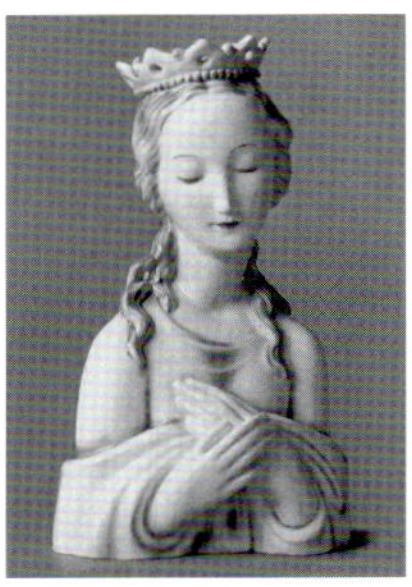

8132F, H. 28,5 cm, B. 17 cm; GSNachf. 6,6a; Fotoarchiv K. Täuber
mit Krone
8132,1,4 Firmenmarke 19, Made in Austria, Hand dec., unsign., H. 29 cm; Quit. 25
8132F/Kr, H. 30 cm, B. 17,5 cm; GSNachf. 6,6b; Fotoarchiv K. Täuber, hier mit der Ang.: H. 25 cm, B. 17 cm
-> vgl. 8842= STEHKOPF
Lit.: GSNachf. 6,6 (m. Abb.), hier mit Ang. des Originaltitels (BÜSTE) „Himmelskönigin".
Quelle: AGM (Kontoblatt der Wiener Manufaktur Josef Schuster für die Firma Porslins Ericsson-Lund, Stockholm?, Schweden vom 7. Januar 1943), hier mit Ang. des Titels BÜSTE „Madonna"; (Bestellung der Firma Wahliss (Wien), vom 24.11.1945 an die „Arbeitsgemeinschaft der Firma Friedrich Goldscheider", hier mit Ang. des Titels „Madonnenbüste"; Bestellungen der Firma Wahliss (Wien), vom 9.11.1945, 19.2.1946 und 27.2.1946, „Madonnenbüste ohne Krone", in Ausführung „Terrakotta"; Rechnung an Maurice Mizrahi, Alexandria, Ägypten, vom 22. April 1953), geliefert in Ausführung „matt"; Fotoarchiv K. Täuber.

8133F

GRUPPE „Kinder" („Kindergruppe") (Tanzendes Kinderpaar) (auf Ovalsockel)
ca. 1937
8133,47,7 gem. M., Firmenmarke 19, Made in Austria, H. 22 cm; WKA 33
8133 H. 22 cm, B. 18 cm; Fotoarchiv K. Täuber
Quelle: AGM (Bestellung von Ferd. Lundquist, Göteborg, Schweden vom 23. August 1950), hier unter dem Titel „Kindergruppe" geführt; Fotoarchiv K. Täuber.

8134F

Herczeg, Claire (Klára)

FIGUR/LAMPE „Tanz" (Tänzerin in Bewegung, ihr langes Kleid mit beiden Händen hochhebend) (auf Rundsockel)
ca. 1937
FIGUR
8134,1,52 Firmenmarke 19, Made in Austria, Hand dec., unsign., Ausformung zw. 1937/38, Originalpick.: „Karl Skala, Wien", H. 35,3 cm; Doro 21.9.1999
8134 (Firmenmarke ohne Ang.), sign. Claire Herczeg, H. 34 cm; Pinhas, S. 155 (m. Abb.)
LAMPE
8134 ohne Ang., da diese Ausführung bislang nur in der Firmenrechnung an den Händler Mizrahi aufscheint. Wahrscheinlich existieren nur wenige Exemplare, die aus der Zeit der Nachkriegsproduktion stammen könnten.
Bem.: Möglicherweise hatte diese Figur / Lampe auch einen noch „griffigeren" Originaltitel. Dieser lässt sich aber nicht mehr verifizieren, so wird jener auf der Rechnung an Mizrahi angegebene allgemeine Titel als gegeben angenommen.
Quelle: AGM (Rechnung an Maurice Mizrahi, Alexandria, Ägypten, vom 22. April 1953), hier unter dem Titel „Tanz Lampe" geführt.

8135

Weiss, Claire (Klára) (d.i. Herczeg, Claire (Klára)

Schreitende elegante Dame mit Hut (auf Ovalsockel)
ca. 1937
8135,11 gem. K, Firmenmarke 19, Made in Austria, sign. Claire Weiss, Ausformung zw. 1945–1953, H. 33 cm; Doro 1.12.2000; WKA 33
81(35),17,10 Firmenmarke 19, Made in Austria, sign. Claire Weiss, Ausformung zw. 1937/38, H. 33,5 cm; GK, S. 24, 131 (m. Abb.); Wien Museum Inv. 182.069
8135,42,11 Firmenmarke 19, Made in Germany, sign. Claire Herczeg, Ausformung zw. 1938–1941, H. 34,5 cm; Kett. 196
Bem.: Die Informationen, dass diese Figur aus der „Übergangszeit, in der Dr. Schwartz die Firma als kommissarischer Verwalter leitete", stammt, sowie die Annahme, dass man unter seiner Ägide wieder beginnend mit der Nr. eins die Modelle durchnummerierte, sind schlichtweg falsch (-> vgl. GK, S. 24). Vielmehr können hier nun mehrere Stücke der Modell-Nr. 8135 nachgewiesen werden. Auch das Exemplar aus dem Wien Museum, welches zu der falschen Annahme und Interpretation führte, kann nun endgültig richtig in das Werkverzeichnis eingereiht werden.
Ausst.: HMW (jetzt Wien Museum) 1985/86.
Lit.: GK, S. 24, 131 (m. Abb.)

8136

Dakon, Stephan

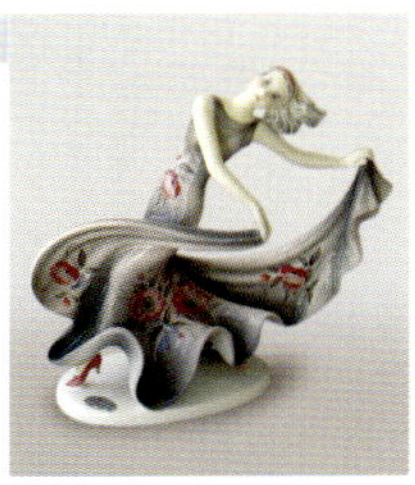

Tänzerin in schwungvoller drehender Bewegung (auf Ovalsockel)
ca. 1935 (für den Entwurf)
8136,6,14 ger. XIIV, gem. SO, S6, Firmenmarke 19, Made in Austria, Pick. 4, H. 22 cm; Sammlung Stressless, Portugal
8136,40,6 ger. XII, gem. N, gem. lig. NO, Firmenmarke 19, Made in Austria, Ausformung zw. 1937/38, H. 22 cm; eBay # 1438389086
-> vgl. 7078

8137F

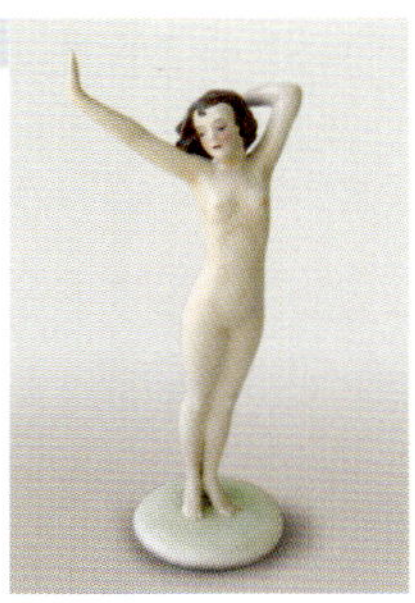

FIGUR „Erwachen" (Stehender Damenakt mit ausgestrecktem Arm) (auf Rundsockel)
ca. 1935/36 (für den Entwurf)
8137,3,53 gem. S1, Firmenmarke 19, Made in Austria, H. 26,5 cm; Sammlung Stressless, Portugal
8137F, H. 27,5 cm, 13 cm; GSNachf. 1,31b
-> vgl. 7431
-> vgl. 7525= ähnl. Modell
Lit.: GSNachf. 1,31 (m. Abb.), hier mit Ang. des Originaltitels „Erwachen".
Quelle: Fotoarchiv K. Täuber

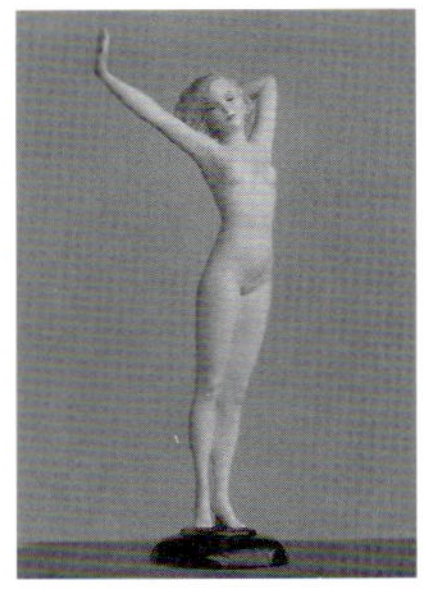

8138F

Weiss, Claire (Klára) (d.i. Herczeg, Claire (Klára)

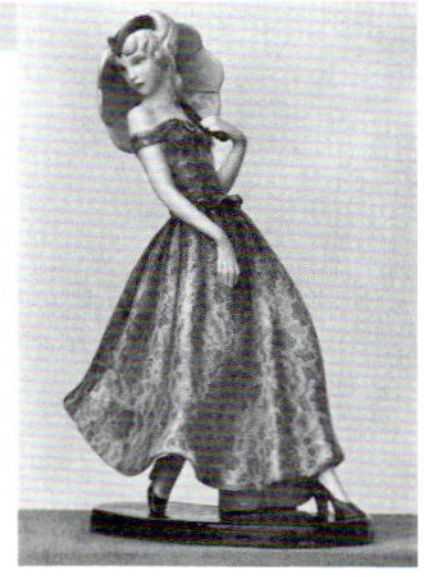

FIGUR „Parisienne" (Schreitende elegante Dame, ihren Schirm in der linken Hand haltend) (auf Ovalsockel)
ca. 1935/36 (für den Entwurf)
8138,3,65 Firmenmarke 19, Made in Austria, Hand dec., unsign., H. 22 cm; Doro 15.3.2005
8138,4,2 Firmenmarke 19, sign. Claire Weiss, vermutl. Ausformung zw. 1937/38, H. 22 cm; Fischer 141
8138,79,2 Firmenmarke 19, Made in Germany, sign. Claire Herczeg, Ausformung zw. 1938–1941, H. 22 cm; BM 57
8138,228,19 gem. M., gem. lig. FDH., Firmenmarke 19, Made in Germany, sign. Claire Herczeg, Ausformung zw. 1938–1941, H. 22 cm; KH
ohne Modell-Nr., gem. lig. HS, Firmenmarke 19, Made in Austria, unsign., vermutl. Ausformung zw. 1945–1953, H. 21,8 cm; Doro 25.4.2000
8138F, H. 22 cm, B. 15,5 cm; GSNachf. 1,16b
-> vgl. 7390; 7723
-> vgl. 8361= seitenverkehrtes Modell
Lit.: GSNachf. 1,16 (m. Abb.), hier mit Ang. des Originaltitels „Parisienne".

8139F

Schwetz-Lehmann, Ida

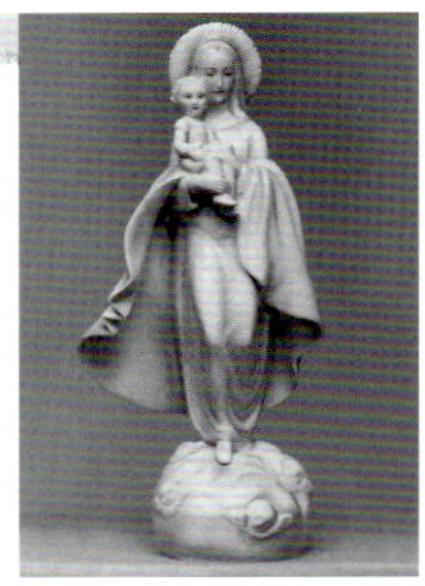

FIGUR „Madonna mit Kind" (Stehende Madonna mit Aureole und Jesuskind) (auf wolkenartigem Sockel)
ca. 1937
8139,16,31 ger. L, XII, S, Firmenmarke 19, Made in Austria, sign. Schwetz-Lehmann, Pick. 6, Ausformung zw. 1937/38, H. 40,6 cm; eBay # 2565491240
8139,77 Firmenmarke 19, Made in Austria, unsign.?, Ausformung zw. 1945–1953, H. 39,3 cm; eBay # 109666922
8139F, H. 40 cm, B. 19 cm; GSNachf. 6,12; Fotoarchiv K. Täuber, hier mit der Ang.: H. 39,5 cm, B. 18 cm
Lit.: GSNachf. 6,12 (m. Abb.), hier mit Ang. des Originaltitels FIGUR „Madonna mit Kind".
Quelle: Fotoarchiv K. Täuber

8140F
-> ÜBERSCHNEIDUNG

Dakon, Stephan

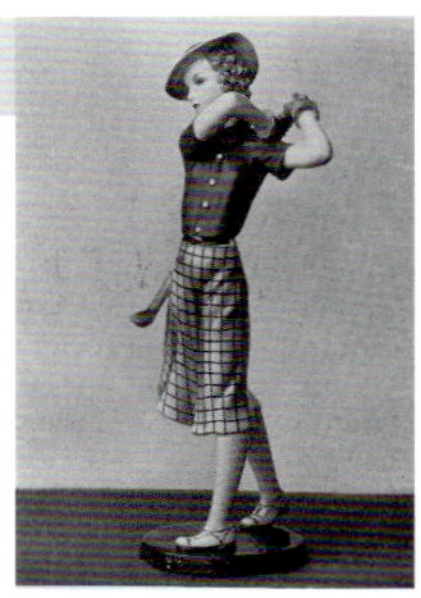

FIGUR „Golf" (Golfspielerin mit Hut und Schläger) (auf Rundsockel)
ca. 1935/36 (für den Entwurf)
8140F, H. 23 cm, B. 9,5 cm; GSNachf. 1,17b
-> vgl. 7416
Lit.: GSNachf. 1,17 (m. Abb.), hier mit Ang. des Originaltitels „Golf".

8140 -> ÜBERSCHNEIDUNG

Bär (Schreitend)
ca. 1937
8140 ger., Firmenmarke 19 (zwei mal), Ausformung zw. 1945–1953, H. 7 cm; Wien Museum Inv. 190.085
Pendant -> vgl. 8147; 8148

8141

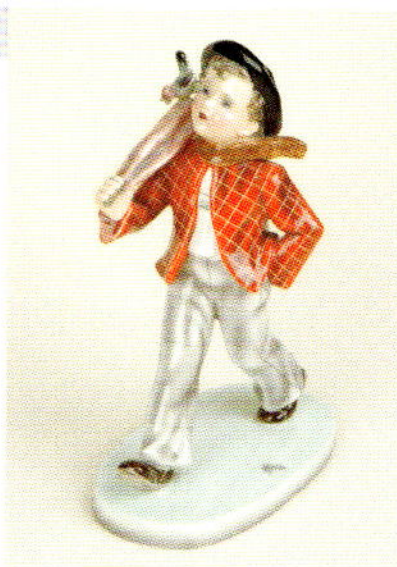

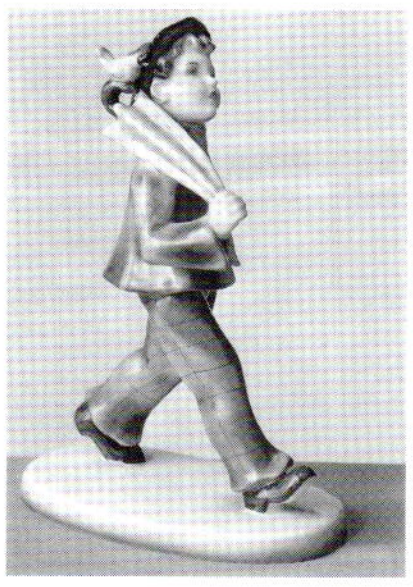

Dakon, Stephan
FIGUR „Wanderbursche" (Schreitender junger Bursche mit Hut und Schirm, darauf ein Vogel sitzend) (auf Ovalsockel)
ca. 1937
8141,22,2 gem. I.M., LS, Firmenmarke 19, Made in Austria, Hand dec., unsign., Ausformung zw. 1937/38, H. 21,6 cm; eBay # 3721486636
8141 ohne Ang., Firmenmarke 19, Made in Austria, unsign., H. 21,5 cm; PB Wien 19
8141 H. 21 cm, B. 10 cm; Fotoarchiv K. Täuber
Bem.: Diese Figur gelangte 1937 erstmalig in den Verkauf (Brief von GS an die Wiener Firma Keramos, vom 25.2.1953).
Quelle: AGM (Brief von der Wiener Manufaktur Goldscheider an die Firma Keramos, vom 25.2.1953); Archiv Mimi Dakon (Originalfoto); Fotoarchiv K. Täuber.

8145F

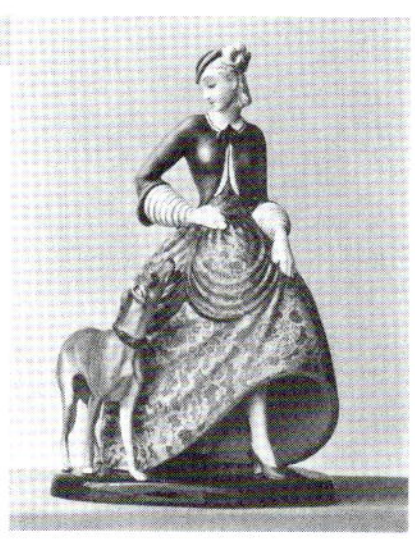

Weiss, Claire (Klára)
(d.i. Herczeg, Claire (Klára)
GRUPPE „Komtesse mit Hund" (Schreitende Tänzerin mit Windhund) (auf Ovalsockel)
ca. 1931 (für den Entwurf)
8145F, H. 23 cm, B. 15 cm; GSNachf. 1,14b
Bem.: Ob es sich bei dem beigestellten Windhund wirklich um einen Entwurf von Claire Herczeg handelt, ist sehr zu bezweifeln. Wurden ihren Tänzerinnen oder Akten doch durch die Manufaktur, teils ohne ihr Einverständnis einzuholen, Hunde beigefügt. Diese Information verdanke ich Herrn Heinz Hoever, der die Künstlerin vor ihrem Tod persönlich befragen konnte und mit ihr länger in Kontakt stand.
-> vgl. 7327
-> vgl. 6454; 6951= jeweils „Komtesse" als EinzelFIGUR
Lit.: GSNachf. 1,14 (m. Abb.), hier mit Ang. des Originaltitels „Komtesse mit Hund".

8147

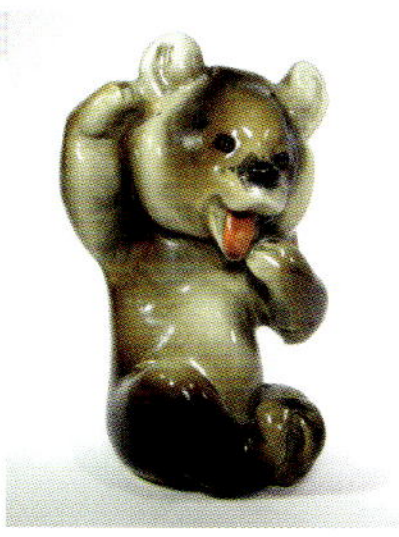

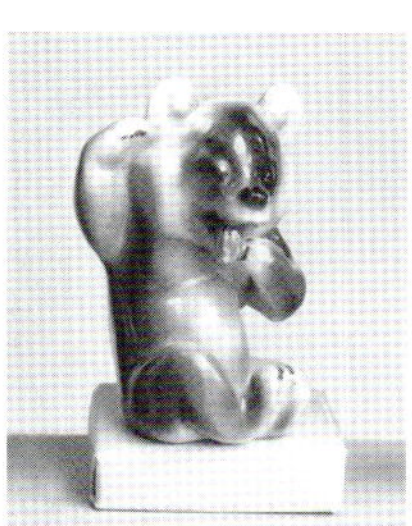

Bär (Sitzend, eine Tatze erhoben)
ca. 1937
8147,1 ohne Firmenmarke, Ausformung zw. 1945–1953, H. 8 cm; Wien Museum Inv. 190.083
8147,7 Firmenmarke 19, Ausformung zw. 1945–1953, H. 8,8 cm; Sammlung a@goldscheider.net
8147 H. 9 cm, B. 5 cm; Fotoarchiv K. Täuber
Pendant -> vgl. 8140; 8148
Quelle: Fotoarchiv K. Täuber

8148F

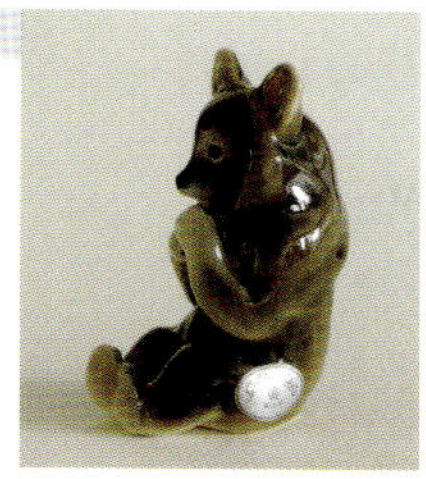

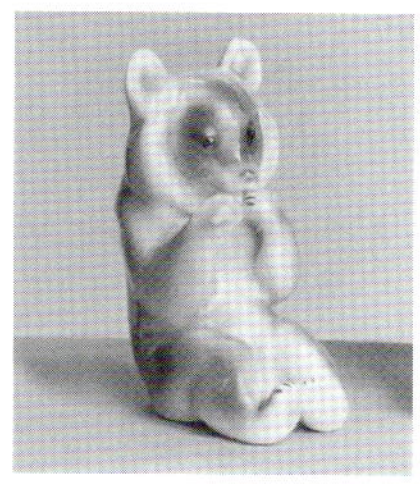

„Bär" (Sitzend, Tatzen am Maul)
ca. 1937
8148,4 ohne Firmenmarke, H. 8,9 cm; Wien Museum Inv. 190.084
8148,6 ger., ohne Firmenmarke, Pick. 1, H. 8,9 cm; Sammlung John Nash
8148 H. 9 cm, B. 4 cm; Fotoarchiv K. Täuber
Pendant -> vgl. 8140; 8147
Quelle: AGM (Kontoblatt der Wiener Manufaktur Josef Schuster für die Firma Porslins Ericsson-Lund, Stockholm?, Schweden vom 7. Januar 1943), hier mit Ang. des Originaltitels „Bär"; Fotoarchiv K. Täuber.
Quelle: Fotoarchiv K. Täuber

8149

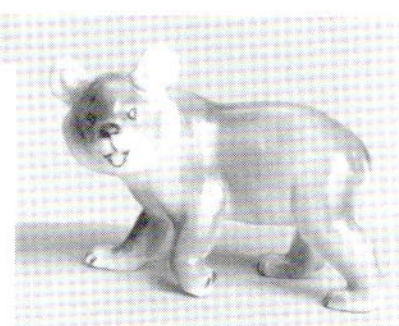

„Bär" (Schreitend)
ca. 1937
8149,22 Firmenmarke 19, Ausformung zw. 1945–1953, H. 8 cm, L. 11 cm; eBay # 3352328627
8149 H. 8 cm, B. 10 cm; Fotoarchiv K. Täuber
Quelle: Fotoarchiv K. Täuber

8151

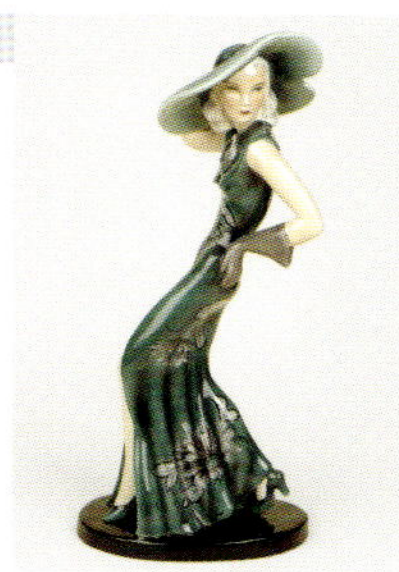

Weiss, Claire (Klára)
(d.i. Herczeg, Claire (Klára)
Stehende elegante Dame mit großem Hut und langen Handschuhen (auf Ovalsockel)
ca. 1935 (für den Entwurf)
8151,55,9 Firmenmarke 19, Made in Germany, unsign., Ausformung zw. 1938–1941, H. 33,5 cm; WKA 38
ohne Ang., Firmenmarke 19, Made in Austria, unsign., Ausformung zw. 1945–1953, H. 33,5 cm; Christie's L. 28.9.1988
-> vgl. 7279
-> vgl. 7367; 8155= jeweils GRUPPE „Maskerade" (mit Barsoi).

8155F

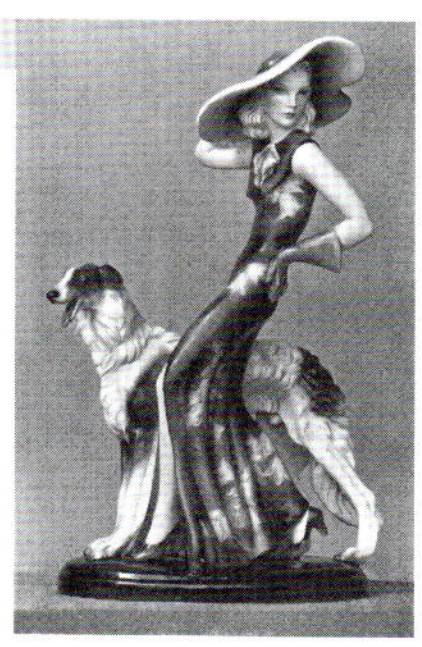

Weiss, Claire (Klára)
(d.i. Herczeg, Claire (Klára)
GRUPPE „Maskerade" (Schreitende elegante Dame mit Barsoi) (auf Ovalsockel)
ca. 1935 (für den Entwurf)
8155,8,11 ger. XIIB, Firmenmarke 19, Made in Austria, sign. Claire Weiss, Pick. 6, Ausformung zw. 1937/38, H. 33,3 cm; Wien Museum Inv. 190.088
8155,18,11 Firmenmarke 19, Made in Austria, sign. Claire Weiss, Ausformung zw. 1937/38, H. 33,5 cm; Doro 20.4.1993
8155,118,19 Firmenmarke 19, unsign., H. 32,8 cm; Doro 21.5.2003
8155,233,13 gem. S., N., Firmenmarke 19, Made in Germany, unsign., Ausformung zw. 1938–1941, H. 34,3 cm; eBay # 7392666
8155F, H. 33,5 cm, B. 23 cm; GSNachf. 1,15b
-> vgl. 7367
-> vgl. 7279; 8151= jeweils Dame ohne Hund
Lit.: GSNachf. 1,15 (m. Abb.), hier mit Ang. des Originaltitels „Maskerade".
Quelle: AGM (Rechnung der Wiener Manufaktur Josef Schuster an Laszlo Bösze, Szeged, Ungarn, vom 5. August 1944); Rechnung an Maurice Mizrahi, Alexandria, Ägypten, vom 22. April 1953), hier jeweils mit Ang. des Originaltitels „Maskerade/Hund".

8156

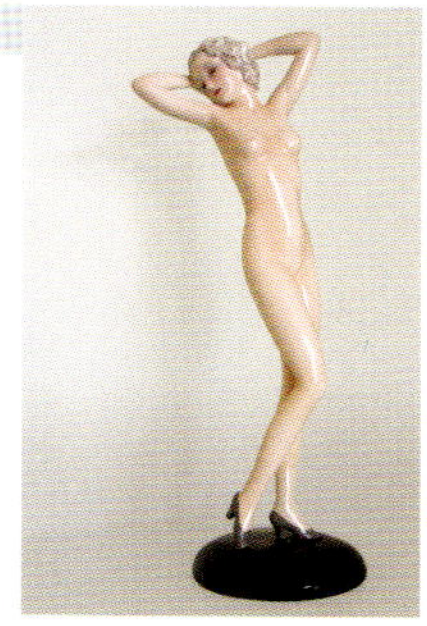

Stehender Mädchenakt (auf Rundsockel)
ca. 1937
8156,50,65 gem. N., gem. lig. NO, Firmenmarke 19, Made in Austria, gepr. „L"= Z6, H. 37,8 cm; Sammlung a@goldscheider.net

8157F

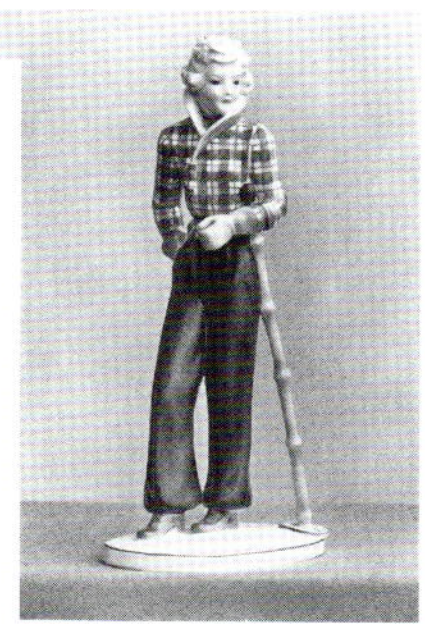

Dakon, Stephan
FIGUR „Skiläuferin" (Stehende Skiläuferin mit Skistock) (auf Ovalsockel)
ca. 1935 (für den Entwurf)
8157,79,11 Firmenmarke 19, Made in Austria, Ausformung zw. 1937/38, H. 22,5 cm; KaM 88
8157,104,10 gepr. G, Firmenmarke 19, Made in Germany, unsign., Ausformung zw. 1938–1941, H. 21 cm; eBay # 660170376
8157F, H. 22,5 B. 10,5 cm; GSNachf. 1,10b
-> vgl. 7149
Lit.: GSNachf. 1,10 (m. Abb.), hier mit Ang. des Originaltitels „Skiläuferin".

8160M
Matula, Viktor
BUCHSTÜTZE „Akt, liegend" (Liegender Damenakt)
ca. 1937
8160M, H. 11 cm, B. 15 cm; GSNachf. 4,6b
Bem.: Die Figuren weisen neben der Seitenverkehrtheit auch Unterschiede in den Arm- und Handstellungen auf. Einer der beiden Akte mit, der andere ohne Tuch.
Lit.: GSNachf. 4,6b (m. Abb.), hier mit Ang. des Originaltitels „Buchstütze Akt, liegend".
Quelle: Nachlass Matula, Wien

8167
Lorenzl, Josef
Schreitende junge Dame, ihr langes Kleid mit der linken Hand nach hinten haltend (auf Ovalsockel)
ca. 1937
8167 (Firmenmarke ohne Ang.), H. 22 cm; Pinhas, S. 151 (m. Abb.)

8169F

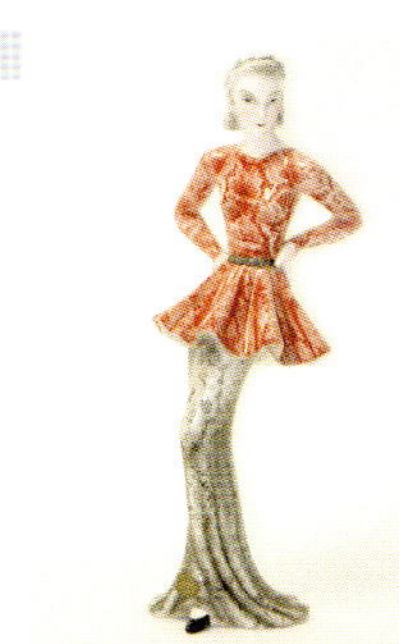

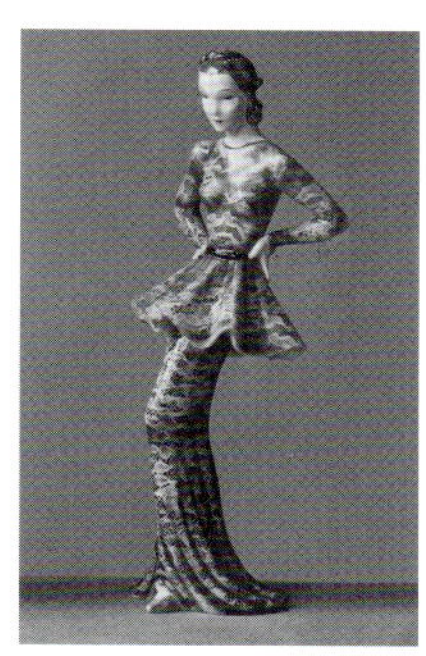

Herczeg, Claire (Klára)
FIGUR „Modedame" (Marlene Dietrich in langem Kleid, ihre Arme in die Hüften gestützt)
ca. 1937
8169,1,12 gem. E, Firmenmarke 19, Made in Austria, Hand dec., unsign., Ausformung zw. 1937/38, H. ca. 31,5 cm; eBay # 1212850331
8169,7 gem. rot XX, 24, Firmenmarke Schuster 1, Made in Germany, unsign., Ausformung zw. 1941–1944, H. 31,5 cm; Wien Museum Inv. 190.153
8169,36 gem. lig. MR, Firmenmarke 19, Made in Austria, Hand dec., unsign., H. 31,5 cm; WKA 35
8169,109,10 ger. XII, F, gem. M., Firmenmarke 26, gepr. „L"= Z6, H. 31,7 cm; eBay # 1254598886; 2015325835
8169,144,4 Firmenmarke 19, Made in Germany, Ausfor-

mung zw. 1938–1941, sign. Claire Herczeg, H. 31,2 cm; Doro 29.5.2001
8169,147,2 (Firmenmarke ohne Ang.), sign. Herczeg, H. 31,5 cm; Christies's A. 18.11.2003
8169,119,2 gem. S.M., Firmenmarke 19, Made in Germany, sign. Claire Herczeg, Ausformung zw. 1938–1941, H. 32 cm; Soth. com. # 1BUI; Bem.: Dieses Stück hat laut AH die Modell-Nr. 8619 eingeprägt. Es muss sich in diesem Fall um ein Versehen des Modelleurs handeln, der offenbar den Paginierstempel falsch eingestellt hatte.
8169,166,66 ger., Firmenmarke 19, unsign., H. 31,5 cm; Doro 21.5.2003
8169,255,? (Firmenmarke ohne Ang.), unsign.?, Pick. ohne Ang., H. 31 cm; Doro 1.6.2005
8169 ohne Ang., (Firmenmarke ohne Ang.), H. 31,5 cm; Privatsammlung, Wien
8169F, H. 31 cm, B. 11 cm; GSNachf. 1,33c
Bem.: Bei der dargestellten Person handelt es sich um die Schauspielerin Marlene Dietrich (27.12.1901, Berlin – 6.5.1992, Paris) –> vgl.: Metten, E. (Hrsg.): Marlene Dietrich, die Geschichte einer Karriere, Wien 1947; unpag. Taf. m. Abb.
Lit.: GSNachf. 1,33 (m. Abb.), hier mit Ang. des Originaltitels „Modedame".
Quelle: AGM (Rechnung an Maurice Mizrahi, Alexandria, Ägypten, vom 22. April 1953), hier unter dem Titel „Abendkleid" geführt, geliefert in Ausführung „gold"; Fotoarchiv K. Täuber.

8173F

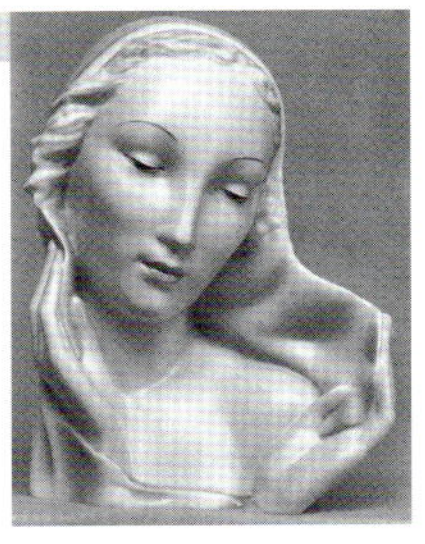

BÜSTE „Madonna"
(Madonna mit offenen Händen)
ca. 1937/38
8173,5,59 Firmenmarke 19, Made in Austria, Hand dec., Pick. 5/a, H. 28 cm; KH
8173F, H. 28 cm, B. 22 cm; GSNachf. 6,7; Fotoarchiv K. Täuber, hier mit der Ang. des Originalpreises RM 35,75
Lit.: GSNachf. 6,7 (m. Abb.), hier mit Ang. des Originaltitels BÜSTE „Madonna".
Quelle: AGM (Rechnung an Maurice Mizrahi, Alexandria, Ägypten, vom 22. April 1953), geliefert in Ausführung „matt"; Fotoarchiv K. Täuber.

8174F

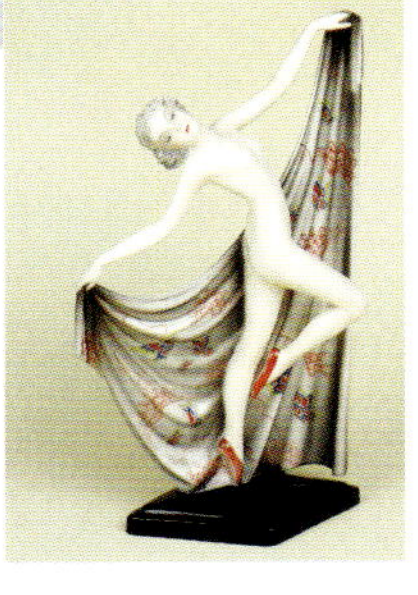

Nieschlag, Karl / Melcher, Josef
FIGUR „Nocturno"
(Stehender Damenakt mit großem Tuch) (auf rautenförmigem Sockel)
ca. 1937/38
8174,2,7 gem. L, Firmenmarke 19, Made in Austria, unsign., H. 40 cm; Wien Museum Inv. 190.177
8174,31,6 ger. L, Firmenmarke 19, Made in Austria, H. 40 cm; Doro 8.3.1994
8174,100,4 ger. 47, Firmenmarke 19, Made in Germany, sign. Nieschlag, Ausformung zw. 1938–1941, H. 40,2 cm, B. 23 cm; PB Wien; Bem.: Die Hautpartien sind in der Art von Biskuitporzellan ausgeführt, also unglasiert.
8174,119,19 ger. X, gem. SG, Firmenmarke 19, Made in Germany, unsign., Ausformung zw. 1941–1944, H. 40,2 cm; Doro 15.3.2005
8174,160,19 gem. rot N.II., Firmenmarke 19, Made in Germany, unsign., Ausformung zw. 1938–1941, H. 40,2 cm; city-antik
8174,192,2 Firmenmarke 19, Made in Germany, unsign., Ausformung zw. 1938–1941, Pick. 6, H. 41 cm; eBay # 475302738
8174,279,19,G Firmenmarke 19, Made in Germany, unsign., Ausformung zw. 1938–1941, H. 40 cm; Stahl 7.6.1997
8174,313,10 gem. B.F., Firmenmarke 19, Made in Germany, unsign., Ausformung zw. 1938–1941, H. 40 cm; city-antik
8174F, H. 40 cm, B. 23,5 cm; GSNachf. 1,34
Bem.: Diese Figur wurde auch noch zw. 1953–1960 bei Carstens (GWM West-Germany) produziert.
–> vgl. 8377= ähnl. Mod.
Lit.: GSNachf. 1,34 (m. Abb.), hier mit Ang. des Originaltitels „Nocturno"; Könecke (2002) belegt auf S. 77 den Künstler „Melcher", der aus einer Preisliste (ca. 1956) entnommen wurde. Einziges mit Melcher sign. Stück der Manufaktur ist ein stehender Mädchenakt (Modell-Nr. 8316). Welcher der beiden Entwerfer schlussendlich für dieses Modell verantwortlich zeichnet, kann zum jetzigen Zeitpunkt nicht geklärt werden.
GS neu: Modell-Nr. 54 „Die Bewegung" (H. 35 cm), Aufl. 3000 Stück

8175F

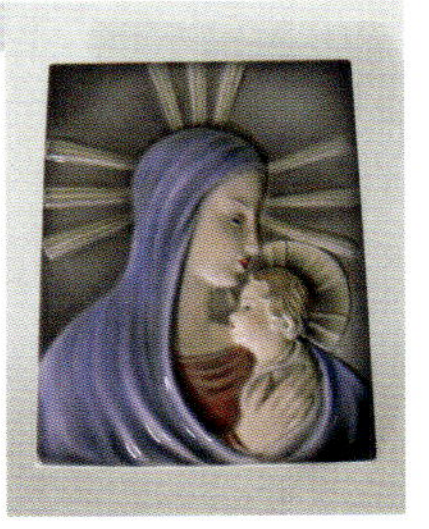

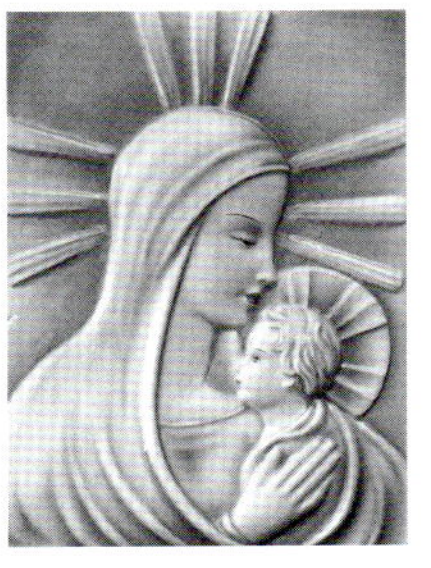

RELIEF „Madonna mit Kind"
(Madonna mit Jesuskind und Aureolen) (Wandrelief)
ca. 1937/38
8175,67,10 Firmenmarke 19, Made in Germany, Ausformung zw. 1938–1941, Originalpick.: „Cl. Attenkofer'sche Buch und Kunsthandlung, Straubing", H. 29 cm, B. 22,6 cm; PB Deutschland
8175F, H. 29 cm, B. 22,5 cm; GSNachf. 6,3a
Lit.: GSNachf. 6,3 (m. Abb.), hier mit Ang. des Originaltitels RELIEF „Madonna mit Kind".
Quelle: Fotoarchiv K. Täuber

8176F
Lorenzl, Josef
FIGUR „Gefangener Vogel" (Der Gefangene Vogel, Niddy Impekoven)
ca. 1922 (für den Entwurf)
FIGUR
8176F, H. fehlt; GSNachf. 1,51
–> vgl. 5054 (BRONZE); 5230; 5960; 5964= Pendant zu 5960; 6204; 6600
–> vgl. 5530 GRUPPE (Trio)
Pendant, bzw. ähnl. Modell –> vgl. 5434
Bem.: s. bei 5230. Leider fehlen in den beiden noch auffindbaren GSNachf. Katalogen die Blätter zur Figur „Gefangener Vogel". Auch sind keine Stücke mit der Modell-Nr. 8176 nachgewiesen. Die Größenfrage muss somit offen bleiben, bis sich neue Quellen hierzu auftun.
Lit.: s. bei 5230
GS neu: Modell-Nr. 82 „Gefangener Vogel" (H. 40 cm), Aufl. 3000 Stück

8178
Matula, Viktor
GRUPPE Zwei stehende Pferde mit Kummet (auf abgerundetem Rechtecksockel)
ca. 1937/38
8178,2,0 gem. L., Firmenmarke 19, Made in Austria, H. 16 cm, B. 24 cm; Sammlung Helmut Riener, St. Florian, Austria
Quelle: Nachlass Matula, Wien

8179
Matula, Viktor
GRUPPE Pferde
ca. 1937/38
8179 H. fehlt
Quelle: Nachlass Matula, Wien

8180
Matula, Viktor
GRUPPE Pferde
ca. 1937/38
8180 H. fehlt
Quelle: Nachlass Matula, Wien

8181M
FIGUR „Akt mit Schale" (Stehender Damenakt, eine Schale haltend) (auf Rundsockel)
ca. 1937/38
8181M, H. 25,5 cm, B. 7,5 cm; GSNachf. 3,5c
Lit.: GSNachf. 3,5 (m. Abb.), hier mit Ang. des Originaltitels „Akt mit Schale".

8182
Prischl, Adolf
WANDMASKE „Maske" (Frauenkopf mit mittellangem Haar, im Halbprofil nach rechts)
ca. 1937/38
8182,3 gem. lig. AH, Firmenmarke 19, Made in Austria, unsign., H. 29,2 cm; eBay # 490325532
Bem.: Diese Maske wurde auch zw. 1953–1960 bei Carstens (GWM West-Germany) unter der Modell-Nr. 526 (H. 18 cm) produziert. Als Entwerfer steht durch die Nummernfolge, beginnend mit der Ziffer 5, Adolf Prischl eindeutig fest.
Quelle: AGM (Bestellung der Firma Wahliss (Wien), vom 24.11.1945 an die „Arbeitsgemeinschaft der Firma Friedrich Goldscheider", hier mit Ang. des (Original-)Titels „Maske"; Bestellung der Firma Wahliss (Wien), vom 27.2.1946 in Ausführung „Terrakotta").

8183F

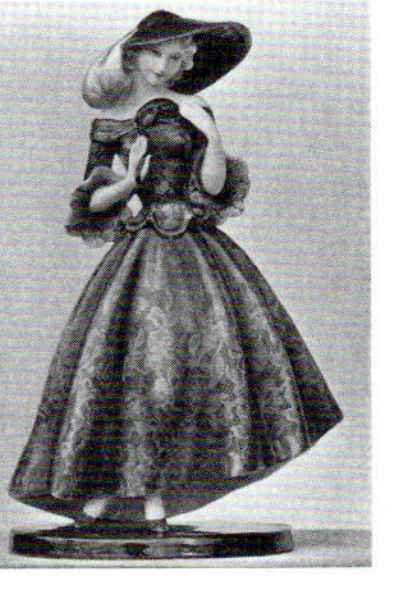

Herczeg, Claire (Klára)
FIGUR „Demaskierung"
(Stehende junge Dame, ihre Maske abnehmend) (auf Ovalsockel)
ca. 1936/37 (für den Entwurf)
8183,7 gem. rot XX., Firmenmarke Schuster 1, Made in Germany, unsign., Ausformung zw. 1941–1944, H. 22 cm; city-antik
8183F, H. 26 cm, B. 16 cm; GSNachf. 1,19b
Bem.: Dieses Modell ist eines jener, welches auch noch in Trenton (USA) produziert wurde.
–> vgl. 7691
Lit.: GSNachf. 1,19 (m. Abb.), hier mit Ang. des Originaltitels „Demaskierung".

8190F

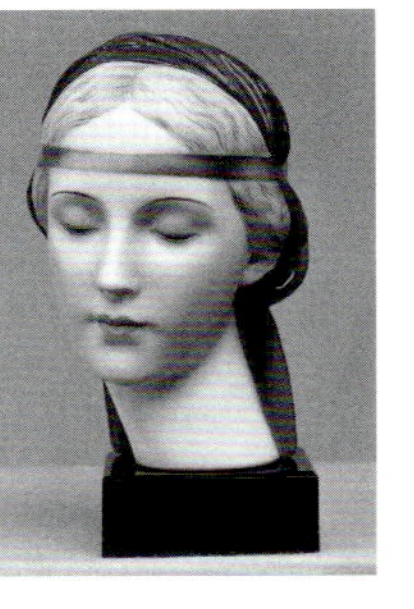

(STEH)KOPF „Kopf ohne/mit Sockel" (Frauenkopf mit geschlossenen Augen, Kopftuch und Stirnband)
ca. 1937/38
8190F, H. 23,5 cm, B. 12 cm; GSNachf. 2,18
Bem.: Dieses Modell scheint einem klassischen, antiken Vorbild nachempfunden zu sein.
Lit.: GSNachf. 2,18 (m. Abb.), hier mit Ang. des Originaltitels „Kopf ohne/mit Sockel".
Quelle: Fotoarchiv K. Täuber

8193M
FIGUR „Akt kniend" (Kniender Damenakt) (auf flachem Rechtecksockel)
ca. 1937/38
8193M, H. 27,5 cm, B. 17,5 cm; GSNachf. 3,5a
Lit.: GSNachf. 3,5 (m. Abb.), hier mit Ang. des Originaltitels „Akt kniend".
Quelle: AGM (Bestellung der Firma Wahliss (Wien), vom 27.2.1946 an die „Arbeitsgemeinschaft der Firma Friedrich Goldscheider", in Ausführung „Terrakotta").

8197F

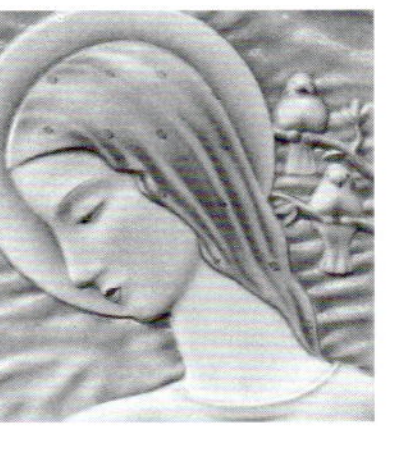

RELIEF „Madonna"
(Madonna mit zwei Tauben auf einem Ast) (Wandrelief)
ca. 1937/38
8197,66,2 Firmenmarke 19, H. 23 cm, B. 23 cm; Fischer 63
8197F, H. 24 cm, B. 23 cm; GSNachf. 6,3b

Lit.: GSNachf. 6,3 (m. Abb.), hier mit Ang. des Originaltitels RELIEF „Madonna".
Quelle: Fotoarchiv K. Täuber

8198M
Prischl, Adolf
WANDMASKE „Maske" (Frauenkopf mit Schal)
ca. 1937 (für den Entwurf)

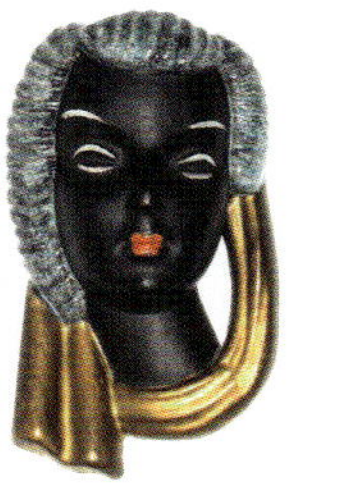

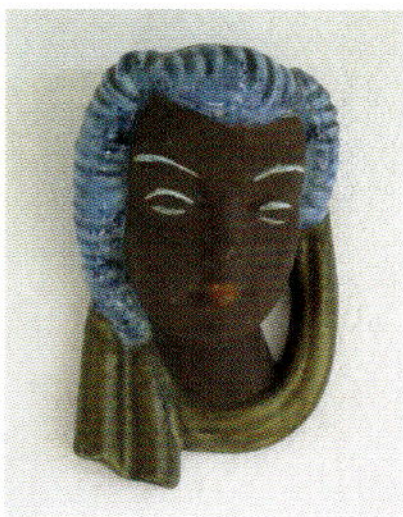

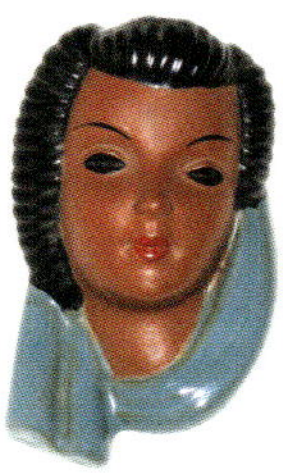

8198,3,96 Firmenmarke 19, Made in Austria, unsign., Ausformung ca. 1937/38, H. 9,2 cm; Wien Museum Inv. 186.653
8198,13,17 ger., Firmenmarke 19 (blau), Made in Austria (blau), Hand dec. (blau), Ausformung ca. 1937/38, H. 9 cm; city-antik
8198,18 gem. lig. AH, Firmenmarke 19, unsign., H. 9,5 cm; Sammlung f@goldscheider.de
8198,19,14 ger., Firmenmarke 19, Made in Austria, Hand dec., unsign., Ausformung ca. 1937/38, H. 10 cm; eBay # 891034488
8198 ger., gem. lig. AH, Firmenmarke 19, Made in Germany, Ausformung zw. 1938–1941, unsign., H. 9,3 cm; Sammlung a@goldscheider.net
8198 (Stempel), Firmenstempel WG 1, Ausformung 50er Jahre, unsign., H. 9,3 cm; PB Deutschland
8198 (Stempel), Firmenstempel WG 1, Ausformung 50er Jahre, unsign., H. 9,4 cm; Sammlung a@goldscheider.net
8198M, H. 9 cm, B. 5,5 cm; GSNachf. 2,10e
Bem.: Diese Maske wurde auch noch zw. 1953–1960 bei Carstens (GWM West-Germany) produziert und wegen der kleinen Ausführung etwas vom Originalentwurf abgewandelt. Der Ohrring, der dem größeren Modell beigefügt ist, wurde hier weggelassen.
-> vgl. 7883
Lit.: GSNachf. 2,10 (m. Abb.), hier mit Ang. des Originaltitels „Maske"; KGSF1.

8199
Prischl, Adolf?
WANDMASKE
ca. 1937/38
8199 ohne Ang., H. 9,5 cm, B. 5,5 cm; Könecke (2002), S. 77
Bem.: Da sich in der angegeben Lit. weder nähere Beschreibung noch Abb. findet, bzw. sich kein Stück nachweisen lässt, kann über das Aussehen der Maske keine nähere Aussage gemacht werden. Der Modellnummernreihenfolge nach ist jedoch anzunehmen, dass es sich um einen Entwurf von Adolf Prischl handeln könnte. Diese Wandmaske wurde auch noch zw. 1953–1960 bei Carstens (GWM West-Germany) produziert.
Lit.: Könecke (2002), S. 77, hier ohne weitere Ang.

8201
Goebel, Kurt
Stehender junger Maler mit Palette (auf Rundsockel)
ca. 1937/38
8201,8 gem. lig. AH, Firmenmarke 19, Made in Germany, unsign., Ausformung zw. 1938–1941, H. 37 cm; eBay # 926323023
8201,65 (Firmenmarke ohne Ang.), unsign., Ausformung zw. 1945–1953, H. 35,5 cm; Christie's SK 12.9.1997; 5.12.1997

8202
Goebel, Kurt
Sitzendes Mädchen mit Hummer am Meer (auf Natursockel mit Stein und Meereswellen)
ca. 1937/38
8202,26,32 ger. 16, Firmenmarke 19, Made in Austria, unsign., H. 21 cm, L. 20 cm; eBay # 1438391824

8204F
Weiss, Claire (Klára)
(d.i. Herczeg, Claire (Klára)
FIGUR „Abendkleid" (Stehende schlanke Dame in langem Kleid und Pelzumhang)
ca. 1937/38

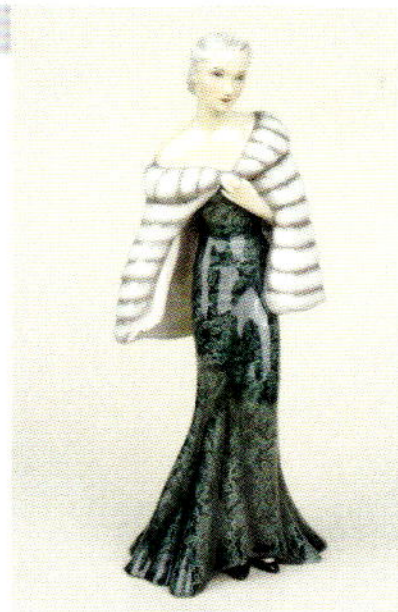

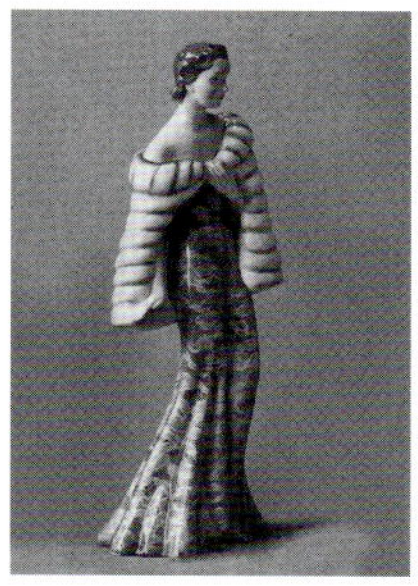

8204,10,14 (Firmenmarke ohne Ang.), sign. Claire Weiss, H. 30,7 cm; Doro 15.5.2002
8204,36,0,41 gem. AN. gem. lig. JR, Firmenmarke 19, Made in Austria, sign. Claire Weiss, H. 31 cm; eBay # 356221882
8204,64,10 gem. AN., SG, Firmenmarke 19, Made in Austria, unsign., gepr. „L"= Z6, H. 31 cm; PB Wien 19
8204,181,? (Firmenmarke ohne Ang.), unsign., Pick. 6, H. 30,4 cm; Doro 1.6.2005
8204F, H. 31 cm, B. 12 cm; GSNachf. 1,33b; Fotoarchiv K. Täuber, hier mit der Ang.: H. 31 cm, B. 13 cm
Lit.: GSNachf. 1,33 (m. Abb.), hier mit Ang. des Originaltitels „Abendkleid".
Quelle: AGM (Rechnung an Maurice Mizrahi, Alexandria, Ägypten, vom 22. April 1953), hier mit Ang. des Originaltitels „Abendkleid", geliefert in Ausführung „matt"; Fotoarchiv K. Täuber.

8205F
Weiss, Claire (Klára)
(d.i. Herczeg, Claire (Klára)
FIGUR „Ivette" (Stehende junge Dame in langem Kleid und Schal)
ca. 1937/38

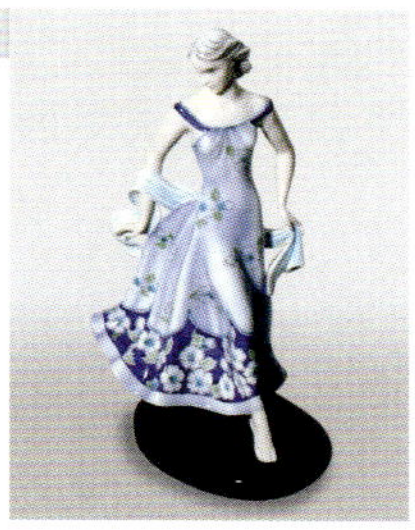

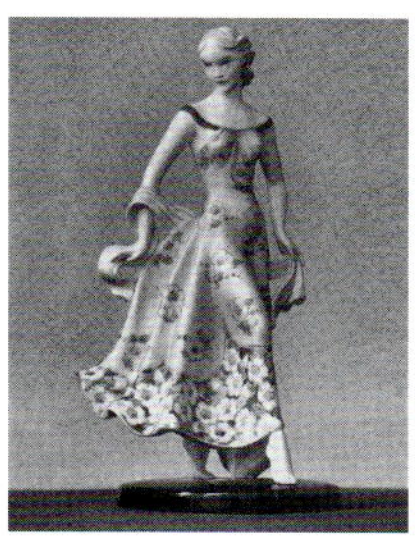

8205,5,11 Firmenmarke 19, Made in Austria, sign. Claire Weiss, H. 35 cm; city-antik
8205,50,4 gem. AN, Firmenmarke 19, Made in Germany, sign. Claire Herczeg, H. 35 cm; Soth./eBay # 1860110227; Christie's SK 22.4.2004
8205,136,67 Firmenmarke 19, unsign., H. 33,5 cm; Doro 12.9.2006
8205F, H. 34 cm, B. 19 cm; GSNachf. 1,35
Lit.: GSNachf. 1,35 (m. Abb.), hier mit Ang. des Originaltitels „Ivette".
Quelle: AGM (Rechnung an Maurice Mizrahi, Alexandria, Ägypten, vom 22. April 1953), hier unter dem Titel „Tanz" geführt.

8207M
Prischl, Adolf
WANDMASKE „Maske" (Frauenkopf mit wehendem Haar)
ca. 1935/36 (für den Entwurf)

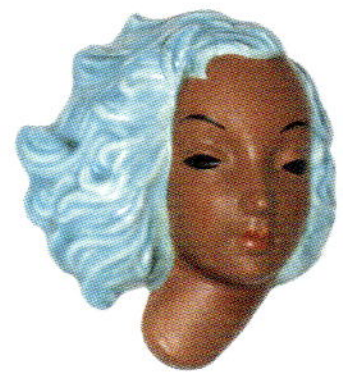

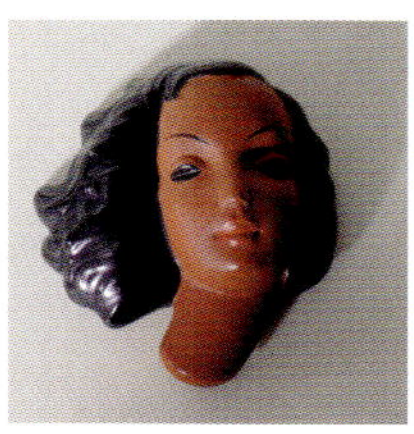

8207,6 Firmenmarke 19, Made in Austria, Ausformung zw. 1945–1953, H. 9,5 cm; Sammlung a@goldscheider.net
8207 (Stempel), Firmenstempel WG 1, unsign., Ausformung 50er Jahre, H. 9,4 cm; PB Deutschland
8207M, H. 9 cm, B. 9,5 cm; GSNachf. 2,3c; 2,10c (hier mit der Ang.: B. 8,5 cm); Fotoarchiv K. Täuber, hier mit der Ang.: H. 9 cm, B. 9 cm
Bem.: Diese Maske wurde auch noch zw. 1953–1960 bei Carstens (GWM West-Germany) produziert.
-> vgl. 7420; 7750
Lit.: GSNachf. 2,3 (m. Abb.), hier mit Ang. des Originaltitels „Maske"; KGSF1; Könecke (2002), S. 77, hier mit Ang. des Künstlers A. Prischl, S. 88 (m. Abb.).
Quelle: Fotoarchiv K. Täuber

8210M
Matula, Viktor
GRUPPE „Steinböcke"(Zwei Steinböcke) (auf Natursockel)
ca. 1937/38
8210M, H. 23 cm, B. 22 cm; GSNachf. 3,26a
-> vgl. 8211; 8212= jeweils EinzelFIGUREN
Lit.: GSNachf. 3,26 (m. Abb.), hier mit Ang. des Originaltitels FIGUR „Steinbock".
Quelle: Nachlass Matula, Wien

8211M
Matula, Viktor
FIGUR „Steinbock" (Äsender Steinbock) (auf Natursockel)
ca. 1937/38
8211M, H. 13 cm, B. 20 cm; GSNachf. 3,26b
Bem.: Diese Figur wurde auch noch zw. 1953–1960 bei Carstens (GWM West-Germany) produziert.
-> vgl. 8210= GRUPPE
Pendant -> vgl. 8212
Lit.: GSNachf. 3,26 (m. Abb.), hier mit Ang. des Originaltitels FIGUR „Steinbock"; Könecke (2002), S. 77, 88 (m. Abb.).
Quelle: Nachlass Matula, Wien

8212M
Matula, Viktor
FIGUR „Steinbock" (mit erhobenem Kopf) (auf Natursockel)
ca. 1937/38
8212M, H. 22,5 cm, B. 17 cm; GSNachf. 3,26c
Bem.: Diese Figur wurde auch noch zw. 1953–1960 bei Carstens (GWM West-Germany) produziert.
-> vgl. 8210= GRUPPE
Pendant -> vgl. 8211
Lit.: GSNachf. 3,26 (m. Abb.), hier mit Ang. des Originaltitels FIGUR „Steinbock"; Könecke (2002), S. 77, 89 (m. Abb.).
Quellen: AGM (Bestellung der Firma Wahliss (Wien), vom 24.11.1945 an die „Arbeitsgemeinschaft der Firma Friedrich Goldscheider"; Bestellung der Firma Wahliss, Wien, vom 27.2.1946 in Ausführung „Terrakotta"; Nachlass Matula, Wien.

8213M
Matula, Viktor
RELIEF „Traube" (Wandrelief)
ca. 1937/38
8213M, H. 23 cm, B. 21 cm; GSNachf. 2,14a
Lit.: GSNachf. 2,14 (m. Abb.), hier mit Ang. des Originaltitels „Relief Traube".
Quelle: Nachlass Matula, Wien

8214
WANDMASKE Madonna mit Schleier und Krone
ca. 1937/38
8214,18,7 gem. XIID, Firmenmarke 19, Made in Austria, H. 28 cm; eBay # 2660980712

8218
Goebel, Kurt
Skifahrerin
ca. 1937/38

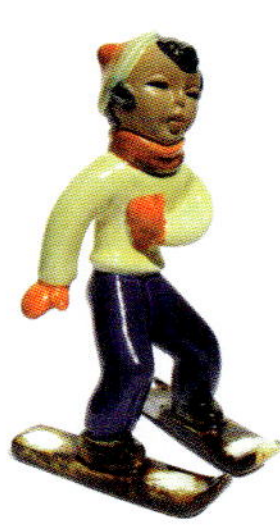

8218,20 Firmenmarke 19, Made in Germany, unsign., Ausformung zw. 1938–1941, H. 14,6 cm; Sammlung a@goldscheider.net

8220F
FIGUR „Skiläuferin"
ca. 1937/38
8220F, H. fehlt
Quelle: AGM (Bestellung von Ferd. Lundquist, Göteborg, Schweden vom 23. August 1950), hier mit Ang. des Originaltitels „Skiläuferin".

8224F/M
Matula, Viktor
GRUPPE „Rehe"
ca. 1937/38
8224F/M, H. fehlt
Quellen: AGM (Kontoblatt der Wiener Manufaktur Josef Schuster für die Firma Porslins Ericsson-Lund, Stockholm?, Schweden vom 7. Januar 1943); (Rechnung der Wiener Manufaktur Josef Schuster an Laszlo Bösze, Szeged, Ungarn, vom 5. August 1944), hier jeweils mit Ang. des Originaltitels GRUPPE „Rehe"; Nachlass Matula, Wien.

8225
Dakon, Stephan (zugeschr.)
GRUPPE Kleines Mädchen, einen Foxterrier am Halsband ziehend (auf Ovalsockel)
ca. 1937/38
8225,15,64 Firmenmarke 19, Made in Austria, Hand dec., unsign., H. 25,5 cm; WKA 38

8226
Dakon, Stephan
Junges Mädchen im Kleid
ca. 1937/38
8226,66 (Firmenmarke ohne Ang.), Ausformung zw. 1945–1953, H. 22 cm; Christie's SK 9.4.1999
Bem.: Da im Auktionskat. keine näheren Angaben gemacht werden bzw. keine Abb. vorhanden ist, kann das Modell hier nicht näher beschrieben werden.

8227M
Matula, Viktor
FIGUR „Lämmchen" (stehend)
ca. 1937/38
8227M, H. 10 cm, B. 14,5 cm; GSNachf. 3,27d
Lit.: GSNachf. 3,27 (m. Abb.), hier mit Ang. des Originaltitels FIGUR „Lämmchen".
Quelle: Nachlass Matula, Wien

8229F
Dakon, Stephan
GRUPPE „Pierrots" (Drei schreitende Kinder, ein kleiner Pierrot mit zwei Mädchen) (auf abgerundetem Rechtecksockel)
ca. 1937/38
8229,20,0 gem. lig. AW, Firmenmarke 19, Made in Germany, gepr. „L"= Z6, Ausformung zw. 1938–1941, H. 25 cm; eBay # 3311793380
8229F, H. 23,5 cm, B. 21 cm; GSNachf. 1,36a; Fotoarchiv K. Täuber, hier mit der Ang.: H. 23 cm, B. 21 cm
-> vgl. 8230= ZweierGRUPPE „Zwei Mädel"
-> vgl. 8232= ZweierGRUPPE „Bub und Mädel"
Lit.: GSNachf. 1,36 (m. Abb.), hier mit Ang. des Originaltitels „Pierrots".
Quelle: Fotoarchiv K. Täuber

8230
Dakon, Stephan
GRUPPE Zwei schreitende Mädchen (auf abgerundetem Rechtecksockel)
ca. 1937/38
8230,50,37 Firmenmarke 19, Made in Germany, unsign., Ausformung zw. 1938–1941, H. 22 cm; eBay # 929589948
Bem.: Diese Mädchengruppe entstammt dem Entwurf der DreierGRUPPE „Pierrots", wobei die beiden den Pierrot flankierenden Mädchen hier als „eigenständiges" Modell angeboten wurden.
-> vgl. 8229= DreierGRUPPE „Pierrots"
-> vgl. 8232= ZweierGRUPPE „Bub und Mädel"

8232F
Dakon, Stephan
GRUPPE „Bub und Mädel" (Zwei schreitende Kinder, ein kleiner Pierrot und ein Mädchen) (auf abgerundetem Rechtecksockel)
ca. 1937/38
8232F, H. 22,5 cm, B. 16 cm; GSNachf. 1,36b
Bem.: Diese beiden Kinder entstammen dem Entwurf der DreierGRUPPE „Pierrots", wobei hier der Pierrot mit einem der Mädchen als „eigenständiges" Modell angeboten wurde.
-> vgl. 8229= DreierGRUPPE „Pierrots"
-> vgl. 8230= ZweierGRUPPE „Zwei Mädel"
Lit.: GSNachf. 1,36 (m. Abb.), hier mit Ang. des Originaltitels „Bub und Mädel"; Könecke (2002), S. 77, 89 (m. Abb.).
Quelle: Fotoarchiv K. Täuber

8233
Dakon, Stephan
Stehende Tänzerin mit nach links geneigtem Kopf, ihr langes Kleid schmetterlingsflügelartig auseinander haltend (auf Ovalsockel)
ca. 1937/38
8233,30,11 (Firmenmarke ohne Ang.), H. 35,5 cm; Christie's SK 21.6.1996

8234F
-> ÜBERSCHNEIDUNG
Dakon, Stephan (zugeschr.)
FIGUR „Tanz" (Schreitende Tänzerin, ihr Kleid seitlich in die Höhe haltend) (auf Ovalsockel)
ca. 1937/38
8234,3,23 Firmenmarke 19 (blau), Made in Austria (blau), Hand dec. (blau), unsign., Pick. 1, Originalpick.: „Mimi Wodak, Edelhausrat, Wien XVI", H. 33 cm, B. 22 cm; Privatsammlung, Wien
8234,11 gem. H., Firmenmarke 19, Made in Austria, unsign., Ausformung zw. 1941–1953, H. 34,3 cm; city-antik
Quelle: AGM (Rechnung an Maurice Mizrahi, Alexandria, Ägypten, vom 22. April 1953), hier mit Ang. des (Original)-Titels „Tanz".

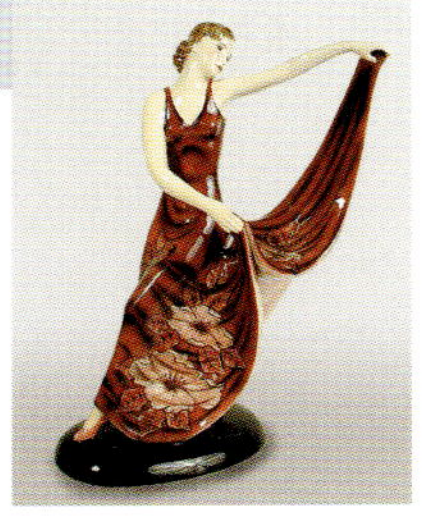

8234 -> ÜBERSCHNEIDUNG
Tänzerin in Pose (ausgebreitete Arme) (auf Rundsockel) mit Blumengebinde)
ca. 1937/38
8234 (Firmenmarke ohne Ang.), H. 33,5 cm; Pinhas, S. 151 (m. Abb.)

8236
Dakon, Stephan (zugeschr.)
Auf einem großen Ball sitzendes Mädchen mit Hut
ca. 1937/38
8236 H. 18 cm, B. 11 cm
Quelle: Fotoarchiv K. Täuber

8239M
Goebel, Kurt
FIGUR „Flötenspieler" (Stehender junger Mann mit Kopfbedeckung, Flöte spielend) (auf Rundsockel)
ca. 1937/38
8239,65 Firmenmarke 19, Made in Austria, unsign., H. 37,3 cm; RM 35; Kett. 86
8239M, H. 38,5 cm, B. 10 cm; GSNachf. 3,10
Lit.: GSNachf. 3,10 (m. Abb.), hier mit Ang. des Originaltitels FIGUR „Flötenspieler".

8241F
Weiss, Claire (Klára)
(d.i. Herczeg, Claire (Klára)
FIGUR „Muffdame" (Stehende Dame in Mantel und Hut mit Muff) (auf Ovalsockel)
ca. 1937/38
8241,1 (Firmenmarke ohne Ang.), unsign.?, Ausformung zw. 1945–1953, H. 38 cm; Christie's SK 9.2.2000

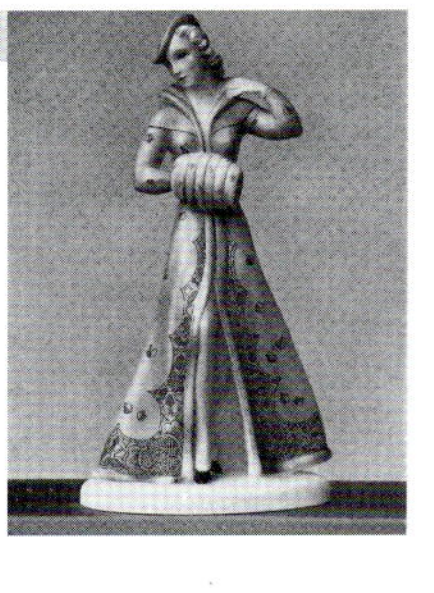

8241,13,14 gem. L, Firmenmarke 19, Made in Austria, sign. C. Weiss, H. 37 cm; Christie's L. 13.12.1989
8241,82,4 (Firmenmarke ohne Ang.), unsign., H. 38,5 cm; Soth. L. 23.10.1987; Christie's SK 27.4.2005
8241,124,31 Firmenmarke Schuster 1, Made in Germany, unsign., Ausformung zw. 1941–1944, H. 37,5 cm; WKA 33
ohne Ang., Firmenmarke Schuster 1, Made in Germany, unsign., Ausformung zw. 1941–1944, H. 37,5 cm; Sigalas 35; 40; 46
8241F, H. 38 cm, B. 20 cm; GSNachf. 1,38
Lit.: GSNachf. 1,38 (m. Abb.), hier mit Ang. des Originaltitels „Muffdame".

8242
Herczeg, Claire (Klára)
Stehende Dame mit Hut und Schirm (auf Rundsockel)
ca. 1937/38
8242,27,19 gem. XII, E, Firmenmarke 26, sign. Claire Herczeg, H. 33 cm; eBay # 486741634

8243
Herczeg, Claire (Klára)
Stehende Dame in langem Kleid mit erhobenem Arm (auf Rundsockel)
ca. 1937/38
8243,23,4 gem. lig. AS, Firmenmarke 26, sign. Claire Herczeg, H. 34,9 cm; eBay # 1200745398
Lit.: Pinhas, S. 160 (m. Abb.), hier mit der falschen Künstlerang. „Dakon" und der falschen Mod.-Nr. 8233. Die richtige Mod.-Nr. 8243 und die Sign. Herczeg konnten selbst geprüft werden.

8245F
Prischl, Adolf
BÜSTE „Madonna" (Madonna mit gekreuzten Händen)
ca. 1936 (für den Entwurf)
8245,3,6 Firmenmarke 19, Made in Austria, Hand dec., unsign., H. 17,8 cm; eBay # 287770146
8245,15,X ger., Firmenmarke 20 (blau), Made in Austria, Hand dec., unsign., Ausformung zw. 1945–1953, H. 15,5 cm; PB
8245,96,6 ger. 47, gem. SG, Firmenmarke 19, Made in Germany, Ausformung zw. 1938–1941, H. 18 cm; Wien Museum Inv. 220.798
8245,98,10 (Firmenmarke ohne Ang.), unsign., H. 20 cm; Doro S. 15.4.1992
8245F, H. 18 cm, B. 12 cm; GSNachf. 6,4b
-> vgl. 7551F
-> vgl. 7985F= RELIEF
Lit.: GSNachf. 6,4 (m. Abb.), hier mit Ang. des Originaltitels BÜSTE „Madonna".
Quelle: Fotoarchiv K. Täuber

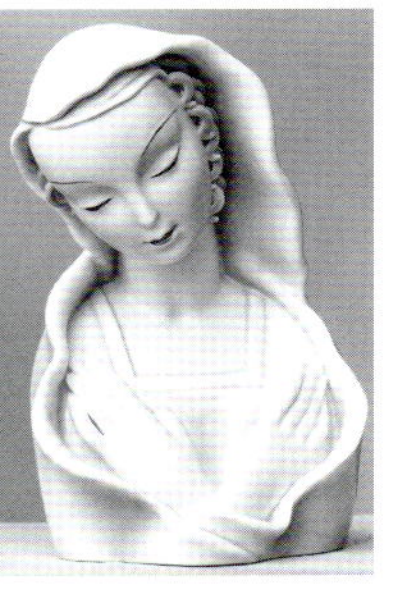

8246
Dakon, Stephan bzw. Goebel, Kurt (zugeschr.)
STEHKOPF Madonna (Frauenkopf mit Kopftuch)
ca. 1937/38
8246 (ger.?), ohne Firmenmarke, Pick. WG 1, Ausformung 50er Jahre, H. 17,1 cm; eBay # 3735255425
8246 ohne Ang., H. 18cm, B. 12,5 cm; Könecke (2002), S. 77, 89 (m. Abb.)
Bem.: Diese Büste (laut Könecke S. 77), vielmehr ein Stehkopf, wurde auch noch zw. 1953–1960 bei Carstens (GWM West-Germany) produziert. Möglicherweise ist die Ang. (oder Zuschreibung) des Künstlers Dakon bei Könecke falsch. Vielmehr nämlich scheint der Entwurf von Kurt Goebel zu sein.
Lit.: KGSF1; Könecke (2002), S. 77, hier mit Ang. des Künstlers S. Dakon, S. 89 (m. Abb.)

8247M
Goebel, Kurt
WANDMASKE „Maske Araberin" (Frauenkopf mit Hand, Ring und Kopftuch)
ca. 1937 (für den Entwurf)
8247,9,12,3 Firmenstempel WG 1, Ausformung 50er Jahre, unsign., H. 9,8 cm; PB Deutschland

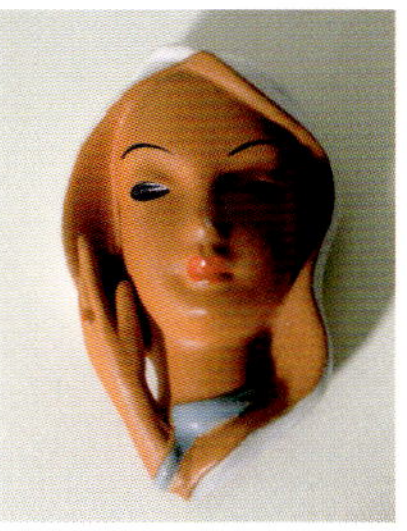

8247 (Stempel), Firmenstempel WG 1, Ausformung 50er Jahre, unsign., H. 9,6 cm; Sammlung a@goldscheider.net
8247M, H. 9,5 cm, B. 6 cm; GSNachf. 2,7b; 2,10f (hier mit der Ang.: H. 9 cm)
Bem.: Diese Maske wurde auch noch zw. 1953–1960 bei Carstens (GWM West-Germany) produziert.
-> vgl. 7813

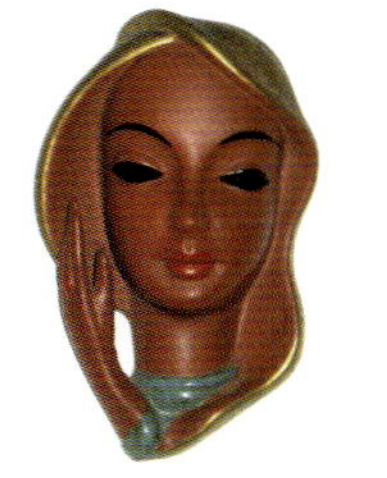

Lit.: GSNachf. 2,7 (m. Abb.), hier mit Ang. des Originaltitels „Maske Araberin"; KGSF1; Könecke (2002), S. 77, hier mit falscher Ang. des Künstlers A. Prischl, S. 89 (m. Abb.).
Quelle: AGM (Auftragszettel für die Firma „Haus der Geschenke", Aachen, vom 16. Januar 1953); Fotoarchiv K. Täuber.

8248M
Prischl, Adolf
WANDMASKE „Maske" (Frauenkopf mit kurzem Kragenansatz)
ca. 1937 (für den Entwurf)
8248,13,X ger., Firmenmarke 19 (blau), Made in Austria (blau), Hand dec. (blau), unsign., Ausformung zw. 1945–1953, H. 9 cm; Wien Museum Inv. 186.652

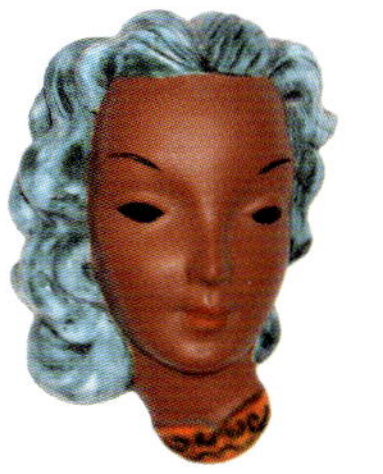

8248,13,X ger., Firmenmarke 19, Ausformung zw. 1945–1953, H. 9 cm; PB Deutschland
8248,13 ger., Firmenmarke 19, Made in Austria, unsign., Ausformung zw. 1945–1953, H. 8,9 cm; eBay # 302224991
8248,16 (Firmenmarke kaum lesbar), Made in Austria, unsign., Ausformung zw. 1945–1953, H. 9,4 cm; Sammlung a@goldscheider.net

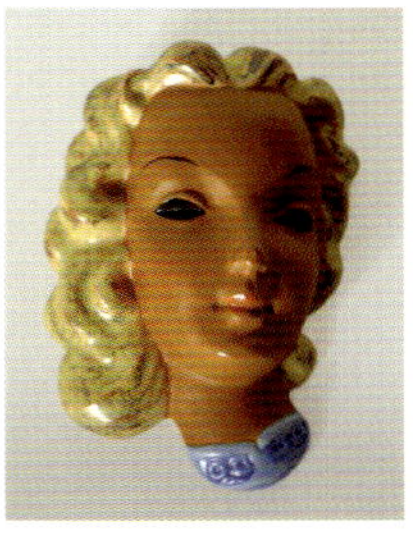

8248,20,16 ger., ohne Firmenmarke, unsign., H. 9 cm; Wien Museum Inv. 186.651
8248,9 Firmenstempel WG 1, Ausformung 50er Jahre, unsign., H. 9,6 cm; Sammlung John Nash
8248M, H. 9 cm, B. 6 cm; GSNachf. 2,10a
Bem.: Diese Maske wurde auch noch zw. 1953–1960 bei Carstens (GWM West-Germany) produziert.
-> vgl. 7816; 7931
Lit.: GSNachf. 2,10 (m. Abb.), hier mit Ang. des Originaltitels „Maske"; KGSF1; Könecke (2002), S. 72, 77, 89 (m. Abb.).

8249F/M
Prischl, Adolf
WANDMASKE „Maske/Goldband" (Frauenkopf mit Haarband und Halstuch)
ca. 1937 (für den Entwurf)
8249,13,G Firmenmarke 19 (blau), Made in Austria (blau), Hand dec. (blau), unsign., Ausformung zw. 1945–1953, H. 9,5 cm; Wien Museum Inv. 186.658

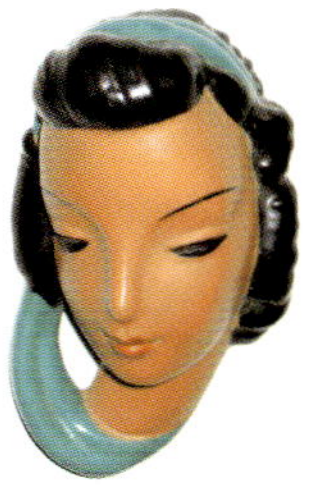

8249,91 ger., gem. lig. AH, Firmenmarke 19, Made in Germany, Ausformung zw. 1938–1941, H. 9,5 cm; PB
8249 (Stempel), Firmenstempel WG 1, Ausformung 50er Jahre, unsign., H. 9,4 cm; Sammlung a@goldscheider.net
8249M, H. 9 cm, B. 5,5 cm; GSNachf. 2,8b; 2,10b
Bem.: Diese Maske wurde auch noch zw. 1953–1960 bei Carstens (GWM West-Germany) produziert.
-> vgl. 8041
Lit.: GSNachf. 2,10 (m. Abb.), hier mit Ang. des Originaltitels „Maske"; KGSF1; Könecke (2002), S. 77, hier mit Ang. des Künstlers A. Prischl, S. 89 (m. Abb.).
Quelle: AGM (Kontoblatt der Wiener Manufaktur Josef Schuster für die Firma Porslins Ericsson-Lund, Stockholm?, Schweden vom 7. Januar 1943), hier mit Ang. des Titels „Kleinmaske"; Fotoarchiv K. Täuber.

8250
Zwei stehende Schafe (auf Rechtecksockel)
ca. 1937/38
8250,16,7 Firmenmarke 19, Made in Austria, H. 22,5 cm; eBay # 732502640

8253F
Prischl, Adolf
FIGUR „Pamela (Akt)" (Kniender Damenakt) (auf Ovalsockel)
ca. 1937/38
8253,1,47 Firmenmarke 19, Made in Austria, Hand dec., unsign., H. 23,5 cm; eBay # 225673227

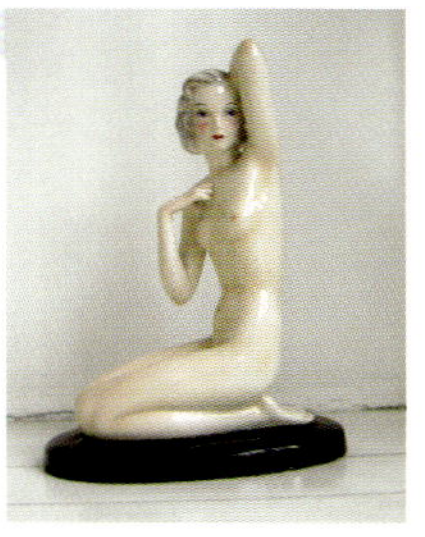

8253,14 ger., gem. W.H., E, G, Firmenmarke 19, Made in Austria, Hand dec., unsign., Ausformung zw. 1945–1953, H. 23 cm; eBay # 457457763
8253,27,3 gem. G, S, Firmenmarke 19, Made in Austria, Hand dec., unsign., H. 24 cm; eBay # 3700830821

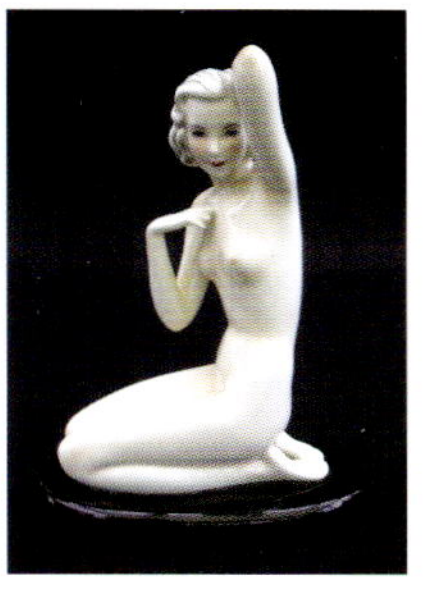

8253,42,6 gem. lig. AS, EW, Firmenmarke 19, Made in Austria, sign. Prischl, H. 23,5 cm; Sammlung Robert Senft
8253,43 gem. GV, XIID, Firmenmarke 19, Made in Austria, Ausformung zw. 1945–1953, H. 24 cm; eBay # 3718715371
8253,121,31 gem. lig. EW, Firmenmarke 19, Made in Germany, sign. Prischl, gepr. „L" = Z6, Ausformung zw. 1938–1941, H. 24,3 cm, B. 17,5 cm; Privatsammlung, Wien

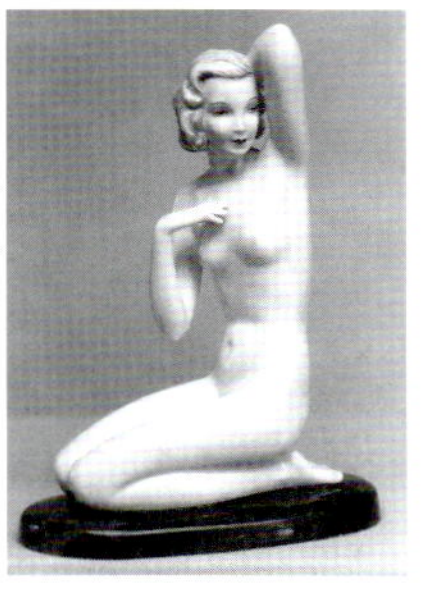

8253,150,1,P gem. B., Firmenmarke 19, Made in Germany, sign. Prischl, Ausformung zw. 1938–1941, Pick. 6, H. 23,5 cm; Wien Museum Inv. 220.746
8253,164,5 Firmenmarke 19, Made in Germany, sign. Claire Herczeg, Ausformung zw. 1938–1941, H. 23,7 cm; Soth. L. 23.6.1989; Bem.: Die Sign. von „Claire Herczeg" ist eindeutig ein Versehen des Modelleurs.
8253F, H. 24 cm, B. 18 cm; GSNachf. 1,39
Bem.: Diese Figur wurde auch noch zw. 1953–1960 bei Carstens (GWM West-Germany) produziert.
Lit.: GSNachf. 1,39 (m. Abb.), hier mit Ang. des Originaltitels „Pamela (Akt)"; KGSF1; KGSF2; Könecke (2002), S. 43 (m. Abb.), 77.
Quelle: AGM (Tantiemenabrechnung 2. Halbjahr 1952 an A. Prischl vom 29.1.1953; Rechnung an Maurice Mizrahi, Alexandria, Ägypten, vom 22. April 1953), hier unter dem Titel „Akt" geführt, geliefert in Ausführung „matt"; Fotoarchiv K. Täuber.

8254
Goebel, Kurt
Sitzendes Liebespaar (auf Ovalsockel)
ca. 1937/38
8254,0 ger., gem. O, Firmenmarke 19, Made in Austria, unsign., Ausformung zw. 1945–1953, H. 24 cm; eBay # 121737922

8258F
Matula, Viktor
FIGUR „Hirsch" (liegend)
ca. 1937/38
8258,2 Firmenmarke 19, Made in Austria, unsign., Ausformung zw. 1945–1953, H. ca. 33 cm; KH

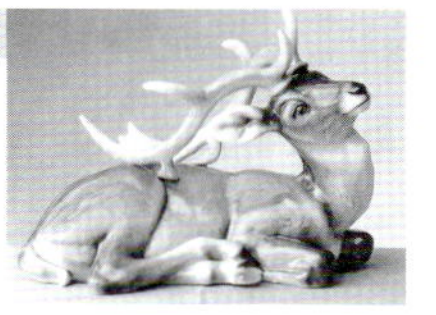

8258,11 Firmenmarke 19, Made in Austria, unsign., Ausformung zw. 1945–1953, H. 30 cm, B. 43 cm; eBay # 3659372225
8258 (Firmenmarke ohne Ang.), unsign., H. 33 cm, L. 41 cm; Doro 9.6.1998
8258F, H. 33 cm, B. 43 cm; GSNachf. 5,27
Bem.: Diese Figur wurde auch noch zw. 1953–1960 bei Carstens (GWM West-Germany) produziert.
Lit.: GSNachf. (2. Ausg. bzw. Erg. Bl.) 5,27 (m. Abb.); Könecke (2002), S. 77, 90 (m. Abb.).
Quellen: AGM (Bestellung der Firma Wahliss (Wien) vom 27.2.1946 an die „Arbeitsgemeinschaft der Firma Friedrich Goldscheider", hier mit Ang. des Originaltitels „Hirsch" (Ausführung Terrakotta); Rechnung an Maurice Mizrahi, Alexandria, Ägypten, vom 22. April 1953), hier mit Ang. des Originaltitels „Hirsch" (Ausführung Fayence); Nachlass Matula, Wien; Fotoarchiv K. Täuber.

8259
Matula, Viktor
Junger Adler (auf Felssockel)
ca. 1937/38
8259,8 Firmenmarke 19, Made in Austria, unsign., Ausformung zw. 1945–1953, H. 21,5 cm; Wien Museum Inv. 190.048

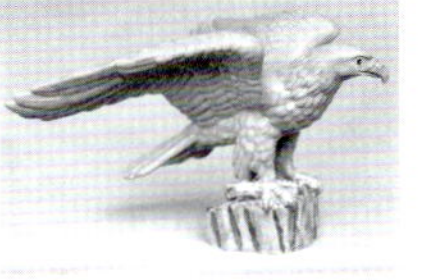

8259 H. 22 cm, B. 33 cm; Fotoarchiv K. Täuber
Quelle: Nachlass Matula, Wien; Fotoarchiv K. Täuber.

8261
Prischl, Adolf
Stehende junge Dame im Mantel mit Pelzkragen, Hut und Pekinese (auf Ovalsockel)
ca. 1937/38
8261,4,2 ger. XII, B, gem. GH., Firmenmarke 19, Made in Austria, H. 31 cm; Zezschwitz 24

8262
Goebel, Kurt
WANDBEHANG Skifahrerin mit Kappe und geschulterten Skiern
ca. 1937/38
8262,44 Firmenmarke 19, Made in Austria, unsign., Ausformung zw. 1945–1953, H. 20,3 cm; eBay # 1316356714

8264M
FIGUR „Sequoia" (Kniender Mädchenakt mit Reh) (auf flachem Rechtecksockel)
ca. 1937/38

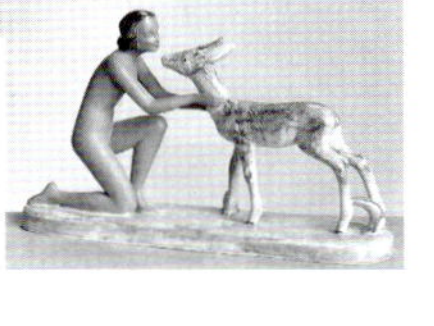

8264M, H. 15 cm, B. 24,5 cm; GSNachf. 3,5b; Fotoarchiv K. Täuber, hier mit der Ang.: H. 17 cm, B. 26 cm
Bem.: Das Wort Sequoia kommt aus dem Indianischen und bedeutet Mammutbaum. Wie es zur Namensgebung bei diesem Modell kam, lässt sich weder nachvollziehen noch durch das Firmenarchiv (AGM) belegen.
Lit.: GSNachf. 3,5 (m. Abb.), hier mit Ang. des Originaltitels „Sequoia".
Quelle: AGM (Kontoblatt der Wiener Manufaktur Josef Schuster für die Firma Porslins Ericsson-Lund, Stockholm?, Schweden vom 7. Januar 1943), hier mit Ang. des Originaltitels „Sequoia"; Fotoarchiv K. Täuber.

8266M
Prischl, Adolf
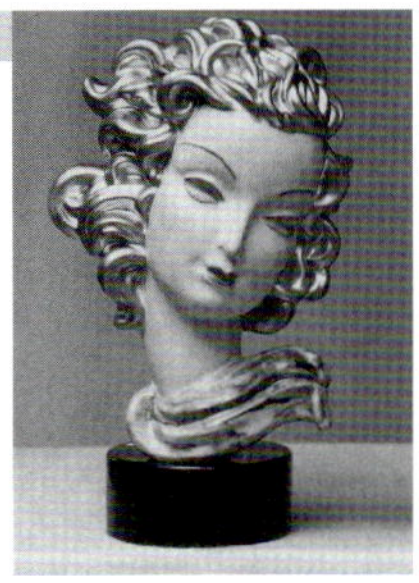
WANDMASKE „Maske"
(Frauenkopf mit Schal)
ca. 1937 (für den Entwurf)
8266 Firmenstempel WG 1, unsign., Ausformung 50er Jahre, H. 9 cm; eBay # 853612623
8266,371/2,10 Firmenstempel WG 1a, unsign., Ausformung 50er Jahre, H. 9 cm; eBay # 2290476140
8266M, H. 9,5 cm, B. 6,5 cm; GSNachf. 2,6b; 2,10d
Bem.: Diese Maske wurde auch noch zw. 1953–1960 bei Carstens (GWM West-Germany) produziert.
-> vgl. 7784
Lit.: GSNachf. 2,6 (m. Abb.), hier mit Ang. des Originaltitels „Maske"; KGSF1; Könecke (2002), S. 77, hier mit Ang. des Künstlers A. Prischl, S. 89 (m. Abb.).
Quelle: Fotoarchiv K. Täuber

8268
WANDMASKE Frauenkopf mit großen Locken (Curls) und Kragen
ca. 1937/38
8268,3 (Firmenmarke ohne Ang.), Pick. ohne Ang., Ausformung zw. 1945–1953, H. 28 cm; Stahl 28.11.1992; 24.4.1993

8269
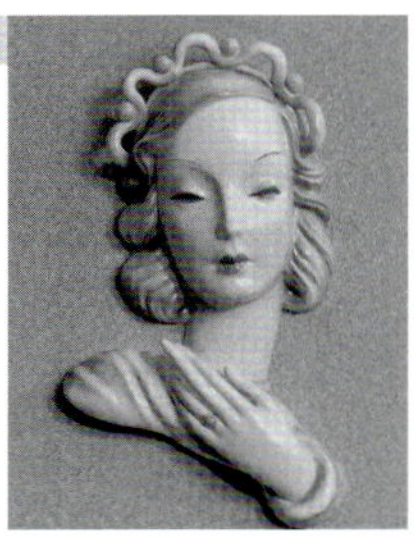
WANDMASKE „Maske mit Golddiadem" (Frauenkopf mit Diadem und Hand)
ca. 1937/38
8269,2 gem. L., ohne Firmenmarke, Ausformung zw. 1945–1953, H. 36 cm; Wien Museum Inv. 186.647
8269,2 gem. S., Firmenmarke 19, Made in Austria, Ausformung zw. 1945–1953, H. 33 cm; Wien Museum Inv. 220.804; Bem.: Dieses Exemplar wurde ohne das mehrfach geschwungene und mit Perlen verzierte Diadem ausgeführt.
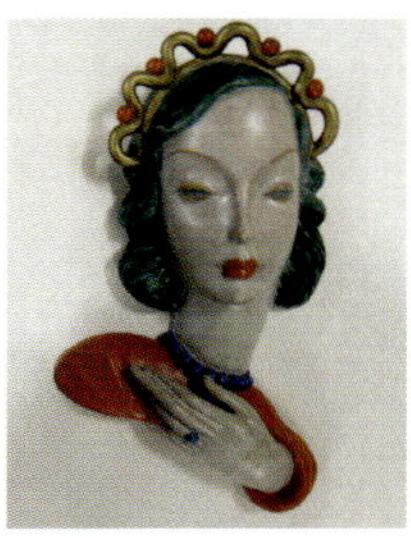
8269,3 gem. X, gem. lig. AH, Firmenmarke 19, Made in ?, H. 33,9 cm; MAK Wien, Inv. Ke 10521-29
8269,9 Firmenmarke 19, H. 34,5 cm; Doro 8.3.2004
8269,33 Firmenmarke 19, Made in Austria, Ausformung zw. 1945–1953, H. 33,5 cm; Soth. L. 20.10.1989
8269,33,19 gem. lig. AH, Firmenmarke 19, Made in Austria, H. 34 cm; eBay # 7301360107
8269,64 Firmenmarke 19a, Made in Germany, Ausformung zw. 1938–1941, Pick. 6, H. 34 cm; eBay # 2598544438
Quelle: AGM (Bestellung der Firma Wahliss (Wien) vom 24.11.1945 an die „Arbeitsgemeinschaft der Firma Friedrich Goldscheider"), hier mit Ang. des (Original-?)Titels „Maske mit Golddiadem"; Bestellung der Firma Wahliss (Wien) vom 27.2.1946 in Ausführung „Terrakotta"); Fotoarchiv K. Täuber.

8270
Goebel, Kurt
Stehendes Mädchen mit Ente (auf Rundsockel)
ca. 1937/38
8270,0 gem. L., Firmenmarke 19, Made in Austria, unsign., Ausformung zw. 1945–1953, H. 36,5 cm; Doro 10.3.2003; PB

8272M
Matula, Viktor
FIGUR „Lämmchen" (Kopf nach links gewandt)
ca. 1937/38
8272M, H. 10,5 cm, B. 13 cm; GSNachf. 3,27e
Lit.: GSNachf. 3,27 (m. Abb.), hier mit Ang. des Originaltitels FIGUR „Lämmchen".
Quelle: Nachlass Matula, Wien

8273M
Matula, Viktor
FIGUR „Lämmchen" (fressend)
ca. 1937/38
8273M, H. 9 cm, B. 13 cm; GSNachf. 3,27f
Lit.: GSNachf. 3,27 (m. Abb.), hier mit Ang. des Originaltitels FIGUR „Lämmchen".
Quelle: Nachlass Matula, Wien

8275
Prischl, Adolf (zugeschr.)
FIGUR Liegende Katze mit Ball
ca. 1937/38
8275 Firmenmarke 19, unsign., H. 4 cm, L. 13 cm; Sammlung mad
Bem.: Für die Zuschreibung an Adolf Prischl danke ich Frau Mag. Uta Matschiner.

8276
Stehender Mädchenakt (in Pose) mit Barsoi (auf Ovalsockel)
ca. 1937/38
8276,6,14 Firmenmarke 19, Made in Austria, Pick. 4, H. 38 cm; WKA 35
8276,102,31 (Firmenmarke ohne Ang.), H. 37,7 cm; Doro 26.11.2002

8277M
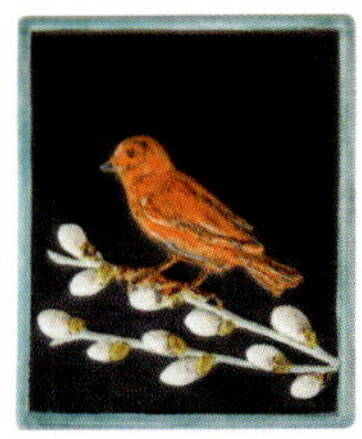
RELIEF „Vogel" (Wandrelief)
ca. 1937/38
8277 ger., gem. W, Firmenmarke 19, Made in Germany, Ausformung zw. 1938–1941, H. 15,5 cm, B. 12,7 cm; Sammlung a@goldscheider.net
8277,14 ger., gem. G, Firmenmarke 19, Made in Austria, Ausformung zw. 1945–1953, H. 15,5 cm, B. 13 cm; eBay # 280019736077
Quelle: AGM (Kontoblatt der Wiener Manufaktur Josef Schuster für die Firma Porslins Ericsson-Lund, Stockholm?, Schweden vom 7. Januar 1943), hier mit Ang. des Originaltitels RELIEF „Vogel".

8278F
Matula, Viktor
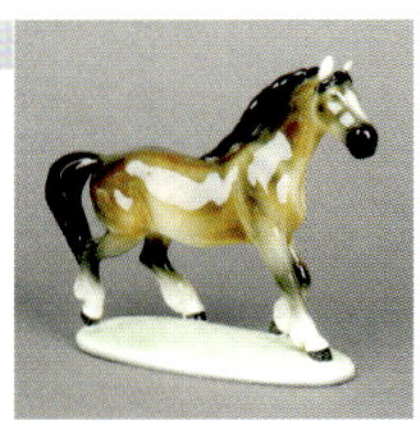
FIGUR „Pferd" (scheuend)
(auf Ovalsockel)
ca. 1937/38
8278,20,8 Firmenmarke 19, Made in Austria, Hand dec., unsign., H. 16 cm; eBay # 1150814678
8278 ohne Ang., H. 17 cm; PB Wien 19
8278F, H. 17 cm, B. 19 cm; GSNachf. 5,19d
Lit.: GSNachf. 5,19 (m. Abb.), hier mit Ang. des Originaltitels FIGUR „Pferd".
Quelle: Nachlass Matula, Wien

8279F
Matula, Viktor
FIGUR „Pferd" (bäumend) (auf Ovalsockel)
ca. 1937/38
8279F, H. 19,5 cm, B. 14 cm; GSNachf. 5,19e
Bem.: GS betitelt das Pferd „bäumend". Der korrekte Ausdruck wäre jedoch „steigend".
Lit.: GSNachf. 5,19 (m. Abb.), hier mit Ang. des Originaltitels FIGUR „Pferd" (bäumend).
Quellen: AGM (Kontoblatt der Wiener Manufaktur Josef Schuster für die Firma Porslins Ericsson-Lund, Stockholm?, Schweden vom 7. Januar 1943); (Rechnung der Wiener Manufaktur Josef Schuster an Laszlo Bösze, Szeged, Ungarn, vom 5. August 1944; Bestellungen der Firma Wahliss (Wien) vom 9. und 24.11.1945 und 27.2.1946 an die „Arbeitsgemeinschaft der Firma Friedrich Goldscheider"; Rechnung an Maurice Mizrahi, Alexandria, Ägypten, vom 22. April 1953), hier jeweils mit Ang. des Originaltitels „Pferd"; Nachlass Matula, Wien.

8280F
Matula, Viktor / Jarl, Karin
FIGUR „Pferd" (trabend) (auf Ovalsockel)
ca. 1937/38
8280,41,19 gem. F., Firmenmarke 19, Made in Germany, sign. Jarl, Ausformung zw. 1938–1941, H. 15,5 cm; eBay # 1455039618; Bem.: Die Sign. „Jarl" muss wohl als Versehen des Modelleurs angesehen werden, da sich im Nachlass Matula ein entsprechendes Foto und Notizen zu diesem Modell finden.
8280F, H. 16,5 cm, B. 18 cm; GSNachf. 5,19b
Lit.: GSNachf. 5,19 (m. Abb.), hier mit Ang. des Originaltitels FIGUR „Pferd".
Quellen: AGM (Kontoblatt der Wiener Manufaktur Josef Schuster für die Firma Porslins Ericsson-Lund, Stockholm?, Schweden vom 7. Januar 1943); (Rechnung der Wiener Manufaktur Josef Schuster an Laszlo Bösze, Szeged, Ungarn, vom 5. August 1944), hier jeweils mit Ang. des Originaltitels „Pferd"; Nachlass Matula, Wien.

8281F
Matula, Viktor
FIGUR „Pferd" (stehend, Kopf nach rechts)
(auf Ovalsockel)
ca. 1937/38
8281,38,66 Firmenmarke 19, unsign., H. 17 cm; eBay # 2224165169
8281F, H. 16,5 cm, B. 18 cm; GSNachf. 5,19a
Lit.: GSNachf. 5,19 (m. Abb.), hier mit Ang. des Originaltitels FIGUR „Pferd".
Quellen: AGM (Rechnung der Wiener Manufaktur Josef Schuster an Laszlo Bösze, Szeged, Ungarn, vom 5. August 1944), hier mit Ang. des Originaltitels „Pferd"; Nachlass Matula, Wien.

8282F / M
Matula, Viktor
FIGUR „Pferd" (wiehernd) (auf Landschaftssockel)
ca. 1937/38
8282F, H. 18,5 cm, B. 19 cm; GSNachf. 5,19f
8282M, ohne Ang.; -> vgl. Quelle: AGM
Bem.: Modellvarianten (Apfelschimmel, bzw. „Hengst", schwarz) in der Staffierung hier belegt durch den in der Quelle genannten Auftragszettel an eine Aachener Firma. Diese Figur wurde auch noch zw. 1953–1960 bei Carstens (GWM West-Germany) produziert.
Lit.: GSNachf. 5,19 (m. Abb.), hier mit Ang. des Originaltitels FIGUR „Pferd"; Könecke (2002), S. 77.
Quellen: AGM (Kontoblatt der Wiener Manufaktur Josef Schuster für die Firma Porslins Ericsson-Lund, Stockholm?, Schweden vom 7. Januar 1943), hier mit Ang. des Originaltitels „Pferd"; (Auftragszettel für die Firma „Haus der Geschenke", Aachen, vom 16. Januar 1953), hier mit dem Vermerk 8282F, „Apfelschimmel"; 8282M „Pferd, schwarz"; Nachlass Matula, Wien.

8283F
Matula, Viktor / Jarl, Karin
FIGUR „Pferd" (fressend) (auf Landschaftssockel)
ca. 1937/38
8283,6,19 ger. XIII, gem. F., Firmenmarke 19, Made in Austria, sign. Jarl, Pick. 6, H. 12 cm; eBay # 3308882016; Bem.: Die Sign. „Jarl" muss wohl als Versehen des Modelleurs angesehen werden, da sich im Nachlass Matula ein entsprechendes Foto und Notizen zu diesem Modell finden.
8283F, H. 12 cm, B. 18 cm; GSNachf. 5,19c
Lit.: GSNachf. 5,19 (m. Abb.), hier mit Ang. des Originaltitels FIGUR „Pferd".
Quelle: Nachlass Matula, Wien

8289M
Matula, Viktor
FIGUR „Lämmchen"
ca. 1937/38
8289M, H. 12 cm, B. 15 cm; GSNachf. 3,27a
Lit.: GSNachf. 3,27 (m. Abb.), hier mit Ang. des Originaltitels FIGUR „Lämmchen".
Quelle: Nachlass Matula, Wien

8290
Drei Birner mit Blattwerk
ca. 1937/38
8290,3 gem. O, Firmenmarke 19, Made in Austria, Besitzerstempel: „Risa Arlt, Konditor, Wien III., Hauptstr. 40", Ausformung zw. 1945–1953, H. 21,5 cm, B. 19,8 cm; Wien Museum Inv. 220.783

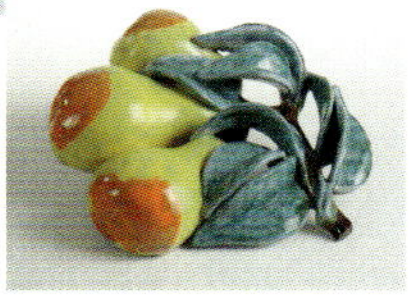

8296
SCHREIBZEUG/ABLAGE
Segelschiff
Inschrift: ÜBERSEE POST LEIPZIG
ca. 1937/38
Form: rechteckig
Dekor: Segelschiff (vollplastisch)
8296,16 ger., Firmenmarke 19, Made in Austria, Ausformung zw. 1945–1953, H. 13,5 cm; Wien Museum Inv. 220.773

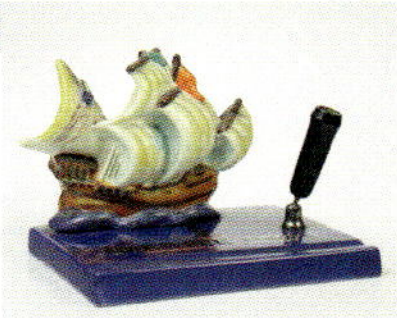

8300
Dakon, Stephan
Schreitende Tänzerin in gerüschtem Ballkleid und Hut (auf Ovalsockel)
ca. 1938
8300,2,3 Firmenmarke 19, Made in Germany, Ausformung zw. 1938–1941, H. 32 cm; eBay # 3721649628
8300,26,31 Firmenmarke 19, Made in Germany, Ausformung zw. 1938–1941, Pick. 6, H. 32 cm; WKA 35 (Lot 1069, außer Kat.)

8301F
FIGUR „Menuette" (Stehende Tänzerin, ihr Kleid mit beiden Händen anhebend) (auf Ovalsockel)
ca. 1938
8301,8 Firmenmarke 19, Made in Germany, Ausformung zw. 1938–1941, H. 31,7 cm; Doro 9.6.1998
8301,11 gem. G, Firmenmarke 19, Made in Austria, Ausformung zw. 1945–1953, H. 31 cm; Doro 15.3.2005
8301,20,6 Firmenmarke 19, Made in Germany, Ausformung zw. 1938–1941, H. 30 cm; Doro K 19.5.2004
Quelle: AGM (Rechnung der Wiener Manufaktur Josef Schuster an Laszlo Bösze, Szeged, Ungarn, vom 5. August 1944), hier mit Ang. des Originaltitels „Menuette".

8302F
Dakon, Stephan
FIGUR „Rumba" (Tänzerin in Bikini, langem Rock/Schleppe und Halstuch) (auf Ovalsockel)
ca. 1938
8302,1,14 gem. HS?, Firmenmarke 19, Made in Austria, Hand dec., unsign., H. 31,5 cm; Sammlung Stressless, Portugal
8302,40,4 gem. F., Firmenmarke 19, Made in Germany, Ausformung zw. 1938–1941, H. 31,8 cm; Doro 24.11.2004
8302,45,66 Firmenmarke 19, Made in Germany, unsign., Ausformung zw. 1938–1941, H. 31,6 cm; Doro 30.11.2001
8302,136,31 Firmenmarke 19, Made in Germany, unsign., Ausformung zw. 1938–1941, H. 31,8 cm; Doro 16.6.1992; 24.5.2000
8302,S gem. M.G., W, Firmenmarke 20, Made in Austria, unsign.?, H. 31,7 cm; city-antik
8302F, H. 32,5 cm, B. 24 cm; GSNachf. 1,40
Bem.: Bei der dargestellten Tänzerin könnte es sich um Margit Symo handeln.
Lit.: GSNachf. 1,40 (m. Abb.), hier mit Ang. des Originaltitels „Rumba".
Quelle: Fotoarchiv K. Täuber

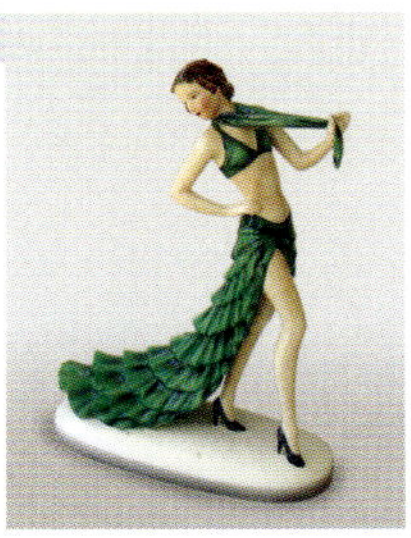

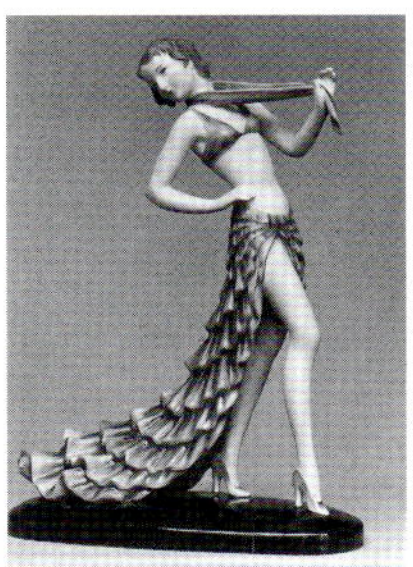

8303F
Dakon, Stephan (zugeschr.)
FIGUR „Skijunge mit Stock" (Kleiner skifahrender Junge mit Zipfelmütze)
ca. 1938
8303F, H. 15 cm, B. 10 cm; GSNachf. 1,41c
Bem.: Diese Figur gelangte 1938 erstmalig in den Verkauf (Brief von GS an die Wiener Firma Keramos, vom 25.2.1953).
Lit.: GSNachf. 1,41 (m. Abb.), hier mit Ang. des Originaltitels „Skijunge mit Stock".
Quelle: AGM (Brief von der Wiener Manufaktur Goldscheider an die Firma Keramos, vom 25.2.1953).

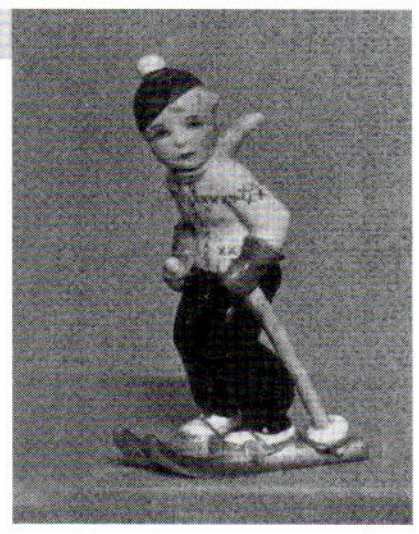

8304F
Dakon, Stephan (zugeschr.)
FIGUR „Skijunge ohne Stock" (Kleiner skifahrender Junge)
ca. 1938
8304,44,31 Firmenmarke 19, Made in Germany, Ausformung zw. 1938–1941, unsign., H. 15,5 cm; eBay # 1486950340
8304F, H. 15,5 cm, B. 11 cm; GSNachf. 1,41a
Bem.: Diese Figur gelangte 1938 erstmalig in den Verkauf (Brief von GS an die Wiener Firma Keramos, vom 25.2.1953).
Lit.: GSNachf. 1,41 (m. Abb.), hier mit Ang. des Originaltitels „Skijunge ohne Stock".
Quelle: AGM (Brief von der Wiener Manufaktur Goldscheider an die Firma Keramos, vom 25.2.1953).

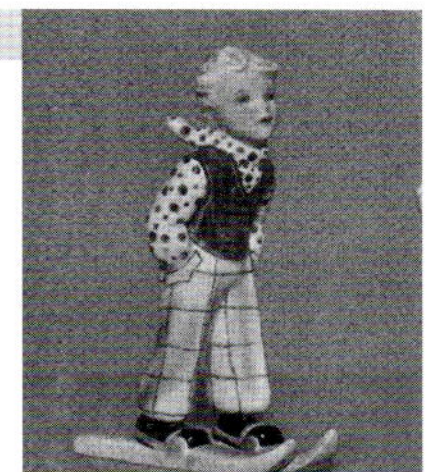

8305F
Dakon, Stephan (zugeschr.)
FIGUR „Skigirl" (Kleines stehendes Mädchen mit Zipfelmütze, ihre Skier haltend)
ca. 1938
8305,122,13 (Firmenmarke ohne Ang.), unsign., H. 17,5 cm; Doro 21.5.1998
8305 Firmenmarke 19, Made in Austria, Hand dec., unsign., H. 17,3 cm; eBay # 1501983253
8305F, H. 18 cm, B. 9,5 cm; GSNachf. 1,41b
Bem.: Diese Figur gelangte 1938 erstmalig in den Verkauf (Brief von GS an die Wiener Firma Keramos, vom 25.2.1953).
Lit.: GSNachf. 1,41 (m. Abb.), hier mit Ang. des Originaltitels „Skigirl".
Quelle: AGM (Brief von der Wiener Manufaktur Goldscheider an die Firma Keramos, vom 25.2.1953).

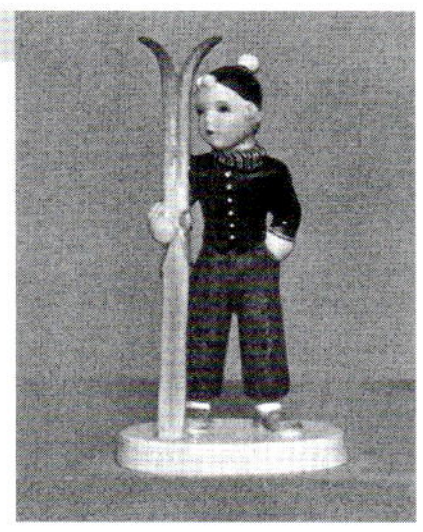

8307
Stehende Madonna mit Aureole und gefalteten Händen (auf halbkugelförmigem Sockel)
ca. 1938
8307 H. 40 cm, B. 14 cm
Quelle: Fotoarchiv K. Täuber

8311
Hahn (auf Ovalsockel)
ca. 1938
8311,7 ger. 71, gem. L., Firmenmarke 19, Made in Austria, Ausformung zw. 1945–1953, H. 11 cm; Wien Museum Inv. 220.761

8312
Stehender Hahn (auf Ovalsockel)
ca. 1938
8312,20 ger. 16A, Firmenmarke 19, Made in Austria, Ausformung zw. 1945–1953, H. 14 cm, B. 18 cm; eBay # 2281366918

8315
Kavalier im Rokokokostüm, seinen Dreispitz haltend (auf Rocaillensockel)
ca. 1938
8315,9 ger. X, gem. H, Firmenmarke 19, Made in Austria, Ausformung zw. 1945–1953, H. 23 cm; Wien Museum Inv. 190.104

8316
Melcher, Josef
Stehender nachdenklicher Mädchenakt mit überkreuzten Armen (auf Rundsockel)
ca. 1938
8316,43,31 Firmenmarke 19, Made in Germany, Ausformung zw. 1938–1941, H. 35,5 cm; Privatsammlung, Wien

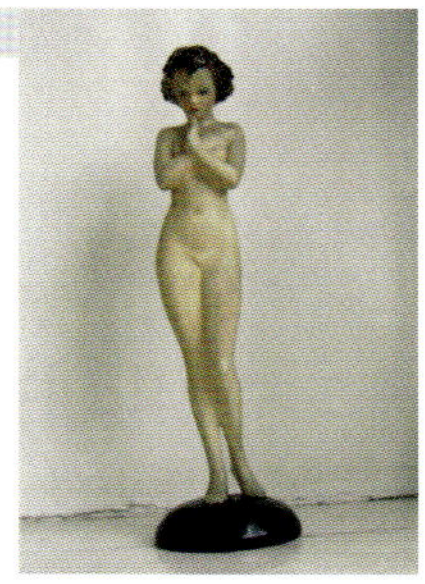

8317F
Dakon, Stephan
FIGUR „Arabella" (Stehende junge Dame, ihr langes Kleid schmetterlingsflügelartig auseinander haltend) (auf Ovalsockel)
ca. 1937 (für den Entwurf)
8317,2,20 gem. lig. KS, Firmenmarke 19, Made in Austria, Hand dec., unsign., Pick. 1, H. 15,5 cm; Sammlung Stressless, Portugal
8317,3,8 gem. MR, Firmenmarke 19, Made in Austria, Hand dec., unsign., H. 15,2 cm; eBay # 254084140
8317,76 Firmenmarke 19, Made in Austria, unsign., Originalpick.: „Rasper & Söhne, Salzburg/Wien", Ausformung zw. 1945–1953, H. 16 cm; city-antik
8317F, H. 16 cm, B. 15,5 cm; GSNachf. 1,20c
-> vgl. 7776; 7961
Lit.: GSNachf. 1,20 (m. Abb.), hier mit Ang. des Originaltitels „Arabella".
Quelle: Archiv Mimi Dakon (Tantiemenabrechnung 1.1.–30.6.1940; Wiener Manufaktur FG, Modellliste Dakon, 7.11.1946; Tantiemenabrechnung 2. Halbjahr 1949; 1. Halbjahr 1950; 2. Halbjahr 1950; (1. Halbjahr 1951, AGM); 2. Halbjahr 1951; 1. Halbjahr 1952), hier jeweils mit dem Originaltitel „Arabella".

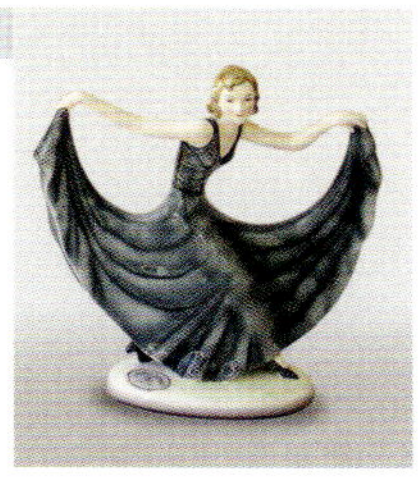

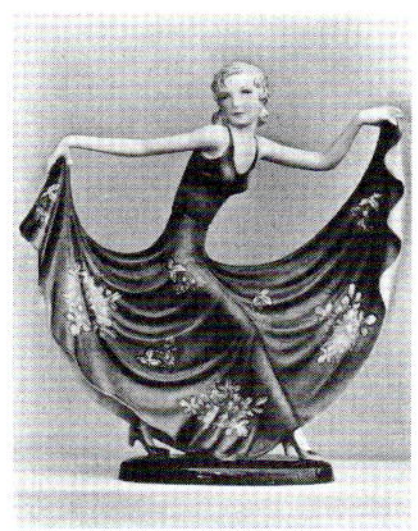

8318F
Dakon, Stephan
FIGUR „Gavotte" (Jessie Matthews als stehende Tänzerin mit Hut, ihr Kleid schmetterlingsflügelartig auseinander haltend) (auf Ovalsockel)
ca. 1935 (für den Entwurf)
8318,S (Firmenmarke ohne Ang.), unsign., Ausformung zw. 1938–1941, H. ca. 16 cm; Christie's SK 12.9.1997
Bem.: Dargestellt ist die engl. Sängerin und Schauspielerin Jessie Matthews (11.3.1907–19.8.1981).
-> vgl. 7257
-> vgl. Lorenzl 7581= ähnl. Modell.
Lit.: DTW, S. 35 (m. Abb.), die hier wiedergegebene zeitgenössische Fotografie von einem Fotografen namens Sascha diente als Vorlage für die Ausführung dieses Modells.
Quelle: Archiv Mimi Dakon (Tantiemenabrechnung 1.1.–30.6.1940; Wiener Manufaktur FG, Modellliste Dakon, 7.11.1946; Tantiemenabrechnung 2. Halbjahr 1949; 1. Halbjahr 1950; 2. Halbjahr 1950; (1. Halbjahr 1951, AGM); 2. Halbjahr 1951, hier auch als „Sorte II" bestellt, 1. Halbjahr 1952), hier jeweils mit Ang. des Originaltitels FIGUR „Gavotte"; AGM.

8319F
Dakon, Stephan
FIGUR „Champagner“ (Nach rechts schreitende junge Dame mit Bikinioberteil, ihren langen Rock schmetterlingsflügelartig auseinander haltend) (auf Ovalsockel)
ca. 1937 (für den Entwurf)

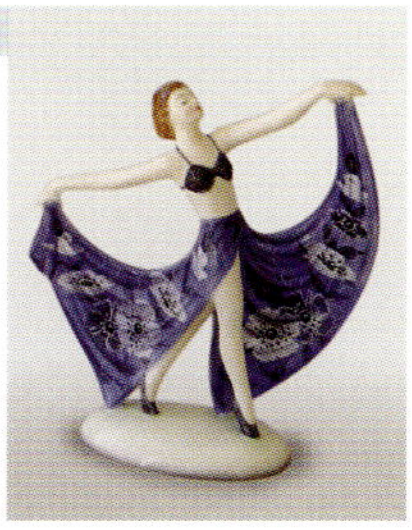

8319,1,3 gem. IS?, Firmenmarke 19, Made in Austria, Hand dec., unsign., H. 18 cm; Sammlung Stressless, Portugal

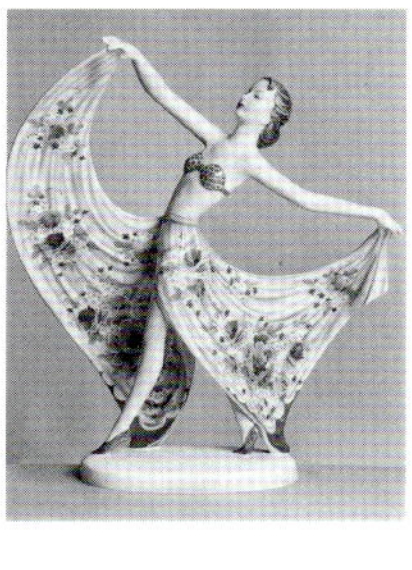

8319,330,31,50 Firmenmarke 19, Made in Germany, Ausformung zw. 1938–1941, H. 18 cm; Mehlis 32
8319F, H. 18 cm, B. 16 cm; GSNachf. 1,21b; Fotoarchiv K. Täuber, hier mit der Ang.: H. 16 cm, B. 13 cm
–> vgl. 7826
Pendant –> vgl. 7857= seitenverkehrtes Modell.
Lit.: GSNachf. 1,21 (m. Abb.), hier mit Ang. des Originaltitels „Champagner“.
Quelle: Fotoarchiv K. Täuber

8320F
Dakon, Stephan
FIGUR „Susette“ (Lilian Harvey, seitlich schreitend) (auf Ovalsockel)
ca. 1938

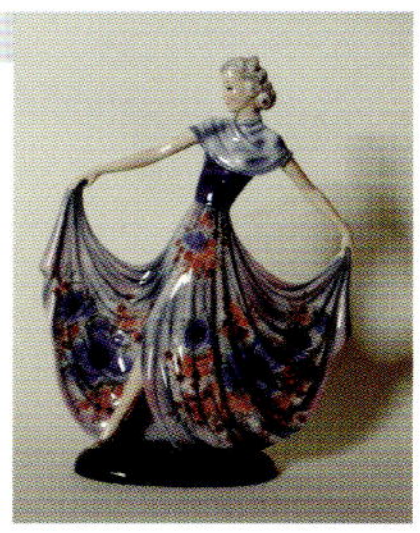

8320,2,42 Firmenmarke 19, Made in Austria, Hand dec., unsign., H. 16,3 cm; eBay # 3746339343
8320,3,85 gem. LS., Firmenmarke 19, Hand dec., unsign., H. 16,5 cm; eBay # 1214616505

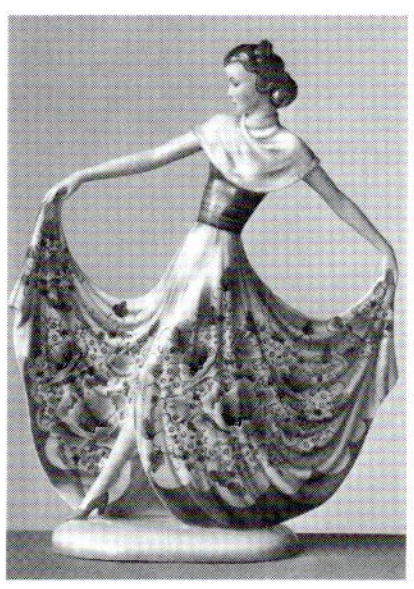

8320,13,7 Firmenmarke 19, Made in Germany, Ausformung zw. 1938–1941, sign. „Lorenzl“ (zwei mal), H. 16,3 cm; Wien Museum Inv. 190.143
8320F, H. 16 cm, B. 13 cm; GSNachf. 1,26b
–> vgl. 8083
Lit.: GSNachf. 1,26 (m. Abb.), hier mit Ang. des Originaltitels „Susette“; DM, 13.1936/37, Nr. 151, S. 98 (m. Abb.), hier eine zeitgenössische Fotografie vom März 1937, die wohl als Vorlage für die Ausführung diente. Dargestellt ist die Schauspielerin Lilan Harvey (eigentl. Lilian Pape, 19.1.1906–27.7.1968), Fotografie: Hämmerer (Ufa).
Quelle: AGM (Rechnung der Wiener Manufaktur Josef Schuster an Laszlo Bösze, Szeged, Ungarn, vom 5. August 1944), hier mit Ang. des Originaltitels „Susette“.

8321
Lorenzl, Josef
„Liane“ (Stehende Tänzerin mit Hut, ihr Kleid schmetterlingsflügelartig auseinander haltend) (auf Ovalsockel)
ca. 1936 (für den Entwurf)
8321,9 (Firmenmarke ohne Ang.), unsign., H. ca. 16 cm; Christie's SK 12.9.1997
8321,22,23 gem. M.G., Firmenmarke 19, Made in Austria, Hand dec., unsign., H. 16,5 cm; eBay # 2661580252
8321,53,31 (Firmenmarke ohne Ang.), unsign., H. 16,5 cm; WKA 38; 44
–> vgl. 7581

8322F
Herczeg, Claire (Klára)
FIGUR „Lydia“ (Schreitende junge Tänzerin, ihr langes Kleid auseinander haltend) (auf Ovalsockel)
ca. 1937 (für den Entwurf)

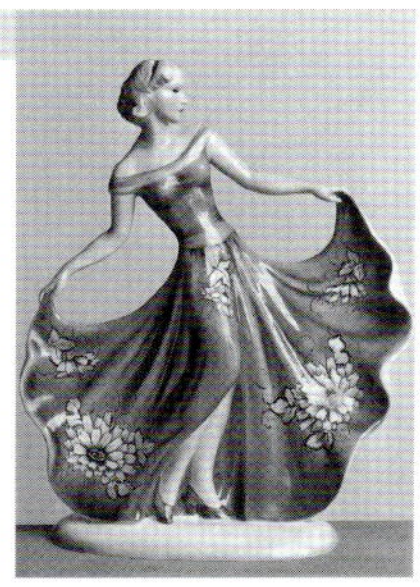

8322,11,14 ger. L, XII D, gem. AN, B.,Firmenmarke 19, Made in Germany, sign Dakon!, Ausformung zw. 1938–1941, H. 16,2 cm; Wien Museum Inv. 220.709;
Bem.: Die Sign. „Dakon“ ist ein Versehen des Modelleurs.
8322,33,4 (Firmenmarke ohne Ang.), sign. Lorenzl!, H. 16 cm; Doro 13.9.2004; Bem.: Die Sign. „Lorenzl“ ist ein Versehen des Modelleurs.
8322,34,0 ger. X,5,B, Firmenmarke 19, Made in Germany, Ausformung zw. 1938–1941, sign. Lorenzl!, H. ca. 16 cm; eBay # 3335059487; Bem.: Die Sign. „Lorenzl“ ist ein Versehen des Modelleurs.
8322F, H. 16 cm, B. 13 cm; GSNachf. 1,22c
–> vgl. 7855; 7963
Lit.: GSNachf. 1,22 (m. Abb.), hier mit Ang. des Originaltitels „Lydia“.
Quelle: Fotoarchiv K. Täuber

8323
RELIEF Sitzendes nacktes Mädchen unter einem Kirschbaum, an einer Kirschblüte riechend (Wandrelief)
ca. 1938
8323,9 gem. ML., Firmenmarke 19, Made in Germany, Ausformung zw. 1938–1941, H. 22 cm, B. 17,4 cm; Sammlung Renée und Otto Schenk, Wien

8325F
Jarl, Karin (Teilentwurf)
GRUPPE „Jagd“ (Reiterin auf scheuendem Pferd und Jagdhund) (auf Ovalsockel)
ca. 1938

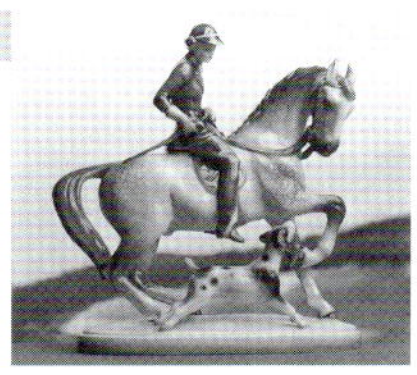

8325,4 Firmenmarke 19, Made in Austria, unsign., Ausformung zw. 1945–1953, H. 25,4 cm; B. 22,5 cm; eBay # 77943664
8325,11,56 gem. G, Firmenmarke 20, Made in Austria, Hand dec., unsign., H. 25 cm; eBay # 2240865059; 3731193722; Bem.: Diese Gruppe wurde Anfang Mai 2004 in Einzelteile zerbrochen, also völlig beschädigt in der eBay-Auktion angeboten. Ende Juni 2004 kam diese Reitergruppe dann erneut in den Handel. Diesmal vermerkt der neue Anbieter „unbeschädigt“. Eindeutig wurde das komplette Stück in der Zwischenzeit vollständig restauriert.
8325F, H. 25,5 cm, B. 22,5 cm; GSNachf. 1,65
Bem.: Die Gesamtkonzeption dieser Gruppe kann nicht ohne weiteres der Künstlerin Karin Jarl zugeordnet werden, denn sie kommt als spezialisierte Tierbildhauerin dafür kaum in Frage. Bei diesem Modell kommt offenbar wieder einmal die Praxis der Manufaktur zum Tragen, aus verschiedenen Entwürfen ein neues Modell zu kreieren. Das Pferd, eindeutig von Jarl, wurde auch noch zw. 1953–1960 bei Carstens (GWM West-Germany) separat produziert wie die Abb. im Firmenkat. KGSF1 zeigt.
Lit.: GSNachf. (2. Ausg., bzw. Erg. Bl.) 1,65 (m. Abb.); KGSF1; Könecke (2002), S. 77, hier mit Ang. der Künstlerin K. Jarl (wohl nur für das Pferd und den Hund), S. 90 (m. Abb.).
Quelle: AGM (Rechnung an Maurice Mizrahi, Alexandria, Ägypten, vom 22. April 1953), hier mit Ang. des Originaltitels „Jagd“; Fotoarchiv K. Täuber.

8327M
Matula, Viktor
FIGUR „Reh“
ca. 1938
8327M, H. 24 cm, B. 20,5 cm; GSNachf. 3,17c
Lit.: GSNachf. 3,17 (m. Abb.), hier mit Ang. des Originaltitels „Reh“.
Quelle: Nachlass Matula, Wien

8328M
WANDMASKE „Maske mit Kopftuch“ (Frauenkopf mit Kopftuch)
ca. 1938

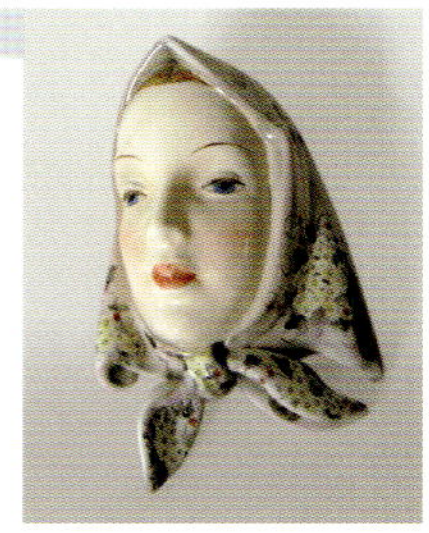

8328,2,29 (Firmenmarke ohne Ang.), H. 25 cm; Christie's SK 18.9.1998
8328,3,51 gem. X, gem. blau G, Firmenmarke 19, Made in Austria, H. 27 cm; eBay # 1228037245
8328,17,29 (Firmenmarke ohne Ang.), H. 24 cm; Christie's SK 12.9.1997

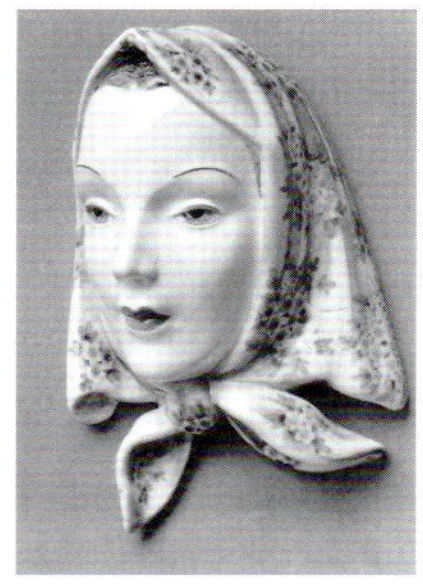

8328,17,76 gem. MR, gem. lig. LR, Firmenmarke 19, Made in Austria, Hand dec., H. 25,7 cm; PB Deutschland
8328,18,G gem. DK., Firmenmarke 19, Made in Germany, Ausformung zw. 1938–1941, H. 26,5 cm; Wien Museum Inv. 186.650
8328,30 Firmenmarke 19, H. 24 cm; Doro Weihn. Auktion 17.12.1994

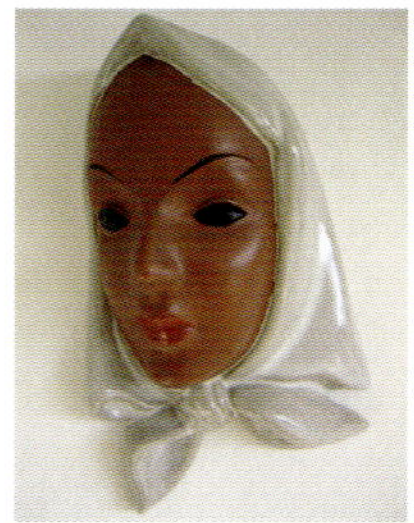

8328M, H. 24 cm, B. 17 cm; GSNachf. 2,5d
Bem.: Diese Maske wurde auch noch zw. 1953–1960 bei Carstens (GWM West-Germany) produziert.
Lit.: GSNachf. 2,5 (m. Abb.), hier mit Ang. des Originaltitels „Maske mit Kopftuch“; Könecke (2002), S. 77, 90 (m. Abb.).
Quelle: AGM (Bestellung der Firma Elisabeth Salle (Wiesbaden) vom 8.1.1951); Fotoarchiv K. Täuber.

8337M
Goebel, Kurt
RELIEF „Malermaske“ (Junge mit Hut, Malerpalette und Pinsel) (Wandrelief)
ca. 1938
8337,2 Firmenmarke 19, Made in Germany, unsign., Ausformung zw. 1938–1941, H. 24 cm; eBay # 2647711144; 2235025652
8337M, H. 24 cm, B. 16 cm; GSNachf. 2,16
Lit.: GSNachf. 2,16 (m. Abb.), hier mit Ang. des Originaltitels „Relief Malermaske“.

8339M
Goebel, Kurt
(STEH)KOPF „Kopf ohne/mit Sockel“ (Frauenkopf mit geschlossenen Augen, Haarschmuck und Kragen)
ca. 1938

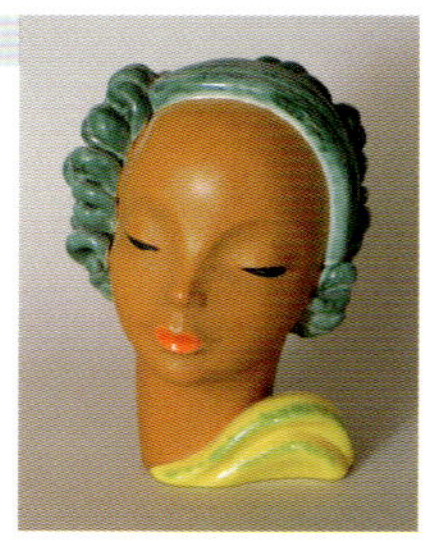

ohne Sockel
8339,1 Firmenmarke 19, unsign., H. 17,8 cm; eBay # 1210432089
8339,3,18 gem. DE, Firmenmarke 19 (blau), Made in Austria, unsign., H. 17 cm; Wien Museum Inv. 186.671
mit Holzsockel
8339 ger., Plak. D, H. 16,5 cm, Sockel 4,5 cm; Sammlung.schmutz@chello.at 72
8339,16 Firmenmarke 19 (blau), unsign., H. 17 cm, Sockel 4,2 cm; Doro 20.10.1997
8339M, H. 16 cm, B. 11 cm; GSNachf. 2,19b
Lit.: N.N.: Neue Keramik, in: ÖK, 7.1936, Nr. 10, S. 15 (m. Abb.), „Man sah auf der Wiener Herbstmesse eine Reihe keramischer Arbeiten, die Originalität und richtiges Erfassen der lebendigen Strömungen in der Entwicklung moderner Keramik zeigten. Besonders die Arbeiten des Keramikers Kurt Goebel, die die Wiener Manufaktur Friedrich Goldscheider mit viel Erfolg herausgebracht hat, fielen vorteilhaft auf“; GSNachf. 2,19 (m. Abb.), hier mit Ang. des Originaltitels „Kopf ohne/mit Sockel“.

8342F
Thomasch, Wilhelm
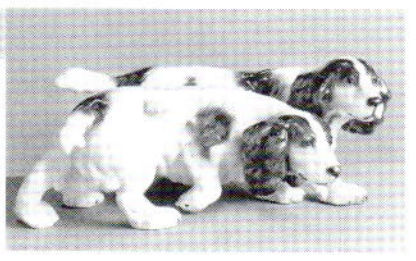
GRUPPE „Spaniels"
(Zwei laufende Spaniel)
ca. 1934 (für den Entwurf)
8342F, H. 10,5 cm,
B. 27,5 cm, GSNachf. 5,8c; Fotoarchiv K. Täuber, hier mit der Ang.: H. 9 cm, B. 25 cm
Bem.: Für die Zuordnung dieser Gruppe und der damit verbundenen Einzeltiere danke ich Herrn Herwig Thomasch, dem Sohn von Wilhelm Thomasch.
-> vgl. 7015F; 7281F
-> vgl. 7061F; 7329F; 8344F= jeweils EinzelFIGUR (vorderer Hund)
-> vgl. 7062F; 7299; 7330F; 8343F= jeweils EinzelFIGUR (rückwärtiger Hund)
Lit.: GSNachf. 5,8 (m. Abb.), hier mit Ang. des Originaltitels GRUPPE „Spaniels".
Quelle: AGM (Bestellung von Ferd. Lundquist, Göteborg, Schweden vom 23. August 1950); Fotoarchiv K. Täuber.

8343F
Thomasch, Wilhelm
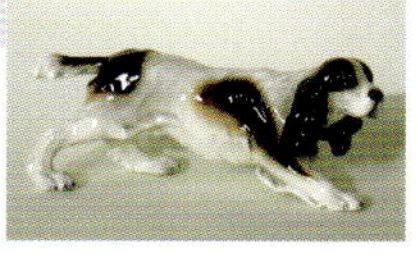
FIGUR „Spaniel"
(Laufender Spaniel)
ca. 1934 (für den Entwurf)
8343,90,68 gepr. G, gem. lig. HM, Firmenmarke 19, unsign., Pick. 6, H. 10 cm, L. 23 cm; Pforzheimer KAH 42
8343F, H. 10 cm, B. 22 cm; GSNachf. 5,8i
Bem.: Für die Zuordnung dieser Figur und der damit verbundenen Gruppe danke ich Herrn Herwig Thomasch, dem Sohn von Wilhelm Thomasch.
-> vgl. 7062F; 7299F; 7330F
-> vgl. 7015F; 7281F; 8342F= jeweils GRUPPE
Pendant -> vgl. 8344
Lit.: GSNachf. 5,8 (m. Abb.), hier mit Ang. des Originaltitels FIGUR „Spaniel" (Rückwärtiger Hund aus der GRUPPE 8342).
Quelle: Fotoarchiv K. Täuber

8344F
Thomasch, Wilhelm
FIGUR „Spaniel" (Laufender Spaniel)
ca. 1934 (für den Entwurf)
8344,14 Firmenmarke 19, Made in Germany, unsign., Ausformung zw. 1938–1941, H. 8,5 cm; WKH
8344F, H. 9 cm, B. 23 cm; GSNachf. 5,8f
Bem.: Für die Zuordnung dieser Figur und der damit verbundenen Gruppe danke ich Herrn Herwig Thomasch, dem Sohn von Wilhelm Thomasch.
-> vgl. 7061F; 7329F
-> vgl. 7015F; 7281F; 8342F= jeweils GRUPPE
Pendant -> vgl. 8343
Lit.: GSNachf. 5,8 (m. Abb.), hier mit Ang. des Originaltitels FIGUR „Spaniel" (Vorderer Hund aus der GRUPPE 8342).

8345F
Dakon, Stephan
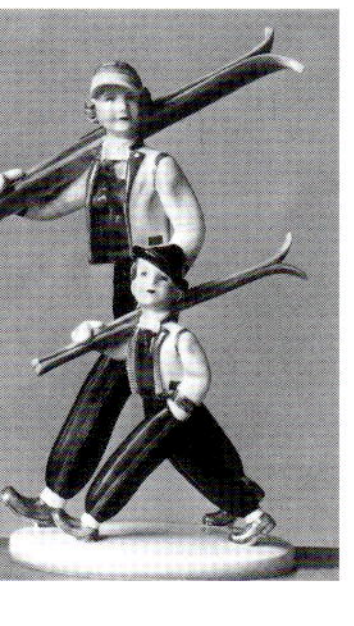
GRUPPE „Tirol" („Skifahrergruppe") (Schreitende Mutter und Kind, ihre Skier geschultert) (auf Ovalsockel)
Inschrift: TYROL
ca. 1938
8345,4,43 gem. LS, Firmenmarke 19, Made in Austria, Hand dec., H. 35 cm; Privatsammlung; Bem.: Die Besonderheit dieses Stückes ist die Beschriftung „Tyrolia Skibindungen" am Sockel. Ob es sich hierbei um eine Sonderanfertigung für die Firma Tyrolia handelt, oder ob der Schriftzug sek. angebracht wurde, lässt sich nicht eindeutig sagen zumal bislang nur dieses eine Exemplar nachweisbar ist und auch nicht geprüft werden konnte ob die Schrift unter der Glasur ist.
8345,32,11 Firmenmarke 19, H. 33,5 cm; Doro 18.11.1997
8345F, H. 33 cm, B. 26 cm; GSNachf. 1,42a; Fotoarchiv K. Täuber, hier mit der Ang.: H. 33 cm, B. 25 cm
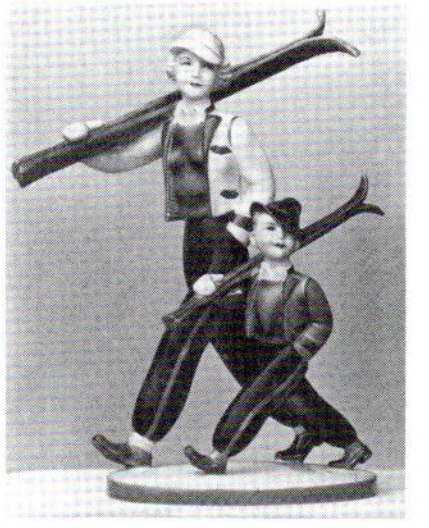
Bem.: Diese Gruppe wurde auch noch zw. 1953–1960 bei Carstens (GWM West-Germany) produziert.
-> vgl. 8346= Skiläuferin allein
-> vgl. 8347= Skibub allein
-> vgl. 8385= RELIEF
Lit.: GSNachf. 1,42 (m. Abb.), hier mit Ang. des Originaltitels „Tirol"; Könecke (2002), S. 77, 90 (m. Abb.).
Quelle: Archiv Mimi Dakon (Tantiemenabrechnung 1.1.–30.6.1939), hier wird die Gruppe unter dem Titel „Skifahrer" geführt; (Modellliste Dakon, undatiert, ca. 1940), hier wird die Gruppe unter dem Titel „Skiläufer" geführt und mit 1938 datiert; (Wiener Manufaktur FG, Modellliste Dakon, 7.11.1946), hier mit Ang. des (Original-)Titels „Skifahrergruppe"; (Tantiemenabrechnung 1. Halbjahr 1949; 2. Halbjahr 1949; 1. Halbjahr 1950; 2. Halbjahr 1950; (1. Halbjahr 1951, AGM); 2. Halbjahr 1951, hier auch als „Sorte II" bestellt; 1. Halbjahr 1952), hier wird die Gruppe jeweils unter dem (Original-)Titel „Tirol" geführt.
Quelle: Fotoarchiv K. Täuber

8346F
Dakon, Stephan
FIGUR „Skifahrerin" („Tirol groß") (Stehende Dame, ihre Skier geschultert) (auf Ovalsockel)
Inschrift: TYROL
ca. 1938
8346,78,68 Firmenmarke 19, Made in Germany, unsign., Ausformung zw. 1938–1941, H. 35 cm; Doro 15.7.1998 (ohne Kat.)
8346F, H. 33 cm, B. 26 cm; GSNachf. 1,42b
-> vgl. 8345= GRUPPE
Pendant -> vgl. 8347= FIGUR „Skibub"
-> vgl. 8385= RELIEF
Lit.: GSNachf. 1,42 (m. Abb.), hier mit Ang. des Originaltitels „Skifahrerin".
Quelle: Archiv Mimi Dakon (Tantiemenabrechnung 1.1.–30.6.1939; Modellliste Dakon, undatiert, ca. 1940), hier mit 1938 datiert; Wiener Manufaktur FG, Modellliste Dakon, 7.11.1946; Tantiemenabrechnung 1. Halbjahr 1950; 2. Halbjahr 1950), hier wird die Figur jeweils unter dem Titel „Skifahrer" geführt; (Tantiemenabrechnung 1. Halbjahr 1949; 2. Halbjahr 1949; 1. Halbjahr 1952), hier wird die Figur jeweils unter dem (Original-)Titel „Tirol (groß)" geführt.

8347F
Dakon, Stephan
FIGUR „Skibub" („Tirol klein") (Stehender Bub, seine Skier geschultert) (auf Ovalsockel)
ca. 1938
8347F, H. 21 cm, B. 18,5 cm; GSNachf. 1,42c
Bem.: Diese Figur wurde auch noch zw. 1953–1960 bei Carstens (GWM West-Germany) produziert.
-> vgl. 8345= GRUPPE
Pendant -> vgl. 8346= Skifahrerin
-> vgl. 8385= RELIEF
Lit.: GSNachf. 1,42 (m. Abb.), hier mit Ang. des Originaltitels „Skibub"; Könecke (2002), S. 77.
Quelle: Archiv Mimi Dakon (Tantiemenabrechnung 1.1.–30.6.1939; Modellliste Dakon, undatiert, ca. 1940), hier mit 1938 datiert; Wiener Manufaktur FG, Modellliste Dakon, 7.11.1946; Tantiemenabrechnung 1. Halbjahr 1950), hier wird die Figur jeweils unter dem Titel „Skifahrer (klein)" geführt; (Tantiemenabrechnung 1. Halbjahr 1949; 2. Halbjahr 1950; (1. Halbjahr 1951, AGM); 2. Halbjahr 1951; 1. Halbjahr 1952), hier wird die Figur jeweils unter dem (Original-)Titel „Tirol (klein)" geführt; (Tantiemenabrechnung 2. Halbjahr 1949), hier unter dem Titel „Skibub" geführt.

8360
Dakon, Stephan
FIGUR „Champagner" (Nach links schreitende junge Dame mit Bikinioberteil, ihren langen Rock schmetterlingsflügelartig auseinander haltend) (auf Ovalsockel)
ca. 1937 (für den Entwurf)
8360,56,4 ger. 5B, gem. lig. ML, Firmenmarke 19, Made in Germany, Ausformung zw. 1938–1941, H. 22 cm; eBay # 2258042601
Bem.: Eine Bestellung (vom 27.2.1946) der Firma Wahliss (Wien) belegt, dass auch das „seitenverkehrte" Modell den Originaltitel „Champagner" trug, wenngleich dort auch diese Order der größeren Version (Modell-Nr. 7857) galt.
-> vgl. 7857
Pendant -> vgl. 7826; 8319= jeweils seitenverkehrtes Modell.
Lit.: GSNachf. 1,21 (m. Abb.), hier das seitenverkehrte Pendant mit Ang. des Originaltitels „Champagner".

8361
Herczeg, Claire (Klára)
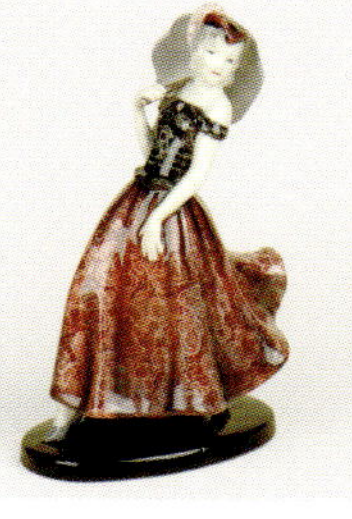
FIGUR „Parisienne" (Schreitende elegante Dame, ihren Schirm mit der rechten Hand haltend) (auf Ovalsockel)
ca. 1938
8361,3,31 Firmenmarke 19, Made in Germany, sign. Claire Herczeg, Ausformung zw. 1938–1941, H. 22 cm; PB Wien 19
8361,10,47 Firmenmarke 19, Made in Germany, sign. Claire Herczeg, Ausformung zw. 1938–1941, H. 22 cm; WKH
8361,11,31,9 gem. XII, E, gem. blau H, Firmenmarke 19, Made in Germany, sign. Claire Herczeg, Ausformung zw. 1938–1941, Pick. 5, H. 21,5 cm; Wien Museum Inv. 190.124
8361,18,31 ger. XII G, gem. blau HG, Firmenmarke 19, Made in Germany, sign. Claire Herczeg, Ausformung zw. 1938–1941, H. 21,2 cm; Wien Museum Inv. 220.701
Bem.: Dieses Modell geht im Ursprung zurück auf die Jahre 1935/36. Zu diesem Zeitpunkt entwarf K. Herczeg (Claire Weiss) eben diese Figur in einer seitenverkehrten Version. Die Dame hält dort ihren Schirm mit der linken Hand. Ob die Firma für beide Modellvarianten den Titel „Parisienne" verwendet hat, lässt sich nicht mit letzter Sicherheit sagen, es ist jedoch anzunehmen.
-> vgl. 7390; 7723; 8138= jeweils seitenverkehrtes Modell.

8363F
Lorenzl, Josef
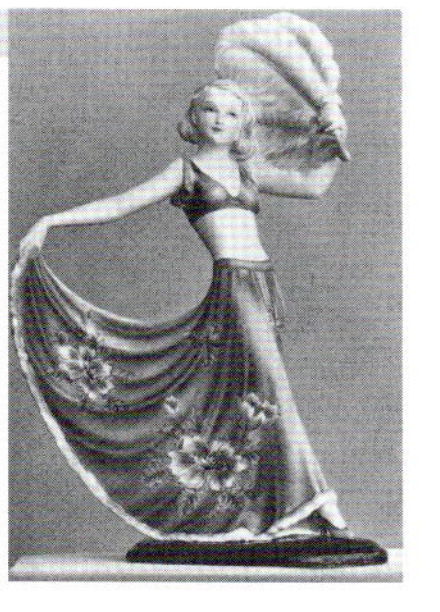
FIGUR „Fächertanz" (Schreitende Tänzerin mit großem Fächer aus Vogelfedern) (auf Ovalsockel)
ca. 1938
8363,43,11 ger. L, 117, Firmenmarke 19, Made in Germany, sign. Lorenzl, Ausformung zw. 1938–1941, H. 43,8 cm; Soth./eBay # 1860110214
8363,133,31 gem. H, Firmenmarke 19, Made in Germany, unsign., Ausformung zw. 1938–1941, H. 42 cm; WKA 35
8363,159,31 Firmenmarke 19, Made in Germany, unsign., Ausformung zw. 1938–1941, H. 43 cm; city-antik
8363F, H. 42,5 cm, B. 27,5 cm; GSNachf. 1,43
Lit.: GSNachf. 1,43 (m. Abb.), hier mit Ang. des Originaltitels „Fächertanz".
Quelle: AGM (Rechnung an Maurice Mizrahi, Alexandria, Ägypten, vom 22. April 1953), hier mit Ang. des Originaltitels „Fächertanz".

8365F

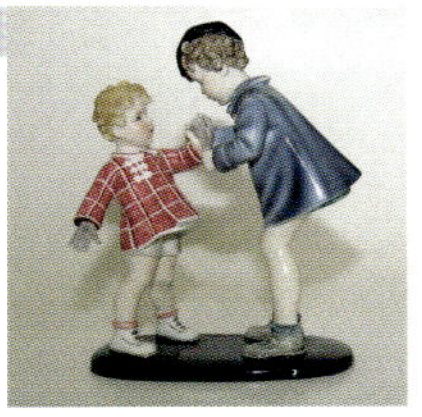

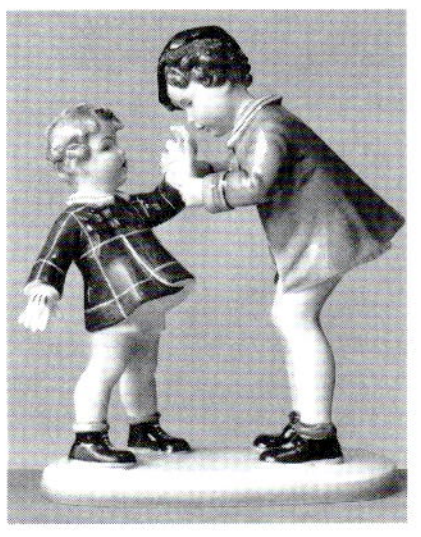

Bouret, Germaine
GRUPPE „Kinder" („Kindergruppe") (Zwei Kleinkinder beim Handschuhe Anziehen) (auf Ovalsockel)
ca. 1938
8365,16,4 gem. B., Firmenmarke 19, Made in Germany, sign. G. Bouret, Ausformung zw. 1938–1941, H. 28 cm; Sammlung Bernard Amigon
8365 H. 24 cm, B. 20 cm; Fotoarchiv K. Täuber
Bem.: Die Entwürfe von Germaine Bouret wurden von Stephan Dakon ausmodelliert.
Quelle: Archiv Mimi Dakon (Tantiemenabrechnung 1.1.–30.6.1939; Modellliste Dakon, undatiert, ca. 1940, hier wird die Gruppe mit 1938 datiert; Tantiemenabrechnung 1. Halbjahr 1950; 2. Halbjahr 1950); (1. Halbjahr 1951, AGM); 2. Halbjahr 1951; 1. Halbjahr 1952), hier jeweils mit Ang. des Originaltitels GRUPPE „Kinder"; (Wiener Manufaktur FG, Modellliste Dakon, 7.11.1946), hier mit Ang. des (Original-)Titels „Kindergruppe"; Fotoarchiv K. Täuber.

8366F
Dakon, Stephan
FIGUR „Golfspielerin" („Golf")
ca. 1938
Quelle: Archiv Mimi Dakon (Tantiemenabrechnung 1.1.–30.6.1939; Modellliste Dakon, undatiert, ca. 1940, hier wird die Figur mit 1938 datiert; Wiener Manufaktur FG, Modellliste Dakon, 7.11.1946); (1. Halbjahr 1951, AGM), hier jeweils mit Ang. des Originaltitels FIGUR „Golfspielerin"; (Tantiemenabrechnung 1. Halbjahr 1949), hier unter dem Titel „Golf" geführt.
Lit.: Pinhas, S. 148 (m. Abb.), hier wird eine elegante stehende Dame (in Schlaghosen und Handschuhen) mit Kappe und einem Spazierstock aus Bambus (auf Rundsockel) gezeigt. Ob es sich um eine Variante zur „Golfspielerin" handelt, die vermutlich einen Golfschläger in Händen hält, kann nur vermutet werden. Bislang gibt es außer diesem nachgewiesenen Stück keine weiteren Belege. Pinhas mutmaßt als dargestellte Person Ginger Rogers.

8367F

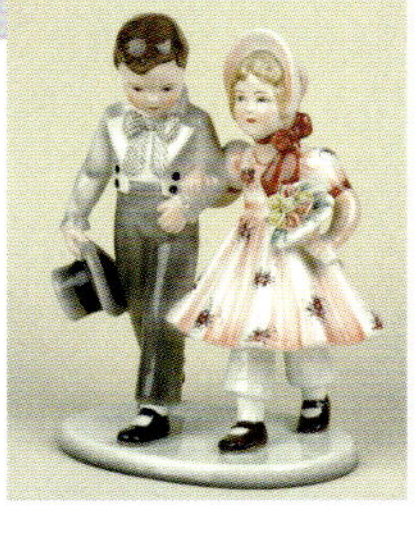

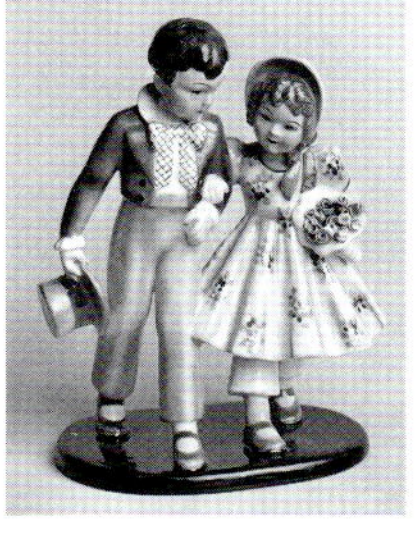

Dakon, Stephan (zugeschr.)
GRUPPE Kinder als Brautpaar im Biedermeiergewand (auf Ovalsockel)
ca. 1938
8367,5 (Firmenmarke ohne Ang.), unsign., H. 22,8 cm; eBay # 1256190576
8367,38,10 gem. M.G.B., Firmenmarke 19, Made in Germany, unsign., Ausformung zw. 1938–1941, H. 22 cm; city-antik
8367 ohne Ang., H. 23 cm; PB Wien 19
8367F, H. 23 cm, B. 17 cm; GSNachf. 1,69
Bem.: Diese Gruppe wurde auch noch zw. 1953–1960 bei Carstens (GWM West-Germany) produziert. Für die Zuschreibung an Stephan Dakon danke ich Frau Mag. Uta Matschiner.
Lit.: GSNachf. (2. Ausg., bzw. Erg. Bl.) 1,69 (m. Abb.); Könecke (2002), S. 77, 90 (m. Abb.).
Quelle: Fotoarchiv K. Täuber

8368M
Dakon, Stephan
RELIEF „Skiläufer" (Aufsteigender Skiläufer mit geschulterten Skiern) (Wandrelief)
ca. 1938
8368,70,3 ger., gem. 12, Firmenmarke 19, Made in Austria, unsign., H. 15,2 cm, B. 12,1 cm; eBay # 1213967690
8368M, H. 15,5 cm, B. 12,5 cm; GSNachf. 2,26c
Pendant -> vgl. 8369M.
Lit.: GSNachf. 2,26 (m. Abb.), hier mit Ang. des Originaltitels „Relief Skiläufer".

8369M
Dakon, Stephan
RELIEF „Skiläuferin" (Abfahrende Skiläuferin) (Wandrelief)
ca. 1938
8369 Firmenmarke 19 (zwei mal), unsign., H. ca. 15,5 cm, B. ca. 12,5 cm; eBay # 894786728
8369M, H. 15,5 cm, B. 12,5 cm; GSNachf. 2,26a
Pendant -> vgl. 8368M.
Lit.: GSNachf. 2,26 (m. Abb.), hier mit Ang. des Originaltitels „Relief Skiläuferin".

8370M
RELIEF „Reh" (Äsendes Reh unter Tanne) (Wandrelief)
ca. 1938
8370,16,62 Firmenmarke 19, Made in Austria, H. 12,8 cm, B. 15,5 cm; WKA 33
8370M, H. 12,5 cm, B. 16 cm; GSNachf. 2,15b
Lit.: GSNachf. 2,15 (m. Abb.), hier mit Ang. des Originaltitels „Relief Reh".

8371F
„Tanz"
ca. 1938
Quelle: AGM (Rechnung der Wiener Manufaktur Josef Schuster an Laszlo Bösze, Szeged, Ungarn, vom 5. August 1944), hier unter dem Titel „Tanz" geführt.

8372M

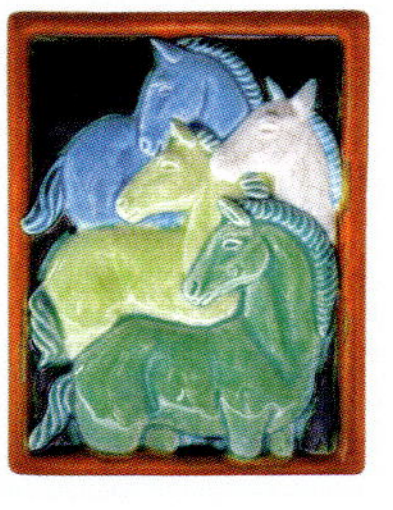

RELIEF „Fohlen" (Wandrelief)
ca. 1938
8372 gem. lig. AH, Firmenmarke 19, Made in Germany, Ausformung zw. 1938–1941, H. 16,1 cm, B. 12,7 cm; Sammlung a@goldscheider.net
8372M, H. 16 cm, B. 12,5 cm; GSNachf. 2,15a
Lit.: GSNachf. 2,15 (m. Abb.), hier mit Ang. des Originaltitels RELIEF „Fohlen".
Quelle: AGM (Kontoblatt der Wiener Manufaktur Josef Schuster für die Firma Porslins Ericsson-Lund, Stockholm?, Schweden vom 7. Januar 1943), hier mit Ang. des Titels RELIEF „Pferde".

8374
WANDMASKE Frauenkopf mit Kopftuch und großem Kragen
ca. 1938
8374,3 Firmenmarke 19, H. 30,3 cm; Doro 19.9.2000
8374,9 gem. G, Firmenmarke 19, Made in Germany, Ausformung zw. 1938–1941, H. 30,5 cm; Wien Museum Inv. 186.662
8374,15 Firmenmarke 19, Made in Germany, Ausformung zw. 1938–1941, H. 30 cm; Soth. L. 20.10.1989

8376F
FIGUR „Versunkenheit (Akt)" (Stehender Mädchenakt) (auf Rundsockel)
ca. 1938
8376,21,14 Firmenmarke 19 (blau), Made in Austria (blau), Hand dec. (blau), H. 35 cm; city-antik
8376,58,19 Firmenmarke 19, Made in Germany, Ausformung zw. 1938–1941, H. 36 cm; WKA 38
8376,127,6 ger. M,56, gem. 3, Firmenmarke 19, Made in Germany, Ausformung zw. 1938–1941, H. 38,1 cm; eBay # 167080590
8376F, H. 35 cm, B. 10 cm; GSNachf. 1,44a
Bem.: Diese Figur wurde auch noch zw. 1953–1960 bei Carstens (GWM West-Germany) produziert.
Lit.: GSNachf. 1,44 (m. Abb.), hier mit Ang. des Originaltitels „Versunkenheit (Akt)"; Könecke (2002), S. 77, 90 (m. Abb.).

8377M -> ÜBERSCHNEIDUNG
RELIEF „Vogel" (Wandrelief)
ca. 1938
8377M, H. 15 cm, B. 12 cm; GSNachf. 2,15v
Lit.: GSNachf. 2,15 (m. Abb.), hier mit Ang. des Originaltitels „Relief Vogel".

8377 -> ÜBERSCHNEIDUNG
Schreitender Damenakt mit großem Tuch (auf Ovalsockel)
ca. 1937
8377 (Firmenmarke ohne Ang.), H. 39 cm; Pinhas, S. 157 (m. Abb.)
Bem.: Ob die bei Pinhas angegebene Mod.-Nr. richtig ist, muss bezweifelt werden. Die Nr. im Firmenkat. GSNachf. ist durch das Relief „Vogel" belegt.
-> vgl. 8174= ähnl. Mod.

8378F
Dakon, Stephan
FIGUR „Mädchen" („Rosy Barsony"/„Mädchen in Sessel") (Auf einem Stuhl sitzend, mit ausgestrecktem Bein)
ca. 1938
8378,21,6 ger. 53, gem. M., Firmenmarke 19, Made in Germany, unsign., Ausformung zw. 1938–1941, H. 26 cm; Wien Museum Inv. 190.147
Quelle: Archiv Mimi Dakon (Tantiemenabrechnung 1.1.–30.6.1939; 2. Halbjahr 1951), hier jeweils mit Ang. des (Original-)Titels FIGUR „Mädchen"; (Modellliste Dakon, undatiert, ca. 1940), hier wird die Figur unter dem Titel „Rosy Barsony" geführt und mit 1938 datiert; (Wiener Manufaktur FG, Modellliste Dakon, 7.11.1946), hier unter dem Titel „Mädchen in Sessel" geführt.

8383F

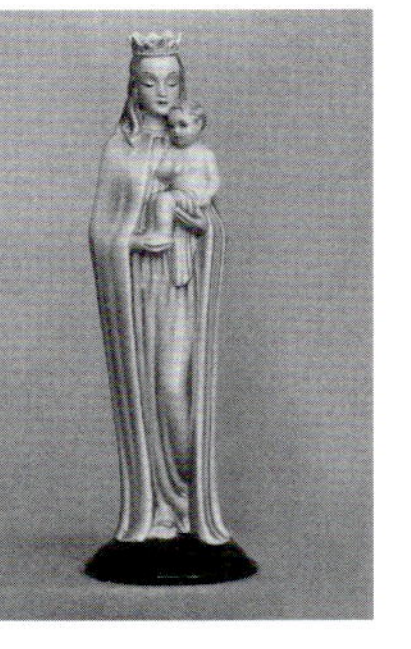

FIGUR „Madonna mit Kind" (Stehende Madonna mit Krone und Jesuskind) (auf Rundsockel)
ca. 1938
8383F, H. 37 cm, B. 12 cm; GSNachf. 6,13
Lit.: GSNachf. 6,13 (m. Abb.), hier mit Ang. des Originaltitels FIGUR „Madonna mit Kind".
Quelle: Fotoarchiv K. Täuber

8384
RELIEF Zwei Rehe (Wandrelief)
ca. 1938
8384 gem. M, Firmenmarke 19, Made in Austria, Ausformung 1938, H. 13,8 cm, B. 15,8 cm; KH Art 1900, Berlin

8385M
Dakon, Stephan
RELIEF „Skiläuferpaar" (Wandrelief)
ca. 1938
8385 Firmenmarke 19, unsign., H. 22,5 cm, B. 16,5 cm; Doro 26.8.1998 (ohne Kat.)
8385M, H. 22 cm, B. 16,5 cm; GSNachf. 2,26b
-> vgl. 8345= GRUPPE
Lit.: GSNachf. 2,26 (m. Abb.), hier mit Ang. des Originaltitels „Relief Skiläuferpaar".

8386
Grasendes Pferd
ca. 1938
8386 ohne Ang., H. fehlt; Könecke (2002), S. 77
Bem.: Diese Figur wurde auch noch zw. 1953–1960 bei Carstens (GWM West-Germany) produziert.
Lit.: Könecke (2002), S. 77, hier ohne weitere Ang.

8388M
Matula, Viktor
RELIEF „Sonnenblumen" (Wandrelief)
ca. 1938
8388M, H. 27 cm, B. 27,5 cm; GSNachf. 2,14b
Lit.: GSNachf. 2,14 (m. Abb.), hier mit Ang. des Originaltitels „Relief Sonnenblumen".
Quelle: Nachlass Matula, Wien

8389F
Prischl, Adolf
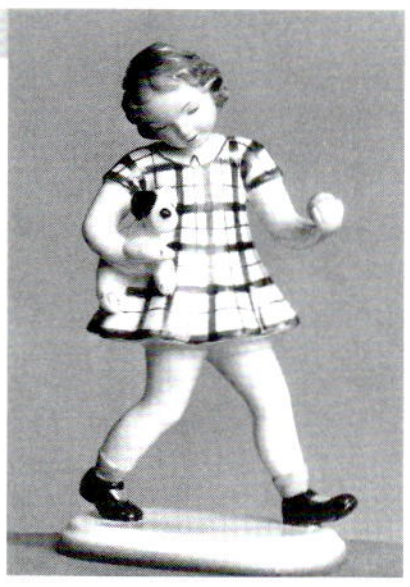
FIGUR „Kind mit Hund" (Schreitendes kleines Mädchen, einen Hund auf dem Arm haltend) (auf Ovalsockel)
ca. 1938
8389F, H. 22 cm, B. 13 cm; GSNachf. 1,45; Fotoarchiv K. Täuber, hier mit der Ang.: H. 22 cm, B. 12 cm
Bem.: Diese Figur gelangte 1938 erstmalig in den Verkauf (Brief von GS an die Wiener Firma Keramos, vom 25.2.1953).
Lit.: GSNachf. 1,44 (m. Abb.), hier mit Ang. des Originaltitels „Kind mit Hund".
Quelle: AGM (Bestellungen der Firma Wahliss (Wien) vom 9. und 24.11.1945 an die „Arbeitsgemeinschaft der Firma Friedrich Goldscheider"; Tantiemenabrechnung 2. Halbjahr 1952 an A. Prischl vom 29.1.1953; Bestellung der Firma Wahliss (Wien) vom 27.2.1946 in Ausführung „Terrakotta"; Brief von der Wiener Manufaktur Goldscheider an die Firma Keramos, vom 25.2.1953); Fotoarchiv K. Täuber.

8390M
Matula, Viktor
FIGUR „Lamm"
ca. 1938
8390M, H 26 cm, B. 30 cm; GSNachf. 3,17a
Lit.: GSNachf. 3,17 (m. Abb.), hier mit Ang. des Originaltitels „Lamm".
Quelle: Nachlass Matula, Wien

8392M
Matula, Viktor
GRUPPE „Pferde"
ca. 1938
Quelle: AGM (Kontoblatt der Wiener Manufaktur Josef Schuster für die Firma Porslins Ericsson-Lund, Stockholm?, Schweden vom 7. Januar 1943), hier mit Ang. des Originaltitels GRUPPE „Pferde".

8396M
Matula, Viktor
FIGUR „Pferd"
ca. 1938
Quelle: AGM (Kontoblatt der Wiener Manufaktur Josef Schuster für die Firma Porslins Ericsson-Lund, Stockholm?, Schweden vom 7. Januar 1943), hier mit Ang. des Originaltitels „Pferd".

8399F
Herczeg, Claire (Klára)
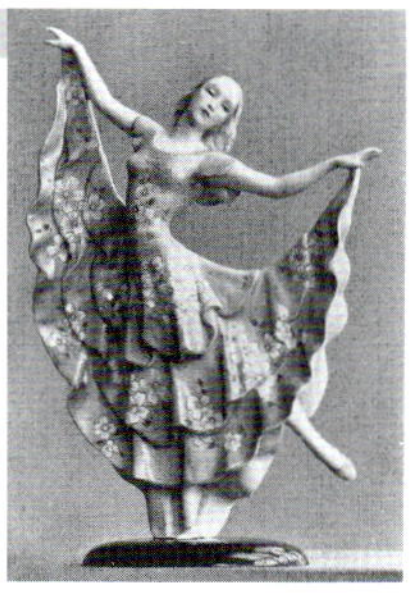
FIGUR „Melodie" (Tänzerin auf einem Bein balancierend, ihr Kleid auseinander haltend) (auf Ovalsockel)
ca. 1938
8399,6,4 Firmenmarke 19, Made in Germany, sign. Claire Herczeg, Ausformung zw. 1938–1941, H. 41,5 cm; Keup 61; Ruef 473; Soth./eBay # 186110223
8399,72,31 (Firmenmarke ohne Ang.), unsign., H. 42 cm; Christie's SK 22.4.2004
8399 ohne Ang., Firmenmarke 19, Made in Germany, Ausformung zw. 1938–1941, unsign., H. 42 cm; Mehlis 21
8399F, H. 42 cm, B. 30,5 cm; GSNachf. 1,46
Lit.: GSNachf. 1,46 (m. Abb.), hier mit Ang. des Originaltitels „Melodie".

8401M
Matula, Viktor
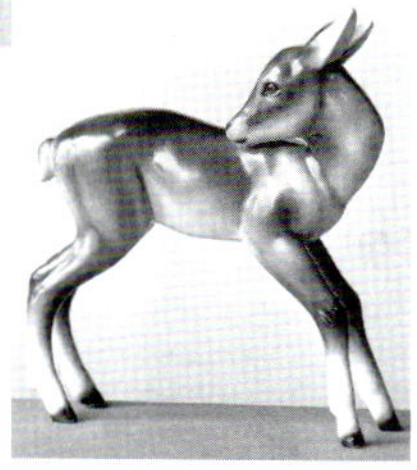
FIGUR „Reh" (Stehendes Reh mit nach hinten gewandtem Kopf)
ca. 1938
8401M, H. 23 cm, B. 21 cm; GSNachf. 3,17b; Fotoarchiv K. Täuber, hier mit der Ang.: H. 22 cm, B. 21 cm
Lit.: GSNachf. 3,17 (m. Abb.), hier mit Ang. des Originaltitels „Reh".
Quellen: AGM (Bestellungen der Firma Wahliss (Wien) vom 9. und 24.11.1945 an die „Arbeitsgemeinschaft der Firma Friedrich Goldscheider"; Bestellung der Firma Wahliss (Wien) vom 27.2.1946 in Ausführung „Terrakotta" und „Fayence"); Nachlass Matula, Wien; Fotoarchiv K. Täuber.

8404
WANDMASKE Frauenkopf mit Kragen (Greta Garbo?)
ca. 1938
8404,13 gem. P, Firmenmarke 19, Made in Austria, Ausformung zw. 1945–1953, H. 27 cm; eBay # 856305346
8404,15 gem. N, Firmenmarke 19, Made in Germany, Ausformung zw. 1938–1941, Sammlungspick.: „Eigentümer Anton Wicho, Wien", H. 26 cm; Wien Museum Inv. 220.806
8404 ohne Ang., H. 27 cm, B. 15 cm; Könecke (2002), S. 77
Bem.: Bei der Dargestellten, könnte es sich um die schwed.-amerikan. Filmlegende Greta Garbo (1905–1990) handeln. Diese Maske wurde auch noch zw. 1953–1960 bei Carstens (GWM West-Germany) produziert.
Lit.: Könecke (2002), S. 77, hier ohne weitere Ang.

8409
WANDMASKE Frauenkopf mit Kragen
ca. 1938
8409,3 Firmenmarke 19, H. 26 cm; Sammlung f@goldscheider.de

8410M
Goebel, Kurt (zugeschr.)
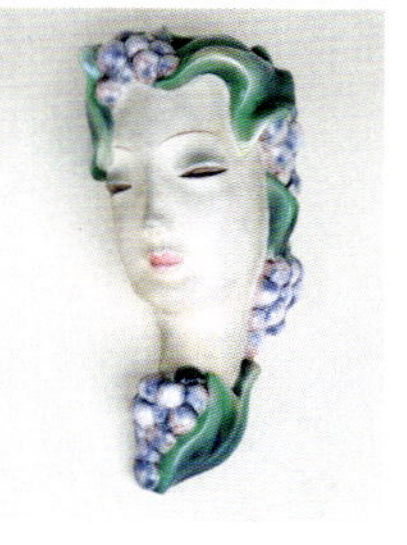
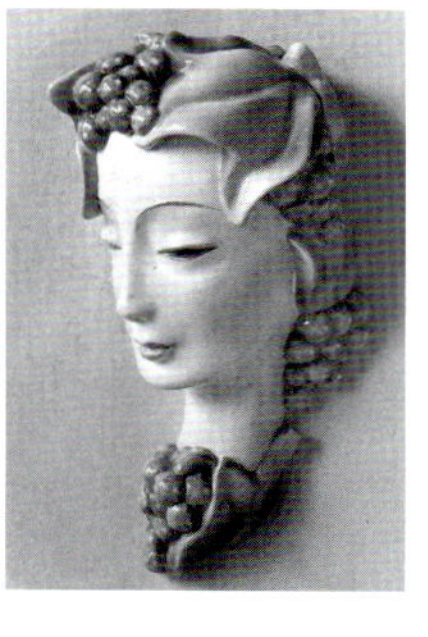
WANDMASKE „Maske „ (Frauenkopf mit Trauben im Haar und am Kragen)
ca. 1938
8410,40 Firmenmarke 19, Made in Austria, unsign., Ausformung zw. 1945–1953, H. 24 cm; Wien Museum Inv. 186.660
8410,66,18 Firmenmarke 19, Made in Austria, Hand dec., unsign., H. 24 cm; PB Wien 19
8410M, H. 23 cm, B. 11 cm; GSNachf. 2,9c; Fotoarchiv K. Täuber, hier mit der Ang.: H. 24 cm, B. 10 cm
Pendant -> vgl. 8411
Lit.: GSNachf. 2,9 (m. Abb.), hier mit Ang. des Originaltitels „Maske".
Quelle: Fotoarchiv K. Täuber

8411M
Goebel, Kurt (zugeschr.)
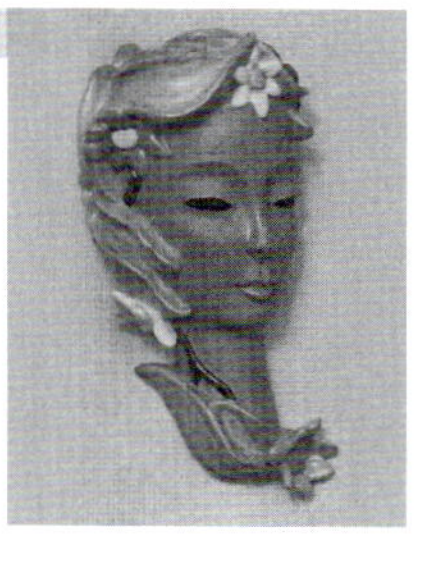
WANDMASKE „Maske „ (Frauenkopf mit Blüten im Haar und Kragen)
ca. 1938
8411,4 gem. G, Firmenmarke 19, unsign., Ausformung zw. 1945–1953, H. 26 cm; eBay # 3721156523
8411M, H. 24 cm, B. 12,5 cm; GSNachf. 2,9a
Lit.: GSNachf. 2,9 (m. Abb.), hier mit Ang. des Originaltitels „Maske".
Pendant -> vgl. 8410

8415
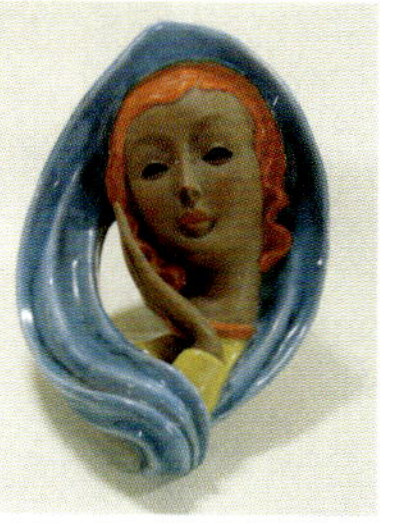
WANDMASKE Frauenkopf mit Kopftuch (Madonna?) mit erhobener Hand
ca. 1938
8415,3 gem. lig. JP, Firmenmarke 19, Made in Germany, Ausformung zw. 1938–1941, H. 25,4 cm; MAK Wien, Inv. Ke 10521-35

8421M
Goebel, Kurt
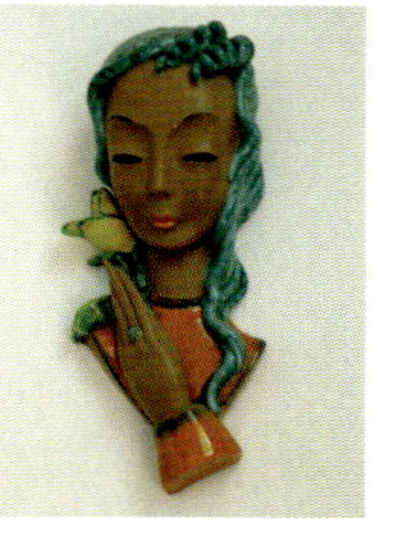
WANDMASKE „Maske „ (Frauenkopf mit langem Haar, Hand und Glockenblume)
ca. 1938
8421,9 ger., gepr. G, Firmenmarke 19, Made in Germany (zwei mal), unsign., Ausformung zw. 1938–1941, H. 32 cm; Sammlung Stressless, Portugal
8421M, H. 30,5 cm, B. 14 cm; GSNachf. 2,9b
Lit.: GSNachf. 2,9 (m. Abb.), hier mit Ang. des Originaltitels „Maske".

8423
WANDMASKE „Maske mit Kopftuch"
ca. 1938
Quelle: AGM (Bestellung der Firma Wahliss (Wien) vom 24.11.1945 an die „Arbeitsgemeinschaft der Firma Friedrich Goldscheider"), hier mit Ang. des (Original-?)Titels „Maske mit Kopftuch"; Bestellung der Firma Wahliss (Wien) vom 27.2.1946 in Ausführung „Terrakotta").

8426
Sitzender Damenakt (mit/ohne Ovalsockel)
ca. 1938
mit Ovalsockel
8426,70,6 Firmenmarke 19, Made in Germany, Ausformung zw. 1938–1941, H. 24,8 cm; Soth. L. 3.5.1991
ohne Sockel
8426,108,19 Firmenmarke 19, Made in Germany, Ausformung zw. 1938–1941, H. 23,5 cm; Doro 21.5.1997

8429F
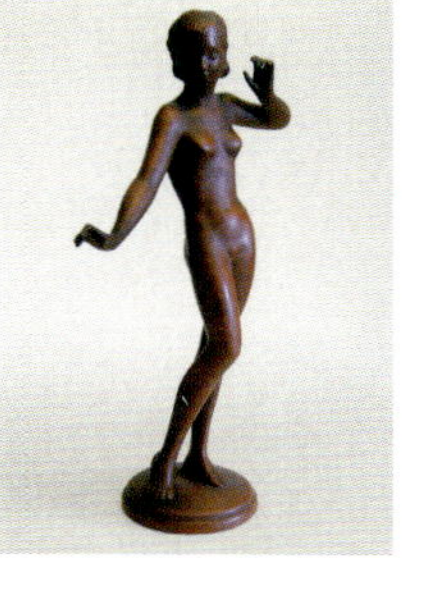
FIGUR „Pikanterie (Akt)" (Stehender Damenakt in Pose) (auf Rundsockel)
ca. 1938
8429,22/IV,1283 ger., ohne Firmenmarke, Ausformung zw. 1953–1960, H. 38,5 cm; Sammlung John Nash
8429F, H. 38,5 cm, B. 23 cm; GSNachf. 1,44b
Bem.: Diese Figur wurde auch noch zw. 1953–1960 bei Carstens (GWM West-Germany) produziert.
Lit.: GSNachf. 1,44 (m. Abb.), hier mit Ang. des Originaltitels „Pikanterie (Akt)"; Könecke (2002), S. 77, 90 (m. Abb.).

8432
RELIEF Bekrönte Madonna mit gekreuzten Händen (Wandrelief)
ca. 1938
8432 gem. blau H., gem. lig. rot GL, Firmenmarke 19, Made in Germany, Ausformung zw. 1941–1944, H. 34 cm, B. 26,5 cm; eBay # 6596394224

8433F

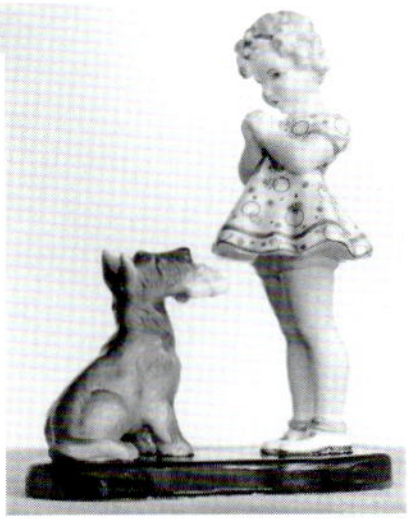

Dakon, Stephan
GRUPPE „Kind mit Hund"
(Kind, Hände gefaltet)
(auf Ovalsockel)
ca. 1938
8433,96,66 Firmenmarke 19, Made in Austria, Hand dec., unsign., H. 21 cm; eBay # 6605899004
8433 H. 20,5 cm, B. 17 cm; Fotoarchiv K. Täuber
Quelle: Archiv Mimi Dakon (Tantiemenabrechnung 1.1.–30.6.1939; Wiener Manufaktur FG, Modellliste Dakon, 7.11.1946; Tantiemenabrechnung 1. Halbjahr 1952), hier jeweils mit Ang. des Originaltitels GRUPPE „Kind mit Hund"; (Modellliste Dakon, undatiert, ca. 1940), hier wird die Gruppe mit der Zusatzbeschreibung „Kind, Hände gefaltet" geführt und mit 1938 datiert; Fotoarchiv K. Täuber.

8437F
Dakon, Stephan
GRUPPE „Akt/Reh" (Kniender Mädchenhalbakt mit kleiner Schale, ein Reh fütternd, „Brüderchen und Schwesterchen") (auf ovalem Natursockel)
ca. 1938
8437,3,83 Firmenmarke 19, Made in Austria, Hand dec., unsign., H. 19,3 cm; city-antik
8437,6,14 Firmenmarke 19, Made in Austria, Hand dec., unsign., H. 19 cm; WKH
8437,20 Firmenmarke 19, Made in Germany, unsign., Ausformung zw. 1938–1941, H. 18 cm; KH
8437,65 ger. N1?, gem. V, Firmenmarke 19, Made in Germany, unsign., Ausformung zw. 1938–1941, H. 18 cm; eBay # 2268509487
8437,65 ger. H, gem. rot H.I., Firmenmarke 19, Made in Germany, sign. Dakon, H. 19,3 cm; KH Ruby Lane (Jane Perlow/Lacremeantik, Great Neck N.Y.)
Lit.: Pinhas, S. 148 (m. Abb.)
Quelle: AGM (Rechnung an Maurice Mizrahi, Alexandria, Ägypten, vom 22. April 1953), hier mit Ang. des (Original-)Titels „Akt/Reh", geliefert in Ausführung „matt".

8438F

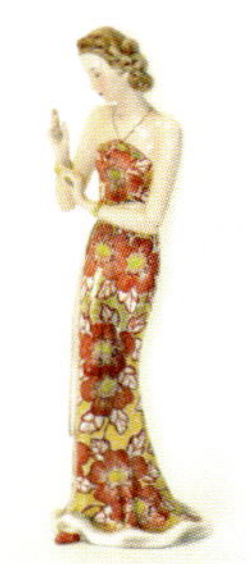

Herczeg, Claire (Klára) (zugeschr.)
FIGUR „Modefigur"
(Stehende schlanke Dame in langem Kleid)
ca. 1938
8438,4,9 Firmenmarke 19, Made in Austria, unsign., Ausformung 1938, Pick. 1, H. 30 cm; Doro 15.3.2005
8438,8,51 Firmenmarke 19, Made in Austria, Hand dec., unsign., Ausformung 1938, H. 29,6 cm; Privatsammlung, Wien

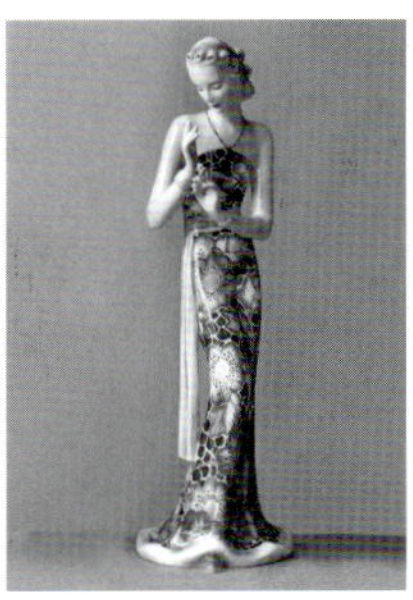

8438,42,31 Firmenmarke 19, Made in Germany, Ausformung zw. 1938–1941, H. 31,5 cm; WKA 38
8438F, H. 33 cm, B. 10 cm; GSNachf. 1,33a; Fotoarchiv K. Täuber, hier mit der Ang.: H. 31 cm, B. 11 cm
Lit.: GSNachf. 1,33 (m. Abb.), hier mit Ang. des Originaltitels „Modefigur".
Quelle: Fotoarchiv K. Täuber

8440F

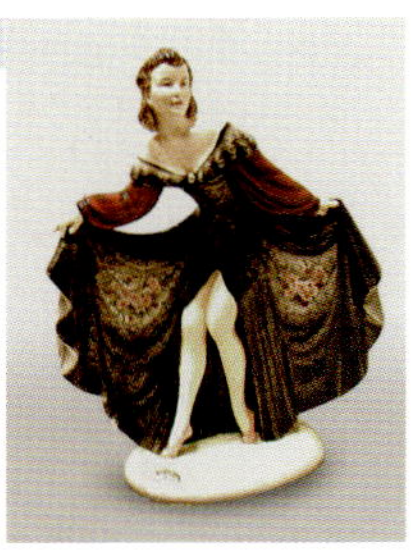

Dakon, Stephan
FIGUR/LAMPE „Jolanthe"
(Tänzerin, ihr Kleid auseinander haltend) (auf Ovalsockel)
ca. 1939
FIGUR
8440,1 Firmenmarke 19, Made in Austria, unsign., Ausformung zw. 1945–1953, H. 32,3 cm; Doro 19.9.2000

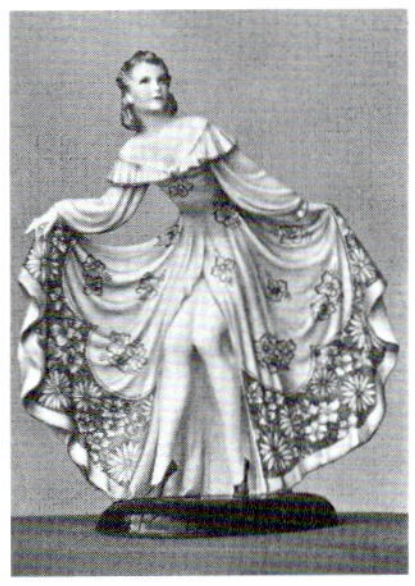

8440,1,2 Firmenmarke 19, Made in Austria, Hand dec., unsign., H. 32 cm; Doro 4.3.2002; city-antik
8440,1,8 gem. lig. HS, Firmenmarke 20, Made in Austria, Hand dec., unsign., H. 32 cm; Doro 14.9.2001
8440,1,12 Firmenmarke 19, Made in Austria, Hand dec., unsign., H. 32,5 cm; Soth. L. 3.3.1989
8440,1,14 ger., gem. rot Ob, Firmenmarke 19, Made in Austria, Hand dec., unsign., H. 33 cm; eBay # 180517634; 399488122; (KH www.tias.com/stores/frozenbeauty)
8440,1,42 gem. MR, VA, Firmenmarke 19, Made in Austria, Hand dec., Pick. 1, H. 32,5 cm; Wien Museum Inv. 190.155
8440,4 gem. L, BW, 1., Firmenmarke 19, Made in Austria, Hand dec., unsign., Ausformung zw. 1945–1953, H. 33 cm; city-antik
8440,4 Firmenmarke 19, Made in Austria (zwei mal), Hand dec., unsign., Ausformung zw. 1945–1953, H. 32 cm; eBay # 1206797076; Christie's SK 11.5.2001
8440,8 gem. M,T, gem. lig. SI, Firmenmarke 19, Made in Austria, Hand dec., unsign., Ausformung zw. 1945–1953, H. 33 cm; eBay # 238974460
8440,10,10,L Firmenmarke 19, Made in Austria, H. 33 cm; eBay # 2231691439
8440,11 Firmenmarke 19, Made in Austria, unsign., Ausformung zw. 1945–1953, H. 32 cm; eBay # 2009370682
8440,13,66 Firmenmarke 19, Made in Austria, Hand dec., unsign., H. 33 cm; eBay # 1232798430
8440,22,45 Firmenmarke 19, Made in Austria, Hand dec., unsign., H. 33,5 cm; WKA 33
8440,66,70 gem. HS, Firmenmarke 19, Made in Austria, unsign., H. 32,5 cm; Quit. 25
8440,69,66 gem. lig. HS, Firmenmarke 19, Made in Austria, unsign., H. 33 cm; eBay # 2500902157
8440,115,19 Firmenmarke 19, H. 32,5 cm; Wendl 30
8440,248,7 (Firmenmarke ohne Ang.), unsign., Pick. 6, H. 32,3 cm; Doro 15.3.2005
8440F, H. 32,5 cm, B. 26,5 cm; GSNachf. 1,47
LAMPE
Es ist kein Exemplar in der Ausführung als Lampe nachweisbar, diese wurde möglicherweise erst nach dem Krieg in nur geringer Stückzahl angefertigt (-> vgl. Quellen).
Bem.: Diese Figur war sowohl zu seiner Entstehungszeit 1939, als auch nach dem Krieg eines der erfolgreichsten Modelle der Manufaktur. Dies bestätigen die Stückzahlen ebenso wie die Verwendung des Motivs als Umschlagabbildung für den GSNachf. Katalog von Schuster. Herr Karl Täuber (ehem. Prokurist der Firma) bestätigte dies auch in einem Interview, wie auch die Tatsache, dass „Jolanthe" das erste, nach Schusters Übernahme der Manufaktur, produzierte Modell war (-> vgl. Kap. zur Firmengeschichte). Offenbar wurde bei Schuster auch überlegt, die Nummerierung neu zu gestalten wie die zahlreichen mit „1" bezeichneten Stücke annehmen lassen. Das erfolgreiche Modell wurde auch in den USA unter der Modell-Nr. 528 produziert (-> vgl. Christie's SK 14.11.2002).
Lit.: GSNachf. 1,47 (m. Abb.), hier mit Ang. des Originaltitels „Jolanthe".
Quelle: Archiv Mimi Dakon (Tantiemenabrechnung 1.1.–30.6.1939; Modellliste Dakon, undatiert, ca. 1940), hier mit 1938 datiert; Wiener Manufaktur FG, Modellliste Dakon, 7.11.1946), hier wird die Figur jeweils unter dem Titel „Tänzerin" geführt; (Tantiemenabrechnung 1. Halbjahr 1949; 2. Halbjahr 1949; 1. Halbjahr 1950; 2. Halbjahr 1950; (1. Halbjahr 1951, AGM); 2. Halbjahr 1951, hier auch als „Sorte II" bestellt), hier wird die Figur jeweils unter dem Titel „Danseuse" geführt; (Tantiemenabrechnung 1. Halbjahr 1952; Rechnung an Maurice Mizrahi, Alexandria, Ägypten, vom 22. April 1953), hier jeweils unter dem Titel „Tanz Lampe" geführt.

8442F

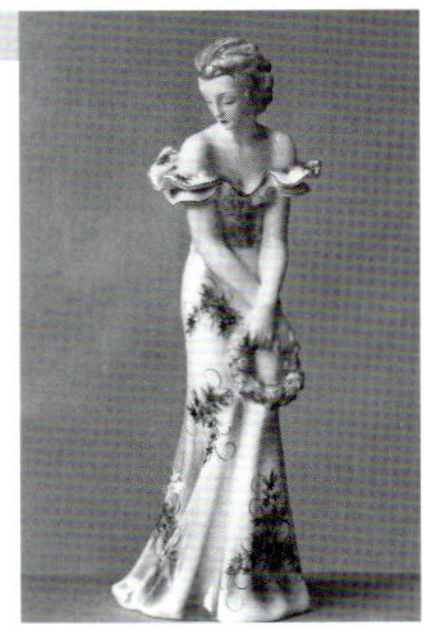

FIGUR „Sommerabend"
(Stehende junge Dame, einen Blumenkranz haltend)
ca. 1939
8442,1,38 gem. lig. HS, Firmenmarke 19, Made in Austria, Hand dec., H. 37,5 cm; PB Wien
8442,4,29 gem. G, gem. rot H, Firmenmarke 19, Made in Austria, H. 36,9 cm; MAK Wien, Inv. Ke 10602-34
8442,17,2 Firmenmarke 19, Made in Austria, Hand dec., H. 31,5 cm; WKA 35
8442,85,31 (Firmenmarke ohne Ang.), H. 36,8 cm; Doro 12.9.2006
8442F, H. 36,5 cm, B. 14 cm; GSNachf. 1,48; Fotoarchiv K. Täuber, hier mit der Ang.: H. 36 cm, B. 12 cm
Lit.: GSNachf. 1,48 (m. Abb.), hier mit Ang. des Originaltitels „Sommerabend". Eine Kommissionsfaktura der „Österreichischen Gruppe-Austria" für die Messe in Utrecht vom 26.2.1953 belegt den Titel „Blumenkranz".
Quelle: Fotoarchiv K. Täuber

8445
Stehende Tänzerin mit Hut und Federfächer (auf Rundsockel)
ca. 1939
8445,22,10 Firmenmarke 19, Made in Germany, Ausformung zw. 1939–1941, H. 32 cm; KH
8445,29,31 gem. rot 4.I, Firmenmarke 19, Made in Germany, Ausformung zw. 1939–1941, H. 31,8 cm; Wien Museum Inv. 190.144
Ausst.: SSH 1992

8446
Herczeg, Claire (Klára)
Junge Dame im Sommerkleid, mit einem Schal tanzend (auf Ovalsockel)
ca. 1939
8446,24,33 gem. AN, Firmenmarke 19, Made in Germany, sign. Claire Herczeg, Ausformung zw. 1939–1941, H. 39,5 cm; WKA 35

8451
Tanzender Damenakt mit langem Tuch (das linke Bein hochhebend) (auf Ovalsockel)
ca. 1939
8451,61,7 Firmenmarke 19, Made in Germany, Ausformung zw. 1939–1941, H. 45,5 cm; Nagel 348; Doro 14.9.2001

8452F

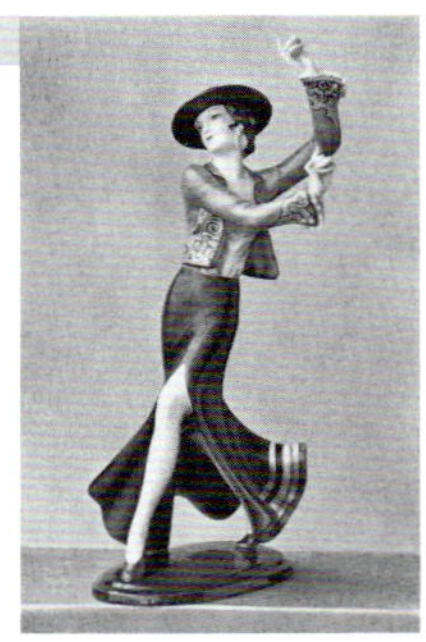

Lorenzl, Josef
FIGUR „Spanischer Tanz"
(„Spanierin" mit Hut und Kastagnetten)
(auf Ovalsockel)
ca. 1939
8452,8,4 Firmenmarke 19, Made in Germany, Ausformung zw. 1939–1941, H. 46 cm; Herr 54
8452F, H. 45 cm, B. 22 cm; GSNachf. 1,49
Bem.: Die in der Quelle angegebene Bestellung von Wahliss (Wien) zeigt, dass offenbar auch der Titel „Spanierin" geläufig war.
Lit.: GSNachf. 1,49 (m. Abb.), hier mit Ang. des Originaltitels „Tänzerin".
Quelle: AGM (Bestellungen der Firma Wahliss (Wien) vom 30.8.1945 und 27.2.1946 an die „Arbeitsgemeinschaft der Firma Friedrich Goldscheider"), hier jeweils mit Ang. des Titels „Spanierin"; Rechnung an Maurice Mizrahi, Alexandria, Ägypten, vom 22. April 1953), hier mit Ang. des (Original-)Titels „Spanischer Tanz".

8453F
Dakon, Stephan
FIGUR „Tanz" („Rüschentanz"/„Colette")
(Stehende Tänzerin mit Beret und Rüschenkleid, dieses mit der rechten Hand anhebend) (auf Ovalsockel)
ca. 1939
8453,1,80 (Firmenmarke ohne Ang.), unsign., Ausformung zw. 1939–1941, H. 32,5 cm; Christie's SK 15.5.2003
8453,68,31 Firmenmarke 19, Made in Germany, unsign., Ausformung zw. 1939–1941, H. 32,7 cm; Christie's SK 15.5.2003
Lit.: Pinhas, S. 151 (m. Abb.)
Quelle: Archiv Mimi Dakon (Tantiemenabrechnung 1.1.–30.6.1939), hier wird die Figur unter dem (Original-)Titel „Tanz" geführt; (Tantiemenabrechnung 1.1.–30.6.1940; Modellliste Dakon, undatiert, ca. 1940), hier mit 1938 datiert; Wiener Manufaktur FG, Modellliste Dakon, 7.11.1946), hier wird die Figur unter dem (Original-)Titel „Rüschentanz" geführt; (Tantiemenabrechnung 1. Halbjahr 1949; 2. Halbjahr 1949; 1. Halbjahr 1950; 2. Halbjahr 1950; (1. Halbjahr 1951, AGM); 2. Halbjahr 1951, hier auch als „Sorte II" bestellt; 1. Halbjahr 1952), hier wird die Figur jeweils unter dem (Original-)Titel „Colette" geführt; AGM.

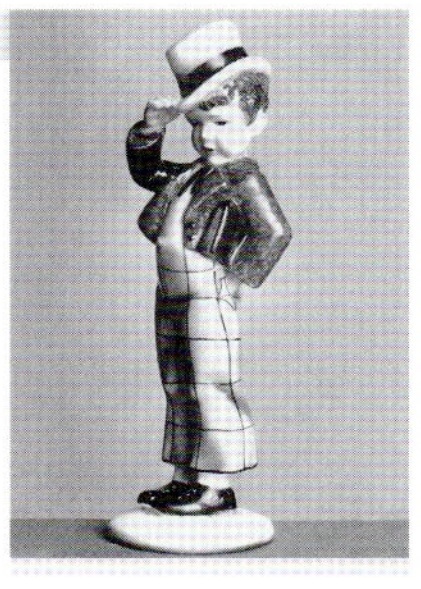

8454F
Dakon, Stephan
FIGUR „Gratulant"
(„Bub mit Zylinder")
(auf Rundsockel)
ca. 1939
8454,1 gem. J., Firmenmarke 19, Made in Austria, unsign., Ausformung zw. 1945–1953, H. 24 cm; eBay # 929900399
8454,4 ger. 32, Firmenmarke 19, Made in Austria, unsign., Ausformung zw. 1945–1953, H. 21 cm; Wien Museum Inv. 220.741
8454 H. 20 cm, B. 9 cm; Fotoarchiv K. Täuber
Pendant –> vgl. 8455
Quelle: Archiv Mimi Dakon (Tantiemenabrechnung 1.1.–30.6.1939; Modellliste Dakon, undatiert, ca. 1940), hier mit 1938 datiert; 1. Halbjahr 1949; 2. Halbjahr 1949; 1. Halbjahr 1950; 2. Halbjahr 1950; (1. Halbjahr 1951, AGM); 2. Halbjahr 1951; 1. Halbjahr 1952), hier wird die Figur jeweils unter dem (Original-)Titel „Gratulant" geführt; (Tantiemenabrechnung 1.1.–30.6.1940; Wiener Manufaktur FG, Modellliste Dakon, 7.11.1946), hier wird die Figur unter dem (Original-)Titel „Bub mit Zylinder" geführt; Fotoarchiv K. Täuber.

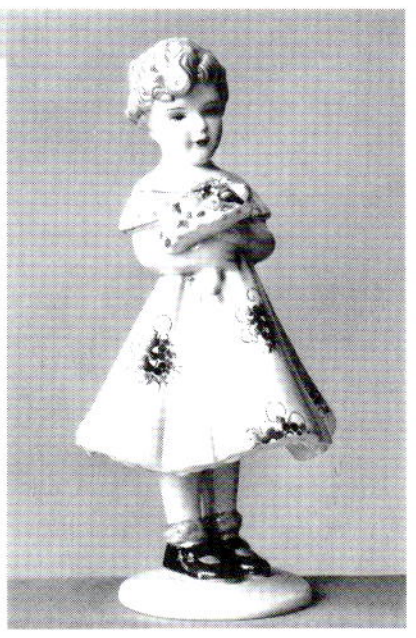

8455F
Dakon, Stephan
FIGUR „Gratulantin"
(„Mädchen mit Blumen")
(auf Rundsockel)
ca. 1939
8455,53 gem. B, Firmenmarke 19, unsign., Ausformung zw. 1939–1953, H. 19 cm; city-antik
8455 H. 18 cm, B. 10 cm; Fotoarchiv K. Täuber
Pendant –> vgl. 8454
Quelle: Archiv Mimi Dakon (Tantiemenabrechnung 1.1.–30.6.1939; Modellliste Dakon, undatiert, ca. 1940), hier wird die Figur unter dem Titel „Mädchen mit Blumen" geführt und mit 1938 datiert; Wiener Manufaktur FG, Modellliste Dakon, 7.11.1946; Tantiemenabrechnung 1. Halbjahr 1949; 2. Halbjahr 1949; 1. Halbjahr 1950; 2. Halbjahr 1950; (1. Halbjahr 1951, AGM); 2. Halbjahr 1951; 1. Halbjahr 1952), hier wird die Figur jeweils unter dem (Original-)Titel „Gratulantin" geführt; AGM; Fotoarchiv K. Täuber.

8457
Lorenzl, Josef
Stehender Mädchenakt mit Blumen (auf Rundsockel)
ca. 1939
8457,63,6 Firmenmarke 19, Made in Germany, Ausformung zw. 1939–1941, H. 33,2 cm; Doro 21.9.1999

8458
Stehendes Mädchen mit Kopftuch und Ring (auf Ovalsockel)
ca. 1939
8458,9 gem. M., gem. rot G, Firmenmarke 19, Made in Austria, Ausformung zw. 1945–1953, H. 29,6 cm; Wien Museum Inv. 220.705

8459
Lorenzl, Josef
„Sonate" (Schreitende Tänzerin mit/ohne Bikinioberteil, ihren Rock schmetterlingsflügelartig auseinander haltend) (auf Dreiecksockel)
ca. 1936 (für den Entwurf)
8459,15,2 Firmenmarke 19, Made in Germany, Ausformung zw. 1939–1941, H. 40,5 cm; BM 54
8459,18, ger. SB, gem. B II, gem. lig. WF, Firmenmarke 19, Made in Germany, Ausformung zw. 1939–1941, H. 41 cm; Wien Museum Inv. 219.187
8459,42,7 Firmenmarke 19, Made in Austria, unsign., Ausformung ca. 1938, H. 40 cm; Christie's SK 4.11.1999; KH
8459,78,31 gem. rot XX, 35, Firmenmarke Schuster 1, Made in Germany, unsign., Ausformung zw. 1941–1944, H. 42 cm; eBay # 170003533764
8459,79,7 gem. rot XX, 37, Firmenmarke Schuster 1, Made in Germany, unsign., Ausformung zw. 1941–1944, H. 42 cm, B. 40 cm; Wien Museum Inv. 219.198
Bem.: Das Modell „Sonate", wurde in dieser Größe in zwei Varianten angeboten: mit dem Bikinioberteil oder ohne.
–> vgl. 7582

8460
Lorenzl, Josef
Posierender Mädchenhalbakt (Revuetänzerin in langem Rock) (auf passigem Sockel)
ca. 1939
8460,9,0 gem. lig. LR., Firmenmarke 19, Made in Germany, sign. Lorenzl, Ausformung zw. 1939–1941, H. 38 cm; eBay # 1456585298

8461
Lorenzl, Josef
Stehender Mädchenhalbakt mit um die Hüften geschlungenem langen Tuch (auf Ovalsockel)
ca. 1939
8461,6,0 (Firmenmarke ohne Ang.), H. 40 cm; Christie's SK 27.4.2005
8461,44,7 Firmenmarke 19, Made in Germany, Ausformung zw. 1939–1941, H. 40 cm; Kastern 74; Herr 34
Lit.: Pinhas, S. 158 (m. Abb.)

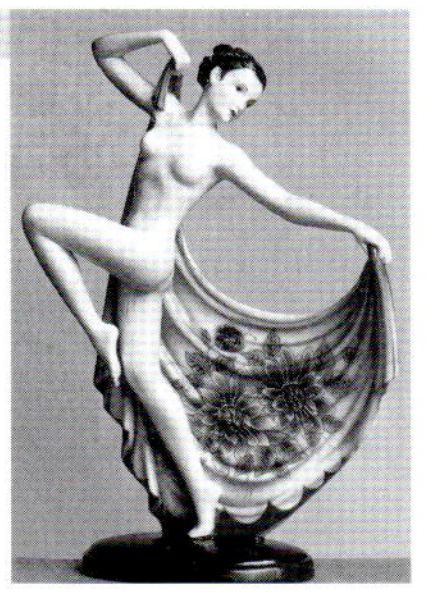

8462
Lorenzl, Josef
„Adagio" (Tanzender Damenakt mit großem Tuch) (auf Ovalsockel)
ca. 1939
8462,7 Firmenmarke 19, Made in Austria, unsign., Pick. 6, Ausformung zw. 1945–1953, H. 41 cm; Christie's SK 15.2.2001
8462,29,19 Firmenmarke 19, Made in Germany, sign. Lorenzl, Ausformung zw. 1939–1941, H. 41,9 cm; eBay # 1423475455
8462,56,66 Firmenmarke 19, unsign., H. 41,2 cm; BM 75
8462 ohne Ang., H. 40 cm, B. 29 cm; Könecke (2002), S. 77
8462, H. 40 cm; B. 28 cm; Fotoarchiv K. Täuber
Bem.: Diese Figur wurde auch noch zw. 1953–1960 bei Carstens (GWM West-Germany) produziert.
Lit.: KGSF1; KGSF2; Könecke (2002), S. 31, 77 (m. Abb.), hier mit der Ang. des (Original-?)Titels „Adagio" und des Künstlers J. Lorenzl; Pinhas, S. 152 (m. Abb.).
Quelle: Archiv Matschiner (Originalfoto); Fotoarchiv K. Täuber.

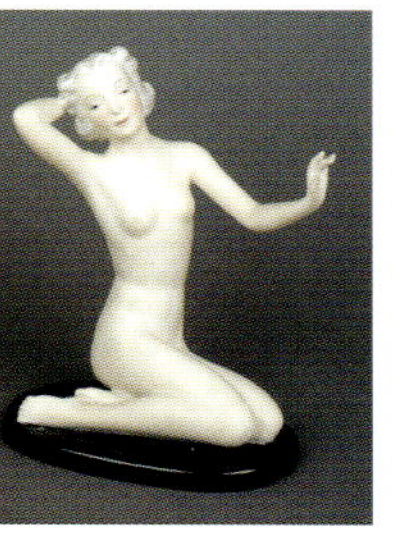

8465
Prischl, Adolf
Sitzender weiblicher Akt mit ausgestrecktem Arm (auf Ovalsockel)
ca. 1939
8465,19,4 Firmenmarke 19, Made in Germany, Ausformung zw. 1939–1941, H. 22 cm; eBay # 293446583; 453545826
8465,34,7,5,C Firmenmarke 19, Made in Germany, unsign., Ausformung zw. 1939–1941, H. 22,2 cm; WKH
8465,45,7,P Firmenmarke 19, Made in Germany, unsign., Pick. 6, Ausformung zw. 1939–1941, H. 22 cm; Schloss Rick. 83
8465 ohne Ang., H. 22 cm; PB Wien 19

8466
FIGUR „Polita" (Tänzerin)
ca. 1939
Quelle: AGM (Bestellungen der Firma Wahliss (Wien) vom 30.8.1945 und 27.2.1946 an die „Arbeitsgemeinschaft der Firma Friedrich Goldscheider"), hier mit Ang. des (Original-?)Titels „Polita".

8469F
Dakon, Stephan
FIGUR „Tänzerin (Dame) mit Herz"
ca. 1939
Quelle: Archiv Mimi Dakon (Tantiemenabrechnung 1.1.–30.6.1939; Wiener Manufaktur FG, Modellliste Dakon, 7.11.1946), hier wird die Figur unter dem (Original-)Titel „Tänzerin mit Herz" geführt; (Tantiemenabrechnung 1.1.–30.6.1940), hier wird die Figur unter dem Titel „Tanz" geführt; (Modellliste Dakon, undatiert, ca. 1940), hier wird die Figur unter dem Titel „Dame mit Herz" geführt und mit 1938 datiert.

8470F
Dakon, Stephan
FIGUR „Tänzerin mit Herz" („Revuegirl")
ca. 1939
Quelle: Archiv Mimi Dakon (Tantiemenabrechnung 1.1.–30.6.1939), hier wird die Figur unter dem (Original-)Titel „Tänzerin mit Herz" geführt; (Modellliste Dakon, undatiert, ca. 1940), hier wird die Figur unter dem (Original-)Titel „Revuegirl" geführt und mit 1938 datiert; (Wiener Manufaktur FG, Modellliste Dakon, 7.11.1946), hier mit Ang. des Titels „Tänzerin".

8471
STEHKOPF Frauenkopf mit Schal, Hut und Feder
ca. 1939
8471,9 ger. III, Firmenmarke 19, Made in Germany, Ausformung zw. 1939–1941, H. 33 cm; Wien Museum Inv. 220.800

8472F
Dakon, Stephan
FIGUR „Tanz" („Tänzerin"/„Adrienne") (Tänzerin mit Bikini und großem Tuch) (auf Ovalsockel)
ca. 1939
8472,75,31 Firmenmarke 19, Made in Germany, unsign., Ausformung zw. 1939–1941, H. 31 cm; Christie's SK 22.4.2004
Lit.: Pinhas, S. 161 (m. Abb.)
Quelle: Archiv Mimi Dakon (Tantiemenabrechnung 1.1.–30.6.1939), hier wird die Figur unter dem (Original-)Titel „Tanz" geführt; (Tantiemenabrechnung 1.1.–30.6.1940; Modellliste Dakon, undatiert, ca. 1940), hier mit 1938 datiert; Wiener Manufaktur FG, Modellliste Dakon, 7.11.1946; Tantiemenabrechnung 2. Halbjahr 1950), hier wird die Figur unter dem (Original-)Titel „Tänzerin" geführt; (Tantiemenabrechnung (1. Halbjahr 1951, AGM); 2. Halbjahr 1951), hier mit dem (Original-)Titel „Adrienne" geführt.

8473F
Dakon, Stephan
FIGUR „Tennisspielerin" („Tennis")
ca. 1939
Quelle: Archiv Mimi Dakon (Tantiemenabrechnung 1.1.–30.6.1939; Modellliste Dakon, undatiert, ca. 1940), hier mit 1938 datiert; Wiener Manufaktur FG, Modellliste Dakon, 7.11.1946), hier wird die Figur jeweils unter dem (Original-)Titel „Tennisspielerin" geführt; (Tantiemenabrechnung 1.1.–30.6.1940; 1. Halbjahr 1949), hier wird die Figur jeweils unter dem (Original-)Titel „Tennis" geführt.

8474F
Dakon, Stephan
FIGUR „Andante" (Stehende Tänzerin mit Bikinioberteil, Shorts und langem Umhang) (auf Ovalsockel)
ca. 1939
8474,3,10 Firmenmarke 19, Made in Germany, Ausformung zw. 1939–1941, H. 34 cm; KH
8474,7 Firmenmarke 19, Made in Austria, unsign., Ausformung zw. 1945–1953, H. 33 cm; WKA 29
8474,11 Firmenmarke 19, Made in Austria, unsign., Ausformung zw. 1945–1953, H. 33 cm; Soth. L. 23.6.1989
Quelle: Archiv Mimi Dakon (Tantiemenabrechnungen 1.1.–30.6.1939; 1. Halbjahr 1949; (1. Halbjahr 1951, AGM), hier wird die Figur jeweils unter dem Originaltitel „Andante" geführt; (Tantiemenabrechnung 1.1.–30.6.1940; Modellliste Dakon, undatiert, ca. 1940), hier mit 1938 datiert; Wiener Manufaktur FG, Modellliste Dakon, 7.11.1946; Tantiemenabrechnung 2. Halbjahr 1950), hier wird die Figur jeweils unter dem (Original-)Titel „Tänzerin" geführt; AGM (Bestellungen der Firma Wahliss (Wien) vom 9. und 24.11.1945 an die „Arbeitsgemeinschaft der Firma Friedrich Goldscheider"; Bestellung der Firma Wahliss (Wien) vom 27.2.1946 in den Ausführungen „Terrakotta" je 37 Stück, und 1 Stück in „Fayence").

8477
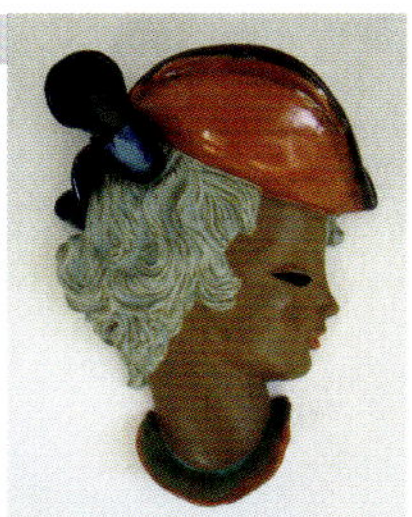
WANDMASKE Frauenkopf (im Profil nach links) mit Hut und Haarschleife
ca. 1939
8477,3, gem. L., Firmenmarke 19, Made in Germany, Ausformung zw. 1939–1941, H. 25 cm; Sammlung f@goldscheider.de

8480
WANDMASKE Frauenkopf mit Hut, Haarspange und Kragen
ca. 1939
8480,3 gem. L., Firmenmarke 19, Made in Germany, Ausformung zw. 1939–1941, H. 33 cm; Soth. L. 23.6.1989; eBay # 2210853073

8482F
Herczeg, Claire (Klára) (zugeschr.)
„Rositta" („Rosita"?) (Tänzerin in drehender Bewegung, mit erhobenem linken Arm) (auf Ovalsockel)
ca. 1939
8482,22,0 Firmenmarke 19, unsign., H. 33,8 cm; Doro 10.3.2003; 13.9.2004
8482,127,31 gem. SP, X, gem. rot XX, 34, Firmenmarke Schuster 1, Made in Germany, unsign., Ausformung zw. 1941–1944, H. 31,5 cm; Wien Museum Inv. 190.150
Bem.: Diese Figur wurde auch noch zw. 1953–1960 bei Carstens (GWM West-Germany) produziert.
Ausst.: SSH 1992
Lit.: SS, S. 22, 23 (m. Abb.); Könecke (2002), S. 77, 91 (m. Abb.).
Quelle: AGM (Rechnung der Wiener Manufaktur Josef Schuster an Laszlo Bösze, Szeged, Ungarn, vom 5. August 1944), hier mit Ang. des Originaltitels „Rositta" (Tippfehler?), eigentlich müsste es wohl „Rosita" heißen.

8483
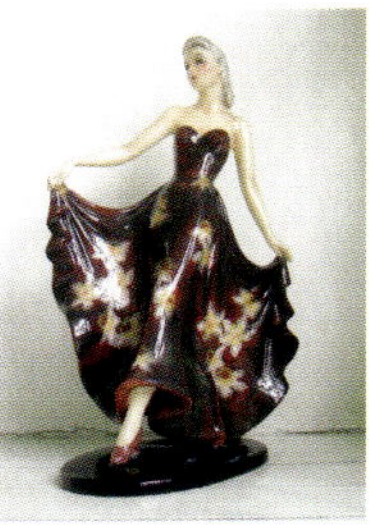
Herczeg, Claire (Klára)
„Phantasie" (Schreitende Tänzerin in langem Kleid) (auf Ovalsockel)
ca. 1939
8483,2,19 gem. rot S., Firmenmarke 19, Made in Germany, sign. Claire Herczeg, Ausformung zw. 1939–1941, H. 36 cm, B. 21 cm; Privatsammlung, Wien
8483,3,2,50 Firmenmarke 19, Made in Germany, sign. Herczeg, Pick. 6, Ausformung zw. 1939–1941, H. 36,5 cm; Wendl 36
8483,6,66 Firmenmarke 19, Made in Austria, unsign., H. 36 cm; WKA 35
8483,11 gem. lig. JT, Firmenmarke 19, Made in Austria, unsign., Ausformung zw. 1945–1953, H. 33,8 cm; Wien Museum Inv. 219.191
8483,65,19,50 Firmenmarke 19, Made in Germany, sign. Claire Herczeg, Ausformung zw. 1938–1941, H. 36 cm; city-antik
8483,139,31,G gem. AN, HE, Firmenmarke 19, Made in Germany, unsign., Ausformung zw. 1939–1941, H. 37 cm; eBay # 3776769745
8483,1341/3 Firmenstempel WG 1, unsign., Pick. WG 1, Ausformung 50er Jahre, H. 36 cm; Mehlis 12
Bem.: Diese Figur wurde auch noch zw. 1953–1960 bei Carstens (GWM West-Germany) produziert.
Lit.: Könecke (2002), S. 77, 91 (m. Abb.).
Quelle: AGM (Bestellung der Firma Wahliss (Wien) vom 27.2.1946 an die „Arbeitsgemeinschaft der Firma Friedrich Goldscheider", hier mit Ang. des (Original-?)Titels „Phantasie", in Ausführung „Terrakotta").

8484F
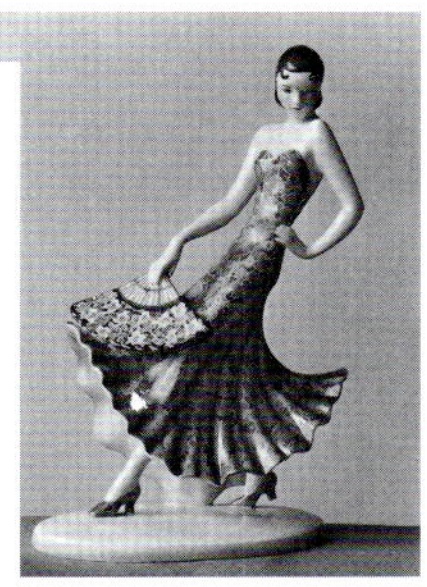
Herczeg, Claire (Klára)
„Tanz" (Spanierin mit Fächer) (auf Ovalsockel)
ca. 1939
8484 H. 39 cm, B. 18 cm; Fotoarchiv K. Täuber
Quelle: AGM (Rechnung an Maurice Mizrahi, Alexandria, Ägypten, vom 22. April 1953), hier unter dem Titel „Tanz" geführt, geliefert in Ausführung „matt"; Fotoarchiv K. Täuber.

8485F
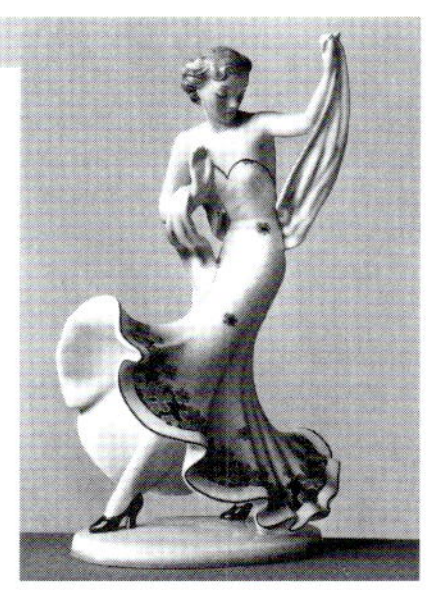
Herczeg, Claire (Klára)
FIGUR/LAMPE „Allegretto" (Tänzerin in langem Kleid und Schal in Bewegung) (auf Ovalsockel)
ca. 1939
FIGUR
8485 ohne Ang., Firmenstempel WG 1, unsign., Ausformung 50er Jahre, H. 35 cm; Ahlden 39
8485 ohne Ang., Firmenstempel WG 1, unsign., Ausformung 50er Jahre, H. 34 cm; BM 38
8485 H. 34 cm, B. 17 cm; Fotoarchiv K. Täuber
LAMPE
Es ist kein Exemplar in der Ausführung als Lampe nachweisbar, diese wurde möglicherweise erst nach dem Krieg in nur geringer Stückzahl angefertigt (-> vgl. Quellen).
Bem.: Diese Figur wurde auch noch zw. 1953–1960 bei Carstens (GWM West-Germany) produziert.
Lit.: KGSF1; Könecke (2002), S. 77, hier mit Ang. des Künstlerin C. Herczeg, S. 91 (m. Abb.).
Quelle: AGM (Bestellungen der Firma Wahliss (Wien) vom 9. und 24.11.1945 an die „Arbeitsgemeinschaft der Firma Friedrich Goldscheider", hier mit Ang. des (Original-?) Titels „Allegretto"; Bestellung der Firma Wahliss (Wien) vom 27.2.1946 in Ausführung „Terrakotta"; Rechnung an Maurice Mizrahi, Alexandria, Ägypten, vom 22. April 1953), hier unter dem Titel „Tanz Lampe" geführt, geliefert in Ausführung „matt"; Fotoarchiv K. Täuber.

8486F
„Tanz" (Schreitende Tänzerin mit Hut und venezianischem Kostüm) (auf Ovalsockel)
ca. 1939
8486,12,11 Firmenmarke 19, H. 33 cm; city-antik
Quelle: AGM (Rechnung an Maurice Mizrahi, Alexandria, Ägypten, vom 22. April 1953), hier mit Ang. des (Original-)Titels „Tanz".

8487
Herczeg, Claire (Klára) (zugeschr.)
Schreitende schlanke Tänzerin mit geschlossenen Augen (auf Ovalsockel)
ca. 1939
8487,25,7 Firmenmarke 19, Made in Germany, unsign., Ausformung zw. 1939–1941, H. 34,5 cm; Phillips L. 25.1.2004
Lit.: Pinhas, S. 161 (m. Abb.), hier mit der falschen Künstlerang. „F. Gattcher". Das Modell kann jedoch mit ziemlicher Sicherheit Claire Herczeg zugeordnet werden.

8489F
GRUPPE „Pferde" (Zwei rivalisierende Hengste) (auf ovalem Natursockel)
ca. 1939
8489,244,9 Firmenmarke 19, Made in Germany, Ausformung zw. 1939–1941, H. 30,8 cm; Doro 19.9.2000
8489 ohne Ang., H. 29 cm, B. 26 cm; Könecke (2002), S. 77, 91 (m. Abb.)
Bem.: Diese Figur wurde auch noch zw. 1953–1960 bei Carstens (GWM West-Germany) produziert.
Lit.: Könecke (2002), S. 77, 91 (m. Abb.).
Quelle: AGM (Kontoblatt der Wiener Manufaktur Josef Schuster für die Firma Porslins Ericsson-Lund, Stockholm?, Schweden vom 7. Januar 1943); (Rechnung an Maurice Mizrahi, Alexandria, Ägypten, vom 22. April 1953), hier jeweils mit Ang. des Originaltitels GRUPPE „Pferde".

8490F/M/T
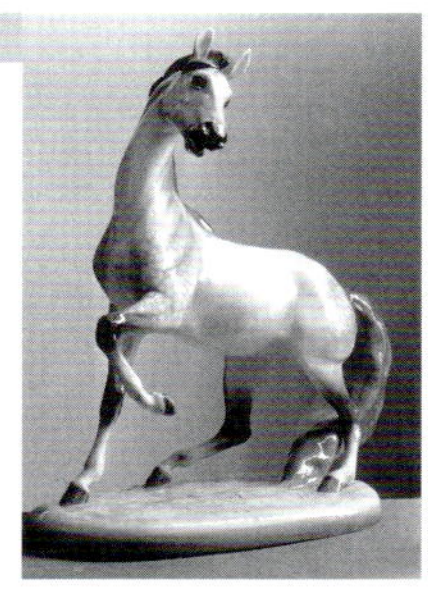
„Pferd" (Scheuender Lipizzaner?) (auf Ovalsockel)
ca. 1939
8490,8 gem. F., Firmenmarke 19, Made in Austria, Ausformung zw. 1945–1953, H. 38 cm, B. 33 cm; eBay # 2228396460
8490F, H. 38 cm, B. 35 cm; GSNachf. 5,24
Pendant -> vgl. 8491
Lit.: GSNachf. (2. Ausg., bzw. Erg. Bl.) 5,24 (m. Abb.)
Quelle: AGM (Kontoblatt der Wiener Manufaktur Josef Schuster für die Firma Porslins Ericsson-Lund, Stockholm?, Schweden vom 7. Januar 1943), hier mit Ang. des Originaltitels „Pferd"; (Bestellung der Firma Wahliss (Wien) vom 27.2.1946 an die „Arbeitsgemeinschaft der Firma Friedrich Goldscheider", hier mit Ang. des Originaltitels „Pferd" (Ausführung Terrakotta); Rechnung der Wiener Manufaktur Josef Schuster an Laszlo Bösze, Szeged, Ungarn, vom 5. August 1944; Rechnung an Maurice Mizrahi, Alexandria, Ägypten, vom 22. April 1953), hier mit Ang. des Originaltitels „Pferd" (Ausführung Fayence); Fotoarchiv K. Täuber.

8491F
Steigendes Pferd (auf ovalem Natursockel)
ca. 1939
8491,75,31 gem. rot XX, 23, Firmenmarke Schuster 1, Made in Germany, unsign., Ausformung zw. 1941–1944, H. 42 cm; Wien Museum Inv. 220.769
8491F, H. 41 cm, B. 34 cm; GSNachf. 5,25
Pendant -> vgl. 8490
Lit.: GSNachf. (2. Ausg., bzw. Erg. Bl.) 5,25 (m. Abb.)

8492
Lorenzl, Josef
Stehendes Mädchen mit Kappe (Béret), drei Herzen tragend (auf Rundsockel)
ca. 1939
8492,5,0 ger. 47A, gem. AN, SG, Firmenmarke 19, Made in Germany, Ausformung zw. 1939–1941, H. 33,8 cm; Soth. A. 30.6.1992

8493

Lorenzl, Josef

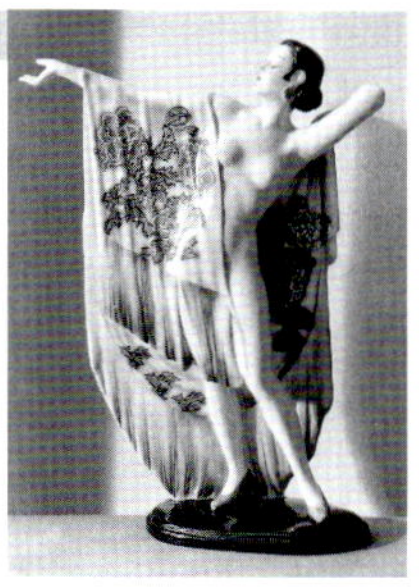

Stehender Damenakt (Spanierin) mit ausgestrecktem Arm und großem Tuch/Schal (auf Ovalsockel)
ca. 1939
8493,22,7 Firmenmarke 19, Made in Germany, sign. Lorenzl, Ausformung zw. 1939–1941, H. 42,8 cm; Doro 24.11.2004
8493,57,31 Firmenmarke 19, Made in Germany, unsign., Ausformung zw. 1939–1941, H. 42,5 cm; city-antik
8493,90,31 gem. WH, Firmenmarke 19, Made in Germany, sign. Lorenzl, Ausformung zw. 1939–1941, H. 42,5 cm; Neumeister 4.6.1994
8493,115,19 gem. rot L, Firmenmarke 19, Made in Germany, unsign., Ausformung zw. 1939–1941, H. 43 cm; city-antik
Quelle: Fotoarchiv K. Täuber

8494F / M

Prischl, Adolf

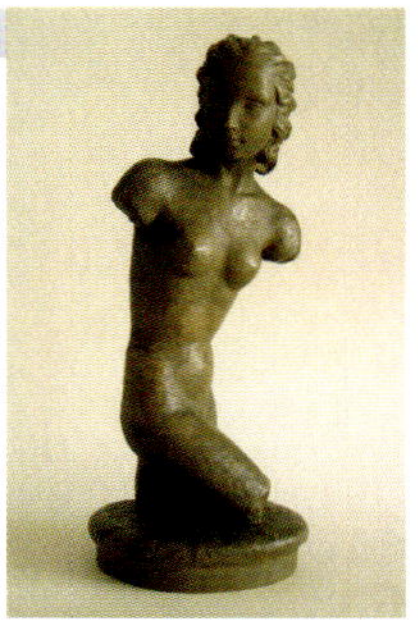

FIGUR „Torso" (Damenakt) (auf Ovalsockel)
ca. 1939
8494,6 Firmenmarke 19a, unsign., Ausformung zw. 1939–1941, H. 43,3 cm; PB Deutschland
8494,7 Firmenmarke 19a, unsign., Ausformung zw. 1939–1941, H. 43,5 cm; WKH
8494,73 Firmenmarke 19a, unsign., Ausformung zw. 1939–1941, H. 45 cm; eBay # 3313194684

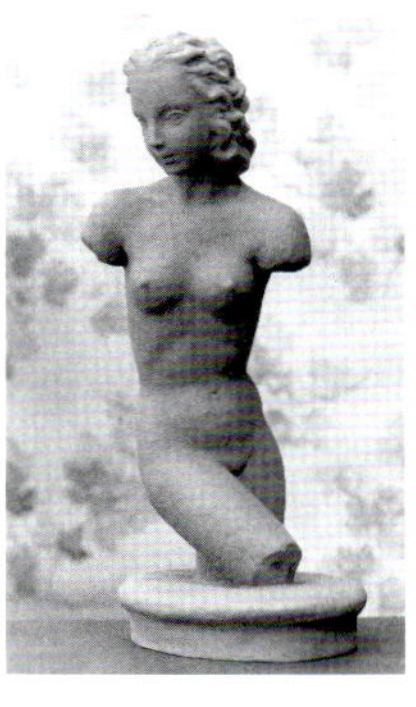

8494 Firmenmarke 19, Made in Austria, unsign., Ausformung zw. 1945–1953, H. 42 cm; Mehlis 17
8494F, H. ca. 43 cm; GSNachf. 1,29
8494M, H. 43 cm, B. 20 cm; GSNachf. 3,30
Lit.: GSNachf. 1,29 (m. Abb.); 3,30 (m. Abb.), hier jeweils mit Ang. des Originaltitels „Torso".
Quelle: AGM (Bestellungen der Firma Wahliss (Wien) vom 9. und 24.11.1945 an die „Arbeitsgemeinschaft der Firma Friedrich Goldscheider"; Bestellung der Firma Wahliss (Wien) vom 27.2.1946 in Ausführung „Terrakotta"; Rechnung an Maurice Mizrahi, Alexandria, Ägypten, vom 22. April 1953), in Ausführung „matt"; Fotoarchiv K. Täuber.

8495F / M

Lorenzl, Josef

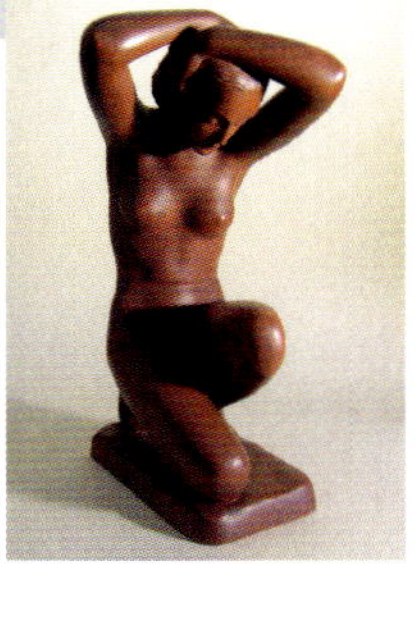

FIGUR „Kniende" (Kniender Damenakt) (auf abgerundetem Rechtecksockel)
ca. 1939
8495,9 (Firmenmarke ohne Ang.), unsign., H. 38 cm; Doro 14.9.2001
8495,14 Firmenmarke 19a, unsign., H. 40 cm; KH
8495,20 Firmenmarke 19, unsign., H. 39 cm; Doro 24.10.1994; 9.6.1998
8495,42,S Firmenmarke 19, Made in Austria, Hand dec., unsign., Ausformung zw. 1945–1953, H. 37 cm; eBay # 3365851350
8495,64 ger. lig. PL, Firmenmarke 19a, H. 38 cm; Doro 15.3.2005
8495,65 Firmenmarke 19, unsign., H. 38,8 cm; Doro 21.5.1997
8495 ger., Firmenmarke 19, unsign., H. 37,8 cm; BM 105
8495 ger., ger. lig. CM, ohne Firmenmarke, H. 36,5 cm; Sammlung John Nash
8495 ger., ohne Firmenmarke, H. 36 cm; Sammlung John Nash
8495F, H. 37 cm; GSNachf. 1,37
8495M, H. 40 cm, B. 20 cm; GSNachf. 3,29
Bem.: Diese Figur wurde auch noch zw. 1953–1960 bei Carstens (GWM West-Germany) produziert.
-> vgl. 8496; 8708
Lit.: GSNachf. 1,37 (m. Abb.); 3,29 (m. Abb.), hier jeweils mit Ang. des Originaltitels „Kniende"; KGSF1; KGSF2; Könecke (2002), S. 37, 77 (m. Abb.), hier mit Ang. des Künstlers J. Lorenzl und des (Original-?)Titels „Akt, Hockende" (für die Produktion bei Carstens?); Matschiner (2006), S. 76 (m. Abb.).
Quelle: AGM (Bestellungen der Firma Wahliss (Wien) vom 9. und 24.11.1945 an die „Arbeitsgemeinschaft der Firma Friedrich Goldscheider"; Bestellung der Firma Wahliss (Wien) vom 27.2.1946 in Ausführung „Terrakotta").

8496

Lorenzl, Josef

FIGUR „Kniende" (Kniender Damenakt) (auf abgerundetem Rechtecksockel)
ca. 1939
8496 ohne Ang., H. 32 cm, B. 16 cm; Könecke (2002), S. 37, 77 (m. Abb.)
Bem.: Diese Figur wurde auch noch zw. 1953–1960 bei Carstens (GWM West-Germany) produziert.
-> vgl. 8495; 8708
Lit.: KGSF1; KGSF2; Könecke (2002), S. 37, 77 (m. Abb.), hier mit Ang. des Künstlers J. Lorenzl und des (Original-?) Titels „Akt, Hockende" (für die Produktion bei Carstens?, zum Originaltitel -> vgl. 8495).

8498

Dakon, Stephan

GRUPPE Zwei Tänzerinnen, eine kniend, die andere stehend (auf Ovalsockel)
ca. 1939
8498 (Firmenmarke ohne Ang.), H. 28 cm; Pinhas, S. 158 (m. Abb.)

8501

Stehende Dame in langem Abendkleid
ca. 1939
8501,12,31 gem. M., Firmenmarke 19, Made in Germany, Ausformung zw. 1939–1941, H. 30,5 cm; eBay # 260056277151

8502F

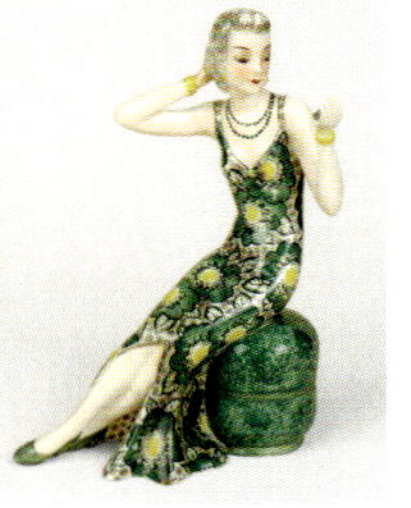

„Abendkleid" (Sitzende Dame, sich im Spiegel ihrer Puderdose betrachtend)
ca. 1939
8502,34,7 Firmenmarke 19, Made in Germany, Ausformung zw. 1939–1941, H. 23 cm; PB Wien 19
8502 H. 20 cm, B. 15 cm; Fotoarchiv K. Täuber
Ausst.: SSH 1992
Lit.: GSN, S. 60, 61 (m. Abb.)

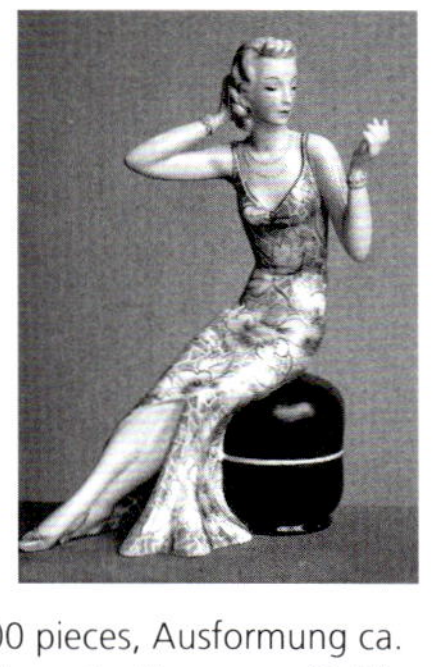

Quelle: AGM (Rechnung an Maurice Mizrahi, Alexandria, Ägypten, vom 22. April 1953), hier mit Ang. des Originaltitels „Abendkleid"; Fotoarchiv K. Täuber.
GS neu: Modell-Nr. 37 „Almost perfect" (H. 20 cm), Aufl. 3000 Stück No. 37-19/3000, GS Stoob 1/a, Töpfernde Frau (gepr.), Worldwide limited Edition 3000 pieces, Ausformung ca. 1988, H. 22 cm; Sammlung Alexandra Herrmann-Goldscheider; Bem.: Dieses Exemplar stammt aus der Produktion Stoob/Burgenland, die in den achtziger Jahren des 20. Jh. versuchte, an die früheren Firmenerfolge anzuknüpfen -> vgl. Firmengeschichte. Der Größenunterschied beruht darauf, dass für die Neuausformung eine Originalfigur der Produktion der 20er Jahre verwendet wurde und durch den Brand bzw. Schwund des Scherbens dann das neu ausgefertigte Stück kleiner ausfällt.

8503F / M

Prischl, Adolf

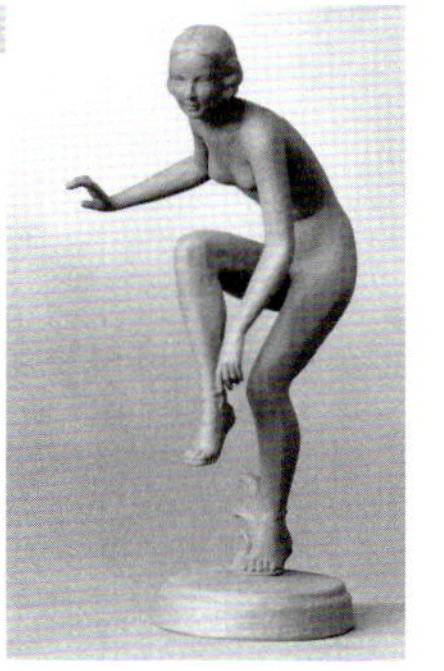

FIGUR „Sandalenbinderin" (Stehender Damenakt, den rechten Fuß hebend, ohne/mit Zöpfe) (auf Rundsockel)
ca. 1939
8503,4,18 (Firmenmarke ohne Ang., blau), unsign., H. 36,3 cm; Doro 4.3.2002
8503,6 (Firmenmarke ohne Ang.), unsign., H. 36,5 cm; Christie's SK 27.6.1997
8503,13,G Firmenmarke 19, Made in Austria, unsign., Ausformung zw. 1945–1953, H. 36 cm; KH

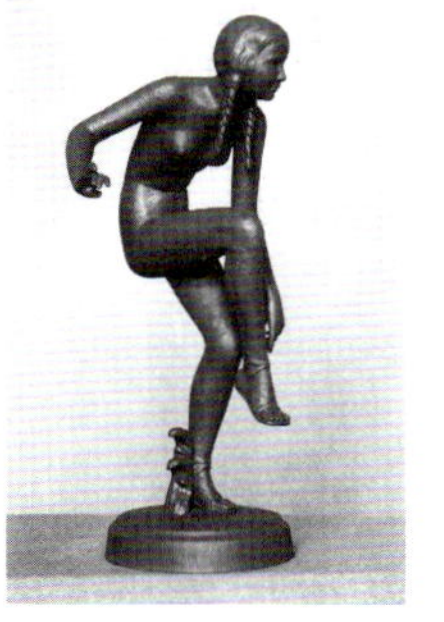

8503,65 Firmenmarke 19, unsign., H. 36 cm; Nagel 381
8503,107,0 ger. 5b, Firmenmarke 19, H. 36,3 cm; BM 69
8503,198,31,56 gem. AN, Firmenmarke 19, Made in Germany, Prischl, Ausformung zw. 1939–1941, H. 35 cm; Wien Museum Inv. 223.731
8503,347,19 Firmenmarke 19, unsign., H. 36 cm; Rammel 65
8503,S Firmenmarke 19, Made in Austria, unsign., Ausformung zw. 1945–1953, H. 36 cm; Wendl 47
8503F, H. 36 cm; GSNachf. 1,6
8503M, H. 36 cm, B. 22 cm; GSNachf. 3,31
Bem.: Diese Figur, die in unterschiedlichen Varianten, was die Oberfläche (unbemalte, sowie strukturierte Terrakotta, mattes Inkarnat, Teilglasur des Beiwerks, oder monochrome Bemalung der ganzen Figur) anlangt, hergestellt wurde, konnte auch noch zw. 1953–1960 bei Carstens (GWM West-Germany) produziert und verkauft werden.
-> vgl. 8702
Ausst.: HMW (jetzt Wien Museum) 1985/86; SSH 1992.
Lit.: GSNachf. 1,6 (m. Abb.), 3,31 (m. Abb.), hier jeweils mit Ang. des Originaltitels „Sandalenbinderin"; KGSF1; GK, S. 24, 132 (m. Abb.); SS, S. 12, 13 (m. Abb.); Könecke (2002), S. 77, 92 (m. Abb.).
Quelle: AGM (Kontoblatt der Wiener Manufaktur Josef Schuster für die Firma Ohlson jr., Malmö, Schweden vom 16. September 1942), hier mit Ang. des Originaltitels „Sandalenbinderin"; Bestellung der Firma Wahliss (Wien) vom 30.8.1945 an die „Arbeitsgemeinschaft der Firma Friedrich Goldscheider"; Bestellungen der Firma Wahliss (Wien) vom 17.10.1945, 9. und 24.11.1945 sowie vom 27.2.1946 jeweils „Sandalenbinderin ohne Zöpfe" in Ausführung „Terrakotta"; Bestellung der Firma Wahliss (Wien) vom 27.2.1946 als „Sandalenbinderin mit Zöpfe" in Ausführung „Terrakotta" und „Fayence"; Rechnung an Maurice Mizrahi, Alexandria, Ägypten, vom 22. April 1953), geliefert in Ausführung „matt"; Archiv Matschiner (Originalfoto); Fotoarchiv K. Täuber.

8505F

Dakon, Stephan

FIGUR „Tanzender Akt" („Figur mit Draperie"/„Schmetterlingtanz")
ca. 1940
8505F, H. 33,5 cm
Quelle: Archiv Mimi Dakon (Tantiemenabrechnung 1.1.–30.6.1940), hier mit Ang. des (Original)-Titels „Tanzender Akt"; (Modellliste Dakon, undatiert, ca. 1940), hier unter dem Titel „Figur mit Draperie" geführt und mit 1938 datiert; (Wiener Manufaktur FG, Modellliste Dakon, 7.11.1946), hier wird die Figur unter dem Titel „Akt tanzend" geführt; (Tantiemenabrechnung 1. Halbjahr 1949; 2. Halbjahr 1949), hier wird die Figur jeweils unter dem (Original)-Titel „Schmetterlingtanz" geführt.

8506 -> ÜBERSCHNEIDUNG

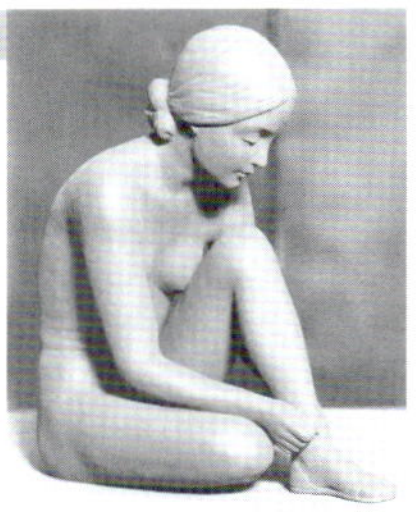

Prischl, Adolf
FIGUR „Sandalenbinderin“ (Sitzender Damenakt mit zurückgekämmtem Haar und Dutt/mit langem lockigem Haar)
ca. 1940
Bem.: Diese Figur wurde auch noch zw. 1953–1960 bei Carstens (GWM West-Germany) produziert.
-> vgl. 8508; 8714
Lit.: KGSF1; KGSF2; Könecke (2002), S. 36, 65, 77 (m. Abb.), hier mit Ang. des (Original?-)Titels „Sandalenbinderin“.
Quelle: Fotoarchiv K. Täuber

8506 -> ÜBERSCHNEIDUNG

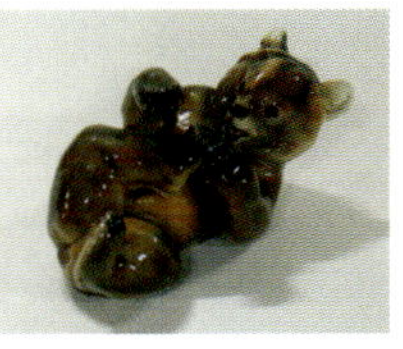

Bär (Liegend)
ca. 1940
8506,9 ger., OX, Firmenmarke 19, Made in Austria, H. 9,8 cm. L. 18,2 cm; MAK Wien, Inv. Ke 10521-16

8508F/M

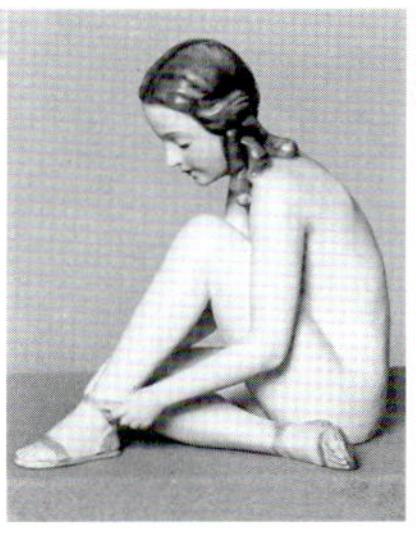

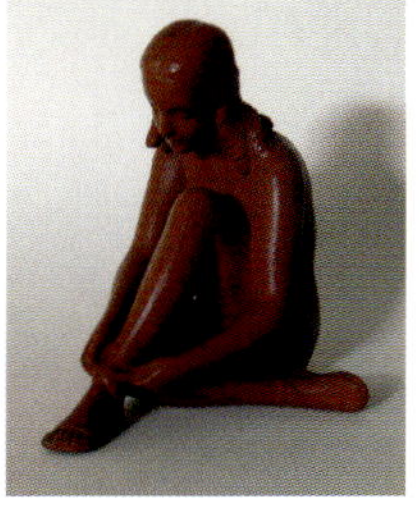

Prischl, Adolf
FIGUR „Sandalenbinderin“ (Sitzender Damenakt mit zurückgekämmtem Haar und Dutt/mit langem lockigem Haar)
ca. 1940
8508,29 Firmenmarke 19a, Made in Austria, unsign., Ausformung zw. 1945–1953, H. 17,8 cm; eBay # 3613228382
8508,49,10 Firmenmarke 19, unsign., Ausformung zw. 1940–1941, H. 17,5 cm; Doro 10.3.2003
8508,66,20 gem. lig. SP, Firmenmarke 19 (blau), Made in Austria (blau), Hand dec. (blau), H. 18 cm; Wien Museum Inv. 190.128
8508,68,G Firmenmarke 19, Made in Germany, Ausformung zw. 1940–1941, H. 18 cm; Nagel 369
8508,263,19 Firmenmarke 19, unsign., H. 17,5 cm; Doro 4.3.2002
8508,1 Firmenstempel WG 1a, Ausformung zw. 1953–1960, H. 18 cm; PB Deutschland
8508,75/IX ger., Firmenstempel WG 1, Ausformung zw. 1953–1960, H. 18 cm; PB Deutschland
8508F, H. 18 cm, B. 15 cm; GSNachf. 1,62
Bem.: Wie die handschriftlichen Notizen auf einem Originalfoto der Manufaktur belegen, wurde die Sandalenbinderin in Fayence (Grundpreis RM 18,–/ Verkaufspreis RM 28,75), in Fayence matt (Grundpreis RM 20,–/ Verkaufspreis RM 32,50) sowie in Terrakotta (Grundpreis RM 16,25/ Verkaufspreis RM 26,–) angeboten. Diese Figur wurde auch noch zw. 1953–1960 bei Carstens (GWM West-Germany) produziert.
Ausst.: SSH 1992
-> vgl. 8506; 8714
Lit.: GSNachf. (2. Ausg., bzw. Erg. Bl.) 1,62 (m. Abb.); KGSF1; KGSF2; SS, S. 12, 13 (m. Abb.); Könecke (2002), S. 36, 65, 67, 77 (m. Abb.), hier mit Ang. des Originaltitels „Sandalenbinderin“.
Quelle: AGM (Kontoblatt der Wiener Manufaktur Josef Schuster für die Firma Porslins Ericsson-Lund, Stockholm?, Schweden vom 7. Januar 1943); (Auftragszettel für die Firma „Haus der Geschenke“, Aachen, vom 16. Januar 1953); (Rechnung an Maurice Mizrahi, Alexandria, Ägypten, vom 22. April 1953), hier mit Ang. des Originaltitels „Sandalenbinderin“, geliefert in Ausführung „matt“; Fotoarchiv K. Täuber.

8510M
Matula, Viktor
GRUPPE „Pferde“
ca. 1940
Quelle: AGM (Kontoblatt der Wiener Manufaktur Josef Schuster für die Firma Porslins Ericsson-Lund, Stockholm?, Schweden vom 7. Januar 1943), hier mit Ang. des Originaltitels GRUPPE „Pferde“.
Quelle: Nachlass Matula, Wien

8511
Matula, Viktor
FIGUR „Pferd“
ca. 1940
8511 H. fehlt
Quelle: Nachlass Matula, Wien

8512
Matula, Viktor
FIGUR „Pferd“
ca. 1940
8512 H. fehlt
Quelle: Nachlass Matula, Wien

8517F
„Madonna mit Kind“ (sitzend)
ca. 1940
8517F, H. 38 cm, B. 24 cm; GSNachf. 6,15
Lit.: GSNachf. (2. Ausg., bzw. Erg. Bl.) 6,15 (m. Abb.)
Quelle: AGM (Bestellungen der Firma Wahliss (Wien) vom 9.11.1945 und 27.2.1946 an die „Arbeitsgemeinschaft der Firma Friedrich Goldscheider“, hier jeweils mit Ang. des (Original?-)Titels „Madonna mit Kind“ in Ausführung „Terrakotta“).

8519F

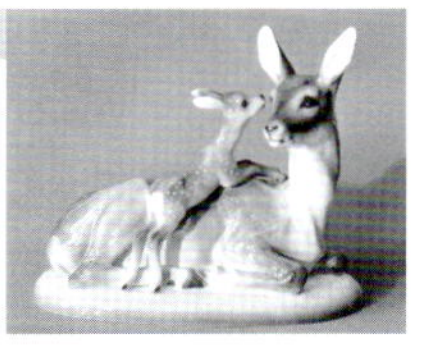

Liegende Ricke mit aufspringendem Rehkitz (auf Ovalsockel)
ca. 1940
8519 H. fehlt
Quelle: AGM (Kommissionsfaktura der „Österreichischen Gruppe-Austria“ für die Messe in Utrecht vom 26.2.1953); Fotoarchiv K. Täuber.

8524
RELIEF Rattenfänger von Hameln (Wandrelief)
ca. 1940
8524,20 Firmenmarke 19, H. 20,8 cm, B. 18,7 cm; BM 38
8524 gem. MT., Firmenmarke 19, H. 21 cm, B. 18 cm; eBay # 256396110

8525M

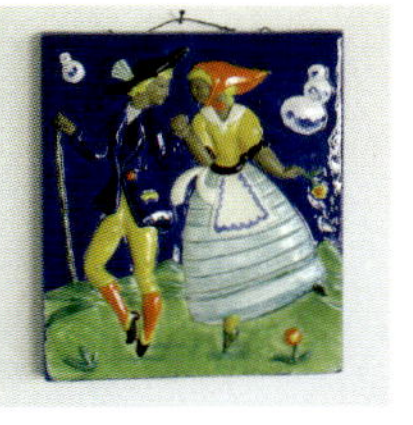

RELIEF Junges Bauernpaar in Tracht, über Wiese tanzend (Wandrelief)
ca. 1940
8525 Firmenmarke 19, Made in Germany, Ausformung zw. 1940–1941, Originalpick.: „Firma Eyck, Heerlen“, H. 21 cm, B. 18,5 cm; Sammlung f@goldscheider.de
8525 Firmenmarke 19, H. 21 cm, B. 19 cm; eBay # 6501820036
Quelle: AGM (Kontoblatt der Wiener Manufaktur Josef Schuster für die Firma Porslins Ericsson-Lund, Stockholm?, Schweden vom 7. Januar 1943), hier mit Ang. des Titels „Relief“.

8531

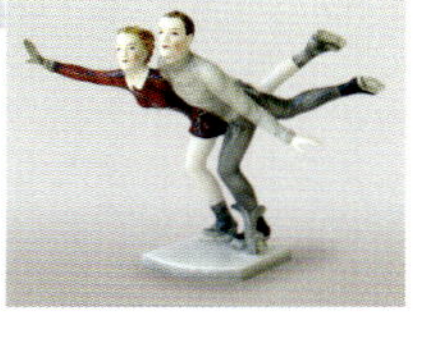

GRUPPE Eisläufer (Eislaufendes junges Paar) (auf Rautensockel)
ca. 1940
8531,28,31 Firmenmarke 19, Made in Germany, Ausformung zw. 1940–1941, H. 21 cm; Sammlung Stressless, Portugal
Lit.: Pinhas, S. 162 (m. Abb.)

8533F

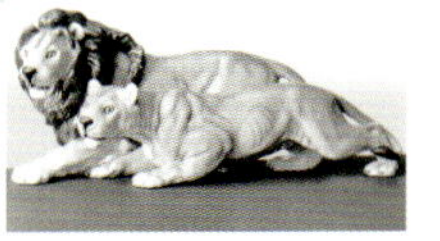

GRUPPE Löwenpaar
ca. 1935 (für den Entwurf)
8533,3,35 Firmenmarke 19, Hand dec., Ausformung zw. 1940–1941 (od. zw. 1945–1953), H. 11 cm; WKA 33
8533F, H. 10,5 cm, B. 27 cm; GSNachf. 5,21b
-> vgl. 7094F
Lit.: GSNachf. (2. Ausg., bzw. Erg. Bl.) 5,21 (m. Abb.)

8538
„Der Liebesbote“ (Stehender Postillion im Biedermeiergewand mit Horn) (auf Ovalsockel mit Pflanze)
Inschrift: DER LIEBESBOTE
ca. 1940
8538,61,7 gem. M., gem. lig. GL, Firmenmarke 19, Made in Germany, Ausformung zw. 1940–1941, H. 38 cm; eBay # 1207307340; 2283848216
8538,67,7 ger. X, gem. M., gem. lig. NO, Firmenmarke 19, Made in Germany, Ausformung zw. 1940–1941, H. 37 cm; eBay # 3746357013
8538,80,31 gem. rot XX, 20, (Firmenmarke ohne Ang.), Ausformung zw. 1940–1945, H. 37 cm; city-antik

8540

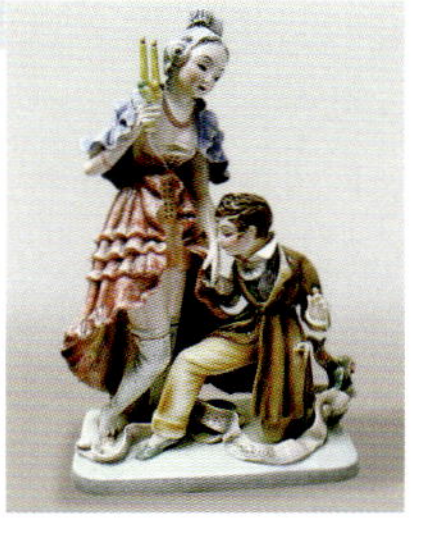

GRUPPE „Der Handkuss“ (Kostümierter Kavalier auf Knien beim Handkuss) (auf Rechtecksockel) mit/ohne Inschrift: DER HANDKUSS
ca. 1940
mit Inschrift
8540,14,31,G gem. AN, HG, Firmenmarke 19, Made in Germany, Ausformung zw. 1940–1941, H. 43 cm; Doro 1.12.2000; city-antik
ohne Inschrift
8540,8,67 gem. S.G, Firmenmarke 19, Made in Austria, H. 44 cm; Wien Museum Inv. 219.165

8561F

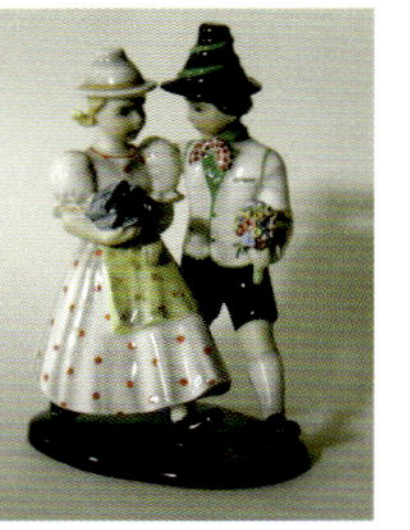

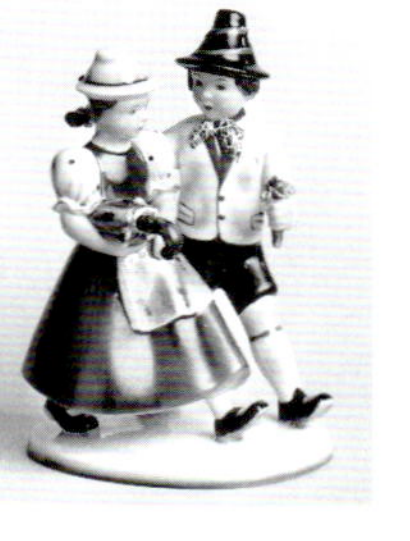

Dakon, Stephan
GRUPPE „Salzburger Kinder“ (auf Ovalsockel)
ca. 1940
8561,1,17 gem. lig. HS, Firmenmarke 20 (blau), Made in Austria, Hand dec., unsign., H. 20,3 cm; eBay # 26025053332
8561,2 gem. F., Firmenmarke 19, unsign., Ausformung zw. 1945–1953, H. fehlt; eBay # 391389592
8561,61,2 Firmenmarke 19, Made in Germany, Ausformung zw. 1940–1941, H. 20,5 cm; Sammlung K. Dechant
8561F, H. 20 cm, B. 15 cm; GSNachf. 1,68
Bem.: Im Stile einer „Hummel-Figur“, -> vgl. Bem. bei Modell-Nr. 8567.
Lit.: GSNachf. (2. Ausg., bzw. Erg. Bl.) 1,68 (m. Abb.)
Quelle: Archiv Mimi Dakon (Tantiemenabrechnung 1.1.–30.6.1940; 1. Halbjahr 1949; 2. Halbjahr 1949; 1. Halbjahr 1950; 2. Halbjahr 1950; 2. Halbjahr 1951; 1. Halbjahr 1952), hier jeweils mit Ang. des Originaltitels „Salzburger Kinder“; (Modellliste Dakon, undatiert, ca. 1940), hier unter dem Titel GRUPPE „Kinder“ geführt und mit 17.5.1939 datiert; Wiener Manufaktur FG, Modellliste Dakon, 7.11.1946), hier mit Ang. des (Original)-Titels „Salzburger Kind“; AGM (Bestellung von Ferd. Lundquist, Göteborg, Schweden vom 23. August 1950; Tantiemenabrechnung 1. Halbjahr 1951); Fotoarchiv K. Täuber.

8563F

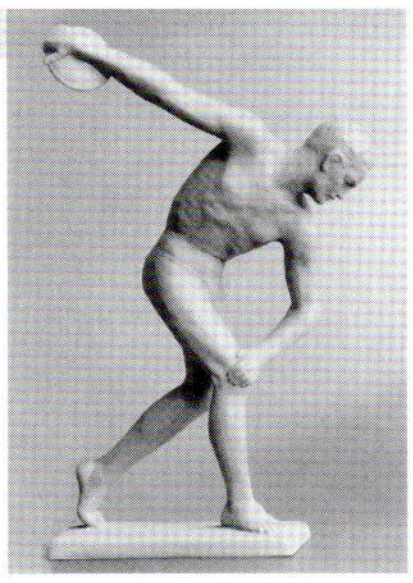

Diskuswerfer beim Wurf (auf flachem Rechtecksockel)
ca. 1940
8563,2,A ger., Firmenmarke Schuster 1a, Ausformung zw. 1941–1944, H. 33,5 cm; eBay # 6517415620
8563F, H. 33 cm, B. 15 cm; GSNachf. 1,66
Bem.: Diese Figur wurde auch noch zw. 1953–1960 bei Carstens (GWM West-Germany) produziert.
Lit.: GSNachf. (2. Ausg., bzw. Erg. Bl.) 1,66 (m. Abb.); Könecke (2002), S. 77, 92 (m. Abb.).
Quelle: Fotoarchiv K. Täuber

8564M

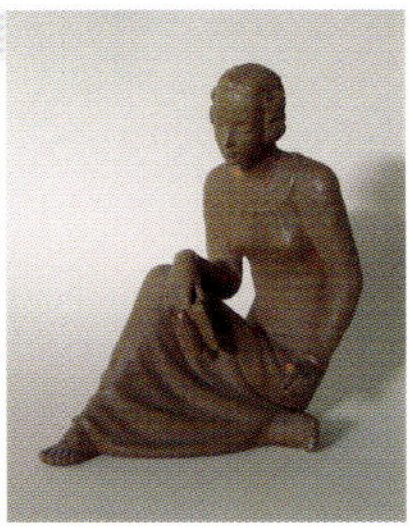

„Lektüre" (Sitzende Frau mit Buch)
ca. 1940
8564,13 Firmenmarke 19a, H. 22 cm; eBay # 6530993486
8564,67 Firmenmarke 19a, Firmenmarke 19 (zwei mal), H. 22 cm; PB Deutschland
-> vgl. 8704
Quelle: AGM (Kontoblatt der Wiener Manufaktur Josef Schuster für die Firma Porslins Ericsson-Lund, Stockholm?, Schweden vom 7. Januar 1943), hier mit Ang. des Originaltitels „Lektüre".

8565

„Madonna mit Kind"
ca. 1940
Quelle: AGM (Bestellungen der Firma Wahliss (Wien) vom 9. und 24.11.1945 an die „Arbeitsgemeinschaft der Firma Friedrich Goldscheider", hier mit Ang. des (Original?-)Titels „Madonna mit Kind"; Bestellung der Firma Wahliss (Wien) vom 27.2.1946, in den Ausführungen „Terrakotta" und „Fayence").

8566

Dakon, Stephan
FIGUR „Bub mit Zylinder" („Bubi")
ca. 1938 (für den Entwurf) Ausführung ab 1940
Quelle: Archiv Mimi Dakon (Modellliste Dakon, undatiert, ca. 1940), hier mit 1938 datiert; Wiener Manufaktur FG, Modellliste Dakon, 7.11.1946), hier mit Ang. des Originaltitels „Bub mit Zylinder"; (Tantiemenabrechnung 1. Halbjahr 1952), hier mit Ang. des Originaltitels „Bubi" (wohl nur eine Abk. für den Bub mit Zylinder).

8567F

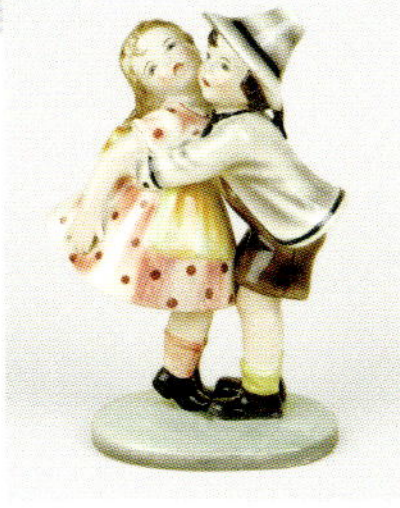

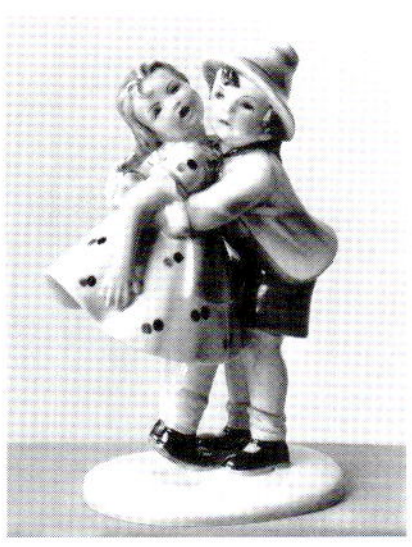

Dakon, Stephan
GRUPPE „Tiroler Kinder" („Gruppe Steirer"/„Widerspenstige") (auf Ovalsockel)
ca. 1939 (für den Entwurf) Ausführung ab 1940
8567,37,7 gem. lig. blau HK., Firmenmarke 19, Made in Germany, unsign., Ausformung zw. 1940–1941, H. 21 cm; eBay # 3328808968
8567 ohne Ang., H. 21 cm; PB Wien 19
8567F, H. 21 cm, B. 14 cm; GSNachf. 1,70
Bem.: Ende der 30er Jahre wurden auch bei Goldscheider, Figuren und Gruppen im Stile der von der Firma Goebel in Oeslau, Deutschland hergestellten „Hummelfiguren" entworfen und erfolgreich auf dem Markt angeboten. Damit sicherte man sich auch in diesem Teilbereich an dem damals wie heute beliebten Genre seinen Marktanteil. Die Modelle Dakons und Sailers für Goldscheider bewahren jedoch zu den Goebel Produkten (die 1935 erstmals auf der Leipziger Messe vorgestellt wurden) ihre Eigenständigkeit, zumal sie häufig als Trachtenfiguren etc. ausgeführt sind. Die Figürchen, die ziemlich nahe der Grenze zum Kitsch angesiedelt sind, stammen fast alle von der Hand Karl Sailers, einem fest angestellten Mitarbeiter bei Goldscheider. Sie sind nur in Ausnahmefällen signiert. Als Beispiele sollen hier die Modelle 8753, 8805–8821, 8823–8826, 8833–8836 dienen.
Lit.: GSNachf. (2. Ausg., bzw. Erg. Bl.) 1,70 (m. Abb.)
Quelle: Archiv Mimi Dakon (Tantiemenabrechnung 1.1.–30.6.1940), hier mit Ang. des Originaltitels „Tiroler Kinder"; (Modellliste Dakon, undatiert, ca. 1940), hier wird die Gruppe unter dem Titel „Steirer" geführt und mit 17.5.1939 datiert; (Wiener Manufaktur FG, Modellliste Dakon, 7.11.1946), hier mit Ang. des (Original)-Titels „Tirolerkind"; (Tantiemenabrechnung 1. Halbjahr 1950; 2. Halbjahr 1950; (1. Halbjahr 1951, AGM); 2. Halbjahr 1951, hier auch als „Sorte II" bestellt), hier jeweils unter dem Titel „Widerspenstige" geführt; AGM (Kommissionsfaktura der „Österreichischen Gruppe-Austria" für die Messe in Utrecht vom 26.2.1953), hier ebenfalls mit dem Titel „Widerspenstige"; Fotoarchiv K. Täuber.

8568

Matula, Viktor
RELIEF Pferde (Wandrelief)
ca. 1940
8568 H. fehlt, B. fehlt
Quelle: Nachlass Matula, Wien

8571F

Dakon, Stephan
FIGUR „Tiroler Bub" (Stehender, essender Bub mit Hut in Tiroler Tracht) (auf Rundsockel)
ca. 1939 (für den Entwurf) Ausführung ab 1940
8571,36 gem. H.D., Firmenmarke 19, Made in Austria, Ausformung zw. 1945–1953, H. 20,3 cm; KH
Bem.: Nach dem Krieg sollte dieses Modell modifiziert und wieder aufgelegt werden. Walter Goldscheider bemängelte bzw. schlug daher zu diesem Modell vor „dass der Tiroler Bub zu kurze Beine hat und ersuche Sie, diese etwas zu verlängern. Der moderne Kopf ist besser, aber ich glaube, dass die Frisur noch immer etwas zu stark ist, so dass sie den Eindruck einer Haube macht. Ich schlage vor, dass Sie vom Hinterkopf entsprechend wegnehmen, damit die Frisur wesentlich leichter wirkt, evtl. könnte man auch die Seitenlocken wegnehmen, doch weiß ich nicht, wie dies wirkt. Wir wollten dies nicht selbst tun, um die Arbeit nicht zu beschädigen, doch könnten Sie es versuchen" (Brief von Walter GS an S. Dakon vom 8.12.1952).
Quelle: Archiv Mimi Dakon (Tantiemenabrechnung 1.1.–30.6.1940; Modellliste Dakon, undatiert, ca. 1940), hier mit 17.5.1939 datiert; Wiener Manufaktur FG, Modellliste Dakon, 7.11.1946; Tantiemenabrechnung 1. Halbjahr 1950; 2. Halbjahr 1950), hier jeweils mit Ang. des Originaltitels „Tirolerbub"; AGM (Brief von Walter Goldscheider an S. Dakon vom 8.12.1952).

8573

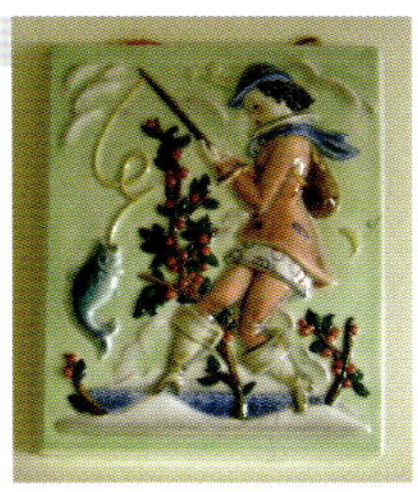

RELIEF Anglerin (Wandrelief)
ca. 1940
8573 gem. AN, Firmenmarke 19, Made in Germany, Ausformung zw. 1940–1941, H. 22,5 cm, B. 18,5 cm; PB Deutschland

8581F

Dakon, Stephan
FIGUR „Kind in (der) Krippe"
ca. 1940
8581F, H. 9 cm
Bem.: Diese Figur ist Teil einer Krippen-Gruppe. Ergänzende Pendants -> vgl. 8585; 8586; 8587; 8588; 8589; 8590; 8592; 8593; 8594; 8595.
Quelle: Archiv Mimi Dakon (Tantiemenabrechnung 1.1.–30.6.1940; Modellliste Dakon, undatiert, ca. 1940; Wiener Manufaktur FG, Modellliste Dakon, 7.11.1946; Tantiemenabrechnung 2. Halbjahr 1950), hier jeweils mit Ang. des Originaltitels „Kind in (der) Krippe".

8582

RELIEF Madonna mit Kind (hochoval)
ca. 1940
8582, H. 35 cm, B. 23 cm
Quelle: Fotoarchiv K. Täuber

8583

Matula, Viktor
RELIEF Zwei Rehe (Wandrelief)
ca. 1940
8583 gem. lig. AH, Firmenmarke 19, unsign., Made in Germany, Ausformung zw. 1940–1941, H. 13,4 cm, B. 15,7 cm; eBay # 1426920732
Quelle: Nachlass Matula, Wien

8584

Matula, Viktor
RELIEF Affe (Wandrelief)
ca. 1940
8584 H. fehlt, B. fehlt
Quelle: Nachlass Matula, Wien

8585F

Dakon, Stephan
FIGUR „Engel (singend)"
ca. 1940
8585,186 Firmenmarke 19, unsign., H. 13 cm; eBay # 6516672438
8585F, H. 13 cm
Bem.: Diese Figur ist Teil einer Krippen-Gruppe. Ergänzende Pendants -> vgl. 8581; 8586; 8587; 8588; 8589; 8590; 8592; 8593; 8594; 8595.
Quelle: Archiv Mimi Dakon (Tantiemenabrechnung 1.1.–30.6.1940; Modellliste Dakon, undatiert, ca. 1940; Wiener Manufaktur FG, Modellliste Dakon, 7.11.1946; Tantiemenabrechnung 2. Halbjahr 1951), hier jeweils mit Ang. des Originaltitels „Engel (singend)".

8586F

Dakon, Stephan
FIGUR „Engel (mit Horn / Trompete)"
ca. 1940
8586F, H. 13 cm
Bem.: Diese Figur ist Teil einer Krippen-Gruppe. Ergänzende Pendants -> vgl. 8581; 8585; 8587; 8588; 8589; 8590; 8592; 8593; 8594; 8595.
Quelle: Archiv Mimi Dakon (Tantiemenabrechnung 1.1.–30.6.1940; Modellliste Dakon, undatiert, ca. 1940), hier jeweils mit Ang. des Originaltitels „Engel (mit Horn)"; (Wiener Manufaktur FG, Modellliste Dakon, 7.11.1946; Tantiemenabrechnung 2. Halbjahr 1950; 2. Halbjahr 1951), hier unter dem Titel „Engel / Trompete" geführt.

8587F

Dakon, Stephan
FIGUR „Engel (mit Laute / Mandoline)"
ca. 1940
8587F, H. 13 cm
Bem.: Diese Figur ist Teil einer Krippen-Gruppe. Ergänzende Pendants -> vgl. 8581; 8585; 8586; 8588; 8589; 8590; 8592; 8593; 8594; 8595.
Quelle: Archiv Mimi Dakon (Tantiemenabrechnung 1.1.–30.6.1940; Modellliste Dakon, undatiert, ca. 1940), hier jeweils mit Ang. des Originaltitels „Engel (mit Laute)"; (Wiener Manufaktur FG, Modellliste Dakon, 7.11.1946; Tantiemenabrechnung 2. Halbjahr 1950), hier unter dem Titel „Engel / Mandoline" geführt.

8588F

Dakon, Stephan
FIGUR „Engel (mit Puppe)"
ca. 1940
8588,40 Firmenmarke 19, Made in Germany, Ausformung zw. 1940–1941, H. 9,5 cm; city-antik
8588F, H. 13 cm
Bem.: Diese Figur ist Teil einer Krippen-Gruppe. Ergänzende Pendants -> vgl. 8581; 8585; 8586; 8587; 8589; 8590; 8592; 8593; 8594; 8595.
Quelle: Archiv Mimi Dakon (Tantiemenabrechnung 1.1.–30.6.1940; Modellliste Dakon, undatiert, ca. 1940; Wiener Manufaktur FG, Modellliste Dakon, 7.11.1946; Tantiemenabrechnung 2. Halbjahr 1950), hier jeweils mit Ang. des Originaltitels „Engel (mit Puppe)".

8589F
Dakon, Stephan
FIGUR „Engel/Josef (mit Schaf/Schäfchen)"
ca. 1940
8589F, H. 24 cm
Bem.: Diese Figur ist Teil einer Krippen-Gruppe. Ergänzende Pendants -> vgl. 8581; 8585; 8586; 8587; 8588; 8590; 8592; 8593; 8594; 8595.
Quelle: Archiv Mimi Dakon (Tantiemenabrechnung 1.1.–30.6.1940), hier mit Ang. des Originaltitels „Engel (mit Schaf)"; (Modellliste Dakon, undatiert, ca. 1940), hier mit dem Titel „Josef mit Schäfchen"; (Wiener Manufaktur FG, Modellliste Dakon, 7.11.1946; Tantiemenabrechnung 2. Halbjahr 1950), hier unter dem Titel „Josef/Schaf" geführt.

8590F
Dakon, Stephan
FIGUR „Madonna (kniend)"
ca. 1940
8590F, H. 24 cm
Bem.: Diese Figur ist Teil einer Krippen-Gruppe. Ergänzende Pendants -> vgl. 8581; 8585; 8586; 8587; 8588; 8589; 8592; 8593; 8594; 8595.
Quelle: Archiv Mimi Dakon (Tantiemenabrechnung 1.1.–30.6.1940; Modellliste Dakon, undatiert, ca. 1940; Wiener Manufaktur FG, Modellliste Dakon, 7.11.1946; Tantiemenabrechnung 2. Halbjahr 1950) hier jeweils mit Ang. des Originaltitels „Madonna (kniend)".

8592F
Dakon, Stephan
FIGUR „Engel (Finger im Mund)"
ca. 1940
8592F, H. 13 cm
Bem.: Diese Figur ist Teil einer Krippen-Gruppe. Ergänzende Pendants -> vgl. 8581; 8585; 8586; 8587; 8588; 8589; 8590; 8593; 8594; 8595.
Quelle: Archiv Mimi Dakon (Tantiemenabrechnung 1.1.–30.6.1940; Modellliste Dakon, undatiert, ca. 1940; Wiener Manufaktur FG, Modellliste Dakon, 7.11.1946; Tantiemenabrechnung 2. Halbjahr 1950), hier jeweils mit Ang. des Originaltitels „Engel (Finger im Mund)".

8593F
Dakon, Stephan
FIGUR „Engel (mit Kerze/mit Leuchter)"
ca. 1940
8593F, H. 13 cm
Bem.: Diese Figur ist Teil einer Krippen-Gruppe. Ergänzende Pendants -> vgl. 8581; 8585; 8586; 8587; 8588; 8589; 8590; 8592; 8594; 8595.
Quelle: Archiv Mimi Dakon (Tantiemenabrechnung 1.1.–30.6.1940; Wiener Manufaktur FG, Modellliste Dakon, 7.11.1946; Tantiemenabrechnung 2. Halbjahr 1950), hier mit Ang. des Originaltitels „Engel (mit Kerze)"; (Modellliste Dakon, undatiert, ca. 1940), hier unter dem Titel „Engel mit Leuchter" geführt.

8594F
Dakon, Stephan
FIGUR „Engel (mit Noten/Notenrolle)"
ca. 1940
8594F, H. 13 cm
Bem.: Diese Figur ist Teil einer Krippen-Gruppe. Ergänzende Pendants -> vgl. 8581; 8585; 8586; 8587; 8588; 8589; 8590; 8592; 8593; 8595.
Quelle: Archiv Mimi Dakon (Tantiemenabrechnung 1.1.–30.6.1940; Wiener Manufaktur FG, Modellliste Dakon, 7.11.1946), hier jeweils mit Ang. des Originaltitels „Engel (mit Noten)"; Modellliste Dakon, undatiert, ca. 1940), hier unter dem Titel „Engel mit Notenrolle".

8595F
Dakon, Stephan
FIGUR „Engel (mit Buch)"
ca. 1940
8595F, H. 13 cm
Bem.: Diese Figur ist Teil einer Krippen-Gruppe. Ergänzende Pendants -> vgl. 8581; 8585; 8586; 8587; 8588; 8589; 8590; 8592; 8593; 8594.
Quelle: Archiv Mimi Dakon (Tantiemenabrechnung 1.1.–30.6.1940; Modellliste Dakon, undatiert, ca. 1940; Wiener Manufaktur FG, Modellliste Dakon, 7.11.1946; Tantiemenabrechnung 2. Halbjahr 1950; 2. Halbjahr 1951), hier jeweils mit Ang. des Originaltitels „Engel (mit Buch)".

8596
Prischl, Adolf
WANDBEHANG/RELIEF „St. Georg" (den Drachen tötend)
ca. 1940
8596 ger., Firmenmarke 19, unsign., H. 30 cm; Fischer 122
Quelle: AGM (Bestellungen der Firma Wahliss (Wien) vom 24.11.1945 und 7.12.1945 an die „Arbeitsgemeinschaft der Firma Friedrich Goldscheider", hier jeweils mit Ang. des Originaltitels „St. Georg"; Bestellung der Firma Wahliss (Wien) vom 27.2.1946 in Ausführung „Terrakotta" je 6 Stück, 1 Stück in „Fayence"); Foto Sammlung mad.

8597
Prischl, Adolf
WANDBEHANG/RELIEF „Betende"
ca. 1940
8597,9 Firmenmarke 19, Made in Germany, unsign., Ausformung zw. 1940–1941, H. 31 cm; eBay # 2222452718
Quelle: AGM (Bestellungen der Firma Wahliss (Wien) vom 9. und 24.11.1945 an die „Arbeitsgemeinschaft der Firma Friedrich Goldscheider", hier mit Ang. des (Original?-)Titels RELIEF „Betende"; Bestellung der Firma Wahliss (Wien) vom 27.2.1946 in Ausführung „Terrakotta"); Foto Sammlung mad.

8598F
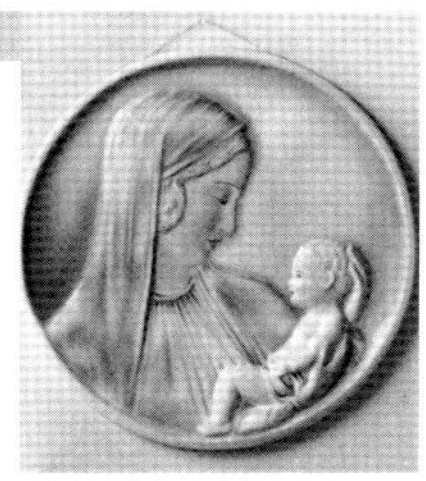
RELIEF/TONDO Madonna mit Kind (Wandrelief)
ca. 1940
8598,128,68,G Firmenmarke 19, Made in Germany, Ausformung zw. 1940–1941, Ø 38,1 cm; eBay # 251964045; 3700182365; 7319571285
8598F, Ø 39 cm; GSNachf. 6,14; Fotoarchiv K. Täuber, hier mit der Ang.: Ø 37 cm
Lit.: GSNachf. (2. Ausg., bzw. Erg. Bl.) 6,14 (m. Abb.)
Quelle: AGM (Kontoblatt der Wiener Manufaktur Josef Schuster für die Firma Porslins Ericsson-Lund, Stockholm?, Schweden vom 7. Januar 1943), hier mit Ang. des Titels „Relief"; Fotoarchiv K. Täuber.

8599
Stehender Damenakt mit Tuch (auf Rundsockel)
ca. 1940
8599,13,31 gem. AN, B, Firmenmarke 19, Made in Germany, Ausformung zw. 1941–1944, H. 34,5 cm; eBay # 65665660195
8599,195/VIII ger., Firmenstempel WG 1, Ausformung 50er Jahre, H. 35 cm; eBay # 372288726
8599 ohne Ang., H. fehlt; Könecke (2002), S. 77, 92 (m. Abb.)
Bem.: Diese Figur wurde auch noch zw. 1953–1960 bei Carstens (GWM West-Germany) produziert.
Lit.: Könecke (2002), S. 77, 92 (m. Abb.), hier ohne weitere Ang.

8600
Tänzerin in Bewegung, ihren Rock mit beiden Armen weit auseinander haltend (auf Ovalsockel)
ca. 1940
8600,42,7 Firmenmarke 19, Made in Germany, unsign., Ausformung zw. 1940–1941, H. 38 cm; Henry's 30.10.1999

8602F
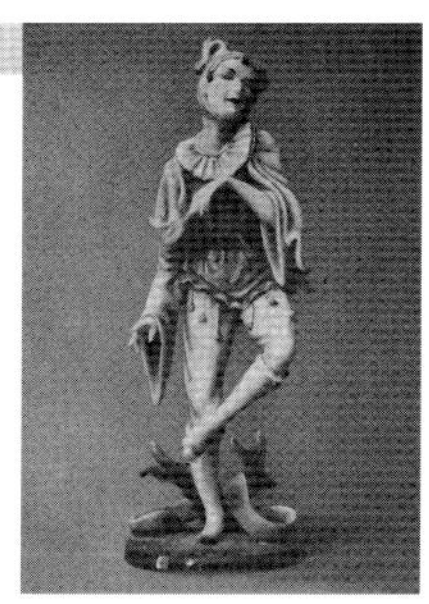
Tanzender Till Eulenspiegel (auf Rundsockel)
ca. 1940
8602,8 gem. S, Firmenmarke 19, Made in Austria, Ausformung zw. 1945–1953, H. 34,3 cm; eBay # 2064159256
8602,11 gem. S., Firmenmarke 19, Made in Germany, Ausformung zw. 1940–1941, H. 35 cm; eBay # 3308466469
8602F, H. 34 cm, B. 13,5 cm; GSNachf. 1,50
Lit.: GSNachf. (2. Ausg., bzw. Erg. Bl.) 1,50 (m. Abb.)

8603
BÜSTE Madonna mit gefalteten Händen
ca. 1940
8603,11,2 Firmenmarke 19, H. 27,5 cm; Wendl 49
8603 ohne Ang., Firmenmarke 19, H. 28 cm; Henry's 30.9.2000

8604
Prischl, Adolf
Schreitende Tänzerin, ihren Rock mit beiden Armen auseinander haltend (auf Ovalsockel)
ca. 1940
8604,5,31 Firmenmarke 19, H. 31,2 cm; Doro 14.9.2001; Fischer 151
8604,6,19 Firmenmarke 19, Made in Germany, Pick. 17 (Entwurf Adolf Prischl, Wien), H. 31,5 cm; Sammlung mad

8606
Pegasus (auf Rechtecksockel)
ca. 1940
8606,8 gepr. G, gem. H., Firmenmarke 19, Made in Germany, Ausformung zw. 1938–1941, H. 16,5 cm; eBay # 6626825755

8607
BUCHSTÜTZE Geflügeltes Pferd (Pegasus?) auf Wolke
ca. 1940
8607 ohne Ang., (Firmenmarke ohne Ang.), H. 16,5 cm; Ruef 473
Bem.: Da sich im Auktionskat. keine Abb. findet, kann die Buchstütze nicht näher beschrieben werden. Der Verdacht, dass es sich um eine Darstellung des Pegasus handelt, liegt jedoch auf der Hand. Es muss aber offen bleiben, ob das Buchstützenpaar unterschiedlich gestaltet wurde bzw. ob beide Ausführungen dieselbe Modell-Nr. tragen.

8608
Dakon, Stephan
„Akt/(mit) Draperie"
ca. 1940
8608, H. 19 cm
Quelle: Archiv Mimi Dakon (Modellliste Dakon, undatiert, ca. 1940), hier mit Ang. des Originaltitels „Akt mit Draperie"; (Wiener Manufaktur FG, Modellliste Dakon, 7.11.1946), hier unter dem Titel „Akt/Draperie" geführt.

8609F
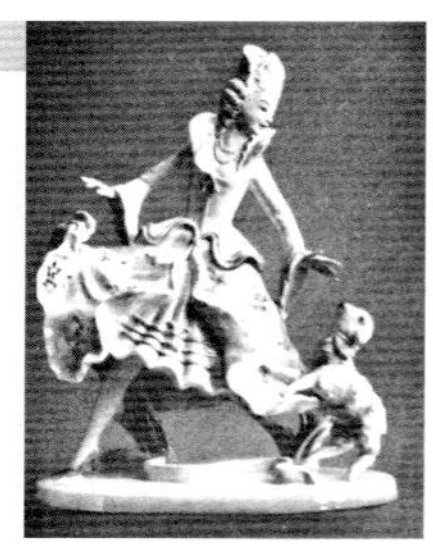
Tanzende Rokokodame mit springendem Lämmchen (auf Ovalsockel)
ca. 1940
8609,14,16 Firmenmarke 19, Made in Germany, Ausformung zw. 1940–1941, H. ca. 33 cm; KH
8609,51,(3?)1 Firmenmarke 19, Made in Germany, Ausformung zw. 1940–1941, H. 31,8 cm; eBay # 2508400225
8609F, H. 33 cm, B. 26 cm; GSNachf. 1,51
Lit.: GSNachf. (2. Ausg., bzw. Erg. Bl.) 1,51 (m. Abb.)

8612 -> ÜBERSCHNEIDUNG
Dakon, Stephan
„Kniende mit Kind"
ca. 1940
8612, H. 48 cm
-> vgl. 8621; 8780= jeweils möglicherweise Höhenvarianten
Quelle: Archiv Mimi Dakon (Modellliste Dakon, undatiert, ca. 1940), hier mit Ang. des Originaltitels „Kniende mit Kind".

8612T
-> ÜBERSCHNEIDUNG
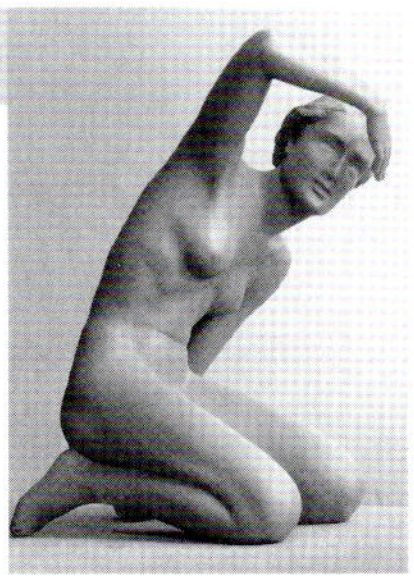
„Schauende" (Kniender Damenakt, die rechte Hand über das Gesicht haltend)
ca. 1940
8612 ohne Ang., H. 35 cm, B. 22 cm; Könecke (2002), S. 77, 92 (m. Abb.)
Bem.: Diese Figur wurde auch noch zw. 1953–1960 bei Carstens (GWM West-Germany) produziert.
Lit.: Könecke (2002), S. 69, 77, 92 (m. Abb.), hier mit Ang. des (Original?-)Titels „Schauende".
Quelle: AGM (Rechnung an Maurice Mizrahi, Alexandria, Ägypten, vom 22. April 1953), in Ausführung „matt"; Fotoarchiv K. Täuber.

8613F
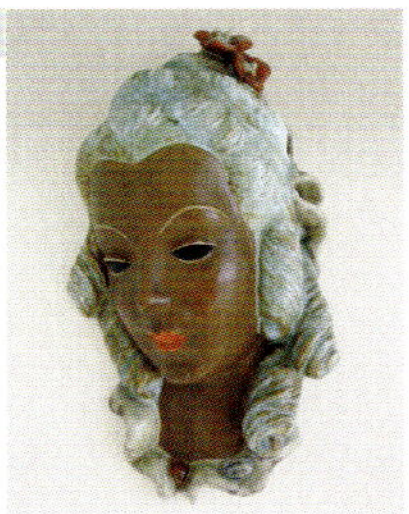
WANDMASKE „Rokoko" (Frauenkopf mit Rokokofrisur und Blumen im Haar, nach rechts blickend)
ca. 1940
8613,3, gem. T, Firmenmarke 25, Pick. 6, H. 26 cm; Sammlung f@goldscheider.de
8613,75 Firmenmarke 19a, H. 26,6 cm;
eBay # 7307541048
Quelle: AGM (Rechnung der Wiener Manufaktur Josef Schuster an Laszlo Bösze, Szeged, Ungarn, vom 5. August 1944, hier mit Ang. des Originaltitels „Maske Rokoko"; Bestellung der Firma Elisabeth Salle (Wiesbaden) vom 8.1.1951).

8614
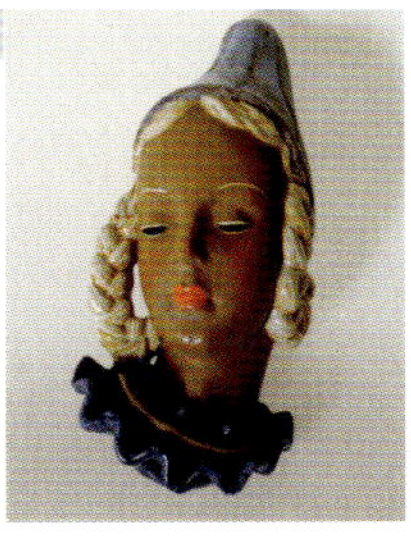
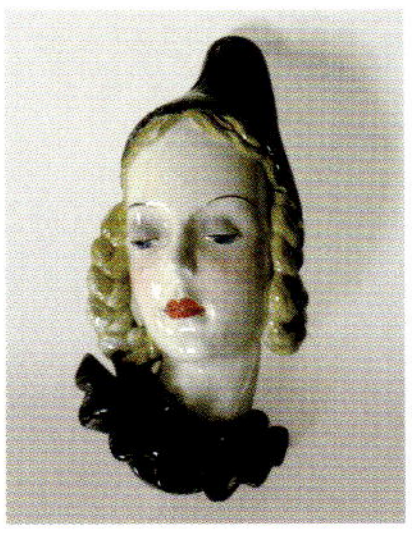
WANDMASKE Frauenkopf mit Trachtenhut und gerüschtem Kragen
ca. 1940
8614,3 gem. lig. AH, Firmenmarke 19a, H. 26 cm; PB Deutschland
8614,4 ger., gem. S., gem. lig. HS., Firmenmarke 19, Pick. 1, H. 26,5 cm; Wien Museum Inv. 220.805
8614,12 Firmenmarke 19, Made in Austria, Hand dec., Ausformung zw. 1945–1953, H. 26 cm; city-antik
8614,15 Firmenmarke 19, H. 26,3 cm; Quit. 58
8614,20 gem. E, Firmenmarke 19, Made in Germany, Ausformung zw. 1940–1941, Pick. 6, H. 26 cm; PB Deutschland
8614,30 Firmenmarke 19, Made in Austria, Ausformung zw. 1945–1953, H. 27 cm; eBay # 2671881422

8615
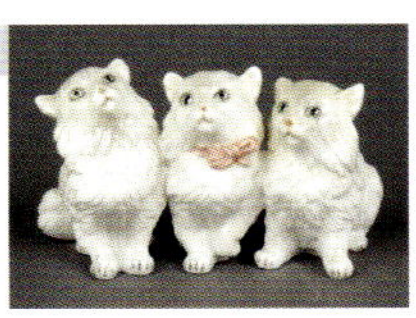
Matula, Viktor
GRUPPE Drei sitzende Kätzchen
ca. 1940
8615,20 Firmenmarke 19, Made in Germany, Ausformung zw. 1940–1941, H. 11 cm; PB Wien 19
-> vgl. 8616= EINZELFigur
Quelle: Nachlass Matula, Wien

8616F
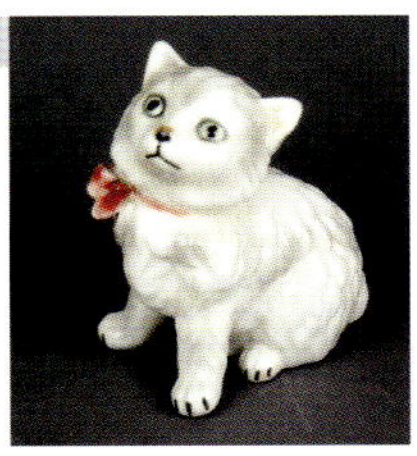
Matula, Viktor
FIGUR Sitzendes Kätzchen
ca. 1940
8616,17,14 Firmenmarke 19, Made in Austria, Hand dec., H. 14 cm; PB Wien 19
8616F, H. 13 cm, B. 13 cm; GSNachf. 5,29; Fotoarchiv K. Täuber, hier mit der Ang.: H. 15 cm, B. 10 cm
-> vgl. 8615= GRUPPE
Lit.: GSNachf. (2. Ausg., bzw. Erg. Bl.) 5,29 (m. Abb.)
Quelle: Nachlass Matula, Wien; Fotoarchiv K. Täuber.

8617
WANDMASKE Frauenkopf mit Haarband (Schleife / Mascherl) und Halstuch
ca. 1940
8617,15 gem. M, Firmenmarke 19a, H. 29,5 cm; eBay # 896117501; 2573773165
8617,15 gem. M, Firmenmarke 19, Pick. 6, H. 31 cm; eBay # 1279415173

8618M
„Frühling"
ca. 1940
Quelle: AGM (Kontoblatt der Wiener Manufaktur Josef Schuster für die Firma Porslins Ericsson-Lund, Stockholm?, Schweden vom 7. Januar 1943), hier mit Ang. des Originaltitels „Frühling".

8619
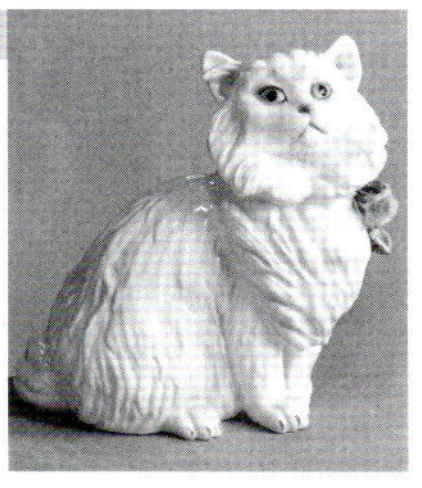
Matula, Viktor
FIGUR Sitzendes Kätzchen
ca. 1940
8619 H. 20 cm, B. 18 cm; Fotoarchiv K. Täuber
-> vgl. 8615= GRUPPE
Quelle: Nachlass Matula, Wien; Fotoarchiv K. Täuber.

8620
BÜSTE Madonna mit Kind
ca. 1940
8620, H. 19 cm, B. 17 cm
Quelle: Fotoarchiv K. Täuber

8621F
Dakon, Stephan
GRUPPE „Kniende („Madonna") mit Kind"
ca. 1940
8621,2 (Malerzeichen blau, ohne Ang.), Firmenmarke 19, Made in Austria, Ausformung zw. 1945–1953, H. 21,4 cm; Quit. 39; Doro 13.9.2004
8621,14,66 gem. rot LS., Firmenmarke 19 (blau), Made in Austria (blau), Hand dec. (blau), unsign., Pick. 5, H. 19 cm; eBay # 211303765
8621,33 gem. ST, Firmenmarke 19, Made in Germany, unsign., Ausformung zw. 1940–1941, H. 22 cm; eBay # 107813207
8621,70,66 Firmenmarke 19, Made in Germany, unsign., Ausformung zw. 1940–1941, H. 22 cm; eBay # 2244533745
-> vgl. 8780= ähnl. Modell
Quelle: Archiv Mimi Dakon (Tantiemenabrechnung 1.1.–30.6.1940), hier wird die Gruppe unter dem (Original)-Titel „Kniende mit Kind" geführt; (Tantiemenabrechnung 1. Halbjahr 1949; 2. Halbjahr 1949), hier wird die Gruppe jeweils unter dem (Original)-Titel „Madonna" geführt; (Wiener Manufaktur FG, Modellliste Dakon, 7.11.1946; Tantiemenabrechnung; 2. Halbjahr 1950), hier jeweils unter dem Titel „Kniende" geführt.

8622F
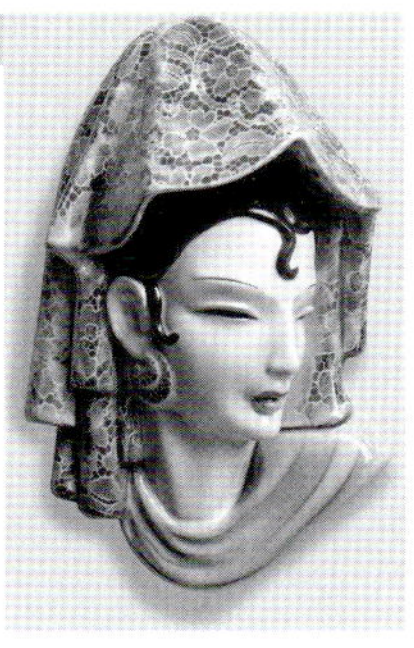
WANDMASKE Frauenkopf mit Mantilla (Spanierin)
ca. 1940
8622,3,G gem. H., Firmenmarke 19, Made in Germany, Ausformung zw. 1940–1941, H. 22,9 cm; eBay # 380357166
8622,4 Firmenmarke 19, Made in Austria, Hand dec., Ausformung zw. 1945–1953, H. 24,5 cm; Herr 45
8622,8 Firmenmarke 19 und 19a, Made in Germany, Ausformung zw. 1940–1941, H. 23 cm; eBay # 405132998
8622,10,10 gem. G, Firmenmarke 19, Made in Austria, Hand dec., Ausformung zw. 1945–1953, H. 23,5 cm; eBay # 2210266649
Quelle: AGM (Rechnung an Maurice Mizrahi, Alexandria, Ägypten, vom 22. April 1953), geliefert in Ausführung „matt"; Fotoarchiv K. Täuber.

8623
WANDMASKE Frauenkopf mit Kopftuch und Doppelkragen (im Profil nach links)
ca. 1940
8623,30 gem. blau N, Firmenmarke 19, Made in Germany, Ausformung zw. 1940–1941, H. 20 cm; city-antik

8624
WANDMASKE Frauenkopf mit Schleier / Halstuch und Lockenfrisur (im Rokokostil)
ca. 1940
8624,16 ger., Firmenmarke 19, H. 24 cm; Doro 7.3.2001
8624,20 gem. M., Firmenmarke 19, Made in Germany, Ausformung zw. 1940–1941, H. 23,9 cm; eBay # 1242845850
8624 gem. M, R, B, Firmenmarke 19, Made in Germany, Ausformung zw. 1940–1941, H. 24 cm; eBay # 2231312111

8625
WANDMASKE Frauenkopf mit Kopfbedeckung und Schleife / Mascherl
ca. 1940
8625,17,63 gem. rot W.E., Firmenmarke 19, Made in Austria, H. 27,5 cm; eBay # 6525059222
8625 gem. F, HG, Firmenmarke 19, Made in Germany, Ausformung zw. 1940–1941, H. 22 cm; eBay # 2231307717; Bem.: Die geringere Höhe rührt von den fehlenden (abgebrochenen) und bei einem unbeschädigten Stück nach unten hängenden Schleifenbändern.

8626
WANDMASKE Frauenkopf mit Schleier und Kragen (vor angedeutetem gotischen Spitzbogen)
ca. 1940
8626,20 gem. LK, Firmenmarke 19, Made in Austria, Ausformung zw. 1945–1953, H. 28 cm; Herr 45
8626,20 Firmenmarke 19, Made in Germany, Ausformung zw. 1940–1941, H. 25,5 cm; Henry's 1.9.2001

8628M
„Sommer"
ca. 1940
Quelle: AGM (Kontoblatt der Wiener Manufaktur Josef Schuster für die Firma Porslins Ericsson-Lund, Stockholm?, Schweden vom 7. Januar 1943), hier mit Ang. des Originaltitels „Sommer".

8629F
Matula, Viktor
FIGUR „Zicklein" (springend) (auf Ovalsockel)
ca. 1940
8629F, H. 37 cm, B. 23 cm; GSNachf. 5,26
Lit.: GSNachf. (2. Ausg., bzw. Erg. Bl.) 5,26 (m. Abb.)
Quellen: AGM (Bestellung der Firma Wahliss (Wien) vom 27.2.1946 in Ausführung „Terrakotta", hier mit Ang. des (Original?-)Titels „Zicklein"; Nachlass Matula, Wien.

8630
Dakon, Stephan
DECKELDOSE „Dose mit Akt"
Form: rund, kugelförmig gedrückt
Knauf: Damenakt (vollplastisch)
ca. 1940
Bem.: Es ist anzunehmen, dass die Form der Dose nicht von Stephan Dakon entworfen wurde, sondern von der Manufaktur wieder, wie bereits mehrfach erwähnt, zwei Entwürfe miteinander kombiniert wurden.
Quelle: Archiv Mimi Dakon (Modellliste Dakon, undatiert, ca. 1940), hier mit der Ang., dass die Dose mit der Modell-Nr. 1094 als Grundelement verwendet wurde; Wiener Manufaktur FG, Modellliste Dakon, 7.11.1946), hier mit Ang. des Originaltitels „Dose mit Akt".
-> vgl. 1094

8632
Dakon, Stephan
Stehender Windhund
ca. 1940
8632, H. 29,5 cm, B. 29 cm
Quelle: Archiv Mimi Dakon (Originalfoto)

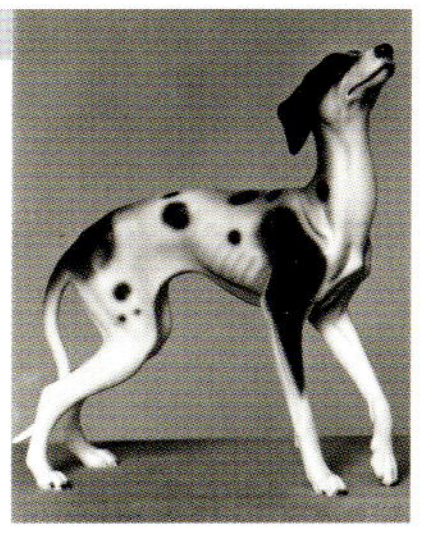

8633
Dakon, Stephan
„Schreitende" (Damenakt) (auf Natursockel)
ca. 1940
8633 ohne Ang., H. 81 cm; Könecke (2002), S. 27, 77 (m. Abb.)
Bem.: Diese Figur, die von der Pose und der Oberflächenbehandlung her einem für Rosenthal entworfenen Akt von Fritz Klimsch ähnelt, wurde auch noch zw. 1953–1960 bei Carstens (GWM West-Germany) produziert. Zum Originaltitel –> vgl. Quelle bei 8740.
–> vgl. 8740
Lit.: KGSF1; KGSF2; Könecke (2002), S. 27, 77 (m. Abb.), hier mit Ang. des Künstlers S. Dakon.

8634
Dakon, Stephan
„Akt"/„Tänzerin" (in Pose) (auf rundem Natursockel)
ca. 1940
8634,0 Firmenmarke 19a, unsign., H. 56 cm; eBay # 6517339050
Quelle: Archiv Mimi Dakon (Modellliste Dakon, undatiert, ca. 1940), hier mit Ang. des (Original)- Titels „Akt"; (Wiener Manufaktur FG, Modellliste Dakon, 7.11.1946), hier unter dem Titel „Tänzerin" geführt.

8635
SCHALE Zwei Singvögel mit drei Küken im Nest (auf einem Baumstamm, der sich am Boden zu einer Schale formt)
ca. 1940
8635,8 ger., Firmenmarke 19, Made in Austria, Ausformung zw. 1945–1953, H. 28 cm; eBay # 3351671585
8635,11 Firmenmarke 19 (blau), Hand dec. (blau), Ausformung zw. 1945–1953, H. 29,5 cm; WKA 33
8635,53,14 ger., Firmenmarke 19, Made in Austria, Hand dec., H. 31,5 cm; city-antik

8637
Stehendes Pferd mit Scheuklappen (auf einem Stapel mit Gesetzbüchern)
ca. 1940
8637,20 Firmenmarke 19, H. 16 cm; Schuler 75

8638
Dakon, Stephan
„Bub/(mit) Flöte"
ca. 1940
8638, H. 22 cm
Quelle: Archiv Mimi Dakon (Modellliste Dakon, undatiert, ca. 1940), hier mit Ang. des Originaltitels „Bub mit Flöte"; Wiener Manufaktur FG, Modellliste Dakon, 7.11.1946), hier unter dem Titel „Bub/Flöte" geführt.

8639
Matula, Viktor
ASCHER
ca. 1940
8639 H. fehlt, B. gehlt
Quelle: Nachlass Matula, Wien

8640
Matula, Viktor
FIGUR Katze
ca. 1940
8640 H. fehlt
Quelle: Nachlass Matula, Wien

8641
Matula, Viktor
FIGUR Katze
ca. 1940
8641 H. fehlt
Quelle: Nachlass Matula, Wien

8644
Prischl, Adolf (zugeschr.)
„Herbst" (Schreitender Damenakt mit Krug und Weintrauben) (auf Landschaftssockel)
ca. 1940
8644,5,28 ger., Firmenmarke 19, unsign., H. 26,4 cm; BM 64; Doro 4.11.1999
8644,9 ger., Firmenmarke 19, Made in Austria, unsign., Ausformung zw. 1945–1953, H. 26 cm; KH
8644,13 Firmenmarke 19a, unsign., H. 26 cm; PB Deutschland
8644,66 ohne Firmenmarke, unsign., Ausformung vermutl. zw. 1945–1953, H. 26 cm; WKH
Bem.: Zum Originaltitel –> vgl. 8735
–> vgl. 8735

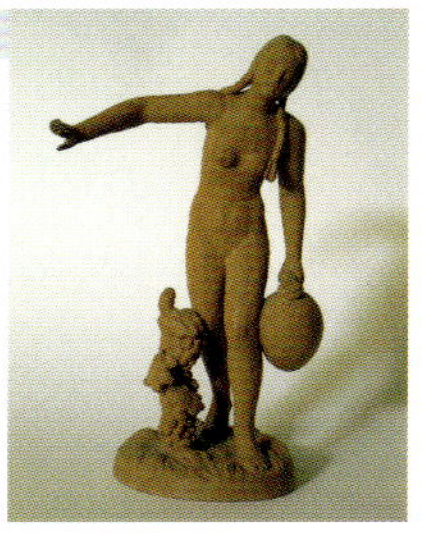

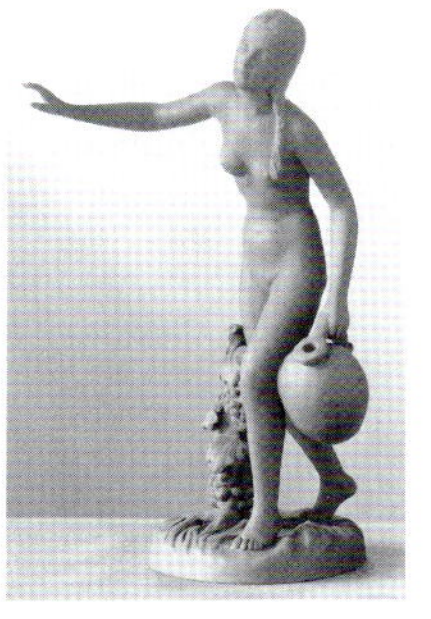

8645
Lorenzl, Josef
Stehender Trommler in Tracht (auf Rundsockel)
ca. 1940
8645,1,7 Firmenmarke 19, H. 33 cm; Sammlung Dr. Peter Hieke, Wien
8645,3,19 Firmenmarke 19, Made in Germany, Ausformung zw. 1940–1941, H. 33 cm; Sammlung Dr. Peter Hieke, Wien

8646
Lorenzl, Josef (zugeschr.)
Stehender Hochzeitslader in Tracht (auf Rundsockel)
ca. 1940
8646,19,7 gem. MM, Firmenmarke 19, Made in Germany, unsign., Ausformung zw. 1940–1941, H. ca. 32 cm; KH

8649
Lorenzl, Josef (zugeschr.)
Stehende Frau in Tracht (auf Rundsockel)
ca. 1940
8649,4,11 gem. AN, Firmenmarke 19, Made in Germany, unsign., Ausformung zw. 1940–1941, H. 31,5 cm; Wendl 31

8651M
RELIEF Liegende Ricke mit Kitz (Wandrelief)
ca. 1940
8651,G gem. A, Firmenmarke 19, Made in Germany, Ausformung zw. 1940–1941, H. 28 cm, B. 28 cm; eBay # 2217353573
Quelle: AGM (Kontoblatt der Wiener Manufaktur Josef Schuster für die Firma Porslins Ericsson-Lund, Stockholm?, Schweden vom 7. Januar 1943), hier mit Ang. des Titels „Relief".

8652F
Lorenzl, Josef
„Rhythmus" (Tänzerin in Pose) (auf Ovalsockel)
ca. 1940
8652,3,19 gem. AN, Firmenmarke 19, Made in Germany, unsign., Ausformung zw. 1940–1941, H. 36 cm; Mehlis 35
8652,11,62 Firmenmarke 19, unsign.?, Ausformung zw. 1940–1941, H. 30 cm; city-antik

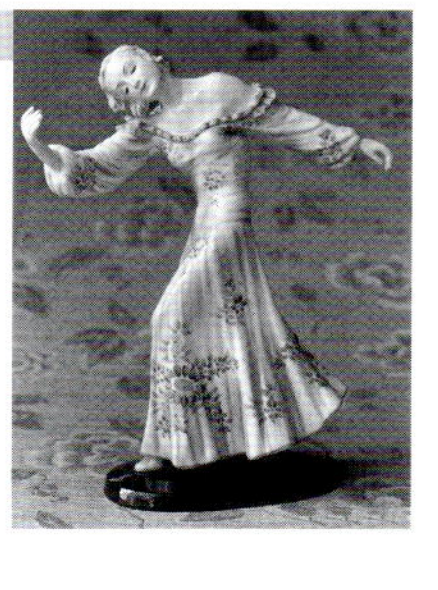

8652,170,7 Firmenmarke 19, Made in Germany, unsign., Ausformung zw. 1940–1941, H. 36 cm; WKA 35
8652,243,19 gem. M.E., Firmenmarke 19, Made in Germany, unsign., Ausformung zw. 1940–1941, Pick. 6, H. 30 cm; Soth./eBay # 1860110210
8652,277,7 gem. lig. CF, gem. rot XI., Firmenmarke 19, Made in Germany, Ausformung zw. 1940–1941, H. 36 cm; WKA 35
8652F, H. 35,5 cm, B. 30 cm; GSNachf. o. Pag
Lit.: GSNachf. (2. Ausg., bzw. Erg. Bl.) o. Pag. (m. Abb.), hier mit Ang. des Originaltitels „Rhythmus".
Quelle: Fotoarchiv K. Täuber

8653F
Dakon, Stephan
GRUPPE „Geheimnis" (Kindergruppe) (Zwei Mädchen mit Puppe und Regenschirm) (auf Ovalsockel)
ca. 1940
8653F, H. 27 cm; B. 18 cm; Fotoarchiv K. Täuber
Quelle: Archiv Mimi Dakon (Tantiemenabrechnung 1.1.–30.6.1940; Modellliste Dakon, undatiert, ca. 1940); Wiener Manufaktur FG, Modellliste Dakon, 7.11.1946; Tantiemenabrechnung 1. Halbjahr 1949; 1. Halbjahr 1950; 2. Halbjahr 1950; (1. Halbjahr 1951, AGM); 2. Halbjahr 1951), hier jeweils mit Ang. des Originaltitels GRUPPE „Geheimnis"; AGM; Fotoarchiv K. Täuber.

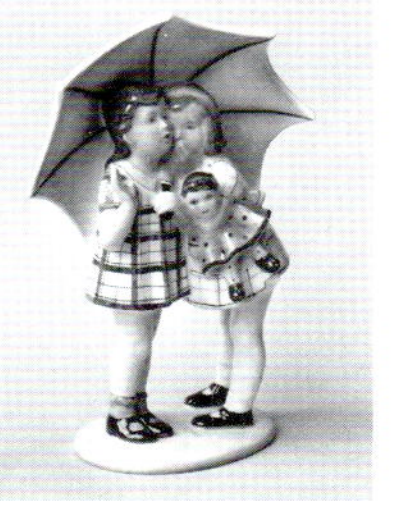

8654F
Lorenzl, Josef
FIGUR/LAMPE „Klassischer Tanz"/„Bewegung" (Tänzerin in Pose, auf einem Bein, ihre Arme ausgestreckt) (auf Ovalsockel)
ca. 1940
FIGUR
8654,12,11 Firmenmarke 19, Made in Germany, unsign., Ausformung zw. 1940–1941, H. 42,7 cm; Christie's SK 15.5.2003; 13.5.2004; Soth. Olympia 1.6.2005
8654,438,7 (Firmenmarke ohne Ang.), Ausformung zw. 1940–1941, H. 43 cm; Fischer Luz. 388
8654,717,31 (Firmenmarke ohne Ang.), unsign., H. 42,3 cm; Doro 12.9.2006
8654,747,19 Firmenmarke 19, Made in Germany, unsign., Ausformung zw. 1940–1941, H. 42,5 cm; Wien Museum Inv. 219.194
8654,S,78 gem. lig. MR, Firmenmarke 19, Made in Austria, Hand dec., unsign., Ausformung zw. 1945–1953, H. 42 cm; eBay # 6528692139
8654F, H. 43,5cm, B. 25 cm
LAMPE
Kein Exemplar in dieser Ausführung nachweisbar, diese wurde möglicherweise erst nach dem Krieg in nur geringer Stückzahl angefertigt (–> vgl. Quellen).
Bem.: Diese Figur wurde auch noch zw. 1953–1960 bei Carstens (GWM West-Germany) produziert.
Lit.: KGSF1; KGSF2; Könecke (2002), S. 32, 77 (m. Abb.), hier mit Ang. des Künstlers J. Lorenzl.
Quelle: AGM (Bestellungen der Firma Wahliss (Wien) vom 9. und 24.11.1945 an die „Arbeitsgemeinschaft der Firma Friedrich Goldscheider", hier mit Ang. des Titels „Klassischer Tanz"; Bestellung der Firma Wahliss (Wien) vom 27.2.1946 in Ausführung „Terrakotta"; Rechnung an Maurice Mizrahi, Alexandria, Ägypten, vom 22. April 1953), hier unter dem Titel „Tanz Lampe" geführt. Archiv Matschiner (Originalfoto), hier mit Ang. des Titels „Bewegung"; Fotoarchiv K. Täuber.

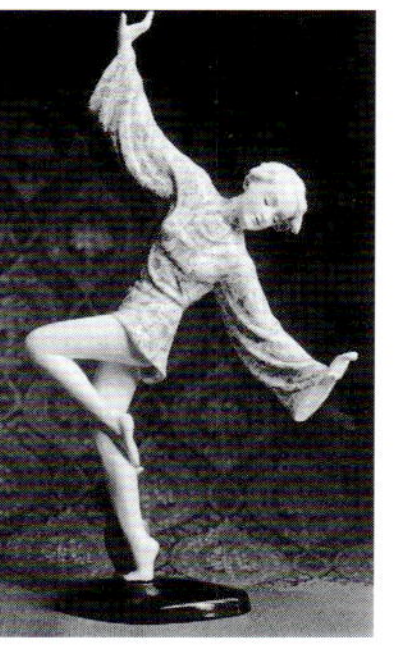

8656F
Dakon, Stephan
GRUPPE „Erzählung" (Kindergruppe)
ca. 1940
8656F, H.(?) 50 cm
Quelle: Archiv Mimi Dakon (Tantiemenabrechnung 1.1.–30.6.1940; Modellliste Dakon, undatiert, ca. 1940), hier mit der Ang. der Größe 50 cm; ob hier die Höhe

oder Breite gemeint ist, kann nicht gesagt werden, da bis Redaktionsschluss kein Stück dieser Gruppe nachgewiesen werden konnte; Wiener Manufaktur FG, Modellliste Dakon, 7.11.1946), hier jeweils mit Ang. des Originaltitels „Erzählung".

8659
Turnender Damenakt mit Medizinball (auf Ovalsockel)
ca. 1940
8659,31 ger. X, gem. lig. CHL, Firmenmarke 19, Made in Germany, Ausformung zw. 1940–1941, H. 28,4 cm; Wien Museum Inv. 190.121
8659,75,7 Firmenmarke Schuster 1, Made in Germany, unsign., Ausformung zw. 1941–1944, H. 18,5 cm; eBay # 358170559
Ausst.: SSH 1992
Lit.: SS, S. 12, 13 (m. Abb.)

8660
Schreitende Dame mit Hut, in kurzem Oberteil und langem Rock (auf Ovalsockel)
ca. 1940
8660,7 gem. rot HA, Firmenmarke 19, Made in Austria, H. 32 cm; KH
8660,48,37 Firmenmarke 19, H. 32,1 cm; Doro 15.9.2003
8660 ohne Ang., H. 33 cm; PB Wien 19
8660F (8661F Druckfehler!), H. 33 cm, B. 18 cm; GSNachf. 1,57
Lit.: GSNachf. (2. Ausg., bzw. Erg. Bl.) 1,57 (m. Abb.);
Bem.: Bei der angegebenen Modell-Nr. 8661 handelt es sich um einen Druckfehler im Firmenkat.

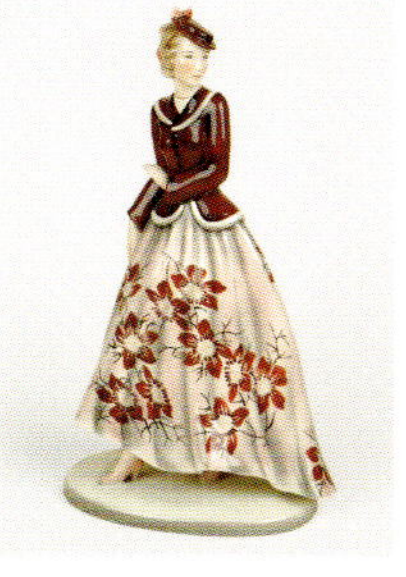

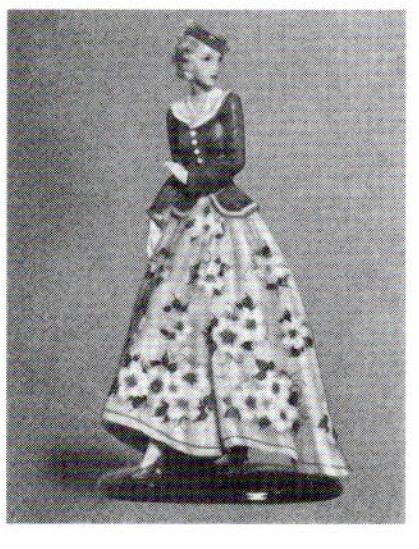

8661
„Oper" (Schlanke Dame in langem Kleid, mit erhobener Hand) (auf Ovalsockel)
ca. 1940
8661,5,2 (Firmenmarke ohne Ang.), H. 33 cm; Doro 8.3.2004
8661,8,64 gem. G, Firmenmarke 19, Made in Austria, Hand dec., H. 33,5 cm; Privatsammlung, Wien
Quelle: AGM (Rechnung an Maurice Mizrahi, Alexandria, Ägypten, vom 22. April 1953), hier mit Ang. des Originaltitels „Oper", geliefert in Ausführung „gold".

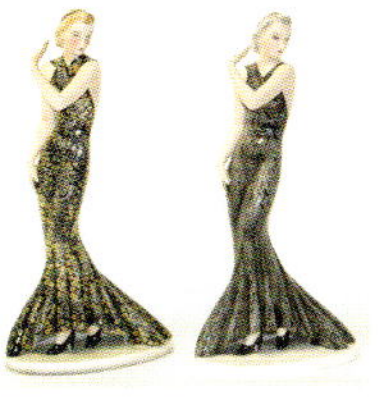

8662
Schreitende schlanke Dame (auf getrepptem Rechtecksockel)
ca. 1940
8662,7 Firmenmarke 19, Made in Germany, Ausformung zw. 1940–1941, H. 33 cm; KaM 93; Christie's SK 15.5.2003
8662,14,10 (Firmenmarke ohne Ang.), Ausformung zw. 1940–1941, H. 33,5 cm; Henry's 27.5.1999
Lit.: Pinhas, S. 162 (m. Abb.)

8663
„Pferd" (Stehendes Fohlen mit nach hinten gewandtem Kopf)
ca. 1940
8663,20 Firmenmarke 19, H. 41 cm; Doro 25.4.2000
Quelle: AGM (Bestellungen der Firma Wahliss (Wien) vom 9. und 24.11.1945 an die „Arbeitsgemeinschaft der Firma Friedrich Goldscheider", hier jeweils mit Ang. des Originaltitels „Pferd"; Bestellung der Firma Wahliss (Wien) vom 27.2.1946 in Ausführung „Terrakotta").

8665M
Lorenzl, Josef
„Sinnende" (Sitzender Damenakt)
ca. 1940
8665,9 Firmenmarke 19a, unsign., Ausformung zw. 1945–1953, H. 33,5 cm; Walldorf 10; Quit. 14
8665,19 (ohne Firmenmarke), unsign., Ausformung zw. 1945–1953, H. 33 cm; PB Deutschland
8665,65 Firmenmarke 19, Ausformung zw. 1945–1953, H. 34 cm; Mehlis 32
8665,66 (ohne Firmenmarke), unsign., Ausformung zw. 1945–1953, H. 33,4 cm; MAK Wien, Inv. Ke 10602-35
8665,66,5 (gem./ger.?) K, Firmenmarke 19a, unsign., H. 35 cm; Ölschläger 10
8665,417/XI,11 ger., (Firmenmarke ohne Ang.), unsign., Ausformung 50er Jahre, H. 34 cm; eBay # 3351893207
Bem.: Diese Figur wurde auch noch zw. 1953–1960 bei Carstens (GWM West-Germany) produziert.
–> vgl. 8726
Lit.: GSNachf. (2. Ausg., bzw. Erg. Bl.) 3,3 (m. Abb.), hier mit Ang. des Originaltitels „Sinnende"; KGSF1; KGSF2; Könecke (2002), S. 35, 77 (m. Abb.), hier mit Ang. des Künstlers J. Lorenzl.
Quelle: AGM (Kontoblatt der Wiener Manufaktur Josef Schuster für die Firma Ohlson jr., Malmö, Schweden vom 16. September 1942), hier mit Ang. des Originaltitels „Sinnende"; Bestellungen der Firma Wahliss (Wien) vom 9. und 24.11.1945 an die „Arbeitsgemeinschaft der Firma Friedrich Goldscheider"; Bestellung der Firma Wahliss (Wien) vom 27.2.1946 in Ausführung „Terrakotta"); Fotoarchiv K. Täuber.

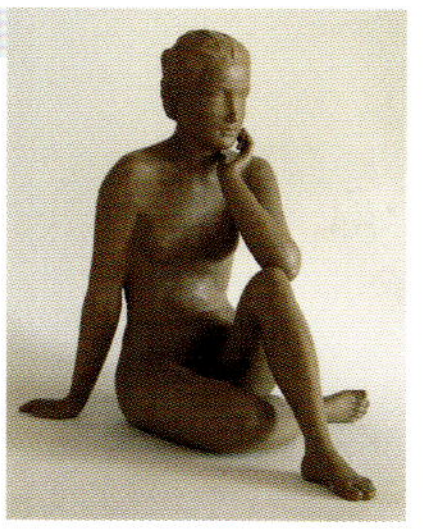

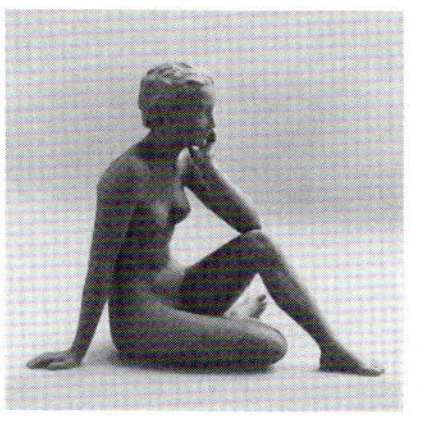

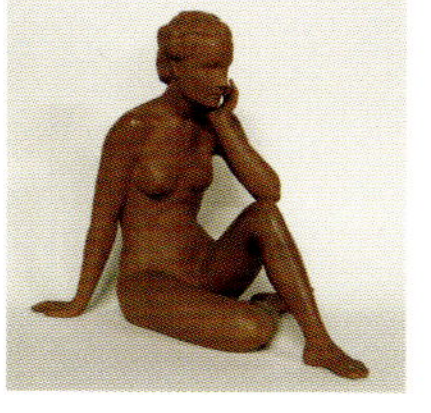

8666
RELIEF „April" (Schäferszene mit Sternzeichen Widder) (Wandrelief)
ca. 1940
8666,5 Firmenmarke 19, Made in Austria, Ausformung zw. 1945–1953, H. fehlt; eBay # 2603371447
8666 (Firmenmarke ohne Ang.), H. 21 cm, B. 20 cm; Fischer 130
8666 (Firmenmarke ohne Ang.), H. fehlt, B. 19,5 cm; Christie's SK 9.4.1999
Bem.: Die Darstellung der nackten Schäferin, die ein kleines Schäfchen unter dem Arm trägt, nur mit einem Hut bekleidet und mit einem Stab zwei weitere Schafe vor sich hertreibend, erinnert an Walter Bosses oder Kurt Goebels Darstellungen. Eine endgültige Zuschreibung an einen der beiden Künstler kann wohl erst dann vorgenommen werden, wenn sich noch weitere dieser Sternzeichen-Darstellungen (hier ein springender Widder in der linken oberen Ecke des Reliefs) nachweisen lassen.
Quelle: AGM (Bestellung der Firma Wahliss (Wien) vom 27.2.1946, hier mit Ang. des (Original?-)Titels RELIEF „April" in Ausführung „Terrakotta").
Pendant –> vgl. 8667

8667
RELIEF „Dezember" (Diana und Reh mit Sternzeichen Schütze) (Wandrelief)
ca. 1940
8667 (Firmenmarke ohne Ang.), H. 21 cm, B. 20 cm; Fischer 130
Bem.: Die Darstellung der Diana mit Pfeil und Bogen, einem Reh nachstellend, erinnert an Walter Bosses oder Kurt Goebels Darstellungen. Eine endgültige Zuschreibung an einen der beiden Künstler kann wohl erst dann vorgenommen werden, wenn sich noch weitere dieser Sternzeichen Darstellungen (hier ein kniender Schütze in der rechten oberen Ecke des Reliefs) nachweisen lassen.
Pendant –> vgl. 8666

8668
RELIEF Heiliger Hubertus (Kniender Hl. Hubertus mit zwei Hunden im Wald) (Wandrelief)
ca. 1940
8668,10,17 Firmenmarke 19, Made in Austria, Hand dec., H. fehlt; eBay # 3349441077

8669
WANDMASKE Frauenkopf mit Hut, Schal und Hand
ca. 1940
8669,2 Firmenmarke 19, Ausformung zw. 1945–1953, H. 28,3 cm; Doro 4.3.2002
8669,3 gem. lig. AH, Firmenmarke 19a, Pick. 6, Ausformung zw. 1945–1953, H. 28,6 cm; eBay # 85880524
8669,9 Firmenmarke 19a, Ausformung zw. 1945–1953, H. 28,4 cm; Doro 25.4.2000
8669,64 Firmenmarke 19a, Ausformung zw. 1945–1953, Pick. 6, H. 28 cm; eBay # 2217732691
8669,72,14 gem. MPI, Firmenmarke 19, Made in Austria, Hand dec., H. 28 cm; eBay # 2235798465
8669,3 ger., ohne Firmenmarke, Pick. WG 1, Ausformung 50er Jahre, H. 27,7 cm; Sammlung John Nash
8669 Firmenstempel WG 1, Ausformung 50er Jahre, H. 28 cm; eBay # 147335224
Bem.: Diese Maske wurde auch noch zw. 1953–1960 bei Carstens (GWM West-Germany) produziert.
Lit.: Könecke (2002), S. 78, 93 (m. Abb.).
Quelle: AGM (Bestellungen der Firma Wahliss (Wien) vom 7.12.1945 und 27.2.1946 in Ausführung „Terrakotta"); Fotoarchiv K. Täuber.

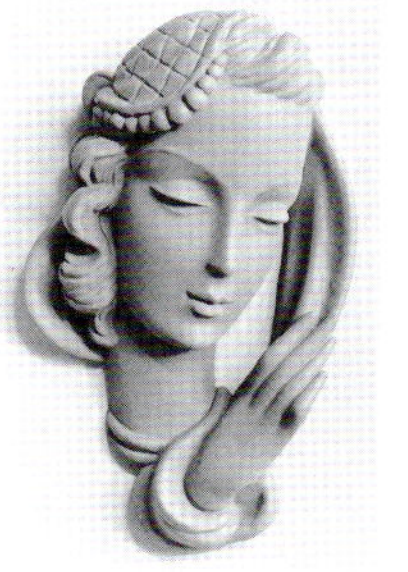

8672
VASE
Form: eiförmig, kurze geweitete Mündung
Dekor: Spitzendekor (Schablone) Waldlandschaft (Rehe, Vögel, Bäume)
ca. 1940
8672,5,16 Firmenmarke 19, H. 40 cm; PB Wien 19
8672,11 ger. K, gem. lig. HS, ohne Firmenmarke, H. 19 cm, Ø 8,3 cm; Wien Museum Inv. 186.691
8672,32 Firmenmarke 19, Made in Austria, Ausformung zw. 1945–1953, H. 42,5 cm, Kett. 39; Bem.: Das AH vermerkt als Dekor „Laufglasur".

8675
VASE
Form: eiförmig, kurze leicht geweitete Mündung
Dekor: Spitzendekor (Schablone) Landschaft (Häuser, Bäume, Brücken)
ca. 1940
8675,11 gem. lig. HS, Firmenmarke 19, Made in Austria, Ausformung zw. 1945–1953, H. 17 cm, Ø 6 cm; Sammlung John Nash
8675,11 Firmenmarke 19, Made in Austria, Ausformung zw. 1945–1953, H. 17 cm, Ø 6 cm; Wien Museum Inv. 186.692

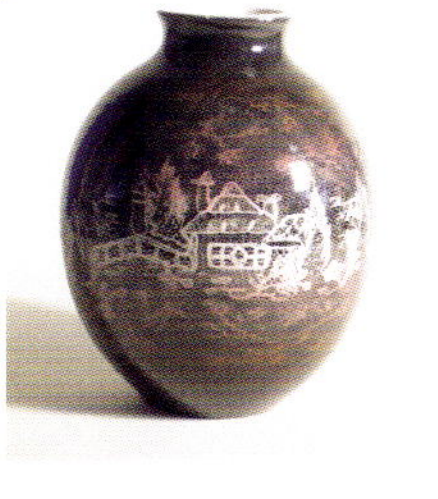

8676
VASE
Form: becherförmig
Dekor: Spitzendekor (Schablone) Landschaft (Häuser und Bäume)
ca. 1940
8676,4 ger., Firmenmarke 19, Ausformung zw. 1945–1953, H. 18 cm; Fischer 61
8676,9 ger., gem. lig. HS, Firmenmarke 19, Made in Austria, Hand dec., Ausformung zw. 1945–1953, H. 14,5 cm; eBay # 653515801
8676 ger., gem. IV, gem. lig. HS, Firmenmarke 19, Made in Austria, Hand dec., Ausformung zw. 1945–1953, H. 17,8 cm, Ø 15 cm; eBay # 1426922086

8678F
VASE
ca. 1940
Quelle: AGM (Rechnung an Maurice Mizrahi, Alexandria, Ägypten, vom 22. April 1953).

8679
VASE
Form: balusterförmig
Dekor: Polospieler zu Pferde, Bäume
ca. 1940
8679,11,50 gem. HS, Firmenmarke 19, Made in Austria, H. 45,5 cm; Doro 6.4.1999
8679,S Firmenmarke 19, Made in Austria, Ausformung zw. 1945–1953, H. 46 cm; Wien Museum Inv. 190.062

8681F
VASE
ca. 1940
Quelle: AGM (Rechnung an Maurice Mizrahi, Alexandria, Ägypten, vom 22. April 1953).

8682M
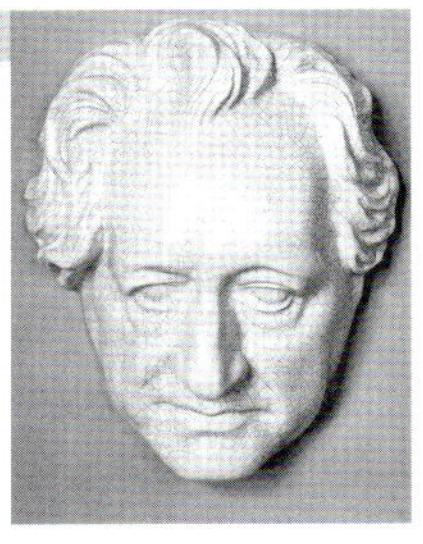
WANDMASKE Johann Wolfgang von Goethe (en Face)
ca. 1940
8682M, H. 28 cm, B. 22 cm; GSNachf. 2,28; Fotoarchiv K. Täuber, hier mit der Ang. des Originalpreises RM 19,50
Bem.: Dargestellt ist der Dichter Johann Wolfgang von Goethe (1749–1832).
Lit.: GSNachf. (2. Ausg., bzw. Erg. Bl.) 2,27 (m. Abb.)
Quelle: Fotoarchiv K. Täuber

8684
WANDMASKE Frauenkopf mit großem Hut und großem Kragen mit Schleife / Mascherl
ca. 1940
8684,20 Firmenmarke 19, Made in Germany, Ausformung zw. 1940–1941, H. 30 cm; eBay # 1496840996
Quelle: AGM (Bestellung der Firma Elisabeth Salle (Wiesbaden) vom 8.1.1951), hier ohne nähere Ang. zur Maske.

8685F
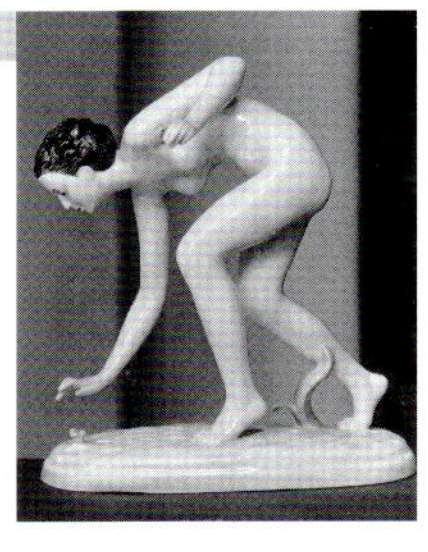
Prischl, Adolf
„Blumenpflückerin"
(Sich bückender Damenakt)
(auf Natursockel)
ca. 1940
8685,8,85 Firmenmarke 19, Made in Austria, unsign., H. 23,5 cm; city-antik
8685,237,1 Firmenstempel WG 1, unsign., Ausformung 50er Jahre, H. 23,5 cm; Doro S 13.10.2004
8685F, H. 25 cm, B. 22 cm; GSNachf. 1,63
Bem.: Diese Figur wurde auch noch zw. 1953–1960 bei Carstens (GWM West-Germany) produziert. Für die Zuschreibung an Adolf Prischl danke ich Frau Mag. Uta Matschiner.
-> vgl. 8738
Lit.: GSNachf. (2. Ausg., bzw. Erg. Bl.) 1,63 (m. Abb.); Könecke (2002), S. 78, 93 (m. Abb.).
Quelle: AGM (Bestellung der Firma Wahliss (Wien) vom 27.2.1946 an die „Arbeitsgemeinschaft der Firma Friedrich Goldscheider", hier mit Ang. des (Original?-)Titels „Blumenpflückerin", in Ausführung „Terrakotta"); Archiv Matschiner (Originalfoto); Fotoarchiv K. Täuber.

8686F
Dakon, Stephan
FIGUR „Akt (tanzend)"
ca. 1940
Quelle: Archiv Mimi Dakon (Tantiemenabrechnung 1.1.–30.6.1940; Wiener Manufaktur FG, Modellliste Dakon, 7.11.1946), hier jeweils mit Ang. des Originaltitels „Akt (tanzend)".

8688F / M
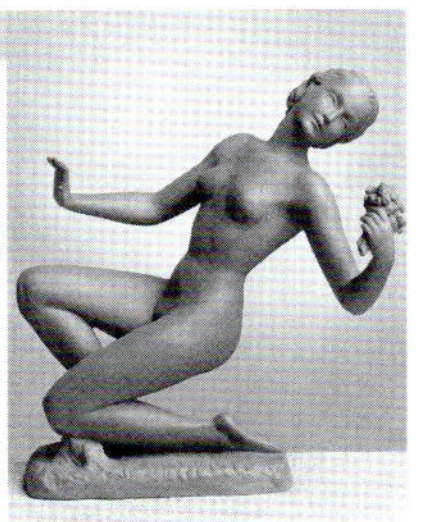
„Freuende" („Akt")
(Kniender Damenakt mit Blumenstrauß)
ca. 1941
8688,15 Firmenmarke 19, H. 33,7 cm; Doro 21.5.1997
8688,65 Firmenmarke 19a, H. 34 cm; KH
8688M H. fehlt; Fotoarchiv K. Täuber
Quelle: AGM (Bestellungen der Firma Wahliss (Wien) vom 9. und 24.11.1945 an die „Arbeitsgemeinschaft der Firma Friedrich Goldscheider", hier jeweils mit Ang. des (Original?-)Titels „Freuende"; Bestellungen der Firma Wahliss (Wien) vom 25.10.1945 und 27.2.1946 in Ausführung „Terrakotta"; Rechnung an Maurice Mizrahi, Alexandria, Ägypten, vom 22. April 1953), hier unter dem Titel „Akt" geführt; Fotoarchiv K. Täuber. hier rückseitig auf dem Originalfoto mit „Freuende" bezeichnet.

8689
WANDMASKE „Maske mit Hand"
ca. 1941
Quelle: AGM (Bestellung der Firma Wahliss (Wien) vom 24.11.1945 an die „Arbeitsgemeinschaft der Firma Friedrich Goldscheider"), hier mit Ang. des (Original?-)Titels „Maske mit Hand".

8690
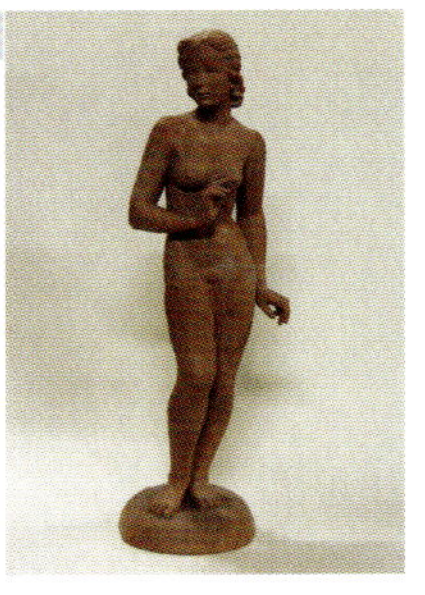
„Bäuerliche Venus"
(Stehender Damenakt)
(auf ovalem Landschaftssockel)
ca. 1941
8690,0 Firmenmarke 19a, H. 60 cm; Schuler 72
8690,67 Firmenmarke 19a, H. 60 cm; KH
8690 Firmenmarke 19a, H. 59,7 cm; MAK Wien, Inv. Ke 10602-37
Quelle: AGM (Bestellungen der Firma Wahliss (Wien) vom 9.11.1945 und 27.2.1946 an die „Arbeitsgemeinschaft der Firma Friedrich Goldscheider", hier mit Ang. des (Original?-)Titels „Bäuerliche Venus", in Ausführung „Terrakotta").

8692
„Grotesker Tanz" (Tanzender Damenakt)
(auf Rundsockel)
ca. 1941
8692,65 Firmenmarke 19a, H. 30 cm; KH
Quelle: AGM (Bestellungen der Firma Wahliss (Wien) vom 9. und 24.11.1945 an die „Arbeitsgemeinschaft der Firma Friedrich Goldscheider", hier jeweils mit Ang. des Originaltitels „Grotesker Tanz"; Bestellung der Firma Wahliss (Wien) vom 27.2.1946 in Ausführung „Terrakotta").

8693
„Akt"
ca. 1941
Quelle: AGM (Kommissionsfaktura der „Österreichischen Gruppe-Austria" für die Messe in Utrecht vom 26.2.1953), hier mit Ang. des Originaltitels „Akt".

8694M
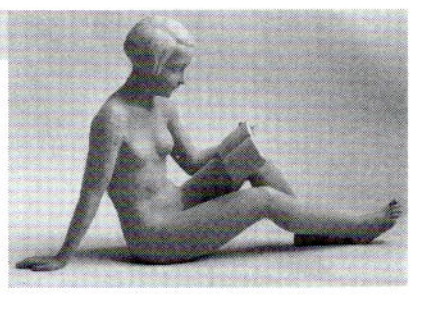
„Lesende" (Sitzender Damenakt mit Buch)
ca. 1941
8694,65 Firmenmarke 19, Ausformung zw. 1945–1953, H. 30 cm; eBay # 914751782
8694 ohne Ang., Firmenmarke 19, H. 33 cm; Spik 565
Quelle: AGM (Bestellungen der Firma Wahliss (Wien) vom 9. und 24.11.1945 an die „Arbeitsgemeinschaft der Firma Friedrich Goldscheider", hier jeweils mit Ang. des (Original?-)Titels „Lesende"; Bestellung der Firma Wahliss (Wien) vom 27.2.1946 in Ausführung „Terrakotta"); Fotoarchiv K. Täuber.

8695
WANDMASKE Mädchenkopf mit Trachtenhut und Kragen
ca. 1941
8695 Firmenmarke 19a, Made in Germany, Ausformung ca. 1941, H. 25 cm; eBay # 45340146

8696
GRUPPE Zwei Rehe (auf ovalem Natursockel)
ca. 1941
8696,8 Firmenmarke 19, Made in Austria, Ausformung zw. 1945–1953, H. 31 cm; eBay # 6520435839

8698T / M
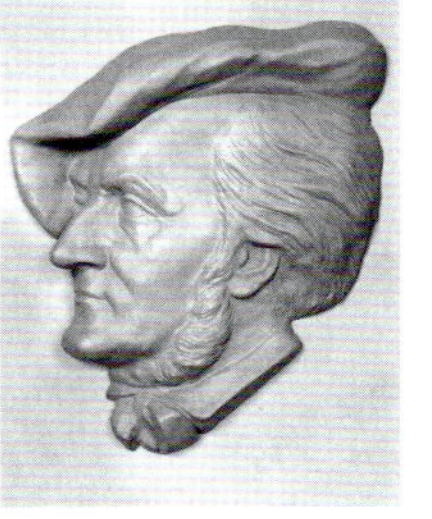
Lorenzl, Josef
WANDMASKE „Wagner"
(Richard Wagner im Profil)
ca. 1941
8698T, H. 27 cm, B. 32 cm; GSNachf. 2,27; Fotoarchiv K. Täuber, hier mit der Ang. des Originalpreises RM 24,25 und bezeichnet mit der Modell-Nr. 8698M (H. 27 cm, B. 25 cm).
Bem.: Dargestellt ist der Komponist Richard Wagner (1813–1883). Diese Maske wurde auch noch zw. 1953–1960 bei Carstens (GWM West-Germany) produziert.
Lit.: GSNachf. (2. Ausg., bzw. Erg. Bl.) 2,27 (m. Abb.); KGSF1; KGSF2; Könecke (2002), S. 56, 78 (m. Abb.), hier mit Ang. des Künstlers J. Lorenzl.
Quelle: AGM (Rechnung an E. Rosenthal, Frankfurt a. M., vom 20. Dezember 1950), hier mit Ang. des Originaltitels MASKE „Wagner".

8699
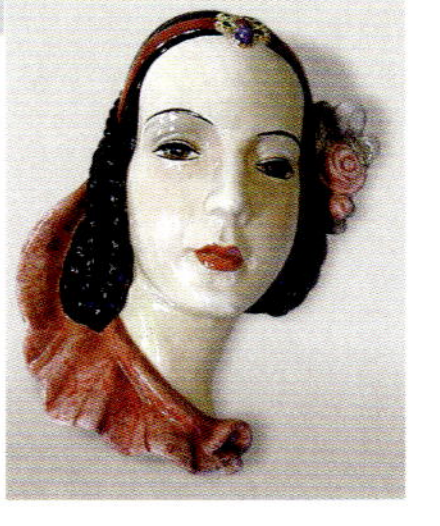
WANDMASKE Frauenkopf mit Kragen und Blume im Haar
ca. 1941
8699,1 Firmenmarke 19, H. 25 cm; Doro 3.2.1998 (ohne Kat.)
8699,1 Firmenstempel WG 1, Ausformung 50er Jahre, H. 26 cm; eBay # 2029744668; 2041094628
8699,20 Firmenmarke 19, Made in Germany, unsign., Ausformung ca. 1941, H. 25 cm; city-antik
8699 ohne Ang., H. 26 cm, B. 16 cm; Könecke (2002), S. 78, 93 (m. Abb.)
Bem.: Diese Maske wurde auch noch zw. 1953–1960 bei Carstens (GWM West-Germany) produziert.
Lit.: Könecke (2002), S. 78, 93 (m. Abb.)
Quelle: AGM (Bestellung der Firma Elisabeth Salle (Wiesbaden) vom 8.1.1951).

8700
WANDMASKE Frauenkopf mit Haarband, Hand und Glockenblumen
ca. 1941
8700,20 gem. GB, Firmenmarke 19, Made in Germany, Ausformung ca. 1941, H. fehlt; eBay # 116526574
Lit.: Pinhas, S. 169 (m. Abb.)
Quelle: AGM (Bestellung der Firma Elisabeth Salle (Wiesbaden) vom 8.1.1951), hier ohne nähere Ang. zur Maske.

8701
WANDMASKE Frauenkopf mit langem Haar und Kopftuch
ca. 1941
8701,20,G gem. N, Firmenmarke 19, Made in Germany, Ausformung ca. 1941, H. 26 cm; Wien Museum Inv. 186.646
8701,20,4 gem. rot 4., Firmenmarke 19, Made in Germany, Ausformung ca. 1941, H. ca. 30 cm (lt. Ang. des Verkäufers); eBay # 1456508038

8702
Prischl, Adolf
FIGUR „Sandalenbinderin (groß)" (Stehender Damenakt, den rechten Fuß hebend) (auf Rundsockel)
ca. 1939 (für den Entwurf)
8702,9 Firmenmarke 19a, H. 47,7 cm; Sammlung a@goldscheider.net
8702,15 Firmenmarke 19a, H. 49,5 cm;
eBay # 122423740

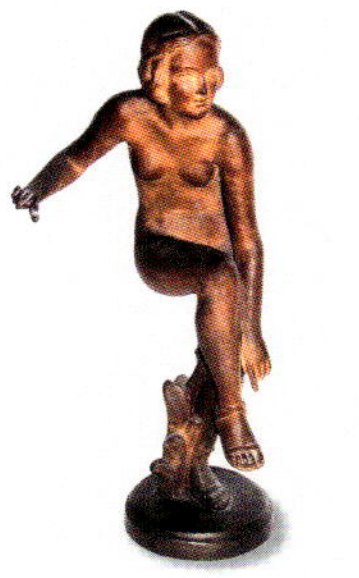

Bem.: Diese Figur wurde auch noch zw. 1953–1960 bei Carstens (GWM West-Germany) produziert.
-> vgl. 8503
Lit.: Könecke (2002), S. 78, hier ohne weiteren Ang.
Quellen: AGM (Bestellung der Firma Wahliss (Wien) vom 27.2.1946 an die „Arbeitsgemeinschaft der Firma Friedrich Goldscheider", hier mit Ang. des Originaltitels „Sandalenbinderin (groß)", in Ausführung „Terrakotta"); Fotoarchiv K. Täuber.

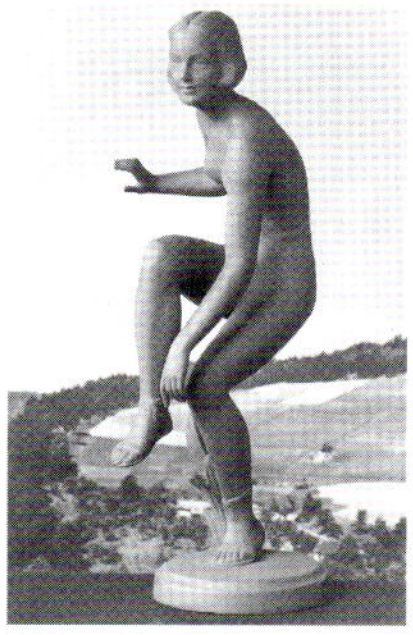

8704
„Lektüre" (Sitzende Frau mit Buch)
ca. 1940 (für den Entwurf)
8704,65 Firmenmarke 19, H. 42 cm;
eBay # 336879705
8704,65 Firmenmarke 19a, H. 40,5 cm;
Doro K 19.5.2004
8704,73 Firmenmarke 19a, H. 42 cm; Wien Museum Inv. 219.145

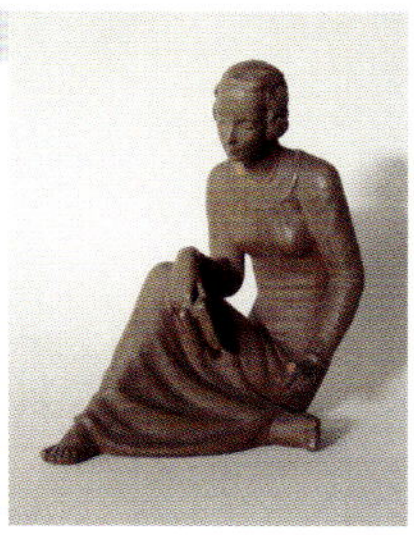

8704 ohne Ang., Firmenmarke 19, H. 42 cm; Hennig 42
Lit.: GSNachf. (2. Ausg., bzw. Erg. Bl.) 3,2 (m. Abb.), hier mit Ang. des Originaltitels „Lektüre".
-> vgl. 8564
Quelle: AGM (Bestellungen der Firma Wahliss (Wien) vom 9. und 24.11.1945 an die „Arbeitsgemeinschaft der Firma Friedrich Goldscheider"; Bestellung der Firma Wahliss (Wien) vom 27.2.1946 in Ausführung „Terrakotta").

8705F
„Papagei"
ca. 1941
Quelle: AGM (Rechnung an Maurice Mizrahi, Alexandria, Ägypten, vom 22. April 1953), hier mit Ang. des Originaltitels „Papagei".

8706F
Lorenzl, Josef
„Feierlicher Rhythmus" (Mary Wigman) (auf Rundsockel)
ca. 1941
8706,29,31,G (ohne Firmenmarke), unsign., H. 37,5 cm; PB Deutschland; Bem.: Ausführung in Terrakotta.

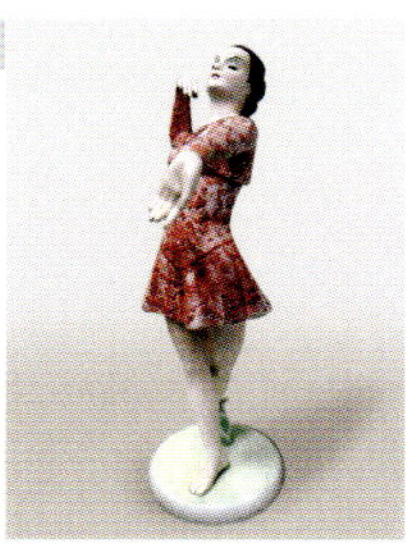

8706,47,7 Firmenmarke 19, Made in Germany, unsign., Ausformung ca. 1941, H. 38 cm; KH
8706,81,31 gem. HG, W., Firmenmarke 19, Made in Germany, unsign., Ausformung ca. 1941, H. 38 cm; Wien Museum Inv. 190.152
8706,213,81 Firmenmarke 19, Made in Germany, Ausformung ca. 1941, unsign., H. 39,5 cm; city-antik

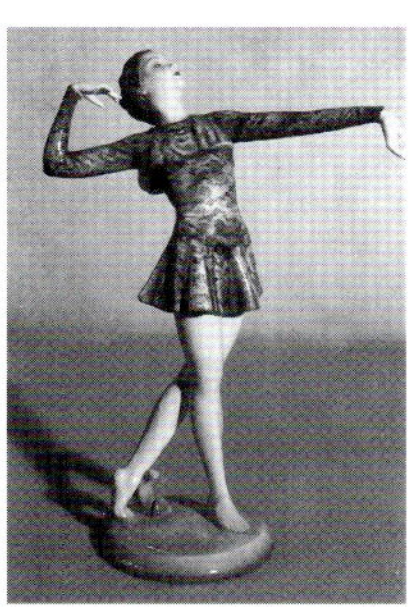

8706F, H. 40 cm, B. 27 cm; GSNachf. 1,60
Bem.: Dargestellt ist die Tänzerin Mary Wigman (13.11.1886, Hannover – 18.9.1973, Berlin), Foto Rudolph.
Lit.: GSNachf. (2. Ausg., bzw. Erg. Bl.) 1,60 (m. Abb.); BT, o. Pag., Bild 120, die hier wiedergegebene zeitgenössische Fotografie von Rudolph diente als Vorlage für die Ausführung dieses Modells.

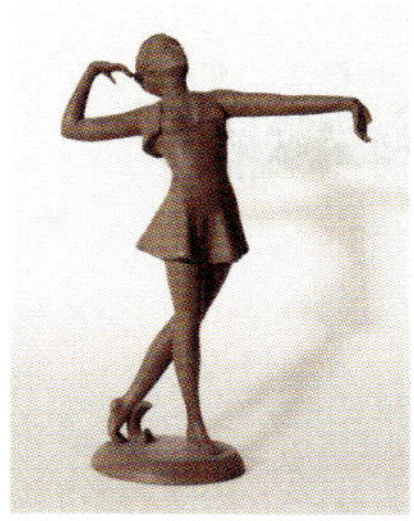

Quelle: AGM (Bestellung der Firma Wahliss (Wien) vom 27.2.1946 an die „Arbeitsgemeinschaft der Firma Friedrich Goldscheider", hier mit Ang. des (Original?-)Titels „Feierlicher Rhythmus", in Ausführung „Terrakotta"); Fotoarchiv K. Täuber.

8707
STEHKOPF Mädchenkopf (vollplastisch) (mit/ohne würfelartigem Sockel)
ca. 1941
mit Sockel
8707 ohne Ang., H. 26 cm; Henry's 26.2.1999
ohne Sockel
8707,71 Firmenmarke 19, H. ca. 20 cm; WKH

8708
Lorenzl, Josef
FIGUR „Kniende" (Kniender Damenakt) (auf abgerundetem Rechtecksockel)
ca. 1941
8708,9 ger., Firmenmarke 19a, unsign., H. ca. 65 cm; Keup 95; eBay # 2892426314; # 5322999817
Bem.: Zum Originaltitel -> vgl. 8495
-> vgl. 8495; 8496
Quelle: AGM (Bestellungen der Firma Wahliss (Wien) vom 9. und 24.11.1945 an die „Arbeitsgemeinschaft der Firma Friedrich Goldscheider", hier jeweils mit Ang. des Titels „Kniende (groß)"; Bestellung der Firma Wahliss (Wien) vom 27.2.1946 in Ausführung „Terrakotta").

8709
„Badende"
ca. 1941
Quelle: AGM (Bestellung der Firma Wahliss (Wien) vom 27.2.1946 an die „Arbeitsgemeinschaft der Firma Friedrich Goldscheider", hier mit Ang. des (Original?-)Titels „Badende" in Ausführung „Terrakotta").

8711F
Lorenzl, Josef
„Dreiklang" (Mary Wigman auf einem Bein tanzend) (auf Rundsockel)
ca. 1941
8711,8,11 Firmenmarke 19, Made in Austria, Hand dec., unsign., Pick. 1, H. 38,5 cm; eBay # 79488965
8711,35,7 gem. HF, Firmenmarke 19, Made in Germany, Ausformung ca. 1941, H. 38 cm;
Doro S 3./4.4.1996

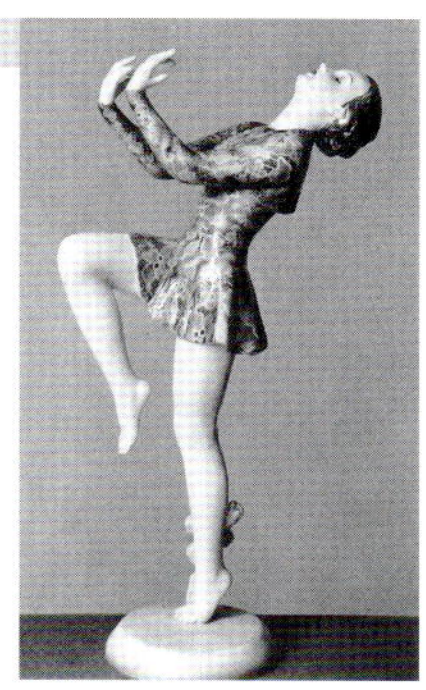

8711,66,5 (Firmenmarke ohne Ang.), unsign., Ausformung ca. 1941, H. 38 cm; Christie's SK 13.5.2004
8711,93,7 Firmenmarke 19, Made in Germany, unsign., Ausformung ca. 1941, H. 38 cm; eBay # 1498292107; 916981264
8711F, H. 38,5 cm, B. 14 cm; GSNachf. 1,58
Bem.: Dargestellt ist die Tanzkünstlerin Mary Wigman (13.11.1886, Hannover–18.9.1973, Berlin) in ihrem Tanz „Festlicher Rhythmus". Diese Figur wurde auch noch zw. 1953–1960 bei Carstens (GWM West-Germany) produziert.
Lit.: GSNachf. (2. Ausg., bzw. Erg. Bl.) 1,58 (m. Abb.); KGSF1; Könecke (2002), S. 77, 93 (m. Abb.), hier mit Ang. des Künstlers J. Lorenzl und des Originaltitels „Dreiklang".
Quelle: AGM (Bestellungen der Firma Wahliss (Wien) vom 9. und 24.11.1945 an die „Arbeitsgemeinschaft der Firma Friedrich Goldscheider", hier jeweils mit Ang. des Originaltitels „Dreiklang"; Bestellung der Firma Wahliss (Wien) vom 27.2.1946 in Ausführung „Terrakotta"); Fotoarchiv K. Täuber.

8712
„Dornauszieherin" (Mädchenakt mit Zöpfen, auf einem Bein stehend) (auf Rundsockel)
ca. 1941
8712,67 Firmenmarke 19, H. 37 cm;
Henry's 27.12.1996
8712 H. fehlt; Fotoarchiv K. Täuber; Bem.: Auf der Rückseite des Fotos handschriftlich bezeichnet „Dornauszieherin".
Quelle: Fotoarchiv K. Täuber

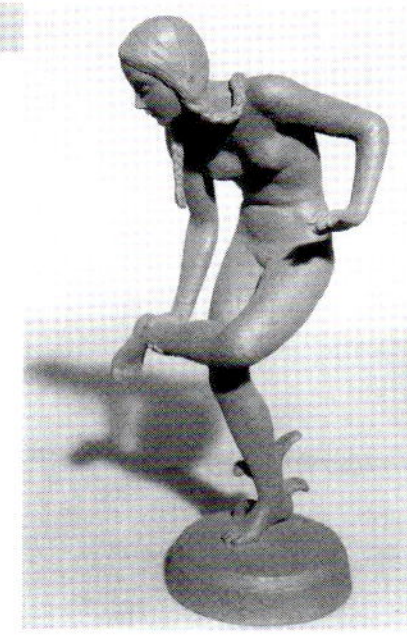

8713F/M
Lorenzl, Josef
„Get(anzte) Harmonie" (Tänzerin in Pose mit ausgestreckten Armen) (auf Rundsockel)
ca. 1941
8713,1,15 gem. lig. HS, Firmenmarke 20 (blau), Made in Austria (blau), Hand dec. (blau), unsign., H. 43 cm;
eBay # 2241305147

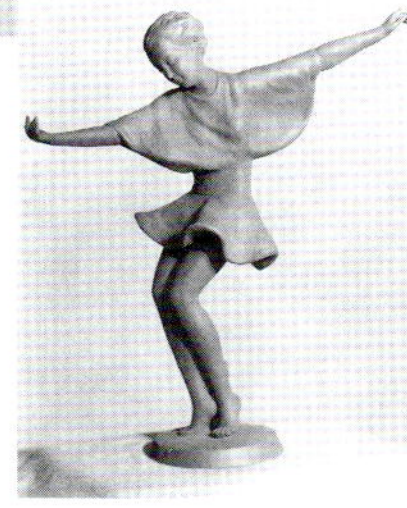

8713,319,7 Firmenmarke Schuster 2, Made in Germany, unsign., Ausformung ca. 1941, H. 41 cm;
Doro 26.8.1998 (ohne Kat.)
8713 ohne Ang. Firmenmarke 19, unsign., H. 41 cm;
Henry's 27.11.1996
8713 ger., ohne Firmenmarke, unsign., H. 41 cm; KH

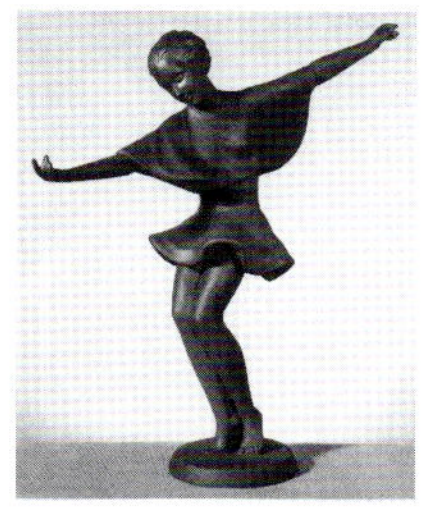

ohne Ang., Firmenmarke 19, unsign., H. 43,5 cm; Nagel Varia 52
8713F, H. 40 cm, B. 38 cm; GSNachf. 1,61; Fotoarchiv K. Täuber, hier mit der Ang.: 8713M, Größenangaben fehlen, rückseitig auf dem Originalfoto mit „Getanzte Harmonie" bezeichnet.
Lit.: GSNachf. (2. Ausg., bzw. Erg. Bl.) 1,61 (m. Abb.); KGSF1; KGSF2; Könecke (2002), S. 67, 78 (m. Abb.), hier mit Ang. des Künstlers J. Lorenzl und des Originaltitels „Harmonie".
Quelle: AGM (Bestellungen der Firma Wahliss (Wien) vom 9. und 24.11.1945 an die „Arbeitsgemeinschaft der Firma Friedrich Goldscheider"), hier jeweils mit Ang. des Originaltitels „Get. Harmonie"; Bestellung der Firma Wahliss (Wien) vom 27.2.1946 in Ausführung „Terrakotta"; Bestellung der Firma Elisabeth Salle (Wiesbaden) vom 8.1.1951); Archiv Matschiner (Originalfoto), hier mit dem Titel „Harmonie", zusätzlich etwas abweichende H. 41 cm, B. 37,5 cm; Fotoarchiv K. Täuber.

8714
Prischl, Adolf
FIGUR „Sandalenbinderin" (Sitzender Damenakt mit zurückgekämmtem Haar und Dutt)
ca. 1941
8714,9 Firmenmarke 19a, unsign., H. 34 cm;
eBay # 6513232068
8714,67 (Firmenmarke ohne Ang.), unsign., H. 32,5 cm;
Christie's SK 4./5.12.1996

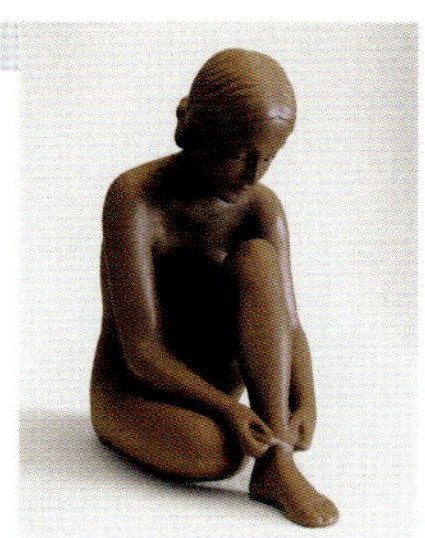

8714 (Firmenmarke ohne Ang.), unsign., H. 33 cm; Doro 4.3.2002
8714 ger., ohne Firmenmarke, unsign., Ausformung 50er Jahre, H. 33,4 cm; Sammlung John Nash
8714 H. 35 cm, B. 30 cm; Fotoarchiv K. Täuber; Bem.: Auf der Rückseite handschriftlich „Sitzende Sandalenbinderin", groß, in ganz Terrakotta, RM 44,50 bezeichnet.

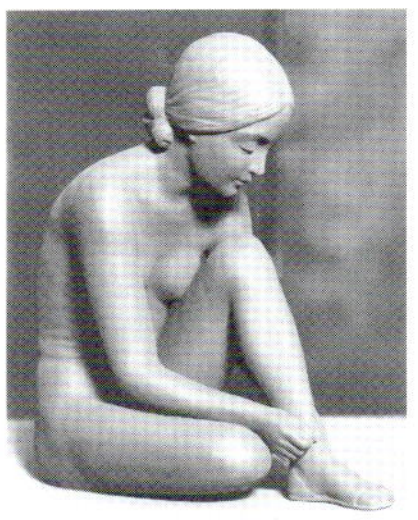

Bem.: Diese Figur wurde auch noch zw. 1953–1960 bei Carstens (GWM West-Germany) produziert.
-> vgl. 8506; 8508
Lit.: KGSF1; KGSF2; Könecke (2002), S. 36, 65, 77 (m. Abb.).
Quelle: AGM (Kontoblatt der Wiener Manufaktur Josef Schuster für die Firma Ohlson jr., Malmö, Schweden vom 16. September 1942), hier mit Ang. des Originaltitels „Sandalenbinderin"; Bestellung der Firma Wahliss (Wien) vom 27.2.1946 an die „Arbeitsgemeinschaft der Firma Friedrich Goldscheider", hier mit Ang. des Originaltitels „Sandalenbinderin (sitzend)", in Ausführung „Terrakotta" und „Fayence"); Fotoarchiv K. Täuber.

8716F

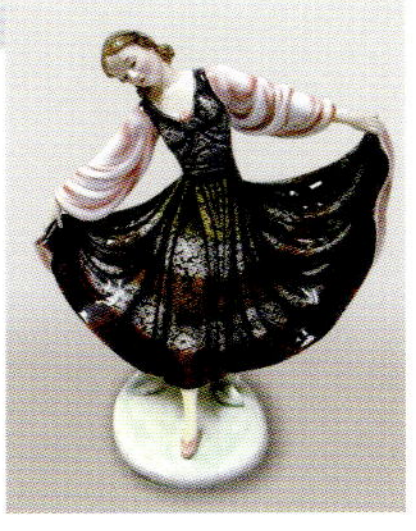

Dakon, Stephan
„Dunja" (Tänzerin, ihr Kleid mit beiden Händen anhebend) (auf Rundsockel)
ca. 1941
8716,8,50 Firmenmarke 19, Made in Germany, unsign., Ausformung ca. 1941, Pick. 6, H. 34,5 cm; Doro 30.11.2001
8716,11 Firmenmarke 19, unsign., Ausformung zw. 1945–1953, H. 34,5 cm; Doro 21.5.1997
8716,S,51 ger., gem. lig. HS, Firmenmarke 19, Made in Austria, Hand dec., unsign., Ausformung zw. 1945–1953, H. 34,5 cm; city-antik
Lit.: Pinhas, S. 163 (m. Abb.)
Quelle: AGM (Tantiemenliste Dakon, ohne Datierung (ca. 1940); Archiv Mimi Dakon (Wiener Manufaktur FG, Modellliste Dakon, 7.11.1946; Tantiemenabrechnung 2. Halbjahr 1950; (1. Halbjahr 1951, AGM); 2. Halbjahr 1951), hier jeweils mit Ang. des Originaltitels „Dunja"; AGM (Rechnung an Maurice Mizrahi, Alexandria, Ägypten, vom 22. April 1953).

8718F

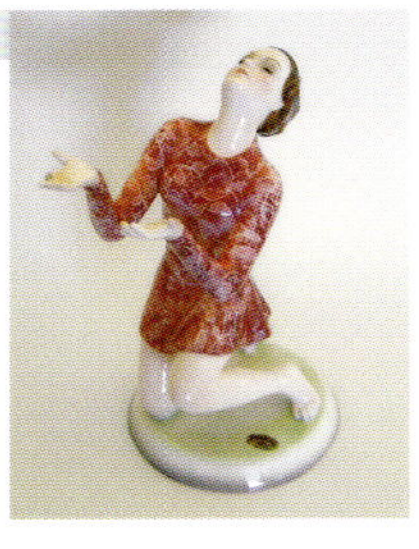

Lorenzl, Josef
„Tanz" (Mary Wigman, in fast kniender hingebungsvoller Pose) (auf Rundsockel)
ca. 1941
8718,14 Firmenmarke 19 und 19a, unsign., Ausformung zw. 1945–1953, H. 26 cm; eBay # 6527154765
8718,90,7 Firmenmarke 19, Made in Germany, Ausformung ca. 1941, H. 26,5 cm; Schopmann 7.5.1998

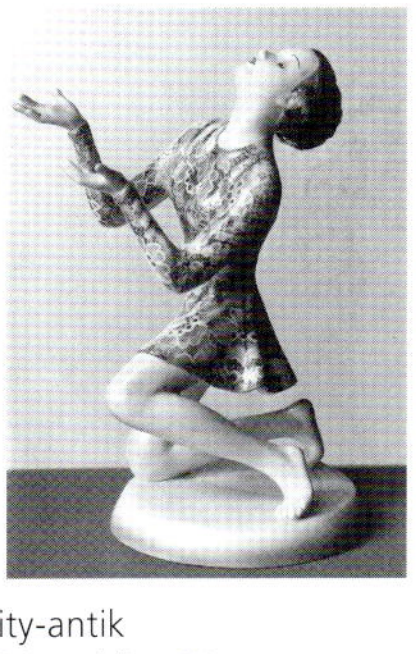

8718,268,7 ger. H, gem. X, Firmenmarke 19, Made in Germany, unsign., Ausformung ca. 1941, Pick. 1, H. 26,5 cm; Sammlung.schmutz@chello.at 47
8718,297,19 Firmenmarke 19, Made in Germany, unsign., Ausformung ca. 1941, Pick. 6, H. 28 cm; Mehlis 23; city-antik
8718F, H. 29 cm, B. 15,5 cm; GSNachf. 1,59
Bem.: Dargestellt ist die Tanzkünstlerin Mary Wigman (13.11.1886, Hannover–18.9.1973, Berlin) in ihrem „Festlichen Rhythmus".
Lit.: GSNachf. (2. Ausg., bzw. Erg. Bl.) 1,59 (m. Abb.); Schaulade, 18.1942, S. 503 (m. Abb.).
Quelle: AGM (Rechnung an Maurice Mizrahi, Alexandria, Ägypten, vom 22. April 1953), hier mit Ang. des (Original)-Titels „Tanz"; Fotoarchiv K. Täuber.

8720

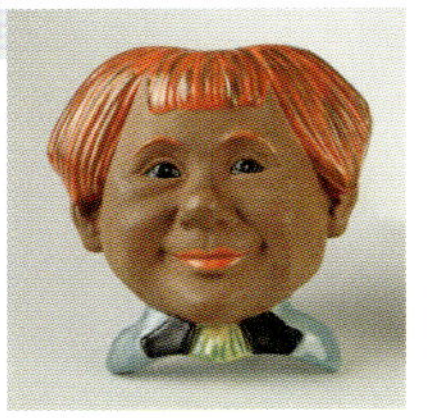

WANDMASKE Max
ca. 1941
8720,81,66 Firmenmarke 19, Made in Austria, H. 22,5 cm; Wien Museum Inv. 190.006/1
Bem.: Diese Maske zeigt den Max, aus der Erzählung „Max und Moritz" von Wilhelm Busch.
Pendant -> vgl. 8721

8721

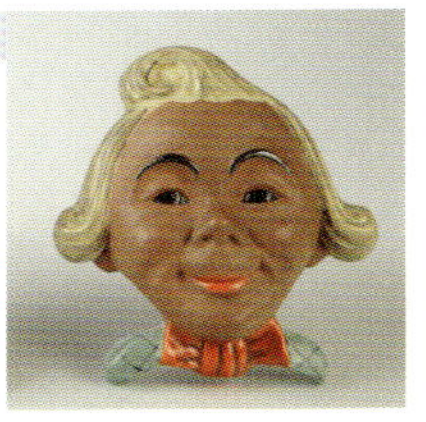

WANDMASKE Moritz
ca. 1941
8721,87,66 Firmenmarke 19, Made in Austria, H. 21,5 cm; Wien Museum Inv. 190.006/2
Bem.: Diese Maske zeigt den Moritz, aus der Erzählung „Max und Moritz" von Wilhelm Busch.
Pendant -> vgl. 8720

8723F

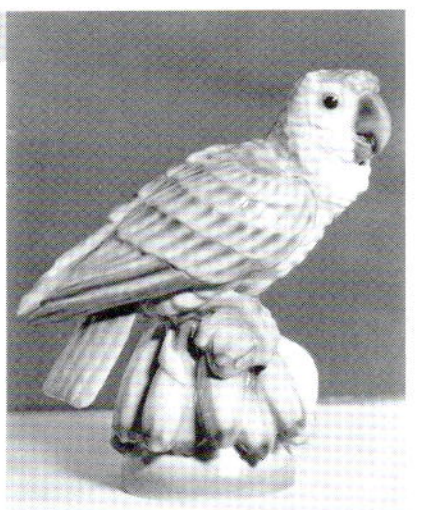

Papagei auf Bananenstaude (auf Rundsockel)
ca. 1941
8723,7 ger., gem. G, Firmenmarke 19, Made in Austria, Ausformung zw. 1945–1953, H. 28,5 cm; Wien Museum Inv. 220.770
8723,18,31 Firmenmarke 19, Made in Austria, H. 32 cm; eBay # 3723905422
Quelle: Fotoarchiv K. Täuber

8725

BUCHSTÜTZE Athleten (linke Buchstütze: Kraftsportler, mit dem Rücken gegen die Bücher drückend, rechte Buchstütze: Kraftsportler, sich mit beiden Händen gegen die Bücher stemmend)
ca. 1941
8725,10 gem. GW., Firmenmarke 19 (blau), Made in Austria (blau), Ausformung zw. 1945–1953, H. 22 cm, B. 20 cm; KH
Bem.: Da es sich um eine Mitt. aus dem KH handelt, konnte nicht ermittelt werden, ob den beiden unterschiedlichen Stützen auch zwei verschiedene Modellnummern zugewiesen wurden.

8726

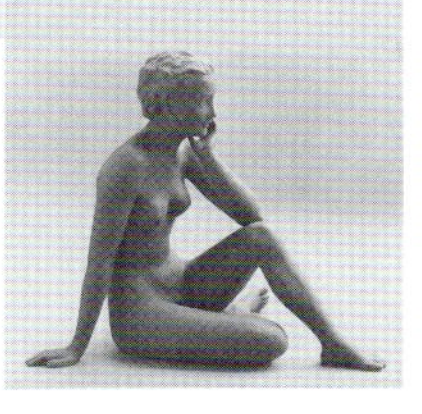

Lorenzl, Josef
„Sinnende" (Sitzender Damenakt)
ca. 1940 (für den Entwurf)
8726 H. 85 cm
-> vgl. 8665
Quelle: AGM (Bestellung der Firma Wahliss (Wien) vom 27.2.1946, hier mit Ang. Originaltitels „Sinnende" in Ausführung „Terrakotta", H. 85 cm).

8727

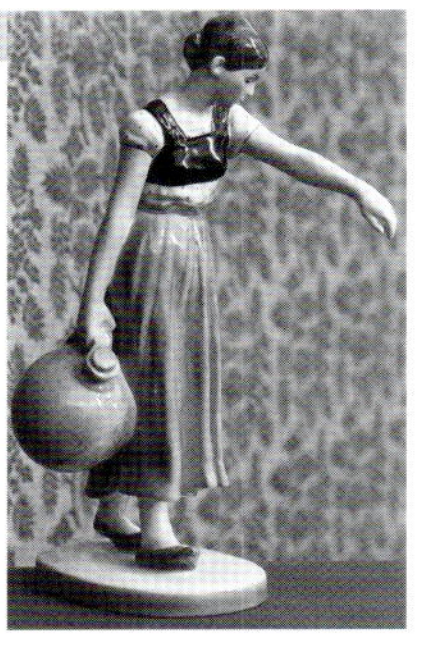

Strasser, Arthur (nach)
„Das Krüglein"/„Krugträgerin" (Schreitendes Bauernmädchen mit Krug) (auf Ovalsockel)
ca. 1898 (für den Entwurf)
8727,67 Firmenmarke 19, unsign., Ausformung zw. 1945–1953, H. 52 cm; eBay # 6545410516; Peege 105
Bem.: Diese Figur stammt im ursprünglichen Entwurf von Arthur Strasser, datiert um das Jahr 1898, und war mehrfach im ÖMKI ausgestellt (-> vgl. Ausst.). Die zeitgenössische Pariser Produktion nahm dieses Modell ebenfalls auf und stellte es in drei Größen her (-> vgl. Lit.).
Ausst.: ÖMKI 1906/07, hier mit dem Originaltitel „Das Krüglein"; ÖMKI 1910/11, hier mit dem Titel „Krüglein".
-> vgl. 7915
-> vgl. 551 (GSB); 576 (GSB); 579 (GSB)
Lit.: GSB, Pl. 18, hier ist die Figur mit dem franz. Originaltitel „Retour de la Fontaine" unter Nr. 551 (H. 90 cm) abgebildet. Zwei weitere Größen mit der Nr. 579 (H. 61 cm) sowie der Nr. 576 (H. 32 cm) finden sich an gleicher Stelle. Ausgeformt wurden diese Objekte in der Pariser Niederlassung. Angaben zum Material werden im Firmenkat. nicht gemacht; KuKHW, 1.1898, S. 354, 362; Neuwirth (1974), S. 74.

8728

„Die Arbeit" („Arbeitsmann")
ca. 1941
Quelle: AGM (Bestellung der Firma Wahliss (Wien) vom 24.11.1945 an die „Arbeitsgemeinschaft der Firma Friedrich Goldscheider", hier mit Ang. des (Original?-)Titels „Die Arbeit"; („Arbeitsmann"); Bestellung der Firma Wahliss (Wien) vom 27.2.1946 in Ausführung „Terrakotta").

8729F

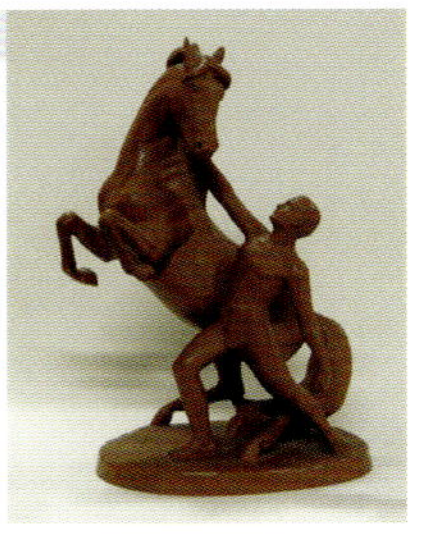

„Rossbändiger" („Pferdebändiger") (Mann, ein sich aufbäumendes Pferd haltend) (auf Natursockel)
ca. 1941
8729,31 Firmenmarke 19, H. 39,5 cm; MAK Wien, Inv. Ke 10521-14
8729F, H. 39 cm, B. 31 cm; GSNachf. 1,67
Lit.: GSNachf. (2. Ausg., bzw. Erg. Bl.) 1,67 (m. Abb.)

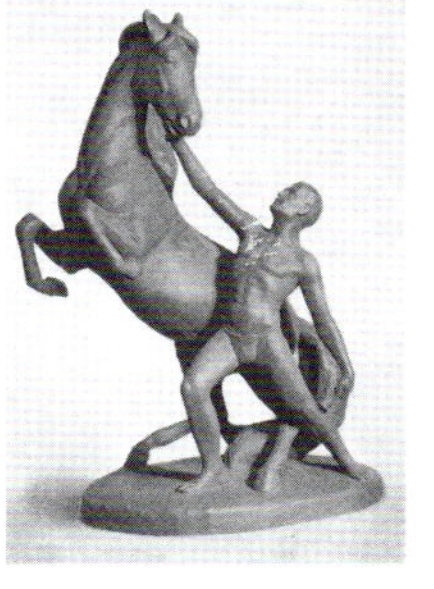

Quelle: AGM (Bestellungen der Firma Wahliss (Wien) vom 9. und 24.11.1945 an die „Arbeitsgemeinschaft der Firma Friedrich Goldscheider", hier jeweils mit Ang. des (Original)-Titels „Rossbändiger"; Bestellung der Firma Wahliss (Wien) vom 27.2.1946 in Ausführung „Terrakotta"; Rechnung an Maurice Mizrahi, Alexandria, Ägypten, vom 22. April 1953), hier mit Ang. des (Original)-Titels „Pferdebändiger".

8730

Auerhahn
ca. 1941
8730,8 (Firmenmarke ohne Ang.), H. 53,3 cm; Doro WA 17.12.1994

8732F

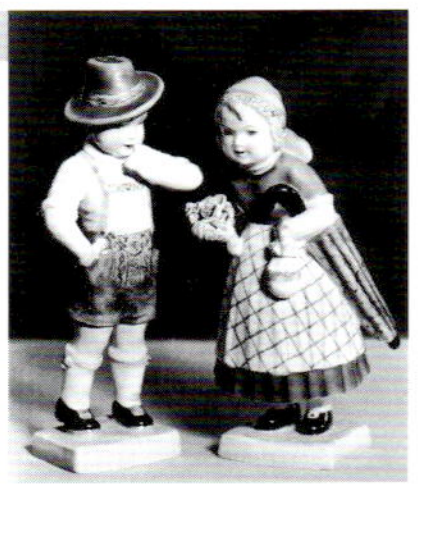

„Sepperl" (Stehender Bub in kurzen Lederhosen und Trachtenhut) (auf Rechtecksockel)
ca. 1941
8732F, H. 22 cm, B. 9 cm; GSNachf. 1,72a
Bem.: Im Stile einer „Hummel-Figur", -> vgl. Bem. bei Modell-Nr. 8567.
Pendant -> vgl. 8733
Lit.: GSNachf. (2. Ausg., bzw. Erg. Bl.) 1,72 (m. Abb.)
Quelle: AGM (Bestellung der Firma Wahliss (Wien) hier mit Ang. des (Original?-)Titels „Sepperl"); Fotoarchiv K. Täuber.

8733F

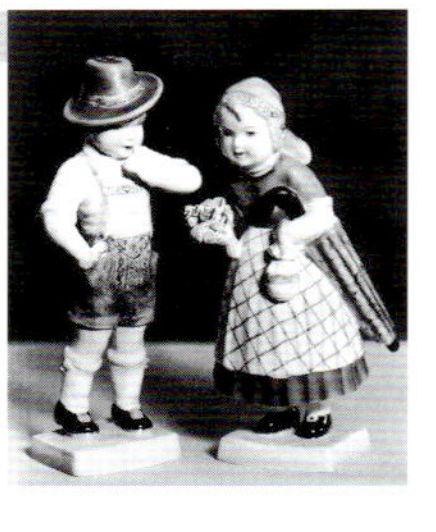

„Mizzerl" (Stehendes Mädchen in Tracht, mit Blumenstrauß und Tasche) (auf Rechtecksockel)
ca. 1941
8733,4 gem. E, Firmenmarke 19, Made in Austria, Ausformung zw. 1945–1953, H. 20 cm; Wien Museum Inv. 219.181

8733,8 gem. lig. ST., LS, Firmenmarke 19, Made in Austria, Hand dec., Ausformung zw. 1945–1953, H. 20,3 cm; eBay # 2622314532
8733,9,I ger., gem. lig. HS.W, Firmenmarke 19, Made in Austria, Ausformung zw. 1945–1953, H. 20 cm; Sammlung Fritz Meyer/Hannover
8733F, H. 20 cm, B. 14 cm; GSNachf. 1,72b
Bem.: Im Stile einer „Hummel-Figur", –> vgl. Bem. bei Modell-Nr. 8567.
Pendant –> vgl. 8732
Lit.: GSNachf. (2. Ausg., bzw. Erg. Bl.) 1,72 (m. Abb.)
Quelle: AGM (Bestellung der Firma Wahliss (Wien), hier mit Ang. des (Original?-)Titels „Mizzerl"); Fotoarchiv K. Täuber.

8734F

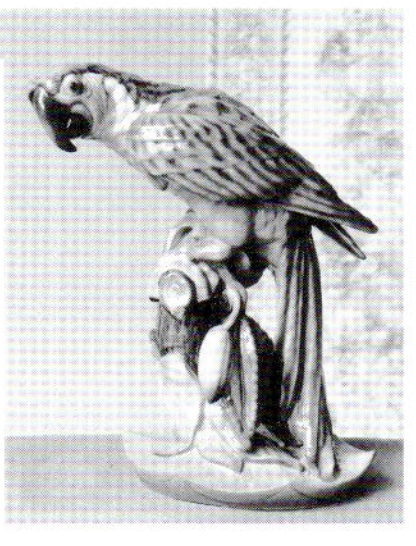

Papagei auf Baumstamm mit Blüten
ca. 1941
8734,4 ger., Firmenmarke 19, Pick. 1, Ausformung zw. 1945–1953, H. 33,5 cm; Doro 20.10.1997
8734,25,S gem. rot. G, Firmenmarke 19, Made in Austria, Ausformung zw. 1945–1953, H. 32 cm; Sammlung Helmut Riener, St. Florian, Austria
8734,S gem. X., 20, Made in Austria, Hand dec., H. 33 cm, B. 26 cm; Sammlung mad
8734F, H. 34 cm, B. 29 cm; GSNachf. 5,23
Lit.: GSNachf. (2. Ausg., bzw. Erg. Bl.) 5,23 (m. Abb.)
Quelle: Fotoarchiv K. Täuber

8735T
Prischl, Adolf (zugeschr.)

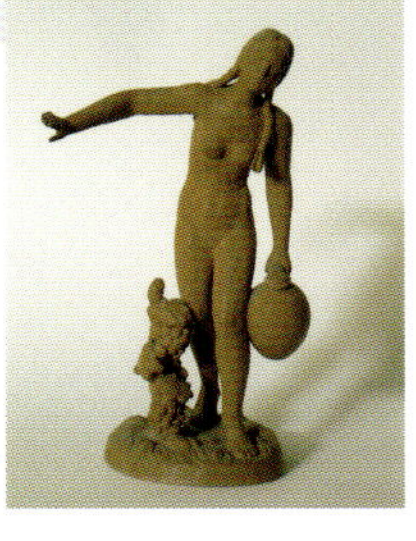

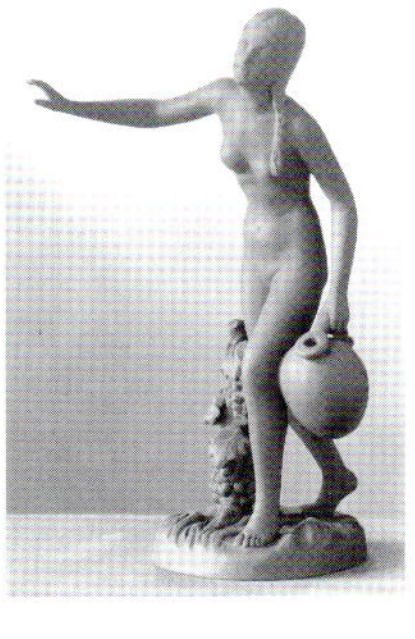

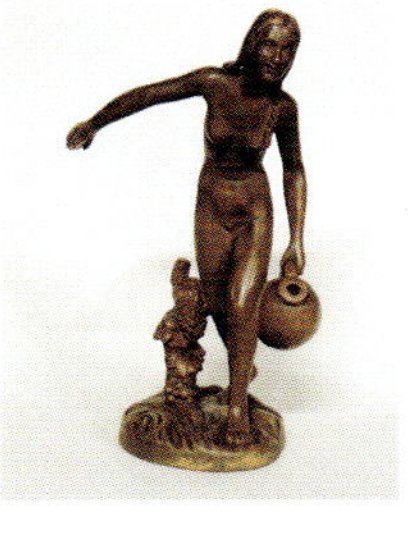

„Herbst" (Schreitender Damenakt mit Krug und Weintrauben) (auf Landschaftssockel)
ca. 1940 (für den Entwurf)
8735,36,9 Firmenmarke 19, Made in Germany, unsign., Ausformung ca. 1941, Pick. 6, H. 33,5 cm; eBay # 488834475
8735,38,19 gem. lig. CZ (oder NU?), Firmenmarke 19, Made in Germany, unsign., Ausformung ca. 1941, Pick. 6, H. 33,5 cm; eBay # 245897258
8735,66 Firmenmarke 19, unsign., Ausformung zw. 1945–1953, H. 32 cm; KH
8735,66 (ohne Firmenmarke), unsign., H. 30,9 cm; MAK Wien, Inv. Ke 10602-33
8735,67 Firmenmarke 19a, unsign., Ausformung zw. 1945–1953, H. 32 cm; Weidler 310; city-antik
–> vgl. 8644
Quelle: AGM (Bestellungen der Firma Wahliss (Wien) vom 9. und 24.11.1945 an die „Arbeitsgemeinschaft der Firma Friedrich Goldscheider", hier jeweils mit Ang. des (Original?-)Titels „Herbst"; Bestellung der Firma Wahliss (Wien) vom 27.2.1946 in Ausführung „Terrakotta"; Rechnung an Maurice Mizrahi, Alexandria, Ägypten, vom 22. April 1953), hier unter dem Titel „Krugträgerin" geführt; Fotoarchiv K. Täuber.

8736F

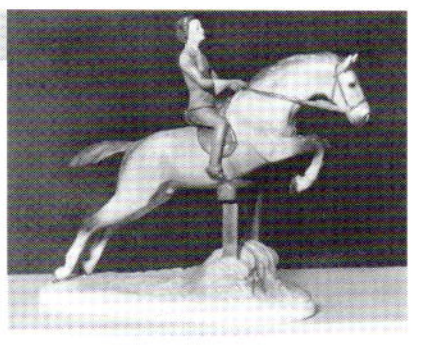

„Reiterin" (auf Pferd über eine Hürde springend) („Jagd") (auf Natursockel)
ca. 1941
8736,6 Firmenmarke 19, H. 37,6 cm, B. 43,2 cm; city-antik
8736 ohne Ang., Firmenmarke 19, Made in Austria, Ausformung zw. 1945–1953, H. 35,5 cm; eBay # 2195811438
8736F, H. 40 cm, B. 42 cm; GSNachf. 1,64; Fotoarchiv K. Täuber, rückseitig auf dem Originalfoto mit „Reiterin auf Hürde" bezeichnet.
Lit.: GSNachf. (2. Ausg., bzw. Erg. Bl.) 1,64 (m. Abb.)
Quelle: AGM (Bestellungen der Firma Wahliss (Wien) vom 9. und 24.11.1945 an die „Arbeitsgemeinschaft der Firma Friedrich Goldscheider", hier jeweils mit Ang. des (Original?-)Titels „Reiterin"; Bestellung der Firma Wahliss (Wien) vom 27.2.1946 in Ausführung „Terrakotta"; Rechnung an Maurice Mizrahi, Alexandria, Ägypten, vom 22. April 1953), hier mit unter dem Titel „Jagd" geführt.

8737F

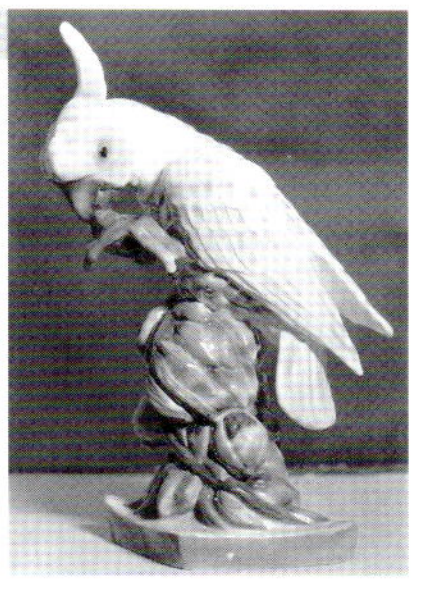

Kakadu auf Ast mit Früchten (auf Rautensockel)
ca. 1941
8737,27,S gem. G, Firmenmarke 19, Made in Austria, Hand dec., Ausformung zw. 1945–1953, H. 33,5 cm; Wien Museum Inv. 190.047
8737 ohne Ang., Firmenmarke 19, H. 34,5 cm; Ahlden 70
8737F, H. 34 cm, B. 24 cm; GSNachf. 5,22
Lit.: GSNachf. (2. Ausg., bzw. Erg. Bl.) 5,22 (m. Abb.)
Quelle: Fotoarchiv K. Täuber

8738
Prischl, Adolf

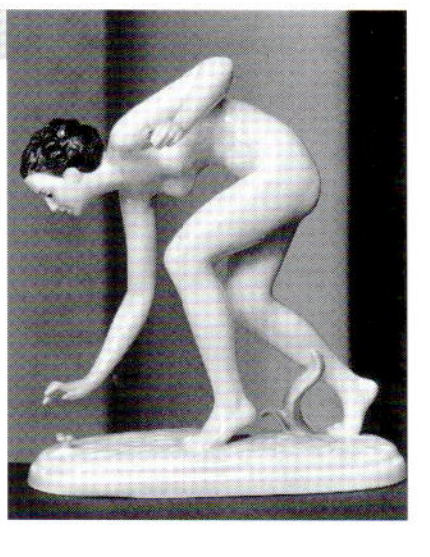

„Blumenpflückerin" (Sich bückender Damenakt) (auf Natursockel)
ca. 1941
8738,15 Firmenmarke 19a, unsign., H. 40 cm; Mehlis 31; eBay # 6514245053
8738,65 Firmenmarke 19, unsign., H. 41 cm; Doro 6.4.1999
–> vgl. 8685

8739F/M
Lorenzl, Josef

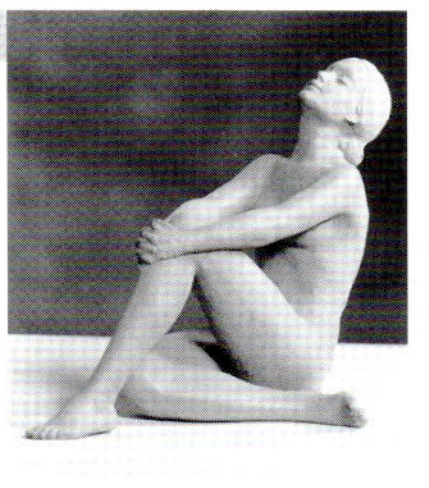

„Sinnende" (Sitzender Damenakt mit geschlossenen Augen)
ca. 1941
8739,11 Firmenmarke 19, Made in Austria, Ausformung zw. 1945–1953, H. 25 cm, B. 25 cm; Sammlung mad
8739,19,31 Firmenmarke Schuster 1, Made in Germany, unsign., Ausformung zw. 1942–1944, H. 25 cm; Doro 26.8.1998 (ohne Kat.)
8739 ohne Ang., H. 25 cm, B. 25 cm; Könecke (2002), S. 37, 78 (m. Abb.)
Bem.: Diese Figur wurde auch noch zw. 1953–1960 bei Carstens (GWM West-Germany) produziert.
Lit.: KGSF1; KGSF2; Könecke (2002), S. 37, 78 (m. Abb.), hier mit Ang. des (Original?-)Titels „Sinnende" und des Künstlers J. Lorenzl.
Quelle: AGM (Rechnung an Maurice Mizrahi, Alexandria, Ägypten, vom 22. April 1953), geliefert in Ausführung „matt"; Fotoarchiv K. Täuber.

8740
Dakon, Stephan

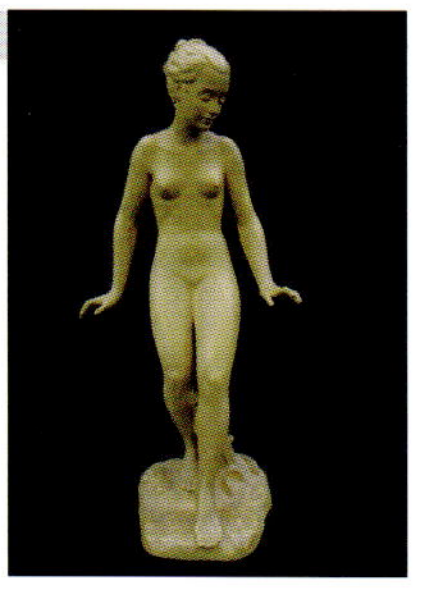

„Schreitende" (Damenakt) (auf Landschaftssockel)
ca. 1940 (für den Entwurf)
8740,13 Firmenmarke 19a, unsign., H. 54,5 cm; Sammlung Robert Senft
8740,67 Firmenmarke 19a, unsign., H. 55 cm; eBay # 1458464612
Bem.: Diese Figur, die von der Pose und der Oberflächenbehandlung her einem für Rosenthal entworfenen Akt von Fritz Klimsch ähnelt, wurde in der größeren Ausformung (81 cm) noch zw. 1953–1960 ausgeformt.
–> vgl. 8633
Lit.: Könecke (2002), S. 27, 77 (m. Abb.), hier für die Höhenvariante (81 cm) und mit Ang. des Künstlers S. Dakon.
Quelle: AGM (Kontoblatt der Wiener Manufaktur Josef Schuster für die Firma Ohlson jr., Malmö, Schweden vom 16. September 1942), hier mit Ang. des Originaltitels „Schreitende".

8742
„Pusztalied" (Ungarische Tänzerin in drehender Bewegung) (auf Rundsockel)
ca. 1941
8742,32,31 Firmenmarke 19, Made in Germany, Ausformung ca. 1941, H. 40 cm; eBay # 2243516948; 3325234661
Quelle: AGM (Bestellungen der Firma Wahliss (Wien) vom 9.11.1945; 24.11.1945 und 27.2.1946 an die „Arbeitsgemeinschaft der Firma Friedrich Goldscheider"), hier jeweils mit Ang. des (Original?-)Titels „Pusztalied" (falsch geschrieben „Pu(s)stalied").

8743F
Knörlein, Rudolf

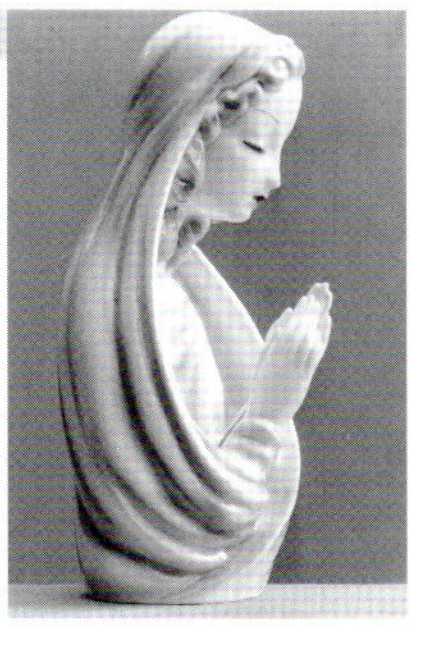

BÜSTE „Ave Maria" (Madonna mit gefalteten Händen)
ca. 1934/35 (für den Entwurf)
8743F, H. fehlt
–> vgl. 7188; 7263; 7858
Lit.: GSNachf. 6,5 (m. Abb.), hier mit Ang. des Originaltitels BÜSTE „Ave Maria".
Quelle: Fotoarchiv K. Täuber

8744
Knörlein, Rudolf

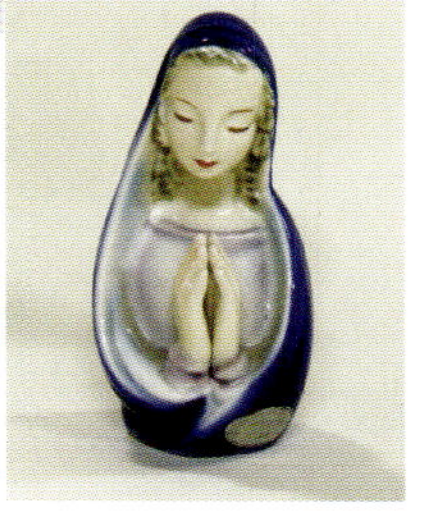

BÜSTE Madonna mit gefalteten Händen
ca. 1941/42
8744,X ger., Firmenmarke 19 (blau), unsign., Pick. 1, H. 11,1 cm; MAK Wien, Inv. Ke 10602-51
–> vgl. 7188; 7263; 7858; 8743= jew. ähnl. Mod.

8746
Dakon, Stephan

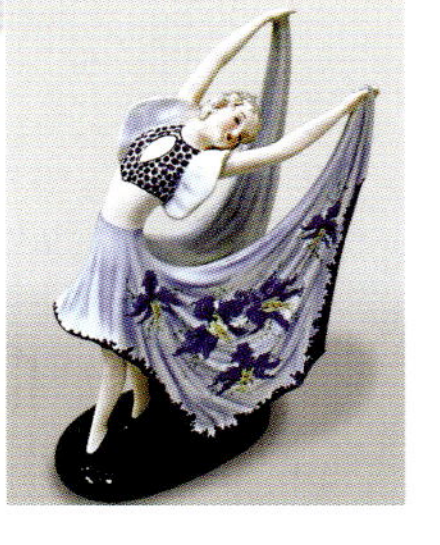

FIGUR „Beauty"/„Schönheit" (Tänzerin, nach links geneigt, ihr Kleid flügelartig zurückhaltend) (auf Ovalsockel)
ca. 1935 (für den Entwurf)
8746,48,7 (Firmenmarke ohne Ang.), unsign., H. 40 cm; city-antik
8746,73,7 Firmenmarke Schuster 1, Made in Germany, unsign., Ausformung zw. 1942–1944, H. ca. 40 cm; KH
Bem.: Dieses Modell ist die „züchtigere" Variante des Goldscheider Erfolgsmodelles „Beauty", welches in zwei Varianten auf den Markt kam. Einmal neigt sich der Körper der Tänzerin nach links, das andere Mal nach rechts. Die Modell-Nr. 8746 weist zusätzlich einige Ab-

wandlungen in Form einer Umgestaltung des Rockes, der sich vorne geschlossen zeigt, sowie an die Ärmel des Oberteils angebrachte Rüschen auf. Alleine dadurch wirkt diese Figur weniger frivol und ist somit dem Zeitgeschmack der 40er Jahre angepasst.
-> vgl. 7195; 7897; 7996
Lit.: Pinhas, S. 163 (m. Abb.)
Quelle: AGM (Bestellungen der Firma Wahliss (Wien) vom 9. und 24.11.1945 an die „Arbeitsgemeinschaft der Firma Friedrich Goldscheider", hier jeweils mit Ang. des Titels „Schönheit"; Bestellung der Firma Wahliss (Wien) vom 27.2.1946 in Ausführung „Terrakotta").

8748
„Südliche Musik" (Tänzerin in Drehung, auf einem Bein) (auf Rundsockel)
ca. 1942
8748,13 Firmenmarke 19a, Ausformung zw. 1945–1953, H. 45,5 cm; Wien Museum Inv. 186.616
8784,24,31 (Firmenmarke ohne Ang.), H. 47 cm; Doro 12.9.2006
Quelle: AGM (Bestellungen der Firma Wahliss (Wien) vom 17.10.1945, 24.11.1945 und 27.2.1946 an die „Arbeitsgemeinschaft der Firma Friedrich Goldscheider", hier jeweils mit Ang. des (Original?-)Titels „Südliche Musik").

8749F
Melcher, Josef / Nieschlag, Karl
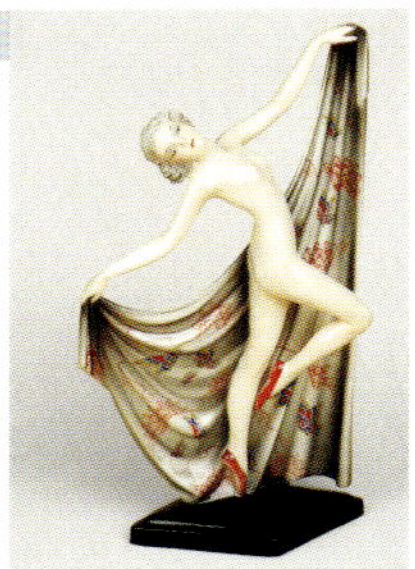
FIGUR „Nocturno" (Stehender Damenakt mit großem Tuch) (auf rautenförmigem Sockel)
ca. 1937/38 (für den Entwurf)
8749 ohne Ang., H. 40 cm; Könecke (2002), S. 78, 94 (m. Abb.)
Bem.: Dieses Modell wird bei Könecke (2002), der sich auf eine Preisliste (ca. 1956) bezieht, unter dem Künstler Melcher geführt. Ein signiertes Exemplar (in anderer Größe, Modell-Nr. 8174), kann für Nieschlag nachgewiesen werden. Einziges mit Melcher sign. Stück der Manufaktur ist ein stehender Mädchenakt (Modell-Nr. 8316). Welcher der beiden Entwerfer schlussendlich für das Modell „Nocturno" verantwortlich zeichnet, kann zum jetzigen Zeitpunkt nicht geklärt werden. Diese Figur wurde auch noch zw. 1953–1960 bei Carstens (GWM West-Germany) produziert.
-> vgl. 8174
Lit.: KGSF1; Könecke (2002), S. 78, 94 (m. Abb.), hier mit Ang. des Originaltitels „Nocturno" und des Künstlers Melcher.
Quelle: AGM (Rechnung der Wiener Manufaktur Josef Schuster an Laszlo Bösze, Szeged, Ungarn, vom 5. August 1944), hier mit Ang. des Originaltitels „Nocturno".
GS neu: Modell-Nr. 54 „Die Bewegung" (H. 35 cm), Aufl. 3000 Stück

8752
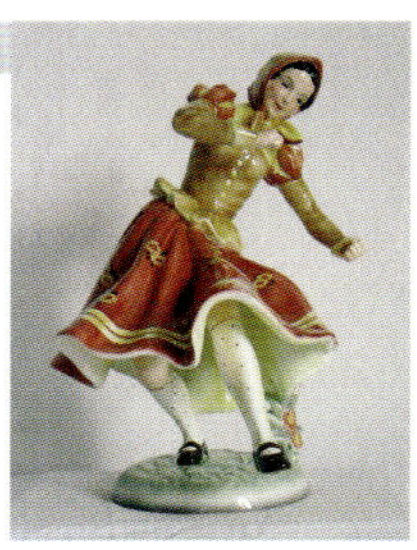
„Bauernpolka" (Tanzende Bäuerin auf rundem Landschaftssockel)
ca. 1942
8752,12,19 ger. O, gem. L8, Firmenmarke 19, Made in Austria, Ausformung zw. 1945–1953, H. 38 cm; Sammlung f@goldscheider.de
Quelle: AGM (Kontoblatt der Wiener Manufaktur Josef Schuster für die Firma Ohlson jr., Malmö, Schweden vom 16. September 1942), hier mit Ang. des Originaltitels „Bauernpolka".

8753F
Sailer, Karl (zugeschr.)
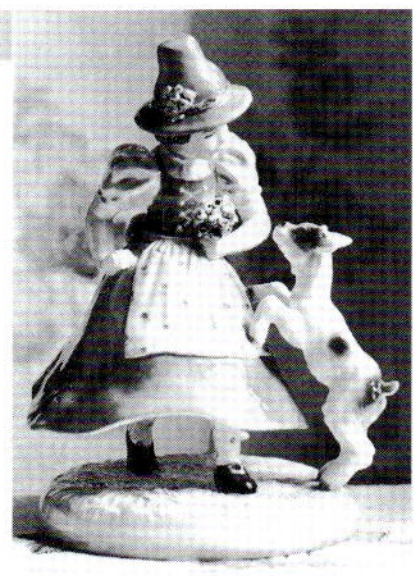
GRUPPE Mädchen in Tracht mit springendem Zicklein (auf Natursockel)
ca. 1942
8753,11,16 Firmenmarke 19, Made in Austria, Ausformung zw. 1945–1953, H. 21 cm; Wien Museum Inv. 190.161
8753,11,19 Firmenmarke 19, Made in Austria, Ausformung zw. 1945–1953, H. 22,9 cm; eBay # 1438468481
8753F, H. 21,5 cm, B. 17 cm; GSNachf. 1,71
Bem.: Im Stile einer „Hummel-Figur", -> vgl. Bem. bei Modell-Nr. 8567.
Lit.: GSNachf. (2. Ausg., bzw. Erg. Bl.) 1,71 (m. Abb.)
Quelle: Fotoarchiv K. Täuber.

8754M
„Träumende"
ca. 1942
Quelle: AGM (Kontoblatt der Wiener Manufaktur Josef Schuster für die Firma Ohlson jr., Malmö, Schweden vom 16. September 1942; Bestellungen der Firma Wahliss (Wien) vom 9. und 24.11.1945 an die „Arbeitsgemeinschaft der Firma Friedrich Goldscheider", hier jeweils mit Ang. des Originaltitels „Träumende"; Bestellung der Firma Wahliss (Wien) vom 27.2.1946 in Ausführung „Terrakotta").

8755M
„Der Wind" (Stehendes Mädchen im Wind, ihren Hut festhaltend) (auf Rundsockel)
ca. 1942
8755,73 Firmenmarke 19a, Ausformung zw. 1945–1953, H. 22,5 cm; Mehlis 31
Quelle: AGM (Kontoblatt der Wiener Manufaktur Josef Schuster für die Firma Porslins Ericsson-Lund, Stockholm?, Schweden vom 7. Januar 1943), hier mit Ang. des Originaltitels „Der Wind".

8756F
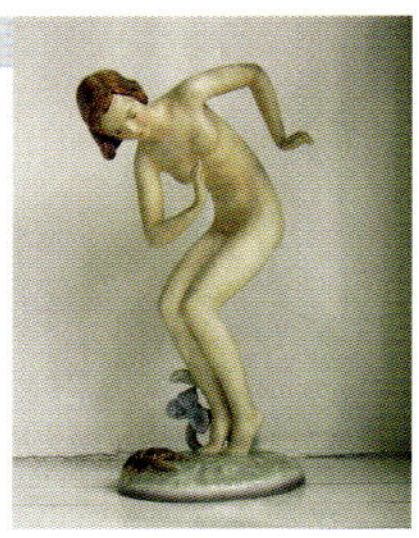
„Hirschkäfer" (Stehender Damenakt mit Hirschkäfer) (auf Landschaftssockel)
ca. 1942
8756,4 gem. G, gem. lig. SP, 20, Made in Austria, Hand dec., Ausformung zw. 1945–1953, H. 39,2 cm, B. 17 cm; Privatsammlung, Wien
8756,5 Firmenmarke 19, Made in Austria, Hand dec., Ausformung zw. 1945–1953, H. 39 cm; Doro 18.5.1999
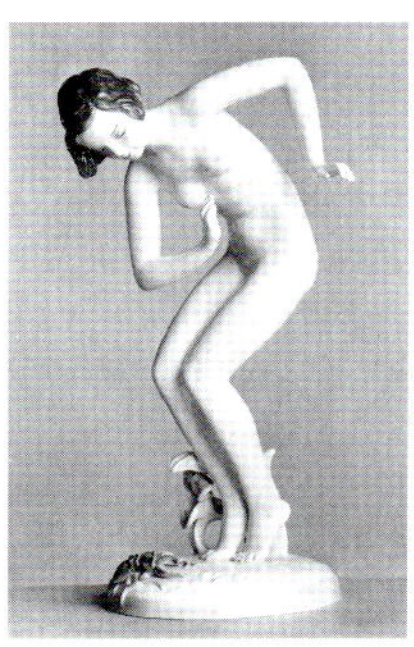
8756,27,9 Firmenmarke 19, Ausformung zw. 1945–1953, H. 39,3 cm; Doro 25.4.2000
8756 H. 39 cm, B. 19 cm; Fotoarchiv K. Täuber
-> vgl. 8770
Quelle: AGM (Rechnung an Maurice Mizrahi, Alexandria, Ägypten, vom 22. April 1953), hier mit Ang. des (Original)-Titels „Hirschkäfer", geliefert in Ausführung „matt"); Fotoarchiv K. Täuber.

8758F
Lorenzl, Josef
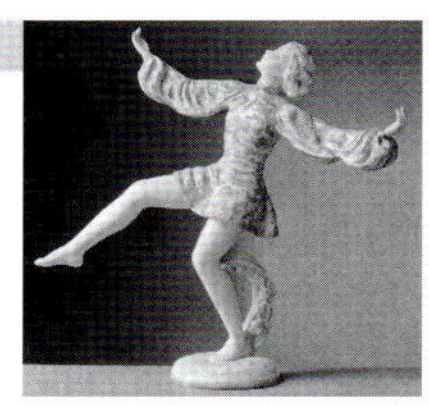
„Scherzo" (Tänzerin auf einem Bein) (auf Rundsockel)
ca. 1942
8758,11 (Firmenmarke ohne Ang.), H. fehlt; Soth. South, Billingshurst 11.10.2000
Quelle: AGM (Bestellung der Firma Wahliss (Wien) vom 27.2.1946 an die „Arbeitsgemeinschaft der Firma Friedrich Goldscheider", hier mit Ang. des (Original?-)Titels „Scherzo" in Ausführung „Terrakotta"; Rechnung an Maurice Mizrahi, Alexandria, Ägypten, vom 22. April 1953), hier unter dem Titel „Tanz" geführt.

8760
Spielendes Mädchen (mit einem kegelförmigen Schlagkreisel und Stab mit Schnur) (auf Rundsockel)
ca. 1942
8760,2,31 gem. lig. SF., gem. rot XX,8, Firmenmarke Schuster 2, Made in Germany, unsign., Ausformung zw. 1942–1944, H. fehlt; eBay # 2243528216

8761F
„Tanz"
ca. 1942
Quelle: AGM (Rechnung an Maurice Mizrahi, Alexandria, Ägypten, vom 22. April 1953), hier unter dem Titel „Tanz" geführt.

8762F
„Tanz"
ca. 1942
Quelle: AGM (Rechnung an Maurice Mizrahi, Alexandria, Ägypten, vom 22. April 1953), hier unter dem Titel „Tanz" geführt.

8763F
„Tanz"
ca. 1942
Quelle: AGM (Rechnung an Maurice Mizrahi, Alexandria, Ägypten, vom 22. April 1953), hier unter dem Titel „Tanz" geführt.

8764
„Blumenbinderin" (Sitzendes Mädchen mit Blumen)
ca. 1942
8764,6 Firmenmarke 19a, H. 28 cm; Weidler 285; eBay # 330032010
Quelle: AGM (Bestellung der Firma Wahliss (Wien) vom 27.2.1946, hier mit Ang. des (Original?-)Titels „Blumenbinderin" in Ausführung „Terrakotta").

8765M
„Der Strumpf" (Laufendes Mädchen) (auf Rundsockel)
ca. 1942
8765,3,19 Firmenmarke 19a, Ausformung zw. 1945–1953, H. 22 cm; eBay # 507609313
8765 (gepr.), 8765 ger.,11,13 Firmenmarke 19, Made in Germany (Z 15), Ausformung ca. 1942, H. 22 cm; Wien Museum Inv. 219.176
Quelle: AGM (Bestellung der Firma Wahliss (Wien), hier mit Ang. des (Original?-)Titels „Der Strumpf").

8769
Laufendes Mädchen mit Haube (auf rundem Natursockel)
ca. 1942
8769,42,7 gem. G, gem. rot XX, 34, Firmenmarke Schuster 1, Made in Germany, Ausformung zw. 1942–1944, H. 24 cm; Wien Museum Inv. 220.744

8770
„Hirschkäfer" (Stehender Damenakt mit Hirschkäfer) (auf Landschaftssockel)
ca. 1942
8770,15,31 gem. G, gem. rot XX, 37, Firmenmarke Schuster 1, Made in Germany, unsign., Ausformung zw. 1942–1944, H. 31,5 cm; Wien Museum Inv. 190.126
Bem.: Diese Figur wurde auch noch zw. 1953–1960 bei Carstens (GWM West-Germany) produziert.
Ausst.: SSH 1992
-> vgl. 8756
Lit.: SS, S. 12, 13 (m. Abb.); Könecke (2002), S. 78, 94 (m. Abb.).

8772
Knabe mit Vogel
ca. 1942
8772 ohne Ang., Firmenmarke Schuster 1?, H. 21 cm; Ahlden 54/55

8778
„Ruhendes Mädchen"
ca. 1942/43
Quelle: AGM (Bestellungen der Firma Wahliss (Wien) vom 9.11.1945 und 27.2.1946 an die „Arbeitsgemeinschaft der Firma Friedrich Goldscheider", hier jeweils mit Ang. des (Original?-)Titels „Ruhendes Mädchen" in Ausführung „Terrakotta").

8780
Dakon, Stephan (zugeschr.)
GRUPPE Kniende Mutter mit Kind
ca. 1942/43
8780,11,9 Firmenmarke 19, Made in Austria, unsign., Ausformung zw. 1945–1953, H. 18 cm;
eBay # 1438746959
-> vgl. 8621= ähnl. Modell

8782
Tanzender Bub in Tracht (Tiroler) (auf Rundsockel mit Bäumchen)
ca. 1942/43
8782,11 gem. H., M., Firmenmarke 19, Made in Austria, Ausformung zw. 1945–1953, H. 22,9 cm;
eBay # 150001093105

8783
„Altdeutscher Tanz"
ca. 1942/43
Quelle: AGM (Bestellungen der Firma Wahliss (Wien) vom 9. und 24.11.1945 an die „Arbeitsgemeinschaft der Firma Friedrich Goldscheider", hier jeweils mit Ang. des (Original?-)Titels „Altdeutscher Tanz"; Bestellung der Firma Wahliss (Wien) vom 27.2.1946 in Ausführung „Terrakotta").

8784
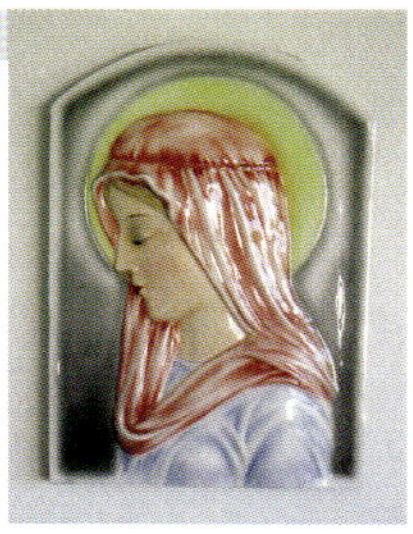
RELIEF Madonna mit Kopftuch und Aureole (im Profil nach rechts) (Wandrelief)
ca. 1942/43
8784, 31 ger., gem. G, O, S, Firmenmarke 19, Made in Austria, Ausformung zw. 1945–1953, H. 33 cm, B. 24 cm; Sammlung f@goldscheider.de

8785
„Leopard" (fauchend)
ca. 1942/43
8785,11 Firmenmarke 19, Made in Austria, Ausformung zw. 1945–1953, H. 14,5 cm, L. 44 cm;
eBay # 220002136028
8785 ohne Ang., H. 13 cm, B. 48 cm; Könecke (2002), S. 78, 94 (m. Abb.)
Bem.: Diese Figur wurde auch noch zw. 1953–1960 bei Carstens (GWM West-Germany) produziert.
Lit.: Könecke (2002), S. 78, 94 (m. Abb.).
Quelle: AGM (Bestellung der Firma Wahliss (Wien), hier mit Ang. des (Original?-)Titels „Leopard").

8792
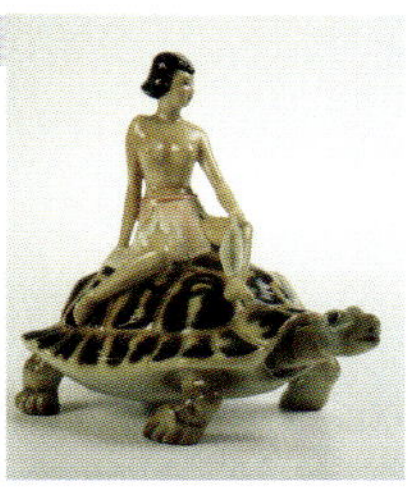
Revuegirl mit Feder, auf Schildkröte reitend/sitzend
ca. 1942/43
8792,11 Firmenmarke 19, Made in Austria, Ausformung zw. 1945–1953, H. 22 cm; Wien Museum Inv. 190.098
Ausst.: SSH 1992

8793
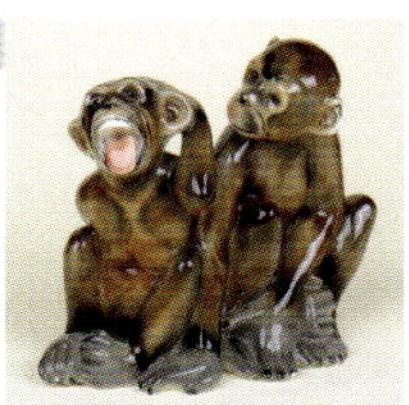
Zwei sitzende Affen
ca. 1942/43
8793,6,25 Firmenmarke 19, Made in Austria, H. 16 cm;
eBay # 250053861040
8793,93,S Firmenmarke 19, Made in Austria, Hand dec., Ausformung zw. 1945–1953, H. 16 cm; eBay # 1303351255
8793 ger. AU, Firmenmarke 19, Made in Austria, Originalpick.: „Rasper & Söhne, Wien", Ausformung zw. 1945–1953, H. 17 cm; PB Wien 19

8796
LAMPE „Akt als Lampe" (Auf einem Bein tanzender Damenhalbakt) (auf Rautensockel)
ca. 1942/43
8796,83,66 gem. G, gem. lig. SP., Firmenmarke 19, Made in Austria, Hand dec., Ausformung zw. 1945–1953, H. 48 cm; Wien Museum Inv. 219.195
Quelle: AGM (Rechnung an Maurice Mizrahi, Alexandria, Ägypten, vom 22. April 1953), hier mit Ang. des (Original)-Titels „Akt als Lampe".

8805
Sailer, Karl
„Skibaby" (Kleine Skifahrerin mit Startnummer „3")
ca. 1948
8805,2,AU ger., gem. H., Firmenmarke 19, Made in Austria, unsign., Ausformung zw. 1948–1953, H. 19 cm; Wien Museum Inv. 220.745
8805,10,I Firmenmarke 19, Made in Austria, unsign., Ausformung zw. 1948–1953, H. 19 cm; eBay # 1141966723
Quelle: AGM (Zession, d. i. der Übertrag der Urheberrechte zwischen der Wiener Manufaktur Friedrich Goldscheider und Karl Sailer vom 25.8.1948), hier mit Ang. des Originaltitels „Skibaby".

8806F
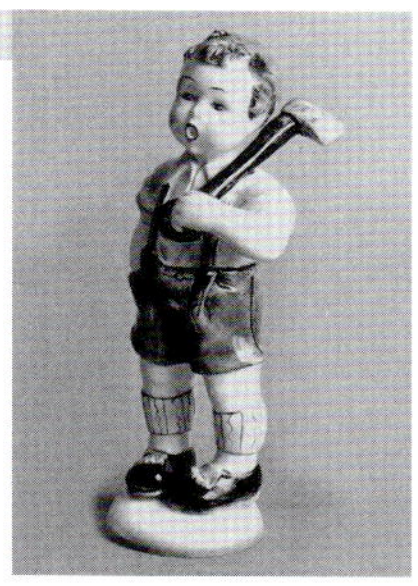
Sailer, Karl
„Holzhackerbub" (Stehender singender Bub mit kurzer Lederhose, eine Axt geschultert) (auf Rundsockel)
ca. 1948
8806F, H. 17,5 cm, B. 8 cm; GSNachf. 1,74b
Bem.: Im Stile einer „Hummel-Figur", -> vgl. Bem. bei Modell-Nr. 8567.
Pendant -> vgl. 8818F
Lit.: GSNachf. (2. Ausg., bzw. Erg. Bl.) 1,74 (m. Abb.)
Quelle: AGM (Zession, d. i. der Übertrag der Urheberrechte zwischen der Wiener Manufaktur Friedrich Goldscheider und Karl Sailer vom 25.8.1948; Tantiemenliste Sailer, ohne Datierung (ca. 1950); Tantiemenabrechnung 2. Halbjahr 1952), hier jeweils mit Ang. des Originaltitels „Holzhackerbub"; Fotoarchiv K. Täuber.

8807
Sailer, Karl
„Fohlen" (springend)
ca. 1948
Quelle: AGM (Zession, d. i. der Übertrag der Urheberrechte zwischen der Wiener Manufaktur Friedrich Goldscheider und Karl Sailer vom 25.8.1948), hier mit Ang. des Originaltitels „Fohlen" (springend).

8808
Sailer, Karl
„Fohlen" (stehend)
ca. 1948
Quelle: AGM (Zession, d. i. der Übertrag der Urheberrechte zwischen der Wiener Manufaktur Friedrich Goldscheider und Karl Sailer vom 25.8.1948), hier mit Ang. des Originaltitels „Fohlen" (stehend).

8809
Sailer, Karl
„Fohlengruppe"
ca. 1948
Quelle: AGM (Zession, d. i. der Übertrag der Urheberrechte zwischen der Wiener Manufaktur Friedrich Goldscheider und Karl Sailer vom 25.8.1948), hier mit Ang. des Originaltitels „Fohlengruppe".

8810F
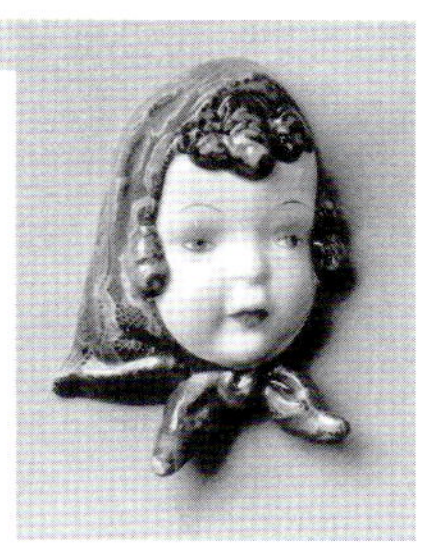
Sailer, Karl
WANDMASKE „Mädi" (Mädchen mit Kopftuch)
ca. 1948
8810,13,0 ger., gem. MR, Firmenmarke 19, H. 9,5 cm; MAK Wien, Inv. Ke 10521-36
8810 Firmenmarke 19, Made in Austria, unsign., Ausformung zw. 1948–1953, H. 10,2 cm; eBay # 126986162
ohne Nr., Firmenstempel WG 1, Ausformung 50er Jahre, H. 10,5 cm; Wien Museum Inv. 220.813
8810F, H. 10 cm, B. 7 cm; GSNachf. 2,29e
Bem.: Diese Maske wurde auch noch zw. 1953–1960 bei Carstens (GWM West-Germany) produziert.
Pendant -> vgl. 8811F
Lit.: GSNachf. (2. Ausg., bzw. Erg. Bl.) 2,29 (m. Abb.); KGSF1; KGSF2; Könecke (2002), S. 57, 78 (m. Abb.), hier mit Ang. des Künstlers Sailer.
Quelle: AGM (Zession, d. i. der Übertrag der Urheberrechte zwischen der Wiener Manufaktur Friedrich Goldscheider und Karl Sailer vom 25.8.1948; Tantiemenliste Sailer, ohne Datierung (ca. 1950); Bestellung von Ferd. Lundquist, Göteborg, Schweden vom 23. August 1950; Tantiemenabrechnung 2. Halbjahr 1952), hier jeweils mit Ang. des Originaltitels „Maske Mädi"; Fotoarchiv K. Täuber.

8811F
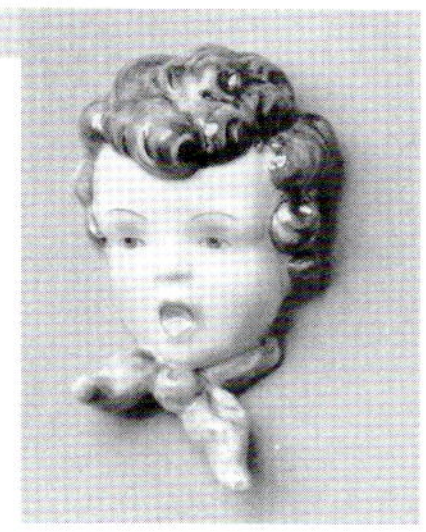
Sailer, Karl
WANDMASKE „Bubi" (Knabe mit Halstuch)
ca. 1948
8811,14,24 Firmenmarke 19, Made in Austria, Hand dec., Ausformung zw. 1948–1953, H. 10 cm; WKH
8811 ger., Firmenmarke 19, Made in Austria, Hand dec., unsign., Ausformung zw. 1948–1953, H. 10,8 cm; eBay # 434409590
ohne Nr., Firmenstempel WG 1, Ausformung 50er Jahre, H. 10,2 cm; Wien Museum Inv. 220.814
8811F, H. 10 cm, B. 7 cm; GSNachf. 2,29f
Bem.: Diese Maske wurde auch noch zw. 1953–1960 bei Carstens (GWM West-Germany) produziert. Wie das Originalfoto der Manufaktur zeigt, wurde hier die falsche Modell-Nr. 8815 eingedruckt.
Pendant -> vgl. 8810F
Lit.: GSNachf. (2. Ausg., bzw. Erg. Bl.) 2,29 (m. Abb.); KGSF1; KGSF2; Könecke (2002), S. 57, 78 (m. Abb.), hier mit Ang. des Künstlers Sailer.
Quelle: AGM (Zession, d. i. der Übertrag der Urheberrechte zwischen der Wiener Manufaktur Friedrich Goldscheider und Karl Sailer vom 25.8.1948; Tantiemenliste Sailer, ohne Datierung (ca. 1950); Bestellung von Ferd. Lundquist, Göteborg, Schweden vom 23. August 1950; Tantiemenabrechnung 2. Halbjahr 1952), hier jeweils mit Ang. des Originaltitels „Maske Bubi"; Fotoarchiv K. Täuber.

8812F
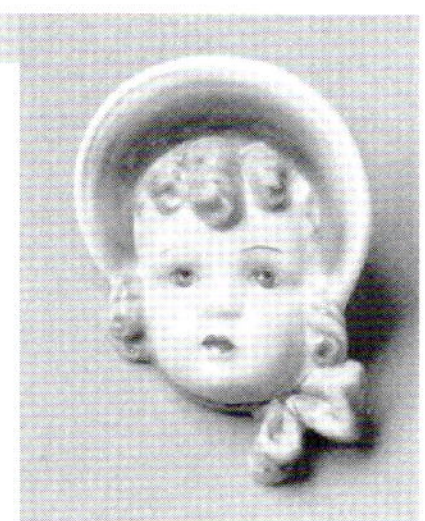
Sailer, Karl
WANDMASKE „Alt Wien" („Mäderl") (Mädchenkopf mit Schutenhut)
ca. 1948
8812,12 Firmenstempel WG 1a, Ausformung 50er Jahre, H. 7 cm; Wien Museum Inv. 220.809
8812F, H. 8 cm, B. 6 cm; GSNachf. 2,29a
Bem.: Diese Maske wurde auch noch zw. 1953–1960 bei Carstens (GWM West-Germany) produziert.
Pendant -> vgl. 8813F
Lit.: GSNachf. (2. Ausg., bzw. Erg. Bl.) 2,29 (m. Abb.); KGSF1; Könecke (2002), S. 78, 94 (m. Abb.), hier mit Ang. des Künstlers Sailer.
Quelle: AGM (Tantiemenliste Sailer, ohne Datierung (ca. 1950), hier mit Ang. des Titels „Mäderl"; Tantiemenabrechnung 2. Halbjahr 1952), hier mit Ang. des Originaltitels „Maske Alt Wien"; Fotoarchiv K. Täuber.

8813F
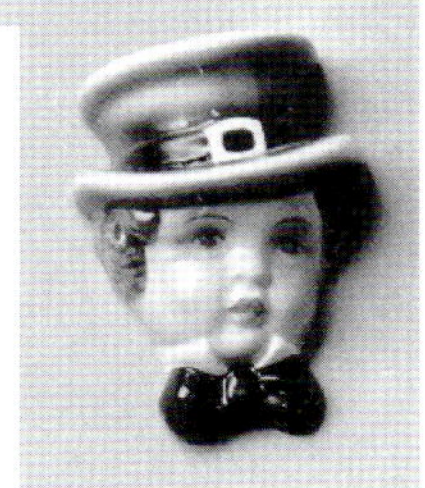
Sailer, Karl
WANDMASKE „Alt Wien" („Buberl") (Knabenkopf mit Biedermeierhut/Zylinder)
ca. 1948
8813,10 ger., gem. B, Firmenmarke 19, Made in Austria, Hand dec., unsign., Ausformung zw. 1948–1953, H. 7,5 cm; eBay # 106677603

8813,12 Firmenstempel WG 1a, Ausformung 50er Jahre, H. 7,3 cm; Wien Museum Inv. 220.810
8813F, H. 8 cm, B. 6 cm; GSNachf. 2,29b
Bem.: Diese Maske wurde auch noch zw. 1953–1960 bei Carstens (GWM West-Germany) produziert.
Pendant –> vgl. 8812F
Lit.: GSNachf. (2. Ausg., bzw. Erg. Bl.) 2,29 (m. Abb.); KGSF1; Könecke (2002), S. 78, 94 (m. Abb.), hier mit Ang. des Künstlers Sailer.
Quelle: AGM (Tantiemenliste Sailer, ohne Datierung (ca. 1950), hier mit Ang. des Titels „Buberl"; Tantiemenabrechnung 2. Halbjahr 1952), hier mit Ang. des Originaltitels „Maske Alt Wien"; Fotoarchiv K. Täuber.

8814F

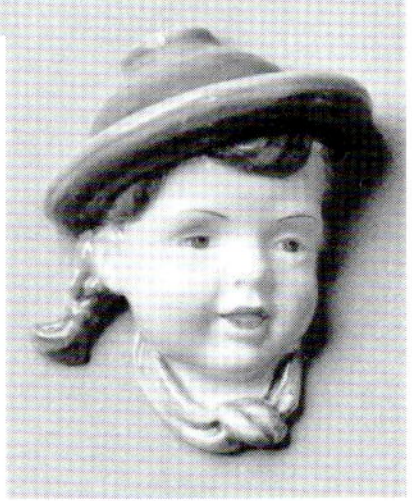

Sailer, Karl
WANDMASKE „Mäderl" („Maske Tirol") (Mädchen mit Trachtenhut, Tirolerin)
ca. 1948
8814,2 gem. RS, Firmenmarke 19, Made in Austria, Hand dec., unsign., Ausformung zw. 1948–1953, H. 10,2 cm; eBay # 46263772
8814,17,44 Firmenmarke 19, Made in Austria, unsign., Ausformung zw. 1948–1953, H. 10 cm; Wien Museum Inv. 186.654
8814,23,X ger., gem. lig. JH, M.G., Firmenmarke 19 (blau), Made in Austria (blau), Hand dec. (blau), unsign., Ausformung zw. 1948–1953, H. ca. 10 cm; eBay # 407677272
8814F, H. 10 cm, B. 7 cm; GSNachf. 2,29c
Bem.: Diese Maske wurde auch noch zw. 1953–1960 bei Carstens (GWM West-Germany) produziert.
Pendant –> vgl. 8815F
Lit.: GSNachf. (2. Ausg., bzw. Erg. Bl.) 2,29 (m. Abb.); KGSF1; KGSF2; Könecke (2002), S. 56, 78 (m. Abb.), hier mit Ang. des Künstlers Sailer.
Quelle: AGM (Tantiemenliste Sailer, ohne Datierung (ca. 1950), hier mit Ang. des Titels „Mäderl"; Bestellung von Ferd. Lundquist, Göteborg, Schweden vom 23. August 1950; Tantiemenabrechnung 2. Halbjahr 1952), hier jeweils mit Ang. des Originaltitels „Maske Tirol"; Fotoarchiv K. Täuber.

8815F

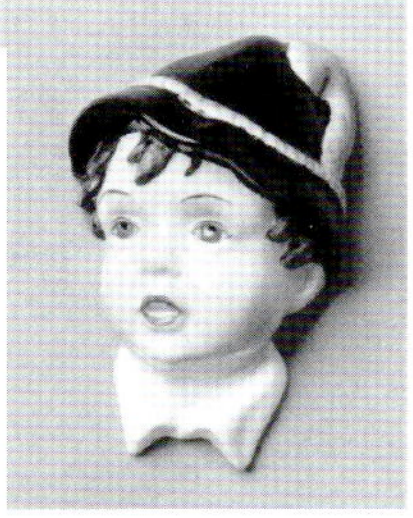

Sailer, Karl
WANDMASKE „Buberl" („Maske Tirol") (Knabe mit Trachtenhut, Tiroler)
ca. 1948
8815,2 gem. RS, Firmenmarke 19, Made in Austria, Hand dec., unsign., Ausformung zw. 1948–1953, H. 10,2 cm; eBay # 46263772
8815,10,23 ger., Firmenmarke 19 (blau), Made in Austria (blau), Hand dec. (blau), unsign., Pick. 1, Ausformung zw. 1948–1953, H. 10,2 cm; eBay # 272966551
8815F, H. 10 cm, B. 7 cm; GSNachf. 2,29d
Bem.: Diese Maske wurde auch noch zw. 1953–1960 bei Carstens (GWM West-Germany) produziert.
Pendant –> vgl. 8814F
Lit.: GSNachf. (2. Ausg., bzw. Erg. Bl.) 2,29 (m. Abb.); KGSF1; KGSF2; Könecke (2002), S. 57, 78 (m. Abb.), hier mit Ang. des Künstlers Sailer.
Quelle: AGM (Tantiemenliste Sailer, ohne Datierung (ca. 1950), hier mit Ang. des Titels „Buberl"; Bestellung von Ferd. Lundquist, Göteborg, Schweden vom 23. August 1950; Tantiemenabrechnung 2. Halbjahr 1952), hier jeweils mit Ang. des Originaltitels „Maske Tirol"; Fotoarchiv K. Täuber.

8816F

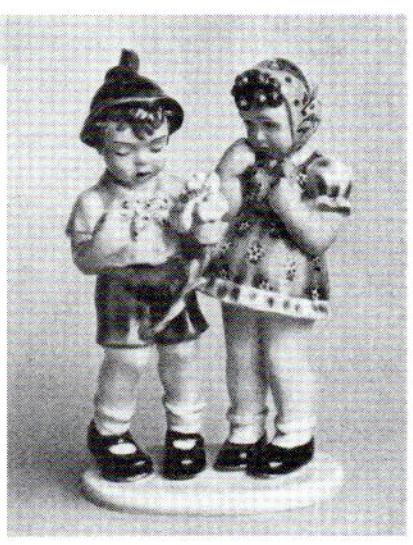

Sailer, Karl
„Sie liebt mich" (Stehender Bub mit Trachtenhut, einer Blume ihre Blütenblätter ausreißend) (auf Rundsockel)
ca. 1948
8816,66 gem. S, gem. lig. HS, Firmenmarke 19, Made in Austria, unsign., Ausformung zw. 1948–1953, H. 21,5 cm; Wien Museum Inv. 190.158
8816F, H. 22,5 cm, B. 8 cm; GSNachf. 1,73b
Bem.: Im Stile einer „Hummel-Figur", –> vgl. Bem. bei Modell-Nr. 8567.
–> vgl. 8821= GRUPPE
Pendant –> vgl. 8817
Lit.: GSNachf. (2. Ausg., bzw. Erg. Bl.) 1,73 (m. Abb.)
Quelle: AGM (Tantiemenliste Sailer, ohne Datierung (ca. 1950); Tantiemenabrechnung 2. Halbjahr 1952), hier mit Ang. des Originaltitels „Sie liebt mich".

8817F

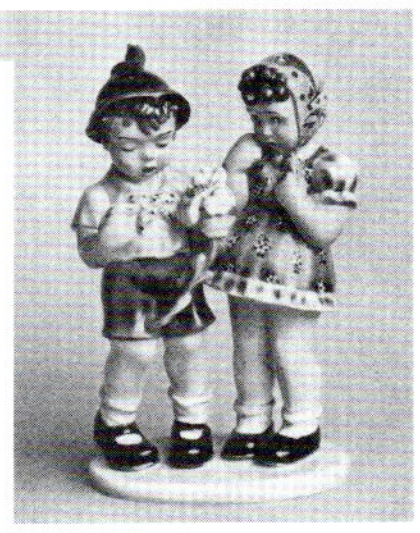

Sailer, Karl
„Liebeskummer" (Stehendes ängstliches Mädchen mit Kopftuch) (auf Rundsockel)
ca. 1948
8817F, H. 22,5 cm, B. 8 cm; GSNachf. 1,73b
Bem.: Im Stile einer „Hummel-Figur", –> vgl. Bem. bei Modell-Nr. 8567.
–> vgl. 8821= GRUPPE
Pendant –> vgl. 8816
Lit.: GSNachf. (2. Ausg., bzw. Erg. Bl.) 1,73 (m. Abb.)
Quelle: AGM (Tantiemenliste Sailer, ohne Datierung (ca. 1950); Tantiemenabrechnung 2. Halbjahr 1952), hier mit Ang. des Originaltitels „Liebeskummer".

8818F

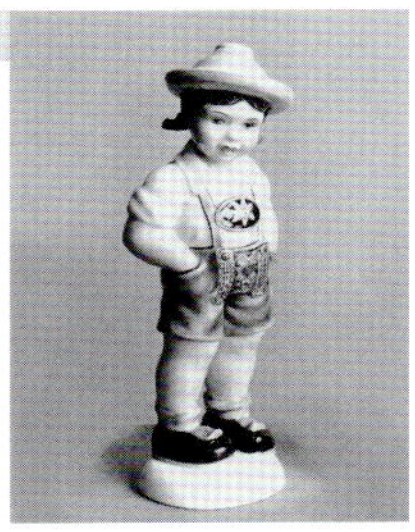

Sailer, Karl
„Mariandl" (Stehendes Mädchen mit Trachtenhut und kurzer Lederhose, Tirolerin) (auf Rundsockel)
ca. 1948
8818F, H. 17,5 cm, B. 8 cm; GSNachf. 1,74a
Bem.: Im Stile einer „Hummel-Figur", –> vgl. Bem. bei Modell-Nr. 8567.
Pendant –> vgl. 8806
Lit.: GSNachf. (2. Ausg., bzw. Erg. Bl.) 1,74 (m. Abb.)
Quelle: AGM (Zession, d. i. der Übertrag der Urheberrechte zwischen der Wiener Manufaktur Friedrich Goldscheider und Karl Sailer vom 25.8.1948; Tantiemenliste Sailer, ohne Datierung (ca. 1950); Tantiemenabrechnung 2. Halbjahr 1952), hier mit Ang. des Originaltitels „Mariandl"; Fotoarchiv K. Täuber.

8819

Sailer, Karl
„Foxterrier" (stehend)
ca. 1948
Quelle: AGM (Zession, d. i. der Übertrag der Urheberrechte zwischen der Wiener Manufaktur Friedrich Goldscheider und Karl Sailer vom 25.8.1948), hier mit Ang. des Originaltitels „Foxterrier" (stehend).

8821F

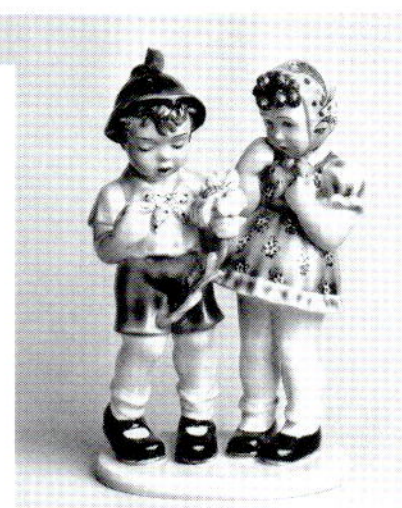

Sailer, Karl
GRUPPE „Sie liebt mich" (Stehendes Kinderpaar) (Bub, einer Blume ihre Blütenblätter ausreißend, daneben ein ängstliches Mädchen) (auf länglichem Ovalsockel)
ca. 1948
8821F, H. 22,5 cm, B. 15 cm; GSNachf. 1,73a
–> vgl. 8816; 8817= jeweils EinzelFIGUR
Lit.: GSNachf. (2. Ausg., bzw. Erg. Bl.) 1,73 (m. Abb.)
Quelle: AGM (Zession, d. i. der Übertrag der Urheberrechte zwischen der Wiener Manufaktur Friedrich Goldscheider und Karl Sailer vom 25.8.1948; Tantiemenliste Sailer, ohne Datierung (ca. 1950); Bestellung von Ferd. Lundquist, Göteborg, Schweden vom 23. August 1950; Tantiemenabrechnung 2. Halbjahr 1952), hier jeweils mit Ang. des Originaltitels GRUPPE „Sie liebt mich"; Fotoarchiv K. Täuber.

8822F

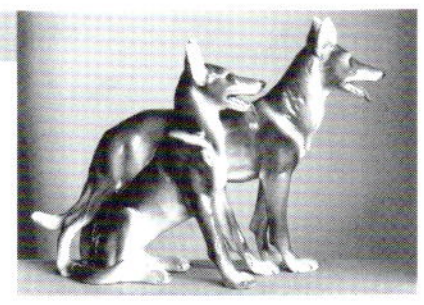

GRUPPE „Wolfshunde" (Zwei Schäferhunde, einer sitzend, der andere stehend)
ca. 1930 (für den Entwurf), ca. 1948 (für die Ausführung)
8822,3,6,2 Firmenmarke 19, Made in Austria, Ausformung zw. 1948–1953, H. 25,5 cm, L. 31 cm; eBay # 3312578610
8822F, H. 28 cm, B. 41 cm; GSNachf. 5,28b
–> vgl. 6128
Lit.: GSNachf. 5,28 (m. Abb.)
Quelle: Fotoarchiv K. Täuber

8823F

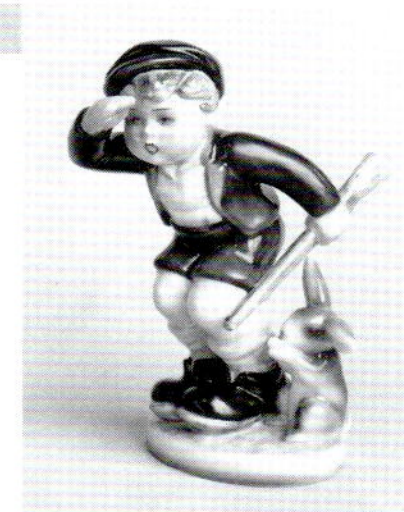

Sailer, Karl
„Waidmanns Heil" („Jäger") (Spähender Bub als Jäger mit Gewehr) (auf Natursockel mit kleinem Hasen)
ca. 1949/50
8823F, H. 15 cm, B. 8 cm; GSNachf. 1,75
Bem.: Im Stile einer „Hummel-Figur", –> vgl. Bem. bei Modell-Nr. 8567.
Lit.: GSNachf. (2. Ausg., bzw. Erg. Bl.) 1,75 (m. Abb.)
Quelle: AGM (Tantiemenliste Sailer, ohne Datierung (ca. 1950); Zession, d. i. der Übertrag der Urheberrechte zwischen der Wiener Manufaktur Friedrich Goldscheider und Karl Sailer vom 13.2.1950), hier jeweils mit Ang. des Originaltitels „Waidmanns Heil"; Bestellung von Ferd. Lundquist, Göteborg, Schweden vom 23. August 1950; Tantiemenabrechnung 2. Halbjahr 1952), hier jeweils unter dem Titel „Jäger" geführt; Fotoarchiv K. Täuber.

8824F

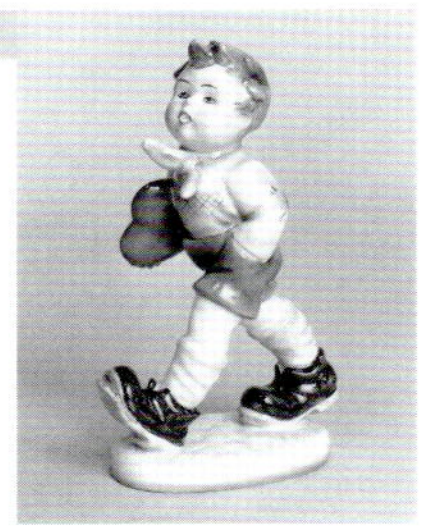

Sailer, Karl
„Herzensdieb" (Schreitender Bub mit Herz unter dem Arm) (auf ovalem Natursockel)
ca. 1949/50
8824,17,41 Firmenmarke 19, unsign., Pick. 1, H. 15,5 cm; Wien Museum Inv. 190.160
8824,17 gem. M1, Firmenmarke 19, Made in Austria, Hand dec., unsign., Pick. 1, H. 15,8 cm; eBay # 3714861672
8824,17 ger., gem. S., gem. lig. HS, Firmenmarke 19, unsign., Pick. 1, H. 15,5 cm; eBay # 3724397685
8824,47,45 Firmenmarke 19, Made in Austria, Hand dec., unsign., Originalpick.: „Karl Skala, XX., Wallensteinstrasse, Wien", H. 15 cm; KH
8824F, H. 15 cm, B. 9 cm; GSNachf. 1,76
Bem.: Im Stile einer „Hummel-Figur", –> vgl. Bem. bei Modell-Nr. 8567.
Lit.: GSNachf. (2. Ausg., bzw. Erg. Bl.) 1,76 (m. Abb.)
Quelle: AGM (Tantiemenliste Sailer, ohne Datierung (ca. 1950); Zession, d. i. der Übertrag der Urheberrechte zwischen der Wiener Manufaktur Friedrich Goldscheider und Karl Sailer vom 13.2.1950; Bestellung von Ferd. Lundquist, Göteborg, Schweden vom 23. August 1950; Tantiemenabrechnung 2. Halbjahr 1952), hier mit Ang. des Originaltitels „Herzensdieb"; Fotoarchiv K. Täuber.

8825F

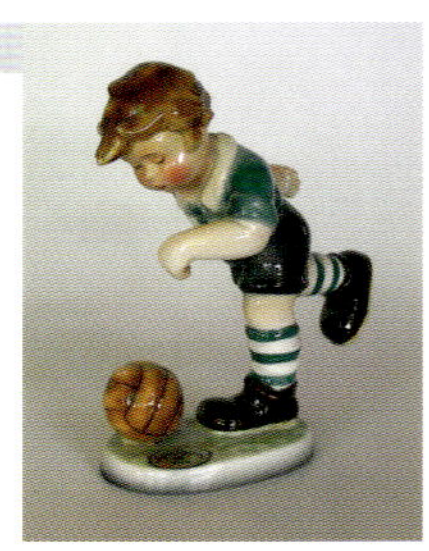

Sailer, Karl
„Fußballer" (Fußball spielender Bub) (auf Natursockel / Ovalsockel)
ca. 1949/50
8825,11,39 Firmenmarke 19, unsign., Pick. 6, H. 14,5 cm; Sammlung.schmutz@chello.at 16
8825F, H. 14,5 cm, B. 12,5 cm; GSNachf. 1,77
Bem.: Im Stile einer „Hummel-Figur", –> vgl. Bem. bei Modell-Nr. 8567.

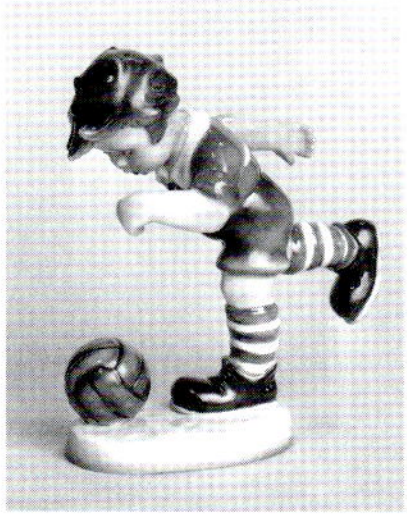

Lit.: GSNachf. (2. Ausg., bzw. Erg. Bl.) 1,77 (m. Abb.)
Quelle: AGM (Tantiemenliste Sailer, ohne Datierung (ca. 1950); Zession, d. i. der Übertrag der Urheberrechte zwischen der Wiener Manufaktur Friedrich Goldscheider und Karl Sailer vom 13.2.1950; Tantiemenabrechnung 2. Halbjahr 1952), hier jeweils mit Ang. des Originaltitels „Fußballer"; Fotoarchiv K. Täuber.

8826F

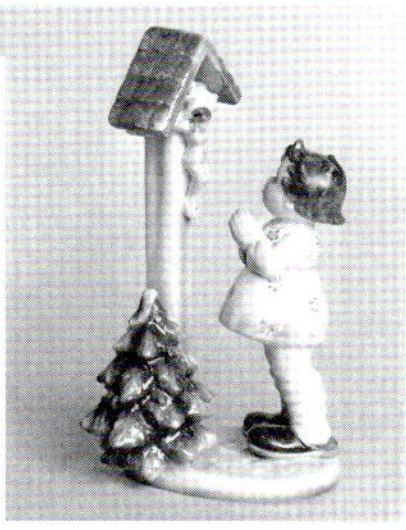

Sailer, Karl
„Waldandacht" („Marterl") (Stehendes Mädchen, vor einem Marterl betend) (auf Natursockel mit kleinem Tannenbaum)
ca. 1949/50
8826,3,12 Firmenmarke 19, Made in Austria, unsign., H. 18,5 cm; eBay # 934613331
8826F, H. 19 cm, B. 10,5 cm; GSNachf. 1,78
Bem.: Im Stile einer „Hummel-Figur", –> vgl. Bem. bei Modell-Nr. 8567.
Lit.: GSNachf. (2. Ausg., bzw. Erg. Bl.) 1,78 (m. Abb.)
Quelle: AGM (Tantiemenliste Sailer, ohne Datierung (ca. 1950); Zession, d. i. der Übertrag der Urheberrechte zwischen der Wiener Manufaktur Friedrich Goldscheider und Karl Sailer vom 13.2.1950); hier jeweils mit Ang. des Originaltitels „Waldandacht"; Tantiemenabrechnung 2. Halbjahr 1952), hier unter dem Titel „Marterl" geführt; Fotoarchiv K. Täuber.

8827
Lorenzl, Josef
„Spanien" („Tänzerin")
ca. 1950
Quelle: AGM (Tantiemenabrechnung Lorenzl 1. Halbjahr 1951), hier mit Ang. des Originaltitels „Spanien"; (Tantiemenabrechnung 10. März 1953), hier unter dem Titel „Tänzerin" geführt.

8828
Lorenzl, Josef
DECKELDOSE „Bonbonniere" (Sitzende Dame mit Ballkleid und Hut)
ca. 1950
8828,11 Firmenmarke 19, Made in Austria, unsign., H. ca. 20 cm; WKH
Quelle: AGM (Tantiemenabrechnung Lorenzl 1. Halbjahr 1951), hier mit Ang. des Originaltitels „Bonbonniere"; (Tantiemenabrechnung 10. März 1953), hier unter dem Titel „Dose" geführt.

8829
Lorenzl, Josef
WANDMASKE „Maske Madonna"
ca. 1950
Quelle: AGM (Tantiemenabrechnungen Lorenzl 1. Halbjahr 1951; 10. März 1953), hier mit Ang. des Originaltitels „Maske Madonna".

8832

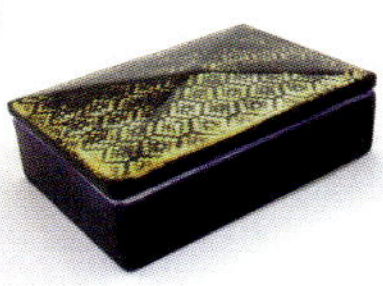

DECKELDOSE
Form: rechteckig, gewölbter Deckel
Dekor: Spitzendekor (Blumen bzw. andere Ornamente)
ca. 1950
8832,11 ger., gem. EM, Firmenmarke 20, Made in Austria, Hand dec., H. 5,5 cm, B. 10,1 cm, L. 14,7 cm; Sammlung a@goldscheider.net; Bem.: Besonderheit dieses Ex. ist die Goldauflage, die für die Staffierung Verwendung fand.
8832,17,38 Firmenmarke 19, Made in Austria, Hand dec., Pick. 1, H. 4,5 cm, B. 9,8 cm, L. 14,5 cm; Wien Museum Inv. 190.004/1+2
Quelle: AGM (Bestellung der Firma Wahliss (Wien), hier unter dem Titel „Dose" geführt).

8833
Sailer, Karl
„Fensterl" (Bub auf kleinen Vogel schauend)
ca. 1950
8833,11,24 Firmenmarke 19 (blau), Made in Austria (blau), Hand dec. (blau), unsign., Pick. 1, H. 13 cm, L. 14,5 cm; Wien Museum Inv. 186.656
Bem.: Das „Fenster", aus dem das Bubengesicht schaut, ist eigentlich nur eine Art „Rahmen", der modelliert und staffiert ist, wie Birkenholz.
Pendant –> vgl. 8834
Quelle: AGM (Tantiemenliste Sailer, ohne Datierung (ca. 1950); Zession, d. i. der Übertrag der Urheberrechte zwischen der Wiener Manufaktur Friedrich Goldscheider und Karl Sailer vom 1.7.1950; Tantiemenabrechnung 2. Halbjahr 1952), hier jeweils mit Ang. des Originaltitels „Fensterl".

8834

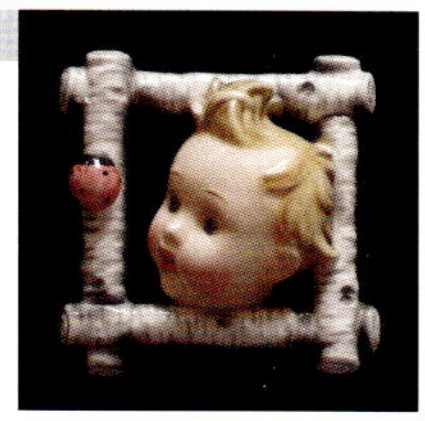

Sailer, Karl
„Fensterl" (Bub auf Marienkäfer schauend)
ca. 1950
8834,3,84 Firmenmarke 19, Made in Austria, unsign., H. 13 cm, B. 13 cm; eBay # 38322170
8834,8,24, 8834 ger., gem. G, Firmenmarke 19 (blau), Made in Austria (blau), Hand dec. (blau), unsign., Pick. 1, H. 13 cm, B. 13 cm; Wien Museum Inv. 186.657
8834,14,53 Firmenmarke 19, Made in Austria, Hand dec., unsign., H. 13,5 cm, B. 13,2 cm; Sammlung John Nash
Bem.: Das „Fenster", aus dem das Bubengesicht schaut, ist eigentlich nur eine Art „Rahmen", der modelliert und staffiert ist, wie Birkenholz.
Pendant –> vgl. 8833
Quelle: AGM (Tantiemenliste Sailer, ohne Datierung (ca. 1950); Zession, d. i. der Übertrag der Urheberrechte zwischen der Wiener Manufaktur Friedrich Goldscheider und Karl Sailer vom 1.7.1950; Tantiemenabrechnung 2. Halbjahr 1952), hier jeweils mit Ang. des Originaltitels „Fensterl".

8835F

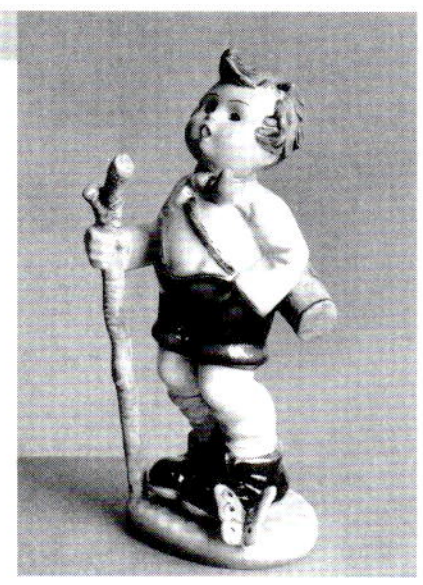

Sailer, Karl
„Botaniker" (Stehender Wandersbub mit Stock und umgehängtem Tornister) (auf Natursockel mit Schmetterling)
ca. 1950
8835,1 (Firmenmarke ohne Ang.), unsign., H. 19 cm; Christie's SK 27.6.1997
8835,14,29 Firmenmarke 19, unsign., H. 17 cm; Doro S 13.10.2004
8835F, H. 17 cm, B. 7 cm; GSNachf. 1,79
Bem.: Im Stile einer „Hummel-Figur", –> vgl. Bem. bei Modell-Nr. 8567.
Lit.: GSNachf. (2. Ausg., bzw. Erg. Bl.) 1,79 (m. Abb.)
Quelle: AGM (Tantiemenliste Sailer, ohne Datierung (ca. 1950); Zession, d. i. der Übertrag der Urheberrechte zwischen der Wiener Manufaktur Friedrich Goldscheider und Karl Sailer vom 1.7.1950; Tantiemenabrechnung 2. Halbjahr 1952), hier jeweils mit Ang. des Originaltitels „Botaniker"; Fotoarchiv K. Täuber.

8836F

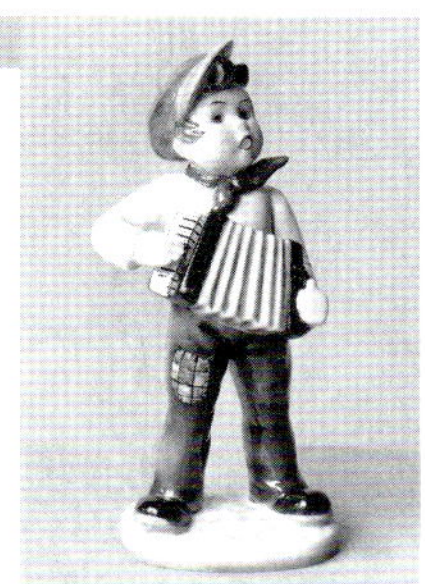

Sailer, Karl
„Harmonika" (Stehender singender und Akkordeon spielender Bub mit Kappe) (auf ovalem Natursockel)
ca. 1950
8836,3 gem. GE, Firmenmarke 19, Made in Austria, unsign., H. 16 cm; Wien Museum Inv. 190.164
8836,14,47 gem. GE, Firmenmarke 19, Made in Austria, unsign., H. 17 cm; Wien Museum Inv. 220.734
8836,17,52 gem. G, gem. lig. SP, Firmenmarke 19, unsign., H. 17 cm; Wien Museum Inv. 220.735
8836,66 Firmenmarke 19, Made in Austria, unsign., Pick. 1, H. 17,8 cm; eBay # 3714855638
8836F, H. 17,5 cm, B. 7 cm; GSNachf. 1,80
Bem.: Im Stile einer „Hummel-Figur", –> vgl. Bem. bei Modell-Nr. 8567.
Lit.: GSNachf. (2. Ausg., bzw. Erg. Bl.) 1,80 (m. Abb.)
Quelle: AGM (Tantiemenliste Sailer, ohne Datierung (ca. 1950); Zession, d. i. der Übertrag der Urheberrechte zwischen der Wiener Manufaktur Friedrich Goldscheider und Karl Sailer vom 1.7.1950; Tantiemenabrechnung 2. Halbjahr 1952), hier jeweils mit Ang. des Originaltitels „Harmonika"; Fotoarchiv K. Täuber.

8842
Schwetz-Lehmann, Ida
STEHKOPF „Himmelskönigin" (Madonna mit Krone) (auf Rechtecksockel)
ca. 1937 (für den Entwurf)
8842, H. 17 cm, B. 10 cm
–> vgl. 8132= BÜSTE
Lit.: GSNachf. 6,6 (m. Abb.), hier mit Ang. des Originaltitels „Himmelskönigin", ausgeführt als Büste.
Quelle: Fotoarchiv K. Täuber.

8843F

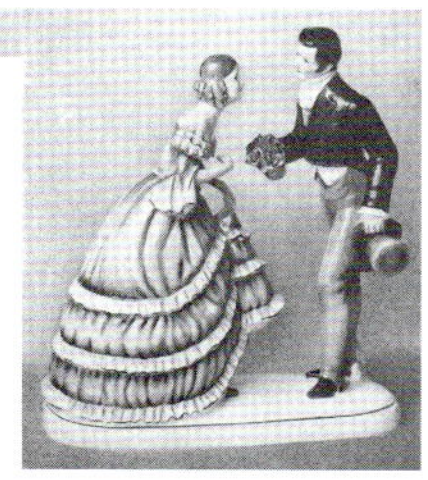

„Verlobung" (Biedermeierdame, einen Blumenstrauß erwartend)
ca. 1950
8843F, H. 28 cm, B. 18 cm; GSNachf. 1,56b
Pendant –> vgl. 8844
–> vgl. 8845= GRUPPE
Lit.: GSNachf. (2. Ausg., bzw. Erg. Bl.) 1,50 (m. Abb.)
Quelle: AGM (Rechnung an Maurice Mizrahi, Alexandria, Ägypten, vom 22. April 1953), hier mit Ang. des Originaltitels „Verlobung".

8844F

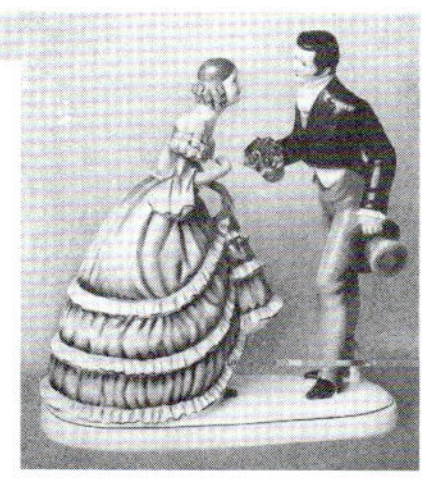

„Verlobung" (Biedermeierkavalier, einen Blumenstrauß überreichend)
ca. 1950
8844F, H. 28 cm, B. 10 cm; GSNachf. 1,56c
Pendant –> vgl. 8843
–> vgl. 8845= GRUPPE
Lit.: GSNachf. (2. Ausg., bzw. Erg. Bl.) 1,50 (m. Abb.)
Quelle: AGM (Rechnung an Maurice Mizrahi, Alexandria, Ägypten, vom 22. April 1953), hier mit Ang. des Originaltitels „Verlobung".

8845F

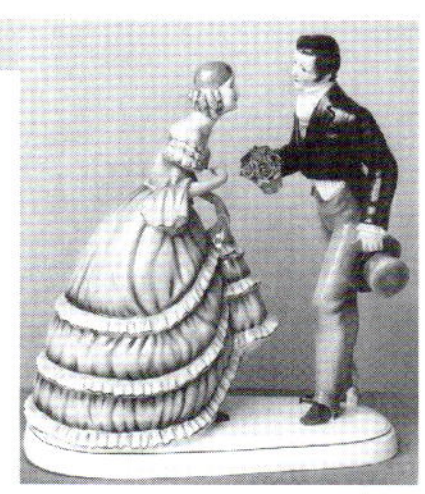

GRUPPE „Verlobung" (Biedermeierkavalier, einen Blumenstrauß an Dame überreichend) (auf länglichem Ovalsockel)
ca. 1950
8845F, H. 28 cm, B. 24 cm; GSNachf. 1,56a
Bem.: Die Vermutung liegt nahe, dass diese Gruppe ebenfalls unter dem Titel „Verlobung" angeboten wurde. Die beiden Einzelfiguren jedenfalls wurden von der Manufaktur so bezeichnet.
–> vgl. 8843; 8844= jeweils EinzelFIGUR
Lit.: GSNachf. (2. Ausg., bzw. Erg. Bl.) 1,50 (m. Abb.)
Quelle: AGM (Rechnung an E. Rosenthal, Frankfurt a. M., vom 15. Januar 1951), hier mit Ang. des Originaltitels GRUPPE „Verlobung"; Fotoarchiv K. Täuber.

8848F

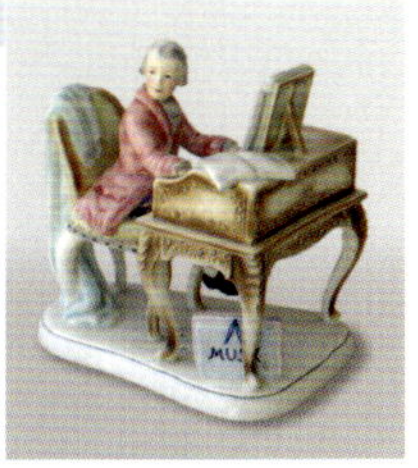

„Mozart" („Mozart/Spinett") (Mozart, am Spinett sitzend) (auf geschwungenem Sockel mit Notenmappe und der Aufschrift MUSIK)
ca. 1950
8848,8,27 Firmenmarke 20, Made in Austria, Hand dec., H. 22 cm; Sammlung Stressless, Portugal
8848,9,S gem. B.W., Firmenmarke 19, Made in Austria, Hand dec., H. 24 cm, B. 23 cm; eBay # 3317128543
Quelle: AGM (Bestellung der Firma Wahliss (Wien), hier jeweils mit Ang. des (Original?-)Titels „Mozart/Spinett"; Rechnung an Maurice Mizrahi, Alexandria, Ägypten, vom 22. April 1953), hier mit Ang. des Originaltitels „Mozart".

8849
Sailer, Karl
„Nachtwächter"
ca. 1950
Bem.: Im Stile einer „Hummel-Figur", -> vgl. Bem. bei Modell-Nr. 8567.
Quelle: AGM (Tantiemenliste Sailer, ohne Datierung (ca. 1950); Tantiemenabrechnung 2. Halbjahr 1952), hier mit Ang. des Originaltitels „Nachtwächter".

8850

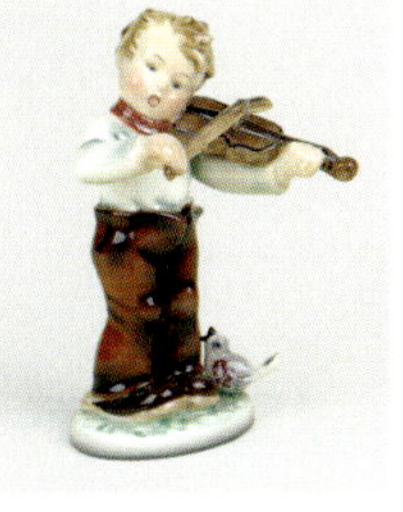

Sailer, Karl
„Geiger" (Stehender Geige spielender Bub) (auf ovalem Natursockel mit kleinem Vogel)
ca. 1950
8850,7,1 ger., (Firmenmarke ohne Ang.), unsign., H. 15,5 cm; Fischer 151
8850,7,14 Firmenmarke 19, Made in Austria, H. 16 cm; Henry's 25.11.1998
8850 ohne Ang., H. 16 cm; PB Wien 19
Bem.: Im Stile einer „Hummel-Figur", -> vgl. Bem. bei Modell-Nr. 8567.
Quelle: AGM (Tantiemenliste Sailer, ohne Datierung (ca. 1950); Tantiemenabrechnung 2. Halbjahr 1952), hier mit Ang. des Originaltitels „Geiger".

8851

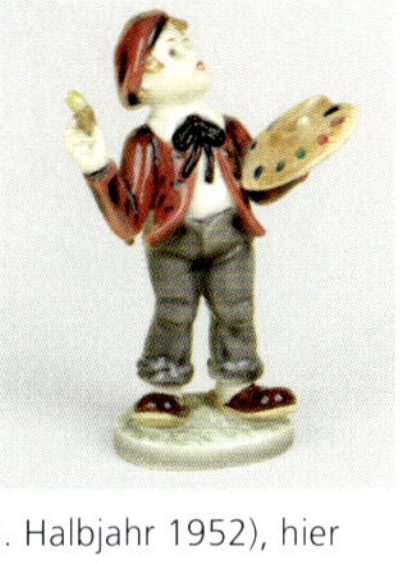

Sailer, Karl
„Maler"
ca. 1950
8851 ohne Ang., H. ca. 16 cm; PB Wien 19
Bem.: Im Stile einer „Hummel-Figur", -> vgl. Bem. bei Modell-Nr. 8567.
Quelle: AGM (Tantiemenliste Sailer, ohne Datierung (ca. 1950); Tantiemenabrechnung 2. Halbjahr 1952), hier mit Ang. des Originaltitels „Maler".

8852
Sailer, Karl
„Wanderlust"
ca. 1950
Bem.: Im Stile einer „Hummel-Figur", -> vgl. Bem. bei Modell-Nr. 8567.
-> vgl. 8854
Quelle: AGM (Tantiemenliste Sailer, ohne Datierung (ca. 1950); Tantiemenabrechnung 2. Halbjahr 1952), hier mit Ang. des Originaltitels „Wanderlust".

8853
Sailer, Karl
„Gitarre"
ca. 1950
Bem.: Im Stile einer „Hummel-Figur", -> vgl. Bem. bei Modell-Nr. 8567.
Quelle: AGM (Tantiemenliste Sailer, ohne Datierung (ca. 1950); Tantiemenabrechnung 2. Halbjahr 1952), hier mit Ang. des Originaltitels „Gitarre".

8854
Sailer, Karl
GRUPPE „Wanderlust"
ca. 1950
Bem.: Im Stile einer „Hummel-Figur", -> vgl. Bem. bei Modell-Nr. 8567.
-> vgl. 8852
Quelle: AGM (Tantiemenliste Sailer, ohne Datierung (ca. 1950); Tantiemenabrechnung 2. Halbjahr 1952), hier mit Ang. des Originaltitels GRUPPE „Wanderlust".

8855F
„Reh" (Liegendes Rehkitz)
ca. 1951
8855,22,35 gem. F., Firmenmarke 20, Made in Austria, Hand dec., H. 13 cm, B. 17 cm; eBay # 1439064910
Quelle: AGM (Rechnung an Maurice Mizrahi, Alexandria, Ägypten, vom 22. April 1953), hier mit Ang. des Originaltitels „Reh".

8856
Dakon, Stephan
„Duo" (Sitzendes nacktes Kind bzw. Putto mit kleinem Vogel spielend) (auf Ovalsockel)
ca. 1951
8856,1,86 Firmenmarke 19, Made in Austria, Hand dec., unsign., H. 10 cm, L. 10 cm; eBay # 2235991283
Quelle: Archiv Mimi Dakon (Tantiemenabrechnung 1. Halbjahr 1952), hier mit Ang. des Originaltitels „Duo".

8857
Dakon, Stephan
„Lampion"
ca. 1951
Quelle: Archiv Mimi Dakon (Tantiemenabrechnung 1. Halbjahr 1952), hier mit Ang. des Originaltitels „Lampion".

8858
Dakon, Stephan
„Trommler"
ca. 1951
Quelle: Archiv Mimi Dakon (Tantiemenabrechnung 1. Halbjahr 1952), hier mit Ang. des Originaltitels „Trommler".

8859
Dakon, Stephan
„Kind"
ca. 1951
Quelle: Archiv Mimi Dakon (Tantiemenabrechnung 1. Halbjahr 1952), hier mit Ang. des Originaltitels „Kind".

8860
Dakon, Stephan
„Kind" (Stehendes, sich waschendes nacktes Kleinkind mit Handtuch, Bürste und kleinem Waschtrog)
ca. 1951
8860,1,8 gem. G, Firmenmarke 19, Made in Austria, Hand dec., unsign., Pick. 1, H. 16,8 cm; city-antik
Quelle: Archiv Mimi Dakon (Tantiemenabrechnung 1. Halbjahr 1952), hier mit Ang. des Originaltitels „Kind".

8861
Dakon, Stephan
„Kind (m. / Schnecke)" (Erschreckter Bub mit Schnecke) (auf Ovalsockel)
ca. 1951
8861,8,29 Firmenmarke 19, Made in Austria, Hand dec., unsign., H. 16 cm; Wien Museum Inv. 220.736
Quelle: Archiv Mimi Dakon (Tantiemenabrechnung 1. Halbjahr 1952), hier mit Ang. des Originaltitels „Kind".

8862F
GRUPPE „Reh(e)"
ca. 1951
Quelle: AGM (Rechnung an Maurice Mizrahi, Alexandria, Ägypten, vom 22. April 1953), hier mit Ang. des Originaltitels GRUPPE „Reh(e)".

8867
„Madonna"
ca. 1951/52
Quelle: AGM (Kommissionsfaktura der „Österreichischen Gruppe-Austria" für die Messe in Utrecht vom 26.2.1953), hier mit Ang. des Originaltitels „Madonna".

8873?
„Madonna"
ca. 1951/52
Quelle: AGM (Kommissionsfaktura der „Österreichischen Gruppe-Austria" für die Messe in Utrecht vom 26.2.1953), hier mit Ang. des Originaltitels „Madonna".

8875

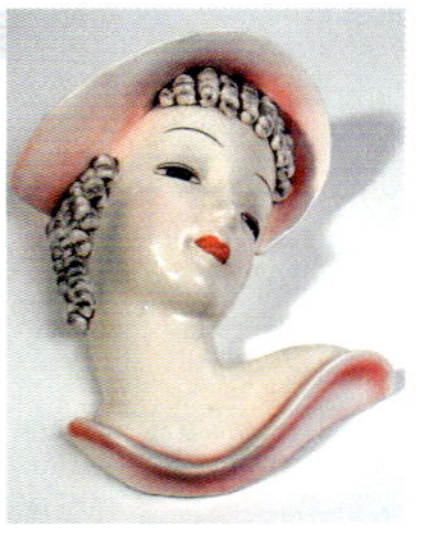

WANDMASKE Frauenkopf mit Hut und Kleidansatz
ca. 1951/52
8875,C,7 ger., Firmenmarke 19, Made in Austria, Hand dec., H. fehlt; city-antik

8877
„Fatima"
ca. 1951/52
Quelle: AGM (Kommissionsfaktura der „Österreichischen Gruppe-Austria" für die Messe in Utrecht vom 26.2.1953), hier mit Ang. des Originaltitels „Fatima".

8878
„Madonna"
ca. 1952
Quelle: AGM (Kommissionsfaktura der „Österreichischen Gruppe-Austria" für die Messe in Utrecht vom 26.2.1953), hier mit Ang. des Originaltitels „Madonna".

8879
VASE
Form: flötenförmig
Dekor: Blumen
ca. 1952
8879,11,34 Firmenmarke 20, Made in Austria, Hand dec., H. 27 cm; eBay # 2230506674; 3776442899

8884

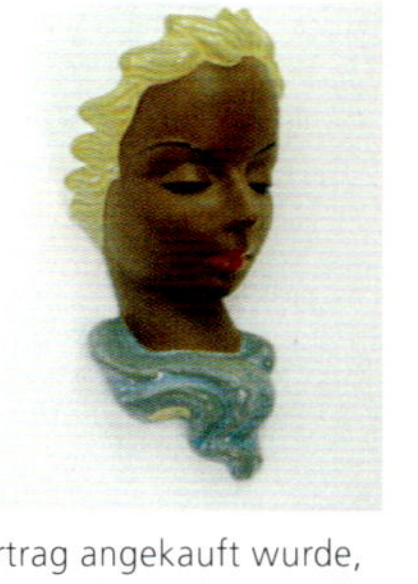

Goebel, Kurt
WANDMASKE Frauenkopf mit Haaransatz und Halstuch
ca. 1952
8884 ger., Firmenmarke 19, Made in Austria, unsign., H. 25,5 cm; Sammlung Stressless, Portugal
Bem.: Diese Maske wurde, bevor sie von der Manufaktur Goldscheider per Zessionsvertrag angekauft wurde, von Kurt Goebel in seiner eigenen Werkstatt hergestellt.
Quelle: AGM (Zession, d. i. der Übertrag der Urheberrechte zwischen der Wiener Manufaktur Friedrich Goldscheider und Kurt Goebel vom 24.10.1952), hier mit Ang. des Titels Wandmaske.

8885
Goebel, Kurt
WANDMASKE Frauenkopf mit Hut und Kleidansatz
ca. 1952
8885,1160/3,6 Firmenstempel WG 1, Pick. WG 1, Ausformung 50er Jahre, H. 25 cm; eBay # 3753591499
8885 ohne Ang., H. fehlt; Könecke (2002), S. 78
Bem.: Diese Maske wurde, bevor sie von der Manufaktur Goldscheider per Zessionsvertrag angekauft wurde, von Kurt Goebel in seiner eigenen Werkstatt hergestellt und auch noch zw. 1953–1960 bei Carstens (GWM West-Germany) produziert.
Lit.: Könecke (2002), S. 78, hier ohne weitere Ang.
Quelle: AGM (Zession, d. i. der Übertrag der Urheberrechte zwischen der Wiener Manufaktur Friedrich Goldscheider und Kurt Goebel vom 24.10.1952), hier mit Ang. des Titels Wandmaske.

8886
Goebel, Kurt
WANDMASKE
ca. 1952
Bem.: Diese Maske wurde, bevor sie von der Manufaktur Goldscheider per Zessionsvertrag angekauft wurde, von Kurt Goebel in seiner eigenen Werkstatt hergestellt.
Quelle: AGM (Zession, d. i. der Übertrag der Urheberrechte zwischen der Wiener Manufaktur Friedrich Goldscheider und Kurt Goebel vom 24.10.1952), hier mit Ang. des Titels Wandmaske.

8887
Goebel, Kurt
WANDMASKE (Frauenkopf?) mit Hut
ca. 1952
8887 ohne Ang., H. fehlt; Könecke (2002), S. 78
Bem.: Diese Maske wurde, bevor sie von der Manufaktur Goldscheider per Zessionsvertrag angekauft wurde, von Kurt Goebel in seiner eigenen Werkstatt hergestellt und auch noch zw. 1953–1960 bei Carstens (GWM West-Germany) produziert.
Lit.: Könecke (2002), S. 78, hier ohne weitere Ang.

8888
Goebel, Kurt
WANDMASKE Madonna mit Kind
ca. 1952
8888,32,S (Firmenmarke ohne Ang.), H. 33 cm; KH
Bem.: Diese Maske wurde, bevor sie von der Manufaktur Goldscheider per Zessionsvertrag angekauft wurde, von Kurt Goebel in seiner eigenen Werkstatt hergestellt.
Quelle: AGM (Zession, d. i. der Übertrag der Urheberrechte zwischen der Wiener Manufaktur Friedrich Goldscheider und Kurt Goebel vom 24.10.1952), hier mit Ang. des Titels Wandmaske.

8889
Goebel, Kurt
WANDMASKE
ca. 1952
Bem.: Diese Maske wurde, bevor sie von der Manufaktur Goldscheider per Zessionsvertrag angekauft wurde, von Kurt Goebel in seiner eigenen Werkstatt hergestellt.
Quelle: AGM (Zession, d. i. der Übertrag der Urheberrechte zwischen der Wiener Manufaktur Friedrich Goldscheider und Kurt Goebel vom 24.10.1952), hier mit Ang. des Titels Wandmaske.

8890
Goebel, Kurt
WANDMASKE Frauenkopf mit Schleier (Kopfschal)
ca. 1952
8890,8 Firmenmarke 20, Made in Austria, Hand dec., H. 26 cm; city-antik
Bem.: Diese Maske wurde, bevor sie von der Manufaktur Goldscheider per Zessionsvertrag angekauft wurde, von Kurt Goebel in seiner eigenen Werkstatt hergestellt.
Quelle: AGM (Zession, d. i. der Übertrag der Urheberrechte zwischen der Wiener Manufaktur Friedrich Goldscheider und Kurt Goebel vom 24.10.1952), hier mit Ang. des Titels Wandmaske.

8891
Goebel, Kurt
WANDMASKE Frauenkopf mit Hut und Halskette
ca. 1952
8891,88,9 ger., Firmenmarke 19, Made in Austria, Hand dec., unsign., H. 21,5 cm; Wien Museum Inv. 186.644
8891,434/2,3 Firmenstempel WG 1, Pick. WG 1, Ausformung 50er Jahre, H. 9,8 cm; Ausformung 50er Jahre, unsign., H. 21,7 cm; Doro 7.3.2001
Bem.: Diese Maske wurde, bevor sie von der Manufaktur Goldscheider per Zessionsvertrag angekauft wurde, von Kurt Goebel in seiner eigenen Werkstatt hergestellt und auch noch zw. 1953–1960 bei Carstens (GWM West-Germany) produziert.
Lit.: KGSF1; Könecke (2002), S. 72, 78, 95 (m. Abb.), hier mit Ang. des Künstlers Goebel.
Quelle: AGM (Zession, d. i. der Übertrag der Urheberrechte zwischen der Wiener Manufaktur Friedrich Goldscheider und Kurt Goebel vom 24.10.1952), hier mit Ang. des Titels Wandmaske.

8892
Goebel, Kurt
WANDMASKE
ca. 1952
Bem.: Diese Maske wurde, bevor sie von der Manufaktur Goldscheider per Zessionsvertrag angekauft wurde, von Kurt Goebel in seiner eigenen Werkstatt hergestellt.
Quelle: AGM (Zession, d. i. der Übertrag der Urheberrechte zwischen der Wiener Manufaktur Friedrich Goldscheider und Kurt Goebel vom 24.10.1952), hier mit Ang. des Titels Wandmaske.

8893
Goebel, Kurt
WANDMASKE
ca. 1952
Bem.: Diese Maske wurde, bevor sie von der Manufaktur Goldscheider per Zessionsvertrag angekauft wurde, von Kurt Goebel in seiner eigenen Werkstatt hergestellt.
Quelle: AGM (Zession, d. i. der Übertrag der Urheberrechte zwischen der Wiener Manufaktur Friedrich Goldscheider und Kurt Goebel vom 24.10.1952), hier mit Ang. des Titels Wandmaske.

8894
Goebel, Kurt
WANDMASKE
ca. 1952
Bem.: Diese Maske wurde, bevor sie von der Manufaktur Goldscheider per Zessionsvertrag angekauft wurde, von Kurt Goebel in seiner eigenen Werkstatt hergestellt.
Quelle: AGM (Zession, d. i. der Übertrag der Urheberrechte zwischen der Wiener Manufaktur Friedrich Goldscheider und Kurt Goebel vom 24.10.1952), hier mit Ang. des Titels Wandmaske.

8895
Goebel, Kurt
WANDMASKE
ca. 1952
Bem.: Diese Maske wurde, bevor sie von der Manufaktur Goldscheider per Zessionsvertrag angekauft wurde, von Kurt Goebel in seiner eigenen Werkstatt hergestellt.
Quelle: AGM (Zession, d. i. der Übertrag der Urheberrechte zwischen der Wiener Manufaktur Friedrich Goldscheider und Kurt Goebel vom 24.10.1952), hier mit Ang. des Titels Wandmaske.

8896
Goebel, Kurt
WANDMASKE
ca. 1952
Bem.: Diese Maske wurde, bevor sie von der Manufaktur Goldscheider per Zessionsvertrag angekauft wurde, von Kurt Goebel in seiner eigenen Werkstatt hergestellt.
Quelle: AGM (Zession, d. i. der Übertrag der Urheberrechte zwischen der Wiener Manufaktur Friedrich Goldscheider und Kurt Goebel vom 24.10.1952), hier mit Ang. des Titels Wandmaske.

8897
Goebel, Kurt
WANDMASKE
ca. 1952
Bem.: Diese Maske wurde, bevor sie von der Manufaktur Goldscheider per Zessionsvertrag angekauft wurde, von Kurt Goebel in seiner eigenen Werkstatt hergestellt.
Quelle: AGM (Zession, d. i. der Übertrag der Urheberrechte zwischen der Wiener Manufaktur Friedrich Goldscheider und Kurt Goebel vom 24.10.1952), hier mit Ang. des Titels Wandmaske.

8899
WANDMASKE Mutter mit Kind
ca. 1952
8899,11,35 Firmenmarke 19, Made in Austria (zwei mal), Pick. 1, H. 35 cm; eBay # 3331947500

8900
Stehendes Rehkitz
ca. 1952
8900 Firmenmarke 19, H. 15 cm; WKA 33

8901F
„Reh"
ca. 1952
Quelle: AGM (Rechnung an Maurice Mizrahi, Alexandria, Ägypten, vom 22. April 1953), hier mit Ang. des Originaltitels „Reh".

8905
Sitzender Damenakt mit ausgestreckten Beinen
ca. 1952
8905 ohne Ang., H. 19 cm, B. 28 cm; Könecke (2002), S. 78, 95 (m. Abb.)
Bem.: Diese Figur wurde auch noch zw. 1953–1960 bei Carstens (GWM West-Germany) produziert.
Lit.: Könecke (2002), S. 78, 95 (m. Abb.).

8909
Lorenzl, Josef
„Suleika"
ca. 1952
Quelle: AGM (Tantiemenabrechnung Lorenzl 10. März 1953), hier mit Ang. des Originaltitels „Suleika".

8913?
Stehendes Rehkitz
1953–1960
8913? ohne Ang., H. 19 cm, B. 16 cm; Könecke (2002), S. 78, 95 (m. Abb.)
Bem.: Diese Figur wurde zw. 1953–1960 bei Carstens (GWM West-Germany) produziert.
Lit.: Könecke (2002), S. 78, 95 (m. Abb.).

8915
Schwarz, (K.?)
„Anmut" (Sitzender Damenakt)
1953–1960
8915 ohne Ang., H. 18 cm, B. 27 cm; Könecke (2002), S. 42, 78 (m. Abb.)
Bem.: Diese Figur wurde zw. 1953–1960 bei Carstens (GWM West-Germany) produziert.
Lit.: KGSF1; KGSF2; Könecke (2002), S. 42, 78 (m. Abb.), hier mit Ang. des Originaltitels „Anmut" und des/der Künstlers/Künstlerin Schwarz.

8917
Pferdekopf
1953–1960
8917 ohne Ang., H. fehlt; Könecke (2002), S. 78
Bem.: Dieser Kopf (eine nähere Beschreibung ist wegen fehlender Abb. bei Könecke und nachweisbarer Stücke im Handel zum jetzigen Zeitpunkt nicht möglich) wurde zw. 1953–1960 bei Carstens (GWM West-Germany) produziert.
Lit.: Könecke (2002), S. 78, hier ohne weitere Ang.

8918
Chludek, Wilhelm
Stehendes Fohlen
1953–1960
8918 ohne Ang., H. 20 cm, B. 13 cm; Könecke (2002), S. 59, 78 (m. Abb.)
Bem.: Diese Figur wurde zw. 1953–1960 bei Carstens (GWM West-Germany) produziert.
Lit.: KGSF1; KGSF2; Könecke (2002), S. 59, 78 (m. Abb.), hier mit Ang. des Künstlers Chludek.

8919
Chludek, Wilhelm
Liegendes Fohlen
1953–1960
8919,4 ger., ohne Firmenmarke, Pick. WG 1, Ausformung 50er Jahre, H. 12,9 cm; eBay # 6501792724
8919 ohne Ang., H. 12 cm, B. 9,5 cm; Könecke (2002), S. 59, 78 (m. Abb.)
Bem.: Diese Figur wurde zw. 1953–1960 bei Carstens (GWM West-Germany) produziert.
Lit.: KGSF1; KGSF2; Könecke (2002), S. 59, 78 (m. Abb.), hier mit Ang. des Künstlers Chludek.

8920
Chludek, Wilhelm
Äsendes Rehkitz
1953–1960
8920 ohne Ang., H. 11,5 cm, B. 15 cm; Könecke (2002), S. 60, 78 (m. Abb.)
Bem.: Diese Figur wurde zw. 1953–1960 bei Carstens (GWM West-Germany) produziert.
Lit.: KGSF1; KGSF2; Könecke (2002), S. 60, 78 (m. Abb.), hier mit Ang. des Künstlers Chludek.

8921
Chludek, Wilhelm
Schreitendes Rehkitz
1953–1960
8921 ohne Ang., H. 18,5 cm, B. 15 cm; Könecke (2002), S. 60, 78 (m. Abb.)
Bem.: Diese Figur wurde zw. 1953–1960 bei Carstens (GWM West-Germany) produziert.
Lit.: KGSF1; KGSF2; Könecke (2002), S. 60, 78 (m. Abb.), hier mit Ang. des Künstlers Chludek.

8922
Chludek, Wilhelm
Liegendes Rehkitz
1953–1960
8922 ohne Ang., H. 12,5 cm, B. 12 cm; Könecke (2002), S. 60, 78 (m. Abb.)
Bem.: Diese Figur wurde zw. 1953–1960 bei Carstens (GWM West-Germany) produziert.
Lit.: KGSF1; KGSF2; Könecke (2002), S. 60, 78 (m. Abb.), hier mit Ang. des Künstlers Chludek.

8923
Matula, Viktor
Stehendes Dromedarfohlen
1953–1960
8923 ohne Ang., H. fehlt; Könecke (2002), S. 78, 95 (m. Abb.)
Bem.: Diese Figur wurde zw. 1953–1960 bei Carstens (GWM West-Germany) produziert.
Lit.: KGSF1; Könecke (2002), S. 78, 95 (m. Abb.), hier mit der Zuschreibung an Chludek. Im Nachlass Matula stellte sich jedoch dessen Autorschaft eindeutig heraus.
Quelle: Nachlass Matula, Wien

8929
WANDMASKE Frauenkopf mit Schal
1953–1960
8929 ohne Ang., H. ca. 24 cm; Könecke (2002), S. 72, 78, 95 (m. Abb.)
Bem.: Diese Figur wurde zw. 1953–1960 bei Carstens (GWM West-Germany) produziert.
Lit.: Könecke (2002), S. 72, 78, 95 (m. Abb.); Pinhas, S. 175 (m. Abb.).

8930
Chludek, Wilhelm
Stehendes Eselsfohlen
1953–1960
8930 ohne Ang., H. 14 cm, B. 15 cm; Könecke (2002), S. 66, 78, 95 (m. Abb.)
Bem.: Diese Figur wurde zw. 1953–1960 bei Carstens (GWM West-Germany) produziert.
Lit.: KGSF1; Könecke (2002), S. 66, 78, 95 (m. Abb.), hier mit Ang. des Künstlers Chludek.

8931
Chludek, Wilhelm
Stehendes Reh
1953–1960
8931 ohne Ang., H. 19,5 cm, B. 16 cm; Könecke (2002), S. 60, 78 (m. Abb.)
Bem.: Diese Figur wurde zw. 1953–1960 bei Carstens (GWM West-Germany) produziert.
Lit.: KGSF1; KGSF2; Könecke (2002), S. 60, 78 (m. Abb.), hier mit Ang. des Künstlers Chludek.

8932
WANDMASKE Frauenkopf mit Kopftuch
1953–1960
8932,921/IV,7 ger., Firmenstempel WG 1, Pick. WG 1, H. 22,9 cm; eBay # 1435534631
8932 ohne Ang., H. fehlt; Könecke (2002), S. 78, 96 (m. Abb.)
Bem.: Diese Maske wurde zw. 1953–1960 bei Carstens (GWM West-Germany) produziert. Könecke belegt, dass diese Modell-Nr. zweimal vergeben wurde.
-> vgl. auch folgenden Eintrag.
Lit.: Könecke (2002), S. 78, 96 (m. Abb.)

8932
Hahn und Henne
1953–1960
8932 ohne Ang., H. fehlt; Könecke (2002), S. 61, 78 (m. Abb.)
Bem.: Diese Figurengruppe wurde zw. 1953–1960 bei Carstens (GWM West-Germany) produziert. Der Firmenkat. KGSF2 belegt, dass diese Modell-Nr. zweimal vergeben wurde.
-> vgl. auch vorherigen Eintrag.
Lit.: KGSF2; Könecke (2002), S. 61, 78 (m. Abb.)

8935
Knörlein, Rudolf
WANDMASKE Frauenkopf mit Hand und Larve (= Gesichtsmaske)
1953–1960
8935 ohne Ang., Firmenstempel WG 1, unsign., H. 21 cm; Kett. 100
8935 (Firmenmarke ohne Ang.), unsign., H. 21 cm; Christie's SK 4./5.12.1996
8935 ohne Ang., H. 19 cm, B. 20 cm; Könecke (2002), S. 54, 78 (m. Abb.)
Bem.: Diese Figur wurde zw. 1953–1960 bei Carstens (GWM West-Germany) produziert und trägt in der Variante ohne Hand und Larve die Modell-Nr. 8941.
-> vgl. 8941= Frauenkopf
Lit.: KGSF1; KGSF2; Könecke (2002), S. 54, 78 (m. Abb.)

8936
RELIEF Christuskopf mit Dornenkrone (montiert auf kreuzförmiger Holzplatte) (Wandbehang)
1953–1960
8936 ohne Ang., H. fehlt; Könecke (2002), S. 78, 96 (m. Abb.)
Bem.: Dieses Relief wurde zw. 1953–1960 bei Carstens (GWM West-Germany) produziert.
-> vgl. 8938 (ohne Holzplatte / Kreuz)
Lit.: Könecke (2002), S. 78, 96 (m. Abb.), hier ohne weitere Ang.

8937
WANDMASKE Madonna (nach rechts blickend)
1953–1960
8937 ohne Ang., H. fehlt; Könecke (2002), S. 78, 96 (m. Abb.)
Bem.: Diese Maske wurde zw. 1953–1960 bei Carstens (GWM West-Germany) produziert.
Pendant -> vgl. 8938
Lit.: Könecke (2002), S. 78, 96 (m. Abb.), hier ohne weitere Ang.

8938
WANDMASKE/RELIEF Christuskopf (nach links blickend)
1953–1960
8938 ohne Ang., H. fehlt; Könecke (2002), S. 78, 96 (m. Abb.)
Bem.: Diese Maske wurde zw. 1953–1960 bei Carstens (GWM West-Germany) produziert.
Pendant -> vgl. 8937
-> vgl. 8936 (auf Holzplatte / Kreuz)
Lit.: Könecke (2002), S. 78, 96 (m. Abb.), hier ohne weitere Ang.

8939
Knörlein, Rudolf
WANDMASKE Engelkopf (mit Flügel) (Cherub)
1953–1960
8939 ohne Ang., H. 10 cm, B. 10 cm; Könecke (2002), S. 78, 96 (m. Abb.)
Bem.: Der von Könecke (aus Firmenunterlagen übernommen) bezeichnete Engelkopf ist eigentlich ein Cherub. Diese Maske wurde zw. 1953–1960 bei Carstens (GWM West-Germany) produziert.
Lit.: Könecke (2002), S. 78, 96 (m. Abb.)

8941
Knörlein, Rudolf
WANDMASKE Frauenkopf mit Kragen
1953–1960
8941 Firmenstempel WG 1, Pick. WG 1, H. 17 cm; Wien Museum Inv. 220.808
8941 ohne Ang., H. 19 cm, B. 20 cm; Könecke (2002), S. 78, 96 (m. Abb.)
Bem.: Diese Maske wurde zw. 1953–1960 bei Carstens (GWM West-Germany) produziert und trägt in der Variante mit Hand und Larve die Modell-Nr. 8935.
-> vgl. 8935= Frauenkopf mit Hand und Larve
Lit.: KGSF1; Könecke (2002), S. 78, 96 (m. Abb.)

8942
WANDMASKE Frauenkopf mit Haarband (im Profil nach rechts)
1953–1960
8942 ohne Ang., H. fehlt; Könecke (2002), S. 78, 96 (m. Abb.)
Bem.: Diese Maske wurde zw. 1953–1960 bei Carstens (GWM West-Germany) produziert.
Lit.: Könecke (2002), S. 78, 96 (m. Abb.)

Goldscheider „Moderne Keramik" und Gebrauchskeramik

Goldscheider "Modern Ceramics" and Utilitarian Ceramics

Ausführung der Modelle zwischen 1928–1938

-> vgl. Kapitel „Populäre und moderne Keramik der dreißiger Jahre unter Walter Goldscheider (1928–1938)"

I M
VASE
Form: kugelförmig, trichterförmige Mündung
Dekor: diverse Dekore, Laufglasur
ca. 1928–1938
I ger., 32 gepr., Firmenmarke 19, Made in Austria, H. 19,5 cm; city-antik
I ger., gem. lig. HS, Firmenmarke 19, Made in Austria, Hand dec., H. 21,5 cm; Wien Museum Inv. 186.690
I M, H. 20 cm, B. 15 cm; GSNachf. 4,2a
Lit.: GSNachf. 4,2a (m. Abb.)

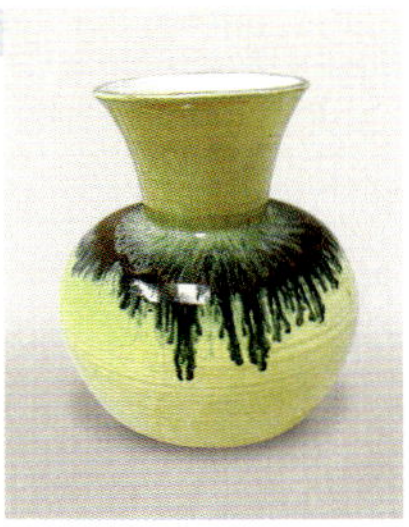

II
Knörlein, Rudolf
VASE
Form: gedrückte Kugelform, hohe trichterförmige Mündung
Dekor: gerillt, Craqueléglasur
ca. 1934
II, H. 20 cm, Ø 16 cm
Bem.: Durch die eigenhändige Datierung Knörleins (1934) auf dem Trägerblatt des Originalfotos kann das Jahr 1934 exakt bestimmt werden.
Quelle: Privatarchiv (Originalfoto)

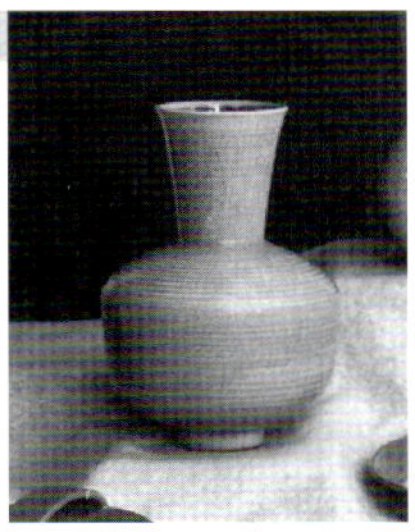

III M
VASE
Form: balusterförmig, trichterförmige Mündung
Dekor: vermutlich diverse Dekore, Craqueléglasur/Laufglasur
ca. 1928–1938
III,37 ger., Firmenmarke 19, Made in Austria, H. 26,7 cm, Ø 13,8 cm; Sammlung a@goldscheider.net
III M, H. 24 cm, B. 12 cm; GSNachf. 4,2b
Lit.: GSNachf. 4,2b (m. Abb.)
Quelle: AGM (Rechnung an Maurice Mizrahi, Alexandria, Ägypten, vom 22. April 1953), geliefert in Ausführung „gold".

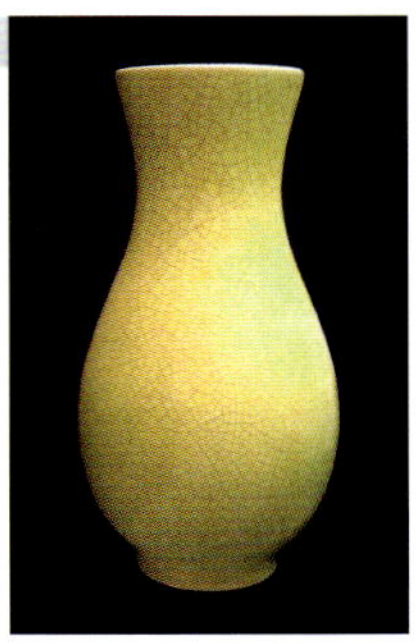

IV M
VASE
Form: eiförmig, hohe trichterförmige Mündung
Dekor: vermutlich diverse Dekore, Laufglasur
ca. 1928–1938
IV M, H. 30 cm, B. 11 cm; GSNachf. 4,2c
Lit.: GSNachf. 4,2c (m. Abb.)
Quelle: AGM (Rechnung an Maurice Mizrahi, Alexandria, Ägypten, vom 22. April 1953), geliefert in Ausführung „gold".

Models made 1928–1938

-> cf chapter "Popular and Modern 1930s Ceramics under Walter Goldscheider (1928–1938)"

V M
VASE
Form: stangenförmig, flache Schulter, kurze geweitete Mündung
Dekor: diverse Dekore
ca. 1928–1938
craqueliert, Laufglasur
5,1 ger., Firmenmarke 19, Made in Austria, H. 33,4 cm, Ø 13,8 cm; Sammlung a@goldscheider.net
Blumenblüten an einem Stängel
V ger.,8 gepr., Firmenmarke 19 (blau), Made in Austria (blau), Hand dec. (blau), H. 35 cm; Doro FVK, Okt. 2006
V M, H. 34 cm, B. 13 cm; GSNachf. 4,2d
Lit.: GSNachf. 4,2d (m. Abb.)
Quelle: AGM (Rechnung an Maurice Mizrahi, Alexandria, Ägypten, vom 22. April 1953).

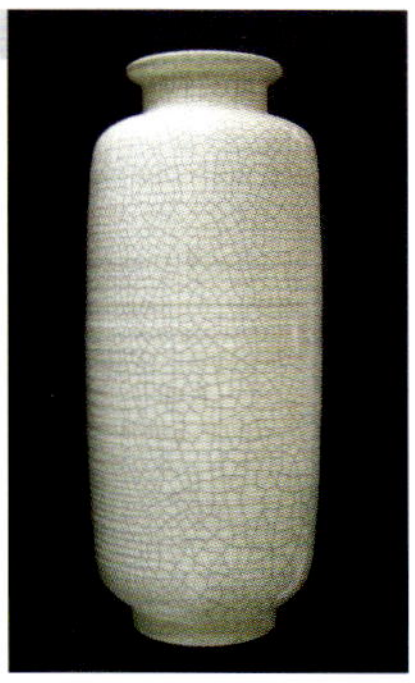

VI M
SCHALE
Form: rund, hochgezogene Wandung, drei (oder vier?) Kugelfüße
Dekor: vermutlich diverse Dekore, Laufglasur
ca. 1928–1938
VI M, H. 8 cm, B. 20 cm; GSNachf. 4,2e
Lit.: GSNachf. 4,2e (m. Abb.)

VII M
SCHALE
Form: rund
Dekor: vermutlich diverse Dekore, Laufglasur
ca. 1928–1938
VII M, H. 3 cm, B. 30 cm; GSNachf. 4,2f
Lit.: GSNachf. 4,2f (m. Abb.)

VIII
Knörlein, Rudolf
VASE
Form: kugelförmig, hohe trichterförmige Mündung
Dekor: Craqueléglasur
ca. 1934
VIII, H. 25 cm, Ø 13 cm
Bem.: Durch die eigenhändige Datierung Knörleins (1934) auf dem Trägerblatt des Originalfotos kann das Jahr 1934 exakt bestimmt werden.
Quelle: Privatarchiv (Originalfoto)

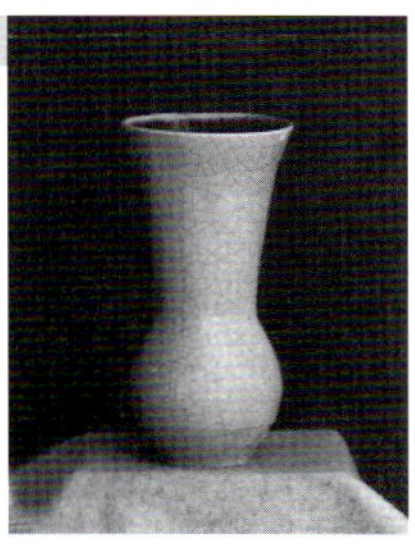

X M
Matula, Viktor
VASE
Form: eiförmig, hohe trichterförmige Mündung
Dekor: vermutlich diverse Dekore, Laufglasur
ca. 1928–1938
X M, H. 25 cm, B. 10 cm; GSNachf. 4,2g
Lit.: GSNachf. 4,2g (m. Abb.)
Quellen: Nachlass Matula, Wien; AGM (Rechnung an Maurice Mizrahi, Alexandria, Ägypten, vom 22. April 1953), geliefert in Ausführung „gold".

XI M
Matula, Viktor
VASE
Form: kugelförmig, trichterförmige Mündung
Dekor: vermutlich diverse Dekore, Laufglasur
ca. 1928–1938
XI M, H. 16 cm, B. 11 cm; GSNachf. 4,2h
Lit.: GSNachf. 4,2h (m. Abb.)
Quelle: Nachlass Matula, Wien

XII
Knörlein, Rudolf
VASE
Form: eiförmig, hohe trichterförmige Mündung
Dekor: Craqueléglasur
ca. 1934
XII, H. 23 cm, Ø 10 cm
Bem.: Durch die eigenhändige Datierung Knörleins (1934) auf dem Trägerblatt des Originalfotos kann das Jahr 1934 exakt bestimmt werden.
Quelle: Privatarchiv (Originalfoto)

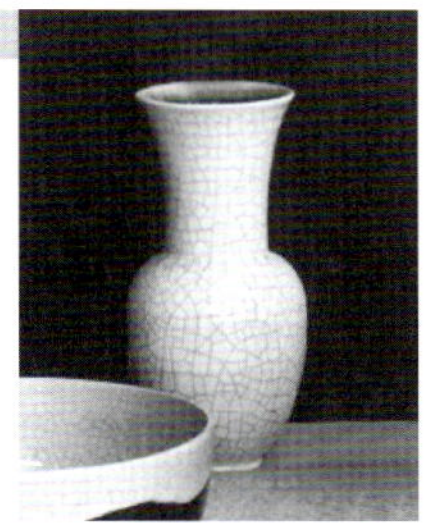

XV
Knörlein, Rudolf
DECKELDOSE
Form: rund
Handhabe: bügelartiger Henkel
Dekor: vermutlich diverse Dekore
ca. 1934
XV, H. 6 cm, Ø 15 cm
Bem.: Durch die eigenhändige Datierung Knörleins (1934) auf dem Trägerblatt des Originalfotos kann das Jahr 1934 exakt bestimmt werden.
Quelle: Privatarchiv (Originalfoto)

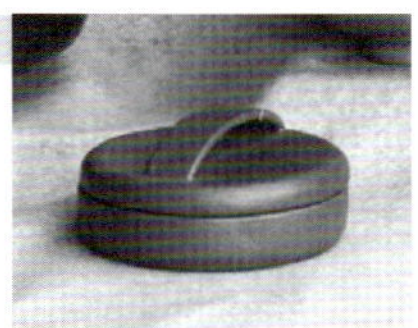

XVI
Knörlein, Rudolf
SCHALE
Form: rund, gewulsteter Stand
Dekor: Craqueléglasur
ca. 1934
XVI, H. 11 cm, Ø 24 cm
Bem.: Durch die eigenhändige Datierung Knörleins (1934) auf dem Trägerblatt des Originalfotos kann das Jahr 1934 exakt bestimmt werden.
Quelle: Privatarchiv (Originalfoto)

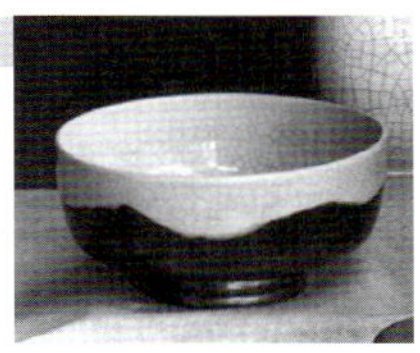

XVIII
Knörlein, Rudolf
SCHALE/TELLER
Form: rund
Dekor: gerillt, abstrakte Malerei (vermutlich diverse Dekore)
ca. 1934
XVIII, H. 5 cm, Ø 27 cm
Bem.: Durch die eigenhändige Datierung Knörleins (1934) auf dem Trägerblatt des Originalfotos kann das Jahr 1934 exakt bestimmt werden.
Quelle: Privatarchiv (Originalfoto)

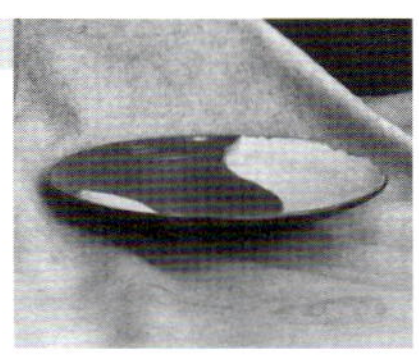

XIX
Knörlein, Rudolf
VASE
Form: stangenartig, leicht gebaucht, etwas geweitete Mündung
Dekor: Blumenblüten, mehrfarbig, Spitzendekor (Schablone)
ca. 1928–1938
XIX ger., 8 gepr., Firmenmarke 19, H. 36,8 cm; Sammlung.schmutz@chello.at 70

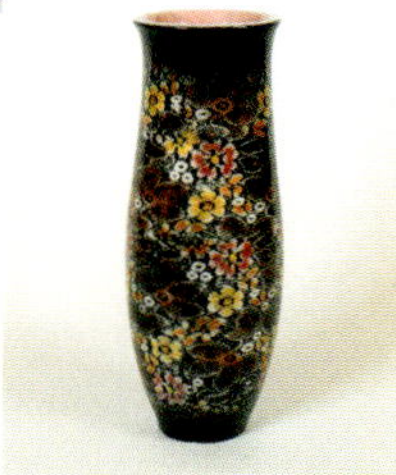

XXV M/F
Knörlein, Rudolf
SCHALE
Form: rund
Dekor: vermutlich diverse Dekore, Laufglasur/Craqueléglasur
ca. 1934
XXV M, H. 10,5 cm, Ø 26 cm; GSNachf. 4,8a
XXV F, H. 10,5 cm, Ø 26 cm
Bem.: Durch die eigenhändige Datierung Knörleins (1934) auf dem Trägerblatt des Originalfotos kann das Jahr 1934 exakt bestimmt werden.
Lit.: GSNachf. 4,8a (m. Abb.)
Quelle: AGM (Rechnung an Maurice Mizrahi, Alexandria, Ägypten, vom 22. April 1953); Privatarchiv (Originalfoto)

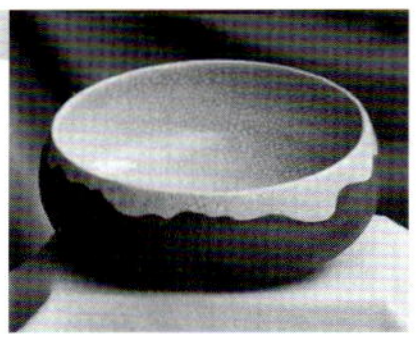

XXIX M/F
Knörlein, Rudolf
SCHALE
Form: blattförmig geschwungen, drei Kugelfüße
Dekor: diverse Dekore, Spitzendekor, Laufglasur/Craqueléglasur
ca. 1934
93,XXIX,66 (Firmenmarke ohne Ang.), H. fehlt, Ø 28 cm; WKA 33
XXIX M, H. 7 cm, Ø 26 cm; GSNachf. 4,8b
XXIX F, H. 7 cm, Ø 26 cm
Bem.: Durch die eigenhändige Datierung Knörleins (1934) auf dem Trägerblatt des Originalfotos kann das Jahr 1934 exakt bestimmt werden.
Lit.: GSNachf. 4,8b (m. Abb.)
Quelle: AGM (Rechnung an Maurice Mizrahi, Alexandria, Ägypten, vom 22. April 1953), geliefert in Ausführung „gold"; Privatarchiv (Originalfoto).

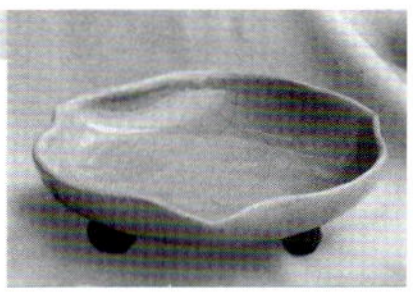

XXX M/F
Knörlein, Rudolf
SCHALE
Form: rund, mehrpassig
Dekor: vermutlich diverse Dekore, Laufglasur/Craqueléglasur
ca. 1934
XXX M, H. 10 cm, Ø 25 cm; GSNachf. 4,8c
XXX F, H. 10 cm, Ø 25 cm
Bem.: Durch die eigenhändige Datierung Knörleins (1934) auf dem Trägerblatt des Originalfotos kann das Jahr 1934 exakt bestimmt werden.
Lit.: GSNachf. 4,8c (m. Abb.)
Quelle: AGM (Rechnung an Maurice Mizrahi, Alexandria, Ägypten, vom 22. April 1953), geliefert in Ausführung „gold"; Privatarchiv (Originalfoto).

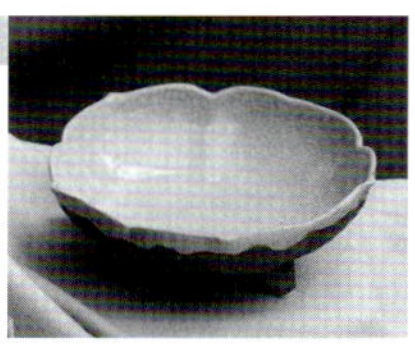

XXXVIII
VASE Kugel
Form: kugelförmig, kurze gewulstete Mündung
Dekor: craqueliert, gelb-orange-braun, Laufglasur
ca. 1928–1938
XXXVIII ger., gem. S, Firmenmarke 19, Made in Austria, H. 20,2 cm, Ø 21,4 cm; MAK Wien, Inv. Ke 10602-76
Ausst.: PSK 1995
Lit.: ADGS, Nr. 29 (m. Abb.)

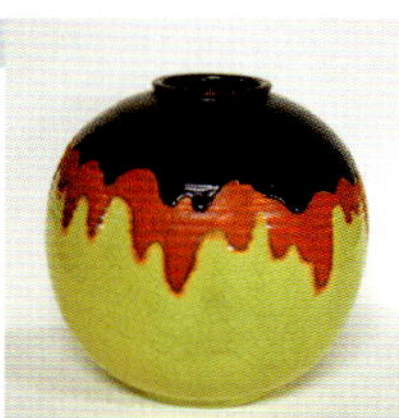

1
Weber, Ernst
LAMPENFUSS Kugel
Form: kugelförmig
Dekor: abstrakte Malerei, Laufglasur
ca. 1928–1938
1 gem. G, Firmenmarke 19, Made in Austria, sign. Ernst Weber, H. 24 cm; Sammlung.schmutz@chello.at 26

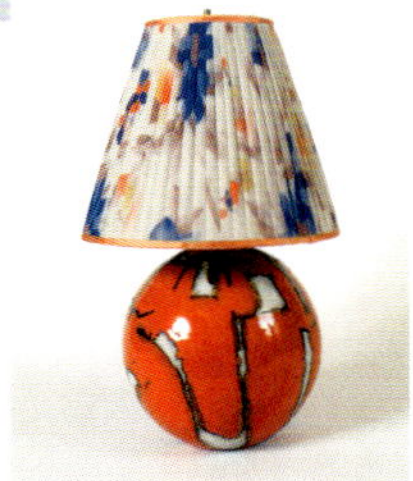

6 s. 69

9F
VASE
ca. 1928–1938
Quelle: AGM (Rechnung an Maurice Mizrahi, Alexandria, Ägypten, vom 22. April 1953).

28
VASE
Form: becherförmig, spitzovaler Querschnitt
Dekor: orange-grün, teils craqueliert, Laufglasur
ca. 1928–1938
28 ger., Firmenmarke 19, Made in Austria, H. 24,3 cm; Doro 4.3.2002

32
DECKELDOSE
Form: rund
Dekor: Kreise
ca. 1928–1938
32 ger. XXVI, A, Firmenmarke 19, Made in Austria, H. 4,5 cm, Ø 13,4 cm; PB Wien 19
32 ger. XXVI, A, Firmenmarke 19, Made in Austria, H. 4,3 cm, Ø 13,7 cm; ADGS, Nr. 28 (m. Abb.); Sammlung Goldscheider, Wien, Inv. G 14
Ausst.: PSK 1995
Lit.: ADGS, Nr. 28 (m. Abb.)
69 s. 669

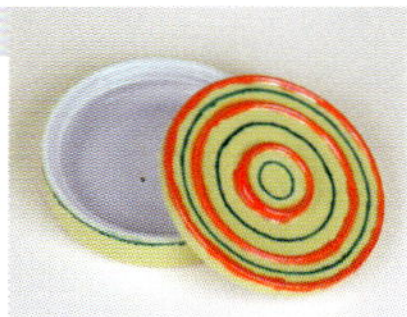

(100)
SERVIETTENHALTER
Form: halbkreisförmig
Handhaben: zwei gerillte Henkel (ohrenartig)
Dekor: mehrfarbig, Laufglasur
ca. 1928–1938
100 gem. rot, ohne Firmenmarke, H. 7,1 cm, L. 14,8 cm; MAK Wien, Inv. Ke 10602-17
Bem.: Dieser Serviettenhalter ist meiner Einschätzung nach kein Erzeugnis der Manufaktur Goldscheider. Es findet sich weder eine eingeprägte Modellnummer noch eine Firmenmarke. Der vollständig unglasierte rote Scherben, sowie die Farbgebung der Glasur sprechen eher gegen die Goldscheiderurheberschaft.
Ausst.: PSK 1995
Lit.: ADGS, Nr. 24 (m. Abb.)

232
Bosse, Walter
VASE/TOPF
Form: konisch
Handhaben: vier Ösen
Dekor: rechteckige Farbfelder mit Blumen, mehrfarbig, Laufglasur
ca. 1928–1938
232 gem., Firmenmarke 19, Made in Austria, unsign., H. 7 cm, L. 13,4 cm, Ø 8,6 cm; PB Wien 19
Bem.: Diese Vase bzw. den Topf wurde auch in einer Höhe von 11,4 cm (Doro 18.11.1997) produziert.
Ausst.: PSK 1995
–> vgl. ohne Nr.
Lit.: ADGS, Nr. 22 (m. Abb.)

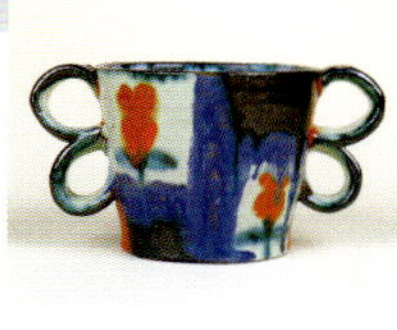

244
ASCHER
Form: zwei, sich überschneidende Kreise
Zigarettenablagen: zwei
Dekor: Farbfelder, mehrfarbig, Laufglasur
ca. 1928–1938
244,448 ger. lig. MK, gem. B, Firmenmarke 19, Made in Austria, H. 3,1 cm, L. 14 cm; Wien Museum Inv. 222.545

246
Bosse, Walter
VASE Blume
Form: becher-/kelchförmig
Handhaben: vier Ösen
Dekor: zwei stilisierte Blumen, rechteckige Farbfelder, mehrfarbig, Laufglasur
246 (gem. mit Bleistift), Firmenmarke 19a, unsign., H. 7 cm, Ø 9 cm; eBay # 3310670315
–> vgl. ohne Nr.

267
Bosse, Walter
VASE/TOPF
Form: konisch
Handhaben: vier Ösen
Dekor: rechteckige Farbfelder mit Herz, Rauten und Wellen, mehrfarbig, Laufglasur
ca. 1928–1938
267,A,27 (Firmenmarke 19?), H. 14,7 cm, Ø 21 cm; Doro 23.5.1991

284
VASE Blatt
Form: kugelförmig, leicht gedrückt, trichterförmige Mündung mit flacher Lippe
Handhaben: zwei Ösen
Dekor: stilisierte Blätter, mehrfarbig
284 (Firmenmarke ohne Ang.), H. 22,5 cm; Pinhas, S. 176 (m. Abb.)

526
Geiger, Benno (zugeschrieben)
VASE
Form: gebaucht, geweitete Mündung
Dekor: stilisierte Blumen (Tulpen und Margariten), mehrfarbig, Laufglasur
ca. 1928–1934
526,7 Firmenmarke 19, Made in Austria, unsign., H. 22,8 cm; Sammlung.schmutz@chello.at 75

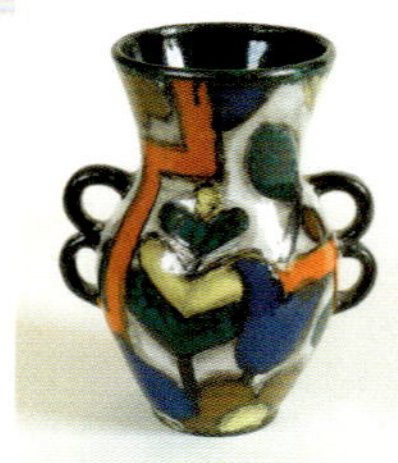

530
VASE
Form: gebaucht, geweitete sechspassige Mündung
Dekor: abstrakte Malerei (Blumenstängel, Blätter, Streifen, Punkte), mehrfarbig, Laufglasur
ca. 1928–1938
530,34 Firmenmarke 19, Made in Austria, H. 30,5 cm; Kett. 196
530 (zweite Nr. durch Pick. überklebt) gem. lig. FP?, Firmenmarke 19, Made in Austria, Originalpick. „V. Schira, Porcelaines, 3. Esc. Gr. Pont, Lausanne", H. 30,5 cm; Schuler 66

531
SCHALE
Form: rund
Dekor: abstrakte Malerei, geometrische Formen, mehrfarbig, Laufglasur
ca. 1928–1938
531,48 gem. B, Firmenmarke 19, Made in Austria, H. 7,5 cm, Ø 24,1 cm; MAK Wien, Inv. Ke 10602-40
531,51 gem. B, Firmenmarke 19, Made in Austria, H. 7,8 cm, Ø 24,5 cm; Sammlung Weilinger, Wien
Ausst.: PSK 1995
Lit.: ADGS, Nr. 3 (m. Abb.)

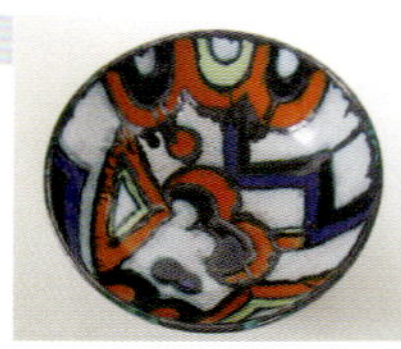

532
SCHALE
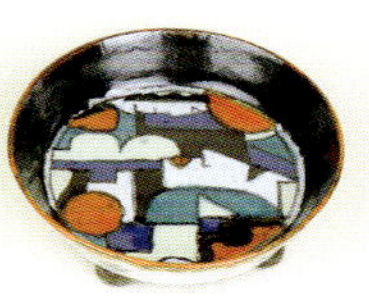
Form: rund, vier Kugelfüße
Dekor: geometrische Formen, mehrfarbig
ca. 1928–1938
0532,3 Firmenmarke 19, Made in Austria, H. 7,2 cm, Ø 23,5 cm; PB Wien 19
Ausst.: PSK 1995
Lit.: ADGS, Nr. 7 (m. Abb.), die hier angegebene Modell-Nr. 582 wurde berichtigt auf 532, da das zweite nachweisbare Exemplar eindeutig mit dieser Nr. gemarkt ist.

549
VASE
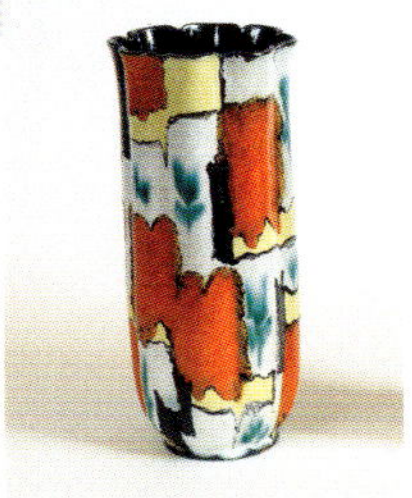
Form: becherförmig, achtpassige Mündung
Dekor: Farbfelder, orange-weiß, Laufglasur
ca. 1928–1938
549,17 gem. N, Firmenmarke 19, Made in Austria, H. 27,8 cm, Ø 12,4 cm; Sammlung.schmutz@chello.at 1
549,68 gem. N, Firmenmarke 19, Made in Austria, H. 27,5 cm; Sammlung Weilinger, Wien

550
VASE
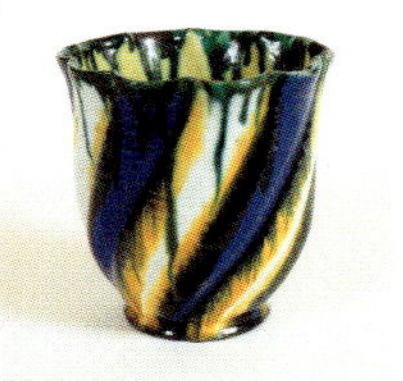
Form: glockenförmig, gerillt, zehnpassige Mündung
Dekor: spiralförmig gedrehte Farbflächen, blau-weiß-gelb, Laufglasur
ca. 1928–1938
550,118 Firmenmarke 19, Made in Austria, H. 18,3 cm, Ø 18 cm; Sammlung.schmutz@chello.at 62

552
VASE
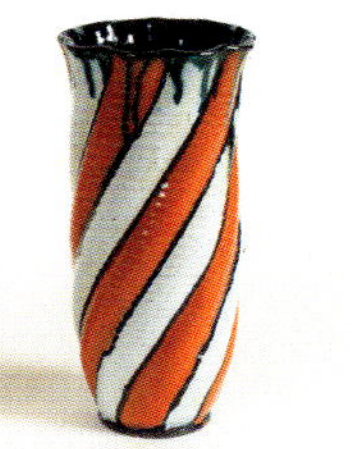
Form: becherförmig, gerillt, zehnpassige Mündung
Dekor: spiralförmig gedrehte Farbflächen, orange-weiß, Laufglasur
ca. 1928–1938
552,172 gem. lig. AR, Firmenmarke 19, Made in Austria, H. 26,1 cm, Ø 13 cm; Sammlung.schmutz@chello.at 76

556
LAMPENFUSS Kugel
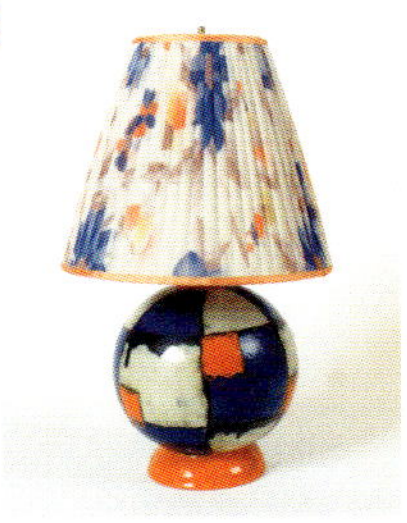
Form: kugelförmig, kurze zylindrische Mündung (für die Lampenfassung), gewulsteter Standring
Dekor: Farbfelder, mehrfarbig, Laufglasur
ca. 1928–1938
556,64 Firmenmarke 19, H. 24,2 cm; Doro 3.5.1995
556,68,9 gem. B, Firmenmarke 19, Made in Austria, H. 23 cm; Sammlung.schmutz@chello.at 17

559
SCHALE / VASE
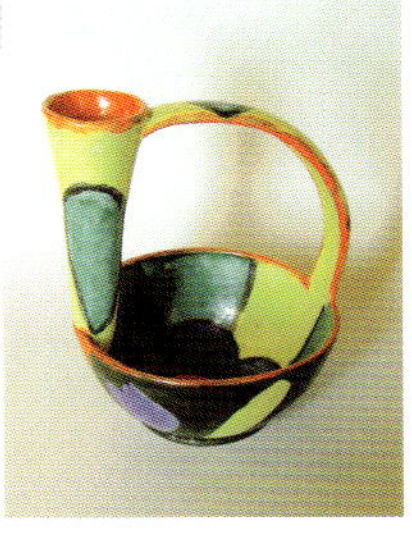
Form: runde Schale, am Rand eingesetzte trichterförmige Vase
Handhabe: halbkreisförmiger Henkel
Dekor: vegetabile Ornamente, Schalenboden mit Blume, mehrfarbig
559,34 gem. K, Firmenmarke 19, Made in Austria, H. 23 cm; Sammlung.schmutz@chello.at 88
559,116 gem. G, Firmenmarke 19, Made in Austria, H. 23,5 cm; Doro 3.5.1995

564
VASE / CACHEPOT / TOPF
Schachbrett
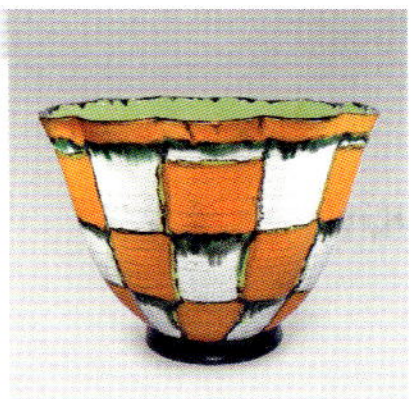
Form: kelchförmig, zwölfpassige Mündung
Dekor: Schachbrettmuster, orange-weiß, abgesetzt mit schwarz-grüner Laufglasur
ca. 1928–1938
564,23 gem. B, Firmenmarke 19, Made in Austria, H. 14,5 cm, Ø 24,2 cm; Doro 20.4.1993
564,88 gem. B, Firmenmarke 19, Made in Austria, H. 17,3 cm, Ø 23,8 cm; Sammlung a@goldscheider.net
564,95 gem. K, Firmenmarke 19, Made in Austria, H. 17,4 cm, Ø 23,7 cm; PB Deutschland
564,102 gem. K, Firmenmarke 19, Made in Austria, H. ca. 18 cm; KH
564,121 gem. G., Firmenmarke 19, Made in Austria, H. ca. 18 cm; KH
564,148 gem. B, 768, Firmenmarke 19, Made in Austria, H. 17,9 cm; MAK Wien, Inv. Ke 10602-80
564,152 ger., gem. B, Firmenmarke 19, Made in Austria, H. 18,5 cm, Ø 24 cm; Sammlung.schmutz@chello.at 28
564,189 Firmenmarke 19, Made in Austria, H. 17,5 cm; Doro 15.9.2003
Ausst.: PSK 1995
Pendant –> vgl. 5710= LAMPE
Lit.: ADGS, Nr. 4 (m. Abb.)

567
VASE
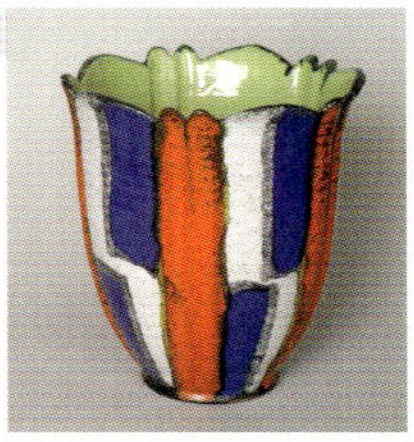
Form: becherförmig, mehrpassige Mündung, gewellte Lippe
Dekor: vertikale Farbfelder, mehrfarbig
ca. 1928–1938
567,100 gem. K, Firmenmarke 19, Made in Austria, H. 18,5 cm, Ø 16,7 cm; Sammlung.schmutz@chello.at 49

568
VASE
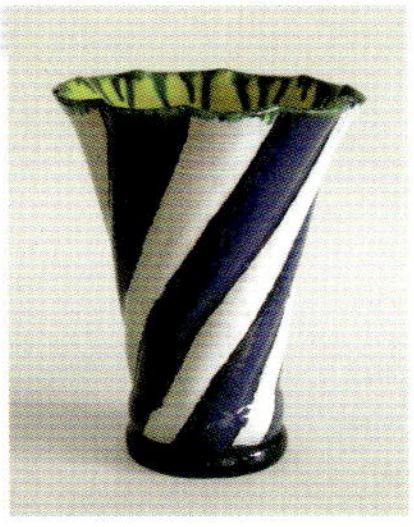
Form: trichterförmig, gerillt, zehnpassige Mündung, zweifach gewulsteter Stand
Dekor: spiralförmig gedrehte Farbflächen, blau-weiß-schwarz, Laufglasur
ca. 1928–1938
568,140 gem. SCH, Firmenmarke 19, Made in Austria, H. 22,7 cm, Ø 18,7 cm; PB Deutschland

569
VASE
Form: becherförmig, sechspassige Mündung
Dekor: abstakte Malerei, Punkte, Laufglasur
ca. 1928–1938
569,119 (Firmenmarke ohne Ang.), H. 22,5 cm; Christie's SK 18.9.1998

574
VASE
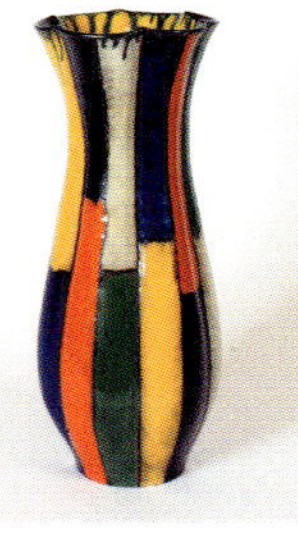
Form: becherförmig, gebaucht, geweitete sechspassige Mündung
Dekor: vertikale Farbfelder, mehrfarbig, Laufglasur
ca. 1928–1938
574,5 gem. lig. AR, Firmenmarke 19, Made in Austria, H. 45,6 cm; Sammlung.schmutz@chello.at 18

576
VASE
Form: gebaucht, geweitete vierpassige Mündung
Dekor: vertikale Farbfelder, mehrfarbig, Laufglasur
ca. 1928–1938
576,12 gem. B, Firmenmarke 19, Made in Austria, H. 23,2 cm; eBay # 1265855641

578
Geiger, Benno
VASE
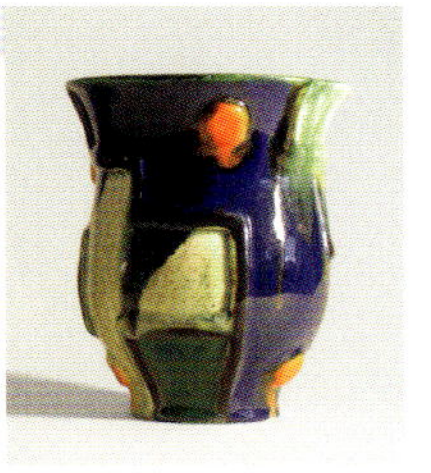
Form: becherförmig, geweitete Mündung
Dekor: unregelmäßige Farbfelder, Kreise, reliefiert, mehrfarbig, Laufglasur
ca. 1928–1934
578,1 gem. W, Firmenmarke 19, Made in Austria, unsign., H. 14,3 cm, Ø 11,9 cm; PB Deutschland
0578,1,20 Firmenmarke 19, Made in Austria, ger. Geiger, H. 20,5 cm, Ø 11,2 cm; Wien Museum Inv. 186.698
Bem.: Zum Vasentypus und zum Künstler –> vgl. Reklame „Moderne Vasen, Fayence von Benno Geiger" in: MW, 11.1929/30, Nr. 18, S. II (m. Abb.).
Ausst.: SSH 1992
Pendants bzw. ähnl. Modelle/Dekore –> vgl. 591; 592; 596
Lit.: SS, S. 32, 33 (m. Abb.).

580
Geiger, Benno
VASE
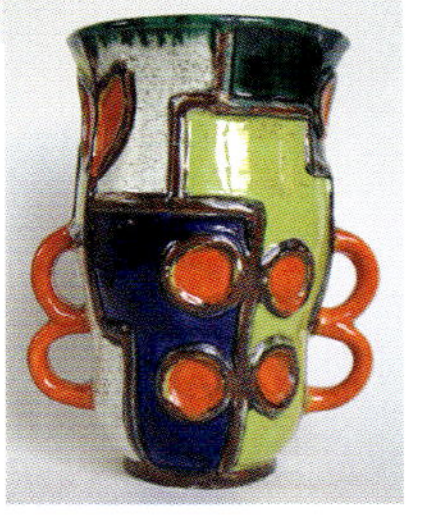
Form: becherförmig, etwas geweitete Mündung
Dekor: unregelmäßige Farbfelder, Ringe, mehrfarbig, Laufglasur
ca. 1928–1934
580 ger., gem. B, Firmenmarke 19, Made in Austria, H. 20 cm, Ø 13 cm; Sammlung Weilinger, Wien

581
VASE
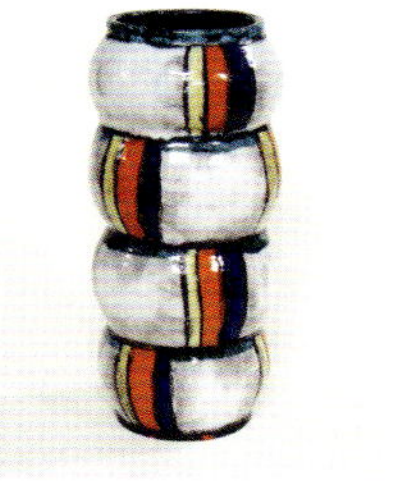
Form: zylinderförmig, vierfach gebaucht, gerade kurze Mündung
Dekor: Farbfelder, Streifen, mehrfarbig
ca. 1928–1938
581,9 gem. B, Firmenmarke 19, Made in Austria, H. 25 cm, Ø 11,6 cm; MAK Wien, Inv. Ke 10602-78
Ausst.: PSK 1995

583
Geiger, Benno (zugeschrieben)
VASE Musikanten
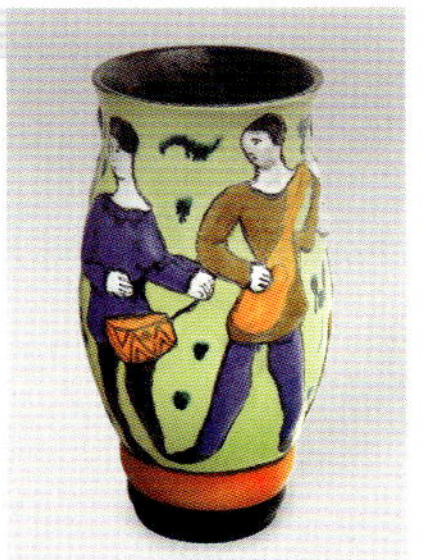
Form: becherförmig, geweitete Mündung
Dekor: vier Figuren (Musikanten)
ca. 1928–1934
583,25 gem. A, Firmenmarke 19, Made in Austria, unsign., H. 30,7 cm, Ø 16 cm; Sammlung a@goldscheider.net
–> vgl. 5739

587
Geiger, Benno (zugeschrieben)
VASE
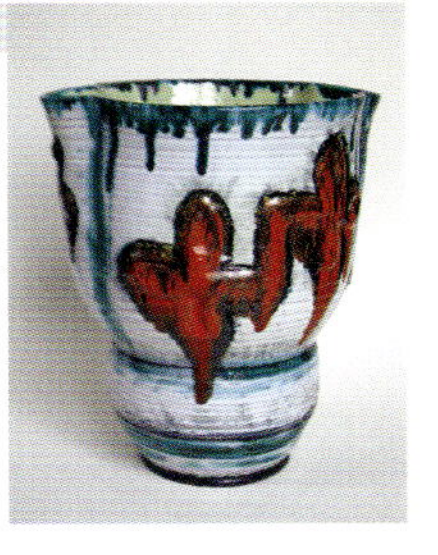
Form: gebaucht, tailliert, vierpassige Mündung
Dekor: stilisierte Blumen, mehrfarbig, Laufglasur
ca. 1928–1938
587,19 gem. B, Firmenmarke 19, Made in Austria, H. 24,8 cm, Ø 22,4 cm; Sammlung Weilinger, Wien

591
Geiger, Benno
VASE
Form: becherförmig, geweitete Mündung, konischer Standring
Dekor: unregelmäßige Farbfelder, (mit/ohne) Kreise, reliefiert, mehrfarbig, Laufglasur
ca. 1928–1934

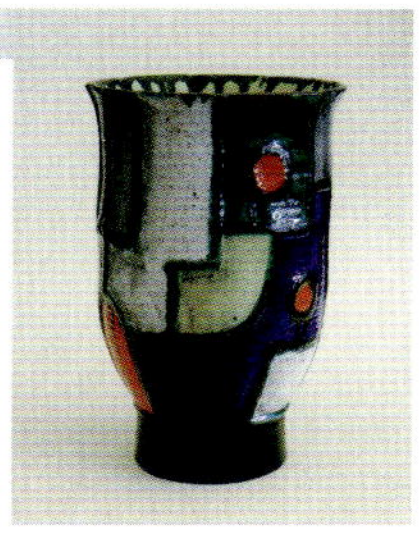

591,65 Firmenmarke 19, Made in Austria, unsign., H. 28,5 cm; Doro 8.11.1994; Bem.: Diese Vase nur mit den Farbfeldern dekoriert.
591,67 gem. lig. AR, Firmenmarke 19, Made in Austria, unsign., H. 28,8 cm, Ø 19,5 cm; PB Deutschland; Bem.: Dieses Exemplar zusätzlich mit Kreisen.
Bem.: Zum Vasentypus und zum Künstler -> Reklame „Moderne Vasen, Fayence von Benno Geiger" in: MW, 11.1929/30, Nr. 18, S. II (m. Abb.).
Pendants bzw. ähnl. Modelle/Dekore -> vgl. 578; 591; 592

592
Geiger, Benno
VASE (Bodenvase)
Form: stangenförmig, hoher konischer Standring
Dekor: unregelmäßige Farbfelder, Kreise, reliefiert, mehrfarbig, Laufglasur
ca. 1928–1934

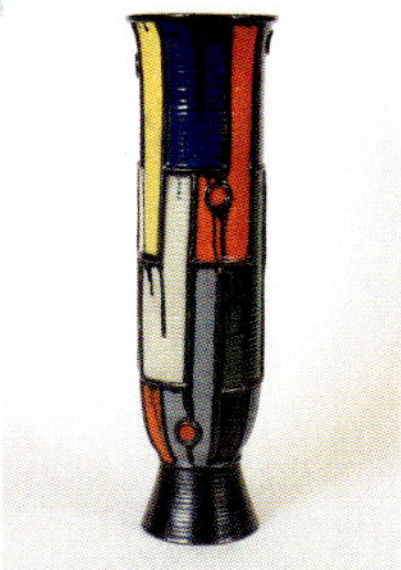

592,33 gem. lig. AR, Firmenmarke 19, Made in Austria, unsign., H. 65 cm, Ø 18,8 cm; Sammlung.schmutz@chello.at 29
Bem.: Zum Vasentypus und zum Künstler -> vgl. Reklame „Moderne Vasen, Fayence von Benno Geiger" in: MW, 11.1929/30, Nr. 18, S. II (m. Abb.).
Pendants bzw. ähnl. Modelle/Dekore -> vgl. 578; 591; 596

594
VASE
Form: zylinderförmig, geweitete vierpassige Mündung, konischer vierpassiger Standring
Dekor: abstrakte Malerei (Blumenstängel, Blätter, Streifen, Punkte), mehrfarbig, Laufglasur
ca. 1928–1938

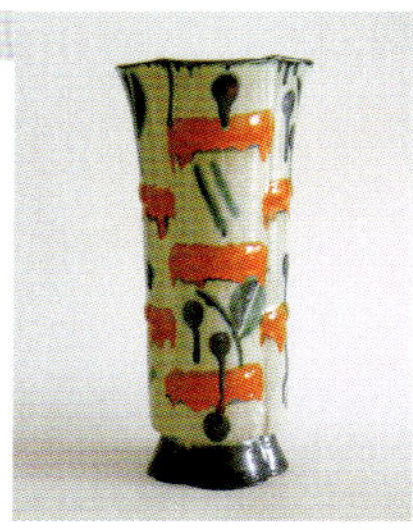

594,13 Firmenmarke 19, Made in Austria, H. 33 cm; eBay # 3652995051
594,28 gem. B, Firmenmarke 19, Made in Austria, H. 30,5 cm, Ø 13 cm; PB Deutschland

595
BUCHSTÜTZE
Form: würfelförmig, abgerundete Ecke
Dekor: Streifen, mehrfarbig, Laufglasur
ca. 1928–1938

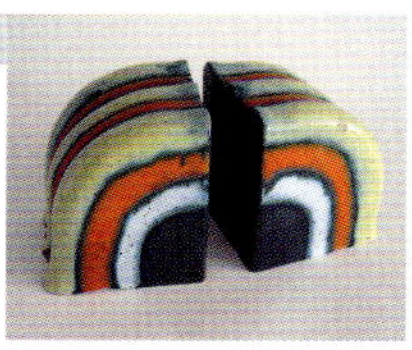

595,30 gem. 594, gem. B, Firmenmarke 19, Made in Austria, H. 11 cm, B. 10,2 cm; Sammlung John Nash
595,33 gem. 594, gem. B, Firmenmarke 19, Made in Austria, H. 11,2 cm, B. 10,2 cm; Sammlung John Nash
595,W ger. gem. T, Firmenmarke 19, Made in Austria, H. 10,1 cm, B. 9,8 cm; eBay # 53316871
595 ger. gem. T, Firmenmarke 19, Made in Austria, H. 10,1 cm, B. 9,8 cm; eBay # 53316871

596
Geiger, Benno
VASE
Form: becherförmig, geweitete Mündung
Dekor: unregelmäßige Farbfelder, reliefert, mehrfarbig, Laufglasur
ca. 1928–1934

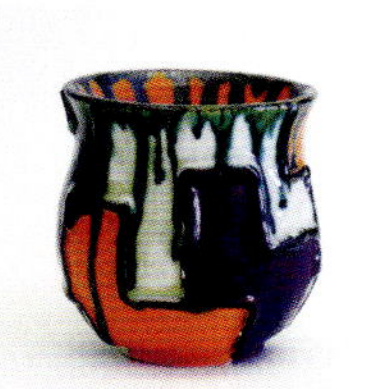

596,4 gem. G., Firmenmarke 19, Made in Austria, unsign., H. 10,5 cm, Ø 10 cm; PB Deutschland
Bem.: Reklame „Moderne Vasen, Fayence von Benno Geiger" in: MW, 11.1929/30, Nr. 18, S. II (m. Abb.).
Pendants bzw. ähnl. Modelle/Dekore -> vgl. 578; 591; 592

606A
Meisinger, Ida
SCHREIBZEUG „Der Modehund" (Sitzender Scotchterrier)
Form: rechteckig, eine Rille als Schreibzeugablage, eine Halterung für den Füller
Dekor: mittig stehender Terrier (vollplastische Fig.)
ca. 1928–1938
606A,22 ger., gem. B, Firmenmarke 19, Made in Austria, H. 15 cm, L. 16,4 cm B. 12 cm; Wien Museum Inv. 220.711
Bem.: Dieser Terrier, der unter den Nummern 5854, 5879 sowie 6026 bei Goldscheider produziert wurde, wird hier durch Beigabe des Schreibzeugs zur angewandten Variante einer Zierfigur. Eine Version als Kerzenhalter findet sich mit der Modell-Nr. 6309, eine weitere gibt es unter Modell-Nr. 6043 als Ascher.

613
VASE Kugel
Form: kugelförmig, kurze trichterförmige Mündung
Dekor: horizontale Streifen, orange-schwarz, Laufglasur
ca. 1928–1938

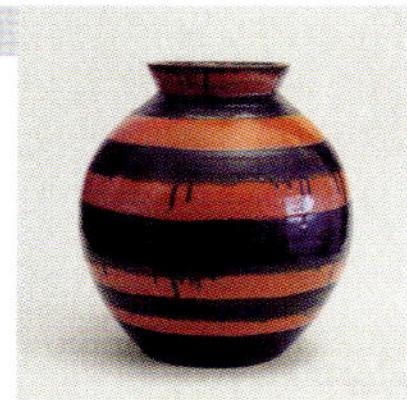

613 ger., ger. lig. MK, gem. lig. AH, Firmenmarke 19, Made in Austria, H. 27,2 cm, Ø 24 cm; PB Deutschland

614
VASE
Form: becherförmig, gebaucht, geweitete mehrpassige Mündung
Dekor: vertikale Farbfelder, mehrfarbig, Laufglasur
ca. 1928–1938

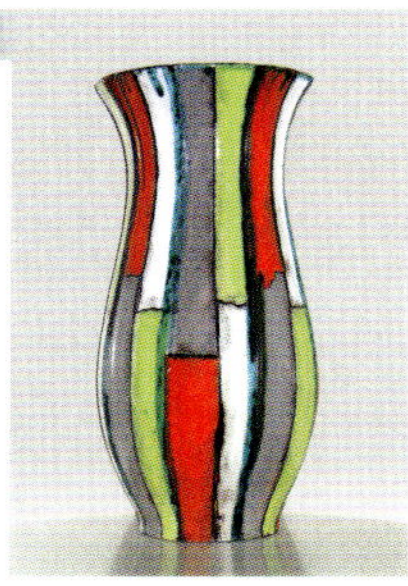

614,35 gem. L, Firmenmarke 19, Made in Austria, H. 35 cm; Sammlung a@goldscheider.net

616
VASE
Form: becherförmig, mittig tailliert, sechspassige geweitete Mündung
Handhaben: sechs Ösen
Dekor: vertikale Farbfelder, mehrfarbig, Laufglasur
ca. 1928–1938

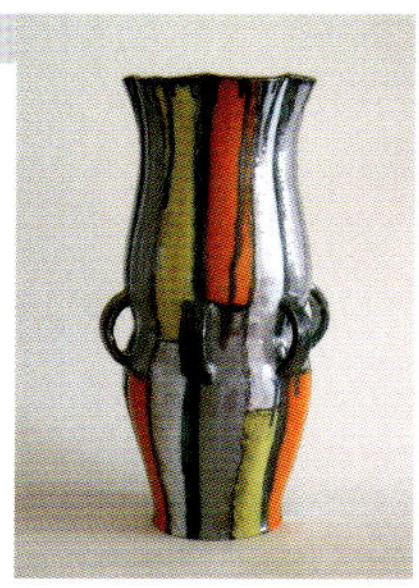

616,2 gem. B, Firmenmarke 19, Made in Austria (zwei mal), H. 35,9 cm, Ø 15,4 cm; PB Deutschland

625
LAMPENFUSS
Form: gebaucht, konzentrisch gerippt
Dekor: weiß-dunkelgrün, Laufglasur
ca. 1928–1938

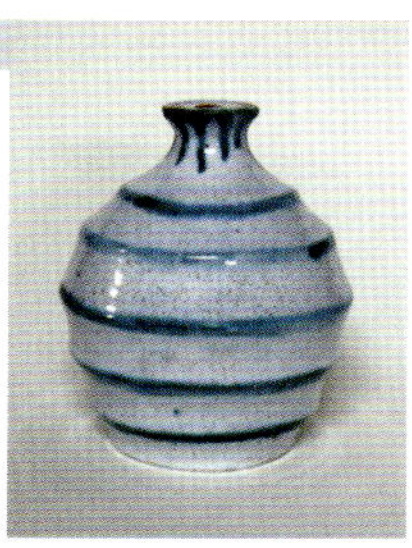

625,6 gem. lig. MR?, Firmenmarke 19, Made in Austria, H. 17,4 cm; MAK Wien, Inv. Ke 10602-75;
Bem.: Die eingepr. Mod.-Nr. 626 wurde vom Staffierer ausgebessert auf 625.
625 gem. lig. MR?, Firmenmarke 19, Made in Austria, H. 17,3 cm; Kett. 100
Lit.: ADGS, Nr. 43 (m. Abb.)
626 s. 625

628
LAMPENFUSS Kugel
Form: kugelförmig, kurze zylindrische Mündung (für die Lampenfassung)
Dekor: geometrische Formen, mehrfarbig, Laufglasur
ca. 1928–1938

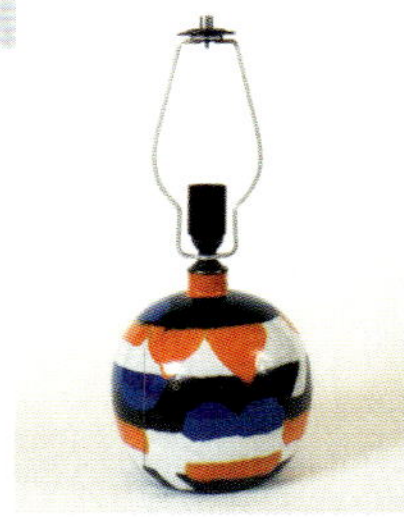

628,11 gem. lig. AR?, Firmenmarke 19, Made in Austria, H. 19,1 cm, Ø 17,6 cm; MAK Wien, Inv. Ke 10602-8
628,13 gem. A, Firmenmarke 19, Made in Austria, H. 19 cm, Ø 19 cm; Sammlung John Nash
628,29 gem. G, (ohne Firmenmarke), H. 19,5 cm, Ø 17 cm; Sammlung.schmutz@chello.at 85
628,33 gem. A, Firmenmarke 19, Made in Austria, H. 22 cm; Neumeister 270
Ausst.: PSK 1995
Lit.: ADGS, Nr. 8 (m. Abb.)

631
VASE Fische
Form: eiförmig, kurze trichterförmige Mündung
Dekor: reliefierte Fische, Punkte, mehrfarbig
ca. 1928–1938

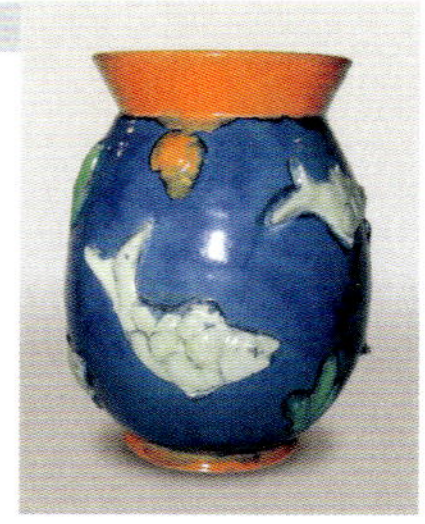

631,2 gem., gem. B, Firmenmarke 19, Made in Austria, H. 23 cm, Ø 18,5 cm; PB Wien

632 s. 637

637
VASE
Form: becherförmig, hoher konischer Stand
Handhaben: vier Ösen
Dekor: abstrakte Malerei, mehrfarbig, Laufglasur
ca. 1928–1938

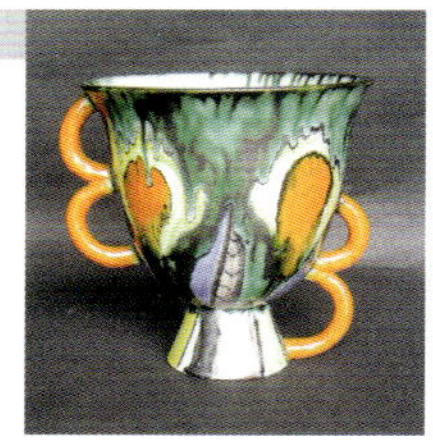

637,76 gem. G., Firmenmarke 19, Made in Austria, H. 16,4 cm, Ø 17 cm; Sammlung Weilinger, Wien
637,101 ger. EW, Firmenmarke 19, Made in Austria, H. 16,5 cm, Ø 18,4 cm; PB Wien
Lit.: Pinhas, S. 176 (m. Abb.), hier mit Ang. der falschen Mod.-Nr. 632.

640
VASE
Form: becherförmig, trichterförmige Mündung
Dekor: stilisierte Blumen, Kreise, mehrfarbig, Laufglasur
ca. 1928–1938
640,126 gem. B, Firmenmarke 19, Made in Austria, H. 17,5 cm; Franzisi 16

656 +A
SCHALE
Form: rund
Handhaben: zwei Ösen
Dekor: abstrakt vegetabil, mehrfarbig, Laufglasur
ca. 1928–1938

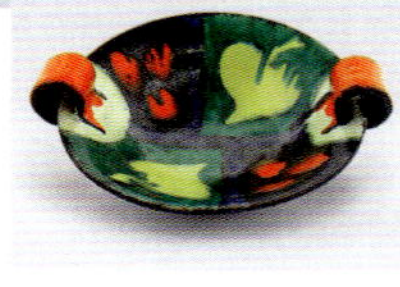

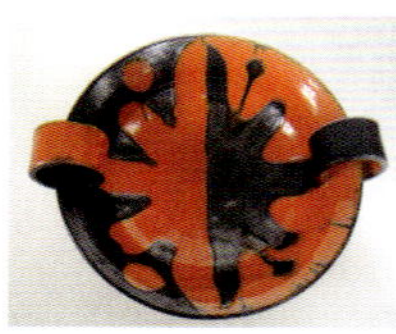

656,24 gem. G., Firmenmarke 19, Made in Austria, H. 6 cm, Ø 26 cm; Sammlung a@goldscheider.net
656,78 gem. G., Firmenmarke 19, Made in Austria, H. ca. 9 cm, Ø 26 cm; eBay # 875815163
656,143 gem. lig. HH, Firmenmarke 19, Made in Austria, H. 8,3 cm, Ø 24,5 cm; PB Wien
656 ger., ger. lig. MK, gem. S?, Firmenmarke 19, Made in Austria, H. 9 cm, Ø 25,1 cm ; Sammlung Weilinger, Wien
656A, ger. lig. MK, gem. B, Firmenmarke 19, Made in Austria, H. 9,4 cm, L. 30,6 cm, Ø 26,7 cm; MAK Wien, Inv. Ke 10602-7
Ausst.: PSK 1995
Lit.: ADGS, Nr. 10 (m. Abb.)

657
SCHALE
Form: oval
Handhaben: zwei Ösen
Dekor: zwei stilisierte Blumen mit acht Blättern
ca. 1928–1938
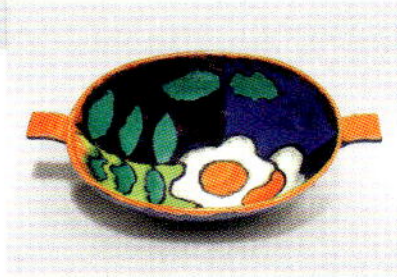
657 ger., gepr. lig. MK, gem. lig. AH, Firmenmarke 19, Made in Austria, H. 7 cm, L. 29,5 cm; PB Wien 19
657,83 Firmenmarke 19, Made in Austria, H. 7 cm, L. 32,3 cm; Sammlung a@goldscheider.net

663
VASE Pferde
Form: gebaucht, geweitete Mündung
Dekor: drei Pferde, mehrfarbig, Laufglasur
ca. 1928–1938
663,24 gem. B, Firmenmarke 19, Made in Austria, H. 20,8 cm, Ø 15,5 cm; Sammlung Weilinger, Wien
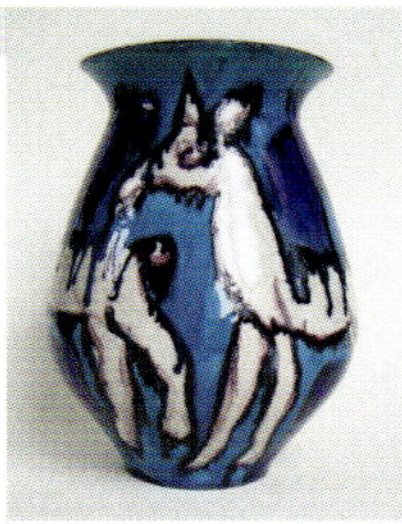

665
SCHALE
Form: rund
Dekor: abstrakte Malerei, mehrfarbig, Laufglasur
ca. 1928–1938
665,18 gem. B, Firmenmarke 19, Made in Austria, H. fehlt, Ø 25 cm; Doro 24.11.1992

666
SCHALE
Form: rund
Dekor: unregelmäßige Farbfelder, mehrfarbig, teils craqueliert, Laufglasur
ca. 1928–1938
666,23 gem. lig. AH, Firmenmarke 19, Made in Austria, Pick. 9, H. 7,2 cm, Ø 20 cm; PB Wien
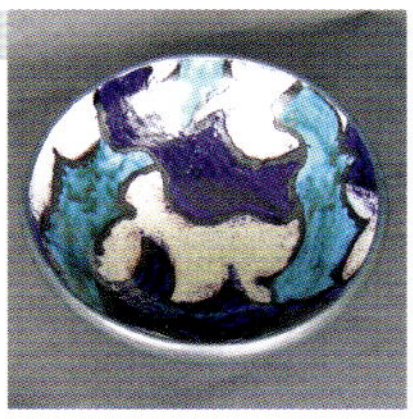

667
Meisinger, Ida
ASCHER Sitzender West Highland Terrier
Form: rund, vier Zigaretten/Zigarrenablagen
Dekor: mittig stehender Terrier (mit heraushängender Zunge) (vollplastische Fig.)
ca. 1928–1938
667,121 ger., Firmenmarke 19, Made in Austria, unsign., H. fehlt, Ø 25 cm; eBay # 2230288007
Bem.: Dieser Terrier, der unter den Nummern 6013; 6025 und 6500 bei Goldscheider produziert wurde, wird hier durch Beigabe der Aschenschale zur angewandten Variante einer Zierfigur. Eine Version als Kerzenhalter findet sich mit der Modell-Nr. 6466, eine weitere gibt es unter Modell-Nr. 606A als Schreibzeug. Offenbar hat die Manufaktur für diesen Aschenbecher neben der Ausführung mit dem Hund auch ein Rehkitz und evtl. weitere Figuren als Motive verwendet, die jedoch nicht ohne weiteres Ida Meisinger zugeordnet werden können.
ASCHER Rehkitz
Form: rund, vier Zigaretten/Zigarrenablagen
Dekor: mittig stehendes Rehkitz (vollplastische Fig.)
667 ger. M, Firmenmarke 19, Made in Austria, H. 14 cm, Ø 25,5 cm; Sammlung a@goldscheider.net
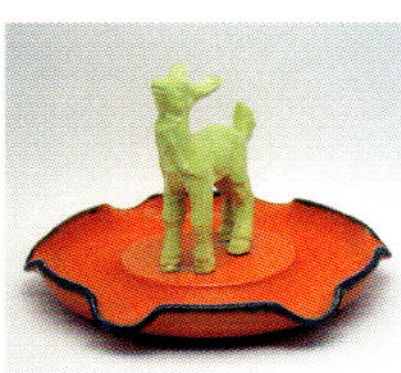

669
Kremser, Lizzi
SCHÜSSEL
Form: rund, hochgezogene Fahne
Dekor: Blumen
ca. 1928–1938
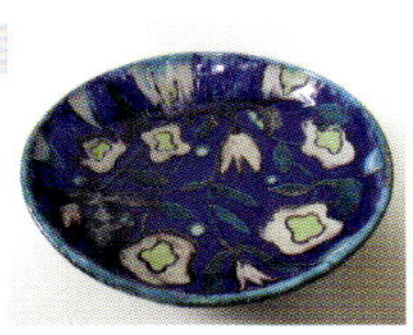
669,20 ger., Firmenmarke 19, Made in Austria, unsign., H. ca. 7,5 cm, Ø 34 cm; Kett. 206 (hier Ang. der Modell-Nr. 69); Doro 10.3.2003
669,54 gem. B., Firmenmarke 19, Made in Austria, unsign., H. 7,5 cm, Ø 34,5 cm; Sammlung Weilinger, Wien
Lit.: Hofmann, E.: Österreichisches Kunsthandwerk II: Keramik, in: ÖK, 3.1932, Nr.1, S. 21 (m. Abb.), hier mit Ang. der Künstlerin „Kremser" und dem Titel „Keramische Schüssel", „Mit besonderer Ambition und Energie hat die Wiener Manufaktur Friedrich Goldscheider den Weg zum neuen Stil gefunden. Ein Stab junger Künstler arbeitet dort und hat die leuchtend bunten Gefäße mit starken, satten Glasuren geschaffen, die besonders das Ausland schnell gewonnen haben. Köstliche Tierfiguren, hübsche blumengeschmückte Schüsseln, Dosen, usw. sind aus dieser Werkstätte hervorgegangen. Die neuen Arbeiten zeigen den Ehrgeiz feiner Craquelés, gedämpfter, vornehmer Glasuren und bemühen sich, mit höheren künstlerischen Werten das heute so schwierige Publikum zu gewinnen".

671
SCHALE
Form: rechteckig
Handhabe: eine stabförmige (über die beiden Längsseiten)
Dekor: rechteckige Farbfelder, mehrfarbig, Laufglasur
ca. 1928–1938
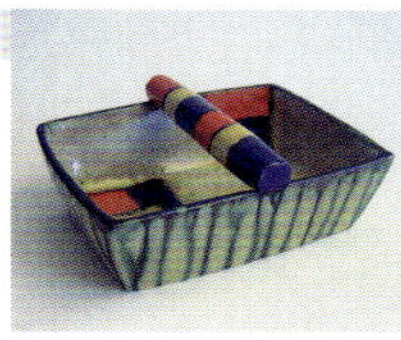
671,13 gem. L, Firmenmarke 19, Made in Austria, H. 9,2 cm, L. 21,3 cm, B. 18 cm; PB Deutschland

674
Geiger, Benno (zugeschrieben)
VASE (Bodenvase)
Form: flaschenförmig, leicht geweitete Mündung, konischer, gewulsteter Standring
Dekor: abstrakte Malerei, mehrfarbig, Laufglasur
ca. 1928–1934
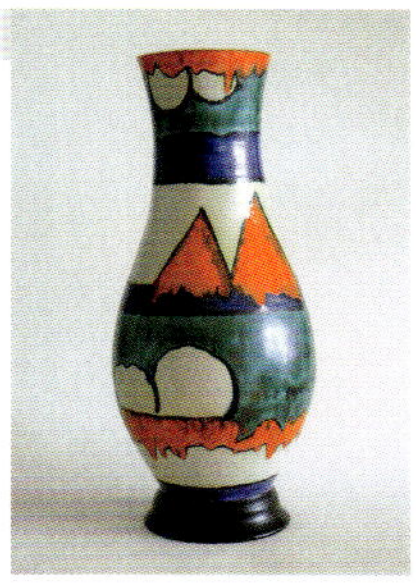
674,79 ger., ger. lig. MK, gem. G., Firmenmarke 19, Made in Austria, unsign., H. 40,3 cm, Ø 17 cm; PB Deutschland

677
Geiger, Benno
SCHALE
Form: vierpassig (oval)
Handhaben: zwei flache, seitlich angesetzte und an den Ecken abgeschrägte Henkel
Dekor: mehrfarbig, Laufglasur
ca. 1928–1938
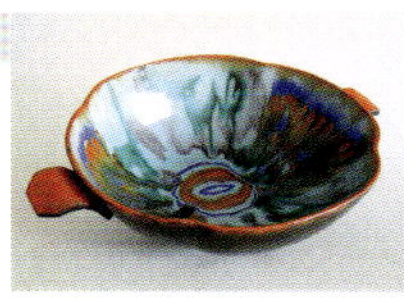
677,MM ger., ger. lig. MK, gem. lig. AH, Firmenmarke 19, Made in Austria, unsign., H. 9,5 cm, L 40 cm; PB Deutschland
Bem.: Für die Zuschreibung –> vgl. 6282.
–> vgl. 6282= ähnl. Modell.

679
VASE
Form: konisch, sich nach oben hin verjüngend, kurze geweitete Mündung
Dekor: vier vertikale, dreifarbige Streifen, graue Farbfelder, mehrfarbig, Laufglasur
ca. 1928–1938
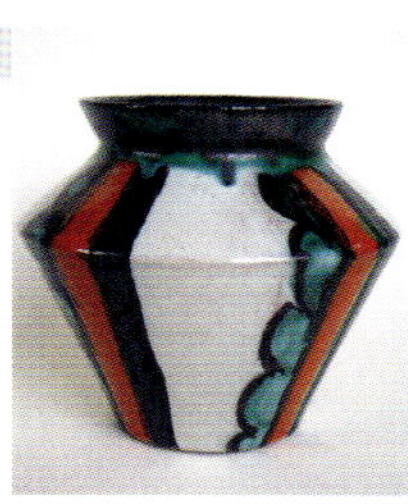
679,154 ger. lig. MK, gem. lig. MK, Firmenmarke 19, Made in Austria, H. 11,4 cm, Ø 13 cm; Sammlung Weilinger, Wien

680
VASE
Form: trichterförmig, leicht konischer Stand
Handhaben: vier Ösen
Dekor: stilisierte Blätter, Blüten, Punkte, mehrfarbig, Laufglasur
ca. 1928–1938
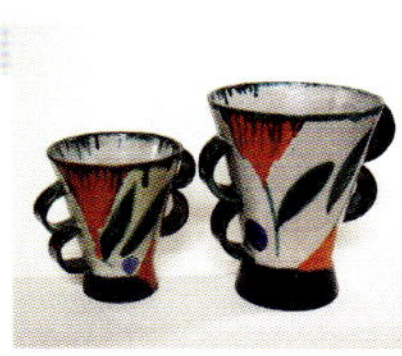
680,64 ger. lig. MK, gem. lig. MK, Firmenmarke 19, Made in Austria, H. 18,6 cm, Ø 15,5 cm; MAK Wien, Inv. Ke 10602-70
680,77 gem. V?, Firmenmarke 19, Made in Austria, H. 13,8 cm, Ø 12,1 cm; MAK Wien, Inv. Ke 10602-67
Bem.: Etwas ungewöhnlich ist die Tatsache, dass beide unter der gleichen Mod.-Nr. nachweisbaren Objekte, um 4,8 cm in der Höhe differieren.
Ausst.: PSK 1995
Lit.: ADGS, Nr. 32, 33 (m. Abb.)

681
SCHALE
Form: rund
Handhaben: zwei Ösen
Dekor: stilisierte Blumen, Blätter, mehrfarbig, Laufglasur
ca. 1928–1938
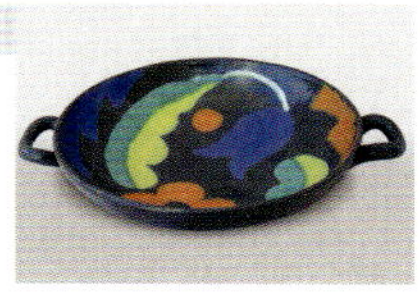
681,77 Firmenmarke 19, Made in Austria, H. fehlt, Ø 21 cm; Doro S. 2./3.12.1998
681 ger., Firmenmarke 19, Made in Austria, H. fehlt, Ø 22,5 cm; Sammlung Stressless, Portugal

688
VASE
Form: becherförmig
Handhaben: zwei (ohrenartig angesetzte)
Dekor: stilisierte Blumen, geometrische Formen, mehrfarbig
ca. 1928–1938
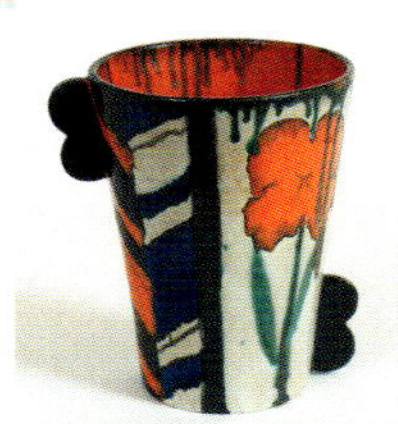
688,222 ger., ger. lig. MK, gem. B, Firmenmarke 19, Made in Austria, H. 15,9 cm; Sammlung.schmutz@chello.at 19
–> vgl. 688A

688A
VASE
Form: becherförmig
Handhaben: zwei (ohrenartig angesetzte)
Dekor: stilisierte Blumen, geometrische Formen, mehrfarbig
ca. 1928–1938
688,A ger., ger. lig. MK, gem. B, Firmenmarke 19, Made in Austria, H. 21 cm; Sammlung Weilinger, Wien
–> vgl. 688

689
VASE
Form: gebaucht (beutelartig), kurze trichterförmige Mündung
Dekor: unregelmäßige Farbfelder, Kreise, Punkte, mehrfarbig, Laufglasur
ca. 1928–1938
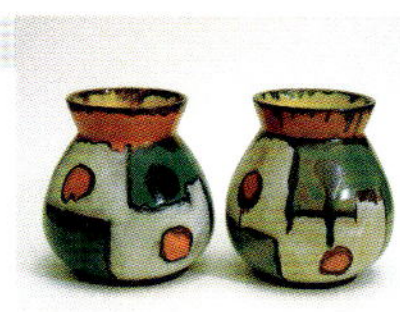
689,9 gem. lig. MK, Firmenmarke 19, Made in Austria, H. 13,1 cm, Ø 13 cm; PB Deutschland
689,26 gem. lig. MK, Firmenmarke 19, Made in Austria, H. 13,3 cm, Ø 13 cm; Sammlung John Nash

690
VASE
Form: kugelförmig, gedrückt, trichterförmige Mündung
Handhaben: zwei Ösen
Dekor: unregelmäßige Farbfelder, mehrfarbig, Laufglasur
ca. 1928–1938
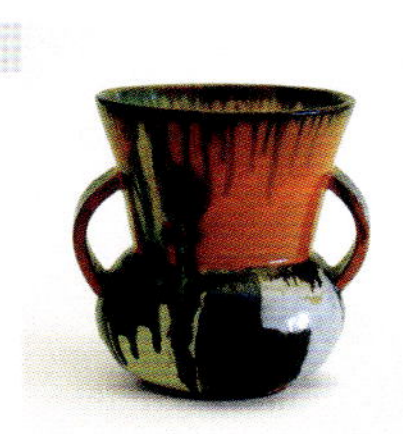
690,13 Firmenmarke 19, Made in Austria, H. 12,7 cm, Ø 14,2 cm; Sammlung.schmutz@chello.at 87
690,46 gem. lig. AH, Firmenmarke 19, Made in Austria, H. 13,9 cm, Ø 14,1 cm; PB Deutschland
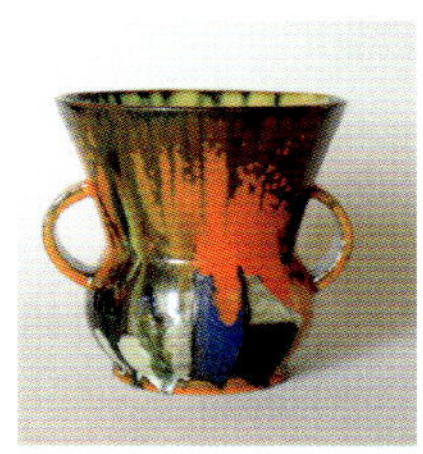

696
DECKELDOSE
Form: zylinderförmig, Deckel kugelig gewölbt
Knauf: konisch
Dekor: horizontale Streifen, mehrfarbig, Laufglasur
ca. 1928–1938
696,13 gem. lig. MK, Firmenmarke 19, Made in Austria, H. 23,7 cm, Ø 12,5 cm; Sammlung.schmutz@chello.at 52

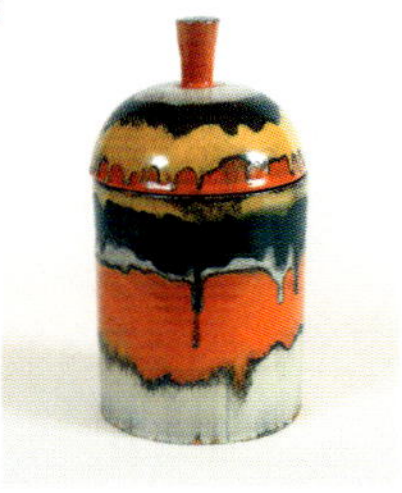

702
DECKELDOSE
Form: rund
Handhaben: vier Ösen
Knauf: eine längliche Öse
Dekor: abstrakte Malerei, mehrfarbig, Laufglasur
702 ger., Firmenmarke 19, Made in Austria, H. 15,5 cm, Ø 13,8 cm; Doro 20.4.1993

703
VASE
Form: balusterförmig, sehr hohe trichterförmige Mündung
Dekor: Laufglasur
ca. 1928–1938
703,4 gem. lig. MK, Firmenmarke 19, Made in Austria, H. 50,1 cm, Ø 20,6 cm; Sammlung.schmutz@chello.at 107

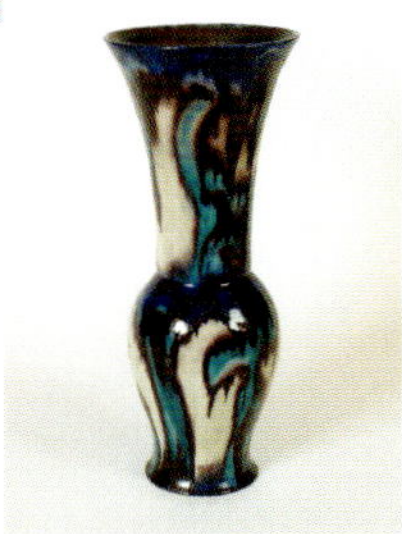

706
VASE
Form: kugelförmig, trichterförmige Mündung
Handhaben: zwei Ösen
Dekor: abstrakte Malerei, mehrfarbig, Laufglasur
ca. 1928–1938
706,4 gem. lig. MK, Firmenmarke 19, Made in Austria, H. 27,2 cm; Sammlung.schmutz@chello.at 30

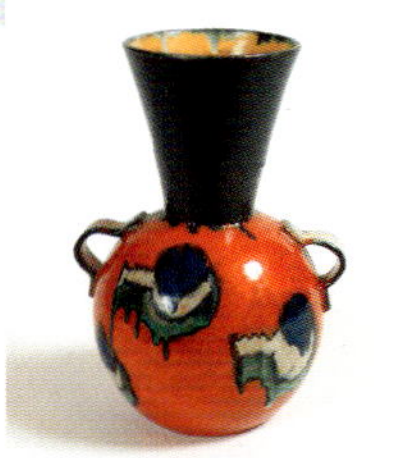

709?
SCHALE / SCHÜSSEL
Form: rund
Handhaben: zwei flache, geschwungene Henkel
Dekor: teils geritzt, abstrakte Malerei, Halbmonde, Kreise, mehrfarbig
709? ger., Firmenmarke 19, Made in Austria, H. fehlt, Ø 25 cm; eBay # 25061233568

713
VASE
Form: kugelförmig, sehr hohe trichterförmige Mündung
Handhaben: vier Ösen
Dekor: Laufglasur
ca. 1928–1938
713,4 Firmenmarke 19, Made in Austria, H. 49,2 cm, Ø 15,6 cm; Sammlung.schmutz@chello.at 106

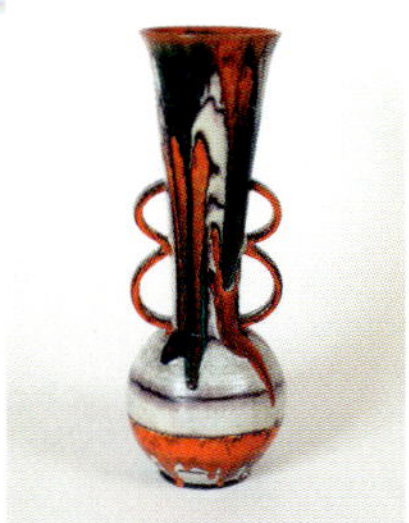

715
KERZENHALTER
Form: zylindrisch, zur Mitte hin eingezogen
Dekor: geometrischer Dekor, Zacken und Punkte, weiß-rot-grün, Laufglasur?
ca. 1928–1938
715,8 Firmenmarke 19, Made in Austria, H. 20 cm, Ø 9,2 cm; Sammlung a@goldscheider.net

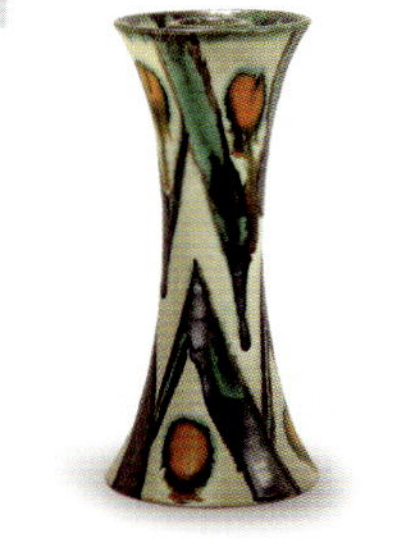

718
BECHER / ZIGARETTEN- bzw. BLEISTIFTHALTER
Form: zylindrisch
Dekor: Frauenkopf im Profil
ca. 1928–1938
718 Firmenmarke 19, Made in Germany, H. 4,8 cm, Ø 5,5 cm; Sammlung a@goldscheider.net
–> vgl. 7719= ähnl. Modell.

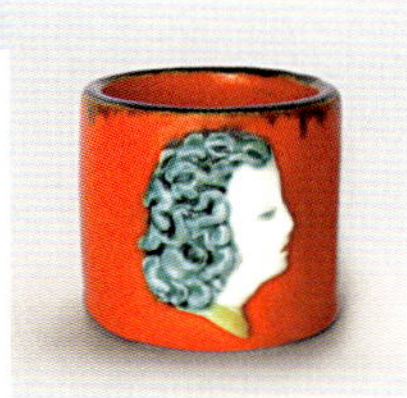

720
VASE
Form: kugelförmig, zylindrische Mündung
Dekor: mehrfarbig, Laufglasur
ca. 1928–1938
720 gem., gem. G., Firmenmarke 19, Made in Austria, H. 14,2 cm; Sammlung.schmutz@chello.at 86
720 Firmenmarke 19, Made in Austria, H. 14 cm; WKA 33
ohne Nr., gem. lig. MK, Firmenmarke 19, Made in Austria, H. 13,9 cm; Sammlung a@goldscheider.net

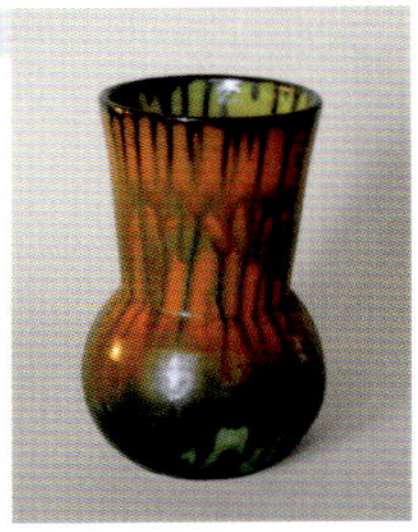

721
VASE
Form: becherförmig, im unteren Drittel tailliert, konische Mündung
Dekor: stilisierte Blätter, Kreise, oder abstrakte Blumen, mehrfarbig, Laufglasur
ca. 1928–1938
721,7 gem. K?, Firmenmarke 19, Made in Austria, H. 14 cm; eBay # 1211244265; Doro S. 27.10.2000
721 ger., gem. B, Firmenmarke 19, Made in Austria, H. 13,4 cm, Ø 9,5 cm; Sammlung a@goldscheider.net

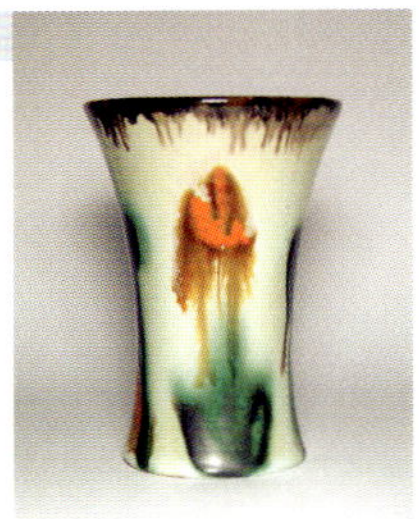

723
VASE
Form: eiförmig, trichterförmige Mündung
Dekor: Rechtecke (verlaufend), mehrfarbig, Laufglasur
ca. 1928–1938
723,83 ger. S., ger. lig. MK, Firmenmarke 19, Made in Austria, H. 13,9 cm; Sammlung Renée und Otto Schenk, Wien

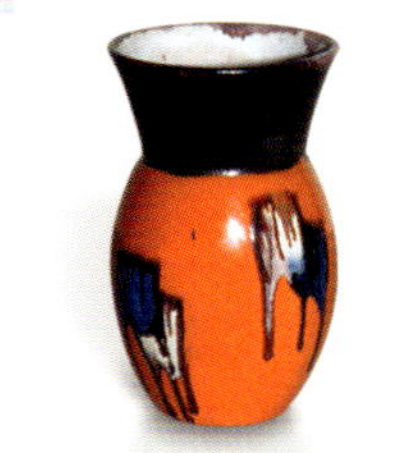

731
DECKELDOSE
Form: rund
Knauf: zwei vertikale ligierte zylinderförmige Stäbe
Dekor: abstrakte Malerei, Laufglasur
ca. 1928–1938
731,2 Firmenmarke 19, Made in Austria, H. 9,5 cm, Ø fehlt; WKA 33

736
DECKELDOSE
Form: rund
Knauf: flache Öse
Dekor: Streifen und geometrische Formen, grau-orange-dunkelgrün, Laufglasur
ca. 1928–1938
736,7 gem. W, Firmenmarke 19 (braun), Made in Austria (braun), H. 7,2 cm, Ø 14,2 cm; Sammlung John Nash

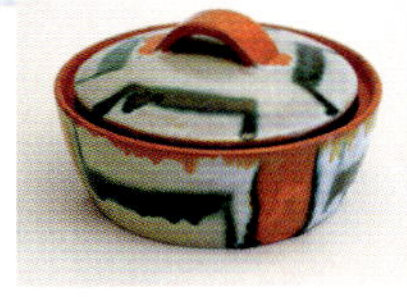

742
VASE
Form: kugelförmig, hohe zylindrische Mündung
Dekor: mehrfarbig, Laufglasur
ca. 1928–1938
742 ger., gem. FK, Firmenmarke 19, H. 26 cm; Sammlung.schmutz@chello.at 20

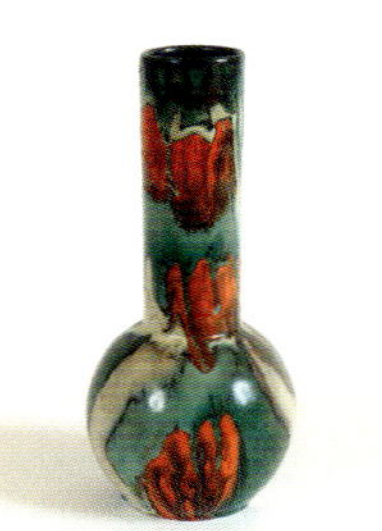

750
VASE
Form: becherförmig, gebaucht, kurze geweitete Mündung
Dekor: stilisierte Blumenblüten, mehrfarbig, Laufglasur
ca. 1928–1938
750, B27 ger., gem. W, Firmenmarke 19, Made in Austria, H. 14,5 cm; Sammlung Weilinger, Wien
750,S ger., gem. lig. TK, Firmenmarke 19, H. 14 cm; Wien Museum Inv. 186.694

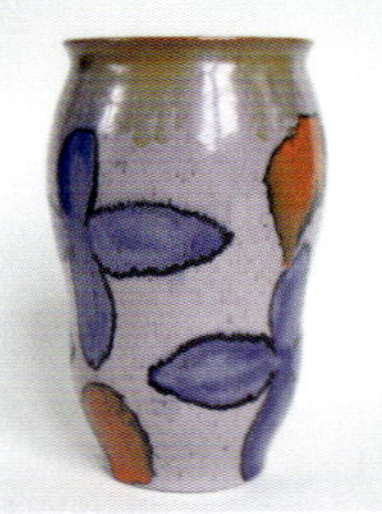

756
VASE Tulpen
Form: becherförmig, geweitete Mündung, flacher, runder Standring
Dekor: stilisierte Tulpen, mehrfarbig
ca. 1928–1938
756 gem., Firmenmarke 21a, (Firmenmarke U), Made in Austria, H. 35,8 cm; Sammlung.schmutz@chello.at 31

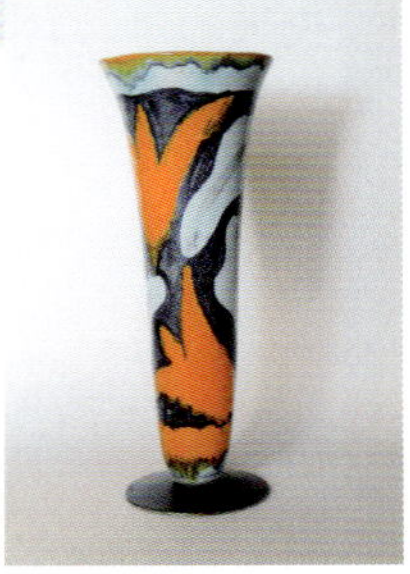

759
Geiger, Benno (zugeschrieben)
VASE
Form: becherförmig, geweitete Mündung, konischer Standring
Dekor: Farbfelder, mehrfarbig, Laufglasur
ca. 1928–1934
759,43,5 ger., gem. M., W, Firmenmarke 19, Made in Austria, unsign., H. 30,5 cm, Ø 13 cm; Wien Museum Inv. 186.697

760
Geiger, Benno
VASE
Form: trichterförmig, konischer Stand
Dekor: mehrfarbig, Laufglasur (diverse Dekore)
ca. 1928–1934
760,48 Firmenmarke 19, Made in Austria, H. 16 cm; WKA 33
760,70 ger. F, gem. lig. AH, Firmenmarke 19, Made in Austria, unsign., H. 17,8 cm; eBay # 1200092600
760,D,150 ger., Firmenmarke 19, Made in Austria, Pick. 11 (Entwurf Benno Geiger, Wien), H. 15 cm; Kett. 15
Bem.: Diese Vase wurde in diversen Dekoren staffiert, wie die beiden hier nachgewiesenen Exemplare belegen.

769
SCHALE
Form: rund
Handhaben: vier Ösen
Dekor: stilisierte Blumenblüten, mehrfarbig
ca. 1928–1938
769 ger., ger. lig. MK, gem. B, Firmenmarke 19, Made in Austria, H. 5,2 cm, L. 27,7 cm, Ø 22,1 cm; PB Wien 19
Ausst.: PSK 1995
Lit.: ADGS, Nr. 16 (m. Abb.)

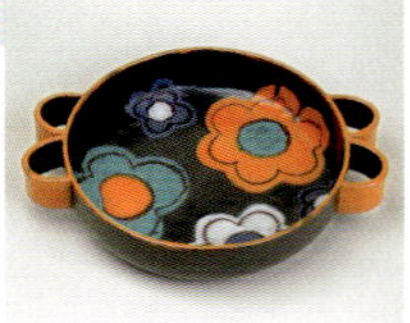

771
DECKELDOSE
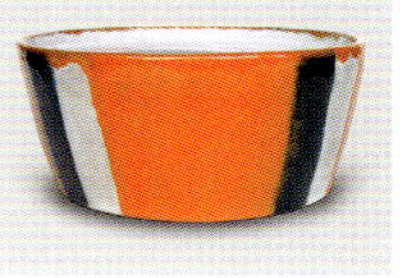
Form, rund, konisch
Dekor: abstrakter Dekor, vertikale Streifen
ca. 1928–1938
771,A Firmenmarke 19, Made in Austria, H. (ohne Deckel) 6,6 cm, Ø 15,3 cm; Sammlung a@goldscheider.net
Bem.: Der Deckel fehlt bei diesem Ex.

777
VASE Fische
Form: becherförmig, geweitete Mündung
Dekor: stilisierte Fische, Pflanzen, mehrfarbig, Laufglasur
ca. 1928–1938
777 ger., ger. lig. MK, Firmenmarke 19, H. 15 cm; Schuler 59

777A
VASE
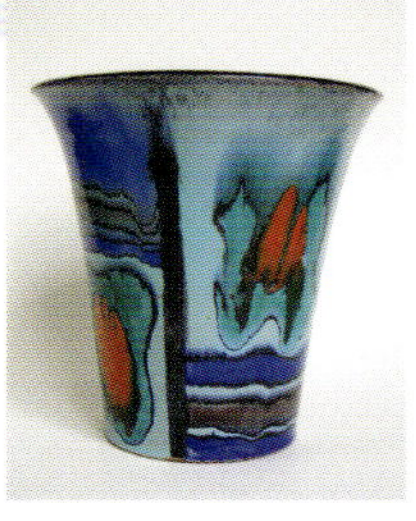
Form: becherförmig, geweitete Mündung
Dekor: vier durch vertikale Streifen getrennte Felder mit abstrakten Formen, mehrfarbig, Laufglasur
ca. 1928–1938
777,A ger., ger. lig. MK, gem. lig. TK, Firmenmarke 19, Made in Austria, H. 17,5 cm, Ø 18 cm; Sammlung Weilinger, Wien

779
VASE Kugel
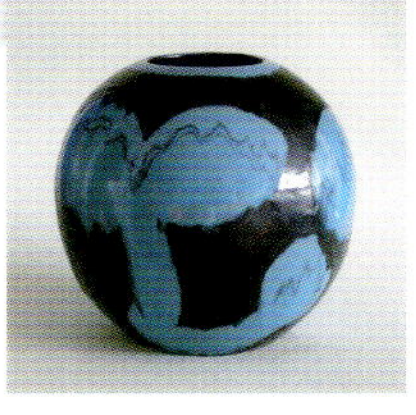
Form: kugelförmig
Dekor: abstrakte Malerei, türkis-schwarz, Laufglasur
ca. 1928–1938
779,26 ger., ger. lig. MK, gem. M (oder W?), Firmenmarke 19, Made in Austria, H. 20,1 cm, Ø 21 cm; PB Deutschland

780
VASE Kugel
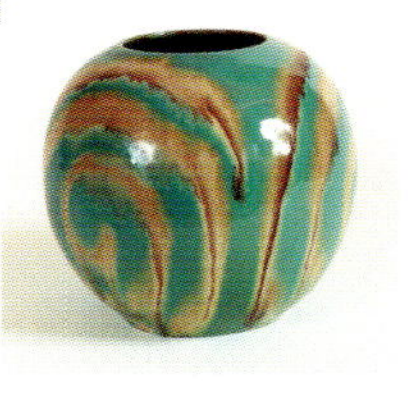
Form: kugelförmig
Dekor: spiralförmig, Laufglasur
ca. 1928–1938
780,26ger., gem. W, Firmenmarke 19, Made in Austria, H. 14,4 cm, Ø 16,5 cm; Sammlung.schmutz@chello.at 77

782
SCHALE
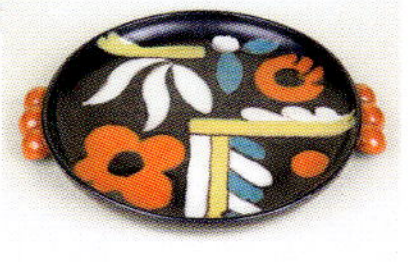
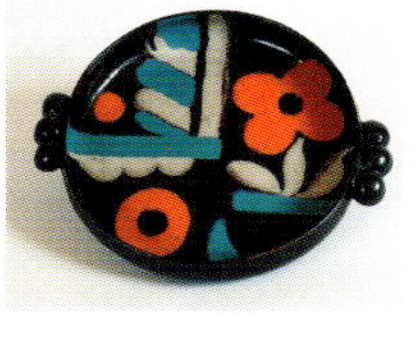
Form: rund
Handhaben: zwei kugelartige Handhaben (ohrenartig angesetzt)
Dekor: stilisierte Blume und abstrakte Formen, rot-türkis-schwarz-weiß
ca. 1928–1938
782,59 gem. W.H., Firmenmarke 19, Made in Austria, Pick. 8, H. 3,5 cm, L. 33 cm, Ø 28 cm; Sammlung.schmutz@chello.at 65
782,W ger., gem. B, Firmenmarke 19, Made in Austria, H. 3,5 cm, L. 32,2 cm, Ø 27,1 cm; PB Wien 19
782 ger. lig. MK, Firmenmarke 19, Made in Austria, H. fehlt, Ø 34 cm; Doro 22.11.1990
Ausst.: PSK 1995
Lit.: ADGS, Nr. 17 (m. Abb.)

784
SCHALE mit Fuß (konischer Standring)
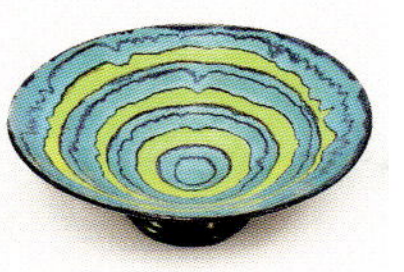
Form: rund
Dekor: Kreise, Laufglasur
ca. 1928–1938
784 Firmenmarke 19, Made in Austria, H. 10,1 cm, Ø 27,5 cm; Sammlung a@goldscheider.net

803
VASE
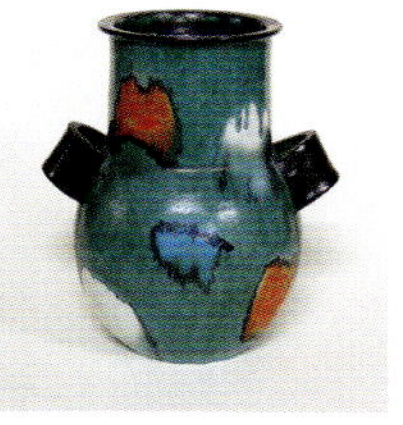
Form: kugelförmig, zylindrische Mündung, flach gedrückte Lippe
Handhaben: zwei Ösen
Dekor: Flecken, mehrfarbig, Laufglasur
ca. 1928–1938
803 ger., ger. lig. MK, gem. G, Firmenmarke 19, Made in Austria, H. 20,3 cm, Ø 14,3 cm; MAK Wien, Inv. Ke 10602-42
Lit.: ADGS, Nr. 25 (m. Abb.)

805
VASE
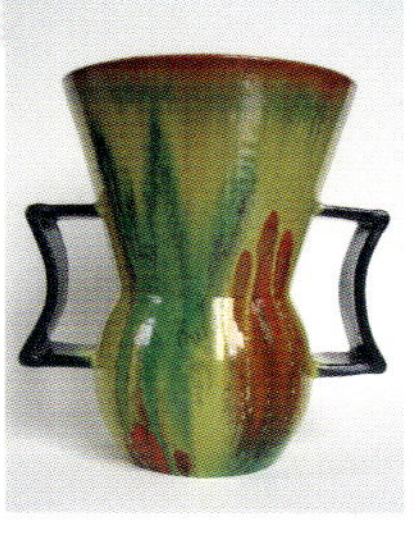
Form: kugelförmig, trichterförmige Mündung
Handhaben: zwei eckige Griffe mit geschwungenem Steg
Dekor: mehrfarbig, Laufglasur (diverse Dekore)
ca. 1928–1938
805,10 ger. lig. MK, Firmenmarke 19, Made in Austria, H. 21 cm; BM 57
805,51 ger., Firmenmarke 19, H. 21 cm; Doro 15.9.2003
805 ger., ger. lig. MK, Firmenmarke 19, Made in Austria, H. 20,1 cm, Ø 17,1 cm; eBay # 2555544954
805 ger., ger. lig. MK, WW, gem. B, Firmenmarke 19, Made in Austria, H. 21 cm, Ø 16,4 cm; Sammlung Weilinger, Wien
Bem.: Diese Vase wurde in diversen Dekoren staffiert, wie die hier nachgewiesenen Exemplare belegen.

810
VASE
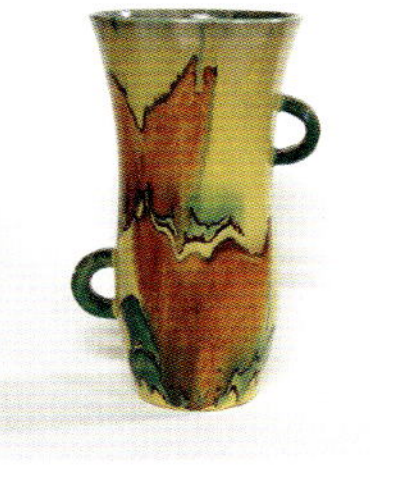
Form: becherförmig, leicht geweitete Mündung
Handhaben: zwei Ösen
Dekor: mehrfarbig, Laufglasur
ca. 1928–1938
810,2,M ger., Firmenmarke 19, Made in Austria, H. 21,4 cm; Kett. 201
810,4,M ger., ger. S, gem. W.H., Firmenmarke 19, H. 21,4 cm, Ø 11,5 cm; MAK Wien, Inv. Ke 10602-68
Ausst.: PSK 1995
Lit.: ADGS, Nr. 26 (m. Abb.)

811
SCHALE
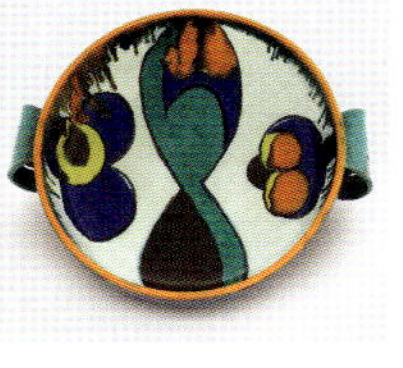
Form: rund
Handhaben: zwei Ösen
Dekor: abstrakt (Blumen, Fisch?), mehrfarbig, Laufglasur
ca. 1928–1938
811 ger., ger. lig. MK, gem. B, Firmenmarke 19, Made in Austria, H. 6,8 cm, Ø 27,5 cm; Sammlung a@goldscheider.net

812
SCHALE
Form: rund
Handhaben: zwei Ösen
Dekor: Blume, mehrfarbig, Laufglasur
ca. 1928–1938
812 (Firmenmarke ohne Ang.), H. fehlt, Ø 24 cm; Pinhas, S. 176 (m. Abb.)

830A
VASE
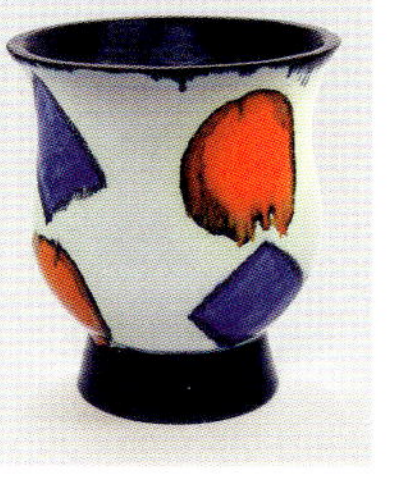
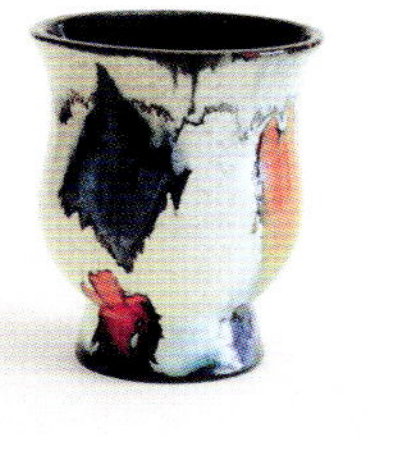
Form: becherförmig, leicht geweitete Mündung, konischer Standring
Dekor: geometrische Formen (Kreise, Quadrate), mehrfarbig, Laufglasur
ca. 1928–1938
830A,S ger., gem. lig. TK, Firmenmarke 19, Made in Austria, H. 10,9 cm, Ø 9,7 cm; PB Wien 19
830A ger., gem. B, Firmenmarke 19, Made in Austria, H. 11,6 cm, Ø 10,2 cm; Sammlung a@goldscheider.net
830A ger., gem. lig. AH, Firmenmarke 19, Made in Austria, H. 11,7 cm, Ø 10,2 cm; Sammlung John Nash
Ausst.: PSK 1995
Lit.: ADGS, Nr. 21 (m. Abb.)

830D
VASE
Form: kugelförmig,
Dekor: mehrfarbig, Laufglasur
ca. 1928–1938
830D ger., Firmenmarke 19, Made in Austria, H. 11 cm; KH

830E
VASE
Form: becherförmig, trichterförmige Mündung, Stand konisch eingezogen
Dekor: diverse Dekore
ca. 1928–1938
830E ger., ger. lig. MK, Firmenmarke 19, Made in Austria, H. 11 cm, Ø 8 cm; Wien Museum Inv. 186.693
830E ger., Firmenmarke 19, Made in Austria, H. 11,5 cm; Doro FVK
830E ger., Firmenmarke 19, Made in Austria, H. 12 cm; WKA 33
Bem.: Diese Vase wurde in diversen Dekoren staffiert, wie die hier nachgewiesenen Exemplare belegen.

830F
VASE
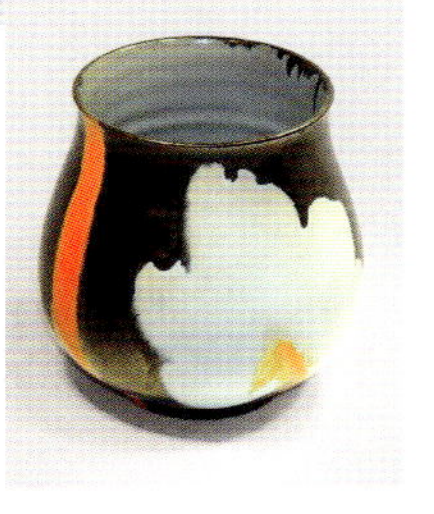
Form: becher-/beutelförmig
Dekor: Farbfelder, mehrfarbig, Laufglasur
ca. 1928–1938
830F ger., gem. W, Firmenmarke 19, Made in Austria, H. 10,3 cm, Ø 11 cm; Wien Museum Inv. 186.699
830F,S ger., gem. lig. AH, Firmenmarke 19, Made in Austria, H. 10,9 cm, Ø 11 cm; Sammlung a@goldscheider.net
Lit.: SS, S. 32, 33 (m. Abb.).

836
SCHALE
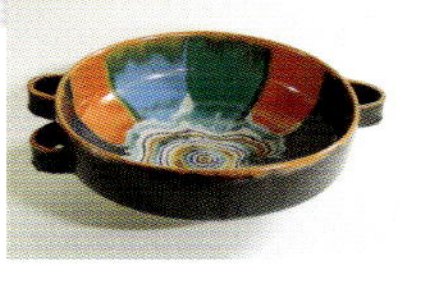
Form: rund
Handhaben: vier Ösen
Dekor: abstrakt, mehrfarbig, Laufglasur
ca. 1928–1938
836 ger., ger. lig. MK, gem. B, Firmenmarke 19, Made in Austria, H. 8,6 cm, L. 34,4 cm, Ø 28,1 cm; MAK Wien, Inv. Ke 10602-10
Ausst.: PSK 1995
Lit.: ADGS, Nr. 13 (m. Abb.)

837
SCHALE
Form: rund
Handhaben: vier Ösen
Dekor: drei Blumenblüten, mehrfarbig, Laufglasur
ca. 1928–1938
837 ger., Firmenmarke 19, H. 12 cm, L. ca. 33,5 cm, Ø 26,5 cm; Doro 20.10.1997
837 (Firmenmarke ohne Ang.), H. 9 cm, L. 33,5 cm, Ø 26,5 cm; Doro 23.5.1991

846
TOPF
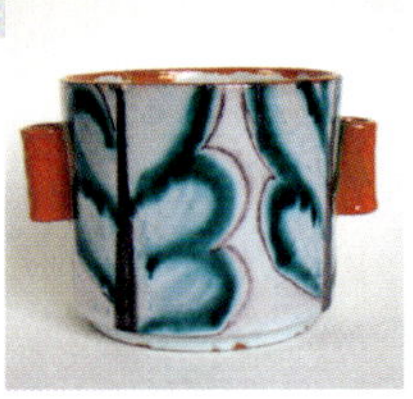
Form: zylinderförmig
Handhaben: zwei Ösen (vertikal)
Dekor: vier stilisierte Blätter/Bäume, mehrfarbig, Laufglasur
ca. 1928–1938
846,2 ger., ger. lig. MK, gem. lig. AH, Firmenmarke 19, Made in Austria, H. 10,8 cm, Ø 12,1 cm; Sammlung Weilinger, Wien

850
Obsieger, Robert (zugeschrieben)
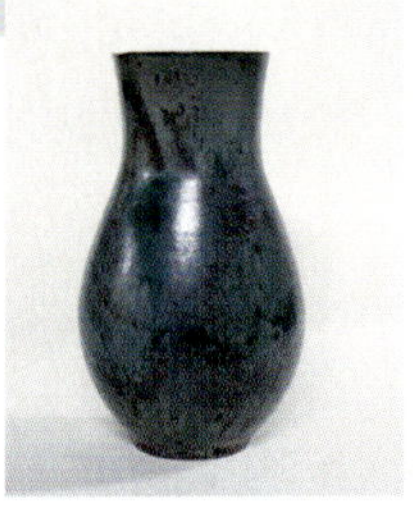
VASE
Form: birnförmig, fast zylindrische Mündung
Dekor: gesprenkelt, Scharffeuerglasur
ca. 1928–1938
850,S,II ger., (ohne Firmenmarke), unsign., Papieretikett S 16.–, H. 25,3 cm, Ø 14,1 cm; MAK Wien, Inv. Ke 7308
Bem.: Das Papieretikett der Zeit zeigt den damaligen Verkaufspreis von 16 Schilling. Die Zuschreibung an Robert Obsieger erfolgt aufgrund vergleichbarer Objekte in: N.N., 50 Jahre Wiener Manufaktur Friedrich Goldscheider, Die Schaulade 11, 1935, S. 152–154 (m. Abb.).
Ausst.: Weihnachtsausst. ÖMKI 1933; Austria in England 1934.

851
Obsieger, Robert (zugeschrieben)
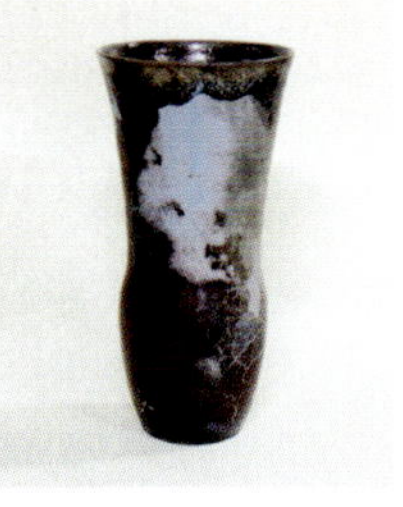
VASE
Form: gebaucht, lang gezogene konische Mündung
Dekor: craqueliert, Scharffeuerglasur
ca. 1928–1938
851,W ger., gem. R4, (ohne Firmenmarke), unsign., H. 23,5 cm, Ø 11,8 cm; MAK Wien, Inv. Ke 7309
Bem.: Die Zuschreibung an Robert Obsieger erfolgt aufgrund vergleichbarer Objekte in: N.N., 50 Jahre Wiener Manufaktur Friedrich Goldscheider, Die Schaulade 11, 1935, S. 152–154 (m. Abb.).
Ausst.: Weihnachtsausst. ÖMKI 1933; Austria in England 1934.
Lit.: Neuwirth (1981), Nr. 41

854
Obsieger, Robert (zugeschrieben)
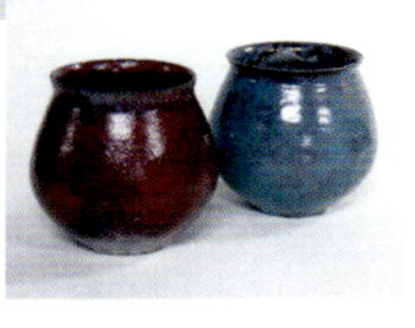
VASE
Form: kugelig, leicht geweitete kurze Mündung
Dekor: gesprenkelt, Scharffeuerglasur (div. Farben)
ca. 1928–1938
854 ger., gem. V, (ohne Firmenmarke), unsign., H. 10,9 cm, Ø 12,2 cm; MAK Wien, Inv. Ke 7314
854,S ger., gem. R94, (ohne Firmenmarke), unsign., H. 10,9 cm, Ø 12,1 cm; MAK Wien, Inv. Ke 7315
Bem.: Die Zuschreibung an Robert Obsieger erfolgt aufgrund vergleichbarer Objekte in: N.N., 50 Jahre Wiener Manufaktur Friedrich Goldscheider, Die Schaulade 11, 1935, S. 152–154 (m. Abb.).
Ausst.: Weihnachtsausst. ÖMKI 1933; Austria in England 1934.
Lit.: Neuwirth (1981), Nr. 40

858
Obsieger, Robert (zugeschrieben)
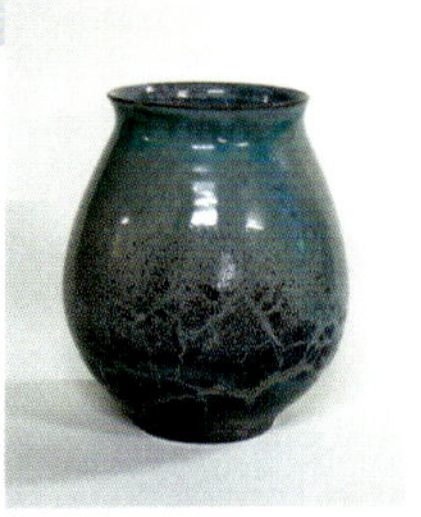
VASE
Form: eiförmig, kurze geweitete Mündung
Dekor: rissig, Scharffeuerglasur
ca. 1928–1938
858,W. ger., gem. R3, (ohne Firmenmarke), unsign., H. 17,2 cm; MAK Wien, Inv. Ke 7312
Bem.: Die Zuschreibung an Robert Obsieger erfolgt aufgrund vergleichbarer Objekte in: N.N., 50 Jahre Wiener Manufaktur Friedrich Goldscheider, Die Schaulade 11, 1935, S. 152–154 (m. Abb.).
Ausst.: Weihnachtsausst. ÖMKI 1933; Austria in England 1934.

860
Obsieger, Robert (zugeschrieben)
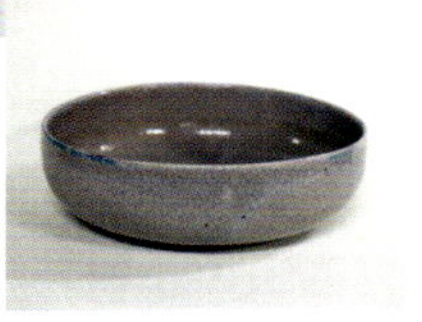
SCHALE
Form: rund
Dekor: craqueliert
ca. 1928–1938
860,S ger., Firmenmarke 19, unsign., Papieretikett S 16.–, H. 6,6 cm, Ø 21 cm; MAK Wien, Inv. Ke 7316
Bem.: Das Papieretikett der Zeit zeigt den damaligen Verkaufspreis von 16 Schilling. Die Zuschreibung an Robert Obsieger erfolgt aufgrund vergleichbarer Objekte in: N.N., 50 Jahre Wiener Manufaktur Friedrich Goldscheider, Die Schaulade 11, 1935, S. 152–154 (m. Abb.).
Ausst.: Weihnachtsausst. ÖMKI 1933; Austria in England 1934.

863
Obsieger, Robert (zugeschrieben)
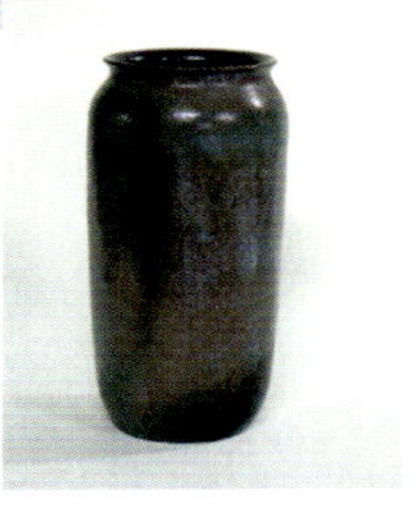
VASE
Form: zylinderförmig, eingezogene Schulter, kurze geweitete Mündung
Dekor: gesprenkelte Scharffeuerglasur
ca. 1928–1938
863 (schwer lesbar), ger. R I, (ohne Firmenmarke), unsign., H. 17,4 cm, Ø 9,2 cm; MAK Wien, Inv. Ke 7311
Bem.: Die Zuschreibung an Robert Obsieger erfolgt aufgrund vergleichbarer Objekte in: N.N., 50 Jahre Wiener Manufaktur Friedrich Goldscheider, Die Schaulade 11, 1935, S. 152–154 (m. Abb.).
Ausst.: Weihnachtsausst. ÖMKI 1933; Austria in England 1934.

874
DECKELDOSE
Form: konisch
Handhaben: zwei zylinderförmige Stifte
Knauf: zwei schräg angesetzte zylinderförmige Stifte
Dekor: Farbfelder, mehrfarbig, Laufglasur
ca. 1928–1938
874,2 ger., ger. lig. MK, gem. B, Firmenmarke 19, Made in Austria, H. 10,8 cm, Ø 16 cm; Wien Museum Inv. 186.695/1+2

875
SCHALE
Form: rund, mehrpassig, gewellter Rand
Dekor: stilisierte Blumen
ca. 1928–1938
875 ger. 1 gepr., gem. lig. TK, Firmenmarke 19, Made in Austria, H. 5,6 cm, Ø 25 cm; GK, S. 21, 114 (m. Abb.); Wien Museum Inv. 182.061
875 gem. A, ger. lig. MK, Firmenmarke 19, Made in Austria, Pick. (ohne Ang.), H. fehlt, Ø 25 cm; Doro 24.11.1992
Ausst.: HMW (jetzt Wien Museum) 1985/86.
Lit.: GK, S. 21, 114 (m. Abb.)

876
Knörlein, Rudolf (zugeschrieben)
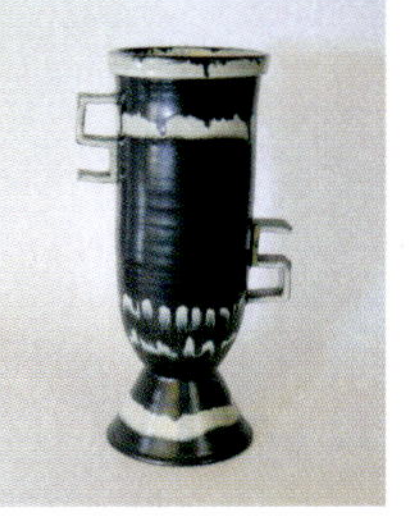
VASE
Form: becherförmig, gerillt, gewulstete Mündung, hoher konischer Standring
Handhaben: zwei S-förmige Handhaben
Dekor: schwarz-weiß, Laufglasur
ca. 1932–1936
876,8 gem. lig. AH, Firmenmarke 19, Made in Austria, unsign., H. 29,5 cm, Ø 11,5 cm, B. 15 cm; Sammlung.schmutz@chello.at 21
876,8 Firmenmarke 19, Made in Austria, unsign., H. 29 cm, Ø 11,5 cm, B. 15 cm; Sammlung.schmutz@chello.at 32
Bem.: Diese Seltenheit findet sich in der Sammlung Rudolf Schmutz. Hier konnte ein Vasenpaar zusammengetragen werden, wobei beide Vasen die gleichen Modellnummern tragen. Für die Zuschreibung an R. Knörlein diente das bei Rochowanski (1930), auf S. 78 abgebildete „Likörservice".

883
Geiger, Benno
VASE
Form: becherförmig, eingezogene flache Schulter, darauf hohe trichterförmige Mündung
Handhaben: vier seitlich angesetzte rechteckige Handhaben
Dekor: abstrakte Malerei, gelb-braun-schwarz, Laufglasur
ca. 1928–1934
883 ger., Firmenmarke 19, (Made in Austria ?), Pick. 11a (Entwurf Benno Geiger), H. 28,4 cm; Doro 18.11.1997
-> vgl. 1325M

886
DECKELDOSE
Form: ??
Handhaben: zwei Henkel
Dekor: abstrakte Malerei?, Laufglasur?
ca. 1928–1938
886 (oder 988?) ger., Firmenmarke 19, (Made in Austria ?), H. 12 cm, Ø 14 cm; Doro 21.12.1993
Bem.: Im Auktionskat. findet sich weder eine exakte Beschreibung der Form noch des Dekors, ebenso fehlt eine Abb. Die Modell-Nr. kann als 886 sowie 988 (umgedreht) gelesen werden. So lange sich kein zweites Stück dieser Dose findet, muss die Frage nach der Nr. offen bleiben.

888
SCHALE Farn
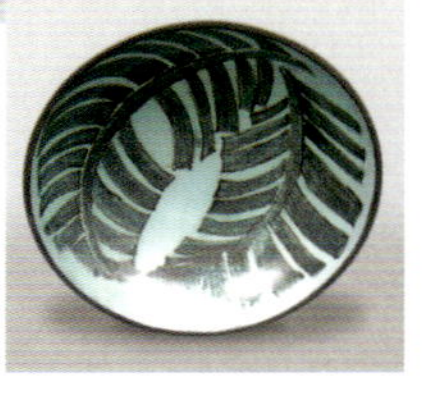
Form: rund
Dekor: zwei Farnblätter, türkis-schwarz
ca. 1928–1938
888 ger., ger. lig. MK, gem. lig. AH, Firmenmarke 19, Made in Austria, H. 7 cm, Ø 27,5 cm; PB Wien

913
Geiger, Benno
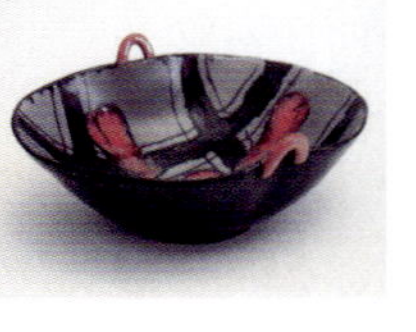
SCHALE
Form: oval
Handhaben: zwei Ösen
Dekor: gitterförmiges Muster mit Punkten, mehrfarbig, Laufglasur
ca. 1928–1934
913,7 ger., Firmenmarke 19, Made in Austria, Pick. 11 (Entwurf Benno Geiger, Wien), H. 8,5 cm, L. 22,7 cm, B. 17,5 cm; PB Deutschland
913 ger., ger. W, gem. lig. TK, Firmenmarke 19, Made in Austria, unsign., H. 8,5 cm, L. 22,5 cm, B. 17,5 cm; Wien Museum Inv. 190.175

924

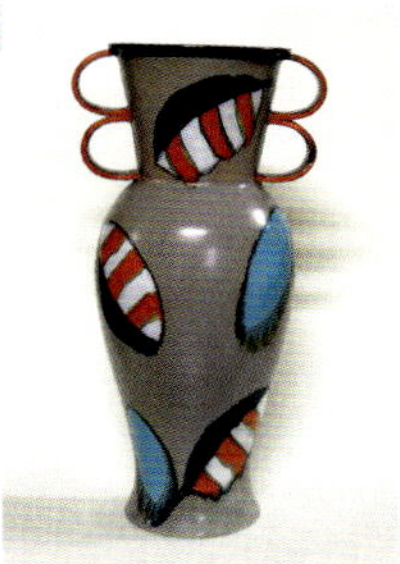

Geiger, Benno (zugeschrieben)
VASE
Form: balusterförmig, trichterförmige Mündung
Handhaben: vier Ösen
Dekor: stilisierte Blätter, mehrfarbig, Laufglasur
ca. 1928–1934
924 ger., ger. lig. MK, gem. B, Firmenmarke 19, Made in Austria, unsign., H. 43 cm, Ø 18,1 cm; MAK Wien, Inv. Ke 10602-26
924,W ger., Firmenmarke 19, Made in Austria, unsign., H. 40,2 cm; Doro 9.11.1993; 8.3.1994
Bem.: Für die Zuschreibung konnte eine signierte, und vom Dekor her sehr ähnliche Vase (ohne Mod.-Nr.), die bei Mehlis in der Auktion 18 (26.5.2000) angeboten wurde, herangezogen werden.
Ausst.: PSK 1995
Lit.: ADGS, Nr. 31 (m. Abb.)

926
Geiger, Benno (zugeschrieben)
SCHALE
Form: rund
Handhaben: zwei spiralförmige Handhaben
Dekor: geometrische Formen (Kreis, Streifen, Oval), mehrfarbig
ca. 1928–1934
926,W ger., Firmenmarke 19, unsign., H. fehlt, Ø fehlt; KH

927

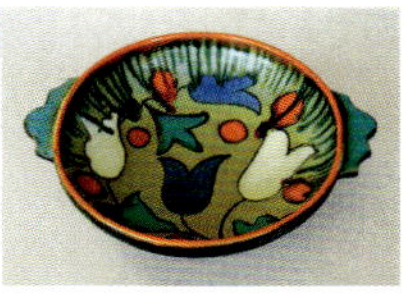

Geiger, Benno
SCHALE Tulpen
Form: rund
Handhaben: zwei flache, mehrfach geschwungene Handhaben (ohrenartig angesetzt)
Dekor: Tulpenblüten, mehrfarbig, Laufglasur
ca. 1928–1934
927 ger., ger. lig. MK, gem. W oder M, Firmenmarke 19, Made in Austria, Pick. 11 (Entwurf Benno Geiger, Wien), H. 5,9 cm, Schalen-Ø 23 cm, Gesamt-Ø 28,5 cm; Sammlung.schmutz@chello.at 63
927 ger., gem. lig. TK, Firmenmarke 19, Made in Austria, unsign., H. 5,9 cm, Schalen-Ø 23 cm, Gesamt-Ø 28,5 cm; Wien Museum Inv. 190.174

937

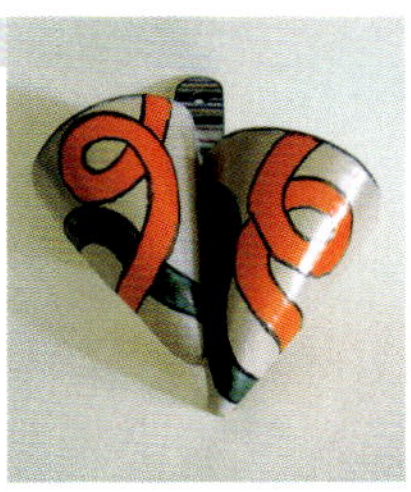

WANDVASE
Form: zwei trichterförmige Gefäße (Spitztüten)
Aufhänger: flacher, rechteckiger Aufhänger mit Loch
Dekor: bandförmig verschlungene Streifen (Luftschlangen), mehrfarbig
ca. 1928–1938
937 ger., ger. lig. MK, Firmenmarke 19, Made in Austria, H. 25,2 cm, B. 23 cm; Sammlung John Nash

938

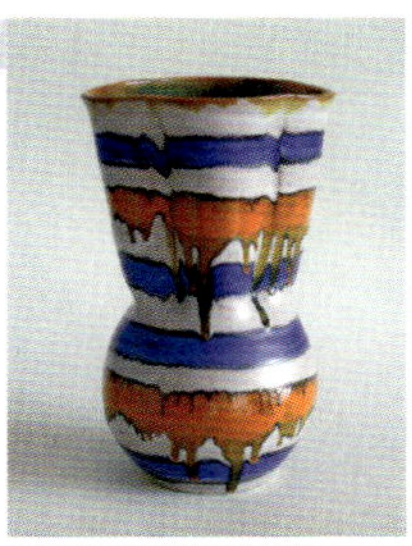

VASE
Form: kugelförmig, trichterförmige, vierpassige Mündung
Dekor: horizontale Streifen, orange-grau-blau-schwarz, Laufglasur
ca. 1928–1938
938 ger., gem. lig. LS, Firmenmarke 19, Made in Austria, H. 22,4 cm, Ø 14,4 cm; PB Deutschland

940
VASE
Form: kugelförmig, blütenartig geöffnete Mündung
Dekor: bandförmig verschlungene Streifen, mehrfarbig, Laufglasur
ca. 1928–1938
940 ger., ger. lig. MK, Firmenmarke 19, Made in Austria, H. 18,8 cm; Doro 18.4.1994

942
VASE
Form: becherförmig, flache Schulter, kurze geweitete Mündung
Dekor: Streifen (V-förmig), mehrfarbig
ca. 1928–1938
942 ger., Firmenmarke 19, H. 23,8 cm; Doro 20.11.2001

945

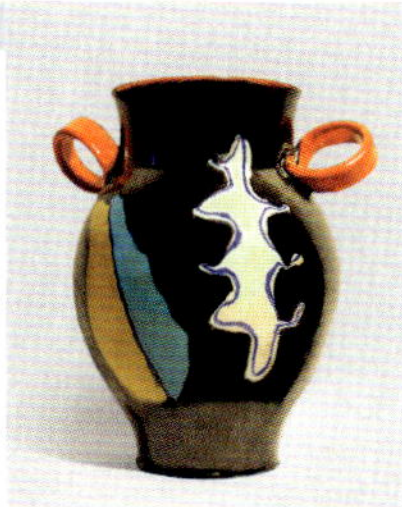

Geiger, Benno (zugeschrieben)
VASE
Form: balusterförmig, kurze zylinderförmige Mündung
Handhaben: zwei gedrehte Ösen
Dekor: abstrakte Malerei (vegetabile Ornamente), mehrfarbig
ca. 1928–1934
945 ger., ger. lig. MK, gem. lig. AH, Firmenmarke 19, Made in Austria, unsign., H. 25,4 cm, Ø 24 cm; Sammlung John Nash
945 ger., ger. lig. MK, Firmenmarke 19, Made in Austria, unsign., H. 26 cm, Ø 24 cm; Sammlung Weilinger, Wien
945 ger., ger. W, Firmenmarke 19, Made in Austria, unsign., H. 25,6 cm, Ø 24 cm; Sammlung a@goldscheider.net

946
Geiger, Benno (zugeschrieben)
DECKELDOSE
Form: eiförmig
Knauf: konischer kurzer Stab
Dekor: abstrakte Malerei (vegetabile Ornamente), mehrfarbig
ca. 1928–1934
946,XV ger., gem. A, Firmenmarke 19, Made in Austria, unsign., H. 15 cm; Mehlis 32

950

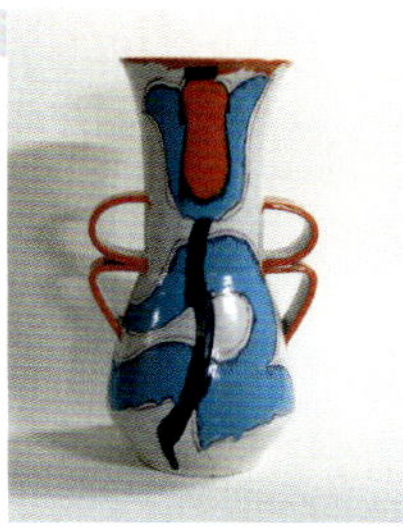

Geiger, Benno
VASE Tulpe
Form: flaschenförmig, geweitete Mündung
Handhaben: vier Ösen
Dekor: stilisierte Tulpe, mehrfarbig, Laufglasur
ca. 1928–1934
950 ger., ger. lig. MK, gem. lig. AH, Firmenmarke 19, Made in Austria, Pick. 11 (Entwurf Benno Geiger, Wien), H. 36,1 cm, Ø 17,3 cm; MAK Wien, Inv. Ke 10602-79
950 Firmenmarke 19, Made in Austria, unsign., H. 35 cm; WKA 33
Ausst.: PSK 1995
Lit.: ADGS, Nr. 35 (m. Abb.)

952

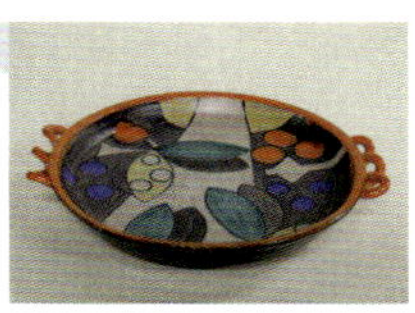

SCHALE
Form: rund
Handhaben: zwei spiralförmige Handhaben
Dekor: stilisierte vegetabile Ornamente (Blätter, Kirschen, Trauben?), mehrfarbig
ca. 1928–1938
952 ger., ger. lig. MK, gem. lig. TK, Firmenmarke 19, Made in Austria, H. 4 cm, L. 31,6 cm, Ø 28 cm; MAK Wien, Inv. Ke 10602-39
Ausst.: PSK 1995
Lit.: ADGS, Nr. 15 (m. Abb.)

955
Matula, Viktor
LEUCHTER
ca. 1928–1938
955 H. fehlt
Quelle.: Nachlass Matula, Wien

958
SCHALE Blumen
Form: rund
Dekor: stilisierter Blumenstängel mit vier Blüten, mehrfarbig
ca. 1928–1938
958 gem. lig. TK, Firmenmarke 19, Made in Austria, H. 8 cm, Ø 22 cm; eBay # 738156743

962
DECKELDOSE
Form: rund, gedrückte Eiform
Handhabe / Dekor: Fisch, vier s-förmig gemalte Wellen
Handhabe / Dekor: Kugel aus Elfenbein (Silbermontierung) / craqueliert
ca. 1928–1938
Fisch
962,1 gem. H, H. 14 cm, Ø 20,7 cm; Sammlung a@goldscheider.net
Kugel
962,11 gem. X, gem. lig. TK, Firmenmarke 19, Made in Austria, Originalpick.: „I. Passerori, Torino“, Ø 20 cm; Sammlung John Jesse, London
Bem.: Wie die beiden Varianten belegen, gab es von dieser Deckeldose mindestens zwei verschiedene Ausführungen. Denkbar wären aber noch weitere Handhaben bzw. Bekrönungen.

965
Geiger, Benno
SCHALE
Form: rund
Dekor: abstrakte Malerei, mehrfarbig, Laufglasur
ca. 1928–1934
965,B ger., Firmenmarke 19, Made in Austria, Pick. 11 (Entwurf Benno Geiger, Wien), H. fehlt cm, Ø fehlt; KH

970
Matula, Viktor
ASCHER
Form: vierpassig, eine Zigarettenablage
Dekor: gitterartige Streifen, mehrfarbig, Laufglasur
ca. 1928–1938
970 ger., Firmenmarke 19, Made in Austria, Pick. 14 (Entwurf Viktor Matula, Wien), H. 5 cm, L. 16,3 cm, B. 16,3 cm; eBay # 1438862832

971
Matula, Viktor
ASCHER
ca. 1928–1938
971 H. fehlt, B. fehlt
Quelle: Nachlass Matula, Wien

972

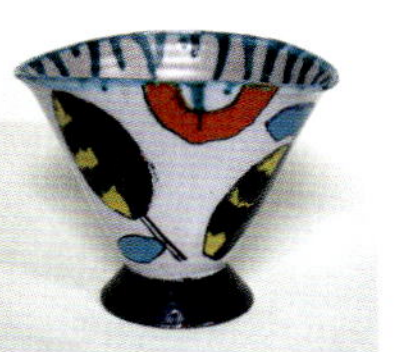

Geiger, Benno (zugeschrieben)
VASE
Form: trichterförmig (zum Oval gedrückt), konischer Stand
Dekor: stilisierte Blätter und Halbkreise, mehrfarbig, Laufglasur
ca. 1928–1934
972 ger., ger. W., gem. lig. AH, Firmenmarke 19, Made in Austria, unsign., H. 16,5 cm, B. 15,9 cm, Ø 20,5 cm; MAK Wien, Inv. Ke 10602-69
972 ger., gem. W, Firmenmarke 19, Made in Austria, unsign., H. 16 cm, B. 15,5 cm, Ø 20,5 cm; Sammlung Weilinger, Wien
Bem.: Für die Zuschreibung konnte eine signierte, und vom Dekor her sehr ähnliche Vase (ohne Mod.-Nr.), die bei Mehlis in der Auktion 18 (26.5.2000) angeboten wurde, herangezogen werden.
Ausst.: PSK 1995
Lit.: ADGS, Nr. 30 (m. Abb.)

988 s. 886

992
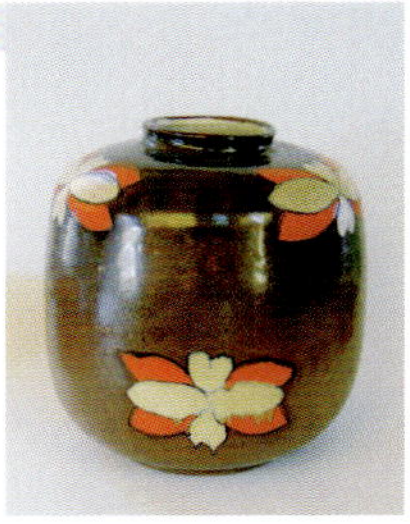
VASE
Form: zylinderförmig, flache Schulter, kurze gewulstete Mündung, kurzer Standring
Dekor: diverse Dekore
ca. 1928–1938
992 ger., ger. (oder gem.?) lig. MK, Firmenmarke 19, Made in Austria, H. 25,5 cm; Doro 10.4.1997
992 ger., ger. lig. MK, gem. lig. AH, Firmenmarke 19, Made in Austria, H. 25,3 cm, Ø 22 cm; Sammlung.schmutz@chello.at 98
992 gem. lig. TK, Firmenmarke 19, Made in Austria, H. 25,8 cm, Ø 24 cm; Sammlung a@goldscheider.net
Bem.: Das im Doro versteigerte Stück war mit gelbem Craquelé glasiert, die Exemplare der Sammlungen Schmutz und a@goldscheider.net sind mit stilisierten Schmetterlingen staffiert. Es ist somit davon auszugehen, dass Goldscheider diese Vase in mindestens zwei verschiedenen Dekoren ausgeführt hat.

997
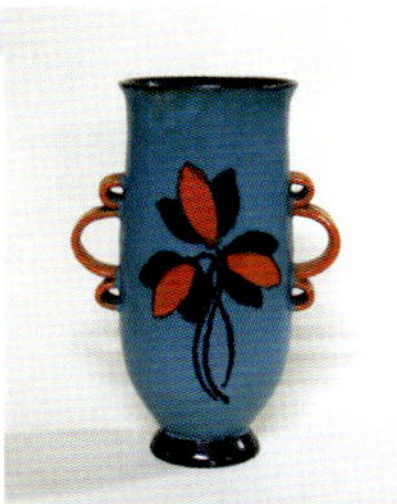
VASE Blumen
Form: becherförmig, geweitete Mündung, kurzer konischer Stand
Handhaben: zwei große Halbkreise (mit je zwei kleinen Halbkreisen)
Dekor: drei stilisierte Blumen (Tulpen?), mehrfarbig, Laufglasur
ca. 1928–1938
997 ger., Firmenmarke 19, H. 24,4 cm, Ø 11,7 cm; MAK Wien, Inv. Ke 10602-77
Ausst.: PSK 1995
Lit.: ADGS, Nr. 20 (m. Abb.)

1003
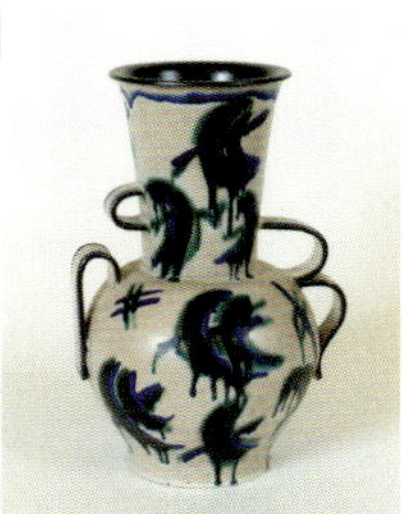
Geiger, Benno (zugeschrieben)
VASE
Form: kugelförmig, geweitete, konische Mündung
Handhaben: vier asymetrische Ösen
Dekor: abstrakte Malerei, mehrfarbig, Laufglasur
ca. 1928–1934
1003 ger. lig. MK, gem B, Firmenmarke 19, Made in Austria, unsign., H. 34 cm; Sammlung.schmutz@chello.at 22

1013
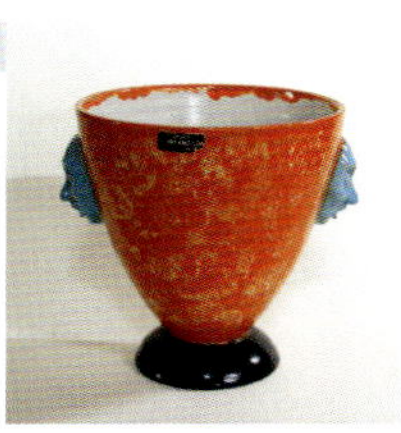
Knörlein, Rudolf
VASE Zwei Masken
Form: becher-/kelchförmig, gewölbter Stand
Dekor/Handhaben: zwei Masken (Maskarons) seitlich als Handhaben (halbplastisch), mehrfarbig
ca. 1932–1936
1013 gem. gelb, gem. B, Malerzeichen undeutlich, Firmenmarke 19, Made in Austria, Pick. 12 (Entwurf Rudolf Knörlein, Wien), H. 17,3 cm, Ø 16,8 cm; MAK Wien, Inv. Ke 10602-41
Ausst.: PSK 1995
Lit.: ADGS, Nr. 40 (m. Abb.)

1021
KERZENHALTER
Form: blütenartige Tülle auf vierpassiger Schale
Dekor: türkis, schwarz, orange, Laufglasur
ca. 1928–1938
1021 ger., ger, lig. MK, Firmenmarke 19, Made in Austria, H. 9,6 cm, Ø 13,2 cm; Sammlung a@goldscheider.net

1022M
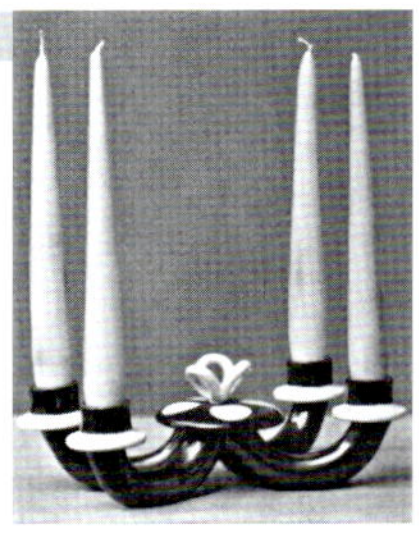
Knörlein, Rudolf
KERZENHALTER „Leuchter" (vierflammig)
Form: u-förmig geschwungene Röhren, darüber mittig eine flache Scheibe mit Bekrönung (drei Ösen)
Dekor: vermutlich diverse Dekore
ca. 1935
1022M, H. 10 cm, Ø 27 cm; GSNachf. 4,9a
Bem.: Durch die eigenhändige Datierung Knörleins (1935) auf dem Trägerblatt des Originalfotos kann das Jahr 1935 exakt bestimmt werden.
Lit.: GSNachf. 4,9a (m. Abb.), hier mit Ang. des Originaltitels „Leuchter".
Quelle: Privatarchiv (Originalfoto)

1025
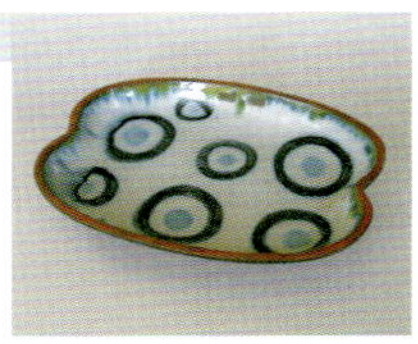
SCHALE
Form: nierenförmig, flach
Handhabe: fehlt, (wohl Korbhenkel)
Dekor: Kreise, Punkte, mehrfarbig, Laufglasur
ca. 1928–1938
1025 gem., gem. X, gem. lig. PL, Firmenmarke 19, Made in Austria, H. 3,5 cm, Ø bzw. L. 27 cm; Sammlung.schmutz@chello.at 89

1026
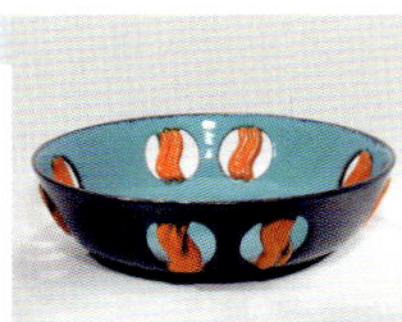
SCHALE
Form: rund, die Wandung achtfach kreisförmig durchbrochen, im Durchbruch jeweils wellenförmige Ornamente
Dekor: dreifarbig, Laufglasur
ca. 1928–1938
1026 gem., 32,37 ger., Firmenmarke 19, Made in Austria, H. 6,8 cm, Ø 24 cm; Sammlung a@goldscheider.net

1027
SCHALE Blumen (Blütenstaude)
Form: rund
Handhaben: zwei, aus drei Bändern gebildete Handhaben
Dekor: drei stilisierte Blumen (Blütenstaude), mehrfarbig
ca. 1928–1938
1027 ger., ger. lig. MK, Firmenmarke 19, Made in Austria, H. ca. 7,5 cm, Ø 25,7 cm, L. ca. 31 cm; Doro 7.3.2001
1027 ger., ger. T, gem. G, Firmenmarke 19, Made in Austria, H. 7,5 cm, Ø ca. 25,7 cm, L. 31 cm; Quit. 9

1034
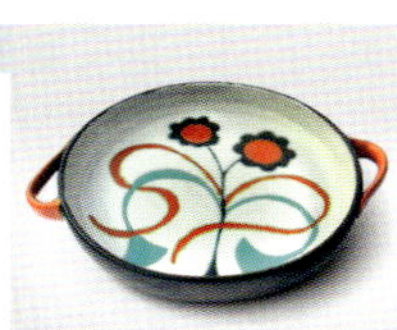
SCHALE Blumen
Form: rund
Handhaben: zwei Henkel (ohrenartig)
Dekor: zwei stilisierte Blumen mit Blattwerk, mehrfarbig
ca. 1928–1938
1034 ger. lig. MK, Firmenmarke 19, Made in Austria, Ø 26 cm; eBay # 160048469874
1034 ger., ger. W, gem. lig. AH, Firmenmarke 19, Made in Austria, H. fehlt, Ø 25 cm; Zof. 25

1039
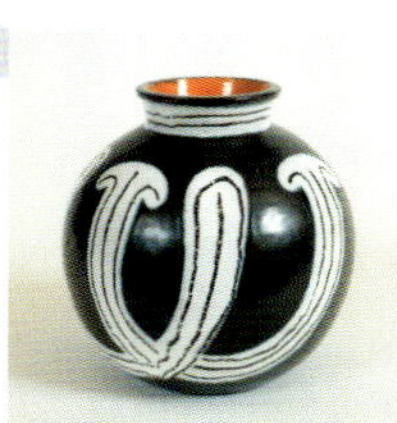
VASE Kugel
Form: kugelförmig, kurze gerade Mündung
Dekor: abstrakte Malerei, Linien
ca. 1928–1938
1039 ger., gem. H, gem. lig. MK, Firmenmarke 19, Made in Austria, H. 21,2 cm, Ø 20 cm; Sammlung.schmutz@chello.at 97

1042
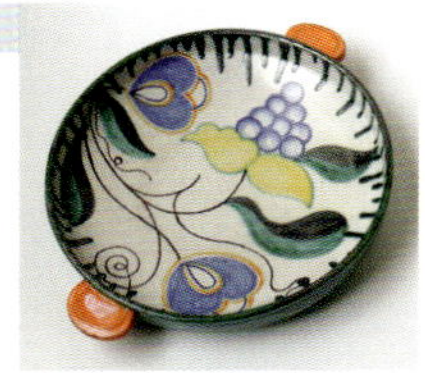
SCHALE Trauben und Blumen
Form: rund
Handhaben: zwei halbkreisförmige Handhaben (ohrenartig angesetzt)
Dekor: stilisierte Trauben, Blumen, mehrfarbig, Laufglasur
ca. 1928–1938
1042 ger., ger. lig. MK, gem. lig. TK, Firmenmarke 19, Made in Austria, H. 6 cm, Ø 29 cm, Gesamtbreite 36 cm; Sammlung R. Krenn, Wien
1042 ger., ger. lig. MK, gem. B, Firmenmarke 19, Made in Austria, H. 6,2 cm, Ø L. 29,4 cm; KH

1043
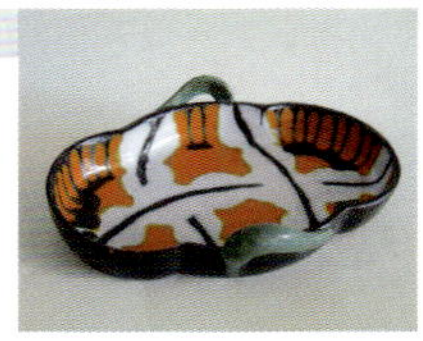
SCHALE
Form: vierpassig
Handhaben: zwei Henkel (ohrenartig)
Dekor: abstrakte Malerei, orange-weiß-grün-schwarz, Laufglasur
ca. 1928–1938
1043 ger., gem. M (oder W?), Firmenmarke 19, Made in Austria, H. 6,2 cm, Ø bzw. L. 29,4 cm; Sammlung John Nash

1050
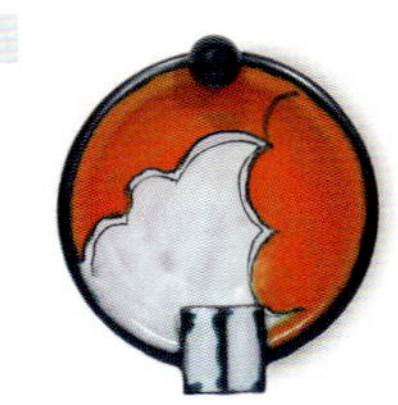
Geiger, Benno
ASCHER
Form: rund, zwei Zigarettenablagen
Handhaben/Knauf: Kugel
Dekor: abstrakte Malerei, mehrfarbig
ca. 1928–1934
1050 ger., gem. lig. TK, Firmenmarke 19, Made in Austria, Pick. 11 (Entwurf Benno Geiger, Wien), H. 3,1 cm; Ø 17,7 cm; Sammlung a@goldscheider.net
–> vgl. 1135= ähnl. Modell.

1054
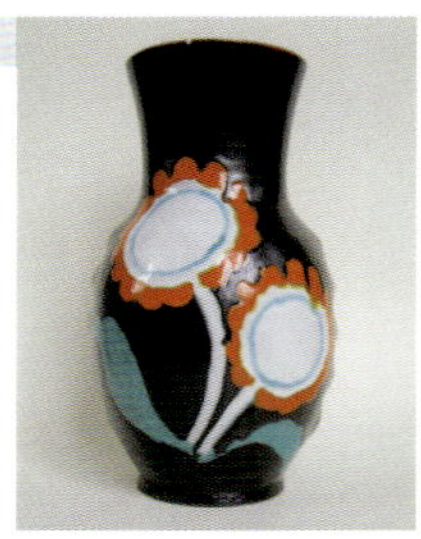
VASE
Form: gebaucht, geweiteter Hals
Dekor: stilisierte Blumen (Tulpen/Sonnenblumen?), mehrfarbig, Laufglasur
ca. 1928–1938
1054 ger. lig. MK, Firmenmarke 19, Made in Austria, H. 27 cm; Doro 9.11.1993; 8.3.1994; 3.5.1995
1054 ger., ger. lig. MK, gem. G., Firmenmarke 19, Made in Austria, H. 26,8 cm; Sammlung Weilinger, Wien
1054 ger., gem. lig. AH, Firmenmarke 19, Made in Austria, H. 26,4 cm; Sammlung.schmutz@chello.at 56

1057
Matula, Viktor
ASCHER
ca. 1928–1938
1057 H. fehlt, B. fehlt
Quelle: Nachlass Matula, Wien

1059
Matula, Viktor
ASCHER
ca. 1928–1938
1059 H. fehlt, B. fehlt
Quelle: Nachlass Matula, Wien

1061
DECKELDOSE
Form: kürbisförmig
Knauf: Kugel
Dekor: Streifen, türkis-schwarz-rot, Laufglasur
ca. 1928–1938
1061 ger., gem. G., Firmenmarke 19, Made in Austria, H. 11,2 cm, Ø 17cm; PB Deutschland

1061
DECKELDOSE
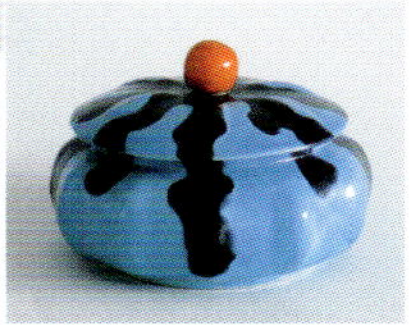
Form: rund, gedrückte Eiform
Knauf: flach, trapezförmig
Dekor: geometrische Formen
ca. 1928–1938
1069 (Firmenmarke ohne Ang.), H. fehlt, Ø 21cm; Pinhas, S. 176 (m. Abb.)

1073
Geiger, Benno
VASE
Form: becherförmig, vierpassige Mündung
Dekor: abstrakt, mehrfarbig, Laufglasur (diverse Dekore)
ca. 1928–1934
1073A gem., ger., W, gem. lig. TK, Firmenmarke 19, Made in Austria, Pick. 11 (Entwurf Benno Geiger, Wien), H. 14,4 cm, Ø 11 cm; Wien Museum Inv. 186.696
1073A ger., Firmenmarke 19, Made in Austria, unsign., H. 15,4 cm; Doro 17.5.1994
Bem.: Diese Vase wurde in diversen Dekoren staffiert wie die beiden hier nachgewiesenen Exemplare belegen.

1074
VASE
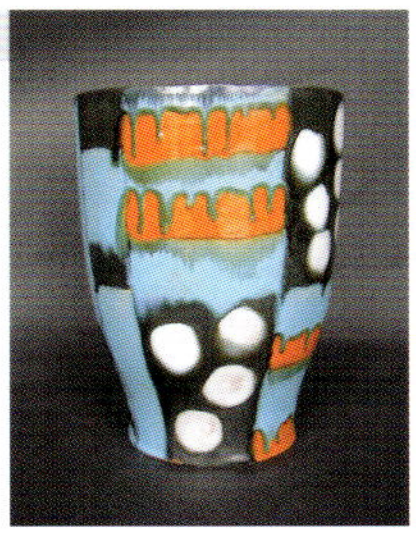
Form: becherförmig
Dekor: abstrakt, mehrfarbig, Laufglasur
ca. 1928–1938
1074,B ger., gem. W, Firmenmarke 19, Made in Austria, H. 19,5 cm, Ø 14,5 cm; PB Wien

1078
SCHALE
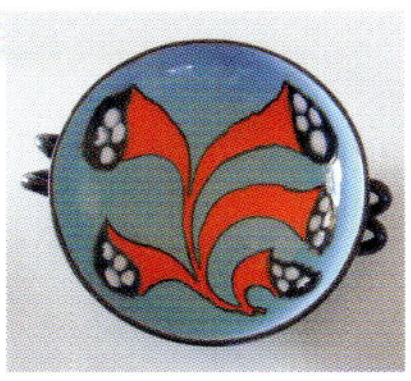
Form: rund
Handhaben: zwei spiralförmige Handhaben
Dekor: stilisierte Blumenblüten
ca. 1928–1938
1078 ger., ger. lig. MK, gem. El., Firmenmarke 19, Made in Austria, H. 6 cm, Ø 27 cm; Sammlung.schmutz@chello.at 23

1082
VASE / TOPF
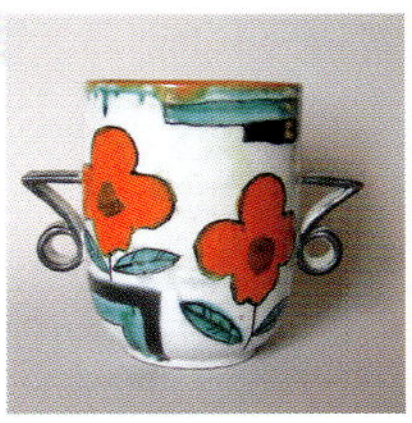
Form: becherförmig
Handhaben: zwei Ösen (an Dreieck)
Dekor: stilisierte Blumen, mehrfarbig, Laufglasur
ca. 1928–1938
1082 ger., Firmenmarke 19, Made in Austria, H. 19,6 cm; Sammlung.schmutz@chello.at 24
Bem.: Bis auf den min. Höhenunterschied und dem Dekor ident. mit Mod.-Nr. 1088.

1086
VASE
Form: becherförmig
Handhaben: zwei Ösen
Dekor: metallische, Kupfer / Schwarz Sprenkelglasur
ca. 1928–1938
1086 ger., ger. TK, gem. lig. AH, Firmenmarke 19, Made in Austria, H. 18 cm; PB Wien 19

1088
VASE / TOPF
Form: becherförmig
Handhaben: zwei Ösen (an Dreieck)
Dekor: Streifendekor, mehrfarbig, Laufglasur
ca. 1928–1938
1088 (Firmenmarke ohne Ang.), H. 19 cm; Pinhas, S. 176 (m. Abb.)
Bem.: Bis auf den min. Höhenunterschied und dem Dekor ident. mit Mod.-Nr. 1082.

1091
VASE
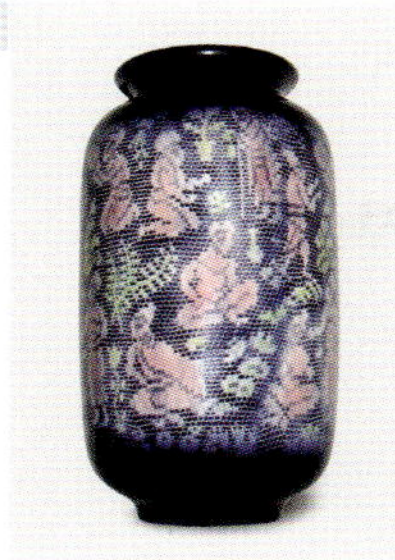
Form: zylinderförmig, flache Schulter, geweitete Mündung
Dekor: Figuren, Spitzendekor (Schablone)
1091,10,I Firmenmarke 19, Made in Austria, Hand dec., H. 20,5 cm; eBay # 2219111766
1091,11,K Firmenmarke 19, Made in Austria, Hand dec., H. 20 cm; Doro FVK 12/1999
1091M, H. 16 cm, B. 12 cm; GSNachf. 4,2i
Bem.: Die Abb. in der genannten Lit. zeigt das Modell in einem anderen Dekor als die nachgewiesenen Stücke. Es ist somit davon auszugehen, dass Goldscheider dieses Modell in mindestens zwei verschiedenen Dekoren ausgeführt hat.
Lit.: GSNachf. 4,2i (m. Abb.)
Quelle: AGM (Rechnung an Maurice Mizrahi, Alexandria, Ägypten, vom 22. April 1953).

1094
DECKELDOSE Frauenkopf
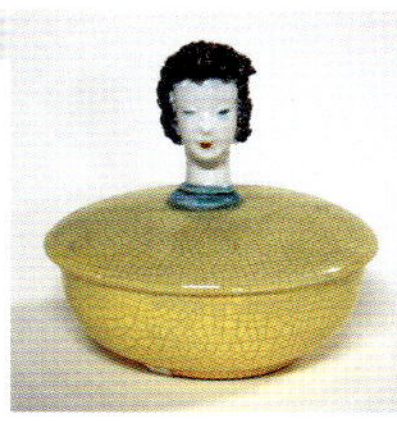
Form: rund, kugelförmig gedrückt
Knauf: Frauenkopf (vollplastisch)
ca. 1938
1094,1 ger., Firmenmarke 19 (blau), Made in Austria (blau), H. 15,2 cm, Ø 16,1 cm; MAK Wien, Inv. Ke 10602-71
1094 Firmenmarke 19, Made in Austria, H. oder Ø 16 cm; PB
-> vgl. 8630 = Variante mit Damenakt (Dakon)
-> vgl. 1219 = ähnl. Modell.

1095
Matula, Viktor
ASCHER
ca. 1928–1938
1095 H. fehlt, B. fehlt
Quelle: Nachlass Matula, Wien

1109
VASE
Form: becherförmig, eingezogene Schulter, geweitete Mündung
Dekor: Kirschblütenzweige
ca. 1928–1938
1109 ger. lig. MK, Firmenmarke 19, Made in Austria, H. 16,6 cm; BM 57

1110
VASE
Form: pokalförmig, frei gedreht
Handhaben: sechs kugelförmige Handhaben (jeweils in einer dreier Gruppe seitlich angesetzt)
Dekor: abstrakte Malerei, vegetabile Formen, mehrfarbig, teils als Laufglasur
ca. 1928–1938
1110 ger., ger. lig. MK, gem. lig. AH, Firmenmarke 19, Made in Austria, H. 27,7 cm, Ø 12 cm; Sammlung Weilinger, Wien

1111F
Knörlein, Rudolf
LAMPE Blumen
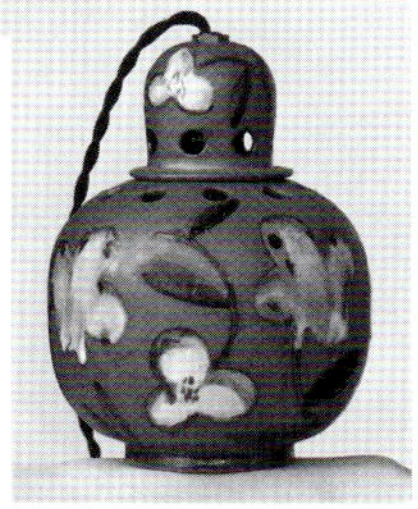
Form: kugelförmig, glockenförmige Bekrönung (für die Lampenfassung)
Dekor: Blumen mit Stängel
ca. 1933
1111F, H. 18 cm, Ø 12 cm
Bem.: Durch die eigenhändige Datierung Knörleins (1933) auf dem Trägerblatt des Originalfotos kann das Jahr 1933 exakt bestimmt werden.
-> vgl. 1112= ähnl. Modell.
Quelle: Privatarchiv (Originalfoto)

1112F
Knörlein, Rudolf
LAMPE Blumen und Blattwerk
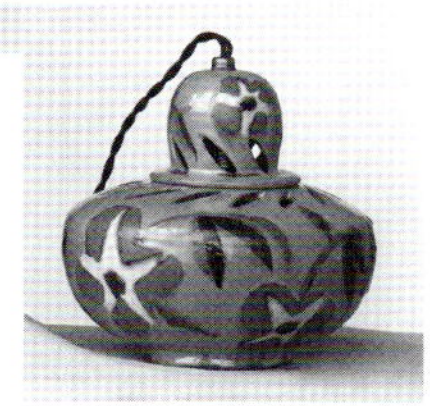
Form: kürbisförmig, glockenförmige Bekrönung (für die Lampenfassung)
Dekor: Blumen mit Stängeln und Blättern
ca. 1933
1112F, H. 16 cm, Ø 18 cm
Bem.: Durch die eigenhändige Datierung Knörleins (1933) auf dem Trägerblatt des Originalfotos kann das Jahr 1933 exakt bestimmt werden.
-> vgl. 1111= ähnl. Modell.
Quelle: Privatarchiv (Originalfoto)

1114
VASE
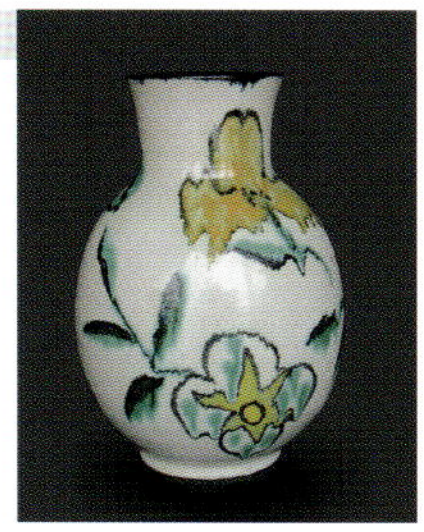
Form: flaschenförmig
Dekor: Blumen, mehrfarbig, Laufglasur
ca. 1928–1938
1114 ger., ger. lig. MK, gem. lig. TK, Firmenmarke 19, Made in Austria, H. 28 cm; PB Wien

1116
DECKELDOSE
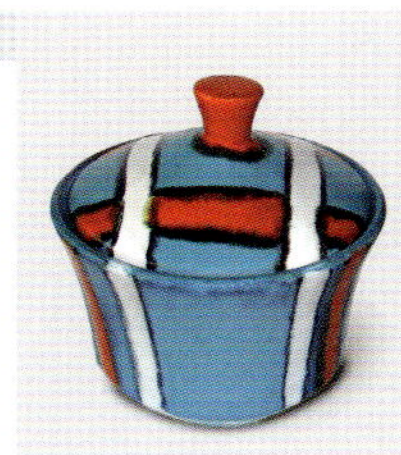
Form: rund
Knauf: konischer Knopf
Dekor: Streifendekor, Laufglasur
ca. 1928–1938
1116,2 Firmenmarke 19, Made in Austria, H. 10,5 cm, Ø fehlt; WKA 33

1118
Knörlein, Rudolf
VASE
Form: balusterförmig, darauf konische Mündung
Handhabe: zwei ohrenförmige Henkel
Dekor: stilisiertes Blattwerk
ca. 1932–1936
1118 Firmenmarke 19, Made in Austria, Pick. 12 (Entwurf Rudolf Knörlein, Wien), H. 42,5 cm; WKA 55

1120
VASE
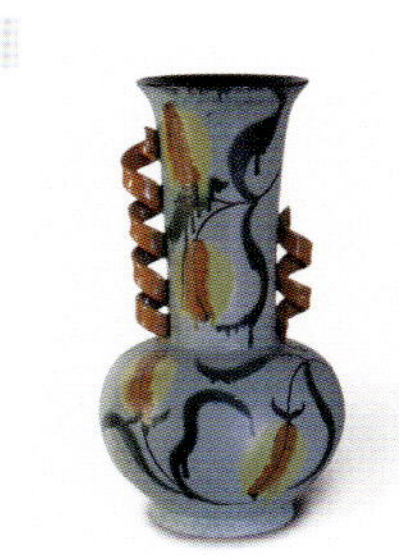
Form: gebaucht auf rundem Stand, lange zylindrische, oben geweitete Mündung
Handhaben: zwei spiralförmige, seitlich angesetzte Handhaben
Dekor: stilisierte Blumen und Blattwerk, mehrfarbig, Laufglasur
ca. 1928–1938
1120 ger., ger. lig. MK, gem. lig. TK, Firmenmarke 19, Made in Austria, H. 39,7 cm, Ø 23 cm; Sammlung a@goldscheider.net

1122
SCHALE
Form: rund
Handhaben: vier blattähnliche Handhaben
Dekor: stilisierte Blumen, Blätter, Kreise, mehrfarbig
ca. 1928–1938
1122 ger., ger. lig. MK, gem. (SH?), Firmenmarke 19, Made in Austria, H. 7 cm, Ø bzw. L. 30,5 cm; eBay # 3160447915

1135
Geiger, Benno (zugeschrieben)
ASCHER
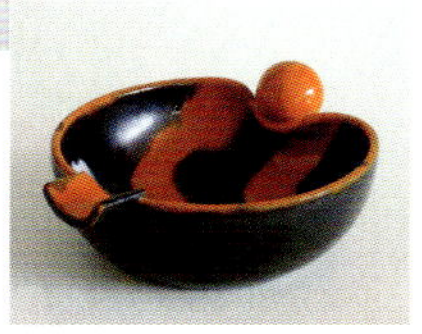
Form: rund, eine Zigaretten-ablage
Handhaben/Knauf: Kugel
Dekor: abstrakte Malerei (Wellen, Punkte), schwarz-orange-weiß
ca. 1928–1934
1135 ger., ger. lig. MK, Firmenmarke 19, Made in Austria, unsign., H. 6,2 cm, Ø 15,2 cm; Sammlung John Nash
Bem.: Für die Zuschreibung –> vgl. 1050= ähnl. Modell.

1137 s. 1139

1139
Knörlein, Rudolf
SCHALE
Form: vierpassig, hochgezogene Wandung
Handhabe: Korbhenkel
Dekor: abstrakte Malerei, mehrfarbig, Laufglasur
ca. 1932–1936
1139 (und 1137) gem., ger. lig. MK, gem. W (oder M?), (ohne Firmenmarke), Pick. 12 (Entwurf Rudolf Knörlein, Wien), H. 6 cm, mit Henkel 18 cm, L. 20,5 cm, B. 16,5 cm; eBay # 2222454888

1141
SCHALE
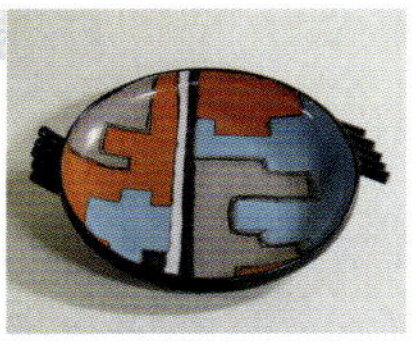
Form: rund
Handhaben: zwei, aus jeweils fünf Stäbchen zusammengesetzte Handhaben (schräg angesetzt)
Dekor: unregelmäßige Farbfelder, Kreuz, Streifen, mehrfarbig
ca. 1928–1938
1141 ger., ger. lig. MK, Firmenmarke 19, Made in Austria, H. 4,5 cm, L. 33 cm, Ø 27,8 cm; MAK Wien, Inv. Ke 10602-72
Ausst.: PSK 1995
Lit.: ADGS, Nr. 18 (m. Abb.)

1158
VASE
Form: flaschenförmig
Dekor: Farbfelder, mehrfarbig, Laufglasur
ca. 1928–1938
1158 Firmenmarke 19, Made in Austria, H. 26,5 cm; Doro 20.4.1993

1165
Bosse, Walter (zugeschrieben)
VASE
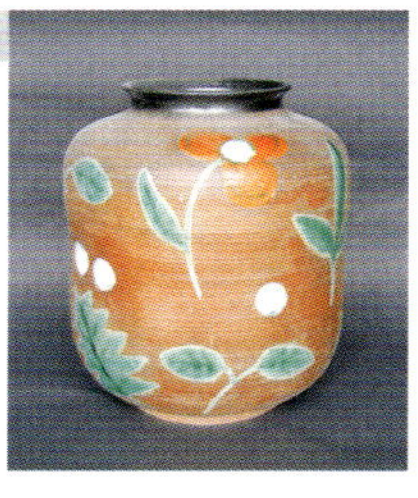
Form: becherförmig, flache Schulter, kurze geweitete Mündung
Dekor: Blumen, Blätter, Punkte, mehrfarbig
ca. 1928–1938
1165 ger., ger. lig. MK, gem. lig. AH, Firmenmarke 19, Made in Austria, unsign., H. 24 cm, Ø 20,5 cm; PB Wien

1166
VASE Tiere Afrikas
Form: balusterförmig
Dekor: Tierdarstellungen (Elefant, Giraffe, Emu, Schlange, Vögel) und Vegetation (Büsche, Palmen)

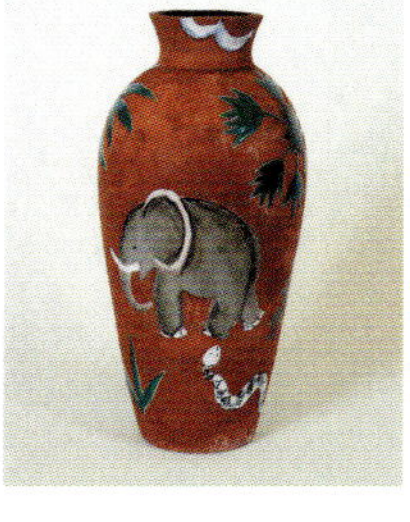

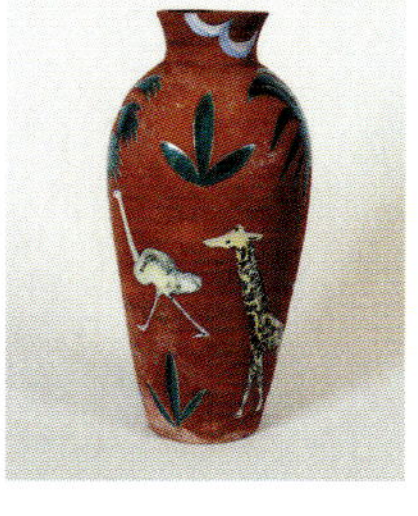

ca. 1928–1938
1166 ger., gem. lig. AH, Firmenmarke 19, Made in Austria, H. 42 cm, Ø 20 cm, Mündung Ø 11,4 cm; Sammlung.schmutz@chello.at 108

1172
Knörlein, Rudolf
KERZENHALTER (einflammig)
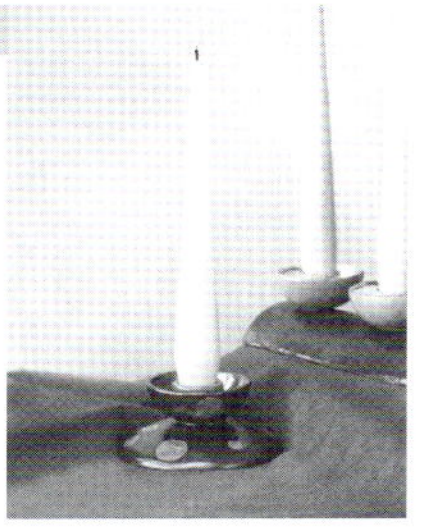
Form: rund, mittig aufgesetzte schüsselartige Kerzentülle
Dekor: Punkte (vermutlich auch Dekorvarianten)
ca. 1934
1172, H. 5 cm, Ø 12 cm
Bem.: Durch die eigenhändige Datierung Knörleins (1934) auf dem Trägerblatt des Originalfotos kann das Jahr 1934 exakt bestimmt werden.
Quelle: Privatarchiv (Originalfoto)

1173
Knörlein, Rudolf
KERZENHALTER (zweiflammig)
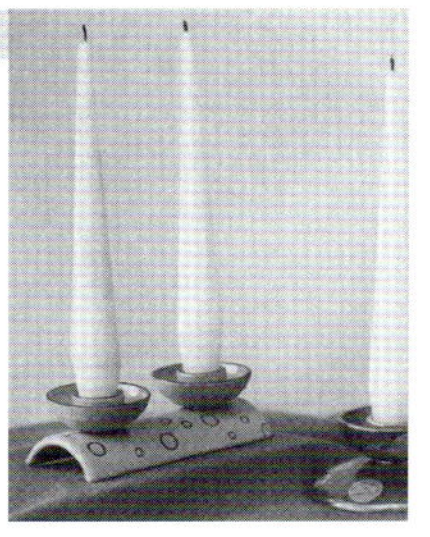
Form: halbröhrenartig, zwei aufgesetzte schüsselartige Kerzentüllen
Dekor: Kreise (vermutlich auch Dekorvarianten)
ca. 1934
1173, H. 5 cm, Ø 16,5 cm
Bem.: Durch die eigenhändige Datierung Knörleins (1934) auf dem Trägerblatt des Originalfotos kann das Jahr 1934 exakt bestimmt werden.
Quelle: Privatarchiv (Originalfoto)

1174
Knörlein, Rudolf
KERZENHALTER (dreiflammig)
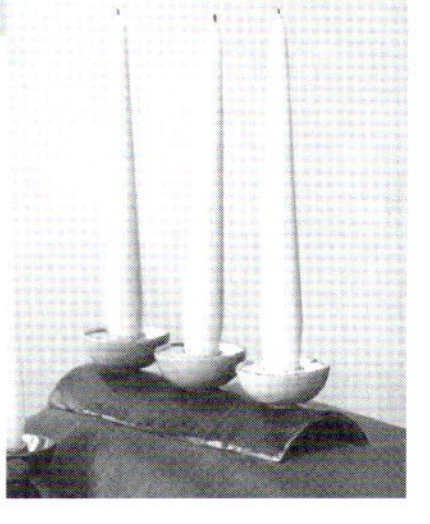
Form: halbröhrenartig, drei aufgesetzte schüsselartige Kerzentüllen
Dekor: Laufglasur (vermutlich auch Dekorvarianten)
ca. 1934
1174, H. 6 cm, Ø 26 cm
Bem.: Durch die eigenhändige Datierung Knörleins (1934) auf dem Trägerblatt des Originalfotos kann das Jahr 1934 exakt bestimmt werden.
Quelle: Privatarchiv (Originalfoto)

1181
SCHALE
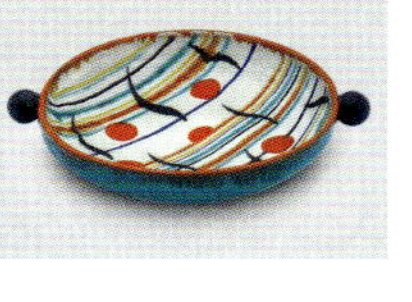
Form: rund
Handhaben: zwei kugelartige Handhaben
Dekor: abstrakte Malerei, Punkte, Streifen, mehrfarbig, Laufglasur
ca. 1928–1938
1181 ger., ger. lig. MK, Firmenmarke 19, Made in Austria, Pick. 2, H. 6,5 cm, Ø 23,8 cm, L. 28,5 cm; Sammlung a@goldscheider.net

1183
VASE
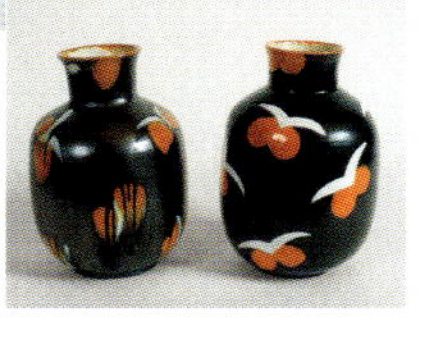
Form: eiförmig, flache Schulter, leicht trichterförmige Mündung
Dekor: stilisierte Früchte? (Kirschen?)
ca. 1928–1938
1183 ger., ger. lig. MK, gem. lig. G., Firmenmarke 19, Made in Austria, Pick. 3/a, H. 28,5 cm; Sammlung.schmutz@chello.at 99
1183 ger., gem. A, Firmenmarke 19, Made in Austria, H. 29 cm; Sammlung.schmutz@chello.at 50

1187
Knörlein, Rudolf
VASE
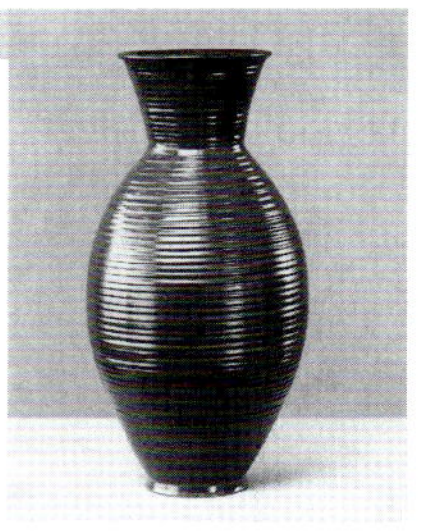
Form: balusterförmig, trichterförmige Mündung
Dekor: gerillt (Dekorvarianten?)
ca. 1933
1187, H. 45 cm, Ø 22 cm
Bem.: Durch die eigenhändige Datierung Knörleins (1933) auf dem Trägerblatt des Originalfotos kann das Jahr 1933 exakt bestimmt werden.
Quelle: Privatarchiv (Originalfoto)

1193
Matula, Viktor
ASCHER
ca. 1928–1938
1193 H. fehlt, B. fehlt
Quelle: Nachlass Matula, Wien

1206
VASE
Form: zylinderförmig, oben und unten eingezogen, mittig zwei erhabene Ringe, kurze gewulstete Mündung
Dekor: Punkte, mehrfarbig, Laufglasur
ca. 1928–1938
1206 ger., ger. lig. MK, gem. Z, Firmenmarke 19, Made in Austria, H. 30,5 cm, Ø 22,8 cm; eBay # 1300094177

1207
VASE Kugel
Form: kugelförmig, kurze trichterförmige Mündung
Dekor: abstrakte Malerei, mehrfarbig, Laufglasur
ca. 1928–1938
1207 Firmenmarke 19, Made in Austria, H. 27,8 cm, Ø ca. 24 cm; Doro 8.3.2004

1209
SCHALE
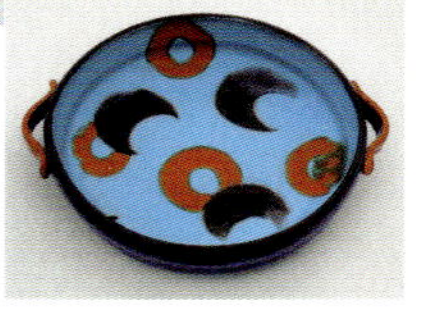
Form: rund
Handhaben: 2 ohrenartig angesetzte Henkel
Dekor: Kreise, Halbmonde, Laufglasur
ca. 1928–1938
1209 ger., ger. PA, gem. lig. AH, Firmenmarke 19, Made in Austria, H. 6,3 cm, Ø 25 cm, L. 29 cm; Sammlung a@goldscheider.net

1219
SCHALE mit vollplastischem Frauenkopf
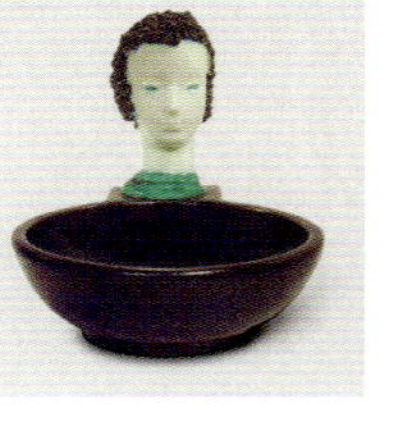
Form: rund
ca. 1928–1938
1219 ger., Firmenmarke 19, Made in Austria, H. 10,4 cm, L. 16 cm; Sammlung a@goldscheider.net
–> vgl. 1094 = ähnl. Modell.

1220
VASE
Form: kugelförmig, hohe trichterförmige Mündung
Dekor: Punkte, Bögen, ziegelrot-braun
ca. 1928–1938
1220 ger. lig. MK, Firmenmarke 19, Made in Austria, H. 34 cm; Doro 22.11.1990

1228
VASE
Form: kugelförmig, hohe trichterförmige Mündung
Dekor: abstrakte Malerei, beige-orange-schwarz
ca. 1928–1938
1228,F ger. lig. MK, gem. 7, Firmenmarke 19, Made in Austria, H. 14,5 cm; Doro 21.9.1999, WKA 27; eBay # 405250555

1230
Matula, Viktor
ASCHER Blume
Form: oktogonal, vier Zigarettenablagen
Dekor: stilisierte Blume, mehrfarbig, Laufglasur
ca. 1928–1938
1230 ger., ger. lig. MK, Firmenmarke 19, Made in Austria, H. 5 cm, Ø 16,7 cm; Sammlung a@goldscheider.net
Quelle: Nachlass Matula, Wien

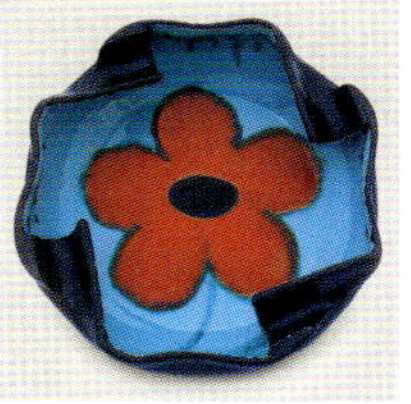

1231
Matula, Viktor
ASCHER
ca. 1928–1938
1231 H. fehlt, B. fehlt
Quelle: Nachlass Matula, Wien

1242
VASE
Form: becherförmig, frei geformt, geweitete Mündung
Dekor: stilisierte Feuerzungen, uranrot-kupfermetallisch, Laufglasur
ca. 1928–1938
1242 ger., gem. S, Firmenmarke 19, Made in Austria, H. 21,3 cm, Ø 16 cm; Sammlung.schmutz@chello.at 25
Ausst.: SSH 1992

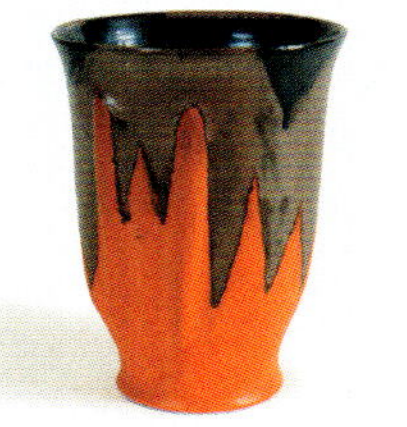

1243
Matula, Viktor
ASCHER
ca. 1928–1938
1243 H. fehlt, B. fehlt
Quelle: Nachlass Matula, Wien

1277M
Matula, Viktor
ASCHER
Form: vierpassig, zwei Zigarettenablagen
Dekor: blau-orange-dunkelgrün, Laufglasur
ca. 1928–1938
1277,I Firmenmarke 19, Made in Austria, unsign., H. ca. 4,5 cm, B. 17,5 cm; Doro 21.5.1997
1277M, H. 4,5 cm, B. 17 cm; GSNachf. 4,1c
Bem.: Die Abb. in der genannten Lit. zeigt das Modell in einem anderen Dekor als das nachgewiesene Stück. Es ist somit davon auszugehen, dass Goldscheider dieses Modell in mindestens zwei verschiedenen Dekoren ausgeführt hat.
Lit.: GSNachf. 4,1c (m. Abb.), hier mit der Ang. „Ascher kann auch in kleinerer Ausführung (ca. 3,5 zu 13 cm) geliefert werden".
Quelle: Nachlass Matula, Wien

1278M
Matula, Viktor
ASCHER
Form: mehrpassig, sechs Zigarettenablagen
Dekor: vermutlich diverse Dekore (Laufglasur, Craquelé)
ca. 1928–1938 (für den Entwurf)
1278 ger., Firmenmarke 19, Made in Germany, Ausformung zw. 1938–1941, H. 4,5 cm, B. 17,5 cm; eBay # 6593393164
1278M, H. 4 cm, B. 15 cm; GSNachf. 4,1e
Lit.: GSNachf. 4,1e (m. Abb.), hier mit der Ang. „Ascher kann auch in kleinerer Ausführung (ca. 3,5 zu 13 cm) geliefert werden".
Quelle: Nachlass Matula, Wien

1279M
Matula, Viktor
ASCHER
Form: rund, zwei Zigarettenablagen
Dekor: Blüten (Sterne?), Spitzendekor (Schablone)
ca. 1928–1938
1279,13 gem. lig. HS, Firmenmarke 19, Made in Austria, Hand dec., unsign., H. 4,4 cm, Ø 18 cm, L. 19 cm; Sammlung a@goldscheider.net
1279M, H. 4 cm, B. 17 cm; GSNachf. 4,1a
Bem.: Die Abb. in der genannten Lit. zeigt das Modell in einem anderen Dekor als das nachgewiesene Stück. Es ist somit davon auszugehen, dass Goldscheider dieses Modell in mindestens zwei verschiedenen Dekoren ausgeführt hat.
Lit.: GSNachf. 4,1a (m. Abb.), hier mit der Ang. „Ascher kann auch in kleinerer Ausführung (ca. 3,5 zu 13 cm) geliefert werden".
Quelle: Nachlass Matula, Wien

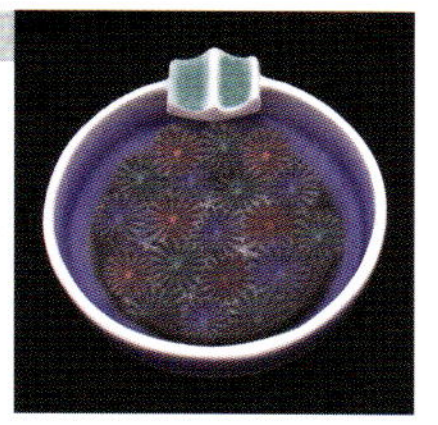

1280M
Matula, Viktor
ASCHER
Form: rund, drei Zigarettenablagen
Dekor: vermutlich diverse Dekore
ca. 1928–1938
1280M, H. 5 cm, B. 15 cm; GSNachf. 4,1f
Lit.: GSNachf. 4,1f (m. Abb.), hier mit der Ang. „Ascher kann auch in kleinerer Ausführung (ca. 3,5 zu 13 cm) geliefert werden".
Quelle: Nachlass Matula, Wien

1281M
Matula, Viktor
ASCHER
Form: sechspassig, drei Zigarettenablagen
Dekor: vermutlich diverse Dekore
ca. 1928–1938
1281,2 Firmenmarke 19, H. 5 cm, B. 18,5 cm; Sammlung Weilinger, Wien
1281M, H. 5 cm, B. 16 cm; GSNachf. 4,1b
Lit.: GSNachf. 4,1b (m. Abb.), hier mit der Ang. „Ascher kann auch in kleinerer Ausführung (ca. 3,5 zu 13 cm) geliefert werden".
Quelle: Nachlass Matula, Wien

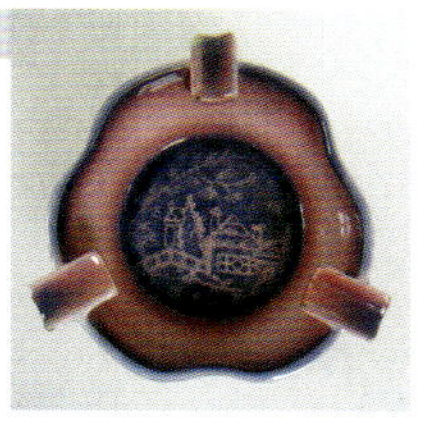

1291M
Matula, Viktor
VASE/TOPF „Vase"
Form: halbkugelförmig, mehrpassig, konischer Standring
Dekor: vermutlich diverse Dekore, Laufglasur
ca. 1928–1938
1291M, H. 16 cm, B. 19 cm; GSNachf. 4,5f
Lit.: GSNachf. 4,5f (m. Abb.), hier mit Ang. des Originaltitels „Vase".
Quelle: Nachlass Matula, Wien

1294F
ASCHER „Ascher, groß"
ca. 1928–1938
Quelle: AGM (Rechnung an Maurice Mizrahi, Alexandria, Ägypten, vom 22. April 1953), hier mit Ang. des (Original)-Titels „Ascher, groß", geliefert in Ausführung „gold".

1297F
ASCHER
ca. 1928–1938
Quelle: AGM (Rechnung an Maurice Mizrahi, Alexandria, Ägypten, vom 22. April 1953), geliefert in Ausführung „gold".

1298M
Matula, Viktor
VASE
Form: flaschenförmig, zylinderförmige, leicht taillierte hohe Mündung
Dekor: vermutlich diverse Dekore, Laufglasur
ca. 1928–1938
1298M, H. 35 cm, B. 16 cm; GSNachf. 4,5g
Lit.: GSNachf. 4,5g (m. Abb.)
Quelle: Nachlass Matula, Wien

1300M
Matula, Viktor
ASCHER
Form: rund, vier Zigarettenablagen
Dekor: diverse Dekore
ca. 1928–1938
1300,2,73 Firmenmarke 19, Made in Austria, unsign., H. 4,7 cm, Ø 18,4 cm, L. 23,3 cm; Sammlung a@goldscheider.net
ohne Nr., Firmenmarke 19, Made in Austria, Hand dec., unsign., H. 4,8 cm; WKA 33
1300M, H. 4,5 cm, B. 18 cm; GSNachf. 4,1g
Bem.: Die Ex. der Sammlung a@goldscheider.net und das bei den WKA angebotene Stück waren mit einem Spitzendekor staffiert, die Abb. im GS Firmenkatalog zeigt das Raucherutensil mit einer Laufglasur (-> vgl. GSNachf. 4,1g). Es ist somit davon auszugehen, dass Goldscheider diesen Ascher in mindestens zwei verschiedenen Dekoren ausgeführt hat.
Lit.: GSNachf. 4,1g (m. Abb.), hier mit der Ang. „Ascher kann auch in kleinerer Ausführung (ca. 3,5 zu 13 cm) geliefert werden".
Quelle: Nachlass Matula, Wien

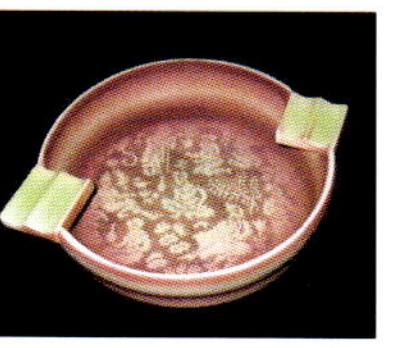

1322M
Matula, Viktor
ASCHER „Ascher, groß"
Form: sechspassig, drei Zigarettenablagen
Dekor: vermutlich diverse Dekore
ca. 1928–1938
1280M, H. 3,5 cm, B. 16 cm; GSNachf. 4,1d
Lit.: GSNachf. 4,1d (m. Abb.), hier mit der Ang. „Ascher kann auch in kleinerer Ausführung (ca. 3,5 zu 13 cm) geliefert werden".
Quelle: AGM (Rechnung an Maurice Mizrahi, Alexandria, Ägypten, vom 22. April 1953), hier mit Ang. des (Original)-Titels „Ascher, groß".
Quelle: Nachlass Matula, Wien

1324M
Matula, Viktor
VASE
Form: kugelförmig, vierpassige, leicht geweitete Mündung
Dekor: vermutlich diverse Dekore, Laufglasur
ca. 1928–1938
1324M, H. 22 cm, B. 13,5cm; GSNachf. 4,5d
Lit.: GSNachf. 4,5d (m. Abb.)
Quelle: Nachlass Matula, Wien

1325M
Geiger, Benno
VASE
Form: becherförmig, eingezogene flache Schulter, darauf hohe trichterförmige Mündung
Handhaben: vier seitlich angesetzte rechteckige Handhaben
Dekor: vermutlich diverse Dekore, Laufglasur
ca. 1928–1934
1325M, H. 29 cm, B. 19 cm; GSNachf. 4,5a
Lit.: GSNachf. 4,5a (m. Abb.)
-> vgl. 883

1326M
Matula, Viktor
VASE
Form: kugelförmig, hohe trichterförmige Mündung, gewulsteter Standring
Dekor: vermutlich diverse Dekore, Laufglasur
ca. 1928–1938
1326M, H. 25 cm, B. 14 cm; GSNachf. 4,5h
Lit.: GSNachf. 4,5h (m. Abb.)
Quelle: Nachlass Matula, Wien

1366M
Matula, Viktor
VASE
Form: eiförmig, flache Schulter, gewulstete Mündung
Dekor: diverse Dekore
ca. 1928–1938
1366 Firmenmarke 19, unsign., H. 26 cm, B. 12 cm, Mündung Ø 12 cm; Wien Museum Inv. 186.689;
Bem.: Hier in der Ausführung „Craquelé".
1366M, H. 25 cm, B. 22 cm; GSNachf. 4,5e
Bem.: Die Abb. in der genannten Lit. zeigt das Modell in einem anderen Dekor als das nachgewiesene Stück. Es ist somit davon auszugehen, dass Goldscheider diese Vase in mindestens zwei verschiedenen Dekoren ausgeführt hat.
Lit.: GSNachf. 4,5e (m. Abb.)
Quelle: AGM (Rechnung an Maurice Mizrahi, Alexandria, Ägypten, vom 22. April 1953).

1367M
Matula, Viktor
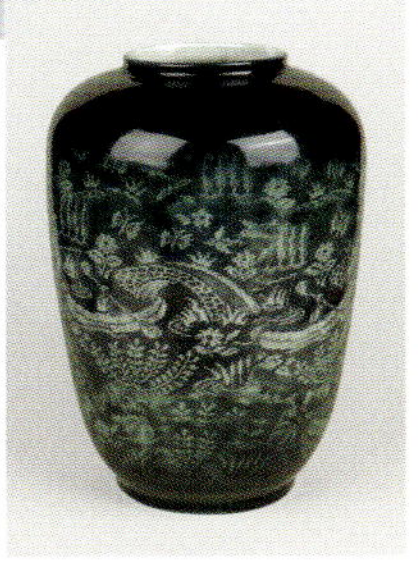
VASE
Form: eiförmig, flache Schulter, enge gewulstete Mündung
Dekor: Spitzendekor (Schablone); Craquelé, (Laufglasur?)
ca. 1928–1938
BLUMEN
1367,11 gem. G, Firmenmarke 19, Made in Austria, Hand dec., unsign., H. 29,6 cm; Wien Museum Inv. 186.688
1367,11,39 Firmenmarke 19, Hand dec., unsign., H. 30 cm; Doro S. 25./26.11.1993
KANUFAHRER IM FLUSS
1367,11,46 gem. lig. SH, Firmenmarke 19, Made in Austria, unsign., H. 29,7 cm; Wien Museum Inv. 182.065
1367,S Firmenmarke 19, Made in Austria, unsign., H. 28,6 cm; Doro 4.3.2002
LANDSCHAFT (BÄUME UND HÄUSER)
1367,5 Firmenmarke 19, Made in Austria, unsign., H. 29, cm; WKA 33
1367M, H. 29 cm, B. 18 cm; GSNachf. 4,5c
Bem.: Die Abb. in der genannten Lit. zeigt das Modell in einem anderen Dekor als die der beiden nachgewiesenen Stücke. Es ist somit davon auszugehen, dass Goldscheider diese Vase in mindestens vier verschiedenen Dekoren ausgeführt hat.
Lit.: GSNachf. 4,5c (m. Abb.)
Quellen: Nachlass Matula, Wien; AGM (Rechnung an Maurice Mizrahi, Alexandria, Ägypten, vom 22. April 1953).

1369M
Matula, Viktor
VASE / DECKELVASE „Vase"
Form: becherförmig, gewulsteter Standring
Handhabe des Deckels: Kugel
Dekor: vermutlich diverse Dekore, Laufglasur
ca. 1928–1938
1369M, H. 31 cm, B. 19,5 cm; GSNachf. 4,5i
Bem.: Die Abb. in der genannten Lit. zeigt eigentlich einen urnenförmigen Deckeltopf. Die Firma bezeichnet dieses Modell in ihren Unterlagen als „Vase".
Lit.: GSNachf. 4,5i (m. Abb.), hier mit Ang. des Originaltitels „Vase".
Quelle: Nachlass Matula, Wien

1370M
SCHALE
Form: rund
Dekor: vermutlich diverse Dekore, Laufglasur
ca. 1928–1938
1370M, H. 10 cm, Ø 26 cm; GSNachf. 4,5b
Lit.: GSNachf. 4,5b (m. Abb.)

1389F
ASCHER „Ascher, klein"
ca. 1928–1938
Quelle: AGM (Rechnung an Maurice Mizrahi, Alexandria, Ägypten, vom 22. April 1953), hier mit Ang. des Originaltitels „Ascher, klein".

1399
Knörlein, Rudolf (zugeschrieben)
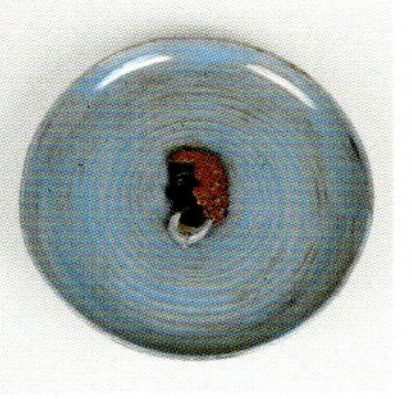
SCHALE Frauenkopf
Form: rund
Dekor: RELIEF Frauenkopf mit Kragen, im Profil nach links
ca. 1932–1936
1399 Firmenmarke 19, Made in Austria, unsign., H. 4,4 cm, Ø 21,5 cm; PB Wien 19
Ausst.: PSK 1995
Lit.: ADGS, Nr. 41 (m. Abb.)

1401
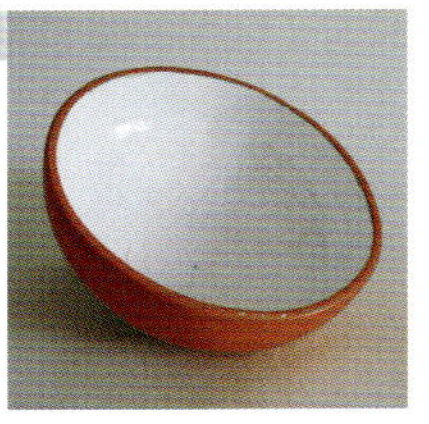
SCHALE
Form: rund
Dekor: innen craqueliert (weiß), außen orange
ca. 1928–1938
1209,8,15 ger. (1)401,B, gem. N (oder Z?), Firmenmarke 19, Made in Austria, H. 5,6 cm, Ø 15,7 cm; Sammlung John Nash
Bem.: Die gepr. Modell-Nr. scheint „ausgebessert" zur Nummer 1401.

1926
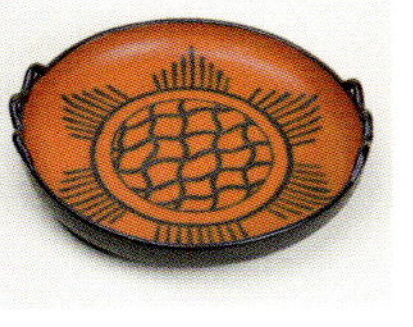
SCHALE
Form: rund
Handhaben: sechs Ösen
Dekor: Netz im Strahlenkranz
ca. 1928–1938
1926 ger., ger. lig. MK, gem. lig. TK, Firmenmarke 19, Made in Austria, H. 6,4 cm, L. 31,2 cm, Ø 30 cm; PB Wien 19
Ausst.: PSK 1995
Lit.: ADGS, Nr. 9 (m. Abb.)

3003M
KRUG
Form: krugförmig, schnabelartiger Ausguss, frei gedreht
Handhabe: ein bandartiger Henkel
Dekor: vermutlich diverse Dekore, Laufglasur
ca. 1928–1938
3003M, H. 30 cm, B. 20 cm; GSNachf. 4,7c
Lit.: GSNachf. 4,7c (m. Abb.)

3016M
VASE
Form: becherförmig, frei gedreht
Dekor: vermutlich diverse Dekore, Laufglasur
ca. 1928–1938
3016M, H. 18 cm, Ø 13 cm; GSNachf. 4,7a
Lit.: GSNachf. 4,7a (m. Abb.)

3028M
SCHALE
Form: oval, leicht gedrückte Wandung, frei geformt
Dekor: vermutlich diverse Dekore, Laufglasur
ca. 1928–1938
3028M, H. 12 cm, Ø 31 cm; GSNachf. 4,7b
Lit.: GSNachf. 4,7b (m. Abb.)

Goldscheider Modelle „ohne Nummerierung"

"Unnumbered" Goldscheider Models

ohne Nr.
Pflasterstein (rötlich)
ca. 1875
ohne Nr., MGE WERK MOLDAU PILSEN, H. 7 cm quadratisch, T. 3,5 cm; Sammlung f@goldscheider.de

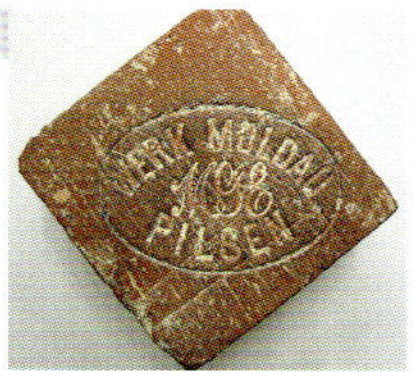

ohne Nr.
Schamotteziegel (beige)
Inschrift: GOLDSCHEIDER PILSEN
ca. 1879
ohne Nr., GOLDSCHEIDER PILSEN, H. 11 cm, B. 20,5 cm, T. 8 cm; Sammlung f@goldscheider.de

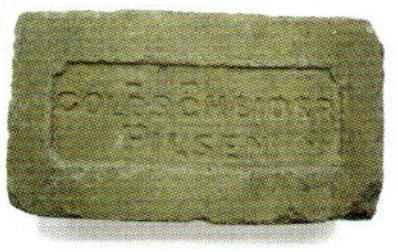

ohne Nr.
KRUG
Form: gebaucht
Handhabe: ein geschwungener Henkel
Dekor: Blumenranken, Kreuze im Vierpass, weiß, braun, grün mit Goldhöhung
ca. 1878
ohne Nr., Firmenmarke 1, H. 28 cm; Sammlung f@goldscheider.de

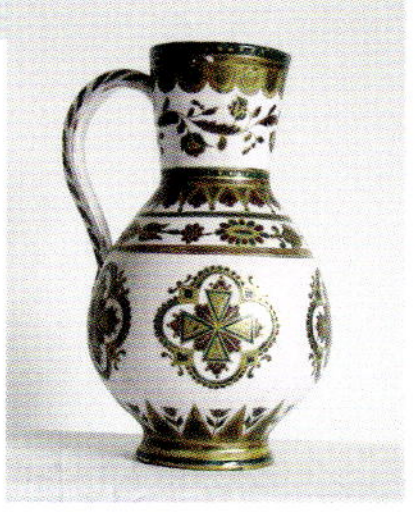

ohne Nr.
VASE Asiatisch reliefiert
Form: gebaucht, kurze Mündung
Dekor: Darstellung asiatischer Kampfszenen, Blumen, Medaillons
ca. 1885/87
ohne Nr., Firmenmarke 4a, H. 20,2 cm, Ø 13,8 cm; PB Deutschland

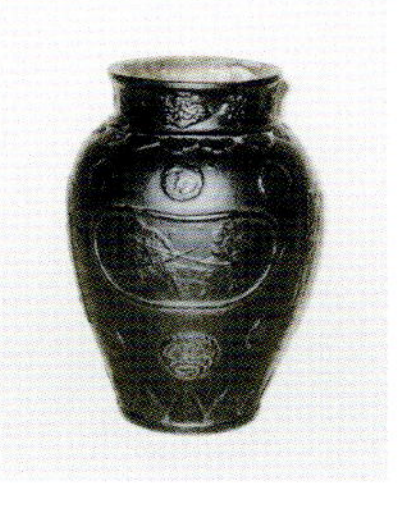

ohne Nr.
WANDBILD Handkuss
Form: unregelmäßig geschwungen, Rahmen mit Rocaillen und Blumenornamenten, goldfarben bronziert
Dekor: Galante Szene, junger Kavalier küsst Dame die Hand (Gemälde, Kaltbemalung)
ca. 1890
ohne Nr., III ger., Firmenmarke 4, Ges. gesch., H. 27 cm, B. 22 cm; Sammlung f@goldscheider.de

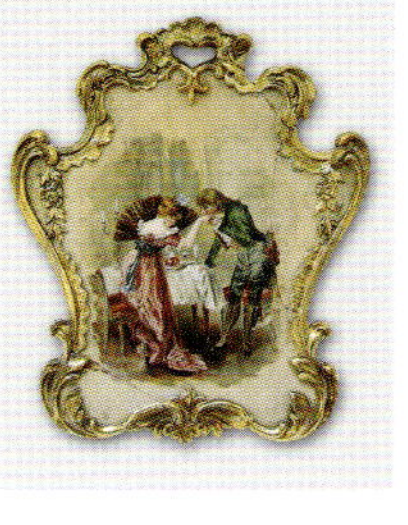

ohne Nr.
WANDBILD Elfen
Form: unregelmäßig geschwungen, Rahmen mit Rocaillen und Blumenornamenten, goldfarben bronziert
Dekor: drei Elfen unter Maiglöckchen stehend (Gemälde, Kaltbemalung)
ca. 1890
ohne Nr., Firmenmarke 4, Ges. gesch., H. 47 cm, B. 25,5 cm; PB Deutschland

ohne Nr.
WANDBILD Pferdewagen in der Puszta
Form: oval, Rahmen mit Rocaillen und Blumenornamenten, goldfarben bronziert
Dekor: Auf Pferdewagen fahrender Ungar, von zwei Hunden begleitet (Gemälde, Kaltbemalung)
ca. 1890
ohne Nr., Firmenmarke 4, Ges. gesch., H. 45,5 cm, B. 36,5 cm; eBay # 3778732713
Bem.: Durch die rückseitig angebrachten Löcher für die Aufhängung erkennt man, dass diese Bilder sowohl für die Hängung als Hochformat, wie auch als Querformat dienten konnten. Die Kaltbemalung erfolgte nach dem Brand. Man entschied also das Motiv und somit das Format erst vor der Bemalung.

ohne Nr.
WANDBILD Reitende Kosaken
Form: oval, Rahmen mit Rocaillen und Blumenornamenten, goldfarben bronziert
Dekor: Durch eine Winterlandschaft reitende Kosaken, von Hunden begleitet (Gemälde, Kaltbemalung)
ca. 1890
ohne Nr., Firmenmarke 4, Ges. gesch., H. 45,5 cm, B. 36,5 cm; eBay # 3778732713
Bem.: Durch die rückseitig angebrachten Löcher für die Aufhängung erkennt man, dass diese Bilder sowohl für die Hängung als Hochformat, wie auch als Querformat dienten konnten. Die Kaltbemalung erfolgte nach dem Brand. Man entschied also das Motiv und somit das Format erst vor der Bemalung.

ohne Nr.
WANDMASKE Lachender Negerjunge mit Strohhut
ca. 1892/95
ohne Nr., ohne Firmenmarke, H. 22 cm; PB Wien 19

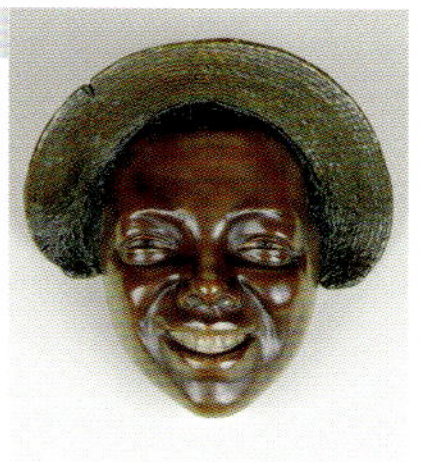

ohne Nr.
WANDTELLER Männerbildnis
Form: rund
Dekor: Männerbildnis (Renaissancetracht), Fahne mit Blumenranken
ca. 1878
ohne Nr., Firmenmarke 1, H. 4 cm, Ø 39,5 cm; PB Wien 19

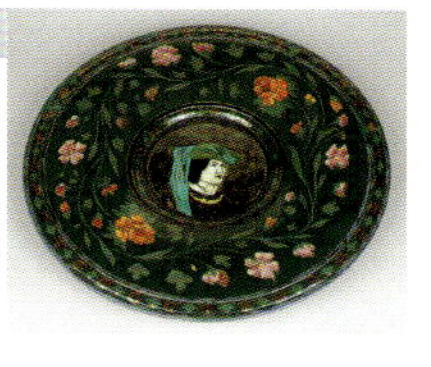

ohne Nr.
WANDTELLER Blumen
Form: rund, Fahne mit Blumenornamenten
Dekor: Amphore, Blumen und Blätter (Gemälde, Kaltbemalung)
ca. 1890
ohne Nr., A, Firmenmarke 3, Vervielf. vorbeh., Ø 24 cm; Sammlung f@goldscheider.de

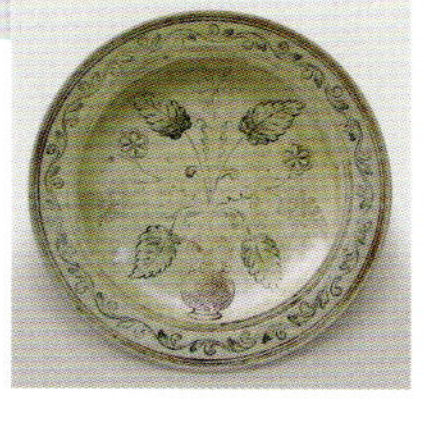

ohne Nr.
WANDTELLER Aschenputtel
Form: rund, Fahne goldfarben bronziert
Dekor: Aschenputtel beim Taubenfüttern (Gemälde, Kaltbemalung)
ca. 1890
ohne Nr., Firmenmarke 4, Ges. gesch., Ø 45,7 cm; eBay # 184685235

ohne Nr.
WANDTELLER Magd mit Kuh
Form: rund, Fahne goldfarben bronziert
Dekor: Magd mit Kuh in ländlicher Umgebung (Gemälde, Kaltbemalung)
ca. 1890
ohne Nr., Firmenmarke 4, Ges. gesch., Ø 48,3 cm; eBay # 3726896758

ohne Nr.
WANDTELLER/RELIEFBÜSTE Araber mit Turban (Fahne mit arab. Inschrift)
Form: rund
Inschrift: undeutbare arab. Schriftzeichen
ca. 1890
ohne Nr., FG (Firmenmarke ohne Ang.), Ø 29 cm; Doro 20.4.1993; SLW 172

ohne Nr.
WANDTELLER/RELIEFBÜSTE „Angeheitert" (Tiroler Bauernbursche mit Hut)
Inschrift: ANGEHEITERT
ca. 1890
ohne Nr., Firmenmarke 4, Vervielf. vorbeh., Ø ca. 23 cm; KH

ohne Nr.
WANDTELLER/RELIEFBÜSTE Götz von Berlichingen
Form: rund
Dekor: Spiegel: Götz, halbplast. Büste, gemalt: 1482 + Wappen, Fahne mit Inschrift: Gottfried Götz von Berlichingen!
ca. 1890
ohne Nr., Firmenmarke 3, Ø 40 cm; Sammlung Perl, Nürnberg
Bem.: Dargestellt ist der Reichsritter Götz von Berlichingen (ca. 1480–1563).
Pendant –> vgl. folgenden Eintrag

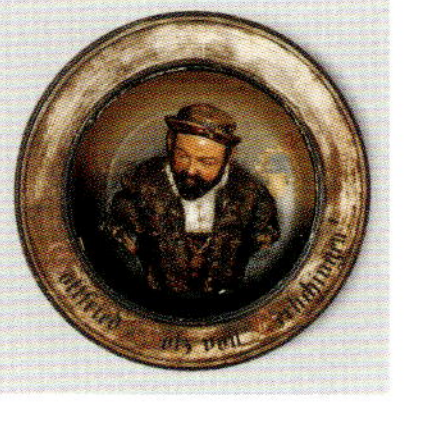

ohne Nr.
WANDTELLER/RELIEFBÜSTE Elisabeth (Frau des Götz von Berlichingen)
Form: rund
Dekor: Spiegel: Elisabeth, halbplast. Büste, gemalt: 1482 + Wappen, Fahne mit Inschrift: Elisabeth Götz' ens Frau!
ca. 1890
ohne Nr., Firmenmarke 3, Ø 40 cm; Sammlung Perl, Nürnberg
Bem.: Dargestellt ist die Frau des Reichsritter Götz von Berlichingen (ca. 1480–1563), Elisabeth.
Pendant –> vgl. vorhergehenden Eintrag

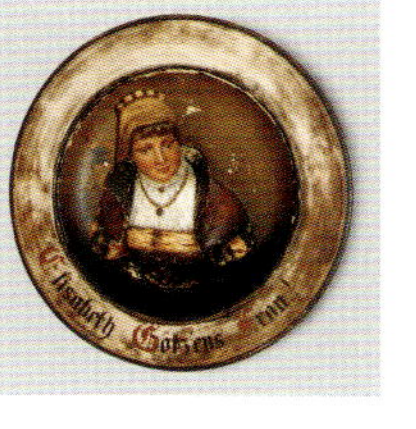

ohne Nr.
Stehende Ägypterin mit Wasserkrug (auf profiliertem Rundsockel)
ca. 1890
ohne Nr., Firmenmarke 11, H. 81 cm; Wien Museum Inv. 190.061

ohne Nr.
Schreitende Holländerin mit zwei Wassergefäßen (auf Landschaftssockel)
ca. 1895
ohne Nr., Firmenmarke 4a, Ges. gesch., H. 68 cm; Wien Museum Inv. 190.060

ohne Nr.
UHR Stehender Mädchenakt an Sonnenblume
ca. 1895/98
ohne Nr., ohne Firmenmarke, Ges. gesch., H. 73,6 cm; MAK Wien, Inv. Ke 10521-12

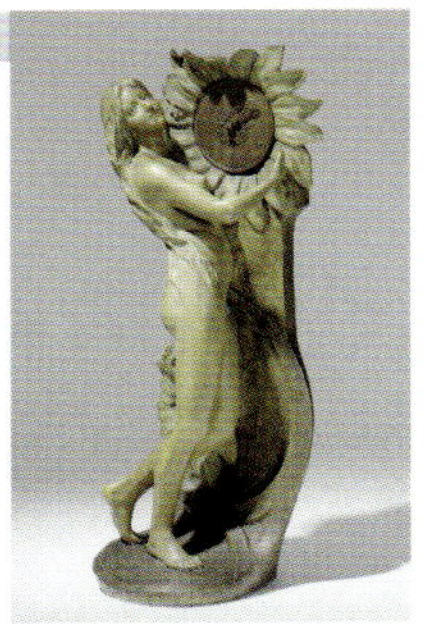

ohne Nr.
Stehende Pierrette mit drei-zipfeliger Mütze
BRONZE / ELFENBEIN (auf Onyxsockel)
ca. 1925–1930
ohne Nr., GS Vienna (Marke Bronze 4), H. 28 cm; Wien Museum Inv. 190.156

ohne Nr.
Sitzender Hase (Kaninchen)
ca. 1938–1941
ohne Nr., 20 Firmenmarke 19, Made in Germany, Ausformung zw. 1938–1941, H. 10,1 cm; Wien Museum Inv. 190.042

ohne Nr.
FIGUR „Spaniel"
(Sitzender Spaniel)
ca. 1934
ohne Nr., Firmenmarke 19, Made in Austria, H. 7,2 cm; Sammlung John Nash

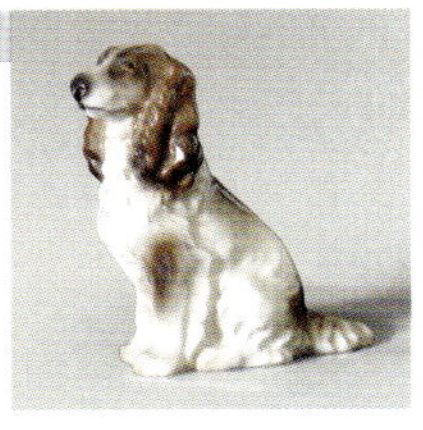

ohne Nr.
FIGUR „Setter"
(Witternder Setter)
ca. 1933
ohne Nr., Firmenmarke 19, H. 4 cm, B. 11 cm; Sammlung John Nash
-> vgl. 6803
-> vgl. 6818= GRUPPE
Lit.: GSNachf. 5,10 (m. Abb.), hier mit Ang. des Originaltitels FIGUR „Setter".

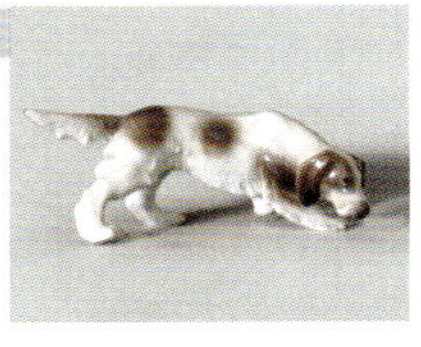

ohne Nr.
Bosse, Walter
„Groteskes Tier" (Hirsch)
ca. 1928/29
ohne Nr., H. fehlt
Lit.: Lehmann, G.: Neuheiten, in: KuKGW, 9.1928/29, S. 126 (m. Abb.), der dargestellte Hirsch wird hier als „Groteskes Tier" bezeichnet und mit dem Attribut „ungequälter keramischer Humor" versehen.

ohne Nr.
Bosse, Walter
Sitzender Affe mit Schale
ca. 1928/29
ohne Nr., Firmenmarke 19, sign. BOSSE AUSTRIA, H. 14,8 cm, B. 15 cm; PB Wien
Lit.: Schreyer-Hartmann, S. 80 (m. Abb.), „zwischen 1924–1936 in Kufstein produziert".

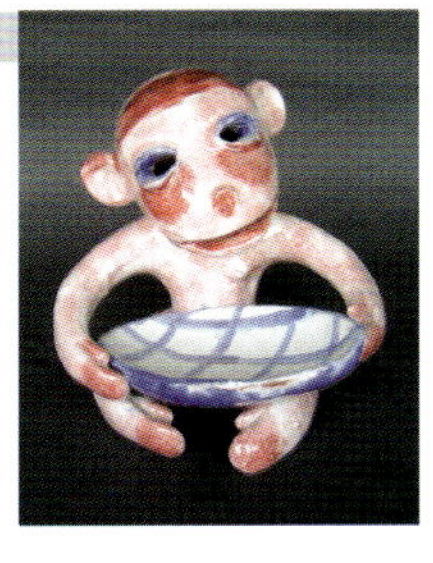

ohne Nr.
Bosse, Walter
Sitzender Knabe mit Kappe und Topf (für Kakteen)
ca. 1928/29
ohne Nr., (Firmenmarke ohne Ang.), H. 12 cm; Pinhas, S. 177 (m. Abb.)

ohne Nr.
Bosse, Walter
Skifahrer mit Kuh
(auf Rechtecksockel)
ca. 1928/29
ohne Nr., Firmenmarke 19, Austria, sign. Bosse, H. 17,6 cm, L. 32,3 cm; PB Wien 19
ohne Nr., Firmenmarke 19, Austria, sign. Bosse, H. 18 cm; Kett. 82
Lit.: Schreyer-Hartmann, S. 48–49 (m. Abb.), dieses Modell wurde von Bosse sowohl in Kufstein, als auch für Goldscheider ausgeführt. Das Kufsteiner Erzeugnis wird im Originalkat. (datiert 1928) von Bosse unter der Nr. 705 geführt.

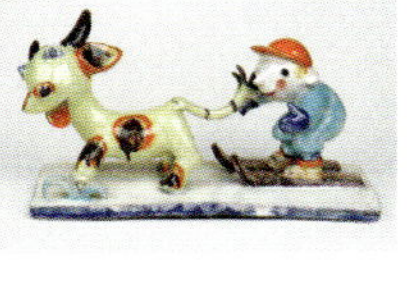

ohne Nr.
Bosse, Walter
„Hochzeitsreise" (groteskes Paar in einem Ruderboot sitzend)
ca. 1928
ohne Nr., H. fehlt, L. 22 cm
Lit.: N.N.: Farbe und Bewegung, Die Fayencen der Wiener Manufaktur Friedrich GS Wien-Währing, in: DB, 5.1928, Nr. 213, S. 38 (m. Abb.) „prächtige Grotesken in ihrer humorvollen Eigenart herzerquickend"; Schreyer-Hartmann, S. 46 (m. Abb.), dieses Modell wurde von Bosse sowohl in Kufstein, als auch für Goldscheider ausgeführt. Das Kufsteiner Erzeugnis wird im Originalkat. (datiert 1928) von Bosse unter der Nr. 549 geführt.

ohne Nr.
Bosse, Walter
BUCHSTÜTZE Igelmann
ca. 1928/29
ohne Nr., Firmenmarke 19a, sign. BOSSE AUSTRIA, H. 15 cm; PB Wien 19
ohne Nr., Firmenmarke 19a, sign. BOSSE AUSTRIA, H. 15,5 cm; PB Wien
ohne Nr., Firmenmarke 19a, Made in Austria (gepr.), Hand dec. (gepr.), unsign., H. 16,5 cm, B. 13 cm; Sammlung mad
Ausst.: HMW (jetzt Wien Museum) 1985/86.
Lit.: N.N.: Farbe und Bewegung, Die Fayencen der Wiener Manufaktur Friedrich GS Wien-Währing, in: DB, 5.1928, Nr. 213, S. 38 (m. Abb.), „prächtige Grotesken in ihrer humorvollen Eigenart herzerquickend"; Lehmann, G.: Neuheiten, in: KuKGW, 9.1928/29, S. 126 (m. Abb.), hier bezeichnet als „Bücherhalter" und mit dem Attribut „ungequälter keramischer Humor" versehen; Schreyer-Hartmann, S. 36 (m. Abb.); GK, S. 21, 43 (m. Abb.).

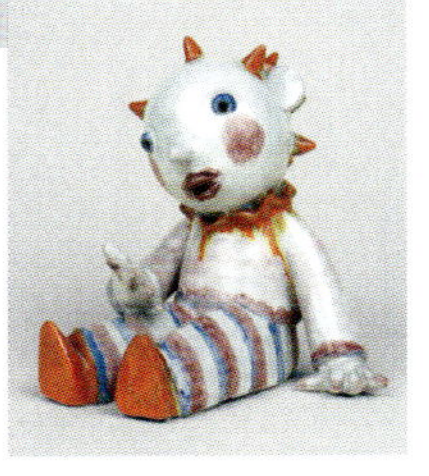

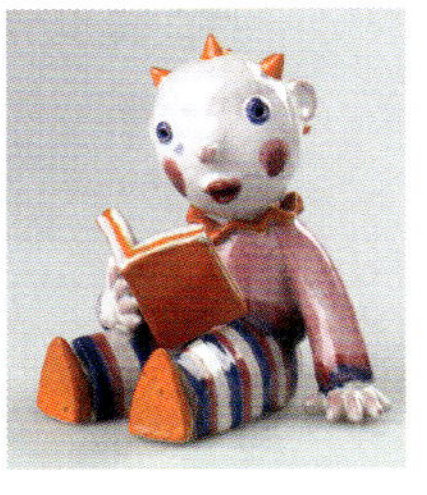

ohne Nr.
Bosse, Walter
„Deckeltopf" in Form eines Kopfes mit Trachtenhut
ca. 1928/29
ohne Nr., H. fehlt
Lit.: Lehmann, G.: Neuheiten, in: KuKGW, 9.1928/29, S. 126 (m. Abb.), der vollplastisch gearbeitete Kopf mit Trachtenhut wird hier nur als „Deckeltopf" bezeichnet und mit dem Attribut „ungequälter keramischer Humor" versehen; Schreyer-Hartmann, S: 28 (m. Abb.).

ohne Nr.
Bosse, Walter
Elefant als Schalenträger
ca. 1928/30
ohne Nr., Firmenmarke 19a, sign. BOSSE AUSTRIA, H. fehlt; PB Wien

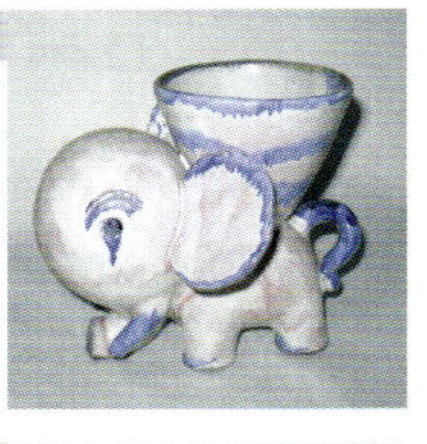

ohne Nr.
Bosse, Walter
KERZENHALTER (zweiflammig) Zwei Dackel
ca. 1928/30
ohne Nr., Firmenmarke 19a, unsign., H. 13 cm, L. 21 cm; PB Wien 19
Lit.: Schreyer-Hartmann, S. 63 (m. Abb.), hier sind ähnl. Dackelfiguren aus der Kufsteiner Zeit, sowie Entwürfe für die Staatl. Majolika-Manufaktur Karlsruhe abgebildet.

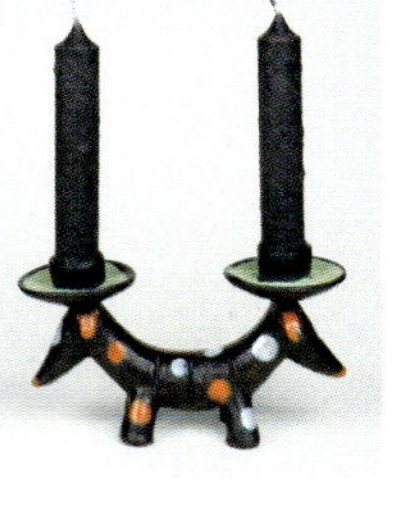

ohne Nr.
Bosse, Walter
SCHALE (Gitterkorb)
Form: trichterförmig, kurzer zylindrischer Stand
Dekor: bogenartig (bzw. dreimal herzförmig) durchbrochene Wandung, Laufglasur
ca. 1937
ohne Nr., Firmenmarke 19a, BOSSE AUSTRIA, H. 8,8 cm, Ø 25,3 cm; PB Wien 19
ohne Nr., Firmenmarke 19a, 7,8 cm, Ø 25,3 cm; Sammlung a@goldscheider.net
ohne Nr., Firmenmarke 19a, BOSSE AUSTRIA, H. 8,8 cm, Ø 25,5 cm; WKA 35 (Lot 1068, außer Kat.)
Bem.: Zwei andere ebenfalls mehrfach durchbrochene, vergleichbare Schalen finden sich in dem Artikel: Der Keramiker Walter Bosse und seine neuen Schöpfungen, in: ÖK, 8.1937, Nr. 12, S. 31 (m. Abb.), hier mit der für Bosse typischen gemalten Blume im Spiegel und der Betitelung „Gitterkörbe". Auch belegt der Vergleich der Objekte hier wieder, dass Bosse für seine eigene Erzeugung und die Entwürfe der Goldscheiderproduktion manches Mal nahezu identische Vorbilder zur Ausführung brachte. Die Datierung der Schale für Goldscheider lässt sich also ziemlich genau in das Jahr 1937 legen.
Ausst.: PSK 1995
Lit.: N.N.: Der Keramiker Walter Bosse und seine neuen Schöpfungen, in: ÖK, 8.1937, Nr. 12, S. 31 (m. Abb.); ADGS, Nr. 39 (m. Abb.)

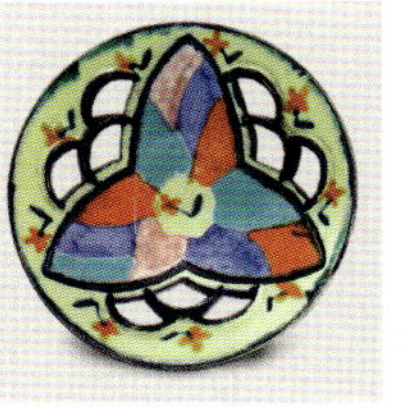

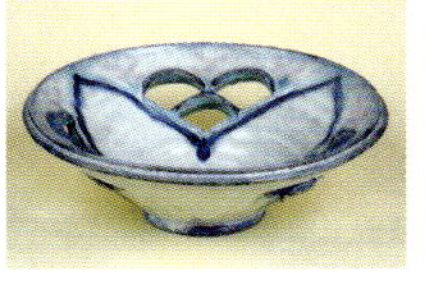

ohne Nr.
Bosse, Walter
SCHALE
Form: sechspassig (teils gezackt)
Handhaben: zwei blattförmige Ösen
Dekor: Blume mit Stängel (appliziert im Spiegel der Schale), mehrfarbig
ca. 1928/30
ohne Nr., Firmenmarke 19a, sign. BOSSE AUSTRIA, H. 7,8 cm, B. 24,2 cm, L. 31,4 cm; Sammlung a@goldscheider.net
ohne Nr., Firmenmarke 19a, unsign., H. 7,7 cm, B. 24,5 cm, L. 30,5 cm; PB Wien
Bem.: Ein ähnlicher, ebenfalls mit Blume im Spiegel und wohl früherer Entwurf für die Wiener Werkstätte, findet sich auch bei Vally Wiesethier -> vgl. ExpKWW, S. 61, Nr. 28 (m. Abb.).
Lit.: Schreyer-Hartmann, S. 84 (m. Abb.), „zwischen 1924–1936 in Kufstein produziert".

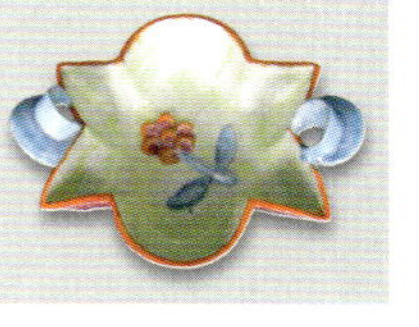

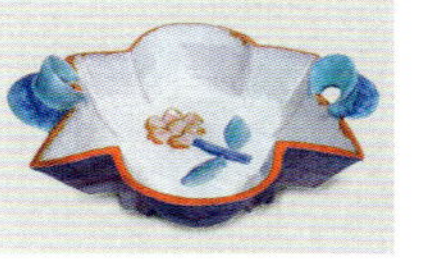

ohne Nr.
Bosse, Walter
VASE mit groteskem Tier (Hund?)
ca. 1928/30
ohne Nr., Firmenmarke 19, sign. BOSSE AUSTRIA, H. 32 cm, B. 21,2 cm; PB Wien
Lit.: Schreyer-Hartmann, S. 57 (m. Abb.), hier die Abb. eines ähnl. grotesken Tieres, welches eine trichterförmige Vase auf dem Rücken trägt welches laut Schreyer-Hartmann zwischen 1924–1936 in Kufstein produziert wurde.

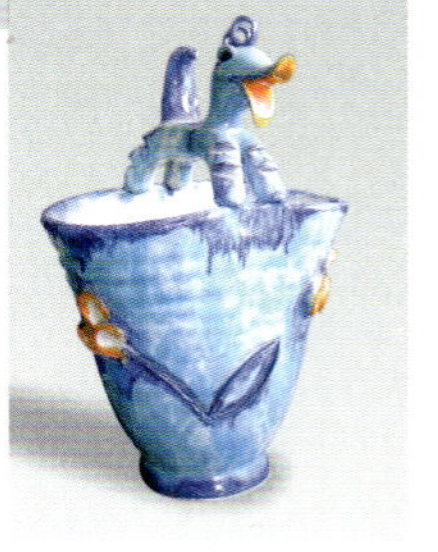

ohne Nr.
Bosse, Walter
VASE Blume
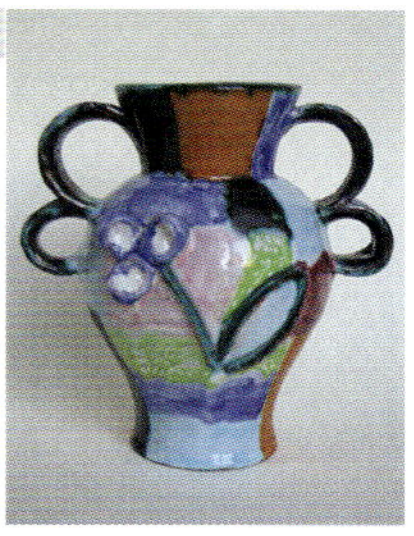
Form: balusterförmig, trichterförmige Mündung
Handhaben: vier Ösen
Dekor: zwei stilisierte, applizierte Blumen mit Blatt, Farbfelder, mehrfarbig, Laufglasur
ca. 1928/30
ohne Nr., Firmenmarke 19a, unsign., H. 14,2 cm; Sammlung Weilinger, Wien
–> vgl. ohne Nr. (H. 30,4 cm)
Lit.: Schreyer-Hartmann, S. 86 (m. Abb.), hier eine ähnl. zwischen 1924–1936 in Kufstein produzierte Vase abgebildet.

ohne Nr.
Bosse, Walter
VASE Blume
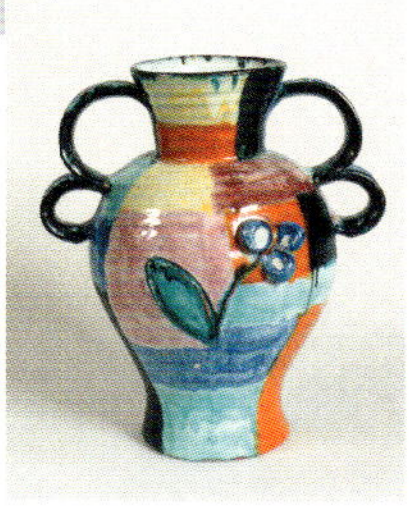
Form: balusterförmig, trichterförmige Mündung
Handhaben: vier Ösen
Dekor: zwei stilisierte, applizierte Blumen mit Blatt, Farbfelder, mehrfarbig, Laufglasur
ca. 1928/30
ohne Nr., Firmenmarke 19a, unsign., H. 30,4 cm; Sammlung.schmutz@chello.at 81
–> vgl. ohne Nr. (H. 14,2 cm)
Lit.: Schreyer-Hartmann, S. 86 (m. Abb.), hier eine ähnl. zwischen 1924–1936 in Kufstein produzierte Vase abgebildet.

ohne Nr.
Bosse, Walter
VASE Blume
Form: balusterförmig, trichterförmige Mündung
Handhaben: zwei ohrenartige Henkel
Dekor: zwei stilisierte, applizierte Blumen mit Blatt, mehrfarbig, Laufglasur
ca. 1928/30
ohne Nr., Firmenmarke 19a, sign. BOSSE AUSTRIA, H. 26,8 cm; Doro 18.11.1997
Lit.: Schreyer-Hartmann, S. 86 (m. Abb.), hier eine Variante dieser Vase abgebildet, die zwischen 1924–1936 in Kufstein produziert wurde.

ohne Nr.
Bosse, Walter
VASE Blume
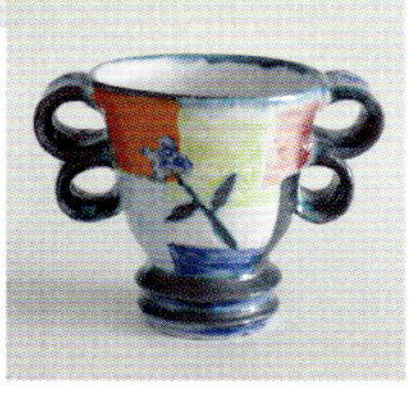
Form: becher-/kelchförmig, zweifach gewulsteter Stand
Handhaben: vier Ösen
Dekor: zwei stilisierte Blumen, rechteckige Farbfelder, mehrfarbig, Laufglasur
ca. 1928/30
ohne Nr., Firmenmarke 19a, sign. BOSSE AUSTRIA, H. 14,1 cm, Ø 19,4 cm; PB Deutschland
ohne Nr., Firmenmarke 19a, sign. BOSSE AUSTRIA, H. 14 cm, Ø fehlt; WKA 35 (Lot 1065, außer Kat.)
ohne Nr., Firmenmarke 19a, sign. BOSSE AUSTRIA, H. 13,5 cm, Ø 18 cm, Wien Museum Inv. 220.775
–> vgl. 246

ohne Nr.
Bosse, Walter
VASE Blume
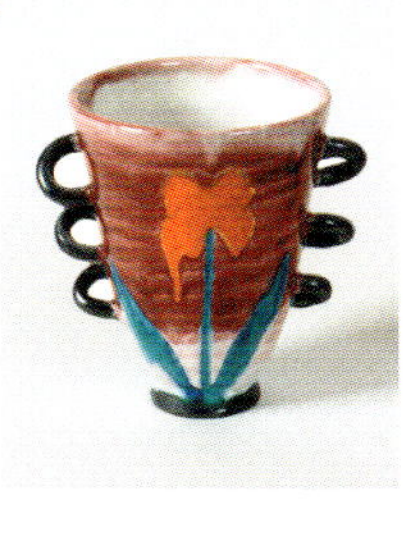
Form: becherförmig
Handhaben: sechs Ösen
Dekor: zwei stilisierte Blumen, mehrfarbig, Laufglasur
ca. 1928/30
ohne Nr., Firmenmarke 19a, unsign., H. 15,2 cm, Ø 12,3 cm; Sammlung.schmutz@chello.at 48
ohne Nr., Firmenmarke 19a, unsign., H. 15,4 cm, Ø 13 cm; Sammlung a@goldscheider.net

ohne Nr.
Bosse, Walter
VASE Blumenblüten
Form: trichterförmig, mittig tailliert (spindelförmig)
Handhaben: zwei Ösen
Dekor: zwei stilisierte Blumenblüten, mehrfarbig, Laufglasur
ca. 1928/30
ohne Nr., Firmenmarke 19a, unsign., H. 23,8 cm, Ø 15,8 cm; MAK Wien, Inv. Ke 10602-18
Bem.: Eine vom Typus her vergleichbare Vase, die mit BOSSE AUSTRIA und der Modell-Nr. 775 gemarkt ist, wurde bei Kett. 153 angeboten. Somit ist das durch Goldscheider ausgeführte Gefäß eindeutig Walter Bosse zuzuordnen.
Ausst.: PSK 1995
Lit.: ADGS, Nr. 12 (m. Abb.)

ohne Nr.
Bosse, Walter
VASE Blume
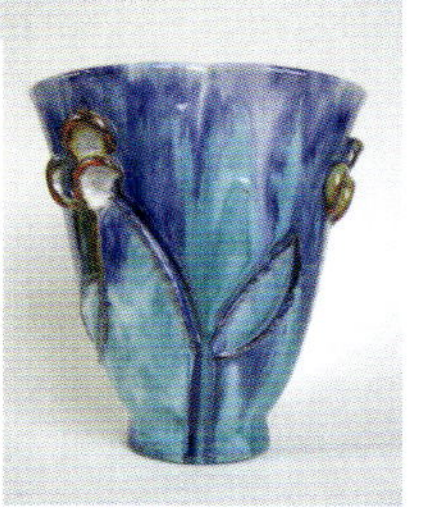
Form: becherförmig, gedrückte Mündung
Dekor: drei stilisierte Blumen, mehrfarbig, Laufglasur
ca. 1928/30
ohne Nr., Firmenmarke 19a, Made in Austria, sign. BOSSE AUSTRIA, H. 20 cm; PB Wien
ohne Nr., Firmenmarke 19a, Made in Austria, sign. BOSSE AUSTRIA, H. 20 cm; Sammlung Weilinger, Wien

ohne Nr.
Bosse, Walter
VASE Blume
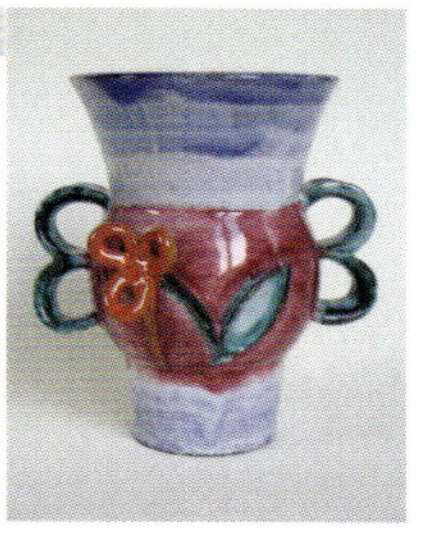
Form: gebaucht, trichterförmige Mündung
Handhaben: vier Ösen
Dekor: zwei stilisierte (ösenförmig), applizierte Blumen mit Blatt, mehrfarbig, Laufglasur
ca. 1928/30
ohne Nr., Firmenmarke 19a, unsign., H. 13,8 cm; Sammlung Weilinger, Wien
ohne Nr., Firmenmarke 19a, unsign., H. 13,9 cm; Sammlung Weilinger, Wien

ohne Nr.
Bosse, Walter
VASE/TOPF
Form: konisch
Handhaben: vier Ösen
Dekor: rechteckige Farbfelder mit Blumen, mehrfarbig, Laufglasur
ca. 1928/30
ohne Nr., Firmenmarke 19a, unsign., H. 11,4 cm; Doro 18.11.1997
–> vgl. 232
Lit.: Schreyer-Hartmann, S. 86–87 (m. Abb.), dieses Modell wurde von Bosse sowohl in Kufstein, als auch für Goldscheider ausgeführt. Das Kufsteiner Erzeugnis wird im Originalkat. (datiert 1928) von Bosse unter der Nr. 765 geführt.

ohne Nr.
Bouval, Maurice
REKLAMESCHILD „Sculpture“ Frederic Goldscheider, Editeur, Paris, Vienne, Terre cuite, Bronze, Marbre, Etain
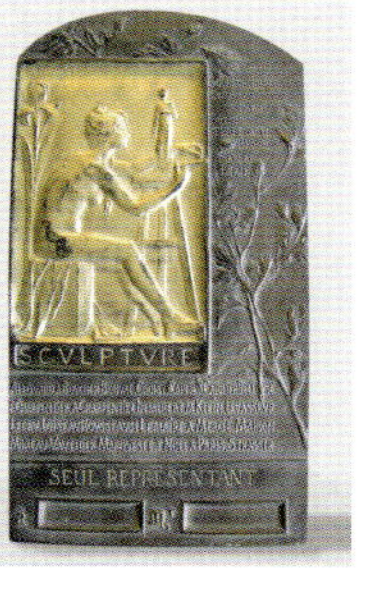

BRONZE
1898
ohne Nr., ohne Firmenmarke, sign. M. Bouval 98, H. 34 cm, B. 18,8 cm; Wien Museum Inv. 182.051
Bem.: Dieses große Reklameschild (für die Werbung an der Hauswand eines Verkaufslokales gedacht), welches Maurice Bouval für Goldscheider als Allegorie der Bildhauerei entwarf, listet neben Rudolf Maison und Arthur Strasser folgende teils sehr bedeutende französische Künstler der Jahrhundertwende auf: Allouard, A. Boucher, Bouval, Croisy, Carlier, Carrièr-Belleuse, F. Charpentier, A. Charpentier, Injalbert, M. Klein, Levasseur, Ledru, Loiseau-Rousseau, H. Lemaire, A. Mercié, Moreau-Vauthier, Marqueste, T. Noël, A. Paris.
Ausst.: HMW (jetzt Wien Museum) 1985/86.
–> vgl. 642; 672; und die Version in Zinn
Lit.: GK, S. 14, 69 (m. Abb.); Kashey, S. 272; Hillenbrand, S. 486; Nagel 354, Herr 22.

ohne Nr.
Bouval, Maurice
REKLAMESCHILD „Sculpture“ Frédéric Goldscheider, Editeur, Paris, Vienne, Terre cuite, Bronze, Marbre, Etain
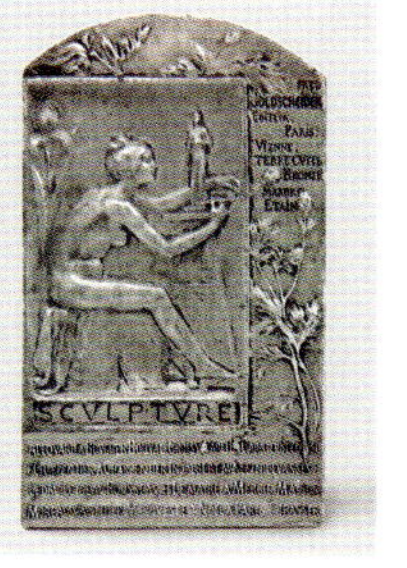

ZINN
1898
ohne Nr., ohne Firmenmarke, unsign., H. 8,6 cm, B. 5 cm; Sammlung a@goldscheider.net
Bem.: Dieses Reklameschild (für die Werbung im Schaufenster, bzw. im Verkaufslokal innen gedacht), welches Maurice Bouval für Goldscheider als Allegorie der Bildhauerei entwarf, listet neben Rudolf Maison und Arthur Strasser folgende teils sehr bedeutende französische Künstler der Jahrhundertwende auf: Allouard, A. Boucher, Bouval, Croisy, Carlier, Carrier-Belleuse, F. Charpentier, A. Charpentier, Injalbert, M. Klein, Levasseur, Ledru, Loiseau-Rousseau, H. Lemaire, A. Mercie, Moreau-Vauthier, Marqueste, T. Noël, A. Paris.
–> vgl. 642; 672 und die Version in Bronze

ohne Nr.
Flora, Francesco
Stehender Amor (an einen Baumstamm gelehnt)
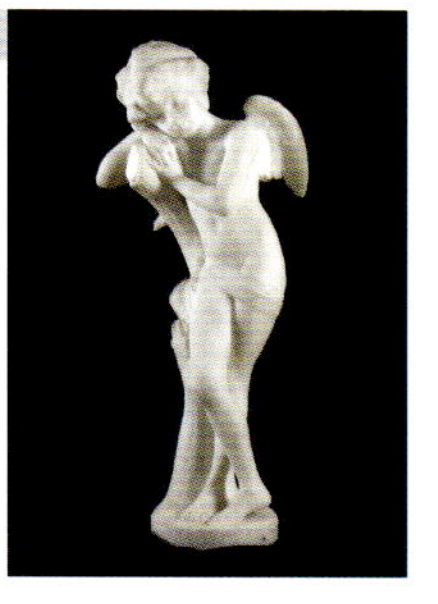
MARMOR
ca. 1900
ohne Nr., Firmenmarke Sculpture Bronze 1, H. 67 cm; Artimo Fine Arts, Brüssel

ohne Nr.
Geiger, Benno
VASE
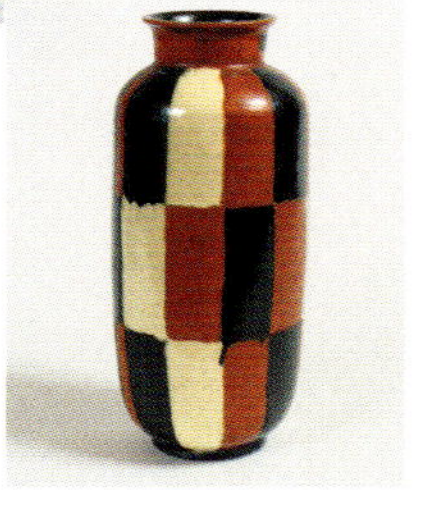
Form: zylinderförmig, flache Schulter, eingezogener Boden, kurze etwas geweitete Mündung
Dekor: schachbrettartig, rot-schwarz-weiß, Laufglasur
ca. 1935
ohne Nr., Firmenmarke 19, Made in Austria, Pick. 11 (Entwurf Benno Geiger, Wien), Pick. 2, H. 33,5 cm; Sammlung.schmutz@chello.at 64

ohne Nr.
Geiger, Benno
VASE
Form: kugelförmig, hohe und weite, leicht eingezogene, trichterförmige Mündung, konischer Standring
Handhaben: zwei gebogene Ösen
Dekor: stilisierte Blätter, mehrfarbig, Laufglasur
ca. 1935
ohne Nr., gem. B, Firmenmarke 19, Pick. 11 (Entwurf Benno Geiger, Wien), H. 25 cm; Mehlis 18
Bem.: Weitere Vasen mit dem gleichen Dekor –> vgl. 924; 972.

ohne Nr.
Goebel, Kurt (zugeschrieben)

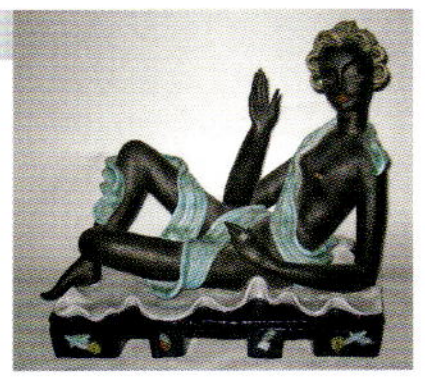

Liegender Damenakt mit Schal, Tuch und erhobener Hand (auf blumenverziertem Rechtecksockel)
ca. 1935
ohne Nr., unleserl. Firmenmarke, unsign., H. 43 cm, L. 47 cm, B. 24 cm; PB Wien
Bem.: Dieser ungewöhnliche und außerordentlich gelungene Akt kann Kurt Goebel zugeordnet werden da Gesicht und Körperlichkeit eindeutig seine Handschrift verraten. Wollte man bei Goldscheider ein vergleichbares Stück finden, so wäre einzig der sitzende Frauenakt von Vally Wieselthier (Modell-Nr. 7789) zu nennen.

ohne Nr.
Goebel, Kurt

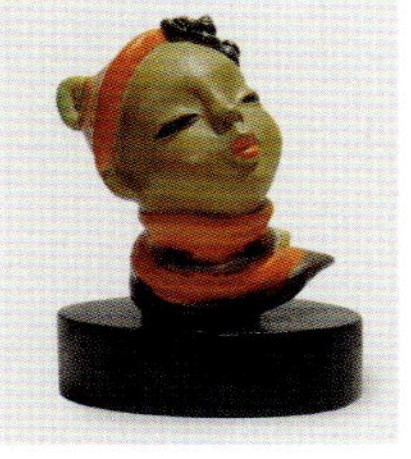

STEHKOPF Frauenkopf mit Mütze und Schal (auf Holzsockel)
ca. 1935
ohne Nr., Plak. D, unsign., H. 8,6 cm, Gesamth. 10 cm; Sammlung a@goldscheider.net
Bem.: Vergleichbare Köpfe dieser Art tragen die Modellnummern 7839 oder 7874.

ohne Nr.
Haniroff
Schreitender Harlekin mit Schirm (auf Ovalsockel)
BRONZE / ELFENBEIN / Onyxsockel
ca. 1905
ohne Nr., Plak. C, sign. Haniroff, H. 26,8 cm; Wien Museum Inv. 190.116
Bem.: Kopf aus Elfenbein oder Bein.
–> vgl. 3131; 3133

ohne Nr.
Hentschel (d.i. Handzel, August)

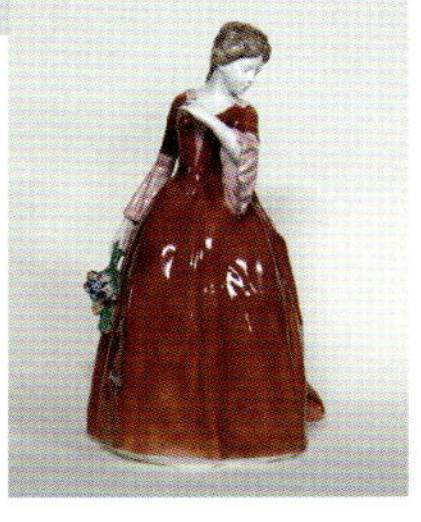

Stehende Dame mit erhobenem Arm (im Watteaukleid) mit Blumenstrauß
ca. 1913
ohne Nr., (ohne Firmenmarke), sign. Hentschel, H. 35,2 cm; MAK Wien, Inv. Ke 10602-46

ohne Nr.
Keller, A.
WANDTELLER „Pfahlbauten Oberbaiern"
Form: rund, Fahne goldfarben bronziert
Dekor: Pfahlbauten im See vor Gebirgslandschaft (Gemälde, Kaltbemalung)
ca. 1890
ohne Nr., Firmenmarke 4, Ges. gesch., sign. A. Keller, Ø 47 cm; eBay # 26231548
Bem.: Der handschriftliche Vermerk „Pfahlbauten Oberbaiern" weißt auf die bayerische Voralpenlandschaft hin.

ohne Nr.
Knörlein, Rudolf

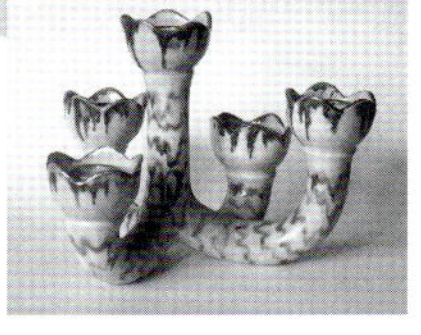

KERZENHALTER (fünfflammig)
Form: röhrenartig, vier zur Mitte hin zusammenlaufende Röhren, blütenartige Kerzentüllen
Dekor: abstrakte Malerei, Laufglasur
ca. 1933
ohne Nr., Maße fehlen
Bem.: Durch die eigenhändige Datierung (1933) Knörleins auf dem Trägerblatt des Originalfotos kann das Jahr 1933 exakt bestimmt werden.
Quelle: Privatarchiv (Originalfoto)

ohne Nr.
Knörlein, Rudolf

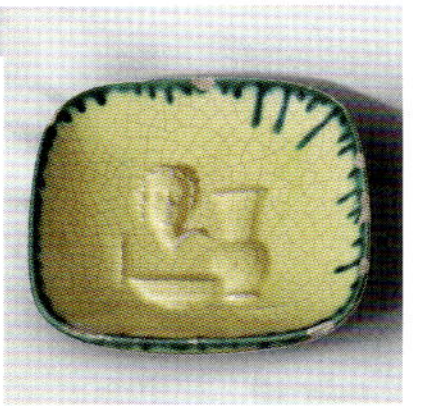

REKLAMESCHALE „50 anni Goldscheider 1885–1935, Vienna"
Dekor: Wandmaske, Schale, Vase, craquelliert (leuchtend gelb / grün)
ca. 1935
ohne Nr., ohne Firmenmarke, unsign., H. 3,6 cm, L. 12,7 cm, B. 12,5 cm; PB Deutschland
Bem.: Diese kleine Reklameschale, die auch als Ascher verwendet werden kann, wurde von Rudolf Knörlein entworfen. Die Motive im Schalenspiegel zeigen Objekte Knörleins wie z.B. die Maske „Turandot" (Modell-Nr. 6427). Hier die Ausführung für den italienischsprachigen Markt.

ohne Nr.
Knörlein, Rudolf

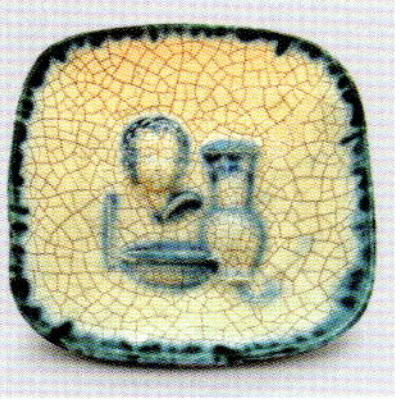

REKLAMESCHALE „50 years Goldscheider 1885–1935, Vienna"
Dekor: Wandmaske, Schale, Vase, craquelliert (weiß / türkis)
ca. 1935
ohne Nr., ohne Firmenmarke, unsign., H. 3,6 cm, L. 12,9 cm, B. 12,8 cm; Sammlung a@goldscheider.net
Bem.: Diese kleine Reklameschale, die auch als Ascher verwendet werden kann, wurde von Rudolf Knörlein entworfen. Die Motive im Schalenspiegel zeigen Objekte Knörleins wie z.B. die Maske „Turandot" (Modell-Nr. 6427). Hier die Ausführung für den englischsprachigen Markt.

ohne Nr.
Knörlein, Rudolf

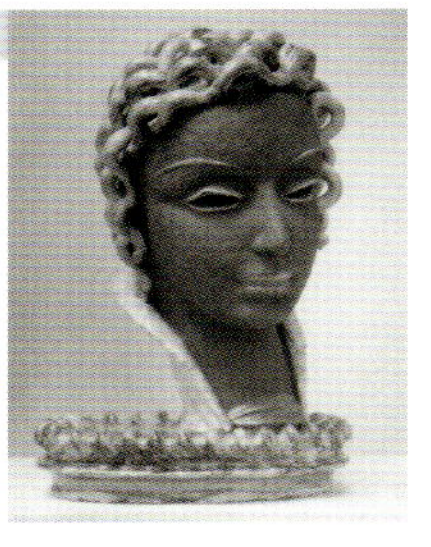

STEHKOPF Mädchen mit gelocktem Haar (auf spiralartig verziertem Rundsockel)
ca. 1933
ohne Nr. H. fehlt
Bem.: Das Original befindet sich im Nationalmuseum Stockholm, Schweden. Belegt wird dies durch ein Originalfoto (Fotoarchiv K. Täuber).
Lit.: ÖK, 5.1934, Nr. 5, S. 17 (m. Abb.), „Mit einigen Frauenköpfen von weichpikantem Ausdruck und lässiger Anmut hat Rudolf Knörlein, einer der wichtigsten keramischen Künstler dieser Erzeugung, Stücke geschaffen, die in London den Wiener Charme mit Glück vertreten".
Quelle: Fotoarchiv K. Täuber.

ohne Nr.
Knörlein, Rudolf (zugeschrieben)

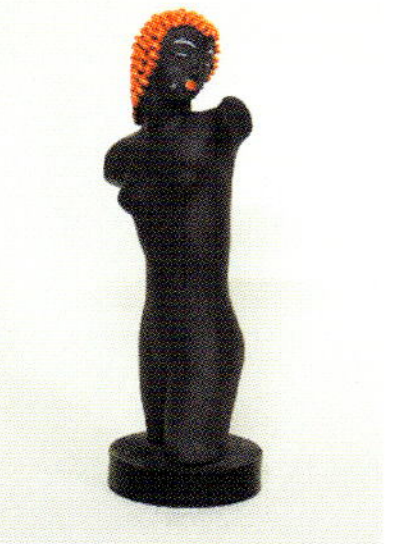

TORSO Mädchenakt mit gelocktem Haar (auf Holzsockel)
ca. 1933
ohne Nr., (ohne Firmenmarke), Gesamth. 27,3 cm; MAK Wien, Inv. Ke 10602-92
Bem.: Trotzdem der Figur die Firmenmarke und Nummerierung fehlt, ist sie eindeutig der Manufaktur Goldscheider zuzuordnen.

ohne Nr.
Laporte-Blairsy (Blaisin?), Léo

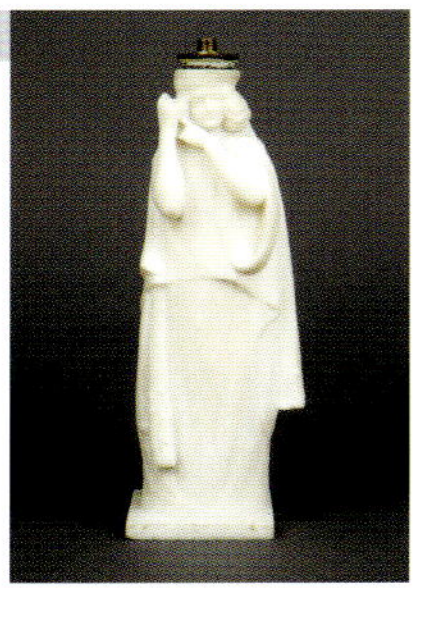

LAMPE Stehende Frauengestalt
ALABASTER / MARMOR?
ca. 1900
ohne Nr., Plak. B, H. 23 cm; PB Wien 19

ohne Nr.
Larsson, Gottfrid Julius Andreas

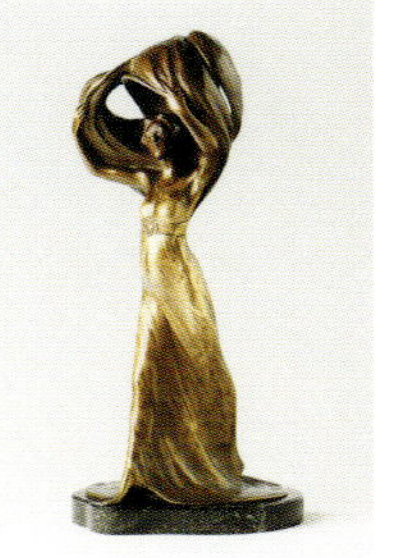

LAMPE Loie Fuller mit großem Tuch (beim Tanz)
BRONZE / Marmorsockel
ca. 1900
ohne Nr., Plak. B, sign. Gottfrid Larsson, H. 43 cm; Sammlung John Jesse, London

ohne Nr.
Meisinger, Ida

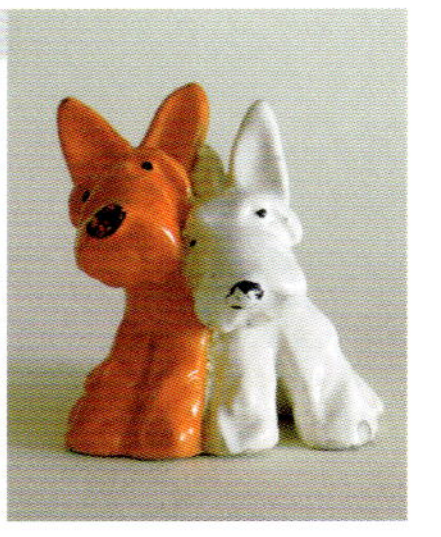

GRUPPE „Scotchterrier" (Zwei sitzende Scotchterrier)
ca. 1928/29
ohne Nr., gem. E, Firmenmarke 19, H. 5,5 cm; Wien Museum Inv. 220.714
ohne Nr., gem. lig. FL, Firmenmarke 19, H. 6 cm; Sammlung John Nash
–> vgl. 5837; 5846

ohne Nr.
Meisinger, Ida

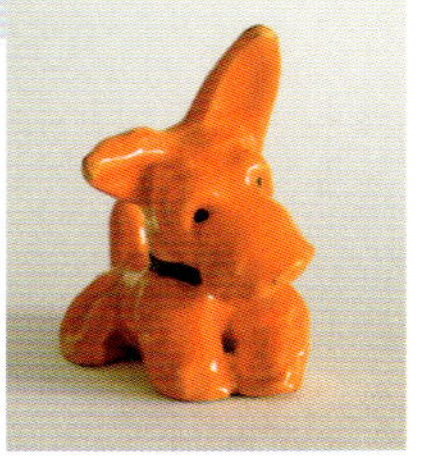

„Der Modehund" (Sitzender Scotch-Terrier)
ca. 1928/29
ohne Nr., Firmenmarke 19, unsign., H. 6,2 cm; PB Deutschland
ohne Nr., Firmenmarke 19, unsign., H. 5,4 cm; PB Wien 19
–> vgl. 5854; 5879; 6026
Lit.: Schilling-Schletter, A.: Hunde-Plastiken, in: MW, 11.1929/30, Nr. 6, S. 12 (m. Abb.); MW, 11.1929/30, Nr. 19, S. 20 (m. Abb.), hier mit Ang. des Originaltitels „Der Modehund" und des Entwerferin Ida Meisinger.

ohne Nr.
Obsieger, Robert (zugeschrieben)

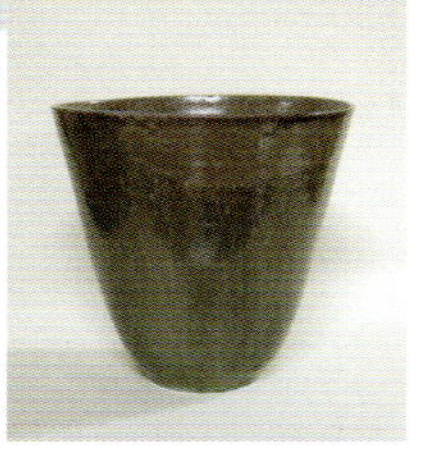

TOPF
Form: trichterförmig
Dekor: grünlichgraue Scharffeuerglasur
ca. 1928–1938
ohne Nr., (ohne Firmenmarke), unsign., H. 19,9 cm, Ø 21,3 cm; MAK Wien, Inv. Ke 7310
Bem.: Die Zuschreibung an Robert Obsieger erfolgt aufgrund vergleichbarer Objekte in: N.N., 50 Jahre Wiener Manufaktur Friedrich Goldscheider, Die Schaulade 11, 1935, S. 152–154 (m. Abb.).
Ausst.: Weihnachtsausst. ÖMKI 1933; Austria in England 1934.

ohne Nr.
Obsieger, Robert (zugeschrieben)

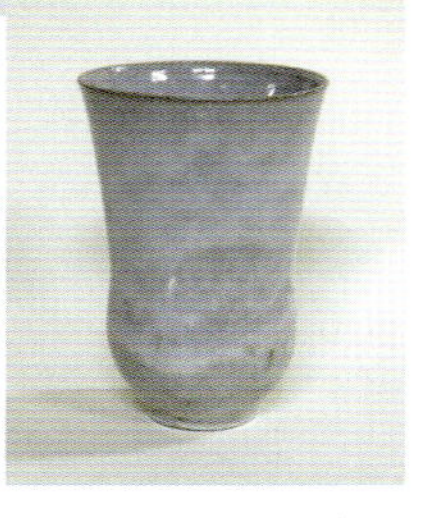

VASE
Form: glockenförmig, trichterförmige Mündung
Dekor: weißliche Scharffeuerglasur
ca. 1928–1938
ohne Nr., ger. R IV, (ohne Firmenmarke), unsign., H. 14,9 cm, Ø 10,7 cm; MAK Wien, Inv. Ke 7313
Bem.: Die Zuschreibung an Robert Obsieger erfolgt aufgrund vergleichbarer Objekte in: N.N., 50 Jahre Wiener Manufaktur Friedrich Goldscheider, Die Schaulade 11, 1935, S. 152–154 (m. Abb.).
Ausst.: Weihnachtsausst. ÖMKI 1933; Austria in England 1934.

ohne Nr.
Sydan
Stehende junge Frau mit wallendem Gewand (auf Felssockel)
ca. 1900
ohne Nr., Firmenmarke Sculpture, H. 115 cm; Wien Museum Inv. 219.109
–> vgl. 2368

ohne Nr.
Thomasch, Wilhelm
FIGUR „Spaniel" (Laufender Spaniel)
ca. 1934
ohne Nr., Firmenmarke 19, H. 4,2 cm, L. 9,5 cm; Sammlung John Nash
Bem.: Für die Zuordnung dieser Figur und der damit verbundenen Gruppe, danke ich Herrn Herwig Thomasch, dem Sohn von Wilhelm Thomasch.
–> vgl. 7061F; 7329F; 8344F
–> vgl. 7015F; 7281F; 8342F= jeweils GRUPPE
Lit.: GSNachf. 5,8 (m. Abb.), hier mit Ang. des Originaltitels FIGUR „Spaniel" (Vorderer Hund aus der GRUPPE 7015).

ohne Nr.
Wagner, A.
WANDTELLER Hütte am See
Form: rund, Fahne goldfarben bronziert
Dekor: Hütte am See (Gemälde, Kaltbemalung)
ca. 1890
ohne Nr., Firmenmarke 4, Ges. gesch., sign. A. Wagner, Ø 47 cm; eBay # 26231910

ohne Nr.
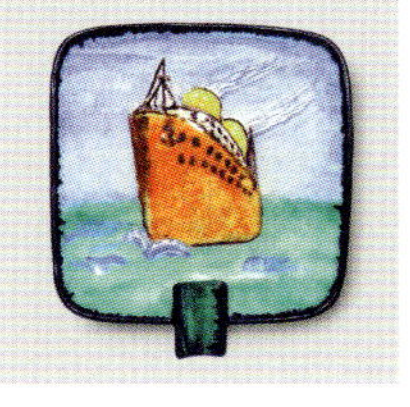
ASCHER Schiff
Form: rechteckig, eine Zigarettenablage
Dekor: Hochseepassagierdampfer
ca. 1935
ohne Nr., Firmenmarke 19, Made in Austria, gem. lig. grün TK, H. 14,6 cm, B. 13,6 cm; Sammlung a@goldscheider.net

ohne Nr.
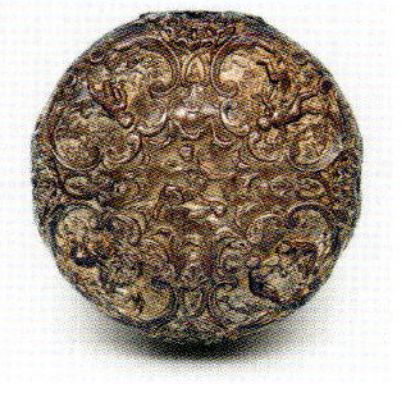
DECKELDOSE Putti
Form: rund
Dekor: spielende Putti auf floralem Dekor, reliefiert
METALL
ca. 1900
ohne Nr., Sculpture Frederic GS, Paris (Firmenmarke Metall 1), H. 2,9 cm, Ø 5,4 cm; Sammlung a@goldscheider.net

ohne Nr.
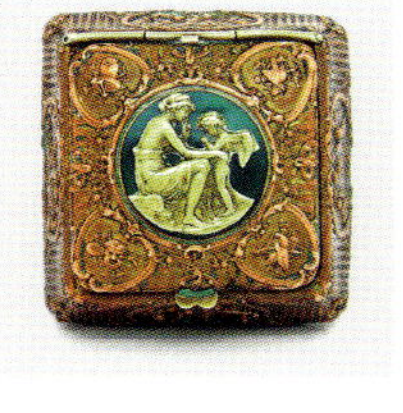
DECKELDOSE Venus und Cupido
Form: quadratisch
Dekor: sitzende Venus mit Cupido, reliefiert
METALL
ca. 1900
ohne Nr., Arthur GS, Paris (Firmenmarke Metall 2), H. 2,8 cm, B. 5,5 cm, L. 5,5 cm; Sammlung a@goldscheider.net

ohne Nr.
DECKELDOSE Kaninchen
Form: rund
Handhabe: Kaninchen (vollplastisch)
Dekor: schwarz, weiß liniert
ca. 1924
ohne Nr., gem. lig. TK, Firmenmarke 19, Made in Austria, H. 13,3 cm, Ø 12,3 cm; Wien Museum Inv. 220.774/1+2

ohne Nr.
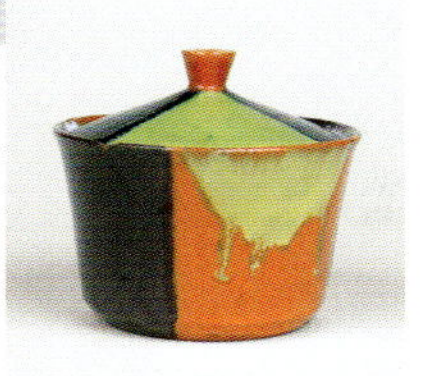
DECKELDOSE
Form: rund
Handhabe: knopfförmig
Dekor: Farbfelder, Laufglasur
ca. 1928–1938
ohne Nr., gem. lig. TK, Firmenmarke 19, Made in Austria, H. 9,2 cm; PB Wien 19
Ausst.: PSK 1995
Lit.: ADGS, Nr. 27 (m. Abb.)

ohne Nr.
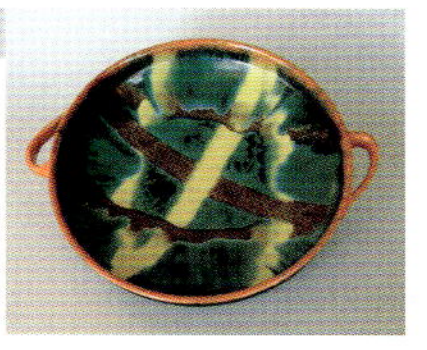
SCHALE
Form: rund
Handhaben: zwei Henkel (ohrenartig)
Dekor: abstrakt, mehrfarbig, Laufglasur
ca. 1928–1938
ohne Nr., ger. WW (oder MM?), ger. lig. MK, Firmenmarke 19, Made in Austria, H. 5,5 cm, Ø 31 cm; Sammlung.schmutz@chello.at 91

ohne Nr.
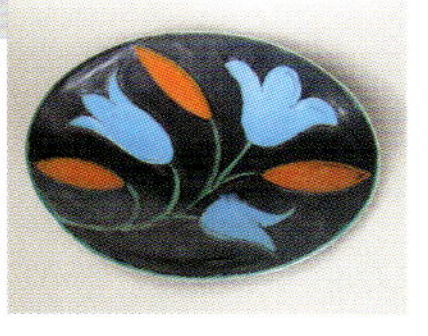
SCHALE
Form: oval, mit Standfuß
Dekor: sechs Tulpen, mehrfarbig
ca. 1928–1938
ohne Nr., gem. lig. MK, Firmenmarke 19, Made in Austria, H. 7 cm, L. 28,8 cm, B. 20,8 cm; Sammlung Weilinger, Wien

ohne Nr.
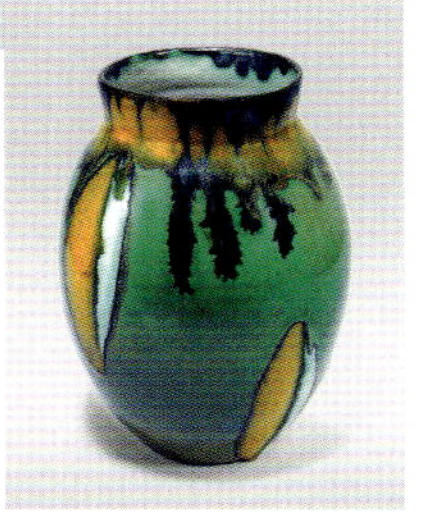
VASE
Form: eiförmig, kurze leicht trichterförmige Mündung
Dekor: stilisierte Blätter, mehrfarbig, Laufglasur
ca. 1928–1938
ohne Nr., gem. lig. KR?, Firmenmarke 19, Made in Austria, H. 15 cm, Ø 10,6 cm; Sammlung a@goldscheider.net

ohne Nr.
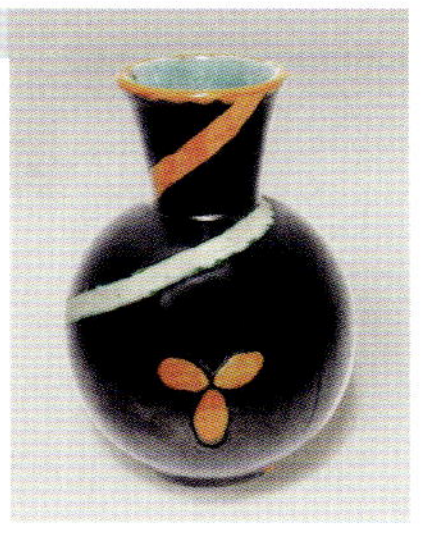
VASE
Form: kugelförmig, trichterförmige Mündung
Dekor: stilisierte Blätter, Streifen / Spiralen, mehrfarbig, Laufglasur
ca. 1928–1938
ohne Nr., Firmenmarke 19, Made in Austria, H. 17 cm, Ø 12,5 cm; Sammlung a@goldscheider.net

ohne Nr.
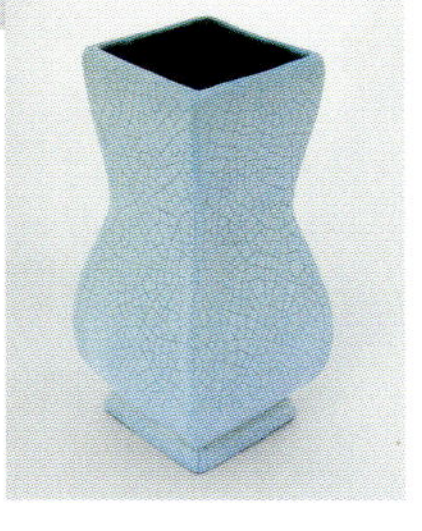
VASE
Form: gebaucht, quadratische Mündung, auf quadratischem Stand
Dekor: craquelliert, vermutlich diverse Dekore
ca. 1928–1938
ohne Nr., Firmenmarke 19, Made in Austria, H. 20,5 cm, Ø 10,8 cm; Sammlung a@goldscheider.net

ohne Nr.
VASE
Form: becherförmig, gewulsteter Stand
Dekor: Osterglocken, geometrische Elemente, mehrfarbig, Laufglasur
ca. 1928–1938
ohne Nr., (ohne Firmenmarke), H. 20,8 cm, Ø 13,5 cm; MAK Wien, Inv. Ke 10602-55
Bem.: Trotz der fehlenden Firmenmarke und der Modell-Nr., ist dieses Gefäß eindeutig ein Erzeugnis der Manufaktur Goldscheider.
Ausst.: PSK 1995
Lit.: ADGS, Nr. 34 (m. Abb.)

ohne Nr.
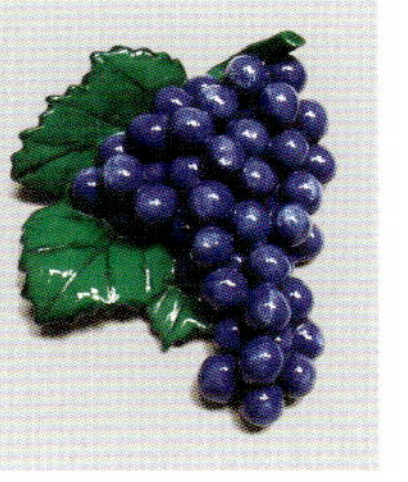
WANDBEHANG
Weintrauben mit Blattwerk
ca. 1935
ohne Nr., Firmenmarke 19, Made in Austria, H. 5,5 cm, L. 24 cm, B. 24 cm; Sammlung a@goldscheider.net

Goldscheider USA

Ausführung der Modelle zwischen 1939–1952
Die meisten Abbildungen stammen aus der Sammlung a@goldscheider.net.

–> vgl. Kapitel „Emigration und der Aufbau von ‚American Goldscheider' (1939–1952)"

Models made 1939–1952
Most illustrations are taken from the collection a@goldscheider.net.

–> cf chapter "Emigration and building up 'American Goldscheider' (1939–1952)"

MODELLNUMMER | MODEL NUMBER

4-USA FIGURINE „Lady Chrysanthemum" (H. 17,8 cm) (Garrett)

8-USA BOX (Lidded Box) (H. 7,8 cm, L. 17,2 cm)

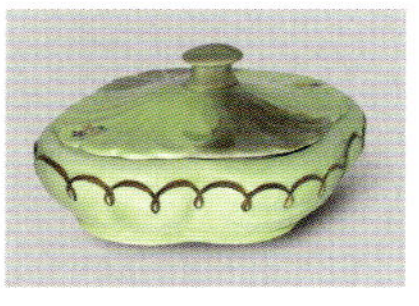

11-USA FIGURINE „Christmas" (H. 16,5 cm) (Porcher)

16-USA FIGURINE „Cotillion" (H. 15,2 cm) (Porcher)

44-USA ASHTRAY / CIGARETTEHOLDER Dutchgirl (H. 17,8 cm)

108-USA FIGURINE Deer (H. 16,5 cm)

110-USA FIGURINE „Quadrille" (H. 15,2 cm) (Porcher)

115-USA BUST „Renaissance Madonna" (with Praying Hands) (H. 32,4 cm) (Funaro)

195-USA FIGURINE Dancer („Alt-Wien") (H. 20,3 cm) (–> vgl. 6458; 7286 im WVZ) (Dakon)

205-USA BUST „Mother of Sorrow" (H. 24,1 cm) (Jacob)

213-USA FIGURINE Rooster (H. 17,8 cm) (–> Group) (Jacob)

215-USA FIGURINE Hen (H. 11,4 cm) (–> Group) (Jacob)

220-USA FIGURINE / MUSICAL CLOCK „Maria Antoinette" (Music: The Blue Danube) (H. 16,5 cm) (Porcher)

221-USA FIGURINE Rococo Dancer (H. fehlt)

224-USA BUST „Javanese Head" (Man) (H. 18,5 cm) Pendant –> vgl. 225-USA (Liedloff)

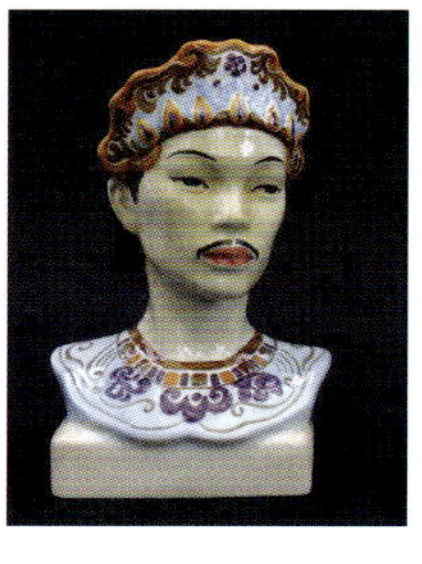

225-USA BUST „Javanese Head" (Woman) (H. 18,5 cm) Pendant –> vgl. 224-USA (Liedloff)

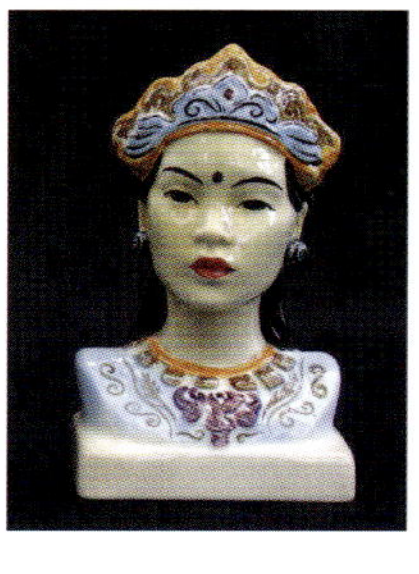

226-USA BUST „Madonna" (H. 21,6 cm) (Funaro)

236-USA FIGURINE „Duchesse Devonshire" (H. 17,5 cm) (Porcher)

246-USA –> „Chinese Teahouse" (Liedloff)

249-USA FIGURINE „Gavotte" (H. 19 cm) (Scott)

254-USA BUST „Madonna" (H. 12,7 cm) (Funaro)

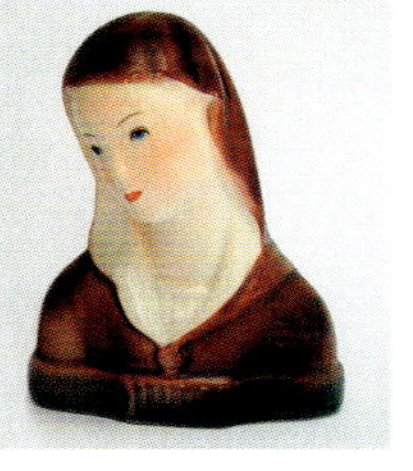

256-USA FIGURINE „Old London" (H. 16,5 cm) (Porcher)

269-USA FIGURINE „Windy Day" (H. 20,3 cm) (Drew)

272-USA FIGURINE / MUSICAL CLOCK „Beautiful Morning" (H. 16,5 cm) (Porcher)

274-USA FIGURINE / MUSICAL CLOCK „Easter Parade" (Music: Easter Parade) (H. 16cm) (Porcher)

279-USA FIGURINE / MUSICAL CLOK Girl with folded Hands (praying) (H. 16,5 cm)

296-USA FIGURINE „Sing-Hi" (Chinese Girl) (H. 18,4 cm) (Liedloff)

301 / 391?-USA BUST Woman with Hand, Fan and Hat (H. 22,8 cm) (Wieselthier)

306-USA FIGURINE „Southern Belle" (H. 25,4 cm) –> vgl. 840-USA; Pendant –> vgl. 882-USA (Porcher)

307-USA BUST „Madonna" (with Child) (H. 17,1 cm) (Funaro)

312-USA BUST „Marquess" (H. 25,4 cm) (Scott)

313-USA FIGURINE Lady with Flowers (H. 16 cm)

315-USA BOX (Lidded Cigarette Box with a Man) (L. 11,4 cm, B. 8,9 cm) (Urbach)

316-USA ASHTRAY (L. 21 cm, B. 11,4 cm) (Urbach)

317-USA BOWL Peasant with Horse (H. 10 cm, Ø 22,8 cm) (Urbach)

318-USA BUST „Praying Madonna" (H. 9 cm)

323-USA FIGURINE Deer (H. 16,5 cm)

325-USA FIGURINE „Spector of the Rose" (Ballerina) (H. 21,5 cm)

329-USA FIGURINE „Tragedy" (Chinese Man Holding a Mask) (H. 35,5 cm) Pendant –> vgl. 330-USA „Comedy" (Woman) (Loveday)

330-USA FIGURINE „Comedy" (Chinese Woman Holding a Mask) (H. 32,4 cm) Pendant –> vgl. 329-USA „Tragedy" (Man) (Loveday)

335-USA FIGURINE Jesus (H. 22,2 cm)

359-USA GROUP Kneeling Girl with Deer (H. 27,9 cm) (Gibbons)

365-USA FIGURINE „Bon Voyage" (Puppy Dog on Suitcase) (H. 15,5 cm) (Jacob)

391 / 301?-USA BUST Woman with Hand, Fan and Hat (H. 22,8 cm) (Wieselthier)

410-USA FIGURINE „Balinese Dancer" (H. 35,5 cm) (Baldwin)

414-USA BOX (Lidded Box with a Woman's Head) (H. fehlt)

416-USA ASHTRAY (with Flower) (H. 2,3 cm, B. 8,6 cm, L. 8,6 cm)

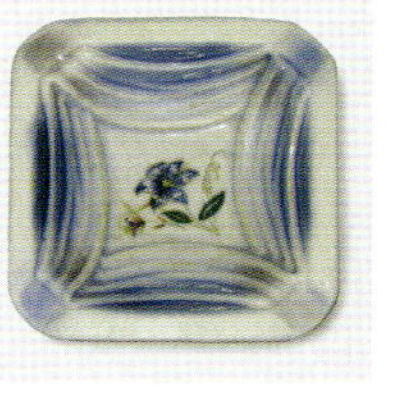

417-USA BOWL (with Flowers) (H. 10,8 cm, L. 39,7 cm)

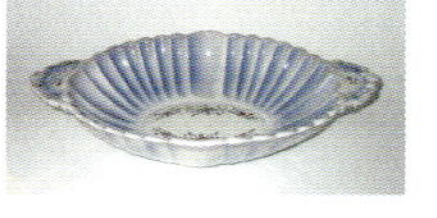

420-USA BOX (Lidded Box with Flowers) (H. 7,7 cm, L. 14,1 cm, B.11 cm)

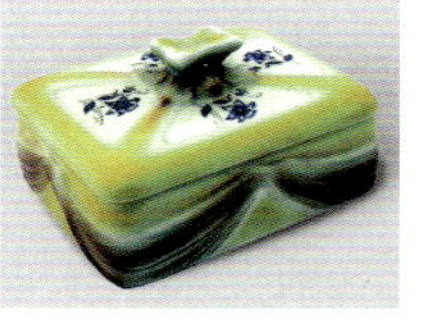

424-USA BOX (H. fehlt)

425-USA DOSE (Lidded Dose with Pattern Reliefs) (Flowers) (H. 26 cm, B. 18 cm)

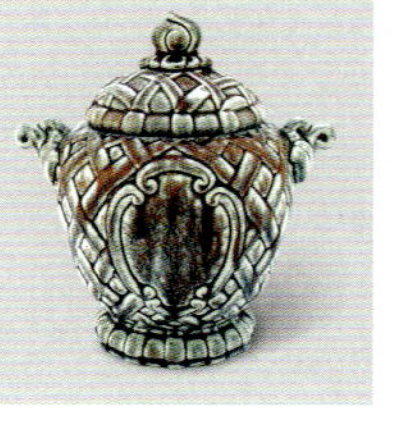

430-USA ASHTRAY Borzoi-Head (Ø ca. 15,9 cm)

440-USA DISH (with Flowers) (L. 8,9 cm, B. 6,3 cm)

504-USA FIGURINE Nude (H. 20,3 cm)

506-USA FIGURINE Puppy Dog (Beagle?) with Suitcase (H. 10,8 cm) (Nr. evtl. 508?)

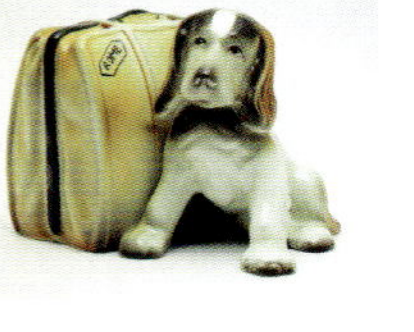

507-USA HEAD Woman with Scarf (H. 30,5 cm)

508 s. 506

510-USA BUST Girl with long Hair (H. 22,9 cm)

515-USA BUST Woman with Hand (H. 26,6 cm) (Collura)

524-USA FIGURINE „Anniversary Girl" (H. 17,1 cm) (Ragow)

530-USA GROUP Cat and Kitten (H. 9 cm)

543-USA FIGURINE Dancer (H. 32,4 cm) (Liedloff)

554-USA FIGURINE Borzoi (H. 35,5 cm)

565-USA FIGURINE Elephant (H. cm; 13,6 cm)

568-USA WALLMASK Woman's Head and Hand (H. ca. 37 cm)

570-USA GROUP Pigmother with two Piglets (H. 6 cm, L. 11,2 cm)

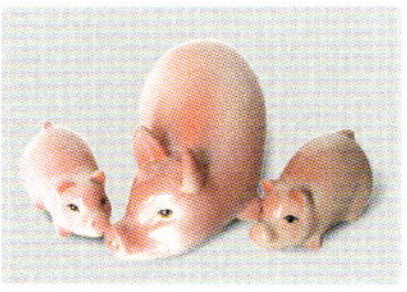

571-USA GROUP Pigmother with two Piglets (H. 5,7 cm, L. 13,6 cm)

574-USA FIGURINE Duckling (Little Duck) (H. 8,8 cm) (Straub)

582-USA ASHTRAY Horse-Head (H. 7,6 cm, Ø ca. 17,1 cm)

570-USA FIGURINE Pig (H. 5,7 cm)

594-USA BUST Girl with a Flower (H. 20,3 cm) (–> vgl. 7435 im WVZ)

598-WALLMASK (–> vgl. 7784 im WVZ)

601-USA FIGURINE Gazelle (H. 26 cm) –> vgl. 708-USA

605-USA FIGURINE Cat (H. 9,5 cm)

609-USA FIGURINE Zebra (H. 19 cm)

610-USA FIGURINE Puppy Dog (H. 14 cm)

621-USA FIGURINE Dog (Spaniel) (H. 23,5 cm)

622-USA FIGURINE Puppy Dog (Beagle?) (H. 10,8 cm)

628-USA FIGURINE Turkey (H. fehlt)

631-USA FIGURINE Dancer with long Skirt (H. 38 cm)

646-USA FIGURINE Dancer with Hat (H. 39,4 cm) (Liedloff)

660-USA BUST Madonna (with Praying Hands) (H. fehlt) –> vgl. 661-USA

661-USA BUST Madonna (with Praying Hands) (H. 22,8 cm) –> vgl. 660-USA

662-USA BUST „Madonna" (with Crown) (H. 19 cm) (Funaro)

662A-USA BUST „Madonna" (without Crown) (H. 16,5 cm) (Funaro)

663-USA BUST „Madonna" (with Crown) (H. 14 cm) (Funaro)

665-USA FIGURINE Dancer (H. 35,5 cm)

669-USA FIGURINE Siamese Cat (H. 24,1 cm)

66(9?)-USA BUST Madonna (H. 17,1 cm)

673-USA FIGURINE Spanish Dancer (H. 43 cm)

685-USA GROUP Mare with a Foal (H. 20,3 cm)

702-USA FIGURINE Pony (H. 12,7 cm)

708-USA FIGURINE Gazelle (H. 17,5 cm) –> vgl. 601-USA

713-USA BUST Madonna (H. 16,5 cm)

714-USA BUST Madonna (H. 20 cm)

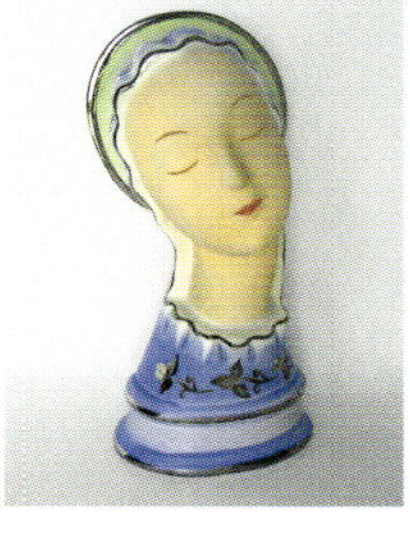

721-USA WALLMASK Woman's Head with Necklace (H. 27,5 cm)

725-USA BUST „Madonna" (H. 11,6 cm)

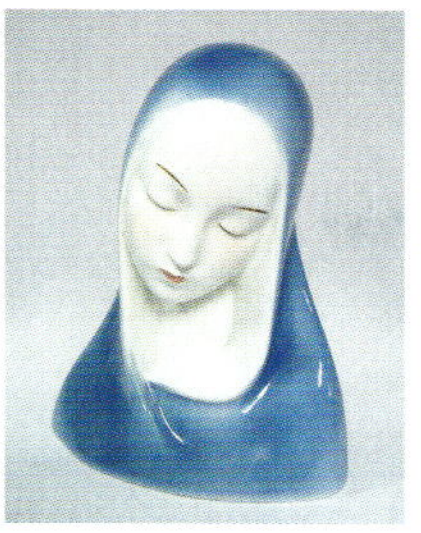

729-USA GROUP Mother Bird Feeding her Baby (H. 16,5 cm)

730-USA FIGURINE Borzoi (H. 22,9 cm) (Schmid)

732-USA WALLMASK (Prischl) (–> vgl. 8041 im WVZ)

737-USA FIGURINE Bird (H. 26,6 cm)

738-USA FIGURINE Bird (H. 26 cm)

739-USA FIGURINE Bird Sitting on a Fir Cone (H. 19,7 cm)

740-USA FIGURINE Bird (H. 21,6 cm)

745-USA FIGURINE Juliet with Doves (H. 31,1 cm)

747-USA HEADS Two Women (H. 17 cm) (Riza)

750-USA WALLMASK Woman with Scarf (H. 14,6 cm)

754-USA FIGURINE Giraffe (H. 26,7 cm)

770-USA FIGURINE Ram (H. 17,8 cm)

772-USA BOWL with a Bird (H. 13,4 cm, Ø 16,2 cm)

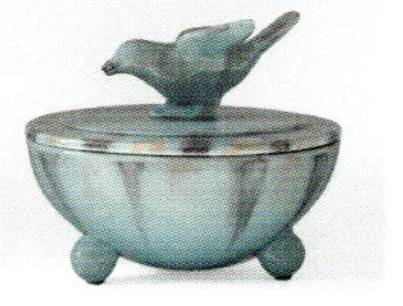

773-USA BOWL (H. 7 cm, Ø 15,9 cm)

777-USA FIGURINE French Bulldog (H. 4 cm, L. 10 cm)

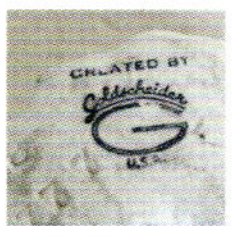

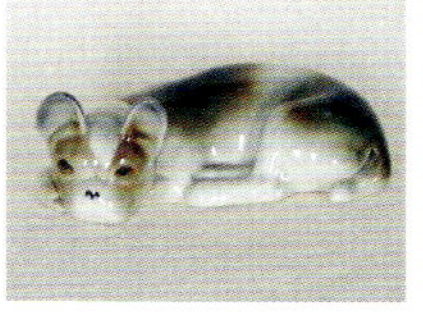

781-USA BUST Madonna (with praying Hands) (H. 22,9 cm) (Funaro)

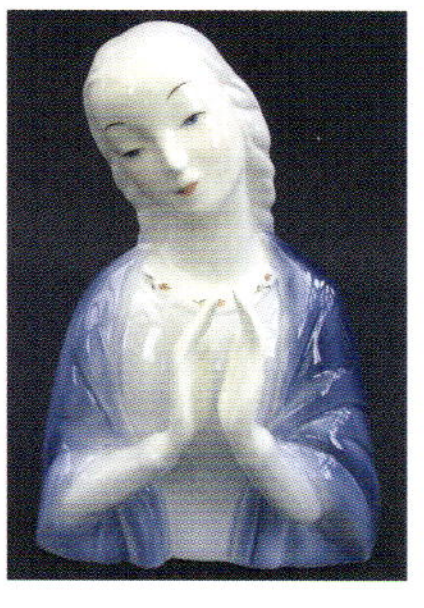

782-USA FIGURINE Rabbit (on back Legs) (H. fehlt)

787-USA FIGURINE Bird (Gull?) (H. 16,5 cm)

795-USA FIGURINE Dancer (H. 19,7 cm) (–> vgl. Dakon 7286 im WVZ, fast ident. Modell)

796-USA FIGURINE Gentleman with Knickerbockers (H. 21,6 cm)

800-USA FIGURINE Woman with Flowers (H. 20 cm)

801-USA FIGURINE „Summer" (H. 20 cm) (Porcher)

802-USA FIGURINE Lady with Hat and Muff (H. 21,5 cm)

805-USA FIGURINE Kitten (Cat) (H. 8,9 cm)

80(5?)-USA BUST/HEAD Dog (H. 15,2 cm)

810-USA FIGURINE „Boudoir" (H. 18,4 cm) (Porcher)

812-USA FIGURINE Rococo Gentleman (H. 23,3 cm) Pendant –> vgl. 813-USA

813-USA FIGURINE Rococo Lady (H. 20,3 cm) Pendant –> vgl. 812-USA

817-USA FIGURINE „Morning in Paris" (H. 22,3 cm) (Liedloff)

822-USA FIGURINE „Asian Musician" (Man) (H. 26,6) Pendant –> vgl. 823-USA (Woman)

823-USA FIGURINE „Asian Dancer" (Woman) (H. 26 cm) Pendant –> vgl. 822-USA (Man)

824-USA BUST Woman with Hat (H. 16 cm)

825-USA BUST/DISH Woman with Hat (H. 17,1 cm)

826-USA BUST Woman with Fan (H. 16 cm) (auch als Büste mit Stoffkleid „Teepuppe")

828-USA FIGURINE Woman with Fan (H. 26,7 cm)

829-USA BUST/HEADS Two Horses (H. 25,4 cm)

831-USA FIGURINE „Blossom Time" (H. 30 cm) (Porcher)

832-USA BUST Woman with Fan (H. 14 cm)

837-USA FIGURINE Cockatoo (H. fehlt)

838-USA BUST „Mongolhead" (Man) (H. 20,3 cm) Pendant –> vgl. 839-USA (Liedloff)

839-USA BUST „Mongolhead" (Woman) (H. ca. 19 cm) Pendant –> vgl. 838-USA (Liedloff)

840-USA FIGURINE „Southern Belle" (H. 20,6 cm) –> vgl. 306-USA (Porcher)

843-USA FIGURINE Rabbit (H. 8,9 cm) Pendant –> vgl. 872-USA

845-USA BUST/HEADS Two Horses (H. 12 cm)

846-USA FIGURINE Lady (H. 23,5 cm)

847-USA FIGURINE Gentleman with Violin (H. 23 cm)

849-USA BUST „Jeanette" (H. 15,9 cm) (Liedloff)

850-USA BUST Woman with a Big Hat (H. 16,5 cm)

854-USA FIGURINE „Chinese Actor" (H. 31,7 cm) (Liedloff)

855-USA FIGURINE „Chinese Actress" (H. 31,7 cm) (Liedloff)

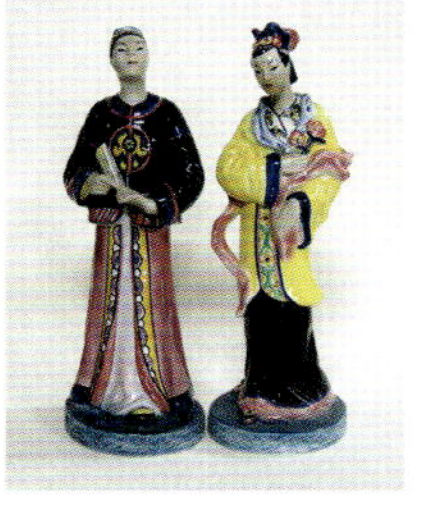

859-USA WALLPLAQUE Woman in Renaissance Costume (H. 14 cm, B. 12 cm)

861-USA FIGURINE Horse (Stallion) (H. 24,1 cm)

862-USA BUST/HEAD Horse (H. 24,1 cm)

865-USA BUST „Madonna" (H. 12 cm) (Funo)

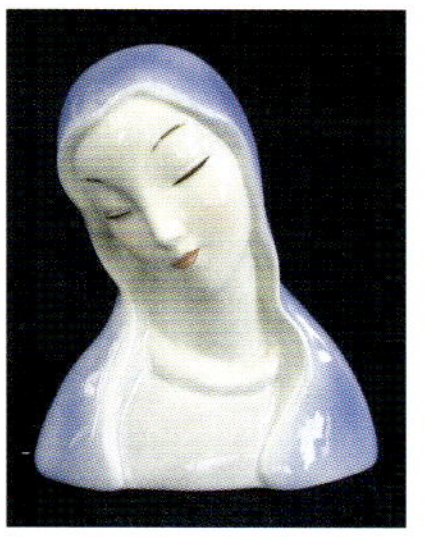

870-USA FIGURINE Young Boy with Flowers (H. 22,5 cm)

872-USA FIGURINE Rabbit (H. 14 cm) Pendant –> vgl. 843-USA

882-USA FIGURINE „Southern Beau" (H. 23,5 cm) Pendant –> vgl. 306-USA (Porcher)

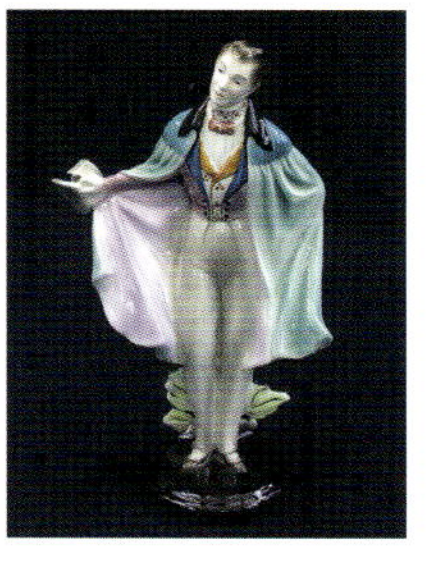

887-USA FIGURINE Cockatoo (H. 35 cm) (Liedloff)

888-USA FIGURINE Woodpecker (H. 16,5 cm)

894-USA FIGURINE „Chinese Prince" (H.29,2 cm) Pendant -> vgl. 895-USA (Liedloff)

895-USA FIGURINE „Chinese Princess" (H.29,2 cm) Pendant -> vgl. 894-USA (Liedloff)

899-USA FIGURINE Nude with Fan (H. 34 cm) (-> vgl. 7544 im WVZ) (Lorenzl)

901-USA VASE (with Feathers) (H. 29,2 cm; Ø 12 cm) (Sakier)

910-USA-VASE (with Feathers) (H. 15,3 cm) (Sakier)

913-USA BOX (Lidded Box for Cigarettes) with Feathers (H. 7,3 cm, L. 10,5 cm, B. 10,5 cm) (Sakier)

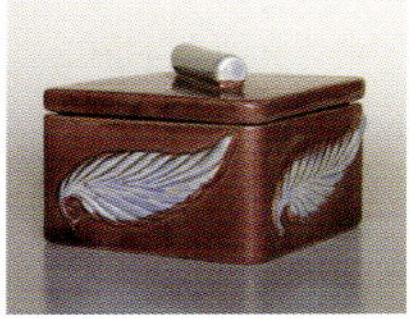

948-USA BOX (H. fehlt)

962-USA BOX (Lidded Box) with Feather (H. 4,6 cm, L. 16 cm, B. 11,3 cm) (Sakier)

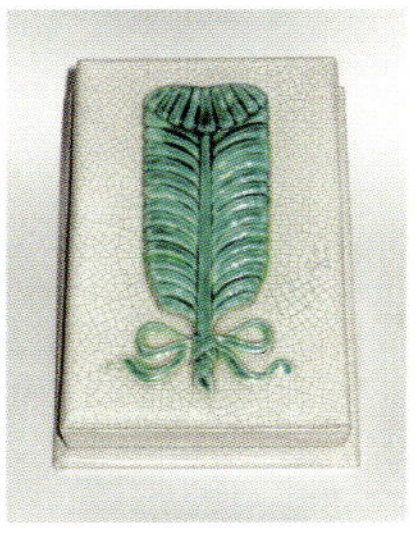

966-USA BOWL (with two Handles) (L. 34,9 cm)

996-USA BOWL (with two Handles as Feathers) (H. 7,5 cm, L. 35,7 cm)

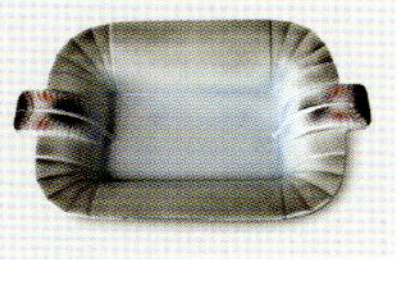

1001-USA BOX (Lidded Box) with Basket Weave Pattern (H. 4,4 cm, L. 14 cm, B. 10,1 cm) (Sakier)

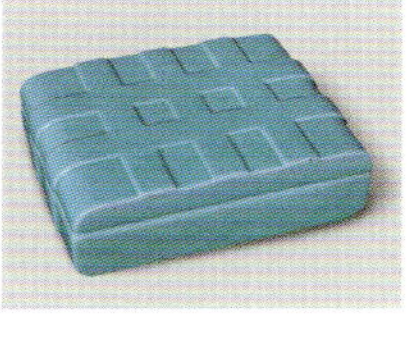

1100-USA FIGURINE „Minuet" (H. 22,2 cm) (Porcher)

1108-USA HEAD „Horsehead" (H. 26,7 cm) (Pippergen/Pippings?)

1114-USA GROUP „Minuet" (Two Roccoco Dancers) (H. 22,8 cm) (Porcher)

1116-USA FIGURINE Sniffing Dog (Spaniel) (H. 9,5 cm)

1120-USA FIGURINE Deer (lying) (H. fehlt, L. 22,8 cm) Pendant -> vgl. 1121-USA (Pippergen/Pippings?)

1121-USA FIGURINE Fawn (lying) (H. fehlt, L. 22,8 cm) Pendant -> vgl. 1120-USA (Pippergen/Pippings?)

1129-USA GROUP „First Party" (H. 22,9 cm) (Porcher)

1130-USA FIGURINE „Mandarin" (H. 27,9 cm) (Porcher)

1132-USA FIGURINE „Scheherezade" (H. 25,4 cm) Pendant -> vgl. „Caliph" (Porcher)

1133-USA BUST „India Prince" (H. 19 cm) Pendant -> vgl. 1134-USA (Liedloff)

1134-USA BUST „Indian Princess" (H. 19 cm) Pendant -> vgl. 1133-USA (Liedloff)

1162 (1169?)- USA FIGURINE Biedermeier Lady (with Hat and Umbrella), Sitting on a Bench (H. 26 cm)

1173-USA FIGURINE „Chinese Boy" (H. 17,1 cm) Pendant -> vgl. 1173-USA (Girl) (Liedloff)

1173-USA FIGURINE „Chinese Girl" (H. 17,1 cm) Pendant -> vgl. 1173-USA (Boy) (Liedloff)

1177-USA BUST „Pierre" (H. 19 cm) Pendant -> vgl. BUST „Denise" (Liedloff)

1181-USA FIGURINE „Countess" (H. 24,1 cm) (Drew)

1189-USA GROUP Madonna (with Child) (H. 24,1 cm) (Funaro)

1204-USA FIGURINE „Lobster Vendor" (H. 22,9 cm) (Baldwin)

1216-USA FIGURINE „Carmencita" (H. 24,1 cm) (Dorier)

1227-USA FIGURINE „Siamese Dancer" (H. 19 cm) Pendant -> vgl. 1228-USA (Woman) (Turner)

1228-USA FIGURINE „Siamese Dancer" (H. 19 cm) Pendant -> vgl. 1227-USA (Man) (Turner)

1243-USA FIGURINE „Chinese Musician" (H. 18,4 cm) (Turner)

1250-USA FIGURINE „Balinese Temple Dancer" (with Straight Head) (H. 26 cm) Pendant -> vgl. 1251-USA (with Leant Head) (Urbach)

1251-USA FIGURINE „Balinese Temple Dancer" (with Leant Head) (H. 24,1 cm) Pendant -> vgl. 1250-USA (with Straight Head) (Urbach)

1253-USA FIGURINE „Tonkinese Theatrical Dancer" (H. 29,2 cm) Pendant -> vgl. „Tonkinese Mandarin" (Urbach)

1254-USA FIGURINE „Chinese Boy" (with Musical Instrument/Lute) (H. fehlt) (Urbach)

1258-USA FIGURINE „Chinese Noble Lady" (H.19 cm) Pendant -> vgl. 1259-USA (Urbach)

1259-USA FIGURINE „Chinese Servant Girl" (H.19 cm) Pendant -> vgl. 1258-USA (Urbach)

1261-USA FIGURINE „Gyamatra" (H. 34,3 cm) (Urbach)

1286-USA FIGURINE „Grecian Dancer" (one Hand Turned Forward) (H. 33 cm) Pendant -> vgl. 2285-USA (Urbach)

2285-USA FIGURINE „Grecian Dancer" (both Hands Turned Sideways) (H. 33 cm) Pendant -> vgl. 1286-USA (Urbach)

2311-USA FIGURINE „Chinese Fan Dancer" (H. 34,9 cm) (Loveday)

235?-USA FIGURINE Horse (H. 12,7 cm) (Pippergen/Pippings?)

2378-USA FIGURINE „Miss Sidens" (H. 17,3 cm) (Porcher)

2388-USA WALLPLAQUE Madonna (with Child) (H. 24,1 cm)

11582-USA FIGURINE „New Bonnet" (H. 15,2 cm) (Porcher)

11592 (11598?)-USA FIGURINE „Lady Caller" (H. 16,5 cm) (Porcher)

11662-USA FIGURINE „Chinese Boy" (H. ca. 37 cm) Pendant -> vgl. 11678-USA „Chinese Fan Dancer" (Loveday)

11678-USA FIGURINE „Chinese Fan Dancer" (H. 34,3 cm) Pendant -> vgl. 11662-USA „Chinese Boy" (Loveday)

11698-USA FIGURINE „Chinese Dagger Dancer" (H. 44,7 cm) Pendant -> vgl. 11702 „Chinese Temple Dancer" (Baldwin)

11702-USA FIGURINE „Chinese Temple Dancer" (H. 45,6 cm) Pendant -> vgl. 11698 „Chinese Dagger Dancer" (Baldwin)

11748-USA FIGURINE „Temptation" (H. fehlt) (Drew)

KÜNSTLER | ARTISTS

Baldwin, Barbara

410-USA FIGURINE „Balinese Dancer" (H. 35,5 cm)

1204-USA FIGURINE „Lobster Vendor" (H. 22,9 cm)

11698-USA FIGURINE „Chinese Dagger Dancer" (H. 44,7 cm) Pendant -> vgl. 11702-USA „Chinese Temple Dancer"

11702-USA FIGURINE „Chinese Temple Dancer" (H. 45,6 cm) Pendant -> vgl. 11698-USA „Chinese Dagger Dancer"

BUST „Balihead" (H. 29,2 cm)

Collura

515-USA BUST Woman with Hand (H. 26,6 cm)

Dakon, Stephan

195-USA FIGURINE Dancer („Alt-Wien") (H. 20,3 cm) (-> vgl. 6458; 7286 im WVZ)

Dorier

1216-USA FIGURINE „Carmencita" (H. 24,1 cm)

Drew, Gloria

269-USA FIGURINE „Windy Day" (H. 20,3 cm)

1181-USA FIGURINE „Countesse" (H. 24,1 cm)

11748-USA FIGURINE „Temptation" (H. fehlt)

FIGURINE „Countess" (H. 20,3 cm)

FIGURINE „Dutch Dancer" (Man) (H. 20,3 cm)

FIGURINE „Dutch Dancer" (Woman) (H. 20,3 cm)

Erika (wohl Pseud.)

FIGURINE „Cotillion" (Man) (H. 22,2 cm)

FIGURINE „Cotillion" (Woman) (H. 19,7 cm)

Funo, P. s. Funaro, P(ena?)
Funaro, P(ena?) oder Fus(s)aro?

115-USA BUST „Renaissance Madonna" (with Praying Hands) (H. 32,4 cm)

226-USA BUST „Madonna" (H. 21,6 cm)

254-USA BUST „Madonna" (H. 12,7 cm)

307-USA BUST „Madonna" (with Child) (H. 17,1 cm)

662-USA BUST „Madonna" (with Crown) (H. 19 cm)

662A-USA BUST „Madonna" (without Crown) (H. 16,5 cm)

663-USA BUST „Madonna" (with Crown) (H. 14 cm)

781-USA FIGURINE „Madonna" (with Praying Hands) (H. 22,9 cm)

865-USA BUST „Madonna" (H. 12 cm)

1189-USA GROUP Madonna (with Child) (H. 24,1 cm)

BUST „Madonna" (with Crossed Hands) (H. 15,9 cm)

Garrett, Blanka

4-USA FIGURINE „Lady Chrysanthemum" (H. 17,8 cm)

FIGURINE „Bachelor Carnation" (H. fehlt)

FIGURINE „Birthday Boy" (H. fehlt)

FIGURINE „Lady Poinsetta" (H. 17,1 cm)

FIGURINE „Lady Rose" (H. 17,1 cm)

FIGURINE „Lady Violet" (H. 17,1 cm)

Gibbons

359-USA GROUP Kneeling Girl with Deer (H. 27,9 cm)

Jacob, A(ugust?)

205-USA BUST „Mother of Sorrow“ (H. 24,1 cm)

213-USA FIGURINE Rooster (H. 17,8 cm) (–> Group)

215-USA FIGURINE Hen (H. 11,4 cm) (–> Group)

365-USA FIGURINE „Bon Voyage“ (Puppy Dog on Suitcase) (H. 15,5 cm)

BUST „Cherub“ (H. 11,4 cm)

FIGURINE Cockatoo (H. 8,2 cm)

FIGURINE Duck (H. 8,2 cm)

GROUP Rooster and Hen (–> 213; 215) (H. 17,8 cm)

GROUP Two Chicks (H. fehlt)

Latour

FIGURINE „Dance Invitation“ (H. 27,9 cm)

Liedloff, Helen

224-USA BUST „Javanese Head“ (Man) (H. 18,5 cm) Pendant –> vgl. 225-USA

225-USA BUST „Javanese Head“ (Woman) (H. 18,5 cm) Pendant –> vgl. 224-USA

246-USA –> „Chinese Teahouse“

296-USA FIGURINE „Sing-Hi“ (Chinese Girl) (H. 18,4 cm)

543-USA FIGURINE Dancer (H. 32,4 cm)

646-USA FIGURINE Dancer with Hat (H. 39,4 cm)

817-USA FIGURINE „Morning in Paris“ (H. 22,3 cm)

849-USA BUST „Jeanette“ (H. 15,9 cm)

838-USA BUST „Mongolhead“ (Man) (H. 20,3 cm) Pendant –> vgl. 839-USA

839-USA BUST „Mongolhead“ (Woman) (H. ca. 19 cm) Pendant –> vgl. 838-USA

854-USA FIGURINE „Chinese Actor“ (H. 31,7 cm)

855-USA FIGURINE „Chinese Actress“ (H. 31,7 cm)

887-USA FIGURINE Cockatoo (H. 35 cm)

894-USA FIGURINE „Chinese Prince“ (H.29,2 cm) Pendant –> vgl. 895-USA

895-USA FIGURINE „Chinese Princess“ (H.29,2 cm) Pendant –> vgl. 894-USA

1133-USA BUST „India Prince“ (H. 19 cm) Pendant –> vgl. 1134-USA

1134-USA BUST „Indian Princess“ (H. 19 cm) Pendant –> vgl. 1133-USA

1173-USA FIGURINE „Chinese Boy“ (H. 17,1 cm)

1173-USA FIGURINE „Chinese Girl“ (H. 17,1 cm)

1177-USA BUST „Pierre“ (H. 19 cm) Pendant –> vgl. BUST „Denise“

BUST „Chinese Head“ (H. 20,9 cm)

BUST „Denise“ (H. 17,8 cm) Pendant –> vgl. 1177-USA BUST „Pierre“

FIGURINE „Butterfly“ (Girl with Butterfly) (H. 29,8 cm)

FIGURINE „Chinese Love“ (Chinese Man in Costume) (H. 29,2 cm)

FIGURINE „Chinese Teahouse“ (Kneeling Chinese Girl with Cup) (H. 17,1 cm)

FIGURINE „Chinese Teahouse“ (Kneeling Chinese Girl with Instrument) (H. 17,8 cm)

FIGURINE „Temple Dancer“ (Man) (H. 20,3 cm) Pendant –> vgl. „Temple Dancer“ (Woman)

FIGURINE „Temple Dancer“ (Woman) (H. 20,3 cm) Pendant –> vgl. „Temple Dancer“ (Man)

FIGURINE Diana with Whippet (H. 33 cm)

GROUP „Twins“ (H. 17,8 cm)

Lorenzl, Josef

899-USA FIGURINE Nude with Fan (H. 34 cm) (–> vgl. 7544 im WVZ)

Loveday, Barbara

329-USA FIGURINE „Tragedy“ (Chinese Man Holding a Mask) (H. 35,5 cm) Pendant –> vgl. 330-USA „Comedy“ (Woman)

330-USA FIGURINE „Comedy“ (Chinese Woman Holding a Mask) (H. 32,4 cm) Pendant –> vgl. 329-USA „Tragedy“ (Man)

2311-USA FIGURINE „Chinese Fan Dancer“ (H. 34,9 cm)

11662-USA FIGURINE „Chinese Boy“ (H. ca. 37 cm) Pendant –> vgl. 11678-USA „Chinese Fan Dancer“

11678-USA FIGURINE „Chinese Fan Dancer“ (H. 34,3cm) Pendant –> vgl. 11662-USA „Chinese Boy“

FIGURINE „Brittany“ (H. 20,3 cm)

FIGURINE „Chinese Princess“ (H. 20,3 cm)

FIGURINE „Mardi Grass“ (Pierrot) (H. 25,4 cm) Pendant –> vgl. „Mardi Grass“ (Pierrette)

FIGURINE „Mardi Grass“ (Pierrette) (H. 25,4 cm) Pendant –> vgl. „Mardi Grass“ (Pierrot)

FIGURINE „Royal Blackamoor“(Man) (H. 38,1 cm) Pendant –> vgl. „ Royal Blackamoor „ (Woman)

FIGURINE „Royal Blackamoor“(Woman) (H. 38,1 cm) Pendant –> vgl. „ Royal Blackamoor „ (Man)

Pipp(ergen oder Pippings?), Robert

235?-USA FIGURINE Horse (H. 12,7 cm)

1108-USA HEAD „Horsehead“ (H. 26,7 cm)

1120-USA FIGURINE Deer (lying) (H. fehlt, L. 22,8 cm) Pendant –> vgl. 1121-USA

1121-USA FIGURINE Fawn (lying) (H. fehlt, L. 22,8 cm) Pendant –> vgl. 1120-USA

FIGURINE „Chinese Haron“ (Bird) (H. 35,5 cm)

Porcher, Peggy

11-USA FIGURINE „Christmas“ (H. 16,5 cm)

16-USA FIGURINE „Cotillion“ (H. 15,2 cm)

110-USA FIGURINE „Quadrille“ (H. 15,2 cm)

220-USA FIGURINE / MUSICAL CLOCK „Maria Antoinette“ (Music: The Blue Danube) (H. 16,5 cm)

236-USA FIGURINE „Duchesse Devonshire“ (H. 17,5 cm)

256-USA FIGURINE „Old London“ (H. 16,5 cm)

272-USA FIGURINE / MUSICAL CLOCK „Beautiful Morning“ (H. 16,5 cm)

274-USA FIGURINE / MUSICAL CLOCK „Easter Parade“ (Music: Easter Parade) (H. 16cm)

306-USA FIGURINE „Southern Belle“ (H. 25,4 cm) –> vgl. 840-USA; Pendant –> vgl. 882-USA

801-USA FIGURINE „Summer“ (H. 20 cm)

810-USA FIGURINE „Boudoir“ (H. 18,4 cm)

831-USA FIGURINE „Blossom Time“ (H. 30 cm)

840-USA FIGURINE „Southern Belle“ (H. 20,6 cm) –> vgl. 306-USA

882-USA FIGURINE „Southern Beau“ (H. 23,5 cm) Pendant –> vgl. 306-USA

1100-USA FIGURINE „Minuet“ (H. 22,2 cm)

1114-USA GROUP „Minuet“ (Two Roccoco Dancers) (H. 22,8 cm)

1129-USA GROUP „First Party“ (H. 22,9 cm)

1130-USA FIGURINE „Mandarin“ (H. 27,9 cm)

1132-USA FIGURINE „Scheherezade“ (H. 25,4 cm) Pendant –> vgl. „Caliph“

2378-USA FIGURINE „Miss Sidens“ (H. 17,3 cm)

11582-USA FIGURINE „New Bonnet“ (H. 15,2 cm)

11592 (11598?)-USA FIGURINE „Lady Caller“ (H. 16,5 cm)

FIGURINE „April Showers“ (H. 15,9 cm)

FIGURINE „Caliph“ (H. ca. 27 cm) Pendant –> vgl. 1132-USA

FIGURINE „Carmen“ (H. 31,5 cm)

FIGURINE „Dancing Lesson“ (H. 24,1 cm)

FIGURINE „Emperor Valse“ (H. 22,9 cm)

FIGURINE „Lady of the Lily Feet“ (H. 26,7 cm)

FIGURINE / MUSICAL CLOCK „Madame Pompadour“ (Music: The Blue Danube) (H. 17,5 cm)

FIGURINE „Old Virginia“ (H. 21,6 cm)

FIGURINE „Prince of Wales“ (H. 18 cm)

FIGURINE „Rendezvous“ (Man) (H. 20,3 cm) Pendant –> vgl. „Rendezvous“ (Woman)

FIGURINE „Rendezvous“ (Woman) (H. 20,0 cm) Pendant –> vgl. „Rendezvous“ (Man)

FIGURINE „Spring“ (H. 20,9 cm)

FIGURINE „Venice“ (H. 29,2 cm)

FIGURINE „Vienna Woods“ (H. 22,9 cm)

FIGURINE / MUSICAL CLOCK „White Christmas“ (Music: White Christmas) (H. 16,5 cm)

FIGURINE „Yankee Doodle Dandy“ (H. 17,1 cm)

Prischl, Adolf

732-USA WALLMASK (–> vgl. 8041 im WVZ)

Ragow, Blanche S.

524-USA FIGURINE „Anniversary Girl“ (H. 17,1 cm)

FIGURINE „Bachelor Carnation“ (H. 19 cm)

FIGURINE „Lady Poinsetta“ (H. 18,4 cm)

FIGURINE „Wedding Bells“ (H. 16,5 cm)

Riza

6??-USA FIGURINE Dancer (H. 36 cm)

747-USA HEADS Two Women (H. 17 cm)

BUST „Gloria“ (H. fehlt)

Sakier

901-USA VASE (with Feathers) (H. 29,2 cm; Ø 12 cm)

910-USA-VASE (with Feathers) (H. 15,3 cm)

913-USA BOX (Lidded Box for Cigarettes) with Feathers (H. 7,3 cm, L. 10,5 cm, B. 10,5 cm)

962-USA BOX (Lidded Box) with Feather (H. 4,6 cm, L. 16 cm, B. 11,3 cm)

1001-USA BOX (Lidded Box) with Basket Weave Pattern (H. 4,4 cm, L. 14 cm, B. 10,1 cm)

Schmid, Ludwig

730-USA FIGURINE Borzoi (H. 22,9 cm)

Scott, Sylvia

249-USA FIGURINE „Gavotte“ (H. 19 cm)

312-USA BUST „Marquess“ (H. 25,4 cm)

FIGURINE „Chinese Guitarist“ (H. 17,8 cm) Pendant –> vgl. „Chinese Poet“ (Man)

FIGURINE „Chinese Poet“ (H. 20 cm) Pendant –> vgl. „Chinese Guitarist“ (Woman)

FIGURINE „Southern Cavalier“ (H. fehlt)

Straub, E.

574-USA FIGURINE Duckling (Little Duck) (H. 8,8 cm)

FIGURINE Horse (H. 29 cm)

Turner, Raymond

1243-USA FIGURINE „Chinese Musician“ (H. 18,4 cm)

FIGURINE „Actress“ (H. 17 cm)

1227-USA FIGURINE „Siamese Dancer“ (H. 19 cm) Pendant –> vgl. 1228-USA (Woman)

1228-USA FIGURINE „Siamese Dancer“ (H. 19 cm) Pendant –> vgl. 1227-USA (Man)

Urbach, Kathi

315-USA BOX (Lidded Cigarette Box with a Man) (L. 11,4 cm, B. 8,9 cm)

316-USA ASHTRAY (L. 21 cm, B. 11,4 cm)

317-USA BOWL Peasant with Horse (H. 10 cm, Ø 22,8 cm)

1250-USA FIGURINE „Balinese Temple Dancer“ (with Straight Head) (H. 26 cm) Pendant –> vgl. 1251-USA (with Leant Head)

1251-USA FIGURINE „Balinese Temple Dancer“ (with Leant Head) (H. 24,1 cm) Pendant –> vgl. 1250-USA (with Straight Head)

1253-USA FIGURINE „Tonkinese Theatrical Dancer“ (H. 29,2 cm) Pendant –> vgl. „Tonkinese Mandarin“

1254-USA FIGURINE „Chinese Boy“ (with Musical Instrument / Lute) (H. fehlt)

1258-USA FIGURINE „Chinese Noble Lady“ (H.19 cm) Pendant –> vgl. 1259-USA

1259-USA FIGURINE „Chinese Servant Girl“ (H.19 cm) Pendant –> vgl. 1258-USA

1261-USA FIGURINE „Gyamatra“ (H. 34,3 cm)

1286-USA FIGURINE „Grecian Dancer“ (One Hand Turned Forward) (H. 33 cm) Pendant –> vgl. 2285-USA

2285-USA FIGURINE „Grecian Dancer“ (Both Hands Turned Sideways) (H. 33 cm) Pendant –> vgl. 1286-USA

FIGURINE „Tonkinese Mandarin“ (H. 30,5 cm) Pendant –> vgl. 1253-USA

PLATE Peasant with Bull (H. 29,2 cm)

Waldron, Katherine

FIGURINE „Chinese Wonderland“ (Man with Turtle) (H. 27,9 cm) Pendant –> vgl. „Chinese Wonderland“ (Woman with Swan)

FIGURINE „Chinese Wonderland“ (Woman with Swan) (H. 29 cm) Pendant –> vgl. „Chinese Wonderland“ (Man with Turtle)

Wieselthier, Vally

301 / 391?-USA BUST Woman with Hand, Fan and Hat (H. 22,8 cm)

WALLMASK / PLAQUE Cherub (H. 12,7 cm) (–> vgl. 6764 im WVZ)

Goldscheider West-Germany

Ausführung der Modelle zwischen 1953–1960
Zitiert nach Könecke (2002), vollständig überarbeitet und ergänzt durch Robert E. Dechant.

–> vgl. Kapitel „Weltmarke der Keramik – West-Germany (1953–1960)“

Models made 1953–1960
Quoted from Könecke (2002), completely revised and enlarged by Robert E. Dechant.

–> cf chapter "World Mark in Ceramics – West Germany (1953–1960)"

14-WG
Bauer, Carli (Carl)
„Chinesin lesend“ (Sitzende lesende Chinesin)

14L-WG
Bauer, Carli (Carl) LAMPE
„Chinesin lesend“ (Sitzende lesende Chinesin)

18-WG
Spiller, Gretl
WANDBEHANG „Chinesischer Träger“ (Chinese mit zwei Gefäßen, für Kakteen/Blumen)

23-WG
Bauer, Carli (Carl)
„Chinesen“ (Sitzende lesende Chinesin)
Pendant –> vgl. 24-WG

23K-WG
Bauer, Carli (Carl)
„Chinesen mit Topf“ (Sitzende lesende Chinesin mit Gefäß, für Kakteen)

23L-WG
Bauer, Carli (Carl)
LAMPE „Chinesin lesend“ (Sitzende lesende Chinesin)

4-WG
Bauer, Carli (Carl)
„Chinesen“ (Sitzende lesender Chinese)
Pendant –> vgl. 25-WG

24K-WG
Bauer, Carli (Carl)
„Chinesen mit Topf“ (Sitzender lesender Chinese mit Gefäß, für Kakteen)

25-WG
Spiller, Gretl
„Chinesen lesend“ (Sitzender lesender Chinese)
Pendant –> vgl. 26-WG

26-WG
Spiller, Gretl
„Chinesen lesend“ (Sitzende lesende Chinesin)
Pendant –> vgl. 25-WG

100-WG
WANDBEHANG Frauenkopf nach rechts blickend (Halbrelief auf Platte)

101-WG
WANDBEHANG Frauenkopf nach links blickend (Halbrelief auf Platte)

102-WG
WANDBEHANG Frauenkopf nach links blickend (Halbrelief auf Platte)

104-WG
WANDBEHANG Frauenkopf nach rechts blickend (Halbrelief auf Platte)

105-WG
WANDBEHANG Frauenkopf im Profil nach links (Halbrelief auf Platte)

106-WG
WANDBEHANG Frauenkopf nach links blickend (Halbrelief auf Platte)

112-WG
WANDBEHANG „Segelschiff“ (Kogge)

200-WG
Sitzender Mädchenakt

202-WG
Goedkte
„Flötenspielerin mit Goldflöte“ (Sitzende Flötenspielerin)

202-WG
Goedkte
„Tanzpaar“ (Balletttänzer)

301-WG
Dakon, Stephan
„Negerin mit Goldschale“ (Sitzendes Negermädchen mit Schale)

302-WG
Dakon, Stephan
„Negerin mit Goldschale“ (Kniendes Negermädchen mit Schale auf dem Kopf)

303-WG
Dakon, Stephan
„Neger“ (Kniender Neger mit Trommel)

304-WG
Dakon, Stephan
„Neger“ (Stehender Neger mit Speer auf Natursockel)

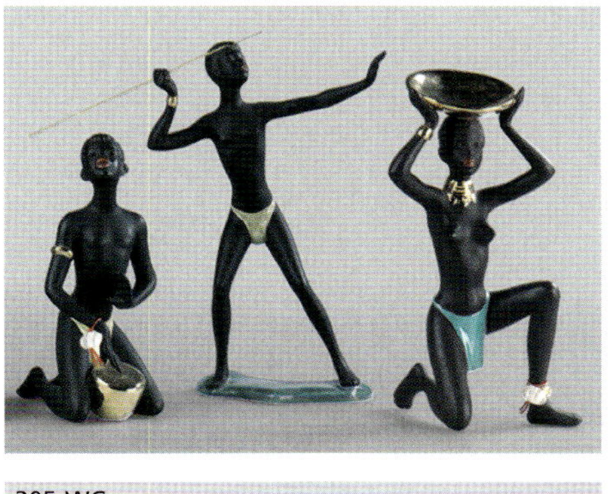

305-WG
Dakon, Stephan
„Akt“ (Liegender Mädchenakt)

306-WG
Dakon, Stephan
„Tanz“ (Tanzender Mädchenakt mit Tamburin)

307-WG
Dakon, Stephan
„Akt“ (Kniender Mädchenakt mit Tuch)

308-WG
Dakon, Stephan
„Neger mit Schild“ (Kniender Neger mit Schild)

309-WG
Dakon, Stephan
Hockende(r?) Neger/in?

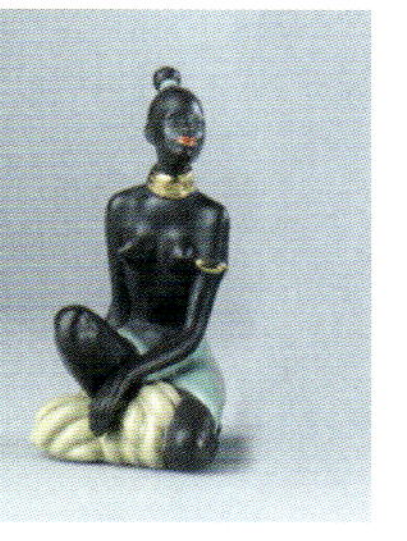

310-WG
Dakon, Stephan
„Negermädchen“ (Kniendes Negermädchen)

311-WG
Dakon, Stephan
„Negermädchen“ (Sitzendes Negermädchen)

312-WG
Dakon, Stephan
„Akt“ (Kniender Mädchenakt)

313-WG
Dakon, Stephan
WANDMASKE „Negerkopf“ (Junge Negerin mit Ohrringen, Kopf- und Halsschmuck)

314-WG
Dakon, Stephan
WANDMASKE „Negerkopf“ (Junge Negerin mit Ohrringen, Kopf- und Halsschmuck)

315-WG
Dakon, Stephan
„Negerin“ (Kniende Negerin mit Kopf- und Halsschmuck)

316-WG
Dakon, Stephan
WANDMASKE „Kindermaske“ (Moritz)
Pendant –> vgl. 317-WG

317-WG
Dakon, Stephan
WANDMASKE „Kindermaske“ (Max)
Pendant –> vgl. 316-WG

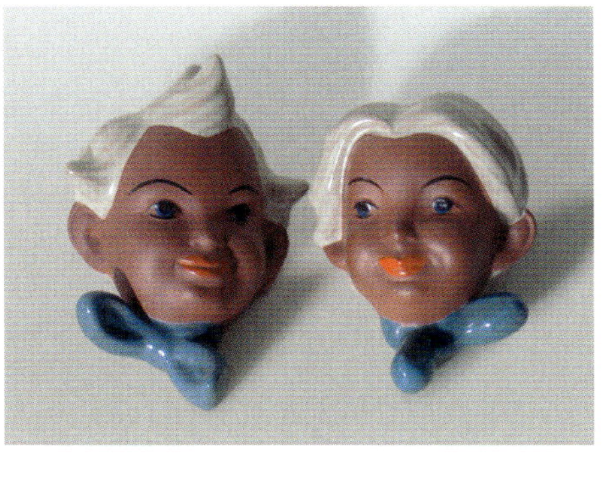

318-WG
Dakon, Stephan
„Neger" (Hockender Neger mit Schale)

320-WG
Dakon, Stephan
„Akt" (Hockender Mädchenakt auf Rundsockel)

321-WG
Dakon, Stephan
WANDMASKE „Maske" (Frauenkopf mit Kopftuch / Schal, geschlossene Augen)

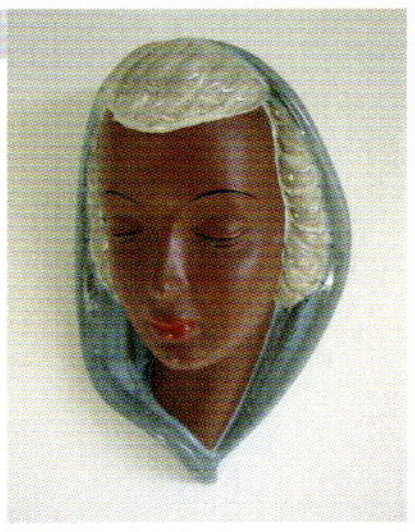

322-WG
Dakon, Stephan
WANDMASKE „Maske" (Frauenkopf mit Kopftuch / Schal, offene Augen)

323-WG
Dakon, Stephan
„Negerin" (Tanzende Negerin auf Natursockel)

324-WG
Dakon, Stephan
WANDMASKE „Maske" (Frauenkopf nach links blickend)

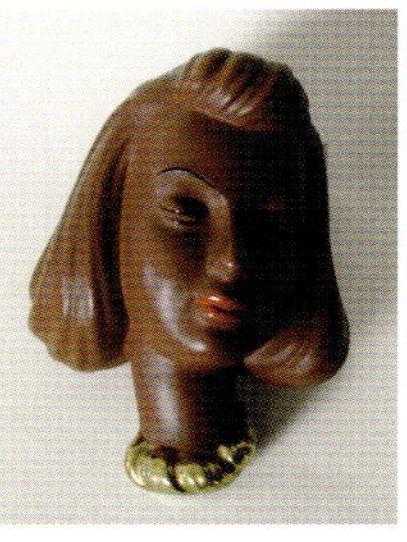

325-WG
Dakon, Stephan
WANDMASKE „Maske" (Frauenkopf mit Kopftuch / Schal)

328-WG
Dakon, Stephan
„Akt" (Kniender Mädchenakt, sich die Sandale richtend)

329-WG
Dakon, Stephan
„Balinesin" (Tanzende Balinesin auf nierenförmigem Sockel)

330-WG
Dakon, Stephan
„Tänzerin" (Tanzendes Mädchen im Ballettkostüm auf Ovalsockel)

331-WG
Dakon, Stephan
„Tänzerin" (Flöte spielende Tänzerin)

332-WG
Dakon, Stephan
„Balinese" (Sitzender Balinese mit Flöte)

333-WG
Dakon, Stephan
„Balinesin" (Kniende Balinesin mit Schale)

334-WG
Dakon, Stephan
„Balinese" (Sitzender Balinese mit Brett?)

335-WG
Dakon, Stephan
„Balinesin" (Tanzende Balinesin mit Fächer, auf nierenförmigem Sockel)

336-WG
Dakon, Stephan
„Tanzfigur" (Flötespielendes Mädchen im Ballettkostüm auf Ovalsockel)

337-WG
Dakon, Stephan
WANDMASKE „Maske Balinesin" (Balinesin mit Ohrringen und Halsschmuck)

338-WG
Dakon, Stephan
„Bogenschütze" (Stehender Bogenschießender Indianer) (auf unregelmäßigem Sockel)

339-WG
Dakon, Stephan
„Akt" (Stehender Mädchenakt mit Tuch, auf einem Bein stehend) (auf Ovalsockel)

340-WG
Dakon, Stephan
„Akt" (Sitzender Mädchenakt, die Arme hinter dem Kopf verschränkt)

342-WG
Dakon, Stephan
„Akt" (Liegender Mädchenakt mit Strohhut)

343-WG
Dakon, Stephan
„Mädchen mit Schale" (Kniender Mädchenhalbakt mit Schale)

345-WG
Dakon, Stephan
„Maske" (Mädchenkopf mit Pferdeschwanz und geschlossenen Augen)

346-WG
Dakon, Stephan
„Maske" (Mädchenkopf mit kurzem Haar nach links blickend)

347-WG
Dakon, Stephan
„Tänzerin" (Balletttänzerin auf einem Bein, auf Ovalsockel)

349-WG
Dakon, Stephan
WANDBEHANG „Sahara-Büste" (Afrikanerin, eine Schale auf dem Kopf tragend)

350-WG
Dakon, Stephan
„Maske" (Mädchenkopf mit Pferdeschwanz im Profil)

351-WG
Dakon, Stephan
„Maske" (Mädchenkopf mit Kragen)

352-WG
Dakon, Stephan
„Tänzerin" (Balletttänzerin, sich ihren Schuh schnürend)

354-WG
Dakon, Stephan
WANDPLASTIK „Negerboot" (Negerin mit Kleinkind im Boot)

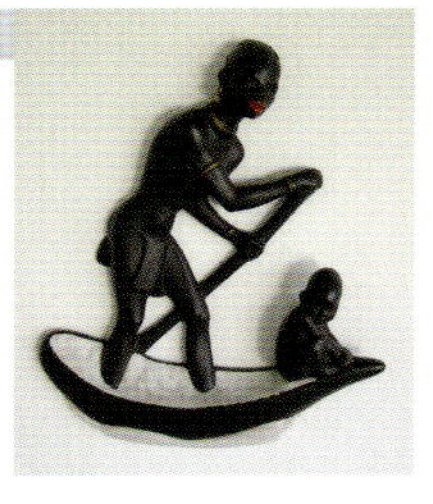

355-WG
Dakon, Stephan
WANDPLASTIK „Negerboot" (Negerin mit Kleinkind auf dem Rücken)

357-WG
Dakon, Stephan
WANDPLASTIK „Schlittschuhläuferin"

358-WG
Dakon, Stephan
„Akt" (Sitzender Mädchenakt mit Buch)

359-WG
Dakon, Stephan
„Akt" (Sitzender Mädchenakt mit Bikini)

360-WG
Dakon, Stephan
„Mädchen mit Koffer" (Mädchen auf Koffer sitzend, daneben ein zweiter Koffer stehend)

361-WG
Dakon, Stephan
„Mädchen mit Aktentasche" (Schreitendes Mädchen mit Aktentasche auf Landschaftssockel)

362-WG
Dakon, Stephan
„Akt" (Kniender Mädchenakt)

363-WG
Dakon, Stephan
„Akt"

364-WG
Dakon, Stephan
„Akt"

400-WG
„Wasserträgerin" (Schreitende Wasserträgerin mit Krug / Vase auf der Schulter) (auf Ovalsockel)

402-WG
Müller
WANDMASKE „Maske" (Frauenkopf mit Kopftuch)

505-WG
Prischl, Adolf
WANDMASKE „Wandmaske Chinesin" (Chinesenmädchen mit Strohhut, nach links blickend)

506-WG
Prischl, Adolf
WANDMASKE „Wandmaske Chinesin" (Chinesenmädchen mit Strohhut, nach rechts blickend)

507-WG
Prischl, Adolf
„Akt" (Sitzender Mädchenakt mit geschlossenen Augen)

508-WG
Prischl, Adolf
WANDMASKE „Maske" (Frauenkopf mit langem Haar, nach rechts geneigt)

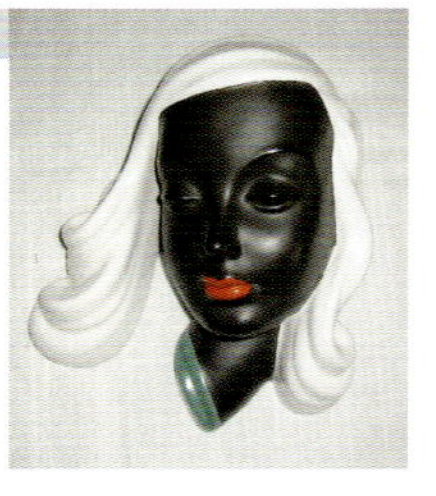

509-WG
Prischl, Adolf
WANDMASKE „Maske" (Frauenkopf mit Halstuch)

510-WG
Prischl, Adolf
WANDMASKE „Maske" (Frauenkopf mit lockigem Haar und Kragen)

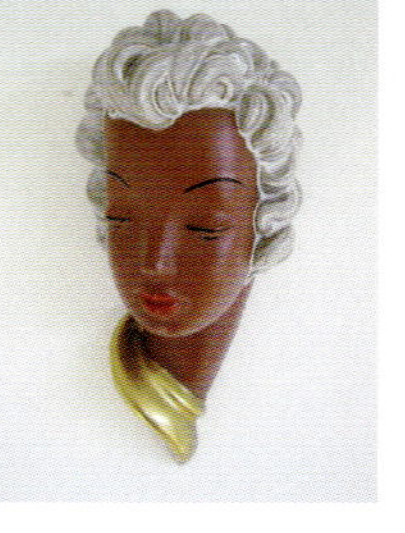

511-WG
Prischl, Adolf
„Akt" (Sitzender Mädchenakt)

512-WG
Prischl, Adolf
WANDMASKE „Neger" (Negerin mit Strohhut und Ohrringen)
Bem.: Der von der Firma im Originaltitel als „Neger" Bezeichnete ist vielmehr eine Frau. Die Ohrringe sind aus Messingdraht gefertigt.

518-WG
Prischl, Adolf
„Akt" (Hockender Mädchenakt mit Ball) (auf Landschaftssockel)

520-WG
Prischl, Adolf
Hockender Mädchenakt mit großem Tuch

525-WG
Prischl, Adolf
„Akt" (Hockender Mädchenakt mit einer Hand zum Boden greifend) (auf Landschaftssockel)

526-WG
Prischl, Adolf
WANDMASKE „Maske" (Frauenkopf mit mittellangem Haar, im Halbprofil nach rechts) (–> vgl. 8182 im WVZ)

527-WG
Prischl, Adolf
WANDMASKE „Maske" (Frauenkopf mit Schal, nach rechts blickend)

529-WG
Prischl, Adolf
WANDMASKE „Maske" (Frauenkopf mit Schal)

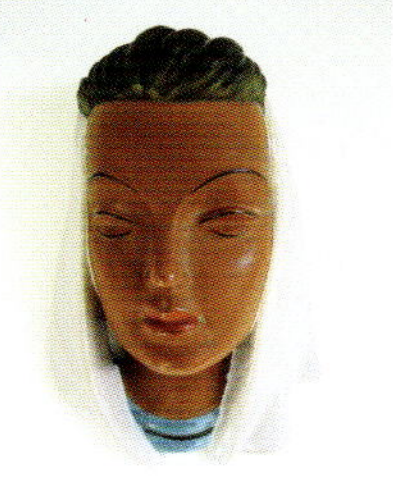

530-WG
Prischl, Adolf
WANDMASKE „Maske" (Frauenkopf mit lockigem Haar)
–> vgl. 531-WG; 532-WG

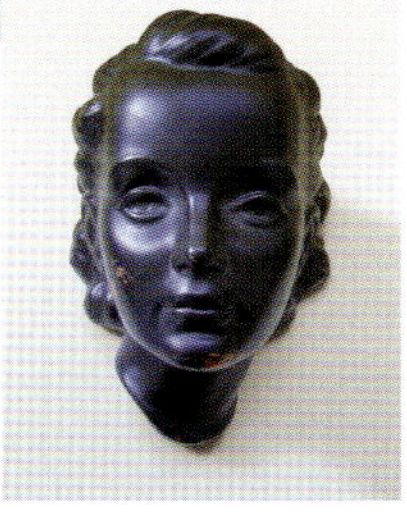

531-WG
Prischl, Adolf
WANDMASKE „Maske" (Frauenkopf mit lockigem Haar)
–> vgl. 530-WG; 532-WG

532-WG
Prischl, Adolf
STEHMASKE „Maske" (Frauenkopf mit lockigem Haar) (auf Rundsockel)

533-WG
Prischl, Adolf
WANDMASKE „Maske" (Frauenkopf mit langem Haar, nach links blickend)

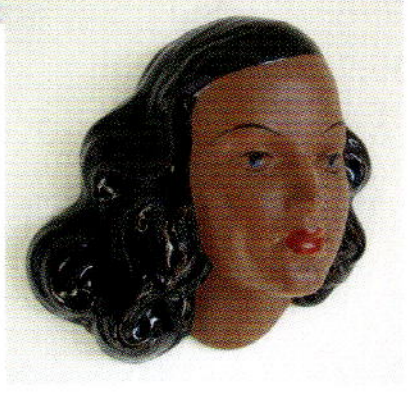

537-WG
Prischl, Adolf
WANDMASKE „Kinderkopf" (Kinderkopf mit Kragen, nach links blickend)
Pendant –> vgl. 538-WG
–> vgl. 542-WG= Variante als „Engel".

538-WG
Prischl, Adolf
WANDMASKE „Kinderkopf" (Kinderkopf mit Kragen, nach rechts blickend)
Pendant –> vgl. 537-WG
–> vgl. 543-WG= Variante als „Engel".

540-WG
Prischl, Adolf
„Akt" (Sitzender Mädchenakt)

542-WG
Prischl, Adolf
WANDMASKE „Engelkopf" (Kinderkopf mit Flügel, nach links blickend)
Pendant –> vgl. 543-WG
–> vgl. 538-WG= Variante als „Kinderkopf".

543-WG
Prischl, Adolf
WANDMASKE „Engelkopf" (Kinderkopf mit Flügel, nach rechts blickend)
Pendant –> vgl. 542-WG
–> vgl. 539-WG= Variante als „Kinderkopf".

544-WG
Prischl, Adolf
WANDMASKE „Maske" (Frauenkopf mit Schal / Halstuch)

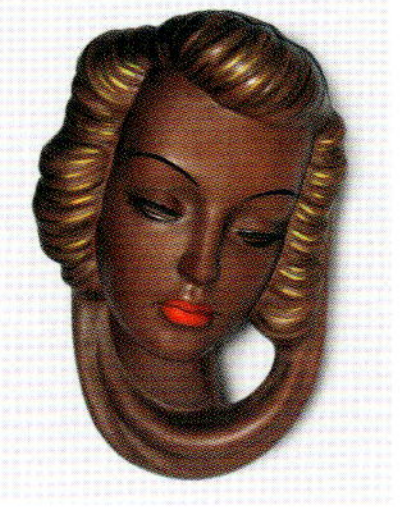

545-WG
Prischl, Adolf
„Akt mit Kugel" (Schreitender Mädchenakt mit Kugel) (auf Ovalsockel mit Kugel)
–> vgl. 545A-WG

545A-WG
Prischl, Adolf
„Akt mit Kugel" (Schreitender Mädchenakt mit Kugel) (auf Ovalsockel mit Kugel)
–> vgl. 545-WG

602-WG
Chludek (nach Dürer, Albrecht)
WANDBEHANG / RELIEF „Betende Hände" (Monogramm: AD)

603-WG
Chludek
„Pudel" (Stehender Pudel)

700-WG
Knörlein, Rudolf
WANDMASKE „Maske" (Negermädchen mit Ohrring und Kopftuch)

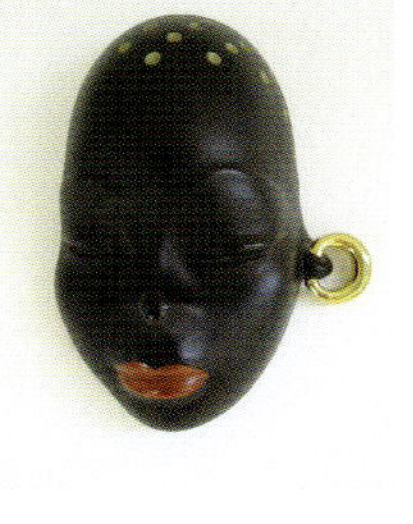

701-WG
Knörlein, Rudolf
WANDMASKE Negerin mit Zöpfen und Halsschmuck

702-WG
Knörlein, Rudolf
BÜSTE Negerin mit Halsschmuck und Ohrringen

703-WG
Knörlein, Rudolf
WANDMASKE Negerin mit goldenem Kopftuch (Kopfschmuck)

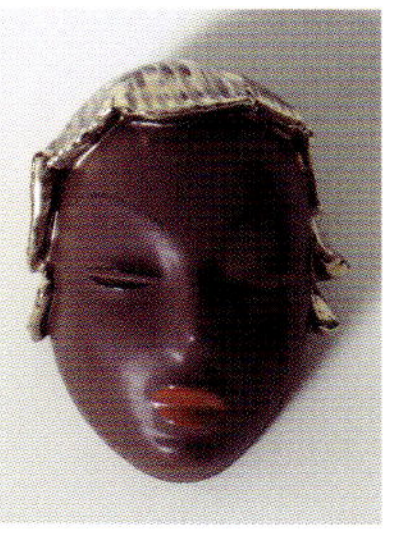

705-WG
Knörlein, Rudolf
„Pinguin" (Stehender Pinguin)

706-WG
Knörlein, Rudolf
BÜSTE „Nofretete"

707-WG
Knörlein, Rudolf
WANDMASKE Negerkind mit Ohrschmuck

750-WG
Knörlein, Rudolf
„Bison" (Stilisiertes Bison beim Angriff)

751-WG
Knörlein, Rudolf
„Sitzende" (Sitzendes Mädchen)

752-WG
Knörlein, Rudolf
„Sitzende" (Sitzendes Mädchen auf Sockel)

753-WG
Knörlein, Rudolf
WANDMASKE „Maske" (Frauenkopf mit Schal/Kopftuch)

754-WG
Knörlein, Rudolf
„Akt" (Sitzender Mädchenakt)

756-WG
Knörlein, Rudolf
„Reh" (Stehendes Reh)

757-WG
Knörlein, Rudolf
„Kniende" (Kniender Mädchenakt auf Rechtecksockel)

758-WG
Knörlein, Rudolf
„Kniende" (Kniendes Mädchen in langem Kleid)

759-WG
Knörlein, Rudolf
„Eule" (Stilisierte Eule)

800-WG
Weiß
WANDMASKE „Maske" (Betende Madonna mit geneigtem Kopf)

801-WG
Weiß
WANDMASKE „Maske" (Frauenkopf mit Kopftuch)

900-WG
WANDMASKE „Maske"

901-WG
Schmidt
WANDMASKE „Maske" (Frauenkopf mit kurzem Haar und Kragen)

1000-WG
Bosse, Walter
„Elefant" (Stilisierter Elefant mit hängendem Rüssel)

1002-WG
Bosse, Walter
„Elefant" (Stilisierter Elefant mit aufgestelltem Rüssel)

1003-WG
Bosse, Walter
WANDBEHANG „Wandpferd" (Stilisiertes Pferd)

1004-WG
Bosse, Walter
WANDBEHANG „Pferd"
(Stilisiertes Pferd)

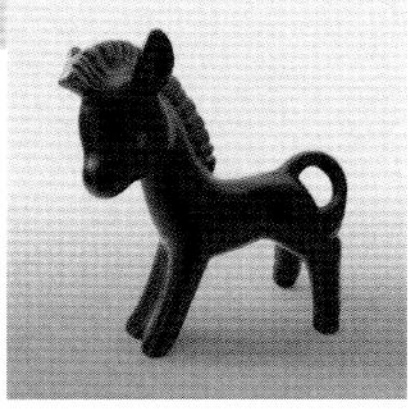

1005-WG
Bosse, Walter
Seehund

1006-WG
Bosse, Walter
„Katze" (Stilisierte sitzende Katze)

1007-WG
Bosse, Walter
WANDBEHANG „Wandhahn" (Stilisierter Hahn)

1008-WG
Bosse, Walter
„Dackel" (Stilisierter sitzender Dackel)

1009-WG
Bosse, Walter
„Giraffe" (Stilisierte Giraffe)

1010-WG
Bosse, Walter
„Häschen" (Stilisiertes sitzendes Häschen)

1011-WG
Bosse, Walter
Kamel / Trampeltier (Stilisiertes Kamel / Trampeltier)
2285-WG s. Werkverzeichnis Wiener Manufaktur FG

2839-WG
Röhrender Hirsch (auf Landschaftssockel)

5891-WG s. Werkverzeichnis Wiener Manufaktur FG
6361-WG s. Werkverzeichnis Wiener Manufaktur FG
6427-WG s. Werkverzeichnis Wiener Manufaktur FG
6816-WG s. Werkverzeichnis Wiener Manufaktur FG
6817-WG s. Werkverzeichnis Wiener Manufaktur FG
6818-WG s. Werkverzeichnis Wiener Manufaktur FG
6905-WG s. Werkverzeichnis Wiener Manufaktur FG
6906-WG s. Werkverzeichnis Wiener Manufaktur FG
6907-WG s. Werkverzeichnis Wiener Manufaktur FG
7025-WG s. Werkverzeichnis Wiener Manufaktur FG
7029-WG s. Werkverzeichnis Wiener Manufaktur FG
7089-WG s. Werkverzeichnis Wiener Manufaktur FG
7420-WG s. Werkverzeichnis Wiener Manufaktur FG
7431-WG s. Werkverzeichnis Wiener Manufaktur FG
7525-WG s. Werkverzeichnis Wiener Manufaktur FG
7552-WG s. Werkverzeichnis Wiener Manufaktur FG
7553-WG s. Werkverzeichnis Wiener Manufaktur FG
7571-WG s. Werkverzeichnis Wiener Manufaktur FG
7579-WG s. Werkverzeichnis Wiener Manufaktur FG
7595-WG s. Werkverzeichnis Wiener Manufaktur FG
7730-WG s. Werkverzeichnis Wiener Manufaktur FG
7739-WG s. Werkverzeichnis Wiener Manufaktur FG
7772-WG s. Werkverzeichnis Wiener Manufaktur FG
7781-WG s. Werkverzeichnis Wiener Manufaktur FG
7784-WG s. Werkverzeichnis Wiener Manufaktur FG
7785-WG s. Werkverzeichnis Wiener Manufaktur FG
7813-WG s. Werkverzeichnis Wiener Manufaktur FG
7816-WG s. Werkverzeichnis Wiener Manufaktur FG
7831-WG s. Werkverzeichnis Wiener Manufaktur FG
7857-WG s. Werkverzeichnis Wiener Manufaktur FG
7936-WG s. Werkverzeichnis Wiener Manufaktur FG
7988-WG s. Werkverzeichnis Wiener Manufaktur FG
8020-WG s. Werkverzeichnis Wiener Manufaktur FG
8025-WG s. Werkverzeichnis Wiener Manufaktur FG
8041-WG s. Werkverzeichnis Wiener Manufaktur FG
8073-WG s. Werkverzeichnis Wiener Manufaktur FG
8083-WG s. Werkverzeichnis Wiener Manufaktur FG
8174-WG s. Werkverzeichnis Wiener Manufaktur FG
8198-WG s. Werkverzeichnis Wiener Manufaktur FG
8199-WG s. Werkverzeichnis Wiener Manufaktur FG
8207-WG s. Werkverzeichnis Wiener Manufaktur FG
8211-WG s. Werkverzeichnis Wiener Manufaktur FG
8232-WG s. Werkverzeichnis Wiener Manufaktur FG
8245-WG s. Werkverzeichnis Wiener Manufaktur FG
8246-WG s. Werkverzeichnis Wiener Manufaktur FG
8247-WG s. Werkverzeichnis Wiener Manufaktur FG
8248-WG s. Werkverzeichnis Wiener Manufaktur FG
8249-WG s. Werkverzeichnis Wiener Manufaktur FG
8253-WG s. Werkverzeichnis Wiener Manufaktur FG
8258-WG s. Werkverzeichnis Wiener Manufaktur FG
8266-WG s. Werkverzeichnis Wiener Manufaktur FG
8282-WG s. Werkverzeichnis Wiener Manufaktur FG
8325-WG s. Werkverzeichnis Wiener Manufaktur FG
8328-WG s. Werkverzeichnis Wiener Manufaktur FG
8345-WG s. Werkverzeichnis Wiener Manufaktur FG
8347-WG s. Werkverzeichnis Wiener Manufaktur FG
8367-WG s. Werkverzeichnis Wiener Manufaktur FG
8376-WG s. Werkverzeichnis Wiener Manufaktur FG
8386-WG s. Werkverzeichnis Wiener Manufaktur FG
8404-WG s. Werkverzeichnis Wiener Manufaktur FG
8429-WG s. Werkverzeichnis Wiener Manufaktur FG
8462-WG s. Werkverzeichnis Wiener Manufaktur FG
8482-WG s. Werkverzeichnis Wiener Manufaktur FG
8483-WG s. Werkverzeichnis Wiener Manufaktur FG
8485-WG s. Werkverzeichnis Wiener Manufaktur FG
8489-WG s. Werkverzeichnis Wiener Manufaktur FG
8495-WG s. Werkverzeichnis Wiener Manufaktur FG
8496-WG s. Werkverzeichnis Wiener Manufaktur FG
8503-WG s. Werkverzeichnis Wiener Manufaktur FG
8506-WG s. Werkverzeichnis Wiener Manufaktur FG
8508-WG s. Werkverzeichnis Wiener Manufaktur FG
8563-WG s. Werkverzeichnis Wiener Manufaktur FG
8599-WG s. Werkverzeichnis Wiener Manufaktur FG
8612-WG s. Werkverzeichnis Wiener Manufaktur FG
8633-WG s. Werkverzeichnis Wiener Manufaktur FG
8654-WG s. Werkverzeichnis Wiener Manufaktur FG
8665-WG s. Werkverzeichnis Wiener Manufaktur FG
8669-WG s. Werkverzeichnis Wiener Manufaktur FG
8685-WG s. Werkverzeichnis Wiener Manufaktur FG
8698-WG s. Werkverzeichnis Wiener Manufaktur FG
8699-WG s. Werkverzeichnis Wiener Manufaktur FG
8702-WG s. Werkverzeichnis Wiener Manufaktur FG
8711-WG s. Werkverzeichnis Wiener Manufaktur FG
8713-WG s. Werkverzeichnis Wiener Manufaktur FG
8714-WG s. Werkverzeichnis Wiener Manufaktur FG
8735-WG s. Werkverzeichnis Wiener Manufaktur FG
8739-WG s. Werkverzeichnis Wiener Manufaktur FG
8749-WG s. Werkverzeichnis Wiener Manufaktur FG
8770-WG s. Werkverzeichnis Wiener Manufaktur FG
8785-WG s. Werkverzeichnis Wiener Manufaktur FG
8810-WG s. Werkverzeichnis Wiener Manufaktur FG
8811-WG s. Werkverzeichnis Wiener Manufaktur FG
8812-WG s. Werkverzeichnis Wiener Manufaktur FG
8813-WG s. Werkverzeichnis Wiener Manufaktur FG
8814-WG s. Werkverzeichnis Wiener Manufaktur FG
8815-WG s. Werkverzeichnis Wiener Manufaktur FG
8885-WG s. Werkverzeichnis Wiener Manufaktur FG
8887-WG s. Werkverzeichnis Wiener Manufaktur FG
8891-WG s. Werkverzeichnis Wiener Manufaktur FG
8905-WG s. Werkverzeichnis Wiener Manufaktur FG
8913-WG s. Werkverzeichnis Wiener Manufaktur FG
8915-WG s. Werkverzeichnis Wiener Manufaktur FG
8917-WG s. Werkverzeichnis Wiener Manufaktur FG
8918-WG s. Werkverzeichnis Wiener Manufaktur FG
8919-WG s. Werkverzeichnis Wiener Manufaktur FG
8920-WG s. Werkverzeichnis Wiener Manufaktur FG
8921-WG s. Werkverzeichnis Wiener Manufaktur FG
8922-WG s. Werkverzeichnis Wiener Manufaktur FG
8923-WG s. Werkverzeichnis Wiener Manufaktur FG
8929-WG s. Werkverzeichnis Wiener Manufaktur FG
8930-WG s. Werkverzeichnis Wiener Manufaktur FG
8932-WG s. Werkverzeichnis Wiener Manufaktur FG
8935-WG s. Werkverzeichnis Wiener Manufaktur FG
8936-WG s. Werkverzeichnis Wiener Manufaktur FG
8937-WG s. Werkverzeichnis Wiener Manufaktur FG
8938-WG s. Werkverzeichnis Wiener Manufaktur FG
8939-WG s. Werkverzeichnis Wiener Manufaktur FG
8941-WG s. Werkverzeichnis Wiener Manufaktur FG
8942-WG s. Werkverzeichnis Wiener Manufaktur FG
8930-WG s. Werkverzeichnis Wiener Manufaktur FG

Arthur Goldscheider, Paris – La Stèle

Ausführung der Modelle zwischen 1924–1928

–> vgl. Kapitel „Dekorative Skulpturen von Arthur Goldscheider in Paris (1900 – ca. 1935)"

ohne Nr.?
Bouraine, A(ndré) Marcel
Seelöwe (auf Felssockel)
BRONZE (versilbert)
ca. 1924–1928
ohne Nr.?, (Firmenmarke La Stèle), sign. M. Bouraine, H. 47,6 cm; Christie's East NY 6.12.2000

ohne Nr.
Dakon, Stephan
Stehendes Mädchen in futuristischem Kostüm
BRONZE (oder Messing?) auf Marmorsockel
ca. 1924–1928
ohne Nr., ohne Firmenmarke, H. 47,5 cm; MAK Wien, Inv. Ke 10602-13
Bem.: Diese Figur stammt aus dem Nachlass von La Stèle.

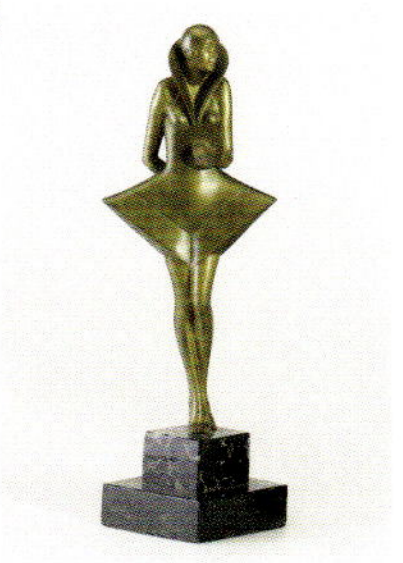

ohne Nr.?
Traverse, Pierre
Laufendes Mädchen mit Reh
BRONZE auf Marmorsockel
ca. 1924–1928
ohne Nr.?, (Firmenmarke La Stèle), France ger., sign. Pierre Traverse, H. 41 cm; Christie's NY 13.6.2003

2
Temporal, Marcel
BÜSTE Beethoven
BRONZE
ca. 1924–1928
2 (Firmenmarke La Stèle), sign. M. Temporal, H. 52,1 cm, B. 29,2 cm; KH

15
Laufendes Mädchen mit Ball
BRONZE auf Marmorsockel
ca. 1924–1928
15 ohne Firmenmarke, H. 33 cm; PB Wien 19
Bem.: Diese Figur stammt aus dem Nachlass von La Stèle.

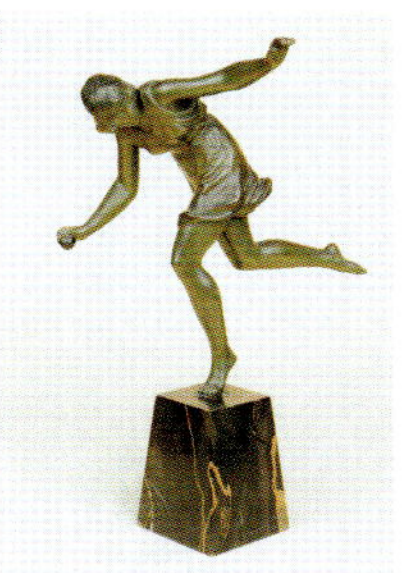

Models made 1924–1928

–> cf chapter "Ornamental Sculpture from Arthur Goldscheider in Paris (1900 – ca 1935)"

21
Philippe, Paul
Stehender Damenakt mit ausgestrecktem Arm
BRONZE auf Marmorsockel
ca. 1924–1928
21 ohne Firmenmarke, sign. P. Philippe, Gesamth. 48,5 cm; PB Wien 19
Bem.: Diese Figur stammt aus dem Nachlass von La Stèle.

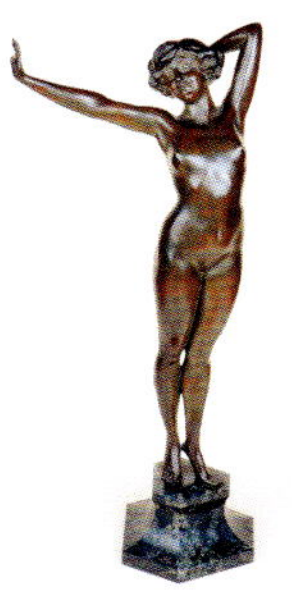

34
Lepage, Céline
„Femme (Mendiante) de Marakech" (Frau / Bettlerin aus Marrakesch) (auf quadratischer Plinthe)
BRONZE
ca. 1921 (für den Entwurf), produziert ca. 1924–1928
34 (Firmenmarke La Stèle), monogr. CD, H. 79,4 cm; MAK Wien, Inv. Ke 10602-15
Bem.: Diese für die Künstlerin und die Manufaktur des Arthur Goldscheider erfolgreiche Figur wurde 1921 entworfen und auf der Pariser Kunstgewerbeausstellung 1925 ausgestellt.
Ausst.: Exposition Internationale des Arts Décoratifs et Industriels Modernes, Paris 1925; Goldscheider, A. (Hrsg.): Artistic Evolution, 1926, Pl. 9 (m. Abb.); PSK 1995.
Lit.: Rouger, G.: Nos Artistes au Maroc, in: L'Art et les artistes, N.S. 6, 1922, S. 36; Exposition Internationale des Arts Décoratifs et Industriels Modernes, Paris 1925, Rapport Général, Paris 1927, Accessoirs du Mobilier, Pl. XXXIV (m. Abb.); 1930: Quand le Meuble Devient Sculpture, Paris 1981, S. 56 (m. Abb.); ADGS, Nr. 2 (m. Abb.).

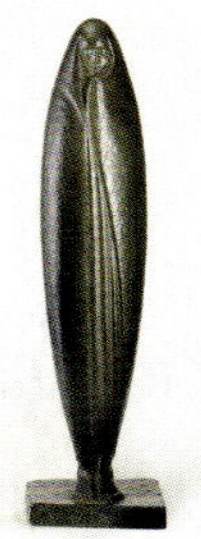

46
Bouraine, A(ndré) Marcel
„Leda" (mit dem Schwan)
BRONZE auf Marmorsockel
ca. 1924–1928
46 (Firmenmarke La Stèle), sign. A. Bouraine, H. 24,2 cm, L. 40,8 cm; PB Wien 19
Lit.: Goldscheider, A. (Hrsg.): Artistic Evolution, 1926, Pl. 11 (m. Abb.), hier in der versilberten Ausführung mit Ebenholz abgebildet; Neuwirth (1974), S. 72.

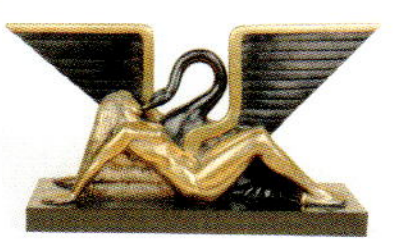

56
Le Faguays, Pierre
Puppenspielerin mit zwei Handpuppen
BRONZE auf Marmorsockel
ca. 1924–1928
56 ohne Firmenmarke, sign. LeFaguays, Gesamth. 49 cm, Figur 38,5 cm; PB Wien
Bem.: Diese Figur stammt aus dem Nachlass von La Stèle.

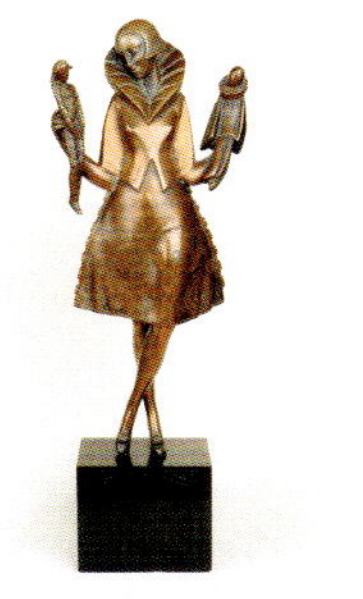

77
Kéléty, Alexandre
Stehendes Mädchen (Halbakt) mit herzförmigem Rock
BRONZE / ELFENBEIN auf Marmorsockel
ca. 1924–1928
77, ohne Firmenmarke, sign. A. Kéléty, H. 29,4 cm; MAK Wien, Inv. Ke 10602-12
Bem.: Diese Figur stammt aus dem Nachlass von La Stèle.

81
Tänzerin mit ausgebreiteten Armen
BRONZE auf Marmorsockel
ca. 1924–1928
81 ohne Firmenmarke, H. 36 cm; PB Wien
Bem.: Diese Figur stammt aus dem Nachlass von La Stèle.

82
Bouraine, A(ndré) Marcel
„Otarie" (Otary) Seelöwe, einen Ball balancierend
BRONZE auf Marmorsockel
ca. 1924–1928
82 (Firmenmarke La Stèle), sign. A. Bouraine, H. 34 cm; PB Wien 19
Bem.: Von dieser Figur gibt es auch eine Ausführung als Lampe, bei den der Leuchtkörper (Argonit) den Ball des Seelöwen ersetzt –> vgl. Goldscheider, A. (Hrsg.): Artistic Evolution, 1926, Pl. 26 (m. Abb.).

106
Bouraine, A(ndré) Marcel
Ballerina auf einem Bein
BRONZE
ca. 1924–1928
106 ohne Firmenmarke, sign. A. Bouraine, H. 28,5 cm; PB Wien 19
Bem.: Diese Figur stammt aus dem Nachlass von La Stèle.

147
Bouraine, A(ndré) Marcel
Tanzender Damenakt auf einem Bein
BRONZE
ca. 1924–1928
147 ohne Firmenmarke, sign. A. Bouraine, H. 30,2 cm; PB Wien 19
Bem.: Diese Figur stammt aus dem Nachlass von La Stèle.

Marcell Goldscheider, Wien

Marcell Goldscheider, Vienna

Ausführung der Modelle zwischen 1928–1939

–> vgl. Kapitel „Marcell Goldscheiders Werkstätten in Wien und Stoke-on-Trent – Die Vereinigten Ateliers für Kunst und Keramik in Wien (1928–1939)"

ohne Nr.
Knörlein, Rudolf
KERZENHALTER mit Behälter für Streichhölzer (einflammig)
Form: s-förmige und rechtwinkelige Konstruktion (auf vierfüßigem Rechtecksockel)
ca. 1928–1932
ohne Nr., (ohne Firmenstempel, Reste eines Pick. Marcell GS), monogr. RK,
H. 20 cm, B. 16,5 cm; PB Wien

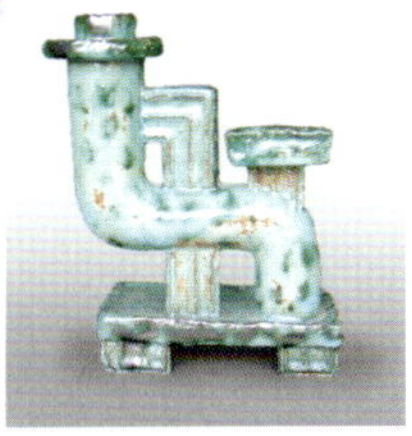

ohne Nr.
Knörlein, Rudolf
RELIEF Tanzendes Mädchen
ca. 1928/29
Lit.: Rochowanski (1929) in: DKuD, 65.1929/30, S. 430 (m. Abb.)

ohne Nr.
Knörlein, Rudolf
RELIEF Frauenbildnis mit Hand und Blume
ca. 1928/29
Lit.: Rochowanski (1930), S. 108 (m. Abb.)

ohne Nr.
Knörlein, Rudolf
SERVICE Likörservice
ca. 1928/29
Lit.: Rochowanski (1930), S. 78 (m. Abb.)

ohne Nr.
Knörlein, Rudolf (zugeschrieben)
FIGUR „Niederblickende" (Stehende junge Dame auf getrepptem Quadrat/Rechtecksockel?)
ca. 1928/29
ohne Ang., H. 100 cm
Lit.: Heizer (1929), S. 714 (m. Abb.)

ohne Nr.
Knörlein, Rudolf (zugeschrieben)
FIGUR „Innehaltende" (Stehende junge Dame auf Quadratsockel)
ca. 1928/29
ohne Ang., H. 40 cm
Lit.: Heizer (1929), S. 714 (m. Abb.)

ohne Nr.
Reinold, Grete
WANDRELIEF/FLIESE Mädchenakt mit Reh
Form: dreieckig
Dekor: Kniender Mädchenakt ein Reh fütternd
ca. 1928–1939
ohne Nr., GS (Firmenstempel Marcell 1), AUSTRIA, sign. G. REINOLD, L. 32,5 cm; eBay # 3352302567

Models made 1928–1939

–> cf chapter "Marcell Goldscheider's Workshops in Vienna and Stoke-on-Trent – United Studios of Art and Ceramic (1928–1939)"

ohne Nr.
Ziffer, Moses (Moshe)
STEHKOPF Frauenkopf
ca. 1928–1939
ohne Nr., GS (Firmenstempel Marcell 2), H. 25,4 cm; eBay # 290150439
Lit.: WIK, 1928, S. 36 (evtl. wurde dort dieses Modell ausgestellt).

ohne Nr.
KERZENHALTER Akrobaten (dreiflammig)
Form: Mann, Hund und Bär in akrobatischer Pose
ca. 1928–1939
ohne Nr., GS (Firmenstempel Marcell 1), H. 26 cm, B. 25 cm; PB Wien
Bem.: Eine groteske Gruppe, die in ihrer Art und der witzigen Darstellung an Walter Bosse erinnert. Sie kann aber auch von einer der Künstlerinnen der Wiener Werkstätte geschaffen worden sein. Eine Zuschreibung ist daher nicht angezeigt.

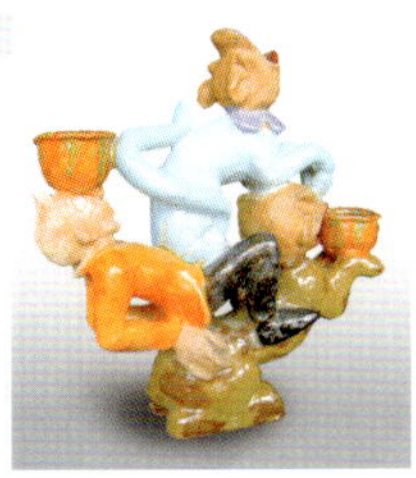

ohne Nr.
FIGUREN/LAMPE Zwei im Kreis laufende Kinder mit fünf Schafen
ca. 1928–1939
FIGUREN
ohne Nr., ohne Firmenstempel, Pick. Marcell 1, H. 17 cm; PB Wien 19
LAMPE
ohne Nr., GS (Firmenstempel Marcell 1), H. 17 cm, Ø 24,6 cm, Gesamth. 59 cm; Wien Museum Inv. 219.167

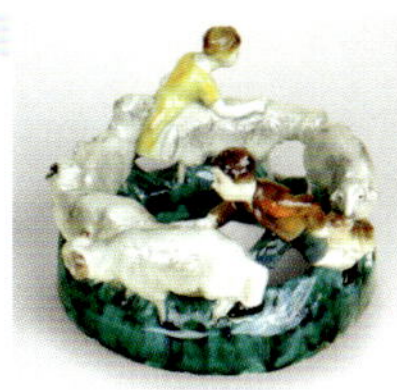

ohne Nr.
TAFELAUFSATZ/TAZZA
Form: Runde Schale auf baumartigem Fuß (auf Rundsockel)
ca. 1928–1939
ohne Nr., (ohne Firmenstempel), Pick. Marcell 1, H. 20 cm; Sammlung Dr. Peter Hieke, Wien

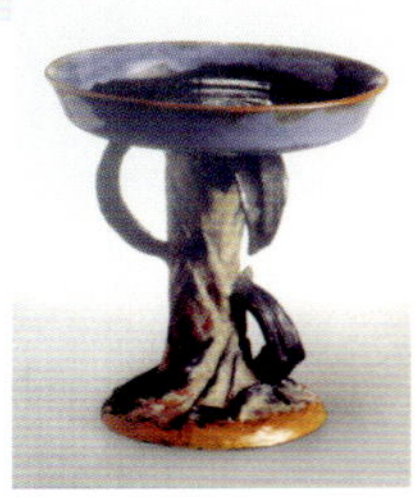

ohne Nr.
WANDMASKE Frauenkopf mit Haarschmuck (grünlich)
ca. 1928–1939
ohne Nr., GS (Firmenstempel Marcell 1), H. 17,6 cm; PB Wien 19

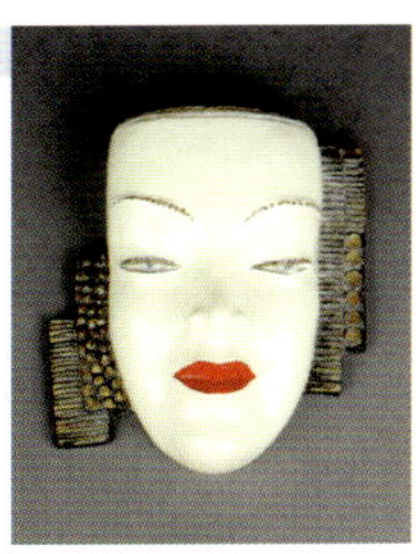

ohne Nr.
WANDMASKE Frauenkopf mit Haarschmuck (gelblich)
ca. 1928–1939
ohne Nr., GS (Firmenstempel Marcell 1), H. 17,2 cm; PB Wien 19

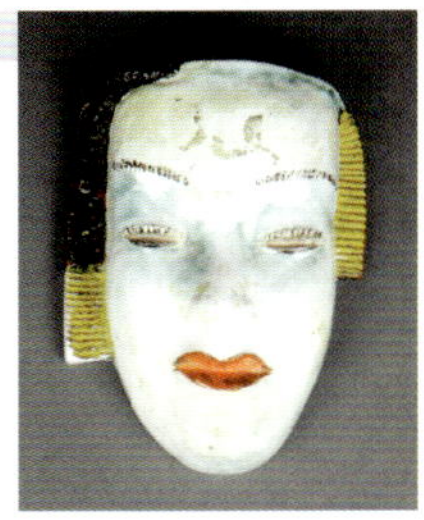

ohne Nr.
Stehendes Küken (auf getrepptem Rechtecksockel)
ca. 1928–1939
ohne Nr., GS (Firmenstempel Marcell 1, blau), H. 11,8 cm; Wien Museum Inv. 190.045

79 –> ÜBERSCHNEIDUNG
Reinold, Grete
SCHALE/TELLER Mädchengesicht
Form: rund
Dekor: Mädchengesicht
ca. 1928
079 GS (Firmenstempel Marcell 1), sign. GRETE REINOLD, Ø 24,6 cm; PB Wien

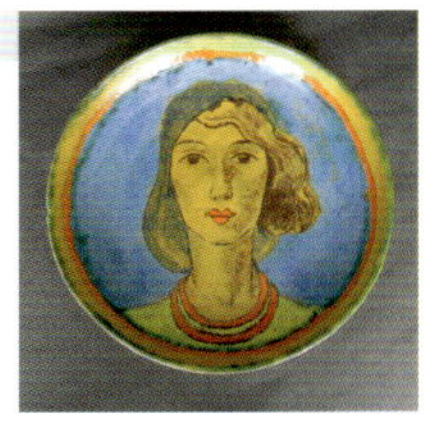

79 –> ÜBERSCHNEIDUNG
TAFELAUFSATZ/SCHALE
Form: rund
Dekor: Kreise
ca. 1928
079 GS (Firmenstempel Marcell 1), H. 8 cm, Ø 25,5 cm; Doro 26.5.2004; 15.3.2005

201
Stuart-Willfort, Irma
SCHALE Verkündigung
Form: rund
Dekor: biblische Verkündigungsszene
ca. 1928
0201,25 GS (Firmenstempel Marcell 1), AUSTRIA, sign. J. Willfort, H. 5,5 cm, Ø 24 cm; Zezschwitz 29

235
RELIEF Asiatin mit Haarschmuck (Wandrelief)
ca. 1928
0235,2,* GS (Firmenstempel Marcell 1), AUSTRIA, H. 33 cm; PB Wien 19

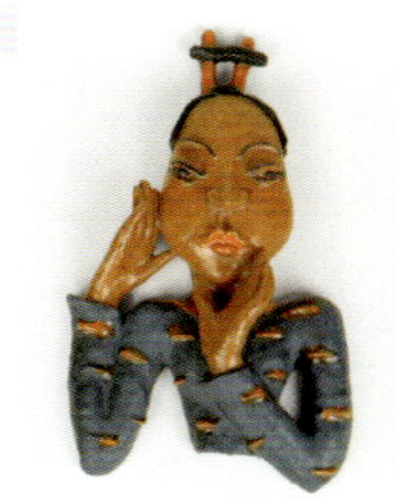

239
BUCHSTÜTZE Stehender lesender Mönch
ca. 1928
0239,16 GS (Firmenstempel Marcell 1), AUSTRIA, H. 17,8 cm; eBay # 1238655120

244
TONDO/RELIEF Madonna mit Kind
ca. 1928
244,6,* GS (Firmenstempel Marcell 1), AUSTRIA, Ø 23 cm; Wien Museum Inv. 223.736
244,8,* GS (Firmenstempel Marcell 1), AUSTRIA, Ø 23 cm; PB Wien 19

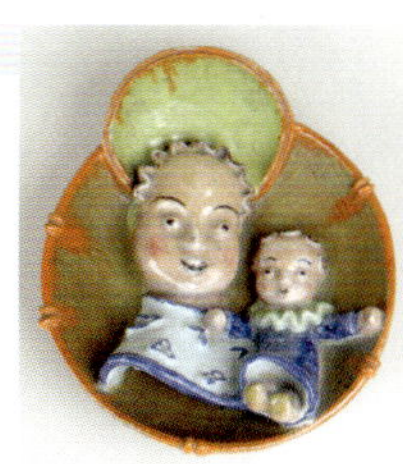

316
Stehender Mädchenakt von Koralle und Seetang umfangen (auf Rundsockel mit Seestern, Muscheln und Seeigel)
ca. 1928
316,1 (ohne Firmenstempel) H. 42,5 cm; MAK Wien, Inv. Ke 10602-89
316,3 GS (Firmenstempel Marcell 1), AUSTRIA, H. 42,7 cm; Wien Museum Inv. 190.079
316,8 GS (Firmenstempel Marcell 1), AUSTRIA, H. 44 cm; PB Wien 19

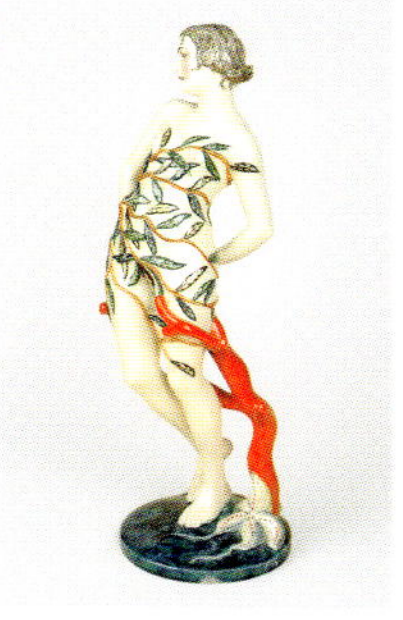

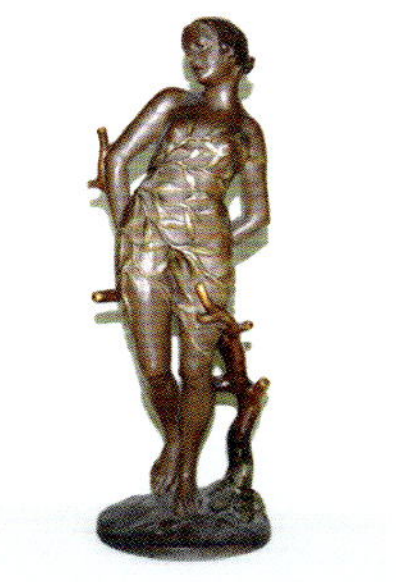

318
Tänzerin mit Schleier und runder Platte darauf Tierdarstellung (auf profiliertem Ovalsockel)
ca. 1928
318,6 GS (Firmenstempel Marcell 1), AUSTRIA, H. 43 cm; Wien Museum Inv. 190.067

319
Carmen (Sitzende Spanierin)
ca. 1928
319,1 gem. S, ohne Firmenstempel, H. 40,5 cm; PB Wien 19

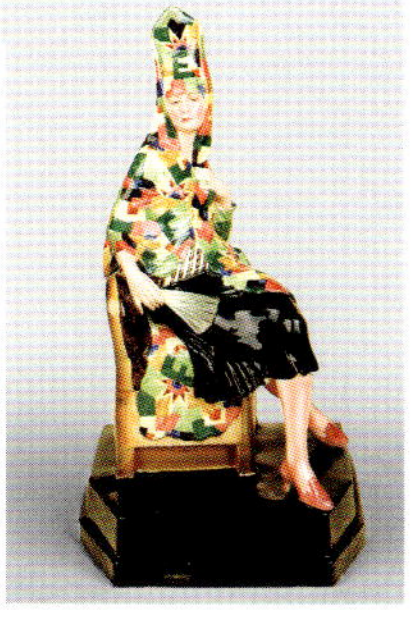

323
Stehende junge Dame mit Hut und Cape (auf Rundsockel)
ca. 1928
323 ger., GS (Firmenstempel Marcell 2), H. 24,8 cm; Wien Museum Inv. 220.710
Bem.: Eines der wenigen in Porzellan ausgeführten Modelle.

325
DECKELDOSE Liegende Dame
Form: oval
Knauf/Bekrönung: liegende, leicht bekleidete Dame (vollplastische Fig.)
Dekor: Blumengirlande
ca. 1928
325 gem. K., gem. lig. TA, GS (Firmenstempel Marcell 1), H. 18 cm; Sammlung Dr. Peter Hieke, Wien

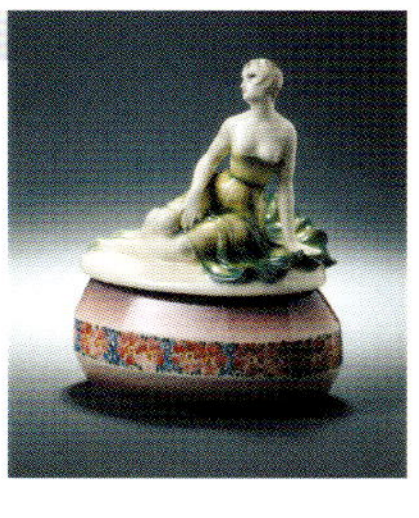

331
KERZENHALTER (zweiflammig)
Form: Stehende Tänzerin in kurzem Kleid, zwei Kerzentüllen haltend
ca. 1928
331 ger., GS (Firmenstempel blau Marcell 2), AUSTRIA, H. 34 cm; GK, S. 27, 135 (m. Abb.); Wien Museum Inv. 182.100
Bem.: Eines der wenigen in Porzellan ausgeführten Modelle.
Ausst.: HMW (jetzt Wien Museum) 1985/86.
Lit.: GK, S. 27, 135 (m. Abb.)

332
GRUPPE / LAMPE Liebespaar (kostümiert) mit Gondel
ca. 1928
GRUPPE
332 GS (Firmenstempel Marcell 1), H. 40,5 cm; Wien Museum Inv. 190.069
332 GS (Firmenstempel Marcell 1), H. 40,5 cm; Doro 20.4.1993
LAMPE
332 GS (Firmenstempel Marcell 1), H. 43 cm; Doro 29.5.2001; 10.3.2003; 25.11.2003

339
Jarl, Karin
Zwei laufende Windhunde (auf Natursockel)
ca. 1928
339 GS (Firmenstempel Marcell 1), H. 16 cm, L. 45 cm; Henry's 26.2.1997; 28.5.1998

358
SCHALE / TAFELAUFSATZ Bär
ca. 1928
358 GS (Firmenstempel Marcell 1), H.31 cm, Ø 26 cm; eBay # 250053860827

371
Bucher, Hertha
„Musikanten" (Sänger und Lautenspielerin)
ca. 1928/29
371,12 ger., ohne Firmenstempel, sign. HB, H. 43 cm; PB Wien 19
Ausst.: BiR, Wien 1929.
Lit.: BiR, 1929, S. 36, 50 (m. Abb.), hier mit Ang. des Originaltitels „Musikanten"; Neuwirth (1974), S. 60.

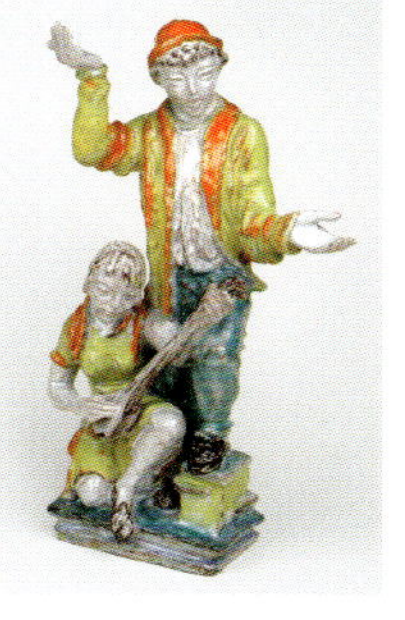

375
KERZENLEUCHTER (dreiflammig)
ca. 1928/29
375 ger., (ohne Firmenstempel), Pick. Marcell 1, H. 16 cm, L. 30 cm; Sammlung Dr. Peter Hieke, Wien

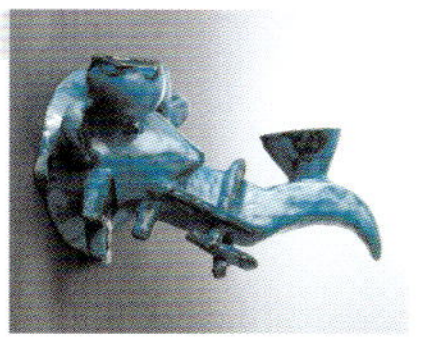

378
Jarl, Karin
Tukan (Pfefferfresser mit exotischen Früchten)
ca. 1928/29
378 ger., blaues ungedeutetes Malerzeichen, GS (Firmenstempel Marcell 1), sign. Karin Jarl, H. 23,8 cm; MAK Wien, Inv. Ke 10602-86

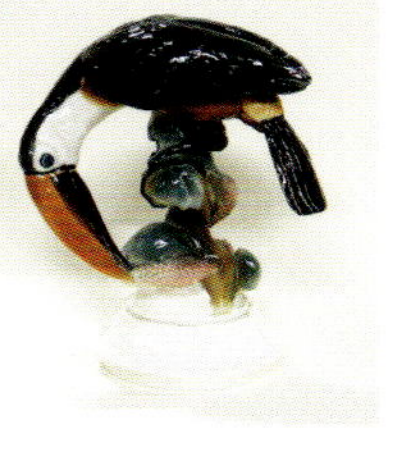

380
Tischer, E.? (unles. sign.)
Schreitende Tänzerin mit großem Kopftuch/Cape, in verträumter Pose (ägyptische Tempeltänzerin?) (auf Ovalsockel)
ca. 1928/29
380 GS (Firmenstempel Marcell 1), unles. sign. E. Tischer?, H. 55,9 cm; eBay # 1503054232
380 (ohne Firmenstempel), unsign., H. 53 cm; Soth. Olympia 3.6.2004

382
GRUPPE Liebespaar mit Kakadu (auf unregelmäßig geschwungenem Sockel)
ca. 1928/29
382,1,0 GS (Firmenstempel Marcell 1), AUSTRIA, Firmenstempel WIENER KUNST-KERAM. WERKST., H. 39 cm; Wien Museum Inv. 219.154

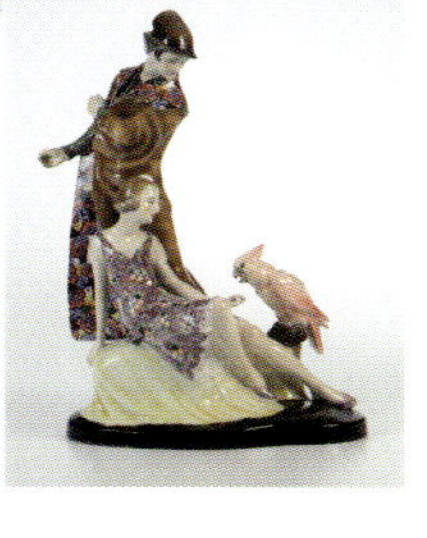

384
Zwei King Charles Spaniels
ca. 1928/29
384 ger., GS (Firmenstempel Marcell 2), H. 11 cm, B. 15 cm; Henry's 29.1.1998
Bem.: Eines der wenigen in Porzellan ausgeführten Modelle.

385
KERZENHALTER (dreiflammig)
Form: zwei zu einem W verschlungene V, quer darüber ein Stab mit den drei Kerzentüllen (auf Ovalsockel)
ca. 1928/29
385,128 GS (Firmenstempel Marcell 1), AUSTRIA, H. 17,7 cm, B. 30,6 cm; PB Wien

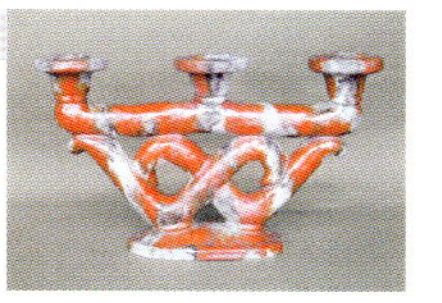

386
Sitzende Geisha mit Reh (auf mehrfach geschwungenem Sockel)
ca. 1928/29
386,3 GS (Firmenstempel Marcell 1), AUSTRIA, H. 25 cm; Sammlung Weilinger, Wien

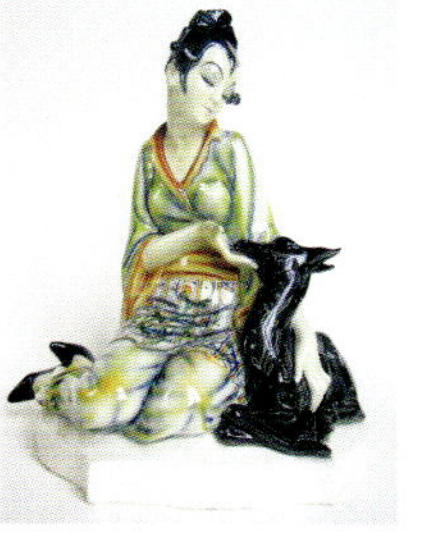

400
SCHALE/BÜSTE Frauenkopf mit flacher Schale auf dem Kopf
ca. 1929
400,79,2 GS (Firmenstempel Marcell 1), H. 18 cm; Wien Museum Inv. 190.171
(400 Nr. fehlt),21 (drei reliefierte Firmenstempel in der Schaleninnenseite), H. 17,4 cm; Doro 20.4.1993;
Bem.: Welche Firmenstempel in der Schaleninnenseite eingeprägt sind, wird im Auktionskat. nicht näher verifiziert.
Ausst.: SSH 1992

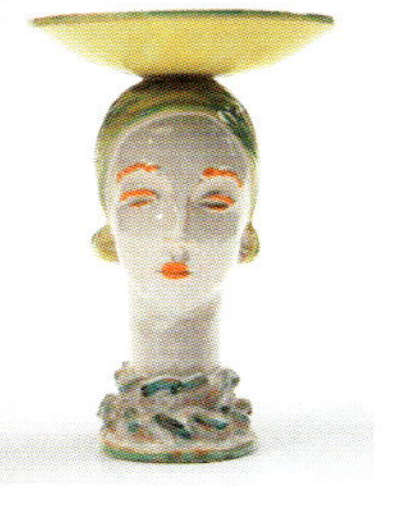

404
SCHALE / TAFELAUFSATZ Zwei Akrobaten (auf Ovalsockel)
Form: oval
Dekor: zwei turnende Akrobaten (vollplastische Figuren)
ca. 1929
404,15 GS (Firmenstempel Marcell 1), H. 14,3 cm; Doro 4.11.1998
404 GS (Firmenstempel Marcell 1), OB GS (Firmenstempel OB), H. 14,5 cm; PB Wien

409
Knörlein, Rudolf
Stehender Mädchenakt mit Reh (Brüderchen und Schwesterchen) (auf profiliertem Quadratsockel)
ca. 1929
409,5,2 GS (Firmenstempel Marcell 1), H. 27 cm; Sammlung Dr. Peter Hieke, Wien
Lit.: Rochowanski (1929) in: DKuD, 65.1929/30, S. 430 (m. Abb.), hier abgebildet als unglasierte Originalkeramik.

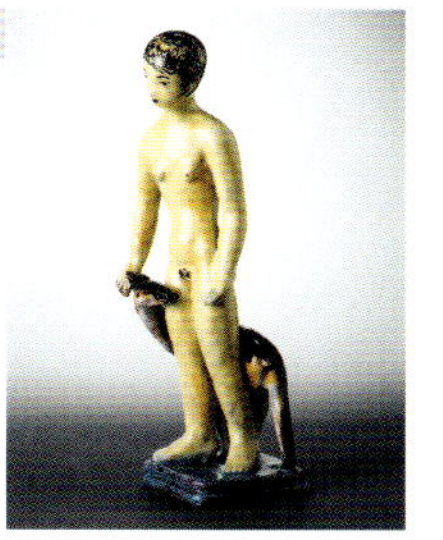

410
Knörlein, Rudolf
„Nische" (Kniendes Mädchen in einer Nische)
ca. 1929 bzw. 1931
410,3 Firmenstempel Marcell OWB, unsign., H. 30 cm, B. 24,5 cm; GK, S. 27, 137 (m. Abb.); Wien Museum Inv. 223.740
Bem.: Die Zuschreibung an

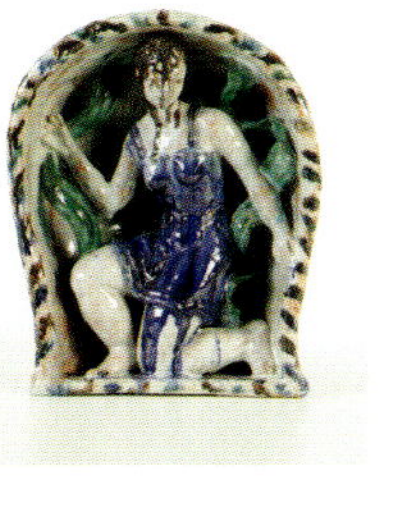

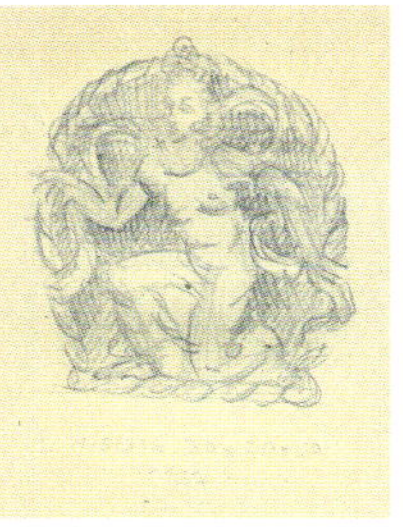

R. Knörlein und die Datierung konnten aufgrund einer originalen Entwurfszeichnung aus der Privatsammlung D vorgenommen werden. Das Blatt im Format 10 x 14 cm ist datiert 1931. Aufgrund des OWB Stempels ist jedoch eher vom Jahr 1929 auszugehen.
Lit.: GK, S. 27, 137 (m. Abb.)
Ausst.: HMW (jetzt Wien Museum) 1985/86.
Quelle: Privatsammlung D

412
Knörlein, Rudolf
STEHKOPF Frauenkopf mit Hut (auf flacher quadratischer Sockelplatte)
ca. 1929
412,37,* GS (Firmenstempel Marcell 1), AUSTRIA, H. 30 cm;
eBay # 2510897170
412,38,* GS (Firmenstempel Marcell 1), AUSTRIA, H. 30 cm; Sammlung.schmutz@chello.at 51
ohne Nr.,43,* GS (Firmenstempel Marcell 1), AUSTRIA, H. 29,5 cm; PB Wien 19
Lit.: Rochowanski (1929) in: DKuD, 65.1929/30, S. 430 (m. Abb.), hier abgebildet in Terrakotta ohne Bemalung sowie ohne die Sockelplatte.

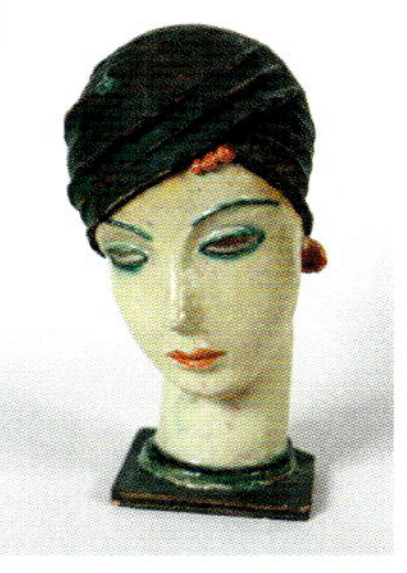

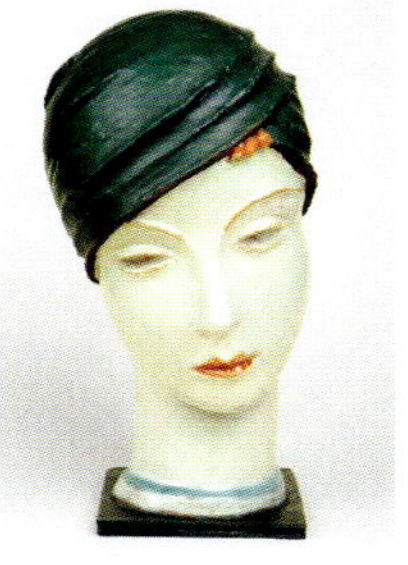

426
Zwei sitzende Scotchterrier (auf Natursockel)
ca. 1929
426 GS (Firmenstempel Marcell 1), H. 23,5 cm; Sammlung a@goldscheider.net

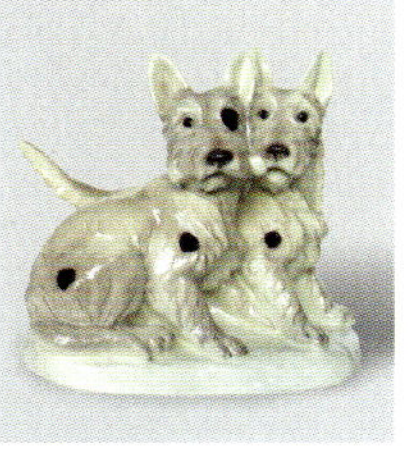

427
Kniende (Japanerin?) mit Tuch
ca. 1929
427,15,1 GS (Firmenstempel Marcell 1), AUSTRIA, H. 43,7 cm; Doro 18.11.1997

428
Sitzender, Männchen machender Yorkshireterrier
ca. 1929
428,8 gem. lig. WK, GS (Firmenstempel Marcell 1), H. 21 cm; eBay # 3316044134

430
Stehendes Kalb (auf Natursockel)
ca. 1929
430,4 ger., GS (Firmenstempel Marcell 1), H. 16,7 cm; Doro 15.9.2003

433
Ente (auf Ovalsockel)
ca. 1929
433,64,1 GS (Firmenstempel Marcell 1), H. 14 cm; Doro 19.9.2000

436
Kostial, Josef
Kniender Pierrot mit Bratsche (Cello?) (auf länglichem geschwungenem und profiliertem Sockel)
ca. 1929
436,8 GS (Firmenstempel Marcell 1), AUSTRIA, H. 28,5 cm; Neuwirth (1974), S. 49 (m. Abb.)
436,10 GS (Firmenstempel Marcell 1), AUSTRIA, H. 31 cm, L. 33,5 cm; eBay # 884984289
Lit.: Neuwirth (1974), S. 49 (m. Abb.)

450
Mann mit zwei Schalen (Schalenträger) (auf flachem Rechtecksockel)
ca. 1929
450,32,1 GS (Firmenstempel Marcell 1), H. 19 cm; PB Wien
450 GS (Firmenstempel Marcell 1), H. 19 cm; WKA 55

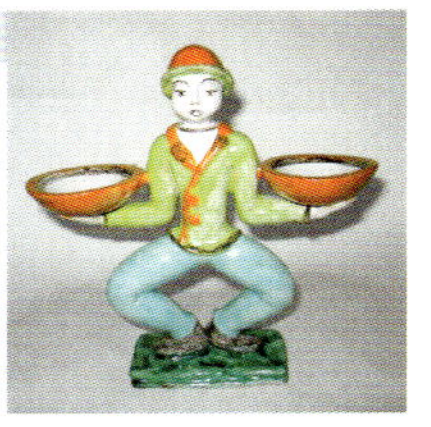

451
Sitzender Airedale Terrier (auf flachem Rechtecksockel)
ca. 1929
451,32,1 GS (Firmenstempel Marcell 1), H. 16 cm; Doro 24.10.1994

458
Sitzender singender Balalaikaspieler
ca. 1929
458,20,1 GS (Firmenstempel Marcell 1), H. 21 cm; Fischer 56; Herr 12
458,21,1 GS (Firmenstempel Marcell 1), H. 20,5 cm; Doro 12.9.2006
Pendant –> vgl. 459
Lit.: Neuwirth (1974), S. 46 (m. Abb.)

459
Tanzende Russin (auf Rechtecksockel)
ca. 1929
459,13,2 GS (Firmenstempel Marcell 1), H. 24 cm; Fischer 56; Herr 12
459,29,1 GS (Firmenstempel Marcell 1), AUSTRIA, H. 23 cm;
eBay # 6549299247
459 ohne Ang., GS (Firmenstempel Marcell 1), AUSTRIA, H. 23 cm; PB Wien
Pendant –> vgl. 458

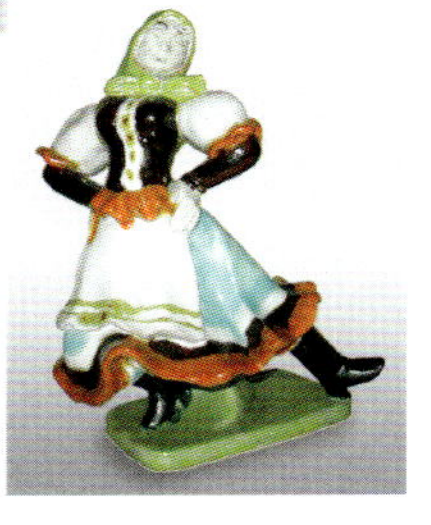

463
Knörlein, Rudolf
RELIEF Frauenkopf im Profil vor Landschaft mit Haus (Wandrelief)
ca. 1929–1932
463,0 (oder Q) GS (Firmenstempel Marcell 1), AUSTRIA, sign. R. Knörlein, H. 35,5 cm, B. 27,9 cm; eBay # 2003092866; 210928086

465
Stehende Frauengestalt im Kostüm „Ballett russe"
ca. 1929–1939
465,5,1 GS (Firmenstempel Marcell 1), AUSTRIA, H. 31,7 cm; Sammlung.schmutz@chello.at 27
Bem.: Der Kostümentwurf geht zurück auf den russ. Maler/Kostümbildner Georgi Poschedajeff. Die originale Zeichnung befindet/befand sich in der Sammlung Jushnij (Jushnij war in den 20er Jahren Leiter des exilrussischen Kabaretts „Der blaue Vogel" in Berlin).

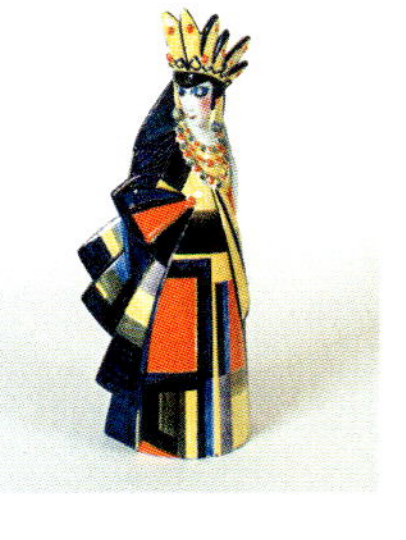

469
Jarl, Karin
Sitzender Kolibri auf Kaktus
ca. 1929–1939
469 (ohne Firmenstempel), sign. Karin Jarl, H. 22 cm; MAK Wien, Inv. Ke 10602-3
Bem.: Durch ein Gegenstück im WKH ließ sich belegen, dass es sich hier um ein Erzeugnis von Marcell GS handelt.

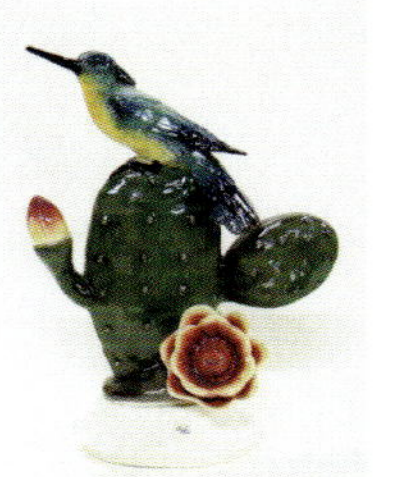

476
Stehendes Mädchen mit Rehbock (Brüderchen und Schwesterchen) (auf gewölbtem und mit Blattwerk verziertem Rundsockel)
ca. 1929–1939
476,5 gem. lig. SS., GS (Firmenstempel Marcell 1), H. 37,5 cm; Wien Museum Inv. 190.097

478
Tanzender Akkordeonspieler (auf Ovalsockel)
ca. 1929–1939
478,3,8 GS (Firmenstempel Marcell 1), AUSTRIA, H. 20,5 cm; PB Wien 19

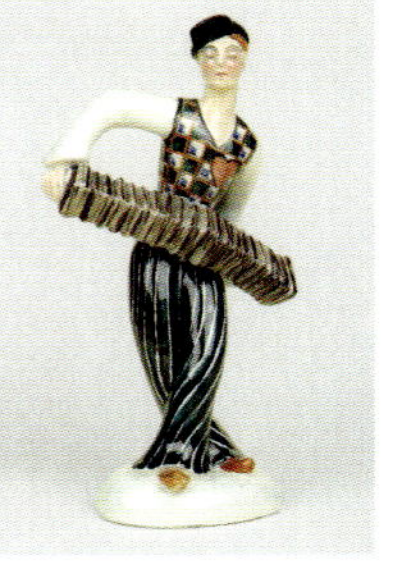

484
Stehender Liliputaner mit Spitzhut und Flöte auf einem Kugelfisch
ca. 1929–1939
484 GS (Firmenstempel Marcell 1), H. 25 cm; Sammlung Dr. Peter Hieke, Wien

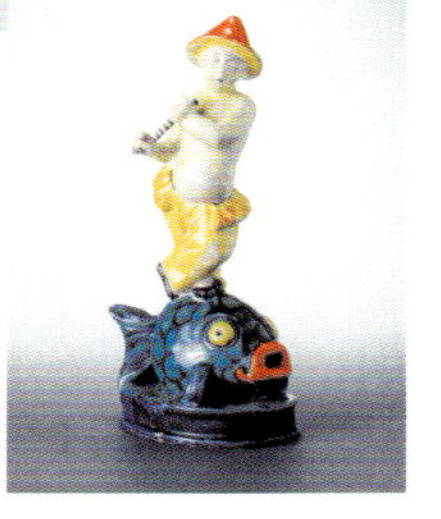

487
BÜSTE Frauenkopf mit Vogel auf der Schulter
ca. 1929–1939
487,10,1 GS (Firmenstempel Marcell 1), AUSTRIA, H. 23 cm; Wien Museum Inv. 186.672
Ausst.: SSH 1992

493
Stehende junge Dame im Wind (ihren Hut mit beiden Händen festhaltend) (auf Landschaftssockel)
ca. 1929–1939
493 (Firmenstempel ohne Ang.), H. 27,3 cm; Doro 13.9.2004

495
Tanzende Mexikanerin
ca. 1929–1939
495,53,1 GS (Firmenstempel Marcell 1), H. 30,5 cm; Doro 12.9.2006
Pendant –> vgl. 496

496
Gitarre spielender Mexikaner
ca. 1929–1939
496,49,1 GS (Firmenstempel Marcell 1), H. 30,5 cm; Doro 12.9.2006
Pendant –> vgl. 495

499
Sitzender, kurzhaariger Dachshund (Dackel)
ca. 1929
499,0 ger. W, GS (Firmenstempel Marcell 1), AUSTRIA, H. 20 cm; Sammlung a@goldscheider.net

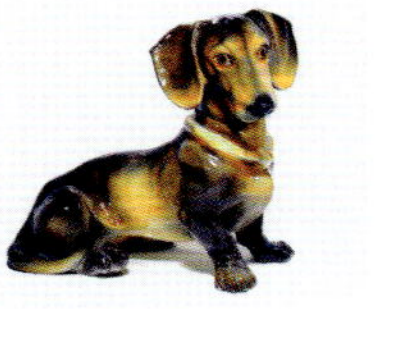

502
Rottenberg, Ena (Emma Helena)
VASE
Form: halbkugelartige Schüssel
Dekor: zwei Mädchen im Blattwerk
ca. 1929–1939
502 GS (Firmenstempel Marcell 1), H. 14,5 cm; Doro 15.5.2002; 10.3.2003

510
BUCHSTÜTZE Jockey
ca. 1929–1939
510,11,2 GS (Firmenstempel Marcell 1), AUSTRIA, H. 23 cm; eBay # 200045120110

514
Stehende Tänzerin (im Kostüm einer Biene) (auf an der Schauseite abgeschrägtem Quadratsockel)
ca. 1929–1939
514,40 GS (Firmenstempel Marcell 1), AUSTRIA, H. 32,5 cm; GK, S. 27, 48 (m. Abb.); Wien Museum Inv. 222.554
Ausst.: HMW (jetzt Wien Museum) 1985/86.
Lit.: GK, S. 27, 48 (m. Abb.); GSN, S. 40, 41 (m. Abb.)
GS neu: Modell-Nr. 26 „Bienenmädchen" (H. 25 cm), Aufl. ohne Ang.

525
Knörlein, Rudolf
WANDMASKE Frauenkopf mit Schleife
ca. 1929–1932
525,36,1 GS (Firmenstempel Marcell 1), AUSTRIA, gem. sign. R. Knörlein, H. fehlt; eBay # 7396692818

535
Reiterin (im Damensitz) auf buntem geflecktem Pferd (auf ovalem Natursockel)
ca. 1929–1939
535,1 GS (Firmenstempel Marcell 1), AUSTRIA, H. 26 cm; Doro 9.11.1993

542
Kostial, Josef
Sitzendes Mädchen (in halblanger Hose) mit Violine (auf Ovalsockel)
ca. 1929–1939
542,3,1 gem. S, GS (Firmenstempel Marcell 1), AUSTRIA, H. 29 cm; Wien Museum Inv. 190.115

584
Sitzender Schottischer Terrier mit Knochen im Maul
ca. 1929–1939
584 GS (Firmenstempel Marcell 1), AUSTRIA, H. 13,5 cm; Sammlung f@goldscheider.de

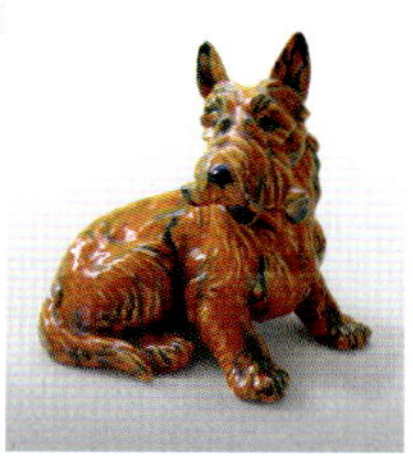

601
Sitzender Yorkshireterrier
ca. 1929–1939
601 ger., GS (Firmenstempel Marcell 1), H. 14,5 cm; WKA 33

602
Tänzerin mit Tuch und Ball (balancierend)
ca. 1929–1939
602 GS (Firmenstempel Marcell 1), AUSTRIA, H. ca. 25 cm; WKH
Bem.: Zu diesem Objekt können keine näheren Ang. gemacht werden, da die Daten als mündliche Mitteilung aus dem WKH übernommen wurden.

610
VASE
Form: kugelförmig, kurze gewulstete Mündung
Dekor: Frauengesichter (im Profil), Blumen, Blüten- und Blattwerk, reliefiert terrakottafarben
ca. 1929–1939

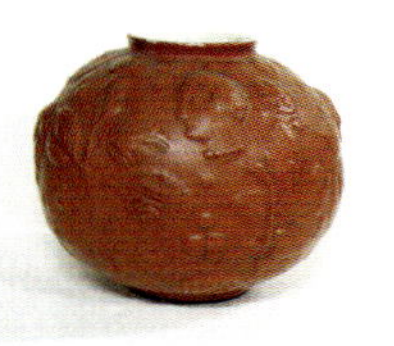

610 (ohne Firmenstempel), H. 22 cm, Ø 27,5 cm; MAK Wien, Inv. Ke 10602-30
rubinrot/goldstaffiert
610 (ohne Firmenstempel), H. 24 cm, Ø 27,5 cm; Sammlung John Nash
Ausst.: PSK 1995.
-> vgl. 1017
Lit.: für die mehrfarbige Variante, GK, S. 27, 136 (m. Abb.); ADGS, Nr. 42 (m. Abb.).

1013
LAMPENFUẞ
Form: sternförmig, kubistisch
ca. 1930–1939
1013 ger., (ohne Firmenstempel), H. 21 cm; Sammlung Dr. Peter Hieke, Wien

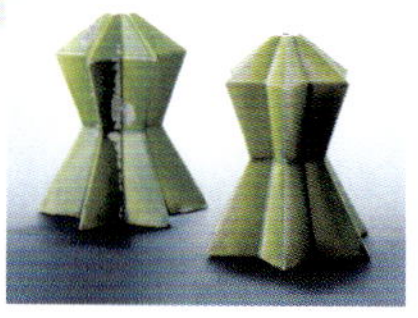

ohne Nr., GS (Firmenstempel Marcell 1), H. 21 cm; Sammlung Dr. Peter Hieke, Wien

1016
LAMPENFUẞ
Form: kugelförmig
Dekor: mehrfarbig, Laufglasur
ca. 1930–1939
1016,113,1 GS (Firmenstempel Marcell 1), AUSTRIA, H. 24 cm; eBay # 260049425876

1017
VASE
Form: kugelförmig, kurze gewulstete Mündung
Dekor: Frauengesichter (im Profil), Blumen, Blüten- und Blattwerk, reliefiert, mehrfarbig
ca. 1930–1939

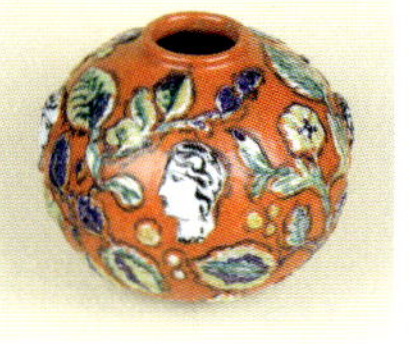

1017,15,2 GS (Firmenstempel Marcell 3), AUSTRIA, H. 24 cm, Ø 26 cm; GK, S. 27, 136 (m. Abb.); Wien Museum Inv. 223.741
ohne Nr., GS (Firmenstempel Marcell 1), H. 23,5 cm, Ø 27,2 cm; MAK Wien, Sammlung GS, Wien, Inv. G 10
Ausst.: HMW (jetzt Wien Museum) 1985/86; PSK 1995.
-> vgl. 610
Lit.: GK, S. 27, 136 (m. Abb.); ADGS, Nr. 42 (m. Abb.).

1024
LAMPENFUẞ
Form: kürbisförmig
Dekor: Landschaft mit Häusern
ca. 1930–1939
1024,10 GS (Firmenstempel Marcell 1), AUSTRIA, H. 24 cm; PB Wien 19

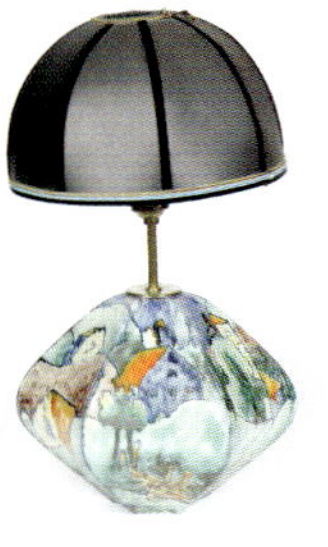

1200
SPIEGEL/BILDERRAHMEN
Form: rechteckig, Fahne mit geometrische Formen (Kreise, Bögen, Zacken)
ca. 1930–1939
1200,8 GS (Firmenstempel Marcell 1), H. 17 cm; PB

2451
BUCHSTÜTZE Sitzender Terrier
ca. 1930–1939
2451,93,1 ger., ger. B, GS (Firmenstempel Marcell 1), AUSTRIA, H. 15 cm; PB Wien
2451,97,1 ger., ger. B, GS (Firmenstempel Marcell 1), AUSTRIA, H. 15 cm; PB Wien

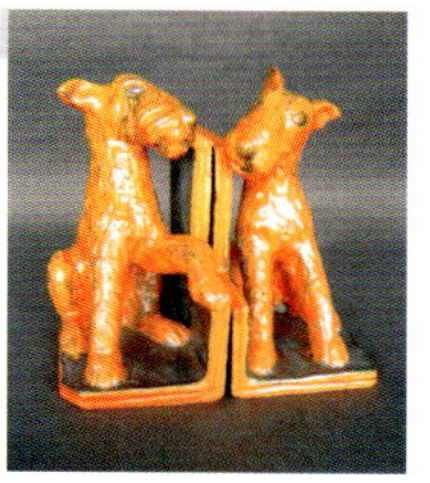

Bem.: Die beiden Stützen unterscheiden sich dadurch, dass einer der Hunde seine linke Vorderpfote hebt.

Marcel Goldscheider, Stoke-on-Trent

Ausführung der Modelle zwischen 1939–1960
Vorwiegend exemplarisch gezeigt an der Sammlung a@goldscheider.net.

–> vgl. Kapitel „Marcell Goldscheiders Werkstätten in Wien und Stoke-on-Trent – Emigration und Art Pottery in England (1939–1960)"

Models made 1939–1960
Most are examples shown at the collection a@goldscheider.net.

–> cf chapter "Marcell Goldscheider's Workshops in Vienna and Stoke-on-Trent – Emigration and Art Pottery in England (1939–1960)"

MODELLNUMMER | MODEL NUMBER

71-UK-Staffordshire GROUP Hares Trio (on a Base) (H. 5,5 cm, Ø 9,1 cm) Sammlung a@goldscheider.net

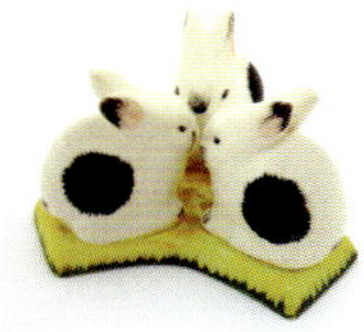

88-UK-Staffordshire TRAY (for Salt and Pepper, with a Girl Bust) (H. 9,5 cm, Ø 15,5 cm) Sammlung a@goldscheider.net

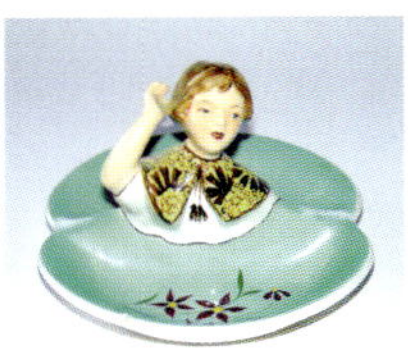

352-UK-Myott VASE (with Antique Figurines) (H. 44,8 cm) Sammlung a@goldscheider.net

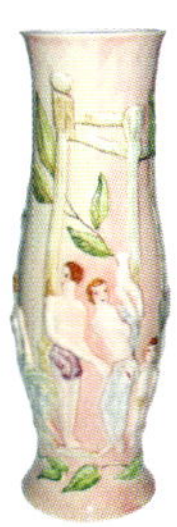

359-UK-Myott FIGURINE Hummingbird (H. 12 cm) eBay # 35077710

429-UK-Staffordshire / Myott FIGURINE „Kiss, Lovers" (H. fehlt) (Marcel Goldscheider)

439-UK-Myott GROUP Rococo Pair Dancing (Oval Base) (H. 33 cm) eBay # 64782480

447-UK-Myott GROUP Puppy Twins (H. 9,2 cm) Sammlung a@goldscheider.net

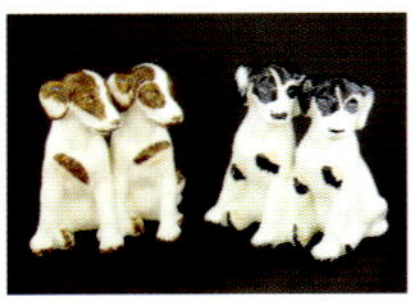

495-UK-Myott FIGURINE George Formby in a Mexican Costume I. (H. 29 cm) Sammlung a@goldscheider.net

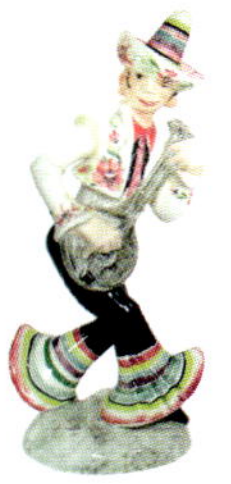

496-UK-Myott FIGURINE George Formby in a Mexican Costume II. (H. 29,6 cm) Sammlung a@goldscheider.net

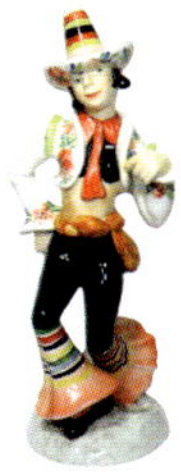

586-UK-Myott BOOKEND Scotch Terrier (H. 14,5 cm; L. 16,5 cm) Sammlung a@goldscheider.net

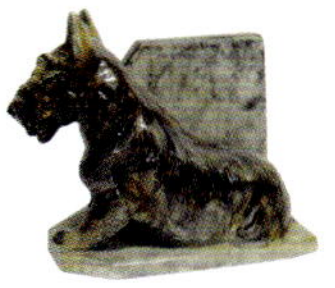

609-UK-Myott FIGURINE Owl (H. 15 cm) Sammlung a@goldscheider.net

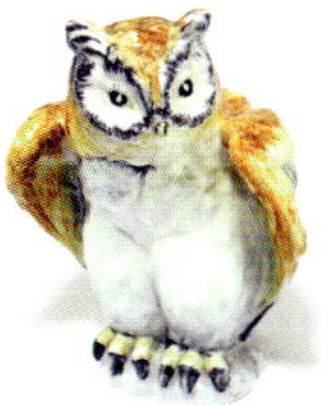

646-UK-Myott BOWL (with Two Kissing Fish) (H. 20,2 cm; L. 20,8 cm) Sammlung a@goldscheider.net

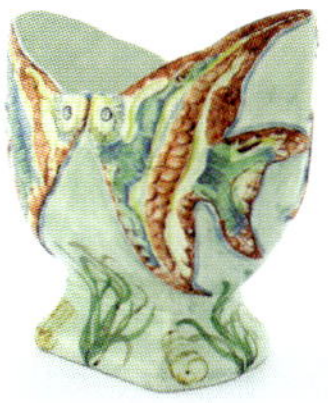

654? s. 854

657? s. 857

668-UK-Myott VASE (with Two Sails) (H. 17,5 cm) Sammlung a@goldscheider.net

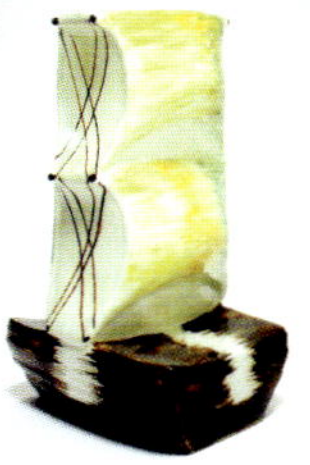

706-UK-Myott FIGURINE Boy in Shorts (H. 11 cm) Sammlung a@goldscheider.net

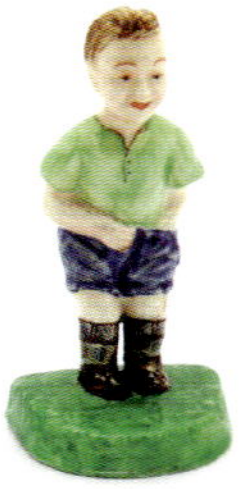

709-UK-Myott HEAD / WALLPLAQUE Girl with a Hat and an Apple (H. 20,3 cm) Sammlung a@goldscheider.net

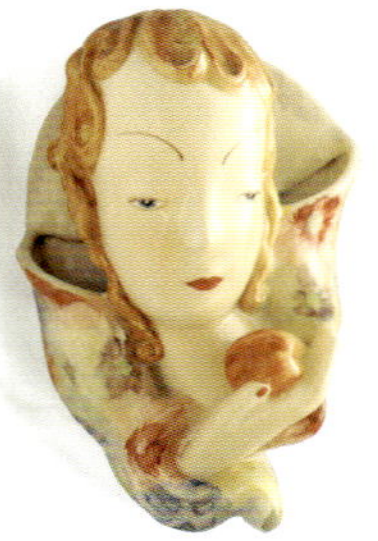

709-UK-Staffordshire HEAD / WALLPLAQUE Girl with a Scarf and an Apple (H. 18,5 cm) Sammlung Anne Myott & Philip Pollitt

710-UK-Staffordshire / Myott FIGURINE „Lady Dancer" (H. fehlt) (Marcel Goldscheider)

727-UK-Staffordshire FIGURINE Girl with a Lamb (oval Base) (H. 15 cm) Sammlung a@goldscheider.net

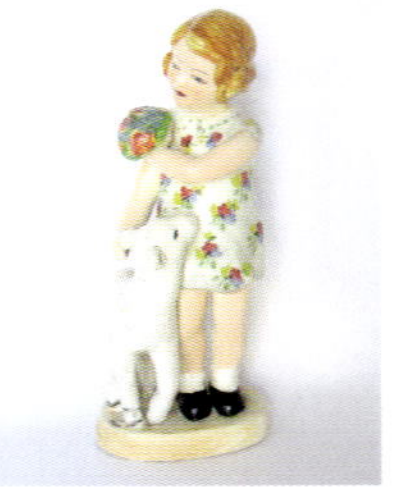

728-UK-MYOTT WALLPLAQUE Winston Churchill with Cigar (H. fehlt)

734-UK-Myott FIGURINE Lady with a Basket (H. 20 cm) Sammlung a@goldscheider.net

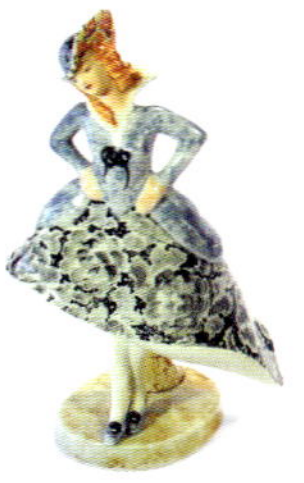

738-UK-Myott VASE with Three Cones (and One Handle) (H. 13,3 cm) Sammlung a@goldscheider.net

753-UK-Myott FIGURINE Boy with a Hat and Suitcase (H. 21,8 cm) Sammlung a@goldscheider.net

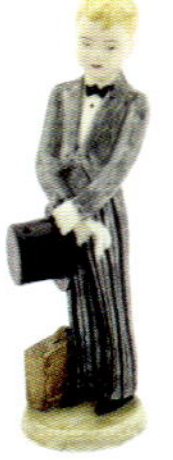

765-UK-Myott TRAY (for Salt and Pepper, with a Girl Bust) (H. 10 cm, Ø 15,5 cm) Sammlung a@goldscheider.net

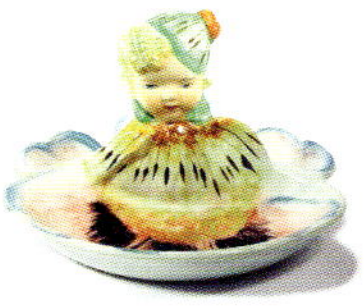

774-UK-Myott VASE (H. 23 cm) Sammlung a@goldscheider.net

780-UK-Myott GROUP „Toucans" (H. 10,7 cm) Sammlung a@goldscheider.net

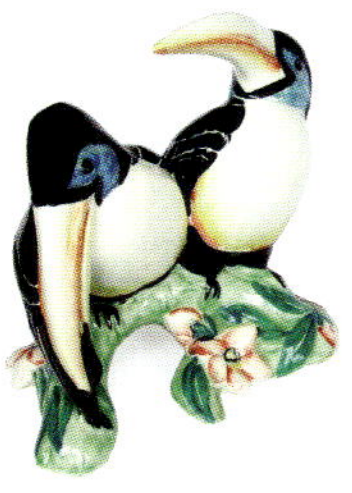

782-UK-Myott TRAY (with a Girls Head Relief) (H. 4,2 cm, L. 12,2 cm) Sammlung a@goldscheider.net

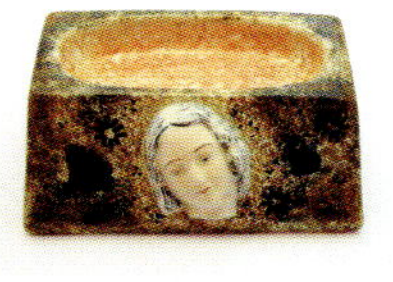

788-UK-Myott VASE (with a Sitting Woman) (H. 15,5 cm, Ø 10,4 cm) Sammlung a@goldscheider.net

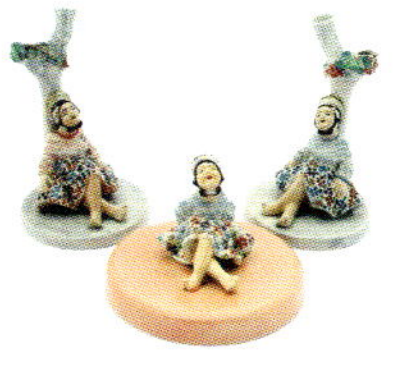

796-UK-Myott WALL-PLAQUE Girl with Scarf (H. 17,5cm) Sammlung Anne Myott & Philip Pollitt

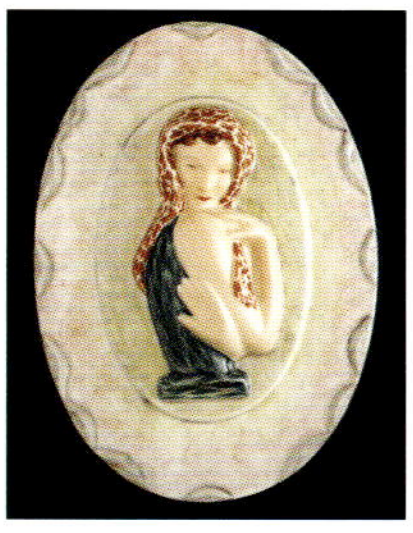

803-UK-Myott BOX (Lidded Box) with a Woman Relief (H. 6,7 cm, L. 14,1 cm, B. 9,4 cm) Sammlung a@goldscheider.net

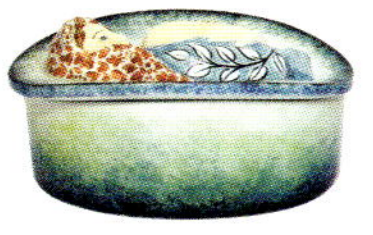

817-UK-Myott FIGURINE Man (after Charles Dickens) (H. 14 cm) Sammlung a@goldscheider.net

818-UK-Staffordshire FIGURINE Duck (H. 14,4 cm) Sammlung a@goldscheider.net

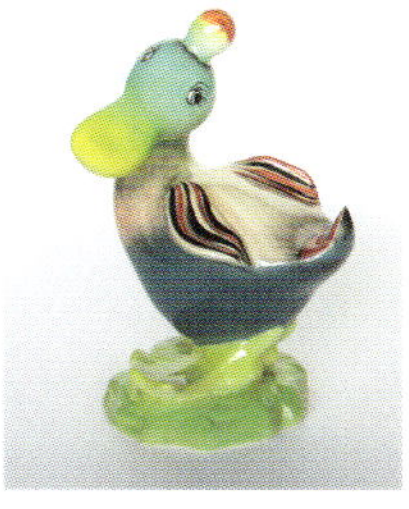

822-UK-Staffordshire FIGURINE Cat (H. 11,5 cm)

834-UK-Staffordshire FIGURINE Nude (H. 32,3 cm) Wien Museum Inv. 190.134

835-UK-Myott BOOKEND Sitting Boy with a Ball (left) (H. 14 cm) Sammlung a@goldscheider.net

835-UK-Myott BOOKEND Sitting Boy with a Ball (right) (H. 14 cm) Sammlung a@goldscheider.net

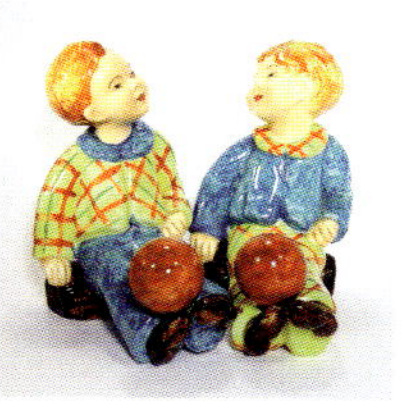

844-UK-Myott FIGURINE Elephant (Oval Stand) (H. 12,1 cm) Sammlung a@goldscheider.net

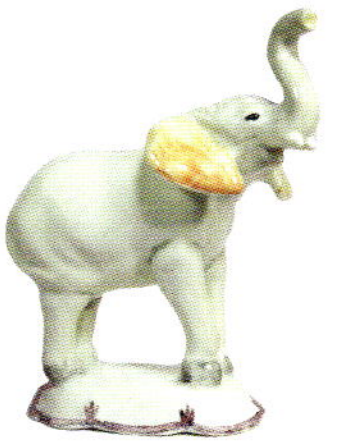

847-UK-Myott FIGURINE Bulldog (H. 12,7 cm) eBay # 31911911

854?-UK-Myott VASE (with a Sail) (H. 16 cm) Sammlung a@goldscheider.net

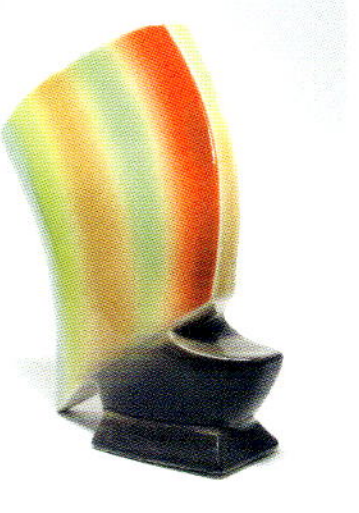

857?-UK-Staffordshire VASE (with a Sail) (H. 22,5 cm) Sammlung a@goldscheider.net

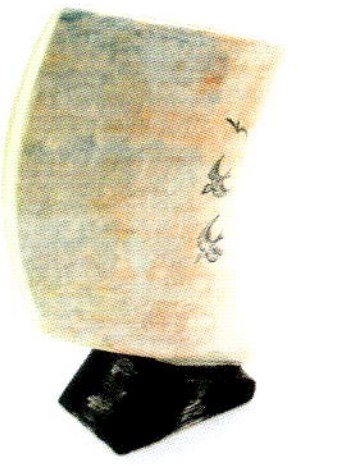

861-UK-Myott FIGURINE Girl Playing Violin (Oval Base) (H. fehlt) eBay # 41966388

872-UK-Staffordshire FIGURINE „Barker" (after Charles Dickens) (H. 12,6 cm) Sammlung a@goldscheider.net

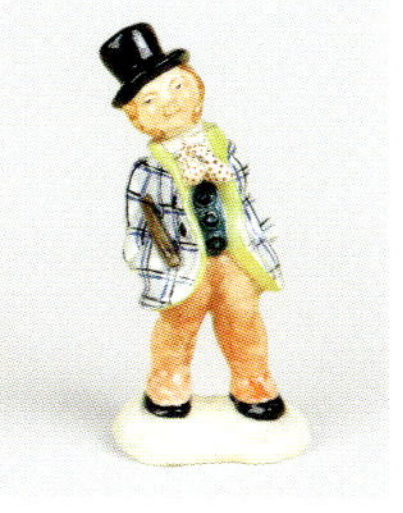

874-UK-Staffordshire FIGURINE Rococo Woman with a Basket (H. 21,6 cm) eBay # 61399411

886-UK-Myott FIGURINE Rabbit (H. 15 cm) eBay # 220023048749

893-UK-Staffordshire FIGURINE „Captured Bird" (H. 27,5 cm)

899-UK-Staffordshire / Myott FIGURINE „Blue Boy" (H. fehlt) (Marcel Goldscheider)

901-UK-Staffordshire FIGURINE Violinist Sitting (H. 16,5 cm) Sammlung a@goldscheider.net

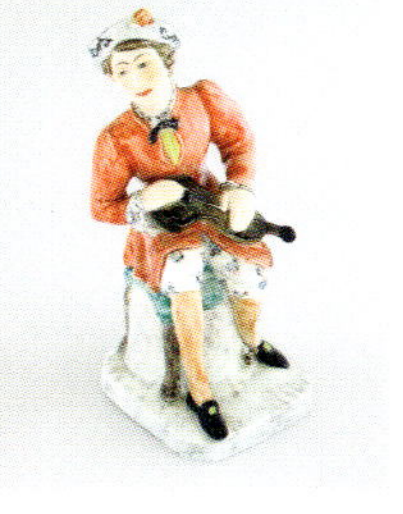

914-UK-Staffordshire FIGURINE „Mr. Squeers" (H. 16,5 cm) Sammlung a@goldscheider.net

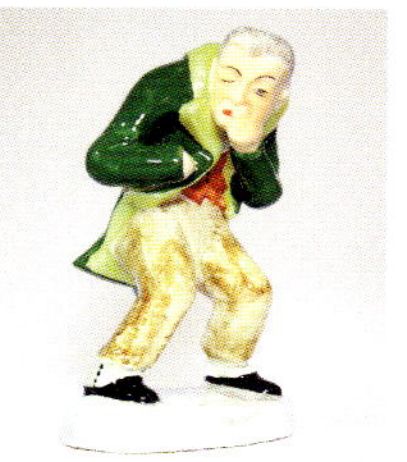

942?-UK-Staffordshire HOUSE „The Old Curiosity Shop immortalised by Charles Dickens". (H. 15,5 cm, L. 19,5 cm, B. 9,5 cm) Sammlung a@goldscheider.net

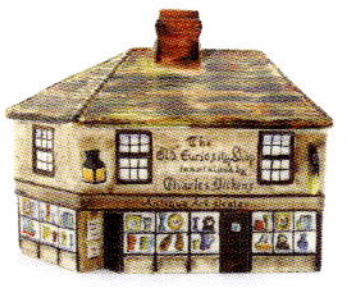

949? s. 942

950-UK-Staffordshire HOUSE „Anne Hathaway's Cottage" (H. 13,5 cm, L. 12,5 cm, B. 9 cm) Sammlung a@goldscheider.net

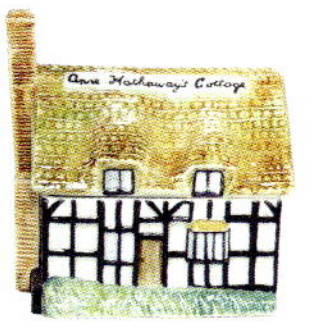

OHNE NUMMERN | WITHOUT NUMBERS

UK-Myott BOWL (Shamrock Shape) (H. 5,1 cm, L. 12 cm; B. 11,7 cm) Sammlung a@goldscheider.net

UK-Myott FIGURINE Pony (Jumping) (H. 20,5 cm) Sammlung a@goldscheider.net

UK-Myott FIGURINE Pony (Rearing) (H. 15 cm) Sammlung a@goldscheider.net

UK-Myott FIGURINE Scotch Terrier (H. 8,1 cm) Sammlung a@goldscheider.net

UK-Myott FIGURINE Young Girl Walking (H. 18 cm) Sammlung Anne Myott & Philip Pollitt

UK-Myott FIGURINE Young Boy with Hat (H. 22,7 cm) Sammlung Anne Myott & Philip Pollitt

UK-Staffordshire FIGURINE Rabbit (H. 6,1 cm, L. 11,5 cm) Wien Museum Inv. 189.043

UK-Staffordshire „St. Paul's Cathedral" (H. fehlt)

UK-Staffordshire „Tower Bridge" (H. fehlt)

UK-Staffordshire „Tower of London" (H. fehlt)

UK-Staffordshire „Windsor Castle" (H. fehlt)

UK-Myott TRAY (Fish Shape) (H. 3 cm, L. 13 cm; B. 8 cm) Sammlung a@goldscheider.net

UK-Staffordshire GROUP Butter Cart (H. 11,5 cm, L. 19,5 cm; B. 10,5 cm) Sammlung a@goldscheider.net

UK-MUG (Horse Head Relief) (H. 11,8 cm, L. 13,3 cm) Sammlung a@goldscheider.net

UK-Myott VASE for Advertising (Oval, with Three Tubes) „Made in England by Goldscheider XX with Myott Son & Co, Staffordshire (H. 11,4 cm) eBay # 26401026

KÜNSTLER | ARTISTS

Goldscheider, Marcel

429 UK-Staffordshire / Myott FIGURINE „Kiss, Lovers" (H. fehlt)

710 UK-Staffordshire / Myott FIGURINE „Lady Dancer" (H. fehlt)

899 UK-Staffordshire/Myott FIGURINE „Blue Boy" (H. fehlt)

Halnon, Frederick James

ohne Nr. UK-Staffordshire / Myott FIGURINE „Nell Gwynne" (H. fehlt)

Lorenzl, Josef (nach)

893-UK-Staffordshire FIGURINE „Captured Bird" (H. 27,5 cm)

Anhang | Appendix

Künstlerverzeichnis | List of Artists

Abschlag
Ein Bildhauer des Namens Abschlag ist in keinem einschlägigen Kunstlexikon nachweisbar.
3251 UHR Der Abschied (Junger Ritter in Rüstung auf Pferd, sich von seinem Mädchen verabschiedend)
3780 s. 3789
3789 Stehender Ziegenbock (auf Rechtecksockel)

Albert-Lefeufre, Louis Etienne Marie
*Louis E.M. Albert-Lefeufre, Bildhauer (*1843 Paris, †13.8.1924 Paris), war Schüler von Dumont und Falguière und bekannt für seine patriotischen Themen sowie für seine Arbeiterdarstellungen. Er stellte zwischen 1875 und 1905 im Salon des Artistes Français, Paris aus. Lit.: TB, Bd. 1, S. 189; Vollmer, Bd. 5, S. 234; Bénézit, Bd. 1, S. 153; AKL, Bd. 2, S. 62.*
„Messidor" Paris, Salon (1902) Lit.: Neuwirth (1974), S. 59

Alloi (Léonce Alloy?)
Nachweisbar ist nur ein franz. Medailleur namens Léonce Alloy. Er stellte zwischen 1898 und 1925 im Salon des Artistes Français, Paris aus. Lit.: TB, Bd. 1, S. 323; Vollmer, Bd. 5, S. 234; AKL, Bd. 2, S. 62.
VASE „Etrusque" ÖMKI (1900/01) Lit.: Neuwirth (1974), S. 59

Allouard, Henri Emile
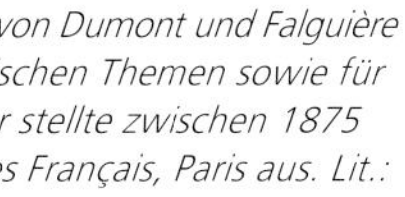
*Henri Emile Allouard, Bildhauer und Aquarellmaler (*11.7.1844 Paris, †12.8.1929 Paris), war Schüler von Lequesne und Schoenewerk. Allouard war ein angesehener und äußerst produktiver Künstler, der auch zahlreiche öffentliche Aufträge ausführte. Lit.: TB, Bd. 1, S. 323; Vollmer, Bd. 5, S. 238; Neuwirth (1974), S. 59; Hillenbrand, S. 7; Bénézit, Bd. 1, S. 223; AKL, Bd. 2, S. 563.*
1441 BÜSTE „Rococo" (Junge Dame nach rechts blickend) (ÖMKI 1898/99) (vgl. 2023; 2079)
2023 BÜSTE „Rococo" (Junge Dame nach rechts blickend) (ÖMKI 1898/99) (vgl. 1441; 2079)
2079 BÜSTE „Rococo" (Junge Dame nach rechts blickend) (ÖMKI 1898/99) (vgl. 1441; 2023)

Alloy s. Alloi

Althan, Josef (d.i. Josef Lorenzl) s. Lorenzl, Josef
Die in der Zeitschrift MW 1928/29 auftauchende falsche Namensform „Althan" für Josef Lorenzl rührt daher, dass er ab 1928 am Althanplatz 6, Wien IX, wohnhaft war. Ob die Firma diesen Namen in der Presse lanciert hat oder ob es sich um einen redaktionellen Fehler des Autors handelte, bleibt ein Rätsel. Es sind jedenfalls bislang keine mit „Althan" signierten Stücke bekannt. Ab 1940 wurde Lorenzls Wohnort umbenannt in Platz der Sudetendeutschen, 1954 bestand dann laut Lehmann noch folgende Adresse: An der unteren alten Donau 441. Lit.: N.N.: Berühmte Frauen und Keramik, in: MW, 10.1929/30, Nr. 6, S. 28 (m. Abb.); Lehmann, 1928–1954.

Amy, Jean Barnabé
*Jean Barnabé Amy, Bildhauer (*11.6.1839 Tarascon, vor dem †24.3.1907 Tarascon), war Schüler von Dumont und Bonnasieux. Er stellte in den Jahren 1868–1907 im Pariser Salon (Salon des Artistes Français) aus und erhielt bei der Weltausstellung in Paris eine lobende Erwähnung für einige seiner ausgestellten Masken. Amy war vor allem Porträtist und schuf in seinen letzten zehn Lebensjahren mehr als 200 Masken von Gestalten aus der Sagenwelt, Mythologie und Religion. Lit.: TB, Bd. 1, S. 425; Bénézit, Bd. 1, S. 281; AKL, Bd. 3, S. 307–308; Neuwirth (1974), S. 59.*
1884 BÜSTE „Defi" (Kleines Mädchen mit Kapuze) ÖMKI (1899/1900; 1900/01) (vgl. 1955= WANDMASKE)
1955 WANDMASKE „Defi" (Kleines Mädchen mit Kapuze) ÖMKI (1899/1900; 1900/01) (vgl. 1884= BÜSTE)

Anglis, J.
J. Anglis war Mitarbeiter/in von Friedrich Goldscheider, Paris. Weitere Daten zu Leben und Schaffen sind nicht bekannt. Lit.: GSB, Pl. 3 (m. Abb.).
503-GSB SCHALE Plateau „Amour Chasseur" (Plateau en vieil Étain)

Anthouard, Magdeleine
Magdeleine Anthouard war Mitarbeiterin von Arthur Goldscheider, Paris. Lit.: Goldscheider, A. (Hrsg.): Artistic Evolution, 1926, T. 1, Pl. 16 (m. Abb.).
VASE (ausgeführt von Edouard Stabi, nach einem Entwurf von M. Anthouard) Metall (lackiert, schwarz/gold) (H. 28,5 cm, B. 12,5 cm) Lit.: Goldscheider, A. (Hrsg.): Artistic Evolution, 1926, T. 1, Pl. 16 (m. Abb.)
VASE (ausgeführt von Edouard Stabi, nach einem Entwurf von M. Anthouard) Metall (lackiert, blau/gold) (H. 18 cm, B. 9 cm) Lit.: Goldscheider, A. (Hrsg.): Artistic Evolution, 1926, T. 1, Pl. 16 (m. Abb.)
LAMPE „Ecrans Lumineux" (Luminous Screens) (Form: quadratisch) BRONZE/Aragonite (H. 30 cm, B. 31,5 cm) Lit.: Goldscheider, A. (Hrsg.): Artistic Evolution, 1926, T. 1, Pl. 25 (m. Abb.)
LAMPE „Ecrans Lumineux" (Luminous Screens) (Form: trapezförmig) BRONZE/Aragonite (H. 33 cm, B. 13,5 cm) Lit.: Goldscheider, A. (Hrsg.): Artistic Evolution, 1926, T. 1, Pl. 25 (m. Abb.)

Arnold, E.
E. Arnold, Bildhauer, war Mitglied der Künstlergruppe „La Stèle" (-> vgl. Kap. „Dekorative Skulpturen von Arthur Goldscheider in Paris"). Lit.: Neuwirth (1974), S. 59.

Arnoux
Arnoux war Mitarbeiter/in von Friedrich Goldscheider, Paris. Weitere Daten zu Leben und Schaffen sind nicht bekannt. Lit.: GSB, Pl. 72 (m. Abb.).
772-GSB „Caresses"
881-GSB „Éléphant"

Auber
3129 Stehende Holländerin mit zwei Körben

Augmüller, Ludwig
*Ludwig Augmüller, Keramiker und Bildhauer (*9.6.1876 Wien, †?), studierte von 1894–1902 unter Zumbusch und von Hellmer in Wien an der Akademie der bildenden Künste und war ab 1908 Lehrer an der K.K. Fachschule für Tonindustrie in Znaim. Augmüller war 1907 Stubenring 2, im I. Bez. und 1908 noch wohnhaft in Wien III., Schützengasse 10. Lit.: Lehmann, 1907–1908; Zühlsdorff, S. 394; AKL, Bd. 5, S. 631.*
3643 JARDINIERE
VASE „Jeu des Sirènes" Paris, Salon (1902) Lit.: Neuwirth (1974), S. 59

Bacqué, Daniel-Joseph
*Daniel-Joseph Bacqué, Bildhauer (*20.9.1874 Vianne/Lot et Garonne, †21.12.1947 Paris), war Mitglied der Künstlergruppe „La Stèle" (-> vgl. Kap. „Dekorative Skulpturen von Arthur Goldscheider in Paris"). Er schuf vor allem Bildnisbüsten und Statuetten sowie weibliche Akte und stellte in den Jahren 1900–1939 im Pariser Salon (Salon des Artistes Français) aus, wo er in den Jahren 1910, 1911 und 1922 jeweils mit Medaillen geehrt wurde. Lit.: Vollmer, Bd. 1, S. 91; Bd. 5, S. 263; Neuwirth (1974), S. 59; Hillenbrand, S. 17; Bénézit, Bd. 1, S. 618; AKL, Bd. 6, S. 189.*

Badin, Jean-Victor
Jean-Victor Badin, Bildhauer (? Toulouse, †?), war Schüler von A. Falguière und M. L. V. Mercier. Badin war Mitarbeiter von Friedrich Goldscheider, Paris. Er war beteiligt an der Weltausstellung 1900 in Paris und stellte in den Jahren 1897–1904 in den Pariser Salons (Salon des Artistes Français) und von 1927–1938 im Salon d' Automne aus. Lit.: TB, Bd. 2, S. 336; Vollmer, Bd. 1, S. 91; Bénézit, Bd. 1, S. 623; AKL, Bd. 6, S. 219.
881-GSB „Éléphant"

Bagge, Eric
*Eric Bagge, Innenarchitekt und Kunstgewerbler (*10.9.1890 Antony/Seine, †?), lehrte an der École des Arts Décoratifs in Paris Architektur und war Mitglied der Künstlergruppe „Évolution" (-> vgl. Kap. „Dekorative Skulpturen von Arthur Goldscheider in Paris"). Er entwarf den Pavillon für „La Stèle" und „Évolution" auf der KGW-Ausst. Paris 1925. Lit.: Goldscheider, A. (Hrsg.): Artistic Evolution, 1926, T. 1, Pl. 1–2 (m. Abb.); Vollmer, Bd. 1, S. 95; Neuwirth (1974), S. 59; Bénézit, Bd. 1, S. 634.*
Vitrine en fer forgé METALL (ausgeführt von Edouard Schenck) (H. 91 cm, B. 76 cm) Lit.: Goldscheider, A. (Hrsg.): Artistic Evolution, 1926, T. 1, Pl. 17 (m. Abb.)

Bakardziev, Georgi Stefanov (Bakardjeff, Georges)
*Georgi S. Bakardziev, Kunstgewerbler (*30.3.1899 Karlovo, †22.1.1972 Sofia). Der bulgarische Keramiker, Glas- und Porzellankünstler lebte in Paris, wo er in den Jahren 1919–1940 seine künstlerische Ausbildung als Glaskünstler erhielt. Bakardziev stellte im Salon 1928 in der Abteilung von Arthur Goldscheider eine Vitrine mit Gläsern (teils emailliert) aus. Er erhielt zahlreiche bulgarische Auszeichnungen, unter anderem den Titel Verdienter Künstler, als er 1940 nach Sofia übersiedelte, sich von seinem „Style moderne" der Pariser Zeit distanzierte und sich ausschließlich seiner heimatlichen Töpfertradition widmete. Lit.: Neuwirth (1974), S. 59; Bénézit, Bd. 1, S. 653; AKL, Bd. 6, S. 347.*

Barabino, Armando (nach)
*Armando Barabino, Maler und Bildhauer (*16.9.1883 Genua, †10.6.1970 Genua), studierte an der Accademia Ligustica Genua Malerei bei Quinzio, seit 1922 Bildhauerei bei Scanzi. Lit.: Vollmer, Bd. 1, S. 109; Neuwirth (1974), S. 59; AKL, Bd. 6, S. 609.*
618-GSB PLAQUE „Vierge et enfant Jésus"
1875 GRUPPE „Madonna mit Kind" (Sitzende Gottesmutter mit Jesuskind, auf getrepptem Steinsockel) (vgl. 1879= BÜSTE)
1879 BÜSTE „Madonna (auf Sockel)" (Rechtecksockel reliefiert mit Krippenszene) (Terrakotta) ÖMKI (1899/1900; 1900/01) (vgl. 1875 Gruppe)

Barfoort
„Pendulette" Paris, Salon (1902) Lit.: Neuwirth (1974), S. 59

Baron, René
René Baron, Bildhauer (? Neuilly-sur-Seine, †?), wird erstmals 1913 und letztmals 1939 erwähnt. Lit.: Bénézit, Bd.1; S. 770; AKL (Archiv, unredigiert).*
505-GSB „STATUE „La Chanson" ZINKGUSS (vgl. 509-GSB) (vgl. 952; 954; 972)
509-GSB „STATUE „La Chanson" BRONZE (vgl. 505-GSB) (vgl. 952; 954; 972)
952 FIGUR „Lied"/„La Chanson" (Stehende Sängerin mit Notenheft) (auf getrepptem Sockel) (vgl. 954; 972) (vgl. 505-GSB; 509-GSB)
972 FIGUR „Lied (mittel)"/„La Chanson" (Stehende Sängerin mit Notenheft) (auf getrepptem Sockel) (vgl. 952; 954) (vgl. 505-GSB; 509-GSB)

954 FIGUR „Lied"/„La Chanson" (Stehende Sängerin mit Notenheft) (auf getrepptem Sockel) (vgl. 952; 972) (vgl. 505-GSB; 509-GSB)

Bate, Firmin
2084 FIGUR „Daphne" Paris, Salon (1902)

Baudry, R. E.
2161 BÜSTE „Silence" ÖMKI (1901/02)
3322 BÜSTE Junge Frau mit Haarband
SPIEGEL Frauenkopf mit langem Haar (Algendekor) Lit.: R., F.: Wiener Kunstkeramik, in: KM, 1902, S. 89 (m. Abb.)

Bauer, Carli (Carl)
*Carli (Carl) Bauer, Keramiker (*10.12.1894 Passau, †17.4.1960 Wien), erlernte das Holzbildhauerhandwerk in Passau und arbeitete in den Jahren 1924–1945 bei der Porzellanmanufaktur Augarten in Wien. Im Jahre 1949 war Bauer als Modelleur bei der Wiener Firma Künstlerkeramik Hermann Heindl tätig und mietete ab 1955 für fünf Jahre die Werkstatt von Gretl Braun (spätere Spiller), wo er ihre erfolgreichen Modelle weiterproduzierte und auch eigene Modelle dem Programm hinzufügte. Bauer erkrankte 1958 an der Keramiker Berufskrankheit Silikose und verstarb zwei Jahre später. Die Werkstatt, die den Namen Carli Bauer bekam, wurde in den 19. Bez., Budinskygasse übersiedelt und konnte noch bis Januar 1970 von seiner Frau weitergeführt werden. Die Erkenntnis, dass die Modelle 14, 23 und 24 von Carli Bauer entworfen wurden, verdanke ich Frau Mag. Uta M. Matschiner. Quelle: Archiv Uta M. Matschiner.*
14-WG „Chinesin lesend" (Sitzende lesende Chinesin)
14L-WG LAMPE „Chinesin lesend" (Sitzende lesende Chinesin)
23-WG „Chinesen" (Sitzende lesende Chinesin mit Gefäß) (Pendant vgl. 24)
23K-WG „Chinesen mit Topf" (Sitzende lesende Chinesin mit Gefäß, für Kakteen)
23L-WG LAMPE „Chinesin lesend" (Sitzende lesende Chinesin)
24-WG „Chinesen" (Sitzender lesender Chinese) (Pendant vgl. 23)
24K-WG „Chinesen mit Topf" (Sitzender lesender Chinese mit Gefäß, für Kakteen)

Beaufils, Armel
*Armel Beaufils, Bildhauer (*28.11.1882 Rennes, †1.5.1978 Paris), war Mitglied der Künstlergruppe „La Stèle" (-> vgl. Kap. „Dekorative Skulpturen von Arthur Goldscheider in Paris"). Er stellte ab 1909 im Pariser Salon (Salon des Artistes Français) aus und erhielt dort 1921 die silberne, im Jahr 1924 die goldene Medaille. Lit.: Vollmer, Bd. 1, S. 144; Neuwirth (1974), S. 59; Bénézit, Bd. 1, S. 446; AKL, Bd. 8, S. 66.*

Beaume (d.i. Bormann, Wilhelm)
Biographie s. Bormann
1486 VASE Putti (Form: zylinder/becherförmig/Dekor: reliefierte Wandung mit Putti und floralen Elementen) (Pendant vgl. 1484) (auch sign. Bormann)
3416 Zwei stehende Windhunde (auf Rechtecksockel) (undeutl. sign. Beaume?) (vgl. 4406= linke Einzelfigur) (Beaume/Bormann zugeschr.) (auch sign. „Lavergne")
3714 GRUPPE „Freundschaft" (Kind mit Hund = Neufundländer) (auch als Deckeldose Modell-Nr. 3714) (vgl. 5908) (vgl. 4014= Hund einzeln)
3878 GRUPPE/LAMPE Zwei stehende Windhunde (Barsois) (auf Rechtecksockel) (vgl. 4407= rechte Einzelfigur) (vgl. 4637= linke Einzelfigur)
4014 Stehender Neufundländer (auf Ovalsockel) (vgl. jeweils GRUPPE 3714; 5908)
4406 Stehender Windhund (auf Ovalsockel) (vgl. 3416= Zweier Gruppe) (Beaume/Bormann zugeschr.)
4407 Stehender Windhund (Barsoi/Greyhound) (auf Ovalsockel) (vgl. 6011) (vgl. 3878= Zweier Gruppe)
4637 Stehender Windhund (Barsoi/Greyhound) (auf Rechtecksockel) (vgl. 3878= Zweier Gruppe)
5908 GRUPPE „Freundschaft" (Kind mit Hund= Neufundländer) (vgl. 3714; vgl. 4014= Hund einzeln)
6011 Stehender Windhund (Barsoi/Greyhound) (auf Ovalsockel) (vgl. 4407) (vgl. 3878= Zweier Gruppe)

Beauvais, Gabriel
Gabriel Beauvais, Bildhauer (Lebensdaten unbekannt, tritt letztmals vor 1930 in Erscheinung), war Mitarbeiter von Arthur Goldscheider, Paris. Lit.: Goldscheider, A. (Hrsg.): Artistic Evolution, 1926, T. 2, Pl. 43 (m. Abb.); Bénézit, Bd. 1, S. 936; AKL (Archiv, unredigiert).
„Chimpanzee" BRONZE (H. fehlt, B. fehlt) Lit.: Goldscheider, A. (Hrsg.): Artistic Evolution, 1926, T. 2, Pl. 43 (m. Abb.)

Beinhofer, Hanns (Johannes)
*Hanns Beinhofer (abweichende Namensform bei Könecke: Johannes Beinhofer), Retoucheur (*4.11.1923 ?, †?) bei Goldscheider in Wien und anschließend bei Carstens auf dem Tönnieshof in Norddeutschland. Er war einer der von Wien aus „übernommenen" Spezialisten, der half, die Facharbeiter auf dem Tönnieshof auszubilden. Ende 1955 kehrte er wieder nach Wien zurück. Lit. und Quelle: AGM, Liste „Adressen von guten und zuverlässigen Arbeitern" (undat., um 1953); Könecke (2002), S. 16.*

Bellaruse? s. Belleuse

Belleuse
4831 Stehender Damenhalbakt (Badende mit großem Tuch) (auf Rundsockel) (undeutl. sign. „Bellaruse"/ „Belleuse"?)

Belville, Jean
Jean Belville, Bildhauer (erwähnt zwischen 1921–1938 in Paris), war Mitglied der Künstlergruppe „La Stèle" (-> vgl. Kap. „Dekorative Skulpturen von Arthur Goldscheider in Paris „). Lit.: Bénézit, Bd. 2, S. 80; AKL (Archiv, unredigiert).

Berger, Josef
*Josef Berger, Architekt (*1898 Wien, †? wird letztmals um 1963 erwähnt), lebte in Wien, wo er in den Jahren 1921–1934 Wohnhausanlagen der Stadt Wien projektierte. Er emigrierte 1934 nach Haifa und übersiedelte 1937 nach London. Berger studierte von 1917–1921 bei Adolf Loos und arbeitete neben Arthur Goldscheider, Paris sowie Friedrich Goldscheider (Weihnachtsschau im Künstlerhaus, Wien 1926), auch für die Bimini Werkstätten-Gesellschaft für angewandte Kunst in Wien. Berger war 1928 wohnhaft in Wien III., Boerhavegasse 7. Lit.: Lehmann, 1928; Neuwirth (1974), S. 59; Zühlsdorff, S. 642; AKL, Bd. 9, S. 356.*
KAMIN (schlangenförmige und florale Ornamente) Lit.: N.N.: Österreich in Paris in: ÖBWK, 1.1924/25, S. 331 (m. Abb.)

Berthou(d), (Paul François)
*(Paul François) Berthoud, (*17.5.1870 Paris, †1939 Paris), war unter anderem Schüler von Antonin Larroux. Er beschickte zwischen 1889 und 1902 den Pariser Salon (Salon des Artistes Français, 1898 „Ehrenvolle Nennung") und zwischen 1907 und 1931 den Salon d' Automne, bei dem er als Gründungsmitglied mitwirkte. Lit.: Neuwirth (1974), S. 59; Bénézit, Bd. 2, S. 218; AKL, Bd. 10, S. 82.*
„Gavroche" (Maske) ÖMKI (1898/99; 1900/01) Lit.: Neuwirth (1974), S. 59

Bertoni, Wander
*Wander Bertoni, Bildhauer (*11.10.1925 Codisotto/ Reggio Emilia), lebt seit 1943 in Wien. Studium in den Jahren 1946–1952 bei Fritz Wotruba. Bertoni ist Gründungsmitglied des „Art Club" (ab 1946), seit 1965 Mitglied der Wiener Sezession und übernahm die Leitung einer Meisterklasse an der Akademie für angewandte Kunst, Wien. Unzählige Ausstellungen und Ausstellungsbeteiligungen u.a. in Wien, Biennale Venedig, Triennale Milano, London usw. Zahlreiche Werke im öffentlichen Raum. Lit.: Vollmer, Bd. 5, S. 301; Bénézit, Bd. 2, S. 227; AKL, Bd. 10, S. 130.*
„Der Büffel" („The Golden Taurus") Fayence in drei Variationen (Aufl. insgesamt 450 Stück) GSG Vienna (um 1991/92) (H. 22 cm) Lit.: Firmen-Kat./Folder: Wander Bertoni, Goldscheider Vienna (Stoob/Bgld. um 1992) o. Pag. (Umschlag Vorderseite, Innenseite jeweils m. Abb.)
„Der reiche Bettler" („The rich Vagrant") Fayence in drei Variationen (Aufl. insgesamt 450 Stück) GSG Vienna (um 1991/92) (H. 52 cm) Lit.: Firmen-Kat./Folder: Wander Bertoni, Goldscheider Vienna (Stoob/Bgld. um 1992) o. Pag. (Innenseite m. Abb.)
„Lingam" („The Lingam") Fayence in drei Variationen (Aufl. insgesamt 450 Stück) GSG Vienna (um 1991/92) (H. 44 cm) Lit.: Firmen-Kat./Folder: Wander Bertoni, Goldscheider Vienna (Stoob/Bgld. um 1992) o. Pag. (Umschlag Rückseite m. Abb.)

Bertrand, A.
Ein Künstler namens A. Bertrand lässt sich in der einschlägigen Literatur nicht nachweisen. Es ist somit, wie häufig bei Goldscheider, an ein Pseudonym zu denken, bzw. ist mit großer Sicherheit der ital. Künstler Ipazzio Antonio Bortone dahinter zu vermuten (weitere Modelle seiner Hand und seine biografischen Daten -> vgl. dort).
1740 s. Bortone
2177 BÜSTE „Sonnenblume" (Junge Frau, nach links blickend) (vgl. 2296; 2458)
2205 FIGUR „Paradieswächter" ÖMKI (1901/02)
2272 FIGUR „Träumerei" (Stehende Frau mit Amphore) (vgl. 2443) (GS neu Modell-Nr. 63 „La Tendresse")
2296 BÜSTE „Sonnenblume" (Junge Frau, nach links blickend) (vgl. 2177; 2458)
2304 BÜSTE „Gebet" ÖMKI (1901/02)
2319 BÜSTE „Constance" (Junge Dame verträumt nach rechts blickend)
2354 BÜSTE „Ziniae"
2443 FIGUR „Träumerei" (Stehende Frau mit Amphore) (vgl. 2272) (GS neu Modell-Nr. 63 „La Tendresse")
2458 BÜSTE „Sonnenblume" (Junge Frau, nach links blickend) (vgl. 2177; 2296)

Besnard, Jean
*Jean Besnard, Keramiker und Maler (*11.6.1889 Paris, †1958 Paris), war der Sohn des Malers Albert Besnard. Er erhielt seine Ausbildung in Paris, wo er auch arbeitete und lebte. Um 1923 schuf er Fayencen für die Pariser Firma Primavera und lieferte um 1925 Modelle für Arthur Goldscheider, Paris. Er war Mitglied der Künstlergruppe „Evolution". Seine Gefäßkeramiken und Figuren sind dem Stile des Art Déco verhaftet. Teilnahme an der KGW-Ausst. in Paris 1925. Besnard beschickte zwischen 1923 und 1946 den Salon des Artistes Décorateurs sowie zwischen 1927 und 1958 den Salon d'Automne, beide in Paris. Lit.: Goldscheider, A. (Hrsg.): Artistic Evolution, 1926, T. 1, Pl. 29 (m. Abb.); Vollmer, Bd. 1, S. 197; Zühlsdorff, S. 131, 396; Bénézit, Bd. 2, S. 242; AKL, Bd. 10, S. 194.*
LAMPE „Lampe en Céramique" (Lamp in Ceramic) (Lampenschirm von Etlin) (Form: keulenförmig) (H. 60,5 cm, B. 37 cm) Lit.: Goldscheider, A. (Hrsg.): Artistic Evolution, 1926, T. 1, Pl. 29 (m. Abb.)

Biško, Josef
Josef Biško, Bildhauer und Kunstgewerbler, absolvierte im Schuljahr 1894/95 die KGW-Schule in Bechyne und wurde bei GS (als Modelleur, Formgießer oder Figurenretoucheur) angestellt. Im Jahr 1907 nennt das AdK seine Kunstterrakottafabrik in Wien XX., Denisgasse 37 und Wien II, Neubaugasse 11. Die Firma produzierte alle möglichen keramischen Waren und exportierte diese in alle Länder. Lit.: Neuwirth (1974), S. 31, 112; Neuwirth (1974a), S. 321; Bechyne 1884–1994, S. 55.

Bläsche
1117 BÜSTE Mädchen mit Lorbeerkranz (Sockel reliefiert mit Sonne und Baum) (vermutlich GS)

Blanc, Pierre
*Pierre Blanc, Bildhauer (*20.10.1902 Lausanne, †3.11.1986 Lausanne), studierte von 1919–1924 in Genf und lebte zwischen 1925 und 1939 in Paris. Der auf Tierbildhauerei spezialisierte Blanc war Mitarbeiter bei Arthur Goldscheider, Paris. Lit.: Goldscheider, A. (Hrsg.): Artistic Evolution, 1926, T. 2, Pl. 42 (m. Abb.); Vollmer, Bd. 1, S. 226; Bénézit, Bd. 2, S. 372; AKL, Bd. 10, S. 380.*
„Cacatoès" (Cockatoo) BRONZE (geschwärzt, teils mit Nickel) (H. 40 cm, B. 12,5 cm) Lit.: Goldscheider, A. (Hrsg.): Artistic Evolution, 1926, T. 2, Pl. 42 (m. Abb.)

Blondat, Max
*Max Blondat, Bildhauer (*30.9.1872 Crain, Yonne, †17.11.1925 Paris), war Mitglied der Künstlergruppe „La Stèle" (-> vgl. Kap. „Dekorative Skulpturen von Arthur*

Goldscheider in Paris"). Er war Mitbegründer der Societé des Artistes décorateurs. Sein Ruhm gründet sich besonders auf dekorative und kunstgewerbliche Arbeiten, bei denen er seine Vielseitigkeit unter Beweis stellt: eine von der Manufacture Nationale in Sèvres hergestellte Miniatur seiner Skulptur (der Brunnen Jeunesse wird zu einem Aschenbecher, einer Ablageschale bzw. einem Tafelaufsatz); Pokale und Trophäen (im Auftrag der Fa. Pommery für ein Luftfahrttreffen); von Edmond Lachenal gefertigte Keramiken (Jeanne d'Arc au bûcher); aber auch Schmuck bis hin zu Kühlerfiguren für Personenkraftwagen gehörten zu seinen Entwürfen. Zahlreiche Objekte im öffentlichen Raum. Lit.: Goldscheider, A. (Hrsg.): Artistic Evolution, 1926, T. 2, Pl. 2 (m. Abb.); TB, Bd. 4, S. 134; Vollmer, Bd. 1, S. 235; Neuwirth (1974), S. 60, 112; Hillenbrand, S. 40; Bénézit, Bd. 2, S. 414; AKL, Bd. 11, S. 579.
„The Child Thinker" (H. fehlt) Lit.: Goldscheider, A. (Hrsg.): Artistic Evolution, 1926, T. 2, Pl. 2 (m. Abb.)

Bofill, Antoine
*Antoine Bofill, Bildhauer (*Barcelona, Lebensdaten unbekannt, zwischen 1894 und 1939 nachweisbar), studierte an der Akademie in Barcelona, war tätig in Paris und stellt zwischen 1894 und 1921 im Pariser Salon des Artistes Français. aus, wo ihm 1902 eine ehrenvolle Erwähnung zu Teil wurde. Lit.: Vollmer, Bd. 1, S. 252; Bénézit, Bd. 2, S. 456; Hillenbrand, S. 43; AKL, Bd. 12, S. 224.*
3042 Stehender Holländerjunge mit Zipfelhaube und Laterne (auf Landschaftssockel)

Bormann, Wilhelm (s. a. Beaume)
*Wilhelm Bormann, Bildhauer und Medailleur (*29.7.1885 Braunschweig, †9.10.1938 Wien), studierte in den Jahren 1898–1905 an der Wiener KGW-Schule unter Josef Breitner (Modellieren). Ab 1908 Mitarbeiter der Wiener Mosaikwerkstätte Leopold Forstner. Bormann war Dozent an der Technischen Hochschule, Wien. Neben seinen, hauptsächlich von Goldscheider ausgeführten Arbeiten, ist er auch bekannt für seine religiösen Darstellungen von Heiligen, auch in Holz, Stein und Bronze. Arbeiten im öffentlichen (Kirchen-)Raum. Bormann war 1908 wohnhaft in Wien III., Gerlgasse 8, 1939 lässt sich seine Wohnung im II. Bez. in der Ybbsstraße 12 und sein Atelier im Prater, Südlicher Amateurpavillon Nr. 246 nachweisen. Lit.: Lehmann, 1908–1939; Neuwirth (1974), S. 60, 113; AKL, Bd. 13, S. 74.*
1484 VASE Putti (Form: quadratisch/becherförmig/Dekor: reliefierte Wandung mit Putti und Blütengirlanden) (Pendant vgl. 1486)
1486 VASE Putti (Form: zylinder-/becherförmig/Dekor: reliefierte Wandung mit Putti und floralen Elementen) (Pendant vgl. 1484) (auch sign. Beaume)
3416 Zwei stehende Windhunde (auf Rechtecksockel) (undeutl. sign. Beaume) (vgl. 4406= linke Einzelfigur)
3622 „Kind mit Hund" (auf Rundsockel) (Fay.) ÖMKI (1910/11)
3714 „Freundschaft" (Kind mit Hund= Neufundländer) (auch als Deckeldose Modell-Nr. 3714) (vgl. 5908; vgl. 4014= Hund einzeln) (auch sign. Beaume)
3878 GRUPPE/LAMPE Zwei stehende Windhunde (Barsois) (auf Rechtecksockel) (vgl. 4407= rechte Einzelfigur) (vgl. 4637= linke Einzelfigur) (sign. Beaume)
4014 Stehender Neufundländer (vgl. Gruppe/Deckeldose Modell-Nr. 3714; 5908) (sign. Beaume)
4195 „Kaninchen" ÖMKI (1912; 1913/14)
4406 Stehender Windhund (auf Ovalsockel) (vgl. 3416= Zweier Gruppe) (Bormann zugeschr.)
4407 Stehender Windhund (Barsoi/Greyhound) (auf Ovalsockel) (vgl. 3878= Zweier Gruppe) (sign. Beaume)
4637 Stehender Windhund (Barsoi) (auf Rechtecksockel) (vgl. 3878= Zweier Gruppe) (sign. Beaume)
5288 „Englischer Greyhound" (auf Ovalsockel)
5457 Stehender Schäferhund (auf Rechtecksockel) (Entwurf evtl. von Hugo Postl?)
5908 „Freundschaft" (Kind mit Hund= Neufundländer) (vgl. 3714; vgl. 4014= Hund einzeln) (auch sign. Beaume)
„Tafelaufsatz, modern" ÖMKI (1912) Lit.: Neuwirth (1974), S. 60

Borrmann s. Bormann, Wilhelm (s.a. Beaume)

Borsi
*Manfredo Borsi, Maler und Keramiker (*15.2.1900, Florenz, †?), war Mitarbeiter/in bei Friedrich Goldscheider, Paris. Lit.: GSB, Pl. 83; Bénézit, Bd. 2, S. 581.*
765-GSB VASE „Nymphe aux Nénuphars"

Bortone, Ipazzio Antonio

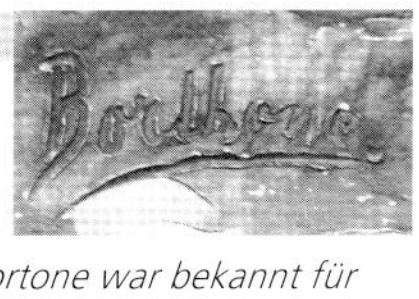

*Ipazzio Antonio Bortone, Bildhauer (*13.6.1844 Ruffano/Lecce, Apulien, †2.4.1938 Lecce), studierte in Neapel bei Tito Angelini. Bortone war bekannt für seine Porträtbüsten und Denkmäler und Brunnen, die sich vorwiegend in Florenz und Lecce befinden. Um seinem Namen einen franz. Klang zu geben, verwendete Goldscheider für ihn auch die Pseudonyme „Bertrand" und „Portou". TB, Bd. 4, S. 380; Bénézit, Bd. 2, S. 583; AKL, Bd. 13, S. 137.*
1258 BÜSTE Dame mit Hut (undeutl. sign.)
1502 BÜSTE „Annita" (Lachendes Bauernmädchen mit Weintrauben) (vgl. 1740; 2130= jew. Variante) (vgl. 2029= WANDMASKE)
1739 BÜSTE „Bummler" (Junger Mann mit Barett) (vgl. 1940= WANDMASKE)
1740 BÜSTE „Annita" (Mädchen mit Kopftuch) (vgl. 2130) (vgl. 1502= Variante) (auch sign. Bertrand)
1940 WANDMASKE „Bummler" (Junger Mann mit Barett) (vgl. 1739= BÜSTE)
2029 WANDMASKE „Annita" ÖMKI (1900/01) (vgl. 1740; 2130= jew. BÜSTE)
2130 BÜSTE „Annita" (Mädchen mit Kopftuch) (vgl. 1740) (sign. Portou) (vgl. 1502= Variante) (vgl. 2029= WANDMASKE)

Bortone s. auch Portou

Boschull

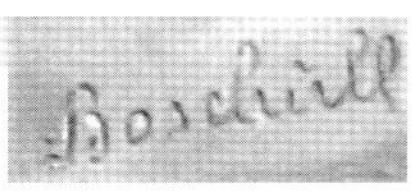

5586 DOPPELBÜSTE Lachendes Kinderpaar

Bosi (Pseud. für Rosé)
„Ruth" (Fay.) Lit.: MW, 11.1929/30, Nr. 11, S. VII (m. Abb.); Neuwirth (1974), S. 60. (vgl. Rosé 3870; 3992; 4141; 5171; 5513)

Bosse, Walter

*Walter Bosse, Bildhauer, Keramiker und Kunstgewerbler (*13.11.1904 Wien, †17.12.1979 Iserlohn/Deutschland), studierte 1919–1922 u.a. bei Franz Cizek und Michael Powolny an der KGW-Schule in Wien. In den Jahren 1921 und 1922 u.a. bei Richard Riemerschmid an der KGW-Schule in München. Er gründete bereits 1920 seine eigene Werkstätte in Kufstein und beschäftigte dort sowie in der Zweigstelle Kitzbühel 1937 bereits 45 Mitarbeiter. Ab 1921 selbständige Arbeit als Keramiker bzw. Mitarbeiter der Wiener Werkstätte unter Josef Hoffmann. Ab 1924 auch für die Porzellanmanufaktur Augarten, Wien sowie ab 1926 bis 1939 Mitarbeiter bzw. Entwerfer für die Firmen Fischer sowie Metzler & Ortloff (beide in Ilmenau). In dieser Zeit nahm er auch an den Leipziger Messen teil. Ab 1939 bis 1969 für Goebel in Oeslau tätig, die 16 seiner Modelle ausführten. Bosse war in den Jahren 1924 bis 1932 Mitglied des Österreichischen Werkbundes und bis 1939 Mitglied des Bayerischen KGW-Vereins in München. Im Jahre 1938 verlegte er seine Werkstätte nach Wien, wo er seine Erzeugnisse in einem eigenen Verkaufslokal (Wien I., Rotenturmstr. 19) anbot. Ein weiteres Wiener Detailgeschäft betrieb Bosse am Ring/Ecke Kärtnerstraße, der sog. „Sirk-Ecke". Nach dem Krieg versuchte Walter Bosse sich mit mehreren Läden in Wien (Tuchlauben, am Ring usw.) sowie mit der Fabrik (zugleich seine Wohnung in dieser Zeit) in der Salzachstraße, Wien XX. Ebenso wurde in Salzburg (Linzer Straße) noch ein Laden und eine weitere Produktionsstätte unterhalten. Bald nach dem Krieg begann auch die Zusammenarbeit mit Herta Baller. Unter ihrem Namen wurden in den 50er Jahren die heute noch sehr gesuchten „Modernen Wiener Bronzen" erzeugt. Vereinzelt wurden diese kleinen Figürchen, Mini-Tiere und Anhänger auch in Silber oder Gold gegossen. Für Goldscheider finden sich seine Entwürfe ab den 20er Jahren für Walter & Marcell, nach dem Krieg dann auch für die West-Germany Produktion sowie für Marcell in England. Seine Kreationen wurden von der Tonindustrie Scheibbs sowie von Achatit Schirmer (Köln), der Porzellanfabrik Altenkunstadt und weiteren deutschen Firmen ausgeführt. Von 1950 bis 1972 übernahm auch die Staatliche Majolikamanufaktur Karlsruhe etliche seiner Modelle (242 Seriennummern!). Im Jahre 1955 übersiedelte Bosse nach Iserlohn, wo er „Wiener Bronzen" bis zum Konkurs 1966 erzeugte. Seine grotesken Figuren, Wandreliefs, Tiere, Leuchter, Schalen usw. zeugen von einer Ideenvielfalt, die in der Keramik ihresgleichen sucht. Zwar sind nur wenige seiner Stücke signiert, selbst die Größeren und Bedeutenderen tragen oft keine Bezeichnung, jedoch sind sein Stil und seine Farbgebung unverkennbar. Waltraud Neuwirth beziffert sein Gesamtwerk auf etwa 8000 Modelle. Bosses erfolgreiche Kreationen, vor allem die Metallarbeiten (wie z.B. seine „Igelascher", Kerzenleuchter und Korkenzieher), werden heute noch nachgeahmt bzw. industriemäßig gefälscht. Lit.: Neuwirth (1974), S. 60, 114–115 (m. Abb.).; Zühlsdorff, S. 398; Schreyer-Hartmann; AKL, Bd. 13, S. 208.*
1000-WG „Elefant" (Stilisierter Elefant mit hängendem Rüssel)
1002-WG „Elefant" (Stilisierter Elefant mit aufgestelltem Rüssel)
1003-WG WANDBEHANG „Wandpferd" (Stilisiertes Pferd)
1004-WG WANDBEHANG „Pferd" (Stilisiertes Pferd)
1005-WG Seehund
1006-WG „Katze" (Stilisierte sitzende Katze)
1007-WG WANDBEHANG „Wandhahn" (Stilisierter Hahn)
1008-WG „Dackel" (Stilisierter sitzender Dackel)
1009-WG „Giraffe" (Stilisierte Giraffe)
1010-WG „Häschen" (Stilisiertes sitzendes Häschen)
1011-WG Kamel/Trampeltier (Stilisiertes Kamel/Trampeltier)
6233 BUCHSTÜTZE Max (als sitzende Figur)
6233 BUCHSTÜTZE Moritz (als sitzende Figur)
6236 BUCHSTÜTZE Blumen (Form: rechtwinkelig, angesetzter Viertelkreisbogen/Dekor: zwei stilisierte Blumen, Laufglasur) (Bosse zugeschr.)
6605 Stehender Elefant (Bosse/Meisinger zugeschr.)
6636 Stehender Clown mit Bandoneon (auf Rundsockel) (Bosse zugeschr.)
6676 Stehender kleiner Engel in Trachtenkostüm (auf Rundsockel) (Bosse zugeschr.)
6696 Stehender vom Notenblatt singender Engel (auf Rundsockel) (Bosse zugeschr.)
6792 Stehender Matrose mit Bandoneon (Bosse zugeschr.)
7838 WANDVASE (Form: zwei becherförmige, miteinander verbundene Gefäße/Dekor: vier stilisierte Blütenkapseln oder Blätter, rot-schwarz-türkis) (Bosse zugeschr.)
232 VASE/TOPF
246 VASE Blume
267 VASE/TOPF
1165 VASE (Bosse zugeschr.)

Bosse, Walter, ohne Nr.
ohne Nr. „Deckeltopf" in Form eines Kopfes mit Trachtenhut
ohne Nr. Elefant als Schalenträger
ohne Nr. „Groteskes Tier" (Hirsch)
ohne Nr. Sitzender Affe mit Schale
ohne Nr. Sitzender Knabe mit Kappe und Topf (für Kakteen)
ohne Nr. Skifahrer mit Kuh (auf Rechtecksockel)
ohne Nr. „Hochzeitsreise" (groteskes Paar in einem Ruderboot sitzend)
ohne Nr. BUCHSTÜTZE Igelmann
ohne Nr. KERZENHALTER (zweiflammig) Zwei Dackel
ohne Nr. SCHALE (Gitterkorb) (Form: trichterförmig, kurzer zylindrischer Stand/Dekor: bogenartig (bzw. drei mal herzförmig) durchbrochene Wandung, Laufglasur)
ohne Nr. SCHALE (Form: sechspassig, teils gezackt/Handhaben: zwei blattförmige Ösen/Dekor: Blume mit Stängel (appliziert im Spiegel der Schale), mehrfarbig)
ohne Nr. VASE mit groteskem Tier (Hund?)
ohne Nr. VASE Blume (Form: balusterförmig, trichterförmige Mündung/Handhaben: vier Ösen/Dekor: zwei stilisierte, applizierte Blumen mit Blatt, Farbfelder,

mehrfarbig, Laufglasur) (2 Varianten: H. 14,2 cm; bzw. 30,4 cm)
ohne Nr. VASE Blume (Form: balusterförmig, trichterförmige Mündung/Handhaben: zwei ohrenartige Henkel/Dekor: zwei stilisierte, applizierte Blumen mit Blatt, mehrfarbig, Laufglasur)
ohne Nr. VASE Blume (Form: becher/kelchförmig, zweifach gewulsteter Stand/Handhaben: vier Ösen/Dekor: zwei stilisierte Blumen, rechteckige Farbfelder, mehrfarbig, Laufglasur) (vgl. 246)
ohne Nr. VASE Blume (Form: becherförmig/Handhaben: sechs Ösen/Dekor: zwei stilisierte Blumen, mehrfarbig, Laufglasur)
ohne Nr. VASE Blume (gebaucht, trichterförmige Mündung/Handhaben: vier Ösen/Dekor: zwei stilisierte (ösenförmig), applizierte Blumen mit Blatt, mehrfarbig, Laufglasur)
ohne Nr. VASE Blumenblüten (Form: trichterförmig, mittig tailliert, spindelförmig/Handhaben: zwei Ösen/Dekor: zwei stilisierte Blumenblüten, mehrfarbig, Laufglasur
ohne Nr. VASE Blume (Form: becherförmig, gedrückte Mündung/Dekor: drei stilisierte Blumen, mehrfarbig, Laufglasur)
ohne Nr. VASE/TOPF (Form: konisch/Handhaben: vier Ösen/Dekor: rechteckige Farbfelder mit Blumen, mehrfarbig, Laufglasur) (vgl. 232)
VASE (Form V-förmige Trichter auf gewulstetem rundem Stand) Lit.: MW, 11.1929/30, Nr. 2, S. IX (m. Abb.)
VASE (Form: V-förmige Trichter auf Kugel mit geradem rundem Stand) Lit.: MW, 11.1929/30, Nr. 2, S. IX (m. Abb.)
VASE (Form: zweifach gebauchter Korpus, geweitete Mündung, kurzer Standring) Lit.: MW, 11.1929/30, Nr. 2, S. IX (m. Abb.)
RAUCHZEUG Elefant Lit.: MW, 11.1929/30, Nr. 22, S. 20 (m. Abb.) (Bosse/Geiger/Meisinger?)
RAUCHZEUG Zigarettenbehälter Lit.: MW, 11.1929/30, Nr. 22, S. 20 (m. Abb.) (Bosse/Geiger/Meisinger?)
KERZENHALTER Lit.: MW, 11.1929/30, Nr. 22, S. 20 (m. Abb.) (Bosse/Geiger/Meisinger?)

Bouchard, Henri

*Henri Bouchard, Bildhauer (*13.12.1875 Dijon, †30.11.1960 Paris), war Mitglied der Künstlergruppe „La Stèle" (-> vgl. Kap. „Dekorative Skulpturen von Arthur Goldscheider in Paris"). Bouchard beschickt von 1895–1951 den Pariser Salon (Société des Artistes Français) und zwischen 1907–1928 den Salon d'Automne. Lit.: Goldscheider, A. (Hrsg.): Artistic Evolution, 1926, T. 1, Pl. 4 (m. Abb.); TB, Bd. 4, S. 425; Vollmer, Bd. 1, S. 279; Neuwirth (1974), S. 60; Bénézit, Bd. 2, S. 616–617; AKL, Bd. 13, S. 279.*
„Jeunesse" („Youth") BRONZE (H. 58 cm, B. 33,5 cm) Lit.: Goldscheider, A. (Hrsg.): Artistic Evolution, 1926, T. 1, Pl. 4 (m. Abb.)
„Game of Chess" BRONZE (H. fehlt, B. fehlt) Lit.: Goldscheider, A. (Hrsg.): Artistic Evolution, 1926, T. 2, Pl. 26 (m. Abb.)

Bounet s. Bouret

Bouraine, (André) Marcel

*(André) Marcel Bouraine, Bildhauer (*1886 Pontoise/Seine-et-Oise, †1948 ?), war Schüler von A. Falguière und Mitglied der Künstlergruppe „La Stèle" (-> vgl. Kap. „Dekorative Skulpturen von Arthur Goldscheider in Paris"). Er beschickte seit 1922 beide Pariser Salons (Salon d'Automne und Société des Artistes Français) mit Figuren, Büsten und Studienköpfen. Zu seinen populären Genres zählen vor allem Harlekins, Amazonen und Tänzerinnen, auch in der Kombination Bronze und Elfenbein. Lit.: Goldscheider, A. (Hrsg.): Artistic Evolution, 1926, T. 1, Pl. 11 (m. Abb.); Vollmer, Bd. 1, S. 284; Neuwirth (1974), S. 60; Hillenbrand, S. 49; Bénézit, Bd. 2, S. 660; AKL, Bd. 13, S. 353.*
ohne Nr.? -LaStèle Seelöwe (auf Felssockel) BRONZE (versilbert)
46-LaStèle „Leda" (mit dem Schwan) BRONZE/Marmorsockel
82-LaStèle „Otarie" (Otary) Seelöwe, einen Ball balancierend BRONZE/Marmorsockel (auch als Lampe mit Argonit)
106-LaStèle Ballerina auf einem Bein BRONZE
147-LaStèle Tanzender Damenakt auf einem Bein BRONZE „Dancing Girl" (Tänzerin in Pose, auf getrepptem Sockel) BRONZE Arthur Goldscheider, Paris (1924) Lit.: L'Illustration, Nr. 4268 vom 20.12.1924, S. 14 (m. Abb.); Lit.: Goldscheider, A. (Hrsg.): Artistic Evolution, 1926, T. 2, Pl. 28 (m. Abb.)
„Jongleuse" (Tänzerin mit Ball) BRONZE/ELFENBEIN Arthur Goldscheider, Paris (1924) Lit.: L'Illustration, Nr. 4268 vom 20.12.1924, S. 14 (m. Abb.)
„Le Boxeurs" BRONZE Paris, Salon (1923) Lit.: Neuwirth (1974), S. 60
„Lanceur de poids et lanceur de javelot" BRONZE Paris, Salon (1923) Lit.: Neuwirth (1974), S. 60
„Sprinter au départ" BRONZE Paris, Salon (1923) Lit.: Neuwirth (1974), S. 60
„Tennis" BRONZE Paris, Salon (1923) Lit.: Neuwirth (1974), S. 60
LAMPE „Otarie" (Otary) Seelöwe, einen Ball balancierend BRONZE/Aragonite (H. 60 cm, B. 34 cm) Lit.: Goldscheider, A. (Hrsg.): Artistic Evolution, 1926, T. 1, Pl. 26 (m. Abb.)
BUCHSTÜTZE „Chien Basset" (Basset Dog) BRONZE (vergoldet), Marmorsockel (H. 11 cm, B. 27,5 cm) Lit.: Goldscheider, A. (Hrsg.): Artistic Evolution, 1926, T. 2, Pl. 20 (m. Abb.)

Bouret, Germaine

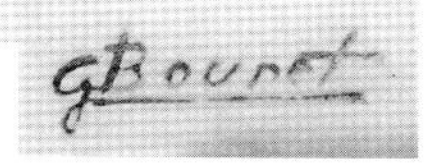

*Germaine Bouret, Zeichnerin, Karikaturistin (*29.5.1907 Paris, †25.1.1953 Paris). Mit Bouret verhält es sich ähnlich wie mit der weitaus bekannteren Künstlerin Berta M. Hummel, deren Figuren nach Zeichnungen bzw. Postkartenmotiven der Klosterschwester gefertigt wurden. Germaine Bouret hat ebenfalls keine bildhauerische Arbeit geleistet, sondern machte sich als Zeichnerin, besser gesagt als Cartoonistin besonders in Frankreich einen Namen. In dem anlässlich ihres fünfzigsten Todestages erschienen Artikel von Serveau (-> vgl. Lit.) werden ihre Entwürfe und deren Umsetzung als Postkarte, Figuren etc. gezeigt. Durch ein Foto im Archiv Mimi Dakon sowie durch die teils „fälschlich" auf den Figuren und Gruppen angebrachte Signatur „Dakon", lässt sich belegen, dass die Entwürfe Bourets von Stephan Dakon für die Manufaktur Goldscheider modelliert wurden. Lit.: Serveau, E.: Hommage à Germaine Bouret, in: La vie du collectionneur, 2003, Nr. 445, S. 1, 13–15 (m. Abb.); AKL (Archiv, unredigiert).*
7039 DreierGRUPPE/ZweierGRUPPE/EinzelFIGUR/SCHREIBTISCHABLAGE Drei schreitende Kinder (zwei Jungen mit einem Mädchen, Arm in Arm gehend) (auf abgerundetem Rechtecksockel)
7273 Stehendes Mädchen mit Hut mit aufspringendem Foxterrier (auf Ovalsockel) (GS neu: Modell-Nr. 30 „Mein Hund und ich")
7594 FIGUR „Joujou" (Stehendes Mädchen mit Ball) (auf Rundsockel)
7663 GRUPPE „Naschkätzchen" (Stehendes Mädchen, einen Terrier fütternd) (auf Ovalsockel)
7664 GRUPPE „Komm mit!" (Stehendes Mädchen mit sitzendem Terrier) (auf Ovalsockel)
7825 GRUPPE Stehendes Mädchen mit Ball und Terrier (auf abgerundetem Rechtecksockel)
7835 GRUPPE Kleines Mädchen, einem Jungen den Schal bindend (auf abgerundetem Rechtecksockel)
8120 GRUPPE Laufendes Mädchen mit zwei Barsois (auf Ovalsockel)
8365 GRUPPE „Kinder" („Kindergruppe") (Zwei Kleinkinder beim Handschuhe Anziehen) (auf Ovalsockel)

Bouret, Pierre-Ernest

*Pierre-Ernest Bouret, Bildhauer und Medailleur (*26.6.1897 Paris, †25.10.1977 Sèvres), war Mitglied der Künstlergruppe „La Stèle" (-> vgl. Kap. „Dekorative Skulpturen von Arthur Goldscheider in Paris"). Lit.: Bénézit, Bd. 2, S. 668; AKL, Bd. 13, S. 369.*

Bouval, Maurice

*Maurice Bouval, Bildhauer und Designer (*9.6.1863 Toulouse, †20.3.1916 Paris?), war Schüler von A. Falguière und entwarf neben Büsten und Statuetten 1898 auch die Goldscheider-Firmenmarke „Sculpture". Bouval war auch für einige andere Modelle der Manufaktur Goldscheider verantwortlich, die in den Jahren 1889–1914 in der Pariser Filiale ausgestellt wurden. Er beschickte von 1891–1914 den Pariser Salon (Société des Artistes Français) und erhielt für seine Teilnahme an der Weltausstellung 1900, ebenfalls in Paris, eine lobende Erwähnung. Seine Arbeiten wurden unter anderem durch die Firmen Colin & Cie., Jollet und Thiébaut Frères, alle Paris, editiert. Lit.: Vollmer, Bd. 1, S. 288; Neuwirth (1974), S. 60; Hillenbrand, S. 50; Bénézit, Bd. 2, S. 690; AKL, Bd. 13, S. 413; Arwas (2002), S. 608.*
642 RELIEF „Sculpture" (umrankt von Blütenzweigen) (Wandrelief) TERRAKOTTA (vgl. Ausführungen in BRONZE und ZINN)
672 REKLAMESCHILD „Sculpture" Fréd. Goldscheider, Editeur, Paris, Vienne, Florence, Bronze, Terre cuite, Marbre, Etain TERRAKOTTA (vgl. 642; Version in Zinn, Version in Bronze)
836-GSB LAMPE „Flambeau électrique" (Blütenstängel in der rechten Hand) (Pendant -> vgl. 853)
853-GSB LAMPE „Flambeau électrique" (Blütenstängel in der linken Hand) (Pendant -> vgl. 836)
860-GSB LAMPE „Femme au Lys, électricité" (Blütenstängel in der rechten Hand) (Pendant -> vgl. 870)
870-GSB LAMPE „Femme au Pavot, électricité" (Blütenstängel in der linken Hand) (Pendant -> vgl. 860)
879-GSB „Le Secret" BRONZE
884-GSB ENCRIER „Pivoine"
2108 BÜSTE „Jvy" (oder „Ivy") (Mädchen mit efeugeschmücktem Haar) ÖMKI (1900/01)
2115 BÜSTE „Meditation" (Frau mit Kopftuch, auf Sockel) (Inschrift: MEDITATION)
„David" mit Steinschleuder (und erlegtem Federvieh) BRONZE (um 1900) (H. 33 cm); Spik 547
o. Nr. REKLAMESCHILD „Sculpture" Frederic Goldscheider, Editeur, Paris, Vienne, Terre cuite, Bronze, Marbre, Etain BRONZE (vgl. 642; 672)
o. Nr. REKLAMESCHILD „Sculpture" Frederic Goldscheider, Editeur, Paris, Vienne, Terre cuite, Bronze, Marbre, Etain ZINN (vgl. 642; 672)

Bozzi, Lorenzo

*Lorenzo Bozzi, Bildhauer und Kunstkritiker (*21.7.1864 San Lorenzo/Maggiore, †?), studierte in Neapel bei Lista und Morelli und debutierte im Salon 1890 in Paris. Bozzi lebte in Paris und war Mitarbeiter von Friedrich Goldscheider, ebenda. Lit.: Bénézit, Bd. 2, S. 711; AKL, Bd. 13, S. 489.*
756-GSB RELIEF Plateau „Mater Purissima"

Bracken, Clio Hinton s. Huneker

Braecke, Pieter Jan (Pierre)

*Pieter Jan (Pierre) Braecke, Bildhauer und Medailleur (*4.10.1858 Nieuwpoort/Belgien, †10.11.1938 Nossegem/Zaventem), studierte an den Akademien in Brügge, Löwen und Brüssel. Einige seiner Entwürfe orientierten sich an den Skulpturen von Constantin Meunier. Braecke entwarf u.a. symbolistisch gehaltene Arbeiten nach Motiven aus Opern von Richard Wagner und war Mitarbeiter von Friedrich Goldscheider, Paris. Ausstellungsbeteiligungen u.a.: Weltausstellung Brüssel 1897, Paris 1900 und Gand 1913. Lit.: Lennep, J. van: Catalogue de la Sculpture Artistes nés entre 1750 et 1882, Bruxelles, 1992, S. 120; Bénézit, Bd. 2, S. 722; AKL, Bd. 13, S. 540.*
857-GSB „Désespoir"

Breitner, Joseph

*Joseph Breitner, Bildhauer (*19.12.1864 Wien, †22.11.1930 Wien), studierte an der KGW-Schule Wien unter Kühne und König, war dann ab 1895 Lehrer an der KGW-Schule des Österreichischen Museums für Kunst und Industrie, Wien und ist in einem Brief von 1914 als Mitarbeiter der Firma Goldscheider erwähnt. Breitner war 1907 wohnhaft in Wien IX, Hernalser Gürtel 24. Lit.: TB, Bd. 4, S. 574; Lehmann, 1907; Neuwirth (1974), S. 60; Zühlsdorff, S. 644; Bénézit, Bd. 2, S. 776; AKL, Bd. 14, S. 91.*

Brenner, (Victor David?)

*(Victor David?) Brenner, Bildhauer und Medailleur (*12.6.1871 Savli/Litauen, †5.4.1924 New York), erhielt seine Ausbildung bei seinem Vater und lebte um 1898 in Paris, später dann in New York. Er nahm an den Weltausstellungen 1900 in Paris sowie 1904 in St. Louis (jeweils*

mit Auszeichnungen) teil. Ob Victor David Brenner wirklich der Entwerfer dieser Gruppe ist, lässt sich nicht mit letzendlich gültiger Sicherheit sagen. Lit.: TB, Bd. 4, S. 581; Vollmer, Bd. 1, S. 309; Bénézit, Bd. 2, S. 783; AKL, Bd. 14, S. 91.
1066 GRUPPE „Adam and Eva" (nach dem Sündenfall auf Stein sitzend)

Brochmann, Edith
*Edith Brochmann, Aufglasurmalerin (*25.12.1924 ?, †?). Quelle: AGM, Liste „Adressen von guten und zuverlässigen Arbeitern" (undat., um 1953).*

Brünelle
Die Signatur „Brünelle" erscheint etwas fragwürdig. Wahrscheinlicher dürfte es sich um einen franz. Künstler namens Brunelle handeln. Nachweisbar ist ein solcher aber ebenso nicht.
2627 Stehende junge Frau mit Lyra (in antikisierendem Kleid) (Sappho?)

Brugier, A.
A. Brugier, Kunstgewerbler, (Lebensdaten unbekannt), schuf Lackarbeiten für Arthur Goldscheider, Paris, und war Mitglied der Künstlergruppe „Évolution" (-> vgl. Kap. „Dekorative Skulpturen von Arthur Goldscheider in Paris"). Lit.: Goldscheider, A. (Hrsg.): Artistic Evolution, 1926, T. 1, Pl. 16 (m. Abb.); Neuwirth (1974), S. 60; AKL (Archiv, unredigiert).
VASE Metall (lackiert, schwarz/gold) (H. 28 cm, B. 20 cm) Lit.: Goldscheider, A. (Hrsg.): Artistic Evolution, 1926, T. 1, Pl. 16 (m. Abb.)

Brunel, H.
H. Brunel, Kunstgewerbler, (Lebensdaten unbekannt), schuf Eisenarbeiten und war Mitglied der Künstlergruppe „Évolution" (-> vgl. Kap. „Dekorative Skulpturen von Arthur Goldscheider in Paris"). Lit.: Neuwirth (1974), S. 60.

Bruyneel, (Victor?)
Victor Bruyneel (franz. Bildhauer, erwähnt zwischen 1881 und 1893), war Mitarbeiter von Friedrich Goldscheider, Paris. Weitere Daten zu Leben und Schaffen sind nicht bekannt. Lit.: GSB, Pl. 44,100 (m. Abb.); AKL (Archiv, unredigiert).
780-GSB STATUE „La Piège" (vgl. 832-GSB)
832-GSB STATUE „La Piège" (vgl. 780-GSB)
898-GSB BUSTE „La Piège"

Brynski / Brynsky
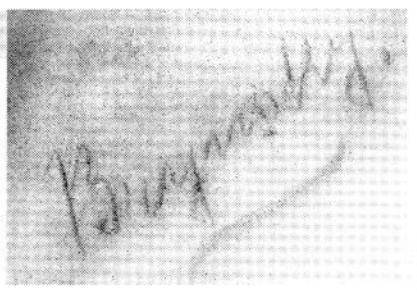
4287 Stehender Knabe mit zwei Enten, die er in den Armen hält (auf Ovalsockel) (sign. „Brynski")
4315 Zwei Putti mit störrischem Widder (auf Rechtecksockel) (sign. „Brynski / Brynsky")
4475 Stehende Dame mit Fächer (auf verziertem Rundsockel) (sign. Brynsky)

Bucher, Edwin
*Edwin Bucher, Tierbildhauer und Goldschmied (*14.3.1879 Luzern, †13.12.1968 Coucy-le-Château), war Schüler bei Bourdelle und Rodin in Paris. Bucher war Mitarbeiter bei Arthur Goldscheider, Paris, und schuf vor allem Tierplastiken, Büsten sowie Statuetten. Lit.: Vollmer, Bd. 1, S. 341; Neuwirth (1974), S. 60; Bénézit, Bd. 2, S. 922; AKL, Bd. 14, S. 665.*
„Taureau argentin" BRONZE Paris, Salon (1923)
Lit.: Neuwirth (1974), S. 60
„Taureau en bataille" BRONZE Paris, Salon (1923)
Lit.: Neuwirth (1974), S. 60
„Vache et veau" BRONZE Paris, Salon (1923)
Lit.: Neuwirth (1974), S. 60

Bucher, Hertha
*Hertha Bucher, Keramikerin (*14.6.1898 Leverkusen, †9.2.1960 Wien), besuchte von 1911 bis 1919 (u.a. bei Franz Cizek und Michael Powolny) die KGW-Schule, Wien. Bucher war auf der Pariser KGW-Ausst. 1925 mit ihrem Modell „Kahnfahrt" (Modell-Nr. 5345) vertreten, welches auch mit rein weißer Glasur auf der „International Exhibition of Ceramic Art" in New York 1928/29 gezeigt wurde. Ihre Gruppe „Musikanten" (ausgeführt durch Marcell Goldscheider) wurde auf der Ausstellung „Das Bild im Raum", Wien 1929, im ÖMKI präsentiert. Bereits ab 1920 unterhielt sie eine eigene Werkstätte. Zahlreiche Ausstellungen folgten bis hin zur Pariser KGW-Ausstellung 1937, wo sie mit Vasen und Terrakotten vertreten war. Hertha Buchers Entwürfe wurden u.a. von der Wiener Werkstätte (ab 1916), Wienerberger, Keramos, Staatliche Majolikamanufaktur Karlsruhe (um 1931) und Rudolf Sommerhuber, Steyr (Oberösterreich), ausgeführt. Für Bauten Josef Hoffmanns gestaltete sie Kaminverkleidungen und Heizungsgitter. Den Alfred-Roller-Preis erhielt sie 1942. Hertha Bucher war 1941/42 wohnhaft in Wien I., Karlsplatz 2 und hatte ihr Atelier in der Mozartgasse 3 im VI. Bez. Lit.: Lehmann, 1941–1942; Vollmer, Bd. 1, S. 341; Neuwirth (1974), S. 41, 60, 116–119 (m. Abb.); Zühlsdorff, S. 398; Fahr-Becker, S. 222; AKL, Bd. 14, S. 668.*
371-MG „Musikanten" (Sänger und Lautenspielerin)
5345 „Kahnfahrt" (Figurengruppe mit Hund in einem Kahn)

Bulier
Lebens- und Schaffensdaten sind unbekannt. Lit.: AKL (Archiv, unredigiert).
2833 BÜSTE Betende Frau (mit großem Kopftuch)
FIGUR „Denker" ÖMKI (1906/07) Lit.: Neuwirth (1974), S. 60

Buß s. Ruß, (Rudolf ? / Willibald?)

Butensky, Jules Léon
*Jules Léon Butensky, Bildhauer (*13.12.1871 Stolivici / Russland, †?), war zuerst Schüler von Hellmer und Zumbusch an der Wiener Akademie. Später in Paris bei Mercié und A. Boucher. Seine Arbeiten für den öffentlichen Raum befinden sich u.a. im Weißen Haus in Washington („Exile"), im Hebrew Institute Chicago („Goliath") und im Metropolitan Museum, New York („Universal Piece"). Butensky wird letztmals 1953 erwähnt. Lit.: Vollmer, Bd. 1, S. 365; Fielding, M.: Dictionary of american Painters, Sculptors and Engravers, 1945; Falk, P. H.: Who is who in american Art, 1985; Bénézit, Bd. 3, S. 53; AKL, Bd. 15, S. 365.*
3800 Nathan der Weise (Stehender Gelehrter mit Turban und weitem Mantel) (vgl. 3839)
3839 Nathan der Weise (Stehender Gelehrter mit Turban und weitem Mantel) (vgl. 3800)

Callot, Jacques H.
Jaques H. Callot, Bildhauer und Medailleur (? Blaru/Seine-et-Oise, †?), war u.a. Schüler von Charles Gauthier. Callot war bekannt für seine Porträtmedaillons. Callot stellte seit 1894 im Pariser Salon aus und errang 1898 den 2. großen Preis. Lit.: TB, Bd. 5, S. 408; Bénézit, Bd. 3, S. 133; AKL, Bd. 15, S. 610.
617-GSB VIDE-POCHE „Médaillon" BRONZE oder ZINN (Pendant vgl. 660)
660-GSB VIDE-POCHE „Fantaisie" BRONZE oder ZINN (Pendant vgl. 617)
686-GSB VASE Etrusque „Eglantine" (vgl. auch 692)
687-GSB VASE „Chrysanthéme" (vgl. auch 693)
692 VASE Etrusque „Eglantine" (reliefiertes Mädchenbildnis nach rechts blickend) (Pendant vgl. 693) (vgl. auch 686-GSB)
693 VASE Etrusque „Chrysanthéme" (reliefiertes Mädchenbildnis nach links blickend) (Pendant vgl. 692) (vgl. auch 687-GSB)

Calm, Lotte (Charlotte Alice Calm-Wierink)
*Lotte Alice Calm, Keramikerin und Kunstgewerblerin (*1.10.1897 Kgl. Weinberge bei Prag, †?, lebte zuletzt in Holland), studierte von 1914 bis 1919 an der KGW-Schule, Wien, unter anderem bei Josef Hoffmann und Michael Powolny. Sie war von 1918–1925 künstlerische Mitarbeiterin der Wiener Werkstätte (Keramik, Graphik, Holzarbeiten sowie Schmuck) und Mitglied des Österreichischen Werkbundes. Teilnahme u.a. an der Modeausstellung 1915, der Kunstschau 1920, beide in Wien, sowie der KGW-Ausst. in Paris 1925. Lit.: Neuwirth (1974), S. 121–122; Zühlsdorff, S. 399; Fahr-Becker, S. 222; AKL, Bd. 15, S. 612.*
5512 Schreitender Orientale mit Turban als Schalenträger (auf Ovalsockel / Schale)

Camus, Jean-Marie
*Jean-Marie Camus, Bildhauer und Maler (*12.11.1877 Clermont-Ferrand, †16.6.1955 Paris), war Schüler von Ernest Barrias und Mitglied der Künstlergruppe „La Stèle" (-> vgl. Kap. „Dekorative Skulpturen von Arthur Goldscheider in Paris"). Lit.: TB, Bd. 5, S. 484; Vollmer, Bd. 1, S. 138; Neuwirth (1974), S. 60; Bénézit, Bd. 3, S. 176; AKL, Bd. 16, S. 80.*

Caniroff s. Haniroff

Capèque s. Czapek, Stanislaus

Cárdenas
2237 WANDMASKE „Odorat" (Frauenkopf mit langem geschwungenem Haar) ÖMKI (1901/02)

Cardona, José
José Cardona, spanischer Bildhauer (Lebensdaten unbekannt), war seit 1908 in Paris ansässig und beschickte bereits seit 1903 dort den Salon der Societé des Artistes Français mit Genrestatuetten und Büsten. Zu seinen Gießern zählte auch Emilio Pinedo, Paris. Lit.: TB, Bd. 5, S. 588; Bénézit, Bd. 3, S. 231; AKL, Bd. 16, S. 364.
42 Stehender, Zigarette rauchender Junge mit Schiebermütze BRONZE

Carier Belteus s. Carrièr-Belleuse

Carlier, Émile-Joseph
*Émile-Joseph Carlier, Bildhauer (*3.1.1849 Cambrai, †11.4.1927 Paris), war Mitarbeiter von Friedrich Goldscheider, Paris. Studium in Paris, u.a. bei François Jouffroy und Henri Chapu. Carlier war Mitglied des Komitees und der Jury der Societé des Artistes Français, wo er in den Jahren 1874–1926 Werke präsentierte. Lit.: TB, Bd. 5, S. 606; Hillenbrand, S. 63; Bénézit, Bd. 3, S. 244; AKL, Bd. 16, S. 416.*
818-GSB GROUPE „Sur la Gréve"

Carlus, Jean
*Jean Carlus, Bildhauer (*21.3.1852 Lavaur/Tarn, †Juni 1930), war Schüler von A. Falguière und A. Mercier. Carlus war ab 1886 Mitglied der Jury der Societé des Artistes Français, wo er bereits ab 1880 Werke präsentierte. Ab 1920 wird er Ritter der Ehrenlegion. Lit.: TB, Bd. 5, S. 12; Bénézit, Bd. 3, S. 251; AKL, Bd. 16, S. 466.*
4646 Stehende französische Bulldogge

Carrièr-Belleuse, Albert-Ernest
*Albert-Ernest Carrièr-Belleuse, Bildhauer (*12.6.1824 Ainsy-le-Château/Ainse, †4.(3.?)6.1887 Sèvres), dessen Werk vor allem antike und zeitgenössische Themen umfasste, war bekannt für seine frivolen Darstellungen. Wegen der Juni-Revolution ging Carrièr zuerst nach London und war ab 1850–1855 in Stoke-upon-Trent für Minton tätig. Napoleon kaufte seine 1863 unter großem Beifall ausgestellte Marmorstatue einer Bacchantin an. Seine Entwürfe wurden u.a. von Sèvres (Manufacture Nationale de Sèvres), wo er als Leiter der künstlerischen Abteilung für einen bedeutenden Aufschwung der Manufaktur verantwortlich war, ausgeführt. Lit.: TB, Bd. 6, S. 73–74; Mackay, S. 59; Zühlsdorff, S. 614; Hillenbrand, S. 64; Bénézit, Bd. 3, S. 292; AKL, Bd. 16, S. 602–603.*
804 Stehender Frauenakt mit Tuch und Katze (auf Rundsockel)
1610 FIGUR „Harmonie"/„Mélodie" (Sitzende junge Dame mit Lyra) (auf Rechtecksockel) (vgl. 2216; 1925= BÜSTE)
1647 BÜSTE „Chriemhilde"/„Frédègonde" (Barbusige Frau mit Löwenkopf und -fell als Kopfbedeckung) (vgl. 1979) (vgl. fast identisches Modell 1772)
1648 BÜSTE „Brunhilde"/(„Brunhaut") (Frau mit Adlerkopf und -schwingen als Kopfbedeckung) (vgl. 1773; 1980) (Pendant vgl. 1772)
1772 BÜSTE „Chriemhild (bekleidet)"/(„Frédègonde") (Bekleidete Frau mit Löwenkopf und -fell als Kopfbedeckung (vgl. 1647= Variante, nackt) (Pendant vgl. 1773; 1980)
1773 BÜSTE „Brunhilde (bekleidet)"/(„Brunhaut") (Frau mit Adlerkopf und -schwingen als Kopfbedeckung) (vgl. 1648; 1980) (Pendant vgl. 1772)
1925 BÜSTE „Harmonie"/„Mélodie" (Sitzende junge Dame mit Lyra) (auf Rechtecksockel) (vgl. 1610; 2216=

jew. Figur) (Pendant vgl. 1941)
1941 BÜSTE „Malerei" (Junge Dame mit Hochsteckfrisur und wallendem Gewand) (nach links blickend) (vgl. 1848; 2104= jew. Figur) (Pendant vgl. 1925)
1979 BÜSTE „Chriemhilde"/„Frédégonde" (Barbusige Frau mit Löwenkopf und -fell als Kopfbedeckung) (vgl. 1647) (vgl. fast identisches Modell 1772)
1980 BÜSTE „Brunhilde"/„Brunhaut" (Frau mit Adlerkopf und -schwingen als Kopfbedeckung) (vgl. 1648; 1773) (Pendant vgl. 1772)
2106 WANDMASKE „Chriemhilde" (vgl. 1772; 1979)
2107 WANDMASKE „Brunhilde" (vgl. 1773; 1980)
2216 FIGUR „Harmonie" (Sitzende junge Dame mit Lyra) (auf Rechtecksockel) (vgl. 1610; 1925= BÜSTE)

Carvin, Louis-Albert

*Louis-Albert Carvin, Bildhauer (*7.8.1875 Paris, †6.1.1951 Clamart/Seine?), war Schüler von Frémiet und Gardet. Carvin stellte zwischen 1891–1946 im Pariser Salon (Société des Artistes Français) aus und erhielt 1894 eine Ehrenvolle Erwähnung. Seine Spezialität war die Tierbildhauerei. Lit.: TB, Bd. 6, S. 97; Mackay, S. 59; Bénézit, Bd. 3, S. 311; AKL, Bd. 17, S. 39.*
1530 Lauernder Luchs auf Baumstamm BRONZE

Caussé, Julien

*Julien (Cadet) Caussé, Bildhauer (*28.6.1869 Bourges, †?), studierte bei A. Falguière und lebte in Paris. Seine allegorischen und mythologischen Statuetten wurden in Gips, Marmor, Bronze und Stein ausgeführt. Seine Werke wurden ab 1888 bei den Ausstellungen der Societé des Artistes Français präsentiert. Lit.: TB, Bd. 6, S. 206; Hillenbrand, S. 65; Bénézit, Bd. 3, S. 374; AKL, Bd. 17, S. 345.*
755-GSB CHOPE „Antique"

Cazaux, Édouard

*Édouard Cazaux, Bildhauer, Keramiker und Kunstgewerbler (*6.9.1889 Caneille, †10.9.1974 La Varenne, Seine), stammt aus einer Töpferfamilie und erlernte früh schon das Töpferhandwerk. Er studierte um 1906 an der Ecole des Beaux-Arts, Paris, und an der Manufacture Nationale de Sèvres. Cazaux war Jury-Mitglied der KGW-Ausst. in Paris 1925, erhielt zahlreiche Auszeichnungen und bekleidete Ehrenämter. Er war Mitglied der Künstlergruppe „Évolution". Auf der Exposition internationale des arts et techniques, Paris 1937, erhielt er eine Auszeichnung. Lit.: Goldscheider, A. (Hrsg.): Artistic Evolution, 1926, T. 1, Pl. 14 (m. Abb.); Vollmer, Bd. 1, S. 420; Neuwirth (1974), S. 60; Zühlsdorff, S. 400; Bénézit, Bd. 3, S. 390.*
VASE (mit Weintrauben und Blättern) (Steingut) (H. 30 cm, B. 28 cm) Lit.: Goldscheider, A. (Hrsg.): Artistic Evolution, 1926, T. 1, Pl. 14 (m. Abb.)
VASE (en faïence émaillée) (H. 17 cm, B. 28,5 cm) Lit.: Goldscheider, A. (Hrsg.): Artistic Evolution, 1926, T. 1, Pl. 17 (m. Abb.)
VASE „Piping dancing girl" (H. 5,2 cm, B. 20,5 cm) Lit.: Goldscheider, A. (Hrsg.): Artistic Evolution, 1926, T. 1, Pl. 18 (m. Abb.)
VASE „The Stag's Bow " (H. 12,5 cm, B. 21,5 cm) Lit.: Goldscheider, A. (Hrsg.): Artistic Evolution, 1926, T. 1, Pl. 18 (m. Abb.)
VASE (mit Blumenfries) (H. 24 cm, B. 21,5 cm) Lit.: Goldscheider, A. (Hrsg.): Artistic Evolution, 1926, T. 1, Pl. 18 (m. Abb.)
FLASCHE/VASE (en faïence décorée) (Form: rund/Dekor: Maskaron) (H. 34,5 cm, B. 28,5 cm) Lit.: Goldscheider, A. (Hrsg.): Artistic Evolution, 1926, T. 1, Pl. 41 (m. Abb.)

Ceccarelli, Ezio

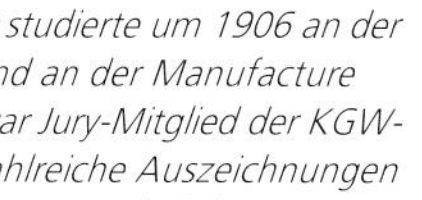

*Ezio Ceccarelli, Bildhauer (*27.7.1865 Montecatini, †Dez. 1927 Volterra), studierte in Florenz bei Bortone und Rivalta und schuf neben Denkmälern für Garibaldi und Columbus 1901 in Florenz das Grabmal Gioacchino Rossini. Neben Ausstellungen im Salon in Paris beteiligte er sich 1909 auch bei der Ausst. im Glaspalast München. Für die Produktion in Paris wurde für Ceccarelli der Name „Fontini" verwendet. Lit.: TB, Bd. 6, S. 250; Neuwirth (1974), S. 61; Bénézit, Bd. 3, S. 394; AKL, Bd. 17, S. 428.*
622-GSB BÜSTE „Automne" (vgl. 734-GSB; 1297)
734-GSB BÜSTE „Automne" (vgl. 622-GSB; 1297)
753-GSB BÜSTE „Coquette" (vgl. 1784)
1297 BÜSTE „Winzerin"/„Automne" (Herbst) (Frau mit großem Hut und Trauben) (vgl. 1619; 1961; 622-GSB; 734-GSB)
1619 BÜSTE „Winzerin"/„Automne" (Herbst) (Frau mit großem Hut und Trauben) (vgl. 1297; 1961; 622-GSB; 734-GSB)
1784 BÜSTE „Coquette" (Lachende, junge Dame mit Haarschleife) (nach rechts oben blickend) (vgl. 753-GSB)
1856 BÜSTE „Jugend"/„Jeunesse" (Lachendes nach rechts blickendes Mädchen mit Tulpenschmuck) (Inschrift: JEUNESSE) ÖMKI (1899/1900) (vgl. 2120; 2412)
1961 BÜSTE „Winzerin" (Frau mit großem Hut und Trauben) (vgl. 1297; 1619; 622-GSB; 734-GSB)
2120 BÜSTE „Jugend"/„Jeunesse" (Lachendes nach rechts blickendes Mädchen mit Tulpenschmuck) (Inschrift: JEUNESSE) ÖMKI (1899/1900) (vgl. 1856; 2412)
2192 BÜSTE „Ave virgo" ÖMKI (1901/02)
2412 BÜSTE „Jugend"/„Jeunesse" (Lachendes nach rechts blickendes Mädchen mit Tulpenschmuck) (Inschrift: JEUNESSE) ÖMKI (1899/1900) (vgl. 1856; 2120)
2948 „Cicero" (Stehende Figur) (auf quadratischer Plinthe) (Inschrift: CICERO) (vermutlich GS)

Cech, Irene

*Irene Cech, Retoucheurin (*6.6.1926, †?), war bei der Firma vom 9.6.1948 bis 29.12.1950 als angelernte Retouscheurin angestellt. Quelle: AGM, Zeugnis vom 29. Dezember 1950.*

Chalon, Louis

*Louis Chalon, Maler und Bildhauer (*15.1.1866 Paris, †1916), bevorzugte als Bildhauer die Kleinplastik (Statutetten, Figurengruppen, Vasen und Lampen), die dem Art Nouveau verhaftet ist und meist die sogenannten „Femme Fleurs" bzw. mythologische Sujets darstellt. Für die in verschiedenen Größen und Materialien editierte „Walküre" erhielt Chalon 1898 in Paris eine „Mention Honorable". Lit.: TB, Bd. 6, S. 341; Hillenbrand, S. 67; Bénézit, Bd. 3, S. 450–451; AKL, Bd. 18, S. 107; Arwas (2002), S. 609.*
35 Walküre (La Walkyrie) (Kämpferin mit Speer auf springendem Pferd) BRONZE/ELFENBEIN/Onyxsockel (vgl. 42)
42 Walküre (La Walkyrie) (Kämpferin mit Speer auf springendem Pferd) BRONZE / Marmorsockel (vgl. 35)
„Printemps de Bretagne" Paris, Salon (1902) Lit.: Neuwirth (1974), S. 61
TAFELAUFSATZ „Eclosion" BRONZE Paris, Salon (1902) Lit.: Neuwirth (1974), S. 61

Charpentier, Alexandre-Louis-Marie

*Alexandre Charpentier, Bildhauer, Medailleur und Kunstgewerbler (*10.6.1856 Paris, †3.3.1909 Neuilly-sur-Seine), studierte in Paris unter Hubert Ponscarme und war befreundet mit Auguste Rodin. Eine Verbindung besteht auch mit Constantin Meunier, da er wie dieser auch demokratische Ideen vertrat und Themen aus der Arbeitswelt bevorzugte. Seit der Gründung 1891 war Charpentier Mitglied der Societé Nationale des Beaux Arts und 1892 Mitbegründer der Abt. KGW im Salon. Gemeinsam mit J. Desbois und Jules Brateau versuchte er die Verwendung von Zinn als kunstgewerbliches Material neu zu beleben. Charpentier war Ritter der Ehrenlegion. Lit.: TB, Bd. 6, S. 405; Neuwirth (1974), S. 61 (hier die falsche Angabe Felix-Maurice); Hillenbrand, S. 68; Bénézit, Bd. 3, S. 507; AKL, Bd. 18, S. 268.*
885-GSB „Christ au Linceul"
886-GSB „Femme au Bain"
887-GSB „La Lutte"
888-GSB „Enfant Musicien" (vgl. 889-GSB)
889-GSB „Enfant Chantant" (vgl. 888-GSB)
„Jeune Provence" Paris, Salon (1898) Lit.: Neuwirth (1974), S. 61

Chauvel, Georges

*Georges Chauvel, Bildhauer (*7.9.1886 Elbeuf, †26.2.1962 Val-Saint-Germain), war Schüler von A.-E. Guilloux und ging mit 25 Jahren nach Paris, wo er Mitglied der Künstlergruppe „La Stèle" war. Er ist bekannt für seine realistisch-expressiven Werke, die ab 1919 im Salon d'Automne ausgestellt wurden. (–> vgl. Kap. „Dekorative Skulpturen von Arthur Goldscheider in Paris"). Lit.: Vollmer, Bd. 1, S. 427; Bénézit, Bd. 3, S. 535; AKL, Bd. 18, S. 344–345; Lit.: Goldscheider, A. (Hrsg.): Artistic Evolution, 1926, T. 2, Pl. 4 (m. Abb.).*
„Woman Archer" (H. fehlt, B. fehlt) Lit.: Goldscheider, A. (Hrsg.): Artistic Evolution, 1926, T. 2, Pl. 4 (m. Abb.)

Chavanne, (Jean-Marie?)

*Jean-Marie Chavanne, Bildhauer, Kupferstecher und Medailleur (*2.1.1797 Lyon, †1860 Lyon), war Schüler seines gleichnamigen Vaters und von André Galle. Chavanne stellte seine Werke von 1822–1852 (1860?) im Salon von Lyon aus. Seine bevorzugten Genres waren Porträtbüsten, Gruppen sowie weibliche Statuetten. Die von Goldscheider ausgeführte Modell-Nr. 1448/1973 wurde offenbar, da er zu diesem Zeitpunkt schon lange verstorben war, seinen Entwurf nachgebildet bzw. ausgeformt. Lit.: TB, Bd. 6, S. 444; Bénézit, Bd. 3, S. 539; AKL, Bd. 18, S. 350.*
1478 FIGUR „Vasenmaler" (Sitzender Römerjüngling mit Krug) (vgl. 1973)
1973 FIGUR „Vasenmaler" (Sitzender Römerjüngling mit Krug) (vgl. 1448)

Cherc

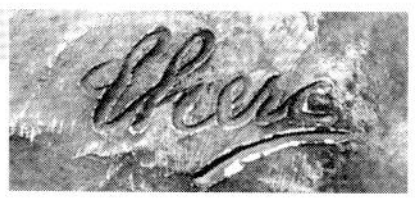

Ein Künstler namens „Cherc" (oder „Cheré" gelesen) lässt sich in der einschlägigen Literatur nicht nachweisen. Es ist somit, wie häufig bei Goldscheider, an ein Pseudonym zu denken. Infrage kommt in diesem Fall der Wiener Bildhauer Hans Schoerk (Lebens- und Schaffensdaten siehe dort), der für Goldscheider auch unter seinem richtigen Namen Modelle entworfen hat. Die mit „Schoerk" (und Abwandlungen sign. Objekte) finden sich dort. Neuwirth führt in ihrem Standardwerk zur Wiener Keramik zwar einige Modelle von „Cherc" auf, die Auflösung des Künstlernamens wird jedoch nicht hergestellt. Bei Pudor (1902), S. 35 wird die Namensform „Choerc" genannt, die eindeutig den Hinweis auf „Schoerk" bietet. Lit.: Neuwirth (1974), S. 61.
571 FIGUR „Die Saat" (Säender Bauer mit Hut) ÖMKI (1899/1900) (auch mit dem Titel „Saemann") (Pendant vgl. 1935; 2220)
606-GSB PLATEAU „Paris et Hélène" BRONZE / Fonte d'Art
757-GSB „Charmeur de Serpents"
766 RELIEF „Printemps" Inschrift: PRINTEMPS (Frühling)
768 RELIEF „Été" Inschrift: ÉTE (Sommer)
770 RELIEF „Autumne" Inschrift: AUTUMNE (Herbst)
772 RELIEF „Hiver" Inschrift: HIVER (Winter)
797 VASE „Lustiger Tanz" (Tanzendes Paar) ÖMKI (1900/01)
811 VASE Sterbender (toter?) Jüngling in der Meeresbrandung (von Frauengestalt mit Fackel und Möwe beobachtet)
966 VASE Zwei stehende Frauen, die eine mit Baum- und Blattwerk reliefierte Kugel in die Höhe halten
1000 FIGUR „Salome" (Stehende Frauengestalt mit lyraähnlichem Instrument) Inschrift: SALOME (vgl. 1900; 2085; 2530)
1459 BÜSTE Frau mit altägyptischem Kopfputz
1470 FIGUR „Negerknabe (fischend)" (Stehender Negerjunge mit Hut, Angel und Fisch) (vgl. 1471; 2134) (vgl. 1472= ähnl. Mod.)
1471 FIGUR „Negerknabe (rauchend)" (Stehender Negerjunge mit Hut und Zündholzschachtel) (vgl. 1470; 2135) (vgl. 1472= ähnl. Mod.)
1472 FIGUR „Negerknabe mit zerbrochenem Krug" (Stehender Negerjunge mit Hut und zerbrochener Vase) (vgl. 2136) (vgl. 1470; 1471= jew. ähnl. Mod.)
1499 VASE Disteln (Form: balusterförmig/Dekor: reliefierte Disteln und Insekten)
1506 BÜSTE „Epheu" (Junge Frau mit Efeuschmuck) (vgl. 1949; 2082)
1559 FIGUR „Fischerknabe (Neger)" (Sitzender Negerjunge mit Strohhut und Angel) (auf Natursockel)
1611 GRUPPE „Kunst bringt Gunst" (Strickender Junge mit Mütze, daneben zwei Mädchen, die seine Künste bewundern) (alle auf einem Mauersockel sitzend)
1620 FIGUR „Raucher" (Stehender lachender Negerjunge mit Zündhölzern)
1628 FIGUR „Dorfprinz" (Stehender rauchender Bursche) (auf quadratischem Landschaftssockel) (vgl. 1928) (vgl. 3823= BÜSTE)
1643 FIGUR „Mignon" (Stehende junge Frau mit langem Haar) (nach rechts blickend) (vgl. 2072; 2436)
1741 FIGUR „Amazonenkönigin" (Stehende Kriegerin, Halbakt auf Felssockel) (vgl. 1880) (vgl. 1923= BÜSTE) ÖMKI (1898/99)
1833 FIGUR „Gladiator" (Römischer Krieger mit Helm, Schild und Schwert) (auf Ovalsockel) (vgl. 2137) (vgl.

2033= BÜSTE)
1834 FIGUR „Bogenspanner" (Römischer Bogenschütze in Rüstung) (auf Ovalsockel) (vgl. 2138) (vgl. 2034= BÜSTE)
1861 FIGUR „Walküre"/„Le Walkyrie" (Stehende Kriegerin mit Lanze) (auf Felssockel)
1880 FIGUR „Amazonenkönigin" (Stehende Kriegerin, Halbakt auf Felssockel) (vgl. 1741) (vgl. 1923= BÜSTE) ÖMKI (1898/99)
1900 FIGUR „Salome" (Stehende Frauengestalt mit lyraähnlichem Instrument) Inschrift: SALOME (vgl. 1000; 2085; 2530)
1902 BÜSTE/LAMPE „Sumpfblume" (Blumengeschmückte junge Frau nach rechts blickend) (vgl. 2265)
1903 FIGUR „Rebecca" (Stehende Frauengestalt neben einer Mauer, ihren Ellenbogen auf eine Amphore stützend) Inschrift: REBECCA (vgl. 2171; 2537)
1920 FIGUR „Der kleine Meisterschütze" (Stehender Bauernjunge mit Armbrust)
1923 BÜSTE „Amazonenkönigin" (mit Kopfputz und Perlenkette) (Kopf nach rechts, geschlossene Augen) (vgl. 1741; 1880= jew. Figur)
1928 FIGUR „Dorfprinz" (Stehender rauchender Bursche) (auf quadratischem Landschaftssockel) (vgl. 1628) (vgl. BÜSTE 3823)
1935 FIGUR „Saat" (Säende Bäuerin) (Pendant vgl. 571; 2220)
1946 BÜSTE „Die Ranke" ÖMKI (1901/02) (vgl. 2195) (vgl. 2159= WANDMASKE)
1949 BÜSTE „Epheu" (Junge Frau mit Efeuschmuck) (vgl. 1506; 2082)
1951 FIGUR/LAMPE „Ex Oriente Lux" (Stehende mit Blüten geschmückte Frauengestalt/mit Schlange) (auf Rundsockel) (Inschrift: EX ORIENTE LUX)
1954 BÜSTE „Secession" (Junges Blumen geschmücktes Mädchen) (nach rechts blickend) (Cherc zugeschr.)
1955 s. Amy 1955
1991 BÜSTE „Boa" (Frau mit Schlange und großem extravagantem Hut) Inschrift: BOA
1993 BÜSTE „Athenais" (Blumengeschmückte junge Frau, nach links blickend) ÖMKI (1901/02) (vgl. 2254) (vgl. 2087= HERME) (vgl. 2087= WANDMASKE)
1994 BÜSTE „Theodora" (mit Helm) (vgl. 2253) (vgl. 2088= HERME)
1996 Muschelträgerin (Stehende, blumengeschmückte Frauengestalt, eine große Muschel haltend) (auf Rundsockel)
2030 FIGUR „Zeitungsständer mit Sockel" (Stehende junge Frau (mit eulengeschmücktem Diadem) auf Bücherstapel, zwei große Papierbögen auffächernd) (auf getrepptem obeliskartigem mit Eulen und Maskarons verziertem Sockel)
2033 BÜSTE „Gladiator"
2034 BÜSTE „Bogenspanner"
2053 FIGUR „Das tägliche Brod" ÖMKI (1900/01) (vgl. 2217)
2072 FIGUR „Mignon" (Stehende junge Frau mit langem Haar) (nach rechts blickend) (vgl. 1643; 2436) (auch sign. Mercadier)
2082 BÜSTE „Epheu" (Junge Frau mit Efeuschmuck) (vgl. 1506; 1949)
2085 FIGUR „Salome" (Stehende Frauengestalt mit lyraähnlichem Instrument) Inschrift: SALOME (vgl. 1000; 1900; 2530)
2087 HERME „Athenais" (vgl. 1993; 2254= jew. BÜSTE) (vgl. 2158= WANDMASKE)
2088 HERME „Theodora" (vgl. 1994; 2253= jew. BÜSTE)
2094 BÜSTE „Flieder" ÖMKI (1900/01) (vgl. 2369; 2370)
2134 FIGUR „Negerknabe (fischend)" (Stehender Negerjunge mit Hut, Angel und Fisch) (vgl. 1470; 1471) (vgl. 1472= ähnl. Mod.)
2135 FIGUR „Raucher" (Stehender Negerjunge mit Hut und Zündholzschachtel) (vgl. 1470; 1471) (vgl. 1472= ähnl. Mod.)
2136 FIGUR „Zerbrochener Krug" (Stehender Negerjunge mit Hut und zerbrochener Vase) (vgl. 1472) (vgl. 1470; 1471= jew. ähnl. Mod.)
2137 FIGUR „Gladiator" (Römischer Krieger mit Helm, Schild und Schwert) (auf Ovalsockel) (vgl. 1833) (vgl. 2033= BÜSTE)
2138 FIGUR „Bogenspanner" (Römischer Bogenschütze in Rüstung) (auf Ovalsockel) (vgl. 1834) (vgl. 2034= BÜSTE)
2158 WANDMASKE „Athenais" (Frauenkopf mit Blumenschmuck im Haar) (vgl. 1993; 2254= jew. BÜSTE) (vgl. 2087= HERME) (Pendant vgl. 2159)
2159 WANDMASKE „Ranke" (Frauenkopf mit Dutt und Blumenschmuck im Haar) (vgl. 1946; 2195= jew. BÜSTE) (Pendant vgl. 2158)
2171 FIGUR „Rebecca" (Stehende Frauengestalt neben einer Mauer, ihren Ellenbogen auf eine Amphore stützend) Inschrift: REBECCA (vgl. 1903; 2537)
2191 FIGUR „Die Ernte" ÖMKI (1902/03) (vgl. 2278) (vgl. 2303= BÜSTE)
2195 BÜSTE „Ranke" ÖMKI (1901/02) (vgl. 1946) (vgl. 2159= WANDMASKE)
2207 BÜSTE „Tulpe" ÖMKI (1901/02; 1902/03) (vgl. 2363; 2367)
2217 FIGUR „Tägliches Brod" ÖMKI (1901/02) (vgl. 2053)
2220 FIGUR „Die Saat" (Säende Bäuerin) ÖMKI (1899/1900) (Pendant vgl. 571; 1935)
2225 BÜSTE Frau mit Kastanien(?) und Blattwerk (Flora?)
2253 BÜSTE „Theodora" (mit Helm) (vgl. 1994) (vgl. 2087= HERME)
2254 BÜSTE „Athenais" (Blumengeschmückte junge Frau, nach links blickend) ÖMKI (1901/02) (vgl. 1993) (vgl. 2086= HERME) (vgl. 2087= WANDMASKE)
2255 BÜSTE „Kastanie" (Junge Frau mit Kastanienzweigen und Blattwerk) (vgl. 2410)
2265 BÜSTE „Sumpfblume" (Blumengeschmückte junge Frau nach rechts blickend) (vgl. 1902)
2278 FIGUR „Ernte" ÖMKI (1902/03) (vgl. 2191) (vgl. 2303= BÜSTE)
2302 BÜSTE „Saat" (Holländerin mit Haube) (nach links blickend)
2303 BÜSTE „Ernte" (vgl. 2191; 2278= jew. FIGUR)
2361 FIGUR „Die Fütterung" ÖMKI (1902/03)
2363 BÜSTE „Tulpe, II. Gr(öße)" ÖMKI (1901/02; 1902/03) (vgl. 2207; 2367)
2367 BÜSTE „Tulpe, III. Gr(öße)" ÖMKI (1901/02; 1902/03) (vgl. 2207; 2363)
2369 BÜSTE „Flieder" ÖMKI (1900/01) (vgl. 2094; 2370)
2370 BÜSTE „Flieder" ÖMKI (1900/01) (vgl. 2094; 2369)
2381 BÜSTE „Dictator" (Kaiser Caligula? mit Lorbeerkranz) (Inschrift: ODERINT DUM METUANT)
2404 BÜSTE „Erdbeere" (Junge Frau nach rechts blickend) (Haar und Sockel geschmückt mit Erdbeeren) (vgl. 2621)
2410 BÜSTE „Kastanie" (Junge Frau mit Kastanienzweigen und Blattwerk) (vgl. 2255)
2436 FIGUR „Mignon" (Stehende junge Frau mit langem Haar) (nach rechts blickend) (vgl. 1643; 2072)
2443 s. Bertrand, A.
2454 FIGUR „Aus der Bretagne" (Bäuerin mit Krug und Schale) (vgl. 2662) (Pendants vgl. 2917; 3840)
2482 Badende mit Tuch (auf Rundsockel)
2519 BÜSTE Junge Frau mit orientalischem (byzantinischem?) Kopfputz
2530 FIGUR „Salome" (Stehende Frauengestalt mit lyraähnlichem Instrument) Inschrift: SALOME (vgl. 1000; 1900; 2085)
2537 FIGUR „Rebecca" (Stehende Frauengestalt neben einer Mauer, ihren Ellenbogen auf eine Amphore stützend) Inschrift: REBECCA (vgl. 1903; 2171)
2572 BÜSTE Junge Frau mit Fuchsienblüten
2621 BÜSTE „Erdbeere" (Junge Frau nach rechts blickend) (Haar und Sockel geschmückt mit Erdbeeren) (vgl. 2404)
2662 FIGUR „Aus der Bretagne" (Bäuerin mit Krug und Schale) (vgl. 2454) (Pendants vgl. 2917; 3840)
2671 BÜSTE Frau mit Blumengirlande
2737 Stehender rauchender Gaucho mit Hut (auf Rechtecksockel)
2770 Wasserträgerin (eine Amphore auf der Schulter tragend) (vgl. F. Gross 2770)
2790 Krugträgerin (Stehende junge Holländerin, ein Gefäß tragend) (auf Landschaftssockel)
3058 GRUPPE „Musique" (Stehendes Kinderpaar, sich verschämt einen Apfel zureichend) (auf Landschaftssockel) (Inschrift: MUSIQUE)
3084 Stehender Junge mit Hut und Schulranzen (auf Felssockel)
3368 Stehende Frau (in antikisierendem Gewand), eine Amphore auf eine Säule stellend
3680 „Animus et vis" (Stehender römischer Krieger mit Helm, Speer und Tiger) (auf passigem Rechtecksockel) (mit oder ohne Inschrift: ANIMUS ET VIS) (auch sign. Lavergne) (vgl. Tiger 3536)
3768 Stehende Frau (daneben große Amphore, reliefiert mit mytholgischen Szenen) (vgl. 3923 Polter)
3823 BÜSTE Rauchender Bursche (auf konischem Sockel) (vgl. FIGUR 1928)
3887 LAMPE „Muschelträgerin" (Stehender Damenakt, eine Muschel über den Kopf haltend) ÖMKI (1899/1900)
4115 Löwe (auf rechteckigem Felssockel)
4194 Römischer Krieger mit Schild (Medusenhaupt) und Speer (auf Rechtecksockel) (Inschrift: ET MIHI SUNT VIRES ET MEA TELA NOCENT)
4747 Stehende Frau in Tracht (mit Haube und Blumenstrauß) (auf Rundsockel)
„Tell" ÖMKI (1898/99) Lit.: Neuwirth (1974), S. 61
„Bretagne" (Auf Felssockel stehende Holländerin, Wasser aus einem Krug in eine Schale gießend) (1902) Lit.: Pudor (1902), S. 35 (m. Abb.)
FIGUR „Winterblume" ÖMKI (1906/07) Lit.: Neuwirth (1974), S. 61

Chéret, Joseph

*Joseph Chéret, Bildhauer und Kunstgewerbler (*12.9.1838 Paris, †13.6.1894 Paris), begann seine Studien bereits mit 13 Jahren und erhielt später seine Ausbildung bei Albert-Ernest Carrière-Belleuse, von dessen reichem Entwurfsschatz einige Modelle bei Goldscheider produziert wurden. Chéret errang eine Ehrenmedaille in Wien und war ab 1887 Mitglied der künstlerischen Direktion der Manufacture Nationale des Sèvres und als freier Mitarb. bei Joseph-Théodore Deck tätig. Lit.: TB, Bd. 6, S. 463; Zühlsdorff, S. 401; Arwas (2002), S. 610; Bénézit, Bd. 3, S. 564.*
1329 BÜSTE „Modern"

Chinard, Joseph (nach)

*Joseph Chinard, Bildhauer (*12.2.1756 Lyon, †20.6.1813 Lyon). Lit.: AKL, Bd. 18, S. 549.*
621 BÜSTE „Madame Récamiere"

Chiparus, Demêtre (Dumitru / Dimitri)

*Demêtre Chiparus, Bildhauer (*16.9.1886 Dorohoi/Rumänien, †22.1.1947 Paris), studierte von 1906–1912 in Italien, wo er in Florenz Schüler von Raffaello Romanelli war. Anschließend bis 1919 Ecole des Beaux Arts in Paris bei Antonin Mercié. Chiparus schuf hauptsächlich dekorative Statuetten, kleine Figurengruppen, Tänzerinnen, u.a. der Ballets Russes. Er wandte sich orientalischen Motiven, Akrobaten und im Spätwerk der Tier-Skulptur auf Marmor- oder Onyx-Sockeln zu. Dabei bevorzugte er die Chryselephantin-Technik. Zu seinen Gießern zählte auch Etling in Paris. Lit.: Neuwirth (1974), S. 61; Hillenbrand, S. 70; Zühlsdorff, S. 582; Bénézit, Bd. 3, S. 597; AKL, Bd. 18, S. 564.*
„Danseuse Ta-Kéo" BRONZE Paris, Salon (1928) Lit.: Neuwirth (1974), S. 61

Chludek, Wilhelm

Wilhelm Chludek, Bildhauer (geb. in Polen), war ab 1953 bei Carstens (Goldscheider West-Germany) in Fredelsloh Mitarbeiter. Modelle von seiner Hand werden nur bei Könecke (2002) nachgewiesen und sind ausnahmslos in den Jahren 1953–1960 erzeugt worden. Lit.: AKL (Archiv, unredigiert); Mitteilung von G. Könecke, 5.9.2004.
602-WG WANDBEHANG/RELIEF „Betende Hände" (monogramiert AD) nach Albrecht Dürer
603-WG „Pudel" (Stehender Pudel)
8918 Stehendes Fohlen
8919 Liegendes Fohlen
8920 Äsendes Rehkitz
8921 Schreitendes Rehkitz
8922 Liegendes Rehkitz
8923 Stehendes Dromedarfohlen s. Matula
8930 Stehendes Eselsfohlen
8931 Stehendes Reh

Choerc s. Cherc

Ciseau

2760 Kniende betende Frau mit Urne/Vase (auf baumartigem Sockel)
2789 BÜSTE Junge Dame mit Kopftuch und Pelzkragen
2854 BÜSTE Jeanne d'Arc (als Kriegerin mit Rüstung und Helm)
2975 BÜSTE Junge Frau mit gesenktem Blick (vgl. 2976)
2976 BÜSTE Junge Frau mit gesenktem Blick (vgl. 2975)
2987 BÜSTE Junge Dame mit Kopftuch
FIGUR „Am Hochofen" ÖMKI (1906/07) Lit.: Neuwirth (1974), S. 61

Clarà, Juan (Joan)
*Juan, eigentl. Joan Clarà (i) Ayats, Bildhauer (*6.12.1875 Olot, Gerona/Spanien, †8.2.1958 Paris), war Schüler von Ernest Barrias und ließ sich 1901 in Paris nieder, wo er sich gemeinsam mit seinem Bruder ein Atelier teilte. Clarà ist vor allem für seine genrehaften Statuetten, besonders Kinderdarstellungen, bekannt. Lit.: TB, Bd. 7, S. 44; ollmer, Bd. 1, S. 446; Neuwirth (1974), S. 61; Hillenbrand, S. 72; Bénézit, Bd. 3, S. 671; AKL, Bd. 19, S. 367.*
61 Mädchen mit Katze spielend (auf Rechtecksockel) BRONZE
250 vgl. 1372?
1308 Zappelndes Kind auf Stuhl BRONZE (wohl Arthur GS)
1372? Kind auf Hocker, sich einen Schuh anziehend BRONZE
1403 „Alé" (Sitzendes, lachendes Kind) (Inschrift: ALÉ) BRONZE (vgl. 1485)
1485 Drei, auf einer Bank sitzende Kinder BRONZE (vgl. 1403)
2719 Mädchen mit Schubkarren (Fay.) (vgl. 2812)
2812 Mädchen mit Schubkarren (Fay.) (vgl. Sydan 2812; vgl. 2719)
„Premier Chausson" Paris, Salon (1905) Lit.: Neuwirth (1974), S. 61
„Nenette" Paris, Salon (1905) Lit.: Neuwirth (1974), S. 61
Auf einen Stuhl kletterndes Kind BRONZE (um 1905) (A/F. GS?) (H. ca. 25 cm); Soth. L. 20.6.1985
Stehendes Kind mit Hund BRONZE (um 1905) (A/F. GS?) (H. ca. 25 cm); Soth. L. 20.6.1985
Sechs lachende Kinder BRONZE/Marmorsockel (um 1905) (A/F. GS?) (H. 18 cm L. 38 cm); Soth. L. 6.6.1997

Cochet, Robert
*Robert Cochet, Bildhauer und Medailleur (*6.5.1903 Paris, †?), stellte im Pariser Salon (Societé des Artistes Français) aus und erhielt 1929 eine Medaille. Cochet war Mitglied der Künstlergruppe „Évolution" (-> vgl. Kap. „Dekorative Skulpturen von Arthur Goldscheider in Paris"). Lit.: Neuwirth (1974), S. 61; Bénézit, Bd. 3, S. 735.*

Colmant, Pierre
Pierre Colmant, Emailleur (nachweisbar zw. 1909–1928 in Paris), war Mitarbeiter von Arthur Goldscheider, Paris. Er erhielt eine Bronzene Medaille bei der KGW-Ausst. in Paris 1925. Lit.: Goldscheider, A. (Hrsg.): Artistic Evolution, 1926, T. 1, Pl. 17 (m. Abb.); Vollmer, Bd. 1, S. 463; Bénézit, Bd. 3, S. 797; AKL (Archiv, unredigiert).
VASE (cuivre gravé avec incrustations d' émail sur argent) (Form von F. La Monaca) (H. 14,8 cm, B. 7 cm) Lit.: Goldscheider, A. (Hrsg.): Artistic Evolution, 1926, T. 1, Pl. 17 (m. Abb.)
VASE (en ciuvre avec incrustations d' émaux sur argent) KUPFER (Form von J.-Charles Goëtz) (H. 18 cm, B. 10 cm) Lit.: Goldscheider, A. (Hrsg.): Artistic Evolution, 1926, T. 1, Pl. 24 (m. Abb.)

Colombier, Amélie
*Amélie Colombier, Bildhauerin (*Paris), studierte bei Franceschi und stellte in den Jahren zwischen 1884–1920 im Pariser Salon aus. Lit.: TB, Bd. 7, S. 247; Bénézit, Bd. 3, S. 800; AKL, Bd. 20, S. 363.*
1222 „Carmencita" BRONZE / Marmorsockel

Congers
„Fischerin" (Stehender Mädchenakt mit Angel) (auf Natursockel) Lit.: B., F.: Keramische Wanderungen durch den Kunstsalon Keller & Reiner Berlin, in: KM, 3.1903, S. 126 (m. Abb.), 128, „Congers kapriziöse kleine Fischerin zeigt bei sorgfältiger Modellierung die absolute Sicherheit, mit der ihr Schöpfer das harmonische Linienspiel in der Bewegung des weiblichen Körpers beherrscht".

Conrat, Ilse (Conrad, J.)
*Ilse Conrat, Bildhauerin (*20.1.1888 Wien, †9.8.1942 München), erhielt ihre erste Ausbildung bei Joseph Breitner und studierte 1898–1901 an der Akademie in Brüssel unter Charles van der Stappen. Anschließend war sie in Wien und München (ab 1928) tätig. Auf dem Wiener Zentralfriedhof findet sich ihr Grabmal für Johannes Brahms. Auf der Internationalen Kunstausst. in München erhält Conrat die Goldmedaille. Lit.: TB, Bd. 7, S. 317; Vollmer, Bd. 4, S. 485; Neuwirth (1974), S. 61 (hier wie auch im Kat. ÖMKI 1906/07 die falsche Angabe „J. Conrad"); AKL, Bd. 20, S. 562.*
2749 BÜSTE „Brahms" ÖMKI (1906/07)

Corbeille
2148 FIGUR „Der Traum" ÖMKI (1901/02) (vgl. 2214= BÜSTE)
2214 BÜSTE „Traum" (vgl. 2148= FIGUR)

Coudray, Georges-Charles
*Georges-Charles Coudray, Bildhauer und Medailleur (*Paris, nachweisbar zwischen 1883–1932), war Schüler von Falguière und Allouard und stellte zwischen 1883–1903 im Salon der Societé des Artistes Français seine meist nach orientalischem Vorbild gefertigten Statuen, Statuetten und Büsten aus. Lit.: TB, Bd. 7, S. 569; Bénézit, Bd. 4, S. 7; AKL, Bd. 21, S. 541.*
665-GSB VASE „Enfant Prisonnier" (Coudray zugeschr.)
666-GSB VASE „Nymphe a la Fontaine" (vgl. 682)
682 VASE „Nymphe a la Fontaine" (Nymphe am Brunnen) (Frauengestalt neben wasserspeiendem Gesicht sitzend) (vgl. 666-GSB)
3169 „Aida" (Stehende junge Frau mit Harfe) (Inschrift: AIDA)

Couturier, (Léon-Philibert)
*Léon-Philibert Couturier, Maler und Schriftsteller (*26.5.1823 Châlon-sur-Saône / Saône-et-Loire, †26.11.1901 Saint-Quentin/Aisne), war bekannt für seine Porträts, Gemälde im historischen Genre, Tierbilder und Stillleben. Die beiden hier nachgewiesenen Modelle, gehen auf ein Bild von Léon-Philibert Couturier zurück. Ein Bildhauer dieses Namens lässt sich nicht nachweisen. Lit.: TB, Bd. 8, S. 7; Bénézit, Bd. 4, S. 43; AKL, Bd. 22, S. 38.*
2820 UHR Napoleon (mit verschränkten Armen) vor Schlachtenszene (Mosaikimitation) stehend (vgl. 2855= FIGUR)
2855 Napoleon (mit verschränkten Armen) stehend (vgl. 2820= UHR)

Coyiegue (d.i. Czapek, Stanislaus)

Croisy, Aristide
*Aristide Croisy, Bildhauer (*31.3.1840 Fagnon/Ardennen, †6.11.1899 Paris), studierte in Paris, stellte in den Jahren 1867–1898 im dortigen Salon aus und errang insgesamt 3 Medaillen. Croisy war Mitarbeiter (Bronzen) von Friedrich Goldscheider, Paris. Lit.: TB, Bd. 8, S. 148; Neuwirth (1974), S. 61; Bénézit, Bd. 4, S. 109–110; AKL, Bd. 22, S. 393.*
791-GSB STATUETTE „Baigneuse"
882-GSB GROUPE „Le Nid" (vgl. 883)
883-GSB GROUPE „Le Nid" (vgl. 882)
VASE „Amour poursuivi" Lit.: R., F.: Wiener Kunstkeramik, in: KM, 2.1902, S. 89 (m. Abb.), „Die Vase, deren Entwurf von Croisy stammt, ist eine stark ausgebauchte Fussurne, um welche oben ein tuchartig arrangierter Schmuck liegt. In diesem Tuch klettert nun ein Amor umher, mit zurückgewandtem Schelmengesicht forschend, ob die schlanke Frauengestalt, welche ihm folgt, ihn noch nicht erreicht hat. Die vortreffliche Modellierung des weiblichen Aktes, die graziöse Komposition der anmutigen Scene vereinigen sich bei dieser Vase vortrefflich mit dem zarten, gelblichgrün-bräunlichen Dekor".

Czapek, Stanislaus (Stanislaw) (Signaturvarianten: Capèque, Czapeque bzw. Czapèque)
*Stanislaus (Stanislaw) Czapek, Bildhauer (*29.6.1874 Jicin/Böhmen, †?), studierte an der KGW-Schule Lwów und anschließend an der Akademie in Wien bei Edmund Hellmer. Ab 1904 war Czapek in Wien ansässig und lange Zeit erster Bildhauer der Manufaktur Friedrich Goldscheider. Die zahlreichen hier aufgelisteten Modelle belegen dies nachhaltig. Czapek war auch in Paris tätig und gestaltete u.a. im Jahre 1900 das Denkmal der Kaiserin Elisabeth in Salzburg. Ab 1908 wurden seine Werke im Salon der Societé des Artistes Français präsentiert. Bei Goldscheider wurden für Czapek vielfältige Namensformen und Pseudonyme verwendet. Die franz. Formen seines Namens (Czapèque bzw. Capèque) machten seine Entwürfe für eine wie Goldscheider stark exportorientierte Firma „griffiger" und somit für den internationalen Markt klangvoller. Das Pseudonym „Podola" ließ sich durch zeitgenössische Reklamen/Annoncen in Magazinen und Zeitungen auflösen. Infrage kommen auch weitere Namen wie Padola, Podolak und Polter. Abschließend kann zu diesen und evtl. noch weiteren möglichen Pseudonymen und Namensformen jedoch noch nichts weiter gesagt werden, zumal sich die Entwürfe und Genres sehr ähneln bzw. größtenteils identisch sind. Die Modelle werden jeweils bei den stark abweichenden Namensformen (zusätzlich) aufgeführt. Lit.: TB, Bd. 8, S. 237–238; Neuwirth (1974), S. 61; Bénézit, Bd. 4, S. 160; AKL, Bd. 23, S. 282.*
330 FIGUR „Melonenverkäufer" (Kniender Araber mit ausgestrecktem Arm) (auf Rechtecksockel) (sign. Capèque)
1176 VASE „Die Quelle" (aus einer Quelle, als wasserspeiendes Gesicht an der Vasenmündung, trinkendes Figurenpaar) (sign. Capèque)
2494 UHR Weinbergschnecke (sign. Capèque)
2560 Diskuswerferin (in antikisierendem Gewand) (vgl. 2793) (sign. Capèque)
2585 UHR/LAMPE „Feenwagen" (Stehende Fee in einem römischen Streitwagen, der von zwei Panthern gezogen wird)
2600 UHR Männliche Figur an Glocke (Gehäuseaufbau mit „byzantinischem Wandmosaik") (Inschrift: VIVOS VOCO MORTVOS PLANGO FVLGVRA FRANGO) (vgl. 2678; 2718) (sign. Capèque)
2678 UHR Männliche Figur an Glocke (Gehäuseaufbau mit „abblätternder Mauer") (Inschrift: VIVOS VOCO MORTVOS PLANGO FVLGVRA FRANGO) (vgl. 2600; 2718) (sign. Capèque)
2718 UHR Männliche Figur an Glocke (Gehäuseaufbau mit „Kirche in weiter Landschaft") (Inschrift: VIVOS VOCO MORTVOS PLANGO FVLGVRA FRANGO) (vgl. 2600; 2678) (sign. Capèque)
2793 Diskuswerferin (in antikisierendem Gewand) (vgl. 2560)
2826 UHR Allegorie des Lebens (vgl. Lavergne 2826!) (sign. Capèque)
3283 Spähende Amazone (mit Schild und Speer) auf Pferd (auf Landschaftssockel) (sign. Capèque)
3288 Spähende Amazone (mit Schild und Dolch) auf Pferd (auf Landschaftssockel) (sign. Capèque)
3424 LAMPE Fackelträgerin (Stehende Frau auf säulenartigem Sockel, zwei Fackeln in die Höhe haltend) (sign. Capèque)
3531 LAMPE „Aquarium" (Zwei Mädchen auf Felsen, in einen Brunnen schauend) (sign. Czapèque)
3625 „Pferde" (Zwei Pferde, bronziert auf marmoriertem Sockel) ÖMKI (1912) (sign. Capèque)
3657 Biedermeiermädchen einen großen Korb auf dem Rücken tragend (sign. Polter)
3822 KERZENHALTER (zweiflammig) Stehender Putto (auf säulenartigem Sockel) (sign. Capèque)
3885 Stehendes Biedermeiermädchen mit Rose (sign. Czapeque) (vgl. 3948)
3894 Amazone mit Schild, vor Pferd stehend (auf gewölbtem Rechtecksockel) (vgl. 4155)
3947 Stehendes Biedermeiermädchen mit Blume FAYENCE / ALABASTER (sign. Capeque)
3948 Stehendes Biedermeiermädchen mit Rose FAYENCE / ALABASTER (sign. Capèque / Czapèque) (vgl. 3885)
4155 Amazone mit Schild, vor Pferd stehend (auf gewölbtem Rechtecksockel) (vgl. 3894)
4222 „Schachmatt" (Sitzendes Rokokopaar beim Schachspiel) (auf girlandenverziertem Ovalsockel) (sign. Podola)
4271 LAMPE Tanzendes Rokokopaar (Dame und Kavalier) (mit auf Blumengirlanden verziertem Rundsockel) (sign. Podola / Polter) (vgl. 4915) (vgl. 4305; 4539= jeweils ähnl. Modell)
4380 Tanzende junge Frau mit Leopard (Bacchantin) (sign. Podola)
4495 „Indische Tänzerin" (auf Rundsockel) (Ruth Saint Denis) (sign. Podola / Rosé)
4539 LAMPE Scherzendes Biedermeierpaar (Dame und Kavalier) (auf mit Blumengirlanden verziertem Rundsockel) (sign. Padola / Podola) (vgl. 4305 und 4915= jeweils ähnl. Modell)
4565 Junge, sich drehende Biedermeiertänzerin (auf Rundsockel) (sign. Padola)
4676 Die Tanzstunde (Violinspieler im Biedermeiergewand mit einem Mädchen tanzend) (auf Rundsockel)

(sign. Padola / Podola / Podolak)
4684 „Ballpause" (Sitzende Krinoline, eine Maske haltend) FAYENCE / ALABASTER (sign. Podola(k)) (vgl. LAMPE 4883; 4884)
4864 BÜSTE Mädchen ALABASTER / MARMOR?
4883 FIGUR / LAMPE „Ballpause" (Sitzende Krinoline, eine Maske haltend) (sign. Podola(k)) (vgl. 4684 Podola(k))
4884 FIGUR / LAMPE „Ballpause" (Sitzende Krinoline, eine Maske haltend) (vgl. 4684 Podola(k))
4915 LAMPE Tanzendes Rokokopaar (Dame und Kavalier) (auf mit Blumengirlanden verziertem Rundsockel) (sign. Podola / Polter) (vgl. 4271) (vgl. 4305; 4539= jeweils ähnl. Modell)
4922 Stehender Harlekin mit Laute und Colombine (auf oktogonalem Sockel) (sign. Podola) (vgl. 4922; 5262; 5263 sign. „Podolak"= Harlekin einzeln)
„Liselotte" (Fay.) ÖMKI (1910/11; 1912) Lit.: Neuwirth (1974), S. 61
„Odette" ÖMKI (1911/12) Lit.: Neuwirth (1974), S. 61
„Willy" ÖMKI (1911/12) Lit.: Neuwirth (1974), S. 61
„Tiger" ÖMKI (1911/12) Lit.: Neuwirth (1974), S. 61
„Windhund" ÖMKI (1912) Lit.: Neuwirth (1974), S. 61
„Aus Großmutters Tagen" ÖMKI (1912) Lit.: Neuwirth (1974), S. 61
„Schneeballen" ÖMKI (1912) Lit.: Neuwirth (1974), S. 61
„Tänzerin" ÖMKI (1912) Lit.: Neuwirth (1974), S. 61
„Chrysis" (Stehender Frauenakt) MARMOR, Salon Paris, (1920/21) Lit.: N.N.: Wiener Kleinplastik einst und jetzt, in: MW, 2.1920/21, Nr. 12, S. 29 (m. Abb.); MW, 3.1921/22, Nr. 4, S. 35 (m. Abb.).
LAMPE Stehende Frauengestalt, einen Arm ausgestreckt, mit dem anderen eine Schale in die Höhe haltend (um 1905) (H. 118,5 cm); Christie's SK 15.5.2003 (hier sign. Capèque)

Czapèque s. Czapek, Stanislaus

Dakon, Stephan (Stefan)

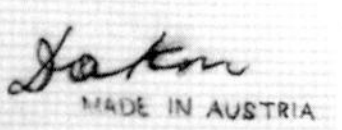

*Stephan Dakon, Bildhauer (*14.11.1904 Wien, †27.2.1992 Wien), besuchte eine Bildhauerschule und absolvierte anschließend eine Lehre in der Kunsterzgießerei im Wiener Arsenal. Dort Bekanntschaft mit Josef Lorenzl, aus der sich dann eine Freundschaft der beiden Künstler entwickelt. Durch Lorenzls Vermittlung erhielt Dakon um 1924 dann eine Anstellung bei der Manufaktur Friedrich Goldscheider. Dakon arbeitete auf Tantiemenbasis (zw. 2–10% vom Verkaufserlös) und wurde im Laufe der Zeit einer der wichtigsten und produktivsten Entwerfer für die Manufaktur. Warum in der Zs. „Moderne Welt" 1928/29 in einem Artikel der Name „A. Stefan" für zwei Modelle Dakons aufscheint, bleibt rätselhaft. Vermutlich handelt sich um einen redaktionellen Fehler und nicht um ein bewusst von der Manufaktur lanciertes Pseudonym. Nach dem Kriegsdienst und Kriegsgefangenschaft in Norwegen nahm er 1945 wieder seine Künstlertätigkeit auf. Hersteller wie Keramos und die Wiener Kunstkeramik wurden auf ihn aufmerksam und führten ebenfalls Modelle von ihm aus. Ebenso war Dakon für die deutschen Porzellanfabriken Hertwig/Katzhütte (-> vgl. eBay # 3728355757) sowie für W. Goebel, Oeslau tätig, für die er zahlreiche figürliche Modelle und Wandmasken entwarf. Dakon-Masken wurden auch bei seinem Freund und Künstlerkollegen Adolf Prischl in dessen Wiener Firma „Kunstkeramik Adolf Prischl" ausgeformt und angeboten. Als Walter Goldscheider 1953 die Firmenrechte an Carstens, Fredelsloh übertrug, wurde auch dort eine stattliche Anzahl von Dakons Masken und Figuren ins Programm aufgenommen. Die Entwürfe der Karikaturistin Germaine Bourets wurden von Dakon nach ihren Zeichnungen ausmodelliert und retouchiert. Aufgelistet sind diese Modelle unter der Künstlerin Bouret (-> vgl.Foto, Archiv Mimi Dakon). Von Dakon stammt nicht nur eines der erfolgreichsten Goldscheidermodelle „Beauty" (Modell-Nr. 7195; 7897; 7996; 8746), sondern man darf ihn als einen der wichtigsten Gestalter für Wandmasken der 30er bis 60er Jahre bezeichnen. Die in der Literatur immer wieder abweichende Namensform „Stefan", konnte über Dakons Frau Mimi aufgelöst werden. Sie bestätigte die Schreibweise „Stephan". Für diesen und zahlreiche weitere Hinweise, die Künstler Dakon, Prischl und Thomasch betreffend, möchte ich an dieser Stelle Frau Mag. Uta M. Matschiner (St. Valentin) recht herzlich danken. Mit ihrer Hilfe gelang es nicht nur, einige Modelle eindeutig an diese Künstler zuzuschreiben, sondern durch Modellfotos, die aus dem Archiv Mimi Dakon stammen, welches ebenfalls von Frau Matschiner verwaltet wird, deren Stellenwert umfassend zu dokumentieren. Lit.: Makus (2000), S. 167, 182; AKL (Archiv unredigiert); Matschiner (2006); Persönliche Mitt. von Frau Mag. Uta M. Matschiner (2002–2005).*
301-WG „Negerin mit Goldschale" (Sitzendes Negermädchen mit Schale)
302-WG „Negerin mit Goldschale" (Kniendes Negermädchen mit Schale auf dem Kopf)
303-WG „Neger" (Kniender Neger mit Trommel)
304-WG „Neger" (Stehender Neger mit Speer auf Natursockel)
305-WG „Akt" (Liegender Mädchenakt)
306-WG „Tanz" (Tanzender Mädchenakt mit Tamburin)
307-WG „Akt" (Kniender Mädchenakt mit Tuch)
308-WG „Neger mit Schild" (Kniender Neger mit Schild)
309-WG Hockende(r?) Neger/in?
310-WG „Negermädchen" (Hockendes bzw. vielmehr kniendes Negermädchen)
311-WG „Negermädchen" (Sitzendes Negermädchen)
312-WG „Akt" (Kniender Mädchenakt)
313-WG WANDMASKE „Negerkopf" (Junge Negerin mit Ohrringen, Kopf- und Halsschmuck)
314-WG WANDMASKE „Negerkopf" (Junge Negerin mit Ohrringen, Kopf- und Halsschmuck)
315-WG „Negerin" (Kniende Negerin mit Kopf- und Halsschmuck)
316-WG WANDMASKE „Kindermaske" (Moritz) (Pendant vgl. 317-WG)
317-WG WANDMASKE „Kindermaske" (Max) (Pendant vgl. 316-WG)
318-WG „Neger" (Hockender Neger mit Schale)
320-WG „Akt" (Hockender Mädchenakt auf Rundsockel)
321-WG WANDMASKE „Maske, Frau mit Tuch" (Frauenkopf mit Kopftuch / Schal, geschlossene Augen)
322-WG WANDMASKE „Maske" (Frauenkopf mit Kopftuch / Schal, offene Augen)
323-WG „Negerin" (Tanzende Negerin auf Natursockel)
324-WG WANDMASKE „Maske" (Frauenkopf nach links blickend)
328-WG „Akt" (Kniender Mädchenakt, sich die Sandale richtend)
329-WG „Balinesin" (Tanzende Balinesin auf nierenförmigem Sockel)
330-WG „Tänzerin" (Tanzendes Mädchen im Ballettkostüm auf Ovalsockel)
331-WG „Tänzerin" (Flöte spielende Tänzerin)
332-WG „Balinese" (Sitzender Balinese mit Flöte)
333-WG „Balinesin" (Kniende Balinesin mit Schale)
334-WG „Balinese" (Sitzender Balinese mit Brett?)
335-WG „Balinesin" (Tanzende Balinesin mit Fächer, auf nierenförmigem Sockel)
336-WG „Tanzfigur" (Flötespielendes Mädchen im Ballettkostüm auf Ovalsockel)
337-WG WANDMASKE „Maske Balinesin" (Balinesin mit Ohrringen und Halsschmuck)
338-WG „Bogenschütze" (Stehender Bogenschießender Indianer auf unregelmäßigem Sockel)
339-WG „Akt" (Stehender Mädchenakt mit Tuch, auf einem Bein stehend) (auf Ovalsockel)
340-WG „Akt" (Sitzender Mädchenakt, die Arme hinter dem Kopf verschränkend)
342-WG „Akt" (Liegender Mädchenakt mit Strohhut)

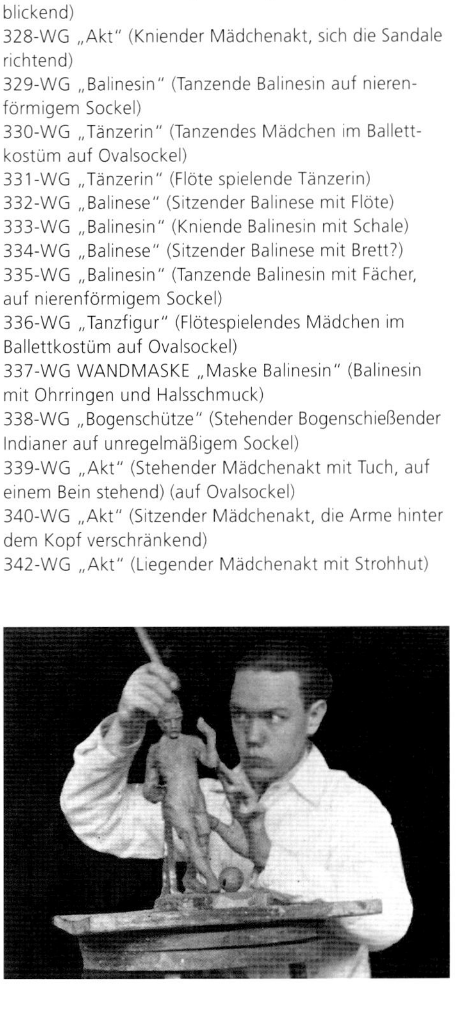

343-WG „Mädchen mit Schale" (Kniender Mädchenhalbakt mit Schale)
345-WG „Maske" (Mädchenkopf mit Pferdeschwanz und geschlossenen Augen)
346-WG „Maske" (Mädchenkopf mit kurzem Haar nach links blickend)
347-WG „Tänzerin" (Balletttänzerin auf einem Bein, auf Ovalsockel)
349-WG WANDBEHANG „Sahara-Büste" (Afrikanerin eine Schale auf dem Kopf tragend)
350-WG „Maske" (Mädchenkopf mit Pferdeschwanz im Profil)
351-WG „Maske" (Mädchenkopf mit Kragen)
352-WG „Tänzerin" (Balletttänzerin, sich ihren Schuh schnürend)
354-WG WANDPLASTIK „Negerboot" (Negerin mit Kleinkind im Boot)
355-WG WANDPLASTIK „Negerboot" (Negerin mit Kleinkind auf dem Rücken)
357-WG WANDPLASTIK „Schlittschuhläuferin"
358-WG „Akt" (Sitzender Mädchenakt mit Buch)
359-WG „Akt" (Sitzender Mädchenakt mit Bikini)
360-WG „Mädchen mit Koffer" (Mädchen auf Koffer sitzend, daneben ein zweiter Koffer stehend)
361-WG „Akt, kniend" (Kniender Mädchenakt)
362-WG „Akt" (Kniender Mädchenakt)
363-WG „Akt"
364-WG „Akt"
5060 Stehende Tänzerin, nach vorne zum Kuss gebeugt (auf trapezförmigem Sockel)
5532 FIGUR / LAMPE Stehende Pierrette mit Maske (auf Rundsockel) FAYENCE / ALABASTER (auch sign. Lorenzl)
5570 FIGUR „Schleiertanz"
5584 „Würde" (Stehende junge Dame in halbtransparentem Kleid und großem Umhang) (auf oktogonalem, abgerundetem Sockel)
5592 „Kugelwerferin"
5595 Sitzende junge Dame mit Fächer (auf Rechtecksockel, mit abgeschrägten Ecken an der Schauseite)
5596 „Pyjama" (Emmy Sturm stehend im Pyjama) (auf querovalem Sockel)
5612 GRUPPE „Revue" (Die „Dolly Sisters" nebeneinander stehend, ihre kurzen Röcke hochhebend) (auf Rechtecksockel) (vgl. 5634; 5649= jeweils Einzelfigur) (vgl. 5819= ViererGRUPPE)
5625 FIGUR „Harvey" (Lilian Harvey als schreitende Tänzerin, ihren Rock mit der rechten Hand anhebend) (auf facettiertem Ovalsockel) (vgl. 6009; 6211)
5629 DOSE „Mädchen"
5634 „Revue" (vgl. 5612; 5649) (vgl. 5819= ViererGRUPPE)
5649 FIGUR / LAMPE „Revue" (Schreitende Einzelfigur der „Dolly Sisters", ihren kurzen Rock hochhebend, vor Straßenlaterne) (auf Rechtecksockel) (LAMPE als Straßenlaterne) (vgl. 5612= ZweierGRUPPE; 5634) (vgl. 5819= ViererGRUPPE) (GS neu Modell-Nr. 6 „Lilli Marlen")
5658 „Dreispitz" (Stehende junge Dame in Karnevalskostüm und Dreispitz) (Lissi Arna?) (auf Dreiecksockel) FAYENCE / ALABASTER
5659 „Banjospiel" (Zerline Balten als stehende Banjospielerin) (auf Rundsockel mit kleiner sitzender Pierrotfigur)
5670 „Bananentanz"
5712 GRUPPE / LAMPE „Turteltauben" (Die Dolly Sisters, Cheek to cheek, zueinander stehend, ihre kurzen Kostüme anhebend) (auf Ovalsockel)
5713 „Zylinder"
5714 „Modern" (Stehende Tänzerin in expressivem Kostüm) (auf Rundsockel)
5776 GRUPPE „Matrosentanz" (Maria Solveg und Katta Sterna im Matrosenkostüm, ZweierGRUPPE) (auf Ovalsockel) (vgl. Einzelfigur 5818; vgl. 5961; vgl. DreierGRUPPE; Soth. N.Y. 14.6.1991)
5781 „Dolly Sister(s)" (Stehend, zum Kuss nach vorne gebeugt, ihr kurzes Kostüm nach hinten weg haltend) (auf getrepptem Trapezsockel)
5808 FIGUR „Clown mit Ball"
5815 „Bubi" (Louise Brooks als stehende junge Dame in kurzen Hosen und Bubikopf) (auf abgerundetem Rechtecksockel) (vgl. 5959) (vgl. 6210= Frisurvarianten: Bubikopf bzw. helles, lockiges Haar)

5818 „Matrosentanz" (Maria Solveg im Matrosenkostüm) (auf Ovalsockel) FAYENCE/ALABASTER (vgl. ZweierGRUPPE 5776; vgl. DreierGRUPPE, Soth. N.Y. 14.6.1991)
5819 „Revue zu 4" (Die „Dolly Sisters" als „Vierer Gruppe" nebeneinander stehend, ihre kurzen Röcke hochhebend) (auf Rechtecksockel) (vgl. 5634; 5649= jeweils Einzelfigur) (vgl. 5612= ZweierGRUPPE)
5863 „Matrosin"
5864 „Maskentanz" (Rita Zabekow in einem Kostüm mit Kapuze, eine Maske haltend) (auf profiliertem Sockel mit Maske)
5865 „Jagdfee"
5876 FIGUR „Fächer(dame)" (Stehende Varietétänzerin mit großem Feder-Fächer) (auf an den Ecken abgeschrägtem, gekehltem Rechtecksockel) (vgl. 6214)
5881 „Frühlingstanz"
5903 „Aschermittwoch" (Stehende Pierrette mit Gitarre) (auf getrepptem rautenförmigem Sockel)
5907 FIGUR „Akt" („Kugelspiel") (Stehender Damenakt mit ausgestreckten Armen) (auf Rundsockel) (vgl. Lorenz 5802= fast ident. Modell)
5914 „Schwestern" (Die Vigo-Drillinge, drei Varietétänzerinnen, als Chorusline hintereinander stehend) (auf abgerundetem Längssockel)
5915 „Sportgirl"
5916 „Tennisspielerin"
5918 „Vogelgezwitscher"
5922 „Coolen more"
5930 „Schottländerin" (Stehende junge Dame mit Hut (Schiffchen) und Kniestrümpfen, ihren (Schotten)-Plisseerock fächerartig auseinander haltend) (auf getrepptem, oktogonalem Sockel)
5932 „Matrosengirl"
5938 „Autogirl" (Stehende junge Dame im Overall mit Kappe und Sportbrille) (auf getrepptem Quadratsockel)
5949 „Donauwalzer"
5950 „Tänzerin"
5959 „Bubi" („Verkl.") (Louise Brooks als stehende junge Dame in kurzen Hosen und Bubikopf) (vgl. 5815)
5961 „Matrosentanz" (vgl. 5776)
5962 „Küsschen" (vgl. 6209)
5975 GRUPPE „Tanz" (Zwei Tänzerinnen, eine davon mit Hut, im Tanzschritt hintereinander posierend) (auf Ovalsockel)
5984 „Kind im Sessel"
6009 FIGUR „Harvey" (Lilian Harvey als schreitende Tänzerin, ihren Rock mit der rechten Hand anhebend) (auf facettiertem Ovalsockel) (vgl. 5625; 6211)
6018 „Skiläuferin"
6019 „Wiener Walzer" („Stern") (Stehende, sich verneigende Pierrette auf Rundsockel mit Spitzhut) (vgl. 6647)
6020 FIGUR „Wiener Walzer" (Stehende Tänzerin, ihr Kleid mit langer schmaler Schleppe, fächerartig ausbreitend) (auf Ovalsockel)
6052 „Libelle"
6067 FIGUR „Hund" („Figur mit Hund")
6068 FIGUR „Ramona" („Tanzfreude")
6070 „Wind"
6071 FIGUR „Cimono" („Kimono")
6074 FIGUR „Daisy" (Stehende junge Dame mit Béret/Pullmannkappe und langen Hosen, eine Gitarre haltend) (auf oktogonalem Sockel) (vgl. 6629)
6075 „Schiffsgirl"
6077 FIGUR „Page im Sessel" (Sitzende junge Dame mit Bubikopf/Pagenschnitt) (auf Rechtecksockel)
6083 „Ryth(mischer) Tanz"
6150 FIGUR „Mary" (Stehende Tänzerin, ihr Kleid mit beiden Händen anhebend) (auf Rundsockel mit 3 kleinen Würfeln)
6170 „Schwimmerin" (Posierende junge Dame im Badedress, ihre Kappe mit der Rechten in die Höhe haltend (auf Rundsockel)
6207 „Tanz" (Schreitende Tänzerin, ihr Kleid fächerartig auseinander haltend) (auf Ovalsockel)
6209 „Küsschen" (vgl. 5962)
6210 „Bubi" (Louise Brooks als stehende junge Dame in kurzen Hosen und Bubikopf bzw. hellem, lockigem Haar) (auf abgerundetem Rechtecksockel) (vgl. 5815; 5959= jeweils Bubikopf)
6211 FIGUR „Harvey" (Lilian Harvey als schreitende Tänzerin, ihren Rock mit der rechten Hand anhebend) (auf facettiertem Ovalsockel) (vgl. 5625; 6009)
6214 FIGUR „Fächer(dame)" (Stehende Varietétänzerin mit großem Feder-Fächer) (auf an den Ecken abgeschrägtem, gekehltem Rechtecksockel) (vgl. 5876)
6277 Ballerina beim Spitzentanz (auf Rundsockel mit Hut)
6330 FIGUR „Kongress" (Lilian Harvey) (auf Rundsockel) (vgl. 6646)
6352 FIGUR „Castanet" (Tanzende Spanierin) (auf Rundsockel)
6360 Spanierin (Carmen) in Pose (auf profiliertem länglichem Sockel)
6361 FIGUR „Sterbender Schwan" (Anna Pawlowa) mit Federputz im Haar (auf Ovalsockel) (vgl. 7426= ähnl. Modell)
6458 FIGUR „Alt-Wien" (Schreitende junge Dame mit Schutenhut und gerüschtem Kleid (auf Rundsockel) (vgl. 7286) (auch für GS USA, Mod.-Nr. 195)
6459 Schreitende junge Dame mit Hut „im Wind" (auf abgerundetem Rechtecksockel)
6460 FIGUR „Innosence" (Stehende junge Dame mit Schutenhut) (auf Rundsockel)
6461 GRUPPE „Tanz" („Tanzgruppe")
6504 „Rosenkönigin" (Tänzerin mit Reifrock) (auf Ovalsockel)
6508 „Fusspitzentanz"
6520 Stehende junge Dame in weiten Hosen (sog. Schlaghosen) mit großem Hut und Ball (auf getrepptem Rechtecksockel?)
6521 Schreitende Tänzerin mit Hut (ihr Kleid mit beiden Händen anhebend) (auf abgerundetem Rechtecksockel)
6535 „Rosenkavalier" s. Lorenzl 6535
6551 FIGUR „Blonder Traum" (Lilian Harvey als schreitende Dame mit großem Hut und Handschuhen, ihr Kleid mit beiden Händen anhebend) (auf Ovalsockel) (vgl. 6606; 7208) (GS neu Modell-Nr. 29 „Noblesse")
6552 Stehende Tennisspielerin (mit Kappe und Schläger) (auf Rundsockel)
6553 Stehende Tänzerin, ihr Kleid auseinander haltend (auf oktogonal facettiertem Rundsockel)
6554 Sitzende junge Dame in langen Kleid (auf Ovalsockel)
6555 „Jane" (Schreitende Tänzerin in Pose , ihr Kleid mit der linken Hand anhebend) (auf Ovalsockel)
6586 Stehende junge Dame mit Blumen (ihr langes Kleid etwas anhebend) (auf Ovalsockel)
6591 FIGUR „Kitty" (Tänzerin mit langem Kleid, in Pose) (auf Ovalsockel)
6606 FIGUR „Blonder Traum" (Lilian Harvey als schreitende Dame mit großem Hut und Handschuhen, ihr Kleid mit beiden Händen anhebend) (auf Ovalsockel) (vgl. 6551; 7208) (GS neu Modell-Nr. 29 „Noblesse")
6629 FIGUR „Daisy" (Stehende junge Dame mit Béret/Pullmannkappe und langen Hosen, eine Gitarre haltend) (auf oktogonalem Sockel) (vgl. 6074)
6630 FIGUR „Walzer" (Schreitende Tänzerin, ihr Kleid schmetterlingsflügelartig auseinander haltend) (auf Ovalsockel) (vgl. 7285)
6631 Sitzendes Mädchen in koketter Pose (auf vierfüßigem Hocker)
6645 FIGUR „Rendezvous" (Stehende Dame im Kostüm mit Hut) (auf Rundsockel)
6646 FIGUR „Kongress" (Lilian Harvey) (auf Rundsockel) (vgl. 6330)
6647 „Wiener Walzer" („Stern") (Stehende, sich verneigende Pierrette auf Rundsockel mit Spitzhut) (vgl. 6019)
6692 Flamencotänzerin in langem Rock (Halbakt) mit Tamburin (auf Ovalsockel) (vgl. 7699)
6702 Stehende junge Dame, an eine Reling gelehnt (auf Segmentalsockel)
6704 FIGUR „Madelaine" („Madeleine"/„Tanz") (Tänzerin auf dem linken Bein tänzelnd, das rechte Bein nach vorne schwingend) (auf Ovalsockel)
6705 FIGUR „Quadrille" (Stehende Tänzerin mit Hut und Handschuhen) (auf Rundsockel)
6707 Schreitende junge Dame (auf Ovalsockel mit Tamburin)
6716 WANDMASKE „Maske Sissy"
6717 „Rel. Blonder Traum"
6737 Ballerina mit ausgestelltem Kleid beim Spitzentanz (auf Rundsockel)
6738 Stehendes Biedermeiermädchen mit Schutenhut und Blumenstrauß (auf Rundsockel)
6755 Stehende junge Dame mit Hut und Schal (auf Rundsockel)
6756 Stehende Dame (Marlene Dietrich?) mit Kappe, Mantel und einem Setter (auf abgerundetem Rechtecksockel)
6757 Stehende Tänzerin mit großem Hut und Handschuhen (ihren langen Rock mit beiden Armen auseinander haltend) (auf Ovalsockel)
6758 Stehende junge Dame ihren Schutenhut mit beiden Händen haltend (auf Rundsockel)
6770 s. Knörlein
6784 Stehendes junges Mädchen in weiten Hosen (sog. Schlaghosen) und Kopftuch (auf Ovalsockel)
6824 Stehende, lesende junge Dame mit Berét (auf Rundsockel)
6825 Schreitende junge Dame im Wind, ihren Hut haltend (auf Ovalsockel) (Dakon zugeschr.)
6842 Stehende junge Dame mit Berét und Handtasche (auf Rundsockel)
6901 Stehende elegante Dame in geschlitztem langen Kleid mit großem Hut und Gitarre (auf Ovalsockel)
6912 FIGUR/LAMPE Stehende junge Dame mit Hut (sombreroartig) und langen Handschuhen (auf profiliertem Rundsockel)
6930 FIGUR „Blonder Traum" (Schreitende Tänzerin mit Hut, ihr Kleid mit beiden Händen seitlich weg haltend) (auf Ovalsockel)
6940 Schreitende junge Dame (einen großen Hut in der Hand haltend) (auf Natursockel)
6942 Stehende Tänzerin (ihr Kleid schmetterlingsflügelartig auseinander haltend) (auf Rundsockel)
6959 Sitzende Dame mit großem Hut (auf zwei Reisekoffern mit Zeitung) (vgl. 7364= ähnl. Modell)
6960 Stehende junge Dame (im Wind) ihren Hut fest haltend (auf profiliertem Rundsockel)
7039 s. Bouret
7050 Stehende junge Dame in einteiligem Hosenanzug mit großem Hut und Ball (auf Ovalsockel)
7059 Tango tanzendes Paar (auf Ovalsockel)
7078 Tänzerin in schwungvoller drehender Bewegung (auf Ovalsockel) (vgl. 8136)
7098 Stehende junge Dame mit großem Hut, ihr Kleid mit der linken Hand anhebend (auf Ovalsockel)
7149 FIGUR „Skiläuferin" (Stehende Skiläuferin mit Skistock) (auf Ovalsockel) (vgl. 8157)
7172 Sitzende junge Dame mit kleinem Hut (auf zwei Reisekoffern)
7194 Stehendes Mädchen mit Hut und Foxterrier (auf Ovalsockel) (GS neu Modell-Nr. 50 „So bin ich")
7195 FIGUR/LAMPE „Beauty" (Tänzerin, nach rechts geneigt, ihr Kleid flügelartig zurückhaltend) (auf Ovalsockel) (vgl. 7897 (nach links); 7996 (nach links); 8746 (nach links))
7199 Tanzende Spanierin mit Kastagnetten (auf Rundsockel) (Dakon zugeschr.)
7200 Surprise (Sitzendes Mädchen mit Hut in einer Dose) (GS neu Modell-Nr. 53 „Die Überraschung")
7208 FIGUR „Blonder Traum" (Lilian Harvey als schreitende Dame mit großem Hut und Handschuhen, ihr Kleid mit beiden Händen anhebend) (auf Ovalsockel) (vgl. 6551; 6606) (GS neu Modell-Nr. 29 „Noblesse")
7226 Sitzendes Mädchen mit Buch (auf Bücherregal/Sockel)
7253 Lesendes Mädchen am Stehpult (auf Rundsockel)
7254 Posierende junge Dame mit Hut (auf Treppe/Sockel)
7256 Elegante Dame mit großem Hut und Foxterrier (in einem Boot sitzend)
7257 FIGUR/LAMPE „Gavotte" (Jessie Matthews als stehende Tänzerin mit Hut, ihr Kleid schmetterlingsflügelartig auseinander haltend) (auf Ovalsockel) (vgl. 8318) (vgl. Lorenzl 7581= ähnl. Modell)
7277 FIGUR „Schaukelpferd" („Kutschpferd")
7285 FIGUR „Walzer" (Schreitende Tänzerin, ihr Kleid schmetterlingsflügelartig auseinander haltend) (auf Ovalsockel) (vgl. 6630)
7286 FIGUR „Alt-Wien" (Schreitende junge Dame mit Schutenhut und gerüschtem Kleid (auf Rundsockel/Ovalsockel) (vgl. 6458)
7287 Stehende Dame mit Hut, langem Kleid und Spitzenumhang (auf Ovalsockel)
7288 „Ninon" (Stehende junge Dame mit Hut und gestuftem Rüschenkleid) (auf Ovalsockel) (Dakon zugeschr.)
7289 Schreitende Tänzerin, ihr Kleid seitlich weg haltend (auf Ovalsockel)
7290 FIGUR „Nora" (Stehende Tänzerin, ihr Kleid schmet-

terlingsflügelartig auseinander haltend) (auf Ovalsockel)
7291 WANDMASKE Kinderkopf (Mädchen mit Hut und Schleife / Mascherl) (vgl. 7413= ähnl. Modell)
7364 Sitzende Dame mit großem Hut (auf zwei Reisekoffern mit Pekinesen) (vgl. 6959= ähnl. Modell)
7393 Tennisspielerin (Stehendes Mädchen mit Tennisracket) (auf Ovalsockel) (Dakon zugeschr.)
7405 GRUPPE Stehender Knabe mit Brezel und Dackel (Dakon zugeschr.)
7406 FIGUR „Triton" (Stehendes Mädchen mit Kappe und Tretroller) (auf abgerundetem Längssockel) (GS neu Modell-Nr. 58 „Rotkäppchen")
7407 Sitzendes Mädchen mit Hut und Schulranzen
7409 WANDMASKE Mädchenkopf mit Hut, Schleife / Mascherl und Kragen
7412 WANDMASKE Mädchenkopf mit Béret und Schal
7413 WANDMASKE Mädchenkopf mit Kappe und Schleife / Mascherl
7415 Schreitende Spanierin mit großem Fächer (auf Ovalsockel)
7416 FIGUR „Golf" (Golfspielerin mit Hut und Schläger) (auf Rundsockel) (vgl. 8140)
7423 Sitzende Dame mit Hut und zwei Foxterriern (auf Backsteinmauer / Sockel)
7424 Stehende Dame mit Hut, Cape (großer Fransenschal), einen kleinen Hund (King Charles?) haltend (auf Rundsockel) (vgl. Lorenzl 7424)
7426 FIGUR „Sterbender Schwan" (mit/ohne Schwanenflügel am Kleid) (auf Ovalsockel) (vgl. 6361= ähnl. Modell)
7500 Schreitende Dame in sportlichem Kostüm und Kappe, zwei Pointer ausführend (auf abgerundetem Rechtecksockel)
7501 Stehender Mädchenakt auf einer Wasserlilienblüte (einen Schmetterling betrachtend)
7514 Schreitende Spanierin mit großem Fächer (auf Ovalsockel)
7560 Junge Golfspielerin beim Abschlag (auf Rundsockel / Ovalsockel)
7561 Sitzende lesende junge Dame (auf Büchersockel)
7562 Schreitender Skibub, seine Skier tragend (auf Ovalsockel)
7563 Stehendes Skifahrerpaar (auf Rundsockel)
7564 Stehendes Mädchen mit Golfsack (auf Rundsockel)
7594 s. Bouret
7595 Schreitender Dudelsackspieler (auf Ovalsockel)
7596 Schreitendes Mädchen in blauer Schuluniform mit Kappe und Schultasche (auf Ovalsockel)
7659 Stehender pausbäckiger Holländerjunge mit Kappe und Clogs, ein Modellsegelboot haltend (auf Rundsockel) (Dakon zugeschr.)
7662 Kinderpaar im Boot
7663 s. Bouret
7664 s. Bouret
7667 Sitzendes lesendes Mädchen mit Foxterrier
7669 „Badende" (Sitzendes Mädchen im Badedress)
7684 Schreitende Tänzerin in mexikanischem Kostüm mit Hut (auf Ovalsockel)
7685 Schreitende Tänzerin, ihr langes Kleid auseinander haltend (auf Ovalsockel)
7688 Stehende junge Dame (im Jagd / Reitoutfit) mit Jagdhund (auf Ovalsockel)
7699 Flamencotänzerin in langem Rock (Halbakt) mit Tamburin (auf Ovalsockel) (vgl. 6692)
7705 WANDMASKE Mädchenkopf mit gerüschtem Kragen (Dakon zugeschr.)
7706 WANDMASKE Kinderkopf (Bub mit Kragen und Mütze)
7774 Kleines Mädchen, das verzweifelt versucht ihren großen Koffer zu schließen (Dakon zugeschr.)
7776 FIGUR „Arabella" (Stehende junge Dame, ihr langes Kleid schmetterlingsflügelartig auseinander haltend) (auf Ovalsockel) (vgl. 7961; 8317)
7777 Schreitende Tänzerin, ihr Kleid mit beiden Armen auseinander haltend (auf Ovalsockel)
7825 s. Bouret
7826 FIGUR „Champagner" (Nach rechts schreitende junge Dame mit Bikinioberteil, ihren langen Rock schmetterlingsflügelartig auseinander haltend) (auf Ovalsockel) (vgl. 8319) (vgl. 7857= seitenverkehrtes Pendant)
7835 s. Bouret
7836 Stehendes kleines Mädchen mit drei Tauben (auf Ovalsockel)
7838 Stehende junge Gitarrespielerin mit Schottenrock und Kappe (auf Ovalsockel)
7849 Stehende Schlittschuhläuferin, ihren Rock fächerartig auseinander haltend (auf Ovalsockel)
7857 FIGUR „Champagner" (Nach links schreitende junge Dame mit Bikinioberteil, ihren langen Rock schmetterlingsflügelartig auseinander haltend) (auf Ovalsockel) (vgl. 8360) (vgl. 7826; 8319= jeweils seitenverkehrtes Pendant)
7868 GRUPPE „Dolly Sisters" (Stehend, mit langen Kleidern, in Tanzpose) (auf Ovalsockel / Rechtecksockel) (vgl. 7991) (vgl. 7958; 7993= jeweils linke EinzelFIGUR) (vgl. 7957; 7992= jeweils rechte EinzelFIGUR)
7897 FIGUR/LAMPE „Beauty" (Tänzerin, nach links geneigt, ihr Kleid flügelartig zurückhaltend) (auf Ovalsockel) (vgl. 7195 (nach rechts); 7996 (nach links); 8746 (nach links))
7957 FIGUR „Dolly" (Stehend, mit langem Kleid, in Tanzpose, rechte Figur) (auf Ovalsockel) (vgl. 7992) (vgl. 7868; 7991= jeweils GRUPPE)
7958 FIGUR „Dolly" (Stehend, mit langem Kleid, in Tanzpose, linke Figur) (auf Ovalsockel) (vgl. 7993) (vgl. 7868; 7991= jeweils GRUPPE)
7960 Sich verbeugende Tänzerin (mit ausgestreckten Armen) (auf Ovalsockel)
7961 FIGUR „Arabella" (Stehende junge Dame, ihr langes Kleid schmetterlingsflügelartig auseinander haltend) (auf Ovalsockel) (vgl. 7776; 8317)
7965 Schreitende Tänzerin mit Bikinioberteil, ihren langen Rock nach vorne weg haltend (auf Ovalsockel)
7991 GRUPPE „Dolly Sisters" (Stehend, mit langen Kleidern, in Tanzpose) (auf Ovalsockel) (vgl. 7868) (vgl. 7958; 7993= jeweils linke EinzelFIGUR) (vgl. 7957; 7992= jeweils rechte EinzelFIGUR)
7992 FIGUR „Dolly" (Stehend, mit langem Kleid, in Tanzpose, rechte Figur) (auf Ovalsockel) (vgl. 7957) (vgl. 7868; 7991= jeweils GRUPPE)
7993 FIGUR „Dolly" (Stehend, mit langem Kleid, in Tanzpose, linke Figur) (auf Ovalsockel) (vgl. 7958) (vgl. 7868; 7991= jeweils GRUPPE)
7996 FIGUR / LAMPE „Beauty" (Tänzerin, nach links geneigt, ihr Kleid flügelartig zurückhaltend) (auf Ovalsockel) (vgl. 7195 (nach rechts); 7897 (nach links); 8746 (nach links))
8006 FIGUR „Lektüre" (Sitzende junge Dame mit Buch) (auf ovalem Sockel, Dekor: Foxterrier)
8026 Junge Dame beim Skiwachsen (auf Ovalsockel)
8069 FIGUR „Großer Akt"
8083 FIGUR „Susette" (Lilian Harvey, seitlich schreitend) (auf Ovalsockel) (vgl. 8320)
8089 Stehendes nacktes Mädchen mit Rehkitz (auf Ovalsockel?)
8117 Schreitendes Mädchen im Badeanzug mit einem Barsoi (auf Ovalsockel)
8120 s. Bouret
8126 FIGUR „Tanzstudie" (Tänzerin mit geneigtem Kopf, ihr langes Kleid mit beiden Händen hochhebend) (auf Ovalsockel)
8127 FIGUR „Frasquita" (Tänzerin mit Bikini und langem Rock, seitlich, in Pose) (auf Ovalsockel)
8129 FIGUR / LAMPE Tänzerin, ihr linkes Bein nach vorne schwingend (auf Ovalsockel)
8136 Tänzerin in schwungvoller drehender Bewegung (auf Ovalsockel) (vgl. 7078)
8140 FIGUR „Golf" (Golfspielerin mit Hut und Schläger) (auf Rundsockel) (vgl. 7416)
8141 FIGUR „Wanderbursche" (Schreitender junger Bursche mit Hut und Schirm, darauf ein Vogel sitzend) (auf Ovalsockel)
8157 FIGUR „Skiläuferin" (Stehende Skiläuferin mit Skistock) (auf Ovalsockel) (vgl. 7149)
8225 GRUPPE Kleines Mädchen, einen Foxterrier am Halsband ziehend (auf Ovalsockel) (Dakon zugeschr.)
8226 Junges Mädchen im Kleid
8229 GRUPPE „Pierrots" (Drei schreitende Kinder, ein kleiner Pierrot mit zwei Mädeln) (auf abgerundetem Rechtecksockel) (vgl. 8230; 8232)
8230 GRUPPE Zwei schreitende Mädel (auf abgerundetem Rechtecksockel) (vgl. 8229; 8230)
8232 GRUPPE „Bub und Mädel" (Zwei schreitende Kinder, ein kleiner Pierrot und ein Mädchen) (auf abgerundetem Rechtecksockel) (vgl. 8229; 8230)
8233 Stehende Tänzerin mit nach links geneigtem Kopf, ihr langes Kleid schmetterlingsflügelartig auseinander haltend (auf Ovalsockel)
8234 FIGUR „Tanz" (Schreitende Tänzerin, ihr Kleid seitlich in die Höhe haltend) (auf Ovalsockel) (Dakon zugeschr.)
8236 Auf einem großen Ball sitzendes Mädchen mit Hut (Dakon zugeschr.)
8246 STEHKOPF Madonna (Frauenkopf mit Kopftuch) (vermutl. eher ein Entwurf von Kurt Goebel)
8300 Schreitende Tänzerin in gerüschtem Ballkleid und Hut (auf Ovalsockel)
8302 FIGUR „Rumba" (Tänzerin in Bikini, langem Rock / Schleppe und Halstuch) (auf Ovalsockel)
8303 FIGUR „Skijunge mit Stock" (Kleiner skifahrender Junge mit Zipfelmütze) (Dakon zugeschr.)
8304 FIGUR „Skijunge ohne Stock" (Kleiner skifahrender Junge) (Dakon zugeschr.)
8305 FIGUR „Skigirl" (Kleines stehendes Mädchen mit Zipfelmütze, ihre Skier haltend) (Dakon zugeschr.)
8317 FIGUR „Arabella" (Stehende junge Dame, ihr langes Kleid schmetterlingsflügelartig auseinander haltend) (auf Ovalsockel) (vgl. 7776; 7961)
8318 FIGUR „Gavotte" (Jessie Matthews als stehende Tänzerin mit Hut, ihr Kleid schmetterlingsflügelartig auseinander haltend) (auf Ovalsockel) (vgl. 7257) (vgl. Lorenzl 7581= ähnl. Modell)
8319 FIGUR „Champagner" (Nach rechts schreitende junge Dame mit Bikinioberteil, ihren langen Rock schmetterlingsflügelartig auseinander haltend) (auf Ovalsockel) (vgl. 7826) (vgl. 7857= seitenverkehrtes Pendant)
8320 FIGUR „Susette" (Lilian Harvey, seitlich schreitend) (auf Ovalsockel) (vgl. 8083)
8345 GRUPPE „Tirol" („Skifahrergruppe") (Schreitende Mutter und Kind, ihre Skier geschultert) (auf Ovalsockel) (Inschrift: TYROL) (vgl. 8346; 8347= jeweils Einzelfigur) (vgl. 8385= RELIEF)
8346 FIGUR „Skifahrerin" („Tirol groß") (Stehende Dame, ihre Skier geschultert) (auf Ovalsockel) (vgl. 8345= GRUPPE; 8347= Pendant) (vgl. 8385= RELIEF)
8347 FIGUR „Skibub" („Tirol klein") (Stehender Bub, seine Skier geschultert) (auf Ovalsockel) (vgl. 8345= GRUPPE; 8346= Pendant) (vgl. 8385= RELIEF)
8360 FIGUR „Champagner" (Nach links schreitende junge Dame mit Bikinioberteil, ihren langen Rock schmetterlingsflügelartig auseinander haltend) (auf Ovalsockel) (vgl. 7857) (vgl. 7826; 8319= jeweils seitenverkehrtes Pendant)
8365 s. Bouret
8366 FIGUR „Golfspielerin" („Golf")
8367 GRUPPE Kinder als Brautpaar im Biedermeiergewand (auf Ovalsockel) (Dakon zugeschr.)
8368 RELIEF „Skiläufer" (Aufsteigender Skiläufer mit geschulterten Skiern) (Wandrelief) (Pendant vgl. 8369)
8369 RELIEF „Skiläuferin" (Abfahrende Skiläuferin) (Wandrelief) (Pendant vgl. 8368)
8378 FIGUR „Mädchen" („Rosy Barsony"/„Mädchen in Sessel") (Auf einem Stuhl sitzend, mit ausgestrecktem Bein)
8385 RELIEF „Skiläuferpaar" (Mutter und Kind mit geschulterten Skiern) (Wandrelief) (vgl. 8345= GRUPPE „Tirol")
8433 GRUPPE „Kind mit Hund" (Kind, Hände gefaltet) (auf Ovalsockel)
8437 GRUPPE „Akt / Reh" (Kniender Mädchenhalbakt mit kleiner Schale, ein Reh fütternd, „Brüderchen und Schwesterchen") (auf ovalem Natursockel)
8440 FIGUR / LAMPE „Jolanthe" (Tänzerin, ihr Kleid auseinander haltend) (auf Ovalsockel)
8453 FIGUR „Tanz" („Rüschentanz"/„Colette") (Stehende Tänzerin mit Beret und Rüschenkleid, dieses mit der rechten Hand anhebend) (auf Ovalsockel)
8454 FIGUR „Gratulant" („Bub mit Zylinder") (auf Rundsockel)
8455 FIGUR „Gratulantin" („Mädchen mit Blumen") (auf Rundsockel)
8469 FIGUR „Tänzerin (Dame) mit Herz"
8470 FIGUR „Tänzerin mit Herz" („Revuegirl")
8472 FIGUR „Tanz" („Tänzerin"/„Adrienne") (Tänzerin mit Bikini und großem Tuch) (auf Ovalsockel)
8473 FIGUR „Tennisspielerin" („Tennis")
8474 FIGUR „Andante" (Stehende Tänzerin mit Bikinioberteil, Shorts und langem Umhang) (auf Ovalsockel)
8498 GRUPPE Zwei Tänzerinnen, eine kniend, die andere stehend (auf Ovalsockel)
8505 FIGUR „Tanzender Akt" („Figur mit Draperie"/ „Schmetterlingtanz")
8561 GRUPPE „Salzburger Kinder" (auf Ovalsockel)

8566 FIGUR „Bub mit Zylinder" („Bubi")
8567 GRUPPE „Tiroler Kinder" („Gruppe Steirer"/ „Widerspenstige") (auf Ovalsockel)
8571 FIGUR „Tiroler Bub" (Stehender essender Bub mit Hut in Tiroler Tracht) (auf Rundsockel)
8581 FIGUR „Kind in (der) Krippe" (vgl. 8585; 8586; 8587; 8588; 8589; 8590; 8592; 8593; 8594; 8595)
8585 FIGUR „Engel" (singend)" (vgl. 8581; 8586; 8587; 8588; 8589; 8590; 8592; 8593; 8594; 8595)
8586 FIGUR „Engel (mit Horn/Trompete)" (vgl. 8581; 8585; 8587; 8588; 8589; 8590; 8592; 8593; 8594; 8595)
8587 FIGUR „Engel (mit Laute/Mandoline)" (vgl. 8581; 8585; 8586; 8588; 8589; 8590; 8592; 8593; 8594; 8595)
8588 FIGUR „Engel (mit Puppe)" (vgl. 8581; 8585; 8586; 8587; 8589; 8590; 8592; 8593; 8594; 8595)
8589 FIGUR „Engel/Josef (mit Schaf/Schäfchen)" (vgl. 8581; 8585; 8586; 8587; 8588; 8590; 8592; 8593; 8594; 8595)
8590 FIGUR „Madonna (kniend)" (vgl. 8581; 8585; 8586; 8587; 8588; 8589; 8592; 8593; 8594; 8595)
8592 FIGUR „Engel (Finger im Mund)" (vgl. 8581; 8585; 8586; 8587; 8588; 8589; 8590; 8593; 8594; 8595)
8593 FIGUR „Engel (mit Kerze/mit Leuchter)" (vgl. 8581; 8585; 8586; 8587; 8588; 8589; 8590; 8592; 8594; 8595)
8594 FIGUR „Engel (mit Noten/Notenrolle)" (vgl. 8581; 8585; 8586; 8587; 8588; 8589; 8590; 8592; 8593; 8595)
8595 FIGUR „Engel (mit Buch)" (vgl. 8581; 8585; 8586; 8587; 8588; 8589; 8590; 8592; 8593; 8594)
8608 „Akt/(mit) Draperie"
8612 „Kniende mit Kind"
8621 GRUPPE „Kniende („Madonna") mit Kind" (vgl. 8780= ähnl. Modell)
8630 DECKELDOSE „Dose mit Akt" (Form: rund, kugelförmig gedrückt/ Knauf: Damenakt, vollplastisch) (vgl. 1094= Dose, ohne Figur)
8632 Stehender Windhund
8633 „Schreitende" (Damenakt) (auf Natursockel) (vgl. 8740)
8634 „Akt"/„Tänzerin" (in Pose) (auf rundem Natursockel)
8638 „Bub/(mit) Flöte"
8653 GRUPPE „Geheimnis" (Kindergruppe) (Zwei Mädchen mit Puppe und Regenschirm) (auf Ovalsockel)
8656 GRUPPE „Erzählung" (Kindergruppe)
8686 FIGUR „Akt (tanzend)"
8716 „Dunja" (Tänzerin, ihr Kleid mit beiden Händen anhebend) (auf Rundsockel)
8740 „Schreitende" (Damenakt) (auf Natursockel) (vgl. 8633)
8746 FIGUR „Beauty"/„Schönheit" (Tänzerin, nach links geneigt, ihr Kleid flügelartig zurückhaltend) (auf Ovalsockel) (vgl. 7195 (nach rechts); 7897 (nach links); 7996 (nach links))
8780 GRUPPE Kniende Mutter mit Kind (Dakon zugeschr.) (vgl. 8621= ähnl. Modell)
8856 „Duo" (Sitzendes nacktes Kleinkind bzw. Putto mit kleinem Vogel spielend) (auf Ovalsockel)
8857 „Lampion"
8858 „Trommler"
8859 „Kind"
8860 „Kind" (Stehendes, sich waschendes nacktes Kleinkind mit Handtuch, Bürste und kleinem Waschtrog)
8861 „Kind (m./Schnecke)" (Erschreckter Bub mit Schnecke) (auf Ovalsockel)
ohne Nr.-LaStéle Stehendes Mädchen in futuristischem Kostüm BRONZE (oder Messing?) (auf Marmorsockel)
„Matrosentanz" (Drei Mädchen im Matrosenkostüm) GWM Wien, Made in Austria, Hand. dec., unsign., (H. 40 cm); Soth. N.Y. 14.6.1991 (vgl. 5776; 5818)

Daubling (d.i. Doblinger, Ferdinand) s. Doblinger, Ferdinand

David, Fernand

*Fernand David, Bildhauer (*3.9.1872 Paris, †13.12.1926 Paris), war Schüler von Barrias und später Mitglied der Künstlergruppe „La Stèle" (-> vgl. Kap. „Dekorative Skulpturen von Arthur Goldscheider in Paris"). David stellte zwischen 1893–1923 im Pariser Salon d'Automne aus. Lit.: TB, Bd. 8, S. 451; Vollmer, Bd. 1, S. 522; Neuwirth (1974), S. 61; Bénézit, Bd. 4, S. 280; AKL, Bd. 24, S. 427.*

Debienne, Noémi(e)

*Noémi(e) Debienne, Bildhauerin (*Moulins/Allier), war tätig in Paris und Schülerin von A. Marquet, Vasselot und Pallez. Debinenne war Mitarbeiterin von Friedrich Goldscheider in Paris und Wien. Sie stellte ab 1894 im Pariser Salon (Salon des Artistes Français) aus und errang 1909 eine Medaille 3. Klasse für ihre Statuen, Statuetten, Porträtbüsten. Lit.: TB, Bd. 8, S. 506; Vollmer, Bd. 1, S. 528; Bénézit, Bd. 4, S. 318; AKL, Bd. 25, S. 64.*
739-GSB „Triomphe de Bacchante" (vgl. 2019)
2019 Stehende, posierende junge Frau in langem Gewand (vgl. 739-GSB)

Defraigne, Gilles

Gilles Defraigne war Mitarbeiter von Arthur Goldscheider, Paris. Lit.: Goldscheider, A. (Hrsg.): Artistic Evolution, 1926, T. 1, Pl. 17 (m. Abb.).
VASE (laqué) (Form von F. La Monaca) (H. 15 cm, B. 15 cm) Lit.: Goldscheider, A. (Hrsg.): Artistic Evolution, 1926, T. 1, Pl. 17 (m. Abb.)

Delagrange, Léon Noël

*Léon Noël Delagrange, Bildhauer (*13.3.1872 Orléans, †4.1.1910 Croix-d' Hains/Bordeaux), war Schüler von Ernest Barrias und beschäftigte sich vorwiegend mit figürlicher Kleinplastik (Ritter, Tänzerinnen) in Bronze und Marmor. Delagrange, der zu den Flugpionieren zählte und 1908 die Bildhauerei deswegen aufgab, kam durch einen Flugzeugabsturz ums Leben. Lit.: Neuwirth (1974), S. 62; Hillenbrand, S. 86; Arwas (2002), S. 611; Bénézit, Bd. 4, S. 371; AKL, Bd. 25, S. 337.*
23 Stehender Deutschordensritter mit Schwert BRONZE/ ELFENBEIN/Marmorsockel
44? BÜSTE Venezianischer Edelmann BRONZE/ELFENBEIN?/ACHATSOCKEL
745 MASKE/JARDINIERE Frauengesicht mit wallendem Haar (Gefäß für Blumen) (vgl. 868) (ÖMKI 1900/01?)
868 MASKE/JARDINIERE Frauengesicht mit wallendem Haar (Gefäß für Blumen) (vgl. 745) (ÖMKI 1900/01?)
2056 FIGUR „Brise" ÖMKI (1900/01)
2059 BÜSTE „Le Secret" ÖMKI (1900/01)
„JARDINIERE" ÖMKI (1900/01) Lit.: Neuwirth (1974), S. 62
Plastiken Paris, Weltausstellung (1900) Lit.: Hecht/Barthel, in: Tonindustrie-Ztg., 24.1900, S. 1584, „Kunstterrakotten bietet FG, Wien an mehreren Stellen der Ausstellung in bekannter sorgfältiger Ausführung dar..., mehrere impressionistische Plastiken von Delagrange".

Delamarre, Raymond

*Raymond Delamarre, Bildhauer (*8.6.1890 Paris, †?), war bis um 1945 in Paris tätig und Mitglied der Künstlergruppe „La Stèle" (-> vgl. Kap. „Dekorative Skulpturen von Arthur Goldscheider in Paris"). Delamarre beteiligte sich 1919 erfolgreich am Wettbewerb um den Rompreis mit der Skulptur „La Gloire ramène le Héros au foyer familial" und erhielt den 2e Premier Grand Prix. 1925 Teilnahme an der Pariser KGW-Ausst. (Goldmedaille). Lit.: Goldscheider, A. (Hrsg.): Artistic Evolution, 1926, T. 2, Pl. 6 (m. Abb.); Vollmer, Bd. 1, S. 535; Bénézit, Bd. 4, S. 375; AKL, Bd. 25, S. 350.*
LAMPE „Femme au Serpent" (Woman and Serpent) (Sitzender Frauenhalbakt mit Schlange, auf Rechtecksockel) BRONZE (vergoldet), Sockel MARMOR/ONYX (H. 19,5 cm, B. 34 cm) Lit.: Goldscheider, A. (Hrsg.): Artistic Evolution, 1926, T. 2, Pl. 6 (m. Abb.)

Della Robbia s. Robbia

Désurmont, (Gil-D.?)

*Gil-D. Désurmont, Maler und Bildhauer (*Tourcoing/Frankreich, †?). Weitere Informationen zu Leben und Schaffen sind nicht bekannt. Lit.: Bénézit, Bd. 4, S. 512.*
904 FIGUR „Kabylin" („Abla") (Inschrift: ABLA) (vgl. 962; 1056) (vgl. 927 BÜSTE)
927 BÜSTE „Kabylin" („Abla") (nach rechts blickend) (vgl. 904; 962; 1056 jeweils als Fig.) (Pendant vgl. 926 Strasser)
935 BÜSTE „Kabylin" („Abla") (nach rechts blickend) (vgl. 904; 962; 1056 jeweils als Fig.) (Pendant vgl. 934 Strasser)
961 s. Strasser
962 FIGUR „Kabylin" („Abla") (Inschrift: ABLA) (vgl. 904; 1056) (vgl. 927 BÜSTE)
1056 FIGUR „Kabylin" („Abla") (Inschrift: ABLA) (vgl. 904; 962) (vgl. 927 BÜSTE)

Dobling s. Daubling (d.i. Doblinger, Ferdinand)

Doblinger, Ferdinand

*Ferdinand Doblinger, Bildhauer, Keramiker und Modelleur (*22.3.1872 Wien, †1935 Wien), studierte 1897–1900 an der Akademie der bildenden Künste, Wien und nahm 1906/07 an der Winterausstellung des ÖMKI, Wien teil. Die Modelle wurden allesamt von Förster & Co. ausgeführt, für die er seit 1902 arbeitete. Er lieferte neben den Figurinen für Goldscheider hauptsächlich Tänzerinnen, Kinder, Paare, Hunde, Jardinieren und Brief- bzw. Postkartenbehältnisse etc. für Förster & Co., Wien, ab 1906 für Amphora bzw. Riessner & Kessel, Turn-Teplitz sowie ab 1910 für die Wiener Kunstkeramische Werkstätte Busch & Ludescher, ab 1935 Entwürfe für Keramos, Wien. Er war Lehrer (später Professor) an der Fachschule Oberleutensdorf (Böhmen). Doblinger war 1908 wohnhaft in Wien XVIII., Herbeckstr. 117. Wie die Auflistung seiner Werke hier belegt, wurden fast alle seine Figuren mit der Sign. „Daubling" versehen. Lit.: Lehmann, 1908; Neuwirth (1974), S. 62, 129–130; Zühlsdorff, S. 405; AKL, Bd. 28, S. 143.*
3301 Stehende Biedermeierdame mit Schultertuch (sign. Daubling)
3501 Schubertlied (sign. Daubling)
3513 Tänzerin (sign. Daubling)
3538 Biedermeierdame mit Hut und zwei Weidenkörben (auf getrepptem Sockel) (sign. Daubling)
3545 Herr mit Zylinder (Biedermeierfigur) (sign. Daubling/ Dobling) (Pendant vgl. 3546) (vgl. Modell-Nr. 4085 „Rundsockel")
3546 Dame mit Hut (Biedermeierfigur) (sign. Daubling) (Pendant vgl. 3545)
3551 Sitzende Biedermeierdame mit lesender Kinderschar (Putti?) (auf getrepptem Sockel) (sign. Daubling)
3565 UHR Rokokopaar (sign. Daubling)
4085 Herr mit Zylinder (Biedermeierfigur) (auf verziertem Rundsockel) (Pendant vgl. 4086) (vgl. Modell-Nr. 3545 „Rechtecksockel") (sign. Daubling)
4086 Dame mit Hut (Biedermeierfigur) (auf verziertem Rundsockel) (Pendant vgl. 4085) (sign. Daubling)
5468 Kamelreitender Afrikanerjunge (auf Landschaftssockel)
„Old England" (Fay.) ÖMKI (1910/11) Lit.: Neuwirth (1974), S. 62
„Biedermeier" ÖMKI (1912) Lit.: Neuwirth (1974), S. 62

Donatello (nach)

Die von Goldscheider editierten Modelle wurden Donatellos Werken nachempfunden bzw. nach diesen reproduziert.
1759 BÜSTE „Nobel-Dame" (Junges Mädchen mit gesenktem Haupt) (auf verziertem Sockel) ÖMKI (1899/1900) (Terrakotta)
1762 BÜSTE „San Lorenzo" (Junger Mann mit Kappe) ÖMKI (1898/99)
2131 BÜSTE Mädchen (Renaissancestil)
„Vornehme Dame" ÖMKI (1898/99) Lit.: Neuwirth (1974), S. 62

Donatello, F(l)agolli A.

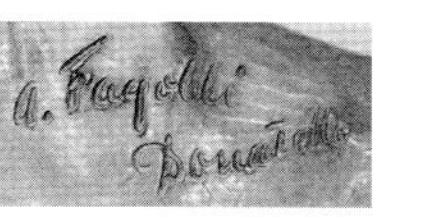

Ein Künstler dieses Namens lässt sich nicht nachweisen. Auch hat er nichts mit „Donatello" zu tun. Wie so oft ist anzunehmen, dass es sich hierbei um ein Pseudonym der Manufaktur Goldscheider handelt.
2090 BÜSTE „Edelweiss" (Frau nach rechts blickend, auf Natursockel mit Blattwerk und Edelweiß) ÖMKI (1902/03) (vgl. 2268)
2097 BÜSTE „Je meurs ou je m'attache"/Lierre (Junge Dame mit Efeu auf Rechtecksockel) (Inschrift: JE MEURS OU JE M'ATTACHE/LIERRE) (sign. Fagoli)
2268 BÜSTE „Edelweiss" (Frau nach rechts blickend, auf Natursockel mit Blattwerk und Edelweiß) ÖMKI (1902/03) (vgl. 2090)

Dreyfus-Stern, (Jean?)

*(Jean?) Dreyfus-Stern, Maler und Kunstgewerbler (*20.12.1890 Paris, †?), war Mitglied der Künstlergruppe „Évolution" (-> vgl. Kap. „Dekorative Skulpturen von Arthur Goldscheider in Paris"). Ob der als Figuren- und Stilllebenmaler bekannte Jean letztendlich infrage*

kommt, konnte bis Redaktionsschluss nicht geklärt werden. Lit.: Vollmer, Bd. 1, S. 594; Neuwirth (1974), S. 62; Bénézit, Bd. 4, S. 734; AKL, Bd. 29, S. 435.

Drouart, Raphaël
*Raphaël Drouart, Kunstgewerbler, Graveur, Maler und Keramiker (*25.12.1884 Choisy-le-Roi, †8.4.1972 Nogent-sur-Marne), war Mitglied der Künstlergruppe „Évolution" (-> vgl. Kap. „Dekorative Skulpturen von Arthur Goldscheider in Paris"). Lit.: Vollmer, Bd. 1, S. 596; Neuwirth (1974), S. 62; Bénézit, Bd. 4, S. 746; AKL, Bd. 29, S. 504.*

Dubret, Henri
*Henri Dubret, Goldschmied und Emailleur (*1872 Dijon †1934), war Mitglied der Künstlergruppe „Évolution" (-> vgl. Kap. „Dekorative Skulpturen von Arthur Goldscheider in Paris"). Lit.: Vollmer, Bd. 5, S. 440; Neuwirth (1974), S. 62; Bénézit, Bd. 4, S. 772; AKL, Bd. 30, S. 99.*

Dupré, J.
1353 FIGUR „Lustiger Zecher"/„Le joyeux Buveur" (Stehender Zecher mit Hut, Humpen und Korbweinflasche) (vgl. 1583)
1583 FIGUR „Lustiger Zecher" (Stehender Zecher mit Hut, Humpen und Korbweinflasche) (vgl. 1352)
1682 BÜSTE „Lustiger Zecher" (Mann mit Hut)
2344 FIGUR „Heuernte" ÖMKI (1902/03) (J. Dupri)

Dupri, J. s. Dupré, J.

Erdös-Meisinger, Ida s. Meisinger, Ida

Escoula, J(ean / Jacques?)
*Jean Escoula, Bildhauer (*26.10.1851 Bagnères-de-Bigorre, †August 1911 Paris) oder Jacques Escoula, Bildhauer (*27.12.1868 Bagnères-de-Bigorre, †?). Von welchem der beiden Bildhauer, die oft verwechselt werden, Goldscheider die Modelle ankaufte, kann wegen der gleichen Vornameninitiale „J" nicht gesagt werden. Zu vermuten ist jedoch wegen seiner zahlreichen Porträts von zeitgenössischen und historischen Persönlichkeiten Jean Escoula. Lit.: Neuwirth (1974), S. 62–63; Bénézit, Bd. 5, S. 179.*
2449 FIGUR „Angelus"/„Angélus" Paris, Salon (1902)
BÜSTE „Diana" ÖMKI (1906/07) Lit.: Neuwirth (1974), S. 63

Etlin
Lit.: Goldscheider, A. (Hrsg.): Artistic Evolution, 1926, T. 1, Pl. 16 (m. Abb.)
LAMPENSCHIRM für die „Lampe en Céramique" (Lamp in Ceramic) (Lampe von Jean Besnard) Lit.: Goldscheider, A. (Hrsg.): Artistic Evolution, 1926, T. 1, Pl. 29 (m. Abb.)

Etoile, A.
1445 GRUPPE „Erste Liebe" (Schäfer und Schäferin in antikisierendem Gewand) (auf Landschaftssockel) (vgl. 1800)
1800 GRUPPE „Erste Liebe" (Schäfer und Schäferin in antikisierendem Gewand) (auf Landschaftssockel) (vgl. 1445)

Ewald, Pierre Albert
*Pierre Ewald, Maler und Keramiker (*5.6.1880 Paris, †?), war in Paris tätig und Mitglied der Künstlergruppe „Évolution" (-> vgl. Kap. „Dekorative Skulpturen von Arthur Goldscheider in Paris"). Lit.: Vollmer, Bd. 2, S. 62; Neuwirth (1974), S. 63; Bénézit, Bd. 5, S. 224; AKL, Bd. 35, S. 463.*

Fagolli s. Donatello, F(l)agolli

Falguière, Alexandre
*Alexandre Falguière, Bildhauer, Maler und Zeichner (*7.9.1831 Toulouse, †19.4.1900 Paris), war zuerst im Atelier von Albert-Ernest Carrier-Belleuse tätig, bevor er ab 1854 an der Ecole des Beaux Arts in Paris bei François Jouffroy Aufnahme fand. Unter Falguières zahlreiche Auszeichnungen sei der im Jahre 1870 verliehene Orden Ritter der Ehrenlegion herausgegriffen. Unzählige Ausstellungen (u.a. im Pariser Salon von 1863–1899) und ein weit verbreitetes Oeuvre zeugen von seiner außerordentlichen Beliebtheit und Schaffenskraft. Viele seiner Skulpturen und Büsten wurden von den Gießereien Thiébaut frères und Hébrard in Bronze ausgeführt. Lit.: TB, Bd. 11, S. 228 ff.; Bénézit, Bd. 5; AKL, Bd. 36, S. 384.*
1256 Balzac (auf einer Bank sitzend) BRONZE

Fargue, Léon(-Paul?)
*Léon(-Paul?) Fargue, Maler und Kunstgewerbler (*4.3.1878 Paris, †1947 Paris), war zwischen 1881 und 1900(? in Paris tätig und dort Mitarbeiter von Arthur Goldscheider. Lit.: Goldscheider, A. (Hrsg.): Artistic Evolution, 1926, T. 1, Pl. 27 (m. Abb.); Bénézit, Bd. 5, S. 302; AKL, Bd. 37, S. 37.*
SCHALE / LAMPE „Coupe Lumineuse" (Luminous Bowl) METALL / Glas (H. 18 cm, B. 49 cm) Lit.: Goldscheider, A. (Hrsg.): Artistic Evolution, 1926, T. 1, Pl. 27 (m. Abb.) Fecler?
2581 BÜSTE Junge Frau mit Kopfputz in Form eines Pfaus

Fischer, Friedrich (s.a. Pecheur)
*Friedrich Fischer, Bildhauer, Keramiker und Hafnermeister (*23.8.1868 Wien, †16.9.1907), studierte an der K.K. Akademie in Wien unter Kundmann und trat ab 1881 mit eigenständigen Arbeiten hervor. Fischer war ab 1897 Mitglied des Wiener Künstlerhauses. Fischer war bis 1908 wohnhaft eingetragen in Wien XIII., Feldkellergasse 7. Goldscheider hat aller bisherigen Erkenntnis nach für ihn die klangvollere franz. Version seines Namens „Pecheur" verwendet. Modelle die mit diesem Namen sign. sind, werden dort aufgeführt. Lit.: Eisenberg, S. 116; Lehmann, 1908; AKL, Bd. 40, S. 332; Quelle: Bildhauerarchiv, Karl Novak.*
2235 BÜSTE „Renaissance" ÖMKI (1903/04)
2440 BÜSTE „Byzanz" ÖMKI (1903/04)
„Edeldame" ÖMKI (1903/04) Lit.: Neuwirth (1974), S. 63
„Ronde des elfes" Paris, Salon (1902)
„Joueuse de flûte" Paris, Salon (1902)
„Le Chant" Paris, Salon (1902)
Vier Modelle Paris, Salon (1905) Lit.: Neuwirth (1974), S. 63

F(l)agolli s. Donatello, F(l)agolli

Fleury
Ein Künstler dieses Namens lässt sich nicht nachweisen. Wie so oft, ist anzunehmen, dass es sich hierbei um ein Pseudonym der Manufaktur Goldscheider handelt.
4313 Stehende Dame (Krinoline) mit Fächer und Kakadu

Flora, Francesco
*Francesco Flora, Bildhauer, Maler und Kunstgewerbler (*30.6.1857 Francavilla, Fontana/Brindisi, †? wird letztmals um 1916 erwähnt), studierte bis 1876 an der Akademie in Venedig und war ab 1885 in Neapel tätig. Flora beschäftigte sich als Bildhauer neben kunstgewerblichen Arbeiten wie Lampen und Schalen auch mit der Skulptur, mit Grabmalen und Reliefs. Lit.: TB, Bd. 12, S. 116; Panzetta; AKL, Bd. 41, S. 297.*
ohne Nr. Stehender Amor (an einen Baumstamm gelehnt) MARMOR

Fontini (d.i. Ceccarelli, Ezio)
Der Name „Fontini" ist ein Pseudonym für Ezio Ceccarelli. Lit.: GSB, Pl. 76.
753-GSB BUSTE „Coquette" (vgl. 1784)

Forster (oder Förster), Karl
*Karl Forster (abweichende Namensform bei Könecke: Förster), Retoucheur (*25.12.1896 †?), zuerst bei Goldscheider in Wien und anschließend bei Carstens in Fredelsloh (Norddeutschland) tätig. Er war einer der von Wien aus „übernommenen" Spezialisten, der half, die Facharbeiter auf dem Tönnieshof auszubilden. Quelle: AGM, Liste „Adressen von guten und zuverlässigen Arbeitern" (undat., um 1953); Könecke (2002), S. 16.*

Freis, Gertrude
*Gertrude Freis, Malerin (*27.3.1923 †?). Quelle: AGM, Liste „Adressen von guten und zuverlässigen Arbeitern" (undat., um 1953).*

Frick, Ferdinand
*Ferdinand Frick, Bildhauer (*1878 Kopparberg/Schweden, †1939 Sao Paolo/Brasilien), studierte in Stockholm und war 1905 in Paris, wo im Salon zwei Werke, die von Goldscheider editiert wurden, ausgestellt waren. Lit.: Vollmer, Bd. 2, S. 159; Neuwirth (1974), S. 63.*
„Louise" Paris, Salon (1905) Lit.: Neuwirth (1974), S. 63
„Froid" Paris, Salon (1905) Lit.: Neuwirth (1974), S. 63

Fried, Nellie
*Nellie Fried (*14.9.1896 Wien, †6.6.1982 London), studierte Innenarchitektur und beschäftigte sich mit Kleinplastiken. Sie bot diese Figuren in einem Kunstgewerbeladen in der Hofburgpassage in Wien an. Für Goldscheider ist vorläufig nur die Modell-Nr. 6392 bekannt. Möglicherweise wurden aber von der Manufaktur noch weitere Entwürfe ihrer Hand ausgeführt. Nellie Fried war die Mutter des bekannten Lyrikers Erich Fried. Lit.: Jessen, Kaukoreit, Wagenbach (Hrsg.): Erich Fried, S. 9–11; Rothschild, T.: Erich Fried (http://oe1.orf.at/highlights/42069.html (vom 5.8.2005).*
6392 Stehende junge Dame „im Wind" (auf Rundsockel, mit Baumstamm)

Fuchs
5650 GEFÄSS Zwei sitzende Affen (jeder mit einer Schale, als Salz- und Pfeffergarnitur)

Fucik-Fischmeister, Grete (Maria Margareta)
*Grete (Maria Margareta) Fucik-Fischmeister, Bildhauerin (*20.4.1910 Wien, †?), besuchte die Graphische Lehr- und Versuchsanstalt, 1930–1933 und 1938/39 Studium an der Wiener KGW-Schule unter B. Löffler, M. Powolny und R. Obsieger. Die Rosenthal Kunstabt. Bahnhof-Selb übernahm 1935/36 den Entwurf für das Relief „Madonna mit Kind" und auch Keramos in Wien produzierte in den Jahren 1938–1940 einige Wandmasken von ihrer Hand. Lit.: Niecol, Bd. 4, S. 152; Neuwirth (1974), S. 63 „Goldscheider Modelle ausgestellt in Paris 1937", S. 149; Makus (2000), S. 169.*

Fuerst, Walter René
*Walter René Fuerst, Maler und Bühnenbildner (*Wien, †Paris?), war ansässig in Paris und dort Mitarbeiter von Arthur Goldscheider. Fuerst beschickte den Salon in Paris (Societé des Artistes Français) ab 1920. Lit.: Goldscheider, A. (Hrsg.): Artistic Evolution, 1926, T. 1, Pl. 14 (m. Abb.); Vollmer, Bd. 2, S. 174.*
TAFELAUFSATZ „Coupe a Fruits" (Fruit Bowl) (H. 31 cm, B. 30 cm) Lit.: Goldscheider, A. (Hrsg.): Artistic Evolution, 1926, T. 1, Pl. 19 (m. Abb.)

Fuka, Franz
Franz Fuka, Bildhauer oder Kunstgewerbler, absolvierte im Schuljahr 1893/94 die KGW-Schule in Bechyne und wurde bei Goldscheider (als Modelleur, Formgießer bzw. Figurenretoucheur) angestellt. Weitere Informationen zu Leben und Schaffen sind nicht bekannt. Lit.: Neuwirth (1974), S. 31; Neuwirth (1974a), S. 320; Bechyne 1884–1994, S. 55.

Galt s. Gult

Gambeauche
2426 FIGUR „Lygia" (vgl. 2639= BÜSTE)
2639 BÜSTE „Lygia" (auf eckigem Sockel mit Fischsymbolen) (Inschrift: LYGIA) (vgl. 2426= FIGUR)
3085 Stehende Römerin mit Fisch (auf Rundsockel)

Gauvenet, Jean-Baptiste
Jean-Baptiste Gauvenet, Kunstgewerbler (Lebens- und Schaffensdaten unbekannt), war neben seiner Tätigkeit für Sèvres (Manufacture Nationale de Sèvres) Mitglied der Künstlergruppe „Évolution" (-> vgl. Kap. „Dekorative Skulpturen von Arthur Goldscheider in Paris"). Lit.: Neuwirth (1974), S. 63; Zühlsdorff, S. 614.

Geiger, Benno
*Benno Geiger, Keramiker und Bildhauer (*5.8.1903 Engelberg/Schweiz, †22.10.1979 Bern), studierte, nach seiner keramischen Vorbildung in Zürich und Lugano, in den Jahren 1925–1927 an der Wiener KGW-Schule unter M. Powolny. 1928 wird Geiger von Walter Goldscheider als künstlerischer Mitarbeiter aufgenommen und später als Abteilungsleiter eingesetzt, wo er sich als einer der wertvollsten und kreativsten Mitarbeiter der Firma zeigt. Geigers Modelle für die Manufaktur werden auf den Jahresausstellungen des Wiener Künstlerhauses gezeigt. Einem Studienaufenthalt in Paris folgt dann 1934 die Rückkehr in die Schweiz, wo er die künstlerische Leitung der Tonwarenfabrik Aedermannsdorf übernimmt. Walter Goldscheider, der um Benno Geigers Qualität als Künstler*

wusste, wollte ihn in den Jahren 1952/53, also in der Zeit des Wiederaufbaues der Firma in Wien, zu einer erneuten Mitarbeit bewegen. Diese abermalige Zusammenarbeit kam jedoch nicht zustande. Geiger sollte „Gebrauchskeramik in guten, einfachen aber effektvollen Glasuren" ausführen. Die neuen Entwürfe sollten „einfach zu reproduzieren und dennoch originell und wirkungsvoll sein". Nach anfänglicher Bereitschaft Geigers schreibt dieser am 3.12.1952 jedoch: „Ich habe mich einige Male mit den Entwürfen für Sie beschäftigt, bin aber zu dem Schluss gekommen, daß diese Arbeit mir nicht liegt. Die vielen Einwände die Ihre Mitarbeiter anlässlich unserer Besprechung gemacht haben, irritieren mich. Eine Kollektion zu machen die modern ist, aber allen Leuten gefallen soll, gut , aber billig und ohne geschulte Leute ausführbar sein soll, das ergibt einen Kompromiss, der hier in der Schweiz im Großen und Ganzen auf dem Gebiete der Kunstkeramik nicht mehr üblich ist". (Briefwechsel zw. Walter Goldscheider und Benno Geiger, vom 18. September 1952 bis 3. Dezember 1952). Geiger, der von 1941 bis 1969 Leiter der keramischen Fachschule Bern war, macht sich durch zahlreiche Ausstellungen in Frankreich, Belgien, Italien und in Österreich einen klangvollen Namen und verfasst zwei Fachbücher zur Keramik: Keramisches ABC (1947), Keramisches Gestalten (1957). Quelle: AGM; Lit.: Neuwirth (1974), S. 63, 150; Schnyder, S. 33 ff.; Benno Geiger 1903–1979, Ausst.-Kat. Aedermannsdorf (1991); AKL Index.

5027 BUCHSTÜTZE Sitzende Frau (Japanerin)
5681 VASE (Form: becherförmig, Mündung mittig gedrückt/Dekor: geometrische Formen, mehrfarbige Laufglasur) (Geiger zugeschr.)
5739 VASE Vier Jahreszeiten (Form: becherförmig, geweitete Mündung/Dekor: vier Figuren, Skifahrerin, Jäger, Mädchen mit Gitarre, Dame mit Tuch) (Geiger zugeschr.)
5742 VASE (Form: balusterförmig / diverse Dekore) (Geiger zugeschr.)
5744 VASE Fünf Reiter (Form: becherförmig, gewulsteter Standring / Dekor: Fünf Reiter auf steigenden Pferden, stilisiertes Blattwerk) (Geiger zugeschr.)
5746 VASE Mann und Frau im Garten (Form: balusterförmig/Dekor: Paar bei der Gartenarbeit) (Geiger zugeschr.)
5767 SCHALE Frauengesicht (Form: rund / Dekor: Frauengesicht vor Stadtlandschaft) (ÜBERSCHNEIDUNG)
5767 SCHALE Reiter (Form: rund / Dekor: Reiter auf springendem Pferd) (ÜBERSCHNEIDUNG)
6017 ASCHER (mit zwölf Ablagen) (Form: rund, zwölf Zigarettenablagen / Dekor: mehrfarbig, Laufglasur)
6282 SCHALE (Form: oval / Dekor: mehrfarbig, Laufglasur) (vgl. 677= ähnl. Mod.)
6283 SCHALE (Form: oval/Handhaben: zwei M-förmige Griffe mit Öse / Dekor: mehrfarbig, Laufglasur) (Geiger zugeschr.)
6375 BUCHSTÜTZE (Form: dreieckig, mit zwei Erhebungen / Dekor: streifenartig, türkis, orange, schwarz) (Geiger zugeschr.)
VASE (groß, mit Staffagen) Lit.: Lehmann, G.: Neuheiten, in: KuKGW, 9.1928/29, S. 127 (m. Abb.), „frische keramische Schilderei von leuchtender Farbigkeit".
VASE (klein, Dekor: Dorf mit Staffagen) Lit.: Lehmann, G.: Neuheiten, in: KuKGW, 9.1928/29, S. 127 (m. Abb.), „frische keramische Schilderei von leuchtender Farbigkeit".
SCHALE (rund) Lit.: Lehmann, G.: Neuheiten, in: KuKGW, 9.1928/29, S. 127 (m. Abb.), „frische keramische Schilderei von leuchtender Farbigkeit".
TELLER (Dekor: Landschaft mit Staffagen) Lit.: Lehmann, G.: Neuheiten, in: KuKGW, 9.1928/29, S. 127 (m. Abb.), „frische keramische Schilderei von leuchtender Farbigkeit".
VASE (Form: becherförmig/Dekor: Farbfelder, reliefartig, Laufglasur) (Fay.) (1930) Lit.: MW, 11.1929/30, Nr. 18, S. II (m. Abb.)
VASE (groß, Dekor: Landschaft und Frauenkopf) (1930) Lit.: MW, 11.1929/30, Nr. 23, S. 20 (m. Abb.)
VASE (mittel, mit Standfuß, Dekor: Landschaft mit Staffagen) Lit.: MW, 11.1929/30, Nr. 23, S. 20 (m. Abb.)
VASE (gebaucht, Dekor: stilisierte Tiere, „Türkisglasur") Lit.: Heizer (1931), S. 236 (m. Abb.)
Pferd (1930) Lit.: MW, 11.1929/30, Nr. 23, S. 20 (m. Abb.)
SCHALE (Form: rund / Dekor: Farbfelder, reliefartig, Laufglasur) (Fay.) (1930) Lit.: MW, 11.1929/30, Nr. 18, S. II (m. Abb.)
SCHALE (Form: rund / Dekor: Pferd) (1930) Lit.: MW, 11.1929/30, Nr. 23, S. 20 (m. Abb.)
SCHALE (Form: oval / Dekor: geometrisch, mit zwei Henkeln) (um 1930) (H. 9 cm, L. 23 cm, B. 18 cm); Herr 23
ASCHER (viereckig) (1930) Lit.: MW, 11.1929/30, Nr. 23, S. 20 (m. Abb.)
BUCHSTÜTZE Sitzender Damenakt mit Bubikopf (nach links blickend) (Fay.) (1930) (Pendant –> vgl. nächsten Eintrag) Lit.: MW, 11.1929/30, Nr. 16, S. VIII (m. Abb.); Neuwirth (1974), S. 63.
BUCHSTÜTZE Sitzender Damenakt mit Bubikopf (nach rechts blickend) (Fay.) (1930) (Pendant –> vgl. vorigen Eintrag) Lit.: MW, 11.1929/30, Nr. 16, S. VIII (m. Abb.); Neuwirth (1974), S. 63.
GRUPPE „Freunde" (Zwei stehende, aneinander geschmiegte Damenakte mit Tüchern) (Fay.) (1930) Lit.: MW, 11.1929/30, Nr. 21, S. VII (m. Abb.); Neuwirth (1974), S. 63.
LAMPENFUSS (1930) Lit.: MW, 11.1929/30, Nr. 24, S. VIII; Neuwirth (1974), S. 63.
WANDBEHANG Madonna (mit Kind) Lit.: Heizer (1931), S. 236 (m. Abb.), „Die Madonna von B. Geiger als Hängefigur mit ausnehmend schönen Farben und schwingenden Linien".
RAUCHZEUG Elefant Lit.: MW, 11.1929/30, Nr. 22, S. 20 (m. Abb.) (Bosse / Geiger / Meisinger?)
RAUCHZEUG Zigarettenbehälter Lit.: MW, 11.1929/30, Nr. 22, S. 20 (m. Abb.) (Bosse / Geiger / Meisinger?)
KERZENHALTER Lit.: MW, 11.1929/30, Nr. 22, S. 20 (m. Abb.) (Bosse / Geiger / Meisinger?)
526 VASE / TOPF (Geiger zugeschr.)
578 VASE
580 VASE
583 VASE Musikanten (Geiger zugeschr.)
587 Vase (Geiger zugeschr.)
591 VASE
592 VASE (Bodenvase)
596 VASE
674 VASE (Bodenvase) (Geiger zugeschr.)
677 SCHALE (vgl. 6282)
759 VASE (Geiger zugeschr.)
760 VASE
883 VASE
913 SCHALE
924 VASE (Geiger zugeschr.)
926 SCHALE
927 SCHALE Tulpen
945 VASE (Geiger zugeschr.)
946 DECKELDOSE (Geiger zugeschr.)
950 VASE Tulpe
965 SCHALE
972 VASE (Geiger zugeschr.)
1003 VASE (Geiger zugeschr.)
1050 ASCHER
1073 VASE
1135 ASCHER (Geiger zugeschr.)
1325 VASE
ohne Nr. VASE (Form: kugelförmig, hohe und weite, leicht eingezogene, trichterförmige Mündung, konischer Standring / Handhaben: zwei gebogene Ösen / Dekor: stilisierte Blätter, mehrfarbig, Laufglasur)
ohne Nr. VASE (Form: zylinderförmig, flache Schulter, eingezogener Boden, kurze etwas geweitete Mündung / Dekor: schachbrettartig, rot-schwarz-weiß, Laufglasur)

Gemignani, Valmore

*Valmore Gemignani, Bildhauer (*1.11.1879 Carrara, †1.5.1958 Florenz), war in Florenz auch als Maler tätig und knüpft in seinen dekorativen Terrakotten an Donatello und Luca della Robbia an. Neben den Entwürfen (überwiegend Windhunde) für Goldscheider, führten auch italienische Firmen wie die Arte della Ceramica, Firenze seine Modelle aus. 1925 erhielt er auf der KGW-Ausst. Paris eine Goldmedaille. Im Jahre 1942 zeigte die Düsseldorfer Kunsthalle zwei seiner Bronze-Arbeiten. Seine bevorzugten Sujets sind Tiere, figürliche und florale Dekore sowie Reliefplatten und Kinder. Lit.: Vollmer, Bd. 2, S. 223; Zühlsdorff, S. 410.*

2895 BÜSTE Junge Frau mit Haube
3858 Sitzender Windhund
3859 Stehender Windhund (auch sign. „Lavergne / Latour") (vgl. 4687) (vgl. auch Latour 4687)
4187 Stehender Windhund (auch sign. „Latour") (vgl. Latour 4374)
4507 Zwei Windhunde
4687 Stehender Windhund (vgl. 3859) (vgl. auch Latour 4687)
„Windhund" ÖMKI (1911/12) (vgl. 3858; 3859; 4187; 4687?) Lit.: Neuwirth (1974), S. 63
Auf einer Schnecke stehendes Mädchen (mit überdimensional großen Schmetterlingsflügeln) (um 1902) (H. 82 cm); Pinhas, S. 61 (m. Abb.)
GRUPPE Geparden (um 1920) (L. 47 cm, H. 24 cm); Doro 31.5.1990

Gennarelli, Amédée (Amédéo)

*Amédée Gennarelli, Bildhauer (*Neapel, †?), war Schüler von Jerace, beschickte in den Jahren 1913–1930 den Salon in Paris (Societé des Artistes Français) und wurde 1923 im Salon von Arthur Goldscheider ausgestellt. Lit.: Vollmer, Bd. 2, S. 224; Neuwirth (1974), S. 63; Hillenbrand, S. 139; Bénézit, Bd. 6, S. 12; AKL Index.*

„Tête d'enfant" Paris, Salon (1923) Lit.: Neuwirth (1974), S. 63
„Enfant à la cornemuse" Paris, Salon (1923) Lit.: Neuwirth (1974), S. 63
„Enfant moqueur" Paris, Salon (1923) Lit.: Neuwirth (1974), S. 63
„La volupté" (BRONZE) Paris, Salon (1923) Lit.: Neuwirth (1974), S. 63; Berman, Bd. 3, S. 2016?
„La passion" (Terre cuite) Paris, Salon (1923) Lit.: Neuwirth (1974), S. 63

Gensoli, Maurice

*Maurice Gensoli, Keramiker (*1892 Oran/Algier, †?), war in den Jahren 1921–1927 Dekorateur und ab 1928–1958 künstlerischer Leiter in Sèvres (Manufacture Nationale de Sèvres), wo er in dem von ihm gegründeten Atelier für Fayencen (um 1925) ebenfalls Leiter wird. Er unterhält zeitgleich eine Werkstatt in Paris, ist Mitglied der Künstlergruppe „Évolution" und liefert für Arthur Goldscheider um 1925 Modelle. Lit.: Zühlsdorff, S. 410.*

Germani, L.-D.

L.-D. Germani, Kunstgewerbler (Lebens- und Schaffensdaten unbekannt), war Mitglied der Künstlergruppe „Évolution" (–> vgl. Kap. „Dekorative Skulpturen von Arthur Goldscheider in Paris").

Gesa, Josep

Josep Gesa, span. Bildhauer und Tischler (Lebens- und Schaffensdaten unbekannt), wird um 1865 in Cardedeu erwähnt. Lit.: AKL Index.

1506 Wassertrinkender Leopard auf Felsen (auf Rechtecksockel)

Giebl, Johann

*Johann Giebl, Spritzer (*11.5.1918, †?). Quelle: AGM, Liste „Adressen von guten und zuverlässigen Arbeitern" (undat., um 1953).*

Goebel, Kurt

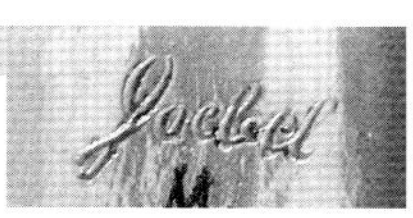

*Kurt Goebel, Bildhauer und Kunstkeramiker (*25.6.1906 Miltsch bei Troppau, †27.6.1991 Wien) besuchte die Schauspielschule und studierte Bildhauerei bei Prof. Gusty Mundt in Wien. Er war langjähriges Mitglied der Wiener Sezession, etablierte sich ab 1936 als freischaffender Künstler und unterhielt mit seiner Frau Erna Bieber eine Kunstkeramikfirma (gegründet in der zweiten Hälfte der 30er Jahre), die unter dem Namen „Goebel-Bieber" kurz nach dem 2. Welt-*

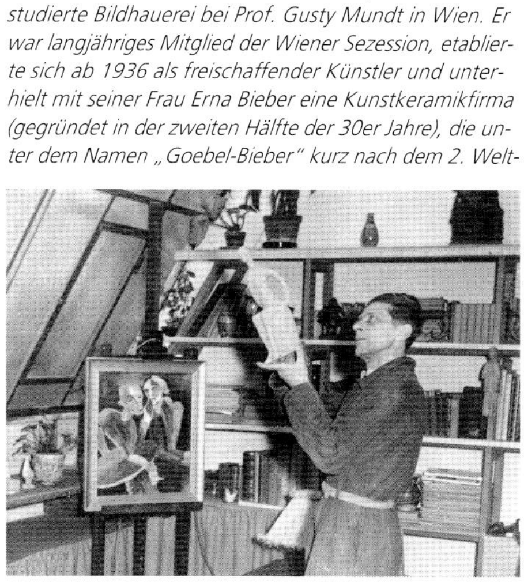

krieg geschlossen wurde. Goebel entwarf, wie die stattliche Liste seiner Modelle zeigt, zahlreiche Modelle für Goldscheider, deren unverkennbarer Charakter auf den ersten Blick zu sehen ist. Seine Wandmasken, Stehköpfe und Figurinen werden gelegentlich mit den Entwürfen Walter Bosses verwechselt. Goebel zeigt jedoch eine eigenständige Handschrift, die in keiner Weise seinem durchaus bekannteren Pendant Bosse nachsteht. Nicht nur wegen der hohen Anzahl seiner Entwürfe für die Manufaktur zählt Goebel zu den wichtigsten Künstlern bei Goldscheider. Einige seiner Arbeiten wurden auch bei Keramos editiert. Goebels kunstkeramisches Atelier (lt. Eintrag bei Lehmann) lässt sich 1937 in der Weimarer Str. 102 und seine Wohnung in derselben Str. auf Nr. 100 nachweisen. Später übersiedelt er in den VI. Bez., Mariahilferstr. 100. Abschließend sei hier aus einer damals führenden österreichischen Kunst-Zs. zit.: „Kurt Goebels Entwicklung erfasst das Bedürfnis der Zeit. Doch wahrt sich der junge, hochbegabte Künstler dabei volle Originalität und seine besonderen Geheimnisse in Brand und Glasuren. Wir begrüßen gerne diese wertvolle Bereicherung österreichischer Keramik", N.N.: Neue Keramik, in: ÖK, 7.1936, Nr. 10, S. 15 (m. Abb.). Lit.: u. Quellen: Lehmann, 1937–1941; Vollmer, Bd. 2, S. 263; Makus (2000), S. 170; AKL Index; Matschiner (2006); Archiv Uta M. Matschiner; Bildhauerarchiv, Karl Novak.

5836 STEHKOPF Frauenkopf mit geschlossenen Augen, Locken (Curls) und Kragenansatz) (Goebel zugeschr.)
6698 GRUPPE Tanzendes Spanierpaar (auf Rundsockel) (Goebel zugeschr.)
6699 Skifahrer (Goebel zugeschr.)
6922 ASCHER Ascher mit Kopf (Mädchenkopf) (Form: rund, mit Standfläche für den Kopf, drei Zigarettenablagen/Dekor: BÜSTE Mädchenkopf mit zwei Mascherln/Schleifen, vollplastisch)
7065 Julia (Stehendes schüchternes Mädchen) (auf Rundsockel) (vgl. 7657)
7448 Sitzendes posierendes Mädchen mit Hut (auf Rundsockel mit Kaktus)
7449 (STEH)KOPF „Kopf ohne/mit Sockel" (Frauenkopf mit geschlossenen Augen, Haarschmuck und Kragen)
7464 Knabe auf Fisch („reitend") (Pendant vgl. 7467)
7465 Stehendes Mädchen im Badeanzug (auf Natursockel mit Baumstamm)
7466 Der Bücherwurm (Stehender, fast glatzköpfiger Mann, ein Buch haltend) (auf Rundsockel mit Büchern)
7467 Stehendes Mädchen mit Fisch (Pendant vgl. 7464)
7518 Skifahrer (Schreitender Skifahrer, seine Skier unter dem rechten Arm tragend) (auf Ovalsockel) (Pendant vgl. 7622; 7887) (vgl. 7655; 7935= jeweils ähnl. GRUPPE)
7522 WANDMASKE Frauenkopf (dunkelhäutig) mit Halstuch/Schal
7523 WANDMASKE Frauenkopf mit Béret/Kappe und Kragen
7583 RELIEF Madonna mit Kind und Sternen (Wandrelief)
7584 WANDMASKE Frauenkopf mit Locken (Curls) und Halstuch
7585 WANDMASKE Frauenkopf mit Hut und Hand
7586 STEHKOPF Frauenkopf mit Kragen
7587 BÜSTE Frauenkopf mit Hand und Halskette
7589 Stehender Knabe mit Hut und Schmetterling (auf Rundsockel)
7604 (STEH)KOPF „Kopf ohne/mit Sockel" (Frauenkopf mit Hand und Blumenblüte)
7605 STEHKOPF Frauenkopf mit Hand und Buch
7607 WANDMASKE Madonna mit Kind
7608 WANDMASKE Frauenkopf mit Hand und Blumenblüte/Trauben
7614 WANDBEHANG Frauenkopf im Profil (gerahmt)
7616 STEHMASKE/WANDMASKE/BUCHSTÜTZE Frauenkopf (auf Rechtecksockel) (vgl. 7879)
7617 TORSO Mädchen mit Halsschmuck, Ohrringen und Haarband
7618 Golfspieler beim Abschlag (auf Rundsockel)
7619 WANDBEHANG Madonna (Frauenkopf) im Profil (gerahmt)
7622 FIGUR „Skiläufer" (Schreitende Skiläuferin mit geschulterten Skiern) (auf Ovalsockel) (Bem.: Die von der Firma als „Skiläufer" bezeichnete Figur ist eindeutig eine „Skiläuferin".) (vgl. 7887; Pendant vgl. 7518) (vgl. 7655; 7935= jeweils ähnl. GRUPPE)
7625 Stehendes Mädchen (an einem Kaktus oder exotischem Baumstamm mit Blüten) (auf Natursockel mit Schnecke)
7653 (STEH)KOPF „Kopf ohne/mit Sockel" (Frauenkopf mit zwei Händen)
7654 WANDMASKE Frauenkopf mit Hand und kleiner Vase
7655 GRUPPE „Skifahrer" (Schreitendes Skifahrerpaar) (auf Ovalsockel) (vgl. 7935; vgl. jeweils EinzelFIGUREN 7518; 7622; 7887)
7656 FIGUR „Romeo" (Stehender junger Mann mit Hut und Kaktus) (auf Rundsockel) (vgl. 7912; 7978) (Pendant vgl. 7657)
7657 Julia (Stehendes schüchternes Mädchen) (auf Rundsockel) (vgl. 7065) (Pendant vgl. 7656)
7686 BUCHSTÜTZEN Beim Fensterln
7813 WANDMASKE/STEHMASKE „Maske Araberin" (Frauenkopf mit Hand, Ring und Kopftuch) (vgl. 8247)
7814 WANDMASKE Frauenkopf mit gelocktem Haar und seitlichem Kragenansatz
7839 STEHKOPF Frauenkopf mit Mütze und Schal
7843 GRUPPE Wandersmann (Bursche) mit springendem Hund (auf Landschaftssockel)
7844 FIGUR „Rendez-vous" (Stehender Bursche mit Stock und Tasche, mit Hut grüßend) (auf flachem Rechtecksockel) (Pendant vgl. 7845) (vgl. 7962= GRUPPE)
7845 FIGUR „Rendez-vous" (Stehendes Mädchen mit Regenschirm und Tasche) (auf flachem Rechtecksockel mit Hund) (Pendant vgl. 7844) (vgl. 7962= GRUPPE)
7859 Stehender Matrose (auf Rundsockel) (Pendant vgl. 7977) (vgl. 7860= GRUPPE, hier ähnl. der EinzelFIGUR)
7860 GRUPPE Stehender Matrose mit Matrosenbraut (auf Ovalsockel) (vgl. 7977= EinzelFIGUR) (vgl. 7859= ähnl. EinzelFIGUR)
7874 STEHKOPF Frauenkopf mit modischer Kappe und Schal
7879 BUCHSTÜTZE Frauenkopf (auf halbkreisförmigem Sockel) (vgl. 7616)
7887 FIGUR „Skiläufer" (Schreitende Skiläuferin mit geschulterten Skiern) (auf Ovalsockel) (Bem.: Die von der Firma als „Skiläufer" bezeichnete Figur ist eindeutig eine „Skiläuferin".) (vgl. 7622) (Pendant vgl. 7518) (vgl. 7655; 7935= jeweils ähnl. GRUPPE)
7889 Stehendes Mädchen mit zwei Hasen (auf Rundsockel)
7890 Stehendes Mädchen mit Puppenwiege (auf abgerundetem Rechtecksockel) (Goebel zugeschr.)
7912 FIGUR „Romeo" (Stehender junger Mann mit Hut und Kaktus) (auf Rundsockel) (vgl. 7656; 7978)
7935 GRUPPE „Skifahrer" (Schreitendes Skifahrerpaar) (auf Ovalsockel) (vgl. 7655) (vgl. ähnl. EinzelFIGUREN 7518; 7622; 7887)
7943 ASCHER „Ascher mit Kopf" (Form: rund (mit Standfläche für den Kopf), vier Zigarettenablagen/Dekor: BÜSTE Frauenkopf, vollplastisch) (vgl. 7944; 7945; 7946= jeweils ähnl. Modell)
7944 ASCHER „Ascher mit Kopf" (Knabenkopf) (Form: rund (mit Standfläche für den Kopf), drei Zigarettenablagen/Dekor: BÜSTE Knabenkopf, vollplastisch) (vgl. 7943; 7945; 7946= jeweils ähnl. Modell)
7945 ASCHER „Ascher mit Kopf" (Mädchenkopf) (Form: rund, mit Standfläche für den Kopf, vier Zigarettenablagen/Dekor: BÜSTE Mädchenkopf, vollplastisch) (vgl. 7943; 7944; 7946= jeweils ähnl. Modell)
7946 ASCHER Mädchenkopf (Form: rund, drei Zigarettenablagen/Dekor: BÜSTE Mädchenkopf mit Schleife/Mascherl im Haar, vollplastisch) (vgl. 7943; 7944; 7945= jeweils ähnl. Modell)
7962 GRUPPE „Rendez-vous" (Stehender Bursche mit Mädchen) (auf flachem Rechtecksockel mit Hund) (vgl. 7844; 7845= jew. EinzelFIGUREN)
7977 Stehendes Mädchen mit Schürze und Schal (Matrosenbraut) (auf Rundsockel) (vgl. 7860= GRUPPE) (Pendant vgl. 7859)
7978 FIGUR „Romeo" (Stehender junger Mann mit Hut und Kaktus) (auf Rundsockel) (vgl. 7656; 7912)
8032 WANDMASKE Frauenkopf mit langem Zopf und Kragen (Goebel zugeschr.)
8033 WANDMASKE Frauenkopf mit wehendem Haar und Hand (Goebel zugeschr.)
8035 WANDRELIEF Mädchenakt mit langem Schal (Tuch) (Goebel zugeschr.)
8036 GRUPPE Brautpaar (Stehendes Paar auf Quadratsockel)
8075 STEHKOPF Frauenkopf mit Locken und Kopftuch (Goebel zugeschr.)
8076 WANDMASKE „Maske" (Frauenkopf mit Hand und Blumenschmuck im Haar)
8102 ASCHER Mandolinenspieler (Form: rund, vier Zigarettenablagen/Dekor: stehender Mandolinenspieler, vollplastisch)
8106 Stehender, aus einem Bierkrug trinkender Bursche in Tracht (auf Ovalsockel)
8201 Stehender junger Maler mit Palette (auf Rundsockel)
8202 Sitzendes Mädchen mit Hummer am Meer (auf Natursockel mit Stein und Meereswellen)
8218 Skifahrerin
8239 FIGUR „Flötenspieler" (Stehender junger Mann mit Kopfbedeckung, Flöte spielend) (auf Rundsockel)
8246 STEHKOPF Madonna (Frauenkopf mit Kopftuch) (–> vgl. Dakon) (Goebel zugeschr.)
8247 WANDMASKE „Maske Araberin" (Frauenkopf mit Hand, Ring und Kopftuch) (vgl. 7813)
8254 Sitzendes Liebespaar (auf Ovalsockel)
8262 WANDBEHANG Skifahrerin mit Kappe und geschulterten Skiern
8270 Stehendes Mädchen mit Ente (auf Rundsockel)
8337 RELIEF „Malermaske" (Junge mit Hut, Malerpalette und Pinsel) (Wandrelief)
8339 (STEH)KOPF „Kopf ohne/mit Sockel" (Frauenkopf mit geschlossenen Augen, Haarschmuck und Kragen)
8410 WANDMASKE „Maske „ (Frauenkopf mit Trauben im Haar und am Kragen) (Pendant vgl. 8411) (Goebel zugeschr.)
8411 WANDMASKE „Maske „ (Frauenkopf mit Blüten im Haar und Kragen) (Pendant vgl. 8410) (Goebel zugeschr.)
8421 WANDMASKE „Maske „ (Frauenkopf mit langem Haar, Hand und Glockenblume)
8884 WANDMASKE Frauenkopf mit Haaransatz und Halstuch
8885 WANDMASKE Frauenkopf mit Hut und Kleidansatz
8886 WANDMASKE
8887 WANDMASKE (Frauenkopf?) mit Hut
8888 WANDMASKE Madonna mit Kind
8889 WANDMASKE
8890 WANDMASKE Frauenkopf mit Schleier (Kopfschal)
8891 WANDMASKE Frauenkopf mit Hut und Halskette
8892 WANDMASKE
8893 WANDMASKE
8894 WANDMASKE
8895 WANDMASKE
8896 WANDMASKE
8897 WANDMASKE
ohne Nr. Liegender Damenakt mit Schal, Tuch und erhobener Hand (auf blumenverziertem Rechtecksockel) (Goebel zugeschr.)
ohne Nr. STEHKOPF Frauenkopf mit Mütze und Schal (auf Holzsockel)
Schreitender Matrose, ein junges Mädchen im Arm haltend (auf Rundsockel) (um 1935/37) Firmenmarke 19, Made in Austria, unsign., (H. 30 cm); Christie's L. 15.4.1987

Goedtke

Der Bildhauer bzw. Keramiker Goedtke wird nur bei Könecke erwähnt. Weitere Hinweise oder Lebensdaten zu diesem Künstler lassen sich nicht finden. Lit.: Könecke (2002), S. 33, 35, 73 (m. Abb.)

202-WG „Flötenspielerin mit Goldflöte" (Sitzende Flötenspielerin)
205-WG „Tanzpaar" (Balletttänzer)

Goëtz, J.-Charles

J.-Charles Goëtz, Kunstgewerbler, Dekorateur (Lebensdaten unbekannt.), war Mitglied der Künstlergruppe „Évolution" (–> vgl. Kap. „Dekorative Skulpturen von Arthur Goldscheider in Paris"). Lit.: Neuwirth (1974), S. 63.

VASE (en dinanderie gravée) MESSING (H. 18 cm, B. 9 cm) Lit.: Goldscheider, A. (Hrsg.): Artistic Evolution, 1926, T. 1, Pl. 20 (m. Abb.)
VASE (métal gravé et laqué) (Form von La Monaco, Lack von Stabi) (H. 11 cm, B. 10 cm) Lit.: Goldscheider, A. (Hrsg.): Artistic Evolution, 1926, T. 1, Pl. 20 (m. Abb.)
VASE (en Faience gravée) (H. 25 cm, B. 32 cm) Lit.: Goldscheider, A. (Hrsg.): Artistic Evolution, 1926, T. 1, Pl. 23 (m. Abb.)
VASE (en dinanderie gravée) MESSING (Form von La Mo-

naca) (H. 28 cm, B. 20 cm) Lit.: Goldscheider, A. (Hrsg.): Artistic Evo ution, 1926, T. 1, Pl. 24 (m. Abb.)
VASE (en ciuvre avec incrustations d'émaux sur argent) KUPFER (Dekor von Colmant) (H. 18 cm, B. 10 cm) Lit.: Goldscheider, A. (Hrsg.): Artistic Evolution, 1926, T. 1, Pl. 24 (m. Abb.)
BRUNNEN „Mosaic Fountain" MOSAIK (H. fehlt, Ø fehlt) Lit.: Goldscheider, A. (Hrsg.): Artistic Evolution, 1926, T. 1, Pl. 37 (m. Abb.)

Goldscheider, Alois
Alois Goldscheider, Prokurist (Geburts- und Sterbedatum unbekannt), Bruder von Friedrich Goldscheider, Prokurist der Firma bis 1920. (-> vgl. Kap. „Die Anfänge – ‚Moritz Goldscheiders Erben' in Pilsen" u. folgende Kap.).

Goldscheider, Arthur
*Arthur Goldscheider, Firmenleiter und später Inhaber der Pariser Niederlassung (*6.6.1874 Pilsen, †?), Sohn von Friedrich Goldscheider, Firmenleiter und später Inhaber der Pariser Niederlassung (-> vgl. Kap. „Dekorative Skulpturen von Arthur Goldscheider in Paris").*

Goldscheider, Camillo
*Camillo Goldscheider, (*28.5.1875 Pilsen, †?), Sohn von Friedrich Goldscheider, Firmeninhaber der „Dekorativen Plastik, Camillo Goldscheider, Friedrich Goldscheider Sohn". (-> vgl. Kap. „Jugendstil – Dekorative Plastik von Camillo Goldscheider in Wien").*

Goldscheider, Eduard
*Eduard Goldscheider, (*1852, †23.5.1892 Pilsen), Bruder von Friedrich Goldscheider, Inhaber der Fabrik in Pilsen. (-> vgl. Kap. „Die Anfänge – Eduard Goldscheider und die ‚Goldscheidrovka' in Pilsen").*

Goldscheider, Erwin Friedrich
*Erwin F. Goldscheider, Prokurist (*11.7.1912 Wien, †10.1.1984 Trenton), Sohn von Walter Goldscheider, Prokurist und Produktionsleiter in Trenton. Quelle: AGM (Gehaltsliste, Mai 1938); Sterbeurkunde vom Office of Registrar of Vital Statistics Township of Ewing. (-> vgl. Kap. „Populäre und moderne Keramik der dreißiger Jahre unter Walter Goldscheider" u. folgende Kap.).*

Goldscheider, Friedrich Siegfried
*Friedrich Goldscheider, (*6.11.1845 Slabetz/Slabce, †19.1.1897 Nizza). Sohn von Moritz Goldscheider, Bruder von Alois Goldscheider, Firmengründer. (-> vgl. Kap. „Friedrich Goldscheider und die Entstehung einer internationalen Manufaktur in der Zeit des Historismus in Wien").*

Goldscheider, Marcell (Marcel)
*Marcell Goldscheider, (*25.5.1887 Wien, †23.4.1964 Worthing/UK), Sohn von Friedrich Goldscheider. Marcell war neben seiner leitenden Funktion für die Firmen (zusammen mit Walter Goldscheider) und die von ihm gegründeten Vereinigten Ateliers für Kunst und Keramik in Wien auch als Künstler (Modelle u.a. „Kiss", „Lovers", „Lady Dancer", „Blue Boy") für seine eigene Manufaktur in Hanley tätig. (-> vgl. Kap. „Marcell Goldscheiders Werkstätten in Wien und Stoke-on-Trent").*

Goldscheider, Moritz
*Moritz Goldscheider, (*1815, †1865 Pilsen), Vater des Firmengründers Friedrich Goldscheider. (-> vgl. Kap. „Die Anfänge – ‚Moritz Goldscheider's Erben' in Pilsen").*

Goldscheider, Peter
*Dr. Peter Goldscheider, (*9.4.1945 Kairo), Enkel von Camillo Goldscheider, Gründer der „Goldscheider Keramik GmbH" in Stoob/Bgld. (-> vgl. Kap. „Die Fortsetzung der Tradition durch Peter Goldscheider in Österreich").*

Goldscheider, Regine
*Regine Goldscheider, Prokuristin und Geschäftsführerin (*19.4.1854 Teplitz/Teplice, † 20.10.1918 Wien), Ehefrau von Friedrich Goldscheider, Prokuristin und Geschäftsführerin. Ab 1901 mit Dr. S. Lewit verheiratet. (-> vgl. Kap. „Die Blüte des Jugendstils unter Regine Goldscheider").*

Goldscheider, Walter
*Walter Goldscheider, Kommerzialrat (*18.3.1879 Wien, †23.3.1962 Rottach), Sohn von Friedrich Goldscheider, war Kommerzialrat und führte das Unternehmen mehrere Jahrzehnte lang, sowohl in Wien wie auch in Trenton. (-> vgl. Kap. „Art-Déco-Meisterstücke der Brüder Walter und Marcell Goldscheider").*

Golter s. Polter

Gori, F. / T. s. Gory, A(ffortunato?)

Gornik, Friedrich
*Friedrich Gornik, Bildhauer (*15.1.1877 Prävali/Slowenien/Kärnten, †26.3.1943 Wien), studierte an der Fachschule Villach und der Wiener KGW-Schule unter Breitner und Strasser (1898–1905). Die künstlerische Mitarbeit bei Goldscheider beginnt etwa 1902/03. Er lieferte Modelle für die Werkstatt von Hugo Kirsch, für Förster & Co. und die Wiener Kunstkeramische Werkstätte Busch & Ludescher. Gornik nahm an zahlreichen Ausstellungen (u.a. Künstlerhaus Wien, Glaspalast München und Große Berliner Kunstausst.) teil und entwarf neben figürlichen Arbeiten in Steinzeug, Fayence und Porzellan auch Bronzen. Ein Schaffensschwerpunkt ist die Tierbildhauerei, welche er durch umfangreiche Studien im Schönbrunner Zoologischen Garten untermauerte. Gornik war 1908 wohnhaft in Wien XIII., Penzingerstr. 126 (später dann Rupertgasse 2, XIV. Bez.), sein Atelier hatte er durchgehend in Wien XIII., Hietzinger Hauptstr. 47. Quelle: TB, Bd. 14, S. 402; Lehmann, 1908–1942; Neuwirth (1974), S. 63, 152; Neuwirth (1974a), S. 156; Zühlsdorff, S. 410–411; Hillenbrand, S. 148; Bénézit, Bd. 6, S. 304.*
1043 JARDINIERE „Kakadu" ÖMKI (1903/04)
2364 FIGUR „Eule" (mit Schreibzeug, auf zwei Büchern sitzend) (vgl. 3853; 3854)
2383 FIGUR „Hahn" ÖMKI (1912) (vgl. 4168)
2423 FIGUR „Kakadu" ÖMKI (1903/04)
2442 FIGUR „Eisbär" (auch als Lampe mit Glassockel)
2655 „Eisbär" ÖMKI (1906/07)
3853 FIGUR „Eule" (mit Schreibzeug, auf zwei Büchern sitzend) (vgl. 2364; 3854)
3854 FIGUR „Eule" (mit Schreibzeug, auf zwei Büchern sitzend) (vgl. 2364; 3853)
3938 Balzender Auerhahn
4168 FIGUR „Hahn" ÖMKI (1912) (vgl. 2383)

Gory, A(ffortunato?)
*Affortunato Gory, Bildhauer (*Florenz, †?), war Schüler von A. Rivalta und ist zwischen 1895 und 1923 nachweisbar. Er stellte im Pariser Salon (Societé des Artistes Français) aus. Lit.: Bénézit, Bd. 6, S. 308; AKL Index.*
2385 BÜSTE „Arôme" (Frau mit Callablütenstengeln) (Inschrift: ARÔME) sign. „F. Gori" (wohl A. Gory) (vgl. 2885= ähnl. Modell)
2474 BÜSTE Junge Frau (nach links blickend)
2885 BÜSTE Flora (Frau mit Callablüten) sign. „T. Gori" (gemeint ist wohl A. Gory) (vgl. 2385= ähnl. Modell)

Gracien, G. / J.
Da sich der Künstler G. bzw. J. Gracien in der einschlägigen Literatur nicht nachweisen lässt, handelt sich offenbar um ein Pseudonym, welches nach dem momentanen Stand der Dinge nicht endgültig aufgelöst werden kann. Der Verdacht liegt jedoch nahe, dass es sich dabei um Arthur Strasser handelt wenn man Sujets und die Ausführung der Figuren mit denen Strassers vergleicht.
374 Wasserverkäufer (Stehender Araber mit Turban, Wasserkrug, Kanne und Schöpfkelle) (sign. G. Gracien) (vgl. 857)
857 s. Strasser, Arthur
1035a FIGUR „Vasenträgerin (halb bekleidet)" (Stehende junge Frau mit Krug auf der Schulter) (auf Rechtecksockel) (sign. J. Gracien) (vgl. 1035b; 1440)
1035b FIGUR „Vasenträgerin (ganz bekleidet)" (Stehende junge Frau mit Krug auf der Schulter) (auf Rechtecksockel) (vgl. 1035a; 1440)
1440 FIGUR „Vasenträgerin" (Stehende junge Frau mit Krug auf der Schulter) (auf Rechtecksockel) (vgl. 1035a; 1035b)

Griser
850 SÄULE / POSTAMENT Drei Frauengesichter mit verflochtenen Haaren (sign. Grißer)
2250 FIGUR „Perle" ÖMKI (1902/03)

Grißer s. Griser

Gross, F. (Karl?)
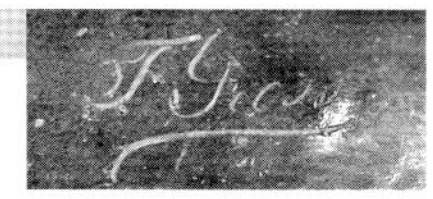
*Da sich ein Künstler namens F. Gross nicht nachweisen lässt, die Stücke für Goldscheider aber in der Regel mit F. Gross sign. sind, muss hier von einem Pseudonym ausgegangen werden. Allenfalls kommt ein Karl Groß, Bildhauer und Kunstgewerbler (*28.1.1869 Fürstenfeldbruck, †4.10.1934 Dresden/ München) infrage. Er war von 1916–1933 Direktor der KGW-Schule in Dresden. Es muss abgewartet werden, bis sich evtl. einmal vollständig mit Vornamen sign. Stücke finden. Lit. (für Karl Groß): TB, Bd. 15, S. 99, Vollmer, Bd. 2, S. 317.*
2346 FIGUR „Orangenpflückerin" ÖMKI (1902/03)
2616 FIGUR/LAMPE Wasserträgerin (einen Krug/Henkeltopf auf der Schulter tragend) (vgl. 2741; 2770 und Modell-Nr. .... H. 39 cm; Zeller 29)
2706 BÜSTE Kriegerin mit Helm
2741 FIGUR/LAMPE Wasserträgerin (einen Krug/Henkeltopf auf der Schulter tragend) (vgl. 2616; 2770 und Modell-Nr. .... H. 39 cm; Zeller 29) (vgl. Cherc 2770)
2770 LAMPE Wasserträgerin (eine(n) Krug/Amphore auf der Schulter tragend) (vgl. 2616; 2741; Modell-Nr. .... H. 39 cm; Zeller 29)
2840 UHR Stehendes Holländermädchen an einem Schiffssteuer drehend (Uhrgehäuse als Felsstück)
2853 BÜSTE Mädchenkopf mit schulterlangem Haar (in der Sockelkartusche die Darstellung von Rotkäppchen mit dem Wolf)
2901 Wasserträgerin (ein Gefäß in die Höhe haltend) (vgl. 3031)
3031 Wasserträgerin (ein Gefäß in die Höhe haltend) (vgl. 2901)

Gross (Karl?) s. Gross, F.

Gruber, (Franz?)
*(Franz?) Gruber, Maler und Bildhauer (*7.1.1878 Wien, †7.6.1945 Wien). Ob dieser, vorwiegend als Maler tätige Künstler, wirklich der Entwerfer der drei unter seinem Namen gelisteten Objekte ist, muss mit einem Fragezeichen versehen bleiben. Gruber war 1908 wohnhaft in Wien III., Landstraßer Gürtel 25. Lit.: Lehmann, 1908; AKL Index.*
4835 Stehende Biedermeierdame mit Schutenhut und Blumenbukett (auf oktogonalem Sockel)
4837 Tanzendes Paar (Biedermeierdame mit Mädchen) (auf Rundsockel)
4914 Tänzerin in Bewegung, ihr Kleid nach hinten hochhaltend (sechseckiger Sockel mit Blumenstrauß)

Grun, Hélène
Hélène Grun, (geb. in Concarneau, Frankreich) war 1928 in Paris wohnhaft und mit einer Vase für Arthur Goldscheider im Pariser Salon vertreten. Lit.: Neuwirth (1974), S. 64.
VASE (métal gravé) Paris, Salon (1928) Lit.: Neuwirth (1974), S. 64

Guénou
Guénou, Kunstgewerbler und Keramiker (Vorname, Lebens- und Schaffensdaten unbekannt), war Mitglied der Künstlergruppe „Évolution" (-> vgl. Kap. „Dekorative Skulpturen von Arthur Goldscheider in Paris"). Lit.: Neuwirth (1974), S. 64.

Guerbe
Guerbe, Bildhauerin (Vorname, Lebens- und Schaffensdaten unbekannt), war Mitarbeiterin bei Arthur Goldscheider, Paris und schuf Kleinbronzen. Lit.: Goldscheider, A. (Hrsg.): Artistic Evolution, 1926, T. 2, Pl. 31 (m. Abb.).
„Bacchic Dance" BRONZE (H. fehlt, B. fehlt) Lit.: Goldscheider, A. (Hrsg.): Artistic Evolution, 1926, T. 2, Pl. 30 (m. Abb.)

Guinard, J.
J. Guinard, Keramiker, (Lebens- und Schaffensdaten unbekannt), war Mitglied der Künstlergruppe „Évolution" und wird erwähnt als künstlerischer Mitarbeiter für keramische Arbeiten bei Arthur Goldscheider, Paris. Lit.: Zühlsdorff, S. 582.

Gult, Johann
Johann Gult, Bildhauer und Keramiker (Lebensdaten unbekannt), war 1930 und 1931 wohnhaft in Wien V.,

Gassergasse 12. Gult war, wie die Auflistung seiner Entwürfe zeigt mit Deckeldosen, Vasen und Schalen, also mit angewandter Kunst für Goldscheider befasst. Lit.: Lehmann, 1930; 1931.
4969 DECKELDOSE Taube (Form: rechteckig/balusterförmig/Knauf: Taube (vollplastisch)/Dekor: Ast- und Blattwerk mit Blüten)
4975 DECKELDOSE Chinese (Form: quaderförmig/Knauf: sitzender Chinese (vollplastische Fig.)/Dekor: vielfarbig, tunkpapierähnl.)
4976 DECKELDOSE Vogel (Form: eiförmig, vier Kugelfüße/Knauf: Liegende Henne (vollplastisch)/Dekor: Durchbrochene Metallarbeit suggerierend (Gult zugeschr.)
4982 VASE (Form: balusterförmig/Dekor: Arabesken, reliefartig aufgelegt)
5108 VASE (Form: stangenförmig, Mündung etwas geweitet, mit umlaufendem Relief, Fuß mit zwei floralen Applikationen/Dekor: Rustika, kobalt, grün)
5109 VASE (Form: becherförmig, leicht geschwungen, Fuß zweifach gewellt/Dekor: Korpus vertikale Blütenranken, Fuß horizontale Blütenranken)
5110 VASE (Form: Mündung und Fuß etwas geweitet, mittig gebaucht/Dekor: florale und Phantasieelemente) (Dekor vgl. 5111)
5111 DECKELVASE (balusterförmig) (Knauf: blütenförmig/Dekor: florale und Phantasieelemente) (Dekor vgl. 5110)
5166 DECKELDOSE
5167 SCHALE

Hafenruckler

5657 Ägyptische Tempeltänzerin (auf abgeschrägtem, mit umlaufendem Reliefdekor (Elefanten, Palmen) verziertem Rechtecksockel)

Hagenauer, Franz

*Franz Hagenauer, (Metall-)Bildhauer und Kunstgewerbler (*22.2.1906 Wien, †26.9.1986 Wien), war Schüler von Franz Cizek und Dagobert Peche, von 1921–1926 bei Anton Hanak an der KGW-Schule in Wien. Hagenauer war Mitglied der Wiener Sezession und des Österreichischen Werkbundes. Er war ab 1931 wohnhaft in Wien VII., Bernardgasse 7. Lit.: Lehmann, 1931 ff.; Vollmer, Bd. 2, S. 353; AKL Index.*
Zigarettendose MESSING sign. Hagenauer (um 1930) (L. 11 cm); Nagel 345

Hahn

5672 Chinesischer Mönch (auf Ovalsockel)

Haidvogel, Florian

*Florian Haidvogel, Spritzer (*19.10.1892, †?). Quelle: AGM, Liste „Adressen von guten und zuverlässigen Arbeitern" (undat., um 1953).*

Halbout, Georges

*Georges Halbout, Bildhauer (*15.12.1895 Paris, †7.6.1986 Bernoux/Drôme), war Mitglied der Künstlergruppe „La Stèle" (–> vgl. Kap. „Dekorative Skulpturen von Arthur Goldscheider in Paris"). Halbout erhielt eine „Mention Honorable" im Pariser Salon 1928 und war zur selben Zeit Mitarbeiter bei Arthur Goldscheider. Lit.: Neuwirth (1974), S. 64; AKL Index.*

Halnon, Frederick James

*Frederick James Halnon, Bildhauer (*8.3.1881 London, †?), war in London ansässig. Er war Schüler von Alfred Drury und Mitarbeiter bzw. als Künstler für Marcell Goldscheider in England tätig. (–> vgl. Kap. „Marcell Goldscheider – Emigration und Art Pottery in England"). In der Literatur wird Halnon letztmalig 1934 im „Who's who in Art" erwähnt und nicht 1903, wie im AKL fälschlich angegeben. Die für Marcell Goldscheider geschaffene Figur „Nell Gwynne" deutet jedoch darauf hin, dass Halnon auch noch um 1940 aktiv war. Lit.: Vollmer, Bd. 2, S. 362; AKL Index.*

Halou, Alfred Jean

*Alfred Jean Halou, Bildhauer (*20.6.1875 Blois/Loir-et-Cher, †1939 Paris), war Schüler von Rodin, Mitbegründer des Salons d' Automne und Mitarbeiter von Friedrich Goldscheider, Paris. Lit.: Vollmer, Bd. 2, S. 363; Bénézit, Bd. 6, S. 696; AKL Index.*
774-GSB PLATEAU „Chasse"

Handzel, August(in)

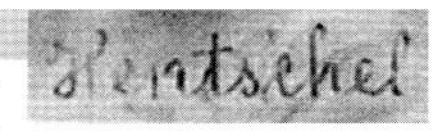

*Der tschech. Künstler Augustin Handzel (*7.6.1896 Mähr. Ostrau/Ostrava, †26.10.1952 Mähr. Ostrau/Ostrava), der an der Prager KGW-Schule unter Sucharda und Drahonovsky seine Ausbildung genoss, unternahm Studienreisen nach Deutschland, Polen und Österreich. Er ist bekannt für seine Architekturplastiken, Porträtbüsten und Reliefs. Lit.: Vollmer, Bd. 2, S. 369. Auch bei Neuwirth (1974), S. 64, gibt es einen Hinweis auf diesen Künstler. Die „deutsche" Namensform „Hentschel" wurde in seinem Fall, wohl wie schon mehrfach angedeutet, bei Goldscheider aus ökonomischen Gründen verwendet. Eindeutige Belege für den Künstler A. Handzel als Entwerfer finden sich in den zeitgenössischen Magazinen: Studio YB, 1914, S. 191 (m. Abb.) und MW, 2.1920/21, Nr. 12, S. 17 (m. Abb.) für die Modelle 4390 und 4757. Die Brüder Konrad (bereits 1907 verstorben und bekannt für seine „Meissen-Kinder") und Rudolf Hentschel (Porzellanmaler und Lehrer an der Porzellanmanufaktur Meissen), die auch gelegentlich in Auktionskatalogen als ausführende Künstler genannt werden, scheiden eindeutig aus. Infrage kommen könnte allenfalls noch ein Bildhauer namens „Eduard" Hentschel, der 1928 bei Lehmann (Wiener Adressbuch) unter den Bildhauern aufgeführt ist. Weitere Hinweise auf ihn fehlen in der einschlägigen Lit. jedoch. Unwahrscheinlich erscheint auch der bei Neuwirth (1974), S. 64 aufgeführte Karl Hentschel. Lit.: Vollmer, Bd. 2, S. 369; AKL Index.*
4362 Sitzende Biedermeierdame mit blumenverziertem Schutenhut (sign. Hentschel)
4376 LAMPE Stehendes Chinesenmädchen einen Topf haltend (auf Rechtecksockel) (sign. Hentschel) (vgl. 5587)
4377 LAMPE Stehendes Chinesenmädchen mit Fächer (auf Rechtecksockel) (sign. Hentschel) (vgl. 5588)
4390 „Im Regen" (Biedermeierdame mit Hut, den Rock anhebend) (sign. Hentschel) ÖMKI (1913/14)
4446 Stehende Rokokodame, ihren Reifrock anhebend (sign. Hentschel)
4450 UHR Zwei stehende Putti (vor klassizistischem, urnenbekröntem Denkmal) (sign. Hentschel)
4482 Stehende Dame (in langem Kleid) mit Blumenstrauß und Buch (Kirchgängerin?) (sign. Hentschel)
4485 Tänzerin in Pose (auf Rundsockel) (sign. Hentschel)
4491 Stehendes Mädchen mit zwei Blumensträußen (auf Rundsockel) (sign. Hentschel)
4612 Stehendes Bauernmädchen in Tracht (Böhmen/Ungarn?) (auf profiliertem Rundsockel) (vgl. 4701= ähnl. Modell/Pendant?) (sign. Hentschel)
4701 Stehendes Bauernmädchen in Tracht (Österreich/Böhmen?) (auf profiliertem Rundsockel) (vgl. 4612= ähnl. Modell/Pendant?) (sign. Hentschel)
4757 „Erwartung" (Sitzendes, kleines Mädchen in Tracht, mit Regenschirm) (Modell-Nr. evtl. auch 4767?)
4767 s. 4757
5587 Stehendes Chinesenmädchen einen Topf haltend (auf Rundsockel) (Fay.; ALABASTER/SPECKSTEIN?) (sign. Hentschel) (vgl. 4376)
5588 Stehendes Chinesenmädchen mit Fächer (auf Rechtecksockel) (sign. Hentschel) (vgl. 4377)
ohne Nr. Stehende Dame mit erhobenem Arm (im Watteaukleid) mit Blumenstrauß
„Stolz" (Stehende Dame mit Hut, Schirm und Schal) Lit.: Studio YB, 1914, S. 193 (m. Abb.), hier mit Ang. des Originaltit. „Stolz".

Haniroff/Xaniroff

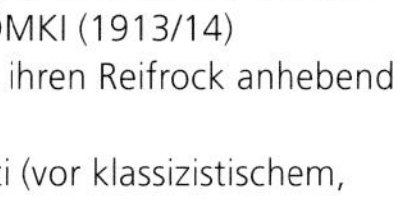

Da sich in der einschlägigen Literatur kein Künstler namens Haniroff nachweisen lässt, für Goldscheider aber zahlreiche Modelle mit diesem, sowie auch der manchmal zu lesenden Form „Xaniroff" sign. sind, muss auch hier wieder von einem Pseudonym ausgegangen werden.
652 VASE Kleines Mädchen mit Frosch an der Vasenwandung stehend
1150 FIGUR „Fanchon" (vgl. 2325= BÜSTE)
1612 WANDMASKE „Araber" (mit Vollbart und Turban) (vgl. 1105= FIGUR; Pendant vgl. 1613)
1614 WANDMASKE „Mahdi" (vgl. 1308; 1423= jew. BÜSTE)
1616 WANDMASKE Nordafrikanerin mit Kopftuch (Haniroff zugeschr.)
1617 WANDMASKE „Negerknabe" (Lachender Negerjunge mit Schiebermütze) (vgl. 1695)
1666 WANDMASKE „Gigerl" (Negerjunge mit Strohhut) (Pendant vgl. 1667)
1667 WANDMASKE „Gigerl" (Neger mit Strohhut, Monokel und Zigarette) (Pendant vgl. 1666) (Haniroff zugeschr.)
1695 BÜSTE Lachender Negerjunge mit Kappe (vgl. 1617)
1811 FIGUR „Säemann" (Säender Bauer) (auf Landschaftssockel) (vgl. 1838) (Pendant vgl. 1812)
1812 FIGUR „Schnitterin" (Mit Sense mähende Bäuerin) (auf Landschaftssockel) ÖMKI (1899/1900, hier Ang. „Xaniroff") (vgl. 1837) (Pendant vgl. 1811)
1813 FIGUR „Marmor-Arbeiter" (Stehender Arbeiter mit großem Hammer/Schlegel) (auf Landschaftssockel)
1837 FIGUR „Schnitterin" (Stehende Bäuerin mit Kopftuch) (Ähren bindend) (auf Landschaftssockel) (vgl. 1812) (Pendant vgl. 1838)
1838 FIGUR „Säemann" (Säender Bauer) (auf Landschaftssockel) (vgl. 1811) (Pendant vgl. 1837)
1969 FIGUR „Wasserträgerin" ÖMKI (1900/01, hier Ang. „Xaniroff")
2000 WANDMASKE „Maske von Knabe mit Leuchter" (Negerjunge mit Fez) (vgl. 1291 Lemassi)
2039 FIGUR „Orchidee" (Stehende Frauengestalt mit Blütenschmuck) (auf Blumen geschmücktem Rundsockel) (vgl. 2277)
2092 FIGUR „Triumph" ÖMKI (1900/01) (vgl. 2201)
2178 BÜSTE „La Fierté" (Inschrift: LA FIERTÈ) ÖMKI (1901/02, hier Ang. „Xaniroff") (vgl. 2371; 2435)
2201 FIGUR „Triumph" ÖMKI (1900/01) (vgl. 2092)
2223 FIGUR „Jagdtrophäe" (Leda besiegt den Schwan) (vgl. 2444)
2234 FIGUR „Der Wind" (Stehendes Mädchen mit zwei Krügen) ÖMKI (1901/02, hier Ang. „Xaniroff") (vgl. 2452)
2238 FIGUR „Vasenträgerin" (Stehende junge Frau eine Amphore auf der rechten Schulter tragend) (auf Landschaftssockel) (vgl. 2409; 2583)
2277 FIGUR „Orchidee" (Stehende Frauengestalt mit Blütenschmuck) (auf Blumen geschmücktem Rundsockel) (vgl. 2039)
2325 BÜSTE „Fanchon" (Junge Dame nach links blickend) (vgl. 1150= FIGUR)
2371 BÜSTE „La Fierté" (Inschrift: LA FIERTÈ) ÖMKI (1901/02, hier Ang. „Xaniroff") (vgl. 2178; 2435)
2405 FIGUR „Harfenspielerin" (Stehende junge Frau mit Harfe) (auf Landschaftssockel)
2409 FIGUR „Vasenträgerin" (Stehende junge Frau eine Amphore auf der rechten Schulter tragend) (auf Landschaftssockel) (vgl. 2238; 2583)
2435 BÜSTE „Stolz" (vgl. 2178; 2371)
2440 BÜSTE Frau mit Zöpfen und Diadem (besetzt mit Perlen und Glassteinen)
2444 FIGUR „Jagdtrophäe" (Leda besiegt den Schwan) (vgl. 2223)
2452 FIGUR „Der Windstoß" (Stehendes Mädchen mit zwei Krügen) (vgl. 2234)
2583 FIGUR „Vasenträgerin" (Stehende junge Frau eine Amphore auf der rechten Schulter tragend) (auf Landschaftssockel) (vgl. 2238; 2409)
2643 BÜSTE Römischer Krieger mit Helm (auf mit Opferszene reliefiertem Sockel)
2759 Schreitendes Bauernmädchen, einen Korb tragend (auf Landschaftssockel)
2980 BÜSTE Biedermeierdame mit Schutenhut (Sockelrelief: Taube auf zwei Rosengirlanden sitzend)
3131 FIGUR/LAMPE Schreitender Harlekin mit Schirm (ohne Nr. BRONZE/ELFENBEIN) (vgl. 3133)
3133 Schreitender Harlekin mit Schirm (auch ohne Nr. BRONZE/ELFENBEIN) (vgl. 3131)
3200 Stehende junge Frau mit großem Hammer und Amboss (auf hohem Rundsockel)
3437 Der erste Kuss (Mädchen mit Weidenkorb, von Jungen geküsst)
3473 LAMPE Stehender Nachtwächter mit Helm (eine eiserne Lampe haltend)
3784 Stehender, nackter Römerjüngling mit Lorbeerkranz und Degen (auf Landschaftssockel)
3810 BÜSTE Frau mit Diadem und Kopftuch (nach links blickend)
4084 Jüngling mit Adler kämpfend (auf Felssockel)
4680 Stehender Knabe, einen Brief lesend (in knielangen Hosen und Strohhut) (auf rechteckigem Landschaftssockel)

4699 Sitzender Pan mit Panflöte (auch sign. Schörck)
4700 Sitzende Nymphe mit Flöte (Pendant vgl. Schoerk 4699)
ohne Nr. Schreitender Harlekin mit Schirm BRONZE / ELFENBEIN / Onyxsockel (vgl. 3131; 3133)
WANDMASKE Gesicht eines lachenden schwarzen Mannes mit hellem Strohhut (um 1895) ohne Nr. Firmenmarke 7, Repr. Res., (H. 25 cm); HMW Inv. 186.642
BÜSTE „Helene" ÖMKI (1906/07) Lit.: Neuwirth (1974), S. 76
FIGUR „Tänzerin" in drehender Bewegung (ihr Kleid mit der rechten Hand am Saum haltend) Lit.: N.N.: Illustrierte Rundschau, Terrakotten von Fr. Goldscheider, Wien, in: VKMH, 17.1902/03, S. 717 (m. Abb.) (hier Ang. Haniroff)
FIGUR „Krugträgerin" Lit.: R., F.: Wiener Kunstkeramik, in: KM, 2.1902, S. 91, „Haniroffs „Krugträgerin" zeigt dunkles Golddekor, d.h. sie hat die schwärzliche Farbe der Eisenbronze und aufgesetzte Lichter wie gelblichem Messington".
Stehender Schmied („mit Schurzfell bekleidet, ein glühendes Hufeisen bearbeitend") (um 1902/03) (vgl. 2427?) Lit.: N.N.: Leipziger Michaelismesse 1904, in: Keram.RS, 12.1904, S. 945; N.N.: Figürliche Terrakotten, in: KM, 4.1904, S. 110 (m. Abb.), hier Künstlerang. „Xaniroff", „Der Oberkörper des Schmiedes ist unbekleidet, das lederne Schurzfell deckt Lenden und Beine gegen die sprühenden Funken, welche der aufschlagende Hammer hervorruft. Modelliert wurde dieses Stück von Bildhauer Xaniroff, dem es vortrefflich gelungen ist, die Tätigkeit des Schmiedes auch in den Muskeln und Adern des Oberkörpers, namentlich dem Arme, zur Erscheinung zu bringen".

Hauchecorne, Gaston

*Gaston Hauchecorne, Bildhauer (*10.9.1880 Le Havre, †8.4.1948 Paris), beschickte den Salon in Paris (Societé des Artistes Français) seit 1922. Hauchecorne arbeitete für Susse Frères und war Mitglied der Künstlergruppe „La Stèle" (-> vgl. Kap. „Dekorative Skulpturen von Arthur Goldscheider in Paris"). Lit.: Vollmer, Bd. 2, S. 390; Neuwirth (1974), S. 64; Bénézit, Bd. 6, S.800.*

Hauf

3078 Sitzender Windhund
5129 Hund (Rasse in der Quelle nicht angegeben) „Englischer Windhund" (liegend) (Fay.) (1920/21) Lit.: N.N.: Wiener Kleinplastik einst und jetzt in: MW, 2.1920/21, Nr. 12, S. 29 (m. Abb.)

Hegenbarth, Ernst

*Ernst Hegenbarth, Bildhauer (*5.3.1867 Ulrichsthal/Böhmen, †14.5.1944 Wien), studierte 1880–1883 an der Fachschule für Glas und Keramik in Steinschönau (Böhmen) dann von 1886–1892 an der KGW-Schule in Wien unter Otto König. Hegenbarth war ab 1907 Mitglied des Wiener Künstlerhauses dem er in den Jahren 1922–1923 als Präsident vorstand. Von seiner Hand stammen auch Denkmale und Figuren im öffentlichen Raum. Hegenbarth war ab 1928 wohnhaft in Wien II., Josef-Gall-Gasse 5, sein Atelier befand sich ebenfalls im II. Bez., Böcklinstr. XIX. Bez. Lit.: TB, Bd. 26, S. 244; Lehmann, 1928 ff.; Bénézit, Bd. 6, S. 854; AKL Index.*
3447 Stehende Frau mit Rosengirlande (auf vorne abgeschrägtem Rechtecksockel)
BÜSTE „Kaiser von Österreich" Paris, Weltausstellung (1900) Lit.: Hecht / Barthel, in: Tonindustrie-Ztg., 24.1900, S. 1584

Heindl, Hermann

*Hermann Heindl, Porträtmaler und Bildhauer (*6.1.1875 Wien, †5.5.1945 Wien), besuchte die KGW-Schule in Wien und entwarf im Jahre 1937 Wandmasken für Goebel, Oeslau. Heindl betrieb in den 40er Jahren eine eigene keramische Werkstatt in Wien XIX., Würthgasse 1. Seine Wohnung lässt sich bis 1942 in Wien XIX., Döblinger Gürtel 21–23 nachweisen. Lit.: Lehmann, 1908–1942; Makus (2000), S. 171.*
6442 BÜSTE Madonna mit Kind
6444 FIGUR „Madonna mit Kind" (Stehende Madonna mit Kind) (vgl. 7730)
7730 FIGUR „Madonna mit Kind" (Stehende Madonna mit Kind) (vgl. 6444)

Heinemann, Fritz

*Fritz Heinemann, Bildhauer (*1.1.1864 Altena/Westfalen, †1.12.1932 Berlin), besuchte 1883–1886 die Nürnberger Kunstschule und von 1886–1889 die Berliner Akademie. Heinemanns Werk ist breit gefächert. In Bronze, Holz und Marmor schuf er zahlreiche Denk- und Grabmäler, Genrefiguren, Porträtbüsten, Tänzerinnen und vieles mehr. Sein „Milchmädchen" (lt. TB im Museum Dessau), könnte die von Goldscheider reproduzierte Holländerin mit den zwei Milchkannen sein. Lit.: TB, Bd. 16, S. 294; Vollmer, Bd. 6, S. 37; Bénézit, Bd. 6, S. 864; AKL Index.*
4242 Schreitende Holländerin mit zwei Milchkannen (auf Rundsockel) (vgl. 4642)
4642 Schreitende Holländerin mit zwei Milchkannen (auf Rundsockel) (vgl. 4242)

Hentschel
s. Handzel, August(in)

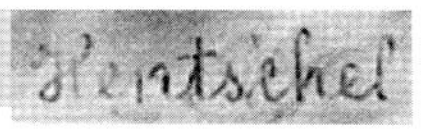

Herczeg, Claire (Klára)

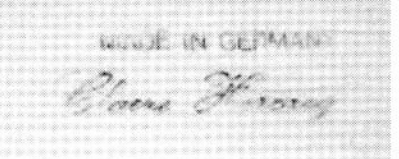

*Klára Herczeg (Claire Weiss, Claire Herczeg), Bildhauerin (*13.10.1906 Budapest, †6.8.1997 Budapest), verlebte ihre Kindheit in Wien. Nach einer Bildhauerlehre bei der Wiener Arsenal-Gießerei studierte sie 1924–1928 an der Budapester Kunsthochschule unter Kisfaludi-Stróbl. Von 1929–1930 Porzellan-Bildhauerei in Berlin, Entwürfe für mehrere Manufakturen, bis 1932 Aufenthalte in Paris und 1939 in London. Ab 1940 hielt sie sich überwiegend wieder in Budapest auf, wo sie rund 30 Bildhauerarbeiten in Stein, Aluminium- und Bronze-Guss für die Hauptstadt und für ungarische Provinzstädte schuf. Ab 1945 erhielt sie weitere staatliche und kommunale Aufträge, die sie zu einer der bedeutendsten Vertreterinnen der sozialistischen ungarischen Bildhauerei werden ließen. Sie entwarf nebenbei und unermüdlich bis ins hohe Alter aber auch noch zahlreiche sozialkritische und religiöse Kleinplastiken, Porträts und Medaillen in Bronze. Die von Klára Herczeg selbst für ihren 90. Geburtstag geplante Ausst. im Petőfi-Museum (Budapest) kam 1996 bedauerlicherweise und auf eigenen Wunsch hin nur mit Exponaten aus Stein, Ton und Bronze zustande. In dieser Schau hätte die Bandbreite dieser vielseitigen Künstlerin noch zu ihren Lebzeiten besonders anhand ihrer besten Vorkriegs-Schöpfungen aus Keramik- und Porzellan gezeigt werden können. In dem Ausst.-Kat. werden diese Arbeiten nicht einmal erwähnt. Die bei Goldscheider verwendeten Signaturen bzw. Schreibformen sind Claire Weis, Weiss sowie ihr späterer Name Herczeg. Ihre Figuren und Gruppen (Anzahlen jew. nach der Ortsangabe) wurden durch Bing & Groendahl (Kopenhagen, 35 Entwürfe), Royal Copenhagen, Hutschenreuther (Selb, 14), Metzler & Ortloff (Ilmenau, 22), Älteste Volkstedter (Volkstedt, 2), Haas & Csjzek A.G. (Schlaggenwald, 1), Royal Doulton (Stoke-on-Trent), Royal Dux (Tschechoslowakei, 1), Herend (mind. 8), Holahazy, VEB Schaubach-Kunst (Lichte, 2), Scheibe (Alsbach, 2), sowie alleine 48 Figuren durch die Firma Rosenthal (Kunstabt. Selb, 23 und Bahnhof Selb, 25) ausgeführt. Lit.: Vollmer, Bd. 2, S. 426, Bd. 6, S. 47; Hoever (1996); Hoever (2005); Niecol, Bd. 3, S. 149–155; Zahlreiche Briefe und persönliche Mitt. vom Weiss-Herczeg-Spezialisten Herrn Heinz Hoever, dem ich an dieser Stelle sehr herzlich für seinen Einsatz und sein großzügiges Entgegenkommen danken möchte.*
6328 „Muff" (Stehende Dame mit Muff und großem Hut) (auf Ovalsockel) (sign. Claire Weiss) (vgl. 7044) (vgl. 7720= GRUPPE)
6447 Posierende junge Dame mit Hut (einen zusammengefalteten Fächer in ihrer rechten Hand haltend) (auf Ovalsockel) (sign. Claire Weiss)
6454 FIGUR „Komtesse" (Stehende Dame mit Hut) (auf Ovalsockel) (sign. Claire Weis(s) (vgl. 6951) (vgl. 7327; 8145= jeweils GRUPPE „Komtesse mit Hund")
6462 Stehende Dame mit Hut, ihr Kleid etwas hochhebend (auf Rundsockel) (sign. Claire Weis)
6478 Stehendes Mädchen mit Hut, einem Terrier eine große Schleife (Mascherl) um den Hals bindend (auf Ovalsockel) (Weiss / Herczeg zugeschr.)
6507 Stehende Dame mit Kappe und Muff (auf Ovalsockel) (sign. Claire Weis)
6747 Stehendes Mädchen mit Hut, ein Schaf betrachtend (auf Ovalsockel) (Weiss / Herczeg zugeschr.)
6769 Sitzendes lesendes Mädchen (sign. Claire Weiss)
6773 Stehendes Mädchen mit Hut, Schirm und Tasche, einen kleinen Hund belehrend (auf Ovalsockel) (sign. Claire Weiss)
6829 Stehendes kleines Mädchen mit Haarschleife und Teddybär (auf Rundsockel) (Weiss / Herczeg zugeschr.)
6830 GRUPPE Tanzendes Paar (Russisches Ballett) (auf Ovalsockel) (sign. Claire Weiss)
6853 Schreitendes Mädchen mit kleinem Terrier, welcher an ihrem Kleid zerrt (auf Ovalsockel) (Weiss / Herczeg zugeschr.)
6900 Tänzerin in Bewegung (auf Ovalsockel) (sign. Claire Weiss)
6951 FIGUR „Komtesse" (Stehende Dame mit Hut) (auf Ovalsockel) (sign. Claire Weiss) (vgl. 6454) (vgl. 7327; 8145= jeweils GRUPPE „Komtesse mit Hund")
6955 Gehendes Kinderpaar (auf Ovalsockel) (Weiss / Herczeg zugeschr.)
7027 GRUPPE Rotkäppchen (mit Korb) und der Wolf (auf Ovalsockel) (sign. Claire Weiss)
7028 Schreitender Junge in Matrosenkostüm (ein Modellsegelschiff hinter sich herziehend) (auf flachem Rechtecksockel) (Weiss / Herczeg zugeschr.)
7035 Stehende Tennisspielerin im Wind (auf Ovalsockel) (sign. Claire Weiss)
7040 Tänzerin in langem Kleid (auf Ovalsockel) (sign. Claire Weiss)
7044 „Muff" (Stehende Dame mit Muff und großem Hut) (auf Ovalsockel) (sign. Claire Weiss) (vgl. 6328) (vgl. 7720= GRUPPE)
7063AF FIGUR „Furchtsame" (Stehendes kleines Mädchen mit Schirm) (auf Ovalsockel)
7063F FIGUR (bzw. GRUPPE) „Furchtsame" (Stehendes kleines Mädchen mit Schirm und Foxterrier) (auf Ovalsockel)
7191 Stehender Junge mit Golfsack (sign. Claire Weiss)
7201 Stehendes Mädchen, ihren Hut haltend (auf Rundsockel mit Kaninchen) (sign. Claire Weiss)
7218 GRUPPE „Jugend" (Nacktes, kniendes Mädchen mit Reh) (auf Ovalsockel)
7222 Stehendes Mädchen mit Haarband und männchenmachendem Pekinesen (auf Ovalsockel)
7279 Stehende elegante Dame mit großem Hut und langen Handschuhen (auf Ovalsockel) (vgl. 8151) (vgl. 7367; 8155= jeweils GRUPPE „Maskerade") (sign. Claire Weiss / Claire Herczeg)
7280 FIGUR „Ballnacht" (Tänzerin mit erhobenen Armen, in Bewegung) (auf Ovalsockel)
7327 GRUPPE „Komtesse mit Hund" (Schreitende Tänzerin mit Windhund) (auf Ovalsockel) (vgl. 8145) (vgl. 6454; 6951= jeweils „Komtesse" als Einzeligur)
7367 GRUPPE „Maskerade" (Schreitende elegante Dame mit Barsoi) (auf Ovalsockel) (vgl. 8155) (vgl. 7279; 8151= jeweils Dame ohne Hund) (sign. Claire Weiss)
7390 FIGUR / LAMPE „Parisienne" (Schreitende elegante Dame, ihren Schirm in der linken Hand haltend) (auf Ovalsockel) (vgl. 7723; 8138) (sign. Claire Weiss)
7505 Stehendes Mädchen mit großer Blume (auf Rundsockel)
7511 „Spanierin" (Stehende Spanierin mit Fächer) (auf Ovalsockel) (sign. Claire Weiss)
7557 Schreitende junge Dame mit Hut, zwei Hutschachteln tragend (auf Ovalsockel) (sign. Claire Weiss)
7648 LAMPE Stehende Dame mit großem Hut und Fächer (auf flachem Rechtecksockel) (sign. Claire Weiss)
7691 FIGUR „Demaskierung" (Stehende junge Dame, ihre Maske abnehmend) (auf Ovalsockel) (sign. Claire Weiss) (vgl. 8183)
7720 GRUPPE Stehende Dame mit Muff, großem Hut und Windhund (Windspiel) (auf Ovalsockel) (vgl. 6328; 7044= jeweils FIGUR Dame) (sign. Claire Weiss)
7723 FIGUR „Parisienne" (Schreitende elegante Dame, ihren Schirm in der linken Hand haltend) (auf Ovalsockel) (vgl. 7390; 8138) (sign. Claire Weiss)
7847 GRUPPE Gänseliesel (Stehendes kleines Mädchen mit Gänsen) (auf Ovalsockel) (vgl. 7861; 8003= SCHALE) (sign. Claire Weiss)
7855 FIGUR / LAMPE „Lydia" (Schreitende junge Tänzerin, ihr langes Kleid auseinander haltend) (auf Ovalsockel) (sign. Herczeg / Weiss) (vgl. 7963; 8322)
7861 GRUPPE Gänseliesel (Stehendes kleines Mädchen mit Gans) (auf Ovalsockel) (vgl. 7847; 8003= SCHALE) (sign. Claire Weiss)
7963 FIGUR „Lydia" (Schreitende junge Tänzerin, ihr

langes Kleid auseinander haltend) (auf Ovalsockel) (vgl. 7855; 8322)
7966 „Ballpause" (Sitzende Tänzerin mit Ballmaske und Fächer) (auf Rundsockel) (sign. Claire Weiss)
8003 SCHALE Gänseliesel (Stehendes kleines Mädchen mit Gans) (vgl. 7847; 7861)
8029 Stehendes Mädchen mit Kappe, Schirm und Skyeterrier (auf Ovalsockel) (sign. Claire Weiss)
8030 FIGUR „Allegretto" (Tänzerin in Pose) (auf Ovalsockel)
8070 FIGUR „Akt" (Stehender Damenakt) (auf Rundsockel)
8134 FIGUR/LAMPE „Tanz" (Tänzerin in Bewegung, ihr langes Kleid mit beiden Händen hochhebend) (auf Rundsockel)
8135 Schreitende elegante Dame mit Hut (auf Ovalsockel) (auch sign. Claire Weiss)
8138 FIGUR „Parisienne" (Schreitende elegante Dame, ihren Schirm in der linken Hand haltend) (auf Ovalsockel) (vgl. 7390; 7723) (auch sign. Claire Weiss)
8145 GRUPPE „Komtesse mit Hund" (Schreitende Tänzerin mit Windhund) (auf Ovalsockel) (vgl. 7327) (vgl. 6454; 6951= jeweils „Komtesse" als Einzelfigur)
8151 Stehende elegante Dame mit großem Hut und langen Handschuhen (auf Ovalsockel) (vgl. 7279) (vgl. 7367; 8155= jeweils GRUPPE „Maskerade")
8155 GRUPPE „Maskerade" (Schreitende elegante Dame mit Barsoi) (auf Ovalsockel) (vgl. 7367) (vgl. 7279; 8151= jeweils Dame ohne Hund) (sign. Claire Weiss)
8169 FIGUR „Modedame" (Marlene Dietrich in langem Kleid, ihre Arme in die Hüften gestützt)
8183 FIGUR „Demaskierung" (Stehende junge Dame, ihre Maske abnehmend) (auf Ovalsockel) (vgl. 7691)
8204 FIGUR „Abendkleid" (Stehende schlanke Dame in langem Kleid und Pelzumhang) (sign. Claire Weiss)
8205 FIGUR „Ivette" (Stehende junge Dame in langem Kleid und Schal) (sign. Claire Weiss)
8241 FIGUR „Muffdame" (Stehende Dame in Mantel und Hut mit Muff) (auf Ovalsockel) (sign. Claire Weiss)
8242 Stehende Dame mit Hut und Schirm (auf Rundsockel)
8243 Stehende Dame in langem Kleid mit erhobenem Arm (auf Rundsockel)
8253 s. Prischl
8322 FIGUR „Lydia" (Schreitende junge Tänzerin, ihr langes Kleid auseinander haltend) (auf Ovalsockel) (vgl. 7855; 7963)
8361 FIGUR „Parisienne" (Schreitende elegante Dame, ihren Schirm mit der rechten Hand haltend) (auf Ovalsockel) (vgl. 7390; 8138= jeweils seitenverkehrtes Modell)
8399 FIGUR „Melodie" (Tänzerin auf einem Bein balancierend, ihr Kleid auseinander haltend) (auf Ovalsockel)
8438 FIGUR „Modefigur" (Stehende schlanke Dame in langem Kleid) (Herczeg zugeschr.)
8446 Junge Dame im Sommerkleid, mit einem Schal tanzend (auf Ovalsockel)
8482 „Rositta" („Rosita"?) (Tänzerin in drehender Bewegung, mit erhobenem linken Arm) (auf Ovalsockel) (Herczeg zugeschr.)
8483 „Phantasie" (Schreitende Tänzerin in langem Kleid) (auf Ovalsockel)
8484 „Tanz" (Spanierin mit Fächer) (auf Ovalsockel)
8485 FIGUR/LAMPE „Allegretto" (Tänzerin in langem Kleid und Schal in Bewegung) (auf Ovalsockel)
8487 Schreitende schlanke Tänzerin mit geschlossenen Augen (auf Ovalsockel) (Herczeg zugeschr.)

Herczeg, Claire (Klára) s. a. Weiss, Claire (dort alle Weiss/Weis sign. Objekte)

Herkner, Friedrich

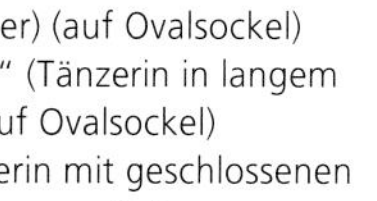

*Friedrich Herkner, Bildhauer (*25.10.1902 Brüx, †1986 Dublin), studierte von 1921–1929 an der Wiener Akademie unter Josef Müllner. Ab 1938 als Professor am National College of Art in Dublin tätig. Lit.: Vollmer, Bd. 6, S. 47; AKL Index.*
4997 Sitzende französische Bulldogge(n) (oder Boston Terrier) (als EinzelFIGUR und GRUPPE)

Herlinger-Schwarz, Ida (Zlata)

*Ida (Zlata) Herlinger-Schwarz, Keramikerin (*19.9.1900 Zagreb/Osijek?, †?), studierte vier Jahre an der Akademie für Kunst und Gewerbe in Zagreb, anschließend 1921–1924 an der Wiener KGW-Schule bei M. Powolny. Sie war auf der Pariser KGW-Ausst. 1925 mit „Amorettengruppen" vertreten, die von Goldscheider editiert wurden. Im Pariser Salon 1928 fand sich vermutlich der in Fayence und Speckstein ausgeführte Kinderkopf, der mit dem „Tête d'enfant" identisch sein dürfte. Lit.: Neuwirth (1974), S. 41, 64; Zühlsdorff, S. 643.*
5433 BÜSTE „Kinderkopf" (nach links blickend) FAYENCE/SPECKSTEIN
Amorettengruppen (1925) Lit.: Neuwirth (1974), S. 64
Amorette mit Kaktus (Fay.) (1925) Lit.: Grell, S. 32f.

Hiller, (Max)

*Höchstwahrscheinlich handelt es sich um Max Hiller, Bildhauer (*28.10.1869 Karlsbad, †23.3.1954 Geislingen). Hiller war zunächst Lehrling (Modelleur?) in der Porzellanfabrik Carl Knollin Fischern bei Karlsbad. Ab 1867 KGW-Schule Wien, ab 1895 freischaffender Bildhauer u.a. für die Galvanoplastische Kunstanstalt in Geislingen/Württemberg. Um 1898 in Wien ansässig, wo er vom Österreichischen Unterrichtsministerium den Auftrag erhielt, eine Erzfigur zu schaffen, die „die Kunst des Landes auf der Weltausstellung 1900 in Paris vertreten sollte". Ab 1921–1934 dann Direktor der Gewerbeschule in Geislingen. Die Porzellanfabrik Bauer, Rosenthal & Co. KG (Kronach) führte elf seiner Entwürfe aus. Lit.: Niecol, Bd. 2, S. 12–15; Träger, G.: Das Kunstschaffen der Egerländer im XX. Jahrhundert, 1984, S. 25; Zühlsdorff, S. 540; AKL Index.*
1299 „Porteuse d'Eau Aveugle" (Schreitende Wasserträgerin mit Krug und Blindenstab)

Hofner, Otto

*Otto Hofner, Bildhauer und Medailleur (*29.3.1879 Wien, †13.3.1946 Wien), besuchte von 1893–1897 die Genossenschaftsschule der Graveure, Gold- und Silberschmiede. In den Jahren 1900–1903 Studium an der Wiener KGW-Schule unter Stefan Schwartz. Ab 1904–1915 war er als Lehrer an der Wiener Genossenschaftsschule tätig. In Hugo Kirschs kunstkeramischer Werkstätte wurden Entwürfe von Hofner ausgeführt, wobei er seine Skulpturen und Medaillen hauptsächlich in Metall und Stein schuf. Entwürfe auch für die Wiener Glasfirma Lobmeyer. Im Jahre 1921 erhielt Hofner den Preis der Stadt Wien. Zahlreiche seiner Arbeiten befinden sich im öffentlichen Raum. Lit.: TB, Bd. 19, S. 292–293; Vollmer, Bd. 2, S. 470; Neuwirth (1974), S. 64, 165; Zühlsdorff, S. 420; Davidson, Bd. 1; Bénézit, Bd. 7, S. 115; AKL Index.*
„Fischer" (Fay.) ÖMKI (1910/11) Lit.: Neuwirth (1974), S. 64
„Windhund" (Fay.) ÖMKI (1910/11) Lit.: Neuwirth (1974), S. 64
„Windhund" (liegend) (Fay.) ÖMKI (1913/14) Lit.: Neuwirth (1974), S. 64

Horsetzky, Melanie von

*Melanie von Horsetzky (von Hornthal), Bildhauerin (*7.4.1852 Wien, †28.4.1931 Wien), war Schülerin von R. Trimmel in Wien und kurze Zeit auch von Auguste Rodin. Für ihre Büste des Admirals Eberau von Eberhorst auf der Londoner Womans Exhibition 1900, erhielt sie eine Goldmedaille. Horsetzkys Domäne waren Porträtbüsten (Marmor, Terrakotta), für Goldscheider können jedoch ausschließlich Figuren nachgewiesen werden. Sie war 1908 wohnhaft in Wien IX., Wasagasse 4, ihr Atelier in Wien I., Salzgries 11–13. Lit.: TB, Bd. 17, S. 530; Lehmann, 1908; Neuwirth (1974), S. 64; Bénézit, Bd. 7, S. 185; Czeike, Bd. 3.*
2666 JARDINIERE Stehende Holländerin mit Korb (vgl. 3591)
3591 JARDINIERE Stehende Holländerin mit Korb (vgl. 2666)
4114 „Lautenspiel" (Sitzende Lautenspielerin) ÖMKI (1912) (vgl. 4580)
4580 „Lautenspiel" (Sitzende Lautenspielerin) ÖMKI (1912) (vgl. 4114)
4880 Kind mit Rabe (auf Fels sitzend)

Hostasch, Anton

Anton Hostasch, Maler (Lebensdaten unbekannt, 1938 wohnhaft Roetzergasse 31, Wien XVII.); Quelle: AGM (Gehaltsliste, Mai 1938; Gehaltsliste, August 1938).

Hourtoni s. Massolini

Houdon, Jean-Antoine (nach)

Die von Goldscheider editierten Modelle Molière und Voltaire, wurden Houdons Büsten nachempfunden bzw. nach diesen reproduziert.
1850 BÜSTE „Molière"
1851 BÜSTE „Voltaire"

Houssin, Edouard Charles Marie

*Edouard Charles Marie Houssin, Bildhauer (*13.9.1847 Douai/Frankreich, †15.5.1917 od. Juli 1919 Paris), war Schüler von Jouffroy und Millet. Von 1873–1914 stellte er im Pariser Salon (Societé des Artistes Français) aus. Ab 1888 leitete Houssin die Wiederherstellung des plastischen Schmucks von Schloß Versailles. Sein Werk umfasst zahlreiche Büsten in Bronze, Marmor, Stein und Terrakotta. Von seinen zahlreichen Ausstellungsbeteiligungen sei die Weltausstellung 1900 in Paris hervorgehoben. Lit.: TB, Bd. 17, S. 568; Bénézit, Bd. 7, S. 201; AKL Index.*
14 BÜSTE Mädchen BRONZE/Marmorsockel (vgl. .... H. 47 cm; Soth. L. 20.5.1994)

Hoyos

4741 Sitzendes Mädchen mit Hasen (Kaninchen) (auf Rechtecksockel)

Huneker (d.i. Bracken, Clio Hinton)

*Clio Hinton Bracken (geb. Huneker), Bildhauerin (*25.7.1870 Rhinebeck/N.Y., †12.2.1925 New York), studierte an der Art Students League, New York, bei A. St.-Gaudens und MacMonnies. In den Jahren 1895–1899 Studien in Paris bei Louis Oury und Auguste Rodin. Huneker war Mitarbeiterin von Friedrich Goldscheider, Paris, und veröffentlichte ihre Arbeiten dort noch unter ihrem Mädchennamen. Lit.: TB, Bd. 4, S. 503; Vollmer, Bd. 1, S. 292; Hillenbrand, S. 50; AKL, Bd. 13, S. 514.*
777-GSB „Cendrier Libellule"
778-GSB „Coque Marine"
793-GSB „Femme au Lotus"
794-GSB „Cendrier Femme a la Rose sauvage"
795-GSB „Cendrier Femme couchée sur feuille"
796-GSB BAS-RELIEF „Naiade au Dauphin" (Pendant vgl. 797-GSB)
797-GSB BAS-RELIEF „Naiade au Cheval Marin" (Pendant vgl. 796-GSB)
802-GSB „Femme au Coquillage"
819-GSB „Cendrier Femme a la Draperie"
820-GSB VIDE-POCHE „Femme a la Coquille"
833-GSB ENCRIER „Repos"
834-GSB ENCRIER „Vigilante"
850-GSB JARDINIÈRE
1047 VASE Frau auf der Vasenmündung sitzend

Iglauer, Fritz

Fritz Iglauer, Betriebsleiter (Lebensdaten unbekannt, 1938 wohnhaft Staudgasse 7–9, Wien XVIII.), bei Walter Goldscheider in den Dreißiger Jahren. Quelle: AGM (Gehaltsliste, Mai 1938; Gehaltsliste, August 1938); Adressbuch der Keram-Industrie 1934, S. 312.

Imenitoff, Nathan

Nathan Imenitoff, Bildhauer (geb. in Dwinsk/Dinaburg/Litauen, lebte in Paris und wird letztmals 1924 erwähnt), war Mitglied der Künstlergruppe „La Stèle" (-> vgl. Kap. „Dekorative Skulpturen von Arthur Goldscheider in Paris") und stellte im Pariser Salon d' Automne ab 1913 aus. Lit.: Vollmer, Bd. 2, S. 547; Neuwirth (1974), S. 64.

Ingels, Domien

*Domien Ingels, Tierbildhauer, Bildhauer und Holzschnitzer (*1881 Gent, †1946 Bachte-Maria-Leerne), studierte an der Königlichen Akademie in Gent wo er später als Prof. tätig war. Er nahm sich vor allem der Pferdeskulptur (hauptsächlich Arbeitspferde) an, aber auch Bildnisbüsten und Kriegsdenkmäler (1914/18) gehören zu seinen bevorzugten Genres. Ingels war auf der Pariser KGW-Ausst. 1925 vertreten. Seine Pferdegruppe war auf der „International Exhibition of Ceramic Art" in New York 1928/29 ausgestellt. Lit.: Vollmer Bd. 2, S. 548; Zühlsdorff, S. 412, 418; Bénézit, Bd. 7, S. 345; AKL Index; Mitt. von Karel Cassiman.*
5238 „Wallonisches Pferd" (Steigendes Pferd auf Rechtecksockel)
5240 Eisbär
5783 Elefant
6143 Pferd (auf Rechtecksockel) (Ingels zugeschr.)

„Cheval debout" Paris, Salon (1928) Lit.: Neuwirth (1974), S. 64
„Wallonisches Pferd" (steigendes Pferd auf quadratischer Plinthe) Lit.: N.N.: Farbe und Bewegung, Die Fayencen der Wiener Manufaktur Friedrich GS Wien-Währing, in: DB, 5.1928, Nr. 213, S. 38 und 40 (m. Abb.), hier mit Ang. des Originaltit. „Wallonisches Pferd", „Von packendem Eindruck sind die Plastiken moderner und modernster Richtung, wie jene prächtigen wallonischen Rosse von Domien Ingels. Hier versteht es die Farbe, in den Hintergrund zu treten, und mit ihrem stillen, graugrünen Ton der breiten Fläche allein die Wirkung zu überlassen".
Pferdegruppe Lit.: Int. Exh. CA 1928/29 (-> vgl. Firmenmarke 19, Made in Austria, H. 49 cm, L. 43 cm; nicht numeriertes Ex. im Wien Museum Inv. 220.771)

Injalbert, Jean-Antoine

*Jean-Antoine Injalbert, Bildhauer (*23.2.1845 Béziers, †3.3.1933 Paris), war Schüler von Auguste Dumont an der École des Beaux-Arts, Paris. Injalbert stellte im Pariser Salon (Societé des Artistes Français) in den Jahren 1872, 1890 und dann wieder ab 1921 aus. Zwischenzeitlich beteiligt er sich fast alljährlich im Salon (Societé Nationale). Aus seinen zahlreichen Auszeichnungen sei der Grand Prix bei der Exposition Universelle 1889 in Paris herausgegriffen. Neben Museumsbesitz in Kopenhagen, Luxemburg, Paris zeugen Denkmale, Monumente und Statuen im öffentlichen Raum von seiner Bedeutung. Injalbert arbeitete neben Goldscheider auch mit Muller & Co, einer Steinzeugfabrik in Ivry Port (bei Paris) zusammen. Lit.: TB, Bd. 19, S. 10; Vollmer, Bd. 6, S. 106; Kashey, S. 272; Hillenbrand, S. 195; Zühlsdorff, S. 527; Bénézit, Bd. 7, S. 352; AKL Index.*
821-GSB VASE „Les Flots"
822-GSB BUSTE „Hippomène"
823-GSB BUSTE „Fierté"
824-GSB BUSTE „Le Dédain"

Jacques, E.

184 JARDINIERE Zwei Frauen, die sich an den Händen halten und deren Kleider die Schale bilden
806 LAMPE Liebespaar
1096 GRUPPE „Römerin im Bade" (Nackte Schönheit mit Kopfputz, die von ihrem Sklaven nach dem Bade abgetrocknet wird) (auf getrepptem Sockel) (vgl. 1579)
1579 GRUPPE „Römerin im Bade" (Nackte Schönheit mit Kopfputz, die von ihrem Sklaven nach dem Bade abgetrocknet wird) (auf getrepptem Sockel) (vgl. 1096)

Jaray, Sándor (Alexander)

*Sándor Jaray, Bildhauer (*11.11.1870 Temesvár/Ungarn, †1916 Berlin?), war neben der Bildhauerei, die er an der Akademie in Wien unter Hellmer und von Zumbusch erlernte, auch Schauspieler. Mit einem Stipendium, das er 1899 erhielt, ging er nach Rom und widmete sich fortan nur mehr der Bildhauerei. 1901 ließ er sich in Berlin (ab 1928 in Wien) nieder und beschickte dort die Große Kunstausst. und den Glaspalast in München 1908. Von seinen Porträtbüsten seien hier Menzel, Schiller und Liszt aufgeführt. Für die Stadt Wien schuf er 1911 das Josef-Kainz-Denkmal in Bronze, welches bei Goldscheider unter der Modell-Nr. 5850 um 1928/29 verkleinert wiedergegeben wurde. Jaray war ab 1928 wohnhaft in Wien III., Weißgerber Lände 38; sein betrieb er Atelier ebenfalls im III. Bez., Landstraßer Hauptstr. 88. Lit.: TB, Bd. 18, S. 427; Lehmann, 1928–1939; Vollmer, Bd. 6, S. 97; Bénézit, Bd. 7, S. 486; AKL Index.*
2306 GRUPPE „Adam und Eva" (als sitzende nackte Kinder) (Inschrift: ADAM UND EVA)
4823 Sitzendes, nacktes Figurenpaar (Adam und Eva?) FAYENCE; ALABASTER / Steinsockel
5850 Josef Kainz als Hamlet (Denkmal)

Jarl, Karin (Karin Jarl-Sakellarios)

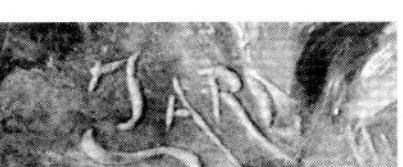

*Karin Jarl-Sakellarios, Bildhauerin (*6.10.1885 Wien, †20.11.1948 Wien), war auf Tierdarstellungen, besonders auf Pferdedarstellungen spezialisiert. Karin Jarl war mit dem Bildhauer Victor Sakellarios verheiratet und arbeitete ab 1925 für die Wiener Porzellanmanufaktur Augarten. Neben zahlreichen Ausstellungsbeteiligungen in Wien war sie auch auf der Pariser KGW-Ausst. 1925 mit Tierplastiken vertreten. Sie präsentierte auf der „International Exhibition of Ceramic Art" in New York 1928/29 das von Goldscheider ausgeführte Nilpferd. Jarl-Sakellarios war 1928 wohnhaft in Wien XVIII., Theresiengasse 16–18. Lit.: Grell, S. 33; Lehmann, 1928; Vollmer, Bd. 2, S. 533; ÖBL, Bd. 3; Neuwirth (1974), S. 41, 64; Zühlsdorff, S. 643.*
339-MG Zwei laufende Windhunde (auf Natursockel)
378-MG Tukan (Pfefferfresser mit exotischen Früchten)
469-MG Sitzender Kolibri auf Kaktus
4797 Stehender Wolfs(Schäfer?)hund (auf abgerundetem Längssockel)
5116 Liegende Katze
5120 Reh (auf rechteckigem Landschaftssockel)
5131 Vogeltränke (Vier sitzende Spatzen auf einem Birkenast)
5133 Tiger
5144 Fauchender Leopard (Jarl zugeschr.)
5147 DECKELDOSE „Schwan" (Form: rund, gebaucht/ Knauf: Schwan (vollplastisch) / Dekor: Seerosenteich)
5148 Stehender Windhund (auf Längssockel)
5149 Liegender Windhund (-> vgl. 5299= ähnl. Modell)
5174 Papagei (auf glockenförmigem Sockel)
5202 Liegender Rehpinscher (auf Rundsockel)
5227 Steigendes Pferd (auf Ovalsockel)
5245 Liegendes Lama
5299 Liegender Windhund FAYENCE; BRONZE (vgl. 5461) (vgl. 5149= ähnl. Modell)
5431 SCHALE Vier Spatzen (an der Tränke sitzend)
5443 Paradiesvogel (auf Ast sitzend) (auf Rundsockel) (Jarl zugeschr.)
5461 Liegender Windhund (vgl. 5299) (vgl. 5149= ähnl. Modell)
5516 GRUPPE / LAMPE Elefant mit reitendem Chinesenpaar (auf Ovalsockel)
5574 Stehender Boxer (Hund) (auf Landschaftssockel)
5575 Sitzende Bulldogge (auf Landschaftssockel)
5604 Liegendes Reh BRONZE
5609 Zwei stehende Terrier (auf Ovalsockel)
6905 GRUPPE „Elefanten" (Zwei stehende Elefanten, groß / klein) (vgl. 6906= Einzelfigur (groß); (vgl. 6907= Einzelfigur (klein)
6906 FIGUR „Elefant" (Stehender Elefant, groß) (vgl. 6905= GRUPPE) (Pendant vgl. 6907)
6907 FIGUR „Elefant" (Stehender Elefant, klein) (vgl. 6905= GRUPPE) (Pendant vgl. 6906)
6929 Stehender Sealyham Terrier
7246 „Zicklein" (liegend)
7247 GRUPPE Zwei Barsois (Einer sitzend, einer liegend)
7248 FIGUR „Spaniel" (sitzend) (vgl. 7578)
7460 GRUPPE „Rehe" (Zwei Rehe auf Ovalsockel) vgl. 7462= Einzelfigur)
7462 FIGUR „Reh" (liegend) (vgl. 7460= GRUPPE)
7492 GRUPPE Zwei Schäferhunde (Einer stehend, einer liegend)
7494 „Wolfshund" (Liegender Schäferhund)
7578 FIGUR „Spaniel" (sitzend) (vgl. 7248)
7924 Pekinese (stehend)
7926 GRUPPE Vier Pekinesen
7987 GRUPPE Zwei Barsois (auf Ovalsockel)
8280 FIGUR „Pferd" (trabend) (auf Ovalsockel) s. Matula
8283 FIGUR „Pferd" (fressend) (auf Landschaftssockel) s. Matula
8325 GRUPPE „Jagd" (Reiterin auf scheuendem Pferd und Jagdhund) (auf Ovalsockel) (Teilentwurf: Pferd und Hund)
„Giraffe mit Palme" (auf Rundsockel) Lit.: Meyer, K.: Arbeitsstoff und Arbeitswert in: KuKGW, 5.1925, SH. 3, S. 9, 10 (m. Abb.), „Die „Giraffe mit Palme", Entwurf von Karin Jarl gibt in dem farbig behandelten, fein glasierten Werkstoff das gesprenkelte, glatte Fell des gewählten Tieres ausgezeichnet wieder. Die Palme ist nicht naturalistisch, sondern nur stilisiert wiedergegeben; von der überhängenden Krone des Baumes wird der durch den überlangen Giraffenhals bedingte allzuscharfe Vorstoß nach der Höhe zu eingedämmt und harmonisch abgerundet; ein für die Komposition des Ganzen recht glücklicher Gedanke".

Jarl-Sakellarios, Karin s. Jarl, Karin

Jonchery, Charles Émile

*Charles Émile Jonchery, Bildhauer (*27.6.1873 Paris, †1937? bzw. vor 1955), arbeitete in Rodins Werkstatt und stellte seine Kleinplastiken, Büsten und Gruppen in Bronze im Pariser Salon (Société des Artistes Français) in den Jahren 1883–1922 aus. Jonchery war Mitarbeiter von Friedrich Goldscheider, Paris. Lit.: TB, Bd. 19, S. 113; Vollmer, Bd. 2, S. 561; Arwas (2002), S. 615; Bénézit, Bd. 7, S. 575; AKL Index.*
726-GSB „Feuille Almée"
BÜSTE Frau mit grünen Haaren BRONZE (H. 80 cm) Mayer, 1992 S. 1714
VASE Mädchen (in ein dünnes Tuch gehüllt) an der mit Ästen umschlungenen Vasenwandung stehend BRONZE (H. 58,7 cm) Christie's SK 27.4.2005

Jozon, Jeanne

*Jeanne Jozon, Bildhauerin, Medailleurin und Kunstgewerblerin (*15.7.1868 Paris, †1946), studierte an der Academie Julian in Paris unter Denis Puech und stellte ihre Arbeiten (vorwiegend Büsten und Statuetten) u.a. im Pariser Salon (Société des Artistes Français) in den Jahren 1893–1911 aus. Dort erhält sie 1897 eine „Mention Honorable". Arbeiten von ihr werden von Lachenal in Keramik- oder Bronzereproduktion vertrieben. Lit.: TB, Bd. 19, S. 216; Neuwirth (1974), S. 64; Hillenbrand, S. 206; Arwas (2002), S. 615; Bénézit, Bd. 7, S. 625; AKL Index.*
„Flambeau" Paris, Salon (1902) Neuwirth (1974), S. 64
Jurjewitsch, Ssergej s. Youriévitch, Serge

Kaniroff s. Haniroff / Xaniroff

Kaye, Y.? s. Wind, Josef (1393)

Kéléty, Alexandre

*Alexandre Kéléty, Bildhauer (*Budapest, †?), war Mitarbeiter von Arthur Goldscheider, Paris. Lit.: Goldscheider, A. (Hrsg.): Artistic Evolution, 1926, T. 2, Pl. 45 (m. Abb.); Lit.: Neuwirth (1974), S. 64; Hillenbrand, S. 212; Bénézit, Bd. 7, S. 740.*
77-La Stèle Stehendes Mädchen (Halbakt) mit herzförmigem Rock BRONZE / ELFENBEIN (auf Marmorsockel)
„Pelicans" BRONZE (H. fehlt, B. fehlt) Lit.: Goldscheider, A. (Hrsg.): Artistic Evolution, 1926, T. 2, Pl. 45 (m. Abb.)
„Hippopotamus" BRONZE (H. fehlt, B. fehlt) Lit.: Goldscheider, A. (Hrsg.): Artistic Evolution, 1926, T. 2, Pl. 48 (m. Abb.)
„Femme à L'èventail" (Bronze gravé et ivoire) BRONZE / ELFENBEIN Paris, Salon (1928) Lit.: Neuwirth (1974), S. 64

Keller, A.

ohne Nr. WANDTELLER „Pfahlbauten Oberbaiern" (Form: rund, Fahne goldfarben bronziert / Dekor: Pfahlbauten im See vor Gebirgslandschaft, Gemälde, Kaltbemalung)

Kienast

4662 Sitzende Kröte (auf Rechtecksockel)

Kinsburger, Sylvain

*Sylvain Kinsburger, Bildhauer (*21.1.1855 Paris, †21.4.1935 Paris?), war Schüler von A. Dumont und G. J. Thomas. Kinsburger stellte im Pariser Salon (Societé des Artistes Français) 1878, 1882–1884 und regelmäßig ab 1887–1926 Porträts, Statuen, Gruppen, Medaillen und Plaketten aus. Lit.: TB, Bd. 20, S. 334–335; Hillenbrand, S. 216; Bénézit, Bd. 7, S. 810; AKL Index.*
663-GSB VIDE-POCHE „Curieux" BRONZE
677-GSB PLATEAU „Danse du Printemps"
684 VASE „Adam" (Pendant vgl. 685) (auch sign. Lemoine)
685 VASE „Eva" (Pendant vgl. 684) (auch sign. Lemoine)

Klakowicz, Casimir (Kazimierz) (H.?)

Casimir (Kazimierz) Klakowicz, Bildhauer (Zablotow, Galizien), wird zwischen 1894 und 1914 erwähnt. Lit.: Neuwirth (1974), S. 64; AKL Index.
SCHALE „Fische" ÖMKI (1903/04), im Ausst.-Kat. Ang. „H." Klakowicz; Lit.: Neuwirth (1974), S. 64

Klein, (Max)

*Max Klein, Bildhauer (*27.1.1847 Gönc/Ungarn, †6.9.1908 Berlin), war in den Jahren 1869–1870 in Rom, anschließend dann in München und Wien. Arbeitete ab 1874 in Berlin an der Akademie bei C. Steffeck und beteiligte sich ab 1877 an zahlreichen Ausstellungen in Berlin, München, Budapest, Paris u.v.a. Orten. Zu seinen Gießern zählte Gladenbeck in Berlin. Seine Denkmäler, Standbilder, Porträts, Statuen, Gruppen und seine deko-*

rativen Plastiken zeugen von seiner ansehnlichen und erfolgreichen künstlerischen Tätigkeit. Lit.: TB, Bd. 19, S. 444–445; Hillenbrand, S. 218; Bénézit, Bd. 7, S. 842; AKL Index.
2524 GRUPPE „Hagar und Ismael" ÖMKI (1903/04)

Klimt, Gustav (nach)

*Gustav Klimt, Maler (*14.7.1862 Baumgarten/Wien, †6.2.1918 Wien). Neben diesem von Goldscheider ausgeführten Mosaik führte auch die Wiener Mosaikwerkstätte etwa ab 1910 Klimt-Entwürfe aus. Lit.: TB, Bd. 19, S. 504–506; Zühlsdorff, S. 647.*
1085 RELIEF Göttin Juno (Wandrelief als Mosaikimitation)

Kloster, M.

M. Kloster, Bildhauer/in (Lebens- und Schaffensdaten unbekannt), war Mitarbeiter von Friedrich Goldscheider, Paris. Lit.: GSB, Taf. 48 (m. Abb.).
652-GSB PLATEAU „Lézards" ZINN/ZINKGUSS

Knörlein, Rudolf

*Rudolf Knörlein, Bildhauer (*23.9.1902 Wien, †7.3.1988 Alt Duvenstedt), absolvierte seine Lehrzeit bei der Gmundner Keramik GmbH (Schleiss, Simandl), Wienerberger Werkstättenschule für Keramik in Wien bei Prof. Obsieger. 1918–1920 (bei Witzmann und Schufinsky) und 1927–1929 (bei Steinhof und Obsieger) Besuch der KGW-Schule Wien. Studium an der Akademie der bildenden Künste Wien bei Prof. Karl Sterrer. Ab Februar 1923 als Werkstattleiter der Scheibbser Keramik, die er Mai 1924 wieder verließ. Mitglied ab 1924 bis zur Auflösung des Österreichischen Werkbundes. Assistent der Deutschen Gesellschaft für christliche Kunst in Neisse O.S. unter Prof. Zutt. Ab etwa Ende 1928 bzw. Anfang 1929 Entwürfe für Marcell Goldscheiders Vereinigte Ateliers für Kunst und Keramik. In seinem eigenhändig geschriebenen Lebenslauf vermerkt Knörlein: „1932 Beginn der Arbeit bei Goldscheider. Von unserem Hochzeitstag, dem 6.12.1934 bis zum 28.2.1936 wohnten wir in Wien, XVIII, Staudgasse 7, das Haus der Firma Goldscheider, bei der ich von 1932 an bis zu unserem Exodus als „Leiter der modernen Abteilung" als „Künstlerischer Leiter" beschäftigt war. Wurde Nachfolger (Benno) Geigers. 1936 im März beginne ich als Betriebsleiter in Gmunden, im Mai werde ich Direktor der Gmundner Keramik. Goldmedaille Weltausst. Paris (1937)". Es folgten Ankäufe u.a. durch das Schwedische Nationalmuseum, Stockholm (Stehkopf Mädchen), das Schlossmuseum Linz sowie die Neue Sammlung – Staatliches Museum für angewandte Kunst in München. Dr. Berdel berief ihn an die Deutsche Meisterschule, staatliche Fachschule für Keramik und Glas (Bunzlau O.S.) als Leiter der Entwurfsklasse. 1948 Meisterprüfung Handwerkskammer Flensburg, anschließend maßgeblich beteiligt am Aufbau der Landesberufsschule für Töpfer in Schleswig-Holstein. Walter Goldscheider attestierte ihm in seinem Zeugnis (Privatarchiv) vom 16. Februar 1953: „Herr R.K. war bei unserer Manufaktur durch eine Reihe von Jahren als künstlerischer Leiter und Bildhauer unserer modernen Abteilung tätig. Wir waren mit seinen Leistungen durchaus zufrieden und hat er eine ganze Reihe von guten künstlerischen figuralen Modellen geschaffen, die zu den besten unserer Kollektion zählen und noch heute als solche anerkannt und verkauft werden. Außerdem hat er während seiner Tätigkeit bei uns verschiedene Gebrauchskeramik geschaffen, die sich damals recht gut verkaufte und auch kunstgewerblich gute Beurteilung fand". R.K. zeichnete neben Goldscheider (Friedrich und Marcell Goldscheider) auch für zahlreiche Entwürfe u.a. der Wiener Werkstätte, für die Gmundner Keramik, sowie E. & C. Carstens verantwortlich. Seit 1962 eigener Betrieb in Alt Duvenstedt, wo Knörlein auch am 7.3.1988 verstarb. Lit.: Rochowanski (1923a); Born, in: DKuD, 69.1931/32; Donau, in: ÖK, 5.1934, Nr. 5, S. 17; Vollmer, Bd. 3, S. 72; Neuwirth (1974), S. 64, 195; R. Knörlein: Keramische Arbeiten 1928–1982, 1982; Zühlsdorff, S. 644; Jahresschau des Schleswig-Holsteinischen Kunsthandwerks 1985, S. 66–67; Deutsches Kunsthandwerk 1986, Verz. der Mitgl., S. 256; Jahresschau des Schleswig-Holsteinischen Kunsthandwerks 1989, S. 130; Triennale des norddeutschen Kunsthandwerks 1991, S. 115; Makus (2000), S. 173, 183; Informationen aus der Privatsammlung D.*

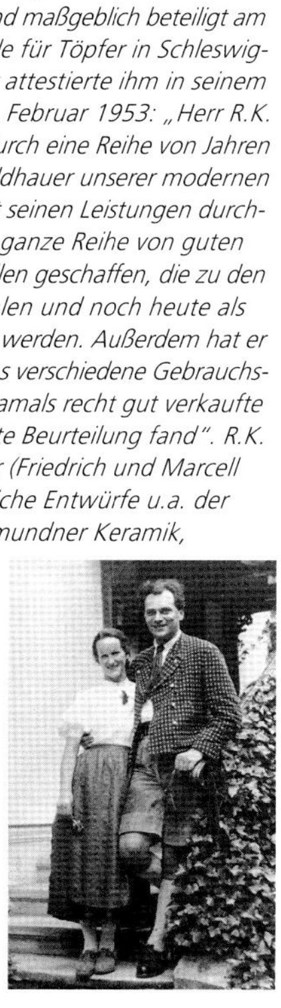

*-MG KERZENHALTER mit Behälter für Streichhölzer (einflammig)
*-MG FIGUR „Niederblickende" (Stehende junge Dame auf getrepptem Quadrat/Rechtecksockel?) (H. 100 cm) Lit.: Heizer (1929), S. 714 (m. Abb.), „Was Tanagra für das alte Hellas war, das ist Wien für das heutige Europa. Einige Proben dieser neuen Tanagrakunst bringen wir hier im Bild. Es sind Erzeugnisse der Firma Marcell GS, Österreichische Werkbundkeramik, Wien. In diesen Arbeiten prägt sich das Wesen des Wienerischen in seiner besten Art charakteristisch aus. Die figürlichen Darstellungen sind von stiller, besinnlicher, fast traumhafter Art, und doch behält die Formgebung das lockere Spielen mit der Form, mit dem Formalen. Spielen kann man aber nur mit einer Sprache, die man versteht, die man beherrscht. Hier ist das Spiel gekonnt, die Eigenart des Materials ist in feiner Weise zur Geltung gebracht. Unvorstellbar sind diese Plastiken in Stein oder Holz; und nur ein leicht formbares Material wie der Ton läßt so ein rieselndes Linienspiel zu, wie es hier bei der großen Plastik der „Niederblickenden" zu sehen ist". (Knörlein zugeschr.)
*-MG FIGUR „Innehaltende" (Stehende junge Dame auf Quadratsockel) (H. 40 cm) Lit.: Heizer (1929), S. 714 (m. Abb.) (Knörlein zugeschr.)
*-MG RELIEF Tanzendes Mädchen Lit.: Rochowanski (1929) in: DKuD, 65.1929/30, S. 430 (m. Abb.)
*-MG RELIEF Frauenbildnis mit Hand und Blume Lit.: Rochowanski (1930), S.108 (m. Abb.)
*-MG SERVICE Likörservice Lit.: Rochowanski (1930), S.78 (m. Abb.)
409-MG Stehender Mädchenakt mit Reh (Brüderchen und Schwesterchen) (auf profiliertem Quadratsockel)
410-MG „Nische" (Kniendes Mädchen in einer Nische)
412-MG STEHKOPF Frauenkopf mit Hut (auf flacher quadratischer Sockelplatte)
463-MG RELIEF Frauenkopf im Profil vor Landschaft mit Haus (Wandrelief)
525-MG WANDMASKE Frauenkopf mit Schleife
700-WG WANDMASKE „Maske" (Negermädchen mit Ohrring und Kopftuch)
701-WG WANDMASKE Negerin mit Zöpfen und Halsschmuck
702-WG BÜSTE Negerin mit Halsschmuck und Ohrringen
703-WG WANDMASKE Negerin mit goldenem Kopftuch (Kopfschmuck)
705-WG „Pinguin" (Stehender Pinguin)
706-WG BÜSTE „Nofretete"
707-WG WANDMASKE Negerkind mit Ohrschmuck
750-WG „Bison" (Stilisiertes Bison beim Angriff)
751-WG „Sitzende" (Sitzendes Mädchen)
752-WG „Sitzende" (Sitzendes Mädchen auf Sockel)
753-WG WANDMASKE „Maske" (Frauenkopf mit Schal/Kopftuch)
754-WG „Akt" (Sitzender Mädchenakt)
756-WG „Reh" (Stehendes Reh)
757-WG „Kniende" (Kniender Mädchenakt auf Rechtecksockel)
758-WG „Kniende" (Kniendes Mädchen in langem Kleid)
759-WG „Eule" (Stilisierte Eule)
6263 KERZENHALTER (zweiflammig) (Form: U-förmig, vier leicht gebogene Standfüße/Dekor: gemalte Linien, Laufglasur) (Knörlein zugeschr.)
6269 KERZENHALTER (fünfflammig) (Form: fünf Kerzentüllen auf rohrartigem Unterbau, über halbkugelförmigem Sockel/Dekor: runde Farbflecken, Laufglasur) (Knörlein zugeschr.)
6305 STEHMASKE/WANDMASKE Frauenkopf mit Haarschmuck mit/ohne Blumenblüte (GS neu: Modell-Nr. 10 „Meine Blüte")
6306 WANDMASKE Frauenkopf mit zurückgekämmten Haaren (Knörlein zugeschr.)
6335 WANDMASKE Frauenkopf mit Halskette und Haarband
6364 STEHKOPF Frauenkopf mit Kragen (offene Augen) (auf flachem Rundsockel) (Pendant -> vgl. 6446) (vgl. 6393= ähnl. Modell)
6393 STEHKOPF Frauenkopf mit Kragen (geschlossene Augen) (auf flachem Quadratsockel) (vgl. 6364; 6446= jeweils ähnl. Modell)
6412 BÜSTE Madonna mit Kind (Knörlein zugeschr.)
6427 WANDMASKE „Turandot" (Frauenkopf mit Schulteransatz) (vgl. 6623; 7453)
6428 WANDMASKE Frauenkopf mit flachem Hut
6445 STEHKOPF Mädchenkopf mit leicht gesenktem Haupt (Lockenfrisur, Curls) (auf Keramik/Holzsockel)
6446 STEHKOPF Frauenkopf mit Kragen (geschlossene Augen) (auf flachem Rundsockel) (Pendant -> vgl. 6364) (vgl. 6393= ähnl. Modell)
6537 WANDMASKE Frauenkopf mit Hut (Kappe)
6542 WANDMASKE Frauenkopf mit Hand und Fächer (Knörlein zugeschr.)
6560 DOPPELSTEHKOPF Zwei Frauenköpfe (einer mit Kappe) (auf flachem Rechtecksockel)
6599 BÜSTE Frauenkopf (nach rechts blickend) mit Schulteransatz (Knörlein zugeschr.)
6623 WANDMASKE „Turandot" (Frauenkopf mit Schulteransatz) (vgl. 6427; 7453)
6654 s. 6599
6658 RELIEF Frauenkopf mit Kappe (Wandrelief)
6660 WANDMASKE Frauenkopf mit Kragen
6661 RELIEF Frauenkopf mit Blume (Wandrelief)
6662 RELIEF Frauenkopf im Profil (nach links) (Wandrelief)
6663 RELIEF Zwei Frauenköpfe vor Fenster (Wandrelief)
6664 WANDMASKE Zwei Frauenköpfe mit Schal
6665 STEHKOPF Frauenkopf mit Hut und Halskette
6712 Stehendes Mädchen mit Reh (auf Spitzovalsockel)
6735 WANDMASKE Frauenkopf mit Halstuch
6739 DOPPELSTEHKOPF Zwei Frauenköpfe (Cheek to cheek) (auf flachem Rechtecksockel)
6763 Stehendes Mädchen mit Taube (auf unregelmäßig geschwungenem Sockel) (Knörlein zugeschr.)
6770 STEHKOPF Kinderkopf mit Kragen (auf Quadratsockel) (vgl. Dakon/Knörlein/Prischl 6770)
6774 WANDMASKE/STEHMASKE „Eva" (Frauenkopf mit Hand und Apfel) (vgl. 7947)
6821 WANDMASKE Frauenkopf mit halblangen Haaren und Schleife/Mascherl (vgl. 6881= TONDO) (Knörlein zugeschr.)
6832 WANDMASKE Frauenkopf mit halblangen Haaren und Kragen (vgl. 6880= ABREISSKALENDER; 6998= DECKELDOSE) (Knörlein zugeschr.)
6859 WANDMASKE Frauenkopf mit/ohne Tulpe (vgl. 7548)
6871 WANDMASKE Frauenkopf mit oben glattem, unten seitlich gelocktem Haar und Halstuch (auch sign. Prischl)
6880 ABREISSKALENDER Frauenkopf mit halblangen Haaren und Kragen (vgl. 6832= WANDMASKE; 6998= DECKELDOSE) (Knörlein zugeschr.)
6881 TONDO Frauenkopf mit halblangen Haaren und Schleife/Mascherl (vgl. 6821= WANDMASKE) (Knörlein zugeschr.)
6882 ABREISSKALENDER Frauenkopf mit Hut (Knörlein zugeschr.)
6883 WANDMASKE Zwei Frauenköpfe (Cheek to cheek) mit Halstuch
6908 SCHALE (Form: rund/Dekor: vier Streifen)
6998 DECKELDOSE Frauenkopf mit halblangen Haaren und Kragen (Form: rechteckig/Dekor: Frauenkopf, reliefiert) (vgl. 6832= WANDMASKE; 6880= ABREISSKALENDER) (Knörlein zugeschr.)
7021 KERZENHALTER (dreiflammig) (Form: astartig, blütenartige Kerzentüllen, sternförmig gezackter Stand) (Sockel)
7025 RELIEF „Madonna mit Kind" (Madonna mit Schleier) (Wandrelief)
7029 RELIEF „Madonna mit Kind" (Madonna und Kind mit Aureole) (Wandrelief)
7081 WANDMASKE/STEHMASKE Frauenkopf mit Hand und Maske (Knörlein zugeschr.)
7121 WANDMASKE „Maske" (Frauenkopf mit Hand und Halskette)
7166 RELIEF Stehendes Mädchen mit zwei Vögeln (Pendant vgl. 7167)
7167 RELIEF Stehender Mädchenakt (Pendant vgl. 7166)
7188 BÜSTE „Ave Maria" (Madonna mit gefalteten Händen) (vgl. 7263; 7858; 8743)
7263 BÜSTE „Ave Maria" (Madonna mit gefalteten Händen) (vgl. 7188; 7858; 8743)
7276 RELIEF „Segler" (Segelschiff nach links) (Wandrelief)
7354 ASCHER (Form: quadratisch, eine Zigarettenablage/

Dekor: Frauenkopf mit Schulteransatz, reliefiert) (Knörlein zugeschr.)
7385 BÜSTE Madonna mit einer Hand
7418 KERZENHALTER (dreiflammig) (Form: röhrenartig, scheibenartige Kerzentüllen) (auf Rund/Ovalsockel)
7430 DECKELDOSE (Form: quadratisch, abgerundete Ecken/Dekor: Frauengesicht, Vase, Schale)
7444 RELIEF Madonna (mit Aureole) (Wandrelief)
7453 WANDMASKE „Turandot" (Frauenkopf mit Schulteransatz) (vgl. 6427; 6623)
7495 WANDMASKE/STEHMASKE „Maske" (Frauenkopf mit zwei Händen und Halstuch)
7527 WANDMASKE Frauenkopf mit Hand und Weintrauben (Knörlein zugeschr.)
7528 Stehender Damenakt (Negerin) (auf Rundsockel)
7542 BÜSTE Damenakt mit Lockenkopf (die Hände an die Brust gelegt) (vgl. 7543= Ausführung mit Tulpe)
7543 BÜSTE „Frühling" (Damenakt mit Lockenkopf, eine Tulpe haltend) (vgl. 7542= Ausführung ohne Tulpe)
7546 BÜSTE Frauenkopf mit gelockten Haaren und Kleid/Blusenansatz (vgl. 7550)
7547 WANDMASKE Frauenkopf mit Kragen (Knörlein zugeschr.)
7548 WANDMASKE Frauenkopf mit Tulpe (vgl. 6859)
7550 BÜSTE Frauenkopf mit gelockten Haaren und Kleid/Blusenansatz (vgl. 7546)
7717 FIGUR „Akt liegend" (mit/ohne Ovalsockel) (Knörlein zugeschr.)
7752 BÜSTE Damenakt mit erhobenen Armen (Knörlein zugeschr.)
7756 STEHKOPF Frauenkopf mit Hut und Kragen (Knörlein zugeschr.)
7854 KERZENHALTER (sechsflammig) (Form: u-förmig geschwungene Röhren, mittig runde Bekrönung mit drei Ösen/Dekor: Craquelé- und Laufglasur)
7858 BÜSTE „Ave Maria" (Madonna mit gefalteten Händen) (vgl. 7188; 7263; 8743)
7947 STEHMASKE „Eva" (Frauenkopf mit Hand und Apfel) (auf Rechtecksockel) (vgl. 6774)
7997 WANDMASKE Frauenkopf mit kurzem Kragenansatz (nach rechts blickend) (Knörlein zugeschr.)
7998 WANDMASKE Frauenkopf mit kurzem Kragenansatz (nach links blickend) (Knörlein zugeschr.)
7999 WANDMASKE Frauenkopf mit Halstuch/Schal (Knörlein zugeschr.)
8063 BÜSTE Damenakt mit Lockenkopf (Knörlein zugeschr.)
8743 BÜSTE „Ave Maria" (Madonna mit gefalteten Händen) (vgl. 7188; 7263; 7858)
8744 BÜSTE Madonna mit gefalteten Händen (vgl. 7188; 7263; 7858, 8743= jew. ähnl. Mod.)
8935 WANDMASKE Frauenkopf mit Hand und Larve (= Gesichtsmaske) (vgl. 8941)
8939 WANDMASKE Engelkopf (mit Flügel) (Cherub)
8941 WANDMASKE Frauenkopf mit Kragen (vgl. 8935)
II VASE
VIII VASE
XII VASE
XV DECKELDOSE
XVI SCHALE
XVIII SCHALE/TELLER
XXV SCHALE
XXIX SCHALE
XXX SCHALE
876 VASE (Knörlein zugeschr.)
1013 VASE Zwei Masken
1022 KERZENHALTER „Leuchter" (vierflammig)
1111 LAMPE Blumen
1112 LAMPE Blumen und Blattwerk
1118 VASE
1139 SCHALE
1172 KERZENHALTER (einflammig)
1173 KERZENHALTER (zweiflammig)
1174 KERZENHALTER (dreiflammig)
1187 VASE
1399 SCHALE Frauenkopf (Knörlein zugeschr.)
ohne Nr. KERZENHALTER (fünfflammig) (Form: röhrenartig, vier zur Mitte hin zusammenlaufende Röhren, blütenartige Kerzentüllen/Dekor: abstrakte Malerei, Laufglasur)
ohne Nr. REKLAMESCHALE „50 anni Goldscheider 1885–1935, Vienna" (Variante -> vgl. nächsten Eintrag)
ohne Nr. REKLAMESCHALE „50 years Goldscheider 1885–1935, Vienna" (Variante -> vgl. vorigen Eintrag)
ohne Nr. STEHKOPF Mädchen mit gelocktem Haar (auf spiralartig verziertem Rundsockel) Bem.: Das Original befindet sich im Nationalmuseum Stockholm, Schweden (Fotoarchiv K. Täuber); Lit.: ÖK, 5.1934, Nr. 5, S. 17 (m. Abb.), „Mit einigen Frauenköpfen von weichpikantem Ausdruck und lässiger Anmut hat Rudolf Knörlein, einer der wichtigsten keramischen Künstler dieser Erzeugung, Stücke geschaffen, die in London den Wiener Charm mit Glück vertreten".
ohne Nr. TORSO Mädchenakt mit gelocktem Haar (auf Holzsockel) (Knörlein zugeschr.)

Kohler

Kohler, Kunstgewerbler/in (Lebens- und Schaffensdaten unbekannt), war Mitglied der Künstlergruppe „Évolution" (-> vgl. Kap. „Dekorative Skulpturen von Arthur Goldscheider in Paris")

Kopp, (Josef?)

*Josef Kopp, Bildhauer, Kunstgewerbler und Maler (*19.4.1877 München, †4.6.1941 München), erhielt seine Ausbildung bei seinem Vater, der Holzschnitzer war. Ab 1885 war Kopp dann auf der Akademie in München. Er war Mitglied der Deutschen Gesellschaft für christliche Kunst und beschäftigte sich erfolgreich mit der Grabplastik. Daneben war er auch kunstgewerblich tätig. Lit.: TB, Bd. 21, S. 299.*
4635 Marabu (auf Felssockel) MARMOR

Korschann, Karl (Korschmann, Charles)

*Karl Korschann, Bildhauer, Kunstgewerbler und Medailleur (*23.7.1872 Brünn/Mähren, †9.3.1943 Brünn), studierte an den Akademien in Wien und Berlin. Ab 1894 verbrachte er die meiste Zeit seines Lebens in Paris und stellte dort u.a. im Salon (Société des Artistes Français) in den Jahren 1894–1905 aus. Auf der Weltausst. in Paris 1900 erhielt er die Bronzemedaille. Die Bandbreite seiner Techniken reicht von polychromierter, vergoldeter und patinierter Bronze über unglasierte Keramik bis zum Zinn. Viele seiner Werke wurden von Louchet editiert. Lit.: TB, Bd. 21, S. 324; Hillenbrand, S. 224; Arwas (2002), S. 615; Bénézit, Bd. 7, S. 951; AKL Index.*
GRUPPE „Fütterung" ÖMKI (1906/07) Lit.: Neuwirth (1974), S. 65

Kostial, Josef

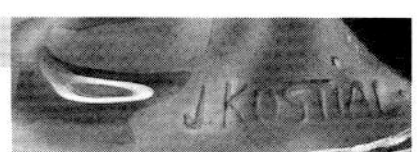

*Josef Kostial, Bildhauer (*17.2.1886 Wien, †9.3.1953 Wien), besuchte die Staatsgewerbeschule und studierte zwei Jahre an der Kunstakad. in Wien. Kostial war Privatschüler bei Erich Swoboda und Josef Kassin. Neben seinen Entwürfen für Goldscheider betätigte er sich auch beim Film für den er Plastiken schuf. Kostials Wohnung in Wien befand sich (nachgeprüft ab 1928) in der Embelgasse 44, Wien V. Lit.: Lehmann, 1928 ff.; Quelle: Bildhauerarchiv, Karl Novak.*
436-MG Kniender Pierrot mit Bratsche (Cello?) (auf länglichem geschwungenem und profiliertem Sockel)
542-MG Sitzendes Mädchen (in halblanger Hose) mit Violine (auf Ovalsockel)
4892 Faun mit Schnecke (vgl. auch Wernekinck 4892)
4903 FIGUR/LAMPE „Carmen" (Sitzende Dame mit Fußkissen) (vgl. 5235; 5326)
4937 „Duett" (Flöte spielender Putto auf Meise sitzend) (auf Natursockel)
4938 „Maskerade" (Lo Hesse und Joachim von Seewitz) (vgl. 5251= Lo Hesse als EinzelFIGUR)
4951 FIGUR/LAMPE Pierrot und Colombine (Halbakt) (auf passigem Sockel) (vgl. Modell „Maskerade")
5094 „Pierrette" (Stehende junge Dame im Harlekinskostüm) (auf Rundsockel) (vgl. 5485)
5102 „Harlekin" (Stehender Till Eulenspiegel mit Flöte) (auf halbovalem Sockel)
5115 „Tanz der Eitelkeit" (Lucie Kieselhausen) (Stehende Pierrette mit Kopfbedeckung) (auf Rundsockel)
5235 „Carmen" (Sitzende Dame mit Fußkissen) FAYENCE/ALABASTER (vgl. 4903; 5326)
5251 „Maskerade" (Stehende Dame mit Muff) (Lo Hesse) (auf Ovalsockel) (vgl. 4938= GRUPPE)
5326 „Carmen" (Sitzende Dame mit Fußkissen) (vgl. 4903; 5235)
5485 „Pierrette" (Stehende junge Dame im Harlekinskostüm) (auf Rundsockel) (vgl. 5094)
5517 FIGUR/LAMPE Galante Rokokoszene (Kavalier mit Dreispitz küsst sie auf die Schulter) (auf abgerundetem, trapezförmigem Sockel)
5549 FIGUR/LAMPE Tänzerin mit großem Turban (ein Bein hebend) (auf Ovalsockel)
„Trunkener Bacchus" (Putto mit Weinbecher) BRONZE Lit.: N.N.: Wiener Kleinplastik einst und jetzt, in: MW, 2.1920/21, Nr. 12, S. 29 (m. Abb.), hier mit Ang. des Originaltitels „Trunkener Bacchus"; Neuwirth (1974), S. 65.

Krajc, Rudolf

*Rudolf Krajc, Maler, Graphiker und Kunstgewerbler (*6.6.1907 Bechyne, †14.3.1934 Bechyne), absolvierte zwischen 1887–1897 die KGW-Schule in Bechyne und wurde bei GS (als Modelleur, Formgießer bzw. Figurenretoucheur) angestellt. Weitere Informationen sind nicht bekannt. Lit.: Neuwirth (1974), S. 31; AKL Index.*

Kratschmer, Josef

Josef Kratschmer, Bildhauer/Kunstgewerbler? (Lebens- und Schaffensdaten unbekannt).
„Perlenkollier" (Stehende Biedermeierdame, ihr Collier betrachtend) ÖMKI (1913/14) Lit.: Studio YB, 1914, S. 191 (m. Abb.); Neuwirth (1974), S. 65; Doro 15.3.2005 (hier ein Ex. ohne Modell-Nr., Firmenmarke Adler 14, Firmenmarke Sculpture (gepr.), H. 27 cm)

Krebs

4958 DECKELDOSE (Form: länglich, hochoval/Knauf: längsoval/Dekor: Blattranken, Zackenfries)
4981 DECKELDOSE (Form: zwölfpassig facettiert, gebaucht, zu Stand und Mündung konisch zulaufend/Dekor: vertikal lig. „Herzen"/Rankenmuster, bzw. stilisierte Blumen)

Kremser, Lizzi (Felicitas)

*Lizzi Kremser, Bildhauerin und Kunstgewerblerin (*23.6.1908 Wien, †?), studierte von 1923–1928 an der KGW-Schule Wien unter Steinhof, Cizek und Powolny. Lit.: Neuwirth (1974), S. 65, 202.*
5953 „Vasenfigur" (Kniendes Mädchen mit balusterförmigem Topf auf der rechten Schulter)
5993 Sitzender Lautenspieler (Kremser zugeschr.)
669 SCHÜSSEL

Krippel, Elisabeth

*Elisabeth Krippel, Keramikerin und Kunstgewerblerin (*18.5.1901 Wien, †?), studierte von 1918–1923 an der KGW-Schule Wien unter Witzmann, Schufinsky, Boehm, Powolny und Selch. Lit.: Neuwirth (1974), S. 202; Zühlsdorff, S. 643.*
LEUCHTER (2-flammig) reliefiertes, florales Dekor (Pendant zu Vase) (um 1925) Lit.: Meyer, K.: Arbeitsstoff und Arbeitswert in: KuKGW, 5.1925, SH. 3, S. 8, 9 (m. Abb.), „Ob sich die Sitte, keramische Leuchter auf den Tisch zu setzen, wieder einbürgern wird (vermutlich kam die Anregung dazu aus Amerika), scheint mir fraglich. Jedenfalls ist der handliche und feste Doppelleuchter mit seiner derben, sicher stilisierten Verzierung ein gelungener Versuch. Ebenso wie die im Dekor ihm verwandte und in der Formengebung ansprechende schlichte Vase auf die Pariser KGW Ausst. geschickt worden".
VASE (becherförmig mit konischem Standring) reliefiertes, florales Dekor (Pendant zu Leuchter) (um 1925) Lit.: Meyer, K.: Arbeitsstoff und Arbeitswert in: KuKGW, 5.1925, SH. 3, S. 8, 9 (m. Abb.), „Ebenso wie die im Dekor ihm verwandte und in der Formengebung ansprechend schlichte Vase auf die Pariser KGW Ausst. geschickt worden".

Kucharzyk, Franz

*Franz Kucharzyk, Bildhauer (*19.3.1880 Rzepienik Strzyzowski/Gorlice/Polen, †?), besuchte die KGW-Schule in Zakopane von 1894–1899. Anschließend an der KGW-Schule in Wien unter Klotz und Strasser. Kucharzyk erhielt 1909 den Rompreis und wird ab 1912 Professor an der KGW-Schule in Wien. Lit.: TB, Bd. 22, S. 40; AKL Index.*
4518 Stehendes Bauernmädchen

Kuen, Walter

Mit Prof. Walter Kuen, Bildhauer (und Prof. an der KGW-Schule Innsbruck) sollte 1952 eine Zusammenarbeit angestrengt werden. Diese kam aber offensichtlich nicht zustande (Briefwechsel zw. Walter Goldscheider und Walter Kuen, vom 4. Juli 1952 bis 27. August 1952). Quelle: AGM.

Kuhn, Dina (Bernhardine)

*Dina (Bernhardine) Kuhn, Keramikerin (*26.4.1891 Wien, †?), studierte von 1912–1920 an der KGW-Schule in Wien (u.a. bei Kolo Moser). Sie war ab 1917 Mitglied der Wiener Werkstätte (entwarf für die Werkstätte zahlreiche Modelle und war damit eine der stilbildenden Keramikerinnen der Zwanziger Jahre) und arbeitete mit den Firmen Iskra, Bimini sowie der Porzellanmanufaktur Augarten, Wien zusammen. In Neutitschein richtet sie sich eine Keramikwerkstatt ein und ist neben zahlreichen Ausstellungen auch auf der Kunstschau Wien 1920 und der Pariser KGW-Ausst. 1925 mit Lampenfüßen, Leuchtern und Vasen vertreten, sowie mit der Gruppe „Adam und Eva" (Modell-Nr. 5248) auf der „International Exhibition of Ceramic Art" in New York 1928/29. Kuhn war ab 1928 wohnhaft in Wien XIX., Kreindlgasse 12. Lit.: Vollmer, Bd. 3, S. 135; Lehmann, 1928–1931; Neuwirth (1974), S. 41, 65, 203; Zühlsdorff, S. 422; Fahr-Becker, S. 228.*

5218 LAMPENFUSS (Form: kegelförmig, darauf eiförmige Bekrönung / Dekor: abstraktes Muster, mehrfarbig)
5248 „Adam und Eva" (als vollplastische Figuren mit Baum und Schlange)
5249 Jäger mit zwei springenden Hirschen
5275 FIGUR / LAMPE „Papagena" (Tänzerin mit exotischem Kostüm und Kopfputz)
5342 VASE (Form: gebaucht, seitlich verschoben mit einer Handhabe / Dekor: abstraktes Muster, mehrfarbig)
5343 KERZENHALTER (zweiflammig)
BLUMENTOPF (klein) (lineare, reliefierte Ornamente) (um 1925) Lit.: Meyer, K.: Arbeitsstoff und Arbeitswert, in: KuKGW, 5.1925, SH. 3, S. 8, 9 (m. Abb.), „Die drei Blumentöpfe (-> vgl. auch die beiden nächsten Einträge, Anm. des Autors) mit ihrem reliefartig auf die dunkelrote Grundfläche aufgesetztem Zierat sind in der Grundform ziemlich einheitlich, sie zeigen ein ganz einfaches, leicht geschwungenes Profil. Mit der Größe der Stücke nimmt auch der Reichtum des Dekors zu, vom gewöhnlichsten, linearen Ornament über das vegetabile bis zu dem reichen Gemisch von Linien-, Blätter- und Figurenschmuck, der bei der dritten Vase mit kühnster Freiheit und beinahe willkürlich über die ganze Fläche verteilt ist. Für diese drei Blumentöpfe wird von der Firma „rote und bunte Majolika" als Werkstoff angegeben. Man könnte natürlich gerade so gut auch „Fayence" dafür sagen. Denn ob man diese ursprünglich aus dem Osten stammende Art von Tonwaren und von Tonbereitung nach dem spanischen Maiorka oder nach dem italienischen Faenza benennen will, ist gleichgültig".
BLUMENTOPF (mittel) (reliefiertes Blattwerk mit Maskarons) (um 1925) Lit.: Meyer, K.: Arbeitsstoff und Arbeitswert, in: KuKGW, 5.1925, SH. 3, S. 8, 9 (m. Abb.)
BLUMENTOPF (groß) (reliefierte Figuren, Putti und Vögel) (um 1925) Lit.: Meyer, K.: Arbeitsstoff und Arbeitswert, in: KuKGW, 5.1925, SH. 3, S. 8, 9 (m. Abb.)
JARDINIERE (Stilisierter Schwan auf konischem Sockel) Lit.: L., A. S.: Vienna, in: The Studio, 1928, S. 367 (m. Abb.), „She (Dina Kuhn, Anm. des Autors) has done some very attractive tea and coffee services, original both in design and form, which have gained much approval. Her art is more austere than that of her sister-artist Hertha Pucher. She has put it more to the service of everyday use. There is simplicity of treatment, a nobility of line, an entire absence of anything pertaining to self-consciousness. All is natural, yet everywhere one senses the thought underlying her work".
VASE (1925) Lit.: Neuwirth (1974), S. 65; Grell S. 33
LAMPENFUSS (1925) Lit.: Neuwirth (1974), S. 65; Grell S. 33
LEUCHTER (1925) Lit.: Neuwirth (1974), S. 65; Grell S. 33
DECKELDOSE / DECKELVASE / Bowle? (mit zwei Handhaben, ösenförmiger Knauf) (1925) Lit.: N.N.: Österreich in Paris, in: ÖBWK, 1.1924/25, S. 341 (m. Abb.)
DECKELDOSE / DECKELVASE / Bowle? (ösenförmiger Knauf) (1925) Lit.: N.N.: Österreich in Paris, in: ÖBWK, 1.1924/25, S. 341 (m. Abb.)

La Monaca, Francesco

*Francesco La Monaca, Bildhauer (*10.2.1882 Catanzaro/Italien, †16.2.1937 Washington), war Schüler von Thomas und Injalbert an der Ecole des Beaux-Arts, Paris, wo La Monaca auch lebte. Er war Mitglied der Künstlergruppe „La Stèle" (-> vgl. Kap. „Dekorative Skulpturen von Arthur Goldscheider in Paris"). Er beteiligte sich an den Ausstellungen des Pariser Salons (Societé des Artistes Français) ab 1908. Lit.: Goldscheider, A. (Hrsg.): Artistic Evolution, 1926, T. 1, Pl. 17 (m. Abb.); Vollmer, Bd. 3, S. 161; Bénézit, Bd. 8, S. 210; AKL Index.*

„Danseur" BRONZE Paris, Salon Société des Artistes Français (1908) Lit.: N.N.: In Bronze lesen, in: AZ, 1984, S. 269
„Siffleur" BRONZE Paris, Salon Société des Artistes Français (1908) Lit.: N.N.: In Bronze lesen, in: AZ, 1984, S. 269
„Tigre en furie" (Fauchender Tiger) BRONZE Arthur Goldscheider, Paris (1924) Lit.: L' Illustration, Nr. 4268 vom 20.12.1924, S. 14 (m. Abb.)
VASE (cuivre gravé) (Dekor von Mergier) (H. 31 cm, B. 21 cm) Lit.: Goldscheider, A. (Hrsg.): Artistic Evolution, 1926, T. 1, Pl. 17 (m. Abb.)
VASE (cuivre gravé avec incrustations d'émail sur argent) (Dekor von Colmant) (H. 14,8 cm, B. 7 cm) Lit.: Goldscheider, A. (Hrsg.): Artistic Evolution, 1926, T. 1, Pl. 17 (m. Abb.)
VASE (laqué) (Dekor von Gilles Defraigne) (H. 15 cm, B. 15 cm) Lit.: Goldscheider, A. (Hrsg.): Artistic Evolution, 1926, T. 1, Pl. 17 (m. Abb.)
VASE (en métal gravé) (Dekor von Mergier) (H. 14,5 cm, B. 6,5 cm) Lit.: Goldscheider, A. (Hrsg.): Artistic Evolution, 1926, T. 1, Pl. 20 (m. Abb.)
VASE (en dinanderie gravée) MESSING (Dekor von J.-Charles Goëtz) (H. 28 cm, B. 20 cm) Lit.: Goldscheider, A. (Hrsg.): Artistic Evolution, 1926, T. 1, Pl. 24 (m. Abb.)
VASE (en dinanderie gravée) MESSING (Dekor von Alexandre Noll) (H. 29 cm, B. 14 cm) Lit.: Goldscheider, A. (Hrsg.): Artistic Evolution, 1926, T. 1, Pl. 24 (m. Abb.)

Labatut, Jules Jacques

*Jules-Jacques Labatut, Bildhauer (*31.7.1851 Toulouse, †?), war ab 1876 Schüler von Joffroy und A. Mercié an der École des Beaux-Arts, Paris. Labatut war namhafter Vertreter der Monumental- und Porträtplastik und erhielt 1881 den Großen Rompreis. Seine Figuren wurden auch von Thiébaut Frères ausgeformt und er war Mitarbeiter der Pariser Bronzewarenfabrikation Goldscheiders. Labatut wird letztmals 1914 erwähnt. Lit.: TB, Bd. 22, S. 163; Neuwirth (1974), S. 34, 65; Hillenbrand, S. 232; Bénézit, Bd. 8, S. 111–112.*

3069 Stehende Frau, eine Mandoline um ihre Schulter tragend (auf Landschaftssockel)
„Junon" Paris, Salon (1902) Lit.: Neuwirth (1974), S. 65

Lacoux, Henri

Henri Lacoux, Maler (Lebensdaten unbekannt), war Schüler von Camille Pissaro. Lit.: Goldscheider, A. (Hrsg.): Artistic Evolution, 1926, T. 1, Pl. 5 (m. Abb.).

VASE „Deer Stalking" (Fayence, Unterglasurmalerei) Lit.: Goldscheider, A. (Hrsg.): Artistic Evolution, 1926, T. 1, Pl. 5 (m. Abb.)

Lamourdedieu, Raoul Eugène

*Raoul Eugène Lamourdedieu, Holz- und Steinbildhauer, Medailleur und Kunstgewerbler (*2.2.1877 Fauguerolles/Lot-et-Garonne, †8.5.1953 Pierrefonds/Oise), lebte in Paris und beschickte den Pariser Salon (Societé des Artistes Français) ab 1905. Er war Mitglied der Künstlergruppe „La Stèle" (-> vgl. Kap. „Dekorative Skulpturen von Arthur Goldscheider in Paris"). Lit.: Goldscheider, A. (Hrsg.): Artistic Evolution, 1926, T. 2, Pl. 33 (m. Abb.), TB, Bd. 22, S. 269; Vollmer, Bd. 3, S. 161; Neuwirth (1974), S. 65; Hillenbrand, S. 235; Bénézit, Bd. 8, S. 213.*

„Amphora" BRONZE (H. fehlt, B. fehlt) Lit.: Goldscheider, A. (Hrsg.): Artistic Evolution, 1926, T. 2, Pl. 40 (m. Abb.)
„Atalanta" BRONZE (H. fehlt, B. fehlt) Lit.: Goldscheider, A. (Hrsg.): Artistic Evolution, 1926, T. 2, Pl. 34 (m. Abb.)
„Awake of Dawn" BRONZE (H. fehlt, B. fehlt) Lit.: Goldscheider, A. (Hrsg.): Artistic Evolution, 1926, T. 2, Pl. 36 (m. Abb.)
„Dancer with Veil" BRONZE (H. fehlt, B. fehlt) Lit.: Goldscheider, A. (Hrsg.): Artistic Evolution, 1926, T. 2, Pl. 38 (m. Abb.)
„Indian Runner" BRONZE (H. fehlt, B. fehlt) Lit.: Goldscheider, A. (Hrsg.): Artistic Evolution, 1926, T. 2, Pl. 35 (m. Abb.)
„Kingfisher" BRONZE (H. fehlt, B. fehlt) Lit.: Goldscheider, A. (Hrsg.): Artistic Evolution, 1926, T. 2, Pl. 33 (m. Abb.)
„Ocella" BRONZE (H. fehlt, B. fehlt) Lit.: Goldscheider, A. (Hrsg.): Artistic Evolution, 1926, T. 2, Pl. 39 (m. Abb.)
„The Pheasant" BRONZE (H. fehlt, B. fehlt) Lit.: Goldscheider, A. (Hrsg.): Artistic Evolution, 1926, T. 2, Pl. 37 (m. Abb.)
„The Turkeycock" BRONZE (H. fehlt, B. fehlt) Lit.: Goldscheider, A. (Hrsg.): Artistic Evolution, 1926, T. 2, Pl. 41 (m. Abb.)

Landowski, Paul-Maximilien

*Paul-M. Landowski, Bildhauer (*1.6.1875 Paris, †April 1961 Boulogne-sur-Seine), war Schüler von Barrias und Mitglied der Künstlergruppe „La Stèle" (-> vgl. Kap. „Dekorative Skulpturen von Arthur Goldscheider in Paris"). Im Jahre 1898 erhielt er den Chenavard-Preis, 1900 dann den Rompreis. 1907 Aufnahme in die Ehrenlegion. Lit.: Goldscheider, A. (Hrsg.): Artistic Evolution, 1926, T. 1, Pl. 3 (m. Abb.); TB, Bd. 22, S. 300; Vollmer, Bd. 6, S. 186; Neuwirth (1974), S. 66; Bénézit, Bd. 8, S. 228.*

„Charmeuse de serpents" („Snake charmer") BRONZE „La Stèle" (H. 50 cm, B. 25 cm) Lit.: Goldscheider, A. (Hrsg.): Artistic Evolution, 1926, T. 1, Pl. 3 (m. Abb.)
„The Joyous Centaur" (H. fehlt, B. fehlt) Lit.: Goldscheider, A. (Hrsg.): Artistic Evolution, 1926, T. 2, Pl. 1 (m. Abb.)
„Bedouin Woman with Jar" BRONZE (H. fehlt, B. fehlt) Lit.: Goldscheider, A. (Hrsg.): Artistic Evolution, 1926, T. 2, Pl. 25 (m. Abb.)

Langer, (Leopold?)

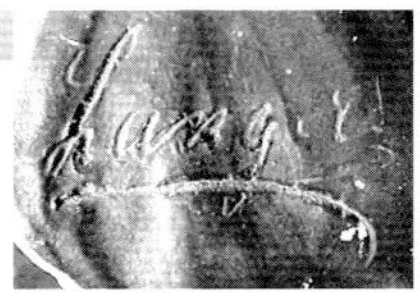

*Leopold Langer, Bildhauer (*15.5.1891 Proßnitz, Mähren, †?), absolvierte die Fachschule für Tonindustrie in Znaim und studierte an der KGW-Schule Wien bei Strnad (1911/12). Seine Arbeiten wurden auch von der Wiener kunstkeramischen Werkstätte ausgeführt. Lit.: Neuwirth (1974), S. 66, 206.*

923 VASE Frauenakt mit Kind und Wassergefäß auf Ast- und Blattwerk
925 BÜSTE Araberin mit Kopftuch und Halsschmuck (vgl. 935)
935 BÜSTE Araberin mit Kopftuch und Halsschmuck (vgl. 925)
1358 BÜSTE „Orientalin" (vgl. 1424)
1360 BÜSTE Afrikaner (nach rechts blickend)
1413 FIGUR „Korbflechter" (Sitzender Knabe einen Korb flechtend) (Sockel als Steinmauer) (vgl. 1650)
1424 BÜSTE „Orientalin" (undeutl. sign. „Lanogé"?) (vgl. 1358)
1650 FIGUR „Korbflechter" (Sitzender Knabe einen Korb flechtend) (Sockel als Steinmauer) (vgl. 1413)
2222 FIGUR (SPIEGEL) „Bei der Toilette" (Sitzender Mädchenakt in einen Spiegel schauend) (auf Landschaftssockel mit Lilien / Callas)
„Zwei Spatzen" ÖMKI (1911/12) Lit.: Neuwirth (1974), S. 66

Lanogé s. Langer, (Leopold?)

Lapigre

2486 FIGUR Mädchen, an heckenumwundenem Mauerdurchbruch stehend (vgl. 2763)
2763 FIGUR / LAMPE Mädchen, an heckenumwundenem Mauerdurchbruch stehend (vgl. 2486)

Laporte-Blairsy (Blaisin?), Léo

*Léo Laporte-Blairsy (Blaisin?), Bildhauer, Graphiker und Kunstgewerbler (*5.4.1865 Toulouse, †1923 Paris), war Schüler von A. Falguière und A. Mercié. Er beschickte den Salon in Paris (Societé des Artistes Français) ab 1888 und schuf vorwiegend Bronzefiguren, die zu Lampen verarbeitet wurden. Von 1902–1907 war er als Modelleur für Sèvres (Manufacture Nationale de Sèvres) sowie u.a. für Susse Frères und Goldscheider Paris tätig. Lit.: TB, Bd. 22, S. 377; Hillenbrand, S. 238; Neuwirth (1974), S. 34; Zühlsdorff, S. 614; Hillenbrand, S. 238; Arwas (2002), S. 615–616.*

ohne Nr. LAMPE Stehende Frauengestalt ALABASTER / MARMOR?
TAFELDEKORATION „Le Déjeuner fleuri" BRONZE (1907); Lit.: Neuwirth (1973), S. 59, 60

Larone s. Laroux

Laroux (d.i. Larroux, Antonin) s. Larroux, Antonin

Larroux, Antonin

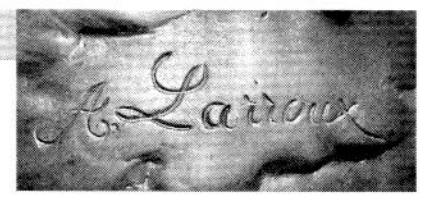

*Antonin Larroux, Bildhauer (*3.6.1859 Toulouse, †11.2.1937 Ort unbekannt), war u.a. Schüler von Falguière und A. Mercié. Larroux war Mitglied der Societé des Artistes Français, deren Salon er ab 1881 beschickte. Larroux erhielt u.a. eine Bronzemedaille auf der Weltausst. in Paris 1900. Lit.: TB, Bd. 22, S. 392; Bénézit, Bd. 8, S. 292; AKL Index.*
908-GSB BUSTE „Méditation"
933 VASE „Masques" (Sechs plastische Gesichter: Nymphen / Faun) (vgl. 1097) (sign. Laroux) (vgl. WANDMASKE 2287; 2288; 2289; 2290; 2291)
1097 VASE „Masques" (Sechs plastische Gesichter: Nymphen / Faun) (vgl. 933) (sign. Laroux) (vgl. WANDMASKE 2287; 2288; 2289; 2290; 2291)
1844 BÜSTE „Modern" ÖMKI (1902/03) (lt. Kat. „Laroux) (vgl. 2292= WANDMASKE)
2287 WANDMASKE „Junger Faun" ÖMKI (1902/03) (Pendants vgl. 2288; 2289; 2290; 2291) (vgl. 933; 1097 VASE „Masques")
2288 WANDMASKE „Bacchantin" (mit Weinblättern im Haar) ÖMKI (1902/03) (Pendants vgl. 2287; 2289; 2290; 2291) (vgl. 933; 1097 VASE „Masques")
2289 WANDMASKE „Faun" (Pendants vgl. 2287; 2288; 2290; 2291) (vgl. 933; 1097 VASE „Masques")
2290 WANDMASKE „Klatschblume" (Frauenkopf mit langem Haar und Mohnblüten) (sign. Leroux) ÖMKI (1902/03) (Pendants vgl. 2287; 2288; 2289; 2291) (vgl. 933; 1097 VASE „Masques")
2291 WANDMASKE „Coquette"/„Kokette" (Frauenkopf mit wallendem Haar) (sign. Laroux) ÖMKI (1902/03) (Pendants vgl. 2287; 2288; 2289; 2290) (vgl. 933; 1097 VASE „Masques")
2292 WANDMASKE „Modern" (vgl. 1844= BÜSTE)
WANDMASKE Faunmaske (mit Blattwerk) Lit.: B., F.: Keramische Wanderungen durch den Kunstsalon Keller & Reiner Berlin, in: KM, 3.1903, S. 125 (m. Abb.), hier mit Ang. des Künstlernamens „Lavroux" und des Titels Faunmaske (vgl. 2287; 2289)
Zwei Model e Paris, Salon (1902) Lit.: Neuwirth (1974), S. 66

Larsson, Gottfrid Julius Andreas

*Gottfrid J. A. Larsson, Bildhauer (*21.11.1875 Narveryd-Vallerstad, Östergötsland, †24.12.1947 Stockholm), studierte in Stockholm, Paris, München usw. Er arbeitete zwischen 1900 und 1910 in Paris und leitete von 1920–1932 eine Kunstschule in Stockholm. Goldmedaillen erhielt Larsson in St. Petersburg (1908), München (1909) sowie in San Francisco (1915). Lit.: Vollmer, Bd. 3, S. 177; Hillenbrand, S. 239.*
ohne Nr. LAMPE Loie Fuller mit großem Tuch (beim Tanz) BRONZE / Marmorsockel
LAMPE „Le Vice fuyant la lumière" GIPS (Salon Paris, 1902) Lit.: N., R.: Keramische Körperkultur, in: AZ, 1992, Nr. 18, S. 565
TINTENFASS Bacchantenstatuette (Salon Paris, 1902) Lit.: N., R.: Keramische Körperkultur, in: AZ, 1992, Nr. 18, S. 565

Latour, (Louis-Marie-Blaise?)

Der bei Neuwirth (1974) auf S. 66 angegebene Künstler Louis-Marie Blaise Latour wird zwar im Pariser Salon (Societé des Artistes Français) letztmalig 1928 erwähnt, ist aber möglicherweise nicht mit jenem identisch, welcher für Goldscheider die zahlreichen Modelle entworfen hat. Vielmehr ist hier wieder von einem „klangvollen" Pseudonym auszugehen, welches die Manufaktur für die internationale Klientel „erfunden" und vermarktet hat. Ein in den USA produziertes Modell mit dem Originaltitel „Dance Invitation" wurde auch mit dem Namen Latour versehen. Diese Tänzerin mit einem Fächer ist keine Nachbildung eines Wiener Modells, sondern eine Neuschöpfung für die amerikanische Produktion. Dies könnte als weiterer Beweis für die o.a. These gelten. Bis aber ein Stück, welches nicht nur mit dem Namen „Latour", sondern auch mit den Vornamen bzw. deren Initialen sign. ist, auftaucht, soll die folgende Künstlerbiografie für Louis-Marie-Blaise Latour an dieser Stelle Raum greifen. Der 1860 in Algier geborene Latour war Schüler von P. Dubois und A. Falguière. Er war tätig in Paris und zwischen 1883–1914 häufig mit Porträts und Genrestatuen im Salon vertreten. Lit.: TB, Bd. 22, S. 422–423; Mackay, S. 224; Bénézit, Bd. 8, S. 316; AKL Index.
2893 LAMPE Stehendes Mädchen mit Blasebalg (auf Rundsockel)
3018 BÜSTE Frau mit Diadem und Halsband
3684 „Pro rude" (Zwei Gladiatoren, gegen eine Löwin kämpfend) (Inschrift: PRO RUDE)
3726 Lucretia (Stehender Frauenhalbakt mit Dolch) (auf Rechtecksockel)
3834 Judith (Stehende bekleidete Frauengestalt mit Schwert) (vgl. 3835 „Halbakt")
3835 Judith (Stehender Frauenhalbakt mit Schwert) (vgl. 3834 „bekleidete Figur") (unleserlich sign.)
3982 Tänzerin (sich nach links drehend)
4094 Sitzender Cellist (im Biedermeiergewand) mit stehendem Biedermeiermädchen (auf Blumengirlanden verziertem, kanneliertem Rundsockel) (vgl. Einzelfiguren 4108; 4109)
4095 Die Liebeserklärung (Sitzendes, händchenhaltendes Rokokopaar an einem Tisch) (auf Blumengirlanden verziertem, kanneliertem Rundsockel)
4108 Sitzender Cellist (im Biedermeiergewand) (vgl. Gruppe 4094)
4109 Stehendes Biedermeiermädchen (mit zwei Rosen am Dekolleté) (vgl. Gruppe 4094)
4158 Biedermeiertänzerin (auf Rundsockel) (vgl. 4252; 4253)
4164 Diana mit Pfeil und Bogen, neben Windhund stehend (auf Vierkantsockel) (vgl. 4310)
4187 Stehender Windhund (auch sign. Gemignani) (vgl. 4374)
4234 Tänzerin (nach links drehend) FAYENCE / ALABASTER (vgl. Latour 4235; 4236 und Rosé 4234!) (auch als Lampe H. 47 cm unsign.)
4235 Tänzerin (nach links drehend) (vgl. Latour 4234; 4236 und Rosé 4234!) (auch als Lampe H. 47 cm unsign.)
4236 Tänzerin (nach links drehend) (vgl. Latour 4234; 4235 und Rosé 4234!) (auch als Lampe H. 47 cm unsign.)
4252 Biedermeiertänzerin (auf Rundsockel) (vgl. 4158; 4253)
4253 Biedermeiertänzerin (auf Rundsockel) FAYENCE / ALABASTER (vgl. 4158; 4252)
4310 Diana mit Pfeil und Bogen, neben Windhund stehend (auf Vierkantsockel) (vgl. 4163)
4347 BÜSTE Biedermeiermädchen mit Schutenhut ALABASTER
4374 Stehender Windhund (vgl. Gemignani / Latour 4187) (ähnl. Modelle vgl. Gemignani 3859; 4687) (Entwurf eindeutig von Valmore Gemignani)
4687 Stehender Windhund s. Gemignani 3859; 4687 (Entwurf eindeutig von Valmore Gemignani)
5393 BÜSTE Junge Biedermeierdame (nach rechts blickend) ALABASTER

Laurana, Francesco (nach)

*Francesco Laurana, Bildhauer, Medailleur und Baumeister (*um 1420/25 Vrana/Zara, †vor dem 12.3.1502 Avignon oder Marseille?). Lit.: TB, Bd. 23, S. 440–442; AKL Index.*
2208 BÜSTE „Isabella de Aragon"

Laurent, Georges Henri

Georges Henri Laurent, Bildhauer (Lebens- und Schaffensdaten unbekannt), war Mitarbeiter bei Arthur Goldscheider, Paris. Lit.: Goldscheider, A. (Hrsg.): Artistic Evolution, 1926, T. 2, Pl. 17 (m. Abb.)
„Capuchin Pigeons" BRONZE (H. fehlt, B. fehlt) Lit.: Goldscheider, A. (Hrsg.): Artistic Evolution, 1926, T. 2, Pl. 44 (m. Abb.)
BUCHSTÜTZE „Canard Mandarin" (Mandarin Duck) BRONZE (versilbert), Marmorsockel (H. 19 cm, B. 20 cm) Lit.: Goldscheider, A. (Hrsg.): Artistic Evolution, 1926, T. 2, Pl. 17 (m. Abb.)
BUCHSTÜTZE „Marabout" BRONZE (vergoldet), Marmorsockel (H. 17 cm, B. 16,5 cm) Lit.: Goldscheider, A. (Hrsg.): Artistic Evolution, 1926, T. 2, Pl. 17 (m. Abb.)

Laurenzl (d.i. Lorenzl, Josef)

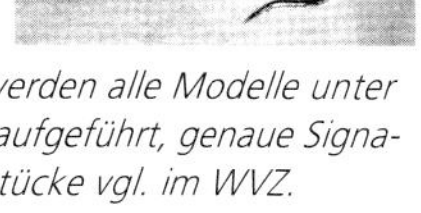

Da eine unübersehbare Anzahl der Modelle sowohl mit der Sign. Lorenzl, als auch mit „Laurenzl" bezeichnet sind, werden alle Modelle unter dem richtigen Namen Lorenzl aufgeführt, genaue Signaturenangaben der einzelnen Stücke vgl. im WVZ.

Lavergne, Adolphe-Jean

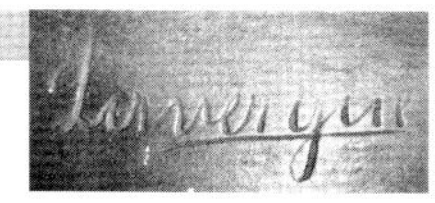

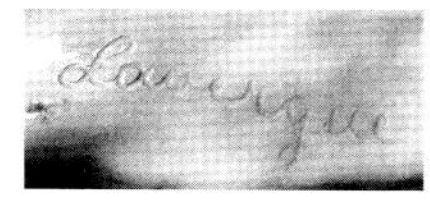

*Adolphe-Jean Lavergne, Bildhauer (*Hautefort, †?), lebte und arbeitete in Paris. Seine Ausstellungstätigkeit im dortigen Salon (Societé des Artistes Français) beginnt 1863 und dauert bis 1876 an. Im Jahre 1928 wird er letztmals erwähnt. Lit.: TB, Bd. 22, S. 474–475; Neuwirth (1974), S. 66; Mackay, S. 225; Hillenbrand, S. 241; Bénézit, Bd. 8, S. 349; AKL Index.*
1210 VASE „Orpheus" mit Lyra, von drei Löwen und zwei Vögeln begleitet (vollplastisch an der Vasenwandung stehend) (vgl. 1251)
1251 VASE „Orpheus" mit Lyra, von drei Löwen und zwei Vögeln begleitet (vollplastisch an der Vasenwandung stehend) (vgl. 1210)
1588 Schreitender Löwe (auf Rechtecksockel)
2385 JARDINIERE Mädchen mit zwei Körben
2758 UHR Am Brunnen (Paar auf säulenartigem Sockel mit Baum)
2826 UHR „Allegorie des Lebens" (vgl. Capèque 2826!)
2896 GRUPPE Reitende indische Schönheit auf einem von Sklaven geführten Elefanten (auf Rechtecksockel)
3207 Stehende junge Frau eine flache Schale in die Höhe haltend (auf Rundsockel) (vgl. 3590)
3256 UHR / LAMPE „Dans Le Arene" (In der Arena) (Römischer Sklave lässt zwei Löwen aus dem Käfig) (Uhrgehäuse als Löwenkäfig, mit zwei auf Säulen stehenden Urnen, welche als Lampenfassungen dienen) (vgl. 3444)
3319 UHR „Historia" (Inschrift: CAVE NE INDIGNUM) (vgl. 3396; 3763)
3369 „Le Grand Condé" (Louis Prince de Condé zu Pferde)
3396 UHR „Historia" (Inschrift: CAVE NE INDIGNUM) (vgl. 3319; 3763)
3416 s. Bormann
3444 UHR / LAMPE „Dans Le Arene" (In der Arena) (Römischer Sklave lässt zwei Löwen aus dem Käfig) (Uhrgehäuse als Löwenkäfig, mit zwei auf Säulen stehenden Urnen, welche als Lampenfassungen dienen) (vgl. 3256)
3534 „Wiener Walzer" (Tanzendes Paar im Biedermeierkostüm)
3526 „Animus et vis" (Stehender römischer Krieger mit Helm, Speer und Tiger) (auf passigem Rechtecksockel) (mit oder ohne Inschrift: ANIMUS ET VIS) (auch sign. Cherc) (vgl. 3680) (vgl. Tiger 3536)
3536 Tiger (auf Rechtecksockel) (vgl. 3988) (vgl. Gruppe „Animus et vis" 3680)
3590 LAMPE Stehende junge Frau eine flache Schale in die Höhe haltend (auf Rundsockel) (vgl. 3207)
3680 „Animus et vis" (Stehender römischer Krieger mit Helm, Speer und Tiger) (auf passigem Rechtecksockel) (mit oder ohne Inschrift: ANIMUS ET VIS) (auch sign. Cherc) (vgl. 3526) (vgl. Tiger 3536)
3718 Stehende Frauengestalt, eine flache Schale in die Höhe haltend (auf zylindrischem, Blumengirlanden verziertem Sockel) (vgl. Tell 3810= fast identisches Modell)
3721 GRUPPE Zwei Zigaretten rauchende Harlekine (vgl. 3766= rechte Einzelfigur; vgl. 3735= LAMPE)
3735 LAMPE Zwei Zigaretten rauchende Harlekine (vgl. 3721= GRUPPE; vgl. 3735= rechte Einzelfigur)
3763 UHR „Historia" (Inschrift: CAVE NE INDIGNUM) (vgl. 3319; 3396)
3766 Zigarette rauchender Harlekin (vgl. 3721= GRUPPE; vgl. 3735= LAMPE)
3859 Stehender Windhund (auch sign. Gemignani/Latour)
3900 Wasserträgerin (ein Henkelgefäß über dem Kopf haltend) (auf Rundsockel)
3988 Tiger (auf Rechtecksockel) (vgl. 3536) (vgl. Gruppe „Animus et vis" 3680)
4123 BÜSTE Chopin (Inschrift: CHOPIN)
4200? LAMPE Salome (Tanzende junge Frau) (auf Rundsockel mit drei Muscheln)
4221 Stehendes Biedermeiermeiermädchen mit Chrysanthemenstrauß
FIGUR „Napoleon" ÖMKI (1906/07) Lit.: Neuwirth (1974), S. 66
GRUPPE „Kleopatra" ÖMKI (1906/07)
VASE Liebespaar mit Blütengirlanden (vollplastisch an der Vasenwandung stehend) (um 1905) (H. 75 cm); Neumeister Dresden 24 (Varia)

Lavergue s. Lavergne, Adolphe-Jean

Laverque s. Lavergne, Adolphe-Jean

Lavroux (d.i. Larroux, Antonin)

Faunmaske (mit Blattwerk) (Wandmaske) Lit.: B., F.: Keramische Wanderungen durch den Kunstsalon Keller & Reiner Berlin, in: KM, 3.1903, S. 125 (m. Abb.), hier mit Ang. des Künstlernames „Lavroux"; (vgl. Antonin Larroux „Maske" oder „Junger Faun" ÖMKI 1902/03).

Le Faguays, Pierre

*Pierre Le Faguays, Bildhauer (*29.8.1892 Nantes, †1935, Paris?), war Schüler von Vibert. Le Faguays stellte ab 1922 im Pariser Salon (Société des Artistes Français) aus und entwarf in den 20er Jahren figürliche Modelle für Arthur Goldscheider, Paris. Er war Mitglied der Künstlergruppen „La Stèle" und „Évolution" und gehörte zu den bedeutenderen und gesuchten Bildhauern des Art Déco. Lit.: Goldscheider, A. (Hrsg.): Artistic Evolution, 1926, T. 1, Pl. 28 (m. Abb.); Vollmer, Bd. 3, S. 199; Neuwirth (1974), S. 66; Zühlsdorff, S. 423; Hillenbrand, S. 242; Bénézit, Bd. 8, S. 426; AKL Index.*

56-LaStèle Puppenspielerin mit zwei Handpuppen BRONZE / Marmorsockel

„Danse Bachique" BRONZE Arthur Goldscheider, Paris (1924) Lit.: L'Illustration, Nr. 4268 vom 20.12.1924, S. 14 (m. Abb.)

LAMPE „Lampe de Parquet" (Parquet Lamp) Stehender Damenakt (Socle en Thuya, Statuette Bronze, Coupe en Alabatre) THUYENHOLZ, BRONZE, ALABASTER (H. 200 cm, B. 34 cm) Lit.: Goldscheider, A. (Hrsg.): Artistic Evolution, 1926, T. 1, Pl. 28 (m. Abb.)

„Egyptian Women" (H. fehlt, B. fehlt) Lit.: Goldscheider, A. (Hrsg.): Artistic Evolution, 1926, T. 2, Pl. 9 (m. Abb.)

„Purity" (H. fehlt, B. fehlt) Lit.: Goldscheider, A. (Hrsg.): Artistic Evolution, 1926, T. 2, Pl. 10 (m. Abb.)

„Agitated Flight" (H. fehlt, B. fehlt) Lit.: Goldscheider, A. (Hrsg.): Artistic Evolution, 1926, T. 2, Pl. 11 (m. Abb.)

„Woman and Puppets" (H. fehlt, B. fehlt) Lit.: Goldscheider, A. (Hrsg.): Artistic Evolution, 1926, T. 2, Pl. 12 (m. Abb.)

„Flower dance, Columbine" (H. fehlt, B. fehlt) Lit.: Goldscheider, A. (Hrsg.): Artistic Evolution, 1926, T. 2, Pl. 13 (m. Abb.)

„Bewitched" (H. fehlt, B. fehlt) Lit.: Goldscheider, A. (Hrsg.): Artistic Evolution, 1926, T. 2, Pl. 14 (m. Abb.)

„Temptation" (H. fehlt, B. fehlt) Lit.: Goldscheider, A. (Hrsg.): Artistic Evolution, 1926, T. 2, Pl. 15 (m. Abb.)

„Girl with the Sandal" (H. fehlt, B. fehlt) Lit.: Goldscheider, A. (Hrsg.): Artistic Evolution, 1926, T. 2, Pl. 16 (m. Abb.)

Le Guluche, Joseph s. WVZ Modell-Nr. 314

Leclancher

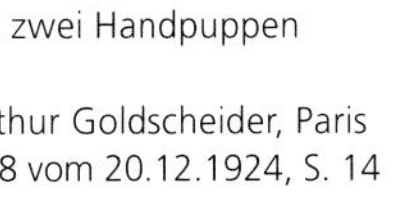

2773 LAMPE Wasserträgerin (Schale in der Rechten)

2859 LAMPE Bergmann (mit Pickel am Bergkristall)

2912 Stehende Frau mit zwei Amphoren (auf Natursockel)

3088 LAMPE Arbeiter, einen Eisblock auf Karren schiebend

3089 BÜSTE Kinderkopf

3156 LAMPE Stehender, an der Feuerstelle (Esse) arbeitender Schmied

3244 Kniender Gärtnerjunge (einen Baum / Rebstock? pflanzend)

3610 Stehender Arbeiter mit Spaten (auf Rechtecksockel)

3620 Mutter mit Kind (auf Felssockel)

3673 Stehender Arbeiter mit Spitzhacke (auf Rechtecksockel)

3969 UHR „Eisengießer" (Zwei Arbeiter mit Handwagen vor dem Ofen) ÖMKI (1906/07)

LeClercq, François-Joseph

3373 s. Rosé

3762 s. Rosé

Ledru, Auguste

*Auguste Ledru, Bildhauer und Kunstgewerbler (*1860 Paris, †5.11.1902 Paris), war Schüler u.a. von Dumont. Ledru stellte seine Arbeiten ab 1883 im Pariser Salon (Société des Artistes Français) aus und errang eine Bronzemedaille auf der Weltausstellung in Paris 1900. Die Firma Susse Frères editierte etwa 40 seiner Bronzen, Silber- und Zinnarbeiten. Zu seinen bevorzugten Sujets zählen Nymphen und Meeresgetier, die er auf Schalen und Schüsseln, Plaketten, Statuen, Vasen und Tellern darstellte. Lit.: TB, Bd. 22, S. 538; Hillenbrand, S. 242; Arwas (2002), S. 616; Bénézit, Bd. 8, S. 414; AKL Index.*

828-GSB Plat „Femme aux Chardons" (Pendant vgl. 929)

829-GSB VIDE-POCHE „Nymphe aux Coquillages"

830-GSB VIDE-POCHE „Ondine"

831-GSB VIDE-POCHE „Grenouilles"

848-GSB PLAT „Naiades"

862-GSB Surtout „Femme au Poisson"

891-GSB „Coquille" No. 1

892-GSB „Coquille" No. 2

929-GSB PLAT „Femme aux Chardons" (Pendant vgl. 828)

930-GSB BAS-RELIEF „Effroi"

VASE „Betrachtung" (Zwei an der Mündung sitzende Mädchen, die in das Innere der Vase blicken) Lit.: R., F.: Wiener Kunstkeramik, in: KM, 2.1902, S. 90, „Helle Türkisdekoration hat eine „Betrachtung" benannte Vase von Ledru erhalten. Die beiden schlanken mädchenhaften Körper, welche auf der oberen Einbiegung der Urne kauern und forschend in das Innere der Vase blicken, haben durch den zarten, leuchtenden, aber etwas kalten Ton des hellen Türkis etwas Herbes, Strenges bekommen, wie es die Idee der forschenden Betrachtung, die sie verkörpern sollen, erheischt. Dieses diskrete helle Türkisdekor eignet sich gut zur Anwendung bei dekorativen Zierstücken, wie es sich durch seine ruhige Wirkung vornehm in jedes Ensemble einfügt".

Lefont

1173 VASE „Der Handkuss" (Amphore mit zwei Henkeln, reliefierte Wandung mit Dekor: Handkuss und Puttenreigen)

2920 Stehendes Biedermeiermädchen mit großem Hut (einen übergroßen Korb tragend)

2991 Stehende junge Frau, eine Lyra spielend (auf Rundsockel)

3022 BÜSTE Junge Frau mit Kopftuch

3025 GRUPPE „Reigen" (Junges Paar einen Reigen tanzend) (auf blumenverziertem mehrpassigem Sockel) ÖMKI (1906/07)

3941 Stehendes Bauernpaar (auf abfallendem Landschaftssockel)

Legastelois, Jules-Prosper

*Jules-Prosper Legastelois, Bildhauer und Medailleur (*24.5.1855 Paris, †1931 Paris?), war Schüler von Levasseur, Carlier und Tonnellier. Legastelois wird im Pariser Salon 1928 erwähnt und war bekannt für seine Medaillen und Porträtplaketten. Lit.: TB, Bd. 22, S. 563; Neuwirth (1974), S. 66; Bénézit, Bd. 8, S. 439.*

„Solfege" ÖMKI (1902/03) Lit.: Neuwirth (1974), S. 66

Leherb (Pseud. für Leherbauer)

*Leherb (eigentlich Helmut Leherbauer), Maler und Graphiker (*14.3.1933 Wien, †28.6.1997 Wien), studierte an der Akademie für angewandte Kunst in Wien, an der Akademie in Stockholm sowie ab 1955 bei A. P. Gütersloh an der Akademie der bildenden Künste in Wien; gehörte in der Anfangszeit zum Kreis der Wiener Schule des Phantastischen Realismus und schuf für die Goldscheider Produktion in Stoob einige surreale Entwürfe. Quelle: Internet: www.aeiou.at/aeiou.encyclop.l/l375000.htm. Lit.: Firmen-Kat., Manufaktur Goldscheider präsentiert: Surrealist Leherb (Stoob/Bgld. um 1991), 12 S. (m. Abb.); AKL Index.*

„Konzert für ein goldenes Telefon" (Vasenkopf) GSG Vienna (um 1990/91) (H. 38 cm) Lit.: Firmen-Kat.: Manufaktur Goldscheider präsentiert: Surrealist Leherb (Stoob / Bgld. um 1991) (Umschlag Vorderseite, Abb.)

„Sonate für eine Träumerin" (Vasenkopf) GSG Vienna (um 1990/91) (H. 40 cm) Lit.: Firmen-Kat.: Manufaktur Goldscheider präsentiert: Surrealist Leherb (Stoob / Bgld. um 1991) S. 9, 11 (m. Abb.); Sammlung.schmutz@chello.at 53

„La Porta della Vita oder: Tür für ein imaginäres Museum" GSG Vienna (um 1990/91) (H. 310 cm) Lit.: Firmen-Kat.: Manufaktur Goldscheider präsentiert: Surrealist Leherb (Stoob / Bgld. um 1991) S. 2–8 (m. Abb.).

Lehmann, Ida (Schwetz-Lehmann, Ida)

*Ida Lehmann (Schwetz-Lehmann), Bildhauerin und Keramikerin (*26.4.1883 Wien, †26.9.1971 Wien), war die Tochter des Bildhauers Bruno Lehmann und studierte von 1904–1911 an der KGW-Schule, Wien (u.a. bei Metzner, Breitner und Powolny). Sie war 1911 Mitbegründerin der „Keramischen Werkgenossenschaft" in Wien, der auch Rosa Neuwirth angehörte. Mitglied des Österreichischen und des Deutschen Werkbundes. Im Jahr 1912 heiratete sie den Künstler Karl Schwetz. Ida Schwetz-Lehmann schuf Modelle für die Wiener Werkstätte, für Keramos, Wien sowie für Gmunden und Epiag, Pirkenhammer. Neben zahlreichen Ausstellungsbeteiligungen, vorwiegend in Wien, nahm sie 1925 an der KGW-Ausst. in Paris teil und erhielt 1926 bei der Hagenbundausstellung den Preis der Stadt Wien. In den Jahren 1952/53, also in der Zeit des Wiederaufbaues der Firma in Wien durch Walter Goldscheider, sollte auch Ida Schwetz-Lehmann, vor allem mit religiöser Thematik (Madonnen etc.), die vorwiegend für den nordamerikanischen Markt gedacht waren, zur Mitarbeit bewogen werden. Dies kam aber wegen anhaltender Differenzen die Bezahlung betreffend nur für drei Modelle zustande (Große Madonna, Kleine Madonnenbüste sowie die Figur des Hl. Franz von Assisi) (Briefwechsel zw. Walter Goldscheider und Ida Schwetz-Lehmann, vom 13. Oktober 1952 bis 27. Januar 1953). Quelle: AGM; Lit.: AGM; TB, Bd. 30, S. 385; Neuwirth (1974), S. 74, 291–292; Neuwirth (1974a), S. 52; Zühlsdorff, S. 444; Fahr-Becker, S. 234; AKL Index.*

2311 GRUPPE „Eitelkeit" ÖMKI (1912)

3815 Stehende Dame mit Blumenbukett (auf quaderförmigem blumenverziertem Sockel) (vgl. auch „Vom Garten" ÖMKI 1912)

3879 Stehende Rokokodame mit Fächer (auf Rundsockel)

3951 BÜSTE Mädchen mit Blumen geschmücktem Haar und Vogerl auf der Schulter

4131 Stehendes Biedermeiermädchen mit Blumenstrauß

4182 Stehende junge Dame mit Hut und Rosenstrauß (auf reliefiertem Ovalsockel)

4192 Stehende, lesende Biedermeierdame (vgl. auch „Lektüre" ÖMKI 1912; 1913/14)

4232 Stehende Dame mit Hut, zwei Blumentöpfe haltend (vgl. auch „Vom Garten" ÖMKI 1912)

4247 Rokokodame mit Rosen s. Meier, Emil 3890 und 4247 Entwurf von Emil Meier!

4761 „Flora" (Tanzendes Mädchen mit blumengeschmücktem Haar) (auf oktogonalem Sockel) (sign. Schwetz) (vgl. auch „Flora" ÖMKI 1910/11)

4798 „Rottraut" (Kniendes Mädchen mit erhobenen Armen) (auf oktogonalem Sockel) (sign. Schwetz)

8132 (BÜSTE) „Himmelskönigin" (Madonna mit gekreuzten Händen, mit / ohne Krone) (sign. Schwetz-Lehmann) (vgl. 8842= STEHKOPF)

8139 FIGUR „Madonna mit Kind" (Stehende Madonna mit Aureole und Jesuskind) (auf wolkenartigem Sockel)

8842 STEHKOPF Himmelskönigin" (Madonna mit Krone) (auf Rechtecksockel) (vgl. 8132= BÜSTE)

„Barock" (Fay.) ÖMKI (1910/11) Lit.: Neuwirth (1974), S. 74

„Flora" (Fay.) ÖMKI (1910/11) Lit.: Neuwirth (1974), S. 74

„Englische Dame" ÖMKI (1911/12) Lit.: Neuwirth (1974), S. 74

„Lektüre" (Fay.) ÖMKI (1912; 1913/14) Lit.: Neuwirth (1974), S. 74

„Vom Garten" ÖMKI (1912) Lit.: Neuwirth (1974), S. 74

„Erwartung" ÖMKI (1912) Lit.: Neuwirth (1974), S. 74

„Hl. Franz von Assisi" (Brief Ida Schwetz-Lehmann an Walter Goldscheider, vom 27. Januar 1953)

Leisek, Georg / Leiser

*Georg Leisek, Bildhauer (*30.6.1869 Wien, †1936 Wien), war Schüler an der Akademie in Wien unter Hellmer und Zumbusch. Er schuf neben Genre- und Tierstatuetten, Denk- und Grabmälern auch figürlich-dekorative Skulpturen für das Hauptportal des Wiener Zentralfriedhofs (1906). Lit.: TB, Bd. 22, S. 598; Bénézit, Bd. 8, S. 466; AKL Index.*

3044 Kleines Mädchen mit Puppe am Waschtrog (undeutl. sign. „Leiser"?)

Lemaire, Hector

*Hector Lemaire, Bildhauer (*15.8.1846 Lille, †Juni 1933 Paris), war Schüler von A. Dumont und A. Falguière. Lemaire erhielt 1865 den Rompreis der Stadt Lille und beschickte den Pariser Salon (Societé des Artistes Français) in den Jahren 1880–1921. Lit.: TB, Bd. 23, S. 16; Vollmer, Bd. 3, S. 208; Bénézit, Bd. 8, S. 481; AKL Index.*

815-GSB „Musique Profane"

816-GSB „Musique Sacrée"
825-GSB „La Vague"
841-GSB BUSTE „Fantaisie"

Lemaître, Eglantine (Robert-Houdin)

*Eglantine Lemaître (geb. Robert-Houdin) Bildhauerin (*6.10.1852 Blois, †26.12.1933 Blois), war spezialisiert auf die Tierplastik und beschickte als Mitglied den Pariser Salon (Societé des Artistes Français) in den Jahren 1884–1913. Vom 10.12.2005–2.4.2006 zeigte man im Chateau Royal de Blois in einer kleinen Sonderausst. etliche ihrer Werke. In dieser Werkschau wurde auch der hier aufgeführte Hund gezeigt, dessen Originalentwurf 1897 datiert (Ausführung bei Goldscheider etwa um 1905). Lit.: TB, Bd. 23, S. 19; Vollmer, Bd. 3, S. 208; Bénézit, Bd. 8; AKL Index; Ausstellungsbroschüre, Blois 2005, unpag.*

Liegender Jagdhund mit Beute (auf rechteckigem Landschaftssockel) BRONZE
„Au coup de fusil" BRONZE (Mitt. von Claire Poumés, Nov. 2005)
„Chasse Gardée" (Hund) BRONZE (Mitt. von Claire Poumés, Nov. 2005)
„Renard défendant sa proie" (Fuchs) BRONZE (Mitt. von Claire Poumés, Nov. 2005)
„Stop!" (Hund) BRONZE (Mitt. von Claire Poumés, Nov. 2005)

Lemassi

1243 FIGUR/LAMPE „Negerknabe mit Leuchter" (Stehender Negerjunge mit Kerzenhalter/Lampe) (Nordafrikaner, vor Hocker stehend) (auf Rundsockel) (vgl. 1291; 1863)
1291 FIGUR „Negerknabe mit Leuchter" (Stehender Negerjunge mit Kerzenhalter) (Nordafrikaner, vor Hocker stehend) (auf Rundsockel) (vgl. 1243; 1863) (vgl. 2000 Haniroff)
1863 FIGUR „Negerknabe mit Leuchter" (Stehender Negerjunge mit Kerzenhalter) (Nordafrikaner, vor Hocker stehend) (auf Rundsockel) (vgl. 1243; 1291)
2035 BÜSTE „Ernte" ÖMKI (1900/01) (vgl. 2315)
2150 WANDMASKE „Römischer Krieger" ÖMKI (1901/02) (vgl. 2152= BÜSTE)
2152 BÜSTE „Römischer Krieger" (Inschrift: CIVIS ROMANUS SUM) ÖMKI (1901/02) (vgl. 2150= WANDMASKE)
2315 BÜSTE „Ernte" ÖMKI (1900/01) (vgl. 2035)
„Tänzerin" ÖMKI (1902/03) Lit.: Neuwirth (1974), S. 66

Lemoine, B.(?)

596-GSB „Prémiere Flèche" (vgl. 598-GSB) (vgl. 1067; 1818)
598-GSB „Prémiere Flèche" (vgl. 596-GSB) (vgl. 1067; 1818)
684 s. Kinsburger
685 s. Kinsburger
1067 FIGUR „Pfeilschnitzer"/„Prémiere Flèche" (Sitzender Araberknabe mit Kopftuch, an einem Pfeil schnitzend) (auf getrepptem Steinsockel) (vgl. 1818) (vgl. 596-GSB; 598-GSB)
1818 FIGUR „Pfeilschnitzer"/„Prémiere Flèche" (Sitzender Araberknabe mit Kopftuch, an einem Pfeil schnitzend) (auf getrepptem Steinsockel) (vgl. 1067) (vgl. 596-GSB; 598-GSB)

Lenoir, Pierre Charles

*Pierre Charles Lenoir, Bildhauer und Medailleur (*22.5.1879 Paris, †9.9.1953 Paris), war Schüler seines Vaters Charles Lenoir und bildete sich u.a. bei A. Falguière weiter. Lenoir war Mitglied der Societé des Artistes Français, deren Salon er seit 1899 mit Porträtbüsten, Figuren und Gruppen sowie Medaillen und Plaketten beschickte. Neben seiner Mitarbeit für Friedrich Goldscheider war er auch Mitglied der Künstlergruppe „La Stèle" (–> vgl. Kap. „Dekorative Skulpturen von Arthur Goldscheider in Paris"). Lit.: Goldscheider, A. (Hrsg.): Artistic Evolution, 1926, T. 1, Pl. 35 (m. Abb.); TB, Bd. 23, S. 57; Vollmer, Bd. 3, S. 211; Neuwirth (1974), S. 66; Bénézit, Bd. 8, S. 516; AKL Index.*

1945 FIGUR „Muschelträgerin" (Stehendes Blumen geschmücktes junges Mädchen, eine große Muschel haltend (auf Ovalsockel) (vgl. 2188; 2221) (vgl. 2063= BÜSTE)
2063 BÜSTE „Felicia" (vgl. 1945; 2221= jew. FIGUR)
2188 FIGUR „Muschelträgerin" (Stehendes Blumen geschmücktes junges Mädchen, eine große Muschel haltend (auf Ovalsockel) (vgl. 1945; 2221) (vgl. 2063= BÜSTE)
2221 FIGUR „Muschelträgerin" (Stehendes Blumen geschmücktes junges Mädchen, eine große Muschel haltend (auf Ovalsockel) (vgl. 1945; 2188) (vgl. 2063= BÜSTE)
BRUNNEN „Confidences" (H. fehlt, Ø fehlt) Lit.: Goldscheider, A. (Hrsg.): Artistic Evolution, 1926, T. 1, Pl. 35 (m. Abb.)
„Dancer putting on her Sendal" (H. fehlt, B. fehlt) Lit.: Goldscheider, A. (Hrsg.): Artistic Evolution, 1926, T. 2, Pl. 3 (m. Abb.)
„Princesse lointaine" BRONZE Paris, Salon (1928) Lit.: Neuwirth (1974), S. 66

Léonard, Agathon

*Agathon Léonard, Bildhauer und Kunstgewerbler (*28.8.1841 Lille, †12.1.1923 Paris), war Schüler an der Akademie in Lille und bei Deplanche. Léonards berühmtestes Werk ist ein Tafelaufsatz mit einer Figurengruppe bestehend aus 15 Porzellanfiguren (Sèvres), die für die Pariser Weltausstellung 1900 entworfen, ausgeführt und mit einer Goldmedaille prämiert wurde. Lit.: TB, Bd. 23, S. 71; Bénézit, Bd. 8, S. 523; AKL Index.*

BÜSTE „Putto" BRONZE GS Paris (um 1895/1900) (H. ca. 15 cm); Henry's 29.10.1996

Leonard, G. (Agathon?)

Höchstwahrscheinlich handelt es sich bei diesem Entwurf auch um den Pariser Bildhauer Agathon Léonard.

„La Vampire" Paris, Salon (1905) Lit.: Neuwirth (1974), S. 66

Lepage, Céline

*Céline Lepage, Bildhauerin (*1882 Warschau, †1928 Paris), war Mitglied der Künstlergruppe „La Stèle". Sie war Mitarbeiterin bei Arthur Goldscheider, Paris und nahm an der KGW-Ausst. Paris 1925 mit der Figur „Femme de Marrakech" teil. Lit.: Goldscheider, A. (Hrsg.): Artistic Evolution, 1926, T. 1, Pl. 9 (m. Abb.); TB, Bd. 23, S. 94; Vollmer, Bd. 3, S. 213; Neuwirth (1974), S. 66; Bénézit, Bd. 8, S. 531; AKL Index.*

34-LaStèle „Femme (Mendiante) de Marrakech" (Frau/Bettlerin aus Marrakesch) (auf quadratischer Plinthe) BRONZE

Lerche s. Stoltenberg-Lerche, Hans-Stefan

Leroux s. Larroux, Antonin

Lesca

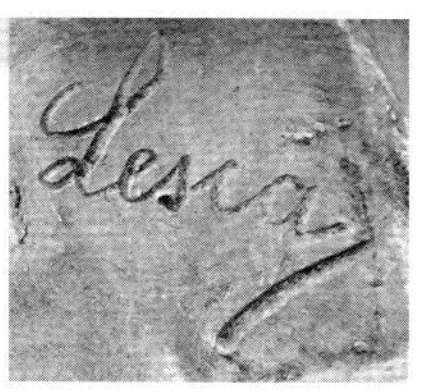

Die mit „Lesca" signierten Modelle dürfen wieder als, von Goldscheider häufig praktiziert, Pseudonym angesehen werden. Es könnte sich hierbei um einen österreichischen oder deutschen Bildhauer handeln, der sich jedoch nicht weiter nachweisen lässt. Auch bei Neuwirth (1974), S. 66 keinerlei Hinweis auf Lebens- und Schaffensdaten. Bei Pudor (1902) wird der Künstlername „Leska" genannt. Infrage kommt auch der hier im Künstlerverzeichnis an anderer Stelle aufgeführte Künstler namens „Lexa".

778 RELIEF „Victoribus" (Junge Frau im Profil, die einen Lorbeerzweig an einen Jüngling reicht) (Inschrift: VICTORIBUS) ÖMKI (1900/01)
952 RELIEF Junge Frau mit Harfe und Blumenschmuck (Wandrelief)
969 RELIEF Junge Bäuerin mit erlegtem Fasan unter Apfelbaumzweig (Wandrelief)
1299 VASE Mädchen in antikisierendem Gewand, neben einer Quelle sitzend (Vasenhals als antike Säule geformt) (große Bodenvase)
1752 BÜSTE „Bébé" (Junger Pierrot, einen Apfel haltend)
2051 WANDMASKE „Chrisis" ÖMKI (1900/01)
2284 BÜSTE „Mispel" ÖMKI (1902/03)
2316 FIGUR „Labor" (Stehende Frau mit Spaten) (Inschrift: LABOR) ÖMKI (1902/03) (vgl. 2573) (vgl. 2387= BÜSTE)
2387 BÜSTE „Labor" (Frau mit langem Haar und Kopftuch) (Sockelrelief: Frau mit Spaten) (vgl. 2316; 2573= jew. FIGUR)
2573 FIGUR „Labor" (Stehende Frau mit Spaten) (Inschrift: LABOR) ÖMKI (1902/03) (vgl. 2316) (vgl. 2387= BÜSTE)
3346 LAMPE Stehendes Mädchen (Tänzerin) in Pose
3454 „Amerika" (Junge Frau mit Fackel, vor Bison stehend)
3455 Stehende Frau mit Kopftuch und Krug (auf Ovalsockel)
3655 GRUPPE „Pierrot und Pierrette" (auf einer Bank sitzend) (auf Ovalsockel) (vgl. 5016; 5327)
3841 Stehendes Mädchen mit Harke (Pendant vgl. 3842)
3842 Stehendes Mädchen eine Garbe tragend (Pendant vgl. 3841)
3868 Fischer mit Netz, Paddel und Krug (auf Rechtecksockel)
5016 GRUPPE/LAMPE „Pierrot und Pierrette" (auf einer Bank sitzend) (auf Ovalsockel) (vgl. 3655; 5327)
5327 „Pierrot und Pierrette" (auf einer Bank sitzend) (auf Ovalsockel) (vgl. 3655; 5016)
„Sisyphus" (Auf Felssockel stehender Jüngling, einen Felsbrocken geschultert) (1902) Lit.: Pudor (1902), S. 36 (m. Abb.)

Leska s. Lesca

Levasseur, Henri-Louis (Le Vasseur)

*Henri-Louis Levasseur, Bildhauer und Kunstgewerbler (*16.4.1853 Paris, †3.11.1934 Paris), war Schüler von Dumont, Thomas und Deplanche. Levasseur war Mitglied der Societé des Artistes Français. Lit.: TB, Bd. 23, S. 149; Neuwirth (1974), S. 66; Hillenbrand, S. 247; Bénézit, Bd. 8, S. 590; AKL Index.*

835-GSB „Diane" No.1; No. 2; No. 3
840-GSB „Déesse de L' Électricité" No 1 (vgl. 921-GSB)
863-GSB „Le Rameaux" No. 1; No. 2; No. 3
871-GSB „Légende des Ruines"
921-GSB „Déesse de L' Électricité" No. 2 BRONZE/Marmorsockel; BRONZE/Elfenbein (vgl. 840-GSB)
JARDINIERE „Surtout" Paris, Salon (1901) Lit.: Neuwirth (1974), S. 66
„Blanche de Castille" Paris, Salon (1905) Lit.: Neuwirth (1974), S. 66

Lévèque, (Auguste Maurice)

*Die Angabe des Künstlers (Auguste-Maurice) Lévèque beruht auf einem Exemplar der Modell-Nr. 1101, das mit einem entsprechenden Metallschild versehen ist und bei Richemond (1999), S. 78, belegt wird. Nachweisen lässt sich aber nur ein belgischer Maler, Bildhauer und Dichter namens Auguste Levêque (*1.3.1864, Nivelles, †22.2.1921, Brüssel). Ob dieser wirklich infrage kommt, steht zu bezweifeln. Möglicherweise handelt es sich um ein Pseudonym, welches Goldscheider kreiert hat. Lit.: TB, Bd. 23, S. 153; Richemond (1999), S. 78; AKL Index.*

684-GSB BUSTE „Esclave" (vgl. 733-GSB) (vgl. 1101; 1338)
733-GSB BUSTE „Esclave" (vgl. 684-GSB) (vgl. 1101; 1338)
766 BÜSTE Junger Afrikaner mit entblößtem Oberkörper und Halsschmuck (Pendant vgl. 768) (Lévèque zugeschr.)
767 BÜSTE Junge Afrikanerin mit entblößtem Oberkörper und Halsschmuck (Pendant vgl. 767) (Lévèque zugeschr.)
1101 BÜSTE „Sklavin"/„Bedouine" (Sklavin, Araberin mit großem Turban) (nach links blickend) (vgl. Thiele 1101; 1338) (Pendant vgl. 1105; 1339) (vgl. 684-GSB; 733-GSB)
1105 BÜSTE „Araber" (mit Turban) (nach rechts blickend) (vgl. 1339; 1899) (Pendant vgl. 1101)
1338 BÜSTE „Sklavin"/„Bedouine" (Sklavin, Araberin mit großem Turban) (nach links blickend) (vgl. Thiele 1101; 1338)
1339 BÜSTE „Araber" (mit Turban) (nach rechts blickend) (vgl. 1105; 1899) (Pendant vgl. 1101)
1341 BÜSTE „Mohrin" (mit Turban) (nach rechts blickend) (auch sign. Thiele)
1877 s. Thiele
1899 BÜSTE „Beduine" (Araber mit Turban) (nach rechts blickend) (vgl. 1105; 1339) (Pendant vgl. 1898)

Lexa, H. (V.?)

H. (V.?) Lexa (Lebens- und Schaffensdaten unbekannt). Ob es sich bei diesem Künstler um ein Pseudonym handelt ist wahrscheinlich, jedoch nicht so offensichtlich wie bei anderen Entwerfern. Auch könnte der Name mit dem hier im Künstlerverzeichnis an anderer Stelle aufgeführten „Lesca" identisch sein. Konkrete Belege für diese Vermutung, gibt es jedoch nicht. Lit.: Neuwirth (1974), S. 66

2401 FIGUR „Erdarbeiter" ÖMKI (1910/11)
„La Gloire" ÖMKI (1903/04) Lit.: Neuwirth (1974), S. 66
„Fischer" (Fay.) ÖMKI (1910/11) Lit.: Neuwirth (1974), S. 66
„Pierrot und Pierrette" = Lesca 3655 Lit.: MW, 2. 1920/21,

Nr. 12, S. 2 (m. Abb.), hier mit Ang. des Originaltitels „Pierrot und Pierrette" und Künstlerang. „V. Lexa"; DD, 50.1922/23, Nr. 10, S. 30; Neuwirth (1974), S. 66. „Fechterin" BRONZE Lit.: Neuwirth (1974), S. 66; MW, 2. 1920/21, Nr. 12, S. 12 (m. Abb.), hier mit Ang. des Originaltit. „Fechterin" und Künstlerang. „V. Lexa".

Lexa, H. (V ?) vgl. Lesca

Lille

BÜSTE Mädchen ÖMKI (1900/01) Lit.: Neuwirth (1974), S. 67

Lindner, Helmut

Bei Neuwirth (1974) auf S. 67 nur die Namensnennung, jedoch keinerlei Hinweis auf Lebens- und Schaffensdaten.
4090 Tänzerin
4511 FIGUR/LAMPE Junge, sich verneigende Biedermeierdame mit drei Rosen (vgl. 4712; 5491)
4573 Diana mit Hund (Stehender Akt mit Köcher, Ausschau haltend) (auf hohem Ovalsockel)
4609 Tänzerin (in drehender Bewegung) (auf Rundsockel) (vgl. 4690; 4749)
4690 Tänzerin (in drehender Bewegung) (auf Rundsockel) (vgl. 4609; 4749)
4712 FIGUR/LAMPE Junge, sich verneigende Biedermeierdame mit drei Rosen (vgl. 4511)
4749 Tänzerin (in drehender Bewegung) (auf Rundsockel) (vgl. 4609; 4690)
4890 Tänzerin
5491 Junge, sich verneigende Biedermeierdame mit drei Rosen (vgl. 4511; 4712)
5616 s. Lorenzl 5616

Linossier, Claudius

*Claudius Linossier, Kunstgewerbler, Silberschmied, Graveur, Metallkünstler (*21.11.1893 Lyon, †8.4.1953 Lyon), war u.a. Mitglied der Künstlergruppe „Évolution" (-> vgl. Kap. „Dekorative Skulpturen von Arthur Goldscheider in Paris"). Linossier erhielt 1928 den Großen Preis von Athen und 1929 von Barcelona. Lit.: Goldscheider, A. (Hrsg.): Artistic Evolution, 1926, T. 1, Pl. 1–2 (m. Abb.); Vollmer, Bd. 3, S. 241; Neuwirth (1974), S. 67; Bénézit, Bd. 8, S. 693; AKL Index.*
Pendule HOLZ/MESSING (für die Figur „Atlas" von Sybille May) (H. 15 cm, B. 47 cm) Lit.: Goldscheider, A. (Hrsg.): Artistic Evolution, 1926, T. 1, Pl. 7 (m. Abb.)

Löw, Jakob

*Jakob Löw, Bildhauer (*4.5.1887 Stanislau/Galizien, †?), war bis 1909 an der Fachschule in Zakopane, anschließend bis 1915 an der KGW-Schule, Wien bei Strnad und Barwig. Löw stellte bei der Weihnachtsschau im Wiener Künstlerhaus 1928 mehrere für Goldscheider entworfene Modelle aus. Seine Spur verliert sich 1939, wo er letztmals erwähnt wird. Lit.: TB, Bd. 23, S. 325; Neuwirth (1974), S. 67; Bénézit, Bd. 8, S. 830; AKL Index.*

Loiseau-Rousseau, Paul Louis Émile

*Paul Louis Émile Loiseau-Rousseau, Bildhauer und Radierer (*20.4.1861 Paris, †1927 Paris), war Schüler von Th. Barrau. Loiseau-Rousseau stellte ab 1891 im Pariser Salon Reliefs und Statuen, Frauenbüsten und Statuetten in Bronze oder Elfenbein sowie Tiergruppen aus. Lit.: TB, Bd. 23, S. 335; Vollmer, Bd. 3, S. 255; Bénézit, Bd. 8, S. 761; AKL Index.*
861-GSB VIDE-POCHE „Repos du Coquillage"
874-GSB VASE „Aquatique"
875-GSB BUSTE „Libellule"

Long

JARDINIERE „Gänse" ÖMKI (1906/07) Lit.: Neuwirth (1974), S. 67

Longer

2222 „Bei der Toilette" (s. Langer, Leopold?) Lit.: B., F.: Keramische Wanderungen durch den Kunstsalon Keller & Reiner Berlin, in: KM, 3.1903, S. 128, „Vor dem Spiegel, der all ihre Reize zurückgibt, steckt sie sich Blüten, die sie von dem Rahmen aufspriessenden Gerank abgebrochen hat, in ihr üppiges Haargewoge. Ihre hingeschmiegten Glieder sind von plastischer Schönheit. Ihre siegesbewußte Haltung verrät den Triumph, den sie darüber empfindet".

Longer vgl. Louger

Longert, F.

Bei Neuwirth (1974) auf S. 67 nur die Namensnennung, jedoch keinerlei Hinweis auf Lebens- und Schaffensdaten.
„Reifenspielerin" ÖMKI (1903/04) Lit.: Neuwirth (1974), S. 67
„Holländische Kugelträgerin" ÖMKI (1903/04) Lit.: Neuwirth (1974), S. 67

Lorenz = Lorenzl

Lorenzl, Josef

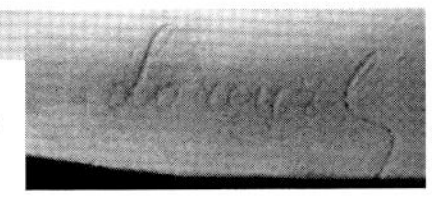

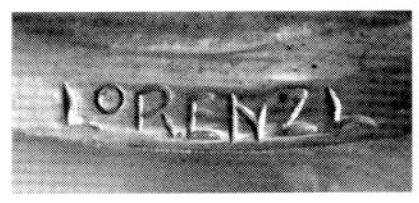

*Josef Lorenzl, Bildhauer (*1.9.1892 Wien, †15.8.1950 Wien). Lorenzl ist nicht nur der Schöpfer eines der bekanntesten, erfolgreichsten und spektakulärsten Modelle der Goldscheiderproduktion, dem „Gefangenen Vogel", sondern lieferte der Manufaktur zahlreiche Modelle. Weitere keramische Arbeiten schuf Lorenzl für die Firma Keramos, die Wiener Kunstkeramik und Porzellanmanufaktur AG, Wien sowie die deutschen Firmen Hertwig und Cortendorf (-> vgl. Schaulade, 13.1937, S. 69 (m. Abb.). Zu seinen Spezialitäten zählten delikate Bronzen (Frauen- und Mädchenakte), die teils durch Elfenbein ergänzt wurden. Diese Bronzen, die bestes Art Déco darstellen, sind unter Sammlern, neben den Figuren für Goldscheider, sehr gesucht. In der Tantiemenabrechnung vom 10. März 1953, in der die Modellnummern 5917, 8827–8829 und 8909 letztmalig genannt werden, verfügt Anna (Njura) Lorenzl, die Frau des Bildhauers, die „übrigen Modelle meines Gatten ausnahmslos zu vernichten". Zu Lorenzls Wiener Wohnungen und zum Pseudonym „Althan" -> vgl. dort. Quellen u. Lit.: AGM, (Gehaltslisten, Mai und August 1938; Tantiemenabrechnung 10. März 1953); Neuwirth (1974), S. 67; Deutsche Elfenbeinkunst vom Fin de Siècle bis zum Art Déco, Abb. 20; Zühlsdorff, S. 425; Hillenbrand, S. 258; Makus (2000), S. 174; Quelle: Bildhauerarchiv, Karl Novak.*
(2285 MASKE „Beethoven" vgl. Bem. im WVZ)
2520 Orientalische Tänzerin mit Kappe und Blumenschmuck im Haar (auf profiliertem Rechtecksockel) (vgl. Bem. im WVZ)
4619 vgl. Meier 4619
5014 Stehender Putto, in ein Horn blasend (auf oktogonalem Blumensockel)
5015 FIGUR/LAMPE Sitzende Tänzerin beim Schuhe schnüren (auf blumenverziertem Rundsockel)
5018 Odaliske (Haremstänzerin mit langem Tuch) (auf facettiertem Ovalsockel) (vgl. 5613; 6063)
5022 GRUPPE/LAMPE Pierrot mit zwei Pierretten (auf vierfüßigem Rundsockel)
5054 „Gefangener Vogel" (Der Gefangene Vogel, Niddy Impekoven) BRONZE/Marmorsockel (auf Ovalsockel) (vgl. 5230; 5960; 5964= Pendant zu 5960; 6204; 6600; 8176) (vgl. 5530= Trio GRUPPE)
5057 GRUPPE/LAMPE Stehender Pierrot mit Laute und Pierrette (als Kinder) (auf Rundsockel)
5058 „Erna Pinner-Puppe" (mit Laute) (auf facettiertem, Blumengirlanden verziertem Rundsockel) (Niddy Impekoven)
5085 „Wilder Tanz" (Tänzerin auf einem Bein) (auf oktogonal facettiertem Sockel) BRONZE/Marmorsockel
5087 Stehende Rokokodame mit Handspiegel (auf Rundsockel)
5088 FIGUR/LAMPE „Carmen" (Stehende Spanierin mit Fächer) (auf Rundsockel) (vgl. 5232; 5277) (vgl. 5534= sitzende Figur)
5092 Stehende Pierrette mit Laute (auf oktogonal facettiertem Sockel)
5178 FIGUR/LAMPE Tänzerin in Pose (mit federverziertem Hut) (auf facettiertem Rundsockel)
5190 Kniende Geisha (sich schminkend) (auf quaderförmigem Sockel)
5193 Sitzender Till Eulenspiegel mit Eule (vgl. 5509)
5196 „Eine kleine Prise" (Schnupfender Kavalier im Rokokokostüm) (auf Rundsockel)
5230 FIGUR/LAMPE „Gefangener Vogel" (Der Gefangene Vogel, Niddy Impekoven) (auf Ovalsockel) (vgl. 5054 (BRONZE); 5960; 5964= Pendant zu 5960; 6204; 6600; 8176) (vgl. 5530= Trio GRUPPE) (Pendant vgl. 5434) GS neu Modell-Nr. 82 „Gefangener Vogel"
5232 „Carmen" (Stehende Spanierin mit Fächer) (auf Rundsockel) (vgl. 5088; 5277) (vgl. 5534= sitzende Figur)
5266 Kartenlegerin (Sitzende Zigeunerin mit Spielkarten)
5267 FIGUR/LAMPE Orientalische Tänzerin in Pose (auf Rundsockel) (vgl. 6023)
5277 „Carmen" (Stehende Spanierin mit Fächer) (auf Rundsockel) (vgl. 5088; 5232) (vgl. 5534= sitzende Figur)
5281 FIGUR/LAMPE „Aida" (Stehende Odaliske) (auf Ovalsockel) (vgl. 6021) (vgl. ALABASTER-Büste Sebök 52)
5282 Stehende junge Tänzerin mit Trommel (auf Rundsockel)
5296 „Herein spaziert!" (Posierender Harlekin mit Akkordeon) (auf Rundsockel)
5316 Stehende, nach vorne gebeugte Pierrette (auf Rundsockel)
5385 BÜSTE Mädchen mit langen Haaren ALABASTER
5397 FIGUR/LAMPE Stehende Tänzerin in russischer Tracht (Mantel und Pelzkappe) (auf facettiertem Rundsockel)
5406 Stehende Sängerin mit Kopftuch und langen Handschuhen (auf Ovalsockel)
5434 Tänzerin (mit breitem Haarband) im Schmetterlingskostüm (nach rechts schreitend) (auf Ovalsockel) (Pendant vgl. 5230)
5435 Tänzerin in Fantasiekostüm mit Kopfputz (auf facettiertem Rundsockel)
5447 Tänzerin (auf einem Bein) im Schmetterlingskostüm (BRONZE/Marmorsockel)
5486 FIGUR/LAMPE Stehende junge Dame in Karnevalskostüm (mit Dreispitz)
5500 FIGUR/LAMPE Stehende Haremstänzerin mit großem Kopfputz (mit ausgestreckten Armen, ihr Kleid auseinander haltend) (Dolores del Rio) (auf Rundsockel)
5501 FIGUR/LAMPE Sitzende Rokokodame mit Buch und Papagei
5504 DECKELDOSE „Akt auf einer Dose" (Kniender Damenakt)
5508 LAMPE Zwei Pierrots mit einer Pierrette (auf facettiertem Rundsockel)
5509 FIGUR/LAMPE Sitzender Till Eulenspiegel mit Eule (vgl. 5193)
5523 Stehende Varietetänzerin mit Zylinder (auf facettiertem Rundsockel)
5525 Schreitende Sklavin, eine Amphore/Henkelkrug auf dem Kopf tragend (auf getrepptem Rechtecksockel)
5528 KERZENHALTER (dreiflammig) Putto mit Blumengirlanden (auf dreifach getrepptem Quadratsockel)
5530 GRUPPE/LAMPE „Gefangener Vogel" (Der Gefangene Vogel, Niddy Impekoven) (als Trio auf Rundsockel) (vgl. jeweils Einzelfigur 5054 (BRONZE); 5960; 5964= Pendant zu 5960; 6204; 6600; 8176) (Pendant vgl. 5434) GS neu Modell-Nr. 82 „Gefangener Vogel"
5532 FIGUR/LAMPE Stehende Pierrette mit Maske (auf Rundsockel) FAYENCE/ALABASTER (auch sign. Dakon)
5534 Carmen (Sitzende Spanierin mit Fächer) (auf halbrundem Sockel) (vgl. 5555; 6666) (vgl. 5088; 5232; 5277= jeweils stehende Figur)
5535 Stehende Haremsdame mit federverziertem Turban (auf Ovalsockel)
5536 Sitzende Haremsdame mit reich geschmückter Kopfbedeckung/Turban (auf podestartigem Sockel, im Schneidersitz)
5537 FIGUR/LAMPE Stehender „Mephisto"
5543 FIGUR/LAMPE Tänzerin mit Kopfputz und großem Schal (auf Ovalsockel)
5547 Stehender Mädchenakt mit Tuch (auf quaderförmigem Sockel) ALABASTER/Marmorsockel
5548 LAMPE Sitzende Spanierin mit Hut und Gitarre (auf halbrundem Sockel)
5555 Carmen (Sitzende Spanierin mit Fächer) (auf halbrundem Sockel) (vgl. 5534; 6666) (vgl. 5088; 5232; 5277= jeweils stehende Figur)
5556 Stehender singender Harlekin mit Akkordeon (auf Rundsockel)
5558 Stehender Harlekin mit Handpuppe (auf Rechtecksockel)
5568 Boxer
5569 GRUPPE/LAMPE „Katta Sterna und Maria Solveg" (hintereinander stehend, mit ausgestreckten Armen) (auf oktogonalem, gekehltem Sockel) (vgl. 5600= Einzelfigur)
5573 Posierende Varietékünstlerin (auf Sockel: Trommel, Saxophon und Tamburin)
5580 Stehende Tänzerin (mit Biedermeierfrisur), ihr langes Kleid anhebend
5581 Kniender Damenakt (Trauben essend) (auf Oval-

sockel)
5582 FIGUR / LAMPE Sitzende Tänzerin im Schmetterlingskleid (auf Rundsockel)
5583 Stehender Ikarus (auf sechseckigem Sockel)
5600 „Katta Sterna" (Einzelfigur stehend, mit ausgestrecktem Arm) (auf oktogonalem, gekehltem Sockel) (vgl. 5569= GRUPPE)
5613 Odaliske (Haremstänzerin mit langem Tuch) (auf facettiertem Ovalsockel) (vgl. 5018; 6063)
5616 „Dolores del Rio als Carmen" (Stehende Spanierin mit Fächer) (auf oktogonalem, gekehltem Sockel) (mit / ohne Inschrift: DOLORES DEL RIO) (vgl. 6208)
5660 Stehende Rokokodame mit Handspiegel (auf Ovalsockel)
5678 ASCHER / RAUCHERGARNITUR „Aschenboy" (Page mit zylindrischem Gefäß für Zigaretten, auf einem Aschenbecher mit vier Ablagen stehend)
5715 Stehende Tänzerin im Schmetterlingskleid (auf Rundsockel)
5724 Schreitender Amor (auf Natursockel)
5726 Stehendes Rokokopaar (auf Ovalsockel)
5732 „Clotilde von Derp" (Schreitende Tänzerin in Kniehosen) (auf Ovalsockel) (vgl. 6248)
5774 FIGUR / LAMPE „Maria Jeritza" (Kammersängerin Maria Jeritza als stehende Figur auf reliefiertem Rundsockel)
5775 GRUPPE „Spanischer Tanz" (Tanzendes Paar in spanischen Kostümen) (auf Ovalsockel) (vgl. 5888= weibliche Einzelfigur)
5780 „Pyjama" (Junge Dame, rittlings auf einem Stuhl sitzend) (auf Rechtecksockel)
5797 Sitzender Damenakt (auf Ovalsockel)
5802 FIGUR „Kokettchen" (Stehender Damenakt mit ausgestreckten Armen) (auf Rundsockel) (vgl. 7227) (vgl. Dakon 5907= fast ident. Modell)
5822 „Spanischer Shawl" (Stehende Tänzerin (Dolores del Rio?) in Pose, mit großem Schal) (auf Rundsockel) (vgl. 6062; 6345)
5839 Stehendes Mädchen in Hosenanzug (auf Rechtecksockel) ALABASTER
5870 „Fächerdame" (Stehende junge Dame in japanischem Kostüm, mit erhobener Hand und großem Fächer) (auf Rundsockel)
5871 GRUPPE Stehende Tänzerin in fächerartig ausgestelltem Kleid, hinter ihr ein Mann mit Geige (auf Rundsockel) (vgl. 5875= ähnl. Modell)
5872 Stehende Tänzerin, ihren langen Rock schmetterlingsflügelartig auseinander haltend (auf Rundsockel) (vgl. 6173) (vgl. 6399= fast ident. Modell)
5875 GRUPPE Stehende Spanierin in fächerartig ausgestelltem Kleid, hinter ihr ein Spanier mit Gitarre (auf Rundsockel) (vgl. 5871= ähnl. Modell)
5882 Stehendes Kind mit Ball und Schubkarren ALABASTER
5888 „Spanischer Tanz" (Spanierin in Tanzpose) (auf Ovalsockel) (vgl. 5775= GRUPPE)
5900 Tänzerin im Schmetterlingskostüm (sich drehend) (auf Rundsockel) (vgl. 6206)
5902 Posierende (tanzende) junge Dame in knielanger Hose, Rüschenbluse und Dreispitz (auf Ovalsockel)
5911 Sitzende spanische Tänzerin, ihren Kopf auf die linke Hand aufgestützt (Carmen?) (auf oktogonalem Sockel)
5912 Stehende Tänzerin in fächerartig ausgestelltem Kleid, Gitarre spielend (auf Rundsockel)
5917 FIGUR / LAMPE Schreitende Tänzerin im Schmetterlingskostüm (ihren Kopf auf die rechte Schulter gelegt) (auf sechseckigem Sockel)
5921 Stehender Till Eulenspiegel mit Laute (auf oktogonalem, getrepptem Sockel)
5923 Stehende kleine Spanierin (Betty Boop) (auf Rundsockel)
5927 Stehende Haremstänzerin (mit gesenktem Kopf) (auf an den Ecken abgeschrägtem Quadratsockel) (vgl. 6245)
5931 Auf Stuhl sitzende junge Dame in kurzen Hosen, ihre Hände keck auf den Knien abstützend (auf abgeschrägtem Quadratsockel)
5939 GRUPPE Harlekin, seiner Columbine die Hand küssend (auf bogenartig facettiertem Rundsockel)
5940 Schreitende Dame in kurzem Kleid mit Schleppe (auf Ovalsockel)
5960 FIGUR / LAMPE „Gefangener Vogel" (Der Gefangene Vogel, Niddy Impekoven, nach links gewandt) (auf Ovalsockel) (Pendant, nach rechts gewandt, vgl. 5964) (vgl. 5054 (BRONZE); 5230; 6204; 6600; 8176) (vgl. 5530= Trio GRUPPE)
5963 Tänzerin in Haremskostüm (hockend, ihr linkes Bein nach hinten gestreckt) (auf Rechtecksockel, mit abgeschrägten Ecken)
5964 FIGUR „Gefangener Vogel" (Der Gefangene Vogel, Niddy Impekoven, nach rechts gewandt) (auf Ovalsockel) (Pendant, nach links gewandt, vgl. 5960) (vgl. 5054 (BRONZE); 5230; 6204; 6600; 8176) (vgl. 5530= Trio GRUPPE)
6003 Tänzerin, auf dem rechten Bein tänzelnd, das linke Bein nach vorne schwingend (auf Rautensockel) (vgl. 6323; 6935)
6004 Schreitender Till Eulenspiegel, die Eule auf dem ausgestreckten Arm haltend (auf getrepptem, hexagonalem Sockel)
6005 Tänzerin (mit Bikini und großem Tuch) auf dem linken Bein tänzelnd, das rechte Bein angewinkelt (auf Rautensockel)
6006 Stehende Japanerin, ein Lampion in die Höhe haltend (auf Quadratsockel)
6021 FIGUR / LAMPE „Aida" (Stehende Odaliske) (auf Ovalsockel) (vgl. 5281)
6023 Orientalische Tänzerin in Pose (auf Rundsockel) (vgl. 5267)
6062 „Spanischer Shawl" (Stehende Tänzerin (Dolores del Rio?) in Pose, mit großem Schal) (auf Rundsockel) (vgl. 5822; 6345)
6063 Odaliske (Haremstänzerin mit langem Tuch) (auf facettiertem Ovalsockel) (vgl. 5018; 5613)
6069 GRUPPE Stehendes Kind, einen sitzenden Schäferhund umarmend (Modell: Lorenzl / Postl) (vgl. 5719= Einzelfigur Hund, Modell von Postl)
6154 Sitzende Dame mit zwei Hunden
6173 Stehende Tänzerin, ihren langen Rock schmetterlingsflügelartig auseinander haltend (auf Rundsockel) (vgl. 5872) (vgl. 6399= fast ident. Modell)
6204 „Gefangener Vogel" (Der Gefangene Vogel, Niddy Impekoven) (auf Ovalsockel) (vgl. 5054 (BRONZE); 5230; 5960; 5964= Pendant zu 5960; 6600; 8176) (vgl. 5530= Trio GRUPPE)
6206 Tänzerin im Schmetterlingskostüm (sich drehend) (auf Rundsockel) (vgl. 5900)
6208 „Dolores del Rio als Carmen" (Stehende Spanierin mit Fächer) (auf oktogonalem, gekehltem Sockel) (vgl. 5616)
6227 Schreitende Tänzerin (etwas nach vorne gebückt), ihr Kleid schmetterlingsflügelartig auseinander haltend (auf Ovalsockel) (vgl. 6693)
6229 Stehende junge Dame mit großem Hut (in Pose) (auf Rundsockel)
6245 Stehende Haremstänzerin (mit gesenktem Kopf) (auf an den Ecken abgeschrägtem Quadratsockel) (vgl. 5927)
6248 „Clotilde von Derp" (Schreitende Tänzerin in Kniehosen) (auf Ovalsockel) (vgl. 5732)
6295 Kniender Mädchenakt mit trinkendem Rehkitz (auf ovalem Landschaftssockel)
6323 Tänzerin, auf dem rechten Bein tänzelnd, das linke Bein nach vorne schwingend (auf Rautensockel) (vgl. 6003; 6935)
6345 „Spanischer Shawl" (Stehende Tänzerin (Dolores del Rio?) in Pose, mit großem Schal) (auf Rundsockel) (vgl. 5822; 6062)
6346 Schreitende Tänzerin, ihr Kleid nach vorne hoch haltend (auf Ovalsockel)
6356 Stehende junge Dame in japanischem Kostüm, mit Blumen geschmücktem Haar und großem Fächer) (auf Rundsockel)
6394 Tänzerin im Schmetterlingskostüm (in Pose) (auf Ovalsockel)
6399 Stehende Tänzerin, ihren langen Rock schmetterlingsflügelartig auseinander haltend (auf Rundsockel) (vgl. 5872; 6173= jeweils fast ident. Modell)
6491 Stehende junge Tänzerin in Pose (auf Rundsockel)
6492 Stehende Tänzerin in Pose, ihr Kleid mit der rechten Hand zurückhaltend (auf Ovalsockel) (vgl. 7732) (vgl. 7193= ähnl. Modell)
6535 „Rosenkavalier" (Octavian als schreitende Figur mit Rose in der rechten Hand) (diverse ovale Sockelformen) (vgl. 6602)
6536 Sitzende elegante junge Dame mit Hut und ausgestrecktem Bein (auf Ovalsockel)
6555 s. 6535
6557 Stehende Dame in venezianischem Kostüm mit sitzendem Windhund (auf Ovalsockel)
6590 Sitzendes Mädchen in langem Kleid mit liegendem Barsoi (auf Ovalsockel)
6600 „Gefangener Vogel" (Der Gefangene Vogel, Niddy Impekoven) (auf Ovalsockel) (vgl. 5054 (BRONZE); 5230; 5960; 5964= Pendant zu 5960; 6204; 8176) (vgl. 5530= Trio GRUPPE)
6602 „Rosenkavalier" (Octavian als schreitende Figur mit Rose in der rechten Hand) (auf Ovalsockel) (vgl. 6535)
6666 Carmen (Sitzende Spanierin mit Fächer) (auf halbrundem Sockel) (vgl. 5534; 5555) (vgl. 5088; 5232; 5277= jeweils stehende Figur)
6672 Stehende, nach vorne gebeugte Rokokodame (auf Ovalsockel)
6690 GRUPPE Biedermeierpaar
6693 Schreitende Tänzerin (etwas nach vorne gebückt), ihr Kleid schmetterlingsflügelartig auseinander haltend (auf Ovalsockel) (vgl. 6227)
6726 GRUPPE Galantes Paar (auf Ovalsockel)
6745 BÜSTE Madonna mit gefalteten Händen (vgl. 7052; 7736) (vgl. 6923= WANDMASKE)
6746 Stehender Damenakt mit großem Tuch (auf gekehltem, geschwungenem Rechtecksockel)
6753 Tänzerin mit Tamburin (La Jana) (auf Ovalsockel) (vgl. 6759)
6759 Tänzerin mit Tamburin (La Jana) (auf Ovalsockel) (vgl. 6753)
6760 Stehender Damenakt mit großem Tuch (auf Rundsockel)
6765 FIGUR / GRUPPE Tänzerin in Pose und liegender Pierrot mit Laute (auf Ovalsockel) (als EINZELFIGUR für die Tänzerin ebenfalls unter Modell-Nr. 6765)
6812 Stehende junge Dame in langem Kleid, Handschuhen und Pekinese (auf Ovalsockel)
6862 Stehende Madonna mit gekreuzten Händen (auf Rundsockel) (vgl. 7735)
6923 WANDMASKE Madonna (vgl. 6745; 7052; 7736= jeweils BÜSTE)
6931 Liegender Damenakt (auf Ovalsockel)
6932 Stehender Damenakt mit ausgestrecktem Arm und großem Tuch (auf Rechtecksockel) (vgl. 7525= Figur ohne Tuch)
6933 Schreitende Dame mit Cape, Hut und Muff (auf Ovalsockel)
6935 Tänzerin, auf dem rechten Bein tänzelnd, das linke Bein nach vorne schwingend (auf Rautensockel) (vgl. 6003; 6323)
7037 „Eifersucht" (Stehende Dame mit Hut und zwei Pekinesen) (auf Rundsockel) (vgl. 7731)
7041 GRUPPE Stehender Damenakt mit sitzendem Schäferhund (auf Rundsockel) (Modell: Lorenzl / Postl) (vgl. 5719= Einzelfigur Hund, Modell von Postl)
7042 Stehende Dame mit Hut, Fuchsstola, Handtasche und Foxterrier (auf Rundsockel)
7051 BÜSTE Flehende Frau mit gefalteten Händen
7052 BÜSTE Madonna mit gefalteten Händen (vgl. 6745; 7736) (vgl. 6923= WANDMASKE)
7053 Tänzerin, das linke Bein abgewinkelt nach oben schwingend (ihren langen Rock mit beiden Händen auseinander haltend) (auf Rautensockel)
7058 Schreitende Tänzerin in Pose (mit Bikinioberteil und langem gerüschtem Rock) (auf Ovalsockel)
7064 Reisende (Stehende junge Dame mit Hut, zwei Koffern und Hutschachtel)
7192 FIGUR „Revue" (Stehende Tänzerin, ihr Kleid auseinander haltend) (auf Ovalsockel) (vgl. 7396)
7193 Stehende Tänzerin in Pose, ihr Kleid mit der rechten Hand zurückhaltend (auf Ovalsockel) (vgl. 6492; 7732= jeweils ähnl. Modell)
7195 Tänzerin (falsche Angabe bei Neuwirth (1974), S. 67, richtig ist Dakon 7195)
7224 „Abendkleid" (Stehende elegante Dame mit Fächer)
7225 „Gaby" (Sich verbeugende Rokokodame) (auf Ovalsockel) (Lorenzl zugeschr.)
7227 FIGUR „Kokettchen" (Stehender Damenakt mit ausgestreckten Armen) (auf Rundsockel) (vgl. 5802) (vgl. Dakon 5907= fast ident. Modell)
7230 Tänzerin in Bewegung (Halbakt mit langem Kleid) (auf Ovalsockel)
7396 FIGUR „Revue" (Stehende Tänzerin, ihr Kleid auseinander haltend) (auf Ovalsockel) (vgl. 7192)
7424 Stehende Dame mit Hut, Cape (großer Fransen-

schal), einen kleinen Hund (King Charles?) haltend (auf Rundsockel) (-> vgl. Dakon 7424, Entwurf von Stephan Dakon)
7454 Schreitende Dame, ihr Kleid mit der linken Hand zurückhaltend (auf Ovalsockel)
7463 Stehender Damenakt mit Barsoi (Marussia und Iwan) (auf Ovalsockel)
7502 FIGUR / LAMPE Stehende schlanke Dame mit Puderdose
7525 Stehender Damenakt mit ausgestrecktem Arm (auf abgeschrägtem Quadratsockel bzw. Rundsockel) (vgl. 7724= GRUPPE) (vgl. 6923= Figur mit Tuch)
7544 FIGUR „Akt mit Fächer" (Stehender Damenakt, einen großen Fächer haltend) (auf Rundsockel) (auch für GS USA, Mod.-Nr. 899)
7581 „Liane" (Stehende Tänzerin mit Hut, ihr Kleid schmetterlingsflügelartig auseinander haltend) (auf abgerundetem rautenförmigem Sockel) (vgl. 8321) (vgl. Dakon 7257= ähnl. Modell)
7582 „Sonate" (Schreitende Tänzerin mit Bikinioberteil, ihren Rock schmetterlingsflügelartig auseinander haltend) (auf Dreiecksockel) (vgl. 8459)
7638 Stehender Mädchenakt mit Tuch / Schal (auf Rundsockel)
7639 „Sibille" (Stehende Tänzerin in Pose, mit nacktem Oberkörper, ihren langen Rock auseinander haltend) (auf Ovalsockel)
7640 „Falter" Stehende nackte Tänzerin mit Schmetterlingskostüm (auf abgerundetem Rautensockel)
7641 Sitzender Damenakt (auf Ovalsockel)
7642 Stehende junge Dame mit Hut und Sonnenschirm
7688 Stehender Damenakt, mit zwei großen Fächer posierend (auf Ovalsockel)
7689 Stehende Tänzerin in Pose, mit erhobenem Arm (auf Rundsockel)
7690 RELIEF Betende Madonna (im Profil nach rechts) (Wandrelief)
7721 GRUPPE Stehende Rokokodame mit Federfächer und Windhund (Windspiel) (auf Ovalsockel)
7722 GRUPPE Stehender Damenakt mit großem Fächer, Hut und Barsoi (auf Ovalsockel)
7724 GRUPPE Stehender Damenakt mit ausgestrecktem Arm und Barsoi (auf Ovalsockel) (vgl. 7524= FIGUR Damenakt)
7731 „Eifersucht" (Stehende Dame mit Hut und zwei Pekinesen) (auf Rundsockel) (vgl. 7031)
7732 Stehende Tänzerin in Pose, ihr Kleid mit der rechten Hand zurückhaltend (auf Ovalsockel) (vgl. 6492) (vgl. 7193= ähnl. Modell)
7735 Stehende Madonna mit gekreuzten Händen (auf Rundsockel) (vgl. 6862)
7736 BÜSTE Madonna mit gefalteten Händen (vgl. 6745; 7052) (vgl. 6923= WANDMASKE)
7827 „Gefangener Vogel" (Nach rechts schreitende Tänzerin im Schmetterlingskostüm) (auf Ovalsockel) (vgl. 5320 u.a.) (Lorenzl zugeschr.)
7916 Sitzender lesender Mädchenakt mit Pumps (auf Polstersessel)
8167 Schreitende junge Dame, ihr langes Kleid mit der linken Hand nach hinten haltend (auf Ovalsockel)
8176 „Gefangener Vogel" (Der Gefangene Vogel, Niddy Impekoven) (vgl. 5054 (BRONZE); 5230; 5960; 5964= Pendant zu 5960; 6600; 8176) (vgl. 5530= Trio GRUPPE)
8321 „Liane" (Stehende Tänzerin mit Hut, ihr Kleid schmetterlingsflügelartig auseinander haltend) (auf Ovalsockel) (vgl. 7581) (vgl. Dakon 7257= ähnl. Modell)
8363 FIGUR „Fächertanz" (Schreitende Tänzerin mit großem Fächer aus Vogelfedern) (auf Ovalsockel)
8452 FIGUR „Spanischer Tanz" („Spanierin" mit Hut und Kastagnetten) (auf Ovalsockel)
8457 Stehender Mädchenakt mit Blumen (auf Rundsockel)
8459 „Sonate" (Schreitende Tänzerin mit / ohne Bikinioberteil, ihren Rock schmetterlingsflügelartig auseinander haltend) (auf Dreiecksockel) (vgl. 7582)
8460 Posierender Mädchenhalbakt (Revuetänzerin in langem Rock) (auf passigem Sockel)
8461 Stehender Mädchenhalbakt mit um die Hüften geschlungenem langen Tuch (auf Ovalsockel)
8462 „Adagio" (Tanzender Damenakt mit großem Tuch) (auf Ovalsockel)
8492 Stehendes Mädchen mit Kappe (Béret), drei Herzen tragend (auf Rundsockel)
8493 Stehender Damenakt (Spanierin) mit ausgestrecktem Arm und großem Tuch / Schal (auf Ovalsockel)
8495 FIGUR „Kniende" (Kniender Damenakt) (auf abgerundetem Rechtecksockel) (vgl. 8496; 8708)
8496 FIGUR „Kniende" (Kniender Damenakt) (auf abgerundetem Rechtecksockel) (vgl. 8495; 8708)
8645 Stehender Trommler in Tracht (auf Rundsockel)
8646 Stehender Hochzeitslader in Tracht (auf Rundsockel) (Lorenzl zugeschr.)
8649 Stehende Frau in Tracht (auf Rundsockel) (Lorenzl zugeschr.)
8652 „Rhythmus" (Tänzerin in Pose) (auf Ovalsockel)
8654 FIGUR / LAMPE „Klassischer Tanz" / „Bewegung" (Tänzerin in Pose, auf einem Bein, ihre Arme ausgestreckt) (auf Ovalsockel)
8665 „Sinnende" (Sitzender Damenakt)
8698 WANDMASKE „Wagner" (Richard Wagner im Profil)
8706 „Feierlicher Rhythmus" (Mary Wigman) (auf Rundsockel)
8708 FIGUR „Kniende" (Kniender Damenakt) (auf abgerundetem Rechtecksockel) (vgl. 8495; 8496)
8711 „Dreiklang" (Mary Wigman auf einem Bein tanzend) (auf Rundsockel)
8713 „Get(anzte) Harmonie" (Tänzerin in Pose mit ausgestreckten Armen) (auf Rundsockel)
8718 „Tanz" (Mary Wigman, in fast kniender hingebungsvoller Pose) (auf Rundsockel)
8739 „Sinnende" (Sitzender Damenakt mit geschlossenen Augen)
8758 „Scherzo" (Tänzerin auf einem Bein) (auf Rundsockel)
8827 „Spanien" („Tänzerin")
8828 DECKELDOSE „Bonbonniere" (Sitzende Dame mit Ballkleid und Hut)
8829 WANDMASKE „Maske Madonna"
8909 „Suleika"
BÜSTE „Aida" ALABASTER (um 1925/26) (sign. Lorenz) (H. 12 cm); Sebök 52 (vgl. Fay.-FIGUR 5281)
FIGUR „Maria Jeritza als Salome" Lit.: N.N.: Farbe und Bewegung, Die Fayencen der Wiener Manufaktur Friedrich GS Wien-Währing, in: DB, 5.1928, Nr. 213, S. 40
FIGUR „Der Sturm" (Frauenakt mit wehendem Haar) BRONZE / Marmorsockel Lit.: N.N.: Farbe und Bewegung, Die Fayencen der Wiener Manufaktur Friedrich GS Wien-Währing, in: DB, 5.1928, Nr. 213, S. 38 und 40 (m. Abb.), hier mit falscher Künstlerang. „Josef Althan".
GRUPPE „Turteltauben" vgl. Dakon, Stephan Modell-Nr. 5712; Lit.: MW, 10.1928/29, Nr. 6, S. 28 (m. Abb.); Nr. 22, S. IV (m. Abb.), hier falsche Künstlerang. „Josef Lorenzl"; Neuwirth (1974), S. 67
FIGUR Nackte Ausdruckstänzerin BRONZE (um 1928/30) Firmenmarke Plak. C (H. 34,5 cm); Ahlden 66
899-USA FIGURINE Nude with Fan (vgl. 7544)

Lorieux, Julien

*Julien (Auguste J. Philibert) Lorieux, Bildhauer und Medailleur (*31.12.1876 Paris, †30.4.1915 Toul/Meurthe-et-Moselle), war Schüler von A. Falguière und Mercié. Lit.: TB, Bd. 23, S. 396; Vollmer, Bd. 3, S. 260; AKL Index.*
571-GSB PLATEAU „Le Cauchemar" ZINN / ZINKGUSS
583-GSB PLATEAU „Offrande a Paris" ZINN / ZINKGUSS

Lornic s. Gornik

Louger vgl. auch Longer

Louger, J.

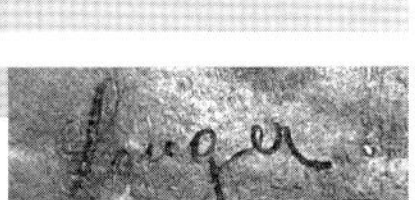

J. Louger, wird bei Pudor (1902) als Mitarbeiter Goldscheiders erwähnt und sein Modell „Kalypso" abgebildet. Weitere Lebens- und Schaffensdaten sind nicht bekannt. Lit.: Neuwirth (1974), S. 67.
896 VASE Kind und Frosch im Schilf mit Libelle (Pendant vgl. 897)
897 VASE Kind und Frosch im Schilf mit Libelle (Pendant vgl. 896)
1075 VASE „Musik" (Damenhalbakt mit Lyra, vollplastisch an der Vasenwandung sitzend, Totenmaske Beethovens an der Vasenmündung) (vgl. 1171; 1172)
1171 VASE „Musik" (Damenhalbakt mit Lyra, vollplastisch an der Vasenwandung sitzend, Totenmaske Beethovens an der Vasenmündung) (vgl. 1075; 1172)
1172 VASE „Musik" (Damenhalbakt mit Lyra, vollplastisch an der Vasenwandung sitzend, Totenmaske Beethovens an der Vasenmündung) (vgl. 1075; 1171)
2491 BÜSTE Holländerin mit Haube (Sockelrelief: Landschaft mit Windmühlen)
2504 „Venetienne" (Junge, am Ruder stehende Gondel fahrende Frau) (Inschrift: VENETIENNE)
2517 UHR Sitzendes Mädchen am Felsen
FIGUR Fischerin mit Netz (Junge Frau, sich nach dem gefülltem Netz bückend) Lit.: N.N.: Figürliche Terrakotten, in: KM, 4.1904, S. 137 (m. Abb.), „Das junge Weib sitzt auf einem am Wasser befindlichen Stein, mit ihrer rechten Hand stützt es seinen weit vorgebeugten Oberkörper auf den Stein, während die Linke das Netz hoch zieht, in dem sich anscheinend reiche Beute gefangen hat. Der Körper des jugendlichen Weibes ist wenig bekleidet, nur ein leichtes Gewand umschließt Hüfte und Oberschenkel, so dass die vollen Formen der Brüste frei bleiben, ebenso wie Arme und Unterschenkel. Das Gesicht des Mädchens, das, weit vorgebeugt, nach dem Netz sieht, verrät Neugierde auf das, was sich gefangen hat, und Freude über den reichen Fang".
BÜSTE „Mädchen" (nach links blickend) Lit.: VKMH, 16.1901/02, S. 592 (m. Abb.)
FIGUR „Die Musik" (Stehendes Mädchen mit kleiner Harfe) Lit.: VKMH, 16.1901/02, S. 592 (m. Abb.)
FIGUR „Kalypso" (Stehender Mädchenhalbakt mit 2 Flöten) (1902) Lit.: Pudor (1902), S. 34 (m. Abb.)
UHR „Am Brunnen" (Stehendes Mädchen an einem Wasserspeier ihren Krug füllend) Lit.: N.N.: Illustrierte Rundschau, Terrakotten von Fr. Goldscheider, Wien, in: VKMH, 17.1902/03, S. 717 (m. Abb.)
VASE Drei Mädchen (Nixen?) im Wasser schwimmend (um 1900) (H. 56 cm); KaM 69

Macadier s. Mercadier, J.

Maier, (Franz?)

Es könnte sich hier um den österreichischen Kunstgewerbler und Porzellanmaler Franz Maier handeln, von dem die Lebens- und Schaffensdaten unbekannt sind. Für ihn ist nur die Mitarbeit bei der Porzellanmanufaktur Josef Böck, Wien belegt. Lit.: Zühlsdorff, S. 426; AKL Index.
„Zum Tee" ÖMKI (1911/12) Lit.: Neuwirth (1974), S. 67
„Englische Dame" ÖMKI (1911/12) Lit.: Neuwirth (1974), S. 67

Maison, Rudolf

*Rudolf Maison, Bildhauer (*29.7.1854 Regensburg, †12.2.1904 München), war Autodidakt und studierte für kurze Zeit am Polytechnikum München Architektur. Er schuf neben Denkmalen (Kaiser Friedrich, Berlin) auch Porträtbüsten, Statuen und Statuetten in Marmor und Bronze. Maison war u.a. beauftragt mit Bildhauerarbeiten für König Ludwig II. von Bayern und die Firma Rosenthal (Kunstabt. Selb) führte zwei seiner Entwürfe aus. Seine Popularität gewann Maison „vielleicht durch seine Vorliebe für die Polychromie der Plastik" (Ostini). Im Besitz des Historischen Museums Regensburg befanden sich ehemals ca. 50 seiner teils lebensgroßen Modelle. Lit.: Ostini; Schmid; TB, Bd. 23, S. 581; Neuwirth (1974), S. 36, 67; Mackay, S. 247; Hillenbrand, S. 265; Niecol, Bd. 3, S. 13; Bénézit, Bd. 9, S. 39; AKL Index.*
732-GSB GROUPE „Le bon Chien" (Deutscher Originaltitel: „Gute Freunde")
1446 FIGUR „Neger" (Stehender Neger mit Schild und Speer) (auf Rundsockel)
1447 FIGUR „Augur" (ÖMKI 1898/99) (vgl. 2147= BÜSTE)
1462 FIGUR „Philosoph" (ÖMKI 1898/99) (vgl. 1888)
1499 GRUPPE „Gänsemädchen" ÖMKI (1898/99)
1888 FIGUR „Philosoph" ÖMKI (1898/99) (vgl. 1462)
2147 BÜSTE „Augur" (Inschrift: AUGUR) ÖMKI (1898/99) (vgl. 1447= FIGUR)
4807 BÜSTE Jeanne D' Arc MARMOR
„Zeitungsträgerin" (Terrakotta) ÖMKI (1899/1900) Lit.: DK, 3.1901, S. 136 (m. Abb.), hier bezeichnet: „Faunmädchen".
„Mohr mit Korb" (Terrakotta) ÖMKI (1899/1900) Lit.: DK, 3.1901, S. 141 (m. Abb.); Neuwirth (1974), S. 67
„Mohrin mit Korb" (Terrakotta) ÖMKI (1899/1900) Lit.: DK, 3.1901, S. 141 (m. Abb.); Neuwirth (1974), S. 67

Malfray, Charles Alexandre

*Charles Malfray, Bildhauer und Zeichner (*19.7.1887 Orléans, †28.5.1940 Dijon), war beeinflusst von Maillol, dessen Nachfolger er an der Akademie Ranson (1931) wurde. Malfray war Mitglied der Künstlergruppe „La Stèle" (-> vgl. Kap. „Dekorative Skulpturen von Arthur Goldscheider in Paris"). Lit.: Vollmer, Bd. 3, S. 306;*

Neuwirth (1974), S. 67; Mackay, S. 247; Bénézit, Bd. 9, S. 103; AKL Index.

Malfrey, J.
J. Malfrey, Kunstgewerbler, Lackarbeiten (Lebens- und Schaffensdaten unbekannt), war Mitglied der Künstlergruppe „Évolution" (-> vgl. Kap. „Dekorative Skulpturen von Arthur Goldscheider in Paris"). Lit.: Neuwirth (1974), S. 67.

Mallener, Theodor
*Theodor Mallener, Bildhauer (*16.7.1872 Olmütz, †?), bildete sich an der Fachschule für Tonindustrie in Znaim, bevor er 1911 an die KGW-Schule nach Wien wechselte und unter Strnad studierte. Weitere Daten zu Leben und Schaffen sind nicht bekannt. Lit.: Neuwirth (1974), S. 67; 216.*
4170 „Spitz" (Liegender Hund) ÖMKI (1912)
4318 Stehender Knabe in Tracht mit Hut, zwei Blumenkörbe tragend (vgl. 5437)
4319 Stehender Knabe in Tracht mit Hut, einen Hasen beim Ohr ziehend (auf Quadratsockel)
4359 „Jeunesse" (Stehendes Biedermeiermädchen mit Rosenstrauß) ÖMKI (1913/14)
5437 SCHALE Stehender Knabe in Tracht mit Hut, zwei Blumenkörbe tragend (vgl. 4318)

Malthner s. Mallener

Mangin, Paul Emile
*Paul Emile Mangin, Goldschmied und Kunstgewerbler (*21.3.1882 Paris, †?), war Mitglied der Künstlergruppe „Évolution" (-> vgl. Kap. „Dekorative Skulpturen von Arthur Goldscheider in Paris"). Lit.: Vollmer, Bd. 3, S. 312; Neuwirth (1974), S. 67.*

Marchand, Harry?
967 VASE Zwei an der Vasenwandung stehende Frauen mit Mohnblumen

Marcuse, Rudolf s. Markuse

Marinko, Werner
*Werner Marinko, Bildhauer (*13.10.1887 Dvorisce/Kroatien, †12.8.1965 Radlbrunn/Niederösterreich), war 1908–1910 Privatschüler bei Franz Barwig. Marinko beteiligte sich 1937 an der Landeskunstausstellung in Krems. Quelle: Bildhauerarchiv, Karl Novak.*
„Bulldoggen" ÖMKI (1912) Lit.: Neuwirth (1974), S. 67

Markuse (d.i. evtl. Rudolf Marcuse?)
*Möglicherweise handelt es sich hier um den Berliner Bildhauer Rudolf Marcuse (*15.1.1878 Berlin, †1928/29?). Marcuse entwarf vorwiegend Kleinplastiken und Denkmäler. Lit.: Zühlsdorff, S. 427.*
„Susanne" (um 1908/09) Lit.: N.N.: Aus Berliner Musterlagern, in: Sprechsaal, 42.1909, S. 308, „... Alles hervorragende Arbeiten. Bei allen diesen Plastiken werden die Körperteile, soweit sie nicht vom Gewand bedeckt sind, ganz matt-glänzend gehalten, während das Gewand sind in glänzender Frische abhebt"; Neuwirth (1974), S. 67

Marquet, Alix
*Alix Marquet, Bildhauer und Maler (*8.1.1875 Oudan/Nièvre, †?), lebte in Paris, war Schüler von Th. Barrau und Mitglied der Societé des Artistes Français. Ausstellungen im Salon ab 1893 mit Erringung der Silbermedaille 1928. Lit.: Vollmer, Bd. 3, S. 330; Neuwirth (1974), S. 67; Bénézit, Bd. 9, S. 252.*
851-GSB BOUGEOIR „Femme sur Feuille"
VASE Paris, Salon (1899) Lit.: Neuwirth (1974), S. 67

Mars-Vallett, Marius
*Marius Mars-Vallett, Bildhauer und Medailleur (*5.11.1869 Leinenc/Chambéry, †15.3.1957 Chambéry), studierte unter A. Falguière und stellte Arbeiten von 1892–1896 im Pariser Salon (Société des Artistes Français) aus. In den Jahren 1897–1922 war er im Salon (Société Nationale des Beaux-Arts) vertreten. Zahlreiche seiner Kleinbronzen wurden durch Siot-Decauville editiert. Lit.: TB, Bd. 24, S. 139; Arwas (2002), S. 617; Bénézit, Bd. 9, S. 256; AKL Index.*
„Akédysérile" BRONZE Lit.: Thomas, A.: Petit Bronzes d'Art, in: L'Art Dec., 3.1900/01, S. 188/189 (m. Abb.)
„Salambo" BRONZE Lit.: Thomas, A.: Petit Bronzes d'Art, in: L'Art Dec., 3.1900/01, S. 188/189 (m. Abb.)

Marschall, (Rudolf?)
*(Rudolf?), Marschall, Medailleur (*3.12.1873 Wien, †24.7.1967 Wien), studierte an der Wiener Akademie (Spezialschule für Medailleurkunst) und wurde 1903 Kammermedailleur. Ab 1905–1938 war er Leiter (Hofrat und Professor) der Akademie der Meisterschule für Medailleurkunst und wurde in einem Brief 1914 als Mitarbeiter bei Goldscheider erwähnt. Lit.: TB, Bd. 24, S. 139; Vollmer, Bd. 3, S. 331; Neuwirth (1974), S. 67; Bénézit, Bd. 9, S. 256; AKL Index.*

Martel, Jan (Jean)
*Jan (Jean) Martel, Bildhauer und Kleinplastiker (*5.3.1896 Le Mollin/Vendée, †?), war Zwillingsbruder von Joël Martel. Beide arbeiteten eng zusammen und schufen ihre Denkmale, Bas-Reliefs, Tier-Kleinplastiken und Dekorationen meist gemeinsam. Jan war, wie sein Bruder, Mitarbeiter bei Arthur Goldscheider, Paris. Lit.: Goldscheider, A. (Hrsg.): Artistic Evolution, 1926, T. 2, Pl. 8 (m. Abb.); Vollmer, Bd. 3, S. 332; Neuwirth (1974), S. 67–68; Bénézit, Bd. 9, S. 264.*
„Le Chat" (The Cat) (in Zusammenarbeit mit Joël Martel) BRONZE (geschwärzt) (H. 39 cm, B. 13 cm) Lit.: Goldscheider, A. (Hrsg.): Artistic Evolution, 1926, T. 2, Pl. 8 (m. Abb.)
„Pigeon" (in Zusammenarbeit mit Joël Martel) BRONZE (H. fehlt, B. fehlt) Lit.: Goldscheider, A. (Hrsg.): Artistic Evolution, 1926, T. 2, Pl. 47 (m. Abb.)

Martel, Joël
*Joël Martel, Bildhauer und Kleinplastiker (*5.3.1896 Le Mollin/Vendée, †?), war Zwillingsbruder von Jan Martel. Beide arbeiteten eng zusammen und schufen ihre Denkmale, Bas-Reliefs, Tier-Kleinplastiken und Dekorationen meist gemeinsam. Joël war, wie sein Bruder, Mitarbeiter bei Arthur Goldscheider, Paris. Lit.: Goldscheider, A. (Hrsg.): Artistic Evolution, 1926, T. 2, Pl. 8 (m. Abb.); Neuwirth (1974), S. 67; Vollmer, Bd. 3, S. 332; Bénézit, Bd. 9, S. 264; AKL Index.*
„Le Chat" (The Cat) (in Zusammenarbeit mit Jean Martel) BRONZE (geschwärzt) (H. 39 cm, B. 13 cm) Lit.: Goldscheider, A. (Hrsg.): Artistic Evolution, 1926, T. 2, Pl. 8 (m. Abb.)
„Pigeon" (in Zusammenarbeit mit Jean Martel) BRONZE (H. fehlt, B. fehlt) Lit.: Goldscheider, A. (Hrsg.): Artistic Evolution, 1926, T. 2, Pl. 47 (m. Abb.)

Masse, (Charles?)
*Charles Masse, Bildhauer (*7.10.1856 (1876?) Génelard/Frankreich, †1913 Troyes), erhielt einen ersten Preis in London (1890) und war Mitarbeiter bei Friedrich Goldscheider, Paris. Lit.: TB, Bd. 24, S. 219; Hillenbrand, S. 272.*
740-GSB „Enfant Scultpteur" (Pendant vgl. 741-GSB)
741-GSB „Enfant Peinture" (Pendant vgl. 740-GSB)

Massolini (s. auch Montini)
4007 Dante (Stehende Figur)

Matula, Viktor
*Viktor Matula, Bildhauer (*16.6.1900 Vichtenstein/Oberösterreich, †19.2.1969), studierte von 1924–1927 an der KGW-Schule in Wien unter Prof. Eugen Steinhof. Matula war spezialisiert auf Tierdarstellungen und zeichnet für die meisten zwischen 1933 und 1939 ausgeformten Tiere bei Goldscheider verantwortlich. Auch gibt es von seiner Hand Entwürfe im Bereich der angewandten Kunst, wie Vasen und Kerzenleuchter. Eine weitere Spezialität Matulas waren die Glasuren. In diesem Bereich entwickelte er zahlreiche Rezepte. Er war 1938 wohnhaft Schopenhauerstraße 19, Wien XVIII. und hatte, da er bis Oktober 1939 fester Mitarbeiter (Bildhauer und Modelleur) bei Goldscheider war, wie auch Adolf Prisch! sein Atelier in der Manufaktur, Staudgasse 7. Durch die Festanstellung bei Goldscheider erklärt sich auch, dass kaum Modelle mit seinem Namen sign. sind, da die Rechte bei der Manufaktur lagen. Lit.: Lehmann, Bd. 1938; Quellen: AGM (Gehaltsliste, Mai 1938; Gehaltsliste August 1938); Nachlass Matula, Wien.*
6353 RELIEF Dackel (Wandrelief)
6354 RELIEF Springendes Reh (Wandrelief)
6451 FIGUR „Reh liegend"
6667 FIGUR Foxterrier
6781 FIGUR „Reh" (auf flachem Rechtecksockel)
6797 FIGUR „Pferd" (Stehendes fressendes Pferd) (auf flachem Rechtecksockel) (vgl. 7386; 7737; 7939= GRUPPE)
6811 FIGUR/BUCHSTÜTZE Rehe
6822 ASCHER Foxterrier
6839 FIGUR Pferd
6866 RELIEF Pferd (Wandrelief)
6888 ASCHER
6890 ASCHER
6892 RELIEF Schiff (Wandrelief)
6894 ASCHER
6917 GRUPPE Enten
6921 ASCHER Fisch
6945 ASCHER Schiff
6946 RELIEF Schiff (Wandrelief)
6979 GRUPPE Rehe
6983 BUCHSTÜTZE Reh
6985 ASCHER
6986 GRUPPE Zwei Pferde (auf flachem Rechtecksockel) (vgl. 6797; 7804)
6992 ASCHER Reh
6994 ASCHER Elefant
7074 FIGUR Pferd
7075 Sitzender Foxterrier (Matula zugeschr.)
7077 Liegender Foxterrier (Matula zugeschr.)
7091 FIGUR Reh
7093 FIGUR Katze
7095 FIGUR Kamel
7106 FIGUR Springbock
7107 FIGUR „Esel"
7109 FIGUR „Pferd"
7110 Stehendes Reh (auf flachem Rechtecksockel)
7122 FIGUR Pferd
7128 ASCHER Elefanten
7154 FIGUR „Rehbock"
7157 FIGUR Pferd
7158 GRUPPE „Pferde" (Zwei stehende Pferde)
7182 ASCHER
7185 RELIEF „Pferd" (Springendes Pferd nach rechts)
7216 ASCHER „Ascher, groß" (Dekor: Vasen)
7235 ASCHER
7243 GRUPPE Rehe
7293 ASCHER Foxterrier
7294 ASCHER Esel
7308 GRUPPE „Rehe" (Drei stehende Rehe) (auf Ovalsockel)
7328 ASCHER
7353 ASCHER Kopf und Vase
7355 ASCHER Reh
7356 ASCHER Fisch
7386 FIGUR „Pferd" (Stehendes fressendes Pferd) (vgl. 6797; 7737; 7939)
7398 Zwei laufende Pferde (auf Rechteckssockel)
7399 FIGUR Pferd
7401 FIGUR „Reh"
7402 FIGUR „Pferd"
7433 LEUCHTER
7442 GRUPPE Pferde
7484 ZIGARETTENDOSE
7486 ZIGARETTENDOSE
7553 FIGUR „Pferd" (Stehendes Fohlen mit nach hinten gewandtem Kopf) (vgl. 8062)
7555 RELIEF „Eichhörnchen" (Eichhörnchen auf Ast sitzend) (Wandrelief)
7559 FIGUR Pferd
7565 Stehendes Mädchen mit Rehkitz (auf flachem Rechtecksockel)
7566 FIGUR Pferd
7567 FIGUR Pferd
7568 „Zicklein" (stoßend)
7569 GRUPPE „Rehe" (Ricke mit säugendem Kitz) (auf flachem Rechtecksockel)
7572 FIGUR Hahn
7670 GRUPPE Rehe
7671 GRUPPE „Rehgruppe" (Zwei Rehe, eines stehend, das andere liegend) (auf Ovalsockel)
7676 RELIEF „Segler" (Segelschiff nach rechts) (Wandrelief)
7713 KERZENHALTER „Leuchter" (vierflammig) (GS neu: Modell-Nr. 80 „Gemeinsam")
7726 RELIEF „Rehe" (Zwei springende Rehe) (Wandrelief)
7737 FIGUR „Pferd" (Stehendes fressendes Pferd) (vgl. 6797; 7386; 7939)
7739 FIGUR „Reh" (Stehendes Reh)
7780 Liegendes Pferd
7781 Stehendes Reh (mit nach hinten gewandtem Kopf)

7795 GRUPPE Rehe (Ricke mit Kitz) (auf Landschaftssockel)
7800 FIGUR/BUCHSTÜTZE „Pferd" (Stehendes Pferd/Fohlen mit nach hinten gewandtem Kopf) (auf flachem Rechtecksockel)
7804 FIGUR „Pferd" (auf flachem Rechtecksockel) (vgl. 6797; 6986)
7805 FIGUR „Pferd" (Schreitendes Pferd/Fohlen) (auf flachem Rechtecksockel)
7810 GRUPPE „Pferde"
7817 FIGUR „Pferd" (Stehendes Pferd/Fohlen) (auf flachem Rechtecksockel)
7818 FIGUR „Reh" (Äsendes Reh) (auf flachem Rechtecksockel) (vgl. 7863)
7819 FIGUR „Pferd" (Stehendes wieherndes Pferd/Fohlen) (auf flachem Rechtecksockel)
7820 FIGUR „Reh" (Stehendes Reh mit nach vorne gerecktem Kopf) (auf flachem Rechtecksockel)
7829 Zwei springende Pferde (auf längsovalem Landschaftssockel)
7834 GRUPPE Pferde
7851 GRUPPE „Zicklein"
7863 FIGUR „Reh" (Äsendes Reh) (auf länglichem Ovalsockel) (vgl. 7818)
7864 FIGUR „Reh" (Stehendes Reh mit nach hinten gewandtem Kopf) (auf flachem Rechtecksockel)
7867 FIGUR „Reh" (auf flachem Rechtecksockel)
7869 FIGUR „Reh" (Stehendes Reh mit erhobenem Kopf) (auf flachem Rechtecksockel)
7875 GRUPPE Zicklein
7900 ASCHER „Ascher mit Schildkröte"
7930 LEUCHTER
7939 GRUPPE „Pferde" (Zwei Pferde) (auf flachem Rechtecksockel) (vgl. 6797; 7386; 7737)
7949 ASCHER „Ascher mit Gänsen"
7974 RELIEF „Fisch" (Wandrelief)
7980 RELIEF „Pferd" (Wandrelief)
7981 FIGUR Pelikan
8020 FIGUR „Pferd", aufbäumend (Steigendes Pferd) (auf Ovalsockel)
8038 RELIEF „Reh" (Wandrelief)
8039 RELIEF „Klee" (Wandrelief)
8052 LEUCHTER
8053 DECKELDOSE (Form: rechteckig/Knauf: röhrenartige, geriffelte Handhabe)
8055 LEUCHTER
8062 FIGUR „Pferd" (Stehendes Fohlen mit nach hinten gewandtem Kopf) (vgl. 7553)
8080 FIGUR Pferd
8081 BUCHSTÜTZE „Elefant" (Stehender, den Rüssel erhebender indischer Elefant)
8091 GRUPPE Zwei „Lämmchen" (Craqueléglasur) (vgl. 8092; 8093)
8092 FIGUR „Lämmchen" (liegend) (vgl. 8091; 8093)
8093 FIGUR „Lämmchen" (stehend) (vgl. 8091; 8092)
8111 GRUPPE Zwei stehende Rehe (auf Ovalsockel)
8112 GRUPPE „Pferde"
8113 FIGUR Zicklein
8116 BUCHSTÜTZE Lämmer
8125 BUCHSTÜTZE Pferd
8160 BUCHSTÜTZE „Akt, liegend" (Liegender Damenakt)
8178 GRUPPE Zwei stehende Pferde mit Kummet (auf abgerundetem Rechtecksockel)
8179 GRUPPE Pferde
8180 GRUPPE Pferde
8210 GRUPPE „Steinböcke"(Zwei Steinböcke) (auf Natursockel) (vgl. 8211; 8212)
8211 FIGUR „Steinbock" (Äsender Steinbock) (auf Natursockel) (vgl. 8210; 8212)
8212 FIGUR „Steinbock" (mit erhobenem Kopf) (auf Natursockel) (vgl. 8210; 8211)
8213 RELIEF „Traube" (Wandrelief)
8224 GRUPPE „Rehe"
8227 FIGUR „Lämmchen" (stehend)
8258 FIGUR „Hirsch" (liegend)
8259 Junger Adler (auf Felssockel)
8272 FIGUR „Lämmchen" (Kopf nach links gewandt)
8273 FIGUR „Lämmchen" (fressend)
8278 FIGUR „Pferd" (scheuend) (auf Ovalsockel)
8279 FIGUR „Pferd" (bäumend) (auf Ovalsockel)
8280 FIGUR „Pferd" (trabend) (auf Ovalsockel)
8181 FIGUR „Pferd" (stehend, Kopf nach rechts) (auf Ovalsockel)
8282 FIGUR „Pferd" (wiehernd) (auf Landschaftssockel)
8283 FIGUR „Pferd" (fressend) (auf Landschaftssockel)
8289 FIGUR „Lämmchen"
8327 FIGUR „Reh"
8388 RELIEF „Sonnenblumen" (Wandrelief)
8390 FIGUR „Lamm"
8392 GRUPPE „Pferde"
8396 FIGUR „Pferd"
8401 FIGUR „Reh" (Stehendes Reh mit nach hinten gewandtem Kopf)
8510 GRUPPE „Pferde"
8511 FIGUR „Pferd"
8512 FIGUR „Pferd"
8568 RELIEF Pferde (Wandrelief)
8583 RELIEF Zwei Rehe (Wandrelief)
8584 RELIEF Affe (Wandrelief)
8615 GRUPPE Drei sitzende Kätzchen (vgl. 8616; 8619= jeweils Einzelfigur)
8616 FIGUR Sitzendes Kätzchen (vgl. 8615= GRUPPE)
8619 FIGUR Sitzendes Kätzchen (vgl. 8615= GRUPPE)
8629 FIGUR „Zicklein" (springend) (auf Ovalsockel)
8639 ASCHER
8640 FIGUR Katze
8641 FIGUR Katze
8923 Stehendes Dromedarfohlen
X VASE
XI VASE
955 LEUCHTER
970 ASCHER
971 ASCHER
1057 ASCHER
1059 ASCHER
1095 ASCHER
1193 ASCHER
1230 ASCHER Blume
1231 ASCHER
1243 ASCHER
1277 ASCHER
1278 ASCHER
1279 ASCHER
1280 ASCHER
1281 ASCHER
1291 VASE
1298 VASE
1300 ASCHER
1322 ASCHER
1324 VASE
1326 VASE
1366 VASE
1367 VASE
1369 VASE

Maurel, E(mile)

*Emile Maurel, Bildhauer (*1873 Marssac/Frankreich, †?) wird letztmalig 1902 erwähnt. Lit.: AKL Index.*
17 VASE mit Heuschrecke BRONZE (poliert)

May, Sybille

Sybille May, Kunstgewerblerin (Lebens- und Schaffensdaten unbekannt), war Mitglied der Künstlergruppe „Évolution" (-> vgl. Kap. „Dekorative Skulpturen von Arthur Goldscheider in Paris"). Lit.: Goldscheider, A. (Hrsg.): Artistic Evolution, 1926, T. 1, Pl. 7 (m. Abb.); Neuwirth (1974), S. 68; Hillenbrand, S. 275.
„Atlas" BRONZE (montiert auf der „Pendule" von Claudius Linossier) (H. 19,5 cm, B. 15 cm) Lit.: Goldscheider, A. (Hrsg.): Artistic Evolution, 1926, T. 1, Pl. 7 (m. Abb.)
„Adoration" (Steingut) (H. 19 cm, B. 21 cm) Lit.: Goldscheider, A. (Hrsg.): Artistic Evolution, 1926, T. 1, Pl. 15 (m. Abb.)
„Poussah" (Steingut) (H. 14 cm, B. 15,5 cm) Lit.: Goldscheider, A. (Hrsg.): Artistic Evolution, 1926, T. 1, Pl. 15 (m. Abb.)
„The Mystery" (Steingut) (H. 18,5 cm, B. 15 cm) Lit.: Goldscheider, A. (Hrsg.): Artistic Evolution, 1926, T. 1, Pl. 15 (m. Abb.)
„Panathene" BRONZE (H. fehlt, B. fehlt) Lit.: Goldscheider, A. (Hrsg.): Artistic Evolution, 1926, T. 2, Pl. 28 (m. Abb.)

Mayre, G.

1645 FIGUR „Ein altes Lied" (Sitzender Flöte spielender Junge als Halbakt) (auf Podest mit sitzendem Löwen) ÖMKI (1898/99)
1689 FIGUR „Flötenspieler" (Sitzender nackter Junge, eine Flöte spielend) (auf getrepptem Sockel) ÖMKI (1899/1900)
1765 BÜSTE „Jason" (Pendant vgl. 1766 Tell)

Méheut, Mathurin

*Mathurin Méheut, Maler, Illustrator, Keramiker und Kunstgewerbler (*21.5.1882 Lamballe/Frankreich, †22.2.1958 Paris), studierte an der Ecole des Beaux Arts in Rennes und war Mitglied der Künstlergruppe „Évolution" (-> vgl. Kap. „Dekorative Skulpturen von Arthur Goldscheider in Paris"). Seine künstlerische Bedeutung erlangte Méheut vor allem als Illustrator. Lit.: TB, Bd. 24, S. 337; Vollmer, Bd. 3, S. 363; Bd. 6, S. 262; Neuwirth (1974), S. 68; Hillenbrand, S. 277; Bénézit, Bd. 9, S. 441; AKL Index.*

Meier, (Emil oder Meier-Michel, Johanna)

3075 LAMPE Stehende Biedermeierdame mit Schutenhut und Blumenkorb (auf Rundsockel)
3963 LAMPE Sitzende Biedermeierdame mit großem Hut und Blumen
4002 Stehende Biedermeierdame mit Schutenhut und Blumenkorb (Sommer)
4068 LAMPE Stehende Biedermeierdame mit federverziertem Hut (vgl. FIGUR, H. 19 cm; RM 25)
4256 Sitzende Dame auf Sofa (mit Blumenvase)
4341 Biedermeiertänzerin im Rosenblütenkleid (auf Rundsockel) (Anna Pawlowa?) (vgl. 4616; 4619)
4408 BÜSTE Mädchen mit leicht geneigtem Kopf ALABASTER
4568 Stehendes Biedermeiermädchen, einen großen, blumengeschmückten Hut haltend (auf Rundsockel) (vgl. Polter 4568)
4569 Mädchen mit Blumenstrauß
4570 „Herbst" (Stehendes Biedermeiermädchen, einen Früchtekorb auf der linken Schulter tragend)
4616 Biedermeiertänzerin im Rosenblütenkleid (auf Rundsockel) (Anna Pawlowa?) (vgl. 4341; 4619)
4619 Biedermeiertänzerin im Rosenblütenkleid (auf Rundsockel) (Anna Pawlowa?) (vgl. 4341; 4616)
4708 Stehendes Mädchen mit Fächer
4846 BÜSTE Knabe mit Schleife/Mascherl (Kronprinz Otto von Habsburg 1916) ALABASTER
5101 GRUPPE/LAMPE Rokokopaar
„Spanisches Mädchen" (um 1910) (H. fehlt) ÖMKI (1911/12); Kett. 100; Lit.: Neuwirth (1974), S. 68
Stehende Biedermeierdame mit federverziertem Hut (um 1911/12) (H. 19 cm); RM 25; (vgl. 4068= LAMPE, H. 40 cm)
Tanzendes Rokokopaar (um 1920) (H. 18 cm); Spik 507; (H. 19 cm) Karrenbauer 127

Meier, Emil

*Emil Meier, Bildhauer und Kunstgewerbler (*22.11.1877 Weckelsdorf/Böhmen, †26.10.1958 Wien), besuchte die Fachschule für Goldschmiede in Turnau und studierte anschließend an der KGW-Schule Wien (1901–1903). Ab 1903 war Meier bei Klinkosch (Hofsilberfabrikant, Wien) beschäftigt. Seit 1910 verheiratet mit der Keramikerin Johanna Meier-Michel, die ebenfalls etliche Modelle für Goldscheider entwarf. Emil Meier lieferte u.a. Entwürfe für die Wiener Kunstkeramische Werkstätte Busch & Ludescher sowie für die „Wiener Keramik". Mitglied des Österreichischen Werkbundes. Emil Meier erhielt 1937 den Österreichischen Staatspreis und 1947 den Goldenen Lorbeer. Emil Meier war ab 1931 wohnhaft in Wien III., Seidlgasse 14. Lit.: TB, Bd. 24, S. 469; Vollmer, Bd. 3, S. 364; Lehmann, 1931–1942; BröPK, S. 147; Neuwirth (1974), S. 68; 218; Neuwirth (1974a), S. 157; Hillenbrand, S. 277; Bénézit, Bd. 9, S. 445; Zühlsdorff, S. 428; AKL Index; Quelle: Bildhauerarchiv, Karl Novak.*
1005 SCHREIBZEUG/TINTENFASS Mädchenkopf mit ausgebreiteten Armen
2320 FIGUR „Heimkehr" ÖMKI (1912)
3861 Stehende Rokokodame mit Blumenkorb (Sommer) (vgl. 4244)
3862 Stehende Rokokodame mit Früchtekorb (Herbst) (vgl. 4052; 4245)
3890 Stehende Rokokodame mit Rosen (vgl. 4247)
4052 LAMPE Stehende Rokokodame mit Früchtekorb (Herbst) (vgl. 3862; 4245)
4243 Stehende Biedermeierdame mit großem Hut
4244 FIGUR/LAMPE Stehende Rokokodame mit Blumenkorb (Sommer) (vgl. 3861)
4245 Stehende Rokokodame mit Früchtekorb (Herbst) (vgl. 3862; 4052)
4246 Stehende, gitarrespielende Rokokodame (Krinoline)
4247 Stehende Rokokodame mit Rosen (vgl. 3890)

4372 „Pensée" (Stehende Rokokodame in Gedanken) (vgl. 4622)
4622 „Pensée" (Stehende Rokokodame in Gedanken) (vgl. 4372)
5407 Stehende Rokokodame mit Rose und Puderquaste „Herbst" (Fay.) ÖMKI (1913/14) Lit.: Neuwirth (1974), S. 68

Meier-Michel, Johanna

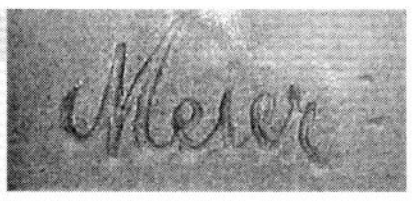

*Johanna Meier-Michel, Bildhauerin und Kunstgewerblerin (*21.6.1876 Böhmisch-Leipa, †12.7.1972 in?), war die Ehefrau von Emil Meier. Sie studierte an der KGW-Schule Wien (1901–1905) und lieferte u.a. Entwürfe für die Wiener Kunstkeramische Werkstätte. Meier-Michel war Mitglied des Österreichischen Werkbundes. Neben ihren figürlichen Arbeiten in Fayence und Porzellan schuf sie auch Werke aus Metall. Johanna Meier-Michel war ab 1931 wohnhaft in Wien III., Seidlgasse 14. Lit.: TB, Bd. 24, S. 469; Lehmann, 1931–1942; BröPK, S. 147; Neuwirth (1974a), S. 157; Zühlsdorff, S. 428; AKL Index; Quelle: Bildhauerarchiv, Karl Novak.*

3876 „Kind mit Löwen" (Putto mit (Raub)katze) ÖMKI (1910/11?)
3977 „Velasquez" (Infantin Margarita) (vgl. 5520)
4365 „Frühling" (Stehende Biedermeierdame mit Blumenkorb) (Pendant vgl. 4366)
4366 „Sommer" (Stehende Biedermeierdame mit Hündchen) (Pendant vgl. 4365)
4563 „Frühling" (Stehende Biedermeierdame mit Blumenkorb)
4720 Stehende Krankenschwester (Rotkreuzschwester) (vgl. 4725)
4725 Stehende Krankenschwester (Rotkreuzschwester) (vgl. 4720)
4763 Stehendes Mädchen mit Fächer (den Rock hochhebend) (auf Rundsockel) FAYENCE / ALABASTER
5258 Stehende Dame mit Fächer (auf Rundsockel)
5520 „Velasquez" (Infantin Margarita) (vgl. 3977)
„Galanterie" ÖMKI (1912) Lit.: Neuwirth (1974), S. 68
„Teddys Frühstück" ÖMKI (1912) Lit.: Neuwirth (1974), S. 68
Stehende Frau mit Schultertuch und Strohhut (auf Rundsockel) (um 1920) (Firmenmarke o. Ang.) (sign. Johanna Meier-Michel / Wien) (H. 24,5 cm); Z 41
Stehende junge Dame mit Hut (Biedermeierkostüm) und Regenschirm (sign. Johanna Meier-Michel) (H. 27 cm); Doro 654 (2.12.1986)

Meisinger, Ida

*Ida Meisinger (Erdös-Meisinger), Bildhauerin und Keramikerin (*9.8.1897 Prachatice / Böhmen, †30.1.1985 Tittmoning / Oberbayern), war zwischen 1924 und 1928 Schülerin der Fachklasse Michael Powolnys an der KGW-Schule in Wien. Ab 1927 verheiratet mit dem in Ungarn geborenen Keramiker Stephan Erdös, mit dem sie 1934 in Znaim eine eigene Werkstätte begründete, die 1946 nach Tittmoning verlegt wurde. Sie war bekannt für ihre unzähligen Terrier und anderen Hunderassen- bzw. Tierfiguren. Sie jedoch auf die reine Tierbildhauerei zu reduzieren, hieße ihr Unrecht tun, wie die umfangreiche Modellliste für Goldscheider zeigt. Sie gehörte neben Bosse, Dakon, Herczeg, Knörlein und Lorenzl zu den bedeutendsten Entwerfern der Manufaktur und war, abgesehen von ihrer Domäne der Tierbildhauerei, nicht nur auf dem Gebiet der angewandten Kunst (Gefäßkeramik) tätig, sondern schuf auch groteske Figuren. Zahlreiche Ausstellungen im In- und Ausland (Einzelausst. u.a. in Stockholm 1937 und im Hetjens-Museum, Düsseldorf 9.3.–7.4.1958, Ausst.-Kat. m. Abb.) und Museumsbesitz (u.a. Düsseldorf, Hamburg, Höhr-Grenzhausen) belegen ihre künstlerische Bedeutung. Lit.: Neuwirth (1974), S. 62, 134; Zühlsdorff, S. 406; AKL, Bd. 34, S. 300.*

5705 Sitzender Yorkshireterrier (Malteser?) (vgl. 5784= ähnl. Modell)
5706 Stehender Airedaleterrier
5729 DECKELDOSE Klarinetten- bzw. Flötenspieler (Form: quaderförmig / Knauf: würfelartig/Dekor (appliziert) Klarinetten- bzw. Flötenspieler (Rattenfänger von Hameln?), Blumendekor)
5784 Liegender Yorkshireterrier (Malteser?) (vgl. 5705= ähnl. Modell)
5787 FIGUR / BUCHSTÜTZE „Elefant"
5811 BUCHSTÜTZE Sitzender Chinese mit Buch
5824 Sitzende Perserkatze
5837 GRUPPE „Scotchterrier" (Zwei sitzende Scotchterrier) (vgl. 5846; ohne Nr.)
5846 GRUPPE „Scotchterrier" (Zwei sitzende Scotchterrier) (vgl. 5837; ohne Nr.)
5854 „Der Modehund" (Sitzender Scotchterrier) (vgl. 5879; 6026; ohne Nr.) (vgl. 6043= ASCHER; 6309= KERZENHALTER; 606A= SCHREIBZEUG)
5856 BUCHSTÜTZE Pferd (kniend)
5879 „Der Modehund" (Sitzender Scotchterrier) (auch als BUCHSTÜTZE) (vgl. 5854; 6026; ohne Nr.) (vgl. 6043= ASCHER; 6309= KERZENHALTER; 606A= SCHREIBZEUG)
5891 BUCHSTÜTZE Elefant
5892 DECKELDOSE (Zigaretten- / Zigarrendose) Rauchende Dame (Form: viereckig, fast quadratisch / Dekor: rauchende Dame, leicht reliefiert)
5937 Stehender Terrier mit Halsband (große Öse) (Meisinger zugeschr.)
5946 SCHALE (Form: rund, sechzehnpassig, vertikale Segmente / Dekor: mehrfarbig, Laufglasur)
5997 SERVIETTENHALTER Dreieck (Form: Dreieck / Dekor: geometrische Formen, bunte Laufglasur)
5998 SERVIETTENHALTER Halbrund (Form: halbrunde Scheibe / Dekor: geometrische Formen, bunte Laufglasur) (Meisinger zugeschr.)
6012 Stehender Scotchterrier
6013 Sitzender West Highland Terrier (mit heraushängender Zunge) (vgl. 6025; 6500= jeweils FIGUR; 6466= KERZENHALTER; 667= ASCHER)
6025 Sitzender West Highland Terrier (mit heraushängender Zunge) (vgl. 6013; 6500= jeweils FIGUR; 6466= KERZENHALTER) (auch als BUCHSTÜTZE)
6026 „Der Modehund" (Sitzender Scotchterrier) (vgl. 5854; 5879) (vgl. 6043= ASCHER; 6309= KERZENHALTER; 606A= SCHREIBZEUG)
6043 ASCHER „Der Modehund" (Sitzender Scotchterrier) (mit vier Ablagen) (vgl. 5854; 5879) (vgl. 6309= KERZENHALTER)
6072 Sitzender Scotchterrier (mit Halsband / Schleife) (vgl. 6101)
6079 Stehender Scotchterrier (Kopf nach links drehend)
6099 Stehende Chinesin mit zwei kleinen Schalen (Kakteentöpfchen / Gewürzschalen?) (Pendant vgl. 6100)
6100 Stehender Chinese mit zwei kleinen Schalen (Rauchzeug / Gewürzschalen?) (Pendant vgl. 6099)
6101 Sitzender Scotchterrier (mit Halsband / Schleife) (vgl. 6072)
6103 BUCHSTÜTZE Springender Scotchterrier (Kopf nach rechts drehend) (Pendant vgl. 6104)
6104 BUCHSTÜTZE Springender Scotchterrier (Kopf nach halb links drehend) (Pendant vgl. 6103)
6239 GRUPPE Hochzeitspaar (stehende Braut mit Bräutigam) (auf Rechtecksockel) (Meisinger zugeschr.)
6252 DECKELDOSE Springendes Reh (Form: rechteckig / Dekor: springendes Reh, reliefiert) (GS neu: Modell-Nr. 31 „Schwarzer Wald, rotes Reh")
6270 Bellender Scotchterrier (in spielerischer Pose)
6284 Sitzendes Mädchen mit Trompete
6285 RELIEF Sitzender Scotchterrier (Wandfliese) (vgl. 6316= DECKELDOSE)
6309 KERZENHALTER (einflammig) „Der Modehund" (Sitzender Scotchterrier auf runder Schale) (vgl. 5854; 5879; 6026) (vgl. 6043= ASCHER)
6315 Sitzende Bulldogge (oder Boston Terrier) (mit Halsband)
6316 DECKELDOSE Sitzender Scotchterrier (Form: rechteckig / Dekor: sitzender Scotchterrier, reliefiert) (vgl. 6285= RELIEF)
6362 „Pat und Patachon" (Zwei sitzende Airedaleterrier) (vgl. 6996) (vgl. 6463= DreierGRUPPE)
6417 Stehender Airedaleterrier
6449 FIGUR „Hundewetter" (Sitzender Terrier mit Regenschirm) (mit / ohne Ovalsockel) (vgl. 6997)
6456 ASCHER Foxterrier (Form: zweifach gemuldet, als sich überschneidende Kreise / Dekor: mittig stehender Foxterrier, vollplastische Fig.)
6463 Drei sitzende Airedaleterrier (vgl. 6362; 6996= jew. ZweierGRUPPE „Pat und Patachon")
6464 Drei sitzende junge Terrier
6466 KERZENHALTER (einflammig) Sitzender West Highland Terrier (mit heraushängender Zunge) (vgl. 6013; 6025; 6500= jeweils FIGUR)
6496 ASCHER Terrier (Form: rund / Dekor: sitzender Terrier, vollplastische Fig.)
6500 FIGUR / BUCHSTÜTZE Sitzender West Highland Terrier (mit heraushängender Zunge) (vgl. 6013; 6025= jeweils FIGUR; 6466= KERZENHALTER)
6533 Zwei stehende, sich beschnuppernde Scotchterrier (auf Rechtecksockel)
6543 WANDMASKE Terrierkopf mit Schleife / Mascherl (vgl. 6611) (vgl. 6610= ZweierGRUPPE)
6545 ASCHER Drei Scotchterrier (Form: rund / Dekor: ein stehender bellender und zwei sitzende Scotchterrier, vollplastische Fig.)
6596 Drei sitzende Terrier beim Kartenspiel (auf flachem Rechtecksockel)
6605 Stehender Elefant (Meisinger / Bosse zugeschr.)
6610 WANDMASKE Zwei Terrierköpfe mit Schleife / Mascherl (vgl. 6611= linker Kopf)
6611 WANDMASKE Terrierkopf mit Schleife / Mascherl (vgl. 6543) (vgl. 6610= ZweierGRUPPE)
6668 GRUPPE Zwei männchenmachende Terrier (Meisinger zugeschr.)
6703 Fressender West Highland Terrier mit Schleife / Mascherl (an der Futterschale) (auf flachem Rechtecksockel)
6708 „Herzensdieb" (Stehender West Highland Terrier, ein Herz mit dem Maul haltend)
6779 Sitzender West Highland Terrier mit Schleife / Mascherl
6780 Sitzender Foxterrier mit Schleife / Mascherl (auf Sockel als Buch / mit kleinem Hufeisen)
6783 Sitzender, jaulender West Highland Terrier mit Schleife / Mascherl
6796 „Pferd" (Springendes Fohlen) (vgl. 6844= DECKELDOSE) (Meisinger zugeschr.)
6844 DECKELDOSE „Pferd" (Springendes Fohlen) (Form: rechteckig / Dekor: springendes Fohlen mit Grasbüschel, reliefiert) (vgl. 6796= FIGUR) (Meisinger zugeschr.)
6861 Foxterrier mit Tennisschläger (auf Quadratsockel)
6873 BUCHSTÜTZE Sitzender Terrier (auf rechtwinklig aufgeschlagenem Buch)
6874 GRUPPE Stehendes Kinderpaar (auf Rechtecksockel) (Meisinger zugeschr.)
6876 GRUPPE Stehendes Mädchen mit Puppe und Terrier (auf Ovalsockel) (Meisinger zugeschr.)
6918 Sitzender West Highland Terrier mit Schottenmütze
6950 GRUPPE Zwei sitzende West Highland Terrier
6977 FIGUR „Hund mit Bassgeige" (Stehender Terrier mit Kontrabass) (vgl. 7878)
6984 GRUPPE Zwei an einem Halsband zerrende Foxterrier (auf Ovalsockel)
6995 „Pat und Patachon" (Zwei sitzende Airedaleterrier) (vgl. 6362) (vgl. 6463= DreierGRUPPE)
6997 FIGUR / BUCHSTÜTZE „Hundewetter" (Sitzender Terrier mit Regenschirm) (vgl. 6449)
7069 Sitzender Terrier mit Damenschuh
7070 „Zahnweh" (Sitzender Terrier mit Kopftuch)
7186 „Zeitungsleser" (Sitzender, in Zeitung lesender Terrier mit Brille)
7237 BUCHSTÜTZE Stehender Pinguin mit Zipfelhaube und Muff (Meisinger zugeschr.)
7282 Zwei Fohlen (auf Landschaftssockel)
7321 FIGUR Hund „Paddler" (Im Ruderboot sitzender Terrier)
7700 FIGUR Hund „Skifahrer" (Skifahrender Terrier)
7701 GRUPPE Hunde „Naschkätzchen" (Zwei in einem Topf sitzende Terrier)
7877 FIGUR „Grippe" (Im Bett liegender Terrier mit Schal)
7878 FIGUR „Hund mit Bassgeige" (Stehender Terrier mit Kontrabass) (vgl. 6977)
7909 GRUPPE Mutter und Kind (Stehende Terrierhündin mit Welpe in der Wiege) (Meisinger zugeschr.)
ohne Nr. GRUPPE „Scotchterrier" (Zwei sitzende Scotchterrier) (vgl. 5837; 5846)
606A SCHREIBZEUG „Der Modehund" (Sitzender Scotchterrier) (vgl. 5854; 5879) (vgl. 6043= ASCHER; 6309= KERZENHALTER)
667 ASCHER Sitzender West Highland Terrier (vgl. 6013; 6025; 6466; 6500)
„Männchen machender" Foxterrier (um 1930) Lit.: Bernd, G.: Heitere Wiener Tierkunst, Ida Meisinger, in: Schaulade, 8.1932, S. 480 (m. Abb.)
Stehender Foxterrier (um 1930) Lit.: Bernd, G.: Heitere Wiener Tierkunst, Ida Meisinger, in: Schaulade, 8.1932, S. 480 (m. Abb.)
Stehender Dackel (um 1930) Lit.: Bernd, G.: Heitere Wiener Tierkunst, Ida Meisinger, in: Schaulade, 8.1932, S. 480 (m. Abb.)
Kleines sitzendes Kätzchen (um 1930) Lit.: Bernd, G.: Heitere Wiener Tierkunst, Ida Meisinger, in: Schaulade,

8.1932, S. 480 (m. Abb.)
RAUCHZEUG Elefant Lit.: MW, 11.1929/30, Nr. 22, S. 20 (m. Abb.) (Bosse / Geiger / Meisinger?)
RAUCHZEUG Zigarettenbehälter Lit.: MW, 11.1929/30, Nr. 22, S. 20 (m. Abb.) (Bosse / Geiger / Meisinger?)
KERZENHALTER Lit.: MW, 11.1929/30, Nr. 22, S. 20 (m. Abb.) (Bosse / Geiger /Meisinger?)

Meisner s. Meisinger, Ida

Meißner, (Max?)

*Max Meißner, Bildhauer (*7.11.1859 Döcklitz/Querfurt, †?). Ob Max Meißner für diese beiden Entwürfe als Künstler infrage kommt, muss mit einem Fragezeichen versehen bleiben so lange kein Exemplar mit dem Vornamen oder mit der Initiale desselben nachgewiesen werden kann. Lit.: TB, Bd. 24, S. 346; Bénézit, Bd. 9, S. 450; AKL Index.*
859 RELIEF Mädchen zwischen zwei Bäumen, an einer Blüte riechend (sign. Meissner)
SCHREIBZEUG (mit Taube) (um 1905/06) Lit.: Pudor (1906), o. S. (m. Abb.)

Melcher, (Josef)

*Josef Melcher, Bildhauer (*27.10.1897 Wien, †?). Ob der Entwerfer bzw. Künstler „Melcher", den Könecke (2002) auf S. 77 für die Modell-Nr. 8174 angibt, mit dem aus dem Bildhauerarchiv Novak identisch ist, konnte letztendlich nicht geklärt werden. Dafür kann eine mit dem Namen „Nieschlag" bezeichnete Figur aus der Zeit zwischen 1938–1941 belegt werden. Einziges mit „Melcher" sign. Objekt der Manufaktur Goldscheider ist ein stehender Mädchenakt (Modell-Nr. 8316). Da es in diesem Zusammenhang bei beiden Künstlern Fragezeichen gibt, werde sie an beiden Stellen gelistet.*
8174 FIGUR „Nocturno" (Stehender Damenakt mit großem Tuch) (auf rautenförmigem Sockel) (vgl. Nieschlag) (vgl. 8749)
8316 Stehender nachdenklicher Mädchenakt mit überkreuzten Armen (auf Rundsockel)
8749 FIGUR „Nocturno" (Stehender Damenakt mit großem Tuch) (auf rautenförmigem Sockel) (vgl. Nieschlag) (vgl. 8174)

Membré, E.

E. Membré, Kunstgewerbler/in (Lebens- und Schaffensdaten unbekannt), war Mitglied der Künstlergruppe „Évolution" (–> vgl. Kap. „Dekorative Skulpturen von Arthur Goldscheider in Paris") und bei Arthur Goldscheider mit Emailarbeiten beschäftigt. Lit.: Neuwirth (1974), S. 68.

Mercadier, J.

1390 BÜSTE „Beppo" (mit gelockten Haaren)
1643 s. Cherc
2072 FIGUR „Mignon" (Stehende junge Frau mit langem Haar) (nach rechts blickend) (vgl. jew. Cherc 1643; 2436)
2436 s. Cherc
Stehender Junge mit Hut, eine Zigarette haltend (um 1900) Firmenmarke Sculpture 13 Repr. Res. (H. 95 cm); Soth. L. 9.6.1989

Mercié, Antonin (Marius Jean)

*(Marius Jean) Antonin Mercié, Bildhauer und Maler (*30.10.1845 Toulouse, †13.(od. 14.)12.1916 Paris), war Schüler von Falguière und Jouffroy. Mercié erhielt 1868 den Rompreis. Werke von seiner Hand finden sich in unzähligen Museen (u.a. in Paris, Chicago, Kopenhagen, Leipzig, Toulouse).–> vgl. auch die bei Mercier aufgeführten Modelle. Lit.: TB, Bd. 24, S. 408; Bénézit, Bd. 9, S. 499; AKL Index.*
717-GSB SERVICE FUMEUR „Narghileh" (Plateau, Cendrier, Porte-Allumettes, Pot a Tabac) (sign. Mercier)
761-GSB JARDINIÈRE „Sphynx" (sign. Mercier)
776-GSB BUSTE „Faunesse"

Mercier

Mercier, Bildhauer, Vorname sowie Lebens- und Schaffensdaten sind nicht bekannt. Infrage kommt auch der Bildhauer Antonin Mercié (–> vgl. dort). Lit.: GSB, Pl. 42 (m. Abb.).
717-GSB SERVICE FUMEUR „Narghileh" (Plateau, Cendrier, Porte-Allumettes, Pot a Tabac)
761-GSB JARDINIÈRE „Sphynx"
776-GSB BUSTE „Faunesse" (sign. Mercié)
2060 FIGUR (LAMPE) „Uhu"

Mergier, Paul-Louis

*Paul-Louis Mergier, Emailleur, Maler und Kunstschriftsteller (*23.8.1891 Orthez/Frankreich, †? wird letztmals 1956 erwähnt), war Mitarbeiter von Arthur Goldscheider, Paris. Lit.: Goldscheider, A. (Hrsg.): Artistic Evolution, 1926, T. 1, Pl. 17 (m. Abb.); Vollmer, Bd. 3, S. 374; Bénézit, Bd. 9, S. 504; AKL Index.*
VASE (cuivre gravé) (Form von F. La Monaca) (H. 31 cm, B. 21 cm) Lit.: Goldscheider, A. (Hrsg.): Artistic Evolution, 1926, T. 1, Pl. 17 (m. Abb.)
VASE (cuivre gravé) (Form von F. La Monaca) (H. 11 cm, B. 10 cm) Lit.: Goldscheider, A. (Hrsg.): Artistic Evolution, 1926, T. 1, Pl. 17 (m. Abb.)

Mérite, Édouard Paul

*Édouard Paul Mérite, Tiermaler und Bildhauer (*7.3.1867 Neubourg/Elsass, †5.2.1941 La Jonchère), war Schüler von Frémiet und Barrias. Mérite stellte in den Jahren 1888–1911 im Pariser Salon (Societé des Artistes Français) überwiegend Tierplastiken aus. Lit.: TB, Bd. 24, S. 415; Bénézit, Bd. 9, S. 508; AKL Index.*
868-GSB „Martin Pécheur"
869-GSB „Souris"

Meunier, Constantin (Émile)

*Constantin Émile Meunier, Bildhauer, Maler und Graphiker (*12.4.1831 Etterbeek/Brüssel, †4.4.1905 Ixelles/Brüssel), war Schüler seines Bruders Jean-Baptiste und des Bildhauers Ch. A. Fraikin. Meunier stellte den Arbeiter in den Mittelpunkt seines Schaffens. Er ist neben Rodin, dem er sich stilistisch und künstlerisch verpflichtet fühlt, ein Hauptvertreter der realistisch-impressionistischen Richtung der Bildhauerei der Jahrhundertwende bis in die zwanziger Jahre. Lit.: TB, Bd. 25, S. 453–455; Hillenbrand, S. 280; Bénézit, Bd. 9, S. 549–550; AKL Index.*
2332 FIGUR „Fischer" (mit Netz) (auf Natursockel)

Miller, Leo s. Müller, Leo

*Leo Miller, Bildhauer (*29.8.1887 Schladming/Steiermark, †1965 Schladming), studierte in den Jahren 1906–1910 an der KGW-Schule Wien unter Klotz und Arthur Strasser, sowie Franz Barwig, wo er einen Großtierkurs besuchte. Miller lebte in Radstadt/Salzburg und gründete 1922 die Radstädter Kunstkeramik. Er schuf neben Tierfiguren hauptsächlich Madonnen, Köpfe und Kachelöfen. Lit.: Neuwirth (1974), S. 68, 221; Quelle: Bildhauerarchiv, Karl Novak.*

Mörtel s. Mörtl, Michael

Mörtl, Michael

*Michael Mörtl (auch Mörth), Bildhauer und Keramiker (*30.3.1878 Fürnitz/Kärnten, †?), studierte an der KGW-Schule Wien (1899–1905) unter Joseph Breitner und Arthur Strasser. Ab 1910 Professor an der Fachschule für Tonindustrie in Znaim. Ab 1923 übersiedelte er nach Karlsbad. Mörtl lieferte u.a. Entwürfe für die Wiener Kunstkeramische Werkstätte, Amphora in Turn sowie für Förster & Co. in Wien. Lit.: Vollmer, Bd. 3, S. 407; Neuwirth (1974), S. 222; Neuwirth (1974a), S. 157; Zühlsdorff, S. 627; AKL Index; Quelle: Bildhauerarchiv, Karl Novak.*
4087 Drei sitzende Eulen (auf Rundsockel mit vier Büchern) (sign. Mörtel)

Molnar, Karl

*Karl Molnar, Gießer (*26.9.1926, †?). Quelle: AGM, Liste „Adressen von guten und zuverlässigen Arbeitern" (undat., um 1953).*

Monder s. Moroder, C.

Mondi

986 RELIEF Mädchen mit Blume (Wandrelief)
1906 BÜSTE „Kleeblatt" (Junge Dame mit Schmetterlingsflügeln) (nach rechts blickend) ÖMKI (1902/03) (vgl. 2328)
2328 BÜSTE „Kleeblatt" (Junge Dame mit Schmetterlingsflügeln) (nach rechts blickend) ÖMKI (1902/03) (vgl. 1906)

Monginot, Charlotte

*Charlotte Monginot, Bildhauerin und Malerin (*18.12.1872 Paris, †? wird letztmals 1962 erwähnt), war Schülerin von Puech, Verlet und Marquestre. Sie beschickte zwischen 1891 und 1930 den Pariser Salon. Lit.: TB, Bd. 25, S. 65; Vollmer, Bd. 6, S. 280; Hillenbrand, S. 287; Bénézit, Bd. 9, S. 755; AKL Index.*
4304 FIGUR / LAMPE Schlafendes Kind auf Bücherstapel (Buchstütze?)
„Jalousie" (Drei auf einer Bank sitzende Kinder) BRONZE / Marmor Arthur Goldscheider, Paris (1924) Lit.: L'Illustration, Nr. 4268 vom 20.12.1924, S. 14 (m. Abb.)

Montenave

Ein Künstler namens Montenave lässt sich in der einschlägigen Literatur nicht nachweisen. Hillenbrand und Kashey belegen einen für Friedrich Goldscheider, Paris, arbeitenden „Monterowe", der sich jedoch auch nicht weiter dokumentieren lässt. Lit.: Kashey, S. 272; Neuwirth (1974), S. 68; Hillenbrand, S. 288.
930 RELIEF Frau mit Buch vor Landschaft (Wandrelief)
931 RELIEF Amazone mit Pferd (nach links) (Wandrelief) (vgl. 1035) (Pendant vgl. RELIEF Walküre mit Pferd, nach rechts)
1035 RELIEF Amazone mit Pferd (nach links) (Wandrelief) (vgl. 931) (Pendant vgl. RELIEF Walküre mit Pferd, nach rechts)
2103 BÜSTE „Lucrezia" (nach links blickend) (auf konischem Sockel) (vgl. 2224; 2434)
2209 BÜSTE „Fides" ÖMKI (1901/02)
2224 BÜSTE „Lucrezia" Inschrift: LUCRECE (undeutl. sign. Montenave?) (vgl. 2103; 2434)
2236 BÜSTE „Sphinx" ÖMKI (1901/02)
2434 BÜSTE „Lucrezia" (nach links blickend) (auf konischem Sockel) (vgl. 2103; 2224)
2448 BÜSTE „Egypterin" (Sockel reliefiert: Sphinx vor aufgehender Sonne u. a.) (vgl. 2595)
2466 BÜSTE Junge Frau (nach links blickend) geschmückt mit großen Blättern (Pendant vgl. 2511)
2511 BÜSTE Junge Frau (nach rechts blickend) geschmückt mit großen Blättern (Pendant vgl. 2466)
2576 BÜSTE Junges Mädchen mit beidseitigem Haarschmuck (Bänder oder Blumenblüten?)
2580 BÜSTE Junge Frau mit gesenktem Haupt (Kleid besetzt mit Farbsteinen / Edelsteinimitation)
2592 BÜSTE Mädchen mit gelockten Haaren und Kappe
2595 BÜSTE „Egypterin" (Sockel reliefiert: Sphinx vor aufgehender Sonne u. a.) (vgl. 2448)
2664 BÜSTE Mädchen mit seitlich geknotetem Haar und Diadem (vgl. 2768; 2842)
2668 BÜSTE Junge Frau mit Schilf (und Lilienblüten?)
2700 BÜSTE Mädchen mit zwei großen Schleifen im Haar
2768 BÜSTE Mädchen mit seitlich geknotetem Haar und Diadem (vgl. 2664; 2842)
2842 BÜSTE Mädchen mit seitlich geknotetem Haar und Diadem (vgl. 2664; 2768)
3401 BÜSTE „Blanca" (Frau mit Diadem und Halskette) Inschrift: BLANCA
SCHALE „Wassernymphe" (Algendekor) Lit.: KM, 2.1902, S. 90, „Die Wassernymphe, eine Arbeit Montenaves, zeigt als Basis Wellen, zu einer Schale stilisiert, aus denen sich unten mit ihnen verfliessend, die schlanke, von reichem Haar umflossene Gestalt der Nymphe hebt, welche mit fremden, neugierigen Augen um sich blickt. Das allmählich Verschwimmende der Konturen passt sich dem gewählten Darstellungsstoffe gut an.
SCHALE „Wellenschlag" (Liegender Frauenakt) (Dekor: „Grès") Lit.: R., F.: Wiener Kunstkeramik, in: KM, 2.1902, S. 91 (m. Abb.), „Dekoration in gelblich-grüner zu schwärzlich-braun übergehender Variation".
RELIEF Walküre mit Pferd (nach rechts) (Wandrelief) Lit.: H., H.: Internationale kunstgewerbliche Ausstellung in Petersburg 1908, in KM, 8.1908, S. 166 (m. Abb.) (Pendant vgl. 981; 1035)

Monterowe s. Montenave

Montini, (Tullio?)

*(Tullio?) Montini (*1878 Verona, †?). Ob es sich hier um den italien. Bildhauer, und Schüler von C. Poli und R. Cristani, Tullio Montini handelt, der in Verona Denkmale und Grabmäler schuf, lässt sich nicht mit letzter Sicherheit sagen. Es könnte sich auch um einen Phantasienamen der Manufaktur handeln. Lit.: TB, Bd. 25, S. 97; Bénézit, Bd. 9, S. 790.*
4007 Dante (stehende Figur) (auch sign. Massolini)
5391 BÜSTE Dante (nach rechts blickend) ALABASTER

Moreau, Hippolyte-François

*Hippolyte-François Moreau, Bildhauer (*1832 Paris, †1927 Neuilly-sur-Seine), war spezialisiert auf Genre-*

figuren und stellte von 1889–1917 im Pariser Salon aus. Hillenbrand weist ihn als Mitarbeiter für Goldscheider, Paris nach. Lit.: TB, Bd. 25, S. 127; Hillenbrand, S. 289; Bénézit, Bd. 9, S. 825; AKL Index.

Moreau-Vauthier, Paul
*Paul (Gabriel Jean) Moreau-Vauthier, Bildhauer und Kunstschriftsteller (*26.11.1871 Paris, †2.2.1936 Poitiers), war Schüler seines Vaters und von G. Thomas. Moreau-Vauthier beteiligte sich u.a. auch an der Weltausst. 1900 in Paris. Lit.: TB, Bd. 25, S. 131; Vollmer, Bd. 6, S. 282; Hillenbrand, S. 289; Bénézit, Bd. 9, S. 831; AKL Index.*
837-GSB VASE „Curieuse"
838-GSB BOUGEOIR „Sommeil"
839-GSB VASE „Triangulaire"
843-GSB STATUE „Peinture"
844-GSB CENDRIER „Perle" BRONZE
Statue „Pariserin" Paris, Weltausstellung (1900) Lit.: Hecht/Barthel, in: Tonindustrie-Ztg., 24.1900, S. 1584

Morice, Léon
*Léon Morice, Bildhauer (*25.1.1868 Angers/Frankreich, †? wird letztmals 1934 erwähnt), war Schüler von Brunclair und F. Charpentier. Beschickte von 1880–1930 den Pariser Salon (Societé des Artistes Français) und errang dort 1910 eine „Mention honourable" sowie 1914 eine Medaille 2. Klasse. Lit.: TB, Bd. 25, S. 152; Bénézit, Bd. 9, S. 853; Christie's SK 9.10.2002; AKL Index.*
1737 „Enfant au Lezard" (Erschreckter Knabe mit Eidechse) (auf Natursockel) BRONZE (Inschrift: ENFANT AU LEZARD)

Moroder, C(hristian?)
In der einschlägigen Literatur lässt sich zwar ein Christian Moroder (nachweisbar zwischen 1903–1956), ein Südtiroler Bildhauer und Holzschnitzer, nachweisen. Ob er aber als Entwerfer für die bei Goldscheider editierte Figur infrage kommt, ist aufgrund der weitverzweigten und bildhauerisch tätigen Familie der Moroders fraglich. Lit.: Vollmer, Bd. 4, S. 425–426; AKL Index.
5276 „Madame Butterfly" (Stehende Japanerin mit Kind) FAYENCE/SPECKSTEIN

Mosch, Hermann
*Hermann Mosch, Gießer (*23.12.1893, †?). Quelle: AGM, Liste „Adressen von guten und zuverlässigen Arbeitern" (undat., um 1953).*

Motto, (Joseph C.)
*Joseph C. Motto, Bildhauer (*6.3.1892 Cleveland/Ohio, †? wird letztmals 1940 erwähnt, South Euclid/Ohio), war Schüler von Matzen, MacNeal und Heber. Motto ist bekannt für seine Porträtbüsten u.a. von W. Shakespeare. Lit.: Vollmer, Bd. 3, S. 431; Who's who in American Art, Vol. 1, 301, Vol. 2, S. 375; AKL Index.*
5773 STEHMASKE Frauenkopf mit langem Haar (auf Rechtecksockel) FAYENCE/BRONZE
6119 Stehende Madonna mit Kind (auf Rundsockel)
7786 WANDMASKE Madonna (Frauengesicht)

Moulin, Eugène (Émile)
*Eugène Émile Moulin, Bildhauer (*9.3.1880 Châlons sur Marne, †(5?).10.1914 Beauvraignes/Somme), war Schüler von Falguière, Mercié und Lemaire. Moulin wird u.a. bei Kashey als Mitarbeiter der Firma Friedrich Goldscheider erwähnt. Lit.: Pudor (1902); TB, Bd. 25, S. 200; Vollmer, Bd. 3, S. .431; Neuwirth (1974), S. 68; Kashey, S. 272; Hillenbrand, S. 291; Bénézit, Bd. 9, S. 910.*
842 VASE ‚Sirenenspiel" (Zwei schwimmende, plastische Nixen, reliefierte Fische)
„Die Musik" (Stehende Frau mit Lyra) Lit.: VKMH, 16.1901/02, S. 592 (m. Abb.)

Mucha, Alphonse (eigentl. Alfons Maria) (in der Art von)
*Alfons Maria Mucha, Maler, Illustrator und Kunstgewerbler (*24.7.(8?).1860 Eibenschütz/Mähren, †14.7.1939 Prag). Lit.: TB, Bd. 25, S. 210; Vollmer, Bd. 3, S. 433; Hillenbrand, S. 292; Bénézit, Bd. 9, S. 923–924.*
766 =Cherc RELIEF „Printemps" (Frühling) (in der Art von) Mucha
768 =Cherc RELIEF „Été" (Sommer) (in der Art von) Mucha
770 =Cherc RELIEF „Autumne" (Herbst) (in der Art von) Mucha
772 =Cherc RELIEF „Hiver" (Winter) (in der Art von) Mucha

Müller
402-WG WANDMASKE „Maske" (Frauenkopf mit Kopftuch)

Müller
Möglicherweise handelt sich bei diesem Entwerfer um den Bildhauer Armin Müller.
4987 DECKELDOSE/FIGUR Mädchenakt (Form: oval/Bekrönung: liegender Mädchenakt (vollplastische Fig.)/Dekor: mehrfarbiger Spritzdekor, Zackenmuster)
5004 Sitzendes Mädchen mit Blumenstrauß (auf einer Steinmauer)

Müller, Armin
3856 Putto, einen großen Pfifferling tragend (Pendant vgl. 4643)
4643 Stehender Putto mit Blumengirlande (auf Rundsockel) (Pendant vgl. 3856)

Müller, Leo
Als Künstler infrage kommt wohl eher der in der Schreibweise des Familiennamens abweichende Bildhauer Leo Miller. Seine biografischen Daten siehe an dieser Stelle.
4132 „Panther" (auf abgerundetem Rechtecksockel) ÖMKI (1912)
„Eichkätzchen" ÖMKI (1912) Lit.: Neuwirth (1974), S. 68
„Moufflon" ÖMKI (1912) Lit.: Neuwirth (1974), S. 68
„Katze" ÖMKI (1912) Lit.: Neuwirth (1974), S. 68

Müllner, Josef
*Josef Müllner, Bildhauer und Medailleur (*1.8.1879 Baden bei Wien, †25.12.1968 Wien), war Schüler von Zumbusch und Hellmer an der Wiener Akademie. Ab 1910 wird er Lehrer (Prof.), ab 1922 Leiter der Meisterschule für Bildhauerei an der KGW-Schule in Wien. Zu seinen Entwürfen zählen u.a. das Lueger Denkmal in Wien, sowie Ehrengräber auf dem Wiener Zentralfriedhof. Müllner wird 1914 in einem Brief als Mitarbeiter der Firma Goldscheider erwähnt. Für diesen Künstler ließen sich bis Redaktionsschluss jedoch keine signierten und somit keine zuweisbaren Modelle belegen. Lit.: TB, Bd. 25, S. 250; Vollmer, Bd. 3, S. 442–443; Neuwirth (1974), S. 68; Bénézit, Bd. 10, S. 8; AKL Index.*

Mur
2240 FIGUR „Rautendelein" (Mädchen, an einem Teich kniend) ÖMKI (1902/03)

Navarre, Henri Edouard
*Henri Navarre, Bildhauer, Medailleur und Glaskünstler (*4.4.1885 Paris, †August 1971 Paris), beschickte alle drei Pariser Salons und war Mitglied der Künstlergruppe „La Stèle" (–> vgl. Kap. „Dekorative Skulpturen von Arthur Goldscheider in Paris"). Lit.: Vollmer, Bd. 3, S. 464; Neuwirth (1974), S. 68; Bénézit, Bd. 10, S. 117; AKL Index.*

Nelson, (Alphonse Henri?)

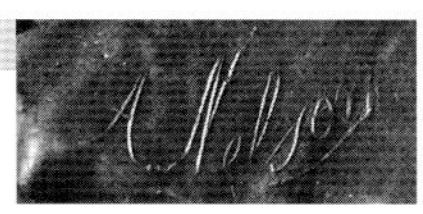

*Alphonse Henri Nelson, Bildhauer (*12.1.1854 Paris, †? wird letztmals 1882 erwähnt), war Mitarbeiter von Friedrich Goldscheider in Paris und Wien. Nelson war Ritter der Ehrenlegion und in Belgien und Frankreich tätig. Lit.: AKL Index.*
775-GSB BUSTE „Femme Fantaisie"
855-GSB BUSTE „Ondine" No. 1 (vgl. 918-GSB; 919-GSB) (vgl. 2091; 2212)
918-GSB BUSTE „Ondine" No. 2 (vgl. 855-GSB; 919-GSB) (vgl. 2091; 2212)
919-GSB BUSTE „Ondine" No. 3 (vgl. 855-GSB; 918-GSB) (vgl. 2091; 2212)
2091 BÜSTE „Ondine"/„Oudine" (Junge Dame mit hochgebundenem Haar) ÖMKI (1900/01; 1902/03) (vgl. 2212)
2212 BÜSTE „Ondine"/„Oudine" (Junge Dame mit hochgebundenem Haar) ÖMKI (1900/01; 1902/03) (vgl. 2091)
2432 BÜSTE „Oudine" ÖMKI (1900/01; 1902/03) (vgl. 2091; 2212)
4522 BÜSTE Frau, nach links blickend MARMOR

Neuwalder-Breuer, Grete
*Grete Neuwalder-Breuer, Bildhauerin und Keramikerin (*28.11.1898 Berlin, † umgekommen 1941 oder 1942 in einem Konzentrationslager), studierte zwischen 1914 und 1919 bei Strnad und Kenner an der KGW-Schule Wien, stellte 1920 das erste Mal in Wien aus und war ab dem selben Jahr Mitarbeiterin der Wiener Werkstätte. Auf der Pariser KGW-Ausst. 1925 war sie mit „Studienköpfen" in Terrakotta vertreten, die von Goldscheider ausgeführt wurden. Auf der Ausstellung „Europäisches Kunstgewerbe" in Leipzig 1927 (Grassimuseum) zeigte sie eine Bronzefigur. Lit.: Grell, S. 33; Vollmer, Bd. 3, S. 475; Neuwirth (1974), S. 41, 68, 230; Zühlsdorff, S. 432; Fahr-Becker, S. 231; Quelle: Bildhauerarchiv, Karl Novak.*
6989 STEHKOPF Mädchenkopf mit Blumenkranz im Haar (Neuwalder-Breuer zugeschr.)
STEHKOPF Frau mit zurückgekämmtem Haar Lit.: N.N.: Österreich in Paris, in: ÖBWK, 1.1924/25, S. 340 (m. Abb.); L'Autriche à L'Exposition Internationale des Arts Décoratifs et Industriels Modernes Paris 1925. Guide, Abb. o.S.-Ang.
WANDMASKE Frau mit Hut (um 1925) (H. 19 cm); Herr 13
BRONZE-Figur (1927) Lit.: Neuwirth (1974), S. 68

Neuwirth, Rosa
*Rosa Neuwirth, Keramikerin, Bildhauerin und Malerin (*25.10.1883 Kgl. Weinberge/bei Prag, †? wird letztmals 1956 erwähnt), studierte zwischen 1899 und 1911 an der KGW-Schule Wien (u.a. bei Kolo Moser und Michael Powolny) und war Schülerin bei Maillol in Paris. Ab 1903/04 Teilnahme an Ausstellungen und ab 1910 Entwurfsausführungen für das Wiener Emailfarbenwerk Schauer & Co. Mitbegründerin der Keramischen Werkgenossenschaft, Wien sowie Mitglied des Österreichischen Werkbundes. Lit.: TB, Bd. 25, S. 423; Vollmer, Bd. 3, S. 475; Neuwirth (1974), S. 68, 232; Zühlsdorff, S. 432; Bénézit, Bd. 10, S. 176; AKL Index.*
„Gimpel" ÖMKI (1911/12) Lit.: Neuwirth (1974), S. 68
„Papagei" ÖMKI (1911/12) Lit.: Neuwirth (1974), S. 68; N.N.: Figürliche Keramik aus zwei Jahrtausenden, Nr. 290 (m. Abb.).

Nieschlag, Karl
*Karl Nieschlag, Bildhauer (*27.5.1909 Matzen/Niederösterreich, †8.1.1975 Wien), war zwischen 1924–1927 an der Wienerberger Werkstättenschule bei Robert Obsieger. Ab 1935 studierte Nieschlag dann an der Akademie in Wien. Durch die Sign. „Melcher" gibt es eine Querverbindung zu diesem Namen –> vgl. auch dort. Quelle: Bildhauerarchiv, Karl Novak.*
8174 FIGUR „Nocturno" (Stehender Damenakt mit großem Tuch) (auf rautenförmigem Sockel) (vgl. 8749) (GS neu Modell-Nr. 54 „Die Bewegung")
8749 FIGUR „Nocturno" (Stehender Damenakt mit großem Tuch) (auf rautenförmigem Sockel) (vgl. Melcher) (vgl. 8174) (GS neu Modell-Nr. 54 „Die Bewegung")

Noël, Tony (Edme Antony Paul)
*Edme Antony Paul Tony-Noël, Bildhauer (*27.1.1845 Paris, †3.10.1909 Palaiseau/Seine et Oise), war Mitarbeiter von Friedrich und Arthur Goldscheider, beide Paris. Lit.: Goldscheider, A. (Hrsg.): Artistic Evolution, 1926, T. 1, Pl. 17 (m. Abb.); TB, Bd. 25, S. 498; Bénézit, Bd. 10, S. 246; AKL Index.*
849-GSB „Querelle D'Amour" No. 2 (vgl. 880-GSB)
852-GSB „Rétaire" No. 1 (vgl. 920-GSB)
880-GSB „Querelle D'Amour" No. 1 (vgl. 849-GSB)
920-GSB „Rétaire" No. 2 (vgl. 852-GSB)
VASE (en grés bleu) (H. 15 cm, B. 15 cm) Lit.: Goldscheider, A. (Hrsg.): Artistic Evolution, 1926, T. 1, Pl. 17 (m. Abb.)

Noll, Alexandre
*Alexandre Noll, Holzbildhauer und Kunstgewerbler (*15.9.1890 Reims, †? Dez. 1970 Fontenay-aux-Roses/Hauts-de-Seine), war Mitglied der Künstlergruppe „Évolution" (–> vgl. Kap. „Dekorative Skulpturen von Arthur Goldscheider in Paris"). Lit.: Goldscheider, A. (Hrsg.): Artistic Evolution, 1926, T. 1, Pl. 24 (m. Abb.); Vollmer, Bd. 3, S. 490; Neuwirth (1974), S. 69; Bénézit, Bd. 10, S. 261.*
VASE (en dinanderie gravée) MESSING (Form von La Monaca) (H. 29 cm, B. 14 cm) Lit.: Goldscheider, A. (Hrsg.): Artistic Evolution, 1926, T. 1, Pl. 24 (m. Abb.)

Obsieger, Robert
*Robert Obsieger, Bildhauer und Keramiker (*23.9.1884 Lundenburg/Böhmen, †27.11.1958 Wien), genoss seine Ausbildung an der Fachschule für Tonindustrie, Znaim. Er studierte von 1909–1914 an der KGW-Schule Wien (u.a. bei Berthold Löffler und Michael Powolny). Obsieger*

hatte 1911 seine erste Ausstellung und war von 1913–1918 Assistent an der Wiener KGW-Schule. Mitglied des Österreichischen Werkbundes. Vom März 1918 bis Juli 1919 war er Professor für keramische Fächer an der Fachschule für Tonindustrie in Znaim. In den Jahren 1921–1932 leitete Obsieger die Werkstättenschule für Keramik an der Wienerberger Ziegelfabriks- und Baugesellschaft. Keramische Arbeiten für die Wiener Werkstätte und zahlreiche Teilnahmen an Ausstellungen auch im Ausland (u.a. an der Pariser KGW-Ausstellung 1925) belegen seine umfangreiche Aktivität als Künstler. Ab 1939 dann Leiter der Wiener KGW-Schule und ab 1945 dann Hochschulprofessur und Vorstand der Meisterklasse für keramische Plastik und Töpferei. Obsieger entwarf hauptsächlich Gebrauchsgegenstände und beschäftigte sich vorwiegend mit Gefäßkeramik, Öfen sowie Heizkörperverkleidungen. Er wurde mit vielen internationalen Auszeichnungen bedacht. Obsieger war 1938 wohnhaft in Wien IV., Phorusgasse 16. Lit.: TB, Bd. 25, S. 553; Vollmer, Bd. 3, S. 503; Bd., 6, S.313; Lehmann, 1938; Neuwirth (1974), S. 69, 234–236 (m. Abb.); Zühlsdorff, S. 433; Fahr-Becker, S. 231; Bénézit, Bd. 10, S. 316; AKL Index.

850 VASE (Obsieger zugeschr.)
851 VASE (Obsieger zugeschr.)
854 VASE (Obsieger zugeschr.)
858 VASE (Obsieger zugeschr.)
860 SCHALE (Obsieger zugeschr.)
863 SCHALE (Obsieger zugeschr.)
ohne Nr. TOPF (Obsieger zugeschr.)
ohne Nr. VASE (Obsieger zugeschr.)
SCHALEN (5 Stück, in diversen Formen) Lit.: N.N.: 50 Jahre Wiener Manufaktur Friedrich Goldscheider, in: Schaulade, 1935, S. 153 (m. Abb.), –> vgl. das Zitat des folgenden Eintrags.
VASEN (7 Stück, in diversen Formen) Lit.: N.N.: 50 Jahre Wiener Manufaktur Friedrich Goldscheider, in: Schaulade, 11.1935, S. 153 (m. Abb.), „... zeigen wir einige der neuesten Goldscheider-Keramiken, nach Entwürfen von Prof. Obsieger. Alle diese Stücke zeichnen sich nicht nur durch handzügige, gut töpferische Formen, sondern auch durch einen fein ausgedrehten, meist noch dazu leicht gerieften Scherben aus. Nimmt man sie in die Hand, so spürt man, daß die „aus der Hand" und nicht aus einer Gußform kommen. Einen besonderen Genuß vermitteln die wie eine feste, dünne Haut aufsitzenden farbigen Mattglasuren. Ihre Tönungen dürfen als wirklich erlesen bezeichnet werden. Wir sehen leuchtendes Rot, tiefes Rotbraun, ein weiches, dunkles Grün und ein sehr eigenes grünliches bis helles Quittengelb, alles dies voller Leben, leicht gefiedert, fein gesprenkelt, oder malerisch breit gefleckt. Man muß es einmal aussprechen; der Raum der wirklichen und sogenannten Kunstkeramik, ist heute weit, sehr weit geworden, eben deshalb bedarf es richtungsweisender Wertarbeiten wie dieser, die Niveau zu schaffen und zu halten vermögen".
VASEN (4, Stück in diversen Formen) Lit.: N.N.: Aus der Wiener Frühjahrsmesse. Die Wiener Edelfirmen I., 50 Jahre Wiener Manufaktur Friedrich Goldscheider, in: ÖK, 6.1935, Nr. 3, S. 15 (m. Abb.), „Schön und erfolgreich zeigen sich auf der Wiener Frühjahrsmesse die neuesten Gefäße nach Entwürfen von Prof. Robert Obsieger. Sie sind durch handzügige gut töpferische Formen und durch einen fein ausgedrehten, meist leicht gerieften Scherben ausgezeichnet. Feine Mattglasuren in erlesenen Tönungen sind ihnen eigen. Sie können als richtungsweisende Leistungen auf dem Gebiete der Kunstkeramik angesprochen werden".

Olesievicz, S.

S. Olesievicz, Kunstgewerbler/in? (Lebens- und Schaffensdaten unbekannt), war Mitglied der Künstlergruppe „Évolution" (-> vgl. Kap. „Dekorative Skulpturen von Arthur Goldscheider in Paris"). Olesievicz war auch Mitarbeiter/in der Firma Primavera, Paris. Lit.: Neuwirth (1974), S. 69; Zühlsdorff, S. 585.

Oury, Louis

Louis Oury, franz. Bildhauer (Lebens- und Schaffensdaten unbekannt), wurde in Montauban geboren und zwischen 1888 und 1899 erwähnt.. Lit.: AKL Index.

2066 BÜSTE „L'Aiglon" (Napoleon II.) (vgl. 2133)
2133 BÜSTE „L'Aiglon" (Napoleon II.) (vgl. 2066)

Ozanne-Cederlund, Frédérique

*Frédérique Ozeanne-Cederlund, Tierbildhauerin (*23.9.1880 Le Havre, †28.7.1966 Antibes), war Schülerin von Valton und Mitglied der Künstlergruppe „La Stèle" (-> vgl. Kap. „Dekorative Skulpturen von Arthur Goldscheider in Paris"). Lit.: Vollmer, Bd. 3, S. 534;Neuwirth (1974), S. 69; Bénézit, Bd. 10, S. 471; AKL Index.*

Padany s. Podany

Padola

Der Name bzw. die Signatur „Padola" muss eindeutig als Pseudonym angesehen werden. Es könnte sich der für Goldscheider überaus produktive Künstler Stanislaus Czapek dahinter verbergen. Biografie siehe dort.

4476 FIGUR/LAMPE Pierrot und Pierrette mit Musikinstrumenten (auf Rundsockel) (vgl. 4773)
4539 LAMPE Scherzendes Biedermeierpaar (Dame und Kavalier) (auf mit Blumengirlanden verziertem Rundsockel) (auch sign. Podola) (vgl. 4305 und 4915= jeweils ähnl. Modell)
4565 Junge, sich drehende Biedermeiertänzerin (auf Rundsockel)
4676 Die Tanzstunde (Violinspieler im Biedermeiergewand mit einem Mädchen tanzend) (auf Rundsockel) (auch sign. Podola/Podolak)
4686 Sitzendes Mädchen, einen blumengeschmückten Hut haltend (auf Rechtecksockel)
4773 Pierrot und Pierrette mit Musikinstrumenten (auf Rundsockel) (vgl. 4476)
4774 Sitzende Pierrette mit Laute (vgl. 4996 und Podany 4774)
4865 Sich verbeugende Tänzerin (mit Schal) (auf Rundsockel)
4935 Stehender Harlekin mit Laute (auf Rundsockel/oktogonalem Sockel) (auch sign. Podolak) (vgl. 5262; 5263 sign Podolak) (vgl. 4922 GRUPPE, sign. Padola)

Padola s.a. Podola

Palma

566-GSB BUSTE „Jeanne D' Arc" ZINN/ZINKGUSS/BRONZE

Palter s. Polter

Paltier s. Polter

Paltir s. Polter

Paris, Auguste

*Auguste Paris, Bildhauer (*1850 Paris-Belleville, †25.3.1915 Colombes/Hauts-de-Seine), war u.a. Schüler von Jouffroy. Auguste Paris debütierte im Pariser Salon 1876 und war Mitarbeiter bei Friedrich Goldscheider. Unter seinen zahlreichen Auszeichnungen sei eine Goldmedaille bei der Weltausst. 1889 besonders hervorgehoben. Lit.: TB, Bd. 26, S. 234; Hillenbrand, S. 314; Bénézit, Bd. 10, S. 573; AKL Index.*

782-GSB STATUE „Éternel Moqueur" (vgl. 783-GSB= BÜSTE; 842-GSB= BÜSTE) (vgl. 1991)
783-GSB BUSTE „Éternel Moqueur" (vgl. 842-GSB) (vgl. 782-GSB= FIGUR)
842-GSB BUSTE „Éternel Moqueur" (vgl. 783-GSB) (vgl. 782-GSB= FIGUR) (vgl. 1991)
1991 BÜSTE „Spottvogel"/„Éternel Moqueur" (Lachender Bursche) (vgl. 783-GSB= BÜSTE; 842-GSB= BÜSTE)

Pecheur (d.i. vermutl. Friedrich Fischer)

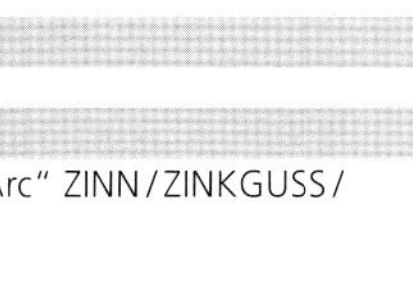

Die mit „Pecheur" signierten Modelle dürfen wieder als, von Goldscheider häufig praktiziert, Pseudonym angesehen werden. Es dürfte sich hier um den österreichischen Bildhauer Friedrich Fischer handeln. Biographische Daten und weitere für den Namen Fischer nachgewiesene Modelle siehe dort. Bei Neuwirth (1974) S. 69 nur die Namensnennung, jedoch keinerlei Hinweis auf Lebens- und Schaffensdaten.

724 VASE Zwei Frauen (Wassernixen?) (an der Vasenwandung) (vgl. 795) (ÖMKI 1900/01?)
795 VASE Zwei Frauen (Wassernixen?) (an der Vasenwandung) (vgl. 724) (ÖMKI 1900/01?)
949 RELIEF Junges Paar mit Buch und Blütenzweig (Wandrelief) (Pendant vgl. 950)
950 RELIEF Schnitter (Junges Paar bei der Feldarbeit) (Wandrelief) (Pendant vgl. 949)
1041 „Schaukel" (Doppel-VASE Junge Frau, zwischen einer Astgabel schwingend) (vgl. 1072)
1072 „Schaukel" (Doppel-VASE Junge Frau, zwischen einer Astgabel schwingend) (sign. Simon) (vgl. 1041)
2372 FIGUR „Poesie" ÖMKI (1901/02)
2457 BÜSTE „Empire" (Junge Dame mit Hochsteckfrisur) (nach rechts blickend) ÖMKI (1906/07)
2506 UHR Stehendes, Seifenblasen pustendes Holländermädchen (Uhrgehäuse reliefiert mit zwei Blumentöpfen/Tulpen)
2891 BÜSTE Junge Dame mit Diadem (Sockelrelief: lyraspielender Jüngling)
4343 BÜSTE Biedermeiermädchen mit Schutenhut ALABASTER
VASE „Wassernixen" (Terrakotta) ÖMKI (1900/01) (vgl. evtl. 795) Lit.: Neuwirth (1974), S. 69
„Prosa" ÖMKI (1901/02) Lit.: Neuwirth (1974), S. 69

Pecquériaux, Henri Paul

*Henri Paul Pecquériaux, Zeichner und Illustrator (*14.10.1889 Avesnelles/Frankreich, †?), war Mitglied der Künstlergruppe „Évolution" (-> vgl. Kap. „Dekorative Skulpturen von Arthur Goldscheider in Paris") und für Arthur Goldscheider als Décorateur tätig. Lit.: Vollmer, Bd. 3, S. 561; Neuwirth (1974), S. 69; AKL Index.*

Pejic, (R.?)

Bei Neuwirth (1974) S. 69 nur die Namensnennung, jedoch keinerlei Hinweis auf Lebens- und Schaffensdaten.

5691 BÜSTE Araber mit Turban (auf quaderförmigem Sockel) MARMOR
5693 KERZENHALTER (dreiflammig) „Leuchter (mit) Fabeltier"
BODENVASE Plastische Frauenbüste, Blumendekor sign. R. Pejic (undeutl. GS Firmenmarke), (um 1920/25) (H. 91 cm); Herr 13

Perigot

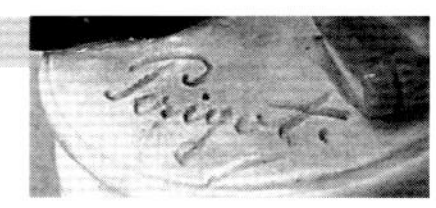

Bei Zühlsdorff nur der allgemeine Hinweis auf eine(n) Kunstgewerbler(in?) und die Mitarbeiterschaft für Goldscheider. Es ist aber sicher davon auszugehen, dass es sich hier, um einen Phantasienamen handelt, der wegen seines (internationalen) Wohlklangs und der damit verbundenen verkaufsfördernden Wirkung verwendet wurde. Lit.: Zühlsdorff, S. 403; Neuwirth (1974), S. 69, hier nur die Nennung des Künstlernamens.

1231 DECKELDOSE Mädchen mit Weihnachtsbaum
2960 Paar an einem Tisch stehend (sich mit Gläsern zuprostend)
3522 vgl. Rosé
3649 JARDINIERE/KONFEKTSCHALE Biedermeiermädchen mit übergroßem quadratischen Korb
3844 UHR Stehendes, sich umarmendes Rokokopaar (auf Blumengirlanden verziertem Sockel/Gehäuse)
3860 Stehende Rokokodame mit Hut (sich auf großen ovalen Korb stützend)
3991 UHR Sitzende Lautenspielerin (Rokokodame)
4017 UHR Dame mit Kavalier
4043 Rokokodame mit großem Korb
4059 Stehendes Biedermeiermädchen mit zwei großen Körben
4079 Krinolinendame mit Blumenvase
4101 UHR Sitzende Biedermeierdame mit Hut, ihren Schuh anziehend
4134? Stehendes lesendes Mädchen (auf Rundsockel)
4163 TAFELAUFSATZ Zwei Putti mit Trauben, eine Schale tragend (Tafelaufsatz/Obstschale)
4227 Stehendes Mädchen mit Schokoladentasse (auf Rundsockel)
4393 „Mozart" (Stehende Figur als Knabe mit Geige, Notenheft und (ohne/mit) Bücherstapel) (vgl. 4610; 5519) (GS neu ohne Modell-Nr. „Mozart")
4451 Stehende Dame im Jagdoutfit mit Barsoi (auf Landschaftssockel)
4557 Stehendes Mädchen in Pumphosen (auf Ovalsockel)
4558 Stehendes Mädchen mit Hut (auf Rundsockel)
4594 Stehendes weinendes Mädchen (auf Rundsockel)

(auch sign. Petri)
4610 „Mozart" (Stehende Figur als Knabe mit Geige, Notenheft und Bücherstapel) FAYENCE / ALABASTER (vgl. 4393; 5519) (GS neu o. Modell-Nr. „Mozart")
4659 Stehendes Biedermeiermädchen mit Blumenstrauß und Schleife im Haar (auf Rundsockel)
4893 Stehendes Mädchen mit Haube und Buch, einen Blumenstrauß hinter dem Rücken versteckend (auf Rundsockel)
5519 „Mozart" (Stehende Figur als Knabe mit Geige, Notenheft und (ohne / mit) Bücherstapel) (vgl. 4393; 5519) (GS neu o. Modell-Nr. „Mozart")
Stehender Knabe mit Fass (um 1905/10) (H. ca. 25 cm); WKH (ohne nähere Ang.)

Perl, Karl

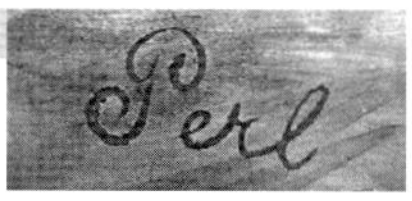

*Karl Perl, Bildhauer und Medailleur (*3.3.1876 Liezen / Steiermark, †2.8.1965 Wien), studierte in Wien unter Hellmer, Kundmann und Zumbusch. Bei der Weihnachtsschau des Wiener Künstlerhauses 1926 wurden Goldscheidermodelle präsentiert. Neben seinen Figurinen und Gruppen für Manufaktur beschäftigte er sich mit dem Entwerfen von Medaillen und Gedenktafeln. Wandmasken von seiner Hand wurden durch die Wiener Firma Keramos veröffentlicht. Perl war ab 1908 wohnhaft in Wien IV., Favoritenstr. 62, später dann im XII. Bez., Zenogasse 5. Lit.: TB, Bd. 26, S. 478; Vollmer, Bd. 3, S. 569; Lehmann, 1908–1941; Davidson, S. 481; Hillenbrand, S. 316; Neuwirth (1974), S. 70; Bénézit, Bd. 10, S. 753; Makus (2000), S. 176; AKL Index.*
5001 Spanisches Paar beim Tanz (auf Rechtecksockel) FAYENCE; ALABASTER (vgl. 5260)
5005 Göttin der Fruchtbarkeit (Ceres / Demeter?) (Stehender Halbakt mit Feldfrüchten, Trauben, Obst und Getreideähren) (auf getrepptem Quadratsockel)
5009 Stehender Frauenakt mit Tuch (auf getrepptem Quadratsockel) ALABASTER / Marmorsockel
5024 Sitzender Araber mit Turban und langer Pfeife (auf Rechtecksockel)
5025 Sitzender Damenakt, mit Leoparden scherzend (auf Landschaftssockel)
5042 König (Stehende posierende Figur mit Krone) (auf quaderförmigem Sockel)
5043 Stehender Chinese mit Kopfbedeckung und zwei erhobenen Zeigefingern (auf zylindrischem Sockel) (Prinz Sou-Chong)
5048 Hänsel und Gretel
5077 Stehender Seemann (auf quaderförmigem Sockel)
5236 „Die sieben Schwaben" (Siebenköpfige Figurengruppe auf rechteckigem Landschaftssockel)
5260 Spanisches Paar beim Tanz (auf Rechtecksockel) (vgl. 5001)
5289 Das letzte Abendmahl (auf Rechtecksockel)
„Seiltänzerin" (Balancierender Damenakt) BRONZE / Marmorsockel (Echt Bronze, Golddekor) Lit.: Ros, L.: Wiener Fayencen und Bronzen, in: KuKGW, 4.1923/24, S. 17 (m. Abb.), hier mit Ang. des Originaltitels „Seiltänzerin", „Wie ein weiches Adagio, das man ganz leise spielen muß, um den Zuhörer nicht zu erschrecken, so mutet diese Figur an. Als etwas rührend Unsicheres und Zartes. Gerade durch diesen schmalen Untersatz, auf dem die noch schmäleren Füßchen zu gehen versuchen, bringt der Künstler die frühlingszarte, knospenhafte, feinnervige Figur mit der tastenden Bewegung erst richtig zur Geltung".

Pernot, Henri

*Henri Pernot, Bildhauer (*10.3.1859 Gent, †1937), war Mitglied der Societé des Artistes Français und wurde in deren Salons 1897, 1904 und 1909 prämiert. Kashey nennt Pernot als Mitarbeiter von Friedrich Goldscheider, Paris. Lit.: TB, Bd. 26, S. 425; Kashey, S. 272; Hillenbrand, S. 317; Bénézit, Bd. 10, S. 759; AKL Index.*

Persan

3091 Gänseliesel (Kleines Mädchen mit Kükenschar) (auf Natursockel)
3840 Holländische Gärtnerin mit Korb und Holzrechen (Pendants vgl. 2662; 2917)

Petri (d.i. Petrides, János (Johann))

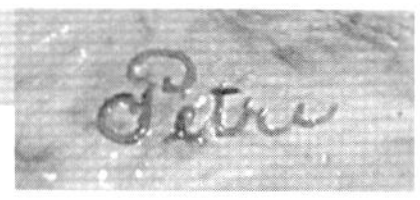

*Da sich eine konkrete Zusammenarbeit zwischen dem in der Lit. (auch bei Neuwirth (1974) S. 70) und von einigen Auktionshäusern immer wieder genannten Berliner Bildhauer Otto Petri (*15.4.1860 Berlin, †1942 Berlin) mit Goldscheider nicht nachweisen lässt, es aber von János Petrides (Biografie siehe dort), einem ungarischen Bildhauer Belege gibt, drängt sich hier die Vermutung auf, dass die Manufaktur den Künstlernamen, wie an anderen Stellen bereits mehrfach erwähnt, „modifizierte". Letztendlich wird jedoch keine Klärung herbeigeführt werden können. Möglicherweise handelt es sich bei dem Namen „Petri" auch nur um ein Pseudonym. Somit werden an dieser Stelle nur die mit „Petri" sign. Modelle aufgelistet und bei Petrides alle Objekte. Die in einigen Auktionskat. angegebene Keramikerin Trude Petri (1906–1968) kann alleine schon wegen ihres Geburtsjahres ausgeschlossen werden.*
1114 VASE Holländisches Mädchen, ihre Schürze hochhebend (vgl. Pendant 1115)
1115 VASE Holländischer (Fischer-)Knabe mit Korb (vgl. Pendant 1114)
1241 VASE Jugendlicher Ballwerfer in antikisierendem Gewand (große Bodenvase)
1461 VASE Diana als Bogenschützin (vollplastisch an der Vasenwandung stehend, säulenartiger Vasenschaft mit Löwenmaskaron) (große Bodenvase)
2345 UHR Stehendes, betendes Mädchen (auf abgerundeter Plinthe, stelen- / quaderförmiger Hintergrund)
2490 BÜSTE Junge Frau mit Kappe (Holländerin?)
2510 Stehender Fischer mit Sturmhaube und Weidenkorb (auf Landschaftssockel) (vgl. Pendant 2535)
2535 Stehende Fischerin (Holländerin) mit Haube, Netz und Korb (auf Landschaftssockel) (vgl. Pendant 2510)
2579 BÜSTE Römischer Krieger mit Helm (bekrönt mit Adler) (auf antiker Säule / Kapitell)
2850 Stehendes Holländermädchen in einem „Teich" angelnd (Fishbowl?)
2875 BÜSTE Ritter mit Helm (bekrönt mit Bourbonenlilie) (auf mit drei Bourbonenlilien verziertem Sockel)
2985 Bauernmädchen, einen Wagen mit Korb ziehend (vgl. 3163)
3095 UHR Sitzender Knabe
3163 Bauernmädchen, einen Wagen mit Korb ziehend (vgl. 2985)
3347 Schreitende Frau, einen flachem Korb haltend (auf Ovalsockel)
3581 Stehendes Biedermeiermädchen mit Korb und Handtasche (Beutel)
3592 Spielendes Kinderpaar (Das Mädchen hält dem Jungen die Augen zu) (auf Ovalsockel)
3629 Putto vor (Grab?)-Denkmal sitzend (Inschrift: 1885–1910)
3724 UHR Stehendes Biedermeiermädchen (auf konischem, mit Blumengirlanden verziertem Sockel)
3782 Schreitender Knabe in Tracht (mit Ledertasche und erlegtem Wild) (auf Landschaftssockel)
3895 LAMPE Blumenmädchen, eine Kugel aus aus Rosen tragend
4026 POSTKARTENHALTER Stehendes Biedermeiermädchen mit Schirm (auf getrepptem Sockel)
4040 Stehender Bauernbub in Tracht mit zwei Vögeln und Umhängetasche (auf Rechtecksockel)
4124 BÜSTE Holländermädchen (auf oktogonalem Längssockel) ALABASTER
4200 Kleines Biedermeiermädchen mit Hut und Schirm
4202 Stehender Jockey (auf quaderförmigem Landschaftssockel) FAYENCE / ALABASTER
4219 Stehendes Biedermeiermädchen mit Korb, Krug und Blumentopf
4264 LAMPE Sitzendes lesendes Biedermeiermädchen
4283 BÜSTE Bub mit Hut ALABASTER
4355 „Die kleine Strickerin" (Stehendes strickendes Biedermeiermädchen mit Hut)
4361 „Vorbereitung zum Tanz" (Stehendes Biedermeiermädchen mit Handspiegel) (auch sign. Petrides)
4458 FIGUR / LAMPE Stehendes Mädchen mit Haube und gefalteten Händen (auf zylindrischem Rundsockel) FAYENCE / ALABASTER; (BRONZE als LAMPE)
4484 Lesende Dame auf Récamiere (Sofa)
4527 BÜSTE Biedermeiermädchen mit Schutenhut ALABASTER
4546 Stehendes Mädchen mit Blumenstrauß ALABASTER / STEINGUSS?
4551 Stehendes Mädchen (Halbakt) mit Gitarre (auf hohem Rundsockel) FAYENCE / ALABASTER
4575 BÜSTE Mädchen mit Haube und gefalteten Händen ALABASTER
4593 BÜSTE Mädchen mit Haube ALABASTER
4594 Stehendes weinendes Mädchen (auf Rundsockel) (auch sign. Perigot) FAYENCE / ALABASTER
4614 Stehender Mädchenhalbakt mit Tuch (auf Rundsockel) FAYENCE; ALABASTER / Marmorsockel
4632 Stehendes kleines Mädchen mit Schiefertafel und Buch (auf Landschaftssockel)
4644 Stehendes kleines Mädchen (auf Landschaftssockel) ALABASTER
4650 Stehendes Mädchen mit Haube und Schiefertafel (auf Rundsockel) FAYENCE / ALABASTER (vgl. auch Polter 4650!)
4651 FIGUR / LAMPE Stehender Knabe mit Haube und Apfel (auf Rundsockel) ALABASTER / STEINGUSS (?) (vgl. auch Polter 4651!)
4655 BÜSTE Kleiner Bub ALABASTER / Marmorsockel
4664 BÜSTE Holländermädchen mit Haube ALABASTER (vgl. 4278= ähnl. Modell)
4678 Stehendes, sich verneigendes Biedermeiermädchen (ihr Kleid etwas hochhebend) (auf Rundsockel)
4713 „Rast" (Sitzendes Bauernmädchen mit Kopftuch) FAYENCE / farbiger MARMOR
4714 Sitzendes Mädchen ALABASTER
4726 BÜSTE Kleines Mädchen (nach links blickend) ALABASTER / Marmorsockel
4727 BÜSTE Biedermeiermädchen mit Schutenhut ALABASTER
4730 BÜSTE Kleines Mädchen mit Hut ALABASTER
4739 BÜSTE Bub mit Kappe ALABASTER (Pendant vgl. 4740)
4740 BÜSTE Holländermädchen ALABASTER (ähnl. Modell 4664) (Pendant vgl. 4739)
4788 BÜSTE Bub (etwas nach rechts blickend) ALABASTER
4850 BÜSTE Bub mit Hut ALABASTER
4854 BÜSTE Kleines Mädchen ALABASTER
5382 BÜSTE Knabe mit Kappe ALABASTER
5386 BÜSTE Kleines Mädchen mit Kappe ALABASTER
5425 Stehendes kleines Mädchen mit Puppe (auf Rechtecksockel) ALABASTER
RELIEF Frauenkopf (im Profil nach rechts) mit Haarschmuck (Dekor „Secession") Lit.: R., F.: Wiener Kunstkeramik, in: KM, 2.1902, S. 91 (m. Abb.), „Das Dekor „Secession" zeigt in der Art seiner Anwendung viel Verwandtes mit den künstlerischen Leder-Reliefarbeiten, welche mit matten diskreten Übermalungen verziert sind. Als Beispiel der in diesem Dekor gehaltenen Arbeiten gibt die Fig. (Abb., Anm. des Autors) 9 ein von Petri entworfenes Relief wieder. In licht gelblicher Kolorierung hebt das gemmenartig fein geschnittene Gesicht von dem lederbraunen Hintergrunde. Das Haar weist eine stumpf-rotbraune Färbung auf, während matte Bronze-Auflagen das Metallmaterial des Halsschmuckes wiedergeben".
VASE Sonnenblumenvase (Dekor: „Secession) Lit.: R., F.: Wiener Kunstkeramik, in: KM, 2.1902, S. 91, „Die Sonnenblumenvase zeigt ebenfalls im Secessionsdekor vollentfaltete stilisierte Sonnenblumen, gelb zu braun schattiert, mit stumpfgrünen Blättern auf lederbraunem, goldübertäubtem Fond. Wir bemerken noch, daß die verschiedenen Kunstgegenstände in mehreren dieser Dekorationen hergestellt werden".
FIGUR „Kokette" ÖMKI (1906/07) Lit.: Neuwirth (1974), S. 70
Biedermeiermädchen auf Stuhl kniend, mit Katze (um 1910/15) (H. 28,8 cm) (vgl. Polter 4317); Doro 20.10.1997
Sitzendes Mädchen mit Hündchen (um 1910) (H. 25 cm); Henry's 9.12.1993
Mädchen mit Hund und Katze auf einer Bank sitzend (um 1910) FG Vienna (Firmenmarke 15?), (H. 23,5 cm, L. 21 cm); Kett. 28

Petri vgl. auch Polter

Petrides, János (Johann)

*János (Johann) Petrides, Bildhauer und Keramiker (*1861 Lemberg, †? wird letztmals 1932 erwähnt), war einer der produktivsten Künstler für die Manufaktur Goldscheider,* *wenn man nicht, wie unter dem Eintrag „Petri" dargestellt, von einem Pseudonym ausgeht. Ein Beleg für die These, dass die Signatur Petri dem Bildhauer Petrides zugeordnet werden kann, liefert die Literatur zur Modell-Nr. 4355 (-> vgl. Studio YB, 1914, S. 190). Seine Modelle sind meist nur mit „Petri" signiert, was wiederum die bei Goldscheider gängige Praxis der „klangvollen" und „griffigen" Namensgebung widerspiegelt. Petrides studierte in Wien und war auch für die Manufaktur Zsolnay in Pécs (Ungarn) tätig. Lit.: TB, Bd. 26, S. 497; Neuwirth (1974), S. 70; Bénézit, Bd. 10, S. 812; AKL Index.*

1114 VASE Holländisches Mädchen, ihre Schürze hochhebend (vgl. Pendant 1115) (sign. Petri)
1115 VASE Holländischer (Fischer-)Knabe mit Korb (vgl. Pendant 1114) (sign. Petri)
1241 VASE Jugendlicher Ballwerfer in antikisierendem Gewand (grosse Bodenvase) (sign. Petri)
1461 VASE Diana als Bogenschützin (vollplastisch an der Vasenwandung stehend, säulenartiger Vasenschaft mit Löwenmaskaron) (große Bodenvase) (sign. Petri)
2345 UHR Stehendes, betendes Mädchen (auf abgerundeter Plinthe, stelen/quaderförmiger Hintergrund) (sign. Petri)
2490 BÜSTE Junge Frau mit Kappe (Holländerin?) (sign. Petri)
2510 Stehender Fischer mit Sturmhaube und Weidenkorb (auf Landschaftssockel) (sign. Petri)
2579 BÜSTE Römischer Krieger mit Helm (bekrönt mit Adler) (auf antiker Säule) (sign. Petri)
2850 Stehendes Holländermädchen in einem „Teich" angelnd (Fishbowl?) (sign. Petri)
2875 BÜSTE Ritter mit Helm (bekrönt mit Bourbonenlilie) (auf mit drei Bourbonenlilien verziertem Sockel) (sign. Petri)
2985 Bauernmädchen, einen Wagen mit Korb ziehend (sign. Petri) (vgl. 3163)
3095 UHR Sitzender Knabe (sign. Petri)
3163 Bauernmädchen, einen Wagen mit Korb ziehend (sign. Petri) (vgl. 2985)
3347 Schreitende Frau, einen flachem Korb haltend (auf Ovalsockel) (sign. Petri)
3581 Stehendes Biedermeiermädchen mit Korb und Handtasche (Beutel) (sign. Petri)
3592 Spielendes Kinderpaar (Das Mädchen hält dem Jungen die Augen zu) (auf Ovalsockel) (sign. Petri)
3629 Putto vor (Grab?)-Denkmal sitzend (Inschrift: 1885–1910) (sign. Petri)
3724 UHR Stehendes Biedermeiermädchen (auf konischem, mit Blumengirlanden verziertem Sockel) (sign. Petri)
3782 Schreitender Knabe in Tracht (mit Ledertasche und erlegtem Wild) (auf Landschaftssockel) (sign. Petri)
3895 LAMPE Blumenmädchen, eine Kugel aus aus Rosen tragend (sign. Petri)
4026 POSTKARTENHALTER Stehendes Biedermeiermädchen mit Schirm (auf getrepptem Sockel) (sign. Petri)
4040 Stehender Bauernbub in Tracht mit zwei Vögeln und Umhängetasche (auf Rechtecksockel) (sign. Petri)
4124 BÜSTE Holländermädchen (auf oktogonalem Längssockel) ALABASTER (sign. Petri)
4200 Kleines Biedermeiermädchen mit Hut und Schirm (sign. Petri)
4202 Stehender Jockey (auf quaderförmigem Landschaftssockel) (sign. Petri)
4219 Stehendes Biedermeiermädchen mit Korb, Krug und Blumentopf (sign. Petri)
4264 LAMPE Sitzendes lesendes Biedermeiermädchen (sign. Petri)
4278 BÜSTE Holländermädchen mit Haube ALABASTER (Petrides zugeschr.) (vgl. 4664= ähnl. Modell)
4283 BÜSTE Bub mit Hut ALABASTER (sign. Petri)
4355 „Die kleine Strickerin" (Stehendes Biedermeiermädchen mit Hut) (sign. Petri)
4361 „Vorbereitung zum Tanz" (Stehendes Biedermeiermädchen mit Handspiegel) (sign. Petri)
4458 FIGUR/LAMPE Stehendes Mädchen mit Haube und gefalteten Händen FAYENCE/ALABASTER/BRONZE (sign. Petri)
4484 Lesende Dame auf Récamiere (Sofa) (sign. Petri)
4527 BÜSTE Biedermeiermädchen mit Schutenhut ALABASTER (sign. Petri)
4546 Stehendes Mädchen mit Blumenstrauß ALABASTER/STEINGUSS? (sign. Petri)
4551 Stehendes Mädchen (Halbakt) mit Gitarre (auf hohem Rundsockel) (sign. Petri)
4575 BÜSTE Mädchen mit Haube und gefalteten Händen ALABASTER (sign. Petri)
4593 BÜSTE Mädchen mit Haube ALABASTER (sign. Petri)
4594 Stehendes weinendes Mädchen (auf Rundsockel) (sign. Petri, auch sign. Perigot) FAYENCE/ALABASTER
4614 Stehender Mädchenhalbakt mit Tuch (auf Rundsockel) FAYENCE; ALABASTER/Marmorsockel (sign. Petri)
4632 Stehendes kleines Mädchen mit Schiefertafel und Buch (auf Landschaftssockel) (sign. Petri)
4644 Stehendes kleines Mädchen (auf Landschaftssockel) ALABASTER (sign. Petri)
4651 Stehender Knabe mit Haube und Apfel ALABASTER/STEINGUSS ? (sign. Petri) (auch sign. Polter 4651!)
4655 BÜSTE Kleiner Bub ALABASTER/Marmorsockel (sign. Petri)
4664 BÜSTE Holländermädchen mit Haube ALABASTER (sign. Petri) (vgl. 4278= ähnl. Modell)
4678 Stehendes, sich verneigendes Biedermeiermädchen (ihr Kleid etwas hochhebend) (auf Rundsockel) (sign. Petri)
4713 „Rast" (Sitzendes Bauernmädchen mit Kopftuch) FAYENCE/farbiger MARMOR
4714 Sitzendes Mädchen ALABASTER (sign. Petri)
4726 BÜSTE Kleines Mädchen (nach links blickend) ALABASTER/Marmorsockel (sign. Petri)
4727 BÜSTE Biedermeiermädchen mit Schutenhut ALABASTER (sign. Petri)
4730 BÜSTE Kleines Mädchen mit Hut ALABASTER (sign. Petri)
4739 BÜSTE Bub mit Kappe ALABASTER (sign. Petri) (Pendant vgl. 4740)
4740 BÜSTE Holländermädchen ALABASTER (sign. Petri) (Pendant vgl. 4739)
4788 BÜSTE Bub (etwas nach rechts blickend) ALABASTER (sign. Petri)
4850 BÜSTE Bub mit Hut ALABASTER (sign. Petri)
4854 BÜSTE Kleines Mädchen ALABASTER (sign. Petri)
5382 BÜSTE Knabe mit Kappe ALABASTER (sign. Petri)
5386 BÜSTE Kleines Mädchen mit Kappe ALABASTER (sign. Petri)
5425 Stehendes kleines Mädchen mit Puppe (auf Rechtecksockel) ALABASTER (sign. Petri)
JARDINIERE „Südfrüchte" ÖMKI (1903/04) Lit.: Neuwirth (1974), S. 70
JARDINIERE „Tannenzapfen" ÖMKI (1903/04) Lit.: Neuwirth (1974), S. 70
FIGUR „Kokette" ÖMKI (1906/07) (hier Angabe „Petri") Lit.: Neuwirth (1974), S. 70
Elektrischer Wandleuchter ÖMKI (1912) Lit.: Neuwirth (1974), S. 70
„Kind mit Eidechse" ÖMKI (1912) Lit.: Neuwirth (1974), S. 70
„Amor" ÖMKI (1912) Lit.: Neuwirth (1974), S. 70
AUFSATZ „Putti" ÖMKI (1912) Lit.: Neuwirth (1974), S. 70
„Kind im Stuhl" ÖMKI (1912) Lit.: Neuwirth (1974), S. 70
RELIEF Frauenkopf (im Profil nach rechts) mit Haarschmuck (Dekor „Sezession") Lit.: KM, 1902, S. 91, (hier mit Ang. des Künstlernamens „Petri").
VASE Sonnenblumenvase Lit.: KM, 1902, S. 91, (hier Ang. des Künstlernamens „Petri").
Biedermeiermädchen auf Stuhl kniend, mit Katze (um 1910/15) (sign. Petri) (H. 28,8 cm) (vgl. Polter 4317); Doro 20.10.1997
Sitzendes Mädchen mit Hündchen (sign. Petri) (um 1910) (H. 25 cm); Henry's 9.12.1993
Mädchen mit Hund und Katze auf einer Bank sitzend (sign. Petri) (um 1910) FG Vienna (Firmenmarke 15?), (H. 23,5 cm, L. 21 cm); Kett. 28

Petrides, Joseph s. Petrides, János (Johann)

Pettarin, A.

„Cäcilie" ÖMKI (1903/04) Lit.: Neuwirth (1974), S. 70

Peyre, Raphael Charles

Raphael Charles Peyre, Bildhauer (1.6.1872 Paris, †3.1.1949 Paris), war Schüler von Falguière, Mercié und Barreau. Lit.: TB, Bd. 26, S. 514; Bénézit, Bd. 10, S. 828; AKL Index.*

„Liliom" (Stehender Mädchenakt mit Lilie im Haar) BRONZE Lit.: Magyar Iparmüvészet, 3.1900, S. 293 (m. Abb.)

Pfeffer, L.

3117 LAMPE Stehendes Mädchen mit zwei Mohnblüten

Pfister, (Émile?)

(Émile) Pfister, Bildhauer (Neuf-Brisach/Haut-Rhin), stellte 1906 im Pariser Salon des Artistes Français aus und erhielt dort eine Mention honorable. Lit.: Bénézit, Bd. 10, S. 839; AKL Index.*

4268 Stehende, sich kämmende junge Frau (Hera?) mit Pfau (auf profiliertem Rundsockel)

Philippe, Paul

Paul Philippe, Bildhauer (Lebensdaten unbekannt), studierte an der Kgl. Akademie in Berlin und in Paris bei A. Larroux. Er war für Friedrich und Arthur Goldscheider tätig und lebte ab 1900 in Paris. Dort stellte er zwischen 1902–1912 u.a. im Salon (Societé des Artistes Français) aus. Seine bevorzugten Materialien waren Bronze, Marmor und Elfenbein und wurden neben Goldscheider auch von Rosenthal & Maeder sowie von Preiss & Kassler editiert. Lit.: N.N.: Bronze-Elfenbein Statuetten der Neuzeit, Künstlerkat. 7. Forts., in: AZ, 1979, S. 581; Bénézit, Bd. 10, S. 849; Internet: www.endymion.co.uk/Details.html (vom 7.6.2005).

21-LaStèle Stehender Damenakt mit ausgestrecktem Arm
111 „Le Réveil" (Stehender Damenakt mit ausgestrecktem Arm) (auf Rundsockel) (Inschrift: LE RÉVEIL) BRONZE/Marmorsockel
3110 Auf Hocker sitzender Mädchenhalbakt mit Katze (auch in BRONZE ausgeführt)
5405 Tänzerin in Pfauenkostüm (auf Rundsockel) (auch sign. „Phillippe")
„Femme au masque" MARMOR/ELFENBEIN (Arthur GS, Paris) (1908) Lit.: N.N.: Bronze-Elfenbein Statuetten der Neuzeit., Künstlerkat. 7. Forts., in: AZ, 1979, S. 581.
„Farniente" BRONZE/ELFENBEIN/MARMOR (Arthur GS, Paris) (1909) Lit.: N.N.: Bronze-Elfenbein Statuetten der Neuzeit., Künstlerkat. 7. Forts., in: AZ, 1979, S. 581.
„Grâce et Force" BRONZE/ELFENBEIN/MARMOR (Arthur GS, Paris) (1909) Lit.: N.N.: Bronze-Elfenbein Statuetten der Neuzeit., Künstlerkat. 7. Forts., in: AZ, 1979, S. 581.

Piémont, T. du s. Prémont, F./J./T. de/du

Piesch

6085 KERZENHALTER (zweiflammig) (Form: zwei Kerzentüllen auf rohrartigem Unterbau, über kegelförmigem Sockel/Handhabe: halbkreisförmig/Dekor: gemalte, offene Ringe auf Craquelé/Laufglasur)
6164 BUCHSTÜTZE Max (als stehende Figur) (Pendant vgl. 6165)
6165 BUCHSTÜTZE Moritz (als stehende Figur) (Pendant vgl. 6164)

Piquard

2394 BÜSTE „Mais" (Bauernmädchen mit Kopftuch) (Sockel reliefiert mit Maiskolben)

Pitra, Josef

Josef Pitra, Bildhauer oder Kunstgewerbler, absolvierte zwischen 1887–1897 die Kunstgewerbeschule in Bechyne und wurde bei GS (als Modelleur, Formgießer bzw. Figurenretoucheur) angestellt. Weitere Informationen sind nicht bekannt. Lit.: Neuwirth (1974), S. 31.

Podany, Rudolf

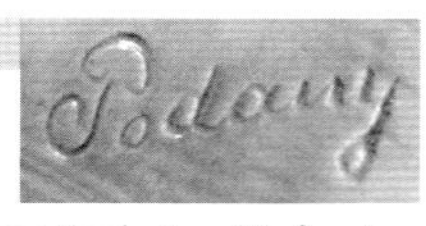

*Rudolf Podany, Bildhauer und Keramiker (*23.8.1876 Wien, †4.6.1963 Wien), war eineinhalb Jahre Schüler von Johannes Benk. Anschließend studierte Podany neun Jahre an der Akademie der bildenden Künste in Wien, war zehn Jahre für die Porzellanmanufaktur Schauer tätig und ab 1920 bei der Wiener Firma Keramos, deren Kommandatist er war, beschäftigt. Podany war ab 1928 wohnhaft in Wien XV., Löschenkohlgasse 40. Lit.: Makus (2000), S. 176; Lehmann, 1928–1942; Quelle: Bildhauerarchiv, Karl Novak.*

4209 LAMPE Stehender Pierrot mit zwei Pierretten (auf Rundsockel mit Blumenstrauß)
4774 Sitzende Pierrette mit Laute (vgl. 4996; und Padola 4774)
4799 Tanzendes Kinderpaar (im Biedermeiergewand) (auf getrepptem Rundsockel)
4814 Adorée-Via Villany („Tanzspiel mit Bändern") (auf oktogonalem Sockel) (vgl. 4908) (GS neu Modell-Nr. 71 „Muse des Tanzes")
4818 Stehende Dame mit entblößter Schulter und langem Kleid (auf Rechtecksockel)

4834 FIGUR/LAMPE „Bei der Toilette" (Sitzende Balldame im Rokokokleid) (auf Ovalsockel) (vgl. 5169; 5329)
4860 BÜSTE Pierrot ALABASTER
4866 BÜSTE Junge Dame mit Haarband ALABASTER
4908 Adorée-Via Villany („Tanzspiel mit Bändern") (auf oktogonalem Sockel) (vgl. 4814) (GS neu Modell-Nr. 71 „Muse des Tanzes")
4996 Sitzende Pierrette mit Laute (vgl. 4774 auch Padola sign.)
5050 Sitzende Pierrette (auf Rechtecksockel)
5169 „Bei der Toilette" (Sitzende Balldame im Rokokokleid) (auf Ovalsockel) (vgl. 4834; 5329)
5329 „Bei der Toilette" (Sitzende Balldame im Rokokokleid) (auf Ovalsockel) (vgl. 4834; 5169)

Podola (d.i. Czapek, Stanislaus)

-> vgl. Bem. bei „Padola" und Biografie bei Czapek.
4222 „Schachmatt" (Sitzendes Rokokopaar beim Schachspiel) (auf girlandenverziertem Ovalsockel)
4380 Tanzende junge Frau mit Leopard (Bacchantin)
4495 „Indische Tänzerin" (auf Rundsockel) (Ruth Saint Denis) (auch sign. Rosé)
4539 LAMPE Scherzendes Biedermeierpaar (Dame und Kavalier) (auf mit Blumengirlanden verziertem Rundsockel) (auch sign. Padola) (vgl. 4305 und 4915= jeweils ähnl. Modell)
4676 Die Tanzstunde (Violinspieler im Biedermeiergewand, mit einem Mädchen tanzend) (auf Rundsockel) (auch sign. Padola/Podolak)
4684 „Ballpause" (Sitzende Krinoline, eine Maske haltend) FAYENCE/ALABASTER (vgl. Czapek/Podolak 4883; 4884)
4915 LAMPE Tanzendes Rokokopaar (Dame und Kavalier) (auf Blumengirlanden verziertem Rundsockel) (auch sign. Polter) (vgl. 4305; 4539= jeweils ähnl. Modell)
4922 Stehender Harlekin mit Laute und Colombine (auf oktogonalem Sockel) (vgl. 4922; 5262; 5263 sign. „Podolak"= Harlekin einzeln)

Podola s.a. Czapek, Padola und Podolak

Podolak, A. / Karl

-> vgl. Bem. bei „Padola".
4676 Die Tanzstunde (Violinspieler im Biedermeiergewand mit einem Mädchen tanzend) (auf Rundsockel) (auch sign. Padola/Podola)
4684 „Ballpause" (Sitzende Krinoline, eine Maske haltend) (vgl. 4883; Czapek/Podola 4883; 4884)
4883 „Ballpause" (Sitzende Krinoline, eine Maske haltend) (vgl. 4684; Czapek/Podola 4883; 4884)
4935 Stehender Harlekin mit Laute (auf Rundsockel/oktogonalem Sockel) (auch sign. Padola) (vgl. 5262; 5263) (vgl. 4922 GRUPPE sign. „Padola")
5262 Stehender Harlekin mit Laute (auf oktogonalem Sockel) (vgl. 4935; 5263) (vgl. 4922 GRUPPE)
5263 Stehender Harlekin mit Laute (auf oktogonalem Sockel) (vgl. 4935; 5262) (vgl. 4922 GRUPPE)
„Der grüne Fächer" (Sitzendes Biedermeiermädchen mit Rundfächer) Lit.: Studio YB, 1914, S. 190 (m. Abb.); hier mit Künstlerang. „Karl Podolak".

Podolsk

SCHALE „Mann mit der Schlange" (Dekor: „Grès") Lit.: R., F.: Wiener Kunstkeramik, in: KM, 2.1902, S. 91 (m. Abb.), „Mit dem Dekor „Grès" sind eine weitere Anzahl künstlerischer Arbeiten geschmückt. Wie aus dem Namen hervorgeht, tragen diese keramischen Produkte die natürliche Farbe des französischen Steinzeuges. Der „Mann mit der Schlange" von Podolsk, der sich am Muschelrand festklammert und, vorsichtig tastend, nach dem sich ringelnden Reptil greift, erscheint in seiner grauschwarzen, leicht geriffelten Tönung wie aus echtem Stein gemeißelt, der dann mit leichter Politur überzogen ist".
VASE „Serpentinentanz" ÖMKI (1900/01) Lit.: Neuwirth (1974), S. 70
SCHALE Muschel „Ein Fang" ÖMKI (1900/01) Lit.: Neuwirth (1974), S. 70

Poirier, B.

1458 Stehende Frau mit langem Kleid und entblößten Brüsten

Polair

2018 FIGUR/(LAMPE) „Berggeist" (Bergsteigender Bursche, eine Blumenstrauß haltend) (auf Felssockel) (elektrifiziert) ÖMKI (1900/01)

Pole

„Sensendengler" ÖMKI (1906/07) Lit.: Neuwirth (1974), S. 70

Pollak, Friedrich (Fritz)

*Friedrich (Fritz) Pollak, Bildhauer (*24.5.1887 Wien, †?), studierte bis etwa 1910 an der KGW-Schule Wien (u.a. bei Michael Powolny). Pollak nahm 1911–1914 regelmäßig an den Ausstellungen des ÖMKI teil und war 1924 bei der Jubiläumsausstellung des Wiener KGW-Vereins sowie 1925 in Paris auf der KGW-Ausst. vertreten. Lit.: Neuwirth (1974), S. 71, 252; Zühlsdorff, S. 437.*
4241 Dame mit Pagen
4387 SCHALE Vier stehende Frauengestalten mit Blumenkränzen (Tafelaufsatz) (GS neu Modell-Nr. 42 „Schale des Frühlings")
„Edeldame" (Fay.) ÖMKI (1913/14) Lit.: Neuwirth (1974), S. 71

Polter (d.i. vermutlich Czapek, Stanislaus)

-> vgl. Bem. bei „Padola".
3657 Biedermeiermädchen einen großen Korb auf dem Rücken tragend
3662 Biedermeierdame mit Hut und zwei quaderförmigen Gefäßen (als Blumenbehälter?)
3923 Stehende Frau (daneben große Amphore, reliefiert mit mythologischen Szenen) (vgl. 3768 Cherc)
4000 Stehendes kleines Mädchen mit großem Schirm und Handtasche (auf profiliertem Rundsockel) (vgl. 4016= UHR)
4016 UHR Stehendes kleines Mädchen mit großem Schirm und Handtasche (auf winkelförmigem Rechtecksockel) (vgl. 4000= FIGUR)
4031 Gänseliesel (Kleines Mädchen mit zwei Gänsen) (auf rechteckigem Landschaftssockel)
4130 Sitzendes nacktes Mädchen auf Scherenstuhl
4250 Stehende Biedermeierdame (daneben kleines, rundes Tischchen mit Blumenstrauß) (sign. Paltir/Paltier)
4267 Sitzende Tänzerin, sich einen Schuh ausziehend (auf Hocker/Stuhl) (vgl. 4897)
4271 LAMPE Tanzendes Rokokopaar (Dame und Kavalier) (auf Blumengirlanden verziertem Rundsockel) (vgl. 4915) (vgl. 4305 und 4539= jeweils ähnl. Modell)
4286 Kaffee trinkendes Mädchen, an einem Tisch sitzend (auf blütenverziertem Rechtecksockel)
4295 Stehender Knabe (Putto) mit Blumenstrauß und Korb auf dem Rücken (auf Rundsockel)
4317 Biedermeiermädchen auf Stuhl stehend, mit Dackel (vgl. Petri Modell-Nr. .... Biedermeiermädchen auf Stuhl kniend, mit Katze)
4401 Stehendes Biedermeiermädchen mit Haube und Blumen (auf mit Blumengirlanden verziertem Rundsockel)
4472 Stehende Rokokodame mit Blumenstrauß
4534 Stehendes, lesendes Mädchen (auf Landschaftssockel mit 4 Büchern) (vgl. Rosé 4534) (vgl. 3849= ähnl. Modell)
4544 Sitzender Knabe beim Schreiben (als „Tisch" dient ein hoher Barockstuhl)
4550 Stehendes Mädchen mit Haube und Blumenstrauß, daneben Hund (Dalmatiner?) (auf ovalem Natursockel)
4568 Stehendes Biedermeiermädchen, einen großen, blumengeschmückten Hut haltend (auf Rundsockel) (vgl. Meier 4568)
4577 FIGUR/LAMPE Sitzende Lautenspielerin mit Hut (Biedermeierfigur) FAYENCE/ALABASTER
4629 Stehende Elsässerin in Tracht (mit Kopfputz/Haube und langem Kleid) (auf quaderförmigem Sockel)
4650 Stehendes Mädchen mit Haube und Schiefertafel (auf Rundsockel) (vgl. auch Petri 4650!)
4651 FIGUR/LAMPE Stehender Knabe mit Haube und Apfel FAYENCE (vgl. auch Petri 4651!)
4677 Stehender Knabe, einen Brief lesend (in knielangen Hosen, mit Hut und Blumenstrauß) (auf Rundsockel)
4895/4897? Sitzende Tänzerin, sich einen Schuh ausziehend (auf Hocker) (vgl. 4267 auf Hocker/Stuhl)
4897 s. auch 4895/4897
4915 LAMPE Tanzendes Rokokopaar (Dame und Kavalier) (auf mit Blumengirlanden verziertem Rundsockel) (auch sign. Podola) (vgl. 4271) (vgl. 4305; 4539= jeweils ähnl. Modell)

Polter vgl. auch Meier, Petri (Petrides) und Rosé

Popineau, François Émile

*François Émile Popineau, Bildhauer (*2.10.1887 St.-Armand-Montrond/Frankreich, †? wurde letztmals 1956 erwähnt), lebte in Paris, war Mitglied der Künstlergruppe „La Stèle" (-> vgl. Kap. „Dekorative Skulpturen von Arthur Goldscheider in Paris") und stellte u.a. im Pariser Salon (Societé des Artistes Français) aus. Lit.: TB, Bd. 27, S. 261–262; Vollmer, Bd. 3, S. 613; Neuwirth (1974), S. 71; Bénézit, Bd. 11, S. 142.*

Portou (d.i. Bortone, Ipazzio Antonio?)

Der Name „Portou" ist ebenso wie im Fall „Bertrand" eine, wie bei Goldscheider öfter praktiziert, Abwandlung (franz. Schreibweise) des Namens Bortone. Biografie siehe unter Bortone.
1258 BÜSTE Dame mit Hut (undeutlich sign.)
1740 s. Bortone (vgl. 2130)
2130 s. Bortone (vgl. 1740)

Postl, Hugo

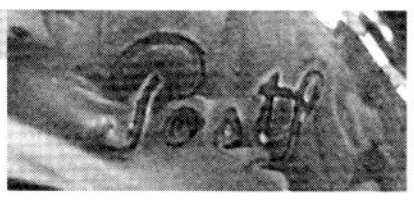

*Hugo Postl, Tierbildhauer (*16.3.1879 Judenburg/Steiermark, †?), war Mitglied zahlreicher Künstlervereinigungen (Künstlerverband österr. Bildhauer, Gemeinschaft bildender Künstler Steiermark, Gemeinschaft bildender Künstler Salzburg usw.). Bereits 1908 wurde vom Landesmuseum Graz die Bronze „Nubischer Wildesel" angekauft. Außer von Goldscheider wurden auch in Hugo Kirschs kunstkeramischer Werkstätte Entwürfe von Postl ausgeführt. Hugo Postl lässt sich bis 1934 in Salzburg, Augustinerstr. 11a nachweisen, jedoch weist ein Sterbekreuz im Artikel, N.N.: Farbe und Bewegung, Die Fayencen der Wiener Manufaktur Friedrich GS Wien-Währing, in: DB, 5.1928, Nr. 213, S. 38 darauf hin, dass er spätestens 1928 schon verstorben sein musste. Lit.: Neuwirth (1974), S. 71; Zühlsdorff, S. 420; Quellen: Bildhauerarchiv, Karl Novak; Kunstarchiv Werner J. Schweiger, Wien.*
4080 Sitzender, kurzhaariger Dachshund (Dackel)
4920 Stehender Schäferhund (auf Ovalsockel)
4944 Liegender Tiger (vgl. 6322= ähnl. Modell)
5052 Stehender Pinguin (auf getrepptem Quadratsockel) (vgl. 5052= GRUPPE)
5053 GRUPPE Zwei stehende Pinguine (auf Rechtecksockel) (vgl. 5052= Einzelfigur)
5457 s. Bormann
5458 FIGUR „Wolfshund" (Liegender Schäferhund) (vgl. 5541; 5542; 5883)
5541 FIGUR „Wolfshund" (Liegender Schäferhund) (vgl. 5458; 5542; 5883)
5542 FIGUR „Wolfshund" (Liegender Schäferhund) (vgl. 5458; 5541; 5883)
5719 „Polizeihund (sitzend)" (Sitzender Schäferhund/Deutscher Schäfer) (vgl. 6069= GRUPPE Stehendes Kind, einen sitzenden Schäferhund umarmend, Modell: Lorenzl/Postl; vgl. 7041= GRUPPE Stehender Damenakt mit sitzendem Schäferhund (auf Rundsockel), Modell Lorenzl/Postl)
5720 Sitzender Foxterrier
5728 Stehender Foxterrier
5814 Sitzender Pointer (nach links blickend)
5858 Pekinese (vgl. 6136; 7320)
5861 Sitzender Pointer (nach oben blickend)
5883 FIGUR „Wolfshund" (Liegender Schäferhund) (vgl. 5458; 5541; 5542)
5957 Stehender Foxterrier BRONZE (Postl zugeschr.)
5974 FIGUR „Windhund" (Liegender Barsoi) (vgl. 6422; 7143)
6069 GRUPPE Stehendes Kind, einen sitzenden Schäferhund umarmend (Modell: Lorenzl/Postl) (vgl. 5719= Einzelfigur Hund)
6116 FIGUR „Skyeterrier" (vgl. 6503; 6802; 6975; 7300; 7519)
6136 Pekinese (vgl. 5858; 7320)
6137 Stehender Dachshund (Dackel)
6322 Liegender Tiger (vgl.4944= ähnl. Modell)
6422 FIGUR „Windhund" (Liegender Barsoi) (vgl. 5974; 7143)
6503 FIGUR „Skyeterrier" (vgl. 6116; 6802; 6975; 7300; 7519)
6525 Stehender Pekinese
6628 „Windhunde" (Zwei laufende Windhunde) (auf Landschaftssockel)

6715 Sitzender Spaniel (vgl. 7144)
6761 GRUPPE / FIGUR Pekinesen (Zwei sitzende Pekinesen) (auch als linke Einzelfigur 6761)
6802 FIGUR „Skyeterrier" (vgl. 6116; 6503; 6975; 7300; 7519)
6975 Skyeterrier (auf Ovalsockel) (vgl. 6116; 6503; 6802; 7300; 7519= jeweils FIGUR)
7143 FIGUR „Windhund" (Liegender Barsoi) (vgl. 5974; 6422)
7144 Sitzender Spaniel (vgl. 6715)
7300 FIGUR „Skyeterrier" (vgl. 6116; 6503; 6802; 6975; 7519)
7320 Pekinese (vgl. 5858; 6136)
7519 FIGUR „Skyeterrier" (vgl. 6116; 6503; 6802; 6975; 7300)
Hund Paris, Salon (1928) Lit.: Neuwirth (1974), S. 71
„Windhund" (liegender Barsoi) (Fay.) (1929/30) Lit.: MW, 11.1929/30, Nr. 17, S. III (m. Abb.); Neuwirth (1974), S. 71
„Schottischer Terrier" (Fay.) (1929/30) Lit.: MW, 10.1928/29, Nr. 24, S. IV (m. Abb.); Neuwirth (1974), S. 71
Liegender Rehbock (auf Ovalsockel) (um 1925) (H. 30,5 cm); eBay #147695290

Potet, Louis (Loys)

*Louis (Loys) Potet, Bildhauer (*1866 Nantes, †? wird letztmals 1933 erwähnt), war Schüler seines Vaters Pierre Potet und von Chapu. Mitarbeiter bei Friedrich Goldscheider, Paris. Potets Spur verliert sich 1933 in Nantes. Lit.: TB, Bd. 27, S. 303; Hillenbrand, S. 328; Bénézit, Bd. 11, S. 171; AKL Index.*
845-GSB „Aux Champs"

Potucek, Ferdinand

Ferdinand Potucek, Retoucheur (15.5.1892–†?); Quelle: AGM, Liste „Adressen von guten und zuverlässigen Arbeitern" (undat., um 1953).

Powolny, Michael

*Michael Powolny, Bildhauer und Keramiker (*18.9.1871 Judenburg, †4.1.1954 Wien), studierte nach einer Hafnerlehre bei seinem Vater an der Fachschule für Tonindustrie in Znaim (1891–1894), dann KGW-Schule Wien (1894–1906), u.a. bei A. Strasser. Powolny begründete 1897 die Wiener Sezession sowie 1905 zusammen mit Berthold Löffler die Wiener Keramik, deren Erzeugnisse von der Wiener Werkstätte vertrieben wurden. In den Jahren 1909–1912 lehrte er an der KGW-Schule in Wien und stand ab 1912 einer eigenen Keramik-Klasse vor. Bis 1936 wirkte er dort als Professor. Powolny war Mitglied im Österreichischen Werkbund und Mitinhaber der Gmundner Keramischen Werkstätte. Auf den bedeutenden Ausstellungen in Paris 1925 (KGW-Ausst.) und der Weltausstellung 1937, ebenfalls in Paris, war er vertreten. Michael Powolny lieferte als vielseitiger Künstler neben der Wiener Werkstätte auch den Firmen Wienerberger, der Wiener Porzellanmanufaktur Josef Böck, Augarten, Wien Entwürfe. Für die Wiener Werkstätte sind Powolnys keramische Arbeiten zu einem Markenzeichen geworden. Durch seine Lehre gingen zahlreiche berühmte Bildhauer und Keramiker. Powolny selber war einer der bedeutenden Bildhauer, Lehrer und Keramiker Wiens, er erhielt 1951 die Ehrenmedaille der Gemeinde Wien. Lit.: TB, Bd. 27, S. 330–331; Vollmer, Bd. 3, S. 619–620; Neuwirth (1974), S. 71, 254–265 (m. Abb.); Zühlsdorff, S. 437; Hillenbrand, S. 328; Fahr-Becker, S. 232; Bénézit, Bd. 11, S. 198.*
2239 UHR „Die Zeit" (Stehende Frauengestalt mit Sanduhr) „Le Temps", Paris, Salon (1902) (vgl. 2301= FIGUR?)
2301 FIGUR „Die Zeit" (vgl. 2239= UHR?)
JARDINIERE Drei stehende Putti als Schalenträger (auf Rundsockel) (unsign., in d. Art Powolny)

Prémont, F. / J. / T. de / du

613 WANDMASKE Nordafrikanerin mit Turban (vgl. 1613) (Pendant vgl. 1612)
1098 BÜSTE „Holländerin" (Junge Frau mit Spitzenhaube) (vgl. 1402)
1102 BÜSTE „Mohrin"/„Negresse au Papillon" (Afrikanerin / Mulattin mit Turban) (vgl. 1878) Bem.: Bis auf den unteren Faltenwurf des Tuches identisches Modell .... (H. 47 cm) A/T 20.4.1994; (H. 48 cm); Richemond (1999), S. 103 (m. Abb.) (sign. T. du Piémont)
1341 s. Thiele
1402 BÜSTE „Holländerin" (Junge Frau mit Spitzenhaube) (vgl. 1098)
1613 WANDMASKE Nordafrikanerin mit Turban (vgl. 613) (Pendant vgl. 1612)
1878 BÜSTE „Mohrin mit Schmetterling"/„Négresse au Papillon" (Afrikanerin / Mulattin mit Turban) (sign. Thiele)

Preyer, Fritz

*Fritz Preyer, Retoucheur (*29.6.1934, †?). Quelle: AGM, Liste „Adressen von guten und zuverlässigen Arbeitern" (undat., um 1953).*

Prince

BÜSTE „Favorite" ÖMKI (1898/99) Lit.: Neuwirth (1974), S. 71

Prischl, Adolf

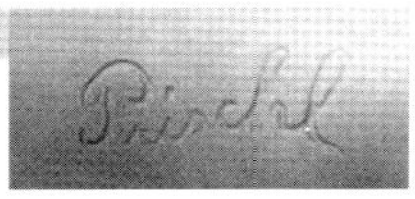

*Adolf Prischl, Bildhauer (*24.1.1912 Wien, †7.6.1970 Wien), war von 1929–1932 bei Prof. Robert Obsieger an der Wienerberger Werkstättenschule. Zahlreiche seiner Modelle wurden neben denen von Goldscheider auch bei der Wiener Kunstkeramik und bei Keramos ausgeführt, bei denen er bis 1946 (nach Kriegsgefangenschaft) als Mitarbeiter verblieb. Im Februar 1947 legte Prischl seine Meisterprüfung als Töpfer und Keramiker ab. Zuerst wohnhaft in der Troststraße 116, Wien X. gründete Prischl eine eigene Werkstätte im 5. Wiener Gemeindebezirk, die später in die Mollardgasse, Wien VI., verlegt wurde. Dort arbeitete auch Karl Sailer mit. Ebenso lieferte Stephan Dakon Entwürfe, beide kannte er bereits von der Goldscheider Manufaktur. Neben eigenen Modellen wurden auch leicht abgewandelte Modelle, die noch für Goldscheider entworfen wurden, in seiner Werkstatt reproduziert. Diese Praxis wurde auch fortgesetzt, als die Firma in Prischl & Sailer umbenannt wurde. Diese Erzeugung wurde 1965 eingestellt und Prischl nahm bei der Wiener Keramikfirma Pawlata eine Anstellung an. Am 7.6.1970 stirbt Adolf Prischl an einem Herzinfarkt. Lit.: Lehmann, 1941; Makus (2000), S. 176, 186 (m. Abb.); Matschiner (2006b) Matschiner (2006c); Quelle: AGM (Gehaltsliste, Mai 1938; Gehaltsliste August 1938); Mitt. von Frau Mag. Uta M. Matschiner.*
505-WG WANDMASKE „Wandmaske Chinesin" (Chinesenmädchen mit Strohhut, nach links blickend)
506-WG WANDMASKE „Wandmaske Chinesin" (Chinesenmädchen mit Strohhut, nach rechts blickend)
507-WG „Akt" (Sitzender Mädchenakt mit geschlossenen Augen)
508-WG WANDMASKE „Maske" (Frauenkopf mit langem Haar, nach rechts geneigt)
509-WG WANDMASKE „Maske" (Frauenkopf mit Halstuch)
510-WG WANDMASKE „Maske" (Frauenkopf mit lockigem Haar und Kragen)
511-WG „Akt" (Sitzender Mädchenakt)
512-WG WANDMASKE „Neger" (Negerin mit Strohhut und Ohrringen)
518-WG „Akt" (Hockender Mädchenakt mit Ball) (auf Landschaftssockel)
520-WG Hockender Mädchenakt mit großem Tuch
525-WG „Akt" (Hockender Mädchenakt mit einer Hand zum Boden greifend) (auf Landschaftssockel)
526-WG WANDMASKE „Maske" (Frauenkopf nach rechts blickend) (vgl. 8182)
527-WG WANDMASKE „Maske" (Frauenkopf mit Schal, nach rechts blickend)
529-WG WANDMASKE „Maske" (Frauenkopf mit Schal)
530-WG WANDMASKE „Maske" (Frauenkopf mit lockigem Haar) (vgl. 531-WG; 532-WG)
531-WG WANDMASKE „Maske" (Frauenkopf mit lockigem Haar) (vgl. 530-WG; 532-WG)
532-WG STEHMASKE „Maske" (Frauenkopf mit lockigem Haar) (auf Rundsockel) (vgl. 530-WG; 531-WG)
533-WG WANDMASKE „Maske" (Frauenkopf mit langem Haar, nach links blickend)
537-WG WANDMASKE „Kinderkopf" (Kinderkopf mit Kragen, nach links blickend) (Pendant vgl. 538-WG) (vgl. 542-WG= Variante als „Engel")
538-WG WANDMASKE „Kinderkopf" (Kinderkopf mit Kragen, nach rechts blickend) (Pendant vgl. 537-WG) (vgl. 543-WG= Variante als „Engel")
540-WG „Akt" (Sitzender Mädchenakt)
542-WG WANDMASKE „Engelkopf" (Kinderkopf mit Flügel, nach links blickend) (Pendant vgl. 543-WG) (vgl. 537-WG= Variante als „Kinderkopf")
543-WG WANDMASKE „Engelkopf" (Kinderkopf mit Flügel, nach rechts blickend) (Pendant vgl. 542-WG) (vgl. 542-WG= Variante als „Kinderkopf")
544-WG WANDMASKE „Maske" (Frauenkopf mit Schal / Halstuch)
545-WG „Akt mit Kugel" (Schreitender Mädchenakt mit Kugel) (auf Ovalsockel mit Kugel) (vgl. 545A-WG)
545A-WG „Akt mit Kugel" (Schreitender Mädchenakt mit Kugel) (auf Ovalsockel mit Kugel) (vgl. 545-WG)
6770 STEHKOPF Kinderkopf mit Kragen (auf Quadratsockel) (vgl. Dakon / Knörlein / Prischl 6770; Entw. eindeutig Knörlein)
6836 WANDMASKE / STEHMASKE Frauenkopf mit Kragen (vgl. 7549) (vgl. 7340= DECKELDOSE)
6871 WANDMASKE Frauenkopf mit oben glattem, unten seitlich gelocktem Haar und Halstuch (auch sign. Knörlein)
7155 WANDMASKE Frauenkopf mit Hut
7312 Schreitender Mädchenakt mit Schmetterling (auf Ovalsockel) (Prischl zugeschr.)
7313 BUCHSTÜTZE Pferd (links / rechts)
7340 DECKELDOSE Frauenkopf mit Kragen (Form: rechteckig / Dekor: Frauenkopf, reliefiert) (vgl. 6836; 7549= jeweils WANDMASKE)
7420 WANDMASKE „Maske" (Frauenkopf mit wehendem Haar) (vgl. 7750; 8207)
7549 WANDMASKE Frauenkopf mit Kragen (vgl. 6836) (vgl. 7340= DECKELDOSE)
7551 BÜSTE „Madonna" (Madonna mit gekreuzten Händen) (vgl. 8245) (vgl. 7985= RELIEF)
7579 WANDMASKE „Madonna" (Madonna mit gefalteten Händen) (vgl. 7865)
7658 Stehende Madonna mit Krone und erhobenen Armen (auf gekehltem Rundsockel)
7672 FIGUR „Akt mit Perlenkette" (auf Rundsockel)
7750 WANDMASKE „Maske" (Frauenkopf mit wehendem Haar) (vgl. 7420; 8207)
7784 WANDMASKE „Maske" (Frauenkopf mit Schal) (vgl. 8266)
7787 BÜSTE Madonna mit erhobenen Händen
7794 WANDMASKE Madonna mit Hand (vgl. 7579= ähnl. Modell)
7797 BÜSTE Frau mit Lockenkopf einen Vogel haltend
7799 „Akt" (Schreitender Damenakt) (auf Ovalsockel)
7816 WANDMASKE „Maske" (Frauenkopf mit kurzem Kragenansatz) (vgl. 7931; 8248)
7831 WANDMASKE „Madonna" (Madonna mit Schleier / Kopftuch)
7862 Stehende Madonna mit gekreuzten Händen (auf Rundsockel)
7865 WANDMASKE „Madonna" (Madonna mit gefalteten Händen) (vgl. 7579)
7883 WANDMASKE Frauenkopf mit Ohrring und Schal (vgl. 8198)
7894 BÜSTE Madonna mit Schleier / Kopftuch
7910 WANDMASKE Frauenkopf mit Kragen und Halsschmuck (Prischl zugeschr.)
7931 WANDMASKE „Maske" (Frauenkopf mit kurzem Kragenansatz) (vgl. 7816; 8248)
7936 Frauenakt als Torso (auf Rundsockel)
7975 „Akt mit Draperie" (Stehender Damenakt, ein langes Tuch mit den Armen hochhaltend) (auf Rundsockel)
7982 STEHKOPF Mädchenkopf mit aufrecht stehender Hand (auf Ovalsockel)
7983 Stehender Mädchenakt mit Tuch (auf Ovalsockel) (Prischl zugeschr.)
7985 RELIEF „Madonna" (Madonna mit gekreuzten Händen) (Wandrelief) (vgl. 7551; 8245= jeweils BÜSTE)
8041 WANDMASKE „Maske / Goldband" (Frauenkopf mit Haarband / Goldband und Halstuch) (vgl. 8249) (auch für GS USA, Mod.-Nr. 732)
8073 FIGUR „Pferd" (Stehendes Fohlen)
8182 WANDMASKE „Maske" (Frauenkopf mit mittellangem Haar, im Halbprofil nach rechts) (vgl. WG 526)
8198 WANDMASKE „Maske" (Frauenkopf mit Schal) (vgl. 7883)
8199 WANDMASKE (Modell vermutlich von A. Prischl)
8207 WANDMASKE „Maske" (Frauenkopf mit wehendem Haar) (vgl. 7420; 7750)
8245 BÜSTE „Madonna" (Madonna mit gekreuzten Händen) (vgl. 7551) (vgl. 7985= RELIEF)
8248 WANDMASKE „Maske" (Frauenkopf mit kurzem Kragenansatz) (vgl. 7816; 7931)

8249 WANDMASKE „Maske/Goldband" (Frauenkopf mit Haarband/Goldband und Halstuch) (vgl. 8041)
8253 FIGUR „Pamela (Akt)" (Kniender Damenakt) (auf Ovalsockel)
8261 Stehende junge Dame im Mantel mit Pelzkragen, Hut und Pekinese (auf Ovalsockel)
8266 WANDMASKE „Maske" (Frauenkopf mit Schal) (vgl. 7784)
8275 FIGUR Liegende Katze mit Ball (Prischl zugeschr.)
8389 FIGUR „Kind mit Hund" (Schreitendes kleines Mädchen, einen Hund auf dem Arm haltend) (auf Ovalsockel)
8465 Sitzender weiblicher Akt mit ausgestrecktem Arm (auf Ovalsockel)
8494 FIGUR „Torso" (Damenakt) (auf Ovalsockel)
8503 FIGUR „Sandalenbinderin" (Stehender Damenakt, den rechten Fuß hebend, ohne/mit Zöpfe) (auf Rundsockel) (vgl. 8702)
8506 FIGUR „Sandalenbinderin" (Sitzender Damenakt mit zurückgekämmtem Haar und Dutt/mit langem lockigem Haar) (vgl. 8508; 8714)
8508 FIGUR „Sandalenbinderin" (Sitzender Damenakt mit zurückgekämmtem Haar und Dutt/mit langem lockigem Haar) (vgl. 8506; 8714)
8596 WANDBEHANG/RELIEF „St. Georg" (den Drachen tötend)
8597 WANDBEHANG/RELIEF „Betende"
8604 Schreitende Tänzerin, ihren Rock mit beiden Armen auseinander haltend (auf Ovalsockel)
8644 „Herbst" (Schreitender Damenakt mit Krug und Weintrauben) (auf Landschaftssockel) (vgl. 8735) (Prischl zugeschr.)
8685 „Blumenpflückerin" (Sich bückender Damenakt) (auf Natursockel) (vgl. 8738)
8702 FIGUR „Sandalenbinderin" (Stehender Damenakt, den rechten Fuß hebend, ohne/mit Zöpfe) (auf Rundsockel) (vgl. 8503)
8714 FIGUR „Sandalenbinderin" (Sitzender Damenakt mit zurückgekämmtem Haar und Dutt) (vgl. 8506; 8508)
8735 „Herbst" (Schreitender Damenakt mit Krug und Weintrauben) (auf Landschaftssockel) (vgl. 8644) (Prischl zugeschr.)
8738 „Blumenpflückerin" (Sich bückender Damenakt) (auf Natursockel) (vgl. 8685)
732-USA WALLMASK (vgl. 8041)

Prozsynski, Henry

*Henri Prozsynski, Bildhauer (*19.4.1887 Pau/Frankreich, †? wird letztmals erwähnt 1956), stellte ab 1910 im Pariser Salon (Societé des Artistes Français) aus, wo er 1921 die Silbermedaille errang. Proszynski lebte in Paris und war Mitarbeiter bei Arthur Goldscheider. Lit.: Vollmer, Bd. 3, S. 630; Neuwirth (1974), S. 71; Bénézit, Bd. 11, S. 279; AKL Index.*
Fontaine „Le Bouc" BRONZE/STEIN Paris, Salon (1928) Lit.: Neuwirth (1974), S. 71

Prunier, D./G.

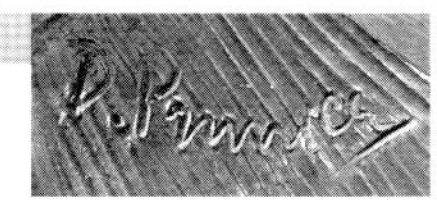

1379 FIGUR „Haremswächter" (Stehender Araber mit Gewehr) (auf Landschaftssockel) (sign. Prunier) (vgl. 1421; 1574) (vgl. 1596= BÜSTE)
1397 FIGUR „Haremswächter" (Stehender Araber mit Gewehr) (auf Landschaftssockel) (vgl. 1379; 1421; 1574) (vgl. 1596= BÜSTE)
1421 FIGUR „Haremswächter" (Stehender Araber mit Gewehr) (auf Landschaftssockel) (sign. G. Prunier) (vgl. 1379; 1397; 1574) (vgl. 1596= BÜSTE)
1574 FIGUR „Haremswächter" (Stehender Araber mit Gewehr) (auf Landschaftssockel) (vgl. 1379; 1397; 1421)
1596 BÜSTE „Haremswächter" (Araber mit Bart und Kopftuch) (sign. D. Prunier) (vgl. 1379; 1397; 1574= jew. FIGUR)

Puchegger, Anton

*Anton Puchegger, Tierbildhauer (*7.4.1878 Payerbach, †25.9.1917 Davos), studierte nach seiner Ausbildung an der Holzschnitzer-Fachschule in Bozen an der Akademie in Wien und Paris. Im Jahre 1905 erhielt er ein Angebot für eine Professur in Znaim, die er aber ablehnte und stattdessen nach Berlin ging. Seine Entwürfe wurden auch von den Firmen Wiener Emailfarbenwerk Schauer, die Kunstkeramische Werkstätte Busch & Ludescher, Wien, der Berliner Porzellanmanufaktur sowie die Schwarzburger Werkstätten für Porzellankunst, Unterweißbach ausgeführt. Lit.: TB, Bd. 27, S. 443; Neuwirth (1974), S. 72, 268; Zühlsdorff, S. 438; Bénézit, Bd. 11, S. 289; AKL Index.*
2388 FIGUR „Bär" ÖMKI (1911/12)
4533 Adler (auf Felssockel) (vgl. 4781; 4782)
4781 Adler (auf Felssockel) (vgl. 4533; 4782)
4782 Adler (auf Felssockel) (vgl. 4533; 4781)
Sitzender Eisbär (um 1905) (H. 27 cm); RM 21

Rab, Herbert

*Herbert Rab, Gipsformengießer (*12.7.1930, †?). Quelle: AGM, Liste „Adressen von guten und zuverlässigen Arbeitern" (undat., um 1953).*

Raby, Ekiel (Rabinovitsch)

*Ekiel Raby, Elfenbeinschnitzer und Dekorateur (*18.10.1893 Lyadi?/Russland, †? wird letztmals erwähnt 1958), lebte in Paris und war Mitglied der Künstlergruppe „Évolution" (-> vgl. Kap. „Dekorative Skulpturen von Arthur Goldscheider in Paris"). Raby nahm teil an der Pariser KGW-Ausst. 1925 und errang dort einen Grand Prix. Lit.: Vollmer, Bd. 4, S. 6, Neuwirth (1974), S. 72; AZ, 1980, S. 209; Bénézit, Bd. 11, S. 374.*

Rapin, Henri

*Henri Rapin, Maler und Dekorateur (*28.2.1873 Paris, †?), war in den zwanziger Jahren bei Sèvres (Manufacture Nationale de Sèvres) als Leiter der künstlerischen Abteilung angestellt. Rapin war Mitglied der Künstlergruppe „Évolution" (-> vgl. Kap. „Dekorative Skulpturen von Arthur Goldscheider in Paris"). Lit.: Zühlsdorff, S. 438; Bénézit, Bd. 11, S. 442; AKL Index.*

Raum, Alfred

*Alfred Raum, Bildhauer (*16.1.1872 Bernau, †?), lebte in Berlin und studierte an der dortigen Akademie. Lit.: TB, Bd. 28, S. 43; Bénézit, Bd. 11, S. 459–460; AKL Index.*
4028 „Dachshund" (Sitzender Dackel) ÖMKI (1912)

Reinold, Grete (Margarete)

*Grete (Margarete) Reinold, Kunstgewerblerin (*30.4.1901 Wien, †?), studierte an der KGW-Schule, Wien (u.a. bei Michael Powolny und Robert Obsieger). Sie war Mitarbeiterin der Wiener Werkstätte und bei Marcell Goldscheider. Neben anderen zahlreichen Ausstellungen war sie auf der KGW-Ausst. 1937 in Paris vertreten. Lit.: Neuwirth (1974), S. 270; Zühlsdorff, S. 439.*
79-MG SCHALE/TELLER Mädchengesicht
ohne Nr.-MG WANDRELIEF/FLIESE Mädchenakt mit Reh

Renard, (Léopold)

*Léopold Renard, Bildhauer (*3.3.1868 Paris, †1.11.1945). Der bei Neuwirth (1974), S. 72 zitierte Marcel Renard (Sohn von Leopold) kommt wegen des Geburtsjahres 1893 nicht infrage. Lit.: Bénézit, Bd. 11, S. 570; AKL Index.*
909 VASE Drei stehende Frauenfiguren
RELIEF „Natura" (Terrakotta) ÖMKI (1900/01) Lit.: Neuwirth (1974), S. 72

Reyam, (David?)

*(David?) Reyam, Maler und Designer (*16.9.1864 New York, †19.5.1943 Wilmington/Delaware), ist der zu vermutende Entwerfer der unten aufgeführten Statue. Ob es sich bei dem amerikanische Maler, der auch in Paris erwähnt wird, wirklich um einen Mitarbeiter von Friedrich Goldscheider, Paris, handelt, kann nicht mit letzter Sicherheit gesagt werden. Lit.: AKL Index.*
738-GSB STATUETTE „Libellule"

Ricard

2800 Biedermeiertänzerin mit langem Kleid (ein großes Tuch über dem Kopf schwingend) (auf Blumen verziertem Rundsockel)
„Tänzerin" ÖMKI (1906/07) Lit.: Neuwirth (1974), S. 72

Richer, Paul Marie Louis Pierre

*Paul Marie Louis Pierre Richer, Bildhauer und Medailleur (*17.1.1849 Chartres, †Dezember 1933), war ursprünglich Arzt. Kashey nennt ihn als Mitarbeiter bei Friedrich Goldscheider, Paris. Lit.: TB, Bd. 28, S. 275; Kashey, S. 272; Bénézit, Bd. 11, S. 665; AKL Index.*
858 Schnitter (Le Faucheur) (Mähender Bauer mit Hut auf Landschaftssockel) BRONZE
KRUG/HUMPEN Wäscherinnen am Fluß (reliefiert) BRONZE (Firmenmarke Sculpture= Plak. B), (H. 12,7 cm); eBay # 2505078966

Richter, Jaroslav

Jaroslav Richter, Bildhauer oder Kunstgewerbler, absolvierte im Schuljahr 1894/95 die KGW-Schule in Bechyne und wurde bei GS (als Modelleur, Formgießer bzw. Figurenretoucheur) angestellt. Weitere Informationen über ihn sind nicht bekannt. Lit.: Neuwirth (1974), S. 31; Neuwirth (1974a), S. 321; Bechyne 1884–1994, S. 55.

Rie-Gomperz, Lucie

*Lucie Rie-Gomperz, Keramikerin (*16.3.1902 Wien, †1.4.1995 London), studierte bei M. Powolny von 1922–1926 an der KGW-Schule in Wien. Lucie Rie-Gomperz floh 1938 mit ihrem Mann (Hans Rie) nach England, wo sie sich in London niederließ und er nach Amerika auswanderte. Bis in ins hohe Alter präsentierte sie in zahlreichen Ausst. in England ihre keramischen Arbeiten. Im Jahr 1989 wurden sogar Ausst. in Osaka und Tokio veranstaltet. Arbeiten von ihr befinden sich u.a. im Victoria und Albert Museum, London. Neuwirth belegt eine Ausstellungsbeteiligung mit Goldscheidermodellen an der Weihnachtsschau im Wiener Künstlerhaus 1928. Modelle von ihr konnten jedoch bis Redaktionsschluss nicht nachgewiesen werden. Lit.: Neuwirth (1974), S. 72; Quellen: www.britannica.com/eb/article?tocid=9063635 (7.7.2005); www.waterman.co.uk/pages/biography/15025.html (1.9.2005).*

Ringuet, Ghislin

*Ghislin Ringuet, Kunstgewerbler (*30.3.1907 Limoges, †?), war Mitglied der Künstlergruppe „Évolution" (-> vgl. Kap. „Dekorative Skulpturen von Arthur Goldscheider in Paris"). Lit.: Vollmer, Bd. 4, S. 72; Neuwirth (1974), S. 72.*

Rivir, Armand

*Armand Rivir, Gold- und Silberschmied (*21.6.1883 Paris, †? wird letztmals erwähnt 1958), war Mitglied der Künstlergruppe „Évolution" (-> vgl. Kap. „Dekorative Skulpturen von Arthur Goldscheider in Paris") und stellte ab 1922 im Pariser Salon d' Automne aus. Dekorative Arbeiten von seiner Hand finden sich in der Abtei Solesmes. Lit.: Vollmer, Bd. 4, S. 78; Neuwirth (1974), S. 72; Bénézit, Bd. 11, S. 748; AKL Index.*

Robbia, ? Della (in der Art/nach)

646 RELIEFPLATTE Madonna mit Kind
2166 BÜSTE Mädchen mit Haarband

Rochella, D.

1409 BÜSTE Araberin mit langem Kopftuch (nach links blickend) (ohne/mit Halskette) (vgl. Thiele 1109; 1340)

Rohel s. Rozel

Rohely s. Rochella und Thiele

Rosca, Archip

Archip Rosca, Kunstgewerbler (? Brasca/Bukowina, †?), studierte an der KGW-Schule in Wien. Lit.: Neuwirth (1974), S. 72.*
„Medizin" ÖMKI (1903/04) Lit.: Neuwirth (1974), S. 72

Rosé

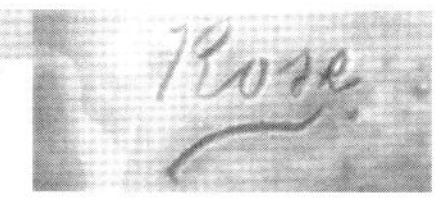

*Bei der Künstlersign. „Rosé" handelt es sich, wie häufig bei Goldscheider, um ein Pseudonym. Es gibt zwar einen Bildhauer namens Albert Dominique Roze (*4.8.1861 Amiens, †17.10.1952 Amiens), von dem bei Bénézit eine Liszt-Büste angeführt wird. Ob hieraus ein Zusammenhang mit diesem franz. Bildhauer hergestellt werden kann, erscheint jedoch mehr als fraglich. Weitere Indizien, dass es sich um einen Scheinnamen handelt, sind eine Reklame für die überaus erfolgreiche Figur „Ruth" in der Zs. MW, 11.1929/30, Nr. 11, S. VII, dort wird als Künstlername „Bosi" angegeben. Bosi lässt sich aber in der einschlägigen Lit. ebenso wenig nachweisen lässt wie „Rosé". Kurz vor Redaktionsschluss tauchte auch noch eine Putten-Gruppe (Mod.-Nr. 3373) auf, die nach dem Werk des belg. Bildhauers François-Joseph LeClercq von Goldscheider editiert wurde und ebenfalls mit „Rosé" sign. ist. Es kann angenommen werden, dass das Pseudonym für einen der für Goldscheider produktiven und wichtigen Bildhauer wie z.B. Stanislaus Czapek gewählt wurde. Lit.: TB, Bd. 29, S. 132; Bénézit, Bd. 12, S. 62; AKL Index.*

2690 UHR „Théodora" (Sarah Bernhardt als Théodora auf einem Thron sitzend) (vgl. 2786)
2786 UHR „Théodora" (Sarah Bernhardt als Théodora auf einem Thron sitzend) (vgl. 2690)
2907 UHR „Der Morgen" (Hahn und drei Hennen)
2934 Mädchen mit Schirm
2955 UHR-LAMPE Frau an verschneitem Glockenturm, die Glocke läutend
2962 UHR Beethoven (mit Notenbuch) auf einem Thron sitzend
2998 LAMPE Sonnenwagen (Römer in zweispännigem Streitwagen)
3077 UHR Biedermeiermädchen mit Uhr und spielender Katze
3116 UHR Drei Mädchen (durch eine geöffnete Tür laufend)
3120 „Heine" (Heinrich Heine mit Buch auf einer Bank mit antikisierender Kratervase sitzend) (Inschrift: HEINE)
3170 LAMPE „Goethe" (lesender Johann Wolfgang von Goethe, an einem Tischchen mit Lampe sitzend) (Inschrift: GOETHE)
3217 BÜSTE „Goethe" (Inschrift: GOETHE)
3234 BÜSTE „Schiller" (Inschrift: SCHILLER)
3248 „Richard Wagner" (sitzende, lesende Figur) (vgl. 3921)
3265 BÜSTE Lächelnde Frau mit Kopftuch
3285 „Mozart" (mit Geige, auf Sessellehne sitzend) (vgl. 3522; 5483)
3295 UHR als Denkmal mit Jüngling und Adler (Inschrift: UT ABSIN(?) CURAE NE OBSIM(?) INJURIOS VIGILO)
3298 BÜSTE „Wagner" (auf quaderförmigem Sockel) (Inschrift: WAGNER)
3330 „Friedrich der Große" (Stehende Figur in Uniform) (auf Landschaftssockel) (vgl. 4020)
3373 Zwei Putti um ein Herz kämpfend (auf Ovalsockel mit Herz, Rosen, Köcher und Bogen) (vgl. 3762) (Modell nach François-Joseph LeClercq)
3377 Zwillinge (Zwei sich umarmende und küssende kleine Mädchen)
3389 UHR Stehende Schönheit, auf einem von vier Hengsten gezogenen Wagen, der aus der Sonne kommt und über den Wolken schwebt (bekrönt von Adler)
3485 BÜSTE „Mozart" (Inschrift: MOZART)
3522 FIGUR / LAMPE „Mozart" (mit Geige, auf Sessellehne sitzend) (teilw. mit Inschrift: MOZART) (auch sign. Perigot) (vgl. 3285; 5483)
3525 Europa auf dem Stier reitend (auf Landschaftssockel)
3606 „Beethoven" (als stehende Figur) (vgl. 4645?)
3682 GRUPPE Kleines Mädchen mit Hund, der einen Korb trägt (vgl. 3683)
3683 GRUPPE Kleines Mädchen mit Hund, der einen Korb trägt (vgl. 3682)
3729 Stehender Knabe mit Büchern (auf Rundsockel)
3762 Zwei Putti um ein Herz kämpfend (auf Ovalsockel mit Herz, Rosen, Köcher und Bogen) (vgl. 3373) (Modell nach François-Joseph LeClercq)
3804 Stehendes Mädchen mit Rosenstrauß (an einer Rose riechend)
3843 Stehendes kleines Mädchen mit Buch (auf Quadratsockel mit Stuhl)
3870 „Ruth" (Tänzerin auf dreipassigem Sockel mit zwei Rosenkörben) (vgl. 3992; 4141; 5171; 5513)
3921 „Richard Wagner" (sitzende, lesende Figur) (vgl. 3248)
3943 Knabe mit Büchern und Schultasche
3980 Stehendes Mädchen mit Cape und großer Schleife (Mascherl) (einen Koffer tragend)
3990 Stehendes Biedermeiermädchen mit Schutenhut und Blumenstrauß (vgl. 4083; 4103)
3992 „Ruth" (Tänzerin auf dreipassigem Sockel mit zwei Rosenkörben) (vgl. 3870; 4141; 5171; 5513)
4020 „Friedrich der Große" (Stehende Figur in Uniform) (auf Landschaftssockel) (vgl. 3330)
4036 LAMPE Stehendes kleines Mädchen an einem kleinen Ofen (sog. „Kanonenofen") kochend (auf Rechtecksockel)
4041 Stehendes Mädchen mit Puppe und Teddybär (auf Rundsockel)
4045 Stehender Mädchenhalbakt
4083 Stehendes Biedermeiermädchen mit Schutenhut und Blumenstrauß FAYENCE / ALABASTER (vgl. 3990; 4103)
4103 FIGUR / LAMPE Stehendes Biedermeiermädchen mit Schutenhut und Blumenstrauß (vgl. 3990; 4083)
4111 BÜSTE „Liszt" (Inschrift: LISZT)
4141 „Ruth" (Tänzerin auf dreipassigem Sockel mit zwei Rosenkörben) (vgl. 3870; 3992; 5171; 5513)
4203 Schreitende Bacchantin (Damenhalbakt mit Kelch, einen Leoparden an blumengeschmücktem Halsband führend) (auf mit Blumengirlanden verziertem Ovalsockel)
4234 Tänzerin (nach links drehend) ALABASTER (vgl. auch Latour 4234 und 4235!)
4412 BÜSTE Beethoven ALABASTER
4463 BÜSTE Mädchen (nach links blickend) ALABASTER
4465 BÜSTE Junges Mädchen mit Schutenhut (nach links blickend) ALABASTER
4495 „Indische Tänzerin" (auf Rundsockel) (Ruth Saint Denis) (auch sign. Podola d.i. Czapek, Stanislaus)
4534 Stehendes lesendes Mädchen (auf Landschaftssockel mit vier Büchern) (auch sign. Polter 4534!) (vgl. 3849= ähnl. Modell)
4535 Liegender Barsoi (vgl. 5796)
4587 BÜSTE Mädchen mit Schutenhut ALABASTER
4588 BÜSTE Mädchen mit Biedermeierfrisur (nach rechts blickend) ALABASTER / Marmorsockel
4645 „Beethoven" (vgl. auch 3606?)
5135 FIGUR / LAMPE Stehende Rokokodame mit Rundfächer (auf Rundsockel)
5171 „Ruth" (Tänzerin auf dreipassigem Sockel mit zwei Rosenkörben) (vgl. 3870; 3992; 4141; 5513)
5483 „Mozart" (mit Geige, auf Sessellehne sitzend) (vgl. 3285; 3522)
5513 „Ruth" (Tänzerin auf dreipassigem Sockel mit zwei Rosenkörben) (vgl. 3870; 3992; 4141; 5171)
5796 Liegender Barsoi (vgl. 4535)
FIGUR „Weltkugel" ÖMKI (1906/07) Lit.: Neuwirth (1974), S. 72

Rosier, (E. od. G.?)

Rosier, Vorname sowie Lebens- und Schaffensdaten sind nicht bekannt. Lit.: GSB, Pl. 49 (m. Abb.).
623-GSB VIDE-POCHE Série III (Pendant vgl. 624-GSB)
624-GSB VIDE-POCHE Série IV (Pendant vgl. 623-GSB)

Rottenberg, Ena (Emma Helena)

*Ena Rottenberg, Kunstgewerblerin (*9.11.1893 Orawiczbanya/Ungarn, †4.6.1952 Wien), studierte (1916/17) und hospitierte bis 1925 an der KGW-Schule Wien (u.a. bei Anton Hanak und Michael Powolny). Sie war ab 1921 Mitarbeiterin der Wiener Werkstätte (Elfenbeinmalereien, geschnittene und geschliffene Glasdekore) und lieferte Modelle für die Porzellanmanufaktur Augarten, Wien. In einem Brief aus dem Jahr 1925 wird sie als Mitarbeiterin für Goldscheider erwähnt. Teilnahme an der KGW-Ausst. Paris 1925. Eine von Goldscheider „Kniende weibliche Figur", ausgeführt in Alabaster, wurde auf der Ausstellung Europäisches KGW 1927 in Leipzig gezeigt, des Weiteren war sie auf der Weihnachtsschau des Wiener Künstlerhauses 1928 mit Modellen der Manufaktur vertreten. Rottenberg war ab 1908 wohnhaft in Wien III., Esteplatz 3. Lit.: TB, Bd. 29, S. 97; Lehmann, 1931 ff.; Neuwirth (1974), S. 72, 272–273 (m. Abb.); Zühlsdorff, S. 440; Fahr-Becker, S. 233; AKL Index.*
502-MG VASE (Form: halbkugelartige Schüssel / Dekor: zwei Mädchen im Blattwerk)
5255 Liegende weibliche Figur (Halbakt) MARMOR / ALABASTER
5464 Sitzender männlicher Akt JADE / SPECKSTEIN?
Kniende weibliche Figur ALABASTER (1927) Lit.: Neuwirth (1974), S. 72
Stehender männlicher Halbakt (mit Tuch) (auf quadratischer Plinthe) Lit.: N.N.: Farbe und Bewegung, Die Fayencen der Wiener Manufaktur Friedrich GS Wien-Währing, in: DB, 5.1928, Nr. 213, S. 38 und 40 (m. Abb.), hier betitelt „Moderne Plastik".

Roy, P.

2903 „Allegorie der Kunst" (Stehende junge Frau in antikisierendem Gewand, eine kleine Statuette betrachtend) (Inschrift: ARS LONGA VITA BREVIS)

Rozel

Der Künstler „Rozel" taucht einzig im Ausst.-Kat. des ÖMKI 1898/99 und vereinzelt bei den beiden weiteren hier angeführten Modellen für Goldscheider auf. Weitere Daten zu Leben und Schaffen sind nicht bekannt.
1062 BÜSTE „Schubert" (vgl. 1145; 1157)
1145 BÜSTE „Schubert" (vgl. 1062; 1157)
1157 BÜSTE „Schubert" (vgl. 1062; 1145)
1687 BÜSTE „Gladiator" (Inschrift: AVE CAESARE! MORITURI TE SALVTANT) ÖMKI (1898/99) (vgl. 2184; 2294)
2184 BÜSTE „Gladiator" ÖMKI (1898/99) (vgl. 1687; 2294)
2294 BÜSTE „Gladiator" ÖMKI (1898/99) (vgl. 1687; 2184)

Rozet, Fan(n)y

*Fan(n)y Rozet, Bildhauerin (*13.6.1881 Paris, †?), war Schülerin von Marqueste und Mitarbeiterin bei Arthur Goldscheider, Paris. Rozet erhielt 1921 eine Bronzemedaille bei der Ausst. im Pariser Salon (Societé des Artistes Français). Lit.: Goldscheider, A. (Hrsg.): Artistic Evolution, 1926, T. 2, Pl. 30 (m. Abb.); Bénézit, Bd. 12, S. 63.*
„On three pink Marble Steps" BRONZE (H. fehlt, B. fehlt) Lit.: Goldscheider, A. (Hrsg.): Artistic Evolution, 1926, T. 2, Pl. 30 (m. Abb.)

Roze, Albert Dominique s. Rosé

Rubelle, A.

Der Künstler „Rubelle" taucht einzig im Ausst.-Kat. des ÖMKI 1900/01 auf. Weitere Daten zu Leben und Schaffen sind nicht bekannt. Lit.: Neuwirth (1974), S. 72, hier nur die Nennung des Künstlernamens.
1977 BÜSTE „Heilige Elisabeth" ÖMKI (1900/01)

Rudolf s. Rudolff

Rudolff

4603 Mädchenakt mit Hündchen (King Charles Spaniel) im Sessel sitzend
4624 Stehendes Biedermeiermädchen mit Schutenhut und Blumenstrauß (Krinoline) (auch sign. Rudolf) (vgl. 4819; 4820)
4819 Stehendes Biedermeiermädchen mit Schutenhut und Blumenstrauß (Krinoline) (auch sign. Rudolf) (vgl. 4624; 4820)
4820 Stehendes Biedermeiermädchen mit Schutenhut und Blumenstrauß (Krinoline) (vgl. 4624; 4819)

Rupp, Herbert

*Herbert Rupp, Gipsgießer (*8.2.1934, †?). Quelle: AGM, Liste „Adressen von guten und zuverlässigen Arbeitern" (undat., um 1953).*

Ruß, N. s. Ruß, (Willibald / Willy)

Ruß, (Russ, Willibald / Willy / Rudolf?)

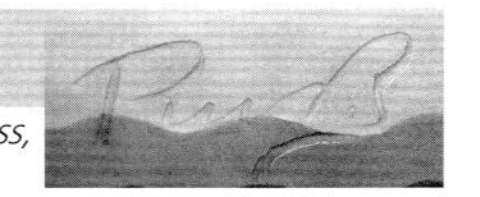

*Willibald (Willy/Willi) Ruß/Russ, Bildhauer und Keramiker (*7.7.1888 Krásné Pole/bei Karlsbad/Böhmen, †? (nach 1942) Merkershausen), genoss seine Ausbildung an der K.K. Fachschule für Keramik und verwandte KGW in Teplitz-Schönau. Er studierte bis 1910 an der KGW-Schule Wien. Ruß war Mitarbeiter bei Ernst Wahliss, Wien, sowie für die Werkstätten Karau und die Wienerberger Ziegelfabriks- und Baugesellschaft, beide in Wien tätig. Im Jahr 1920 gründete er eine kunstkeramische Werkstatt in Schönfeld bei Karlsbad und stellte 1927 in Berlin und Essen aus. Ruß war ab 1908 wohnhaft in Wien VII., Kaiserstraße 47–49, später dann im III. Bez., Baumgasse 69. Lit.: Neuwirth (1974), S. 274–275 (m. Abb.); Lehmann, 1908–1942; Zühlsdorff, S. 441; AKL Index; Cerný.*
4483 FIGUR / LAMPE Juno mit Pfau (Stehender Damenhalbakt mit großem Tuch) (auf längsovalem Sockel) (GS neu Modell-Nr. 74 „Göttin")

Rust, L.

1428 BÜSTE „Triumphator" (Römer mit Lorbeerkranz) (Inschrift: IO TRIVMPHE) (vgl. 2123)
2123 BÜSTE „Triumphator" (Römer mit Lorbeerkranz) (Inschrift: IO TRIVMPHE) (vgl. 1428)

Sailer, Karl

*Karl Sailer, Retoucheur und Modelleur (*12.11.1905, †?), scheint von seinem Beruf als Retoucheur her kommend in die Bildhauerei hineingewachsen zu sein, wie die späteren zahlreichen Modelle belegen. Sailer war von 1944–1946 bei der Wiener Porzellanmanufaktur Augarten und ab 1953 vorübergehend*

Leiter des kunstkeramischen Betriebes von Wilhelm Thomasch, Sierndorf. Mit Adolf Prischl führte er dann gemeinsam bis 1965 die Firma Kunstkeramik Prischl & Sailer. Quelle & Lit.: AGM, Liste „Adressen von guten und zuverlässigen Arbeitern" (undat., um 1953); Makus (2000), S. 178, 186.
5840 FIGUR/LAMPE Stehende Tänzerin im Schmetterlingskostüm (auf Rundsockel) (vgl. 6366)
6366 Stehende Tänzerin im Schmetterlingskostüm (auf Rundsockel) (vgl. 5840)
8753 GRUPPE Mädchen in Tracht mit springendem Zicklein (auf Natursockel) (Sailer zugeschr.)
8805 „Skibaby" (Kleine Skifahrerin mit Startnummer „3")
8806 „Holzhackerbub" (Stehender singender Bub mit kurzer Lederhose, eine Axt geschultert) (auf Rundsockel) (Pendant vgl. 8818)
8807 „Fohlen" (springend)
8808 „Fohlen" (stehend)
8809 „Fohlengruppe"
8810 WANDMASKE „Mädi" (Mädchen mit Kopftuch) (Pendant vgl. 8811)
8811 WANDMASKE „Bubi" (Knabe mit Halstuch) (Pendant vgl. 8810)
8812 WANDMASKE „Alt Wien" („Mäderl") (Mädchenkopf mit Schutenhut) (Pendant vgl. 8813)
8813 WANDMASKE „Alt Wien" („Buberl") (Knabenkopf mit Biedermeierhut/Zylinder) (Pendant vgl. 8812)
8814 WANDMASKE „Mäderl" („Maske Tirol") (Mädchen mit Trachtenhut, Tirolerin) (Pendant vgl. 8815)
8815 WANDMASKE „Buberl" („Maske Tirol") (Knabe mit Trachtenhut, Tiroler) (Pendant vgl. 8814)
8816 „Sie l ebt mich" (Stehender Bub mit Trachtenhut, einer Blume ihre Blütenblätter ausreißend) (auf Rundsockel) (Pendant vgl. 8817) (vgl. 8821= GRUPPE)
8817 „Liebeskummer" (Stehendes ängstliches Mädchen mit Kopftuch) (auf Rundsockel) (Pendant vgl. 8816) (vgl. 8821= GRUPPE)
8818 „Mariandl" (Stehendes Mädchen mit Trachtenhut und kurzer Lederhose, Tirolerin) (auf Rundsockel) (Pendant vgl. 8806)
8819 „Foxterrier" (stehend)
8821 GRUPPE „Sie liebt mich" (Stehendes Kinderpaar) (Bub, einer Blume ihre Blütenblätter ausreißend, daneben ein ängstliches Mädchen) (auf länglichem Ovalsockel) (vgl. 8816; 8817= jeweils EinzelFIGUR)
8823 „Weidmanns Heil" („Jäger") (Spähender Bub als Jäger mit Gewehr) (auf Natursockel mit kleinem Hasen)
8824 „Herzensdieb" (Schreitender Bub mit Herz unter dem Arm) (auf ovalem Natursockel)
8825 „Fußballer" (Fußball spielender Bub) (auf Natursockel)
8826 „Waldandacht" („Marterl") (Stehendes Mädchen, vor einem Marterl betend) (auf Natursockel mit kleinem Tannenbaum)
8833 „Fensterl" (Bub auf kleinen Vogel schauend) (Pendant vgl. 8834)
8834 „Fensterl" (Bub auf Marienkäfer schauend) (Pendant vgl. 8833)
8835 „Botaniker" (Stehender Wandersbub mit Stock und umgehängtem Tornister) (auf Natursockel mit Schmetterling)
8836 „Harmonika" (Stehender singender und Akkordeon spielender Bub mit Kappe) (auf ovalem Natursockel)
8849 „Nachtwächter"
8850 „Geiger" (Stehender Geige spielender Bub) (auf ovalem Natursockel mit kleinem Vogel)
8851 „Maler"
8852 „Wanderlust" (vgl. 8854= GRUPPE)
8853 „Gitarre"
8854 GRUPPE „Wanderlust" (vgl. 8852)

Sain, Marius Joseph

*Marius-Joseph Sain, Bildhauer (*18.10.1877 Avignion, †1908), war u.a. Schüler von Felix Charpentier und Mitglied der Künstlergruppe „La Stèle" (-> vgl. Kap. „Dekorative Skulpturen von Arthur Goldscheider in Paris"). Lit.: TB, Bd. 29, S. 316; Neuwirth (1974), S. 72; Bénézit, Bd. 12, S. 181; AKL Index.*

Saint-Paul, Jean

*Jean Saint-Paul, Kunstgewerbler, Maler, Illustrator (*4.9.1897 Paris, †?), stellte u.a. im Pariser Salon d' Automne aus und war Mitglied der Künstlergruppe „Évolution" (-> vgl. Kap. „Dekorative Skulpturen von Arthur Goldscheider in Paris"). Lit.: Vollmer, Bd. 4, S. 145; Bénézit, Bd. 12, S. 194; AKL Index.*

Sakellarios, Karin Jarl s. Jarl, Karin

Sala, Jean

*Jean Sala, Maler und Glaskünstler (*7.1.1895 Barcelona, †? wird letztmals um 1935 erwähnt), war Mitglied der Künstlergruppe „Évolution" (-> vgl. Kap. „Dekorative Skulpturen von Arthur Goldscheider in Paris"). Lit.: Goldscheider, A. (Hrsg.): Artistic Evolution, 1926, T. 1, Pl. 8 (m. Abb.); TB, Bd. 29, S. 332; Vollmer, Bd. 4, S. 146; Neuwirth (1974), S. 72; Bénézit, Bd. 12, S. 203; AKL Index.*
Fisch (Glas) (H. 31 cm; B. 25 cm) Lit.: Goldscheider, A. (Hrsg.): Artistic Evolution, 1926, T. 1, Pl. 8 (m. Abb.)
VASE (en Pate de Verre) (Glas) (H. 28 cm; B. 19 cm) Lit.: Goldscheider, A. (Hrsg.): Artistic Evolution, 1926, T. 1, Pl. 22 (m. Abb.)

Salzmann

VASE „Affen und Krokodil" Lit.: R., F.: Wiener Kunstkeramik, in: KM, 2.1902, S. 89 (m. Abb.), „Der Entwurf der Vase „Affen und Krokodil" ist von Salzmann. Einbiegend verjüngt sie sich zu einem Palmenstamme, der seine schön geschwungenen Blätter entfaltet. An Fuß und Stamm dieser Palme hocken, in lebhafter Bewegung festgehalten, Affen, welche mit einem drolligen Gemisch von Neugier und Entsetzen auf ein Krokodil starren, das sich von der Basis heraufwindet. Das Dekor dieser Vase ist bräunlich Algen".

Sardon

1497 s. Tell, E.
1913 s. Tell, E.

Sarrabezolles, Charles (Carlo)

*Charles (Carlo) Sarabezolles, Bildhauer (*27.12.1888 Toulouse, †11.2.1971 Paris), studierte unter A. Mercié und L. Marqueste, beschickte regelmäßig den Pariser Salon (Societé des Artistes Français) und entwarf zahlreiche Kriegerdenkmäler. Sarrabezolles erhielt zahlreiche Preise (u.a. Rompreis 1914, zwei goldene Medaillen bei der KGW-Ausst. Paris 1925) und war Mitarbeiter bei Arthur Goldscheider, Paris. Lit.: TB, Bd. 29, S. 469; Vollmer, Bd. 4, S. 161; Neuwirth (1974), S. 74; Bénézit, Bd. 12, S. 300; AKL Index.*
„Ame de France" (Statuette) BRONZE (Cire perdue) (Paris, Salon (1928) Lit.: Neuwirth (1974), S. 74

Sartorio, Antoine

*Antoine Sartorio, Bildhauer (*27.1.1885 Menton, †18.2.1988 Jouques), war Schüler von Injalbert und Hannaux. Sartorio stellte im Pariser Salon (Societé des Artistes Français) aus und erhielt dort 1934 die Silbermedaille. Er war vertreten auf der KGW-Ausst. Paris 1925 und Mitglied der Künstlergruppe „La Stèle" (-> vgl. Kap. „Dekorative Skulpturen von Arthur Goldscheider in Paris"). Lit.: Vollmer, Bd. 4, S. 161; Bénézit, Bd. 12, S. 306; AKL Index.*

Sauger s. Louger, J.

Savard

Savard, Kunstgewerbler (Lebens- und Schaffensdaten unbekannt), war Mitglied der Künstlergruppe „Évolution". Lit.: Goldscheider, A. (Hrsg.): Artistic Evolution, 1926, T. 1, Pl. 10 (m. Abb.).
VASE Metall (Kupfer? mit Einlegearbeiten) (H. 30 cm, B. 24 cm) Lit.: Goldscheider, A. (Hrsg.): Artistic Evolution, 1926, T. 1, Pl. 10 (m. Abb.)

Savine, Léopold Pierre Antoine

*Léopold Pierre Antoine Savine, Bildhauer (*6.3.1861 Paris, †7.10.1834 Paris), war Schüler von Dumont und Injalbert und Mitarbeiter bei Friedrich Goldscheider, Paris. Savine erhielt bei der Weltausst. in Paris 1900 eine Bronzemedaille. Lit.: TB, Bd. 29, S. 509; Hillenbrand, S. 362; Bénézit, Bd. 12, S. 338; AKL Index.*
758-GSB „Moissonneur" (Pendant vgl. 759-GSB)
759-GSB „Moissonneuse" (Pendant vgl. 758-GSB)

Schadt, Wilhelmine

Wilhelmine Schadt (Lebensdaten unbekannt, 1938 wohnhaft Gentzgasse 72, Wien XVIII.), war Mitarbeiterin bei Friedrich Goldscheider, Wien. Quelle: AGM (Gehaltsliste, Mai 1938; Gehaltsliste August 1938).

Scherk s. Schoerk, Hans

Schimelpfenig s. Schimmelpfennig

Schimmelpfennig, (Oswald)

Oswald Schimmelpfennig (Lebens- u. Schaffensdaten unbekannt), war ein in Berlin tätiger Bildhauer der 1902 und 1904 an der Großen Berliner Kunst-Ausst. beteiligt war und dessen Brunnenfigur „Gänsejunge" (Bronze) noch heute im Krankenhaus (Berlin, Eberswalderstraße) steht.
4262 Sitzender Frauenakt auf Stuhl (mit jungem Löwen)
4332 UHR „Herzensfrage" ALABASTER/Marmorsockel (Inschrift: HERZENSFRAGE)

Schmid

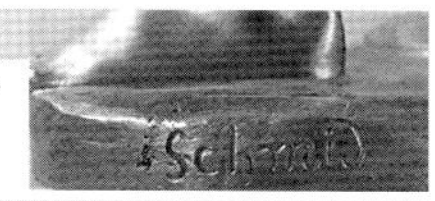

1486 Stehender, nackter Knabe BRONZE/Marmorsockel

Schmidt

901-WG WANDMASKE „Maske" (Frauenkopf mit kurzem Haar und Kragen)

Schmidt, Henri

3191 Stehendes Mädchen, mit fünf Katzen spielend (auf Rundsockel)

Schmidt-Kestner, Erich

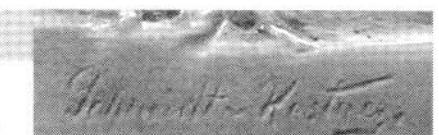

*Erich Schmidt-Kestner, Bildhauer (*15.1.1877 Berlin, †?), war bis etwa 1913 Mitarbeiter der Schwarzburger Werkstätten für Porzellankunst. Seine Bronzen wurden von Hermann Noack, Berlin, ediert. Lit.: TB, Bd. 30, S. 169; Hillenbrand, S. 370; Zühlsdorff, S. 630; Neuwirth (1974), S. 74; Bénézit, Bd. 12, S. 469; AKL Index.*
3024 FIGUR „Setter" (vorne) (mit/ohne Landschaftssockel) (auch sign. Strobl) (vgl. 6135) (vgl. 6659; 7100= jeweils GRUPPE)
3615 Brunst-Hirsch (Röhrender Hirsch)
3630 Tänzerin in Bewegung (Halbakt) (auf Rechtecksockel)
3785 Europa mit dem Stier (auf Rechtecksockel)
3846 „Hirschkuh mit Kalb" ÖMKI (1912)
4237 Reh
4695 Zwei Pudel (auf Ovalsockel)
6135 FIGUR „Setter" (vorne) (mit/ohne Landschaftssockel) (vgl. 3024) (vgl. 6659; 7100= jeweils GRUPPE)
6648 FIGUR „Setter" (rückwärts) (vgl. 7066) (vgl. 6659; 7100= jeweils GRUPPE)
6659 GRUPPE „Setter" (vgl. 7100) (vgl. 3024; 6135= jeweils Einzelfigur vorne) (vgl. 6648; 7066= jeweils Einzelfigur rückwärts)
7066 FIGUR „Setter" (rückwärts) (vgl. 6648) (vgl. 6659; 7100= jeweils GRUPPE)
7100 GRUPPE „Setter" (vgl. 6659) (vgl. 3024; 6135= jeweils Einzelfigur vorne) (vgl. 6648; 7066= jeweils Einzelfigur rückwärts)
GRUPPE „Tänzerinnen" (um 1908/09) Lit.: N.N.: Aus Berliner Musterlagern, in: Sprechsaal, 42.1909, S. 308, „... Alles hervorragende Arbeiten. Bei allen diesen Plastiken werden die Körperteile, soweit sie nicht vom Gewand bedeckt sind, ganz matt-glänzend gehalten, während das Gewand sind in glänzender Frische abhebt"; Neuwirth (1974), S. 74
„Brunst-Hirsch" (vermutlich Modell-Nr. 3615) (um 1908/09) Lit.: N.N.: Aus Berliner Musterlagern, in: Sprechsaal, 42.1909, S. 308; Neuwirth (1974), S. 74

Schoedl

Schoedl (aus Tetschen), Vorname sowie Lebens- und Schaffensdaten sind nicht bekannt. Lit.: Neuwirth (1974), S. 74.
„Dogge" ÖMKI (1912) Lit.: Neuwirth (1974), S. 74

Schoerck s. Schoerk, Hans

Schoerk, Hans

*Hans Schoerk, Bildhauer (*6.12.1849 Wien, †20.9.1899 Wien?), war Schüler der Wiener Akademie und spezialisiert auf die Porträtbildhauerei. Goldscheider verwendete für diesen ungemein produktiven Künstler das Pseudonym „Cherc". Auf die Umformung des Namens ins Französische für den internationalen Markt wurde an anderer Stelle schon mehrfach hingewiesen. Weiters finden sich auch die Namensformen Scherk bzw. Schörck und*

Schoerck sowie lt. TB auch Schörg. Bei Pudor (1902), S. 35 wird zusätzlich die Namensform „Choerc" genannt, die eindeutig den Hinweis auf den Wiener Bildhauer Hans Schoerk bietet. Dieser war nachweislich 1907 und 1908 wohnhaft gemeldet in Wien IV., Kreuzgasse 14, sein Atelier befand sich im XVIII. Bez., Martinstr. 50. Die Unstimmigkeit zwischen dem Sterbedatum (Bildhauerarchiv Karl Novak) und Lehmanns Wohnungsanzeiger lässt sich mit dem Weiterbestand der Wohnung durch die Witwe Schoerks einfach erklären. Lit.: Eisenberg, S. 502; Lehmann, 1908; TB, Bd. 30, S. 239; Neuwirth (1974), S. 74; Bénézit, Bd. 12, S. 517; AKL Index; Quelle: Bildhauerarchiv, Karl Novak.
4301 BÜSTE Frau mit Kopftuch (nach links blickend) ALABASTER (sign. „Scherk")
4302 BÜSTE Frau mit Kopftuch (nach rechts blickend) ALABASTER (sign. „Scherk")
4653 BÜSTE Mädchen mit Haube (nach links blickend) ALABASTER
4699 Sitzender Pan mit Panflöte (sign. „Schörck", auch sign. Haniroff) (Pendant vgl. Haniroff 4700)
4745 Stehender Knabe mit großen Augen (Google Eyes) (auf Rundsockel) (sign. „Schörck")
4769 Stehender Putto mit Blumengirlande (auf Rundsockel) (sign. „Schörck")
4793 BÜSTE Frau mit Haube / Hut ALABASTER (sign. „Scherk")
4882 BÜSTE Knabe ALABASTER (sign. „Scholz"?)
4984 Damenakt auf einer Sessellehne sitzend (auf rechteckigem Podest) (sign. „Schörck")
5376 BÜSTE Mädchen MARMOR (sign. „Scherk")
„Byzanz" ÖMKI (1903/04) Lit.: Neuwirth (1974), S. 74

Scholz s. Schoerk, Hans

Schram, Alois Hans

*Alois Hans Schram, Maler, Bildhauer und Plakatkünstler (*20.8.1864 Wien, †8.4.1919 Wien), studierte an der Wiener Akademie unter Griepenkerl und Makart. Schram befasste sich hauptsächlich mit Historien- und Genrebildern sowie mit der Porträtmalerei. Lit.: Eisenberg, S. 502; TB, Bd. 30, S. 275; Bénézit, Bd. 12, S. 523; AKL Index.*
1156 RELIEF Lautenspielerin (Wandrelief als Mosaikimitation)
1964 BÜSTE „Ilka" (Junge Frau mit Blumenschmuck) ÖMKI (1900/01) (vgl. 1968)
1968 BÜSTE „Ilka" (Junge Frau mit Blumenschmuck) (auf hohem Sockel mit Blumenranken) ÖMKI (1900/01) (vgl. 1964)
2868 BÜSTE Frau mit Kopftuch

Schuller, L (?)

L. Schuller, Bildhauer oder Kunstgewerbler (Lebens- und Schaffensdaten nicht bekannt), war Mitarbeiter bei Friedrich Goldscheider, Paris und schuf eine komplette Garnitur für Schreibtisch und Salon (Ascher, Tintenfässer, Kerzenhalter etc.) im antiken Stil. Lit.: GSB, Pl. 46 (m. Abb.).
697-GSB GARNITURE ANTIQUE „Encrier" No. 1 (Weitere Teile dieser Garnitur vgl. 698-GSB bis 702-GSB)
698-GSB GARNITURE ANTIQUE „Encrier" No. 2 (Weitere Teile dieser Garnitur vgl. 697-GSB; 699-GSB bis 702-GSB)
699-GSB GARNITURE ANTIQUE „Plumier" (Weitere Teile dieser Garnitur vgl. 697-GSB; 698-GSB; 700-GSB bis 702-GSB)
700-GSB GARNITURE ANTIQUE „Cendrier" (Weitere Teile dieser Garnitur vgl. 697-GSB bis 699-GSB; 701-GSB; 702-GSB)
701-GSB GARNITURE ANTIQUE „Flambeau" (Weitere Teile dieser Garnitur vgl. 697-GSB bis 700-GSB; 702-GSB)
702-GSB GARNITURE ANTIQUE „Bougeoir" (Weitere Teile dieser Garnitur vgl. 697-GSB bis 701-GSB)

Schulz

2497 Stehende Elsässerin mit Blumenkorb, Blumengirlande und Gans (auf oktogonalem Sockel)

Schuster, Josef

Josef Schuster, Kaufmann und Firmeninhaber der Wiener Manufaktur Friedrich Goldscheider nach der Arisierung im Dezember 1938. Geboren um 1900, stammte aus Dinkelsbühl, Deutschland. Zu seiner Biografie –> vgl. Kap. „Die Manufaktur Josef Schuster in den Kriegsjahren".

Schuster, Lilly

Josef Schusters Ehefrau Lilly war in der „Wiener Manufaktur Friedrich Goldscheider Nachf., Inhaber Josef Schuster" für die künstlerische Leitung zuständig (–> vgl. Kap. „Die Manufaktur Josef Schuster in den Kriegsjahren").

Schwarz, (K.?)

Zum Bildhauer (bzw. Modelleur?) (K.?) Schwarz können an dieser Stelle keine Lebens- und Schaffensdaten geliefert werden. Es lässt sich nicht einmal sagen, ob es sich um einen Künstler oder eine Künstlerin handelt. Könecke (2002), S. 78 macht keine näheren Ang., sondern verzeichnet nur einen sitzenden Damenakt (Modell-Nr. 8915). Weiterhin gibt es noch die Schubert-Büste (Modell-Nr. 5677), wobei aber beide Künstler nicht identisch sein müssen.
5677 BÜSTE Schubert
8915 „Anmut" (Sitzender Damenakt)

Schwarzbart, Franz

Franz Schwarzbart, Prokurist (Lebensdaten unbekannt, 1938 wohnhaft Staudgasse 7–9, Wien XVIII.). Quelle: AGM (Gehaltsliste, Mai 1938).

Schwetz / Schwetz-Lehmann, Ida s. Ida Lehmann

Ida Schwetz-Lehmann, Lebens- und Schaffensdaten sowie die Auflistung sämtlicher Modelle siehe unter ihrem Mädchennamen Lehmann, Ida.

Seeger, Ernst s. Seger, Ernst

Seger, Ernst

*Ernst Seger, Bildhauer und Medailleur (*19.9.1868 Neurode/Niederschlesien, †August 1939 Berlin), studierte 1884–1886 an der Kunstschule Breslau bei Prof. Christian Behrens, danach in Paris (Atelier Rodin). Seit 1892 beteiligte er sich u.a. an den Berliner Akademieausst., der Großen Berliner Kunstausst. und den Ausst. im Glaspalast München. Ab 1894 als Bildhauer in Berlin. Seine Entwürfe wurden u.a. von der Staatlichen Majolikamanufaktur Karlsruhe, von Rosenthal (Kunstabt. Selb, und Bahnhof-Selb) und Lauchhammer ausgeführt. Lit.: TB, Bd. 30, S. 441; Vollmer Bd. 4, S. 253; Bénézit, Bd. 12, S. 634; Niecol, Bd. 3, S. 123; AKL Index.*
3697 „Helena"/„Helene" (Stehender Frauenakt mit Tuch) (vgl. 3698) (GS neu Modell-Nr. 7 „Schönheit")
3698 „Helena"/„Helene" (Stehender Frauenakt mit Tuch) (vgl. 3697) (GS neu Modell-Nr. 7 „Schönheit")

Seidan, Thomas

*Thomas (Tomás) Seidan, Bildhauer (*5.9.1830 Prag, †2.12.1890 Wien), lebte ab 1850 in Wien. Die Namensformen Sidan und Sydan sind wohl Pseudonyme für Seidan. Eine weitere Namensvariante könnte auch der in der einschlägigen Lit. nicht nachweisbare Künstler Simon sein, mit dessen Namen zahlreiche Modelle für Goldscheider sign. sind. Lit.: TB, Bd. 30, S. 455; Bénézit, Bd. 12, S. 644; AKL Index.*
RELIEF BRONZE (Sitzende Nymphe, Haare kämmend) Lit.: Neuwirth (1973), S. 54–63, hier S. 56 (m. Abb.)
„Toilette de la Nymphe" Paris, Salon (1901) Lit.: Neuwirth (1974), S. 74
JARDINIERE „Songeuse" Paris, Salon (1902) Lit.: Neuwirth (1974), S. 74

Sevcik, Hanns

*Hanns Sevcik (abweichende Namensform bei Könecke: Hans Sefzik), Retoucheur (*5.12.1905, †?) bei Goldscheider in Wien und anschließend bei Carstens auf dem Tönnieshof in Norddeutschland. Er hatte dort die Leitung des Raumes der Retoucheure inne und war von Sommer 1953 bis April 1956 als einer der von Wien aus „übernommenen" Spezialisten tätig. Nach dem April 1956 kehrte er wieder nach Wien zurück. Lit. und Quelle: AGM, Liste „Adressen von guten und zuverlässigen Arbeitern" (undat., um 1953); Könecke (2002), S. 16, 79.*

Sidan vgl. auch Simon und Sydan

Sidan

–> vgl. Biografie Thomas Seidan.
2317 FIGUR „Das Gehör" (Stehendes Mädchen an Muschel hörend)

Simart, Marie-Louise

Marie-Louise Simart, Bildhauerin (Lebens- und Schaffensdaten unbekannt), war Mitglied der Künstlergruppe „La Stèle" (–> vgl. Kap. „Dekorative Skulpturen von Arthur Goldscheider in Paris"). Lit.: Goldscheider, A. (Hrsg.): Artistic Evolution, 1926, T. 2, Pl. 46 (m. Abb.).
„Le Cheval de Troie" (The Trojan Horse) BRONZE (vergoldet) auf Marmorsockel (H. 64 cm, B. 60 cm) Lit.: Goldscheider, A. (Hrsg.): Artistic Evolution, 1926, T. 2, Pl. 46 (m. Abb.)

Simon vgl. auch Sidan und Sydan

Simon

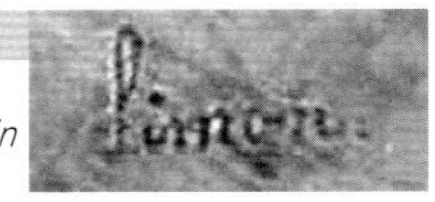

Eindeutig lässt sich sagen, dass die Signatur „Simon" ein von Goldscheider verwendetes Pseudonym ist. Wer sich schlussendlich dahinter verbirgt, konnte letztendlich nicht ermittelt werden.
712 VASE Acht Mädchen, Ringelreigen tanzend ÖMKI (1901/02 „Frühling"?)
726 SCHALE Liegender Frauenakt
878 VASE „Am Brunnen" (Kniende, trinkende Frauengestalt, Vasenmündung als Wasserspeier)
1006 VASE Mädchen mit Vogel spielend (reliefiert)
1072 „Schaukel" (Doppel-VASE Junge Frau zwischen einer Astgabel schwingend) (sign. Simon) (vgl. 1041 sign. Pecheur)
1340 VASE Stehendes Paar (sie reicht ihm eine Schale mit frischem Wasser)
1575 FIGUR „Aschenbrödel" (Tauben fütternd) (vgl. 1936)
1869 FIGUR „Fischer" ÖMKI (1899/1900) (Pendant vgl. 1870)
1870 FIGUR „Fischerin" ÖMKI (1899/1900) (Pendant vgl. 1869)
1874 FIGUR „Eva" (ohne / mit elektr. Broncezweig) ÖMKI (1902/03)
1924 WANDMASKE „Poppaea" ÖMKI (1899/1900) (Pendant vgl. 1952)
1932 HERME „Kind mit Schmetterling" ÖMKI (1899/1900)
1936 FIGUR „Aschenbrödel" (Tauben fütternd) (vgl. 1575)
1943 GRUPPE „Phantasie" (Nackte Frauengestalt (mit Harfe) auf Adler thronend) ÖMKI (1899/1900) (vgl. 2180)
1952 WANDMASKE „Nero" ÖMKI (1899/1900) (Pendant vgl. 1924)
1956 BÜSTE „Lorely" (Junge Frau mit Adler als Kopfputz) (Inschrift: LORELY) (vgl. 2226 sowie folgenden Eintrag)
1956 BÜSTE „Harmonia" (Junge Frau mit Schwan als Kopfputz) (Inschrift: HARMONIA) ÖMKI (1899/1900) (vgl. 2119) (vgl. vorher gehenden Eintrag)
2017 FIGUR „Ruth" (Stehende Frau mit Kornähren) (auf Natursockel) (Inschrift: RUTH) (vgl. 2110= Variante)
2083 FIGUR „Orpheus" ÖMKI (1900/01)
2110 FIGUR „Ruth" (Stehendes, sich bückendes Bauernmädchen mit Kornähren) (Inschrift: RUTH) ÖMKI (1902/03) (vgl. 2017= Variante)
2119 BÜSTE „Harmonia" ÖMKI (1899/1900) (vgl. 1956)
2180 GRUPPE „Phantasie" (Nackte Frauengestalt (mit Harfe) auf Adler thronend) ÖMKI (1899/1900) (vgl. 1943)
2187 BÜSTE „Pax" (Junge Frau mit Diadem aus Lorbeer) (Inschrift: PAX) ÖMKI (1901/02) (vgl. 2280; 2433)
2226 BÜSTE „Loreley" (Junge Frau mit Adler als Kopfputz) (vgl. 1956)
2229 UHR „Frühling" (Stehende Frauengestalt vor stilisiertem Kirschbaum) ÖMKI (1901/02) (vgl. 2299)
2231 FIGUR „Carmen" ÖMKI (1901/02)
2242 UHR „Psyche" (Inschrift: PSYCHE)
2280 BÜSTE „Pax" ÖMKI (1901/02) (vgl. 2187; 2433)
2299 UHR „Frühling" (Stehende Frauengestalt vor stilisiertem Kirschbaum) ÖMKI (1901/02) (vgl. 2229)
2305 UHR „Amicitia" (Stehendes Liebespaar reicht sich die Hände) (Inschrift: AMICITIA VINCIT HORAS) (vgl. 2523; 2736)
2433 BÜSTE „Pax" ÖMKI (1901/02) (vgl. 2187; 2280)
2523 UHR „Amicitia" (Stehendes Liebespaar reicht sich die Hände) (Inschrift: AMICITIA VINCIT HORAS) (vgl. 2305; 2736)
2575 Stehendes Holländermädchen mit Korb und Krug (auf Landschaftssockel)
2605 BÜSTE Junge Frau mit orientalischer Kopfbedeckung und Halsschmuck
2633 „Neugierde" (Stehendes Holländermädchen

neben einer Standuhr) (vgl. 2693)
2693 „Neugierde" (Stehendes Holländermädchen neben einer Standuhr) (vgl. 2633)
2698 BÜSTE Holländerin (in der Sockelkartusche Landschaftsdarstellung)
2702 FIGUR/LAMPE „Topfgucker in" (Kleines Holländermädchen in einen großen Topf schauend) (vgl. 2807; vgl. Sydan 2699) FAYENCE/BRONZE
2736 UHR „Amicitia" (Stehendes Liebespaar reicht sich die Hände) (Inschrift: AMICITIA VINCIT HORAS) (vgl. 2305; 2523)
2807 „Topfguckerin" (Kleines Holländermädchen in einen großen Topf schauend) (vgl. 2702; vgl. Sydan 2699)
2857 UHR „Das Rendevouz" (Biedermeierkavalier mit Taschenuhr auf seine Geliebte wartend, diese versteckt sich hinter einer Mauer/Uhrgehäuse)
3027 GRUPPE Stehendes Biedermeiermädchen, ein Kleinkind im Arm haltend
3945 Sitzendes Biedermeiermädchen (auf Scherenstuhl) mit geigenartigem Instrument
UHR „Weltall" (In Sonne Mond und Sternen schwebende Frauengestalt mit großem Tuch) (1902) Lit.: Pudor (1902), S. 33 (m. Abb.)
RELIEF „Chloë" (Terrakotta) ÖMKI (1899/1900) Lit.: Neuwirth (1974), S. 74
GRUPPE „Frühlingswehen" ÖMKI (1906/07) Lit.: Neuwirth (1974), S. 74
BÜSTE „Ruth" (Frau mit Kopftuch und Kornähren) (Inschrift: RUTH) Lit.: Pinhas, S. 58 (m. Abb.)

Skal / Skall (d.i. vermutlich Skála, Jaroslav)

*Jaroslav Skála, Bildhauer (*27.12.1881 Pardubice/Böhmen, †11.1.1919, Pardubice/Böhmen), war Schüler von Sucharda und Myslbek. Lit.: TB, Bd. 31, S. 109; Bénézit, Bd. 12, S. 864; AKL Index.*
2318 BÜSTE „Beethoven" (halbplastischer Kopf auf quaderförmigem Sockel) (Inschrift: LUDWIG VAN BEETHOVEN) ÖMKI (1902/03)
2888 BÜSTE „Mignon" (Junge Dame (Burgunderin?) mit Haube) (auf Kartuschensockel mit Landschaftsdarstellungen) ÖMKI (1901/02)

Skalák, Jindřich

Jindřich Skalák, Bildhauer oder Kunstgewerbler, absolvierte zwischen 1887–1897 die KGW-Schule in Bechyne und wurde bei GS (als Modelleur, Formgießer bzw. Figurenretoucheur) angestellt. Weitere Informationen sind nicht bekannt. Lit.: Neuwirth (1974), S. 31.

Spiller, Gretl (Margarete / Grete / Gretl Braun)

*Gretl (Margarete/Grete/Gretl Braun) Spiller, Keramikerin und Malerin (*1.2.1922 Wien, † Sept. 2002 Wien), war Schülerin der KGW-Schule in Wien unter Robert Obsieger. Nach ihrer Meisterprüfung im Töpfer- und Keramikergewerbe meldete sie 1947 ihr eigenes Gewerbe in Wien IV., Viktorgasse 18 an und stellte im gleichen Jahr auf der Wiener Messe aus. Später, nachdem sie nach Südamerika auswanderte, aber bereits nach einem halben Jahr wieder zurückkehrte, vermietete sie ihre Werkstatt an Carl (Carli) Bauer der ihre Modelle weiter produzierte und auch eigene Modelle ins Programm hinzufügte. Als sie ihre Arbeit in Wien wieder aufnahm, kamen neben den Keramiken auch Metallobjekte und Bronzen hinzu. Quelle: Archiv Uta M. Matschiner.*
18-WG WANDBEHANG „Chinesischer Träger" (Chinese mit zwei Gefäßen, für Kakteen/Blumen)
25-WG „Chinesen lesend" (Sitzender lesender Chinese) (Pendant vgl. 26)
26-WG „Chinesen lesend" (Sitzende lesende Chinesin) (Pendant vgl. 25)

Spörk, Marie

*Marie Spörk, Dessineurin (*5.9.1920, †?). Quelle: AGM, Liste „Adressen von guten und zuverlässigen Arbeitern" (undat., um 1953).*

Spuler, Erwin s. Spuller

Spuller, (R.)

R. Spuller, weitere Daten bzw. der Vorname lassen sich nicht ermitteln (keinerlei Einträge für R. Spuller bei TB, Vollmer oder im Wiener Adressbuch). Spuller ist durch die bislang bei Grell, S. 33 einzig nachgewiesenen „Exotischen Figuren" (1925) als Maler sowie als Bildhauer einzustufen. Bei einem der unter Modell-Nr. 5060 (Thomasch) ausgeführten Modelle findet sich die Malersign. „R. Spuller". Unter größtem Vorbehalt könnte auch der erst 23.3.1906 in Augsburg geborene und für die Staatliche Majolika-Manufaktur Karlsruhe tätige Keramiker, Maler und Lithograph Erwin Spuler infrage kommen. -> vgl. N.N.: Neue künstlerische Majolika-Arbeiten, in: Schaulade, 9.1933, S. 87 (m. Abb.).
Exotische Figuren (1925) („in bemaltem, rotem Ton") Lit.: Grell, S. 33; Neuwirth (1974), S. 74

Srisser? s. Griser

Stabi, Edouard

Edouard Stabi war Mitarbeiter von Arthur Goldscheider, Paris. Lit.: Goldscheider, A. (Hrsg.): Artistic Evolution, 1926, T. 1, Pl. 16 (m. Abb.)
VASE (nach einem Entwurf von Madeleine Anthouard) Metall (lackiert, schwarz/gold) (H. 28,5 cm, B. 12,5 cm) Lit.: Goldscheider, A. (Hrsg.): Artistic Evolution, 1926, T. 1, Pl. 16 (m. Abb.)
VASE (nach einem Entwurf von Madeleine Anthouard) Metall (lackiert, blau/gold) (H. 18 cm, B. 9 cm) Lit.: Goldscheider, A. (Hrsg.): Artistic Evolution, 1926, T. 1, Pl. 16 (m. Abb.)

Stankovits, Christa

Von Christa Stankovits, einer Bildhauerin oder Kunstgewerblerin (wohnhaft 1952 Huschkagasse 20, Wien XIX.), sollten ab 1953 Vasen in Produktion gehen. Diese wurden nach 36 Stück eingestellt, da zu diesem Zeitpunkt die Manufaktur nach Deutschland verkauft wurde. Die bis dahin in „weißer Fayence" erzeugten Stücke wurden mit der Entwerferin aufgeteilt und die Modell- und Gipsformen an sie übergeben. (Brief von Walter Goldscheider an Christa Stankovits, vom 13. April 1953). Quelle: AGM.

Stefan, A. (d.i. Stephan / Stefan Dakon)

Ob das Pseudonym „A. Stefan" von Goldscheider wirklich auf Figuren verwendet wurde oder ob es sich hier um einen redaktionellen Fehler der Zs. „Moderne Welt" handelt, kann nicht nachvollzogen werden, da bislang keine mit „A. Stefan" signierten Stücke nachweisbar sind. Stephan Dakons Modelle werden umfassend unter seinem richtigen Namen aufgeführt -> vgl. Dakon, Stephan.
„Modern" (Stehende Tänzerin in expressivem Kostüm) (auf Rundsockel) (um 1928) Lit.: N.N.: Berühmte Frauen und moderne Keramik, in: MW, 10.1928/29, Nr. 6, S. 29 (m. Abb.); Neuwirth (1974), S. 74
LAMPE „Die Dolly Sisters" identisch mit FIGUREN-Gruppe „Turteltauben" von Josef Lorenzl (Fay.) (um 1928) Lit.: MW, 10.1928/29, Nr. 6, S. 28 (m. Abb.) (Bem.: hier der falschen Ang. „A. Stefan"); Neuwirth (1974), S. 74

Steiner, Leo

*Leo Steiner, Spritzer (*10.9.1931, †?). Quelle: AGM, Liste „Adressen von guten und zuverlässigen Arbeitern" (undat., um 1953).*

Steinhof, Eugen

*Eugen Steinhof, Maler, Architekt, Bildhauer und Prof. (*5.10.1880 Wien, †10.7.1952 Los Angeles), studierte Plastik und Architektur in Wien und München. Steinhof war selbst ab 1923 Lehrer an der KGW-Schule in Wien und wurde 1925 in einem Brief als Mitarbeiter der Manufaktur Friedrich Goldscheider genannt. Lit.: Neuwirth (1974), S. 74, 304–305; Zühlsdorff, S. 643; AKL Index.*
STEHKOPF/BÜSTE Weiblicher Kopf BRONZE (1927) Lit.: L., A. S.: Vienna, in: The Studio, 92.1926, T. 1, S. 438–439 (m. Abb.); Europäisches KGW, Ausstellung Leipzig, 1927; Neuwirth (1974), S. 74.

Stellmacher, (Eduard?)

*Eduard Stellmacher, Porzellanmodelleur und Bildhauer (*7.6.1868 Volkstedt bei Rudolstadt, †? wird letztmals 1937 erwähnt), lebte in Turn-Teplitz. Lit.: TB, Bd. 31, S. 584; Neuwirth (1974), S. 74; Bénézit, Bd. 13, S. 220; AKL Index.*
2218 BÜSTE „Erwachen" ÖMKI (1901/02) (vgl. 2343)
2343 BÜSTE „Erwachen" ÖMKI (1901/02) (vgl. 2218)

Stepan, A(dolf)

Adolf Stepan, als Bildhauer bzw. Kunstgewerbler Autodidakt. Er lebt heute in Hinterbrühl/Niederösterreich. Lebens- und Schaffensdaten zu seiner Person sind nicht bekannt. Quelle: Mitteilung von Frau Elfi Sommer, Stoob/Bgld.
VASE/OBJEKT „Florida Flamingos" GSG Vienna (um 1991) (H. 32 cm) Lit.: Firmen-Kat./Folder: Geo-Art, Goldscheider Vienna (Stoob/Bgld. um 1992) o. Pag. (Innenseite 3, m. Abb.)
VASE/OBJEKT „Goldzahn-Elefanten" GSG Vienna (um 1991) (H. 26 cm) Lit.: Firmen-Kat./Folder: Geo-Art, Goldscheider Vienna (Stoob/Bgld. um 1992) o. Pag. (Innenseite 4, m. Abb.)
VASE/OBJEKT „Ikarus" GSG Vienna (um 1991) (H. 20 cm) Lit.: Firmen-Kat./Folder: Geo-Art, Goldscheider Vienna (Stoob/Bgld. um 1992) o. Pag. (Innenseite 3, m. Abb.)
VASE/OBJEKT „Lichthüter" GSG Vienna (um 1991) (H. 36 cm) Lit.: Firmen-Kat./Folder: Geo-Art, Goldscheider Vienna (Stoob/Bgld. um 1992) o. Pag. (Innenseite 3, m. Abb.)
VASE/OBJEKT „Mexiko-Kaktus" GSG Vienna (um 1991) (H. 40 cm) Lit.: Firmen-Kat./Folder: Geo-Art, Goldscheider Vienna (Stoob/Bgld. um 1992) o. Pag. (Innenseite 4, m. Abb.)
VASE/OBJEKT „Sonntags-Krawatte" GSG Vienna (um 1991) (H. 33 cm) Lit.: Firmen-Kat./Folder: Geo-Art, Goldscheider Vienna (Stoob/Bgld. um 1992) o. Pag. (Innenseite 3, m. Abb.)
VASE/OBJEKT „Spielende Raben" GSG Vienna (um 1991) (H. 40 cm) Lit.: Firmen-Kat./Folder: Geo-Art, Goldscheider Vienna (Stoob/Bgld. um 1992) o. Pag. (Innenseite 4, m. Abb.)
VASE/OBJEKT „Stadtpark-Pfau" GSG Vienna (um 1991) (H. 30 cm) Lit.: Firmen-Kat./Folder: Geo-Art, Goldscheider Vienna (Stoob/Bgld. um 1992) o. Pag. (Innenseite 3, m. Abb.)
VASE/OBJEKT „Tanzende Mädchen" GSG Vienna (um 1991) (H. 29 cm) Lit.: Firmen-Kat./Folder: Geo-Art, Goldscheider Vienna (Stoob/Bgld. um 1992) o. Pag. (Innenseite 4, m. Abb.)
VASE/OBJEKT „Wiegende Papageien" GSG Vienna (um 1991) (H. 33 cm) Lit.: Firmen-Kat./Folder: Geo-Art, Goldscheider Vienna (Stoob/Bgld. um 1992) o. Pag. (Innenseite 4, m. Abb.)
VASEN „Sternzeichen" (12 einzelne Vasen mit den verschiedenen Sternzeichen) GSG Vienna (um 1991) (H. fehlt) Lit.: Firmen-Kat./Folder: Geo-Art, Ihr Sternzeichen in Keramik/Your Sign of the Zodiac in Ceramic, Goldscheider Vienna (Stoob/Bgld. um 1992) o. Pag. (Leporello m. Abb.)
ASCHER/OBJEKT „Zigaretten-Plaza" GSG Vienna (um 1991) (B. 30 cm) Lit.: Firmen-Kat./Folder: Geo-Art, Goldscheider Vienna (Stoob/Bgld. um 1992) o. Pag. (Außenseite 5, m. Abb.)
LEUCHTER/OBJEKT „Lichterkranz" GSG Vienna (um 1991) (H. 20 cm, B. 30 cm) Lit.: Firmen-Kat./Folder: Geo-Art, Goldscheider Vienna (Stoob/Bgld. um 1992) o. Pag. (Außenseite 5, m. Abb.)
OBJEKT „Brieftheke" GSG Vienna (um 1991) (H. 16 cm, B. 29 cm) Lit.: Firmen-Kat./Folder: Geo-Art, Goldscheider Vienna (Stoob/Bgld. um 1992) o. Pag. (Außenseite 5, m. Abb.)
OBJEKT „Solo-Lipizzaner" GSG Vienna (um 1991) (H. 26 cm, B. 29 cm) Lit.: Firmen-Kat./Folder: Geo-Art, Goldscheider Vienna (Stoob/Bgld. um 1992) o. Pag. (Außenseite 5, m. Abb.)

Steuer, Bernard Adrien

*Adrien Bernard Steuer, Bildhauer (*21.6.1853 Paris, †1913 Paris), war Schüler von Jouffroy, Lequesne und A. Millet. Steuer stellte im Pariser Salon (Société des Artistes Français) aus und erhielt in den Jahren 1882–1884 und 1886 dort Auszeichnungen. Bei der Weltausst. Paris 1900 erhielt er ebenfalls eine Mention honorable. Lit.: TB, Bd. 32, S. 21; Bénézit, Bd. 13, S. 238; AKL Index.*
792-GSB PLATEAU „Femme Moyen Age"

Stoltenberg-Lerche, Hans Stephan

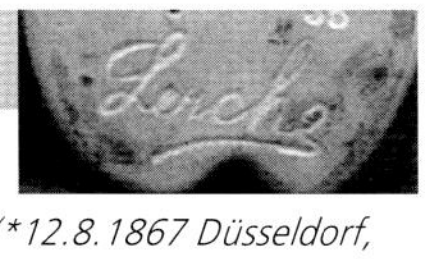

*Hans Stephan Stoltenberg-Lerche, Maler, Bildhauer, Medailleur und Kunstgewerbler (*12.8.1867 Düsseldorf, †17.4.1920 Rom), war Schüler von Ernest Carrière. Er schuf zahlreiche Modelle in Bronze und Keramik für Pariser Verleger. Ab 1900 war er in Rom ansässig, stellte auf*

der Weltausstellung Paris aus (Goldmedaille) und nahm 1902 an der 1. Internationalen Ausstellung für moderne dekorative Kunst in Turin teil. Seine künstlerische Tätigkeit war sehr breit gefächert: Arbeiten mit farbigen Tonschlickern, Metallgeräte, Schmuck, figürliche Keramik und ab 1915 unternahm er sogar Experimente mit farbigen Gläsern bei Fratelli Toso, Murano. Lit.: TB, Bd. 32, S. 106; Hillenbrand, S. 395; Zühlsdorff, S. 446; Bénézit, Bd. 8, S. 550–551; Arwas (2002), S. 616; AKL Index.

510 VASE „Chasse aux Papillons" (Schmetterlingsjagd) (Zwei Putti, einer plastisch, der andere reliefiert, auf Schmetterlingsjagd) (vgl. 561-GSB)
518 VASE Zwei Putti auf der Vasenwandung
519-GSB VASE „Enfants jouant" ZINN (Stoltenberg-Lerche zugeschr.)
521-GSB VASE „Enfant au Tambourin" ZINN (Pendant vgl. 679-GSB) (Stoltenberg-Lerche zugeschr.)
523-GSB VASE „Enfant Négrillon" ZINN (Stoltenberg-Lerche zugeschr.)
524-GSB VASE „Les deux Amis" ZINN (Stoltenberg-Lerche zugeschr.)
525-GSB VASE „Enfants Grimpant" ZINN (Stoltenberg-Lerche zugeschr.)
526-GSB VASE „Ibsen" (Stoltenberg-Lerche zugeschr.)
527-GSB VASE „Saut de Gamin" ZINN (Stoltenberg-Lerche zugeschr.)
528-GSB PLATEAU „Turc" ZINN (Pendant vgl. 564-GSB)
529-GSB PLATEAU „Nayade" ZINN
532-GSB PLATEAU „Printemps" ZINN/ZINKGUSS
554-GSB VASE „Précoce Vigneron" ZINN/ZINKGUSS (Stoltenberg-Lerche zugeschr.)
557-GSB VASE „Enfant a Califourchon" ZINN/ZINKGUSS (Stoltenberg-Lerche zugeschr.)
560-GSB VASE „Enfants Sauveteurs" ZINKGUSS (Stoltenberg-Lerche zugeschr.)
561-GSB VASE „Chasse aux Papillons" ZINN/ZINKGUSS (vgl. 510)
564-GSB PANNEAU „Turc" ZINN/ZINKGUSS (Pendant vgl. 528-GSB)
629-GSB VIDE-POCHE „Grenouille" ZINN
668 VASE An der Vasenwandung sitzender lachender Knabe
678-GSB JARDINIÈRE „Rocaille" (Stoltenberg-Lerche zugeschr.)
679-GSB VASE „Enfant joueur de Flute" ZINN (Pendant vgl. 521-GSB) (Stoltenberg-Lerche zugeschr.)
690 VASE An der Vasenmündung sitzender Knabe (in die Mündung blickend)
4770 Sitzender, Flöte spielender Pan (vgl. 5032 DECKELDOSE)
5032 DECKELDOSE Sitzender, Flöte spielender Pan (vgl. 4770 FIGUR)
VASE „egyptisch" ÖMKI (1898/99) Lit.: Neuwirth (1974), S. 66
VASE ÖMKI (1898/99) Lit.: Neuwirth (1974), S. 66
VASE „Les Enfants" ÖMKI (1900/01) Lit.: Neuwirth (1974), S. 66
VASE Drei Putti (um 1898/1900) (H. ca. 52 cm) (evtl. identisch mit: „Les Enfants"); Henry's 27.11.1996

Straeten, Georges van der

*Georges van der Straeten, Bildhauer (*21.12.1856 Gent, †1928 Paris), war bis 1882 Jurist und genoss ab 1883 dann eine Bildhauerausbildung bei Gustave Kasteleyn und Jef Lambeaux. Van der Straeten war von 1883–1928 in Paris tätig und stellte seine Arbeiten von 1885–1912 im Pariser Salon (Société des Artistes Français) aus. Zu seinen bevorzugten Sujets zählten Bildnisbüsten, Figuren von Straßenjungen, Komödianten, Pierrots und Statuetten im Rokokostil, die u.a. auch von M. Plot und der Societé de Bronzes, beide Paris gegossen wurden. Lit.: TB, Bd. 32, S. 150–151; N.N.: Bronze-Elfenbein Statuetten der Neuzeit, Künstlerkat., in: AZ, 1980, S. 459; Mackay, S. 361; Hillenbrand, S. 395; Bénézit, Bd. 13, S. 292; AKL Index.*

Stehende, sich verbeugende Biedermeierdame mit Haube BRONZE/ELFENBEIN/Onyxsockel A. Goldscheider Paris= Marke Bronze 5 (um 1905) (H. 22,5 cm); Stahl 4.9.1993; (H. 22,7 cm); Soth. com. # 42PR; (ohne Firmenmarke?) (H. 22 cm); Soth. L. 14.11.1979
„Mimi" (Statuette 1830) BRONZE/ELFENBEIN Paris, Salon Société des Artistes Francais (1909) Lit.: N.N.: Bronze-Elfenbein Statuetten der Neuzeit., Künstlerkat., in: AZ, 1980, S. 459
„Rêverie" (Statuette style Empire) Paris, Salon Société des Artistes Francais (1909) Lit.: N.N.: Bronze-Elfenbein Statuetten der Neuzeit. Künstlerkat., in: AZ, 1980, S. 459
„Pierrette" (stehend) BRONZE/ELFENBEIN Arthur Goldscheider, Paris (1924) Lit.: L'Illustration, Nr. 4268 vom 20.12.1924, S. 14 (m. Abb.)

Strakatý, Anton

Anton Strakatý, Bildhauer oder Kunstgewerbler, absolvierte im Schuljahr 1892/93 die KGW-Schule in Bechyne und wurde bei GS (als Modelleur, Formgießer bzw. Figurenretoucheur) angestellt. Weitere Informationen sind nicht bekannt. Lit.: Neuwirth (1974), S. 31; Neuwirth (1974a), S. 320; Bechyne 1884–1994, S. 55.

Strasser, Arthur

*Arthur Strasser, Bildhauer und Maler (*8.4.1854 Adelsberg/Krain, †8.11.1927 Wien), studierte von 1871–1875 an der Akademie der bildenden Künste in Wien bei Viktor Tilgner und ging mit einem Stipendium 1881–1883 nach Paris. Er war von 1899–1919 Lehrer an der KGW-Schule in Wien und zählt zu den Gründungsmitgliedern der Wiener Sezession. Unter dem Einfluss eines Ägyptenaufenthaltes schuf Strasser zahlreiche Büsten und Figuren zu nordafrikanischen und orientalischen Themen. Unter seinen zahlreichen Auszeichnung seien hervorgehoben: Wien 1871: Neulingspreis der Akad., München 1888: die kleine Goldmedaille, 1896: der Kaiserpreis, Wien 1898: die große Staatsmedaille und der Reichelpreis, Paris 1900: die große Goldmedaille. Im gleichen Jahr errichtete er im Auftrag der österr. Regierung für die Pariser Weltausst. die Bronzegruppe „Triumphwagen des Marc Anton", die anschließend vor der Wiener Secession Aufstellung fand und sich noch heute dort befindet. Ab 1901 war Strasser Ritter der franz. Ehrenlegion. 1928 fand eine Gedächtnisausst. im Künstlerhaus Wien statt. Strassers besonderer Verdienst war die naturalistische Polychromierung in der Keramik, mit der er sich schon sehr früh beschäftigte und sie perfektionierte. Strasser war ab 1908 wohnhaft in Wien III., Metternichgasse 7. Lit.: Eisenberg, S. 553; Lehmann, 1908; TB, Bd. 32, S. 157; Neuwirth (1974), S. 74, 306–308 (m. Abb.); Mackay, S. 361; Kleindel, S. 524; Bénézit, Bd. 13, S. 296; AKL Index.*

211 BÜSTE Orientale mit Vollbart, Turban und Fransenschal (Strasser zugeschr.)
212 BÜSTE Orientale mit Kopfputz (Fez) (Strasser zugeschr.)
217 BÜSTE Orientale mit Kopfputz (Fez) (Strasser zugeschr.)
331a FIGUR „Schlangenbändiger" (Stehender Araber mit Turban, Schlange und Flöte) (Strasser zugeschr.)
331b FIGUR „Wasserträger" (Strasser zugeschr.)
336 BÜSTE Araberin mit Kopftuch (nach rechts blickend) (Strasser zugeschr.)
524 BÜSTE Araber mit Vollbart und Turban
525 BÜSTE Jüngling mit Kappe in Renaissancetracht
537-GSB BUSTE „Nègre" BRONZE (ohne Sockel) ZINKGUSS (mit Sockel)
539-GSB BUSTE „Fumeur" (Téte d'étude) (Rauchender Bursche) (vgl. 1219) (Strasser zugeschr.)
545-GSB BUSTE „Antique" (ohne Sockel) (vgl. 545-GSB „Femme Fellah")
545-GSB BUSTE „Femme Fellah" (mit Sockel) (vgl. 545-GSB „Antique")
549-GSB BUSTE „Soudanais" (mit/ohne Sockel) (Strasser zugeschr.)
551-GSB STATUE „Retour de la Fontaine" No. 1 (vgl. 576-GSB; 579-GSB; 7915)
576-GSB STATUE „Retour de la Fontaine" No. 3 (vgl. 551-GSB; 579-GSB; 7915)
579-GSB STATUE „Retour de la Fontaine" No. 2 (vgl. 551-GSB; 576-GSB; 7915)
608-GSB STATUE „Porteur D'Eau" No. 1 ZINKGUSS (vgl. 609-GSB; 1174; 1600)
609-GSB STATUE „Porteur D'Eau" No. 2 BRONZE (vgl. 608-GSB; 1174; 1600)
610-GSB STATUE „Marchand D'Eau" No. 1 ZINKGUSS (vgl. 611-GSB)
611-GSB STATUE „Marchand D'Eau" No. 2 BRONZE (vgl. 610-GSB)
857 Wasserverkäufer (Stehender Araber mit Turban, Wasserkrug, Gefäß und kleiner Schale) ÖMKI (1898/99?) (Strasser zugeschr.) (vgl. 371 Gracien; 1175; 1601)
872 FIGUR „Egyptischer Krieger" (auf passigem Sockel mit arab. Schriftzeichen) (vgl. 884 BÜSTE) (Pendant vgl. 873) (Strasser zugeschr.)
873 FIGUR „Vasenträgerin" (auf passigem Sockel mit arab. Schriftzeichen) (vgl. 885 BÜSTE) (Pendant vgl. 872) (Strasser zugeschr.)
884 BÜSTE Ägypter mit Bart (nach links blickend) (Pendant vgl. 885) (vgl. 872 FIGUR) (Strasser zugeschr.)
885 BÜSTE Ägypterin (nach rechts blickend) (Pendant vgl. 884) (vgl. 873 FIGUR) (Strasser zugeschr.)
903 FIGUR „Kabyle" („Antar") (vgl. 961; 1055) (vgl. 926 BÜSTE) (Pendant vgl. 904; 962; 1056 Désurmont)
926 BÜSTE „Kabyle" („Antar") (nach links blickend) (vgl. 934) (vgl. 903; 961; 1055 FIGUR) (Pendant vgl. 927 Désurmont)
959 BÜSTE Junger Orientale (Araber) mit Vollbart und Turban (Strasser zugeschr.)
934 BÜSTE „Kabyle" („Antar") (nach links blickend) (vgl. 926) (vgl. 961; 1055 FIGUR) (Pendant vgl. 935 Désurmont)
961 FIGUR „Kabyle" („Antar") (vgl. 903; 1055) (vgl. 926 BÜSTE) (Pendant vgl. 904; 962; 1056 Désurmont)
1009 BÜSTE Araber mit Vollbart und Turban (Strasser zugeschr.)
1010 BÜSTE Junger Araber mit Bart und Turban
1026 Stehender Araber (auf Quadratsockel) (Strasser zugeschr.)
1055 FIGUR „Kabyle" („Antar") (Inschrift: ANTAR, in arab. Schriftzeichen) (vgl. 903; 961) (vgl. 926 BÜSTE) (Pendant vgl. 904; 962; 1056 Désurmont)
1060 BÜSTE Schwarzer Araber mit Turban (Strasser zugeschr.)
1142 „Haruspex" (Stehender römischer Wahrsager) (auf Rechtecksockel) (Inschrift: CAES. HARUSPEX) (vgl. 1262)
1174 FIGUR „Wasserträger" (Stehender Nordafrikaner mit großem Wasserschlauch/tierischer Wasserbehälter) (auf Natursockel) (ÖMKI 1898/99) (vgl. 1600; 608-GSB; 609-GSB)
1175 FIGUR „Wasserverkäufer" ÖMKI (1898/99) (vgl. 857; 1601) (vgl. 610-GSB; 611-GSB)
1176 FIGUR „Afrikanischer Sprecher" ÖMKI (1899/1900)
1193 BÜSTE „Berber" ÖMKI (1899/1900)
1219 BÜSTE „Raucher"/„Fumeur" (Téte d'étude) (Rauchender Bursche) (vgl. 539 GSB) (Strasser zugeschr.)
1232 FIGUR „Othello" ÖMKI (1899/1900)
1262 „Haruspex" (Stehender römischer Wahrsager) (auf Rechtecksockel) (Inschrift: CAES. HARUSPEX) (vgl. 1142)
1314 FIGUR „Stoffverkäufer" (Stehender Nordafrikaner, Stoffballen auf Kopf und Schulter tragend) (auf Rechtecksockel) ÖMKI (1898/99) (vgl. 1609)
1334 „Wollverkäufer" (Stehender Nordafrikaner einen großen Sack Wolle auf der Schulter tragend) (auf Natursockel) ÖMKI (1898/99) (vgl. 1608)
1564 FIGUR „Araber mit Säbel" (auf felsigem Landschaftssockel) (vgl. 1824)
1600 FIGUR „Wasserträger" (Stehender Nordafrikaner mit großem Wasserschlauch/tierischer Wasserbehälter) (auf Natursockel) (ÖMKI 1898/99) (vgl. 1174; 608-GSB; 609-GSB)
1601 FIGUR „Wasserverkäufer" ÖMKI (1898/99) (vgl. 857; 1175) (vgl. 610-GSB; 611-GSB)
1604 Stehender Nordafrikaner mit Turban, breitem Gürtel und langem Stab (auf oktogonalem Sockel)
1608 „Wollverkäufer" (Stehender Nordafrikaner einen großen Sack Wolle auf der Schulter tragend) (auf Natursockel) ÖMKI (1898/99) (vgl. 1334)
1609 FIGUR „Stoffverkäufer" (Stehender Nordafrikaner, Stoffballen auf Kopf und Schulter tragend) (auf Rechtecksockel) ÖMKI (1898/99) (vgl. 1314)
1652 FIGUR „Janitschare" (mit Turban, Schild, Säbel und langem Stab) (auf Landschaftssockel) (vgl. 2043; 2225) (vgl. 1921= BÜSTE)
1824 FIGUR „Araber mit Säbel" (auf felsigem Landschaftssockel) (vgl. 1564)
1876 FIGUR „Araber mit Krügen" (mit Turban und zwei Wassergefäßen) (auf Landschaftssockel) (Strasser zugeschr.)
1921 BÜSTE „Janitschare" (vgl. 1652; 2043; 2225= jew. FIGUR)
2043 FIGUR „Janitschare" (mit Turban, Schild, Säbel und langem Stab) (auf Landschaftssockel) (vgl. 1652; 2225) (vgl. 1921= BÜSTE)
2225 FIGUR „Janitschare" (mit Turban, Schild, Säbel und

langem Stab) (auf Landschaftssockel) (vgl. 1652; 2043) (vgl. 1921= BÜSTE)
7915 „Das Krüglein"/„Krugträgerin" (Schreitendes Bauernmädchen mit Krug) (auf Ovalsockel) ÖMKI (1906/07) (Bem.: Ausformung 30er Jahre) (vgl. 8727; 551-GSB; 576-GSB; 579-GSB)
8727 „Das Krüglein"/„Krugträgerin" (Schreitendes Bauernmädchen mit Krug) (auf Ovalsockel) ÖMKI (1906/07) (Bem.: Ausformung 30er Jahre) (vgl. 7915; 551-GSB; 576-GSB; 579-GSB)
„Mohr" (Terrakotta) ÖMKI (1899/1900) Lit.: Neuwirth (1974), S. 74
„Retour de la fontaine" BRONZE ÖMKI (1899/1900) s. Modell-Nr. 7915
„Soudanais" BRONZE ÖMKI (1899/1900) Lit.: Neuwirth (1974), S. 74
„Jeune Japonais" BRONZE ÖMKI (1899/1900) Lit.: Neuwirth (1974), S. 74
„Krug" ÖMKI (1903/04) Lit.: Neuwirth (1974), S. 74

Strobl (d.i. Schmidt-Kestner, Erich)

3024 Setter (auf Landschaftssockel) (auch sign. Schmidt-Kestner) (vgl. unles. Modell-Nr. Knabe, Setter am Halsband führend)
unles. Modell-Nr. GRUPPE Knabe, Setter am Halsband führend (auf Landschaftssockel) Firmenmarke Sculpture 13, Fabriqué en Autriche, gepr. „S"= Z2, H. 38,7 cm; Wien Museum Inv. 190.055

Stuart-Willfort, Irma

*Irma Stuart-Willfort., Bildhauerin (*10.5.1882 Wien, †1969 Wien), studierte 1905–1909 an der KGW-Schule Wien unter A. Strasser. 1909/10 Studium an der École des Beaux Arts in Paris. 1927 wieder in Wien unter M. Powolny („arbeitete Studien in keramischer Plastik") an der KGW-Schule. Ausstellungsbeteiligung 1930 „Buch und Raum der Gegenwart" im Künstlerhaus Wien, 1931 Teilnahme an der 52. Jahresausstellung im Künstlerhaus, Wien. Lit.: Neuwirth (1974), S. 309; Quelle (für das Todesjahr): Kunstarchiv Werner J. Schweiger, Wien.*
201-MG SCHALE Verkündigung
7211 STEHKOPF Frauenkopf mit gelockten Haaren und Kopftuch (halbplastisch)

Studenovsky, Johann

Johann Studenovsky, Bildhauer oder Kunstgewerbler, absolvierte im Schuljahr 1890/91 die KGW-Schule in Bechyne und wurde bei GS (als Modelleur, Formgießer bzw. Figurenretoucheur) angestellt. Weitere Informationen sind nicht bekannt. Lit.: Neuwirth (1974), S. 31; Neuwirth (1974a), S. 320; Bechyne 1884–1994, S. 55.

Suchan (d. i. Suchanek, Vacslav)

VASE „Hirschkäfer" ÖMKI (1902/03) Lit.: Pudor (1902), S.109 (m. Abb.); Neuwirth (1974), S. 75

Suchanek, Vacslav

Vacslav Suchanek, Kunstgewerbler (geb. in Krein bei Neustadt a.d. Mettau/Böhmen, †?), arbeitete neben Goldscheider auch für die Wiener Kunstkeramische Werkstätte Busch & Ludescher. Er war spezialisiert auf Tierplastiken. Lit.: Neuwirth (1974a), S. 159; Zühlsdorff, S. 447.
990 VASE Zwei Chamäleons
2355 FIGUR „Falke" (auf Natursockel) ÖMKI (1911/12)
2365 FIGUR „Adler" (mit Beute) (auf Natursockel) ÖMKI (1902/03)
2366 FIGUR „Auerhahn" ÖMKI (1902/03; 1911/12)
VASE „Hirschkäfer" ÖMKI (1902/03) Lit.: Pudor (1902), S.109 (m. Abb.); Neuwirth (1974), S. 75

Sydan vgl. auch Sidan und Simon

Sydan

–> vgl. Biografie Thomas Seidan.
ohne Nr. Stehende junge Frau mit wallendem Gewand (auf Felssockel) (vgl. 2368)
2176 FIGUR „Le Parfum" (Stehende Frauengestalt mit wallendem Tuch und großer (Mohn?-) Blume) ÖMKI (1901/02) (auch sign. Tell)
2199 FIGUR „Sonnenaufgang" (Stehende junge Frau auf Felssockel) (vgl. ohne Nr.; 2199b)
2199b FIGUR „Sonnenaufgang" (Stehende junge Frau mit wallendem Tuch auf Felssockel) (vgl. ohne Nr.; 2199; 2368)
2352 FIGUR „Edelstein" (Stehende Frauengestalt mit Kopfputz und gesenktem Haupt, in ihren Händen Blumenblüten und einen Kristall haltend) ÖMKI (1902/03) (auch sign. Tell)
2368 FIGUR „Sonnenaufgang, II. Gr(öße)" (Stehende junge Frau mit wallendem Tuch auf Felssockel) (auf sechseckiger Plinthe) (vgl. ohne Nr.; 2199b)
2492 SPIEGEL Stehender Mädchenhalbakt in einen Spiegel schauend (auf Landschaftssockel mit Baum)
2597 Wasserträgerin (Junge Bäuerin mit geschultertem Krug) (auf Landschaftssockel)
2699 BÜSTE Holländermädchen (ihre Schürze zu einer Vase/Vide-Poche? geformt) (vgl. Simon 2702; 2807)
2727 Stehendes Bauernmädchen mit Korb auf dem Rücken und Schale (auf Rundsockel)
2771 LAMPE Stehendes Biedermeiermädchen eine Schatulle öffnend (auf Rundsockel)
2812 Mädchen mit Schubkarren (Terrakotta) (vgl. Clara 2719; 2812)
2897 s. Tell, E.
2905 Stehendes Mädchen mit großem Weidenkorb
3312 Bäuerin mit Futtertrog und Kuh (auf Natursockel)
3487 JARDINIERE Ovaler Korb mit darauf sitzendem Mädchen und Knaben
3679 LAMPE Scherzendes Rokokopaar (auf Ovalsockel)
SPIEGEL „Malerei" (Vor einem Spiegel sitzende Malerin mit Pinsel und Palette) (H. m. Spiegel ca. 80 cm) (vgl. auch Tell 2897) Lit.: N.N.: Figürliche Terrakotten, in: KM, 4.1904, S. 138, 139 (m. Abb.), „Eine sehr eigenartige Darstellung „Malerei" , modelliert von Bildhauer Sydan. Ein junges Mädchen sitzt auf einem Felsblock und malt scheinbar ihr eigenes Bild; die Täuschung wird dadurch hervorgebracht, dass an die Terrakottenfigur ein Planspiegel anstösst, welcher die Form des sitzenden Mädchens im Spiegelbild wiedergibt. Die linke Hand des Mädchens hält die Palette, während die Rechte den Pinsel führt. Das linke Bein ist etwas hochgezogen, so dass das leichte Gewand, das sich an den Oberkörper eng anschmiegt, auch die bekleideten Körperteile erkennen lässt. Ein Lorbeerzweig ist noch vor dem Sitzplatz angebracht und deutet an, dass die „Malerin", in diesem Falle der Spiegel, die Natur trefflich wiederzugeben versteht. Dieses Bildwerk, das mit Spiegel eine Höhe von rund 80 cm hat, ist nicht bloss als Dekorationsgegenstand bestimmt, sondern will auch als Gebrauchsgegenstand benutzt werden, etwa als Aufsatz auf einen Wandschrank, wo es dann nicht nur die Figur der Malerin wiedergegeben wird, sondern auch die Gesichter der Beschauerinnen, die mit neugierigem Blick das Bildwerk und ihre eigenen hübschen Gesichter in dem Spiegel erblicken werden".
FIGUR „Spaziergang" ÖMKI (1906/07) Lit.: Neuwirth (1974), S. 75

Szeiler, Joseph

*Joseph Szeiler, Bildhauer (*1924 Ungarn, †1986, Burslem/UK), studierte ab 1944 an der Budapester Universität (Tierarzt). Im Jahre 1947 ging er nach Österreich, um dann bereits im Juli 1948 nach Großbritannien zu übersiedeln, wo er sich als Minenarbeiter versuchte. Szeiler wurde 1949 Lehrling bei Marcell Goldscheider und blieb für ca. 6 Monate. Seine eigene Firma gründete Szeiler dann 1951 in Hanley. Er vergrößerte sein Geschäft 1955 und zog um nach Burslem, wo er neben Vasen, Figuren, Konfektdosen und anderen keramischen Erzeugnissen hauptsächlich Tierfiguren produzierte. Nach seinem Tod im Jahre 1986 wurde in seiner „Moorland Pottery" ein kleines Museum eingerichtet, wo heute noch zahlreiche Formen seiner Modelle aufbewahrt werden. Lit.: PGGTR, 1957, S. 1042.*

Szwarc, Marek

*Marek Szwarc, Bildhauer, Metallplastiker, Maler und Kunstschriftsteller (*9.5.1892 Zgierc/Polen, †Dez. 1958 Paris od. Zgierc?), war Mitarbeiter von Arthur Goldscheider, Paris. Szwarc absolvierte zwischen 1910–1913 sein Studium an der École des Beaux-Arts in Paris unter A. Mercié sowie an der Akademie in Krakau. Ab 1914–1920 vorwiegend mit der Malerei um sich dann ab 1922 wieder der Bildhauerei zu widmen. Szwarcs Kunst fußt auf der Tradition des jüd. Kunsthandwerks und ist beeinflusst durch die afrikan. und assyr. Plastik. Lit.: Goldscheider, A. (Hrsg.): Artistic Evolution, 1926, T. 1, Pl. 21 (m. Abb.); TB, Bd. 32, S. 383; Vollmer, Bd. 4, S. 408; Bd. 6, S. 441; Bénézit, Bd. 13, S. 414; AKL Index.*
VASE (en cuivre repoussé) (Dekor: Liegender Rehbock) KUPFER (H. 27 cm, B. 16,5 cm) Lit.: Goldscheider, A. (Hrsg.): Artistic Evolution, 1926, T. 1, Pl. 21 (m. Abb.)
VASE (en Faience Décorée sous couverte) (H. 17,3 cm; B. 14,3 cm) Lit.: Goldscheider, A. (Hrsg.): Artistic Evolution, 1926, T. 1, Pl. 8 (m. Abb.)

Tacke

4634 Stehender Königspinguin (auf Felssockel) FAYENCE/ ALABASTER
4660 Stehender Collie (auf Rechtecksockel)
4689 Perlhuhn (vgl. 4744)
4744 Perlhuhn (vgl. 4689)
4810 Adler (auf Felssockel)

Take s. Tacke

Tallichet, S.

S. Tallichet, Bildhauer (Lebens- und Schaffensdaten unbekannt), war Mitarbeiter bei Arthur Goldscheider in Paris und Mitglied der Künstlergruppe „La Stèle" (–> vgl. Kap. „Dekorative Skulpturen von Arthur Goldscheider in Paris"). Lit.: Neuwirth (1974), S. 75.

Tegner, Rudolph Christopher Puggaard

*Rudolph Tegner, Bildhauer (*12.7.1873 Kopenhagen, †5.6.1950 Esboenderup/Dänemark), studierte von 1890–1893 an der Akademie in Kopenhagen, anschließend bis 1897 in Paris. Danach hielt sich Tegner in Italien, Spanien, Griechenland und Ägypten auf. Tegner war beeinflusst von Rodin und war Mitarbeiter bei Arthur Goldscheider, Paris. Bei der Weltausst. Paris 1900 erhielt er eine Bronzemedaille. Lit.: Goldscheider, A. (Hrsg.): Artistic Evolution, 1926, T. 2, Pl. 32 (m. Abb.). Lit.: TB, Bd. 32, S. 503–504; Vollmer, Bd. 4, S. 424; Bénézit, Bd. 13, S. 515; AKL Index.*
„Leda" BRONZE (H. fehlt, B. fehlt) Lit.: Goldscheider, A. (Hrsg.): Artistic Evolution, 1926, T. 2, Pl. 32 (m. Abb.)

Teichgrab, Alois

*Alois Teichgrab, Spritzer, Maler (*30.5.1906 Zierl/Tirol, †?), war vom 21.3.1922 bis 30.6.1953 bei Goldscheider beschäftigt. In den Jahren 1922–1925 erhielt er eine Ausbildung als Kerammaler. Er wurde im Jahr 1939 unter Josef Schuster zum Abteilungsleiter für die Spritzerei, Dessinage und den Brennhausbetrieb ernannt. Teichgrab entwarf Dekore und Glasurzusammenstellungen und entwickelte neue technische Arbeitsvorgänge. Quellen: AGM Liste „Adressen von guten und zuverlässigen Arbeitern" (undat., um 1953); Zeugnis von Walter Goldscheider vom 30.6.1953.*

Telcs, Ede (Eduard)

*Ede Telcs, Bildhauer und Medailleur (*12.5.1872 Baja/ Ungarn, †19.7.1948 Budapest), studierte in Wien unter Zumbusch und war tätig in Budapest. Anfänglich beschäftigte sich Telcs mit ungar. Volkstypen, die er in Terrakotta ausführte. Später dann zahlreiche Medaillen und Plaketten sowie Grab- und Denkmäler u.a. von Munkácsy und Kossuth. Telcs erhielt bei der Weltausst. Paris 1900 eine Bronzemedaille. Weitere Namensformen für Ede Telcs sind Telsch und Telzec. Das von Goldscheider verwendete Pseudonym „E. Tell" ist mit großer Wahrscheinlichkeit auch diesem ungar. Bildhauer zuzuordnen. Lit.: TB, Bd. 32, S. 510; Vollmer, Bd. 4, S. 426; Bénézit, Bd. 13, S. 518; AKL Index.*
3913 Stehendes Mädchen mit Rosen und Korb (Ausführung auch in ALABASTER, ohne Korb) (auch sign. Telzec, Telzer)
1255 GRUPPE „Der kleine Durstige" (Terrakotta) ÖMKI (1899/1900)
1410 BÜSTE „Raucher" BRONZE ÖMKI (1899/1900)
„Veilchen" (Fay.) ÖMKI (1910/11) Lit.: Neuwirth (1974), S. 75

Tell, E.

Da sich ein Bildhauer namens E. Tell in der einschlägigen Literatur nicht nachweisen lässt, muss man hier auch wieder an ein Pseudonym denken. Ähnlich wie bei den Namen Padola oder Polter, hinter denen sich der „erste Bildhauer" bei Goldscheider, Stanislaus Czapek, verbirgt, könnte man diesen auch hier vermuten. Infrage kommt aber wegen der großen Namensähnlichkeit eher der Ungar Ede Telcs. Wer schlussendlich mit dem Pseudonym E. Tell anonymisiert wurde, wird sich nicht klären lassen.

659 VASE Stehender Mädchenakt (am Vasenkorpus) (vgl. 853) (Pendant vgl. 675)
673 SCHALE Nymphe am Teich
675 VASE Stehender Mädchenakt (am Vasenkorpus) (Pendant vgl. 659)
676 SCHALE Nymphe in eine Muschel kletternd
796 VASE (Form: gebaucht, hochgezogene Mündung/Dekor: liegender Frauenakt (vollplastische Fig., Libellendekor)
841 VASE Stehende, nach vorne gebeugte junge Frau (deren Kleid sich zur Vase formt) (GS neu Modell-Nr. 1 „Melodie")
853 VASE Stehender Mädchenakt (am Vasenkorpus) (vgl. 659)
1497 FIGUR „Capriesertin"/„Fille de Capri" (Stehende Wasserträgerin mit Strohhut) (auf Natursockel) (am Metallschild sign. „Sardon") (vgl. 1913) (vgl. 1761= BÜSTE)
1565 FIGUR „Caprieser" (vgl. 1760= BÜSTE)
1655 FIGUR „Caprieser" (Pendant vgl. 1656) (vgl. 1760= BÜSTE)
1656 FIGUR „Capriesertin" (Pendant vgl. 1655) (vgl. 1761= BÜSTE)
1760 BÜSTE „Caprieser" (Junger Mann mit Strohhut (auf ährenverziertem Sockel) (Pendant vgl. 1761)
1761 BÜSTE „Capriesertin" (Junge Frau mit Strohhut) (auf ährenverziertem Sockel) (Pendant vgl. 1760)
1766 BÜSTE „Medea"/„Medee" (Frauenbüste) (Inschrift: MEDEE) (Pendant vgl. 1765 Mayre)
1767 GRUPPE „Geschwisterkuss" (Stehendes sich küssendes Kinderpaar) (auf Ovalsockel)
1768 GRUPPE „Der kleine Doktor" (Biedermeierkinderpaar) (auf Rechtecksockel mit Schemel) (Tell zugeschr.)
1769 FIGUR „Krugträgerin" (Wasserträgerin mit Gefäß auf dem Kopf) (vgl. 2274)
1790 s. 796
1907 Stehende junge Frau mit Strohhut und zwei Wassergefäßen (auf Landschaftssockel)
1913 FIGUR „Capriesertin (groß)" (Stehende Wasserträgerin mit Strohhut) (auf Natursockel) (vgl. 1497) (vgl. 1761= BÜSTE)
2040 FIGUR (SPIEGEL) „Spiegelträgerin" (Stehender Frauenakt, einen Spiegel haltend und in diesen schauend) (vgl. 2359) (vgl. 2109= Variante)
2046 FIGUR „Muschelträgerin" (Stehende, nach vorne gebeugte nackte Frauengestalt, eine große Muschel haltend) (vgl. 2111= bekleidet)
2064 FIGUR „Serpentintänzerin" (Loie Fuller) (auf Natursockel) (vgl. 2428)
2109 FIGUR (SPIEGEL) „Spiegelträgerin" (Stehende Frau, einen Spiegel haltend und in diesen schauend) (vgl. 2040; 2359= jew. Variante, nackt)
2111 FIGUR „Muschelträgerin (bekleidet)" (Stehende, nach vorne gebeugte bekleidete Frauengestalt, eine große Muschel haltend) (vgl. 2046= nackt)
2149 FIGUR „Die Nacht" (Junge Frau in wallendem Gewand das sich zu einer Schale formt)
2176 LAMPE „Le Parfum" (Stehende Frauengestalt mit wallendem Tuch und großer (Mohn?-) Blume) ÖMKI (1901/02) (vgl. auch Sydan)
2211 FIGUR „Sonnenblume" (Stehender Frauenhalbakt, umgeben von wucherndem Blattwerk und Sonnenblume)
2215 FIGUR „Schmetterling" (Nackte Frau mit Schmetterlingsflügeln, an einem Blütenkelch stehend)
2274 FIGUR „Krugträgerin" (Wasserträgerin mit Gefäß auf dem Kopf) (vgl. 1769)
2259 FIGUR (SPIEGEL) „Beim Spiegel" (Stehende, sich an einen Spiegel lehnende Frau, umgeben von Blattwerk)
2283 FIGUR „Kugelträgerin" ÖMKI (1902/03)
2335 UHR „Patientia" ÖMKI (1902/03)
2352 FIGUR „Edelstein" (Stehende Frauengestalt mit Kopfputz und gesenktem Haupt, in ihren Händen Blumenblüten und einen Kristall haltend) ÖMKI (1902/03, hier mit der Künstlerang. „Sydan")
2357 FIGUR „Rückkehr vom Brunnen" (Frau mit Gefäß, sich nach Krug bückend) (ÖMKI 1902/03; 1903/04) (vgl. 2447; 2876)
2358 FIGUR „Kugelträgerin mit Glaskugel"
2359 FIGUR (SPIEGEL) „Spiegelträgerin" (Stehender Frauenakt, einen Spiegel haltend und in diesen schauend) (vgl. 2040) (vgl. 2109= Variante)
2428 FIGUR „Serpentintanz" (Loie Fuller) (auf Natursockel) (vgl. 2064)
2447 FIGUR „Rückkehr vom Brunnen" (Frau mit Gefäß, sich nach Krug bückend) (ÖMKI 1902/03; 1903/04) (vgl. 2357; 2876)
2486 UHR Der Kuss (Inschrift: FUGIAT AETAS MANEAT AMOR)
2711 Tanzendes Biedermeier-Paar
2876 FIGUR „Rückkehr vom Brunnen" (Frau mit Gefäß, sich nach Krug bückend) (ÖMKI 1902/03; 1903/04) (vgl. 2357; 2447)
2897 SPIEGEL „Malerei" (Vor einem Spiegel sitzende Malerin mit Palette) (vgl. auch Sydan „Malerei" = identisches Modell)
2982 Stehendes Mädchen, einen Hasen haltend (auf Landschaftssockel) (STEINGUSS?)
3009 Stehendes Biedermeiermädchen (ihr Kleid etwas hochhebend) (nach links blickend) FAYENCE / ALABASTER (vgl. 4656) (Pendant vgl. 3067)
3067 FIGUR / LAMPE Stehendes Biedermeiermädchen (ihr Kleid etwas hochhebend) (nach rechts blickend) (vgl. 4657) (Pendant vgl. 3009)
3427 Sitzende Dame und stehender Kavalier mit Dreispitz in Rokokokostümen
3497 Sitzende junge Frau (Akt) mit Lyra (Sappho?) (vgl. 3576; 3577; 4625)
3511 Sitzender, sich bückender Frauenakt (auf Felssockel) (vgl. 3512)
3512 Sitzender, sich bückender Frauenakt (auf Felssockel) FAYENCE;ALABASTER / Marmorsockel (vgl. 3511)
3563 Stehende Wäscherin mit Trog (auf Rechtecksockel)
3576 Sitzende junge Frau (Akt) mit Lyra (Sappho?) (vgl. 3497; 3577; 4625; 4697)
3577 Sitzende junge Frau (Akt) mit Lyra (Sappho?) (vgl. 3497; 3576; 4625; 4697)
3716 GRUPPE „Mutterglück" (Mutter mit Kind auf reliefiertem Sockel) ÖMKI (1906/07)
3810 LAMPE Stehende Frauengestalt, einen geschmückten Kranz (?) in die Höhe haltend (auf zylindrischem, mit Blumengirlanden verziertem Sockel) (vgl. Lavergne 3718= fast identisches Modell)
4625 Sitzende junge Frau (Akt) mit Lyra (Sappho?) FAYENCE / ALABASTER (vgl. 3497; 3576; 3577; 4697)
4656 Stehendes Biedermeiermädchen (ihr Kleid etwas hochhebend) (nach links blickend) (vgl. 3009) (Pendant vgl. 3067)
4657 Stehendes Biedermeiermädchen (ihr Kleid etwas hochhebend) (nach rechts blickend) (vgl. 3067) (Pendant vgl. 4656)
4697 Sitzende junge Frau (Akt) mit Lyra (Sappho?) (vgl. 3497; 3576; 3577; 4625)
4753 BÜSTE Dame mit entblößter Schulter ALABASTER
5105 „Judith" (Stehender Frauenhalbakt mit Stirnband und Gürtel) (auf oktogonalem facettiertem Sockel)
VASE „Nymphentanz" ÖMKI (1900/01) Lit.: Neuwirth (1974), S. 75
„Wasserträgerin" (mit zwei umgehängten Krügen) (auf Felssockel) Lit.: N.N.: Figürliche Terrakotten, in KM, 4.1904, S. 137 (m. Abb.), „Die junge Frau hat die Wassergefäße mit einem Strick zusammen gebunden, den sie über die linke Schulter gelegt hat, ihr linker Arm umfasst das kürbisförmige, an der rechten Brust anliegende Gefäß, während die rechte Hand einen zweiten Strick hält, der am Halse des zweiten auf dem Rücken hängenden Gefäßes befestigt ist. Durch diese Befestigung erreicht die Frau, dass das auf der Schulter hängende Gefäß fest liegt und nicht etwa während des Gehens herabrutscht. Die Stellung der Frau entspricht derjenigen, welche dieselbe beim Gehen und Tragen der Last einzunehmen gezwungen ist. Die Unterarme und Füße der Frau sind unbekleidet; der Oberrock ist hochgeschürzt, ein kurzes Leibchen umschließt die Büste, während eine weitere, von vorn offene Jacke die Kleidung vervollständigt".

Telsch (d.i. Telcs, Ede / Eduard)

1255 GRUPPE „Der kleine Durstige" ÖMKI (1899/1900)
1410 BÜSTE „Raucher" BRONZE ÖMKI (1899/1900)

Telzec (d.i. Telcs, Ede / Eduard)

„Veilchen" (Fay.) ÖMKI (1910/11) Lit.: Neuwirth (1974), S. 75

Temporal, Marcel

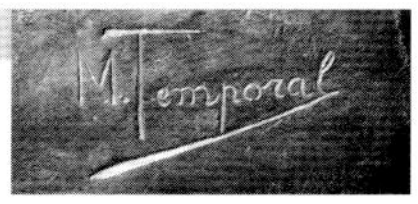

Marcel Temporal, Bildhauer (Lebens- und Schaffensdaten unbekannt), war Mitarbeiter bei Arthur Goldscheider in Paris und Mitglied der Künstlergruppe „La Stèle" (-> vgl. Kap. „Dekorative Skulpturen von Arthur Goldscheider in Paris"). Lit.: Goldscheider, A. (Hrsg.): Artistic Evolution, 1926, T. 1, Pl. 6 (m. Abb.); Neuwirth (1974), S. 75.

2-LaStèle BÜSTE Beethoven BRONZE
„Madrigal" (Fayence) (H. 31 cm, B. 23 cm) Lit.: Goldscheider, A. (Hrsg.): Artistic Evolution, 1926, T. 1, Pl. 6 (m. Abb.)

Thiele, (P.?)

684-GSB BUSTE „Esclave" (vgl. 733-GSB) (vgl. 1101; 1338)
733-GSB BUSTE „Esclave" (vgl. 684-GSB) (vgl. 1101; 1338)
1101 BÜSTE „Sklavin"/„Bedouine" (Sklavin / Araberin mit großem Turban, nach links blickend) (vgl. Lévèque 1101; 1338; 1877) (vgl. 684-GSB; 733-GSB)
1109 BÜSTE „Bulgarin" (mit langem Kopftuch) (nach links blickend) (Rohely ? sign.) (vgl. 1340; 1898) (vgl. 1615= WANDMASKE)
1338 BÜSTE „Sklavin"/„Bedouine" (Sklavin, Araberin mit großem Turban) (nach links blickend) (vgl. Lévèque 1101)
1341 BÜSTE „Mohrin" (mit Turban) (nach rechts blickend) (auch sign. Lévèque)
1615 WANDMASKE „Bulgarin" (vgl. 1109; 1340; 1898= jew. BÜSTE)
1877 BÜSTE „Sklavin"/„Bedouine" (Sklavin / Araberin mit großem Turban, nach links blickend) (vgl. Lévèque 1101; 1338) (vgl. 684-GSB; 733-GSB)
1878 BÜSTE „Mohrin mit Schmetterling"/„Négresse au Papillon" (Afrikanerin / Mulattin mit Turban) (auch sign. Prémont) (vgl. 1102)
1898 BÜSTE „Bulgarin" (mit langem Kopftuch) (nach links blickend) (vgl. 1109; 1340)
2195 BÜSTE „Die Ranke" vgl. Cherc 1946
FIGUR „Zephir" (um 1912) Lit.: N.N.: Die Leipziger Messe, Ostern 1913 in: DER, 13.1913/14, S. 83.

Thom, A.

1862 BÜSTE „Römischer Krieger" ÖMKI (1903/04)

Thomasch, Wilhelm

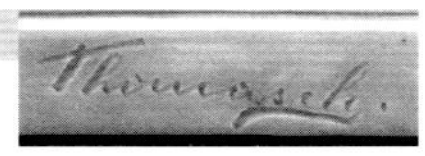

*Wilhelm Thomasch, Bildhauer (*22.8.1893 Teplitz-Schönau, †15.12.1964 Sierndorf), besuchte die KGW-Schule in Teplitz. Anschließend Studium an der Akademie der Bildenden Künste, Wien. Betriebsleiter bei Goldscheider. Er erwarb bereits 1926 eine seit 1857 bestehende Töpferei in Sierndorf, wo anfangs vorwiegend Blumentöpfe, Schalen und Vasen produziert wurden. Seine figuralen Entwürfe sind nicht nur für die Produktion von Goldscheider von Bedeutung. Auch Repräsentationsbauten in Wien, wie z. B. das ehemalige Kriegsministerium, verzierten seine Objekte. In den 50er Jahren stellte er die Produktion um. Zu dieser Zeit ist auch Karl Sailer, ein früherer Mitarbeiter bei Goldscheider, bei ihm beschäftigt. Es wurden dem damaligen Zeitgeist entsprechend ähnliche Entwürfe wie bei Goldscheider ausgeführt. Wandmasken, Afrikanerdarstellungen und Hundefiguren konnten nicht nur in Europa, sondern auch bis nach Übersee vertrieben werden. Sein Betrieb wurde, nachdem er bereits 1964 verstarb, noch bis 1970 von der Familie weitergeführt und dann geschlossen. Lit. u. Quelle: Makus (2000), S. 180; Matschiner (2006); Archiv Uta M. Matschiner.*

4978 Putto mit Vase auf der Schulter (sich auf einen Blumentopf kniend) (profilierter Rundsockel)
5030 FIGUR / LAMPE Sitzende Pierrette mit Laute FAYENCE / ALABASTER
5060 „Faszination" (Stehender weiblicher Halbakt mit großem Tuch und Kopfputz) (auf facettiertem Rundsockel) (GS neu Modell-Nr. 89 „Das Versprechen")
5117 Sitzender Pierrot mit Laute (auf abgerundetem Längssockel)
5136 Tanzender Pierrot mit Dame (auf facettiertem Ovalsockel)
5151 DECKELDOSE / RAUMLUFTERFRISCHER Elefant (Form: kugelförmig / Knauf: Elefant, vollplastische Figur / Dekor: Mündung und Deckel länglich oval durchbrochen)
5187 Tänzerin (Halbakt auf dem rechten Bein tanzend) (auf passigem Sockel mit umlaufendem Perlstab) BRONZE vermutlich GS
5211 TAFELAUFSATZ (Form: rund, flache Schale über zylinderförmigem, rautenreliefiertem Stand, perlenstabverzierte Halbkugelfüße / Dekor: mehrfarbige Laufglasur)

5214 GRUPPE Pierrot und Pierrette (Pierrette mit Blumengirlande, rücklings auf Pierrot) (div. Sockelformen)
5314 Stehende Tänzerin mit Kopfputz (einen langen Schal hinter sich haltend) (auf Rundsockel)
5354 DECKELDOSE Kugel (Form: kugelförmig mit drei Kugelfüßen / Knauf: Kugel/ Dekor: Blüten)
5356 DECKELDOSE Kugel (Form: kugelförmig mit zwei Handhaben / Knauf: drei Zacken/Dekor: Blüten- und Blattwerk) (Thomasch zugeschr.)
5366 KERZENHALTER (dreiflammig) Tanzender Frauenakt in rundem Reifen (auf Ovalsockel) (GS neu Modell-Nr. 43 „Perpetuum mobile")
5427 Stehender Knabe mit Kappe und Tragebeutel (auf oktogonalem Sockel)
5493 TELLER (Form: rund / Dekor an der Fahne: Blumen und Blattwerk)
7015 GRUPPE „Spaniels" (Zwei laufende Spaniels) (vgl. 7281; 8342) (vgl. 7061; 7329; 8344; ohne Nr.= jeweils EinzelFIGUR, vorderer Hund) (vgl. 7062; 7330; 8343= jeweils EinzelFIGUR, hinterer Hund)
7061 FIGUR „Spaniel" (Laufender Spaniel) (vgl. ohne Nr.) (Pendant vgl. 7062) (vgl. 7015; 7281; 8342= jeweils GRUPPE)
7062 FIGUR „Spaniel" (Laufender Spaniel) (vgl. 7299; 7330; 8343) (Pendant vgl. 7061) (vgl. 7015; 7281; 8342= jeweils GRUPPE)
7187 GRUPPE Zwei Musikanten (Akkordeonspieler und Sängerin als bekleidetes Terrierpärchen) (auf flachem Rechtecksockel) (Thomasch zugeschr.)
7281 GRUPPE „Spaniels" (Zwei laufende Spaniels) (vgl. 7015; 8342) (vgl. 7061; 7329; 8344= jeweils EinzelFIGUR, vorderer Hund) (vgl. 7062; 7330; 8343= jeweils EinzelFIGUR, hinterer Hund)
7299 FIGUR „Spaniel" (Laufender Spaniel) (vgl. 7062; 7330; 8343) (Pendant vgl. 7061) (vgl. 7015; 7281; 8342= jeweils GRUPPE)
7329 FIGUR „Spaniel" (Laufender Spaniel) (Pendant vgl. 7330) (vgl. 7015; 7281; 8342= jeweils GRUPPE)
7330 FIGUR „Spaniel" (Laufender Spaniel) (vgl. 7062; 7299; 8343) (Pendant vgl. 7329) (vgl. 7015; 7281; 8342= jeweils GRUPPE)
8342 GRUPPE „Spaniels" (Zwei laufende Spaniels) (vgl. 7015; 7281) (vgl. 7061; 7329; 8344= jeweils EinzelFIGUR, vorderer Hund) (vgl. 7062; 7330; 8343= jeweils EinzelFIGUR, hinterer Hund)
8343 FIGUR „Spaniel" (Laufender Spaniel) (vgl. 7062; 7299; 7330) (Pendant vgl. 8344) (vgl. 7015; 7281; 8342= jeweils GRUPPE)
8344 FIGUR „Spaniel" (Laufender Spaniel) (Pendant vgl. 8343) (vgl. 7015; 7281; 8342= jeweils GRUPPE)

Tischer, E. (?)

380-MG Schreitende Tänzerin mit großem Kopftuch / Cape, in verträumter Pose (ägyptische Tempeltänzerin?) (auf Ovalsockel)

Toinko, Heinrich

*Heinrich Toinko, Maler (*3.6.1888, †?). Quelle: AGM, Liste „Adressen von guten und zuverlässigen Arbeitern" (undat., um 1953).*

Tomaczewsky

RELIEF „Chrysanthème" (Terrakotta) ÖMKI (1900/01) Lit.: Neuwirth (1974), S. 75

Traverse, Pierre

*Pierre Traverse, Bildhauer und Kleinplastiker (*1.4.1892 Saint-André-de-Cubzac/Dordogne, †? wird 1958 letztmals erwähnt), war Schüler von Injalbert. Traverse lebte in Paris, stellte u.a. im Salon (Societé des Artistes Français) aus und erhielt nicht nur dort mehrere Auszeichnungen. Traverse war Mitglied der Künstlergruppe „La Stèle" (-> vgl. Kap. „Dekorative Skulpturen von Arthur Goldscheider in Paris"). Lit.: Goldscheider, A. (Hrsg.): Artistic Evolution, 1926, T. 2, Pl. 5 (m. Abb.); TB, Bd. 33, S. 361; Vollmer, Bd. 4, S. 465–466; Neuwirth (1974), S. 75; Hillenbrand, S. 407; Bénézit, Bd. 13, S. 778; AKL Index.*
ohne Nr.-LaStèle Laufendes Mädchen mit Reh BRONZE/ Marmorsockel
„Mother and Child" (Kniende Mutter mit Kleinkind) (H. fehlt, B. fehlt) Lit.: Goldscheider, A. (Hrsg.): Artistic Evolution, 1926, T. 2, Pl. 5 (m. Abb.)

Ullmaier s. Ullmann

Ullmann, Robert

*Robert Ullmann, Bildhauer (*18.7.1903 Mönchengladbach, †19.3.1966 Wien), studierte zwischen 1911–1917 bei F. Cizek an der Jugendschule in Wien sowie 1919–1927 an der Akademie bei J. Müllner. Ullmann arbeitete u.a. für die Wiener Porzellanmanufaktur Augarten und erhielt zahlreiche Preise und u.a. die Goldmedaille des Künstlerhauses Wien. Ullmann schuf zahlreiche Skulpturen, Grabdenkmäler und Gedenktafeln für Wien. Lit.: TB, Bd. 33, S. 556; Rutter, S. 61, 63; Vollmer, Bd. 4, S. 491; Hillenbrand, S. 410; Bénézit, Bd. 13, S. 901; AKL Index.*
7054AF FIGUR „Windhund" (Stehender Barsoi) (vgl. 7054F= GRUPPE) (vgl. 7571; 7972)
7054BF FIGUR „Windhund" (Liegender Barsoi) (vgl. 7054F= GRUPPE) (vgl. 7571; 7972)
7054F GRUPPE „Windhunde" (Ein stehender und ein liegender Barsoi) (vgl. 7054AF / BF= jeweils EinzelFIGUR) (vgl. 7571; 7972)
7571AF FIGUR „Windhund" (Stehender Barsoi) (vgl. 7571F= GRUPPE) (vgl. 7054; 7972)
7571BF FIGUR „Windhund" (Liegender Barsoi) (vgl. 7571F= GRUPPE) (vgl. 7054; 7972)
7571F GRUPPE „Windhunde" (Ein stehender und ein liegender Barsoi) (vgl. 7571AF / BF= jeweils EinzelFIGUR) (vgl. 7054; 7972)
7972AF FIGUR „Windhund" (Stehender Barsoi) (vgl. 7972F= GRUPPE) (vgl. 7054; 7571)
7972BF FIGUR „Windhund" (Liegender Barsoi) (vgl. 7972F= GRUPPE) (vgl. 7054; 7571)
7972F GRUPPE „Windhunde" (Ein stehender und ein liegender Barsoi) (vgl. 7972AF / BF= jeweils EinzelFIGUR) (vgl. 7054; 7571)

Valentin, Max

*Max Valentin, Bildhauer und Architekt (*14.7.1875 Fürstenwalde, †vor 1929), studierte von 1894–1903 an der Akademie der Künste in Berlin bei Prof. Ludwig Manzel. Entwürfe von ihm wurden von KPM Berlin und von der Firma Rosenthal (Kunstabt. Selb) ausgeführt. Lit.: TB, Bd. 34, S. 65; Neuwirth (1974), S. 75; Niecol, Bd. 3, S. 142; AKL Index.*
3599 „Badende" (Stehender Damenakt mit Tuch) (auf Felssockel) (vgl. 3681= ähnl. Mod.)
3681 „Badende" (Stehender Damenakt mit Tuch) (auf achteckigem Sockel) (vgl. 3599= ähnl. Mod.)
4496 Schreitende Tänzerin, ihr langes Kleid seitlich hochhebend (auf getrepptem Ovalsockel) (vgl. 4794)
4794 Schreitende Tänzerin, ihr langes Kleid seitlich hochhebend (auf getrepptem Ovalsockel) (vgl. 4496)

Vaňek, František

František Vaňek, Bildhauer oder Kunstgewerbler, absolvierte zwischen 1887–1897 die KGW-Schule in Bechyne und wurde bei GS als Modelleur, Formgießer bzw. Figurenretoucheur angestellt. Weitere Informationen sind nicht bekannt. Lit.: Neuwirth (1974), S. 31.

Vansteen

Da sich ein Künstler namens Vansteen in der der einschlägigen Lit. nicht nachweisen lässt, ist wieder an ein Pseudonym zu denken. Durch das holl. Genre der Figuren verwendete Goldscheider einen dem niederländ. ähnl. Namen.
2398 BÜSTE „Holländisches Kind" (Mädchen mit Spitzenhaube) (nach rechts blickend) (vgl. 2635)
2635 BÜSTE „Holländisches Kind" (Mädchen mit Spitzenhaube) (nach rechts blickend) (vgl. 2398)
Holländerin (Stehendes Holländermädchen mit Blumentopf) (1902) Lit.: Pudor (1902), S. 106 (m. Abb.); Neuwirth (1974), S. 75

Verbanck, Georges (Géo)

*Georges (Géo) Verbanck, Bildhauer und Medailleur (*28.2.1881 Gent, †1961 Aartselaar), war Schüler von L. van Biesbroeck an der Akademie in Gent sowie bei van der Stappen an der Akademie in Brüssel. Verbanck war später dann selbst Prof. an der Akademie in Gent. Lit.: TB, Bd. 34, S. 224; Vollmer, Bd. 5, S. 23; Neuwirth (1974), S. 75; Mackay, S. 382; Hillenbrand, S. 414; Bénézit, Bd. 14, S. 121; AKL Index.*
5242 Stehender Putto mit Früchten (auf Rundsockel) Paris, Salon (1928) (-> vgl. Wernekinck 5242)

Verschneider, Jean

*Jean Verschneider, Bildhauer (*29.8.1872 Lyon, †20.12.1943), war Schüler von J. Perrin und Injalbert sowie Mitglied der Künstlergruppen „La Stèle" und „Évolution" (-> vgl. Kap. „Dekorative Skulpturen von Arthur Goldscheider in Paris"). Verschneider beschickte u.a. den Pariser Salon (Societé des Artistes Français). Lit.: Goldscheider, A. (Hrsg.): Artistic Evolution, 1926, T. 1, Pl. 33 (m. Abb.); Neuwirth (1974), S. 75; Hillenbrand, S. 415; Bénézit, Bd. 14, S. 180.*
„Le saut" BRONZE Paris, Salon (1923) Lit.: Neuwirth (1974), S. 75
„La lecon" BRONZE Paris, Salon (1923) Lit.: Neuwirth (1974), S. 75
SCHALE / LAMPE „Coupe Lumineuse" (Luminous Bowl) METALL / Glas (H. 18 cm, B. 49 cm) Lit.: Goldscheider, A. (Hrsg.): Artistic Evolution, 1926, T. 1, Pl. 27 (m. Abb.)
BRUNNEN „Fountaine de Parc" (Parc Fountain) (Form: rund) (Mosaique de Pate de Verre, bleu-verte-or) (H. über Wasserobfläche 25 cm, Ø 90 cm) Lit.: Goldscheider, A. (Hrsg.): Artistic Evolution, 1926, T. 1, Pl. 33 (m. Abb.), Bem.: „im Besitz des Maharadja von Kapurthala".
BRUNNEN „Fountain in Wrought-Iron and Mosaic" SCHMIEDEEISEN / MOSAIK (H. fehlt, Ø fehlt) Lit.: Goldscheider, A. (Hrsg.): Artistic Evolution, 1926, T. 1, Pl. 34 (m. Abb.)
BRUNNEN „Mosaic Fountain with Wrought-Iron Pedestal" SCHMIEDEEISEN / MOSAIK (H. fehlt, Ø fehlt) Lit.: Goldscheider, A. (Hrsg.): Artistic Evolution, 1926, T. 1, Pl. 36 (m. Abb.)
BRUNNEN „Park Fountain with double Basin" SCHMIEDEEISEN / MOSAIK (H. fehlt, Ø fehlt) Lit.: Goldscheider, A. (Hrsg.): Artistic Evolution, 1926, T. 1, Pl. 38 (m. Abb.)
BRUNNEN „Fountain in Stone and Wrought-Iron" SCHMIEDEEISEN/STEIN (H. fehlt, Ø fehlt) Lit.: Goldscheider, A. (Hrsg.): Artistic Evolution, 1926, T. 1, Pl. 39 (m. Abb.)
BRUNNEN „Fountain with Jardinière" (H. fehlt, Ø fehlt) Lit.: Goldscheider, A. (Hrsg.): Artistic Evolution, 1926, T. 1, Pl. 40 (m. Abb.)

Vez, Constant

*Constant Vez, Bildhauer (*Paris, †?). Lit.: Neuwirth (1974), S. 75.*
„Foot-balleurs" BRONZE Paris, Salon (1923) Lit.: Neuwirth (1974), S. 75
„Coureur à l'arrivée" Paris, Salon (1923) Lit.: Neuwirth (1974), S. 75

Vidal, Henri

*Henri Vidal, Bildhauer (*4.5.1864 Charenton / Seine, †1918 Paris), war Schüler von M. Moreau und beschickte zwischen 1882–1913 den Pariser Salon (Societé des Artistes Français). Unter seinen zahlreichen Auszeichnungen sei die Silbermedaille bei der Weltausst. Paris 1900 hervorgehoben. Lit.: TB, Bd. 34, S. 332; Hillenbrand, S. 415; Bénézit, Bd. 14, S. 212; AKL Index.*
2009 FIGUR „Diana" ÖMKI (1900/01) (vgl. 2114)
2114 FIGUR „Diana" ÖMKI (1900/01) (vgl. 2009)

Vigoureux, Pierre Octave

*Pierre Octave Vigoureux, Bildhauer (*4.4.1884 Avallon / Yonne, †24.10.1965 Nogent-sur-Marne), war Schüler von H. Lemaire. Vigoureux lebte in Dijon und war dort ab 1935 der Direktor der École des Beaux-Arts. Er beschickte seit 1906 den Pariser Salon (Societé des Artistes Français), erhielt zahlreiche Auszeichnungen auf diversen Ausst. und war Mitglied der Künstlergruppe „La Stèle" (-> vgl. Kap. „Dekorative Skulpturen von Arthur Goldscheider in Paris"). Lit.: Goldscheider, A. (Hrsg.): Artistic Evolution, 1926, T. 1, Pl. 13 (m. Abb.); TB, Bd. 34, S. 358–359; Vollmer, Bd. 5, S. 34; Neuwirth (1974), S. 75; Hillenbrand, S. 416; Bénézit, Bd. 14, S. 235–236; AKL Index.*
„Young girl with deer" BRONZE (H. 61 cm, B. 58 cm) Lit.: Goldscheider, A. (Hrsg.): Artistic Evolution, 1926, T. 1, Pl. 13 (m. Abb.)

Villanis, Emmanuel

*Emmanuel Villanis, Bildhauer (*1858 Lille, †1920), war in den Jahren zwischen 1880 und 1910 ein viel beschäftigter Künstler. Sein bevorzugtes Genre waren Frauenbildnisse als Figuren und Büsten, häufig im orientalischen Stil. Sie wurden in Bronze, aber auch in anderen Materialien wie Terrakotta, von verschiedenen Firmen (u.a. durch die Societé de Bronzes, Paris) ausgeführt. Villanis beteiligte sich an zahlreichen Ausstellungen, u. a. in Mailand 1881, Rom*

1883, Turin 1884 und auch an der Weltausstellung 1889. Seine Bronzen sind heute auf Auktionen sehr gesucht. Lit.: Hillenbrand, S. 416; Bénézit; Pinhas, S. 196; AKL Index.
3208 „Captive" (Stehende Sklavin mit Handfesseln) (auf Mauersockel)

Villieu

1887 BÜSTE „Beatrice" (Junge Dame mit Diadem und Kopftuch) (vgl. 2122; 2593) (vgl. 2200= WANDMASKE)
1981 BÜSTE „Lilie"/(„Pureté") (Junge Frau in antikisierendem Gewand mit Blumenkranz im Haar, Dekolleté mit Lilienblüten geschmückt) (vgl. 2213; 2411)
2122 BÜSTE „Beatrice" (Junge Dame mit Diadem und Kopftuch) (auf konischem Sockel) (vgl. 1887; 2593) (vgl. 2200= WANDMASKE)
2200 WANDMASKE „Beatrice" (Frauenkopf mit Diadem und Kopftuch) ÖMKI (1901/02; 1902/03) (vgl. 1887; 2122; 2593= jeweils BÜSTE)
2213 BÜSTE „Lilie"/„Pureté" (Junge Frau in antikisierendem Gewand mit Blumenkranz im Haar, Dekolleté mit Lilienblüten geschmückt) (Inschrift: PURETÉ) ÖMKI (1900/01) (vgl. 1981; 2411)
2411 BÜSTE „Lilie" (Junge Frau in antikisierendem Gewand mit Blumenkranz im Haar, Dekolleté mit Lilienblüten geschmückt) (vgl. 1981; 2213)
2593 BÜSTE „Beatrice" (Junge Dame mit Diadem und Kopftuch) (Inschrift: BEATRICE) ÖMKI (1901/02; 1902/03) (vgl. 1887) (vgl. 2122= ähnl. Modell) (vgl. 2200= WANDMASKE)

Vondrák, Hugo

Hugo Vondrák, Bildhauer oder Kunstgewerbler, absolvierte zwischen 1887–1897 die KGW-Schule in Bechyne und wurde bei GS als Modelleur, Formgießer bzw. Figurenretoucheur angestellt. Weitere Informationen sind nicht bekannt. Lit.: Neuwirth (1974), S. 31.

Waderé, Heinrich

*Heinrich Waderé, Bildhauer (*2.7.1865 Colmar/Elsaß, †27.2.1950 München), war 1884–1891 Meisterschüler unter Syrius Eberle an der Akademie in München. Ab 1896 Prof., zwischen 1900 und 1933 Professur an der Kgl. KGW-Schule, München. Waderé errang eine Bronzemedaille auf der Weltausst. Paris 1900 und erhielt hohe Orden. Zahlreiche seiner bedeutenden Werke befinden sich in den Museen von Colmar, Mulhouse (beide Frankreich) und in München. Lit.: TB, Bd. 35, S. 13–14; Vollmer, Bd. 5, S. 62; Neuwirth (1974), S. 75–76, Mackay, S. 388; Zühlsdorff, S. 450; Bénézit, Bd. 14, S. 382; AKL Index.*
615-GSB BUSTE „Grecque" BRONZE (vgl. 1268; 2692)
1268 BÜSTE „Venuskopf"/„Venus Grecque" ÖMKI (1898/99; 1899/1900; 1902/03; 1906/07) (vgl. 615-GSB; 2692)
1561 BÜSTE „Elsässerin" (Lachendes Mädchen mit großer Haarschleife) ÖMKI (1900/01)
1789 BÜSTE „Madonnenkopf" (Frau mit geneigtem Haupt) (auf Rundsockel) ÖMKI (1900/01)
1860 BÜSTE „Psyche" ÖMKI (1906/07) (vgl. 2055; 2081)
2055 BÜSTE „Psyche" ÖMKI (1906/07) (vgl. 1860; 2081)
2081 BÜSTE „Psyche" ÖMKI (1906/07) (vgl. 1860; 2055)
2154 WANDMASKE „Madonnenkopf" (Frauenkopf mit Kopftuch) ÖMKI (1898/99; 1901/02)
2692 BÜSTE „Venus" ÖMKI (1898/99; 1899/1900; 1902/03; 1906/07) (vgl. 615-GSB; 1268)
2196 BÜSTE „Madonna" ÖMKI (1906/07)
„Elsässerin" ÖMKI (1906/07) (vgl. 1561) Lit.: Neuwirth (1974), S. 76

Wagner

Stehendes Mädchen in Wintergarderobe und Muff (auf Rundsockel) ALABASTER/ELFENBEIN (H. 27,5 cm) Doro 15.3.2005

Wagner, A.

ohne Nr. WANDTELLER Hütte am See (Form: rund, Fahne goldfarben bronziert/Dekor: Hütte am See, Gemälde, Kaltbemalung)

Waldmann, (Oscar?)

*Oscar Waldmann, Bildhauer (*25.6.1856 Genf, †15.3.1937 Paris), war Schüler von A. Kreling in Nürnberg und von H. Bovy in Genf. Waldmann lebte in Paris (hier Studien bei Gardet) und beschickte ab 1885 regelmäßig den Salon (Societé des Artistes Français). Auf der Weltausst. in Paris 1900 erhielt er eine Silbermedaille. Lit.: TB, Bd. 35, S. 74; Bénézit, Bd. 14, S. 402; AKL Index.*
2331 FIGUR „Löwe" ÖMKI (1902/03)

Waldmüller

5737 FIGUR Katze (vgl. 5769= BUCHSTÜTZE)
5769 BUCHSTÜTZE Katze (vgl. 5737= FIGUR)
5791 FIGUR Strauß (auf Rechtecksockel)

Walter, A. s. Walter, Victor Amalric

Walter, Victor Amalric

*Victor Amalric Walter, Glaskünstler, Kunstgewerbler und Maler (*19.5.1870 Sévres, †9.11.1959, Lury-sur-Arnon), war Mitglied der Künstlergruppe „Évolution" (-> vgl. Kap. „Dekorative Skulpturen von Arthur Goldscheider in Paris"). Ab 1885 Ausbildung an der Ecole Nationale de Céramique und der Manufacture Nationale des Sévres. In den zwanziger Jahren wurden seine Arbeiten u.a. durch Arthur Goldscheider, Paris verkauft. Im Jahre 1919 gründete er seine eigene Werkstatt in Nancy, die bis 1932 Bestand hatte. Seine Berühmtheit erlangte Walter durch seine Glaskunst (Pâte de Verre). Lit.: Zühlsdorff, S. 457.*

Warzel, Franz

Franz Warzel, Bildhauer oder Kunstgewerbler, absolvierte im Schuljahr 1894/95 die KGW-Schule in Bechyne und wurde bei GS (als Modelleur, Formgießer bzw. Figurenretoucheur) angestellt. Weitere Informationen sind nicht bekannt. Lit.: Neuwirth (1974), S. 31; Neuwirth (1974a), S. 321; Bechyne 1884–1994, S. 55.

Watzka, Friedrich Roland (Bedrich)

*Prof. Friedrich Roland Watzka, Bildhauer und Retoucheur (*31.3.1906 Ústi nad Labem/Böhmen, †? wird letztmals um 1929 erwähnt), absolvierte die keramische Fachschule in Teplitz, besuchte die Kunstakademie Prag und war bei GS als Retoucheur angestellt. Er leitete seit 1947 die Fachabteilung Industriekeramik an der Wiesbadener Werkkunstschule und war bei weiteren (nicht näher bezeichneten) Porzellanfirmen tätig. In einem Brief an die Manufaktur ersuchte er um erneute Anstellung mit den Worten „Wenn Sie einen Arbeitsplatz für mich hätten und eine Wohnung, dann würde ich gern wieder bei der Firma Goldscheider schaffen und meinen derzeitigen Arbeitsplatz kündigen, denn meine Sehnsucht nach Wien ist groß." (Brief von F.R. Watzka an die Manufaktur GS vom 16. Juli 1949). Quelle: AGM; Lit.: AKL Index.*
5663 Papagei, auf Kugel sitzend

Weber, Ernst

*Ernst Weber, Bildhauer (*18.4.1905 Steyr, †?), besuchte vor seinem Eintritt in die KGW-Schule Wien (Studium bei Cizek und M. Powolny) die Fachschule für Tonindustrie in Znaim. In den Jahren 1920–1922 war er an der Schleiss-Schule in Gmunden. Weber war neben Goldscheider auch für die Firma Rudolf Sommerhuber, Steyr (Oberösterreich) tätig. Lit.: Neuwirth (1974), S. 76; 338; Zühlsdorff, S. 618; Quelle: Kunstarchiv Werner J. Schweiger, Wien.*
5790 „Aschenkindl" (Sitzender Junge mit Aschenschale und Streichholztopf)
5820 GRUPPE Zwei Pferde (auf Ovalsockel)
6086 WANDMASKE/WANDVASE Mädchenkopf mit Locken
6288 WANDMASKE Frauenkopf mit Hand, eine Maske haltend (vgl. 6787)
6787 WANDMASKE Frauenkopf mit Hand, eine Maske haltend (vgl. 6288)
LAMPENFUSS (trichterförmig auf gewölbtem Stand) (um 1930) Lit.: MW, 11.1929/30, Nr. 24, S. VIII (m. Abb.)
VASE (balusterförmig, stark craquelierter Grund, Laufglasur) Lit.: Heizer (1931), S. 236 (m. Abb.)
1 LAMPENFUSS Kugel

Weiher

975 FIGUR „Tiger" (liegend) (auf Rechtecksockel)

Weis, Claire s. Weiss, Claire

Weiß

Ob es sich bei dem Künstler(in?) Weiß um Claire Weiß (Herczeg) handelt, lässt sich nicht sagen. Es ist aber eher davon auszugehen, dass eine Zusammenarbeit nach dem Krieg nicht mehr zustande kam.
800-WG WANDMASKE „Maske" (Betende Madonna mit geneigtem Kopf)
801-WG WANDMASKE „Maske" (Frauenkopf mit Kopftuch)

Weiss, Claire (Klára) (d.i. Herczeg, Claire (Klára))

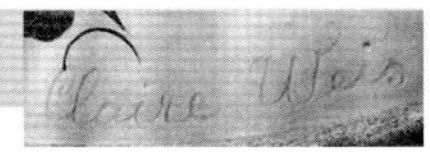

Lebens- und Schaffensdaten -> vgl. Herczeg, Claire (Klára)

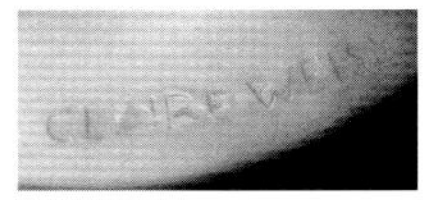

6328 „Muff" (Stehende Dame mit Muff und großem Hut) (auf Ovalsockel) (vgl. 7044) (vgl. 7720= GRUPPE)
6447 Posierende junge Dame mit Hut (einen zusammengefalteten Fächer in ihrer rechten Hand haltend) (auf Ovalsockel) (sign. Claire Weiss)
6454 FIGUR „Komtesse" (Stehende Dame mit Hut) (auf Ovalsockel) (sign. Weis(s), Claire) (vgl. 6951) (vgl. 7327; 8145= jeweils GRUPPE „Komtesse mit Hund")
6462 Stehende Dame mit Hut, ihr Kleid etwas hochhebend (auf Rundsockel) (sign. Weis, Claire)
6507 Stehende Dame mit Kappe und Muff (auf Ovalsockel)
6747 Stehendes Mädchen mit Hut, ein Schaf betrachtend (auf Ovalsockel) (Weiss/Herczeg zugeschr.)
6769 Sitzendes lesendes Mädchen
6773 Stehendes Mädchen mit Hut, Schirm und Tasche, einen kleinen Hund belehrend (auf Ovalsockel)
6829 Stehendes kleines Mädchen mit Haarschleife und Teddybär (auf Rundsockel) (Weiss/Herczeg zugeschr.)
6830 GRUPPE Tanzendes Paar (Russisches Ballett) (auf Ovalsockel)
6853 Schreitendes Mädchen mit kleinem Terrier, welcher an ihrem Kleid zerrt (auf Ovalsockel) (Weiss/Herczeg zugeschr.)
6900 Tänzerin in Bewegung (auf Ovalsockel)
6951 FIGUR „Komtesse" (Stehende Dame mit Hut) (auf Ovalsockel) (sign. Weihs, Claire) (vgl. 6454) (vgl. 7327; 8145= jeweils GRUPPE „Komtesse mit Hund")
6955 Gehendes Kinderpaar (auf Ovalsockel) (Weiss/Herczeg zugeschr.)
7027 GRUPPE Rotkäppchen (mit Korb) und der Wolf (auf Ovalsockel)
7028 Schreitender Junge in Matrosenkostüm (ein Modellsegelschiff hinter sich herziehend) (auf flachem Rechtecksockel) (Weiss/Herczeg zugeschr.)
7035 Stehende Tennisspielerin im Wind (auf Ovalsockel)
7040 Tänzerin in langem Kleid (auf Ovalsockel)
7044 „Muff" (Stehende Dame mit Muff und großem Hut) (auf Ovalsockel) (vgl. 6328) (vgl. 7720= GRUPPE)
7191 Stehender Junge mit Golfsack
7201 Stehendes Mädchen, ihren Hut haltend (auf Rundsockel mit Kaninchen)
7222 Stehendes Mädchen mit Haarband und Männchen machendem Pekinesen (auf Ovalsockel)
7279 Stehende elegante Dame mit großem Hut und langen Handschuhen (auf Ovalsockel) (vgl. 8151) (vgl. 7367; 8155= jeweils GRUPPE „Maskerade") (sign. Claire Weiss/Claire Herczeg)
7367 GRUPPE „Maskerade" (Schreitende elegante Dame mit Barsoi) (auf Ovalsockel) (vgl. 8155) (vgl. 7279; 8151= jeweils Dame ohne Hund)
7390 FIGUR/LAMPE „Parisienne" (Schreitende elegante Dame, ihren Schirm in der linken Hand haltend) (auf Ovalsockel) (vgl. 8138) (vgl. 8361= seitenverkehrtes Modell)
7511 „Spanierin" (Stehende Spanierin mit Fächer) (auf Ovalsockel)
7557 Schreitende junge Dame mit Hut, zwei Hutschachteln tragend (auf Ovalsockel)
7648 LAMPE Stehende Dame mit großem Hut und Fächer (auf flachem Rechtecksockel)
7691 FIGUR „Demaskierung" (Stehende junge Dame, ihre Maske abnehmend) (auf Ovalsockel) (vgl. 8183)
7720 GRUPPE Stehende Dame mit Muff, großem Hut und Windhund (Windspiel) (auf Ovalsockel) (vgl. 6328; 7044= jeweils FIGUR Dame)
7847 GRUPPE Gänseliesel (Stehendes kleines Mädchen mit Gänsen) (auf Ovalsockel) (vgl. 7861; 8003= SCHALE)
7855 FIGUR/LAMPE „Lydia" (Schreitende junge Tänzerin, ihr langes Kleid auseinander haltend) (auf Ovalsockel) (vgl. 7963; 8322) (sign. Herczeg/Weiss)
7861 GRUPPE Gänseliesel (Stehendes kleines Mädchen mit Gans) (auf Ovalsockel) (vgl. 7847; 8003= SCHALE)
7966 „Ballpause" (Sitzende Tänzerin mit Ballmaske und

Fächer) (auf Rundsockel) (sign. Weiss, Claire)
8003 SCHALE Gänseliesel (Stehendes kleines Mädchen mit Gans) (vgl. 7847; 7861)
8029 Stehendes Mädchen mit Kappe, Schirm und Skyeterrier (auf Ovalsockel)
8135 Schreitende elegante Dame mit Hut (auch sign. Herczeg)
8138 FIGUR „Parisienne" (Schreitende elegante Dame, ihren Schirm mit der linken Hand haltend) (auf Ovalsockel) (vgl. 7390) (vgl. 8361= seitenverkehrtes Modell) (auch sign. Herczeg, Claire)
8151 Stehende elegante Dame mit großem Hut und langen Handschuhen (auf Ovalsockel) (vgl. 7279) (vgl. 7367; 8155= jeweils GRUPPE „Maskerade")
8155 GRUPPE „Maskerade" (Schreitende elegante Dame mit Barsoi) (auf Ovalsockel) (vgl. 7367) (vgl. 7279; 8151= jeweils Dame ohne Hund)
8183 FIGUR „Demaskierung" (Stehende junge Dame, ihre Maske abnehmend) (auf Ovalsockel) (vgl. 7691)
8204 FIGUR „Abendkleid" (Stehende, schlanke Dame in langem Kleid und Pelzumhang)
8205 FIGUR „Ivette" (Stehende junge Dame in langem Kleid und Schal)
8241 FIGUR „Muffdame" (Stehende Dame in Mantel und Hut mit Muff) (auf Ovalsockel)

Wernekinck, Sigismund

*Sigismund Wernekinck., Bildhauer (*18.3.1872 Bromberg/Posen, †8.11.1921 Berlin), studierte 1891–1899 an der Preußischen Akademie der Künste in Berlin bei Gerhard Janensch, Ernst Herter und Peter Breuer. Er war Meisterschüler von Reinhold Begas und stellte ab 1901 im Glaspalast München sowie ab 1903 auf der Großen Berliner Kunstausst. usw. aus. Seine Tierfiguren, Bildnisbüsten und Entwürfe für Porzellanstatuetten, wurden u.a. durch Merkelbach & Wick (Grenzhausen), Metzler & Ortloff (Illmenau), Heubach (Lichte) und die Firma Rosenthal (Kunstabt. Selb) ausgeführt, Bronzen u.a. durch Gladenbeck, Berlin. Lit.: TB, Bd. 35, S. 401; Vollmer, Bd. 5, S. 114; Neuwirth (1974), S. 36, 37, 76; Niecol, Bd. 3, S. 163; Zühlsdorff, S. 450–451; AKL Index.*
2408 s. WVZ
4120 Ballspielende Katze (auf Ovalsockel)
4144 „Eisbär" ÖMKI (1912)
4145 „Ente und Wildkatze" (auf Ovalsockel) ÖMKI (1912)
4364 Gans (auf Ovalsockel)
4383 Zwei Wüstenfüchse
4386 Haubentaucher mit zwei Jungen (vgl. 5081 DECKELDOSE)
4477 BÜSTE Mädchen
4685 Fuchs
4828 Eule (auf Felssockel)
4873 Eule (auf Felssockel)
4892 Faun mit Schnecke (vgl. auch Kostial 4892)
5081 DECKELDOSE Haubentaucher mit zwei Jungen (Form: eiförmig / Knauf / Bekrönung: Haubentaucher mit zwei Jungtieren / Dekor: stilisierte Blütenmalerei) (vgl. 4386 FIGUR)
5242 Stehender Putto mit Früchten (–> vgl. Verbanck 5242)
„Meerschweinchen" ÖMKI (1911/12) (H. 15 cm); Ahlden 36; Lit.: Neuwirth (1974), S. 76

Werner, Franz

*Franz Werner, Bildhauer (*11.6.1872 Breziny/Birkigt/(Tetschen)/Böhmen, †4.9.1910 Breziny/Birkigt), studierte zwischen 1891 und 1894 an der Akademie in Wien unter E. Hellmer und von 1897–1901 an der KGW-Schule Wien bei Otto König und A. Strasser. Lit.: TB, Bd. 35, S. 405; Bénézit, Bd. 14, S. 551; AKL Index.*
2340 FIGUR „Affe als Philosoph" (sitzend auf Buch) (Inschrift: PHILOSOPHIE) (vgl. „Affe", ÖMKI 1902/03)
„Affe" ÖMKI (1902/03) (vgl. 2340) Lit.: Neuwirth (1974), S. 76
„Marabou" ÖMKI (1903/04) Lit.: Neuwirth (1974), S. 76

Wiegand, Martin

*Martin Wiegand, Bildhauer, Maler und Zeichner (*27.11.1867 Ilmenau, †8.8.1961 München), studierte an der Kunstakademie Dresden und an der Akademie in München. Er war ab 1888 als freischaffender Künstler tätig und in den Jahren 1906–1908 künstlerischer Leiter der Kgl. Sächsischen Porzellan-Manufaktur, Meissen. Ab 1910 ließ sich Wiegand in München nieder, und unterhielt ein eigenes Atelier mit Brennofen in Nymphenburg. Seine keramischen Arbeiten umfassen hauptsächlich allegorische Gruppen und Figuren (aus der Mythologie, Märchen etc.). Lit.: TB, Bd. 35, S. 529;Zühlsdorff, S. 457; AKL Index.*
5061 TAFELAUFSATZ / OBSTSCHALE Vier, eine runde Schale tragende Putti (auf getrepptem Quadratsockel)

Wieselthier, Vally

*Vally (Valerie) Wieselthier, Bildhauerin, Keramikerin, Glas- und Textilkünstlerin (*25.5.1895 Wien, †1.9.1945 New York), besuchte 1912–1914 die Kunstschule für Frauen und Mädchen in Wien und studierte anschließend bis 1920 an der KGW-Schule, unter anderem bei Kolo Moser, Josef Hoffmann sowie bei Michael Powolny, wo sie in der Werkstätte für Keramik Hospitantin war. Ab 1917 war sie dann Mitarbeiterin der Wiener Werkstätte. Diese Zusammenarbeit endete vorläufig 1922. Wieselthier nahm an der Pariser KGW-Ausst. 1925 teil und wurde dort auch prämiert. Die Zusammenarbeit mit Goldscheider (Austria) lässt sich bis in die 30er Jahre nachweisen, wobei Wieselthier auch unter anderem mit den Firmen Gmundner Keramik und Schleiss (beide Gmunden), den Porzellanmanufakturen Augarten und Josef Böck, Lobmeyer (alle Wien) und dem Österreichischen Werkbund kooperierte. Ab 1927 war sie dann bis zur Schließung 1932 wieder mit der Wiener Werkstätte verbunden. Im Jahr 1933 zog sie nach Chicago, dann bald nach New York. Im Sommer 1937 kehrte sie noch einmal nach Wien zurück, wo offenbar auch der „Sitzende Damenakt" (Modell-Nr. 7789) für Goldscheider fixiert wurde. Sicher zeichnet Wieselthier noch für etliche Entwürfe verantwortlich, die von der Wiener Manufaktur ausgeführt und vertrieben wurden. Zuschreibungen wurden aber wegen der teils stilistischen Nähe zu den Entwürfen Walter Bosses (besonders bei Schalen und Wandmasken) und zu Rudolf Knörlein (Köpfe und Masken, die ebenfalls die typischen Haarlocken aufweisen) nicht vorgenommen. Für Goldscheider USA lassen sich bislang mindestens zwei Modelle nachweisen. Lit.: TB, Bd. 35, S. 541; Vollmer, Bd. 5, S. 131; Neuwirth (1974), S. 76, 447–451 (m. Abb.); Zühlsdorff, S. 451; Fahr-Becker, S. 236; Hörmann; AKL Index.*
5351 TEEGARNITUR Zuckerdose, Sahnekännchen, Tasse
6764 WANDMASKE Cherub (Engelskopf mit Flügeln) (auch für GS USA)
7789 Sitzender Frauenakt
301 / 391?-USA BUST Woman with Hand, Fan and Hat

Willfort, Irma s. Stuart-Willfort, Irma

Wind, Josef

*Josef Wind, Bildhauer (*16.6.1864 München, †? wird letztmals um 1900 erwähnt), studierte an der Akademie in München und stellte im Glaspalast München ab 1889 sowie bei der Internationalen Kunstausst. im Künstlerhaus Wien 1894 aus. Lit.: TB, Bd. 36, S. 49; Bénézit, Bd. 14, S. 645; AKL Index.*
40 „Schlangenbändigerin" / „Charmeuse des Serpents (nue)" (Stehende nackte Orientalin mit Turban, Flöte und Schlange) BRONZE Paris (vgl. 644-GSB; 1399)
644-GSB STATUETTE „Charmeuse des Serpents (nue)" (vgl. 40 BRONZE; 1399)
1119 FIGUR „Gefesselte Sklavin" (Stehende, an einen Baumstamm mit Handfesseln gebundene, nackte Araberin mit Turban) (vgl. 1393) (Wind zugeschr.)
1285 FIGUR „Krugträgerin" / „Krughändlerin" (Schreitende Frau mit Turban und zwei / drei Krügen) (auf Rundsockel) ÖMKI (1898/99) (vgl. 1727; 1997)
1295 FIGUR „Schlangenbändigerin" (Stehender Akt einer Orientalin mit Tuch, Turban, Flöte und Schlange), (Terrakotta) ÖMKI (1898/99) (vgl. 40) (vgl. 1399= Variante)
1393 FIGUR „Gefesselte Sklavin" (Stehende, an einen Baumstamm mit Handfesseln gebundene, nackte Araberin mit Turban) (vgl. 1119) (Wind zugeschr.)
1399 FIGUR „Schlangenzauberin" (Stehender Halbakt einer Orientalin mit Tuch, Turban, Flöte und Schlange), (Terrakotta) ÖMKI (1898/99) (vgl. 40) (vgl. 1295= Variante)
1727 FIGUR „Krugträgerin" / „Krughändlerin" (Schreitende Frau mit Turban und zwei Krügen) (auf Rundsockel) ÖMKI (1898/99) (vgl. 1285; 1997)
1849 BÜSTE „Heilige Cäcilie" (auf Rechtecksockel mit sechs reliefierten Porträts und einer Orgel spielenden Mädchenfigur) ÖMKI (1900/01)
1997 FIGUR „Krugträgerin" / „Krughändlerin" (Schreitende Frau mit Turban und zwei Krügen) (auf Rundsockel) (vgl. 1285; 1727)
„Hl. Cäcilie" (Terrakotta) ÖMKI (1899/1900; 1900/01) (vgl. 1849) Lit.: Neuwirth (1974), S. 76

Wiskot, Karl

*Karl Wiskot, Maler (*27.5.1906, †?). Quelle: AGM, Liste „Adressen von guten und zuverlässigen Arbeitern" (undat., um 1953).*

Xaniroff s. Haniroff

Youriévitch, Serge

*Serge Youriévitch, Bildhauer und Psychologe (*1876, Paris, †1969), studierte in Rom. Er war durch Auguste Rodin beeinflusst, mit dem er jahrelang (1903–1914) zusammenarbeitete. Ab 1909 stellte Youriévitch in den verschiedenen Pariser Salons seine vorwiegend figürlichen Bronzen aus. Seine von Goldscheider editierte Figur der Tänzerin Nattowa wurde 1924 vom franz. Staat erworben. Lit.: TB, Bd. 36, S. 361–362; Vollmer, Bd. 5, S. 189; Bénézit, Bd. 14, S. 814; AKL Index. Quelle: http://www.artnet.de/artwork/424014901/serge-yourievitch-diana.html (vom 14.6.2005).*
„La Danseuse Nattowa" (Nattowa the Dancer) BRONZE Arthur Goldscheider, Paris (1923) Lit.: L'Illustration, Nr. 4268 vom 20.12.1924, S. 14 (m. Abb.); Goldscheider, A. (Hrsg.): Artistic Evolution, 1926, T. 2, Pl. 27 (m. Abb.)

Zago, Emil (Emile / Emilio)

*Emilio Zago, Bildhauer (*23.7.1877 Triest, †?), war Schüler an der KGW-Schule in Wien und nannte sich nach der Ausst. der KGW-Schule 1901 dann Emilio. Lit.: Neuwirth (1974), S. 76; Quelle: Kunstarchiv Werner J. Schweiger, Wien.*
2551 FIGUR „Dante" (Schreitender Dante Alighieri, auf Felssockel) ÖMKI (1903/04; 1906/07)
FIGUR Sitzender Schmied (mit Hammer und Amboss auf Rechtecksockel) Lit.: N.N.: Figürliche Terrakotten, in: KM, 4.1904, S. 138 (m. Abb.), „Die kräftige Gestalt hat sich auf dem Amboss niedergelassen, der rechte Arm, der sich mit dem Ellenbogen auf das Knie stützt, hält den vorgebeugten Kopf, während die linke Hand lässig auf dem linken Oberschenkel ruht; der schwere Schmiedehammer steht zwischen den Beinen, und liegt der Stiel desselben gegen das Schurzfell des Schmiedes an. Bildhauer Zago, der diesen Schmied modelliert hat, hat den menschlichen Körper und die Muskulatur desselben gut studiert, er hat aber auch die Schmiede bei der Arbeit beobachtet und deren Gewohnheiten in Kleidung und Haltung vortrefflich wiederzugeben verstanden. Auch das Beiwerk, der Amboss, das zerbrochene Wagenrad, sowie der Hammer sind in der Wiedergabe richtig getroffen worden".

Zailler s. Zeiller

Zardier?

1915 BÜSTE „Angelica" (Junge Dame mit Haarschmuck, nach links blickend) (auf Ovalsockel) (Inschrift: ANGELICA)

Zeiller, Paul

*Paul Zeiller, Tierbildhauer (*22.10.1888 München, †25.6.1915 Bad Ems), war Schüler von W. von Rümann an der Akademie in München. Zeiller ging 1904 nach Berlin und arbeitete auch für die Manufaktur Heubach (Lichte) sowie für die Porzellanfirma Gebr. Metzler & Ortloff in Ilmenau. Lit.: TB, Bd. 36, S. 437; AKL Index.*
4445 Geier (auf Felssockel) ALABASTER
4688 (oder 4888?) Hahn (sign. „Zailler")

Zezelitz, Hanns

*Hanns Zezelitz, Retoucheur (*3.7.1921, †?). Quelle: AGM, Liste „Adressen von guten und zuverlässigen Arbeitern" (undat., um 1953).*

Ziffer, Moses (Moshe)

*Moses (Moshe) Ziffer, Bildhauer (*26.4.1902 Przemysl, †1989 Tel Aviv?), studierte 1924–1928 an der KGW-Schule in Wien unter Prof. Eugen Steinhof und wurde als Mitarbeiter der Vereinigten Ateliers für Kunst und Keramik, Marcell Goldscheider geführt. Ziffer lebte in Tel Aviv und stellte 1928 bei der Weihnachtsschau im Wiener Künstlerhaus aus. Lit.: Vollmer, Bd. 5, S. 206; Neuwirth (1974), S. 76, Zühlsdorff, S. 644; AKL Index.*
ohne Nr.-MG STEHKOPF Frauenkopf

Markenverzeichnis I
Marks Table

Wiener Manufaktur Friedrich Goldscheider

Firmenstempel I Firm stamps

GS 1
ca. 1878

GS 1a
ca. 1895

GS 2
ca. 1878–1888

GS 2a
ca. 1878–1888

GS 2b
ca. 1878–1888

GS 3
ca. 1887–1897

GS 4
ca. 1890–1910

GS 4a
ca. 1887–1897

GS 5
ca. 1895

GS 6
ca. 1895–1910

GS 7
ca. 1895–1910

GS 7a
ca. 1895–1910

GS 8
ca. 1895–1898

GS 8a
ca. 1895–1898

GS 8b
ca. 1895–1898

GS 8c
ca. 1905–1910

GS 9
ca. 1900–1917

GS 10
ca. 1900–1912

GS 11
ca. 1890–1899

GS 11a
ca. 1890–1899

GS 12
ca. 1890–1895

GS 12a
ca. 1890–1895

GS 12b
ca. 1890–1895

GS 13 Sculpture
ca. 1900–1920

GS 13a Sculpture
ca. 1912–1920

GS Adler 14
ca. 1910–1918

GS Adler 15
ca. 1905–1910

GS Adler 15a
ca. 1900–1912

GS Adler 16
ca. 1900–1918

GS Adler 17
ca. 1902–1910

GS 18
ca. 1917–1922

GS 19 (gestempelt I stamped), ca. 1922–1941

GS 19a (geprägt I impressed), ca. 1936–1941

GS 19b
ca. 1922–1925

GS 20
ca. 1936–1952

GS 21
ca. 1925–1930

GS 21a
ca. 1925–1930

GS 22
ca. 1922

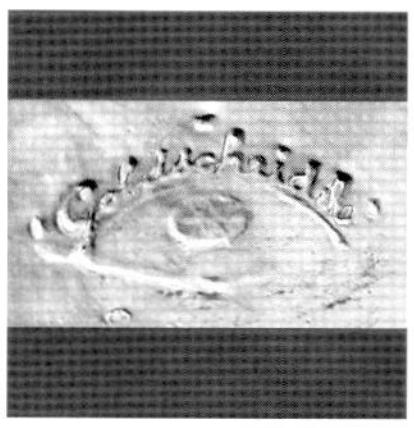

GS 22a (geprägt | impressed), ca. 1925–1930

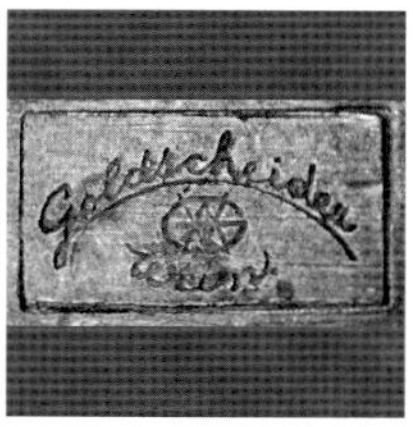

GS 23
ca. 1922

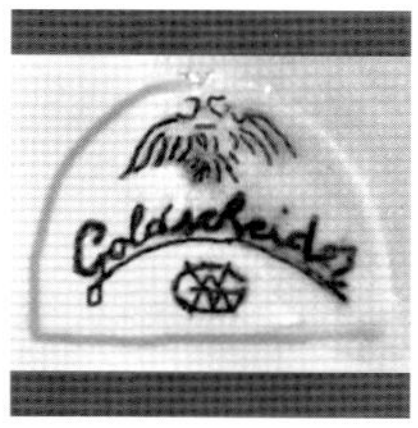

GS 24
ca. 1922

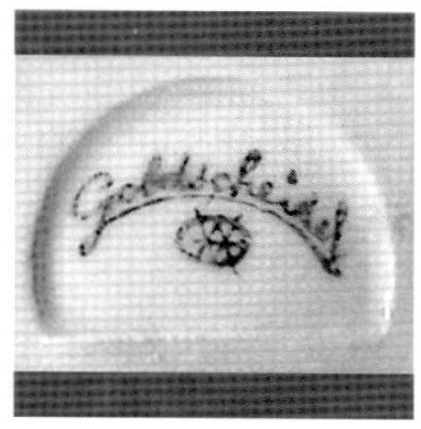

GS 25
ca. 1922–1940

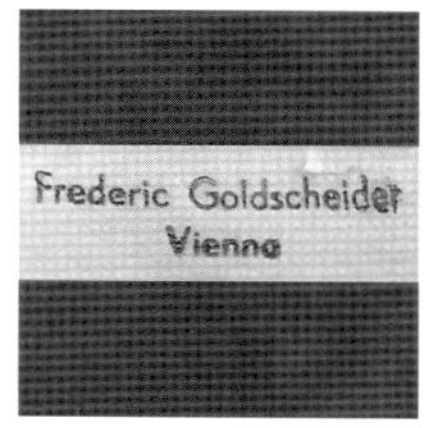

GS 26
ca. 1936–38

Wiener Manufaktur Friedrich Goldscheider

Plaketten und Pickerl (inkl. Künstlerpickerl) | Badges and stickers (incl. stickers of artists)

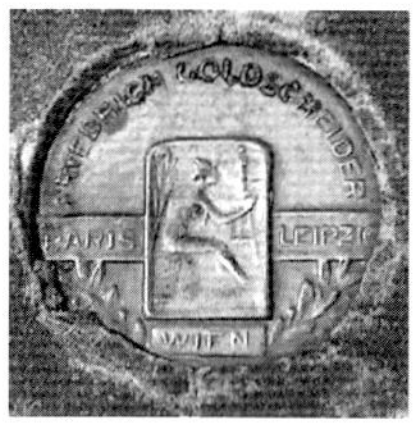

GS PLAK A
ca. 1900–1913

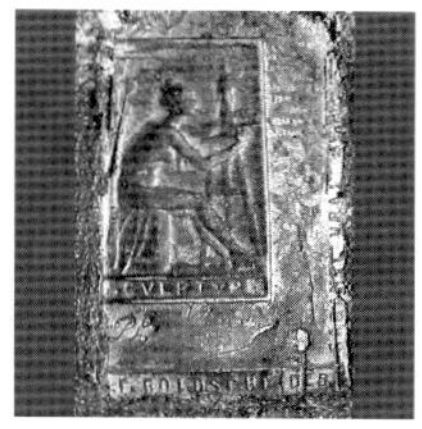

GS PLAK B
ca. 1900–1913

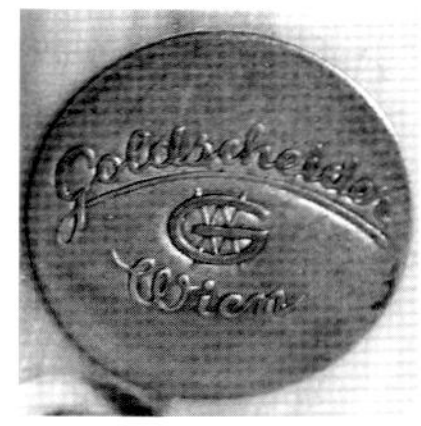

GS PLAK C
ca. 1914–1930

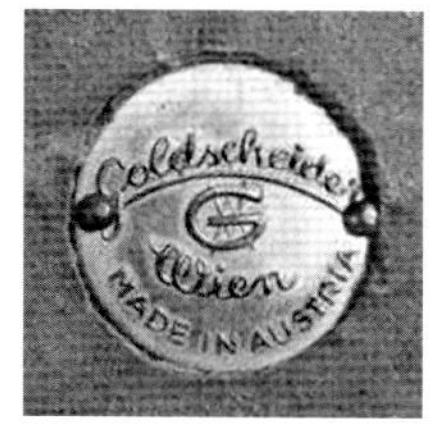

GS PLAK D
ca. 1938

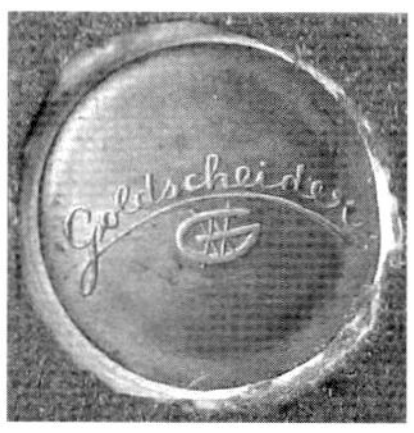

GS PLAK E
ca. 1922

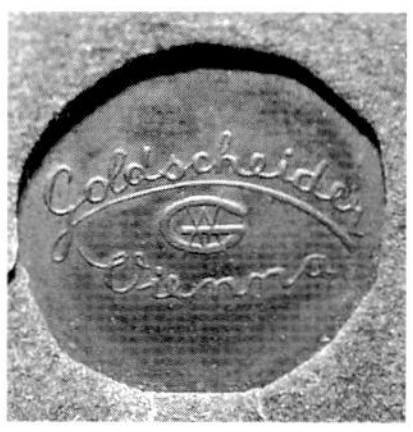

GS PLAK F
ca. 1925

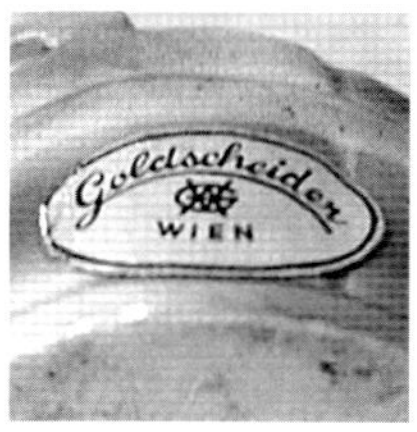

GS PICK 1
ca. 1922–1942

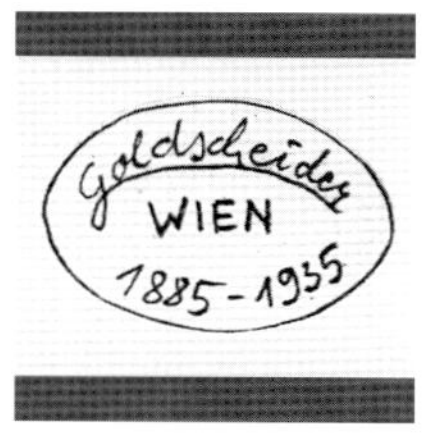

GS PICK 2
ca. 1922–1942

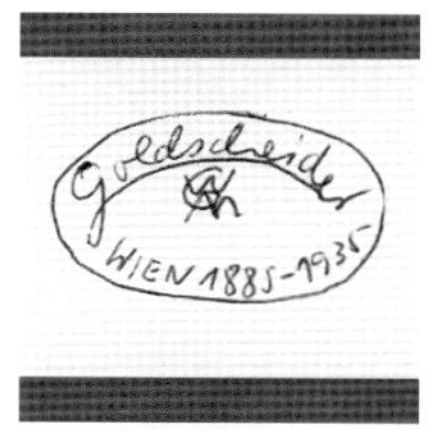

GS PICK 2a
ca. 1922–1942

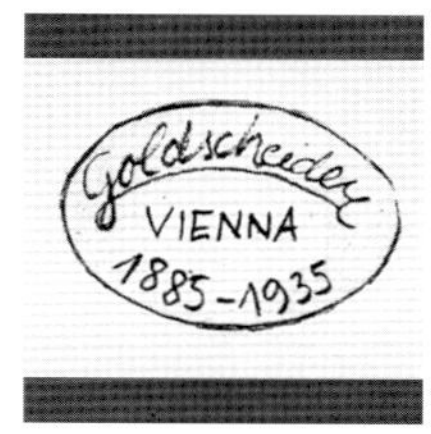

GS PICK 3
ca. 1922–1942

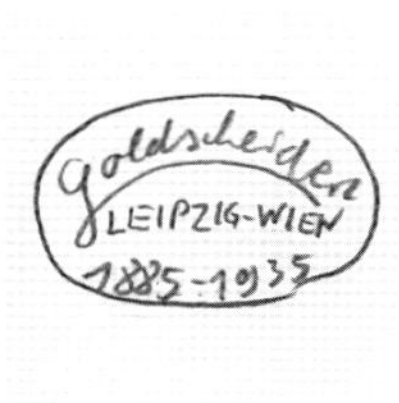

GS PICK 3a
ca. 1922–1942

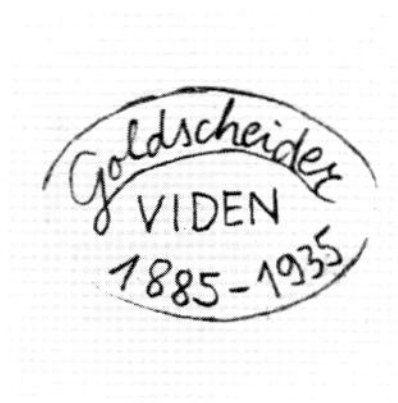

GS PICK 4
ca. 1922–1942

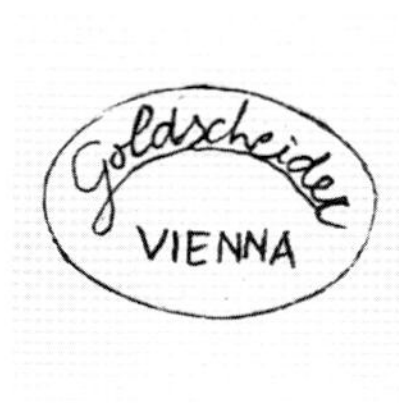

GS PICK 5
ca. 1922–1942

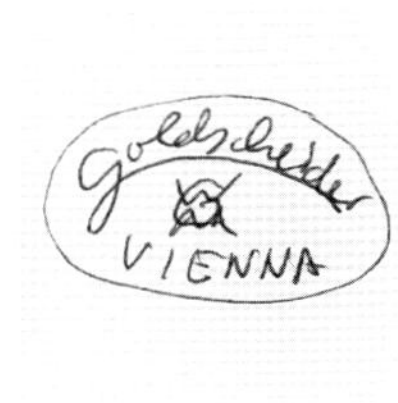

GS PICK 5a
ca. 1922–1942

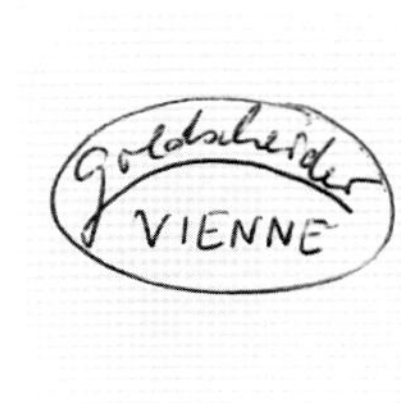

GS PICK 5b
ca. 1922–1942

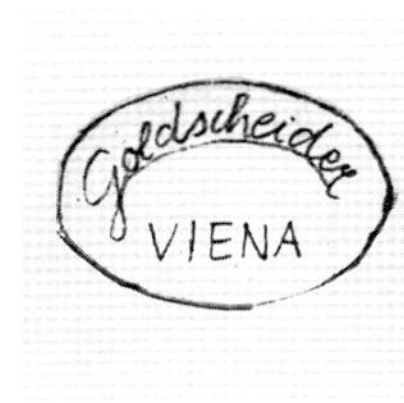

GS PICK 5c
ca. 1922–1942

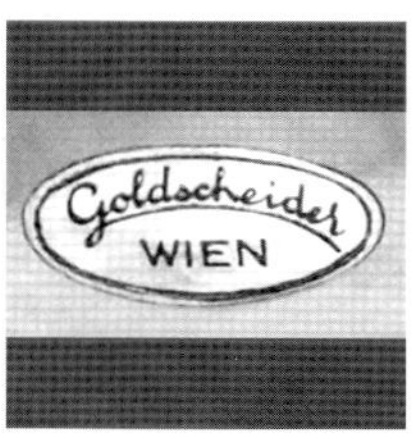

GS PICK 6
ca. 1922–1942

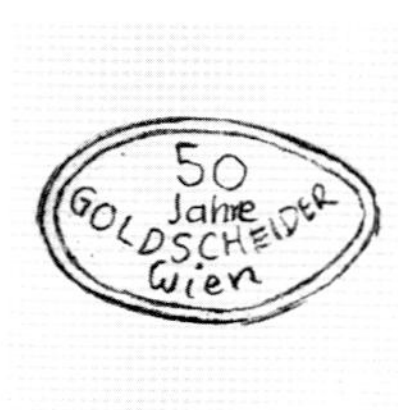

GS PICK 7
ca. 1922–1942

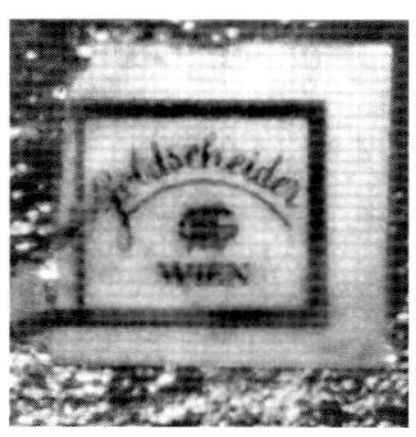

GS PICK 8
ca. 1922–1942

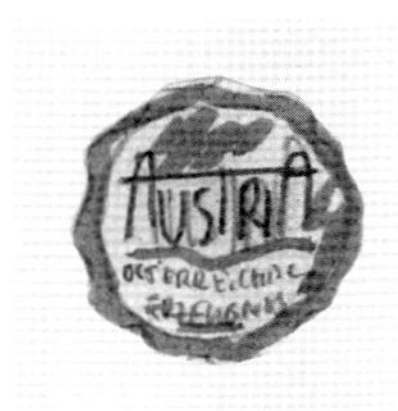

GS PICK 9
ca. 1922–1942

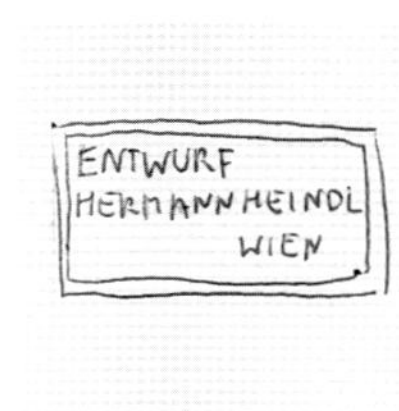

GS PICK 10
ca. 1928–1935

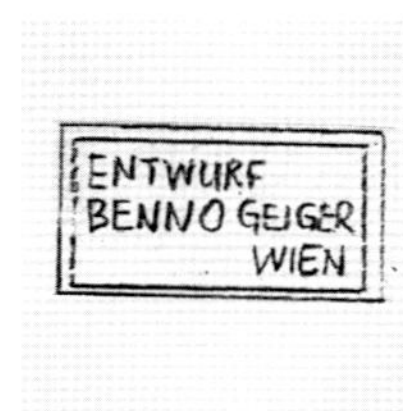

GS PICK 11
ca. 1928–1935

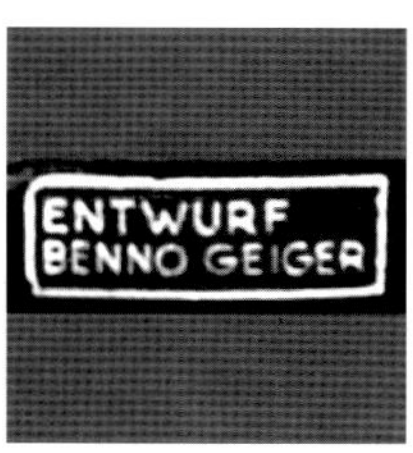

GS PICK 11a
ca. 1928–1935

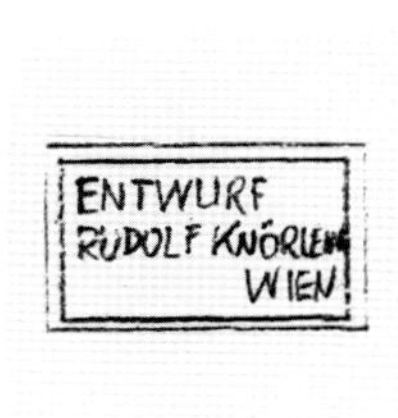

GS PICK 12
ca. 1928–1935

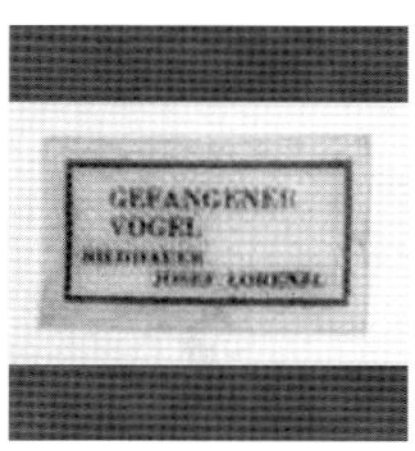

GS PICK 13
ca. 1928–1935

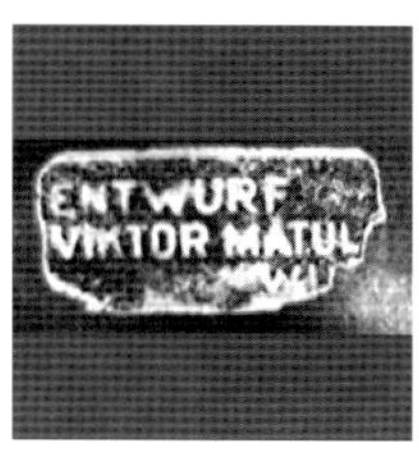

GS PICK 14
ca. 1928–1935

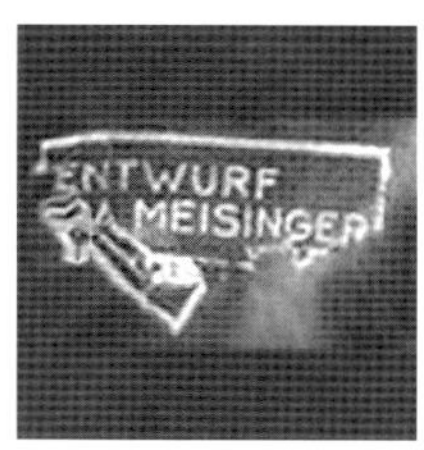

GS PICK 15
ca. 1928–1935

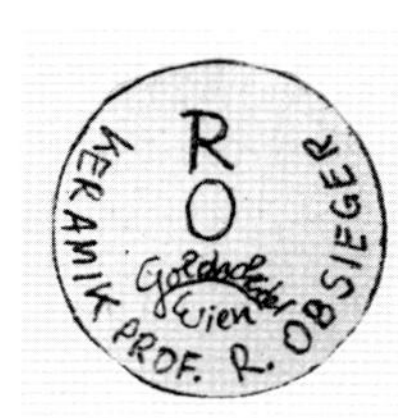

GS PICK 16
ca. 1928–1935

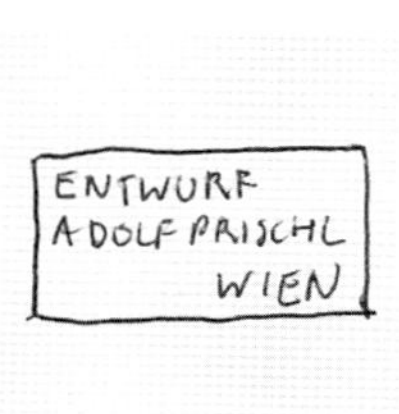

GS PICK 17
ca. 1928–1935

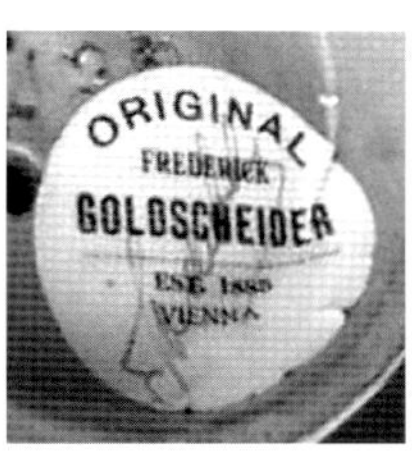

GS PICK 18
ca. 1928–1935

Wiener Manufaktur Friedrich Goldscheider

Zeichen und Zusätze zu den Firmenstempeln I Signs and additional marks to firm stamps

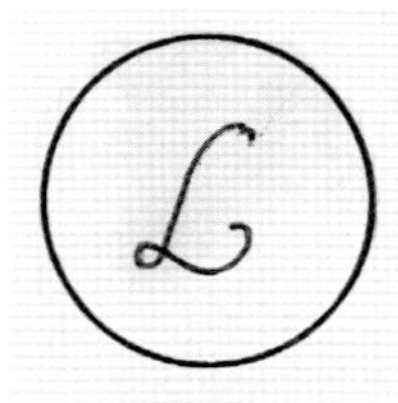

GS Z 1
ca. 1885–1922

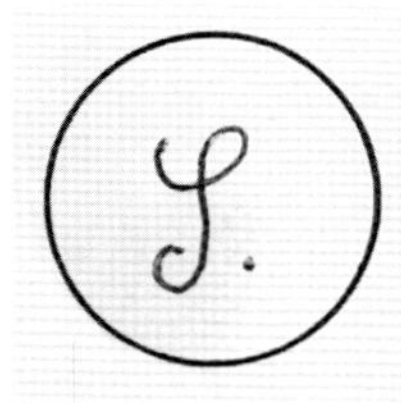

GS Z 2
ca. 1885–1922

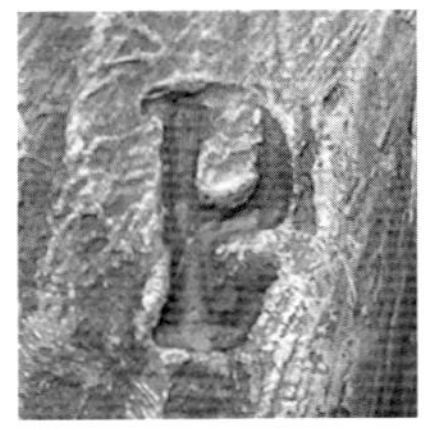

GS Z 3
ca. 1885–1922

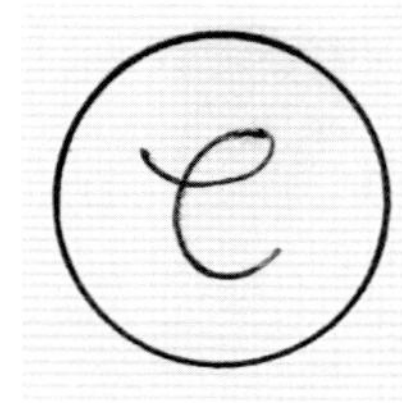

GS Z 4
ca. 1885–1922

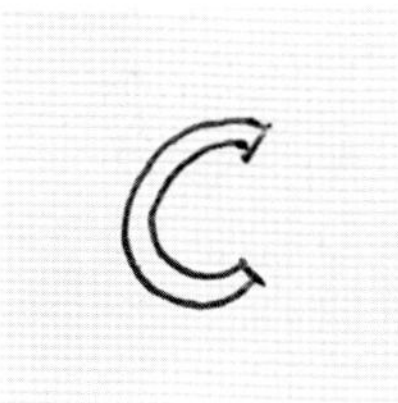

GS Z5
ca. 1885–1922

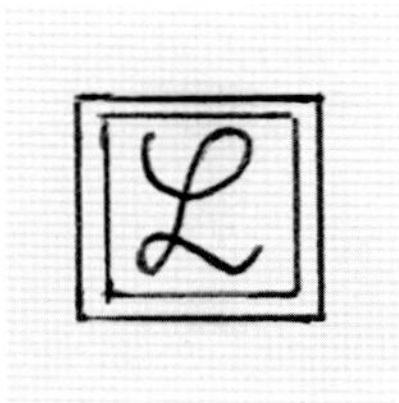

GS Z 6
ca. 1885–1922

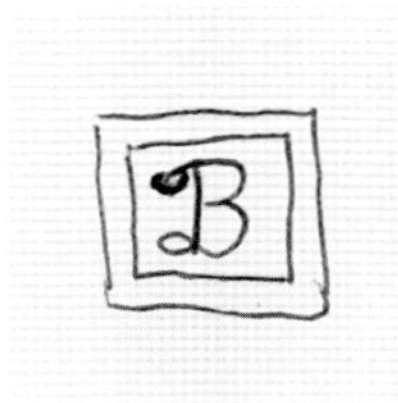

GS Z 6a
ca. 1885–1922

GS Z 7
ca. 1885–1922

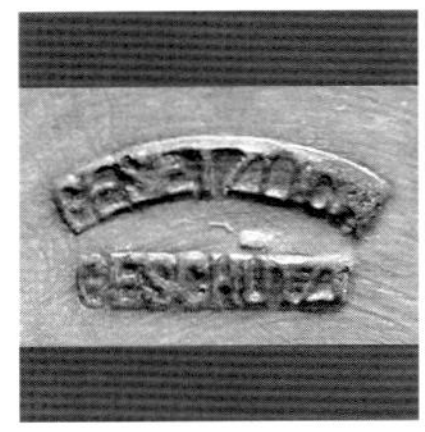

GS Z 8
ca. 1885–1922

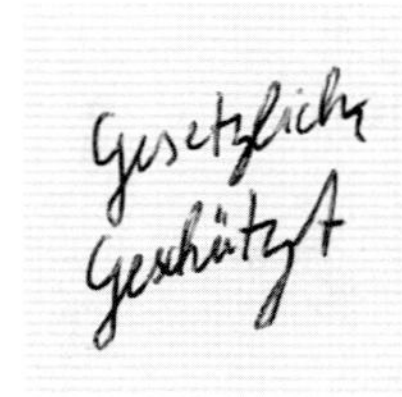

GS Z 8a
ca. 1885–1922

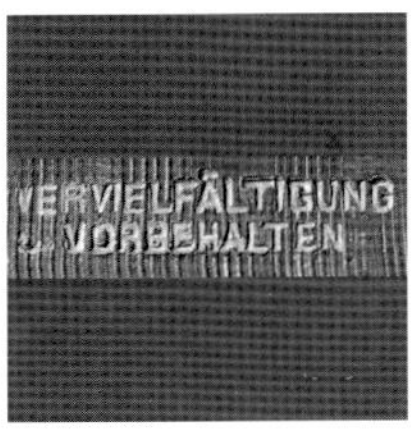

GS Z 9
ca. 1885–1922

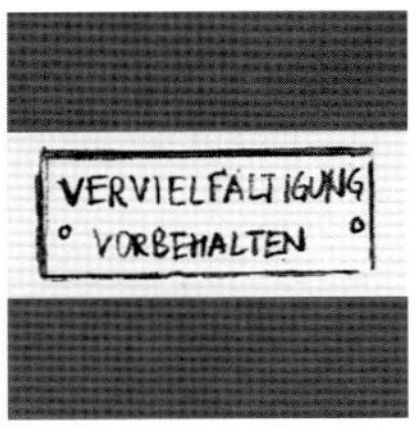

GS Z 10
ca. 1885–1922

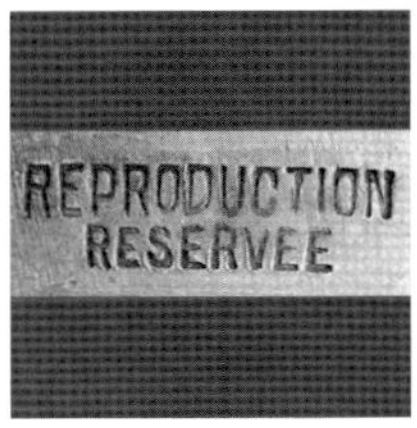

GS Z 11
ca. 1885–1922

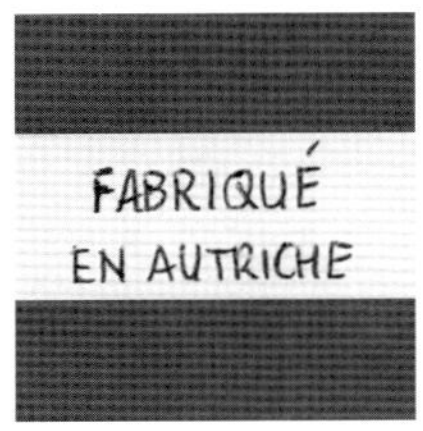

GS Z 12
ca. 1885–1922

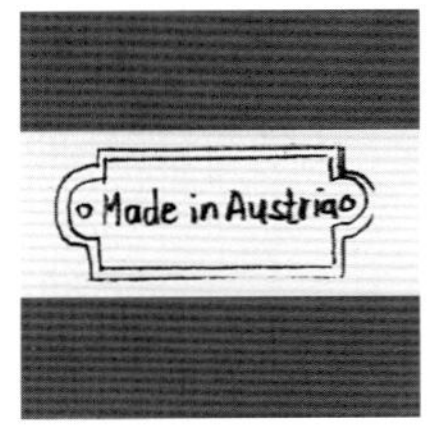

GS Z 13
ca. 1885–1922

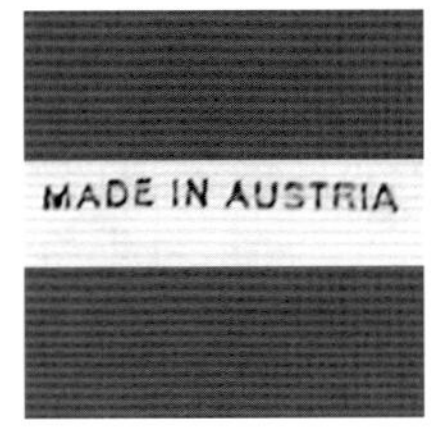

GS Z 14

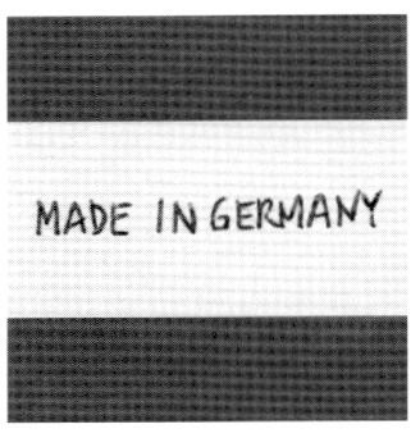

GS Z 15

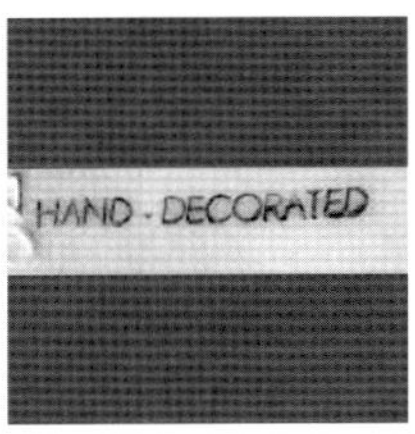

GS Z 16

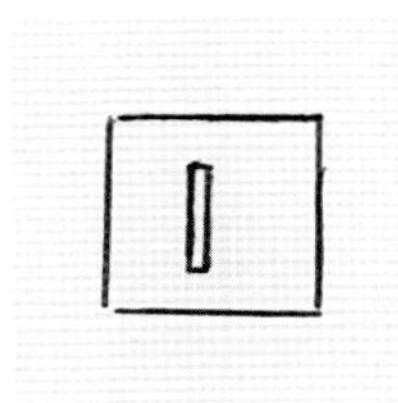

GS Z 17
ca. 1885–1922

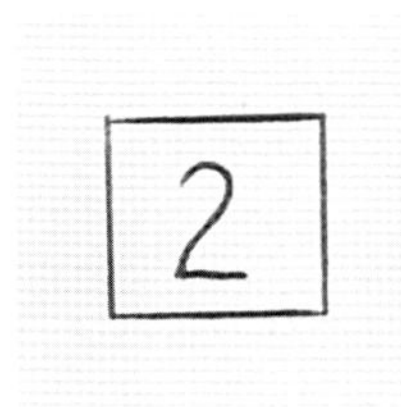

GS Z 18
ca. 1885–1922

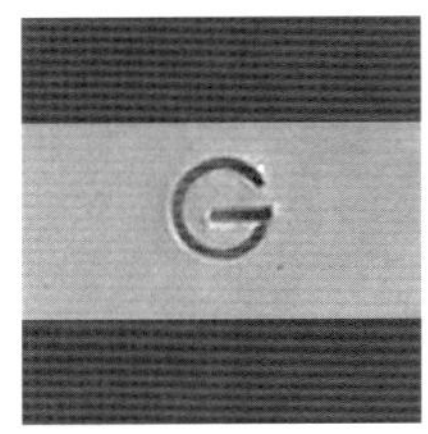

GS Z 19
ca. 1925

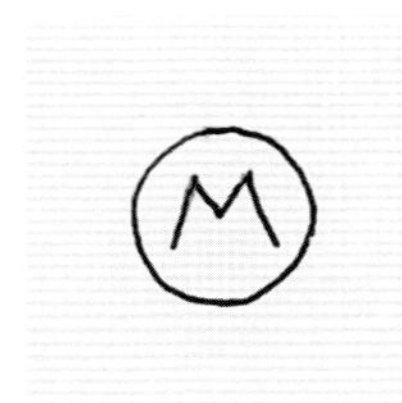

GS Z 20
ca. 1925

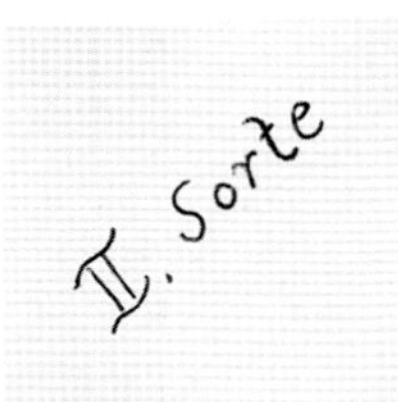

GS Z 21
ca. 1925

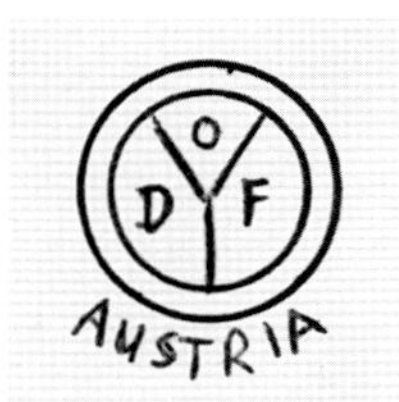

GS Extra-Stempel I extra Stamp 1, ca. 1930

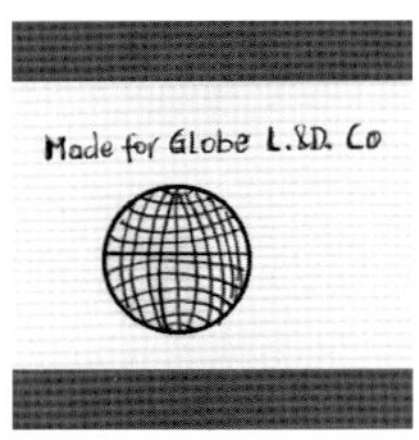

GS Extra-Stempel I extra Stamp 2, ca. 1930

Wiener Manufaktur Friedrich Goldscheider

Bronzen und Alabaster I Bronze and alabaster

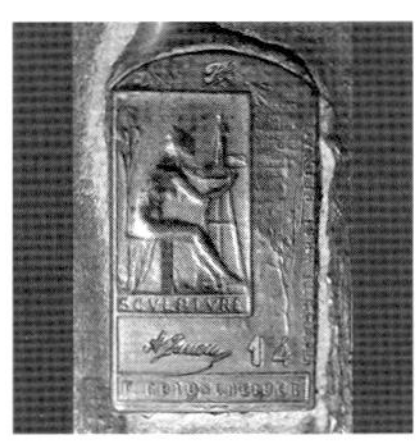

GS Bronze 1 und I and Alabaster, ca. 1900

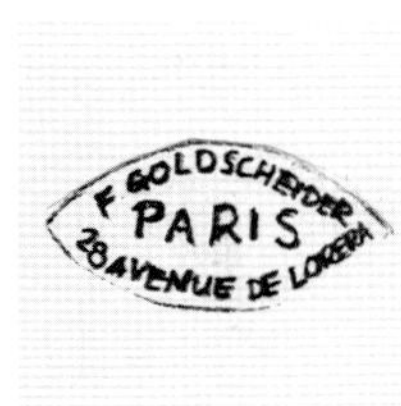

GS Bronze 2
ca. 1900

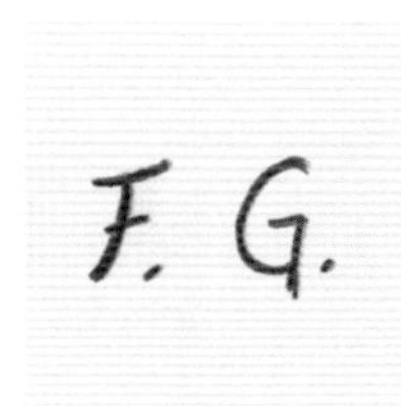

GS Bronze 3
ca. 1915

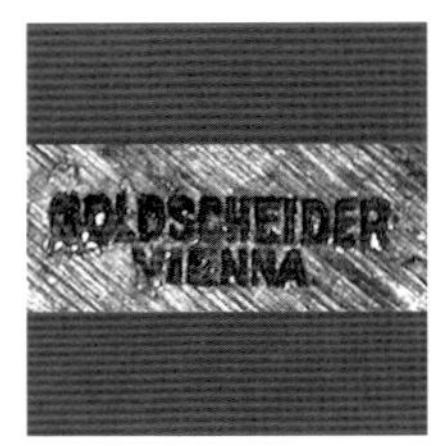

GS Bronze 4
ca. 1925

Wiener Manufaktur Josef Schuster, vorm. Friedrich Goldscheider

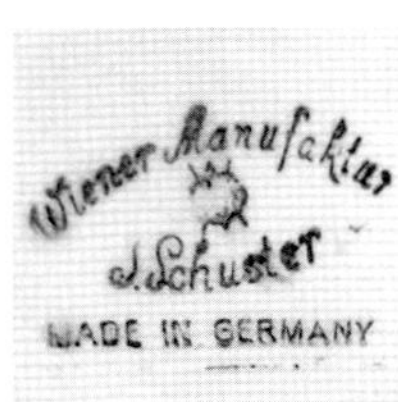

Schuster 1
ab I from 1941

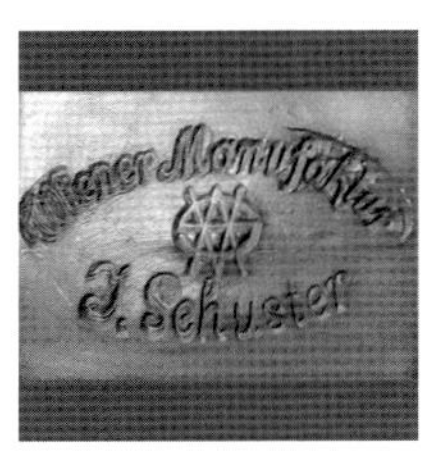

Schuster 1a (geprägt I impressed), ab I from 1941

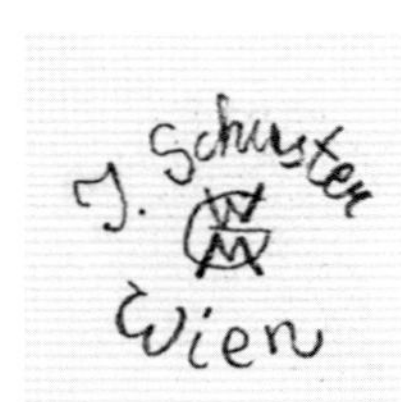

Schuster 2
ab I from 1944

Goldscheider USA & Goldcrest

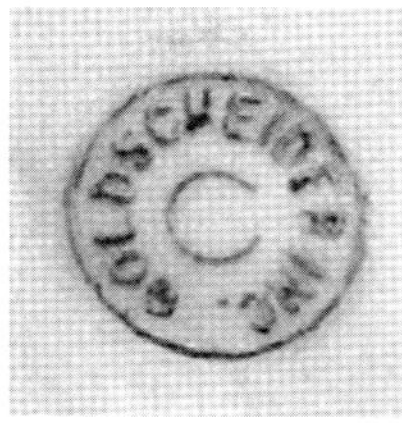

GS USA 1, ab I from 1940 + ab I from 1950

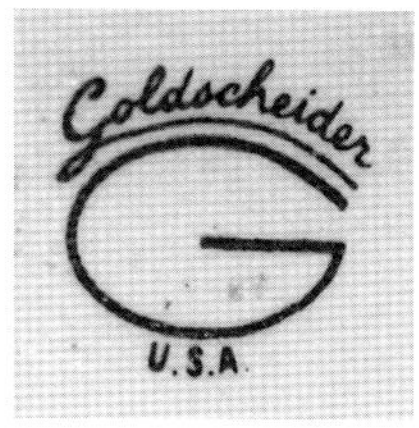

GS USA 2 ab I from 1941

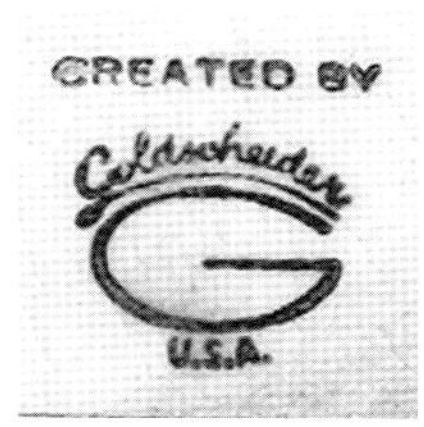

GS USA 2a

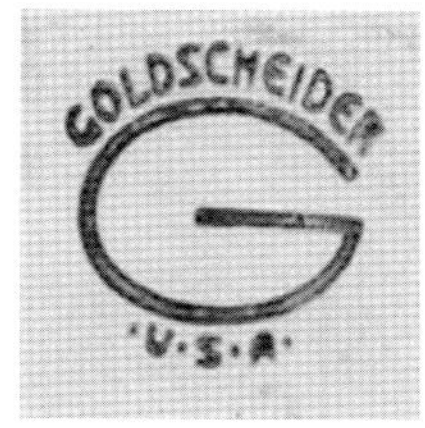

GS USA 3 ab I from 1944

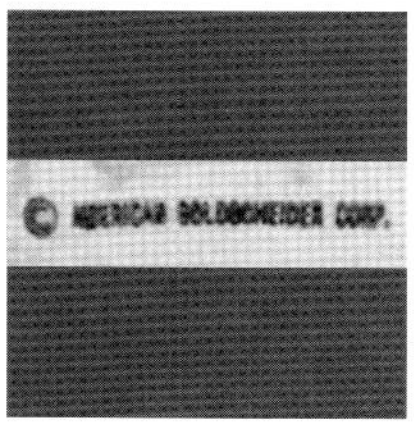
GS USA 4

GS USA 5 ab I from 1942

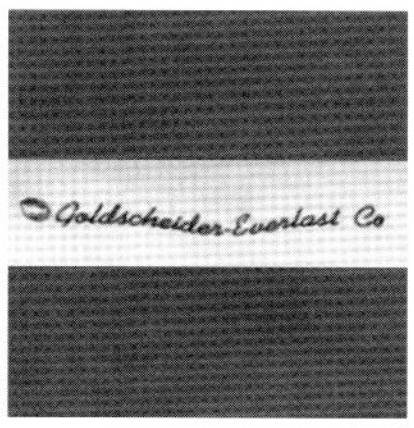

GS USA 6a ab I from 1946

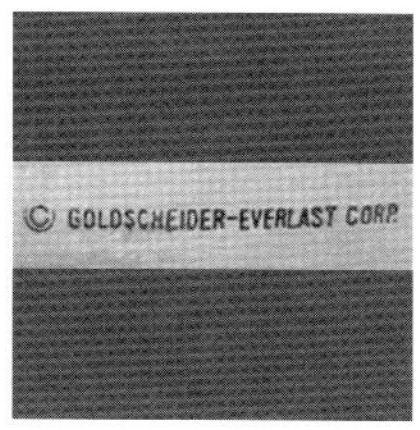

GS USA 6b ab I from 1946

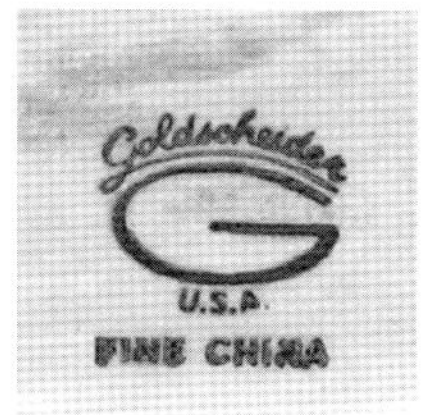

GS USA 7 ab I from 1946

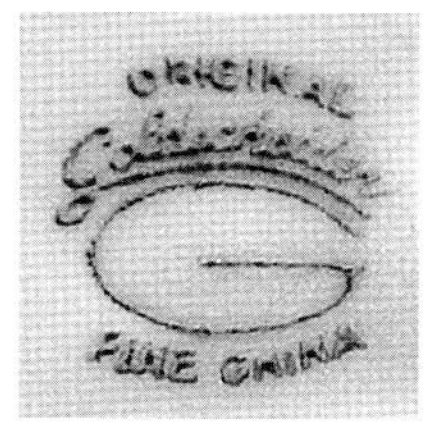
GS USA 8 ab I from 1946

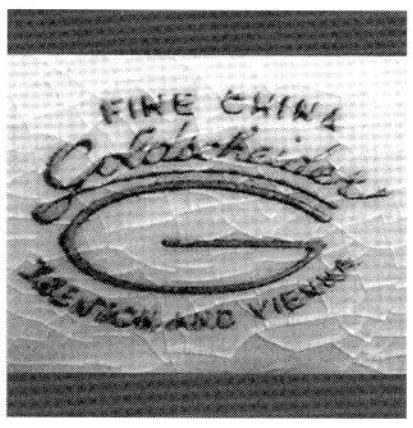
GS USA 9 ab I from 1950

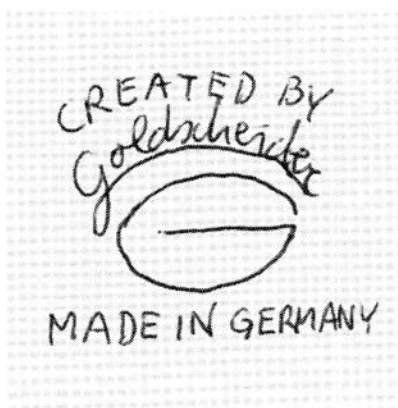

GS USA 10 (Import) ab I from 1953

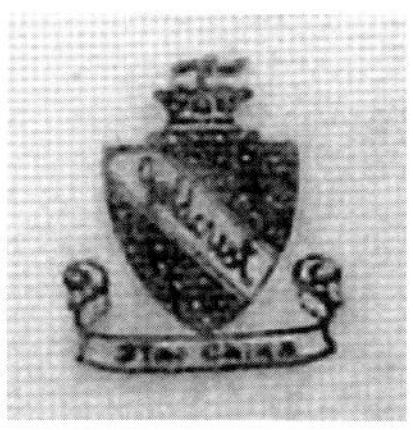
GC USA I 1948

GC USA II 1948

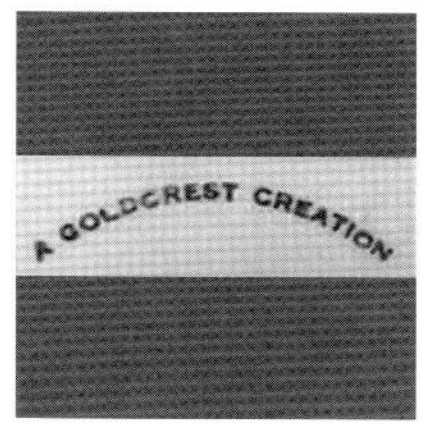

GC USA III

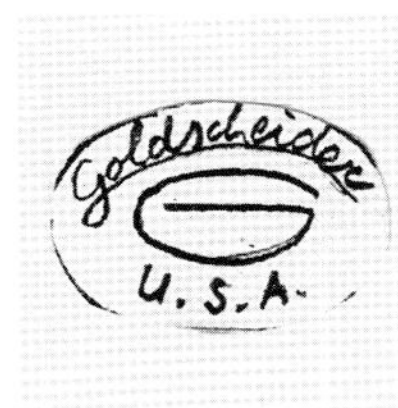

GS USA 1 (Pickerl I Sticker), ab I from 1946

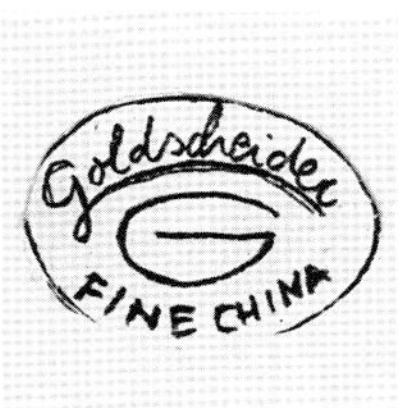

GS USA 2 (Pickerl I Sticker), ab I from 1946

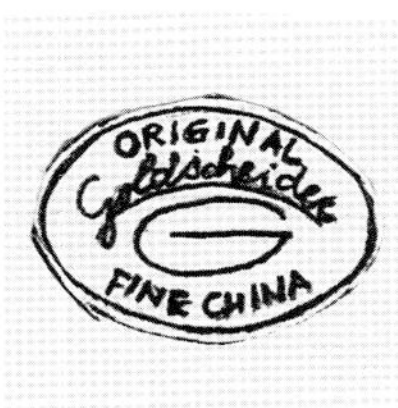

GS USA 3 (Pickerl I Sticker), ab I from 1948

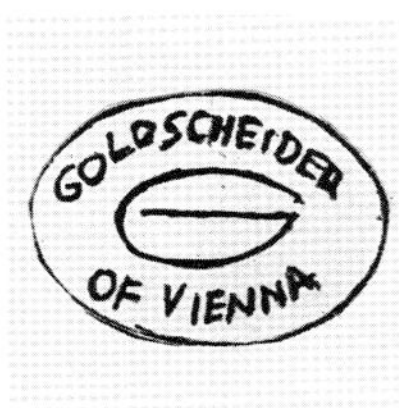

GS USA 4 (Pickerl I Sticker), ab I from 1952

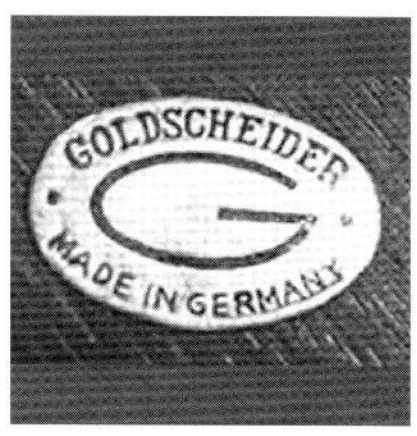

GS USA (Import) I, ab I from 1953 (Pickerl I Sticker)

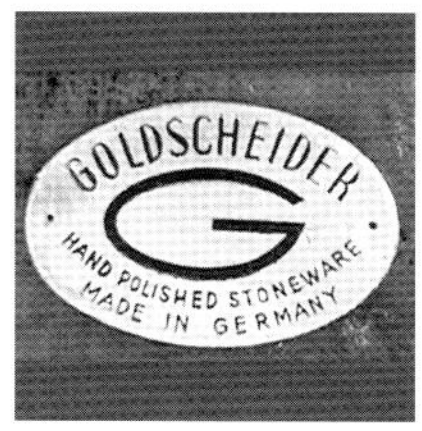

GS USA (Import) II, ab I from 1953 (Pickerl I Sticker)

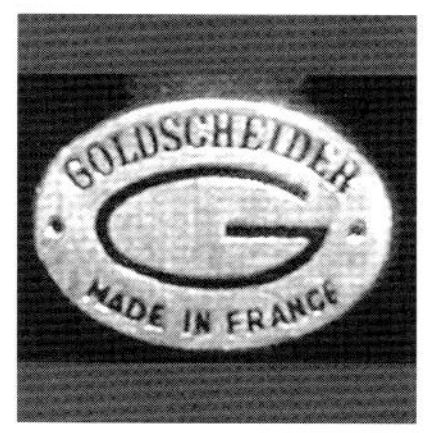

GS USA (Import) III, ab I from 1953 (Pickerl I Sticker)

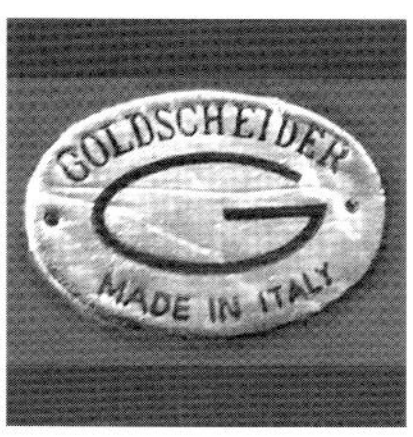

GS USA (Import) IV, ab I from 1953 (Pickerl I Sticker)

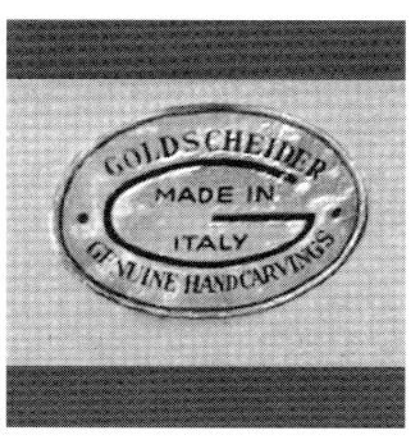

GS USA (Import) V, ab I from 1953 (Pickerl I Sticker)

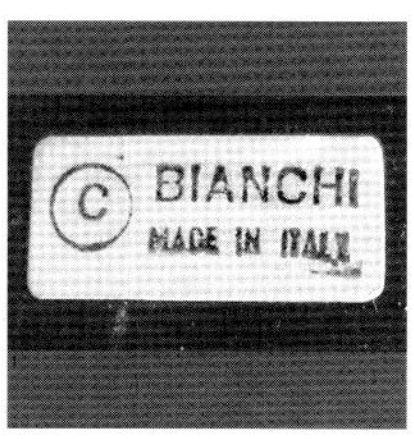

GS USA (Import) VI, ab I from 1953 (Pickerl I Sticker)

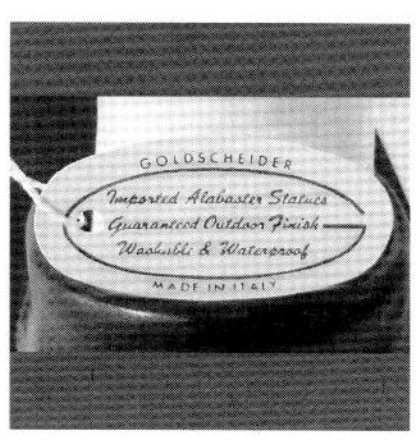
GS USA (Import) VII, ab I from 1953 (Anhänger I Tag)

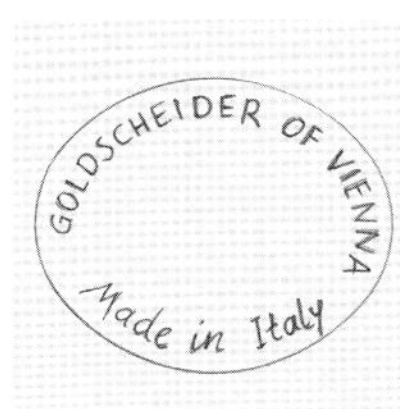

GS USA (Import) VIII, ab I from 1953 (Anhänger I Tag)

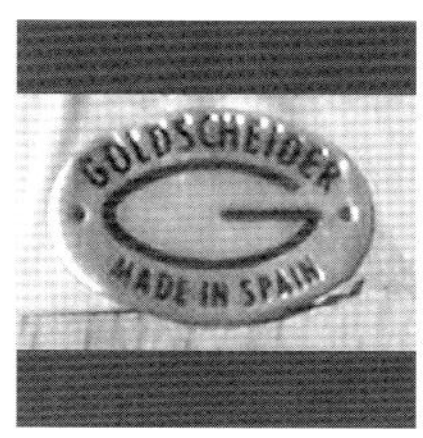

GS USA (Import) IX, ab I from 1953 (Pickerl I Sticker)

Goldscheider West-Germany (Carstens)

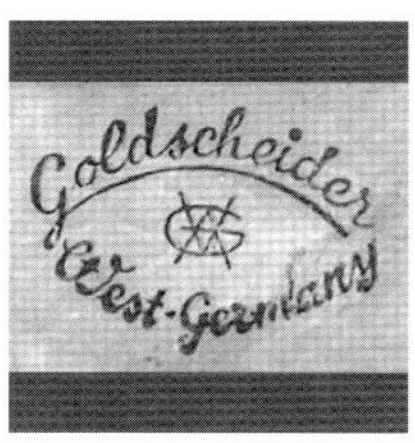

GS WG 1 1953–1960

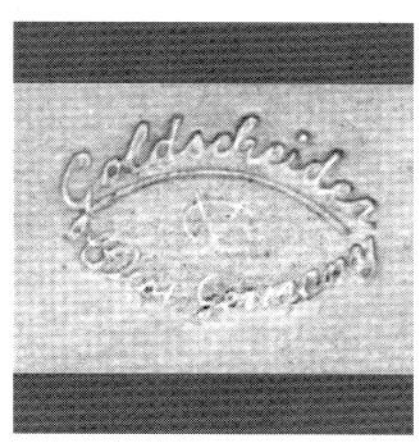
GS WG 1a (geprägt I impressed), 1953–1960

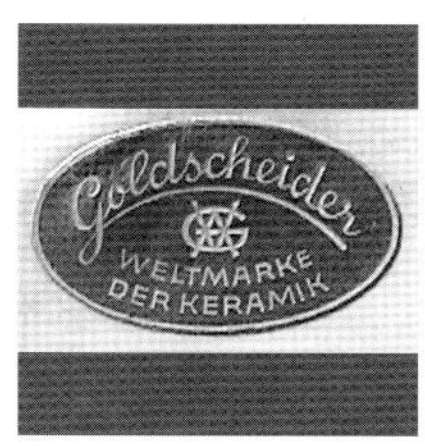

GS WG (Pickerl I Sticker) 1 Gold, 1953–1960

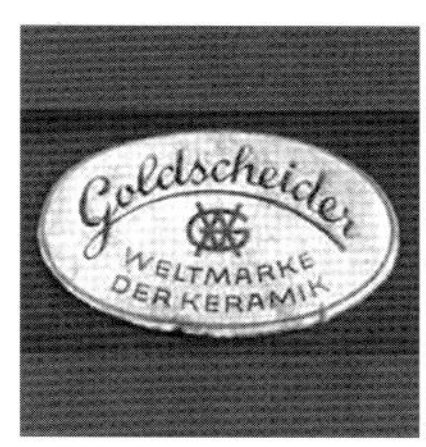

GS WG (Pickerl I Sticker) 1a Silber I Silver, 1953–1960

Goldscheider & Österreichischer Werkbund

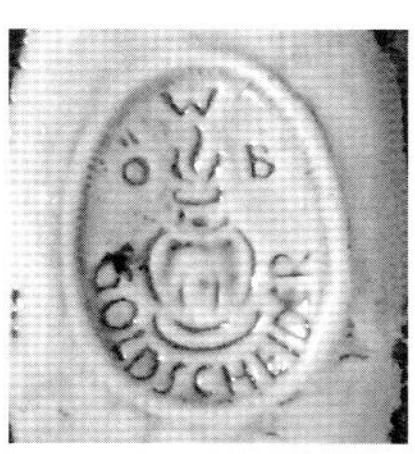

OWB (Österreichischer Werkbund)

Eduard Goldscheider

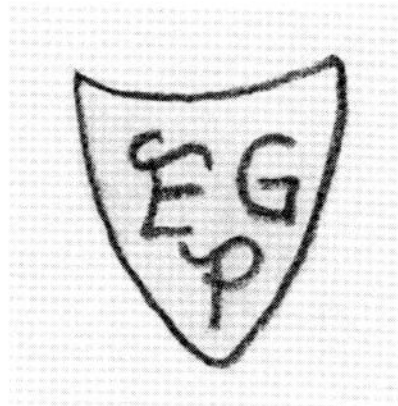

GS Eduard Pilsen 1

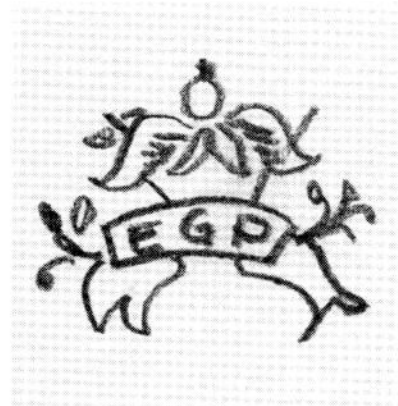

GS Eduard Pilsen 2

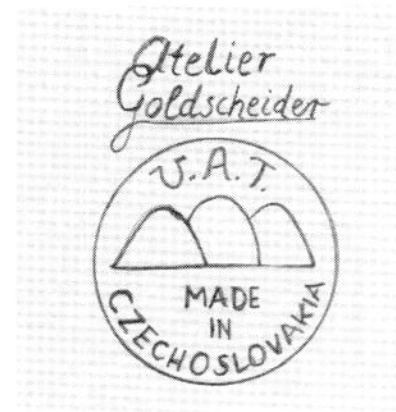

Marke I Mark CZ

Camillo Goldscheider

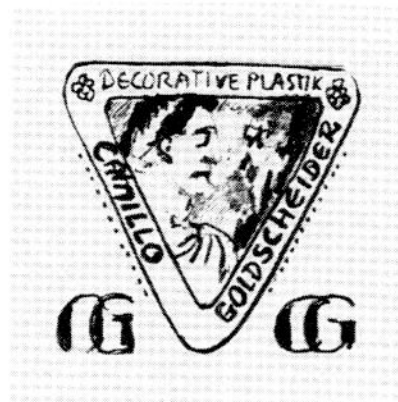

GS Camillo Wien

Arthur Goldscheider, Paris – La Stèle

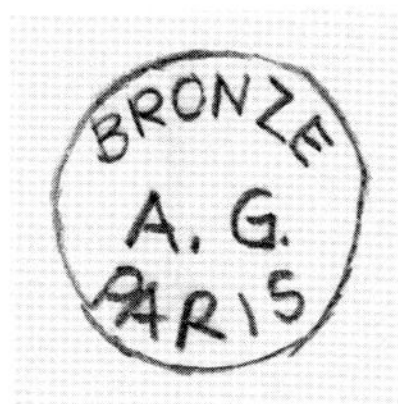

GS Arthur 1
1914–1935

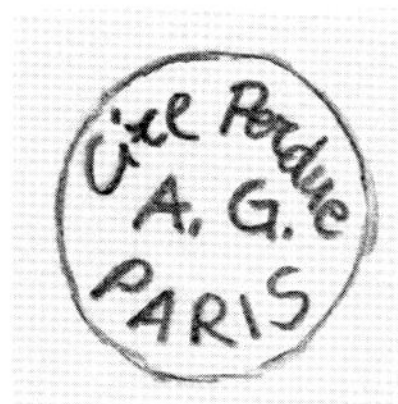

GS Arthur 2
1914–1935

La Stèle
1924–1928

Marcell Goldscheider, Wien I Vienna

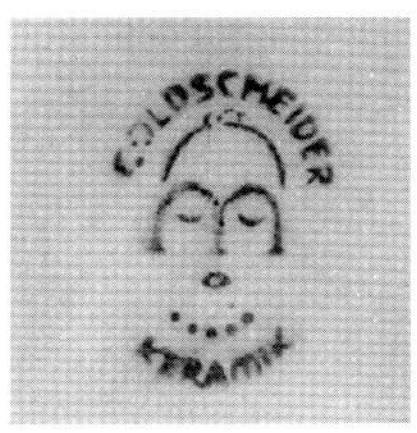

GS Marcell Wien 1
1928–1939

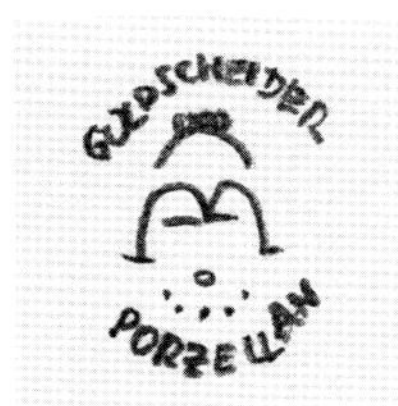

GS Marcell Wien 2
1928–1939

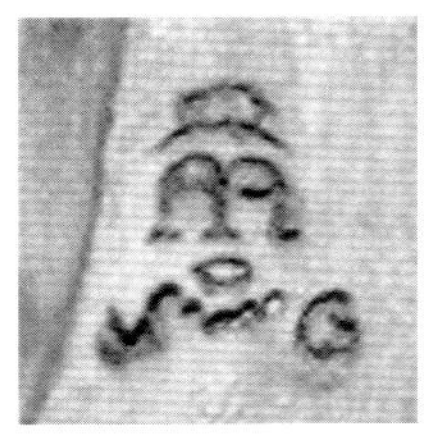
GS Marcell Wien 3
1928–1939

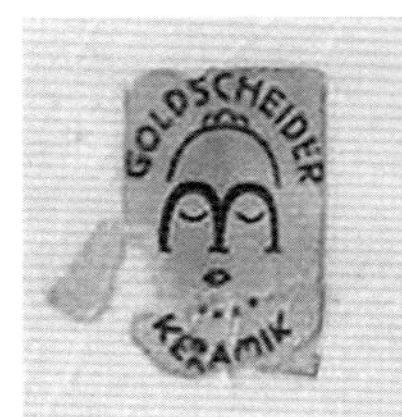

GS Marcell Wien (Pickerl I Sticker) 1, 1928–1939

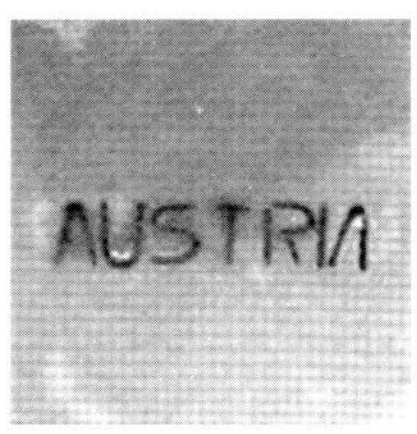

GS Marcell Wien, (Zeichen I Sign) 1, 1928–1939

Marcel Goldscheider, Stoke-on-Trent

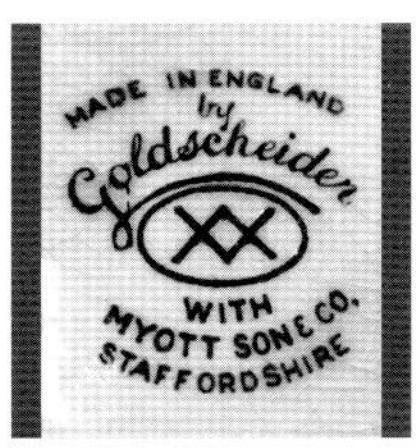

GS UK 1
ab I from 1940

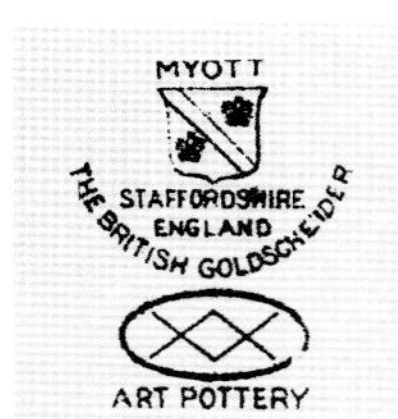

GS UK 1a

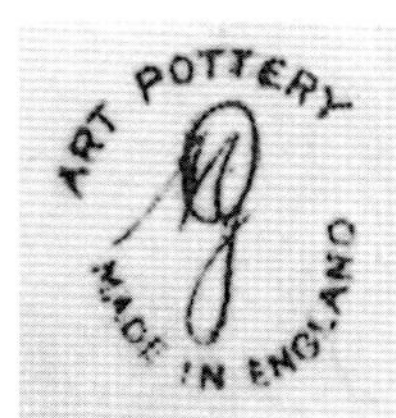

GS UK 2
ab I from 1945

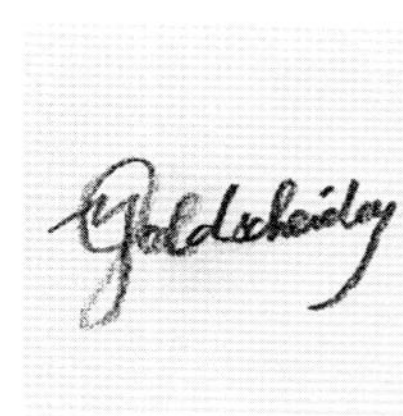

GS UK 3
ab I from 1950

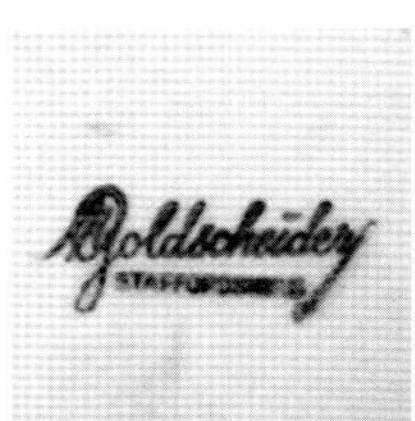

GS UK 4
ab I from 1953

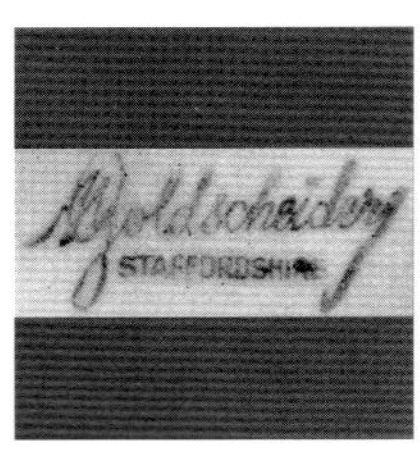

GS UK 5
ab I from 1955

Peter Goldscheider, Stoob

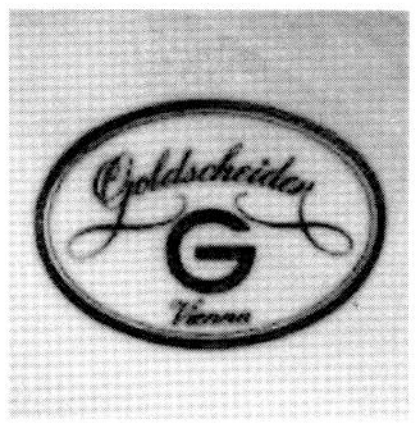

GS Stoob 1
1987–1994

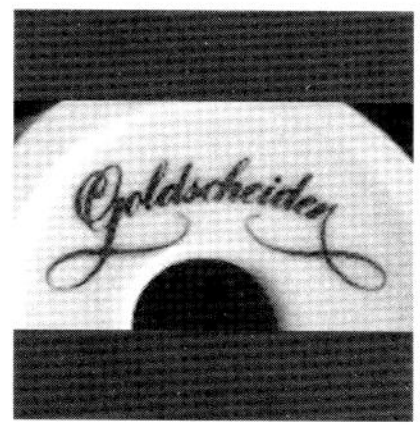

GS Stoob 1a
1987–1994

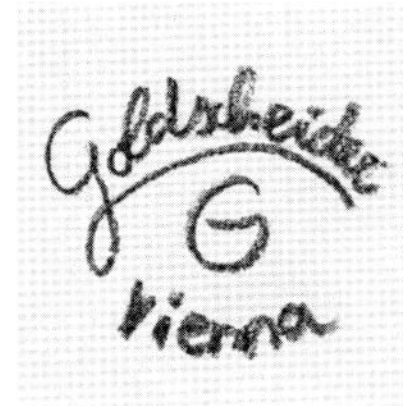

GS Stoob 2 (geprägt | impressed), 1987–1994

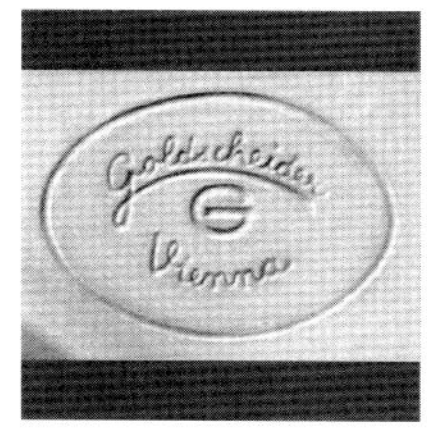
GS Stoob 3 (geprägt | impressed), 1987–1994

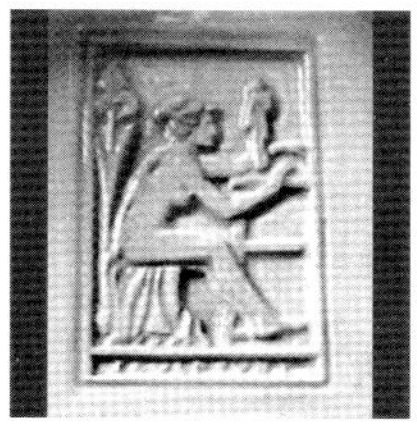
GS Stoob 4 (geprägt | impressed), 1987–1994

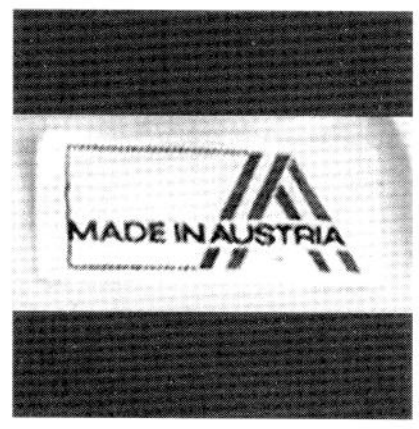

GS Stoob (Pickerl | Sticker) 1

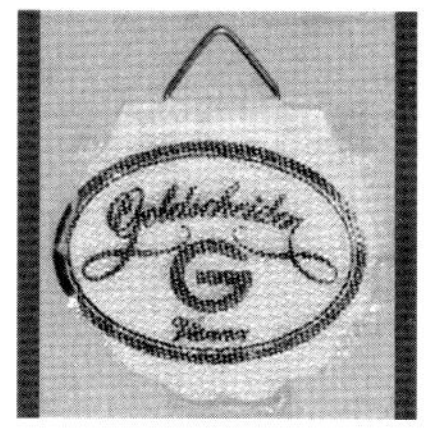

GS Stoob (Aufhänger | Hanger) 1

Tipps für Sammler I

Pointers for Collectors

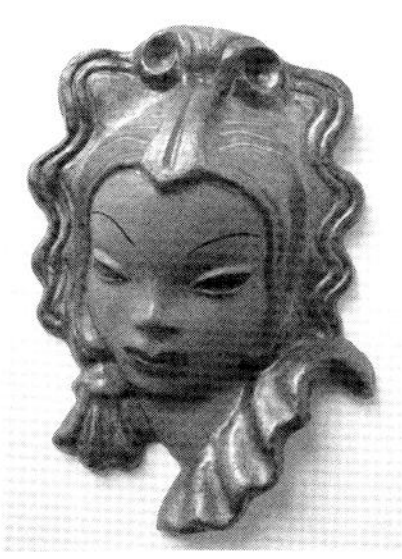

Diese Wandmaske von Stephan Coneye oder Paul Kinsler, die auch stilistisch nicht den Goldscheider-Modellen ähnelt, trägt rückseitig einen nachträglich aufgemalten Goldscheider-Stempel und lässt sich als absichtliche Fälschung enttarnen. (Sammlung f@goldscheider.de)

This wall mask by either Stephan Coneye or Paul Kinsler, which does not even stylistically resemble Goldscheider models, bears on the back a Goldscheider stamp that was painted on later and is thus easily exposed as a deliberate counterfeit. (Collection f@goldscheider.de)

Das abgebildete Paar trägt zwar den Schriftzug „Goldscheider", ist aber keiner Goldscheider-Produktionsstätte zuordenbar. Es dürfte sich eher um ein böhmisches Erzeugnis handeln, vermutlich Royal Dux. (Sammlung a@goldscheider.net)

The couple shown is marked "Goldscheider" in lettering but the piece cannot be assigned to any Goldscheider place of production. It is probably a product made in Bohemia, presumably by Royal Dux. (Collection a@goldscheider.net)

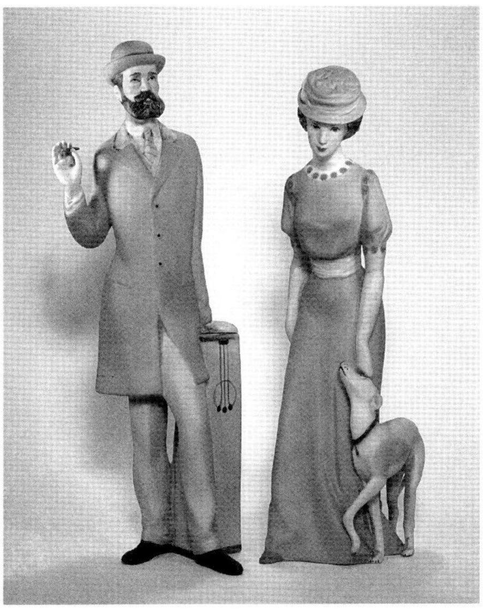

Der weltweite Vertrieb der Goldscheider Keramik seit Ende des 19. Jahrhunderts führte nicht nur zum Begriff der „Weltmarke der Keramik", sondern machte Goldscheider-Erzeugnisse einem großen und internationalen Kundenkreis zugänglich. Von Sammlern konnten Goldscheider-Objekte bis in die 1990er Jahre hinein über den Kunsthandel sowie bei größeren internationalen Auktionshäusern wie Christie's, Sotheby's oder dem Dorotheum erworben werden; oft wurden sie auch bei kleineren Häusern gehandelt. Dies hat sich in den letzten Jahren mit dem Aufkommen des Internets grundsätzlich geändert.

Sammler und Händler aus der ganzen Welt können auf einmal direkt Kontakt miteinander aufnehmen und Teile ihrer Sammlungen beispielsweise über das Internet-Aukionshaus ebay einem vielfach größeren Kundenkreis anbieten. Von preiswerten englischen und amerikanischen Goldscheider-Erzeugnissen, über die derzeit weniger beliebten Biedermeier- und Klassizismus-Modelle der Wiener Manufaktur, die stärker nachgefragten Objekte der Angewandten Modernen Keramik und Wandmasken der 1930er Jahre, bis hin zu den hochpreisigen glamourösen Tänzerinnen, ist „online" eine Fülle an Goldscheider-Modellen erhältlich.

Dabei sind die Verkäufer nicht nur erfahrene Kunsthändler, sondern sehr oft auch Privatpersonen, die ein Erbstück auf diese Weise verkaufen wollen. Einige dieser Verkäufer sind relativ professionell und nutzen beispielsweise den kostenlosen Expertise-Service auf www.goldscheider.de, um richtige Datierungen und Künstler-Zuweisungen zu erfahren. Andere wiederum geben versehentlich oder auch bewusst falsche Angaben, um das Modell interessanter zu machen und den Preis in die Höhe zu treiben. Es gibt auch einige schwarze Schafe, die Exponate aus anderen Manufakturen als Goldscheider ausgeben und sogar Goldscheider-Stempel auf dem Objekt nachzeichnen. Dies ist vor allem bei den Tänzerinnen der 1920er und 30er Jahre von Manufakturen wie Hertwig (Katzhütte), Royal Dux und österreichischen Manufakturen wie Schauer, Thomasch, Prischl oder Keramos oft der Fall. Die Krönung der

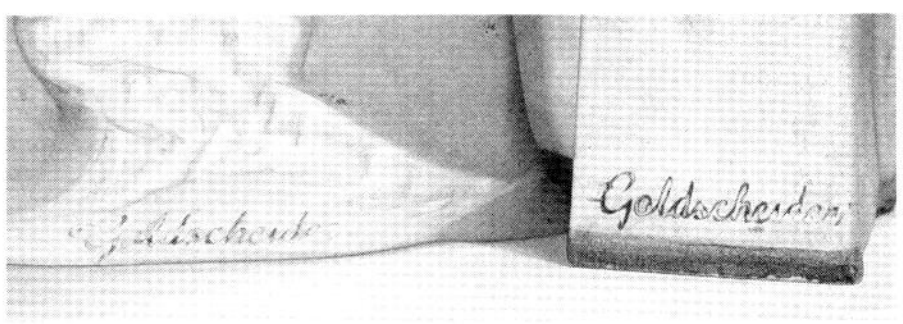

The fact that Goldscheider ceramics were sold worldwide from the late 19th century not only led to the "World Mark in Ceramics" being a concept to be reckoned with but also made Goldscheider products available to a large international clientele. Collectors could acquire Goldscheider objects on into the 1990s through antiques dealers as well as the larger international auction houses such as Christie's, Sotheby's and the Dorotheum; they were also sold by small auction houses. All this has fundamentally changed in recent years, thanks to the internet.

Now collectors and dealers from all over the world can contact each directly and offer parts of their collections for sale to a vastly larger clientele, for instance, via the online auction ebay. A wealth of Goldscheider models, ranging from affordable English and American Goldscheider products through the currently less sought-after Biedermeier and Neo-Classical models from the Viennese manufactory, the more popular applied Modern Ceramics and 1930s wall masks to the glamorous dancers, that fetch top prices, is available online.

The vendors of these objects are not just experienced art dealers but are often individuals who want to sell an inherited piece in this way. Some of these vendors adopt a relatively professional attitude and make good use of the appraisal service offered free of charge at www.goldscheider.de to ensure correct dating and attribution to artists. Others, however, provide erroneous data, either inadvertently or deliberately to make models more interesting and drive up prices. There are also some black sheep who try to pass off inferior products from other factories as Goldscheider pieces and are not above drawing the Goldscheider stamp on them. This is particularly often the case with 1920s and 1930s dancers from such factories as Hertwig (Katzhütte) and Royal Dux as well as the Austrian factories Schauer, Thomasch, Prischl and Keramos. The ultimate in clumsy Goldscheider fakes was to be admired at a German auction house: a numbered Spanish dancer from Keramos, on which someone had painted the "Goldscheider GWM"stamp. However, they forgot the "d" and wrote "Golscheider" instead and "Wien" was replaced by "Wein" – perhaps consumption of that beverage wine at the same time led to such an amateurish yet amusing forgery.

Some erroneous attributions can, however, probably also be credited to the similarities between the stamps used by several factories, which made their own stamps closely resemble the Goldscheider stamp to take advantage of the enormous popularity of Goldscheider ceramics. It is very easy to mistake the stamp used by the Cologne factory König & Lengsfeld for the Goldscheider "Sculpture" stamp. The products made by the Thomasch factory are also often confused with Goldscheider pieces because of their stamp and on stylistic grounds.

However, objects by other makers from the Jugendstil era, especially such Bohemian factories as Amphora, Bernhard Bloch and Johann Maresch, whose products resemble Goldscheider designs and whose number sequences are virtually identical in structure, are also praised erroneously as Goldscheider pieces to raise prices. The similarity of the models also consists in the fact that many artists who designed for Goldscheider (including Thomasch, Bosse, Prischl, Goebel and Dakon) worked for other factories in later years and that in some cases very similar models were produced at such factories.

In the case of the Orientalizing models, complete plaster of Paris casts of Goldscheider originals are increasingly encountered at French flea markets. Many of them are of astonishingly good quality and at first glance not recognizable as copies. The bust of an Arab shown (46 cm high; copy of Model 1339) is a counterfeit of this kind. Even though it is quite accurately reproduced as far as external details are concerned, several spurious features make it easy to spot as a fake.

schlechtesten Goldscheider-Fälschung war in einem deutschen Auktionshaus zu bewundern: Eine nummerierte spanische Tänzerin von Keramos, der von jemandem der „Goldscheider GWM"-Stempel aufgemalt worden war. Allerdings wurde das „d" vergessen und nur „Golscheider" daraufgeschrieben und anstatt „Wien" hieß es „Wein" – vermutlich führte der gleichzeitige Konsum desselben zu dieser dilettantischen und amüsanten Fälschung.

Einige falsche Zuweisungen sind aber auch der Ähnlichkeit der Stempel verschiedener Manufakturen zuzuordnen, die aufgrund der hohen Beliebtheit der Goldscheider-Keramik das Aussehen ihres eigenen Stempels sehr nahe an dem des Goldscheider-Stempels ausgerichtet haben. So ist der Stempel der Kölner Manufaktur König & Lengsfeld dem Goldscheider-Stempel „Sculpture" zum Verwechseln ähnlich. Die Erzeugnisse der Manufaktur Thomasch werden ebenfalls öfters wegen des Stempels und Stils mit Goldscheider verwechselt.

Aber auch Objekte anderer Hersteller aus der Zeit des Jugendstils, gerade von böhmischen Manufakturen wie Amphora, Bernhard Bloch oder Johann Maresch, deren Produkte den Goldscheider-Entwürfen ähnlich sehen und deren Nummernfolgen fast identisch aufgebaut sind, werden fälschlicherweise als Goldscheider angepriesen, um höhere Preise zu erzielen. Die Ähnlichkeit der Modelle liegt teilweise aber auch daran, dass viele der für Goldscheider entwerfenden Künstler (wie Thomasch, Bosse, Prischl, Goebel, Dakon) in den Folgejahren für andere Manufakturen tätig waren und dort teilweise sehr ähnliche Modelle produzierten.

Bei den Modellen des Orientalismus werden verstärkt auf französischen Flohmärkten komplette Nachgüsse aus Gips angeboten, die teilweise sogar eine erstaunlich gute Qualität erreichen und auf den ersten Blick nicht als Kopien zu enttarnen sind. Bei der abgebildeten Büste eines Arabers (46 cm hoch; Kopie des Modells 1339) handelt es sich um solch eine Fälschung. Obwohl äußerlich recht detailgetreu reproduziert, ist sie aufgrund mehrerer Aspekte als Fälschung zu erkennen.

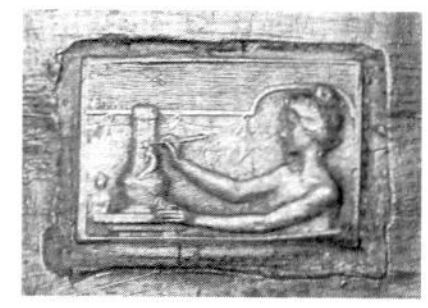

Die Marke der Manufaktur König & Lengsfeld sieht der „Sculpture"-Marke von Goldscheider zum Verwechseln ähnlich und unterscheidet sich lediglich darin, dass sie die Figur der Bildhauerin nicht ganzkörperlich abbildet. (PB Deutschland)

The mark used by König & Lengsfeld so closely resembles the Goldscheider "Sculpture" mark that it is often mistaken for it; the only difference between the two marks is that the König & Lengsfeld mark does not reproduce the entire figure of the sculptor. (PB Deutschland)

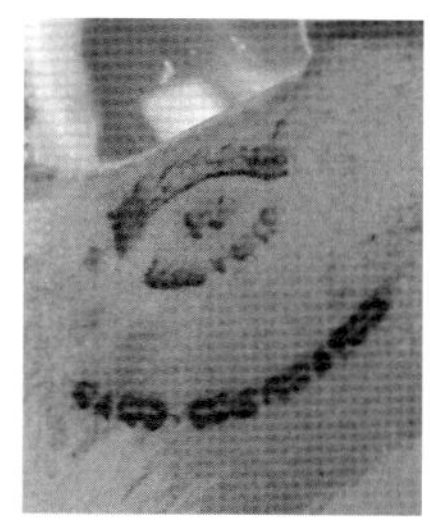

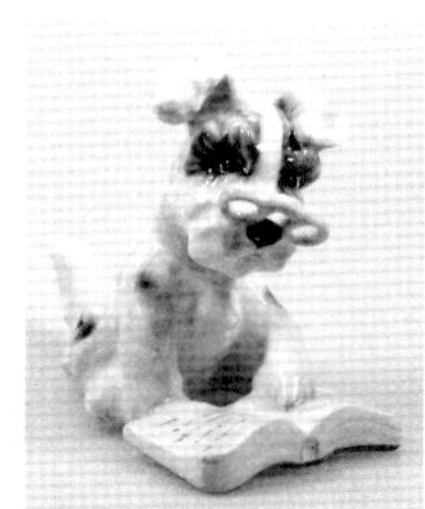

Die Manufaktur des ehemaligen Mitarbeiters von Goldscheider, Wilhelm Thomasch, produzierte ähnliche Modelle; die Gestaltung ihres Firmenstempels ist dem von Goldscheider zum Verwechseln ähnlich. (Sammlung Karl Aumann, Österreich; Foto des Stempels aus PB Deutschland)

The factory founded by former Goldscheider employee Wilhelm Thomasch produced similar models; the design of the Thomasch firm stamp is easily mistaken for the Goldscheider stamp. (Karl Aumann Collection, Austria; photo of the stamp from PB Deutschland)

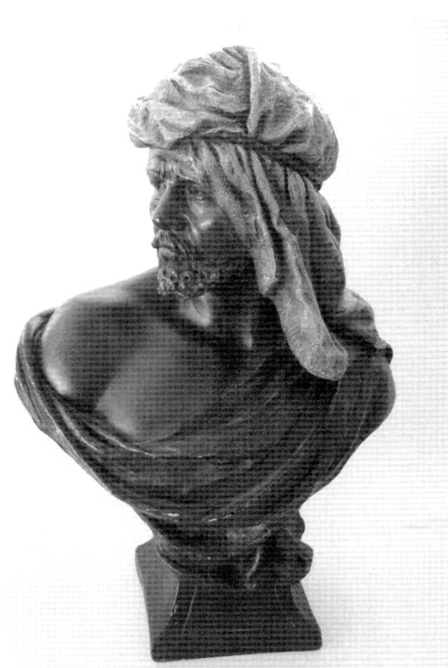

Fälschung einer Araber-Büste aus Gips.

Counterfeit bust of an Arab in plaster of Paris.

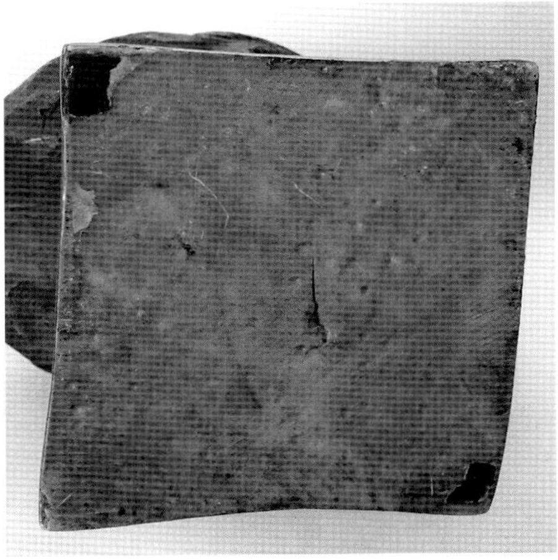

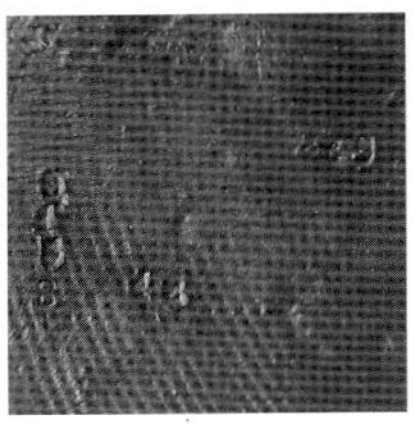

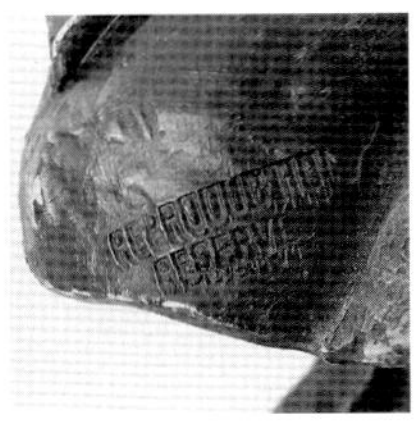

Unterseite des Sockels, gefälschte Modellnummer und Reproduktions-Hinweis der Araber-Büste aus Gips.

Underside of the pedestal, faked model number and reproduction notification from the plaster of Paris bust of an Arab.

Die hier abgebildete kleine, blaue Vase wurde beispielsweise im Handel als Goldscheider deklariert, obwohl sie weder in ihrer Form noch Qualität mit Goldscheider vergleichbar ist.

The little blue vase shown here, for instance, was advertised on the market as Goldscheider although it does not stand comparison with Goldscheider in respect of either form or quality.

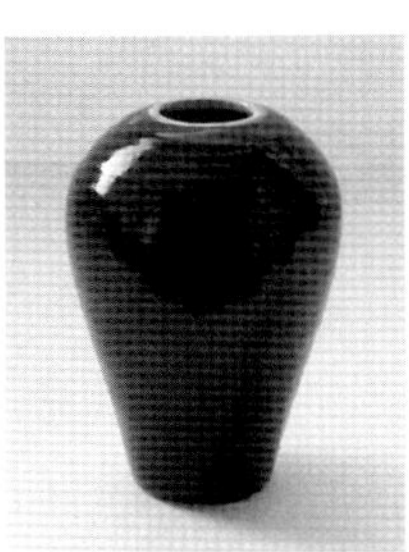

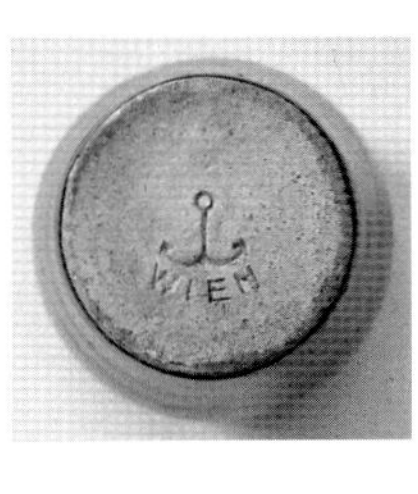

Zunächst ist die Araber-Büste viel schwerer als das Original, bedingt durch das Material Gips und die Tatsache, dass sie innen nicht hohl ist. Der Sockel der Fälschung ist zudem mit einem modernen Harz versiegelt. Originale Goldscheider-Büsten aus dieser Zeit sind zumeist aus Terrakotta und innen hohl. Neben dem Material geben vor allem die Goldscheider-Stempel und die Nummerierung Aufschluss über eine Figur. Die vorliegende Fälschung weist zwar eine Kombination von drei Ziffernfolgen auf (0158,419,43), die hinten auf dem Sockel versetzt abgebildet sind. Diese sind jedoch für solch ein Modell zu klein und in ihrer Schriftart zu modern. Dass sie versetzt erscheinen ist unüblich, taucht aber bei einigen Goldscheider-Modellen aus dieser Zeit auf.

Für den französischen Markt produzierte Goldscheider-Objekte wurden mit dem Stempel „Reproduction Reservée" markiert (oft auch „Reservé" geschrieben), im deutschsprachigen Raum war stattdessen „Vervielfältigung vorbehalten" zu lesen. Auch die gefälschte Araberbüste trägt diesen französischen Stempel, allerdings ist der Stempel stärker verzerrt als sonst. Dieser Stempel wäre auch eher am Sockel, d.h. bei den Zahlen, platziert worden und nicht am Umhang des Mannes. Zudem fehlt auch der Goldscheider-Firmenstempel. Es gibt zwar Goldscheider-Objekte ohne einen Firmenstempel, allerdings trifft man auf solche Modelle eher selten.

Oft kommt es auch vor, dass Modelle als Goldscheider ausgegeben werden, sobald im Stempel die Initialen FG, EG, MG oder AG vorkommen – entsprechend den Namen Friedrich, Eduard, Marcell oder Arthur Goldscheider. Auch der so genannte Anker-Stempel, der in einigen Büchern als Goldscheider katalogisiert worden ist und dem ganz frühen Goldscheider-Anker-Stempel oder dem Wiener Firmenzeichen von Marcell Goldscheider (in Form eines M bzw. eines lachenden Gesichts) mit viel Phantasie ähnelt, setzt fälschlicherweise Objekte mit Goldscheider in Verbindung, die stilistisch weit von Goldscheider entfernt sind.

Wie kann man also eine Goldscheider-Figur richtig datieren, den Zustand erkennen und bewerten? Das vorliegende Werkverzeichnis bietet hier eine hervorragende Grundlage, denn anhand der eingeprägten oder eingeritzten Modellnummern (jeweils die erste Ziffernfolge) kann man zunächst nachschauen, ob ein ähnliches Exemplar bereits mit Provenienz dokumentiert ist. Sollte dies nicht der Fall sein – bei insgesamt über 10.000 entworfenen Goldscheider-Modellen, die in hunderttausendfacher Ausführung mit unterschiedlichen Dekoren produziert worden sind, wird dies öfter vorkommen – lässt sich anhand der Stempel-Übersicht im Markenverzeichnis zumindest die Produktionszeit bestimmen. Wie gesagt gibt es nur sehr wenige Goldscheider-Objekte, die keinen

First of all, the bust of an Arab is much heavier than the original because it is made of plaster of Paris and also because it is not hollow. Further, the pedestal of the fake piece is coated with a modern sealing resin. Original Goldscheider busts from this period are usually made of terracotta and are hollow. Quite apart from the material, the Goldscheider stamp and the numbering of a piece are instructive on a figure. The present piece is marked with a combination of three staggered number sequences (0158,419,43) on the back of the pedestal. These numerals, however, are too small for a model of this kind and the script used for them is too modern. It would not be usual for them to be staggered but they do appear in this arrangement on some Goldscheider models from this period.

Goldscheider objects produced for the French market were marked with the stamp "Reproduction Reservée" (often also spelled "Reservé"). The corresponding mark in German-speaking countries is "Vervielfältigung vorbehalten" ["Reproduction rights reserved"]. The fake bust of an Arab bears this Goldscheider French stamp but here it is more distorted than is usual. This stamp would have been more likely to have been put on the pedestal, ie, with the numerals, and not on the figure's cloak. Moreover, the Goldscheider firm stamp is missing. Admittedly, there are Goldscheider objects without the firm stamp but such models tend to be rare.

It also frequently happens that models are passed off as Goldscheider when the stamp contains the initials "FG," "EG," "MG" or "AG" – corresponding to the names Friedrich, Eduard, Marcell and Arthur Goldscheider. Also what is known as the anchor stamp, which is catalogued in some publications as Goldscheider and which, by a stretch of the imagination might appear to resemble the very early Goldscheider anchor stamp or the Viennese firm mark used by Marcell Goldscheider (in the form of an "M" or a laughing face), erroneously links objects with Goldscheider that are, however, stylistically at a far remove from the authentic pieces.

How, then, can one date a Goldscheider figure correctly, recognize the condition it is in and appraise it? The present catalogue of works provides an excellent basis because one can first check on the grounds of the impressed or incised model numbers (in each case the first sequence of numerals) whether a similar piece has been documented with provenance. Should this not be the case – this will frequently occur since more than 10000 Goldscheider models were designed and produced in the hundreds of thousands when all the different decoration schemes are taken into consideration – the survey of stamps in the marks index makes it possible to ascertain at least when a piece was made. As has been said, there are very few Goldscheider objects indeed that bear no stamp at all: such models are usually fastened to a pedestal or have a protecting mount at the bottom that conceals the model numbers.

Another important clue is the artist signature, which occurs on so many Goldscheider models. The index of artists makes it possible to ascertain whether a particular artist worked for Goldscheider and which models of his are known to date. However, mistakes with signatures were very often made at Goldscheider so that sometimes several artists or, alternatively, varying spellings of a name can be traced to a single model number (also see the Catalogue of Works on this).

Authentication can be outlined as follows, using the sculpture of a boy as an example (photos: Kunsthaus Wiesinger, Wels): the naturalistic style and the painting suggest a model from the period before 1900 because figures of children worked up in terracotta were a common motif at that time. Since the original metal sign on the base has survived, we learn the title of this piece of sculpture: "Coq du village," by the sculptor Cherc. The designer's signature is also incised in the base.

When this piece is viewed from the side, the sculptural dimension of this work, which is almost a meter high, becomes apparent. Apart from the firm stamp "Friedrich Goldscheider Wien," the base also bears the mark that indicates manufacture for the French market: "Reproduction Reservée." On the back of the base, the model number, the piece number and the modeler's mark are discernible. The design for model 1628 should be dated around 1898: this is the number 69. The number 55 indicates the piece and the modeler's number. The painter's mark is XV.

It can be definitively stated that not everything bearing the marks "Made in Austria," "Reproduction Reservée" or sequences of numbers automatically turns out to be a Goldscheider figure. Moreover, the material and the casting of a piece should also always be examined. In the case of the

Stempel tragen: Meistens sind diese Modelle auf einem Sockel befestigt oder tragen unten einen Schutz, der die Modellnummern verdeckt.

Auch die Künstler-Signatur, die bei Goldscheider auf vielen Modellen zu finden ist, dient als wichtiger Anhaltspunkt. So lässt sich anhand des Künstlerverzeichnisses feststellen, ob dieser Künstler für Goldscheider gearbeitet hat und welche Modelle bislang von ihm bekannt sind. Allerdings wurden bei Goldscheider sehr oft Fehler bei der Signierung gemacht, so dass manchmal mehrere Künstler oder unterschiedliche Namensschreibweisen einer Modellnummer nachweisbar sind (siehe hierzu auch das Werkverzeichnis).

Die Echtheits-Beurteilung lässt sich anhand der Skulptur eines Knaben (Fotos: Kunsthaus Wiesinger, Wels) wie folgt skizzieren: Der naturalistische Stil und die Bemalung deuten auf ein Modell aus der Zeit vor 1900 hin, denn in Terrakotta ausgearbeitete Kinderfiguren waren damals ein gängiges Sujet. Dank des noch erhaltenen Metallschilds am Sockel erfahren wir den Titel dieser Skulptur: „Coq du village", vom Bildhauer Cherc. Die Signatur des Entwerfers findet sich zudem noch eingeritzt im Sockel.

Seitlich erkennt man die plastische Dimension dieses fast einen Meter hohen Werkes. Am Sockel ist neben dem Firmenstempel „Friedrich Goldscheider Wien" auch der Hinweis für den französischen Markt zu lesen: „Reproduction Reservée". Auf der Hinterseite des Sockels sind die Modellnummer, Stücknummer und Modelleur-Kennziffer erkennbar. Der Entwurf des Modells 1628 ist um das Jahr 1898 anzusiedeln, es ist das 69. Exemplar und die Modelleur-Nummer ist die 55. Der Maler hat die Kennziffer XV.

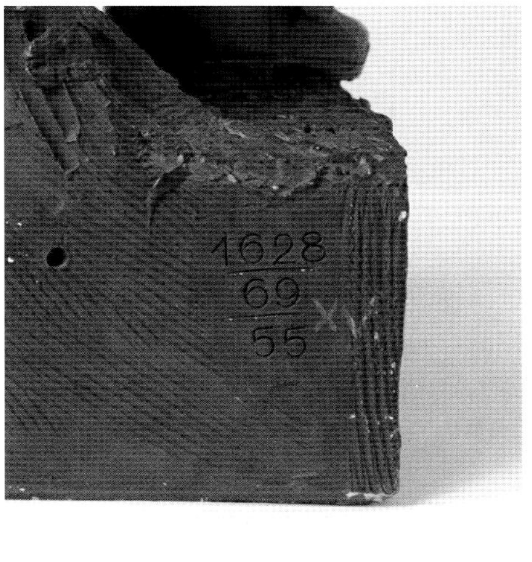

Anhand der Figur dieses Knaben lässt sich, wie im Text beschrieben, der Prozess der Echtheits-Beurteilung nachvollziehen.

Using this sculpture of a boy as an example, readers can follow the authentication procedure as outlined in the text.

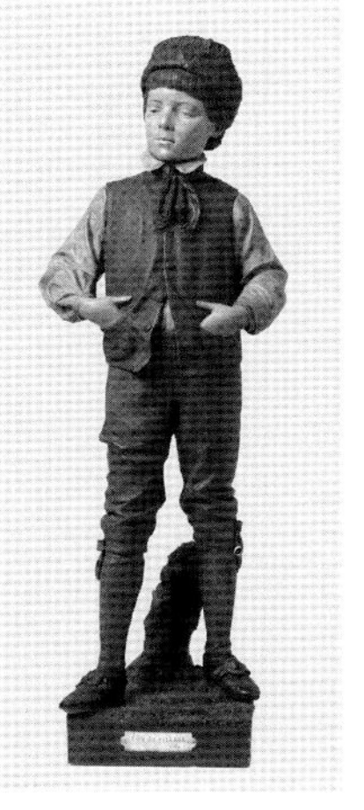

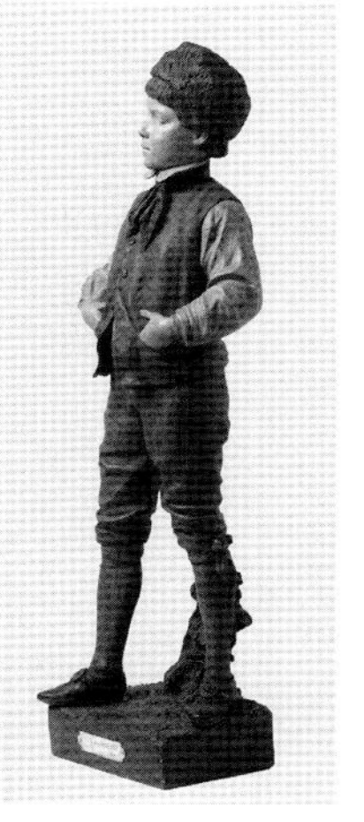

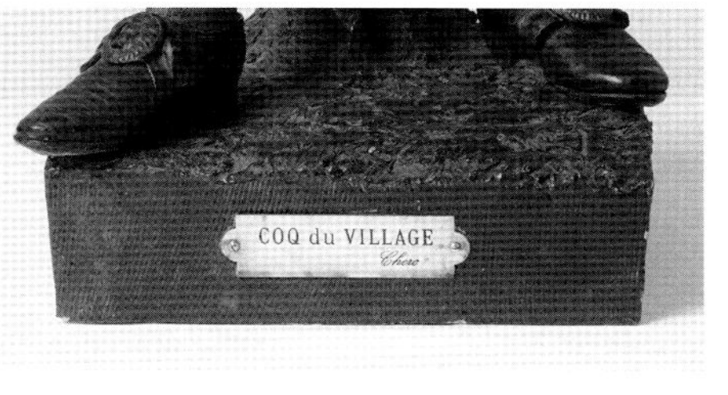

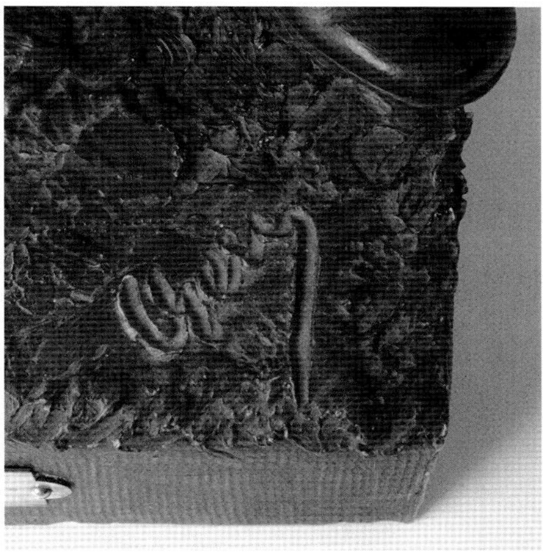

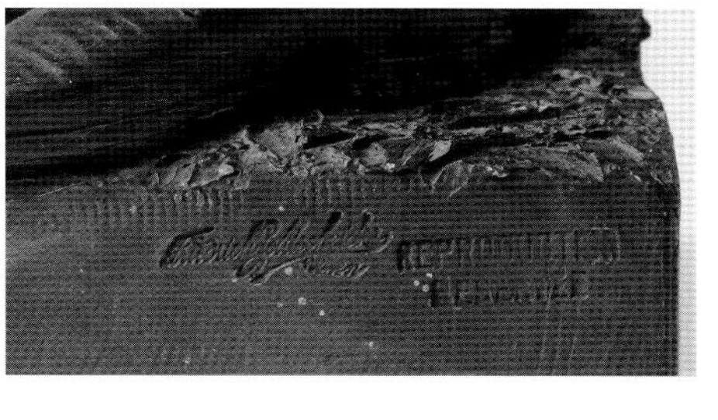

Grundsätzlich kann man festhalten, dass nicht alles, wo „Made in Austria", „Reproduction Reservée" oder Nummernfolgen zu sehen sind, automatisch Goldscheider-Figuren sind. Außerdem sollte immer auch das Material und die Ausformung untersucht werden. Bei Wandmasken muss man beispielsweise wissen, dass ab den 1920er Jahren fast alle Goldscheider-Wandmasken zunächst aus weißem, Ende der 30er Jahre auch aus rotem Scherben produziert worden sind und ausgehöhlt waren. In den 50er Jahren wurde nur der minderwertigere rote bzw. bräunliche Scherben benutzt. Auf dem Markt auftauchende Kopien der populären „Eva" (ModNr. 6774) sind meistens aus Gips und innen nicht hohl.

Interessant sind auch die seit den 20er Jahren eingesetzten Pickerl (Aufkleber), die in den 50er Jahren um den Slogan „Weltmarke der Keramik" ergänzt worden sind. Sie tauchen manchmal auch bei Objekten auf, die nicht von Goldscheider sind und missbrauchen den Qualitätsnamen zur Erzielung höherer Verkaufspreise.

Zum Alltag des Sammlerlebens gehören aufgrund des „hohen Alters" und der „Anfälligkeit" der Figuren auch Restaurierungen. Beim Kauf sollte man exponierte und somit bruchgefährdete Stellen wie Hände, Arme, Schuhe, Kopfschmuck und Accessoires untersuchen. Am besten betastet man hierbei die Figuren, da nachträglich aufgetragene Farbe dadurch am leichtesten zu erkennen ist. Manche Kenner benutzen hierfür auch eine Stecknadel, die dann in der neuen Farbmasse kurz stecken bleibt (nicht jeder Kunsthändler freut sich allerdings über diese Behandlung seiner Verkaufsobjekte). Selbstverständlich ist auch die Betrachtung unter UV-Licht hilfreich, wobei heutzutage professionelle Restaurierungen auch unter Schwarzlicht nicht sichtbar sind.

Aufgrund des hohen künstlerischen Anspruchs und dem damit verbundenen Produktionsaufwand sind aber generell spezifische Goldscheider-Fälschungen sehr selten. Es ist eher das Unwissen, das zu falschen Interpretationen verleitet. Das vorliegende Buch soll anhand der detaillierten Firmengeschichte und des umfassenden, bebilderten Werkverzeichnisses als Nachschlagewerk dienen und die Freude an dem schönen und vielfältigen Sammelgebiet der Goldscheider-Keramik entfachen. Sollten Sie spezielle Fragen haben, können Sie gerne die Autoren unter info@goldscheider.de kontaktieren. Wir wünschen Ihnen viel Freude beim Sammeln.

wall masks, it is necessary to known that from the 1920s almost all Goldscheider wall masks were made from white clay body and, from the late 1930s, they were also produced from red clay body and were by then hollow. In the 1950s, only the inferior red or brownish clay body was used. Most copies of the popular "Eva" ["Eve"] (ModNr. 6774) that appear on the market are made of plaster of Paris and are not hollow.

Another interesting aspect is the glued on labels used from the 1920s, to which the slogan "Weltmarke der Keramik" ["World Mark in Ceramics"] was added in the 1950s. Such labels occasionally occur on objects that are not by Goldscheider and represent abuse of a name that stands for quality to raise prices.

Restoration necessitated by the "great age" and "fragility" of the figures is part and parcel of the collector's life. When buying, one should investigate exposed places that are subject to chipping, breaking and cracking. The best thing to do is to gently palpitate the figures since painted added later can be most easily spotted in this way. Some knowledgeable collectors use a pin when doing this, which briefly catches in the new coat of paint (not all dealers, however, rejoice at the objects they are selling being subjected to such hands-on treatment). Of course, looking at a piece under ultraviolet light is helpful. The catch here is that professional restoration work done nowadays does not even show up under black light (Wood's light: the near ultraviolet range).

Since Goldscheider pieces are of such high quality and attempting to reproduce them is invariably linked with an elaborate production process, deliberate counterfeits of them are on the whole quite rare. It is in fact lack of knowledge that has usually led to faking in such cases. The present book is intended as a handbook to be used for its detailed history of the firm and comprehensive illustrated catalogue of works to kindle a delight in collecting in such a fine and diverse field. If you have specific questions, you are welcome to contact the authors at info@goldscheider.de. We hope you will really enjoy collecting Goldscheider ceramics.

Bibliografie I
Bibliography

Bücher – Aufsätze – Reklame I
Books – Essays – Advertising

Acatos
Acatos, Sylvio, Ceramics and Glass. Intern. Auctions, New York 1990–1991

ADEC
ADEC. Annuaire des Cotes. Art Price Annual, Paris, 1994 ff.

AKL
Allgemeines Künstlerlexikon: die bildenden Künstler aller Zeiten und Völker, Bd. 1 ff., München u.a. 1992 ff.

Alt-Wien
W. Kos u. C. Rapp (Hrsg.), Alt-Wien. Die Stadt, die niemals war, 2. überarb. Aufl., Wien 2005

Amigon, Bernard (Interview), Germaine plaît beaucoup par la fraicheur et l'humor de ses dessins, Le Chineur 2003, No. 71, Paris 2003

Amort
Amort, Andrea u. Mimi Wunderer-Gosch (Hrsg.), Österreich tanzt. Geschichte und Gegenwart, Wien, Köln, Weimar 2001

AdA
Annuel des Arts, 1990 ff.

ARGOS
ARGOS Glcbal Price guide of Art & Antiques, Den Haag 1995/96

API
Art Price Index International, Madison, Conn. 1994–1995

ASI
The Art Sales Index Vol. 19–30, Weybridge 1986/87–1997/98

Arwas (1987)
Arwas, Victor, Art Deco, New York 1987, S. 163, 292

Arwas (1992)
Arwas, Victor, Art Deco Sculpture, London 1992

Arwas (2002)
Arwas, Victor, Art Nouveau, London 2002

Atterbury
Atterbury, Paul u.a., Miller's Twentieth Century Ceramics, London 1999

Aubel
Aubel, Hermann u. Marianne, Der künstlerische Tanz unserer Zeit, Königstein i. Ts., Leipzig 1935

Baarova, Lida, Die süße Bitterkeit meines Lebens, Koblenz 2C01

Battenberg
Battenberg. Kunst-Auktionen-Preise (später: Battenberg Antiquitätenführer), Augsburg 1991–1995/96

Battenberg
Battenberg Antiquitätenführer (Forts. von Battenberg. Kunst-Auktionen-Preise), Augsburg 1996/97 ff.

Bauer-Paulfranz, L., Fabrikmarken-ABC. Die Fabrikmarken d. Keram- u. Glasindustrie, Coburg 1930

Bayer, A. H., Worlds Fair Chicago (1893), Wien 1894, S. 75

Bechyně 1884–1994. 110 let Středni průmyslové školy keramické, Bechyně 1994

Becker, Kurt, Lilian, das Mädchen mit den 100 Gesichtern, Uhu 4, 1928/29, S. 54 (m. Abb.)

Bénézit
Bénézit, E., Dictionnaire critique et documentaire des Peintres, Sculpteurs, Dessinateurs et Graveurs, Neuauflage, Paris 1999

Benno Geiger 1903–1979, Ausst. Kat., Solothurn 1991

Berdel, Eduard, Keramik und Glas auf der Weltausstellung Brüssel 1910, Sprechsaal 43, 1910, S. 641

Berman
Berman, Harold, Bronzes. Sculptors & Founders 1800–1930, Vol. 1–4, Chicago 1977

Bernd
Bernd, Gottfried, Heitere Wiener Tierkunst: Ida Meisinger, Die Schaulade 8, 1932, S. 480 (m. Abb.)

BT
Berühmte Tänzerinnen. Garbáty Zigarettenbilderalbum, Berlin, um 1930

Bie, Oskar, Schnackenberg. Kostüme, Plakate und Dekorationen, München 1920

Bloch (1872)
Bloch, J., Pilsen und sein Kammerbezirk. Firmen und Adressbuch, Pilsen 1872

Bloch (1876)
Bloch, J., Adressenbuch für Pilsen und das westliche Böhmen, Pilsen 1876

Bloit-Fournier-Richemond
Bloit, Michel u. J-C. Fournier u. Stéphane Richemond, Réalisme et Art Nouveau. Deux Siecles de Ceramique de L'Isle-Adam et Parmain (Val-D'Oise) 1815–1950, 2. durchgesehene und korrigierte Auflage, L'Isle-Adam 1996

Blühm, Andreas u.a.: The Colour of Sculpture, 1840–1910, Zwolle 1996

Böhme, Fritz, Tanzkunst, Dessau 1926

Bonynge, Richard, A Collectors guide to Theatrical Postcards, London 1988

Born
Born, Wolfgang, Figürliche Keramik von Rudolf Knörlein – Wien, DKuD 69, 1931/32, S. 112–115 (m. Abb.)

Braig, Rosewith, Bröhan-Museum Berlin, Braunschweig 1984, S. 40–42

Brö 2,1
Bröhan Museum, Bd. 2,1: Kunst der Jahrhundertwende und der 20er Jahre, Berlin 1976, S. 399–407 (m. Abb.)

Brö 2,2
Bröhan Museum, Bd. 2,2: Kunst der Jahrhundertwende und der 20er Jahre, Berlin 1977, S. 351

Brö PK
Bröhan-Museum, Porzellan-Kunst, Berlin 1969, S. 146–147

Brühl, Georg (Mitarb.), Kunst um 1900 – Sammlung Georg Brühl, Gera 1985

Brunnbauer
Brunnbauer, Heidi, Im Cottage von Währing/Döbling, Bd. 2: Interessante Häuser, interessante Menschen, Gösing 2006

Bük (1888)
Bük, Julius von, Von der Wiener Jubiläumsgewerbeausstellung (1888), Sprechsaal 21, 1888, S. 635

Bük (1891)
Bük, Julius von, Die Jubiläumsausstellung in Prag (1891), Sprechsaal 24, 1891, S. 695

Bük (1891a)
Bük, Julius von, Die Thonindustrieausstellung in Budapest 1891, Sprechsaal 24, 1891, S. 819

Bük (1908)
Bük, Julius von, Wiener Keramik, KeramRS 16, 1908, S. 581–583

Burger, W., Die Deutsche Gewerbeschau in München, Keramik u. Glas, Dekorative Kunst 25, 1921/22, S. 265, 267 (m. Abb.)

Buxbaum, Gerda, Mode aus Wien 1815–1938, Salzburg u.a. 1986, S. 255–259 (m. Abb.)

Carter, Julie, On the up, The Goldscheider figurine..., Antiques and collectables for pleasure and profit, 2005, Nr. 8, S. 49 (m. Abb.)

Casteels, Maurice, Die Sachlichkeit in der modernen Kunst, Paris, Leipzig 1930

Catley (1987)
Catley, Bryan, Art Deco and other figures, Woodbridge 1987

Catley (1981)
Catley, Bryan, Art-Déco- und Jugendstilfiguren, München 1981

Černý
Černý, Zbyněk, Willy Russ, Bildhauer und Keramiker, Ausst.-Kat. Cheb/Marktredwitz, Cheb 2004

Cushion, John P., Manuel de la céramique européenne, Fribourg 1987, S. 567

Czeike
Czeike, Felix, Historisches Lexikon Wien, Bd. 1–6 und Erg. Bd., Wien 1992–2004

Dal Vecchio, Dekorative Plastik, Internationales Zentralblatt für Baukeramik und Glasindustrie 24, 1909, S. 3601

Davidson
Davidson, Mortimer G., Kunst in Deutschland 1933–1945, Bd. 1: Skulpturen, Tübingen 1988

Deutsches Kunsthandwerk 1986, Ausst. Kat., Düsseldorf 1986, S. 256

Dictionnaire du Ballet moderne, Paris 1957

Dolinschek, Ilse, Die Bildhauerwerke in den Ausstellungen der Wiener Sezession von 1898–1910, München 1989

Donau
Donau, Grete, Österreichische Kunst in London (1934), ÖK 5, Nr. 5, 1934, S. 17 (m. Abb.)

Dressler (1909)
Dressler, Dresslers Kunstjahrbuch, Rostock, Leipzig 1906 ff.

Dry, Graham, Modell 3913 Telzec (d.i. Telcs), Sammler-Journal 24, Nr. 12, 1995, S. 1822

Dry, Graham, Modell 3184 A. Péche BÜSTE, Sammler-Journal 32, Nr. 10, 2003, S. 7

Dufrene, Maurice, Authentic Art Deco Interiors, Woodbridge 1989, 223 S. (m. Abb.)

Duncan (1978)
Duncan, Alastair, Art Nouveau Sculpture, London 1978, 95 S.

Duncan (1998)
Duncan, Alastair, The Paris Salons 1895–1914, Vol. 4: Ceramics and Glass, Woodbrigde 1998, 463 S. (m. Abb.)

Edenhofer (2001)
Edenhofer, René, Deutsch-Wagramer Kunst-Keramik 1932–1940, Deutsch-Wagram 2001

Eisenberg
Eisenberg, Ludwig, Das geistige Wien. Künstler- u. Schriftsteller-Lexikon, Wien 1893

Eisler, Max, Österreichische Werkkultur, hrsg. v. ÖWB, Wien 1916

Enkelmann, S., Tänzer unserer Zeit, München 1937

Etzersdorfer
Etzersdorfer, Irene, Arisiert. Eine Spurensuche im gesellschaftl. Untergrund d. Republik, Wien 1995, S. 149–178 (m. Abb.)

Europäische Elfenbeinkunst vom Fin de Siècle bis zum Art Déco (1880–1940), Ausst. Kat., Erbach 1978, Kat. Nrn. 43, 44

Faber
Faber, Monika, Die Frau, wie du sie willst. Kult und korrigierte Körper, Atelier Manassé 1922–1938, Wien, München 1998

Fahr-Becker
Fahr-Becker, Gabriele, Wiener Werkstätte 1903–1932, Köln 1994

Falk, Peter H. (Hrsg.), Who was who in American Art, New York 1985

Feuerstein, G. u.a., Moderne Kunst in Österreich, Wien u.a. 1965

Fielding, Mantle, Dictionary of American Painters, Sculptors & Engravers, New York 1945

Franzke
Franzke, Irmela, Jugendstil (Battenberg Antiquitäten-Kataloge), München 1987, S. 149, 175

Franzke 1
Franzke, Irmela, Jugendstil, 2. Aufl., München 1995

Franzke 2
Franzke, Irmela, Jugendstil (Bad. Landesmuseum Karlsruhe), Karlsruhe 1987, S. 338, 339, 361

Frotscher
Frotscher, Sven, dtv-Atlas Keramik und Porzellan, München 2003

Frottier
Frottier, Elisabeth, Michael Powolny, Wien, Köln 1990

SS
Fuchs-Nebel, Susanne (Red.), Schöner Schein. Goldscheider Keramik u. andere Kunststücke. Schloss Halbturn 1992, Ausst.-Kat., Halbturn 1992

G., P., Illustrierte Rundschau, Wiener Terrakotten, VKMH 16, 1901/02, S. 591–592 (m. Abb.)

Gabor, Mark, Die Geschichte des Pin-up, München 1975

Gansterer, Helmut A., Erfolg: Die kunstvolle Wiederauferstehung der Goldscheiders, Trend 21, 1990, S. 149

Garner (1978)
Garner, Philippe (Hrsg.), The Encyclopedia of Decorative Arts 1890–1940, New York 1978, S. 221

Garner (1981)
Garner, Philippe (Hrsg.), Dekorative Kunst 1890–1940, Wiesbaden 1981, S. 221

Geiger, Benno, Keramisches ABC, Bern 1947

Geiger, Benno, Keramisches Gestalten, Ravensburg 1957

Glüber, Wolfgang, Von Jugendstil bis Art Déco. Schönheit in Elfenbein, Erbach 1998

Gmelin
Gmelin, L., Die 1. internationale Ausstellung für moderne dekorative Kunst in Turin 1902, KuHW 52, 1901/02, S. 328–334 (m. Abb.)

Godden, Geoffrey, Encyclopedia of British Pottery on Porcelain Marks, Nachdruck, London 1979

Goldscheider, Filipp, Unternehmerische Aktivitäten der Manufaktur Goldscheider aus Pilsen im Kontext der Jahre 1867 bis 1963, Diss. Prag, 2005

Goldscheider, Arthur (Hrsg.), Artistic Evolution, Texte von Yvanhoé Rambosson, franz. Ausg. unter dem Titel.: L' Évolution Artistique. Revue destinée a la diffusion des idées d'art moderne dans l'industrie et contenant des reproductions d'œuvre d'artistes des groupes „La Stèle" et „Évolution", Bd. 1 u. 2, Paris 1926

Goldscheider, Marcell, Modern oder zeitgemäß? Stimmen zur Intern. Kunstgewerbeausstellung in Paris (1925), Das Handelsmuseum 40, 1925, S. 5–6

Goldscheider, Marcel(l), Authentic Report of its Origin and Development, Firmen-Katalog/Prospekt, Hanley, um 1951, 6 gez. Bl.

KGSF1
Goldscheider (Fredelsloh), Kunstkeramische Manufaktur Goldscheider, Katalog (Herstellungsprogramm), ca. 30x20 cm, Fredelsloh/Tönnieshof, um 1955/56, o. Pag. (21 Bl.)

KGSF2
Goldscheider (Fredelsloh), Kunstkeramische Manufaktur Goldscheider, Katalog (Herstellungsprogramm), ca. 20x20,5 cm, Fredelsloh/Tönnieshof, um 1956/57, o. Pag. (ca. 113 Bl.)

Goldscheider (of Vienna, USA), Goldscheider of Vienna, Inc., Firmen-Katalog, Trenton 2005, 36 S. m. Abb.

GSB
Goldscheider (Paris), Frederic Goldscheider Fabrique de Bronzes, Rue de Paradis 45, Firmen-Katalog, Paris, ca. 1895/1900, 101 Tafeln

GSN
Goldscheider (Stoob), Goldscheider. The Royal Family of Porcelain. Since 1885, Firmen-Katalog, Stoob 1988, 66 S.

GSNF
Goldscheider (Stoob), Goldscheider. The Royal Family of Porcelain. Since 1885, Firmen-Katalog, Faltblatt/Folder, Stoob 1988, o. Pag. (5 S.)

Goldscheider (Stoob), Wander Bertoni, Firmen-Katalog, Faltblatt/Folder, Stoob um 1992, o. Pag.

Goldscheider (Stoob), Manufaktur Goldscheider präsentiert: Surrealist Leherb, Firmen-Katalog, Stoob, um 1991, o. Pag. (17 S.)

Goldscheider (Stoob), Geo-Art. (Künstler: A. Stepan), Firmen-Katalog, Faltblatt/Folder, Stoob, um 1992, o. Pag.

Goldscheider (Stoob), Geo-Art. Ihr Sternzeichen in Keramik. (Künstler: A. Stepan), Firmen-Katalog, Faltblatt/Folder, Stoob, um 1992, Leporello, o. Pag.

GSF
Goldscheider (Wien), Friedrich Goldscheider, Kunstgewerbliches Etablissement Kunst-Terracotta-Fabrik Wien XVIII. Staudgasse 7–9, Generaldepot für Frankreich, England, Belgien, Franz. Schweiz, Frédérich Goldscheider 45 Rue de Paradis: Paris; Niederlage – Leipzig – Kaufhaus, Katalog (Herstellungsprogramm), Wien, 1901, (34 S., ca. 80 Bl.); ÖNB, Sign. 1,830.344-D Neu Sond.

Goldscheider (Wien), 25 Jahre Wiener Fayence, Firmen-Katalog, Wien 1929

GSNachf.
Goldscheider (Wien), Wiener Manufaktur Friedrich Goldscheider Nachf., Firmen-Katalog, Wien, um 1940, o. Pag. (lose Bl. in Mappe)

GSNachf. (2. Ausg., bzw. Erg. Bl.)
Goldscheider (Wien), Wiener Manufaktur Friedrich Goldscheider Nachf., Firmen-Katalog, 2. Ausg. bzw. Erg. Blätter, Wien, um 1941/42, o. Pag. (lose Bl. in Mappe)

Grell
Grell, Johanna, Die Pariser Kunstgewerbe Ausstellung 1925, KeramRS 33., Nr. 38, 1925 (Sonderdr.), S. 31–33

Greven's Adressbuch von Köln und Umgebung..., 1904, Jg. 50. 1904, Teil II, S. 327

Grove
„Grove", The Dictionary of Art, Vol. 12, London 1996, S. 872

Samml. Gruber
Gruber, H. R. u. Stolte (W. Bearb.), Jugendstil. Glas-Keramik-Zinn-Möbel. Sammlung Gruber, Ausst. Kat., Kaiserslautern 1983, S. 104, 105, 180 (m. Abb.)

H., H., Internationale Kunstgewerbliche Ausstellung in Petersburg 1908. Keramik u. Glas, KM 8, 1908, S. 166–168 (m. Abb.)

Haenel, E., Wirtschaft und Kunst, Die Kunst 22, 1910, S. 43–44 (m. Abb.)

Hakenjos, Bernd (Bearb.), Europäische Keramik des Jugendstils. Modern Style Art Nouveau, Ausst. Kat., Düsseldorf 1974

ExpKWW
Hamacher, Bärbel (Bearb.), Expressive Keramik der Wiener Werkstätte 1917–1930, München 1992

Hargrove
Hargrove, June E., The life and work of Albert Ernest Carrier-Belleuse, New York, London 1977

Hartmann
Hartmann, P. W., Kunstlexikon, Sersheim 1997, 1663 S.

Haslam (1977)
Haslam, Malcom, Marks and monograms of the modern movement 1875–1930, Guildford, London 1977

Haslam (1987)
Haslam, Malcom, Collector's Style Guide ART DECO, New York 1987, S. 40 ff., 154 ff.

Hecht/Barthel
Hecht, H. u. P. Barthel, Die Thonindustrie auf der Pariser Weltausstellung 1900. Die Thonwaren Österreich-Ungarns, Thonindustrie-Zeitung 24, 1900, S. 1584

Heizer (1929)
Heizer, Wilhelm, Wiener Fayencen, Die Schaulade 5, 1929, S. 714–715 (m. Abb.)

Heizer (1931)
Heizer, Wilhelm, Edle Fayencen, Die Schaulade 7, 1931, S. 236 (m. Abb.)

Heller
Heller, Hermann (Hrsg.), Die Österreichische Ausstellung in London 1906 in Wort und Bild, Wien, London 1910, S. 66, 102

Hennig, Wolfgang, Kunsthandwerk um 1900 – Schenkung Brühl 1966 und 1986, Berlin 1986, S. 184

Hevesi (1898)
Hevesi, Ludwig, Arthur Strasser, KuKHW 1, 1898, S. 349–365 (m. Abb.)

Hevesi (1899)
Hevesi, Ludwig, Die Winterausstellung im Österreichischen Museum (1899/1900), KuKHW 2, 1899, S. 15–16

Hildenbrandt, Fred, Tänzerinnen der Gegenwart, Zürich, Leipzig 1931

Hillenbrand
Hillenbrand, Gunnar, Bronzesignaturen und Gießereimarken, Augsburg 1992

Hillier, Bevis, The world of art deco, New York 1971, S. 116, 117

Himmelheber
Himmelheber, Georg, Kunst des Biedermeier 1815–1835, München 1988, S. 288

Hirsch
Hirsch, ?, Pariser Kunstgewerbeausstellung (1925), KeramRS 33, 1925, S. 448

Hirschfeld, Heinz, Von der Laute zum Banjo, EW 14, Nr. 16, S. 16–18, 43

Hislop, siehe Art Sales Index

Högl (1993)
Högl, Helga, Die wiedergegründete Manufaktur Goldscheider Stoob, Sammler-Journal 22, Nr. 7, 1993, S. 1092–1094

Högl (1998)
Högl, Helga, Pflegeleicht und treu, Sammler-Journal 27, Nr. 9, 1998, S. 124–125

Hörmann
Hörmann, Marianne, Vally Wieselthier 1985–1945, Wien, Köln, Weimar 1999

Hoever (1996)
Hoever, Heinz, Porzellan von Claire Weiss, TuS Nr. 199, 1996, S. 36–40 (m. Abb.)

Hoever (2005)
Hoever, Heinz, Claire Weiss, CD-Rom, Friedrichshafen, 2005

Hofmann, Else, Österreichisches Kunsthandwerk II: Keramik, ÖK 3, 1932, S. 20–23 (m. Abb.)

Hottenroth (1994)
Hottenroth, Hans-Hagen, Die Tonindustrie Scheibbs (1923–1933) Scheibbser Keramik (seit 1937), Scheibbs 1994

Hottenroth (2001)
Hottenroth, Hans-Hagen u. Johanna, Die Linzer Keramik 1933–1982. Eine Dok., Scheibbs 2001

Hottenroth (2002)
Hottenroth, Hans-Hagen u. Johanna, Die Radstätter Keramik. Ofen-, Bau- u. Kunstkeramik, Scheibbs 2002

Hruška
Hruška, M., Kniha Pamětní Královského města Plzně od roku 775 až 1870, Pilsen 1883

Huth
Huth; Walter (Hrsg.), Handbuch des Spezialhandels in Porzellan, Glas, Haus- und Küchengeräten, Bamberg 1927

Industrial Directory of New Jersey, Hudson Dispatch 1946

Jacquemin, Dominique, L'Œuvre de Germaine Bouret: Un univers de fraicheur au prix existence besogneuse, Le Chineur 2003, No. 71, S. 116–123 (m. Abb.)

„... jeder Mensch ist ein Tänzer." Ausdruckstanz in Deutschland zwischen 1900 und 1945, Ausst. Kat., Gießen 1993

Jessen, Kaukoreit, Wagenbach (Hrsg.), Erich Fried. Eine Chronik: Das biograph. Lesebuch, Berlin 1998

Kashey
Kashey, Elisabeth/Robert, 19th Century French and Western European Sculpture in Bronze..., New York 1985, S. 272, 296

Kattinger, Matthäus, Goldscheider-Keramiken. „Aussteiger" will Vergangenheit überholen, Parnass Nr. 6, 1987, S. 42–45

Keramiken von Stephan Erdös und Ida Erdös-Meisinger, Ausst. Kat., Düsseldorf 1958

Der Keramiker Walter Bosse und seine neuen Schöpfungen, ÖK 8, Nr. 12, 1937, S. 31 (m. Abb.)

Kiessig (1899)
Kiessig, Paul, Die keramische Branche auf der diesjährigen Oster-Vormesse in Leipzig, Thonindustrie-Zeitung 23, 1899, S. 383

Kiessig (1902)
Kiessig, Paul, Leipziger Michaelismesse 1902, DER 2, 1902/03, S. 393

Kiessig (1904)
Kiessig, Paul, Leipziger Ostervormesse (1904), DER 3, 1903/04, S. 760–761, Abb. auf S. 740

Kjellberg
Kjellberg, Pierre, Les Bronzes du XIXe siecle. Dictionnaire des Sculpteurs, Paris 1987, S. 660

Klein (1986)
Klein, Dan u. M. Bishop, Decorative Art. Die Entwickl. des mod. Designs 1880–1980, dt. Ausg., Oxford 1988, S. 145, 152

Klein u.a. (1986)
Klein, Dan N. McClelland u. M. Haslam, „In the deco style", New York 1986, 288 S.

Klein u.a. (1991)
Klein, Dan N. McClelland u. M. Haslam, Art Déco, dt. Ausg., dt. Ausg., Stuttgart 1991, 288 S.

Kleindel
Kleindel, Walter, Das große Buch der Österreicher, Wien 1987, S. 149, 524

Klinge
Klinge, Ekkart, Deutsche Keramik des 20. Jahrhunderts, Bd. 1, Düsseldorf 1975, 329 S.

Kluge, Liane, Nicht wieder Modelle für Porzellanfiguren (Klari Herczeg), Sonntag (Berlin) 07.04.1957

Knörlein, Rudolf, Keramische Arbeiten von 1928–1982, Alt Duvenstedt 1982, o. Pag. (m. Abb.)

Knowles
Knowles, Eric, 100 years of Decorative Arts, London 1993, 254 S.

Koch, Alexander (Hrsg.), 1. Internationale Ausstellung für moderne decorative Kunst in Turin 1902, Darmstadt, Leipzig 1902

Koegler
Koegler, Horst, Friedrichs Ballettlexikon, Velber 1972

Könecke (2000)
Könecke, Gerald, Goldscheider West-Germany. Figuren u. Wandmasken 1953–1960, Göttingen 2000

Könecke (2002)
Könecke, Gerald, Goldscheider West-Germany. Figuren u. Wandmasken 1953–1960, 2. geänd. Aufl., Göttingen 2002

Körting
Körting, W. (Hrsg.), GLAPO. Nachweis f. Glas, Porzellan, Keramik Tafelgeräte, 4. Aufl, Bamberg 1954

Körting 1
Körting, W. (Hrsg.), Glas-Porzellan-Keramik-Waren, Tafelgeräte, 3. Aufl., Lüneburg o.J (um 1956)

Koller
Koller, Gabriele, Die Kunstgewerbeschule des K.K. Österreichischen Museums für Kunst und Industrie, Wien, Dissertation, Wien 1983

Konečný
Konečný, Milan, Česká keramika (1748–1948), Prag 2002

Kovel (1986)
Kovel, Ralph/Terry, Kovels new Dictionary of Marks, New York 1986

Kovel (1999)
Kovel, Ralph/Terry, Kovels Price List, New York 1999

Krause
Krause, Uwe, Die Liste, Dortmund 1996

DKT
Der künstlerische Tanz. Eckstein-Halpaus Zigarettenbilderalbum, Dresden um 1934/35

KPJb
Kunstpreis-Jahrbuch, München 1988 ff.

Kurinsky, Samuel, Gold and Silver Smithing; A Judaic Tradition. Part II – Europe and America. Hebrew History Federation, Factpaper 17-II, www.hebrewhistory.info

Kuzmany, Karl M., Wiener Keramik, Die Kunst 22, 1910, S. 45–47 (m. Abb.)

L., A. S., Vienna (Prof. Eugen Steinhof), Studio 92, 1926, S. 438–439 (m. Abb.)

L., A. S., Vienna (Hertha Bucher, Dina Kuhn), Studio 95, 1928; S. 367 (m. Abb.)

Labek, Alfred, Zur Ausstellung „Das befreite Handwerk" (1934), ÖK 5, Nr. 11, 1934, S. 20 (m. Abb.)

Lambert
Lambert, Th. (Hrsg.), Die moderne Dekoration auf der Pariser Weltausstellung 1900, Stuttgart 1900, Taf. 17, 18

Lang, Gordon, Pottery & Porcelain Marks, London 1995

Lehmann, Gustav, Neuheiten, KuKGW 9, 1928/29, S. 125–127 (m. Abb.)

Lehner
Lehner, Lois, Lehner's Encyclopedia of U.S. Marks on Pottery, Porcelain and Clay, Paduca 1988

Leisching
Leisching, Eduard, Die Winterausstellung des k. k. Österreichischen Museums (1903/04), KuKHW 7, 1904, S. 26 m. Abb., 47 m. Abb., 50

Lennep, Jacques van, Catalogue de la Sculpture Artistes nés entre 1750 et 1882, Brüssel 1992

Leopold
Leopold, Silke und Robert Maschka, Who's who in der Oper, Kassel u.a. 1997

Leuchs, siehe Zeitschriften

Kaisersl. 1987
Loth, Heidel (Bearb.), Jugendstil. Kunsthandwerk u. Bilder. Eine private Sammlung, Ausst. Kat., Kaiserslautern 1987, S. 69

Ludwigg, Heinz (Hrsg.), 40 (Vierzig) Jahre Wintergarten. Festschrift zur Wiedereröffnung, Berlin 1928

M., W., Buch und Raum der Gegenwart, Ausstellung der Genossenschaft der Bildenden Künstler Wiens, DKuD 67, 1930/31, S. 67–71 (m. Abb.)

Mackay
Mackay, James, The Dictionary of Western Sculptors in Bronze, Woodbridge 1977

Makus (1998)
Makus, Horst, 50er Jahre Keramik. Der Alltag der Moderne, Stuttgart 1998

Makus (2000)
Makus, Horst, 50er Jahre Wandmasken, Schönheit und Exotik (Deutschland u. Österreich), Stuttgart 2000

Malkowsky, Georg, Der Bildhauer H. St. Lerche, Die Pariser Weltausstellung in Wort und Bild, Berlin 1900, S. 194–197 (m. Abb.)

Martin
Martin, J., Nos Peintres et Sculpteurs, Graveurs, Dessinateurs, Paris 1897

Matschiner (2006)
Matschiner, Uta M., Nackt – Ästhetik der Nacktheit in den 50er Jahren, Linz 2006

Matschiner (2006a)
Matschiner, Uta M., Nackt – Ästhetik der Exoten in den 50er Jahren, Linz 2006

Matschiner (2006b)
Matschiner, Uta M., Prischl Kunstkeramik Wien, Das Lebenswerk von Adolf Prischl (1912–70), Linz 2006

Matschiner (2006c)
Matschiner, Uta M., Prischl Wien, Sammler-Journal, 2006, Nr. 12, S. 44–50 (m. Abb.)

Mayer
Mayer, E. (Hrsg.), International Auction Records, Vol. 22–36, Lausanne 1987–1997

Mayr-Oehring, Erika u. Erika Doppler, Orientalische Reise, Malerei und Exotik im späten 19. Jahrhundert, Wien 2003, S. 252–253 (m. Abb.)

McClinton (1972)
McClinton, Katherine M., Art deco. A guide for collectors, New York 1972

McClinton (1986)
McClinton, Katherine M., Art deco. A guide for collectors, 2. Auflage, New York 1986

McCready, Karen, Art deco and modernist ceramics, London 1995, S. 150, 151

McGlauflin, Alice C. (Hrsg.), Who is who in American Art, Vol. 1–2 (1935/36–1937/38), Washington 1935, Vol. 1 S. 301, Vol. 2 S. 375

Meisenbach, J.A., Putten mit Tieren, Keramos 2, 1923, S. 305–309 m. Abb.

Messerli Bolliger, Barbara, Keramik in der Schweiz. Von den Anfängen bis heute, Zürich 1993

Metten, E. (Hrsg.), Marlene Dietrich, die Geschichte einer Karriere, Wien 1947

Meyer
Meyer, Konrad, Arbeitsstoff und Arbeitswert. Eine Betracht. über Wiener Fayenceplastiken, KuKGW 5, 1925 SH. 3, S. 8–12 (m. Abb.)

Midant, Jean-Paul, L'Art Nouveau. Jugendstil in Frankreich, Köln, Eltville 1999

Miller
Miller, Judith/Martin, Art Déco. Ein Handbuch f. Sammler, Rastatt 1992, S. 94–95

Miller's A
Millers, Miller's Antiques Price Guide, Vol. 13, Augsburg 1992

Miller's A&A
Millers, Miller's Art Nouveau & Art Deco Buyers Guide, London 1995

Miller's C
Millers, Miller's Collectables Price Guide, Vol. 3, London 1991/92

Minkus, Fritz, Die Erste Internationale Ausstellung für moderne dekorative Kunst in Turin (1902), KuKHW 5, 1902, S. 27, 402–406

Müller, Renate, Jugendstil. Möbel, Glas, Keramik..., München 1999, S. 117, 155

Myott, Anne u. Philip Pollitt, The Mystery of Myott. A history of the Myott Pottery, Stoke on Trent 1891–1991, Stafford 2003

Nagel, Gert W., Modell 4458 Petri, TuS 1995, Nr. 185, S. 6

Nagy, L. v., Die Frau im Pyjama, Uhu 4, 1925/26, S. 51 (m. Abb.)

Neuwirth (1973)
Neuwirth, Waltraud, Die Wiener Manufaktur Friedrich Goldscheider, Keramos 59, 1973, S. 54–63 + Markentaf.

Neuwirth (1974)
Neuwirth, Waltraud, Wiener Keramik. Historismus-Jugendstil-Art Déco, Braunschweig 1974, S. 25–77, 472–473

Neuwirth (1974a)
Neuwirth, Waltraud, Österreichische Keramik des Jugendstils, München 1974, S. 45–52, 80, 156–159, 192, 291, 320–337

Neuwirth (1981)
Neuwirth, Waltraud (Bearb.), Österreichische Keramik 1900–1980 Linz, Graz, Salzburg, Ausst. Kat., Linz u.a. 1981

Neuwirth (1981a)
Neuwirth, Waltraud, Die Keramik der Wiener Werkstätte, Bd. 1: Originalkeramiken 1920–1931, Wien 1981

Niecol
Niecol, Emmy, Rosenthal Kunst- und Zierporzellan 1897–1945, Bd. 1–5, Wolnzach 2001–2003

Nikolaus, Paul, Tänzerinnen, München 1919

N.N., Ausstellung des Kunstgewerbevereins in Wien, Sprechsaal 24, 1891, S. 63

N.N., Die keramische Ausstellung in Dresden 1891, Sprechsaal 24, 1891, S. 816, 839

N.N., Plenar-Versammlung Freitag 5.1.1894 (des Niederösterreich. Gewerbevereins), WNÖG 55, 1894, S. 1

N.N., Protokoll der (1526.) Plenar-Versammlung des Niederösterreich. Gewerbevereins am 5.1.1894, WNÖG 55, 1894, S. 14

N.N., Protokoll der (1528.) Plenar-Versammlung des Niederösterreich. Gewerbevereins am 19.1.1894, WNÖG 55, 1894, S. 38

N.N., o.T. (Ausst. im Gewerbeverein Dezember 1893), WNÖG 55, 1894, S. 74

N.N., Friedrich Goldscheider (Weltausst. Antwerpen 1894), Sprechsaal 27, 1894, S. 1111

N.N., o.T. (Ausst. im Gewerbeverein Dezember 1894), WNÖG 55, 1894, S. 469

N.N., Aus unserem Ausstellungszimmer, WNÖG 56, 1895, S. 67

N.N., Protokoll der ausserordentlichen General-Versammlung des Niederösterreich. Gewerbevereins am 3.1.1896, WNÖG 57, 1896, S. 13

N.N., Plenar-Versammlung Freitag 14.2.1896 (des Niederösterreich. Gewerbevereins), WNÖG 57, 1896, S. 53

N.N., Bericht über die Ausstellung der Abtheilung für Kunstgewerbe (des Niederösterreich. Gewerbevereins), WNÖG 57, 1896, S. 125

N.N., General-Versammlung Freitag 7.5.1897 (des Niederösterreich. Gewerbevereins), WNÖG 58, 1897, S. 204

N.N., General-Bericht des Verwaltungsrathes über das 57. Vereinsjahr des Niederösterreich. Gewerbevereins, WNÖG 58, 1897, S. 220

N.N., Ausserordentliche General-Versammlung Freitag 3.12.1897 (des Niederösterreich. Gewerbevereins), WNÖG 58, 1897, S. 493

N.N., Aus unserem Ausstellungszimmer, WNÖG 59, 1898, S. 1

N.N., Protokoll der Plenar-Versammlung des Niederösterreich. Gewerbevereins am 21.1.1898 (zu Alois GS), WNÖG 59, 1898, S. 26

N.N., Plenar-Versammlung Freitag 9.12.1898 (des Niederösterreich. Gewerbevereins), WNÖG 59, 1898, S. 419

N.N., Bericht über die Ausstellung der Abtheilung für Kleingewerbe (des Niederösterreich. Gewerbevereins), WNÖG 60, 1899, S. 20

N.N., Weihnachtsausstellung (1.12.1899–6.1.1900 des Niederösterr. Gewerbevereins), WNÖG 60, 1899, S. 579

N.N., Bericht über die Weihnachtsausstellung im Niederösterr. Gewerbevereine (1.12.1899–6.11.1900), WNÖG 61, 1900, S. 63

N.N., Bericht über die von der Abtheilung für Kunstgewerbe anlässl. der Generalversammlung veranstaltete Ausst., WNÖG 61, 1900, S. 313

N.N., Patentanmeldung: FG, Wien: Verfahren der Decorierung plastischer Gegenstände, ÖPB 2, 1900, S. 850

N.N., Die Thonwaren-Industrie Deutschlands, Österreichs und der Schweiz auf der Weltausstellung in Paris, KM 1, 1901, S. 92–93

N.N., Patenterteilung: FG, Wien: Verfahren der Decorierung plastischer Gegenstände (Patent Nr. 3855), ÖPB 3, 1901, S. 243

N.N., „Aufgebot"= Patentanmeldung: Arthur GS, Kunstgewerbetreibender Paris: Verfahren zur Herstellung einer, durchscheinendes und undurchsichtiges Email ersetzende Masse, ÖPB 3, 1901, S. 796

N.N., Patenterteilung: Arthur GS, Atelier f. Kunstwerke Paris: Verfahren zur Herstellung einer, durchscheinendes und undurchsichtiges Email ersetzende Masse (Patent Nr. 7052), ÖPB 4, 1902, S. 187

N.N., Unsere keramische Ausstellung (= Reichenberger keramische Ausst. 1902), Mitt. d. Nordböhm. GWMus. 20, 1902, S. 85–88

N.N., Von der ersten Internationalen Ausstellung für dekorative Kunst in Turin 1902, KM 2, 1902, S. 162–163

N.N., Illustrierte Rundschau, Terrakotten von Fr. Goldscheider, Wien, VKMH 17, 1902/03, Nr. 6, S. 715–718 (m. Abb.)

N.N., Plenar-Versammlung Freitag 23.1.1903 (des Niederösterreich. Gewerbevereins), WNÖG 64, Nr. 4, 1903, Titelseite

N.N., Patent-Erlöschung durch Zeitablauf: Arthur GS in Paris: Verfahren zur Herstellung einer, durchscheinendes und undurchsichtiges Email ersetzende Masse (Patent Nr. 7052), ÖPB 5, 1903, S. 343

N.N., Erste Kärntnerische Kunstgewerbliche Ausstellung in Klagenfurt 1903, KuKHW 6, 1903, S. 526–528 (m. Abb.)

N.N., o. T. (zu einer Lampe von Maurice Bouval), Le Style Moderne, L' Art Appliqué, 1903/04, Taf. 5 (m. Abb.)

N.N., Figürliche Terrakotten, KM 4, 1904, S. 110–112 (m. Abb.)

N.N., Leipziger Messführer, Ostervormesse 7.–12. März 1904, DER 3, 1903/04, S. 740 (m. Abb.)

N.N., „Aufgebot"= Patentanmeldung: FG, Firma in Wien: Mattlack, einem bekannten weingeisthältigen Lack wird Stärke beigemischt, ÖPB 6, 1904, S. 197

N.N., Patenterteilung: FG, Wien: Mattlack (Patent Nr. 16882), ÖPB 6, 1904, S. 499

N.N., Figürliche Terrakotten, KM 4, 1904, S. 136–140 (m. Abb.)

N.N., Leipziger Ostermesse 1904, KeramRS 12, 1904, S. 378, 379

N.N., Leipziger Michaelismesse 1904, KeramRS 12, 1904, S. 945

N.N., Neuheiten in Terrakotta- und Tonfiguren u.s.w., KeramRS 12, 1904, S. 642

N.N., Leipziger Ostervormesse (1905), DER 4, 1904/05, S. 746

N.N., o. T. (zur Ausstellung bei d. Fa. Heckert in Halle a. S.), KM 5, 1905, S. 25 (m. Abb.)

N.N., „Aufgebot"= Patentanmeldung: FG, in Wien: Mosaikimitation und Verfahren zur Herstellung derselben, ÖPB 7, 1905, S. 59

N.N., Patenterteilung: FG, Wien: Mosaikimitation und Verfahren zur Herstellung derselben (Patent Nr. 20387), ÖPB 7, 1905, S. 335

N.N., Michaelismesse 1905, KeramRS 13, 1905, S. 915–917

N.N., Österreichische Kunsttonwaren, KeramRS 14, 1906, S. 1253–1254

N.N., Patent-Erlöschung durch Zeitablauf: FG in Wien: Verfahren der Dekorierung plastischer Gegenstände (Patent Nr. 3855), ÖPB 8, 1906, S. 346

N.N., Patent-Erlöschung durch Zeitablauf: FG in Wien: Mosaikimitation und Verfahren zur Herstellung derselben (Patent Nr. 20387), ÖPB 8, 1906, S. 346

N.N., Patent-Erlöschung durch Zeitablauf: FG in Wien: Mattlack (Patent Nr. 16882), ÖPB 9, 1907, S. 681

N.N., Aus Berliner Musterlagern (1907), Sprechsaal 40, 1907, S. 684

N.N., Aus Berliner Musterlagern 1909, Sprechsaal 42, 1909, S. 308

N.N., Friedrich Goldscheider, Kunstgewerbliche Etablissements, Kunst-Terrakotta-Fabrik, Atelier künstlerischer Fayencen, Jahrbuch der Gesellschaft österreichischer Architekten, 1909, S. 73, 1 Bl. o. Pag. (m. Abb.)

N.N., o. T. (zur Frühjahrsmesse Lpz. 1909; Erfindung: „Pierre de Vienne" Stein(guss)masse), KeramRS 17, 1909, S. 177

N.N., Moderne Plastik (Camillo GS, Messe Lpz. 1909), DER 9, 1909/10, S. 596 (m. Abb.)

N.N., Leipziger Ostervormesse 1910, kurzweg Frühjahrsmesse genannt, DER 9, 1909/10, S. 1030

N.N., Verwaltungsrat (zu Walter Goldscheider), WNÖG 71, 1910, S. 34

N.N., o. T. (zu Walter Goldscheider), KeramRS 18, 1910, S. 291, 441

N.N., „Aufgebot"= Patentanmeldung: FG, Erste Wiener Terrakottafabrik..., Verfahren zur Dekorierung keramischer Gegenstände durch Tränken oder Durchfärben mit Metallsalzlösungen, ÖPB 13, 1911, S. 91

N.N., Patenterteilung: FG, Erste Wiener Terrakottafabrik...: Verfahren zur Dekorierung keramischer Gegenstände durch Tränken oder Durchfärben mit Metallsalzlösungen (Patent Nr. 49972), ÖPB 13, 1911, S. 885

N.N., Kunstkeramik (FG), DER 11, 1911/12, S. 566

N.N., Austrian architecture and decoration, StudioYB, 1911, S. 250 (m. Abb.)

N.N., Illustrierte Rundschau über die Leipziger Ostervormesse 1912, DER 12, 1912/13, S. 112

N.N., Die Leipziger Herbstmesse 1912, DER 12, 1912/13, S. 674

N.N., „Aufgebot"= Patentanmeldung: FG, Erste Wiener Terrakottafabrik..., Verfahren zur Herstellung kraquelierter Verzierungen auf Gegenständen, ÖPB 15, 1913, S. 225

N.N., Zurückziehungen und Versagungen nach erfolgten Aufgeboten: FG, Erste Wiener Terrakottafabrik..., Verfahren zur Herstellung kraquelierter Verzierungen auf Gegenständen (aufgeboten am 1.3.1913), ÖPB 15, 1913, S. 852

N.N., Goldene Medaille auf der Internationalen Baufach-Ausstellung in Leipzig (1913), KeramRS 21, 1913, S. 464

N.N., Keramik und Glas auf der Internationalen Baufach-Ausstellung in Leipzig (1913), KeramRS 21, 1913, S. 473

N.N., Die Leipziger Messe, Ostern 1913 (Marmorwaren, Porzellan, Glas u. Steingut), DER 13, 1913/14, S. 83

N.N., Patent-Erlöschung durch Zeitablauf: FG in Wien: Verfahren zur Dekorierung keramischer Gegenstände durch Tränken oder Durchfärben mit Metallsalzlösungen (Patent Nr. 49972), ÖPB 16, 1914, S. 470

N.N., Austria, StudioYB, 1914, S. 189–193, 197 (jew. m. Abb.)

N.N., o. T. (zu Walter Goldscheider), KeramRS 24, 1916, S. 284

N.N., o. T. (zu Walter Goldscheider), KeramRS 25, 1917, S. 29

N.N., o. T. (zur Produktionspalette), KeramRS 25, 1917, S. 128

N.N., Unsere Toten: Frau Lewit-Goldscheider, KeramRS 26, 1918, S. 216

N.N., (Modell: V. Lexa (d.i. Lesca) BRONZE/Marmorsockel „Fechterin"), MW 2, 1920/21, Nr. 12 S. 12 (m. Abb.)

N.N., Wiener Kleinplastik einst und jetzt, MW 2. 1920/21, Nr. 12 S. 28–29 (m. Abb.)

N.N., o. T. (zur Namensänderung), KeramRS 29, 1921, S. 395

N.N., o. T. (zur Namensänderung), KeramRS 30, 1922, S. 230

N.N., Messräume der feinkeramischen Industrie, Keramos 2, 1923, S. 216–217 m. Abb.

N.N., „Aufgebot"= Patentanmeldung: FG, Wiener Manufaktur FG, Verfahren zur Erzeugung farbiger Muster durch Aufspritzen, ÖPB 25, 1923, S. 382

N.N., Patenterteilung: FG, Wiener Manufaktur FG, Verfahren zur Erzeugung farbiger Muster durch Aufspritzen (Patent Nr. 96834), ÖPB 26, 1924, S. 68

N.N., Österreich in Paris (1925), ÖBWK 1, 1924/25, S. 331, 340, 341 (m. Abb.)

N.N., Ehrenpreis/Hauptpreis (für ein Preisausschreiben der Zs.), Modell 5537 Lorenzl „Mephisto" KuKGW 8, 1927/28, o. Pag.

N.N., Leipziger Messe 1928, KuKGW 8, 1927/28, S. 309 (m. Abb.)

N.N., Farbe und Bewegung. Die Fayencen der Wiener Manufaktur FG Wien-Währing, DB 5, 1928, S. 38 (m. Abb.) und 40

N.N., Berühmte Frauen und moderne Keramik. Sämtl. Modelle aus d. Wiener Manufaktur FG, MW 10, 1928/29, Nr. 6 S. 28–29 (m. Abb.)

N.N., Hunde und Katzen. Zur Porzellantierausstellung Ernst Wahlis, Wien, Kärntnerstrasse, MW 11, 1929/30, Nr. 20 S. 8

N.N., Wiener Keramik. (Marcell GS), WNÖG 90, 1929, S. 686

N.N., Verband der feinkeramischen Betriebe Österreichs im N.-Ö. Gewerbeverein, WNÖG 90, 1929, S. 893

N.N., Die Wiener Plastikerin Grete Neuwalder, DKuD 66, 1930, S. 246–250 (m. Abb.)

N.N., 50 Jahre Wiener Manufaktur Friedrich Goldscheider, Die Schaulade 11, 1935, S. 152–154 (m. Abb.)

N.N., Aus der Wiener Frühjahrsmesse. Die Wiener Edelfirmen I. 50 Jahre Wiener Manufaktur FG, ÖK 6, Nr. 3, 1935, S. 20 (m. Abb.)

N.N., Neue Keramik. (Wiener Herbstmesse 1937), ÖK 7, Nr. 10, 1936, S. 25 (m. Abb.)

N.N., Neue Keramik. (Wiener Manufaktur FG), ÖK 8, Nr. 7, 1937, S. 22 (m. Abb.)

N.N., Goldscheidrovka – Plzeňská Hrnčírna, Nová Doba, 24.12.1939, Pilsen 1939

N.N., Making Goldscheider Ware in America, Ceramic Industry, Aug. 1942, S. 43–44 (m. Abb.)

N.N., The Goldscheiders greet the Canadian Market, Gift Buyer, Dez. 1945, o. Pag.

N.N., Goldscheiders American Venture in 8th Year of Prosperity, Ceramic Industry, Sept. 1947, S. 98–100 u. Umschlag (m. Abb.)

N.N., Goldscheiders open new plant, Ceramic Industry, Juni 1948, S. 55

N.N., Goldscheider Vienna Plant Going Full Tilt Once More, Trenton Evening Times 22.2.1952, o. Pag.

N.N., Goldscheider Inc., Ceramic Industry, März 1952, S. o. Pag. (m. Abb.)

N.N., Figürliche Keramik aus 2. Jahrtausenden (Museum f. KHW Frankfurt), Ausst. Kat., Frankfurt 1963

N.N., Das Preisbild (63) Figürliche Keramik, 20. Jahrhundert, AZ 1979, S. 514 (m. Abb.)

N.N., Bronze-Elfenbein Statuetten der Neuzeit. Künstlerkat. Forts. 2 ff., AZ 1979, S. 483, 501–503, 528–529, 581 jew. (m. Abb.)

N.N., Bronze-Elfenbein Statuetten der Neuzeit. Künstlerkat. Forts. 17 ff., AZ 1980, S. 209, 378, 379, 459, 587 teils (m. Abb.)

N.N., In Bronze lesen (Francesco La Monaca), AZ 1984, S. 296

N.N., Salon-Skulptur, Verlag Goldscheider, Wien, Paris, Leipzig, Florenz..., AZ 1986, Nr. 9, S. 245 (m. Abb.)

N.N., Goldscheider (Ausstellungsbericht Halbturn), TuS 1992, Nr. 154, S. 130

N.N., (Modell: Seger „Schönheit" (= 3697 „Helena/ Helene"), AZ 1992, S. 400 (m. Abb.)

N.N., Modell 7286 Dakon, TuS 1996, Nr. 205, S. 149

N.N., Modell 3170 Rosé, AZ 1996, S. 339 (m. Abb.)

N.N., Modell 631? Meisinger, Sammler-Journal 28, 1999, S. 157

N., R., Keramische Körperkultur. Schöner Schein, Ausst.-Bericht Halbturn, AZ 1992, S. 565 (m. Abb.)

ÖBL
Österreichisches Biographisches Lexikon, Bd. 1 ff., Graz, Köln, Wien 1957 ff.

Officieller Bericht der K.K. Centralcommission für die Weltausstellung in Chicago 1893, Wien 1894, S. 16, 17

Orientirungsschematismus der kön. Kreisstadt Pilsen, Pilsen 1879

Ostini
Ostini, Fritz v., Rudolf Maison, DK 3, 1901, S. 131–141 (m. Abb.)

Panzetta
Panzetta, Alfonso, Dizionario degli scultori italiani dell'ottocento e del primo novecento, Turin 1994, 311 S.

Pazaurek (1904)
Pazaurek, Gustav E., Alte und neue Beleuchtungskörper, Mitt. d. Nordböhm. GWMus. 22, 1904, S. 90–91 (m. Abb.)

Pazaurek (1921)
Pazaurek, Gustav E., Steingut. Formgebung u. Geschichte, Stuttgart 1921, Spalte 53

Pelichet
Pelichet, Edgar, La Céramique Art Déco, Paris 1988, S. 9, 103, 180, 188

Pelichet 1
Pelichet, Edgar, Jugendstil Keramik, Anierès, Genf 1976, S. 181

Pelka (1924)
Pelka, Otto, Keramik der Neuzeit, Leipzig 1924

Pelka (1925)
Pelka, Otto, Deutsche Keramik, Reutlingen 1925

Pendl
Pendl, Erwin, Österreich auf der Weltausstellung Paris 1900, Wien 1900

Peter (1992)
Peter, Frank-Manuel, Das tänzerische Lichtbild. Hugo Erfurth als Dokumentarist des frühen Ausdruckstanzes, Hugo Erfurth, Photograph zwischen Tradition und Moderne, Köln 1992, S. 45–52 (m. Abb.)

Peter (2002)
Peter, Frank-Manuel, Die Sacharoffs. Zwei Tänzer aus dem Umkreis des Blauen Reiters, Köln 2002

Pinhas
Pinhas, Ora, Goldscheider, A Catalogue of Selected Models, Shepton Beauchamps, 2006

Planer
Planer, Franz (Hrsg.), Das Jahrbuch der Wiener Gesellschaft 1929, Wien 1929

Pohl, Walter, Schönheiten aus Porzellan, AZ/Tagblatt, Nov. 1987

Prager, Wolfgang, Goldscheider-Keramik, Weltkunst, 1986, Nr. 5, S. 660–663

Préaud
Préaud, Tamara, Ceramics of the 20th Century, New York 1982

Pudor (1902)
Pudor, Heinrich, Dokumente des modernen Kunstgewerbes, Serie A, Bd. 1, Berlin 1902, S. 13, 68–73, 108–109

Pudor (1902a)
Pudor, Heinrich, Erste Internationale Ausstellung für dekorative Kunst in Turin 1902, KM 2, 1902, S. 119

Pudor (1906)
Pudor, Heinrich, Erziehung zum Kunstgewerbe, Berlin 1906, o. Pag.

Pudor (1907)
Pudor, Heinrich (Red.), Dokumente des modernen Kunstgewerbes. Die Leipziger Messe 1907 Nr. 1, Leipzig 1907, S. 15

Pudor (1908)
Pudor, Heinrich (Red.), Dokumente des modernen Kunstgewerbes. Michaelismesse 1908 Nr. 2, Leipzig 1908, S. 50, 102, 104

R., F., Wiener Kunstkeramik, KM 2, 1902, S. 88–91 (m. 9 Abb.)

R., F., Keramische Wanderungen durch den Kunstsalon von Keller & Reiner in Berlin, KM 3, 1903, S. 125–128 (m. 5 Abb.)

Rambosson, siehe Goldscheider, Arthur

Rauch, Alexander, Goldscheider-Wien, Pan Nr. 9, 1987, S. 74 (m. Abb.)

Richemond (1995)
Richemond, Stéphane, Terre cuites orientalistes de XIXe et Xxe siècles, L'Estampille L'Objet d'Art No. 290, 1995, S. 54–62 (m. Abb.)

Richemond (1996)
Richemond, Stéphane, Terre cuites de Goldscheider, L'Estampille L'Objet d'Art No. 301, 1996, S. 66–73 (m. Abb.)

Richemond (1999)
Richemond, Stéphane, Terres cuites orientalistes et africanistes : 1860–1940, Paris 1999, 238 S.

Rochowanski (1923)
Rochowanski, L. W., Der tanzende Schwerpunkt, Zürich, Leipzig, Wien 1923

Rochowanski (1923a)
Rochowanski, L. W., Wiener Keramik, Leipzig, Wien, 1923

Rochowanski (1925)
Rochowanksi, L. W., Österreich auf der Pariser Ausstellung (1925), DKuD 57, 1925/26, S. 69–76

Rochowanski (1929)
Rochowanski, L. W., Wiener Wohnungskunst und Kunstgewerbe, DKuD 65, 1929/30, S. 418–430 (m. Abb.)

Rochowanski (1930)
Rochowanski, L. W. (Hrsg.), Ein Führer durch das österreichische Kunstgewerbe, Leipzig, Wien 1930, S, 78, 108 (m. Abb.)

Röntgen
Röntgen, Robert E., Marks on German, Bohemian and Austrian Porcelain, Exton, Pa. 1981

Rössler, Arthur, Österreichisches Kunsthandwerk im Wiener Künstlerhaus (1927), ÖBWK 3, 1926/27, S. 74

Ros
Ros, Lilli, Wiener Fayencen und Bronzen. Begleitworte zu den Abb. über neue Werke v. FG, Wien, KuKGW 4, 1923/24, S. 15–19 (m. Abb.)

Rosenzweig
Rosenzweig, Rudolf, Geschichte der Juden in Brennporitschen, in: Gold, Hugo: Die Juden und Judengemeinden Böhmens, Brünn u.a. 1934, 735 S.

Rouger, Gustave, Nos Artistes au Maroc, L'Art et les artistes, N.S. 6, Paris 1922

Rudoe, Judy, Decorative Arts 1859–1950. A Catalogue of the British Museum Collection, überarbeitete Auflage, London 1994

Rutter
Rutter, Josef, Kunst in Österreich, Leoben 1934

Schilling-Schletter, Alexander, Hunde-Plastiken, MW 11, 1929/30, Nr. 6 S. 12–13 (m. Abb.)

Scheffler
Scheffler, Wolfgang, Werke um 1900, Berlin 1966, S. 198, 224

Scheffran, Barbara u.a., Prager Jugendstil, Heidelberg 1992

Schilling, W. u. S. Schneemann, Zerbrechliche Familien. Eine motivgeschichtl. Betrachtung europ. Porzellane..., Hohenberg/Eger 1994, 212 S. (m. Abb.)

Schmid
Schmid, Max, Rudolf Maison, VKMH 1, 1898/99, S. 1–12 (m. Abb.)

Schmidt, A. (1890)
Schmidt, Alexander, Von der Universal-Ausstellung Paris 1889, Sprechsaal 23, 1890, S. 77–78

Schmidt, A. (1894)
Schmidt, Alexander, Eine Meßausstellung (Leipzig Ostermesse 1894), Sprechsaal 27, 1894, S. 435

Schmidt, A. (1897)
Schmidt, Alexander, Nachruf auf Friedrich Goldscheider, Sprechsaal 30, 1897, S. 121

Schmidt, Otto, Wiener Typen und Strassenbilder, Wien 2000, 107 S.

Schnyder
Schnyder, Rudolf, Vier Berner Keramiker. Werner Burri, Benno Geiger..., Bern 1985, S. 32–51 (m. Abb.)

Schreyer-Hartmann
Schreyer-Hartmann, Cherica u. Hans-Hagen (und Johanna) Hottenroth, Walter Bosse. Leben, Kunst und Handwerk 1904–1979, Wien 2000, 168 S. (m. Abb.)

Schulz
Schulz, Arthur, Deutsche Sculpturen der Neuzeit, Ser. 1–3, Berlin, New York 1900

Schwarzburger Werkstätten für Porzellankunst, Hohenberg/Eger 1993, 420 S.

Schweiger (1982)
Schweiger, Werner, Wiener Werkstätte. Kunst u. Handwerk 1903–1932, Wien 1982

Serveau, Erwan, Hommage à Germaine Bouret, La vie du collectionneur, Nr. 445, 2003, S. 1, 13–15 (m. Abb.)

SOTH
Sothebys, Sotheby's Guide. Antiques and their Prices worldwide, Vol. 3–5, Harmondsworth 1988–1990

SOTH
Sothebys, Sotheby's Kunst-Auktionen-Preise, Bd. 1–2, Augsburg 1986–1987

Thiemann
Spielmann, Hans (Hrsg.), Europäische Keramik seit 1950. Sammlung Hans Thiemann, Hamburg 1979

Špirková
Špirková, Věra, Židovská komunita v Plzni, Domažlice 2000

Stehlik's, siehe Zeitschriften

Suhr, Werner, Der künstlerische Tanz, Leipzig 1922

DTW
Die Tanzbühnen der Welt, Eckstein-Halpaus Zigarettenbilderalbum, Dresden, um 1934/35

TB
Thieme, U. u. F. Becker (Hrsg.), Allgemeines Lexikon der bildenden Künstler von der Antike bis zur Gegenwart, Studienausgabe, Bd. 1–37, Leipzig 1999

Thomas, Albert, Petits Bronzes D'Art, L'Art Décoratif 3, 1900/01, S. 181–189 (m. Abb.)

Träger, Gertrud, Das Kunstschaffen der Egerländer im XX. Jahrhundert, Bd. 2, Stuttgart 1984

Trenkwald, H., Keramik, Glas, Textilien, MW 2, 1920/21, Nr. 12, S. 17–18 (m. Abb.)

Triennale des norddeutschen Kunsthandwerks 1991, Ausst. Kat., Schleswig 1991, S. 115

TuSA
Trödler & Sammeln Auktionspreise, Bd. 1 ff., Reichertshausen 1994/95 ff.

Venzmer
Venzmer, Wolfgang, Jugendstil (Bestandskat. Landesmuseum Mainz), Mainz 1990, S. 341

Verzeichnis der Mitglieder des Niederösterreichischen Gewerbevereines 1930, Wien 1930, S. 73, 196, 198–199, 201, 213, 235

Villa, Maria Cristina, La Saga die Goldscheider: una Passione di Famiglia, Ceramic Antica, 2005, Nr. 11, S. 39–55 (m. Abb.)

Villany, Adorée-Via, Tanzreform und Pseudo-Moral. Krit.-satyr. Gedanken aus meinem Bühnen- und Privatleben, Paris 1912

Vollmer
Vollmer, Hans (Hrsg.), Allgemeines Lexikon der bildenden Künstler des XX. Jahrhunderts, Studienausgabe, Bd. 1–6, Leipzig 1999

Währing
Arbeitsgem. Währinger Heimatkunde (Hrsg.), Währing. Ein Heimatbuch des 18. Wiener Gemeindebezirkes, T. 3, Wien 1925, S. 707

Wallner, Susanne (Hrsg.), Schwarzburger Werkstätten für Porzellankunst, Hohenberg/Eger 1993

GK
Walther, Susanne (Hrsg.), Goldscheider Keramik Wien. Historismus-Jugendstil-Art déco, Ausst.-Kat., Wien 1985

Walzer/Templ
Walzer, Tina u. Stephan Templ, Unser Wien. „Arisierung" auf österreichisch, Berlin 2001

Ware, George, Deutsches und österreichisches Porzellan, dt. Ausg. (maschinenschr. Vervielf.), Frankfurt a. M. 1950, o. Pag.

Ware, George, German and Austrian Porcelain, Frankfurt a. M. 1964

Weidmann, Dieter: Goldscheider, TuS Nr. Juli, 2006, S. 36–44 (m. Abb.)

Weiß, Gustav, Ullstein Porzellanbuch, Berlin 1994

Wenzel
Wenzel, Alfred, Neue Terrakotten von Vally Wieselthier, DKuD 67, 1930/31, S. 390–398 (m. Abb.)

ADGS
Wenzl, Monika, Art Déco Keramik Goldscheider, Ausst.-Kat. 1995, hrsg. von d. Österr. Postsparkasse unter Mitarb. v. Waltraud Neuwirth, Wien 1995, o. Pag.

Westerman
Westerman, Gerhart von u. Karl Schumann, Knaurs Operführer, München, Zürich 1975

Witek
Witek, Hans u. Hans Safrian, Und keiner war dabei. Dokumente des alltägl. Antisemitismus in Wien 1938, Wien 1988, S. 145

Witkiewicz
Witkiewicz, Jan S., Ta' nczace figurki. Z kolekcji Alaina Bernarda, Warschau 2004

Witzmann, Reingard, Wiener Typen. Histor. Alltagsfotos aus d. 19. Jh., Dortmund 1982

Zühlsdorff
Zühlsdorff, Dieter, Marken-Lexikon. Porzellan und Keramikreport 1885–1935, Stuttgart 1988

Ausstellungen I Expositions

Barc.
BARCELONA 1929: Exposición internatcional de Barcelona 1929, Ausst. Kat., Barcelona 1929, S. 156

CHICAGO 1893: Amtlicher Special-Katalog der österr. Abth. Auf der Weltausstellung Chicago 1893, Wien 1893, S. 54, 70

HANNOVER 1951: Deutsche Industriemesse. Amtl. Messekatalog 1951, Hannover 1951

HANNOVER 1952: Deutsche Industriemesse. Amtl. Messekatalog 1952, Hannover 1951

HANNOVER 1954: Deutsche Industriemesse. Amtl. Messekatalog 1954, Hannover 1954

HANNOVER 1955: Deutsche Industriemesse. Amtl. Messekatalog 1955, Hannover 1955

HANNOVER 1956: Deutsche Industriemesse. Amtl. Messekatalog 1956, Hannover 1956

HANNOVER 1957: Deutsche Industriemesse. Amtl. Messekatalog 1957, Hannover 1957

LMA (1897)
LEIPZIG 1897: Leipziger Mess-Adressbuch. Oster-Vormesse 1897, Leipzig 1897, S. 21, 96, 255, 292

LMA (1898)
LEIPZIG 1898: Leipziger Mess-Adressbuch. Oster-Vormesse und Oster-Hauptmesse 1898, Leipzig 1898, S. 29

LMA (1899)
LEIPZIG 1899: Leipziger Mess-Adressbuch. Oster-Vormesse und Oster-Hauptmesse 1899, Leipzig 1899, S. 34

LMA (1900)
LEIPZIG 1900: Offizielles Leipziger Mess-Adressbuch. Oster-Vormesse und Oster-Hauptmesse 1900, Bd. 1, Leipzig 1900, S. 41, 198, 493

LMA (1903)
LEIPZIG 1903: Offizielles Leipziger Mess-Adressbuch. Oster-Vormesse und Oster-Hauptmesse 1903, Leipzig 1903, S. 38

LMA (1904)
LEIPZIG 1904: Offizielles Leipziger Mess-Adressbuch. Oster-Vormesse und Oster-Hauptmesse 1904, Leipzig 1904, S. 43, 461

LMA (1910)
LEIPZIG 1910: Offizielles Leipziger Mess-Adressbuch. Oster-Vormesse 1910, Leipzig 1910, S. 55, 56, 296, 570

Guide LF
LEIPZIG 1912: Guide to the Leipzig Fair. Directory of Exhibitors..., Leipzig 1912, S. 66, 148

LMA (1917)
LEIPZIG 1917: Herbst Mustermesse Leipzig. Amtl. Aussteller Verz. 1917, Leipzig 1917, S. 48

LMA (1918)
LEIPZIG 1918: Herbst Mustermesse Leipzig. Mess-Adressbuch, Amtl. Aussteller Verz. 1918, Leipzig 1918, S. 48

LMA (1924)
LEIPZIG 1924: Herbstmesse, Leipzig 1924, S. 480

LMA (1926)
LEIPZIG 1926: Frühjahrsmesse 1926. Messeweiser für Kunstgewerbe u. Wertarbeit, KuKGW 6, 1925/26, o. Pag.

Europ.KGW
LEIPZIG 1927: Europäisches Kunstgewerbe 1927, Ausst. Kat., Leipzig 1927

LMA (1937)
LEIPZIG 1937: Amtliches Leipziger Messadressbuch. Frühjahr 1937, Leipzig 1937, S. 20

LONDON 1902: Catalogue of the Exhibition of Austrian fine art and decor. Furnishing, London 1902, S. 27

AiL 1934
LONDON 1934: Austria in London. Austrian national Exhib. of Industry-Art-Travel-Sport, Dorland Hall 1934, Ausst. Kat., London 1934, S. 67, 70

MÜNCHEN 1922: Deutsche Gewerbeschau München 1922. München u.a. 1922, S. 39, 277

MÜNCHEN 1922: Deutsche Gewerbeschau München 1922. Amtlicher Führer, München 1922

Int. Exh. CA
NEW YORK u.a. 1928/29: International Exhibition of Ceramic Art. Catalogue 1928/29, New York u.a. 1928

PARIS 1900: Exposition Internationale Universelle de 1900. Catalogue General Officiel. Bd. 14. Groupe XII, Paris 1900, S. 52

KÖAbt.
PARIS 1900: Weltausstellung Paris 1900. Kat. d. Österr. Abtheilung Nr. 8, Gruppe XII u. XV, Wien 1900, S. 188

PARIS 1925: Exposition Internationale des Arts décoratifs et industriels modernes Paris 1925, Ausst. Kat., Paris 1925, S. 18, 37, Taf. 12, 34

AutrP 1925
PARIS 1925: L'Autriche à l'Exposition Internationale des Arts Décoratifs et Industriels Modernes 1925. Guide, Wien 1925, S. 39–49, Abb. o. Pag.

ÖBWK
PARIS 1925: Künstler und Werkstätten, die an der Österreichischen Ausstellung in Paris beteiligt sind, ÖBWK 1, 1924/25, Nr. Sept. S. XV

TURIN 1902: 1ère Exposition Internationale d'Art décoratif moderne. Cat. de la section Autrichienne, Turin 1902, S. 18

TURIN 1902: vgl. Koch, Alexander

WIEN 1873: Officieller General-Catalog, Weltausstellung 1873 in Wien, Wien 1873

WIEN 1888: Katalog der Jubiläumsgewerbeausstellung (NÖ Gew. Verein), Wien 1888

ÖMKI 1898/99
WIEN 1898/99: K.u.K Österreichisches Museum für Kunst und Industrie, Winterausstellung 1898/99, Ausst. Kat., Wien 1898, Kat. Nr. 134–158

ÖMKI 1899/1900
WIEN 1899/1900: K.u.K Österreichisches Museum für Kunst und Industrie, Winterausstellung 1899/1900, Ausst. Kat., Wien 1899, Kat. Nr. 189–219

ÖMKI 1900/01
WIEN 1900/01: K.u.K Österreichisches Museum für Kunst und Industrie, Winterausstellung 1900/01, Ausst. Kat., Wien 1900, Kat. Nr. 180–210

ÖMKI 1901/02
WIEN 1901/02: K.u.K Österreichisches Museum für Kunst und Industrie, Winterausstellung 1901/02, Ausst. Kat., Wien 1901, Kat. Nr. 221–247

ÖMKI 1902/03
WIEN 1902/03: K.u.K Österreichisches Museum für Kunst und Industrie, Winterausstellung 1902/03, Ausst. Kat., Wien 1902, Kat. Nr. 345–381

ÖMKI 1903/04
WIEN 1903/04: K.u.K Österreichisches Museum für Kunst und Industrie, Winterausstellung 1903/04, Ausst. Kat., Wien 1903, Kat. Nr. 305–325

ÖMKI 1906/07
WIEN 1906/07: K.u.K Österreichisches Museum für Kunst und Industrie, Winterausstellung 1906/07, Ausst. Kat., Wien 1906, Kat. Nr. 218–245

ÖMKI 1909/10
WIEN 1909/10: K.u.K Österreichisches Museum für Kunst und Industrie, Ausst. Österr. KGW 1909/10, Ausst. Kat., Wien 1909, S. 75 (Interieurs Nr. 9, 12, 22)

ÖMKI 1910/11
WIEN 1910/11: K.u.K Österreichisches Museum für Kunst und Industrie, Ausst. Österr. KGW 1910/11, Ausst. Kat., Wien 1910, Kat.Nr. 527–538, 2089, 2090

ÖMKI 1911/12
WIEN 1911/12: K.u.K Österreichisches Museum für Kunst und Industrie, Ausst. Österr. KGW 1911/12, Ausst. Kat., Wien 1911, Kat.Nr. 601–616

ÖMKI 1912
WIEN 1912: K.u.K Österreichisches Museum für Kunst und Industrie, Frühjahrsausst. 1912, Ausst. Kat., Wien 1912, Kat.Nr. 97–132

ÖMKI 1913/14
WIEN 1913/14: K.u.K Österreichisches Museum für Kunst und Industrie, Ausst. Österr. KGW 1913/14, Ausst. Kat., Wien 1913, Kat.Nr. 284–291

KSW (1920)
Wien 1920: Kunstschau Wien 1920, Katalog, Wien 1920

HmesseW (1921)
WIEN 1921: Herbstmesse, Bd. 1 u. 2, Ausst. Kat., Wien 1921

FmesseW (1923)
WIEN 1923: Frühjahrsmesse, Ausst. Kat., Wien 1923

WIEN 1924: Wiener Kunstgewerbe Verein 1924. Jubiläumsausstellung 1884–1924, Ausst. Kat., Wien 1924, S. 32

FmesseW (1926)
WIEN 1926: Frühjahrsmesse, Ausst. Kat., Wien 1926

WIK (1926)
WIEN 1926: Weihnachtsschau im Künstlerhaus. Kunst u. KHW 1926, Ausst. Kat., Wien 1926, S. 56, 61

WIK (1928)
WIEN 1928: Weihnachtsschau im Künstlerhaus. Kunst u. KHW 1928, Ausst. Kat., Wien 1928, S. 36

BiR 1929
WIEN 1929: Das Bild im Raum. Führer durch d. Ausst. d. Wiener Frauenkunst 1929, Wien 1929, S. 36, 37, 50, 78 (Reklame Marcell)

WIK (1929/30)
WIEN 1929/30: Weihnachtsschau im Künstlerhaus 1929/30, Ausst. Kat., Wien 1929, S. 31, 35, 36

FmesseW (1930)
WIEN 1930: Frühjahrsmesse, Ausst. Kat., Wien 1930

HmesseW (1930)
WIEN 1930: Herbstmesse, Ausst. Kat., Bd. 1 u. 2, Wien 1930

BuR 1930
WIEN 1930: Buch und Raum der Gegenwart 1930, Wien 1930, S. 46–48

BuP 1931
WIEN 1931: Blume und Plastik 1931, Ausst. Kat., Wien 1931, S. 7, 25, 26, 28, 41

WIEN 1934: 50 Jahre Wiener Kunstgewerbe-Verein – Jubiläumsausst. 1934, Ausst. Kat., Wien 1934, S. 11

WIEN 1934: Das befreite Handwerk. Geschmack u. Wohnkultur 1934, KGW Ausstellung, Ausst. Kat., Wien 1934, S. 26, 30

FmesseW (1938)
WIEN 1938: Frühjahrsmesse, Ausst. Kat., Wien 1938

HmesseW (1938)
WIEN 1938: Herbstmesse, Ausst. Kat., Bd. 1 u. 2, Wien 1938

WIM (1951)
WIEN 1951: 53. Wiener Internationale Messe, Ausst. Kat., Wien 1951

WIM (1952)
WIEN 1952: 55. Wiener Internationale Messe, Ausst. Kat., Wien 1952

WIM (1952a)
WIEN 1952: 56. Wiener Internationale Messe, Ausst. Kat., Wien 1952

WIM (1953)
WIEN 1953: 57. Wiener Internationale Messe, Ausst. Kat., Wien 1953

Reklame | Advertising

Annoncen, Muster-Register (Modelle mit Musterschutz)

Friedrich Goldscheider: Terracottafabrik in Wien, Sprechsaal 20, 1887, S. 608

Eduard Goldscheider: Reklame „Specialitäten: hochfeuerfestes Sanitäts-Kochgeschirr“, Sprechsaal 21, 1888, S. 134

Friedrich Goldscheider: Erste Wiener Terracottafabrik, Sprechsaal 22, 1889, S. 733

Friedrich Goldscheider: Erste Wiener Terracottafabrik, Sprechsaal 23, 1890, S. 261

Friedrich Goldscheider: Erste Wiener Kunst-Terracottafabrik, Sprechsaal 23, 1890, S. 753

Friedrich Goldscheider: Erste Wiener Kunst-Terracotta-Fabrik und Atelier für Kunstgewerbe, Sprechsaal 24, 1891, S. 247

Friedrich Goldscheider: Erste Wiener Kunst-Terracotta-Fabrik und kunstgewerbliche Ateliers, Sprechsaal 24, 1891, S. 765

Eduard Goldscheider: Stellenanzeige „sucht tüchtigen Formengießer“, Sprechsaal 24, 1891, S. 867

Friedrich Goldscheider: Muster-Register (Juli 1893), KeramRS 1, 1893, S. 321

Friedrich Goldscheider: Muster-Register (Sept. 1893), KeramRS 1, 1893, S. 531

Friedrich Goldscheider: Kunstgewerbliche Etablissements Wien-Paris-Leipzig-Berlin, Sprechsaal 27, 1894, S. 797

Friedrich Goldscheider: Muster-Register (Febr. 1895), KeramRS 3, 1895, S. 153

Friedrich Goldscheider: Muster-Register (April 1895), KeramRS 3, 1895, S. 345

Friedrich Goldscheider: Muster-Register (Juni 1895), KeramRS 3, 1895, S. 545

Friedrich Goldscheider: Muster-Register (Sept. 1895), KeramRS 3, 1895, S. 803, 805

Friedrich Goldscheider: Muster-Register (Sept. 1896), KeramRS 4, 1896, S. 863

Friedrich Goldscheider: Muster-Register (April 1897), KeramRS 5, 1897, S. 347

Friedrich Goldscheider: Muster-Register (April 1898), KeramRS 6, 1898, S. 353

Friedrich Goldscheider: Muster-Register (Mai 1898), KeramRS 6, 1898, S. 421

Friedrich Goldscheider: Muster-Register (Aug. 1898), KeramRS 6, 1898, S. 757

Friedrich Goldscheider: Muster-Register (Juli 1899), KeramRS 7, 1899, S. 881

Friedrich Goldscheider: Kunst-Terracotta-Fabrik, KeramRS 10, 1902, S. 137

J. Pilar: Kunstterracotta-Fabrik Wien-Oberlaa, KeramRS 11, 1903, S. 200

Frédéric Goldscheider: Frédéric Goldscheider Editeur d'Art, Paris, Art et Décoration 13, 1903, S. 2 (m. Abb.)

Friedrich Goldscheider: Kunst-Terrakotta-Fabrik in Wien, KeramRS 12, 1904, S. 158, 182, 210, 236

Friedrich Goldscheider: Kunst-Terrakotta-Fabrik, Atelier künstlerischer Fayencen, KeramRS 12, 1904, S. 798, 825, 853

Friedrich Goldscheider: Kunst-Terrakotta-Fabrik, Atelier künstlerischer Fayencen, KeramRS 13, 1905, S. 157, 185, 230

Friedrich Goldscheider: Kunst-Terrakotta-Fabrik, Atelier künstlerischer Fayencen, KeramRS 13, 1905, S. 793, 820

Friedrich Goldscheider: Fabrik künstlerischer Terrakotten und Fayencen, KeramRS 14, 1906, S. 180, 208, 762, 792, 820

Friedrich Goldscheider: Fabrik künstlerischer Terrakotten und Fayencen, KeramRS 15, 1907, S. 146, 208

Friedrich Goldscheider: Kunstterrakotta-Fabrik, Ateliers künstlerischer Fayencen, Leuchs Adressbuch aller Länder der Erde der Kaufleute, Fabrikanten..., Bd. 14, Nürnberg 1908, S. 1308

Friedrich Goldscheider: Erste Wiener Terrakotta-Fabrik und Atelier für künstlerische Fayencen, KeramRS 17, 1909, S. 150

Camillo Goldscheider: Camillo Goldscheider, FG Sohn, Dekorative Plastik, DER 9, 1909/10, S. 628, 902, 990 (m. Abb.)

Camillo Goldscheider: Camillo Goldscheider, FG Sohn, Dekorative Plastik, DER 9, 1909/10, Nr. 19 (Umschlag, m. Abb.)

Friedrich Goldscheider: Kunstsalon FG Leipzig, Heimatkalender für Thüringen und das Osterland 1914, Dresden 1913, S. 216 (m. Abb.)

Friedrich Goldscheider: Kunststeinwerkstätten FG Leipzig, Heimatkalender für Thüringen und das Osterland 1914, Dresden 1913, S. 217 (m. Abb.)

Friedrich Goldscheider: Wiener Manufaktur FG, MW 2, 1920/21, Nr. 12 S. 2 (m. Abb.)

(Modell: V. Lexa (d.i. Lesca) BRONZE/Marmorsockel „Fechterin"), MW 2, 1920/21, Nr. 12 S. 12 (m. Abb.)

Friedrich Goldscheider: Wiener Manufaktur FG (Wien–Leipzig–Paris), MW 3, 1921/22, Nr. 1 S. 39 (m. Abb.)

Friedrich Goldscheider: Wiener Manufaktur FG (Wien–Leipzig–Paris), MW 3, 1921/22, Nr. 3 S. 39 (m. Abb.)

Friedrich Goldscheider: Wiener Manufaktur FG (Wien–Leipzig–Paris), MW 3, 1921/22, Nr. 4 S. 35 (m. Abb.)

Friedrich Goldscheider: Wiener Manufaktur FG (Wien–Leipzig–Paris), MW 3, 1921/22, Nr. 5 S. 37 (m. Abb.)

Friedrich Goldscheider: Wiener Manufaktur FG (Wien–Leipzig–Paris), MW 3, 1921/22, Nr. 6 S. 39 (m. Abb.)

Friedrich Goldscheider: Wiener Manufaktur FG, Deutsche Gewerbeschau München 1922, amtl. Führer, Anzeigen-Anhang

Walter & Marcell Goldscheider: GWM Wien, DD 50, 1922/23, Nr. 4 S. 32 (m. Abb.)

Walter & Marcell Goldscheider: GWM Wien–Leipzig (Wiener Manufaktur Friedrich Goldscheider), DD 50, 1922/23, Nr. 5 S. 26 (m. Abb.)

Walter & Marcell Goldscheider: GWM Wien–Leipzig (Wiener Manufaktur Friedrich Goldscheider), DD 50, 1922/23, Nr. 6 S. 29 (m. Abb.)

Walter & Marcell Goldscheider: GWM Wien–Leipzig, DD 50, 1922/23, Nr. 7 S. 38 (m. Abb.)

Walter & Marcell Goldscheider: GWM, DD 50, 1922/23, Nr. 8 S. 38 (m. Abb.)

Walter & Marcell Goldscheider: GWM, DD 50, 1922/23, Nr. 9 S. 39 (m. Abb.)

Walter & Marcell Goldscheider: GWM Fayencen, eigene Niederlassung: Leipzig, DD 50, 1922/23, Nr. 10 S. 30 (m. Abb.)

Arthur Goldscheider: Arthur Goldscheider Paris, L'Illustration, Nr. 4268 (20.12.1924), S. 14 (m. Abb.)

Friedrich Goldscheider: Wiener Manufaktur FG 1885–1925, Modell .... „Hirschen-Gruppe", KuKGW 5, 1924/25, o. Pag.

Friedrich Goldscheider/Walter & Marcell Goldscheider: Wiener Manufaktur FG, ÖBWK 1, 1924/25, Nr. März S. III (m. Abb.)

Friedrich Goldscheider/Walter & Marcell Goldscheider: Wiener Manufaktur FG, ÖBWK 1, 1924/25, Nr. April S. III (m. Abb.)

Friedrich Goldscheider/Walter & Marcell Goldscheider: Wiener Manufaktur FG, ÖBWK 1, 1924/25, Nr. Mai u. Juni jew. S. III (m. Abb.)

Friedrich Goldscheider/Walter & Marcell Goldscheider: Wiener Manufaktur FG, ÖBWK 1, 1924/25, Nr. Juli/Aug. S. III (m. Abb.)

Friedrich Goldscheider/Walter & Marcell Goldscheider: Wiener Manufaktur FG, ÖBWK 1, 1924/25, Nr. Sept. S. III (m. Abb.)

Friedrich Goldscheider: Wiener Manufaktur FG 1885–1925, Modell .... „Hirschen-Gruppe", KuKGW 6, 1925/26, Nr. 12, o. Pag.

Friedrich Goldscheider: Wiener Manufaktur FG 1885–1925, Modell 5345 Bucher „Kahnfahrt", KuKGW 6, 1925/26, Nr. 13, o. Pag.

Friedrich Goldscheider: Modell 5433 Herlinger-Schwarz „Kinderkopf", ÖBWK 3, 1926/27, S. 104 (m. Abb.)

Friedrich Goldscheider: Modell 5705 Meisinger Sitzender Hund (Malteser?), ÖBWK 5, 1928/29, S. 239 (m. Abb.)

Friedrich Goldscheider: Wiener Manufaktur FG gegr. 1885, KuKGW 9, 1928/29, S. 127 (m. Abb.)

Friedrich Goldscheider/Walter & Marcell Goldscheider: Wiener Manufaktur FG, MW 10, 1928/29, Nr. 21 S. IV (m. Abb.)

Friedrich Goldscheider/Walter & Marcell Goldscheider: Wiener Manufaktur FG, MW 10, 1928/29, Nr. 22 S. IV (m. Abb.)

Friedrich Goldscheider/Walter & Marcell Goldscheider: Wiener Manufaktur FG, MW 10, 1928/29, Nr. 23 S. V (m. Abb.)

Friedrich Goldscheider/Walter & Marcell Goldscheider: Wiener Manufaktur FG, MW 10, 1928/29, Nr. 24 S. IV (m. Abb.)

Friedrich Goldscheider/Walter & Marcell Goldscheider: Wiener Manufaktur FG, MW 10, 1928/29, Nr. 25 S. IV (m. Abb.)

Friedrich Goldscheider/Walter & Marcell Goldscheider: Wiener Manufaktur FG, MW 10, 1928/29, Nr. 26 S. V (m. Abb.)

Friedrich Goldscheider/Walter & Marcell Goldscheider: Wiener Manufaktur FG, MW 11, 1929/30, Nr. 1 S. VI (m. Abb.)

Friedrich Goldscheider/Walter & Marcell Goldscheider: Wiener Manufaktur FG, MW 11, 1929/30, Nr. 2 S. IX (m. Abb.)

Friedrich Goldscheider/Walter & Marcell Goldscheider: Wiener Manufaktur FG, MW 11, 1929/30, Nr. 3 S. IX (m. Abb.)

Friedrich Goldscheider/Walter & Marcell Goldscheider: Wiener Manufaktur FG, MW 11, 1929/30, Nr. 4 S. IV (m. Abb.)

Friedrich Goldscheider/Walter & Marcell Goldscheider: Wiener Manufaktur FG, MW 11, 1929/30; Nr. 5 S. XI (m. Abb.)

Friedrich Goldscheider/Walter & Marcell Goldscheider: Wiener Manufaktur FG, MW 11, 1929/30, Nr. 6 S. VI (m. Abb.)

Friedrich Goldscheider/Walter & Marcell Goldscheider: Wiener Manufaktur FG, MW 11, 1929/30, Nr. 7 S. VIII (m. Abb.)

Marcell Goldscheider: Vereinigte Ateliers für Kunst u. Keramik, WNÖG 90, 1929, S. 837

Friedrich Goldscheider/Walter & Marcell Goldscheider: Wiener Manufaktur FG, MW 11, 1929/30, Nr. 8 S. V (m. Abb.)

Friedrich Goldscheider/Walter & Marcell Goldscheider: Wiener Manufaktur FG, MW 11, 1929/30, Nr. 9 S. II (m. Abb.)

Friedrich Goldscheider/Walter & Marcell Goldscheider: Wiener Manufaktur FG, MW 11, 1929/30, Nr. 10 S. V (m. Abb.)

Friedrich Goldscheider/Walter & Marcell Goldscheider: Wiener Manufaktur FG, MW 11, 1929/30, Nr. 11 S. VII (m. Abb.)

Friedrich Goldscheider/Walter & Marcell Goldscheider: Wiener Manufaktur FG, MW 11, 1929/30, Nr. 12 S. VI (m. Abb.)

Friedrich Goldscheider/Walter & Marcell Goldscheider: Wiener Manufaktur FG, MW 11, 1929/30, Nr. 13 S. VI (m. Abb.)

Friedrich Goldscheider/Walter & Marcell Goldscheider: Wiener Manufaktur FG, MW 11, 1929/30, Nr. 15 S. VI (m. Abb.)

Friedrich Goldscheider/Walter & Marcell Goldscheider: Wiener Manufaktur FG, MW 11, 1929/30, Nr. 16 S. VIII (m. Abb.)

Friedrich Goldscheider/Walter & Marcell Goldscheider: Wiener Manufaktur FG, MW 11, 1929/30, Nr. 17 S. III (m. Abb.)

Friedrich Goldscheider/Walter & Marcell Goldscheider: Wiener Manufaktur FG, MW 11, 1929/30; Nr. 18 S. II (m. Abb.)

Friedrich Goldscheider/Walter & Marcell Goldscheider: Wiener Manufaktur FG, MW 11, 1929/30, Nr. 19 S. 20 (m. Abb.)

Friedrich Goldscheider/Walter & Marcell Goldscheider: Wiener Manufaktur FG, MW 11, 1929/30, Nr. 20 Innenseite d. vorderen Umschlages (m. Abb.)

Friedrich Goldscheider/Walter & Marcell Goldscheider: Wiener Manufaktur FG, MW 11, 1929/30, Nr. 21 S. VII (m. Abb.)

Friedrich Goldscheider/Walter & Marcell Goldscheider: Wiener Manufaktur FG, MW 11, 1929/30, Nr. 22 S. 20 (m. Abb.)

Friedrich Goldscheider/Walter & Marcell Goldscheider: Wiener Manufaktur FG, MW 11, 1929/30, Nr. 23 S. 20 (m. Abb.)

Friedrich Goldscheider/Walter & Marcell Goldscheider: Wiener Manufaktur FG, MW 11, 1929/30, Nr. 24 S. VIII (m. Abb.)

Friedrich Goldscheider/Walter & Marcell Goldscheider: Wiener Manufaktur FG, MW 11, 1929/30, Nr. 25 S. III (m. Abb.)

Friedrich Goldscheider: Wiener Manufaktur FG Wien–Leipzig. Fayence – Bronze – Moderne Keramik, Die Schaulade 6, 1930, S. 243 (m. Abb.)

Friedrich Goldscheider: Wiener Manufaktur FG Fayence – Keramik, Die Schaulade 8, 1932, S. 926

Friedrich Goldscheider: Modell 5963 Kremser „Vasenfigur", ÖK 3, 1932, S. 35 (m. Abb.)

Raddatz & Co (Berlin), Schenken macht Freude, Raddatz & Co (Verkaufsprospekt), Berlin, um 1932–35, 8 S. (m. Abb.)

Friedrich Goldscheider: Wiener Manufaktur FG, Adressbuch von Österreich 1950, Wien 1950, S. 765

Marcel Goldscheider: Marcel Goldscheider Hanley, Stoke-on-Trent, UK, The Pottery Gazette and Glass Trade Review 1950, London 1950, S. 1855

Patente | Patents

Friedrich Goldscheider: Österreichische Patentschrift 3855: Verfahren der Decorierung plastischer Gegenstände (Angemeldet am 30.3.1900), 2 Seiten

Friedrich Goldscheider: Österreichische Patentschrift 16882: Mattlack (Angemeldet am 13.3.1903), 1 Seite

Friedrich Goldscheider: Österreichische Patentschrift 20387: Mosaikimitation und Verfahren zur Herstellung derselben (Angemeldet am 21.1.1904), 1 Seite

Erste Wiener Terrakottafabrik...: Österreichische Patentschrift 49972: Verfahren zur Dekorierung keramischer Gegenstände durch Tränken oder Durchfärben mit Metallsalzlösungen (Angemeldet am 8.10.1909), 2 Seiten

Wiener Manufaktur FG: Österreichische Patentschrift 96834: Verfahren zur Erzeugung farbiger Muster durch Aufspritzen (Angemeldet am 3.3.1923), 1 Seite

Internet

www.aeion.at: Goldscheider, Friedrich (Wiener Manufaktur), Ö-Lexikon

www.aluminet.org/manufacturers/everlast.html: 28.09.2004, Goldscheider Everlast Corp.

www.familysearch.co.uk: Goldscheider, Marcel

Zeitschriften | Magazines

AdK
Adressbuch der Keram-Industrie, Jg. 2, 8, 12, 17–21, 23, 24, Coburg 1887, 1904, 1913, 1930–1941, 1954/55, 1959

AdO
Adressbuch der Ostmark (Forts. v.: Adressbuch von Österreich für Industrie, Handel, Gewerbe und Landwirtschaft; 1942/43 u.d.T.: Adressbuch der Alpen- u. Donaugaue für Industrie, Handel und Gewerbe), Wien 1939–1942/43

AvÖ
Adressbuch von Österreich für Industrie, Handel, Gewerbe und Landwirtschaft (Forts.: Adressbuch der Ostmark 1939–1942/43, ab 1950 wieder AvÖ), Wien 1924, 1925/26, 1929, 1931, 1937, 1938, 1950, 1951

Allgemeines Adressen-Buch der königl. Hauptstadt Prag und vollständiges Verzeichnis handelsgerichtlich protokollierter Firmen, Prag 1875

American Ceramic Society Bulletin, Vol. 25, 1946

AZ
Antiquitäten-Zeitung, Jg. 1979–1996, München 1979–1996

L'Art Déc.
L'Art Décoratif, Jg. 3, Paris 1900/01

Art et Décoration, T. 11–13, Paris 1902–1903

Bloch, siehe Bücher – Aufsätze – Reklame

Blom's Engros- und Export-Adressbuch von Berlin und Vororten, 3. Aufl., Berlin 1907/08

DB
Die Bühne. Wochenschrift für Theater und Gesellschaft, Nr. 1–393, Wien 1924–1933

CMReg.
Central-Marken-Register 1900, Wien 1901

Ceramic Industry, Jg. 1942–1952, Chicago 1942–1952

Compass
Compass. Industrielles Jb., Protektorat Böhmen u. Mähren... (ab 1939/40 u.d.T.: Industrie-Compass. Protektorat Böhmen u. Mähren), Prag 1939–1942/43

Crockery and Glass Journal Directory, 1952

DD
Die Dame, Jg. 50–58, Berlin 1922/23–1930/31

Dekor. Kunst
Dekorative Kunst, Jg. 25, München 1921

DER
Deutsche-Export-Revue, Jg. 1–13, Berlin 1901/02–1913/14

DKuD
Deutsche Kunst und Dekoration, Bd. 57–69, Darmstadt 1925/26–1929/30

EW
Elegante Welt, Jg. 14, Berlin 1925

L'Estampille
L'Estampille L'Objet d'Art, Dijon, 1995–1996

ExpAÖ
Exportadressbuch von Österreich 1952, Wien 1952

Firmenbuch Österreich, siehe Handelsregister Österreich

Das Handelsmuseum, Jg. 40, Wien 1925

HRÖ
Handelsregister Österreich (später u.d.T.: Firmenbuch Österreich), Wien 1987, 1990–1995

High-Life
High-Life-Almanach, Jg. 5, 7, 9–10, Wien 1909–1914/15

IC
Industrie-Compass (Deutschösterreich../Österreich/später: Deutsches Reich: Ostmark, Hrsg: R. Hanel), Wien 1920/21, 1924/25–1931/32, 1934/35, 1937/38–1941/42

Industrie-Compass. Protektorat Böhmen u. Mähren, siehe Compass

IZB
Internationales Zentralblatt für Baukeramik und Glasindustrie, Jg. 24, Wien, Berlin 1909

Jahrbuch der Österreichischen Industrie, Hrsg.: R. Hanel, Wien 1910, 1913, 1914, 1916, 1918

K-EM
K-E Magazin, Jg. 1–3, Berlin 1925–1927

KM
Keramische Monatshefte, Jg. 1–8, Halle 1901–1908

KeramRS
Keramische Rundschau, Jg. 1–33, Berlin 1893–1925

Keramos
Keramos, Jg. 2, München 1923

DK
Die Kunst. Monatshefte f. freie u. angewandte Kunst, Jg. 3, 22, München 1901, 1910

KfA
Kunst für Alle, Jg. 1–58, München 1886–1943

KuHW
Kunst und Handwerk, Zs. des Bayer. KGW-Vereins zu München, Jg. 32, München 1901/02

KuKGW
Kunst und Kunstgewerbe, Jg. 4–9, Nürnberg 1923–1929

KuKHW
Kunst und Kunsthandwerk, Jg. 1–7, Wien 1898–1904

Das Leben, Jg. 1–12, Leipzig 1923–1934

Lehmann
Lehmann's Allgemeiner Wohnungs-Anzeiger nebst Handels- und Gewerbeadressbuch für ... Wien. Später u.d.T.: Wiener Adressbuch, Lehmanns Wohnungsanzeiger, Adressbuch von Wien, Wien 1880–1954

Leuchs Adressbuch aller Länder der Erde der Kaufleute, Fabrikanten..., Bd. 14, 10. Ausg., Nürnberg 1908

DM
Das Magazin, Jg. 1–13, Berlin 1924/25–1936/37

Magyar Iparmüvészet, Jg. 3–4, Budapest 1900–1901

Mitt. d. Nordböhm. GWMus.
Mitteilungen des Nordböhmischen Gewerbemuseums, Jg. 20–23, Reichenberg (Böhmen) 1902–1905

MW
Moderne Welt, Jg. 1–11, Wien, Leipzig 1918/19–1929/30

Nová Doba, Pilsen 1939

Nové Plzeňské Noviny, Pilsen 1892

ÖK
Österreichische Kunst, Jg. 1–8, Wien 1929/30–1937

ÖZK
Österreichischer Zentralkataster, Bd. 1, 2. verm. u. verb. Aufl., Wien 1908, S. 1651

ÖPB
Österreichisches Patent-Blatt, Jg. 1–26, Wien 1899–1924

ÖBWK
Österreichs Bau- und Werkkunst, Jg. 1–6, Wien 1924/25–1929/30

Pilsner Bote, Pilsen 1863

Pilsner Reform, Pilsen 1874

Plzeňské Listy, Pilsen 1892

PGGTR
The Pottery Gazette and Glass Trade Review, 1940–1960

The Pottery Gazette and Glass Trade Review, Reference Book and Directory, 1953–1960

Pottery and Glass Record, London 1956

RDM
Revue des Monats, Jg. 1–8, Berlin 1926/27–1933

Sammler-Journal, Schwäbisch Hall, 1993–2003

Schaulade
Die Schaulade, Jg. 5–18, Bamberg 1929–1942

SM
Scherls Magazin, Jg. 1–9, Berlin 1924/25–1933

Sprechsaal
Sprechsaal, Jg. 20–43, Coburg 1887–1910

Stehlik's Handels-Adressbuch der Kaufleute und Fabrikanten von Österreich-Ungarn, Jg. 7, Wien, Pest, Leipzig 1885/86

Studio
The Studio. An illustr. Magazine of fine and applied Art, Vol. 92–95, London 1926–1928

Studio YB
The Studio Yearbook of Decorative Art, 1911–1926, 1936, London, Paris, New York 1911–1926, 1936

Der Tanz, Jg. 1–17, Berlin 1927/28–1944

Taschenbuch für Keramiker 1924, Berlin 1924

Tonindustrie-Ztg
Tonindustrie-Zeitung und Fachblatt für Zement, Beton, Gips..., Jg. 23–54, Berlin 1899–1904, 1927–1930

Tonwaren-, Ofen- und Töpfer-Zeitung, Jg. 46, Leipzig 1922, S. 54

Trenton Evening Times, Trenton 1952

TuS
Trödler & Sammeln, Reichertshausen, 1995 ff.

Uhu
Uhu. Das neue Monats-Magazin, Jg. 1–9, Berlin 1924/25–1932/33

VKMH
Velhagen und Klasings Monatshefte, Jg. 1; 16–17, Bielefeld, Leipzig 1898/99; 1901/02

WZBl.
Waarenzeichenblatt, Jg. 8, Berlin 1901, S. 41

Weltkunst, 1995 ff.
WBZ
Wiener Bauindustrie-Zeitung und Wiener Bauten-Album, Jg. 19, Wien 1901/02

WNÖG
Wochenschrift des Niederösterreichischen Gewerbevereins, Jg. 55–90, Wien 1894–1929

ZMA
Zentral-Marken-Anzeiger 1909, S. 225

Abkürzungsverzeichnis | Abbreviations

Abb.	Abbildung
AeD	Art et Décoration
ähnl.	ähnlich
AGM	Archiv Goldscheider
AH	Auktionshaus
Ahlden	Auktionshaus Schloß Ahlden
AKL	Allgemeines Künstlerlexikon
Ang.	Angabe
A/P/T	Auktionshäuser Ader/Picard/Tajan (Paris)
arab.	arabisch
Arnold	Auktionshaus Arnold (Frankfurt)
Art.	Artikel
A/T	Auktionshaus Ader/Tajan (Paris)
Aufl.	Auflage
Ausst.	Ausstellung
Ausst.-Kat.	Ausstellungskatalog
AZ	Antiquitäten-Zeitung
B.	Breite
Bamberger	AH Bamberger (Nürnberg)
Bd.	Band
Bem.	Bemerkung
Bergmann	AH Bergmann (Nürnberg)
Berlinghof	AH Berlinghof (Heidelberg, Walldorf/Baden)
Bezeichn.	Bezeichnung
Bl.	Blatt
Blank	AH Blank (Bad Vilbel)
Bloss	AH Bloss (Merzenhausen)
BM	AH Bolland & Marotz (Bremen)
Bode	AH Bode (Pforzheim)
Bödiger	AH Bödiger (Bonn)
böhm.	böhmisch
Bonhams	AH Bonhams (London u.a.)
Bukowskis	AH Bukowskis (Stockholm)
Burchard	AH Burchard Galleries, (St. Peterburg/Florida)
bzw.	beziehungsweise
ca.	circa
Cheffins	AH Cheffins, Grain & Comins (Cambridge/UK)
Christie's A.	AH Christie's (Amsterdam)
Christie's East	AH Christie's East (New York)
Christie's G.	AH Christie's (Genf)
Christie's L.	AH Christie's (London)
Christie's SK.	AH Christie's (South Kensington)
Christie's Sy.	AH Christie's (Sydney)
cm	Zentimeter
CuxAH	AH Cuxhavener Auktionshalle (Cuxhaven)
Dawo	AH Dawo (Saarbrücken)
DB	Die Bühne
DD	Die Dame
Dekor. Kunst	Dekorative Kunst
dec.	decorated
DER	Deutsche Export-Revue
DGW	Deutsche Gewerbeschau (München)
d.i.	das ist (der ist)
DK	Die Kunst
DKuD	Deutsche Kunst und Dekoration
DL	Das Leben
DM	Das Magazin
Doro	AH Dorotheum (Wien)
Doro B	AH Dorotheum (Filiale Bregenz)
Doro F	AH Dorotheum (Filiale Favoriten)
Doro G	AH Dorotheum (Filiale Graz)
Doro K	AH Dorotheum (Filiale Klagenfurt)
Doro L	AH Dorotheum (Filiale Linz)
Doro P	AH Dorotheum (Filiale Prag)
Doro S	AH Dorotheum (Filiale Salzburg)
Doro SP	AH Dorotheum (Filiale St. Pölten)
Doro WA	AH Dorotheum (Wien, Weihnachtsauktion)
Doro WN	AH Dorotheum (Filiale Wiener Neustadt)
Dreweatt Neate	AH Dreweatt Neate (Donnington/UK)
DT	Der Tanz
D&K	AH Deutsch & Keil (Wien)
Dumousset-Debureaux	AH Dumousset-Debureaux (Paris)
eBay	AH eBay (Internet-AH: www.ebay.com)
Ed.	Edition/Editeur
ehem.	ehemalig
eigentl.	eigentlich
Erg.	Ergänzung
europ.	europäisch
evtl.	eventuell
EW	Elegante Welt
Falkkloo's	AH Falkkloo's (Malmö)
Fay.	Fayence
FG	Friedrich Goldscheider
Fischer	AH Fischer (Jürgen Fischer, Heilbronn)
Fischer Luz.	AH Fischer (Luzern)
franz.	französisch
Franzisi	AH Franzisi (Nürnberg)
Fs.	Festschrift
FVK	Freiverkauf
G 68 Auctions	AH Gallery Sixtyeigth Auctions (Toronto)
geb.	geboren (*)
Geb.	Geburtstag
gem.	gemalt
gepr.	geprägt
ger.	geritzt
Ges. gesch.	Gesetzlich geschützt
Gesamth.	Gesamthöhe
gest.	gestorben (†)
GK	Goldscheider Keramik. Ausst.-Kat., Wien 1985/86
griech.	griechisch
Gros/Del	AH Gros/Delettrez (Paris)
GS	Goldscheider
GSB	Goldscheider Firmenkatalog „Fabrique de Bronzes" (Fréderic GS Paris, 45 Rue de Paradis)
GSF	Goldscheider Firmenkatalog „Friedrich Goldscheider, Kunstgewerbliches Etablissement Kunst-Terracotta-Fabrik Wien..."
GSG Vienna	Goldscheider Stoob/Bgld.
GSNachf.	Goldscheider Firmenkatalog „Wiener Manufaktur Friedrich GS Nachf."
GSN	Goldscheider Firmenkatalog „The Royal Family of Porcelain, since 1885"
GSNF	Goldscheider Firmenkatalog (Faltbl./Folder) „The Royal Family of Porcelain, since 1885"
GWM	Manufakturstempel der zwanziger Jahre (-> vgl. Markenübersicht)
GWMus.	Gewerbemuseum
H.	Höhe
Hack. Ill.	Hackebeils Illustrierte
Hampel	AH Hampel (München)
Hand dec.	Hand decorated
handschr.	handschriftlich
Heickmann	AH Heickmann (Chemnitz)
Hennig	AH Hennig (Leipziger Kunstauktionshaus)
Henry's	AH Henry's (Mutterstadt)
Herr	AH Herr (Köln)
Hrsg.	Herausgeber
HMW	Historisches Museum (Wien), jetzt Wien Museum
holl.	holländisch
HR	Hofrat
HSAK	Hochschule für angewandte Kunst
IC	Industrie Compass
Ilko	KH Ilko (Salzburg)
IZB	Internationales Zentralblatt für Baukeramik und Glasindustrie
Inv.	Inventar
Inv.-Nr.	Inventarnummer
Jg.	Jahrgang
Jgg.	Jahrgänge
KAHS	AH Schwabing
KaM	AH Kunsthaus am Museum (Carola van Ham, Köln)
Kap.	Kapitel
Karrenbauer	AH Karrenbauer (Konstanz)
Kastern	AH Kastern (Hannover)
Kat.	Katalog
Kaupp	AH Kaupp (Staufen)
KE-M	KE-Magazin
KeramRS	Keramische Rundschau
Keramos	Keramos (Zs.)
Kett.	AH Ketterer (München)
Keup	AH Keup (Regensburg)
KfA	Die Kunst für Alle
KGSF1	Firmenkat. Kunstkeramische Manufaktur Goldscheider (Carstens/Fredelsloh), o.Pag. (21 Bl.)
KGSF2	Firmenkat. Kunstkeramische Manufaktur Goldscheider (Carstens/Fredelsloh), o.Pag. (113 Bl.)

KGW	Kunstgewerbe
KH	Kunsthandel
KM	Keramische Monatshefte
KNS	AH Kunstauktionen nächst St. Sebald (Nürnberg)
König	AH König (Nürnberg)
KollerZ	AH Koller (Zürich)
KPJb	Kunstpreisjahrbuch
Kühling	AH Kühling (Kempten)
Künstlerang.	Künstlerangabe
KuHW	Kunst und Handwerk
KuKGW	Kunst- und Kunstgewerbe
KuKHW	Kunst und Kunsthandwerk
L.	Länge
L'Art Déc.	L'Art Décoratif
L'Estampille	L'Estampille L'Objet d'Art
Lankes	AH Lankes (Hof/Nürnberg)
lat.	lateinisch
Lempertz	AH Lempertz (Köln)
lig.	ligiert
Lit.	Literatur
Lpz.	Leipzig
Lüffe	AH Lüffe (Bamberg)
Lyon/T.	AH Lyon & Turnbull (Edinburgh)
Mehlis	AH Mehlis (Plauen)
Metz	AH Metz (Heidelberg)
mexikan.	mexikanisch
MG	Marcell Goldscheider
Mitt.	Mitteilung
MKAH	AH Meerbuscher Kunstauktionshaus (Meerbusch)
Modell-Nr.	Modellnummer
monogr.	monogrammiert
Mskr.	Manuskript
Müller	AH Müller (Landshut)
Mus.	Museum
MW	Moderne Welt
Nachf.	Nachfolger
Nagel	AH Nagel (Stuttgart)
Nagel D.	AH Nagel (Dresden)
Nagel L.	AH Nagel (Leipzig)
Neumeister	AH Neumeister (München)
niederländ.	niederländisch
Nr.	Nummer
Nrn.	Nummern
Nusser	AH Nusser (München)
o. J.	ohne Jahr
o. Pag.	ohne Paginierung (Seitenzählung)
ÖBWK	Österreichs Bau- und Werkkunst
ÖK	Österreichische Kunst
ÖMAK	Österreichisches Museum für Kunst und Industrie
ÖP	Österreichisches Patentamt
ÖPB	Österreichisches Patent-Blatt
ÖPS	Österreichische Patentschrift
ÖWB	Österreichischer Werkbund
ÖZK	Österreichischer Zentralkataster
Pat.	Patent
PB	Privatbesitz (Private Collection)
Peege	AH Peege (Freiburg i.Br.)
Peretz	AH Peretz (Saarbrücken)
PGGTR	The Pottery Gazette and Glass Trade Review
Phillips E.	AH Phillips, Son and Neale (Edinburgh)
Phillips L.	AH Phillips, Son and Neale (London)
Phillips S.	AH Phillips, Son and Neale (Sydney)
Phillips Z.	AH Phillips, Son and Neale (Zürich)
Pick.	Pickerl (österr. für Aufkleber, Sticker)
Pickford	AH Pickford (Hertford/UK)
Plak.	Plakette
Pseud.	Pseudonym
Quit.	Auktionshaus Quittenbaum (München)
Quit. H.	Auktionshaus Quittenbaum (Hamburg)
Rammel	AH Rammel (Bamberg)
RDM	Revue des Monats
Rehm	AH Rehm (Augsburg)
Repr. Res.	Reproduction Reservé
Ribeyre-Baron	AH Ribeyre-Baron (Paris)
Rieber	AH Rieber (Stuttgart)
RKAH	AH Rheingauer Kunstauktionshaus (Eltville-Erbach)
RM	AH Reimann & Monatsberger (Ludwigsburg)
Robert	AH Marie-François Robert (Paris)
röm.	römisch
Ruef	AH Ruef (München)
s.	siehe
s.a.	siehe auch
S.	Seite
Schaulade	Die Schaulade
Schloß Hag.	AH Schloß Hagenburg (Hagenburg)
Schloß Rick.	AH Schloß Ricklingen (Garbsen bei Hannover)
Schlüter	AH Schlüter (Hamburg)
Schmidt	AH Schmidt (Regensburg)
Schopmann	AH Schopmann (Hamburg)
Schuler	AH Schuler (Zürich)
Sebök	AH Sebök (Bamberg)
seitl.	seitlich
sek.	sekundär
SH.	Sonderheft
Sigalas	AH Sigalas (Hildrizhausen)
Sign.	Signatur
sign.	signiert
SJ	Sammler-Journal
Slawinski	AH Slawinski (Felton/Ca, USA)
SM	Scherls Magazin
sog.	sogenannt
Sonderdr.	Sonderdruck
Soth. A.	AH Sotheby's (Amsterdam)
Soth. Ama.	AH Sotheby's (http://www.sothebys.amazon.com)
Soth. Arcade	AH Sotheby's Arcade Auctions (New York)
Soth. com.	AH Sotheby's (http://www.sothebys.com)
Soth. G.	AH Sotheby's (Genf)
Soth. L.	AH Sotheby's (London)
Soth. Mon.	AH Sotheby's (Monaco)
Soth. Monrepos	AH Sotheby's (Monrepos/Ludwigsburg)
Soth. NY.	AH Sotheby's (New York)
Soth. Olympia	AH Sotheby's (Olympia/London)
Soth. Rgbg.	AH Sotheby's (Thurn & Taxis/Regensburg)
Soth. Z.	AH Sotheby's (Zürich)
Spik	AH Spik (Berlin)
Stahl	AH Stahl (Hamburg)
Studio	The Studio
Studio YB	The Studio Yearbook of Decorative Art
Sworders	AH Sworders (Stansted/UK)
Tajan	AH Etude Tajan (Paris)
teilw.	teilweise
Tonindustrie-Ztg.	Tonindustrie-Zeitung und Fachblatt für Zement, Beton, Gips...
Tradera	AH Tradera (Internet-AH: www.tradera.de)
TuS	Trödler & Sammeln
u.a.	und andere
u.d.T.	unter dem Titel
undeutl.	undeutlich
ungar.	ungarisch
unsign.	unsigniert
Urauff.	Uraufführung
verb.	verbessert
Vervielf.	Vervielfältigung
vgl.	vergleiche
VKMH	Velhagen und Klasings Monatshefte
vollplast.	vollplastisch
vorbeh.	vorbehalten
vorm.	vormals
Walldorf	AH Walldorf (Walldorf)
WBZ	Wiener Bauindustrie-Zeitung
Weichmann	AH Weichmann (Wiesbaden)
Weidler	AH Weidler (Nürnberg)
WeltM	Welt-Magazin
Wendl	AH Wendl (Rudolstadt)
WG	West-Germany
Wilbois	AH Wilbois (Münster/Westf.)
WKA	AH Wiener Kunstauktionen/Kinsky (Wien)
WKH	Wiener Kunsthandel
WM	Wiener Magazin
WW	Wiener Werkstätte
WZBl.	Wa(a)renzeichenblatt
Z	AH Zeller (Lindau)
zahlr.	zahlreiche
Zengen	AH Zengen (Bonn)
Zezschwitz	AH Zezschwitz (München)
Zof.	AH Zofingen (Zofingen/Schweiz)
zugeschr.	zugeschrieben
zusätzl.	zusätzlich
zw.	zwischen

Register | Index

Die Ziffern verweisen auf die Seitenzahlen des deutschen Texts. **Fette Ziffern** bedeuten Abbildungsseiten innerhalb der Firmengeschichte. Eine Ziffer mit „f" verweist auch auf die nächstfolgende Seite; „ff" verweist auf die zwei oder mehr Seiten, die folgen.

The numerals refer to the page numbers of the German text. **Bold numerals** indicate pages with illustrations. A numeral with "f" refers also to the next following page; "ff" refers to the two or more pages that follow.

Abschlag 370, 375, 554
Accademia Liguistica Genua 554
Accius 358
Adler-Stempel **19**, 72, **72**
Älteste Volkstedter, Manufaktur 571
Aerograph 88, **89**, 142
Akademie Barcelona 556
Akademie Berlin 571, 575, 590, 593, 601, 603
Akademie Brüssel 561, 601
Akademie Dresden 603
Akademie für Kunst und Gewerbe Zagreb 572
Akademie Gent 572, 601
Akademie Kopenhagen 599
Akademie Krakau 599
Akademie Lille 579
Akademie München 575, 602f
Akademie Prag 602
Akademie Stockholm 578
Akademie Wien 92, 554f, 558, 561, 565f, 572ff, 578, 583, 587, 590, 593, 595f, 598, 600
Albert-Lefeufre, Louis 554
Alexander Potteries 262
Alloi, Léonce 554
Allouard, Henri Emile 30, 229, 321, 342, 344, 533, 554, 561
Almost Perfect (Marke) **275**
Althan, Josef 415, 554, 580
American Goldscheider Inc. 182, **182**, 184f, 603
Amphora, Manufaktur 57, 565, 586
Amy, Jean Barnabé 336, 339, 554
Angelini, Tito 556
Anglis, J. 292, 554
Anker-Stempel **22**, 23
Anthouard, Magdeleine 241, **241**, 554
Aragon, Isabella de 350
Arna, Lissi 412
Arnold, E. 235, 554
Arnoux 301, 304, 554
Art Pottery 269, **269**
Art Students League New York 572
Arte della Ceramica Firenze 567
Arthur Goldscheider, Éditeur d'Art 231
Artistic Evolution 237, 239, 241, 245
Auber 369, 554
Augarten, Manufaktur 147, 160, 275, 447, 555f, 573, 576, 592, 603
Augmüller, Ludwig 373, 554
Ausschuss des österreichischen Kunstgewerbes 113
Austria in London, Ausst. 134, 260

Baarova, Lida 483
Bacqué, Daniel-Joseph 235, 242, 554
Badin, Jean-Victor 554
Bagge, Eric 235, 240, 554
Bakardziev, Georgi Stefanov 554
Baldwin, Barbara 188, 536, 538
Baller, Herta 556
Ballets Russes 560
Balten, Zerline 412
Balzac, Honoré de 316
Bara, Theda 408
Barabino, Armando 290, 336, 554
Barfoort 554
Barner 403
Baron, René 292, 307f, 554f
Barrau, Th. 580, 583
Barreau 590
Barrias 409, 558, 560, 565, 576, 586
Barwig 580, 583, 586
Bate, Firmin 344, 555
Baudry, R.E. 52, 348, 371, 555
Bauer, Carli 541, 555, 597
Bauer, Rosenthal & Co. KG Kronach 572
Bayerischer Kunstgewerbeverein 556
Beaufils, Armel 235, 555
Beaume 323, 371, 374, 377, 379, 386, 390, 419, 422, 555
Beauvais, Gabriel 245, 555
Bechstein, Ludwig 345
Beethoven, Ludwig van 57, 298, 309f, 341, 344, 349, 353, 355, 358, 368, 373, 386, 390, 545
Begas, Reinhold 603
Behrens, Christian 596
Beinhofer, Hanns 221, 555
Belleuse 394, 555
Belville, Jean 244, 555
Benirschke, Karl August 159
Berger, Josef 98, 101, 555
Bergerac, Cyrano de 330, 332, 336, 344
Berlichingen, Elisabeth von **24**
Berlichingen, Götz von **24**, 531
Berlin, Kommandantenstraße 86 42, 46
Bernhard Bloch, Manufaktur 34, 371
Bernhard, Sarah 28, 364, 366, 370
Berthou(d) 555
Bertoni, Wander 279, **279**, 555
Bertrand, A. 278, 348, 350, 353ff, 357, 360f, 555
Besnard, Jean 235, 241, 555
Beswick 187
Bieber, Erna 567
Biesbroeck, L. van 601
Billroth, Theodor **288**
Bimini Werkstätten Wien 555, 576
Bing & Groendahl, Manufaktur 571
Biško, Josef 56, 555
Bismarck, Otto von 332, 337
Blaas'sche Gemälde 28
Blackpool, International Gifts & Fancy Goods Fair 271
Bläsche 313, 555
Blanc, Pierre 245, **245**, 555
Blondat, Max 235, 242, 555f
Blot, E. 229
Blume und Plastik, Ausst. 260
Boehm 575
Bofill, Antoine 368, 556
Boleyn, Anne **25**, **289**, 290
Bonnasieux 554
Boop, Betty 420
Borghèse, Pauline 298
Borgia, Lucrezia 345
Bormann, Wilhelm 99, 323, 371, 373f, 377, 379, 382, 386, 390, 405, 407, 419, 422, 556
Borsi 300, 556
Bortone, Ipazzio Antonio 323, 330, 339, 342, 346, 556, 559
Boschull 411, 556
Bosi 379, 381, 401, 409, 556, 593
Bosse, Gretl 161
Bosse, Walter 106ff, **106ff**, 160ff, **163**, 167, 223, **224**, 254, 415, 426f, 435ff, 439, 441, 505, 518, 528, 532f, 543f, 556f, 568, 585
Bouchard, Henri 235, 244, 557
Boucher, A. 229, 533, 558
Bouraine, Marcel 232, 235, **235**, **238f**, 240f, **240**, 244, **244**, 545, 557
Bourdelle, Antoine 558
Bouret, Germaine **141**, **145**, **264f**, 446, 454, 464, 467, 472f, 483, 494, 557, 562
Bouret, Pierre-Ernest 244, 557
Bouval, Maurice 30, 56, **56**, 229, **229**, 296f, 302ff, 345f, 533, 557
Bovy, H. 602
Bozzi, Lorenzo 300, 557
Braecke, Pieter Jan 303, 557
Brahms, Johannes 365, 561
Brateau, Jules 559
Braun, Gretl 555, 597
Breitner, Josef 556, 569, 578
Breitner, Joseph 65, 557, 561, 586
Brenner 311, 557f
Breuer, Peter 603
Brévété S.G.D.G (Marke) 24
British Industry Fair 267
Brochmann, Edith 558
Brooks, Louise 416, 420, 426,
Brugier, A. 235, 240, 558
Brünelle 363, 558
Brunclair 587
Brunel, H. 235, 558
Brunner, Walter 154
Bruyneel 301f, 305, 558
Brynski 384, 387, 558
Buch und Raum der Gegenwart, Ausst. 259, **260**
Bucher, Edwin 558
Bucher, Hertha 98, **98**, 101, 254, 257, **257**, 406, 547, 558
Bulier 366, 558
Busch & Ludescher, Wiener Kunstkeramische Werkstätte 565, 569, 584, 593, 599
Busch, Wilhelm **290**, 425f, 508
Butensky, Léon Jules **65**, 375f, 558

Caligula 358
Callot, Jacques **28**, 295f, 298, 558
Calm, Lotte 408, 558
Camillo Goldscheider, Friedrich Goldscheider Sohn 78, 230
Camus, Jean-Marie 235, 558
Capèque 31, **57**, **67**, 314, 362ff, 366, 370ff, 375, 377f, 381, 558
Cárdenas 351, 558
Cardona, José 289, 558
Carlier, Émile-Joseph 229, 302, 533, 558, 578
Carlus, Jean 390, 558
Carrièr-Belleuse, Albert-Ernest 30, **51**, 229, 301, 326f, 331, 338ff, 345, 350, 558ff, 566
Carrière, Ernest 597
Carstens, Christian 224, 574
Carstens, Ernst 224, 574
Carstens, Manufaktur 191, 214, 220ff, **220f**, **223**, 224, **224**, 270, 342, 419, 429, 431, 441, 446, 448, 458f, 461ff, 469ff, 480ff, 486ff, 492, 477, 493, 494f, 497ff, 504ff, 514ff, 555, 560, 562, 566, 574, 596
Carvin, Louis-Albert 324, 559
Caussé, Julien 300, 559
Cazaux, Édouard 235, 240f, 242, **242**, 559
Cazzolla, Luciano 192
Ceccarelli, Ezio 295, 299f, 318, 326, 332, 335, 340, 346, 349, 359, 367, 559
Cech, Irene 559
Cederlund, Frédérique Ozeanne 235
Chalon, Louis 229, 288f, 559
Chaplet, Ernest 242
Chapu 592
Charpentier, Alexandre 30, 229, 305, 533, 559
Charpentier, F. 30, 229, 533, 587, 595
Chauvel, Georges 235, 242, 559
Chavanne **48**, 323, 340, 559
Cherc 31, **32f**, **37**, **50**, **52**, **54**, 294f, 300f, 308f, 322f, 324, 326f, 330, 334ff, 338f, 341ff, 344ff, 354, 357ff, 362ff, 368f, 371, 375ff, 381f, 392, 559f
Chéret, Joseph 30, 318, 560
Chicago Gift Show 188
Chinard, Joseph 295, 560
Chiparus, Demêtre 245, 560
Chludek, Wilhelm 223, 515f, 543f, 560
Churchill, Winston **183**, 264
Cicero 367
Ciseau **54**, 365f, 368, 560
Cizek, Franz 106f, 556, 558, 570, 575, 601
Clarà, Juan **289**, 318, 320, 323, 365f, 561
Cleopatra 335, 338, 370
Clown Grock 436
Cochet, Robert 561
Colditz, Sachsen 198
Colin & Cie 229, 557
Collura 536, 538

Colmant, Pierre 241, **241**, 561
Colombier, Amélie 315, 561
Columbus 559
Congers 561
Conrat, Ilse 365, 561
Corbeille 347, 350, 5619
Cortendorf, Manufaktur 580
Coudray, Georges-Charles 297, 370, 561
Couturier 366, 561
Cries of Old London 265, **265**
Cristani, R. 586
Croisy, Aristide 51, 229, 301, 533, 561
Czapek, Stanislaus 31, 290, 363, 365f, 370ff, 375, 377f, 381f, 385, 387f, 391, 394f, 405, 561f, 588, 591, 593, 599

Dakon, Stephan **87**, 93, **97**, **99**, **101**, **116ff**, **134**, 135f, **135**, **141**, **143**, 180, 205, 207, 218, **222f**, 223, **231**, **264**, 399, 409ff, 415ff, 432ff, 447ff, 461, 463ff, 467ff, 472ff, 478, 480ff, 488, 491ff, 496ff, 506, 508f, 511, 514, 536ff, 541f, 545, 557, 562ff, 585, 592, 597
Dante Alighieri 300, 336, 346, 349, 362f, 379f, 406
Das befreite Handwerk, Ausst 134, 261
Das Bild im Raum, Ausst. 257
Daubling 370, 372f, 380, 565
David, Fernand 235, 565
Debienne, Noémi(e) 299, 342, 565
Debut, M. 409
Deck, Joseph-Théodore 560
Decor Chromo nouveau 49
Decor Ebène chromé 49
Defoe, Daniel 318
Defraigne, Gilles 241, 565
Dekor Algen 51, **51**, 555
Dekor Grès 52
Dekor Heliotrop 52
Dekor Holz 52
Dekor Secession 52
Dekor Türkis 52
Delagrange, Léon Noël **30**, 53, 229, 288f, 300, 304, 343, 565
Delamarre, Raymond 242, 565
Denecke, Carl 46
Denk am Graben, Fachgeschäft 147
Deplanche 579
Derp, Clotilde von 414, 427
Desbois, J. 559
Désurmont **41**, 305ff, 311, 565
Deutsche Gesellschaft für christliche Kunst 574f
Deutsche Industrie-Messe Hannover 211, **220**, 221
Deutscher Werkbund 578
Dickens, Charles 269, **270f**, 551
Dietl, Siegfried 160
Dietrich, Marlene 119, **135**, 439, 486
Disney, Walt 212, **213**
Doblinger, Ferdinand 370, 372f, 380, 408, 565
Dolly Sisters **87**, 116, 399, 410ff, 415f, 475, 479
Donatello 330f, 346, 565, 567
Donatello, F(l)agolli 345, 353, 565
Dorier 538f
Drahonovsky 570
Drew, Gloria 186, 536, 538
Dreyfus-Stern 235, 565f
Drouart, Raphaël 235, 566
Drury, Alfred 570
Dubarry, Marie-Jeanne Bécu, Comtesse 335, 369
Dubois, P. 577
Dubret, Henri 235, 566
Dürer, Albrecht 543
Dumont, August 554, 572f, 578f, 595
Dupré, J. 319, 325, 328, 356, 566

Eberle, Syrius 602
Eberth, Anny 403
École des Arts Décoratifs Paris 554
École des Beaux-Arts Dijon 601
École des Beaux-Arts Paris 558, 560, 566, 572, 576, 599
École des Beaux-Arts Rennes 584
Elisabeth, Kaiserin 561
Energieministerium 195
Epiag, Manufaktur 578
Erbteilungsvertrag 46, **46**
Erdös, Stephan 585
Erika 538
Ernst Wahliss, Fachgeschäft 147, 195, 213f, 465, 493, 496, 594
Erste Kärntnerische Ausstellung Klagenfurt 58
Erste Wiener Terracottafabrik 23, 42
Erste Wiener Terrakottafabrik und Atelier für künstlerische Fayencen Friedrich Goldscheider **64**, 65, 68, 230
Escoula, J. 361, 566
Etlin 241, 566
Etling 560
Etoile, A. 321, 332, 566
Everlast Corporation 185, 190f, 387, 392, 408, 432, 451, 454, 462, 474
Évolution 235, 240, **240**, 245, 554f, 558f, 561, 565ff, 569, 575, 578, 580, 583f, 586ff, 593, 595, 601f
Ewald, Pierre Albert 566
Exhibition of Austrian Art and Decorative Furnishing London 56f, 63
Exposiciòn Internacional Barcelona 134
Exposition Internationale des Arts Décoratifs et Industriels Modernes Paris (KGW-Ausst.) 98, 151, 232, 402, 5542f, 558ff, 565, 567, 572f, 576, 578f, 587f, 591ff, 603

F. Goldscheider (Marke) 23
F.G. (Marke) 22
Fachschule für Glas und Keramik Steinschönau 571
Fachschule für Goldschmiede Turnau 584
Fachschule für Keramik und Glas Bunzlau 574
Fachschule für Tonindustrie Znaim 554, 576, 583, 586ff, 592, 602
Fachschule Oberleutensdorf 565
Fachschule Villach 569
Falguière, Alexandre 316, 554, 557ff, 561, 566, 576ff, 582f, 586f, 590
Fargue, Léon 241, **241**, 566
Fecler 363
Filialen 46, 47, 101, 113, 231
Fine China U.S.A (Marke) 189, 191
Fischböck, Hans 163
Fischer, Friedrich 31, 156, 351, 360, 566, 588
Fischer, Ilmenau 556
Fleischer, Max 420
Fleury 384, 566
Flora, Francesco **56**, 533, 566
Florenz, Atelier für Marmor Sculpturen Rotonde Barbetti 42, 46, 47, 228
Förster & Co. Wien, Manufaktur 565, 569, 586
Förster, Karl 221, 564
Fontini 300, 559, 566
Formby, George **262**, 552
Fraikin, Ch.A. 586
Franceschi 561
Franz von Assisi, Hl. 578
Freis, Gertrude 566
Frémiet 559, 586
Frick, Ferdinand 566
Fried, Erich 566
Fried, Nellie 430, 566
Friedrich (Frédéric) Goldscheider, Paris 30, **47**, 228, 230, 533, 554, 556ff, 561, 565, 572f, 579, 583, 586ff, 592f, 595f
Friedrich der Große 371, 379
Friedrich Goldscheider in Wien, Berlin und Leipzig 48
Friedrich Goldscheider Porzellan-Manufactur Carlsbad 23, **23**
Friedrich Goldscheider, kunstgewerbliche Werkstatt 57
Fuchs 412, 566
Fucik-Fischmeister, Grete 151, 566
Fuerst, Walter René 241, 566
Fuka, Franz 56, 566
Fuller, Loie **62**, 344, 360
Funaro, P. (o. Funo) 536ff

Galle, André 559
Galvanoplastische Kunstanstalt, Geislingen/Württ. 572
Gambeauche 360, 364, 369, 566
Garbo, Greta 495
Gardet 559
Garibaldi 559
Garrett, Blanka 536, 538
Gauthier, Charles 558
Gauvenet, Jean-Baptiste 235, 566
Geiger, Benno **109f**, 112, **112f**, 210, 254, 398, 413f, 422, 427, 430, 518ff, 524ff, 533, 566f, 574
Gemeinschaft bildender Künstler Salzburg 591
Gemeinschaft bildender Künstler Steiermark 591
Gemignani, Valmore 65, **71**, 367, 376, 381, 385, 387, 391, 567
Generali 160
Gennarelli, Amédée 567
Genossenschaftsschule der Graveure, Gold- und Silberschmiede Wien 572
Gensoli, Maurice 235, 567
GEO-ART 279f, **281**
Georg Roth, jun., Fachgeschäft 146
Germani, L.-D. 235f
Gesa, Josep 323, 567
Gesetzlich geschützt (Marke) 24
Gibbons 536, 538
Giebl, Johann 567
Gigerl 25, 320ff, 328, 330, 333
Gladenbeck, Bronzegießerei 603
Glaspalast-Ausstellung München 559, 569, 573, 596, 603
Gluck, Christoph Willibald Ritter von 329
Gmundner Keramik, Manufaktur 214f, 572, 578, 592, 603
Goebbels, Joseph 483
Goebel, Kurt 135f, **136f**, **140**, 145, 205f, 218, 416, 437, 444, 448, 459ff, 464ff, 468, 472ff, 487ff, 492, 495, 505, 514f, 534, 567f
Goebel, Porzellanfabrik 216, 501, 562, 567f, 571
Goedtke 223, 541, 568
Göring, Hermann 175
Goethe, Johann Wolfgang von 65, 289, 329, 355, 370, 506
Goëtz, J.-Charles 236, 241f, **241**, 568f
Goldcrest (Marke) 189
Goldcrest Ceramics Corp. 189f
Golddekor 52
Goldscheider (Staffordshire) Pottery, Ltd. 269
Goldscheider Ceramics **190**, 191
Goldscheider Inc. 191
Goldscheider Keramik GmbH 274, 280
Goldscheider of Vienna Incorporated **190**, 191f, **192**
Goldscheider Pottery 184
Goldscheider U.S.A. (Marke) 184
Goldscheider, Alois 15ff, 24, 42, 46, 65, 68, 76, 84f, 569
Goldscheider, Alphonse 17
Goldscheider, Arthur 17, 23, 46f, **46f**, 57, 65, 98, 113, 228ff, **228**, **230ff**, 235ff, **237f**, 239, **240ff**, 241f, 244f, **244f**, 252, 545, 554f, 557ff, 565ff, 569f, 573, 577f, 583, 586f, 590, 593ff, 597, 599f
Goldscheider, Brigitte 274
Goldscheider, Camillo 17, 23, **46**, 47, 49, 57, 64f, **76**, 78f, **78f**, 167, 204, 228, 569
Goldscheider, Charlotte 191f
Goldscheider, Eduard 15, 17ff, **18f**, 569
Goldscheider, Emanuel 15, 17
Goldscheider, Erwin 68, 76, 148, 151, 159, 182, **182**, **187**, 189, 191f, 200, 212, 569
Goldscheider, Friedrich 15, 17, 23, 26ff, **26**, 30ff, 34, 42f, **42f**, 46, 64, **76**, 151, 183, 201, 204, 274, 280, 569f
Goldscheider, Fritz 204, **204**, 214f, 274
Goldscheider, Heinrich 15
Goldscheider, Henriette 15
Goldscheider, Inge 262
Goldscheider, Irving 183
Goldscheider, Jolanthe (geb. Hecht) 64, 79
Goldscheider, Julie (geb. Thieben) 15, 18f
Goldscheider, Klara 15
Goldscheider, Lilly (geb. Kuschnitzky) 68, 76, 113, 154, 156, 164ff, 178, 182f, **187**, 199ff, 204, 215
Goldscheider, Marcel(l) 23, 46, **46**, 65, 71, 74, 76, 84, 93ff, 96, 101, 104, 113, 122, 187, 204, **204**, 228, 248f, **248**, 251f, 257, 259ff, **259f**, 262ff, **262**, **264**, **266f**, **268f**, 403, 467, 546, 550f, 558, 570, 574, 593, 599, 603
Goldscheider, Margot **76**
Goldscheider, Moritz 14ff, 19, 569
Goldscheider, Peter F. 191f

Goldscheider, Peter 274f, 280, 569
Goldscheider, Regine (geb. Schwarz) 17, 22, 42, 46, **46**, 48f, 56, 60, 64f, 68, 74, 76, **76**, 569
Goldscheider, Rosemarie 248, 260f, 262, 267
Goldscheider, Samuel 14f
Goldscheider, Siegfried Friedrich 15ff, 22, 569
Goldscheider, Walter 23, 42, 46, **46**, 68, 74, 76, 84f, 102, 104, 106ff, 112f, **113**, 122f, 138, 141, 145f, 148f, 151, 154, 156ff, **156f**, 164ff, **171**, 178f, 182f, **182f**, 185, **187**, 189f, 192, 194f, 199ff, 203f, **204**, 206ff, **209**, 212f, **213**, 221, 224, 228, 248, 254, 259, **259f**, **264**, 270, 274, 431, 451, 462, 562, 566, 569, 572, 575, 578, 599
Goldscheider'sche Porzellan-Manufaktur und Majolica-Fabrik 23
Goldscheider'sche Porzellan-Manufaktur und Porzellanmalerei 22
Goldscheider-Everlast Corporation 185, **187**, 189
Goldscheider-Nippes 101, 120, **122**, 431
Goldscheidrovka 18, 20, **20**, 569
Gollhammer, F., Manufaktur 200, 214
Gornik, F. 358, 376, 378, 381, 569
Gornik, Friedrich 58, 310, 357f, 360, 364, 569
Gory, A. 358, 361, 367, 569
Gracien 290, 310, 321, 569
Grand Hotel Bad Hofgastein 199
Griepenkerl 596
Grimm, Gebrüder 446
Griser **29**, 303, 352, 569
Gross, F. **58**, 356, 363, 365ff, 569
Gruber 394f, 569
Grun, Hélène 569
Guénou 236, 569
Guerbe 244, 569
Gütersloh, A.P. 578
Guilloux, A.-E. 559
Guinard, J. 236, 569
Gult, Johann 92, 396, 400f, 569f
Guse 207
Gutenberg (Johannes Gensfleisch) 293, 311
GWM (Monogram) 84, **84**, 177, **177**, 212, 221, 223, 277

Haas & Czjzek, Fachgeschäft 147, 201
Habsburg, Otto von 394
Hafenruckler 412, 570
Hagenauer, Franz 570
Hagenbundausstellung 578
Hahn 413, 570
Haidvogel, Florian 570
Halbout, Georges 235, 570
Halnon, Frederick James 267, 551, 570
Halou, Alfed Jean 301, 570
Hanak, Anton 570, 594
Handzel, August(in) 385ff, 389, 392, 411, 534, 570f
Haniroff 52, **54f**, **58**, **96**, 296, 313, 326, 328f, 333, 334, 340f, 345, 348f, 351, 353, 355, 358ff, 363ff, 368ff, 375, 380, 391, 534, 570f
Hanley, Stoke-on-Trent 198, 262, **267**, 268
Hannaux 595
Hanne, Alphonse 290
Hardegg, Graf 159
Harlow, Jean 401
Harrison & Son 198
Harrogate, Messe 271
Hartl, Bruno 403
Harvey, Lilian **118**, 412, 422, 426, 429, 434ff, 452, 492
Hauchecorne, Gaston 235, 571
Hauf 369, 400, 571
Hauptmann, Gerhart 351
Haydn, Joseph 289
Heber 587
Hébrard 566
Hecht, Jakob 64
Hegenbarth, Ernst 53, 372, 571
Heim, Jacques 483
Heindl, Hermann 431, 469, 555, 571
Heine, Heinrich 369,
Heinemann, Fritz 383, 390, 571
Heinzle, Josef 160
Hellmer 554, 558, 561, 573, 578, 587, 589, 603
Henry VIII. **25**, **267**, **289**
Hentschel 385ff, 389, 392, 411, 534, 570f

Herczeg, Klára 135, **135**, **139**, **150**, 428, 431ff, 439f, 442ff, 446ff, 451f, 454ff, 461, 463, 466, 468f, 473ff, 478, 480f, 484ff, 492f, 495f, 498, 571f, 602f
Herend, Manufaktur 571
Herkner, Friedrich 85, 397, 572, 585
Herlinger-Schwarz, Ida 98, 101, 407, 572
Hermann Noack, Bronzegießerei 595
Herter, Ernst 603
Hertwig 85
Hertwig/Katzhütte, Manufaktur 562, 580
Hesse, Lo **71**, 396, 404
Heubach, Manufaktur 603
Hiller 318, 572
Hitler, Adolf 178, 483
Höpler, Ernst 165f
Hoffmann, Josef 556, 558
Hofner, Otto 572
Holzer-Defanti, Constantin 396
Horsetzky, Melanie von 364, 373, 380, 388, 394, 572
Hostasch, Anton 572
Hotel Tyrol, Innsbruck 212
Houdebine 229
Houdon, Jean-Antoine 334, 572
Houssin, Edouard Charles Marie 288, 572
Hoyos 392, 572
Hugo Lippmann & Co 65, 67
Hummel, Berta M. 557
Huneker 301ff, 310, 572
Hutschenreuther, Manufaktur 403, 571

Iglauer, Fritz 134, 150, 572
Imenitoff, Nathan 235, 572
Impekoven, Niddy 89, **89**, 116, 398, 402f, 420f, 425, 435, 473
Industrieausstellung Teplitz 18
Ingels, Domien 403f, 415, 425, 572f
Injalbert, Jean-Antoine 229, 302, 533, 573, 576, 595, 601
Internationale Baufach-Ausstellung Leipzig 73
Iskra 576

Jacob, A. 186, 536, 539
Jacques, E. 289, 301, 312, 325, 573
Janensch, Gerhard 603
Jaray, Sándor 355, 394, 417, 573
Jarl, Karin **95**, 98, 135, 141, **249**, 393, 400ff, 404f, 407, 409, 411, 443f, 453, 459f, 464, 477, 479, 490, 492, 547f, 573
Jeanne d'Arc (Jungfrau von Orléans) **76**, 294, 314, 322, 325, 334, 338, 366, 393
Jerace 567
Jeritza, Maria 116, 415
Joffroy 576
Johann Maresch, Manufaktur 34
Jollet 557
Jonchery, Charles Émile 299, 573
Josef Böck, Manufaktur 57, 582, 592, 603
Jouffroy, François 566, 572, 586, 588, 597
Jozon, Jeanne 573
Jubiläumsausstellung Wien 24, 96
Jugendschule Wien 601

Kainer, Ludwig 400
Kainz, Josef 417
Kaiser Franz Joseph 183
Kassin, Josef 575
Kasteleyn, Gustave 598
Katz, Morris 216
Kéléty, Alexandre **231**, 244f, 545, 573
Keller & Reiner 58
Keller, A. 534, 573
Keramische Fachschule Bern 567
Keramische Werkgenossenschaft Wien 578, 587
Keramos, Manufaktur 147, 447, 464, 467, 485, 491, 495, 558, 562, 565f, 568, 578, 580, 589f, 592
Kienast 391, 573
Kieselhausen, Lucie 400
Kinsburger, Sylvain 297f, 573
Kirsch, Hugo 569, 572, 591
Kisfaludi-Stróbl 571
Klakowicz, Casimir 573
Klein 229, 362, 533, 573f
Klimsch, Fritz 504, 509

Klimt, Gustav **59**, 312, 574
Klinko Hofsilberfabrikant Wien 584
Kloster, M. 297, 574
Klotz 573, 586
Knörlein, Rudolf 122f, **123ff**, 135, 143, 180, 209f, **209f**, 215, 218, 221, **221**, 223, 254, **257**, 259, **259**, 264, **264**, 427ff, 433ff, 445f, 448ff, 453f, 456, 458ff, 469f, 474, 478, 480f, 509, 516ff, 524, 526ff, 530, 534, 543, 546ff, 574f, 585
Köhler, Karl 180
König, Otto 557, 571, 603
Kohler 575
Kopp 390, 575
Korschann, Karl 575
Kossuth 599
Kostial, Josef **84**, 85, 86, 92, 101, 395f, 399f, 403ff, 408ff, 548f, 575
Kowacz, Anna 178,
KPM Berlin, Manufaktur 601
Krajc, Rudolf 56, 575
Kramer 213
Kratschmer, Josef 575
Kraus, Theodor Ludwig 160
Krebs 396, 575
Kreling, A. 602
Kremser, Lizzi 420f, 521, 575
Krippel, Elisabeth 575
Kucharzyk, Franz 387, 575
Kühne 557
Kuen, Walter 575
Künstlerhaus Wien 252, 259, 566, 569, 571, 580, 589, 593f, 598, 601, 603
Künstlerverband österr. Bildhauer 591
Kuhn, Dina 98f, **99**, 101, 402, 404, 406, 576
Kundmann 566
Kunsterzgießerei im Wiener Arsenal 93
Kunstgewerbeausstellung Paris 98, 151, 232, 402, 554f, 558ff, 565, 567, 572f, 576, 578f, 587f, 591ff, 603
Kunstgewerbeschule Bechyne 555, 566, 575, 590, 593, 597ff, 601f
Kunstgewerbeschule Bern 210
Kunstgewerbeschule Dresden 569
Kunstgewerbeschule Lwów 561
Kunstgewerbeschule München 556, 602
Kunstgewerbeschule Prag 570
Kunstgewerbeschule Wien 71, 106, 112f, 556ff, 566, 569ff, 574ff, 578, 580, 583ff, 591ff, 597ff, 602f
Kunstgewerbeschule Zakopane 575, 580
Kunstgewerbliche Fachschule Teplitz 57, 71, 92, 594, 600, 602
Kunsthochschule Budapest 571
Kunstkeramische Manufaktur Friedrich Goldscheider 220f, 224
Kunstschule Breslau 596
Kunstschule für Frauen und Mädchen Wien 603
Kunstschule Nürnberg 571
Kunze 160
Kuschnitzky, Richard 183, 199, 211
Kush, Rosa 156, 165, 179

L'Évolution Artistique 237, 241f
La Jana 439
La Monaca, Francesco 230, 232, 235, 241, **241**, 565, 576
La Stèle 232, **232**, 235, **236**, 242, 245, 554f, 557ff, 565, 570ff, 576, 578f, 582, 587f, 595f, 599ff
Labatut, Jules Jacques 369, 576
Lachenal, Edmond 556, 573
Lacoux, Henri 239, 576
Länderbank 159
Lambeaux, Jef 598
Lamourdedieu, Raoul 235, 244f, 576
Landowski, Paul-Maximilien 235, **236**, 242, 244, 576
Langer 306f, 319ff, 327, 350, 576
Lapigre 361, 365, 576
Laporte-Blairsy, Léo 229, 534, 576
Larroux, Antonin **56**, 58, **58**, 305f, 312, 334, 353f, 555, 577, 590
Larsson, Gottfrid Julius Andreas **62**, 229, 534, 577
Latour 367f, 374ff, 380ff, 389, 406, 539, 577
Lauchhammer, Manufaktur 596
Laurana, Francesco 350, 577
Laurent, Georges Henri 244f, **244**, 577

Lavergne, Adolphe-Jean **49**, **57**, **67**, **69**, **84**, 315f, 325, 358, 365ff, 370ff, 377f, 381f, 396, 577
Le Faguays, Pierre 232, **238**, 241f, **243**, 244, 545, 578
Le Guluche, J. 290
Leclancher **60**, 365, 367, 369f, 373f, 378, 578
LeClercq, François-Joseph **77**, 371, 375, 593
Ledru, Auguste 229, 302ff, 533, 578
Lefont 314, 367f, 378, 578
Legastelois, Jules-Prosper 578
Lehar, Franz 398
Leherb **278**, 279, 578
Lehmann, Bruno 578
Lehmann, Ida 65, 209f, 355, 375, 377f, 381f, 393, 484, 513, 578, 596
Lehucher, Marguerite 242
Leipzig, Neumarkt 31, 42, 46, 60, 73, **73**, 78, 96, **96**, 101, 146, 231
Leipziger Frühjahrsmesse 65
Leipziger Herbstmesse 62, 71, 230
Leipziger Messe 32, 99, 146, **146**, 158, 556
Leipziger Michaelis-Messe 24, 62
Leipziger Ostermesse 58, 60, 71, 79, 230
Leisek, Georg 368, 578
Leiser 368
Leitich, Hermann 160
Lemaire, Hector 229, 302, 533, 578f, 587, 601
Lemaître, Eglantine 579
Lemassi 316f, 335, 342, 347, 355, 579
Lemoine, B. **48**, 294, 311, 333, 579
Lemoyne, Jean-Baptiste 369
Lenoir, Charles 579
Lenoir, Pierre Charles 235, 242, 339, 343, 349f, 579
Lenzkirch 351f, 354ff, 368, 371, 373, 384
Léonard, Agathon 579
Lepage, Céline 231f, **231**, 235, 240, 545, 579
Lequesne 554, 597
Les Neveux de J. Lehmann 245
Lesca 301, 307f, 318, 330, 343, 353, 355, 358, 363, 372, 374, 376f, 397, 406, 579
Levasseur, Henri-Louis 30, 229, 302, 304, 306, 533, 578f
Lévèque, August Maurice **38**, 297, 299f, 312, 318f, 336f, 579
Lewit, Regine -> Goldscheider Regine (geb. Schwarz)
Lewit, S. 56, 569
Lexa, H. 359, 374, 406, 579f
Lichtenberg, von 155
Liebig, Baron Gisbert von 160
Liebig, Baron Theodor von 160
Liedloff, Helen 184, 536ff
Lille 580
Lindner, Helmut **73**, 380, 387ff, 391f, 395, 408, 580
Link, Otto 155
Linossier, Claudius 236, **237**, 580
Lista 557
Liszt, Franz 295, 309f, 380
Lobmeyer, Manufaktur 160, 572, 603
Loeffler, Alice 180
Loeffler, Arthur Georg 180, 211
Löffler, Berthold 566, 587, 592
Löw, Jakob 254, 580
Löwidt, Leopold 20
Loiseau-Rousseau, Paul 229, 304, 533, 580
Long 580
Longert, F. 580
Loos, Adolf 555
Lorenzl, Josef **85**, **88ff**, 89, 92f, **94ff**, 96, 99, **101ff**, **116**, **118**, 135, **138**, **144**, 151, 207, 278, 362, 397ff, 401ff, 425ff, 432ff, 437ff, 441f, 444, 446ff, 451ff, 457ff, 464, 466, 468f, 473, 477, 485f, 492f, 496ff, 504ff, 513, 515, 538f, 551f, 562, 577, 580ff, 585
Lorieux, Julien 294, 582
Louchet 575
Louger, J. **31**, 305, 311, 314, 361f, 396, 582
Louis, Prince de Condé 371
Loveday, Barbara 536, 538f
Ludwig II. 582
Ludwig XV. 335, 369
Luis, Bart 483
Luise, Königin, Herzogin zu Mecklenburg 313, 316, 332, 334

M (Marke) **248**, 249f
M. Goldscheider's Erben 16f, **16f**
M. Goldscheider's Erben, Thon- und Kunststein-manufactur, Werk Moldau 17
M. Plot, Bronzegießerei 598
MacMonnies 572
MacNeal 587
Maier 582
Mailänder Messe 146
Maillol 582, 587
Maison, Rudolf **54**, **76**, 229, 298, 321ff, 336, 347, 393, 533, 582
Makart, Hans 596
Malfray, Charles 235, 582f
Malfrey, J. 236, 583
Mallener, Theodor 65, **68**, **72**, 381, 384f, 407, 583
Mangin, Paul Emile 236, 583
Manzel, Ludwig 601
Marcell Goldscheider Keramik 260
Marchand, Harry 308, 583
Margarita, Infantin 378, 409
Marinko, Werner 583
Markuse 583
Marqueste 229, 533, 594f
Marquestre 586
Marquet, Alix 303, 565, 583
Marschall 65, 583
Marshal Pétain **236**
Mars-Vallett, Marius 583
Martel, Jean 242, **242**, 245, 583
Martel, Joël 235, 242, **242**, 245, 583
Masse 299, 583
Massolini 583
Matray, Ernst 410, 415f
Matthews, Jessie 491
Mattlack 58
Matula, Viktor **115**, 151, **173**, 218, 429, 432, 437, 440ff, 448ff, 452f, 455ff, 459f, 463f, 467, 469ff, 485ff, 492, 494f, 500f, 503f, 516f, 525ff, 583f
Matzen 587
Maurel, E. 288, 584
May, Sybille 236, **237**, 240, 244, 584
Mayer, Arthur 154, 160f, 167
Mayre, G. 327f, 331, 584
Méheut, Mathurin 236, 584
Meier, Emil 309, 355, 369, 377ff, 383ff, 388f, 392ff, 399, 407, 584f
Meier-Michel, Johanna **74**, 86, 369, 377, 379f, 383ff, 388f, 392ff, 399, 404, 409, 584f
Meisinger, Ida **104**, **113f**, 120, **120ff**, 413ff, 428ff, 437f, 440ff, 448, 450, 453ff, 468, 475f, 520f, 534, 585f
Meissen Porzellanmanufaktur 570, 603
Meißner 303, 586
Melcher, Josef **138**, **151**, 486, 491, 510, 586f
Membré, E. 236, 586
Mercadier, J. 320, 344, 586
Mercié, Antonin 229, 301, 343, 533, 558, 560, 576f, 582, 586f, 590, 595, 599
Mercier 298, 300, 554, 558, 586
Merck, Finck & Co Bank 168
Mergier, Paul-Louis 241, 586
Mérite, Édouard Paul 304, 586
Merkelbach & Wick, Manufaktur 603
Metzler & Ortloff Ilmenau, Manufaktur 556, 571, 603
Metzner 578
Meunier, Constantin 356, 557, 559, 586
Meyer, Konrad 99
Miller, Leo 586
Millerand **236**
Millet 572, 597
Ministerium für Wirtschaft und Arbeit 176
Mitzi Pinget Fachgeschäft 197f
Moderne Keramik 106, **107**, 114, **115**, 143, 149, **149**
Mörtl, Michael 380, 586
Molière 334
Molnar, Karl 586
Mondi 309, 337, 356, 586
Monginot, Charlotte 232, 384, 586
Montenave 52, **54**, 306, 310, 345, 350f, 360ff, 371, 586
Montini 379, 406, 586
Moreau, Hippolyte-François 586f

Moreau, M. 601
Moreau-Vauthier, Paul 53, 229, 302f, 533, 587
Morelli 557
Morice, Léon 330, 587
Moroder, C. 404, 587
Mosaikimitation 58, **59**, 60, 62
Mosch, Hermann 587
Moser, Kolo 576, 587
Moszkowski, Moritz 396
Motto, Josef C. **104**, 414, 424, 471, 587
Moulin, Eugène Emile **31**, 302, 587
Mozart, Wolfgang Amadeus 144, **144**, 309f, 344, 356, 370, 372, 386, 389, 408f, 514
Mucha, Alphonse 300f, 587
Müller 223, 397, 542, 587
Müller, Armin **69**, 376, 390, 587
Müller, Leo 587
Müller, P. 178
Müllner, Josef 65, 351, 572, 587, 601
München, Deutsche Gewerbeschau für Keramik und Glas 86, 96
Muller & Co, Steinzeugfabrik 573
Mundt, Gusty 567
Munkácsy 599
Mur 587
Museum für angewandte Kunst Wien 207
Myott, Ashley 263
Myott, Son & Co. 262ff, **262**, **266f**, 267f
Myott, Sydney 263
Myslbek 597

Napoléon **53**, 294, 311, 313, 315, 324, 329, 344, 346, 366f, 558
National College of Art Dublin 572
Nationalbibliothek 207
Nattowa 603
Navarre, Henri Edouard 235, 587
Nelson 301, 303, 306, 345, 350, 360, 387, 587
Nero Kaiser 339
Neukomm 53
Neuwalder, Grete 98, 101, 445, 587
Neuwirth, Rosa 578, 587
Nieschlag, Karl **138**, 486, 510, 586f
Noël, Tony 229, 241, 303f, 306, 533, 587
Noll, Alexandre 236, 240, **241**, 587

O'Connell, Leo 192
O'Connell, Timothy 192
Oberösterreichisches Landesmuseum Linz 207
Obsieger, Robert 134, **134**, 149, 151, 210, 524, 534, 566, 574, 587f, 592f, 597
Österreichische Kontrollbank 161f, **163**, 164, 166f, 174ff, **174**
Österreichischer Werkbund 252, **257**, 556, 558, 570, 574, 578, 584f, 587f, 592, 602
Österreichisches Exportförderungsinstitut 113
Olesievicz, S. 236, 588
Opéra-Comique Paris 230
Original Goldscheider (Marke) 188
Ornstein, Jack 185, 189
Oury, Louis 344, 346, 572, 588
Ozanne-Cederlund, Frédérique 588

Padola 387f, 391, 393f, 561, 588
Pajou, Augustin 369
Pallez 565
Palma 294, 588
Palti(e)r 383, 588
Paris, Auguste 301, 303, 341, 533, 588
Paris, Avenue de l'Opera 28 42, 46f, **47**, 60, 228, **230**
Paris, Rue de Paradis 29 269
Paris, Rue de Paradis 34 72, 101
Paris, Rue de Paradis 45 30, 42, 46f, 60, 65, 228, 230f
Parivienna 248
Patent 50f, 58, 65, 88, **88**, 228, **228**
Paul, Josef 113, 194f
Pawlata, Manufaktur 592
Pawlowa, Anna 429
Pazaurek 62
Peche, Dagobert 570
Pecheur 31, **48**, 298, 301, 307, 310f, 358, 361f, 367, 384, 566, 588
Pechmann, Fachgeschäft 199

Pecquériaux, Henri 236, 588
Pejic 413, 588
Pension Arenberg 204
Perigot 316, 367, 372f, 376ff, 386, 388ff, 395, 409, 588f
Perl, Karl 92, 99, 101, 397ff, 403ff, 589
Pernot, Henri 589
Perrin, J. 601
Persan 369, 376, 589
Petrazit 78f, **79**
Petri(des), János 52, 312, 316, 322, 356, 361ff, 366ff, 373, 375, 377, 379, 381f, 385ff, 406f, 589f
Petri, Otto 589
Petri, Trude 589
Pettarin, A. 590
Peyre, Raphael Charles 590
Pfeffer, L. 369, 590
Pfeiffer, Rudolf 160
Pfister 383, 590
Philippe, Paul **94**, 229, **232**, 289, 369, 407, 545, 590
Philippovich, von 177
Piémont, T. du 326, 590
Pierre de Vienne 65, 71
Piesch 424f, 590
Pilar, J., Wiener Kunstterrakotta-Fabrikant 58
Pilsen, Zweite Gewerbe, Industrie, Land und Forstwirtschaft Ausstellung 17
Pinedo, Emilio 558
Pippergen (o. Pippings), Robert 538f
Piquard 288, 359
Pirkenhammer, Manufaktur 578
Pissaro, Camille 239, 576
Pítra, Josef 56, 590
Plichta, J. 134
Podany, Rudolf **74f**, 382, 393ff, 397f, 401, 406, 590f
Podola(k) **76**, 382, 385, 387f, 391, 395, 404, 561, 591
Podolsk 52, 591
Poertzel, Otto 403
Poirier, B. 322, 591
Polair 342, 591
Pole 591
Poli, C. 586
Pollak, Fritz 65, **69**, 383, 385, 591
Polter **71f**, 374, 378f, 381, 383f, 387f, 390f, 395, 561, 591
Ponscarme, Hubert 559
Popineau, François Emile 235, 591
Porcher, Peggy 186, 536ff
Portinari, Beatrice 336, 346, 349, 363
Portou 316, 346, 556, 591
Porzellanfabrik Carl Knollin 572
Poschedajeff, Georgii 548
Postl, Hugo **119**, 380, 395f, 398, 407, 410, 414, 416ff, 420f, 423ff, 428, 430, 433, 436, 438f, 441, 445, 449, 455, 461, 591f
Potet, Louis 303, 592
Potet, Pierre 592
Pottery Poets 188
Potucek, Ferdinand 592
Pouyat, Guerin 113
Powolny, Michael 58, 106, 112, 351, 354, 376, 556, 558, 566, 572, 575, 578, 585, 587, 591ff, 599, 602f
Preiss & Kassler, Manufaktur 590
Prémont 295, 312, 320, 592
Preyer, Fritz 592
Prima Esposizione d'Arte Decorativa Moderna Turin **56**, 57
Primavera 555
Prince 592
Prischl Kunstkeramik **223**, 562, 592
Prischl, Adolf 135f, 138, 151, **170**, 180, 207, **208**, 218, 221, 223, **223**, 440, 442f, 450, 455f, 458, 462, 464, 466f, 470ff, 475ff, 479, 481, 486ff, 495, 497, 499f, 502, 504, 506f, 509, 537, 539, 542f, 562, 583, 592f
Privilegium **15**, 18, 22, **22**, **26**, 27
Proszynski, Henri 245, 593
Prüfungsstelle Keramische Industrie 155, 174
Prunier 320f, 325, 593
Psychiatrische Anstalt Am Steinhof 167
Puchegger, Anton 358, 387, 393, 593
Puech, Denis 573, 586

Quinzio 554

Rab, Herbert 593
Raby, Ekiel 236, 593
Ragow, Blanche S. 536, 539
Rambosson, Yvanhoé 236ff, 241, 244f
Rapin, Henri 236, 593
Rappová, Adéla 20, **20**
Rasper, Fachgeschäft 147
Rath, Stefan 160
Raum, Alfred **67**, 379, 593
Récamier, Julie 295, 329
Reichskunstkammer 162
Reichsschriftumskammer 162
Reichstreuhandstelle 158
Reichswirtschaftsministerium 175, 178
Reinold, Grete 546, 593
Renard 305, 593
Reproduction Réservée (Marke) 24
Reyam 299, 593
Ricard 366, 593
Richer, Paul 303, 593
Richter, Jaroslav 56, 593
Rie-Gomperz, Lucie 254, 593
Riemerschmid, Richard 556
Riessner, Manufaktur 34, 57, 565
Ringuet, Ghislin 236, 593
Rio, Dolores del **95**, 116, 408, 412, 416, 426, 429
Rivalta 559, 569
Rivir, Armand 236, 593
Riza 537, 539
Robbia, Luca Della 295, 348, 567, 593
Rochella, D. 320, 593
Rodin, Auguste 558f, 570, 572f, 586, 596, 599, 603
Romanelli, Raffaello 560
Roosevelt, Franklin D. 183, **183**
Rosca, Archip 593
Rosé **60**, **64ff**, **69**, **72f**, **144**, 364, 366, 368ff, 386ff, 390, 400f, 408f, 415, 593f
Rosenthal & Maeder, Manufaktur 417, 590
Rosenthal 85, 211, 504, 509, 566, 571, 582, 596, 601, 603
Rosier 295, 594
Rossini, Gioacchino 559
Rottenberg-Henny, Ena 98, 101, 404, 408, 549, 594
Roy, P. 367, 594
Royal Copenhagen, Manufaktur 571
Royal Doulton, Manufaktur 187, 571
Royal Dux, Manufaktur 571
Rozel 311, 313f, 328, 349, 354, 594
Rozet, Fan(n)y 244, 594
Rubelle, A. 340, 594
Rubens, Peter Paul 313, 317
Rudolf Sommerhuber Steyr, Manufaktur 558, 602
Rudolff 389, 394, 594
Rümann, W. von 603
Rupp, Herbert 594
Ruß, Willibald **70**, 387, 594
Rust, L. 321, 594

Sacharoff, Alexander 414, 427
Sachs, Hans 403
Sailer, Karl 135, 205ff, **205f**, 218, 254, 430, 501, 510ff, 592, 594f, 600
Sain, Marius Joseph 235, 595
Saint Denis, Ruth **66**, 377, 379, 409
Saint-Paul, Jean 236, 595
Sakellarios, Victor 573
Sakier 538f
Sala, Jean 236, **237**, 241, 595
Salzmann 51, 595
Samson, Ernesto 208, 270
Santini, Ado 192
Santini, Simona 192
Sardon 323, 337, 595
Sardou, Victorien 326, 342f, 364, 366, 370
Sarrabezolles, Charles 595
Sartorio, Antoine 235, 595
Savard 240, 595
Savine, Léopold Pierre Antoine 300, 595
Scanzi 554
Schadt, Wilhelmine 595
Scheibbser Keramik, Manufaktur 574
Scheibe-Alsbach, Manufaktur 571
Schenk, E. 236
Schifffahrtsgesellschaften 212
Schiller, Friedrich 57, 65, 289, 329, 355, 370
Schimmelpfennig 383f, 595
Schleiss-Schule Gmunden 602
Schmid 323, 595
Schmid, Ludwig 537, 539
Schmidt 223, 543, 595
Schmidt, Alexander 25, 43
Schmidt, Henri 370, 595
Schmidt-Kestner, Erich 65, 368, 373, 375f, 383, 391, 425, 436, 448f, 595
Schnackenberg, Walter 396
Schnitt- & Thonwaarengeschäft 17f
Schoedl 595
Schoenewerk 554
Schoerk, Hans 31, 384, 390ff, 395, 397, 406, 559, 595f
Schram, Alois Hans **59**, 313, 340, 367, 596
Schubert, Franz 311, 313f, 372, 398, 413
Schufinsky 574f
Schuller, L. 298, 596
Schulz 362, 596
Schumann, Robert 414, 427
Schumetz, Georg 160, 167
Schuster, Josef 161ff, 170f, **170ff**, 174ff, **174ff**, **179**, 194, 199, 201, 203, 206, 208, 596, 599
Schuster, Karl 162
Schuster, Lilly 164, 170, **171**, 179, 194, 596
Schwartz, Stefan 154ff, 158f, 161, 167, 484, 572
Schwarz 221, 413, 515, 596
Schwarz, Heinrich 49
Schwarzbart, Franz 113, 150, 154f, 183, 596
Schwarzburger Werkstätten für Porzellankunst 593, 595
Schwedisches Nationalmuseum **209**
Schwetz, Karl 578
Schwetz-Lehmann, Ida –> Lehmann, Ida
Scott, Sylvia 191, 536, 539
Sculpture (Marke) 56, **56**, 72, **72**, **229**, 277, 557
Seewitz, Joachim von **71**, 396
Sefzik, Hans 221
Seger, Ernst **66**, 374, 596
Seidan, Thomas 596
Seifert, Elisabeth 178
Selch 575
Sevcik, Hanns 596
Sèvres, Manufacture Nationale 556, 558ff, 567, 576, 593
Shakespeare, William 291, 302, 311ff, 316, 321, 341, 346, 353, 587
Sidan 355, 596
Simart, Marie-Louise 245, **245**, 596
Simon **28**, **30**, **48**, 51, **51**, 57, 62, 298f, 304, 309, 311, 318, 325f, 338f, 342, 344, 346, 348f, 351, 353f, 360, 362ff, 368, 378, 596f
Singer Nähmaschinen 191
Siot-Decauville 583
Skal(l) 355, 367, 597
Skalák, Jindřich 56, 597
Soc. Reima Paris 113
Societé de Bronzes 598, 601
Societé des Artistes décorateurs 556
Societé des Artistes Français 229f, 245, 579
Société Nationale des Beaux-Arts 229
Solveg, Maria 410, 415f
Sommer, Michael 274
Spiller, Gretl 541, 597
Spörk, Marie 597
Spuller 98, 597
St.-Gaudens, A. 572
Staatliche Majolikamanufaktur Karlsruhe 556, 558, 596f
Staatsamt für Industrie, Handel, Gewerbe und Verkehr 194
Stabi, Edouard 240, 554, 597
Staffordshire 262f, 552ff
Stankovits, Christa 597
Stappen, Charles van der 561, 601
Stefan, A. 597
Steffeck, C. 573
Stein, Eduard 113, 183
Steiner, Leo 597

Steinhof, Eugen 574f, 583, 597, 603
Stellmacher & Kessel, Manufaktur 34, 57
Stellmacher 350, 356, 597
Stepan, A. 280, 597
Sterna, Katta 410f, 415
Sterrer, Karl 574
Steuer, Bernard Adrien 301, 597
Stoke-on-Trent Art School 266
Stoltenberg-Lerche, Hans Stephan 229, 292f, 296ff, 393, 398, 597f
Stoob 274, **275f**, 277, 279, 569, 578
Stoob, Keramikschule 274
Straeten, Georges van der 230, 232, 598
Strakatý, Anton 56, 598
Strasser 569, 575
Strasser, Arthur 34, **35**, **38**, **40f**, 65, 229, 289f, 292ff, 303ff, 313ff, 325ff, 333, 336, 338, 343, 351, 476, 508, 533, 569, 586, 592, 598f, 603
Straub, E. 537, 540
Strauss, Richard 433
Strnad 576, 580, 583, 587
Strobl 368, 599
Stuart-Willfort, Irma 452, 546, 599
Studenovský, Johann 56, 599
Sturm, Emmy 411
Suchan(ek), Vacslav 309, 357, 599
Sucharda 570, 597
Susse Frères 571, 576, 578
Swoboda, Erich 575
Sydan 348, 357, 362ff, 367, 370, 372, 374, 535, 599
Symo, Margit 491
Szeiler, Joseph 266, 599
Szwarc, Marek 241, 599

Tacke 390ff, 599
Täuber, Karl 138, 147, 151, 194, 204
Tallichet, S. 235, 599
Tegner, Rudolph 244, 599
Teichgrab, Alois 170, 599
Telcs, Ede 377f, 599
Tell, E. 52, **53f**, 58, **63**, 296f, 301ff, 323, 325, 327, 331, 337, 343ff, 350, 352f, 356f, 360f, 365, 367f, 371ff, 389ff, 400, 599f
Tell, Wilhelm 327
Telsch 316, 320, 600
Telzec 600
Temple, Shirley 457
Temporal, Marcel 235ff, 242, 244, 545, 600
Terracotta und Majolikafabrik 23, 42
Theatre National de l'Opéra Paris 230, **230**
Thiébaut Frères 557, 566, 576
Thieben, Emanuel 18
Thieben, Jakob 15
Thiele **38**, 297, 299, 312, 318f, 326, 336f, 349, 600
Thom, A. 335, 600
Thomas 573, 576, 579, 587
Thomasch, Wilhelm **86f**, 92, **93**, 215, 218, 278, 396, 398, 400ff, 405ff, 431, 445, 447, 450, 454ff, 493, 535, 562, 595, 600f
Thon.- und Porzellanwaarenfabrik 42
Tilgner, Viktor 598
Tischer, E. 547, 601
Tönnieshof 220f, **221f**, 223f, 555, 596
Toinko, Heinrich 601
Tomaczewsky 601
Tonnellier 578
Tonwarenfabrik Aedermannsdorf 566
Traverse, Pierre 235, 242, 545, 601
Trenton, New Jersey 182ff, **184**, 186, **187**, 188ff, **190**, 192, 432, 439, 459, 461f, 468, 471, 481, 486, 569
Trimmel, R. 572
Turner, Raymond 538, 540
Tutanchamun 87
Tutter, Karl 403

Uffert, F.R. 161
Ullmann, Robert 145, 447, 463, 478f, 601
United Studios of Art and Ceramic 260, **260**
Universal Pictures 188
Unterglasur-Dekoration 88, 142, 170, 188, 198
Urbach, Kathi 190, 536, 538, 540

Valentin, Max 373f, 387, 393, 601
Valton 588
Vaňek, František 56, 601
Vansteen 359, 364, 601
Vasselot 565
VEB Schaubach-Kunst, Manufaktur 571
Velasquez 378, 409
Verbanck, Georges 404, 601
Verband der feinkeramischen Betriebe Österreichs 113
Verband deutscher Fabriken für Gebrauchs- Zier- und Kunstporzellan und Keramik 156
Vereinigte Ateliers für Kunst und Keramik 248, 251f, 260, 569, 574
Verlet 586
Vermögensverkehrsstelle 156ff, **156f**, 163, 174ff, **176**
Verschneider, Jean 235f, 241f, **242**, 244, 601
Vervielfältigung vorbehalten (Marke) 24
Vez, Constant 601
Vibert 578
Vidal, Henri 342, 346, 601
Vigée-Lebrun, Marie Louise Élisabeth 322
Vigo-Drillinge **117**, 419
Vigoureux, Pierre 235, 240, 601
Villanis, Emmanuel 370, 601f
Villany, Adorée-Via **75**, 393, 395
Villieu 336, 341, 346, 349f, 359, 363, 602
Vindabona (Marke) 22
Vondrák, Hugo 56, 602

Wacha 167, 177
Waderé, Heinrich 295, 317, 325, 332, 335, 343f, 347, 349, 364, 602
Wagner 602
Wagner, A. 535, 602
Wagner, Richard 65, 295, 308, 310, 337, 370, 378, 506, 557
Waldmann 356, 602
Waldmüller 414f, 602
Waldron, Katherine 540
Walsh, Raoul 412, 426
Walter, Victor Amalric 236, 602
Warzel, Franz 56, 602
Watzka, Friedrich Roland 413, 602
Weber, Ernst **109**, **113**, **130**, **132**, 415f, 424, 428, 441, 518, 602
Weiher 308, 602
Weiß 223, 543, 602
Weiss, Claire –> Herczeg, Klára
Weltausstellung Brüssel 557
Weltausstellung Chicago 31
Weltausstellung Paris 24, 53, 554, 557f, 573ff, 577ff, 587, 592, 595, 597ff, 601f
Weltausstellung St. Louis 558
Weltausstellung Wien 17
Weltmarke der Keramik 220, **220**, 223, 274
Werkkunstschule Wiesbaden 602
Werkstätten Karau 594
Werkstätten 46
Wernekinck, Sigismund 65, 73, **84**, 381, 385, 387, 391, 394f, 399, 603
Werner, Franz 356, 603
Wiederhielm, O. 179, 195, 198, 201, 203
Wiegand, Martin 399, 603
Wien, Staudgasse 7/9 42f, 46, 49, 60, 65, 85, 104, 123, 149f, 159f, 162, 165, 179, 215
Wiener Arsenal-Gießerei 571
Wiener Emailfarbenwerk Schauer & Co 587, 590, 593
Wiener Fayence 89, 135, 149, **149**
Wiener Frühjahrsmesse 158, 199, 259
Wiener Herbstmesse 136, 195, 199, **201**, 214, 259
Wiener Kunstgewerbeverein 27, 113, 591
Wiener Kunstkeramische Werkstätte 547, 576, 585f
Wiener Manufaktur Friedrich Goldscheider 84, **84**, 87, 92, 104, 122, 134, 149, 154, 161, 163ff, 200f, **201**, 203, 205, 211, 215, 274
Wiener Manufaktur Friedrich Goldscheider, Nachf. 170, 174
Wiener Manufaktur Friedrich Goldscheider's Nachfolger 164f
Wiener Manufaktur Josef Schuster, vorm. Friedrich Goldscheider 176f, 179, 194
Wiener Messe 113, 146, 158, 211
Wiener Mosaikwerkstätte Leopold Forstner 556

Wiener Porzellanmanufaktur 22
Wiener Sezession 567, 570, 592, 598
Wiener Typen 24, **25**, 289f, 318
Wiener Werkbundsiedlung 134
Wiener Werkstätte **71**, 86, 252, 556, 558, 576, 578, 587f, 592ff, 603
Wienerberger Werkstättenschule für Keramik, Wien 574, 587, 592, 594, 601
Wienerberger, Manufaktur 558, 588, 594
Wieselthier, Vally 184, 406, 439, 471, 534, 536, 540, 603
Wigman, Mary 507f
Wind, Josef **41**, 289, 296, 313, 317, 320, 329, 334, 341, 603
Wiskot, Karl 603
Witzmann 574f
Wotruba, Fritz 555
Wranovsky 155

Yourievitch, Serge 232, 244, 603

Zabekow, Rita 417
Zago, Emil 362, 603
Zametzer 403
Zardier 337, 603
Zeiller, Paul 386, 391, 603
Zezelitz, Hanns 603
Ziffer, Moses 254, 546, 603
Zsolnay, Manufaktur 590
Zumbusch, K. 378, 554, 558, 573, 578, 587, 589, 599

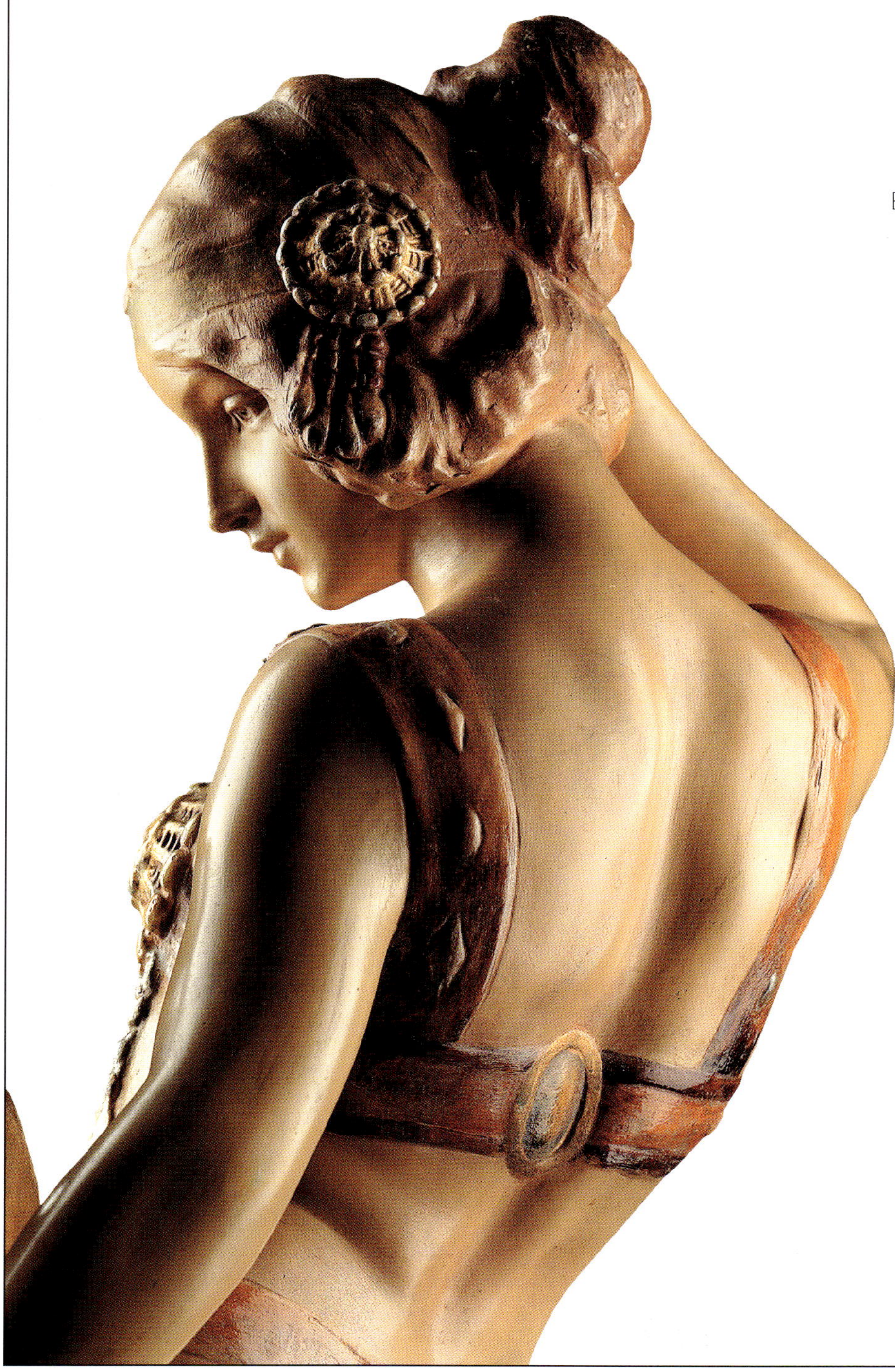

Pfister, Girl combing her hair and peacock, designed ca. 1913, manufactured by Goldscheider, Vienna, ceramics, height 90.5 cm

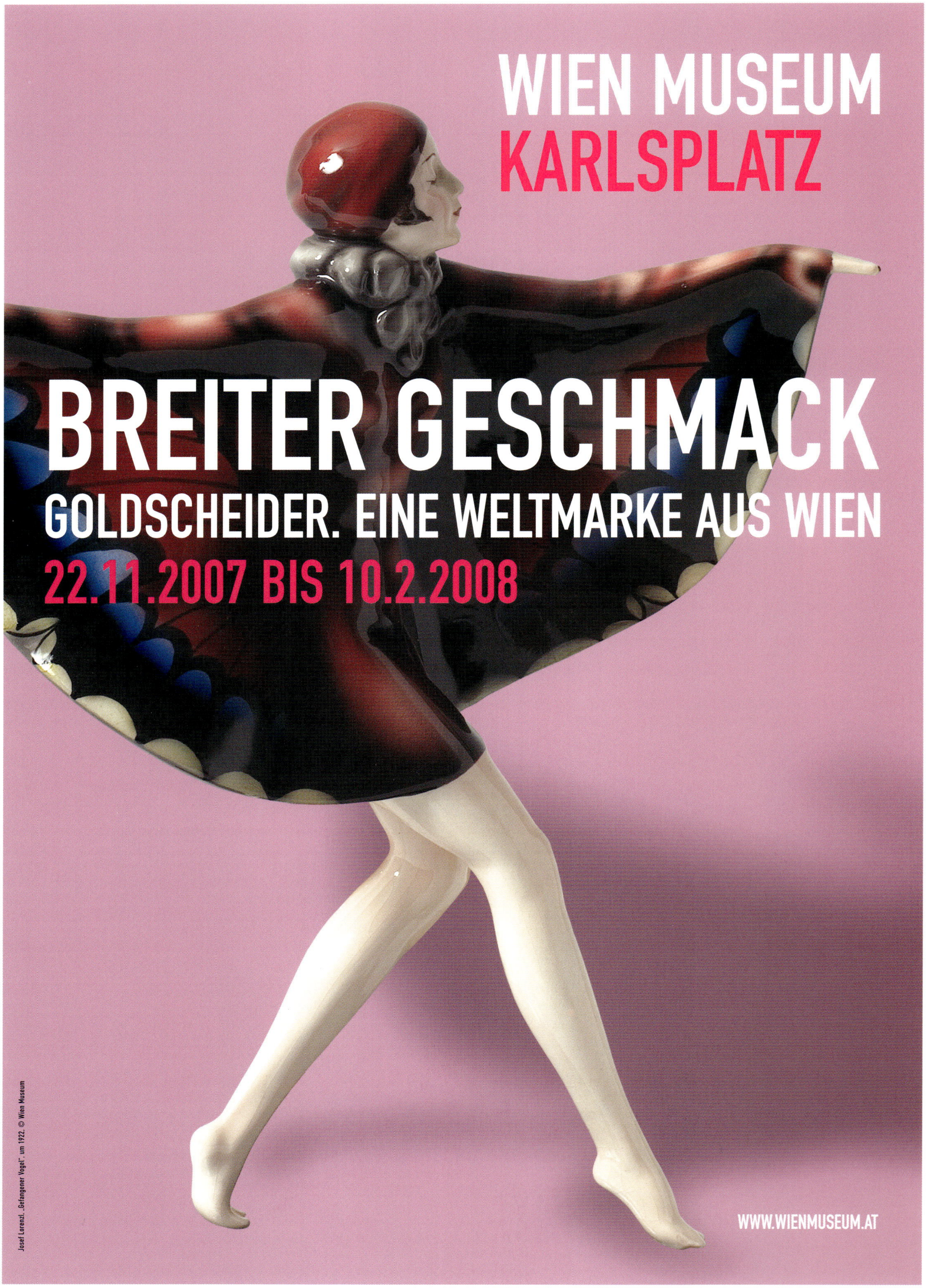
WIEN MUSEUM
KARLSPLATZ
BREITER GESCHMACK
GOLDSCHEIDER. EINE WELTMARKE AUS WIEN
22.11.2007 BIS 10.2.2008
WWW.WIENMUSEUM.AT
Josef Lorenzl, „Gefangener Vogel“, um 1922. © Wien Museum

Mit freundlicher Unterstützung von | Generously supported by
Ceramica-Stiftung, Basel / Switzerland

Autoren | Authors
Robert E. Dechant, Regensburg / Germany
Filipp Goldscheider, München / Germany

Lektorat | Editorial Office
Text & Kunst KONTOR
Elke Thode, Bodman / Germany

Übersetzung | Translation
Joan Clough, Castallack / Great Britain

Grafische Gestaltung | Layout
B.M. Design
Gabi Koch, Stuttgart / Germany

Offset-Reproduktionen | Offset Reproductions
Repromayer, Reutlingen / Germany

Druck | Printed by
Everbest Printing Co Ltd

Dieses Buch wurde gedruckt auf 100 % chlorfrei gebleichtem Papier und entspricht damit dem TCF-Standard.
This book has been printed on paper that is 100 % free of chlorine bleach in conformity with TCF standards.

Bibliografische Information der Deutschen Nationalbibliothek:
Die Deutsche Nationalbibliothek verzeichnet diese Publikation in der Deutschen Nationalbibliografie; detaillierte bibliografische Daten sind im Internet über http://dnb.ddb.de abrufbar.
Bibliographic information published by Deutsche Nationalbibliothek (German National Library):
The Deutsche Nationalbibliothek lists this publication in the Deutsche Nationalbibliografie (German National Bibliography); detailed bibliographic data is available on the internet at http://dnb.ddb.de.

ISBN 978-3-89790-216-9

Made in 2007

Bildnachweis | Photo Credits
Alle Bildrechte bei den Autoren, Sammlungen und Fotografen wie folgt:
All photo rights by the authors, collections and photographers as following:

Artimo Fine Arts, Brüssel: © Fotos Xavier Breuls de Tiecken
city-antik: © Fotos Dr. Barbara Kramer-Egghart
Im Kinsky Kunstauktionen GmbH: © Fotos Im Kinsky Kunstauktionen GmbH
MAK – Österreichisches Museum für angewandte Kunst / Gegenwartskunst, Wien: Fotos © MAK / Georg Mayer
Sammlung PB Wien 19: © Fotos Atelier Doris Kucera
Sammlung John Jesse, London: © Fotos Mike Bruce Gate Studios
Sammlung Anne Myott & Philip Pollitt: © Fotos Phil Myott
Sammlung Hansl Rudolf, St. Florian: © Fotos Fotostudio Manfred Lang GmbH
Sammlung.Schmutz@chello.at: © Fotos Robert Zahornicky, Pressbaum
Wien Museum: © Wien Museum, Fotos Hervé Massard, Gisela Stiegler
ModNr. 5230,2131,14 „Gefangener Vogel" | "Captured Bird", vgl. S. 89, Collection f@goldscheider.de: © Foto www.artwerk-online.de

© VG Bild-Kunst, Bonn 2007, für
Wander Bertoni, S. 279
Édouard Cazaux, S. 242
Kurt Goebel, S. 136, 137, 140 und | and Werkverzeichnis | Catalogue of Works S. 460–461, 464–466, 468, 473–477, 480–483, 487, 489, 492, 495, 514, 534,
Claudius Linossier, S. 237
Jan und Joël Martel, S. 242
Alexandre Noll, S. 241

Titelabbildung | Cover illustration
ModNr. 5714 „Modern", vgl. S. 413
Collection Weilinger, Wien;
© Foto Winfried Stürzl, Stuttgart

Umschlag-Rückseite (von links nach rechts) | Back cover (from left to right)
ModNr. 2664 Büste | Bust, vgl. S. 54 und Werkverzeichnis | Catalogue of Works S. 364
ModNr. 5230 „Gefangener Vogel" | "Captured Bird", vgl. S. 89–92 und Werkverzeichnis | Catalogue of Works S. 402
ModNr. 592 Vase, vgl. S. 112 und Werkverzeichnis | Catalogue of Works S. 520
ModNr. 7261 Maske | Mask, vgl. S. 131 und Werkverzeichnis | Catalogue of Works S. 453

Frontispiz | Frontispiece
Markenschild der Goldscheider-Manufaktur | A Goldscheider manufactory sign (PB Wien 19), vgl. S. 147

Kontaktieren Sie bitte die Autoren unter www.goldscheider.de bzw. info@goldscheider.de, wenn Sie ein neues Modell entdeckt haben oder Fragen zu Goldscheider-Objekten haben.
Please contact the authors at www.goldscheider.de or info@goldscheider.de if you have discovered a new model or have any questions about Goldscheider objects.

ARNOLDSCHE art books are available internationally at selected bookstores and from the following distribution partners:
USA – ACC/USA, Easthampton, MA, info@antiquecc.com
CANADA – NBN Canada, Toronto, lpetriw@nbnbooks.com
UK – ACC/GB, Woodbridge, Suffolk, sales@antique-acc.com
FRANCE – Fischbacher International Distribution, Paris, libfisch@wanadoo.fr
BENELUX – Coen Sligting Bookimport, Amsterdam, sligting@xs4all.nl
SWITZERLAND – OLF S.A., Fribourg, Information@olf.ch
JAPAN – UPS United Publishers Services, Tokyo
THAILAND – Paragon Asia Co., Ltd, Bangkok, info@paragonasia.com
AUSTRALIA / NEW ZEALAND – Bookwise International, Wingfield, customer.service@bookwise.com.au

For general questions, please contact ARNOLDSCHE Art Publishers directly at art@arnoldsche.com, or visit our homepage at **www.arnoldsche.com** for further information.

Genealogical Table

(Deutsche Version der Stammtafel:
siehe Vorsatz dieses Buchs)

Moritz Goldsch
1815 – 186

(Siegfried) Friedrich Goldscheider ⚭ Regine Schwarz
1845 – 1897 | 1854 – 1918

Alois Goldscheider
? – ?

Eduard Goldsch
1853 – 189

PILSEN (community of heirs)

from 1865 M. Goldscheider's Erben
from 1877 M. Goldscheider's Erben, Thon- und Kunststeinmanufaktur, Werk Moldau
from 1878 Brüder Goldscheider

Arthur Goldscheider
1874 – ?

Camillo Goldscheider ⚭ Jolanthe Hecht
1875 – ? | ? – ?

Alphons Goldsch
1877 – 187

PARIS & Studio in FLORENCE (Arthur)

from 1900 Frédéric Goldscheider, Éditeur d'Art
from 1918 Arthur Goldscheider, Éditeur d'Art
ca. 1935 firm closed down

VIENNA (Camillo)

1909–1910 Camillo Goldscheider, Friedrich Goldscheider Sohn

Dr. Friedrich "Fritz" Goldscheider
1906 – 1971

Dr. Peter Goldscheider
born 1945

STOOB (Peter)

from 1987 Goldscheider Keramik G.m.b.H